BIOLOGY
Understanding Life

BIOLOGY
Understanding Life

Sandra Alters, Ph.D

Adjunct Associate Professor of Biology and Education
Departments of Biology and Educational Studies
University of Missouri—St. Louis

with 1183 *illustrations*

Mosby

St. Louis Baltimore Boston Carlsbad Chicago Naples New York Philadelphia Portland
London Madrid Mexico City Singapore Sydney Tokyo Toronto Wiesbaden

Mosby
Dedicated to Publishing Excellence

A Times Mirror
Company

Editor: John E. Fishback
Associate Developmental Editor: Kathleen M. Naylor
Vice President and Publisher: James M. Smith
Project Manager: Linda McKinley
Senior Production Editor: Rich Barber
Manufacturing Supervisor: Theresa Fuchs
Designer: Elizabeth Fett
Art: Precision Graphics, and Pagecrafters, Inc.
Photographers: Stewart Halperin; Richard H. Gross, Motlow State College
Cover Photograph © by Masterfile

Printed in the United States of America
Composition by Graphic World, Inc.
Printing/binding by Von Hoffmann Press, Inc.

Mosby-Year Book, Inc.
11830 Westline Industrial Drive
St. Louis, Missouri 63146

International Standard Book Number 0-8151-3846-6

24519

96 97 98 99 00 / 9 8 7 6 5 4 3 2 1

To Brian

About the Author

Eighteen years of teaching experience and sixteen years of textbook writing experience by Dr. Sandra Alters (formerly, Gottfried) have come together in this writing of *Biology: Understanding Life.* Dr. Alters received the Bachelor's degree in biology from St. Joseph College and the Master's degree in biology from Wesleyan University. She continued on to earn the Ph.D. degree in science education at the University of Connecticut. Her research interests in science education include a focus on the use of textbooks in biology classrooms and in college biology teaching. Her science education research has resulted in publications in *The Journal of Research in Science Teaching, The American Biology Teacher,* and the 1993 Yearbook of the Association for the Education of Teachers in Science, *Excellence in Education Teachers of Science.*

Dr. Alters has diverse teaching experience. After starting out as a high school biology teacher, Dr. Alters spent 5 years teaching at the community college level and then 3 years teaching pre-clinical sciences to nursing students in a hospital-based program. For the last 6 years she has been a professor of biology and education at the University of Missouri—St. Louis. There, Dr. Alters' teaching responsibilities included teaching general biology in both large lecture hall and small classroom settings.

In 1995, Dr. Alters took a leave of absence from the University, spending a year working on the revision of *Biology: Understanding Life.* She is now adjunct Associate Professor of Biology and Education at the University of Missouri—St. Louis. Dr. Alters is married and lives in Boston.

PREFACE

WOW! A new title, a new design, and a meticulously reviewed and revised second edition are here in *Biology: Understanding Life,* second edition. Our title has changed from our first edition title, *Biology Today*. My name has changed too—in fact, so much has changed with our second edition that you may only recognize the clear, concise, and accurate writing for which so many of you complimented us. We thank you for those compliments.

This textbook is written for nonmajors college students or for mixed groups of majors and nonmajors students who may have no background in biology. Therefore, I have organized and written the text presupposing no prior knowledge. That is also the reason that this general biology book has a human focus—some or most of your students may be "science-shy" but will still find the human body inherently interesting. But *Biology: Understanding Life* does not stop there—it includes comprehensive attention to all the topics general biology courses most often address.

An Overview of the Revision

The many professors we surveyed and met with commented that the writing approach and style of the first edition worked well for their students. They appreciated the complete explanation of concepts, which were placed in a context with which students could relate and construct meaning. I have retained that approach and style and, working in conjunction with an outstanding Mosby development team, have added other features to expand our student focus and help students learn in ways in which *they* suggest will help. Two new student-oriented features are:

- *In-text glossary with pronunciation guide of bold-faced terms.* All bold faced terms are defined, and all difficult-to-pronounce scientific words have a pronunciation guide. In addition to this important new feature, special attention has been given to defining all new and unfamiliar words within the narrative to aid student understanding. Often, tables or charts help by listing important words, such as the names of hormones or sexually-transmitted diseases, and then listing their definitions or descriptions along with other pertinent information. A glossary of important terms and their definitions is also included in an appendix at the back of the book.
- *Answers to student questions.* This special feature was developed by surveying over 1000 students across the country enrolled in introductory biology courses for nonmajors. We asked them to list questions that bothered, intrigued, or otherwise interested them. From their many responses, we extracted the most frequently asked questions that fit within the content areas of our 38 chapters, including questions on evolution, AIDS, drug testing, birth control, ethical guidelines in genetic engineering, and the use of animals in research. We titled this feature "Just Wondering . . . " and think that students will be drawn to the questions of their peers and will be interested in the answers.

While talking to and surveying students, we did not forget the professors! They told us that we needed to expand our discussions of biotechnology and genetic engineering, and should include more information on sexually transmitted diseases, cancer, and drugs. They asked that we move the information on cell division from its position at the beginning of the book to a later chapter dealing with DNA. Chapter 18 (DNA, Gene Expression, and Cell Reproduction) reflects this move. Also, they asked that we increase our focus on the methods of science, emphasizing to students, as Harvard paleontologist Stephen J. Gould once put it, that "scientists reach conclusions for the damndest of reasons: intuitions, guesses, and redirections after wild goose chases . . . " For this reason, we have modified our approach to scientific methodologies in Chapter 1, retaining our explanation of classical scientific methods, but pointing out that step-wise methods are only one way in which scientists do their work. In addition to these revisions in Chapter 1, in response to professors' comments, we have included the following:

- *Twenty essays on "How Science Works."* These essays focus on the applications, processes, and methodologies of science as well as on the tools and discoveries of scientists. The topics are diverse and include "Is There Bias in Science?" "Tracking the HIV Epidemic," "DNA Sequencing," "The Evolution of Food Guides," and "The Merging of Sciences: Using the Theory of Plate Tectonics to Understand Evolution."

- *A new chapter titled "Biotechnology and Genetic Engineering."* Months of work went into the development of this chapter, and included the help of molecular biologists and biological artists to help provide a visually instructive chapter that included accurate and up-to-date content written for the nonmajor. This chapter begins with the history of classical biotechnology so that students can better understand the context of the work that is being done today at the molecular level. After describing natural gene transfer among bacteria, the chapter describes human-engineered gene transfer using eukaryotes. The chapter ends with an extensive discussion of the applications of molecular biotechnology, such as the production of proteins for therapeutic use, the development of genetically engineered vaccines, gene therapy, and food biotechnology.

- *Inclusion of material on sexually transmitted diseases, cancer, and drugs.* First, a listing of common sexually transmitted diseases (STDs) and pertinent information students should know about them is presented as a table in Chapter 21, Sex and Reproduction. Then, descriptive material about STDs, including discussions of the organisms that cause each disease, is incorporated into Chapter 25 (Viruses and Bacteria), Chapter 26 (Protists and Fungi), and Chapter 29 (Invertebrates: Patterns of Structure, Function, and Reproduction). Mutations and the molecular biology of cancer is included in Chapter 20 (Human Genetics). In addition, Chapter 13 (Nerve Cells and How They Transmit Information) now includes the following topics: How Drugs Affect Neurotransmitter Transmission, Drug Addiction, and Psychoactive Drugs and Their Effects.

Revision of the Art Program

Many professors commented that a revision of our textbook would benefit by a careful review of the art program. Therefore, we and our expert reviewers analyzed each illustration for accuracy, coherence and flow with the manuscript, continuity in the use of color, and accuracy and helpfulness of the accompanying legends. As a result, over 80% of the illustrations in this book have been re-rendered. In addition to a dedication to content accuracy, the following criteria were used as guiding principles as the art program was reviewed, modified, and reconceptualized:

- Illustrations and photographs should be used to make abstract concepts more concrete and to enhance student understanding.
- The content of the narrative and the art should match—concepts, processes, or structures should not be presented in the art that are not discussed in the narrative. Additionally, the narrative and the art should be integrated in a meaningful way.
- Illustrations that depict processes should use a clear, simple layout. For complex processes, numbered steps should be used.
- Legends should be a helpful guide to the illustration. The legend should not introduce material that is not in the text.
- Art should be visually appealing so that students are prompted to "take a look."
- Color should be consistent among structures. For example, students should always see cytoplasm depicted in a particular color, DNA in another, and cell membranes, in a third. This consistency

helps students to visually recognize structures and to more easily understand the illustrations.

We hope that you are as pleased with our new art program as are we and our reviewers. Not only do we think that the art program in this textbook is an excellent learning tool, we think that the new design of *Biology: Understanding Life* creates an exceptionally inviting look for students.

Organization of the Book

We have made only slight modifications in the organization of *Biology: Understanding Life* based on feedback from professors. In general, this textbook is organized in a traditional "micro" to "macro" approach, and is divided into seven parts encompassing 38 chapters. Although the chapters have been placed in a particular sequence, special care has been taken to write the chapters so each can stand alone, affording the professor the opportunity to restructure the sequence of the book to meet the needs of his or her students and classes.

Part One introduces students to scientific methodologies and the unifying themes of biology. A chapter on the basics of chemistry completes this introductory part. *Part Two* discusses the structure of cells and the various ways in which they capture, use, and store energy. *Part Three* uses the human body as a vertebrate example to discuss systems. *Part Four* begins with a discussion of DNA, gene expression, and cell reproduction. This molecular foundation leads to the study of Mendelian genetics and human genetics. *Part Four* ends with a discussion of sex, reproduction, and development. Students then study evolution in *Part Five,* beginning their study with the scientific evidence of evolution and concluding it with a discussion of origins and the evolution of the five kingdoms of life. After describing viruses, *Part Six* discusses the anatomy and physiology of bacteria, protists, fungi, plants, and animals. It concludes with a discussion of biotechnology and the roles these organisms (including humans) play in genetic engineering. The concluding section of the book, *Part Seven,* begins with two chapters that delve into animal behavior. The next four chapters describe populations, communities, ecosystems, and biomes, respectively. The final chapter of the book approaches the topic of "environmentalism" and helps students understand the interactions between humans and the environment that are so crucial today.

Within this organization, the chapter changes from the first edition are:

- The first edition Chapter 3 (Cell Structure and Function) and Chapter 4 (Cell Membranes) have been merged, resulting in the second edition Chapter 3 (Cell Structure and Function). Overlapping material regarding cell membranes has been revised and condensed in the new chapter.
- The first edition Chapter 5 (Cell Division) has been incorporated with Chapter 24 (The Molecular Basis of Inheritance), resulting a single chapter in the second edition: Chapter 18 (DNA, Gene Expression, and Cell Reproduction). This chapter has also been relocated.
- Chapter 13 (Nerve Cells and How They Transmit Information) now includes a discussion of psychoactive drugs.
- The sequence of chapters in Part 4, How Humans Reproduce and Pass on Biological Information (Part 5 in the first edition) has been changed. Chapter 18 (DNA, Gene Expression, and Cell Reproduction) now opens this part (rather than closes it) to lay groundwork for the rest of the chapters. Also, Chapter 19 (Patterns of Inheritance) and Chapter 20 (Human Genetics) are placed before Chapter 21 (Sex and Reproduction) and Chapter 22 (Development Before Birth) rather than after these chapters. Chapter 20 now includes a discussion of the molecular biology of cancer in the context of human genetics.
- Chapter 26 (The Evolution of the Five Kingdoms of Life) and Chapter 27 (Human Evolution) now comprise a single chapter in the second edition: Chapter 24 (The Evolution of the Five Kingdoms of Life). The overlapping content of the two first-edition chapters has been revised and merged, resulting in a more coherent single chapter. No content has been sacrificed, however, in this new, longer chapter.
- Chapter 25 (Viruses and Bacteria), Chapter 26 (Protists and Fungi), and Chapter 29 (Invertebrates: Patterns of Structure, Function, and Reproduction) have added information regarding sexually transmitted diseases. This information is included within these chapters so that students can first gain an understanding of the types of organisms that cause these diseases before studying the diseases themselves.
- Chapter 31 (Biotechnology and Genetic Engineering) is a new chapter. It is included at the end of Part 6, which describes viruses and the

five kingdoms of organisms, some of which are used in genetic engineering. In this way, students first study bacterial structure, for example, before they are introduced to the use of bacterial plasmids in gene transfer. They also study plants and animals before reading information regarding the ways in which they are involved in applications of genetic engineering.

All these chapter changes result in a textbook that is now composed of 38 chapters instead of the former 40 chapters but that includes the topics addressed in the first edition and more. Overlap and redundancy have been omitted, but additions of new features such as the in-text glossary, "How Science Works" and "Just Wondering . . . " boxes, and additional information has resulted in some chapters being lengthened.

Chapter Organization and Features

The chapters of *Biology: Understanding Life* have an organization that is designed to engage student interest and help students learn. Each chapter opens with a photo and accompanying vignette that are eye-catching, thought-provoking, and interesting. Together, the photo and vignette lead the student directly into the chapter content. Sometimes these vignettes explain something that may already be familiar to the student in a context that relates to the chapter material. For example, Chapter 17 (Hormones) begins with a vignette about the dangers of anabolic steroids. Other vignettes focus on unusual phenomena that nevertheless have a biological explanation. Chapter 12 (Excretion) opens with a discussion about why some species of turtles "cry" and how this crying is related to salt and water balance.

Opposite the opening photo page are a listing of four key concepts discussed within the chapter, and a short outline (using the first-level and second-level headings only) of the chapter content. The key concepts and outlines act as advance organizers for students, providing them with a roadmap and conceptual framework from which to begin their reading and study.

Within each chapter are concept summaries that provide synopses of the content as key points in the chapter. They are short (usually no more than three sentences long) and effectively review preceding material. Students will find them valuable when reviewing for examinations.

Two types of boxed essays are incorporated within chapters of *Biology: Understanding Life.* A student-oriented "Just Wondering . . . " box is included in each chapter. As mentioned previously, these boxes were developed from questions that general biology students asked when we surveyed them. Each box is placed near the content with which it fits or expands upon best. The topics are diverse and interesting. We thank the more than 1000 students who helped with this feature.

In addition to "Just Wondering . . . " boxes, we have included 20 "How Science Works" boxes. These essays are an outgrowth of professors' desire for us to focus on scientific processes and methodologies. Again, the topics are diverse and interesting, and focus on the applications, processes, and methodologies of science as well as on the tools and discoveries of scientists. We hope this feature reflects well what you asked for . . . and more.

Each chapter closes with a summary that lists the key concepts in the preceding content. Two categories of questions follow. Knowledge and comprehension questions comprise the first category. Knowledge-level questions center on recall of information. Comprehension-level questions go a step further and ascertain whether a student *understands* the information. Critical thinking questions make up the second category of questions. Critical thinking questions are of four types: application, analysis, synthesis, and evaluation. The Instructor's Manual identifies each question in the textbook by number and lists its specific question type within each category. The answers to the questions are found in Appendix A of the textbook.

Along with Appendix A, three other appendices are found in the back of the textbook. Appendix B is a table of classification of the five kingdoms of life. Because of the human focus of this book, animals are classified through class, and the vertebrates are classified through order. Appendix C and Appendix D are new to this edition. Appendix C is the Periodic Table of Elements (which has been deleted from Chapter 2). Appendix D is Units of Measurement and will help students with metric conversions.

The glossary provides the pronunciation, definition, and derivation for key terms in the book and also gives the chapter in which each term appears. This glossary is complete and easy to use and will help students immeasurably as they make their way through the text.

Ancillaries

Outstanding supplements enhance the *process* of teaching and learning.

Artpack

(0-8151-0858-3)

Packaged free with the text, full-color reproductions of the overhead transparencies are provided to help students follow along in lecture and comprehend detailed artwork.

Study Guide to Accompany Biology: Understanding Life, 2nd Edition

(0-8151-0837-0)

This guides provides exercises and summaries that enable students to link and apply what they have learned in the textbook to their everyday lives.

Instructor's Manual with Test Bank

(0-8151-0829-X)

This valuable teaching tool is a useful and comprehensive aid to instructors teaching biology. Providing you with full classroom support, the manual includes:

Chapter Overviews that correspond to those in the text.

Sample lecture outlines based on student objectives.

Topics for class discussion suitable for distribution to students.

Supplementary material sources, including films, slides, and other references.

Relevant barcodes for the videodiscs are conveniently organized by chapter.

Test bank with over 1000 objective test items.

Computerized Test Bank*

(Macintosh 0-8151-0831-1; Windows 0-8151-1830-3)

This state-of-the-art test-generation software offers an impressive array of user-friendly features:

Two-track design (Easy Test for the novice and Full-Test for the expert).

Full editing capabilities.

Ability to import text and graphics from conventional word-processing and graphics programs.

Test generation by review, question number, criteria, or formula.

Support for special symbols and characters, support for over 700 printers, and full mouse support.

The most complete set of digitally editable programs available.

Transparency Acetates*

(0-8151-0832-X)

Featuring key illustrations from the text, 198 full-color transparencies are an excellent supplement for lectures. Lettered callouts are consistently large and bold so they can be viewed easily, even from the back of a large lecture hall.

The Human Body Videodisc*

(0-8016-6680-5)

Provide your students with outstanding visual reinforcement and let them make first-hand observations about the human body with this customized videodisc. An indispensable resource to help further explain concepts from *Biology: Understanding Life,* this videodisc features an overview of the structure and function of each system in the human body and includes anatomical artwork, photographs of gross anatomy, micrographs of relevant tissues, and moving sequences to show the system in action or to explain how a process works.

Mechanisms of Life: Stability and Change Videodisc

(0-8016-4193-4)

This customized resource combines high-resolution artwork with engaging film clips on dynamic biological processes to provide outstanding visual reinforcement for classroom presentation. The four sections of the videodisc are Physiology, Genetics, Evolution, and Ecology.

View Study Disc*

(0-8151-0834-6)

Easy to use and implement as an instructional tool, this CD-ROM contains all the illustrations from the text. Images are arranged by chapter and by concepts. A slide show tool enables selection of prearranged images. Images can be printed in full size for use as acetates. Images may be exported for use with other programs and applications, such as the computerized Test Bank. Each disc contains start-up files for both Windows and Macintosh.

Exploring Biology Today Laboratory Manual

Ann Wilke
(0-8151-9341-6)

This laboratory manual features 30 two-hour experiments based on the learning cycle. This method of investigation allows students to manipulate objects, draw conclusions, and apply knowledge based on observation.

Instructor's Manual to Accompany Exploring Biology Today

(0-8151-9342-4)

This resource clearly shows lab instructors and teaching assistants *how* to foster critical-thinking skills in their students. In addition, it offers direction in setting up materials for successful completion of experiments.

Biology Newsletter*

(0-8151-1833-8)

Offering great material to supplement lectures, this valuable bi-annual update features expert opinions on late-breaking and controversial topics in biology.

ACKNOWLEDGMENTS

A textbook is not developed by an author alone. So many people play a role in the review, reconceptualization, design, layout, art program, and editorial aspects of a textbook; my writing is only a small part of all this work. The Mosby team as well as outside experts with whom we all interact are crucial to publishing a book that will be a dynamic and integral addition to your teaching. My thanks to everyone involved in the publication of this exciting second edition.

In particular, heartfelt thanks goes to John Fishback, acquisitions editor, who continually strove for excellence as *Biology: Understanding Life* was being developed. His fresh ideas coupled with his unceasing work on this book were significant factors that resulted in a quality publication. A key figure in the day-to-day development of this book, Kathy Naylor, has my deepest gratitude. Kathy, whose "official" title is developmental editor, wore many hats on this project: artist, biologist, editor, manager, and perfectionist. I sincerely appreciate her long hours, attention to detail, and personal commitment. On top of that, she was an incredible pleasure to work with.

The production team at Mosby did an outstanding job with this book, and I am appreciative of their work. Linda McKinley managed the production of the text; Rich Barber, the production editor, did an excellent job with the page makeup but more importantly kept the book on schedule despite difficult circumstances. Liz Fett produced the elegant design of the text, its cover, and all the ancillary covers. Theresa Fuchs, the manufacturing supervisor, also did a wonderful job in keeping this book on schedule.

The marketing team at Mosby, led by Cathy Bailey and Colleen Murray, developed an exciting program for presenting the book to the educational community, including an informative brochure, salesperson training sessions, and a prospectus of a sample chapter.

I would like to thank all the students who participated in the survey; I had fun answering your questions and reviewing your comments on using the text. I owe a special debt of gratitude to those professors who participated in the focus group; their contributions of time and experience added greatly to the relevance and accuracy of the text:

Susan Bard, Howard Community College
Clyde Bottrell, Tarrant County Junior College
Salman Elawad, Pensacola Junior College
Rick Firenze, Broome Community College
Richard Peirce, Pasadena City College
Michael Postula, Parkland College
Michael Wood, Del Mar College

In addition to the focus group members, the following scientific professionals played important roles as technical reviewers and consultants in the project:

Brian Alters, University of Southern California
Ed Anderson, University of Arkansas
Carl Colson, W. Virginia Wesleyan College
Warren Ehrhardt, Daytona Beach Community College

Merrill Emmett, University of Colorado
Rick Firenze, Broome Community College
Keith Morrill, S. Dakota State University
Dean Nelson, University of Wisconsin
Kenneth Raymond, E. Washington University, Cheney
Steven Salaris, Maryville University
Martin Spalding, Iowa State University
George Spomer, University of Idaho
Steve Taber, University of Austin
Sam Tarsitano, S.W. Texas State University
Todd Thuma, Macon College, GA
Fred Wasserman, Boston University
Carl Thurman, University of Northern Iowa
Bill Cockerham, Fresno Pacific College
Wayland Ezell, St. Cloud State University
Kristi Sather-Smith, Hinds Community College
Michael Postula, Parkland College
Richard Peirce, Pasadena City College
John Conroy, University of Winnipeg
Salman Elawad, Pensacola Community College
Richard Thomas, Illinois Central College
John Clausz, Carroll College
Gregory Gillis, Bunker Hill Community College
Cynthia J. Moore, Washington University, Missouri

Special thanks to the following art reviewers:

Salman H. Elawad, Pensacola Junior College
H.W. Elmore, Marshall University
Robert Galbraith, Crafton Hills College
John Phythyon, Saint Norbert College

We hope that *Biology: Understanding Life,* second edition will be an even better learning tool for students than the first edition, *Biology Today.* And we hope that students and professors alike will continue their dialogue with us. These interactions, we feel, have helped make this textbook a useful, informative, and helpful adjunct to any college-level nonmajors or "mixed" introductory biology course.

Sandra Alters

LEARNING WITH

BIOLOGY: UNDERSTANDING LIFE

CHAPTER 3

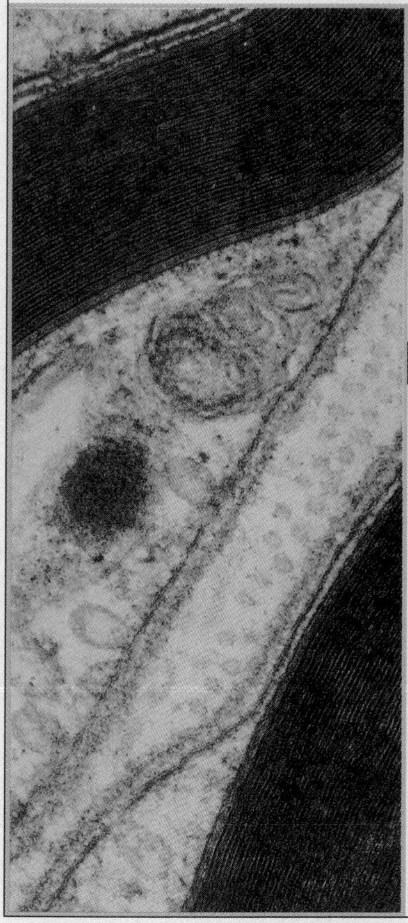

CELL STRUCTURE AND FUNCTION

THE PHOTO may not look like any highway you've ever seen, but that's exactly what it is. This highway and billions more like it make up a transportation network within your body. The threads, seen within the area bounded by the darkened layers, are called microtubules, and they shuttle particles within your body's cells. Amazingly, they can move particles in two directions at the same time!

This photo is of a nerve cell axon, magnified 17,000 times by a powerful electron microscope. Axons of nerve cells bring messages away from nerve cells to other nerve cells, organs, or muscles. The darkened areas are layers of fatty insulation that help nerves like this one transmit impulses quickly. Because axons are long (some extend from the base of your spinal cord to your toes, for example), micro-

Chapter opening photos and vignettes spark students' interest and draw then into the chapter

n transporting
cleus of the cell
axon—some-
axon tip.
one type of
d within cells.
structures per-
ding energy,
es, and direct-
s. But what are
s called "cells"?
than the empty
ish microscop-
mid-1600s. Sci-
cells are the
t can exist in-
take in nutri-
to release en-
tes. They can
uli, and main-
nment differ-
dings. In multi-
as humans,
naintain life. In
such as the
3-1, each cell
survives independently. This chapter will help you become familiar with the structure of cells and how they work. By studying their structure and function, you will also see their tremendous diversity and complexity (Figure 3-1) and will begin to understand the cellular level of organization of the human body.

The Cell Theory

In 1839 botanist Matthias Schleiden and zoologist Theodor Schwann formulated the theory that all living things are made up of cells. In other words,

KEY CONCEPTS

- All organisms on Earth are cells or are made up of cells.
- Most cells are microscopically small, providing sufficient surface area with respect to volume for cell survival.
- All cells can be grouped into two broad categories: prokaryotic cells and eukaryotic cells.
- Prokaryotic cells, which are characterized by the bacteria, are simpler in structure than eukaryotic cells, which include all other types of cells.

OUTLINE

51

- *Key Concepts help direct students' attention to the most important topics in the chapter.*
- *Chapter Outlines act as an advanced organizer for students.*

▶ *These essays focus on the applications, discoveries, tools of scientists, and processes and methodologies.*

The Tools of Scientists

HOW Science WORKS

Exploring the Microscopic World

A world invisible to the naked eye had its roots of discovery in the late 1600s with the work of two pioneers in the field of microscopy: Anton van Leeuwenhoek and Marcello Malpighi. Microscopy is the use of a *microscope*—an optical instrument consisting of a lens or a combination of lenses for magnifying things that are too small to see clearly or at all. Leeuwenhoek's and Malpighi's inventions allowed them to observe such things as plant and animal tissues, blood cells, and sperm; little escaped their observant and technologically aided eyes.

Microscopes improved somewhat over the next 200 years, but it was not until the beginning of the 20th century that this technology began to advance in sophistication at a fast pace, resulting in the array of light microscopes and electron microscopes that scientists routinely use today. Photographs prepared from three types of microscopes—the compound light microscope, the transmission electron microscope, and the scanning electron microscope—are the ones most often used in this book.

The *compound light microscope* (Figure 3-A) is an instrument that uses two lenses (therefore, compound) to magnify an object. A mirror focuses light from the room up to the eye from beneath the specimen, or a lamp is used for illumination (therefore, light). The compound light microscope visualizes eukaryotic cells (such as plant and animal cells) and prokaryotic cells (bacteria) and can magnify them up to 1500 times. Special stains are often used to visualize particular structures, as is shown in Figure 3-B. *Brightfield microscopy,* the technique used to visualize this stained specimen, passes light directly through the specimen. Unstained organisms can be seen quite well using a technique called *phase-contrast microscopy.* This technique allows the organism to be viewed while still alive (staining kills cells) and uses direct and indirect lighting to intensify the variations in density within the cell as shown in Figure 3-C. A vari-

3-A

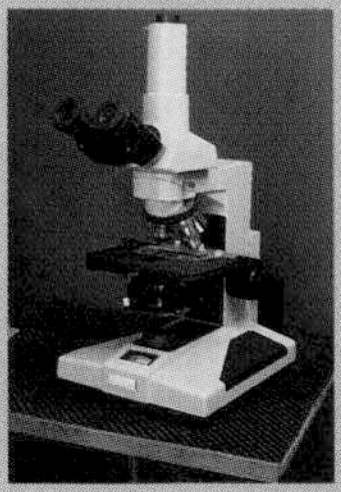

3-B

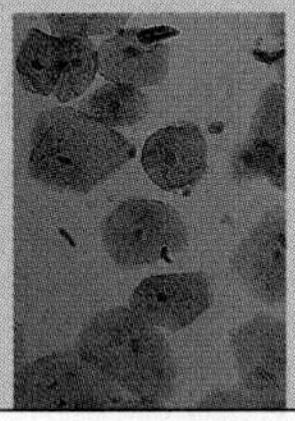

3-C

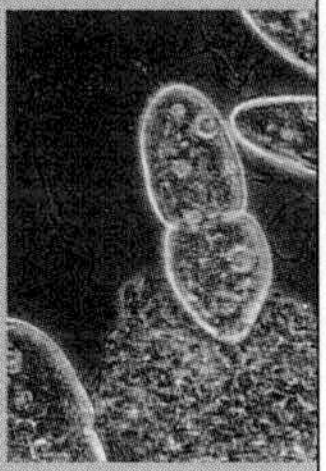

Tables help students access and organize important information.

TABLE 3-1

Eukaryotic Cell Structures and Their Functions

STRUCTURE	DESCRIPTION	FUNCTION
Exterior Structures		
Cell wall	Outer layer of cellulose or chitin, or absent	Protection, support
Plasma membrane	Lipid bilayer in which proteins are embedded	Regulation of what passes in and out of cell, cell-to-cell recognition
Cytoskeleton	Network of protein filaments, fibers, and tubules	Structural support, cell movement
Flagella (cilia)	Cellular extensions with 9 + 2 arrangement of pairs of microtubules	Motility or moving fluids over surfaces
Interior Structures and Organelles		
Endoplasmic reticulum	Network of internal membranes	Formation of compartments and vesicles; modification and transport of proteins; synthesis of carbohydrates and lipids
Ribosomes	Small, complex assemblies of protein and RNA, often bound to ER	Sites of protein synthesis
Nucleus	Spherical structure bounded by double membrane, site of chromosomes	Control center of cell
Chromosomes	Long threads of DNA associated with protein	Sites of hereditary information
Nucleolus	Site within nucleus of rRNA synthesis	Synthesis and assembly of ribosomes
Golgi bodies	Stacks of flattened vesicles	Packaging of proteins for export from cell
Lysosomes	Membranous sacs containing digestive enzymes	Digestion of various molecules
Mitochondria	Bacterialike elements with inner membrane highly folded	"Power plant" of the cell
Chloroplasts	Bacterialike elements with inner membrane forming sacs containing chlorophyll, found in plant cells and algae	Site of photosynthesis in plant cells and algae

prokaryotic cells (pro **kare** ee **ot** ick **cells**) cells that have a relatively simple structure and no membrane-bounded nucleus or other membrane-bounded organelles; all bacteria are prokaryotes.

eukaryotic cells (yoo **kare** ee **ot** ick **cells**) cells that have a more complicated structure than prokaryotic cells; eukaryotic cells have a membrane-bounded nucleus and a variety of cell organelles.

▶ *Mini glossaries with pronunciation guides assist students with difficult and unfamiliar terminology right where they appear.*

▶ *Concept summaries synthesize key concepts and help students prepare for exams.*

These vesicles eventually fuse with the membranes of another organelle called the **Golgi body.**

Smooth ER has no ribosomes attached to its surfaces. Therefore it does not have the grainy appearance of rough ER and does not manufacture proteins. Instead, smooth ER has enzymes bound to its inner surfaces that help build carbohydrates and lipids. Cells specialized for the synthesis of these molecules, such as animal cells that produce male or female sex hormones, have abundant smooth ER.

> *The endoplasmic reticulum (ER) is an extensive system of membranes that divides the interior of eukaryotic cells into compartments and channels. Rough ER makes and transports proteins destined to leave the cell. Smooth ER helps build carbohydrates and lipids.*

Golgi Bodies

First described by the physician Camillo Golgi in the last half of the nineteenth century, Golgi bodies look like microscopic stacks of pancakes in the cytoplasm. Animal cells each contain 10 to 20 sets of these flattened membranes. Plant cells may contain several hundred because the Golgi bodies are involved in the synthesis and maintenance of plant cell walls (a structure animal cells do not have). Collectively, the Golgi bodies are referred to as the **Golgi complex.**

Molecules come to a Golgi body in vesicles pinched off from the ER. The membranes of the vesicles fuse with the membranes of a Golgi body. Once inside the space formed by the Golgi membranes, the molecules may be modified by the formation of new chemical bonds or by the addition of carbohydrates. For example, mucin, which is a protein with attached carbohydrates that forms a major part of the mucous secretions of the body, is put together in its final form in the Golgi complex. This refining of molecules occurs in stages in the Golgi complex, with different parts of the stack of membranes containing enzymes that do specific jobs. When molecular products are ready for transport, they are sorted and pinched off in separate vesicles (Figure 3-12). Each vesicle travels to its destination and fuses with another membrane. The vesicles containing those molecules that are to be secreted from the cell fuse with the plasma membrane. In this way the contents of the vesicle are liberated from the cell (Figure 3-13). Other vesicles fuse with the membranes of organelles such as lysosomes, delivering the new molecules to their interiors.

> *The Golgi complex is the delivery system of the eukaryotic cell. It collects, modifies, packages, and distributes molecules that are made at one location within the cell and used at another.*

Lysosomes

The new molecules delivered to **lysosomes** are digestive enzymes—molecules that help break large molecules into smaller molecules. Lysosomes are, in fact, membrane-bounded bags of many different digestive enzymes (Figure 3-14). Several hundred of these organelles may be present in one cell alone.

Lysosomes and their digestive enzymes are extremely important to the health of a cell. They help cells function by aiding in cell renewal, constantly breaking down old cell parts as they are replaced

plasma membrane a thin, flexible lipid bilayer that encloses the contents of a cell.
hydrophilic (hi droe **fil** ick) a term referring to polar molecules that form hydrogen bonds with water.
hydrophobic (hi droe **fo** bic) a term referring to nonpolar molecules that cannot form hydrogen bonds with water.
fluid mosaic model the most well-accepted theoory regarding the nature of the cell membrane; it describes the fluid nature of a lipid bilayer studded with a mosaic of proteins.
cytoplasm (sye toe **plaz** um) a viscous fluid within a cell that contains all cell organelles *except* the nucleus.
endoplasmic reticulum (en doe **plaz** mik ri **tik** yuh lum) an extensive system of interconnected membranes that forms flattened channels and tubelike canals within the cytoplasm of a cell.
rough ER a kind of endoplasmic reticulum that is covered with ribosomes and hence resembles long sheets of sandpaper; in tandem with its ribosomes, rough ER manufactures and transports proteins designed to leave the cell.
smooth ER a kind of endoplasmic reticulum that does not have ribosomes attached to its surface and hence manufactures no proteins; instead, smooth ER helps produce carbohydrates and lipids.
ribosomes (**rye** bow **somes**) the organelles at which proteins are manufactured.
Golgi body (**gol** gee **body**) an organelle that collects, modifies, and packages molecules that are made at different locations within the cell and prepares them for transport.
Golgi complex a term collectively referring to all of the Golgi bodies within a cell.
lysosomes (lye so somes) membrane-bounded organelles that are essentially bags of many different digestive enzymes. Lysosomes break down old cell parts or materials brought into the cell from the environment and are extremely important to the health of a cell.

▶ *Illustrations make abstract concepts more concrete to enhance student learning. Multi-part illustrations show the sequence of events and how processes take place.*

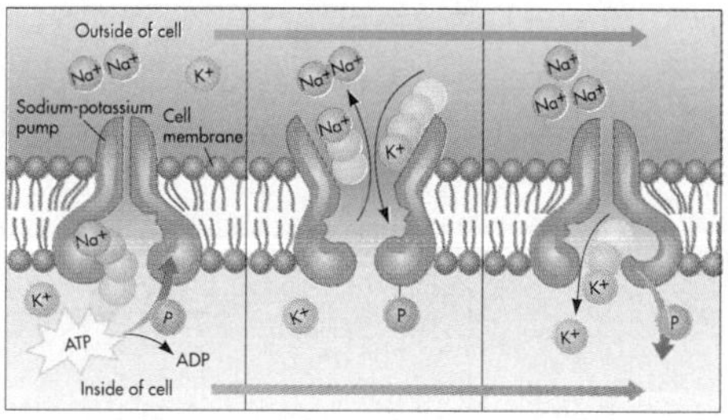

FIGURE 3-25 The sodium-potassium pump. The pump uses energy to move sodium ions (Na^+) out of the cell and potassium ions (K^+) into the cell. Three sodium ions are moved out for every two potassium ions that are moved in by this transport channel. The relative concentrations of sodium and potassium ions within and outside of the cell are important for many functions of the human body.

portant functions of any cell. Without it, the cells of your body would be unable to maintain the proper concentrations of substances they need for survival.

> *Active transport is the movement of a solute across a membrane against the concentration gradient with the expenditure of chemical energy. This process requires the use of a transport protein specific to the molecule(s) being transported.*

More than one third of all the energy expended by a cell that is not dividing is used to actively transport sodium (Na^+) and potassium (K^+) ions. The type of channel by which *both* ions are transported across the cell membrane *in opposite directions* is a *coupled channel,* which has binding sites for both molecules on one membrane transport protein. This remarkable coupled channel is called the **sodium-potassium pump,** and it uses energy to move these ions across the cell membrane. Sodium ions are moved out of the cell to maintain a low internal concentration relative to the concentration outside the cell. Conversely, potassium ions are moved into the cell to maintain a high internal concentration relative to the concentration outside the cell (Figure 3-25). (Three sodium ions are moved out for every two potassium ions that are moved in by the channel.) This transport mechanism is important in most cells of the human body but is particularly important for the proper functioning of muscle and nerve cells.

Endocytosis

Certain types of cells transport particles, small organisms, or large molecules such as proteins into their cells. In humans, for example, white blood cells police the body fluids and ingest substances as large as invading bacteria. In nature, some single-celled organisms often eat other single-celled organisms whole. How can cells move such large substances into their interiors?

Cells such as these ingest particles or molecules that are too large to move across the membrane by a process called **endocytosis,** which literally means "into the cell." Even if the substance is not

facilitated diffusion a type of transport process in which molecules move across the cell membrane by means of a carrier protein but down the concentration gradient without an input of energy by the cell.
active transport the movement of a solute across a membrane against the concentration gradient with the expenditure
sodium-potassium pump the coupled channel that uses energy to move sodium (Na^+) and potassium (K^+) ions across the cell membrane. Three sodium ions are moved out of the cell for every two potassium ions that are moved in.
endocytosis (**enn** doe sye **toe** sis) a process in which cells engulf large molecules or particles and bring these sub-

Just Wondering ***boxes discuss the most frequently asked questions from a survey of over 1,000 students.***

JUST WONDERING. . . .

How do scientists know what is in the nucleus of a cell?

The story of scientists' quests to understand the structure of the nucleus began in the early 1700s. Microscopes at that time, although primitive by today's standards, were sophisticated enough to allow scientists to see the nuclei of plant cells and some animal cells. However, it wasn't until 1833 that a nucleus was thought to be in every cell. Prior to that time, scientists held the view that some cells had no nuclei because they could not see them. Today we realize that scientists were simply observing some cells during cell division, when the nucleus loses its membrane and the nuclear material looks much different from that of a nondividing cell.

As time went on, microscope technology improved with the invention of the high-power oil immersion lens in 1870. The microtome was invented in 1866, which could slice individual cells into thin sections. In addition, new methods of fixing and staining cells were developed. Some dyes were produced that were highly specific for staining certain cell parts, such as the nucleus. All of these factors allowed scientists to view the nuclei of cells with much greater magnification, detail, and clarity.

As the quality of optical equipment and cell preparation techniques improved, the precision increased with which cells "at rest" and undergoing nuclear division could be studied. The nucleus no longer looked like a granule-filled compartment that split indiscriminately during cell division. Scientists could now see a nucleus filled with threads and bands of mate- ... *tin* ... ice ... lor ... *ne,* ... ro-

During the 1800s many scientists were also studying the principles of heredity and the nature of fertilization as well as the events of cell division. So investigations into the function of the nuclear material paralleled the study of its structure. By 1866, Gregor Mendel published his theories of inheritance. (His work is described in Chapter 19.) In the late 1870s, scientists described the process of nuclear division and the process of nuclear reduction-division that occurs in sex cells. By 1884, scientists realized that fertilization in both animals and plants consists of the fusion of maternal and paternal sex cells. Taking all of this information into account, an American graduate student named Walter Sutton hypothesized that the hereditary factors were on the chromosomes. At the turn of the century, scientists began to look toward the chemical nature of this hereditary material. Then, for more than half a century, chemists joined biologists in their quest to understand the structure of nuclear material.

The rest of this fascinating story of discovery is continued later in this book. Chapter 18, "DNA, Gene Expression and Cell Reproduction," chronicles investigations that experimentally determined that the hereditary material is located in the nucleus. It goes on to discuss the discovery of the molecular makeup of the hereditary material, culminating with the Nobel prize-winning work of two young scientists, James Watson and Francis Crick. As you can see, your question took scientists from many fields over 100 years to answer. Today, using electron microscopes and highly sophisticated laboratory techniques, molecular biologists and others are still asking and answering questions about the nucleus and the nature of the hereditary material.

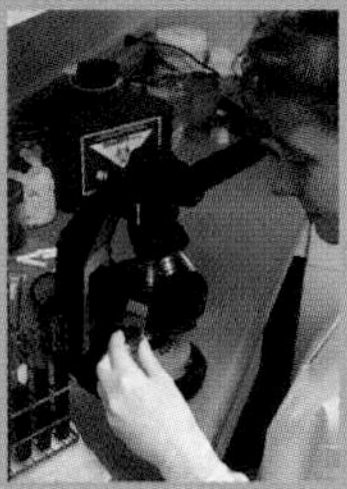

... re nucleoli. In ... e nucleus" and ... s a closed com- ... he. Other than ..., this compart- ... lls. ... the nucleus is ... ner of the two membranes actually forms the boundary of the nucleus, and the outer membrane is continuous with the ER. At various spots on its surface, the double membrane fuses to form openings called *nuclear pores.* Figure 3-15 shows the surface of the nuclear membrane of a cell that was frozen and then cracked during a special type of preparation for the electron microscope. The pores look like pock-

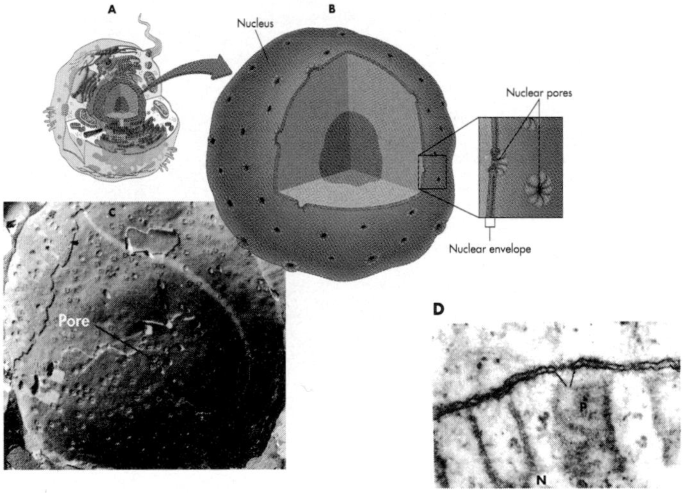

FIGURE 3-15 The nucleus. A, The nucleus, or controlling center of the cell. **B,** The outer, double membrane of the nucleus encloses the chromatin and one or more nucleoli. This double membrane is dotted with openings called nuclear pores. The nuclear membrane is fused at the edges of each pore. **C,** Electron micrograph showing nuclear pores perforating the nuclear envelope (*36,000x*). These pores are visible in **D,** designated by *P. C* is the cytoplasm and *N* is the nucleus.

marks on the surface of the membrane. These ringlike holes are lined with proteins and serve as passageways for molecules entering and leaving the nucleus.

Within the nucleus is the hereditary material **deoxyribonucleic acid, or DNA.** DNA determines whether your hair is blond or brown or whether a plant flowers in pink or white. It controls all activities of the cell. To accomplish these amazing feats, DNA performs one job: it directs the synthesis of ribonucleic acid, or RNA, which in turn directs the synthesis of proteins. (The structure, interactions, and roles of both DNA and RNA are discussed further in Chapter 18.)

DNA is bound to proteins in the nucleus, forming a complex called *chromatin.* In a cell that is not dividing, the chromatin is strung out, looking like strands of microscopic pearls. But as a cell begins to divide, the DNA coils more tightly around the

deoxyribonucleic acid (DNA) (de **ok** see rye boh new klay ick) the hereditary material; DNA controls all cell activities and determines all of the characteristics of organisms.

chromosomes (krow muh **somes**) shortened, thickened structures consisting of DNA coiled tightly around proteins.

nucleolus (noo **klee** oh lus) a darkly staining region within the nucleus of a cell that contains a special area of DNA that directs the synthesis of ribosomal ribonucleic acid, or rRNA.

Reference illustrations show students how the detail they are studying relates to the larger structure.

- *Chapter summaries consolidate key concepts and provide an effective review for exams.*

- *Knowledge-level questions center on recall of information, and Comprehension-level questions ascertain whether students* understand *chapter content.*

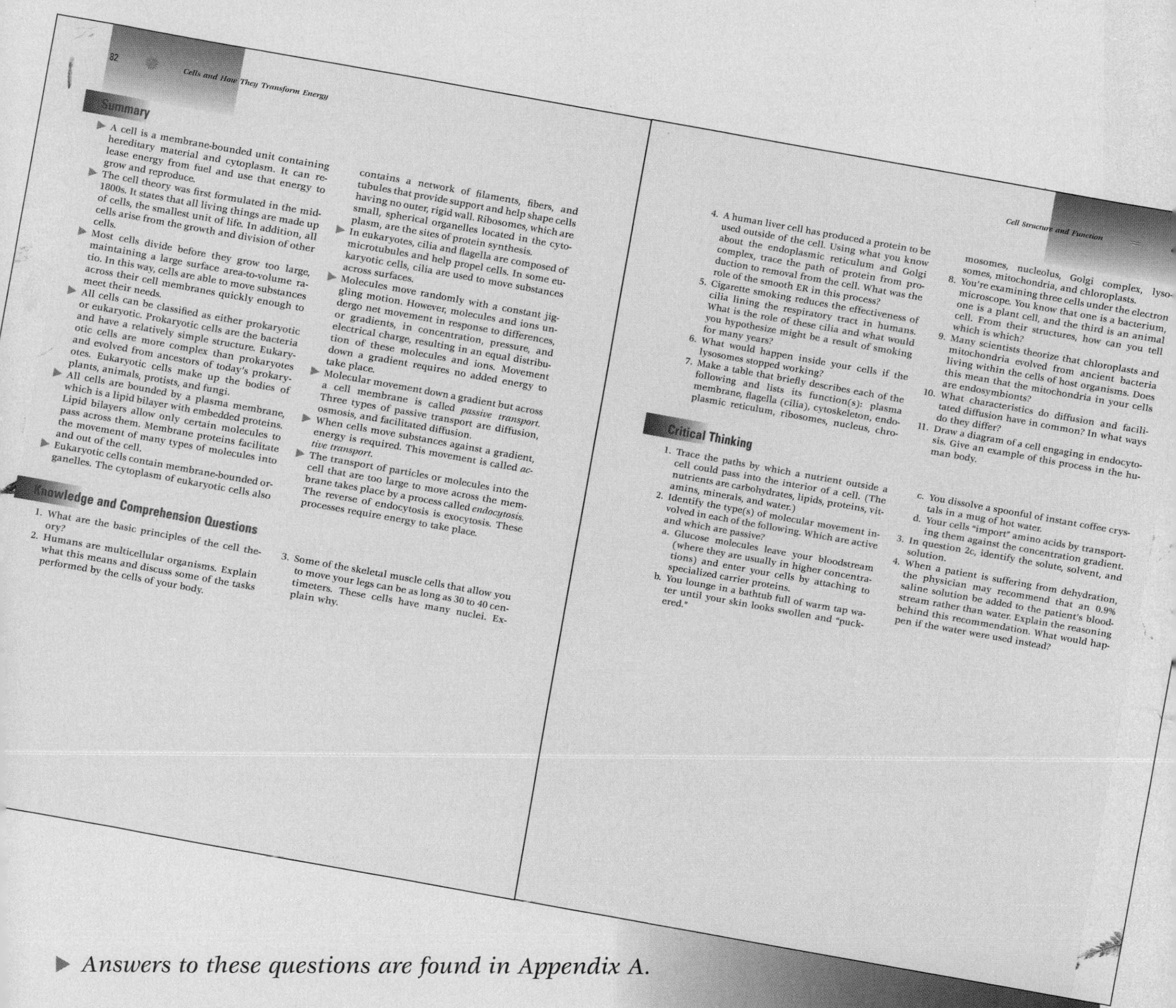
82 Cells and How They Transform Energy

Summary

- A cell is a membrane-bounded unit containing hereditary material and cytoplasm. It can release energy from fuel and use that energy to grow and reproduce.
- The cell theory was first formulated in the mid-1800s. It states that all living things are made up of cells, the smallest unit of life. In addition, all cells arise from the growth and division of other cells.
- Most cells divide before they grow too large, maintaining a large surface area-to-volume ratio. In this way, cells are able to move substances across their cell membranes quickly enough to meet their needs.
- All cells can be classified as either prokaryotic or eukaryotic. Prokaryotic cells are the bacteria and have a relatively simple structure. Eukaryotic cells are more complex than prokaryotes and evolved from ancestors of today's prokaryotes. Eukaryotic cells make up the bodies of plants, animals, protists, and fungi.
- All cells are bounded by a plasma membrane, which is a lipid bilayer with embedded proteins. Lipid bilayers allow only certain molecules to pass across them. Membrane proteins facilitate the movement of many types of molecules into and out of the cell.
- Eukaryotic cells contain membrane-bounded organelles. The cytoplasm of eukaryotic cells also contains a network of filaments, fibers, and tubules that provide support and help shape cells having no outer, rigid wall. Ribosomes, which are small, spherical organelles located in the cytoplasm, are the sites of protein synthesis.
- In eukaryotes, cilia and flagella are composed of microtubules and help propel cells. In some eukaryotic cells, cilia are used to move substances across surfaces.
- Molecules move randomly with a constant jiggling motion. However, molecules and ions undergo net movement in response to differences, or gradients, in concentration, pressure, and electrical charge, resulting in an equal distribution of these molecules and ions. Movement down a gradient requires no added energy to take place.
- Molecular movement down a gradient but across a cell membrane is called *passive transport.* Three types of passive transport are diffusion, osmosis, and facilitated diffusion.
- When cells move substances against a gradient, energy is required. This movement is called *active transport.*
- The transport of particles or molecules into the cell that are too large to move across the membrane takes place by a process called *endocytosis.* The reverse of endocytosis is exocytosis. These processes require energy to take place.

Knowledge and Comprehension Questions

1. What are the basic principles of the cell theory?
2. Humans are multicellular organisms. Explain what this means and discuss some of the tasks performed by the cells of your body.
3. Some of the skeletal muscle cells that allow you to move your legs can be as long as 30 to 40 centimeters. These cells have many nuclei. Explain why.
4. A human liver cell has produced a protein to be used outside of the cell. Using what you know about the endoplasmic reticulum and Golgi complex, trace the path of protein from production to removal from the cell. What was the role of the smooth ER in this process?
5. Cigarette smoking reduces the effectiveness of cilia lining the respiratory tract in humans. What is the role of these cilia and what would you hypothesize might be a result of smoking for many years?
6. What would happen inside your cells if the lysosomes stopped working?
7. Make a table that briefly describes each of the following and lists its function(s): plasma membrane, flagella (cilia), cytoskeleton, endoplasmic reticulum, ribosomes, nucleus, chromosomes, nucleolus, Golgi complex, lysosomes, mitochondria, and chloroplasts.
8. You're examining three cells under the electron microscope. You know that one is a bacterium, one is a plant cell, and the third is an animal cell. From their structures, how can you tell which is which?
9. Many scientists theorize that chloroplasts and mitochondria evolved from ancient bacteria living within the cells of host organisms. Does this mean that the mitochondria in your cells are endosymbionts?
10. What characteristics do diffusion and facilitated diffusion have in common? In what ways do they differ?
11. Draw a diagram of a cell engaging in endocytosis. Give an example of this process in the human body.

Critical Thinking

1. Trace the paths by which a nutrient outside a cell could pass into the interior of a cell. (The nutrients are carbohydrates, lipids, proteins, vitamins, minerals, and water.)
2. Identify the type(s) of molecular movement involved in each of the following. Which are active and which are passive?
   a. Glucose molecules leave your bloodstream (where they are usually in higher concentrations) and enter your cells by attaching to specialized carrier proteins.
   b. You lounge in a bathtub full of warm tap water until your skin looks swollen and "puckered."
   c. You dissolve a spoonful of instant coffee crystals in a mug of hot water.
   d. Your cells "import" amino acids by transporting them against the concentration gradient.
3. In question 2c, identify the solute, solvent, and solution.
4. When a patient is suffering from dehydration, the physician may recommend that an 0.9% saline solution be added to the patient's bloodstream rather than water. Explain the reasoning behind this recommendation. What would happen if the water were used instead?

Cell Structure and Function

- *Answers to these questions are found in Appendix A.*

- *Critical Thinking questions are of four types: application, analysis, synthesis and evaluation.*

BRIEF CONTENTS

PART ONE
AN INTRODUCTION TO BIOLOGY AND CHEMISTRY

PART TWO
CELLS AND HOW THEY TRANSFORM ENERGY

PART THREE
HUMAN BIOLOGY: THE STRUCTURE AND FUNCTION OF THE BODY

PART FOUR
HOW HUMANS REPRODUCE AND PASS ON BIOLOGICAL INFORMATION

PART FIVE
EVOLUTION: HOW ORGANISMS CHANGE OVER TIME

PART SIX
THE UNITY AND DIVERSITY OF ORGANISMS

PART SEVEN
HOW ORGANISMS INTERACT WITH EACH OTHER AND WITH THE ENVIRONMENT

APPENDICES

DETAILED CONTENTS

PART ONE

AN INTRODUCTION TO BIOLOGY AND CHEMISTRY

1 The Themes of Biology Today, 2

2 The Chemistry of Life, 22

PART *Two*

CELLS AND HOW THEY TRANSFORM ENERGY

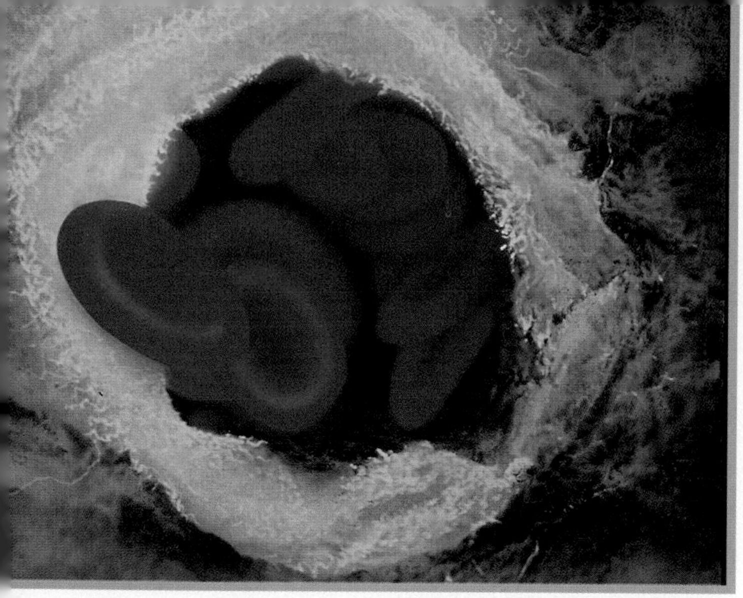

PART THREE

HUMAN BIOLOGY: THE STRUCTURE AND FUNCTION OF THE BODY

15 Senses, 304

16 Protection, Support, and Movement, 322

17 Hormones, 344

PART FOUR
HOW HUMANS REPRODUCE AND PASS ON BIOLOGICAL INFORMATION

18 DNA, Gene Expression, and Cell Reproduction, 366

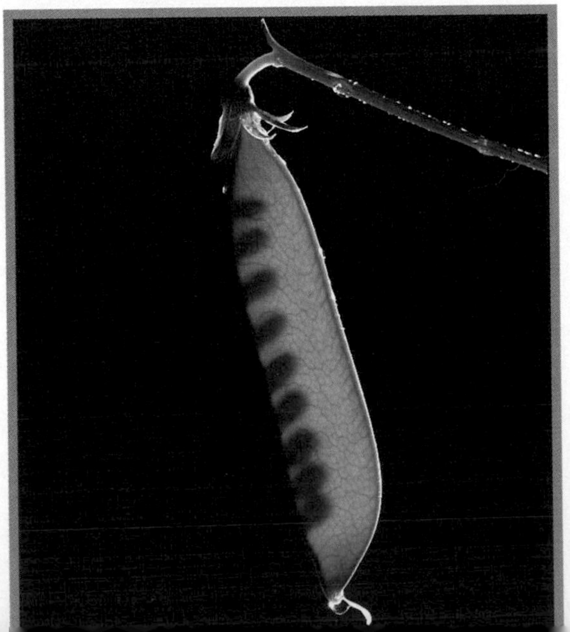

22 Development Before Birth, 466

PART FIVE

EVOLUTION: HOW ORGANISMS CHANGE OVER TIME

23 The Scientific Evidence of Evolution, 488

24 The Evolution of the Five Kingdoms of Life, 510

PART SIX
THE UNITY AND DIVERSITY OF ORGANISMS

27 Plants: Reproductive Patterns and Diversity, 592

28 Plants: Patterns of Structure and Function, 614

29 Invertebrates: Patterns of Structure, Function, and Reproduction, 634

30 Chordates and Vertebrates: Patterns of Structure, Function, and Reproduction, 658

31 Biotechnology and Genetic Engineering, 680

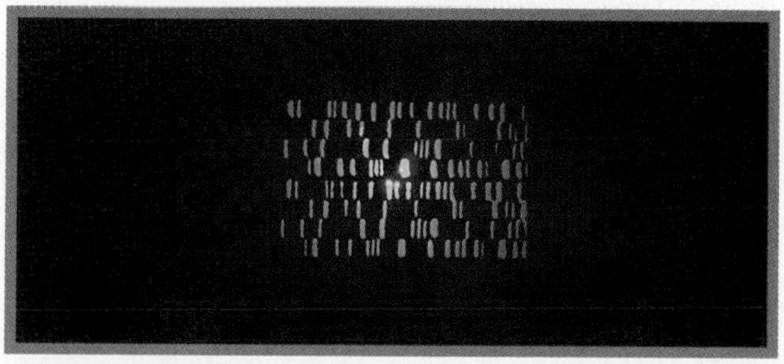

PART SEVEN

HOW ORGANISMS INTERACT WITH EACH OTHER AND WITH THE ENVIRONMENT

35 Interactions Within Communities of Organisms, 760

36 Ecosystems, 780

37 Biomes and Life Zones of the World, 798

38 The Biosphere: Today and Tomorrow, 824

APPENDICES

BIOLOGY
Understanding Life

PART ONE

AN INTRODUCTION TO BIOLOGY AND CHEMISTRY

CHAPTER 1

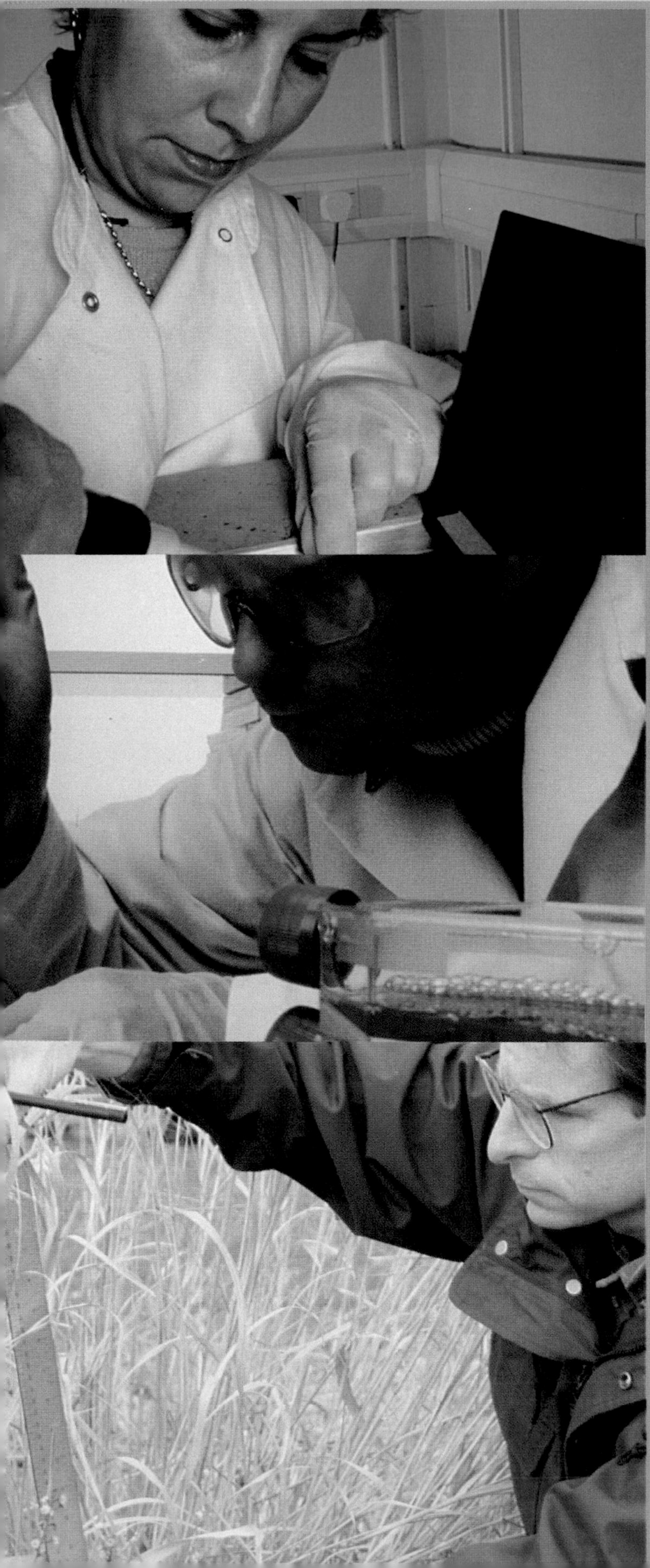

THE THEMES OF BIOLOGY TODAY

BIOLOGY TODAY has many faces. Molecular biologists study life at the chemical level, probing the workings of the hereditary material and the molecular "chain of command" within cells. Cell biologists study individual cells or groups of cells, often by growing them outside of organisms in cultures, and ask questions regarding cell-to-cell interactions and the effects of the environment on cells. Working at the organism level, biologists study animals, plants, and other multicellular organisms. Other biologists study populations—individuals of the same species occurring together at one place and at one time—and are interested in interactions among them. Population biologists study topics such as the changes in population sizes and the causes of these fluctuations. Some biologists work with a global orientation and are therefore interested in questions having worldwide impact. These biologists study the effects of burning billions of acres of tropical rain forest on global weather patterns and trends, for example.

Today, the study of biology includes these focuses and myriad others. Biologists ask questions that probe the intricacies of life from the perspectives of both distinct subdisciplines and blended subdisciplines of biology, calling on the knowledge and techniques of related fields. They answer questions that add to our knowledge base and probe areas such as the transfer of hereditary material from organism to organism, the relationship between diet and disease, and the development of new food plants. This probing advances knowledge in applied fields such as medicine, agriculture, and industry and results in the creation of products that enhance and lengthen lives. As you read *Biology: Understanding Life,* you will find out more about these and other topics that are a part of biology today. As the ancient Greeks put it, you will have a *bios logos*—a discourse on life.

KEY CONCEPTS

- Scientists study the natural world by using scientific methods, which most often include making observations, developing hypotheses to explain those observations, and then testing those hypotheses.
- Scientific explanations that are supported by overwhelming evidence over time are called theories—powerful concepts that help scientists make dependable predictions about the world.
- Rising out of the abundance of scientific explanations are many themes regarding the natural world.
- Modern taxonomy, the classification of the diverse array of living things, categorizes organisms based on their common ancestry.

OUTLINE

Scientific Process: The Unifying Theme Among All Sciences

Although scientists and "their" science may differ in their focuses, scientists—biologists, chemists, physicists, and so forth—study the *natural world* and share scientific methods that usually include the following key features: *observation, hypothesis development,* and *hypothesis testing.* A **hypothesis** is a tentative explanation or prediction that guides inquiry. Scientists test their hypotheses in a variety of ways, producing results that are open to verification by others in their disciplines. Because a hallmark of science is the testable hypothesis, science does not try to explain philosophical or religious questions such as "What is the meaning of life?" or "Does life exist after death?" within the framework of scientific inquiry.

As just noted, an important process of science is observation, but observation alone does not constitute science. The predecessors of scientists—the ancient Greeks and Romans such as Plato and Aristotle—were skilled observers, but they are usually referred to as *naturalists* rather than scientists. Their "science" was called *natural history.* Until the mid-1800s, the study of life (the emerging science of biology) continued to have a natural history perspective, focusing on descriptions and comparisons of organisms and providing explanations to questions based on observation alone. At this time, inquiry in the physical sciences often involved hypothesis testing, but many naturalists felt that the living world transcended this mode of inquiry. A critical shift in biological investigation occurred during the nineteenth century: a preoccupation with form moved to an investigation of function and process. Biologists went beyond observation to put forth hypotheses and test them.

> *Scientists study the natural world using a variety of scientific methods. These methods usually include observation, hypothesis formation, and hypothesis testing.*

Developing Hypotheses

The pathways of thinking that scientists follow in developing hypotheses vary greatly. One way scientists formulate hypotheses is by induction, or **inductive reasoning,** a pattern of thought in which a person develops generalizations from specific instances. For example, before formulating hypotheses, scientists may gather information regarding their questions or observations by searching the scientific literature and synthesizing the information they find. They develop their hypotheses based on information and observations of specific instances of a phenomenon. For example, you may observe that many of your friends grow healthy and robust houseplants. Because you struggle to keep yours alive, you might ask them their "secret." Many may answer that one key to growing healthy houseplants is giving them regular doses of Brand X fertilizer. You may have also read advertisements and other literature stating that Brand X fertilizer causes houseplants to grow 30% taller and produce healthier, greener plants. From these observations and testimonials, you might make the generalization that Brand X fertilizer helps houseplants grow well. This generalization may be stated in an "If . . . then . . ." format. Your hypothesis might be the following: *If* houseplants are given Brand X fertilizer, *then* they will grow at least 30% taller and be greener than plants with no fertilizer. As you can see, a hypothesis can be a *prediction.*

Scientists often develop hypotheses in ways other than by inductive reasoning. (In fact, most philosophers of science would say that scientists rarely if ever develop hypotheses by induction.) Often, hypothesis formation is simply the result of a hunch, a guess, or an idea springing to mind for no apparent reason. Creativity and imagination often spawn hypotheses, as do accidents. A classic accident that resulted in a scientific breakthrough was the discovery of the antibiotic, penicillin, in 1928 by Sir Alexander Fleming. While performing experiments with *Staphylococcus* bacteria, Fleming noted that some of his bacterial cultures had become contaminated with mold. Surrounding the mold colonies were clear zones where no bacteria grew. Fleming hypothesized that the mold produced an antibacterial substance. By the early 1940s other scientists had purified penicillin, demonstrated its potency, and developed preparations that could be injected into humans to fight certain bacterial infections.

> *A hypothesis is a tentative explanation or prediction that guides the investigation of a question. Scientists develop hypotheses in a variety of ways.*

Testing Hypotheses

Scientists develop hypotheses, predictions, and explanations of observed phenomena in myriad ways and test them in myriad ways also. Testing a hypothesis involves producing data or evidence that either supports or refutes the prediction made in the hypothesis. This "testing" process involves the use of deduction, or **deductive reasoning.** Deduction begins with a general statement (the prediction about Brand X fertilizer and its effects on houseplants, for example) and proceeds to a specific statement (the effects of Brand X fertilizer on *your* houseplants).

Although induction and deduction may seem to be separate patterns of reasoning, they are not. While scientists are formulating hypotheses, they also think about ways to test their hypotheses and what might happen as a result. Then, as scientists test their hypotheses, they compare their mental arguments for or against their hypotheses with what their testing actually shows.

The ways in which a hypothesis is tested and the ways in which data and evidence are collected depend on the hypothesis itself and on the scientist doing the testing. For example, paleontologists often test their hypotheses about prehistoric life by gathering evidence from living ancestors of prehistoric species or by gathering evidence from fossils, the preserved remains (or other traces) of organisms. Some population biologists test their hypotheses regarding interactions among populations using computer (mathematical) models. Biomedical researchers may test hypotheses by comparing two or more groups of subjects with regard to an intervention that occurs naturally, such as comparing the percent of unimmunized persons contracting the flu with the percent of immunized persons contracting the disease.

Many scientists test hypotheses by means of *controlled experiments.* In a controlled experiment, a researcher manipulates (changes) a single factor and observes other factors that change in response to the manipulated factor. All other factors are kept constant (consistent) throughout the experiment; that is, they are controlled. In analyzing the simplified hypothesis in our previous houseplant scenario (hypotheses can be much more complex), notice that the factor you are manipulating (the fertilizer) is stated after the "if" part of the hypothesis. This factor is called the **independent variable,** or manipulated variable. (The word *variable* refers to a factor that changes, or varies.) To test your hypothesis, you will need to add different amounts of Brand X fertilizer to groups of houseplants. Also notice that the factor stated after the "then" part of the hypothesis depends on the manipulated (independent) variable. This factor is called the **dependent variable,** or responding variable. It varies in response to changes in the independent variable. In your experiment, plant growth and color are dependent variables.

The variables to be kept constant include any factors that can change in the experimental setup. In this case, they include the type of soil, the amount of time the plants are exposed to the light, the temperature in which the plants are grown, and the type of plant tested. Can you think of any other variables that may need to be controlled?

As you design a controlled experiment to test your hypothesis, you should first outline how you would manipulate the independent variable. In this case, you may decide to give one group of plants the "dose" of fertilizer suggested on its container and to give some plants slightly more and others slightly less. If the suggested dose is 30 milliliters (ml) of properly diluted liquid fertilizer, you may decide to give a second group of plants 15 ml, a third group of plants 45 ml, and a fourth group of plants 60 ml. Each group of plants should be the same type of plant, such as all periwinkle plants of approximately the same size. You should pot them all in the same soil in the same type and size of pot, place them approximately the same distance from the same light source, and water them all with the same amount of water at the same time. These plants will receive the *treatment*—in this case, the fertilizer. They are the experimental plants. A fifth group of plants should be treated the same as the others *except* that it will not receive the treatment (fertilizer). This group serves as the **control** (Figure 1-1). The control is the standard against which the treatment plants can be compared. Any changes in the control are the result of factors other than the treatment.

Over a preselected time, you should observe your plants and collect *data* regarding leaf color and plant height. (You could also choose other indica-

hypothesis (hi **poth** uh sis) a tentative explanation or prediction that guides scientific inquiry.

inductive reasoning a pattern of thought in which a person develops generalizations from specific instances.

deductive reasoning a pattern of thought in which a person begins with generalizations and proceeds to specific instances.

independent variable the factor that is manipulated during a controlled experiment.

dependent variable the factor that varies in response to changes in the independent variable during a controlled experiment.

control the standard against which experimental observations or conclusions may be checked to establish their validity.

FIGURE 1-1 A controlled experiment to determine how fertilizer influences plant growth. A, The independent or manipulated variable is the amount of Brand X fertilizer added to each plant. The dependent or responding variable is the amount of plant growth. All other variables such as temperature, type of soil, amount of light, type of plant, and amount of water will remain constant. **B,** Setup of a controlled experiment showing the amount of fertilizer (independent variable) added to each plant except for the control plant.

tors of plant growth such as number of leaves.) The data regarding the height of your plants are called *quantitative data* because these data are based on numerical measurements (centimeters, for example). The data regarding leaf color are called *qualitative data* because these data are descriptive and not based on numerical measurements.

At the beginning of your experiment, you should measure the height of each plant in each group and record these data in a data table. Find the *mean* (average) height of the plants in each group by summing the heights of the plants and dividing each sum by the number of plants. Record your data for each group. At the end of your predetermined growing period, 6 weeks for example, you measure plant height again, determine the mean heights for each group, and record these data. Your summary data table may look like this:

MEAN PLANT HEIGHT IN CENTIMETERS (cm)

Am't of Fertilizer	0 ml	15 ml	30 ml	45 ml	60 ml
0 weeks	145	150	148	151	149
6 weeks	166	180	181	179	184
Increase in plant height	21	30	33	28	35

Regarding the collection of qualitative data, you should determine whether the color of the leaves of each plant is yellow, light green, medium green, or dark green. Record these data in a data table. Find the *mode* (the value that occurs most often) for each group. Record the mode for each group in a summary data table. Do the same at the end of your predetermined growing period. Your summary data table may look like this:

DESCRIPTION OF LEAF COLOR (MODE FOR EACH GROUP)

Am't of Fertilizer	0 ml	15 ml	30 ml	45 ml	60 ml
0 weeks	mg	mg	mg	mg	mg
6 weeks	lg	mg	dg	dg	dg
Change in leaf color	neg	nc	pos	pos	pos

KEY: y = yellow; mg = medium green; lg = light green; dg = dark green; neg = turned lighter (negative change); pos = turned darker (positive change); nc = no change

Do your data support your hypothesis? If I add Brand X fertilizer to the soil of my houseplants, they will grow at least 30% taller and be greener than plants with no fertilizer. Looking at the qualitative data, the plants with fertilizer were greener than the control with no fertilizer. In this experiment, 30 ml to 60 ml produced the greenest plants.

Looking at the quantitative data, can you determine if the plants were at least 30% taller than the control? First, let's compare the plants treated with 15 ml fertilizer with the control plants. The control plants grew 21 centimeters and the test plants grew 30 centimeters—9 centimeters more than the control. Is this difference at least 30% more (30% greater than the control growth)? This question is really asking "What percent of 21 is 9?" To find out, you divide 9 by 21, which yields 0.43. Multiplying by 100 to change this number into a percentage yields 43%. In other words, the plants receiving 15 ml of Brand X fertilizer grew 43% taller than the control! Using the same reasoning (1) the plants receiving 30 ml of Brand X fertilizer grew 57% taller, (2) the plants receiving one and 45 ml grew 33% taller, and (3) the plants receiving 60 ml grew 67% taller. In summary, these data do support your hypothesis; in fact, your test data suggest that Brand X fertilizer may help houseplants grow even greener and longer than the manufacturer suggests!

Why do you think the manufacturer of Brand X fertilizer is not making greater claims regarding this product? Your experiment certainly shows much greater than a 30% increase in growth. The answer lies in the limitations of your experiment. First, you experimented with one type of houseplant. Brand X fertilizer is probably sold for use on a variety of houseplants. Not all plants may grow as vigorously in response to the fertilizer as your plants. Second, you probably experimented on five small groups of plants. Scientists often perform experiments over and over again, or *repeated trials,* before drawing conclusions from their data. Using groups of plants for each test dose of fertilizer is considered repeated trials, but re-running the experiment with larger numbers of plants would yield more reliable data. The Brand X fertilizer manufacturing company probably performs hundreds of experiments on a variety of houseplants before drawing conclusions from their data and generalizing to the population of houseplants grown throughout the United States (or possibly throughout the world).

Evidence that contradicts a hypothesis shows it to be false; that is, it disproves the hypothesis. However, evidence that supports a hypothesis does *not* establish that further testing will also produce supporting evidence. Unforeseen factors may affect the outcome of a future experiment. In your experiment, for example, the data collected using your plants supported your hypothesis. But data collected using other plants in someone else's home may not support your hypothesis. In summary, then, scientists cannot actually *prove* hypotheses (or anything!); they can only *disprove* hypotheses or *support* hypotheses with evidence.

Theory Building

As scientists repeat, or replicate, each other's work, their data may uphold a hypothesis again and

Processes and Methodologies

HOW *Science* WORKS

Is There Bias in Science?

Do you think the following statements are true or false?

(1) The work of the scientist is objective and controlled.

(2) Scientists use methodologies to remove any personal biases they may have.

(3) Scientists have rules that govern how they make decisions, and they abide by those rules flawlessly.

Are the above statements a true rendition of how science works? No! Are scientists charlatans, then, out to pull the wool over each one of our eyes? No! That statement is untrue also. But in science, as in investigation in any discipline, one fact has an impact on what humans do and how they go about their work: their past.

Everything we observe, whether it be a sight, sound, smell, or touch, is processed through the cognitive filter of our past. No one can make an observation that is not biased by the context of their education, their prior work, and other factors. And by its very nature, scientific inquiry starts with expectations of the outcome; scientists call these expectations hypotheses. As scientists develop their hypotheses and design studies to test them, they give relevance to certain observations and choose methods that appear appropriate based on their prior knowledge and experience. As data are collected and analyzed, a scientist makes decisions regarding how the data relate to one another, or which data are useful. You may be thinking that these types of decisions are simply subjective decisions based on personal judgment. You may be right . . . but when does subjectivity become bias in science?

Stephen Jay Gould, a research scientist and Harvard professor, provides an interesting example that probes this question in his book *The Mismeasure of Man*. He relates the story of Paul Broca, one of the many 19th century scientists who were engaged in craniometry: the study of the brains of human races. Broca, a professor of clinical surgery in Paris, hypothesized that the size of a brain was related to its degree of intelligence. Broca gathered data to test his hypothesis by pouring lead shot into skulls to measure their cranial capacities. He used meticulous techniques and repeated them with each measurement. In addition, he collected data by weighing brains himself directly after an autopsy. Ostensibly, he conducted his studies with careful controls of his procedures so that bias and error would not affect his measurements. However, as Dr. Gould points out, "Broca's cardinal bias lay in his assumption that human races could be ranked in a linear scale of mental worth" (p. 86). Gould goes on to suggest that Broca's bias influenced the type of data he collected as he *subconsciously* searched for those data that would support his hypothesis. In addition, Gould asserts that Broca subconsciously manipulated the data he did collect so that they would support his hypothesis. If the data clearly did not support his hypothesis, he would develop other criteria that would supersede the importance of the data that didn't "fit." A recent re-evaluation of Broca's data suggests that his data are reliable, but his interpretations of the data are unsubstantiated. In other words, brain size is not related to intelligence.

Scientists today usually try to eliminate as much bias in their work as is possible. To evaluate the effects of a new drug, for example, medical researchers often choose a double-blind methodology. In such a study, half of the participants are randomly chosen to receive the new drug. The other half of the participants receive a placebo, a pill that resembles the drug but contains no medication. Neither the researchers nor the participants know which pill is the drug (thus, double-blind); the pills are stamped with code letters by the drug company and the code is revealed only after the study ends.

Recently, the American Association for the Advancement of Science published the *Benchmarks for Science Literacy*, a set of guidelines for teachers of science. This document approaches the topic of bias in science in this way:

> Scientists in any one research group tend to see things alike, so even groups of scientists may have trouble being entirely objective about their methods and findings. For that reason, scientific teams are expected to seek out the possible sources of bias in the design of their investigations and in their data analysis. Checking each other's results and explanations helps, but that is no guarantee against bias. (p. 13)

Processes and Methodologies—cont'd

The cold fusion debate of recent years is tangible evidence that there is no guarantee against bias in science. In the spring of 1989, B. Stanley Pons of the University of Utah and Martin Fleischmann of Southampton University announced that they created a room-temperature (cold) nuclear fusion reaction—one that produced more energy than was initially needed to power the reaction. If what they claimed was true, Pons and Fleischmann could be the recipients of a Nobel Prize and this energy-producing reaction could be a boon to humankind. However, after their work was published, it was challenged by other researchers who could not reproduce their results. Nevertheless, the National Cold Fusion Institute (NCFI) opened in Salt Lake City in August of 1989, supported by state funds. By the fall of 1990, an external scientific review of the NCFI concluded "that neither the institute's various experiments nor those anywhere else have firmly established the existence of cold fusion." In January of 1991, Pons resigned from the University of Utah and in June of that year the NCFI closed. Cold fusion still has enthusiastic supporters, however, who attend conferences on the topic and present data they say support the cold fusion claim. So the cold fusion debate goes on and keeps the scientific community ever aware that they, too, must continually evaluate their work and others' in light of the threats of subjectivity, sloppy science, fraud . . . and bias in science.

again. As the explanations of such consistently supported and related hypotheses are woven together, "grand explanations" that account for existing data and that consistently predict new data are developed. Interwoven hypotheses upheld by overwhelming evidence are called theories. The word *theory* in a scientific context has a much stronger meaning than the everyday use of the term. A **scientific theory** is a synthesis of hypotheses and is a powerful concept that helps scientists make predictions about the world. Theories are supported by such an overwhelming weight of evidence that they are accepted as scientifically valid statements. However, analyses of data from different scientific perspectives can result in opposing theories based on the same evidence. Often, continued analyses over time result in one scientific theory prevailing over opposing scientific theories. In addition, the possibility always remains that future evidence will cause a theory to be revised, since scientific knowledge grows and changes as new data are collected, analyzed, and then synthesized with previous information.

Scientific theories *never* become scientific laws. **Scientific laws** describe natural phenomena. For example, the law of inertia states that an object in motion tends to stay in motion and an object at rest tends to stay at rest. In general, theories explain what laws describe.

The Unifying Themes of Biology Today

Biology is often viewed by the nonscientist as an accumulation of facts, but it is much more than that. As the beginning of this chapter shows, biology (as well as all sciences) is a way, or process, of understanding the world. As a result of observing living things, their environments, and the interactions between them, scientists have posed questions, have put forth hypotheses based on those observations, and have tested their predictions of the living world. Rising out of this abundance of tested predictions are the themes, or accepted explanations, that permeate the science of biology. Although the specific details of these themes may be updated or changed as biologists modify their hypotheses, the themes themselves transcend time.

Living Things Display Both Diversity and Unity

The diversity, or variety, of living things is astounding. Biologists estimate that from 5 million to 30 million *different* species exist on Earth. As you can see from this range of numbers, scientists are unsure how many species exist; they have discovered, described, and catalogued probably less

scientific theory a synthesis of hypotheses; a powerful concept that helps scientists make predictions about the world.

scientific laws descriptions of natural phenomena.

than half of them. Unfortunately, many species are becoming extinct, or dying out, as their habitats are destroyed, before scientists can even study and classify them.

A **species** is a population of organisms that interbreed freely in their natural settings and do not interbreed with other populations. Put simply, members of one species are reproductively isolated from members of other species. Organisms that do not reproduce sexually such as bacteria and certain plants, animals, protists, and fungi are designated as species by means of their morphological characteristics (those relating to their form and structure) and biochemical characteristics (those relating to their chemical composition and processes).

Only a sampling of the array of the species humans have seen and categorized is shown in Figure 1-2. Although these organisms are very different from one another, each has characteristics that are common to all species. (These shared characteristics, however, may or may not be visible to the naked eye.) The remaining themes describe how and why organisms are diverse in their types but unified in their patterns.

Living Things Are Composed of Cells and Are Hierarchically Organized

A cell is a microscopic mass of protoplasm, a chemically active mixture of complex substances suspended in water that is bounded by a membrane and contains hereditary material. Until the invention of the microscope around 1600, naturalists could not probe this invisible level of organization of living things. And it was not until the mid-1800s when advances in the technology of the microscope allowed botanist Matthias Schleiden and zoologist Theodor Schwann to determine (through repeated lines of inquiry) that the unit of structure of all living things is the **cell.** Shortly thereafter, the German medical microscopist Rudolf Virchow argued that all cells can arise only from preexisting cells. The cells, he wrote, are "the last constant link in the great chain of mutually subordinated formations

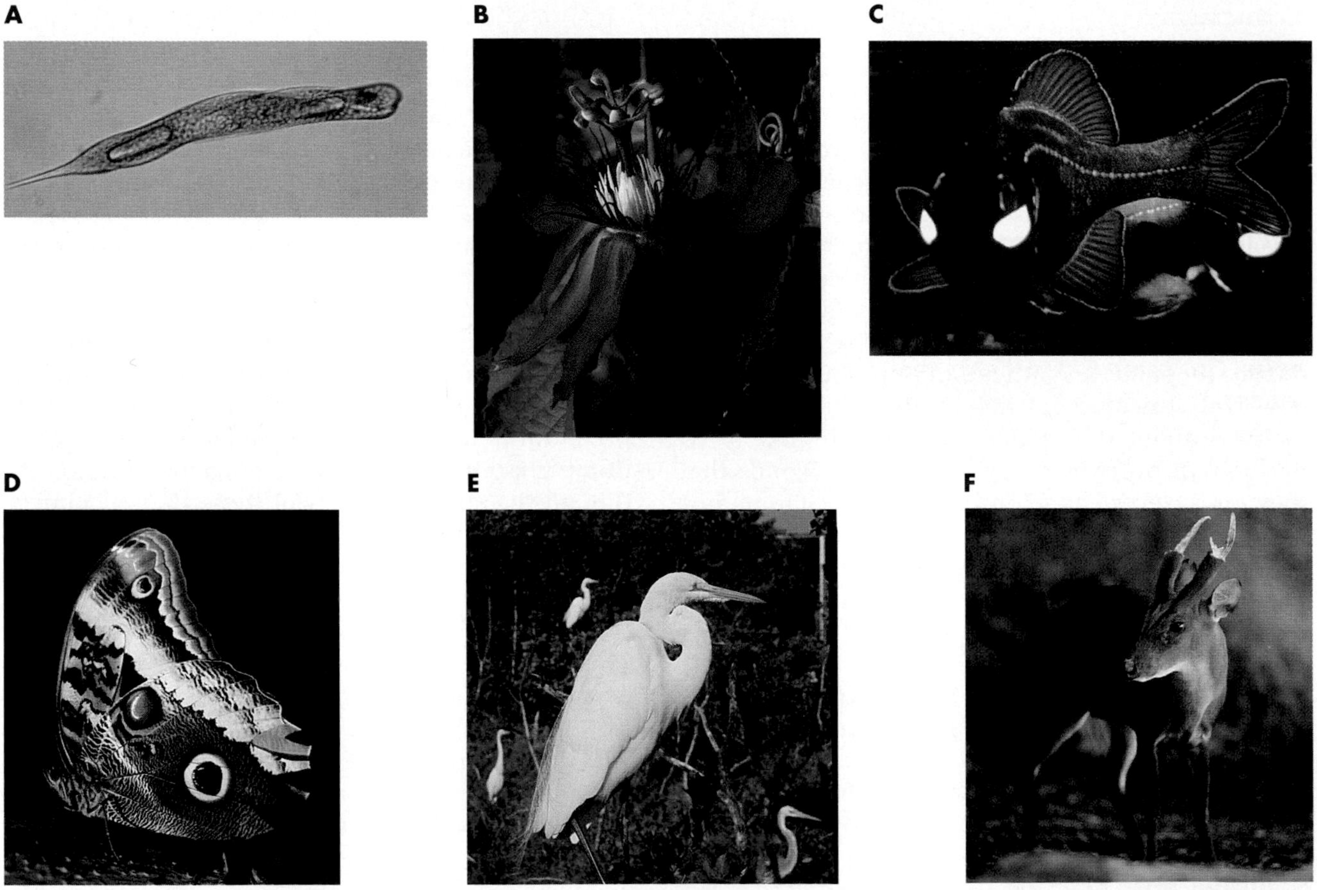

FIGURE 1-2 The diversity of species. A sampling of the variety of species found on Earth. **A,** Euglena, a type of protozoan. **B,** Red passion flower. **C,** Flashlight fish. **D,** Owl butterfly. **E,** Egret. **F,** Muntjac, "barking deer" from southeastern Asia and the East Indies.

that form tissues, organs, systems, the individual. Below them is nothing but change."

Virchow's statement refers to the *hierarchy,* or levels, of organization seen in all living things. Living things, or **organisms,** are either multicellular (composed of many cells) or unicellular (composed of a single cell). Some organisms called *colonial organisms* are single-celled organisms that work and live together as a team. But whether unicellular, multicellular, or colonial, every organism has the cell as its simplest level of structure and function. Smaller units, such as atoms and molecules, make up cells but are not *living* units.

As shown in Figure 1-3, cells in multicellular organisms are organized to form the structures and perform the functions of the organism. The next more complex, or more inclusive, level of organization is the tissue. **Tissues** are groups of similar cells that work together to perform a function. Grouped together, tissues form a structural and functional unit called an **organ.** An **organ system** is a group of organs that function together to carry out the principal activities of the organism—its most complex level of organization.

Interactions take place within an organism among its levels of organization. For example, when a fox sees and then chases a mouse, a complex series of events occur within the fox. First, the nerve *cells* embedded in the back wall of its eyes (*organs*) conduct impulses to its brain (an *organ*). The vision center within the brain (*nervous tissue*) interprets these impulses, resulting in the fox seeing the mouse. Impulses speed to other brain centers (*nervous tissues*), which integrate this visual message with messages from olfactory *cells* in its nose. The brain coordinates these impulses, then sends out a response in the form of nerve impulses to the muscles (*organs*). Skeletal muscles contract, allowing the animal to chase after its prey.

Living Things Interact with Each Other and with Their Environments

Interactions occur not only among the levels of organization within organisms but also between organisms and their external environments. Biologists usually classify the interactions between organisms and their environments into the following hierarchy of levels of organization: populations, communities, ecosystems, and the biosphere. These levels of organization build on the levels of organization of individual organisms and are depicted in Figure 1-3.

A **population** consists of the individuals of a given species that occur together at one place and at one time. To continue the example, the foxes living in a small forest make up a population. The foxes interact with one another in a variety of ways. Sometimes they compete for the same limited resources and for mates. Conversely, individuals within animal populations may also work cooperatively for common purposes. Foxes often hunt together, for example, working with one another to overtake and trap prey.

Populations of different species that interact with one another make up a **community** of organisms. A forest community may be made up of populations of bacteria, fungi, earthworms, plant-eating and animal-eating insects, mice, deer, salamanders, foxes, snakes, hawks, trees, and grasses. These populations within the forest community compete with one another for resources as do organisms within populations. The foxes, snakes, and hawks, for example, compete with one another to capture the mice for food. Other types of interactions may also exist within the community, such as mutualistic relationships in which two different species live in a close association for the benefit of both. Fungi and plants are often interdependent on one another. The fungi live near the roots of many plants, such as trees in a forest. The fungi envelop the roots and send billions of minute cell extensions into the soil. These microscopic "fingers" of fungus absorb water and nutrients better than the roots could alone, and they pass the substances to the plant. In turn, the fungus uses certain products that the plant makes by photosynthesis.

An **ecosystem** is a community of plants, animals, and microorganisms that interact with one

species (**spee** shees *or* **spee** sees) a population of organisms that interbreed freely in their natural settings and do not interbreed with other populations. The most narrow and specific taxonomic classification; a taxonomic subcategory of **family.**

cell a microscopic mass of protoplasm; the unit of structure of all living things.

organisms (**or** guh **niz** ums) living things. Organisms can be either multicellular or unicellular.

tissues groups of similar cells that work together to perform a function.

organ a group of tissues that forms a structural and functional unit.

organ system a group of organs that function together to carry out the principal activities of an organism.

population a group that consists of the individuals of a given species that occur together at one place and at one time.

community populations of different species that interact with one another in a particular place.

ecosystem a community of plants, animals, and microorganisms that interact with one another and their environments and that are interdependent.

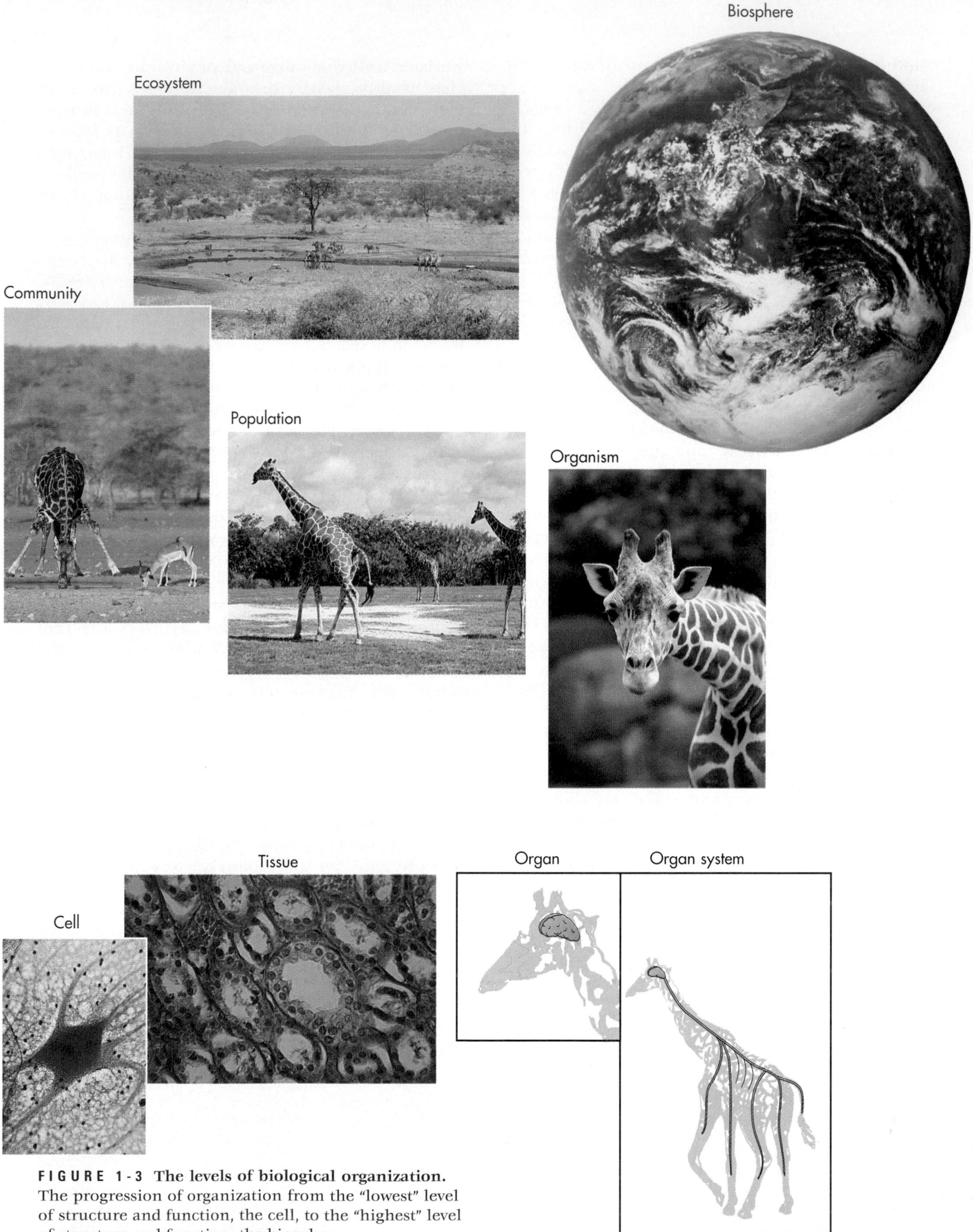

FIGURE 1-3 The levels of biological organization. The progression of organization from the "lowest" level of structure and function, the cell, to the "highest" level of structure and function, the biosphere.

another and their environments and that are interdependent. The forest ecosystem includes all the organisms previously discussed as well as nonliving components of the environment such as air and water, which contribute substances needed for the ecosystem to function.

The **biosphere** is the part of the Earth where biological activity exists. Most people refer to the biosphere as simply "the environment." Within this global environment, living things interact with each other and with nonliving resources in myriad ways. Organisms other than humans use only *renewable* (replaceable) *resources.* When they die, decomposers return the nutrients held within the dead organisms to the soil and air. Humans, however, use many *nonrenewable resources* and fill the land with wastes. Scientists are testing and refining alternative renewable energy sources such as solar, wind, and water energy to help curb the use of nonrenewable energy sources such as coal, oil, and natural gas. The most severe crisis of natural resource destruction is occurring today in the tropical rain forests. Species are dying out as their rain forest habitats, or homes, are lost. In addition, humans are overpopulating the land and polluting the water and the air. The destiny of future generations of all living things depends on our solving the problems humans have created and our learning to live on this planet without harming it and everyone's future existence.

Living Things Transform Energy and Maintain a Steady Internal Environment

Organisms need energy to do the work of living. This work involves many processes, such as movement, cell repair, reproduction, and growth. It also involves maintaining a stable internal environment in spite of a differing external environment. Energy drives the chemical reactions that underlie all these activities.

What is the source of the energy that fuels life processes and helps organisms maintain an inner equilibrium? Ultimately, it comes from the sun. For example, certain organisms within our forest ecosystem such as the trees and grasses make their own food by capturing energy from the sun in a process called **photosynthesis.** These organisms are called **producers.** During photosynthesis, producers (primarily green plants) convert the energy in sunlight into chemical energy by locking it within the bonds of the food molecules they synthesize. Organisms that cannot make their own food, such as the insects, mice, deer, salamanders, foxes, snakes, and hawks in the forest, are called **consumers.** They feed on the producers and on each other, passing energy along that was once captured from the sun (Figure 1-4). Both the producers and the consumers release the stored energy in food by breaking down its molecules bit by bit. As food molecules are broken down, much of the energy that is released is used to do work, but some of it is lost as heat and is therefore unusable. Consequently, organisms need a continual input of energy to fuel the chains of chemical reactions that move, store, and free energy needed to perform the activities of life. **Decomposers** such as many types of bacteria and fungi break down the organic molecules of dead organisms, serving as the last link in the flow of energy through an ecosystem and contributing to the recycling of nutrients within the environment.

Living Things Exhibit Forms That Fit Their Functions

Would it make sense to try to turn a screw with a hammer or eat soup with a knife? Of course not—tools and kitchen utensils are structured in specific ways to do specific jobs. Just as the shape and structure of a spoon or a screwdriver fits its function, the structures of living things fit their functions. Biologists sum up this idea with the phrase "form fits function." By analyzing form, inferences can be made regarding function. Conversely, knowing function gives insights into form.

For example, look at the variety of bird feathers shown in Figure 1-5. By analyzing their forms, determine which is used to insulate the bird against the cold. First, determine the characteristics of a good insulator from your everyday experience. Are you warmer in cold weather when you wear one thick sweater, or are you warmer when you wear many thinner layers that together may have the same thickness as the sweater? The answer is that

biosphere the part of the Earth where biological activity exists.

photosynthesis (**fote** oh **sin** the sis) a process by which certain organisms make their own food by capturing energy from the sun.

producers organisms that make their own food using the process of photosynthesis.

consumers organisms that cannot make their own food and must feed on producers or other consumers.

decomposers organisms that break down the organic molecules of dead organisms.

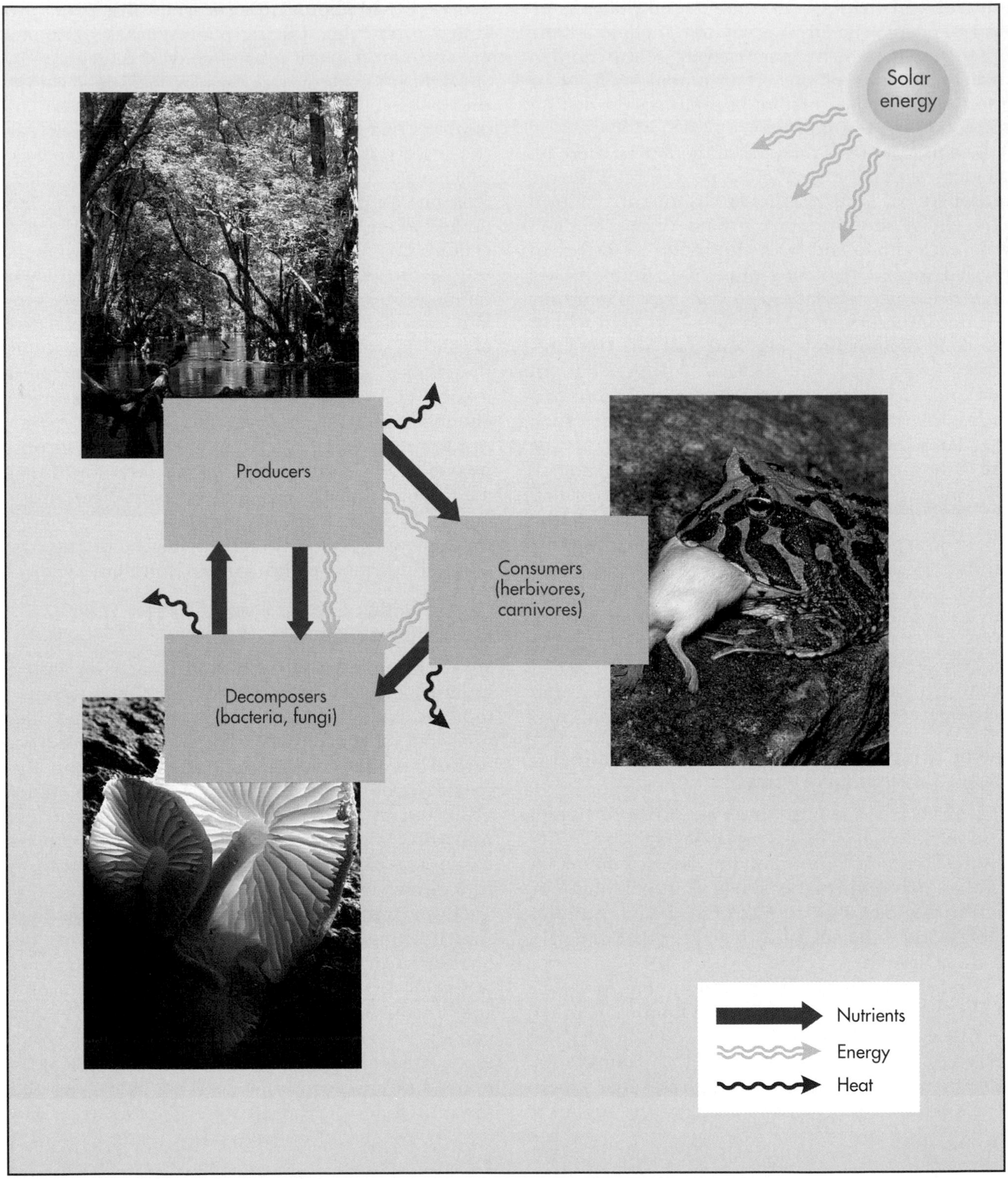

FIGURE 1-4 Flow of energy through an ecosystem. Producers transform the sun's energy (*yellow wavy lines*) to chemical energy by means of photosynthesis. Nutrients (*blue arrows*) are then transferred from producer to consumer and from consumer to consumer. Decomposers such as fungi break down the organic molecules of dead organisms, making these nutrients available for re-use. Some energy is lost as heat (*brown wavy lines*); therefore organisms need a constant input of energy to perform the activities of life.

A

B

C

FIGURE 1-5 Different types of feathers: form fits function. By examining the feathers shown here, can you determine which feather is the best insulator? The feather in **A** is used for flight. The feather in **B** is a display feather and is used by peacocks to attract females (peahens). The feather in **C** is a down feather and is used for insulation. Arranged in layers, these feathers trap air that is warmed by the bird's body.

many layers provide better insulation against the cold because they trap air between the layers. This trapped air is then warmed by the heat radiating from your body. In essence, you create a blanket of warm air around your body. Looking at the bird feathers, which feather might trap air within its structure?

Living Things Reproduce and Pass on Biological Information to Their Offspring

All organisms give rise to organisms of their same kind—they reproduce. They reproduce both sexually and asexually. In sexual reproduction, two parents give rise to offspring; in asexual reproduction, only one parent gives rise to offspring. In either case, genes—the units of heredity—are passed from one generation to the next.

Genes are made up of molecules of **DNA,** or deoxyribonucleic acid. In these molecules of DNA lies the "code of life," instructions that are translated into a working organism. Interestingly, these instructions take the form of molecular subunits of the DNA molecule called *nucleotides.* Using only four different nucleotides, DNA codes for all the structural and functional components of an organism.

The secret to DNA's ability to carry information regarding the variety of structural and functional components lies in its code. The four nucleotides of DNA are used like code letters to produce code words, each composed of three letters. These code words are then sequenced to produce code sentences, which translated in a living organism results in the formation of a specific molecule designed to do a specific job. Thus the diversity of life is produced by the same code letters that produce code words and code sentences unique to each of the many species of organisms of the living world.

Living Things Change Over Time, or Evolve

All organisms have common characteristics because they are related to one another. Just as you have a history and a family tree, so does the Earth's family of organisms. Yet all organisms have differences because as they changed over time, they diverged from one another. The Earth itself has changed from its beginnings some 4.6 billion years ago.

Scientists know very little about what "incubator Earth" was like nearly 4 billion years ago, but they

deoxyribonucleic acid (de ok see **rye** boh new **klay** ick) **(DNA)** molecules that make up genes and constitute the "code of life"; DNA codes for all the structural and functional components of an organism.

do agree that it was a harsh environment different from today's. Under these conditions, many scientists hypothesize that the elements and simple compounds of the primitive atmosphere reacted with one another, forming complex molecules. In a way that scientists can only hypothesize, biochemical change took place over time and resulted in the appearance of single-celled organisms approximately 3.5 billion years ago. The remains of these early cells (and of any organisms) preserved in rocks are called *fossils.* Fossils provide scientists with a record of the history of life and document the changes in living things that have taken place over billions of years. By 1.4 to 1.2 billion years ago, the fossil record documents the existence of cells more complex than the first cells, and by 500 million years ago, an abundance of multicellular organisms—with members of groups similar to those that exist today—had appeared.

The fossil record is only one piece of evidence suggesting that living things have changed over time. Using this and other types of evidence, Charles Darwin, a nineteenth century English naturalist, developed what was then a hypothesis of organismal change over time or, as he put it, "descent with modification." He termed his hypothesis **evolution.** Darwin also proposed a mechanism by which evolution took place. Since Darwin's time, his hypothesis has been consistently supported by an overwhelming amount of scientific data and has therefore become a well-accepted theory. This theory embodies the ideas that organisms alive today are descendants of organisms that lived long ago and that organisms have changed and diverged from one another over billions of years. Scientists still ponder, examine, and develop hypotheses regarding details of the mechanisms of evolution but agree that evolution has and is taking place.

Classification: A Reflection of Evolutionary History

As in any family tree, some organisms are more closely related than others. The species of the world are no exception. In the family tree of life, organisms having a common ancestor in the not-too-distant past are said to be *closely related.* Organisms having a common ancestor farther down the family tree are said to be *distantly related.* Modern **taxonomy,** the classification of the diverse array of species, categorizes organisms based on their common ancestry (their evolutionary history). Therefore, organisms that are close relatives and resemble one another in a variety of ways are placed in common groupings reflecting their similarities and closeness; organisms not closely related are placed in groupings separate from one another that reflect their differences.

Taxonomists group all living things into broad categories called **kingdoms.** However, taxonomists are divided into various schools of thought regarding their methods. The taxonomic scheme used in this book recognizes five kingdoms of life: Monera, Protista, Plantae, Fungi, and Animalia (Figure 1-6). Kingdom Monera includes the bacteria. This kingdom differs from the other four kingdoms in that its cells lack membrane-bounded intracellular structures called *organelles.* In addition, the hereditary material of the cell is not bounded by a membrane. Such cells are called *prokaryotes.* Finding these organisms is an easy task—bacteria live almost everywhere. If you took a moist cotton swab and drew it across any surface (except one that had been sterilized), you will collect an array of bacteria that is visible under a microscope. The organisms of the other four kingdoms are made up of cells that do have organelles and a membrane-bounded nucleus. Such cells are called *eukaryotes.*

Ancestors of the monerans gave rise to organisms in the kingdom Protista. The protist kingdom consists of primarily single-celled eukaryotes, whose cells are much more complex and much larger than the bacteria. By collecting a small amount of pond water and looking at it under a microscope, you can see some of the organisms in this kingdom. Some of the differences between the single-celled protists and the bacteria can be easily seen in Figure 1-6. This kingdom also includes the algae, which include both multicellular and unicellular organisms that seem to be more closely related to the protists than to the plants, animals, or fungi.

The remaining three kingdoms originated from the protists. The plants are multicellular organisms that live on land and make their own food by using carbon dioxide from the air and light energy from the sun. Examples of plants are mosses, ferns, pine trees, and flowering plants. Animals are multicellular organisms that cannot make their own food and that obtain food by eating and digesting other organisms. Fungi are decomposers (as are many species of bacteria) and survive by breaking down substances and absorbing the breakdown products. Many of these multicellular eukaryotes live off organisms that are no longer alive, such as fallen trees and leaves and dead animals. Some fungi, however, attack living organisms, causing human diseases or conditions such as athlete's foot and ringworm or plant diseases such as potato blight. The kingdoms of organisms are further sub-

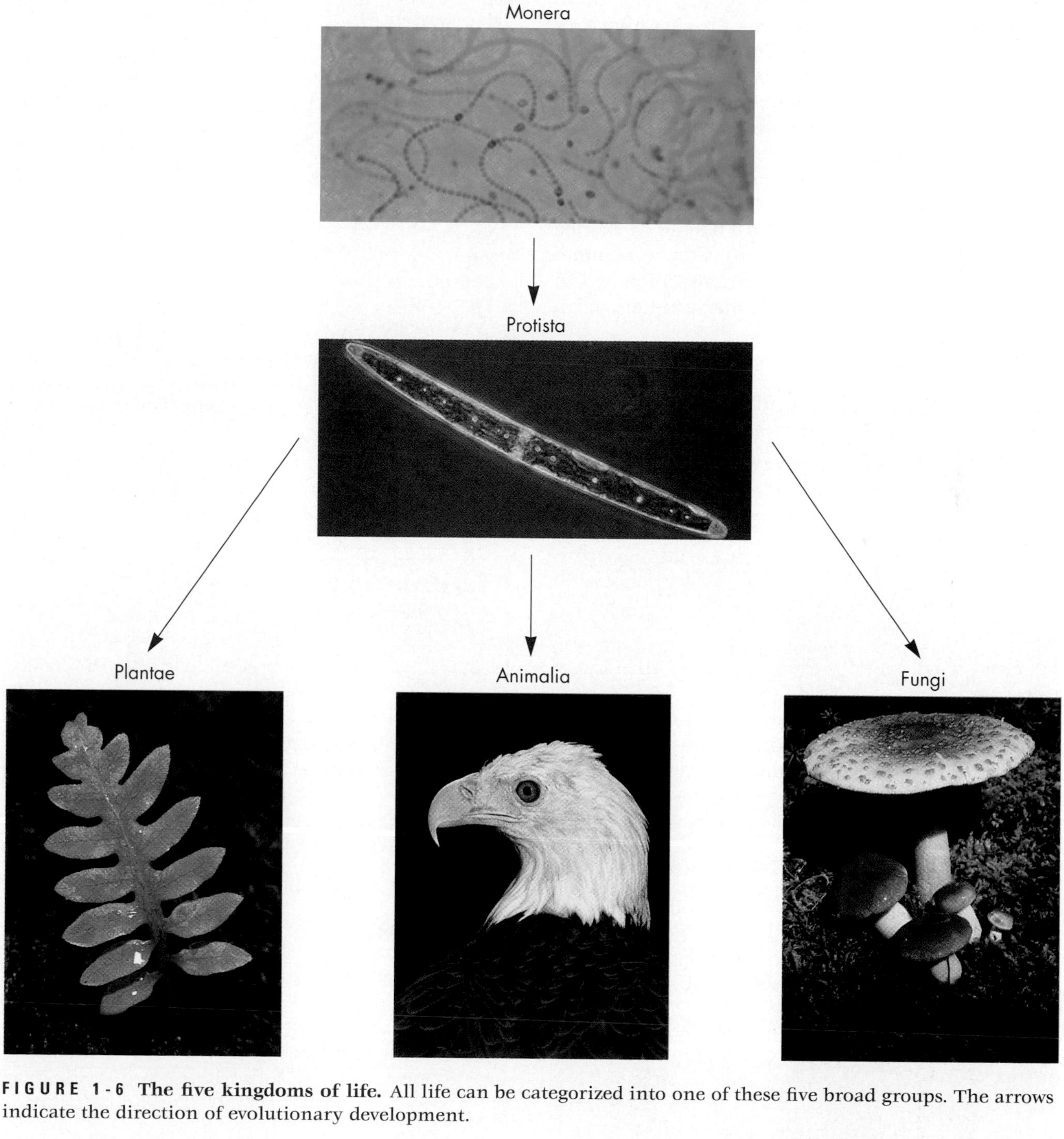

FIGURE 1-6 The five kingdoms of life. All life can be categorized into one of these five broad groups. The arrows indicate the direction of evolutionary development.

divided into groupings that reflect an increasing closeness in evolutionary history as determined by many types of evidence, including organisms' similarities in their existing behavioral (in the case of animals) and biochemical characteristics. Kingdoms Monera, Plantae, and Fungi are each subdivided into **divisions.** Kingdoms Protista and Animalia are each subdivided into **phyla** (singular, phylum). (These two subdivisions are taxonomic equals.) The animal kingdom, for example, has approximately 19 phyla depending on the particular classification system that is referenced. Figure 1-7 shows some representatives of the animal kingdom. One of the phyla within this kingdom is the phylum Chordata: animals with a backbone. Notice that the starfish and the butterfly shown as members of the kingdom Animalia are not included in phylum Chordata, yet all the other animals origi-

FIGURE 1-7 Classification scheme for a squirrel. Moving from the largest taxonomic category, kingdom, to the smallest subcategory, species. Fewer and fewer organisms are included in each subcategory as the organisms are separated into groups reflecting their evolutionary closeness. Many types of animals are in the animal kingdom. Only squirrels are in the genus *Tamiasciurus.*

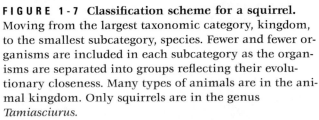

JUST WONDERING. . . .

I usually find scientific words hard to understand. Why are they always so long and hard to pronounce?

The scientific names of organisms are usually descriptive Latin or Greek names. In this way, scientists worldwide have a common language to refer to the myriad species that have been described. In addition, these languages are "dead" languages; they are not used conversationally today. Therefore, the meanings of the descriptors do not change with the times as do certain words in the English language. Latin and Greek are also often the languages from which the names of structures, processes, and diseases are derived. Students not familiar with these languages often find names and terms based on these languages unfamiliar. Learning the meanings of prefixes and suffixes and parts of words that are used in many contexts is often helpful. In *Biology: Understanding Life,* the derivation of many unusual and difficult words is explained to help students understand the meanings behind the names.

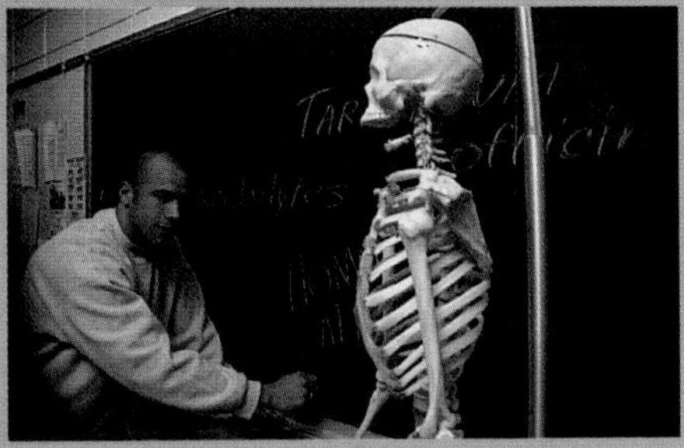

nally shown are included. These animals are more closely related to one another than to the starfish or butterfly.

The next subcategory is **class.** The class Mammalia is a subgroup of phylum Chordata. The snake and the fish are not included in this group because they are not mammals. Mammals are characterized by having skin with hair and by nourishing their young with milk secreted by mammary glands. The next subcategory is **order.** One order of class Mammalia is the order Rodentia, which includes most gnawing mammals. Humans are not gnawing animals and are not included in this subgroup. The last three subgroupings, in order of increasing relatedness, are **family, genus,** and **species.** Of all the organisms shown, which one is most closely related to squirrels?

The last two categories in the hierarchy of classification, genus and species, provide the scientific name of an organism. This system is known as **binomial nomenclature**—literally, "two-name naming." Scientifically, humans are *Homo sapiens,* meaning "wise man." House cats are *Felis domesticus,* meaning "domesticated cat." Notice that the genus name is capitalized and the species name begins with a lowercase letter. Both are italicized. What is the scientific name for the squirrel in our example?

Biology Today and You

Learning about biology today is more than learning the scientific names of organisms around you or learning biological "facts." It also involves learning about problem solving, how scientists go about their work, and worlds you may never have known existed, such as the world of microbes, the world of overpopulation, and the world of modern gene technology. Learning about biology today is learning how to protect the planet and how to take care of your body. But most of all, learning about biology today is learning about yourself—your evolutionary roots and your connectedness to all things living on this Earth.

evolution (ev oh **loo** shun) a theory of organismal change over time originally developed by Charles Darwin; it embodies the ideas that organisms alive today are descendants of organisms living long ago, and that organisms have changed and diverged from one another over billions of years.

taxonomy (tack **sawn** uh mee) a method of classification of the diverse array of species based on their common ancestry.

kingdoms broad categories in which taxonomists group all living things; the taxonomic system used in this book recognizes five kingdoms of life: *Monera, Protista, Plantae, Fungi,* and *Animalia.*

divisions subcategories of the kingdoms Monera, Plantae and Fungi into which taxonomists place organisms having similar evolutionary histories; taxonomically equivalent to **phyla.**

phyla (**fye** luh) subcategories of the kingdoms Protista and Animalia into which taxonomists place organisms having similar evolutionary histories; taxonomically equivalent to **divisions.**

class a taxonomic subcategory of **divisions** and **phyla;** as organisms appear to be increasingly related to each other, they are more and more narrowly classified within taxonomic subcategories.

order a taxonomic subcategory of **class.**

family a taxonomic subcategory of **order.**

genus (**gee** nus) a taxonomic subcategory of **family.**

binomial nomenclature (bye **no** mee uhl **no** men **clay** chur) literally, "two-name naming"; the system of determining the scientific name of an organism using its genus and species classifications. Thus, an organism's genus becomes its first name, and its species becomes its last name.

Summary

- Scientists study the natural world by using scientific methods. Scientific methods are processes used to answer the questions scientists ask and usually include observation, hypothesis formation, and hypothesis testing. Scientists develop hypotheses and test them in various ways, producing data (evidence) that either support or refute a prediction (hypothesis).
- As they weave together explanations of consistently supported and related hypotheses, scientists develop grand explanations that account for existing data and that consistently predict new data. Theories are interwoven hypotheses upheld by overwhelming evidence over time and are powerful concepts that help scientists make dependable predictions about the world.
- Although scientific knowledge grows and changes, certain themes, or accepted explanations, rise out of the abundance of tested predictions and permeate the science of biology. Although the specific details of these themes may be updated or changed as biologists modify their hypotheses, the themes themselves transcend time.
- Seven themes of life are described in this textbook:
  a. Living things display both diversity and unity.
  b. Living things are composed of cells and are hierarchically organized.
  c. Living things interact with each other and their environments.
  d. Living things transform energy and maintain a steady internal environment.
  e. Living things exhibit forms that fit their functions.
  f. Living things reproduce and pass on biological information to their offspring.
  g. Living things change over time, or evolve.
- Modern taxonomy, the classification of the diverse array of species, categorizes organisms based on their common ancestry (their evolutionary history). Organisms that are close relatives and resemble one another in a variety of ways are placed in common groupings reflecting their similarities and closeness. Organisms that are not closely related are placed in groupings separate from one another reflecting their differences.
- Taxonomists group all living things into broad categories, or kingdoms. This book uses a five-kingdom system of classification: Monera, Protista, Plantae, Fungi, and Animalia. Ancestors of the monerans (the bacteria) gave rise to organisms in the kingdom Protista (single-celled eukaryotes and the algae). Ancestors of the protists gave rise to the remaining three kingdoms of organisms: plants, animals, and fungi. The kingdoms of organisms are further subdivided into groupings that reflect an increasing closeness in evolutionary history as determined by a variety of criteria.

Knowledge and Comprehension Questions

1. Describe three topics that biologists might study.
2. The text states that scientists cannot prove hypotheses. Explain this statement.
3. What is a scientific theory? What is the difference between a scientific theory and the everyday word *theory*?
4. List the seven unifying themes of biology described in this textbook.
5. List and describe the four levels of internal organization in multicellular organisms.
6. Name and describe the different levels of interactions that occur between organisms and their environments.
7. How do humans' interactions with the biosphere differ from those of all other organisms?
8. From what ultimate source do you obtain the energy that keeps you alive? Explain how you obtain energy from that source.
9. What is DNA? Explain its role in living things.
10. Briefly state the theory of evolution. What are fossils and how do they relate to this theory?
11. Name the five kingdoms of living things and give an example of an organism from each. To which kingdom do you belong?
12. Place the following terms in their correct sequence, starting with the term that refers to the group containing the largest number of organisms: phylum, species, genus, order, family, class, kingdom.

Critical Thinking

1. You've been hearing about how regular exercise (say, walking for 1 hour 4 days a week) could help people lose weight. You decide to do a series of controlled experiments to test this idea and persuade 50 students at your school to participate. State your hypothesis and identify the independent and dependent variables.
2. What is a control? How would you set up the control in the preceding experiment?
3. Using the information that two organisms are members of the same order, determine whether the following statements are true or false. Explain your answer.
   a. The organisms must belong to the same class.
   b. The organisms must belong to the same genus.
   c. The organisms must have the same "first name" using binomial nomenclature.
4. If all living things are composed of one or more cells, it follows that a computer-driven machine cannot be alive. And yet such a machine can transform energy, do work, make copies of itself, and evolve over time to better suit the challenges of its environment. Can machines be considered living? From a biological perspective, what constitutes life?

CHAPTER 2

THE CHEMISTRY OF LIFE

WATER . . . WITHOUT IT you would die. In fact, life on Earth could not go on if deprived of this amazing liquid, and evolution could never have taken place without it. What is it about water that makes it so important to life?

The clues needed to answer this question lie in the chemical structure of water. Put another way, water is important because of its characteristics, which in turn depend on its parts and how they are put together. In the photo, you can easily observe some of these characteristics. Notice how the water sticks together to form a droplet. Notice, too, how the droplet adheres to the blade of grass in spite of the opposing force of gravity. The story of how the parts of water relate to its characteristics and the answer to why water is important to you will soon be told—as you begin your study of the chemistry of life.

KEY CONCEPTS

- The chemical interactions of atoms, the smallest units of elements, are largely determined by the distribution of their electrons.
- Atoms may combine by means of chemical bonds to form molecules and compounds.
- Water has properties that are extremely important to life.
- The main groups of biological molecules are carbohydrates, lipids, proteins, and nucleic acids.

OUTLINE

Atoms

Chemistry is the science of matter, the physical material that makes up everything in the universe. Matter is anything that takes up space and has a measurable amount of substance, or mass. All matter (including water) is made up of submicroscopic parts called **atoms.** Although scientists know a great deal about atoms, a simplified explanation provides a good starting point in understanding their complex structures.

Every atom is made up of particles tinier than the atom itself. These subatomic particles are of three types: **protons, neutrons,** and **electrons.** Protons and neutrons are found at the core, or nucleus, of the atom. Electrons surround the nucleus.

Protons have mass and carry a positive (+) charge. Neutrons, although similar to protons in mass, are neutral and carry no charge. Electrons have very little mass and carry a negative (−) charge. For this reason, the *atomic mass* of an atom is defined as the combined mass of all its protons and neutrons without regard to its electrons.

The number of protons in an atom, called the *atomic number,* is the same as the number of electrons in that atom. Atoms are therefore electrically neutral; the positive charges of the protons are balanced by the negative charges of the electrons. The number of neutrons in an atom, however, may or may not equal the number of protons. Atoms that have the same number of protons but different numbers of neutrons are called **isotopes** (Figure 2-1). Isotopes of an atom differ in atomic mass but have similar chemical properties because they have the same number of electrons. Electrons primarily determine the chemical properties of atoms because atoms interact by means of their electrons and not their protons or neutrons.

> *An atom is a core (nucleus) of protons and neutrons surrounded by electrons. The electrons largely determine the chemical properties of an atom.*

The key to the chemical behavior of atoms lies in the arrangement of their electrons. Although scientists cannot precisely locate the position of any individual electron at a particular time, they

Applications

HOW *Science* WORKS

Using Isotopes to Detect and Treat Disease

Isotopes are elements that have the same number of protons and electrons but have different numbers of neutrons. For example, hydrogen has three isotopes: hydrogen (one proton and no neutrons), deuterium (one proton and one neutron), and tritium (one proton and two neutrons) (see Figure 2-1, *B*). Because only the number of neutrons is different, all isotopes of an element have similar chemical properties. In fact, tritium can "substitute" for hydrogen in chemical reactions.

Some isotopes are unstable and can emit subatomic particles: neutrons, protons, electrons, alpha particles (high-energy helium nuclei), and gamma rays (a form of electromagnetic radiation). These unstable isotopes are termed *radioactive.* When a radioactive isotope emits subatomic particles, a new, more stable element is eventually formed (although this process may take a long time). This process is called *decay.*

Because of isotopes' radioactivity and the fact that isotopes of a particular atom are nearly indistinguishable from one another to organisms, doctors use them in many ways to detect and treat disease.

When used in diagnosis, isotopes are called *tracers* or *labels* because they can be used to follow the fate of certain substances in the body. For example, suppose your doctor wants to investigate the rate at which your thyroid gland is using iodine, a substance in your diet that your body uses to make thyroid hormone. To determine this rate, your doctor injects you with a tracer—a solution of radioactive iodine, which your body will use in the same way it uses the "normal" iodine you acquire in your food. Equipment that counts the emission of particles from the nuclei of the radioactive iodine is placed over your thyroid gland and measures how fast the tracer iodine is entering your thyroid. The doctor can then see *exactly* how fast your thyroid is using iodine by simply reading the information on the detector (Figure 2-A).

Another way isotopes are used is in the treatment of cancer, especially those cancers that are present on the surface of the body, such as skin cancers. A patch containing powerful radioactive isotopes is taped over the cancerous tumor. The particles emitted from the isotope bombard the tumor and destroy the cancerous tissue. Treatment with isotopes does have some drawbacks, however. Isotope therapy causes nausea and breakdown of the normal tissue surrounding the tumor. In addition, treatment with isotopes is not a cure for cancer—the isotopes can only destroy localized tumors. Cancer cells that have migrated to other parts of the body are unaffected by isotope therapy. Still, treatment with isotopes offers hope to those in the early stages of cancer.

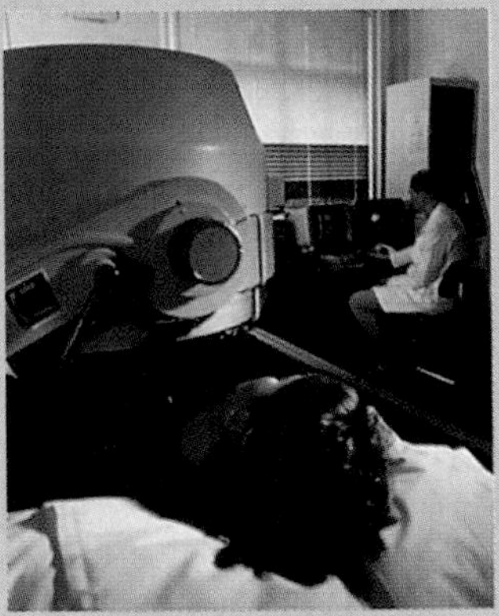

FIGURE 2-A Isotopes are used to diagnose and treat illness.

can predict where an electron is most likely to be. This volume of space around a nucleus where an electron is most likely to be found is called the **shell** of that electron.

Atoms can have many electron shells. Atoms with more than two electrons have more than one electron shell. In Figure 2-2, *A*, the nucleus is shown as a small circle surrounded by spheres. These spheres represent electron shells. In Figure 2-2, *B*, electrons are shown within the shells, which are depicted here two dimensionally as circles. Notice that the atom of nitrogen pictured in Figure 2-2, *B*, has electrons occupying two shells. The innermost shell contains two electrons. The second (outermost) shell has five electrons.

Because the energy of electrons increases as their distance from the attractive force of the nucleus increases, the various electron shells of atoms are also called *energy levels.* Electrons occupying increasingly distant shells from the nucleus have a stepwise increase in their levels of energy. The energy of electrons at the various energy levels is also

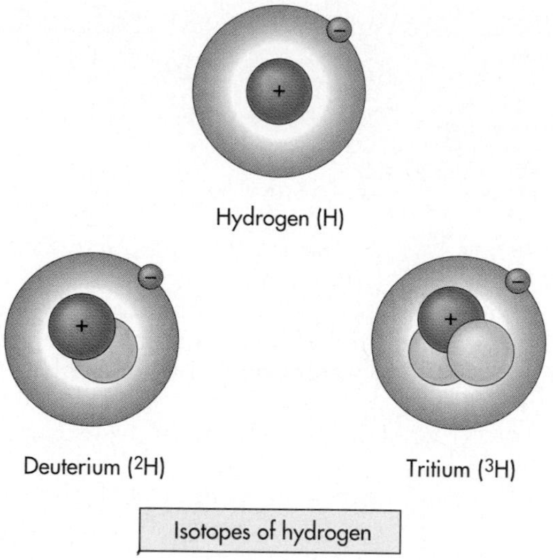

FIGURE 2-1 Isotopes. Isotopes of an atom differ in atomic mass but have similar chemical properties. Three naturally occurring isotopes of hydrogen exist. Each has a single proton in its nucleus but has different numbers of neutrons. Deuterium, tritium, and hydrogen therefore differ in their atomic masses but have the same chemical properties as one another.

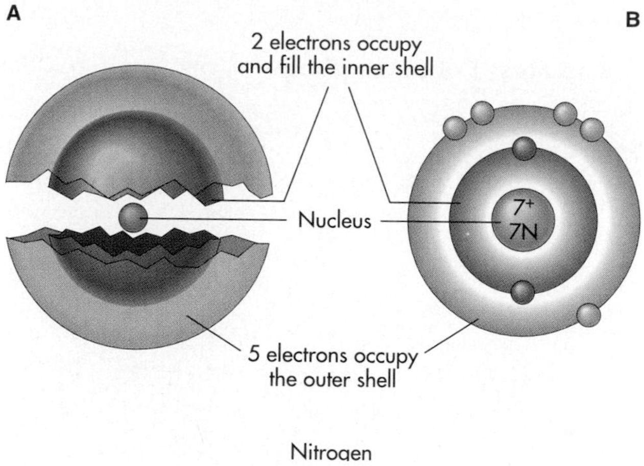

FIGURE 2-2 Representation of an atom. A, The nucleus of the atom is surrounded by electrons occupying electron shells or energy levels. **B,** A two-dimensional representation of a nitrogen atom showing the number of electrons found in each shell and the number of protons and neutrons in the nucleus.

influenced by the number of protons in the nucleus, since the attractive force of the nucleus increases as its number of protons increases.

> *The farther an electron is from the nucleus, the more energy it has.*

Molecules and Compounds

The identity of an atom is determined by its number of protons. For example, an atom containing two protons is helium, a gas you have probably seen used to blow up balloons. An atom possessing seven protons is nitrogen (Figure 2-2, *B*).

Elements are pure substances that are made up of a single kind of atom and cannot be separated into different substances by ordinary chemical methods. The atoms of most elements interact with one another. Two or more atoms held together by sharing electrons are **molecules.** Molecules can be made up of atoms of the same element or atoms of different elements. Molecules made up of atoms of different elements are called **compounds.** (Compounds are also atoms of different elements held together by electrostatic attraction [ionic bonds] see p. 27).

The oxygen in the air consists of molecules made up of pairs of atoms of the same element—oxygen. These pairs are represented by the chemical formula O_2. A *chemical formula* is a type of "shorthand" used to describe a molecule. The atoms are represented by symbols, such as *O* for oxygen. (Chemical symbols are shown in Table 2-1 on p. 26.) A subscript shows the number of these atoms present in the molecule. An example of a com-

atoms submicroscopic particles that make up all matter.
protons particles found at the core of an atom. Protons carry a positive electrical charge.
neutrons (noo trons) particles found at the core of an atom. Neutrons carry no electrical charge.
electrons particles that surround the core of an atom. Electrons carry a negative electrical charge.
isotopes (eye suh topes) atoms that have the same number of protons but different numbers of neutrons.
shell a term for the volume of space around an atom's nucleus where an electron is most likely to be found.
elements pure substances that are made up of a single kind of atom and cannot be separated into different substances by ordinary chemical methods.
molecules two or more atoms held together by sharing electrons.
compounds molecules or ionically bonded substances made up of atoms of different elements.

TABLE 2-1

The Most Common Elements on Earth and Their Distribution in the Human Body

ELEMENT	SYMBOL	ATOMIC NUMBER	PERCENT OF HUMAN BODY BY WEIGHT	APPROXIMATE PERCENT OF EARTH'S CRUST BY WEIGHT	IMPORTANCE OR FUNCTION
Oxygen	O	8	65.0	46.6	Necessary for cellular respiration, component of water
Carbon	C	6	18.5	0.03	Backbone of organic molecules
Hydrogen	H	1	9.5	0.14	Electron carrier, component of water and most organic molecules
Nitrogen	N	7	3.3	Trace	Component of all proteins and nucleic acids
Calcium	Ca	20	1.5	3.6	Component of bones and teeth, trigger for muscle contraction
Phosphorus	P	15	1.0	0.07	Backbone of nucleic acids, important in energy transfer
Potassium	K	19	0.4	2.6	Principal positive ion in cells, important in nerve function
Sulfur	S	16	0.3	0.03	Component of most proteins
Chlorine	Cl	17	0.2	0.01	Principal negative ion bathing cells
Sodium	Na	11	0.2	2.8	Principal positive ion bathing cells, important in nerve function
Magnesium	Mg	12	0.1	2.1	Critical component of many energy-transferring enzymes
Iron	Fe	26	Trace	5.0	Critical component of hemoglobin in the blood
Copper	Cu	29	Trace	0.01	Key component of many enzymes
Zinc	Zn	30	Trace	Trace	Key component of some enzymes
Molybdenum	Mo	42	Trace	Trace	Key component of many enzymes
Iodine	I	53	Trace	Trace	Component of thyroid hormone
Silicon	Si	14	Trace	27.7	—
Aluminum	Al	13	Trace	6.5	—
Manganese	Mn	25	Trace	0.1	—
Fluorine	F	9	Trace	0.07	—
Vanadium	V	23	Trace	0.01	—
Chromium	Cr	24	Trace	0.01	—
Boron	B	5	Trace	Trace	—
Cobalt	Co	27	Trace	Trace	—
Selenium	Se	34	Trace	Trace	—
Tin	Sn	50	Trace	Trace	—

pound with which you are all familiar is water, H_2O. In this chemical formula, the symbol *H* stands for the element hydrogen.

Three factors influence whether an atom will interact with other atoms. In addition, these factors influence the type of interactions likely to take place. These factors are (1) the tendency of electrons to occur in pairs, (2) the tendency of atoms to balance positive and negative charges, and (3) the tendency of the outer shell, or energy level, of electrons to be full. This third factor is often called the **octet rule.**

The word *octet* means "eight objects" and refers to the fact that eight electrons is a stable number for an outer shell; the outer electron shell of many atoms contains a maximum of eight electrons. (The first energy level is an exception to this rule because it contains a maximum of two electrons.) The octet rule states that an atom with an unfilled outer shell has a tendency to interact with another atom or atoms in ways that will complete this outer shell. Although the octet rule does not apply to all atoms, it does apply to all biologically important ones—those involved in the structure, energy needs, and information systems of living things. These atoms and the molecules they make up are discussed later in this chapter.

> *An atom is identified by its number of protons. Its interaction with other atoms is determined by the number and arrangement of its electrons.*

Atoms of elements that have equal numbers of protons and electrons and have full outer-electron energy levels are the only ones that exist as single atoms. Atoms with these characteristics are called **noble gases,** or inert gases, because they do not react readily with other elements. Many stories explain why they are called noble, but they all center around the concept of nobility—those who have everything (in this case a full outer shell), need nothing, and interact little with others. Most of the noble gases—helium, neon, argon, krypton, xenon, and radon—are rare. In addition, they are relatively unreactive and unimportant in living systems. Helium and radon are probably the most well known and the least rare noble gases. Radon, in fact, has gained more notice in recent years. Formed in rock or soil particles from the radioactive decay of radium, radon gas can seep through cracks in basement walls and remain trapped in homes that are not well ventilated. Prolonged exposure to radioactive radon in levels greater than those normally found in the atmosphere is thought to lead to lung cancer.

Nature of the Chemical Bond

An atom having an incomplete outer shell can satisfy the octet rule in one of three ways:

1. It can gain electrons from another atom.
2. It can lose electrons to another atom.
3. It can share one or more electron pairs with another atom.

Such interactions among atoms result in **chemical bonds,** forces that hold atoms together. If the force is caused by the attraction of oppositely charged particles formed by the gain or loss of electrons, the bond is called **ionic.** If the force is caused by the electrical attraction created by atoms sharing electrons, the bond is called **covalent.** Other weaker kinds of bonds also occur.

Ions and Ionic Bonds

Electrons stay in their shells because they are attracted to the positive charge of the nucleus. However, electrons far from the nucleus are not held as tightly as electrons closer to the nucleus. In addition, they have more energy and they interact with other atoms more easily than close, tightly held, energy-poor electrons. Therefore, atoms typically interact with other atoms by means of the electrons in their outermost (highest) energy levels, or shells. The types of interactions that occur tend to result in atoms with completed outer shells, thus satisfying the octet rule.

An atom tends to lose electrons if it has an outer shell needing any electrons to be complete. An atom tends to gain electrons if it has a nearly completed outer shell. Interestingly, the former type of atom tends to interact with the latter type, resulting in a completed outer shell for each. For example, an atom with a nearly completed outer shell tends to "take" enough electrons to complete its outer shell from an atom having only one or two electrons in an outer shell that needs eight for completion. Once these outer electrons are gone from the atom giving up electrons, the next shell in becomes its new, complete outer shell. As a result of this interaction, neither atom is electrically neu-

octet rule (ock **tet rool**) one of three factors that influence whether an atom will interact with other atoms; the octet rule states that an atom with an unfilled outer shell has a tendency to interact with another atom or atoms in ways that will complete this outer shell.

noble gases atoms of elements that have equal numbers of protons and electrons and have full outer-electron energy levels and thus do not react readily with other elements.

chemical bonds forces that hold atoms together.

ionic (eye **on** ick) a type of chemical bond between atoms that is caused by the attraction of oppositely charged particles formed by the gain or loss of electrons.

covalent (ko **vay** lent) a type of chemical bond that is caused by the electrical attraction created by atoms sharing electrons.

tral; the number of protons no longer equals the number of electrons in either atom. The atom taking on electrons acquires a negative charge. The atom giving up electrons acquires a positive charge. They are not called atoms because they are no longer electrically neutral. Such charged particles are called **ions.**

Figure 2-3, *A*, illustrates the formation of ions with a specific example of electron "give and take." Sodium (Na) is an element with 11 protons and 11 electrons. It is a soft, silver-white metal that occurs in nature as a part of ionic compounds. One familiar compound is sodium chloride, or table salt. Of sodium's 11 electrons, 2 are in its innermost energy level (full with 2 electrons), 8 are at the next level, and 1 is at the outer energy level. Because of this distribution of electrons, its outer energy level is not full, and therefore the octet rule is not satisfied.

Chlorine (Cl) is an element with 17 protons and 17 electrons. In its molecular form (Cl_2), it is a greenish-yellow gas that is poisonous and irritating to the nose and throat. In the ionic compound sodium chloride, however, it does not have these characteristics. Of chlorine's 17 electrons, 2 are at its innermost energy level, 8 at the next energy level, and 7 at the outer energy level. Chlorine, like sodium, has an outer energy level that is not full. Sodium and chlorine atoms can interact with one another in a way that results in both having full outer energy levels.

When placed together, the metal sodium and the gas chlorine react explosively. The single electrons in the outer energy levels of the sodium atoms are lost to the chlorine. The result is the production of Na^+ and Cl^- ions. These ions come together as their opposite charges attract one another. This type of attraction is called *electrostatic attraction* and results in ionic bonding of the sodium and chloride ions. As these ions are drawn to one another, they form geometrically perfect crystals of salt (Figure 2-3, *B* and *C*).

> *An ionic bond is an attraction between ions of opposite charge.*

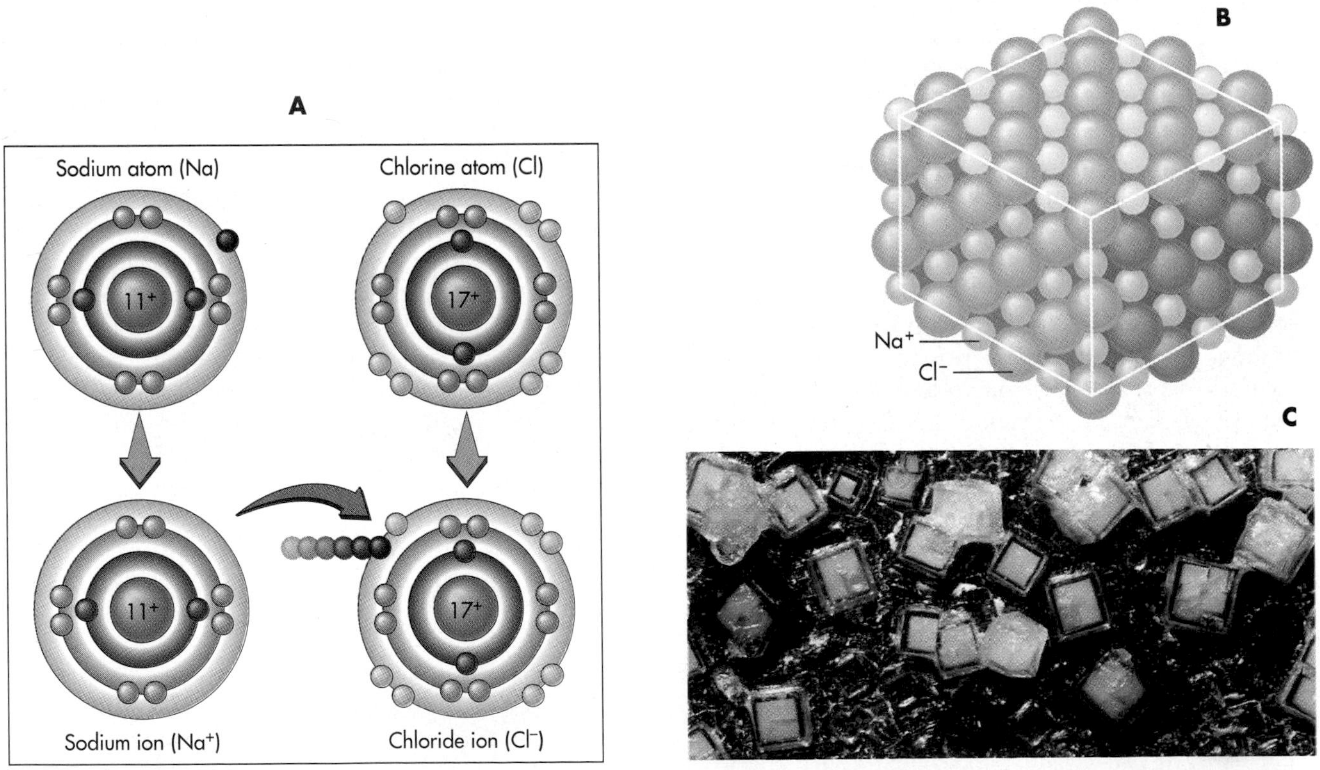

FIGURE 2-3 The formation of an ionic bond. A, The sodium atom (*Na*) loses an electron, or negative charge, from its unfilled outer shell. Lacking this negative charge, it becomes a positively charged ion. The chlorine atom (*Cl*) gains an electron or negative charge, filling its outer electron shell. Therefore, it becomes a negatively charged ion. **B,** These ions come together forming salt (*NaCl*) as their opposite charges attract one another. **C,** Salt assumes a crystalline shape because of the ionic bonding and regular atomic arrangement between sodium and chlorine.

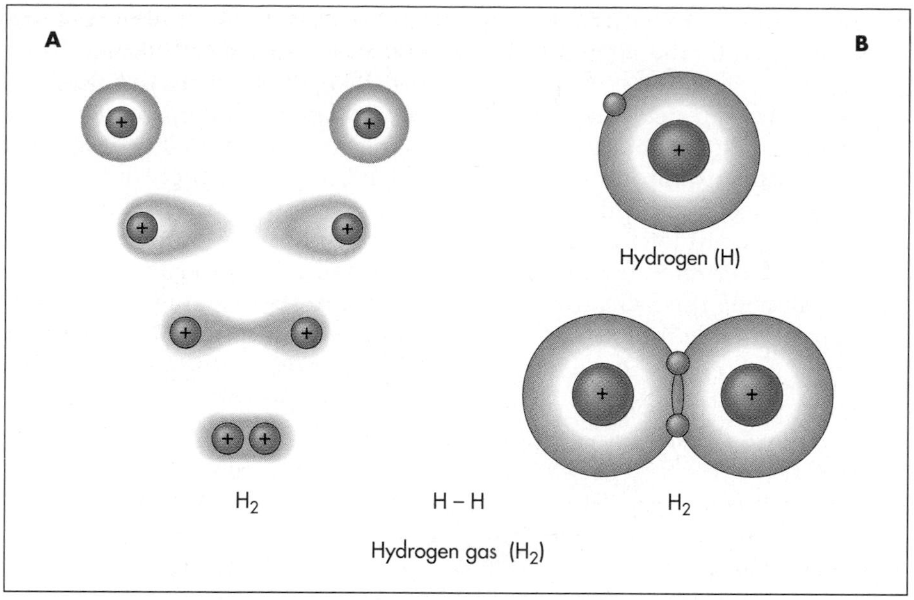

FIGURE 2-4 Covalent bonds. Covalent bonds form when two atoms share electrons. Hydrogen has only one electron in its outer shell but is electrically balanced with two electrons. When hydrogen atoms get close enough together, they form pairs by sharing their electrons. These paired atoms are called diatomic molecules. **A,** The two atoms of hydrogen shown here depict the single electron shell of each as a cloudlike area surrounding the nucleus. The electron clouds first overlap and then unite. **B,** The atoms of hydrogen shown here depict the single electron shell of each as a ring surrounding the nucleus, with the electrons shown as dots.

The transfer of electrons between atoms is an important chemical event—one type of chemical reaction. In fact, this type of chemical interaction has a special vocabulary. When an atom loses an electron, it is said to be *oxidized.* The reaction is called an **oxidation,** a term that comes from the original meaning (an alternative definition still valid today), which is "to combine with oxygen." For example, when iron combines with oxygen in the presence of moisture, it becomes oxidized. The product of this oxidation is commonly known as rust. Conversely, when an atom gains an electron (or loses oxygen), it is said to be *reduced.* The reaction is called a **reduction.** Oxidations and reductions always occur together because an electron lost from an atom cannot exist by itself—it will be transferred to another atom. Therefore the shortened term **redox reaction** is often used to describe this paired transfer.

The loss of an electron by an atom is an oxidation. The gain of an electron is a reduction. Together, these paired electron transfers between atoms are called redox reactions.

Covalent Bonds

Covalent bonds form when two atoms share electrons. Hydrogen is a simple example of an atom that usually shares electrons with other atoms. As you can see in Figure 2-4, a hydrogen atom has a single electron and an unfilled outer electron shell. A filled outer shell at this energy level requires only two electrons. When hydrogen atoms are close enough to one another, an interesting thing happens. They form pairs, with each of their single electrons moving around the two nuclei.

ions (eye ons) charged particles formed by atoms that have either acquired or lost electrons and have therefore developed a negative or positive charge.

oxidation (ok si **day** shun) a reaction that involves an atom losing an electron.

reduction a reaction that involves an atom gaining an electron.

redox reaction a term used to describe the occurrence of oxidation and reduction reactions, which always happen together.

These paired atoms of hydrogen are called *diatomic molecules* and are represented by the chemical formula H_2.

As a result of this sharing of electrons, the diatomic hydrogen gas molecule is electrically balanced because it now contains two protons and two electrons. In addition, each hydrogen atom has two electrons in its outer shell, completing this shell. This relationship also results in the pairing of two free electrons. Thus, by sharing their electrons, the two hydrogen atoms form a stable molecule.

> *A covalent bond is formed by the sharing of one or more pairs of electrons.*

Covalent bonds can be very strong, that is, difficult to break. *Double bonds,* those bonds in which two pairs of electrons are shared, are stronger than *single bonds,* covalent bonds sharing only one pair of electrons. As you might expect, *triple bonds,* those bonds in which three pairs of electrons are shared, are the strongest of these three types of covalent bonds. In chemical formulas that show the structure of covalently bonded molecules, single bonds are represented by a single line between two bonded atoms, double bonds by two lines, and triple bonds by three lines. For example, the structural formula of hydrogen gas is $H—H$, oxygen gas is $O=O$, and nitrogen gas is $N \equiv N$.

An atom can also form covalent bonds with more than one other atom. Carbon (C), for exam-

JUST WONDERING . . . .

How do scientists know what atoms look like if they cannot see them?

The scientist's description of atoms is called the *atomic theory* and is an explanation of chemical phenomena scientists *can* see. The atomic theory explains observable phenomena such as the differences among elements, the ability of elements to combine to form compounds, and the differences that exist between compounds and mixtures. Although philosophers pondered questions regarding the makeup of matter as far back as the ancient Greeks, John Dalton, an English schoolteacher, developed an atomic theory in the early 1800s, which was based on his experimentation. The basic tenets of his theory are the following:

1. Elements are made up of small particles called *atoms.*
2. The atoms of an element are all the same, but the atoms of different elements are different from one another and have different properties.
3. When atoms of more than one element combine in chemical reactions, they form compounds, but the atoms themselves do not change. Within a given compound, the relative number and kind of atoms remain constant.

Surprisingly, his theory has undergone little revision since he proposed it.

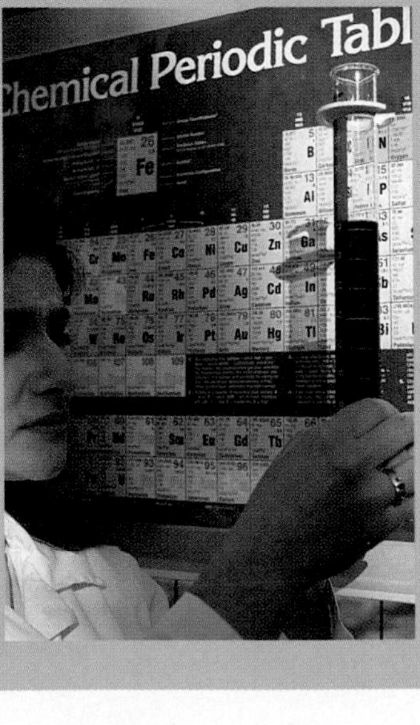

After Dalton's time, a variety of investigators added to our knowledge about the structure of atoms. In the mid-1800s, researchers studying radiation gathered evidence suggesting that it consisted of a stream of negatively charged particles, which they termed "electrons." Further investigation throughout the late 1800s and into the early 1900s regarding the nature and behavior of radiation and radioactive particles by such noted scientists as J.J. Thompson, Robert Millikan, Henri Becquerel, Marie Curie, and Ernest Rutherford led to the understanding of other subatomic particles.

Today, particle physicists and nuclear physicists use a wide range of sophisticated instruments to study atoms. Most recently, particle physicists have determined by hurtling subatomic particles into one another that matter is made up of two kinds of particles: quarks and leptons. Physicists have described how quarks form into protons and neutrons and how these two subatomic particles congregate to create atomic nuclei. However, most chemists still live in the world of protons, neutrons, and electrons rather than in the world of quarks and leptons, for only these three subatomic particles have bearing on chemical behavior.

ple, contains six electrons: two in the inner shell and four in the outer shell. To satisfy the octet rule, it must gain four additional electrons by sharing its four outer-shell electrons with another atom or atoms, forming four covalent bonds. Because there are many ways that four covalent bonds may form, carbon atoms are able to participate in many different kinds of molecules in living systems, such as proteins and carbohydrates.

The strength of a covalent bond refers to the amount of energy needed to make or break that bond. The energy that goes into making the bond is held within the bond and is released when the bond is broken. Therefore covalent bonds are actually a storage place for energy as well as a type of chemical "glue" that holds molecules together. Living things store and use energy by means of making and breaking covalent bonds, thereby using molecules as a type of energy currency.

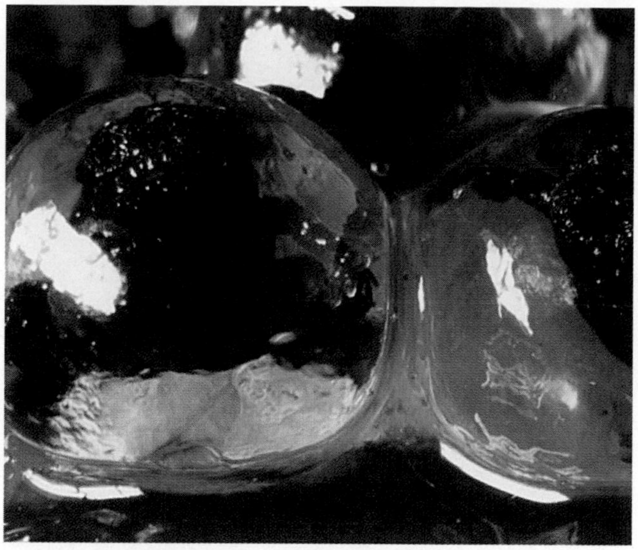

FIGURE 2-5 Water is the cradle of life. Many kinds of organisms, like these small frogs seen through the transparent walls of their eggs, begin life in water.

The Cradle of Life: Water

One covalently bonded molecule that plays a major role in living systems is water—H_2O. In fact, water is the most abundant molecule in your body, making up about two thirds of your body weight. Although it seems to be a simple molecule, water has many surprising properties. For example, of all the common molecules on Earth, only water exists as a liquid at the Earth's surface. When life on Earth was beginning, this liquid provided a medium in which other molecules could move around and interact. Life evolved as a result of these interactions. And life, as it evolved, maintained these ties to water (Figure 2-5). Three fourths of the Earth's surface is covered by water. Where water is plentiful, such as in the tropical rain forests, the land abounds with life. Where water is scarce, such as in the desert, the land seems almost lifeless except after a rainstorm. No plant or animal can grow and reproduce without some amount of water.

The chemistry of life, then, is water chemistry. Water has a simple molecular structure: one oxygen atom bonded by single covalent bonds to two hydrogen atoms. The resulting molecule satisfies the octet rule, and its positive and negative charges are balanced. Because the oxygen atom in each water molecule contains eight protons and each hydrogen atom contains only one proton, the electron pair shared in each covalent bond is more strongly attracted to the oxygen nucleus than to either of the hydrogen nuclei. Although the electrons surround both the oxygen and hydrogen nuclei, the negatively charged electrons are far more likely to be found near the oxygen nucleus at a given moment than near one of the hydrogen nuclei. Because of this situation, the oxygen end of the water molecule has a partial negative charge. The hydrogen end has a partial positive charge (Figure 2-6). Molecules such as water that have opposite partial charges at different ends of the molecule are called **polar molecules.** Water is one of the most polar molecules known.

The polarity of water contributes to its ability to attract other molecules and form special types of chemical bonds with them. These bonds, called **hydrogen bonds,** have approximately 5% to 10% the strength of covalent bonds. Hydrogen bonds are weak electrical attractions between the H of an NH group or an OH group, and other oxygen or nitrogen atoms. In fact, water forms bonds with other water molecules in just this way: the hydrogen atoms of some water molecules are attracted to the oxygen atoms of other water molecules. The ability of water molecules to form weak bonds among themselves and with other molecules is the reason for much of the organization and chemistry of living things. Although hydrogen bonds are weak, these bonds are constantly made and broken. (Each lasts only 1/100,000,000,000 of a second!) The cumulative effect of very large numbers of hydrogen bonds is responsible for the many impor-

polar molecules molecules (two or more atoms sharing electrons) that have opposite partial charges at either end because electrons are shared unequally.

hydrogen bonds weak electrical attractions between the H of an NH group or an OH group, and other oxygen or nitrogen atoms.

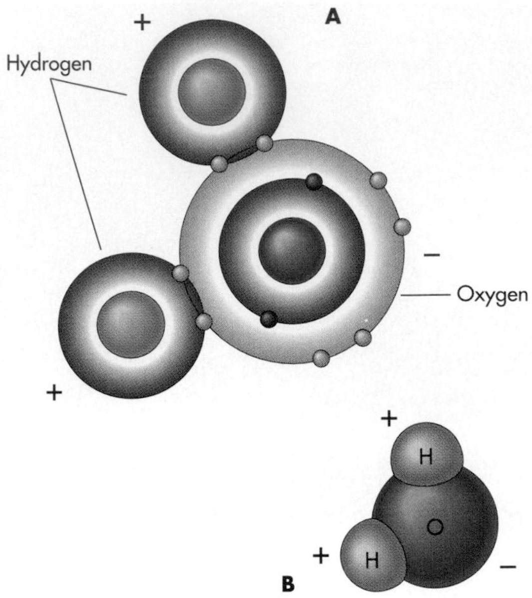

FIGURE 2-6 The structure of water. A, Water is a polar covalently bonded molecule. One oxygen atom (*O*) shares a pair of electrons with two hydrogen atoms (*H*). **B,** In this model the + and - represent the partial charges at opposite sides of the water molecule.

tant physical properties of water (Figure 2-7) and is the answer to why water sticks together as seen in the chapter opener photograph.

> *Much of the biologically important behavior of water results because its oxygen atom attracts electrons more strongly than its hydrogen atoms do. As a result, water molecules each have electron-rich (−) and electron-poor (+) regions, giving them positive and negative poles.*

Water Is a Powerful Solvent

Water molecules gather closely around any particle that exhibits an electrical charge, such as ions and polar molecules. For example, sodium chloride (table salt) is made up of the positively charged sodium (Na^+) and negatively charged chloride (Cl^-) ions. These ions are attracted to one another and cluster in a regular pattern, forming crystals. When you put salt in water, some ions break away from the crystals because the positive ends of some water molecules are attracted to the Cl^- ions, while the negative ends of other water molecules are attracted to the Na^+ ions; these attractions are stronger than the attraction between the ions that keeps the crystal together. Therefore, the ions are pulled from their positions in the crystal. Water molecules then surround each ion, forming a *hydration shell,* which keeps the ions apart. The salt is said to be *dissolved* (Figure 2-8).

Similarly, hydration shells form around all polar molecules and ions. Compounds that dissolve in water this way are said to be **soluble** in water. Chemical interactions readily take place in water because so many kinds of compounds are water soluble and therefore move among water molecules as separate molecules or ions.

Water Organizes Nonpolar Molecules

Remember the old saying that "oil and water don't mix?" This statement is true because oil is a nonpolar molecule and cannot form hydrogen bonds with water. Instead, the water molecules form hydrogen bonds with each other, causing the water to exclude the nonpolar molecules. It is almost as if nonpolar molecules move away from contact with the water. For this reason, nonpolar molecules are referred to as being **hydrophobic.** The word *hydrophobic* comes from Greek words meaning "water" (*hydros*) and "fearing" (*phobos*). This tendency for nonpolar molecules to band together in a water solution is called *hydrophobic bonding.* Hydrophobic forces determine the three-dimensional shapes of many biological molecules, which are usually surrounded by water within organisms.

Water Ionizes

The covalent bonds of water molecules sometimes break spontaneously. When this happens, one hydrogen atom nucleus (a proton) dissociates from the rest of the water molecule, leaving behind its electron. Because its positive charge is no longer balanced by an electron, it is a positively charged hydrogen ion, H^+. The remaining part of the water molecule now has an extra electron. It is therefore a negatively charged hydroxyl ion, OH^-. This process of spontaneous ion formation is called **ionization:**

$$H_2O \rightarrow OH^- + H^+$$

Only very few water molecules are ionized at a single instant in time. Scientists calculate that the fraction of water molecules dissociated (ionized) at any given time in pure water is 0.0000001. This tiny number can be written another way by using exponential notation. This is done by counting the number of places to the right of the decimal point. Because there are seven places, this number is writ

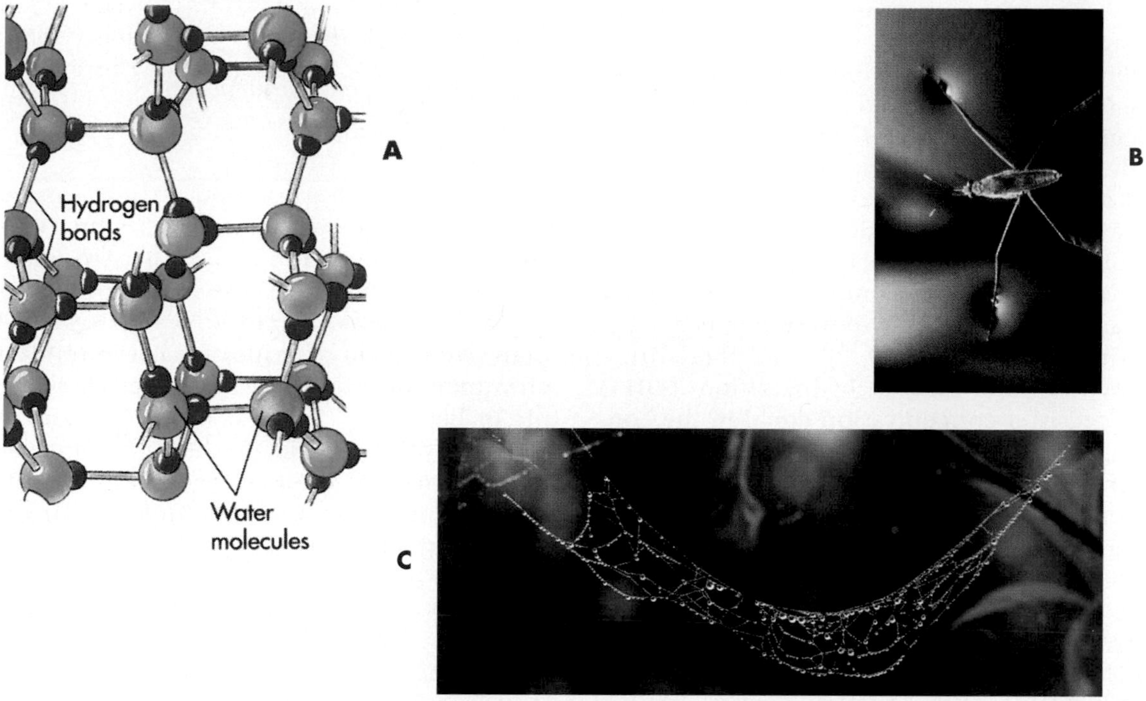

FIGURE 2-7 Hydrogen bonds are responsible for many important physical properties of water. **A,** An illustration showing the many hydrogen bonds that are constantly made and broken between water molecules. Hydrogen bonds form because of weak electrical attractions between the water molecules. **B,** As a result of hydrogen bonding, water molecules "stick together" with such force at the water's surface that they can bear the weight of organisms such as this water strider. **C,** Water molecules also stick or adhere to other substances with charged surfaces such as this web.

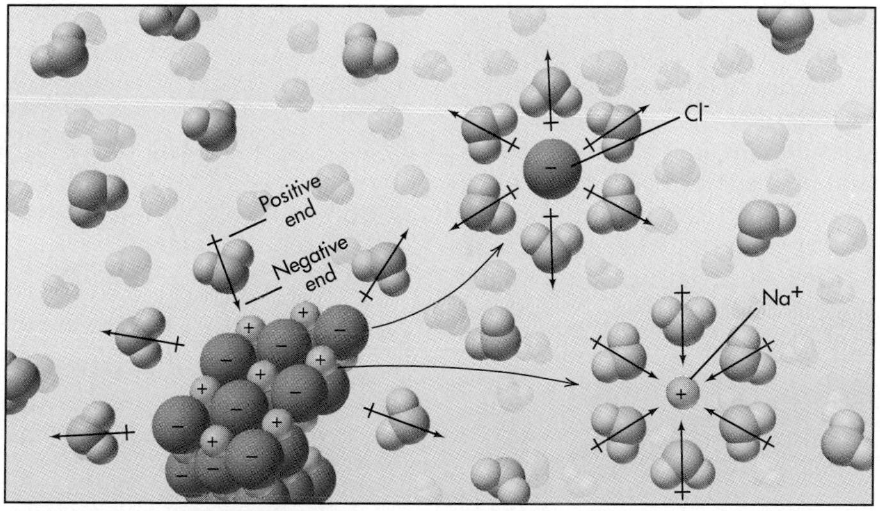

FIGURE 2-8 How salt dissolves in water. Water molecules are attracted to other charged particles. Sodium chloride (*NaCl*), or table salt, is made up of positively charged sodium (Na^+) ions and negatively charged chloride (Cl^-) ions. The negative ends of water molecules orient toward the positive sodium ions. The positive ends of water molecules orient toward the negative chloride ions. These attractions are strong enough to pull the ions from the crystal.

ten as 10^{-7}. The minus sign means that the number is less than 1.

To indicate the concentration of H^+ ions in a solution, scientists have devised a scale based on the slight degree of spontaneous ionization of water. This scale is called the **pH scale.** (The letters *pH* stand for the *p*ower of the hydrogen ion [H^+].) The pH values of this scale generally range from 0 to 14. The pH of a solution is determined by taking the negative value of the exponent of its hydrogen ion concentration. For example, pure water has a hydrogen ion concentration of 10^{-7} and therefore a pH of 7. When water ionizes, hydroxyl ions (OH^-) are produced in a concentration equal to the concentration of hydrogen ions. In pure water the concentrations of H^+ and OH^- ions are equal to 10^{-7}. Because these ions join spontaneously, water, at pH 7, is neutral.

Any substance that dissociates to form H^+ ions when it is dissolved in water is called an **acid.** The more hydrogen ions an acid produces, the stronger an acid it is. Although an acid produces a higher concentration of H^+ ions than pure water (0.00001 as opposed to 0.0000001, for example), its pH is lower. Using the above numbers to illustrate: the first number (0.00001 or 10^{-5}) represents the hydrogen ion concentration of an acid. The negative value of its exponent results in a pH of 5. The second number (0.0000001 or 10^{-7}) represents the hydrogen ion concentration of water. The negative value of its exponent results in a pH of 7. Each one unit decrease in pH, however, does not correspond to a onefold increase in acidity. In fact, depending on the acid and how completely it ionizes, a change of one unit may correspond to as much as a tenfold increase in acidity.

Figure 2-9 shows many common acids and their pH values. The pH of champagne, for example, is about 4. This low pH is due to the dissolved carbonic acid that causes champagne to bubble. Some bodies of water are acidic, such as peat bogs (pH 4 to 5). The hydrochloric acid (HCl) of your stomach ionizes completely to H^+ and Cl^-. It forms a strong acid with a pH of 2 to 3. Stronger acids are rarely found in living systems.

pH refers to the relative concentration of H^+ ions in a solution. Low pH values indicate high concentrations of H^+ ions (acids), and high pH values indicate low concentrations.

Any substance that combines with H^+ ions, as OH^- ions do, is said to be a **base.** Any increase in the concentration of a base lowers the H^+ ion concentration. Bases therefore have pH values higher than water's neutral value of 7. For example, the environment of your small intestine is kept at a basic pH of between 7.5 and 8.5. Strong bases such as sodium hydroxide (NaOH) have pH values of 12 or more. As with acids, a change of 1 in the pH value of a base may reflect up to a tenfold change in pH, depending on the base.

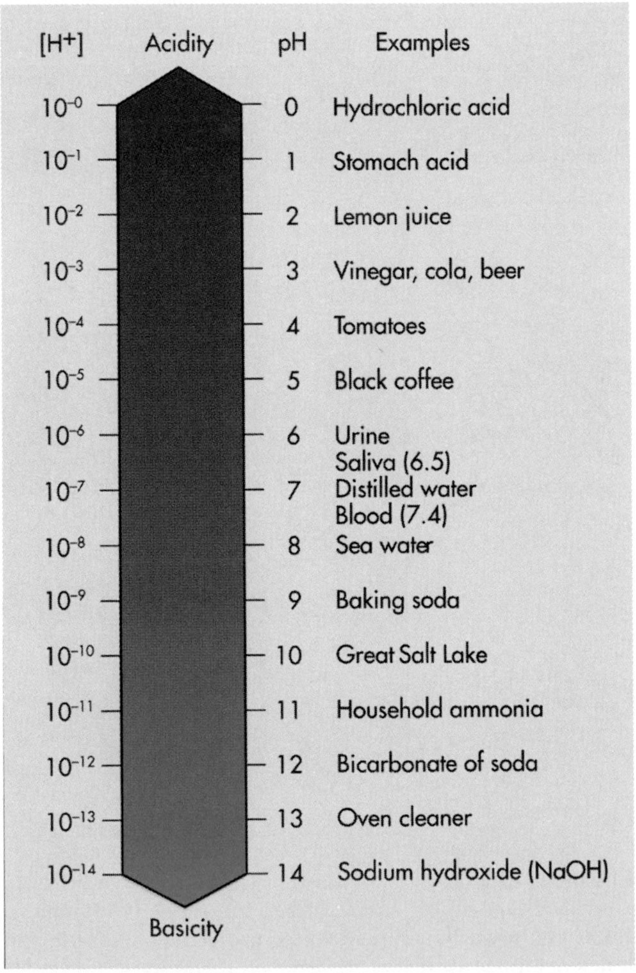

FIGURE 2-9 The pH scale. pH refers to the relative concentration of H^+ ions in a solution. Low pH values indicate high concentrations of H^+ ions (acids), and high pH values indicate low concentrations of H^+ (bases).

Inorganic Versus Organic Chemistry

With the exception of water, most molecules that are formed by living organisms and make up their structures contain the element *carbon.* This fascinating element lends itself to being the basis of living material because of its ability to form four bonds with other atoms and therefore interact with other molecules in myriad ways. Silicon, an atom with similar bonding properties, is abundant in the Earth's crust but not in living things. However, silicon is used by some organisms in building outer skeletons or shells. An example of a group of such organisms is the diatoms, one of the most abundant protists on this planet.

The carbon-containing molecules that make up living things are called **organic compounds.** Organic compounds are often very large and are usually held together by covalent bonds. The study of organic molecules and their interactions is called *organic chemistry.* The study of molecules and substances not containing carbon is called *inorganic chemistry.* Inorganic compounds are often quite small and are usually held together by ionic bonds. A few carbon-containing, covalently bonded inorganic molecules exist (such as carbon dioxide and carbonic acid), but they are considered inorganic molecules because they do not make up the structure of living things.

Although carbon is the underlying component of biological molecules, 10 other elements are common in living organisms. These elements are listed in Table 2-1 and are compared according to their frequency in the Earth's crust. Notice that the great majority of atoms in living things (96.3% in fact) are either oxygen, carbon, hydrogen, or nitrogen.

The Chemical Building Blocks of Life

It is often helpful to think of an organic molecule as a carbon-based core with other special parts attached. Each of these special parts is really a group of atoms called a **functional group** and has definite chemical properties. A hydroxyl group (−OH), for example, is a functional group. The most important functional groups are illustrated in Figure 2-10. These groups are important because most chemical reactions that occur within organisms involve the transfer of a functional group from one molecule to another. Other frequent chemical reactions involve the breaking of carbon-carbon bonds.

Some of the molecules of living things are simple organic molecules, often having only one functional group. Other molecules, called *macromolecules,* are far larger and contain thousands of atoms. Macromolecules often have many functional groups. There are four major groups of biologically important macromolecules: complex carbohydrates, lipids, proteins, and nucleic acids. The four major classes of biologically important macromolecules are presented in Table 2-2.

Although these classes of macromolecules are each composed of different building blocks, the process by which their building blocks are put together is the same. All except the lipids are polymers, macromolecules that are built by forming covalent bonds between similar building blocks, or monomers, to form long chains. **Dehydration synthesis** is a process by which monomers (and the building blocks of lipids) are put together. During dehydration synthesis, one molecule of water is removed (dehydration) from each two monomers that are joined (synthesis). One monomer loses its hydroxyl group (−OH), and the other loses an atom of hydrogen (H). Having lost electrons they were sharing in covalent bonding, both monomers bond covalently with one another (see Figure 2-19). The process of dehydration synthesis uses energy, which is stored in the bond that is made, and takes place with the help of special molecules called *enzymes* (see Chapter 4).

soluble (sol you ble) able to be dissolved in water.

hydrophobic (hi dro fo bic) a term referring to nonpolar molecules that cannot form hydrogen bonds with water.

ionization (eye uh ni zay shun) the spontaneous formation of charged particles (ions) very typical of water molecules, caused by the breaking of the molecules' covalent bonds.

pH scale a succession of values (scale) based on the slight degree of spontaneous ionization of water that indicates the concentration of H^+ ions in a solution.

acid any substance that dissociates to form H^+ ions when it is dissolved in water.

base any substance that combines with H^+ ions when it is dissolved in water.

organic compounds the carbon-containing molecules that make up living things.

functional group a group of atoms with definite chemical properties that is attached to the carbon-based core of an organic molecule.

dehydration synthesis the process by which monomers are put together to form polymers.

Polymers are disassembled in an opposite process called **hydrolysis** (see Figure 2-19). During hydrolysis the bonds are broken between monomers with the addition of water (and in the presence of enzymes). In fact, the term *hydrolysis* literally means "to break apart" (lysis) "by means of water" (hydro). The hydroxyl group of a water molecule bonds to one monomer, and the hydrogen atom bonds to its neighbor. The energy held in the bond is released.

Carbohydrates

Carbohydrates are molecules that contain carbon, hydrogen, and oxygen, with the concentration of hydrogen and oxygen atoms in a 2:1 ratio. Abundant energy is locked in their many carbon-hydrogen bonds. Plants, algae, and some bacteria produce carbohydrates by the process of photosynthesis. Most organisms use carbohydrates as an

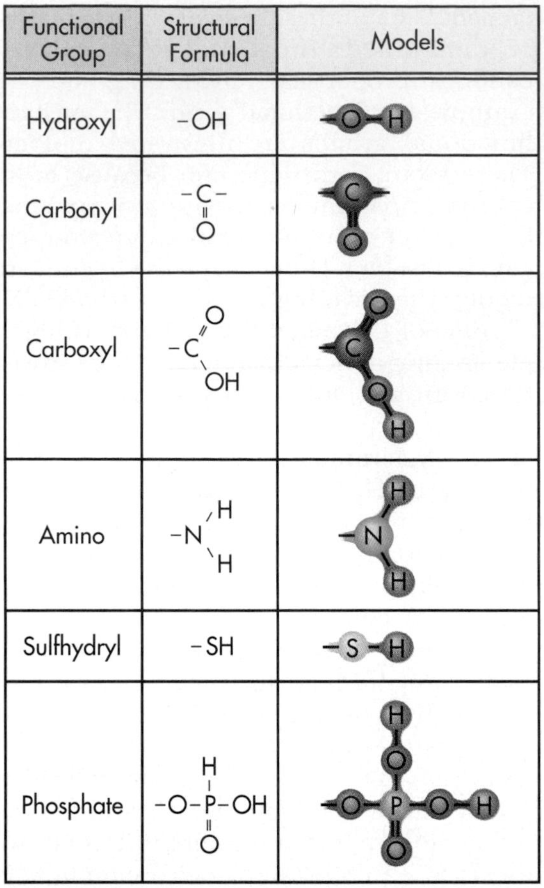

FIGURE 2-10 The chemical building blocks of life, functional groups. Six of the most important functional groups involved in chemical reactions.

TABLE 2-2

Biologically Important Macromolecules

MACROMOLECULE	SUBUNIT	FUNCTION	EXAMPLE
Carbohydrates			
Starch, glycogen	Glucose	Stores energy	Potatoes
Cellulose	Glucose	Makes up cell walls in plants	Paper
Chitin	Modified glucose	Makes up the exterior skeleton in some animals	Crab shells
Lipids			
Fats	Glycerol + three fatty acids	Store energy	Butter
Phospholipids	Glycerol + two fatty acids + phosphate	Make up cell membranes	All membranes
Steroids	Four carbon rings	Act as chemical messengers	Cholesterol, estrogen
Proteins			
Globular	Amino acids	Help chemical reactions take place	Hemoglobin
Structural	Amino acids	Make up tissues that support body structures and provide movement	Muscle
Nucleic Acids			
DNA	Nucleotides	Helps code hereditary information	Chromosomes
RNA	Nucleotides	Helps decode hereditary information	Messenger RNA

important fuel, breaking these bonds and releasing energy to sustain life.

Among the least complex of the carbohydrates are the simple sugars or **monosaccharides.** This word comes from two Greek words meaning "single" (monos) and "sweet" (saccharon) and reflects the fact that monosaccharides are individual sugar molecules. Some of these sweet-tasting sugars have as few as three carbon atoms. The monosaccharides that play a central role in energy storage, however, have six. The primary energy-storage molecule used by living things is glucose ($C_6H_{12}O_6$), a six-carbon sugar with seven energy-storing carbon-hydrogen bonds. Notice in Figure 2-11 that glucose, like other sugars, exists as a straight chain or as a ring of atoms.

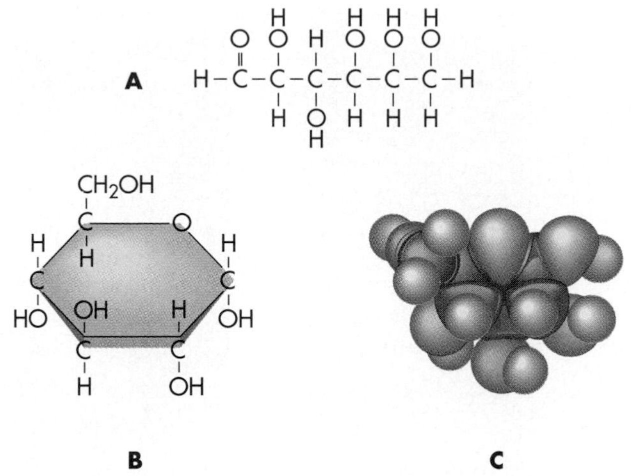

FIGURE 2-11 Structure of a glucose molecule. The structural formula of glucose in its linear form **(A)** and as a ring structure **(B)**. Three-dimensional model of glucose **(C)**. (Hydrogen, blue; Oxygen, red; Carbon, black).

> *Sugars are among the most important energy-storage molecules in living things.*

Glucose is not the only sugar with the formula $C_6H_{12}O_6$. Other monosaccharides having this same formula are fructose and galactose. Because these molecules have the same molecular formula as glucose but are put together slightly differently, they are called *isomers,* or alternative forms, of glucose (Figure 2-12). Your taste buds can tell the difference: fructose is much sweeter than glucose.

In living systems, two of these three sugars are often found covalently bonded to one another. Two monosaccharides linked together form a **disaccharide.** Many organisms, such as plants, link monosaccharides together to form disaccharides that are less readily broken down while being transported within the organism (Figure 2-13). Sucrose (table sugar) is a disaccharide formed by linking a molecule of glucose to a molecule of fructose. It is the common transport form of sugar in plants. Lactose, or milk sugar (glucose + galactose), is a disaccharide produced by many mammals to feed their young.

Not only do organisms unlock, use, and transport the energy within carbohydrate molecules, they store this energy. To do this, however, organisms must convert soluble sugars such as glucose to an insoluble form to be stored. Sugars are made insoluble by joining them together into long polymers called **polysaccharides.** Plants store energy in polysaccharides called **starches.** The starch amylose, for example, is made up of hundreds of glucose molecules linked together in long, unbranched chains. Most plant starch is a branched version of amylose called *amylopectin.* Animals store glucose in highly branched polysaccharides called **glycogen** (Figure 2-14).

> *Starch and glycogen, both types of polysaccharides, are storage forms of glucose. Plants store sugar as starch, whereas animals store sugar as glycogen.*

The chief component of plant cell walls is a polysaccharide called *cellulose.* Cellulose is chemically similar to amylose but is bonded in a way that

hydrolysis (hi **drol** ul sis) the process by which polymers are disassembled into monomers.
carbohydrates molecules that contain carbon, hydrogen, and oxygen, with the concentration of hydrogen and oxygen atoms in a 2:1 ratio.
monosaccharides (**mon** o **sack** uh rides) simple sugars.
disaccharide (dye **sack** uh ride) two monosaccharides linked together.
polysaccharides (**pol** ee **sack** uh rides) long, insoluble polymers composed of sugar.
starches polysaccharides that are the storage form of sugar in plants.

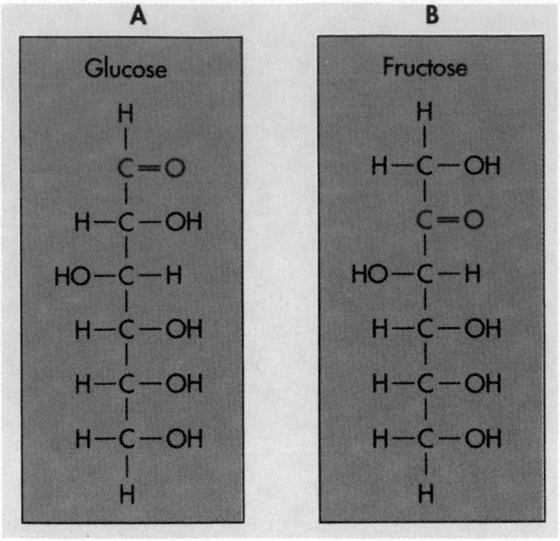

FIGURE 2-12 Isomers. Glucose (A) and fructose (B) have the same molecular formula of $C_6H_{12}O_6$, but their atoms are arranged differently within the molecule, making them alternative forms (isomers).

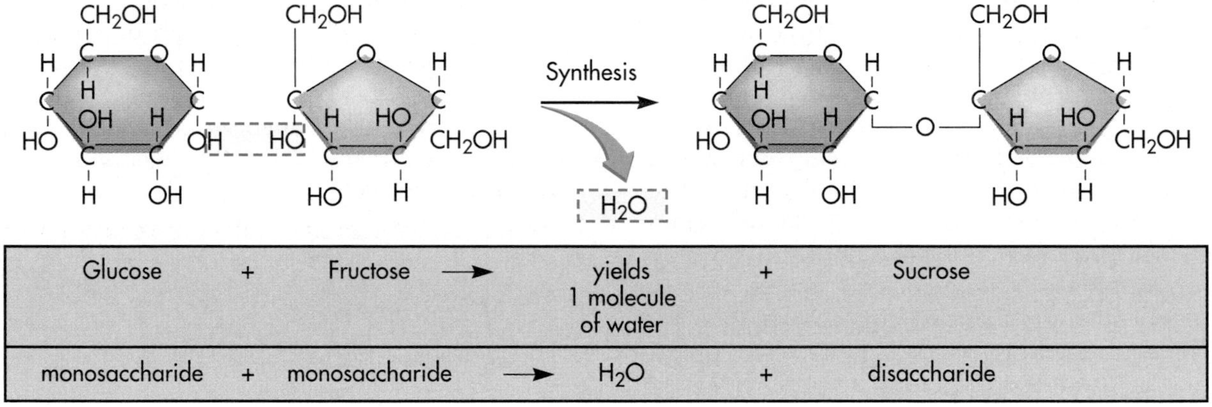

FIGURE 2-13 Disaccharides. Sucrose is a disaccharide that is formed from the linkage of two monosaccharides, glucose and fructose. Sucrose is a common transport form of sugar in plants.

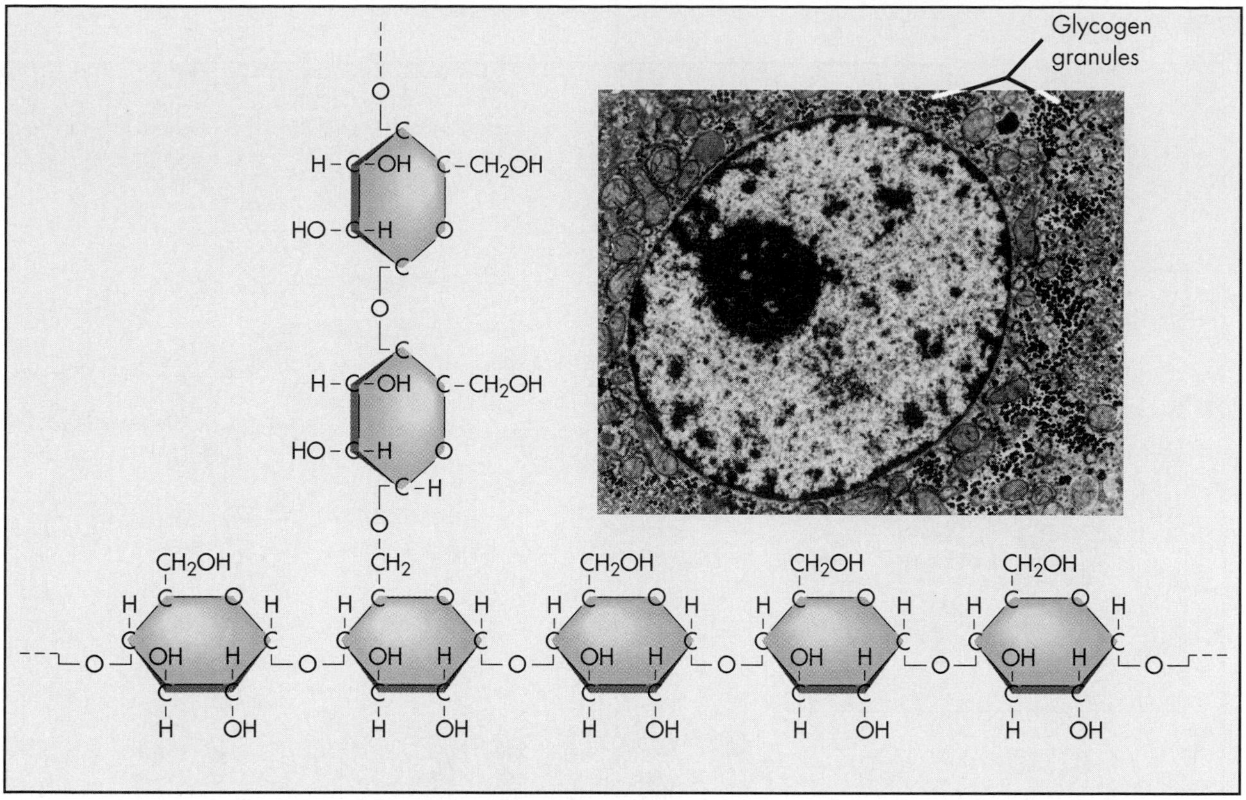

FIGURE 2-14 Glycogen, a starch. Animals store energy by converting glucose into a highly branched polysaccharide called glycogen. The electron micrograph shows a liver cell surrounded by this storage molecule (*8700x*).

most organisms cannot digest (Figure 2-15). For this reason, cellulose works well as a biological structural material and occurs widely in this role in plants. The structural material in insects, many fungi, and certain other organisms is a modified form of cellulose called *chitin*. Chitin is a tough, resistant surface material that is also relatively indigestible.

Fats and Lipids

When organisms store glucose molecules for long periods, they usually store them as **fats** rather than as carbohydrates. Fats are large molecules made up of carbon, hydrogen, and oxygen, as are the carbohydrates, but their hydrogen-to-oxygen ratio is higher than 2:1. For this reason, fats contain more energy-storing carbon-hydrogen bonds than carbohydrates. In addition, fats are nonpolar, insoluble molecules, so they work well as storage molecules.

Fats are only one kind of **lipid.** Lipids include a wide variety of molecules, all of which are soluble in oil but insoluble in water. This insolubility is because almost all the bonds in lipids are nonpolar carbon-carbon or carbon-hydrogen bonds (see p. 32). Three important categories of lipids are (1) oils, fats, and waxes; (2) phospholipids; and (3) steroids.

Lipids are composite molecules; that is, they are made up of more than one component. Oils and fats are built from two different kinds of subunits:

1. Glycerol: Glycerol is a three-carbon molecule with each carbon bearing a hydroxyl (−OH) group. The three carbons form the backbone of the fat molecule.
2. Fatty acids: Fatty acids have long *hydrocarbon* chains (chains consisting only of carbon and hydrogen atoms) ending in a carboxyl (−COOH) group. Three fatty acids are

glycogen (**glye** ko jen) highly branched polysaccharides that are the storage form of sugar in animals.
fats large molecules made up of carbon, hydrogen, and oxygen, with a hydrogen-to-oxygen ratio higher than 2:1; one type of lipid.
lipid composite molecules made up of glycerol and fatty acids (in the case of oils and fats) or carbon rings (in the case of steroids).
triglyceride (try **gliss** er ide) a fat molecule in which three fatty acids are attached at each of the three carbon atoms of a glycerol molecule.

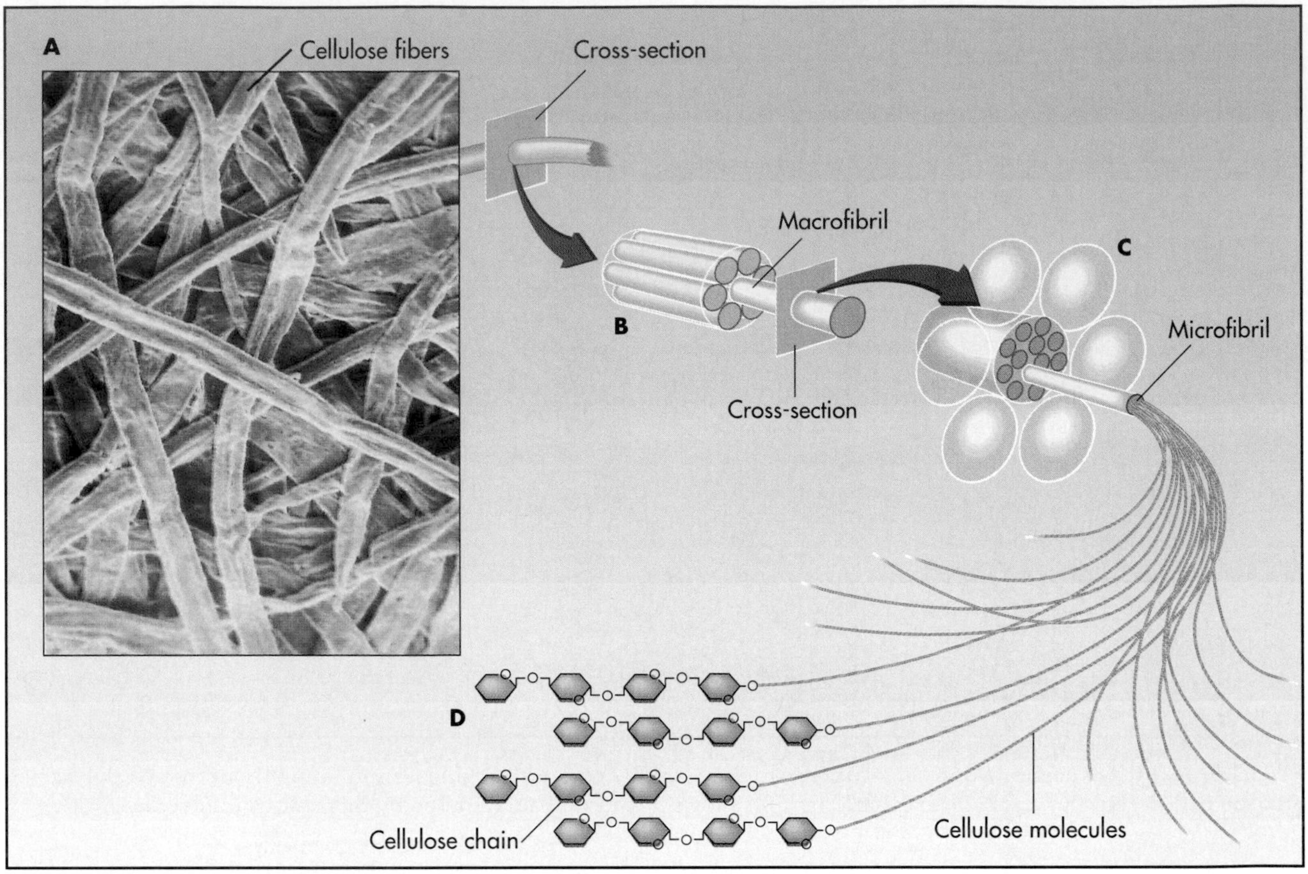

FIGURE 2-15 Structure of cellulose. Cellulose fibers from a ponderosa pine (**A**). Macrofibrils compose each fiber (**B**). Each macrofibril is composed of bundles of microfibrils (**C**). Microfibrils, in turn, are composed of bundles of cellulose chains (**D**). Cellulose fibers can be very strong; this is one reason why wood is such a good building material.

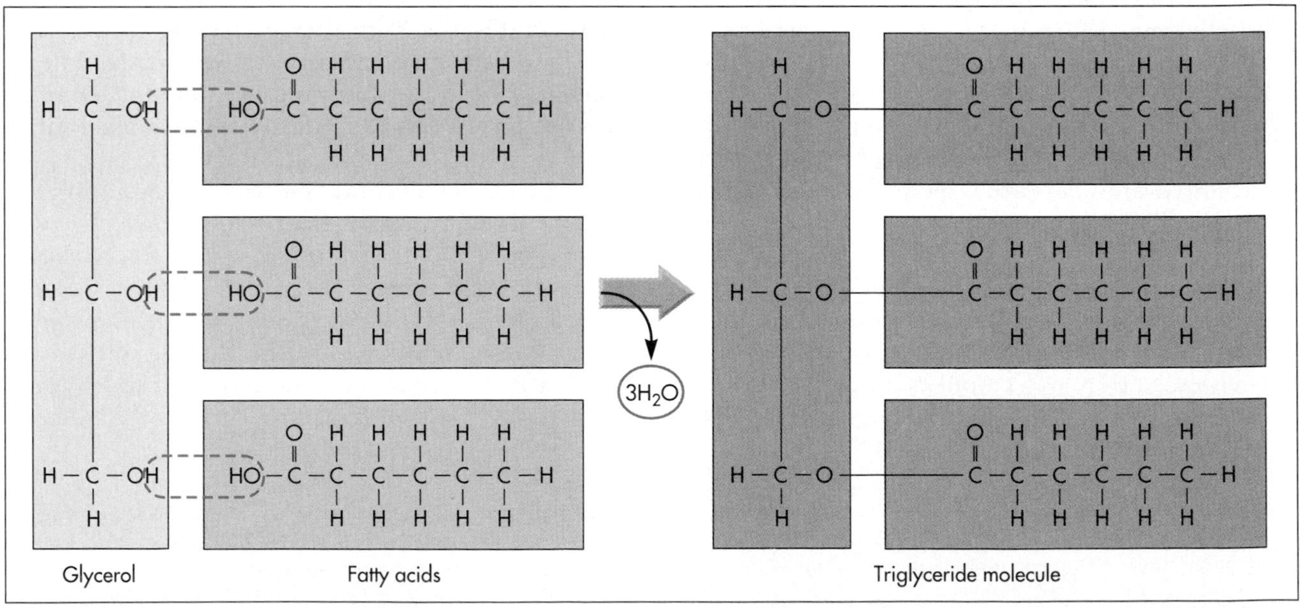

FIGURE 2-16 Structure of a triglyceride. Triglycerides are composite molecules, made up of three fatty acid molecules bonded to a single glycerol molecule with the loss of water molecules.

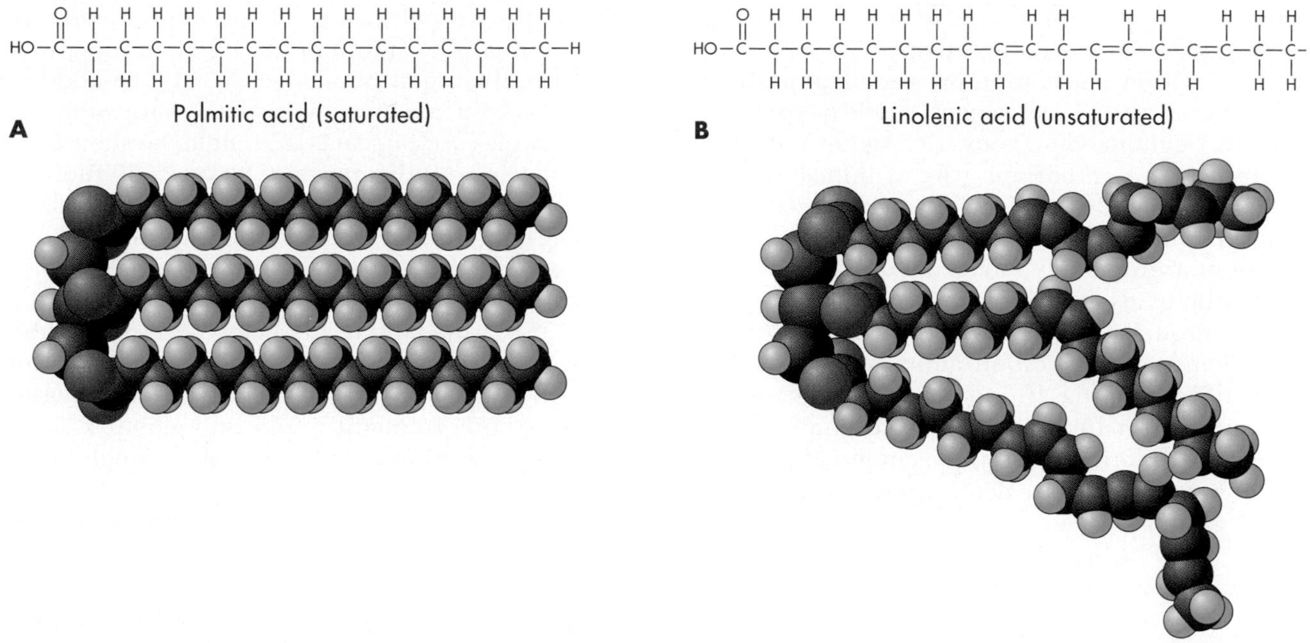

FIGURE 2-17 Saturated and polyunsaturated fats. A, Palmitic acid, a fatty acid with only single bonds between its carbon atoms, has a maximum of hydrogen atoms and is a saturated fat. **B,** Linolenic acid, with three double bonds and thus fewer than the maximum number of hydrogen atoms bonded to the carbon chain, is an unsaturated fatty acid.

attached to each glycerol backbone (Figure 2-16). Because there are three fatty acids, the resulting fat molecule is called a **triglyceride:**

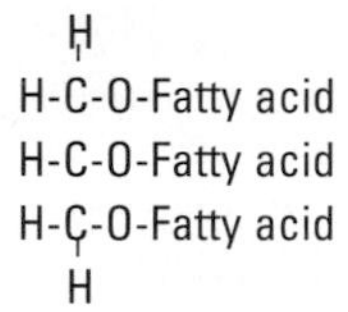

The difference between fats and oils has to do with the number of double bonds in their fatty acids. As Figure 2-17 shows, a fatty acid with only single bonds between its carbon atoms can hold more hydrogen atoms than a fatty acid with double bonds between its carbon atoms. A fatty acid that carries as many hydrogen atoms as possible, such as the fatty acid in Figure 2-17, *A,* is saturated. Fats composed of fatty acids with double bonds are unsaturated because the double bonds replace some of the hydrogen atoms. If a fat has more than one double bond, it is polyunsaturated. Polyunsaturated fats (Figure 2-17, *B*) have low melting points and are therefore liquid fats, or oils. The fatty acids of most plant triglycerides such as vegetable oils are unsaturated. (Exceptions are the tropical oils.) Animal fats, in contrast, are often saturated and occur as hard fats. Human diets with large amounts of saturated fats appear to upset the normal balance of fatty acids in the body, a situation that can lead to diseases of the circulatory system.

> *Fats are important energy-storing molecules. They are made up of three fatty acid chains attached to a glycerol backbone.*

Waxes, which are used by land plants and some animals as a waterproofing material, differ from fats and oils by having a chemical backbone slightly different from glycerol. Phospholipids are also similar to oils except that one of their fatty acids is replaced by a phosphate group attached to a nitrogen-containing group. Phospholipids play a key role in the structure of cell membranes (see Chapter 3). Membranes often contain steroids, a lipid having a structure very different from oils. Steroids are composed of four carbon rings. Most of your cell membranes contain the steroid cholesterol. Male and female sex hormones (discussed in Chapter 21) are also steroids.

Proteins

Proteins are the third major group of macromolecules that make up the bodies of organisms. Proteins play diverse roles in living things. Perhaps

> **proteins** long, complex chains of amino acids linked end to end.

the most important proteins are **enzymes,** proteins capable of speeding up specific chemical reactions. Other short proteins called **peptides** are used as chemical messengers within your brain and throughout your body. Collagen, a structural protein, is an important part of bones, cartilage, and tendons. Despite their varied functions, all proteins have the same basic structure: a long chain of amino acids linked end to end.

Amino acids are small molecules containing an amino group ($-NH_2$), a carboxyl group (–COOH), a hydrogen atom, a carbon atom, and a *side chain* that differs among amino acids. In a generalized formula for an amino acid, the side chain is shown as *R*. The identity and unique chemical properties of each amino acid are determined by the nature of the R group.

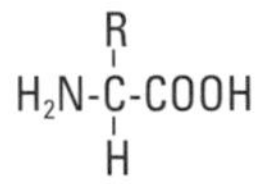

Only 20 different amino acids make up the diverse array of proteins found in living things. Each protein differs according to the amount, type, and arrangement of amino acids that make up its structure. These 20 "common" amino acids are illustrated in Figure 2-18. They are grouped according to the chemical nature of their side chains.

Each amino acid has a free amino group ($-NH_2$) at one end and a free carboxyl group (–COOH) at the other end. During dehydration synthesis, each of these groups on separate amino acids loses a molecule of water between them, forming a covalent bond that links the two amino acids (Figure 2-19, *A*). This bond is called a *peptide bond.* A long chain of amino acids linked by peptide bonds is a *polypeptide.* Proteins are long, complex polypeptides. The great variability possible in the sequence of amino acids in polypeptides is perhaps the most important property of proteins, permitting tremendous diversity in their structures and functions.

The sequence of amino acids that makes up a particular polypeptide chain is termed the *primary structure* of a protein (Figure 2-20). This sequence determines the further levels of structure of the protein molecule resulting from bonds that form between these groups. Having the proper sequence of amino acids, then, is crucial to the functioning of a protein. Put simply, if the protein does not assume its correct shape, it will not work properly or at all. Because different amino acid functional groups have different chemical properties, the shape of a protein may be altered by a single amino acid change.

The functional groups of the amino acids in a polypeptide chain interact with their neighbors, forming hydrogen bonds (see p. 31). In addition, portions of a protein chain with many nonpolar side chains (see Figure 2-18) tend to be shoved into the interior of the protein because of their hydrophobic properties. Because of these interactions, polypeptide chains tend to fold spontaneously into sheets or wrap into coils. This folded or coiled shape is called its *secondary structure.* Proteins made up largely of sheets often form fibers such as keratin fibers in hair, fibrin in blood clots, and silk in spiders' webs. Proteins that have regions forming coils frequently fold into globular shapes such as the globin subunits of hemoglobin in blood.

Hydrogen bonding can also result in proteins with more complex shapes than the secondary structure. The next level of structure is called *tertiary structure.* For proteins that consist of subunits (separate polypeptide chains), the way these subunits are assembled into a whole is called the *quaternary structure.*

> *Approximately 20 different amino acids are used in various sequences and combinations to make up a wide variety of proteins. These amino acids have different properties from one another. These differences and the sequence in which the amino acids are arranged determine the shape of a protein, which is crucial to its proper functioning.*

Nucleic Acids

Organisms store information about the structures of their proteins in macromolecules called **nucleic acids.** Nucleic acids are long polymers of repeating subunits called *nucleotides.* Each nucleotide is made up of three smaller building blocks (Figure 2-21):

1. A five-carbon sugar
2. A phosphate group ($-PO_4^{=}$)
3. An organic, nitrogen-containing molecule called a *base*

To form the nucleic acid chain, the sugars and phosphate groups making up the nucleotides are linked; a nitrogenous base protrudes from each sugar (Figure 2-22). The order in which the nucleotides are linked together forms a code that ultimately specifies the order of amino acids in a particular protein.

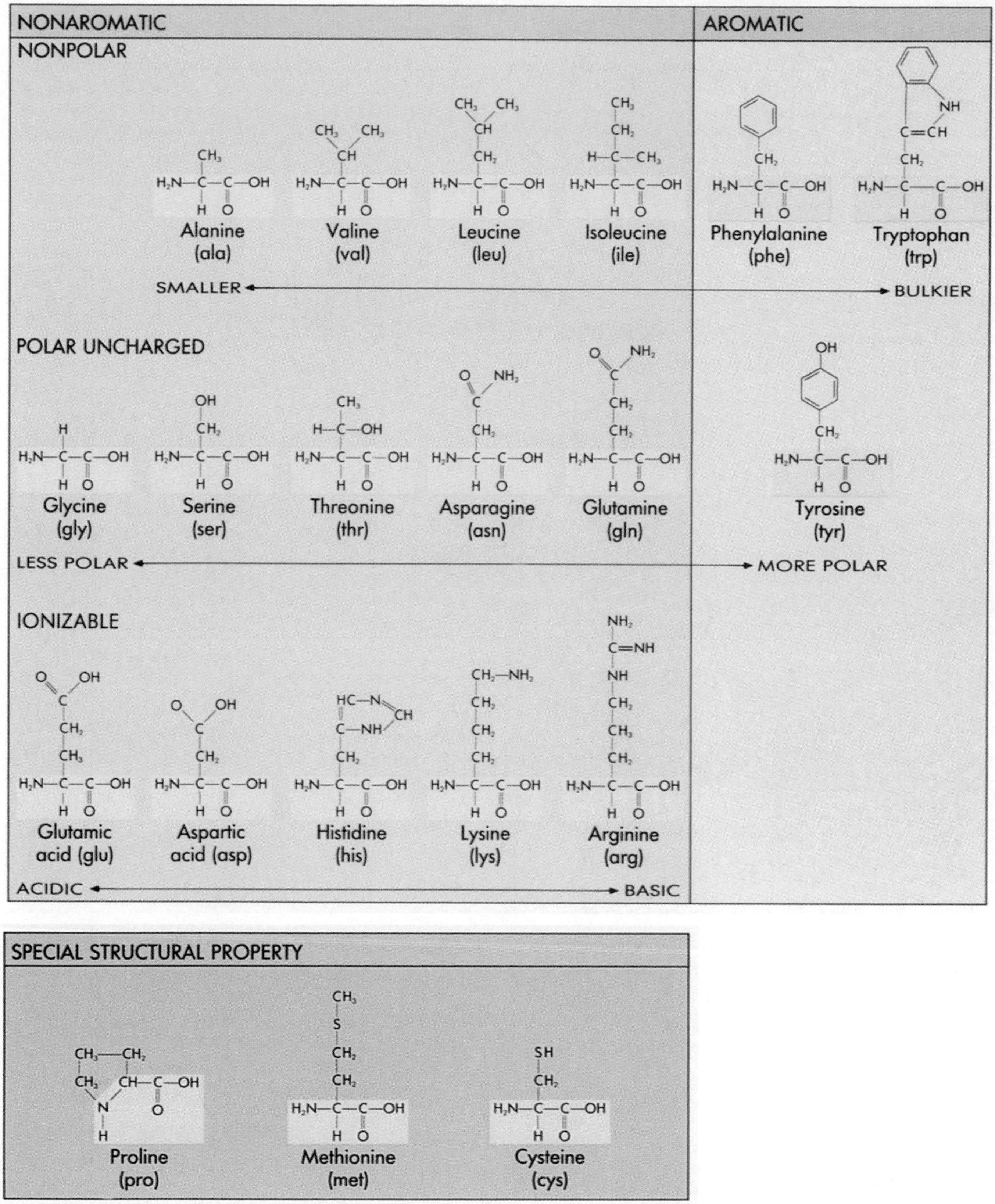

FIGURE 2-18 The 20 common amino acids. Each amino acid has the same chemical backbone (*yellow box*) but differs from the others in the side chain, or R group. Six of the amino acid R groups are nonpolar. Those forming ring structures are called aromatic amino acids. Another six are polar but do not ionize to charged forms. Five more are also polar but capable of ionizing to charged forms. The remaining three have special chemical (structural) properties that play important roles in forming links between protein chains or forming kinks in their shapes.

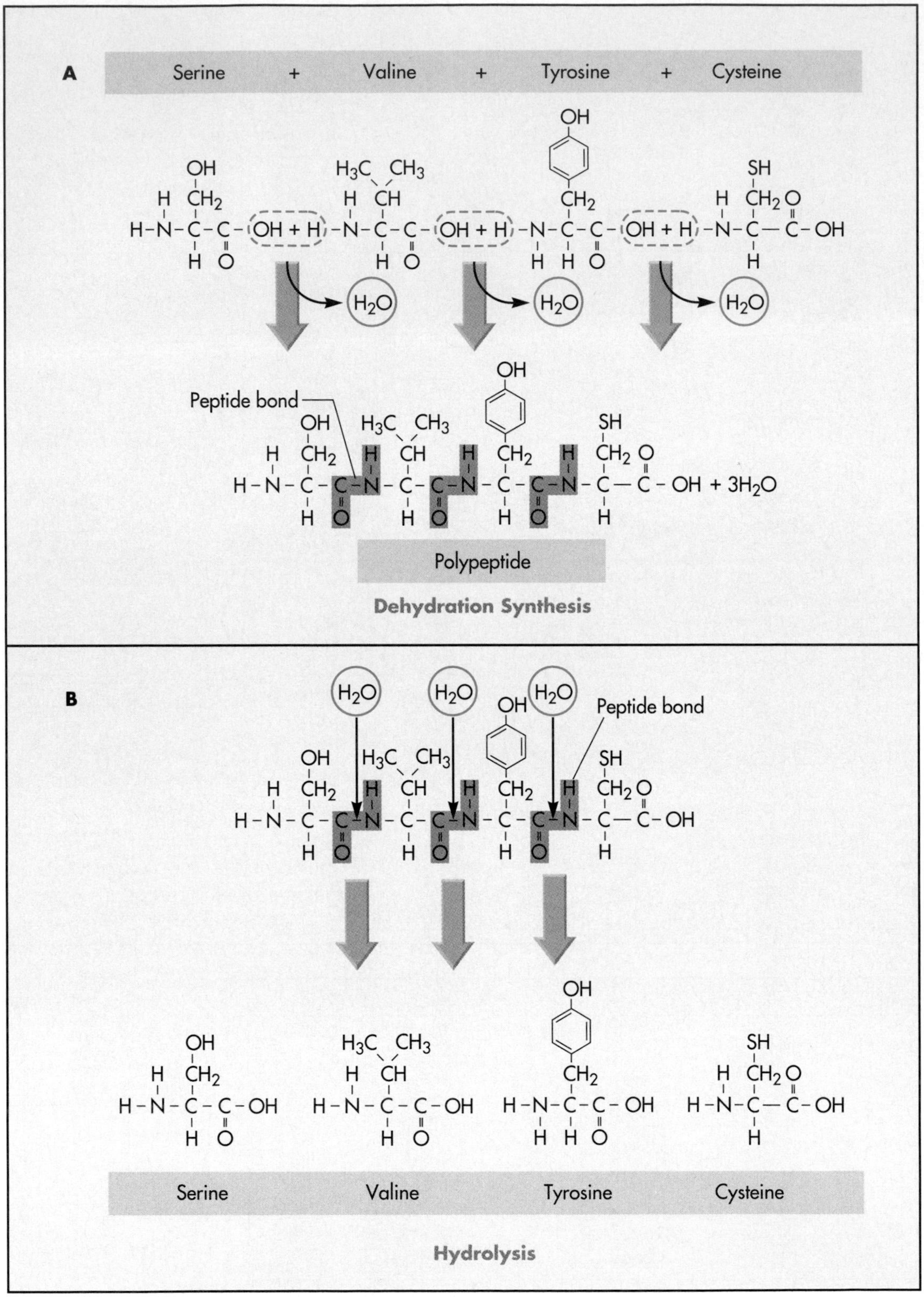

FIGURE 2-19 How a polypeptide chain is formed and broken. **A,** During dehydration synthesis, peptide bonds are formed between adjacent amino acids. A chain of amino acids is called a polypeptide. **B,** During hydrolysis, a molecule of water is added to each peptide bond that links adjacent amino acids, breaking the bond between them. This separates the molecules into individual amino acids.

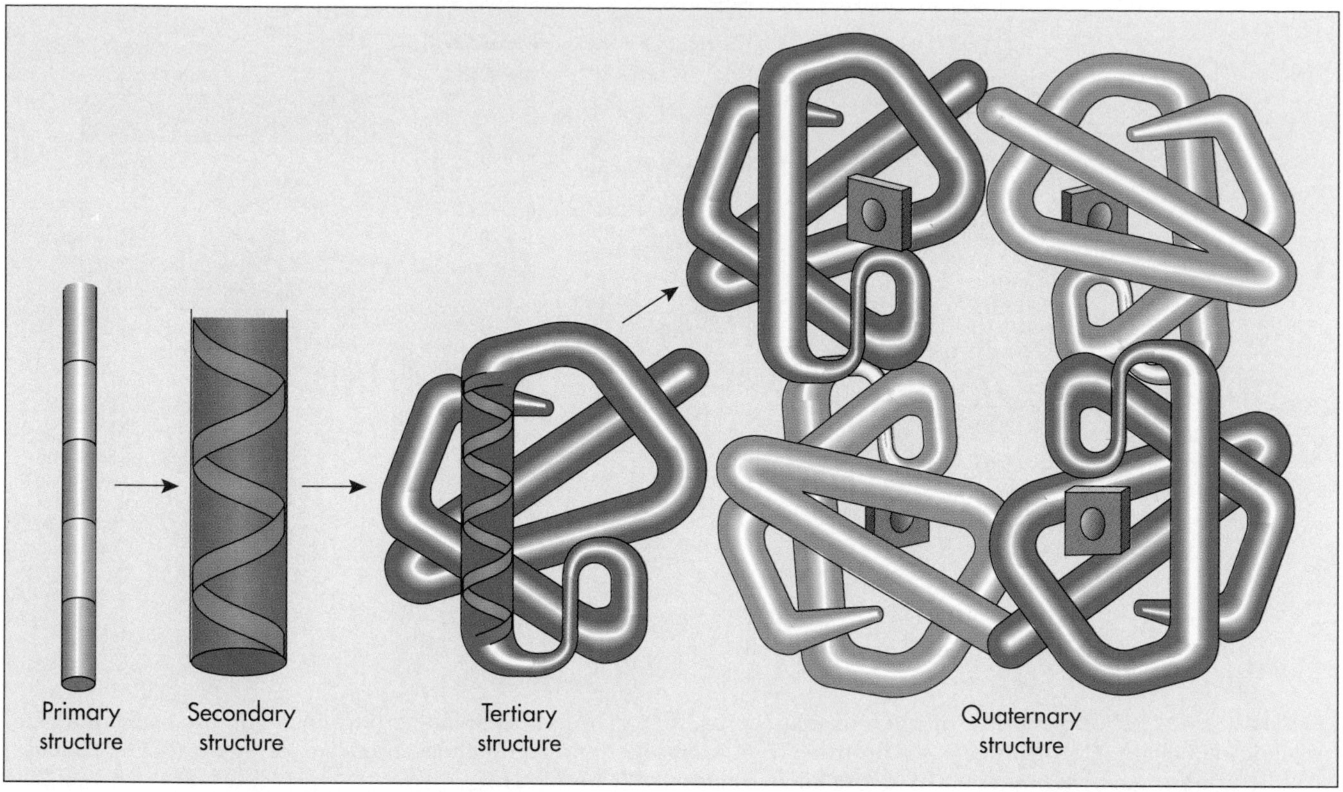

FIGURE 2-20 How primary structure determines a protein's shape. The sequence of amino acids that makes up a particular polypeptide chain of a protein is known as the primary structure. This sequence determines the further levels of structure of the protein molecule resulting from bonds that form along the chain. The proper sequence of amino acids, as well as shape, is crucial to the functioning of a protein.

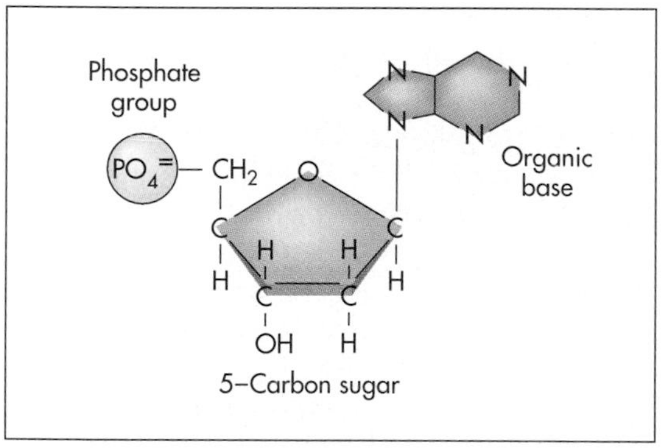

FIGURE 2-21 The structure of a nucleotide. A nucleotide is composed of a five-carbon sugar, a phosphate group, and an organic nitrogen base.

Hereditary information is a code composed of a sequence of nucleotides that signifies the sequence of amino acids in the proteins of living things.

Organisms have two forms of nucleic acid. One form, deoxyribonucleic acid (DNA), stores the information for making proteins. The other form, ribonucleic acid (RNA), directs the production of proteins. Details of the structure of DNA and the ways it interacts with RNA are presented in Chapter 18.

enzymes proteins capable of speeding up specific chemical reactions.
peptides short chains of amino acids.
nucleic acids (new **klay** ick) long polymers made of repeating subunits called nucleotides; nucleic acids store information about the structure of proteins.

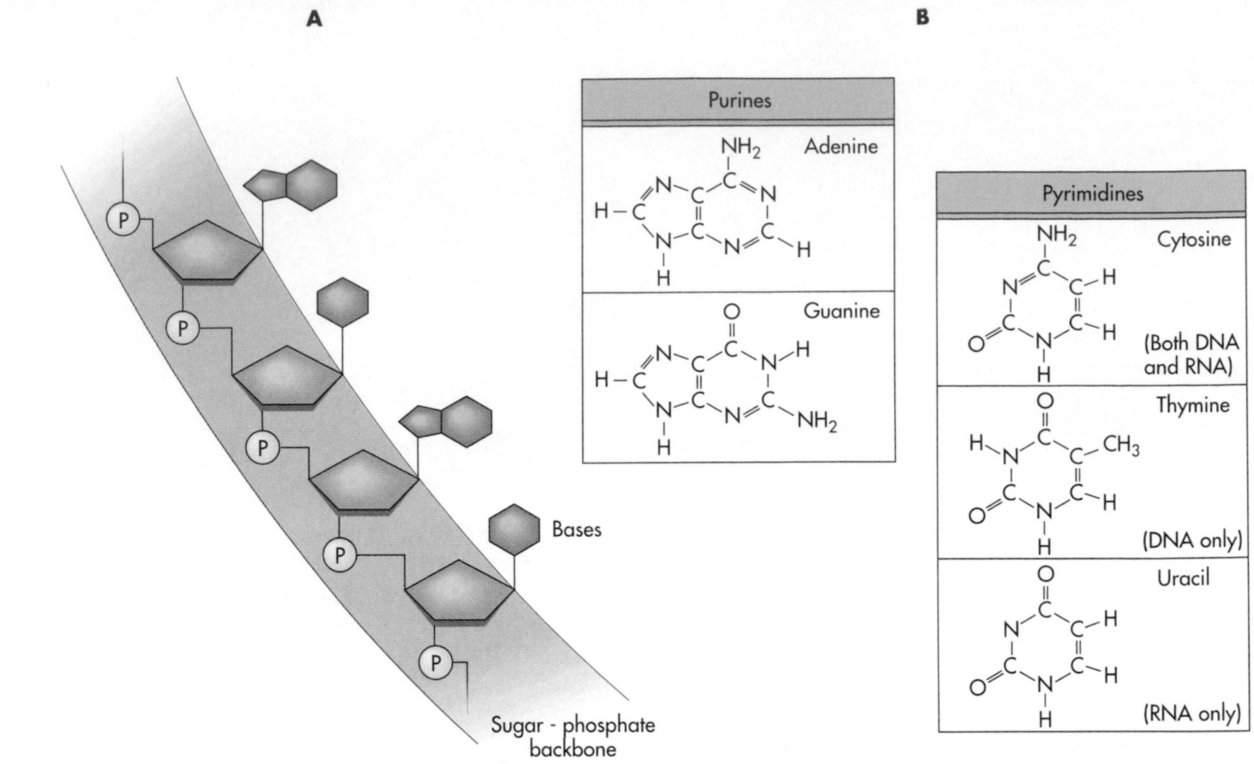

FIGURE 2-22 The formation of nucleic acid chains. A, Sugar and phosphate groups make up the backbone of a nucleic acid chain. Nitrogenous bases protrude from each sugar and complete each nucleotide subunit of the chain. **B,** The five nitrogen bases that occur in the nucleic acids of DNA and RNA.

Summary

- Atoms make up all matter and are composed of three types of subatomic particles: protons, neutrons, and electrons. Protons and neutrons are found at the core of the atom, and electrons surround this core.
- The chemical behavior of an atom is largely determined by the distribution of its electrons, particularly the number of electrons in its outermost energy level. Atoms are most stable when their outer energy levels are filled. Electrons are lost, gained, or shared until this condition is reached.
- The atoms of most elements interact with one another. Two or more atoms held together by sharing electrons are molecules. Molecules can be made up of atoms of the same element or atoms of different elements. Molecules made up of atoms of different elements are called compounds. Compounds are also atoms of different elements held together by electrostatic attraction.
- Water is an extremely important molecule for living things. It has one outstanding chemical property: it is a polar molecule, having positive and negative ends, and attracts other charged particles.
- Water dissolves polar substances and excludes nonpolar substances. Chemical interactions readily take place in water because so many kinds of molecules are water soluble and therefore move among water molecules as separate molecules or ions.
- The chemistry of life is concerned with the interactions among biological molecules, or organic molecules. Organic molecules have a carbon core with various functional groups attached to the carbon atoms.
- Most organisms use carbohydrates as an important fuel. The most important of the energy-storing carbohydrates is glucose, a six-carbon sugar.
- Certain lipids—the fats—are important in the long-term storage of energy. Phospholipids are a major component of cell membranes.
- Proteins are linear polymers of amino acids. Because the 20 amino acids that occur in proteins have side chains with differing chemical properties, the shape and therefore the functioning of a protein are critically affected by its particular sequence of amino acids.
- Hereditary information is stored as a sequence of nucleotides in the nucleotide polymer DNA. A second form of nucleic acid, RNA, directs the production of proteins.

Knowledge and Comprehension Questions

1. a. Draw a diagram of an atom with an atomic number of 1. Label the nucleus and the subatomic particles, and show the electrical shape of each particle.
   b. Draw an isotope of the same atom.
2. List and explain three factors that influence how an atom interacts with other atoms. What is the significance of the octet rule?
3. Nitrogen gas is formed via a triple bond between two nitrogen atoms. What type of bonding is this? What does this bond have in common with the bonds forming water molecules?
4. What characteristics of water make it so unusual? How do these traits affect the interactions of other molecules dissolved or suspended in it? Relate these properties to water's molecular structure.
5. Distinguish between organic and inorganic molecules. Which would you primarily study if you wanted to learn more about the human body?
6. The pH of the digestive juices within the human small intestine is between 7.5 and 8.5. Would you describe this environment as acidic, neutral, or basic? Would the concentration of H^+ ions in this material be high or low with respect to a neutral pH?
7. Both carbon and water are discussed in this chapter as the essential components of life on Earth. How are the bonding characteristics of each important aspects of their biological importance?
8. Distinguish among monosaccharides, disaccharides, and polysaccharides. To what group of biological macromolecules do they belong, and why are they important?
9. Why do plants and animals not store glucose as is for future use? What forms of storage are most common in plants and animals?
10. Structurally and functionally, compare and contrast RNA and DNA.
11. Discuss the three classes of macromolecules taken in as food energy by humans. Which chemical process, dehydration synthesis or hydrolysis, do you think is essential in the digestion of food?

Critical Thinking

1. Based on what you learned in this chapter about the interactions of oil and water, suggest at least one reason why it is necessary to clean up an oil spill in the ocean.
2. While doing your grocery shopping, you find this label on a product:

Nutritional Information Per Serving	
Calories 270	Calories from fat 45
Total fat	5 grams
Saturated fat	1.5 grams
Total carbohydrates	48 grams
Sugar	4 grams
Protein	8 grams

   What macromolecules are present in this food? Would this food be part of a heart-healthy diet? Defend your answer with evidence.

3. Name two functions of proteins in living systems. How might an error in the DNA of an organism affect protein function?

PART TWO

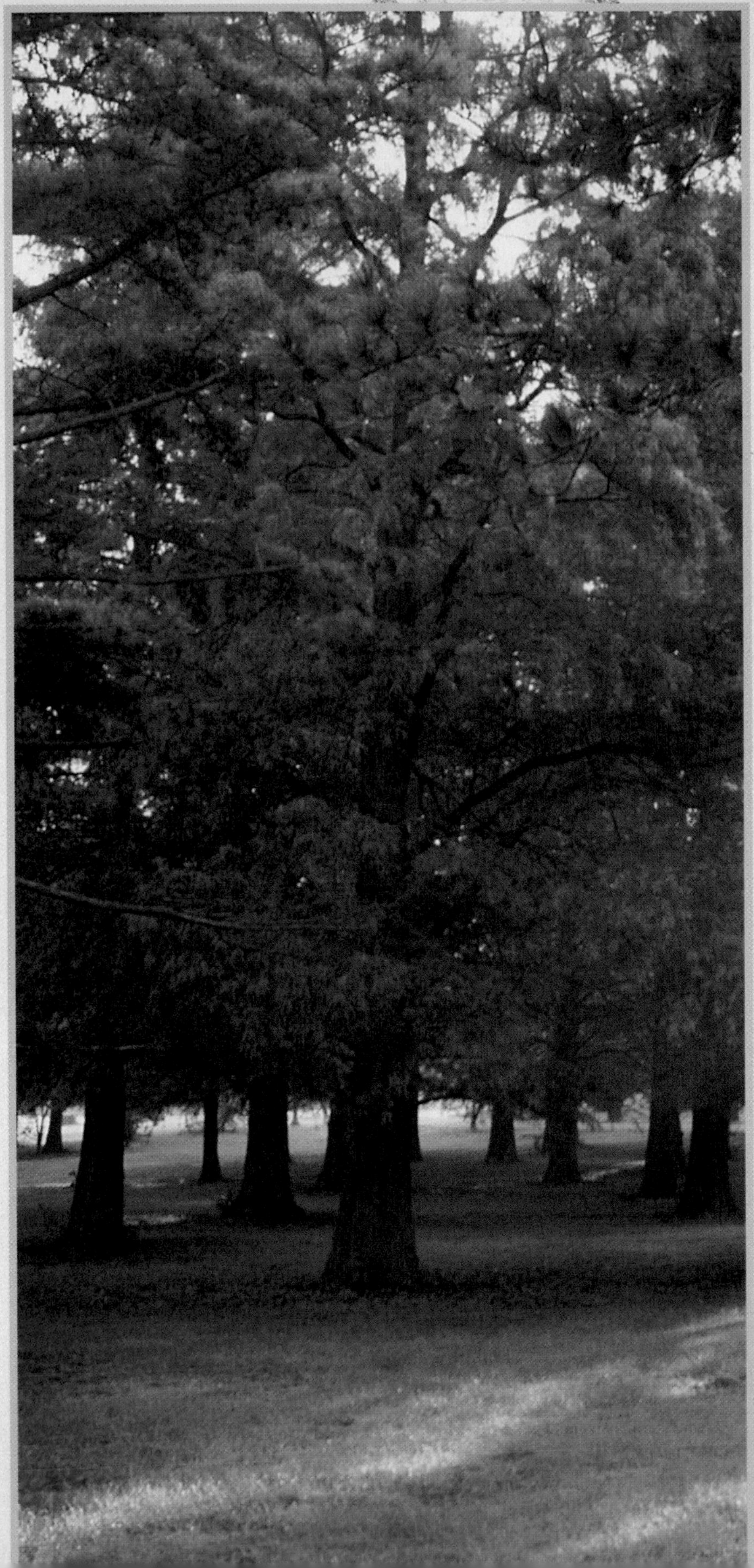

CELLS AND HOW THEY TRANSFORM ENERGY

CHAPTER 3

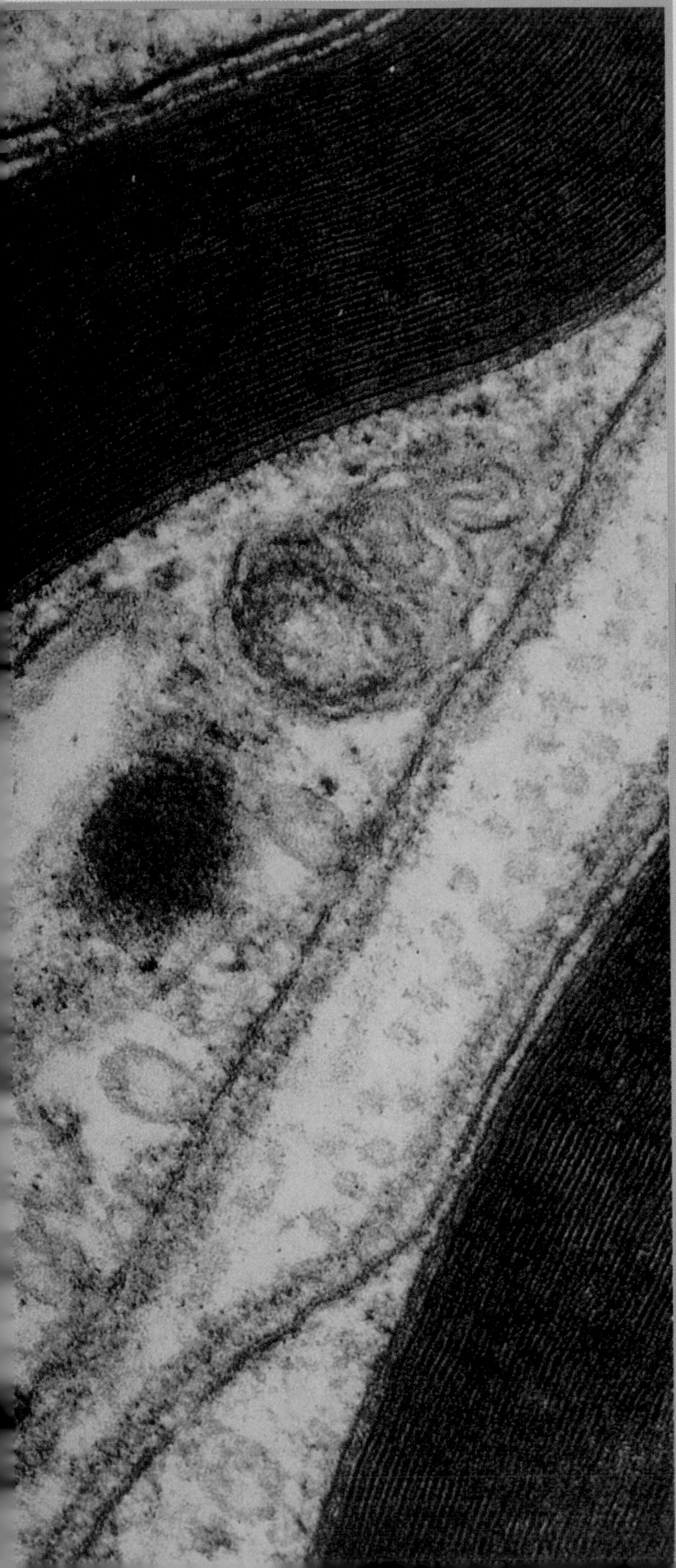

CELL STRUCTURE AND FUNCTION

THE PHOTO may not look like any highway you've ever seen, but that's exactly what it is. This highway and billions more like it make up a transportation network within your body. The threads, seen within the area bounded by the darkened layers, are called microtubules, and they shuttle particles within your body's cells. Amazingly, they can move particles in two directions at the same time!

This photo is of a nerve cell axon, magnified 17,000 times by a powerful electron microscope. Axons of nerve cells bring messages away from nerve cells to other nerve cells, organs, or muscles. The darkened areas are layers of fatty insulation that help nerves like this one transmit impulses quickly. Because axons are long (some extend from the base of your spinal cord to your toes, for example), microtubules are important in transporting molecules from the nucleus of the cell along the length of the axon—sometimes all the way to the axon tip.

Microtubules are only one type of myriad structures found within cells. Many other subcellular structures perform jobs such as providing energy, manufacturing molecules, and directing the activities of cells. But what are these complex structures called "cells"?

Cells are much more than the empty "boxes" seen by the British microscopist Robert Hooke in the mid-1600s. Scientists today know that cells are the smallest unit of life that can exist independently. Cells can take in nutrients, break them down to release energy, and get rid of wastes. They can reproduce, react to stimuli, and maintain an internal environment different from their surroundings. In multicellular organisms such as humans, cells work together to maintain life. In single-celled organisms such as the paramecium in Figure 3-1, each cell survives independently. This chapter will help you become familiar with the structure of cells and how they work. By studying their structure and function, you will also see their tremendous diversity and complexity (Figure 3-1) and will begin to understand the cellular level of organization of the human body.

KEY CONCEPTS

- All organisms on Earth are cells or are made up of cells.
- Most cells are microscopically small, providing sufficient surface area with respect to volume for cell survival.
- All cells can be grouped into two broad categories: prokaryotic cells and eukaryotic cells.
- Prokaryotic cells, which are characterized by the bacteria, are simpler in structure than eukaryotic cells, which include all other types of cells.

OUTLINE

The Cell Theory

In 1839 botanist Matthias Schleiden and zoologist Theodor Schwann formulated the theory that all living things are made up of cells. In other words,

The Tools of Scientists

HOW *Science* WORKS

Exploring the Microscopic World

3-A

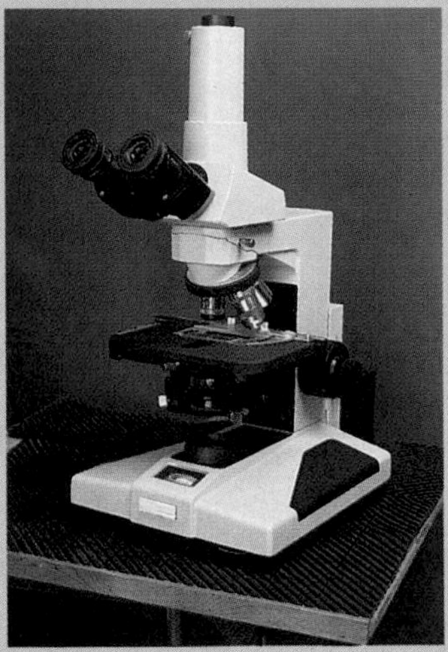

3-B

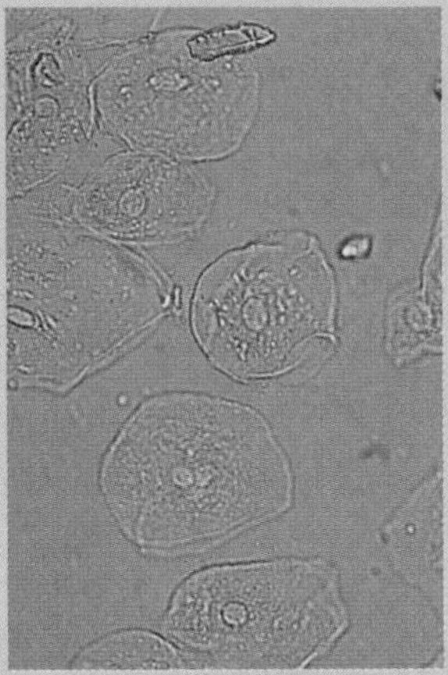

A world invisible to the naked eye had its roots of discovery in the late 1600s with the work of two pioneers in the field of microscopy: Anton van Leeuwenhoek and Marcello Malpighi. Microscopy is the use of a *microscope*—an optical instrument consisting of a lens or a combination of lenses for magnifying things that are too small to see clearly or at all. Leeuwenhoek's and Malpighi's inventions allowed them to observe such things as plant and animal tissues, blood cells, and sperm; little escaped their observant and technologically aided eyes.

Microscopes improved somewhat over the next 200 years, but it was not until the beginning of the 20th century that this technology began to advance in sophistication at a fast pace, resulting in the array of light microscopes and electron microscopes that scientists routinely use today. Photographs prepared from three types of microscopes—the compound light microscope, the transmission electron microscope, and the scanning electron microscope—are the ones most often used in this book.

The *compound light microscope* (Figure 3-A) is an instrument that uses two lenses (therefore, compound) to magnify an object. A mirror focuses light from the room up to the eye from beneath the specimen, or a lamp is used for illumination (therefore, light). The compound light microscope visualizes eukaryotic cells (such as plant and animal cells) and prokaryotic cells (bacteria) and can magnify them up to 1500 times. Special stains are often used to visualize particular structures, as is shown in Figure 3-B. *Brightfield microscopy,* the technique used to visualize this stained specimen, passes light directly through the specimen. Unstained organisms can be seen quite well using a technique called *phase-contrast microscopy.* This technique allows the organism to be viewed while still alive (staining kills cells) and uses direct and indirect lighting to intensify the variations in density within the cell as shown in Figure 3-C. A vari-

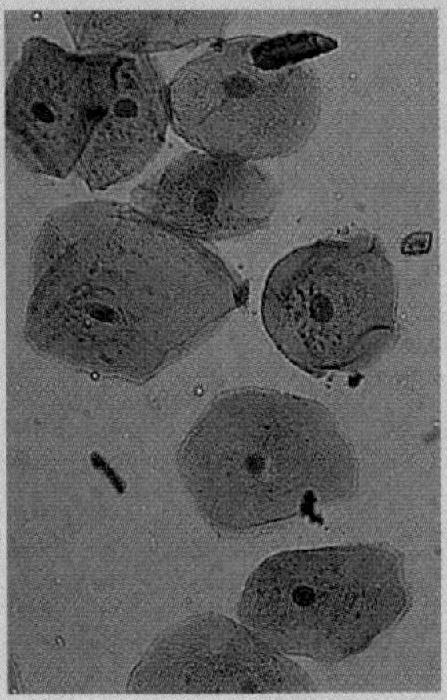

3-C

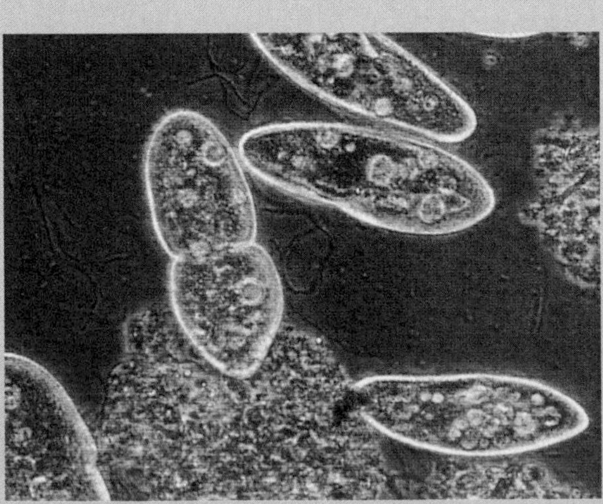

The Tools of Scientists—cont'd

3-D

3-E

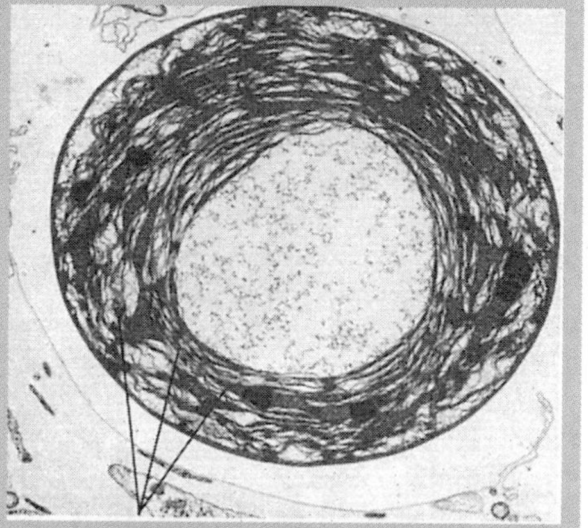

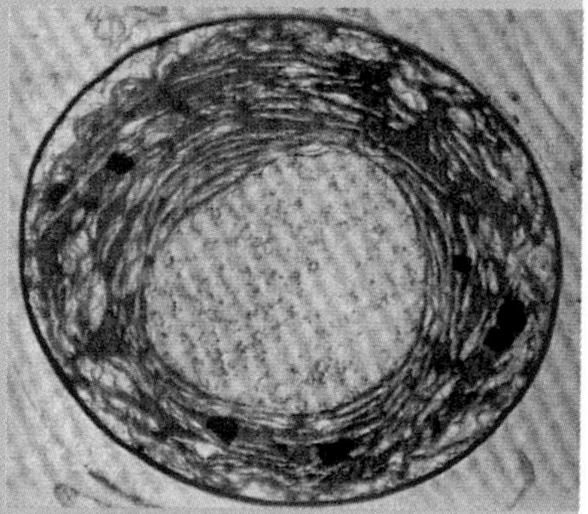

3-F

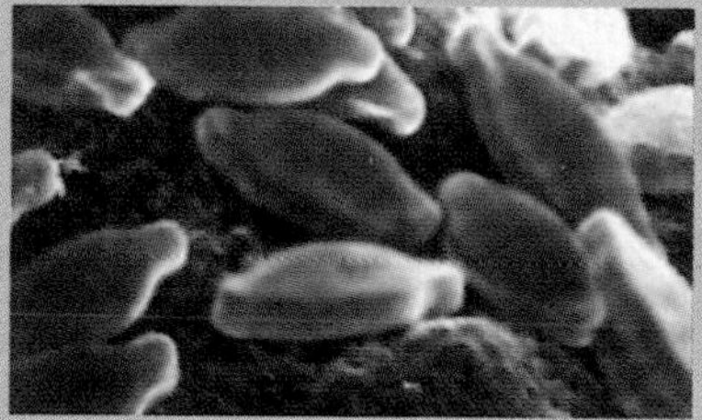

ety of other techniques can be used with the compound microscopes to view organisms, but these two are most commonly used.

The *electron microscope* (Figure 3-D) does not use light to visualize specimens but instead uses a fine beam of electrons transmitted to a specimen in a vacuum. The transmitted electrons, after partial absorption by the object, are focused by magnets to form the image of the specimen. Electrons have a shorter wavelength than visible light; this difference gives the electron microscope a greater resolving power than the compound light microscope. Resolving power is the ability to distinguish two points as being separate from one another. Therefore, specimens can be magnified more highly with the electron microscope and still transmit a clear image. As with light microscopy, a variety of techniques can be used to prepare specimens. In addition, there are two types of electron microscopes: the *transmission electron microscope (TEM)* and the *scanning electron microscope (SEM)*.

The transmission electron microscope is usually used to visualize slices of cells that have been specially prepared. The images are therefore of the interiors of cells, as is shown in Figure 3-E. In these photographs (called electron micrographs), you can see cell organelles. The photograph at the top is a typical electron micrograph. The one on the bottom has been colorized: color has been added by artists after the electron micrograph was taken to help point out specific cell structures. Colorized electron micrographs are used in this book.

The scanning electron microscope is usually used to visualize surfaces of specimens that have been glued or taped to a metal slide and then covered with a microscopically thin layer of a metal, usually gold or platinum. The metal gives off secondary electrons when excited by a beam of electrons scanned across its surface. These secondary electrons are collected on a screen, which results in an image of the surface of the specimen. Figure 3-F is a scanning electron micrograph of diatoms, a form of algae.

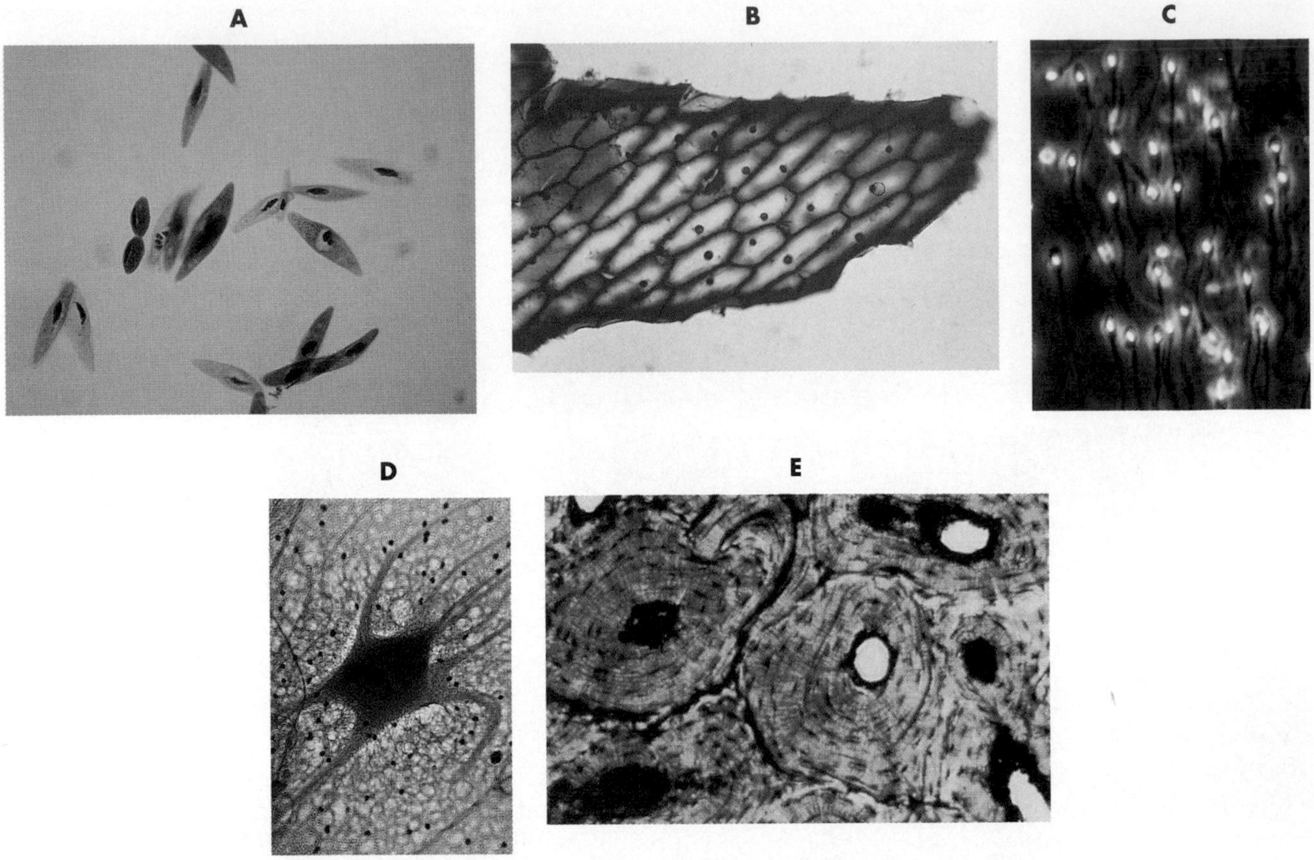

FIGURE 3-1 Cells diverse in shape, structure, and function. A, *Paramecium,* a single-celled protozoan. **B,** Onion cells. **C,** Sperm cells. **D,** Nerve cell. **E,** Fat cells.

they realized that cells make up the structure of such things as houseplants and people—but not of rocks or soil. It took another 50 years and the work of Rudolf Virchow for scientists to understand another basic concept about cells. Living cells can only be produced by other living cells. As Virchow put it, "all cells from cells." Together, these concepts are called the **cell theory.**

The cell theory is a profound statement regarding the nature of living things. It includes three basic principles:

1. All living things are made up of one or more cells.
2. The smallest *living* unit of structure and function of all organisms is the cell.
3. All cells arise from preexisting cells.

When these statements were formulated in the mid-1800s, scientists discarded the idea of *spontaneous generation:* that living things could arise from the nonliving. This theory suggested that frogs could be born of the mud in a pond and that rotting meat could spawn the larvae of flies. After

FIGURE 3-2 The marine green alga *Acetabularia.* *Acetabularia* is a large, single-celled organism with clearly differentiated parts, such as the stalks and elaborate "hats" visible here. Each cell is a different individual several centimeters tall.

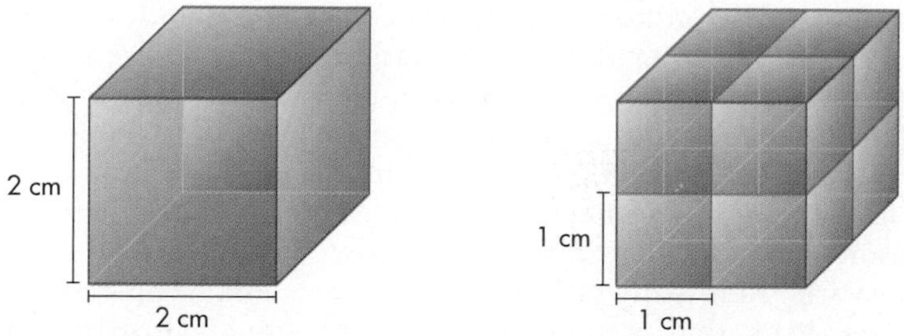

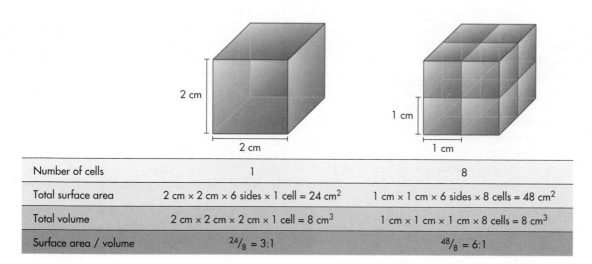

Number of cells	1	8
Total surface area	2 cm × 2 cm × 6 sides × 1 cell = 24 cm^2	1 cm × 1 cm × 6 sides × 8 cells = 48 cm^2
Total volume	2 cm × 2 cm × 2 cm × 1 cell = 8 cm^3	1 cm × 1 cm × 1 cm × 8 cells = 8 cm^3
Surface area / volume	$^{24}/_{8}$ = 3:1	$^{48}/_{8}$ = 6:1

FIGURE 3-3 Cells maintain a large surface-to-volume ratio. The smaller the cell, the larger its surface area–to-volume ratio. Multicellular organisms are made up of many microscopic cells. This organization increases the total surface area of the cell membranes within living things.

the development of the cell theory, scientists recognized that life arose directly from the growth and division of single cells. Today, scientists think that life on Earth represents a continuous line of descent from the first cells that evolved on Earth.

> *All organisms on Earth are cells or are made up of groups of cells. In addition, all organisms are descendants of the first cells.*

Why Aren't Cells Larger?

Most animal cells are extremely small, ranging in diameter from about 10 to 30 micrometers (μm). There are 1000 μm in 1 millimeter (mm)—the width of a paper clip's wire. So as you might expect, most cells are invisible to the naked eye without the aid of a microscope. Your red blood cells, for example, are so small that it would take a row of about 2500 of them to span the diameter of a dime. Only a few kinds of cells are large. Individual cells of the marine alga *Acetabularia,* for example, are up to 5 centimeters long (Figure 3-2). If you had eggs for breakfast, you were eating single cells! Other kinds of cells are long and thin, such as nerve cells that run from your spinal cord to your toes or fingers. But very few cells are as large as a hen's egg or as long as a nerve cell.

To understand why cells are so small, you must first realize that most cells are constantly working, doing such jobs as breaking down molecules for energy, producing substances that cells need, and getting rid of wastes. Each cell must move substances in and out across its boundary—the cell membrane—quickly enough to meet its needs. Therefore the amount of membranous surface area a cell has in relationship to the volume it encloses is crucial to its survival.

To illustrate, suppose that a cell is a cube like the one pictured on the left in Figure 3-3, with sides 2 centimeters (cm) in length. The surface area of each side of the cell is 2 cm × 2 cm, or 4 cm^2. The total surface area of the six sides is 6 × 4 cm^2, or 24 cm^2. Now let's compare that surface area to the volume of this cell. Its volume is calculated by multiplying its length × width × height: 2 cm × 2 cm × 2 cm, or 8 cm^3. Our hypothetical cell, then, has 24 cm^2 of surface area for 8 cm^3 of volume, or 3 cm^2 of surface area for each cm^3 of volume—a 3:1 ratio.

Suppose, however, that most cells were never able to grow this large. Suppose that as their sides grew to 1 cm, the cells were stimulated to divide. The right upper corner of Figure 3-3 shows eight cells having sides of 1 cm each. These eight cells comprise the same volume as the single cell 2 cm on a side. But how does the surface area compare? The surface area of each cell is 1 cm × 1 cm × 6 sides, or 6 cm^2. The total surface area of the 8 cells is 48 cm^2. These eight smaller cells, then, have 48 cm^2 of surface area for 8 cm^3 of volume, or 6 cm^2 of surface area for each cm^3 of volume—a 6:1 ratio.

> **cell theory** a statement regarding the nature of living things, which holds that all living things are made up of cells, that the smallest living unit of structure and function of all organisms is the cell, and that all cells arise from preexisting cells.

The smaller cells have twice as much surface area for the same amount of volume as the larger cell!

In general, then, the smaller the cell, the larger its surface area-to-volume ratio. Cells that are too large have a surface area-to-volume ratio that is too small to move substances in and out (across their membranes) fast enough to meet their needs.

Another reason that cells generally do not grow large has to do with their controlling centers, or **nuclei.** Scientists think that a nucleus could not control all the activities of an active cell if it grew too large. In fact, some large complex cells, such as the unicellular paramecia, have two nuclei. Large cells with only one nucleus, such as unfertilized egg cells, are usually inactive cells.

> *Most cells are microscopically small, resulting in a surface area-to-volume ratio sufficient to move substances across their cell membranes quickly enough to meet their needs.*

Two Kinds of Cells

All cells can be grouped into two broad categories: **prokaryotic cells** and **eukaryotic cells.** Cells are placed into one of these categories based on their types of structure. Prokaryotic cells, or prokaryotes, have a simpler structure than eukaryotes and were the first type of cell to exist as life arose on Earth billions of years ago. Eukaryotic cells evolved from these simpler cells.

You know the prokaryotes as bacteria. The cells of organisms within the other four kingdoms of life—plants, animals, protists, and fungi—are eukaryotes. Although the structure of cells in either group may vary among species, prokaryotes and eukaryotes each have distinctive features common to each group. In addition, because members of both groups are living cells, they also have some of the same features. Almost all cells have the following four characteristics:

1. A surrounding membrane
2. A thick fluid enclosed by this membrane that, along with the other cell contents, is called *protoplasm*
3. Organelles, or "little organs," located within the protoplasm (of eukaryotes) that carry out certain cellular functions
4. A control center called a *nucleus* (in eukaryotes) or *nucleoid* (in prokaryotes) that contains the hereditary material DNA

The structure of bacterial cells is discussed in Chapter 25, but Table 3-2 located on p. 75 will help you clarify and classify the differences among three cell types: plant cells and animal cells (both eukaryotes) and bacterial cells (prokaryotes). Also, Figure 3-4 (a bacterial cell), Figure 3-5 (an animal

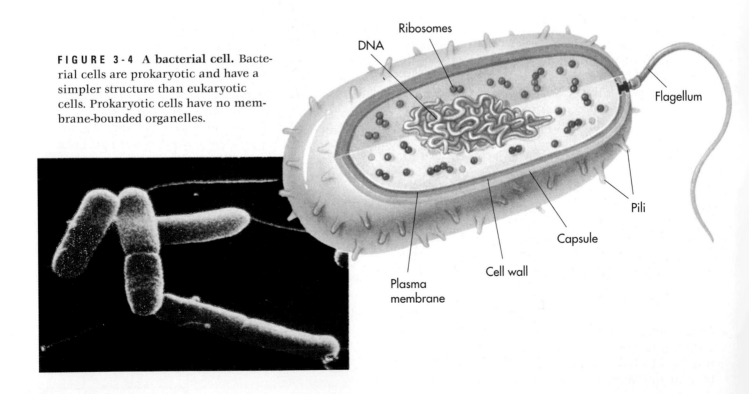

FIGURE 3-4 A bacterial cell. Bacterial cells are prokaryotic and have a simpler structure than eukaryotic cells. Prokaryotic cells have no membrane-bounded organelles.

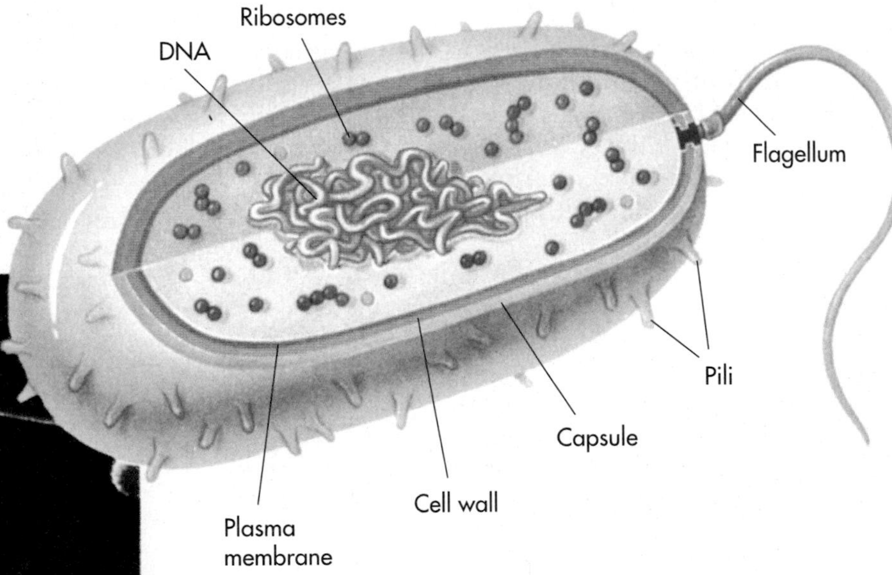

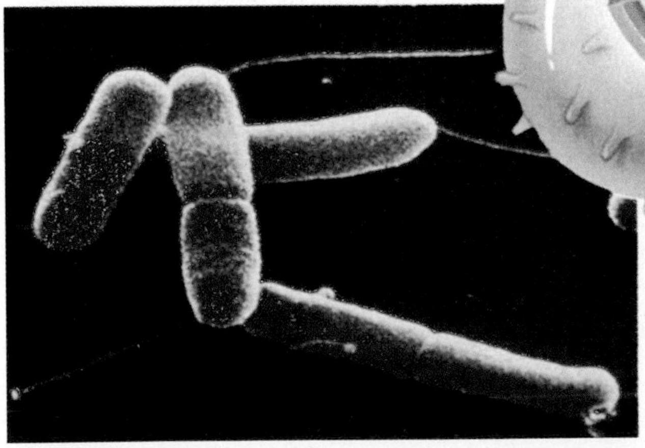

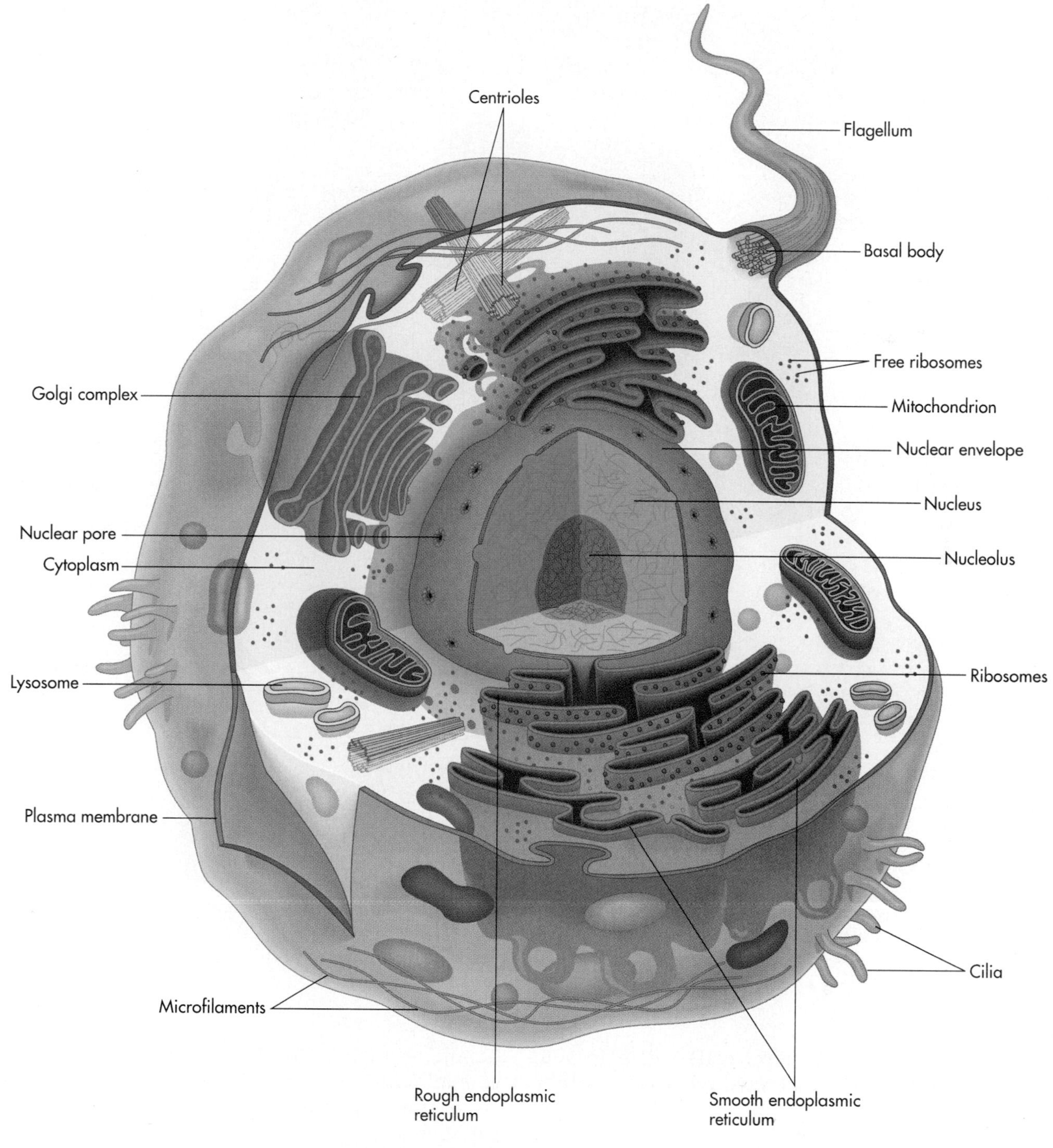

FIGURE 3-5 An animal cell. A generalized representation of an animal cell showing the major organelles.

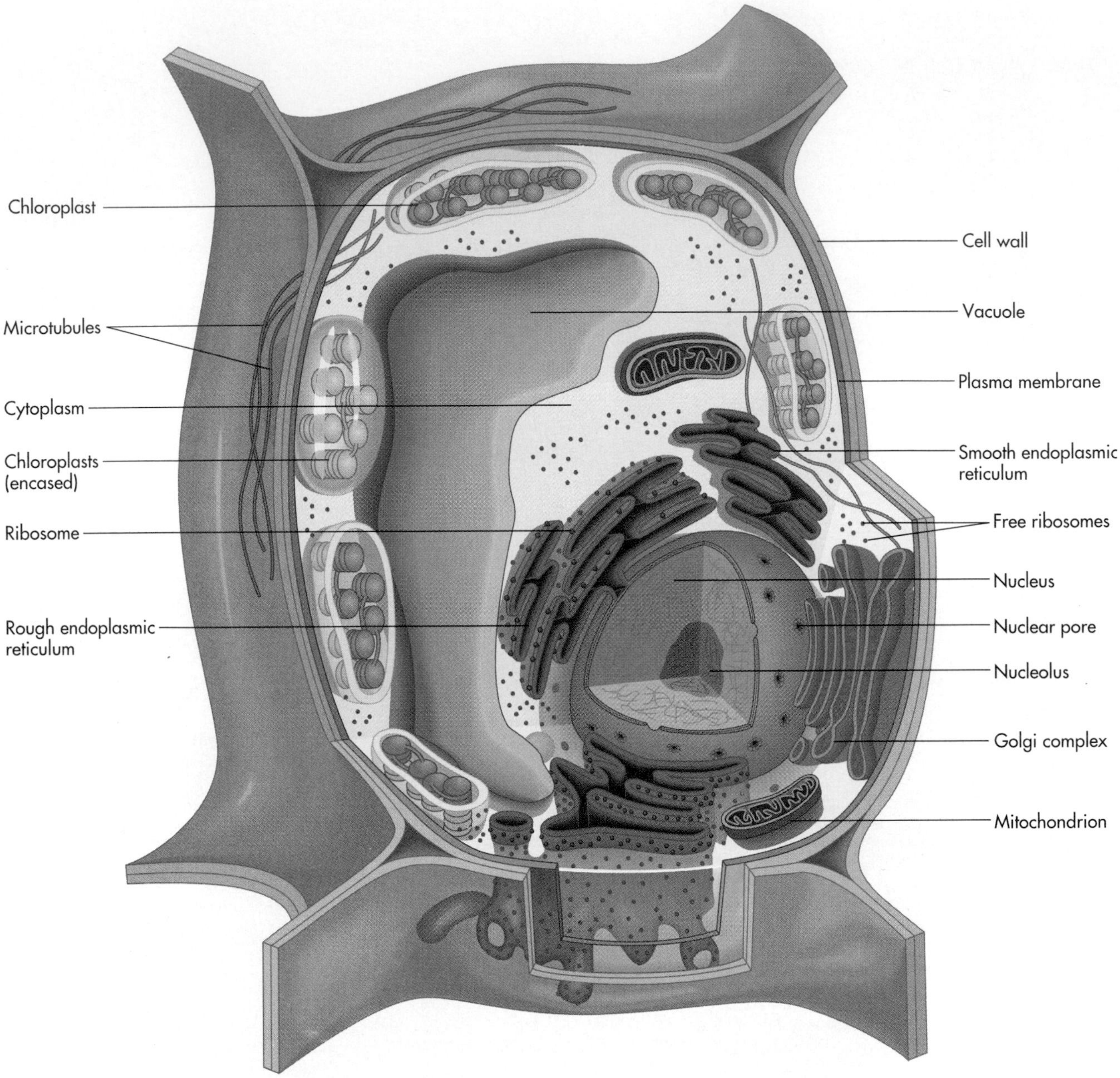

FIGURE 3-6 A plant cell. A generalized representation of a plant cell showing the major organelles. Plant cells differ from animal cells in that they have a rigid cell wall, contain chloroplasts, and frequently have large, water-filled vacuoles occupying a major part of the cell volume.

cell), and Figure 3-6 (a plant cell) will help you visualize prokaryotic and eukaryotic cell structures.

Eukaryotic Cells: An Overview

Although eukaryotic cells are quite a diverse group, they share a basic architecture. They are all bounded by a membrane that encloses a semifluid material crisscrossed with a supporting framework of protein. All eukaryotic cells also possess many organelles (Table 3-1). These organelles are of two general kinds: (1) membranes or organelles derived from membranes and (2) bacterialike organelles.

Most biologists agree that the bacterialike organelles in eukaryotes were derived from ancient *symbiotic bacteria.* The word *symbiosis* means that two or more organisms live together in a close association. An organism that is symbiotic within another is called an *endosymbiont.* The major endosymbionts that occur in eukaryotic cells are mi-

TABLE 3-1

Eukaryotic Cell Structures and Their Functions

STRUCTURE	DESCRIPTION	FUNCTION
Exterior Structures		
Cell wall	Outer layer of cellulose or chitin, or absent	Protection, support
Plasma membrane	Lipid bilayer in which proteins are embedded	Regulation of what passes in and out of cell, cell-to-cell recognition
Cytoskeleton	Network of protein filaments, fibers, and tubules	Structural support, cell movement
Flagella (cilia)	Cellular extensions with 9 + 2 arrangement of pairs of microtubules	Motility or moving fluids over surfaces
Interior Structures and Organelles		
Endoplasmic reticulum	Network of internal membranes	Formation of compartments and vesicles; modification and transport of proteins; synthesis of carbohydrates and lipids
Ribosomes	Small, complex assemblies of protein and RNA, often bound to ER	Sites of protein synthesis
Nucleus	Spherical structure bounded by double membrane, site of chromosomes	Control center of cell
Chromosomes	Long threads of DNA associated with protein	Sites of hereditary information
Nucleolus	Site within nucleus of rRNA synthesis	Synthesis and assembly of ribosomes
Golgi bodies	Stacks of flattened vesicles	Packaging of proteins for export from cell
Lysosomes	Membranous sacs containing digestive enzymes	Digestion of various molecules
Mitochondria	Bacterialike elements with inner membrane highly folded	"Power plant" of the cell
Chloroplasts	Bacterialike elements with inner membrane forming sacs containing chlorophyll, found in plant cells and algae	Site of photosynthesis in plant cells and algae

prokaryotic cells (pro **kare** ee **ot** ick **cells**) cells that have a relatively simple structure and no membrane-bounded nucleus or other membrane-bounded organelles; all bacteria are prokaryotes.

eukaryotic cells (yoo **kare** ee **ot** ick **cells**) cells that have a more complicated structure than prokaryotic cells; eukaryotic cells have a membrane-bounded nucleus and a variety of cell organelles.

tochondria, which occur in all but a very few eukaryotic organisms, and chloroplasts, which occur in algae and plants.

The Plasma Membrane

Every cell is bounded by a **plasma membrane,** so named because it encloses the *protoplasm*—the semifluid cell contents, including the cell organelles and nucleus (control center) of the cell. The plasma membrane has two main components: *phospholipids* and *proteins.* Lipids, as you may recall from Chapter 2, are made up of a glycerol backbone attached to three fatty acid chains. Phospholipids have only two fatty acid chains; phosphoric acid takes the place of the third (Figure 3-7). Phosphoric acid is a phosphorylated alcohol: an alcohol molecule bonded to a phosphate group.

Lipids are nonpolar molecules and will not dissolve in (form hydrogen bonds with) water. Phosphorylated alcohols, however, *are* polar and *do* interact with (form hydrogen bonds with) water. Therefore, phospholipids have a portion of the molecule termed **hydrophilic,** meaning "water loving," and a portion of the molecule that is **hydrophobic,** meaning "water fearing."

When a collection of phospholipid molecules is placed in water, their hydrophobic fatty acid chains are pushed away by the water molecules that surround them. The water molecules move toward the molecules that form hydrogen bonds with them—the phosphoric acid. As a result of these orientations of the two parts of phospholipids to water, they form two layers of molecules, with their hydrophobic "tails" oriented inward (away from the water) and their hydrophilic "heads" oriented outward (toward the water) (see Figure 3-7). This double-layered structure is called a lipid bilayer and is the foundation of the membranes of all living things.

> *The basic foundation of biological membranes is a lipid bilayer. Lipid bilayers form spontaneously, with the nonpolar tails of the phospholipid molecules pointed inward and the polar heads pointed outward.*

Did you ever blow soap bubbles when you were a child? Take a moment to think about how the bubbles looked. Can you remember how the soap film forming the bubble seemed to swirl and move as the bubble floated in the air? Lipid bilayers are "fluid" much like the film of the soap bubbles you remember. The fluid nature of the bilayer is caused by movement of the phospholipid molecules. Although water constantly forces the phospholipid molecules to form a bilayer, the hydrogen bonds between the water molecules and the hydrophilic heads of the phospholipids are constantly breaking and re-forming. Therefore individual molecules move about somewhat freely.

> *A lipid bilayer forms a fluid, flexible covering for a cell and keeps the watery contents of the cell on one side of the membrane and the watery environment on the other.*

If cells were encased with pure lipid bilayers, however, they would be unable to survive because the interior of the lipid bilayer is completely nonpolar. It repels most water-soluble molecules that attempt to pass through it. In fact, as Figure 3-8 shows, the lipid bilayer allows only a few uncharged polar molecules, such as individual water (H_2O) and ammonia (NH_3) molecules, to pass through. (Ammonia is a common cellular waste product formed from the breakdown of proteins.) In addition, substances that are soluble in oil, such as vitamins A, D, and E, can also move across the lipid bilayer. It prevents, however, any water-soluble substances such as sugars, amino acids, and proteins as well as charged particles (important ions) such as calcium (Ca^{2+}) and potassium (K^+) from entering or leaving the cell. Such molecules pass through the cell membrane by means of proteins that are embedded in the lipid bilayer (see below). This model of the cell membrane is widely accepted today. It is called the **fluid mosaic model,** a name that describes the fluid nature of a lipid bilayer studded with a mosaic of proteins.

The variety of proteins that penetrate the lipid layer control the interactions of the cell with its environment. Some of these proteins, called *channels,* act as doors that let specific molecules into and out of the cell. Other proteins, called *receptors,* recognize certain chemicals such as hormones that signal cells to respond in particular ways. Receptor proteins cause changes within the cell when they come in contact with such chemical "messengers" or signals. A third type of protein identifies a cell as being of a particular type. These cell surface markers help cells identify one another, an ability that helps cells function correctly in a multicellular organism (such as your body).

> *Proteins embedded within the plasma membrane regulate interactions between the cell and its environment.*

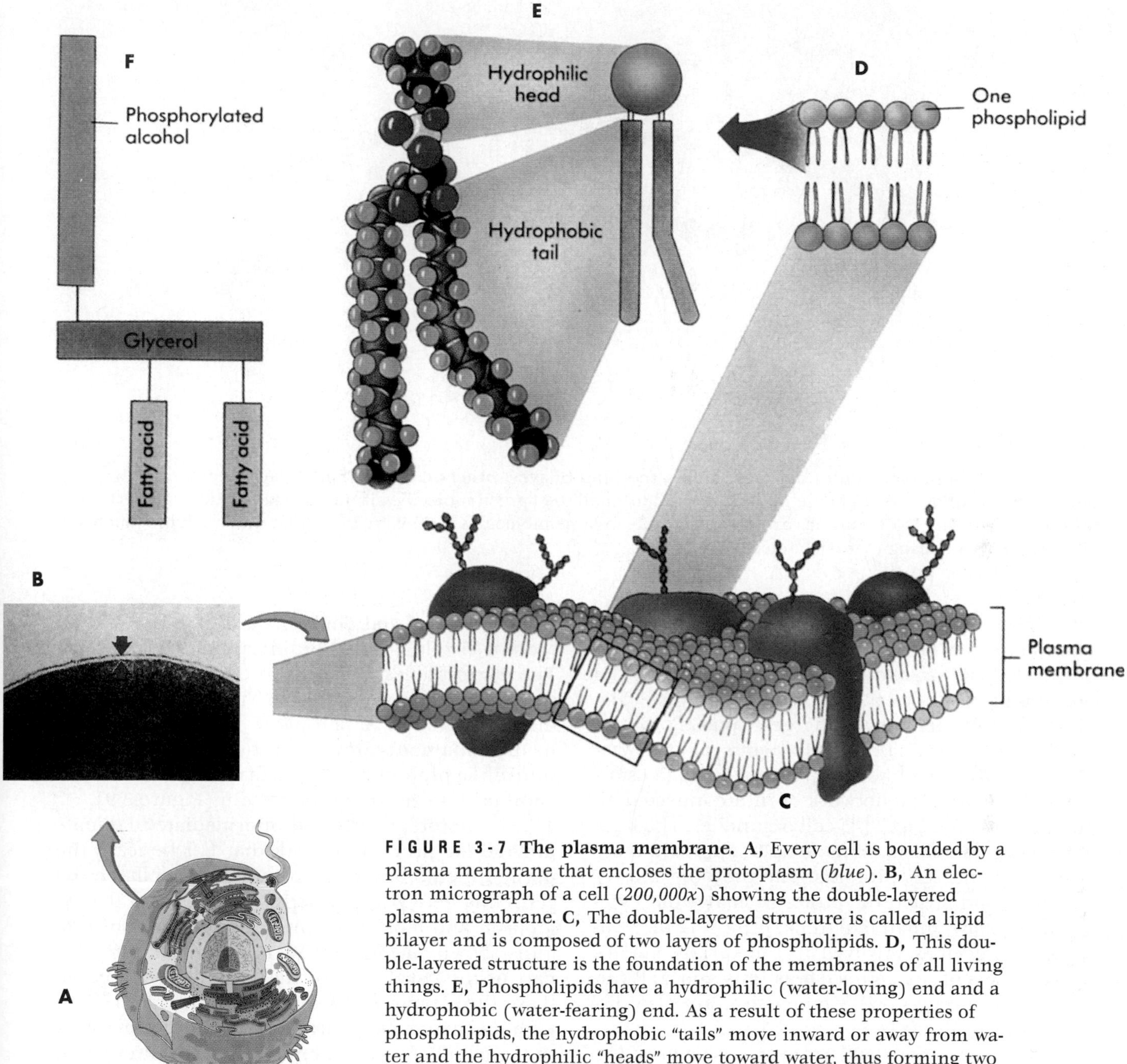

FIGURE 3-7 The plasma membrane. A, Every cell is bounded by a plasma membrane that encloses the protoplasm (*blue*). **B,** An electron micrograph of a cell (*200,000x*) showing the double-layered plasma membrane. **C,** The double-layered structure is called a lipid bilayer and is composed of two layers of phospholipids. **D,** This double-layered structure is the foundation of the membranes of all living things. **E,** Phospholipids have a hydrophilic (water-loving) end and a hydrophobic (water-fearing) end. As a result of these properties of phospholipids, the hydrophobic "tails" move inward or away from water and the hydrophilic "heads" move toward water, thus forming two layers of molecules. **F,** Each phospholipid is composed of a glycerol molecule attached to two chains of fatty acids. The phosphorylated alcohol is the phosphate functional group for this particular phospholipid.

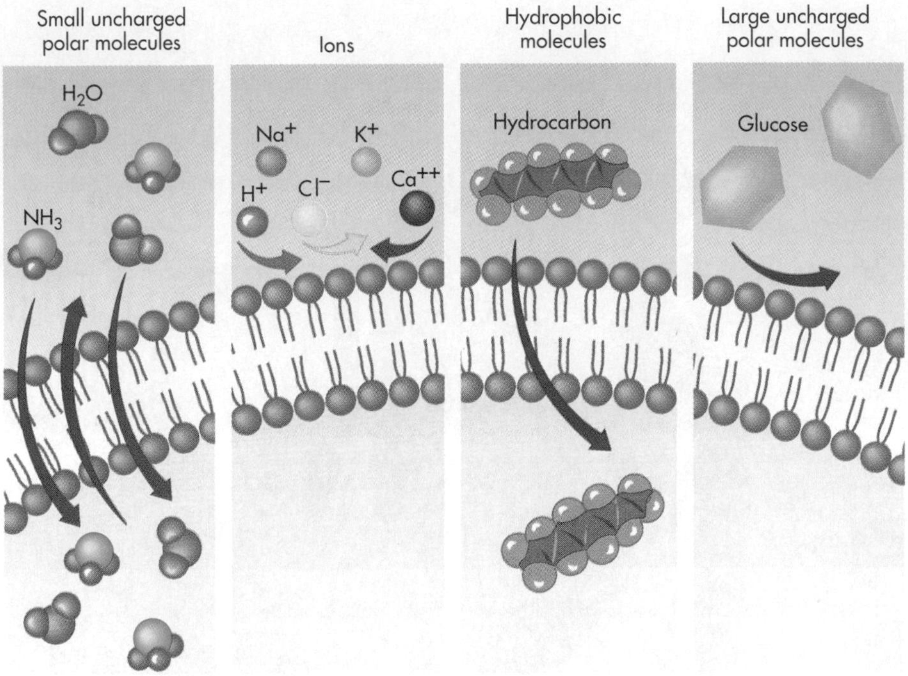

FIGURE 3-8 Some molecules can pass across the lipid bilayer; others cannot. Lipid bilayer membranes are permeable to lipids (hydrophobic molecules) and to small uncharged molecules if they are polar like water (H_2O) and ammonia (NH_3). Membranes are not permeable to large molecules if they are polar and water soluble (such as glucose) or to anything that is charged, such as ions.

The Cytoplasm and Cytoskeleton

The **cytoplasm** of the cell is the viscous fluid containing all cell organelles *except* the nucleus. (The protoplasm includes the nucleus.) The word *cytoplasm* literally means "living gel" (plasm) "of the cell" (cyto). The major components of the cytoplasm are (1) a gel-like fluid (the cytosol), (2) storage substances, (3) a network of interconnected filaments and fibers, and (4) cell organelles. The fluid part of the cytoplasm is made up of approximately 75% water and 25% proteins. The proteins, mostly enzymes and structural proteins, make the cytoplasm viscous—much like thickening gelatin. Nonprotein molecules involved in the various chemical reactions of the cell, such as ions and ATP (an important energy-storing molecule), are also dissolved in this fluid.

Storage substances vary from one type of cell to another. For example, the cytoplasm of liver cells contains large molecules of glycogen, a storage form of glucose. Fat cells contain a large lipid droplet.

The cytoplasm also contains a *cytoskeleton*—a network of filaments and fibers that do many jobs. The cytoskeleton is made up of three different types of fibers: *microfilaments, microtubules,* and *intermediate filaments.* The microfilaments are the thinnest—a twisted double chain of protein—and are not visible with an ordinary light microscope. The microtubules, a chain of proteins wrapped in a spiral to form a tube, are the thickest members of the cytoskeleton. As the name suggests, the intermediate filaments are an in-between size. These are threadlike protein molecules that wrap around one another to form "ropes" of protein (Figure 3-9).

The microfilaments and intermediate filaments provide the protein network that lies beneath the plasma membrane to help support and shape the cell. This network of proteins extends into the cytoplasm and looks something like the web of proteins shown in Figure 3-10. Microtubules are also part of this "skeleton" of the cytoplasm. Together, these three types of protein fibers provide the cell with mechanical support and help anchor many of the organelles—mitochondria are an exception. They also help move substances from one part of the cell to another.

The cytoskeleton is a network of filaments and fibers within the cytoplasm that helps maintain the shape of the cell, move substances within cells, and anchor various structures in place.

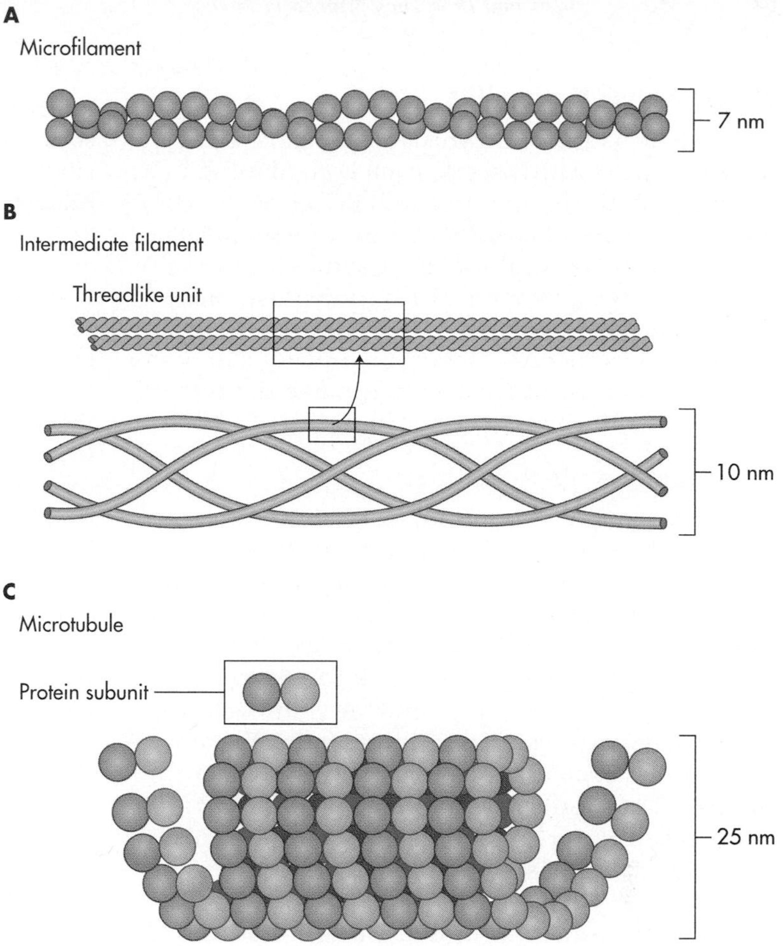

FIGURE 3-9 Types of fibers found in the cytoskeleton. **A,** Microfilaments are the thinnest filaments making up the cytoskeleton and are composed of a twisted double chain of protein. **B,** Intermediate filaments are threadlike protein molecules that wrap around each other to form "ropes" of protein. **C,** Microtubules are the thickest filaments of the cytoskeleton and are composed of a chain of proteins wrapped in a spiral that forms a tube.

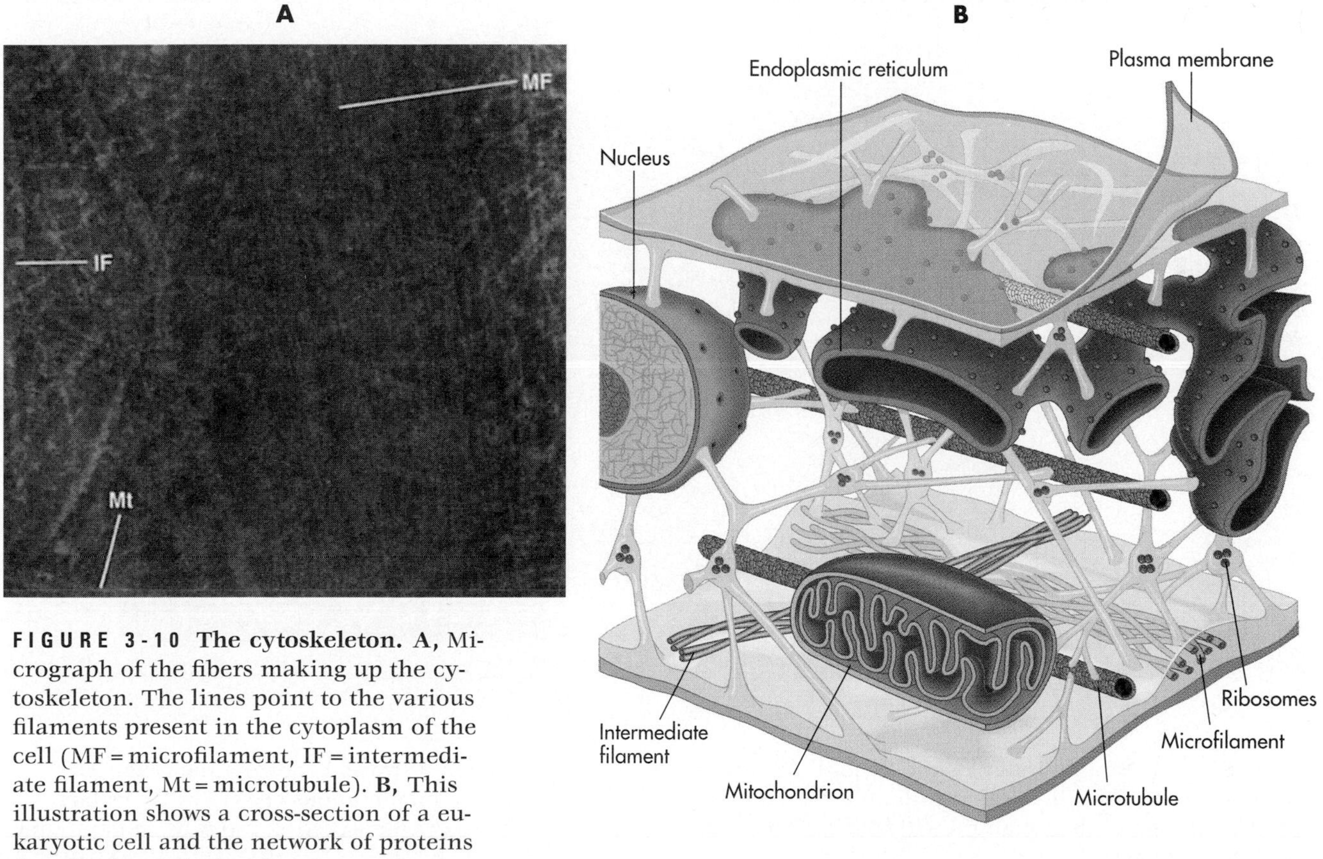

FIGURE 3-10 The cytoskeleton. **A,** Micrograph of the fibers making up the cytoskeleton. The lines point to the various filaments present in the cytoplasm of the cell (MF = microfilament, IF = intermediate filament, Mt = microtubule). **B,** This illustration shows a cross-section of a eukaryotic cell and the network of proteins that lies beneath the plasma membrane and helps support and shape the cell.

Membranous Organelles

Enmeshed within the cytoskeletal fibers, the cell organelles constantly work, each contributing in a special way to the life and well-being of the cell. Most organelles of eukaryotic cells are bounded by membranes. Many different enzymes—molecules that help specific chemical reactions take place—are attached to these membranes. In this way, these subcellular structures form organized compartments within the cytoplasm and make the cell an efficiently running living machine.

The Endoplasmic Reticulum

The **endoplasmic reticulum,** or ER, is an extensive system of interconnected membranes that forms flattened channels and tubelike canals within the cytoplasm, almost like a cellular subway system. The channels are used to help move substances from one part of the cell to another. The name *endoplasmic reticulum* may sound very strange. It is, however, descriptive of its location and appearance. The word *endoplasmic* means "within the cytoplasm." The word *reticulum* comes from a Latin word meaning "a little net." And that is exactly what the ER looks like—a net within the cytoplasm.

There are two types of endoplasmic reticula in cells: **rough ER** and **smooth ER.** The *rough* refers to the minute, spherical structures called **ribosomes** covering the surface of one type of ER. With ribosomes dotting their surfaces, rough ER membranes look like long sheets of sandpaper (Figure 3-11).

Ribosomes are the places where proteins are manufactured. Some ribosomes are attached to the rough ER, but others are in the cytoplasm bound to cytoskeletal fibers. The cytoplasmic ribosomes help produce proteins for use in the cell, such as those making up the structure of the cytoskeletal fibers or certain organelles. The proteins made at the ribosomes of the rough ER are most often destined to leave the cell. Cells specialized for secreting proteins, such as the pancreatic cells that manufacture the hormone insulin, contain large amounts of rough ER. After the proteins are manufactured at the ribosomes on the surface of the ER, they enter the inner space of the ER. Within this channel, the proteins may be changed by enzymes bound to the inner surface of the ER membrane. Carbohydrate molecules are often added to them.

When the proteins reach the end of their journey in the ER, they are encased in tiny membrane-bounded sacs called *vesicles.* These vesicles are formed at sections of smooth ER that are continuous with the rough ER. "Buds" of the smooth ER pinch off as the newly formed proteins reach them.

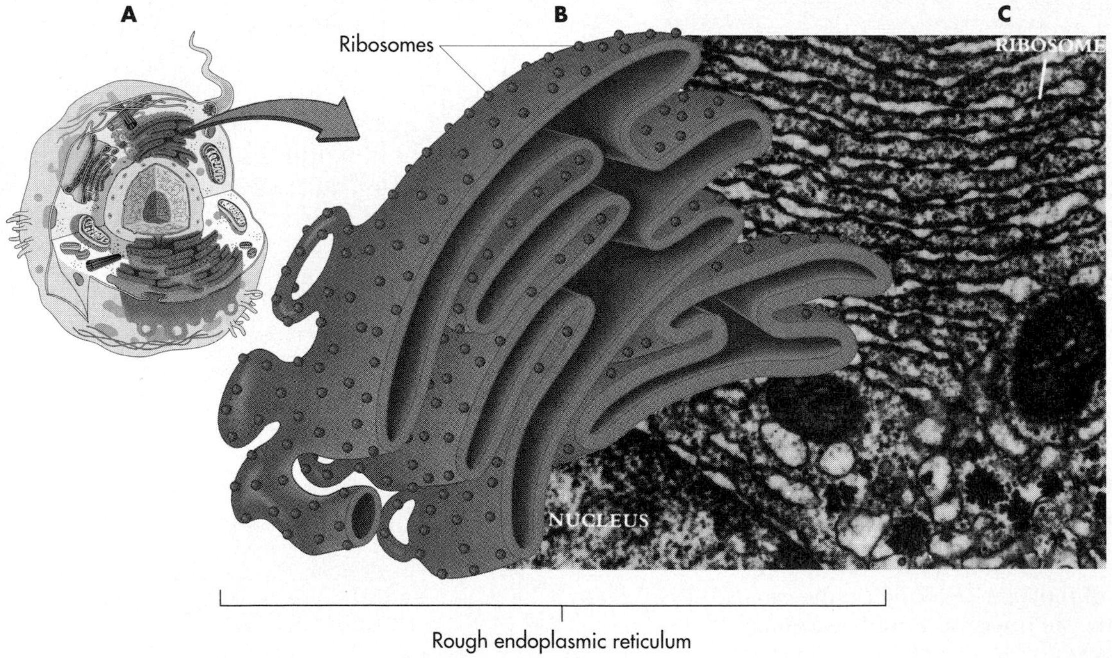

FIGURE 3-11 Rough endoplasmic reticulum (rough ER). A, A reference illustration showing the rough ER within a cell. **B,** An illustration of rough ER. Note the ribosomes covering the membranous channels. **C,** An electron micrograph of the rough ER within a rat liver cell (*2300x*). Cells specialized for secreting proteins contain large amounts of rough ER.

These vesicles eventually fuse with the membranes of another organelle called the **Golgi body.**

Smooth ER has no ribosomes attached to its surfaces. Therefore it does not have the grainy appearance of rough ER and does not manufacture proteins. Instead, smooth ER has enzymes bound to its inner surfaces that help build carbohydrates and lipids. Cells specialized for the synthesis of these molecules, such as animal cells that produce male or female sex hormones, have abundant smooth ER.

> *The endoplasmic reticulum (ER) is an extensive system of membranes that divides the interior of eukaryotic cells into compartments and channels. Rough ER makes and transports proteins destined to leave the cell. Smooth ER helps build carbohydrates and lipids.*

Golgi Bodies

First described by the physician Camillo Golgi in the last half of the nineteenth century, Golgi bodies look like microscopic stacks of pancakes in the cytoplasm. Animal cells each contain 10 to 20 sets of these flattened membranes. Plant cells may contain several hundred because the Golgi bodies are involved in the synthesis and maintenance of plant cell walls (a structure animal cells do not have). Collectively, the Golgi bodies are referred to as the **Golgi complex.**

Molecules come to a Golgi body in vesicles pinched off from the ER. The membranes of the vesicles fuse with the membranes of a Golgi body. Once inside the space formed by the Golgi membranes, the molecules may be modified by the formation of new chemical bonds or by the addition of carbohydrates. For example, mucin, which is a protein with attached carbohydrates that forms a major part of the mucous secretions of the body, is put together in its final form in the Golgi complex. This refining of molecules occurs in stages in the Golgi complex, with different parts of the stack of membranes containing enzymes that do specific jobs. When molecular products are ready for transport, they are sorted and pinched off in separate vesicles (Figure 3-12). Each vesicle travels to its destination and fuses with another membrane. The vesicles containing those molecules that are to be secreted from the cell fuse with the plasma membrane. In this way the contents of the vesicle are liberated from the cell (Figure 3-13). Other vesicles fuse with the membranes of organelles such as lysosomes, delivering the new molecules to their interiors.

> *The Golgi complex is the delivery system of the eukaryotic cell. It collects, modifies, packages, and distributes molecules that are made at one location within the cell and used at another.*

Lysosomes

The new molecules delivered to **lysosomes** are digestive enzymes—molecules that help break large molecules into smaller molecules. Lysosomes are, in fact, membrane-bounded bags of many different digestive enzymes (Figure 3-14). Several hundred of these organelles may be present in one cell alone.

Lysosomes and their digestive enzymes are extremely important to the health of a cell. They help cells function by aiding in cell renewal, constantly breaking down old cell parts as they are replaced

plasma membrane a thin, flexible lipid bilayer that encloses the contents of a cell.

hydrophilic (hi droe **fil** ick) a term referring to polar molecules that form hydrogen bonds with water.

hydrophobic (**hi** droe **fo** bic) a term referring to nonpolar molecules that cannot form hydrogen bonds with water.

fluid mosaic model the most well-accepted theory regarding the nature of the cell membrane; it describes the fluid nature of a lipid bilayer studded with a mosaic of proteins.

cytoplasm (**sye** toe **plaz** um) a viscous fluid within a cell that contains all cell organelles *except* the nucleus.

endoplasmic reticulum (en doe **plaz** mik ri **tik** yuh lum) an extensive system of interconnected membranes that forms flattened channels and tubelike canals within the cytoplasm of a cell.

rough ER a kind of endoplasmic reticulum that is covered with ribosomes and hence resembles long sheets of sandpaper; in tandem with its ribosomes, rough ER manufactures and transports proteins designed to leave the cell.

smooth ER a kind of endoplasmic reticulum that does not have ribosomes attached to its surface and hence manufactures no proteins; instead, smooth ER helps produce carbohydrates and lipids.

ribosomes (rye bow **somes**) the organelles at which proteins are manufactured.

Golgi body (**gol** gee **body**) an organelle that collects, modifies, and packages molecules that are made at different locations within the cell and prepares them for transport.

Golgi complex a term collectively referring to all of the Golgi bodies within a cell.

lysosomes (lye so somes) membrane-bounded organelles that are essentially bags of many different digestive enzymes. Lysosomes break down old cell parts or materials brought into the cell from the environment and are extremely important to the health of a cell.

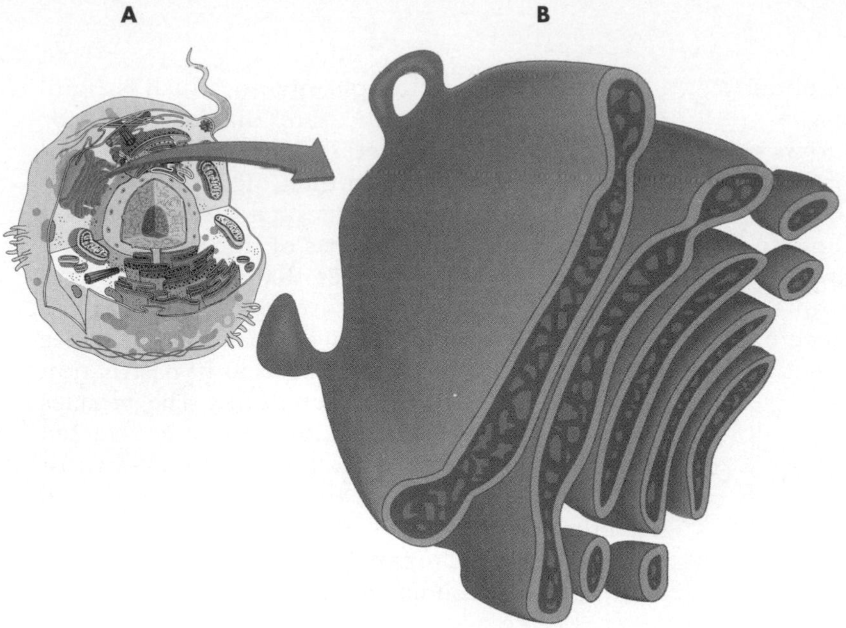

FIGURE 3-12 A Golgi body. A, Reference photo showing a Golgi body within a cell. **B,** A Golgi body is a structure that collects, modifies, packages, and distributes molecular products. When these molecular products are ready for transport, they are pinched off into vesicles that move either to the cell membrane to exit the cell or to locations within the cell.

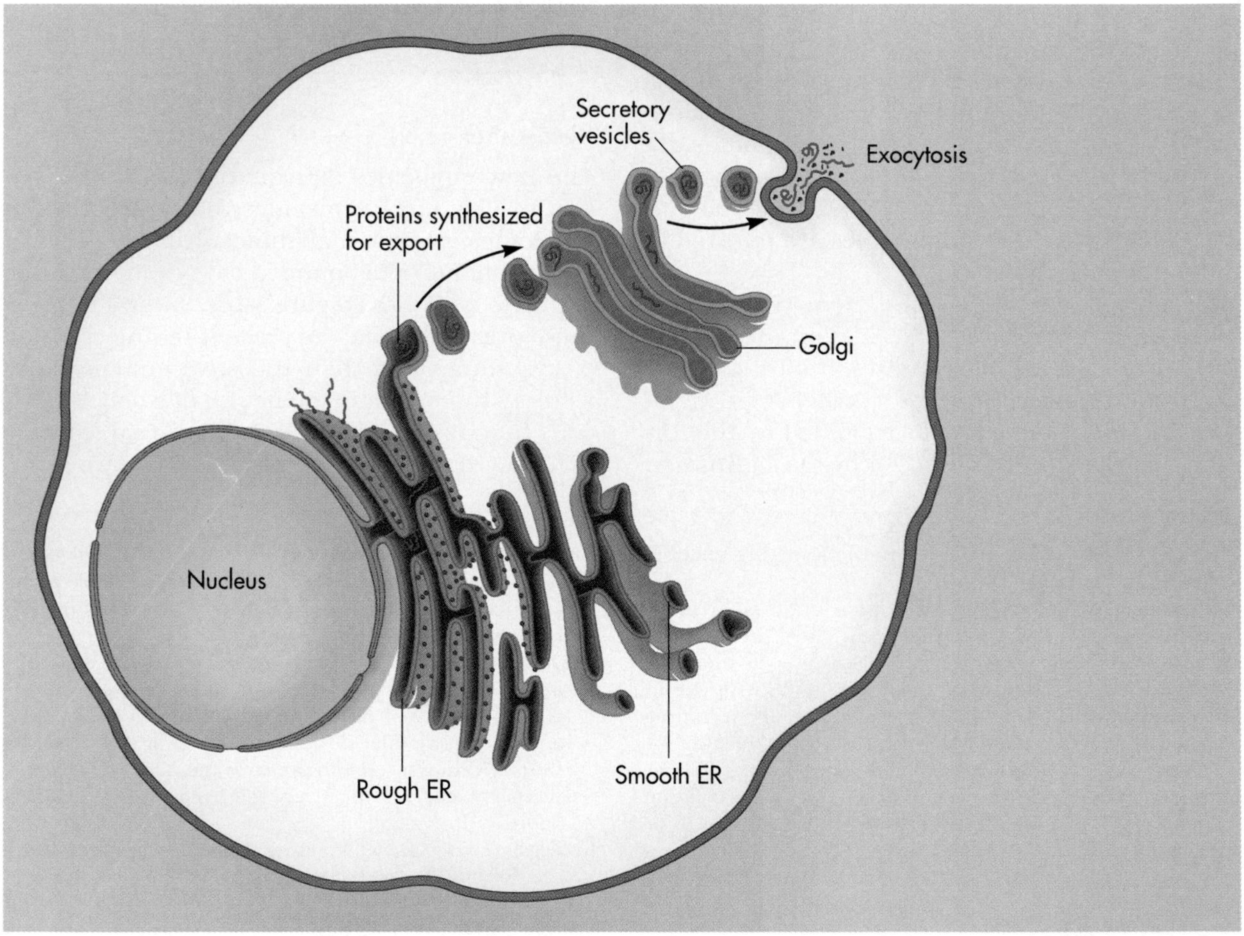

FIGURE 3-13 How a molecule is packaged for transport in the Golgi body. Proteins used to build cell membranes or used outside the cell are usually synthesized at the ribosomes attached to the ER. These proteins move through the Golgi complex, are modified there, and are finally encased in secretory vesicles. These vesicles then fuse with the cell's plasma membrane. Those destined for export are secreted in a process known as exocytosis.

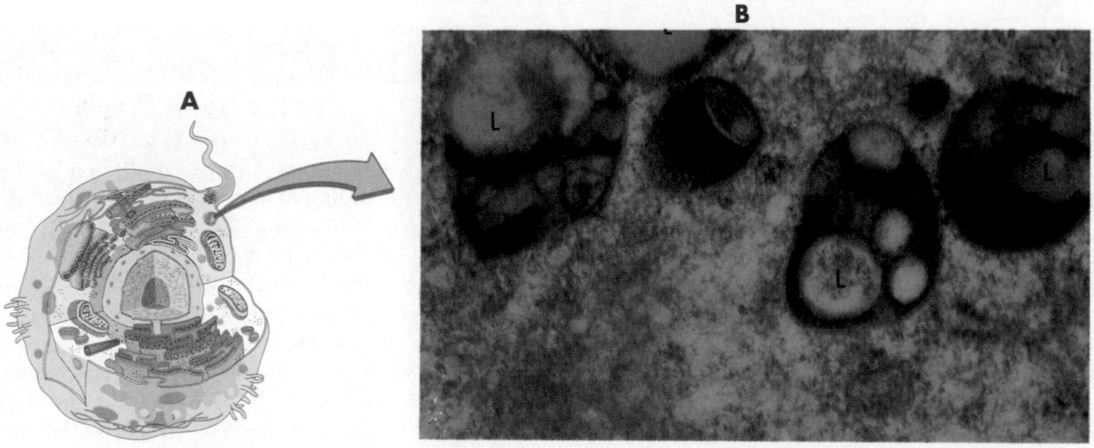

FIGURE 3-14 **Lysosomes. A,** Reference illustration of a lysosome within a cell. **B,** Electron micrograph of lysosomes (*L*). The blackened areas within the lysosomes are digestive enzymes. These enzymes break down old cell parts and other materials brought into the cell (*45,000x*).

with new cell parts. During development, lysosomes help re-model tissues, such as the reabsorption of the tadpole tail as the tadpole develops into a frog. In some cells, lysosomes also break down substances brought into the cell from the environment. For example, one job of certain white blood cells is to get rid of bacteria invading the body. The cytoplasm of these cells flows around their prey, engulfing them in a membrane-bounded sac called a *vacuole.* A lysosome then fuses with this vacuole, and digestion of the invader begins. In the past, lysosomes were thought to be active during the aging process, breaking open and destroying whole cells. Scientists now know this idea to be false, since lysosome enzymes work only in the acid environment of the lysosome and become inactive at the relatively neutral pH of the cytoplasm.

What happens to the digestion products of lysosomes? Substances such as parts of bacteria are packaged in a vesicle, are transported to the plasma membrane, and exit the cell. Other molecules, such as the breakdown products of old cell parts, may simply be released into the cytoplasm. These cellular building blocks can be recycled—used once again to build new cell parts.

> *Lysosomes are vesicles that contain digestive enzymes. These enzymes break down old cell parts or materials brought into the cell from the environment.*

Vacuoles

The word **vacuole** comes from a Latin word meaning "empty." However, these membrane-bounded storage sacs only look empty. Within them can be found such substances as water, food, and wastes. Their number, kind, and size vary in different kinds of cells.

Vacuoles are most often found in plant cells. These giant water-filled sacs play a major role in helping plant tissues stay rigid. In mature plant cells, the vacuole is often so large that it takes up most of the interior of the cell. Certain single-celled eukaryotes (protists) and some yeasts also contain vacuoles. Some fresh water protists contain contractile vacuoles. These organelles take up extra water that tends to flow into these organisms. Periodically, these vacuoles fuse with the plasma membrane and expel their contents, keeping the cell from swelling and bursting. Many protists have food vacuoles, places where food is stored and eventually digested by fusion with lysosomes.

The Nucleus

The **nucleus,** or control center of the cell, is made up of an outer, double membrane that encloses

> **vacuole** (**vack** yoo **ole**) membrane-bounded storage sacs within eukaryotic cells that hold such substances as water, food, and wastes; most often found within plant cells.
>
> **nucleus** (**noo** klee us) the control center of the cell, which is made up of an outer, double membrane that encloses the cell's chromosomes and one or more nucleoli.

JUST WONDERING. . . .

How do scientists know what is in the nucleus of a cell?

The story of scientists' quests to understand the structure of the nucleus began in the early 1700s. Microscopes at that time, although primitive by today's standards, were sophisticated enough to allow scientists to see the nuclei of plant cells and some animal cells. However, it wasn't until 1833 that a nucleus was thought to be in every cell. Prior to that time, scientists held the view that some cells had no nuclei because they could not see them. Today we realize that scientists were simply observing some cells during cell division, when the nucleus loses its membrane and the nuclear material looks much different from that of a nondividing cell.

As time went on, microscope technology improved with the invention of the high-power oil immersion lens in 1870. The microtome was invented in 1866, which could slice individual cells into thin sections. In addition, new methods of fixing and staining cells were developed. Some dyes were produced that were highly specific for staining certain cell parts, such as the nucleus. All of these factors allowed scientists to view the nuclei of cells with much greater magnification, detail, and clarity.

As the quality of optical equipment and cell preparation techniques improved, the precision increased with which cells "at rest" and undergoing nuclear division could be studied. The nucleus no longer looked like a granule-filled compartment that split indiscriminately during cell division. Scientists could now see a nucleus filled with threads and bands of material, which they named *chromatin* after the Greek word for color, since this material readily took on color when stained. The word *chromosome,* meaning "colored body," was proposed in 1888.

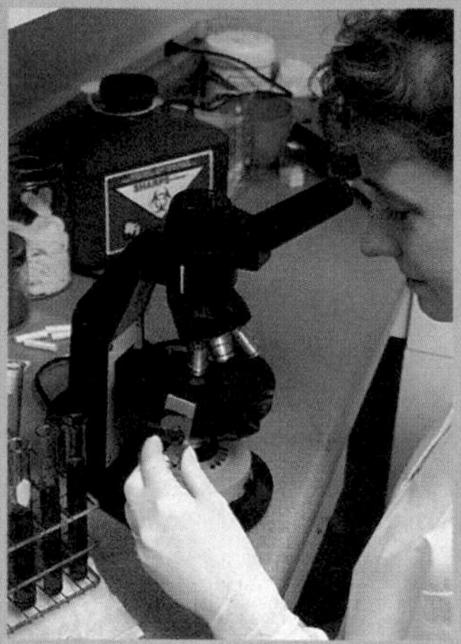

During the 1800s many scientists were also studying the principles of heredity and the nature of fertilization as well as the events of cell division. So investigations into the function of the nuclear material paralleled the study of its structure. By 1866, Gregor Mendel published his theories of inheritance. (His work is described in Chapter 19.) In the late 1870s, scientists described the process of nuclear division and the process of nuclear reduction-division that occurs in sex cells. By 1884, scientists realized that fertilization in both animals and plants consists of the fusion of maternal and paternal sex cells. Taking all of this information into account, an American graduate student named Walter Sutton hypothesized that the hereditary factors were on the chromosomes. At the turn of the century, scientists began to look toward the chemical nature of this hereditary material. Then, for more than half a century, chemists joined biologists in their quest to understand the structure of nuclear material.

The rest of this fascinating story of discovery is continued later in this book. Chapter 18, "DNA, Gene Expression and Cell Reproduction," chronicles investigations that experimentally determined that the hereditary material is located in the nucleus. It goes on to discuss the discovery of the molecular makeup of the hereditary material, culminating with the Nobel prize–winning work of two young scientists, James Watson and Francis Crick. As you can see, your question took scientists from many fields over 100 years to answer. Today, using electron microscopes and highly sophisticated laboratory techniques, molecular biologists and others are still asking and answering questions about the nucleus and the nature of the hereditary material.

the chromosomes and one or more nucleoli. In fact, the word *eukaryote* means "true nucleus" and refers to the fact that the nucleus is a closed compartment bounded by a membrane. Other than large fluid-filled vacuoles in plants, this compartment is the largest in eukaryotic cells.

The outer, double membrane of the nucleus is called the *nuclear envelope.* The inner of the two membranes actually forms the boundary of the nucleus, and the outer membrane is continuous with the ER. At various spots on its surface, the double membrane fuses to form openings called *nuclear pores.* Figure 3-15 shows the surface of the nuclear membrane of a cell that was frozen and then cracked during a special type of preparation for the electron microscope. The pores look like pock-

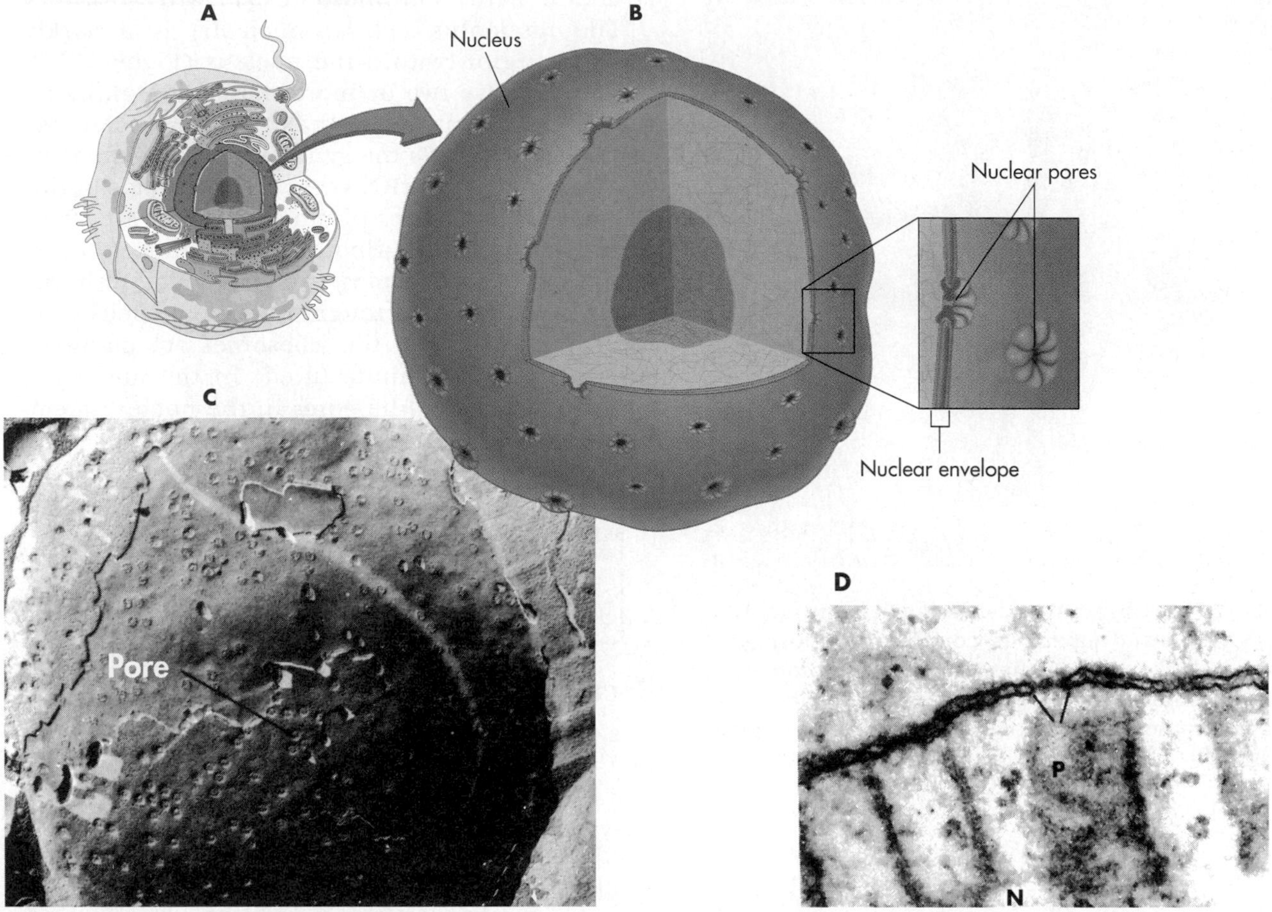

FIGURE 3-15 The nucleus. A, The nucleus, or controlling center of the cell. **B,** The outer, double membrane of the nucleus encloses the chromatin and one or more nucleoli. This double membrane is dotted with openings called nuclear pores. The nuclear membrane is fused at the edges of each pore. **C,** Electron micrograph showing nuclear pores perforating the nuclear envelope (*36,000x*). These pores are visible in **D,** designated by *P. C* is the cytoplasm and *N* is the nucleus.

marks on the surface of the membrane. These ringlike holes are lined with proteins and serve as passageways for molecules entering and leaving the nucleus.

Within the nucleus is the hereditary material **deoxyribonucleic acid,** or **DNA.** DNA determines whether your hair is blond or brown or whether a plant flowers in pink or white. It controls all activities of the cell. To accomplish these amazing feats, DNA performs one job: it directs the synthesis of ribonucleic acid, or RNA, which in turn directs the synthesis of proteins. (The structure, interactions, and roles of both DNA and RNA are discussed further in Chapter 18.)

DNA is bound to proteins in the nucleus, forming a complex called *chromatin.* In a cell that is not dividing, the chromatin is strung out, looking like strands of microscopic pearls. But as a cell begins to divide, the DNA coils more tightly around the

deoxyribonucleic acid (DNA) (de **ok** see **rye** boh new **klay** ick) the hereditary material; DNA controls all cell activities and determines all of the characteristics of organisms.

chromosomes (**krow** muh **somes**) shortened, thickened structures consisting of DNA coiled tightly around proteins.

nucleolus (noo **klee** oh lus) a darkly staining region within the nucleus of a cell that contains a special area of DNA that directs the synthesis of ribosomal ribonucleic acid, or rRNA.

FIGURE 3-16 Chromosomes. As a cell begins to divide, DNA coils and supercoils, condensing to form shortened, thickened structures called chromosomes. The chromosomes in this cell are ready to separate from one another in a way that will provide a complete set of hereditary material to each new daughter cell.

proteins, condensing to form shortened, thickened structures called **chromosomes** (Figure 3-16).

The **nucleolus** (plural nucleoli) is a darkly staining region within the nucleus (Figure 3-17). Most cells have two or more nucleoli. Making up the bulk of the nucleolar material is a special area of DNA that directs the synthesis of **ribosomal ribonucleic acid,** or **rRNA.** As rRNA is made at this DNA, it forms clumps of molecules that are structural components of ribosomes. Proteins, another component of ribosomes, can also be found in the nucleolus. They are brought into the nucleus from the cytoplasm. After the ribosomes are manufactured (or partly manufactured) in the nucleolus, they move through the pores in the nuclear membrane and pass into the cytoplasm.

> *The nucleus of a eukaryotic cell contains the hereditary material, or DNA. When the cell divides, the DNA condenses into compact structures called chromosomes.*

Bacterialike Organelles

The idea of the symbiotic origin of mitochondria and chloroplasts has had a controversial history, and a few biologists still do not accept it. The endosymbiotic hypothesis is, however, the most widely agreed-on model at this time regarding the origin of mitochondria and chloroplasts in eukaryotic cells.

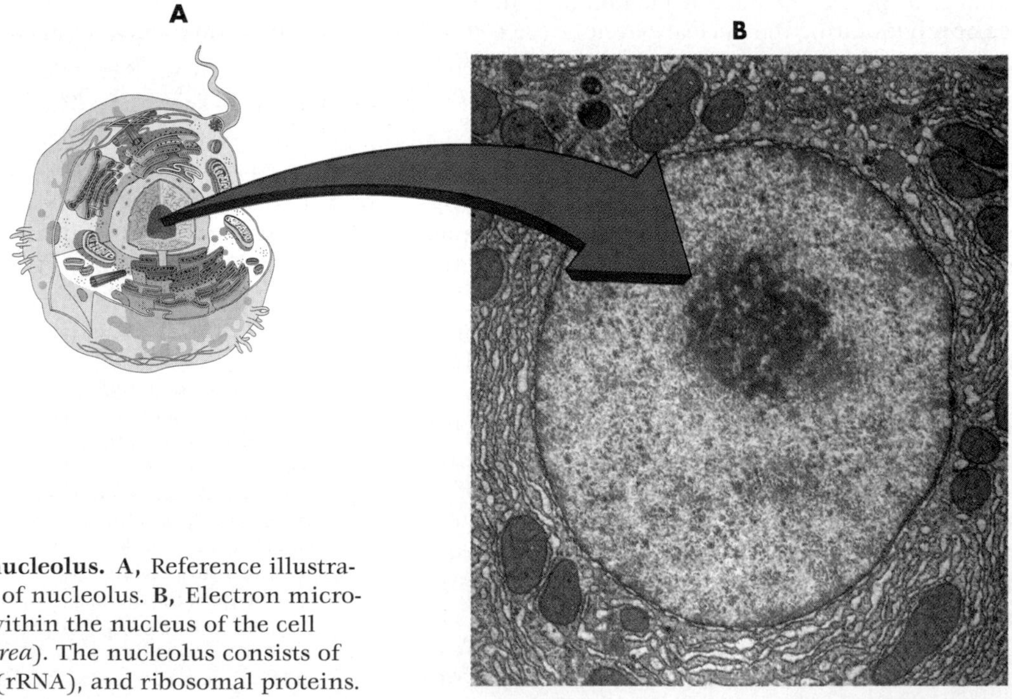

FIGURE 3-17 The nucleolus. A, Reference illustration showing location of nucleolus. **B,** Electron micrograph of a nucleolus within the nucleus of the cell (*darkly stained green area*). The nucleolus consists of DNA, ribosomal RNA (rRNA), and ribosomal proteins.

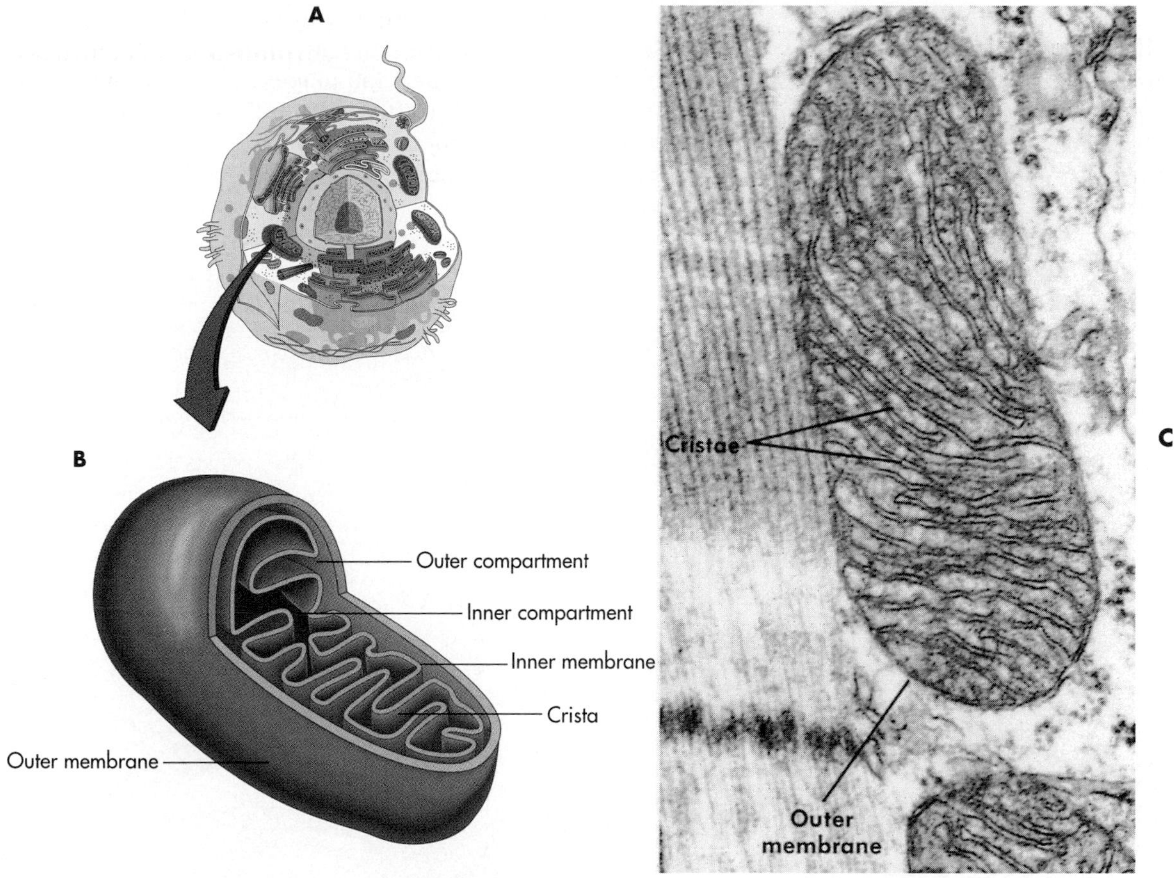

FIGURE 3-18 The mitochondrion. A, Reference illustration of a cell showing sausage-shaped mitochondria. **B,** This illustration shows the outer membrane and extensively folded inner membrane. These infoldings are called cristae. **C,** Electron micrograph of mitochondria, which break down fuel molecules and release energy for cell work.

Mitochondria

The **mitochondria** that occur in most eukaryotic cells are thought to have originated as symbiotic bacteria. According to this hypothesis, the bacteria that became mitochondria were engulfed by eukaryotic cells early in their evolutionary history. Before they had acquired these bacteria, the host cells were unable to carry out chemical reactions necessary for living in an atmosphere that had increasing amounts of oxygen. The engulfed bacteria were able to carry out these reactions *using oxygen* and are considered to be the precursors to mitochondria.

Mitochondria are oval, sausage-shaped, or threadlike organelles about the size of bacteria that *have their own DNA* (which suggests that they were once free-living bacteria). They are bounded by a double membrane. The outer of the two membranes is smooth and defines the shape of the organelle. The inner membrane, however, has many folds called *cristae* that dip into the interior of the mitochondrion. These cristae resemble the folded membranes that occur in various groups of bacteria. Notice in Figure 3-18, *B* that this arrangement of membranes forms two mitochondrial compartments.

The job of the mitochondria is to break down fuel molecules, releasing energy for cell work. The two most important fuels of cells are glucose and fatty acids. Some organisms, such as plants, make their fuel (glucose) using the raw materials of carbon dioxide, water, and sunlight. Other organisms, such as animals, eat food and digest it to produce glucose

ribosomal ribonucleic acid (rRNA) (rye buh **so** mull **rye** boh new **klay** ick) a type of ribonucleic acid that is manufactured by the DNA within a eukaryotic cell's nucleolus; rRNA molecules are structural components of ribosomes.

mitochondria (**mite** oh **kon** dree uh) oval, sausage-shaped, or threadlike cellular organelles approximately the size of bacteria that have their own DNA; these eukaryotic cell organelles break down fuel molecules, releasing energy for cell work.

and fatty acids, which can then be transported to the cells. In both plants and animals, these fuels are broken down by cells, releasing energy by means of a series of oxygen-requiring reactions called *cellular respiration*. The energy released during these reactions is stored for later use in special molecules called ATP. (See Chapters 4 and 5 for a detailed description of ATP and cellular respiration.) Using chemical symbols, the reactions of cellular respiration can be summarized as follows:

$$\underset{\text{Glucose}}{C_6H_{12}O_6} + \underset{\text{Oxygen}}{6O_2} \rightarrow \underset{\text{Carbon dioxide}}{6CO_2} + \underset{\text{Water}}{6H_2O} + \text{Energy}$$

Cellular respiration begins in the cytoplasm, but most of the energy from the breakdown of glucose and fatty acids is generated in the mitochondria. The enzymes that are used in this breakdown are bound to the membranes of the mitochondrion. Within this closed compartment, the complex series of reactions needed to break apart glucose and to capture the liberated energy and store it are accomplished in an orderly and efficient way separated from the rest of the cell.

> *The mitochondria apparently originated as endosymbiotic bacteria. In a complex series of reactions using oxygen, cell fuel is broken down to release the energy within mitochondria.*

Chloroplasts

Symbiotic events similar to those postulated for the origin of mitochondria also seem to have been in-

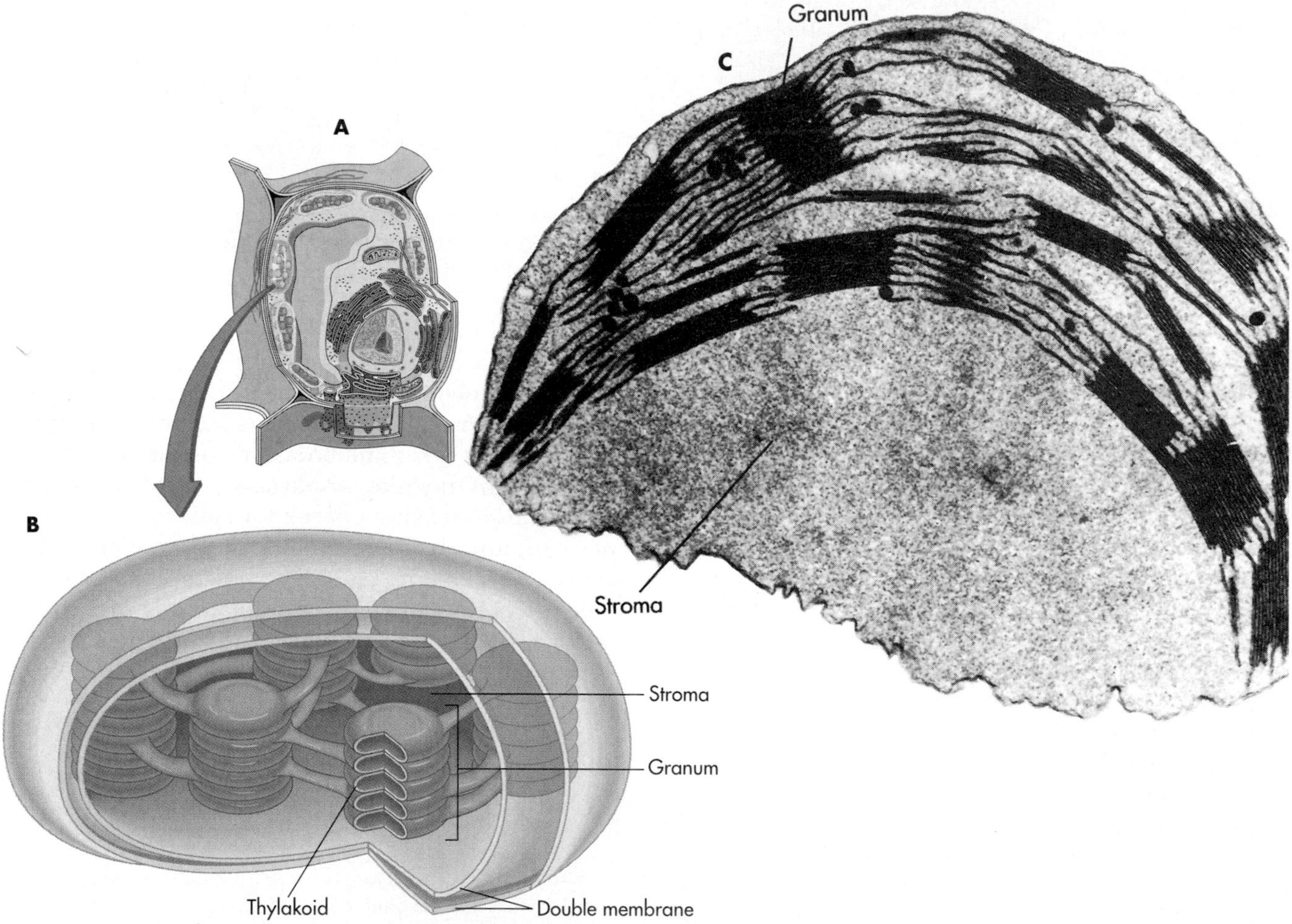

FIGURE 3-19 The chloroplast. **A,** Energy-capturing chloroplast shown here in a plant cell. **B,** Chloroplasts have a double outer membrane and a system of interior membranes that form sacs called thylakoids. A stack of thylakoids is called a granum. **C,** Electron micrograph of a chloroplast showing the grana (*2,700x*).

volved in the origin of **chloroplasts.** They are thought to be derived from symbiotic photosynthetic bacteria. These energy-capturing organelles are found in the cells of plants and algae. In these organelles the energy in sunlight is used to power the reactions that make the cellular fuel—glucose—by using molecules of carbon dioxide from the air. Together, the complex series of chemical reactions that perform these tasks is known as *photosynthesis.* Using chemical symbols, this series of reactions can be summarized as follows:

$$6CO_2 + 6H_2O + \text{Sun's energy} \longrightarrow C_6H_{12}O_6 + 6O_2$$

Carbon dioxide, Water, Glucose, Oxygen

The glucose can then be broken down in the mitochondria to release its energy to ATP for immediate use or short-term storage, or the glucose can be stored as complex carbohydrates for later use.

Chloroplasts have a structure similar to the mitochondria. Like mitochondria, chloroplasts have their own DNA and are bounded by a double membrane. The inner of these membranes forms an extensive array of saclike invaginations called *thylakoids.* Stacks of thylakoids are called *grana* (Figure 3-19). Chlorophyll, a chemical that can absorb light energy from the sun and that allows photosynthesis to take place, is found within the thylakoids. As with the mitochondria, chloroplasts provide an orderly, closed compartment within the cell in which a series of reactions can occur.

Chloroplasts, which are located within the cells of plants and algae, apparently originated as endosymbiotic photosynthetic bacteria. Within chloroplasts, carbon dioxide, water, and light energy are used to produce the cell fuel glucose during a series of reactions called photosynthesis.

Cilia and Flagella

Single-celled eukaryotic cells are often motile—able to move within their environments. One way in which these cells move is by means of cell extensions that look somewhat like hairs. These structures, called **cilia,** are short and often cover a cell. In human cells, cilia are found only on sections of certain cells and are used to move substances across their surfaces. Figure 3-20 shows the cilia that line the trachea, or windpipe. They help sweep invading particles and organisms up the trachea and away from the lungs. Some cells have whiplike extensions called **flagella.** Used strictly for movement, these structures are longer than cilia, but fewer are usually present on a cell. Sperm are the only human cells that have flagella. Normal human sperm have a single flagellum, as you can see in Figure 3-1, *C.*

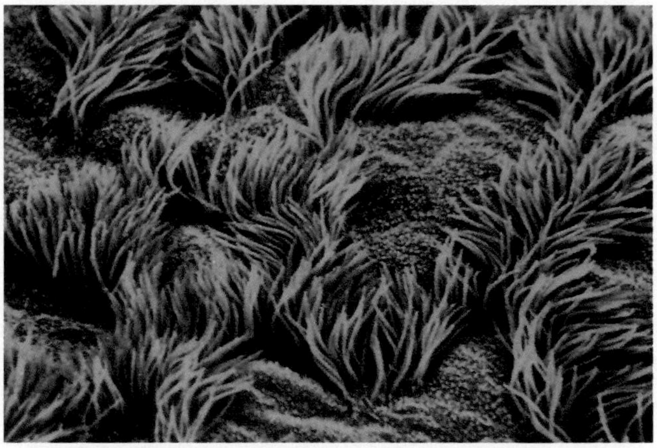

FIGURE 3-20 Cilia lining the trachea, or windpipe. These cilia help move particles and bacteria away from the lungs and upward toward the oral cavity.

Although cilia and flagella differ in length, they have the same structure: they are bundles of microtubules covered with the plasma membrane of the cell. As you can see in Figure 3-21, nine pairs of microtubules surround a single, central pair. As these microtubules dip into the cell beneath the level of the plasma membrane, they connect with another structure called the *basal body.* Also composed of microtubules, a basal body serves to anchor a cilium or flagellum to the cell.

Cilia and flagella in eukaryotes beat in a whiplike fashion because of microtubules sliding past each other in a manner similar to the sliding of actin and myosin filaments in muscle contraction (see Chapter 16). The structure of the flagella in bacteria differs from eukaryotic flagellar structure. In addition, bacteria swim by rotating their flagella rather than by whipping them as eukaryotes do.

Cilia and flagella are whiplike organelles of motility that protrude from some eukaryotic cells.

chloroplasts (**klor** oh plasts) energy-capturing organelles that are found in the cells of plants and algae; these eukaryotic cell organelles manufacture carbohydrates using energy from the sun and carbon dioxide.

cilia (**sill** ee uh) hairlike extensions that often cover the surface of some eukaryotic cells and allow them to move or to move substances across their surfaces.

flagella (fluh **jell** uh) whiplike extensions from some cells that allow them to move.

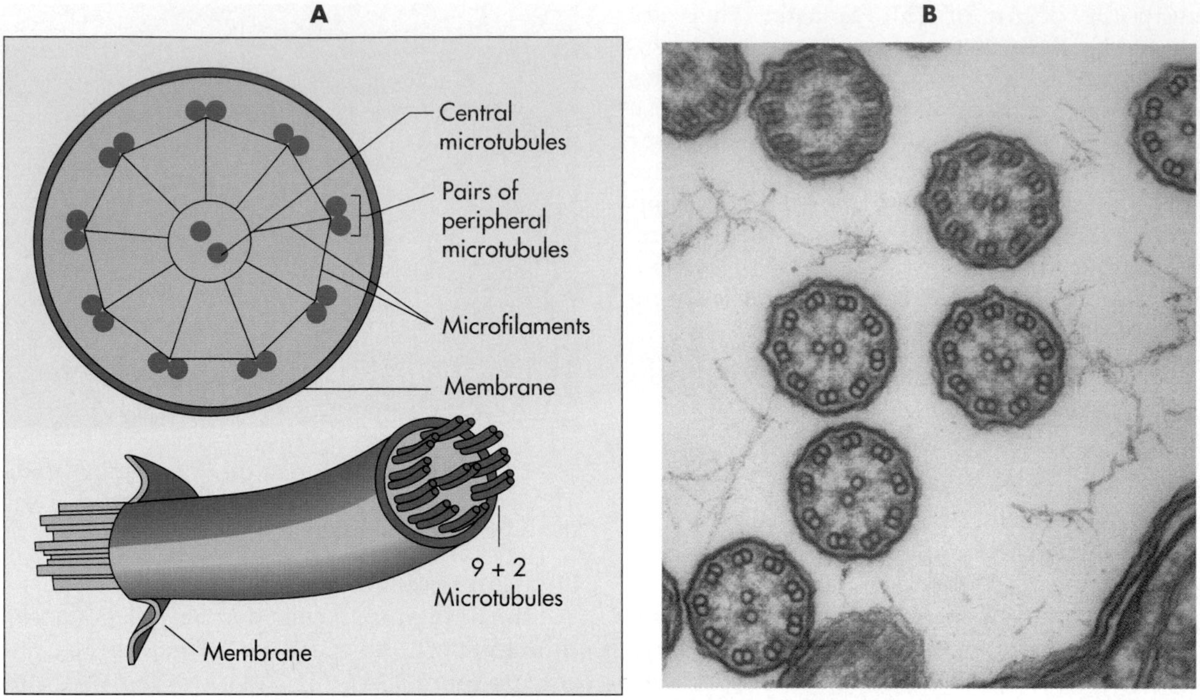

FIGURE 3-21 The structure of cilia and flagella. A, Cilia and flagella have the same structure. Each has nine pairs of peripheral microtubules surrounding a central pair of microtubules. The microtubules are connected by microfilaments. **B,** Colorized micrograph showing a cross-section of cilia (*74,500x*).

Cell Walls

Many plants, algae, and fungi have a rigid structure called a **cell wall** that surrounds the plasma membrane. Almost all plant cell walls are made up of cellulose, large molecules formed by the linking of glucose units. Many single-celled eukaryotes have a similar type of cell wall. However, other single-celled eukaryotes, such as *Paramecium* and *Euglena* species, have an outer structure called a *pellicle* that is made up of protein. Fungi have cell walls that contain chitin, the same substance that is found in the shells of organisms such as grasshoppers and lobsters.

Cell walls perform many jobs for cells. In plants and fungi, they help impart a stiffness to the tissues. They also provide some protection from a drying environment. In single-celled organisms, cell walls give shape to the organisms and help protect them. Pellicles are even more specialized in their function, often serving as anchors for defense mechanisms such as organelles that discharge sticky or toxic threads.

Summing Up: How Bacterial, Animal, and Plant Cells Differ

Table 3-2 summarizes the differences among bacterial, animal (including human), and plant cells. As the table shows, prokaryotes (bacterial cells) and eukaryotes (plant and animal cells) differ greatly in structure and complexity. The most distinctive difference between these two cell types is the extensive subdivision of the interior of eukaryotic cells by membranes. (Prokaryotic cells are described in more detail in Chapter 25.) The rest of this chapter will explain how the substances cells use move across their plasma membranes.

How Substances Move into and out of Cells

Cells usually live in an environment where they are bathed in water. When you consider that bacteria live on your body and that protists live in the soil, this fact may not seem to be true. However, all cells must move substances across their membranes to survive, and water is the liquid in which their molecules are dissolved. In addition, water provides a fluid environment within which molecules can move. Therefore water must surround cells, even in microscopic amounts that may not be readily apparent.

How do molecules get where they are going? To answer this question, you must understand that molecules cannot move *purposefully* from one spot to another. Molecules move randomly. All molecules and small particles have a constant, inherent

TABLE 3-2

A Comparison of Bacterial, Animal, and Plant Cells

	BACTERIUM	ANIMAL	PLANT
Exterior Structures			
Cell wall	Present (protein-polysaccharide)	Absent	Present (cellulose)
Plasma membrane	Present	Present	Present
Flagella (cilia)	Flagella sometimes present	Sometimes present	Sperm of a few species possess flagella
Interior Structures and Organelles			
Endoplasmic reticulum	Absent	Usually present	Usually present
Microtubules	Absent	Present	Present
Centrioles	Absent	Present	Absent
Golgi bodies	Absent	Present	Present
Nucleus	Absent	Present	Present
Mitochondria	Absent	Present	Present
Chloroplasts	Absent	Absent	Present
Chromosomes	A single circle of naked DNA	Multiple units, DNA associated with protein	Multiple units, DNA associated with protein
Ribosomes	Present	Present	Present
Lysosomes	Absent	Usually present	Present
Vacuoles	Absent	Absent or small	Usually a large single vacuole in mature cell

"jiggling" motion called *Brownian movement.* And because molecules are always jiggling, they tend to bump into things, such as other molecules. A bump may push a molecule in a particular direction until it bumps into another molecule and gets pushed again. This random motion in all directions often results in a net movement of molecules in a particular direction in response to differences in concentration, pressure, or electrical charge. These differences are referred to as a **gradient.** The term *net movement* means that although individual members of a group of molecules are moving in different directions, the resulting movement of the group is in one direction. This movement results in a uniform distribution of molecules.

Movement That Does Not Require Energy

The net movement of molecules often requires no input of energy. In this type of movement, molecules move from regions of high concentration to those of low concentration or from regions of high pressure to regions of low pressure. Ions, because they each carry an electrical charge, move toward unlike charges or away from like charges.

As molecules or ions move from regions of high concentration to regions of low concentration, they are referred to as going *down* a concentration, pressure, or electrical gradient. Some molecules move into and out of cells as they move down gradients. Molecular movement down a gradient but across a cell membrane is called **passive transport.** There are three types of passive transport: diffusion, osmosis, and facilitated diffusion.

Diffusion

Everyone has had experience with molecules in the air moving down a concentration gradient. Did you ever wake up to the smell of fresh brewed coffee? Did you ever spill a bottle of a liquid that had a strong odor, such as ammonia or perfume? The molecules that reached the smell receptors in your nose moved from an area of high concentration (the spill) to an area of low concentration (your nose). They moved down the concentration gradi-

cell wall a rigid structure that surrounds some cells' plasma membranes.

gradient differences in concentrations, pressures, and electrical charges that result in the net movement of molecules in a particular direction.

passive transport molecular movement down a gradient but across a cell membrane; three types of passive transport are diffusion, osmosis, and facilitated diffusion.

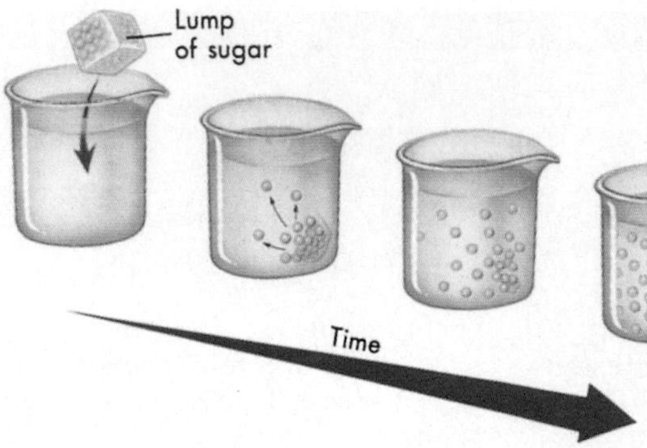

FIGURE 3-22 Process of diffusion. Sugar molecules move from an area of higher concentration (*the cube*) to an area of lower concentration (*the water in the beaker*); eventually there will be an even distribution of sugar molecules throughout the water.

ent, eventually becoming evenly spread. This process is illustrated in Figure 3-22 with a lump of sugar.

> *Diffusion is the net movement of molecules from a region of higher concentration to a region of lower concentration, eventually resulting in a uniform distribution of the molecules. This movement is the result of random, spontaneous molecular motions.*

Osmosis

Osmosis is a special form of diffusion in which water molecules move from an area of higher concentration to an area of lower concentration across a differentially permeable membrane. A *differentially permeable membrane* allows only certain types of molecules to pass through it, or permeate it, freely. Cell (plasma) membranes are differentially permeable membranes (and are also called *semipermeable* or *selectively permeable membranes*). In fact, most types of molecules that occur in cells cannot pass freely across the plasma membrane (see p. 60).

The cytoplasm of a cell consists of many different types of molecules and ions dissolved in water. The mixture of these molecules and water is called a *solution.* Water, the most common of the molecules in the mixture, is called the *solvent.* The other kinds of molecules dissolved in the water are called *solutes.*

Because of diffusion, both solvent and solute molecules in a cell move from regions where the concentration of each is greater to regions where the concentration of each is less. When two regions are separated by a membrane, however, what happens depends on whether the molecule can pass freely through that membrane.

> *Osmosis is the diffusion of water molecules across a differentially permeable membrane.*

The cytoplasm of living cells contains approximately 1% dissolved solutes. If a cell is immersed in pure water, interesting things begin to happen. Because the water has a lower concentration of solutes than the cell, water molecules begin to move into the cell. The reason is simple: water moves down the concentration gradient, or from a region of higher concentration of water molecules to a region of lower concentration of water molecules. The pure water is said to be **hypotonic** (Greek *hypo,* "under") with respect to the cytoplasm of the cell because its concentration of solutes is less than the concentration of solutes in the cell.

As you can see in Figure 3-23, a cell in a hypotonic solution begins to blow up like a balloon. Likewise, if a cell is placed in a **hypertonic** (Greek *hyper,* "over") solution, one with a solute concentration higher than the cytoplasm of the cell, water will move out of the cell, and the cell will shrivel. If a cell is placed in a solution with the same concentration of solutes as its cytoplasm, an **isotonic** (Greek *isos,* "equal") solution, water will move into and out of the cell, but no net movement will take place.

> *A solution with a solute concentration lower than that of another fluid is said to be hypotonic to that solution. A solution with a solute concentration higher than that of another fluid is said to be hypertonic to that solution. Solutions having equal solute concentrations are isotonic to one another.*

Intuitively you may think that as "new" water molecules diffuse into a cell placed in a hypotonic solution, the pressure of the cytoplasm pushing against the cell membrane builds. This is indeed what happens. At the same time, however, the water molecules that continue to diffuse into the cell also exert a pressure—**osmotic pressure.** Because the pressure of the cytoplasm within the cell opposes osmotic pressure in this case, diffusion of wa-

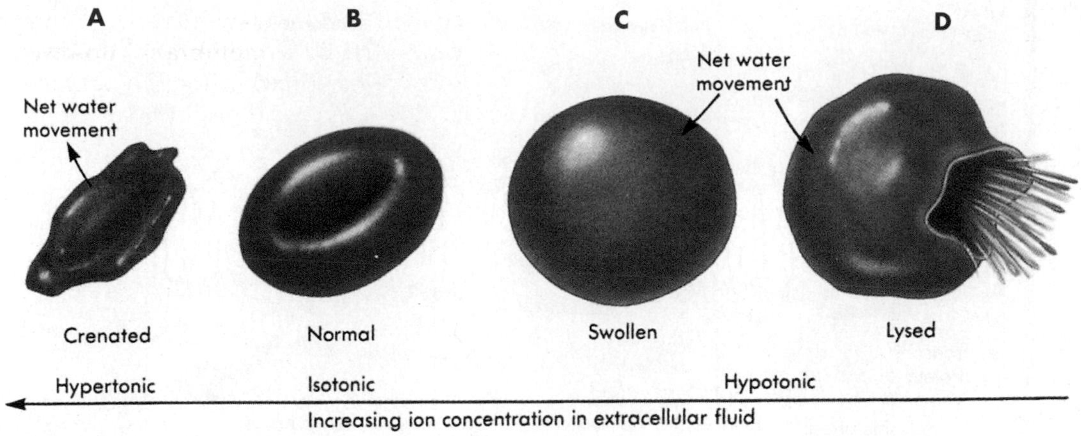

FIGURE 3-23 Effects of hypertonic, isotonic, and hypotonic solutions on red blood cells. A, When the outer solution is hypertonic with respect to the cell, water moves out of the cell and it shrinks. The cell is said to be crenated (shrunken). **B,** A cell in an isotonic solution will retain its shape because the amount of water entering the cell equals the amount of water leaving the cell. **C,** When the outer solution is hypotonic with respect to the cell, water moves into the cell and it swells. **D,** As water continues to enter a cell from a hypotonic solution, it will eventually cause the cell to burst.

ter molecules into the cell will not continue indefinitely. The cell will eventually reach an equilibrium—a point at which the osmotic force driving water inward is counterbalanced exactly by the pressure outward of the cytoplasm. However, the pressure within the cell may become so great that the cell bursts like a balloon. The cell is said to *lyse.*

> *Water moves into cells placed in hypotonic solutions and out of cells placed in hypertonic solutions.*

Single-celled and multicellular organisms have various mechanisms that work to keep their cells from swelling and bursting or shriveling like prunes. Single-celled organisms that live in fresh water, for example, battle a constant influx of water. Some of these organisms have one or more organelles that collect water from the cell's interior and transport it to the cell surface. Plant cells have cell walls that support their membranes, so plant cell membranes press outward against cell walls in hypotonic solutions and pull in away from cell walls in hypertonic solutions.

Many multicellular animals, such as humans, circulate a fluid through their bodies that bathes cells in isotonic liquid. By controlling the composition of its circulating body fluids, a multicellular organism can control the solute concentration of the fluid bathing its cells, adjusting it to match that of the cells' interiors. Your blood, for example, contains a high concentration of a protein called *albumin,* which serves to elevate the solute concentration of the blood to match that of your tissues so that the movement of water molecules is regulated.

Facilitated Diffusion

One of the most important properties of any cell is its ability to move substances necessary for survival into its interior and to get rid of unnecessary or harmful substances. Your cells, for example, move glucose into the cytoplasm from the bloodstream to be used for fuel. Without the ability to move glucose into your cells, you would die.

Cells can perform these feats because they have differentially permeable membranes. The cell membrane can select which molecules are to enter and leave and in which concentrations because most molecules must enter or leave the cell

osmosis (os **moe** sis) a special form of diffusion in which water molecules move from an area of higher concentration to an area of lower concentration across a differentially permeable membrane.

hypotonic (**hi** poe **tawn** ick) refers to a solution with a solute concentration lower than that of another fluid.

hypertonic (**hi** per **tawn** ick) refers to a solution with a solute concentration higher than that of another fluid.

isotonic (**eye** so **tawn** ick) refers to solutions having equal solute concentrations to one another.

osmotic pressure the pressure that water exerts on a cell as it diffuses into the cell.

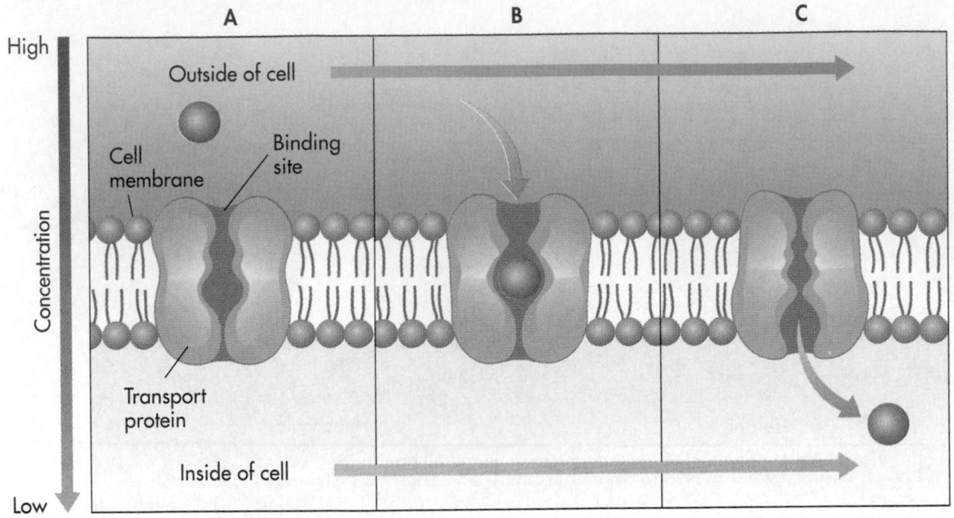

FIGURE 3-24 Facilitated diffusion. Facilitated diffusion is similar to diffusion in that it is a passive process in which molecules move from a region of higher concentration to a region of lower concentration. However, in facilitated diffusion, molecules move across a cell membrane through specific membrane-bound protein channels.

through protein doors, or channels. These doors are not open to every molecule that presents itself; only a particular molecule can pass through a given kind of door. Some of the channels in the cell membrane help certain molecules and ions enter or leave the cell and speed their movement by providing them with a passageway. These passageways are most likely transport proteins that extend from one side of the membrane to the other. After the transport protein binds with a solute molecule, its shape changes, and the molecule moves across the membrane. This type of transport process, in which molecules move down the concentration gradient without an input of energy by the cell, is called **facilitated diffusion** (Figure 3-24). Facilitated diffusion helps rid the cell of certain molecules present in high concentrations and moves molecules into the cell that are present on the outside in high concentrations.

Facilitated diffusion has two essential characteristics: (1) it is *specific,* with only certain molecules being able to traverse a given channel, and (2) it is *passive,* the direction of net movement being determined by the relative concentrations of the transported molecule inside and outside the membrane.

> *Facilitated diffusion is the movement of selected molecules across the cell membrane by specific transport proteins along the concentration gradient and without an expenditure of energy.*

Movement That Does Require Energy

Cells often move substances into or out of the cell *against* the gradients of concentration, pressure, and electrical charge. A cell uses energy to move molecules against a gradient much like you might use energy to move something against gravity. For example, if you are driving downhill, you can put your car into neutral and coast (although that is not the safest way to drive downhill). The car will continue to move without a push from the engine. As soon as you come to a hill, however, you must put the car in gear and press on the accelerator or the car will soon come to a stop. Cells, too, need to use energy to move molecules "uphill," or against a gradient. Cells also need to expend energy to move large molecules or particles into the cell that cannot move across the cell membrane.

Active Transport

There are many molecules that a cell takes up or eliminates against a concentration gradient. Some, such as sugars and amino acids, are molecules the cell needs to use for energy or to build new cell parts. Others are ions such as sodium and potassium that play a critical role in such functions as the conduction of nerve impulses. These many kinds of molecules enter and leave cells by way of a variety of selectively permeable transport channels. In all these cases, a cell must expend energy to transport these molecules against the concentration gradient and maintain the concentration difference. This type of transport is called **active transport.** Active transport is one of the most im-

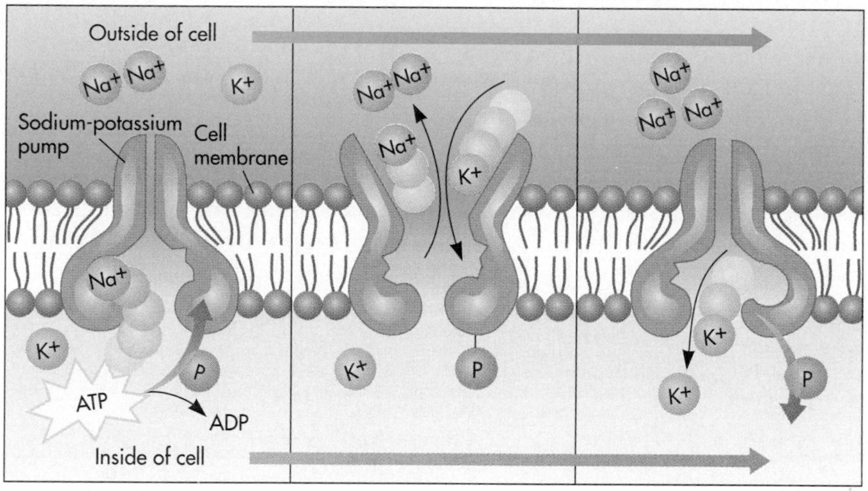

FIGURE 3-25 The sodium-potassium pump. The pump uses energy to move sodium ions (Na^+) out of the cell and potassium ions (K^+) into the cell. Three sodium ions are moved out for every two potassium ions that are moved in by this transport channel. The relative concentrations of sodium and potassium ions within and outside of the cell are important for many functions of the human body.

portant functions of any cell. Without it, the cells of your body would be unable to maintain the proper concentrations of substances they need for survival.

> *Active transport is the movement of a solute across a membrane against the concentration gradient with the expenditure of chemical energy. This process requires the use of a transport protein specific to the molecule(s) being transported.*

More than one third of all the energy expended by a cell that is not dividing is used to actively transport sodium (Na^+) and potassium (K^+) ions. The type of channel by which *both* ions are transported across the cell membrane *in opposite directions* is a *coupled channel,* which has binding sites for both molecules on one membrane transport protein. This remarkable coupled channel is called the **sodium-potassium pump,** and it uses energy to move these ions across the cell membrane. Sodium ions are moved out of the cell to maintain a low internal concentration relative to the concentration outside the cell. Conversely, potassium ions are moved into the cell to maintain a high internal concentration relative to the concentration outside the cell (Figure 3-25). (Three sodium ions are moved out for every two potassium ions that are moved in by the channel.) This transport mechanism is important in most cells of the human body but is particularly important for the proper functioning of muscle and nerve cells.

Endocytosis

Certain types of cells transport particles, small organisms, or large molecules such as proteins into their cells. In humans, for example, white blood cells police the body fluids and ingest substances as large as invading bacteria. In nature, some single-celled organisms often eat other single-celled organisms whole. How can cells move such large substances into their interiors?

Cells such as these ingest particles or molecules that are too large to move across the membrane by a process called **endocytosis,** which literally means "into the cell." Even if the substance is not

facilitated diffusion a type of transport process in which molecules move across the cell membrane by means of a carrier protein but down the concentration gradient without an input of energy by the cell.

active transport the movement of a solute across a membrane against the concentration gradient with the expenditure of energy. This process requires the use of a transport protein specific to the molecule(s) being transported.

sodium-potassium pump the coupled channel that uses energy to move sodium (Na^+) and potassium (K^+) ions across the cell membrane. Three sodium ions are moved out of the cell for every two potassium ions that are moved in.

endocytosis (enn doe sye **toe** sis) a process in which cells engulf large molecules or particles and bring these substances into the cell packaged within vesicles.

being transported against a concentration gradient, the cell must use energy for this process to occur. During endocytosis, a cell envelops the particle with fingerlike extensions of the membrane-covered cytoplasm (Figure 3-26). The edges of the membrane eventually meet on the other side of the particle. Because of the fluid nature of the lipid bilayer, the membranes fuse together, forming a vesicle around the particle.

> *Endocytosis is a process in which cells engulf large molecules or particles and bring these substances into the cell packaged within vesicles.*

If the material that is brought into the cell contains an organism (Figure 3-26, *A* and *C*) or some other fragment of organic matter, the endocytosis

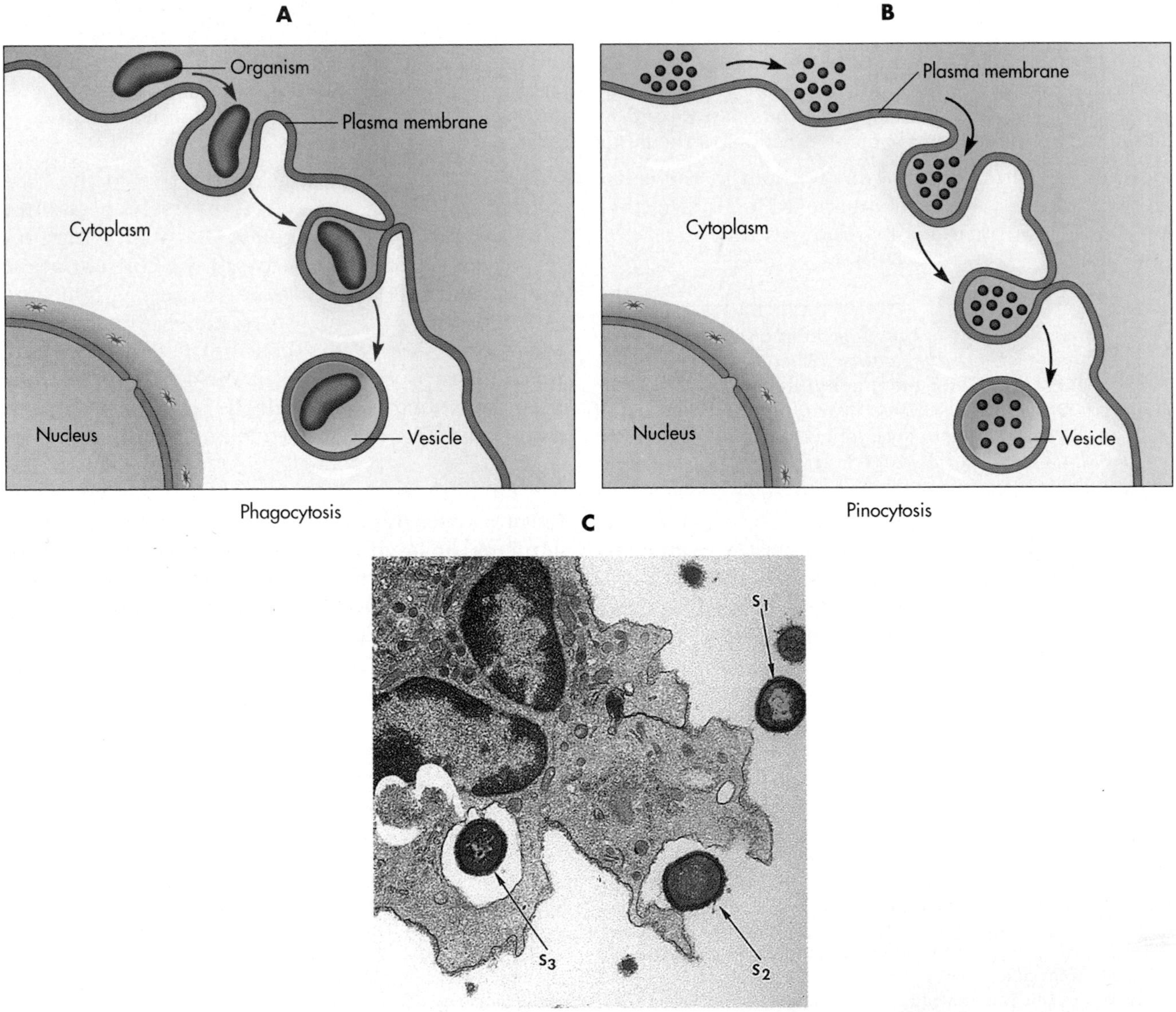

FIGURE 3-26 Phagocytosis and pinocytosis. A, Phagocytosis is the ingestion of an organism or particulate matter. **B,** Pinocytosis is the ingestion of dissolved materials (*liquids*). **C,** Colorized micrograph showing the phagocytosis of a bacterium (*14,000x*). One bacterial cell is free (S_1), one is in the process of being phagocytized (S_2), and one has been phagocytized (S_3).

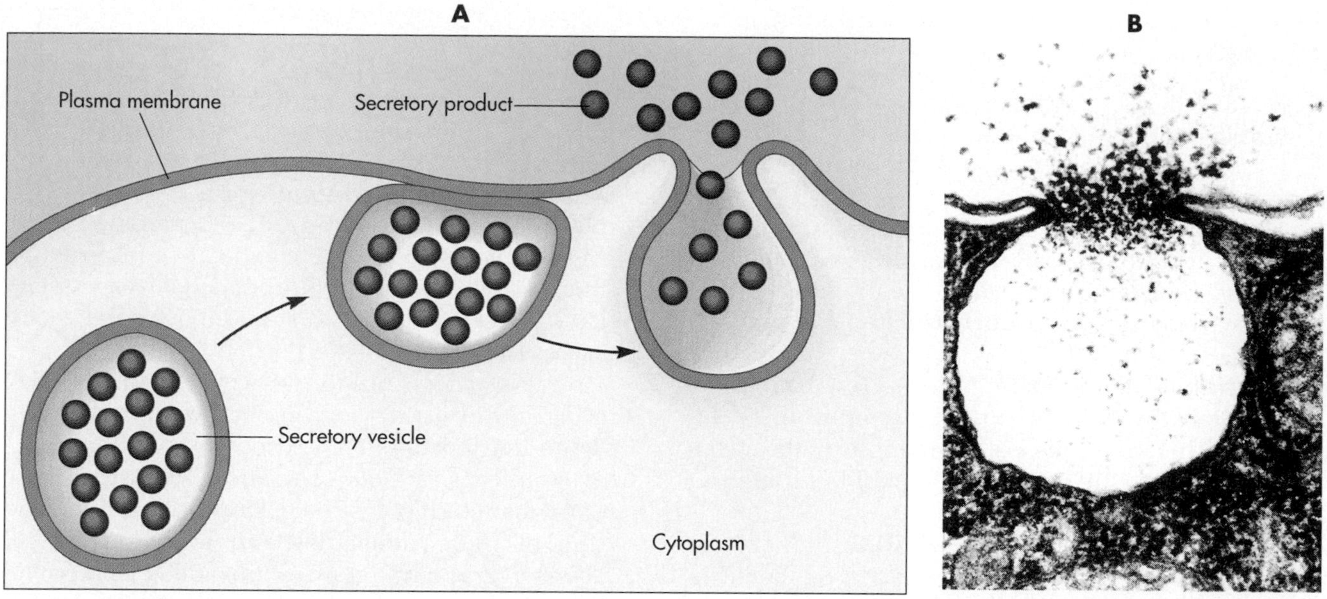

FIGURE 3-27 Exocytosis. A, Proteins and other molecules are secreted from cells in small pockets called vesicles, whose membranes fuse with the cell membrane. When this fusion occurs, the contents of the vesicles are released to the cell surface. **B,** An electron micrograph (*33,000x*) showing contents of a vesicle being expelled from a cell.

is called **phagocytosis** (Greek *phagein,* "to eat," and *cytos,* "cell"). If the material brought into the cell is liquid—contains dissolved molecules as in Figure 3-26, *B*—the endocytosis is referred to as **pinocytosis** (Greek *pinein,* "to drink"). Pinocytosis is common among the cells of multicellular animals. Human egg cells, for example, are "nursed" by surrounding cells that secrete nutrients the maturing egg cell takes up by pinocytosis.

Virtually all animal cells are constantly carrying out endocytosis, trapping extracellular fluid in vesicles and ingesting it. Within the cell, these vesicles fuse with lysosomes, tiny cellular bags of digestive enzymes, to break down these large particles into molecules usable to the cell. Rates of endocytosis vary from one cell type to another but can be surprisingly large. Some types of white blood cells, for example, ingest 25% of their cell volume each hour!

Exocytosis

The reverse of endocytosis is **exocytosis.** As with endocytosis, a cell must expend energy for exocytosis to occur. During exocytosis, a cell discharges material by packaging it in a vesicle and moving the vesicle to the cell surface. The membrane of the vesicle fuses with the cell membrane, and the contents are expelled (Figure 3-27). In plants, exocytosis is the main way that cells move the materials from the Golgi body and out of the cytoplasm to construct the cell wall. In animals, many cells are specialized for secretion using the mechanism of exocytosis, such as cells that produce and secrete digestive enzymes or hormones.

phagocytosis (**fag** oh sye **toe** sis) a type of endocytosis in which a cell ingests an organism or some other fragment of organic matter.

pinocytosis (**pie** no sye **toe** sis) a type of endocytosis in which a cell ingests liquid material containing dissolved molecules.

exocytosis (**ek** so sye **toe** sis) the reverse of endocytosis; the discharge of material from a cell by packaging it in a vesicle and moving the vesicle to the cell surface.

Summary

- A cell is a membrane-bounded unit containing hereditary material and cytoplasm. It can release energy from fuel and use that energy to grow and reproduce.
- The cell theory was first formulated in the mid-1800s. It states that all living things are made up of cells, the smallest unit of life. In addition, all cells arise from the growth and division of other cells.
- Most cells divide before they grow too large, maintaining a large surface area-to-volume ratio. In this way, cells are able to move substances across their cell membranes quickly enough to meet their needs.
- All cells can be classified as either prokaryotic or eukaryotic. Prokaryotic cells are the bacteria and have a relatively simple structure. Eukaryotic cells are more complex than prokaryotes and evolved from ancestors of today's prokaryotes. Eukaryotic cells make up the bodies of plants, animals, protists, and fungi.
- All cells are bounded by a plasma membrane, which is a lipid bilayer with embedded proteins. Lipid bilayers allow only certain molecules to pass across them. Membrane proteins facilitate the movement of many types of molecules into and out of the cell.
- Eukaryotic cells contain membrane-bounded organelles. The cytoplasm of eukaryotic cells also contains a network of filaments, fibers, and tubules that provide support and help shape cells having no outer, rigid wall. Ribosomes, which are small, spherical organelles located in the cytoplasm, are the sites of protein synthesis.
- In eukaryotes, cilia and flagella are composed of microtubules and help propel cells. In some eukaryotic cells, cilia are used to move substances across surfaces.
- Molecules move randomly with a constant jiggling motion. However, molecules and ions undergo net movement in response to differences, or gradients, in concentration, pressure, and electrical charge, resulting in an equal distribution of these molecules and ions. Movement down a gradient requires no added energy to take place.
- Molecular movement down a gradient but across a cell membrane is called *passive transport.* Three types of passive transport are diffusion, osmosis, and facilitated diffusion.
- When cells move substances against a gradient, energy is required. This movement is called *active transport.*
- The transport of particles or molecules into the cell that are too large to move across the membrane takes place by a process called *endocytosis.* The reverse of endocytosis is exocytosis. These processes require energy to take place.

Knowledge and Comprehension Questions

1. What are the basic principles of the cell theory?
2. Humans are multicellular organisms. Explain what this means and discuss some of the tasks performed by the cells of your body.
3. Some of the skeletal muscle cells that allow you to move your legs can be as long as 30 to 40 centimeters. These cells have many nuclei. Explain why.

4. A human liver cell has produced a protein to be used outside of the cell. Using what you know about the endoplasmic reticulum and Golgi complex, trace the path of protein from production to removal from the cell. What was the role of the smooth ER in this process?
5. Cigarette smoking reduces the effectiveness of cilia lining the respiratory tract in humans. What is the role of these cilia and what would you hypothesize might be a result of smoking for many years?
6. What would happen inside your cells if the lysosomes stopped working?
7. Make a table that briefly describes each of the following and lists its function(s): plasma membrane, flagella (cilia), cytoskeleton, endoplasmic reticulum, ribosomes, nucleus, chromosomes, nucleolus, Golgi complex, lysosomes, mitochondria, and chloroplasts.
8. You're examining three cells under the electron microscope. You know that one is a bacterium, one is a plant cell, and the third is an animal cell. From their structures, how can you tell which is which?
9. Many scientists theorize that chloroplasts and mitochondria evolved from ancient bacteria living within the cells of host organisms. Does this mean that the mitochondria in your cells are endosymbionts?
10. What characteristics do diffusion and facilitated diffusion have in common? In what ways do they differ?
11. Draw a diagram of a cell engaging in endocytosis. Give an example of this process in the human body.

Critical Thinking

1. Trace the paths by which a nutrient outside a cell could pass into the interior of a cell. (The nutrients are carbohydrates, lipids, proteins, vitamins, minerals, and water.)
2. Identify the type(s) of molecular movement involved in each of the following. Which are active and which are passive?
   a. Glucose molecules leave your bloodstream (where they are usually in higher concentrations) and enter your cells by attaching to specialized carrier proteins.
   b. You lounge in a bathtub full of warm tap water until your skin looks swollen and "puckered."
   c. You dissolve a spoonful of instant coffee crystals in a mug of hot water.
   d. Your cells "import" amino acids by transporting them against the concentration gradient.
3. In question 2c, identify the solute, solvent, and solution.
4. When a patient is suffering from dehydration, the physician may recommend that an 0.9% saline solution be added to the patient's bloodstream rather than water. Explain the reasoning behind this recommendation. What would happen if the water were used instead?

CHAPTER 4

THE FLOW OF ENERGY WITHIN ORGANISMS

DID YOU KNOW that your body takes constant "rollercoaster rides" that you can't even feel? At the beginning of a rollercoaster ride, the cars are pulled up a high hill by a cable. For the cable to pull the cars, it must be run by an engine that supplies energy for the cable to do its work. Energy is, in fact, the ability to do work. In the case of the rollercoaster, energy is needed to pull the cars up the hill; otherwise, they would go nowhere. After the cars reach the top and are freed of the cable, they swoop down the hill, releasing this input of energy and coasting around the rails at breakneck speed.

All the chemical reactions within your body—your metabolism—take place in a similar way. Chemical reactions are changes in molecules in which one substance is changed into another. However, for chemical reactions to take place, they need an input of energy to get started. This input of energy helps molecules "climb" an energy "hill" and interact with one another.

Starting Chemical Reactions

Figure 4-1 is a graph of the varying energy states of the molecules in a chemical reaction. Doesn't this graph look like the beginning hill of a rollercoaster? The **substrates** are the molecules entering into the chemical reaction—the rollercoaster cars. The energy needed to pull the substrates up the energy hill is called the *free energy of activation.* This energy input destabilizes the bonds of the substrates. Molecules with unstable bonds interact with one another more easily than molecules with stable bonds do. In this way, the free energy of activation allows the chemical reaction to "go." The substrates undergo a chemical change resulting in new bonding arrangements between the molecules. This chemical change may involve the breaking of bonds and the making of new bonds. The changed substrates are called **products.**

> *Free energy of activation is needed to start a chemical reaction.*

Sometimes substrates are chemically broken down to yield products. When a substrate is broken down, energy is released from the chemical bonds that were holding it together. This type of reaction is called **exergonic,** meaning "energy out." In an exergonic reaction, the products contain less energy than the substrate. The excess energy is released (Figure 4-2). Exergonic reactions happen *spontaneously;* that is, they absorb free energy from their surroundings and occur with no additional input of energy from another chemical reaction.

You may have seen an example of an exergonic reaction if you have ever used baking soda and water to remove the light blue substance that can build up on the terminals of a car's battery. This substance is formed when battery acid seeps from inside the battery and reacts with the lead on the battery terminals. As you pour a baking soda solution onto the battery, it fizzes and bubbles and the material that has built up disappears. What happened? The baking soda and acid reacted to form

KEY CONCEPTS

- Energy, the capacity to do work, continually flows throughout living systems to maintain life.
- Energy can never be destroyed, but as it flows throughout living systems, much of it is changed to heat, becoming unusable for performing work.
- Metabolic pathways, which are regulated by enzymes, are the means by which energy is moved, stored, carried, and freed in living systems.
- ATP is a molecule used universally by living cells to capture and release energy in chemical reactions.

OUTLINE

Starting chemical reactions
Energy flow and change in living systems
Regulating chemical reactions
Enzymes: Biological catalysts
Storing and transferring energy
ATP: The energy currency of living things
Carrying and transferring energy using ATP

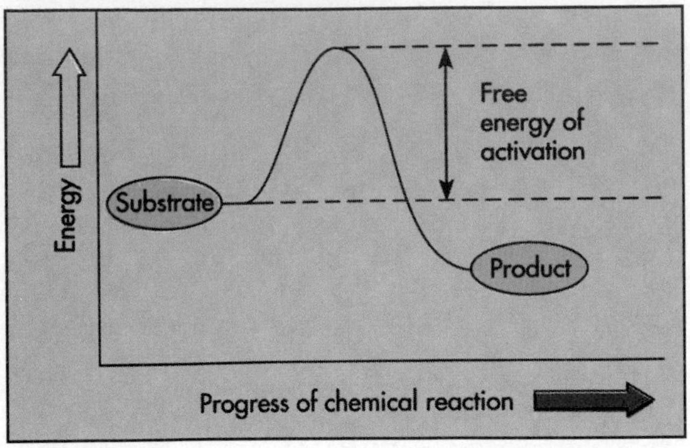

FIGURE 4-1 A chemical reaction. Before a chemical reaction occurs, energy must be supplied to the substrates to destabilize existing chemical bonds. This energy is called the free energy of activation.

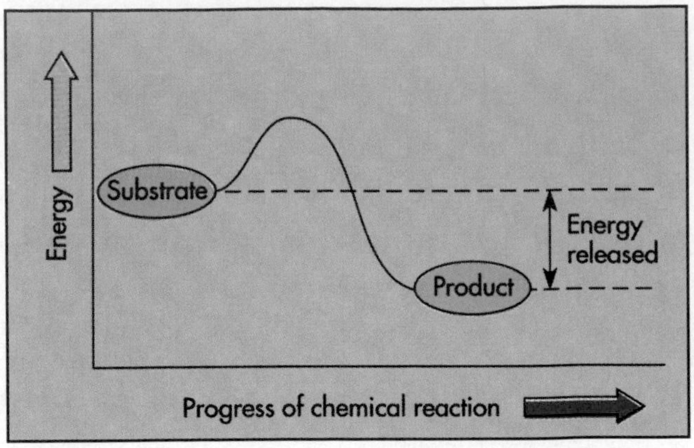

FIGURE 4-2 An exergonic reaction. In an exergonic reaction when a substrate is broken down, energy is released from its chemical bonds. The reaction product(s) contain less energy than the substrate.

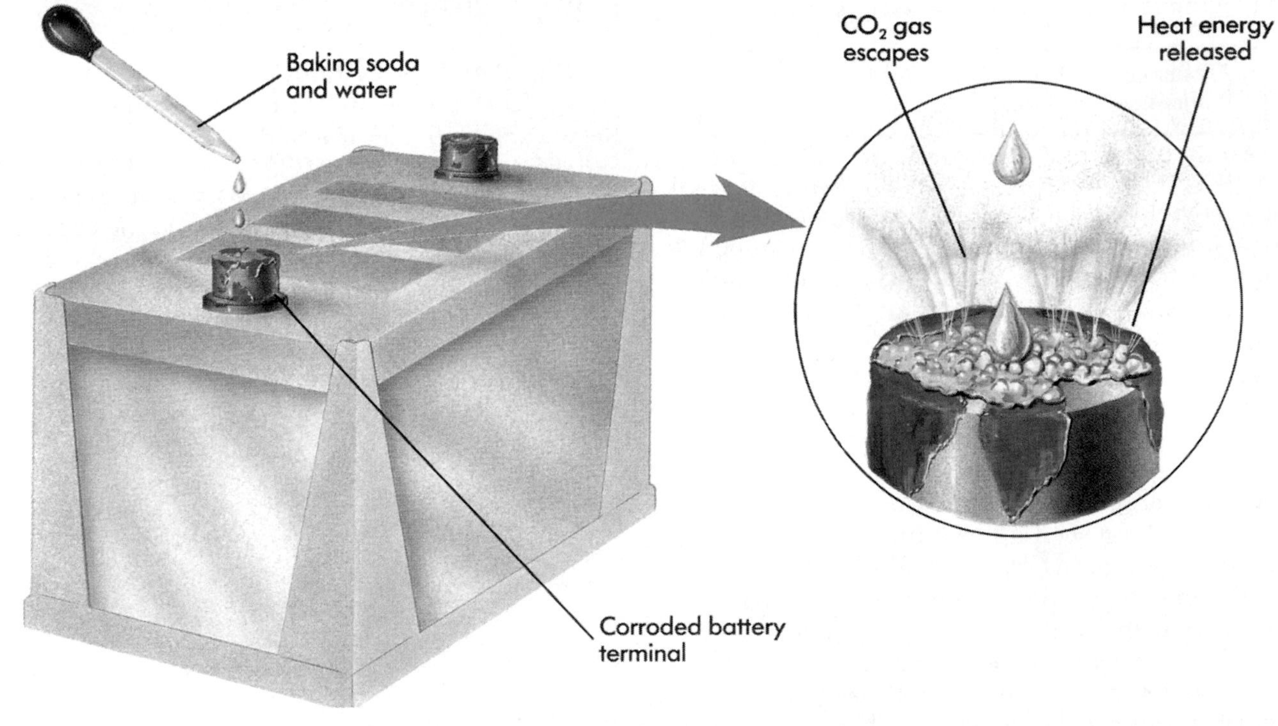

FIGURE 4-3 An example of an exergonic reaction. When you clean a car battery with baking soda and water, these two substances react with the acid on the battery terminals, producing carbon dioxide gas (CO_2) and water. Energy is released as heat.

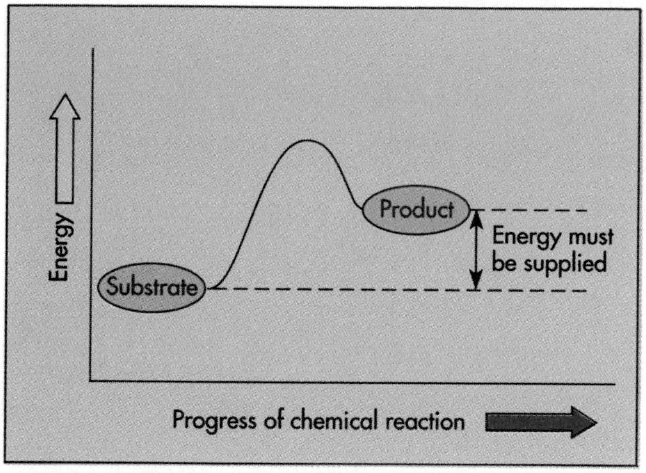

FIGURE 4-4 An endergonic reaction. In an endergonic reaction, the products of the reaction contain more energy than the substrate(s). Extra energy must be supplied for the reaction to proceed.

two new substances: carbon dioxide gas and water. In addition, energy was released as heat (Figure 4–3). Exergonic reactions take place in your body too. In fact, a series of exergonic reactions breaks down the glucose in your cells to supply your body with energy. The heat released from these reactions helps keep your body warm. Reactions such as this, which release energy by breaking down complex molecules into simpler molecules, are called **catabolic reactions.**

In some chemical reactions, substrates are chemically joined to yield a product. When substrates are joined, energy is used to build the chemical bonds holding the product together. This type of reaction is called **endergonic,** meaning "energy in." In an endergonic reaction, the product contains more energy than the substrates (Figure 4-4). Endergonic reactions *do not occur spontaneously;* that is, they require more free energy of activation to drive the reaction than the substrates are able to absorb from their surroundings. Endergonic reactions play important roles in your body, such as putting molecules together to build muscles and bones. Reactions like this, which use energy to build complex molecules from simpler molecules, are called **anabolic reactions.**

In living things, exergonic and endergonic reactions are *coupled reactions,* meaning that they occur in conjunction with one another. In this way the energy released when molecules are split (exergonic reactions) is used to power the combining of molecules (endergonic reactions) (Figure 4-5).

Reactions that break apart substrates release energy. These reactions are exergonic. Reactions that bond substrates together store energy. These reactions are endergonic. In living systems, exergonic reactions are coupled with endergonic reactions, supplying the energy for endergonic reactions to take place.

Energy Flow and Change in Living Systems

Life can be viewed as a constant flow of energy that is channeled by organisms to do the work of living. As you learned at the beginning of this chapter, energy is the ability to do work, and work is many things, such as the pull of a cable on a rollercoaster car or the swift dash of a horse. It is also heat, such as the blast from an explosion or a warming fire. Energy can exist in many forms, including mechanical force, heat, sound, electricity, light, radioactivity, and magnetism. All these forms of energy are able to create change—to do work.

Energy exists in two states. Energy not actively doing work but having the capacity to do so is called **potential energy.** The rollercoaster cars perched atop that first hill, for example, possess this type of stored energy. They have the capacity to roll downhill, having stored the energy from their pull up the hill. Then, as the cars roll down the hill, they are actively engaged in doing work. This form of energy is called **kinetic energy,** or the energy of motion. Much of the work performed by living organisms involves changing potential en-

substrates substances entering into a chemical reaction.

products substances obtained as the result of chemical reaction; substrates that have undergone a chemical change.

exergonic reaction (ek sur **gon** ick) a chemical reaction (the reciprocal action of chemical agents upon each other) in which energy is released and, therefore, the products contain less energy than the substrate.

catabolic reactions chemical reactions that release energy by breaking down complex molecules into simpler molecules.

endergonic reaction (enn der **gon** ick) a chemical reaction in which the products contain more energy than the substrates; energy must be supplied for this kind of reaction to proceed.

anabolic reactions chemical reactions that use energy to build complex molecules from simpler molecules.

potential energy energy not actively doing work but having the capacity to do so.

kinetic energy (kuh **net** ick) energy actively doing work; the energy of motion. (Energy is the ability to do work.)

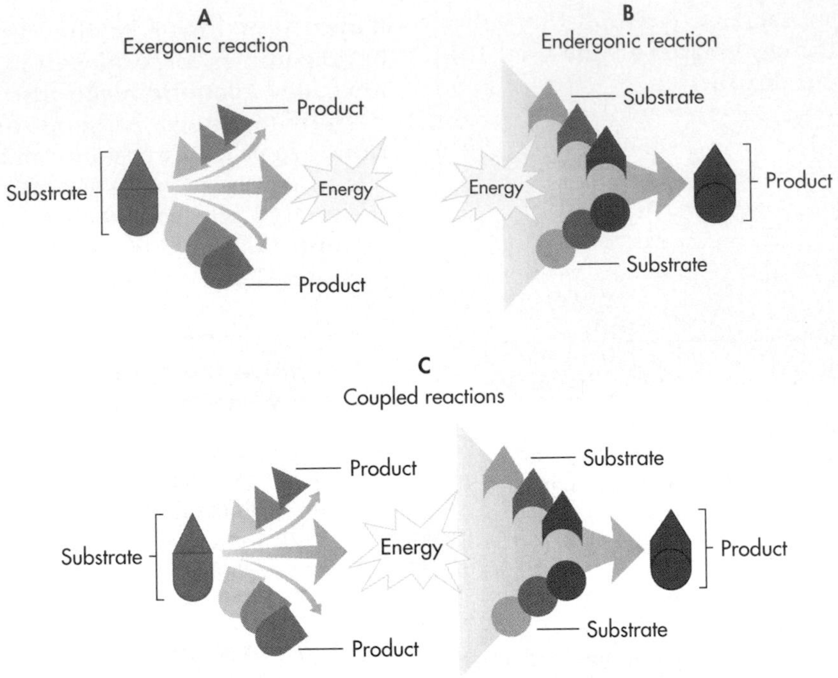

FIGURE 4-5 Coupled reactions. In coupled reactions, the energy released in an exergonic reaction is used to drive an endergonic reaction.

ergy to kinetic energy (Figure 4-6). The exergonic reactions that supply your body with energy, for example, change the potential energy stored in the food you eat to kinetic energy for cellular work.

> *Energy is the capacity to do work, or bring about change. It can exist in many forms, such as mechanical force or heat. Energy actively doing work is called kinetic energy; stored energy is called potential energy.*

All the changes in energy that take place in the universe, from nuclear explosions to the buzzing of bees, are governed by two laws called the **laws of thermodynamics.** The **first law of thermodynamics** states that energy can change from one form to another and from one state to another, but it can never be lost. New energy cannot be made. The total amount of energy in the universe remains constant.

> *The first law of thermodynamics states that energy cannot be created or destroyed; it can only be changed from one form or state to another.*

FIGURE 4-6 The energy of motion, kinetic energy. Transforming potential energy into kinetic energy, this young athlete has gone from a stationary position to executing a somersault in midair.

The **second law of thermodynamics** states that all objects in the universe tend to become more disordered and that the disorder in the universe is continually increasing. Stated simply, disorder is more likely than order. You can relate to this concept as you try to keep your personal environment in order. Your bedroom continually becomes messy, for example, unless you make an effort to keep it neat. Interestingly, molecules also become increasingly disordered.

Disorder with respect to molecules refers to their random motion. As you may recall from Chapter 3, this random motion, or disorder, is an inherent jiggling that is characteristic of all molecules and small particles. Their disorder increases as energy is added to molecules in the form of heat. This heat is the energy that is lost as energy transfers are made between molecules during chemical reactions. This lost heat energy is often called the *kinetic energy of molecular motion*. Although the total amount of energy in the universe does not change, the amount of useful energy available to do work decreases as progressively more energy is degraded to heat. With the addition of this heat, molecules move more quickly, becoming more disordered.

> *The second law of thermodynamics states that disorder in the universe constantly increases. Energy spontaneously converts to less organized forms.*

The energy lost to disorder is referred to as **entropy.** In fact, entropy is a measure of the disorder of a system. Sometimes the second law of thermodynamics is simply stated as "entropy increases." So although energy cannot be destroyed, the universe is constantly moving toward increasing entropy. Eventually, the universe will have wound down like a forgotten clock. Scientists speculate that this will occur approximately 100 billion years from now.

Although energy cannot come into or go out of the universe, the Earth is constantly receiving "new" energy from the sun. Much of this energy heats up the oceans and continents. Some of it is captured by photosynthetic organisms such as green plants. In photosynthesis, the energy from sunlight is changed to chemical energy, combining small molecules into more complex molecules by means of endergonic reactions. This stored energy can be shifted to other molecules by forming different chemical bonds. In addition, this stored energy can be changed into kinetic energy: motion, light, electricity—and heat. Thus energy continuously flows into and through the biological world, with new energy from the sun constantly flowing in to replace the energy that is lost as heat.

> *Photosynthetic organisms such as green plants change energy from the sun to other forms of energy that drive life processes. This energy is never destroyed, but as it is exchanged in chemical reactions, much of it is changed to heat, a form of energy that is not useful for performing work.*

Regulating Chemical Reactions

The flow of energy within an organism like yourself consists of a long series of coupled reactions. Energy is moved from one molecule to the next by means of exergonic and endergonic reactions. Some of this energy is stored in the bonds of molecules that make up the structure of your body. Some is freed to do cellular work. And some is lost as heat. These chains of reactions that move, store, and free energy are called *metabolic pathways*. Metabolic pathways accomplish jobs such as obtaining energy from the food you eat and repairing tissues that are worn out or damaged. By means of such pathways, your body and the bodies of all living things work to maintain order and avoid increasing entropy. Put simply, all the metabolic pathways in your body work to help you survive. But for these chemical reactions to occur, they must be pushed over the hill of activation energy. In addition, they must be controlled so that they occur on schedule.

laws of thermodynamics two laws that govern all of the changes in energy that take place in the universe.

first law of thermodynamics a law stating that energy cannot be created or destroyed; it can only be changed from one form or state to another.

second law of thermodynamics a law stating that disorder in the universe constantly increases. Energy spontaneously converts to less organized forms.

entropy (enn truh pee) the energy lost to disorder; entropy is a measure of the disorder of a system.

Enzymes: Biological Catalysts

As you read earlier in this chapter, chemical reactions require activation energy to get started. This energy is needed for various reasons. In some chemical reactions, it is used to help break old bonds so that new bonds can be formed. In others, it is used to excite electrons—to help them achieve a higher energy level or shell—so that they will pair up in covalent bonds. And in others, this energy helps molecules overcome the mutual repulsion of their many electrons so that they can get close enough to react. In any chemical reaction, the free energy of activation performs one or more of these jobs.

If molecules are moving very quickly and bump into one another forcefully, the kinetic energy of the bump can provide enough energy for activation. This situation rarely occurs in living organisms, however, because the chemical reactions of living things take place in the moderate temperatures of living cells. But living systems contain proteins called **enzymes** that lower (or lessen) the free energy of activation that is needed, allowing chemical reactions to take place.

Enzymes are proteins that act as *catalysts*. A catalyst is a substance that increases the rate of a chemical reaction but is not chemically changed by the reaction. It lowers the barrier of the free energy of activation—reduces the amount of energy needed for the reaction to occur—by bringing substrates together so that they can react with one another or by placing stress on the bonds of a single substrate, making it more reactive (Figure 4-7). Enzymes are biological catalysts and control all the chemical reactions making up the metabolic pathways in living things. Life is therefore a process regulated by enzymes.

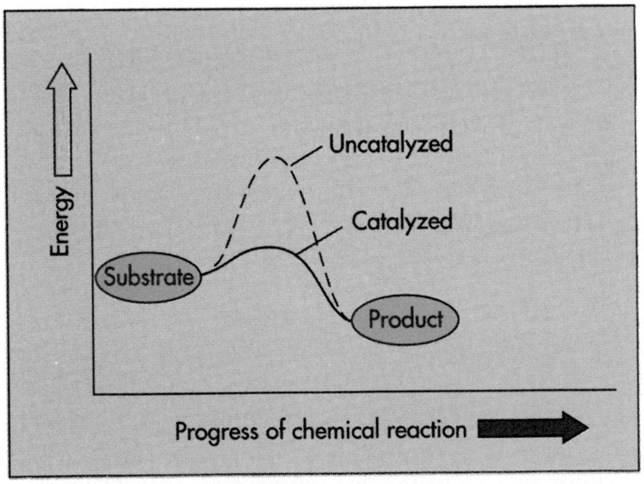

FIGURE 4-7 The function of enzymes. Enzymes are able to catalyze particular reactions because they lower the amount of activation energy required to initiate the reaction.

> *Enzymes are biological catalysts that reduce the amount of free energy of activation needed for a chemical reaction to take place, thus speeding up the reaction.*

How Enzymes Work

Enzymes are proteins that have one or a few grooves or furrows on their surfaces. These surface depressions are called **active sites** and are determined by the side groups of the amino acids making up the protein. Nonpolar side groups of amino acids, for example, tend to be shoved into the interior of the protein because of their hydrophobic (water-fearing) interactions, thereby imparting a particular shape to that portion of the protein (see Chapter 2).

The active sites are the locations on the enzyme where a reaction is catalyzed—where *catalysis* takes place. For catalysis to occur, a substrate must fit into the surface depression of an enzyme so that many of its atoms nudge up against atoms of the enzyme. The fit between a substrate and an enzyme is much like putting your foot into a tight-fitting shoe. However, the binding of a substrate in some cases causes the enzyme to adjust its shape slightly, allowing a better fit called an *induced fit*. Just as your feet will only fit into certain shoes, substrates can only fit into, or bind with, the active site of certain enzymes. Therefore enzymes typically catalyze only one or a very few different chemical reactions. Because of this specificity, each cell in your body contains from 1000 to 4000 different types of enzymes.

Enzymes catalyze both endergonic and exergonic reactions. Figure 4-8 shows how an enzyme catalyzes an exergonic reaction. In this example, a molecule of sucrose binds to the active site of an enzyme. After binding takes place, certain atoms within the active site of the enzyme chemically interact with the sucrose. This interaction causes a slight change in the shape of both the enzyme and the sucrose molecule. This change in shape places stress on the bonds joining the glucose and fructose subunits, reducing the amount of free energy of activation needed to be absorbed by the molecules for them to react. The bond then breaks, and the products are released from the enzyme.

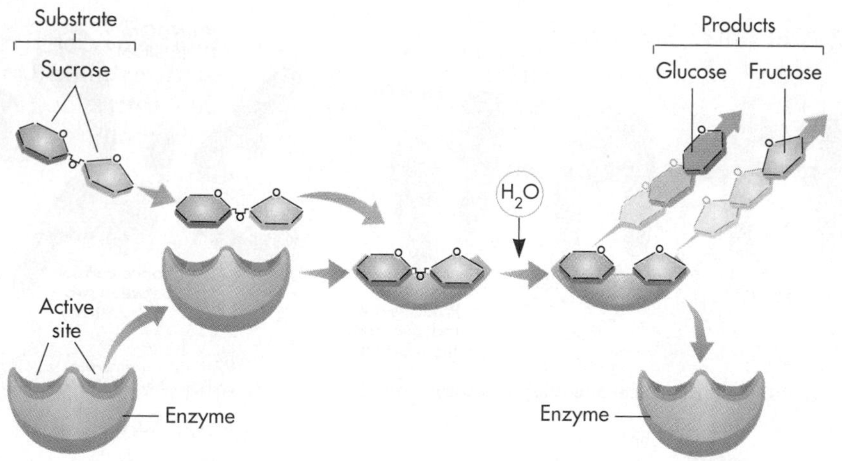

FIGURE 4-8 How an enzyme catalyzes an exergonic reaction. In this reaction, sucrose binds to the active site of an enzyme. A reaction takes place between the active site atoms and the sucrose atoms, causing a change in shape of the enzyme and the sucrose molecule. As a result of this change in shape, stress is placed on the bonds holding the glucose and fructose subunits together in the sucrose molecule. This stress lowers the amount of energy needed to break the bonds. When the bond breaks, the glucose and fructose are released from the enzyme.

Enzymes also catalyze reactions that bind two substrates together (Figure 4-9). Both substrates bind at the active site on the enzyme. The binding of the substrates to the enzyme distorts and weakens their chemical bonds. In addition, their interaction with the enzyme orients the substrates so that their reactive sites are in contact with one another and react. Put simply, the enzyme is like a dating service. It gets the substrates together so that their meeting does not simply occur by chance. Interestingly, the Chinese characterize enzymes in a similar way, calling them *tsoo mei,* or "marriage brokers." Moreover, enzymes do not get used up after one interaction takes place. They are only intermediaries in chemical reactions and are then released, available to catalyze yet another reaction.

> *Enzymes work by bringing substrates together so that they react more easily and by placing stress on bonds, which lowers the amount of free energy of activation that must be absorbed by the substrates for them to react.*

Factors That Affect Enzyme Activity

The activity of an enzyme is affected by anything that changes its three-dimensional shape. If it loses its shape, or is denatured, an enzyme cannot bind with a substrate. Three environmental conditions that can affect enzyme activity in this way are pH, temperature, and the binding of specific chemicals. You can see the protein in an egg become denatured when you cook it. The heat of the cooking process permanently changes the shape of the egg proteins, which results in the egg becoming firmer and more opaque.

Most human enzymes function best between 36° C and 38° C, close to body temperature. Likewise, the enzymes of other living things work best at temperature ranges specific to the organism. Bacteria that live in hot springs, for example, have enzymes that function at temperatures between 74°C and 76°C (Figure 4-10, *A*). At temperatures colder or warmer than the enzyme's optimum range, the bonds between the amino acids that are responsible for the enzyme's three-dimensional shape become weak or rigid, changing the shape of the active site and causing chemical reactions to stop.

Enzymes also work best within a particular range of pH values. As previously stated, pH is a measure of the hydrogen ion concentration of a solution (see Chapter 2). Most enzymes work best within the range of pH 6 to 8. Some enzymes, however, function in environments with a high hydrogen ion concentration (low pH). For example, the

> **enzymes** (enn zymes) biological catalysts that reduce the amount of free energy of activation needed for a chemical reaction to take place, thus speeding up the reaction.
>
> **active sites** the grooved or furrowed locations on the surface of an enzyme where reactions are catalyzed.

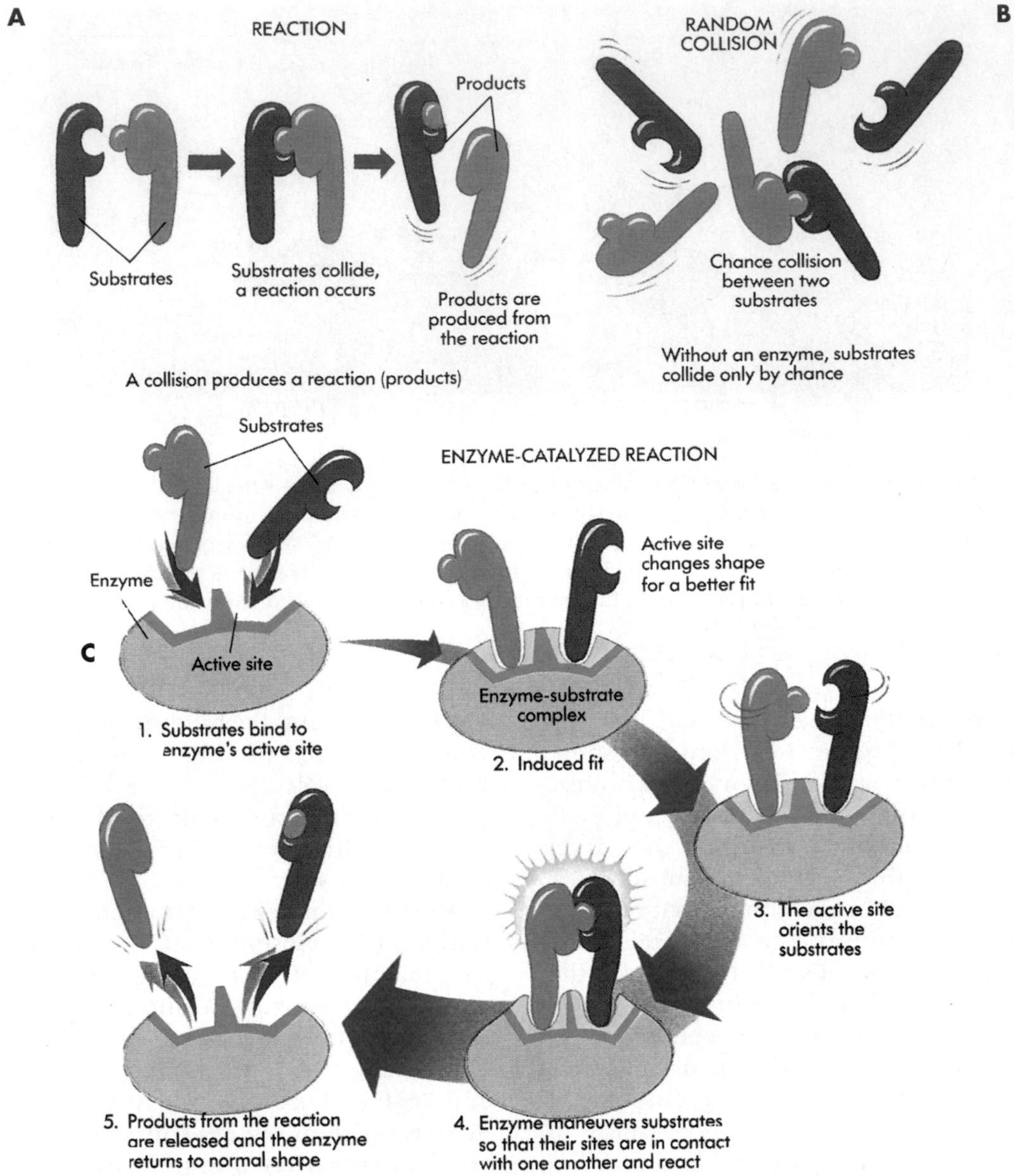

FIGURE 4-9 Enzymes also catalyze reactions between two substrates. A, A collision between two substrates causes a reaction to occur, resulting in products. **B,** Without an enzyme, substrates only collide by chance. **C,** With an enzyme, a reaction is more likely to occur. The binding of the substrates to the active site of the enzyme weakens their chemical bonds so that less energy is needed for a reaction to occur. The enzyme also orients the substrates in the best possible position for a productive reaction.

enzyme pepsin digests proteins in your stomach, a highly acidic environment, at pH 2 (Figure 4-10, *B*).

The activity of an enzyme is sensitive not only pH and temperature but also to the presence of specific chemicals that bind to the enzyme and cause changes in its shape. By means of these specific chemicals, a cell is able to turn enzymes on and off. When the binding of a chemical changes the shape of the protein and the active site so that it can catalyze a chemical reaction, the chemical is called an *activator.* When the binding causes a change in the active site that shuts off enzyme activity, the chemical is called an *inhibitor.* Enzymes usually have special binding sites for the activator and inhibitor molecules that affect them, and these binding sites are different from their active sites.

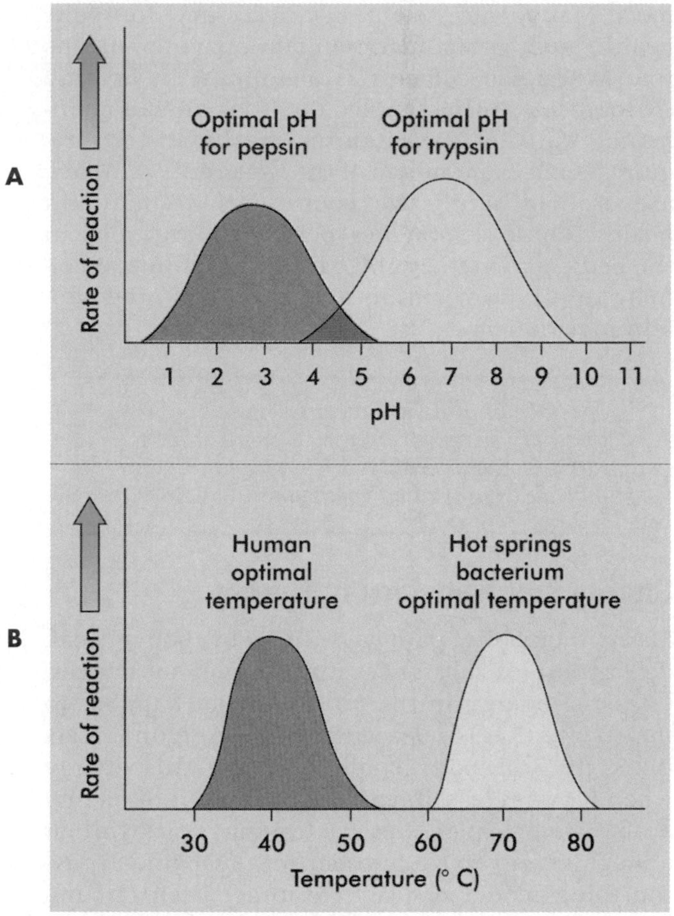

FIGURE 4-10 Enzymes are sensitive to the environment. The activity of an enzyme is influenced by pH **(A)** and temperature **(B).** Human enzymes tend to work best within a pH range of 6 to 8 and at a temperature of about 37 °C (body temperature).

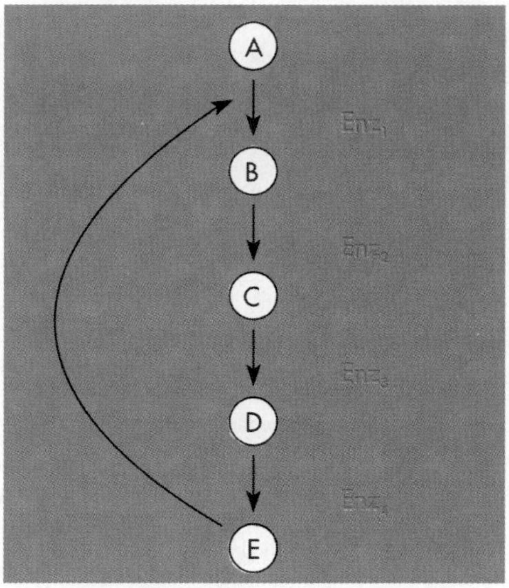

FIGURE 4-11 End-product inhibition. Product E controls the rate of its own synthesis by acting on ENZ_1, the enzyme that catalyzes the first reaction in the pathway.

Enzyme action is often regulated by inhibitors in a process called **negative feedback.** In this process the enzyme catalyzing the first step in a series of chemical reactions has an inhibitor binding site to which the end product of the pathway binds. As the concentration of the end product builds up in the cell, it begins to bind to the first enzyme in the metabolic pathway, shutting off that enzyme. In this way, the end product is feeding information back to the first enzyme in the pathway, shutting the pathway down when an additional end product is not needed (Figure 4-11). Such *end-product inhibition* is a good example of the way many enzyme-catalyzed processes within cells are self-regulated. For example, the enzyme threonine deaminase catalyzes the first reaction in a series that converts the amino acid threonine to another amino acid, isoleucine. The cells of the body need both amino acids in proper concentrations to make proteins. The presence of an adequate amount of isoleucine for protein manufacture shuts off the enzyme threonine deaminase and stops any further conversion of threonine to isoleucine.

> *The activity of enzymes is regulated by changes in enzyme shape; these changes result when activator and inhibitor molecules bind to specific enzymes.*

Cofactors and Coenzymes

Enzymes often have additional parts in their structures that are made up of molecules other than proteins (Figure 4-12). These additional chemical parts are called **cofactors.** Cofactors help enzymes catalyze chemical reactions. For example, many enzymes have metal ions such as zinc, iron, or copper locked into their active sites. These ions help draw electrons from substrate molecules. One of your digestive enzymes, carboxypeptidase, breaks down proteins in foods by using a zinc ion to draw electrons away from the bonds being broken in the

negative feedback a mechanism by which the products of a process regulate the process.
cofactors nonprotein molecules that bind to enzymes and help them catalyze chemical reactions.

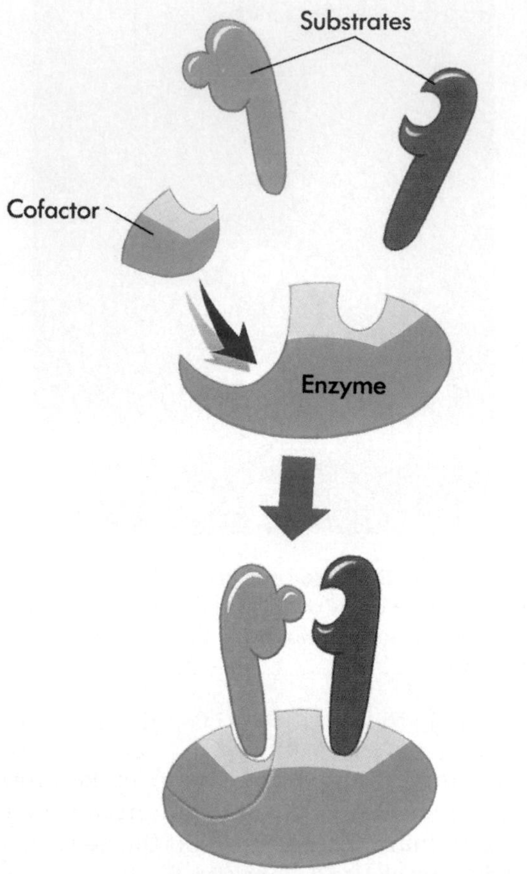

FIGURE 4-12 Cofactors. Cofactors assist enzymes in catalyzing chemical reactions. They are an integral part of some enzymes and must be present for these enzymes to work.

food. Many trace elements necessary for your health, such as manganese, help enzymes in this way. When the cofactor is a nonprotein organic (carbon-containing) molecule, it is called a **coenzyme.** Many of the vitamins that your body requires, such as members of the B-vitamin complex, are used to synthesize coenzymes to maintain health. These coenzymes perform many jobs in the body, playing key roles in the reactions of cellular respiration, amino acid synthesis, and protein metabolism.

> *Special nonprotein molecules called* cofactors *help enzymes catalyze chemical reactions. Cofactors that are nonprotein organic molecules are called coenzymes.*

Storing and Transferring Energy

Many metabolic pathways in your body break down complex substances into simpler substances, releasing energy in the process. What happens to the energy that is released in these exergonic reactions? If your body could only use this energy when it was released, you might run out of energy a few hours after lunch, leaving you without enough energy to eat another meal. Obviously, living things could not survive under such circumstances. They have evolved ways to capture energy when it is released so that it can be used to do cell work.

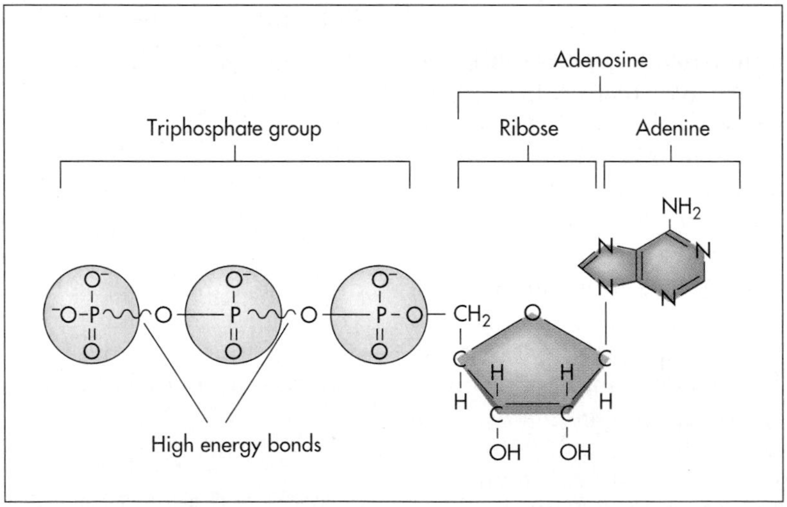

FIGURE 4-13 Adenosine triphosphate (ATP) is the primary energy currency of the cell. ATP has three subunits: a five-carbon sugar called ribose, a double-ringed structure called adenine, and three phosphate groups called a triphosphate group. The phosphate groups are linked together by high-energy bonds (*wavy lines*).

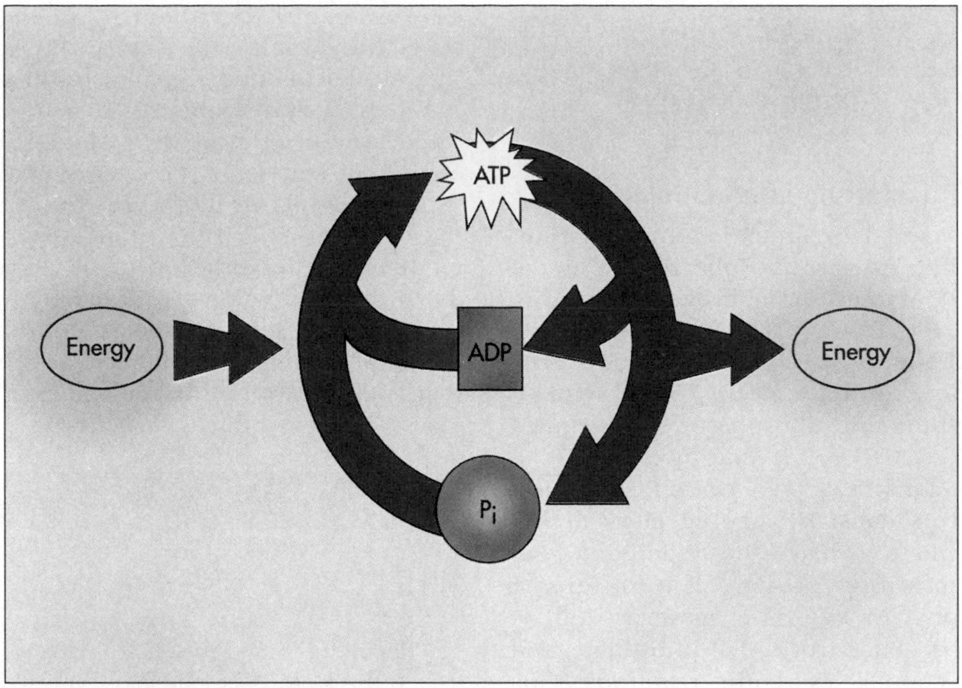

FIGURE 4-14 The ATP-ADP cycle. When ATP is used to drive the energy-requiring activities of living things, the high-energy bond that links the last phosphate group to the ATP molecule is broken, releasing energy. Cells always contain a pool of ATP, ADP, and phosphate (P_i), so ATP is continually being made from ADP, phosphate, and the energy released from exergonic reactions.

ATP: The Energy Currency of Living Things

One way your body stores energy is by converting it to fat. In fact, many of us think we have too much of this stored energy. Fat, however, cannot serve our cells' immediate energy needs. It is used for long-term storage.

Glycogen is another molecule used by the body to store energy. Although glycogen can be more readily converted into energy than body fat can, it is still unable to meet immediate energy demands. The primary molecule used by cells to capture energy and supply it at a moment's notice is a molecule called **adenosine triphosphate,** or **ATP.**

Each ATP molecule is made up of three subunits (Figure 4-13):

1. A five-carbon sugar called *ribose*
2. A double-ringed molecule called *adenine*
3. Three phosphate groups (PO_4) linked in a chain called a *triphosphate group*

Together, the ribose sugar and the adenine rings are called *adenosine.* The "working end" of the molecule, however, is the triphosphate group.

Notice in the diagram that the bonds linking the phosphate groups in ATP are shown as wavy lines. These symbols stand for special bonds called *high-energy bonds.* These bonds are special because they are fairly unstable and require very little free energy of activation to be broken. They are also special because a great deal of free energy is released when ATP is broken down into ADP and Pi.

ATP is often referred to as the energy currency of the cell because it is used much like money. When cells break down molecules in exergonic reactions, the energy that is released can be captured in molecules of ATP and carried there until it is needed. Likewise, when cells need energy to drive endergonic reactions, the cell can "spend" ATP to provide this energy.

coenzyme (ko enn zyme) cofactor that is a nonprotein organic molecule.
adenosine triphosphate (ATP) (uh **den** o **seen** try **fos** fate) the primary molecule used by cells to capture energy and later release it during chemical reactions.
adenosine diphosphate (ADP) (uh **den** o **seen** dye **fos** fate) the molecule that remains after ATP has been used to drive an endergonic reaction.

ATP, the universal energy currency of all cells, can capture energy in its high-energy bonds and later release this energy.

Carrying and Transferring Energy Using ATP

When ATP is used to drive an endergonic reaction, the bond that links the last phosphate group to the rest of the ATP molecule is broken, releasing needed energy (Figure 4-14). The molecule that remains is called **adenosine diphosphate,** or **ADP.** In addition, the phosphate group (often symbolized as P_i, or inorganic phosphate) exists on its own. Thus ATP $\rightarrow$ ADP + P_i + energy.

Cells always contain a pool of ATP, ADP, and phosphate. ATP is constantly being cleaved into ADP and phosphate to drive the endergonic, energy-requiring processes of the cell by means of coupled reactions. In addition, however, ATP is continually being made from ADP, phosphate, and energy during coupled exergonic reactions. This recycling happens quickly. In fact, if you could mark every ATP molecule in your body at one instant in time and then watch them, they would be gone in a flash. Most cells maintain a particular molecule of ATP for only a few seconds before using it.

ATP, ADP, and phosphate are continually being recycled within living cells, thus capturing, carrying, and releasing energy.

ATP is used to fuel a variety of cell processes. It is used, for example, when fireflies and deep water fishes produce light. It is used to build larger, more complex molecules like proteins from smaller, simpler molecules like amino acids. It also provides energy for you to move, fueling the reactions that take place in your muscles causing muscle fibers to contract. Cells also use ATP to help move substances against a gradient—it is the fuel of active transport. Finally, the energy in ATP is changed into electrical energy, primarily in nerves.

Where does the energy come from that is captured in molecules of ATP? The ultimate source of energy, of course, is the energy of the sun that is captured by plants during the process of photosynthesis. Plants and the animals that eat them are food sources for other animals. Your body breaks down the food you eat into molecules such as glucose that are usable for fuel by your cells. Your cells break down these fuel molecules, releasing energy that is then captured in molecules of ATP.

JUST WONDERING . . . .

I don't get it. What does phosphate have to do with energy?

Phosphate (PO_4) is a functional group—an arrangement of atoms that tends to act as a unit in chemical reactions. The phosphate group itself has nothing *directly* to do with energy. However, its bonding characteristics, as a part of certain molecules, result in high-energy bonds that release a large amount of free energy when the bonds are broken. This energy can be used to drive other chemical reactions.

Summary

- All the chemical reactions in a living organism are referred to as its *metabolism.*
- Chemical reactions involve changing one substance into another. The substances entering the reaction are substrates. The changed substrates are products.
- Energy is needed to initiate a chemical reaction. This energy is the free energy of activation.
- In exergonic reactions, substrates are broken down to yield products. The products have less energy than the substrates; the excess energy is released. So little free energy of activation is needed to drive these reactions that this energy is absorbed from the environment and the reaction occurs spontaneously.

- In endergonic reactions, substrates are chemically joined to yield a product. The product has more energy than the substrates. So much free energy of activation is needed to drive these reactions that it cannot be absorbed from the surroundings. In living systems, endergonic reactions are coupled with exergonic reactions to provide the energy needed to drive the endergonic process.
- Energy exists in two states. Stored energy, or energy not actively doing work, is potential energy. Energy of motion, or energy actively doing work, is kinetic energy. Energy can be changed from one form to another during chemical reactions.
- Energy cannot be created or destroyed; it can only be changed from one form or state to another. However, energy is continually being lost as heat during chemical reactions. Heat is a form of energy that is not useful for performing work.
- Chains of chemical reactions that move, store, carry, and release energy in living systems are called *metabolic pathways*. Special proteins called *enzymes* regulate these series of reactions.
- Enzymes lower the free energy of activation and speed up reactions by bringing substrates together so that they can react or by stressing chemical bonds that must break before a reaction can take place. Enzyme activity is affected by temperature, pH, and the binding of certain chemicals.
- Adenosine triphosphate, or ATP, is the energy currency of the cell in that it can capture energy from an exergonic reaction, carry this energy, and then "spend" it when needed

Knowledge and Comprehension Questions

1. What is the role of free energy of activation in a chemical reaction?
2. Cells within your bone marrow are continuously producing blood cells for the body. Reactions occur that involve creating complex molecules from simple molecules. Are these processes dependent on anabolic or catabolic reactions? Bone marrow cells also receive nutrients, such as glucose, to fuel the cell. What type of reaction would you hypothesize occurs as glucose is broken down, releasing energy?
3. We have studied how the body stores lipids for later use as a fuel source. Similarly, glycogen is the human short-term storage form of sugars. The liver stores glycogen until energy is needed, and it is converted to glucose for quick use. What form of energy is represented by these lipid and glycogen stores?
4. State the first and second laws of thermodynamics.
5. When asked about his messy room, a friend who's studied science simply replies, "Entropy increases." Explain what this means.
6. Explain the role played by photosynthetic organisms in the biological flow of energy. Why is it important?
7. What are metabolic pathways? Explain their significance.
8. If a catalyst is not changed by a reaction, what purpose does it serve? Why do the cells of your body require so many enzymes (1000 to 4000 per cell)?
9. What do enzyme activators and inhibitors have in common? Which plays a key role in the process of negative feedback?
10. What are cofactors and coenzymes? Name a biologically important example of each.
11. ATP has been called the "energy currency of the cell." Explain what this means.

Critical Thinking

1. Our cells exhibit a continual cycle of ATP buildup and breakdown involving exergonic and endergonic reactions. What do you think might be the biological advantage of such a simple energy exchange system?
2. Industrial pollution can change the pH of a pond or river to make the water more acidic. How can this affect the metabolic pathways of the plants that live in the water?
3. The chapter states that "life is a process regulated by enzymes." What might be the sources of these enzymes? If a particular enzyme is not available in a person's cells, what sequence of events might result to produce it?
4. The cellular cycle of ATP $\rightarrow$ ADP + P_i + energy $\rightarrow$ ATP is occurring continuously in all of our cells. Why is the input of fuel (food energy) needed to run this cycle?

CHAPTER 5

CELLULAR RESPIRATION

DID YOU KNOW that you and the steam locomotive in the photo have a lot in common? Both you and a train speeding down a track are complex machines, and you both burn fuel for energy. But you and a steam engine burn fuel in ways that have important differences.

In a steam engine, coal or oil is burned in a firebox in the presence of oxygen. During exergonic reactions, molecules of this fuel are broken down to carbon dioxide and water, releasing energy as heat. The heat from the firebox warms water in an adjacent boiler. The steam created in the boiler is fed, under high pressure, to a reciprocating (back-and-forth) piston-cylinder assembly. When the piston is to the left within the cylinder, high pressure steam pushes it to the right. This movement of the piston pushes steam previously in the right side of the cylinder out an exhaust port. (The locomotive's "puffing" from its smoke stack is the release of exhaust steam.) With the piston now on the right side of the cylinder, high pressure steam is admitted that pushes the piston to the left and so forth. Connecting rods from the pistons move the driving wheels, which propel the train. Because the train engine has no way to store the energy liberated from these reactions, this energy is used immediately, and the train speeds ahead.

In the engines of each of your cells—mitochondria—the products of glucose breakdown are burned in the presence of oxygen. During a series of exergonic reactions, molecules of glucose are broken down to carbon dioxide and water, releasing energy. Most of this energy is released as heat and helps maintain your body temperature. But about 38% of it is captured in molecules of ATP—the energy currency of your cells. In fact, you do a better job of converting the energy in food to useful energy for your body—the train engine can convert only about 25% of the energy in coal into useful energy. And by using ATP, your body can capture, carry, and spend energy when it is needed, an ability that this old steam locomotive will never have.

KEY CONCEPTS

- All the activities that organisms perform require energy to power the endergonic chemical reactions that underlie their processes.
- Most living things release the stored energy in food molecules needed to drive endergonic reactions by an oxygen-dependent process called *cellular respiration.*
- Cellular respiration captures energy in molecules of adenosine triphosphate (ATP), an energy-carrying molecule, during three complex series of reactions: glycolysis, the Krebs cycle, and the electron transport chain.
- Organisms that do not use oxygen, such as certain microorganisms, make ATP by a process known as *fermentation.*

OUTLINE

Using Chemical Energy to Drive Metabolism

All the activities that organisms perform—bacteria swimming, a cat purring, you reading these words—use energy. These activities require energy because endergonic chemical reactions underlie all life activities. For example, energy-requiring chemical reactions cause muscles to contract and are involved in the active transport of molecules across nerve cell membranes, maintaining the crucial balance of various ions necessary for their functioning. In addition, living things constantly use energy to build larger molecules from smaller ones. Chains of reactions move, store, and free the energy needed to perform these activities of life. As you may recall from Chapter 4, such chains of reactions are called *metabolic pathways.*

In metabolic pathways, exergonic reactions—those that release energy—occur spontaneously with the help of enzymes, absorbing enough free energy as heat from their surroundings to "go." Endergonic reactions—those that require an input of energy—do not occur spontaneously because they are unable to absorb enough free energy from their environment to drive the reaction. So where does the energy come from to drive endergonic reactions? All living things get this energy in the same way: they *couple* endergonic reactions with exergonic reactions—those that release energy from molecules such as ATP. In this way, the energy of ATP is spent to enable an endergonic reaction to go (Figure 5-1).

Chemical energy powers metabolism by driving endergonic reactions. Chemical energy from exergonic reactions is captured in molecules of ATP, and the splitting of ATP is coupled to endergonic reactions, providing the energy necessary to drive them.

How Cells Make ATP: Variations on a Theme

The energy used to make ATP is actually energy from the sun, captured by plants, algae, and certain bacteria in a process called **photosynthesis.** During this process, the energy from the sun powers the synthesis of glucose molecules, using the raw materials of carbon dioxide and water (Figure 5-2). Glucose is a sugar that plants use as a source of energy and as building blocks to construct larger molecules such as the cellulose found in plant cell walls. Plants also manufacture starch from glucose, a food that can be stored in plant cells and then converted to glucose when energy is needed. By eating plants and by eating animals that eat plants, humans and other animals can harvest the energy from the sun that was originally captured and stored by plants.

Living things, including photosynthetic organisms, make ATP by breaking down nutrient molecules such as glucose to release the energy stored in its bonds. The complex series of chemical reactions during which ATP is made from nutrient molecules in the presence of oxygen is known as **cellular respiration.** The term *cellular* refers to this process taking place in the cells of living things. The term *respiration* describes the process: the breakdown of fuel molecules in the presence of oxygen, with a resulting release of energy. Cellular respiration is called an *aerobic* process because this series of reactions is oxygen dependent. Do not confuse this process with the breathing of oxygen gas that your body carries out, which is also called respiration. However, the process of cellular respiration does depend on the oxygen you breathe. *All*

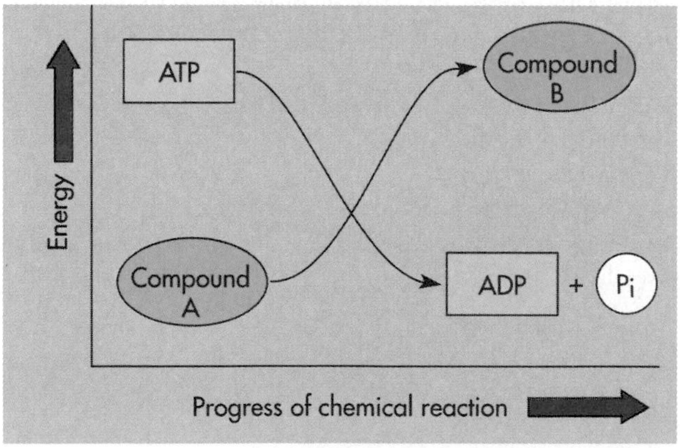

FIGURE 5-1 A coupled reaction. The energy that is released from the conversion of ATP to ADP+ P_i (an exergonic reaction) is used to drive the conversion of compound A to compound B (an endergonic reaction).

FIGURE 5-2 Photosynthesis. This orange tree captures energy from the sun and produces carbohydrates from carbon dioxide (CO_2) and water (H_2O).

organisms that take in oxygen—including plants, which produce oxygen as a byproduct of photosynthesis—break down nutrient molecules by means of cellular respiration.

Organisms that do not "breathe" oxygen, such as certain microorganisms, make ATP by a process known as **fermentation.** This process consists of glycolysis (an initial series of reactions of cellular respiration) and one or two additional reactions that take place **anaerobically**—literally, "without" (*an*) "oxygen" (*aerobically*). Baker's yeast, for example, can generate ATP by fermentation. During this process the yeast produces carbon dioxide gas, an end product that causes bread dough to rise. Certain yeasts produce alcohol during fermentation and are used in the production of beers, wines, and other alcoholic beverages.

All organisms use either cellular respiration or fermentative pathways to generate ATP, suggesting a strong evolutionary relationship among all living things. The only organisms that use neither of these two ATP-generating processes are single-celled parasites called *Chlamydia.* One species of *Chlamydia* causes certain diseases of the eyes and also one of the most prevalent sexually transmitted diseases in the United States: nongonococcal urethritis. Another species causes diseases in animals. These organisms, among the smallest of the bacteria, can grow only within other cells, and they obtain their ATP from the cells they infect. For this reason, scientists call chlamydia "energy parasites." *Chlamydia* are thought to have lost the ability to generate ATP during their evolution.

> *All cells that make their own ATP do so by means of cellular respiration or fermentation.*

How Oxygen-Using Organisms Release ATP from Food Molecules

An Overview of Cellular Respiration

The process of cellular respiration is described chemically in the following formula:

$$C_6H_{12}O_6 + 6O_2 \xrightarrow{36\,ADP + 36P_i \;\rightarrow\; 36\,ATP\ (energy)} 6CO_2 + 6H_2O$$

$$\text{Glucose} + \text{Oxygen} \longrightarrow \text{Carbon dioxide} + \text{Water}$$

where: ADP is adenosine diphosphate and P_i is inorganic phosphate

Although the above formula is stated as if glucose were broken down to carbon dioxide and water in a single step, this is not the case. The formula is a summary of a complex series of reactions, much as you might summarize the details of a book in a single paragraph. This simplified formula shows that as cellular respiration occurs, the net effect is the breakdown of one molecule of the substrate glucose in the presence of six molecules of oxygen to yield end products of six molecules of carbon dioxide and six molecules of water. During this process, enough energy is liberated from the glucose as it is being cleaved to power 36 endergonic reactions: 36 molecules of ATP are bonded to 36 atoms of inorganic phosphate—an energy yield of 36 ATPs. Therefore the breakdown of glucose during cellular respiration releases energy (is exergonic), which helps to produce molecules of ATP, a type of endergonic reaction.

> *Cellular respiration is a series of chemical reactions that capture energy in molecules of ATP. During cellular respiration, glucose is broken down in the presence of oxygen, yielding carbon dioxide and water and capturing energy in molecules of ATP.*

The complex series of reactions of cellular respiration can be divided into three parts: **glycolysis,** the **Krebs cycle,** and the **electron transport chain.** The word *glycolysis* comes from Greek words meaning "to break apart" (*lysis*) a "sugar" (*glyco*). Glycolysis takes place in the cytosol of the cell and is the first stage of extracting energy from glucose. No oxygen is needed for it to take place. It is

photosynthesis (fote oh **sin** thuh sis) the process whereby energy from the sun is captured by organisms and used to produce carbohydrates and other organic molecules.

cellular respiration the complex series of chemical reactions during which ATP is generated from the chemical breakdown of nutrient molecules in the presence of oxygen.

fermentation a process whereby organisms that do not "breathe" oxygen generate ATP from the chemical breakdown of nutrient molecules.

anaerobically (**an** er **oh** bick ly) occurring without oxygen. Fermentation reactions, for example, occur anaerobically.

glycolysis (glye **kol** uh sis) the first of the three series of chemical reactions of cellular respiration, in which glucose is broken down to pyruvate..

Krebs cycle the second series of chemical reactions of cellular respiration, in which pyruvate, the end product of glycolysis, is oxidized to carbon dioxide. Also called the citric acid cycle.

electron transport chain a group of electron carriers located on the inner mitochondrial membrane where the third series of chemical reactions of cellular respiration takes place.

a metabolic pathway in which ATP is generated, but the total yield of ATP molecules is small—only two ATPs for each original glucose molecule. When glycolysis is completed, the six-carbon glucose has been cleaved in half, yielding two three-carbon molecules called *pyruvate,* or pyruvic acid. The two pyruvate molecules still contain most of the energy that was present in the one original glucose molecule.

The *Krebs cycle* takes place in the mitochondria of the cell and is the second stage of extracting energy from glucose. Named after the biochemist Sir Hans Krebs, this cycle of reactions begins with a two-carbon molecule that is produced by the removal of one carbon dioxide molecule (CO_2) from each pyruvate molecule formed by glycolysis. Each of these two-carbon molecules combines with a four-carbon molecule to form a six-carbon molecule called *citric acid.* For this reason, the Krebs cycle is also often called the **citric acid cycle.** For every two pyruvate molecules entering the Krebs cycle as the result of the glycolytic breakdown of one glucose molecule, two more ATP molecules are made and a large number of electrons are removed from the substrates in the cycle. These electrons are passed to special molecules also located in the mitochondria called *electron carriers.* These electron carriers are part of the electron transport chain. By means of chemical reactions and processes that take place along the chain, 32 ATPs are produced. The process of cellular respiration is summarized in Figure 5-3. You may not fully understand this summary illustration now; refer to Figure 5-3 as you continue reading this chapter.

> *Three series of reactions make up the process of cellular respiration: glycolysis, the Krebs cycle, and the electron transport chain.*

Oxidation-Reduction

As chemical reactions take place in metabolic pathways such as cellular respiration, energy stored in chemical bonds is transferred to new chemical bonds, with the electrons shifting from one energy level to another. In some (but not all)

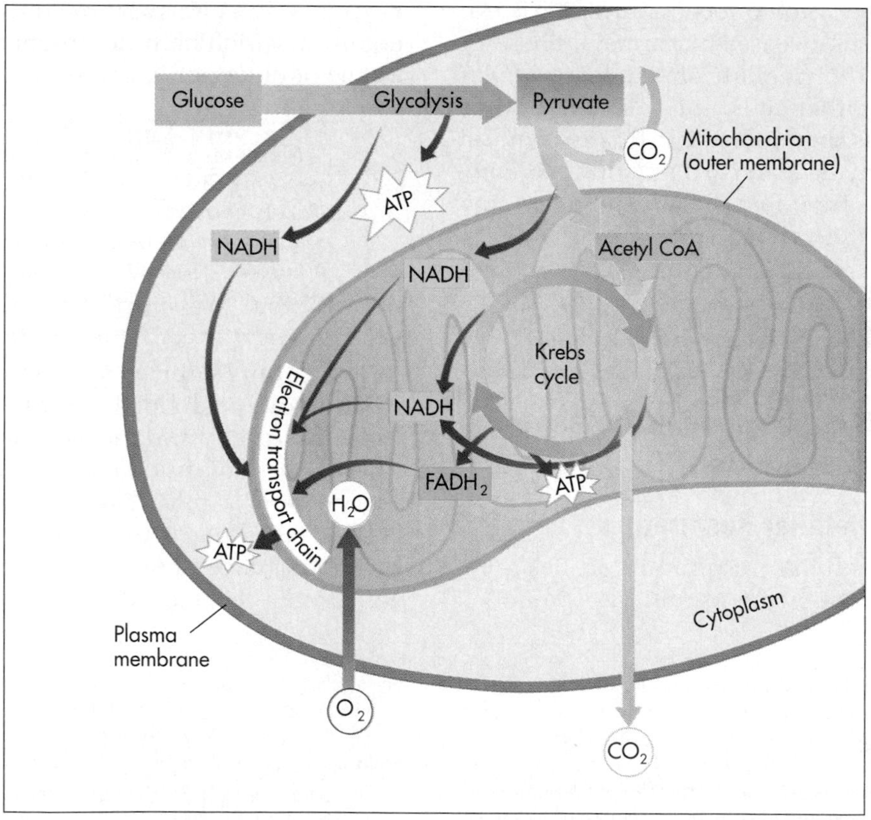

FIGURE 5-3 An overview of cellular respiration. Living things, including plants, use cellular respiration to capture and store energy in molecules of ATP. During this series of chemical reactions, glucose is broken down in the presence of oxygen producing carbon dioxide (CO_2) and water (H_2O) and capturing energy in the form of ATP. (36 ATPs are produced for every one glucose molecule.)

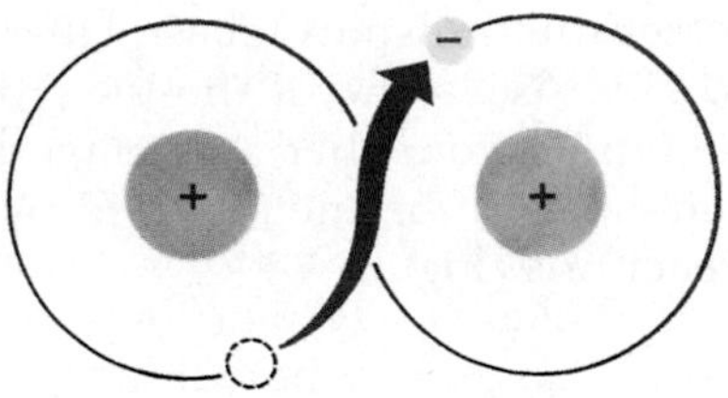

FIGURE 5-4 Oxidation-reduction. Oxidation is a loss of an electron; reduction is the gain of one.

chemical reactions, electrons actually pass from one atom or molecule to another. As explained in Chapter 2, this class of chemical reaction is called an **oxidation-reduction reaction.** Oxidation-reduction reactions are critically important to the flow of energy through living systems and are essential to the flow of energy in cellular respiration.

When an atom or a molecule gives up an electron, it is *oxidized.* The process by which this occurs is called *oxidation.* The name reflects that in biological systems, oxygen, which strongly attracts electrons, is the most frequent electron acceptor. Therefore, atoms that give up electrons to oxygen are "acted upon" by oxygen, or oxidized.

When an atom or molecule gains an electron, it becomes *reduced.* The process is called *reduction.* This name reflects that the addition of an electron reduces the charge by one. For example, if a molecule had a change of +2, the addition of an electron (−1) would reduce the molecule's charge to +1.

Oxidation and reduction always take place together because every electron that is lost by one atom (oxidation) is gained by some other atom (reduction) (Figure 5-4). Together they are therefore called **redox reactions.** In a redox reaction the charge of the oxidized atom is increased, and the charge of the reduced atom is lowered.

> *Oxidation is the loss of an electron by an atom. Reduction is the gain of an electron by an atom.*

Electron Carriers

In biological systems, electrons often do not travel alone from one atom to another but instead take along a proton. As you may recall from Chapter 2, a proton and an electron together make up a hydrogen atom. In living things, special electron carriers transfer electrons as hydrogen atoms from one molecule to another. One such carrier is the molecule *nicotinamide adenine dinucleotide,* or *NAD^+*, a molecule synthesized from the vitamin niacin. The + sign indicates that NAD is the oxidized form of the molecule. It can accept two electrons and one proton (H^+) to become reduced to NADH. Notice that this molecule (NADH) no longer bears a positive charge because it has been reduced. In addition, the extra H^+ and electron (hydrogen atom) are shown by the addition of one H to NAD.

Flavine adenine dinucleotide, or *FAD,* is another special electron carrier. It can accept *two* electrons and *two* protons to be reduced to $FADH_2$. The subscript 2 indicates the number of hydrogen atoms that FAD has accepted.

Both NADH and $FADH_2$ can pass on the electrons and protons they accept to other carriers. Although not a part of the electron transport chain, NADH and $FADH_2$ bring the electrons and protons they have accepted during the processes of glycolysis (NADH only) and the Krebs cycle (both NADH and $FADH_2$) to the electron transport chain to be accepted by carriers in the chain. By passing electrons and protons in a series of redox reactions, the molecules in this chain create a flow of energy that works in special ways to produce most of the ATP of cellular respiration.

A More Detailed Look at Cellular Respiration

Food is a complex mixture of sugars, lipids, proteins, and other molecules. The first thing that happens in food's journey toward ATP production is that digestive system enzymes break down complex molecules to simple ones. Complex sugars like those found in vegetables and pasta, for example, are split into simple sugars such as glucose or into sugars that are changed to glucose. Proteins, which are found in foods such as meats and nuts, are broken down to amino acids. Complex lipids like those found in oil-based salad dressings are split into fatty acids and glycerol. These steps taken by the digestive system yield no usable energy, but

citric acid cycle another name for the Krebs cycle.
oxidation-reduction reaction a chemical reaction in which electrons pass from one atom or molecule to another; one atom or molecule loses an electron (oxidation) while the other gains an electron (reduction).
redox reaction another name for an oxidation-reduction reaction.

they change a diverse array of complex molecules into a small number of simpler molecules that can be used by your cells as fuel.

The primary metabolic pathways in your body that break apart these simple molecules to release their stored energy and capture it in molecules of ATP are glycolysis, the Krebs cycle, and the electron transport chain. Once they are changed to glucose, carbohydrates are completely metabolized this way. After the digestion of proteins to amino acids, their carbon portions are chemically modified and are then metabolized by the Krebs cycle and the electron transport chain, skipping over glycolysis. Likewise, some of the breakdown products of the fats in your diet are converted to substances that can also be metabolized by these two series of reactions (Figure 5-5).

Glycolysis

Scientists think that glycolysis was one of the first metabolic processes to evolve. One reason is that this process uses no molecular oxygen and therefore occurs readily in an environment devoid of oxygen—a characteristic of the atmosphere of the

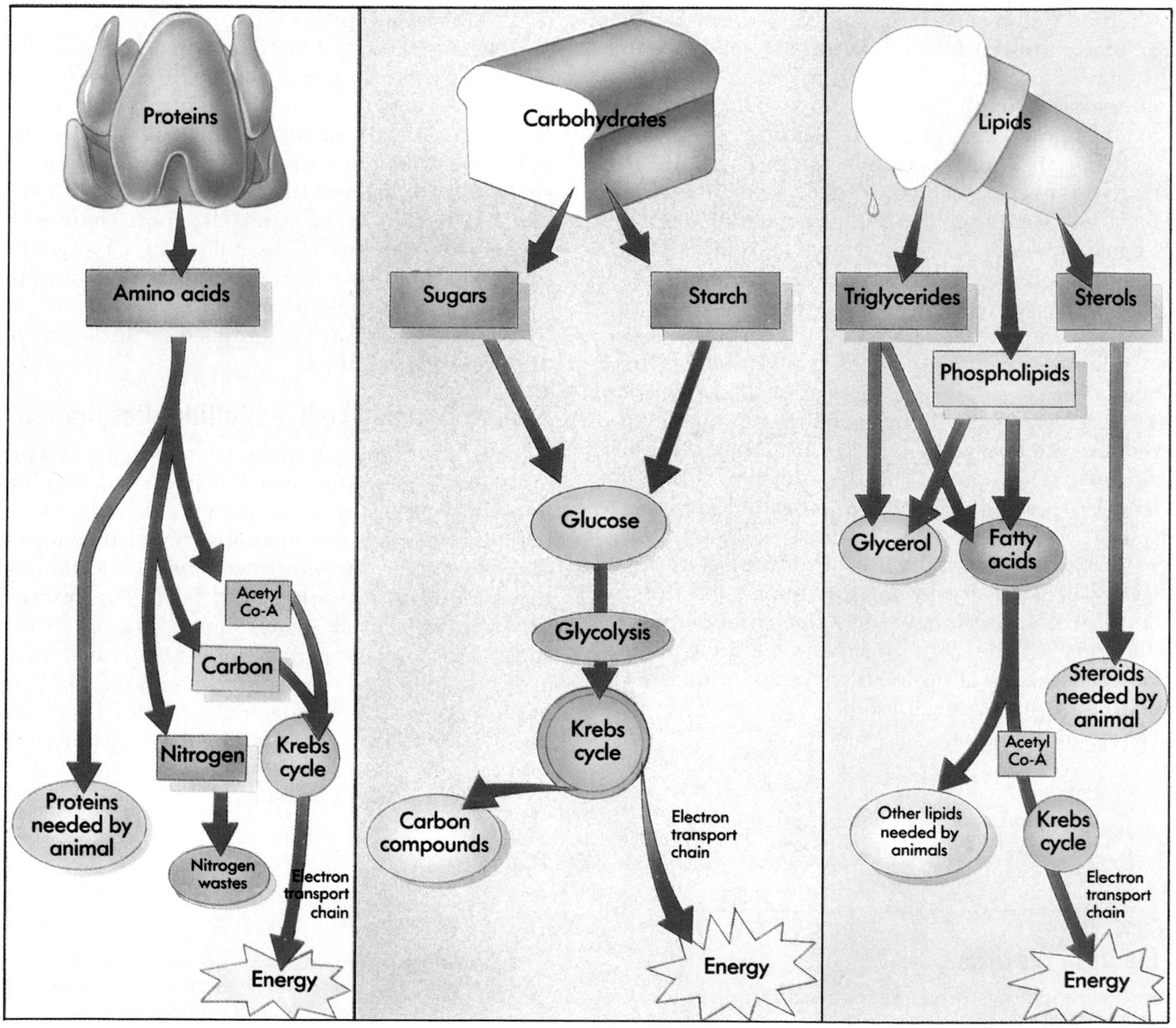

FIGURE 5-5 The complex molecules in food must be broken down into simpler molecules before they can be used by cells as fuel. Proteins are broken down into amino acids. Carbohydrates are broken down into simpler sugars. Lipids are broken down into fatty acids and glycerol. Note that proteins and lipids, once broken down into their simpler subunits, enter the Krebs cycle without undergoing glycolysis.

JUST WONDERING. . . .

I work out a lot and take protein supplements. Do they really help me develop my muscles, or are they just broken down in my body?

You probably take protein or amino acid-containing powders, drinks, and/or wafers made from milk, soy, yeast, or gelatin proteins. As was discussed in this chapter, the proteins we ingest are disassembled into their component amino acids during digestion. Eventually, these amino acids are reassembled into human proteins such as enzymes, hormones, or antibodies that fight infection; the amino acids can also be chemically modified and processed through the Krebs cycle and the electron transport chain, yielding energy. Protein *is* needed for a healthy body, but will protein supplements produce a healthier, more muscular body than can be attained by a normal diet alone?

The protein needs for most healthy people have been well-documented by research. The Recommended Dietary Allowance (RDA) for adults is 0.8 gram of protein per kilogram of body weight. That means that the RDA for a 138-pound adult is about 50 grams per day. Protein needs do increase during growth periods, pregnancy, and breast-feeding. If additional protein is needed during these periods, then it might seem that athletes such as body builders would need additional protein to enlarge their muscles. However, researchers disagree as to whether athletes' needs are significantly greater than those of other healthy adults.

Some researchers believe that because athletes already eat more food than most other people to meet their energy needs, their protein intake is already likely to be adequate. Others suggest that athletes need to double their protein intake to meet actual increases in need. Studies demonstrate, however, that eating extra protein does not stimulate muscle cells to divide and make more muscle. Instead exercise, together with the normal recommended intake of protein, increases the size of individual muscle cells.

Other studies indicate that most Americans already consume about twice the recommended amount: about 100 grams of protein per day. By supplementing their diets with extra protein, athletes may be consuming 125 grams or more of protein each day. And it is possible that too much may not be better than just enough. The reason for this is the following: when protein molecules are broken down during digestion, a portion is either used for energy or converted to body fat and stored for future energy needs. The remainder of the molecule is a waste product, and elimination of these wastes makes additional metabolic work for the body. This may not be much of a problem for healthy people, but it can be an additional strain for those with liver or kidney disease. Even in healthy people, elimination of unneeded protein wastes often leads to increased loss of calcium and body water.

Most nutrition experts agree that body builders and other athletes do not need protein supplements. Instead, during training they should exercise appropriately and follow the same dietary guidelines recommended for everyone.

primitive Earth. In addition, all of the reactions of glycolysis occur free in the cytoplasm; none are associated with any organelle or membrane structure. "Early" cells most certainly had no specialized structures such as organelles; thus metabolic processes that evolved independently of membrane systems are probably evolutionarily older than those associated with membrane systems.

All living things (except the chlamydia) possess glycolytic enzymes. However, most present-day organisms are able to extract considerably more energy from glucose molecules than glycolysis does. For example, of the 36 ATP molecules you obtain from each glucose molecule that you metabolize, only 2 are obtained by glycolysis. Why, then, is glycolysis still maintained even though its energy yield is comparatively meager?

This simple question has an important answer: evolution is a step-by-step process. Change occurs during evolution by improving on past successes. In catabolic metabolism, glycolysis was an improvement. A *catabolic process* is one in which complex molecules are broken down into simpler ones. Cells that could not carry out the catabolic reactions of glycolysis were at a competitive disadvantage; cells that were capable of glycolysis survived the early competition of life. Later improvements in catabolic metabolism built on this success. Glycolysis was not discarded during the course of evolution but rather was used as the starting point for the further extraction of chemical energy.

The catabolic pathway of glycolysis is shown in Figure 5-6. Its reactions can be divided into four stages:

1. Stage A—*Glucose mobilization.* During reactions 1 and 2, glucose is changed into a compound that can be split into two molecules. During each of these two reactions, ATP molecules have their outermost phosphate groups enzymatically removed. The bonds holding these phosphate groups release a high amount of free energy when broken. The energy in this case is used to attach the phosphate group to the glucose molecule. Before energy can be released from glucose, your body must spend two ATP to prime the glucose pump.

2. Stage B—*Cleavage.* During reaction 3, the six-carbon product of glucose mobilization is split into two three-carbon molecules.

3. Stage C—*Oxidation.* During reaction 4, one H^+ and two electrons are removed from each three-carbon molecule and are donated to NAD^+. NAD^+ acts as an electron carrier, forming two molecules of NADH for each original glucose molecule. As this chemical change takes place, an inorganic phosphate molecule is bonded to each resulting three-carbon molecule (BPG) by means of high-energy bonds.

4. Stage D—*ATP generation.* During reactions 5 through 7, the two high-energy phosphate groups on each three-carbon molecule (BPG) are removed by enzymes and bonded to two ADP molecules, producing four ATP molecules for each original glucose molecule and two 3PG molecules. A molecule of water is removed from each 3PG molecule producing two PEP molecules. The end product of these reactions after the removal of two high-energy phosphate groups is two three-carbon molecules called pyruvate.

Because each glucose molecule is split into *two* three-carbon molecules, the overall net reaction sequence yields two ATP molecules, as well as two molecules of pyruvate:

$$\begin{array}{c} -\ 2\ \text{ATP Stage A} \\ +\ 2\ (2\text{ATP})\ \text{Stage D} \\ \hline +\ 2\ \text{ATP} \end{array}$$

Although this is not a great amount of energy (only 2% of the energy in the glucose molecule), glycolysis does generate ATP. During the first anaerobic, or airless, stages of life on Earth, this reaction sequence was the only way for living things to extract energy from food molecules.

> *The sequence of reactions of glycolysis generates a small amount of ATP by reshuffling the bonds of glucose molecules. Glycolysis is a very inefficient process, capturing only about 2% of the available chemical energy of glucose.*

The three most significant changes that take place during glycolysis are as follows:

1. Glucose is converted to pyruvate.
2. ADP + P_i is converted to ATP.
3. NAD^+ is converted to NADH.

These three products can continually be formed as long as the substrates used to produce them are available. The glucose is supplied by the food you eat. ADP and P_i continually become available as ATP is broken down to do cellular work. But cells contain only a small amount of NAD^+, and the supply is quickly depleted unless NADH passes along its hydrogen atom to another electron carrier. In this way NADH is oxidized to form NAD^+ once again.

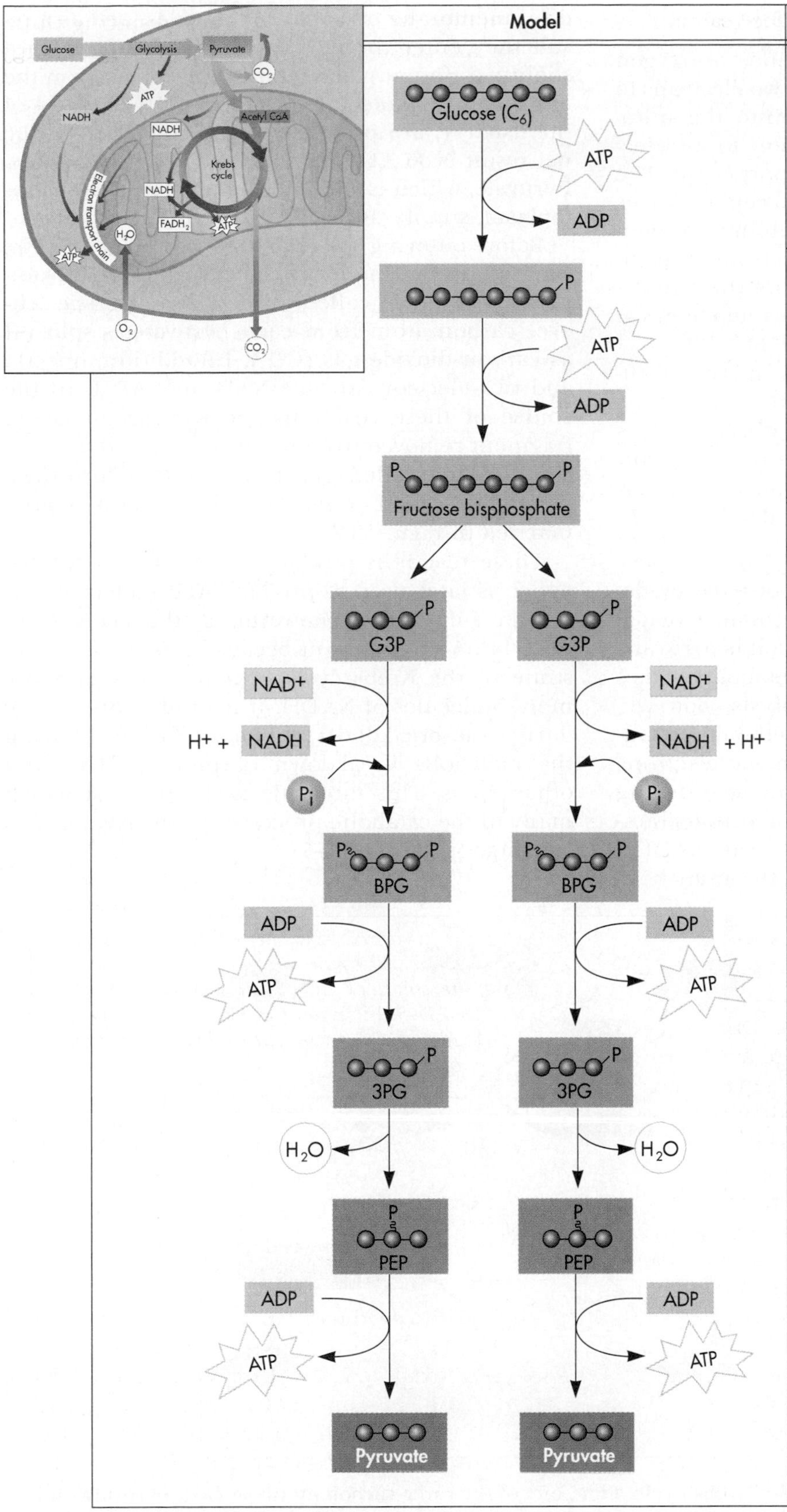

Description	Stage
1. A phosphate group is added to glucose by ATP.	Stage A Glucose mobilization (1,2)
2. Rearrangement, followed by a second addition of a phosphate group by ATP, yields Fructose bisphosphate.	
3. The six-carbon molecule is split into two three-carbon molecules, ultimately yielding 2 molecules of G3P (Glyceraldehyde 3–Phosphate))	Stage B Cleavage (3)
4. Oxidation produces two NADH molecules and gives two molecules of BPG (Bisphosphoglycerate) each with one high–energy phosphate bond.d	Stage C Oxidation (4)
5. Removal of high–energy phosphate by two ADP molecules produces two ATP molecules and gives two 3PG (3–Phosphoglycerate) molecules.	Stage D ATP generation (5–7)
6. Removal of water gives two PEP (Phosphoenol pyruvate) molecules each with a high energy phosphate bond.	
7. Removal of high–energy phosphate by two ADP produces two ATP molecules and gives two pyruvate molecules.	

FIGURE 5-6 Glycolysis. Glycolysis produces a net yield of two ATP molecules for each original glucose molecule, as well as two molecules of NADH and two molecules of pyruvate.

Cells recycle NADH in one of two ways:

1. *Aerobic metabolism.* In the presence of oxygen gas, NADH passes its proton and two electrons to a molecule that shuttles them into the mitochondria, ultimately passing them to an electron carrier in the electron transport chain. The oxygen, an excellent electron acceptor, forms the last link in the chain. Each atom of oxygen accepts two protons and two electrons to form one molecule of water. In addition, three molecules of ATP are formed during various chemical processes that take place (see p. 111).
2. *Anaerobic metabolism.* When oxygen is not available, another organic molecule must accept the hydrogen atom instead. This process occurs when organisms such as bacteria or yeast grow without oxygen and is called *fermentation.* Fermentation is described in more detail later in this chapter.

In all aerobic organisms—like you—the oxidation of glucose continues with the further oxidation of the product of glycolysis, which is pyruvate. In eukaryotic organisms, aerobic metabolism takes place in the mitochondria. Glycolysis, however, takes place in the cytoplasm outside the mitochondria. Therefore the cell must move the electrons from the two NADH molecules produced during glycolysis into the mitochondrion. (The mitochontrial membrane is impermeable to the NADH.) These electrons are carried across the mitochondrial membrane by carrier molecules, and as a result they enter the electron transport chain in a slightly different place than the electrons from the other NADH molecules. Only 2 ATPs (instead of the usual 3) are produced from each NADH. The net result is 36 ATPs for each molecule of glucose. Pyruvate, which is in high concentration in the cytoplasm, simply diffuses into the mitochondrion.

Before entering the reactions of the Krebs cycle, each pyruvate molecule is cleaved into a two-carbon molecule called an *acetyl group.* The left-over carbon atom from each pyruvate is split off as carbon dioxide gas (CO_2). In addition, one H^+ and two electrons reduce NAD^+ to NADH. In the course of these reactions the two-carbon acetyl fragment removed from pyruvate is added to a carrier molecule called *coenzyme A,* or *CoA,* forming a compound called *acetyl-CoA.* This process is summarized in Figure 5-7.

These reactions produce a molecule of NADH, which is later used to produce ATP molecules. Of greater importance, however, is the acetyl-CoA. Acetyl-CoA is important because it is the first substrate in the Krebs cycle, which generates many more molecules of NADH. It is produced not only during the breakdown of glucose but also during the metabolic breakdown of proteins, fats, and other lipids. This molecule is the point at which many of the catabolic processes of eukaryotic cells converge (see Figure 5-5).

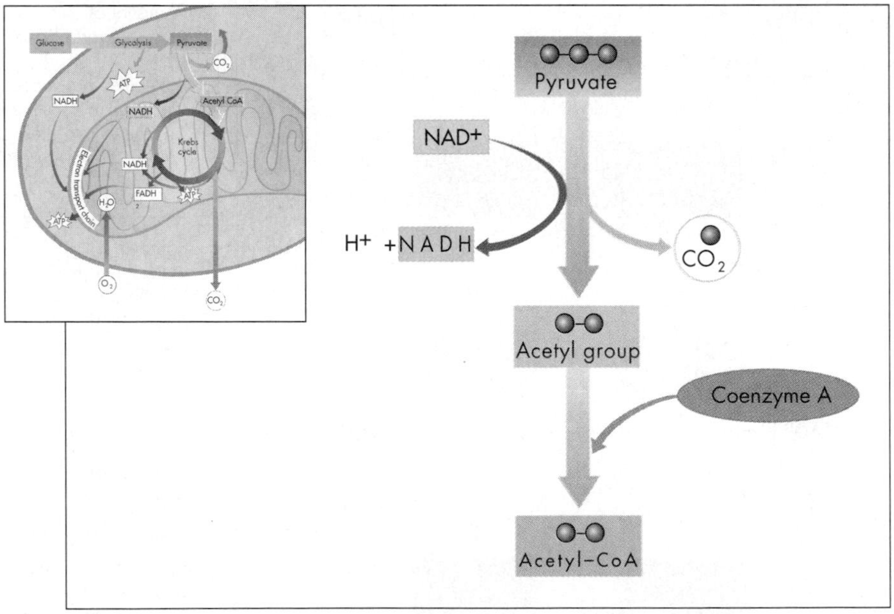

FIGURE 5-7 How pyruvate enters the Krebs cycle. First, one of the three carbon atoms of each pyruvate molecule is released in the form of carbon dioxide (CO_2). Left over is a two-carbon molecule called an acetyl group and a hydrogen and two electrons, which combine with NAD^+ to produce NADH. Coenzyme A, a carrier molecule, is added to the acetyl group, forming acetyl-CoA, which is the starting compound for the Krebs cycle.

The Krebs Cycle

The Krebs cycle, which oxidizes acetyl-CoA, consists of eight enzyme-mediated reactions. These reactions are shown in Figure 5-8. The cycle has two stages:

1. Stage A—*Preparation reactions.* These two reactions set the scene. In the first reaction, acetyl-CoA joins a four-carbon molecule from the end of the cycle to form the six-carbon molecule citric acid. In the next reaction, chemical groups are rearranged.
2. Stage B—*Energy extraction.* Four of the remaining six reactions are oxidations in which hydrogen ions and electrons are removed from the intermediate compounds in the cycle to form three NADH molecules for each acetyl-CoA, six for each original glucose molecule. In addition, one molecule of $FADH_2$ is formed, two for each original glucose. During the cycle a molecule called *GTP* is produced, which is quickly used to produce an ATP; two ATP are generated for each original molecule of glucose.

Together, the eight reactions make up a cycle that begins and ends with the same four-carbon molecule. At every turn of the cycle, acetyl-CoA enters and is oxidized to CO_2 and H_2O, and the hydrogen ions and electrons are donated to electron carriers.

In the process of aerobic respiration the glucose molecule has been consumed entirely. Its six carbons were first split into three-carbon units during glycolysis. One of the carbons of each three-carbon unit was then lost as CO_2 in the conversion of pyruvate to acetyl-CoA, and the other two were lost during the oxidations of the Krebs cycle. Part of the glucose molecule's energy and its electrons, which are preserved in four ATP molecules and the reduced state of 12 electron carriers, are all that is left.

> *The breakdown of the two pyruvate molecules produced from one glucose molecule during glycolysis generates 2 ATP molecules, 10 molecules of NADH, and 2 molecules of $FADH_2$. NADH and $FADH_2$ can be used to generate ATP.*

The Electron Transport Chain

Embedded within the inner mitochondrial membrane are many series of electron carrier proteins. Each series is known as an electron transport chain (Figure 5-9). Here, electrons are passed from one carrier protein to another. The NADH molecules formed during glycolysis and the subsequent oxidation of pyruvate carry their electrons to this membrane. As you may recall, each NADH contains a pair of electrons and a proton gained when NADH was formed from NAD^+. The $FADH_2$ molecules are already attached to this membrane, each containing two protons and two electrons gained when $FADH_2$ was formed from FAD.

Three of the electron carrier proteins in the electron transport chain act as proton pumps. A proton pump is an active transport mechanism in cell membranes that forces hydrogen ions (protons) out of a cell. As each of these carriers accepts electrons in turn, the electrons fall to lower energy levels, releasing energy in the process. In addition, these carriers change shape as they accept the electrons, forming a channel from the inner compartment, or matrix, of the mitochondrion to the outer compartment—the space between the mitochondrial membranes. The energy released from the electrons being passed along the chain pumps protons through these channels and against the concentration gradient into the outer compartment of the mitochondrion (Figure 5-10). As a high concentration of protons builds up in the outer compartment, the protons begin to diffuse back across the membrane through the membrane protein *ATP synthase,* visible in electron micrographs as projections on the inner surface of the membrane (see Figure 5-10). As the protons travel inward through these channels, proteins associated with the channels couple the energy from the movement of the protons to the reaction of bonding P_i to ADP, forming ATP. Oxygen is the final acceptor for both the electrons and protons. Water is therefore an end product of cellular respiration, with one molecule of water produced for each NADH or $FADH_2$. Because the electron transport chain uses oxygen as its terminal electron acceptor, this series of reactions is also termed the *respiratory assembly.*

> *The electrons harvested from glucose and transported to the mitochondrial membrane by NADH drive protons out across the inner membrane. The return of the protons through the membrane protein ATP synthase results in the generation of ATP.*

Because their electrons activate three pumps, most NADH molecules produced during cellular respiration ultimately cause the production of *three* ATP molecules. Each $FADH_2$, which activates two of the pumps, leads to the production of *two* ATP molecules. However, eukaryotes carry out gly-

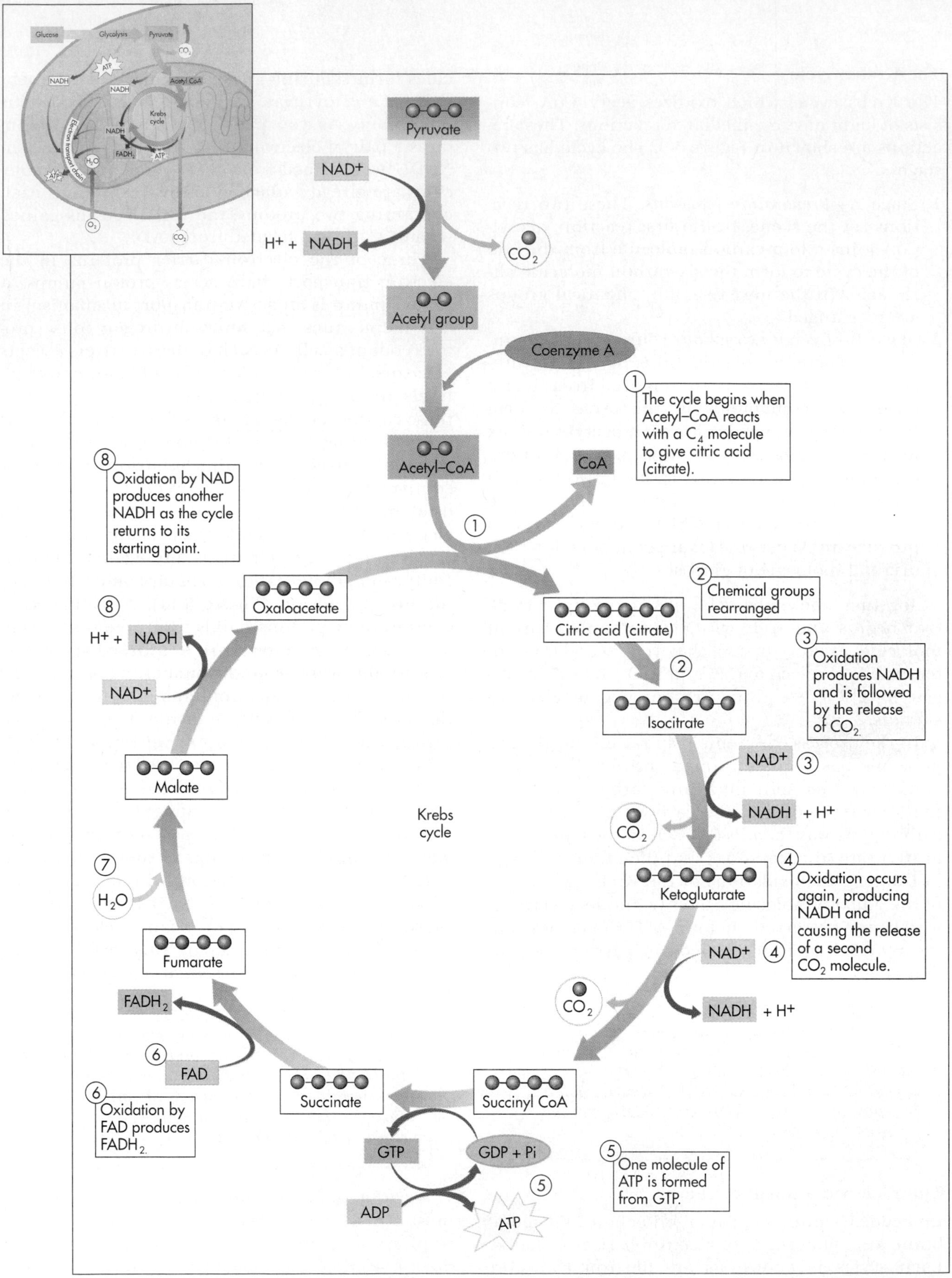

FIGURE 5-8 **The Krebs cycle.** The Krebs cycle produces three molecules of NADH, one molecule of $FADH_2$, and one molecule of ATP for each acetyl-CoA molecule that enters the cycle.

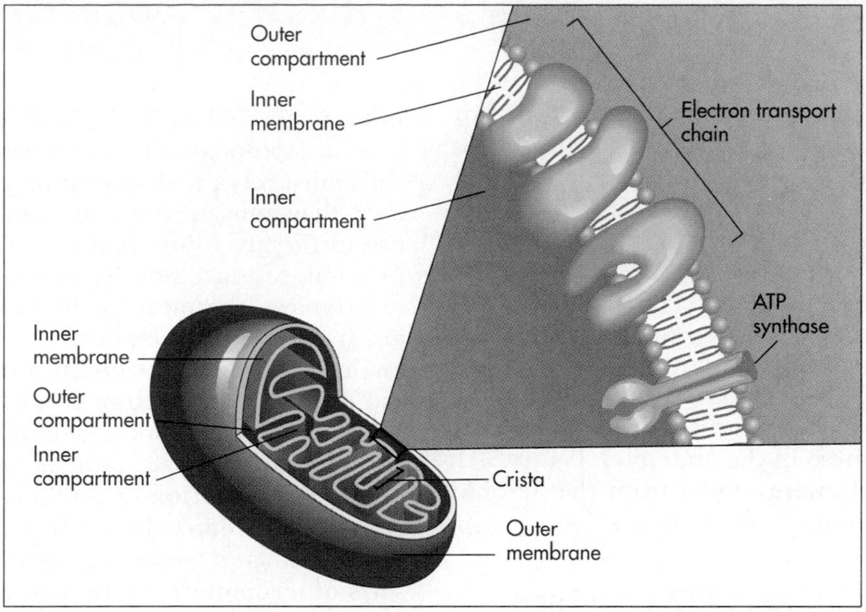

FIGURE 5-9 Site of the electron transport chain. Located deep within the inner mitochondrial membrane, electron carrier proteins accept electrons from the NADH and $FADH_2$ formed during glycolysis and the Krebs cycle.

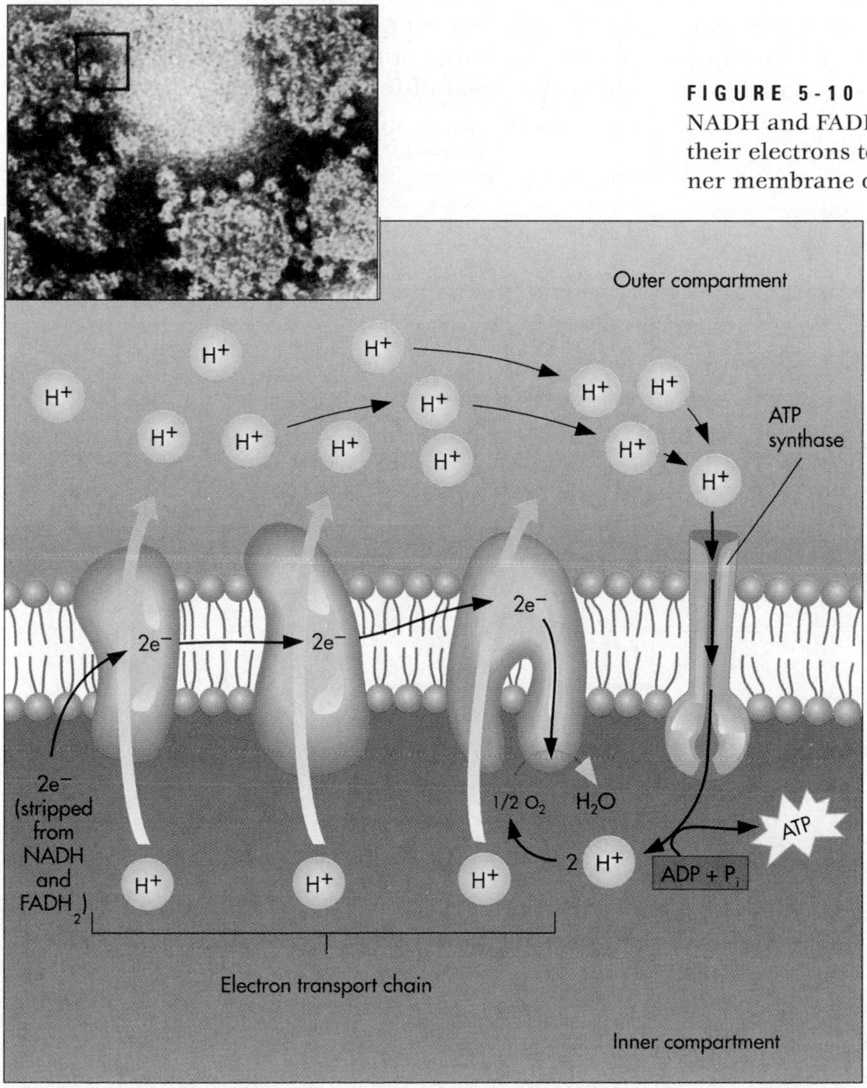

FIGURE 5-10 The electron transport chain. The NADH and $FADH_2$ formed during the Krebs cycle pass their electrons to electron receptors located in the inner membrane of the mitochondrion.

colysis in the cytoplasm and the Krebs cycle within the mitochondria. This separation of the two processes within the cell requires transporting the electrons of the NADH created during glycolysis across the mitochondrial membrane. In fact, only the electrons of NADH are shuttled across the membrane, not the entire molecule. The electron transfers that are made as these electrons are shuttled into the mitochondria result in the formation of only 2 ATP, not 3 as is usually the case. Thus each glycolytic NADH produces only two ATP molecules instead of three in the final total. Figure 5-11 describes the total energy yield from the aerobic metabolism of one molecule of glucose.

How Organisms Release ATP from Food Molecules Without Using Oxygen

Aerobic metabolism cannot take place in the absence of oxygen because a final electron acceptor is missing from the electron transport chain. Not only will the reactions of the electron transport chain come to a halt, but the Krebs cycle reactions will not take place because NADH molecules will not be recycled to NAD^+ molecules, needed electron acceptors in the cycle. In such a situation, cells must rely on glycolysis to produce ATPs. However, they must regenerate NAD^+ from NADH for use in the glycolytic pathway. During aerobic respiration, cells accomplish this task by means of the reactions of the electron transport chain. During anaerobic respiration, bacteria and yeasts produce an organic molecule from pyruvate that will accept the hydrogen atom from NADH and thus re-form NAD^+. These reactions, including the glycolytic pathway, are termed *fermentation*. The end products of fermentation depend on the organic molecule that is produced from pyruvate.

Bacteria and yeasts carry out many different sorts of fermentations. In fact, more than a dozen fermentative processes have evolved among bacteria, each process using a different organic molecule as the hydrogen acceptor. Often the resulting reduced compound is an acid. In some organisms, such as yeasts, an organic molecule called *acetaldehyde* is formed from pyruvate and then reduced, producing an alcohol (Figure 5-12, *A*). This particular type of fermentation is of great interest

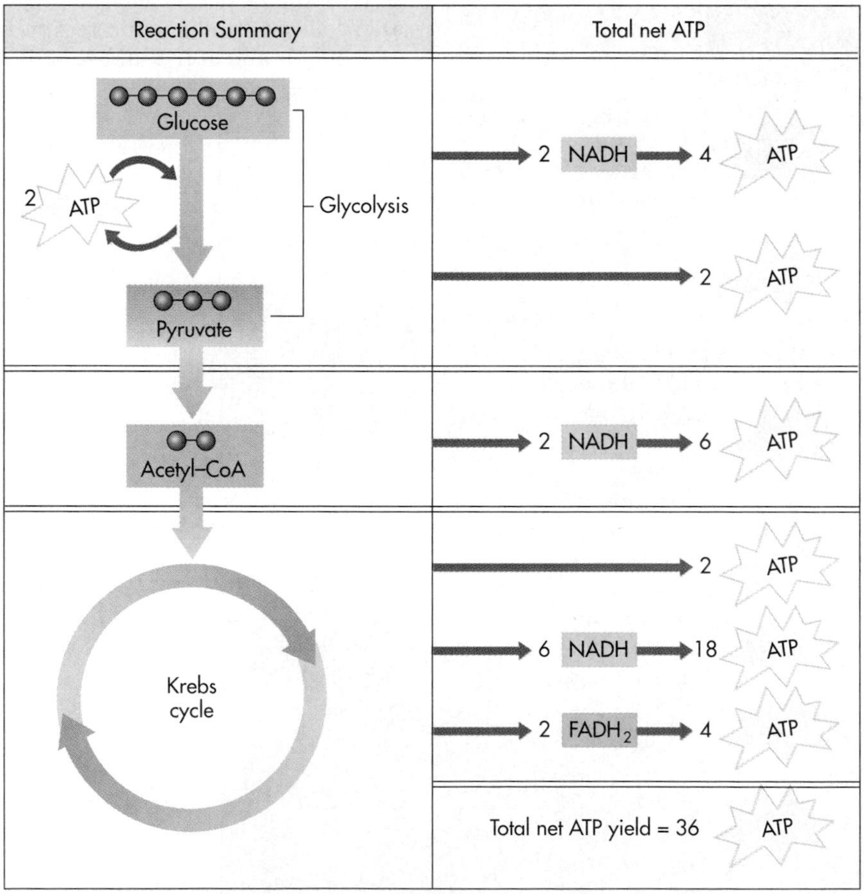

FIGURE 5-11 An overview of the energy extracted from the oxidation of glucose. A net total of 2 ATP are produced during glycolysis. Two NADPH molecules are also produced, which each yield only 2 (rather than 3) molecules of ATP after their electrons are shuttled from the cytoplasm to the inner mitochondrial membrane. Two NADH, yielding 6 ATP, are produced as pyruvate is oxidized to acetyl-CoA. In the Krebs cycle, 2 ATP are produced directly; 6 NADH are produced, yielding 18 ATP; and 2 $FADH_2$ are produced, yielding 4 ATP.

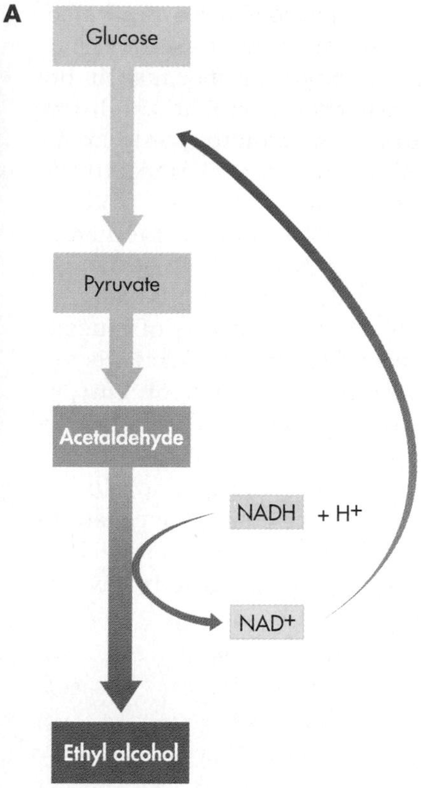

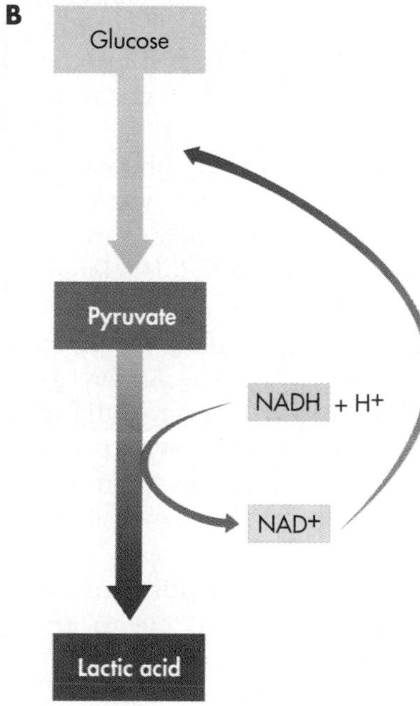

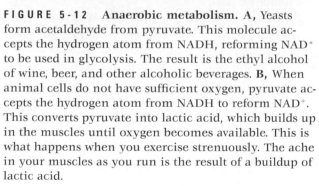

FIGURE 5-12 Anaerobic metabolism. A, Yeasts form acetaldehyde from pyruvate. This molecule accepts the hydrogen atom from NADH, reforming NAD^+ to be used in glycolysis. The result is the ethyl alcohol of wine, beer, and other alcoholic beverages. **B,** When animal cells do not have sufficient oxygen, pyruvate accepts the hydrogen atom from NADH to reform NAD^+. This converts pyruvate into lactic acid, which builds up in the muscles until oxygen becomes available. This is what happens when you exercise strenuously. The ache in your muscles as you run is the result of a buildup of lactic acid.

to people because it is the source of the ethyl alcohol in wine and beer. However, ethyl alcohol is an undesirable end product for yeast because it becomes toxic to the yeast when it reaches high levels. That is why natural wine contains only about 12% alcohol—12% is the amount it takes to kill the yeast fermenting the sugars.

Muscle cells undergo a similar fermentative process when they produce ATP without sufficient oxygen (Figure 5-12, *B*). During strenuous exercise, muscle cells break down large amounts of glucose to produce ATP. However, because muscle cells are packed with mitochondria, the pace of glucose breakdown far outstrips the blood's ability to deliver oxygen for aerobic respiration. Therefore the muscle cells switch from aerobic respiration to fermentation. Muscle cells do not change pyruvate to another organic molecule to be reduced as the bacteria and yeasts do but instead directly reduce pyruvate to lactic acid. This acid builds up in the muscle and tends to produce a sensation of muscle fatigue. Gradually, however, the lactic acid is carried by the blood to the liver, where it is broken down aerobically. This need for oxygen to break down the lactic acid produces what is termed an *oxygen debt* and is the reason a person continues to pant after completing strenuous exercise.

> *In fermentations, which are anaerobic processes, pyruvate molecules or organic compounds produced from pyruvate are reduced, accepting the electrons of NADH generated in the glycolytic breakdown of glucose. These last reactions of the fermentative process produce the NAD^+ needed for the anaerobic breakdown of glucose to continue.*

Summary

- All the activities performed by living things involve chains of reactions that use, move, carry, store, and free energy. These chains of reactions are called *metabolic pathways.* In metabolic pathways, organisms couple exergonic reactions that release energy from molecules such as ATP with endergonic reactions to enable them to take place.
- Almost all living things, including photosynthetic organisms, make ATP by a process called *cellular respiration.* Certain organisms make ATP by an anaerobic process called *fermentation.*
- Cellular respiration is composed of three series of chemical reactions: glycolysis, the Krebs cycle, and the electron transport chain. During these reactions, glucose is broken down in the presence of oxygen, capturing energy in molecules of ATP. The byproducts of this process are carbon dioxide and water.
- The energy stored in chemical bonds can be transferred from one atom or molecule to another. In some chemical reactions, electrons pass from one atom or molecule to another. This class of chemical reaction is called an *oxidation-reduction reaction.*
- NAD^+ and FAD are important electron carriers in living systems. Both become reduced by accepting electrons and protons and can then pass on the electrons and protons they accept to other carriers. Such carriers play important roles in the production of ATP.
- In the human body, most carbohydrates are changed to glucose and completely metabolized by cellular respiration. Parts of protein and lipid molecules are also metabolized by part of this metabolic pathway.
- Cellular respiration yields 36 ATP molecules from glycolysis, the Krebs cycle, and the electron transport chain.
- Organisms that metabolize nutrient molecules anaerobically by the process of fermentation use pyruvate or a molecule derived from pyruvate to accept the electrons produced during the glycolytic breakdown of glucose. This process produces end products that are frequently acids or alcohols.

Knowledge and Comprehension Questions

1. What is ATP, and why is it important to life?
2. Distinguish between cellular respiration and fermentation. What purpose do they serve?
3. Explain what this formula means. What process does it describe?

$$C_6H_{12}O_6 + 6O_2 \longrightarrow 6CO_2 + 6H_2O + \text{Energy}$$

4. Does oxidation always involve the acceptance of an electron by oxygen? Define the term *oxidation.*
5. What do NAD^+ and FAD have in common? How do they differ?

6. You stop at a fast-food restaurant to eat a burger and fries. Explain how your body converts these foods into usable energy.
7. Summarize the main events of glycolysis.
8. The chapter says that glycolysis is inefficient; explain why. If it is inefficient, why do our cells still carry out the process?
9. Summarize what happens during the Krebs cycle.
10. Create a table that summarizes the output of aerobic metabolism for glycolysis, the oxidation of pyruvate, and the Krebs cycle.
11. We have discussed before how the chemistry of life is integrally involved with the nature of water. What is the link between water and the electron transport chain, also called the "respiratory assembly"?
12. By filling in the diagram below, summarize the stages, products, and byproducts of cellular respiration. Show which events occur outside the mitochondrion and which take place inside the organelle. Include the number of ATP molecules produced at each stage.

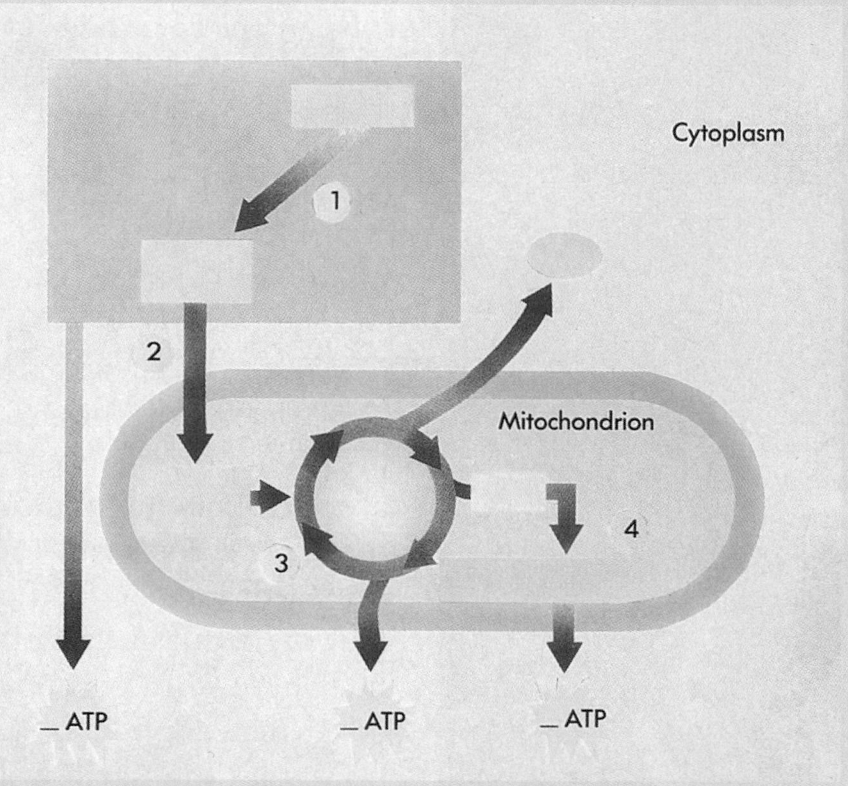

Critical Thinking

1. Is the following statement true or false? The process of glucose breakdown during cellular respiration has both endergonic and exergonic elements. Support your answer with evidence.
2. Decorative ponds often have a fountain in which water is sprayed into the air. What might be a reason, other than aesthetic appeal, that this is done?
3. Chapter 3 briefly described the endosymbiont theory, which suggests that mitochondria originated as bacteria that were engulfed by pre-eukaryotic cells. What have you learned in this chapter that would lead you to more firmly hold to that theory or to discard it as false?

CHAPTER 6

Photosynthesis

HOW PLANTS CAPTURE AND STORE ENERGY FROM THE SUN

An eclipse of the sun is an awesome phenomenon to observe. But did you ever think about what might happen if the light of the sun were blocked from the Earth forever? You might predict that the temperature of the Earth's atmosphere would drop, and you would be correct. Obviously, you might also think that the Earth would become shrouded in darkness. And that is true too. But more devastating consequences would occur. Life on the surface of the Earth—as we know it today—would totally and completely end.

The living organisms of Earth depend on the sun and have for billions of years. Your life depends on a process called *photosynthesis,* a process powered by the sun. By means of photosynthesis, energy from the sun is captured and stored in carbohydrate molecules. This food ultimately builds and fuels the bodies of all living things.

KEY CONCEPTS

- Virtually all living things on the surface of the Earth depend on the sun for their existence.
- Photosynthesis is the process whereby plants, algae, certain single-celled eukaryotes, and some bacteria produce organic molecules from carbon dioxide by harvesting energy from sunlight.
- Some of the complex events of photosynthesis require light energy and some do not.
- The summary equations of photosynthesis and cellular respiration are the reverse of one another, although the individual steps do not make up reverse pathways.

OUTLINE

Producers and Consumers of Food

Plants, algae, certain single-celled eukaryotes, and some bacteria are the members of the living world able to carry out photosynthesis. Photosynthetic organisms harvest light energy from the sun and change it into stored chemical energy within food. Organisms that produce their own food by photosynthesis, along with a few others that use chemical energy in a similar way, are called **autotrophs.** The word *autotroph* literally means "self-feeder." All organisms live on the food produced by autotrophs, including the autotrophs themselves.

Organisms that cannot produce their own food are called **heterotrophs,** literally "other-feeders." Heterotrophs consume other organisms and, ultimately, the food produced by autotrophs. At least 95% of the species of organisms on Earth—all animals, all fungi, and most protists and bacteria—are heterotrophs. They live by feeding on the chemical energy that is fixed (incorporated into carbohydrates) by photosynthesis. Therefore all living things—plants, bacteria, and you—share the same ultimate dependency on the sun (Figure 6-1).

> *Most autotrophs manufacture their own food by converting the energy in sunlight into chemical energy by means of the process of photosynthesis. Heterotrophs cannot make their own food and ultimately depend on autotrophs—and the sun—for survival.*

The Energy in Sunlight

How do photosynthetic organisms "harvest" light energy from the sun? And how do they use this energy to create chemical bonds in food molecules? First, it is important to understand what light energy is.

As you may recall from Chapter 4, energy can exist in many forms, such as mechanical force, heat, sound, electricity, light, radioactivity, and magnetism. All these forms of energy are able to create change—to do work. Sunlight is a form of energy known as *electromagnetic energy,* also called *radiation.*

FIGURE 6-1 Autotrophs and heterotrophs. A, Photosynthesizer (autotroph). The grass under these trees grows actively during the hot, rainy season, and after capturing energy from the sun, the grass converts the energy to molecules of sucrose that are stored as starch. **B,** Herbivore (heterotroph). These wildebeests consume the grass and incorporate some of its stored energy into their bodies. **C,** Carnivore (heterotroph). This lion feeds on other animals, incorporating part of the prey's stored energy into their own bodies.

Electromagnetic energy travels as waves. The waves can be envisioned as ripples on the surface of a still pond after a pebble has been thrown into the water. Electromagnetic "ripples," however, are disturbances in electric and magnetic fields moving through space.

The array of electromagnetic waves coming from the sun vary in length. Their lengths are measured from the crest of one wave to the crest of the next and are expressed in meters or in billionths of meters (nanometers, or nm). The shortest wavelengths are gamma rays; the longest are radio waves.

The full range of electromagnetic radiation in the universe is called the *electromagnetic spectrum* (Figure 6-2). Visible light is only a small part of this spectrum. But it too is made up of many different wavelengths. Remember the last rainbow you saw? Its array of colors was caused by the separation of the various wavelengths of visible light as they passed through tiny droplets of water in the air.

Although electromagnetic energy travels as waves, it also behaves like individual particles—discrete "packets" of energy—called *photons.* The energy content of each photon is inversely proportional to the wavelength of the radiation. Put simply, short-wavelength radiation (such as gamma rays and x-rays) contains photons of a high energy level; long-wavelength radiation (such as microwaves and radio waves) contains photons of a low energy level. Within the range of visible light, then, the shorter wavelengths of violet light carry a greater "energy punch" than the longer wavelengths of red light.

Sunlight can be described both as waves, or disturbances in electric and magnetic fields, and as discrete packets of energy called photons.

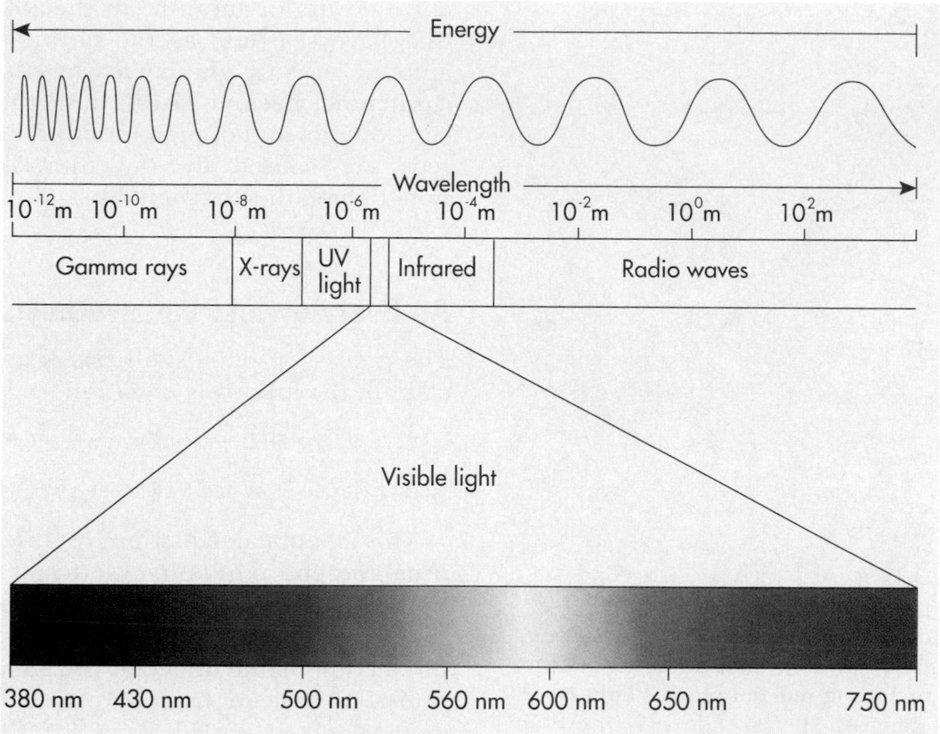

FIGURE 6-2 The electromagnetic spectrum. Light is a form of electromagnetic energy and is conveniently thought of as a wave. The shorter the wavelength of light, the greater the energy. Visible light represents only a small part of the electromagnetic spectrum, that between 380 and 750 nanometers (nm).

Capturing Light Energy in Chemical Bonds

A molecule "captures" the energy of sunlight when a photon of energy boosts an electron of the molecule to a higher energy level than it already occupies. This electron is then said to be "excited." (See Chapter 2 for a description of the energy levels of electrons.) Boosting an electron to a higher energy level requires just the right amount of energy—no more and no less. For example, when you climb a ladder, you must raise your foot just so far to climb a rung, not 1 centimeter more or less. Likewise, specific atoms can absorb only certain photons of light—those that correspond to available energy levels. Therefore a given atom or molecule has a characteristic range, or *absorption spectrum,* of photons it is capable of absorbing depending on the electron energy levels that are available in it.

Molecules that absorb some visible wavelengths and transmit (allow them to pass through) or reflect others are called **pigments.** Pigments are responsible for all the colors you see. For example, a car looks red because it is painted with a pigment that absorbs variable amounts of different wavelengths, including red, but reflects light enriched in red wavelengths. These red wavelengths bounce off the pigment and are reflected back to your eyes. Pigments located in special cells within your eyes then absorb the photons of this red light. The resulting electron excitations ultimately generate nerve impulses that are sent to the brain. The brain interprets these impulses as red (Figure 6-3). Likewise, white objects absorb no light and reflect all the wavelengths of visible light to your eye. A black object absorbs all wavelengths, reflecting none.

autotrophs (aw toe trofes) organisms that produce their own food by photosynthesis or chemosynthesis.
heterotrophs (het ur oh trofes) organisms that cannot produce their own food. Heterotrophs must consume other organisms and, ultimately, the food produced by autotrophs.
pigments molecules that absorb some visible wavelengths of light and transmit or reflect others.

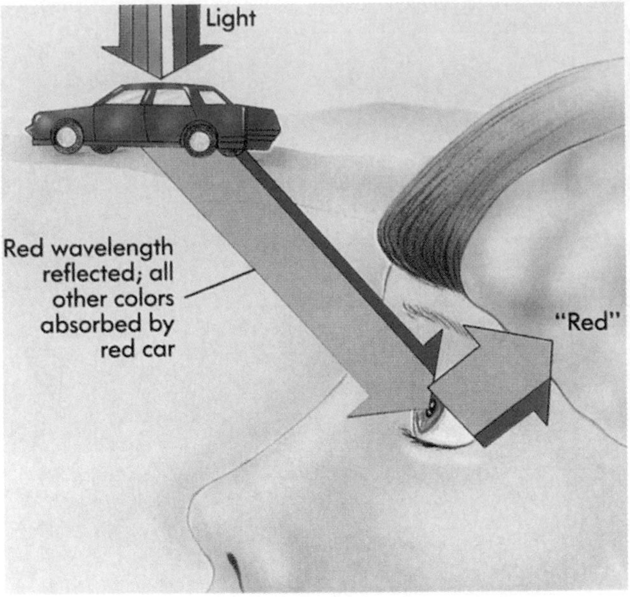

FIGURE 6-3 How you see color. A red car is painted with a pigment that absorbs variable amounts of different wavelengths, including red, but reflects light enriched in red wavelengths. Special cells in your eyes absorb the red light. The brain then interprets impulses sent to these cells as red.

> *A pigment is a molecule that absorbs some wavelengths of light and transmits or reflects others. The wavelengths absorbed by a particular pigment depend on the energy levels available in the molecule to which light-excited electrons can be boosted.*

Organisms have evolved a variety of pigments. Two groups of pigments that are important to the process of photosynthesis are the **carotenoids** and the **chlorophylls.** Carotenoids absorb photons of green, blue, and violet wavelengths and reflect red, yellow, and orange. Chlorophylls absorb photons of violet-blue and red wavelengths and reflect green and yellow (Figure 6-4). Chlorophyll is used as the primary light gatherer in all plants and algae and in certain photosynthetic bacteria—the cyanobacteria. The carotenoids are important, however, because they absorb wavelengths of light that chlorophyll cannot, and then pass the energy to chlorophyll, thereby increasing the spectrum of light that can be absorbed. Chlorophyll masks the presence of the carotenoids, which become visible only when chlorophyll breaks down. This breakdown occurs in autumn (in the areas of the world that have seasons) as the days shorten and the amount of sunlight decreases. As a result, the carotenoids become visible, providing the magnificent yellows and oranges of fall foliage. The bright reds are caused by the unmasking of other non-light-gathering pigments.

An Overview of Photosynthesis

The process of photosynthesis is described chemically in the following equation:

$$6CO_2 + 6H_2O + \text{Light energy} \longrightarrow C_6H_{12}O_6 + 6O_2$$

$$\text{Carbon dioxide} + \text{Water} + \text{Light energy} \longrightarrow \text{Glucose} + \text{Oxygen}$$

This chemical equation represents the beginning and end points in a series of chemical reactions and merely summarizes complex events. It shows that the net effect of photosynthesis is the production of one molecule of glucose and six molecules of oxygen from six molecules of carbon dioxide and six molecules of water in the presence of sunlight. (The main products of photosynthesis are actually the carbohydrates sucrose and starch, both of which contain glucose. Glucose is used in our discussion to show the relationship between photosynthesis and respiration.)

A more accurate summary of the chemical reactions of photosynthesis is represented by the following equation:

$$6CO_2 + 12H_2O + \text{Light energy} \longrightarrow C_6H_{12}O_6 + 6H_2O + 6O_2$$

Here, water appears on both sides of the equation—as a substrate and as a product. Writing the equation in this way points out that the water molecules on the left side of the equation and those on the right side are *not* the same water molecules. The six molecules of oxygen (12 atoms) produced by photosynthesis are derived from those in the water molecules on the left side of the equation. Plants, algae, and the cyanobacteria produce oxygen in this way. In addition, the water molecules on the right side of the equation derive their oxygen atoms from the molecules of carbon dioxide. Glucose (an organic molecule) is constructed from the carbon and oxygen in carbon dioxide (an inorganic molecule) and the hydrogen in water. During the process of photosynthesis, inorganic molecules are broken apart and their atoms are reshuffled and put back together again to form organic molecules—molecules of living things. Photosynthesis, then, is a process during which the living and nonliving worlds meet.

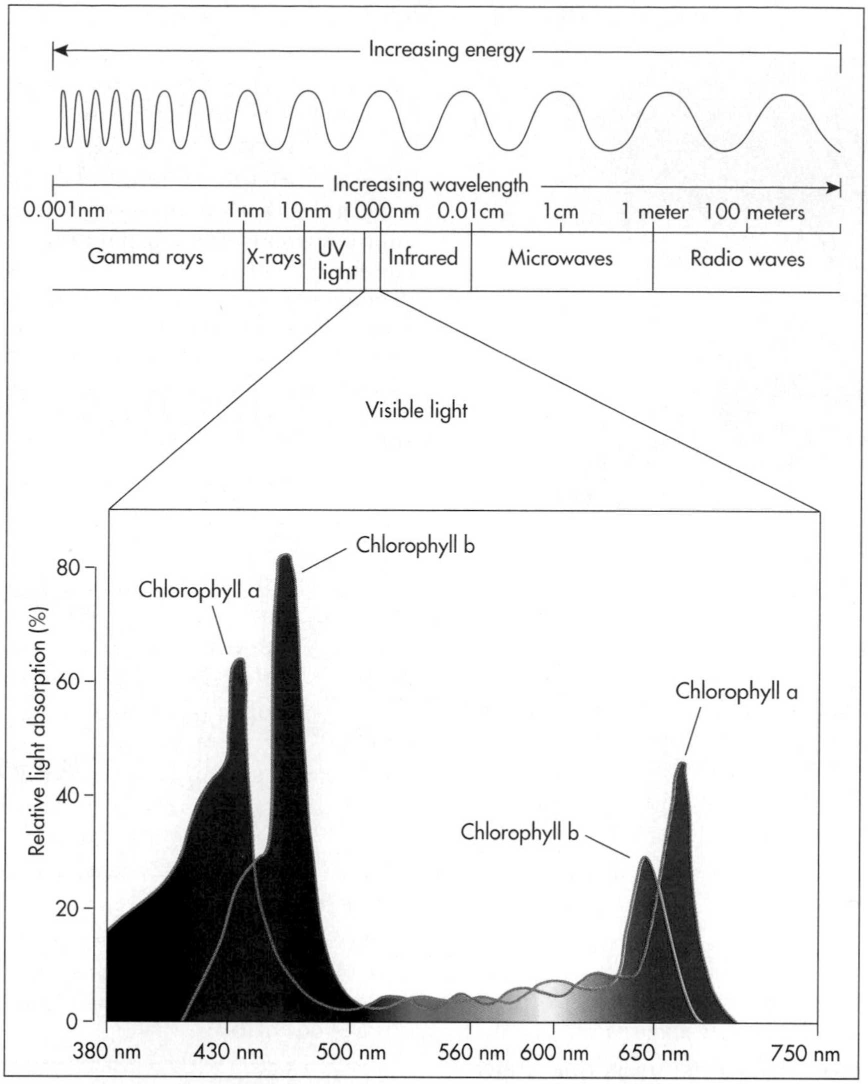

FIGURE 6-4 Absorption spectra for chlorophylls. Chlorophylls, such as chlorophyll *a* and chlorophyll *b*, absorb predominantly violet-blue and red light in two narrow bands of the spectrum.

Photosynthesis, an energy-requiring series of reactions, produces carbohydrates from carbon dioxide and water. In addition, plants, algae, and cyanobacteria release oxygen as a byproduct of photosynthesis.

The complex events of photosynthesis involve two sets of chemical reactions (Figure 6-5). During the first set of reactions, adenosine triphosphate (ATP) and the reduced form of nicotinamide adenine dinucleotide phosphate (NADPH) are formed using energy captured from sunlight. As you learned in Chapter 4, ATP is the primary molecule used by cells to capture and carry energy and to supply energy at a moment's notice. NADPH is an electron carrier. It behaves much like the NADH of cellular respiration (see Chapter 5). Its oxidized form, $NADP^+$, can accept two electrons and one proton (H^+) to become reduced to NADPH. The electrons that $NADP^+$ accepts contain a great deal of energy, thus producing molecules that carry en ergy in the form of energized electrons. The rea tions of photosynthesis that produce ATP and NADPH are called the **light-dependent reactions** because they take place only in the presence of light.

carotenoids (kuh **rot** uh noids) pigments that absorb photons of green, blue, and violet wavelengths and reflect red, yellow, and orange; they are second only to the chlorophylls in importance in photosynthesis.

chlorophylls (**klor** uh fils) pigments that absorb photons of violet-blue and red wavelengths and reflect green and yellow; chlorophylls are the primary light gatherer in all plants and algae and in almost all photosynthetic bacteria.

light-dependent reactions the reactions of photosynthesis that produce ATP and NADPH, which can occur only in the presence of light.

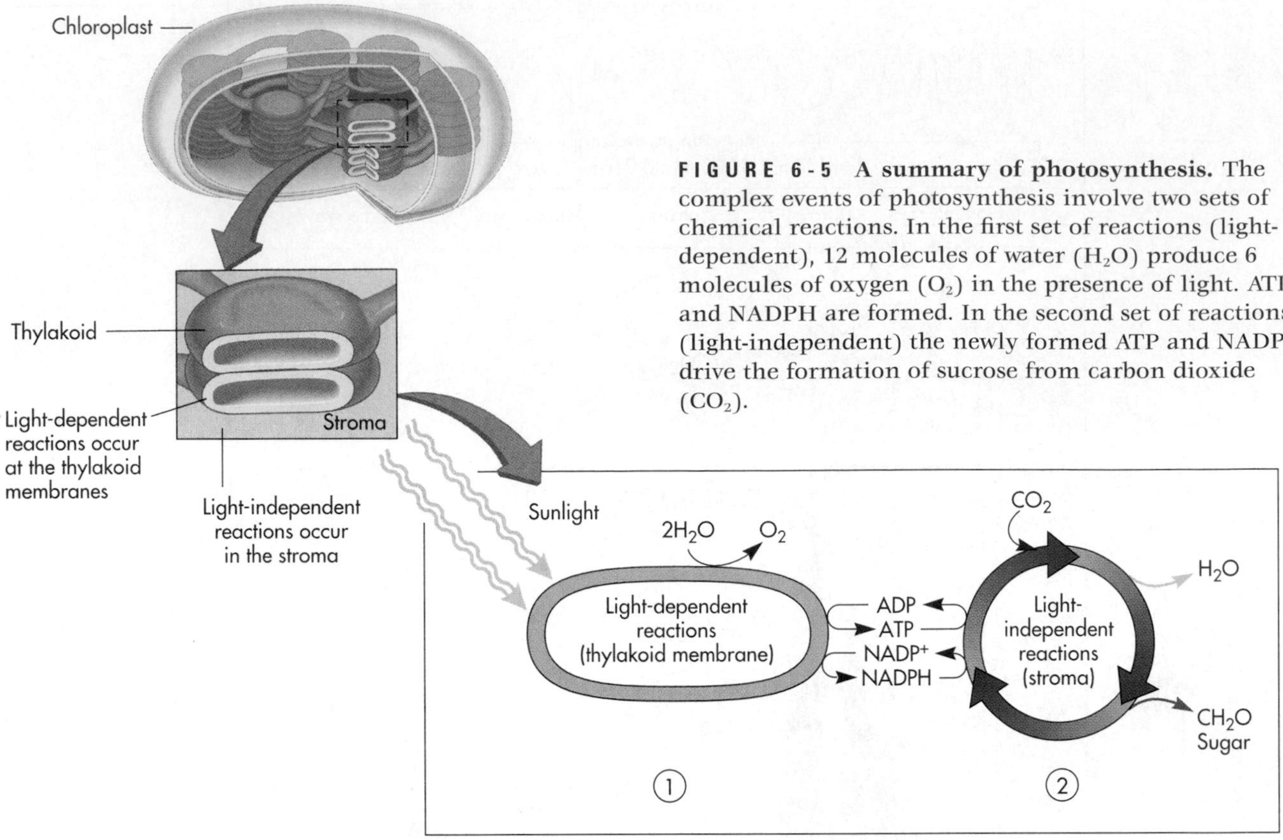

FIGURE 6-5 A summary of photosynthesis. The complex events of photosynthesis involve two sets of chemical reactions. In the first set of reactions (light-dependent), 12 molecules of water (H_2O) produce 6 molecules of oxygen (O_2) in the presence of light. ATP and NADPH are formed. In the second set of reactions (light-independent) the newly formed ATP and NADPH drive the formation of sucrose from carbon dioxide (CO_2).

A second series of reactions uses this newly formed ATP and NADPH to provide energy for the formation of sucrose from carbon dioxide. These reactions are called the **light-independent reactions** because there is no direct involvement of light in the reactions.

The reactions of photosynthesis take place on *photosynthetic membranes.* In photosynthetic bacteria these membranes are infoldings of the cell membrane. In plants, algae, and the photosynthetic single-celled protists called *euglenoids,* all the reactions of photosynthesis are carried out within cell organelles known as **chloroplasts** (Figure 6-6). Chloroplasts, which were probably derived evolutionarily from photosynthetic bacteria (the cyanobacteria), have a system of internal membranes. These membranes are organized into flattened sacs called **thylakoids.** Stacks of thylakoids are referred to as **grana.** Each thylakoid is a closed compartment, a fact that plays an important role in the formation of ATP. Surrounding the thylakoids is a fluid called the **stroma.** It contains the enzymes of the light-independent reactions.

> *Photosynthesis takes place in the chloroplasts of photosynthetic eukaryotic cells. It involves two series of reactions: the synthesis of ATP and NADPH, and the use of ATP and NADPH to produce sucrose.*

Light-Dependent Reactions: Making ATP and NADPH

The light-dependent reactions of photosynthesis perform the important job of changing the energy in sunlight into usable chemical energy: ATP and NADPH. These molecules are used in the light-independent reactions to make sucrose from carbon dioxide. The first event leading to ATP and NADPH production is the capturing of a photon of light by a pigment.

How Photosystems Capture Light

In all but the most primitive bacteria, light is captured by individual networks of chlorophylls, carotenoids, and other pigment molecules. These

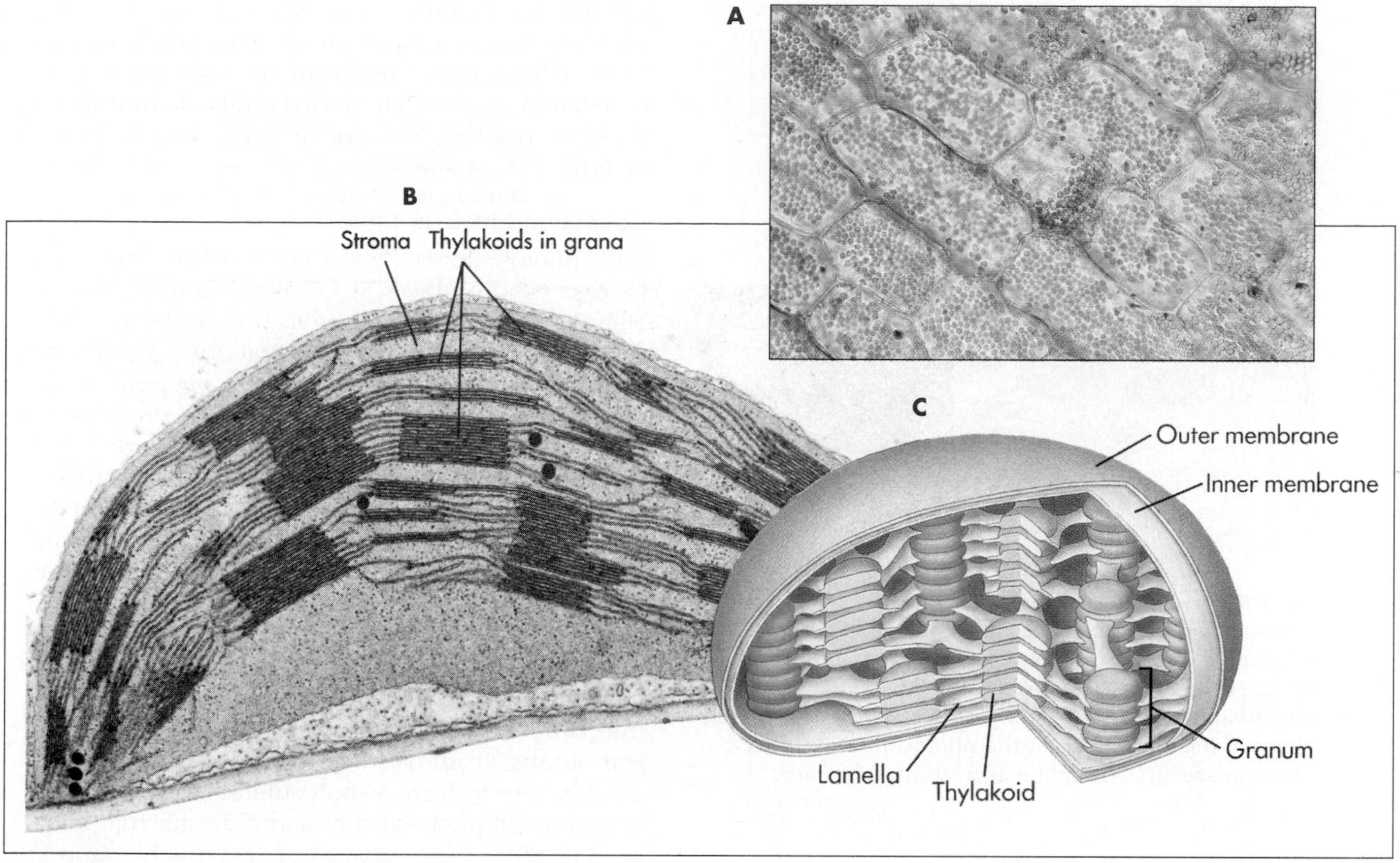

FIGURE 6-6 Structure of a chloroplast. A, Chloroplast in a living plant cell. **B,** An electron micrograph (*2700x*) of a chloroplast showing the photosynthetic membranes or thylakoids; these thylakoids are stacked into structures called grana (singular, granum). The fluid surrounding the grana is called the stroma. **C,** An illustration of a chloroplast with the double membrane cut away to show the interior of this organelle.

networks, located in the thylakoid membranes of the chloroplasts, are called *photocenters.* The pigment molecules within the photocenters are arranged in such a way that they act like sensitive "antennae." These antennae capture and funnel photon energy to special molecules of chlorophyll called **chlorophyll *a*.** The variety of pigments in the photocenters allows a plant to capture energy from a range of wavelengths of light and use it to power photosynthesis.

When light of the proper wavelength (photon) strikes any pigment molecule of the photocenter, one of its electrons is boosted to a higher energy level. This pigment molecule passes the energy to a neighboring molecule within the photocenter. The energy is passed from one pigment molecule to the next, until it reaches the molecule of chlorophyll *a* that will participate in photosynthesis. This molecule is called *P700* or *P680.* The *P* stands for pigment. The number refers to the wavelength at

light-independent reactions the reactions of photosynthesis that use ATP and NADPH to provide energy for the formation of sucrose from carbon dioxide; to occur, these reactions do not require light.

chloroplasts (**klor** oh plasts) energy-producing organelles found in the cells of plants and algae; the sites of photosynthesis in plants.

thylakoids (**thigh** luh koids) flattened, saclike membranes within chloroplasts. These membranes contain the enzymes necessary for the light-dependent reactions.

grana (**gra** nuh) stacks of thylakoid membranes within chloroplasts.

stroma (**stroh** muh) within a chloroplast, a fluid that surrounds the thylakoids and contains the enzymes of the light-independent reactions.

chlorophyll *a* the name given to special molecules of chlorophyll that absorb wavelengths of light in the 680 to 700-nanometer range, in the far-red portion of the light spectrum.

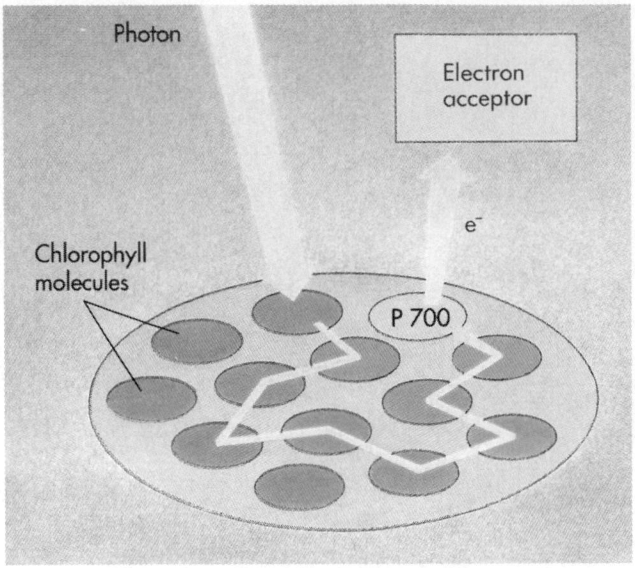

FIGURE 6-7 How photocenters work. When light of a proper wavelength strikes any chlorophyll molecule within a photocenter, the photon's energy is absorbed by it. The energy passes from one chlorophyll molecule to another until it encounters P700 or P680, which channel the energy out of the photocenter to the first electron acceptor in an electron transport chain.

which this molecule of chlorophyll *a* absorbs light. Both P700 and P680 absorb wavelengths in the far-red portion of the light spectrum (Figure 6-7).

A simple analogy to the form of energy transfer in the photocenter is the initial break in a game of pool. If the cue ball squarely hits the point of the triangular array of 15 pool balls, the two balls at the far corners of the triangle fly off. None of the central balls move at all. The energy is transferred through the central balls to the most distant ones. In a similar way, the pigment molecules of the photocenter channel energy to a molecule of chlorophyll *a* in the form of excited electrons.

> *A photocenter is an array of pigment molecules within the thylakoid membranes of a chloroplast. It acts like a light antenna, capturing and directing photon energy toward a single molecule of chlorophyll* a *that will participate in photosynthesis.*

Plants and algae have two different types of photosystems that play a role in photosynthesis. These photosystems are named after the order in which they evolved in photosynthetic organisms: **photosystem I** and **photosystem II**. In photosystem I, energy is transferred to a molecule of chlorophyll *a*, P700 (see Figure 6-7). In photosystem II, energy is transferred to a molecule of chlorophyll *a* called *P680,* whose name refers to its absorption peak of 680 nanometers. These two kinds of chlorophyll *a* act as reaction centers in each type of photosystem.

Noncyclic Electron Flow

Both photosystems absorb light at the same time (Figure 6-8). However, it is easiest to first describe what happens in photosystem II to understand the events of the light-dependent reactions. When four photons of light are absorbed by pigment molecules in photosystem II, the energy is funneled to P680. This energy causes four electrons to be ejected from the P680 molecule. The electrons are captured sequentially by an electron acceptor located within the thylakoid membrane. However, an electron "hole" is now left in the P680 molecule. The P680 pulls four electrons from two molecules of water (H_2O), normally a hard task to accomplish. The P680 strongly attracts these electrons, however, causing each molecule of water to be split into two hydrogen ions and an oxygen atom. Oxygen atoms from two split molecules of water quickly join to form a molecule of oxygen (O_2)—a byproduct of photosynthesis and the gas that helps keep us alive. (This process occurs one electron at a time; only a single electron is or can be energized in a P680 [or P700] molecule. Only a single electron can be ejected, and the electron hole that is formed must be filled before another electron can be energized and ejected.)

Meanwhile, the electron acceptor passes the four energized electrons from the P680 to the next in a series of electron acceptors. This series of electron acceptors or carrier proteins is called an *electron transport chain* as in cellular respiration (see Chapter 5). Here, electrons are passed from one carrier protein to another. As each of these carriers accepts electrons in turn, the electrons fall to lower energy levels, releasing energy in the process. The energy released from the electrons being passed along the chain pumps the protons (H^+) from the stroma through the thylakoid membrane and into the interior of the thylakoid compartment (not shown in Figure 6-8). As a high concentration of protons builds up within this thylakoid space, the protons begin to diffuse back across the membrane through the membrane protein ATP synthase. As the protons travel outward through these channels, enzymes couple the energy from the movement of the protons to the reaction of bonding inorganic phosphate (P_i) to ADP, forming ATP.

When the electrons finish their journey through the electron transport chain, they are then accepted

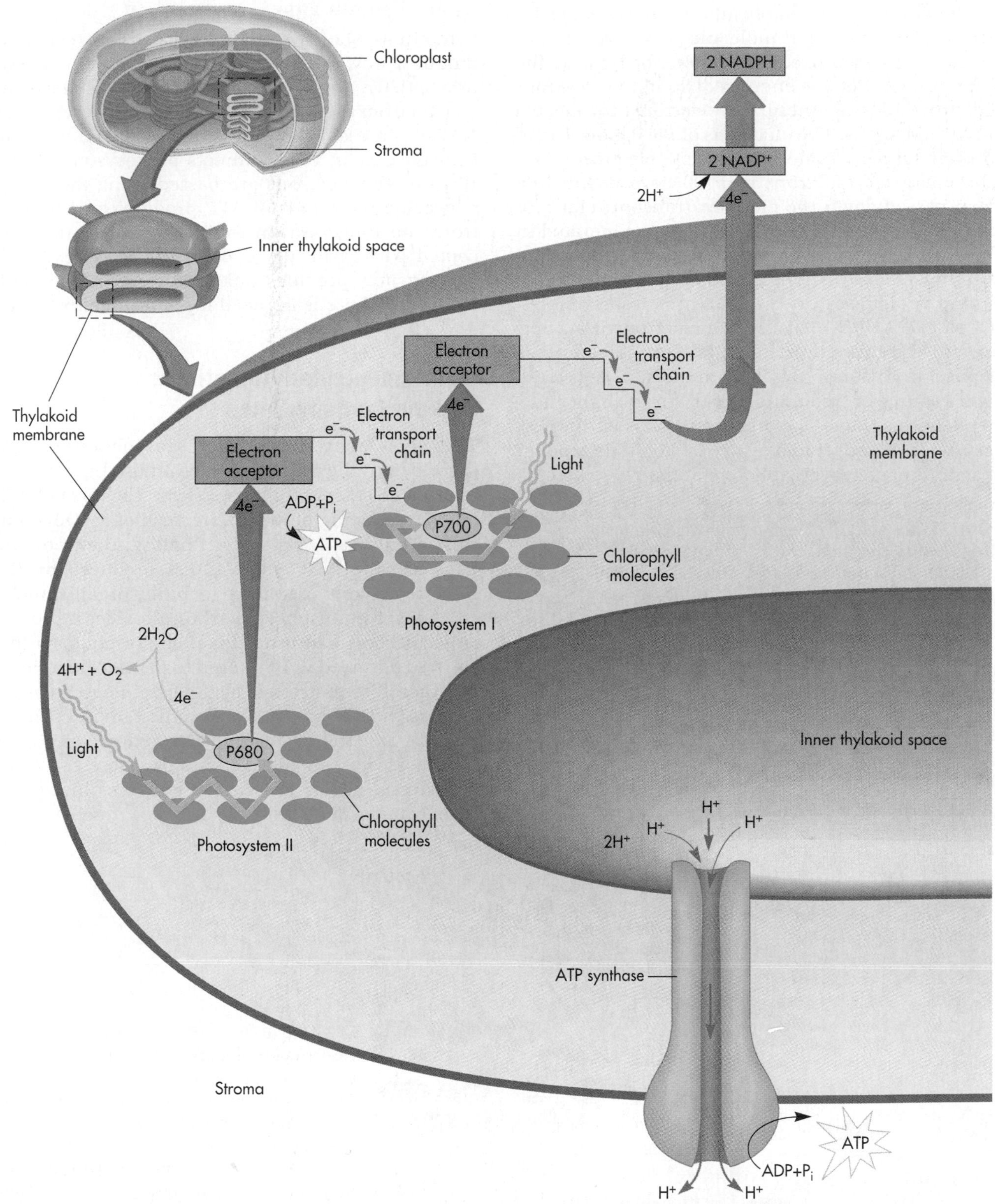

FIGURE 6-8 Light-dependent reactions: noncylic electron flow. Noncyclic electron flow harvests energy in a process that involves photosystems I and II working together to produce NADPH and ATP. These two products are used as the energy source in the light-independent reactions to power the generation of sucrose from carbon dioxide.

by the chlorophyll *a* molecule of photosystem I—P700. This chlorophyll molecule, as you recall, has its electrons energized by photons of light at the same time as P680 is energized. So it, too, develops electron holes as energized electrons are ejected from their shells. The electrons of P680, now devoid of their "boost" of energy, fill these electron holes. The energized electrons of P700 have meanwhile been passed down the electron transport chain of photosystem I. ATP molecules are not generated in this transport chain. Instead, the last acceptors are two molecules of $NADP^+$, each of which picks up a proton in addition to two electrons to become reduced to NADPH. Because the electrons it accepts are still highly energized, NADPH serves as a storage depot for this energy. This molecule is considered a source of "reducing power" for the light-independent reactions. The processes that occur in photosystems I and II make up the light-dependent reactions of eukaryotic photocenters.

Plants and algae use a two-stage photosystem to carry out the light-dependent reactions of photosynthesis. During noncyclic electron flow, energized electrons ejected from photosystem II are passed down an electron transport chain, triggering events of ATP production. The electron hole in photosystem II is filled by electrons from the breakdown of water, a process that releases oxygen as a byproduct. Energized electrons ejected from photosystem I, along with hydrogen ions, reduce $NADP^+$ to NADPH.

Cyclic Electron Flow

Sometimes photosystem I works on its own in a modified way, independent of photosystem II (Figure 6-9). In this situation, the excited electrons ejected from the P700 molecule are not accepted by $NADP^+$. Instead, they are shunted to the electron transport chain that connects photosystems I and II. Here, the electrons are passed along the chain, triggering the events of ATP production. The electrons, no longer in an excited state, are then accepted back by P700 to complete a cycle. *Cyclic electron flow* provides eukaryotic organisms with ATP when there is no need for additional NADPH.

Light-Independent Reactions: Making Carbohydrates

The light-dependent reactions of photosynthesis described in the preceding section use light energy to produce (1) metabolic energy in the form of ATP and (2) reducing power in the form of NADPH. But this is only half the story. Photosynthetic organisms use the ATP and NADPH produced by the light-dependent reactions to build organic molecules from atmospheric carbon dioxide, a process called **carbon fixation.** This phase of photosynthesis does not need light energy to drive its reactions. It is therefore referred to as the *light-independent reactions*. In addition, it is called the **Calvin cycle** after the researcher who discovered its steps. In plants, algae, and the euglenoids, these reactions are carried out by a series of enzymes located in the stroma of the chloroplast.

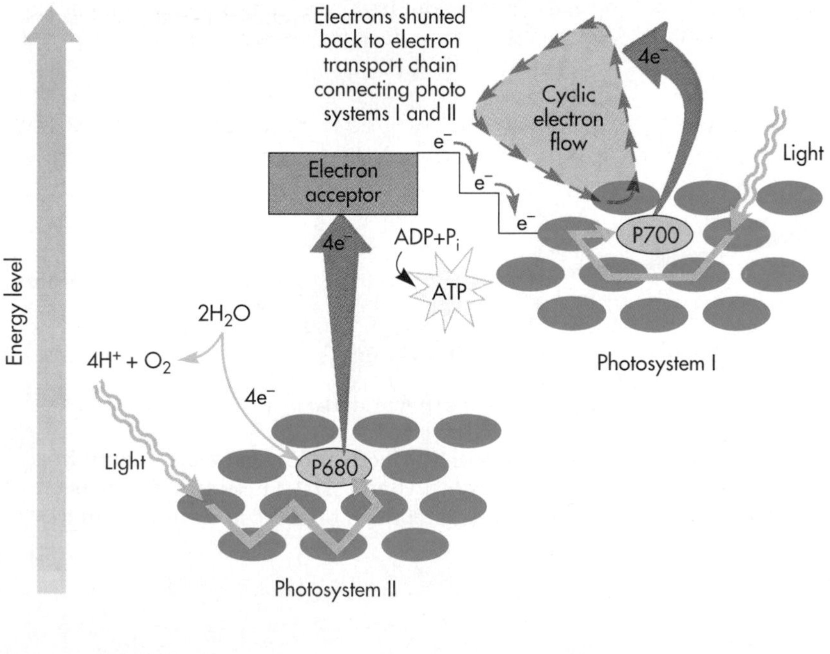

FIGURE 6-9 Light-dependent reactions: cyclic electron flow. Cyclic electron flow harvests energy in a process that involves photosystem I working independently of photosystem II. This method of generating energy produces ATP, but it does not directly produce NADPH as does noncyclic electron flow.

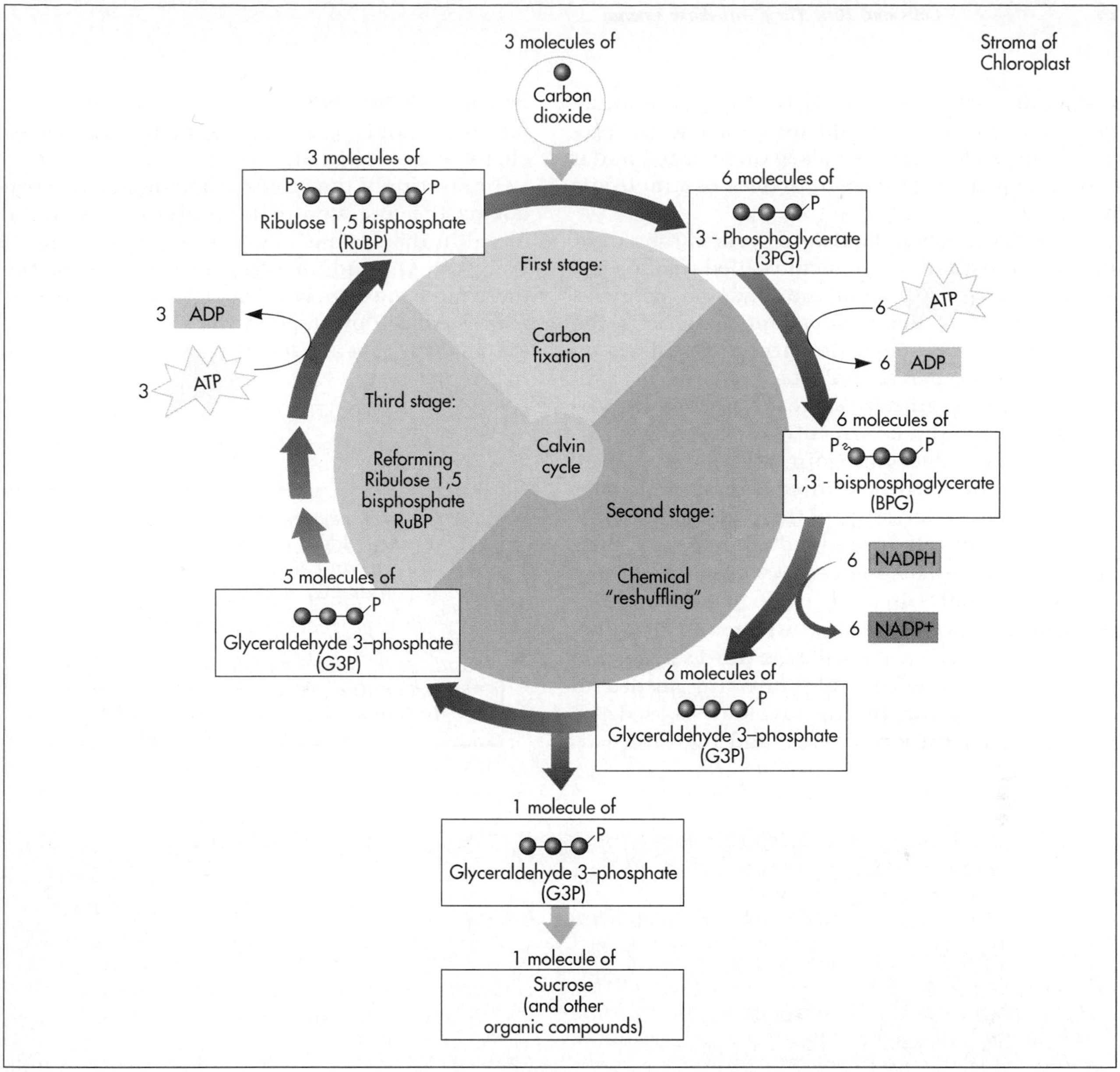

FIGURE 6-10 **The Calvin cycle.** For every three molecules of carbon dioxide that enter the cycle, one molecule of the three-carbon compound glyceraldehyde 3-phosphate is produced. Notice that the process requires energy, supplied as ATP and NADPH. ATP and NADPH are generated by the light-dependent reactions.

The Calvin cycle has three stages that produce one three-carbon sugar, *glyceraldehyde 3-phosphate,* from three molecules of carbon dioxide (Figure 6-10). Two molecules of this sugar are used to manufacture one sucrose molecule (and other carbohydrates). Referring back to the equation of photosynthesis, it takes six molecules of carbon dioxide to produce one molecule of sucrose (shown as glu-

photosystem I the first photosynthetic pathway to evolve; in it, energy is transferred to a molecule of chlorophyll *a* called *P700,* which then acts as the photosystem's reaction center. Photosystem I, along with hydrogen ions, reduces $NADP^+$ to NADPH.

photosystem II a photosynthetic pathway in which energy is transferred to a molecule of chlorophyll *a* called *P680,* which then acts as the photosystem's reaction center. Energized electrons ejected from photosystem II are passed down an electron transport chain, triggering events of ATP production.

carbon fixation a process by which organisms use the ATP and NADPH produced by light-dependent photosynthetic reactions to build organic molecules from atmospheric carbon dioxide.

Calvin cycle another name for the light-independent reactions that take place during photosynthesis.

cose in the equation on p. 120). For one molecule of glyceraldehyde 3-phosphate (G3P) to be produced, the cycle must take place three times, and it must take place six times to produce one molecule of sucrose.

In the first stage of the Calvin cycle, carbon dioxide joins a five-carbon molecule called *ribulose 1, 5 bisphosphate (RuBP)* from the "end" of the cycle. This process is called *carbon fixation.* That is, the inorganic carbon molecule carbon dioxide is attached to a biological molecule, making it available for use to synthesize sucrose and complex carbohydrates. For three turns of the cycle, three carbon dioxide molecules join with three RuBP molecules and form six three-carbon molecules (3PG). In the second stage of the cycle, these molecules are "chemically reshuffled" to produce glyceraldehyde 3-phosphate molecules. Only one of these six G3P molecules is used in the manufacture of carbohydrates. The other five are used to resynthesize three molecules of RuBP (the third stage of the cycle). These molecules of RuBP are needed to keep the cycle going. In this way, RuBP is used as a "handle" to fix carbon dioxide, and the three-carbon molecules produced are used to build sugars for energy and storage and to reconstruct or recycle molecules of RuBP.

Interestingly, the three-carbon molecules formed during the first step of the Calvin cycle are also found in the process of glycolysis (see Chapter 5). Using the ATP and NADPH from the light reactions, these molecules are bonded together and are chemically reduced by steps in the second stage of the Calvin cycle that are actually a reversal of the steps in glycolysis.

Three main events take place during the light-independent reactions of photosynthesis: (1) carbon is fixed with the attachment of carbon dioxide to an organic molecule, (2) sucrose and other carbohydrates are produced from molecules of carbon dioxide, and (3) the organic molecule that attaches to carbon dioxide is re-formed to begin the cycle again.

JUST WONDERING . . .

How come the weeds and crabgrass in my lawn grow during the hot summer yet the rest of my lawn dies?

The answer to your question lies in the fact that not all plants undergo photosynthesis in the same way. Kentucky bluegrass and similar grasses use C_3 photosynthesis, the type of photosynthesis described in this chapter. It is termed C_3 because molecules of CO_2 are used to form a three-carbon compound, the first stable product of photosynthesis produced in the Calvin cycle (see Figure 6-10). The weeds in your lawn and crabgrass, which thrive during a hot, dry summer, use a C_4 photosynthetic pathway.

During C_4 photosynthesis, CO_2 is incorporated into a four-carbon molecule *before* it enters the Calvin cycle. The four-carbon molecule is then decarboxylated (carbon is dropped off) to form a three-carbon molecule, which then enters into the normal Calvin cycle pathway. Although C_4 photosynthesis seems like a cumbersome way to generate three carbon molecules, it is actually an advantage to plants growing in hot, dry, sunny regions. Why? C_4 plants use water more efficiently and use less energy than C_3 plants in high temperatures because of their unique leaf anatomy and their ability to fix CO_2 by mechanisms and under conditions that C_3 plants cannot. In essence, C_4 plants are able to grow and thrive while C_3 plants wither and die in dry heat and intense sunlight. But not only pesky weeds have this adaptation. Corn and sugarcane, two extremely important crops, are also C_4 plants.

Relationships: Photosynthesis and Cellular Respiration

Chapter 5 told the story of cellular respiration: the breakdown of glucose and the capture of its energy in molecules of ATP. Interestingly, the by-products of these series of reactions are carbon dioxide and water—the substrates of photosynthesis. Conversely, oxygen gas, which is the by-product of eukaryotic photosynthesis, is an essential substrate of cellular respiration. The photosynthetic products, carbohydrates, are the initial substrate for cellular respiration (Figure 6-11).

Within the chloroplast, sunlight drives the production of sucrose, using up carbon dioxide and water while generating oxygen. Oxygen is converted to water in the mitochondria of nonphotosynthetic plant cells (such as the roots) or in photosynthetic cells in the dark by accepting electrons harvested from glucose molecules after the electrons have been used by the electron transport system to drive the synthesis of ATP. Nonphotosynthetic organisms (such as you) require an outside source of carbohydrates and oxygen, and these organisms generate carbon dioxide and water. This "outside source" of carbohydrates is the plant material you eat or the organisms you eat that have eaten plants. Plants store the carbohydrates they do not need for cellular respiration and biosynthesis as starch and complex sugars. We and other organisms consume these starches and sugars when we eat fruits and tubers of plants as well as other plant parts.

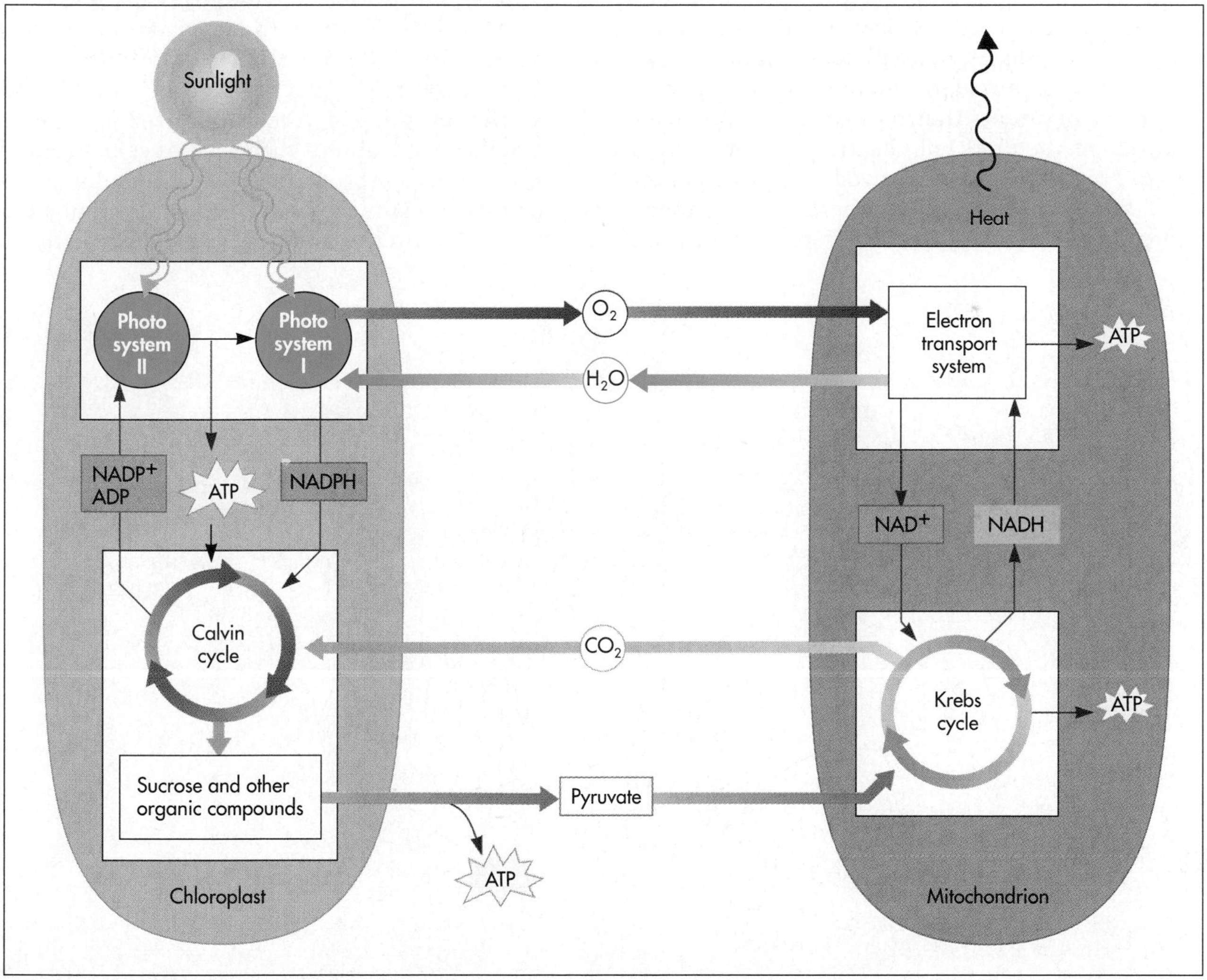

FIGURE 6-11 The metabolic machine. The diagram shows how the processes of photosynthesis and respiration are connected. The products of photosynthesis (sucrose and other organic compounds) are converted to pyruvate and used by the nonphotosynthesizing parts of the plant to obtain energy. Leaf cells (the primary photosynthesizing parts of the plant) use the products of photosynthesis to obtain energy in the dark (when the light-dependent reactions are not taking place).

Summary

- Photosynthesis is a process whereby energy from the sun is captured by living organisms and used to produce molecules of food. Organisms capable of making their own food—plants, algae, euglenoid protists, and some bacteria—are called *autotrophs.* Organisms not capable of making their own food, called *heterotrophs,* depend on the food produced by autotrophs.
- The energy from the sun is electromagnetic radiation and behaves as both waves and discrete packets of energy called *photons.* Visible light is only a part of the full range of the electromagnetic spectrum.
- Special molecules called *pigments* absorb some wavelengths of light and reflect others. When a molecule of a pigment absorbs this energy, one of its electrons is boosted to a higher energy level than it already occupies. Two important types of pigments that play a role in photosynthesis are the chlorophylls and the carotenoids. These pigments are organized into photocenters and serve to capture light energy and change it into chemical energy useful in photosynthesis.
- During photosynthesis, atmospheric carbon dioxide is used to produce molecules of sugar. The events of this synthesis involve two series of chemical reactions. During the first series of reactions, ATP and NADPH molecules are produced and oxygen is given off as a byproduct. During the second series of reactions, ATP and NADPH are used to power the synthesis of sucrose and other carbohydrates.
- The light-dependent reactions of photosynthesis are those that produce ATP and NADPH. These molecules are used during the light-independent reactions to produce sucrose.
- Cellular respiration and photosynthesis are intimately connected biochemical pathways (although they generally do not take place in the same cells). The byproducts of cellular respiration—carbon dioxide and water—are the substrates of photosynthesis. Conversely, the byproduct of photosynthesis—oxygen—and its product of carbohydrates are the substrates of cellular respiration. These major metabolic pathways are the energy crossroads of cells.

Knowledge and Comprehension Questions

1. Compare autotrophs and heterotrophs. Which are you?
2. You hold a prism in front of a window and the incoming sunlight forms a rainbow on the floor. Explain what happened.
3. Fill in the blanks: Sunlight is a form of energy known as ________________, also called ________________. Sunlight travels as ________________ and as discrete packets of energy called ________________.
4. Why does someone's blue sweater look blue?
5. Explain what the following equation means. What process is being described, and why is it important?

$$6CO_2 + 6H_2O + \text{Light energy} \longrightarrow C_6H_{12}O_6 + 6O_2$$

6. In which part of photosynthesis is sugar produced?
7. What are the products of each of the two series of reactions involved in photosynthesis? What is meant by the terms *light-dependent reactions* and *light-independent reactions?*
8. Define *photocenter* and *chlorophyll* a, and explain their significance.
9. Describe the two-stage photosystem used by plants and algae to carry out light reactions.
10. Summarize the main events that take place during the light-independent reactions of photosynthesis.
11. Some scientists think that a large nuclear explosion could blanket the Earth with a thick haze of atmospheric dust lasting for years. Explain how this would affect life on Earth.

Critical Thinking

1. Imagine you place a potted plant in your classroom window after weighing it carefully. Pot and all, it weighs just 3 pounds. Watering it properly, you let the plant grow for a year and then reweigh it. Pot and all, it now weighs $4^1/_2$ pounds. Where precisely did the $1^1/_2$ pounds of extra plant mass come from?
2. The roots of plants are not exposed to sunlight—they are under the ground. How do they manufacture ATP?
3. Study the absorption spectrum for chlorophyll in Figure 6-4. Are all wavelengths equally effective for photosynthesis? Support your answer using evidence gleaned from Figure 6-4.

PART THREE

HUMAN BIOLOGY: THE STRUCTURE AND FUNCTION OF THE BODY

CHAPTER 7

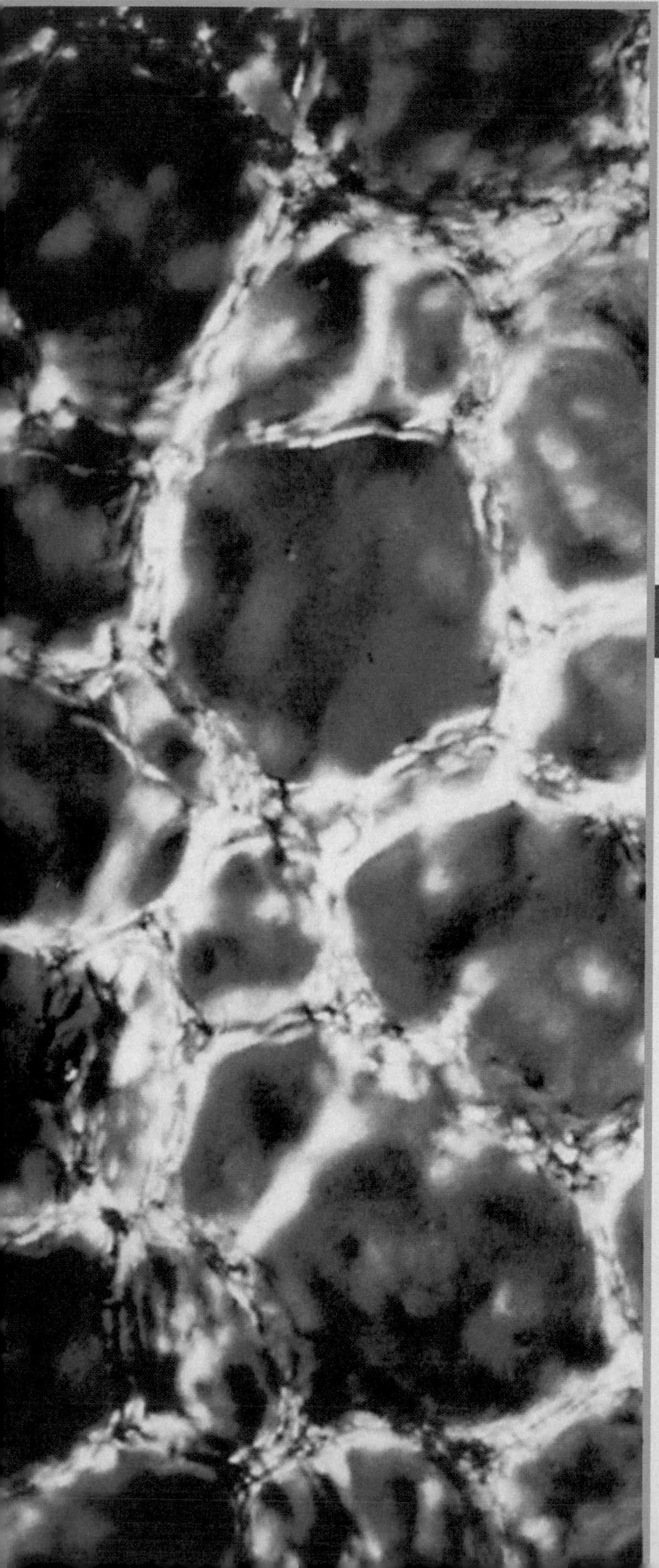

LEVELS OF ORGANIZATION IN THE HUMAN BODY

DID YOU EVER blow soap bubbles through a straw when you were a child or see a bubble-blowing machine at a science museum? The bubbles form one on top of the other, looking much like the pattern in the photo. But although the "bubbles" in the photo are filled with air, they are not bounded by soap. They are enclosed by cells and are filled with the air in your lungs. In fact, these bubbles are not bubbles at all. They are microscopic air sacs that help deliver the oxygen in the air you breathe to your bloodstream.

Although several hundred million air sacs can be found in each of your lungs, your lungs are not made up solely of air sacs. They include many other structures, such as blood vessels and air passageways. Together, these structures and others help the lungs do their job—transporting oxygen to the bloodstream and ridding the blood of the waste gas carbon dioxide. A complex structure such as the lung is called an *organ* and forms only one of the levels of organization of the human body.

KEY CONCEPTS

- There are several levels of organization in the human body: atoms and molecules are organized into cells, cells are organized into tissues, tissues are organized into organs, and organs are organized into organ systems.
- Four basic types of tissues make up the organs: epithelial, connective, muscle, and nervous.
- The human body can be divided into 11 different organ systems: digestive, respiratory, circulatory, immune, urinary, nervous, skeletal, integumentary, muscular, endocrine, and reproductive.
- The organ systems interact with one another to maintain a stable internal environment, or "steady state," called *homeostasis.*

OUTLINE

How the Human Body Is Organized

The human body, like the bodies of all other multicellular animals, is made up of many different types of cells. In fact, the human body contains more than 100 different kinds of cells! These cells are not distributed randomly but are organized to form the structures and perform the functions of the human body, just as workers in a factory may be organized to manufacture a product.

The cells of the body are organized into **tissues.** Tissues are groups of similar cells that work together to perform a function. Traditionally, tissues are divided into four basic types based on their function: **epithelial, connective, muscle,** and **nervous** (Figure 7-1). For example, the cells making up the walls of the air sacs of your lungs are a type of epithelial tissue—tissue that covers body surfaces and lines its cavities. All four tissue types are described later in this chapter.

Two or more tissues grouped together to form a structural and functional unit are called **organs.** Your heart is an organ. It contains cardiac muscle tissue wrapped in connective tissue and "wired" with nerves. All of these tissues work together to pump blood through your body. Other examples of organs are the stomach, skin, liver, and eyes.

An **organ system** is a group of organs that function together to carry out the principal activities of the body. For example, the digestive system is composed of individual organs concerned with the breaking up of food (teeth), the passage of food to the stomach (esophagus), the storage and partial digestion of food (stomach), the digestion and absorption of food and the absorption of water (intestine), and the expulsion of solid waste (rectum). The human body contains 11 principal organ systems (Table 7-1), which are discussed in later chapters.

The organ systems are all encased in the human body (Figure 7-2), which is an incredible living machine. The human body has the same general body architecture that all animals with a backbone have (Figure 7-3). It includes a long tube that travels from one end of the body to the other, from mouth

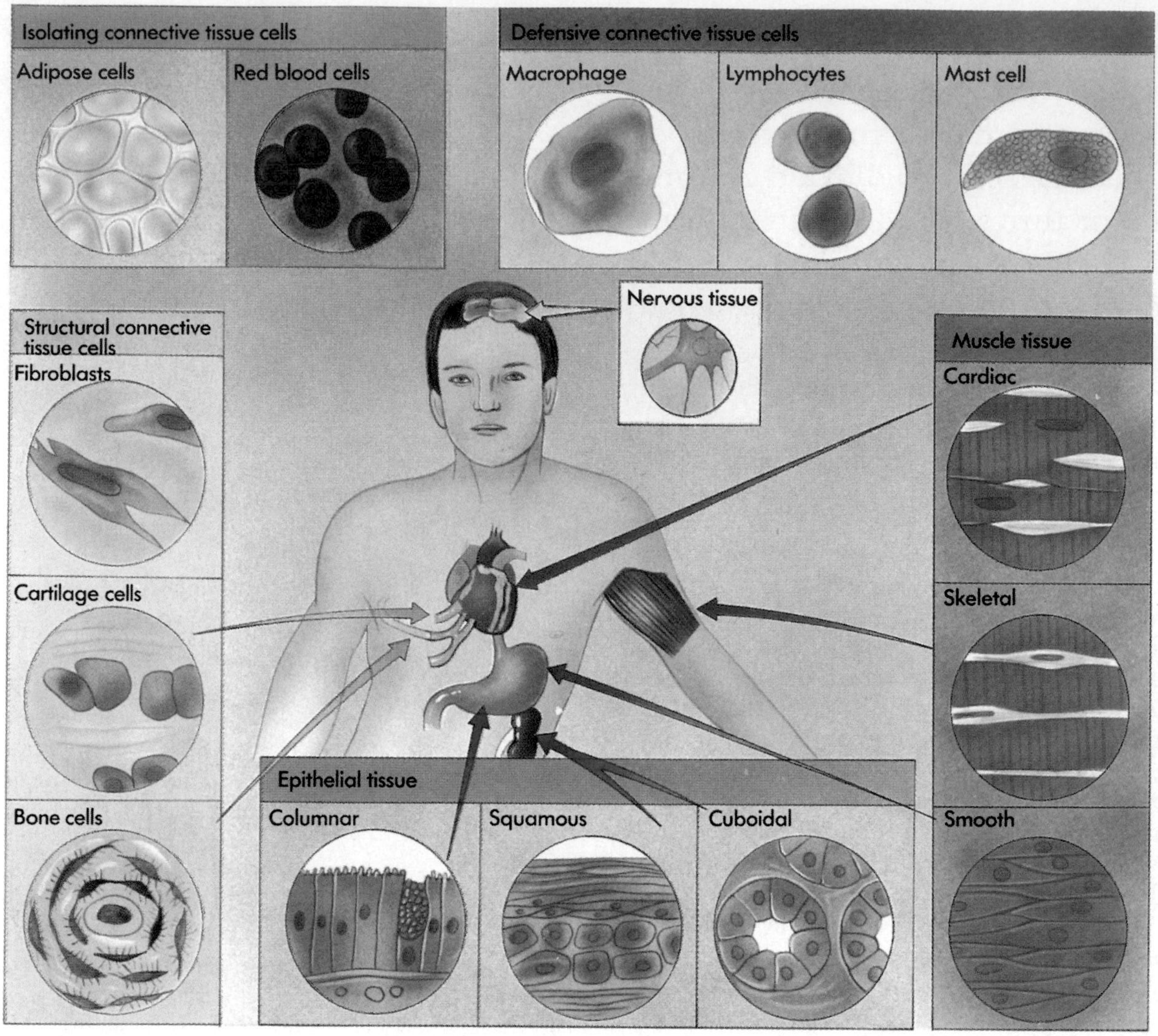

FIGURE 7-1 The basic tissue types and their cells. The tissues in the body are divided into four basic types according to function: epithelial, connective, muscle, and nervous.

to anus. This tube is suspended within an internal body cavity called the **coelom.** In humans the coelom is divided into two main parts: (1) the thoracic cavity, which contains the heart and lungs, and (2) the abdominal cavity, which contains organs such as the stomach, intestines, and liver. The lower portion of the abdominal cavity is often referred to as the *pelvic cavity.* The body is supported by an internal scaffold, or skeleton, made up of bones that grow as the body grows. A bony skull surrounds the brain, which is located in a cavity separate from the coelom, called the *cranial cavity.* In addition, a column of bones called the *vertebral column* forms the backbone, or spine, of the human body. These bones surround a nerve cord, the spinal cord, that relays messages between the brain and other parts of the body.

Tissues

Epithelial Tissue

Epithelial cells guard and protect your body; these cells are really the doors to the inner you. Put simply, epithelial cells cover and line the surfaces of the body, both internal and external. Remember, you can think of your body as a tube, with your skin forming the outside of the tube and your digestive system forming the inside of the tube. Therefore the surfaces of your inner tube, as well as your skin, are made up of epithelial cells. And, because epithelial cells cover the surfaces of the body, they determine which substances enter and which do not.

All epithelial cells, collectively called the *epithelium,* are broadly similar in form and function.

TABLE 7-1

The Major Human Organ System

SYSTEM	FUNCTIONS	COMPONENTS	CHAPTER
Digestive	Breaks down food and absorbs breakdown products	Mouth, esophagus, stomach, intestines, liver, and pancreas	8
Respiratory	Supplies blood with oxygen and rids it of carbon dioxide	Trachea, lungs, and other air passageways	9
Circulatory	Brings nutrients and oxygen to cells and removes waste products	Heart, blood vessels, blood, lymph, and lymph structures	10
Urinary	Removes wastes from the bloodstream	Kidney, bladder, and associated ducts	12
Nervous	Receives and helps body respond to stimuli	Nerves, sense organs, brain, and spinal cord	13–15
Endocrine	Coordinates and regulates body processes and functions	Pituitary, adrenal, thyroid, and other ductless glands	17
Skeletal	Protects the body and provides support for locomotion and movement	Bones, cartilage, and ligaments	16
Muscular	Produces body movement	Skeletal, cardiac, and smooth muscles	16
Integumentary	Covers and protects the body	Skin, hair, nails, and sweat glands	16
Immune	Helps defend the body against infection and disease	Lymphocytes, macrophages, and antibodies	11
Reproductive	Produces sex cells and carries out other reproductive functions	Testes, ovaries, and other associated reproductive structures	21

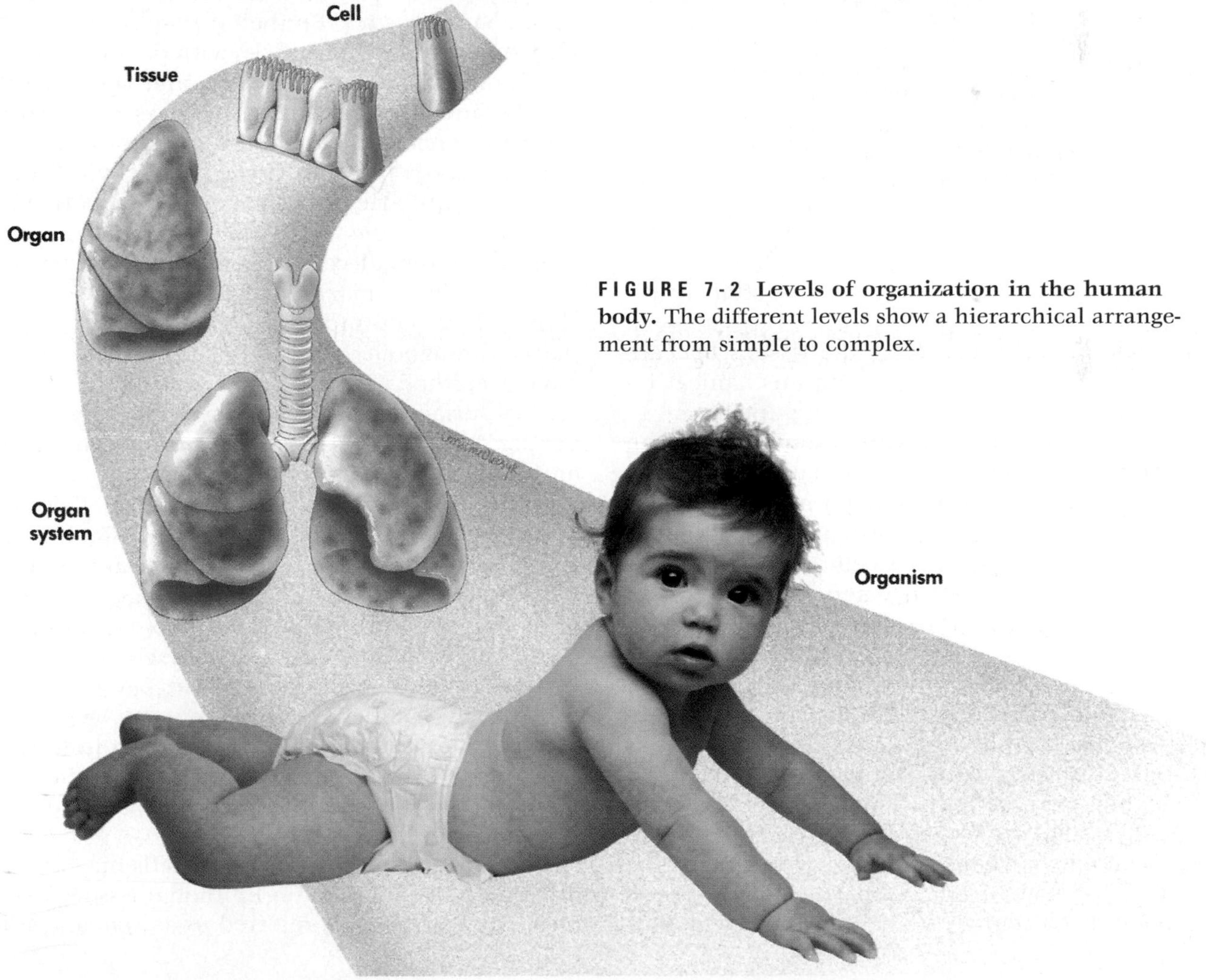

FIGURE 7-2 Levels of organization in the human body. The different levels show a hierarchical arrangement from simple to complex.

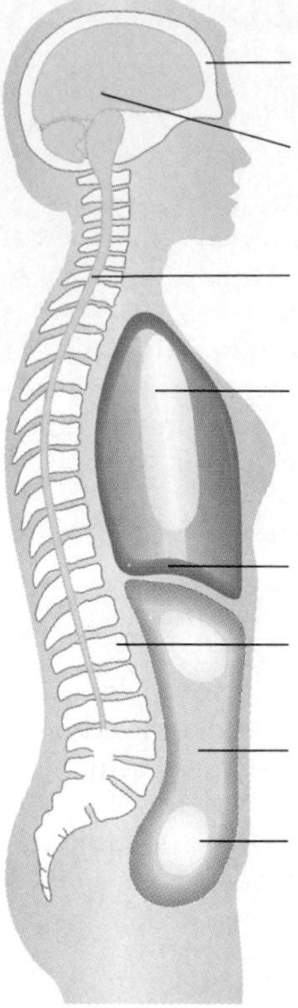

FIGURE 7-3 Architecture of the human body. Humans, like all vertebrates, have a spinal cord and brain enclosed in the vertebral column and skull. In mammals a muscular diaphragm divides the coelom into the thoracic cavity and the abdominal cavity.

The epithelial layers of the body function in six different ways:

1. *Protection.* They protect the tissues beneath them from drying out, sustaining mechanical injury, and being invaded by microorganisms.

2. *Absorption.* They provide a barrier that can help or hinder the movement of materials into the tissue beneath. Because epithelium encases all of the body's surfaces, every substance that enters or leaves the body must cross an epithelial layer.

3. *Sensation.* Many sensory nerves end in epithelial cell layers. The epithelium therefore provides a sensory surface.

4. *Secretion.* Certain epithelial cells are specialized to produce and discharge substances; they are called *glands.* Glands may be single cells, such as the mucus-secreting cells lining the intestine. Most glands, however, are multicellular structures, such as the thyroid gland located in your neck or the pituitary gland that hangs like a tiny pea from the underside of your brain. Even your sweat glands are made up of many cells. Because many glands lie deep within your body and do not cover or line a surface, they may seem inappropriately described as being composed of epithelial tissue. However, glands form during embryological development from infoldings of epithelial cell layers, so they fit in this category.

5. *Excretion.* Specialized epithelial cells in the kidney excrete waste products during the formation of urine. Also, the epithelium forming the air sacs of the lungs excretes the waste gas carbon dioxide.

6. *Surface transport.* Some epithelial cells have hairlike projections called *cilia.* These cilia beat in unison, causing a wavelike movement in the thin film of mucus that bathes the cells' surfaces, sweeping particles along in the process. The cells lining parts of your respiratory passageways, for example, use this technique to keep foreign particles from entering the lungs. Cigarette smoking damages these cilia, resulting in accumulations of mucus in the throat and respiratory passageways that lead to "smoker's cough" and, over time, to serious upper respiratory problems.

Structurally, epithelial tissues share some common characteristics. Epithelial tissues are usually only one or a few cells thick (with the exception of the skin), are packed together and stacked very tightly, and have very few blood vessels running through them. The circulation of nutrients, gases, and wastes in epithelial tissue occurs by diffusion from the capillaries of neighboring tissue. In addition, although the chemical reactions of many types of epithelial cells take place at a very slow rate, many have amazing regenerative powers. Your skin cells, for example, are continually being replaced throughout your lifetime. Your liver, a gland having epithelial tissue on its absorptive and secretory surfaces, can readily regenerate substantial portions of tissue if parts are surgically removed or are damaged by certain diseases.

There are three main shapes of epithelial cells: *squamous, cuboidal,* and *columnar* (Table 7-2). Squamous cells are thin and flat. They are found in places such as the air sacs of the lungs, the lining of blood vessels, and the skin. Cuboidal cells have complex shapes but look like cubes when the tissue is cut at right angles to the surface. These cells are found lining tubules in the kidney and the ducts of glands. Columnar cells look like tiny columns (as their name suggests) when they are viewed from the side. Much of the digestive tract is lined with columnar epithelium.

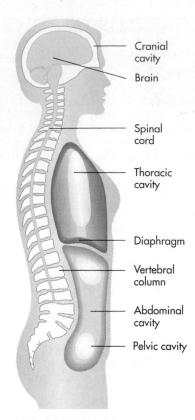

The various types of epithelial cells may also be arranged in various ways. Epithelial tissue that is only one cell thick is referred to as *simple epithe-*

TABLE 7-2

Epithelial Tissue

TISSUE	LOCATION	FUNCTIONS	DESCRIPTION
Simple squamous epithelium 	Lining of blood vessels, air sacs of lungs, kidney tubules, and lining of body cavities	Diffusion, filtration, and passage of materials where little protection is needed	The simplest of all epithelial tissues, simple epithelium consists of a single layer of thin, flat cells.
Cuboidal epithelium 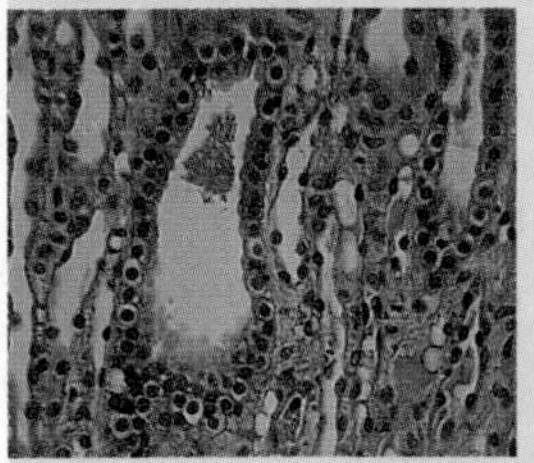	Kidney tubules, glands and their ducts, terminal bronchioles of lungs, and surface of ovaries and retina	Secretion, absorption, and movement of substances	The structure is a single layer of cube-shaped cells. Some have microscopic extensions called microvilli. Some have cilia that protrude from their surfaces.
Columnar epithelium 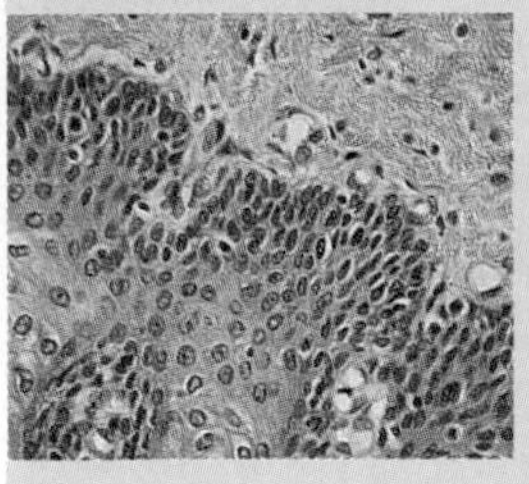	Lining of the digestive and upper part of respiratory tracts, auditory and uterine tubes	Secretion of mucus	The structure is a single layer of tall, narrow cells. Some have microvilli or cilia.

tissues groups of similar cells that work together to perform a function.

epithelial tissue (ep uh **thee** lee uhl) groups of similar cells that cover body surfaces and line body cavities.

connective tissue groups of similar cells that provide a framework for the body, join its tissues, help defend it from foreign invaders, and act as a storage site for specific substances.

muscle tissue groups of similar cells that are capable of contraction.

nervous tissue groups of similar cells that are specialized to conduct electrical impulses, and other supporting cells.

organs two or more tissues grouped together to form a structural and functional unit.

organ system a group of organs that function together to carry out the principal activities of the body.

coelom (**see** lum) an internal body cavity common to all animals with a backbone. In humans the coelom is divided into two parts: the thoracic cavity and the abdominal cavity.

TABLE 7-2

Epithelial Tissue—cont'd

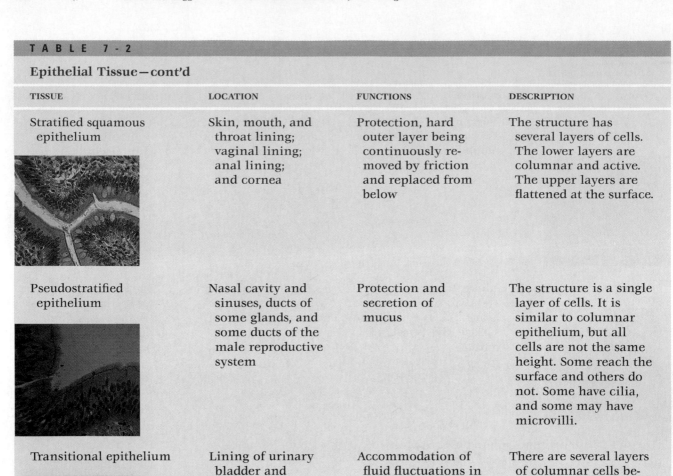

TISSUE	LOCATION	FUNCTIONS	DESCRIPTION
Stratified squamous epithelium 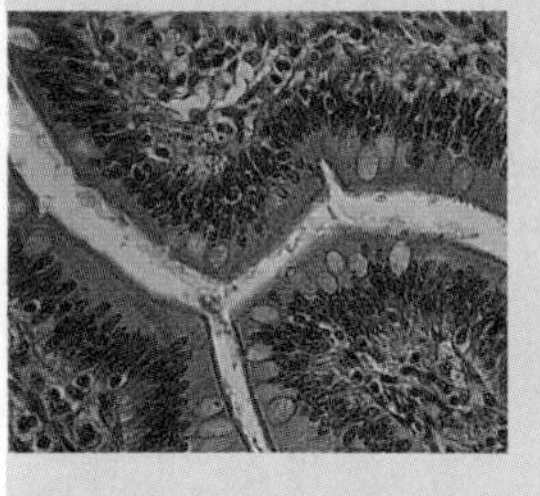	Skin, mouth, and throat lining; vaginal lining; anal lining; and cornea	Protection, hard outer layer being continuously removed by friction and replaced from below	The structure has several layers of cells. The lower layers are columnar and active. The upper layers are flattened at the surface.
Pseudostratified epithelium 	Nasal cavity and sinuses, ducts of some glands, and some ducts of the male reproductive system	Protection and secretion of mucus	The structure is a single layer of cells. It is similar to columnar epithelium, but all cells are not the same height. Some reach the surface and others do not. Some have cilia, and some may have microvilli.
Transitional epithelium 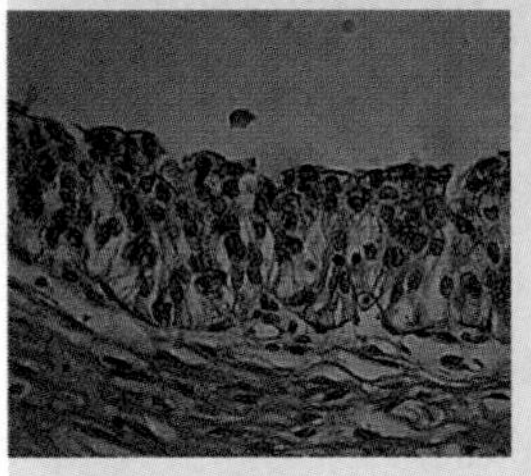	Lining of urinary bladder and ureters	Accommodation of fluid fluctuations in an organ or tube by stretching easily	There are several layers of columnar cells beneath layers of surface cells. Cells are flattened when tissue is is stretched.

lium. It is usually found in areas where substances diffuse through the tissue and where substances are secreted, excreted, or absorbed. The cells lining your blood vessels are simple squamous epithelium, for example, whereas much of your digestive tract is lined with simple columnar epithelium. *Stratified epithelium* is made up of two or more layers. These layers are usually protective. The surface of your skin is stratified squamous epithelium. *Pseudostratified epithelium* only looks as though it is layered. In reality, this tissue is made up of only one layer of cells—some tall, some short. The sides of the top portions of the tall cells rest on top of the shorter cells, looking like two or more layers of cells. Some of your airways are lined with pseudostratified epithelium. *Transitional epithelium* is tissue that can stretch. The cells of transitional epithelium are cubelike; when stretched, the cells appear thin and flat. Transitional epithelium is found in the bladder and ureters of the urinary tract, an organ and tubes (respectively) that accommodate fluid fluctuations.

JUST WONDERING . . .

How come some animals can regenerate certain tissues and body parts and we can't?

The idea of humans regenerating limbs is, of course, science fiction. But, as you suggested, many kinds of animals can regenerate parts. Sponges, for example (see p. 640), are capable of extensively regenerating and replacing lost parts. Sometimes their regeneration results in a new organism and is considered a form of asexual reproduction. This type of asexual reproduction also occurs in echinoderms such as sea stars (see p. 654). When one arm is removed from a sea star, the animal will grow a new arm; however, if the cut arm has a piece of the central disk attached, it too will grow into a new individual. Vertebrates (a subphylum of animals to which humans belong) show some capacity for tissue regeneration. The best examples are the amphibians; in fact, salamanders have been extensively studied to try to answer just the question you raised.

Developmental biologists have been intrigued by the regeneration question for years. They have found that the regenerated limbs of salamanders arise from epidermal tissue that surrounds a mass of rapidly growing cells capable of differentiating into connective tissue, bone, cartilage, vascular tissue, and lymphatic vessels. These cells apparently arise from the stump and eventually grow and develop into a new limb.

But what controls the growth and differentiation of these cells? And what controls their patterning—the ability of the proper type of cell to develop in the appropriate place, forming a fully functional and anatomically correct limb? Understanding the mechanisms by which these processes occur is central to understanding the process of regeneration. Scientists are a long way from having complete answers to these questions. Researchers know that hormones play a role in influencing the normal course of growth during regeneration. But they are also studying the role of one class of regulatory genes called *homeobox genes,* which scientists know control patterns of development. However, the exact role of these genes in pattern formation remains unclear.

Humans are capable of regenerating many of their tissues except mature nerve cells (when their cell bodies are damaged) and muscle cells. Injuries outside of the brain and spinal cord do not involve the cell bodies of nerve cells; thus some regeneration takes place if a wound is not too serious. Most serious injuries of the human body (and some not so serious) result in the development of scar tissue. This tissue is formed by fibroblasts (see p. 142), which divide rapidly in the injured area and secrete large quantities of the protein collagen. This substance "fills in" the area of lost cells . . . certainly not a process nearly as complex or useful as that of salamander limb regeneration!

Epithelial cells cover and line the internal and external surfaces of the body and compose the glands. These cells are of three shapes and are arranged in ways that best suit their functions.

Connective Tissue

The cells of connective tissue provide the body with structural building blocks and potent defenses. In addition, connective tissue joins the other tissues of the body. Connective tissue performs an assortment of important jobs for the body. The varied composition of the different types of connective tissue reflects this diversity. However, connective tissue is generally made up of cells that are usually spaced well apart from one another and are embedded in a nonliving substance called a *matrix.* In fact, connective tissue is made up of a great deal more matrix than cells. This matrix varies in consistency among the different types of connective tissue—from a fluid to a gel to crystals.

The different types of connective tissues and cells are categorized in many different ways. One way to group them is by function: *defensive, structural,* and *isolating connective tissue.* Defensive cells, those that protect the body from attack, float in a matrix of blood plasma. They roam the circulatory system, hunting invading bacteria and foreign substances. An example of this type of cell is the lymphocyte, a special type of white blood cell (Figure 7-4). Structural connective tissue cells, such as bone and cartilage cells, stay in one place, secreting proteins into the empty spaces between them. These proteins provide structural connective tissue with a fibrous matrix, giving it the strength it needs to support the body and provide connections among tissues. Isolating connective tissue cells act as storehouses, accumulating specific substances such as fat and hemoglobin.

> *Connective tissue and its cells provide a framework for the body, join its tissues, help defend it from foreign invaders, and act as storage sites for specific substances.*

Cells That Defend

The three principal defensive cell types are *lymphocytes* (see Figure 7-4), *macrophages,* and *mast cells.* All of these cells are dispersed throughout the body either in the blood or among other tissues.

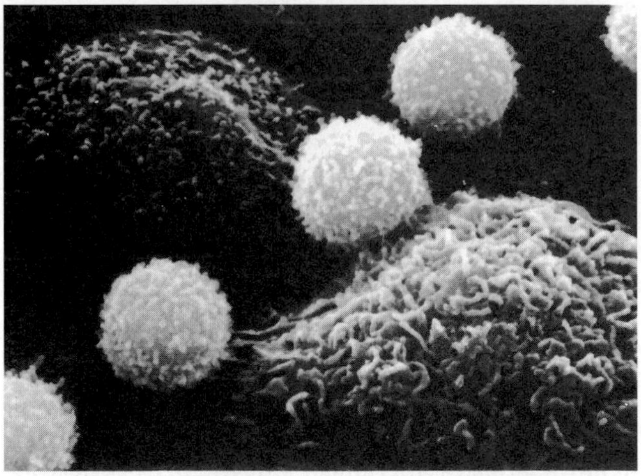

FIGURE 7-4 Lymphocytes and macrophages. The lymphocytes are small and spherical; the macrophages are larger and more irregular in form. Both types of cells play key roles in the body's defense against disease (*1500x*).

Lymphocytes are a type of white blood cell that circulates in the blood or resides in the organs, vessels, and nodes of the lymphatic system. Both these cells and the lymphatic system itself play complex, key roles in the body's defense against infection. Your body has an amazing trillion or so lymphocytes ready to attack foreign cells or viruses that enter the body or to produce specific antibodies that can act against specific substances. (How lymphocytes and antibodies function in the body's defense against disease is described in Chapter 11.)

Macrophages are abundant in the bloodstream and also in the fibrous mesh of many tissues, such as the lungs, spleen, and lymph nodes. They develop, or differentiate, from white blood cells called *monocytes.* Usually macrophages move about freely, but sometimes they stay in one place, attached to fibers. These cells may be thought of as the janitors of the body, cleaning up cellular debris and invading bacteria by a process known as *phagocytosis*—an engulfing and digesting of particles.

Mast cells produce substances that are involved in the body's inflammatory response to physical injury or trauma. One important substance produced by mast cells is histamine, a chemical that causes blood vessels to dilate, or widen. As more blood then flows through the vessels, it brings added oxygen and nutrients and dilutes any toxins, or poisons. The increased blood flow also aids the movement of defensive leukocytes coming to the area. Mast cells, although important in the inflammatory response, also play a role in allergic reactions (see Chapter 11).

> *Protective connective tissue cells defend the body against foreign invaders. Lymphocytes attack foreign cells or viruses that enter the body or produce specific antibodies that can act against specific substances. Macrophages then engulf and digest the invader. Mast cells enlarge the blood vessels in response to trauma, speeding the healing process.*

Cells and Tissues That Shape and Bind

The three principal types of cells found in structural connective tissue are *fibroblasts,* cartilage cells *(chondrocytes),* and bone cells *(osteocytes).* These cells produce substances that cause the tissues of which they are a part to have distinctive characteristics.

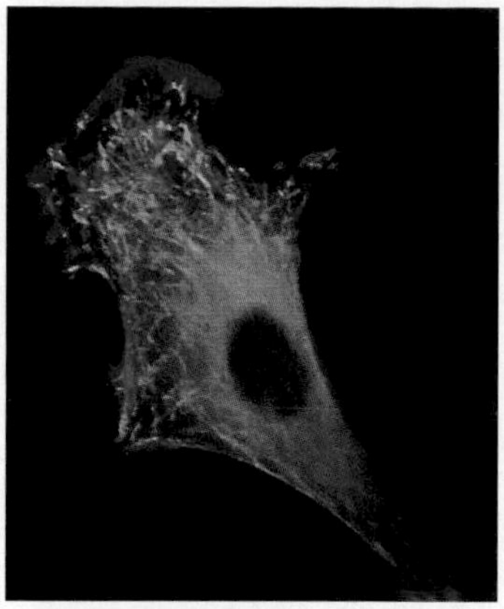

FIGURE 7-5 A fibroblast. Note the flat, irregular, branching shape. Fibroblasts secrete fibers of protein, constructing a web that supports other fibroblasts (*700x*).

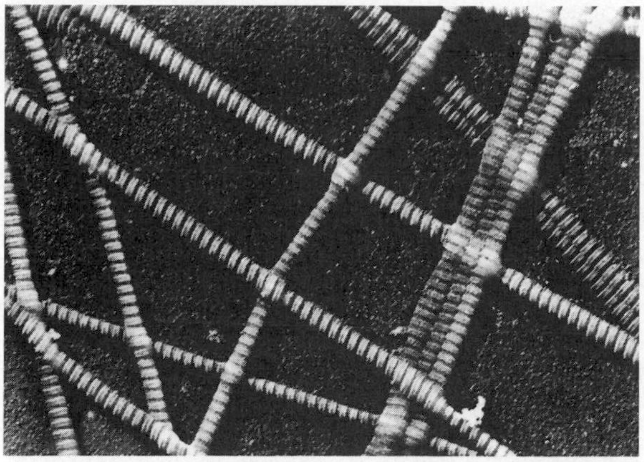

FIGURE 7-6 Collagen fibers. Each fiber is composed of many collagen strands and is very strong (*45,000x*).

Fibroblasts

Of all the connective tissue cells, fibroblasts are the most numerous. They are flat, irregular, branching cells that secrete fibers into the matrix between them (Figure 7-5). These fibers are of three basic types: collagen, reticulin, and elastic.

Both the collagen fibers and the reticulin fibers are made up of the protein collagen, the most abundant protein in the human body (Figure 7-6). Collagen fibers are strong and wavy. These properties allow the connective tissues that they compose to be somewhat flexible and stretchy, without the fibers themselves stretching. *Dense fibrous connective tissue* is primarily made up of collagen fibers. In Table 7-3 you can see bundles of collagen fibers with widely spaced rows of fibroblasts. This type of connective tissue is very strong and is found as tendons connecting muscles to bones; it composes the lower layer of the skin and makes strong attachments between organs.

Reticulin is a fine branching fiber that forms the framework of many glands such as the spleen and the lymph nodes. It also makes up the junctions between many tissues. The tissue formed by fibroblasts and reticulin alone is called *reticular connective tissue* (see Table 7-3). Elastic fibers, as the name suggests, act much like rubber bands. They are not made of collagen but of a protein called *elastin,* a "stretchy" protein. *Elastic connective tissue* is made up of branching elastic fibers with fibroblasts interspersed throughout (see Table 7-3). This type of tissue is found in structures that must expand and then return to their original shape, such as the lungs and large arteries.

Loose connective tissue contains various connective tissue cells and fibers within a semifluid matrix (see Table 7-3). Fibroblasts and macrophages are the most common cells in loose connective tissue. This tissue also contains loosely packed elastic and collagen fibers; it is therefore a somewhat strong but very flexible tissue. Loose connective tissue is distributed widely throughout the body and is found wrapping nerves, blood vessels, and tissues; filling spaces between body parts; and attaching the skin to the layers beneath it. If you have ever skinned chicken before cooking, for example, you have seen the loose connective tissue that binds the skin to the muscle beneath.

Cartilage Cells

Chondrocytes are the cells that produce cartilage, a specialized connective tissue that is hard and strong. Cartilage is found in many places such as the ends of long bones, the airways of the respiratory system, and the spaces between the vertebrae. Cartilage is made up of cells that secrete a matrix consisting of a semisolid gel and fibers. The fibers are laid down along the lines of stress in long parallel arrays (groups or arrangements). The result of this process is a firm and flexible tissue that does not stretch.

As the cartilage cells secrete the matrix, they wall themselves off from it and eventually come to lie in tiny chambers called *lacunae.* In Table 7-3, you can see that although there are three different types of cartilage—*hyaline cartilage, elastic cartilage,* and *fibrocartilage*—they all have cartilage cells within lacunae. Their differences lie in the matrix.

TABLE 7-3

Connective Tissue

TISSUE	LOCATION	FUNCTIONS	DESCRIPTION
Dense fibrous connective tissue 	Tendons (attach muscle to bone) and ligaments (attach bones to bones), attachments between organs and dermis of the skin	Support, ability to withstand great pulling forces in the direction in which the fibers are oriented	The structure consists of mostly collagen fibers with occasional rows of collagen-producing cells.
Reticular connective tissue 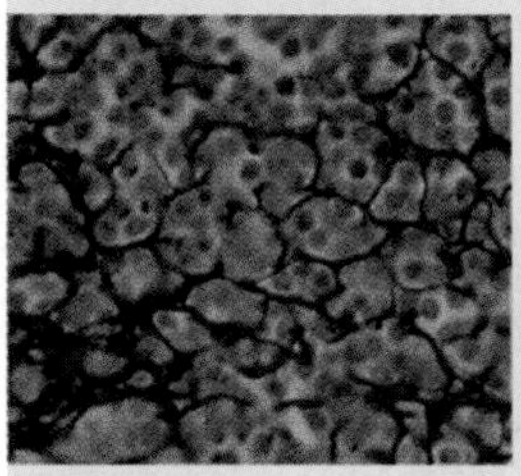	Liver, lymph nodes, spleen, and bone marrow	Support	The structure is intertwined reticular fibers.
Elastic connective tissue 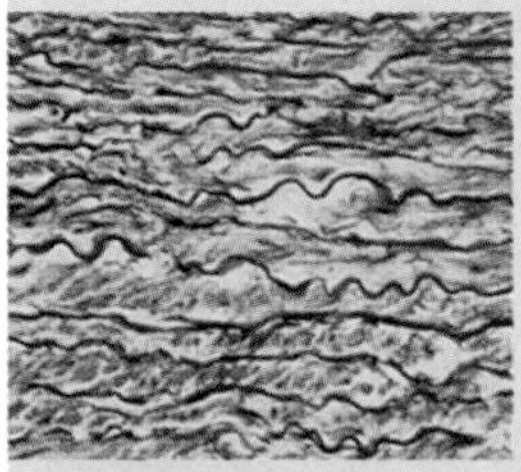	Lung tissue, arteries	Strength with stretching and recoil	The structure consists of elastic fibers dotted with cells.
Loose connective tissue 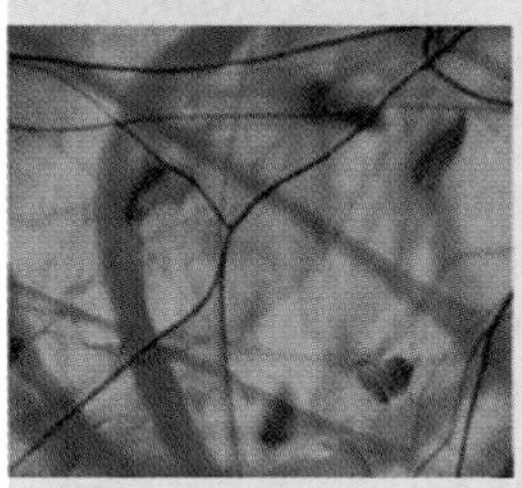	Widely distributed throughout body; packing between glands, muscles, and nerves; attachments between the skin and underlying tissue	Support, loose packing	The structure consists of cells within a fine network of fibers (mostly collagen), which they produce. The cells and fibers are separated from each other by fluid-filled spaces.
Hyaline cartilage 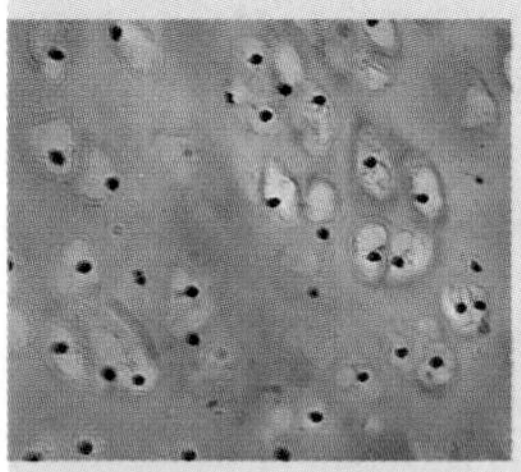	Ends of long bones, joints, respiratory tubes, costal cartilage of ribs, nasal cartilage, and embryonic skeleton	Flexible support, reduction of friction between movable bones	Cartilage cells are found in lacunae within a rigid, transparent matrix. Collagen fibers are small and not visible.

TABLE 7-3

Connective Tissue—cont'd

TISSUE	LOCATION	FUNCTIONS	DESCRIPTION
Elastic cartilage 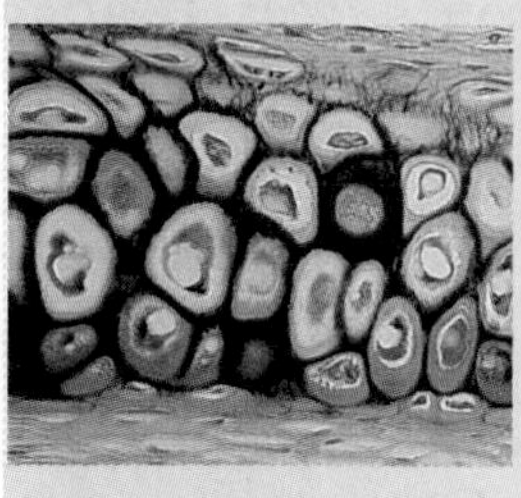	Auditory tube, external ear, epiglottis	Rigidity with flexibility, returning to original shape after being stretched	The structure resembles hyaline cartilage but has elastic fibers.
Fibrocartilage 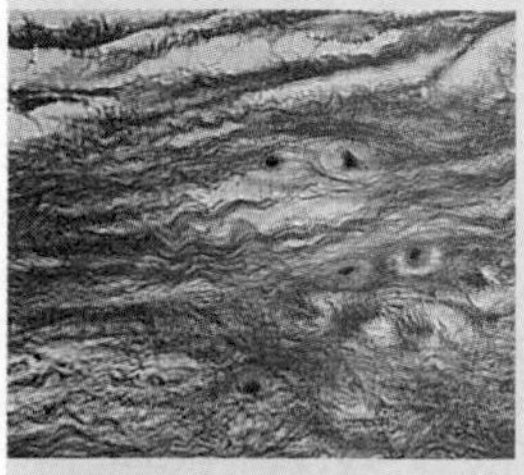	Connection between pubic bones, intervertebral disks	Support, connection, shock absorption, and ability to withstand considerable pressure	The structure resembles hyaline cartilage but has thick bundles of collagen fibers.
Bone 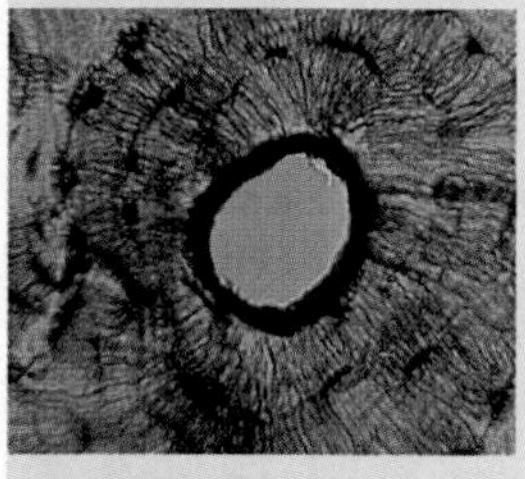	Bones of skeleton	Strength, support, and protection of internal organs; storage of calcium; attachment for muscles	Bone-making cells (osteocytes) are found in lacunae. In compact bone (as shown) the lacunae are arranged in circles around the Haversian canals, which contain blood vessels and nerves. The structure is a hard, mineralized matrix.
Adipose tissue 	Under the skin insulation of organs such as the heart, kidneys, and breasts	Storage, insulation, energy, support of organs	The structure consists of lipid-filled, ring-shaped cells packed together.
Blood 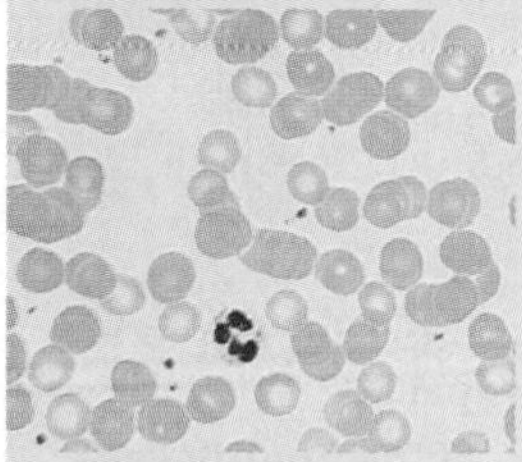	Blood vessels, heart	Protection of body from infections; transportation of oxygen, nutrients, wastes, and other materials; regulation of body temperature	The structure consists of blood cells in a fluid matrix.

Hyaline cartilage has very fine collagen fibers in its matrix that are almost impossible to see under the light microscope. During your development before birth, most of your skeleton was composed of hyaline cartilage. As an adult, you have hyaline cartilage on the ends of your long bones, cushioning the places where these bones meet. Hyaline cartilage also rings the windpipe, keeping this airway propped open, and makes up parts of your ribs and nose. Elastic cartilage, as the name suggests, has elastic fibers embedded in its matrix. It is found where support with flexibility is needed, such as in the external ear. Fibrocartilage has collagen fibers embedded in its matrix. It is therefore a very tough substance and is used in places of the body where shock absorbers are needed. It is found, for example, as discs between the vertebrae and in the knee joint.

Bone Cells

Osteocytes are the cells that produce bone. As in cartilage, these cells are isolated in lacunae. They lay down a matrix of collagen fibers that becomes coated with small, needle-shaped crystals of calcium. The calcium makes the bone rigid, whereas the fibers keep the bone from being brittle. Bone makes up the adult skeleton, which supports the body and protects many of the organs. Bone is also a storehouse for calcium.

The bones of your skeleton have two types of internal structure: *spongy bone* and *compact bone.* Spongy bone makes up the ends of long bones and the interiors of long bones, flat bones, and irregular bones (Figure 7-7). It is composed of an open lattice of bone that supports the bone just as beams support a building (Figure 7-8). It also helps keep bones somewhat lightweight. The spaces within the latticework of bone are filled with red bone marrow, the substance that produces most of the body's blood cells.

Compact bone is denser than spongy bone and gives the bone the strength to withstand mechanical stress. In compact bone, the cells lay down matrix in thin concentric rings, forming tubes of bone around narrow channels or canals (see Table 7-3).

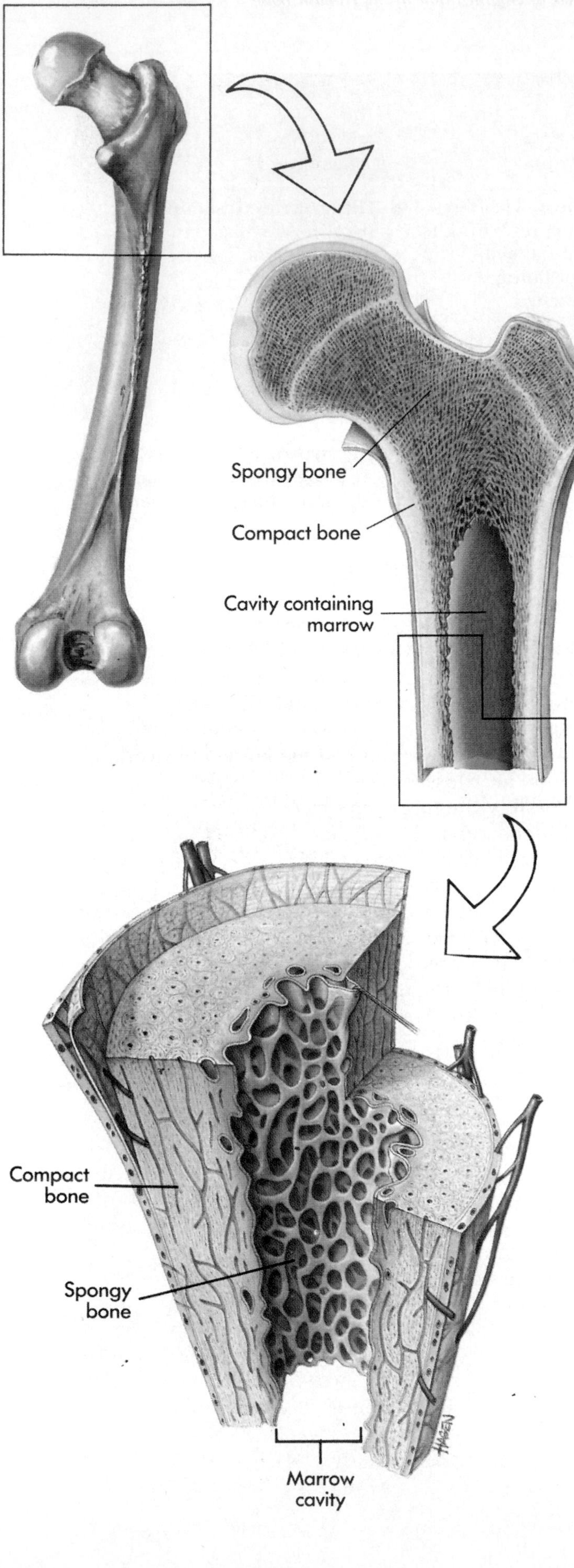

FIGURE 7-7 Structure of a long bone. A long bone, such as the bone located in your upper leg, is composed of spongy bone and compact bone; spongy bone adds support and (along with the hollowed-out shaft of compact bone) keeps the bone somewhat lightweight. Compact bone adds strength.

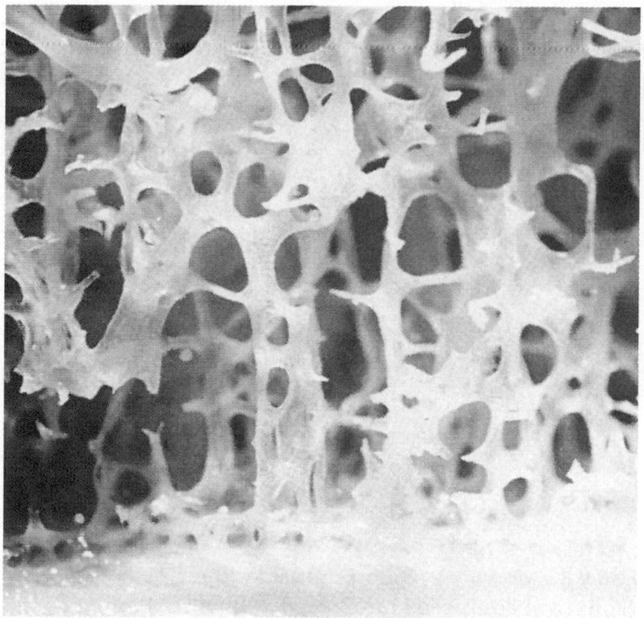

FIGURE 7-8 Spongy bone. Although you may think of bone as solid, spongy bone is composed of a delicate latticework. Bone, like most tissues in your body, is a dynamic structure, constantly renewing itself (*10x*).

These canals run parallel to the length of the bone, are interconnected, and contain blood vessels and nerves. The blood vessels provide a lifeline to the bone-forming cells, and the nerves control the diameter of the blood vessels and thus the flow through them.

> *Each of the three different types of structural connective tissue cells produces substances that cause the tissues of which they are a part to have distinctive characteristics. Fibroblasts produce fibers of different types that are found in dense and loose connective tissue, reticular connective tissue, and elastic connective tissue. Cartilage cells lay down a matrix of gel and fibers. Bone cells lay down a matrix of fibers that becomes coated with calcium.*

Cells and Tissues That Isolate

The third general class of connective tissue is composed of cells that specialize in accumulating and transporting particular molecules. Isolating connective tissues include the fat cells of adipose tissue, as well as pigment-containing cells.

As you can see in Table 7-3, groups of fat cells as seen under a microscope bear a striking resemblance to chicken wire. The "wire" is really the cytoplasm and nucleus of each cell pushed against the cell membrane by a large fat droplet. The fat takes up much of the cell and is released when the body needs it for fuel. Although many people think that they have more fat cells than they need for fuel emergencies, fat serves other purposes too. Adipose tissue helps shape and pad the body and insulates against heat loss.

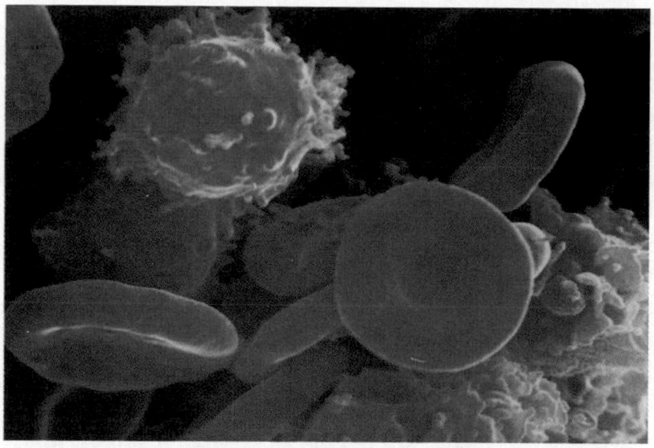

FIGURE 7-9 Blood cells. White blood cells, or leukocytes, are roughly spherical and have irregular surfaces. Red blood cells, or erythrocytes, are disks with depressed centers (*3000x*).

Possibly the most important "isolating" cells are red blood cells, one of the solids that float in the fluid connective tissue called *blood*. Blood cells are classified according to their appearance (Figure 7–9), either as *erythrocytes* (red blood cells) or *leukocytes* (white blood cells). Some types of white blood cells were described earlier in this chapter; they include the macrophages and lymphocytes that defend the body. The role of red blood cells is very different. Red blood cells act as mobile transport units, picking up and delivering gases. Cell fragments called *platelets* are also present in the blood. These cell pieces play an important role in the clotting of blood.

Red blood cells are the most common of the blood cells. There are about 5 billion in every milliliter of blood. During their maturation in the red bone marrow, they lose their nuclei and mitochondria and their endoplasmic reticula dissolve. As a result of these processes, red blood cells are relatively inactive metabolically, but they still perform the life-sustaining job of carrying oxygen to your tissues. This oxygen is carried by the iron-containing pigment hemoglobin. This pigment imparts the color to red blood cells. Hemoglobin is produced within the red bone marrow as the red blood cells are formed. An amazing 300 million molecules of hemoglobin become isolated within each red blood cell.

Blood cells float in a fluid intercellular matrix, or plasma. This fluid is both the banquet table and the refuse heap of your body because practically every substance used and discarded by cells is found in the plasma. These substances include the sugars, lipids, and amino acids that are the fuel of the body, as well as the products of metabolism such as the waste gas carbon dioxide. The plasma also contains minerals such as calcium used to form bone; fibrinogen, which helps the blood to clot; albumin, which gives the blood its viscosity; and antibody proteins produced by lymphocytes. Every substance secreted or discarded by cells is also present in the plasma.

> *Many connective tissues are specialized for accumulating particular classes of molecules, such as fats and pigments. Red blood cells accumulate the oxygen-carrying pigment hemoglobin.*

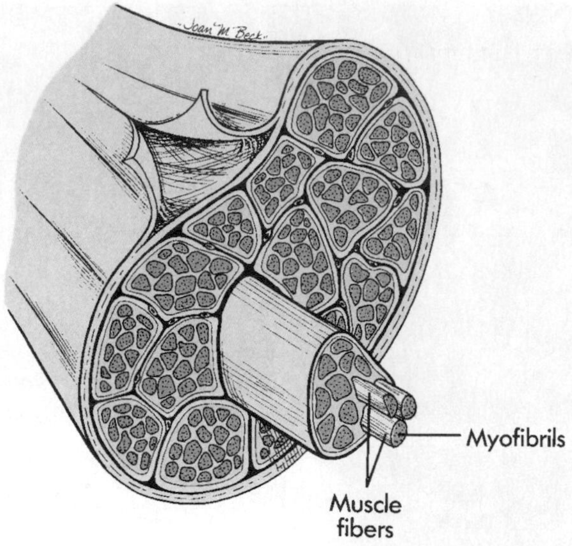

FIGURE 7-10 A skeletal muscle fiber. Muscle fibers (cells) are composed of many myofibrils. Myofibrils, in turn, are made up of microfilaments, which are responsible for muscle contraction.

Muscle Tissue

Muscle cells are the workhorses of your body. The distinguishing characteristic of muscle cells is the abundance of special thick and thin microfilaments. These microfilaments are highly organized to form strands called *myofibrils.* Each muscle cell is packed with many thousands of these myofibril strands (Figure 7-10). The myofibrils shorten when the microfilaments slide past each other, causing the muscle to contract. Table 7-4 shows the three different kinds of muscle cells of the human body: *smooth muscle, skeletal muscle,* and *cardiac muscle.*

Smooth Muscle

Smooth muscle cells are long, with bulging middles and tapered ends and a single nucleus. The cells are organized into sheets, forming smooth muscle tissue. This tissue contracts involuntarily—you cannot consciously control it. Because it is found in the organs, or viscera, smooth muscle tissue is also often called *visceral muscle tissue.*

Some smooth muscle contracts when it is stimulated by a nerve or hormone. Examples of smooth muscle that contract in this way are the muscles found lining your blood vessels and those that make up the iris of your eye. But nerves do not reach each muscle cell. In many cases, impulses may be able to pass directly from one smooth muscle cell to another so that a wave of contraction can pass through a layer of some kinds of smooth muscle all by itself. In other smooth muscle tissue, such as that found in the wall of the intestines, individual cells may contract spontaneously when they are stretched, leading to a slow, steady squeeze of the tissue.

Skeletal Muscle

Skeletal muscles are attached to your bones and allow you to move your body. These muscles are called *voluntary muscles* because you have conscious control over their action. They are also called *striated muscles* because the tissue has "stripes"—microscopically visible bands, or striations (see Table 7-4). These striations result from the organization of thick and thin microfilaments within the myofibrils and the alignment of the myofibrils with one another. The myofibrils are organized so that all the myofibrils within a single cell contract at the same time when the muscle cell is stimulated by a nerve.

Striated muscle cells are extremely long. A single muscle cell, or *fiber,* may run the entire length of a muscle. Each fiber has many nuclei that are pushed to the edge of the cell and lie just under the cell membrane. Bundles of these muscle cells are wrapped with connective tissue and joined with other bundles to form the muscle itself.

TABLE 7-4

Muscle Tissue

TISSUE	LOCATION	FUNCTIONS	DESCRIPTION
Smooth muscle 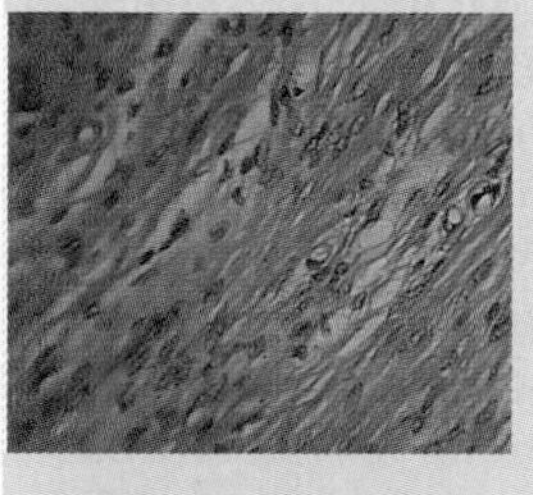	Walls of hollow organs, pupil of eye, skin (attached to hair), and glands	Regulation of size of organs, forcing of fluid through tubes, control of amount of light entering eye, and production of "goose-flesh" in the skin; under involuntary control	The tissue is not striated. The spindle-shaped cells have a single, centrally located nucleus
Skeletal muscle 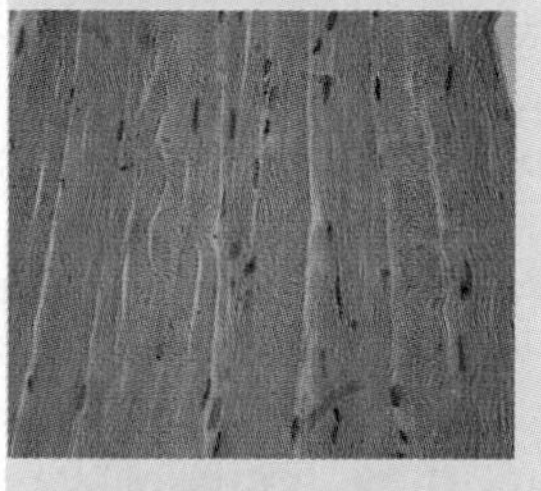	Attachment to bone	Movement of the body, under voluntary control	The tissue is striated. Cells are large, long, and cylindrical with several nuclei.
Cardiac muscle 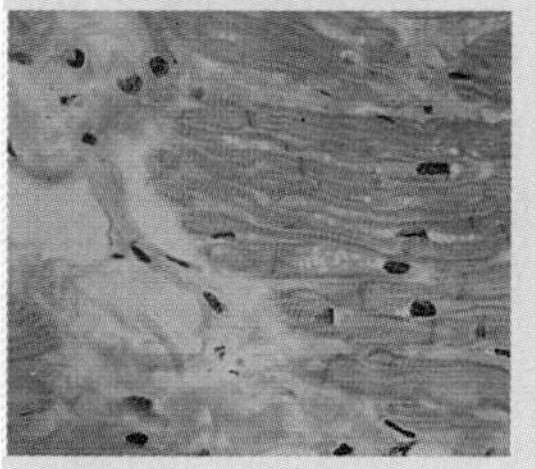	Heart	Pumping of blood, under involuntary control	The tissue is striated. Cells are cylindrical and branching with a single centrally located nucleus

Cardiac Muscle

The heart is composed of striated muscle fibers, but these fibers are arranged differently from their arrangements in skeletal muscle. Instead of very long cells running the length of the muscle, heart muscle is composed of chains of single cells. These chains of cells are organized into fibers that branch and interconnect, forming a latticework (see Table 7-4). This lattice structure is critical to how heart muscle functions, and it allows an entire portion of the heart to contract at one time.

Muscle cells contain microfilaments that are capable of contraction. Smooth muscle contracts involuntarily and is located in the walls of certain internal structures such as blood vessels and the stomach. Skeletal muscle is connected to bones and allows you to move your body. Cardiac muscle makes up the heart, acting as a pump for the circulatory system.

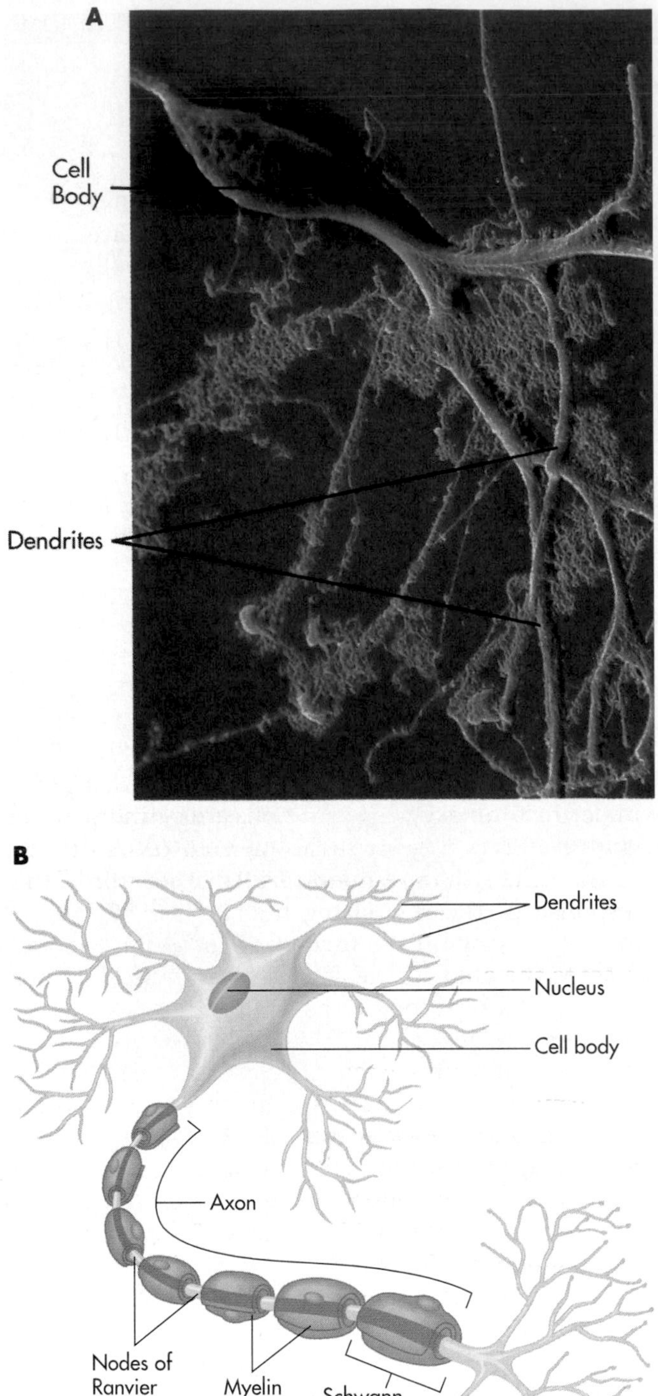

FIGURE 7-11 A human neuron. A, Electron micrograph of a neuron (*7500x*). The cell body is at the upper left, with the axon extending up out of view. The branching network of fibers extending down from the cell body is made up of dendrites, which carry signals to the cell body. **B,** This diagram illustrates the generalized structure of a vertebrate neuron.

Nervous Tissue

The fourth major class of tissue in humans is nervous tissue. It is made up of two kinds of cells: (1) *neurons,* which transmit nerve impulses, and (2) *supporting cells,* which nourish and protect the neurons.

Neurons are cells specialized to conduct an electro-chemical "current" (Figure 7-11). The cell body of a neuron contains the nucleus of the cell. Two different types of projections extend from the cell body. One set of projections, the *dendrites,* act as antennae for the reception of nerve impulses and conduct these impulses toward the cell body. A single projection called an *axon* conducts impulses away from the cell body. When axons or dendrites are long, they are referred to as *nerve fibers.* Some nerve fibers are so long, in fact, that they can extend from your spinal cord all the way to your fingers or toes. Single neurons over a meter in length are common.

The *nerves,* which appear as fine white threads when they are viewed with the naked eye, are actually composed of clusters of axons and dendrites. Like a telephone trunk cable, they include large numbers of independent communication channels—bundles of hundreds of axons and dendrites, each connecting a different nerve cell with a different muscle fiber or sensory receptor. In addition, the nerve contains numerous supporting cells bunched around the nerve fibers. In the brain and spinal cord, which together make up the central nervous system, these supporting cells are called *glial cells.* The supporting cells associated with nerve fibers of all other nerve cells, which make up the peripheral nervous system, are called *Schwann cells.*

> *Neurons are cells that are specialized to conduct electrical signals. Nerve tissue is made up of these cells and supporting cells.*

Organs

The four major classes of tissues that we have discussed in this chapter (see Figure 7-1) are the building blocks of the human body. These tissues form the organs of the body. Each organ contains several different types of tissue coordinated to form the structure of the organ and to perform its function. A muscle, for example, is composed of muscle cells that together make up the muscle tissue. Bundles of this tissue are wrapped in connective tissue and wired with nervous tissue. Muscles can help you walk, pump your blood, and digest your food. Different combinations of tissues are found in different organs that perform different functions.

Organ Systems

An organ system is a group of organs that function together to carry out the principal activities of the body. Figure 7-12 shows the 11 major organ systems of the human body. The skeletal system supports and protects your body. It is moved by the large, voluntary muscles of the muscular system. Other muscles in this system help move internal fluids throughout your body. The nervous system regulates most of the organ systems. It can sense conditions in both your internal and external environments and help your body respond to this environmental information. The organs of your endocrine system secrete chemicals called *hormones* that also regulate body processes and functions. The circulatory system is the transportation system of the body. It brings nutrients and oxygen to your cells and removes the waste products of metabolism. Along with the immune and integumentary (skin) systems, it also helps defend the body against infection and disease. The respiratory system works hand in hand with the circulatory system, supplying the blood with oxygen and ridding it of the waste gas carbon dioxide. The food you eat is broken down by the digestive system and is absorbed through the intestinal walls into the bloodstream. Solid wastes are also eliminated from the body by this organ system. Liquid wastes are eliminated by the urinary system after it collects waste materials and excess water from the bloodstream. And to ensure the continuance of the human race, the reproductive system produces gametes, or sex cells, that can join in the process of fertilization to produce the first cell of a new individual.

The Organism: Coordinating It All

As you can see from their descriptions, the organ systems interact with one another to keep the organism—you—alive and well. This state of "wellness" is called **homeostasis.** Put another way, homeostasis is the maintenance of a stable internal environment despite what may be a very different external environment. To maintain this internal equilibrium, all your molecules, cells, tissues, organs, and organ systems must work together.

Your body maintains a steady state by means of feedback systems, or **feedback loops.** Feedback loops are mechanisms by which information regarding the status of a physiological situation or system is fed back to the system so that appropriate adjustments can be made. The thermostat in your house or apartment works by means of a feedback loop and is a good analogy to use to understand feedback loops in your body. If you set the temperature in your house at 68° F, for example, the furnace will run until that temperature is reached. A sensor in the thermostat monitors the temperature in the room, and this information triggers the electrical stimulation of the furnace. As soon as the temperature near the thermostat reaches 68° F, the thermostat sends an electrical signal to the furnace to turn off. In this way the temperature fluctuates within a small range. Such a feedback loop is called a **negative feedback loop** because the change that takes place is negative—the furnace turns off—to counter the rise in temperature. In other words, the response of the regulating mechanism (the thermostat turning off the furnace) is negative with respect to the output (the heat of the furnace). Figure 7-13 shows examples of negative feedback loops.

Most regulatory mechanisms of your body work by means of negative feedback loops. The secretion of hormones is regulated in this way (see Chapter 17). Sufficient levels of specific hormones in the blood trigger mechanisms that result in the shutdown of their secretion. Likewise, your body temperature is maintained by means of negative feedback, much as the temperature of your home is maintained. When your body temperature rises too high, your thermostat (a special portion of the brain called the *hypothalamus*) senses this rise and counters it. Messages race along your nerve fibers to your sweat glands, triggering the release of

homeostasis (hoe mee oh **stay** sis) the maintenance of a stable internal environment despite what may be a very different external environment.

feedback loops mechanisms by which information regarding the status of a physiological situation or system is fed back to the system so that appropriate adjustments can be made.

negative feedback loop a feedback loop in which the response of the regulating mechanism is negative with respect to the output. Most of the body's regulatory mechanisms work by means of negative feedback loops.

positive feedback loop a feedback loop in which the response of the regulating mechanism is positive with respect to the output.

sweat, which evaporates and cools the body. Blood vessels near the surface of the skin dilate, or widen, bringing blood near the surface, where the heat is dissipated. Likewise, when your body temperature falls, the hypothalamus triggers tiny muscles under the skin to contract, causing a shiver that generates heat. Blood vessels near the surface of the skin constrict, lessening the flow of blood there and the amount of heat that radiates from the body.

Very few body mechanisms are regulated by means of **positive feedback loops.** In positive feedback loops the response of the regulating mechanism is positive with respect to the output. You can see the problems inherent with this type of regulatory mechanism—a situation continues to intensify rather than shutting down. For example, in a positive feedback loop, your furnace would be stimulated to stay on when the temperature in the house reached a particular setting. Positive feedback loops most often disrupt the steady state and can even lead to death. In circulatory shock, for example, a severe loss of blood can result in such a decreased blood volume that the body cannot compensate—homeostasis cannot be maintained. The blood pressure drops and the blood flow is reduced so much that the heart and brain do not receive enough oxygen. As the heart weakens, the blood flow decreases more, weakening the heart even further. As the body becomes damaged from the loss of oxygen and nutrients, the shock worsens and can result in death. There are, however, some examples of and good uses for positive feedback in the body, including oxytocin release during childbirth.

> *The molecules, cells, tissues, organs, and organ systems that make up an organism must all work together to maintain homeostasis—a stable internal environment. Homeostasis is maintained by means of negative feedback loops, regulatory mechanisms that slow down or shut down output systems when they reach certain levels. For example, a high internal body temperature triggers mechanisms that lower the temperature.*

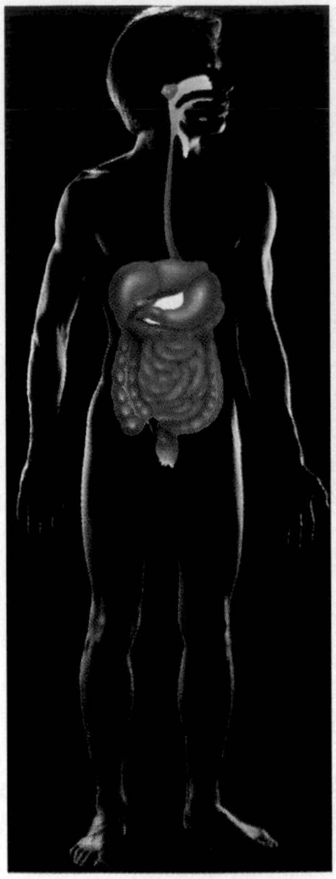
Digestive system

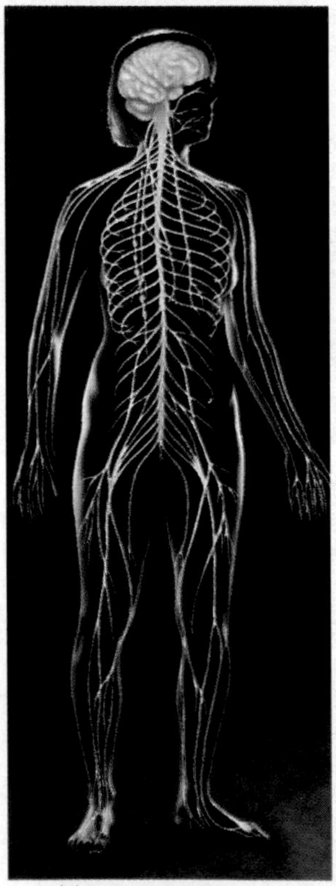
Nervous system

FIGURE 7-12 The major organ systems of vertebrates. The body is composed of combinations of the four types of tissue, assembled in various ways. The many organs working together to carry out the principal activities of the body are traditionally grouped together as organ systems.

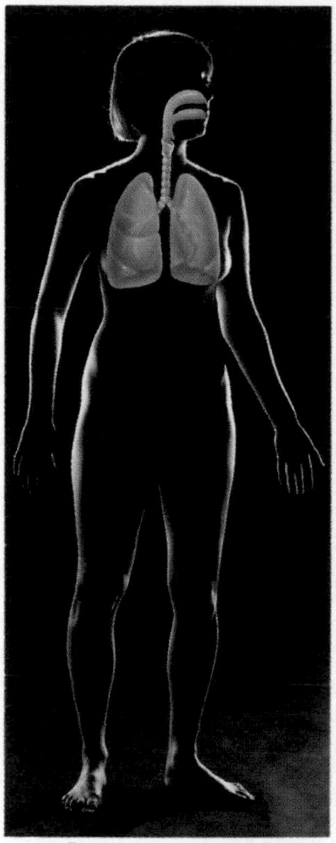
Respiratory system

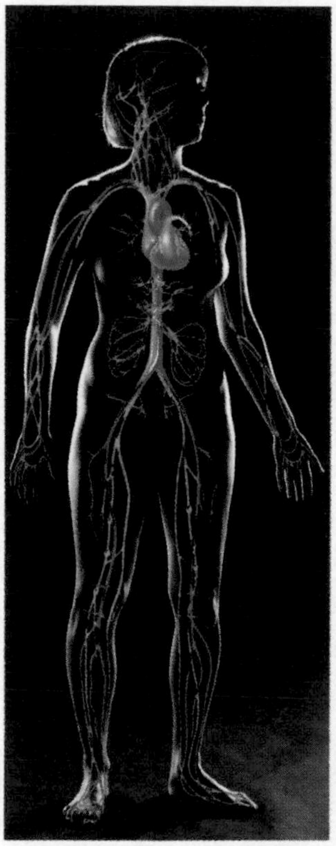
Circulatory system

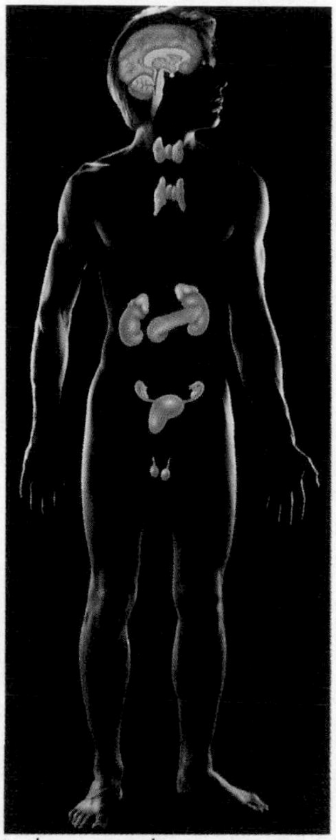
Endocrine and immune systems

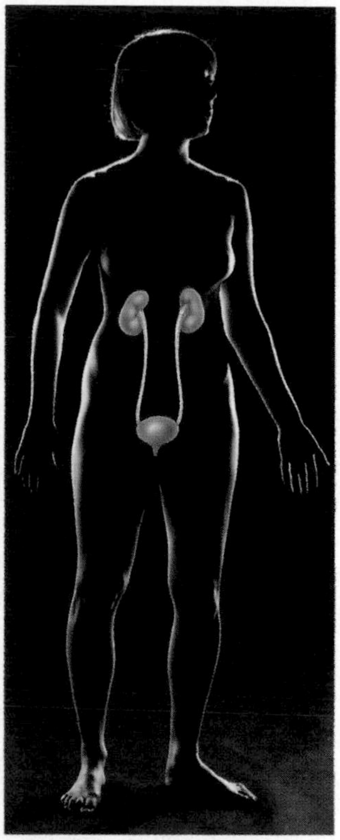
Urinary system

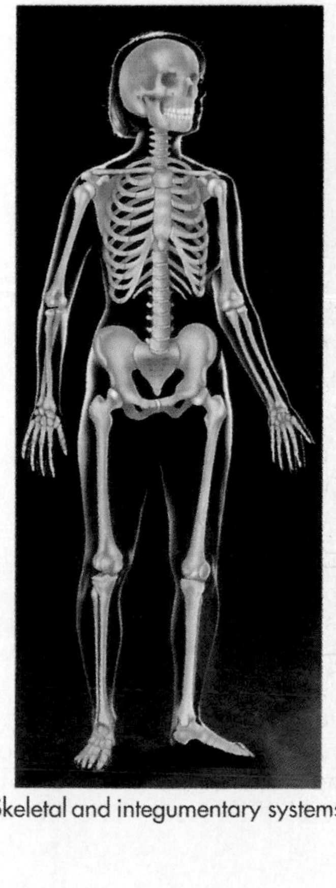
Skeletal and integumentary systems

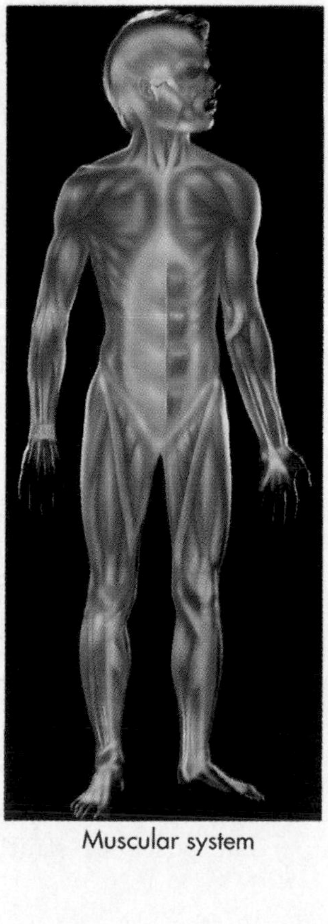
Muscular system

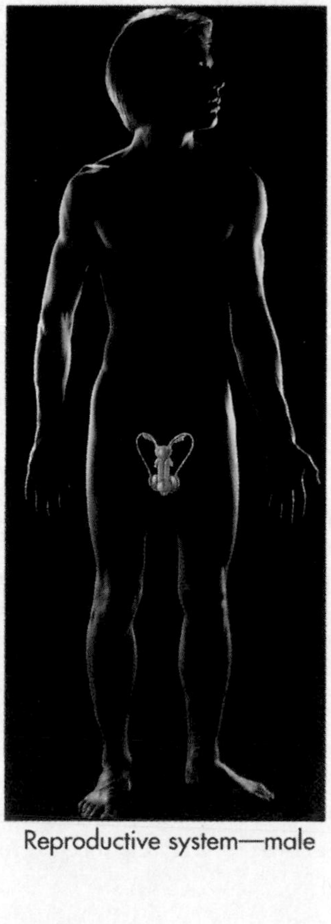
Reproductive system—male

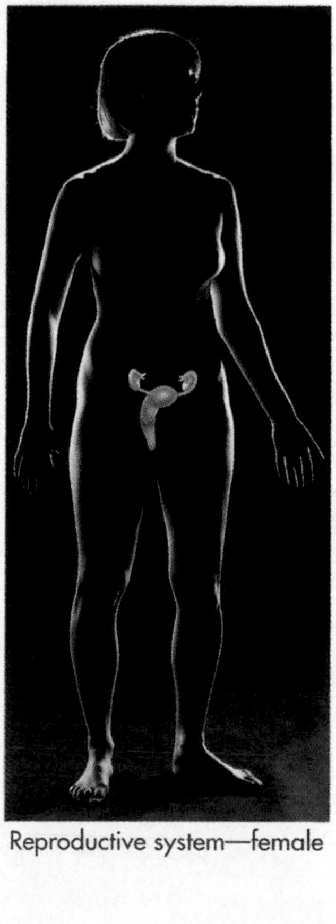
Reproductive system—female

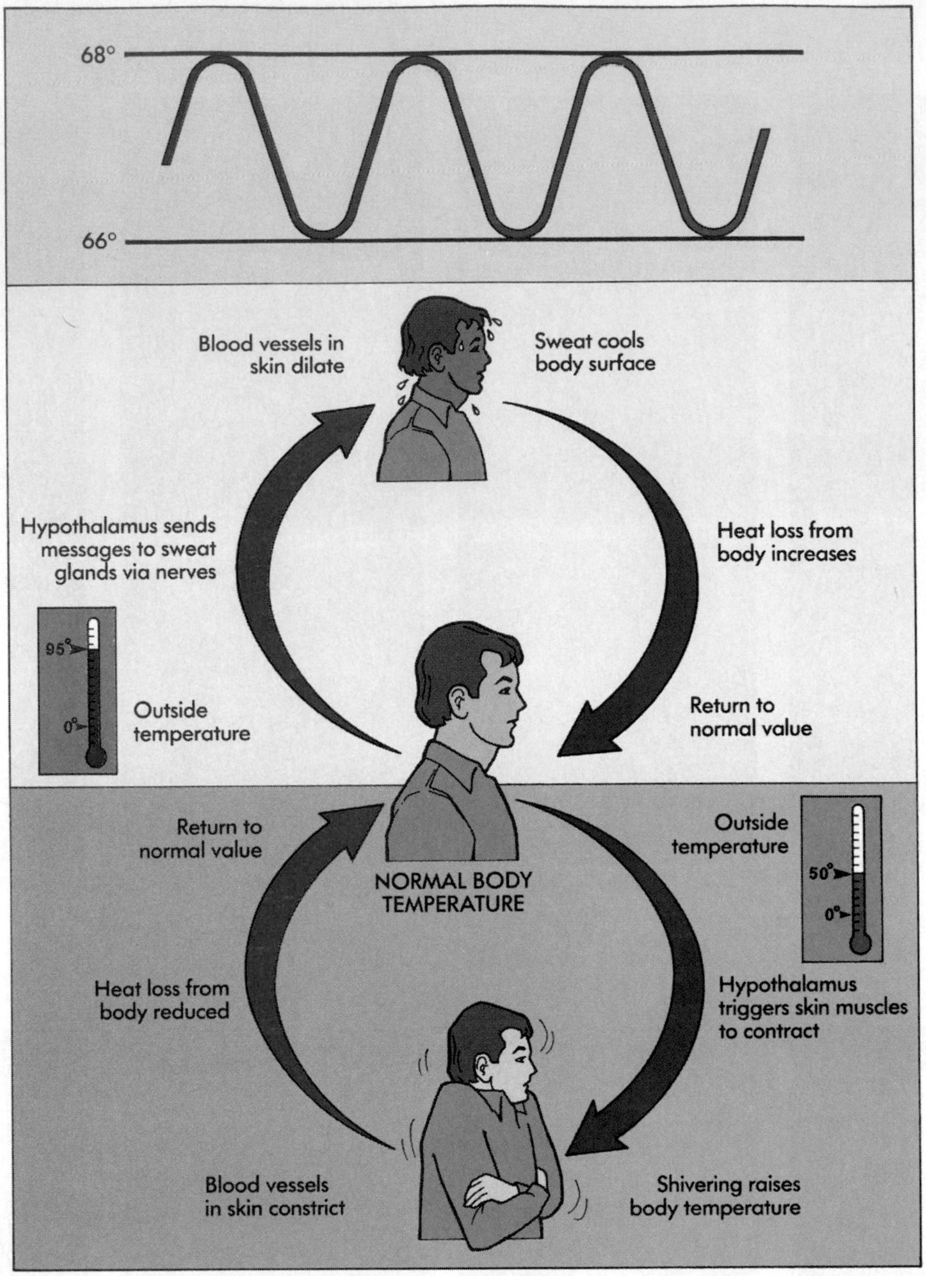

FIGURE 7-13 A negative feedback loop. In a negative feedback loop, the response is negative with respect to the output. Heat generated by the furnace (the output) triggers it to shut off (a negative response) when the temperature reaches a set point (68° F, for example). The temperature falls with no output until it reaches another point (66° F, for example), a point at which the furnace is triggered to go on. (Turning on is negative with respect to no output.) A similar process occurs in your body and is described in the text.

Summary

- The human body is organized hierarchically; that is, small units of structure and function together form larger units of structure and function, eventually forming the whole organism. Specifically, the cells of the body are organized into tissues. Several different tissues grouped together in a structural and functional unit make up an organ. Organs that work together to carry out particular body activities are called *organ systems.* The human body is an organism made up of 11 major organ systems.
- The four basic types of tissue are epithelial tissue, connective tissue, muscle tissue, and nerve tissue.
- Epithelium covers the surfaces of the body. The lining of the major body cavities is composed of simple epithelium, whereas the exterior skin is composed of stratified epithelium. The major

gland systems are also derived from epithelium.

- The defensive connective tissue cells are macrophages, which engulf foreign bacteria and antibody-coated cells or particles; lymphocytes, certain of which produce antibodies; and mast cells, which release inflammation-producing chemicals at sites of trauma.
- The structural connective tissue cells include cartilage cells (chondrocytes), bone cells (osteocytes), and fibroblasts.
- The isolating connective tissue cells include fat cells and pigment-accumulating cells, such as red blood cells.
- Muscle contraction provides the force for mechanical movement of the body. There are three kinds of muscle tissue: smooth muscle, which is found in internal organs; skeletal muscle, which moves the body parts; and cardiac muscle, which makes up the heart—the pump of the circulatory system.
- Nerve cells provide the body with a means of rapid communication. There are two kinds of nerve cells: neurons and supporting cells. Neurons are specialized for the conduction of electrical impulses.
- Organs are body structures composed of several different tissues grouped into a structural and functional unit. An organ system is a group of organs that function together to carry out the principal activities of the body.
- The human body is organized so that its parts form an integrated whole. These parts work together at each level of organization to maintain homeostasis—a stable internal environment within the human organism.

Knowledge and Comprehension Questions

1. Distinguish among cells, tissues, organs, and organ systems. Give an example of each.
2. What are the four basic types of tissue in the human body? Give an example of each.
3. Draw and label the three shapes of epithelial cells. Give an example of where each is found in your body.
4. What do lymphocytes, erythrocytes, and fibroblasts have in common? How do they differ?
5. Red blood cells provide the life-sustaining role of carrying oxygen to your body cells. For what cellular process is this oxygen necessary?
6. Everybody hates fat cells. Do they serve a purpose and if so, what purpose(s)?
7. Which type of muscle tissue is under voluntary control and may be specifically strengthened by such exercises as lifting weights and doing push-ups? Which type may be strengthened by aerobic exercises such as running and cross-country skiing?
8. What is the nature of the signals carried on nerve fibers? Which is quicker—communication by neurons or by hormones?
9. Create a table that lists the major function(s) and at least two components of each of the principal organ systems in your body.
10. What is homeostasis? How is it achieved?

Critical Thinking

1. Our bones generally have reached their full growth and mature length by the age of 20 years. How then is the femur, a leg bone, of a 30-year-old man able to heal fully when broken in an accident?
2. Your cat scratches your finger, and the skin surrounding the scratch turns reddish and feels warm. Explain why.
3. The text states that connective tissues are generally made up of cells spaced well apart in a nonliving matrix. Using your knowledge of cells and tissues, suggest what the functions of a matrix might be.
4. Certain tissues in your body do not repair themselves if damaged (for example, nervous tissue within the brain and spinal cord), whereas other tissues such as skin rapidly repair themselves. Based on what you learned in this chapter, suggest a reason why nerve tissue in your central nervous system does not repair itself as readily as skin.

CHAPTER 8

DIGESTION

WHAT IS WRONG with this picture? Take a critical look at the photo and solve this nutrition puzzle. Fish is certainly a healthy choice, you might be thinking, and you are correct. In fact, evidence suggests that a group of substances in fish oil protects the human body against heart attack. In addition, fish is generally lower in calories than beef or poultry and has other healthy qualities. The problem is that the typical American dinner usually contains double, triple, or sometimes quadruple the amount of protein needed for the entire day! The half-pound of fish here is too hefty a portion of protein.

Vegetables and a potato are also healthy food choices. However, the butter that is saturating these foods is not so good a choice. This fat, plus the fat in the salad dressing and ice cream, means that at least half the calories in this meal come from fat—far too many when you realize that one tablespoon of vegetable oil satisfies your daily requirement.

The sugary soda and dessert probably caught your eye. As you may know, refined sugar, or sucrose, is a relatively simple carbohydrate that provides only calories—something most Americans already consume in excess. Sugar is therefore often called a source of "empty" calories, and it also promotes tooth decay.

How would you change the composition of this meal to make it healthier? Do you know which types of foods to add and which to cut down on or omit? To do this task, you may need to know the answers to a few questions. What are nutrients? Why do we need them to survive? What amounts and types of nutrients do we need? And how do our bodies get these nutrients from the food we eat? The answers to some of these questions begin the story of digestion. By the time you finish reading this chapter, you should be able to answer all the questions.

KEY CONCEPTS

- Carbohydrates, fats, and proteins supply the body with energy and the raw materials to build the substances it needs.
- Vitamins and minerals help the chemical reactions of the body take place; water provides the medium for these reactions.
- Digestion breaks down the energy nutrients—carbohydrates, fats, and proteins—into smaller molecules that can be used to produce ATP by means of cellular respiration.
- Parts of the digestive system are specialized to perform both the mechanical and chemical processes of digestion.

OUTLINE

The Nutrition-Digestion Connection

Humans are heterotrophs, organisms that cannot produce their own food. Heterotrophs must ingest, or take in, food. At least 95% of the species of organisms on Earth—all animals, all fungi, and most protists and bacteria—are heterotrophs. Food provides these "other-feeders" with two things: (1) energy and (2) the raw materials to build the substances they need. The energy in food is described in units called *calories,* or more properly kilocalories (thousands of calories, or kcal). The raw materials are called **nutrients.**

There are six classes of nutrients: *carbohydrates, fats* (lipids), *proteins, vitamins, minerals,* and *water.* These are the substances, in fact, that make up most of your body. Assuming that you are a proper weight for your height, your body is made up of about 60% water and about 20% fat. The other 20% is mostly protein, carbohydrate, combinations of these two substances, and two major minerals found in your bones: calcium and phosphorus. Other minerals and vitamins make up less than 1% of you.

As you may recall from Chapter 2, carbohydrates, lipids, and proteins are organic (carbon-containing) compounds and are used by your body as a source of energy, or kilocalories. They are therefore often referred to as the energy nutrients. (Proteins, however, are not a preferred energy source.) In addition, these three organic compounds are used as building blocks for growth and repair, as well as to produce other substances your body may need.

Your body does not obtain energy from vitamins, minerals, or water. Vitamins are organic molecules that perform functions such as helping your body to form red blood cells and to unlock the energy in carbohydrates, fats, and proteins. Although required only in small amounts, each of the 13 different vitamins plays a vital role in your body (Table 8-1). Many, such as the B vitamins, are used for the synthesis of coenzymes, small molecules that help enzymes work. For example, niacin and riboflavin, two of the B vitamins, are used to make the coenzymes nicotinamide adenine dinucleotide (NAD) and flavin adenine dinucleotide (FAD), which function in cellular respiration (see Chapter 5). The roles and interactions of some other vitamins, such as vitamin C, are not as well understood.

Minerals are inorganic substances and are transported around the body as ions dissolved in the blood and other body fluids. Your body uses a variety of minerals that perform a variety of functions (see Table 8-1). Calcium, for example, does many jobs, including making up a part of the structure of your bones and teeth and helping your blood to clot. Sodium plays a key role in regulating the fluid balance within your body. Magnesium is an important player in the process of releasing energy from carbohydrates, fats, and proteins.

Water provides the medium in which all the body's reactions take place. Along with dissolved substances and (sometimes) suspended solids, it bathes your cells, makes up much of the cytosol within your cells, and courses through your arteries and veins. It even lubricates your joints and cushions organs such as the brain and spinal cord. The atoms that make up water molecules are also held within the bonds of the energy molecules. These atoms are key players when bonds are broken (by hydrolysis) or formed (by dehydration synthesis) in these molecules. (See Chapter 2 for a description of the processes of hydrolysis and dehydration synthesis.)

Your body obtains the energy and raw materials it needs to survive from six classes of nutrients. Carbohydrates, fats, and proteins provide energy and building blocks. Vitamins, minerals, and water help body processes take place. Some minerals are also incorporated into body structures.

To obtain these nutrients and energy from the food you ingest, your body must first digest the food. **Digestion** is a process in which food particles are broken down into small molecules that can be absorbed by the body. Digestion is also carried out mechanically in the mouth as food is crushed by the teeth and in the stomach as food is churned by the stomach's muscular walls. Digestion is carried out chemically in three ways: (1) by hydrochloric acid (HCl), which denatures, or unfolds, protein molecules and disrupts the protein glue that holds cells together; (2) by bile salts, which emulsify, or separate, large lipid droplets into much smaller lipid droplets; and (3) by a variety of highly specific enzymes that help cleave certain chemical bonds (Table 8-2). As you read in Chapter 4, enzymes are proteins that speed up the rate of chemical reactions in living things. In fact, without enzymes, you would die before needed chemical reactions took place!

The enzymes that help digest the energy nutrients—proteins, carbohydrates, and lipids—are of three basic types:

1. **Proteases,** which break down proteins to smaller polypeptides and polypeptides to amino acids
2. **Amylases,** which break down starches and glycogen to sugars (Both starch and glycogen are storage forms of polysaccharides)
3. **Lipases,** which break down the triglycerides in lipids to fatty acids and glycerol

Proteins, carbohydrates, and lipids, along with their breakdown products, are described in Chapter 2.

TABLE 8-1

Major Vitamins and Minerals

VITAMINS AND MINERALS	MAJOR FUNCTION	RICH FOOD SOURCES	RDA* (MG)	SIGNS AND SYMPTOMS OF DEFICIENCY	SIGNS AND SYMPTOMS OF TOXICITY
Vitamins					
A	Formation of visual pigments, maintenance of epithelial cells	Green or yellow fruits and vegetables, milk products, liver	0.8-1.0	Night blindness, dry skin, growth failure	Headache, nausea, birth defects
Thiamin	Coenzyme in CO_2 removal during cellular respiration	Pork, whole grains, seeds, and nuts	1.1-1.5	Mental confusion, loss of muscular coordination	None from food
Riboflavin	Part of coenzymes FAD and FMN	Liver, leafy greens, dairy products, eggs, meats	1.3-1.7	Dry skin, cracked lips	None from food
Niacin	Part of coenzymes NAD^+ and $NADP^‡$	Wheat bran, tuna, chicken, beef, enriched breads and cereals	15-19	Skin problems, diarrhea, depression, death	Skin flushing
Pantothenic acid	Part of coenzyme A, energy metabolism	Widespread in foods	4-7†	Rare	None from food
B_6	Protein metabolism	Animal proteins, spinach, broccoli, bananas	1.6-2.0	Anemia, headaches, convulsions	Numbness, paralysis
B_{12}	Coenzyme for amino acids and nucleic acid metabolism	Animal foods	0.002	Pernicious anemia	None from food
Biotin	Coenzyme in carbohydrate and fat metabolism, fat synthesis	Liver, peanuts, cheese, egg yolk	0.03-0.10†	Rare	Rare
Folate	Nucleic acid and amino acid synthesis	Green leafy vegetables, orange juice, liver	0.18-0.20	Anemia, embryonic neural tube defects	None from food
C	Collagen synthesis antioxidant	Citrus fruits, broccoli, greens	60	Scurvy, poor wound healing bruises	Diarrhea, kidney stones
D	Absorption of calcium and phosphorus, bone formation	Vitamin D fortified milk, fish	0.005-0.010	Rickets (bone deformities)	Calcium deposits in soft tissues, growth failure
E	Antioxidant	Vegetable oils	8-10	Rare	Muscle weakness, interference with vitamin K metabolism
K	Sythesis of blood clotting substances	Green leafy vegetables, liver	0.06-0.08	Hemorrhage	Anemia

*Recommended dietary allowances (1989) values are for adults.
†Estimated minimum requirement for adults (no RDA established).
‡Estimated safe and adequate daily dietary intake for adults (no RDA established).

TABLE 8-1

Major Vitamins and Minerals—cont'd

VITAMINS AND MINERALS	MAJOR FUNCTION	RICH FOOD SOURCES	RDA* (MG)	SIGNS AND SYMPTOMS OF DEFICIENCY	SIGNS AND SYMPTOMS OF TOXICITY
Major Minerals					
Calcium (Ca)	Component of bone and teeth, blood clotting, nerve transmission, muscle action	Dairy products, canned fish	800-1200	Osteoporosis	Kidney stones
Phosphorus (P)	Component of bone and teeth, energy transfer (ATP), component of nucleic acid	Dairy products, meat, soft drinks	800-1200	None	Bone loss if calcium intake low
Magnesium (Mg)	Bone formation, muscle and nerve function	Wheat bran, green vegetables, nuts	280-350	Weakness	Rare
Sodium (Na)	Osmotic pressure, nerve transmission	Salt, seafood, processed food	500‡	Nausea, vomiting, muscle cramps	Possible hypertension
Chlorine (Cl)	HCl synthesis, osmotic pressure, nerve transmission	Salt, processed food	700‡	Rare	Possible hypertension
Potassium (K)	Osmotic pressure, nerve transmission	Fruits and vegetables	2000‡	Heart irregularities, muscle cramps	Slowed heart rate
Iron (Fe)	Hemoglobin synthesis, oxygen transport	Liver, meat, enriched breads and cereals	10-15	Anemia	Constipation, death from overdose of children's iron supplements
Zinc (Zn)	Component of many enzymes, including those involved in growth, sexual development, and immune function	Seafood, liver, meats, whole grains	12-15	Skin rash, poor growth, hair loss	Diarrhea, depressed immune function
Iodine (I)	Thyroid hormone production	Iodized salt, seafood	0.15	Goiter	Interference with thyroid function
Fluorine (F)	Strengthener of teeth	Fluoridated water, tea	1.5-4.0	Increased risk of dental caries	Stained teeth during development
Copper (Cu)	Iron metabolism, component of many enzymes	Liver, cocoa, whole grains	1.5-3.0†	Anemia	Rare, vomiting

TABLE 8-2

Digestive Enzymes

SOURCE	ENZYME	SUBSTRATE	DIGESTION PRODUCT
Salivary gland	Amylase	Starch, glycogen	Disaccharides
Stomach	Pepsin	Proteins	Short polypeptides
Small intestine	Peptidases	Short peptides	Amino acids
	Lactase Maltase disaccharidases Sucrase	Disaccharides	Glucose, monosaccharides
Pancreas	Pancreatic amylase	Starch, glycogen	Disaccharides
	Trypsin Chymotrypsin Carboxypeptidase	Proteins	Polypeptides
	Lipase	Triglycerides	Fatty acids and glycerol

During digestion, proteins are unfolded by hydrochloric acid and are then cleaved into peptides and individual amino acids by protease enzymes. Starch and glycogen are digested to sugars by amylases. Triglycerides are digested to fatty acids and glycerol by lipases.

Where It All Begins: The Mouth

Does your mouth ever water when you think about your favorite food? Does the smell of some of your favorite foods also evoke this reaction? Do you know why this reaction occurs?

The water in your mouth is really a secretion from a set of glands called **salivary glands.** The locations and names of each of these glands are shown in Figure 8-1. Their secretion, *saliva,* is a solution that consists primarily of water, mucus, and the digestive enzyme **salivary amylase.** Other substances can be found in smaller amounts, such as antibodies and a bacteria-killing enzyme.

Salivary amylase breaks down starch into molecules of the disaccharide maltose. During this reac-

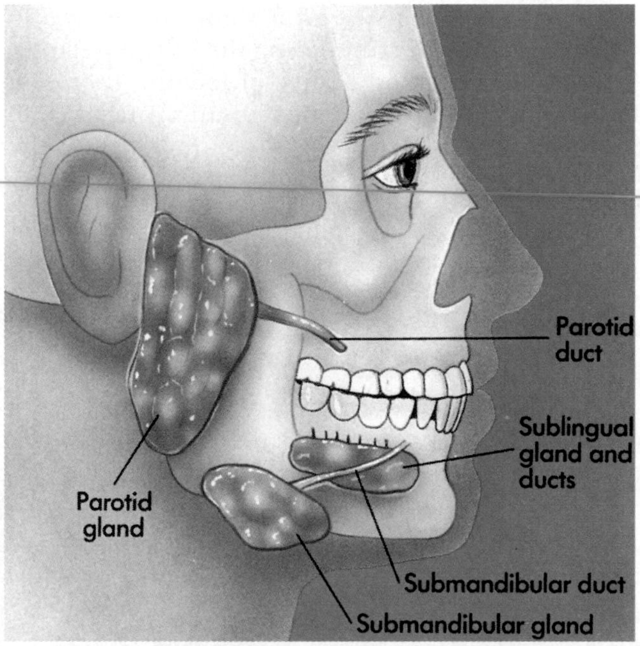

FIGURE 8-1 The salivary glands. These paired glands secrete saliva, which contains salivary amylase, an enzyme that breaks down starch to maltose.

nutrients carbohydrates, fats, proteins, vitamins, minerals, and water.

digestion a process in which food particles are broken down into small molecules that can be absorbed by the body.

proteases (**pro** tee **ace** es) digestive enzymes that break down proteins to smaller polypeptides, and polypeptides to amino acids.

amylases (**am** uh lace es) digestive enzymes that break down starches and glycogen to sugars.

lipases (lye pays es) digestive enzymes that break down the triglycerides in lipids to fatty acids and glycerol.

salivary glands (sal ih ver ee **glandz**) the paired glands of the mouth that secrete saliva.

salivary amylase (**sal** i ver ee **am** uh lace) a digestive enzyme found in saliva that breaks down starch into molecules of the disaccharide maltose.

tion, the starch is broken down on the addition of water—a process called *hydrolysis.*

$$\text{Starch} + \text{Water} \xrightarrow{\text{Salivary amylase}} \text{Maltose}$$

Enzymes that break down substances by hydrolysis are called *hydrolyzing enzymes.* Along with playing this role in digestion, the saliva also moistens and lubricates food so that it is swallowed easily and does not scratch the tissues of the throat.

The secretions of the salivary glands are controlled by the nervous system. Although the nervous system works to maintain a constant secretion of saliva in the mouth, it speeds up the secretion when it is stimulated by the presence (or sometimes the sight, smell, or thought) of food. When food is in the mouth, nerve endings called *chemoreceptors,* which are sensitive to the presence of certain chemicals, send a signal to the brain, which responds by stimulating the salivary glands. Did you ever suck on a slice of lemon? Then you know that the most potent stimuli are acid solutions; lemon juice can increase the rate of salivation eightfold.

Mechanical digestion begins when the teeth tear food apart into tiny pieces. In this way the surface area of the food is increased, allowing the digestive enzymes to mix with the food and break it down more quickly and completely. The teeth of humans, as well as those of other organisms, are specialized in different ways. This specialization depends on the type of food an organism eats and how it obtains its food. Humans are *omnivores.* An omnivore eats both plant and animal foods. As a result, human teeth are structurally intermediate between the pointed, cutting teeth characteristic of *carnivores,* or meat eaters, and the flat, grinding teeth characteristic of *herbivores,* or plant eaters. In fact, the teeth of humans are like carnivores in the front and herbivores in the back.

The four front teeth in the upper and lower jaws of humans are *incisors* (Figure 8-2, *A*). These teeth are sharp and chisel shaped and are used for biting. On each side of the incisors are pointed teeth

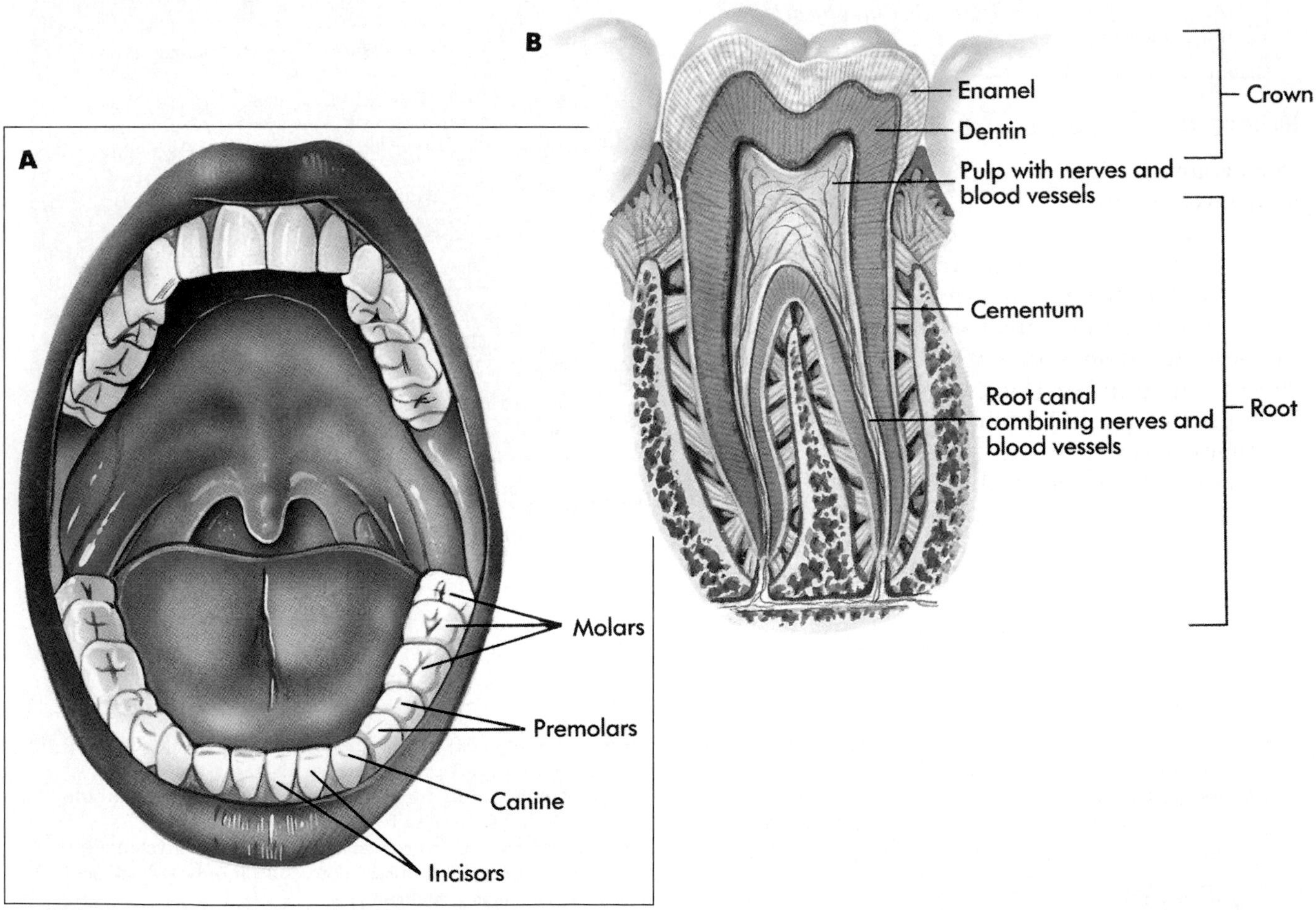

FIGURE 8-2 Teeth in humans. A, The four front teeth are incisors, used for biting. Canines are used for tearing food. The premolars and molars are used for grinding and crushing. **B,** Although you may forget until you are in the dentist's chair, each tooth in your mouth is alive and contains nerves and blood vessels.

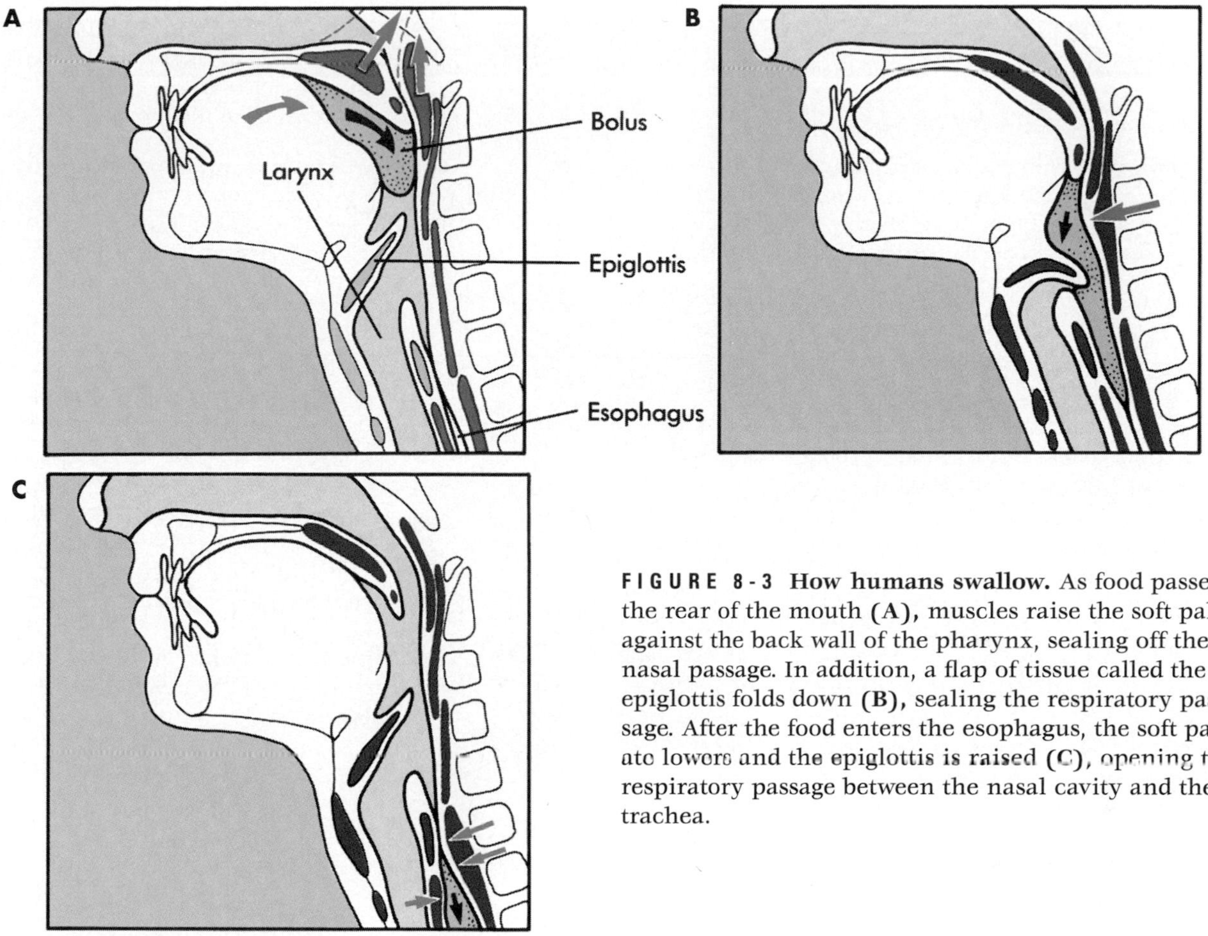

FIGURE 8-3 How humans swallow. As food passes the rear of the mouth **(A)**, muscles raise the soft palate against the back wall of the pharynx, sealing off the nasal passage. In addition, a flap of tissue called the epiglottis folds down **(B)**, sealing the respiratory passage. After the food enters the esophagus, the soft palate lowers and the epiglottis is raised **(C)**, opening the respiratory passage between the nasal cavity and the trachea.

called *canines,* which are used in tearing food. Behind each canine, on each side of the mouth and along both top and bottom jaws, are two *premolars* and three *molars,* all of which have flattened ridged surfaces for grinding and crushing food. In early childhood, however, humans do not have these 32 adult teeth but only 20 "baby teeth." These first teeth are lost during childhood and are replaced by the 32 adult teeth.

Each tooth is alive and is rooted in the bones of the upper and lower jaw. The gums cover this bone; the portion of the tooth protruding above the gumline is called the *crown* (Figure 8-2, *B*). The portion of the tooth that extends into the bone is called the *root.* Inside lies a central, nourishing *pulp* that contains nerves, blood vessels, and connective tissue. The nerves and blood vessels pass out of the tooth through holes at the root. The actual chewing surface of the tooth is made up of enamel—the hardest substance in the body. It is layered over the softer dentin that forms the body of the tooth.

Teeth shred and grind the plant and animal material that humans ingest as food. Saliva secreted into the mouth moistens the food, which aids its journey into the rest of the digestive system and begins its enzymatic digestion.

The Journey of Food to the Stomach

As food is chewed and moistened, the tongue forms it into a balllike mass called a *bolus* and pushes it into the **pharynx.** The pharynx is the upper part of the throat (Figure 8-3). As this happens, the soft palate raises up, sealing off the nasal cavity and preventing any food from entering this chamber. The soft palate is the tissue at the back of the roof of the mouth. The pressure of the food in the pharynx stimulates nerves in its walls that begin the swallowing reflex, an involuntary action. As part of

pharynx (**fair** inks) the upper part of the throat.

JUST WONDERING . . . .

I get heartburn a lot. What causes it?

Heartburn results when the acid contents of the stomach splash back into the esophagus at the cardiac opening. The acid burns the esophagus—but it feels like your heart is on fire! You're not alone in your pain: 10% of adults report experiencing heartburn about once a week, and more than 30% experience it once a month.

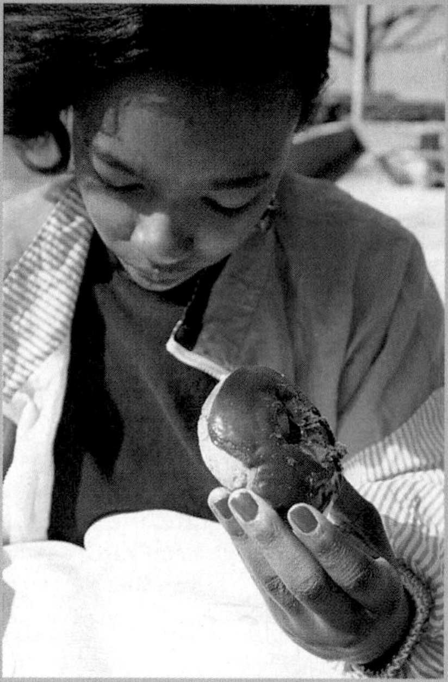

Occasional heartburn is not a serious condition. However, frequent heartburn can irritate the delicate lining of the esophagus. In severe cases, stomach acid injures the esophagus so much that it bleeds. If your heartburn is severe and frequent, you should consult a physician.

There are a few things you can do to avoid or relieve heartburn. If you smoke, understand that cigarette smoke relaxes the lower esophageal sphincter, the muscle that closes the cardiac opening. Therefore, quitting smoking will reduce your heartburn or get rid of it altogether. In addition, try not to overeat, which stuffs your stomach too full of food and promotes reflux (movement of stomach contents into the esophagus). Avoid eating close to bedtime, because lying down after eating with food in your stomach also promotes reflux. Try elevating the head end of your bed about six inches. Tight clothing around the middle causes some people trouble, as does obesity. Both conditions put extra pressure on the abdominal organs and may force food into the esophagus. Last, avoid the following foods and medications if you can; they all relax the lower esophageal sphincter: alcohol, chocolate, fats, peppermints, birth control pills, antihistamines, antispasmodics, some heart medications, and some asthma medications. As a last resort, many persons find relief by taking over-the-counter liquid antacids.

this reflex action, the voice box, or **larynx,** raises up to meet the **epiglottis,** a flap of tissue that folds back over the opening to the larynx. With this action the epiglottis acts much like a trapdoor, closing over the **glottis,** the opening to the larynx and trachea (your windpipe), so that food will not go down the wrong way. If you place your hand over your larynx (Adam's apple), you can feel it move up when you swallow.

After passing into the pharynx and then bypassing the windpipe, the food enters the **esophagus,** a food tube that connects the pharynx to the stomach (Figure 8-4). The esophagus, which is about 25 centimeters long (a bit less than a foot), pierces the diaphragm before it connects to the stomach. The diaphragm is a sheetlike muscle that forms the floor of the chest cavity; it is a muscle of breathing (see Chapter 8). The opening in the diaphragm that allows the esophagus to pass is called the *esophageal hiatus* (opening).

The esophagus ends at the door to the stomach. This door is called the *cardiac opening* and is ringed by a circular muscle called the *lower esophageal sphincter.* When the ring of muscle contracts, or tightens, it closes the cardiac opening. When it relaxes, or loosens, it opens this stomach door. Although it is in the abdominal cavity, the cardiac opening is located very close to the heart, which lies just above it in the thoracic cavity.

The esophagus does not take part in digestion but instead acts like an escalator, moving food down toward the stomach. The circular muscles in its walls produce successive waves of contractions called **peristalsis** (Figure 8-5). The movement of food to the stomach is therefore not dependent on gravity, so even astronauts in zero gravity—or you standing on your head—can swallow without difficulty.

> *When the tongue pushes food into the pharynx, nerves stimulate the swallowing reflex. The food then enters the esophagus, or food tube, and is moved to the stomach by rhythmic muscular contractions called peristalsis.*

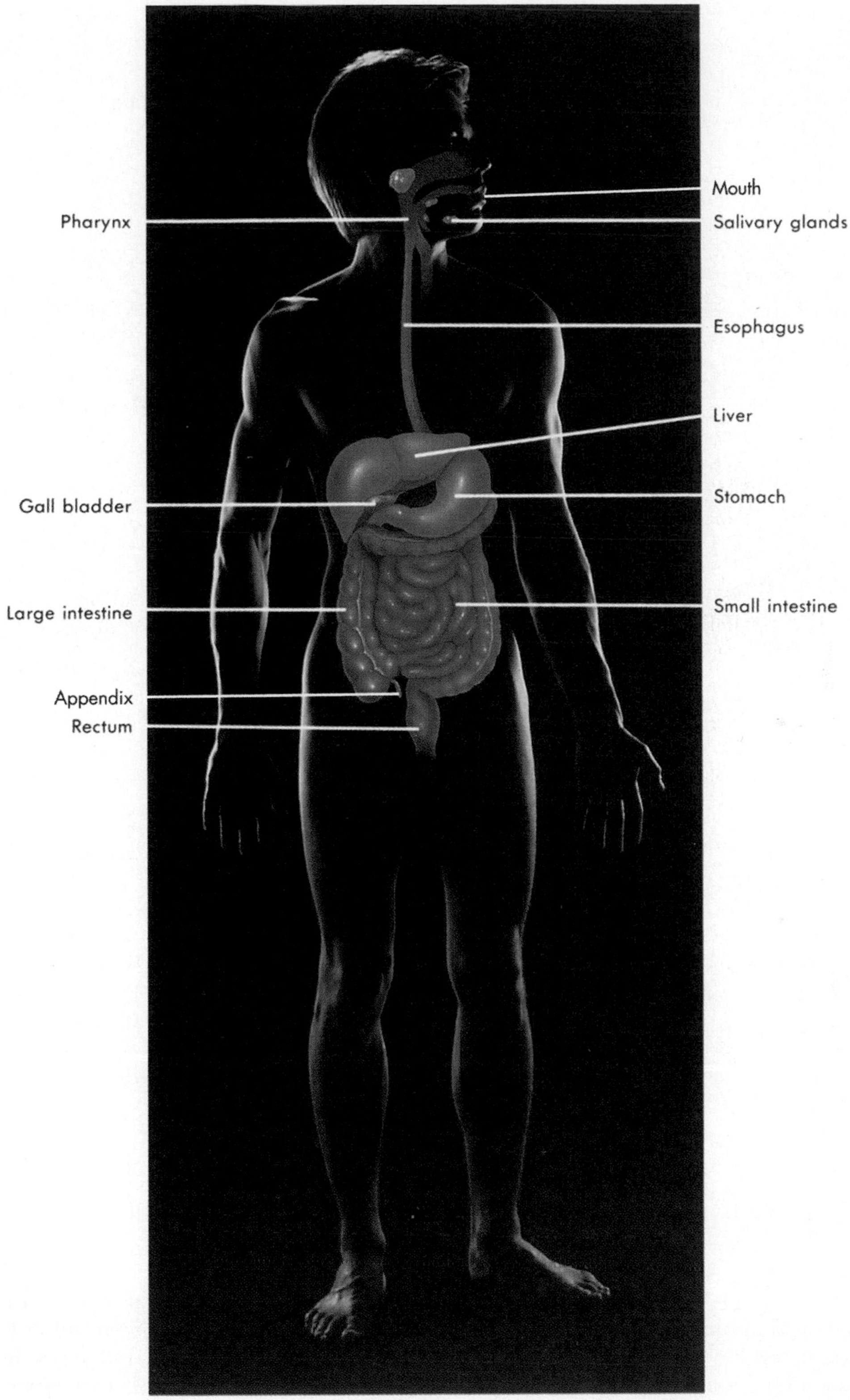

FIGURE 8-4 The human digestive system. The digestive system as located in the body.

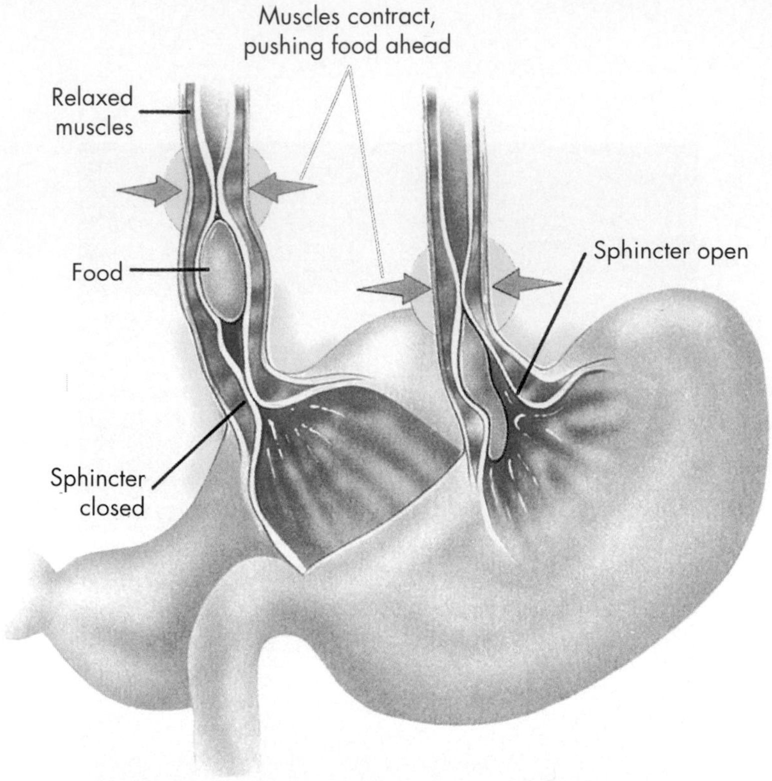

FIGURE 8-5 Peristalsis. Successive waves of contraction move food down the esophagus to the stomach.

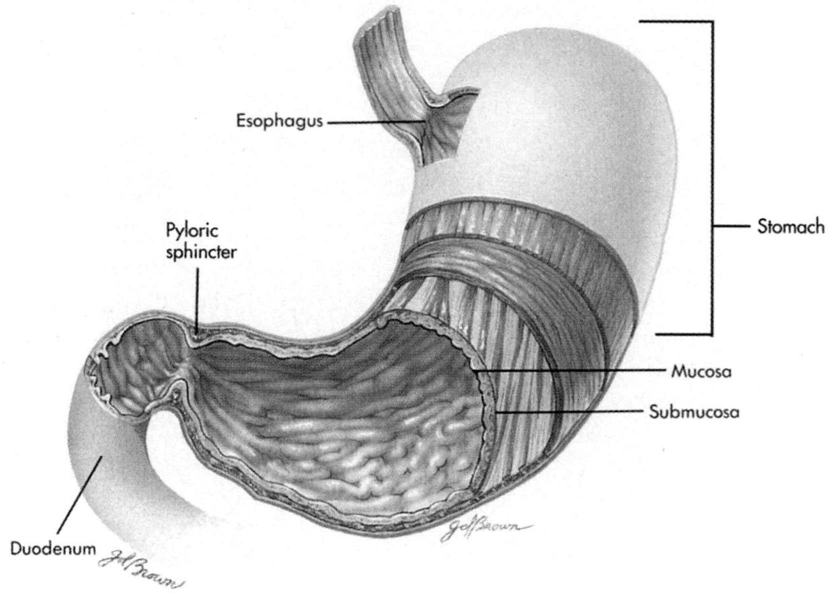

FIGURE 8-6 The stomach. The stomach stores food, partially digests it with hydrochloric acid and proteases, and moves the food to the small intestine.

Preliminary Digestion: The Stomach

The lower esophageal sphincter relaxes when food reaches it, allowing food to enter the **stomach.** The stomach is a muscular sac in which the food is collected and partially digested by hydrochloric acid and proteases (Figure 8-6). The stomach then "feeds" this food, little by little, to the primary organ of digestion, the small intestine.

The stomach and intestinal tract have the same basic structural plan (Figure 8-7). Their interiors are lined with a layer of tissue called the *mucosa.* The mucosa consists of epithelial cells, blood and lymph vessels, and a thin layer of muscle. It covers a deeper, thicker layer of connective tissue, the *submucosa,* which is rich in blood vessels and nerves. Surrounding the connective tissue are layers of smooth muscle tissue—three in the stomach and two in the intestines. An envelope of tough connective tissue called *serosa* serves as the outer covering of the digestive tract (as well as the other ab-

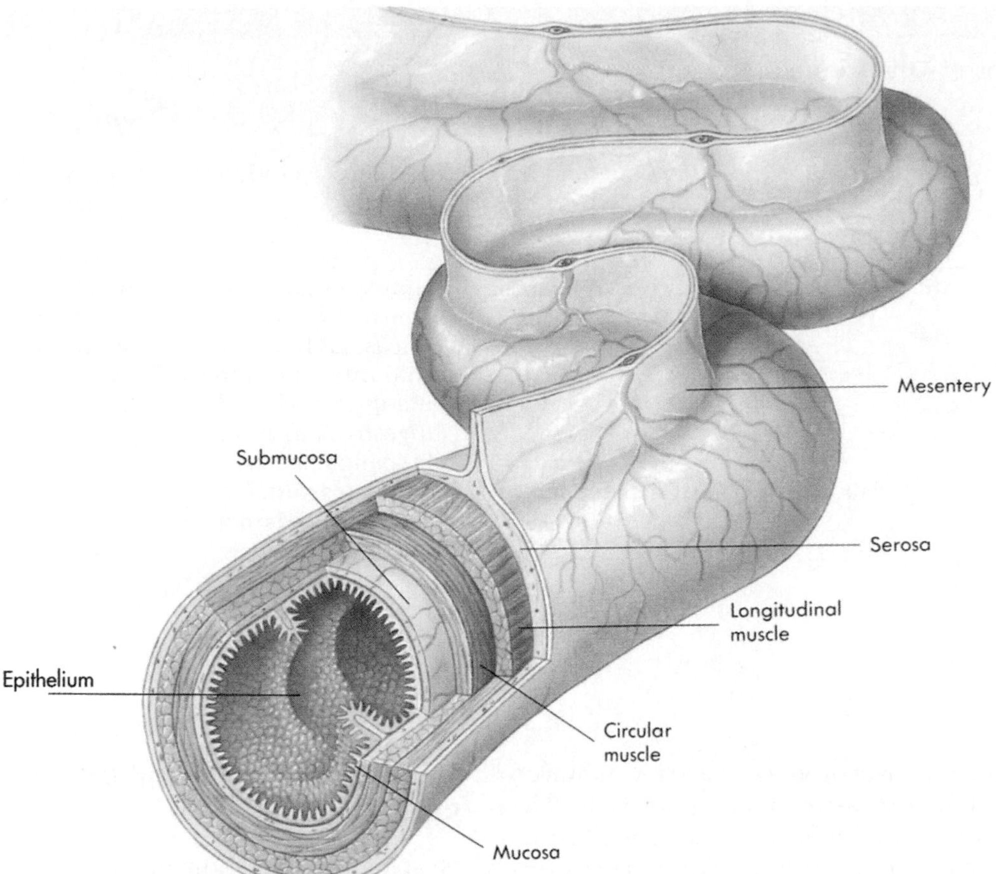

FIGURE 8-7 Digestive tract. Many layers of tissue characterize the digestive tract.

dominal organs). Thin sheets of connective tissue called *mesentery* are attached to the serosa along most of the intestinal tract, holding it in place and serving as a highway for blood vessels and nerves.

The inner surface of the stomach is dotted with **gastric glands** that extend from the epithelium but dip deeply into the mucosa. Two different kinds of cells in these glands secrete a gastric juice made up of hydrochloric acid and the protein pepsinogen. After the secretion of both, the acid chemically interacts with the pepsinogen, converting it to the protein-hydrolyzing enzyme **pepsin.** The hydrochloric acid also softens the connective tissue in foods; denatures, or unfolds, large protein molecules; and kills most bacteria that may have been ingested with the food. The pepsin digests only proteins, breaking them down into short polypeptides. Other epithelial cells are specialized for the secretion of mucus. This mucus, produced in large quantities, lubricates the stomach wall and protects the stomach from digesting itself.

The stomach controls the production of gastric juice by means of a digestive hormone called **gastrin** (Table 8-3). Hormones are regulating chemicals that are made at one place in the body and work in another. Gastrin is produced by endocrine (hormone-secreting) cells that are scattered throughout the epithelium of the stomach. Some

larynx (**lair** inks) the voice box, which is located at the upper end of the human windpipe.
epiglottis (**ep** i **glot** is) a flap of tissue that folds back over the opening to the larynx, thus preventing food or liquids from entering the airway.
glottis (**glot** is) the opening to the larynx and trachea.
esophagus (ih **sof** uh gus) the food tube that connects the pharynx to the stomach.
peristalsis (**pear** i **stal** sis) successive waves of contractions of the esophagus and small intestine that move food along these parts of the digestive system.
stomach a muscular sac in which food is collected and partially digested by hydrochloric acids and proteases.
gastric glands glands dotting the inner surface of the stomach that secrete a gastric juice of hydrochloric acid and pepsinogen.
pepsin an enzyme of the stomach that digests only proteins, breaking them down into short peptides.
gastrin a digestive hormone of the stomach that controls the production of gastric juice.

TABLE 8-3

Hormones of Digestion

HORMONE	SOURCE	STIMULUS	ACTION	NOTE
Gastrin	Stomach	Entry of food into stomach	Regulates secretion of HCl	This hormone is unusual in that it acts on the same organ that secretes it.
Cholecystokinin (CCK)	Duodenum	Arrival of food in small intestine	Stimulates gallbladder contraction and release of bile salts into intestine, stimulates secretion of digestive enzymes by pancreas	CCK bears a striking structural resemblance to gastrin.
Secretin	Duodenum	HCl in duodenum	Stimulates pancreas to secrete bicarbonate, which neutralizes stomach acid	Secretin was the first hormone to be discovered (1902).

stomachs greatly overproduce gastrin, however, which results in excessive acid production. The causes of this overproduction include such factors as stress, heredity, diet, and smoking. This extra acid may attack the walls of the first portion of the small intestine, or duodenum, burning holes through the wall. These holes are called *duodenal ulcers* (see essay in Chapter 25). Because the contents of the small intestine are not normally acidic, this organ is much less able to withstand the disruptive actions of stomach acid than the wall of the stomach is. For this reason, more than 90% of all ulcers are duodenal, although other ulcers sometimes occur in the stomach.

Food stays in the stomach for approximately 2 to 6 hours. During this time the contractions of the muscular wall of the stomach churn the food, mixing it with the gastric juice and mucus. By the time the food is ready to leave the stomach as a substance called *chyme,* it has the consistency of pea soup. The gate to the small intestine, the *pyloric sphincter,* opens to allow just a bit of the chyme to pass. When the acid in this chyme is neutralized and the food is digested, the pyloric sphincter is signaled by the nervous system to open again, allowing the next bit of chyme to pass.

In the stomach, concentrated acid breaks up connective tissue and unfolds proteins. These proteins are digested by pepsin into short polypeptides. Starches and lipids are not digested in the stomach.

Terminal Digestion and Absorption: The Small Intestine

Most of the digestion of food takes place in the small intestine. Within this organ, starches and glycogen are broken down to sugars, proteins to amino acids, and lipids to fatty acids and glycerol. These products of digestion then pass through the cells of the intestinal mucosa and diffuse into the blood in underlying blood vessels. The triglycerides move into the lymph in neighboring lymphatic vessels.

All of this activity takes place in the **small intestine,** the tubelike portion of the digestive tract that begins at the pyloric sphincter and ends at its T-shaped junction with the large intestine. Although the small intestine is approximately 6 meters long—long enough to stretch from the ground to the top of a two-story building—only the first 25 centimeters (8 inches) is actively involved in digestion. This initial portion of the small intestine is called the **duodenum.** The other portions of the small intestine, the **jejunum** and the **ileum,** are highly specialized (along with the duodenum) to aid in the absorption of the products of digestion by the blood and lymph.

Accessory Organs That Help Digestion

Some of the enzymes necessary for digestion are secreted by the salivary glands, epithelial cells of the stomach, and epithelial cells of the duodenum. The others are secreted by the **pancreas,** a long gland that lies beneath the stomach and is surrounded on one side by the curve of the duodenum (see Figure 8-4). A tiny duct runs from the pancreas to the small intestine and serves as the passageway for the pancreatic juice. As you can see from Table 8-2, this secretion of the pancreas includes a number of digestive enzymes.

The **liver** is another organ that works with the duodenum to digest food. This organ, which weighs over 3 pounds, lies just under the diaphragm (see Figure 8-4). It is one of the most complex organs of the body and performs more than 500 functions! Although the liver produces no digestive enzymes, it does help in the digestion of lipids by secreting a collection of molecules called **bile.** One of the many components of bile is bile pigments, breakdown products of the hemoglobin from old, worn-out red blood cells. These pigments give bile its greenish color. The bile also contains bile salts, which are substances that act much like detergents, breaking lipids up into minute droplets of triglycerides, similar to droplets of cream suspended in milk. The liver manufactures bile salts from cholesterol. Excess bile is stored and concentrated in the **gallbladder** on the underside of the liver. A bile duct brings bile from the liver and gallbladder to the small intestine.

> *The liver and pancreas, although not organs of the digestive system, help digestion take place. The pancreas secretes a number of digestive enzymes. The liver produces bile as one of its numerous and diverse functions. One component of bile, bile salts, aids in lipid digestion.*

Digestion

As the food enters the small intestine, some of it has already been partially digested. Salivary amylase has broken some of the bonds in the starches and glycogen, producing the disaccharide maltose. However, much starch and glycogen remain undigested. In the small intestine, **pancreatic amylase** breaks down these carbohydrates to maltose. Maltose, sucrose (table sugar), and lactose (milk sugar) are digested to the monosaccharides glucose, fructose, and galactose by enzymes called **disaccharidases** that are produced by specialized epithelial cells of the small intestine. Some people do not produce the enzyme to digest lactose, however, and are therefore unable to digest milk. These persons cannot drink milk without experiencing cramps and, in some cases, diarrhea.

Some of the proteins in the food entering the small intestine have also been partially digested. The hydrochloric acid of the stomach has unfolded these proteins, and pepsin has cleaved some of them to shorter polypeptides. Three other enzymes produced by the pancreas complete the digestion of proteins: **trypsin, chymotrypsin,** and **carboxypeptidase.** These enzymes work as a team with **peptidases** produced by cells in the intestinal epithelium, breaking down polypeptides into shorter chains and then to amino acids. These enzymes are secreted in an inactive form and become active in the presence of a particular enzyme secreted by cells in the intestinal epithelium. This way, they do not digest their way down the pancreatic duct to the small intestine.

small intestine the tubelike portion of the digestive tract that begins at the pyloric sphincter and ends at its T-shaped junction with the large intestine.

duodenum (**doo** oh **dee** num *or* doo **odd** un um) the initial short segment of the small intestine that is actively involved in digestion and absorption of nutrients.

jejunum (ji **joo** num) the second portion of the small intestine, which is highly specialized for absorption of nutrients and extends from the duodenum to the ileum.

ileum (**ill** ee um) the third and final part of the small intestine which is highly specialized for absorption of nutrients. The ileum follows the jejunum.

pancreas (**pang** kree us *or* **pan** kree us) a long gland that lies beneath the stomach and is surrounded on one side by the curve of the duodenum; it secretes a number of digestive enzymes and the hormones insulin and glucagon.

liver a large, complex organ weighing over 3 pounds, lying just under the diaphragm, that performs more than 500 functions in the body, including the secretion of bile, which aids in the digestion of lipids.

bile a collection of molecules secreted by the liver that help in the digestion of lipids.

gallbladder a sac attached to the underside of the liver, where bile is stored and concentrated.

pancreatic amylase (**pang** kree **at** ick *or* **pan** kree **at** ick **am** uh lace) a digestive enzyme secreted by the pancreas that works within the small intestine to break down starch and glycogen to maltose.

diaccharidases (dye **sack** uh rye **days** is) enzymes, produced by specialized epithelial cells of the small intestine, that break down the disaccharides maltose, sucrose, and lactose to the monosaccharides glucose, fructose, and galactose.

trypsin (**trip** sin) an enzyme produced by the pancreas, which, together with chymotrypsin and carboxypeptidase, completes the digestion of proteins in the small intestine.

chymotrypsin (**kye** moe **trip** sin) an enzyme produced by the pancreas, which, together with trypsin and carboxypeptidase, completes the digestion of proteins in the small intestine.

carboxypeptidase (kar **bok** see **pep** ti **dace**) an enzyme produced by the pancreas, which, together with trypsin and chymotrypsin, completes the digestion of proteins in the small intestine.

peptidases (**pep** teh **dace** is) enzymes produced by cells in the intestinal epithelium that work as a team with trypsin, chymotrypsin, and carboxypeptidase to break down polypeptides into shorter chains and then to amino acids.

The lipids are not digested until they reach the small intestine. Because they are insoluble in water, they tend to enter the small intestine as globules. Before these lipids can be digested, they are emulsified, or made soluble, by bile salts. When the lipid globules are in the form of triglycerides, **pancreatic lipase** cleaves them into fatty acids and glycerol.

In the first part of the small intestine, the duodenum, digestion is completed. Undigested starch and glycogen are digested to disaccharides and then to monosaccharides. Undigested proteins and other polypeptides are digested to shorter peptides and then to amino acids. Triglycerides are made soluble and are then digested to fatty acids and glycerol.

As you can see, the digestion of food involves so many players that the digestive team could use a manager! In fact, the key players in digestion—the liver, gallbladder, pancreas, stomach, and small intestine—have more than one manager. These managers of digestion are hormones (see Table 8-3). Earlier in this chapter you read how the hormone gastrin controls the release of hydrochloric acid in the stomach. Two other hormones, **secretin** and **cholecystokinin (CCK),** control digestion in the small intestine (Figure 8-8).

When chyme enters the small intestine, the acid in it stimulates cells in the intestinal mucosa to produce secretin. This hormone does two things: first, it stimulates the release of an alkaline fluid called *sodium bicarbonate* from the pancreas. This solution neutralizes the acid in the chyme so that it will not damage the wall of the small intestine

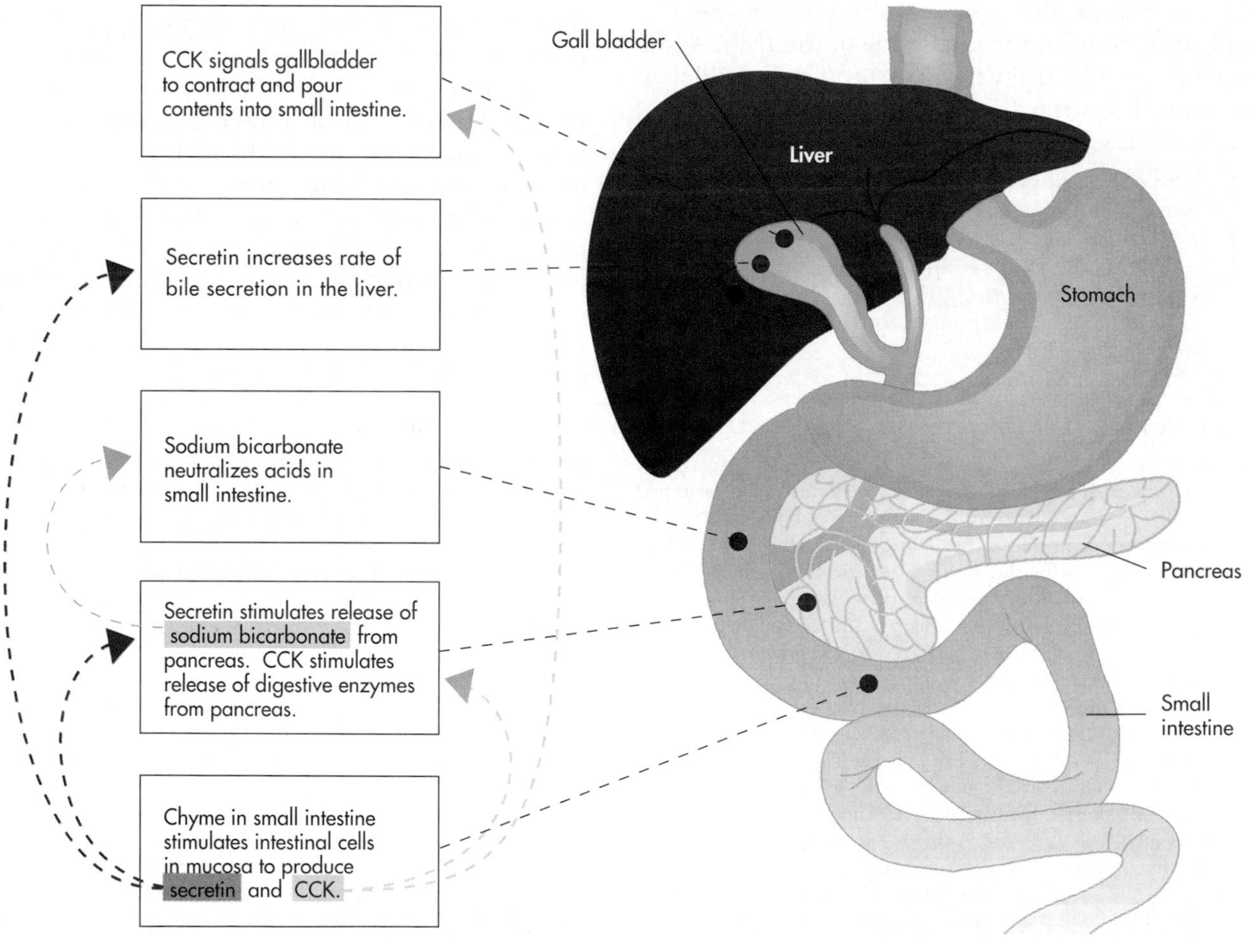

CCK signals gallbladder to contract and pour contents into small intestine.
Secretin increases rate of bile secretion in the liver.
Sodium bicarbonate neutralizes acids in small intestine.
Secretin stimulates release of sodium bicarbonate from pancreas. CCK stimulates release of digestive enzymes from pancreas.
Chyme in small intestine stimulates intestinal cells in mucosa to produce secretin and CCK.
Gall bladder
Liver
Stomach
Pancreas
Small intestine

FIGURE 8-8 Digestive hormones manage many digestive enzymes. Secretin and cholecystokinin control and integrate digestive processes in the small intestine. The boxes describe events that take place during digestion. The line from each box to an organ of the digestive system shows the approximate location of the events. The arrows indicate how one event triggers another in the digestive process.

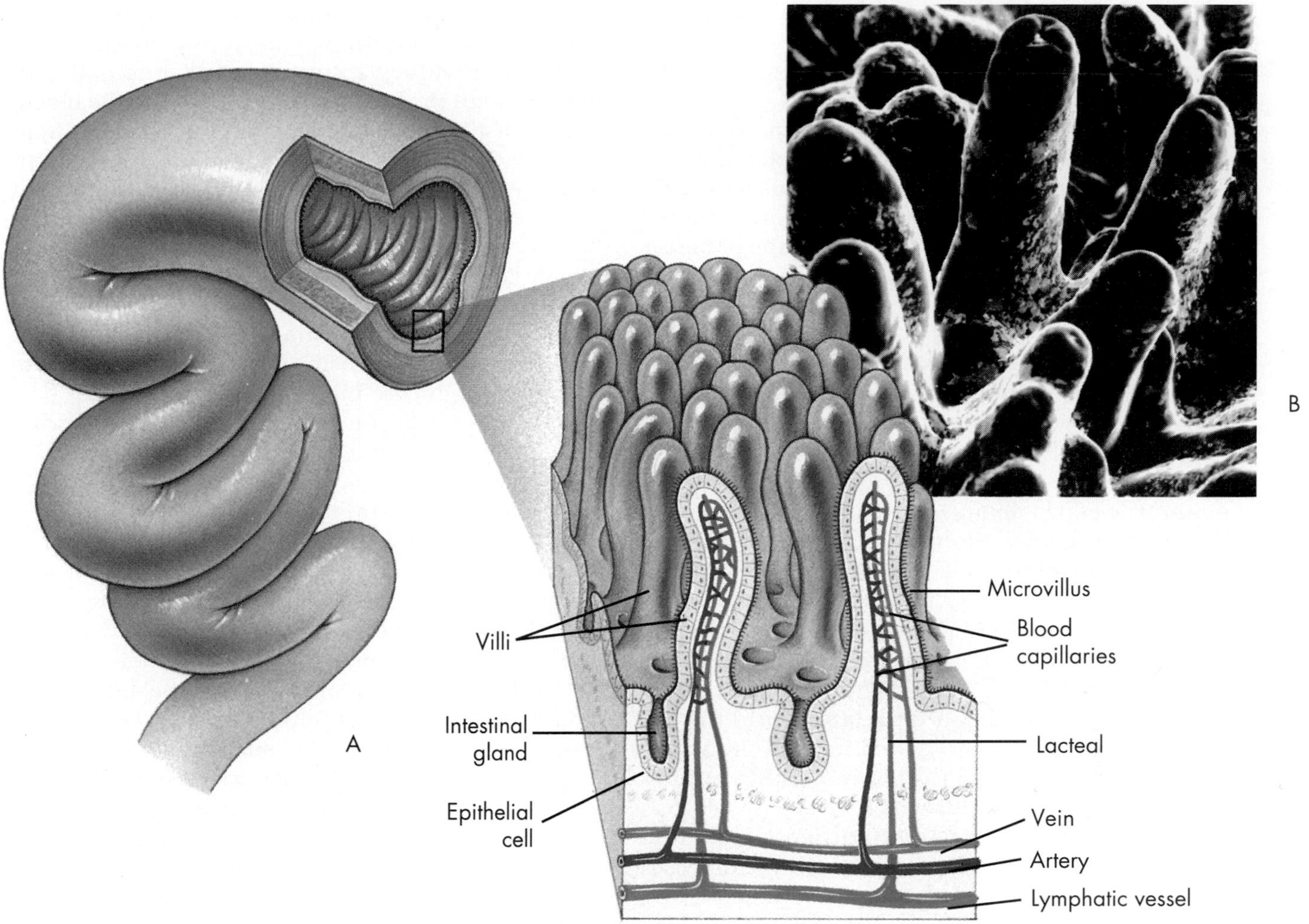

FIGURE 8-9 Structure of villi. A, Villi are fingerlike projections of tissue located on the inner surface of the small intestine. The villi, and microvilli on the surface of the villi, greatly increase the surface area of the small intestine, providing more room for absorption. **B,** Electron micrograph of intestinal villi *(150x)*.

pancreatic lipase (**pang** kree **at** ick **lye** pace *or* **lip** ace) a digestive enzyme secreted by the pancreas that works in the small intestine to break down triglycerides to fatty acids and glycerol.

secretin (si **kre** tin) one of the hormones that control digestion in the small intestine.

cholecystokinin (CCK) (**kol** uh **sis** tuh **kine** un) one of the hormones that control digestion in the small intestine.

villi (**vil** eye) fine, fingerlike projections of the epithelium of the small intestine.

and produces the proper pH in which the pancreatic and intestinal enzymes will work. In addition, secretin increases the rate of bile secretion in the liver.

The presence of fatty acids and partially digested proteins in the chyme stimulates the mucosa to produce CCK. This enzyme signals the gallbladder to contract and pour its contents into the small intestine. It is also a stimulus to the pancreas to release its digestive enzymes.

Absorption

The amount of material passing through the small intestine is startlingly large. An average person consumes about 800 grams of solid food and 1200 milliliters of fluid each day, for a total volume of about 2 liters, or a little more than 2 quarts. To this amount is added about 7 liters of fluid secreted by the salivary glands, stomach, pancreas, liver, and the small intestine itself. The total is a remarkable 9 liters, or almost $2^1/_2$ gallons! Of the 800 grams of solid food and 9 liters of liquid that enter the digestive tract, only about 50 grams of solid and 100 milliliters of liquid leave the intestinal tract as waste, or *feces.* The small intestine absorbs the 750 grams of nutrients per day and most of the water; the large intestine absorbs the remaining water—approximately 750 milliliters.

To aid in the absorption of nutrients, the internal surface area of the small intestine increases dramatically in three ways. First, the mucosa and submucosa of the small intestine are thrown into folds; they do not have a smooth inner surface like a garden hose. This folded surface, in turn, is covered by fine, fingerlike projections of the epithelium. These projections are called **villi** (singular villus) and are so small that it takes a microscope to see them (Figure 8-9). In addition, the epithelial cells of the villi are covered on their exposed surfaces by cytoplasmic projections called *microvilli.* The infoldings, the villi, and the microvilli provide the small intestine with a surface area of about 300 square meters, or 2700 square feet, an area greater than the floor space in many homes!

Within each villus is a network of capillaries and a lymphatic vessel called a *lacteal.* Although the nutrients from digested food do not have far to travel—through a single layer of epithelial cells and a single layer of cells forming the wall of the capillary or lacteal—a great deal of work must be done. Each monosaccharide and amino acid must catch a ride on a carrier molecule to get into an epithelial cell. Energy is needed to ferry many of these molecules across cell membranes—a process called *active transport* (see Chapter 3). Others are taken up by facilitated diffusion. Once the monosaccharides and amino acids are in the epithelial cells, however, they accumulate and eventually diffuse through the base of the cell and into the blood. When in the bloodstream, they are quickly swept away to the liver for processing and storage. When blood levels of glucose are sufficient to supply all your cells with this fuel of cellular respiration (see Chapter 5), the liver stores glucose as glycogen. When more glucose is needed, such as between meals, the liver readily converts the glycogen back to glucose. In this way, the liver is your metabolic bank, accepting deposits and withdrawals in the currency of glucose molecules.

The absorption of lipids takes place somewhat differently. After lipids are broken down into fatty acids and glycerol, they become surrounded by bile salts. Packaged in this way, they move to the cell membranes of the villi. As you may recall from Chapter 3, one of the main ingredients in cell membranes is lipid. Therefore when the fatty acids and glycerol from digestion come into contact with these cell membranes, they discard their shell of bile salts and easily move across the membrane and into the cell. Short-chained fatty acids are absorbed directly into the bloodstream. Longer-chained fatty acids are reassembled into triglycerides by the endoplasmic reticulum and are then encased in protein. After this processing, they pass out of the epithelial cells and into the lacteal. These protein-coated triglycerides are then transported in the lymphatic fluid through a system of vessels that drains the lymph into the blood at the left subclavian vein, a major blood vessel at the base of the neck.

The internal surface area of the small intestine is increased by the presence of inner folds and projections, which results in a more efficient absorption of nutrients and water. Monosaccharides and amino acids are absorbed into the intestinal epithelium by means of facilitated diffusion and active transport and then diffuse into the bloodstream. Epithelial cells re-form fatty acids and glycerol to triglycerides and shuttle them to the blood by means of the lymphatic system. Water is absorbed by osmosis.

Concentration of Solids: The Large Intestine

The large intestine, or **colon,** is much shorter than the small intestine—only about a meter and a half, or 5 feet, long. It is wide, however, having a diameter slightly less than the width of your hand. In

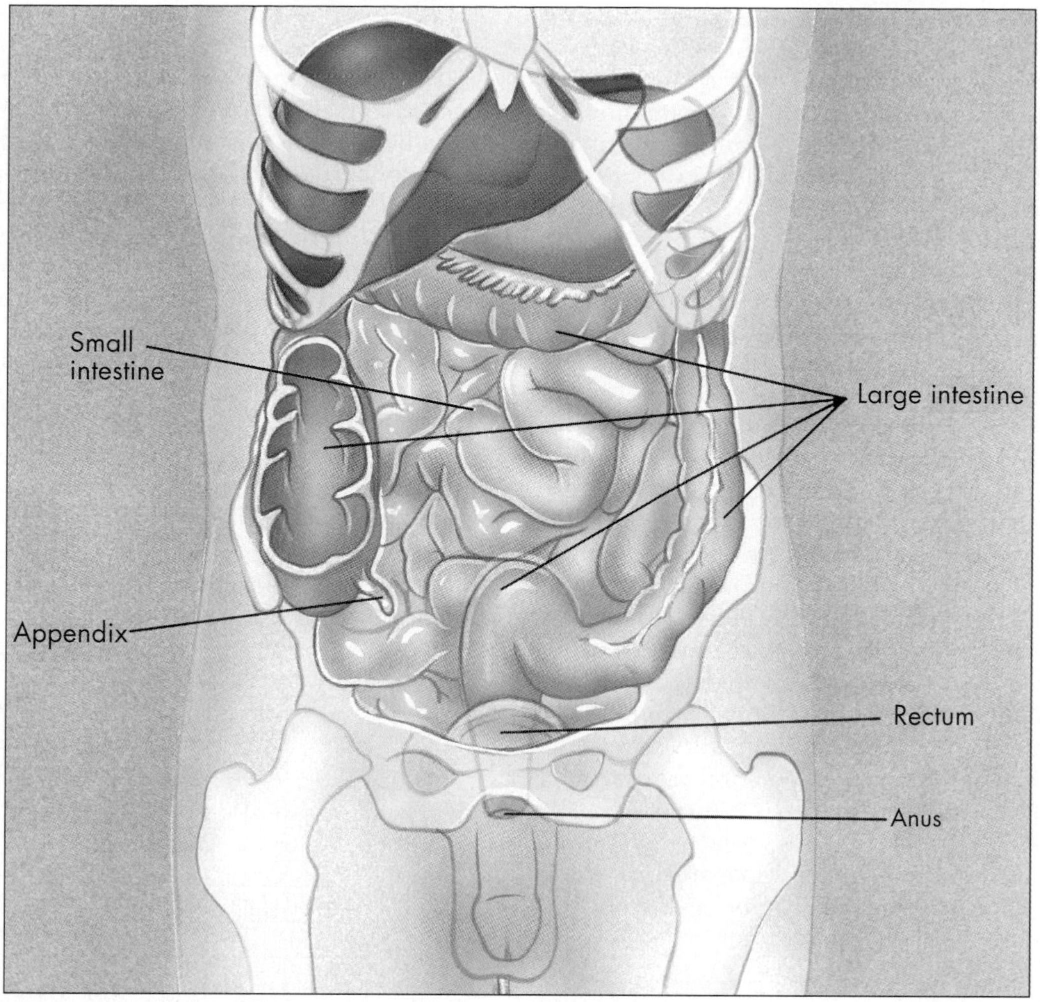

FIGURE 8-10 Location of the large intestine. The large intestine, or colon, terminates at the rectum. The appendix hangs from a blind (dead-end) pouch where the small intestine meets the colon. The appendix serves no function in humans.

contrast, the small intestine has a diameter only slightly larger than the width of two of your fingers. The small intestine joins the large intestine about 7 centimeters up from its end, creating a blind pouch out of the beginning of the large intestine (Figure 8-10). Hanging from this pouch is the **appendix,** a structure that serves no essential purpose in humans. An infection of the appendix is called *appendicitis* and can be quite serious and painful. In certain animals the blind pouch is more fully developed and is called the *cecum.* It serves as a place where the cellulose of plant cell walls is digested by the activity of intestinal bacteria and other microorganisms. Humans cannot digest cellulose, so it becomes a digestive waste. This waste, however, is important to the regular movement of the feces through the large intestine. Also called *dietary fiber,* undigested plant material provides bulk against which the muscles of the large intestine can push.

The junction of the small and large intestines is in the lower right side of the abdomen. From there the large intestine goes up the right side of the abdomen to the liver. It then turns left, crossing the abdominal cavity just under the diaphragm. On the left side of the abdominal cavity, it turns downward, ending at a short portion of the colon called

colon the large intestine; its function is to absorb sodium and water, and to eliminate wastes.

appendix (uh **pen** diks) a pouch that hangs from the beginning of the large intestine and serves no purpose in humans.

the **rectum.** The rectum terminates at the **anus,** the opening for the elimination of the feces.

No digestion takes place within the large intestine. Its job is to absorb sodium and water, to eliminate wastes, and to provide a home for friendly bacteria. These bacteria help keep out disease-causing microbes and produce certain vitamins, especially vitamin K.

Waste materials move slowly along the smooth interior of the large intestine as water and sodium are slowly reabsorbed. As they move along, the wastes become more compacted. If the wastes move too slowly through the colon, too much water may be reabsorbed, leading to a difficulty in elimination called *constipation.* Conversely, if the wastes move too quickly, as happens with certain intestinal illnesses, not enough water may be removed, resulting in diarrhea. Eventually, the solids within the colon pass into the rectum as a result of the peristaltic contractions of the muscles encasing the large intestine. From the rectum the solid material passes out of the anus through two anal sphincters. The first of these is composed of smooth muscle. It opens involuntarily in response to a pressure-generated nerve signal from the rectum. The second sphincter, in contrast, is composed of skeletal muscle. It is subject to voluntary control from the brain, thus permitting a conscious decision to delay *defecation,* or the elimination of waste.

The large intestine serves primarily to reabsorb water and sodium from digestive wastes and eliminate the remainder, or feces.

Diet and Nutrition

The digestion of food yields no usable energy but changes a diverse array of complex molecules into a small number of simpler molecules that can be used by your cells as fuel for cellular respiration. During cellular respiration, adenosine triphosphate (ATP) molecules are produced—the energy currency of your body. As you read in Chapter 5, glucose is completely broken down by the processes of glycolysis, the Krebs cycle, and the electron transport chain to yield 36 usable molecules of ATP. In addition, the carbon portions of amino acids and the glycerol backbones of triglycerides are converted to substances that can also be metabolized by two of these pathways. Fatty acids are metabolized by another metabolic pathway to yield ATP molecules.

Any intake of food in excess of that required to maintain the blood sugar (glucose) level and the glycogen reserve in the liver results in one of two consequences. Either the excess glucose is metabolized by the muscles and other cells of the body or it is converted to fat and stored within fat cells. Only when all the body's energy needs have been met—including the energy needed to run chemical reactions, move muscles, and digest food—will kilocalories be stored as fat. Think of your body as a giant scorecard, keeping track of kilocalories eaten and kilocalories used. If you eat more than you use, you will gain weight. Unfortunately, 10% to 25% of all teenagers and 25% to 50% of all adults in the United States are obese. In other words, they weigh at least 20% more than the average weight for their height (Figure 8-11).

As the digestive process breaks down food into molecules usable in cellular respiration, it also unlocks the vitamins and minerals from these foods.

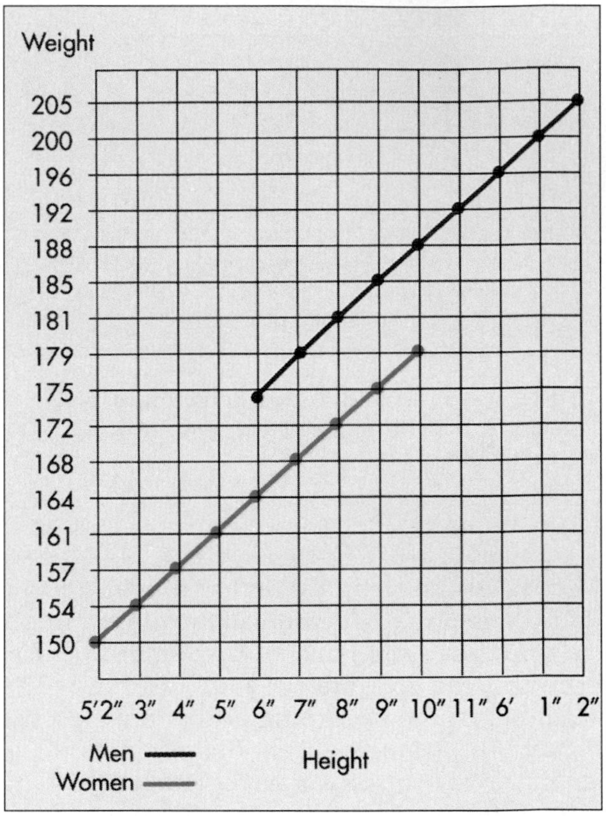

FIGURE 8-11 Obesity ranges for men and women. Obesity is usually characterized as the state of being more than 20% heavier than the average person of the same sex and height. The graph charts height versus weight, showing the lowest weights at which a person of an indicated height is considered obese by most physicians.

HOW *Science* WORKS

Applications

The Evolution of Food Guides

Today, more Americans than ever before are concerned about nutrition. This interest in a healthy diet has caught the attention of advertisers and the media, and food fads and fashions have proliferated. Oat bran, fish oils, and "good" cholesterol have all received acclaim for their healthy nutritional properties, but it seems that as soon as one food or nutrient is hailed as the key to a healthy diet, another quickly takes its place. Given all the conflicting studies and claims about what we should and should not eat, how can we be sure that our diet is really nutritious?

A practical approach is to use a food guide, an easy-to-use chart developed by nutritionists to help people monitor the quality of their diets. A food guide groups foods of similar nutrient composition together. Guides also recommend how much of each food should be eaten to meet nutrient needs. For example, the guide presently in use in the United States, the food pyramid, groups fruits together because they contribute a major portion of vitamins A, C, and folate to the diet. This guide recommends eating two to four servings of fruit daily. And because not all foods in a grouping are equally nutritious, nutritionists recommend eating a variety of foods from each group.

In 1943, in an effort to promote nutrition education, the U.S. Department of Agriculture (USDA) introduced the basic seven food guide. This plan remained in use until 1956, when it was replaced by the basic four guide. The basic seven guide divided foods into seven groups: (1) green and yellow vegetables; (2) oranges, tomatoes, and raw salads; (3) potatoes; (4) milk and cheese; (5) meat, poultry, fish, and eggs; (6) bread, flour, and cereals; and (7) butter and margarine. The basic four guide combined the basic seven's three fruit and vegetable groupings into one group and eliminated the butter and margarine group. The basic four plan was not revised for more than 20 years, and it formed the basis for most nutrition education during that time.

By the late 1970s, critics charged that the basic four guide did not reflect current scientific findings about the role of such nutrients as fats in the development of such diseases as heart disease and some cancers. Responding to these concerns, the USDA presented the *Hassle-Free Guide to a Better Diet* in 1979. This guide, sometimes referred to as the "basic five," added a fifth food group (fats, sweets, and alcoholic beverages) and recommended limiting intake of foods rich in these substances. Critics still were not satisfied, however, that these recommendations went far enough. For instance, recent studies have shown that a diet high in fiber was instrumental in the prevention of colon cancer. Other researchers had uncovered a link between a diet high in fat and an increased incidence of breast cancer. Nutritionists wanted a food guide that stressed the necessity of including high-fiber foods and limiting fat intake.

Again, the USDA responded with a revised food guide. The *Food Wheel: A Daily Pattern for Food Choices* recommended eating more fiber-rich fruits, vegetables, and whole grains and limiting dairy products and meats (foods that are high in fat) in the diet. By this time the development and use of food guides had become a political and economic issue in addition to one of health. Meat and dairy producers, for example, were not pleased with the latest food guide that recommended the limitation of their products in the American diet. The USDA realized that on this issue, it could not yield to the concerns of these producers and still present a scientifically valid food guide.

The latest food guide was introduced in 1992. Called the *Food Pyramid,* this guide continues to reflect health concerns. In some ways it is a return to the basic seven concept with more food groups, but it also adds recent recommendations on healthy eating by providing a visual model of the optimal diet. Grouped together at the base of the pyramid are cereals and other grain products. These foods have the highest number of recommended servings. Fruits and vegetables, grouped separately, form the next tier. Fewer servings of these foods are needed for a healthy diet, so the tier is smaller. Because they include dietary fat, meats, meat alternatives, and dairy products form the next, still smaller tier. At the top are high-sugar or fat foods and alcoholic beverages, which offer few vitamins and minerals in relation to calories.

Applications—cont'd

Food guides can be a convenient way to assess the nutritional quality of one's diet, but they are not perfect. Combination foods such as casseroles and ethnic foods such as Oriental or Mexican dishes may still be difficult to classify without checking recipes. As scientific research uncovers new information, food guides will continue to change, and nutrition educators will have to revise their teachings. It seems unlikely that any new food guide will be used for as long as was the now obsolete "Basic Four."

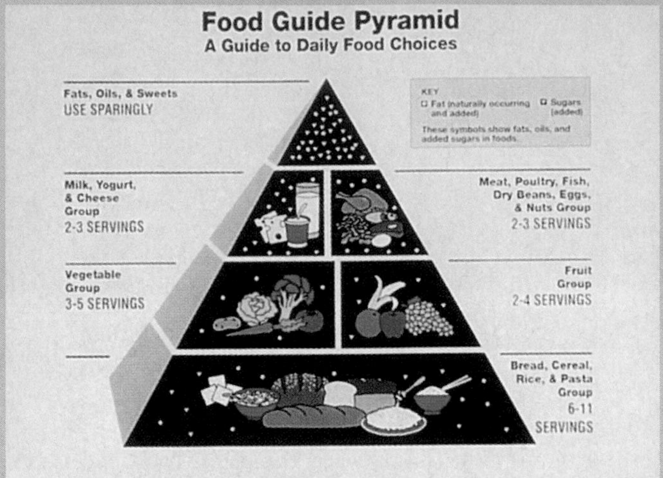

Vitamins fall into two general categories: water soluble and fat soluble (see Table 8-1). The water-soluble vitamins enter the cells of the intestinal mucosa and move into the bloodstream. The fat-soluble vitamins are carried across the membranes of the intestinal cells associated with the fatty acids and glycerols. They are also transported to the bloodstream by means of the lymphatic system. Minerals are absorbed into the bloodstream as ions. (The recommended dietary allowances of the various vitamins and minerals are listed in Table 8-1.) Most people who eat a sufficient amount and variety of foods get the vitamins they need in the food they eat.

Proteins, lipids, and carbohydrates provide more than energy for the diet—they provide the raw materials to build the substances the body needs. In fact, proteins are used primarily as building blocks and not as a source of energy. Of the 20 different amino acids that make up proteins, humans can manufacture only 12. Therefore 8 of the amino acids, called **essential amino acids,** must be obtained by humans from proteins in the food they eat (Figure 8-12). Protein foods that contain the essential amino acids in amounts proportional to the body's need for them are called high-quality, or complete, proteins. Unfortunately, many high-quality protein foods such as meat, cheese, and eggs are high in animal fat and cholesterol as well. A high percent of animal fat and cholesterol in the diet has been shown to result in weight gain and increase the risk of heart attack because of the clogging of blood vessels with cholesterol deposits. To reduce fat intake yet get the amino acids you need, you should combine any legume (dried peas, beans, peanuts, or soy-based food) with any grain, nut, or seed, or you should combine any grain, legume, nut, or seed with small amounts of milk, cheese, yogurt, eggs, red meat, fish, or poultry.

So how much protein, fat, and carbohydrate *should* you have in your diet? The U.S. Senate Select Committee on Nutrition and Human Needs recommends that 12% of your daily intake of kilocalories come from proteins, 30% *or less* come from fats, and 58% from carbohydrates (Figure 8-13). You may want to approximate the number of kilocalo-

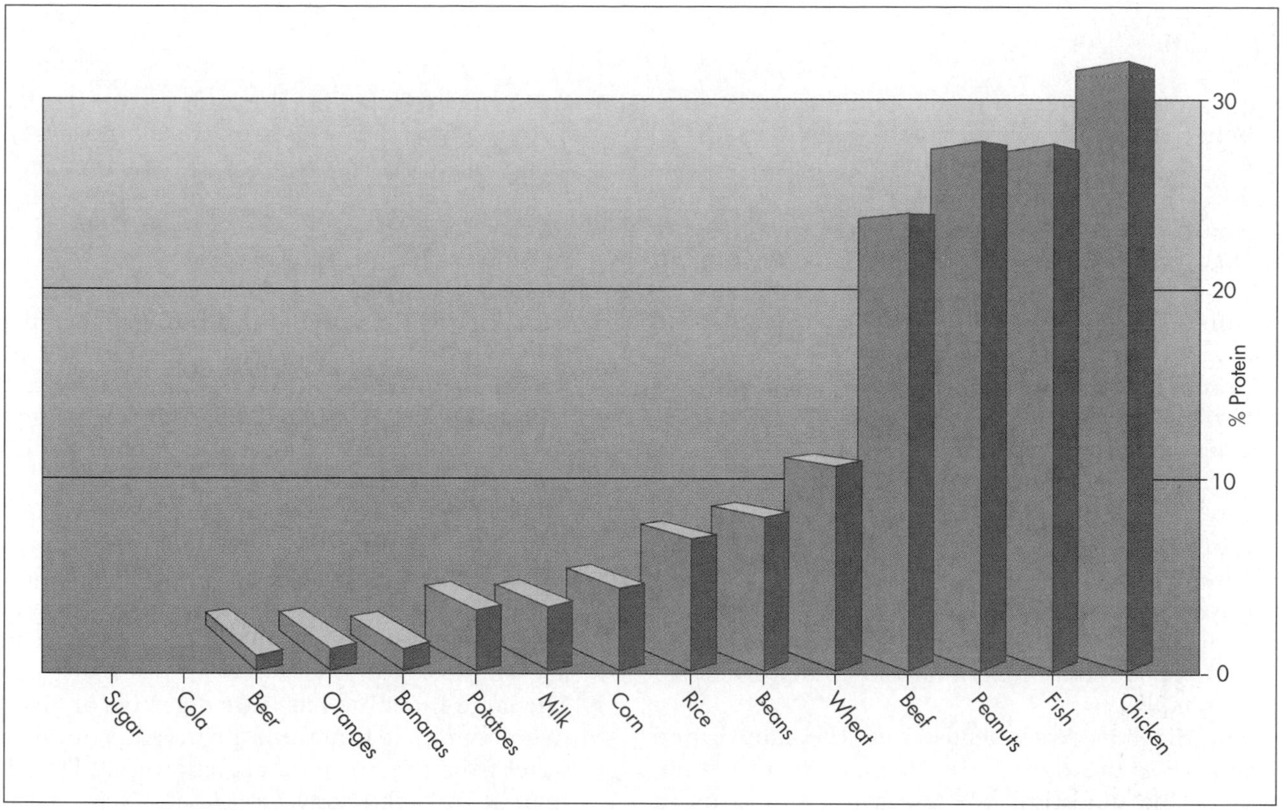

FIGURE 8-12 The protein content of a variety of common foods. Eight of the essential amino acids humans need must be obtained from complete proteins in the diet. Chicken, fish, and beef contain complete proteins.

FIGURE 8-13 Representative proteins, fats, and carbohydrates. The foods shown in each of the photographs are representative dietary proteins, fats, and carbohydrates. Notice that eggs and cheese are shown in both the protein group and the fat group. Measured in grams, these foods have approximately equal amounts of fat and protein. However, fats yield 9 calories per gram, whereas proteins and carbohydrates yield 4 calories per gram. Therefore, eggs and cheese yield more calories from fat than from protein.

ries in the meal shown at the beginning of the chapter. How many of its kilocalories come from proteins, fats, and carbohydrates? Although you may not be analyzing all of its nutritional aspects, what recommendation can you make for it to be a healthier meal?

rectum the lower portion of the colon, which terminates at the anus.

anus the opening of the rectum for the elimination of feces.

essential amino acids the eight amino acids that humans cannot manufacture and therefore must obtain from proteins in the food they eat.

Summary

- Nutrients are substances in food that are used to help the body grow and repair and sustain itself. There are six classes of nutrients: carbohydrates, fats (lipids), proteins, vitamins, minerals, and water.
- Digestion is the process whereby food is broken down into a form that is usable by the body's cells. Carbohydrates are digested to monosaccharides, lipids to fatty acids and glycerol, and proteins to amino acids. Amino acids are primarily used as building blocks. Monosaccharides and glycerol are primarily used as the fuel of cellular respiration and other metabolic pathways that produce ATP molecules, the energy currency of the body.
- Vitamins and minerals are not broken down during digestion but are released from the food of which they are a part. These nutrients help various chemical reactions and bodily processes take place.
- The digestive tract leads from the mouth and pharynx, through the esophagus, to the stomach. The digestion of carbohydrates begins in the mouth with the action of salivary amylase. The stomach juices contain a concentrated acid in which the protein-digesting enzyme pepsin is active.
- Food passes from the stomach to the small intestine where the pH is neutralized. A variety of enzymes, many synthesized in the pancreas, act to complete digestion. Most digestion occurs in the first 25 centimeters of the small intestine, a portion called the *duodenum*.
- The products of digestion are absorbed across the walls of the small intestine, which possess numerous villi and achieve a large surface area. Amino acids and monosaccharides pass into the bloodstream. Fatty acids and glycerols pass into the lymphatic system and, coated with proteins, are transported to the bloodstream.
- Glucose and other metabolic products of digestion do not enter the general circulation directly but instead flow to the liver. The liver removes and stores any excess metabolic products and maintains blood glucose levels within narrow bounds.
- The large intestine has little digestive or absorptive activity; it functions principally to absorb water from the waste that is left over and to eliminate it from the body.
- The U.S. Senate Select Committee on Nutrition and Human Needs recommends that 12% of the daily intake of kilocalories come from proteins, 30% or less from fats, and 58% from carbohydrates.

Knowledge and Comprehension Questions

1. Name the six classes of nutrients, and explain why you need each one in your diet.
2. Describe the digestive fates of proteins, lipids, and carbohydrates. What three types of enzymes are involved?
3. Where does the process of human digestion begin?
4. You bite off a mouthful from a crunchy apple and chew it thoroughly. Describe the role played by each type of tooth in your mouth.
5. Is the muscle activity of peristalsis under voluntary control or is it an involuntary process? Does digestion occur in the esophagus as peristalsis is occurring?
6. What important tasks does the hydrochloric acid in the human stomach accomplish? What protects the stomach wall from potential damage by this chemical product?
7. What is the duodenum? Summarize the digestive activities that take place there.
8. Identify and give a function for each of the following: gastrin, secretin, and cholecystokinin. What do they have in common?
9. Discuss the features of the small intestine that increase its internal surface area. Why is this increase important?
10. Summarize how your body produces ATP from glucose, amino acids, triglycerides, and fatty acids.

Critical Thinking

1. Many people of other countries live a vegetarian lifestyle. This has become more common in the United States in recent years also. Discuss how a vegetarian could create a balanced diet while eliminating meat and animal products.
2. While on a hike, a student finds the dry skull of a small mammal. How could the student determine whether that animal ate only plant products?
3. Why do you think that the liver and pancreas are not called organs of the digestive system although they serve important accessory roles in digestion?
4. For years, baking soda (sodium bicarbonate) has been a popular home remedy for indigestion and heartburn. Explain why.
5. Keep a record of everything you eat and drink for 3 days. Then (this could be painful!) analyze your diet. Compare it with the recommendations in Figure 8-13. How could you improve your nutritional intake?

CHAPTER 9

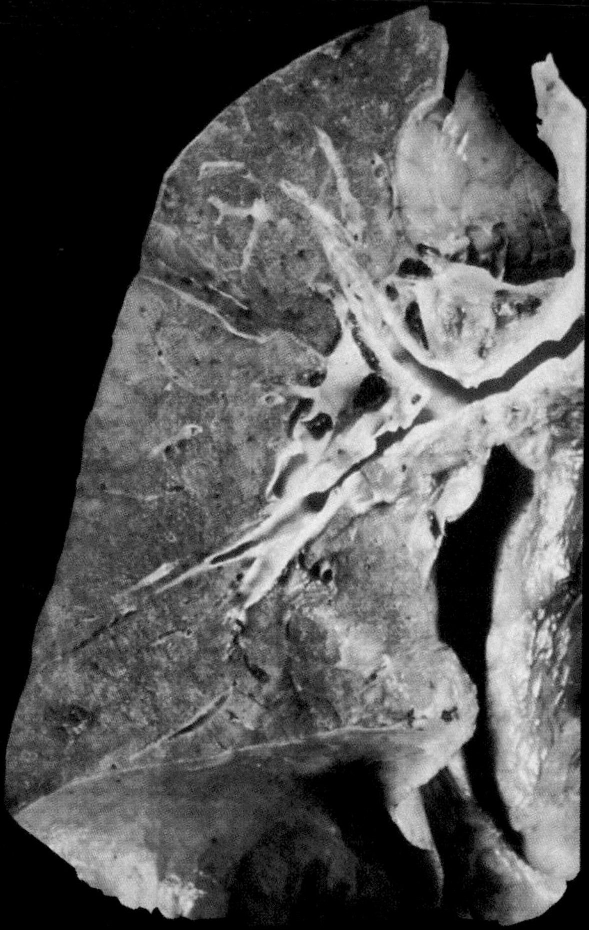

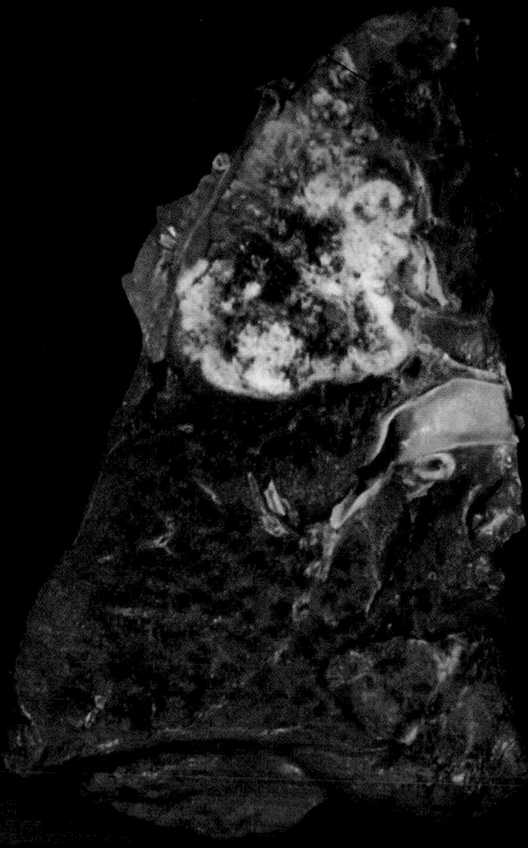

RESPIRATION

WHICH LUNG WOULD YOU CHOOSE? Probably the healthy, red lung in the top photo. Unbelievably, however, more than 50 million Americans choose to have lungs similar to those in the bottom photo. Why? Because these people choose to smoke cigarettes.

When people smoke cigarettes, they inhale many particles and gases. Carbon monoxide is one of the colorless, odorless gases present in cigarette smoke (and in your car's exhaust). It binds to the oxygen-carrying molecules of hemoglobin in your blood, reducing their oxygen-carrying capability by as much as 15%. Tobacco also contains nicotine, a drug that is physically addicting. In addition, nicotine stimulates the nervous system in a way that causes the heart to race and the blood pressure to rise. The blood vessels also narrow—including the arteries that supply the heart with blood. Both the nicotine and carbon monoxide work in different ways to decrease the amount of oxygen that can get to the heart. If the decrease is severe, a part of the heart muscle can die, an experience commonly referred to as a *heart attack*.

Tar is another substance found in cigarette smoke. It is the residue of smoke—what is left over after the nicotine and moisture have been removed. In the bottom photo, you can easily see how tar blackens lung tissue. The whitish areas are no longer functional—they are cancerous tumors caused by the tar. This person did not survive the addiction to cigarettes.

KEY CONCEPTS

- In its journey to the lungs, air passes through the nostrils, nasal cavities, pharynx, larynx, trachea, bronchi, and bronchioles.
- In the process of breathing, a negative pressure within the thoracic cavity causes air to be pulled into the lungs; a positive pressure causes air to be pushed out of the lungs.
- Oxygen diffuses into the blood and carbon dioxide diffuses out of the blood at the lungs; the reverse exchange of gases occurs at the body cells..
- Cigarette smoking damages respiratory structures and can lead to such disorders as lung cancer, chronic bronchitis, and emphysema.

OUTLINE

Respiration

Your lungs play an important role in the process of **respiration** (Figure 9-1). The term *respiration* is used in many ways when speaking about the processes of the human body; therefore its definition can be confusing. Used alone, respiration refers to gas exchange—the uptake of oxygen (O_2) and the release of the waste gas carbon dioxide (CO_2) by the whole body. The phrase *cellular respiration* uses the term respiration in a more specific way: it is the chemical process by which cells break down fuel molecules to release energy, using oxygen and producing carbon dioxide. Cellular respiration is a process that uses the oxygen you breathe in and produces the carbon dioxide you breathe out. (This process is described in detail in Chapter 5.)

At the millions of tiny air sacs, or **alveoli,** that make up most of your lungs, oxygen enters the blood and carbon dioxide leaves. The exchange of these gases between the blood and the alveoli is known as **external respiration.** The movement of air into and out of the lungs is **breathing.**

After the blood picks up oxygen at the lungs and gets rid of carbon dioxide, it travels to the heart and gets a push out to the rest of the body. The cells of

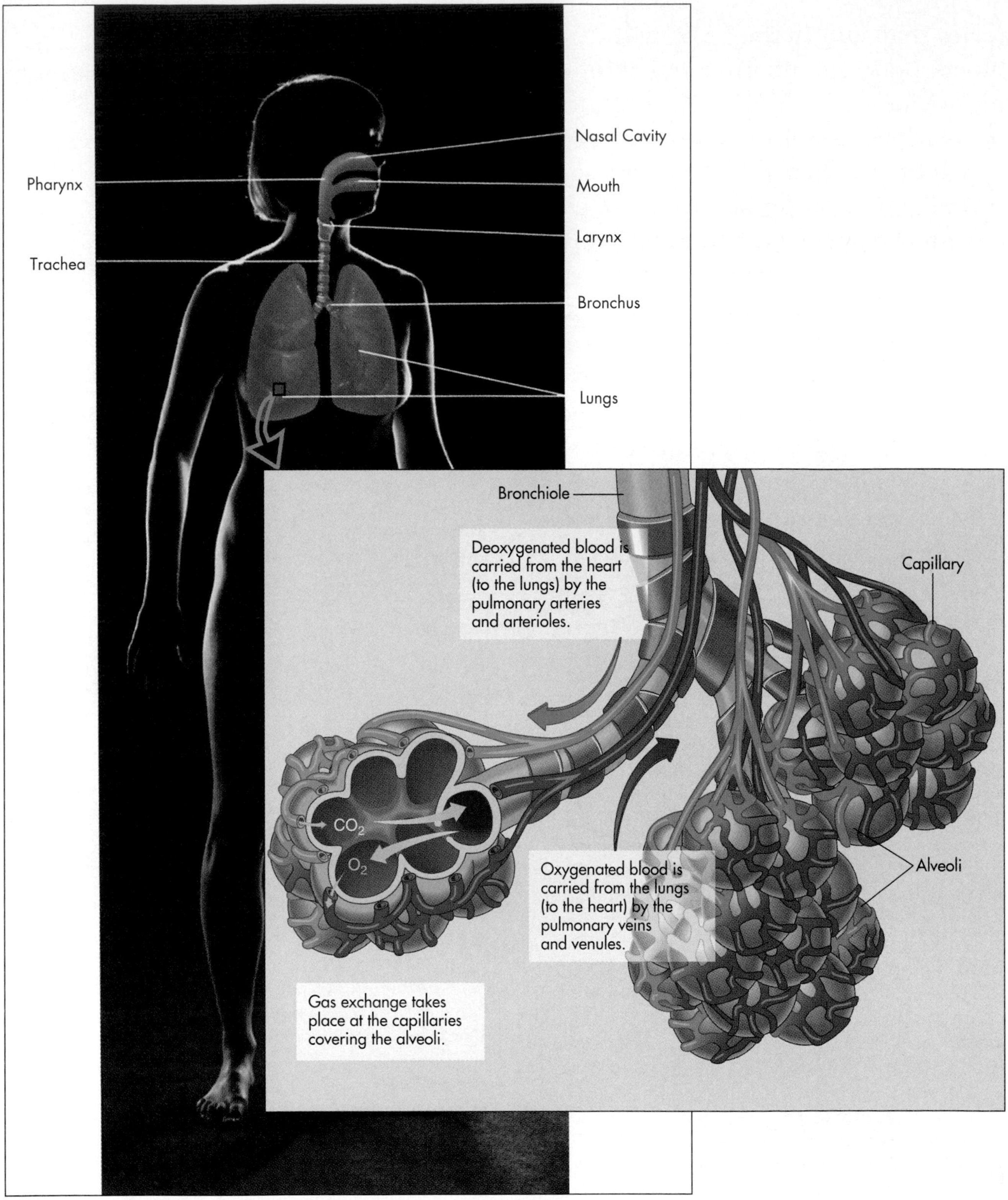

FIGURE 9-1 The human respiratory system. A section of the lung is enlarged to show the exchange of the respiratory gases, oxygen (O_2) and carbon dioxide (CO_2), between the blood and alveoli.

the body receive the oxygen they need and get rid of the waste gas carbon dioxide as the blood moves past the cells within microscopic blood vessels. At the cells, carbon dioxide moves by diffusion from the tissue fluid into the blood. Tissue fluid (also called *interstitial fluid* or *intercellular fluid*) is a waterlike fluid derived from the blood that bathes all the cells of the body. Oxygen moves in the opposite direction—from the blood into the tissue fluid. This process is called **internal respiration**, the exchange of oxygen and carbon dioxide between the blood and the tissue fluid.

> *Respiration is the uptake of oxygen and the release of carbon dioxide by the body. The processes of cellular respiration, internal respiration, and external respiration are all part of the general process of respiration.*

The Pathway of Air into and out of the Body

How many times per minute do you breathe? A breath consists of taking air into the lungs, or *inspiration,* and expelling air from the lungs, or *expiration.* Try to breathe normally, count the number of breaths you take in 15 seconds, and multiply by 4. Is your breath rate within the average of 14 to 20 breaths per minute?

Nasal Cavities

As you breathe in, air first enters your body through the nostrils. The nostrils are lined with hairs that filter out dust and other particles from the air. The air is warmed and moistened as it swirls around in the **nasal cavities.** These cavities, located above your oral cavity and behind your nose, are bordered by projections of bone covered with epithelial tissue. This tissue stays moist with mucus secreted by its many mucous glands. This sticky fluid helps trap dirt and dust that you breathe in. The epithelium is also covered with tiny, hairlike projections called *cilia* (see Figure 9–3). The word *cilia* comes from a Latin word meaning "eyelashes." These "cell eyelashes" beat in unison, creating a current in the mucus that carries the trapped particles toward the back of the nasal cavity. From here, the mucus drips into the throat and is swallowed—at a rate of over a pint per day!

> *As air moves about within the nasal cavities, it is warmed and moistened.*

The Pharynx

After passing through the nasal cavities, air enters the **pharynx,** or throat. The pharynx extends from behind the nasal cavities to the openings of the esophagus and larynx. The esophagus, as you may recall from Chapter 7, is the food tube, a passageway for food to the stomach. The **larynx,** or voice box, lies at the beginning of the trachea. The **trachea** is the air passageway that runs down the neck in front of the esophagus and that brings air to the lungs.

The Larynx

The larynx is a cartilaginous box shaped somewhat like a triangle. Stretched across the upper end of the larynx are the **vocal cords** (Figure 9-2). The vocal cords are two pieces of elastic tissue covered with a mucous membrane. Muscles within the larynx pull on the cartilage, which in turn pulls or relaxes the tension on the vocal cords. When the vocal cords are stretched tightly, the space between them—the **glottis**—is closed. In this way the vocal cords provide a "backup" for the epiglottis that flaps

respiration a term that, when used alone, refers to the uptake of oxygen and the release of carbon dioxide by the whole body.
alveoli (al vee uh lye) microscopic air sacs in the lungs where oxygen enters the blood and carbon dioxide leaves.
external respiration the exchange of carbon dioxide and oxygen gases at the alveoli in the lungs.
breathing the movement of air into and out of the lungs.
internal respiration the exchange of carbon dioxide between the blood and the tissue fluid.
nasal cavities two hollow areas, located above the oral cavity and behind the nose, that are bordered by projections of bone covered with moist epithelial tissue.
pharynx (fair inks) the upper part of the throat that extends from behind the nasal cavities to the openings of the esophagus and larynx.
larynx (lair inks) the voice box, which is located at the beginning of the trachea.
trachea (tray kee uh) the windpipe; the air passageway that runs down the neck in front of the esophagus and that brings air to the lungs.
vocal cords two pieces of elastic tissue, covered with a mucous membrane, that are stretched across the upper end of the larynx and are involved in the production of sound.
glottis (glot is) the space between the vocal cords; the opening to the larynx and trachea.

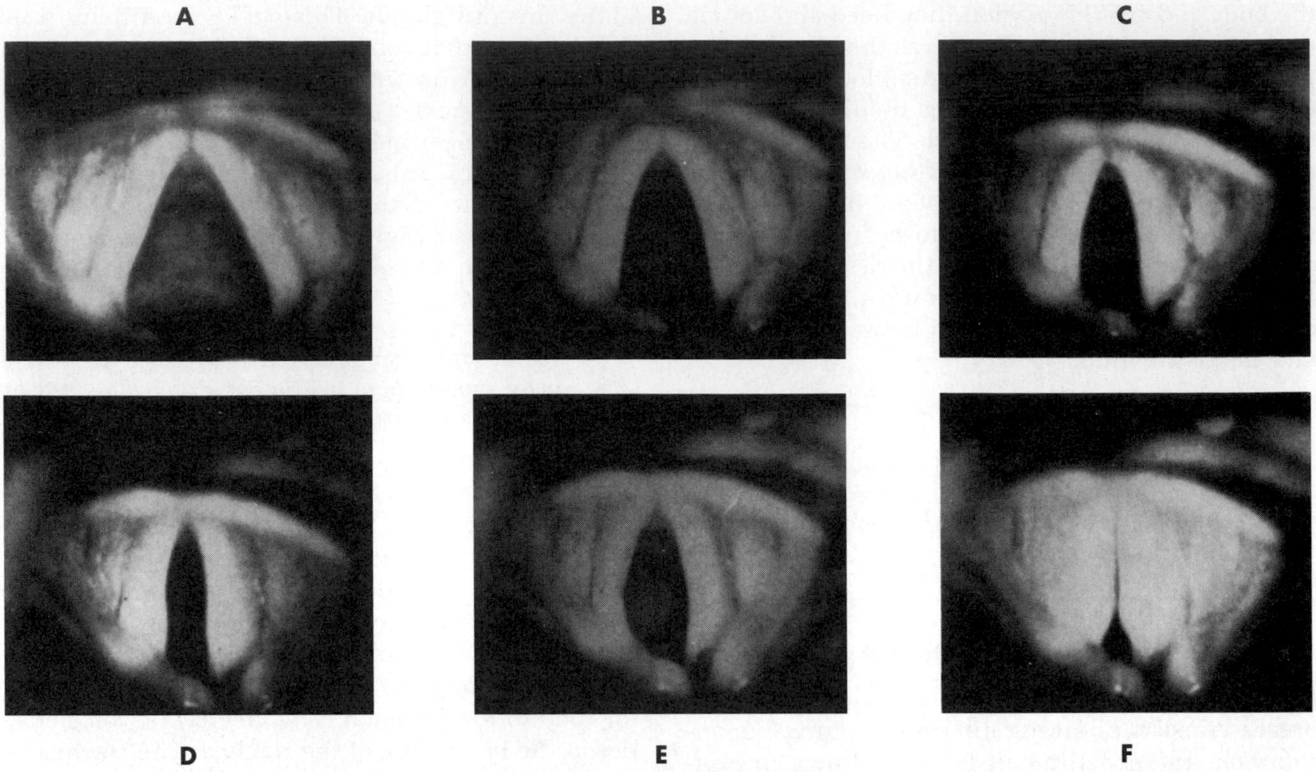

FIGURE 9-2 Movement of the vocal cords. The space between the two folds of tissue is the glottis. In this series of photographs, **A-F**, the vocal cords are shown in the process of closing off the glottis.

over the glottis during swallowing. Both structures work to prevent food and drink from going down the wrong way. The vocal cords also produce the sounds you make as air rushes by and causes them to vibrate. You can illustrate this principle by stretching a rubber band between your fingers. Have someone repeatedly pluck the band. At the same time increase and then decrease the stretch on the band. What happens? Your vocal cords work in much the same way to produce a variety of pitches of sound. But you also have a mouth with lips and a tongue to form the sounds into words. Your lungs add a power supply and volume control to your personal musical instrument—your voice.

The Trachea

Put your hand on your larynx, or Adam's apple, and then picture about 4 or 5 inches of garden hose attached to its bottom end. A garden hose is about the diameter of your trachea, or windpipe, which extends downward from your larynx toward your lungs. The trachea has thin walls, similar to the thickness of those in the hose. Garden hoses are reinforced with materials such as rubber or vinyl to keep them from collapsing; your trachea is reinforced with rings of cartilage. The cartilage wraps around the trachea only part way, forming C shapes that begin and end where the windpipe lies next to the esophagus. Press gently on your windpipe just below your Adam's apple, and you can feel some of these cartilaginous rings.

The inner walls of the trachea are lined with ciliated epithelium (Figure 9-3). Certain cells in the epithelium secrete mucus. Together, the cilia and the mucus provide your windpipe with an up escalator for any particles or microbes you may have inhaled. This escalator brings substances up to the pharynx where they are swallowed and eliminated through the digestive tract. In the trachea of a cigarette smoker, however, action of the cilia is impaired, causing mucus to build up in the airway. The result is that the tars in cigarettes are not caught and expelled with the action of cilia. They move easily into the lungs and settle there. A chronic cough, often called *smoker's cough,* is triggered by accumulations of mucus below the larynx.

> *Air passes from the nasal cavities and into the throat, or pharynx. From the pharynx, it passes over the vocal cords that are stretched across the larynx. From this voice box, air moves into the trachea, or windpipe, on its way to the lungs.*

FIGURE 9-3 Colorized scanning electron micrograph of cilia. Cilia such as these line the trachea and sweep trapped particles out of the respiratory tract.

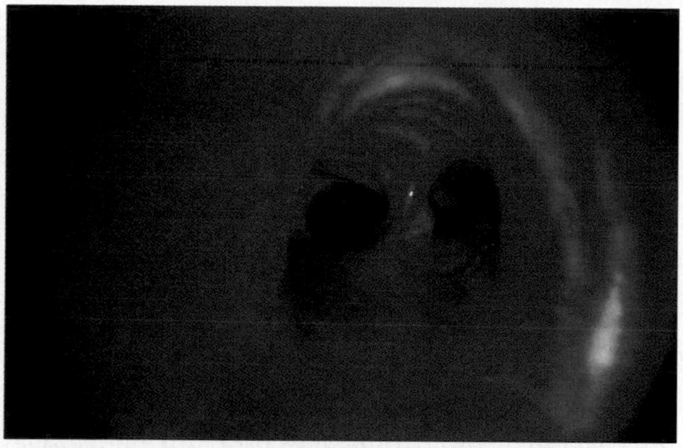

FIGURE 9-4 The trachea. The two holes that you see are the primary bronchi. One bronchus leads to each lung.

The Bronchi and Their Branches

If you could look down your trachea as in Figure 9-4, the two black holes you would see would be your **primary bronchi.** These airways are structured much like the trachea but are smaller in diameter. One bronchus goes to each lung, branching into three right and two left **secondary bronchi** serving the three right and two left lobes of the lungs. The heart is nestled into the left side of the lungs, taking up some of the space that a third left lobe might occupy.

The secondary bronchi divide into smaller and smaller branches, looking much like an upside-down tree (Figure 9-5), until they end in thousands of passageways called **respiratory bronchioles.** These airways have a diameter less than that of a pencil lead. Their walls have clusters of tiny pouches that, along with the respiratory bronchioles, are the sites of gas exchange. These pouches, or air sacs, are called *alveoli.*

FIGURE 9-5 Secondary bronchioles. Secondary bronchi divide into smaller and smaller branches, resembling an upside-down tree or a system of roots.

> *Air moves from the trachea into the two bronchi that supply each lung. The bronchi divide into smaller and smaller branches ending in respiratory bronchioles having outpouchings called alveoli, the sites of gas exchange.*

primary bronchi (**bronk** eye) the two airways branching immediately from the trachea into each lung.

secondary bronchi outbranchings of each of the two primary bronchi. Three secondary bronchi serve the three right lobes of the lungs, whereas two secondary bronchi serve the two left lobes of the lungs.

respiratory bronchioles (**bron** kee olz) the last division of bronchi. Bronchioles have thousands of tiny air passageways whose walls have clusters of tiny pouches, or alveoli.

The Alveoli: Where Gas Exchange Takes Place

The alveoli provide a perfect place for carbon dioxide and oxygen to diffuse between the air in the lungs and the blood. These clusters of microscopic air sacs are bounded by membranes made up of a single layer of epithelial cells. A network of capillaries tightly clasps each alveolar sac. The capillary walls are also only one cell thick and press against the alveolar epithelium. These two adjacent membranes provide the thinnest possible barrier between the blood in the capillaries and the air in the alveoli.

The alveoli also provide another important component of efficient gas exchange: a large surface area. In fact, if the epithelial membrane of all your alveoli was spread out flat, it would cover a tennis court! The capillaries cover this enormous surface, creating patterns much like tightly woven spider webs, providing nearly a continuous sheet of blood over the alveolar surface.

Large white blood cells called *macrophages* are also found at the alveoli. These cells work to remove any particles or microbes that have escaped the other defenses of the airways. The macrophages transport the invaders to the bronchioles or to the lymphatic system. Sometimes the job is too big, however, and particles remain in the lungs. In smokers the action of the macrophages is impaired, making the lungs more susceptible to disease and injury.

The alveoli provide a thin, enormous surface area over which gas exchange can take place.

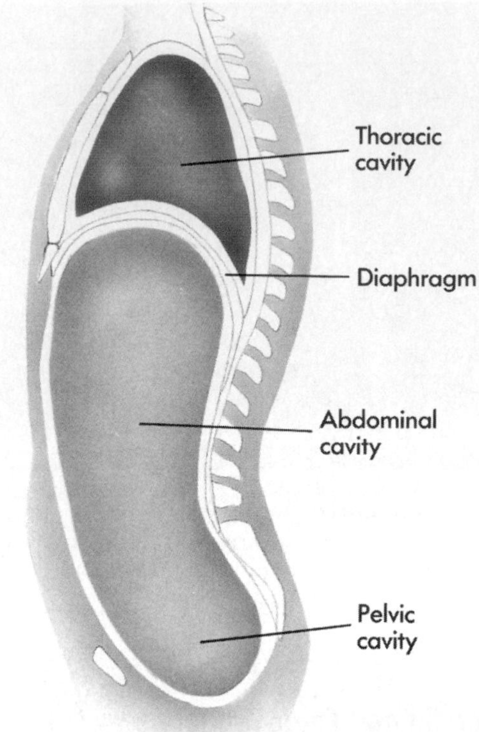

FIGURE 9-6 Location of the diaphragm. The diaphragm forms a partition between the thoracic and abdominal cavities.

The Mechanics of Breathing

Air moves into and out of your lungs as the volume of your thoracic cavity is made larger and smaller by the action of certain muscles. The **thoracic cavity,** or chest cavity, is within the trunk of your body above your diaphragm and below your neck. The **diaphragm** is a sheet of muscle that forms the horizontal partition between the thoracic cavity and the abdominal cavity. Various blood vessels and the esophagus puncture it as they traverse these two body cavities. The position of the diaphragm is shown in Figure 9-6.

The diaphragm is assisted by other muscles of breathing. These muscles extend from rib to rib—from the lower border of each rib to the upper border of the rib below—and are called **intercostal muscles.** The word *intercostal* literally means "between" (inter) "the ribs" (costal). You have two sets of intercostals: the *internal intercostals* and the *external intercostals.* The internal intercostals are those that lie closer to the interior of the body (as their name suggests) and that have fibers extending obliquely downward and backward (from front to back). The external intercostals extend from back to front, having fibers that are directed downward and forward.

Inspiration

When you are breathing quietly—not exerting yourself physically—your diaphragm and external intercostals alone are responsible for the change in the size of your thoracic cavity. This change in size results in the movement of air into and out of your lungs. During inspiration the dome-shaped diaphragm contracts, flattening somewhat and thereby lowering the floor of the thoracic cavity. The external intercostals contract, raising the rib cage. Notice in Figure 9-7 how these two actions increase the size of the thoracic cavity.

The interior walls of the thoracic cavity are lined with a thin, delicate, sheetlike membrane called the *pleura.* The pleura folds back on itself to cover each lung. These two parts of the membrane are close to one another, separated only by a thin film of fluid. As the volume of the thoracic cavity increases during inspiration, the lungs also expand, held to the wall of the thoracic cavity by co-

HOW *Science* WORKS

Discoveries

Asthma—New Findings about an "Old" Ailment

Asthma is a chronic lung disease affecting about 10 million people in the United States. *Chronic* means that a disease persists for a long time. In fact, people with asthma usually have the disease for their lifetimes, although some people affected with asthma in childhood "outgrow" it. So far there is no cure. However, new drugs and preventive measures can lessen the severity of asthma attacks and allow sufferers to lead normal lives. In addition, scientists have recently discovered that some persons with hard-to-treat asthma have underlying viral infections of the lungs. This new discovery may lead to new treatments for these sufferers.

The underlying cause of asthma attacks is hypersensitive airways. The airways (bronchi and bronchioles) of a person with asthma are much more sensitive to certain stimuli, such as dust, pollen, or animal dander, than are the airways of other people. In an asthma attack or episode, the lining of the airways swells. The bronchial muscles contract. Mucus production increases. This combination of swelling, increased mucus production, and muscle contraction narrows the airways. The person with asthma then has trouble bringing air into the lungs and also has trouble getting air out. Forcing air through the narrowed airways can produce a whistling or wheezing sound.

It was once thought that asthma was a psychosomatic illness, with no physical cause. Some physicians believed that the symptoms were all in a person's mind or that the person could somehow control the attacks. Others believed that asthma was a bad habit, like a temper tantrum, that showed a person's extreme need for attention. Today, the physiological mechanism of asthma is well known. It is understood that asthma is indeed a physical illness, although emotions can influence the frequency and severity of asthma attacks. Stimuli or triggers that are known to cause asthma attacks include allergens such as pollen, dust, or certain foods; physical factors such as coughing, sneezing, rigorous exercise, or cold temperatures; viral infections; and chemical irritants such as cigarette smoke.

Asthma is treated by teaching the asthma patient to recognize the triggers that bring on attacks and to prevent attacks before they can occur. Asthma patients are encouraged to measure their lungs' vital capacity by exhaling into a device called a *peak flow meter.* The measurements shown by this device can help the physician and the patient to develop an individualized treatment plan.

Treatment can involve several types of drugs. *Antiinflammatory agents* lessen the swelling of the airways—these *drugs* are taken to *prevent* an attack. *Bronchodilators* that are inhaled are taken during an attack—these fast-acting drugs work to open constricted airways. However, bronchodilators must be used sparingly. Recent studies have shown that overuse of bronchodilator inhalers can lead to a worsening of asthma and even to death in a few cases. The inhaler itself does not cause the death. Instead, the asthma patient comes to rely on the fast-acting nature of the drug and does not notice that her asthma might be worsening. A severe attack of asthma can take such a person by surprise, leading to serious complications and death. That is why asthma patients must participate as much as possible in monitoring and treating their condition. This participation can lessen the severity of the disease, giving asthma patients some control over the episodes and the opportunity to lead a normal life.

The recent discovery that the lungs of some young asthma patients are chronically infected with the virus that causes bronchitis suggests that antiviral therapy might be a useful treatment for some children. Researchers are unsure whether low-grade viral infections also plague certain adult asthma sufferers. Research is currently under way to answer these questions and determine new therapies.

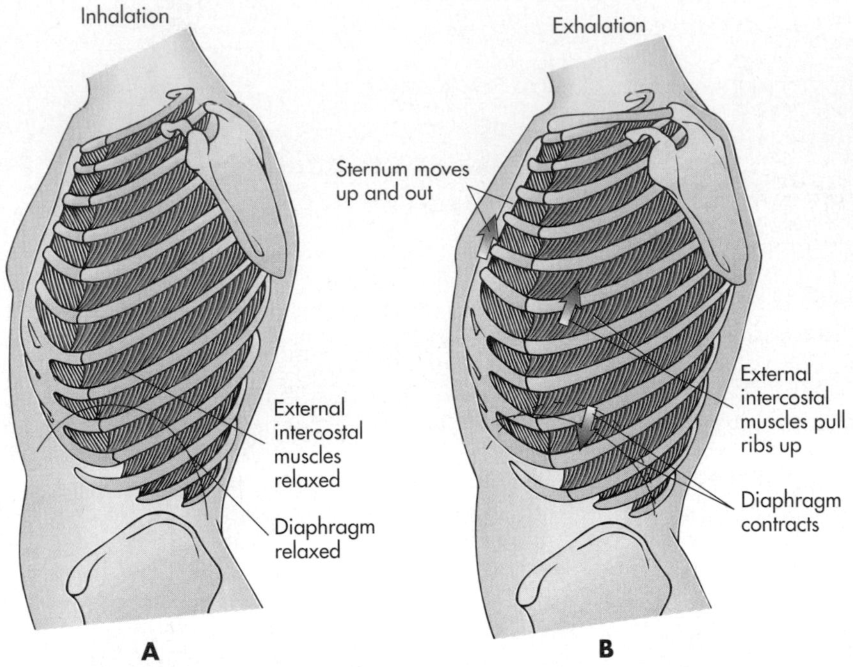

FIGURE 9-7 How a human breathes. A, Expiration: At the end of an expiration, the diaphragm and other respiratory muscles are relaxed. The diaphragm takes on a dome shape and the ribs fall, reducing the volume of the chest cavity and forcing air outward. **B,** Inspiration: The diaphragm contracts, becoming flattened, and the ribs are raised, increasing the volume of the chest cavity. As a result of the increased volume, air rushes inward.

hesion of the water molecules between the two membranes. Put simply, the lungs stick to the thoracic wall and move with it.

As the lungs expand in volume, the air pressure within the lungs decreases because there are fewer air molecules per unit of volume. As a result, air from the environment outside the body is pulled into the lungs, equalizing the pressure inside and outside the thoracic cavity. By means of this process, you breathe in 13,638 liters (more than 3000 gallons) of air every day!

Expiration

The lungs contain special nerves called *stretch receptors.* When the lungs are stretched to their normal inspiratory capacity, these receptors send a message to a respiratory center in the brain. This respiratory center is located in parts of the brainstem called the *medulla* and the *pons* (see Chapter 14). The respiratory center stops sending "contract" messages to the muscles of breathing, which causes them to relax—a passive process in contrast to the active process of inspiration. As the diaphragm relaxes, it assumes its domelike shape, reducing the volume of the thoracic cavity. Likewise, as the external intercostals relax, the rib cage drops, reducing the volume of the thoracic cavity further. The volume of the lungs, in turn, decreases, aided by the recoil action of the lungs' elastic tissue. The reduced volume of the lungs results in an increase in the air pressure within them. Air is forced out of the lungs, equalizing the pressure outside and inside the thoracic cavity. The cycle of inspiration and expiration is shown in Figure 9-8.

The brainstem also contains areas sensitive to changes in the levels of carbon dioxide in the blood that control the rate of the cycle of inspiration and expiration. Hydrogen ions (H^+) that are produced from a series of chemical reactions involving carbon dioxide also play an extremely important role in influencing the activity of the respiratory center (see Internal Respiration, p. 192). So when you have a high level of carbon dioxide in your blood, the level of hydrogen ions in your blood increases (blood pH decreases), and your breathing rate is "stepped up."

Deep Breathing

When you breathe deeply, such as during physical exercise, the internal intercostals as well as other muscles in the chest and abdomen help out the muscles of respiration. During inspiration, certain muscles attached to the breastbone—the bone in the center of your chest to which your ribs are

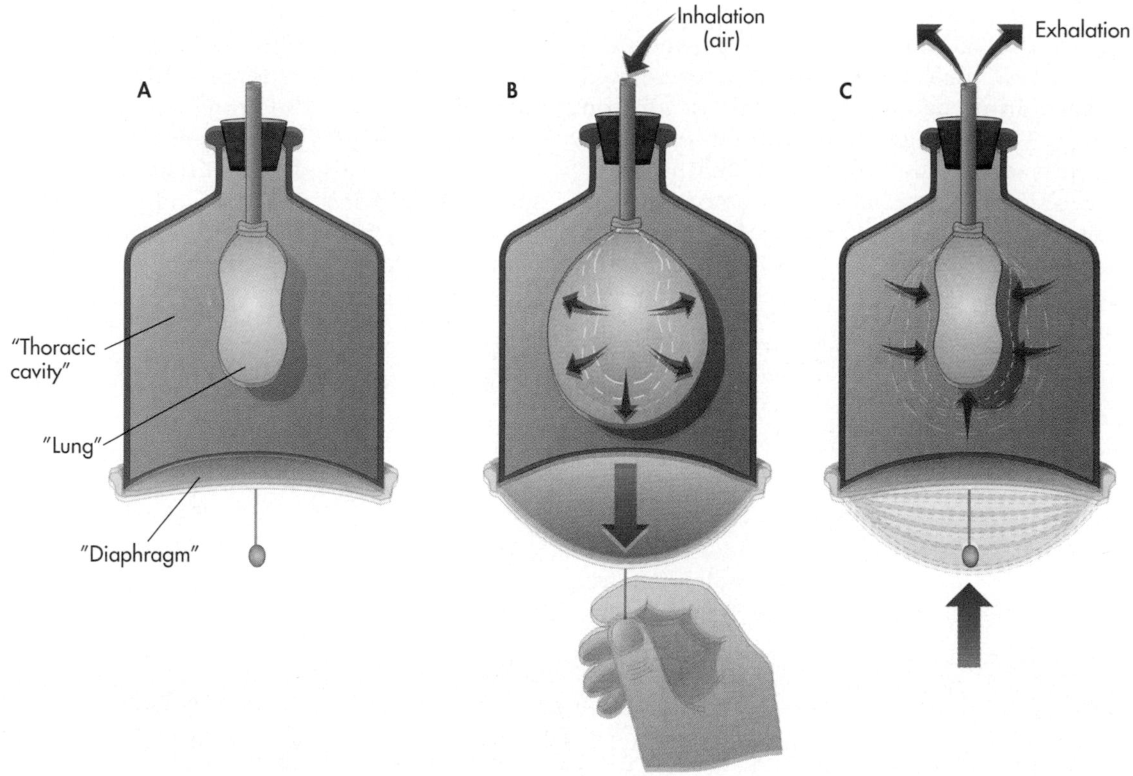

FIGURE 9-8 A simple experiment that shows how you breathe. In the jar is a balloon **(A)**. When the diaphragm is pulled down, as shown in **B**, the balloon expands. When it is relaxed **(C)**, the balloon contracts. In the same way, air is taken into your lungs when your diaphragm contracts and flattens, increasing the volume of your chest cavity. When your diaphragm relaxes and resumes its dome shape, the volume of your chest cavity decreases and air is expelled.

attached—pull up on it. In addition, other muscles pull up on the first two ribs. This action increases the volume of the chest cavity more than during quiet breathing, so more air flows into the lungs. By deep breathing and increasing the rate of breathing, world champion runners have been shown to increase their air intake fifteenfold.

Expiration, a passive process during quiet breathing, becomes an active process during deep breathing. The internal intercostals pull down on the rib cage. Abdominal muscles also contract, pulling down on the lower ribs and compressing the abdominal organs, which results in a push up on the diaphragm. These actions additionally decrease the volume of the thoracic cavity and cause more air to be expelled than during quiet breathing.

Inspiration occurs when the volume of the thoracic cavity is increased and the resulting negative pressure causes air to be pulled into the lungs. Expiration occurs when the volume of the thoracic cavity is decreased and the resulting positive pressure forces air out of the lungs.

Lung Volumes

How much air do you move into and out of your lungs during inspiration and expiration? You can find out by performing this simple procedure: Fill a large jar with water and invert it in a pan of water. Be careful not to let any of the water seep out of the jar while you are turning it upside down. Mark the level of the water in the jar with tape or a wax pencil. Take a piece of rubber tubing or garden hose about a foot long and put one end up into the jar. Put the other end into your mouth and exhale normally. The air you breathe out will displace the water. Mark the new level of the water. Remove the jar from the water and fill it to this line with water.

thoracic cavity (thu **rass** ick) the chest cavity in humans, located within the trunk of the body and extending above the diaphragm and below the neck.

diaphragm (**dye** uh **fram**) a sheet of muscle that forms the horizontal partition between the thoracic cavity and the abdominal cavity.

intercostal muscles (inn ter **kos** tul) literally, "between-the-rib muscles"; muscles that extend from rib to rib and assist the diaphragm in the breathing process.

Now measure the volume of the water between the two lines with a measuring cup or a graduated cylinder. Its volume equals the volume of air you breathed out. The average adult male breathes out 500 milliliters of air during quiet breathing, or slightly more than 1 pint. This volume of air—the amount inspired or expired with each breath—is called the **tidal volume.**

Of the 500 milliliters of air you normally breathe in, only about 350 milliliters reaches the alveoli. The other 150 milliliters is either on its way into or out of the lungs, occupying space in the nose, pharynx, larynx, trachea, and bronchial tree. This space is called *dead air space* because it serves no useful purpose in gas exchange. Some of this air, in fact, will never reach the lungs. And some air, called *residual air,* remains in the lungs—even during deep breathing.

JUST WONDERING . . . .

I'm a smoker thinking about quitting. Does my smoking really affect the health of my children?

Cigarette smoking damages the lungs of children in more ways than people realize, and the damage can start in the womb. Recent reports from the Harvard Six Cities Study, a 20-year investigation into the effects of air pollution and lung hazards, conclude that smoking during pregnancy can detrimentally affect a child's lung function *for the rest of the child's life!* Lung function refers to the vital capacity of the lungs—the maximum volume of air that can be pushed out of the lungs during one exhalation after a single, deep inhalation. The reason for the reduced lung function in these children is unclear. However, many investigators hypothesize that it is due to their developing smaller lungs that produce less elastic tissue, a protein that helps the lungs recoil during exhalation. There is also a positive correlation between mothers who smoke and the incidence of asthma in their children. In addition, questions have been raised about the incidence of maternal smoking and hearing defects in their children, as well as the occurrence of sudden infant death syndrome (SIDS).

In addition to these conclusions, researchers conducting other parts of the Six Cities Study showed that children exposed to tobacco smoke before the age of 6 had a 2 to 3 percent decrease in lung function—a decrease that normally does not occur until the late 20s or early 30s. Children growing up in households with smokers also develop more lower respiratory conditions such as shortness of breath, wheezing, coughing, and bronchitis. Statistics from the American Heart Association state that secondhand smoke causes up to 300,000 lower respiratory tract infections such as pneumonia and bronchitis in children younger than the age of $1^1/_2$. In 5% of these cases the infections are so severe that the child must be hospitalized. Quitting smoking will not only benefit your health but the health of your children as well.

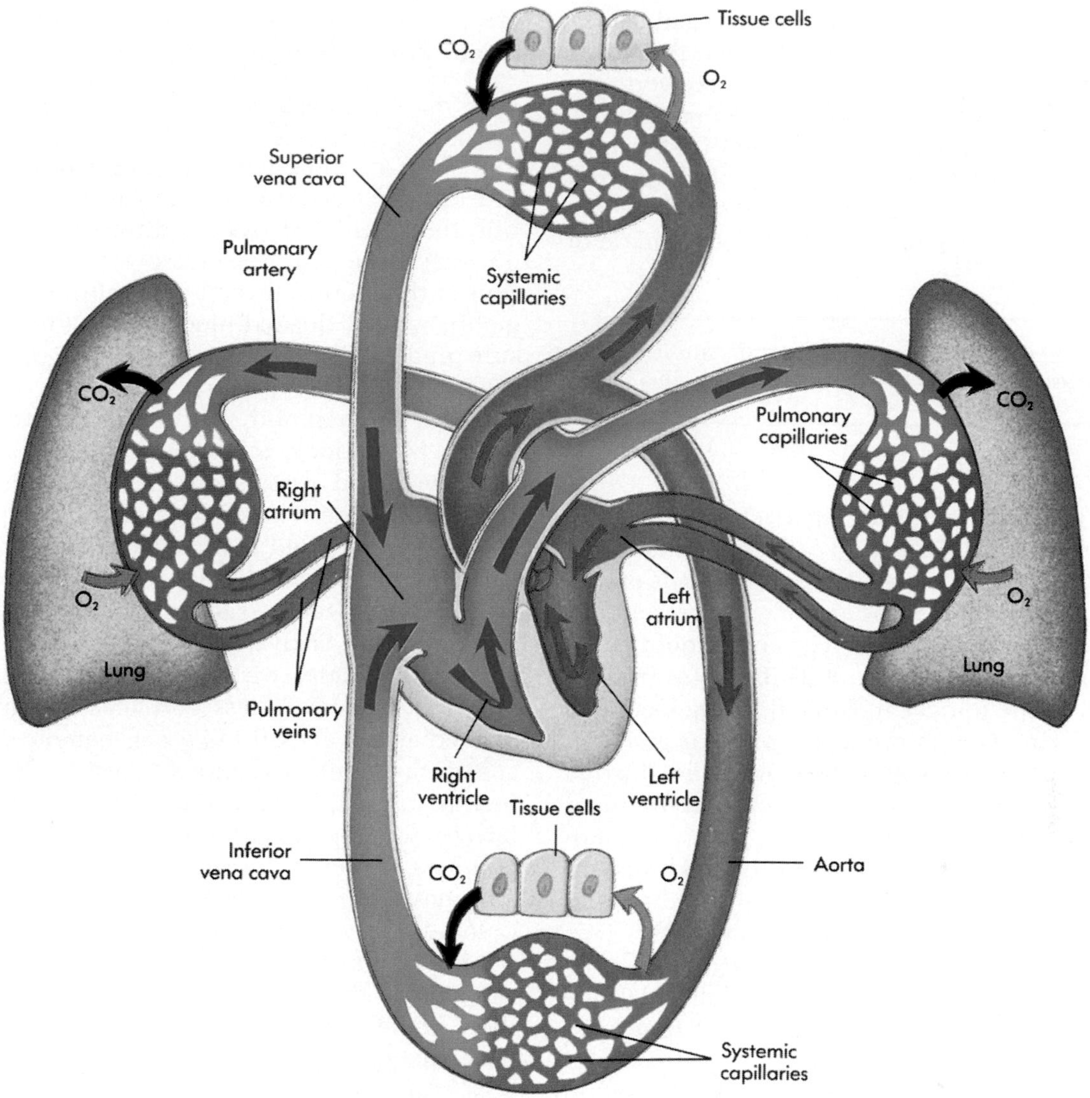

FIGURE 9-9 The route of blood during gas transport. As blood moves through the body, it gives up its supply of oxygen (O_2) to the tissues and collects carbon dioxide (CO_2). This deoxygenated blood is then pumped through the heart to the lungs, where CO_2 in the blood and O_2 in the alveoli are exchanged. The reoxygenated blood flows to the heart and is pumped to the rest of the body tissues.

Gas Transport and Exchange

The transport of gases throughout the body is assisted by the circulatory system. Without the help of a "highway" of blood, scientists estimate that it would take a molecule of oxygen 3 years to diffuse from your lung to your toe! Humans could not survive if gas transport were this slow.

Although the pathway of blood throughout the body will be described in detail in the next chapter, it is helpful to understand the basic routing of blood when discussing gas transport. Notice in Figure 9-9 that the upper right chamber of the heart collects incoming blood from the upper and lower body. This blood has given up its supply of oxygen to the tissues, so it is oxygen poor. Along its route, however, it collected the waste product of cellular respiration, which is carbon dioxide. Deoxygenated blood such as this (more carbon dioxide than oxygen) is shown as blue in the diagram. The deoxygenated blood passes from the upper to the lower right chamber of the heart, where it is pumped to the lungs. Some of this blood goes to the right lung, and some goes to the left lung.

At the lungs, carbon dioxide within the blood of the capillaries surrounding the alveoli and the oxy-

tidal volume the amount of air inspired or expired with each breath.

gen in the air of the alveoli are exchanged. External respiration, the exchange of gases at the lungs, works by the process of diffusion. It converts deoxygenated blood to oxygenated blood (more oxygen than carbon dioxide). This blood is shown as red in Figure 9-9. The oxygenated blood flows from the lungs to the left side of the heart, where the lower left chamber pumps it out to other parts of the body.

The circulatory system aids in respiration by transporting gases throughout the body.

External Respiration

Air is made up of many different kinds of molecules, such as oxygen, nitrogen, and carbon dioxide. Each of these gases exerts a pressure that depends on the number of molecules of the specific gas present per unit of volume. Molecules of a gas in a liquid (such as blood) also exert a pressure. Differences in pressures of a particular gas in two adjoining locations within a living system produce a *pressure gradient.* Molecules of a gas such as oxygen or carbon dioxide tend to move from an area of higher pressure to an area of lower pressure, or down the pressure gradient. Each gas moves according to its own pressure gradient, unaffected by the pressure gradients of other gases with which it might be mixed. This movement is a type of diffusion.

As deoxygenated blood (blue in Figure 9-9) arrives at the lungs, the pressure of the carbon dioxide in the blood is greater than the pressure of carbon dioxide in the air within the alveoli. Therefore carbon dioxide diffuses out of the blood and into the alveoli. Likewise, the pressure of the oxygen in the air within the alveoli is greater than the pressure of the small amount of oxygen in the blood. Therefore oxygen diffuses out of the alveoli and into the blood.

Most of the carbon dioxide (70%) in the blood travels around bound to water molecules in the fluid portion of the blood. The carbon dioxide and water combine chemically to form molecules of carbonic acid (H_2CO_3) but quickly break apart, or dissociate, into bicarbonate ions (HCO_3^-) and hydrogen ions (H^+). About one fourth (23%) of the carbon dioxide is carried in the red blood cells, bound to the oxygen-carrying molecule **hemoglobin (Hb)**. It is carried, however, by a different portion of the molecule than oxygen is. A small amount (7%) of the carbon dioxide is simply dissolved in the blood. At the alveoli, the dissolved carbon dioxide first moves out of the blood. This decrease in its concentration triggers a reversal of the chemical reactions just described:

$$H^+ + HCO_3^- \rightarrow H_2CO_3 \rightarrow H_2O + CO_2$$

Carbon dioxide also dissociates from hemoglobin, and the freed carbon dioxide molecules diffuse into the alveoli. As oxygen diffuses from the alveoli, very little of it (3%) dissolves in the fluid portion of the blood. Instead, it combines with hemoglobin within the red blood cells. When the pressure gradient is high, as in the alveoli, hemoglobin binds with large amounts of oxygen. When oxygenated, hemoglobin turns bright red, which makes blood look red. You will notice, however, that the blood vessels on the underside of your wrists look blue. These blood vessels are veins carrying deoxygenated blood. Deoxygenated blood is dark red but appears blue through layers of skin.

At high altitudes, such as in mountainous regions, the air is thinner and the pressure of the oxygen molecules within the air is lower than at sea level. Therefore less of a pressure gradient is created at the alveoli. As a consequence, less oxygen diffuses into the blood, causing shortness of breath, nausea, and dizziness. This condition is referred to as *high-altitude sickness.* Mountain climbers, athletes working out or playing at high altitudes (such as Mile High Stadium in Denver, Colorado), and tourists visiting an area of high altitude (such as Mexico City, Mexico) need to slowly work up to their normal levels of activity to give their bodies time to adjust to the lower oxygen pressure.

During external respiration, carbon dioxide moves down a pressure gradient, diffusing from the blood in capillaries surrounding the alveoli to air in the alveoli. Likewise, oxygen moves down a pressure gradient, diffusing from the alveolar air to the blood in the capillaries surrounding the alveoli. Carbon dioxide is carried in the fluid portion of the blood primarily as bicarbonate ions; oxygen is carried within the red blood cells by the oxygen-carrying molecule hemoglobin.

Internal Respiration

Oxygenated blood is pumped out to the body by the left ventricle, or lower left chamber, of the heart. It travels through the aorta to other large arteries that soon branch into smaller and smaller arteries. Eventually the blood vessels become so small that only one red blood cell at a time can pass. These

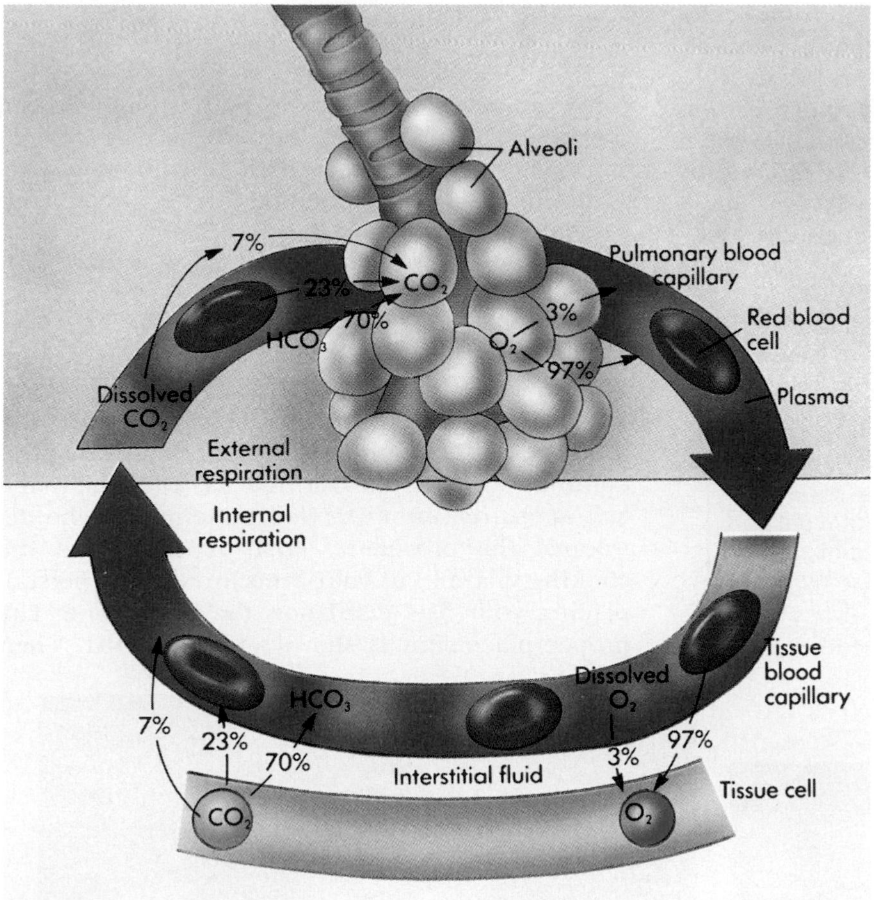

FIGURE 9-10 How gases are transported during external and internal respiration. External respiration takes place in the lungs, as CO_2 leaves the blood, diffusing into the alveoli, and O_2 enters the blood, diffusing from the alveoli. During internal respiration, oxygenated blood travels to the tissues. Oxygen diffuses from the blood into the tissues while carbon dioxide diffuses from the tissues into the blood.

vessels are the capillaries and are the sites of internal respiration.

Internal respiration is the exchange of oxygen and carbon dioxide between the blood within capillaries at the tissues and the tissue fluid that bathes the cells. As this exchange takes place, oxygenated blood becomes progressively deoxygenated. In essence, the processes of internal respiration are the reverse of the processes of external respiration. As the blood begins its journey from the lungs around the body, it is oxygen rich. The oxygen molecules in the blood exert a higher pressure than those in the cells because oxygen is continuously being used by the cells in the process of cellular respiration. Consequently, oxygen molecules begin to dissociate from the hemoglobin, diffuse into the tissue fluid, and from there diffuse into the cells. Conversely, the levels of carbon dioxide are higher in the cells than in the blood because carbon dioxide is continuously being produced during cellular respiration. Consequently, carbon dioxide diffuses from the cells into the tissue fluid and then into the blood.

As gas exchange at the capillaries continues, an interesting thing happens. The oxygen supply of the blood decreases, thereby decreasing the pressure of oxygen in the blood. Diffusion of oxygen into the tissue fluid does not slow down, however, because the pH at the capillaries lowers as carbon dioxide diffuses into the blood. Remember, carbon dioxide is carried in the blood primarily as bicarbonate ions. As these ions are formed, hydrogen ions are also produced. The buildup of these hydrogen ions makes the blood increasingly acidic as more carbon dioxide diffuses in. (See Chapter 2 for a discussion of the relationship between hydrogen ions and pH.) This acid environment helps split more oxygen from hemoglobin, thereby enhancing oxygen's diffusion into the tissue fluid. (Figure 9-10 shows the processes of internal and external respiration.)

hemoglobin (Hb) (hee muh **glow** bin) an oxygen-carrying molecule within red blood cells that also carries some carbon dioxide.

Temperature also helps split oxygen from hemoglobin. During exercise or any type of exertion, your body needs more oxygen delivered to its cells. However, active cells also produce more carbon dioxide and heat, thereby helping to split oxygen from hemoglobin and meet their own needs. This is one example of how your body works to maintain homeostasis, a state of internal equilibrium.

> *During internal respiration, carbon dioxide moves down a pressure gradient, diffusing from the tissue fluid surrounding the body cells to the blood. Likewise, oxygen moves down a pressure gradient, diffusing from the blood within capillaries to the tissue fluid. As carbon dioxide diffuses into the blood, the resultant increase in hydrogen ions increases the dissociation of oxygen from hemoglobin.*

Choking: A Common Respiratory Emergency

Have you ever been eating with someone who started to choke? Did you know what to do? Choking is caused when food or a foreign object becomes lodged in the windpipe. When you are with someone who is choking, first notice whether the person can talk, breathe, or cough. If so, stay with the person until the airway is cleared by coughing. Do not try to slap the person on the back. The slapping may only cause the food to become more deeply lodged in the windpipe.

If a person cannot talk or cough and appears not to be breathing, administer several short, quick abdominal thrusts. This technique is called the *Heimlich maneuver,* after Dr. Henry Heimlich who developed the procedure. First, stand behind the choking victim. Put your arms around the person, placing your fists just below the breastbone. The proper placement is shown in Figure 9-11. Then

A B C

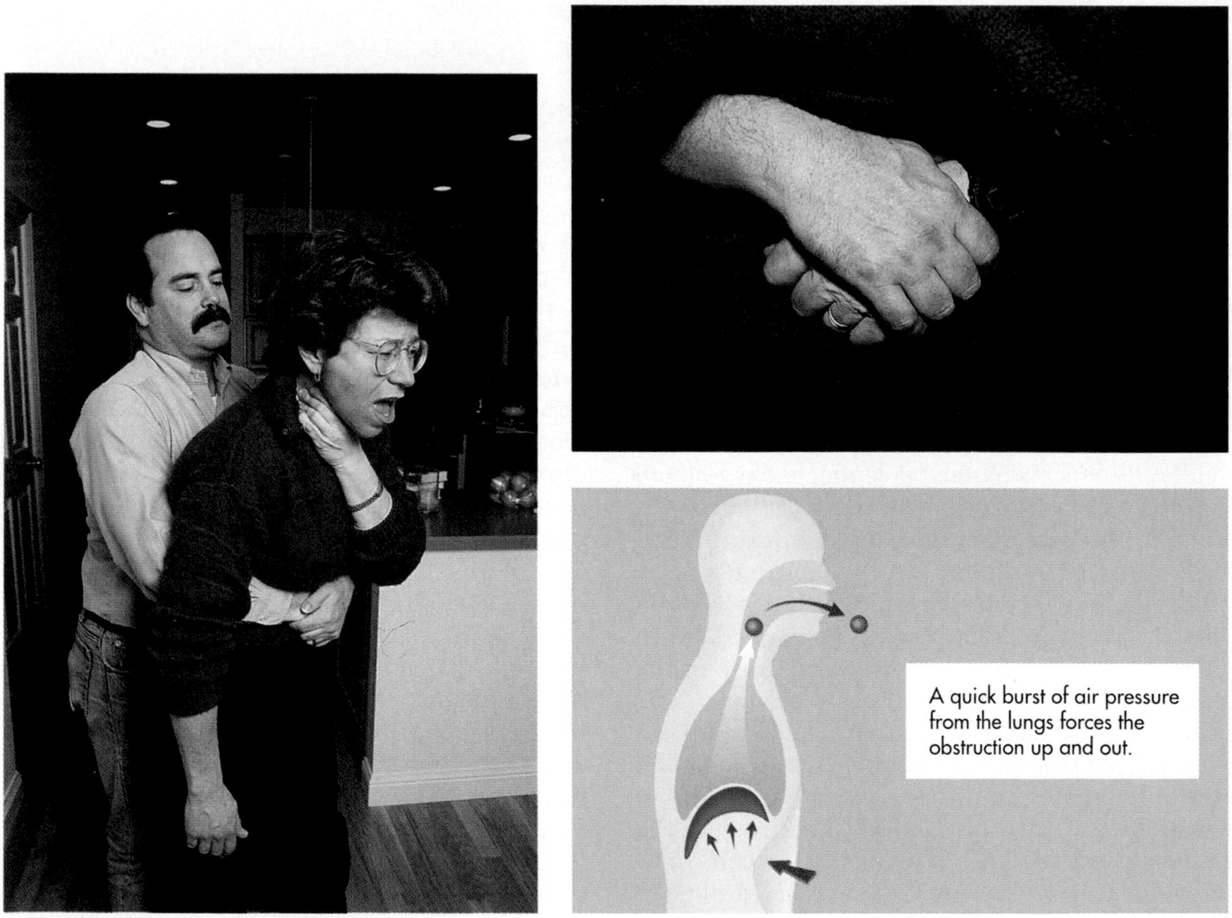

A quick burst of air pressure from the lungs forces the obstruction up and out.

FIGURE 9-11 The Heimlich maneuver. A procedure where air in a person's lungs is used to expel or pop out an obstruction in the airway. A person stands behind the victim **(A)**, positions a fist just below the victim's sternum **(B)**, and firmly and quickly pulls up and back under the ribs **(C)**. This action forces air out of the lungs, dislodging the obstruction from the airway.

give a series of quick, sharp, upward and inward thrusts. These thrusts push in on the diaphragm and the thoracic cavity, suddenly decreasing its volume. This sudden decrease creates a surge in air pressure below the obstruction, which usually projects it forcefully from the windpipe.

Chronic Obstructive Pulmonary Disease

The term **chronic obstructive pulmonary disease (COPD)** is used to refer to disorders that block the airways and impair breathing. COPD affects 25 million people in the United States alone and is responsible for at least 50,000 deaths per year. Most doctors agree that it is one of the fastest growing health problems in this country. Two disorders commonly included in COPD are **chronic bronchitis** and **emphysema.**

The term *chronic* refers to something that occurs over an extended time. Chronic bronchitis, then, does not refer to an isolated, one-time infection or inflammation. It refers to an inflammation of the bronchi and bronchioles that lasts for at least 3 months each year for 2 consecutive years with no accompanying disease as a cause. The primary cause is cigarette smoking. Air pollution and occupational exposure to industrial dust are much less frequent causes.

Cigarette smoking paralyzes ciliated epithelial cells so that they can no longer effectively remove incoming particles and microbes. It also causes increased mucus production by cells lining the trachea. A continued buildup of mucus provides food for bacteria, and infection can result. The mucus also plugs up the respiratory "plumbing."

Normally, bronchioles widen, or dilate, during inspiration; they narrow, or constrict, during expiration. If mucus is plugging various bronchioles, some air may therefore be able to get to the alveoli

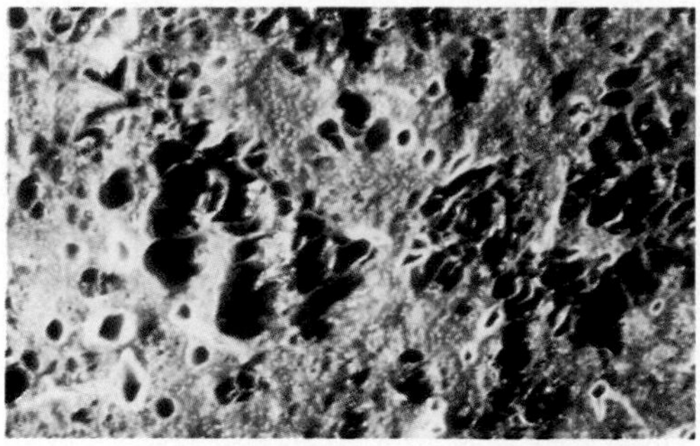

FIGURE 9-12 Alveoli ruptured as a result of emphysema. Emphysema permanently damages lungs and, along with cancer, is one of the dangerous risks of smoking.

beyond the plugged bronchioles but may not be able to get out. Coughing spells produce pressure within these continuously inflated alveoli that ruptures their walls, decreasing the surface area over which gas exchange can take place (Figure 9-12). In addition, the lungs lose their elasticity and the ability to recoil during exhalation, staying filled with air. This disorder is aptly called *emphysema,* meaning "full of air." A person with emphysema has to work voluntarily to exhale.

People with COPD find that they have more respiratory infections than they did before the disorder and that these infections last longer. They have a morning cough or may cough all day. They may also tire easily and become short of breath with minimal physical exertion. Some people with COPD feel as though they cannot breathe at times. As the disorder progresses, some find it difficult to do a day's work or accomplish the daily activities of living. Periods of breathlessness increase. Some persons suffer bouts of respiratory failure and must be hospitalized. COPD is serious and deadly—but in most cases, it is avoidable.

chronic obstructive pulmonary disease (COPD) a term used to refer to disorders that block the airways and impair breathing.

chronic bronchitis (bron **kye** tis) an inflammation of the bronchi and bronchioles that lasts for at least 3 months each year for 2 consecutive years with no accompanying disease as a cause; one of the disorders commonly included in chronic obstructive pulmonary disease.

emphysema (em fi **see** muh) a chronic obstructive pulmonary disease in which mucus plugs various bronchioles, trapping air within alveoli and often causing them to rupture. In this condition, the lungs lose their elasticity and the ability to recoil during exhalation, and instead stay filled with air.

Summary

- Respiration is the uptake of oxygen and the release of carbon dioxide by the body. It includes the processes of cellular respiration, internal respiration, external respiration, and breathing.
- Breathing consists of taking air into the lungs, or inspiration, and expelling air from the lungs, or expiration. The average rate of breathing is 14 to 20 breaths per minute.
- Air passes through many respiratory structures on its way to the lungs. The nasal cavities, located behind the nostrils, warm and moisten incoming air. The air then passes down the throat, or pharynx, and into the larynx. The larynx, or voice box, sits on top of the windpipe, or trachea. The trachea divides into two bronchi, which further divide into smaller and smaller bronchioles, ending in microscopic air sacs called *alveoli.*
- The epithelium of the trachea, the bronchi, and some of the bronchioles produces mucus that helps trap foreign particles. In addition, cilia beat to produce a current in the mucus that brings particles up to the throat to be swallowed.
- During inspiration, the diaphragm, a sheetlike muscle that forms the floor of the thoracic cavity, flattens somewhat as it contracts, increasing the volume of the thoracic cavity. Muscles between the ribs pull up on the rib cage, also increasing this volume. The negative pressure that is created pulls air into the lungs. During expiration, these muscles relax, decreasing the volume of the thoracic cavity. The positive pressure that is created pushes air out of the lungs.
- Oxygen in the air and carbon dioxide in the blood are exchanged at the alveoli. This exchange is called *external respiration.*
- Deoxygenated blood from the body is pumped by the heart to the lungs. Here it is oxygenated during external respiration and is then returned to the heart to be pumped to the rest of the body.
- At the capillaries in the body tissues, oxygen in the blood diffuses out of the blood and into the tissue fluid surrounding the cells and then into the cells for use during cellular respiration. The waste of cellular respiration, carbon dioxide, diffuses out of the cells, into the tissue fluid, and then into the blood. This process is called *internal respiration.*
- To assist a person who is choking and cannot talk or breathe, administer several short, quick, abdominal thrusts with your arms around the victim and your fists just below the breastbone.
- *Chronic obstructive pulmonary disease (COPD)* refers to disorders that impair movement of the air in the respiratory system. Chronic bronchitis and emphysema are two common COPD disorders that are most frequently caused by cigarette smoking.

Knowledge and Comprehension Questions

1. Distinguish among respiration, cellular respiration, internal respiration, and external respiration.
2. Your biology instructor poses a problem in class, and you suggest a solution. Explain how you produced the necessary sounds.
3. Explain how differences in air pressure help you to breathe.
4. Explain what happens during gas exchange at the alveoli. What gases are exchanged, and what forces "drive" this exchange?
5. A normal hemoglobin level is an essential circulatory function. What is carried by the hemoglobin?
6. Deoxygenated blood appears blue while in the veins. Why then, when you cut a vein and look at your injury, does your blood appear red as it leaves the wound?
7. Explain the process of gas exchange at the capillaries. What gases are exchanged, and why does an exchange of gases occur?
8. What symptoms indicate that a person is choking and it would be essential to apply the Heimlich maneuver to the person?
9. While on vacation in the mountains, you find that you feel lightheaded and short of breath. Explain why.

Critical Thinking

1. Hiccups occur when the diaphragm begins to contract spasmodically, causing a sudden inhalation. What do you hypothesize causes the sound effect that results?
2. How does the circulatory system assist the process of respiration? Why is this help important?
3. Why do you think that people with COPD and emphysema tire easily?

CHAPTER 10

CIRCULATION

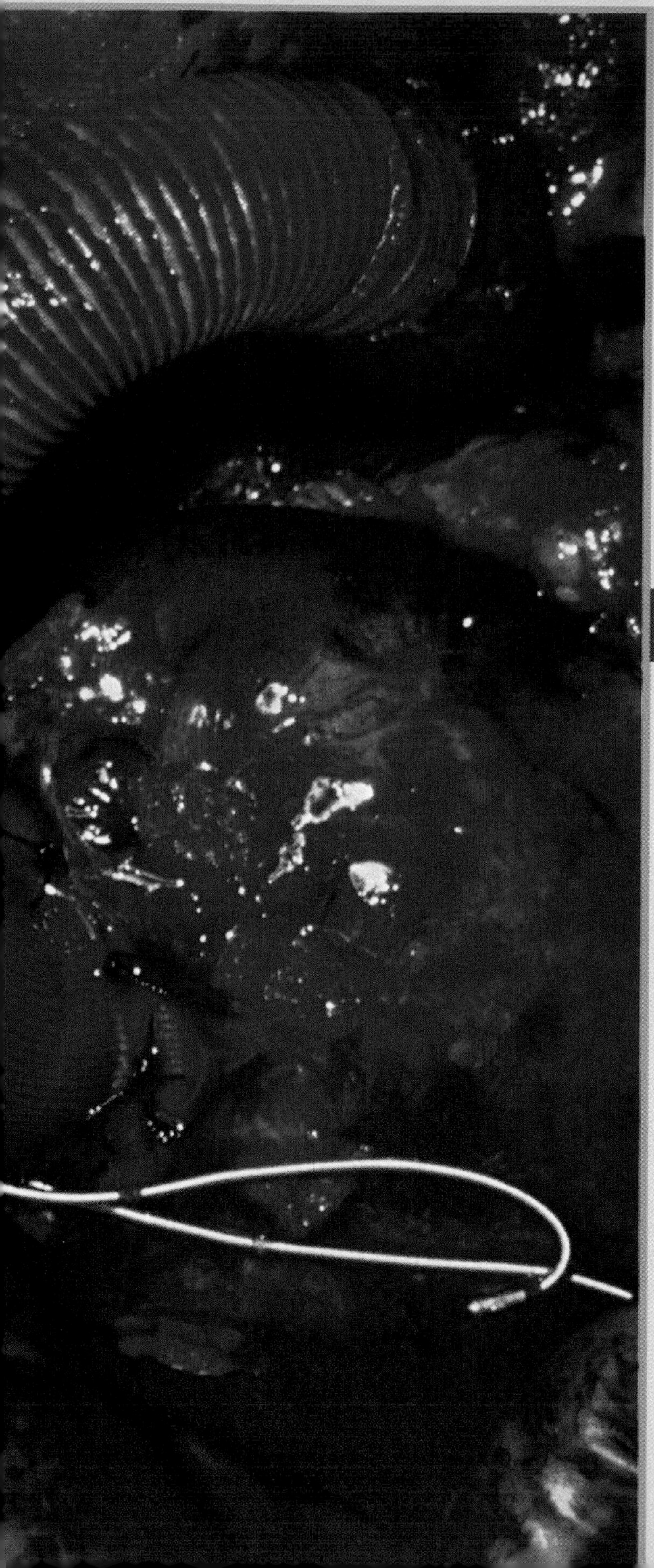

OPEN HEART SURGERY, formerly dangerous and rarely performed, is now commonplace. Although this type of surgery certainly carries risks that vary from patient to patient, it can save and extend lives in ways that were impossible only decades ago. The photo shows aorta replacement surgery. The aorta, the largest artery in the body, is the vessel that receives the rush of blood pumped by the heart to the body. As you can see from the hoselike artificial aorta that is stitched in place, this vessel extends straight up from the heart beneath, and then curves down to supply blood to the lower body. Branches from the aorta bring blood to the upper body as well as the arms, legs, and torso.

Physicians have not only constructed replacement parts for the vessels and valves of the heart, they have designed replacements for the heart itself. History was made in 1982 when a 61-year-old dentist from Seattle, Washington, had his diseased heart replaced with a total artificial heart (TAH). This bridge to life—the Jarvik-7—was named for the doctor who designed it. A product of over three decades of research on mechanical replacements for the heart, the Jarvik-7 was constructed primarily of polyurethane and a Dacron polyester mesh and was actually held together by Velcro.

Since 1982, the Jarvik-7 has been implanted in approximately 180 patients, but it is no longer approved for use by the Food and Drug Administration. Air tubes that supplied strong pulses of air to the artificial heart, "pushing" the blood from its chambers, caused problems; the place where they entered the body became infected easily. Also, the artificial heart often produced blood clots that traveled to the brain, resulting in strokes. A new, smaller artificial heart has taken its place—the Jarvik 2000.

Called a *left ventricular assist device,* this new design helps a person's diseased heart; it does not replace it. The small, egg-shaped pump sits in the left lower chamber of the patient's heart. It connects to a battery pack worn outside the body and is regulated by microprocessors (computer chips). As amazing as these "high tech" replacements and assist devices are, scientists still have not been able to duplicate the body's own remarkable pump—the heart.

KEY CONCEPTS

- The circulatory system is made up of the heart, the blood vessels, and the blood; the heart, a muscular pump, circulates the blood throughout a system of vessels that permeates the body.
- The circulatory system is the transportation system of the body, carrying nutrients, oxygen, and hormones to the cells and removing metabolic wastes.
- The lymphatic system helps the circulatory system by collecting, filtering, and returning the fluid that moves out of the blood and into the tissues and that does not directly return to the blood by itself.
- Cardiovascular diseases, or diseases of the heart and blood vessels, are the leading causes of death in the United States.

OUTLINE

Functions of the Circulatory System

Your circulatory system is made up of three components: (1) the **heart**, a muscular pump; (2) the blood vessels, a network of living tubelike vessels that permeate the body; and (3) the blood, which circulates within these vessels (see Figure 10-3). Together, the heart and blood vessels—the "plumbing" of the circulatory system—are known as the **cardiovascular system.** The prefix *cardio* refers to the heart, and the suffix *vascular* refers to blood vessels. However, the terms *cardiovascular system* and *circulatory system* are often used interchangeably.

Your circulatory system is like a roadway that connects the various muscles and organs of your body with one another. It serves four principal functions:

1. Nutrient and waste transport
2. Oxygen and carbon dioxide transport
3. Temperature maintenance
4. Hormone circulation

In addition, proteins and ions in the fluid portion of the blood regulate the movement of water between the blood and the tissues. Specialized cells defend the body against invading microorganisms and other foreign substances. (These two functions are discussed later in the chapter.) Many biologists argue that the functions of the circulatory system are truly functions of the blood alone; however, the blood cannot perform these functions without circulating throughout the body within vessels, pushed by a powerful pump.

Nutrient and Waste Transport

The nutrient molecules that fuel cell metabolism are transported to the cells of the body by the circulatory system. Sugars and amino acids pass into the bloodstream at the small intestine, diffusing into a fine net of blood vessels below the mucosa. Most fatty acids are absorbed by the epithelial cells lining the small intestine. After being resynthesized into fats and packaged for transport, these fats pass into lymphatic vessels and are transported to the blood by the lymphatic system. The blood takes all these digestive products to the liver for processing. There, some of these molecules are converted to glucose, which is released into the bloodstream. Others, such as essential amino acids and vitamins, pass through the liver unchanged. Still others, such as excess energy molecules and excess amino acids, are used for the synthesis of glycogen and body fat and are stored for later use.

From the liver, the blood carries glucose and other energy molecules to all the body's cells. In addition, the blood also brings molecules such as amino acids to the cells, which are used as building blocks to produce other substances. The cells, in turn, release the waste products of metabolism into the bloodstream. The blood carries most of these wastes to a cleansing organ, the kidney, which captures and concentrates them for excretion in the urine (see Chapter 12). The cleansed blood then passes back to the heart.

Oxygen and Carbon Dioxide Transport

The cells of the body carry out cellular respiration and need oxygen for this series of reactions to take place. Oxygen is transported to the cells of the body by the circulatory system. Within the lungs, oxygen molecules diffuse into the circulating blood through the walls of capillaries, which are very fine blood vessels (see Chapter 9). This oxygen passes into red blood cells suspended in the liquid portion of the blood. From the lungs the blood carries its cargo of oxygen to all cells of the body. At the same time, it picks up a waste product of cellular respiration, carbon dioxide. The blood then returns to the lungs, where the carbon dioxide is released and a fresh supply of oxygen is captured.

Temperature Maintenance

As you read in Chapter 4, energy is continuously being lost as heat during the chemical reactions that take place in the cells of your body. The blood distributes this heat as it circulates, helping to maintain your body temperature. As the blood circulates, it passes through delicate, microscopic networks of blood vessels that lie under your skin. As the blood passes through these vessels, it gives up heat because the environment is usually cooler than the body's temperature of 98.6°F. The body works to balance the amount of heat produced and the amount of heat lost to maintain a stable internal temperature.

To maintain this balance, a regulatory center in the brain, the hypothalamus, acts like your own personal thermostat, constantly monitoring body temperature and stimulating regulatory processes. If your body temperature drops, for example, signals from this center cause surface blood vessels to narrow, or constrict. Constriction of these blood vessels limits blood flow to the surface of the skin and lessens heat loss. Conversely, if your body temperature rises, signals from this center cause surface blood vessels to widen, or dilate. Dilation increases blood flow to the surface of the skin and increases heat loss (Figure 10-1).

Hormone Circulation

The chemical reactions and other activities of the body are coordinated by nerve signals and hormones. Hormones are the body's chemical mes-

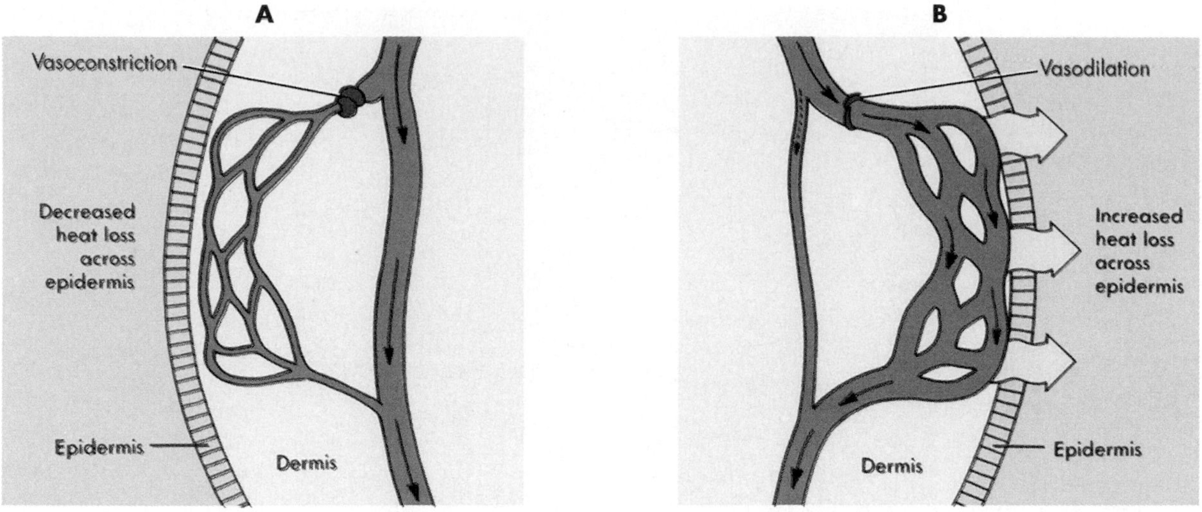

FIGURE 10-1 Regulation of heat loss. A, Vasoconstriction. The amount of heat lost from the body can be regulated by controlling the flow of blood to the skin surface. Constriction of the surface blood vessels limits blood flow and lessens heat loss. **B,** Vasodilation. Dilation of the surface blood vessels increases blood flow and thus increases heat loss.

sengers; they are produced in one place and produce an effect in another. The circulatory system is the highway within which hormones travel throughout the body from their site of production and, in this way, reach the target tissues that are capable of responding to them.

> *The circulatory system transports oxygen and nutrients to cells, transports carbon dioxide and metabolic wastes away from cells, helps maintain a stable internal temperature, and carries chemical messengers called hormones throughout the body. Substances within the blood regulate the movement of water between the blood and the tissues and defend the body against disease.*

The Heart and Blood Vessels

As the blood is pushed along by the heart to begin its journey throughout the body, it leaves the heart within vessels known as **arteries** (Figure 10-2). From the arteries, the blood passes into a network of **arterioles,** or small arteries. From these, it eventually is forced through the **capillaries,** a fine latticework of very narrow tubes, which get their name from the Latin word *capillus,* meaning "a hair." As blood passes through these capillaries, gases are exchanged between the blood and the tissues, nutrients are delivered to the tissues, and wastes are picked up from the tissues. After its journey through the capillaries, the blood passes into a third kind of vessel: the **venules,** or small **veins.** A network of venules and larger veins collects the circulated blood and carries it back to the heart.

Arteries and Arterioles

Arteries are vessels that carry blood away from the heart. The adult human heart pumps about 70 milliliters of blood—a little over 2 ounces—into the arteries with each beat. On average, your heart beats about 75 times per minute, pumping more than 5 liters, or $5^1/_2$ quarts, of blood into your arteries each minute of your life. The vessels leading out from the heart expand slightly in diameter and then recoil before the next heartbeat in response to the pressure of blood surging into them.

The walls of the arteries are made up of three layers of tissue (Figure 10-3, *A*) and have a hollow core called the *lumen* through which blood flows. *Endothelial cells* (the inner epithelium of blood vessels and the heart) line arteries and are in contact with the blood. Surrounding these cells is a thick layer of smooth muscle and elastic fibers. The elas-

heart the muscular pump of the circulatory system.
cardiovascular system (**kar** dee oh **vas** kyuh lur) the heart and blood vessels, the "plumbing" of the circulatory system.
arteries (**art** uh rees) blood vessels that carry blood away from the heart.
arterioles (are **teer** ee oles) small arteries that lead from arteries to capillaries.
capillaries (**kap** uh **lare** ees) fine latticeworks of microscopic blood vessels that permeate tissues.
venules (**vayn** yooles) small veins that connect capillaries to larger veins.
veins (**vayns**) blood vessels that carry blood to the heart.

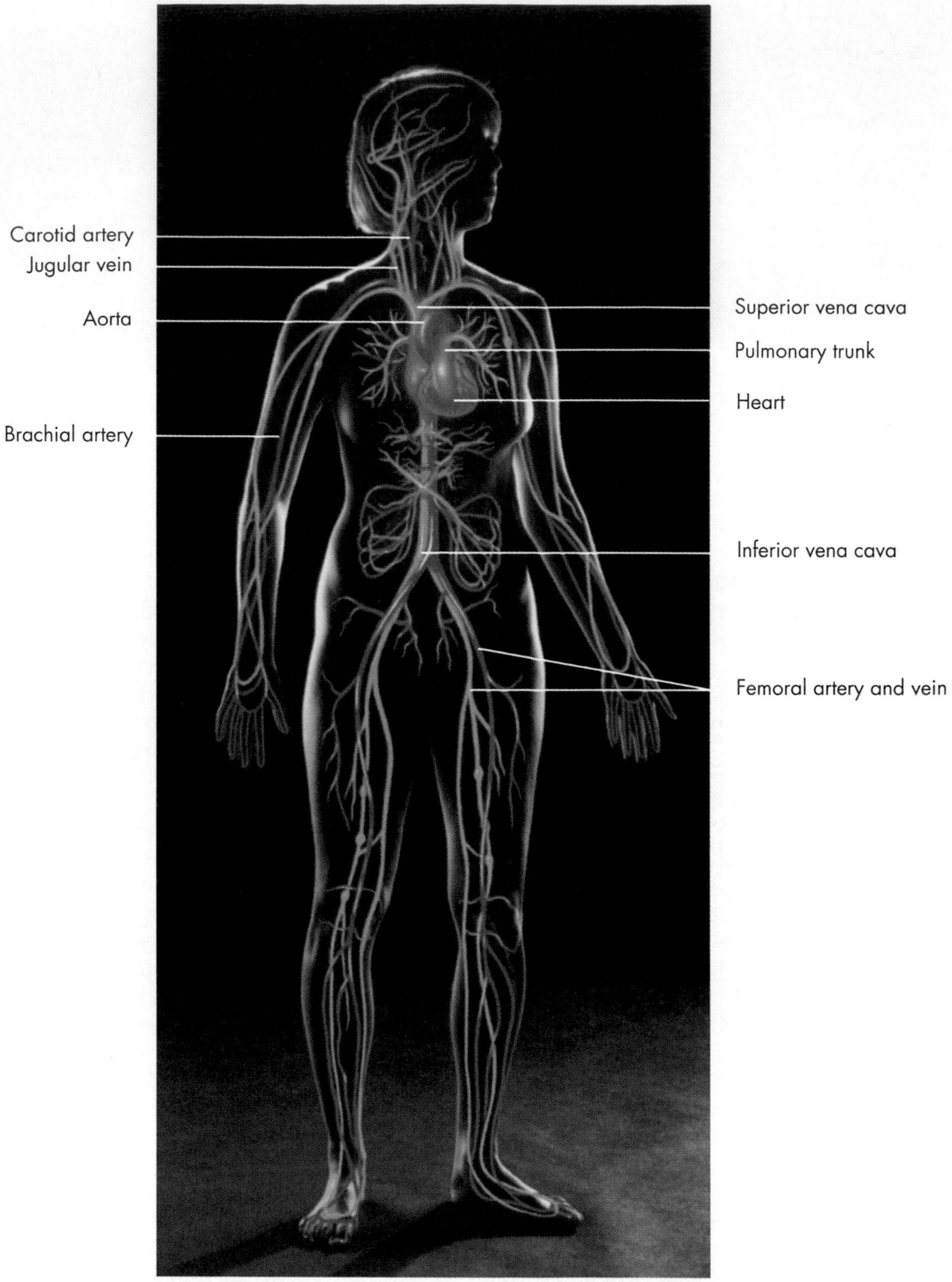

FIGURE 10-2 The human circulatory system. The circulatory system is composed of the heart, blood vessels, and blood.

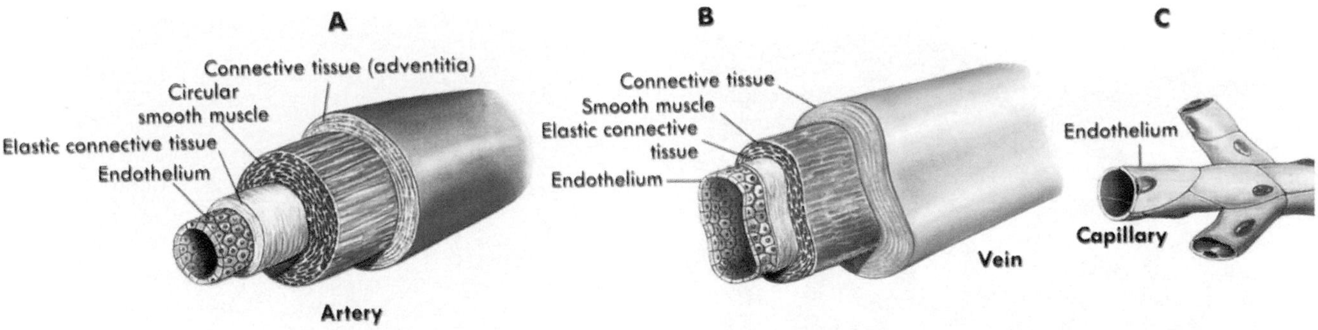

FIGURE 10-3 The structure of some important blood vessels. Blood leaves the heart through arteries **(A)** and returns to the heart through veins **(B)**. The capillaries **(C)** are the tubes through which the blood is forced in the tissues.

tic tissue allows the artery to expand and recoil in response to the pulses of blood. The steady contraction of the muscle layer strengthens the wall of the vessel against overexpansion. This muscle and elastic layer is encased within an envelope of protective connective tissue.

Arterioles differ from arteries in that they are smaller in diameter. The walls of the largest arterioles are constructed very much like those of arteries. However, as arterioles near the capillaries, their diameter decreases until they consist of nothing but a layer of endothelium wrapped with a few scattered smooth muscle cells.

The muscle cells within the walls of arterioles tighten or relax in response to messages from nerves and hormones. When these muscles relax, the arteriole dilates and the blood flow through it increases. This response is exactly what is happening when you blush. When you are embarrassed (or when you become overheated), signals from the nerve fibers connected to muscles surrounding the arterioles are inhibited, which relaxes the smooth muscle and causes the arterioles in the skin to dilate. The increased blood flow brings heat to the surface for escape and causes you to have a red face. Conversely, when you are scared or cold, the muscles in the walls of the arterioles in your skin contract. When the muscles contract, the arteriole constricts and the blood flow through it decreases, conserving your body heat. When you are scared, the body constricts these vessels to route more blood to other tissues such as your skeletal muscles.

Capillaries

Capillaries are microscopic blood vessels that connect arterioles with venules. They have a simple structure (Figure 10-3, *C*) and are little more than tubes with walls one cell thick and with a length that would barely stretch across the head of a pin. The internal diameter of the capillaries is about the same as that of red blood cells, causing these cells to squeeze through the capillaries single file (Figure 10-4). The closeness between the walls of the capillaries and the membranes of the red blood cells facilitates the diffusion of gases, nutrients, and wastes between them—a swap of oxygen and nutrients for carbon dioxide and other metabolic waste products.

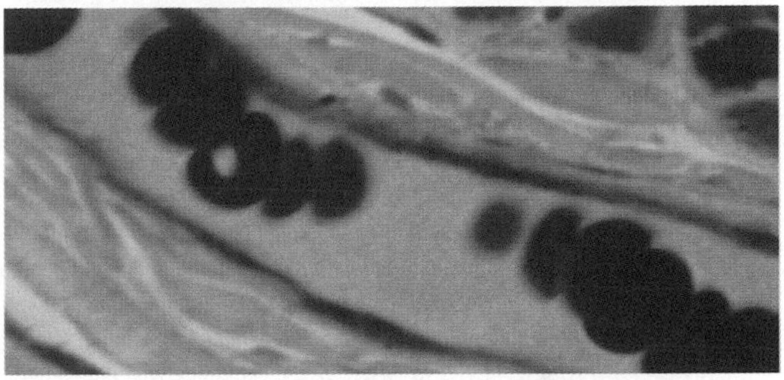

FIGURE 10-4 Red blood cells moving through a capillary in single file. Many capillaries are even smaller than those shown here from the bladder of a monkey *(2,500x)*. Red blood cells will even pass through capillaries narrower than their own diameter, pushed along by the pressure generated by a pumping heart.

Your entire body is permeated with a fine mesh of capillaries, networks that amount to several thousand kilometers in overall length. In fact, if all the capillaries in your body were laid end to end, they would extend across the United States! These networks of capillaries are called **capillary beds.** In a capillary bed, some of the capillaries connect arterioles and venules directly (Figure 10-5) and are called **thoroughfare channels.** From these channels, loops of true capillaries—those not on

capillary beds (kap uh lare ee) networks of capillaries.
thoroughfare channels capillaries within capillary beds that connect arterioles and venules directly.

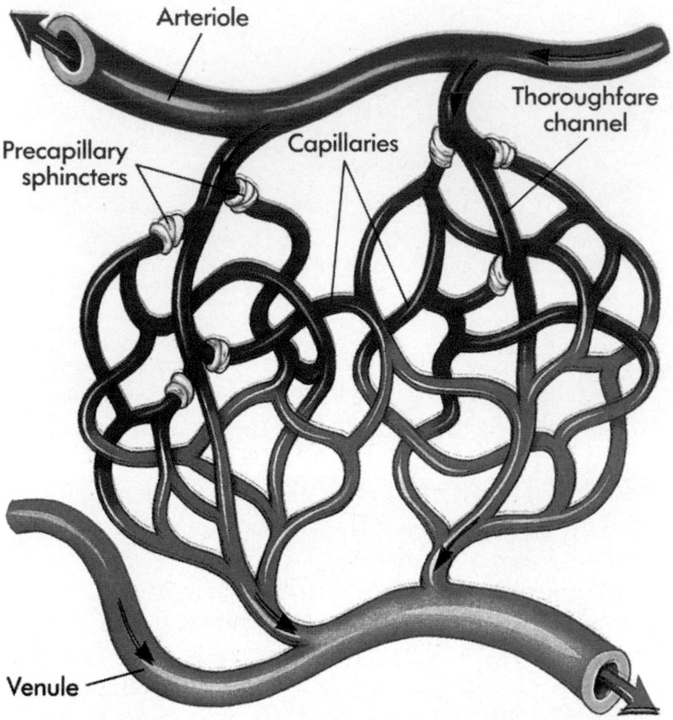

FIGURE 10-5 The capillary bed connects arterioles with venules. The most direct connection is via thoroughfare channels that connect arterioles directly to venules. Branching from these thoroughfare channels is a network of finer capillary channels. Most of the exchange of gases, nutrients, wastes, and ions between body cells and red blood cells occurs while the red blood cells are in this capillary network.

the direct flow route from arterioles to venules—leave and return. Almost all exchanges between the blood and the cells of the body occur through these loops. The entry to each loop is guarded by a ring of muscle that, when contracted, blocks flow through the capillary. Restricting blood flow in capillary beds near the surface of the skin is a powerful means by which the body can limit heat loss in addition to the constriction of surface arterioles. The body can also cut down on the flow within a capillary bed when heavy flow is not needed—to your muscles, for example, when you are resting. Likewise, the body can increase the flow within a capillary bed when the need increases—to your small intestine, for example, after a meal. Interestingly, you do not have enough blood to fill all your capillary beds if they were all open at the same time. If such a situation occurred, you would faint because of lack of sufficient blood to the brain.

Veins and Venules

Venules are small veins that collect blood from the capillary beds and bring it to larger veins (Figure

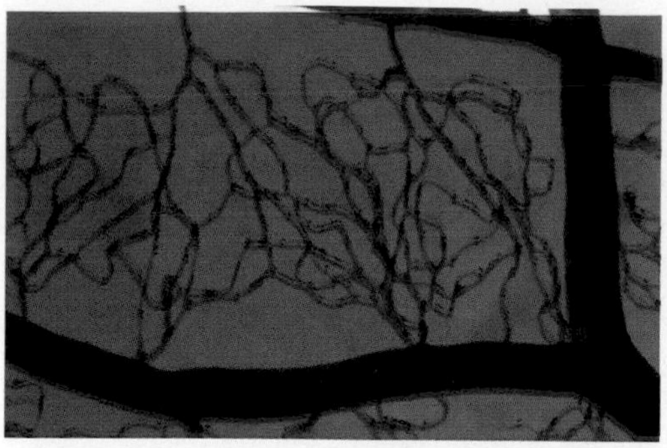

FIGURE 10-6 Venules. Venules collect blood from the capillary bed and deliver it to the larger veins *(10x)*.

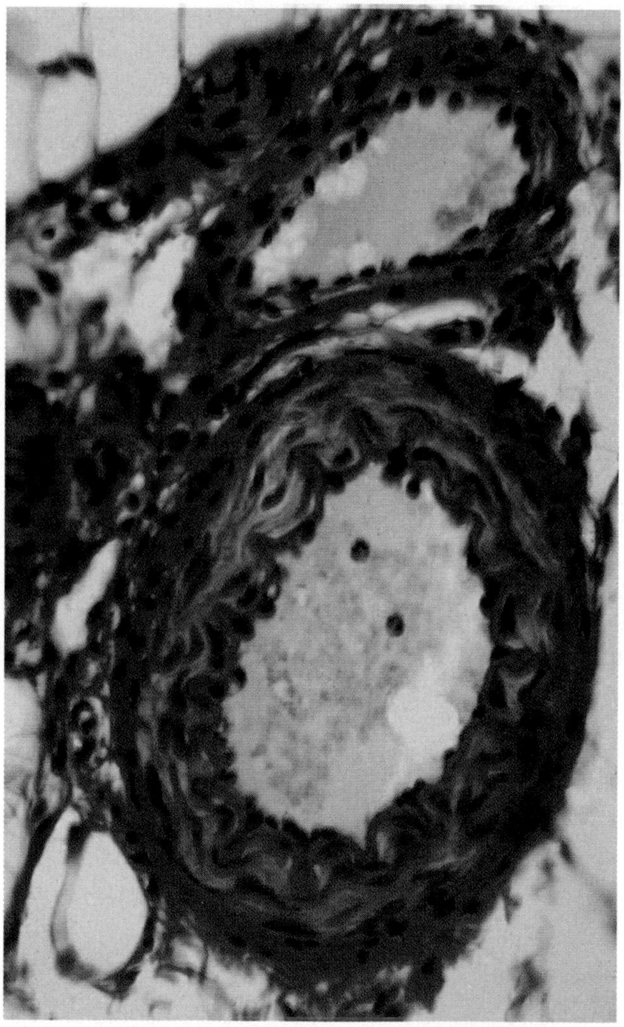

FIGURE 10-7 A closer look at blood vessels. The vein *(top)* has the same general structure as an artery *(bottom)* but has much thinner layers of muscle and elastic fiber. An artery will retain its shape when empty, but a vein will collapse *(700x)*.

10-6) that carry it back to the heart. The force of the heartbeat is greatly diminished by the time the blood reaches the veins, so these vessels do not have to accommodate pulsing pressures as arteries do. Therefore the walls of veins, although similar in structure to those of the arteries, have much thinner layers of muscle and elastic fiber (Figures 10-3, *B,* and 10-7). Lacking much of this supportive tissue, the walls of veins collapse when they are empty (although this situation would never occur in the living body). Empty arteries, on the other hand, stay open like tiny pipes. In addition, the lumen of veins is larger than that of arteries. This difference in size is related to the lower pressure of blood flowing within veins back toward the heart—large vessels present less resistance to the flow than smaller vessels do.

The pathway of blood back to the heart from much of the body is an uphill struggle. The pressure pushing the blood upward in the veins of your legs, for example, approximately equals the force of gravity pulling it down. As skeletal muscles contract, they press on veins and help move blood along. In addition, many veins have one-way valves that help blood move back toward the heart by preventing its backflow (Figure 10-8). If some of the valves in a vein are weak, however, gravity can force blood back through these valves, overloading a portion of a vein and pushing its walls outward. These "stretched out" veins, called *varicose veins,* are often seen in the legs at the surface of the skin.

Arteries and veins are the major vessels of the circulatory system and have walls composed of three layers: (1) an innermost, thin layer of endothelial tissue that is in contact with the blood; (2) a layer of elastic fibers and smooth muscle serviced by nerves; and (3) a layer of protective connective tissue. The walls of arteries have more elastic tissue and smooth muscle than veins do, which helps them accommodate pulses of blood pumped from the heart. Capillaries are microscopic blood vessels that have walls only one cell thick. The exchange of substances between the cells and the blood takes place at the capillaries.

The Heart: A Double Pump

The human circulatory system is referred to as a *closed circulatory system* because blood flows throughout the body confined within blood vessels. The blood is therefore separated from the rest of the body's fluids and does not mix freely with

FIGURE 10-8 A one-way valve in a vein. These valves assist the blood in its movement toward the heart by preventing the blood from moving backwards *(20x).*

them. The plumbing of a closed circulatory system requires not only a system of vessels through which the blood can continuously flow throughout the body but also a pump to push the blood along. The pump of the human circulatory system is the heart, but it is two pumps in one. Figure 10-9 is a diagram of a frontal section through the heart and shows the organization of this double pump. The left side (one pump) has two connected chambers, as does the right side (the other pump). But the two sides, or pumps, of the heart are not directly connected with one another.

Circulatory Pathways

The journey of blood around the body starts with the entry of oxygenated blood into the heart from the lungs. Oxygenated blood from the lungs enters the left side of the heart, emptying directly into the upper left chamber of the heart, the **left atrium,** through large vessels called the **pulmonary veins.** These veins are unusual in that they carry oxygenated blood; other veins, because they carry blood back to the heart from the body tissues, carry deoxygenated blood. The word *pulmonary* refers to the lungs. The circulation of blood to and from the lungs is therefore called the **pulmonary circulation** (Figure 10-10).

left atrium (**left ay** tree um) the upper left chamber of the heart, into which oxygenated blood enters from the lungs.

pulmonary veins (**puhl** muh **nare** ee **vaynz**) large blood vessels that carry oxygenated blood from the lungs to the left atrium of the heart.

pulmonary circulation (**puhl** muh **nare** ee sur kyuh **lay** shun) the part of the human circulatory system that carries blood to and from the lungs.

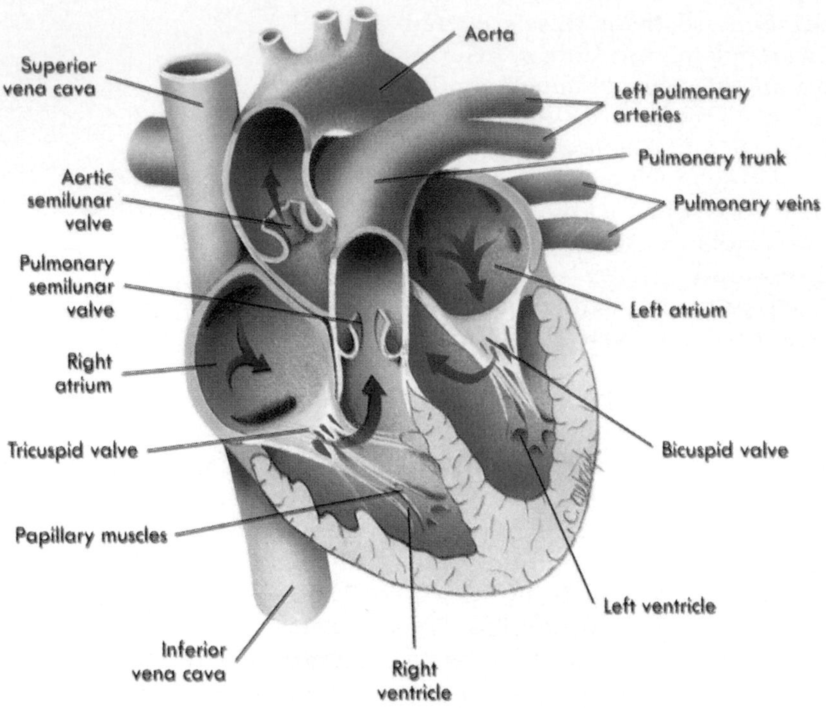

FIGURE 10-9 Blood flow through the human heart. Frontal section of the heart revealing the four chambers and the direction of blood flow through the heart. The great vessels bringing blood to or away from the heart are also shown.

From the left atrium of the heart, blood flows through a one-way valve, the **bicuspid valve,** into the lower, adjoining chamber, the **left ventricle.** Most of this flow (roughly 80%) occurs while the heart is relaxed. The atrium then contracts, pushing the remaining 20% of its blood into the ventricle. After a slight delay the ventricle contracts. The walls of the ventricle are far more muscular than those of the atrium, and as a result, this contraction is much stronger. It forces most of the blood out of the ventricle in a single strong pulse. The blood is prevented from going back into the atrium by the bicuspid valve, whose flaps are pushed shut as the ventricle contracts. Strong fibers that prevent the flaps from moving too far when closing are attached to their edges. If the flaps did move too far, they would project out into the atrium.

Prevented from reentering the atrium, the blood takes the only other way out of the contracting left ventricle: an opening that leads into the largest artery in the body—the **aorta.** The aorta is closed off from the left ventricle by a one-way valve, the **aortic semilunar valve.** It is oriented to permit the flow of the blood out of the ventricle. As the blood is pushed forcefully out of the left ventricle to make its trip around the body, it rushes into the aorta, causing the elastic, muscular walls of this artery to bulge slightly outward. Quickly, however, the aorta walls recoil as a stretched elastic band does when released. This recoil action pushes on the blood; some of it is pushed backward against the valve. The valve is constructed in such a way that it snaps shut in response to this backflow. The other one-way valves found in the heart and blood vessels are constructed in a similar manner, preventing the backflow of blood.

Many arteries branch from the aorta, carrying oxygen-rich blood to all parts of the body. The pathway of blood vessels to the body regions and organs other than the lungs is called the **systemic circulation** (see Figure 10-10), with the aorta being the first and largest vessel in the circuit. The first arteries to branch off the aorta are the coronary arteries, which carry freshly oxygenated blood to the heart itself; the muscles of the heart do not obtain their supply of blood from directly within the heart. From the arch of the aorta, the carotid arteries branch off and bring blood to networks of vessels in the neck and head. The subclavian arteries bring blood to the shoulders and arms. The aorta then descends down the trunk of the body, with arteries branching off to supply various organs such

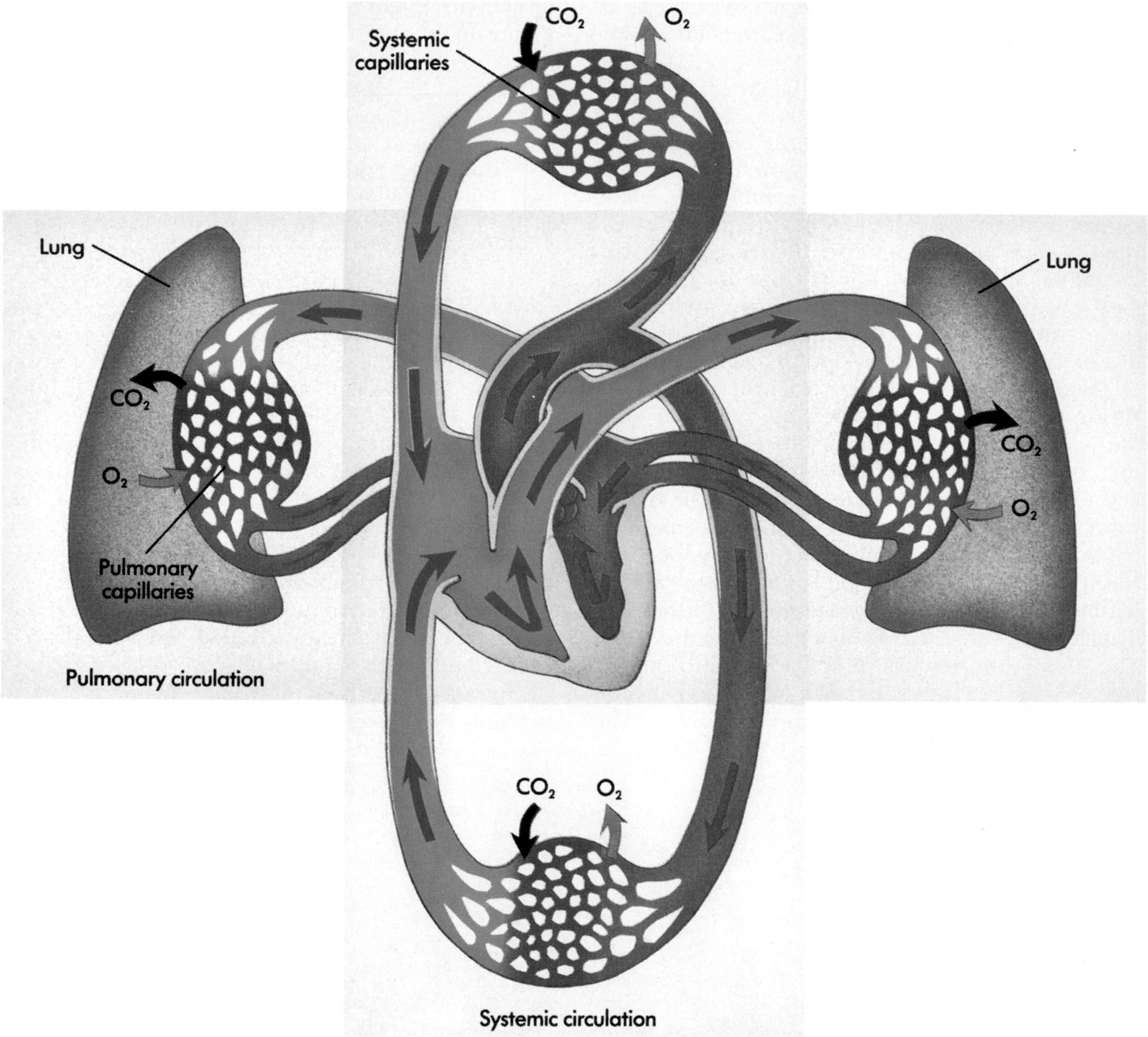

FIGURE 10-10 Pulmonary circulation versus systemic circulation. The circulation of the blood to and from the lungs is called the pulmonary circulation. The circulation of the blood to the body regions and organs other than the lungs is called the systemic circulation. This illustration diagrams those pathways.

bicuspid valve (bye **kus** pid **valv**) a one-way valve in the heart through which blood flows from the left atrium to the lower, adjoining left ventricle.

left ventricle (**left ven** truh kul) a chamber in the heart into which blood flows from the left atrium. The left ventricle then pumps blood through the aorta into the arteries.

aorta (ay **ort** uh) the largest artery in the body. It receives blood from the left ventricle and has many vessels branching from it that bring blood throughout the body (with the exception of the lungs).

aortic semilunar valve (ay **ort** ik **sem** ee **loo** ner **valv**) a one-way valve that permits blood flow from the left ventricle of the heart to the aorta.

systemic circulation the pathway of blood vessels to the body regions and organs other than the lungs.

as the kidneys, liver, and intestines. The aorta divides into two major vessels at the level of the lower back, one traveling to each leg.

The blood that flows into the arterial system eventually returns to the heart after delivering its supply of oxygen to the cells of the body and picking up the waste gas carbon dioxide. This exchange takes place at the capillaries. In returning, blood passes through a series of veins, eventually entering the right side of the heart. Two large veins collect blood from the systemic circulation. The **superior vena cava** drains the upper body, and the **inferior vena cava** drains the lower body. These veins dump deoxygenated blood into the right atrium.

The right side of the heart is similar in organization to the left side. However, the muscular walls of the right ventricle are not as thick as those of the left. Blood passes from the right atrium into the right ventricle through a one-way valve, the **tricuspid valve.** It passes out of the contracting right ventricle through a second valve, the **pulmonary semilunar valve,** into a single **pulmonary artery,** which subsequently branches into two pulmonary arteries that carry deoxygenated blood to the lungs. The blood then returns from the lungs to the left side of the heart, replenished with oxygen and cleared of much of its load of carbon dioxide. (The human circulatory system as a whole is outlined in Figure 10-2.)

> *The circulation of the blood to and from the lungs is called the pulmonary circulation. The circulation of the blood through the parts of the body other than the lungs is called the systemic circulation.*

How the Heart Contracts

The contraction of the heart depends on a small cluster of specialized cardiac muscle cells that is embedded in the upper wall of the right atrium (Figure 10-11). This cluster of cells, called the **sinoatrial (SA) node,** automatically and rhythmically sends out impulses that initiate each heartbeat. The SA node is therefore nicknamed the "pacemaker" of the heart.

The impulse initiated by the SA node causes both atria to contract simultaneously and also excites a bundle of cardiac muscle cells located at the base of the atria. These cells are known as the **atrioventricular (AV) node.** The AV node conducts the impulse to a strand of specialized muscle in the **septum,** the tissue that separates the two sides of

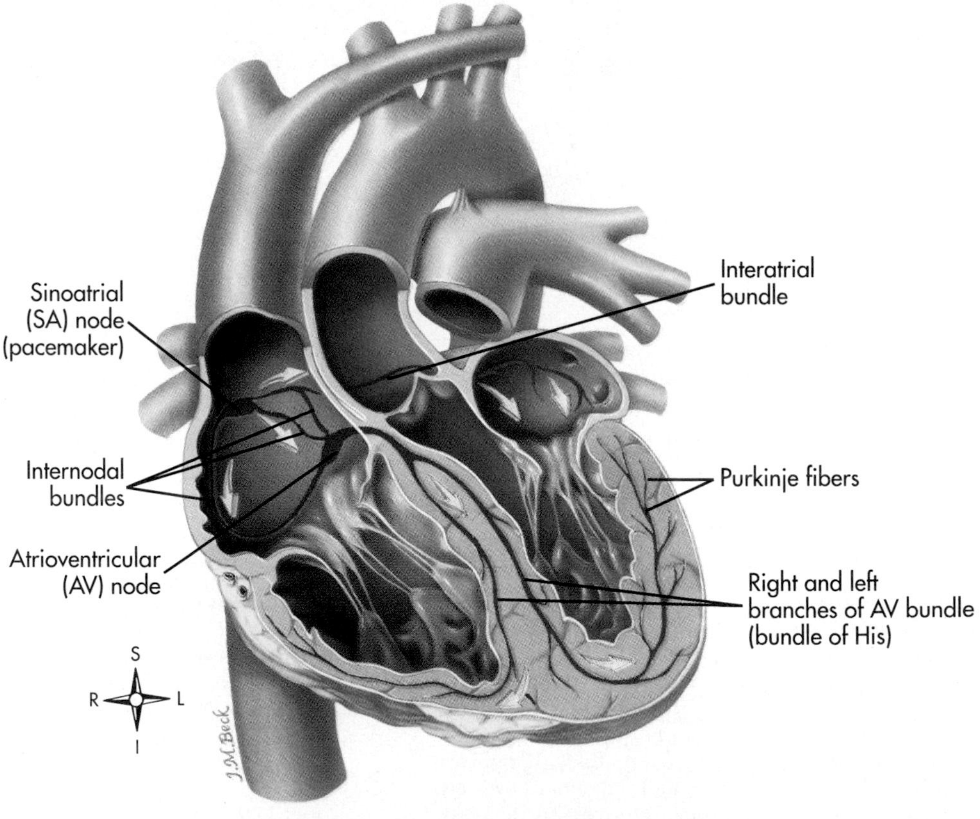

FIGURE 10-11 The conduction system of the heart. Four structures compose the conduction system of the heart—sinoatrial (SA) node, atrioventricular (AV) node, AV bundle, and Purkinje fibers. These specialized groups of cardiac muscle cells initiate an electrical impulse throughout the heart. The impulse begins in the SA node and spreads to the AV node. The AV node then initiates a signal that is conducted through the ventricles by way of the AV bundle and Purkinje fibers.

the heart. This strand of impulse-conducting muscle, known as the **bundle of His,** has branches that divide to the right and left. On reaching the apex (lower tip) of the heart, each branch further divides into conducting fibers called **Purkinje fibers,** which initiate the almost simultaneous contraction of all the cells of the right and left ventricles. The conduction of the impulse from the atria to the ventricles takes approximately 0.2 second. During this delay the atria finish emptying their contents into the corresponding ventricles before the ventricles start to contract.

An excitatory impulse is initiated in the SA node of the heart, which causes the atria to contract and the impulse to be passed along to other specialized groups of cardiac cells: the AV node, bundle of His, bundle branches, and Purkinje fibers. The passage of this excitatory impulse causes the almost simultaneous contraction of all cells of the right and left ventricles.

Monitoring the Heart's Performance

The heartbeat is really a series of events that occurs in a predictable order. A physician can gain information about the health of the heart and events occurring during the heartbeat by making several different kinds of observations. The simplest is to listen to the heart at work.

The first sound heard, a low-pitched *lubb,* is caused by the turbulence in blood flow created by the closing of the bicuspid and tricuspid valves at the start of ventricular contraction. A little later a higher-pitched *dupp* can be heard, signaling the closing of the pulmonary and aortic semilunar valves at the end of ventricular contraction. If the valves are not closing fully or if they open too narrowly, a slight backflow of blood occurs within the heart. This backflow can be heard as a sloshing sound; this condition is known as a *heart murmur.*

A second way to examine the events of the heartbeat is to monitor the blood pressure, the force exerted by the blood on blood vessel walls. During the first part of the heartbeat the atria are filling and contracting. At this time the pressure in the arteries leading from the left side of the heart out to the tissues of the body decreases slightly because blood is not being forced into them by the left ventricle. This period of relaxation (with respect to the ventricles) is referred to as the **diastolic period.** The force the blood exerts on the blood vessels at this time is the diastolic pressure. During the contraction of the ventricles, a pulse of blood is forced into the systemic arterial system by the left ventricle, immediately raising the blood pressure within these vessels. This period of contraction (with respect to the ventricles) ends with the closing of the aortic semilunar valve and is referred to as the **systolic period.** The force the blood exerts on the blood vessels at this time is the systolic pressure.

superior vena cava (vee nuh **kay** vuh *or* **vay** nuh **kah** vuh) a large vein that drains blood from the upper body and returns it to the right atrium of the heart.

inferior vena cava a large vein that drains blood from the lower body and returns it to the right atrium of the heart.

tricuspid valve (try **kus** pid **valv**) a one-way valve within the heart, through which blood passes from the right atrium to the right ventricle.

pulmonary semilunar valve (**puhl** muh **nare** ee **sem** ih **loo** ner **valv**) a one-way valve that permits blood flow from the right ventricle of the heart to the pulmonary artery.

pulmonary artery an artery that extends from the right ventricle of the heart; it branches into two smaller pulmonary arteries that carry deoxygenated blood to the lungs.

sinoatrial (SA) node (sye no ay tree uhl **node**) a small cluster of specialized cardiac muscle cells that are embedded in the upper wall of the right atrium and automatically and rhythmically send out impulses that initiate each heartbeat.

atrioventricular (AV) node (ay tree oh ven **trik** yuh lur **node**) a group of specialized cardiac muscle cells located in the base of the atria that receives the impulses initiated by the sinoatrial node and conducts them to the bundle of His in the heart septum.

septum tissue that separates the two sides of the heart.

bundle of His a strand of impulse-conducting muscle located in the septum of the heart that conducts heartbeat impulses from the AV node to the ventricles of the heart.

Purkinje fibers (pur **kin** jee **fye** burs) conducting fibers that branch from the bundle of His and initiate the almost simultaneous contraction of all the cells of the right and left ventricles.

diastolic period (**dye** uh **stol** ik) the time of ventricular relaxation during the first part of a heartbeat, when the atria are filling.

systolic period (sis **tol** ik) the time of ventricular contraction during a heartbeat, in which a pulse of blood is forced into the systemic arterial system by the left ventricle.

Blood pressure is measured in units called millimeters (mm) of mercury (Hg). These units refer to the height to which a column of mercury is raised in a tube by an equivalent pressure. On the basis of this system of measurement, normal blood pressure values are 70 to 90 (mm of Hg) diastolic and 110 to 130 systolic. Blood pressure is expressed as the systolic pressure over the diastolic pressure, such as 110 over 70. When the inner walls of the arteries accumulate fats, as they do in the condition known as **atherosclerosis,** the diameters of the passageways are narrowed. Such narrowing is one cause of elevated systolic and diastolic blood pressures.

A third way to monitor the events of a heartbeat is to measure the electrical changes that take place as the heart's chambers both contract and relax. Because so much of the human body is made up of water, it conducts electrical currents rather well. Therefore as the impulses initiated at the SA node pass throughout the heart as an electrical current, this current passes in a wave throughout the body. Although the magnitude of this electrical pulse is tiny, it can be detected with sensors placed on the skin. A recording made of these impulses (Figure 10-12) is called an **electrocardiogram (ECG).**

Three successive electrical pulses are recorded in a normal heartbeat. The first pulse occurs when the atria contract; this electrical event is called the *P wave* on the ECG (Figure 10-12, *D*). There is a much stronger pulse (*the QRS complex*) $^{2}/_{10}$ of a second later, reflecting both the contraction of the ventricles and the relaxation of the atria (Figure 10-12, *F*). Finally, a third pulse (*the T wave*) occurs caused by the relaxation of the ventricles (Figure 10-12, *G*).

The heartbeat is a series of events that occurs in a predictable order. A physician can gather data about the health of the heart by monitoring the heartbeat in various ways: by listening to the sounds the heart makes, by monitoring the blood pressure, and by measuring the electrical changes that take place as it contracts and relaxes.

The Blood

Your blood makes up about 8% of your body. It is a viscous, or thick, fluid made up of two parts: a liquid portion called **plasma** and a portion consisting of **formed elements**—several different kinds of cells and cell parts suspended within the plasma.

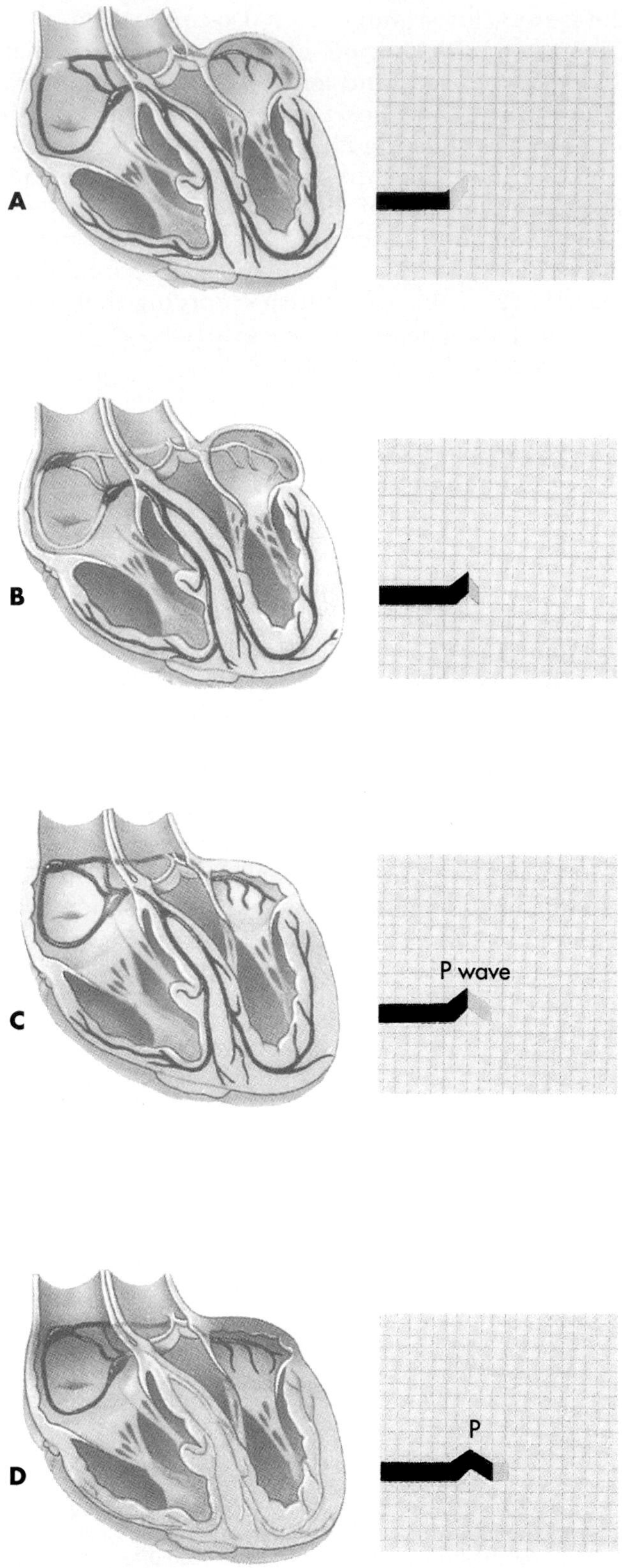

FIGURE 10-12 Events represented by the electrocardiogram (ECG). A through **D,** The P wave represents the depolarization of cardiac muscle tissue in the SA node and atrial walls. Before the QRS complex **(E, F)** is observed, the AV node and AV bundle depolarize. **G,** The T wave is observed as the ventricular walls repolarize. Depolarization triggers contraction in the affected muscle tissue. Thus cardiac muscle contraction occurs *after* depolarization begins.

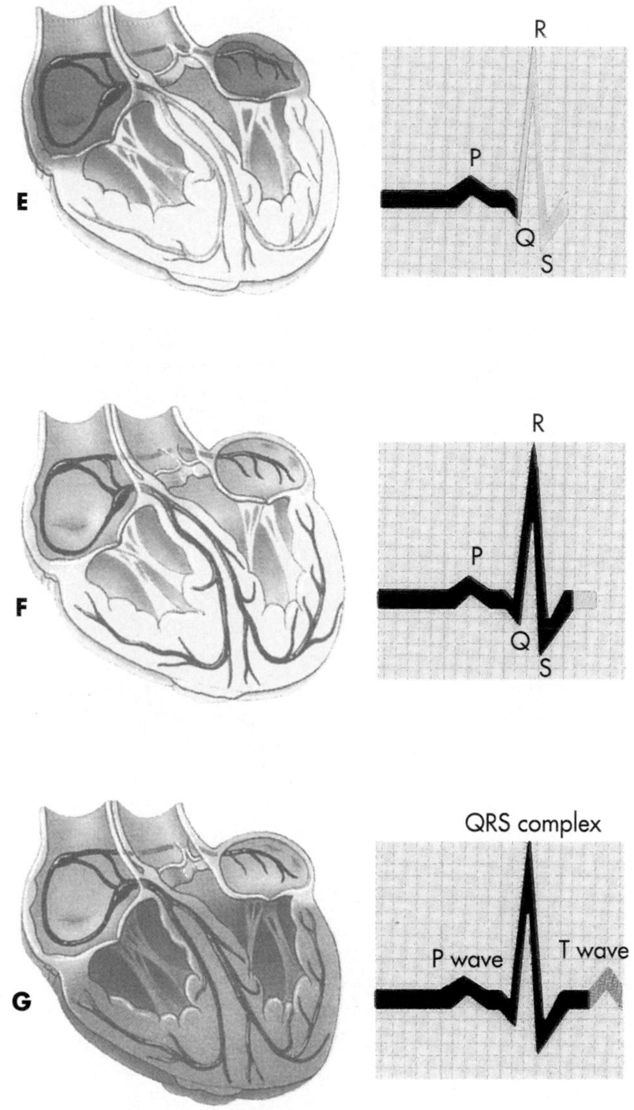

FIGURE 10-12, cont'd.

Blood Plasma

Blood plasma is a straw-colored liquid made up of water and dissolved substances. The dissolved substances can be grouped into three categories:

1. *Nutrients, hormones, respiratory gases, and wastes.* These substances can be thought of as the traffic on the highway of blood. Dissolved within the plasma are substances that move from one place to another in the body and that are used or produced by the metabolism of cells. These substances include glucose, lipoproteins (a soluble form of lipid), amino acids, vitamins, hormones, and the respiratory gases.
2. *Salts and ions.* Plasma is a dilute salt solution. Chemically, the word *salt* refers to more than just table salt. It is a general term applied to any substance composed of positively and negatively charged ions. In water, salts dissociate into their component ions. An ion is a charged particle—an atom that has an unequal number of protons and electrons (see Chapter 2). The chief plasma ions are sodium (Na^+), chloride (Cl^-), and bicarbonate (HCO_3^-). In addition, there are trace amounts of other ions, such as calcium (Ca^{2+}), magnesium (Mg^{2+}), copper (Cu^{2+}), potassium (K^+), and zinc (Zn^{2+}). In living systems these ions are called *electrolytes.* Electrolytes serve three general functions in the body. First, many are essential minerals. Second, they play a role in the movement of water—osmosis—between various compartments within the body. Third, they help maintain the acid-base (pH) balance required for normal cellular activities.
3. *Proteins.* Blood plasma is approximately 90% water. As the plasma circulates past the cells of the body, water moves between the tissue fluid and the blood from an area of higher

atherosclerosis (**ath** uh **row** skluh **row** sis) a disease in which the inner walls of the arteries accumulate fat deposits, narrowing the passageways.
electrocardiogram (ECG) (ih **lek** trow **kard** ee uh **gram**) a recording of the electrical impulses that pass throughout the heart as it contracts and relaxes.
plasma (**plaz** muh) a straw-colored fluid that forms the liquid portion of blood. Plasma is made up of water and dissolved substances such as nutrients, hormones, respiratory gases, wastes, salts, ions, and proteins.
formed elements the solid portion of blood that is suspended in blood plasma; composed principally of erythrocytes, leukocytes, and platelets.

Applications

HOW *Science* WORKS

Creating Fat Replacements and Substitutes for "Heart Healthy" Foods

Cheeseburgers, french fries, ice cream . . . most Americans love fatty foods. One problem with the fats in our diets is that they provide 9 calories per gram—more than twice the amount of energy in a gram of carbohydrate or protein. Therefore, eating fats can easily result in the consumption of too many calories and bulges around our middles. A more serious problem with high fat consumption is that too much cholesterol in the bloodstream contributes to the formation of plaque in the arteries—masses of fats and other substances that block arteries and impair their proper functioning. This buildup most often occurs in the arteries that supply the heart and brain with blood, and it can result in coronary heart disease, heart attacks, and strokes.

Nutritional experts currently recommend that no more than 25% to 30% of total calories in the diet of a healthy person of normal weight come from the various kinds of fats. But 37% of the typical American diet consists of calories from fat. Clearly, Americans need to find ways to reduce their fat consumption.

Food scientists have found ways to make many of our favorite, fat-laden foods "heart healthy" by using fat replacements or fat substitutes. *Fat replacements* are created by taking normal food ingredients, such as proteins from eggs and milk, water, or various carbohydrates, and processing and combining them to produce an ingredient that has some of the properties of fats without their calories or artery-clogging characteristics. Polydextrose, a starch-based fat replacement, is currently used in frozen desserts, puddings, and cake frostings. Maltodextrin, another starch-based product, is used in salad dressings and margarines.

Fat substitutes differ from fat replacements in that they are a single, unique ingredient and usually have even fewer calories than fat replacements. A fat substitute already available is marketed under the brand name Simplesse. This fat substitute is used in the manufacture of such foods as ice cream, salad dressings, and cheese products. Simplesse has a fatlike consistency and texture because it is composed of proteins from either milk or egg whites that have been heated and blended to create nearly microscopic spheres of protein. These mistlike particles mimic the "mouth feel" of real fat globules—that smooth, creamy texture we enjoy in a chocolate bar or a scoop of ice cream.

Simplesse has one drawback. It does not retain its fatlike consistency and taste when it is heated. Olestra, a fat substitute not yet on the market, can withstand heat. Olestra is a sucrose polyester: a sucrose molecule to which six, seven, or eight fatty acid chains are attached. Triglycerides, or natural fats, are also composed of fatty acids; thus Olestra comes closer than Simplesse to mimicking their molecular structure. Because fat-digesting enzymes do not break down Olestra, it bypasses the body's digestive and absorptive mechanisms (and has no calories). This ability, combined with Olestra's textural resemblance to natural fat, makes it a very convincing and effective fat substitute, which can be used in a variety of products from oils to ice cream. Studies are presently under way to test Olestra's safety, and the preliminary reports are positive; FDA approval is pending.

Some nutritionists argue that the best way to lower fat consumption is to develop a taste for foods that are naturally low in fat rather than searching for substitutes for fatty foods. The manufacturers of Olestra caution that it should not be used to replace a balanced diet low in natural fat. Rather, fat substitutes offer the opportunity to eat a once-forbidden treat occasionally without increasing fat consumption above healthy levels, as well as to modify the level of fat in the diet using replacements or substitutes in moderation. Fat substitutes are also useful for those on no-fat diets for health reasons.

concentration of water to an area of lower concentration of water. Blood contains a concentration of proteins that balances that of the cells, thereby balancing the concentration of water between the cells and the blood. For this reason, water is not osmotically sucked out of the blood. The types of proteins in blood vary. Some of these proteins are antibodies and other proteins are active in the immune system. Others are fibrinogen and prothrombin, key players in blood clotting. Taken together, however, these proteins make up less than half of the amount of protein that is necessary to balance the protein content of the other cells of the body. The rest consists of a protein called *serum albumin,* which circulates in the blood as an osmotic counterforce.

Blood plasma contains nutrients, hormones, respiratory gases, wastes, and a variety of ions and salts. It also contains high concentrations of the protein serum albumin, which functions to keep the blood plasma in osmotic equilibrium with the cells of the body.

Types of Blood Cells

Although blood is liquid, 45% of its volume is actually occupied by cells and pieces of cells, collectively called *formed elements.* There are three principal types of formed elements in the blood: erythrocytes, leukocytes, and platelets (Figure 10-13).

Erythrocytes

In only one teaspoonful of your blood, there are about 25 billion **erythrocytes,** or **red blood cells.** Each erythrocyte is a flat disk with a central depression (Figure 10-14), something like a doughnut with a hole that does not go all the way through. Almost the entire interior of each cell is packed with the oxygen-carrying molecule hemoglobin (Figure 10-15). One estimate, in fact, is that each erythrocyte can carry 280 million molecules of hemoglobin!

Mature erythrocytes do not have nuclei or the ability to manufacture proteins. Red blood cells are therefore unable to repair themselves and consequently have a rather short life span. Erythrocytes live only about 4 months. Old, worn-out erythrocytes are processed by macrophages in the spleen, liver, and bone marrow. New erythrocytes are constantly being synthesized and released into the blood by cells within the soft interior marrow of bones at the amazing rate of 2 million per second!

Leukocytes

Less than 1% of the cells in human blood are **leukocytes,** or **white blood cells.** In fact, there are only about 1 or 2 leukocytes for every 1000 erythrocytes. Leukocytes are larger than red blood cells; they contain no hemoglobin, have nuclei, and are essentially colorless. There are several kinds of leukocytes, and each has a different function. All functions, however, are related to the defense of the body against invading microorganisms and other foreign substances (see Chapter 11).

There are two major groups of leukocytes: **granulocytes** and **agranulocytes.** Granulocytes are circulating leukocytes and get their name from the tiny granules in their cytoplasm. In addition, they have lobed nuclei (see Figure 10-13). The granulocytes are classified into three groups by their staining properties. About 50% to 70% of them are **neutrophils,** cells that migrate to the site of an injury and stick to the interior walls of the blood vessels. They then form projections that enable them to push their way into the infected tissues, where they engulf, or *phagocytize,* microorganisms and other foreign particles. The term *phagocytosis* comes from a Latin word meaning "cell eating." **Baso-**

erythrocytes (ih **rith** row sites) or **red blood cells,** one type of formed element of the blood that resembles flat disks with central depressions and are packed with the oxygen-carrying molecule hemoglobin.

leukocytes (**loo** ko sites) or **white blood cells,** one type of formed element of the blood that is larger than red blood cells and is essentially colorless. There are several kinds of leukocytes, including macrophages and lymphocytes, but all of them function in defending the body against invading microorganisms and foreign substances.

granulocytes (**gran** yuh low sites) one of the two major groups of leukocytes distinguished by their cytoplasmic granules and lobed nuclei.

agranulocytes (ay **gran** yuh low sites) one of the two major groups of leukocytes; they have neither cytoplasmic granules nor lobed nuclei.

neutrophils (**noo** truh filz) one of three kinds of granulocytes; neutrophils migrate to the site of an injury and engulf microorganisms and other foreign particles.

		Blood cell type	Description	Function	Life Span
Red blood cells		Erythrocyte 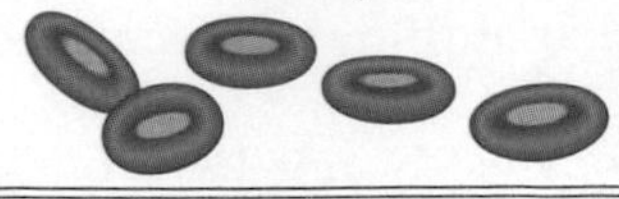	Flat disk with a central depression, no nucleus, contain hemoglobin.	Transport oxygen (O_2) and carbon dioxide (CO_2).	About 120 days.
White blood cells (leukocytes)	Granulocytes	Neutrophil 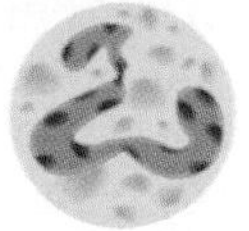	Spherical; with many lobed nucleus, no hemoglobin, pink-purple cytoplasmic granules.	Cellular defense-phagocytosis of small microorganisms.	Hours to 3 days.
		Eosinophil 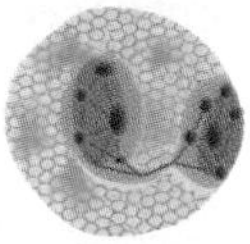	Spherical; two-lobed nucleus, no hemoglobin, orange-red staining cytoplasmic granules.	Cellular defense- phagocytosis of large microorganisms such as parasitic worms, releases anti-inflammatory substances in allergic reactions.	8 to 12 days.
		Basophil 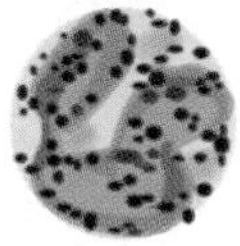	Spherical; generally two-lobed nucleus, no hemoglobin large purple staining cytoplasmic granules.	Inflammatory response - contain granules that rupture and release chemicals enhancing inflammatory response.	Hours to 3 days.
	Agranulocytes	Monocyte 	Spherical; single nucleus shaped like kidney bean, no cytoplasmic granules, cytoplasm often blue in color.	Converted to macrophage which are large cells that entrap microorganisms and other foreign matter.	Days to months.
		B-lymphocyte 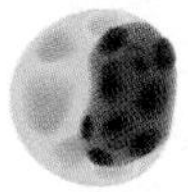	Spherical; round singular nucleus, no cytoplasmic granules.	Immune system response and regulation, antibody production sometimes cause allergic response.	Days to years.
		T-lymphocyte 	Spherical; round singular nucleus, no cytoplasmic granules.	Immune system response and regulation; cellular immune response.	Days to years.
Platelets		Platelets 	Irregularly shaped fragments, very small pink staining granules.	Control blood clotting or coagulation.	7 to 10 days.

FIGURE 10-13 Types of blood cells. Different blood cells perform different functions and have varying lifespans, as this chart shows.

phils, a second kind of granulocyte, contain granules which rupture and release chemicals that enhance the body's response to injury or infection. They play a role in causing allergic responses. The third kind of granulocyte, the **eosinophils,** is also believed to be involved in allergic reactions. In addition, they act against certain parasitic worms.

Agranulocytes have no cytoplasmic granules, nor are their nuclei lobed. One group of agranulocytes, the **monocytes,** circulate as the granulocytes do. Monocytes are attracted to the sites of injury or infection where they are converted into **macrophages**—enlarged, amoebalike cells that entrap microorganisms and particles of foreign matter by phagocytosis. They usually arrive after the neutrophils and clean up any bacteria and dead cells. **Lymphocytes,** the other type of agranulocyte, recognize and react to substances that are foreign to the body, sometimes producing a protective immunity to disease. Occasionally, however, these cells produce an inflammation or an allergic response (see Chapter 11).

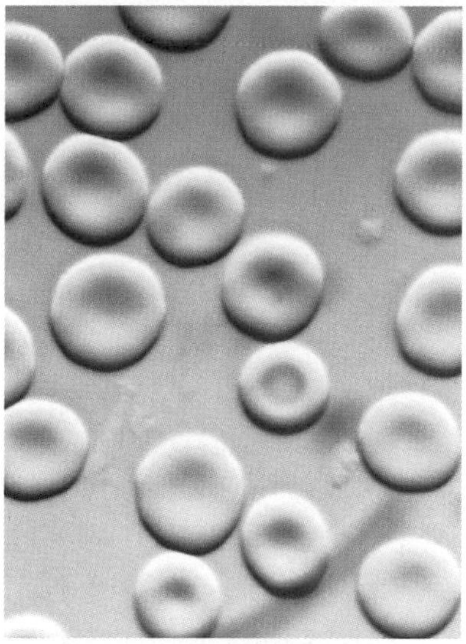

FIGURE 10-14 Human erythrocytes. Mature human erythrocytes have a characteristic collapsed appearance, like a pillow on which someone has sat *(1000x)*.

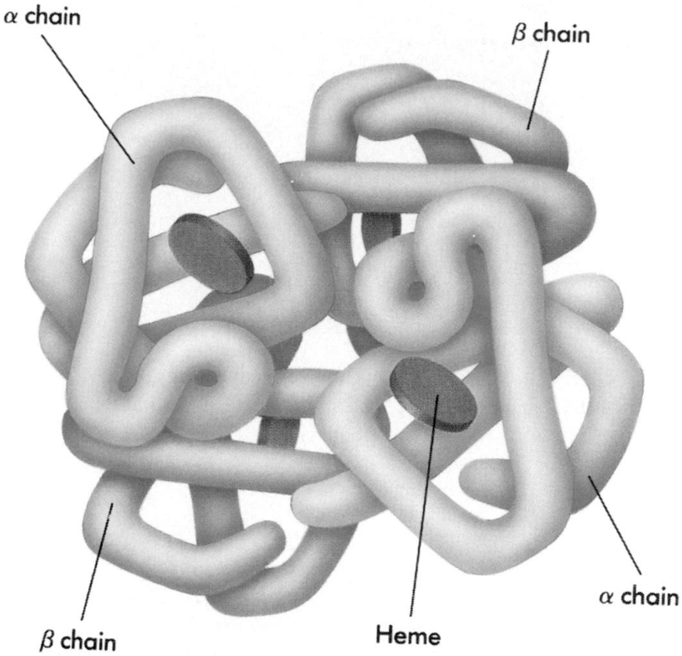

FIGURE 10-15 A hemoglobin molecule. The interior of a red blood cell is packed with hemoglobin. Each hemoglobin molecule is composed of four polypeptide chains. At the center of each chain is a heme group that contains iron. Oxygen binds to the iron for transport.

Platelets

Certain large cells within the bone marrow called *megakaryocytes* regularly pinch off bits of their cytoplasm. These cell fragments, called **platelets,** enter the bloodstream and play an important role in controlling blood clotting, or coagulation. The clotting of blood is a complicated process initiated by damage to blood vessels or tissues.

When an injury occurs, platelets clump at the damaged area, temporarily blocking blood loss. The damaged tissues and platelets release a complex of substances called *thromboplastin.* The release of this complex begins a cascade of events, one dependent on the other. A simplified explanation is as follows: thromboplastin interacts with calcium ions, vitamin K, and other clotting factors to form an enzyme called *prothrombin activator.* Prothrombin activator brings about the conversion of prothrombin to thrombin. Thrombin converts fibrinogen, a soluble protein, to fibrin, an insoluble, threadlike protein. Fibrin threads, along with trapped red blood cells (Figure 10-16), form the clot—a plug at the damaged area so that blood cannot escape.

basophils (**bay** soh filz) one of three kinds of granulocytes; basophils contain granules that rupture and release chemicals that enhance the body's response to injury or infection; they also play a role in causing allergic responses.

eosinophils (ee oh **sin** oh filz) one of three kinds of granulocytes; eosinophils are believed to be involved in allergic reactions and also act against certain parasitic worms.

monocytes (**mon** oh sites) a group of agranulocytes that circulates as the granulocytes do; monocytes are attracted to the sites of injury or infection, where they mature into macrophages and engulf any bacteria or dead cells that neutrophils may have left behind.

macrophages (**mak** row **fay** djus) enlarged, amoebalike cells that entrap microorganisms and particles of foreign matter by phagocytosis.

lymphocytes (**lim** foh sites) a type of agranulocyte that recognizes and reacts to substances that are foreign to the body, sometimes producing a protective immunity to disease. Lymphocytes are cells that make up the immune system, the specific resistance to disease.

platelets cell fragments present in blood that play an important role in blood clotting.

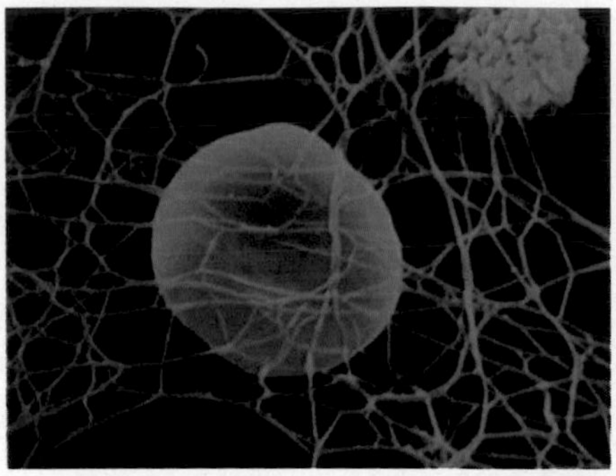

FIGURE 10-16 Fibrin threads have caught a red blood cell. Eventually, many red blood cells will become caught in this net, forming a clot (*4000x*).

> *The three principal types of formed elements in the blood are erythrocytes, leukocytes, and platelets. Erythrocytes, or red blood cells, carry oxygen; leukocytes, or white blood cells, defend the body against invading microorganisms and other foreign substances; and platelets play an important role in blood clotting.*

The Lymphatic System

Although the blood proteins and electrolytes help maintain an osmotic balance between the blood and the tissues, the blood loses more fluid to the tissues than it reabsorbs from them. Of the total volume of fluid that moves from the blood into and around the tissues, about 90% reenters the cardiovascular system. Where does the other 10% go? The answer is to the body's one-way, passive circulatory system, the **lymphatic system** (Figure 10-17). This system counteracts the effects of net fluid loss from the blood.

By osmosis and diffusion, blind-ended lymphatic capillaries (see Figure 10-17) fill with tissue fluid (including small proteins that have diffused out of the blood). From these capillaries, the tissue fluid—now called **lymph**—flows through a series of progressively larger vessels to two large lymphatic vessels, which structurally resemble veins. These vessels drain into veins of the circulatory system near the base of the neck through one-way valves. Although the lymphatic system has no heart to pump lymph through its vessels, the fluid is driven through them when the vessels are squeezed by the movements of the body's muscles. The lymphatic vessels contain a series of one-way valves (Figure 10-18) that permits movement only in the direction of the neck.

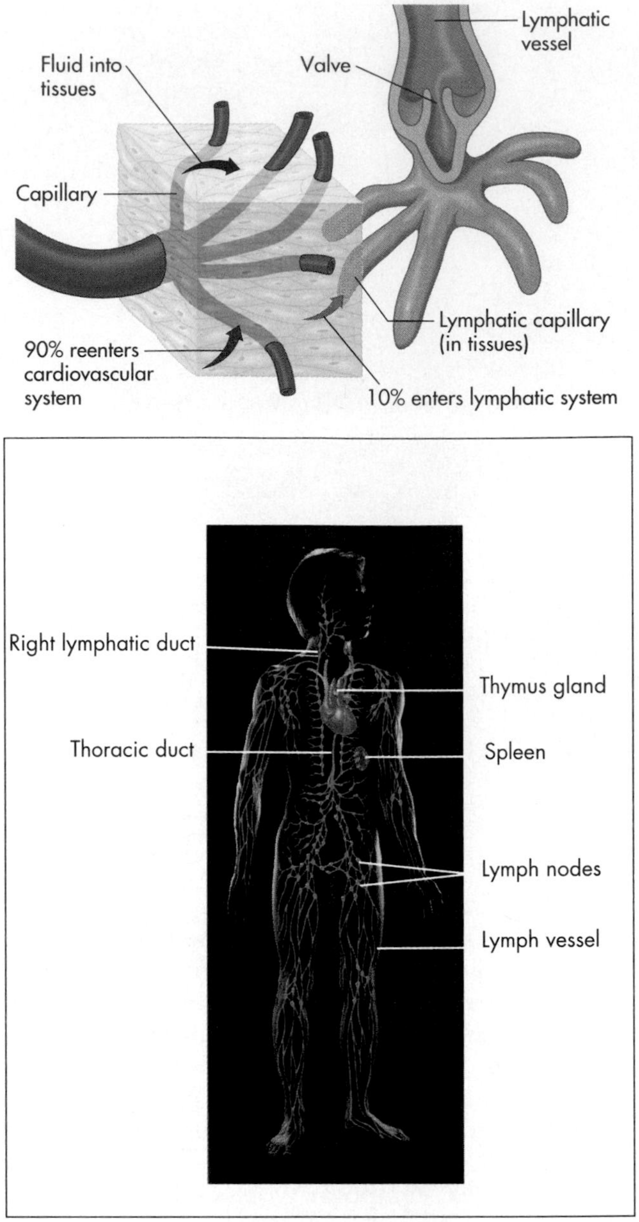

FIGURE 10-17 The human lymphatic vessels and nodes. Other organs containing lymphatic tissue include the spleen, thymus, appendix, tonsils, and bone marrow. The enlarged diagram at the top shows the path of the excess fluid that leaves the arteriole end of a capillary bed, enters the adjacent tissue spaces, and is absorbed by lymphatic capillaries.

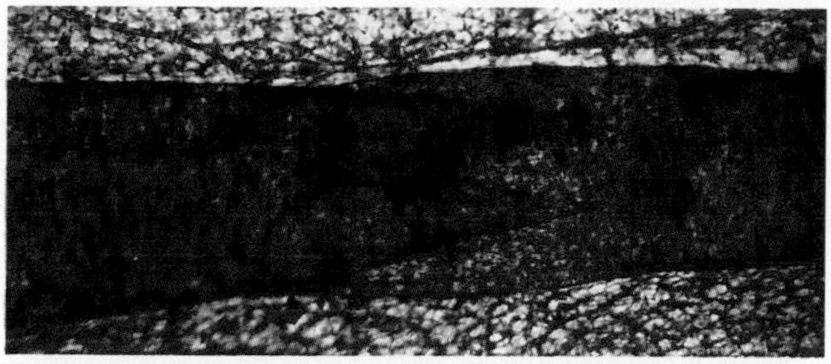

FIGURE 10-18 A lymphatic vessel. Flow from left to right is not impeded because this flow tends to force open the valve. Flow from right to left is prevented because "backward" flow tends to force the one-way valve closed *(25x)*.

Small, ovoid, spongy structures called **lymph nodes** are located in various places of the body along the route of the lymphatic vessels. They are clustered in areas such as the groin, armpits, and neck, filtering the lymph as it passes through. Some lymphocytes, the cells that activate the immune response, reside in the lymph nodes. (These cells are discussed in more detail in Chapter 11.) In addition, the lymphatic system has two organs: the **spleen** and the **thymus.** The spleen stores an emergency blood supply and also contains white blood cells. Specific types of white blood cells in the spleen destroy old red blood cells, filter microorganisms out of the blood as it passes through, and initiate an immune response against the foreign microbes. The thymus plays an important role in the maturation of certain lymphocytes called *T cells,* which are an essential part of the immune system.

Approximately 10% of the fluid that moves out of the blood and into the cells and cell spaces at the capillaries does not return to the blood directly. It is collected and then returned to the blood by a system of one-way, blind-ended vessels called the lymphatic system.

Diseases of the Heart and Blood Vessels

Cardiovascular diseases are the leading cause of death in the United States. More than 42 million people in this country have some form of cardiovascular disease—about one person in five.

Heart attacks, the most common cause of death in the United States, result from an insufficient supply of blood reaching an area of heart muscle. Heart attacks may be caused by a blood clot forming in one of the vessels that supplies the heart with blood, thereby blocking its blood supply. They may also result if a vessel is blocked sufficiently by fatty deposits, especially cholesterol and triglycerides. Recovery from a heart attack is possible if the segment of damaged heart tissue is small enough that other blood vessels in the heart can supply the damaged tissues. *Angina pectoris,* which literally means "chest pain," occurs for reasons similar to those that cause heart attacks, but it is not as severe. In this case, a reduced blood flow to the heart muscle weakens the cells but does not kill them. The pain may occur in the heart and often also in the left arm and shoulder.

The amount of heart damage associated with a small heart attack may be relatively slight and thus difficult to detect. It is important that such damage be detected, however, so that the overall condition of the heart can be evaluated properly. Electrocardiograms are very useful for this purpose because they reveal abnormalities in the timing of heart contractions, abnormalities that are associated with the presence of damaged heart tissue. Damage to the atrioventricular (AV) node, for example, may delay as well as reduce the second, ventricular pulse. Unusual conduction routes may lead to continuous disorganized contractions called *fibrillations.* In many fatal heart attacks, ventricular fibrillation is the immediate cause of death.

Many factors contribute to heart disease and heart attacks. One factor is heredity. You may inherit the predisposition to heart disease from your parents. Other factors involve eating too much saturated fat and cholesterol, not exercising, and being overweight. Cigarette smoking greatly in-

lymphatic system (lim **fat** ik) the body's one-way, passive circulatory system, which collects and returns tissue fluid to the blood that does not return to the blood directly from the body's tissues.

lymph (**limf**) the name given tissue fluid when in the vessels of the lymphatic system.

lymph nodes (**limf nodes**) small, ovoid, spongy structures located in various places in the body along the routes of the lymphatic vessels, which filter lymph as it passes through them.

spleen an organ of the lymphatic system that stores an emergency blood supply and also contains white blood cells.

thymus (**thigh** muss) a small gland located in the neck that plays an important role in the maturation of certain lymphocytes called *T cells,* which are an essential part of the immune system. The thymus is active in children.

Just Wondering . . . .

I often hear about people who die suddenly of a heart attack even though they exercised and ate right. So why should I bother?

As you have discovered, there is no guarantee that if you follow health recommendations regarding coronary heart disease, you will be free of this disease and the threat of heart attacks. Following health recommendations helps you *reduce* your risk of heart attack, however. In addition, researchers have discovered new information that sheds light on a possible reason why some people die of heart attacks despite their heart-healthy lifestyles.

Cardiovascular disease is the number one killer of Americans, so assessing your risk level and doing something about it *is* important. Next time you're in class, look around the room and realize that two out of every five students in that class will, according to today's statistics, die of heart and blood vessel disease. The major risk factors of this disease are physical inactivity, elevated serum cholesterol (a fat that often sticks to artery walls), cigarette smoking, obesity, chronically high blood pressure, and a family history of heart and blood vessel disease. The more of these risk factors you have, the *more likely* you are to have a heart attack or stroke.

So why does someone die from a heart attack when the chances are low according to the risk factors noted here? In some cases, persons are discovered (during autopsy perhaps) to have congenital heart abnormalities—problems of heart structure or function with which they were born and were unaware. Suddenly, under some physical or emotional time of stress, their heart gives out. But new information has come to light that may help us understand many more of these seemingly paradoxical cases. The presence of high blood concentrations of a newly discovered, cholesterol-carrying molecule: lipoprotein(a).

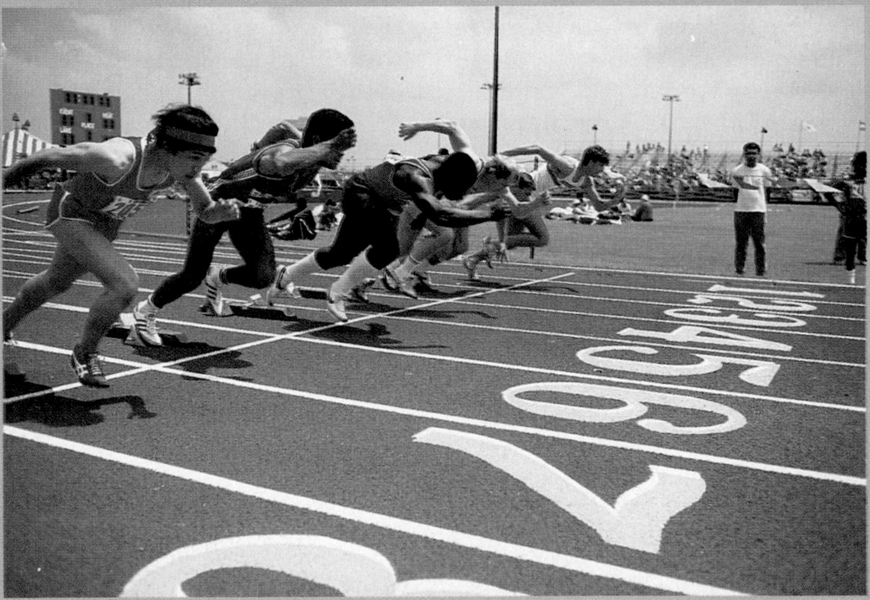

As scientists' understanding developed of how the arteries become clogged with fat, they came to understand that the blood levels of two cholesterol-carrying molecules, high-density lipoproteins (HDL) and low-density lipoproteins (LDL), were possibly more important than the blood level of total cholesterol. LDL, the so-called bad cholesterol, carries cholesterol to the cells (including those lining the blood vessels). HDL, the so-called good cholesterol, carries cholesterol away from the cells and facilitates its removal from the body. In fact, the preceding risk factors tend to correlate in significant ways with LDL and HDL levels. For example, cigarette smokers tend to have lower HDL levels than do nonsmokers.

Recently scientists discovered a new cholesterol-carrying lipoprotein called lipoprotein(a). Shown to be an important factor in the development of heart and blood vessel disease, lipoprotein(a) has been found in high concentrations in the blood of persons with a low risk of this disease. Researchers are currently studying this "new" lipoprotein to understand its role in cardiovascular disease. With this understanding may come modifications or additions to the recommendations of "heart healthy" lifestyles.

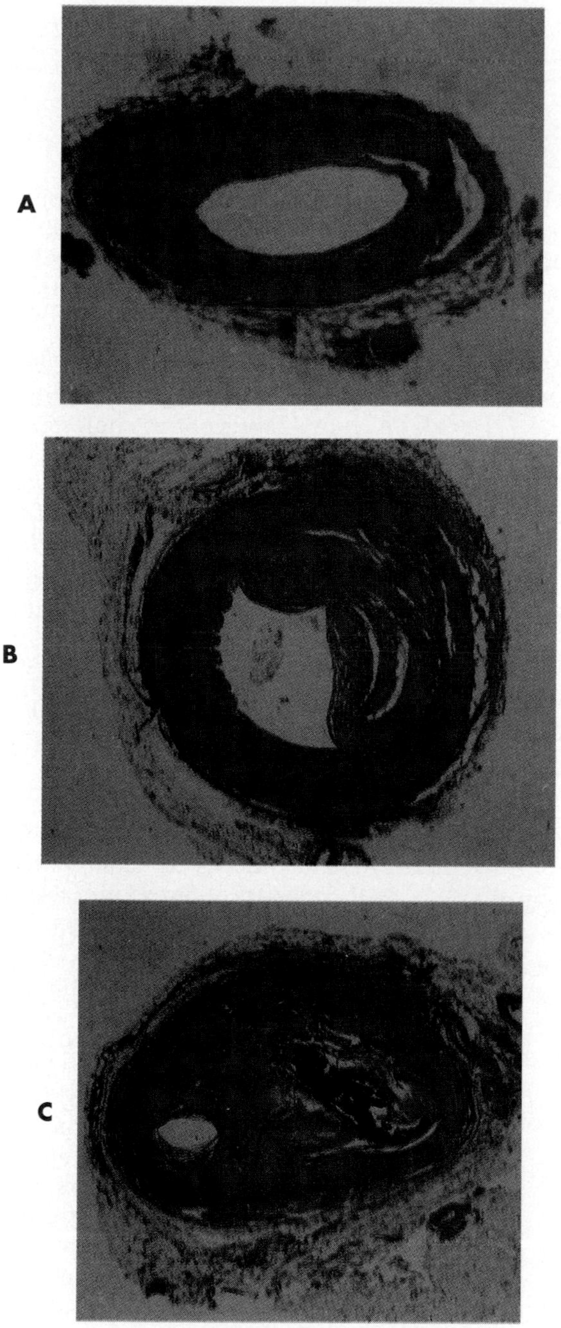

FIGURE 10-19 The path to a heart attack. A, The coronary artery shows only minor blockage. **B,** The artery exhibits severe atherosclerosis—much of the passage is blocked by buildup of cholesterol and other lipids on the interior walls of the artery. **C,** The coronary artery is almost completely blocked *(30x)*.

creases a person's risk for heart attacks and strokes, although the physiologic mechanisms involved are not totally understood at this time.

Strokes are caused by an interference with the blood supply to the brain and often occur when a blood vessel bursts in the brain. A stroke may also be caused by a blood clot, or *thrombus,* that forms in these vessels or by a blood clot that has traveled to the brain from another location. This type of clot is called an **embolus.** Blood clots may be caused by cancer or other diseases. The effects of strokes depend on the severity of the damage and the specific location of the stroke.

Arteriosclerosis is a thickening and hardening of the walls of the arteries. Blood flow through such arteries is restricted, and they lack the ability to dilate and so have difficulty accommodating the volume of blood pumped out by the heart. The narrowing of these vessels forces the heart to work harder. *Atherosclerosis* is a form of arteriosclerosis in which masses of cholesterol and other lipids build up within the walls of large and medium-sized arteries (Figure 10-19). These masses are referred to as **plaques.** The accumulation of plaques impairs the arteries' proper functioning. When this condition is severe, the arteries can no longer dilate and constrict properly and the blood moves through them with difficulty. The accumulation of cholesterol is thought to be the prime contributor to atherosclerosis, and diets low in cholesterol are now prescribed to help prevent this condition. Atherosclerosis contributes to both heart attacks and strokes.

embolus (**em** bo lus) a travelling blood clot.
arteriosclerosis (ar **teer** ee oh skluh **row** sis) a thickening and hardening of the walls of arteries.
plaques (**plaks**) masses of cholesterol and other lipids that can build up within the walls of large and medium-sized arteries in a disease called *atherosclerosis;* the accumulation of such masses impairs the arteries' proper functioning.

Summary

- The *circulatory system* is made up of three components: the heart, the blood vessels, and the blood. Together, the heart and blood vessels—the "plumbing" of the circulatory system—are known as the *cardiovascular system*. These terms are often used interchangeably.
- The circulatory system transports nutrients, wastes, respiratory gases, and hormones and plays an important role in temperature maintenance of the body.
- The heart is a double pump, pushing both pulmonary (lung) circulation and systemic (general body) circulation. Because the two circulations are kept separate within the heart, the systemic circulation receives only fully oxygenated blood.
- The general flow of blood circulation through the body is a circuit starting from the heart, which pumps blood out via muscled arteries to the capillary networks that interlace the tissues of the body; the blood returns to the heart from these capillaries via the veins.
- The contraction of the heart is initiated at the SA node, or pacemaker, as a periodic spontaneous impulse. The impulse spreads across the surface of the two atrial chambers, causing all of these cells to contract.
- The passage of the impulse to the ventricles is briefly delayed by tissue that insulates the two segments of the heart from one another. Only a narrow channel of specialized conducting cells is able to carry the impulse from the atria to the ventricles. The delay in the passage of the impulse permits the atria to empty completely into the ventricles before ventricular contraction occurs.
- The heartbeat can be heard, or monitored, by tracking changes in blood pressure through the period of filling and contracting of the atria (the diastolic period) and the contraction of the left ventricle (the systolic period). The impulses can also be measured directly as an electrical current that passes in a wave throughout the body. A recording of this electrical wave is called an *electrocardiogram* (ECG).
- The liquid portion, or plasma, of the circulating blood contains the proteins and ions that are necessary to maintain the blood's osmotic equilibrium with the surrounding tissues.
- The formed elements of the blood are the red blood cells (erythrocytes), white blood cells (leukocytes), and platelets. The red blood cells transport oxygen, the white blood cells defend the body against disease, and the platelets are essential to the process of blood clotting.
- The lymphatic system gathers fluid from the body that has been lost from the circulatory system by diffusion and returns it via a system of lymphatic capillaries, lymphatic vessels, and two large lymphatic ducts to veins in the lower part of the neck.
- Cardiovascular diseases, diseases of the heart and blood vessels, are the leading cause of death in the United States. Heart attacks, strokes, and atherosclerosis are all serious cardiovascular diseases.

Knowledge and Comprehension Questions

1. What are the principal functions of the circulatory system?
2. You have just finished a great lunch of a chicken sandwich and a tossed salad with oil and vinegar dressing. Discuss what happens to the molecules of the food after they pass into your bloodstream.
3. It is a hot summer day, and as you ride your bike, you start feeling very warm. Explain how your circulatory system helps to keep you from overheating.
4. Distinguish between the terms *circulatory system* and *cardiovascular system.*
5. What significant change occurs in the blood from the time it enters the capillaries until it enters the venules?
6. Mystery novels sometimes describe people who "turn pale with fear." Describe the circulatory reason for this.
7. How does the body regulate blood flow within the capillary beds? Why is this important?
8. Compare the structure of arteries, capillaries, and veins. Relate any differences to their respective functions.
9. The text states that the heart is actually two pumps in one; explain this duality. Can you make a hypothesis as to why these two sides do not make contact?
10. The explanation of heart function indicates the tremendous muscular strength of this organ. What type of muscle is found in the heart? Do you think that this muscle can be strengthened voluntarily?
11. Why are valves necessary in heart function?
12. What are the SA node and the AV node? Explain their significance.
13. What is the range of normal blood pressure? Have you ever experienced high blood pressure? How was it treated?
14. Explain how a high-fat diet could place you at risk for cardiovascular disease.

Critical Thinking

1. List the risk factors in your family history and life-style for cardiovascular disease. Which factors can be changed? Which cannot? What can you do to lower your risk of heart disease?
2. The opening vignette to this chapter discussed the left ventricular assist device that has replaced the total artificial heart (TAH). What factors in heart anatomy and physiology allow this smaller, more specific pump to replace the TAH?

CHAPTER 11

DEFENSE AGAINST DISEASE

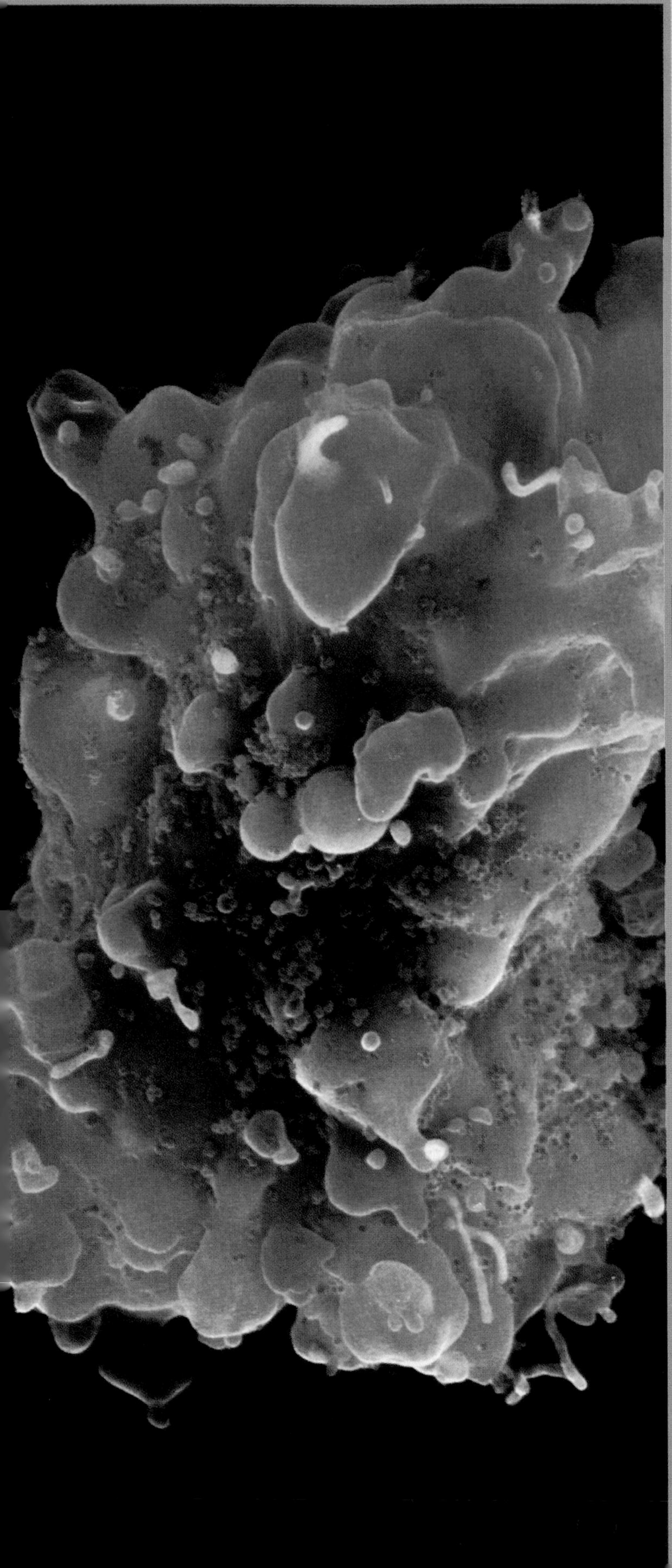

EVERY DAY you fight for your life. Some of your attackers swarm over your skin. Some enter your mouth, nose, eyes—any place where they may be able to gain a foothold in your tissues and spread. The battles between you and these invaders are silent and deadly. The fighting is one on one; chemical warfare is commonplace.

It is your immune system that fights and wins these thousands of battles, yet the war within you rages on continually. Each day thousands of people lose their inner wars, succumbing to invasions of the body by viruses, bacteria, fungi, or protists. People whose immune system is impaired, such as those infected with the deadly virus that causes acquired immunodeficiency syndrome (AIDS), have the hardest time winning battles against infectious diseases because the AIDS virus attacks the immune system itself, lowering its defenses. Therefore, AIDS patients contract many other diseases quite easily.

Looking at the photo, you can see the AIDS virus doing some of its dirty work, attacking one of the most important cells of the immune system—a helper T cell. Helper T cells identify foreign invaders and stimulate the production of other cells to fight an infection. Although thousands of times larger than the AIDS virus particles (colored blue for this photograph), the helper T cell is inactivated and sometimes killed by this virus. The body is then an easy mark for other invaders to enter—unnoticed—and win the war between life and death.

Nonspecific Versus Specific Resistance to Infection

Your body works in many ways to keep you healthy and free of infection. An *infection* results when microorganisms or viruses enter the tissues, multiply, and cause damage. Your body has one set of defenses against infection called **nonspecific defenses.** These defenses work to keep out *any* foreign invader, just as the walls and roof of your home protect you from any of the elements of the environment—rain, hail, wind, the sun's rays, insects, or the neighbor's dog, for example. But your body has another set of defenses, too. This set of defenses are called **specific defenses** because they work against *each particular type* of microbe that may invade your body. A specific defense in your home, for example, may be a certain chemical you use to rid your kitchen of ants in the summer. Your nonspecific defenses consist of mechanical and chemical barriers and cells that attack invaders in general. Your specific defense is called **immunity,** and it consists of cellular and molecular responses to particular foreign invaders.

KEY CONCEPTS

- The body has two defense systems: nonspecific defenses working to keep out any foreign invader and specific defenses working against each particular type of foreign cell or virus that may invade the body.
- The nonspecific defenses of the body are the skin and mucous membranes, chemicals that kill bacteria, and the inflammatory response.
- The specific defense of the body is the immune response, a two-branched response that works in a complex way to rid the body of a foreign invader and to develop a future resistance within the body to that particular invader.
- When the immune system is attacked, as during infection with the deadly AIDS virus, infections normally defeated by the immune system may overpower the body.

OUTLINE

The defenses of the body are both nonspecific—acting against any foreign invaders—and specific—acting against particular invaders.

Nonspecific Resistance

Unbroken skin is an effective mechanical barrier to any foreign substance entering the body. Most microorganisms cannot penetrate the skin of a healthy individual. However, if the skin is injured, many types of microorganisms will grow and multiply at the injured site, causing an inflammation or infection.

An *inflammation* occurs when cells are damaged by microbes, chemicals, or physical injuries. Characterized by redness, pain, heat, and swelling, an inflammation consists of a series of events that removes the cause of the irritation, repairs the damage that was done, and protects the body from further invasion and infection. During this nonspecific response, blood vessels dilate, or widen, bringing an increased supply of blood to the injured area. This increased flow brings defensive substances to the site of injury as fluid and phagocytic white blood cells pass out of the capillaries. Lymphatic drainage removes dissolved poisonous substances that may accumulate there. Blood clots wall off the area, preventing the spread of microbes or other injurious substances to other parts of the body. Phagocytes, both neutrophils and monocytes, migrate to the area and ingest microbes and other foreign substances. Nutrients stored in the body are also released to the area to support these defensive cells.

The mucous membranes of the body, such as those lining your mouth, nose, throat, eyelids, urinary tract, and genital tract, also defend the body against invaders. These membranes are sticky and trap microbes and foreign particles much as flypaper traps flies. Some of the mucous membranes of the body also have other physical or chemical aids for combating infection. For example, the membranes covering the eyes and eyelids are constantly washed in tears, a fluid that contains a chemical called lysozyme, which is deadly to most bacteria. The ciliated mucous membranes of the respiratory passageways have a comparatively thick coating of mucus that not only traps invaders but efficiently moves them from the respiratory passageways to be swallowed by the digestive system. In the digestive system, microorganisms are killed by the acid environment of the stomach. The environments of the urinary and genital tracts are somewhat acidic too—an unfavorable situation for many foreign bacteria. Foreign bacteria are also kept from invading the body by bacteria that are normal inhabitants of the body, colonizing such areas as the mouth, throat, colon, vagina, and skin. Figure 11-1 summarizes some of the nonspecific defense mechanisms of the body.

The nonspecific defenses of the body include the skin and mucous membranes, chemicals that kill bacteria, and the inflammatory process.

Specific Resistance

Sometimes invaders get by your nonspecific defenses. This situation may occur if your body is invaded by many organisms at one time or if the invaders are extremely *virulent,* that is, if they are good at establishing an infection and damaging the body. In addition, the defenses of a person ill with certain diseases, such as diabetes mellitus or cancer, may not be as strong as those of persons who are free of disease.

Invaders that get into the blood, lymph, or tissues encounter a second line of defense—the **immune system,** your body's specific defense. Have you ever had measles, chickenpox, or mumps? All these infections are caused by specific viruses that enter the body through the respiratory system, invade cells, multiply, and spread. As you may know from experience, the body is usually effective in combating these diseases. In addition, having had one of these diseases confers an immunity, or resistance, to getting that same disease again. The mechanism that provides you with this resistance is the immune system—the backbone of your health—protecting you not only from measles, mumps, and chickenpox but also from myriad other diseases. And although the immune system is not always able to protect a person from infection, the immune system of a healthy person staves off most infections and usually overcomes those that take hold.

The specific defense of the body is the immune response.

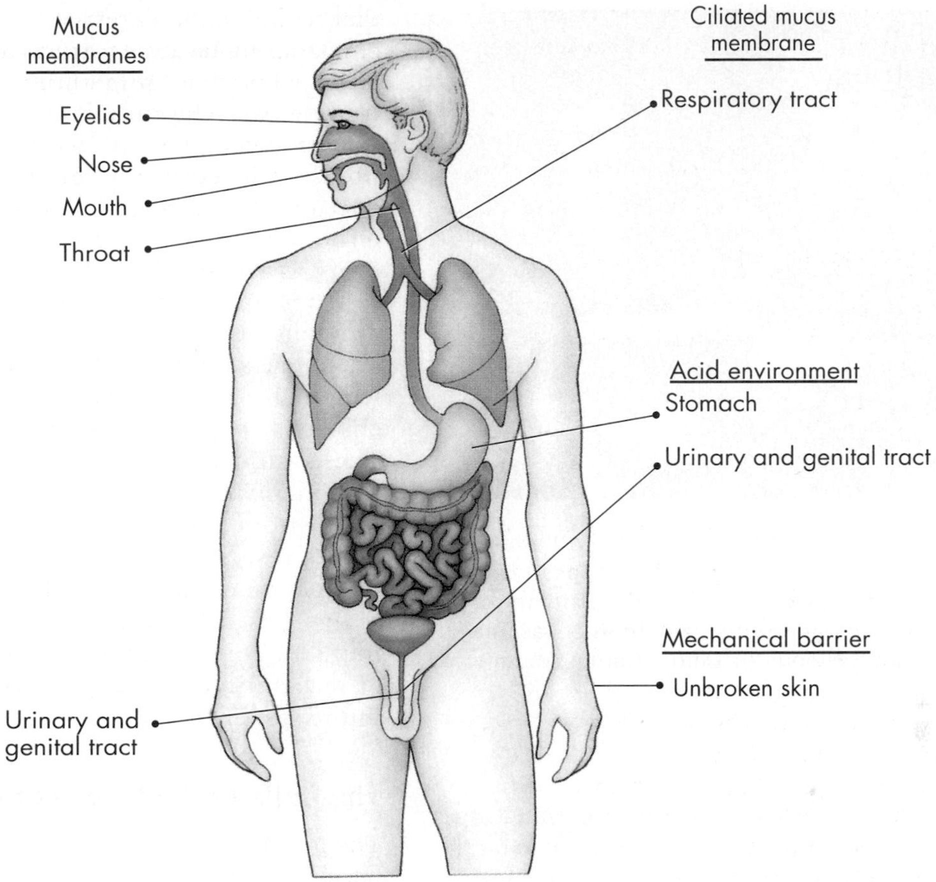

FIGURE 11-1 Nonspecific defenses in the human body. These nonspecific defenses are the body's first line of defense against foreign invaders.

Discovery of Specific Resistance: The Immune Response

In 1796 an English country doctor named Edward Jenner (Figure 11-2) carried out experiments that marked the beginning of immunology, the study of the responses of the body when it is challenged by molecules foreign to it. His experiments had to do with a disease called *smallpox,* a common and frequently deadly disease in Jenner's day. Although thousands died from smallpox, many survived. Jenner noted that these survivors were now immune from becoming reinfected. The only other group of people immune to smallpox were milkmaids who had caught another, much milder, form of the pox called *cowpox.* Jenner hypothesized that cowpox somehow conferred protection against smallpox.

To test his hypothesis, Jenner deliberately infected healthy persons with fluid he removed from the pox vesicles of sick milkmaids, causing these healthy individuals to develop cowpox. Many of the people Jenner infected with cowpox became immune to smallpox, just as he had predicted. Scientists now know that smallpox is caused by the variola virus and that cowpox is caused by a different but similar virus called the *vaccinia virus*. The word vaccinia is derived from the Latin *vacca* meaning "cow." Jenner's patients who were injected with fluid that contained cowpox virus mounted a defense against the cowpox infection, a defense that was also effective against infection by the similar

nonspecific defenses a set of defenses the body has to keep out any foreign invader; these defenses include the skin and mucous membranes.

specific defenses a set of cellular and molecular defenses that protect the body from each particular microbe that may invade it.

immunity (ih **myoon** ih tee) resistance to foreign antigens.

immune system the body's specific defenses: populations of specialized white blood cells that combat foreign antigens.

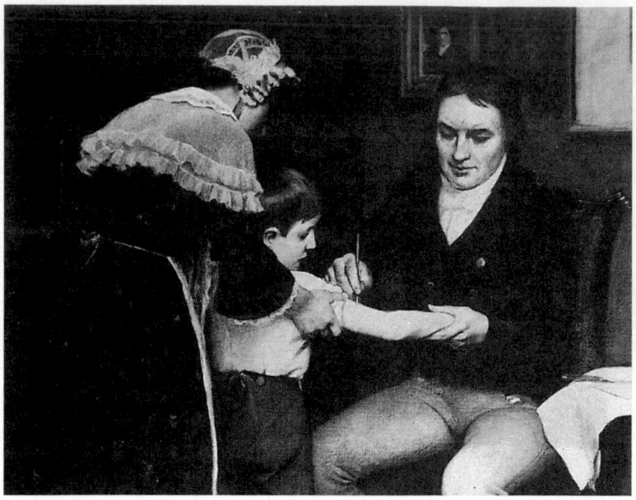

FIGURE 11-2 The birth of immunology. Edward Jenner (1749-1823) vaccinated James Phipps in the early 1800s with cowpox, protecting him from the disease smallpox. The underlying principles of vaccination were not understood until more than a century later.

smallpox virus. Jenner's procedure of injecting a harmless microbe into a person or animal to confer resistance to a dangerous one is called **vaccination.**

It was a long time after Jenner's experiments before scientists understood how one microbe conferred resistance to another. Further important information was added about 85 years after Jenner by Louis Pasteur of France. Pasteur was studying fowl cholera, a serious bacterial disease of chickens. From diseased chickens, Pasteur isolated bacteria that brought on the disease when these organisms were injected into healthy birds. These birds became seriously ill and died.

During a period of experimentation, Pasteur injected healthy birds with an old culture of fowl cholera—one that had sat on his benchtop for several weeks—because he had no fresh cultures. Interestingly, the injected birds became only slightly ill and then recovered. Pondering this outcome, Pasteur reasoned that the culture had become weakened as it sat for many days at room temperature. Trying to get his experiment to work, he then injected the recovered birds with massive doses of fresh cultures—active virulent fowl cholera bacteria. These birds did *not* develop fowl cholera! Chickens receiving the same injections but that had not previously recovered from a mild form of the illness all died. Pasteur began to realize that his old culture acted on the birds in a unique way. Clearly, these old, weakened bacteria possessed something that could confer resistance yet not cause the chickens to become seriously ill. Scientists now know what that "something" was: molecules protruding from the surface of the bacterial cells.

Every cell has proteins, carbohydrates, and lipids on its surface (Figure 11-3). It was the presence of molecules on the surface of the cholera bacteria—molecules different from any of the birds' own—to which the chickens were responding. Nonself, or foreign, molecules such as these are called **antigens.** Chickens injected with weakened or killed fowl cholera bacteria are immune to later infection because the bacterial antigens cause the chickens to produce proteins called **antibodies.** These antibodies are able to recognize any future cholera invaders, weakened or normal, and prevent them from causing disease (Figure 11-4). The body's response to foreign molecules, such as the production of antibodies directed against a specific antigen, is called an **immune response.** The immune response is a result of the complex activities of recognition and defense carried out by the immune system.

The Cells of the Immune System

The immune system is different from the other body systems in that it cannot be identified by interconnected organs. In fact, the immune system is not a system of organs, nor does it have a single controlling organ. The immune system is, instead, made up of cells that are scattered throughout the body but have a common function: reacting to specific foreign molecules.

The cells of the immune system are white blood cells, or leukocytes (see Chapter 10). Two types of white blood cells are involved in the immune system: **phagocytes** and **lymphocytes.** Two classes of lymphocytes play roles in specific resistance: *T cells* and *B cells.* These cells arise in the bone marrow, circulate in the blood and lymph, and reside in the lymph nodes, spleen, liver, and thymus (Figure 11-5). Although not bound together, these white blood cells exchange information and act in concert as a functional, integrated system—your "army" of 200 billion defenders—that is called into action when antigens invade your body.

The immune system is composed of white blood cells, or leukocytes. Two types are involved: phagocytes and lymphocytes (T cells and B cells).

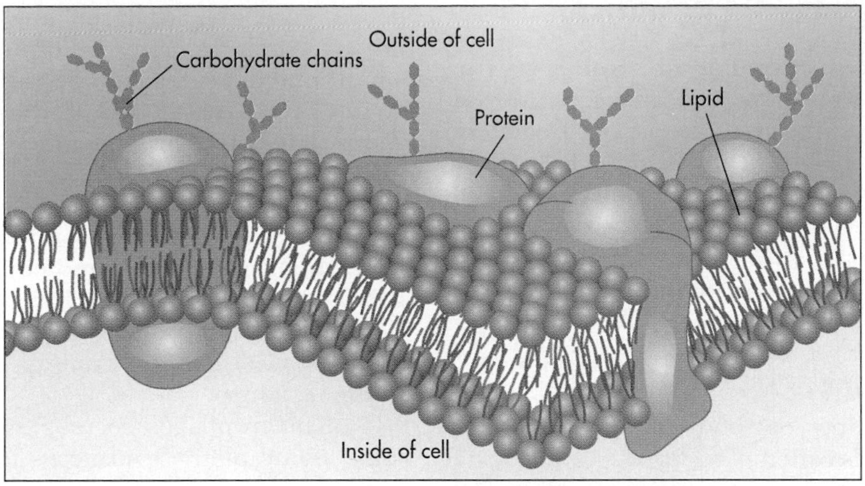

FIGURE 11-3 The outer surface of a cell is imbedded with proteins, carbohydrates, and lipids. The glycoproteins (carbohydrate-protein complexes) often serve as highly specific cell surface markers that identify particular cell types and also identify the cell as "self."

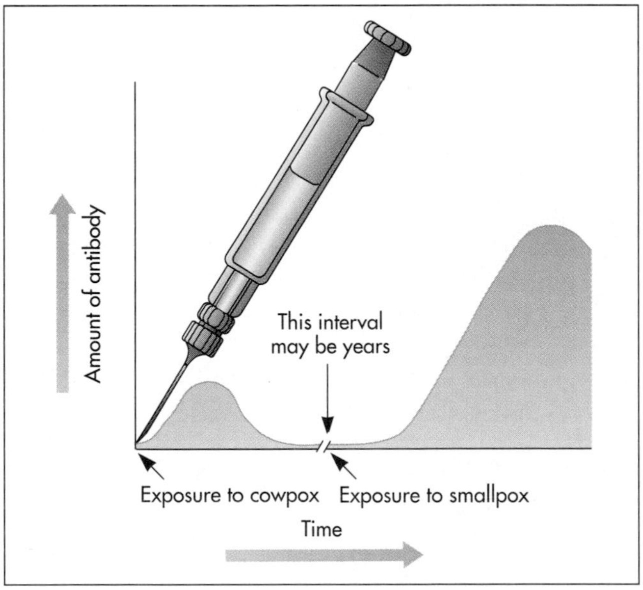

FIGURE 11-4 Immunity to smallpox. The immunity to smallpox that Jenner's patients acquired was a direct result of their cowpox inoculation. The cowpox inoculation caused Jenner's patients to develop antibodies to the antigen. An exposure to smallpox later stimulated them to produce the antibody in much larger amounts than before, resulting in immunity to smallpox.

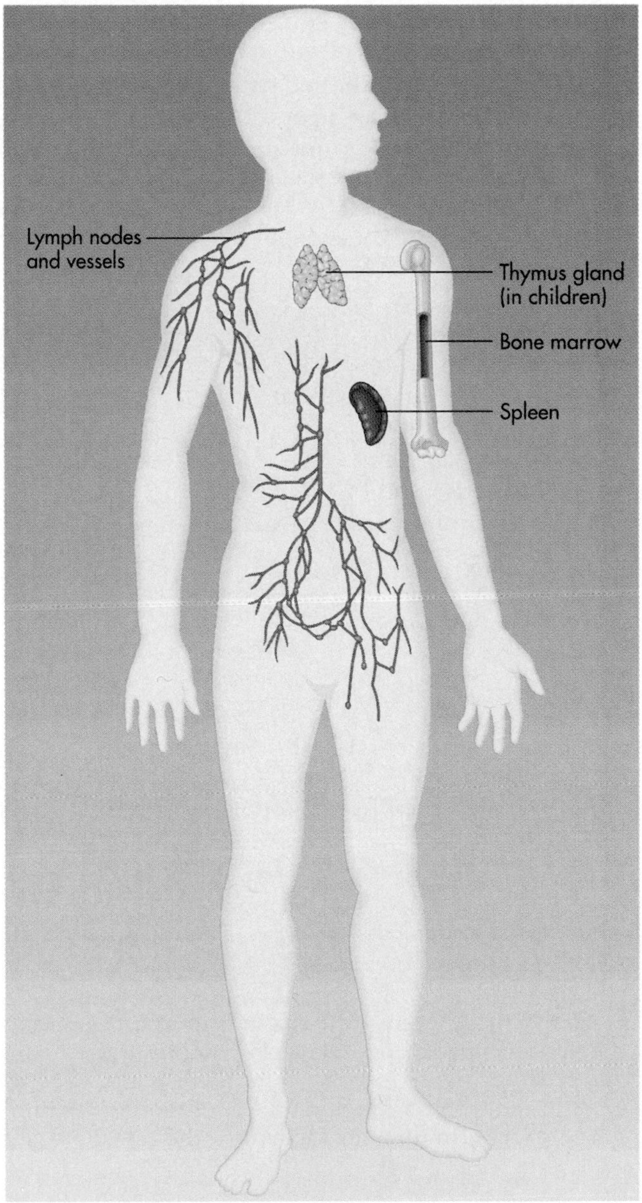

FIGURE 11-5 Organs in which immune cells reside. The principal organs in which the cells of the immune system mature and reside include the thymus, bone marrow, spleen, lymph nodes, and lymph vessels.

The Immune Response: How It Works

The immune system responds in several ways to defend your body against foreign invaders. Some of the cells of the immune system react immediately to invasion by foreign antigens. Other cells work to protect your body against future attack. The immune response is a complex yet coordinated effort, and it involves several types of immune protection. An overview of the alarm-sounding activation of the immune response is shown in Figure 11-6.

Sounding the Alarm

Constantly patrolling your body are roving armies of white blood cells called *phagocytes.* A phagocyte is a cell that destroys other cells by engulfing and ingesting them. This process is called *phagocytosis.* Two types of white blood cells are phagocytes: neutrophils and macrophages. *Macrophages* are the phagocytes that play a key role in the body's immune response.

Large, irregularly shaped cells, the macrophages act as the body's scavengers. Macrophages phagocytize anything that is not normal, including cell debris, dust particles in the lungs, and invading microbes. When the body is not under attack, only a small number of macrophages circulate in the body's bloodstream and lymphatic system. In response to infection, precursors of macrophages called *monocytes* develop into mature macrophages in large numbers.

When macrophages encounter foreign microbes in the body, they attack and engulf them and then display parts of these microbes on their surfaces. This display, as well as proteins secreted by the macrophages, is important in activating the immune response.

The macrophages secrete many different proteins. Some of these proteins trigger the maturation of monocytes into macrophages, thereby increasing their numbers. Another protein, *interleukin-1,* signals the brain to raise the body temperature, producing a fever. The higher temperature aids the immune response and inhibits the growth of invading microorganisms.

At the onset of the infection, specialized white blood cells other than the monocytes also go to work. Reacting to cell surface changes that occur

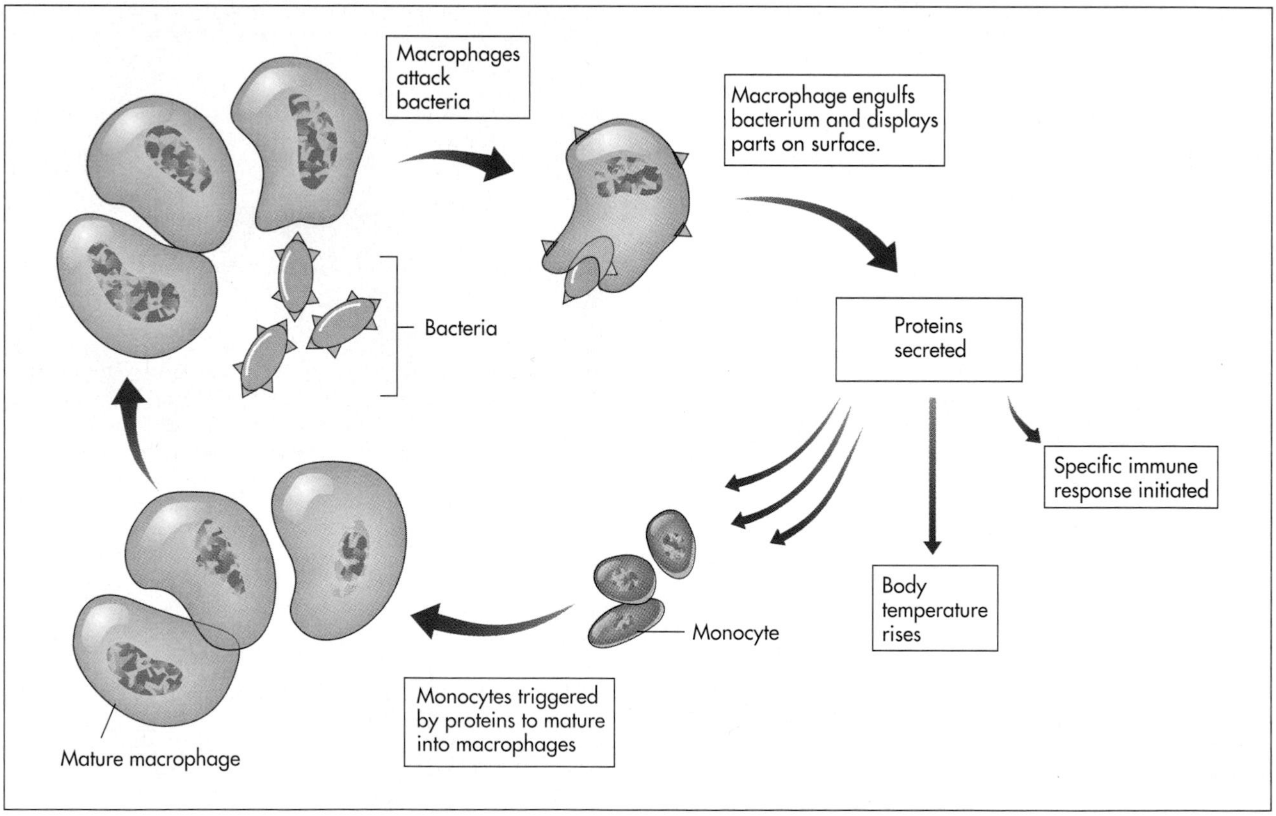

FIGURE 11-6 Activation of the immune response. When macrophages encounter microbes such as bacteria, they phagocytize them and display their antigenic parts on their cell surfaces. The macrophages also secrete various proteins that trigger the immune response, cause the body temperature to rise (thereby inhibiting microbial growth), and trigger the maturation of monocytes into macrophages. When mature, these macrophages will engulf microbes, continuing this cycle until the infection subsides.

on cancer cells or virally infected cells, *natural killer cells* attack them, breaking them apart. Rather than ingesting infected or diseased cells as macrophages do, natural killer cells secrete proteins that create holes in the membranes of the cells they attack. In addition, natural killer cells secrete toxins that poison these cells. The cells under attack burst open and die. This initial defense by macrophages and natural killer cells peaks within a day or two of the infection. This response buys time for the immune system to respond.

The Two Branches of the Immune Response

In addition to its other interactions mentioned previously, interleukin-1 activates helper T cells that have been presented with antigens by the macrophages, triggering the immune response. **T cells,** or **T lymphocytes,** develop in the bone marrow but then migrate to the thymus. The thymus, a small gland located in the upper chest and extending upward toward the neck, is the place where T (thymus) cells mature. This organ is large and active in children; it dwindles in size throughout childhood until, by puberty, it is quite small. In the thymus, T cells develop the ability to identify invading bacteria and viruses by the foreign molecules (antigens) on their surfaces. Tens of millions of different T cells are produced, each specializing in recognizing one particular foreign antigen. No invader can escape being recognized by at least a few T cells. There are five principal kinds of T cells:

1. *Helper T cells,* which initiate the immune response
2. *Cytotoxic T cells,* which break apart cells that have been infected by viruses and break apart foreign cells such as incompatible organ transplants
3. *Inducer T cells,* which oversee the development of T cells in the thymus
4. *Suppressor T cells,* which limit the immune response
5. *Memory T cells,* which respond quickly and vigorously if the same foreign antigen reappears

When they are triggered by interleukin-1 and the presentation of antigens, the helper T cells can stimulate both branches of the immune response: the **cell-mediated immune response** and the **antibody-mediated immune response.**

The Cell-Mediated Immune Response

The activation of helper T cells by interleukin-1 and the binding of antigen to these activated helper T cells unleash a chain of events known as the *cell-mediated immune response.* The main event of this response is that cytotoxic T cells ("cell-poisoning" cells) recognize and destroy infected body cells.

When a helper T cell has been activated, it produces a variety of chemical substances collectively called *lymphokines.* One type of lymphokine attracts macrophages to the site of infection, and another inhibits their migration away from it. Another of the lymphokines stimulates T cells that are bound to foreign antigens to undergo cell division many times. This cell division produces enormous quantities of T cells capable of recognizing

vaccination (vak suh **nay** shun) a procedure that involves the injection (or oral administration, in some cases) of a weakened or killed microbe into a person or animal in order to confer resistance to a disease-causing microbe.

antigens (**ant** ih jens) foreign molecules that induce the formation of antibodies, which specifically bind to the foreign substance and mark it for destruction.

antibodies proteins produced by plasma cells (B lymphocytes) that recognize foreign antigens and prevent them from causing disease.

immune response the body's specific response to foreign molecules, such as the production of antibodies directed against a particular antigen. The immune response is a result of the complex activities of recognition and defense carried out by the immune system.

phagocytes (**fag** oh sites) white blood cells that patrol the body and destroy foreign cells by engulfing and ingesting them.

lymphocytes (**lim** foe sites) white blood cells that are formed in the bone marrow, circulate in the blood and lymph, and reside in lymph tissue such as the lymph nodes, spleen, and thymus.

T cells or T lymphocytes white blood cells that develop in the bone marrow but migrate to the thymus to mature. As they mature, T cells develop the ability to identify body cells that have been infected with fungi, some viruses, and certain bacteria. They also recognize foreign body cells. T cells provide the cell-mediated immune response.

cell-mediated immune response one of the two branches of the immune response; it is initiated by helper T cells that have been activated by interleukin-1 and the presence of antigens, and results in cytotoxic T cells recognizing and destroying body cells.

antibody-mediated immune response one of the two branches of the immune response; it is initiated by helper T cells that have been activated by interleukin-1 and the presence of antigens, and results in B cells producing antibodies. The antibodies bind to the antigens they encounter and mark them for destruction.

the antigens specific to the invader. Each type of activated T cell does a specific job.

The activated inducer T cells trigger the maturation of immature lymphocytes in the thymus into mature T cells. Activated cytotoxic T cells kill the body's own cells that have been infected with fungi, some viruses, and bacteria that produce slowly developing diseases such as tuberculosis. In addition, they act to kill cells that have become cancerous. Cytotoxic T cells bind to these infected or abnormal cells by means of molecules on their surfaces that specifically fit antigens, much as a key fits a lock. Because the entire cell binds to the abnormal cells (by means of specific cell-surface proteins), this response is called *cell mediated.* The cytotoxic T cells then secrete a chemical that breaks apart the foreign cell (Figure 11-7). Unfortunately for patients undergoing organ transplantation, cytotoxic T cells also recognize foreign body cells in a similar way. Because of this, the immune system attacks transplanted tissue, leading to the rejection of transplanted organs.

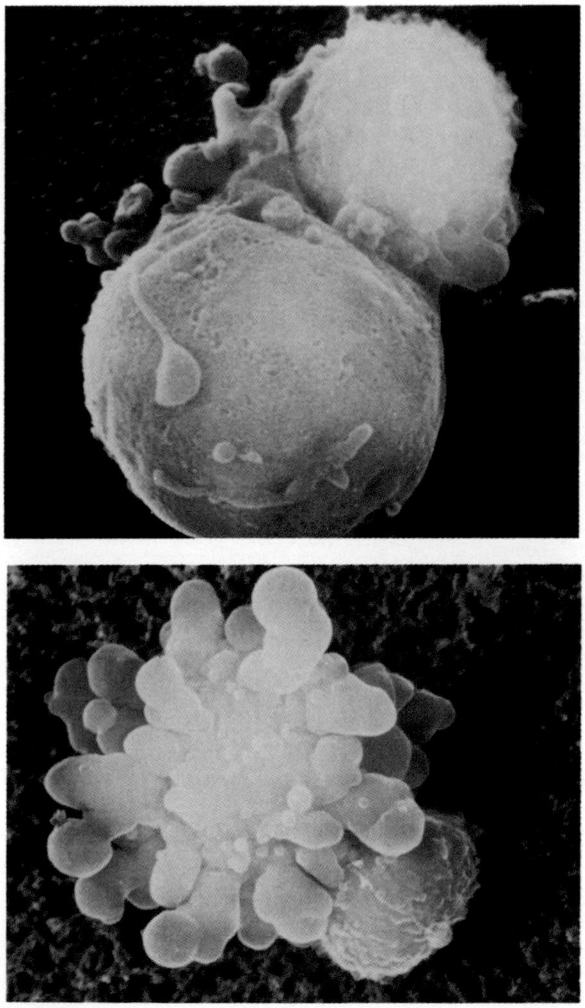

FIGURE 11-7 Cytotoxic T cell: a killer cell. The cell-mediated response involves cytotoxic T cells binding to antigens on the surfaces of infected cells and secreting a chemical that breaks the infected cells apart. The top photo shows a cytotoxic T cell binding with a tumor cell; at bottom, the tumor cell is disintegrating.

After the cytotoxic cells and macrophages do their jobs, the cell-mediated immune response begins to shut down. The cells in charge of shutdown are the suppressor T cells. The number of suppressor T cells slowly begins to rise after activation by the helper T cells. However, it takes about 1 to 2 weeks for their numbers to increase to a point where they are able to suppress the cytotoxic T cell response.

After suppression, or shutdown, a population of T cells persists, probably for the life of the individual. Referred to as memory T cells, these helper and cytotoxic T cells provide an accelerated and larger response to any later encounter with the same antigens. A summary of the cell-mediated immune response is shown in Figure 11-8.

T cells carry out the cell-mediated immune response, during which cytotoxic T cells recognize and destroy body cells infected with certain bacteria, viruses, and fungi. In addition, they destroy transplanted cells and cancer cells. Helper T cells initiate the response, activating cytotoxic T cells, macrophages, inducer T cells, and as the infection subsides, suppressor T cells.

The Antibody-Mediated Immune Response

When helper T cells are stimulated to respond to foreign antigens, they activate the cell-mediated immune response as described and activate a second, more long-range defense called the *antibody-mediated immune response.* Depending on the types of antigens present, the helper T cells may stimulate either or both of these branches of the immune response. The key players in antibody-mediated immunity are lymphocytes called **B cells,** or **B lymphocytes.** The B cells are named after a digestive organ in birds called the *bursa of Fabricius* in which these lymphocytes were first discovered. However, B cells mature in the bone marrow of humans, which may be a convenient way for you to remember these cells. The antibody response is sometimes called the *humoral response,* which refers to the fact that B cells secrete antigen-specific chemicals into the bloodstream—one of the body fluids called "humours" long ago.

On their surfaces, B cells each have about 100,000 copies of a protein receptor that binds to

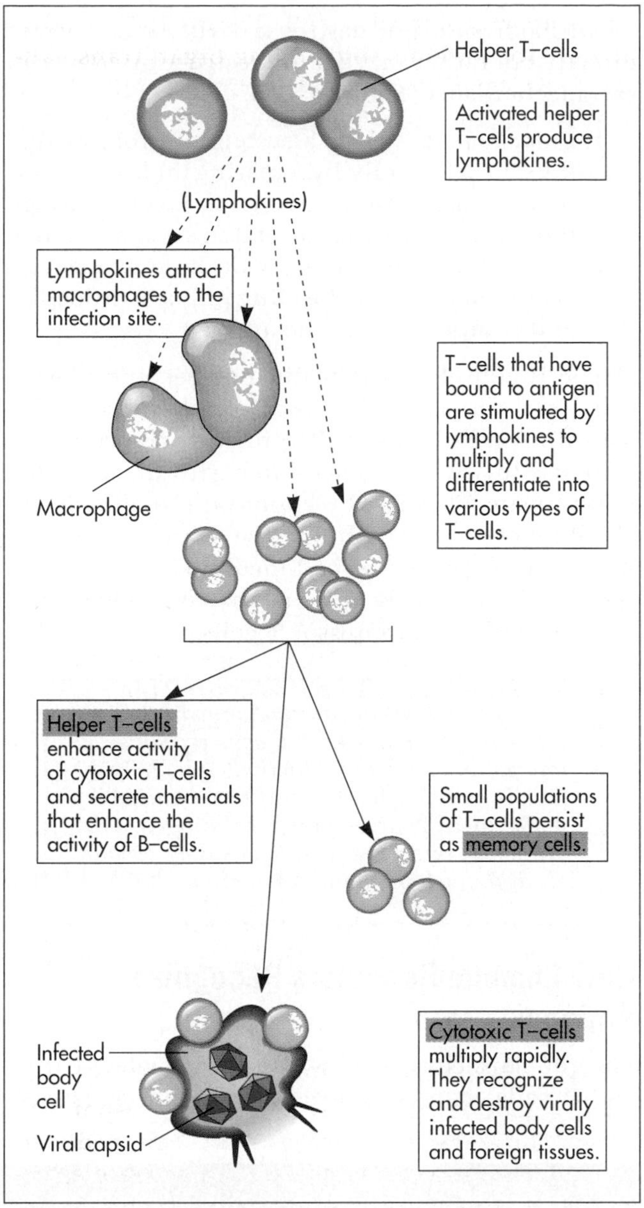

FIGURE 11-8 Summary of the antibody-mediated response. Interleukin-1, a protein secreted by macrophages that triggers the immune response, activates helper T cells. Activated helper T cells secrete lymphokines, proteins that attract macrophages to the infection site. Lymphokines also stimulate T cells that have recognized the foreign antigen to differentiate into helper T cells, memory cells, and cytotoxic T cells. The cytotoxic T cells multiply rapidly, and recognize and destroy the invaders—virally infected body cells and/or foreign tissues.

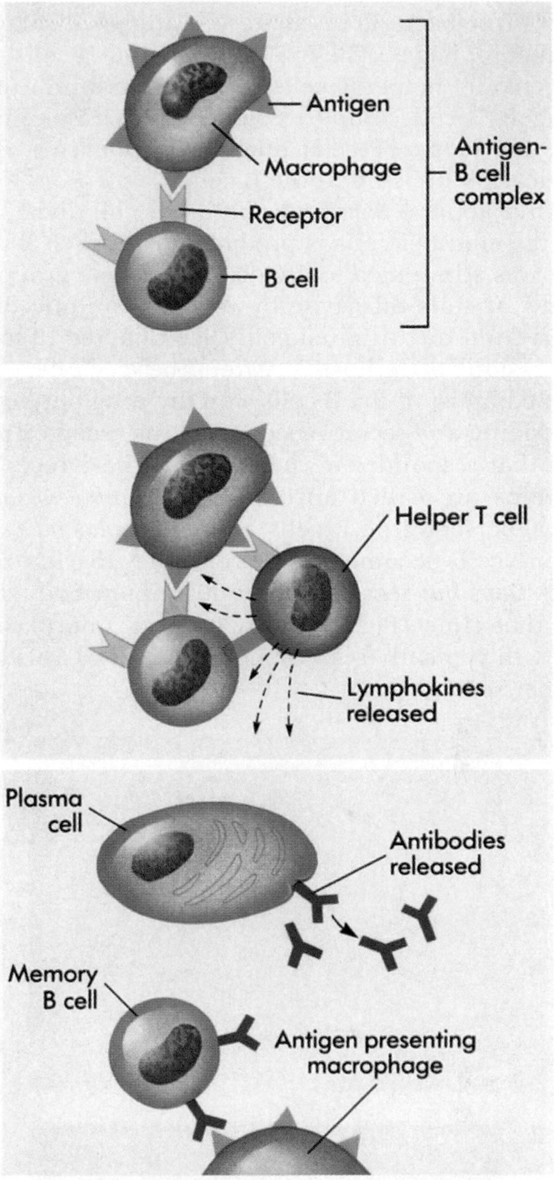

FIGURE 11-9 Summary of the antibody-mediated immune response. The antibody-mediated immune response involves the participation of three immune cell types: B cells, helper T cells, and antigen-presenting macrophages.

antigens. Because different B cells bear different protein receptors, each recognizes a different, specific antigen. At the onset of a bacterial infection, for example, the receptors of one or more B cells bind to bacterial antigens. The B cells may bind to

> **B cells or B lymphocytes** (lim foe sites) white blood cells that develop and mature in the bone marrow. As they mature, B cells develop the ability to identify bacterial antigens and respond by developing into plasma cells, which are specialized to produce antibodies. B cells provide the antibody-mediated immune response.

either free bacteria or bacterial antigens displayed by macrophages. These antigen-bound B cells are detected by helper T cells, which then bind to the antigen–B cell complex (Figure 11-9). After binding, the helper T cells release lymphokines that trigger cell division in the B cell.

After about 5 days and numerous cell divisions, a large clone of cells is produced from each B cell that was stimulated to divide. A *clone* is a group of identical cells that arise by repeated mitotic divisions from one original cell. (See Chapter 18 for a discussion of mitosis.) After these clones are formed, most of the B cells stop dividing but begin producing and secreting copies of the receptor protein that responded to the antigen. These receptor proteins are called **antibodies,** or *immunoglobulins.* The secreting B cells are called *plasma cells.* After B cells become plasma cells, they live for only a few days but secrete a great deal of antibody during that time (Figure 11-10). In fact, one plasma cell will typically secrete more than 2000 antibodies per second!

A

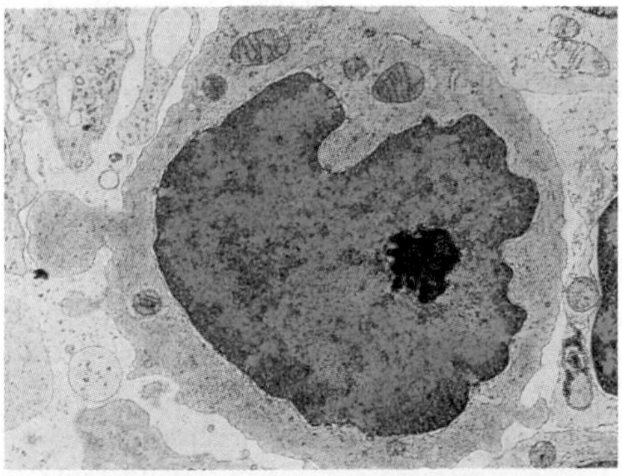

B

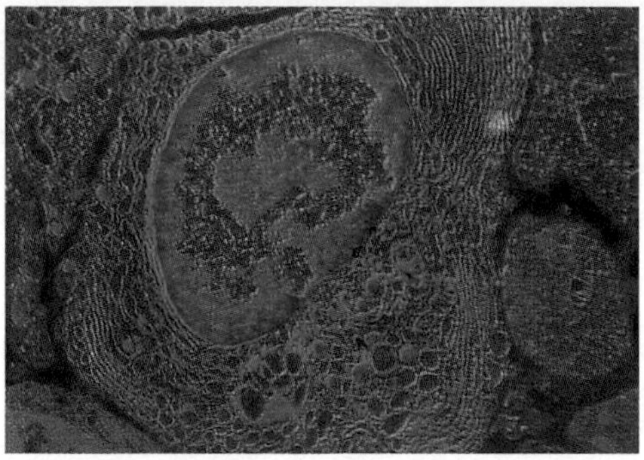

FIGURE 11-10 A plasma cell (bottom) and a B cell (top). A, Colorized micrograph of a B cell *(5400x).* These cells are key players in antibody-mediated immunity. **B,** Colorized micrograph of a plasma cell *(4700x).* Plasma cells are formed from B cells and secrete large amounts of antibodies. The large amount of endoplasmic reticulum on which antibodies are made, can be seen in the cytoplasm as green lines. ER cannot be seen in **(A).**

Antibodies do not destroy a virus or bacterium directly but rather mark it for destruction by one of two mechanisms (Figure 11-11):

1. *Complement.* This is a system of proteins that kills foreign cells by creating holes in their membranes. Water floods the cell through these holes, causing the cell to swell and burst.
2. *Macrophages.* The activity of these phagocytes is enhanced by the binding of both antibodies and complement to the antigen.

The B cell clones that did not become plasma cells live on as circulating lymphocytes called *memory B cells.* These cells provide an accelerated response to any later encounter with the stimulating antigen. This is why immune individuals are able to mount a prompt defense against infection. As in the case of the cell-mediated immune response, the antibody response is shut down after several weeks by suppressor T cells.

B cells carry out the antibody-mediated immune response in which B cells recognize foreign antigens and, if activated by helper T cells, produce large quantities of antibody molecules directed against the antigen. The antibodies bind to the antigens they encounter and mark them for destruction.

How Immune Receptors Recognize Antigens

The cell surface receptors of lymphocytes (T cells and B cells) can recognize specific antigens with

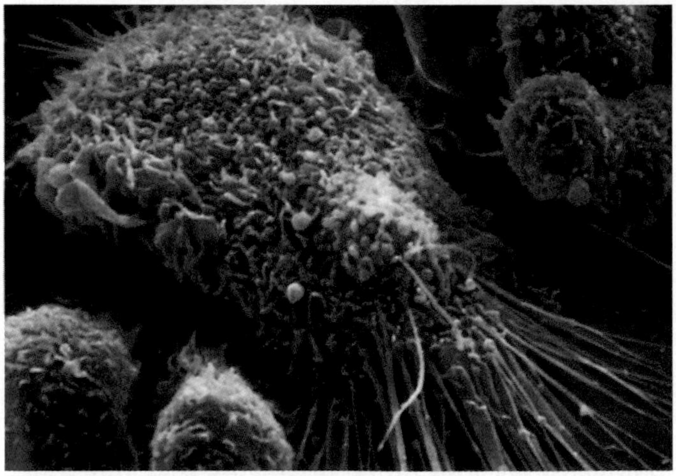

FIGURE 11-11 A macrophage ingesting cells. Macrophages engulf "marked" antigens by phagocytosis. This macrophage is snaring its "prey" with strandlike cell extensions. Electron micrograph *(3200x).*

great precision. Even single amino acid differences between proteins can often be discriminated, with a receptor recognizing one form and not the other. This high degree of precision is a necessary property of the immune system because without it, the identification of foreign antigens would not be possible in many cases; the differences between self and foreign (nonself) molecules can be very subtle.

A typical antibody, or immunoglobulin molecule, consists of four polypeptide chains. A polypeptide is a chain of amino acids linked end-to-end by peptide bonds (see Chapter 2). There are two identical short strands, called *light chains,* and two identical long strands, called *heavy chains.* The four chains are held together by disulfide (-S-S-) bonds, forming a Y-shaped molecule (Figure 11-12). The two "arms" of the Y determine which antigen will bind to the antibody. Antibodies recognize, or lock onto, antigens by means of special binding sites in their arms. These binding sites are made up of specific sequences of amino acids that determine the shape of the site. The specificity of the antibody molecule for an antigen depends on this shape. An antigen fits into the binding site on the antibody like a hand into a glove. Changes in the amino acid sequence of an antibody can alter the shape of this region and, by doing so, change the antigen that can bind to that antibody, just as changing the size of a glove will alter which hand can fit into it.

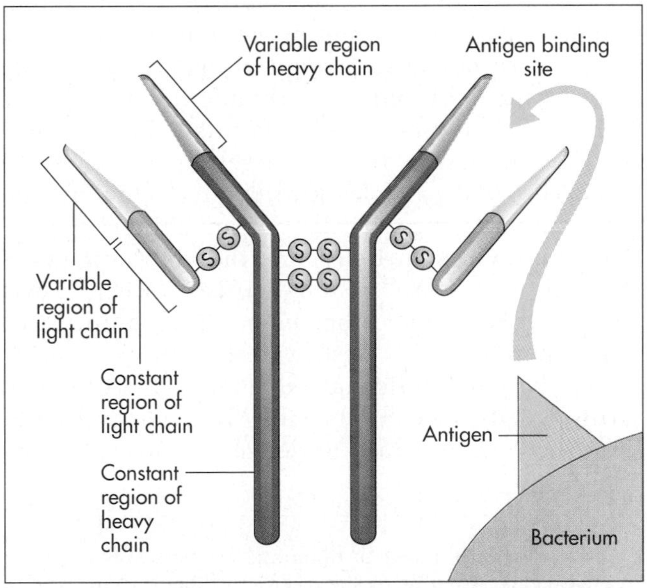

FIGURE 11-12 The structure of an antibody molecule. Each antibody molecule is composed of two identical light chains and two identical heavy chains. Disulfide bonds hold the chains together. The antibody binds the antigen to which it is specific at the antigen binding site (the variable region of the molecule).

> *An antibody molecule recognizes a specific antigen because it possesses binding sites into which an antigen can fit, much as a key fits into a lock. Changes in the amino acid sequence at the binding sites alter the antibody's shape and thus change the identity of the antigen that is able to fit into them.*

The stem of the Y determines what role the antibody plays in the immune response. An antibody can have one of five different stems and therefore be a member of one of the five classes of antibodies. For example, the antibodies produced in the first week of an infection are primarily class M antibodies and are called *IgM* (immunoglobulin M). After the first week, IgG antibodies are primarily produced.

Immunization: Protection Against Infection

Although scientists have unraveled many of the mysteries of the immune response since Jenner's day, scientists still use the same basic idea of vaccination that Jenner used to help people become resistant to specific diseases. A vaccine is made up of disease-causing microbes or toxins (poisons) that have been killed or changed in some way so as not to produce the disease. Injection with these antigens causes an antibody (B cell) response, with the production of memory cells. A booster shot induces these memory cells to differentiate into antibody-producing cells and more memory cells. So vaccination causes your body to build up antibodies against a particular disease without getting the disease. This type of immunity is called **active immunity.**

antibody a protein produced in the blood by a B lymphocyte that specifically binds circulating antigen and marks cells or viruses bearing antigens for destruction.

active immunity a kind of immunity in which a vaccination or the contraction of a disease causes the body to build up antibodies against that particular disease.

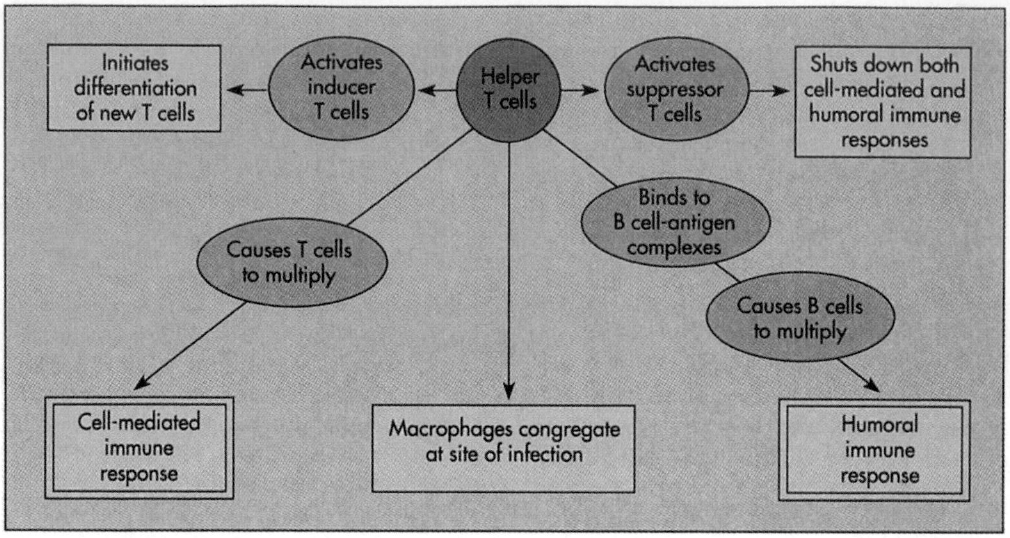

FIGURE 11-13 Helper T cells are the key to the immune response. This diagram outlines the many roles of the helper T cell.

Scientists can now use the techniques of genetic engineering to produce vaccines. Instead of growing cultures of an agent of infection such as a bacterium or a virus, scientists can, in the laboratory, produce large quantities of antigenic proteins from disease-causing microbes. These antigens can then be used as vaccines.

Vaccination to produce active immunity works well for many diseases but is a somewhat slow process. After the injection of the antigen, it takes weeks for the body to develop sufficient antibodies and memory cells to combat the disease. However, another type of immunity, **passive immunity,** can be used when protection is needed quickly. One example is the use of *antitoxin* when a person has been bitten by a snake.

When a poisonous snake bites a person or other animal, it injects a toxin (poison) into the body. This toxin circulates in the bloodstream. Certain toxins damage the nerve cells they reach and may result in death. An antitoxin is a preparation of antibodies specific for a toxin. These antibodies are injected into the victim and they bind with the toxin, preventing the toxin from causing damage. Although effective, this type of borrowed immunity lasts for only a short time. The body soon uses or eliminates these antibodies, so they will not be available if another snakebite occurs.

Newborns have a type of passive immunity borrowed from their mothers. Antibodies that pass from mother to fetus through the placenta provide babies with a short-term immunity that subsides by the time the baby has produced its own system of antibodies. In addition, if a mother nurses her baby, antibodies in the mother's milk provide the baby with some protection.

Defeat of the Immune System: AIDS

Helper T cells are the key to the entire immune response because they initiate the proliferation of both T cells and B cells (Figure 11-13). Without helper T cells, the immune system is unable to mount a response to foreign antigens. The AIDS virus, called **human immunodeficiency virus (HIV),** is deadly because it mounts a direct attack on T lymphocytes, mainly targeting the helper T cells. The HIV particles kill helper T cells by entering them, using the cells' own machinery to produce more virus particles, and then bursting open the cells to release the new viruses (see p. 548). These viruses infect other helper T cells until the entire population is destroyed. In a normal individual, helper T cells make up 60% to 80% of circulating T cells; in patients with AIDs, helper T cells often become too rare to detect. Because the im-

passive immunity a kind of immunity conferred on an individual when he or she receives antibodies from another source. An example of passive immunity is a baby receiving antibodies during breastfeeding.

human immunodeficiency virus, or HIV (im yoo no de **fish** un see **vye** rus) a deadly virus that weakens the ability of the immune system to mount a defense against infection because it attacks and destroys helper T cells.

HOW *Science* WORKS

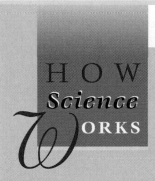

Discoveries: Tracking the HIV Epidemic

Lucy Bradley-Springer, PhD, RN*

In 1981, public health officials in the United States recognized a new disease that eventually became known as the *acquired immunodeficiency syndrome (AIDS).* We now know that AIDS is the final phase of a chronic immune function disorder caused by the human immunodeficiency virus (HIV). By the time HIV infection was making its presence known in the United States, it had already gained a solid foothold in Africa, Eastern Europe, the Caribbean, and South America.

There is evidence that HIV had infected people in Africa as early as 1959, where it was called "slim disease" because of the severe weight loss that accompanied advancing illness. Although we may never know the exact origin of HIV, many scientists believe that it was introduced into the human population by one of the primate species found in Africa. The disease was contained in small areas of Africa for several decades until urbanization, changing work habits, and improved transportation brought infected people into contact with others in widely dispersed geographic regions. It was then only a matter of time before HIV spread throughout the world.

There are currently 10 to 12 million HIV-infected people living in the world. HIV and an associated epidemic of tuberculosis continue to spread rapidly around the world, with Southeast Asia currently showing the greatest increases. One global estimate predicts that 100 million people will be infected with HIV by the year 2000.

HIV and What It Does

HIV is an RNA virus that was discovered in 1983. RNA viruses are also known as *retroviruses.* Like all viruses, HIV is an obligate parasite: it cannot survive and replicate unless it is inside a living cell.

HIV can enter a cell by binding to specific (CD4) receptor sites on the cell's surface (see Figure 11-A). Once the viruses' RNA enters a cell, it is transcribed into a single strand of viral DNA with the assistance of an enzyme called *reverse transcriptase.* This strand then replicates itself and becomes double-stranded viral DNA. At this point, the viral DNA can splice itself into genetic material in the nucleus of the host cell, becoming a permanent part of the cell's genetic structure. Since all of the cell's genetic materials are replicated during cellular division, all of this cell's daughter cells will also be infected. And since the genetic material now contains viral DNA, the cell's genetic codes can direct the cell to make more HIV.

Although HIV can infect several types of human cells, immune dysfunction results predominantly from the destruction of helper T lymphocytes. These lymphocytes are targeted because they have more CD4 receptor sites on their surfaces than other cells. The helper T lymphocyte is a white blood cell that plays a pivotal role in the

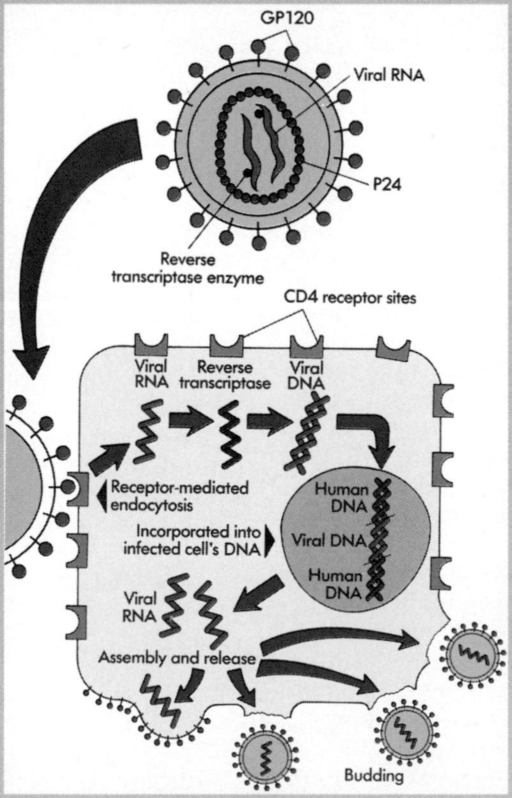

FIGURE 11-A How HIV replicates within a cell.

*Co-Director, New Mexico AIDS Education and Training Center Assistant Professor, University of New Mexico School of Medicine

ability of the immune system to recognize and defend against foreign invaders. When HIV replicates within an infected helper T cell, newly formed viruses bud out from the cell's membrane. This process not only releases new HIV to infect other cells, it also eventually kills the host cell.

Humans normally have 800 to 1200 helper T lymphocytes/mm^3 of blood. In HIV infection, a point is eventually reached where so many helper T cells are destroyed that there are not enough to direct and coordinate immune responses. This sets the stage for the development of life-threatening illnesses.

Transmission

HIV is transmitted from human to human through infected blood, semen, vaginal secretions, and breast milk. If these infected fluids are introduced into an uninfected person's body, the potential for transmission occurs. Sexual contact with an HIV-infected partner is the most common method of transmission. Sexual activity provides an opportunity for contact with semen, vaginal secretions, and/or blood. Although male homosexuals were the initial targets of HIV in this country, heterosexual transmission is becoming more prevalent and is the most common method of infection for women.

Sharing equipment to inject drugs is a major means of transmission to both sexes in many large metropolitan areas and is becoming more common in smaller cities and rural areas. HIV has also been transmitted in blood products. This is now rare in the United States because of a strict program of donor screening, blood testing, and heat or chemical treatment of hemophilia clotting-factor products.

Transmission from HIV-infected mothers to their infants can occur during pregnancy, at the time of delivery, or after birth through breast feeding. Studies have found that 9% to 30% of infants born to HIV-infected women will be born infected. This means that 70% to 91% of these infants will *not* be infected. *All* infants born to HIV-infected mothers will be positive on the HIV antibody test, however, because maternal antibodies cross the placental barrier. It may take 15 months before it is known whether an infant is infected with HIV or not. Breast feeding is a rare, but well-documented, HIV transmission risk. Because of this, HIV-infected mothers in the United States are encouraged not to breastfeed their infants.

Prevention

Although advances are being made, there is still no evidence that a cure or a vaccine will be developed soon. The only defense against HIV is prevention through education about risky behaviors and risk-reduction practices. The good news is that behaviors that create a risk for transmission of HIV have been clearly identified and are avoidable.

Drug Use

Risks for HIV infection related to injecting drugs can be *eliminated* by not injecting drugs or by only using equipment that is never shared. Risk *reduction* during drug use involves cleaning equipment between use with bleach and water. Individuals who use drugs of any kind (including alcohol and drugs that are smoked, snorted, or swallowed rather than injected) may put themselves at risk of sexual transmission if they make unwise sexual decisions while under the influence. Remember that HIV infection is only one of many health risks associated with drug and alcohol use.

Sexual Activity

Elimination of HIV risk related to sexual activity includes abstinence, sexual activity in a mutually monogamous relationship where neither partner is infected, and sexual activity in which the penis, vagina, mouth, or rectum does not touch someone else's penis, vagina, mouth, or rectum. Risk *reduction* includes correct and consistent use of condoms and refraining from unprotected anal intercourse, the most risky form of sexual activity.

Perinatal Transmission

Elimination of risk for infants can be accomplished by preventing HIV infection in women of childbearing age, providing birth control to infected women, or terminating the pregnancies of infected women. It is important to remember that abortion is an unacceptable choice for many women. HIV-infected women who wish to have a baby can be counseled in ways to *decrease* the risk to the infant. Recent studies have shown that babies born to HIV-infected women who are in the early phases of the disease are infected at a significantly lower rate than babies born to mothers with severe immune suppression. There is also evidence that the use of antiretroviral medication during pregnancy can significantly reduce the risk to the infant.

Spectrum of HIV Infection

Early Infection

Initial infection with HIV causes an immune reaction during which HIV-specific antibodies are produced. These antibodies can be measured in a person's blood at a median of 2 months after infection (Figure 11-B). Some people will have measurable HIV antibodies as early as 3 weeks after infection, whereas others may take as long as 6 months. The most accurate tests that we currently have for HIV detect these antibodies and not the virus itself. This means that some people can be infected (and able to transmit HIV to others) for as long as 6 months before the test would show that

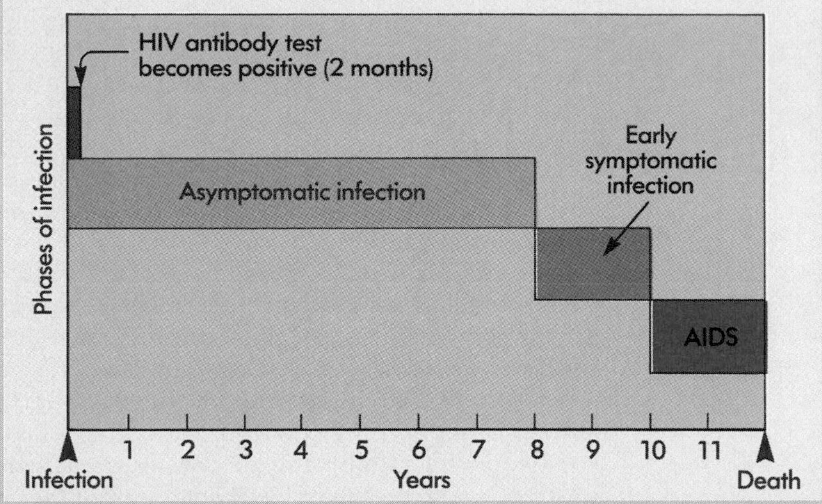

FIGURE 11-B The spectrum of HIV infection: Median times for progression of HIV disease.

they were infected. In some people, a flulike syndrome of fever, fatigue, nausea, and diarrhea accompanies the development of HIV antibodies.

The median time between HIV infection and a diagnosis of AIDS is about 10 years. HIV-infected people remain generally healthy during this time, but vague symptoms, including fatigue, headaches, low-grade fevers, and night sweats may occur. Because the symptoms are vague, people may not be aware that they are infected during this "asymptomatic" phase. During this time, infected people are able to continue their usual activities, which may include risky sexual and drug-using behaviors. This creates a public health problem because infected people can transmit HIV to others even if they have no symptoms.

Early Symptomatic Disease

Toward the end of the asymptomatic phase and before a diagnosis of AIDS, early symptomatic disease develops. Early symptoms can include persistent fevers, night sweats, and fatigue. These may be severe enough to interrupt normal routines. Swollen lymph glands and localized infections may also occur. The most common problem at this point is thrush, a fungal infection of the mouth that rarely causes problems in healthy adults. Other infections that can occur at this time include shingles (caused by the *Varicella zoster* virus), persistent vaginal yeast infections, and outbreaks of oral or genital herpes.

AIDS

A diagnosis of AIDS cannot be made until the HIV-infected person meets criteria established by the Centers for Disease Control and Prevention (CDC). These criteria are more likely to occur when the immune system becomes severely compromised. AIDS is diagnosed when an individual with HIV develops at least one of these additional conditions:

1. The helper T lymphocyte count drops below 200/mm^3.
2. One of 24 specific opportunistic diseases develops. Opportunistic diseases include cancers and infections that would not occur if the immune system were working properly. The most common HIV-related opportunistic diseases are *Pneumocystis carinii* pneumonia (PCP), *Mycobacteriun avium* complex, and Kaposi's sarcoma (KS).
3. Wasting syndrome occurs. Wasting syndrome is defined as a loss of 10% or more of ideal body mass.
4. Dementia develops. Symptoms of AIDS-related dementia include decreased ability to concentrate, apathy, depression, social withdrawal, and slowed response rates. AIDS dementia can progress to coma.

The median time for survival after a diagnosis of AIDS is 2 years, but this varies greatly. Some people with AIDS live for 6 or more years, but others survive for only a few months. Advances in the treatment and diagnosis of HIV infection have increased survival times, but AIDS fatality rates remain high.

Impact of HIV in the United States

By the end of 1993, more than 361,000 cases of AIDS and more than 217,000 AIDS-related deaths had been reported in the United States. In addition, HIV is estimated to infect at least 1 million people in this country. In the United States, the fastest growing groups of people with HIV and AIDS are women and adolescents. Of the more than 44,000 cases of AIDS that have been reported in women, 38% were reported in 1993. During that same year, 29% of the more than 15,000 AIDS cases in people aged 13 to 25 were reported. The face of AIDS in the United States is not only changing along the lines of gender and age, it is also becoming an increasing problem for people of color and people who live in poverty.

Conclusion

By all accounts, the global HIV epidemic will continue for the foreseeable future. People who are already infected will be using health care services more frequently over the next 10 to 15 years. Currently, prevention of new cases is the only way to decrease the devastation caused by HIV. Unfortunately, these efforts are often hampered by social, cultural, and governmental policies that restrict education about risk-reduction measures. If we cannot overcome these barriers, the HIV epidemic will continue to expand, causing human suffering, premature death, and strains on limited health care resources.

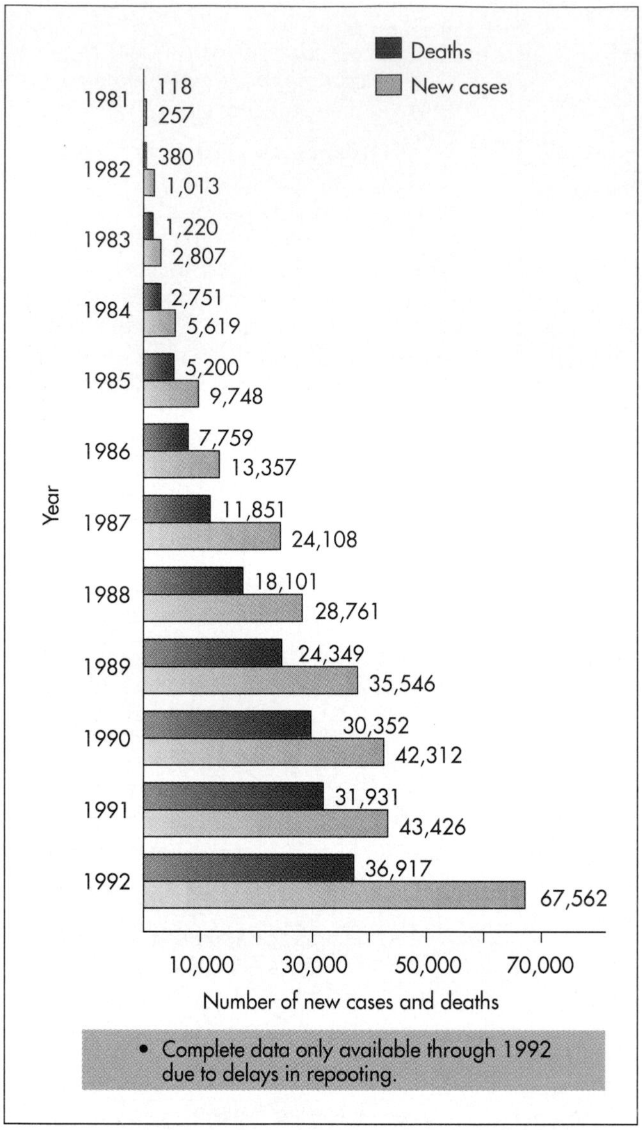

FIGURE 11-14 AIDS cases and deaths since 1981. By the end of 1993, more than 361,000 cases of AIDS and more than 217,000 AIDS-related deaths had been reported in the United States. One global estimate predicts that 100 million people will be infected with HIV by the year 2000.

mune response cannot be initiated, any one of a variety of infections proves fatal. This is the primary reason that AIDS is a particularly devastating disease (Figure 11-14).

> *The AIDS virus weakens the ability of the immune system to mount a defense against infection because it attacks and destroys helper T cells.*

The AIDS virus was discovered to be a cause of human disease in the early 1980s; it is clear that it is a serious disease. The fatality rate of AIDS is virtually 100%, but some AIDS patients live for many years after diagnosis. The disease is *not* highly infectious; that is, it is not transmitted from person to person by casual contact. It is transmitted only by the direct transfer of body fluids, typically in semen or vaginal fluid during sexual activity, in blood during transfusions, or by contaminated hypodermic needles. It also crosses the placenta to infect babies within the womb of an infected mother.

Efforts to develop a vaccine against AIDS continue, both by splicing portions of the AIDS surface protein gene (gp 120) into a harmless virus (Figure 11-15) and by attempting to develop a harmless strain of AIDS. In 1994, scientists reported that human studies of two gp 120 vaccines showed that the vaccines stimulated the production of antibodies in uninfected persons, which could then be used to inactivate several laboratory strains of HIV. Unfortunately, however, the vaccine did not work against HIV taken from the bloodstream of AIDS patients. All approaches to developing an AIDS vaccine are limited by the fact that different strains of AIDS virus seem to possess different surface antigens.

Allergy

Although the human immune system provides effective protection against viruses, bacteria, parasites, and other microorganisms, sometimes it does its job too well, mounting a major defense against a harmless antigen. Such immune responses are called **allergic reactions.** Hay fever, the sensitivity that many people exhibit to proteins released from plant pollen, is a familiar example of an allergy. Many other people are sensitive to proteins released from the feces of a minute house-dust mite (Figure 11-16). This microscopic insect lives in the house dust present on mattresses and pillows and eats the dead epithelial tissue that everyone sheds from their skin in large quantities daily. Many people sensitive to feather pillows are actually allergic to the feces of mites that are residents of the feathers.

What makes an allergic reaction uncomfortable and sometimes dangerous is involvement of class E (IgE) antibodies. The binding of antigens to these antibodies initiates an inflammatory response; powerful chemicals cause the dilation of blood vessels and a host of other physiological changes. Sneezing, runny nose, and fever often result. In some instances, allergic reactions can be far more dangerous, resulting in anaphylactic shock, a severe and life-threatening response of the immune system.

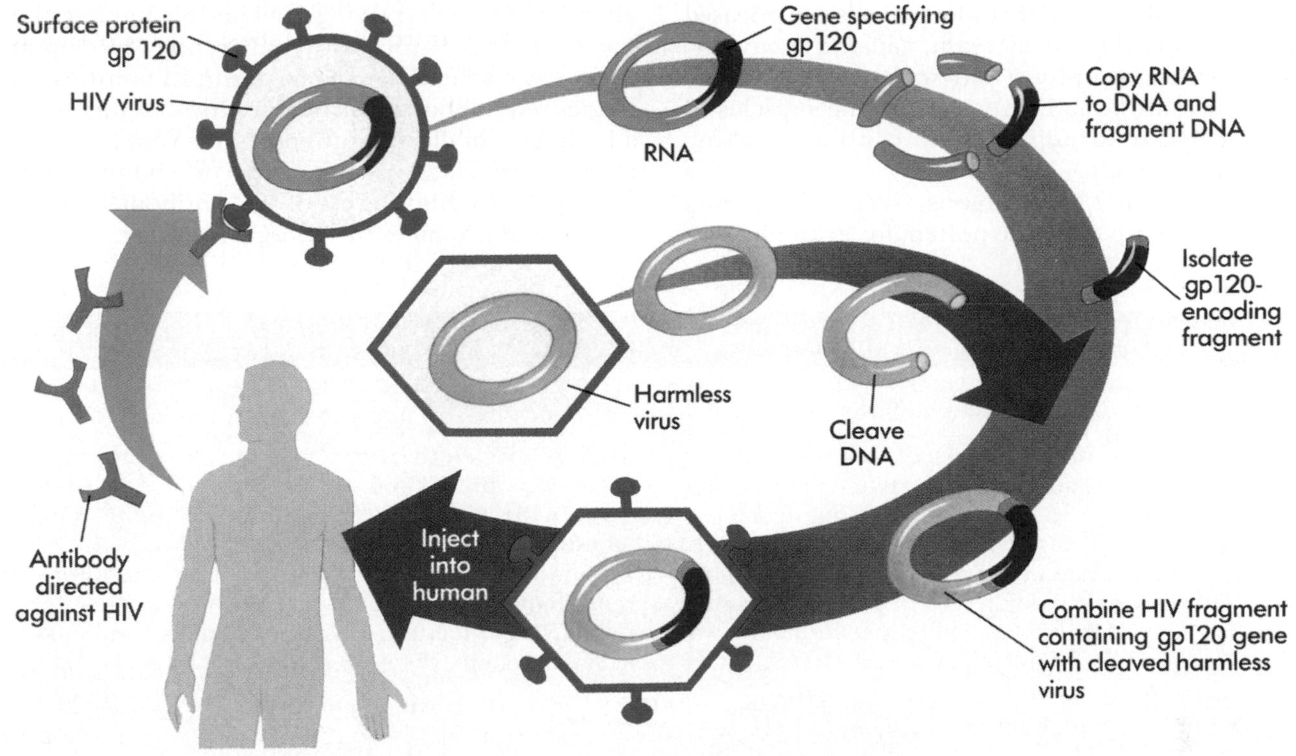

FIGURE 11-15 How scientists are constructing a vaccine for AIDS. Of the several genes of the human immunodeficiency virus (HIV), one is selected (gp120) that directs the synthesis of a certain surface feature on the virus. All of the other HIV genes are discarded. This one gene is not in itself harmful to humans. This one gene or fragment of it is inserted into the DNA of a harmless vaccinia cowpox virus, resulting in a harmless virus whose surface imitates the AIDS virus. Persons injected with the vaccinia virus do not become ill, but they do develop antibodies directed against the infecting virus surface. At this time a vaccine genetically-engineered in this way has been developed; it attacks several strains of HIV. It is currently being tested on human subjects.

FIGURE 11-16 The house-dust mite *Dermphagoides.* Mites like this live in the dust in your house. Many people are allergic to the proteins released in these mites' feces.

allergic reactions an immune response that results from the immune system mounting a major defense against a harmless antigen.

allergens antigens that are initiators of strong immune (allergic) responses.

The chemicals that dilate blood vessels are released suddenly into the bloodstream, causing many vessels to widen abruptly. Their widening causes the blood pressure to fall. In addition, the muscles of the trachea, or windpipe, may contract, making breathing difficult.

Not all antigens are **allergens,** initiators of strong immune responses. Nettle pollen, for example, is as abundant in the air as ragweed pollen, but few people are allergic to it. Nor do all people develop allergies; the sensitivity seems to run in families. Allergies require both a particular kind of antigen and a high level of class E antibodies. The combination of an appropriate antigen on the one hand and inappropriately high levels of IgE antibodies on the other hand produces the allergic response.

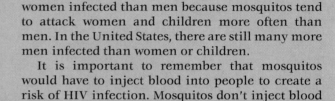

JUST WONDERING . . .

Can I get AIDS from an insect bite?

No. Mosquitos and other insects do not transmit HIV infection. The main way that we know that this doesn't happen is through looking at the statistics for HIV. If mosquitos spread HIV, we would expect to see more HIV infection in areas where mosquitos live. Although HIV does occur in areas such as Florida and Texas where mosquitos live, we tend to see higher concentrations of HIV in places like San Francisco and New York, which do not have a mosquito problem.

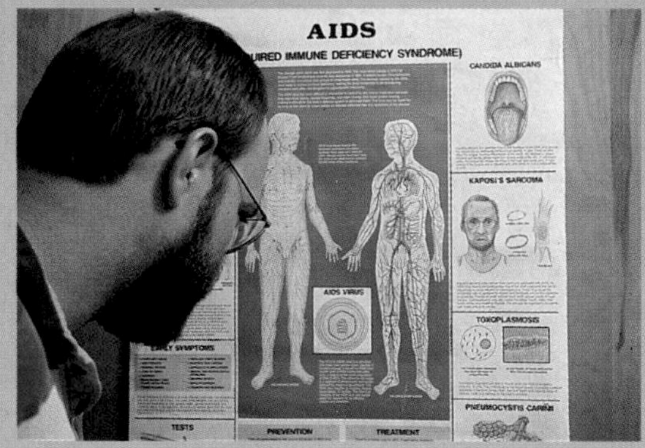

In addition, we would expect to see HIV more equally distributed among men, women, and children. This is because insects don't discriminate according to a person's sex, they just want blood. In fact, if mosquitos transmitted HIV, we'd probably see more children and women infected than men because mosquitos tend to attack women and children more often than men. In the United States, there are still many more men infected than women or children.

It is important to remember that mosquitos would have to inject blood into people to create a risk of HIV infection. Mosquitos don't inject blood when they bite, they inject saliva. And since HIV doesn't get into the saliva of mosquitos, they do not inject HIV. Also, mosquitos would not only have to inject infected blood, they would have to inject a whole lot of it. Mosquitos simply do not carry enough blood around to cause an infection.

Summary

- Although immunity was discovered almost 200 years ago, it has only recently been understood that resistance to disease is achieved by populations of white blood cells collectively called the *immune system.*
- The human body has both nonspecific and specific defenses against disease. The nonspecific defenses, those acting against any foreign invader, include the skin and mucous membranes, chemicals that kill bacteria, and the inflammatory process. The specific defense of the body is the immune response, a complex series of processes carried out by the cells of the immune system.
- There are two types of white blood cells involved in the immune system: phagocytes and lymphocytes. Two classes of lymphocytes play roles in specific resistance: T cells and B cells. These lymphocytes are capable of recognizing specific foreign substances (usually proteins) called *antigens.*

- Lymphocytes recognize specific antigens by means of receptor proteins on the lymphocyte cell surface.
- The immune response is initiated by groups of T cells called *helper T cells.* When stimulated by macrophages and antigens, helper T cells can activate the two branches of the immune response: the cell-mediated immune response (T cell response) and the antibody-mediated immune response (B cell response).
- In the cell-mediated immune response, cytotoxic T cells attack infected body cells. In the antibody-mediated immune response, B cells are converted to plasma cells that secrete proteins called *antibodies.* These antibodies specifically bind circulating antigen and mark cells or viruses bearing antigens for destruction.
- The high level of specificity of an antibody for a particular antigen is caused by the three-dimensional shape of the ends of the arms, or antigen-binding sites, of the molecule. Slight changes in the amino acid sequence alter the shape of these sites and thus the identity of molecules able to fit into it.
- Vaccination, which is an injection with disease-causing microbes or toxins that have been killed or changed in some way so as to be harmless, causes the body to build up antibodies against a particular disease. Scientists can now use laboratory-made antigenic proteins as vaccines.
- The AIDS virus is deadly because it primarily attacks helper T cells, the key to the entire immune response. Because helper T cells are no longer present to initiate the immune response, any one of a variety of otherwise commonplace infections proves fatal.
- An allergic reaction is an immune response against a harmless antigen. It is produced by the combination of a particular kind of antigen and a high level of a certain class of antibody.

Knowledge and Comprehension Questions

1. Distinguish between nonspecific and specific defenses.Which could be considered the first line of defense and why?
2. Name your body's nonspecific defenses and summarize how each protects you against infection or injury.
3. How is your immune response involved if you get a particle of foreign matter, such as dust, in the mucous membranes of your eyelid?
4. How is the immune system different from other body systems discussed so far?
5. In earlier chapters, we discussed how a cell membrane may have proteins or carbohydrates on its surface to act as "cell markers" was discussed. How do you think these would be interpreted if this cell were placed in the tissues of another organism?
6. What are the five principal types of T cells? Summarize the function of each.
7. Summarize the events of the cell-mediated immune response. What cells are involved?
8. Summarize the events of the antibody-mediated immune response. What cells are involved?
9. How is an antibody molecule able to discriminate between millions of cells in a tissue and bind to only one specific antigen?
10. A nurse vaccinates a child against the tetanus bacterium. Later, the child receives a booster shot. Describe how these two vaccinations affect the child's immune response during subsequent exposures to tetanus.
11. Which virus is responsible for AIDS? Explain why this virus is so deadly.
12. Many people—including you, perhaps—are allergic to something. Explain what an allergic reaction is and why it occurs. What causes its uncomfortable symptoms?

Critical Thinking

1. Why do you think it is important that there are phagocytes constantly circulating in the bloodstream and in body tissues?
2. You had measles as a child. Now your younger brother has measles, but you don't catch it from him. Explain why.
3. Pediatricians often encourage mothers to breast-feed infants because maternal antibodies circulate in milk and provide the baby with some protection against infection. What sort of immune response is involved here? Why is this immunity only temporary?

CHAPTER 12

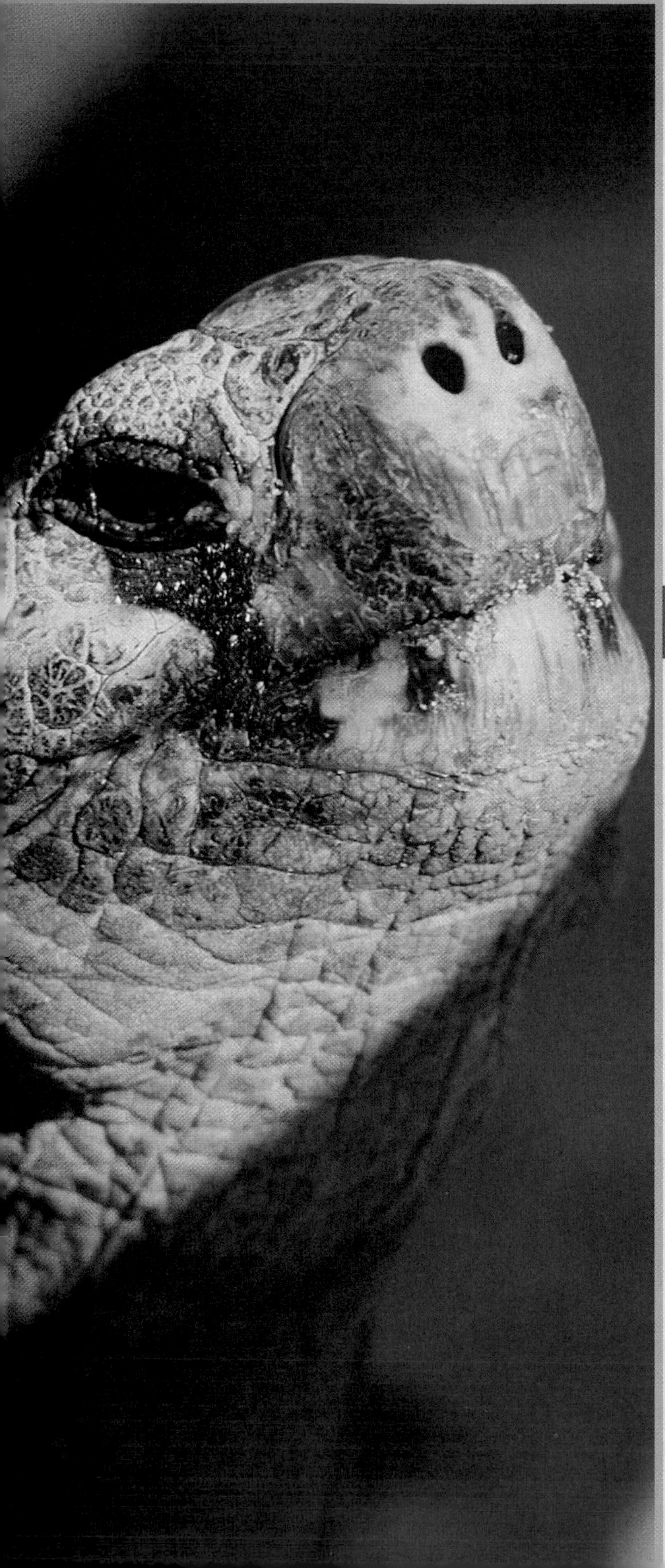

Excretion

The story of excretion is, to a large extent, a story about water and salt. And that story is the reason this sea turtle is crying.

Animals live not only on land but in salt water, in fresh water, in marshes . . . in a wide array of environments. In any environment, an animal needs water to live. Water bathes its cells and helps move substances within its body. Water is the medium in which substances are dissolved and in which chemical reactions take place. The amount of water in the body is crucial to survival—having too much water can cause as many problems as having too little.

In animal bodies, water is separated within compartments—any space bounded by the selectively permeable membranes of cells. Cells themselves are compartments, as are such structures as the interiors of the blood vessels and the heart. Water moves among the compartments within the body by osmosis, moving from areas of a high concentration of water to areas of a lower concentration of water. The concentration of salts—compounds that form ions in water—affects the concentration of water in the various body compartments and therefore water's movement. To regulate the water in the various body compartments, animals must regulate the amount of salt as well—a balancing act of sorts.

Animals have evolved various mechanisms that deal with their water and salt problem. The production of salt water "tears" is only one mechanism. Because the turtle lives in a salty environment, water tends to leave its less salty body. To compensate, the turtle drinks a lot of seawater but then gets rid of the salt and keeps the water. Interestingly, the sea turtle has evolved excretory organs located in the corners of its eyes that eliminate salt. So shedding salty tears is really one way of accomplishing its balancing act. Animals that live on land do not have the problem of water moving into or out of their bodies through the skin in response to the concentration of water in their environments. Therefore land animals can regulate the balance of water and salt in their bodies without shedding a sea of salty tears.

KEY CONCEPTS

- The major organs of excretion in the human body are the lungs and the kidneys.
- The formation of urine by the kidneys not only rids the body of metabolic wastes but controls the balance of water and salts, processes that are important to the maintenance of homeostasis.
- The kidneys work by removing most of the water and dissolved substances from the blood and then replacing what is needed by the body at that time.
- Other organs of the urinary system—the ureters, bladder, and urethra—temporarily store and then excrete the urine produced by the kidneys.

OUTLINE

What Substances Does the Body Excrete?

Although the story of **excretion** is, to a large extent, a story about water and salt balance, this process also regulates the concentrations of other ions and molecules and removes metabolic wastes from the body. The metabolic wastes of excretion, however, are not the wastes of the digestive process removed during **elimination.** Unabsorbed digestive wastes are not involved in cellular metabolism—the chemical reactions of the body. Technically, they never pass *into* the body's cells or blood but remain outside as they pass down the pipeline of the digestive tract. Metabolic wastes, on the other hand, are substances produced within the body by various chemical processes such as cellular respiration.

The only true excretory products that are removed from the body during elimination are bile pigments and the salts of certain minerals. **Bile pigments** are produced by the liver from the breakdown of old, worn-out red blood cells. These pigment molecules are combined with other substances to form bile, which is carried by a duct from the liver, to the gallbladder (bile's storage pouch), and then to the small intestine. You may recall from Chapter 8 that bile plays a role in the emulsification of fat globules during the digestion of food. The bile pigments leave the body mixed with the digestive wastes and are the cause of the characteristic color of the feces. Because of its role in the excretion of the bile pigments, the liver is considered an organ of excretion. The large intestine is also considered an organ of excretion because cells lining its walls excrete the salts of some minerals such as calcium and iron (Figure 12-1).

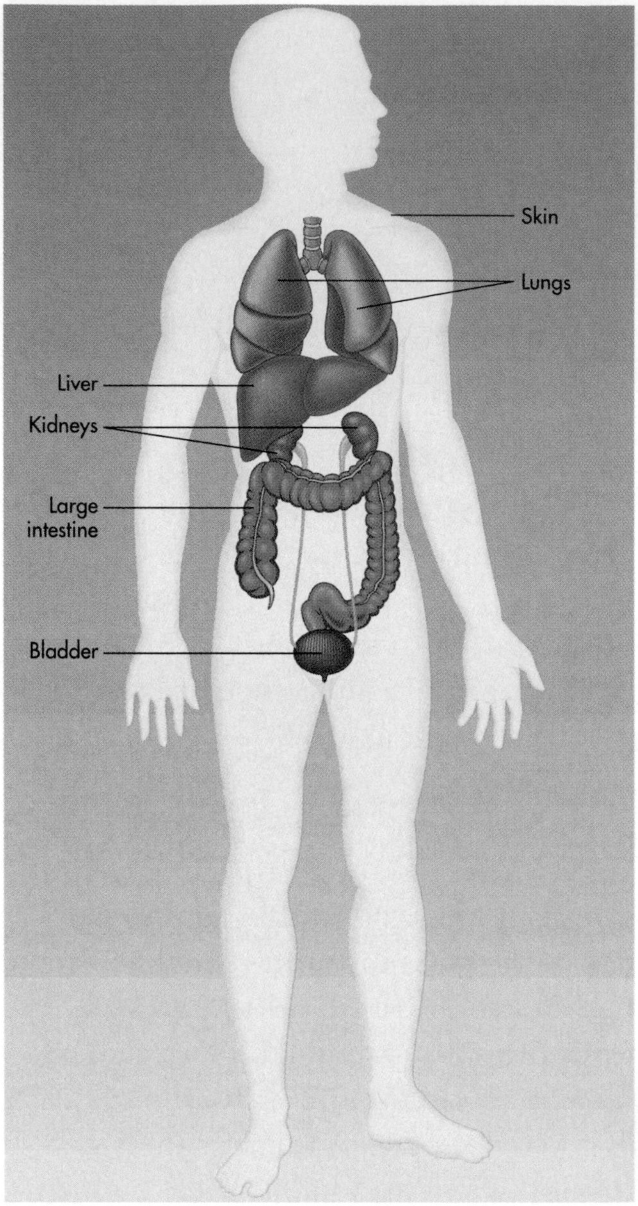

FIGURE 12-1 Organs that carry out excretion. In addition to the kidneys, the skin, lungs, liver, and large intestine all play a role in the excretion of metabolic wastes from the body.

Bile pigments excreted by the liver and the minerals excreted by the large intestine make up only a tiny fraction of the excretory products of the human body. The major products of excretion are water, salts (primarily sodium chloride [NaCl]—table salt), carbon dioxide, and nitrogen-containing molecules. Carbon dioxide (CO_2) is a gas produced during cellular respiration and is excreted primarily by the lungs (see Chapter 9). The lungs also excrete water in the form of water vapor—which is the reason you can see your breath on a cold day as the water vapor condenses to droplets. A small amount of salt and often a large amount of water also leave the body by means of glands in the skin (see Chapter 16).

Nitrogen-containing molecules, or **nitrogenous wastes,** are produced from the breakdown of proteins and nucleic acids. As discussed in Chapter 8, humans need to take in protein as a source of amino acids. These amino acids are used to construct body-building proteins. The liver breaks down any extra amino acids you may eat—those not needed for building new body parts or repairing old ones. In the liver, enzymes break down these amino acids by removing their amino groups ($-NH_2$), a process called *deamination.* The molecules that result can then be used to supply the body with energy. The amino group is a leftover that cannot be used.

> *The primary metabolic waste products of the human body are carbon dioxide, water, salts, and nitrogen-containing molecules.*

In the liver, amino groups ($-NH_2$) are chemically converted to ammonia (NH_3). Ammonia is quite toxic, or poisonous, to all cells, and the body must get rid of it quickly, not allowing its concentration to increase. A bit of chemical reshuffling takes care of this problem, as the ammonia combines with carbon dioxide to form **urea,** the primary excretion product from the deamination of amino acids. In fact, approximately 90% of all nitrogenous wastes are eliminated from the human body in the form of urea. Two other nitrogenous wastes found in small amounts in the urine are uric acid and creatinine. **Uric acid** is formed from the breakdown of nucleic acids (DNA and RNA) found in the cells of the food you eat and from the metabolic turnover of your nucleic acids and adenosine triphosphate (ATP). **Creatinine** is derived primarily from a nitrogen-containing molecule called *creatine* found in muscle cells.

In humans, most nitrogenous (nitrogen-containing) waste products are excreted as urea.

The Organs of Excretion

The skin, liver, and large intestine play a minor role in the excretion of metabolic wastes from the body. The primary organs of excretion are the lungs (see Chapter 9) and the **kidneys** (see Figure 12-1). As mentioned previously, the lungs excrete carbon dioxide and water vapor. The kidneys excrete the ions of salts, such as Na^+ (sodium), K^+ (potassium), Cl^- (chloride), Mg^{2+} (magnesium), and Ca^{2+} (calcium). In addition, they excrete the nitrogenous wastes urea, creatinine, and uric acid, along with small amounts of other substances that may vary depending on diet and general health. These substances, together with water, form the excretion product called **urine.**

The kidneys produce urine by first filtering out most of the molecules dissolved in the blood, then selectively reabsorbing useful components, and finally secreting a few other waste products into this remaining filtrate, which becomes urine. As the kidneys process the blood in this way, they regulate its chemical composition and water content and, in turn, the chemical and fluid environment of the body. For example, although almost no amino acids are removed from the blood by the kidneys, almost half its urea is removed and then excreted. In addition, the kidneys maintain the concentrations of all ions within narrow boundaries. This strict maintenance of specific levels of ion concentrations keeps the blood's pH at a constant value, maintains the proper ion balances for nerve conduction and muscle contraction, and affects the amount of water reabsorbed into the bloodstream from the filtrate.

The primary organs of excretion are the lungs and the kidneys. The kidney is a regulatory as well as an excretory organ.

An Overview of How the Kidney Works

Your body has two kidneys, each about the size of a small fist, located in the lower back region and partially protected by the lower ribs (see Figure 12-1). Each is a living filtration plant that balances the concentrations of water and salts (ions) in the blood and, at the same time, excretes wastes. Put simply, kidneys work in the following way: As blood flows through the millions of microscopic filtering systems, or **nephrons,** within each kidney, the fluid portion of the blood is forced through cap-

excretion (ex **skree** shun) a process whereby metabolic wastes and excess water and salt are removed from the blood and passed out of the body during urination.
elimination a process whereby unabsorbed digestive wastes leave the body during defecation.
bile pigments substances produced by the liver from the breakdown of old,worn-out red blood cells; bile pigments enter the small intestine with the bile and are the cause of the characteristic color of the feces.
nitrogenous wastes (nye **troj** uh nus) nitrogen-containing molecules that are produced as waste products from the body's breakdown of proteins and nucleic acids.
urea (yoo **ree** uh) the primary excretion product from the deamination of amino acids.
uric acid (**yoor** ik) a nitrogenous waste found in small amounts in the urine that is formed from the breakdown of nucleic acids found in the cells of food and from the metabolic turnover of nucleic acids and ATP.
creatinine (kree **at** uh neen) a nitrogenous waste found in small amounts in the urine that is formed from a nitrogen-containing molecule in muscle cells called creatine.
kidneys along with the lungs, the primary organs of excretion. Kidneys excrete the ions of salts and the nitrogenous wastes urea, creatinine, and uric acid, along with small amounts of other waste products.
urine an excretion product composed of water and the substances excreted by the kidneys.
nephrons (**nef** rons) the microscopic filtering systems of the kidneys in which urine is formed.

illary membranes into the nephron tubules. The formed elements of the blood (red blood cells, white blood cells, and platelets), along with most proteins, stay within the blood. Much of the fluid portion of the blood, including ions and molecules other than large proteins, passes through the membrane. The fluid that passes through the membrane is called the **filtrate.** Most of the water, as well as selected ions and molecules, is then reabsorbed from the filtrate back into the blood as the filtrate passes through the rest of the nephron.

The human kidney maintains the proper balance of water in the body, retains substances the body needs, and eliminates metabolic wastes by first filtering most substances out of the blood and then reabsorbing what is needed.

The fluid that passes through the membrane filters of each nephron contains many molecules that are of value to the body, such as glucose, amino acids, and various salts and ions. Humans (and all vertebrates) have evolved a means of selectively reabsorbing these valuable molecules without absorbing the waste molecules that are also dissolved in the filtered waste fluid, or urine. Such selective reabsorption gives the vertebrates great flexibility because the membranes of different groups of animals have evolved specific transport channels that reabsorb a variety of different molecules. This flexibility is a key factor underlying the ability of different vertebrates to function in many diverse environments. They can reabsorb small molecules that are especially valuable in their particular habitats and not absorb wastes. In all vertebrates the kidney carries out the processes of filtration, reabsorption, and excretion. It can function with modifications in fresh water, in the sea, and on land (Figures 12-2 and 12-3).

A Closer Look at How the Kidney Works

The Anatomy of the Kidney

The kidneys look like two, gigantic, reddish-brown kidney beans. Substances enter and leave the kidneys through blood vessels that pierce the kidney near the center of its concave border. If you were to slice a kidney vertically, dividing it into front (ventral) and back (dorsal) portions, you would see an open area called the *renal pelvis* near the concave border. The urine produced by the nephrons within the kidneys is carried by *collecting ducts*

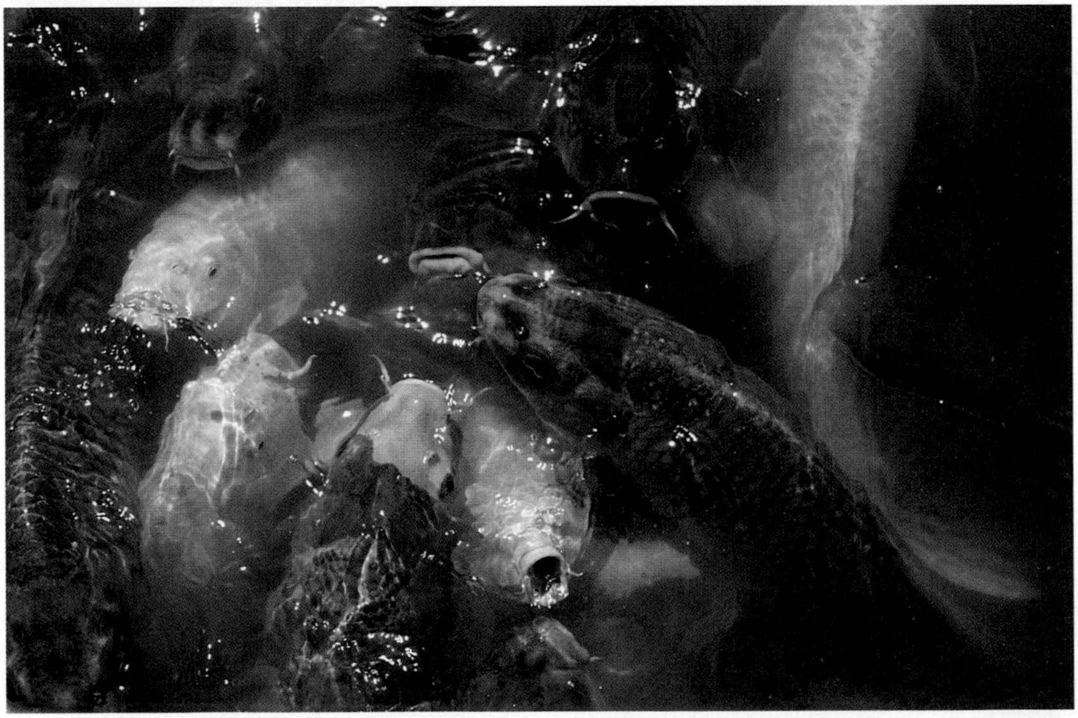

FIGURE 12-2 Freshwater fishes produce large quantities of dilute urine. Because these fish live in fresh water, they take on water by osmosis. To maintain a proper water balance, they drink very little and do not reabsorb water in the nephron.

FIGURE 12-3 Kangaroo rats produce small quantities of concentrated urine. In contrast to freshwater fishes, the kangaroo rat can concentrate urine to a high degree by reabsorbing water. As a result, it avoids losing any more moisture than necessary. Because kangaroo rats live in dry habitats such as deserts, this feature is extremely important to them.

that empty into this space. From here, urine exits the kidney via a tube called a **ureter.**

The kidney itself is made up of outer, reddish tissue called the *cortex* and inner, reddish-brown tissue called the *medulla.* Lined up within the medulla are triangles of tissue called the *renal pyramids* (Figure 12-4). (The word *renal* means "kidney.") The cortex and pyramids are made up of approximately 1 million nephrons, their collecting ducts, and the blood vessels that surround them.

Figure 12-5 shows the plan of the major blood vessels within a kidney. Blood enters each kidney via the right and left *renal arteries.* Branches split off these major arteries, travel up the sides of the pyramids, and meet at the interface of the cortex

filtrate the water and dissolved substances that are filtered out of the blood during the formation of urine.
ureters (**yoor** uh ters) the tubes that carry urine from the kidney to the bladder.

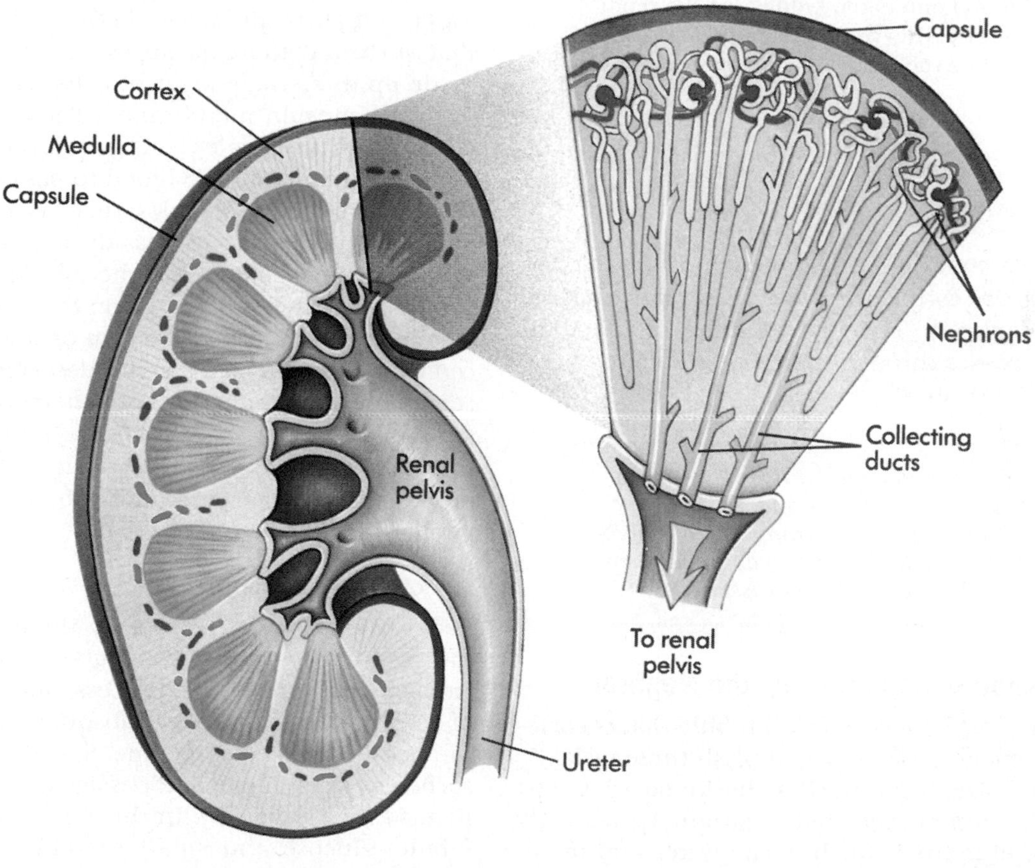

FIGURE 12-4 Structure of the kidney. The enlarged portion of the diaphragm (right) shows a detailed section of a renal pyramid and the cortex above. The renal pyramids are composed of nephrons, their collecting ducts, and surrounding blood vessels.

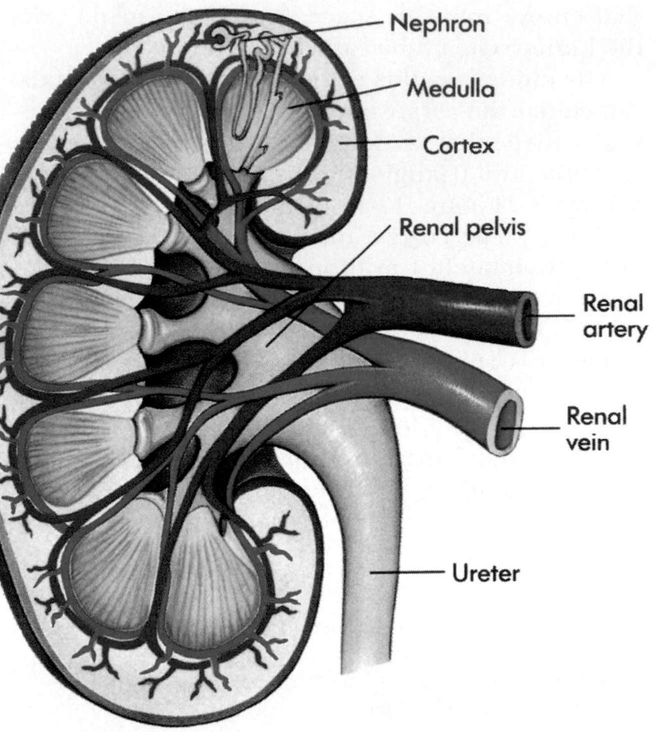

FIGURE 12-5 Plan of the major blood vessels in the kidney. Blood enters the kidney via the renal artery, which splits into branches that travel into the cortex. Blood exits the kidney through the renal vein.

and medulla. Smaller branches extend into the cortex, giving rise to the arterioles that enter each individual nephron. The total volume of blood in your body passes through this network of blood vessels every 5 minutes!

> *The kidney is made up of two main types of tissue: an outer cortex and an inner medulla. These tissues are composed of microscopic filtering units called nephrons, their collecting ducts, and blood vessels.*

The Workhorse of the Kidney: the Nephron

Each nephron (Figure 12-6) is a tubule that is structured in a special way to accomplish three tasks:

1. *Filtration.* During filtration the blood is passed through membranes that separate blood cells and proteins from much of the water and molecules dissolved in the blood. Together the water and dissolved substances are referred to as the *filtrate.*
2. *Selective reabsorption.* During reabsorption, desirable ions and metabolites and most of the water from blood plasma are recaptured from the filtrate, leaving nitrogenous wastes, excess water, and excess salts behind for later elimination. The filtrate is now called *urine.*
3. *Tubular secretion.* During secretion the kidney *adds* materials such as potassium and hydrogen ions, ammonia, and potentially harmful drugs to the filtrate from the blood. These secretions rid the body of certain materials and control the blood pH.

Filtration

At the front end of each nephron tube is a filtration apparatus called *Bowman's capsule.* The capsule is shaped much like a caved-in tennis ball and surrounds a tuft of capillaries called the *glomerulus.* The glomerular capillaries branch off an entering arteriole termed the *afferent arteriole* (named after the Latin verb *affero,* which means "going toward"). The walls of the glomerular capillaries, along with specialized cells of the capsule, act as filtration devices. The pressure of the incoming blood forces the fluid within the blood through the capillary walls and through spaces between capsular cells that surround the capillaries. Capillary walls are made up of a single layer of cells having differentially permeable membranes. These walls do not allow large molecules such as proteins and the formed elements of the blood to pass through. Water and smaller molecules such as glucose, ions, and nitrogenous wastes pass through easily.

This filtrate passes into the tubule of the nephron at Bowman's capsule. From there, it passes into a coiled portion of the nephron called the *proximal convoluted tubule.* This name describes the coiled (convoluted) tubule as that portion closest to (proximal to) Bowman's capsule. Both these structures lie in the cortex of the kidney. As the filtrate passes through the proximal tubule, reabsorption begins.

Selective Reabsorption

Reabsorption is carried out by the epithelial cells throughout the length of the tubule. During this extremely discriminating process, specific amounts of certain substances are reabsorbed depending on the body's needs at the time. Substances are reabsorbed by both active and passive transport mechanisms (see Chapter 3). In the proximal convoluted tubule, glucose and small proteins are put back into the blood. These substances move out of the tubule and into the blood within surrounding capillaries (Figure 12-7).

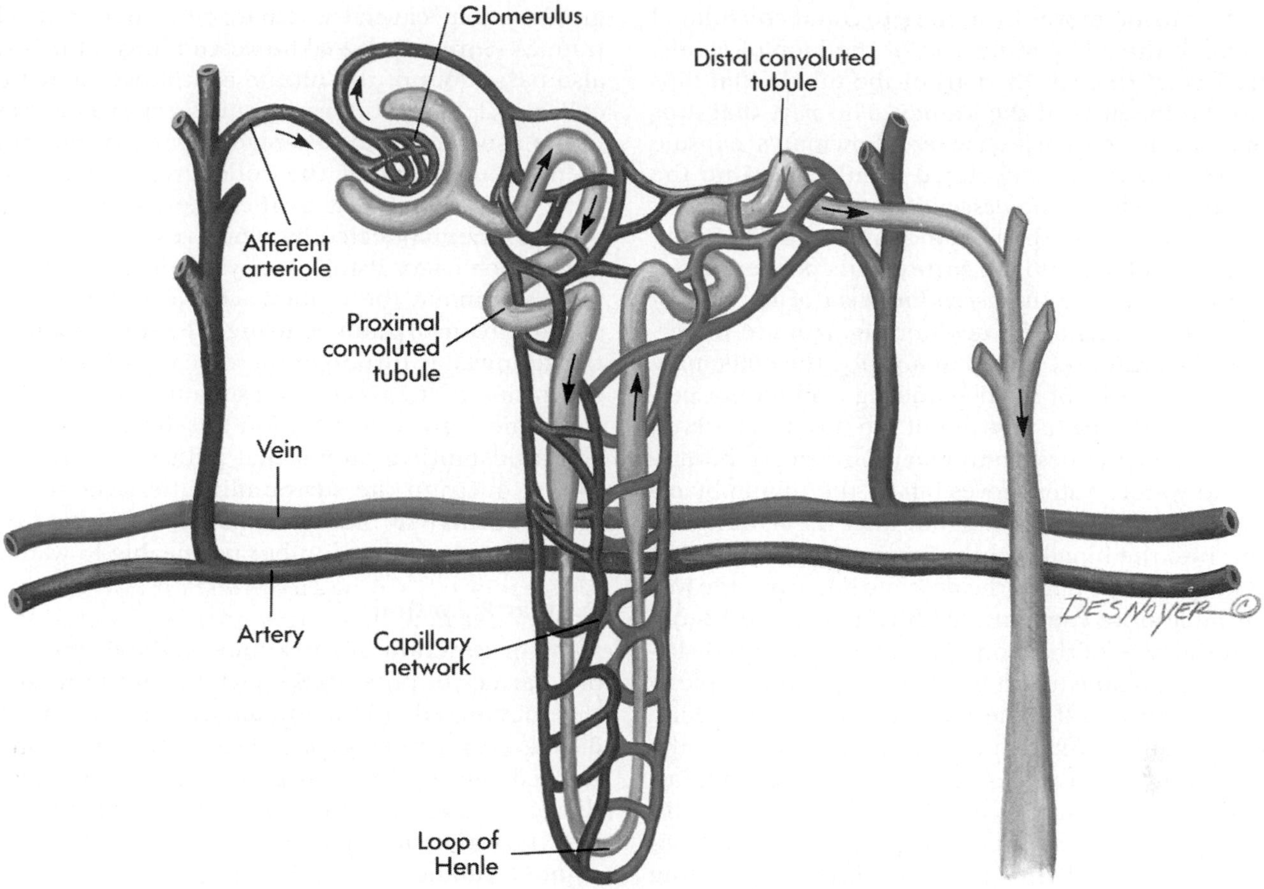

FIGURE 12-6 Structure of a nephron. Part of the nephron is located in the cortex, and part is located in the medulla. Nephrons filter the blood, reabsorb substances to the blood from the blood filtrate, and secrete certain ions and potentially harmful substances into the filtrate. These processes take place along specific portions of the nephron tubule.

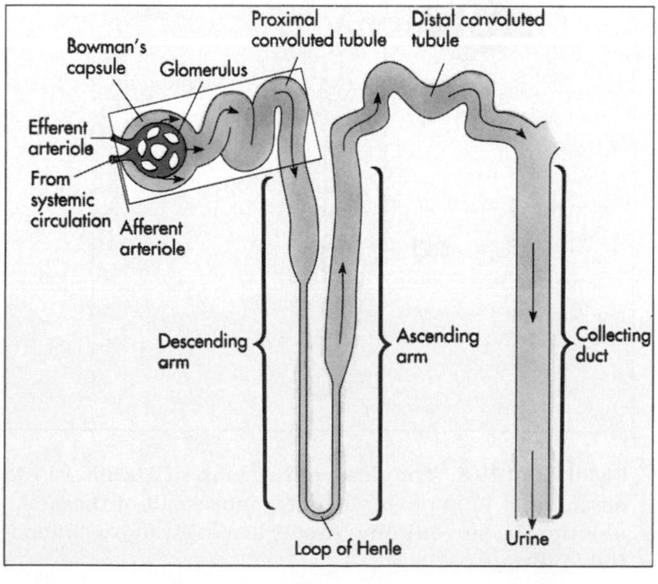

FIGURE 12-7 Path of the filtrate through the nephron. In this figure and the five that follow here and on pages 250-251, the filtrate is traced through the nephron. In the first stage of filtration, the filtrate passes into the nephron at Bowman's capsule. The filtrate then enters the proximal convoluted tubule, where glucose and small proteins are reabsorbed by the blood.

The filtrate moves from the proximal convoluted tubule to the descending arm of the loop of Henle. The *loop of Henle* is the part of the tubule that dips into the medulla of the kidney. The part that dips down from the cortex (where Bowman's capsule and the proximal convoluted tubule are) into the medulla is called the *descending arm of the loop of Henle.* The tubule then extends in the opposite direction, back up into the cortex. This portion of the nephron is called the *ascending arm of the loop of Henle.* The walls of the descending arm are impermeable to salt and urea. Put simply, the cells making up the walls of the descending arm do not permit salt and urea to pass out of the tubule. However, the walls of the descending arm are freely permeable to water. Water moves out of the tubule by osmosis, passing into the surrounding tissue fluid and then into the blood within surrounding capillaries. As water passes out of the descending arm, it leaves behind a more concentrated filtrate (Figure 12-8).

At the turn of the loop, the walls of the tubule become permeable to salt but much less permeable to water (Figure 12-9). Therefore as the filtrate (which became quite concentrated as it moved down the descending arm) passes up the ascending arm, salt passes out into the surrounding tissue fluid by diffusion. This movement of salt produces a high concentration of salt in the tissue fluid surrounding the bottom of the loop.

Higher in the ascending arm, the walls of the tubule contain active transport channels that pump even more salt out of the filtrate within the kidney tubule (Figure 12-10). This active removal of salt from the ascending arm causes water to diffuse outward from the filtrate just above the ascending arm at the *distal convoluted tubule.* The walls of this portion of the kidney tubule are permeable to water, which is not the case in the ascending arm. Left behind in the filtrate are some water and the urea that initially passed through the glomerulus as nitrogenous waste; eventually the urea concentration becomes very high in the tubule.

Finally, the filtrate empties into a *collecting duct* that passes back into the medulla (Figure 12-11). The collecting ducts of all the nephrons bring the urine to the renal pelvis, an open area in which the urine collects before it flows out of the kidney. Unlike other parts of the kidney tubule, the lower part of the collecting duct is permeable to urea. During this final passage, some of the concentrated urea in the filtrate diffuses into the surrounding tissue fluid, which has a lower urea concentration than the filtrate. A high urea concentration in the tissue fluid surrounding the loop of Henle results. This high concentration of urea produces the osmotic gradient that caused water to move out of the filtrate as it passed down the descending arm. Water also passes out of the filtrate as it moves down the collecting duct. This movement occurs as a result of the osmotic gradient created by both the movement of urea out of the collecting duct and the movement of salt out of the ascending arm. The water is then collected by blood vessels in the kidney, which carry it into the systemic circulation.

In summary, the kidney achieves a high degree of water reabsorption by using the salts and urea in the glomerular filtrate to increase the osmotic concentration of the tissue fluid surrounding the nephron. This high concentration of salts and urea creates an osmotic gradient that pulls water from the filtrate out into the surrounding tissue fluid. It is collected there by blood vessels impermeable to the high urea concentration but permeable to water.

Tubular Secretion

A major function of the kidney is the elimination of a variety of potentially harmful substances that you may eat, drink, or inhale or that may be produced during metabolism. The human kidney has evolved the ability to detoxify the blood through the process of tubular secretion. During this process the cells making up the walls of the distal convoluted tubule take substances from the blood within surrounding capillaries and from within the surrounding tissue fluid and put them into the filtrate (urine) within the kidney tubule (Figure 12-12). Ammonia (NH_3) from the deamination of

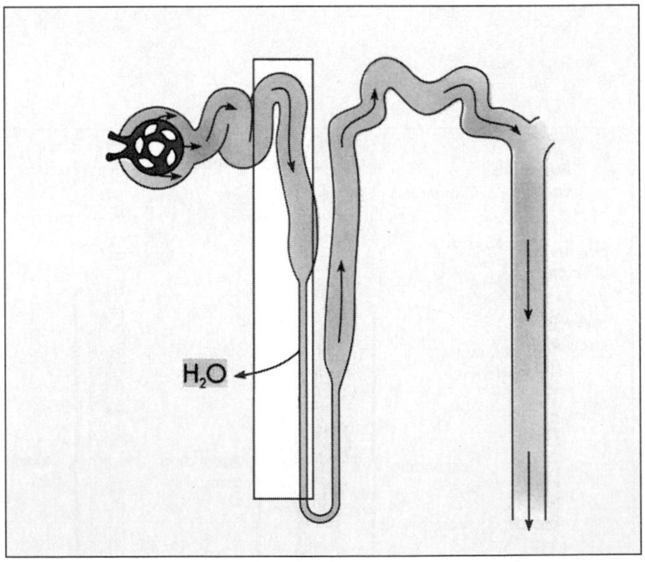

FIGURE 12-8 The descending loop of Henle. In the descending loop of Henle, water moves out of the tubule into the surrounding tissue, leaving a more concentrated filtrate.

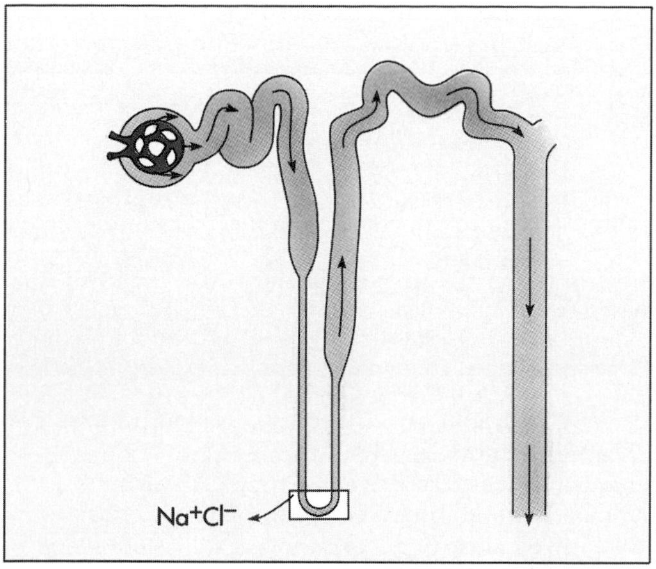

FIGURE 12-9 The turn of the loop of Henle. At the turn of the loop of Henle, the walls of the tubule become more permeable to salt (Na^+ and Cl^- ions) but less permeable to water. As the concentrated filtrate moves through the turn of the loop, salt diffuses out of the tubule into the surrounding tissues. The area surrounding the bottom of the loop then acquires a high concentration of salt.

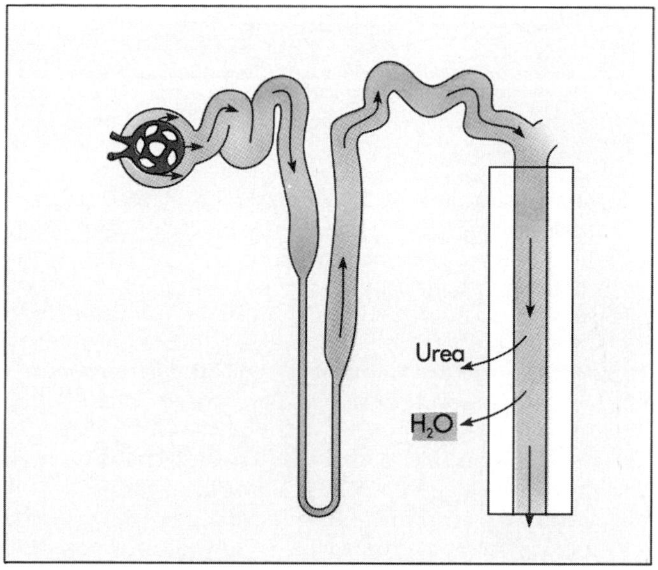

FIGURE 12-11 The collecting duct. In the collecting duct, urea and water pass out of the filtrate. The filtrate, or urine, moves out of the collecting duct to the renal pelvis, in which urine collects before it flows from the kidney.

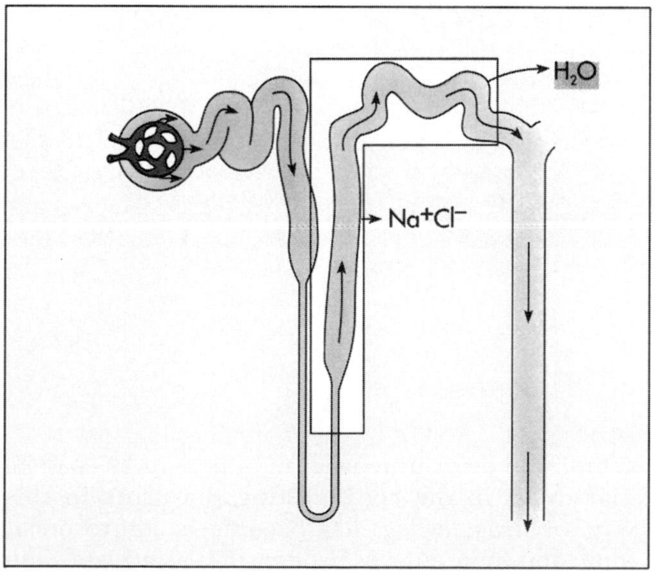

FIGURE 12-10 The ascending loop of Henle. In the ascending loop, channels in the cell membranes of the tubule cells actively pump salt (Na^+ and Cl^-) out of the kidney tubule. At the distal convoluted tubule, water diffuses out of the filtrate. The filtrate now contains only some water and urea.

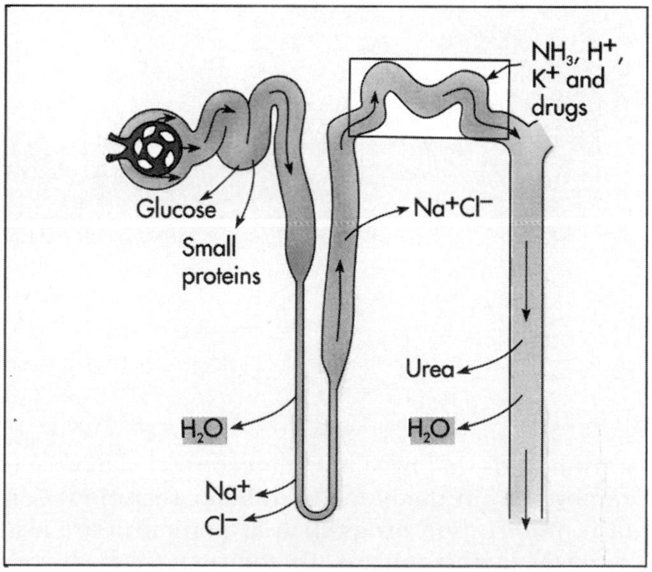

FIGURE 12-12 The distal convoluted tubule. Another important function of the kidney is tubular secretion. The distal convoluted tubule secretes substances such as ammonia, drugs, and ions into the filtrate. This illustration also summarizes the process of selective reabsorption.

JUST WONDERING. . . .

Where I work, they do drug testing. Are these tests accurate? I worry they'll accuse me of something I didn't do.

You are not alone in being tested for drugs at your place of employment. Statistics from the early 1990s show that approximately 5% of businesses of all sizes test their employees for drugs. Large companies tend to follow this practice more than small companies, as do manufacturing, utility, and transportation-related companies. The most common type of screening used in the United States is preemployment screening, although many companies test their employees periodically, randomly, or if they find cause. The purpose for drug testing in the workplace is to prevent and reduce harm or loss to the public, the employer, fellow employees, and employees themselves.

Drug testing was first used in the United States by the Department of Defense in the 1960s and early 1970s to screen servicemen and women returning from the war in Vietnam. In the 1980s, drug testing methods became more reliable, and private companies, especially those concerned about public safety (such as some in the transportation industry), began testing employees. Many laws were passed in the late 1980s that established drug-free workplaces within the federal government and random drug-screening programs in transportation industries such as aviation, rail, mass transit, and trucking.

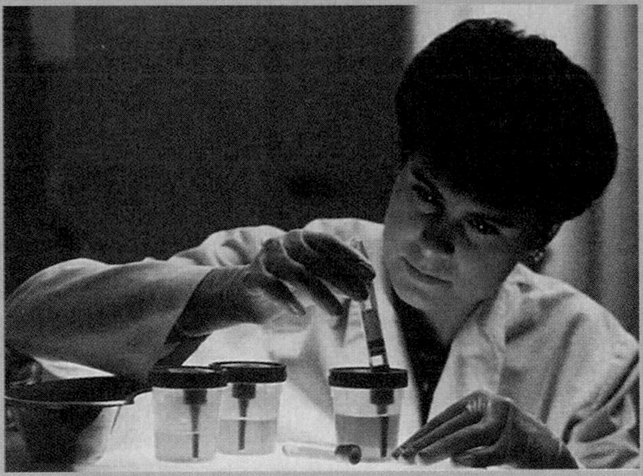

Major improvements have come with the introduction of tests that use recombinant DNA techniques to amplify trace chemicals in urine (see Chapter 31). The enzyme-multiplied immunoassay technique (EMIT) has become the standard approach for screening large numbers of urine samples. EMIT can be used in a robot tester that employs a light sensor to read urine samples and prints out a value for each of five or six drugs present—it can churn out 18,000 results an hour. The detection period depends on the type and dose of the drug. EMIT is more than 98% accurate, with error biased toward nondetection.

Because EMIT responds to a broad range of opiate and amphetamine compounds, it sometimes kicks out a "positive" for harmless prescription drugs such as ibuprofen and decongestants, as well as for foods such as herbal teas and poppy seeds. For this reason, samples that show a positive result on EMIT are then retested by more cumbersome but 100% accurate procedures (gas chromatography–mass spectrometry). For example, eating poppy seeds may trigger a positive signal on EMIT, but the confirmation test tells that a key heroin breakdown product (6-0-acetylmorphine) is not present. So don't worry; the tests are highly reliable and you shouldn't be wrongly accused.

amino acids that have not been converted to urea is removed from the blood by tubular secretion. Certain prescription drugs such as penicillin are also removed by this process. In addition, the body rids itself of harmful drugs such as marijuana, cocaine, heroin, and morphine by tubular secretion. These drugs are processed by the liver, and their breakdown products are secreted into the filtrate within the kidney tubules. Specific tests for the breakdown products of these drugs can be performed on urine samples to determine whether a person has particular drugs in the body. Testing the urine in this way, or *drug testing,* has become a controversial topic among groups of people such as athletes and employees involved in certain types of work, for example, airline pilots.

Tubular secretion can also be thought of as a fine-tuning mechanism. By removing specific amounts of hydrogen ions (H^+) from the blood and secreting them into the filtrate, the kidney can keep the pH of the blood at a constant level (7.35 to 7.45). (See Chapter 2 for an explanation of pH.) Likewise, the potassium ion (K^+) concentration of the blood is fine tuned by tubular secretion. The proper concentration of potassium ions is important to the proper functioning of muscles, including the heart.

A Summary of Urine Formation

The roughly 2 million nephrons that form the bulk of the two human kidneys receive a flow of approximately 2000 liters of blood per day—enough to fill about 20 bathtubs! The nephron first filters the blood, removing most of its water and all but its largest molecules and its cells. The nephron then selectively reabsorbs substances back into the blood. This process is driven by two factors: (1) the development of a high osmostic gradient surrounding the loop of Henle and (2) the varying permeability of the membranes of the cells lining the kidney tubule.

Because of the varying permeability of kidney tubule cells, substances move from the filtrate back into the blood by diffusion and active transport mechanisms at specific places along the length of the tubule. Each substance will be reabsorbed only to a particular threshold level, however, with the rest remaining in the filtrate. For example, glucose is reabsorbed at the proximal convoluted tubule to a threshold level of 150 milligrams per 100 milliliters of blood. Any glucose above this threshold will be excreted in the urine. The amount of glucose in the blood of a healthy person does not normally exceed this limit. However, persons with diabetes mellitus (sugar diabetes) excrete glucose because they fail to produce a hormone called *insulin* that promotes glucose uptake by the cells. Urea, on the other hand, has a very low threshold that is reached quickly. Therefore most urea stays in the filtrate (except the urea that diffuses from the collecting duct, adding to the osmotic gradient surrounding the loop of Henle).

The high osmotic gradient surrounding the loop of Henle causes most of the water filtered from the

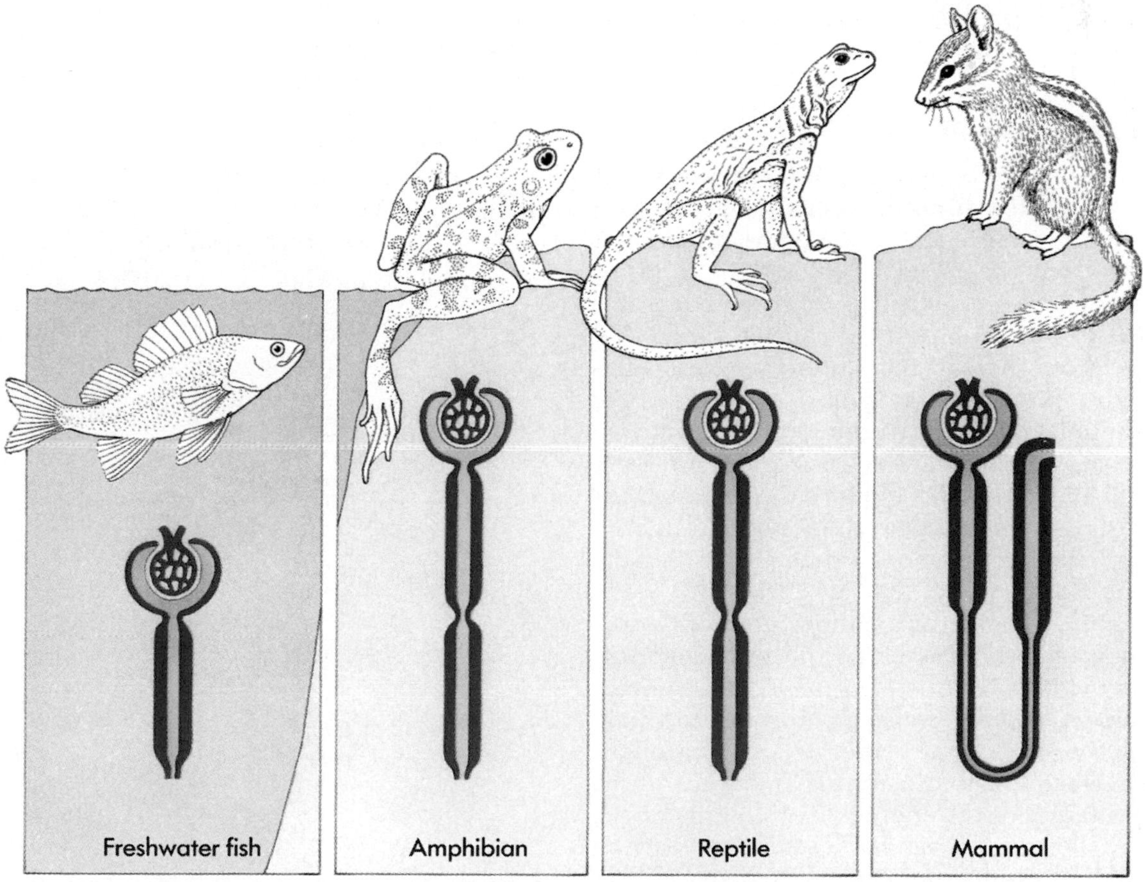

FIGURE 12-13 The function of the loop of Henle is the conservation of water. Animals that do not need to conserve water such as freshwater fishes do not have loops of Henle. Only birds and mammals have a loop in the nephron tubule that allows the reabsorption of water.

blood to be reabsorbed, or conserved, in the descending arm of the loop. Water conservation is, in fact, the job of this loop. Vertebrates that do not need to conserve water have no loop. For example, a freshwater fish drinks little water and produces large amounts of urine. Because the body fluid of a freshwater fish is hyperosmotic (contains more dissolved substances and therefore has a lower concentration of water) to the water in which it lives, water is not reabsorbed in its nephrons. The excess water that enters its body passes instead through the relatively straight nephron tubes to be excreted as urine (Figure 12-13). In general, the longer the loop of Henle, the more water can be reabsorbed. Animals such as desert rodents that have highly concentrated urine have exceptionally long loops of Henle.

The nephron also controls blood pH, fine tunes the concentrations of certain ions and molecules, and removes potentially harmful drugs from the blood by the process of tubular secretion. In these ways the kidney contributes to the chemical and water balance of the body, thus functioning as a major organ of internal equilibrium, or homeostasis. The resulting fluid is urine, a waste that is excreted from the body.

The Urinary System

The kidneys are only one part of the **urinary system**—a set of interconnected organs (Figure 12-14) that not only remove wastes, excess water, and excess ions from the blood but store this fluid, or urine, until it can be expelled from the body. The urine exits each nephron by means of the collecting duct. From there, it flows into the renal pelvis (see Figure 12-4). This area narrows into a tube called the *ureter* that leaves the kidney on its concave border, near where the renal artery and vein enter and exit. By means of muscular contractions (peristalsis), these muscular tubes—one from each kidney—bring the urine to a storage bag called the **urinary bladder.**

The urinary bladder is a hollow muscular organ that, when empty, looks much like a deflated balloon. As the bladder fills with urine, it assumes a pear shape. On the average, the bladder can hold 700 to 800 milliliters of urine, or almost a quart. However, when less than half this amount is in the bladder, special nerve endings in the walls of the bladder, called *stretch receptors,* send a message to the brain that results in the desire to urinate.

The urinary bladder empties into a tube called the **urethra.** This tube leads from the underside of the bladder to the outside. The urethra in men plays a role in the reproductive system, carrying semen to the outside of the body during ejaculation. However, the urinary and reproductive systems have no connection in women. In addition, the urethra in women is much shorter than the urethra in men. For this reason, women contract urinary tract (bladder) infections more easily than men.

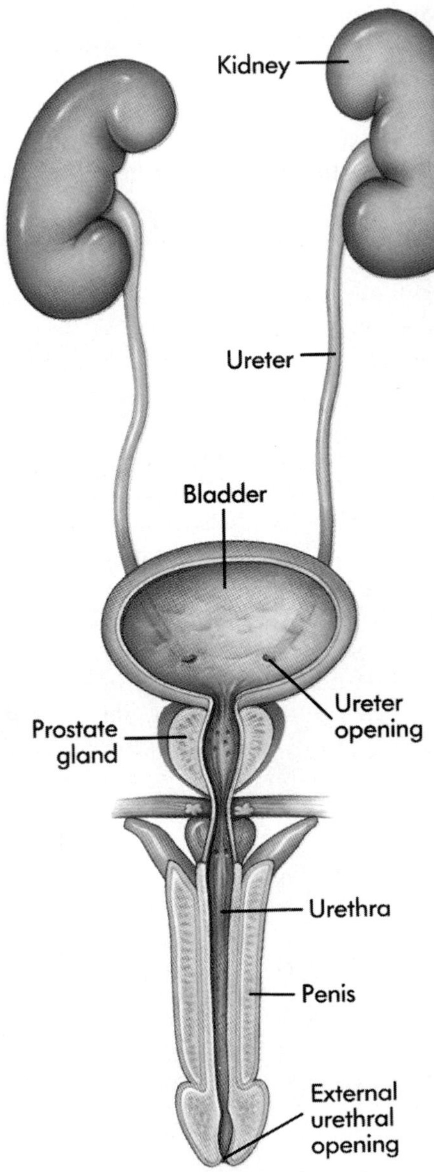

FIGURE 12-14 The human urinary system. The urinary system in males and females includes the kidneys, ureters, urinary bladder, and urethra. This diagram shows the male urinary system.

The Kidney and Homeostasis

The kidney is an excellent example of an organ whose principal function is homeostasis, the maintenance of constant physiological conditions within the body. It is concerned with both water balance and ion balance. To maintain homeostasis, however, the urinary system depends in part on the endocrine (hormone) system.

A *hormone* is a chemical messenger sent by a gland to other cells of the body. Hormones are produced in one part of the body and have an effect in another (see Chapter 17). The hormone that regulates the rate at which water is lost or retained by the body is **antidiuretic hormone (ADH).** It is secreted by the pituitary gland at the base of the brain. Its effect on the excretion of water is easy to remember if you know that a *diuretic* is a substance that helps remove water from the body. Because ADH is an *anti*diuretic, it works to conserve, or retain, water.

ADH works by controlling the permeability of the distal convoluted tubules and collecting ducts to water (Figure 12-15, *A*). When ADH levels increase, the permeability of the collecting ducts to water increases. Water therefore moves out of the ducts by osmosis and back into the blood within surrounding capillaries. When ADH levels decrease, the permeability of the collecting ducts to water decreases. Therefore less water is reabsorbed from the urine (filtrate). Instead, it remains in the ducts and is excreted from the body. Alcohol inhibits the release of ADH, resulting in decreased reabsorption of water and therefore increased urination, which causes dehydration.

Another hormone regulates the level of sodium ions (Na^+) and potassium ions (K^+) in the blood. This hormone is called **aldosterone** (Figure 12-15, *B*). When aldosterone levels increase, the kidney tubule cells increase their reabsorption of sodium ions from the filtrate and decrease their reabsorption of potassium ions. Put simply, aldosterone promotes the retention of sodium and the excretion of potassium. In addition, because the concentration of sodium in the blood affects the reabsorption of water, the increase in sodium in the blood causes water to move by osmosis from the filtrate into the blood.

The principal function of the kidney is homeostasis—the maintenance of a constant internal environment. Two hormones help the kidney maintain homeostasis. The hormone ADH regulates the amount of water reabsorbed at the collecting duct. The hormone aldosterone promotes the conservation of sodium and therefore water and promotes the excretion of potassium.

Problems with Kidney Function

Kidney Stones

During the formation of urine, salts and other wastes to be excreted from the body are normally dissolved in the water of the filtrate and pass with it out of the kidney as urine. Sometimes, however, certain salts do not stay dissolved—most notably calcium salts or uric acid—but instead form crystals called **kidney stones.** Each year more than 300,000 Americans form these stones for reasons such as diminished water intake, diets high in protein and calcium, genetic disorders, infections, and the misuse of medications. Kidney stone attacks are excruciatingly painful; the stone moves through the kidney and then blocks the flow of urine. Small stones—those with diameters less than 5 millimeters (about one fifth of an inch)—pass through the ureters, bladder, and urethra within a few days. Larger stones require certain medical procedures to remove them.

Since 1984, a treatment for kidney stones called *shock-wave lithotripsy* has been available. Before this time, surgery was the only means of relief. Surgery is still used, however, to remove stones larger than 2 centimeters in diameter or stones that are causing infection. During one form of surgery a tube is inserted into the renal pelvis. An ultrasonic probe within the tube first bombards the

urinary system (**yoor** uh **nare** ee **sis** tem) a set of interconnected organs that not only remove wastes, excess water, and excess ions from the blood but store this fluid until it can be expelled from the body.

urinary bladder a hollow muscular organ in the pelvic cavity that acts as a storage pouch for urine.

urethra (yoo **ree** thruh) a muscular tube that brings urine from the urinary bladder to the outside; in men, the urethra also carries semen to the outside of the body during ejaculation.

antidiuretic hormone, or ADH (**anti** dye yoo **ret** ik **hor** mone) the hormone that regulates the rate at which water is lost or retained by the body; it is secreted by the pituitary gland at the base of the brain.

aldosterone a hormone that regulates the level of sodium ions and potassium ions in the blood; it promotes the retention of sodium and the excretion of potassium.

kidney stones crystals of certain salts that can develop in the kidney and block urine flow.

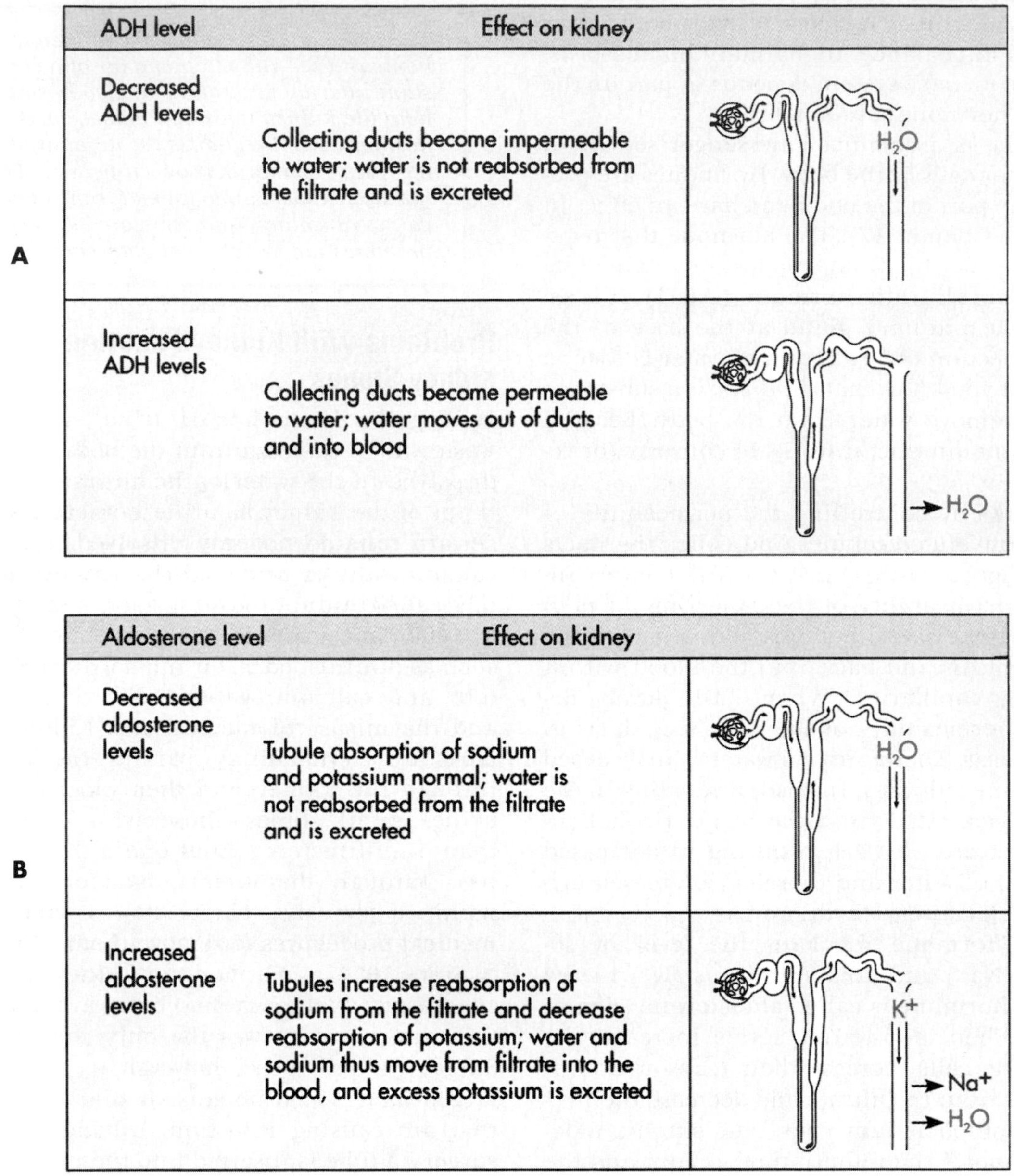

ADH level
Effect on kidney
Decreased ADH levels
Collecting ducts become impermeable to water; water is not reabsorbed from the filtrate and is excreted
H_2O
A
Increased ADH levels
Collecting ducts become permeable to water; water moves out of ducts and into blood
H_2O
Aldosterone level
Effect on kidney
Decreased aldosterone levels
Tubule absorption of sodium and potassium normal; water is not reabsorbed from the filtrate and is excreted
H_2O
B
Increased aldosterone levels
Tubules increase reabsorption of sodium from the filtrate and decrease reabsorption of potassium; water and sodium thus move from filtrate into the blood, and excess potassium is excreted
K^+
Na^+
H_2O

FIGURE 12-15 How hormones regulate kidney function. A, ADH is a hormone that helps the body conserve water by acting on the permeability of the distal convoluted tubules and collecting ducts to water. When ADH levels decrease, more water is excreted from the body. When ADH levels increase, more water is conserved. **B,** Aldosterone regulates the levels of sodium and potassium in the body by acting on the permeability of the distal convoluted and collecting tubules to these ions and water. When Na^+ levels are low and K^+ levels are high, aldosterone acts on the nephron tubules to correct the imbalance. Aldosterone causes Na^+ and water to be reabsorbed in the blood and excess K^+ to be excreted.

stone with sonic waves, crushing it. Tiny forceps then protrude from the tube and are used to collect the pieces of the stone. In another type of surgical procedure the kidney is cut open and forceps are used to remove the stones.

Shock-wave lithotripsy involves a procedure quite different from these two. The patient is lowered up to the neck into a tank of water. Guided by x-ray monitors, intense sound waves are directed at the stone, shattering it. Once the stone is broken apart, the pieces are excreted in the urine.

> *Kidney stones are crystals of certain salts that develop in the kidney and block urine flow. They may be surgically removed or broken apart with shock-wave therapy.*

Renal Failure

Renal failure occurs when the filtration of the blood at the glomerulus either slows or stops. In acute renal failure, filtration stops suddenly. Acute renal failure can have many causes, such as a decreased flow of blood through the kidneys as a result of problems with the heart or blockage of a blood vessel, damage to the kidney by disease, or the presence of a kidney stone blocking urine flow.

In chronic renal failure, the filtration of the blood at the glomerulus slows gradually. Unfortunately, this condition is usually irreversible because it is most commonly caused by injury to the glomerulus. These injuries have many causes, such as the deposit of toxins, bacterial cell walls, or molecules produced by the immune system within the glomerulus; the coagulation of the blood within the glomerulus; or the presence of a disease such as diabetes.

Treatments for Renal Failure

If the kidneys become unable to excrete nitrogenous wastes, regulate the pH of the blood, and regulate the ion concentration of the blood because of renal failure, the individual will die unless the blood is filtered. This job is accomplished by a machine called an *artificial kidney.* The process of filtering blood in this way is called **dialysis.** During dialysis, small amounts of the patient's blood are pumped through tubes to one side of a selectively permeable membrane (Figure 12-16). This membrane is a derivative of cellulose, very similar to material that makes up the cell walls of plants. On the other side of the membrane is a fluid called the *dialysate.* The dialysate contains the same concentration of ions as that normally found in the bloodstream. Because small molecules can pass across the membrane, any extra ions in the patient's blood move by diffusion into the dialysate until their concentrations on both sides of the membrane are equal. In addition, the dialysate contains no wastes, so the wastes in the patient's blood also diffuse into the dialysate.

Unfortunately, dialysis is a slow process, usually taking 6 to 8 hours. Recently, however, researchers have created a new synthetic membrane made of a plastic copolymer said to be 25% more efficient than cellulose membranes. Other scientists are also currently developing tailor-made artificial cells to replace or help kidney cells. The artificial cells currently being developed are designed to change the toxic urea and ammonia that build up in patients with renal failure into useful amino acids.

Each year in the United States, approximately 7000 patients with renal failure receive kidney transplants. During a transplantation operation a patient's nonfunctioning kidney is removed and a donor kidney is implanted. Although two thirds of these transplant patients suffer serious problems of organ rejection, many of their problems are alleviated with the use of immunosuppressive drugs. Recently, treating patients with monoclonal antibody therapy has been shown to reverse rejection in a significant number of patients.

Monoclonal antibodies are special proteins produced in the laboratory that are all descendants of a single cell. These proteins are therefore all alike. In addition, they are antibodies and will attack a specific antigen (see Chapter 11). The monoclonal antibodies developed for use with patients undergoing kidney transplantation target the specific immune cells involved in the rejection of the donor kidney. The use of monoclonal antibodies and new drug therapies hold much promise for recipients of kidney transplants.

> *Renal failure is a reduction in the filtration rate of blood in the glomerulus. This condition has many possible causes and may be treated by dialysis or kidney transplantation.*

renal failure a disease that occurs when the filtration of the blood at the glomerulus either slows or stops; in acute renal failure, filtration stops suddenly, while in chronic renal failure, the filtration of blood at the glomerulus slows gradually.

dialysis (dye **al** uh sis) a method of treating renal failure in which blood is filtered through a machine called an *artificial kidney.*

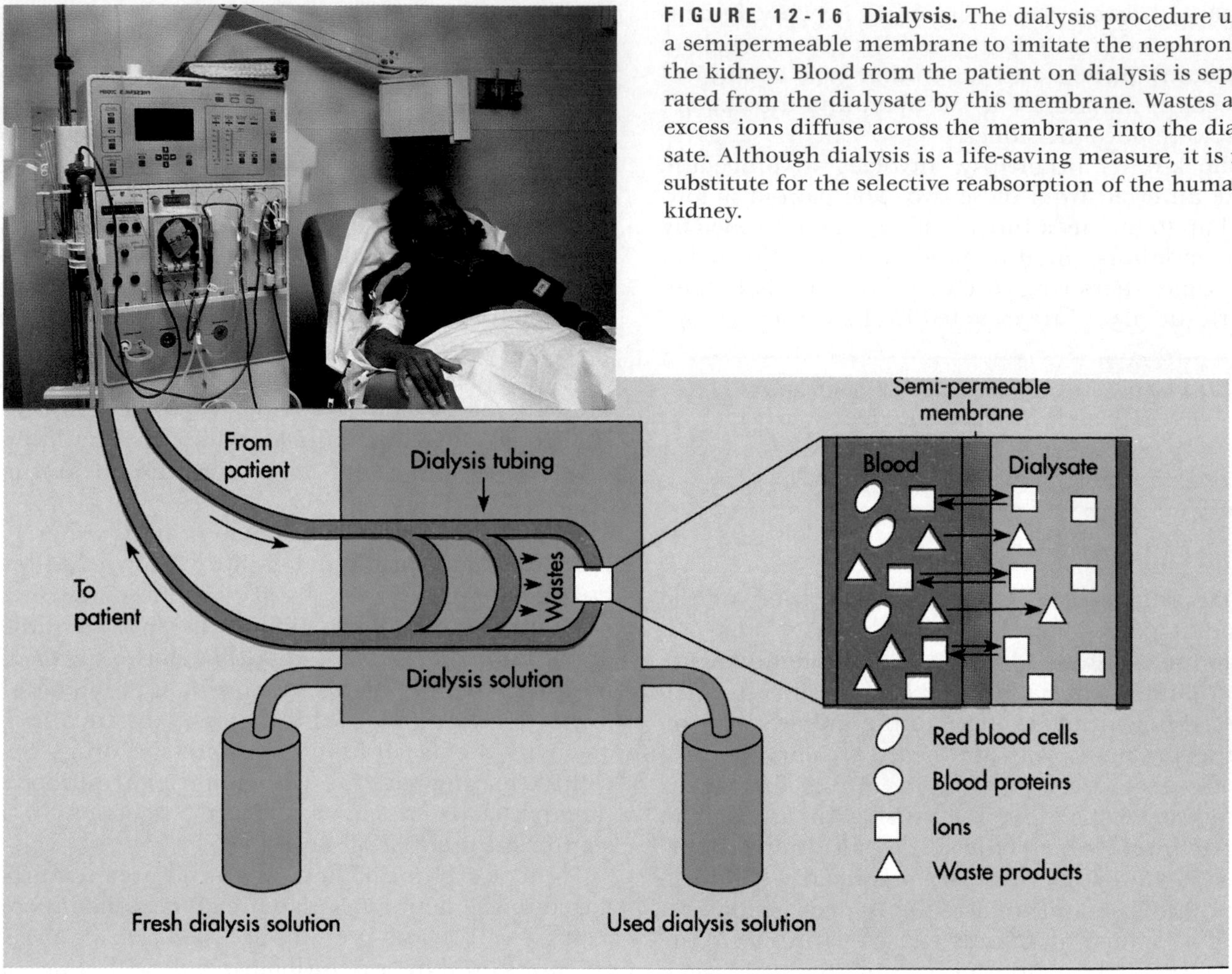

FIGURE 12-16 Dialysis. The dialysis procedure uses a semipermeable membrane to imitate the nephron of the kidney. Blood from the patient on dialysis is separated from the dialysate by this membrane. Wastes and excess ions diffuse across the membrane into the dialysate. Although dialysis is a life-saving measure, it is no substitute for the selective reabsorption of the human kidney.

Summary

- Excretion is a process whereby metabolic wastes, excess water, and excess salts are removed from the blood and passed out of the body. Elimination is a process that takes place as digestive wastes leave the body during defecation.
- The excretory products of the body are bile pigments, nitrogen-containing molecules (nitrogenous wastes), carbon dioxide, water, and salts. The bile pigments are excreted by the liver and passed out of the body by means of the digestive system. Most carbon dioxide is excreted by the lungs. The kidneys excrete most of the nitrogenous wastes along with excess water and salts.
- The kidneys excrete wastes as urine. Urine is formed within microscopic tubular units of the kidney called *nephrons.* During the formation of urine, most of the water and other small molecules are first filtered out of the blood. These substances are called the *filtrate.*
- As the filtrate flows through the nephron tubule, substances are selected for reabsorption into the blood within surrounding capillaries. Various substances are reabsorbed along sections of the tubule because the permeability of its walls varies along its length. In addition, most of the water removed from the blood is reabsorbed from the filtrate because of a high osmotic gradient that surrounds certain sections of the tubule.
- Different groups of vertebrates can function in a wide variety of environments because of adaptations in their nephrons that allow selective reabsorption of molecules valuable to their particular habitats.

- By means of tubular secretion, the kidney also excretes the breakdown products of a variety of potentially harmful substances from the blood, such as marijuana, cocaine, heroin, morphine, and prescription drugs. The breakdown products of these substances can be detected in the urine.
- The kidney is made up of an outer region called the *cortex* and an inner region called the *medulla*. Certain portions of the nephron lie in the cortex, and another portion dips into and out of the medulla. Because of the high osmotic gradient created within the medulla, the reabsorption of water takes place there.
- Urine leaves the kidneys by means of muscular tubes called *ureters*. It is conveyed to a storage pouch called the *urinary bladder*. A tube called the *urethra* brings urine to the outside. These interconnected organs make up the urinary system.
- The principal function of the kidney is homeostasis. Two hormones help the kidneys control the balance of water and ions in the blood, thus helping maintain homeostasis.
- Crystals of salts that sometimes form within the kidney are called *kidney stones*. If these stones do not pass out of the body on their own, they may need to be removed. Surgery and shockwave lithotripsy are two ways in which kidney stones are removed.
- When the filtration of the blood at the glomerulus is seriously impaired, the blood must be filtered by means of dialysis. In severe cases of renal failure, a patient may require kidney transplantation.

Knowledge and Comprehension Questions

1. What is excretion? Explain its importance to life.
2. What are the primary metabolic waste products that you excrete?
3. What do urea, urine, and uric acid have in common? How do they differ?
4. The kidneys are integrally linked to blood pressure control in the body by their regulation of water and dissolved salts. What part of the blood contains these components?
5. How are our muscles dependent on the kidney for proper function?
6. Draw a diagram of a human kidney showing the location of the renal pelvis, collecting ducts, ureter, cortex, medulla, renal pyramids, and renal arteries.
7. Summarize how your kidneys help your body conserve water.
8. Explain why urine can reveal the use of certain drugs.
9. Briefly summarize the process by which urine is formed.
10. What is homeostasis? Explain how ADH and aldosterone are important in maintaining homeostasis.
11. Describe two problems that can occur with kidney function.
12. The text states that more than two thirds of kidney transplantations are somewhat unsuccessful because of organ rejection. What system is involved in such rejection

Critical Thinking

1. Where do you think the carbon dioxide used in the formation of urea comes from? Where does the remainder of excess carbon dioxide go to be excreted?
2. Imagine you are the sole survivor of a ship that sank in the ocean, and you have managed to crawl onto a raft containing nothing but bottles of whiskey. After several days, you are very thirsty. Without water you may die. Should you drink the whiskey or the ocean water, or neither?

CHAPTER 13

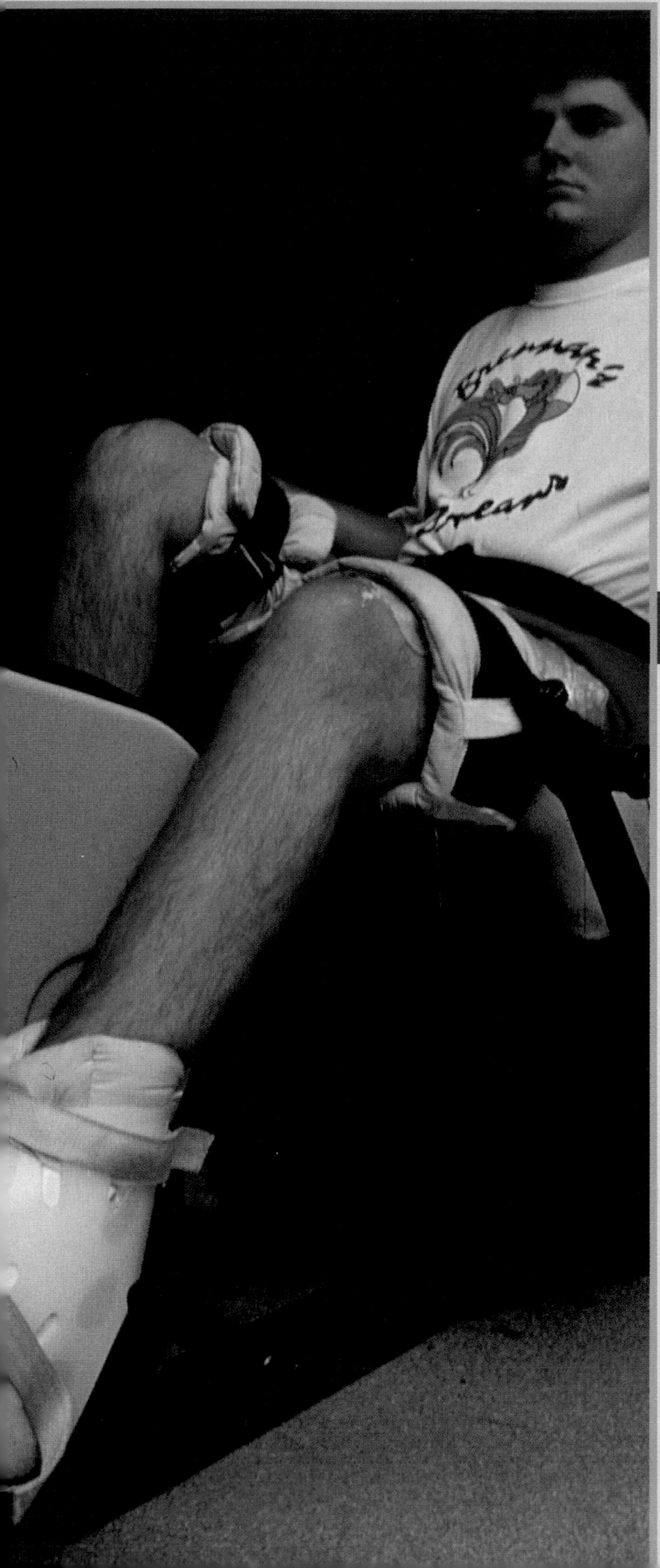

NERVE CELLS AND HOW THEY TRANSMIT INFORMATION

YOU WON'T FIND this machine among the Nautilus equipment at your local health club. It wasn't designed for physical fitness buffs but for paraplegics and quadriplegics—people who cannot move their legs. Impossible as it sounds, this machine can take over the job of nerves that tell leg muscles to move.

Nerves bring messages from the brain to other parts of the body by means of electrical impulses that sweep along their membranes. This stationary bicycle with its attached computer mimics this nervous activity. The computer generates pulses of electrical current and sends them to the patient's hamstrings, quadriceps, and buttocks by means of electrodes. These electrical pulses "fire" each leg muscle for an instant and in a specific order. The result is what you see in the photo—a pedaling motion that helps keep a patient's muscles, metabolism, and even mood in shape.

In the past decade, scientists have discovered that cells in the peripheral nervous system—those outside the spinal cord and brain—are capable of limited regeneration. Some researchers are now experimenting with ways to enhance this process. For example, researchers in California have designed computer chips riddled with microscopic holes through which they are trying to get the severed ends of nerve cells to grow. They hope that the chips will not only help cut nerve cells reconnect but also relay messages along damaged nerve cells.

Scientists have had a more difficult time finding ways to encourage cells in the central nervous system (the brain and spinal cord) to repair themselves. Within the past decade, researchers have learned ways to stimulate nerve cell growth and regeneration in the brains of rats. Unfortunately, these techniques do not work in humans; the regeneration of human brain and spinal cord tissue is merely science fiction at this time. However, science and technology are entering arenas in which many paralysis victims, like the one in the photo, may be able to take a walk—or pedal—in ways they never before thought possible.

KEY CONCEPTS

- The tissues within your body communicate with one another by means of slow messengers called hormones and fast messengers called nerve impulses.
- Nerve impulses are electrical signals transmitted along the membranes of nerve cells, or neurons.
- The nerve impulse is initiated by a stimulus and is spread because of electrical and chemical forces at work along the length of the neuron.
- Nerve impulses are transmitted to other neurons, organs, or muscles by means of a variety of chemicals called neurotransmitters; drugs affect body functions by interfering with the normal activity of neurotransmitters.

OUTLINE

The Communication Systems of Your Body

The cells of your body communicate with one another in several ways. Your nervous system is only one means of communication. A simple way in which cells communicate with one another is by direct contact: open channels between adjacent cells allow ions and small molecules to pass freely from one cell to another. This method of communication allows cells to interact with their neighbors but is too slow and inefficient a method for interactions among cells that are far apart, such as those in your toes and those in your fingers. This process can be likened to running a large corporation with only face-to-face interactions between the people who sit next to each other! It would be better, in terms of distant communication, if managers of the various departments within a corporation sent memos to their staff members, instructing them what to do.

The varied tissues within the body communicate in a similar way by means of chemical instructions called *hormones.* The hormone "memos" are produced by one of several different *endocrine glands,* secreted into the bloodstream, and are carried around the body by the circulatory system. Each hormone, however, interacts only with certain target tissues, just as a memo that is sent through interoffice mail is delivered only to those persons to whom it is addressed. (Hormones are discussed in more detail in Chapter 17.)

Although hormones are an important means of communication within your body, they do not serve all the communication needs. For example, if the message to be delivered to your leg muscles is "Contract quickly, we are about to be hit by a car," hormones are too slow a message system. That is, a person in an emergency does not send a memo to the police but uses the telephone to get help quickly. That, in effect, is what your nervous system does.

The quick message system of your body is the **nervous system.** Your nervous system, as well as the nervous systems of all complex animals, is made up of nerve cells, or **neurons.** Neurons are specialized cells that transmit signals throughout the body (Figure 13-1). Bundled in groups, the long cell extensions of neurons make up nerves. Their message signals are called **nerve impulses.** Like the dots and dashes of Morse code, all nerve impulses are the same, differing only in their frequencies, their points of origin, and their destinations.

The command center of your nervous system is the **brain,** a precisely ordered but complicated maze of interconnected neurons—a large biological computer. The brain is connected by a network of neurons to both the hormone-producing glands and the individual muscles and other tissues. This dual channel of command permits great flexibility: the signals can be slow and persistent (hormones), fast and transient (nerve signals), or any combination of the two.

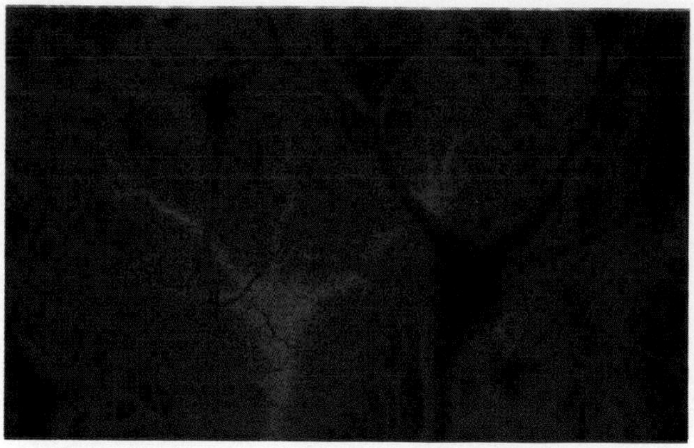

FIGURE 13-1 A network of neurons. Nerve cells, or neurons, transmit nerve impulses throughout the body. Highly specialized, neurons are extremely sensitive to conditions within the body and to external conditions to which the body is subjected.

Two forms of communication integrate and coordinate body functions in humans, as well as in all other vertebrate animals and some invertebrates. Neurons are specialized cells that transmit rapid signals called nerve impulses, reporting information or initiating quick responses in specific tissues. Hormones are chemical messengers that trigger widespread prolonged responses, often in a variety of tissues.

The Nerve Cell, or Neuron

Despite the fact that individual neurons vary widely in size and shape, all neurons have the same basic parts: a cell body and cell extensions called **dendrites** and **axons.** Some neurons have only a few cell extensions; others are bushy with these projections. Many of the neurons in your body, such as those extending from the base of your spinal cord to your toes, have extensions that are a meter or more long!

Figure 13-2 illustrates some of the structural differences that exist among neurons. These differences are intimately related to the varying functions of nerve cells. **Sensory neurons** transmit information to the central nervous system (Figure

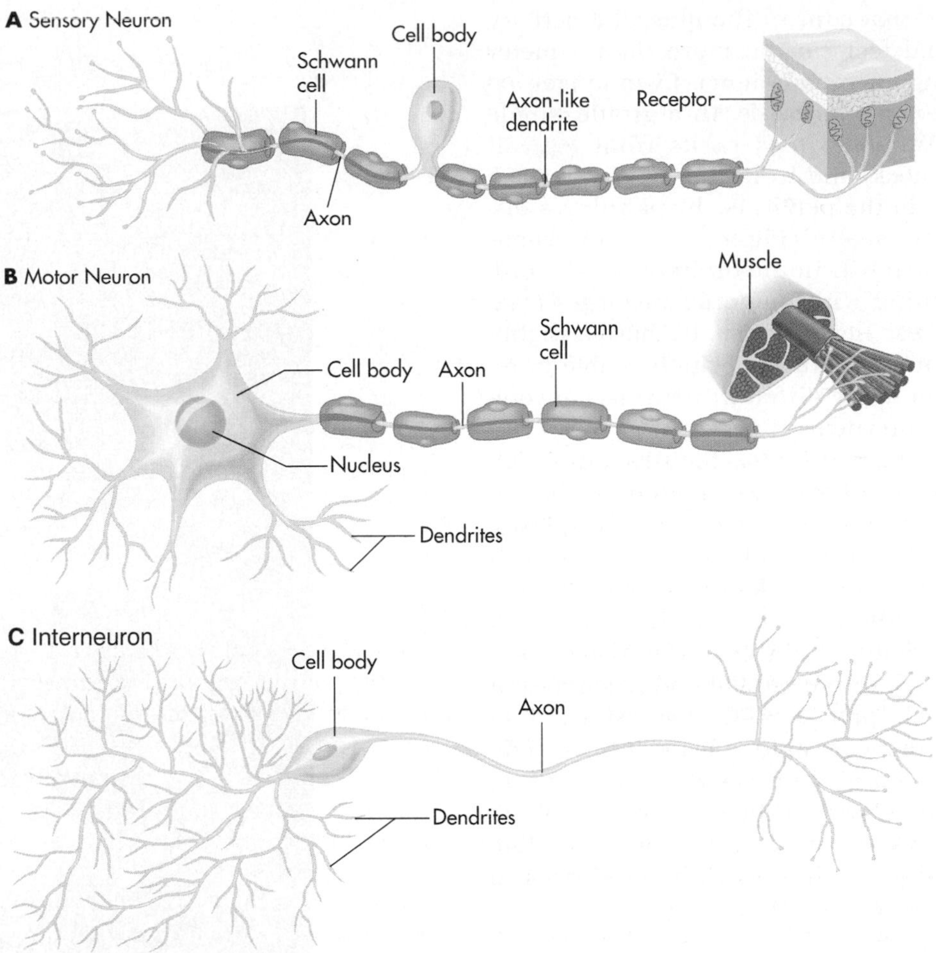

FIGURE 13-2 Types of vertebrate neurons. A, Sensory neurons carry signals from sense organs to the brain and spinal cord. **B,** The axons of motor neurons carry commands from the brain to muscles and glands. **C,** Interneurons within the brain and spinal cord integrate sensory and motor impulses.

13-2, *A*). Each of these neurons has one long dendrite bringing messages from all parts of the body. The cell bodies of sensory neurons lie near the central nervous system. **Motor neurons** transmit commands away from the central nervous system (Figure 13-2, *B*). Each of these neurons has one long axon bringing messages to muscles and glands. The cell bodies of most motor neurons lie in or near the central nervous system. Axons carry messages from the cell body and are usually quite long. Many axons, for example, extend from the spinal cord to muscles or glands. As mentioned previ-

nervous system the quick message system of the body; the nervous system in all complex animals is made up of nerve cells that transmit electrical signals throughout the body.

neurons (**noor** ons) nerve cells.

nerve impulses the electrochemical message signals sent by nerves.

brain the command center of the nervous system; a precisely ordered but complicated maze of interconnected neurons connected to the hormone-producing glands, the muscles, and other tissues.

dendrites projections extending from neuron cell bodies that act as antennae for the reception of nerve impulses and conduct these impulses toward the neuron cell body.

axons (**ak** sons) single projections extending from neuron cell bodies that conduct impulses away from the cell body.

sensory neurons neurons that are specialized to receive information and transmit it to the central nervous system.

motor neurons neurons that are specialized to transmit information from the central nervous system to the muscles and glands.

ously, the axons that control the muscular activity in your legs and feet can be more than a meter long (depending on your height). Even longer axons occur in larger mammals. In a giraffe, single axons extend from the toes in its front legs all the way to the neck, and from the toes in its back legs all the way to the pelvis, each spanning a distance of several meters (Figure 13-3). **Interneurons,** those located within the brain or spinal cord, integrate incoming with outgoing messages (Figure 13-2, *C*). These neurons usually have a highly branched system of dendrites, which is able to receive input from many different neurons converging on a single interneuron.

The *cell body* of a neuron has the structures typical of a eukaryotic cell, such as a membrane-bound nucleus. Surrounding the nucleus is cytoplasm, which contains the various cell organelles. The cell body is responsible for producing substances that are necessary for the nerve cell to live. The rough endoplasmic reticulum, for example, makes proteins that are used for the growth and regeneration of the nerve cell processes, or extensions. Substances such as these are able to leave the cell body and travel down the cell processes of the neuron, riding on "currents" within the cytoplasm or along "tracks" made by microtubules and microfilaments. In addition, some materials travel back to the cell body to be degraded or recycled.

A nerve cell, or neuron, is made up of a cell body that contains a nucleus and other cell organelles and two types of cellular projections: axons and dendrites. Dendrites bring messages to the nerve cell body; axons carry messages from the nerve cell body. Structural differences exist among neurons; these differences relate to functional differences among neurons.

Most neurons do not exist alone but have companion cells nearby. These companion cells are called *neuroglia* and provide nutritional support to neurons. Special neuroglial cells called **Schwann cells** are wrapped around many of the long cell processes of sensory and motor neurons of the peripheral nervous system. Such long cell extensions are nerve fibers. The fatty wrapping created by multiple layers of many Schwann cell membranes is called the **myelin sheath.** The myelin sheath insulates the axon; however, the Schwann cells are wrapped around the axon in such a way that uninsulated spots occur at regular intervals (Fig-

FIGURE 13-3 Some nerve cells are quite long. From the neck of this giraffe, axons of neurons extend all the way to the toes of the front legs—a distance of about 5 meters.

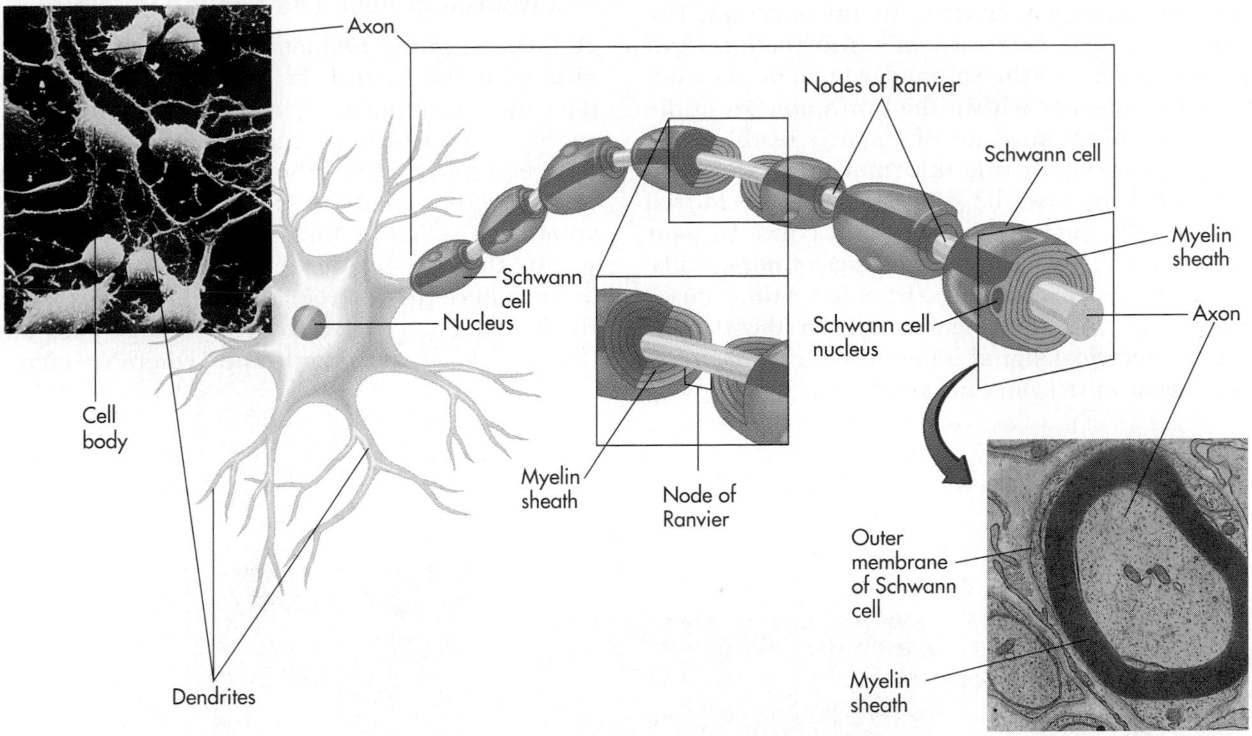

FIGURE 13-4 The insulation of neurons. Long cell extensions of many neurons, particularly the dendrites of sensory neurons and the axons of motor neurons, are encased at intervals by Schwann cells, which form the myelin sheath. The myelin sheath has many layers, visible in the electron micrograph, that insulate the axon or axonlike dendrite. The nerve impulse jumps from uninsulated node to uninsulated node and so moves very rapidly down the nerve fiber.

ure 13-4). These uncovered spots are called **nodes of Ranvier.** These nodes and the myelin sheath (discussed in more detail later in this chapter) create conditions that speed the nerve impulse as it is conducted along the surface of the axon.

The Nerve Impulse

How do nerve cells conduct nerve impulses? The answer is not obvious from examining the amazingly complex network of nerve cells that extends throughout the body. Looking at this network and trying to understand how the nervous system operates is a little like gazing at a telephone wire and trying to understand how it transmits your voice to a receiver far away—nothing can be seen moving along the wires. The key to understanding how your telephone (and your nerve cells) transmits information is an understanding of the abstract concept of electricity.

Electricity is a form of energy—an invisible form—and you can only see, hear, or feel its effects. Electricity is invisible because it is a flow of electrons, subatomic particles so minute that they cannot be visualized even with a high-powered microscope. In a telephone line a flow of electrons carries information from your telephone to the receiver of the person whom you called. The telephone changes the sound energy of your voice into electrical energy, transmits it over the telephone lines, and then changes it back to sound energy when it reaches its destination.

interneurons nerve cells located within the brain or spinal cord that integrate incoming information with outgoing information.

Schwann cells special neuroglial cells that are wrapped around many of the long cell processes of sensory and motor neurons of the peripheral nervous system.

myelin sheath (**mye** uh lin **sheeth**) the fatty wrapping surrounding some axons and long dendrites of sensory neurons created by multiple layers of many Schwann cell membranes; the myelin sheath insulates the nerve fiber.

nodes of Ranvier (**nodes of** ron **vyay**) uninsulated spots on the myelin sheath between two Schwann cells.

Neurons carry information in a similar way. The job of nerve cells is to transmit information from the environment to the spinal cord and brain, from one cell to another within the brain, and from the brain and spinal cord to other parts of the body. Nerve cells transmit this information in the form of electrical signals. The stimulation of specialized receptor cells, such as the rods and cones in your eyes, or specialized receptor endings of nerve cells, such as pressure receptors in your skin, causes electrical signals to be generated in these cells. Once an electrical signal is generated in a receptor, it can travel in the nervous system.

The Neuron at Rest: The Resting Potential

How do neurons conduct electricity? The story begins with the neuron at rest—a neuron not conducting an impulse. A resting neuron, interestingly, is electrically charged. This electrical charge has been measured in the laboratory (by means of microelectrodes) to be approximately −70 millivolts (mV) (Figure 13-5, *A*). The negative charge means that the inside of the cell (near the membrane where the microelectrode is placed) is negatively charged relative to extracellular fluid along the outside of the membrane (where a reference

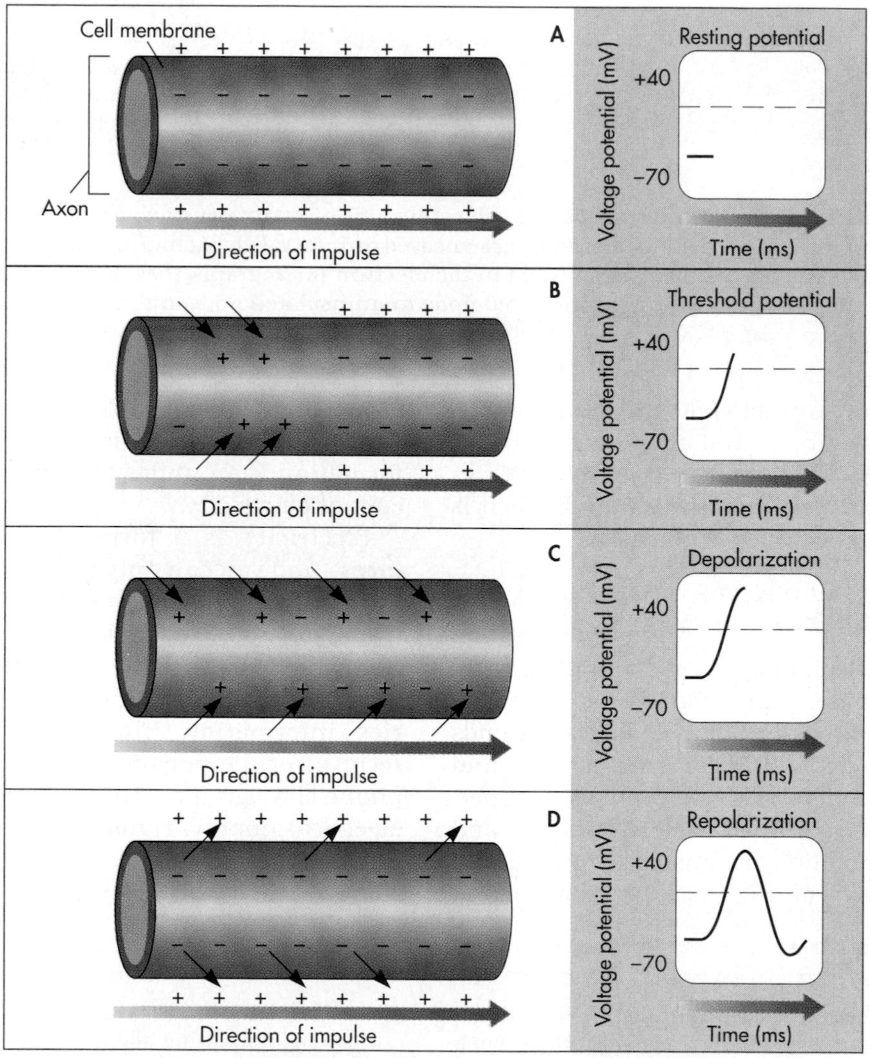

FIGURE 13-5 The stages of membrane depolarization. A, When a neuron is at rest, there is a net negative charge along the inside of the nerve cell membrane of −70 mV. **B,** When a neuron is sufficiently stimulated, Na^+ ions enter the cell, abolishing the voltage difference. **C,** Enough positive ions enter to establish a net interior positive charge. **D,** The exit of K^+ ions then returns the interior to a net negative charge.

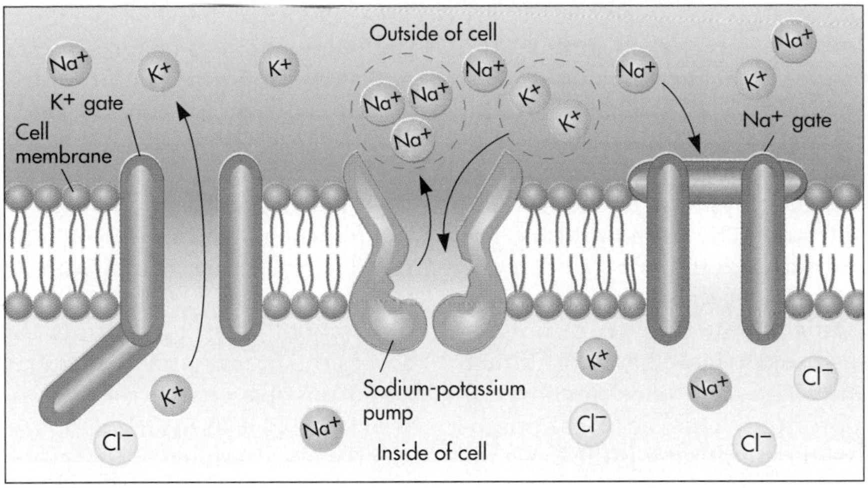

FIGURE 13-6 The sodium-potassium pump helps maintain the resting potential in a neuron. The sodium-potassium pump consists of enzymes that are embedded in the cell membranes of neurons (and other cells). The pump maintains the interior negative charge and exterior positive charge of the neuron as it pumps Na^+ ions outside the cell and K^+ ions inside the cell. However K^+ ions are constantly diffusing back out of the cell and Cl^- ions remain within the cell, blocked from diffusing out because the nerve cell membrane is only slightly permeable to them. Due to these events, the interior of the cell becomes slightly negative. This separation of negative and positive charges across the cell membrane of a neuron is called polarization. Polarization characterizes the resting potential of the neuron.

electrode is placed). A millivolt is one thousandth of a volt, the unit measure of electrical potential. The term *electrical potential* refers to the amount of potential energy created by a separation of positive and negative charges (in this case, along the inside and outside of the cell membrane of the neuron). This potential energy comes from the work that is done to separate these charges. For example, work is necessary to roll a boulder up a hill. Some of the actively working energy (kinetic energy) used to roll the boulder uphill is stored in the boulder (potential energy) as it sits at the top of the hill. In the same way, positive and negative charges that are separated from one another possess potential energy. The energy it takes to separate these oppositely charged particles is released if they rejoin.

The charges that are separated from one another along the nerve cell membrane are carried as ions—atoms with unequal numbers of protons and electrons (see Chapter 2). The ions that play the principal role in the development of the electrical potential along the membrane of the neuron (and that are also important to many other body processes) are sodium ions (Na^+) and potassium ions (K^+).

Embedded within the cell membranes of neurons are enzymes called **sodium-potassium pumps** that actively transport these ions across the cell membrane (see Chapter 3 and Figure 13-6). These transmembrane proteins (enzymes) use the energy stored in molecules of adenosine triphosphate (ATP) to move potassium ions into the neuron at the same time that they move sodium ions out of the neuron. However, the potassium ions can simply diffuse back out of the cell through channels or tunnels open to them. Put simply, potassium ions actively enter the cell through one type of "door" but then exit by another. The situation with the sodium ions is different. These ions *cannot* move back into the cell even though they would normally do so by diffusion. There are no open inward channels in the neuron membrane through which the sodium ions can pass. In other words, they have no open door to get back into the cell. Sodium specific channels exist, but they are closed at −70 mV (the resting potential).

As a result of constantly pumping sodium ions out, the cell lowers the concentration of positively charged sodium ions within its interior. In this way a chemical gradient of sodium ions is established, with a higher concentration of sodium ions outside the cell than inside the cell. Negatively charged

sodium-potassium pumps enzymes embedded within cell membranes of neurons (and other cells) that transport sodium ions out of the cell and potassium ions into the cell.

chloride ions (Cl^-) are attracted to the positively charged sodium but are unable to follow the sodium ions out of the cell because the neuron cell membrane is only slightly permeable to them. Therefore an electrical gradient is also established; the inside of a nerve cell along its membrane is more negatively charged than the outside of the cell along the membrane. The membrane of the neuron is said to be *polarized;* there is a difference in charge on the two sides of the membrane. Resting neurons—ones not conducting an impulse—have this difference in electrical charge on either side of the membrane. The difference in charge is called the **resting potential.** This electrical potential difference across the membrane is the basis for the transmission of signals by nerves.

The action of sodium-potassium transmembrane pumps and ion-specific membrane channels produces conditions that result in the separation of positive and negative ions along the inside and outside of the nerve cell membrane. This separation of charged particles creates an electrical potential difference, or electrical charge, along the membrane of the resting neuron. This electrical potential difference is defined as the resting potential.

Conducting an Impulse: The Action Potential

A neuron transmits a nerve impulse when it is excited by an internal or external environmental change called a *stimulus* (pl., stimuli). Examples of stimuli are pressure, chemical activity, sound, and light. Specialized *receptors* detect stimuli. For example, the rods and cones in the retina of your eye are sensitive to light, pressure receptors in your skin allow you to feel a hug, and certain cells in your nose allow you to smell your favorite dessert baking in the oven.

Stimuli cause a nerve impulse to be transmitted by initiating events that change the electrical potential difference of the receptor cell or nerve cell membrane. This change in electrical potential difference is called *depolarization.* The first event to occur when a neuron is sufficiently stimulated, depolarizing the membrane to a level called the **threshold potential** (see Figure 13-5, *B*), is the opening of the sodium-specific channels in the membrane. You can think of these channels as being operated by electricity. A change in voltage across the nerve cell membrane opens and closes them. What actually happens is that the sudden change in voltage changes the shape of the proteins forming the channels. As their shape changes, these channels become permeable to sodium ions and allow them to diffuse into the cell.

As the sodium-specific channels open, a few sodium ions move rapidly into the cell, diffusing from where there are more sodium ions (outside the cell) to where there are fewer (inside the cell). This permeability for sodium ions across the cell membrane wipes out the local electrochemical gradient (the polarization) and further depolarizes the nerve cell membrane. Amazingly, although only about 1 in 10 million ions actually moves across the membrane, the interior of the cell develops a positive charge relative to the outside of approximately +40 mV (see Figure 13-5, *C*), a 110-mV electrical difference from the −70 mV resting potential. This electrical difference occurs because a new electrical potential develops when the inwardly directed permeability for sodium ions develops across the cell membrane. This rapid change in the membrane's electrical potential is called the **action potential.**

Although it may seem hard to believe, the depolarization of the cell membrane (the action potential) lasts for only *a few thousandths of a second* (milliseconds) because the sodium channels close quickly. They cannot reopen until after the resting potential is reestablished and another depolarization occurs, triggering them again. During this inactive state of the sodium channels, a nerve impulse cannot be conducted. This period, which is only milliseconds long, is called the *refractory period.*

When the sodium channels close, potassium ions move outward as they usually do through potassium channels. Many neurons contain voltage-sensitive potassium channels that open as the membrane depolarizes and the sodium channels shut down. It is this event—the movement of potassium ions out of the cell—that repolarizes the membrane (see Figure 13-5, *D*). The sodium-potassium pump (see Figure 13-6) works to maintain the resting potential on a more long-term basis. The whole process of stimulation, depolarization, and recovery happens in the blink of an eye (Figure 13-7). In fact, 500 such cycles could occur in the time it takes you to say the words "nerve impulse."

Transmission of the Nerve Impulse: A Propagation of the Action Potential

An action potential at one point on the nerve cell membrane is a stimulus to neighboring regions of the cell membrane. The change in membrane potential causes sodium channels to open, depolarizing the adjacent section of membrane. Put simply, depolarization at one site produces depolarization at the next. In this way the initial depolarization passes outward over the membrane, spreading out

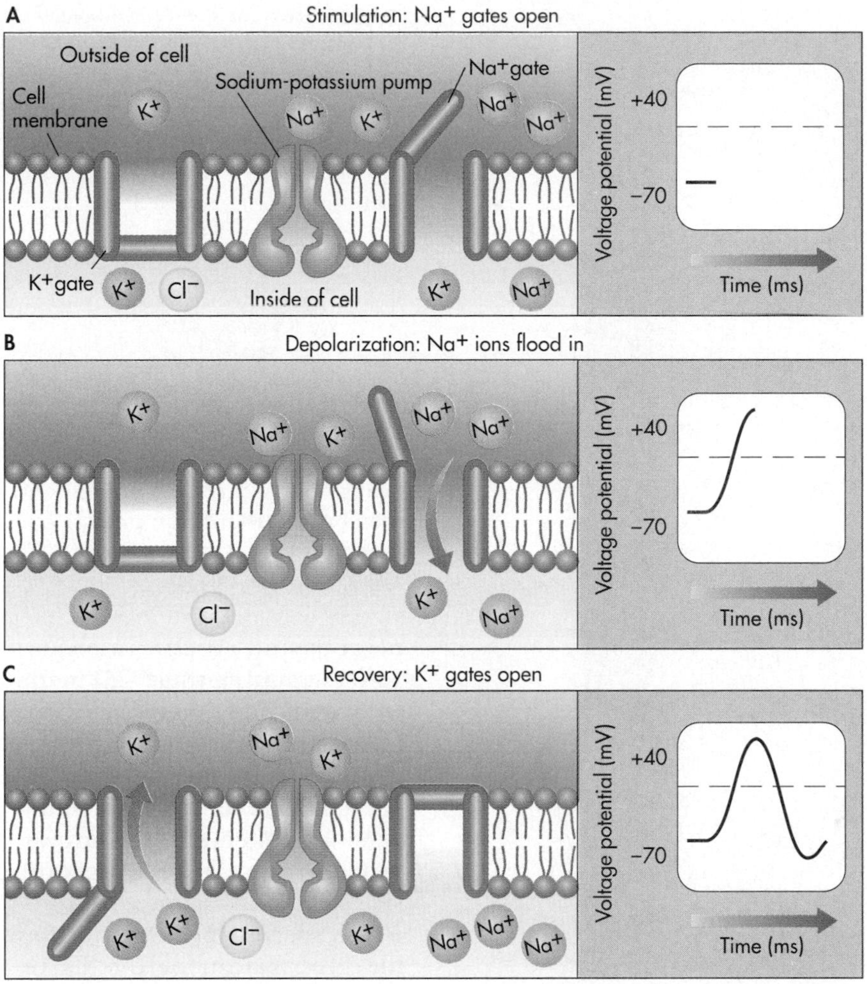

FIGURE 13-7 The action potential. The stages of the action potential are: **A,** Stimulation of a neuron, which causes Na^+ gates in the neuron to open; **B,** Depolarization, in which the Na^+ ions flood into the neuron, depolarizing the electrochemical gradient established during the resting potential; and **C,** Recovery, in which K^+ gates open, allowing K^+ ions to leave, thus repolarizing the membrane.

in all directions from the site of stimulation (Figure 13-8). Like a burning fuse, the signal is usually initiated at one end and travels in one direction, but it would travel out from both directions if it were lit in the middle. The self-propagating wave of depolarization that travels along the nerve cell membrane is the nerve impulse.

> *A nerve impulse arises when a receptor is stimulated. This stimulation results in a rapid change in the electrical potential difference along the nerve cell membrane. This transient disturbance, or depolarization, has electrical consequences to which nearby transmembrane proteins respond, spreading the depolarization in wavelike fashion.*

The nerve impulse is an *all-or-nothing response.* The amount of stimulation applied to the receptor or neuron must be sufficient to open enough sodium channels to generate an action potential. Otherwise, the cell membrane will simply return

resting potential an electrical potential difference, or electrical charge, along the membrane of a resting neuron; it is the basis for the transmission of signals by nerves.

threshold potential a level of nerve membrane depolarization in which sodium-specific channels in the membrane open, allowing sodium ions to diffuse into the cell. The level at which a stimulus results in an action potential along the nerve cell membrane.

action potential the rapid change in a membrane's electrical potential caused by the depolarization of a neuron to a certain threshold.

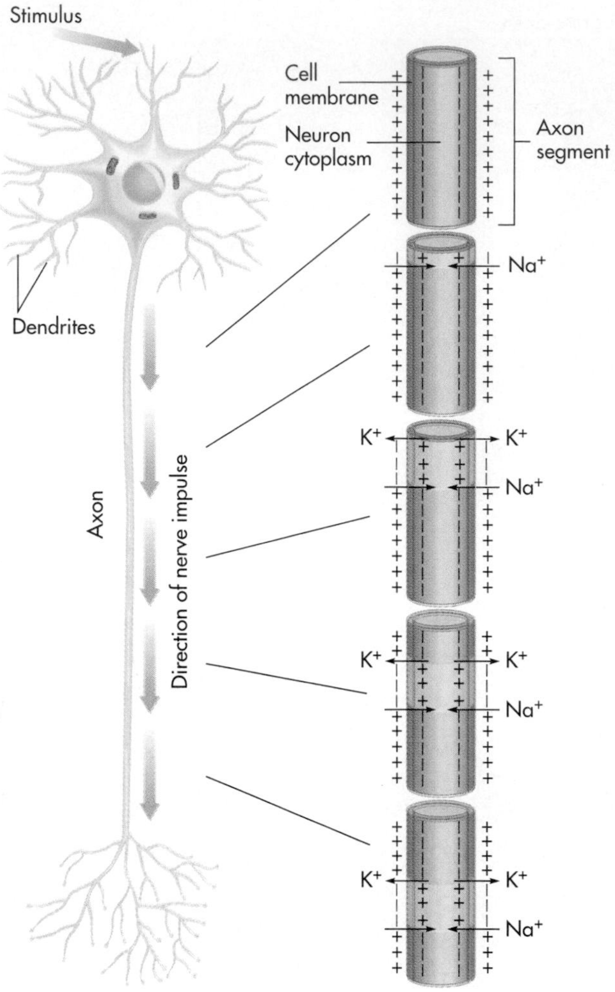

FIGURE 13-8 Transmission of a nerve impulse. Depolarization moves along a neuron in a self-propagating wave. The process of depolarization can be compared with the burning of a fuse because the signal usually starts at one end and travels in one direction. However, if the signal were able to start in the middle of the neuron, depolarization would travel in both directions outward.

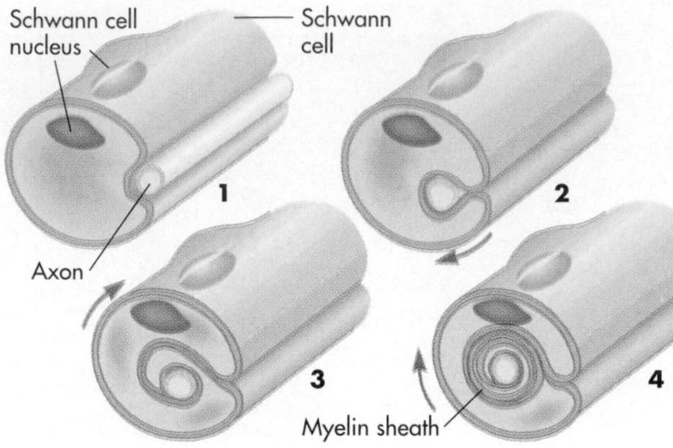

FIGURE 13-9 Schwann cells. Schwann cells wrap around nerve cell processes forming stacks of lipid-rich layers. The progressive growth of the Schwann cell membrane around the process contributes the many membrane layers characteristic of myelin sheaths. These layers act as an excellent electrical insulator.

to the resting potential. For any one neuron the action potential is always the same. But any stimulus intense enough to open a sufficient number of sodium channels will depolarize the membrane to the threshold level and initiate an impulse. The neuron is said to have fired. This wave of depolarization has a constant amplitude (the height or strength of the wave). The speed at which impulses are conducted, however, varies among nerves.

Speedy Neurons: Saltatory Conduction

As mentioned earlier, many neurons have axons and axonlike dendrites that are covered by neuroglial Schwann cells. These cells envelop many of the long cell processes of sensory and motor neurons and certain interneurons as well (Figure 13-9), wrapping their cell membranes around them so many times that they form stacks of lipid-rich layers. In fact, some neurons have as many as 100 membrane layers surrounding them! This "cell wrapper" is the myelin sheath.

Schwann cells wrap around the length of an axon or axonlike dendrite, one after the other, with spaces separating one from the next (Figure 13-10). These spaces, the nodes of Ranvier, are critical to the propagation of the nerve impulse in myelinated cells. The myelin sheath is an insulator; it prevents the transport of ions across the neuron membrane beneath it. However, within the small gaps between Schwann cells, the surface of the axon is exposed to the intercellular fluid surrounding the nerve. An action potential can be generated only at these gaps. In fact, the pumps and channels that move ions across the neuron membrane are concentrated at the nodes of Ranvier, enhancing ion movements at these spots.

The action potential moves along the nerve cell membrane differently in myelinated cells than in unmyelinated cells. The wave of membrane depolarization that travels down the axon of unmyelinated neurons is impossible in myelinated axons due to the myelin sheath, which acts as an electrical insulator. Instead, the action potential jumps from one node to the next (much as mountain climbers rappel down a cliff [Figure 13-11]), causing a depolarization there. This depolarization opens the voltage-sensitive sodium channels at that node, resulting in the production of an action potential. This very fast form of nerve impulse conduction is known as *saltatory conduction,* from the Latin word *saltare,* meaning "to jump."

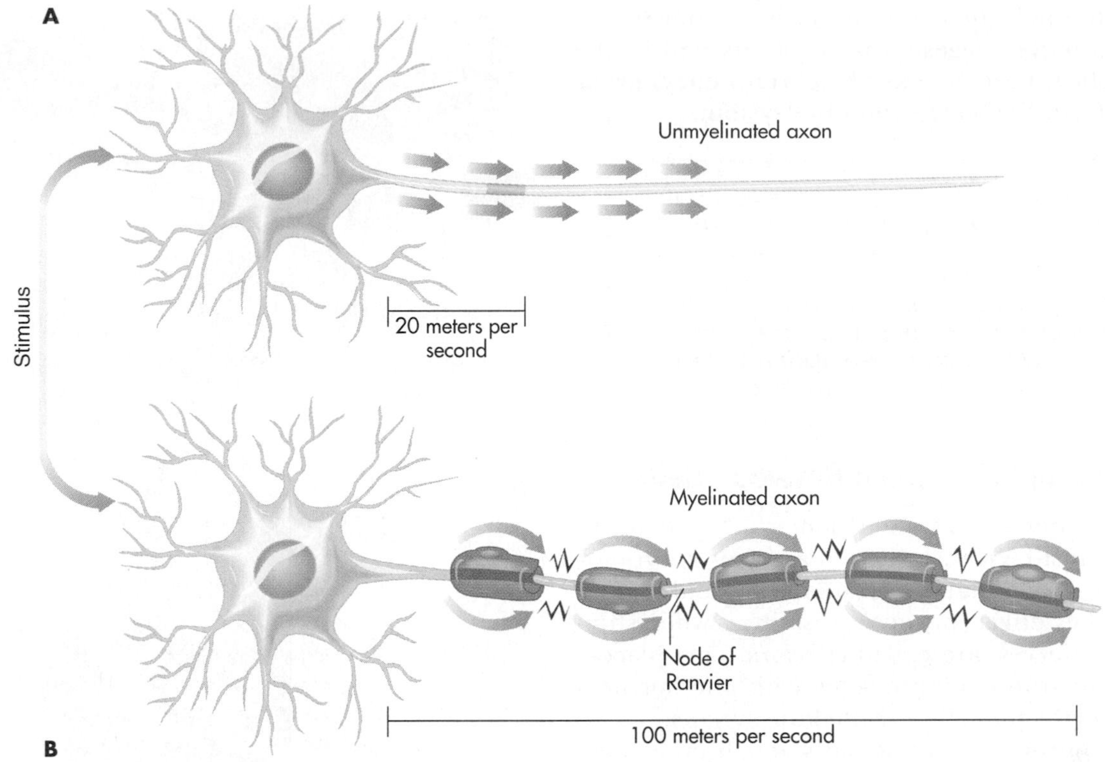

FIGURE 13-10 Saltatory conduction. A, In a fiber without Schwann cells, each portion of the membrane becomes depolarized in turn, like a row of falling dominoes. **B,** In neurons of the same diameter as in **A,** the nerve impulse moves faster along a myelinated fiber because the wave of depolarization jumps from node to node without ever depolarizing the insulated membrane segments between nodes.

FIGURE 13-11 A mountain climber rappeling down a cliff. The way in which a mountain climber rappels down a cliff—pushing off the rock face, landing a distance away, and then pushing off again—is analogous to saltatory condition, the movement of ions from node to node along a myelinated nerve fiber.

Impulses conducted by myelinated neurons travel much faster than impulses conducted by nerve fibers of the same diameter without this insulation. In fact, impulses conducted in large-diameter myelinated neurons travel up to 270 miles per hour (120 meters per second). These myelinated neurons can transmit a signal from your toes to your brain in less than half a second!

Multiple sclerosis (MS), one of the leading causes of serious neurological disease in adults, affects approximately 1 out of every 2000 people in the United States. This disease results in the destruction of large patches of the myelin sheath around neurons of the brain and spinal cord. Left behind are hardened scars (scleroses) that interfere with the transmission of nerve impulses. The resulting slowed transmission of signals in the nervous system results in a gradual loss of motor activity. Although the cause of MS is unknown, evidence strongly suggests that proteins within the myelin sheath are attacked by certain enzymes or by cells from the body's immune system.

> *Not all nerve impulses propagate as a continuous wave of depolarization spreading along the neuronal membrane. Along myelinated neurons, impulses travel by jumping along the membrane, leaping over insulated portions. Impulses travel much more quickly along these myelinated neurons than along unmyelinated ones.*

Transmitting Information Between Cells

When the nerve impulse reaches the end of an axon, it must be transmitted to another neuron or to muscle or glandular tissue. Muscles and glands, because they effect (or cause) responses when stimulated by nerves, are called *effectors*. This place—where a neuron communicates with another neuron or an effector cell—is called the **synapse.**

Most neurons do not actually touch other neurons or cells with which they communicate. Instead, there is a minute space (*billionths* of a meter across) separating these cells (Figure 13-12). This space or gap is called the **synaptic cleft.** The nerve impulse must cross this gap and does so by the direct passage of electrical current or by changing an electrical signal to a chemical signal. Chemical synapses are the prevalent type of synapse in humans (and all vertebrates).

The membrane on the axon side of the synaptic cleft is called the *presynaptic membrane*. In chemical synapses, when a wave of depolarization reaches the presynaptic membrane, it stimulates the release of organic molecules called **neurotransmitters** into the cleft. These molecules are stored in thousands of small, membrane-bound sacs located at the tips of the axon. Each sac contains from 10,000 to 100,000 molecules of neurotransmitter. These chemicals diffuse to the other side of the gap. Once there, they combine with receptor molecules in the *postsynaptic membrane* (associated with either a dendrite or a cell body) of the target cell. When they do, they cause ion channels to open.

> *A synapse is a junction between an axon tip and another cell, usually including a narrow gap separating the two cells. Passage of the impulse across the gap is by an electrical current or, more likely in vertebrates, a chemical signal from the axon.*

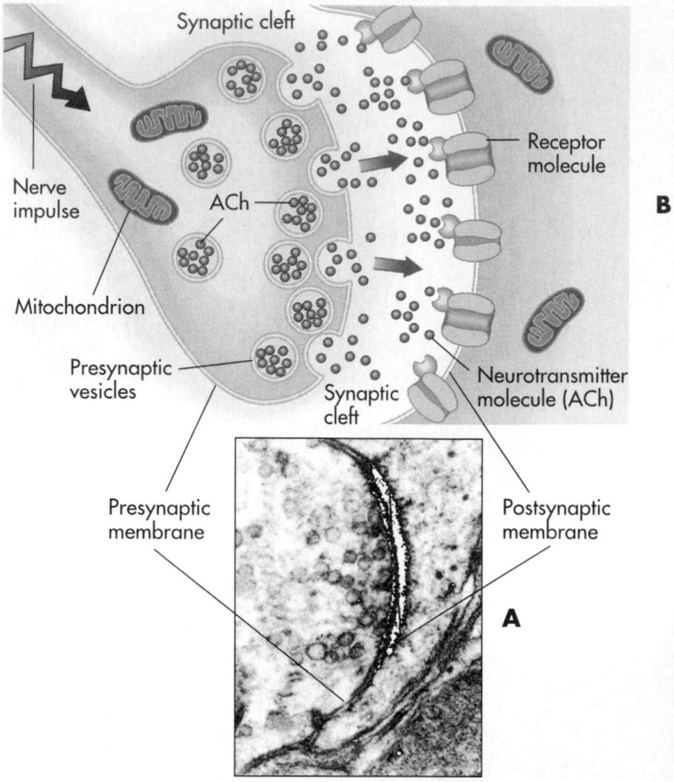

FIGURE 13-12 A synaptic cleft between neurons. **A,** An electron micrograph (*50,000x*) of a synaptic cleft. **B,** Nerve signals cross the synapse by changing from an electrical signal to a chemical signal.

Chemical junctions between neurons and other neurons or effector cells have a distinct advantage over direct electrical connections—flexibility. The chemical transmitters can be different in different junctions. Just as you might take an aspirin to stop headache pain or cough syrup to subdue a cough, different neurotransmitters result in different kinds of responses. In fact, more than 60 different chemicals have been identified that act as neurotransmitters or that act to modify the activity of neurotransmitters.

Neuron-to-Muscle Cell Connections

Synapses between neurons and skeletal muscle cells are called **neuromuscular junctions.** The neurotransmitter found at neuromuscular junctions is *acetylcholine*. Passing across the gap, the acetylcholine molecules bind to receptors in the postsynaptic (muscle cell) membrane, opening sodium channels. This influx of sodium ions depolarizes the muscle cell membrane, which initiates a wave of depolarization that passes down the muscle cell (Figure 13-13). This wave of depolarization releases calcium ions, which in turn trigger muscle contraction.

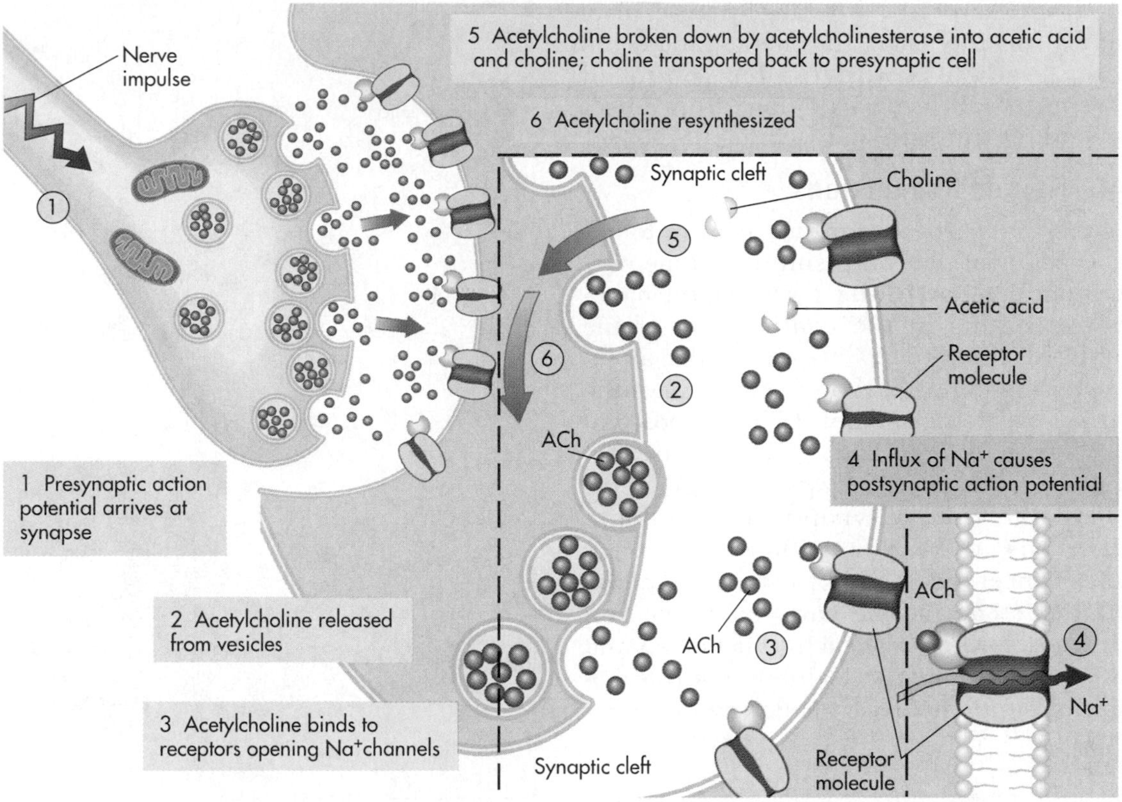

FIGURE 13-13 The sequence of events in synaptic transmission. This diagram illustrates in detail how a nerve impulse travels across the synaptic cleft and stimulates an adjacent neuron. *ACh* is an abbreviation for acetylcholine.

> *At a neuromuscular junction, acetylcholine released from an axon tip depolarizes the muscle cell membrane, releasing calcium ions that trigger muscle contraction.*

After an impulse has been transmitted across the synaptic cleft, the neurotransmitter must be broken down or the postsynaptic membrane will remain depolarized. The breakdown products of the neurotransmitter then diffuse or are actively transported from the postsynaptic cell. In general, some or all of the breakdown products are transported back to the presynaptic cell to be reused. For example, the neurotransmitter acetylcholine is broken down to acetic acid and choline by an enzyme called *acetylcholinesterase.* Choline is transported back to the presynaptic cell, where it is used to make molecules of acetylcholine. Acetylcholinesterase is one of the fastest-acting enzymes in the body, breaking down one acetylcholine molecule every 40 microseconds. The fast work of acetylcholinesterase permits as many as 1000 impulses *per second* to be transmitted across the neuromuscular junction.

Interestingly, nerve gases and the agricultural insecticide parathion work by blocking the action of acetylcholinesterase. These chemicals can cause death because they produce continuous neuromuscular transmission, which results in a continuous muscular contraction of vital muscles such as those involved in breathing and the circulation of

synapse (**sin** aps) a place where a neuron communicates with another neuron or effector cell.

synaptic cleft (sin **nap** tik **kleft**) the space or gap between two adjacent neurons, which the nerve impulse must cross.

neurotransmitters (**noor** oh **trans** mit urs) chemicals released when nerve impulses reach the axon tip of a nerve cell; neurotransmitters then cross the synaptic cleft to combine with receptor molecules on the target cell.

neuromuscular junctions (**noor** oh **mus** kyuh ler **jungk** shuns) synapses between the terminal endings of axons and skeletal muscle cells.

blood. Many drugs work by affecting synapses also. For example, cocaine, local anesthetics, and some tranquilizers work by destroying the control of neurotransmitter release. (See "How drugs affect neurotransmitter transmission.")

Neuron-to-Neuron Connections

Impulses are transmitted from one neuron to another by a variety of neurotransmitters. Some neurotransmitters depolarize the postsynaptic membrane, which results in the continuation of the nerve impulse. This type of synapse is called an *excitatory synapse.* Other neurotransmitters have the reverse effect, reducing the ability of the postsynaptic membrane to depolarize. This type of synapse is called an *inhibitory synapse.* A single nerve cell can have *both* kinds of synaptic connections to other nerve cells. And as you might expect, excitatory signals cancel out inhibitory signals, modifying each other's effects. The postsynaptic neuron keeps score as the impulses reach its dendrites and cell body, and it responds accordingly. For this reason the postsynaptic neuron is called an *integrator* (Figure 13-14).

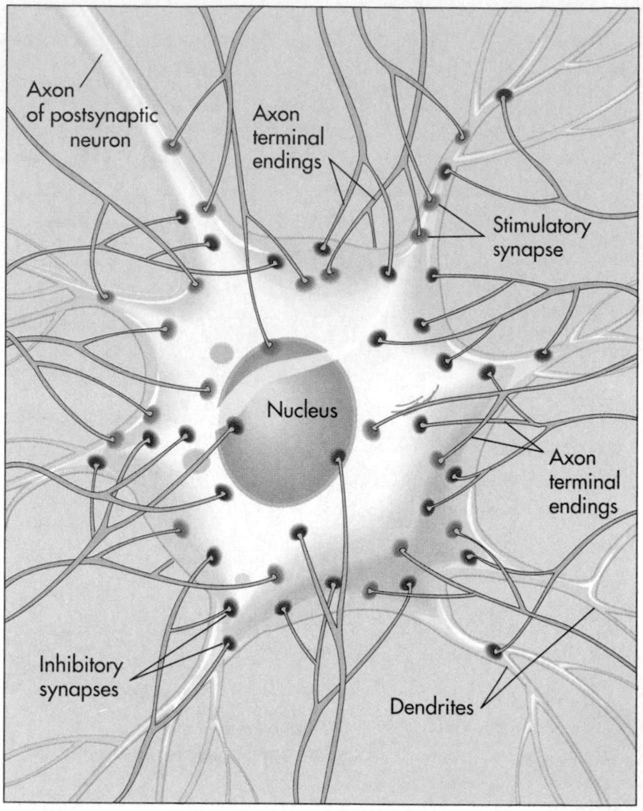

FIGURE 13-14 Integration of nerve impulses takes place on the neuron cell body. The synapses made by some axons are inhibitory, tending to counteract depolarization of the postsynaptic membrane; these synapses are indicated in *reddish-brown.* The synapses made by other axons are stimulatory, tending to depolarize the postsynaptic membrane; these synapses are indicated in *green.* The summed influences of these inputs determine whether the postsynaptic membrane will be sufficiently depolarized to initiate a nerve impulse.

> *The dendrites and cell body of a postsynaptic neuron integrate the information they receive from presynaptic neurons. The summed effect of both excitatory and inhibitory signals either facilitates depolarization or inhibits it.*

Within your body, synapses are organized into functional units with definite patterns, similar to electrical circuits. Just as your house is wired in a definite pattern to provide electricity to various appliances and light fixtures, your body is wired in a specific manner. But the manner in which your body is wired is much more complex than that in your home, allowing you to maintain internal homeostasis as well as to adapt to the outside environment.

How Drugs Affect Neurotransmitter Transmission

Drugs are substances that affect normal body functions. Although the specific actions of drugs vary, they all work by interfering with the normal activity of neurotransmitters. In this way, drugs affect communication between neurons or between neurons and muscles or glands.

One way in which drugs work is to decrease the amount of neurotransmitter that is released from a presynaptic neuron. Some drugs directly block neurotransmitter release. Other types of drugs work more indirectly, causing neurotransmitter molecules to leak out of their storage vesicles and then to be degraded by enzymes. Reserpine, a tranquilizer that also lowers blood pressure, is such a drug. It interferes with the storage of the neurotransmitter norepinephrine (also called noradrenaline). This neurotransmitter is one of two neurotransmitters active in the sympathetic nervous system, a branch of the autonomic nervous system that controls involuntary body functions (see p. 299). Among norepinephrine's functions within the sympathetic nervous system is that it acts to constrict blood vessels, which increases blood pressure. Decreasing the amount of norepinephrine decreases the constriction of the blood vessels and lowers the blood pressure.

Some drugs increase the amount of a neurotransmitter or its effects at the synapse. These types of drugs either enhance the release of neurotransmitter molecules, inhibit the action of enzymes that degrade neurotransmitter molecules at the postsynaptic neuron, or chemically resemble the neurotransmitter and mimic its effects at the postsynaptic neuron. In all cases, the postsynaptic neuron becomes or remains stimulated by the neurotransmitter or its mimic. *Amphetamines,* drugs that stimulate the brain, are of this type and are described in more detail in the next section.

Some drugs that chemically resemble specific neurotransmitters act in still another way. These mimics occupy receptor sites but do not stimulate postsynaptic neurons. They simply block the neurotransmitter molecules from the sites. Therefore these drugs block the effects of the neurotransmitter.

> *Drugs are substances that affect normal body functions by interfering with the normal activity of neurotransmitters.*

Drug Addiction

Psychoactive drugs are chemical substances that affect neurotransmitter transmission in specific parts of the brain. Some psychoactive drugs are used medically to alter moods or to treat diseases or disorders. For example, the drug diazepam, commonly known as Valium, is used to control anxiety. Imipramine is a psychoactive drug that works as an antidepressant. Morphine is sometimes used to control pain after surgery. However, many psychoactive drugs (such as morphine) are abused. That is, they are used for nonmedical reasons, are taken in doses that may cause damage to the body, and often result in personally destructive, antisocial, and crime-related behaviors.

The chronic use of psychoactive drugs results in **drug addiction:** a compulsive urge to continue using the drug, physical and/or psychological dependence on the drug, and a tendency to increase the strength (dosage) of the drug. Persons physically dependent on a drug show symptoms of this dependence when they stop taking the drug. These symptoms are called *withdrawal symptoms;* their effects are usually opposite to the effects caused by the drug. For example, if a drug relieves pain, those physically addicted to it become hypersensitive to pain when they stop taking the drug. Drugs that cause euphoria, a feeling of intense well-being, will result in depression on withdrawal. Unfortunately, these feelings are often the ones addicts try to relieve by taking the drug in the first place. Symptoms of withdrawal become a new stimulus for the drug-taking response, resulting in a cycle of addiction that is hard for the addict to break.

Chronic drug users also tend to increase the dosage of the drug they take because they become drug tolerant. **Drug tolerance** is a decrease in the effects of the same dosage of a drug in a person who takes the drug over time. Therefore, the chronic user of a drug must take increasingly higher dosages of a drug for it to continue to elicit the same response. One reason behind the phenomenon of drug tolerance is that the chronic use of a drug stimulates liver enzymes to degrade, or break down, a drug with increasing swiftness. Another metabolic reason for drug tolerance is that brain cells become less responsive to a drug over time. As drug users increase the amount of the drug they take, dangerous side effects appear more often and become stronger.

Compulsive drug-seeking behavior and psychological dependence have sociological and psychological dimensions. Unfortunately, the biological factors involved in drug addiction, other than those already mentioned, are not clear. However, recent research on alcoholism (an addiction described in the next section) points to the conclusion that this addiction is a disease. Twin and adoptive studies, animal studies, and physiological studies of addicts all provide evidence supporting the hypothesis that a gene may exist that predisposes those inheriting this gene to alcohol addiction. It is hoped that future research will reveal further information and understanding regarding drug addiction that could lead to more effective treatments and preventive measures.

Psychoactive Drugs and Their Effects

Table 13-1 lists the major classes of psychoactive drugs. Their actions relate to the neurotransmitters to which they are chemically similar and to the parts of the brain or spinal cord that have receptors for those neurotransmitters. Just as neurons communicate with skeletal muscles by secreting acetylcholine at neuromuscular junctions, neurons in specific parts of the brain and spinal cord commu-

drug addiction a compulsive urge to continue using a psychoactive drug, physical and/or psychological dependence on the drug, and a tendency to increase the dosage of the drug.

drug tolerance a decrease in the effects of the same dosage of a drug in a person who takes the drug over time.

JUST WONDERING. . . .

What happens to your body when you quit using drugs?

When a person is addicted to a drug, his or her body becomes used to the drug's activity instead of the natural neurotransmitter activity the drug replaces. Under these conditions, the body actually *suppresses* its normal neurotransmitter function. In addition to causing these effects in the brain, drugs affect various organ systems of the body, often damaging tissues and making the organs more susceptible to disease. Therefore, when a person stops taking a drug, the body must adjust to the absence of the drug's activity and begin to use its own mechanisms once again. The bodily responses to these adjustments are referred to as *withdrawal.* In addition, the body begins healing the damage done to organs.

Withdrawal can be described as a rebound response. When a drug is no longer used, the neurotransmitter activity it supplied is also gone. Meanwhile, however, the body's normal neurotransmitter functions are still suppressed. In essence, the body swings from a high level of drug-related neurotransmitter activity to no activity or low levels of activity. The body reacts with symptoms such as tension, anxiety, anger, pain, and panic. The severity of these withdrawal symptoms is related to the amount of drug use and, consequently, the amount of suppression. Withdrawal effects also vary among individuals and may vary for a particular individual at different times in his or her life.

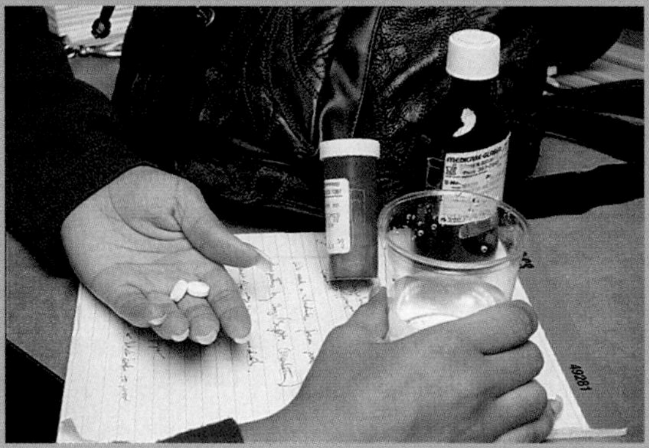

In addition to the generalized symptoms already mentioned, withdrawal may cause various other effects depending on the drug. For example, withdrawal from opiates results in certain physiological changes in the body, such as changes in blood pressure, whereas withdrawal from cocaine results in mood and behavioral changes. Withdrawal from alcohol and other depressants can be life-threatening and can result in convulsions; tremors of the hands, tongue, and eyelids; and other symptoms such as vomiting, heart palpitations, elevated blood pressure, and anxiety. One goal of drug withdrawal programs, therefore, is to stop the drug in a safe manner and allow the body's natural chemistry to return. Sometimes drug withdrawal must be managed with other drugs to achieve this goal.

nicate by means of specific neurotransmitters. After reading Chapter 14, which will help you understand brain anatomy, you may choose to reread this discussion of drugs.

Depressants

Depressants are drugs that slow down the activity of the central nervous system (brain and spinal cord). The **sedative-hypnotics** are central nervous system depressants that induce sleep (sedatives) and reduce anxiety (hypnotics). This group of drugs interacts with gamma-aminobutyric acid (GABA), an inhibitory brain neurotransmitter. The interaction takes place at receptor sites found close together on postsynaptic neurons located in the amygdala of the limbic system and throughout the cerebral cortex. When GABA binds to these postsynaptic neurons, it slows the neurons' rate of firing. When a sedative-hypnotic drug binds to the postsynaptic neurons, it enhances GABA's inhibitory effects and results in calmness. Scientists are unsure, however, why the binding of benzodiazepine reduces anxiety whereas the binding of a barbiturate produces sedation. However, barbiturates and benzodiazepines bind at separate sites near the GABA receptor site.

Sedative-hypnotics are used to treat sleep disorders, epileptic convulsions, and anxiety and are sometimes used as anesthetics for dental surgery. However, they are addictive drugs that can have dangerous side effects from either large single doses or

TABLE 13-1

Major Classes of Psychoactive Drugs

CLASS	TYPE	EXAMPLES	STREET NAMES
Depressants	Sedative hypnotic	Barbiturates (Seconal, Nembutal, Phenobarbital) Bensodiazepines (Valium)	Downers, dolls
		Methaqualone	Ludes, sopors
	Alcohol	Beer, Wine	
Opiates		Opium	
		Morphine	Mexican brown, black tar
		Codeine	
		Heroin	
Stimulants		Amphetamines	Uppers, speed, bennies
		Cocaine	Crack, ice
		Nicotine	
		Caffeine	
Hallucinogens		LSD	
		Marijuana (cannabis)	Pot, weed, grass
		PCP (phencyclidine)	Angel dust, trank
		Psychedelic mushrooms	

prolonged use. Some of these side effects are the inability to think clearly, regression of the personality to childlike characteristics, emotional instability and irritability, unsteadiness when walking, and an indifference to personal hygiene. Because they suppress body functions, a sedative-hypnotic drug overdose can be fatal.

Another type of depressant drug is *alcohol* (ethanol). The amount of ethanol in an alcoholic beverage is designated by the word "proof"; the percentage of alcohol in the beverage is half that of the proof. Eighty-proof whiskey, for example, is 40% ethanol. About 20% of the alcohol a person consumes is absorbed from the stomach. It immediately enters the bloodstream and travels all over the body. The rest passes into the small intestine and is then absorbed. Once in the bloodstream, ethanol, like other psychoactive drugs, is able to pass through the cells that make up the walls of the blood vessels within the brain. These cells, because of the way they are joined together, screen out many substances harmful to the brain. However, lipid-soluble drugs (all psychoactive drugs) are able to pass through the cells themselves. Once in the brain, ethanol acts at the same binding site as barbiturates and has sedating effects as do barbiturates. These close interactions may account for the potentially lethal interactions of alcohol or barbiturates and benzodiazepines. In other words, it is extremely dangerous to take benzodiazepines (such as Valium) with alcohol or a barbiturate (sleeping pill) because one enhances the activity of the other and at relatively low doses the combination can lead to death.

The effects of alcohol include increased heart rate, loss of alertness, blurred vision, and decreased coordination. Persons who use alcohol on a regular basis can proceed through the stages of addiction (use/tolerance/dependency/abuse) and display the characteristics of addicts described previously.

depressant (dih **press** unt) drugs that slow down the activity of the central nervous system.

sedative-hypnotics (**sed** uh tiv hip **not** iks) central nervous system depressants that induce sleep (sedatives) and reduce anxiety (hypnotics).

Long-term use of alcohol can result in liver damage, ulcers, inflammation of the pancreas, nutritional disorders (due to both the behavior of the alcoholic and the metabolic changes that result from alcoholism), heart disease, and, in pregnant women, children who exhibit fetal alcohol syndrome (see Just Wondering box in Chapter 22).

Opiates

The **opiates** are compounds derived from the milky juice of the poppy plant *Papaver somniferum* or their synthetic (human-made) derivatives. These drugs are *narcotic analgesics.* An analgesic is a drug that stops or reduces pain without causing a person to lose consciousness. A narcotic is a drug that produces sedation and euphoria; originally it referred only to the opiates. Today, however, the term *narcotic* is used more loosely and refers to any addictive drug that produces narcosis, a deadened or dazed state. It is used incorrectly when referring to stimulant drugs, which are described later in this chapter.

The opiates work by mimicking naturally occurring morphinelike neurotransmitters called *endorphins* (*end*ogenous m*orphine*like substances). Endorphins help us cope with pain and help modulate our response to emotional trauma. They bind to opiate receptors concentrated in areas of the central nervous system that include (1) the portion of the thalamus that conveys sensory input associated with deep, burning, aching pain; (2) portions of the midbrain and spinal cord involved in integrating pain information; (3) portions of the limbic system that govern the emotions; and (4) portions of the brainstem that control respiratory reflexes, pupil constriction, cough suppression, and gastric secretion and motility. This distribution of opiate receptors is shown in a guinea pig brain in Figure 13-15.

Some opiates have therapeutic uses. Dextromethorphan, an ingredient in some cough medicines, is a synthetic opiate that stimulates brainstem receptors that control coughing. Paregoric, a drug given to infants and small children to control diarrhea and accompanying cramps, contains a small amount of opium that acts to control pain and decrease gastric secretion and motility. Codeine is sometimes prescribed for the control of pain. Taken in an uncontrolled way, however, the opiates are highly addictive and dangerous drugs.

Stimulants

Stimulants include the *amphetamines* and *cocaine.* Stimulant drugs enhance the activity of two neurotransmitters: norepinephrine and dopamine.

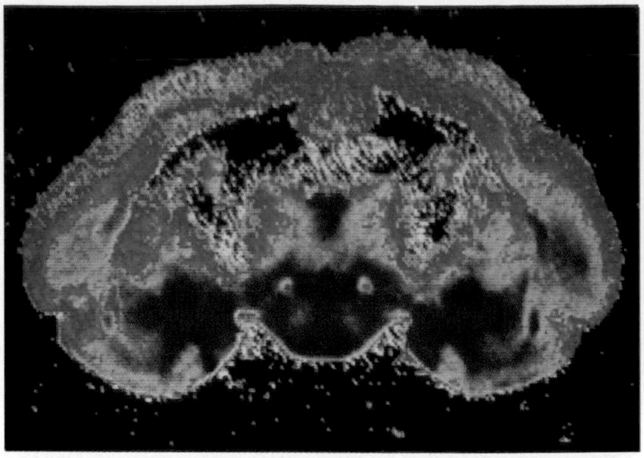

FIGURE 13-15 Opiate receptors in the guinea pig brain. This scan of guinea pig brain shows the concentration of opiate receptors; the highest density is indicated in red; yellow indicates moderate density; blue indicates low density; purple and white indicate very low densities.

These neurotransmitters function in brain pathways that regulate emotions, sleep, attention, and learning. The alerting, stimulating effects produced by these drugs relate to their action in the cerebral cortex and the action of norepinephrine in the sympathetic division of the autonomic nervous system (see p. 299). The euphoria users feel relates to the action of these neurotransmitters in the limbic area of the brain. Stimulant drugs act by moving into presynaptic neurotransmitter storage vesicles, which causes norepinephrine and dopamine to move into the synaptic cleft. The drugs then block the breakdown and recycling of the neurotransmitter molecules, causing a depletion of the neurotransmitters. When their supply of neurotransmitters is gone, drug users must take more and more of the stimulant to achieve their "high." In fact, the nervous system eventually becomes so depleted of neurotransmitter that drug users cannot get through a day without the stimulant because their nervous systems are no longer working properly.

If a user begins to take in large amounts of a stimulant drug such as cocaine, the drug can cause a schizophrenialike mental disorder characterized by paranoia, the hearing of voices, and irrational thought. This disorder appears to be caused by an overstimulation of dopamine receptors. Chronic cocaine use can also lead to long-lasting and severe physical changes in the brain, as well as severe damage to the tissues of the nose and the lungs, heart disease, epileptic seizures, and respiratory failure.

Despite the fact that cocaine and other stimulants can be extremely dangerous drugs, they have important therapeutic uses. The drug methylphenidate (Ritalin) is used in the treatment of attention deficit disorder. The seemingly paradoxical effects of this drug (hyperactive children are slowed down) are attributed to the drug's attention-focusing effects, allowing hyperactive children to focus their attention for longer periods of time and, therefore, resulting in their not moving quickly from one activity or place to the next. Another important therapeutic use of stimulant drugs is the use of cocaine as a local anesthetic. When injected into peripheral nerves, cocaine blocks conduction on nerve fibers that transmit sensation. For this reason, synthetic cocaine derivatives are used for local anesthesia in most dental and eye surgery.

Nicotine is also a stimulant drug, although it produces much milder effects than the amphetamines and cocaine. Along with stimulating the release of norepinephrine and dopamine, nicotine also affects the nicotinic receptors of the autonomic nervous system by affecting the release of acetylcholine in the presynaptic neurons. Depending on the dosage of nicotine, it may either increase or decrease the release of this neurotransmitter. Nicotinic receptors are found in both divisions of the autonomic nervous system. Because the divisions of the autonomic nervous system oppose one another in their activities and because the effect on acetylcholine release is variable, the effects of nicotine are often variable. For example, nicotine may cause the heart rate to increase or decrease. The result is that the heart rhythm becomes somewhat irregular. But the effects of nicotine on the brain and body are only one reason not to smoke cigarettes. The carbon monoxide in cigarette smoke replaces some of the oxygen in a smoker's red blood cells, thus interfering with the delivery of oxygen to the heart, brain, and other vital organs. In addition, smoking impairs the functions of the respiratory system. Cigarette smoke also contains a number of known cancer-causing agents and has been directly linked to cancers of the lungs, throat, and mouth. Cigarette smoking may also result in heart disease, stroke, high blood pressure, ulcers, earlier onset of menopause, excessive wrinkling of the skin, and sleep problems.

Hallucinogens

Hallucinogens, or psychedelic drugs, cause sensory perceptions that have no external stimuli; that is, a person hears, sees, smells, or feels things that do not exist. These drugs bear a close chemical resemblance to the neurotransmitters norepinephrine, dopamine, and serotonin (a transmitter involved with mood, anxiety, and sleep induction). However, scientists have not been able to pinpoint the exact mechanisms by which psychedelic drugs might affect the transmission of these neurotransmitters. The best hypothesis at this time is that hallucinogenic drugs act on two small nuclei in the brainstem (containing serotonin, norepinephrine, and dopamine neurons) that act as filtering stations for incoming sensory stimuli. These two nuclei are part of the reticular formation. Although small, these two nuclei give rise to axons that branch out to interact with billions of neurons in the cerebral cortex and cerebellum. Scientists think that hallucinogens may disrupt the sorting process and allow a surge of sensory data to overload the brain.

Hallucinogens can have devastating effects. Regular use of marijuana, the psychoactive plant *Cannabis,* can result in impaired eye-hand coordination, increased heart rate, panic attacks, anxiety, paranoia, depression, immune system impairment, upper respiratory system damage, and decreased levels of sex hormones. Regular use of this drug can also result in two syndromes: (1) *acute brain syndrome,* a condition marked by perceptual distortions, sleep and memory problems, disorientation with regard to time and place, and the inability to concentrate or sustain attention to important stimuli in the environment, and (2) *amotivational syndrome,* characterized by apathy, fatigue, poor judgment, loss of ambition, and diminished ability to carry out plans. Other hallucinogens such as LSD, PCP, peyote, and mescaline can result in psychotic behavior (derangement of the personality and loss of contact with reality). These episodes can trigger chronic mental health problems or can result in suicide.

The only hallucinogen shown to have possible therapeutic effects is marijuana. It has been found to relieve nausea in some patients undergoing chemotherapy treatment for cancer. The courts have approved the use of this drug in specific cases of this nature.

opiates (**oh** pee uts) compounds derived from the milky juice of the poppy plant that act as narcotic analgesics, which stop or reduce pain without causing a person to lose consciousness.

stimulants drugs that enhance the activity of two neurotransmitters, norepinephrine and dopamine, resulting in an alerting and euphoric effect.

hallucinogens (huh **loo** suh nuh jens) psychedelic drugs that cause sensory perceptions that have no external stimuli; they cause a person to see, hear, smell, or feel things that do not exist.

Summary

- The human body has two primary means of internal communication: hormones and nerve impulses. Nerve impulses are quick and transient electrical signals. Hormones are slower, persistent chemical signals.
- Nerve cells are called *neurons.* These cells have a cell body that contains structures typical of a cell and cellular extensions called *axons* and *dendrites.* Dendrites receive incoming messages and bring them toward the cell body. Axons carry messages from the cell body.
- Nerve cells transmit information in the form of electrical signals. A resting neuron (one not conducting an impulse) has an electrical potential difference across its cell membrane. This electrical potential, or resting potential, occurs because of the separation of positively and negatively charged ions along the inside and outside of the nerve cell membrane.
- A neuron transmits an impulse when it is excited by an environmental change, or stimulus. Specialized receptors detect stimuli and can initiate events that change the electrical potential difference along the nerve cell membrane. This change in electrical potential is called *depolarization.*
- If a neuron is depolarized to a certain threshold, sodium channels in the membrane open. This permeability for sodium ions across the cell membrane wipes out the local electrochemical gradient and further depolarizes the nerve cell membrane. This rapid change in the membrane's electrical potential is called the *action potential.*
- Within a few thousandths of a second, the sodium channels close. When these channels close, ion permeabilities are changed once again, and potassium tends to move out of the cell. This change in the electrochemical gradients at the cell membrane changes the electrical potential of the membrane once again, resulting in a return to the resting potential.
- An action potential at one point on the nerve cell membrane causes depolarization of the adjacent section of membrane. A "wave" of depolarization continues along the nerve membrane. This self-propagating wave of depolarization is the nerve impulse.
- The long cell processes of many neurons are wrapped in cells in which lipid-rich membranes act as insulation. Spaces between the cells that make up the neuron cell wrapper help speed the nerve impulse as the action potential jumps from one node to the next. This very fast form of nerve impulse conduction is known as *saltatory conduction.*
- When a nerve impulse reaches the axon tip of most vertebrate nerve cells, it causes the release of chemicals called *neurotransmitters.* These chemicals pass across a space called the *synaptic cleft* to the next cell. In neuron-to–muscle cell connections the neurotransmitter triggers muscle cell contraction. In neuron-to-neuron connections the neurotransmitter may cause either an excitatory response or an inhibitory response.
- The integration of nerve impulses occurs on the cell body membranes of individual neurons, which receive both excitatory and inhibitory signals. These signals tend to cancel each other out, with the final amount of depolarization depending on the mix of the signals received.

- Drugs affect body functions by interfering with the normal activity of neurotransmitters. In this way, drugs affect communication between neurons or between neurons and muscles or glands.
- Psychoactive drugs are chemical substances that affect neurotransmitter transmission in specific parts of the brain. Some psychoactive drugs are used medically to alter the mood or to treat diseases or disorders. However, many psychoactive drugs are used for nonmedical reasons, are taken in doses that may cause damage to the body, and often result in personally destructive, antisocial, and crime-related behaviors. The chronic use of psychoactive drugs results in drug addiction, which is a compulsive urge to continue using the drug, physical and/or psychological dependence on the drug, and a tendency to increase the strength (dosage) of the drug.
- Major classes of psychoactive drugs are depressants (sedative-hypnotics, alcohol), opiates (opium, morphine), stimulants (amphetamines, cocaine), and hallucinogens (LSD, marijuana).

Knowledge and Comprehension Questions

1. What do neurons and hormones have in common? How are they different?
2. If you spill hot coffee on your hand, you quickly jerk your hand away. What type of neuron responded to the hot coffee first?
3. Explain the term *resting potential.* What creates it, and why is it important?
4. Explain how a nerve impulse travels through the body.
5. What is saltatory conduction? Explain how it works and what advantages it offers.
6. You decide to move your finger, and it moves. Explain how your nervous system communicated your decision to the muscles in your finger.
7. In question 6, explain why your finger muscles stopped contracting when you wanted them to stop.
8. Why are neurons of the central nervous system called *integrators?* What do they integrate?
9. A drug may cause decreased communication between neurons or between neurons and muscles or glands. By what common action do drugs cause such varied effects?
10. Susan takes sleeping pills regularly with no ill effects. She assumes that having a few drinks at a party is not overly dangerous either and plans to still use her regular bedtime sedative. Do you agree that this behavior is safe?
11. How do cocaine and amphetamines affect neurotransmitters such as norepinephrine and dopamine?

Critical Thinking

1. Is the nerve impulse a chemical reaction? Explain the nature of such an impulse as you understand it.
2. Basing your answer on drug action, how would you explain the fact that psychoactive drugs are among the most commonly abused type of drug?
3. Can you hypothesize a link between drug tolerance and increased criminal or antisocial behavior among long-term addicts?
4. Some people claim that marijuana is not an addictive psychoactive drug because it does not cause physical addiction to the substance. How might you refute this argument?

CHAPTER 14

THE NERVOUS SYSTEM

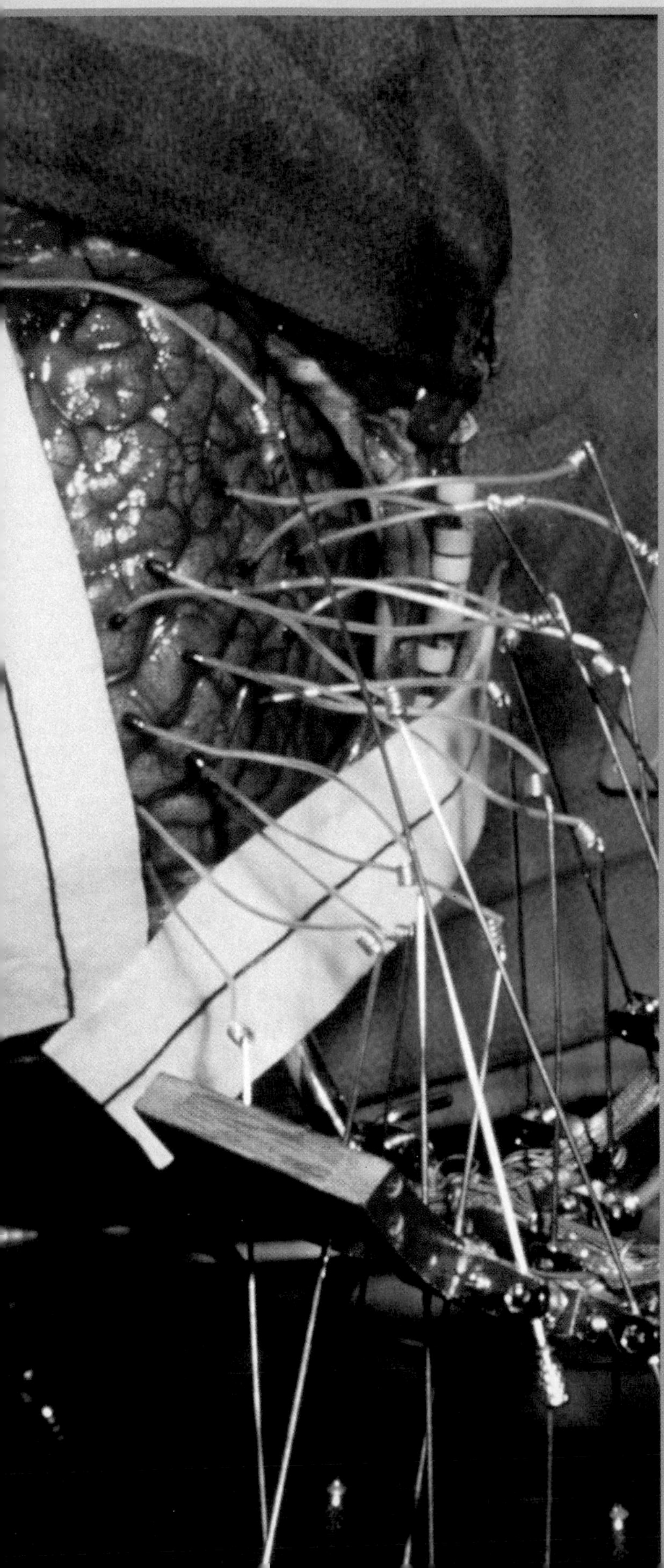

THIS INCREDIBLE PHOTOGRAPH is of an exposed human brain undergoing a procedure known as *electrocorticography*. This procedure is used to map the electrical activity of the outer, highly-active layer of the brain called the *cortex*—the part of the brain that allows you to speak, smell, see, move, and remember. Sixteen electrodes are usually used in this mapping procedure, their tips fashioned into carbon balls supported by flexible wires connected to a sensing and recording device.

Medical researchers have developed a wide range of techniques to describe the electrical activity of the human brain, including noninvasive techniques that can collect and record information without surgery, and invasive techniques such as electrocorticography. Although these techniques are used to help diagnose and treat patients with various brain-related disorders and psychiatric diseases, electrocorticography is routinely used during epilepsy surgery.

Epilepsy is a disorder in which patients experience recurrent episodes of sudden, high frequency discharges of the brain's neurons. These discharges result in the patient experiencing seizures, or convulsions. During a seizure, the patient may lose consciousness. Various behaviors may also occur such as staring into space, falling to the floor, and experiencing involuntary muscular contractions. Scientists think that seizures occur because of an imbalance between excitatory and inhibitory neurotransmission.

Although 5 percent of the general population will have a seizure at some time in their lives, less than 1 percent will have chronic seizures classified as epilepsy. Of these persons, most will be able to manage their medical condition well with anticonvulsant drugs. However, some persons develop an epilepsy that is difficult to manage and eventually decide to have their disorder treated with surgery. For these patients, it is important that brain surgeons determine the focal area of their seizure activity so that this part of the brain can be removed without taking tissue essential for important functions. Using various techniques to map brain function (including electrocorticography), brain surgeons are successful in about 80% of all epilepsy surgeries and help patients control otherwise unmanageable seizures.

KEY CONCEPTS

- The nervous system is made up of the central nervous system, which consists of the brain and spinal cord, and the peripheral nervous system, which consists of nerves that extend from the brain and spinal cord to the muscles and glands.
- The cerebrum, the dominant part of the human brain, is covered with gray matter, or unmyelinated neurons, that perform "higher" cognitive functions such as perceiving, thinking, remembering, and feeling.
- Other parts of the brain and spinal cord perform jobs of integration such as controlling subconscious movements of the skeletal muscles and coordinating complex reflexes such as breathing.
- The sensory pathways of the peripheral nervous system bring messages from receptors to the brain and spinal cord; the motor pathways relay commands to the skeletal muscles by means of the somatic nervous system and to smooth and cardiac muscles and glands by means of two opposing branches of the autonomic nervous system.

OUTLINE

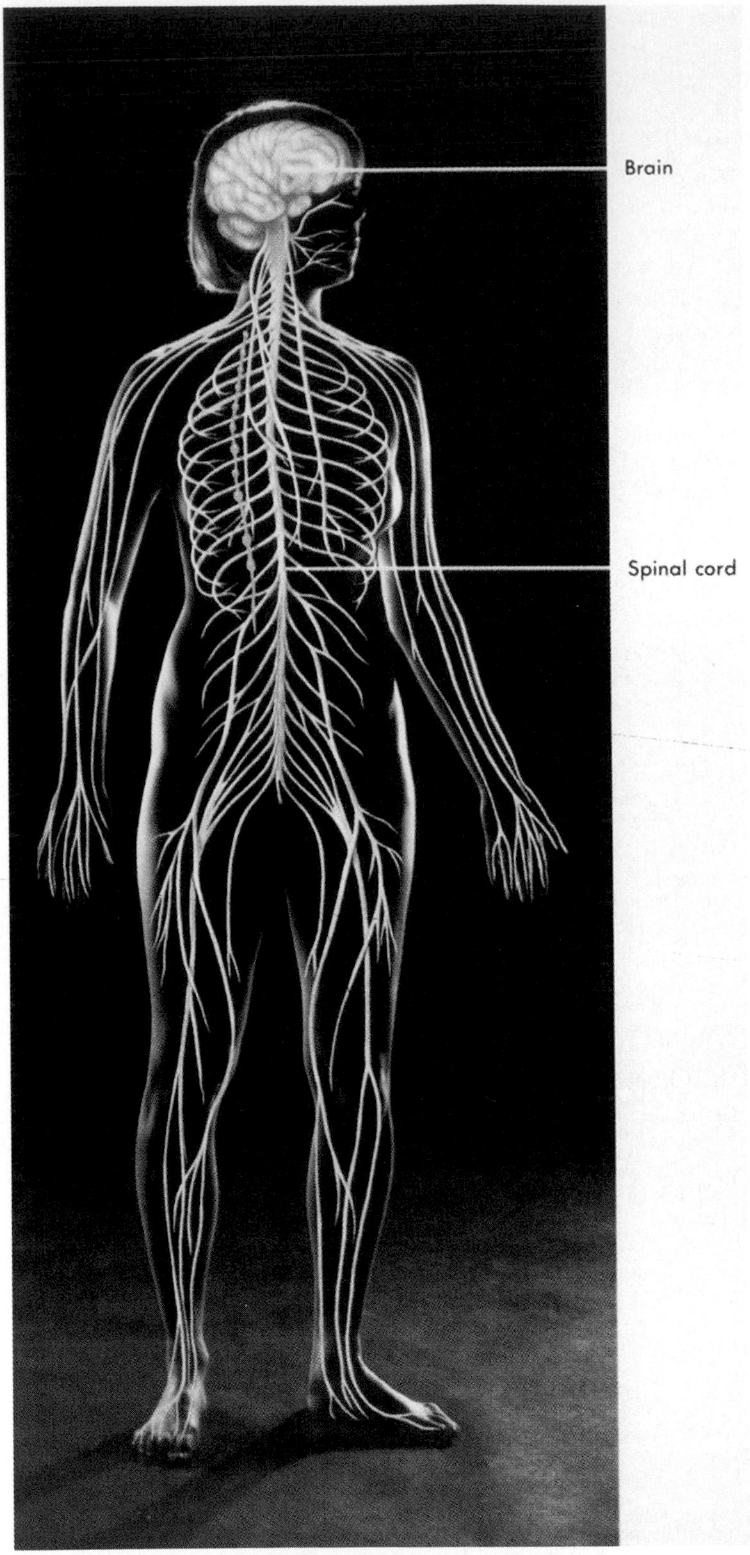

FIGURE 14-1 The human nervous system. The nervous system is comprised of the central nervous system (the brain and spinal cord) and the peripheral nervous system.

The Organization of the Nervous System

The brain is only one part of the complex network of neurons known as the **nervous system** (Figure 14-1). One job of the nervous system is to gather information about the body's internal and external environments. The nervous system then performs its other job—processing and responding to the information it has gathered. These nervous system responses are nerve impulses—commands sent out to the body's muscles and glands, directing them to react in an appropriate way.

Structurally, the nervous system can be divided into two main parts (Figure 14-1): (1) the **central nervous system**—the site of information processing within the nervous system, which is made up of the **brain** and **spinal cord,** and (2) the **peripheral nervous system**—an information highway made up of nerves that bring messages to and from the brain and spinal cord. The nerves of the peripheral nervous system are made up of the long cell processes of nerve cells (nerve fibers), support cells (see Chapter 13), connective tissue, and blood vessels. The nerves of the peripheral nervous system contain the nerve fibers of two different types of neurons: **sensory neurons,** which transmit information to the central nervous system, and **motor neurons,** which transmit commands away from the central nervous system (Figure 14-2). Sensory neurons are also called *afferent neurons,* from the Latin prefix *affero* meaning "going toward." Motor neurons are also called *efferent neurons,* from the Latin prefix *effero* meaning "going away from."

The human nervous system is made up of the central nervous system, consisting of the brain and the spinal cord, and the peripheral nervous system. Within the peripheral nervous system, sensory pathways transmit information to the central nervous system, and motor pathways transmit commands from it.

One group of motor neurons controls voluntary responses, such as coordinating the movement of muscles in your legs so that you can cross the street. These responses are called *voluntary* because you consciously choose whether or not to do this activity. Motor neurons that control voluntary responses make up the **somatic nervous system.** The word *somatic* means "body" and refers to the fact that these neurons carry messages to your skeletal muscles—those that move the parts of your body. In addition, certain voluntary activities that may seem somewhat out of your control, such as blinking and breathing, are also directed by the somatic

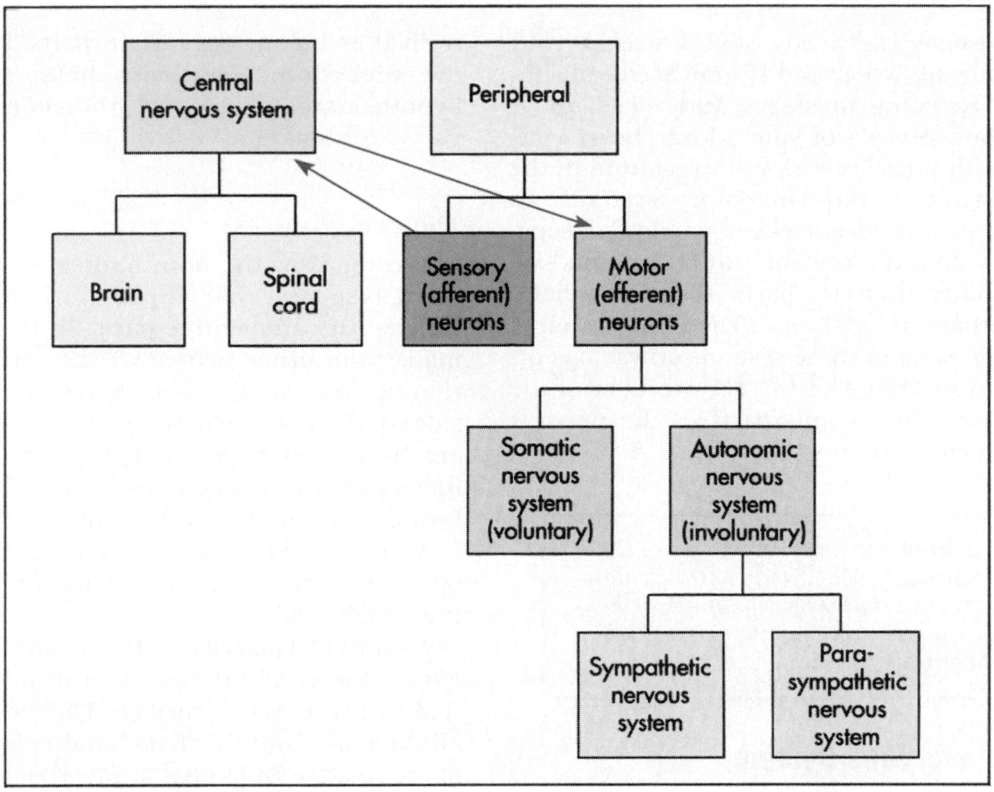

FIGURE 14-2 The organization of the human nervous system. The nervous system is divided into two parts: the central nervous system and the peripheral nervous system. The central nervous system is composed of the brain and spinal cord. The peripheral nervous system is composed of sensory pathways that transmit information to the central nervous system and motor pathways that transmit information away from the central nervous system.

nervous system. Such activities are actually *reflexes,* automatic responses to stimuli that are mediated by the spinal cord or lower portions of the brain—those closest to the spinal cord.

Another group of motor neurons carries messages that control *involuntary responses.* These responses include such activities as mixing the food in your stomach with acid and enzymes after a meal and pumping adrenaline into your bloodstream when you dodge an oncoming car. Motor neurons that carry messages about involuntary activities make up the **autonomic nervous system.** The word *autonomic* comes from Greek words meaning "self" *(auto)* "governing" *(nomos).* This portion of the nervous system, then, literally takes care of you—by itself! In general, it works to pro-

nervous system the body's complex network of neurons, which gathers information about the body's internal and external environments, and processes and responds to that information.

central nervous system one of the two main parts of the nervous system; the site of information processing within the nervous system. The central nervous system is made up of the brain and spinal cord.

brain one of the two main parts of the central nervous system. The brain consists of four main parts: the cerebrum, the cerebellum, the diencephalon, and the brainstem.

spinal cord one of the two main parts of the central nervous system; the spinal cord runs down the neck and the back. It receives information from the body, relays this information to the brain, and sends information from the brain to the body. Some simple reflexes are also integrated there.

peripheral nervous system (puh **rif** uh rul) one of the two main parts of the nervous system; it consists of an information highway of nerves that bring messages to and from the brain and spinal cord.

sensory neurons neurons that are specialized to change specific stimuli into nerve impulses and transmit this information to the central nervous system.

motor neurons neurons that are specialized to transmit nerve impulses from the central nervous system to muscles and glands.

somatic nervous system (soe **mat** ik) the branch of the peripheral nervous system consisting of motor neurons that send messages to the skeletal muscles and control voluntary responses.

autonomic nervous system (awe tuh **nom** ik) the branch of the peripheral nervous system consisting of motor neurons that control the involuntary and automatic responses of the glands and the nonskeletal muscles of the body.

mote homeostasis (a "steady state") within your body. The autonomic nervous system accomplishes this feat by carrying messages that speed up or slow down the activities of your glands, heart muscle, and smooth muscles such as those found in the digestive, excretory, and circulatory systems. In fact, these opposing messages are carried on separate neurons, dividing the autonomic nervous system functionally into two parts: the sympathetic and parasympathetic systems. (The effect of nerve impulses from each of these systems on various organs is shown in Figure 14-14.) An overview of the relationships of the various parts of the nervous system is shown in Figure 14-2.

> *One group of motor neurons makes up the somatic nervous system and controls voluntary responses. Another group of motor neurons makes up the autonomic nervous system and controls involuntary responses.*

The Central Nervous System

In many ways, your central nervous system can be compared to the central processing unit (CPU) of your computer. Without your CPU, input (what you type on the keyboard, for example) is not processed by any of the programs in your computer. Therefore you will have no output either on your monitor or from your printer. In a similar but much more complex way, your brain and spinal cord make sense of incoming sensory information and then produce outgoing motor impulses. This function of the central nervous system is called *integration.* Each part of the central nervous system plays its own unique role in integration. Many simple nervous reactions, such as pulling your hand away from a hot stove, are integrated in the spinal cord. Other more complex nervous system reactions, such as breathing, are controlled in a lower portion of the brain called the *brainstem.* And the most complex or highest functions of the nervous system, such as thinking, remembering, and feeling, are all integrated in a portion of the brain called the *cerebrum* . . . a part of the brain that can do things a computer will *never* be able to do!

The Brain

Your brain weighs about 3 pounds and contains an amazing 100 billion (100,000,000,000) neurons, a number that does not include neuroglial (supporting) cells. Both types of cells work together to form the intricate structure of the brain. Although a complicated whole, the human brain can be described as having four main parts: the **cerebrum,** the **cerebellum,** the **diencephalon** (thalamus and hypothalamus), and the **brainstem.** These four parts are shown in Figure 14-3.

The Cerebrum

The cerebrum, the dominant part of the human brain, is so large that it appears to wrap around and envelop the other three parts. In the brains of humans (and other primates) the cerebrum is split into two halves, or hemispheres, the right and left sides of the brain. These two sides of the cerebrum are connected by a single, thick bundle of nerve fibers called the **corpus callosum.** The corpus callosum is a communication bridge that allows information to pass from one side of the brain to the other, so each side "knows" what the other is thinking and doing.

Most of the activity of the cerebrum takes place within the **cerebral cortex,** a thin layer of tissue that forms its outer surface. The word *cortex* actually means "rind" or "bark" and helps give a sense of the relative thickness of this tissue to the underlying cerebral tissue. Your cortex cap is made up of unmyelinated neurons (see Chapter 13 for a discussion of myelin) and is densely packed with neuron cell bodies. It therefore appears gray and is referred to as *gray matter.*

Although the cerebral cortex is only a few millimeters thick, it has a tremendous surface area. It gains this surface area by lying in deep folds called *convolutions,* making the surface of the cerebrum look like a jumble of hills and valleys. These convolutions increase the surface area of the cerebral cortex 30 times. If laid out flat, it would cover an area of 5 square feet!

The cerebral cortex of each hemisphere of the brain is divided into four sections by deep grooves among its many convolutions. These sections are the **frontal lobe, parietal lobe, temporal lobe,** and **occipital lobe** (Figure 14-4). By examining the effect of injuries to particular sites on the cerebrum, scientists were first able to determine the approximate location of the various activities of each lobe of the cerebral cortex. More recently, using positron emission tomography (PET) scans, researchers have been able to determine more specifically the areas of the cerebral cortex that are used during various activities. Figure 14-5 shows a series of PET scan photos of the brain of a person performing a series of intellectual tasks related to words. These photos show the increase in blood flow that occurs at the part of the brain performing the task.

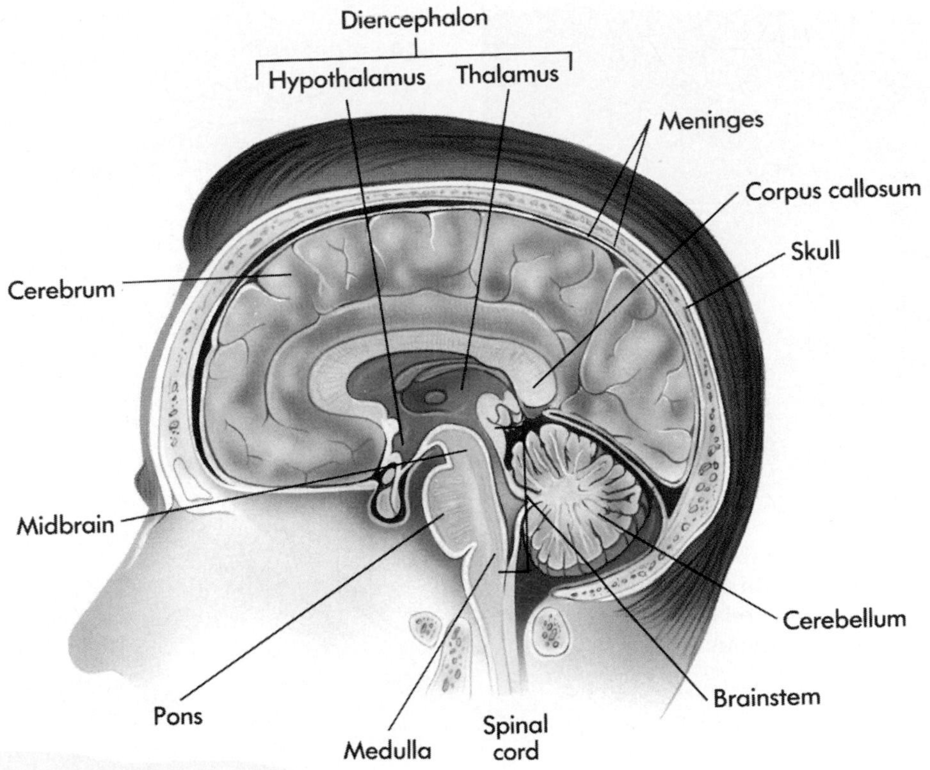

FIGURE 14-3 The human brain. The brain has four main parts, each with different functions: the cerebrum, the cerebellum, the diencephalon, and the brainstem.

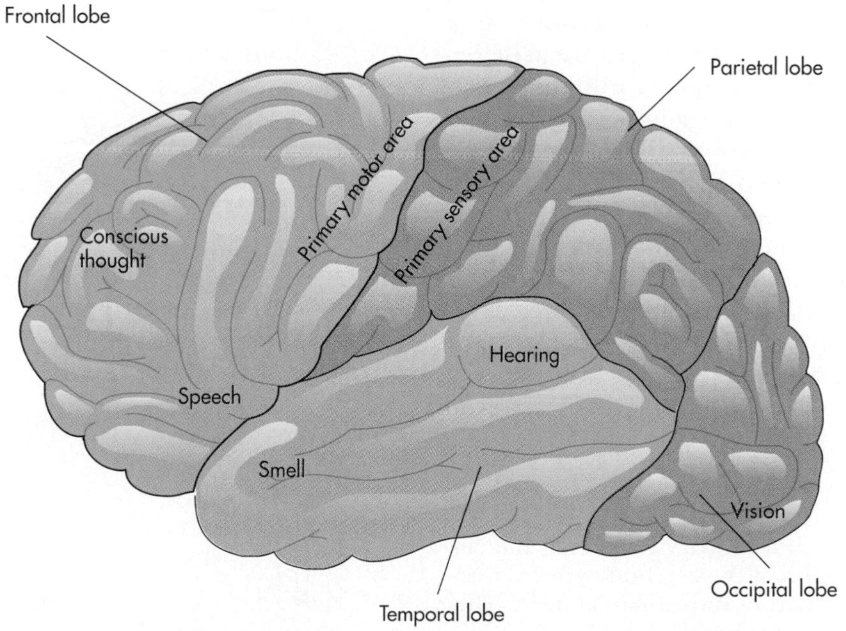

FIGURE 14-4 The cerebral cortex. The cerebral cortex has four sections: the frontal, parietal, temporal, and occipital lobes. Each of these lobes is responsible for various motor, sensory, and associative activities.

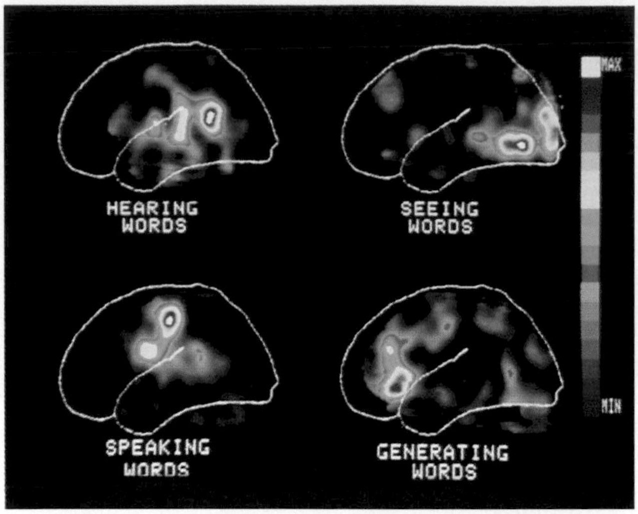

FIGURE 14-5 Visualizing brain activity. These PET scans show the areas of the brain that are most active during varied intellectual tasks related to words. Red indicates the most intense activity; blue the least.

There are three major types of activities that take place within the lobes of the cerebral cortex: motor, sensory, and association activity (see Figure 14-4). The **motor area,** the part of the brain that sends messages to move your skeletal muscles, straddles the rearmost portion of the frontal lobe. Each point on its surface is associated with the movement of a different part of the body (Figure 14-6). Right behind the motor area, on the leading edge of the parietal lobe, lies the **sensory area.** Each point on the surface of the sensory area represents sensory receptors from a different part of the body, such as the pressure sensors of the fingertips and the taste receptors of the tongue. Other sensory areas are located on other lobes. For example, the auditory area lies within the temporal lobe; different surface regions of this area correspond to different tonal patterns and rhythms. The visual area lies on the occipital lobe, with different sites corresponding to different positions on the retina.

The remaining areas within the cerebral cortex are referred to as **association areas.** These areas ap-

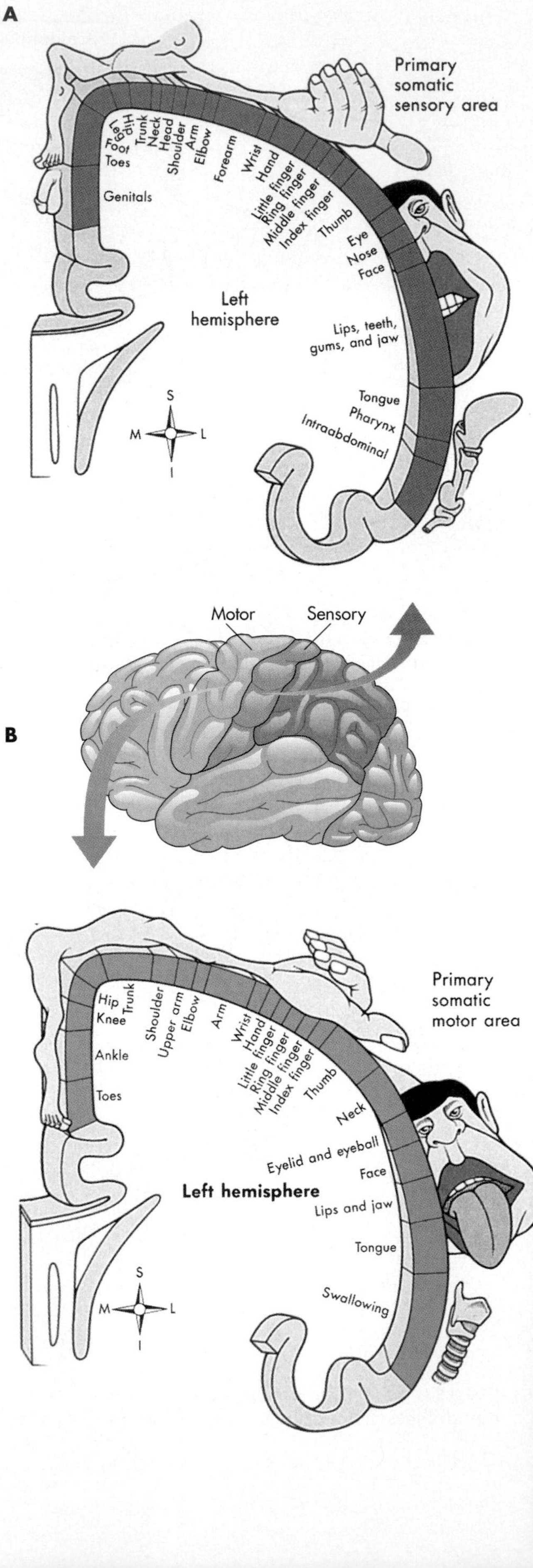

FIGURE 14-6 Motor and sensory regions of the cerebral cortex. Each region of the cerebral cortex is associated with a different part of the human body. **A,** This diagram represents the location of sensory regions in the left hemisphere. **B,** This diagram represents points in the left hemisphere that control motor functions.

pear to be the site of higher cognitive activities, such as planning and contemplation. The associative cortex represents a far greater portion of the total cortex in primates than it does in any other mammal and reaches its greatest extent in humans. In a mouse, for example, 95% of the surface of the cerebral cortex is occupied by motor and sensory areas. In humans, only 5% of the surface is devoted to motor and sensory functions; the remainder is associative cortex.

> *The cerebral cortex is the major site of higher cognitive processes such as sense perception, thinking, learning, and memory. It makes up a thin layer on the surface of the cerebrum, the largest and most dominant part of the brain.*

Although each hemisphere of the cerebrum contains motor, sensory, and association areas, the hemispheres are responsible for different associative activities. The right side of the cerebrum controls spatial relationships, musical and artistic ability, and expression of emotions. You might think of it as your "artistic" and "visual" side. The left side controls speech, writing, logical thought, and mathematical ability. It is your "logical" or "verbal" side. Injury to the left hemisphere of the cerebrum, for example, often results in the partial or total loss of speech, but a similar injury to the right side does not. Also, several speech centers control different aspects of speech. An injury to one speech center produces halting but correct speech; injury to another speech center produces fluent, grammatical, but meaningless speech; and injury to a third center destroys speech altogether. Injuries to other sites on the surface of the brain's left hemisphere result in impairment of the ability to read, write, or do arithmetic. Comparable injuries to the right hemisphere have very different effects, resulting in impairment of three-dimensional vision, musical ability, and the ability to recognize patterns and solve inductive problems. The significance of this clustering of associative activities in different areas of the brain is not clear and remains a subject of much interest.

> *The right side of the cerebrum controls spatial relationships, musical and artistic ability, and expression of emotions. The left side controls speech, writing, logical thought, and mathematical ability.*

In addition to the gray matter of the cerebral cortex, other masses of gray matter are located deep within the cerebrum. These islands of gray

cerebrum (suh **ree** brum) the largest and most dominant part of the human brain, which is divided into two halves, or hemispheres, connected by the corpus callosum. The cerebrum governs motor, sensory, and association activity.

cerebellum (**ser** uh **bell** um) the part of the brain located below the occipital lobes of the cerebrum that coordinates subconscious movements of the skeletal muscles.

diencephalon (**dye** un **sef** uh lon) the part of the brain consisting of the thalamus and hypothalamus.

brainstem the part of the brain consisting of the midbrain, pons, and medulla that brings messages to and from the spinal cord and controls important body reflexes such as the rhythm of the heartbeat and rate of breathing.

corpus callosum (**kor** pus kuh **low** sum) a single, thick bundle of nerve fibers that connects the two sides of the cerebrum in humans and primates.

cerebral cortex (suh **ree** brul **kor** tecks) a thin layer of tissue, called gray matter, that forms the outer surface of the cerebrum, and within which most of the activity of the cerebrum occurs, including higher cognitive processes such as learning and memory.

frontal lobe one of the four sections of the cerebral cortex of each hemisphere of the brain; the frontal lobe integrates motor activity and certain aspects of speech.

parietal lobe (puh **rye** uh tul) one of the four sections of the cerebral cortex of each hemisphere of the brain; the parietal lobe integrates sensory activities and stores information required for speech.

temporal lobe one of the four sections of the cerebral cortex of each hemisphere of the brain; the temporal lobe receives and interprets nerve impulses regarding hearing and smell.

occipital lobe (ock **sip** uh tul) one of the four sections of the cerebral cortex in each hemisphere of the brain; the occipital lobe receives and interprets nerve impulses regarding vision.

motor area the part of the brain that sends messages to move your skeletal muscles; it straddles the rearmost portion of the frontal lobes.

sensory area the area of the brain, located right behind the motor area, on the leading edge of the parietal lobe, that deals with information regarding stimuli; each point on the surface of the sensory area represents receptors from a different part of the body.

association areas the areas of the brain that appear to be the sites of higher cognitive activities, such as planning and contemplation.

matter, shown in Figure 14-7, are often collectively referred to as the **basal ganglia.** Ganglia are groups of nerve cell bodies; these areas of gray matter are therefore described by their name as groups of nerve cell bodies located at the base, or lowest level, of the cerebrum. (Generally, however, groups of nerve cell bodies having a similar function are called *nuclei* when they are located within the central nervous system, and they are called *ganglia* when they are located within the peripheral nervous system.) The basal ganglia are connected to one another, to the cerebral cortex, and to parts of the brain not yet described—the thalamus and the hypothalamus—by bridges of nerve fibers. These ganglia play important roles in the control of large, subconscious movements of the skeletal muscles, such as swinging the arms while walking. Injury to the basal ganglia can result in an array of uncontrolled muscular activity, such as shaking, aimless movements, sudden jerking, and muscle rigidity. Parkinson's disease is associated with the degeneration of parts of the basal ganglia.

The cerebral nuclei, also called the basal ganglia, are groups of nerve cell bodies located at the base of the cerebrum that control large, subconscious movements of the skeletal muscles.

The cerebral cortex, or gray matter, covers an underlying solid white region of myelinated nerve fibers. These nerve fibers are the highways of the brain, bringing messages from one part of the brain to another. These highways, like the message highways in all parts of the nervous system, are composed of individual axons and long dendrites all bundled together like the strands of a telephone cable. Within the central nervous system, bundles of nerve fibers are called *tracts.* In the peripheral nervous system, they are called *nerves.* The tracts that make up the white matter of the cerebrum run in three principal directions: (1) from a place in one hemisphere of the brain to another place in the same hemisphere, (2) from a place in one hemisphere to a corresponding place in the other hemisphere, and (3) to and from the cerebrum to other parts of the brain and spinal cord.

The cerebral white matter, which lies beneath the cerebral cortex, consists of myelinated nerve fibers that send messages from one part of the central nervous system to another.

The Thalamus and Hypothalamus

At the base of the cerebrum but not part of it, lying close to the basal ganglia, are paired oval masses of

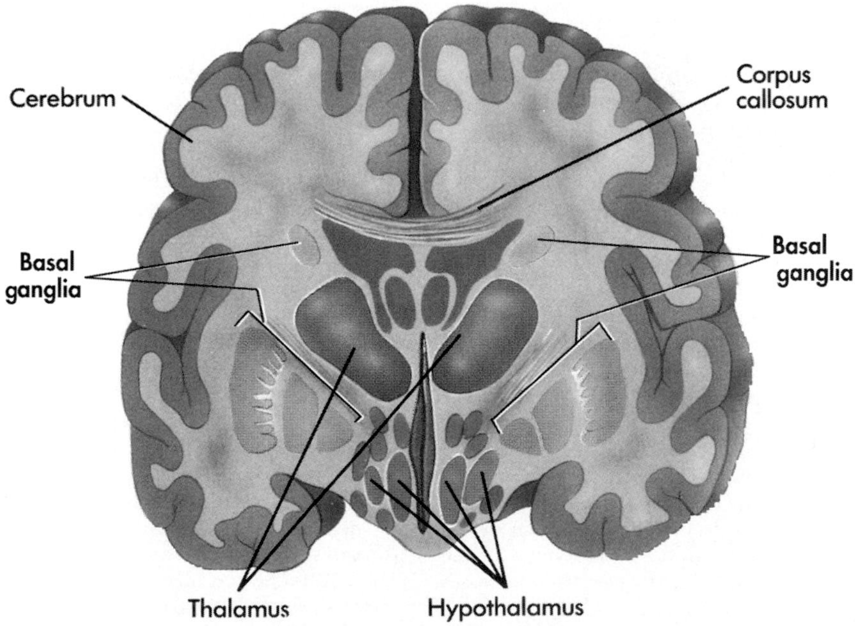

FIGURE 14-7 The basal ganglia. The basal ganglia, shaded in bright green and olive green, are groups of nerve cell bodies. The ganglia are connected to the thalamus and hypothalamus and play a part in controlling many subconscious behaviors.

gray matter called the **thalamus.** A view of these nuclei in Figure 14-7 shows them connected by a bridge of gray matter and indicates how the thalamus—meaning "inner chamber"—got its name. This structure acts as a relay station for most sensory information. This information comes to the thalamus from the spinal cord and certain parts of the brain. The thalamus then sends these sensory signals to appropriate areas of the cerebral cortex. In addition, the thalamus interprets certain sensory messages such as pain, temperature, and pressure.

The **hypothalamus,** located beneath the thalamus (*hypo* means "under"), controls the activities of various body organs. Its position is shown in Figure 14-7. The hypothalamus works to maintain homeostasis, a steady state within the body, by means of its various activities such as the control of body temperature, respiration, and the heartbeat. It also directs the hormone secretions of the pituitary, which is located at the base of the brain.

The hypothalamus is linked by a network of neurons to the cerebral cortex. This network, together with the hypothalamus, is called the **limbic system.** The term *limbic* is derived from a Latin word meaning "border." As shown in Figure 14-8, the neurons of the limbic system form a ringlike border around the top of the brainstem. The operations of the limbic system are responsible for many of the most deep-seated drives and emotions of vertebrates, including pain, anger, sexual drive, hunger, thirst, and pleasure.

> *The thalamus and hypothalamus are masses of gray matter that lie at the base of the cerebrum. The thalamus receives sensory stimuli, interprets some of these stimuli, and sends the remaining sensory messages to appropriate locations in the cerebrum. The hypothalamus controls the activities of various body organs and the secretion of certain hormones.*

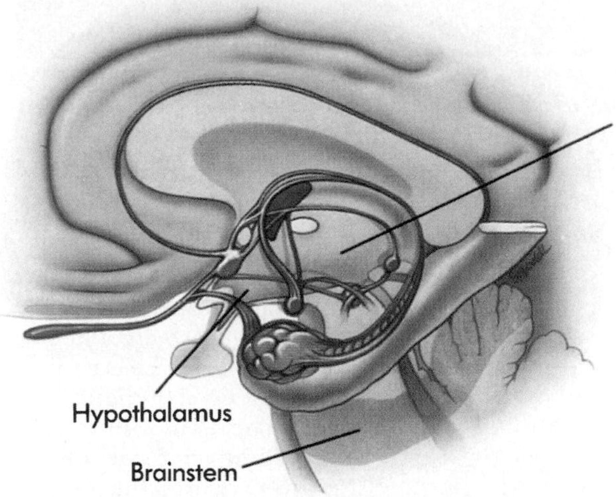

FIGURE 14-8 The limbic system. Shaded in blue in this diagram, the limbic system is responsible for many of your most deep-seated drives and emotions.

The Cerebellum

The cerebellum is a relatively large part of the brain, weighing slightly less than half a pound. It is located below the occipital lobes of the cerebrum, as shown in Figure 14-3. Although its name means "little cerebrum," the cerebellum does not perform cerebral functions. Instead, it coordinates subconscious movements of the skeletal muscles. Sensory nerves bring information to the cerebellum about the position of body parts relative to one another, the state of relaxation or contraction of the skeletal muscles, and the general position of the body in relation to the outside world. These data are gathered in the cerebellum and synthesized. The cerebellum then issues orders to motor neurons that result in smooth, well-coordinated muscular movements, contributing to overall muscle tone, posture, balance, and equilibrium.

> *The cerebellum coordinates subconscious movements of the skeletal muscles.*

basal ganglia (**bay** suhl **gang** lee uh) groups of nerve cell bodies located at the base of the cerebrum that control large, subconscious movements of the skeletal muscles.

thalamus (**thal** uh muss) paired oval masses of gray matter at the base of the cerebellum, lying close to the basal ganglia, that receive sensory stimuli, interpret some of these stimuli, and send the remaining sensory messages to appropriate locations in the cerebrum.

hypothalamus (hye poe **thal** uh muss) a mass of gray matter located beneath the thalamus that regulates vital body functions such as respiration and the heartbeat and directs the hormone secretions of the pituitary gland.

limbic system a network of neurons, which together with the hypothalamus, forms a ringlike border around the top of the brainstem; the operations of the limbic system are responsible for many deep-seated drives and emotions of vertebrates, including pain, anger, sexual drive, hunger, thirst, and pleasure.

JUST WONDERING . . . .

I've often seen adults on talk shows discussing repressed childhood memories they only recently recalled. How can people repress memories and then suddenly recall them?

Many people have had the experience of a memory suddenly coming into consciousness, but the validity of such experiences gained credence when Vietnam veterans began to describe suddenly vivid memories of forgotten experiences they had during the war. These long-buried memories that had been repressed came out when a later emotional event triggered their release. In addition, the publicity given to the sexual abuse of children within the last decade has prompted many victims of this trauma to remember and relate their repressed experiences in "flashbacks" of memories of their abuse, which are activated by an emotional or physical stimulus years later.

Although the exact physiological mechanism is unknown, researchers do have some idea of how a person can repress and suddenly recall memories. A portion of the limbic system (see Figure 14-8) called the *hippocampus* (not specifically pointed out in the illustration), acts as the brain's "file clerk." To store a memory, the hippocampus directs the parts of an experience, such as the sights, smells, and sounds, to different sectors of the brain through a network of neurons. Thus, different parts of memory are stored in different places, and the neuron network, if it is fired often, allows access to the memory. If the neuron network is not fired, the memory can be forgotten—or simply repressed. If the proper stimulus can cause the network of neurons to fire, a long-buried memory can come to the surface, often with fine details still intact. It is not unusual for people experiencing such memory flashbacks to be able to describe the time of day, the objects in the room, and the clothes they were wearing during the experience they are recalling. Researchers continue to be fascinated with the brain's capacity to store and retrieve information and are still actively researching the factors involved in memory.

The Brainstem

If you think of the brain as being shaped somewhat like a mushroom, the cerebrum, thalamus, and hypothalamus would be its cap. The brainstem would be the mushroom's stalk (see Figure 14-3). Its 3 inches of length consist of three parts—the midbrain, pons, and medulla. Each part makes up about an equal length of the brainstem and contains tracts of nerve fibers that bring messages to and from the spinal cord. In addition, each portion of the brainstem contains nuclei that govern important reflex (automatic) activities of the body. Many of the cranial nerves, nerves that enter the brain rather than the spinal cord, enter at the brainstem. These cranial nerves bring messages to and from the regulatory centers of the brainstem or use the brainstem as a relay station. These nerves can be seen in Figure 14-9 and are discussed in more detail later in the chapter.

At the top of the brainstem sits the **midbrain,** extending down from the lower portion of the thalamus and hypothalamus about an inch—approximately the distance from the tip of your thumb to its first knuckle. If you were to cut open the midbrain, you would see both white and gray matter. The white matter consists of nerve tracts that connect the upper parts of the brain (cerebrum, thalamus, and hypothalamus) with lower parts of the brain (pons and medulla). In addition, the midbrain contains nuclei that act as reflex centers for movements of the eyeballs, head, and trunk in response to sights, sounds, and various other stimuli. For example, if a plate falls off the counter behind you, you probably turn around quickly and automatically. That is your midbrain at work.

The term **pons** means "bridge," and it is actually two bridges. One bridge consists of horizontal tracts that extend to the cerebellum, connecting this part of the brain to other parts and to the spinal cord. The other bridge consists of longitudinal tracts that connect the midbrain and structures above to the medulla and spinal cord below. In addition, the gray matter of the pons contains nuclei that work with certain nuclei in the medulla to help control respiration.

The **medulla,** the lowest portion of the brainstem, is continuous with the spinal cord below. Because of its location, a large portion of the medulla consists of tracts of neurons that bring messages up from the spinal cord and others that take messages down to the spinal cord. Interestingly, most of these tracts cross over one another within the medulla. Therefore sensory information from the right side of the body is perceived in the left side of

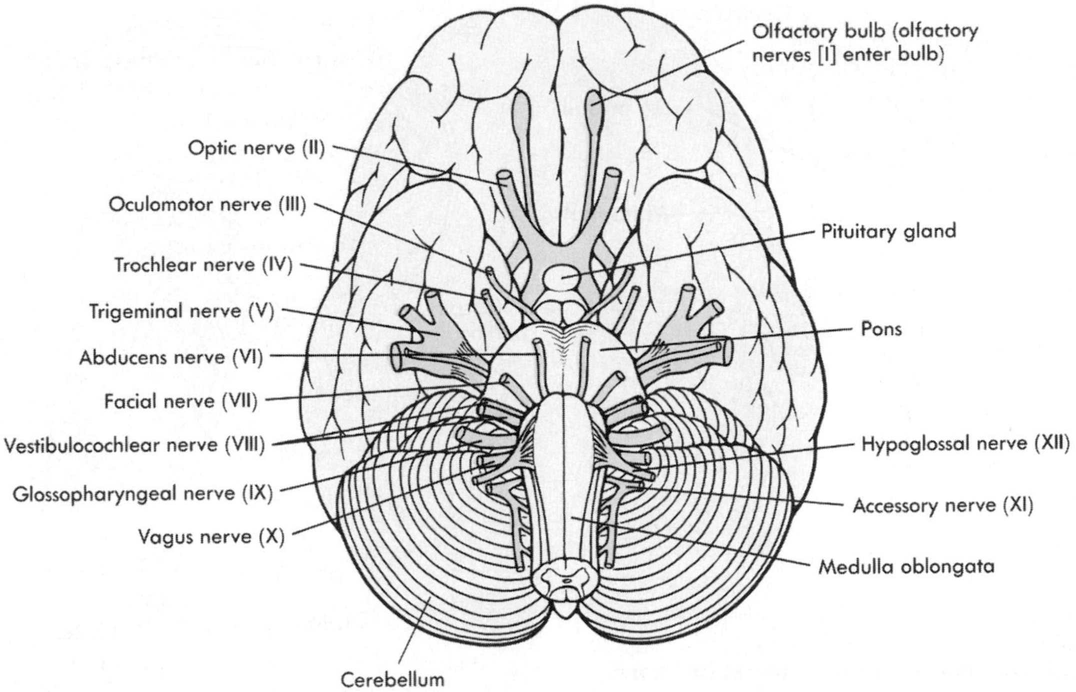

FIGURE 14-9 A view of the brain from its underside. In this view, the origins of the cranial nerves can be clearly seen.

the brain and vice versa. Likewise, the right side of the brain sends messages to the left side of the body; the left side of the brain controls the right side. In addition to these tracts, the medulla contains reflex centers that regulate heartbeat, control the diameter of blood vessels, and adjust the rhythm of breathing. Centers there also control less vital functions such as coughing, sneezing, and vomiting.

Throughout the entire length of the brainstem but concentrated in the medulla weaves a complex network of neurons called the **reticular formation.** All of the sensory systems have nerve fibers that feed into this system, which serves to "wiretap" all of the incoming and outgoing communication channels of the brain. In doing so, the reticular formation monitors information concerning the incoming stimuli and identifies important ones. The reticular formation also plays a role in consciousness, increasing the activity level of many parts of the brain when aroused and decreasing in activity during periods of sleep. Interestingly, this is the part of the brain that causes a knockout during boxing. When the head is twisted sharply and suddenly during a punch to the jaw, the brainstem (and reticular formation) is twisted, resulting in unconsciousness.

The brainstem, consisting of the midbrain, pons, and medulla, contains tracts of nerve fibers that bring messages to and from the spinal cord. In addition, nuclei located in various parts of the brainstem control important body reflexes such as the rhythm of the heartbeat and rate of breathing.

midbrain the top part of the brainstem; it contains nerve tracts that connect the upper and lower parts of the brain, and nuclei that act as reflex centers.

pons (**ponz**) a part of the brainstem that consists of bands of nerve fibers that act as bridges, connecting various parts of the brain to one another; the pons also brings messages to and from the spinal cord.

medulla (muh **dull** uh) the lowest portion of the brainstem, continuous with the spinal cord below. The medulla is the site of neuron tracts, which cross over one another, delivering sensory information from the right side of the body to the left side of the brain and vice versa.

reticular formation (ri **tik** yuh ler) a complex network of neurons woven throughout the entire length of the brainstem, but concentrated in the medulla, that monitors information concerning incoming stimuli and identifies important ones.

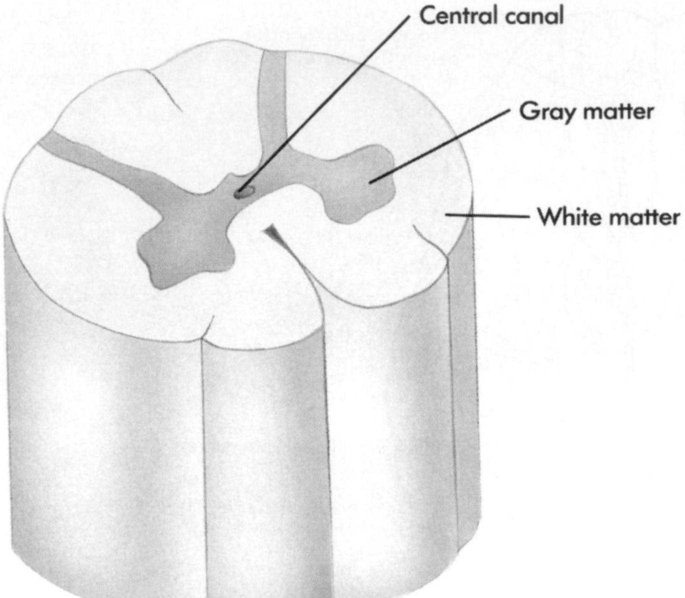

FIGURE 14-10 Section through the spinal cord. The location of the gray matter, the white matter, and the central canal is shown here.

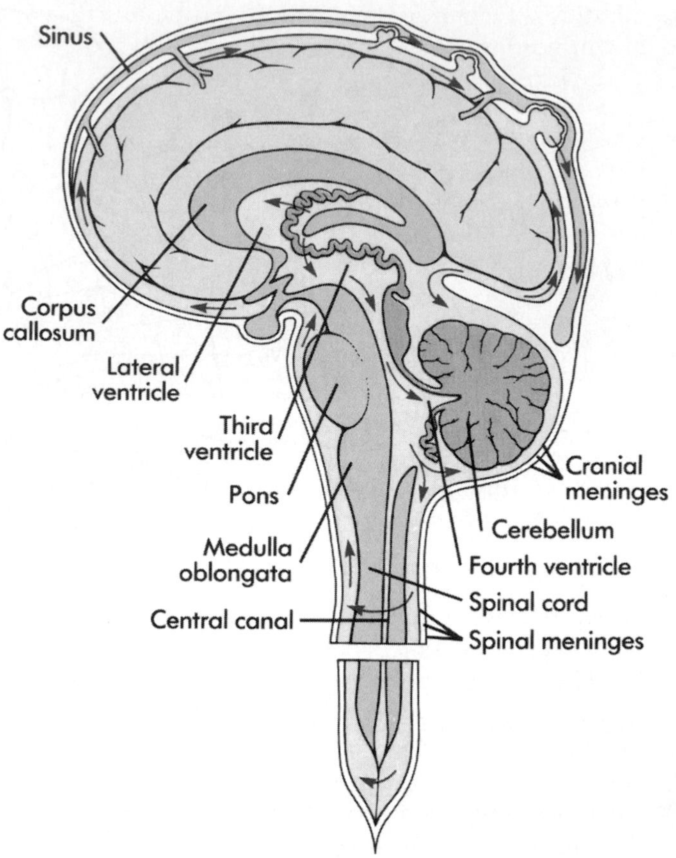

FIGURE 14-11 Flow of the cerebrospinal fluid. Arrows indicate the direction of flow. Notice that the fluid travels around the cerebrum and into the ventricles within the brain. The cerebrospinal fluid acts like a shock absorber for the brain and is also the medium in which gases such as CO_2 dissolve. The amount of gases in the cerebrospinal fluid triggers the cells in the brain to adjust such factors as the heartbeat and breathing rates.

The Spinal Cord

The central nervous system also includes the brain's extension, the spinal cord. The spinal cord runs down the neck and back within an inner "tunnel" of the vertebral column, or spine. This tunnel is created by the stacking of vertebrae one on another, with their central spaces, or *foramina* (singular *foramen*) in alignment with one another. This bony casing protects the spinal cord from injury, just as the bones of the skull protect the brain.

The spinal cord receives information from the body by means of **spinal nerves** (see Figure 14-12). It carries this information to the brain along organized tracts of myelinated nerve fibers (white matter) and similarly sends information from the brain out to the body. In addition, the gray matter of the spinal cord integrates responses to certain kinds of stimuli. These integrative pathways are called **reflex arcs** and are discussed in more detail later in the chapter.

The spinal cord, the part of the central nervous system that runs down the neck and back, receives information from the body, carries this information to the brain, and sends information from the brain to the body.

The white matter tracts of the spinal cord, unlike the white matter of the brain, are located on the exterior of the spinal cord with gray matter in the center (Figure 14-10). (The brain, as discussed earlier, is covered with gray matter, which surrounds the white matter hidden below.) In addition, the spinal cord has a tiny central canal that pierces its length. This tubelike space is filled with **cerebrospinal fluid** and is continuous with fluid-filled spaces, or *ventricles,* in the brain. (The arrows in Figure 14-11 show the direction of flow of this fluid.) The cerebrospinal fluid is a filtrate of the blood, formed by specialized structures in the ventricles. This fluid acts as a shock absorber, cushioning the brain and the spinal cord. In fact, your brain is actually *floating* in this fluid, although its volume would barely fill an average-sized glass. In addition, the cerebrospinal fluid brings nutrients, hormones, and white blood cells to different parts of the brain.

As well as being encased in the skull and vertebrae and cushioned by the cerebrospinal fluid, both the brain and spinal cord are protected by three layers of membranes called the **meninges.** The outermost of these layers is called the *dura mater,* from the medieval Latin meaning "tough mother." Surrounding the brain, the dura mater is made up of two parts: one that adheres to the inner side of the bones of the cranium and another that lies closer to the brain. Surrounding the spinal cord, however, the dura mater is only a single layer. In between it and the bones of the vertebrae is a space filled with fat, connective tissue, and blood vessels called the *epidural* (outside the dura) *space.* At the level of the lower back, the epidural space is where anesthetics are injected to numb the lower portion of the body for certain operations or childbirth. Another space exists between the middle (arachnoid layer) and inner (pia mater) membranes of the meninges. The cerebrospinal fluid circulates within this space, called the *subarachnoid space.* The subarachnoid space is continuous with the central canal of the vertebral column and the ventricles of the brain (see Figure 14-11).

> *The brain and spinal cord are protected by a cushion of fluid called the cerebrospinal fluid, layers of membranous coverings called the meninges, and bones of the skull and vertebral column.*

The Peripheral Nervous System

Just as your central nervous system can be compared with the CPU of your computer, your peripheral nervous system can be compared with the cables that connect the CPU to the other pieces of hardware—the peripherals—that make up your computer. In your computer, connector cables carry information to the monitor to be visualized on the screen or to the printer to be printed. So too the nerve "cables" of your peripheral nervous system bring information to your muscles, instructing them to contract and relax in patterns that result in the integrated and flowing movements of your body. For example, food in your stomach (the input) sends signals along your peripheral nervous system to the medulla to be interpreted, resulting in the release of gastric juice and the churning of the muscular stomach walls (the output). These are only two examples of how the peripheral nervous system works to connect the integrative portion of your nervous system with its *receptors* (structures that detect stimuli) and *effectors* (muscles and glands that respond to that stimuli).

Sensory Pathways

Seeing a sunset, hearing a symphony, experiencing pain—all these stimuli travel along sensory neurons to the central nervous system and arrive there in the same form: as nerve impulses. Put simply, every nerve impulse is identical to every other one. How, then, does the brain distinguish between pleasure and pain or sight and sound? Interestingly, the information that the brain derives from sensory input is based solely on the source of the impulse and its frequency. Thus if the auditory nerve is artificially stimulated, the central nervous system perceives the stimulation as sound. If the optic nerve is artificially stimulated in exactly the same manner and degree, the stimulation is perceived as a flash of light. Increasing the intensity of the stimulation of either receptor will result (within limits) in an increase in the frequency of nerve impulses and therefore produce a perception of a brighter light or a louder noise.

Many kinds of receptors have evolved among vertebrates, with each receptor sensitive to a different aspect of the environment. Sensory receptors are able to change specific stimuli into nerve impulses by having low thresholds for specific types of stimuli and high thresholds for others. Receptors in the retina of your eye, for example, have a low threshold for light. Appropriate stimuli open ion channels within the membrane of the receptor cell, thus depolarizing it (see Chapter 13). This depolarization is called a *generator potential.* When the generator potential reaches the threshold level, it initiates an action potential in the sensory neu-

spinal nerves nerves by which the spinal cord receives and transmits information; spinal nerves include sensory and motor neurons.

reflex arcs pathways that integrate responses to certain types of stimuli within the spinal cord.

cerebrospinal fluid (suh **ree** brow **spy** nul) a liquid filtrate of blood, formed by specialized structures in the ventricles of the brain, that acts as a shock absorber, cushioning the brain and spinal cord.

meninges (muh **nin** jeez) the three layers of protective membranes covering both the brain and the spinal cord.

rons with which the receptor synapses. Simple receptors, such as pain receptors in the skin, may be simply the dendrite endings of sensory neurons. The generator potential then initiates an action potential along that same neuron.

Sensory nerve fibers that travel directly to the brain are axons, not dendrites as in the spinal nerves. The cell bodies and dendrites of these sensory neurons are located outside the brain in the sense organ. Their axons are bundled into groups with other sensory nerve fibers arising from the same area to form certain **cranial nerves.** The cranial nerves that conduct only sensory stimuli are those associated with sight, sound, smell, and equilibrium, such as cranial nerve I (see Figure 14-9), the olfactory nerve, which brings in "smell" messages from your nose; cranial nerve II, the optic nerve, which brings in "sight" messages from your eyes; and cranial nerve VIII, which brings impulses from your ears regarding hearing and balance. Other cranial nerves called **mixed nerves** are made up of both sensory (incoming or afferent) and motor (outgoing or efferent) nerve fibers, such as cranial nerve VII, which shuttles messages to and from your salivary and tear glands, and cranial nerve X (the vagus nerve), which regulates the function of your heart rate, respiration rate, and digestive activities. The motor nerve fibers of cranial nerves are also axons; the cell bodies of these motor neurons are located within the brain. Cranial nerve XII, which stimulates movement in your tongue, carries outgoing (motor) messages only.

Sensory and motor nerve fibers that travel directly to the spinal cord make up the spinal nerves (Figure 14-12). Nerve fibers that serve the same general area of the body, whether of sensory or motor neurons, are bundled together to form these nerves. Thus all spinal nerves are mixed nerves. Close to the spinal cord, however, motor and sensory fibers separate from one another. The sensory (afferent) fibers are myelinated axonlike dendrites that extend from the source of stimulation (your fingers or toes, for example) to a swelling near their entrance to the dorsal (back) side of the spinal cord. This swelling is a ganglion that contains the nerve cell bodies of these sensory neurons. The short axons of these neurons extend from this ganglion to the gray matter of the spinal cord. Here, the axon ends of these sensory neurons synapse with neurons that play a role in integrating incoming messages with outgoing responses.

Integration

"Integrating" neurons, or **interneurons,** receive incoming messages and send appropriate outgoing messages in response. Interneurons extend from the spinal cord to the brain, make up part of the brain tissue itself, and in the case of certain reflex pathways, are located within the gray matter of the spinal cord. The pathway of a nerve impulse from stimulus to response may involve a single interneuron, as in the case of certain reflexes. In a few of the simplest reflexes, a sensory neuron may synapse directly with a motor neuron, with no interneuron(s) playing a role. Most pathways, however, involve many interneurons that, in sequence, receive the incoming message, direct it to other interneurons for interpretation (such as specific areas of the cerebrum), and bring the outgoing message to the appropriate motor neurons.

Motor Pathways

The short dendrites of spinal motor neurons, located in anterior portions of the gray matter of the spinal cord, synapse with interneurons. These dendrites then conduct the impulses picked up from the interneurons to their cell bodies located within the same area of gray matter of the spinal cord. The nerve impulse sweeps along the membrane of the axon of each motor neuron to effectors: muscles or glands that produce a response. Motor pathways to skeletal muscles contain a single motor neuron; motor pathways to smooth and cardiac muscles and to glands are made up of a series of two motor neurons.

The bundles of axons leaving the spinal cord join with the dendrites of sensory neurons entering the same level of the cord, forming spinal nerves. A total of 31 pairs of spinal nerves brings messages to and from specific areas of the body. Figure 14-1 shows the approximate location of the spinal nerves (those emerging from the spinal cord) and, in general, the areas of the body that they serve.

Sensory receptors change stimuli into nerve impulses. The nerve fibers of sensory neurons then bring this information to the brain or spinal cord where, except in the case of very simple reflexes, interneurons interpret them and send appropriate outgoing messages. The axons of motor neurons conduct these impulses to muscles and glands, the structures that effect a response.

The Somatic Nervous System

The central nervous system directs different types of responses in different ways. Movements of the skeletal muscles are controlled by messages from

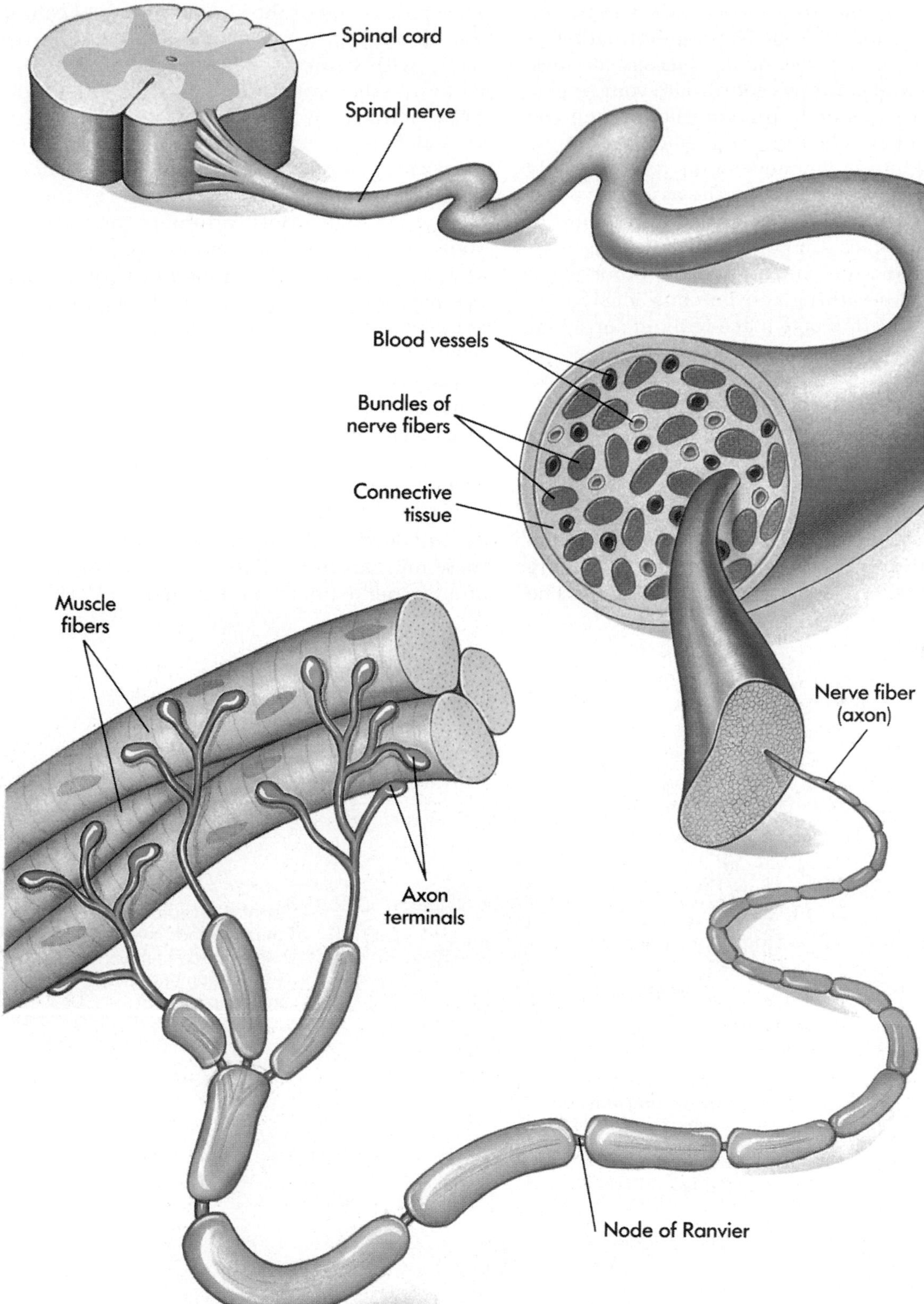

FIGURE 14-12 Structure of a nerve. A nerve is composed of bundles of nerve fibers and various layers of connective tissue.

the brain and spinal cord via pathways that contain a single motor neuron. These motor pathways are referred to as the *somatic nervous system.* Messages along these pathways coordinate your fingers when you grasp a pencil, spin your body when you dance, and outstretch your arms when you hug a friend. These muscular movements are primarily subject to conscious control by the associative cortex of the cerebrum. However, not all movements of the skeletal muscles are conscious. Blinking your eyes, putting one foot in front of the other when you walk, and breathing can be consciously controlled but most often take place without conscious thought.

The somatic branch of the peripheral nervous system consists of motor fibers that send messages to the skeletal muscles.

A *reflex* is an automatic response to nerve stimulation. Very little, if any, integration (and certainly no thinking) takes place during reflex activity. The knee jerk is one of the simplest types of reflexes in the human body; a sensory neuron synapses directly with a motor neuron (Figure 14-13). This pathway—the pathway an impulse follows during reflex activity—is called a *reflex arc.* Simple reflex arcs such as the knee jerk are called *monosynaptic reflex arcs* (literally, "one synapse"). A monosynaptic reflex arc is not subject to control by the central nervous system. Most voluntary muscles within your body possess such monosynaptic reflex arcs, although usually in conjunction with other more complex reflex pathways. It is through these more complex paths that voluntary control is established. In the few cases in which the monosynaptic reflex arc is the only feedback loop present, its function can be clearly seen, such as in the knee jerk reaction. If the ligament just below the kneecap is struck lightly by the edge of your hand or by a doctor's rubber mallet, the resulting sudden pull stretches the muscles of the upper leg, which are attached to the ligament. Stretch receptors in these muscles immediately send an impulse along afferent nerve fibers to the spinal cord, where these

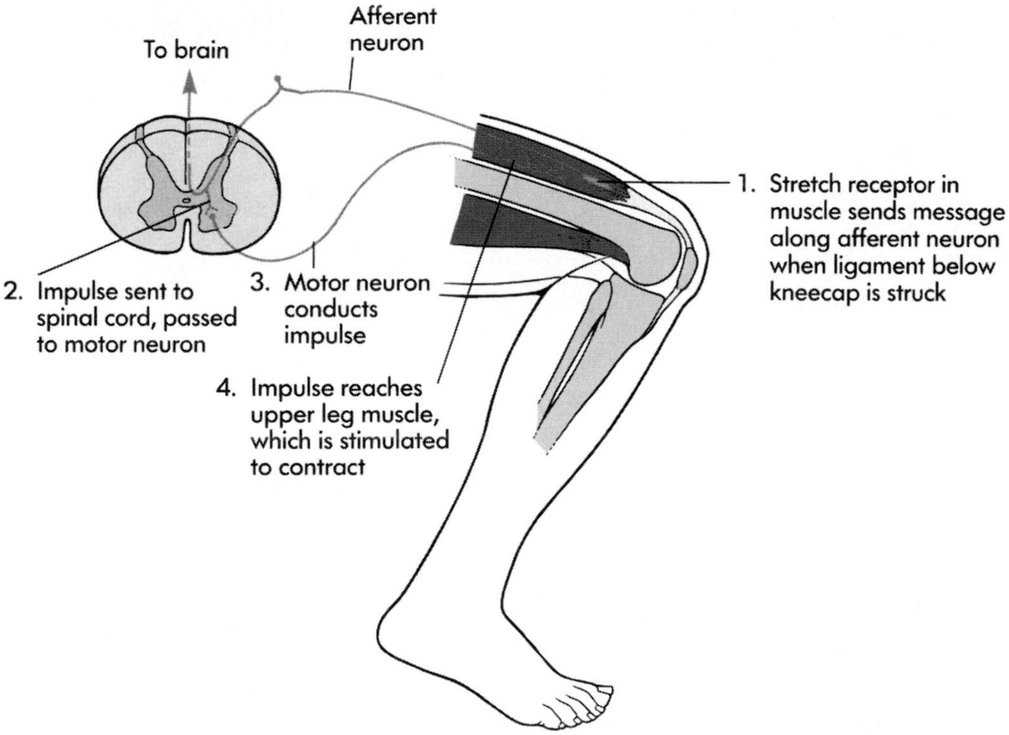

FIGURE 14-13 The knee jerk reflex. When your kneecap is tapped, the muscle in your leg is stretched, and this stimulation of the stretch receptor travels through afferent (sensory) neurons to the spinal cord, where it synapses directly with a motor neuron, which stimulates the leg muscle to contract.

fibers synapse directly with motor neurons that extend back to upper leg muscles, stimulating them to contract and the leg to jerk upward. Such reflexes play an important role in maintaining posture.

> *A reflex is an automatic response to nerve stimulation. The pathway of nervous activity in a reflex is called a reflex arc. Impulses travel along sensory neurons to the spinal cord, which sends a message out via motor neurons without involving the brain.*

The Autonomic Nervous System

When you enter a dark movie theater, the pupils of your eyes dilate, or become larger. When you are frightened, adrenaline is pumped into your bloodstream and your heart beats faster. Neither of these reactions is governed by conscious thought—they occur automatically and involve the action of smooth muscle, cardiac muscle, or glands. The motor pathways that control such involuntary and automatic responses of the glands and nonskeletal muscles of the body are referred to as the *autonomic nervous system*. The autonomic nervous system takes your temperature, monitors your blood pressure, and sees to it that your food is properly digested. The body's internal physiological condition is thus fine-tuned within relatively narrow bounds, a regulatory process called *homeostasis*.

The autonomic nervous system is made up of two divisions, the **parasympathetic nervous system** and the **sympathetic nervous system,** that act in opposition to each other, speeding up or slowing down certain bodily processes (Figure 14-14). Each of these motor pathways consists of a series of two motor neurons. In the parasympathetic system the axons of the motor neurons leaving the spinal cord extend to ganglia located near the muscles or organs they affect. In the sympathetic system the axons leaving the spinal cord are much shorter and extend only to ganglia located near the vertebral column. The axons of second motor neurons in each system extend from these ganglia to their targets.

The neurotransmitters used by each of these two branches of the autonomic nervous system differ at the axon ends of the second motor neuron—at the synapse with the target organ. The actions of these different neurotransmitters oppose each other. Thus because each gland (except the inner portion of the adrenal gland), smooth muscle, and cardiac muscle is "wired" to *both* systems, an arriving signal will either stimulate or inhibit the organ. For example, the sympathetic system speeds up the heart and slows down digestion, whereas the parasympathetic system slows down the heart and speeds up digestion. In general, the two opposing systems are organized so that the parasympathetic system stimulates the activity of normal body functions such as the churning of the stomach, the contractions of the intestine, and the secretions of the salivary glands. The sympathetic system, on the other hand, generally mobilizes the body for greater activity, as in increased respiration or a faster heartbeat. The "decision" to stimulate or inhibit a muscle, organ, or gland is "made" by the central nervous system.

> *The autonomic nervous system is a branch of the peripheral nervous system that consists of two antagonistic sets of motor fibers. Messages sent along these fibers control smooth and cardiac muscle as well as glands.*

cranial nerves any of the 12 pairs of nerves that enter the brain through holes in the skull.

mixed nerves nerves that are made up of both sensory (incoming or afferent) and motor (outgoing or efferent) nerve fibers.

interneurons nerve cells found in the spinal cord and brain that are situated between other neurons and receive incoming messages and send outgoing messages in response.

parasympathetic nervous system (pare uh sim puh thet ik) a subdivision of the autonomic nervous system that generally stimulates the activities of normal internal body functions and inhibits alarm responses; acts in opposition to the sympathetic nervous system.

sympathetic nervous system (sim puh thet ik) a subdivision of the autonomic nervous system that generally mobilizes the body for greater activity; acts in opposition to the parasympathetic nervous system.

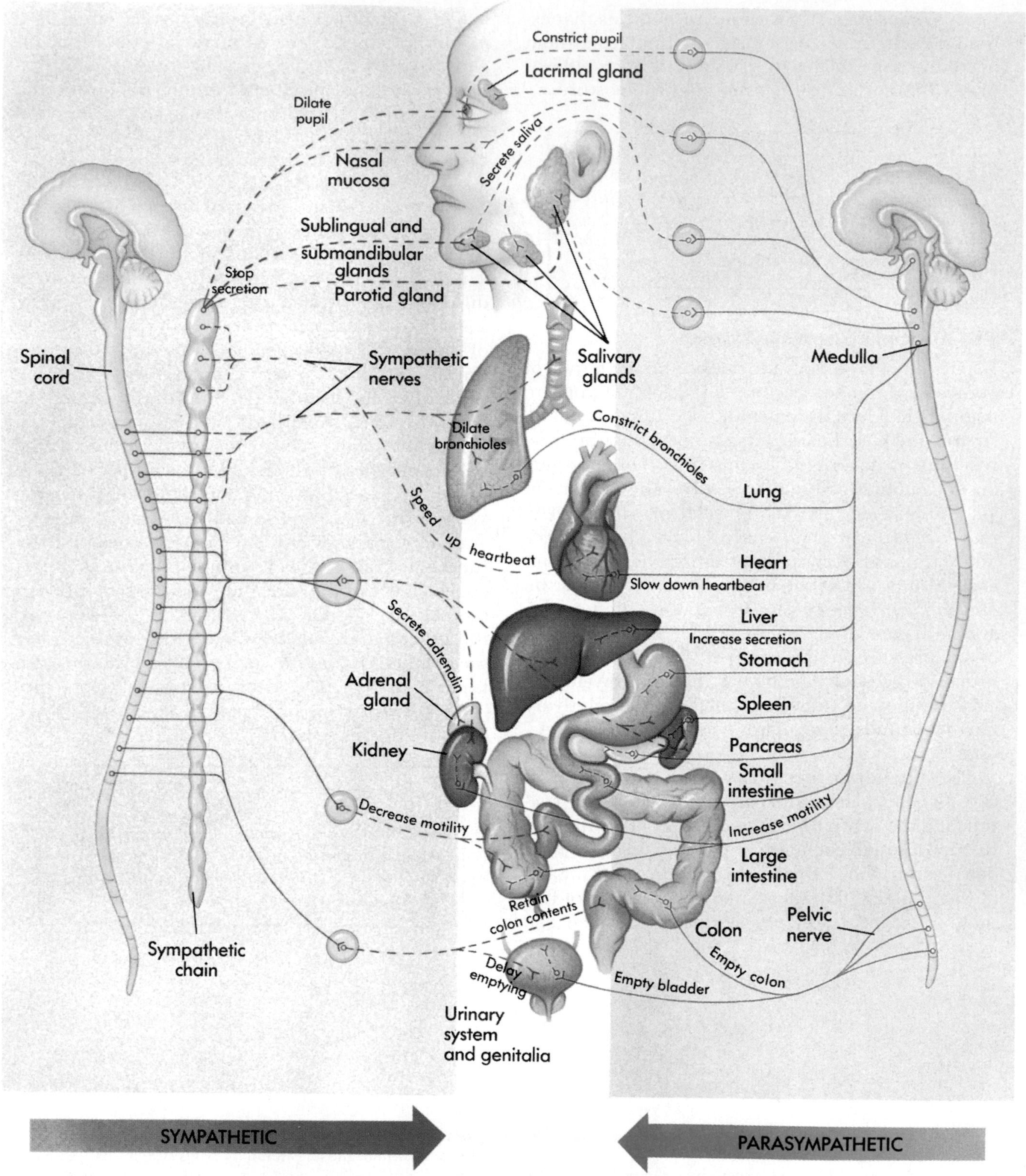

FIGURE 14-14 The sympathetic and parasympathetic nervous systems. The ganglia of sympathetic nerves are located near the spine, and the ganglia of parasympathetic nerves are located far from the spine, near the organs they affect. A nerve runs from both of the systems to every organ indicated, except the adrenal gland.

HOW *Science* WORKS

Discoveries

Solving the Mystery of Alzheimer's Disease

Alzheimer's disease was first described in 1907 by a Swiss psychiatrist, Dr. Alois Alzheimer. One of Dr. Alzheimer's patients, aged 51, was having substantial memory difficulties. She continued to deteriorate, forgetting words and losing her ability to reason. She died 4 years later. At the autopsy, the woman's brain was found to be much smaller than average for her age and sex, and the ventricles (cavities) of the brain were much larger. Her brain was also riddled with waxy patches where nerve cells had once been.

Currently affecting an estimated 2 to 3 million Americans, Alzheimer's disease is characterized by a gradual loss of memory and reasoning. Affected individuals cannot remember things that they heard or saw just a few minutes previously. They have trouble finding their way around and eventually forget how to talk, to feed themselves, even to swallow. There is no cure.

Because Alzheimer's is most common among people older than 65 years of age, doctors at first believed Alzheimer's to be a normal part of old age as the brain simply wore out. Evidence soon mounted, however, that it was a genetic disorder—Alzheimer's showed up most often in particular families, for example. Studying how it is inherited, researchers soon focused on three human chromosomes, 14, 19, and 21. Many if not all Alzheimer's cases could be linked to one of these three chromosomes.

In 1991, researchers identified the normal function of the "Alzheimer's gene" on chromosome 21. The gene at fault encodes a component of nerve cell membranes called *amyloid protein.* Amyloid is found in every nerve cell, but in Alzheimer's patients its synthesis goes awry, spitting out pieces of amyloid that pile up in sticky masses on the cell surface and eventually kill the nerve cell. The mutation appears to make the amyloid protein more fragile and liable to be chopped into fragments that accumulate in the brain.

As the population ages and people begin to live long enough for defective amyloid genes to begin their deadly work, Alzheimer's disease is becoming more common (Figure 14-A). It now affects 10% of the population older than 65 and up to 50% of persons 85 and older. Researchers are now focusing their efforts on finding ways to block the formation of amyloid fragments or their attachment to the surface of brain cells. For example, researchers are searching for ways to inhibit the enzyme that chops off the amyloid fragments. A cure will not be available tomorrow, but researchers seem to be on the road to success.

FIGURE 14-A Solving the mystery of Alzheimer's disease. The number of people over age 65 is steadily increasing. Therefore Alzheimer's disease will affect an increasing number of people unless a cure is found.

Summary

- The nervous system consists of the peripheral nervous system and the central nervous system. The peripheral nervous system, made up of the nerves of the body, gathers information about the internal and external environments and brings it to the central nervous system via sensory nerve cells. The central nervous system, made up of the brain and spinal cord, then processes and responds to that information. These responses are sent out to the body via the motor nerve cells of the peripheral nervous system.
- The cerebrum, the dominant part of the human brain, is split into two hemispheres connected by a nerve tract called the *corpus callosum*. Each hemisphere is divided further by deep grooves into four lobes: the frontal, parietal, temporal, and occipital lobes.
- Various cognitive activities take place in the outer gray matter of the cerebrum, or cerebral cortex. Specific sensory areas are located in the various lobes of the cortex. One large sensory area lies on the leading edge of the parietal lobe. Other sensory areas, the auditory and visual areas, lie within the temporal and occipital lobes, respectively. The motor region is mostly a part of the frontal lobe. Association areas connect all parts of the cerebral cortex and govern such functions as memory, reasoning, intelligence, and personality.
- The cerebral white matter, myelinated nerve fibers that lie beneath the cerebral cortex, sends messages from one part of the central nervous system to another.
- The basal ganglia, groups of nerve cell bodies located at the base of the cerebrum, control large, subconscious movements of the skeletal muscles. Additional groups of nerve cell bodies also located near the base of the cerebrum but not part of it are the thalamus and the hypothalamus. These parts of the brain receive and interpret certain sensory stimuli and control the activities of various body organs and glands, respectively.
- The cerebellum is the part of the brain that coordinates unconscious movements of the skeletal muscles.
- The brainstem consists of three parts: the midbrain, pons, and medulla. All three parts contain nerve fibers that bring messages to and from the spinal cord. In addition, the brainstem contains groups of nerve cell bodies that control important body reflexes.
- The spinal cord brings messages to the brain and sends messages from the brain out to the body. It also integrates the incoming and outgoing information of reflex arcs, with the exception of certain simple reflexes.
- In the peripheral nervous system, sensory receptors change stimuli into nerve impulses. These impulses travel along the nerve fibers of sensory neurons to the brain or spinal cord. There, integrating neurons called *interneurons* interpret this sensory information and send appropriate outgoing messages. The axons of motor neurons conduct these impulses to muscles and glands, the structures that effect a response.
- No matter what the stimulus, a nerve impulse takes the same form: depolarization of a nerve fiber membrane.
- The motor pathways are divided into somatic pathways, which relay commands to skeletal muscles, and autonomic pathways, which stimulate the glands and other muscles of the body. The autonomic pathways consist of nerve pairs having antagonistic neurotransmitters, one of which stimulates while the other inhibits. In general, the parasympathetic nerves stimulate the activity of normal internal body functions and inhibit alarm responses, and the sympathetic nerves do the reverse.

Knowledge and Comprehension Questions

1. Distinguish among the nervous system, central nervous system, and peripheral nervous system. What are the functions and components of each?
2. Distinguish between the somatic and autonomic nervous systems. What type of response does each control? Give an example of each type.
3. A friend accidentally steps on your foot and you quickly pull it away. Where in the nervous system do you think that this reflex is integrated?
4. A sentimental movie makes you remember your beloved grandparents as tears come to your eyes. Where in the nervous system would the connection between the movie (visual) and emotional reaction be integrated?
5. What is the brainstem? Name its three major components and summarize its functions.
6. You're studying in a quiet library. Suddenly someone drops an armful of books; you hear the loud crash and turn your head toward the noise. Explain how your sensory and motor pathways allowed you to detect and respond to this stimulus.
7. In question 6, why did you perceive this stimulus as a loud noise (rather than, say, as a bright light)?
8. The organs of your body are controlled by two branches of the autonomic nervous system, the sympathetic and parasympathetic. Each organ is reached by neurons from each system. Why *two* systems? Why use two neutrons to communicate with an organ rather than one?

Critical Thinking

1. Looking at Figure 14-6, can you create a hypothesis explaining why a greater proportion of the sensory and motor area of the cerebral cortex is devoted to hands and facial structures rather than an equal distribution according to body parts? Use your own body as a model.
2. Doctors have warned parents to avoid applying their own low-fat diet to small toddlers, because a higher fat level is essential for the body's production of myelin. Why do you think this may be so important in prenatal care and in the diet of infants and toddlers?
3. Which would you hypothesize to be more similar in function: the limbic systems of a human being and a lion or the associative cerebral cortex of these vertebrates? Why?
4. If the spinal cord is severed in an injury, the closer the point of damage to the brain the more severe the disability. For example, a break near the neck may result in paralysis from the neck down. Why do you think this occurs?

CHAPTER 15

Senses

CAN THE MIND block out pain? These firewalkers appear not to feel the hot coals they are walking on. In 1965, Ronald Melzack and Patrick Wall described a means by which the body could short-circuit its own warning system. They called this revolutionary idea the *gate-control theory of pain.* Melzack and Wall hypothesized that signals coming down from the brain could put a stop to signals going up the spinal cord. In this way, the "gate" was shut to certain incoming sensory messages before they got to the brain. In the same way, they explained, people tune out background noise to focus on a task, or football players continue to play the game with a broken leg, not realizing their situation. Current physiological research supports the idea that various parts of the brain exert control over the perception of pain. In fact, psychologists who treat chronic pain sufferers use the brain's ability to focus on certain sensory information and screen out other data to help patients develop strategies for the control of pain.

Pain receptors are distributed widely throughout the body and are very important to your well-being; they are the body's way of knowing when it is in danger of being harmed. Simple in structure, pain receptors are the branching ends of dendrites of certain sensory neurons. These neurons bring the pain message to the spinal cord, where it is transferred to tracts of neurons that travel up the cord to the brain.

The Nature of Sensory Communication

Pain is not a type of environmental stimulus itself but can be caused by various types of stimuli such as heat, cold, and pressure. Sensory receptors can change environmental stimuli such as these into nerve impulses. Some receptors are composed of nervous tissue; others are not but are capable of initiating a nerve impulse in an adjacent neuron. Many kinds of receptors have evolved among vertebrates, with each receptor sensitive to a different aspect of the environment (Figure 15-1).

For you to be aware of, or to sense, your internal or external environment, certain events must take place. First, a change in the environment (the stimulus) must be of sufficient magnitude to open ion channels within the membrane of the receptor cell, thus depolarizing it (see Chapter 13). This depolarization is called a *generator potential.* When the generator potential reaches the threshold level, it initiates an action potential—a nerve impulse—in the sensory neurons with which the receptor synapses. In simple receptors (the dendrite endings of sensory neurons) the generator potential initiates an action potential along that same neuron. The impulse is conducted by nerve fibers to either the spinal cord or the brain.

KEY CONCEPTS

- Sensory receptors are the body's "window on the world," detecting stimuli from both the external and the internal environments.
- Stimuli, or changes in the environment, are detected when they are of a sufficient magnitude to open ion channels within the membranes of receptor cells, depolarizing them and thus initiating action potentials along those neurons or adjacent neurons.
- The body has a mosaic of receptors referred to as the general senses, which detect stimuli such as pain, pressure, touch, and temperature within the body's internal and external environments.
- The body has a few specialized receptors referred to as the special senses, which detect smell, taste, sight, hearing, and balance.

OUTLINE

A B C

FIGURE 15-1 Types of sensory information processed by sensory receptors. A, A person who feels sick is relying on information about the body's internal environment. A headache, for example, causes pain because of the activation of pain receptors located within the body. **B,** This child can ride his bike without hands by using information about his position in space. **C,** This child is reading, using her eyes to perceive patterns of light. In this way, the reader is sensing information about the external environment.

Impulses conducted to specific sensory areas of the cerebral cortex produce conscious sensations (see Chapter 14). Only the cerebral cortex can "see" a flower, "hear" a symphony, or "feel" a paper cut. Although not a part of the cortex, the thalamus can sense pain, but it is unable to distinguish its source or intensity. Impulses that end at the spinal cord or brainstem do not produce conscious sensations. They may, however, result in reflex activity such as the rhythmic contraction of the muscles of breathing or the turning of your head toward a startling noise.

Receptors provide the body with three types of information:

1. Information about the body's internal environment
2. Information about the body's position in space
3. Information about the external environment

Receptors that sense the internal environment are located deep in the body within the walls of blood vessels and organs. They tell you when you are hungry, thirsty, sick, or tired. Receptors sensitive to stimuli outside the body are located at or near the body surface. They allow you to see, hear, taste, and feel various stimuli in the environment. Receptors that provide information about body position and movement are located in the muscles, tendons, joints, and inner ear. They tell you whether you are lying down or standing up and where the various parts of your body are in relationship to one another.

Sensory receptors are cells that can change environmental stimuli into nerve impulses. Specific receptors detect certain stimuli that inform the body about both its external and internal environments and its position in space.

Sensing the Body's Internal Environment

Many of the neurons that monitor body functions are simply nerve endings that depolarize in response to direct physical stimulation—to temperature, to chemicals such as carbon dioxide diffusing into the nerve cell, or to a bending or stretching of the neuron cell membrane. Among the simplest of these neurons are those that report on changes in body temperature and blood chemistry.

Temperature-sensitive neurons in the hypothalamus, a tiny portion of the brain that lies above the brainstem, act as your body's thermostat. These neurons constantly take your temperature by monitoring the temperature of your blood. If the temperature of the blood rises, such as when you are sick or vigorously exercising, neurons in the hypothalamus trigger your body's heat loss mechanisms. Such mechanisms include dilating, or widening, the blood vessels closest to the skin so that excess heat can be lost to the environment. Likewise, if the temperature of the blood drops, such as when you are in a cold environment, neurons in the hypothalamus trigger your body's heat production mechanisms. Such mechanisms include shivering, a cycle of contraction and relaxation of your skeletal muscles producing body heat as a waste product of the cellular respiration that generates adenosine triphosphate (ATP) for muscle contraction.

Other receptors are sensitive to the levels of carbon dioxide and oxygen in your blood as well as its pH. These receptors are embedded within the walls of your arteries at several locations in the circulatory system. Bathed by the blood that flows through the arteries, these chemical receptors provide input to respiratory centers in the medulla and pons, which use this information to regulate the rate of respiration. When carbon dioxide and pH levels in the blood rise and the oxygen level falls, the respiratory centers respond by increasing the respiration rate.

Various other receptors that sense the environment within your body have membranes whose ion channels open in response to mechanical force. Put simply, twisting, bending, or stretching these nerve endings results in depolarization of their membranes, causing these nerves to "fire." These receptors differ from one another primarily in their locations and in the way that they are oriented with respect to the stimulus. Pain receptors, for example, are widely distributed throughout the body. They respond to many different types of stimuli when these stimuli reach a level that can endanger the body. Pain receptors deep within the body detect such internal environmental stresses as inadequate blood flow to an organ, excessive stretching of a structure, and spasms of muscle tissue.

Specialized mechanical receptors sensitive to pressure changes are attached to the walls of three major arteries of your body (the carotids and the aorta) and constantly monitor your blood pressure. These highly branched networks of nerve endings detect the stretching of the walls of these arteries caused by the "push" of the blood as it is pumped out of the heart. These neurons fire at a low rate at all times, sending a steady stream of impulses to the cardiac (heart) center in the medulla. A rise in blood pressure stretches the walls of these vessels, resulting in a higher frequency of nerve impulses, which inhibits the cardiac center. A drop in blood pressure lessens the stretch, resulting in a lower frequency of nerve impulses, which stimulates the cardiac center.

Receptors that sense the body's internal environment are located within the walls of blood vessels and organs.

Sensing the Body's Position in Space

Your body uses receptors called **proprioceptors** to sense its position in space. Proprioceptors tell you how much your arms and legs are bent, where they are in relationship to your body, and where your head is relative to the ground even if you cannot see!

Proprioceptors buried deep within your muscles keep track of the degree to which your muscles are contracted (Figure 15-2). These receptors are actually specialized muscle cells called *muscle spindles.* Wrapped around each spindle is the end of a sensory neuron called a **stretch receptor.** When a muscle is stretched, the muscle spindle gets longer and stretches the nerve ending, repeatedly stimulating it to fire. Conversely, when the muscle contracts, the tension on the fiber lessens and the stretch receptor ceases to fire. The frequency of stretch receptor discharges along the nerve fiber tells the body the degree to which a muscle is contracted at any given moment. The central nervous system uses this information to control those movements that involve the combined action of several muscles, such as those involved in breathing and walking. Other proprioceptors are found in your tendons, which is the connective tissue that joins your muscles to your bones, and in the tissue surrounding your joints. These receptors are also stimulated when they are stretched and help protect your muscles, tendons, and joints from excessive tension and pulling.

Gravity and motion receptors are other types of proprioceptors. These receptors, both located in the inner ear, help you maintain your equilibrium, or balance. They are discussed later in the chapter.

Receptors called proprioceptors, located within the skeletal muscles, tendons, and inner ear, give the body information about the position of its parts relative to each other and to the pull of gravity.

proprioceptors (**pro** pree oh **sep** turz) receptors located within skeletal muscles, tendons, and the inner ear that give the body information about the position of its parts relative to each other and to the pull of gravity.

stretch receptor sensory neuron endings that are stimulated to fire when tension is placed on them.

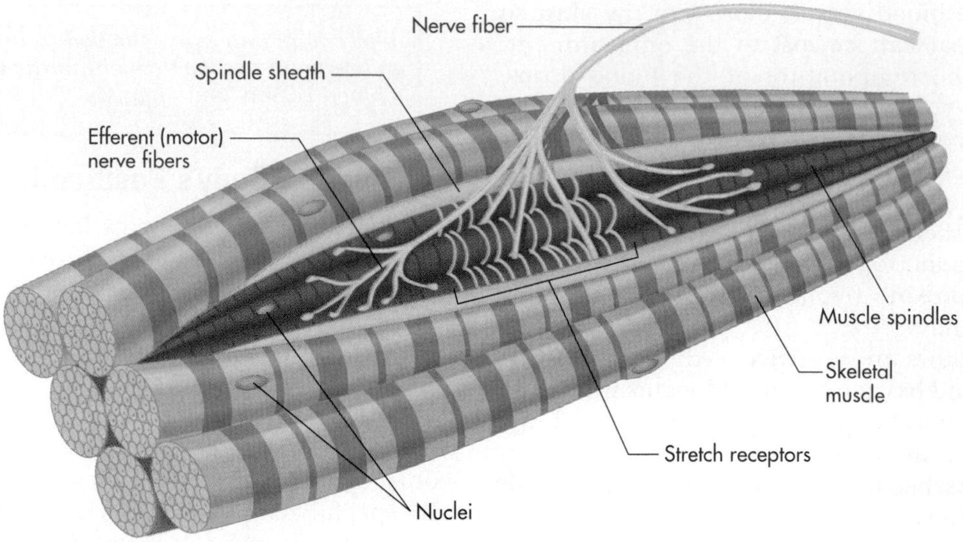

FIGURE 15-2 A stretch receptor embedded within skeletal muscle. Stretching of the muscle elongates the spindle fibers, which deforms nerve endings, causing them to fire and send a nerve impulse out along the nerve fiber.

Sensing the External Environment

Various receptors are located at or near the body surface that sense changes in the external environment. Many types of simple, microscopic receptors are located within the skin. These receptors detect touch, pressure, pain, and temperature—the so-called general senses. Other, more structurally complex receptors are located in specific places at or near the body surface. These receptors detect smell, taste, sight, and hearing and help you keep your balance. They are your special senses.

The General Senses

The body has a mosaic of mechanical receptors in the skin for touch, pressure, and pain. These receptors are the same as those found deeper in the body, but because they are located close to the surface of the skin, they provide information about the external rather than the internal environment. Figure 15-3 is a diagram of a cross section of human skin showing these specialized nerve endings.

Disk-shaped dendrite endings called *Merkel's disks* and egg-shaped receptors called *Meissner's corpuscles* are two types of touch receptors. (The word *corpuscle* means "little body.") Both are widely distributed in the skin but are most numerous in the hands, feet, eyelids, tip of the tongue, lips, nipples, clitoris, and tip of the penis. In addition, free nerve endings wrap around the roots of hairs and detect any stimulus, such as the wind or an insect, that moves body hair. Pressure receptors called *Pacinian corpuscles* are located deeper within the skin. As shown in Figure 15-3, Pacinian corpuscles look layered, much like an onion. The layers are made up of connective tissue with dendrites sandwiched between. These receptors are most numerous in the nipples and external geni-

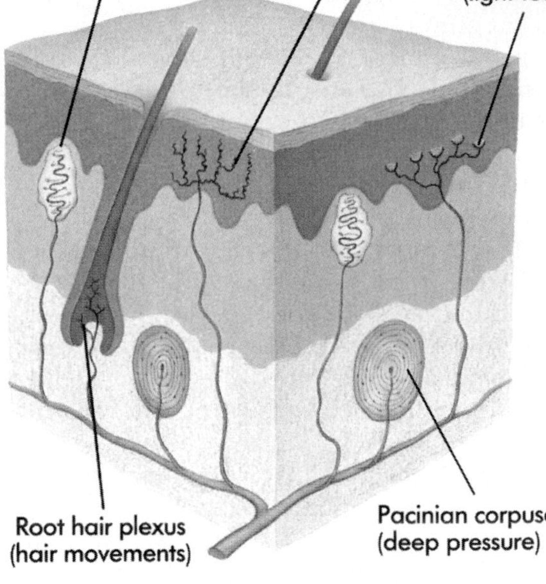

FIGURE 15-3 Receptors in the human skin. These receptors, although they are found within the body, sense the external environment because they are located close to the surface of the skin.

tals of both sexes. Pacinian corpuscles are also a part of the body's internal sensing system because of their locations around joints and tendons, in muscles, and in certain organs.

Two populations of nerve endings in the skin are sensitive to changes in temperature. One set is stimulated by a lowering of temperature (the cold receptors) and the other by a raising of temperature (the heat receptors). Scientists are unsure how these receptors work. However, evidence suggests that changes in the temperature cause changes in the shapes of the proteins that make up the sodium channels of their cell membranes, leading to the depolarization and firing of these neurons.

> *Various receptors in the skin sense touch, pressure, pain, and temperature stimuli from the external environment.*

The Special Senses

The special senses—smell, taste, sight, hearing, and balance—have receptors that are much more specialized and complex than the receptors for the general senses. Smell and taste receptors sense chemicals, whereas hearing receptors sense sound waves. Other receptors in the ears detect motion and the effect of gravity on the body. Sight receptors sense wavelengths of electromagnetic energy known as *visible light.*

Smell

Although smell receptors are structurally the simplest of your special senses, they can distinguish several thousand different odors. Smell receptors are also called **olfactory receptors,** a name derived from Latin words meaning "to make" *(facio)* "a smell" *(oleo).* They are *not* located (surprisingly) in your nostrils but in a $^{1}/_{2}$-inch square of tissue in the roof of the nasal cavity—just behind the bridge of your nose (Figure 15-4).

Olfactory receptors consist of neurons whose cell bodies are embedded in the nasal epithelium, the tissue that lines the nasal cavity. Dendrites extend from these cell bodies. Cilia, microscopic projections of the dendrites, poke out of the epithelium like minute tufts of hair and are bathed by the mucus that covers this tissue. Gases in the air dissolve in the mucus and come into contact with the cilia.

Scientific evidence suggests that the olfactory receptors detect different smells because of specific binding of airborne gases with receptor chemicals located within the cilia. This interaction opens ion channels within the membrane of the receptor so

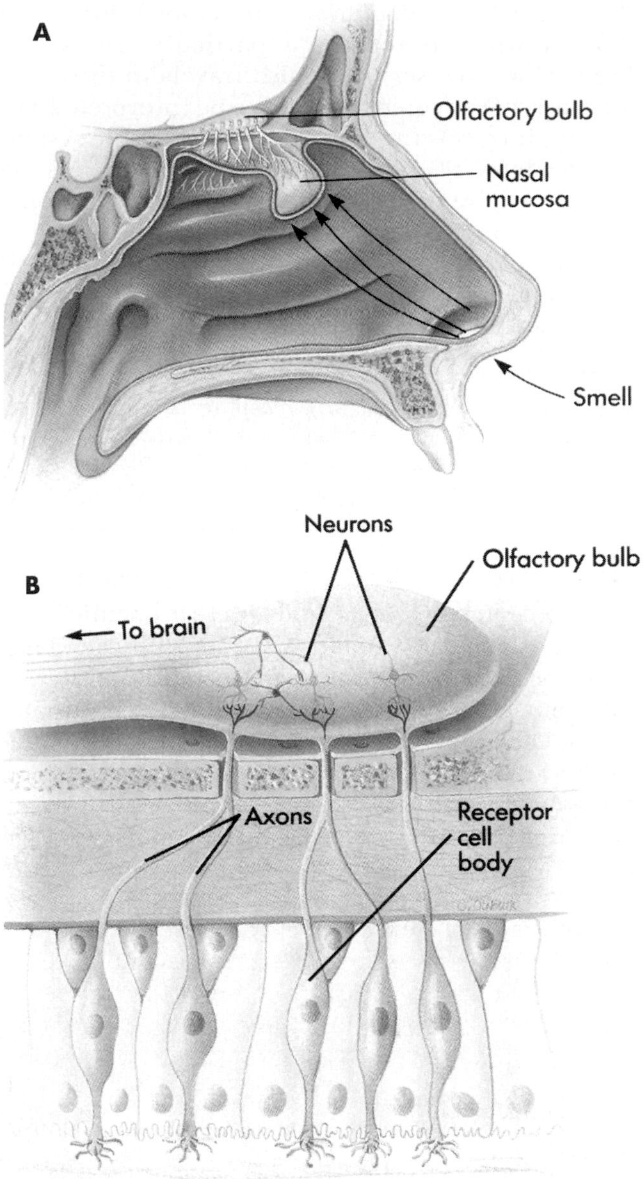

FIGURE 15-4 The sense of smell. A, Humans smell by means of olfactory receptor cells located in the lining of the nasal cavity. The receptor cells are neurons. Airborne gases combine with chemicals within the cilia on each neuron's dendrites, and the neuron is depolarized. **B,** Axons from these sensory neurons carry the impulse back through the olfactory nerve directly to the cerebral cortex in the brain.

> **olfactory receptors** (ole **fak** tuh ree) neurons whose cell bodies are embedded in the nasal epithelium; they detect smells when different airborne chemicals bind with receptor chemicals in their ciliated dendrite endings.

that a generator potential is developed, firing the neuron when it reaches a particular threshold. From here, the nerve impulse travels to the olfactory area of the cerebral cortex to be interpreted. On its way, it travels through the brain's limbic system (see Chapter 14), the area of the brain responsible for many drives and emotions. Does the odor of baking cookies please you? Does the smell of rotting garbage cause you to turn your head in disgust? That is your limbic system at work—in conjunction with your cerebrum and your sense of smell.

> *The smell, or olfactory, receptors are located in the nasal epithelium and detect airborne chemicals as they bind with receptor chemicals.*

Taste

Taste receptors, or **taste buds,** detect chemicals in the foods you eat. Humans have four kinds of taste buds located on the tongue that each respond to a different class of stimulus: salty, sweet, sour, and bitter. Each type of taste bud is concentrated on different areas of the tongue, with sweet and salty on the front, sour on the sides, and bitter at the back (Figure 15-5). Humans, of course, perceive a rich and diverse array of tastes that are much more complex than simply salty, sweet, sour, and bitter. Interestingly, these tastes are composed of different combinations of impulses from just these four types of chemoreceptors. In addition, the sense of taste interacts with the sense of smell to produce the taste sensation, as you may have noticed when you had a "stuffy nose" and could perceive only the four basic tastes—those not dependent on smell for their perception.

Nearly 9000 taste buds are packed within the short projections on your tongue known as *papillae.* Figure 15-5 shows a drawing of a typical papilla on the tongue and points out the location of taste buds deep within the papilla. Taste buds are microscopic structures shaped like tiny onions. Each is made up of 30 to 80 receptor cells bound together by support cells. Hairlike projections of the receptor cells poke through an opening in the taste bud called a *taste pore.* The receptor cells are stimulated by the various chemicals in food as these chemicals dissolve in the saliva and come into contact with the cellular "hairs." A certain threshold of stimulation produces generator potentials within the receptor cells, firing neighboring sensory nerve fibers. These nerve fibers travel from the tongue and throat to the cerebrum, medulla, and thalamus. Some of these nerve fibers also travel to the reticular formation (see Chapter 14), the part of

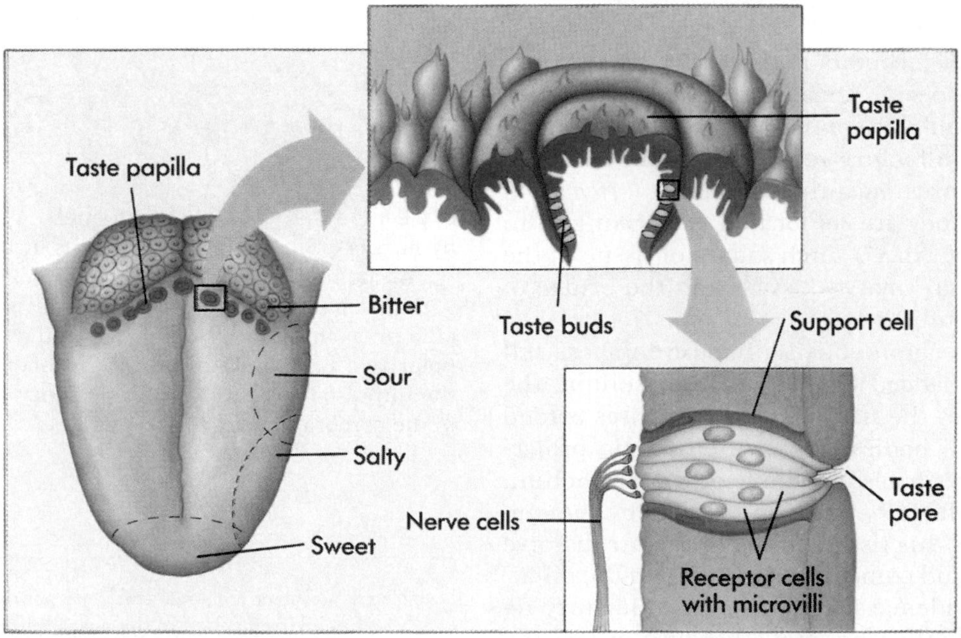

FIGURE 15-5 Taste. Humans have four kinds of taste buds (bitter, sour, salty, and sweet) located on different regions of the tongue. Groups of taste buds are typically organized in sensory projections called papillae. Individual taste buds are bulb-shaped collections of chemical receptor cells that open into the mouth through a pore.

the brain that monitors incoming stimuli, identifying important information.

> *The taste receptors, or taste buds, are microscopic chemoreceptors embedded within the papillae of the tongue. They work with the olfactory receptors to produce the taste sensation.*

Sight

The eye is the only human sense organ that can detect wavelengths of electromagnetic energy, or radiation, known as *visible light.* Visible light is only a part of the full range of electromagnetic radiation, which is called the electromagnetic spectrum (see Figure 6-2). Electromagnetic energy is repeating disturbances in electrical and magnetic fields in the atmosphere and can be thought of as a wave, much like the repeating disturbances or tiny waves caused by a stone thrown in a still pond.

The electromagnetic waves coming from the sun vary in length. Their lengths are measured from the crest of one wave to the crest of the next and are expressed in meters or in billionths of meters (nanometers, or nm). The shortest wavelengths are gamma rays; the longest are radio waves. Visible light has wavelengths in between these two extremes, but it, too, is made up of many different wavelengths. Remember the last rainbow you saw? Its array of colors was caused by the separation of the various wavelengths of visible light as they passed through tiny droplets of water in the air.

The receptors of the eye sensitive to the various wavelengths of visible light are called **rods** and **cones.** These receptors are located in the back of the eye and act somewhat like film in a camera. Light that falls on the eye is focused by a lens onto these receptors, just as the lens of a camera focuses light on film.

It is important to discuss how the eye's sensory receptors work before discussing the structure of the eye in greater detail. A primary event is the absorption of a photon of light by a pigment. Photons are units or packets of electromagnetic energy. One complexity of electromagnetic energy is that it travels as waves but also behaves like individual units. In the eye, photons of light are absorbed by the pigment *rhodopsin.* This pigment is located in the tips of rod and cone cells that make up tissue at the back of the eye called the **retina** (see Figure 15-9). Rhodopsin consists of *retinal,* a derivative of vitamin A, coupled to a membrane protein called *opsin.* As each rhodopsin complex absorbs a photon of light, the molecular shapes of both retinal and opsin change. As a result, retinal separates from opsin. The activated opsin triggers a series of events that are responsible for *closing* sodium channels. As discussed in Chapter 13, stimuli usually lead to an *opening* of sodium channels, allowing sodium ions to rush into the cell, *depolarizing* the receptor membrane. Sodium channels close in rod and cone cells when they are stimulated. The result is *hyperpolarization*—the interior of the rod becomes even more negatively charged than before. Hyperpolarization of the cell in turn activates a mechanism that *lowers* the amount of neurotransmitter released by the cell. This decrease in neurotransmitter release lessens the frequency of firing in adjacent neurons—nerve signals that eventually reach the brain and are interpreted as patterns of light and dark.

> *Light receptors are rod and cone cells located within the retina at the back of the eye. The pigment within these cells absorbs photons of light, which causes a series of events leading to a hyperpolarization of the receptor cells and a subsequent firing of adjacent neurons.*

Rod cells are able to detect low levels of light because even one photon can stimulate a rod cell. In addition, they detect only white light; thus vision with rod cells alone is black, white, and shades of gray. Did you ever try to see color in extremely dim light? Your rods alone are at work, and they cannot see color. Color vision is achieved by your cone cells, which function in bright light. There are three kinds of cone cells (although they all look the same under the electron microscope). Each type of cone cell possesses rhodopsin molecules that have

taste buds microscopic receptors embedded within the papillae of the tongue that work with olfactory receptors to produce taste sensations; humans have taste buds specialized to respond to four different kinds of sensations: salty, sweet, sour, and bitter.

rods light receptors located within the retina at the back of the eye that function in dim light and detect white light only.

cones light receptors located within the retina at the back of the eye that function in bright light and detect color.

retina (ret uh nuh) tissue at the back of the eye that is composed of rod and cone cells and that is sensitive to light.

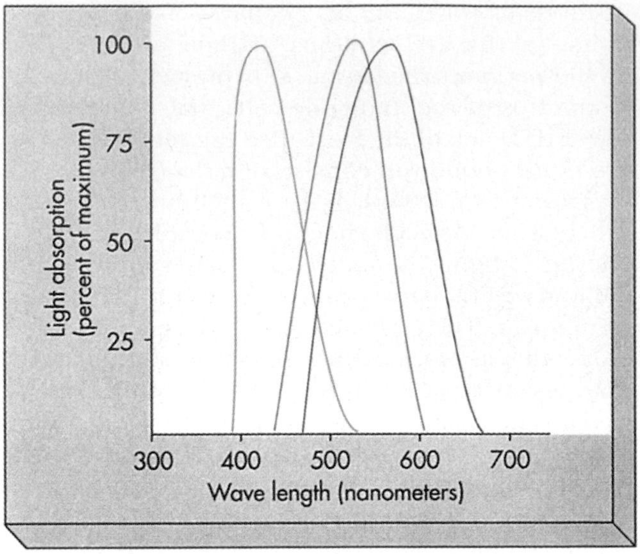

FIGURE 15-6 The absorption spectrum of human vision. The wavelength of light absorbed by the visual pigment of cone cells depends on the opsin proteins to which the pigment is bound. There are three such proteins, producing cones that absorb at 455 nanometers *(blue)*, 530 nanometers *(green)*, and 625 nanometers *(red)*.

slightly different shapes from one another because they have different types of opsin. These differences determine which wavelengths of light the pigment (and therefore the cone cell) will absorb. One type absorbs wavelengths of light in the 455-nanometer (blue-absorbing) range. Another type absorbs in the 530-nanometer (green-absorbing) range, and the third in the 625-nanometer (red-absorbing) range. The color you perceive depends on how strongly each group of cones is stimulated by a light source. The pigment in rod cells, on the other hand, absorbs in the 500-nanometer range. As shown in Figure 15-6, this range encompasses the three primary colors (blue, green, and red), which together make up white light.

Rod cells function in dim light and detect white light only, whereas cone cells function in bright light and detect color.

The basic structure of the human eye is shown in Figure 15-7. Your eyes are each about 1 inch in diameter and are covered and protected by a tough outer layer of connective tissue called the **sclera.**

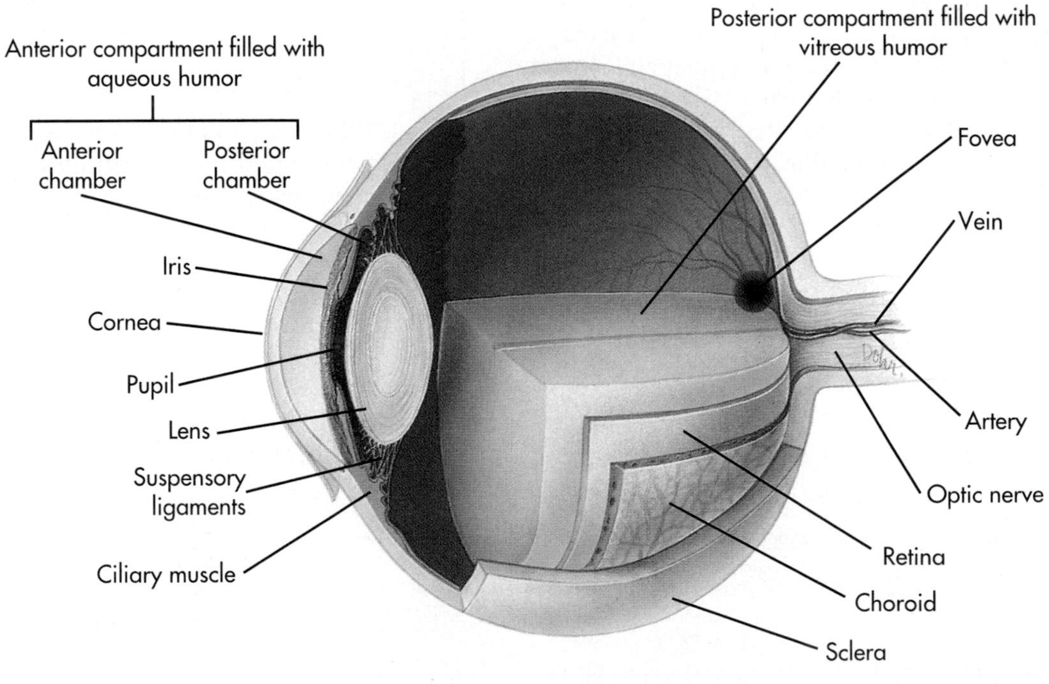

FIGURE 15-7 Structure of the human eye. Light passes through the transparent cornea and is focused by the lens on the retina at a particular location called the fovea. The retina is rich in rods and cones.

The front of the eye is transparent, allowing light to enter the eye. This portion of the eye's outer layer is called the **cornea.** Because the cornea is rounded, it not only allows light to enter the eye but bends it as well. This bending, or refraction, of light occurs as the light waves slow down as they move from the air and pass through the tissue of the cornea. You can see this phenomenon if you put a straw in a glass of water. The parts of the straw both outside and inside the water do not appear connected. If the cornea were flat, the light waves or rays would exit the cornea traveling in a path parallel to their original path. However, the cornea is rounded; the light waves that enter above, below, or to the side of the center of the cornea are bent slightly toward one another. Therefore the cornea is the first structure of the eye that begins to focus incoming light onto a point at the rear of the eye.

A compartment behind the cornea is filled with a watery fluid called the **aqueous humor.** The word *humor* refers to "fluid within the body," and the word *aqueous* means "water." Fresh aqueous humor is continually produced as old fluid is drained into the bloodstream. This fluid nourishes both the cornea and the lens, because neither structure has a supply of blood vessels. In addition, the aqueous humor (along with fluid further back in the eye called the *vitreous* [Latin for "glass"] *humor*) creates a pressure within the eyeball that maintains the eyeball's shape and keeps the retina pressed properly against the back of the eyeball.

The **lens** lies just behind the aqueous humor and plays a major role in focusing the light that enters the eye onto the retina at the back of the eye. It looks much like a lemon drop candy or a somewhat flattened balloon. If you were to cut a lens in half, it would look similar to an onion because it is made up of layer upon layer of protein fibers. The lens is encircled by ligaments that suspend it within the eye. These ligaments are attached to a tiny circular muscle called the **ciliary muscle** that, by contracting or relaxing, slightly changes the shape of the lens. The greater the curve of the lens, the more sharply it bends light rays toward one another.

In persons with normal vision (Figure 15-8, *A*), this bending of the light results in its being focused on the retina. Some people, however, may have an

JUST WONDERING. . . .

What is radial keratotomy and how does it work?

Radial keratotomy is a surgical procedure to treat myopia, or nearsightedness. In myopia, the eye focuses the images of distant objects in front of the retina rather than on the retina. In other words, you cannot see objects clearly that are far away. Either the lens or the cornea of the eye are too curved, focusing the image at too short a distance within the eye, or the eyeball is too long. Radial keratotomy corrects this condition by flattening out the cornea, which brings the focal point farther back in the eye.

The term *keratotomy* means cutting the cornea. Radial keratotomy refers to cutting symmetrical incisions that emanate from the center of the cornea (but that do not reach its center). These incisions look somewhat like the spokes on a wheel. Cutting the cornea flattens it because the incisions relax the tissue at right angles to the incisions (Figure 15-*A*); the spaces created eventually become filled in with scar tissue.

This procedure has evolved since its introduction in the late 1970s. Initially, 32 incisions were used, but research has shown that far fewer incisions give nearly the same results if great care is given to their depth and placement. The use of lasers to develop corneal flattening without incisions is currently in clinical trials, and will most likely be in use by the time you read this.

15-A

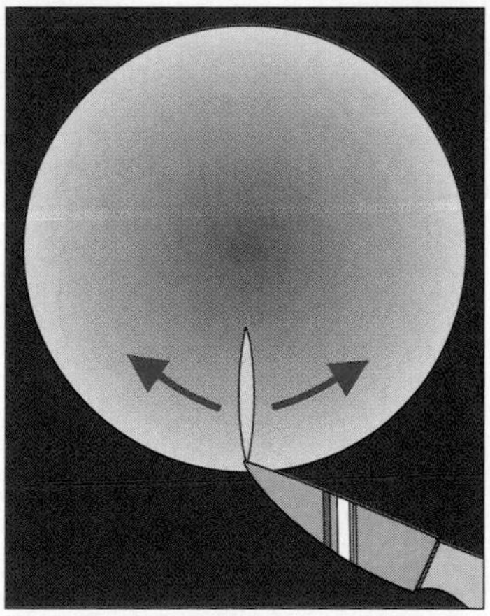

sclera (**sklear** uh) the tough outer layer of connective tissue that covers and protects the eye.

cornea (**kor** nee uh) the rounded, transparent portion of the eye's outer layer that permits light to enter the eye.

aqueous humor (**ayk** wee us **hyoo** mur) the watery fluid that fills the chamber behind the cornea.

lens a body in the eye lying just behind the aqueous humor that plays a major role in focusing the light that enters the eye onto the retina at the back of the eye.

ciliary muscle (**sill** ee err ee) a tiny circular muscle that slightly changes the shape of the eye's lens by contracting or relaxing.

elongated eyeball or a thickened lens. In either case, the light is focused on a spot in front of the retina. Such people are nearsighted. As Figure 15-8, *B* shows, this condition can be corrected with a lens that spreads the light rays entering the eye so that they focus on the retina. Some people are farsighted, having either a shortened eyeball or a thin lens. Light is focused on a spot behind the retina. This condition can be corrected with a lens that converges entering light rays so that they focus on the retina.

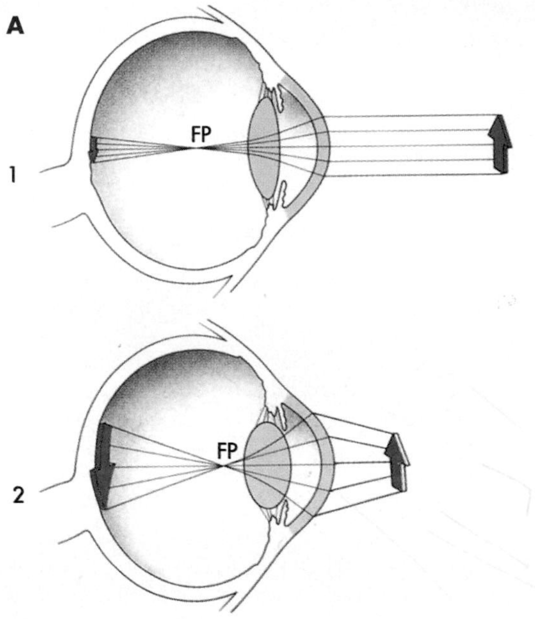

FIGURE 15-8 Focusing the human eye. Contraction of the ciliary muscles pulls on suspensory ligaments and changes the shape of the lens, which alters the focal point forward or backward. The focal point (*FP*) is where light rays cross. **A,** In focusing a distant object, the lens is flattened, focusing the image on the retina. As for focusing on close objects, the lens is more rounded. **B,** Visual disorders and their correction by various lenses. Upper left, nearsightedness; upper right, correction of nearsightedness with a concave lens; lower left, farsightedness; lower right, correction of farsightedness with convex lens.

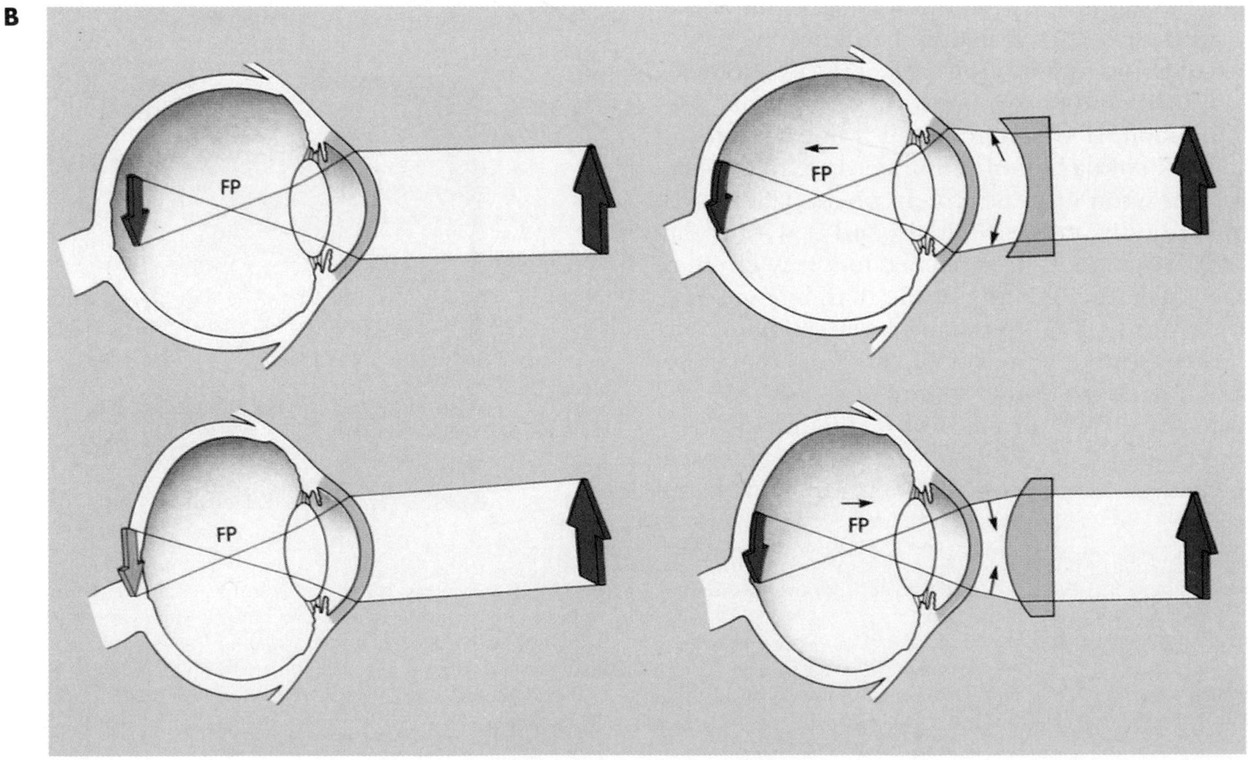

The amount of light entering the eye is controlled by a diaphragm called the **iris,** which lies between the cornea and the lens. The iris reduces the size of the transparent zone, or **pupil,** of the eye (what you see as a black dot) through which the light passes. Together, the ciliary muscle and the iris make up two of the structures of the middle layer of the eye. The third structure is the **choroid.** As shown in Figure 15-7, the choroid extends toward the back of the eye as a thin, dark-brown membrane that lines the sclera. It contains vessels carrying blood that nourishes the retina and a dark pigment that absorbs light rays so that they will not be reflected within the eyeball.

The inner layer of the eye is the retina. The retina lines the back of the eye and contains the rods and cones. The human retina contains about 7 million cones, most of which are located at a central region of the retina called the **fovea.** The lens focuses incoming light on this spot, the area of sharpest vision because of this high concentration of cones. There are no rods in the fovea. Some are located just outside the fovea and increase in concentration as the distance from it increases.

Each foveal cone cell makes a one-to-one connection with a special kind of neuron called a *bipolar cell* (Figure 15-9). Each of the bipolar cells is connected in turn to a ganglion cell, whose axon is

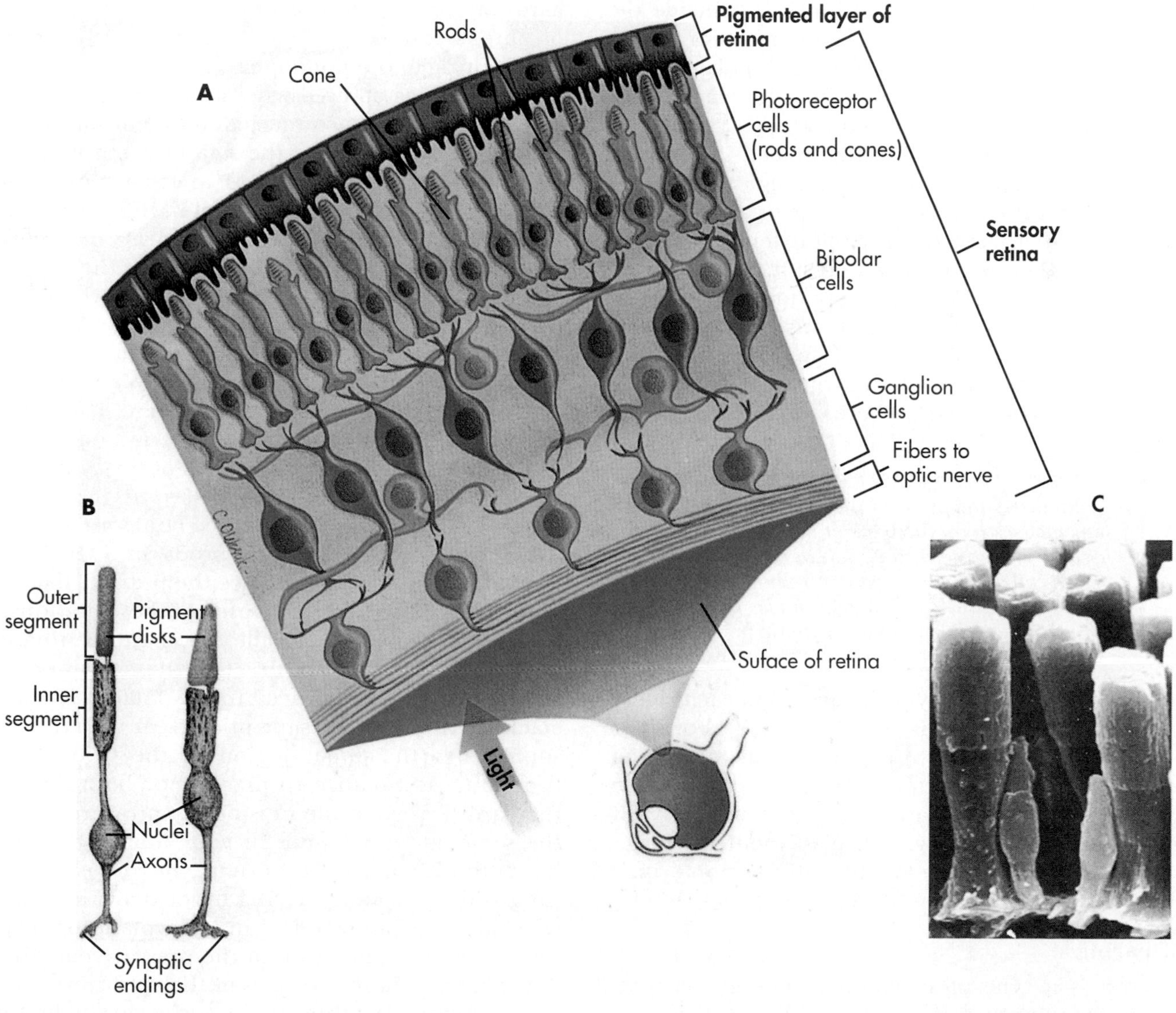

FIGURE 15-9 Structure of the retina. A, Pigmented and sensory layers of the retina. Note that the rods and cones are at the rear of the retina. Bipolar cells receive the hyperpolarization stimulus from the cone cells and transmit a depolarization stimulus to the ganglion cells. The ganglion cells, whose axons are part of the optic nerve, transmit the impulse to the brain. **B,** Rod and cone cells. Rods detect white light only, cone cells detect color. **C,** Electron micrograph of rod and cone cells *(7000x).*

part of the **optic nerve.** The bipolar cells receive the hyperpolarization stimulus from the cone cells and transmit a depolarization stimulus to the ganglion cells. The axons of the ganglion cells transmit the impulses to the brain. The frequency of pulses transmitted by any one receptor provides information about light intensity. The pattern of firing among the different foveal axons provides a point-to-point image. The different cone cells provide information about the color of the image.

The relationship of receptors to bipolar cells to ganglion cells is one-to-one-to-one within the fovea. Put simply, each receptor (cone) cell in the fovea synapses with its own bipolar cell that in turn synapses with its own ganglion cell. Outside the fovea, however, the output of many receptor cells is channeled to one bipolar cell. Many bipolar cells, in turn, synapse with one ganglion cell. In fact, in the outer edge of the retina, more than 125 receptor cells feed stimuli to each ganglion cell in the optic nerve. In this outer, or peripheral, region, many additional neurons cross-connect the ganglion cells with one another and carry out extensive processing of visual information. As a result, this portion of the retina does not transmit a point-to-point image as the fovea does but transmits instead a processed version of the visual input that may be interpreted simply as movement. Think of the periphery of your retina as a detector, whereas your fovea is an inspector.

Light enters the pupil of the eye, and the lens focuses it on the back of the eye, on an area of the retina that is rich in cone cells. These cells, along with surrounding rod and cone cells, initiate nervous impulses that travel to the cerebrum via the optic nerve.

The positioning of your two eyes on each side of the head sends two slightly different sets of information to the brain about what you are seeing. Each set of information is slightly different because each eye views an object from a slightly different angle. This slight displacement of images gives *depth perception*—a three-dimensional quality—to your sight.

Hearing

Sound is a type of mechanical energy resulting from the vibration of an object. This vibration disturbs the air around the object, pushing the molecules in the air closer to one another, compressing them. These areas of compression are followed by areas of molecules that are not compressed. If you could visualize these disturbances, they would look much like the concentric circles of waves on the surface of water that occur when you throw a pebble into a still pond. Sound waves are these wavelike disturbances of molecules in the air (or in certain other substances such as water, glass, iron, or wood) caused by a vibrating object. Two parts of the ear—the outer and middle ear—work together to transmit sound waves to the inner ear, where sound stimuli are changed into nerve impulses (Figure 15-10).

The flaps of skin on the outside of the head that are called ears are only one part of the *outer ear.* Each flap, or **pinna,** funnels sound waves into an **auditory canal.** The auditory canal, which is about 1 inch long, leads directly to the eardrum. The eardrum, or **tympanic membrane,** is a thin piece of fibrous connective tissue that is stretched over the opening to the **middle ear.** On its external side the eardrum is covered with skin. On its internal side the eardrum is connected to one of the smallest bones in the body—the *hammer,* or **malleus.** The malleus is connected to another tiny bone called the *anvil,* or **incus.** The anvil, in turn, is connected to a third bone called the *stirrup,* or **stapes.** These three bones are the structures of the middle ear that pick up sound vibrations from the outer ear and transfer them to the inner ear.

Sound waves entering the outer ear beat against the tympanic membrane, causing it to vibrate like a drum. Vibrations of this membrane cause the malleus to move with a rocking motion because it is attached to the internal surface of the tympanic membrane. This rocking is transferred, in turn, to the incus and stapes. These three bones are hinged to one another in a way that produces a lever system, a mechanism that causes them to act like an amplifier and increase the force of the vibrations.

Also located in the middle ear is an opening to the eustachian tube (a structure named after an Italian physician who lived in the 1500s). The **eustachian tube** connects the middle ear with the nasopharynx (the upper region of the throat). Its function is to equalize air pressure on both sides of the eardrum when the outside air pressure is not the same as the pressure in your middle ear. You have probably had the experience of your ears popping while you were driving up or down a mountain, taking off and landing in an airplane, or deep-sea diving; it is the result of the pressure equalization between these two sides of the eardrum.

The stirrup, the third in the series of middle ear bones, is attached to a membrane that separates the middle ear from the inner ear. This membrane covers the **oval window,** the entrance to the *inner ear,* the part of the ear in which hearing actually takes place. In addition, other parts of the inner ear de-

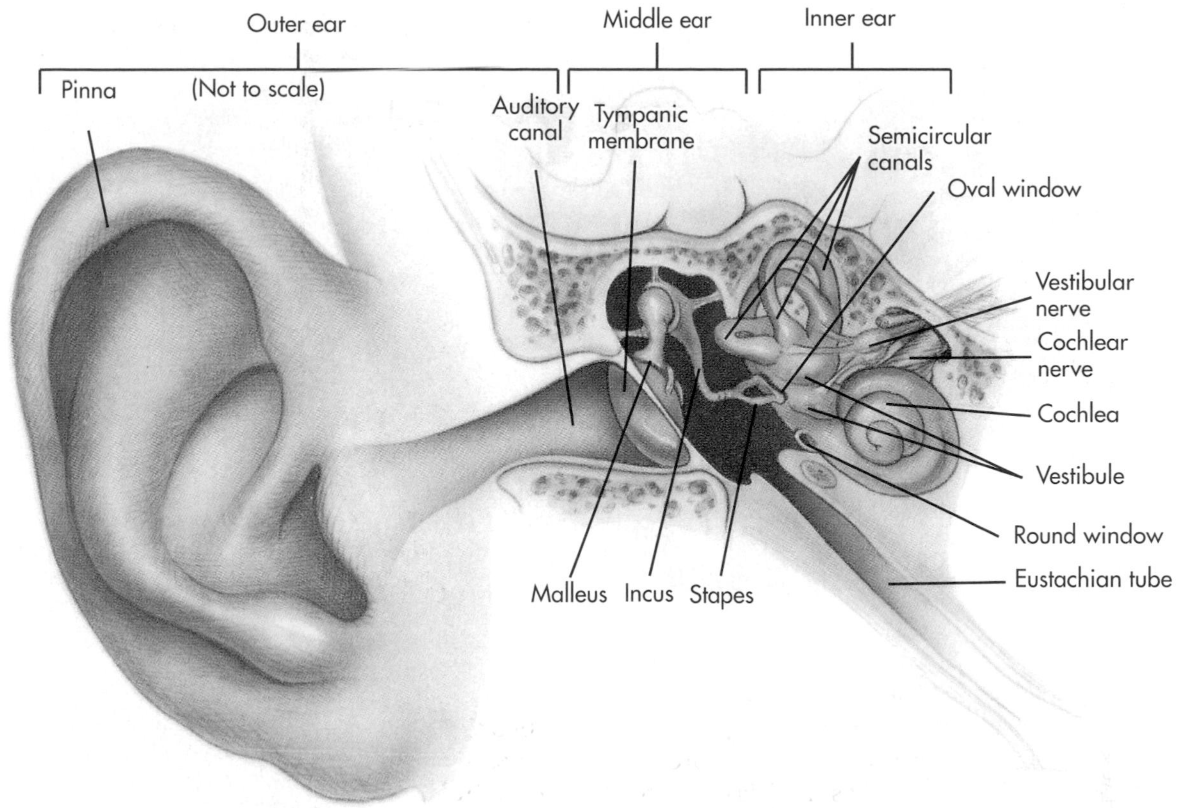

FIGURE 15-10 Structure of the human ear. The human ear is composed of outer, middle, and inner sections. The outer ear extends from the pinna to the tympanic membrane. The middle ear contains bones that transmit sound vibrations from the tympanic membrane to the cochlea. The cochlea contains the organ of hearing and makes up part of the inner ear.

iris (**eye** rus) a diaphragm lying between the cornea and lens that controls the amount of light entering the eye.

pupil the opening in the center of the iris through which light passes.

choroid (**kor** oyd) a thin, dark-brown membrane that lines the sclera of the eye and contains blood vessels that nourish the retina, and a dark pigment that absorbs light rays so they will not be reflected within the eyeball.

fovea (**foe** vee uh) a spot on the retina that has the highest concentration of cones; the lens focuses images on this spot, resulting in sharp vision.

optic nerve the nerve that carries impulses from the retina to the brain.

pinna (**pin** uh) the ear flap located on the outside of the head.

auditory canal (awd uh **tore** ee) a 1-inch long canal that receives sound waves funneled from the pinna and carries them directly to the eardrum.

tympanic membrane (tim **pan** ik) a thin piece of fibrous connective tissue that is stretched over the opening to the middle ear; the tympanic membrane is also known as the *eardrum*.

middle ear the portion of the ear that contains three bones that amplify the force of sound vibrations as they conduct them from the tympanic membrane to the oval window of the inner ear.

malleus (**mal** ee us) a very small bone, connected to the internal side of the eardrum, that works with two other small bones, the incus and stapes, to amplify sound vibrations and carry them from the outer ear to the inner ear. The malleus is also known as the *hammer*.

incus (**ing** kus) one of the three bones of the middle ear that amplify sound vibrations and carry them from the outer ear to the inner ear. The incus is also known as the *anvil*.

stapes (**stay** peez) one of the three bones of the middle ear that amplify sound vibrations and carry them from the outer ear to the inner ear. The stapes is also known as the *stirrup*.

eustachian tube (yoo **stay** kee un *or* yoo **stay** shun) a structure that connects the middle ear with the nasopharynx (the upper throat); it equalizes air pressure on both sides of the eardrum when the outside air pressure is not the same as the pressure in the middle ear.

oval window the entrance to the cochlea. The stirrup fits into this membrane-covered opening.

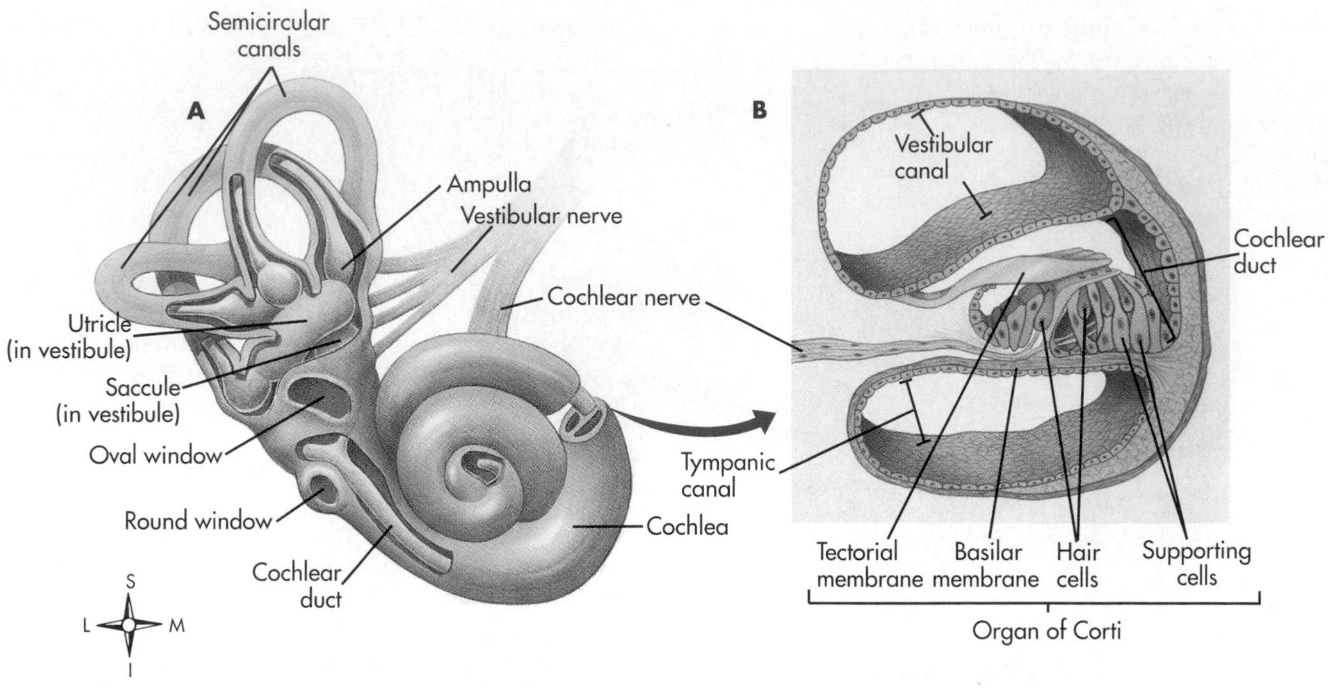

FIGURE 15-11 Structure of the inner ear. A, The inner ear contains the organs of hearing and the organs that are responsible for maintaining the body's equilibrium. **B,** The inset shows a section of the cochlea. Hair cells in the organ of Corti detect sound and send the information to the brain via the cochlear nerve. The vestibular and cochlear nerves join to form the eighth cranial nerve.

tect motion and the effect of gravity on the body.

The oval window is the entrance to the fluid-filled **cochlea**—the part of the inner ear that contains the organ of hearing (Figure 15-11). To understand the cochlea's structure and function, imagine a tapering, blind-ended tube having two membrane-covered holes at its wider end. The upper hole is the oval window, and the lower hole is called the **round window.** Also imagine that another, smaller membranous tube runs down its center, serving as a partition between the oval and round windows and bisecting the large tube into upper and lower channels, or canals. The upper channel is called the *vestibular canal;* the lower channel is called the *tympanic canal.* The inner tube stops short of the blind, tapered end of the outer tube, however, so the two canals connect with one another there. Imagine this tube rolled up like a jellyroll, and you have the basic structure of the cochlea. In fact, the word *cochlea* comes from a Latin word meaning "snail" and describes the rolled-up shape of this structure extremely well.

The inner tube of the cochlea is called the *cochlear duct,* and it contains specialized cells that are the receptors of hearing. These cells are called *hair cells* and are embedded in the floor of the inner tube, the basilar membrane. The hairs that project from the hair cells stick up into the cochlear duct and are covered by a roof called the *tectorial membrane.* Together, the hair cells with supporting cells of the basilar membrane and the overhanging tectorial membrane are called the **organ of Corti**—the organ of hearing.

How does the organ of hearing detect sound? As the stapes rocks in the oval window, it sets the fluid within the vestibular canal in motion. As the oval window membrane rocks inward, it pushes on the fluid in the vestibular canal. This fluid moves forward in waves that pass into the tympanic canal. When the waves reach the membrane-covered round window, it bulges outward. Meanwhile, the stapes rocks backward in the oval window, setting up an opposite motion of the fluid. As the waves then flow backward through the tympanic canal on the way to the vestibular canal, they push on the basilar membrane. The basilar membrane bends, causing the hairs of the receptor cells to be pressed against the tectorial membrane. As the hairs bend, they develop generator potentials that fire adjacent neurons. These impulses then pass to the brain to be interpreted.

The brain is able to interpret pitch (the highness or lowness of a sound) because sounds of different pitches produce sound waves of different frequencies (numbers of waves per second). These differences set up differing wave patterns within the fluid of the inner ear that cause specific regions of the basilar membrane to vibrate more intensely than others. In addition, louder sound waves cause greater vibrations of the basilar membrane than can be interpreted by the brain. Repeated exposure to extremely loud noises such as gunshots, jet engines, and loud music can damage the hairs of the receptor cells and cause partial but permanent hearing loss.

> *The ear has three parts: the outer, middle, and inner ear. The outer ear funnels sound waves in toward the eardrum, which changes these waves into mechanical energy. This energy is then transmitted to the bones of the middle ear, which increase its force. The inner ear, a complex of fluid-filled canals, contains the organ of hearing. The receptor cells of this organ change the mechanical sound energy into nerve impulses.*

Balance

The inner ear has two other portions that are each structured as a fluid-filled tube within a tube like the cochlea, but their shapes are different from that of the cochlea. As shown in Figure 15-11, the cochlea makes up one side of the inner ear. The bulge in the midsection of the inner ear is called the **vestibule,** and it contains structures that sense whether you are upside down or right side up. In other words, it detects the effects of gravity on the body.

Inside this bulge are two sacs called the **utricle** and the **saccule** (both words mean "a little bag or sac") (Figure 15-12, A). Within each of these sacs is a flat area composed of both ciliated and nonciliated cells. Spread over the surfaces of these cells is a layer of jellylike material. Embedded within the jelly are small pebbles of calcium carbonate called *otoliths* (literally, "ear stones") (Figure 15-12, B). The ciliated cells are the receptor cells and have long cilia and thin cell extensions that stick up into the jellylike microscopic tufts of hair (Figure 15-12, C). When the head is moved, the otoliths slide within the jelly just as a hockey puck slides on ice. As the otoliths move, they pull on the jelly, which pulls on the cilia, bending them (Figure 15-12, D). Any shift in the position of the otoliths results in different cilia being bent and cilia being bent to a specific degree and in a specific direction. These stimuli initiate generator potentials that are transmitted to neurons that make up a branch of cranial nerve VIII (see Chapter 14). The brain interprets these messages, resulting in your perception of "up" with respect to the pull of gravity.

Positioned above the saccule and utricle are three fluid-filled **semicircular canals** (see Figure 15-11). These loops are oriented at right angles to one another in three planes. At the base of each loop is a group of ciliated sensory cells that are connected to neurons. Lying above these cells is a mass of jellylike material. When the head moves, the fluid within the semicircular canals moves and pushes the jelly (and the cilia) in a direction opposite to that of the motion. (You have experienced this phenomenon when you accelerate suddenly in your car and your head is thrown backward.) This movement initiates the depolarization of the cell membrane, which triggers a nerve impulse. Because the three canals are each oriented differently, movement in any plane is sensed by at least one of them. Complex movements are analyzed in the brain as it compares the sensory input from each canal.

> *The vestibule of the inner ear detects your position with respect to gravity, whereas the semicircular canals detect the direction of your movement.*

cochlea (**kock** lee uh) a winding, snail-shaped tube that contains the organ of hearing and forms a portion of the inner ear.
round window a membrane-covered hole at the wider end of the cochlea.
organ of Corti (**kort** ee) the organ of hearing; the collective term for the hair cells, the supporting cells of the basilar membrane, and the overhanging tectorial membrane.
vestibule (**ves** tuh byool) a structure within the inner ear that detects the effects of gravity on the body.
utricle (**yoo** trih kul) the larger of two membranous sacs within the vestibule, containing both ciliated and nonciliated cells, that functions in the maintenance of bodily equilibrium and coordination.
saccule (**sak** yool) the smaller of two membranous sacs within the vestibule, containing both ciliated and nonciliated cells, that functions in the maintenance of bodily equilibrium and coordination.
semicircular canals (sem ih **sur** kyuh lur) three looped fluid-filled canals in the inner ear that detect the direction of the body's movement.

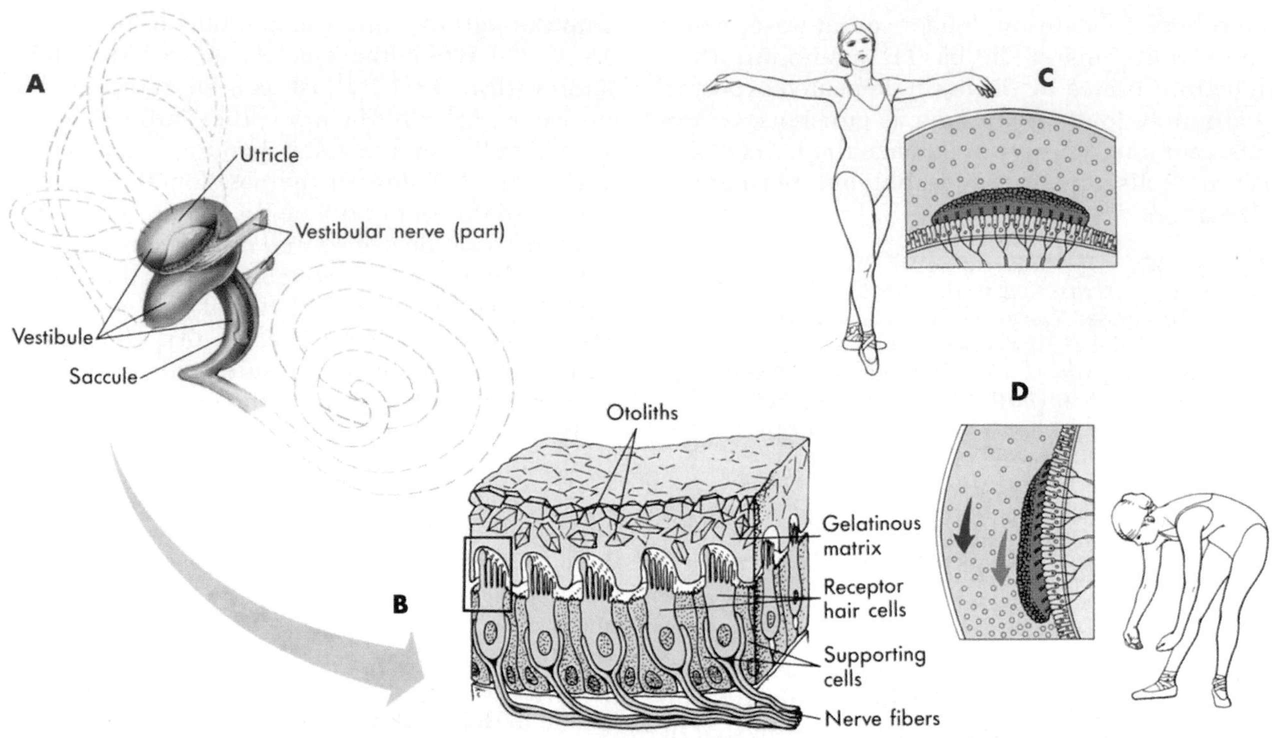

FIGURE 15-12 Balance is maintained by the semicircular canals. A, Structure of vestibule showing placement of utricle and saccule. **B,** Section showing otoliths. **C,** Cilia and otoliths in an upright position. **D,** Cilia and otoliths displaced by gravity as a person bends over.

Summary

- Sensory receptors change environmental stimuli into nerve impulses. These stimuli, or changes in the environment, must be of a sufficient magnitude to open ion channels within the membranes of receptor cells, depolarizing them. This initial depolarization is called a *generator potential* and initiates an action potential, or nerve impulse, that is carried to either the spinal cord or brain.
- Some sensory receptors monitor body functions. Temperature-sensitive neurons in the hypothalamus, for example, monitor the temperature of the blood and trigger mechanisms to regulate this temperature. Other receptors embedded in the walls of certain arteries are sensitive to the carbon dioxide, oxygen, and pH levels of the blood and provide input to respiratory centers in the medulla and pons. Still others monitor the blood pressure and provide feedback to a cardiac center that regulates heartbeat.
- Sensory receptors called *proprioceptors* gather information about the body's position in space. Proprioceptors buried deep within the muscles keep track of the degree to which they are contracted. Gravity is detected by structures in the inner ear called the *utricle* and *saccule.* Motion is sensed by the semicircular canals, also located in the inner ear. Together, these two structures help you maintain your balance and know which way is up.
- Various receptors located at or near the body surface sense changes in the external environment. Many simple receptors located in the skin detect general environmental stimuli such as touch, pressure, pain, and temperature. More specialized senses—smell, taste, sight, hearing, and balance—have more complex receptors located only at certain locations.
- The olfactory receptors, which are located in the nasal epithelium just behind the bridge of the nose, detect various smells. Scientific evidence suggests that the olfactory receptors detect these different smells because of specific binding of airborne gases with receptor chemicals located within the cilia of the olfactory receptor cells.
- Taste receptors, or taste buds, are located within the short projections on the tongue known as *papillae.* The receptor cells of the taste buds de-

tect sweet, sour, salty, and bitter tastes in food as food molecules dissolve in the saliva and come into contact with cellular "hairs" that poke out of the taste pores.

- The human eye detects wavelengths of electromagnetic energy known as *visible light*. Electromagnetic energy is generated by charged subatomic particles moving through space emitting waves of electrical and magnetic energy. The human eye contains receptor cells called *rods* and *cones* that contain pigments capable of absorbing units of electromagnetic energy of certain wavelengths. The absorption of light causes a change in the shape of the pigment, which in turn triggers a series of events resulting in a hyperpolarization of the receptor cell membrane. This hyperpolarization leads to an adjustment in the frequency of the firing of adjacent neurons—nerve signals that eventually reach the brain and are interpreted as patterns of light and dark.
- Structurally, the human eye is similar to a camera. Light entering the eye is focused by a lens onto the receptors located at the back of the eye, which act like film in a camera.
- Sound waves are disturbances of molecules in the air (and in certain other substances) that result from the vibration of an object. The ear collects sound waves and changes them into other forms of mechanical energy that stimulate the organ of hearing within the inner ear. The receptor cells of this organ propagate nerve impulses that are carried to the cerebral cortex.

Knowledge and Comprehension Questions

1. The text states that the perception of pain involves an individual's reaction to particular sensory stimuli. Would you consider pain, therefore, an objective or subjective state of being?
2. Explain the events that occur at the cellular level that allow you to "sense" something.
3. Which type of receptor would be most highly stimulated by an experience of riding on an amusement park ride such as a roller coaster? Why?
4. Researchers have found the hypothalamus to be very involved in the sense of satiation or "feeling full" after a meal and thus integrally involved in weight control. What type of information would you theorize is being sent to the hypothalamus?
5. John has chronic allergies that cause him to suffer from constant sinus congestion. What special senses may be highly affected by this condition?
6. Explain how you can smell the difference between your favorite perfume and your least favorite food.
7. You may have noticed that certain smells are able to evoke specific memories and emotions—both good and bad. Explain why.
8. What are taste buds? Explain their role in allowing you to taste the flavors in your favorite ice cream. What other sense (besides taste) is involved?
9. How are you able to see different colors?
10. Explain the process that allows you to see objects.
11. Draw a diagram of the human eye. Label the sclera, cornea, aqueous humor, lens, iris, pupil, and fovea.
12. Is there any difference in the manner in which you see in a well-lit environment as compared with your vision in a darkened room? What type of cells are involved specifically?
13. Mr. Garcia undergoes minor eye surgery and must wear a patch over his left eye for 3 days. During this time, he finds that he has trouble grasping things because he keeps underreaching and overreaching. Explain why.
14. You switch on the radio to your favorite program. Explain how your ears enable you to hear the sounds.
15. Even with your eyes closed, you can tell when your head tilts to one side. Explain how.

Critical Thinking

1. In a very interesting series of experiments, migrating song birds were placed in a cage lined with carbon paper similar to that used in credit card slips. The birds would peck at the wall of the cage, attempting to get out, but not at random—by far the greatest density of pecks is in the direction they were migrating when captured! Now when a large magnet is placed in the room, they peck in the direction of the magnet, eventhough they cannot see it. What sort of sensory system do you imagine the birds possess that lets them orient in this way?
2. Some children are born with the inability to feel pain. Do you think that this is an advantage or a disadvantage with respect to the well-being of the person? Explain the reasons for your answer.

CHAPTER 16

PROTECTION, SUPPORT, AND MOVEMENT

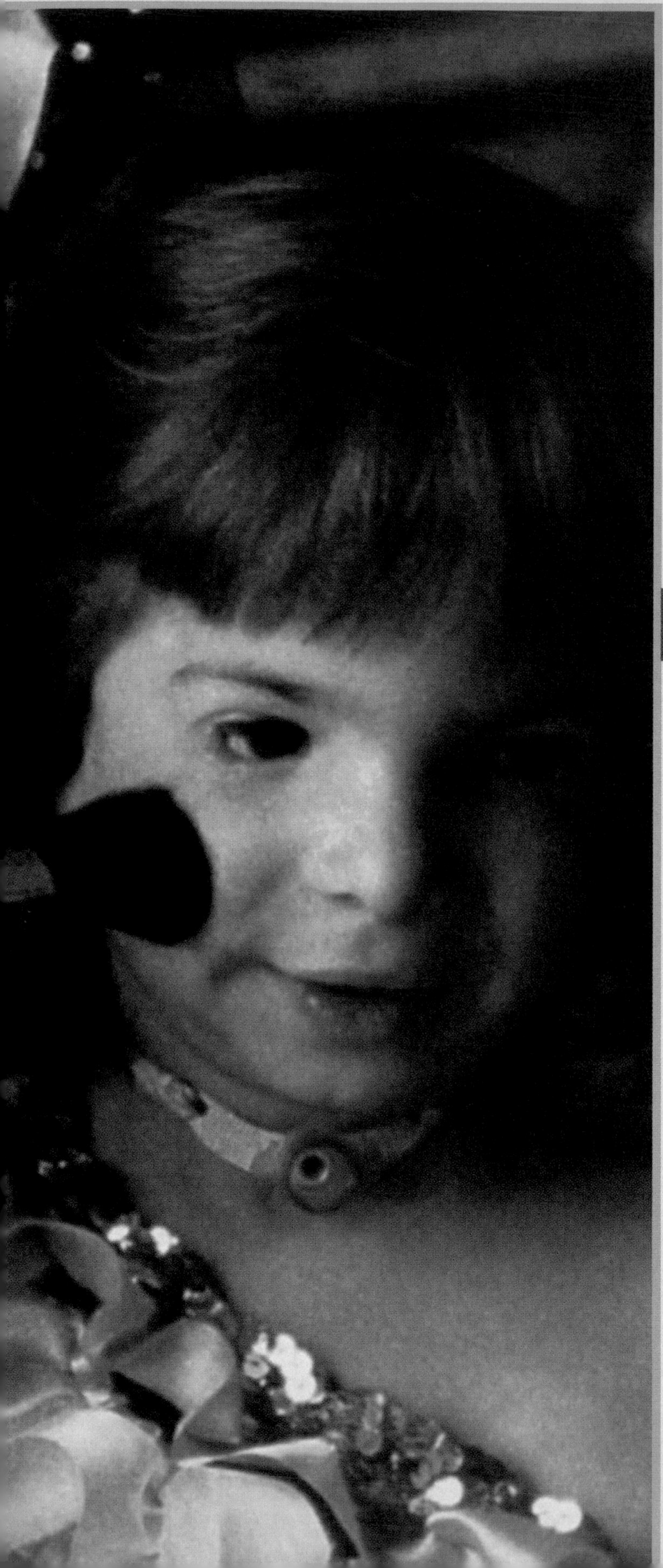

FASHIONING FACES USING COMPUTERS is one type of medical miracle taking place at Johns Hopkins Children's Facial Rehabilitation Center in Baltimore, Maryland. Erin Williams, born with a genetic disorder called Treacher Collins' syndrome, knows all about this medical miracle. She was born without a jaw, cheekbones, or fully formed ears. Although she still faced surgery at the time this photo was taken, Erin does not look like a child who had to struggle just to eat and breathe during her infancy.

Doctors at the Children's Center performed "phantom surgery" to plan the reconstruction of Erin's face. Using three-dimensional computer images of her skull, doctors simulated Erin's surgery before they actually operated on her. Phantom surgery can be used to experiment with cutting and fitting bone pieces from other parts of a patient's body, or the computer can be used to design an artificial bone. This technology is called computer-aided design, or CAD for short.

To help Erin, doctors recreated her skull as a computer graphic using a stack of computerized tomography (CT) scan images. Each CT scan image pictures a thin slice of the body. A CAD computer stacks these images to produce a three-dimensional graphic that can be rotated and viewed from any angle. Using this image, doctors designed prostheses—artificial bones—for Erin. The instructions to build these prostheses were put on tape and then fed into a computer specialized in computer-aided manufacturing (CAM).

One company used a CAM computer attached to a milling machine to make Erin a plastic jaw. During surgery, a real jaw was fashioned for Erin using the plastic model as a template. To do this, physicians used bits of Erin's ribs and pieces of a chemically treated South Pacific coral that mimics human bone. Another company used CAD/CAM to construct titanium cheekbones for her. This lightweight, durable, biocompatible metal will hold bone grafts that will be added to Erin's face as she grows. After her teenage years, Erin will no longer need reconstructive surgery. The growth and development of her face, fashioned with modern technology, will be complete.

KEY CONCEPTS

- Movement of the human body is possible because your tissues have a flexible outer covering and are supported by a scaffolding of bone that is powered by skeletal muscles.
- The skin is a dynamic organ that flexes with body movement, protects the body from water loss and invasion by microbes, provides the body with a sensory surface, and helps control the body's internal temperature.
- The rigid yet flexible bones of the body form an internal framework called the skeletal system, which supports the soft tissues of the body; provides places for the attachment of the muscles; and acts as a lever system against which muscles can pull.
- The skeletal muscles—the muscles of body movement—are made up of long, cylindrical cells packed with organized arrangements of thick and thin microfilaments capable of causing muscular contraction.

OUTLINE

How Skin, Bones, and Muscles Work Together

Bones not only give shape to the body and protect its delicate inner structures but also help you move. Your skeletal muscles are attached to your bones, pulling on one while anchored to another. Rigid yet flexible, bones are able to bear a considerable amount of weight. If you have ever tried to lift a heavy object, you are familiar with this basic problem of movement: gravity exerts a force, or "pull," on objects. For any motion to take place, the force exerted on an object—including your body parts—must be greater than the opposing force of gravity. The manner in which muscles are attached to bones allows bones to be used as levers to increase the strength of a movement, similar to the way a claw hammer increases the force you can exert on a nail to pull it out of a piece of wood.

To supply the force needed for movement, your body uses the chemical energy of adenosine triphosphate, or ATP (see Chapter 4). By splitting an ATP molecule into adenosine diphosphate (ADP) and inorganic phosphate (P_i), energy is made available to do the work of movement. Your body uses this energy to move certain microfilaments within your muscle cells, resulting in the shortening of those cells. When a large group of muscle cells shortens all at once, the muscle cells exert a great deal of force. For a muscle to use this force to produce movement, it must direct its force against an object. In humans, muscles pull on bones, so for example, muscle contraction results in the lifting of a leg or the bending of a finger. And although your movement is determined largely by your muscles and their attachments to bones, your skin plays an important role too. It is a soft and flexible covering, stretching to accommodate the myriad movements of which your body is capable.

> *Movement results from the contraction of skeletal muscles anchored to bones. These muscles use bones like levers to direct force against an object. When a body part is pulled to a new position, the skin stretches to accommodate the change.*

Skin

Your skin is far more than simply an elastic covering of epithelial cells encasing your body's muscles, blood, and bones. Instead, it is a dynamic organ that performs many functions:

1. Skin is a protective barrier. It keeps out microorganisms that would otherwise infect the body. Because skin is waterproof, it keeps the fluids of the body in and other fluids out. Skin cells also contain a pigment called *melanin*, which absorbs potentially damaging ultraviolet radiation from the sun.
2. Skin provides a sensory surface. Sensory nerve endings in skin act as your body's pressure gauge, telling you how gently to caress a loved one and how firmly to hold a pencil. Other sensors embedded in the skin detect pain, heat, and cold (see Chapter 15). Skin is the body's point of contact with the outside world.
3. Skin compensates for body movement. Skin stretches when you reach for something and contracts quickly when you stop reaching. It grows when you grow.
4. Skin helps control the body's internal temperature. When the temperature is cold, the blood vessels in the skin constrict, so less of the body's heat is lost to the surrounding air. When it is hot, these same vessels dilate, giving off heat. In addition, glands in the skin release sweat, which then absorbs body heat and evaporates, cooling the body surface.

Your skin is the largest organ of your body. In an adult human, 15% of the total body weight is skin. Much of the multifunctional role of skin reflects the fact that its tissues are made up of a variety of specialized cells. In fact, one typical square centimeter of human skin contains 200 nerve endings, 10 hairs with accompanying microscopic muscles, 100 sweat glands, 15 oil glands, 3 blood vessels, 12 heat receptors, 2 cold receptors, and 25 pressure-sensing receptors (Figure 16-1). Together with the hair and nails, the skin is called the **integument** (meaning "outer covering"), or the **integumentary system.**

> *The skin is the largest organ of the body and is made up of myriad specialized cells. It covers the surface of the body, protecting it while helping to control the body's internal temperature, providing a sensory surface, and compensating for body movement.*

Bones

Bone is a type of connective tissue (see Chapter 7) consisting of widely separated bone cells embedded in a matrix of collagen fibers and mineral salts. The bones' mineral salts are needle-shaped crystals of calcium phosphate and calcium carbonate. The collagen fibers (see Figure 7-6) are coated and sur-

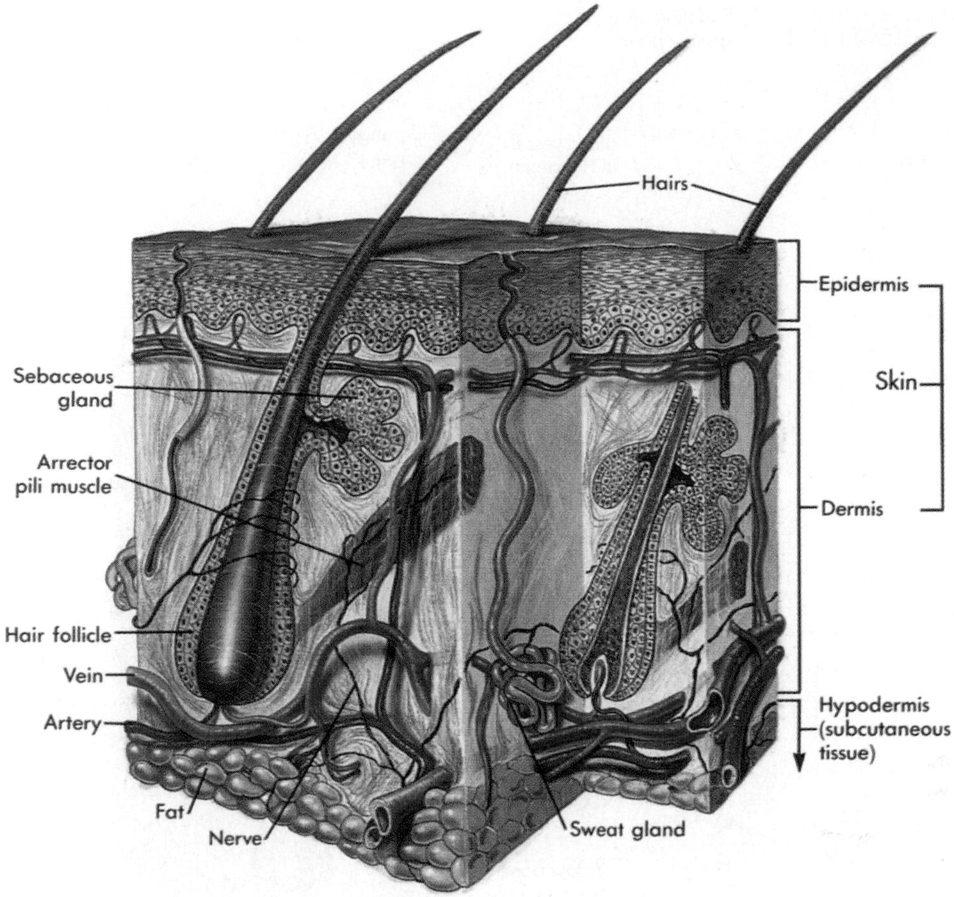

FIGURE 16-1 The structure of human skin. Human skin is composed of three layers: the epidermis, the dermis, and the subcutaneous layer. Within these layers are specialized tissues and cells that perform many specialized functions.

rounded by these mineral salts. Interestingly, these two components give bone a structure strikingly similar to that of fiberglass, producing bones that are rigid yet flexible. You can see this interplay of flexibility and hardness by soaking a chicken bone in vinegar overnight. The acetic acid of the vinegar dissolves the mineral salts in the bone. Without the hardness of the mineral salts, the collagen fibers leave the bone so flexible that you can tie it in a knot!

Living cells are as important a component of bone as the collagen fibers and mineral salts are. New bone is formed by cells called *osteoblasts.* These cells secrete the collagen fibers on which the body later deposits mineral salts. The osteoblasts lay down bone in thin, concentric layers called *lamellae* (Figure 16-2), like layers of insulation wrapped around an old pipe. You can think of the osteoblasts as being within the insulation and the open tube of the pipe as a narrow channel called the **Haversian canal.** Haversian canals run parallel to the length of the bone and contain blood vessels and nerves. The blood vessels provide a lifeline to living bone-forming cells, whereas the nerves control the diameter of the blood vessels and thus the flow through them. Nutrients and oxygen diffuse

integument (in **teg** you ment) in vertebrates, the skin, hair, and nails.

integumentary system (in **teg** you **men** tuh ree) another name for the vertebrate body's integument: the skin, hair, and nails.

bone a type of connective tissue consisting of widely separated bone cells embedded in a matrix of collagen fibers and mineral salts; it forms the vertebrate skeleton.

Haversian canal (huh **vur** shun) in long bones, a narrow channel that runs parallel to the length of the bone and contains blood vessels and nerves. Some Haversian canals run crosswise to the length of the bone to connect the lengthwise canals.

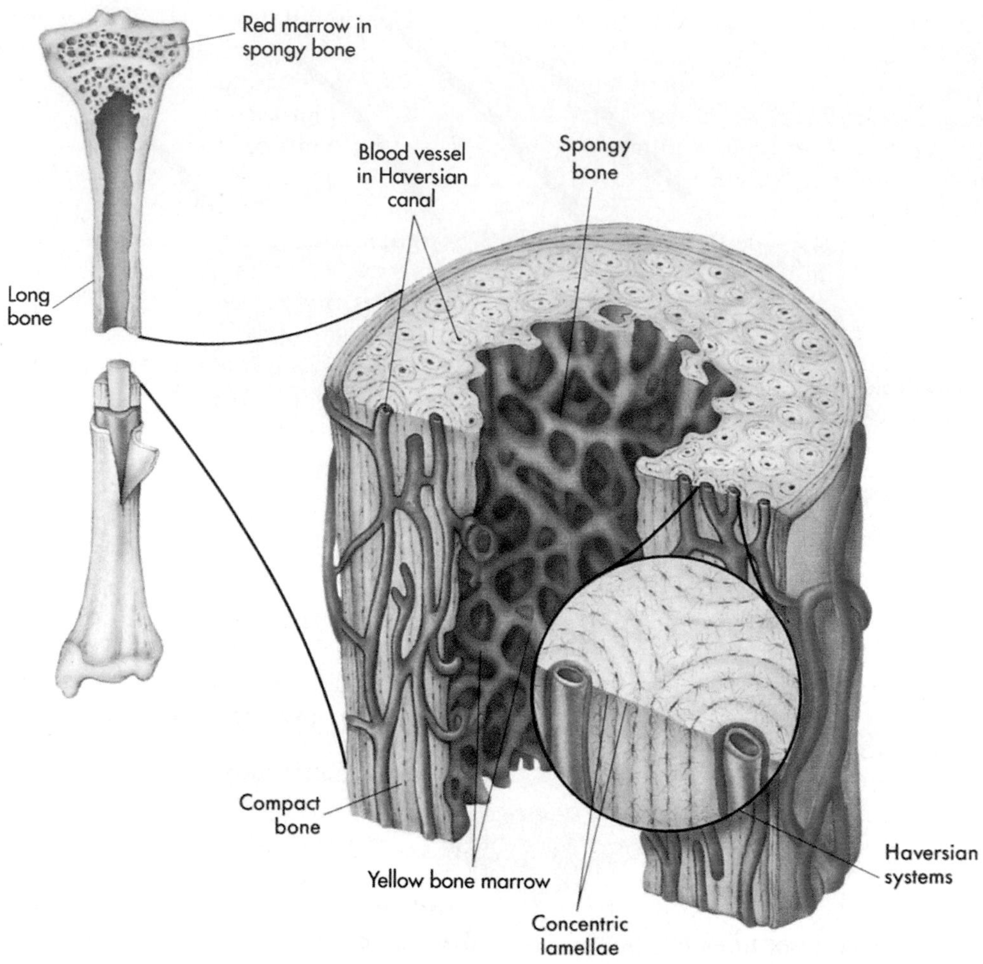

FIGURE 16-2 The organization of bone, shown at three levels of detail. Some parts of bone are dense and compact, giving the bone strength. Other parts are spongy, with a more open lattice; it is here (within red bone marrow) that most red blood cells are formed. Inset shows a magnified Haversian system with its concentric lamellae.

from the bloodstream into thin cellular processes of the osteoblasts, and metabolic wastes diffuse out.

> *Bone is a type of connective tissue made up of widely separated living cells that secrete collagen fibers into their surrounding matrix. The body deposits mineral salts on these fibers. The salts give bone hardness, whereas the collagen fibers contribute flexibility. This bone tissue appears under the light microscope as concentric rings surrounding a central canal containing blood vessels and nerves.*

The bones of the human skeleton are composed of spongy bone and compact bone. Microscopically, *compact bone* has the concentric ring structure described previously. As shown in Figure 16-2, it runs the length of long bones, such as those in your arms and legs. However, if bones were completely made up of compact bone, your body would be very heavy, and your arms and legs would be nearly impossible to move! Instead, the central core of long bones is a hollow cylinder, and this cavity is filled with a soft, fatty connective tissue called *yellow bone marrow.*

Spongy bone makes up most of the ends of long bones and most of the bone tissue of short bones (like your wrist and ankle bones), flat bones (like your ribs), and irregularly shaped bones (like some of your facial bones). Spongy bone is an open latticework of thin plates, or bars, of bone. Microscopically, its structure does not show a regular concentric ring structure like compact bone; it has a somewhat more irregular organization. The spaces within its bony latticework are filled with *red bone marrow*. Here, most of the body's blood cells are formed. Surrounding the spongy bone tissue are layers of compact bone (see Figure 16-2). The compact bone gives bones the strength to withstand mechanical stress, and spongy bone provides some support and a storage place for the red bone marrow while helping to lighten bones.

> *The bones of the skeleton contain two kinds of tissue: spongy bone and compact bone. Compact bone runs the length of long bones and has no spaces within its structure visible to the naked eye. Spongy bone is found in the ends of long bones and within short, flat, and irregularly shaped bones. It looks like interweaving bars of bone with the intervening spaces filled with red bone marrow.*

The Skeletal System

The 206 bones of the body make up a working whole called the **skeleton,** or **skeletal system.** Together, the bones of the skeleton perform many important functions, some of which have already been mentioned. To summarize, (1) The skeletal system provides support for the body—otherwise you would be a shapeless blob of skin-covered tissues and organs. (2) It also provides for movement, with individual bones serving as points of attachment for the skeletal muscles and acting as levers against which muscles can pull. (3) Delicate internal structures are protected by the skeleton. The brain, for example, lies within a strong, bony casing called the *skull;* the rib cage surrounds and protects the heart and lungs. (4) The bones of the skeleton are a storehouse for minerals such as calcium and phosphorus. Amazingly, bone tissue is continually broken down and re-formed, the mineral salts being transported to other parts of the body on demand. In fact, scientists have estimated that your body completely replaces your skeleton over a period of 7 years! (5) Last, certain inner portions of bones contain red marrow and are your internal factory of blood cells. Without this factory you would die because red blood cells have a short life span of approximately 120 days. Without these cells your body would have no efficient means to transport life-giving oxygen to your tissues. White blood cells are also produced in the red bone marrow. One of the main jobs of certain classes of these cells is to ingest bacteria and debris. White blood cells also form the cells of your immune system (see Chapter 11). It is essential to your survival that the red marrow continue to manufacture these cells, which have varying life spans and help protect your body from infection and disease.

Scientists divide your internal scaffolding of bones into two parts: the axial skeleton and the appendicular skeleton (Figure 16-3). The **axial skeleton** is your axis, or central column, of bones off which the appendages (arms and legs) of your **appendicular skeleton** hang. The appendicular skeleton also includes the bones that serve to attach the appendages to the axial skeleton.

The Axial Skeleton

The 80 bones of the axial skeleton include the bones of the skull, vertebral column, and rib cage. The **skull** contains 8 cranial and 14 facial bones. Find the following skull bones in Figure 16-4: 1 frontal bone that forms the forehead, 2 parietal bones that form a large portion of the sides and top of the skull, 2 temporal bones that form the sides of the skull toward the back, and 1 occipital bone that forms the lower back portion of the skull. The irregularly shaped ethmoid and sphenoid bones are not easy to locate in a diagram because only a tiny portion of these bones lies on the surface of the skull. The ethmoid bone sits between the eyes; only a fraction of it can be seen forming a part of each orbit, or eye socket. The rest of this bone forms part of the nasal cavity and the floor of the cranial cavity. The sphenoid bone makes up part of the back and sides of the eye sockets. Looking somewhat like wings, it extends from one side of the skull to

skeleton the collective term for the body's 206 bones. The skeleton provides support for the body, helps the body move (along with the muscles), and protects delicate internal structures.

skeletal system another term for the skeleton.

axial skeleton (ak see uhl) the central column of the skeleton, from which the appendages (arms and legs) of the appendicular skeleton hang. The axial skeleton consists of 80 bones, including the skull, vertebral column, and rib cage.

appendicular skeleton (ap en **dik** you lur) the portion of the human skeleton, consisting of 126 bones, that forms the bones of the appendages (arms and legs) and the bones that help attach the appendages to the axial skeleton.

skull the framework of the head, which in humans consists of 8 cranial and 14 facial bones.

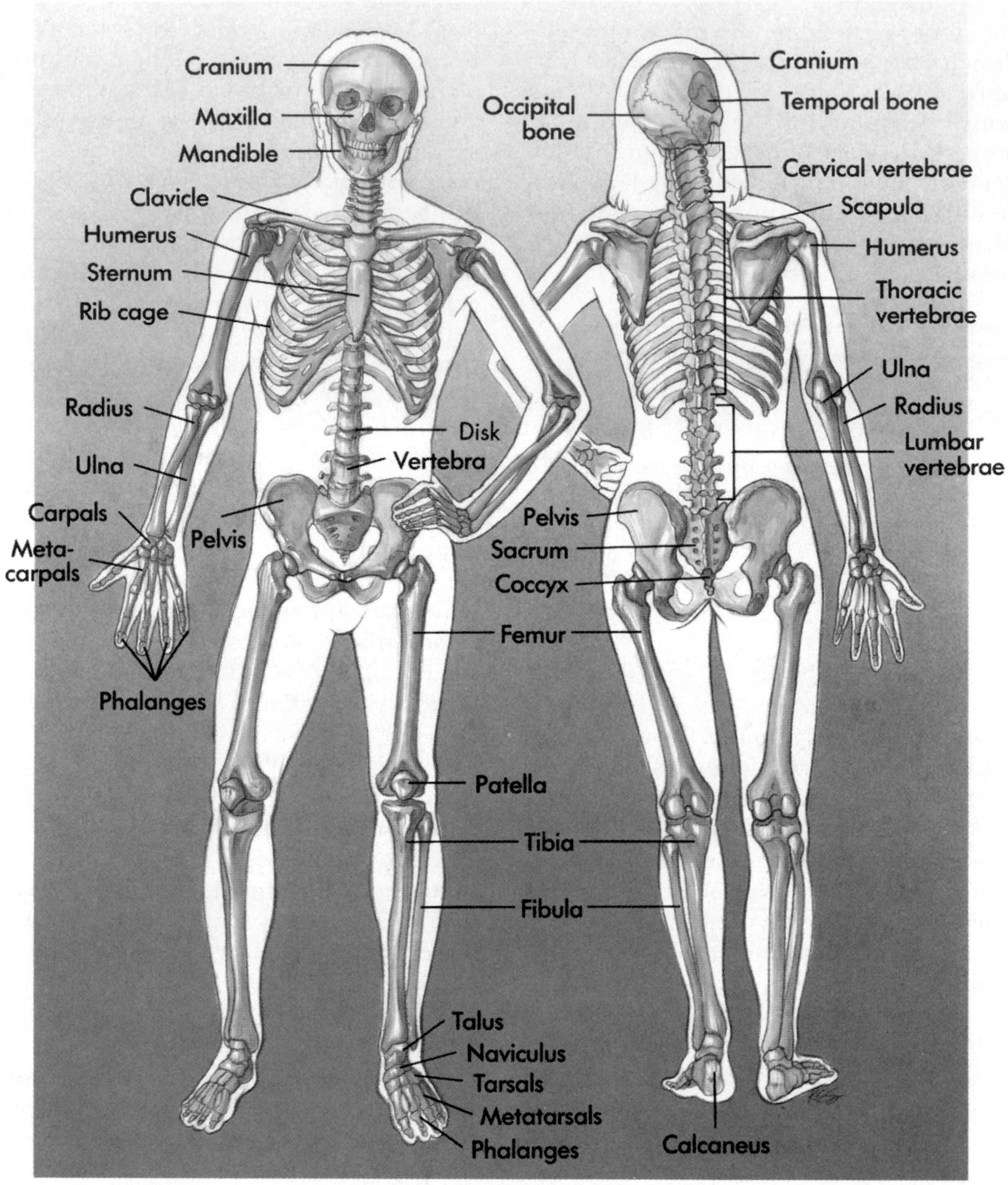

FIGURE 16-3 The human axial and appendicular skeletons. The axial skeleton is made up of the bones of the head, face, rib cage, and vertebral column. The appendicular skeleton includes the bones of the legs and feet, arms and hands, clavicles, scapula, and pelvis.

the other, forming the middle portion of the base of the skull.

Some of the 14 facial bones are easy to see in Figure 16-4: 1 mandible, or lower jawbone; 2 maxillae, which unite to form the upper jawbone; 2 nasal bones, which form the bridge of the nose; and 2 zygomatic bones, or cheekbones. Look below the part of the ethmoid bone that forms part of the nasal septum to find the vomer. Along with the ethmoid and cartilage, the vomer helps divide the nose into right and left nostrils. The paired lacrimal bones are tiny and, therefore difficult to see and are located at the inner corners of the orbits near the nose. Each of these bones has a groove forming part of the tear ducts, canals that drain excess fluid bathing the eyes. The two L-shaped palatine bones are hidden from view because they form the back portion of the roof of your mouth and extend upward to form part of the floor and sides of the nasal cavity. The two inferior nasal conchae are "swirled" bones and project into the nasal cavity. The structure of these bones causes the air that enters the nose to be circulated and warmed before being breathed into the lungs.

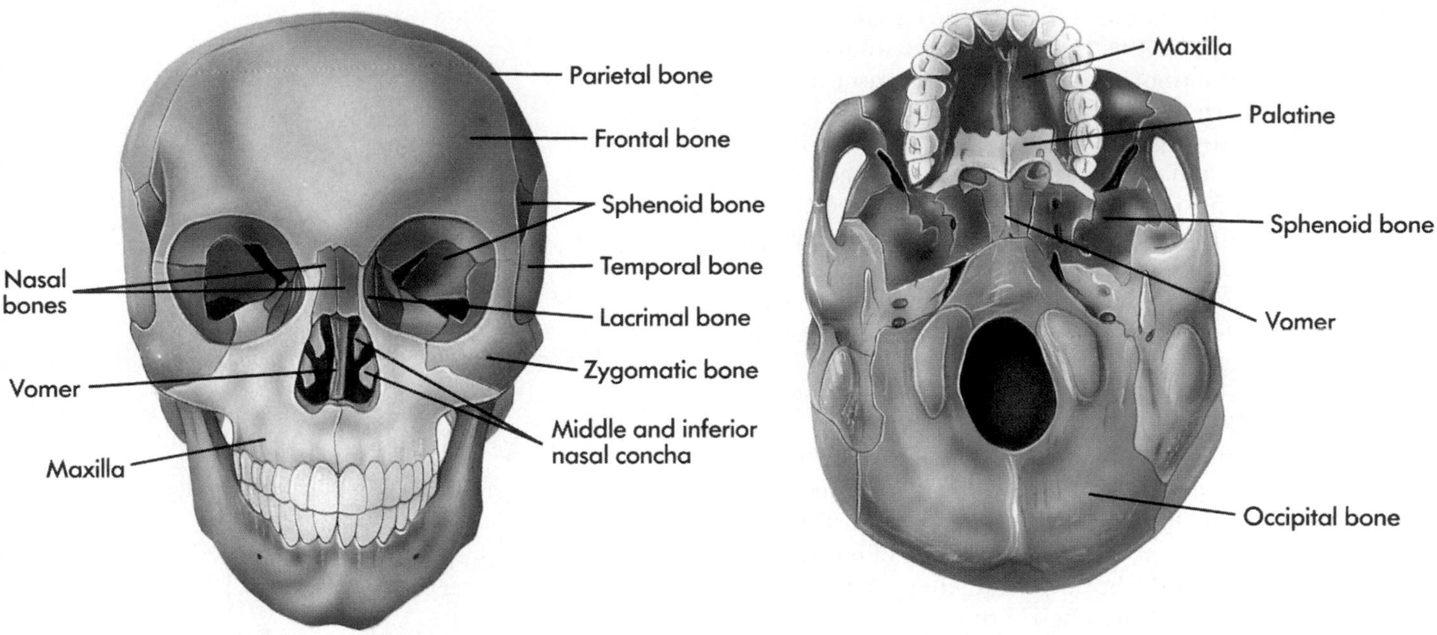

FIGURE 16-4 Frontal view of the human skull. The skull contains 8 cranial and 14 facial bones.

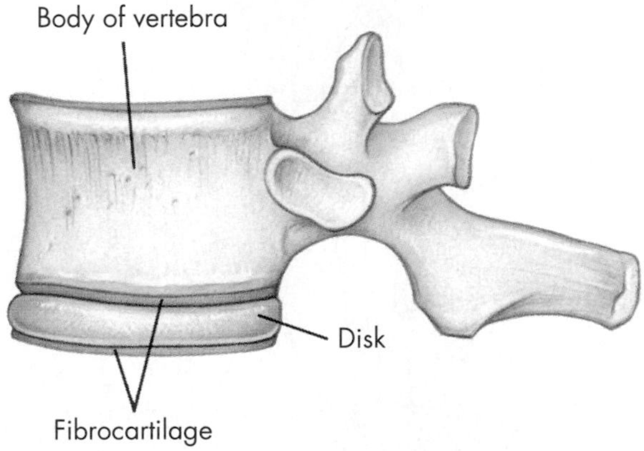

FIGURE 16-5 A vertebra with disk attached. The vertebral column is composed of 26 vertebrae. Between each vertebra and the next is a disk composed of an outer layer of fibrocartilage and an inner, more elastic layer.

The **vertebral column** is made up of 26 individual bones called *vertebrae* (Figure 16-5). Stacked one on top of the other, these bones act like a strong yet very flexible rod that supports the head. In addition, some of the vertebrae serve as points of attachment for the ribs. Each vertebra has a central hole, or foramen. Lined up, the vertebrae form a bony canal protecting the spinal cord, which runs down much of its length.

The 7 vertebrae closest to the head are the cervical (neck) vertebrae. Next are 12 thoracic (chest) vertebrae. Following these are 5 lumbar (lower back) vertebrae. The sacrum and the coccyx, or tailbone, are the last vertebrae in the column. Positioned between the vertebrae (except the sacrum and coccyx) are disks of fibrocartilage called **intervertebral disks.** These disks act as shock absorbers, provide the means of attachment between one vertebra and the next, and permit movement of the vertebral column.

Attached to the 12 thoracic vertebrae are 12 pairs of ribs. The ribs curve around to the front of the thoracic (chest) cavity, producing a bony cage that protects the heart and lungs. The upper 7 pairs of ribs directly connect to a flat bone that lies at the midline of the chest called the *breastbone,* or sternum. These 14 ribs each connect to the sternum by means of a strip of hyaline cartilage and are therefore called *true ribs.* The remaining five pairs of ribs, called *false ribs,* do not directly connect to the sternum. Instead the cartilages of ribs 8 through 10

vertebral column (ver tuh brul) the collection of 26 vertebrae, stacked one on top of the other along the midline of the back, that acts as a strong, flexible rod and supports the head in a human skeleton. The vertebral column is also known as the backbone.

intervertebral disks (in tur ver tuh brul) disks of fibrocartilage, positioned between all vertebrae except the sacrum and coccyx, that act as shock absorbers, provide the means of attachment between one vertebra and the next, and permit movement of the vertebral column.

attach to each other and then to the cartilage of the seventh ribs pair. Ribs 11 and 12 do not attach to the sternum but hang free, supported by muscle tissue. These ribs are therefore called *floating ribs.*

The remaining bone of the axial skeleton is the *hyoid bone,* a name that means "U-shaped." This bone is supported by ligaments in the neck. It is the only bone in the body that does not form a joint with other bones. The tongue is attached to the *hyoid bone.* Because this bone is often broken when a person is strangled, the hyoid can provide important evidence in certain murder cases.

The Appendicular Skeleton

The appendicular skeleton is made up of the bones of the appendages (arms and legs), the **pectoral (shoulder) girdle,** and the **pelvic (hip) girdle.** The pectoral girdle is made up of two pairs of bones: the clavicles, or collarbones, and the scapulae, or shoulder blades. Find these bones in Figures 16-3 and 16-6, and then locate the edges of your own clavicles and scapulae. The bones of the pectoral girdle support and articulate with (form joints with) the arms. Both arms, or upper extremities, contain 60 bones.

As shown in Figure 16-6, the bone of the upper arm is called the *humerus.* This bone articulates with both the clavicle and the scapula. The other end of this long bone articulates with the two bones of the forearm: the radius and the ulna. The wrist is made up of eight short bones called *carpals,* lined up in two rows of four. These bones are held together by ligaments and articulate with the bones of the hand, or metacarpals. The five metacarpals articulate with the bones of the fingers, or phalanges (singular phalanx). Each finger has three phalanges; the thumb has only two.

The pelvic girdle is made up of two bones called *coxal bones.* You know these bones as pelvic bones or hip bones. Find these bones in Figures 16-3 and 16-7. Male and female skeletons can be distinguished from one another by their pelvic bones. Females have a wider pelvis with a large, oval opening in the center for giving birth. The male pelvic girdle, in contrast, is narrower and has a smaller, heart-shaped opening in the center.

The bones of the pelvis support and articulate with the legs. Together, both legs, or lower extremities, contain 60 bones as the arms do. As shown in Figure 16-7, the bone of the upper leg, or thigh, is called the *femur.* This bone has a rounded, ball-shaped head that articulates with a depression, or socket, in the hip bone. The femur, the longest and heaviest bone in the body, articulates at its lower end with the two bones of the lower leg: the tibia

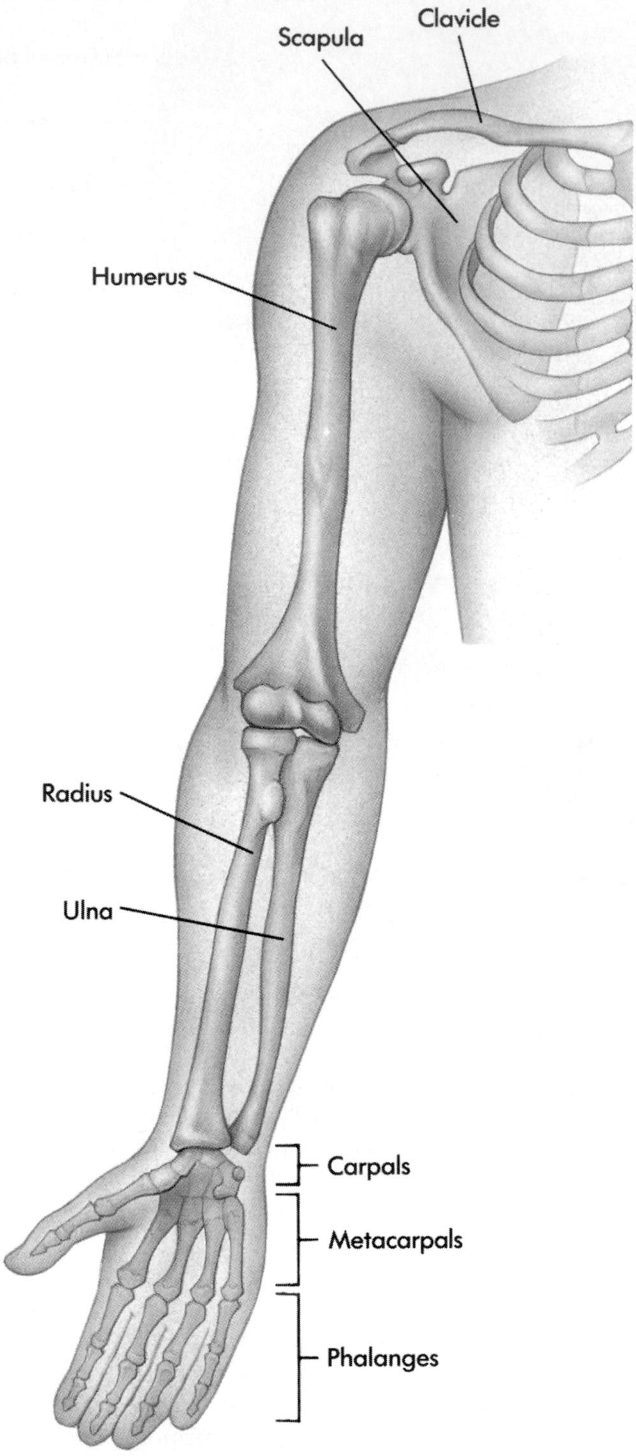

FIGURE 16-6 Bones of the pectoral girdle and arm, wrist, and hand. Notice that the bones of the pectoral girdle support and form joints with the bones of the arm.

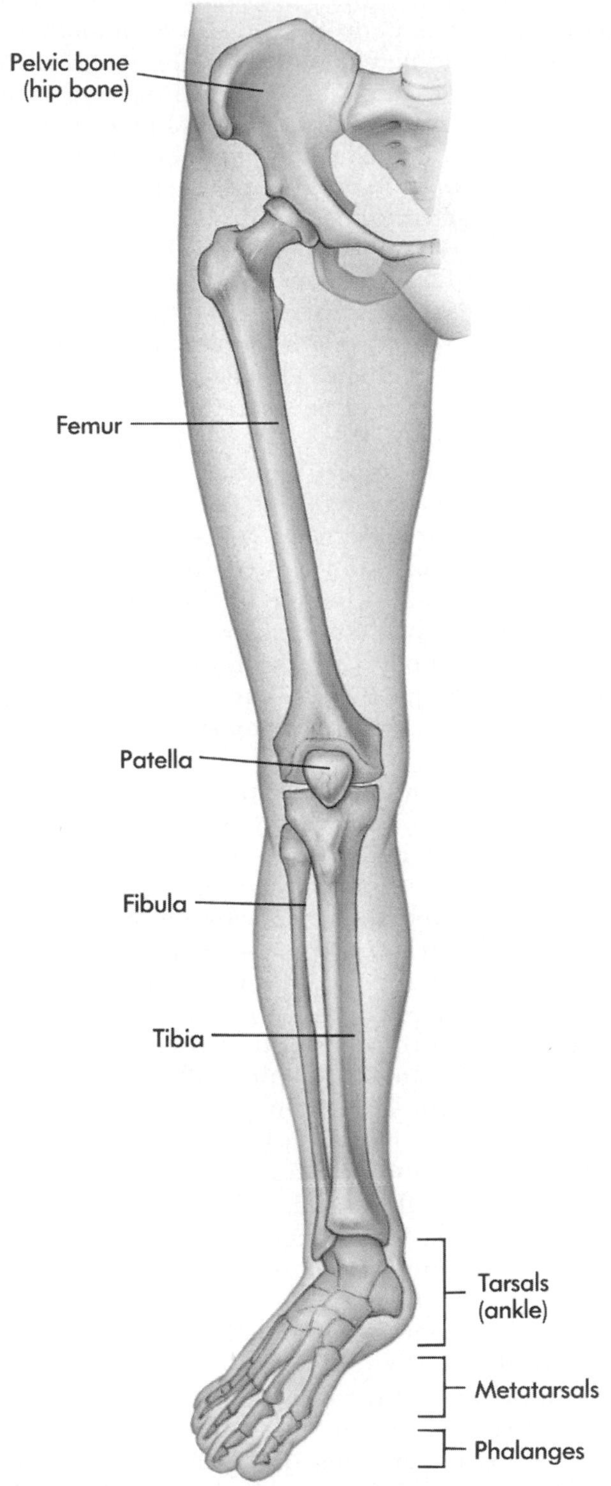

FIGURE 16-7 Bones of the pelvic girdle and leg, ankle, and foot. The two pelvic bones (also called coxal or hip bones) articulate with the head of each femur, or thigh bone.

JUST WONDERING....

I work all day at a computer and have developed a pain in my wrist that shoots up my arm. I've heard a lot of talk about carpal tunnel syndrome, but I'm not sure what it is. Could that be my problem?

As you can see in Figure 16-6, the carpals are the wrist bones. The carpal tunnel is a narrow passageway within your wrist through which the median nerve passes on its way to your middle, ring, and little fingers. Some types of activities that use the wrists and hands repetitively, such as typing, golfing, carpentry, and certain types of factory work, can cause the tissues around the tunnel to become inflamed and swollen. When this happens, the median nerve becomes compressed, resulting in numbness and/or tingling in the hand and fingers and pain in the wrist that may shoot up or down the arm.

Your physician can perform tests to determine whether you have carpal tunnel syndrome. Often the cure for this condition is as easy as resting the affected wrist and hand or wearing a splint designed to inmobilize the wrist. In more serious cases, injections of steroid drugs relieve the symptoms, with surgery being a last resort.

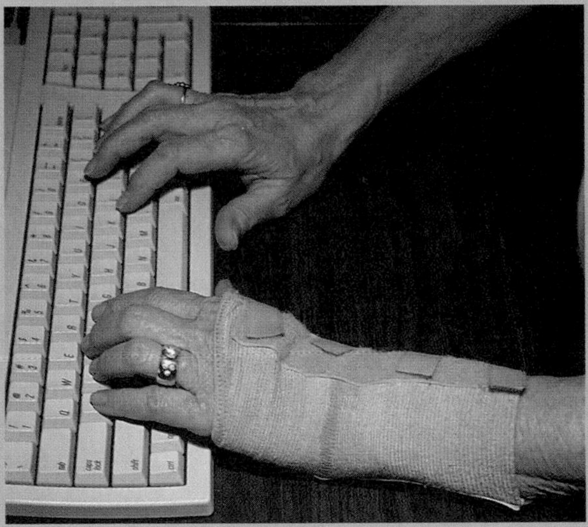

pectoral girdle (pek tuh rul) the part of the appendicular skeleton that is made up of two pairs of bones: the clavicles, or collarbones, and the scapulae, or shoulder blades. The pectoral girdle is also known as the shoulder girdle.

pelvic girdle the part of the appendicular skeleton that is made up of the two bones called coxal bones, pelvic bones, or hip bones. The pelvic girdle is also known as the hip girdle.

and the fibula. The tibia is the bone commonly referred to as the *shinbone* and can be felt at the front of the lower leg. If you have ever had shin splints, you are well aware of this bone. Shin splints are an inflammation of the outer covering of the tibia caused by repeated tugging by muscles and tendons. Vigorous walking, running, or other types of exercise can sometimes result in shin splints, which can be very painful.

The patella, or kneecap, is a small, triangular bone that sits in front of the joint formed by the femur, tibia, and fibula. The patella is a sesamoid bone, one that is formed within tendons where pressure develops.

Each ankle is made up of seven short bones called *tarsals.* One of these bones forms the heel and is the largest and strongest ankle bone. The tarsals are held together by ligaments and articulate with the bones of the foot, or metatarsals. The five metatarsals articulate with the bones of the toes, or phalanges. Like the fingers, each toe has three phalanges; like the thumb, the "big toe" has only two.

> *The 206 bones of the body make up the skeletal system, which can be divided into the axial skeleton and the appendicular skeleton. The axial skeleton includes your central axis—the skull, vertebral column, and rib cage. The appendicular skeleton is made up of the appendages (arms and legs) and the bones that help attach the appendages to the axial skeleton.*

Joints

If you have ever broken a leg or an arm and had a cast covering a joint, you have learned firsthand just how important joints are to movement. A **joint,** or **articulation,** is a place where bones, or bones and cartilage, come together. All joints are not alike; some permit little or no movement (immovable joints), some permit limited movement (slightly movable joints), and some permit a considerable amount of movement (freely movable joints). The amount of movement afforded by a joint is a direct result of how tightly the bones are held together at that location.

The bones of the skull articulate with one another in a type of immovable joint called a **suture.** The edges of each skull bone are ragged but fit tightly with the ragged edges of adjoining bones—just as puzzle pieces fit together (see Figure 16-3). A layer of dense connective tissue helps keep the bones from separating (see Chapter 7). Another example of immovable joints is the articulation of your teeth with the mandible and maxillae. Here, peg-shaped roots fit into cone-shaped sockets in the jawbones. A ligament lies between each tooth and its socket, holding each tooth in place. **Ligaments** are bundles, or strips, of dense connective tissue that hold bones to bones. A third example of an immovable joint is the articulation between the first pair of ribs and the breastbone. Each of these two ribs is connected to the sternum by a strip of hyaline cartilage, connections that change to bone during adult life.

As you twist your forearm to the right and left or twist your lower leg in the same manner, imagine the movement taking place between the shafts of the radius and ulna or between the tibia and fibula at their lower ends. Dense connective tissue is present at these locations, holding these bones together while permitting some flexibility. These two articulations are examples of slightly movable joints. The articulation between one vertebra and the next is also an example of a slightly movable joint. Here, a broad, flat sheet of fibrocartilage covers the top and bottom of each intervertebral disk (see Figure 16-5), which is sandwiched between adjacent vertebrae, creating a somewhat flexible connection.

In slightly movable and immovable joints, there is no space between the articulating bones. The bones forming these joints are held together by dense connective tissue or cartilage. In freely movable joints, there is a space between the articulating bones. Such joints are common in your body and are called **synovial joints.** Figure 16-8 diagrams the components of a synovial joint. Hyaline cartilage covers the ends of the articulating bones, which are separated by a fluid-filled space, or joint cavity. The entire joint is encapsulated in a double layer of connective tissue, with the joint cavity and its fluid lying between these two layers. The outer layer of the capsule is made up of dense connective tissue that holds the bones of the joint together. Some of the connective tissue fibers are arranged in bundles forming strips of tissue, or ligaments. Although this outer capsule holds the bones of the joint in place, it permits a wide range of motion. The inner layer of the capsule is made up of loose connective tissue, including elastic fibers and adipose tissue, or fat. This inner layer of tissue secretes the fluid of the joint cavity, called *synovial fluid.* In fact, the synovial joint gets its name from this "egg white–like" fluid; the word *synovial* comes from a Greek word and a Latin word meaning "with" *(syn)* "an egg" *(ovum).* Synovial fluid lubricates the joint, provides nourishment to the cartilage covering the bone, and contains white blood cells that battle infection. Table 16-1 summarizes

TABLE 16-1

Locations and Motions of Various Joints in the Human Body

TYPE OF JOINT	LOCATION	TYPE OF MOVEMENT
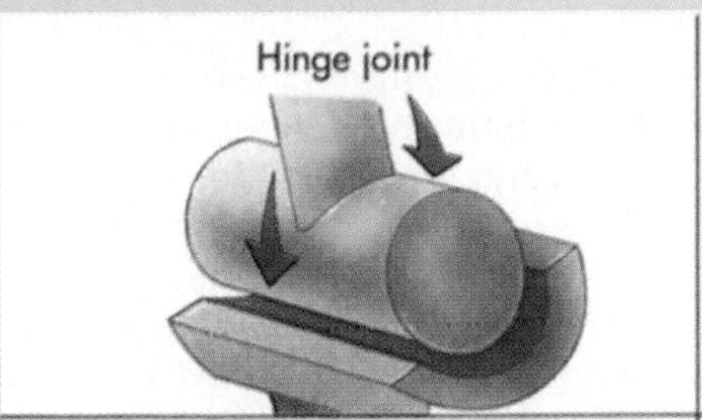	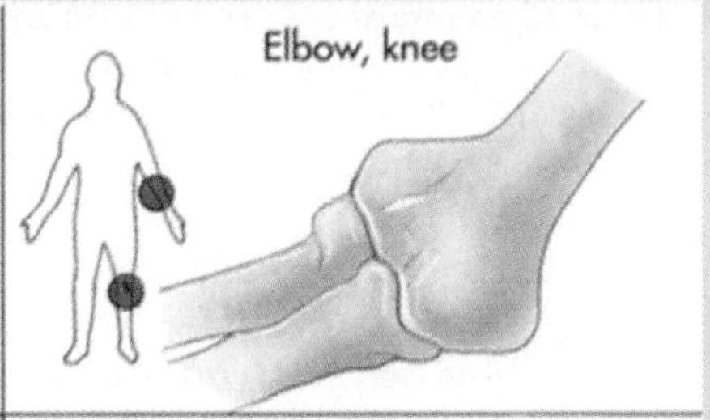	Hinge movement
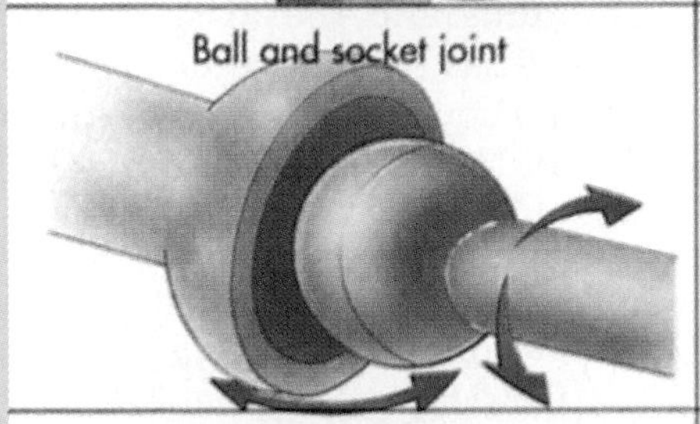	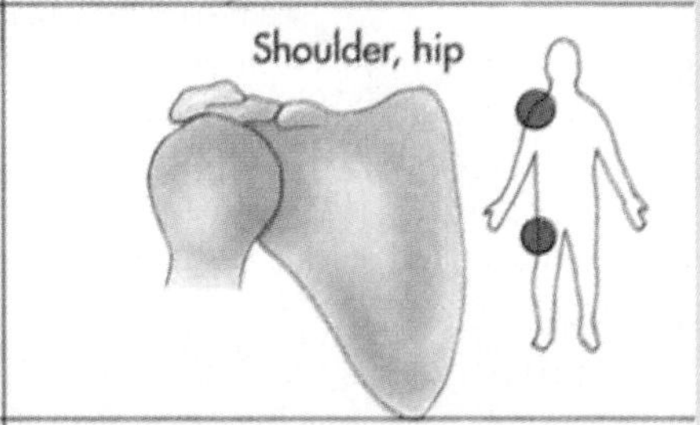	Wide range of movement in almost any direction
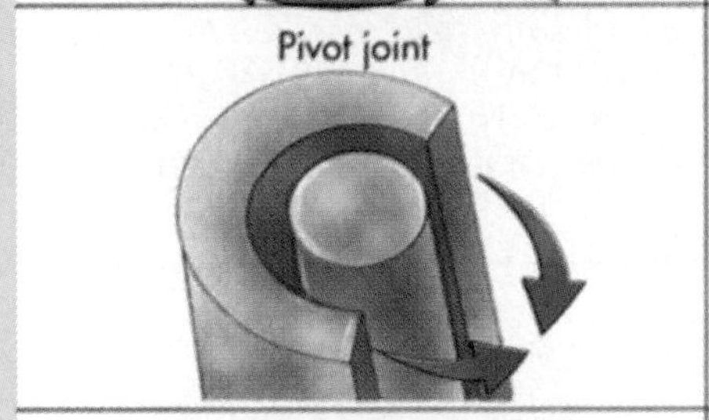	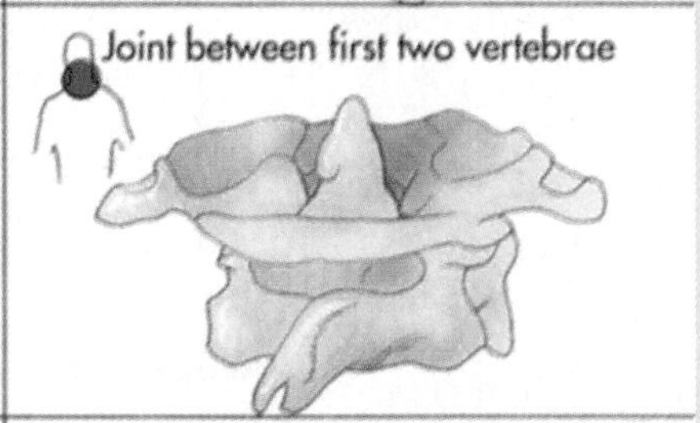	Side-to-side movement
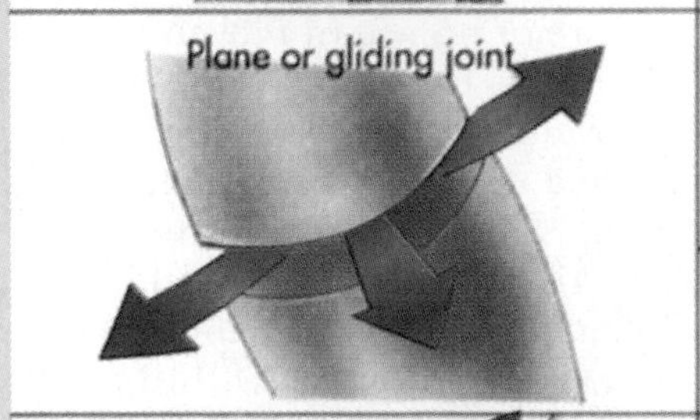	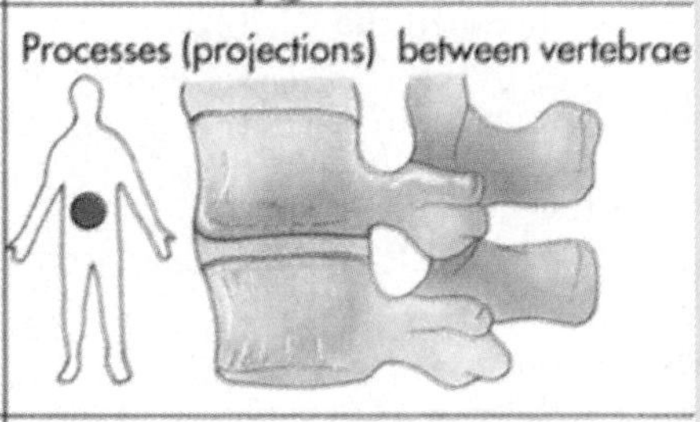	Sliding in many different directions
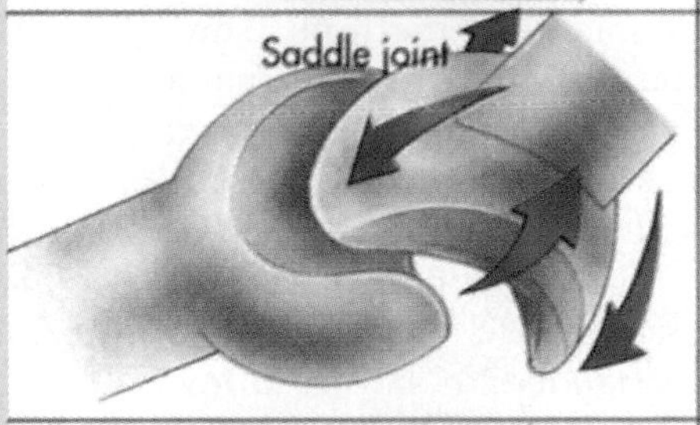	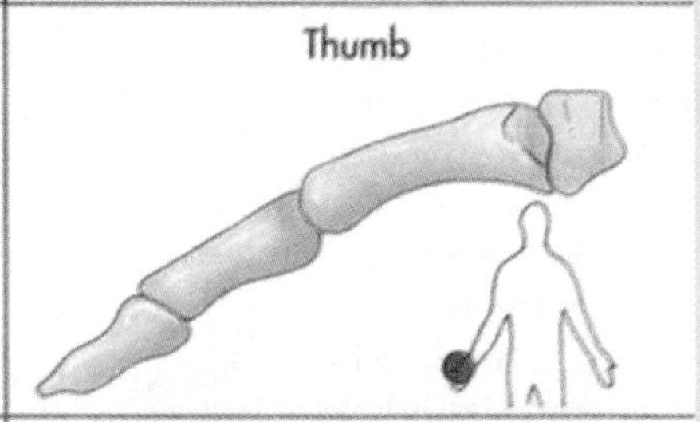	Movement in right angles
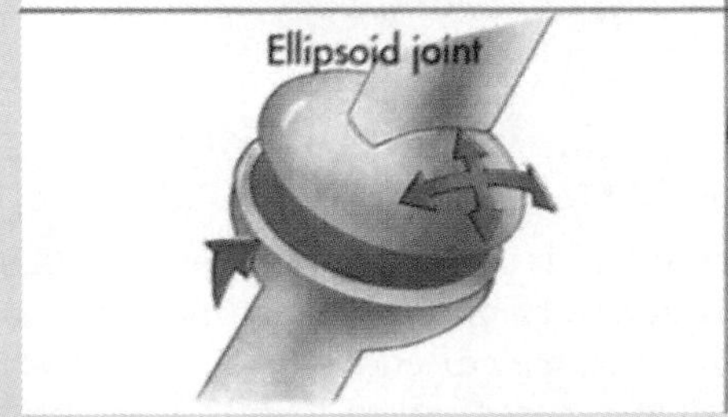	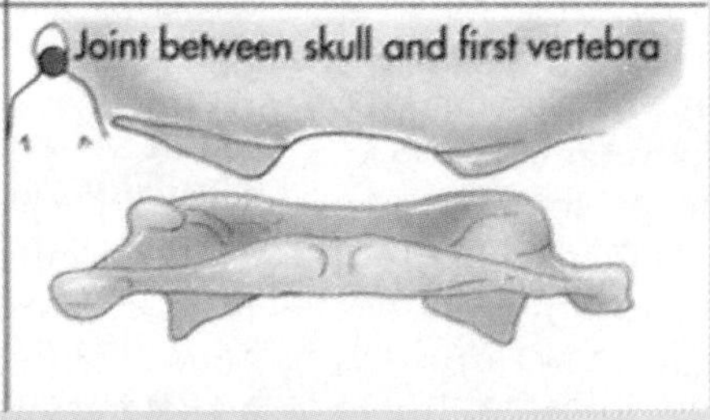	Nearly hinge movement, restriction of rotation

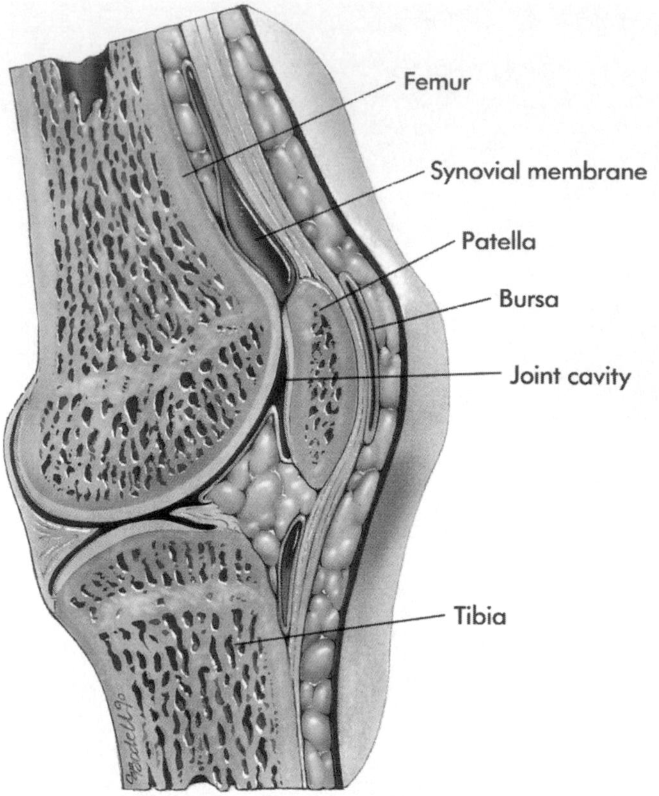

FIGURE 16-8 **A synovial (freely movable) joint.** The articulation of the kneecap (patella) with the femur and tibia is an example of a synovial joint. In the knee joint a sac filled with synovial fluid (a bursa) lies on top of the kneecap, providing a cushion between the bone and skin.

the types of movements possible at synovial joints and gives other examples of joints.

Although synovial joints permit a wide range of movement, this movement is limited by several factors, including the tension, or tightness, of the ligaments and muscles surrounding the joint and the structure of the articulating bones. You can see the various types of movement of synovial joints on your own body. For example, move your hand at the wrist or your foot at the ankle. During this movement the carpals or tarsals slide over one another to produce gliding movements. Other joints allow angular movements, which as the name suggests, increase or decrease the angle between two bones. When body builders flex their biceps, for example, they bring their forearms closer to their upper arms, decreasing the angle between the radius/ulna and the humerus. The elbow joint permits this movement and is called a **hinge joint.** A hinge joint allows movement in one plane only, similar to how a door opens and closes by means of its hinges. Another type of movement is rotation, the movement of a bone around an axis. You see this type of movement when a pitcher winds up and throws a baseball or when you move your head from side to side.

The type of joint the pitcher is using at the shoulder is called a **ball-and-socket joint.** In this joint, the rounded head or ball of the humerus articulates with the concavity or socket formed by the ends of the clavicle and scapula. When you move your head from side to side, the type of joint you are using is a **pivot joint,** formed by the first two vertebrae, the atlas and the axis. Your skull rests on the ringlike atlas (named after the mythological Greek god who could lift the Earth) and pivots on a projection of the axis rising from below.

A joint is a place where bones and cartilage come together. The amount of movement allowed between two bones depends on how tightly the bones are held together. Freely movable joints, those in which a space exists between articulating bones, are common in the body and include such joints as the elbow and knee.

Muscles

Bones and joints are of no use in movement unless muscles are attached to the bones. Muscles provide the power for movement and are made up of specialized cells packed with intracellular fibers capable of shortening. Your body has three different kinds of muscle tissue: smooth muscle, cardiac muscle, and skeletal muscle (see Table 7-4). Each type is found in certain locations within the body and performs specific functions. Smooth muscle tissue is made up of sheets of cells that are found in many organs, doing such jobs as mixing food in your stomach and narrowing the interior of your arteries to restrict blood flow. Cardiac muscle tissue makes up the heart—the pump of your circulatory system. Skeletal muscle tissue makes up the muscles that are attached to bones, allowing you to move your body; these muscles are the nearly 700 muscles of the muscular system.

The Muscular System

The muscular system is shown in Figure 16-9. Its muscles are attached to bones by means of cords of dense connective tissue called **tendons.** One end of a skeletal muscle is attached to a stationary bone; this attachment is called the *origin.* The other end

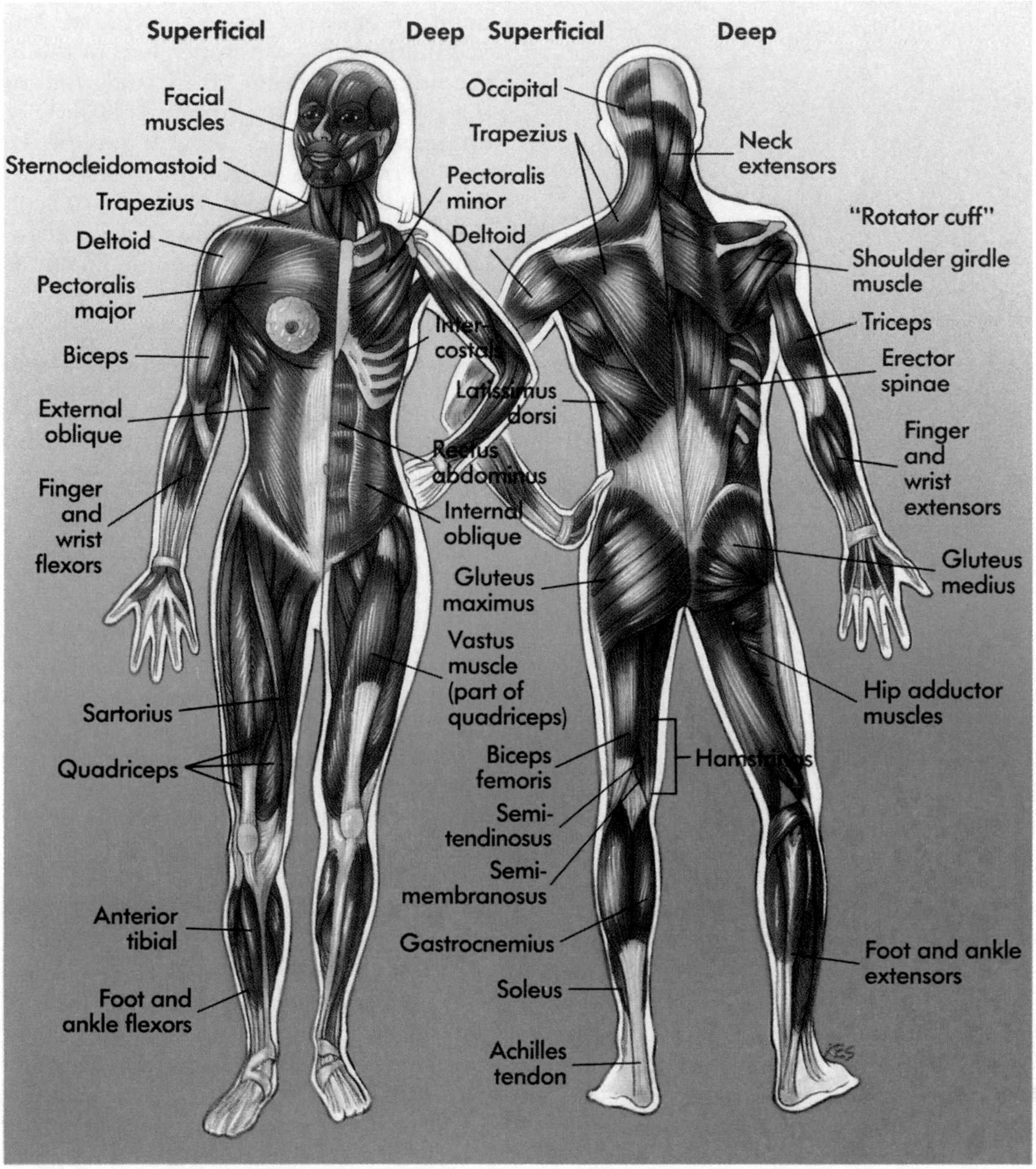

FIGURE 16-9 The human muscular system. Body movements are accomplished by the contraction of the skeletal muscles, which are attached to the bones.

joint a place within the skeletal system where bones, or bones and cartilage, come together; also known as an articulation.

articulation (ar **tik** you **lay** shun) another term for a joint.

suture (**soo** chur) a type of immovable joint.

ligaments (**lig** uh munts) bundles or strips of dense connective tissue that hold bones to bones.

synovial joints (suh **no** vee uhl) a freely movable joint in which a fluid-filled space exists between the articulating bones.

hinge joint a kind of joint that allows movement in one plane only.

ball-and-socket joint a kind of joint that allows rotation, or the movement of a bone around an axis.

pivot joint a kind of joint that allows side-to-side movement.

tendons cords of dense connective tissue that attach muscles to bones.

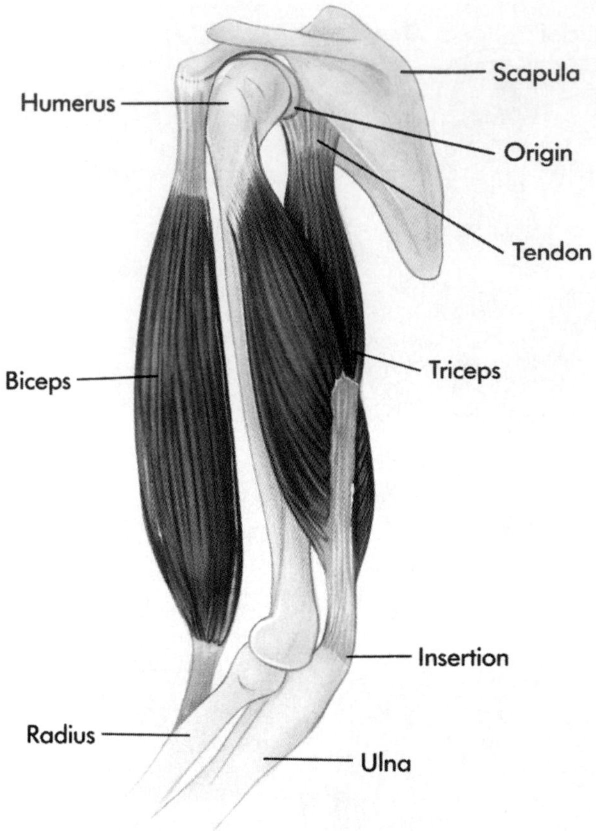

FIGURE 16-10 The attachments of skeletal muscles to bones. The biceps is attached to the stationary scapula at the origin and to the movable radius at the insertion with straps of tissue called tendons. The origin anchors the muscle as it pulls at the insertion.

is attached to a bone that will move; this attachment is called the *insertion*. The origin serves to anchor the muscle as it pulls at the insertion. As an example, Figure 16-10 shows the biceps and triceps muscles and their origins and insertions. The biceps brachii is located on the front of the upper arm. In fact, the word *brachii* means "arm." This muscle has two upper ends, or heads (hence, *biceps,* meaning "double headed"). These two heads have origins on the edges of the scapula near the front side of the humerus. Their tendons pass from the scapula over the upper end of the humerus and then blend into the fattened midsection, or belly, of the biceps. The insertion of the biceps is by means of flattened tendons into the radius. When the biceps shortens, the insertion is brought closer to the origin of the muscle, an action that pulls the forearm to the upper arm. To lower the forearm, the triceps brachii goes into action. Skeletal muscles oppose each other in this way; such opposing muscle pairs are called *antagonists.* In this case, the action of the triceps opposes the action of the biceps. The triceps sits on the back of the arm with one of its three heads originating on edges of the scapula near the back side of the humerus. Another head originates along the upper half of the back side of the humerus and the third along the lower half of the back side of the humerus. The triceps inserts on the ulna near the elbow. When the triceps shortens, its insertion is brought closer to its origin and the forearm is brought downward. The distance between the origin and insertion is measured as the length of the muscle and tendons, not as a straight line between the two points.

The nearly 700 muscles of the skeletal system are attached to bones by means of connective tissue called tendons. One end of a muscle is attached to a stationary bone (the origin) and the other to a movable bone (the insertion). The origin anchors the bone as it pulls at the insertion.

How Skeletal Muscles Contract

Skeletal muscles are called *striated muscles* because their cells appear marked with striations, or lines, when viewed under the light microscope (see Figure 16-12, *A*). These striations are caused by an orderly arrangement of microfilaments within skeletal muscle cells. Special groupings of these microfilaments are the contractile units of muscle cells.

Skeletal muscle cells are extremely long cells formed by the end-to-end fusion of shorter cells during embryonic development. These long muscle cells are called *muscle fibers.* Each muscle fiber contains all the nuclei of the fused cells pushed out to the periphery of the cytoplasm (Figure 16-11). To see the relationship between muscle fibers and muscles, study Figure 16-11, noting that a muscle is made up of bundles of muscle fibers, bound together with connective tissue, and nourished by blood vessels. As you study the diagram, you will notice that the names of certain parts of muscle cells have the prefix *sarco,* which comes from a Greek word meaning "flesh." Table 16-2 lists these differences in terminology.

Each muscle fiber is packed with *myofibrils,* which are cylindrical, organized arrangements of special thick and thin microfilaments capable of shortening the muscle fiber (see Chapter 3). In muscle cells, microfilaments are called *myofilaments* (*myo* means "muscle"). The so-called thin myofilaments are made up of the protein **actin.** The thick myofilaments are made up of a much larger protein, **myosin.** All the myofilaments within a muscle fiber are lined up in such a way that the cells appear to have interior bands of light and dark lines, a feature readily observable in Figures 16-11 and 16-12.

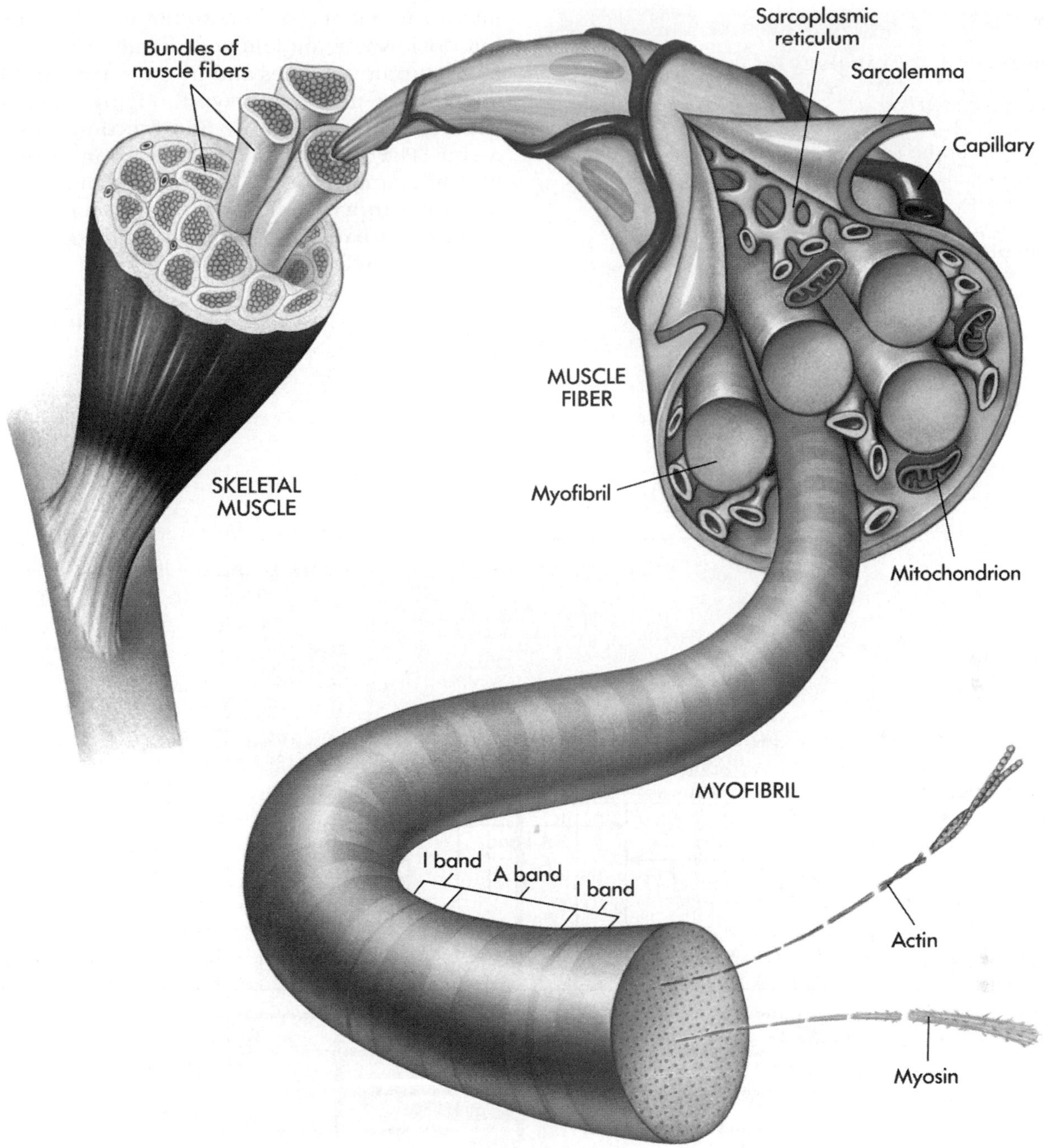

FIGURE 16-11 Structure of a skeletal muscle. This diagram shows progressively magnified views of a skeletal muscle.

Figure 16-12 is a close-up view of the banding pattern in a muscle fiber as shown in an electron micrograph. This pattern is repeated throughout each muscle fiber. Below the electron micrograph in Figure 16-12, the diagram shows how the actin and myosin filaments are arranged, forming this banding pattern. The thin actin filaments are attached to plates of protein that appear as dark lines called *Z lines.* The actin filaments extend from the Z lines (plates) equally in two directions, perpendicular to the plates. In a resting muscle the actin filaments that extend from two sequential Z lines are not long enough to reach each other. Instead, they are joined to one another by interdigitating

actin (ak tin) a protein that makes up the thin myofilaments in a muscle fiber.

myosin (my uh sin) a protein that makes up the thick myofilaments in muscle fiber.

TABLE 16-2

Muscle Cell Terminology

CELL COMPONENT	MUSCLE CELL COMPONENT
Cell membrane (or plasmalemma)	Sarcolemma
Cytoplasm	Sarcoplasm
Endoplasmic reticulum	Sarcoplasmic reticulum

myosin filaments, similar to the way your fingers interlock when you fold your hands. This arrangement of protein plates and myofilaments produces the banding patterns shown in Figure 16-12. The I band contains the Z line and the thin actin filaments. The A band contains a portion of the thin filaments and the thick myosin filaments. The H zone appears as a light zone running down the center of the A band; it contains only thick filaments held in place by a series of fine threads called the

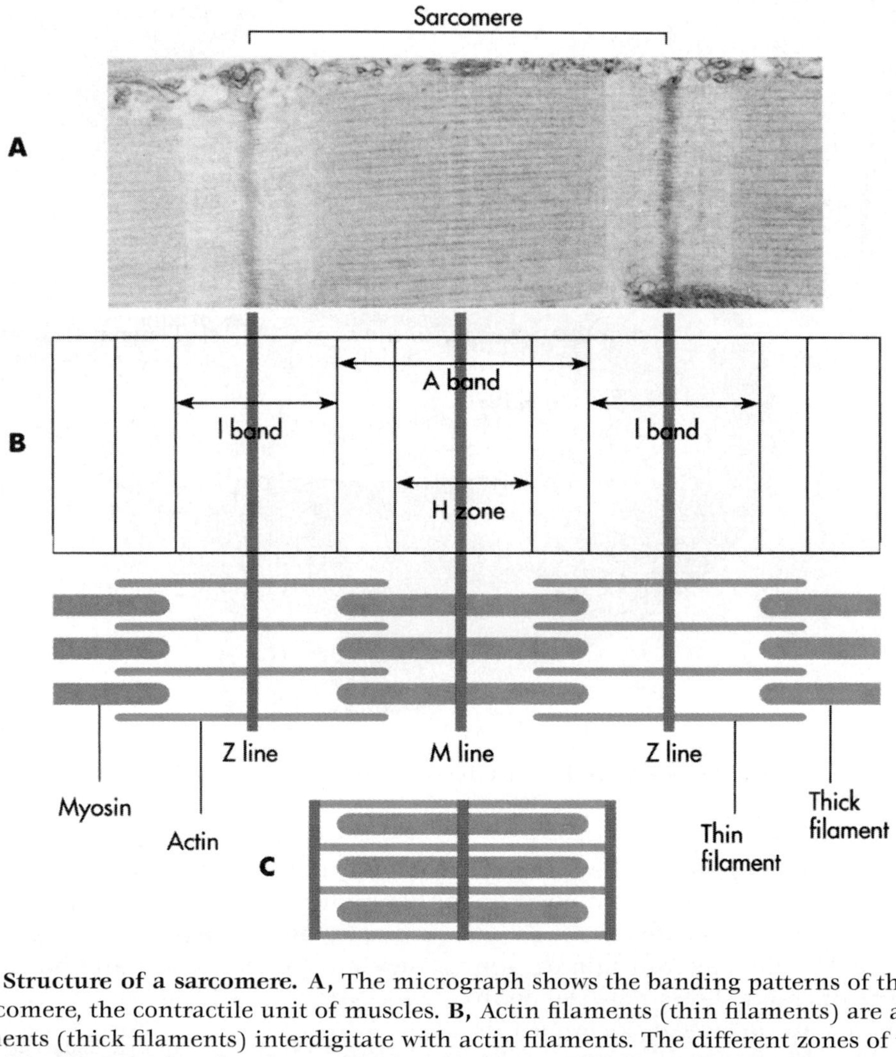

FIGURE 16-12 Structure of a sarcomere. A, The micrograph shows the banding patterns of the thick and thin filaments in a sarcomere, the contractile unit of muscles. **B,** Actin filaments (thin filaments) are attached to the Z line. Myosin filaments (thick filaments) interdigitate with actin filaments. The different zones of the muscle are also illustrated in this diagram; the I band contains thin filaments and Z line, the A band contains portions of thin filaments and the thick filaments, and the H zone contains only thick filaments that have been anchored in place by the M line. **C,** In a contracting sarcomere, the thin filaments move toward the M line. As this movement occurs, the I band disappears.

Applications

HOW Science WORKS

Can certain types of running shoes help you run faster than others?

In the 1963 Hollywood comedy *Son of Flubber,* an absent-minded professor invented a magical bouncy substance that, when inserted into the boots of the hapless football team players, gave them a lift and helped them to win their big game. What was fantasy then is the talk of runners today.

Shoes are the most important tool of any runner. They must both cushion the impact of the foot against the running surface and prevent exaggerated sideways movements of the foot that can lead to injury. In modern shoes, the foot is protected from unwanted sideways movements by a strong flexible plastic cup around the heel that holds it firmly in place. To cushion a shoe, manufacturers commonly use a midsole material made of springy ethyl vinyl acetate (EVA) encapsulated in various ways within polyurethane.

In the last few years, shoe manufacturers have added a new twist to shoe cushioning. Noting experiments carried out at Harvard University, which revealed that runners ran 3% faster on tracks made slightly elastic, some shoe manufacturers reasoned that these conclusions should also apply to the shoes running on the track. Manufacturers therefore designed an "elastic" shoe that uses an "energy-return" system.

In a typical energy-return shoe, the cushion is provided by hollow tubes of flexible thermoplastic that quickly spring back to their original shape after being compressed. The idea is that the springback action of the tubes imparts energy, aiding liftoff of the runner's foot.

Do they work? Shoe manufacturers say yes, but some researchers disagree. They argue that a functional energy-return system would have to capture the energy from the rear of the foot after it hit the pavement and transfer it to the front of the foot, not simply sending it back up the rear of the foot as energy-return shoes do. Perhaps more to the point, many energy-return shoes are not as flexible as their "less technologically advanced" counterparts, resulting in a more jarring run, just as a stiff suspension in a car produces a jarring ride. In addition, the goal for most recreational runners is to expend energy and get a good aerobic workout—not to minimize energy expenditure using special shoes.

M line. The part of the myofibrils lying between adjacent Z lines is called the **sarcomere,** the contractile unit of muscles.

> *Skeletal muscle cells are long, multinucleated cells called muscle fibers. A muscle is made up of bundles of muscle fibers. Each muscle fiber is packed with organized arrangements of microfilaments that are capable of contracting.*

Molecularly, each actin filament (that is, thin filament) consists of two strings of proteins wrapped around one another, like two strands of loosely wound pearls. The result is a long, thin, helical filament. Myosin has an unusual shape: one end of the molecule is a coil of two chains that forms a

sarcomere (sar koe mere) the repeating bands of actin and myosin myofilaments that appear between two Z lines in a muscle fiber; the contractile unit of muscles.

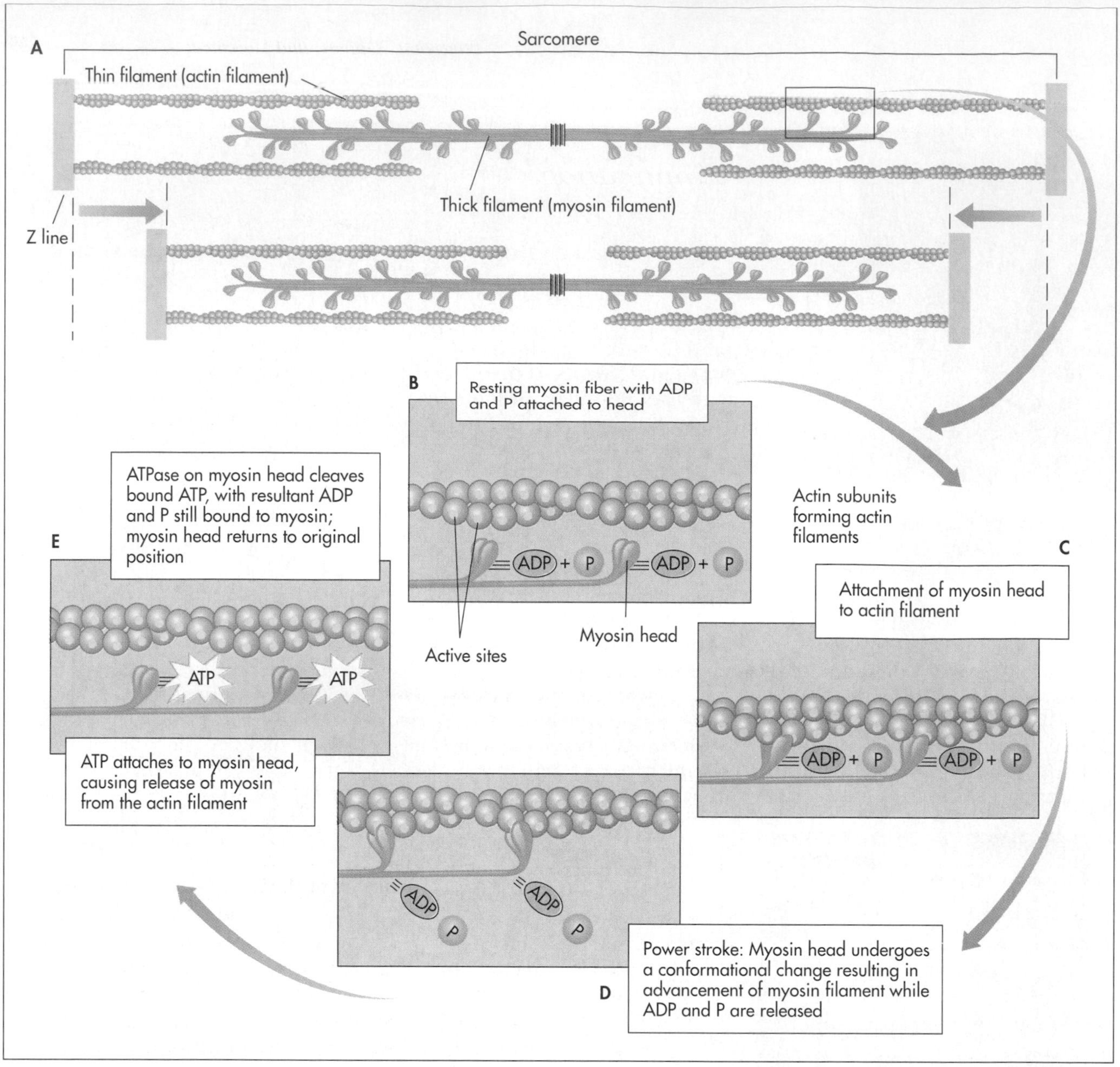

FIGURE 16-13 How the myofilaments move during muscle contraction. A, The sliding of myosin filaments past the actin filaments pulls the actin filaments toward the center. **B,** An illustration of a microscopic model of how myosin slides past actin. Myosin moves along the actin filament by first binding to it and then moving forward as the result of a change in the shape of the myosin head.

very long rod, whereas the other end consists of a double-headed globular region. In electron micrographs, a myosin molecule looks like a two-headed snake. The contraction of myofilaments occurs when the heads of the myosin filaments change shape. This causes the myosin filaments to slide past the actin filaments, their globular heads "walking" step by step along the actin (Figure 16-13).

How does myofilament sliding lead to muscle contraction? Because actin myofilaments are anchored to the Z lines, the Z lines are pulled closer to one another when actin slides along the myosin, contracting the sarcomere. All the sarcomeres of a myofibril contract simultaneously, shortening the myofibril. In addition, all the myofibrils of a muscle fiber usually contract at the same time. However, all the muscle fibers within a muscle *do not* contract simultaneously. The forcefulness, speed, and degree of a muscle contraction depend on the number of muscle fibers that contract, their positions in relation to one another, and the frequency of nerve stimuli.

Muscle cells contract as a result of the sliding of actin and myosin filaments past one another. Changes in the shape of the ends of the myosin molecules, which are located between adjacent actin filaments, cause the myosin molecule to move along the actin, producing contraction of the myofilament.

How Nerve Impulses Signal Muscle Fibers to Contract

The contraction of skeletal muscles is initiated by a nerve impulse that causes a series of events ultimately resulting in the interaction of actin and myosin. The nerve impulse arrives as a wave of depolarization along the nerve fiber (see Chapter 13). The end of the nerve fiber, the *motor endplate,* is microscopically close to the surface of the muscle fiber, forming a **neuromuscular junction** (Figure 16-14). When the wave of depolarization reaches the motor endplate, it causes the release of the neurotransmitter acetylcholine from the nerve cell into the junction. The acetylcholine passes across to the muscle fiber membrane and opens the ion channels of that membrane, depolarizing it. Infoldings of the membrane of the muscle fiber called *transverse (T) tubules* carry this action potential deep into the muscle fiber, reaching each myofibril. In turn, the depolarization of the muscle fiber membrane opens calcium ion channels in the membrane of the *sarcoplasmic reticulum,* which is a tubular, branching latticework of endoplasmic reticulum that wraps around each myofibril like a sleeve (see Figure 16-11). In resting muscle, calcium ions are actively pumped through these channels, concentrating calcium ions within the spaces or sacs of the sarcoplasmic reticulum and thus polarizing the sarcoplasmic reticulum membrane with respect to calcium ions. When the ion channels of the sarcoplasmic reticulum membrane open, calcium ions stored within the sarcoplasmic reticulum move across its membrane and into the sarcoplasm (cytoplasm) (Figure 16-15). The presence of calcium ions triggers the series of chemical reactions of contraction.

In resting muscle, myosin filaments are not free to interact with actin because the myosin binding sites on the actin (called *cross bridge binding sites*) are not available. These binding sites are blocked by a threadlike molecule called *tropomyosin* (see Figure 16-15). Along this tropomyosin thread lie other globular proteins called *troponin.* The troponin molecules bind the calcium molecules when they are released from their storage sacs in the sarcoplasmic reticulum. When this binding occurs, the troponin molecules change their shape. With this change, the tropomyosin thread is repositioned to a new location where it does not interfere with myosin interaction. The globular heads of the myosin filament bind to the actin, forming cross bridges that exert tension on the actin threads, pulling them toward the center of the sarcomere. As in a chain reaction, the myosin then cleaves a molecule of ATP (see Chapter 4) and uses the energy that is released to break the cross bridge. Spontaneously, the now-freed head of the myosin forms a new cross bridge at another actin binding site, once again pulling the actin thread closer to the center of the sarcomere. This cycle of formation, breaking, and re-formation of the cross bridges results in the speedy and simultaneous contraction of the sarcomeres within a muscle fiber. After the passage of the stimulating nerve impulse, the membrane's calcium channels close. The sarcoplasmic reticulum repolarizes by the active transport of calcium ions back out of the cytoplasm and into the reticular spaces. As the level of calcium in the cytoplasm decreases, the tropomyosin-troponin complexes again block the myosin binding sites on actin, and the muscle relaxes. Thus the release of calcium by the nerve fiber's stimulation of the sarcoplasmic reticulum releases the troponin that, together with an input of energy, results in the contraction of the myofibril.

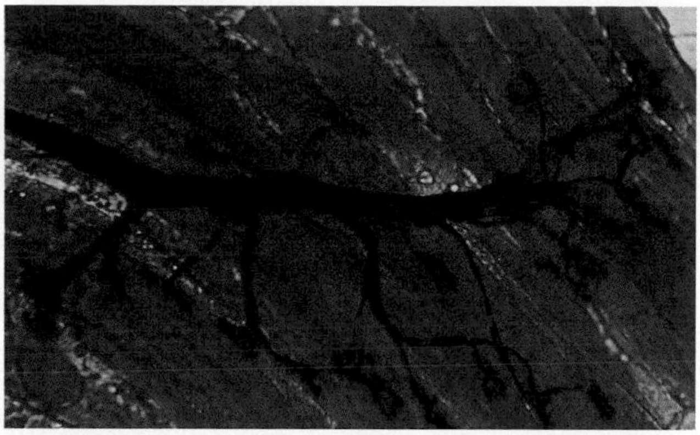

FIGURE 16-14 A neuromuscular junction. The thick central portion is the body of an axon, with small branches each terminating in a motor endplate. This special synapse between a motor nerve fiber and a muscle fiber is the neuromuscular junction. *(460x)*

> *In skeletal muscle, contraction is initiated by a nerve impulse. Acetylcholine passes across the neuromuscular junction from the nerve to the muscle, initiating the process that causes the muscle to contract. This process is driven by the energy from ATP.*

neuromuscular junction (ner oh **mus** kyuh lur) a synapse between a neuron and a skeletal muscle cell.

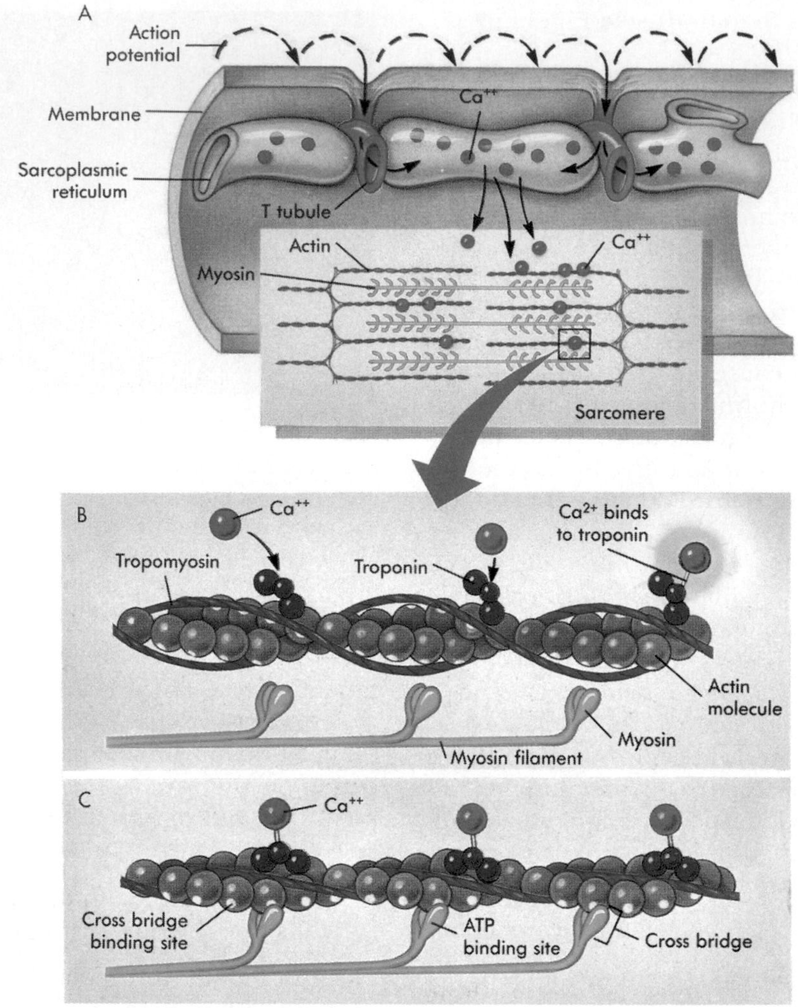

FIGURE 16-15 How nerve impulses signal muscle fibers to contract. A, When a nerve impulse reaches a motor endplate, acetylcholine causes a depolarization in the muscle fiber membrane. The T tubules in the membrane transmit this action potential to each myofibril, causing calcium ions to be released from their storage sites in the sarcoplasmic reticulum. **B,** For muscle contraction to occur, myosin must be able to bind to the actin at specific sites called cross bridge binding sites. However, in resting muscle, these sites are obstructed by tropomyosin threads. When a muscle is signaled to contract, the calcium ions released from the sarcoplasmic reticulum bind to troponin, a molecule that studs the tropomyosin threads. **C,** This binding causes the tropomyosin to move out of the way so that myosin can bind to actin at the cross bridge binding sites.

Summary

- The human body is able to move from place to place because it is supported by a rigid internal framework of bones powered by skeletal muscles. The skeletal muscles are attached to the bones and use them as levers to move the body itself or other objects. The skin provides a flexible covering for the body, accommodating each movement, no matter how large or small.
- In addition to providing an elastic covering for the body, the skin serves as a protective barrier against water loss and invading microbes, provides a surface by which you sense your environment, and helps control your body's internal temperature.
- Although bones may not look like living tissue, they contain living cells embedded in a matrix of collagen fibers and mineral salts. The collagen provides flexibility, whereas the mineral salts provide hardness. Bone is nourished by blood that flows in blood vessels permeating its interior.
- Compact bone is found along the length of long bones, surrounding a central core of fatty yellow marrow. A lighter, bony latticework called

spongy bone is found within the ends of long bones and within short, flat, and irregularly shaped bones.

- Together, all the bones of the body make up the skeletal system. Along with providing support for the body and places for the attachment of skeletal muscles, the skeletal system protects delicate internal structures such as the heart, lungs, and brain; stores minerals such as calcium and phosphorus; and produces blood cells in its red marrow.
- Scientists divide the skeletal system into two parts: the axial skeleton and the appendicular skeleton. The axial skeleton is your central column of bones and consists of the skull, vertebral column, and rib cage. The appendicular skeleton consists of the bones of the appendages (the arms and legs), and the bones off which the appendages hang (the collarbones, shoulder blades, and pelvic bones). A joint, or articulation, is a place where bones, or bones and cartilage, come together.
- There are three kinds of muscle tissue: smooth muscle tissue, which is found in many body organs and contracts spontaneously; cardiac muscle tissue, which makes up the heart and may initiate contraction spontaneously; and skeletal muscle tissue, which is organized into trunks of long fibers and contracts only when stimulated by a nerve. Skeletal muscles are attached to bones and move the body. These are the muscles of the muscular system.
- Skeletal muscles move bones by pulling on one while anchored to another. They are made up of bundles of extremely long muscle cells called *muscle fibers.* Each muscle fiber is packed with myofibrils, which are cylindrical, organized arrangements of special thick and thin filaments called *myofilaments,* capable of shortening the muscle fiber as they slide past one another.
- The contraction of skeletal muscles is initiated by a nerve impulse, which causes a series of events that ultimately result in the interaction of actin and myosin. Acetylcholine passes across the neuromuscular junction from the nerve to the muscle, triggering the events leading to this interaction.

Knowledge and Comprehension Questions

1. You raise your hand to answer a question in class. Explain the roles played by your bones and skeletal muscles in this movement.
2. Human babies learn extensively by touching and manipulating toys and objects. What physiological role does skin serve in this process?
3. What is bone, and what are its functions? Explain its importance.
4. Leukemia patients sometimes receive red bone marrow transplants as treatment for their disease. From what kind of bone tissue would this marrow be obtained?
5 What type of bone tissue composes most of the bones of the axial skeleton?
6. What do you hypothesize happens to excess calcium taken in by the body yet not needed at the moment?
7. Mrs. Gorman has injured her back; her doctor tells her she has a "slipped disk." What type of "disk" is involved, and what is its function?
8. What do hinge, pivot, and ball-and-socket joints have in common? How are they different?
9. What connective tissue connects muscles to bones? What connects muscles to each other?
10. Many older people suffer from osteoarthritis, which causes painful inflammation and decreased mobility in the knees, fingers, shoulders, and elbows. What skeletal structures seem, therefore, to be the targets for this disease?
11. Define actin, myosin, and myofilament. Explain their roles in the contraction of a muscle fiber.
12. Explain how a nerve impulse can lead to a muscle contraction.

Critical Thinking

1. From your observations, describe the changes that occur to skin and bone as people age.
2. In recent years, there has been tremendous progress made in the development and use of metallic artificial joints and bone replacement. What activities and bone functions would be easily served by such artificial structures? Are there functions of the bone that could not be filled by such replacements?

CHAPTER 17

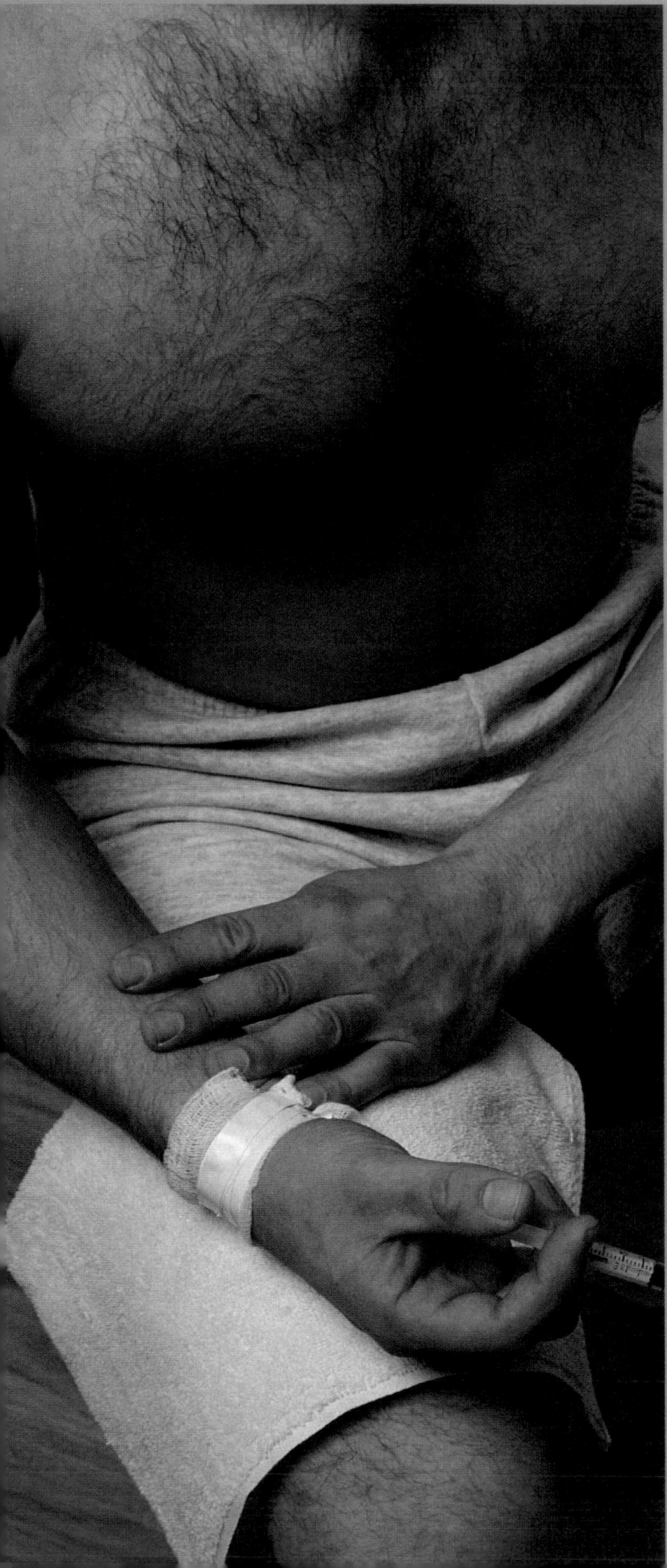

Hormones

Shooting up with anabolic steroids has been the downfall of many winners. These controversial drugs are really synthetic hormones, chemicals that affect the activity of specific organs or tissues. The various hormones of the body all affect their "target" tissues in unique ways. Anabolic steroids affect the body in ways similar to the male sex hormone testosterone and stimulate the buildup of muscles. But along with building a championship body, anabolic steroids strikingly change the body's metabolism.

Female athletes on steroids experience side effects such as shrinking breasts, a deepening voice, and an increase in body hair. Male athletes find that their testicles shrink. Some users also experience life-threatening kidney and liver damage. Youngsters who take these drugs risk stunting their growth because anabolic steroids cause bones to stop growing prematurely. And a great deal of controversy still surrounds claims that anabolic steroids can cause psychological effects such as "steroid rage," a state of mind in which users attack people and things around them.

Today, scientists still lack scientific data regarding all aspects and consequences of anabolic steroid use. However, it is clear that their use is risky at the least—and may put users in the cemetery rather than in the winner's circle.

Endocrine Glands and Their Hormones

A **hormone** is a chemical messenger sent by a gland to other cells of the body. Traditionally, animal hormones have been described by scientists as the chemical products of glands that travel within the bloodstream to all parts of the body, causing an effect on specific cells, or target organs, far removed from that gland. Glands are individual cells or groups of cells that secrete substances. Their secretory portions are made up of specialized epithelial cells. The glands that secrete hormones spill these chemicals directly into the bloodstream and are called **endocrine,** or ductless, **glands.** Glands that secrete other substances such as digestive enzymes or sweat route their secretions to specific destinations by means of ducts. For example, the digestive enzyme pancreatic amylase flows directly from the pancreas to the small intestine and goes nowhere else. Glands having this kind of associated ductwork are called **exocrine glands.**

Today, most scientists have expanded their definition of hormones to include any chemical produced by one cell that causes an effect in another. Included in this description, then, are substances such as neurotransmitters—chemicals produced by the axon end of a nerve cell that travel to and bind with the dendrite end of an adjacent nerve cell, contributing to the propagation of the nerve impulse along that neuron (see Chapter 13). Such chemicals are often called *local hormones* because

KEY CONCEPTS

- Hormones are chemical messengers produced by cells; they regulate other cells of the body.
- Hormones produced by glands and transported to cells via the bloodstream are known as endocrine hormones, whereas hormones that are secreted into the tissue fluid to affect nearby cells are known as local hormones.
- Hormones affect cells in one of four ways: by regulating their secretions and excretions, by helping them to respond to changes in the environment, by controlling activities related to reproductive processes, and by influencing their proper growth and development.
- The production of hormones is controlled by a feedback mechanism that works in much the same way as a thermostat in a house.

OUTLINE

they affect neighboring target cells. The human body produces many local hormones; they are described later in this chapter.

Hormones are chemical messengers secreted by cells that affect other cells. Hormones that travel within the bloodstream and affect cells in another part of the body are called endocrine hormones. Hormones that do not travel within the bloodstream but only affect cells lying near the secretory cells are called local hormones.

The 10 different endocrine glands of the human body make over 30 different hormones. Together, these glands are called the **endocrine system** (Figure 17-1). The endocrine system works with the nervous system to integrate the functioning of the various tissues, organs, and organ systems of the body. The nervous system sends messages to muscles and glands, regulating muscular contraction and glandular secretion. The hormones of the endocrine system, on the other hand, carry messages to virtually any type of cell in the body. The messages of the endocrine hormones are varied but can be grouped into four categories:

1. *Regulation:* Hormones control the internal environment of the body by regulating the secretion and excretion of various chemicals in the blood, such as salts and acids.
2. *Response:* Hormones help the body respond to changes in the environment and cope with physical and psychological stress.
3. *Reproduction:* Hormones control the female reproductive cycle and other reproductive processes essential to conception and birth and control the development of sex cells, the reproductive organs, and secondary sexual characteristics (those that make men and women different) in both sexes.
4. *Growth and development:* Hormones are essential to the proper growth and development of the body from conception to adulthood.

Once molecules of a hormone are released into the bloodstream, they travel throughout the body. Although hormone molecules may pass billions of cells, specific hormones only affect specific cells called *target cells.* Hormones recognize target cells because they bind to receptor molecules embedded within the cell membrane or located within the cytoplasm of the cell. The binding of a hormone molecule to a receptor molecule activates a chain of events in the target cell that results in the effect of the hormone being expressed.

Two major classes of endocrine hormones work within the human body: **peptide hormones** and **steroid hormones.** Peptide hormones are made of amino acids, but the amino acid chain length varies greatly from hormone to hormone. The smallest are actually modifications of the single amino acid tyrosine. Somewhat larger are short peptide hormones that are several amino acids in length. Polypeptide hormones have chain lengths of several dozen or more amino acids, such as the hor-

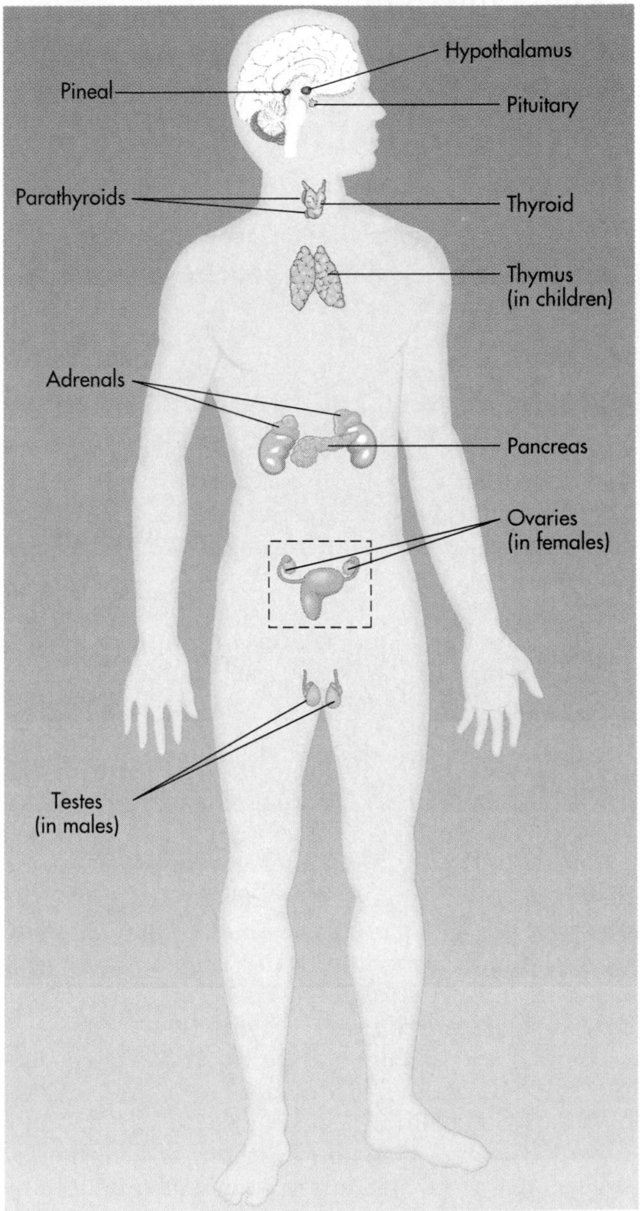

FIGURE 17-1 The human endocrine system. The endocrine glands pictured in this diagram secrete chemical "messengers" that travel through the bloodstream to affect other cells in the body.

mone insulin. Even larger are protein hormones that may have over 200 amino acids with carbohydrates attached at several positions.

Unable to pass through the cell membrane, peptide hormones bind to receptor molecules embedded in the cell membrane of target cells. The binding of a hormone to a receptor triggers an increase in that cell's production of a compound referred to as a *second messenger.* A second messenger triggers enzymes that cause the cell to alter its functioning in response to the hormone (Figure 17-2). For example, prolactin stimulates cells of the mammary glands to produce milk. Target cells respond as enzymes go into action catalyzing reactions that produce the components of mother's milk. Other types of hormone responses include the secretion of substances from target cells and the closing or opening of certain protein doors within target cell membranes. Cyclic adenosine monophosphate (cyclic AMP for short), a cousin of ATP (see Chapter 4), acts as a second messenger to many cells. Besides cyclic AMP, other second messenger molecules have been discovered.

Steroid hormones are all made from cholesterol, a lipid synthesized by the liver. You know cholesterol as that dietary devil present in certain foods such as eggs, dairy products, and beef. A characteristic of steroid hormones is their set of carbon rings. Steroid hormones, being lipid soluble, pass freely through the lipid bilayer of the cell membrane. Once inside a cell, these hormones bind to receptor molecules located within the cytoplasm of target cells. Together, the hormone-receptor complex moves into the nucleus of the cell, causing the cell's hereditary material, or DNA, to trigger the production of certain proteins (Figure 17-3). In response to the sex hormones estrogen and testosterone, for example, the proteins produced are those involved in such processes as the development and maintenance of female and male sexual characteristics.

Two main classes of endocrine hormones are peptide hormones and steroid hormones. Both travel within the bloodstream to all parts of the body but affect only certain target cells. Peptide hormones bind to receptors on the cell membrane of target cells and ultimately trigger enzymes that alter cell functioning. Steroid hormones bind to receptors within the cytoplasm of target cells and ultimately cause the hereditary material of the cell to produce specific proteins.

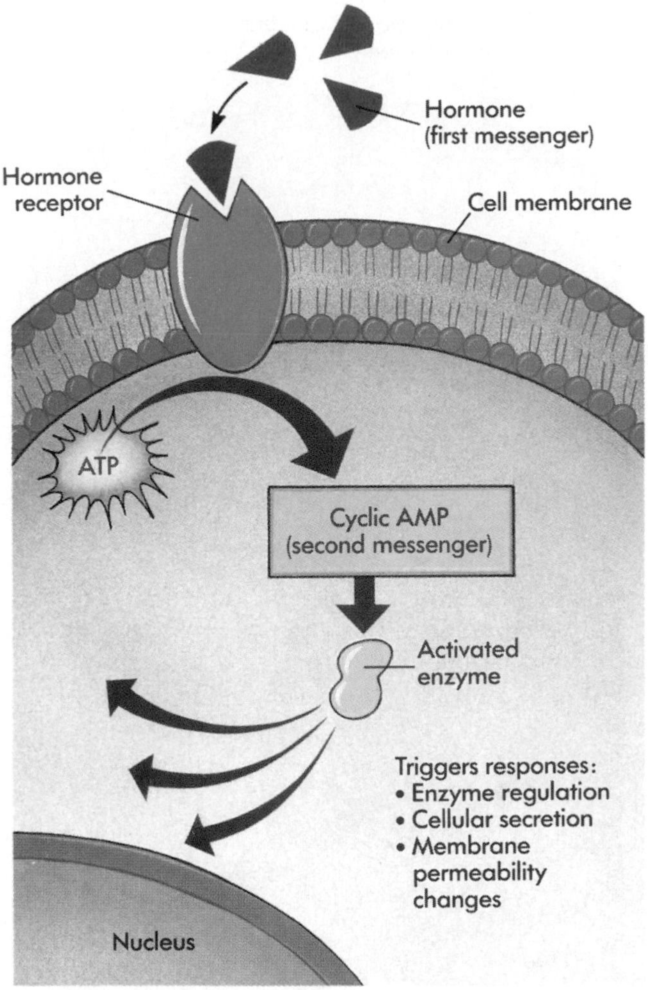

FIGURE 17-2 How peptide hormones work. Peptide hormones bind to target cell membranes and trigger an increase of second messenger compounds, such as cyclic AMP, within these cells. The second messenger in turn activates enzymes that alter cell function in response to the hormonal message.

hormone a chemical messenger secreted and sent by a gland to other cells of the body.

endocrine glands (**enn** doe krin) ductless glands that secrete hormones and spill them directly into the bloodstream.

exocrine glands (**ek** so krin) glands whose secretions reach their destinations by means of ducts.

endocrine system the collective term for the 10 different endocrine glands of the human body, which secrete over 30 different hormones.

peptide hormones one of the two main classes of endocrine hormones. Peptide hormones are made of amino acids and are unable to pass through cell membranes; instead, they bind to receptor molecules embedded in the membranes of target cells.

steroid hormones one of the two main classes of endocrine hormones. Steroid hormones are made of cholesterol and are able to pass through cell membranes.

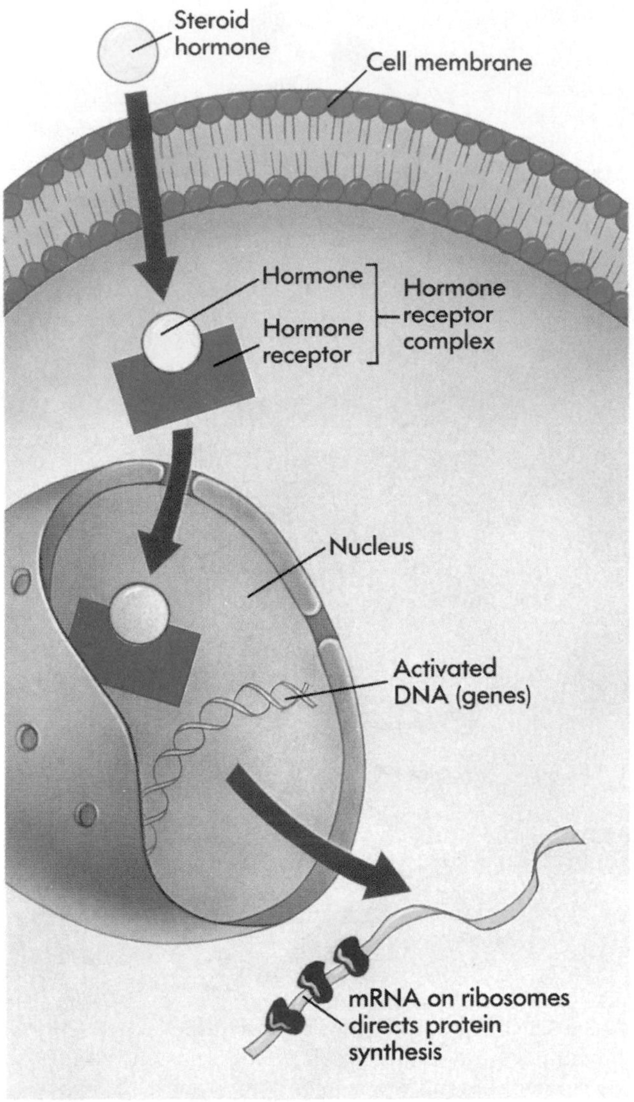

FIGURE 17-3 How steroid hormones work. Steroid hormones are able to pass through the cell membrane without the aid of a receptor molecule. Inside the cell, they bind with receptor molecules. The hormone-receptor complex then enters the nucleus of the cell, where it acts on DNA to produce proteins. These proteins control physiological processes such as growth and development.

The production of hormones is regulated by a mechanism called a *feedback loop.* In general, hormonal feedback loops work in the following way: endocrine glands are initially stimulated to release hormones. Stimulation of an endocrine gland occurs in one of three ways:

1. *Direct stimulation by the nervous system:* The sensation of fear, for example, can cause the autonomic nervous system to trigger the release of the hormone adrenaline from the adrenal medulla.
2. *Indirect stimulation by the nervous system by means of releasing hormones:* The hypothalamus is a specialized portion of the brain that produces and secretes releasing hormones. Some releasing hormones stimulate the release of other hormones; some prevent the release.
3. *The concentration of specific substances in the bloodstream:* The blood level of a substance such as glucose or calcium, for example, may signal an endocrine gland to turn on or turn off.

After an endocrine gland secretes its hormone into the bloodstream, the hormone travels throughout the body via the circulatory system and interacts with target tissues. The target tissues cause the desired effect to be produced. This effect acts as a new stimulus to the endocrine gland (Figure 17-4). Put simply, the body feeds back information to each endocrine gland after it releases its hormone. In a positive feedback loop, the information that is fed back causes the gland to produce more of its hormone. In a **negative feedback loop,** the feedback causes the gland to slow down or to stop the production of its hormone. Most hormones work by means of negative feedback loops. (Specific examples of feedback mechanisms and interactions are discussed throughout this chapter.)

The Pituitary Gland

The **pituitary** is a powerful gland that secretes nine different hormones. Although it secretes so many hormones, it is amazingly tiny—about the size of a marble. The pituitary gland hangs from the underside of the brain, supported and cradled within a bony depression of the sphenoid bone.

Controlling the Pituitary: The Hypothalamus

The pituitary secretes seven major hormones from its larger front portion, or lobe, the anterior pituitary. It secretes two from its rear lobe, the posterior pituitary. The secretion of these hormones is regulated by a mass of nerve cells that lies directly above the pituitary, making up a small part of the floor of the brain. This regulatory nervous tissue, the **hypothalamus,** is connected to the pituitary by a stalk of tissue (Figure 17-5, *A*). The hypothalamus uses information it gathers from the peripheral nerves and other parts of the brain to stimulate or inhibit the secretion of hormones from the anterior pituitary. In this way, the hypothalamus acts like a production manager, receiving information about the needs of the company's customers and regulating the production of products to satisfy

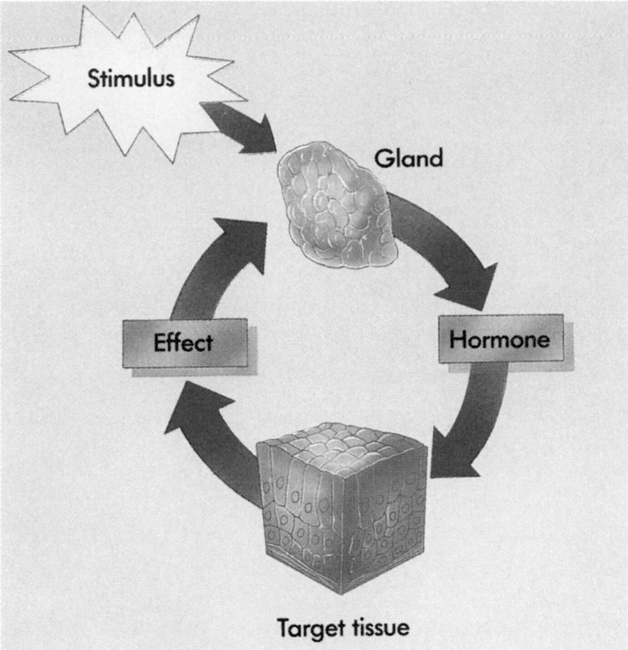

FIGURE 17-4 A simple feedback loop. In response to a stimulus, an endocrine gland releases a specific hormone that acts on a specific target tissue. The effect of the hormone on the target tissue either causes the gland to release more of the hormone (positive feedback) or causes the gland to slow or stop its production of the hormone (negative feedback).

those needs. The hypothalamus accomplishes its management job by producing *releasing hormones* that affect the secretion of specific hormones from the anterior pituitary. The hypothalamus also produces two hormones that do not regulate hormonal release in the pituitary. When they are needed by the body, the hypothalamus signals the pituitary to release them.

> *The pituitary is a tiny gland that hangs from the underside of the brain. The secretion of its many diverse hormones is controlled by a mass of nerve cells lying directly above it called the hypothalamus. The hypothalamus stimulates or inhibits the secretion of hormones from the pituitary by means of releasing hormones. In addition, the hypothalamus produces two hormones that it stores in the pituitary.*

The Anterior Pituitary

The seven hormones produced by the anterior pituitary regulate a wide range of bodily functions (Figure 17-5, *B*). Four of these hormones are called **tropic hormones.** The word *tropic* comes from a Greek word meaning "turning" and refers to the ability of tropic hormones to turn on or stimulate other endocrine glands. Of the four tropic hormones, two are **gonadotropins.**

The gonads are the male and female sex organs, the testes and the ovaries. The gonadotropins are hormones that affect these sex organs (considered endocrine glands because they secrete sex hormones). The two gonadotropins are *follicle-stimulating hormone (FSH)* and *luteinizing hormone (LH).* In females, FSH targets the ovaries and triggers the maturation of one egg each month. In addition, it stimulates cells in the ovaries to secrete female sex hormones called *estrogens.* In men, FSH targets the testes and triggers the production of sperm. LH stimulates cells in the testes to produce the male sex hormone testosterone. In females, a surge of LH near the middle of the menstrual cycle stimulates the release of an egg. In addition, LH triggers the development of cells within the ovaries that produce another female sex hormone—progesterone. (See Chapter 21 for a description of the organs and processes of the reproductive system.)

The two other tropic hormones are *adrenocorticotropic hormone (ACTH)* and *thyroid-stimulating hormone (TSH).* ACTH triggers the adrenal cortex to produce certain steroid hormones. The adrenal glands are located on top of the kidneys (see Figure 17-1). Each of these two glands has two distinct parts: an outer cortex and an inner medulla. ACTH stimulates the adrenal cortex to produce hormones that regulate the production of glucose from noncarbohydrates such as fats and proteins. Others regulate the balance of sodium and potassium ions in the blood. Still others contribute to the development of the male secondary sexual characteristics.

negative feedback loop within the endocrine system, a mechanism that causes a gland to slow down or stop the production of its hormone.

pituitary (puh **too** ih tare ee) a small but powerful gland hanging from the underside of the brain that is controlled by the hypothalamus and secretes nine major hormones.

hypothalamus (hye poe **thal** uh muss) a mass of nerve cells lying directly above the pituitary that regulates the pituitary's secretion of hormones, based on information it receives from peripheral nerves and parts of the brain.

tropic hormones (**trope** ik) four of the seven hormones produced by the anterior pituitary gland; the tropic hormones turn on, or stimulate, other endocrine glands.

gonadotropins (go **nad** oh **tro** pins) two of the four tropic hormones; gonadotropins affect the male and female sex organs.

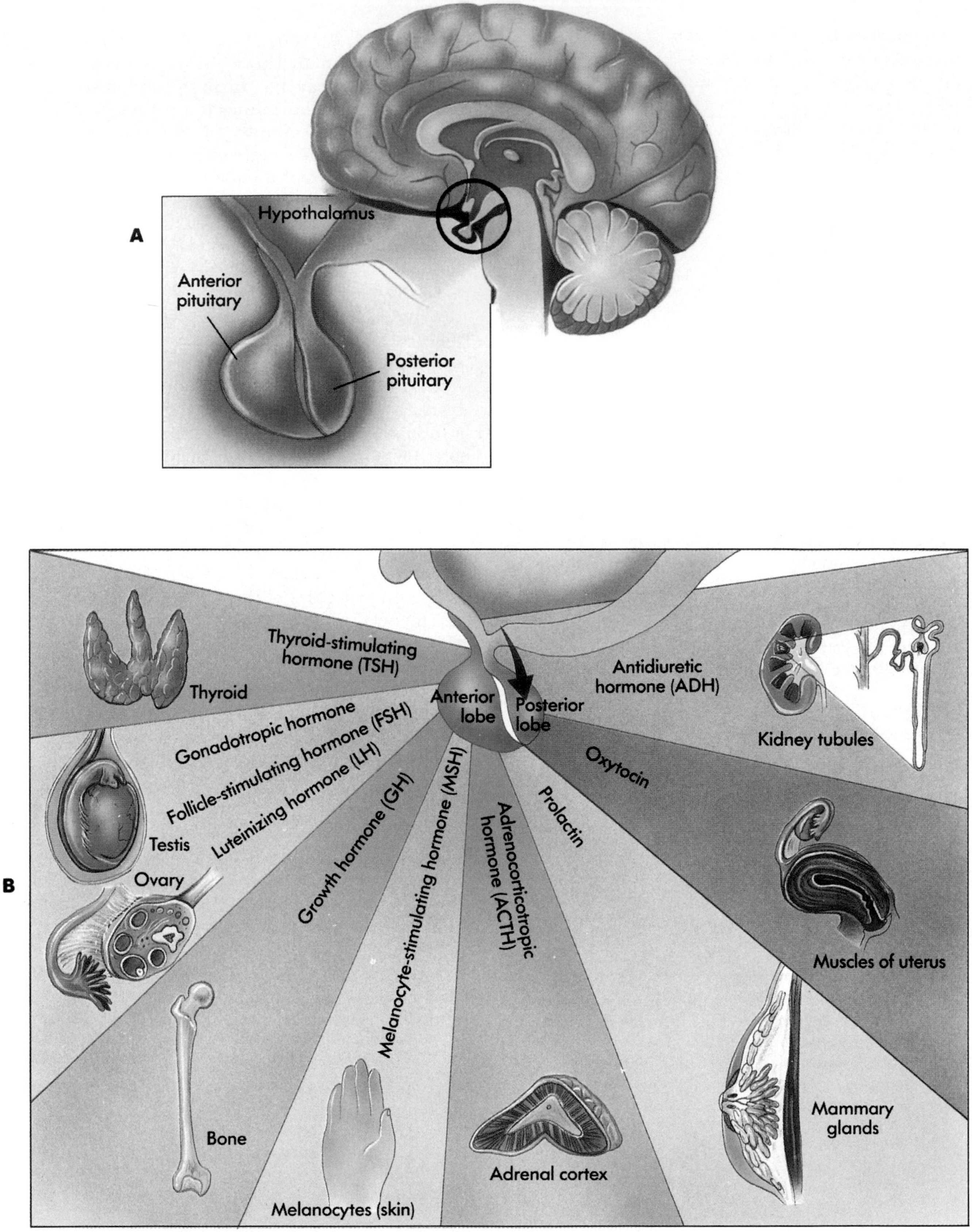

FIGURE 17-5 The role of the pituitary. A, In this diagram, the stalk of connecting tissue between the hypothalamus and the pituitary can be clearly seen. **B,** The hormones secreted by the pituitary can be divided into anterior pituitary hormones and posterior pituitary hormones.

TSH triggers the thyroid gland to produce the three thyroid hormones. This endocrine gland is located on the front of the neck, just below the voice box (see Figure 17-1). Its hormones control normal growth and development and are essential to proper metabolism. (Further discussion of ACTH is on p. 355, and further discussion of TSH is on p. 352.)

The front portion of the pituitary, the anterior pituitary, secretes seven hormones. Of these seven, four stimulate other endocrine glands and are called tropic hormones.

Growth hormone (GH) is produced by the anterior pituitary and works with the thyroid hormones to control normal growth. GH increases the rate of growth of the skeleton by causing cartilage cells and bone cells to reproduce and lay down their intercellular matrix. In addition, GH stimulates the deposition of minerals within this matrix. GH also stimulates the skeletal muscles to grow in both size and number. In the past, children who did not produce enough GH did not grow to an average height; this condition is called *hypopituitary dwarfism.* However, in the past decade, scientists have been able to use the techniques of genetic engineering to insert the human GH gene into bacteria to produce human GH. Currently, this laboratory-made hormone is being used successfully for treating growth disorders caused by hyposecretion (underproduction) of GH in children. The opposite problem may also occur: during the growth years, some children produce too much GH. This hypersecretion (overproduction) can cause the long bones to grow unusually long (Figure 17-6) and result in a condition known as *giantism.* In adults, hypersecretion of GH causes the bones of the hands and face to thicken, resulting in a condition known as *acromegaly* (Figure 17-7).

FIGURE 17-6 The world's tallest woman. Sandy Allen is shown here with her mother, brother, and dog. Giantism is caused by the oversecretion of growth hormone.

FIGURE 17-7 Effects of acromegaly. These photographs of a woman with acromegaly, at ages 16 and 33, show the thickening of the facial bones that results from the oversecretion of growth hormone.

Prolactin is another hormone secreted by the anterior pituitary. Prolactin works with estrogen, progesterone, and other hormones to stimulate the mammary glands in the breasts to secrete milk after a woman has given birth to a child. During the menstrual cycle, milk is not produced and secreted because prolactin levels in the bloodstream are very low. Late in the menstrual cycle, however, as the levels of progesterone and estrogen fall, the pituitary is stimulated by the hypothalamus to secrete some prolactin. This rise in prolactin, although not sufficient to cause milk production, does cause the breasts of some women to feel sore before menstruation. After menstruation, estrogen levels begin to rise, and prolactin secretion is once again inhibited.

Melanocyte-stimulating hormone (MSH) acts on cells in the skin called *melanocytes,* which synthesize a pigment called *melanin.* This pigment is taken up by epidermal cells in the skin, producing skin colorations from pale yellow (in combination with another pigment called *carotene*) to black. Variations are caused by the amount of pigment the melanocytes produce; this variation is genetically determined and is an inherited characteristic.

The Posterior Pituitary

The posterior lobe of the pituitary stores and releases two hormones that are produced by the hypothalamus: *antidiuretic hormone (ADH)* and *oxytocin*. ADH helps control the volume of the blood by regulating the amount of water reabsorbed by the kidneys. For example, receptors in the hypothalamus can detect a low blood volume by detecting when the solute concentration of the blood is high. When the hypothalamus detects such a situation, it triggers its specialized neurosecretory cells to make ADH. This hormone is transported within axons to the posterior pituitary, which releases the hormone into the bloodstream. ADH binds to target cells in the collecting ducts of the nephrons of the kidneys, increasing their permeability. More water then moves out of these ducts and back into the blood, resulting in a more concentrated urine. ADH also acts on the smooth muscle surrounding arterioles. As these muscles tighten, they constrict the arterioles, an action that helps raise the blood pressure. Alcohol suppresses ADH release, which is why excessive drinking leads to the production of excessive quantities of urine and eventually to dehydration.

Oxytocin is another hormone of the posterior pituitary: it is produced in the hypothalamus and transported within axons to the posterior pituitary for secretion. In women, oxytocin is secreted during the birth process, triggered by a stretching of the cervix of the uterus at the beginning of the birth process. Oxytocin binds to target cells of the uterus, enhancing the contractions already taking place. The mechanism of oxytocin secretion is an example of a **positive feedback loop** in which the effect produced by the hormone enhances the secretion of the hormone. For this reason, oxytocin is used by physicians to induce uterine contractions when labor must be brought on by external means. Oxytocin also targets muscle cells around the ducts of the mammary glands, allowing a new mother to nurse her child. The suckling of the infant triggers the production of more oxytocin, which aids in the nursing process and helps contract the uterus to its normal size.

> *The rear lobe of the pituitary, or posterior pituitary, stores and releases two hormones, ADH and oxytocin, which are produced by the hypothalamus.*

The Thyroid Gland

Sitting like a large butterfly just below the level of the voice box, the thyroid gland can be thought of as your metabolic switch. This gland secretes hormones that determine the rate of the chemical reactions of your body's cells. Put simply, thyroid hormones determine how fast bodily processes take place.

The thyroid hormones are *thyroxine (T4)* and *triiodothyronine (T3)*. These hormones are called *amines:* single, modified amino acids. They are not considered to be true peptide hormones, however, because they act on the DNA of target cells as steroid hormones do. They are also unique because an inorganic ion—iodine—is part of their structures.

Your body uses iodine in the food you eat to help make the thyroid hormones; the *3* or *4* in each hormone name refers to the number of atoms of iodine in each hormone. Foods such as seafood and iodized salt are good sources of dietary iodine. If the diet contains an insufficient amount of iodine, the thyroid gland enlarges. This condition is called a *hypothyroid goiter* (Figure 17-8).

The hypothalamus and the thyroid gland work together to keep the proper level of thyroid hormone circulating in the bloodstream. This level is detected by the hypothalamus. A low level of thyroid hormones stimulates the hypothalamus to secrete a releasing factor—a chemical message—to the anterior pituitary. This message tells the pituitary to release more TSH. The thyroid responds, thereby raising the blood level of T_3 and T_4 back to normal (Figure 17-9). This mechanism of action is

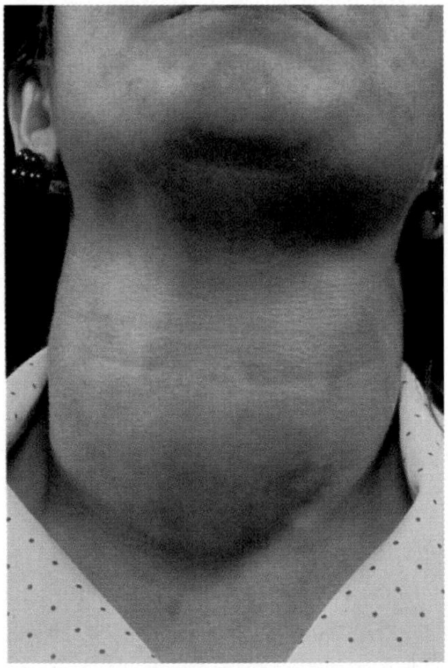

FIGURE 17-8 A goiter. The thyroid gland becomes enlarged when the diet lacks iodine. Iodine is used in the production of the thyroid hormones; when not enough iodine is available, the thyroid cannot produce thyroid hormones, and the thyroid swells from overstimulation by the anterior pituitary.

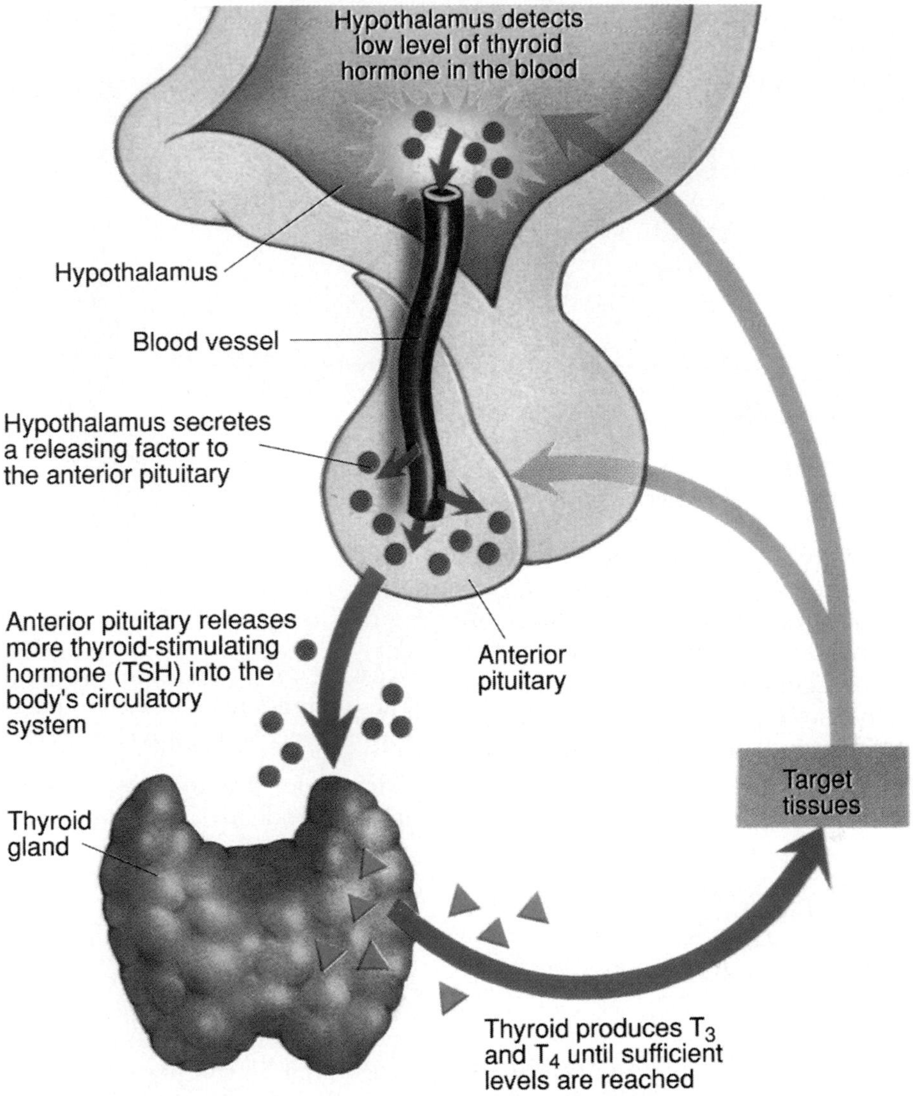
Hypothalamus detects low level of thyroid hormone in the blood
Hypothalamus
Blood vessel
Hypothalamus secretes a releasing factor to the anterior pituitary
Anterior pituitary releases more thyroid-stimulating hormone (TSH) into the body's circulatory system
Anterior pituitary
Target tissues
Thyroid gland
Thyroid produces T_3 and T_4 until sufficient levels are reached

FIGURE 17-9 A negative feedback loop is used in the release of thyroid hormone. When enough thyroid hormone has been produced, the hypothalamus stops producing thyroid-releasing hormone.

an example of a negative feedback loop in which the effect produced by stimulation of a gland shuts down the stimulus. Shutdown occurs when a sufficient effect has been produced, similar to the mechanism of a thermostat. In your home, your furnace is triggered to go on when the temperature goes below the thermostat setting. The furnace stays on until the house heats up to the desired level. The thermostat then signals the furnace to turn off.

In certain disease conditions the amount of thyroid hormones in the bloodstream cannot be regulated properly. If the thyroid produces too much of the thyroid hormones, a person may feel as though the "engine is racing," with such symptoms as a rapid heartbeat, nervousness, weight loss, and protrusion of the eyes (Figure 17-10). This condition is called *hyperthyroidism.* On the other hand, if the thyroid produces too little of the thyroid hormones, a person may feel run down, with such symptoms as weight gain and slow growth of the hair and fingernails. This condition is called *hypothyroidism.* Various factors can be the underlying cause of such problems; often medication or surgery can correct the situation.

positive feedback loop within the endocrine system, a mechanism in which the effect produced by the hormone enhances the secretion of the hormone; the mechanism of oxytocin secretion works this way.

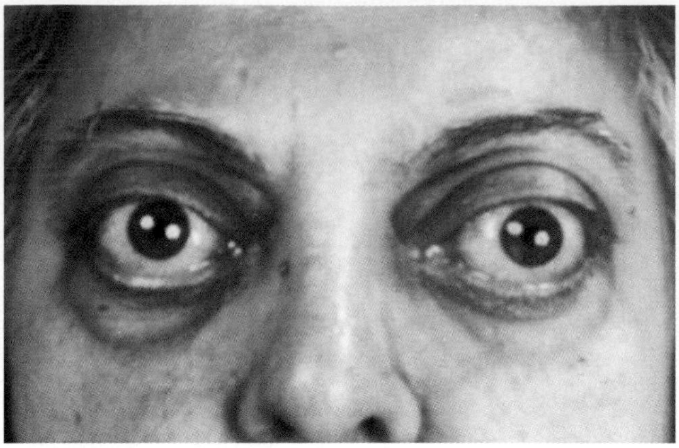

FIGURE 17-10 Hyperthyroidism. One of the symptoms of an overactive thyroid is protrusion of the eyes.

In addition to secreting the thyroid hormones, the thyroid gland secretes a hormone called *calcitonin,* or *CT.* This hormone works to balance the effect of another hormone called *parathyroid hormone,* or *PTH.* PTH regulates the concentration of calcium in the bloodstream. Calcium is an important structural component in bones and teeth and aids in the proper functioning of nerves and muscles.

> *The thyroid gland, located in the neck near the voice box, produces hormones that regulate the body's metabolism.*

The Parathyroid Glands

Embedded in the posterior side of the thyroid are the **parathyroid glands.** Most people have two parathyroids on each of the two lobes of the thyroid, as shown in Figure 17-11. These are the glands that secrete PTH, which works antagonistically to CT to help maintain the proper blood levels of various ions, primarily calcium. Two of the many problems related to abnormal calcium levels in the blood are kidney stones and osteoporosis. If calcium levels in the blood remain high, tiny masses of calcium may develop in the kidneys. These masses, called *kidney stones,* can partially block the flow of the urine from a kidney. If calcium levels in the blood remain low, calcium may be removed from the bones, a disorder known as *osteoporosis.* Osteoporosis is most common in middle-aged and elderly women, who have stopped secreting estrogen at menopause (see Chapter 21). Estrogen stimulates bone cells to take calcium from the blood to build bone tissue.

PTH and CT work in the following way to keep calcium at an optimum level in the blood: If the

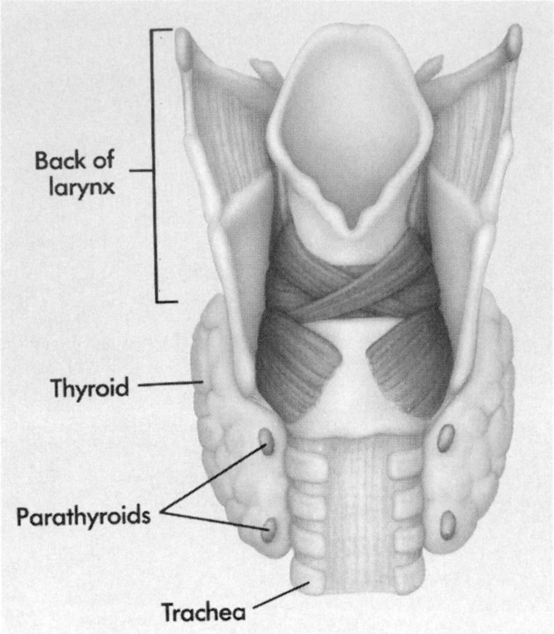

FIGURE 17-11 Location of the parathyroid glands. These glands are embedded in the underside of the thyroid and secrete parathyroid hormone.

calcium level is too low, PTH stimulates the activity of osteoclasts, or bone-destroying cells. These cells liberate calcium from the bones and put it into the bloodstream. PTH also stimulates the kidneys to reabsorb calcium from urine that is being formed and stimulates cells in the intestines to absorb an increased amount of calcium from digested food. CT acts antagonistically to PTH. When the level of calcium in the blood is too high, less PTH is secreted by the parathyroids and more CT is secreted by the thyroid. The CT inhibits the release of calcium from bone and speeds up its absorption, decreasing the levels of calcium in the blood. These interactions of PTH and CT are an example of a negative feedback loop that does not involve the hypothalamus or pituitary gland. The level of calcium in the blood directly stimulates the thyroid and parathyroid glands (Figure 17-12).

> *The parathyroid glands, embedded in the posterior side of the thyroid, secrete a hormone that helps maintain the proper levels of various ions such as calcium in the bloodstream.*

The Adrenal Glands

The two **adrenal glands** are named for their position in the body: above (*ad* meaning "near") the kidneys (*renal* meaning "kidney"). Each of these triangular glands has two parts with two different

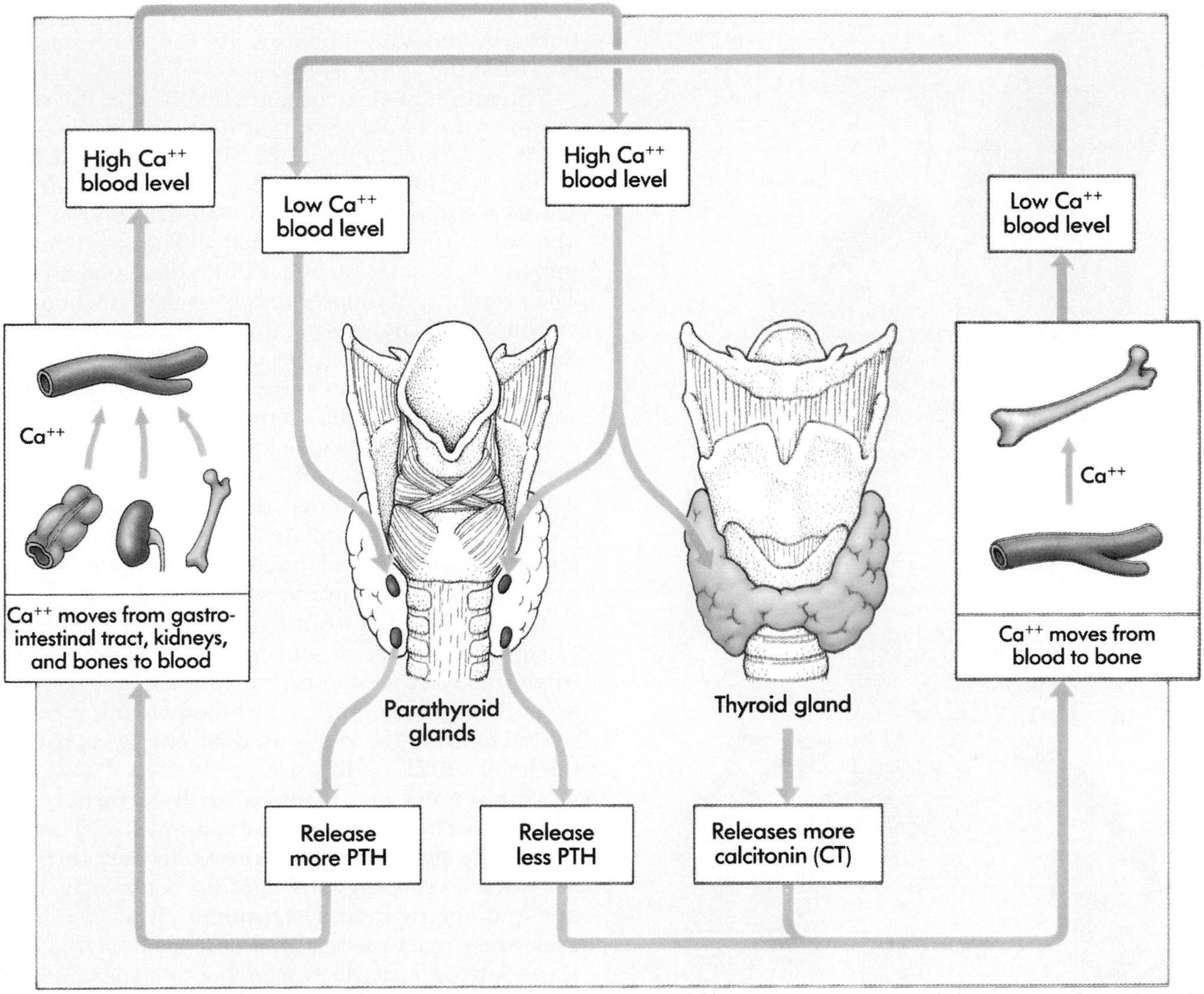

FIGURE 17-12 How parathyroid hormone and calcitonin work to maintain proper calcium levels in the blood. Parathyroid hormone stimulates the removal of calcium from the bones, the reabsorption of calcium from the urine, and the absorption of calcium from digested food when calcium levels are low. When calcium levels in the blood are high, calcitonin works to inhibit calcium release from bone and speed up its absorption.

functions (Figure 17-13). The **adrenal cortex** is the outer, yellowish portion of each adrenal gland. The word *cortex* comes from a Latin word meaning "bark" and is often used to refer to the outer covering of a tissue, organ, or gland. The **adrenal medulla** is the inner, reddish portion of the gland and is surrounded by the cortex. Not surprisingly, the word *medulla* comes from a Latin word meaning "marrow" or "middle."

The Adrenal Cortex

As you may recall, the anterior pituitary gland secretes the hormone ACTH, adrenocorticotropic hormone. This hormone, as its name implies, stimulates the adrenal cortex to secrete a group of hormones known as **corticosteroids.** These steroid hormones act on the nucleus of target cells, triggering the cell's hereditary material to produce certain

parathyroid glands (**pare** uh **thigh** royd) four small glands embedded in the posterior side of the thyroid that produce parathyroid hormone.

adrenal glands (uh **dree** nul) two triangular glands located above the kidneys; each adrenal gland consists of two parts: the adrenal cortex and the adrenal medulla.

adrenal cortex (uh **dree** nul **kor** tecks) the outer, yellowish portion of each adrenal gland that secretes a group of hormones known as corticosteroids in response to the hormone ATCH, or adrenocorticotropic hormone.

adrenal medulla (uh **dree** nul muh **dull** uh) the inner, reddish portion of each adrenal gland. The adrenal medulla is surrounded by the cortex and secretes the hormones adrenaline and noradrenaline.

corticosteroids (**kort** ik oh **stare** oyds) a group of hormones, secreted by the adrenal cortex in response to ACTH, that act on the nuclei of target cells, causing the cell's hereditary material to produce certain proteins.

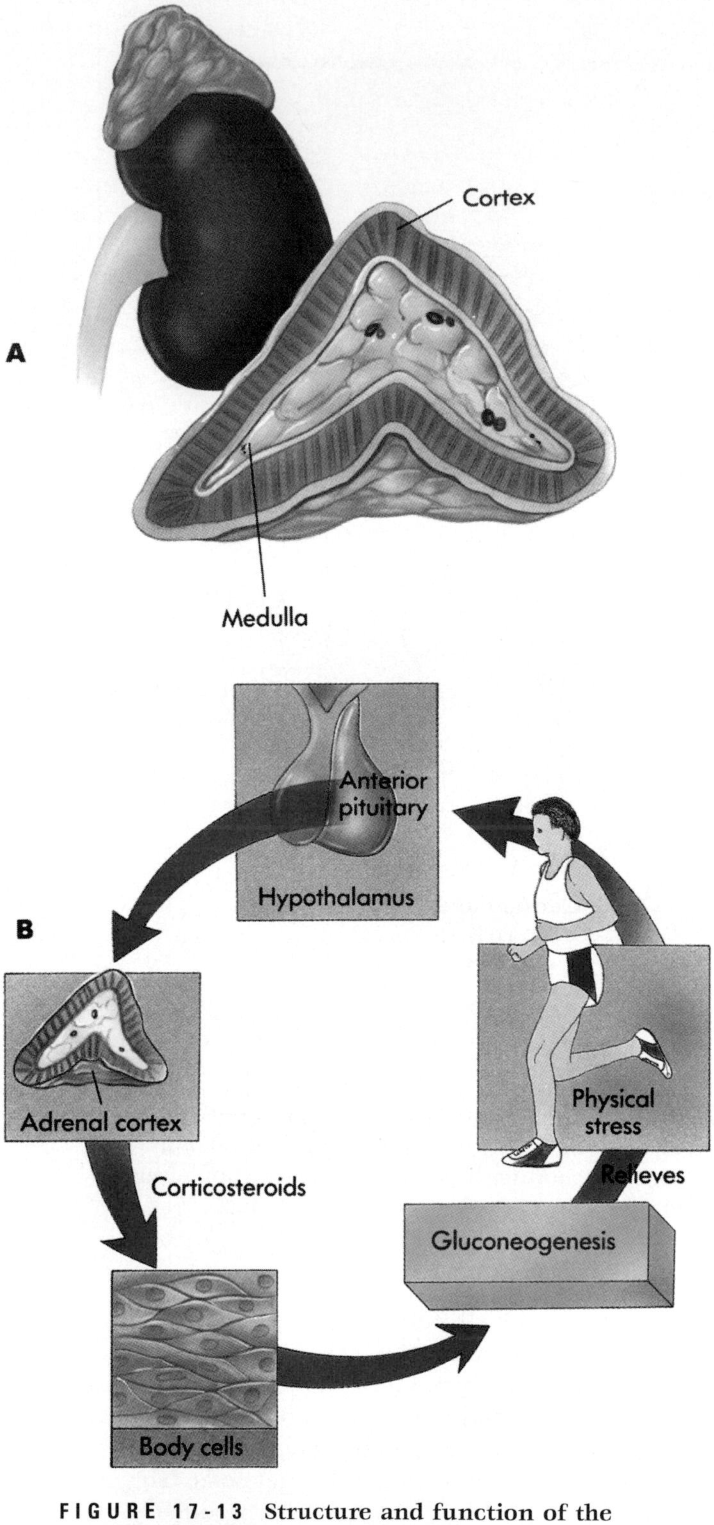

FIGURE 17-13 Structure and function of the adrenal glands. A, The adrenal glands are located on top of each kidney. Each gland has a cortex (the outer portion) and a medulla (the inner portion). **B,** The adrenal cortex plays a role in stress reduction. When the body is under stress, the hypothalamus, through the action of releasing hormone, induces the anterior pituitary to secrete ACTH. ACTH stimulates the adrenal cortex to produce corticosteroids. These corticosteroids cause the body to make glucose from noncarbohydrates such as protein, providing extra energy and thus reducing stress.

proteins. The two main types of corticosteroids produced by the adrenal cortex are the *mineralocorticoids* and the *glucocorticoids.*

The mineralocorticoids are involved in the regulation of the levels of certain ions within the body fluids. The most important of this group of hormones is *aldosterone.* It affects tubules within the kidneys, stimulating them to reabsorb sodium ions and water from the urine that is being produced, putting these substances back into the bloodstream. The secretion of aldosterone is triggered when the volume of the blood is too low, such as during dehydration or blood loss. Special cells in the kidneys monitor the blood pressure. When the blood pressure drops, these cells secrete an enzyme that begins a chain of reactions ending with the secretion of aldosterone. Conversely, when the blood pressure is within a normal range, the cell detectors in the kidneys are not stimulated, the release of aldosterone is not triggered, and the kidney tubules are not stimulated to conserve sodium and water.

The glucocorticoids affect glucose metabolism, causing molecules of glucose to be manufactured in the body from noncarbohydrates such as proteins. This glucose enters the bloodstream, is transported to the cells, and is used for energy as part of the body's reaction to *stress.*

Almost everyone is familiar with the term *stress.* And almost everyone can give examples of stressful situations: their boss "chewing them out" in front of co-workers, their kids fighting constantly with one another, or their sustaining a physical injury. The stress reaction was first described in 1936 by Hans Selye, a researcher who has since become the acknowledged authority on stress. Dr. Selye explained how the body typically reacted to stress—any disturbance that affects the body—and called this reaction the *general adaptation syndrome.* Over a prolonged period of stress, the body reacts in three stages: (1) the alarm reaction, (2) resistance, and (3) exhaustion. Contrary to maintaining homeostasis within the body, the general adaptation syndrome works to help the body gear up to meet an emergency.

During the alarm reaction, the body goes into quick action. Imagine entering your place of work and having your boss confront you, accusing you—in front of the office staff—of making a costly mistake. Your body reacts with a quickening pulse, increased blood flow, and an increased rate of chemical reactions within your body. Why does your body react in this way? Although the adrenal cortex is involved in the stress reaction, the beginning of the story lies in an understanding of the middle section of the adrenal glands, the adrenal medulla.

The Adrenal Medulla

The adrenal medulla is different from most other endocrine tissue in that its cells are derived from cells of the peripheral nervous system and are specialized to secrete hormones. These cells are triggered by the autonomic nervous system, which controls involuntary or automatic responses. The other major nervous tissues with direct endocrine function are the secretory portion of the hypothalamus in the brain and the posterior pituitary just under the hypothalamus.

The two principal hormones made by the adrenal medulla are *adrenaline* and *noradrenaline* (also called *epinephrine* and *norepinephrine*). These two hormones are primarily responsible for the alarm reaction. The hypothalamus is responsible for sending the alarm signal (via the autonomic nervous system) to the adrenal medulla. The hypothalamus picks up the alarm signal as it monitors changes in the emotions and carries it as nerve signals on tracts of neurons that connect the hypothalamus with the emotional centers in the cerebral cortex. It can therefore sense when the body perceives an emotional stress. It can also sense physical stress, such as cold, bleeding, and poisons in the body. The hypothalamus reacts to stress by readying the body for fight or flight; it first triggers the adrenal medulla to dump adrenaline and noradrenaline into the bloodstream. These hormones cause the heart rate and breathing to quicken, the rate of chemical reactions to increase, and glucose (stored in the liver) to be dumped into the bloodstream. In general, the actions of adrenaline and noradrenaline increase the amounts of glucose and oxygen available to the organs and tissues most used for defense: the brain, heart, and skeletal muscles.

A summary of the stress reaction is shown in Figure 17-14. The diagram also shows that the hypothalamus triggers the pituitary to relase ACTH, which is also involved in the stress reaction (see "The Adrenal Cortex").

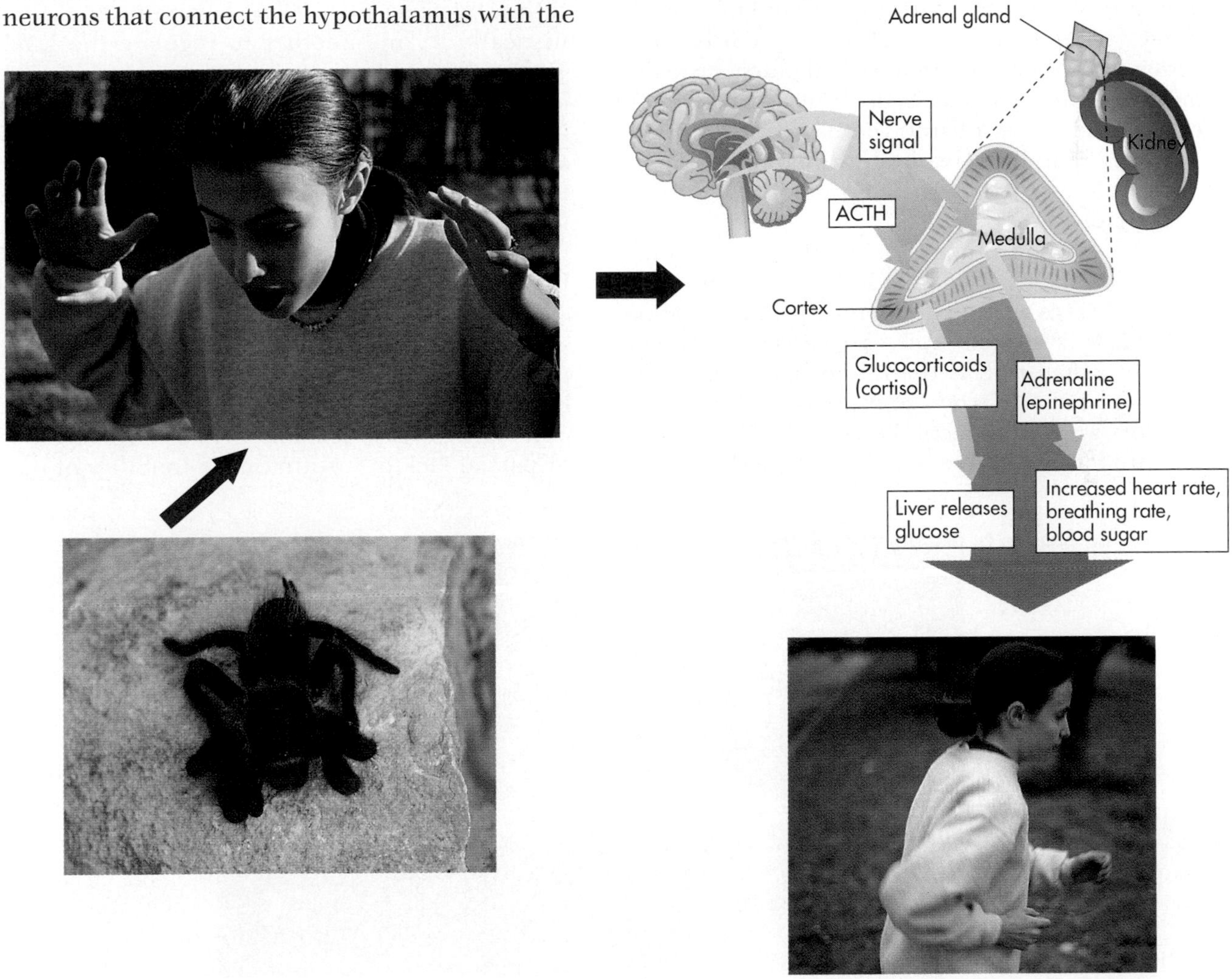

FIGURE 17-14 Summary of the stress reaction. This diagram shows the events that take place during the alarm stage.

The adrenal glands, located on top of each kidney, are divided into two secretory portions: an outer cortex and an inner medulla. The hormones of the adrenal cortex, the corticosteroids, primarily regulate the level of sodium and consequently water in the bloodstream and stimulate the liver to produce glucose from stored carbohydrates. The hormones of the adrenal medulla, adrenaline and noradrenaline, ready the body to react to stress.

During the second stage of stress, the resistance stage, the hypothalamus triggers continuing responses by releasing regulating factors. These factors stimulate the pituitary to release ACTH, GH, and TSH. TSH stimulates the thyroid to secrete T_4, which stimulates the liver to break down stored carbohydrates (glycogen) to glucose. GH also stimulates the liver to produce glucose from glycogen, providing the body with an abundant energy source. ACTH stimulates the adrenal cortex to secrete both mineralocorticoids and glucocorticoids. The mineralocorticoids cause the body to retain sodium and water, raising the blood pressure and providing more blood volume in the case of blood loss. The glucocorticoids also promote the production of glucose.

If a person continues to be highly stressed over a long time, the body may lose the fight and enter the third stage of the stress reaction: exhaustion. This stage is serious and can cause death. One cause of exhaustion is the loss of potassium ions, which are excreted when sodium ions are retained. This loss severely affects the cell's ability to function properly. Another cause is depletion of the glucocorticoids, resulting in a sharp drop in the blood glucose level. The organs also become weak and may cease to function. To combat chronic stress, people can learn ways to psychologically handle their stress and can work toward a level of health and fitness that will help their bodies cope with the physical effects of stress.

The Pancreas

The pancreas, located alongside the stomach (see Figure 17-1), is two glands in one: an exocrine gland and an endocrine gland. As an exocrine gland, it secretes the digestive enzymes discussed in Chapter 8. As an endocrine gland, it secretes the hormones *insulin* and *glucagon.*

The endocrine portion of the pancreas consists of separated clusters of cells that lie among the exocrine cells. For this reason, these cells are called *islets*—the **islets of Langerhans** (Figure 17-15). Separate types of cells within the islets produce insulin and glucagon. These hormones act antagonistically to one another to regulate the level of glucose in the bloodstream. Glucagon increases the blood glucose level by triggering the liver to convert stored carbohydrates (glycogen) into glucose and to convert other nutrients such as amino acids into glucose. Insulin decreases the blood glucose level by helping body cells transport glucose across their membranes. In addition, insulin acts on the liver to convert glucose into glycogen and fat for storage (Figure 17-16).

Diabetes mellitus is a set of disorders in which a person tends to have a high level of glucose in the blood. There are many variations of the disorder and many causes. The underlying cause is the lack or partial lack of insulin or the inability of tissues

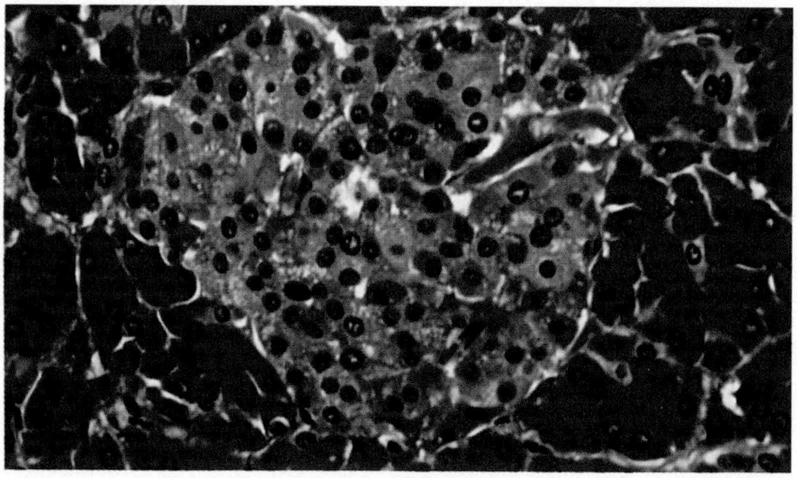

FIGURE 17-15 Islets of Langerhans. Glucagon and insulin are produced by the islets of Langerhans, which are more lightly stained than other types of cells in this preparation.

JUST WONDERING . . .

My sister has juvenile diabetes. Will she grow out of it?

As you may know from living with your sister, *diabetes mellitus* is a disorder in which the pancreas does not produce adequate amounts of the hormone insulin or the body becomes insensitive to it. There are two major types of *diabetes mellitus:* insulin-dependent diabetes (type I) and non-insulindependent diabetes (type II). The names of these diseases have changed over the years. Your sister has what's now called *insulin-dependent (type I) diabetes.*

In type I diabetes, which strikes chiefly in childhood (thus, juvenile diabetes), the body loses its ability to produce insulin. Affecting about 5% of diabetics, type I diabetes is thought to be a hereditary disease in which the body's immune system attacks the pancreas, destroying its ability to produce insulin. As a result, muscles, fat, and liver cells are prevented from absorbing sugar from the blood; the sugar is excreted in urine while the undernourished cells literally starve. The high levels of blood glucose cause the thickening of capillary and artery walls, constricting blood flow and damaging critical organs. Researchers are working to uncover a way to prevent the immune system from attacking the pancreas.

Luckily, you and your sister are living at a time when management techniques of type I diabetes have been discovered that help people live long, healthy lives. (This was not the case just a few decades ago.) Diabetes is treated by adding insulin directly to the bloodstream by injection (as a protein, it would be destroyed in the stomach if administered as a pill), usually twice a day. But taking insulin injections is only part of the treatment for diabetes. Your sister should also be on a personal management plan for her illness, which will help her avoid the complications of diabetes. Such a plan should include nutrition and exercise recommendations, glucose monitoring instructions, and a medical plan to monitor organs and body systems that may be affected by her diabetes. For more information, you and your family should contact the branch of the American Diabetes Association in your area.

to respond to insulin, which leads to increased levels of glucose in the blood. The high levels of glucose in the blood result in water moving out of the body's tissues by osmosis. The kidneys remove this excess water through increased urine production, which causes excessive thirst and dehydration.

The primary types of diabetes mellitus are termed *type I* and *type II diabetes.* Type I diabetes is described in the student question box, "Just Wondering." Persons with type II diabetes, or maturity onset diabetes, have some effective insulin but not enough to meet their needs. Others have sufficient insulin, but the cells in many parts of the body lack enough insulin receptors. These persons can often treat their disorder with diet alone. Research on the possibility of transplanting islets of Langerhans for persons who do not produce enough insulin holds much promise as a lasting treatment for this disease.

> *The pancreas secretes two hormones, insulin and glucagon, that act antagonistically to one another, regulating the level of glucose in the bloodstream.*

islets of Langerhans (**eye** lets of **lang** ur **hanz**) the separate types of cells within the exocrine cells of the pancreas that produce the hormones insulin and glucagon.

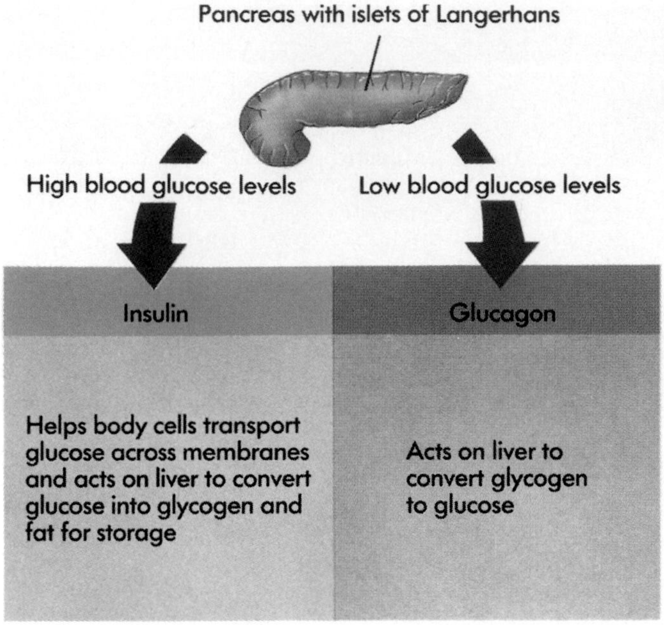
Pancreas with islets of Langerhans
High blood glucose levels
Low blood glucose levels
Insulin
Glucagon
Helps body cells transport glucose across membranes and acts on liver to convert glucose into glycogen and fat for storage
Acts on liver to convert glycogen to glucose

FIGURE 17-16 Functions of insulin and glucagon. Glucagon and insulin work antagonistically. Insulin helps cells transport glucose across their membranes and acts on the liver to convert the excess glucose to glycogen and fat. These activities lower blood glucose levels. When blood glucose levels are low, glucagon triggers the liver to convert glycogen to glucose.

The Pineal Gland

The **pineal gland** gets its name because it looks like a tiny pine cone embedded deep within the brain (see Figure 17-1). Although a great deal of research has been conducted regarding the workings of this gland, it still remains somewhat of a mystery. Interestingly, the pineal is the possible site of your biological clock, the control center that regulates your daily rhythms such as sleeping and waking. It may also stimulate such activity as the onset of puberty.

The Thymus Gland

The thymus gland is a small gland located in the neck a few inches below the thyroid (see Figure 17-1). In the thymus, certain immune system cells called *T lymphocytes* develop the ability to identify invading bacteria and viruses (see Chapter 11). The thymus produces a variety of hormones to promote the maturation of these cells. This gland is quite active during childhood but is replaced by fat and connective tissue by the time a person reaches adulthood.

The Ovaries and Testes

The ovaries produce female sex cells, or eggs, and the testes produce male sex cells, or sperm. In addition, the ovaries produce the hormones estrogen and progesterone, and the testes produce testosterone. The detailed structure of these endocrine glands and the specific roles their hormones play are discussed in Chapter 21.

Nonendocrine Hormones

The principal endocrine glands and their hormones are listed in Table 17-1. However, certain cells in the body produce chemical messengers that regulate nearby cells without traveling in the bloodstream and are therefore not considered endocrine hormones. These intercellular chemical messengers are often called *local hormones* and include a wide variety of substances. Certain cells in the walls of the stomach and small intestine, for example, secrete hormones that regulate the release of digestive juices by various cells. The ends of axons secrete a variety of transmitter substances that stimulate the dendrites or cell body of a neighboring neuron, allowing a nerve impulse to be conducted from neuron to neuron. Cells throughout the body secrete substances called *prostaglandins* that are derived from the cell membranes. These local hormones accumulate in regions of tissue injury or disturbance. They stimulate smooth muscle contraction and the dilation and constriction of blood vessels. Overproduction of prostaglandins swells cerebral blood vessels, so their walls press against nerve tracts in the brain, causing pain. Aspirin relieves this pain, commonly known as a headache, because it inhibits prostaglandin production.

In recent years, a group of local hormones has been identified within the brain. The most commonly known local brain hormones are the *enkephalins* and *endorphins*. Both have pain-killing properties similar to the drug morphine. It has been suggested that these chemical messengers are the body's natural painkillers, and they have been linked to a feeling of well-being referred to as *runner's high* experienced by some joggers.

Certain cells in the body produce hormones such as prostaglandins, enkephalins, and endorphins that regulate nearby cells. Because these chemicals do not travel in the bloodstream, they are not considered endocrine hormones.

pineal gland (pin ee uhl) a gland resembling a tiny pine cone embedded deep within the brain; though its exact function remains a mystery, it is believed to be the site of the control center that regulates the body's daily rhythms. It may also stimulate the onset of puberty.

TABLE 17-1

Endocrine glands and their hormones

ENDOCRINE GLAND AND HORMONE	TARGET TISSUE	PRINCIPAL ACTIONS
HYPOTHALAMUS		
Releasing hormones	Other endocrine glands	Stimulate the release of hormones by other endocrine glands
POSTERIOR PITUITARY		
Oxytocin	Uterus Mammary glands	Stimulates contraction of uterus and milk production
Antidiuretic hormone (ADH)	Kidneys	Stimulates reabsorption of water by the kidneys
ANTERIOR PITUITARY		
Follicle-stimulating hormone (FSH)	Sex organs	Stimulates ovarian follicle, spermatogenesis
Luteinizing hormone (LH)	Sex organs	Stimulates ovulation and corpus luteum formation in females
Adrenocorticotropic hormone (ACTH)	Adrenal cortex	Stimulates secretion of adrenal cortical hormones
Thyroid-stimulating hormone (TSH)	Thyroid	Stimulates secretion of T_3 and T_4
Growth hormone (GH)	Cartilage and bone cells, skeletal muscle cells	Stimulates division of cartilage and bone cells, growth of muscle cells, and deposition of minerals
Prolactin	Mammary glands	Stimulates milk production
Melanocyte-stimulating hormone (MSH)	Melanocytes	Stimulates production of melanin
THYROID GLAND		
Thyroxine (T_4) and triiodothyrone (T_3)	General	Regulates metabolism
Calcitonin	Bone	Regulates calcium levels in the blood
PARATHYROID GLAND		
Parathyroid hormone (PTH)	Bone, kidney, small intestine	Regulates calcium levels in the blood
ADRENAL CORTEX		
Aldosterone	Kidney	Increases sodium and water reabsorption and potassium excretion
Glucocorticoids	General	Stimulate manufacture of glucose
ADRENAL MEDULLA		
Adrenaline and noradrenaline	Heart, blood vessels, liver, fat cells	Regulate fight or flight response: increase cardiac output, blood flow to muscles and heart, conversion of glycogen to glucose
PANCREAS (ISLETS OF LANGERHANS)		
Insulin	Liver, skeletal muscle, fat	Decreases blood glucose levels by stimulating movement of glucose into cells
Glucagon	Liver	Increases blood glucose levels by converting glycogen to glucose
OVARY		
Estrogens	General, female reproductive organs	Stimulate development of secondary sex characteristics in females, control monthly preparation of uterus for pregnancy
Progesterone	Uterus	Completes preparation of uterus for pregnancy
	Breasts	Stimulates development
TESTIS		
Testosterone	General	Stimulates development of secondary sex characteristics in males and growth spurt at puberty
	Male reproductive structures	Stimulates development of sex organs, structures spermatogenesis

Summary

- Hormones are the chemical products of cells, used as messengers to affect other cells within the body. Hormones that are secreted into the bloodstream are called *endocrine hormones.* The 10 different endocrine glands make over 30 different hormones. Together, the glands that secrete the endocrine hormones are known as the *endocrine system.*
- Hormones that are not secreted into the bloodstream but move within the tissue fluid to affect cells near their sources are called *local hormones.* These intercellular chemical messengers include a wide variety of substances.
- Hormones affect cells in one of four ways: by regulating their secretions and excretions, by helping them to respond to changes in the environment, by controlling activities related to reproductive processes, and by influencing their proper growth and development.
- The two primary types of endocrine hormones are the peptide hormones and the steroid hormones. Peptide hormones work by binding to receptor molecules embedded in the cell membranes of target cells and ultimately triggering enzymes that cause cells to alter their functioning. Steroid hormones work by binding to receptor molecules within the cytoplasm of cells and ultimately causing the hereditary material to produce certain proteins.
- The production of hormones is regulated by feedback mechanisms in the following way: an endocrine gland is initially stimulated directly or indirectly by the nervous system or by the concentration of various substances in the bloodstream. The gland secretes a hormone that interacts with target cells, and the effect produced by the target cells acts as a new stimulus to increase or decrease the amount of hormone produced.
- The pituitary gland, which hangs from the underside of the brain, secretes nine different hormones: seven from its anterior lobe and two produced by the hypothalamus that are stored in its posterior lobe. The hypothalamus is a small mass of brain tissue lying above the pituitary. The hypothalamus controls the pituitary by means of releasing hormones.
- Four of the hormones produced by the anterior pituitary control other endocrine glands. In addition, the anterior pituitary secretes growth hormone, which works with the thyroid hormones to control normal growth; prolactin, which works with other female sex hormones to stimulate the mammary glands to secrete milk after childbirth; and melanocyte-stimulating hormone, which affects certain pigment-producing cells of the body.
- The posterior pituitary secretes two hormones produced by the hypothalamus. Antidiuretic hormone helps control the volume of the blood by regulating the amount of water reabsorbed by the kidneys. Oxytocin is secreted during the birth process and enhances uterine contractions.
- The parathyroids, located on the underside of the thyroid gland in the neck, secrete parathyroid hormone. This hormone works with one of the thyroid hormones, calcitonin, to help maintain the proper blood levels of various ions, primarily calcium.
- The adrenal glands, located on top of the kidneys, have two different secretory portions: the outer cortex and the inner medulla. The cortex secretes a group of hormones known as *corticosteroids* that regulate the levels of certain mineral ions, water, and glucose in the bloodstream. The adrenal medulla secretes the hormones adrenaline and noradrenaline, which ready the body for action during times of stress.

Knowledge and Comprehension Questions

1. What are hormones, and why are they important?
2. How do the "messages" sent by the endocrine system differ from those carried by the nervous system?
3. Why is cyclic AMP required to act as a "second messenger" in peptide hormone-related processes?
4. Puberty is the physiological period when adolescents quickly mature and develop secondary sex characteristics (such as facial hair, breast development, menses). This is a period of great increase in hormonal levels. What class of hormones would you hypothesize are responsible for many of the physiological changes in puberty?
5. Describe how a feedback loop regulates the production of a hormone. What is the difference between a negative feedback loop and a positive feedback loop?
6. If the fluid content of the blood is low, where would this information initially be received and processed? What hormone would be released and from which endocrine gland would it be released?
7. Explain the term *tropic hormone.* Identify the four tropic hormones produced by the anterior pituitary and summarize the function(s) of each.
8. After playing softball on a hot sunny day, you are sweating hard and feeling very thirsty. A friend suggests getting a beer. Will the beer help your body replace the fluids it has lost? Explain.
9. Describe two hormones that together regulate calcium levels in your blood and discuss how they do it.
10. Describe the three stages of the general adaptation syndrome. What hormones are involved?
11. Which hormones regulate the level of glucose in your blood? Describe how they do this. Where are they produced?

Critical Thinking

1. In the past, people living in remote inland areas often suffered from goiter. Can you assess the reasons for increased goiter problems in such places?
2. Chronic emotional stress has been linked to a greater than normal risk of various diseases. How might this link be explained?

PART FOUR

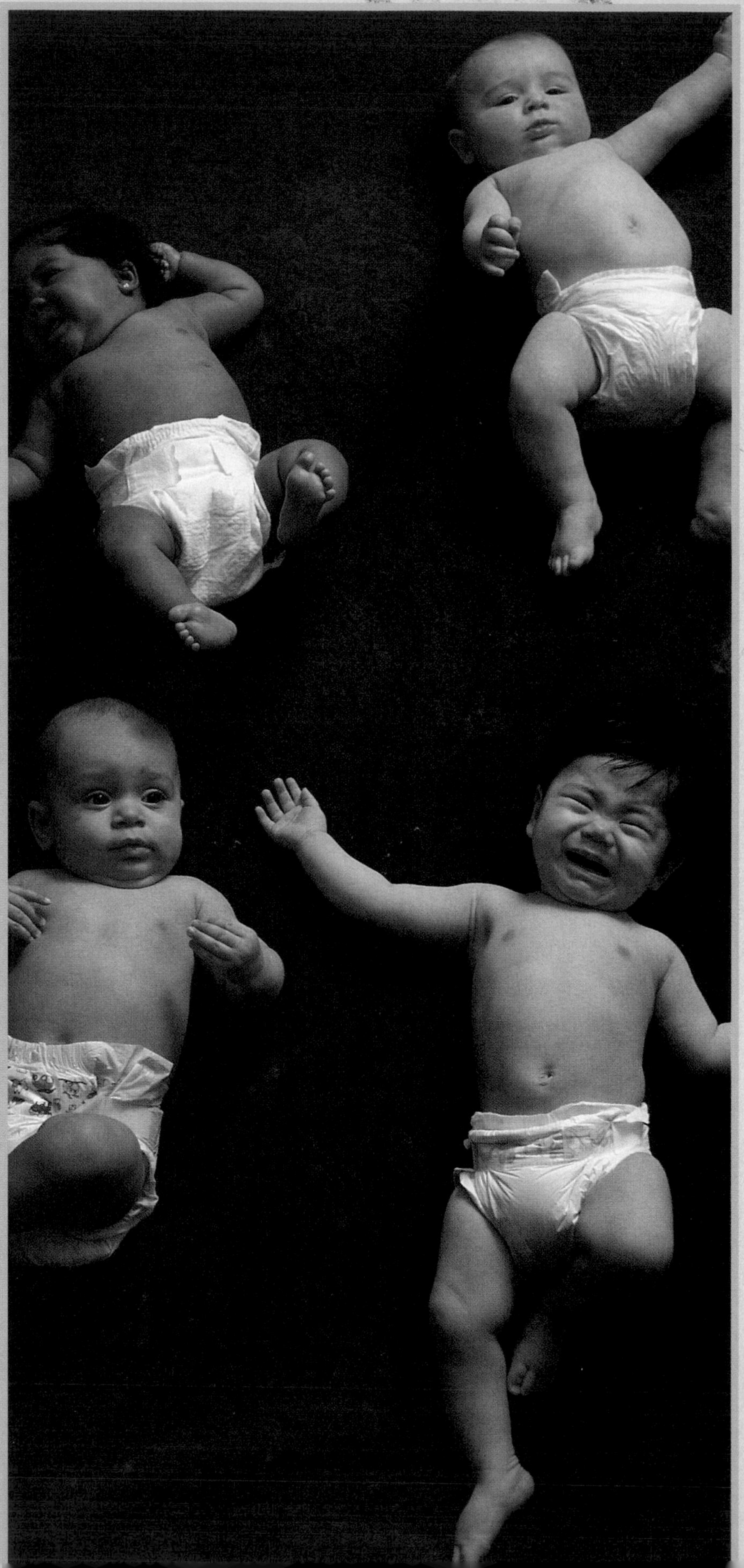

HOW HUMANS REPRODUCE AND PASS ON BIOLOGICAL INFORMATION

CHAPTER 18

DNA, GENE EXPRESSION, AND CELL REPRODUCTION

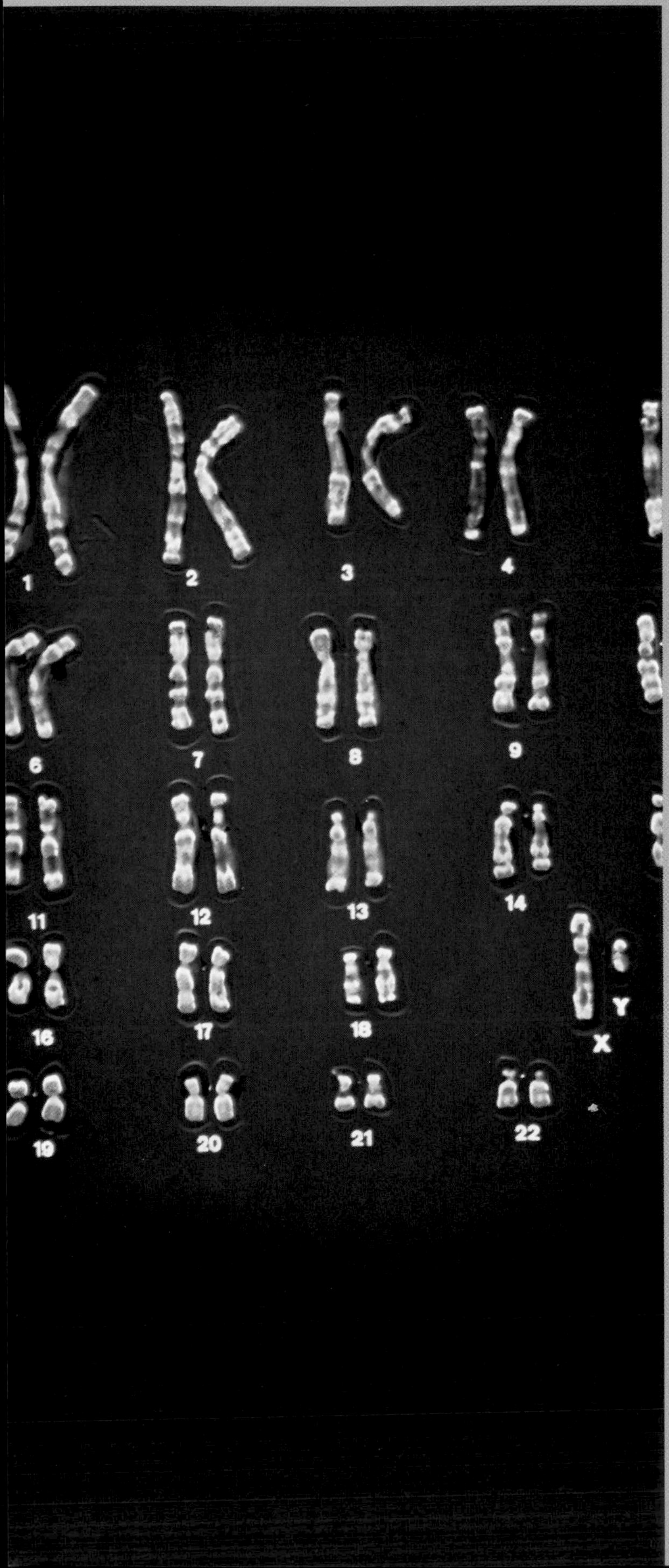

WHEN YOU WERE BORN, did you come with instructions? You may not know it, but you did. The brightly stained bodies in the photo are human "instructions" called *chromosomes,* but they cannot be read in the usual way. Instead, using a special code that can be understood by other special molecules, they direct all the chemical reactions that take place in your body. Instructions in the chromosomes determine such things as whether your hair is red or brown and whether you are male or female. But most important, they hold the key to the mystery of what makes *you.*

Chromosomes

The chapter opening photo is a *karyotype,* a particular array of chromosomes that belongs to an individual. Scientists make this type of picture by specially treating a cell just as it is about to divide. The cell is then photographed, and the chromosomes are cut out of the photograph and arranged on a piece of paper. Each chromosome has a mate, so they are paired and arranged in order of decreasing size.

Since the discovery of chromosomes, scientists have learned a great deal about their structure and function. Scientists know that the chromosomes of eukaryotes are made up of a complex of deoxyribonucleic acid (DNA) and protein. This complex is called **chromatin.** The DNA is the part of the chromatin that contains hereditary information, commonly called the *code of life.* DNA exists as one very long, double-stranded molecule that extends unbroken through the chromosome's entire length. The relationship of the amount of DNA to protein in a single chromosome can be seen in Figure 18-1.

To fit into cells, DNA strands are coiled, much as you might coil up your garden hose. The coiling of DNA within cells, however, is much more complicated. Figure 18-2 shows how this coiling is accomplished to provide organization and orderliness. First, the double strand of DNA winds like thread on a spool around groups of tiny proteins called *histones.* In this form the DNA looks similar to a string of pearls (see Figure 18-2). This string of pearls then wraps up into larger coils called *supercoils.* These supercoils are looped and packaged with other proteins to form chromatin. And as Figure 18-2 shows, the chromatin can be condensed by further coiling to form the chromosomes, a process that takes place just before and during cell division. (Cell division is described later in this chapter.)

> *Proteins and the hereditary material DNA make up the chromatin of the cell. When coiled and condensed, chromatin forms chromosomes.*

The Location of the Hereditary Material, DNA

Today, scientists understand (in considerable detail) the way that information in the DNA of a developing organism is translated into its eyes, arms, and brain. This understanding is the result of more than a half century of work by a succession of scientific investigators. Gregor Mendel's work, done in the late 1800s, suggested that traits are inherited as discrete packets of information. At the turn of the century, Walter Sutton suggested that these packets of information were on the chromosomes, but he had no direct evidence to support this hypothesis. In 1910, however, Thomas Hunt Morgan's experiments provided the first clear evidence upholding

KEY CONCEPTS

- Most traits reflect the action of enzymes; the hereditary message is a molecular code in DNA, located in the nucleus of eukaryotic cells, which results in the production of these enzymes.
- RNA plays key roles in the translation of the hereditary message of DNA.
- Before cell division, the hereditary material of cells condenses to form chromosomes; in this form, the hereditary material can be distributed in an orderly fashion during cell division.
- Mitotic cell division produces new cells during the growth and repair of human cells; meiotic cell division produces sex cells in humans.

OUTLINE

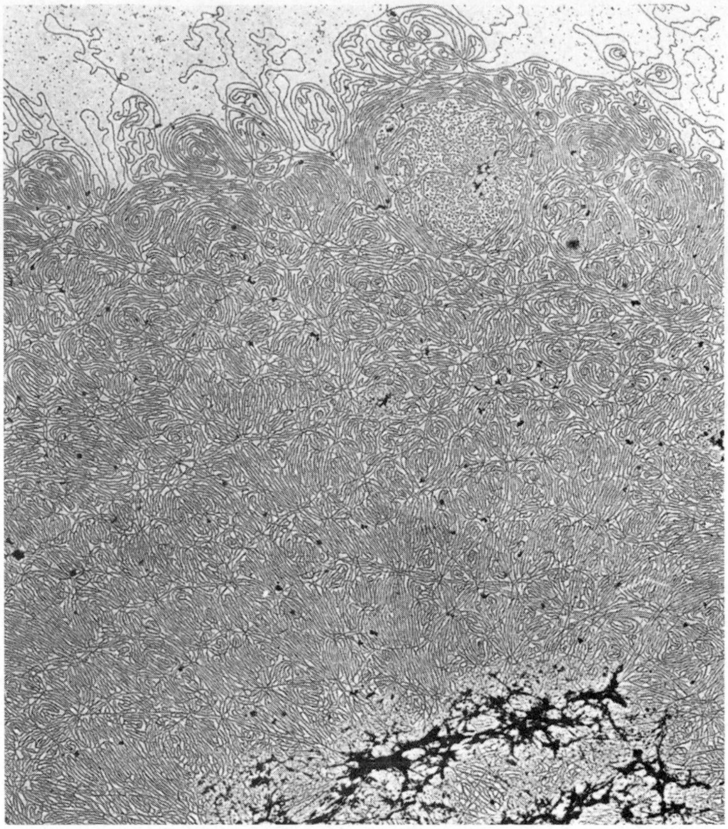

FIGURE 18-1 A human chromosome contains an enormous amount of DNA. The dark element at the bottom of the photograph is the protein component of a single chromosome. All of the surrounding material is the DNA of that chromosome.

Sutton's theory. (The work of Mendel, Sutton, and Morgan is described in Chapter 19.) Beginning in the 1930s, experiments by biologists such as Joachim Hammerling further probed the question of where the hereditary information is stored within the cell. Hammerling's experiments identified the nucleus as the likely place of the hereditary material.

Identifying the nucleus as the most likely source of hereditary information again focused attention on the chromosomes, which are located within the nucleus. Scientists had been studying the chromosomes since the late 1800s and had suspected them to be the vehicles of inheritance. In 1869 a German scientist named Friedrich Miescher isolated the DNA contained within the chromosomes of various types of cells. He did not call it DNA (this designation would not come until the 1920s). Miescher called this chromosomal material **nucleic acid** because it seemed to be specifically associated with the cell nucleus and was slightly acidic.

The Chemical Nature of the Hereditary Material: Nucleic Acids

In the 1920s the American biochemist, P. A. Levene, discovered that two types of nucleic acid were located within cells. Today scientists understand that one type of nucleic acid contains the hereditary message, whereas the other type helps express that message. Levene found that both types were nearly identical in their structures and contained three molecular parts (Figure 18-3): (1) phosphate ($-PO_4^-$) groups, (2) five-carbon sugars, and (3) four types of nitrogen-containing bases. Although nucleic acids as a whole are acidic and tend to form hydrogen ions when in solution, the nitrogen-containing bases tend to accept hydrogen ions. For this reason, these portions of nucleic acids are called *bases.*

The Nucleic Acids: DNA and RNA

Levene found that nucleic acids are composed of roughly equal portions of these three molecular

FIGURE 18-2 How DNA is coiled within chromosomes. A, An electron micrograph of rat liver DNA showing how DNA wrapped around histones resembles a string of pearls. **B,** The DNA double helix is wrapped around special proteins called histones. The DNA-histone package is then coiled into larger coils appropriately called supercoils. The supercoils are themselves coiled and packaged with other protein to form chromatin. Just before cell division, the chromatin is further coiled to form visible chromosomes.

B

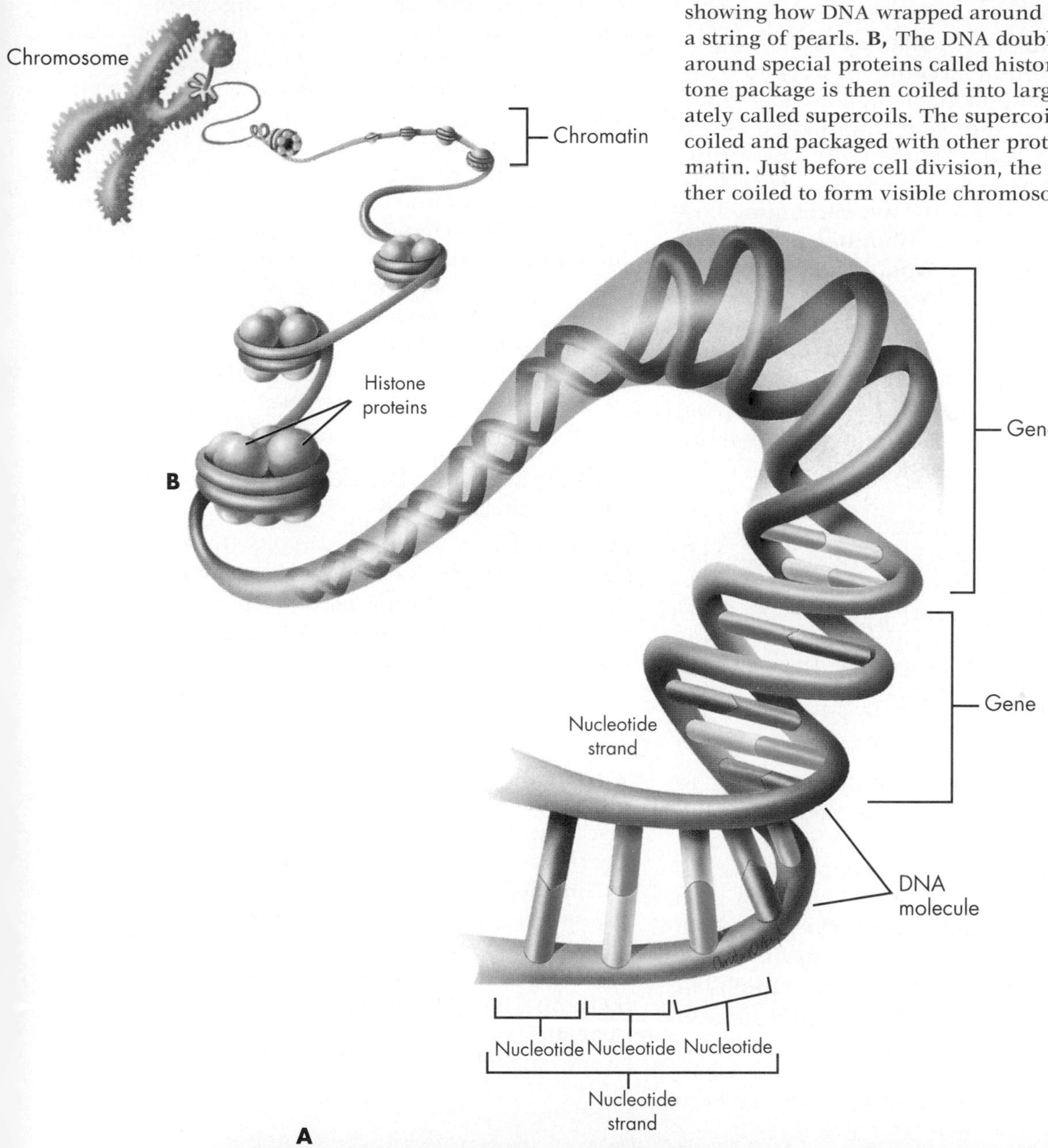

A

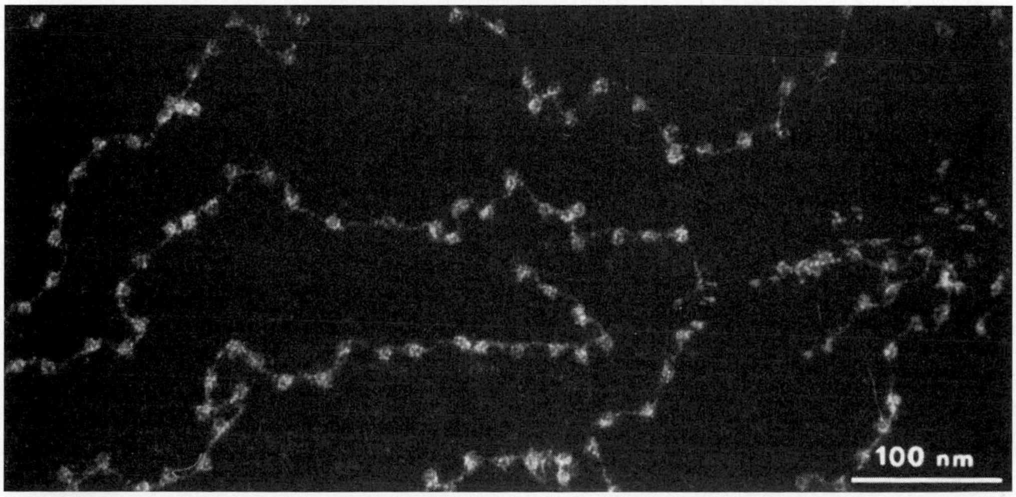

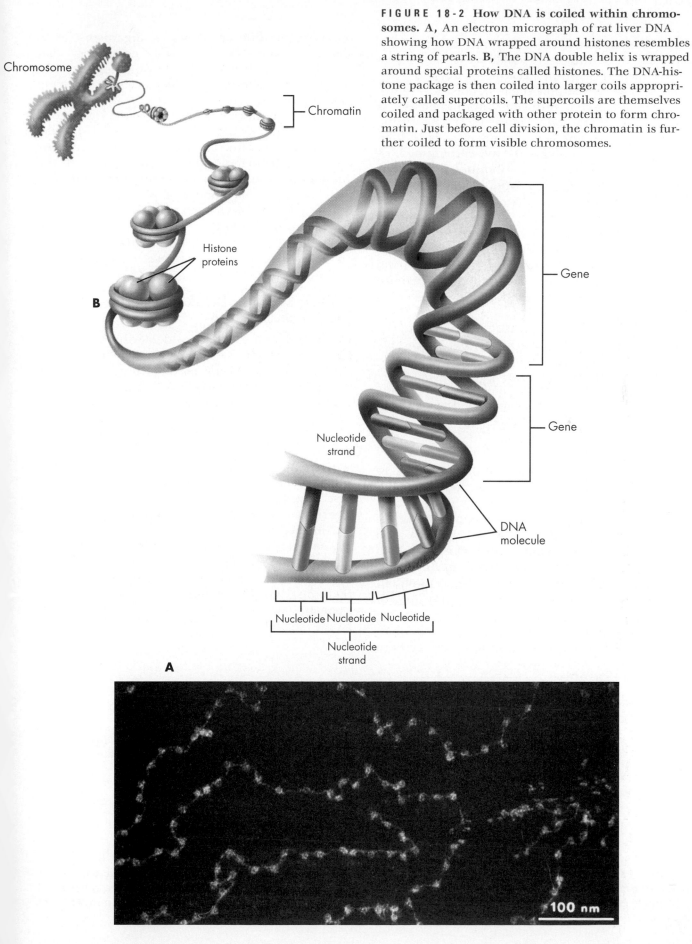

parts. He concluded (correctly) from this information that these three molecular parts, bonded together, form the unit of structure of nucleic acids. Each unit, a five-carbon sugar bonded to a phosphate group and a nitrogen-containing base, is called a **nucleotide** (see Figure 18-3). One of the two types of nucleic acids found in cells is composed of units that contain the five-carbon sugar ribose. It is therefore called **ribonucleic acid,** or **RNA** for short. The other type of nucleic acid is composed of units containing a five-carbon sugar similar to ribose but having one less oxygen atom. This sugar is called *deoxyribose,* and the nucleic acid is called **deoxyribonucleic acid (DNA).** Figure 18-4 shows the structure of both ribose and deoxyribose sugars. Can you find the difference between them? How do their names reflect this difference?

There are two types of nucleic acids located within cells: deoxyribonucleic acid (DNA) and ribonucleic acid (RNA). Both types of nucleic acid are composed of units called nucleotides, which are made up of three molecular parts: a sugar, a phosphate group, and a base.

Each nucleotide making up the structure of DNA and RNA contains one of four different bases. DNA and RNA both contain the bases adenine and guanine, which are double-ring compounds called **purines** (Figure 18-5). DNA also contains the bases thymine and cytosine, which are single-ring compounds called **pyrimidines.** RNA contains the pyrimidine cytosine as well but contains the single-ring base uracil instead of thymine.

In the late 1940s, the experiments of Erwin Chargaff showed that the proportion of bases varies in the DNA of different types of organisms, as is shown in Table 18-1. This evidence suggests that DNA has the ability to be used as a molecular code. Its base composition varies as its code varies from organism to organism.

Along with the variations among the DNA molecules of different organisms, Chargaff also noted an important similarity: the amount of adenine present in DNA molecules is always equal to the amount of thymine, and the amount of guanine is always equal to the amount of cytosine (A = T and G = C). (Notice the roughly equal percentages of A and T and of G and C in each organism listed in Table 18-1.) Therefore, DNA molecules always have an equal proportion of purines (A and G) and pyrimidines (C and T).

A Major Scientific Breakthrough: The Structure of DNA

In 1952, Alfred Hershey and Martha Chase performed now-famous experiments showing that the DNA in chromatin, not chromatin protein, is the hereditary material. They reasoned that if DNA carried hereditary information, then bacteria infected with viruses would contain viral DNA inside their cells, which would direct the formation of new virus particles. If, on the other hand, the protein carried the hereditary message, then virally infected bacteria would contain viral proteins within their cells. The experiments of Hershey and Chase showed clearly that viral DNA, not viral protein, was inside the infected bacteria.

The next puzzle to solve was how the nucleotides of DNA were put together. The significance of the regularities in the proportion of pyrimidines and the proportion of purines in DNA pointed out by Chargaff became clear through the work of two later experiments.

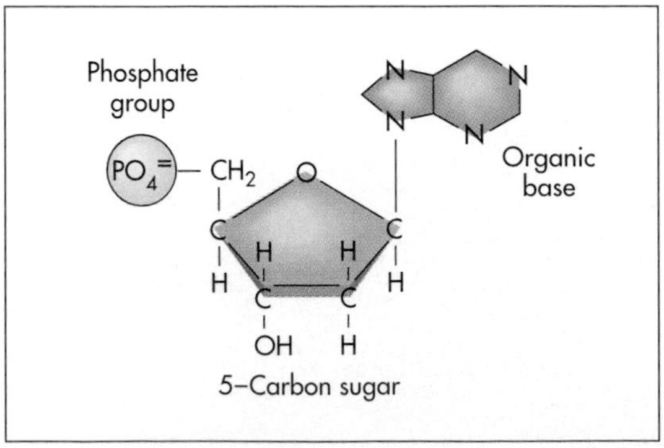

FIGURE 18-3 Structure of a nucleotide. A nucleotide contains three different molecular components: a phosphate group (PO_4), a five-carbon sugar, and a nitrogen-containing base.

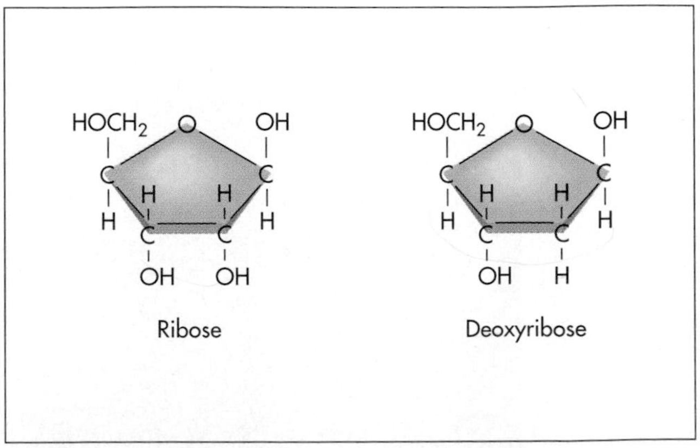

FIGURE 18-4 Comparison of ribose and deoxyribose. Both sugars have five carbons, but deoxyribose has one fewer oxygen atom. Ribose is found in RNA and deoxyribose is found in DNA.

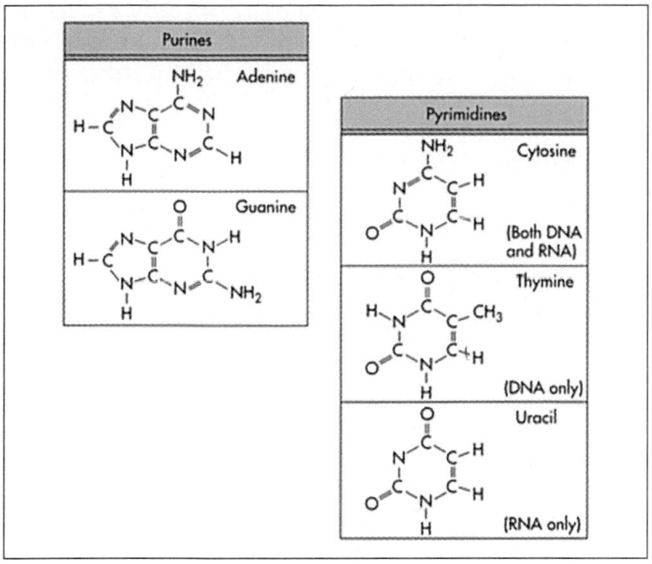

FIGURE 18-5 Purine and pyrimidines. Both DNA and RNA contain the purines as well as one of the pyrimidines, cytosine. The pyrimidine thymine is found only in DNA, and the pyrimidine uracil only in RNA.

Two British biophysicists, Rosalind Franklin and Maurice Wilkins, carried out x-ray diffraction analysis of fibers of DNA. In this process the DNA molecule is bombarded with an x-ray beam. When individual rays encounter atoms, each ray's path is bent or diffracted; the pattern created by these diffractions can be captured on photographic film. With careful analysis of the diffraction pattern, it is possible to develop a three-dimensional image of the molecule causing the diffractions of the x-rays. The diffraction patterns Franklin and Wilkins obtained (Figure 18-6) suggested that the DNA molecule was a helical coil. Put simply, this complex molecule was shaped like a spring.

Learning informally of Franklin and Wilkins' results before they were published in 1953, James Watson and Francis Crick (Figure 18-7), two young scientists at Cambridge University in England, quickly worked out a likely structure of the DNA molecule. They built models of the nucleotides, assembled them into molecular structures, and then tested each to see whether its structure fit with what they knew from Chargaff's and Franklin and Wilkins' work. They finally hit on the idea that the molecule might be a double helix (two springs twisted together) in which the bases of the two strands pointed inward toward one another (Figure 18-8). Pairing a purine (which is large) with a pyrimidine (which is small) resulted in a helical "ladder" with "rungs" of uniform length. In fact, always pairing adenine with thymine and cytosine with guanine yielded a molecule in which A = T and C = G—a molecule consistent with Chargaff's observations. In Watson and Crick's model, the

TABLE 18-1

Chargaff's Analysis of DNA Nucleotide Base Compositions

ORGANISM	BASE COMPOSITION (MOLE PERCENT) A	T	G	C
Escherichia coli	26.0	23.9	24.9	25.2
Streptococcus pneumoniae	29.8	31.6	20.5	18.0
Mycobacterium tuberculosis	15.1	14.6	34.9	35.4
Yeast	31.3	32.9	18.7	17.1
Sea Urchin	32.8	32.1	17.7	18.4
Herring	27.8	27.5	22.2	22.6
Rat	28.6	28.4	21.4	21.5
Human	30.9	29.4	19.9	19.8

chromatin (**kro** muh tin) the complex of deoxyribonucleic acid (DNA) and protein that makes up the chromosomes of eukaryotes.

nucleic acid (noo **klay** ick) a long polymer of repeating subunits called nucleotides; the two types of nucleic acid within cells are deoxyribonucleic acid (DNA) and ribonucleic acid (RNA).

nucleotide (**noo** klee o tide) a single unit of nucleic acid, consisting of a five-carbon sugar bonded to a phosphate group and a nitrogen-containing base.

ribonucleic acid (RNA) (**rye** boh noo **klay** ick) one of two types of nucleic acid found in cells; the nucleotides of RNA contain the 5-carbon sugar ribose.

deoxyribonucleic acid (DNA) (de **ok** see **rye** boh noo **klay** ick) one of two types of nucleic acid found in cells; the nucleotides of DNA contain the 5-carbon sugar deoxyribose.

purines (**pyoor** eens) double-ring compounds that are components of nucleotides. Both DNA and RNA contain the purines adenine and guanine.

pyrimidines (pye **rim** uh deens *or* pih **rim** uh deens) single-ring compounds that are components of nucleotides. Both DNA and RNA contain the pyrimidine cytosine; in addition, DNA contains the pyrimidine thymine, and RNA contains the pyrimidine uracil.

Discoveries

HOW *Science* WORKS

Is DNA or Are Chromosomal Proteins the Genetic Material?

Students studying biology today sometimes take the information presented in textbooks for granted. For instance, it seems a *given* that all the genetic information about you—your appearance, how your cells function, your gender—is contained in your DNA. But as little as 45 years ago, many scientists thought that the genetic information was contained in and passed along to offspring by proteins located in chromosomes. A series of important experiments that helped prove DNA and not protein was the genetic material was performed in 1952 by Alfred Hershey and Martha Chase. Building on an increasing body of experimental evidence that pointed to DNA, Hershey and Chase used viruses to carry out a clear either/or test and showed that an infection by a virus involved an injection of the virus DNA, not protein, into a cell.

Hershey and Chase reasoned that if DNA was indeed the genetic carrier, then bacteria infected with a virus would contain DNA from the virus inside their cells. They reached this conclusion because they knew that when viruses infect cells, they somehow direct the cell to manufacture more virus particles. On the other hand, if protein were the material of which genes are made, then virus-infected bacteria would contain viral proteins within their cells.

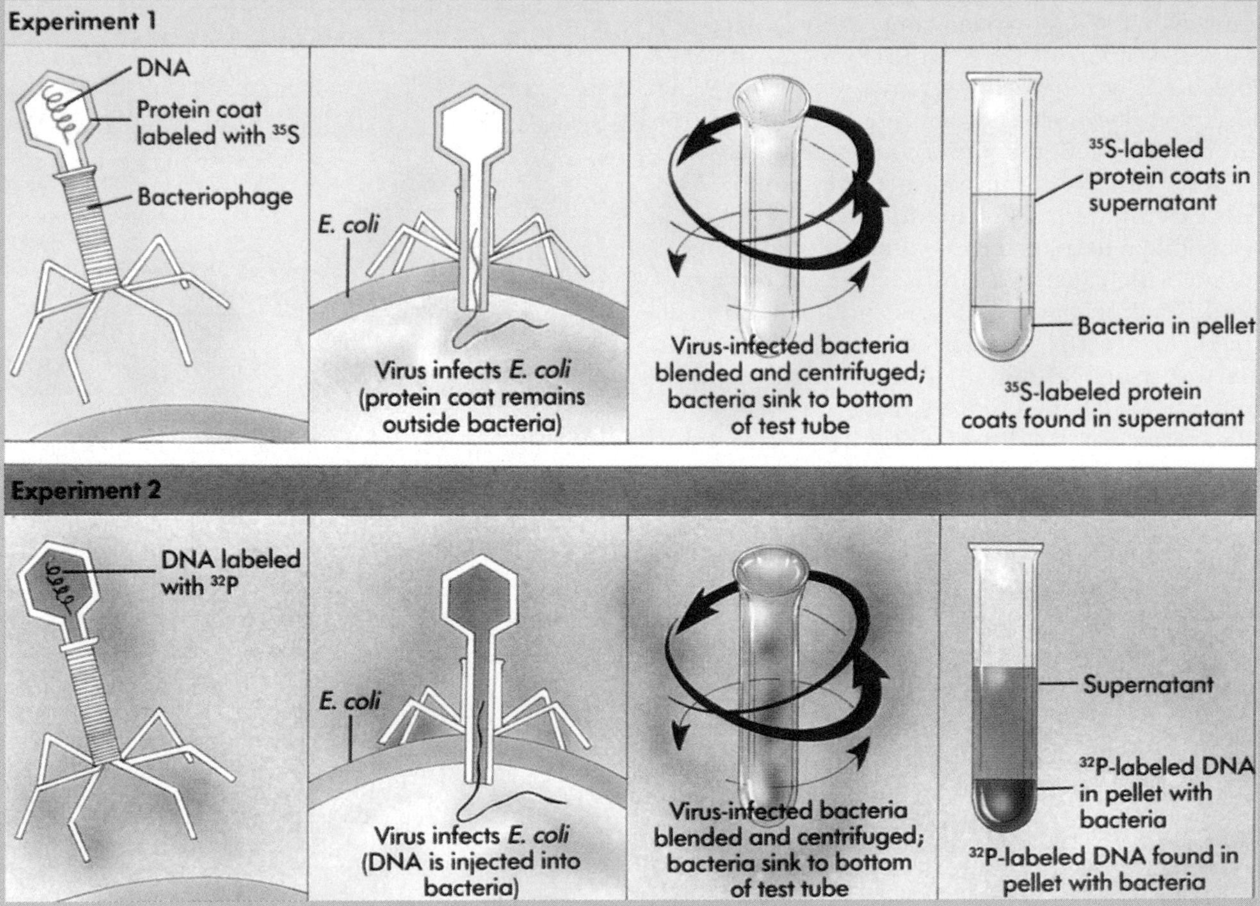

FIGURE 18-A How the Hershey-Chase experiments were performed.

Discoveries—cont'd

The challenge for Hershey and Chase was to find a way to look inside an infected cell and see what parts of the infecting virus could be found. Hershey and Chase chose to work with a bacteriophage, a type of virus that infects bacteria. Bacteriophages (like all viruses) have a nucleic acid core surrounded by a protein coat. Bacterial viruses also have "legs" attached to the protein coat that allow them to "dock" with a bacterium.

A key part of Hershey and Chase's experiments was the way they told DNA and protein apart—they used radioactive "name tags." Proteins contain sulfur but DNA does not, so radioactive sulfur served as a label for protein. DNA contains phosphorus but protein does not, so radioactive phosphorus served as a label for DNA.

Hershey and Chase performed two experiments (Figure 18-A). In experiment 1, they labeled the protein coat of the bacteriophage with ^{35}S, an isotope of sulfur. (See Chapter 2 for a discussion of isotopes.) Then they mixed the labeled bacteriophage with *Escherichia coli* bacteria (which are found in abundance in your digestive tract) and allowed the viruses to infect the bacteria. To examine the bacteria after infection, Hershey and Chase had to separate the protein coats that did not enter the bacteria from the infected bacteria. They did this by blending their mixture and then centrifuging it (spinning it rapidly), which pulled the heavier bacteria to the bottom of a test tube, leaving the lighter protein coats in the solution on top called the *supernatant.* Hershey and Chase found all the ^{35}S-labeled protein in the supernatant, not in the bacteria. Thus the bacteriophage protein did not enter the bacteria during infection.

In experiment 2, Hershey and Chase labeled the bacteriophage DNA with an isotope of phosphorus, ^{32}P, and performed the same mixing, blending, and centrifuge procedure as they did in experiment 1. The ^{32}P was found in the bottom of the test tube, with the bacteria. Clearly, ^{32}P-labeled DNA had entered the bacteria.

These experiments went a long way toward convincing the scientific community that DNA was the genetic material. Within the year, Watson and Crick's discovery of DNA's structure showed how easily DNA could fulfill its suggested genetic role. The protein hypothesis, unsubstantiated by experimentation, was no longer considered tenabl

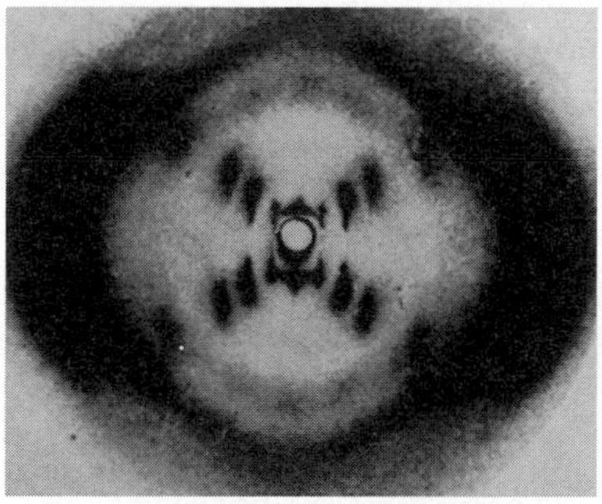

FIGURE 18-6 Evidence for the helical structure of DNA. This x-ray diffraction photograph of crystals of DNA was made in 1953 by Rosalind Franklin in the laboratory of Maurice Wilkins. It suggested to Watson and Crick that the DNA molecule was a helix, like a winding staircase.

FIGURE 18-7 Watson and Crick and the double helix. Watson and Crick examine their first model of the DNA double helix.

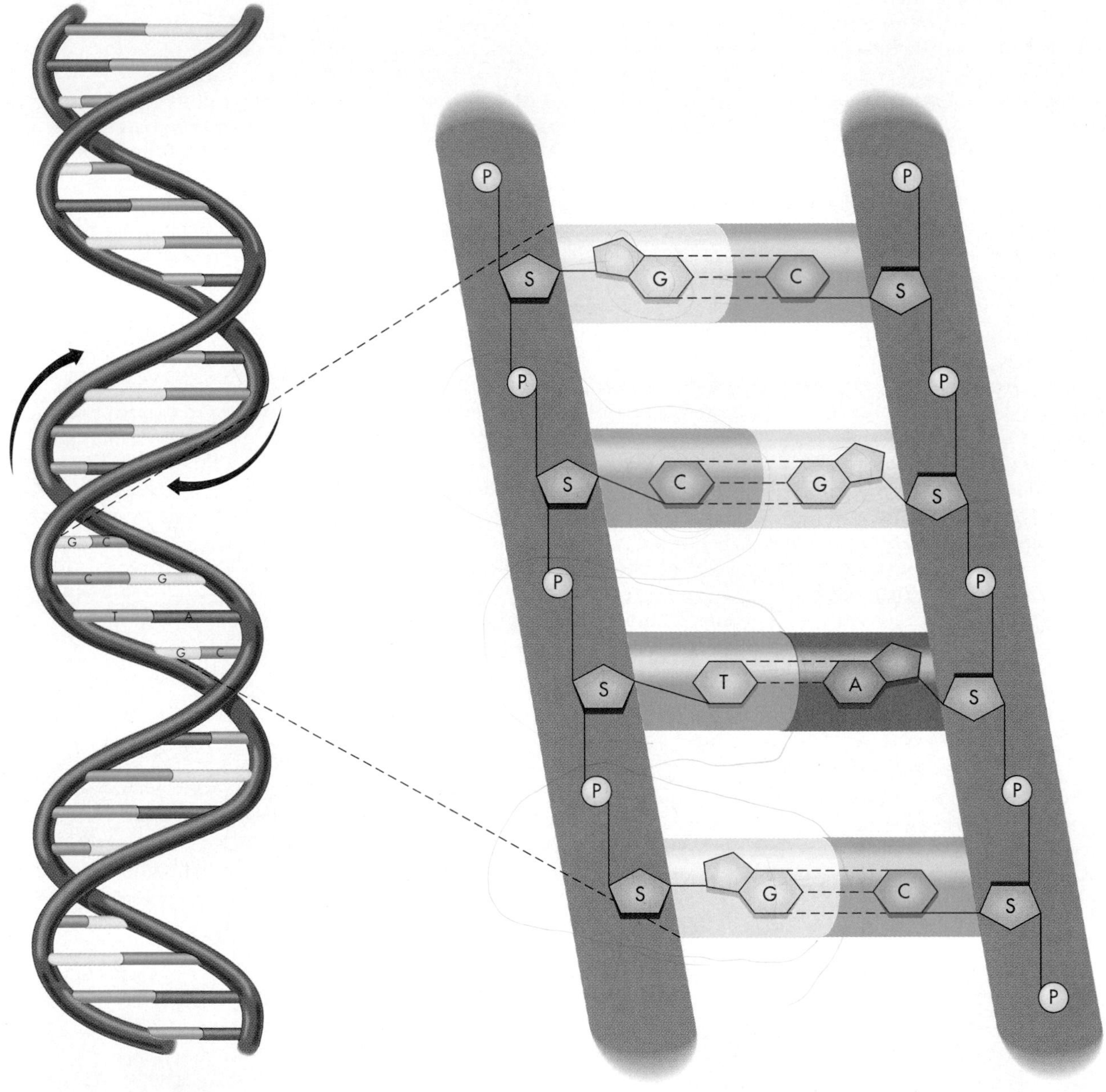

FIGURE 18-8 Structure of DNA. The diagram on the left shows the location of the nucleotide bases. The arrows indicate the direction of replication of the DNA strands, which run in opposite, antiparallel directions. In a DNA molecule, only two base pairs are possible: Adenine (*A*) binds with thymine (*T*), and guanine (*G*) pairs with cytosine (*C*). The diagram on the right shows that a G-C base pair is bonded by three hydrogen bonds, and an A-T base pair is bonded by two hydrogen bonds. The sugar and phosphate units are linked in an alternating fashion to form the sides of the helix.

sugar and phosphate units linked in an alternating fashion to form the sides of this twisted ladder (the springs in the earlier analogy). For their groundbreaking theory, Watson and Crick shared the Nobel Prize in 1962.

Chargaff determined that although the proportion of the bases in DNA varied among organisms, the amount of adenine always equaled the amount of thymine and guanine always equaled cytosine. This information, coupled with the work of Rosalind Franklin, Maurice Wilkins, James Watson, and Francis Crick, led to the determination that the DNA molecule is a double-stranded, helical, molecular ladder. The bases of each strand of DNA form rungs of uniform length, and alternating sugar-phosphate units form the ladder uprights.

How DNA Replicates

The Watson-Crick model immediately suggested that the basis for copying the genetic information is the complementarity of its bases. One side of the DNA ladder may have any base sequence, resulting in a particular code that the body understands. This sequence of bases then determines the sequence of bases making up the other side of the ladder. If the sequence on one side is ATTGCAT, for example, the sequence of the other side would have to be TAACGTA—its complementary image.

Scientists have also learned how DNA replicates, or makes more DNA. Before a cell divides, the bonds between the complementary bases break in short sections of the double-stranded DNA molecules, and the complementary strands separate from one another. This process is commonly referred to as *unwinding,* or *unzipping.* Each split in the molecule is called a *replication fork.* At these forks, each separated strand serves as a template for the synthesis of a new complementary strand. Figure 18-9 shows a diagram of the replication fork of DNA during the replication process. Notice the Y-shaped replication point. In a process directed by enzymes, free nucleotide units that are present in the nucleus link to complementary bases on each of the DNA strands. The sugars and phosphates of the new nucleotides bond together to form the backbones of both new strands. Thus the process of DNA replication begins with one double-stranded DNA molecule and ends with two double-stranded DNA molecules. Each double strand contains one strand from the parent molecule plus a new complementary strand assembled from free nucleotides. Each of the new double-stranded molecules is identical to the other and is also identical to the original parent molecule.

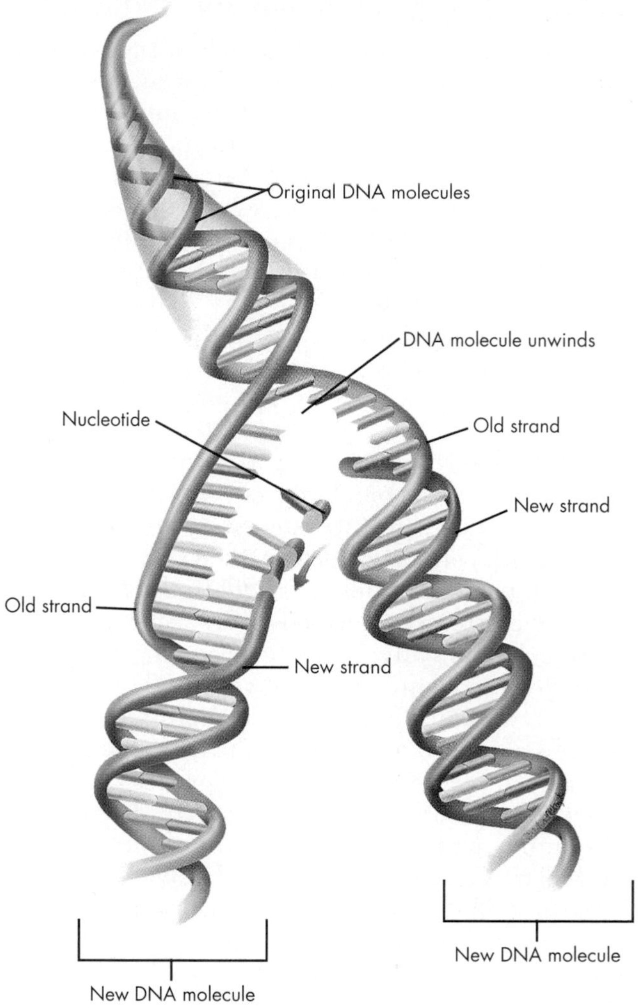

FIGURE 18-9 A replication fork. The double-stranded DNA molecule separates at its bases, forming a split, or fork. At this fork, each separated strand acts as a template for the synthesis of a new complementary strand.

DNA begins replication by unwinding at intervals along its helix before cell division. In a process initiated and coordinated by enzymes, free nucleotides bond to the exposed bases, producing two new DNA strands, each identical to one another and to the original double strand from which they were replicated.

Genes: The Units of Hereditary Information

Before the discovery of the structure of DNA by Watson and Crick, scientists were working to determine how DNA directs the growth and development of an organism. They asked the question, "What is the hereditary message that DNA carries?" This question was answered by means of experiments conducted in the late 1940s and early 1950s by two geneticists—George Beadle and Edward Tatum. These researchers, whose work officially marked the beginning of molecular genetics, studied biochemical characteristics as the expression of genes. They worked with the common red bread mold *Neurospora.*

At the same time Beadle and Tatum were doing their work, various biochemists were also studying the manufacture and breakdown of organic molecules within cells. They determined that cells build and degrade molecules by sequences of steps in which each step is catalyzed by a specific enzyme. These enzymatically controlled sequences of steps are called biochemical pathways.

Beadle and Tatum studied mold cultures they had exposed to x-rays. The x-rays induced mutations, or changes, in the DNA of the organism, resulting in a variety of mutants, each unable to manufacture certain amino acids. Using known information about the biochemical pathways of *Neurospora,* they hypothesized that specific enzymes must be involved in the manufacture of these amino acids and that some or all of the enzymes in each biochemical pathway must not be doing their jobs. To test their hypothesis, Beadle and Tatum chose mutants unable to synthesize the amino acid arginine. They supplied each mutant with various compounds intermediate in the arginine pathway and observed whether or not the mutant was then able to synthesize arginine. Using this method, Beadle and Tatum were able to infer the presence of defective enzymes in this biochemical pathway (Figure 18-10).

From these studies, Beadle and Tatum proposed the **one gene–one enzyme theory,** which states that the production of a given enzyme is under the control of a specific gene. If the gene mutates, the enzyme will not be synthesized properly or will not be made at all. Therefore the reaction it catalyzes will not take place, and the product of the reaction will not be produced. Today scientists have refined this hypothesis to say that the production of a given polypeptide (a portion of an enzyme or other protein) is under the control of a single gene. In addition, their conclusions have been upheld by other biochemical evidence. For their groundbreaking work, Beadle and Tatum were awarded the Nobel Prize in Physiology and Medicine in 1958.

These experiments and other related ones have brought a clear understanding of what the unit of heredity is. A unit of heredity—a **gene**—is a sequence of nucleotides that codes for (scientists say encodes) the amino acid sequence of an enzyme or other protein. Although most genes code for a string of amino acids (polypeptides), there are also genes devoted to the production of special forms of RNA, which play important roles in protein synthesis.

> *DNA carries a code that directs the synthesis of polypeptides—pieces of enzymes and other proteins that orchestrate and regulate the growth, development, and daily functioning of cells. Each hereditary unit, or gene, codes for one polypeptide.*

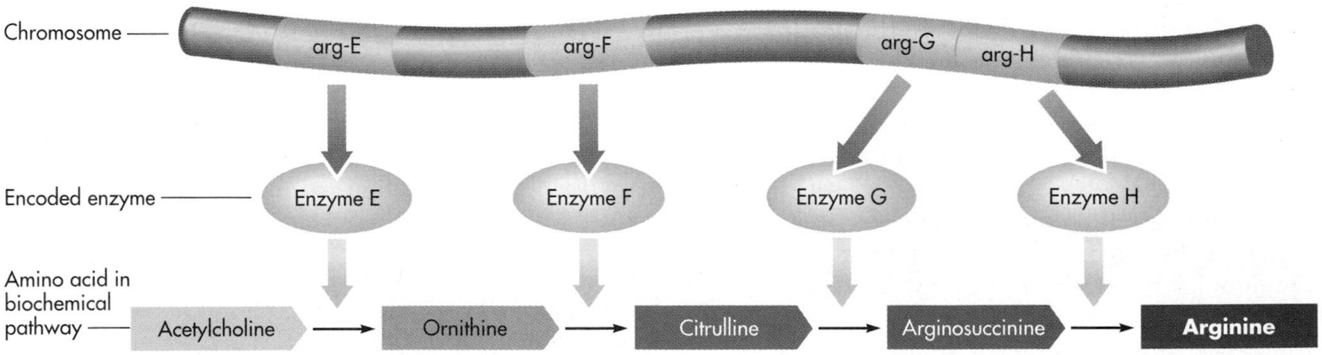

FIGURE 18-10 The one gene–one enzyme theory. This theory states that the production of a given enzyme is under the control of a specific gene. This theory was postulated by Beadle and Tatum after their work with *Neurospora.* This illustration shows the biochemical pathway of the synthesis of the amino acid arginine, which Beadle and Tatum studied. The genes that direct the synthesis of the enzymes that catalyze specific steps in this synthetic pathway are depicted at the top of the illustration.

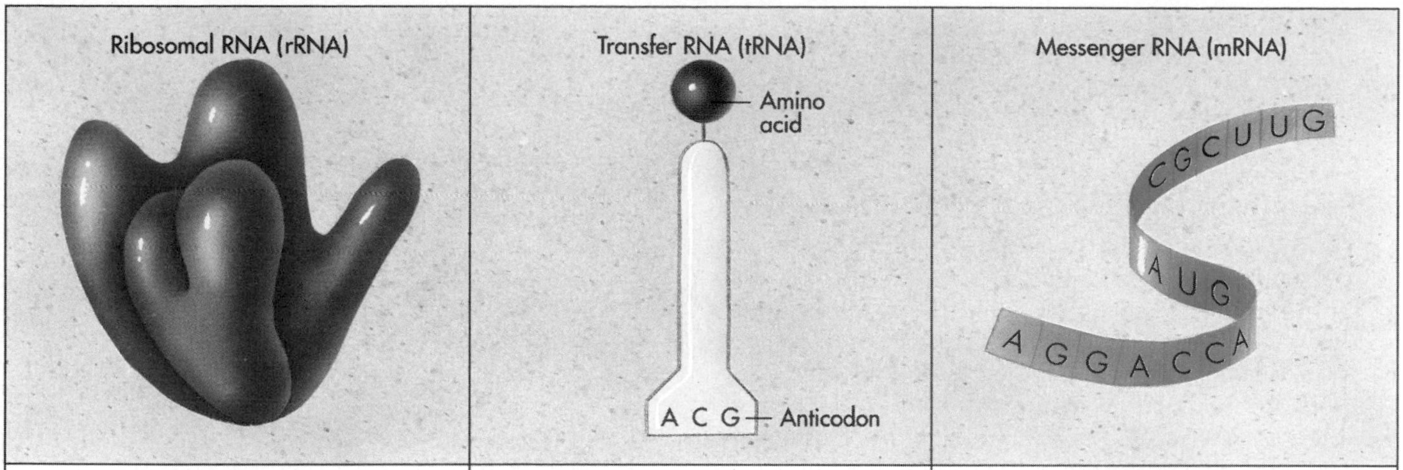

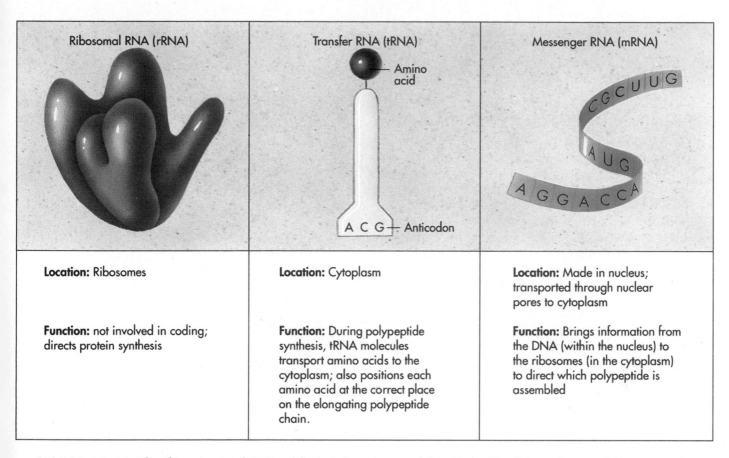

Ribosomal RNA (rRNA)	Transfer RNA (tRNA)	Messenger RNA (mRNA)
Location: Ribosomes	**Location:** Cytoplasm	**Location:** Made in nucleus; transported through nuclear pores to cytoplasm
Function: not involved in coding; directs protein synthesis	**Function:** During polypeptide synthesis, tRNA molecules transport amino acids to the cytoplasm; also positions each amino acid at the correct place on the elongating polypeptide chain.	**Function:** Brings information from the DNA (within the nucleus) to the ribosomes (in the cytoplasm) to direct which polypeptide is assembled

FIGURE 18-11 The three types of RNA with their locations and functions. The illustrations and diagrammatic representations of the shapes of these molecules and are used to depict each type of RNA throughout this chapter.

Gene Expression: How DNA Directs the Synthesis of Polypeptides

An Overview

Interestingly, polypeptides are not made in the nucleus where DNA is located. Instead, they are made in the cytoplasm at the **ribosomes.** These complex polypeptide-making factories contain more than 50 different proteins in their structure. Along with proteins, ribosomes are also made up of RNA—a hint that RNA molecules play an important role in polypeptide synthesis. In fact, as shown in Figure 18-11, cells contain three types of RNA, and each plays a special role in the manufacture of polypeptides.

The type of RNA found in ribosomes is called **ribosomal RNA (rRNA).** A second type of RNA, called **transfer RNA (tRNA),** is found in the cytoplasm. During polypeptide synthesis, tRNA molecules transport amino acids (used to build the polypeptide) to the ribosomes. In addition, tRNA molecules position each amino acid at the correct place on the elongating polypeptide chain. A third

one gene-one enzyme theory a scientific explanation that states that the production of a given enzyme is under the control of a specific gene. If the gene mutates, the enzyme will not be synthesized properly or will not be made at all. Therefore, the reaction it catalyzes will not take place, and the product of the reaction will not be produced.

gene a unit of heredity formed of a sequence of nucleotides that codes for the amino acid sequence of a polypeptide.

ribosomes (rye buh somes) minute, round structures found on the endoplasmic reticulum of eukaryotic cells or free in the cytoplasm of both prokaryotes and eukaryotes; ribosomes are the places where proteins are manufactured.

ribosomal RNA (rRNA) (rye buh so mull) a type of ribonucleic acid found in ribosomes that plays a role in the manufacture of polypeptides.

transfer RNA (tRNA) a type of ribonucleic acid found in the cytoplasm that transports amino acids, used to build polypeptides, to the ribosomes; tRNA molecules also position each amino acid at the correct place on the elongating polypeptide chain.

type of RNA, called **messenger RNA (mRNA),** brings information from the DNA (within the nucleus) to the ribosomes (in the cytoplasm) to direct which polypeptide is assembled.

In the first step of polypeptide synthesis, mRNA is constructed from free RNA nucleotides by the use of a single strand of DNA as a template. After noncoding sequences are removed from the mRNA strand, the mRNA leaves the nucleus and travels to the ribosomes, where polypeptide synthesis takes place.

Transcribing the DNA Message to RNA

The first step in the process of polypeptide synthesis and gene expression is the copying of the gene into a strand of messenger RNA. Scientists call this copying process **transcription.** Transcription begins when a special enzyme, called an *RNA polymerase,* binds to a particular sequence of nucleotides on a single DNA strand. This sequence of nucleotides is located at the beginning of the gene that is being expressed. You can think of such a sequence of nucleotides as an enzyme code that says, "Start here." Scientists call this place on a DNA strand a *promoter site.*

First, the DNA base-pair bonds break, no longer holding the double-stranded helix together. As the DNA strands separate, one strand begins to function as a template as RNA polymerase binds to a promoter site on the DNA molecule. RNA nucleotides bind with the now-exposed DNA bases in a sequence complementary to that of the DNA. For example, RNA nucleotides having the base adenine pair with DNA nucleotides having the base thymine. Likewise, RNA nucleotides having the base uracil pair with DNA nucleotides having the base adenine. Figure 18-12 shows the process of RNA transcription. Which DNA nucleotides bond with RNA nucleotides having the base guanine?

As the RNA polymerase moves along the DNA strand encountering each DNA nucleotide in turn, it adds the corresponding complementary RNA nucleotide to the growing single strand of mRNA. When the enzyme arrives at a special stop sequence (frequently a loop in the DNA) located at the end of the gene, it disengages from the DNA and releases the newly assembled mRNA strand. A processing step then removes segments of RNA from the original transcript that are not used in polypeptide synthesis. These extra sequences of nucleotides that intervene between the polypeptide-specifying portions of the gene are called *introns* (they *intrude* into the gene but are not expressed). The remaining segments of the gene—the nucleotide sequences that encode the amino acid sequence of the polypeptide—are called *exons* (*expressed* portions of a gene). Cutting out introns and splicing together exons results in the final mRNA strand. After cutting and splicing occurs, the mRNA strands leave the nucleus through the nuclear pores and travel to the ribosomes in the cytoplasm of the cell (Figure 18-13).

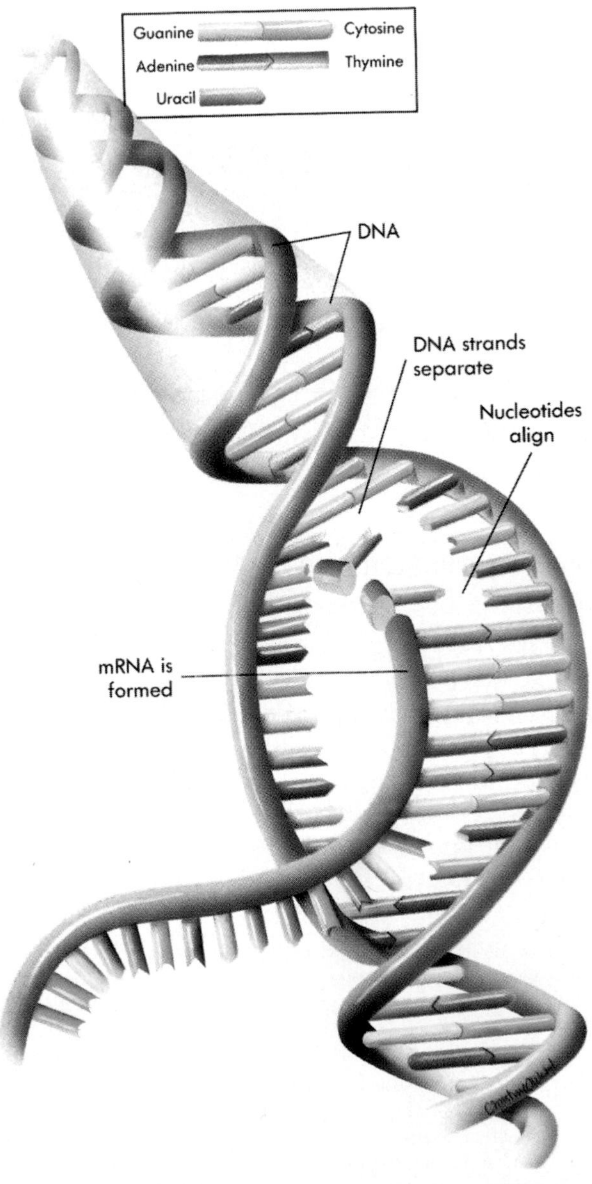

FIGURE 18-12 Transcription. One of the strands of DNA functions as a template on which nucleotide building blocks are assembeled by RNA polymerase (not shown here) into RNA.

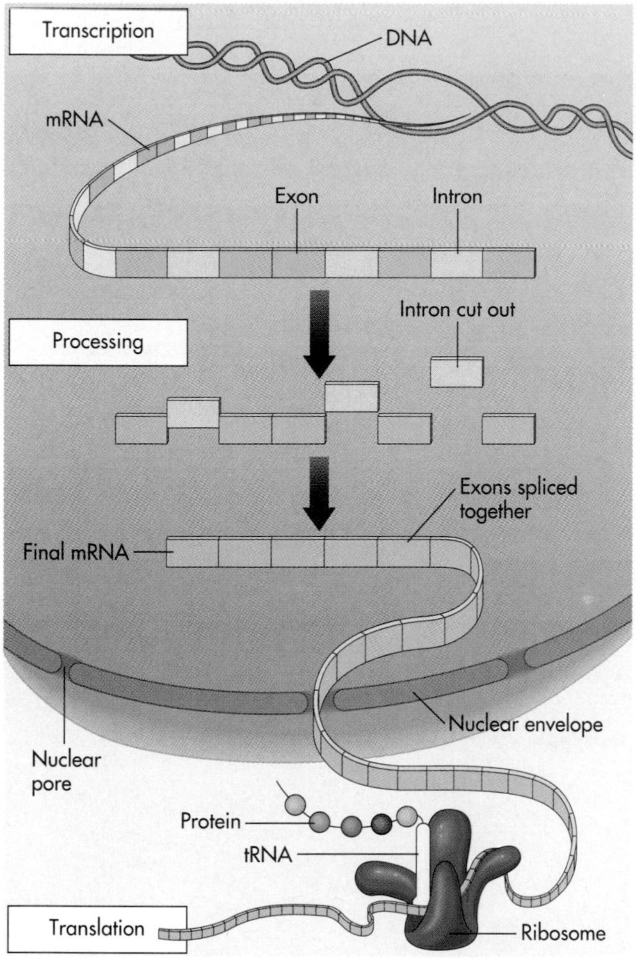

FIGURE 18-13 Polypeptide synthesis in eukaryotes. Genes in eukaryotes are long and contain extra sequences of nucleotides called introns, which are not used in protein synthesis. These introns must be cut out of the mRNA transcript of the gene, leaving exons, before protein synthesis can proceed.

Translating the Transcribed DNA Message into a Polypeptide

In the second step of gene expression, the mRNA—using its *copied* DNA code—directs the synthesis of a polypeptide. Scientists call this decoding process **translation.** In 1961, as a result of experiments led by Francis Crick (one of the researchers who worked out the structure of DNA), scientists learned that the DNA code is made up of sequences of three nucleotide bases. These triplet-nucleotide code words, as they appear on the transcribed mRNA, are called **codons.** Researchers soon broke the code and learned which codons stand for which amino acids, the building blocks of polypeptides.

Table 18-2 is a list of the mRNA codons and the amino acids for which they code. The table is structured so that every triplet-base sequence of the four mRNA bases is listed. As you can see from the table, using a three-base sequence of four different bases produces 64 possible codon combinations ($4 \times 4 \times 4$, or 4^3). Approximately 20 amino acids are used in polypeptide production, so extra codons exist, providing alternative code words for many of the amino acids. Also notice that certain codons do not code for amino acids but act as stop signals for the process of translation.

During the process of translation, the genetic code (the sequence of codons) is deciphered. First, the initial portion of the mRNA transcribed from DNA in the nucleus binds to an rRNA molecule interwoven in the ribosome (Figure 18-14). The mRNA lies on the ribosome in such a way that only the three-nucleotide portion of the mRNA molecule—the codon—is exposed at the polypeptide-making site. As each bit of the mRNA message is exposed in turn, a molecule of transfer RNA (tRNA) binds to the mRNA.

Each tRNA molecule has an **anticodon** loop—a portion of the molecule with a sequence of three base pairs complementary to a specific mRNA codon (Figure 18-15). On the opposite end of the anticodon loop, each transfer RNA molecule carries an amino acid specific to its anticodon sequence. For example, a tRNA molecule having the anticodon AGA carries the amino acid serine. A special family of enzymes links the amino acids to the tRNA molecules. Using the information in Table 18-2 and your understanding of the relationship between codons and anticodons, can you determine which amino acid a tRNA molecule with the anticodon GGG carries?

One by one, as tRNA molecules bind to codons at the mRNA, amino acids are lined up in an order determined by the sequence of codons. As each amino acid is added to the chain, a bond is formed. When an amino acid bonds to the forming polypeptide, its

messenger RNA (mRNA) a type of RNA that brings information from the DNA within the nucleus to the ribosomes in the cytoplasm and directs polypeptide synthesis.

transcription (trans **krip** shun) the first step in the process of polypeptide synthesis and gene expression, in which a gene is copied into a strand of messenger RNA.

translation (trans **lay** shun) the second step of gene expression, in which the mRNA, using its copied DNA code, directs the synthesis of a polypeptide.

codons sequences of three nucleotide bases in transcribed mRNA that code for specific amino acids, which are the building blocks of polypeptides.

anticodon a portion of a tRNA molecule with a sequence of three base pairs complementary to a specific mRNA codon.

TABLE 18-2

The Genetic Code

FIRST LETTER	SECOND LETTER: U	C	A	G	THIRD LETTER
	Phenylalanine	Serine	Tyrosine	Cysteine	U
U	Phenylalanine	Serine	Tyrosine	Cysteine	C
	Leucine	Serine	Stop	Stop	A
	Leucine	Serine	Stop	Typtophan	G
	Leucine	Proline	Histidine	Arginine	U
C	Leucine	Proline	Histidine	Arginine	C
	Leucine	Proline	Glutamine	Arginine	A
	Leucine	Proline	Glutamine	Arginine	G
	Isoleucine	Threonine	Asparagine	Serine	U
A	Isoleucine	Threonine	Asparagine	Serine	C
	Isoleucine	Threonine	Lysine	Arginine	A
	(Start); Methionine	Threonine	Lysine	Arginine	G
	Valine	Alanine	Aspartate	Glycine	U
G	Valine	Alanine	Aspartate	Glycine	C
	Valine	Alanine	Glutamate	Glycine	A
	Valine	Alanine	Glutamate	Glycine	G

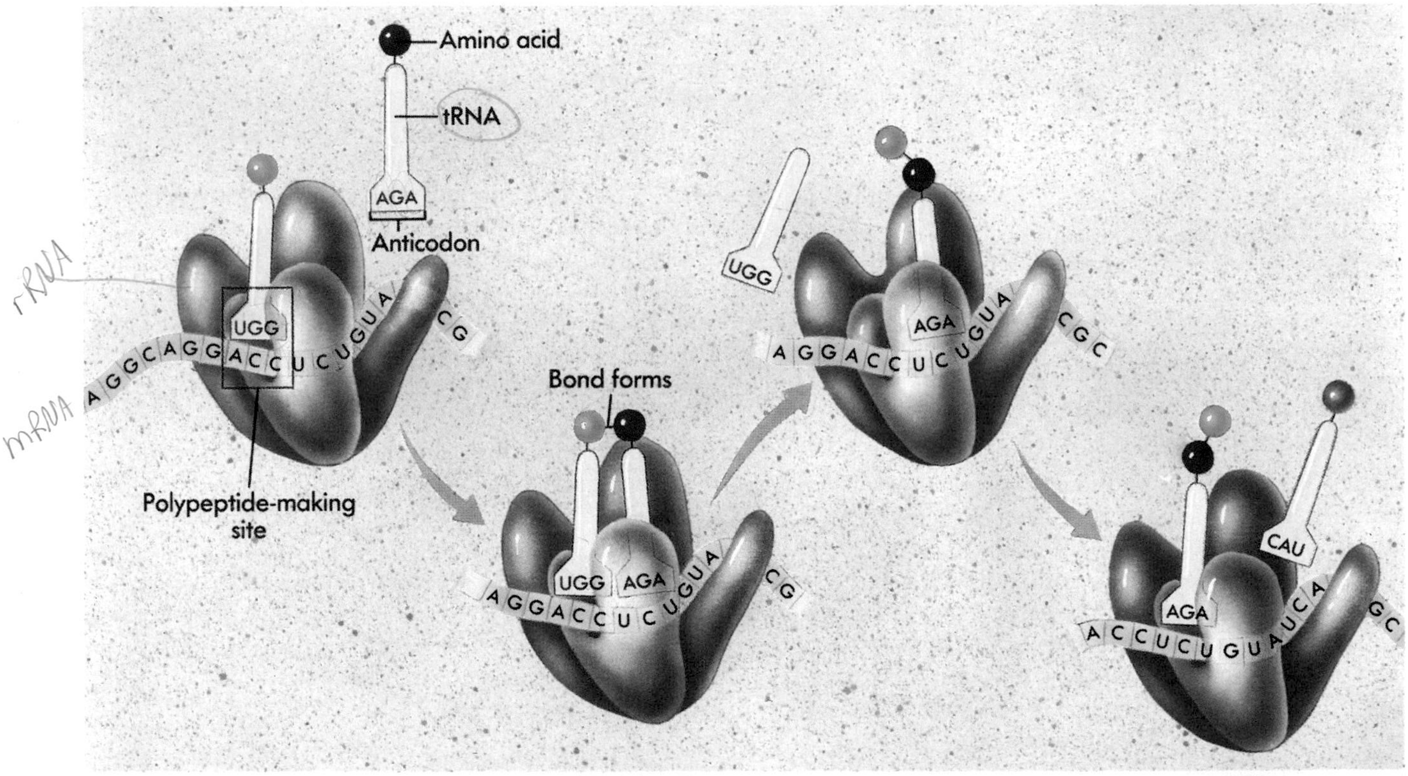

FIGURE 18-14 Translation. The steps of translation are as follows:

1. mRNA binds to rRNA.
2. tRNA with an attached amino acid approaches the mRNA-rRNA complex.
3. The tRNA attaches to the mRNA at a special binding site. The amino acid at the end opposite to the anticodon loop forms a bond with the amino acid attached to the adjacent tRNA molecule.
4. The adjacent tRNA molecule breaks its bond with the mRNA and falls from its site.
5. The rRNA moves along the mRNA, exposing the next codon of mRNA.

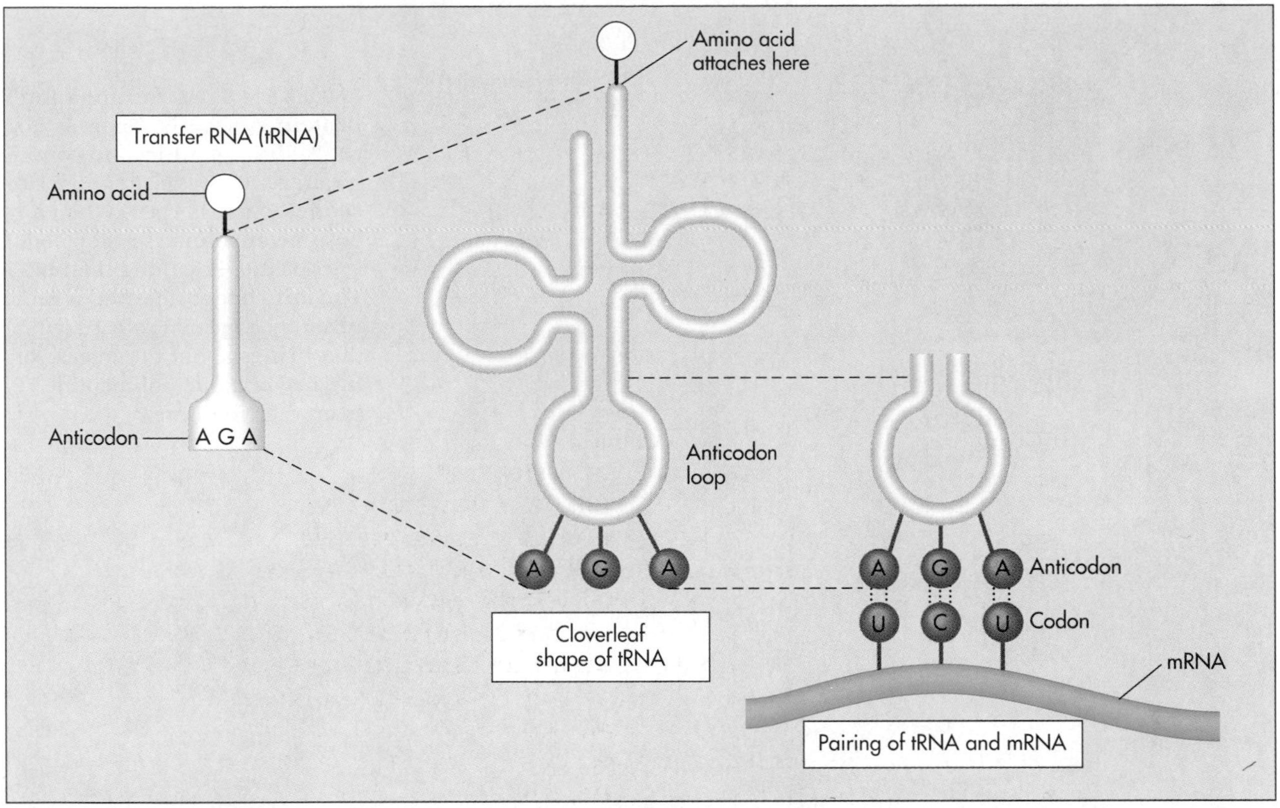

FIGURE 18-15 Structure of a tRNA molecule. A tRNA molecule is shaped like a cloverleaf. The anticodon is at the bottom of the cloverleaf, and the amino acid specific to the anticodon is at the top. The cloverleaf structure is not shown throughout this chapter; instead a more simplified tRNA molecule is used. It is symbolized by a bar as shown to the left. On the right, the anticodon is shown pairing with a complementary codon on the mRNA molecules.

bond to its carrier tRNA molecule breaks. The tRNA falls from its site on the ribosome, leaving that site vacant. The ribosome then moves along the mRNA strand, exposing the next codon of the mRNA. When a tRNA molecule that recognizes this next codon appears, its anticodon bonds to the codon, adding a new amino acid to the growing chain. When a stop codon is encountered (see Table 18-2 and Figure 18-16), no tRNA exists to bind to it. Instead, it is recognized by special release factors, proteins that bring about the release of the newly made polypeptide from the ribosome. Figure 18-17 shows an overview of protein synthesis.

> *Polypeptide synthesis takes place at the ribosomes. Ribosomes bind to sites at one end of a mRNA strand and then move down the strand, exposing the codons one-by-one. At each step of a ribosome's progress, it exposes a codon to binding by a tRNA molecule having a three-base sequence complementary to the exposed mRNA codon. The amino acid carried by that particular tRNA molecule is then added to the end of the elongating polypeptide chain.*

Figure 18-18 shows that more than one ribosome at a time reads and translates the mRNA message to synthesize a polypeptide. Here, a group of many ribosomes (called a *polyribosome*) is reading along an mRNA molecule taken from a cell of a fly. The polypeptides being made can be seen dangling behind each ribosome.

Regulation of Eukaryotic Gene Expression

Cells not only know how to make particular polypeptides but also when to make them. For example, during your fetal development, specific enzymes played crucial roles at certain times, directing the series of biochemical reactions that resulted in your growth. As this growth and development took place, your genes were transcribed in a specific order, each gene for a specified time. Likewise, the cells in your body that produce digestive enzymes know when to manufacture these enzymes. Your red blood cells—not other types of cells—synthesize hemoglobin to carry oxygen to all parts of your body. How are these specific genes activated?

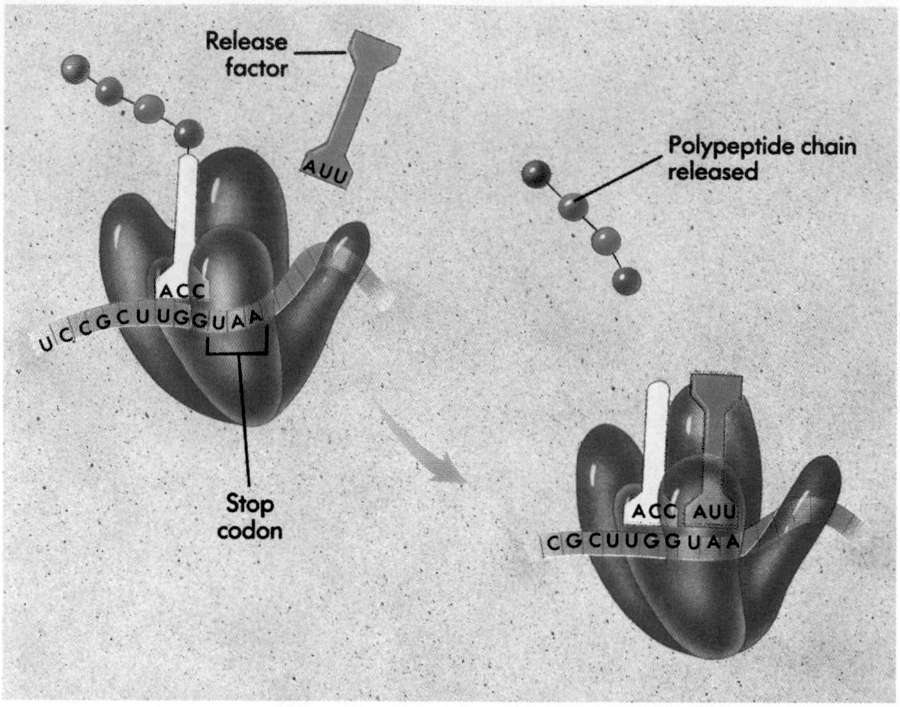

FIGURE 18-16 Termination of protein synthesis. There is no tRNA with an anticodon complementary to any of the three stop codons, such as UAA. When a ribosome encounters a stop codon, it stops moving along the mRNA. The anticodon, however, *is* recognized by a special release factor that brings about the release of the newly made polypeptide from the ribosome.

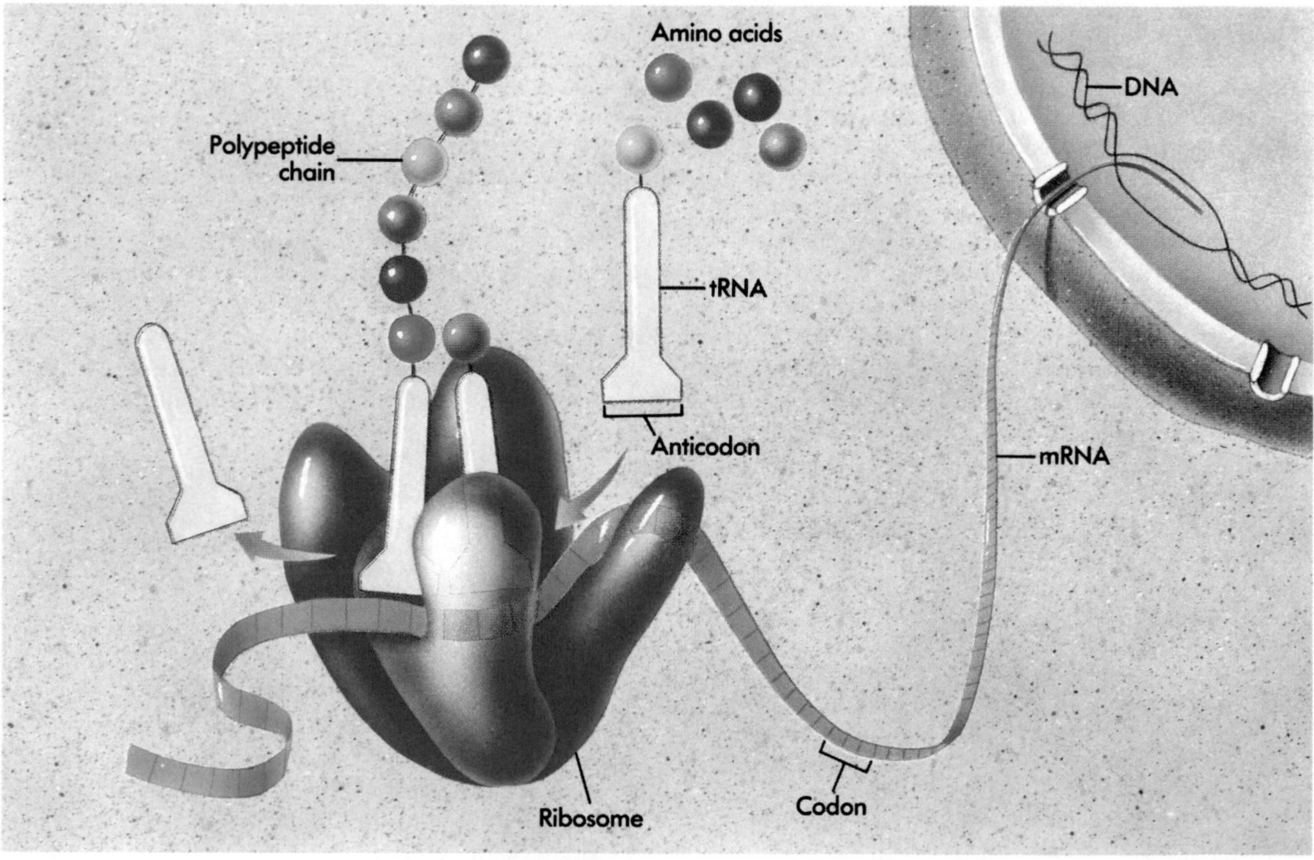

FIGURE 18-17 Overview of protein synthesis. This diagram presents a summary of protein synthesis, beginning in the nucleus where DNA is transcribed into mRNA. mRNA leaves the nucleus and enters the cytoplasm where it binds with a ribosome. The ribosome exposes condons on the mRNA enabling tRNA molecules with complementary anticodons to bind. These tRNA molecules carry amino acids. In this manner, a chain of amino acids—a polypeptide—is formed.

A

B

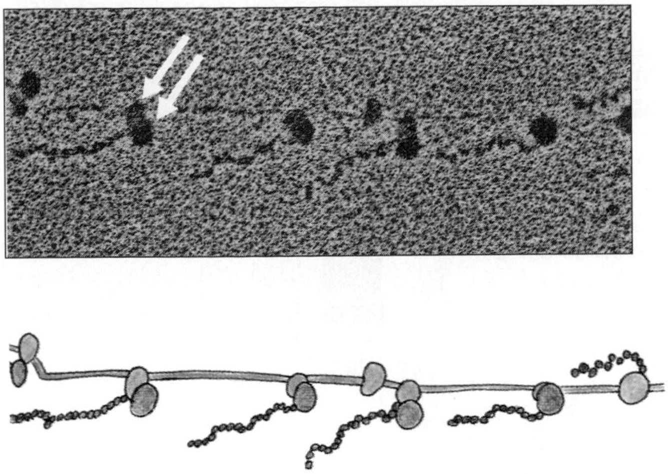

FIGURE 18-18 Protein synthesis in a fly. A, An electron micrograph of ribosomes in the fly *Chironomus tentans.* These ribosomes are reading along an mRNA molecule from left to right, assembling polypeptides that dangle behind them like tails. Visible are the two subunits *(arrows)* of each ribosome translating the mRNA. **B,** An illustration depicting the same process described above. The mRNA molecule is shown in green, the ribosome subunits are shown in two different shades of brown, and the polypeptides are shown in purple.

Molecular geneticists are now beginning to understand these mechanisms in eukaryotic cells. Interestingly, cells have gene switches. These switches are really proteins that interact with specific nucleotide sequences called **regulatory sites.** Regulatory sites control the transcription of genes. Control can be positive and result in turning a gene on, or it can be negative and result in turning a gene off. In eukaryotes, control is generally positive because eukaryotic genes are usually inactive—so genes must be turned on. In one type of gene switch, steroid hormones, along with protein receptor molecules from the cytoplasm, activate transcription by binding to regulatory sites on the DNA. Scientists are probing the details of this regulatory process.

Eukaryotic cells also regulate gene expression by the number of identical genes they transcribe at one time. This use of multiple copies of genes is a way to control the amount of a particular protein that is produced at a particular time. Some genes have related, but not identical, nucleotide sequences. Transcribing these genes at the same time produces related products. Such multiple copies of genes in eukaryotes, called *gene families,* are derived from a common ancestral gene and are a reflection of the evolutionary process.

> *Organisms control the expression of their genes by selectively inhibiting the transcription of some genes and facilitating the transcription of others. Regulatory proteins control the transcription process.*

Cell Division, Growth, and Reproduction

The highly ordered and orchestrated process of DNA replication, transcription, and translation results in the expression of your heredity—an expression of the uniqueness of you. The fact that DNA is packaged neatly in chromosomes not only allows a tremendous amount of information to reside in each cell of your body but also allows your cells to separate this genetic material in an organized way during cell division so that each cell receives the appropriate complement of DNA.

Cell division occurs in your body (as well as in other eukaryotes) as a means of cellular growth, repair, and reproduction. To illustrate, look at the human life cycle diagrammed in Figure 18-19. A **life cycle** is the progression of stages an organism passes through from its conception until it conceives another similar organism. The human life cycle is representative of the life cycle of all animals. The baby in the diagram represents that part of the life cycle during which a new individual has been produced by the fusion of **gametes,** or sex cells, from a male and a female of the same species. The female gamete is the egg, and the male gamete is the sperm.

After a person (or other animal) grows to sexual maturity, the sex organs begin to produce gametes by a type of cell division called **meiosis.** During meiosis, one parent cell produces four sex cells, but these cells are *not* identical to the parent cell. Each

regulatory sites in eukaryotic cells, specific nucleotide sequences that control the transcription of genes.

life cycle the progression of stages an organism passes through from its conception until it conceives another similar organism.

gametes (**gam** eets) sex cells; the female gamete is the egg, and the male gamete is the sperm.

meiosis (my **oh** sis) a type of cell division by means of which the sex organs of a mature animal produce gametes. During meiosis, one parent cell produces four sex cells; each gamete produced during this process contains half the number of chromosomes of the original parent cell.

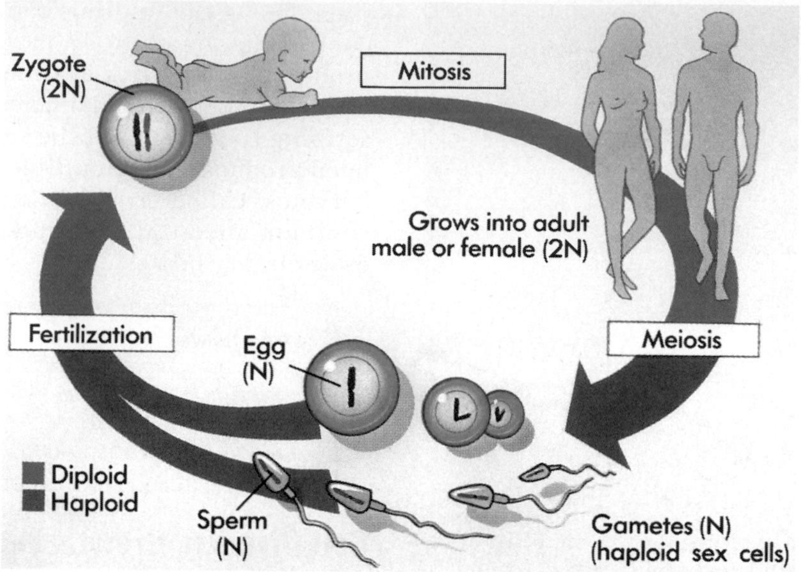

FIGURE 18-19 The human life cycle. Mitosis, which produces identical diploid daughter cells, is responsible for the growth of a human into an adult. Meiosis, on the other hand, produces haploid gametes (sex cells) that are *not* identical—each gamete contains half the hereditary material of the original cell. Meiosis in human females produces one functional gamete and three nonfunctional gametes. Meiosis in human males produces four functional gametes. During fertilization, one gamete from each of the two parents comes together to form a new individual.

sex cell is **haploid;** that is, it contains *half* the amount of hereditary material of the original parent cell. It is a single set of genetic information. Because of this reduction in chromosome number, one sex cell from each of two parent organisms can join together in a process called **fertilization** to form the first cell of a new individual that has a full complement of hereditary material. This new cell is **diploid.** That is, it contains double the haploid amount—a double set of the genetic information. This type of reproduction, which involves the fusion of gametes to produce the first cell of a new individual, is called **sexual reproduction.**

The first cell of any multicellular organism divides, and those cells divide, and so on. This type of cell division is called **mitosis** and produces two daughter cells from one parent cell. These daughter cells have genetic information that is identical with one another and with the original parent cell. As the growth and development of the individual continues, certain cells are triggered by other developmental processes to begin to differentiate, or become different, from one another while continuing to divide. The animal life cycle begins again when the growth and development of the individual is complete and sexual maturity is reached.

> *Mitosis is a process of cell division that produces two identical cells from an original parent cell. Meiosis is a process of cell division that produces four cells from one parent cell. Each of these four cells has one set of genetic information rather than two sets like the parent cell.*

Plants and animals are similar in that both use the process of mitosis to grow. Also, the life cycles of both plants and animals have two phases: a haploid phase and a diploid phase. In animals, however, the diploid phase of the life cycle predominates. The haploid phase consists only of single-celled gametes. In most organisms, these gametes live only 1 or 2 days after meiosis is complete unless fertilization occurs. During fertilization, two haploid gametes join to form the first cell in the diploid phase of the life cycle. Plant life cycles differ from animal life cycles because most plants have distinct *multicelled* haploid as well as diploid phases. One phase usually dominates over the other, but both phases are always present.

In plant life cycles, the haploid phase is the **gametophyte** (gamete-plant) generation and the diploid phase is the **sporophyte** (spore-plant) gen-

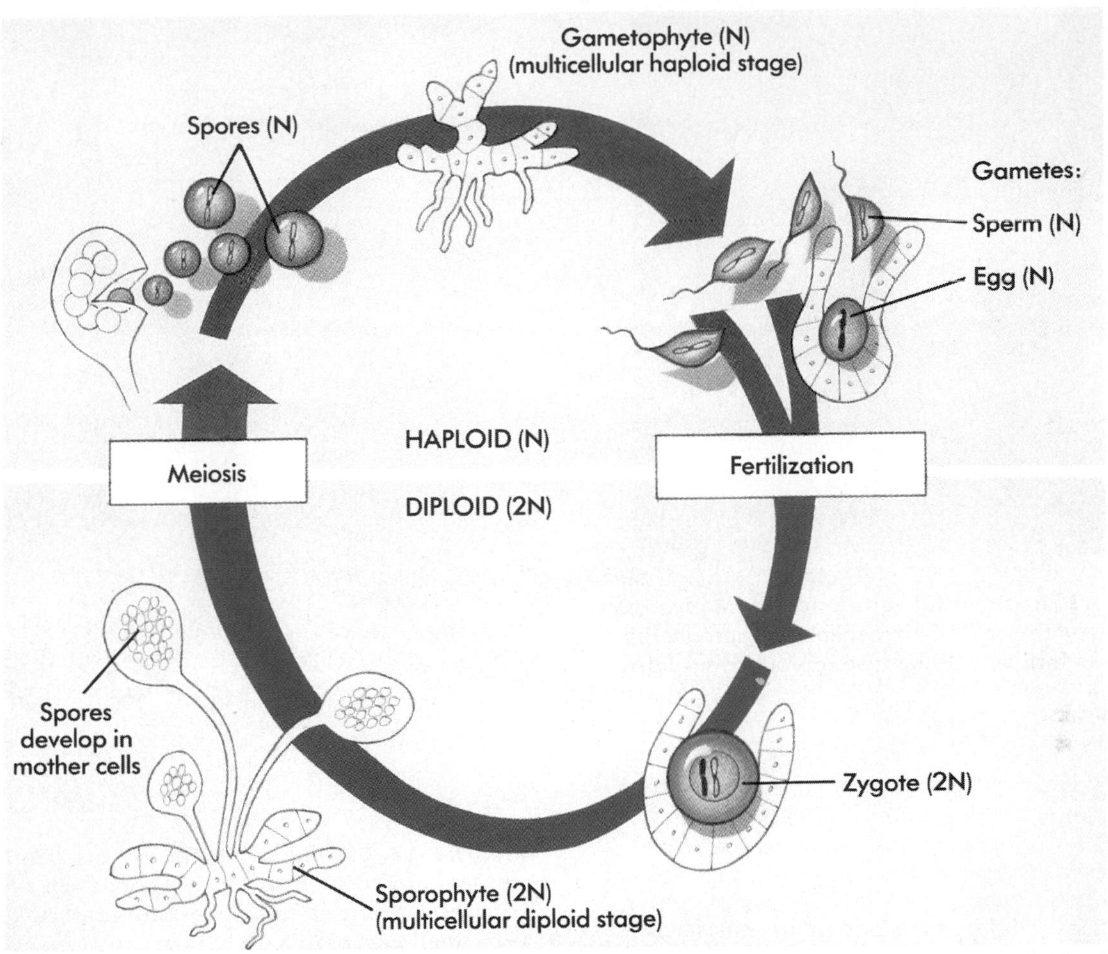

FIGURE 18-20 The life cycle of plants. Notice how the sporophyte phase of a plant, which is diploid, alternates with the gametophyte phase, which is haploid.

eration. Compare the generalized plant life cycle in Figure 18-20 with the animal (human) life cycle in Figure 18-19. Although haploid plants are common, haploid animals are rare. This results in some striking differences in the life cycles of animals and plants. In animals, gametes are usually produced in diploid individuals by the process of meiosis. In plants, gametes are produced in haploid individuals by the process of mitosis. The product of fertilization in plants is the first cell of a diploid spore-producing plant, not a diploid gamete-producing organism as in animals. (Spores are haploid, single-celled reproductive bodies capable of growing into the haploid gametophyte plant.) Because sporophytes are diploid, they produce spores by meiosis, reducing the number of chromosomes by half during this process.

haploid (**hap** loyd) a term describing a sex cell that contains half the amount of hereditary material of the original parent cell.

fertilization the union of a male gamete (sperm) and a female gamete (egg).

diploid a term describing a cell that contains double the haploid amount of genetic information; the full complement of genetic information of an organism.

sexual reproduction a type of reproduction that involves the fusion of gametes to produce the first cell of a new individual.

mitosis (my **toe** sis *or* mih **toe** sis) a process of cell division that produces two identical cells from an original parent cell.

gametophyte (geh **mee** toe fite) the haploid phase of a plant life cycle that alternates with the diploid phase; also known as the gamete-plant generation.

sporophyte (**spor** ih fite) the diploid phase of a plant life cycle that alternates with the haploid phase; also known as the spore-plant generation.

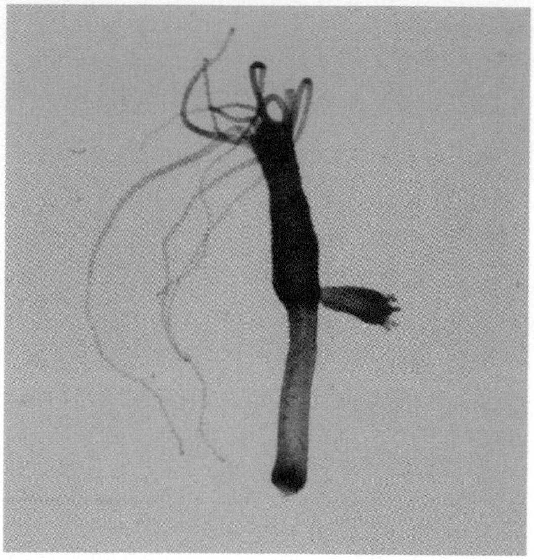

FIGURE 18-21 Asexual reproduction. Many organisms, such as this hydra, are capable of reproducing asexually by mitosis. Hydra also reproduce sexually.

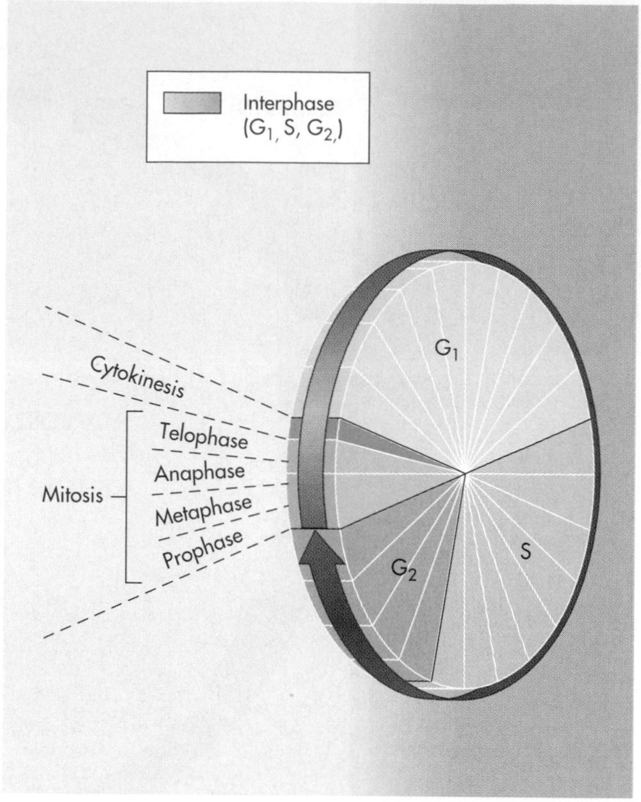

FIGURE 18-22 The cell cycle. In this diagram, each wedge represents 1 hour of the 24-hour cell cycle of a "typical" eukaryotic cell. Note that mitosis takes up a very small part of this cycle.

> *In plants, unlike animals, mitosis occurs during both growth and reproduction. Haploid plants produce gametes by mitosis, whereas diploid plants produce spores by meiosis. The life cycle of a plant consists of the alternation of the growth and reproduction of each of these plant types.*

Single-celled eukaryotes (such as many of the protists) generally reproduce by mitosis; this process plays no role in the growth of single-celled organisms but plays a role in the growth of populations of single-celled organisms. During reproduction, the parent organism divides by mitosis, producing two identical organisms. This type of reproduction is called **asexual reproduction.** Some simple animals such as hydra pictured in Figure 18-21 and many plants are capable of reproducing asexually by mitosis.

Mitosis

Notice in Figure 18-22 that only a small portion of a cell's "life," called the **cell cycle,** is involved with mitosis. The activities of **interphase** occupy most of the cell cycle.

In the past, interphase was called the *resting stage.* Scientists now know that a cell in interphase is far from resting. In fact, it doubles its size and replicates its DNA, as well as carrying out its normal life functions. During the G_1 stage of interphase (the time gap between the last mitosis and the start of DNA synthesis), the cell is growing. For many organisms, this growth period occupies the major portion of the cell's life span—possibly days, months, or years. During the S (synthesis) stage of interphase, a complete replica of the cell's DNA is synthesized. The cell now contains two complete, identical copies of hereditary information. These two copies of hereditary information can be seen as sister chromatids when the chromosomes become visible.

During the G_2 stage of interphase (the time gap between the end of DNA synthesis and the beginning of mitosis), the supercoils of DNA, normally diffuse or strung out within the cell nucleus, begin the long process of *condensation.* During this process the complex of DNA and proteins coil into more tightly compacted bodies that become visible as chromosomes during mitosis. In addition, during the G_2 phase the cell begins to assemble the "machinery" (microtubules that make up the spindle) that it will later use to move and divide the chromosomes. In animal cells the centrioles replicate. These organelles are surrounded by microtubule-

JUST WONDERING . . .

How does a cell "know" when to divide or reproduce?

Your question is an important one because cell reproduction is vital to such processes as growth, reproduction, immunity, and tissue repair. Scientists have been working on finding the answer to this question for more than 20 years. In doing so, they have also investigated how the cell coordinates the processes that take place during interphase, such as the replication of cell organelles and chromosomes. Although they are still refining and expanding their understandings, scientists found many answers to these questions by the late 1980s.

The cell cycle of eukaryotes appears to be controlled by chemical reactions that take place in the cytoplasm, but this control is modulated by events in the nucleus. Mitosis is delayed until DNA is replicated, for example, or until any damage to DNA has been repaired.

Three proteins are key players in the regulation of the cell cycle: *cdc2* (cell division cycle) protein, *cyclin,* and *MPF* (maturation producing factor). MPF is a complex of cyclin and cdc2 that has been activated by enzymes. MPF (with its key functional component cdc2) is a *protein kinase,* an enzyme that transfers phosphate groups from adenosine triphosphate (ATP) to proteins.

As discussed in Chapter 4, ATP is the "energy currency" of the cell. The bonding of phosphate groups to molecules of ATP is a means of storing energy released from exergonic reactions; likewise, the removal of phosphate groups from ATP is a way to make energy available for endergonic reactions. The reactions catalyzed by MPF (and thus cdc2 proteins) control the cell cycle by triggering molecular interactions that result in the events of mitosis. For example, one such set of reactions causes certain proteins in the nuclear envelope to dissociate, thus leading to the disintegration of the nuclear envelope. MPF also activates enzymes that degrade cyclin, resulting in its own loss of activity at the end of mitosis. Enzymes called *phosphatases* then catalyze reactions leading to the reformation of the nuclear membrane and inactivating cyclin-degrading enzymes so that the cycle can begin once again.

The story of cell cycle regulation is a complicated one, but in its simplest form, it is a story of regulatory enzymes that pervade all life. Scientists hope their work into the details of these regulatory mechanisms will lead to ways to induce cells to regenerate—a capability most human tissues do not have (see Just Wondering, Chapter 7). Also, this work might provide insight into why cancer cells grow out of control and how to stop this growth.

organizing centers that play a key role in the mitotic process; most plants and fungi lack centrioles.

> *Interphase is the portion of the cell cycle in which the cell grows and carries out normal life functions. During this time the cell also produces an exact copy of the hereditary material DNA as it prepares for cell division.*

Mitosis is a continuous sequence of events that occurs just after interphase and that results in the division of the chromosomes duplicated during its S phase. To more easily understand this process, scientists divide its events into the following phases: prophase, metaphase, anaphase, and telophase (Figure 18-23).

asexual reproduction the production of a new individual without the union of gametes.

cell cycle the time from the generation of a new cell until it reproduces. The cell cycle includes interphase, nuclear division, and cytoplasmic division.

interphase (in tur faze) the portion of the cell cycle preceding mitosis in which the cell grows and carries out life functions. During this time the cell also doubles in size and produces an exact copy of its heredity material, DNA, as it prepares for cell division.

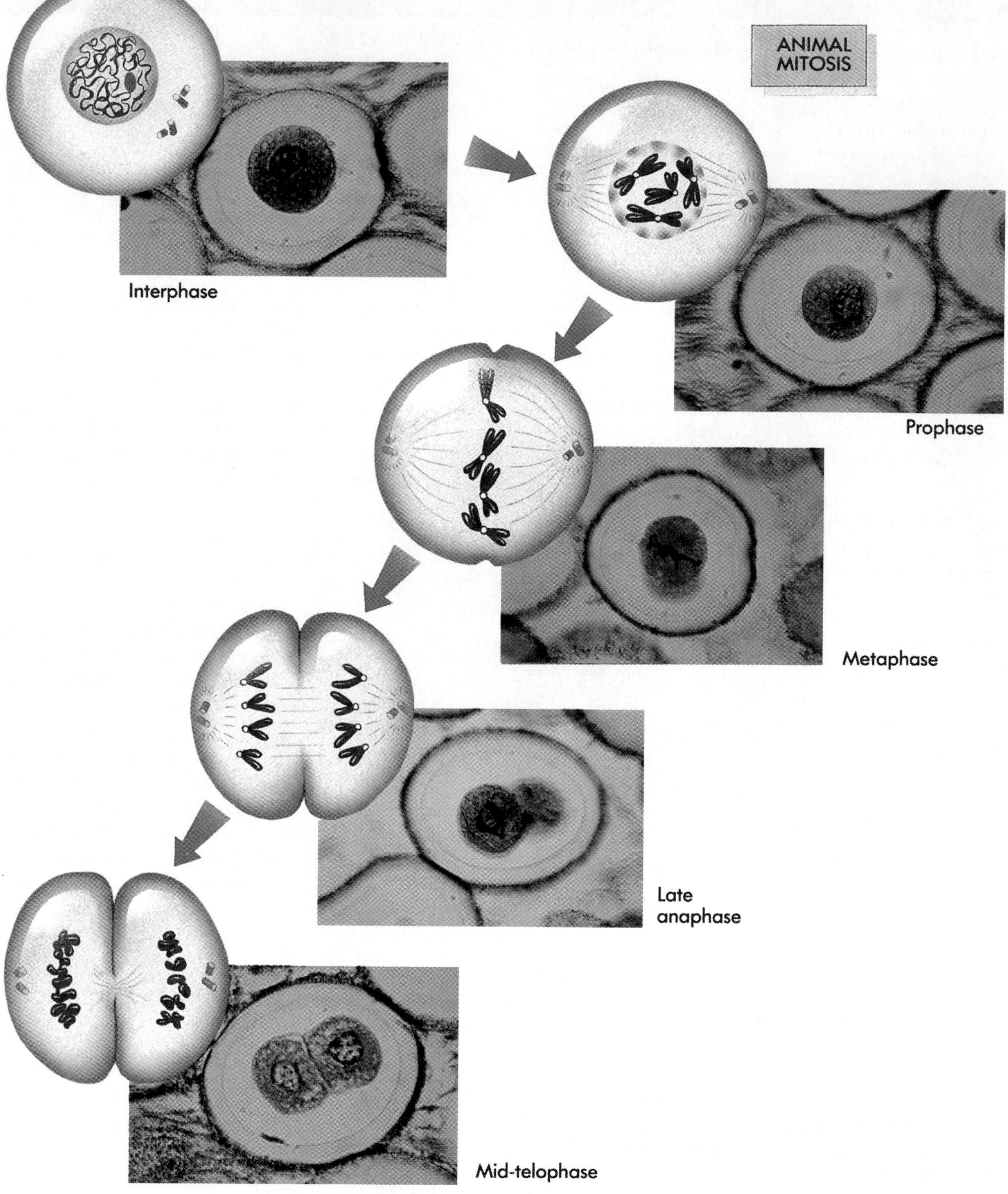

FIGURE 18-23 The stages of mitosis in an animal cell. The stages of mitosis are shown here in photographs and drawings of a dividing *Ascaris* (roundworm) cell. Although they are barely noticeable in the photographs, centrioles play a key role in animal cell mitosis. Although interphase is shown here to complete the cell cycle, it is not a stage of mitosis.

Prophase

The first stage of mitosis, **prophase,** begins when the chromosomes have condensed to the point where they become visible under a light microscope. As prophase continues, the chromosomes continue to shorten and thicken, looking much bulkier at the end of prophase than at its beginning. The nucleolus, which was previously conspicuous, disappears. This disappearance is due to the nucleolus being unable to make ribosomal RNA (rRNA) when the part of the chromosome bearing the rRNA genes is condensed. And it is rRNA that makes up most of the substance of the nucleolus (see Chapter 3).

Prophase is the stage of mitosis characterized by the appearance of visible chromosomes.

While the chromosomes are condensing, another series of equally important events is also occurring: special microtubules (thin, tubelike, protein structures [see Chapter 3]) called the *spindle fibers* are being assembled. In animal cells, these spindle fibers extend from a pair of related microtubular structures called *centrioles* (see Figure 18-23). Although the centrioles were once thought to play a role in forming the spindle, recent evidence suggests that this is not the case. Instead, the microtubules of the spindle appear to form from granules surrounding the centrioles called the *microtubule-organizing center (MTOC)*. Interestingly, plant cells do not contain centrioles, but spindle fibers form in plant cells.

In early prophase the centrioles of animal cells begin to move away from one another. By the end of prophase, each member of the pair has moved to an opposite end, or pole, of the cell. As the spindle fibers form, the nuclear envelope breaks down, forming small vesicles that disperse in the cytosol. Without the nucleus in the way, the spindle fibers form a bridge between the centrioles, spanning the distance from one pole to the other. Some spindle fibers extending from each pole attach to their side of the centromere of each sister chromatid. By this time, each chromatid has developed a *kinetochore* at its centromere. A kinetochore is the structure to which several pole-to-centromere microtubules will attach. The effect is to attach one sister chromatid to one pole and the other sister chromatid to the other pole. During later stages of mitosis the sister chromatids separate and move to opposite poles of the cell, pulled by the spindle fibers. The proper attachment of each spindle fiber to the kinetochore is therefore critical to the process of mitosis; a mistake is disastrous. The attachment of two sister chromatids to the same pole, for example, results in their not separating, so they end up in the same daughter cell. Such a situation can result in a cell with an extra chromosome. Depending on the type of cell, it may be abnormal or it may die.

By the end of prophase, spindle fibers that radiate from opposite poles of the cell attach to each kinetochore at the centromere.

When the centrioles reach the poles of the cell in animal mitosis, they radiate an array of microtubules outward (toward the cell membrane) as well as inward (toward the chromosomes). This arrangement of microtubules is called an *aster.* The function of the aster is not well understood. Evidence suggests that the aster probably acts as a support during the movement of the sister chromatids.

Metaphase

The second phase of mitosis, **metaphase,** begins when the chromatid pairs line up in the center of the cell. In reality, the chromosomes are not in a line but form a circle. They only appear to form a line when viewed two dimensionally with a light microscope as in Figure 18-23. Figure 18-24 is a diagram of a three-dimensional view of an animal cell at metaphase. As you can see, the chromosomes form a circle perpendicular to the direction of the spindle fibers. Positioned by the microtubules attached at their centromeres, all the chromosomes are equidistantly arranged between the two poles at the "equator" of the cell. The region of this circular arrangement, called the *metaphase plate,* is not a physical structure but indicates approximately where the future axis of cell division will be.

Metaphase is the stage of mitosis characterized by the alignment of the chromosomes, equidistant from the two poles of the cell.

prophase (pro faze) the first stage of mitosis. Prophase begins when the chromosomes have condensed; they become visible under a light microscope.

metaphase (met uh faze) the second phase of mitosis. Metaphase begins when the chromosomes align themselves equidistantly from the two poles of the cell.

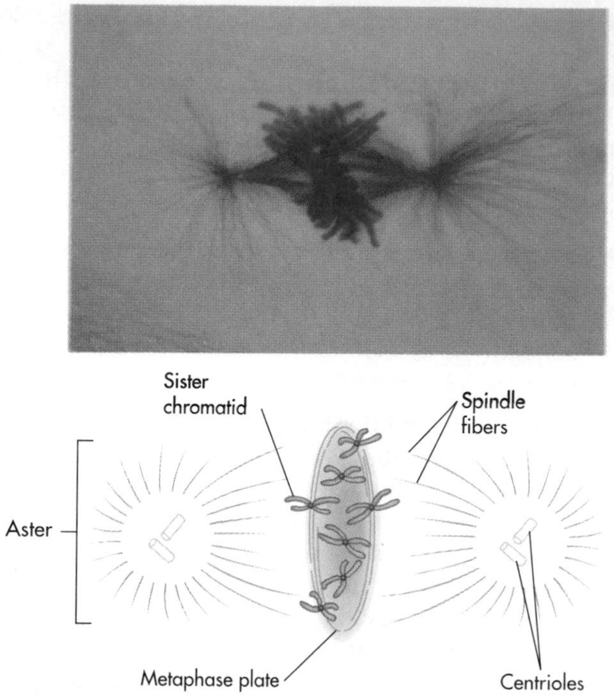

FIGURE 18-24 The metaphase plate. In metaphase, the chromosomes array themselves around the spindle midpoint.

Anaphase

At the beginning of **anaphase** the sister chromatids separate at the centromere, freeing them from their attachment to each other. Before this split, the chromatids are tugged in two directions at once by opposing microtubules—somewhat like a cellular tug-of-war. With the separation of the chromatids (now called chromosomes), they move rapidly toward opposite poles of the cell, each pulled at its kinetochore by attached, shortening microtubules.

> *Anaphase is the stage of mitosis characterized by the physical separation of sister chromatids and their movement to opposite poles of the cell.*

Telophase

The separation of the sister chromatids in anaphase equally divides the hereditary material that replicated just before mitosis. Therefore each of the two new cells that are forming receive a complete, identical copy of the chromosomes. This partitioning of the genetic material is the essence of the process of mitosis.

The events of this last phase of mitosis, **telophase,** ready the cell for **cytokinesis,** or division of the cytoplasm. The spindle fibers are chemically disassembled and therefore disappear. The nuclear envelope re-forms around each set of what were sister chromatids, now chromosomes. These chromosomes begin to uncoil, returning to their normally strung out, more diffuse state. The DNA of the nucleolus begins making ribosomal RNA once again; this rRNA is visible, resulting in the reappearance of the nucleolus. In summary, the events of telophase are very much like the reverse order of the events of prophase.

> *Telophase is the stage of mitosis during which the mitotic apparatus assembled during prophase is disassembled, the nuclear envelope is reestablished, and the normal use of the genes present in the chromosomes is reinitiated.*

Cytokinesis

At the end of telophase, mitosis is complete. The eukaryotic cell has divided its duplicated hereditary material into two nuclei that are positioned at opposite ends of the cell. While this process has been going on, the cytoplasmic organelles, such as the mitochondria, have been somewhat equally distributed to each side of the cytoplasm. These organelles replicate throughout interphase.

At this point the process of cell division is still not complete. The division of the cytoplasm—that portion of the cell outside the nucleus—has not yet begun. The stage of the cell cycle at which cell division actually occurs is called *cytokinesis.* Cytokinesis generally involves the division of the cell into approximately equal halves.

> *Cytokinesis is the physical division of the cytoplasm of a eukaryotic cell into two daughter cells.*

Cytokinesis in Animal Cells

In human cells and in the cells of other eukaryotes that lack cell walls, cytokinesis occurs by a pinching of the cell in two. This pinching is accomplished by a belt of microfilaments that encircles the cell at the metaphase plate. These microfilaments contract, forming a *cleavage furrow* around the circumference of the cell (Figure 18-25). As contraction proceeds, the furrow deepens until the opposing edges of the membrane make contact with one another. Then the membranes fuse, separating the one cell into new cells.

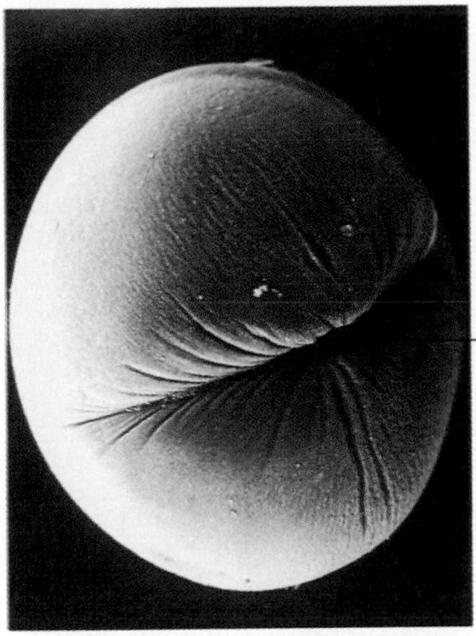

FIGURE 18-25 Cytokinesis in an animal cell. A cleavage furrow is forming around this dividing sea urchin egg.

Cytokinesis in Plant Cells

In plants and some algae, a rigid wall surrounds the cell. Therefore cytokinesis involves the laying down of a new cell wall between the two daughter cells rather than a pinching in of the cytoplasm. Plants manufacture new sections of membrane and wall from tiny vesicles most likely derived from the Golgi complex. These membrane "parts" accumulate at the metaphase plate and fuse, beginning the formation of a partition called the *cell plate.* A new cell wall forms between the two membranes of the cell plate. Figure 18-26 shows cytokinesis in a plant cell and the formation of a cell plate.

Meiosis

Most animals, plants, algae, fungi, and certain protists reproduce sexually. In sexual reproduction, gametes of opposite sexes or mating types (sometimes just termed + *cells* and − *cells*) unite in the process of fertilization, producing the first cells of new individuals (see Figure 18-19).

Humans have 46 chromosomes in all of their body cells—the diploid amount. If the two cells that joined in fertilization each contained 46 chromosomes, however, the first cell of the future offspring would have 92 chromosomes. An individual born after 10 generations would have more than 47,000 chromosomes! Such a continuing addition of chromosomes to each new individual is obviously an unworkable situation. Even early investigators realized that there must be some mechanism during the course of gamete formation to reduce the number of chromosomes. They reasoned that if sex cells were formed with half the number of chromosomes characteristic of the cells of that species, then the fusion of these cells during fertilization would produce cells of new individuals with the proper number of chromosomes. Investigators soon observed this special type of cell division and named it *meiosis,* from a Greek word meaning "less."

During meiosis the diploid number of chromosomes is reduced by half, forming haploid cells. For this reason, meiosis is often called *reduction-division.* Four daughter cells result from two divisions of one original parent cell, although in human females, only one of the four cells is functional (see Chapter 21). In animals, meiosis occurs in the cells that produce gametes.

> *Meiosis is a process of nuclear division in which the number of chromosomes in cells is halved, forming gametes in most animals and spores in plants.*

anaphase (**ann** uh **faze**) the third phase of mitosis. During anaphase, sister chromatids separate and move to opposite poles of the cell.

telophase (**tel** uh **faze**) the last phase of mitosis. During telophase, the mitotic apparatus assembled during prophase is disassembled, the nuclear envelope is reestablished, and the normal use of the genes present in the chromosomes is reinitiated.

cytokinesis (**sye** toe kuh **nee** sis) the physical division of the cytoplasm of a cell into two daughter cells.

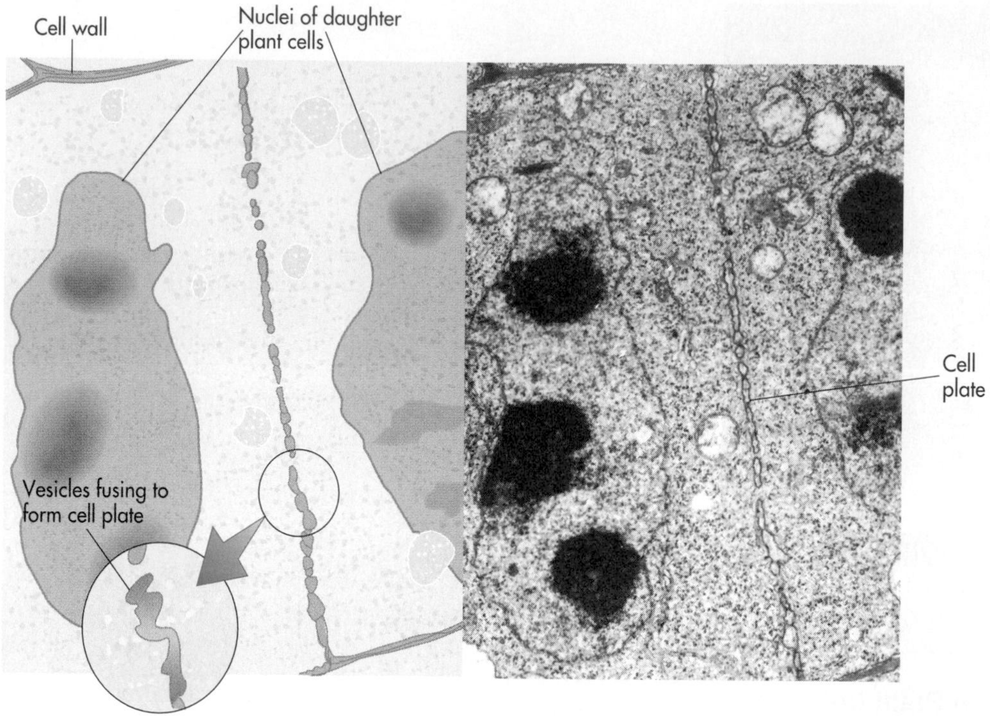

FIGURE 18-26 Cytokinesis in plant cells. In this photograph and companion drawing, a cell plate is forming between daughter nuclei. The cell plate is composed of pieces of cell membrane (vesicles) that fuse together.

The Stages of Meiosis

Although meiosis is a continuous process, scientists divide it into stages as they do the process of mitosis. The two forms of cell division have much in common. They are both special forms of nuclear division (although these processes are often referred to as forms of cell division for convenience). Meiosis consists of two sets of divisions called **meiosis I and meiosis II.** Each set is divided into prophase, metaphase, anaphase, and telophase, just as in mitosis. In meiosis, however, prophase I is much more complicated.

Meiosis is preceded by an interphase that is similar to the interphase of mitosis. During interphase the chromosomes are replicated, resulting in each chromosome consisting of two genetically identical sister chromatids held together at the constricted centromere. The centrioles also replicate.

Meiosis I

In prophase I, the individual chromosomes condense as their DNA coils more tightly, thus becoming visible under a light microscope. Because this DNA replicated before meiosis, each chromosome consists of two sister chromatids joined at the centromere.

Chromosomes, as you can see in the chapter opener photo, come in pairs. Each pair is formed during fertilization as an egg and sperm fuse. Each of these gametes contributes one chromosome to each pair of chromosomes. Thus, you received half your chromosomes (one set of genetic information) from your father and half (a second set of genetic information) from your mother. Likewise, all organisms produced by sexual reproduction receive half their chromosomes from one parent organism and half from the other. These pairs of chromosomes are called **homologous chromosomes,** or **homologues** (Figure 18-27).

Each homologous chromosome contains genes that code for the same inherited traits, such as eye color and hair color. During prophase I, homologous chromosomes line up side by side in a process called **synapsis.** Because each chromosome is made up of two sister chromatids, the paired homologous chromosomes together have four chromatids. These synapsed homologues are therefore called *bivalents* (meaning "too strong"), or *tetrads* (meaning "groups of four").

The process of synapsis begins a complex series of events called **crossing over.** During crossing over, homologous *nonsister* chromatids actually cross over one another. These crossed-over pieces break

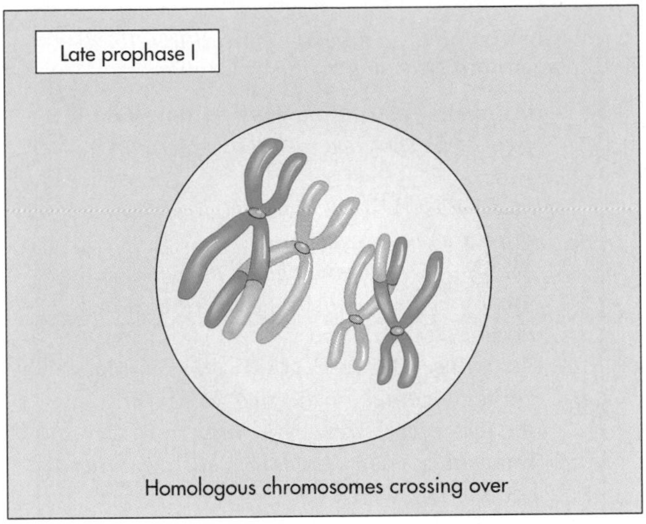

FIGURE 18-27 Late prophase in meiosis I. Chromosomes are replicated during interphase. Each chromosome is paired (homologous chromosomes). During prophase I, they line up side by side (called synapsis) and exchange genetic material (called crossing over), which is shown in this illustration.

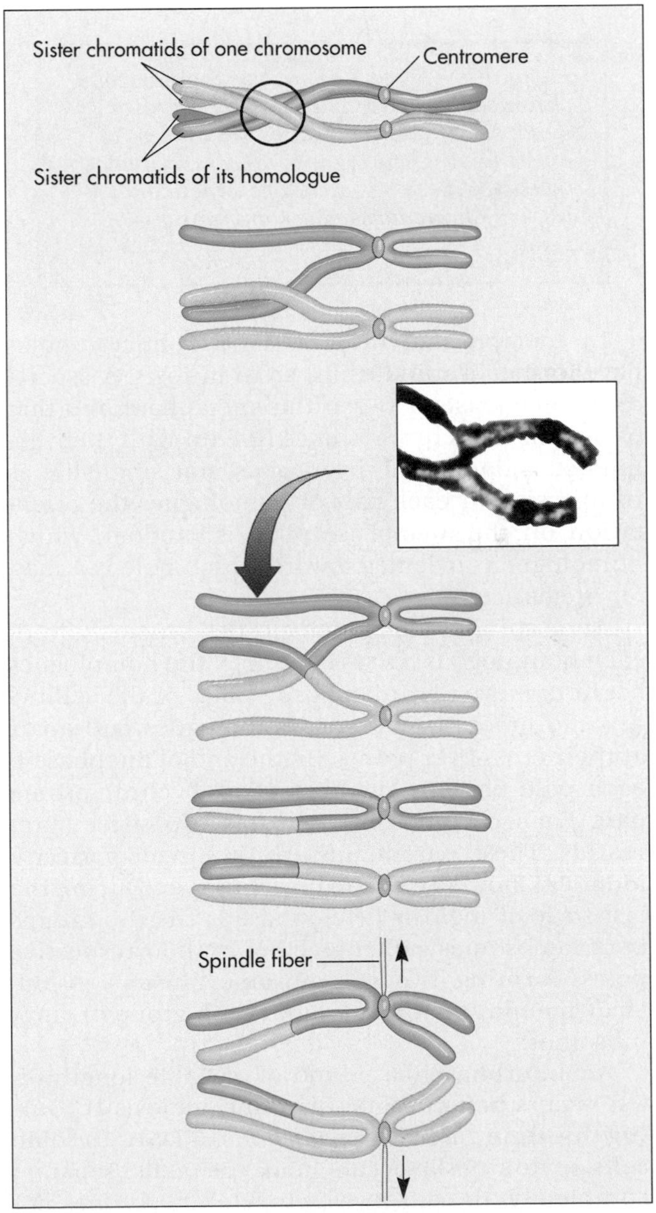

FIGURE 18-28 Crossing over. The circle highlighting the upper chromosome indicates one region where homologous chromosomes are being held together tightly. Crossing over occurs when the paired chromosomes exchange genetic material at such locations. Once crossing over is complete, chromosomes move apart slightly, forming an X-shaped structure called a chiasmata (photograph). Crossing over produces new combinations of genes.

away from the chromatids to which they are attached and reattach to the nonsister chromatid—literally exchanging parts (Figure 18-28). Once crossing over is complete, the nuclear envelope dissolves and the homologues begin to move apart. These four chromatids cannot separate from one another, however, because (1) the sister chromatids are held together at their centromeres and (2) the paired homologues are held together at the points where crossing over occurred. As the chromosomes move apart somewhat, the points of crossing over can be seen (under a light microscope) as X-shaped structures called **chiasmata** (see Figure 18-28). Crossing over is a significant event in meiosis because it produces new combinations of genes. This process provides one way in which offspring have a genetic makeup different from either parent.

> *Early in prophase I of meiosis, homologous chromosomes pair up in a process called synapsis. Synapsis initiates the process of crossing over. During this process, homologous chromosomes exchange genetic material, which produces new combinations of genes.*

In metaphase I of meiosis, the microtubules have formed a spindle, just as in mitosis. A crucial difference exists between this metaphase and that of mitosis: the chromosomes line up with their homologues double file in meiosis, not single-file as in mitosis. For each pair of homologues the orientation on the metaphase plate is random; which homologue is oriented toward which pole is a matter of chance.

After spindle attachment to the kinetochore of each homologous pair is complete, the homologues begin to move toward opposite poles of the cell. As this movement occurs, the homologues pull apart at their cross-over points. By the end of anaphase I, each pole has one member of each chromosome pair. Each chromosome still has two sister chromatids. These chromatids are no longer perfectly identical, however, due to crossing over. During the last stage of meiosis I (telophase I) the two groups of chromosomes gather together at their respective poles, forming two chromosome clusters. A nuclear membrane forms around each group of chromosomes.

An interphaselike period of variable length often occurs between meiosis I and meiosis II. During this time, *there is no replication of DNA*. In some cells, cytokinesis occurs and the cells separate completely. In cells in which cytokinesis does not occur, telophase I simply merges with prophase II.

> *The first meiotic division is traditionally divided into four stages plus interphase:*
>
> 1. *Prophase I. Homologous chromosomes pair and exchange pieces of genetic material.*
> 2. *Metaphase I. Homologous chromosomes align on a central plane.*
> 3. *Anaphase I. Homologous chromosomes move toward opposite poles. Chromatids do not separate.*
> 4. *Telophase I. Individual chromosomes gather together at the two poles and the nuclear membranes re-form.*
> 5. *Interphase. The haploid cells separate completely.*

Meiosis II

At the end of anaphase I, each pole of the original cell has a haploid complement of chromosomes. That is, each pole has *half* the normal amount of chromosomes—only one member of each homologous pair. Each chromosome, however, still has two sister chromatids. Meiosis II separates these sister chromatids. Because of crossing over in the first phase of meiosis, these sister chromatids are *no longer identical* to one another.

The two haploid cells formed during meiosis I divide during meiosis II. The nuclear envelopes disappear, and the spindle fibers form. The chromosomes line up, their sister chromatids separate, and they move to the opposite poles of each cell. At this point the nucleoli reorganize and nuclear envelopes form around each set of chromosomes (Figure 18-29).

Meiosis II results in the production of four daughter cells, each with a haploid number of chromosomes. The cells that contain these haploid nuclei function as gametes for sexual reproduction in most animals and as spores in plants. A comparison of mitotic and meiotic cell division is shown in Figure 18-30.

> *During meiosis II, four haploid daughter cells are produced from the two haploid daughter cells of meiosis I. These two haploid cells were produced during meiosis I from one original diploid parent cell.*

The Importance of Meiotic Recombination

The reassortment of genetic material that occurs during meiosis generates variability in the hereditary material of the offspring. To understand why this is true, remember that most organisms have

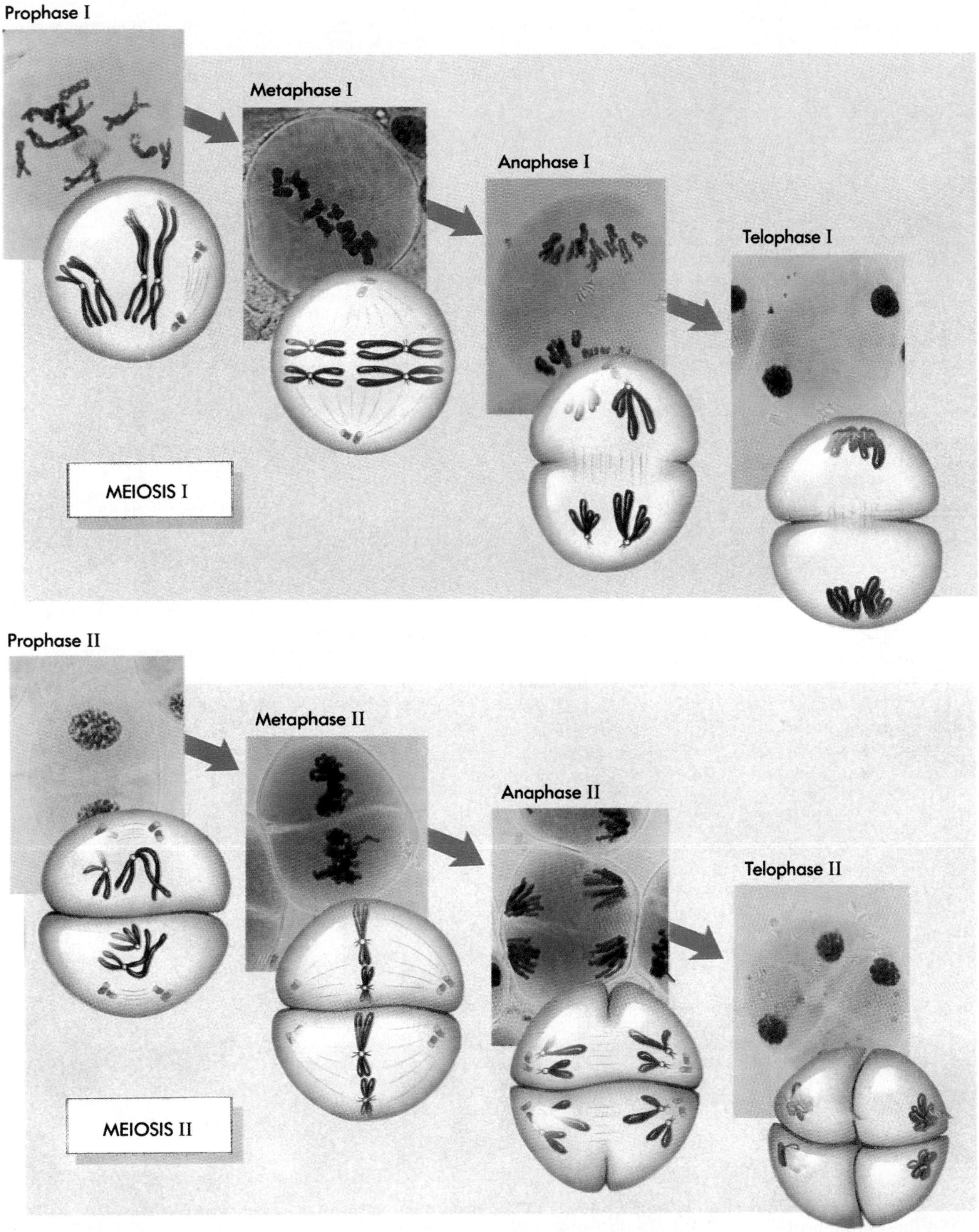

FIGURE 18-29 Meiosis. Meiosis ensures the reassortment of genetic material. It is preceded by interphase, during which the chromosomes are replicated. An interphaselike period of variable length often occurs between meiosis I and meiosis II. During this time there is no replication of chromosomes (DNA).

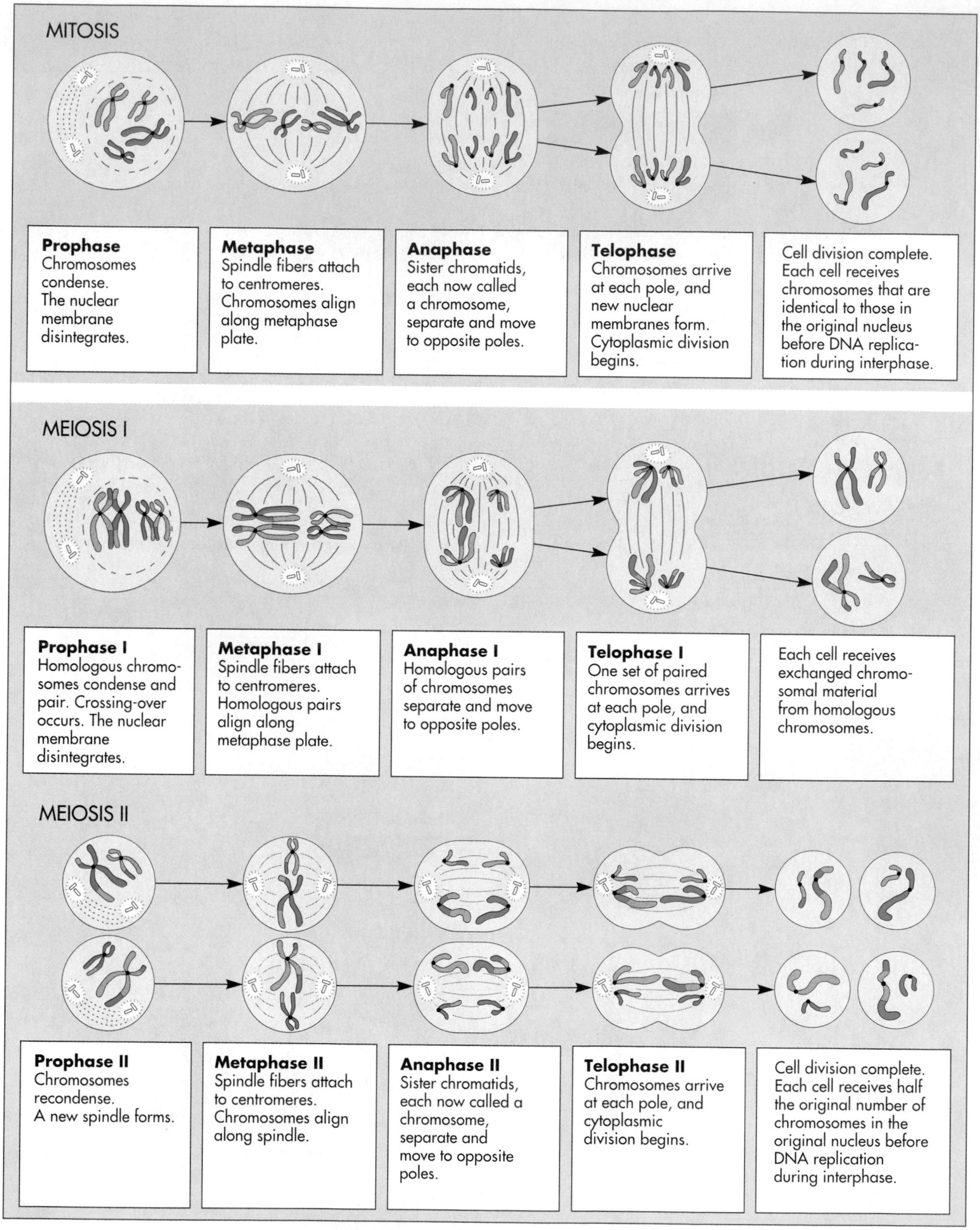
MITOSIS
Prophase
Chromosomes condense.
The nuclear membrane disintegrates.
Metaphase
Spindle fibers attach to centromeres.
Chromosomes align along metaphase plate.
Anaphase
Sister chromatids, each now called a chromosome, separate and move to opposite poles.
Telophase
Chromosomes arrive at each pole, and new nuclear membranes form.
Cytoplasmic division begins.
Cell division complete.
Each cell receives chromosomes that are identical to those in the original nucleus before DNA replication during interphase.
MEIOSIS I
Prophase I
Homologous chromosomes condense and pair. Crossing-over occurs. The nuclear membrane disintegrates.
Metaphase I
Spindle fibers attach to centromeres.
Homologous pairs align along metaphase plate.
Anaphase I
Homologous pairs of chromosomes separate and move to opposite poles.
Telophase I
One set of paired chromosomes arrives at each pole, and cytoplasmic division begins.
Each cell receives exchanged chromosomal material from homologous chromosomes.
MEIOSIS II
Prophase II
Chromosomes recondense.
A new spindle forms.
Metaphase II
Spindle fibers attach to centromeres.
Chromosomes align along spindle.
Anaphase II
Sister chromatids, each now called a chromosome, separate and move to opposite poles.
Telophase II
Chromosomes arrive at each pole, and cytoplasmic division begins.
Cell division complete.
Each cell receives half the original number of chromosomes in the original nucleus before DNA replication during interphase.

FIGURE 18-30 A comparison between mitosis and meiosis. Meiosis involves two nuclear divisions with no DNA replication between them. Meiosis therefore produces four daughter cells, each with half the original amount of DNA. Mitosis produces two identical daughter cells, each with the same number of chromosomes as the original mother cell.

more than one chromosome. Humans, for example, have 23 different pairs of homologous chromosomes, one of each pair from the father and one of each pair from the mother. Each human gamete receives one of the two copies of each of the 23 different chromosomes, but which copy of a particular chromosome it receives is random. For example, the copy of chromosome 14 that a particular human gamete receives has no influence on which copy of chromosome 5 that it will receive. Each of the 23 pairs of chromosomes goes through meiosis independently of all the others, so there are 2^{23} (more than 8 million) different possibilities for the kinds of gametes that can be produced, and no two of them are alike. In addition, crossing over adds even more variability to the random assortment of chromosomes. The subsequent union of two gametes thus creates a unique individual, a new combination of 23 chromosomes that probably has never occurred before and probably will never occur again. When you study evolution in Chapters 23 and 24, you will see that variability within species is essential to the process of evolution by natural selection.

meiosis I and meiosis II (my **oh** sis) the two-staged process of nuclear division in which the number of chromosomes in cells is halved during gamete formation. Both meiosis I and meiosis II can be further divided into four stages: prophase, metaphase, anaphase, and telophase.

homologous chromosomes (homologues) (hoe **mol** uh gus **kro** muh **somes**/**hom** uh logs) pairs of chromosomes that all organisms produced by sexual reproduction receive. Half of these chromosomes are from one parent organism and half from the other. Homologous chromosomes each contain genes that code for the same inherited trait.

synapsis (suh **nap** sis) a process during prophase I of meiosis in which homologous chromosomes line up side-by-side, initiating the process of crossing over.

crossing over a complex series of events during meiosis in which homologous nonsister chromatids cross over one another. These crossed-over pieces break away from the chromatids to which they are attached and reattach to the nonsister chromatid.

chiasmata (keye **az** muh tuh) in meiosis, the point of crossing over where parts of chromosomes have been exchanged during synapsis; under a light microscope, a chiasma appears as an X-shaped structure.

Summary

- The hereditary material is deoxyribonucleic acid, or DNA. In eukaryotes, DNA is combined with protein to form a complex called *chromatin.* DNA is made up of nucleic acids. Ribonucleic acid, or RNA, plays key roles in the translation of the hereditary message of DNA. It too is made up of nucleic acids.
- The nucleic acids of all organisms are made up of nucleotides. Each nucleotide contains three molecular parts: one phosphate group, one five-carbon sugar, and one nitrogen-containing base. The relative proportion of the bases found in DNA varies from species to species.
- DNA has the structure of a double helix, two molecular "chains" held to each other by hydrogen bonds between its nucleotides. Its four bases are adenine (A), thymine (T), cytosine (C), and guanine (G). The nucleotide A always bonds with T and G with C. These bases are complementary to one another.
- Before cell division, DNA replication is orchestrated by a battery of enzymes that "unzip" the DNA, cause it to unwind, and then use each of the single strands as a template to assemble a complementary new strand.
- The hereditary message of DNA is actually a code within its sequence of bases. A gene, or unit of heredity, is a sequence of bases that codes for a specific polypeptide, or portion of an enzyme. Most hereditary traits reflect the actions of enzymes.
- The expression of hereditary information in all organisms takes place in two stages. First, in the process of transcription, a portion of the DNA message is copied onto a single strand of mRNA by using the DNA as a template. Second, in the process of translation, an amino acid chain is assembled by a ribosome and tRNA, using the mRNA base sequence to direct the sequence of amino acids.
- The control of gene expression is exercised largely by regulating transcription. Cells have "on" and "off" gene transcription switches. Generally, in eukaryotic cells, control is positive; cellular signals are required before transcription can start.
- The packaging of DNA into chromosomes allows a tremendous amount of information to reside in each cell of a eukaryote and also allows this hereditary material to be separated in an organized way during cell division. Eukaryotic cells divide during the growth, repair, and reproduction of organisms.
- Mitosis is the process of nuclear division that distributes a complete, identical set of chromosomes to each of two daughter cells. This type of nuclear division is used in the growth and repair of multicellular organisms and in the asexual reproduction of single-celled eukaryotes and some simple animals. It produces cells that are identical to the parent cell.
- Meiosis is the process of nuclear division that distributes half the complement of chromosomes to each of four daughter cells. This type of nuclear division is used to produce gametes for sexual reproduction in humans.
- Crossing over is an essential element of meiosis. This process produces sister chromatids that are not identical to each other. Crossing over and the reassortment of genetic material that occurs during meiosis provides for new combinations of hereditary material during sexual reproduction.

Knowledge and Comprehension Questions

1. Explain how a unique characteristic of yours, such as your hair color, is a result of specific chemical instructions. Where do these instructions originate?
2. What are the two types of nucleic acids found in your cells? Describe the structures of each.
3. Explain the term *double helix,* and describe its structure.
4. What characteristics of DNA did Chargaff's experiments reveal? Why was this significant?
5. Distinguish among rRNA, tRNA, and mRNA. What does each abbreviation stand for, and what are the respective functions of each?
6. Is mitosis or meiosis responsible for the tremendous genotypic and phenotypic variation among humans? Explain.
7. Place these steps in the correct sequence and label each stage. What process is being described?
   a. The chromosomes line up in the center of the cell.
   b. The nuclear envelope forms and the chromosomes uncoil.
   c. The chromosomes grow shorter and thicker, and spindle fibers form.
   d. The sister chromatids separate and move to opposite poles of the cell.
8. Place these steps in the correct sequence and label each stage. What process is being described?
   a. Homologous chromosomes move toward opposite poles of the cell; chromatids do not separate.
   b. Chromosomes gather together at the two poles of the cell, and the nuclear membranes re-form.
   c. Homologous chromosomes pair and exchange segments.
   d. Homologous chromosomes align on a central plane.
   e. The haploid cells separate completely.
9. Compare the cells that result from mitosis with those that result from meiosis. How are they different?
10. Explain the importance of the genetic recombination that occurs during meiosis.
11. Compare the cells resulting from meiosis with their mother cell. What functions do these cells serve?

Critical Thinking

1. What do you hypothesize might occur if there is an error made in base arrangement during DNA transcription? Suggest a possible sequence of subsequent events.
2. There is often a tremendous amount of repetition in the genetic code for enzyme production. Why do you suppose such a backup system is necessary to an organism?

CHAPTER 19

PATTERNS OF INHERITANCE

THE SAME YET DIFFERENT—how often have you heard that phrase? And what does it really mean? Scientists would answer the last question with a single word: variation.

Variation can be seen in the group of flamingos in the photograph. Although the flamingos are all the same—with long legs and beaks and pink feathers covering their bodies—they are all different. Look closely at the same characteristic in each. One has longer legs than the others. Another has a slightly different shade of pink. What other differences do you see?

Clearly, each species of living things exhibits variation. Humans, for example, all have characteristics you recognize as human, but each person (except for identical twins) looks different from all others. What is the source of this variation? And how are these differences distributed among populations of living things.

Inheritance and Variation Within Species

Today it is common knowledge that organisms inherit characteristics from their parents. During sexual reproduction, parents pass on traits to their offspring by means of genetic material within the eggs of the mother and sperm of the father. The intermingling of parental genes that takes place at fertilization, the union of the egg and sperm, is the material of variation. Organisms produced by asexual reproduction exhibit less variation for just this reason. Their genetic makeup is derived from one parent only, so offspring are genetically identical to that parent, with the exception of mutations. For example, a plant produced asexually by rooting a cutting will have the same genes as the "mother" plant, but a plant produced by cross-pollination—a type of sexual reproduction—will have certain characteristics of both its parents.

> *Sexual reproduction introduces variation within a species because offspring inherit characteristics from both their parents. Asexual reproduction does not introduce variation because the offspring is an exact duplicate of its parent.*

Historical Views of Inheritance

Although genetic inheritance seems obvious today, this fact was not always obvious to scientists and philosophers. Hippocrates (460-377 BC) believed that a child inherited traits from "particles" given off by all parts of the bodies of the father and mother. These particles, he suggested, travel to the sex organs. During intercourse, the father's particles merge with the mother's particles to form the child. This idea of inheritance was held by many until the mid-1800s. Most theories of the direct transmission of hereditary material assumed that the male and female traits blended in the offspring. Thus a parent with red hair and a parent with brown hair would be expected to produce children with reddish-brown hair, and a tall parent and a short parent would produce children of intermediate height. However, taken to its logical conclusion, this theory suggests that all individuals within a species would eventually look like one another as their traits continually blended together.

KEY CONCEPTS

- Over the centuries, the nature of heredity was misunderstood by scientists and philosophers until Mendel, Sutton, Morgan, and others in the late 1800s began to uncover its mechanisms.
- Working with pea plants that exhibited alternative forms of various characteristics (traits), Mendel established that traits are inherited as discrete packets of information, which scientists today call genes.
- Through his experiments with pea plants, Mendel also discovered that alternative forms of genes (alleles) separate from one another during the formation of gametes and are distributed independently of other traits.
- Mendel and other early geneticists took the first real steps toward solving the puzzles of inheritance and laid the foundation for one of the great revolutions in thinking of this century: an understanding of the nature of genetic material and how it is transmitted from generation to generation.

OUTLINE

Other ideas regarding inheritance were formulated after the invention of a simple, handheld microscope. Anton van Leeuwenhoek (1632-1723) observed sperm for the first time with the microscope, drew pictures of them, and developed hypotheses regarding inheritance based on his observations. A widely held notion at the time was that each sperm contained a tiny but whole human (Figure 19–1). *Preformationists* (as this group was called) were separated into two camps. The spermists thought that sperm encased microscopic humans. Because the human was fully formed, the body contained sperm that each encased another preformed individual, and so forth . . . ad infinitum! Ovists, on the other hand, held that it was the eggs of the mother that contained minute humans.

Not until the mid-1800s with the work of Schleiden, Schwann, and Virchow (see Chapter 3) did scientists realize that new life arose from old life in the form of new cells arising from old cells. By means of the growth and division of these cells, new organisms developed from individual cells of parent organisms. Scientists of the late 1800s studied the nuclei of cells to uncover the mysteries of cell growth and division. They observed the complex process of mitosis (see Chapter 18) and wondered why cells went through this intricate process. Would it not be more efficient for a cell to simply pinch in two along its middle? By 1883, scientists knew that the complex process of cell division ensured an equal distribution of the nuclear material, to two daughter cells. And not only did each daughter cell receive the same amount of nuclear material; each received a complete amount of the nuclear material.

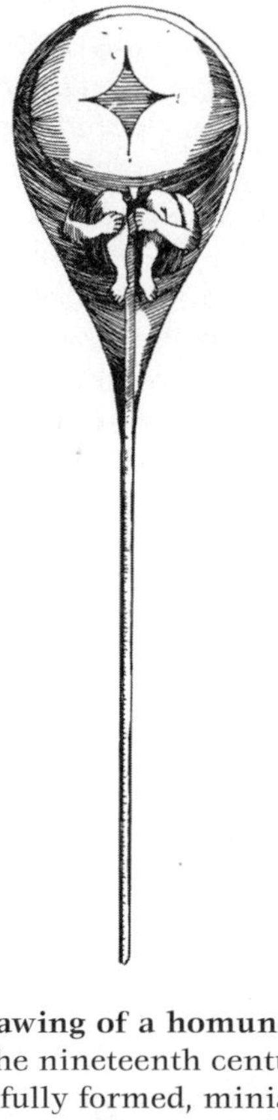

FIGURE 19-1 Drawing of a homunculus. A widely held belief before the nineteenth century was that sperm contained a fully formed, miniature human called a homunculus. It was supposedly implanted in the uterus during fertilization, where it grew to maturity.

At this same time, scientists also observed that an even more complex series of nuclear events (now known as *meiosis* [see Chapter 18]) preceded the formation of eggs and sperm. By 1885, several scientists independently concluded that this nuclear material was the physical bond that linked generations of organisms. But how nuclear material regulated the development of fertilized eggs and how it was related to heredity and variation were still mysteries. About 1900, scientists began to answer this question by piecing together research of the day with research that had long been ignored: the work of Gregor Mendel. Mendel was an Austrian monk trained in botany and mathematics at the University of Vienna.

The Birth of the Study of Inheritance

Approximately 25 years before scientists had discovered a link between heredity and the complex processes of mitosis and meiosis, Mendel began his work with the garden pea. Mendel chose the pea plant because it was an annual plant that was small and easy to grow and had a short generation time. Therefore he could conduct experiments involving numerous plants and obtain results relatively quickly.

Pea plants are well suited to studies of inheritance because each pea flower contains both female parts (stigma, style, and ovary) and male parts (filaments that support anthers). Both are enclosed and protected by the petals (Figure 19-2). The gametes produced within each flower—pollen grains within the anthers and eggs within the ovary—are able to fuse and develop into seeds of new plants. Fertilization of this sort, called *self-fertilization,* takes place naturally within individual pea flowers if they are not disturbed. As a result, the offspring of self-fertilizing garden peas are derived from one pea plant, not two. After generations of self-fertilization, some plants produce offspring consistently identical to the parent with respect to certain defined characteristics; these plants are said to be **true-breeding.**

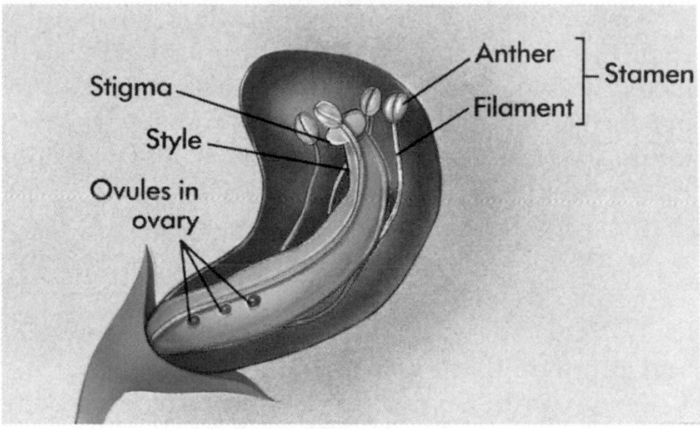

FIGURE 19-2 Anatomy of a pea plant. In the flower of a pea plant, the petals enclose the male (anther) and female (stigma) parts, ensuring that self-fertilization will take place.

Mendel selected a number of different true-breeding *varieties* of pea plants with which to work. Differing varieties, or strains, of an organism each belong to the same species, but each has one or more distinctive characteristics that are passed from parent to offspring. To study the inheritance patterns of these characteristics, Mendel took true-breeding plants and artificially cross-fertilized them. *Cross-fertilization* occurs when the pollen of one plant fertilizes the egg cells of another plant. To do this, Mendel removed the anthers of a flower before they shed pollen and then dusted this flower with pollen from another plant. In this way, Mendel was able to perform experimental crosses between two different true-breeding varieties of pea plants that exhibited differences regarding particular traits. The offspring, or progeny, of the cross between two different varieties of plants of the same species are called *hybrids.* (In other contexts, the word *hybrid* may also refer to the cross between two different species of organisms. A mule, for example, is the hybrid offspring of a horse and a donkey.)

Gregor Mendel's Experiments to Determine Inheritance Patterns

Mendel first designed a set of experiments that involved crossing varieties of pea plants differing from one another in a single characteristic. Mendel chose seven different characteristics, or **traits,** to study (Figure 19-3). Although various other varieties of pea plants exist, Mendel chose only those varieties that differed from one another clearly and distinctly.

The purpose of Mendel's experiments was to observe the offspring from the crossing of each pair of plants and to look for patterns in the transmission of single traits. Although he was not the first to perform such experiments, he was the first to count and classify the peas that resulted from his crosses and compare the proportions with mathematical models. Mendel planned to use these data to try to deduce the laws by which these traits are passed from generation to generation. First, Mendel crossed plants having contrasting forms of the single traits listed in Figure 19-3 by artificially fertilizing one with the other. Mendel called these plants the **parental (P) generation** and called their hybrid offspring the **first filial (F_1) generation.** (The word *filial* is from Latin words meaning "son" and "daughter.") These progeny are called *monohybrids* because they are the product of two plants that differ from one another in a single trait.

When Mendel crossed two contrasting varieties, such as purple-flowered plants with white-flowered plants, the hybrid offspring that he obtained were not intermediate in flower color, as the theory of blending inheritance predicted. Instead, all hybrid offspring in each case resembled *only one* of their parents. Thus in a cross of white-flowered plants with purple-flowered plants, *all* the F_1 offspring had purple flowers. In a cross of tall plants and short plants, *all* the F_1 offspring were tall plants. Mendel referred to the form of a trait that was expressed in the F_1 plants as *dominating,* or **dominant,** and to the alternative form, which was not expressed in the F_1 plants, as **recessive.** He chose the word *reces-*

true-breeding a term referring to organisms that produce offspring consistently identical to the parent with respect to certain defined characteristics after generations of self-fertilization.

traits distinguishing features or characteristics.

parental (P) generation the members of a cross between pure-breeding organisms, giving rise to offspring called the F_1.

first filial (F_1) generation (fil ee uhl) the hybrid offspring of the parental (P) generation.

dominant in an organism carrying a pair of contrasting alleles for a particular trait, the form of the trait (the allele) that will be expressed.

recessive in an organism carrying a pair of contrasting alleles for a particular trait, the form of the trait (the allele) that recedes or disappears entirely.

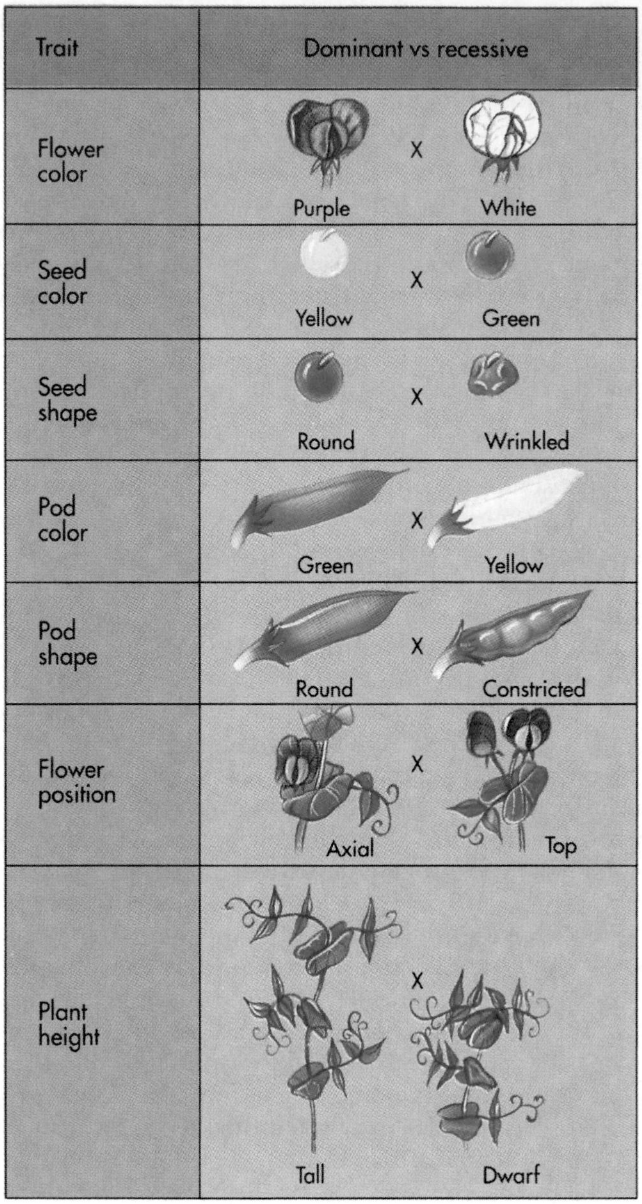
Trait
Dominant vs recessive
Flower color
Purple
X
White
Seed color
Yellow
X
Green
Seed shape
Round
X
Wrinkled
Pod color
Green
X
Yellow
Pod shape
Round
X
Constricted
Flower position
Axial
X
Top
Plant height
Tall
X
Dwarf

FIGURE 19-3 The seven pairs of contrasting traits in the garden pea studied by Mendel. To determine inheritance patterns, Mendel crossed plants with contrasting forms of each of seven traits. The parent plants were the P (parental) generation, and their offspring were the F_1 (first filial) generation. Among the progeny of these crossing experiments, Mendel found that some traits dominated other traits.

sive because this form of the trait receded, or disappeared entirely, in the hybrids. For each of the seven contrasting pairs of traits that Mendel examined, one member of each pair was dominant; the other was recessive. Figure 19-3 shows which traits Mendel found to be dominant and which traits he found to be recessive.

Then Mendel went a step further. He allowed each F_1 plant to mature and self-pollinate. He collected and planted seeds from each plant, which produced the **second filial (F_2) generation.** He found that most of the F_2 plants exhibited the dominating form of the trait and looked like the F_1 plants, but some exhibited the recessive form. None of the plants had blended characteristics. Therefore Mendel was able to count the numbers of each of the two contrasting varieties of F_2 progeny and compare these results.

Earlier in history, scientists had carried out hybridization experiments. However, the plants these scientists chose often produced hybrids that differed in appearance from both their parents because they exhibited blended traits. Also, scientists had not quantified the results of their experiments. Mendel's change in experimental design—counting the progeny—was a key component of his ability to unravel the mystery of certain inheritance patterns.

When Mendel quantified the traits he observed in the F_2 generation, he discovered that for every three plants exhibiting the dominant form of the trait, one exhibited the recessive trait. Put another way, both contrasting forms of the parental characteristics reappeared in the F_2 generation in an approximate 3:1 ratio: three fourths of the plants exhibited the dominant form (determined in the F_1 generation), and one fourth exhibited the recessive form. Figure 19-4, *A*, illustrates Mendel's experiments using the trait of flower color; Figure 19-4, *B*, lists the results Mendel obtained in the F_2 generation for each of the contrasting forms of the seven traits. Notice that these numbers reflect a 3:1 ratio of dominant-to-recessive forms for each trait. Notice also that in his experiments, Mendel used hundreds and sometimes thousands of plants. Why do you think that such a procedure is part of a good experimental design?

Mendel went on to examine what happened when each F_2 plant was allowed to self-pollinate. He found that the recessive one fourth were always true-breeding. For example, self-fertilizing white-flowered F_2 plants reliably produced only white-flowered offspring. By contrast, only one third of the dominant purple-flowered F_2 individuals (one fourth of the total offspring) proved true-breeding, whereas two thirds were not true-breeding. This last class of plants produced dominant and recessive F_3 individuals in a ratio of 3:1. The ratio of individuals in the entire F_2 population was the following (Figure 19-5):

1 true-breeding dominant: **2** not true-breeding dominant:
1 true-breeding recessive

A

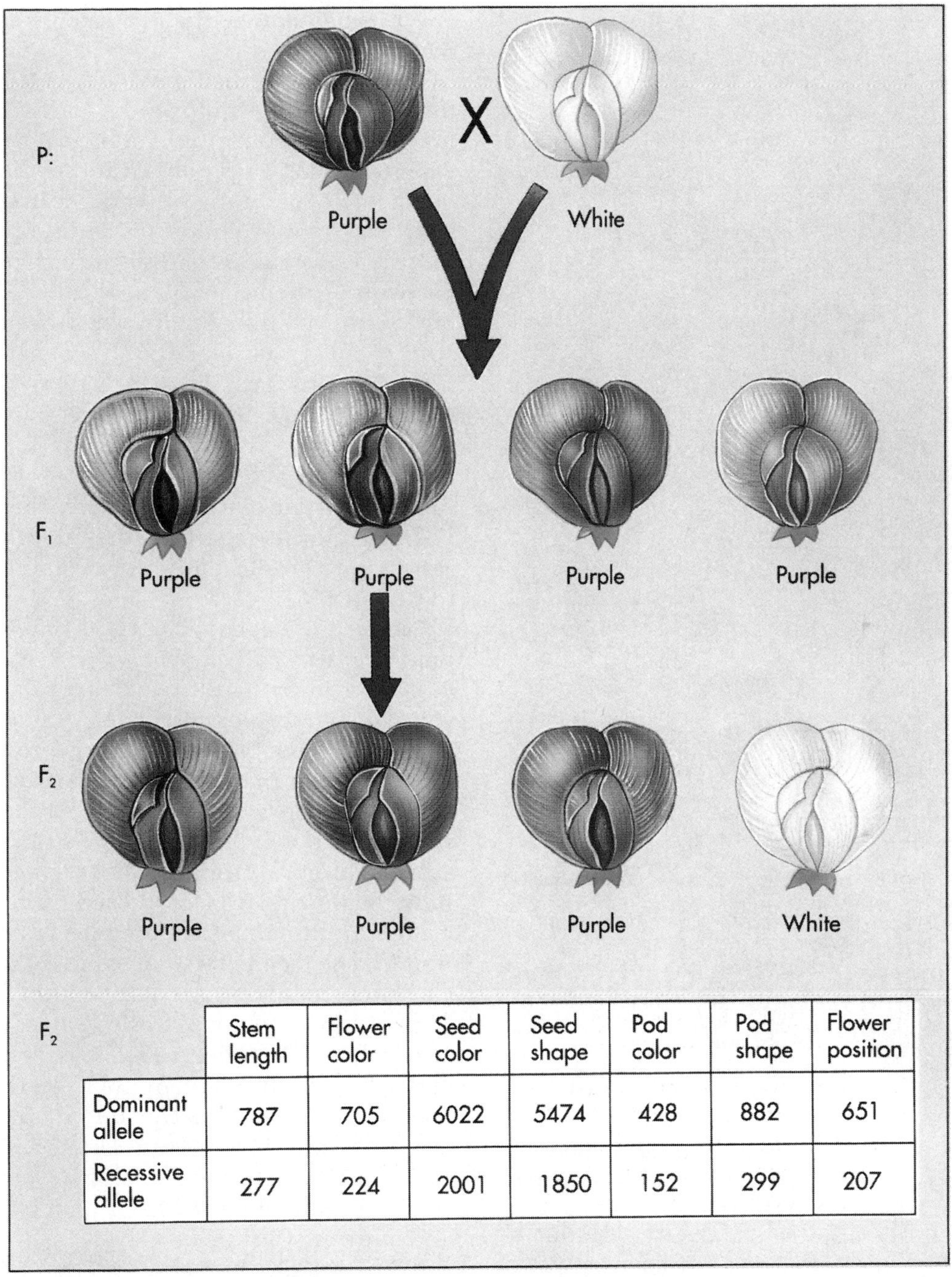

B

F2

	Stem length	Flower color	Seed color	Seed shape	Pod color	Pod shape	Flower position
Dominant allele	787	705	6022	5474	428	882	651
Recessive allele	277	224	2001	1850	152	299	207

FIGURE 19-4 Mendel's experiments refute the theory of blending inheritance. A, When crossing a purple flower with a white flower, Mendel observed that all the F_1 generation flowers were purple. When the F_1 plants were allowed to self-pollinate, three quarters of the F_2 generation exhibited the dominant trait (purple) and one-quarter the recessive trait (white). There were no intermediate or blended flowers. **B,** Mendel found the approximate 3:1 ratio of dominant to recessive forms to be exhibited in each of the seven traits he studied.

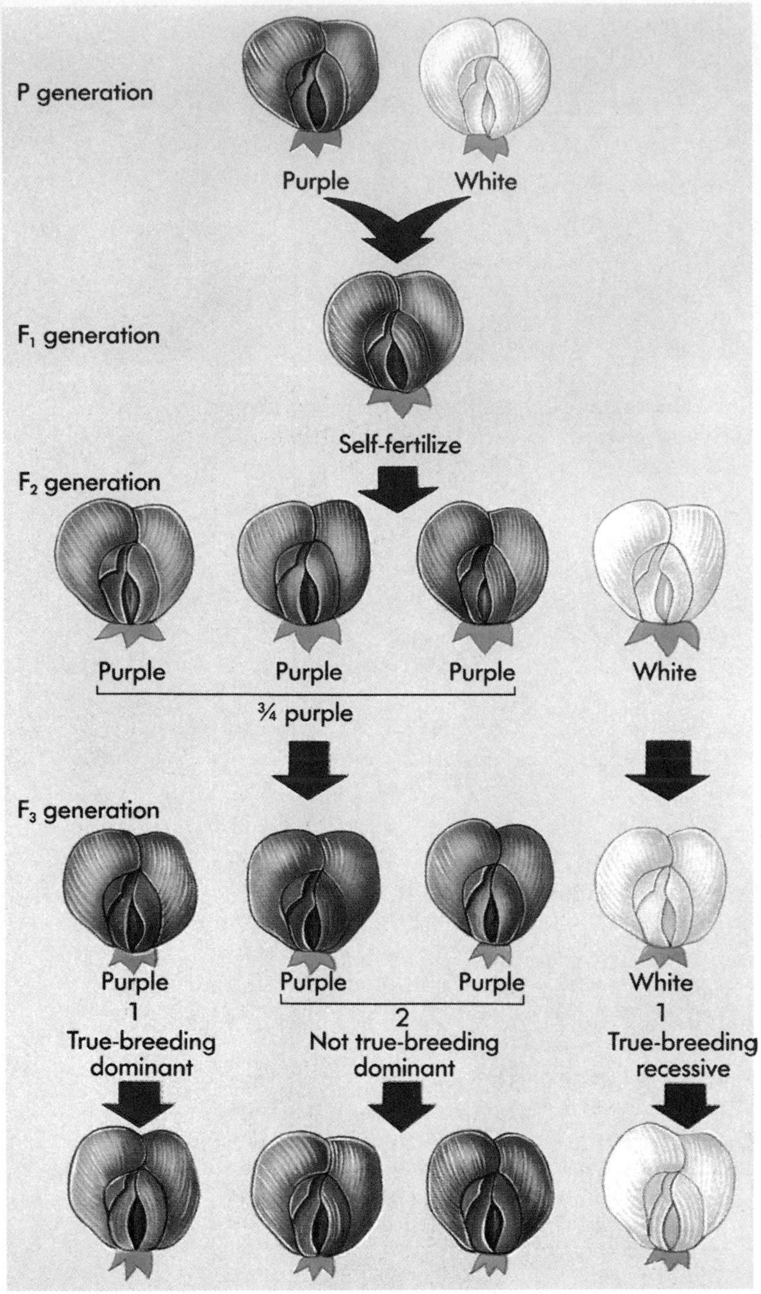

FIGURE 19-5 The F_3 generation allowed Mendel to determine which of the F_2 plants were true-breeding. By allowing the F_2 generation to self-pollinate, Mendel reasoned from the F_3 offspring that the F_2 generation exhibited the ratio of one true-breeding dominant, to two not-true-breeding dominant, to one true-breeding recessive.

Conclusions Mendel Drew from His Experiments

From his experimental data, Mendel drew conclusions regarding the nature of heredity—conclusions that have withstood tests of time and further experimentation. In fact, Mendel's work is historically looked upon as the birth of **genetics,** the branch of biology dealing with the principles of heredity and variation in organisms. The statements that follow regarding Mendel's conclusions are considered to be the first established principles of genetics.

First, as mentioned previously, Mendel observed that the plants exhibiting the traits he studied did not produce progeny of intermediate appearance when crossed. These observations did not uphold the theory of blending inheritance but suggested instead that traits are inherited as discrete "packets" of information that are either present or absent in a particular generation. Mendel called these discrete bits of information *factors.* These factors, hypothesized Mendel, act later in the offspring to produce the trait. Today, scientists call these factors **genes,** the units of transmission of hereditary characteristics in an organism. In addition, scientists know that a gene is a segment, or piece, of DNA occupying a particular place on a particular chromosome. (These concepts regarding the molecular nature of genes and chromosomes are explained more fully in Chapter 18.)

Second, for each pair of traits that Mendel examined, one alternative form was not expressed in the F_1 hybrids, although it reappeared in some F_2 individuals. As mentioned earlier, Mendel referred to the form of the trait that was expressed in the F_1 plants as "dominating"; today the preferred term is dominant. He referred to the alternative form, which was not expressed in the F_1 plants, as recessive. He inferred from these observations that each individual, with respect to each trait, contains two factors. Each pair of factors may contain information for (be a code for) the same form of a trait, or each member of the pair may code for an alternative form of a trait. Today, scientists call each member of a factor pair an **allele.** An allele is a particular form of a gene. Each human, for example, receives one allele for each gene from the mother's egg and one allele from the father's sperm. The 46 chromosomes of each human cell are actually 23 paired chromosomes, one member of each pair from the sperm and one from the egg.

Third, Mendel hypothesized that because (1) the two alleles that coded for a trait remained "uncontaminated," not blending with one another as the theory of blending inheritance predicted and (2) the results obtained from experiments on various traits showed similar results, then pea hybrids must form egg cells and pollen cells (gametes) in which the alleles for each trait separate from one another "in equal shares" during gamete formation. This concept is referred to as **Mendel's law of segrega-**

tion. Put in today's terms with today's understandings: each gamete receives only one of an organism's pair of alleles. Chance determines which member of a pair of alleles becomes included in a gamete. This random segregating process takes place during the process of meiosis, or reduction division of the nuclei of cells destined to be sex cells.

Fourth, Mendel realized that plants exhibiting the dominant trait in his monohybrid crosses of pea plants, when self-fertilized, would breed true or would produce plants exhibiting either the dominant or the recessive form of the characteristic in a 3:1 ratio, respectively. He observed that plants exhibiting the recessive trait, when self-fertilized, would always breed true. These data suggested to Mendel that true-breeding plants receive *only* the dominant factors or the recessive factors from each of their parents and that non–true-breeding plants were hybrids, which received the dominant and the recessive factors in equal shares. Today, scientists call an individual having two identical alleles for a trait **homozygous** for that trait. The prefix *homo* means "the same"; the suffix *zygous* refers to the zygote, or fertilized egg. An individual having two different alleles for a trait is said to be **heterozygous** (*hetero* means "different") for that trait.

> *Mendel's first law of inheritance, the law of segregation, states that each gamete receives only one allele of each pair of alleles in an organism's genetic makeup.*

Scientists understand that Mendel's results were well defined because he was studying traits that exhibited complete dominance. This is not always the case. Incomplete dominance is a situation in which neither member of a pair of alleles exhibits dominance over the other. In fact, the "blending" of genes observed by some early investigators may have been visible expressions of incomplete dominance. Since Mendel's time, many examples of incomplete dominance have been found for various traits in both plants and animals (see Chapter 20).

Analyzing Mendel's Experiments

Looking back to Mendel's experiments, you can use the information from the conclusions he drew to further analyze and understand his experiments and data. Mendel used letters to represent alleles, with the dominant allele commonly denoted by an uppercase letter and the recessive allele by the lowercase of the same letter. In the parental generation, the cross of true-breeding (therefore homozygous) pea plants with purple flowers (the dominant trait) and true-breeding pea plants with white flowers (the recessive trait) can be represented as PP × pp (Figure 19-6). The × denotes a cross between two plants. Because the purple-flowered parent can produce only P gametes and the white-flowered parent can produce only p gametes, the union of an egg and a pollen grain from these parents can produce only heterozygous Pp offspring in the F_1 generation. But because the P allele is dominant, all of the F_1 individuals have purple flowers. The p allele, although present, is not visibly expressed.

To distinguish between the presence of an allele and its expression, scientists use the term **genotype** to refer to an organism's allelic makeup and the term **phenotype** to refer to the expression of those genes. The phenotype—the organism's outward appearance—is the end result of the functioning of the enzymes and the proteins coded by an organism's genotype.

The genotype of the F_1 generation of pea plants from the cross of true-breeding plants with purple flowers (PP) and true-breeding plants with white flowers (pp) is Pp, but the phenotype of the flowers of the hybrids is purple because P is the dominant allele. When these F_1 plants are allowed to self-fertilize, the P and p alleles segregate randomly during gamete formation. Their subsequent union at

second filial (F_2) generation (fil ee uhl) the offspring of the F_1 generation.
genetics (juh net iks) the branch of biology dealing with the principles of heredity and variation in organisms.
genes the units of transmission of hereditary characteristics in an organism; each gene is a segment of DNA that occupies a particular place on a particular chromosome.
allele (uh leel) each member of a factor pair containing information for an alternative form of trait that occupies corresponding positions on paired chromosomes.
Mendel's law of segregation the concept that each gamete receives only one of an organism's pair of alleles. Chance determines which member of a pair of alleles becomes included in a gamete.
homozygous (hoe muh zye gus) a term referring to an individual who has two identical alleles for a trait.
heterozygous (het uhr uh zye gus) a term referring to an individual who has two different alleles for a trait.
genotype (jeen uh type) an organism's allelic (genetic) makeup.
phenotype (fee nuh type) the outward appearance or expression of an organism's genes.

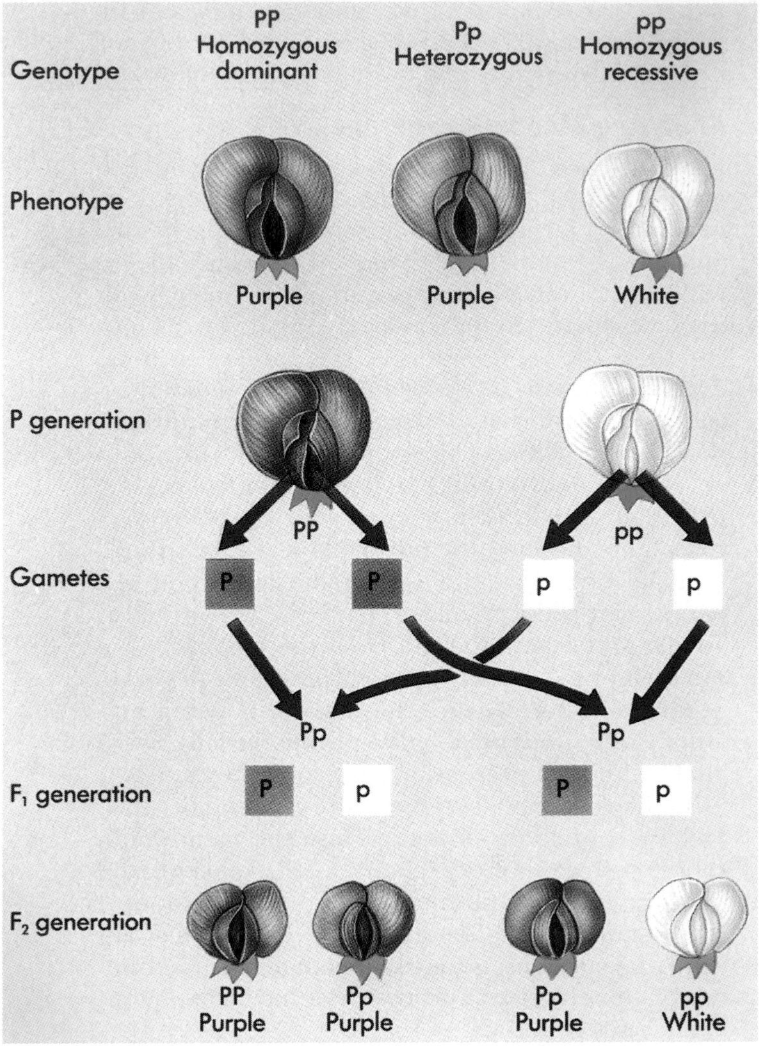

FIGURE 19-6 Analysis of Mendel's experiments. Mendel determined that the physical characteristics of his plants were a reflection of alleles, which he represented with letters. In this diagram, *P* represents purple flower color, and *p* represents white flower color. Because true-breeding purple flowers can produce only P gametes and white flowers only p gametes, a cross between these two types of flowers (the P generation) can yield only purple flowers (Pp) in the F_1 generation. The p allele is recessive and thus is not expressed in the F_1 generation. The P and p alleles segregate randomly during gamete formation and randomly recombine in the F_2 generation to produce plants in an approximate ratio of 1 PP:2 Pp:1 pp.

fertilization to form F_2 individuals is also random. Figure 19-6 shows the possible combinations of the gametes formed by the F_1 plants. Their random combination produces plants in an approximate ratio of 1 PP:2 Pp:1 pp. From these genotypes, can you determine the phenotypes of these plants?

Using a simple diagram called a **Punnett square** is another way to visualize the possible combinations of genes in a cross. Named after its originator, the English geneticist Sir Reginald Punnett, the Punnett square is used to align possible female gametes with possible male gametes in an orderly way. The male gametes are shown along one side of the square; the female gametes are shown along the other. As in Figure 19-7, the square is divided into smaller squares as columns are drawn vertically and rows are drawn horizontally, providing a "cell" for each possible combination of gene pair.

Whether using a Punnett square or visualizing the gametes and their recombinations as in Figure 19-6, you can see the expected ratios of the three kinds of F_2 plants: one fourth are true-breeding pp white-flowered plants, two fourths are heterozygous Pp purple-flowered plants, and one fourth are pure-breeding PP purple-flowered individuals. The 3:1 phenotypic ratio is really a 1:2:1 genotypic ratio.

The outward appearance of an individual (the expression of the genes) is referred to as its phenotype. The genetic makeup of an individual is referred to as its genotype. The genotype represents the alleles for given genes that are present. The use of a Punnett square is one way to visualize the genotypes of progeny in simple Mendelian crosses and to illustrate their expected ratios.

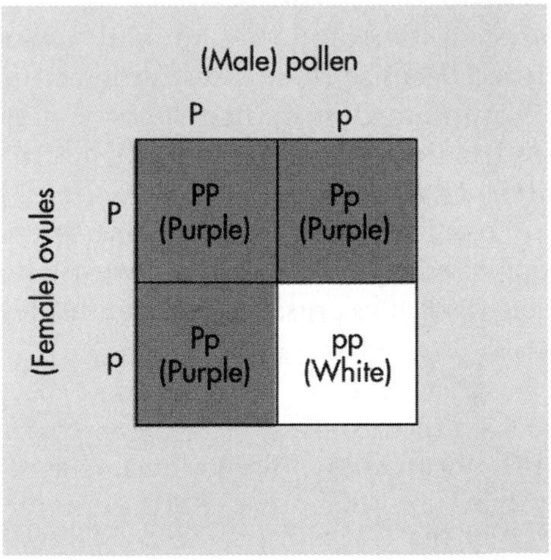

FIGURE 19-7 Punnett square analysis of Mendel's F_1 generation. Each smaller cell within the square contains the possible F_2 phenotypic combinations from this cross of F_1 plants. The phenotypic ratio is 3:1, but the genotypic ratio is 1:2:1. Look back at Figure 19-5 and see that the genotypic ratio matches the expected F_2 ratio of true-breeding dominant to non-true-breeding dominant to true-breeding recessive.

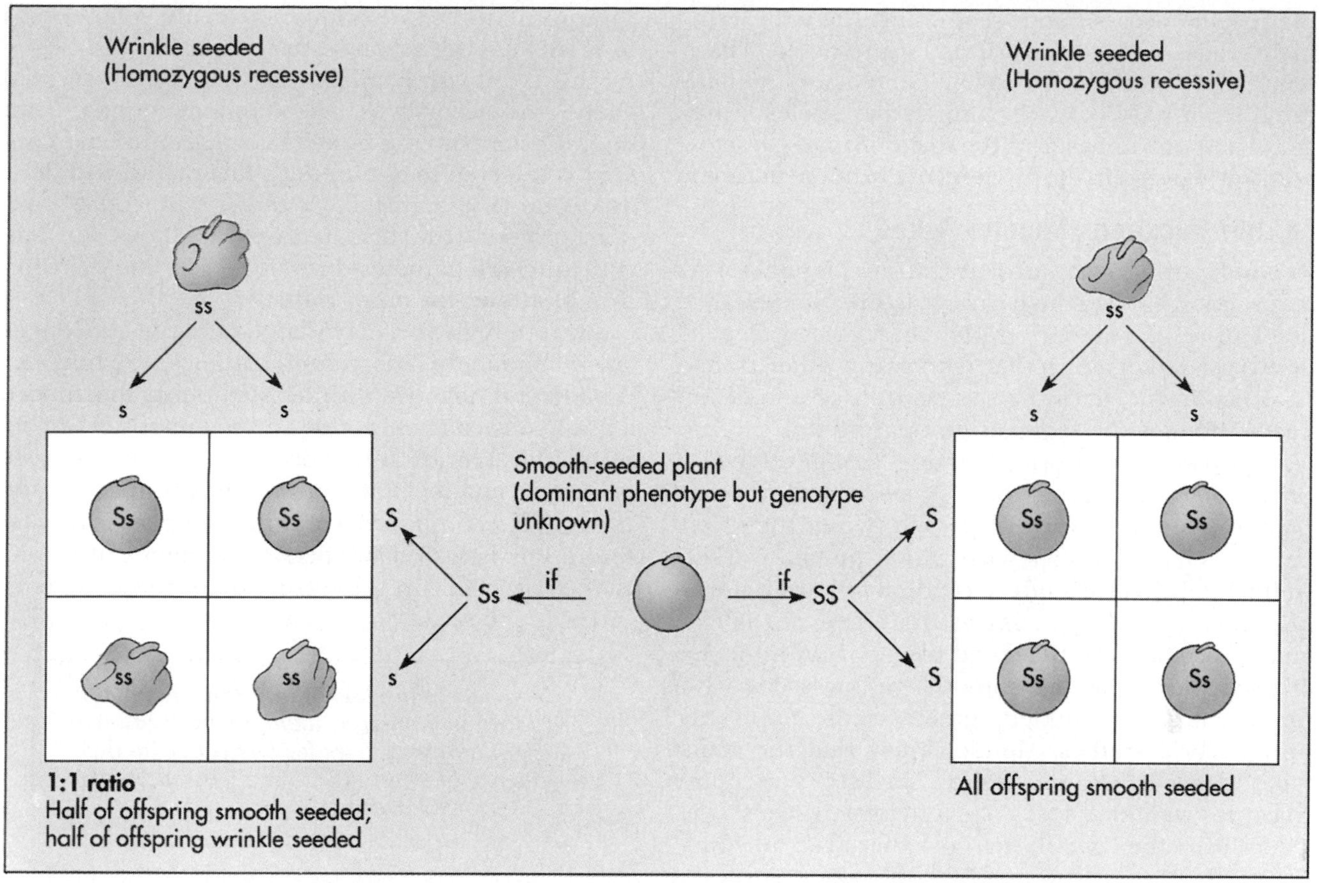

FIGURE 19-8 A testcross. The phenotypically dominant test plant, which has smooth seeds but an unknown genotype, is in the center, crossed with a known homozygous recessive plant having wrinkled seeds. The Punnett square *(left)* shows the expected genotypes and phenotypes if the test plant is heterozygous dominant. The Punnett square on the right shows the expected genotypes and phenotypes if the test plant is homozygous dominant.

How Mendel Tested His Conclusions

To test his conclusions further and to distinguish between homozygous dominant and heterozygous phenotypes, Mendel devised two procedures. The first test, already mentioned, was to self-fertilize the plant having a dominant phenotype. If the plant is homozygous dominant, it breeds true: purple-flowered plants produce progeny with purple flowers, tall plants produce tall plants, and so forth. If the plant is heterozygous, on the other hand, it produces dominant and recessive offspring in a 3:1 ratio when self-fertilized.

The other procedure Mendel used is called a *testcross.* In this procedure, Mendel crossed the phenotypically dominant test plant with a known homozygous recessive plant. He predicted that (1) if the test plant is homozygous for the dominant trait, the progeny will all be hybrids, Pp for example, and will therefore look like the test plant and that (2) if the test plant is heterozygous, then *half* the progeny will be heterozygous and look like the test plant, but half the progeny will be homozygous recessive and therefore will exhibit the recessive characteristic. Figure 19-8 shows the two possible testcross scenarios, using the characteristics that Mendel actually used in his first testcross experiments.

What did Mendel's data reveal? When he performed the testcross using hybrid F_1 plants having smooth seeds (Ss) and crossed them with a variety of plant having wrinkled seeds (ss), he obtained 208

Punnett square (pun et) a simple diagram that provides a way to visualize the possible combinations of genes in a cross and that illustrates their expected ratios. Male gametes are shown along one side of the square and female gametes are shown along the other. Each possible combination of gametes is shown in cells within the square.

plants: 106 with smooth seeds, and 102 with wrinkled seeds—a 1:1 ratio just as he predicted. These data confirmed the primary conclusion Mendel drew from earlier work: alternative alleles segregate from one another in the formation of gametes, coming together in the progeny in a random manner.

Further Questions Mendel Asked

Expanding on his law of segregation, Mendel asked a new question. Do the pairs of factors (alleles) that determine particular traits segregate independently of factor pairs that determine other traits? In other words, does the segregation of one factor pair influence the segregation of another?

To answer his question, Mendel first developed a series of true-breding lines of peas that differed from one another with respect to two of the seven pairs of characteristics with which he had worked in his monohybrid studies. He then crossed pairs of plants that exhibited contrasting forms of the two characteristics and that bred true. For example, he crossed plants having smooth, yellow seeds with plants having wrinkled, green seeds. From his monohybrid studies, Mendel knew that the traits "smooth seeds" (S) and "yellow seeds" (Y) are dominant to "wrinkled seeds" (s) and "green seeds" (y). Therefore the genotypes of the true-breeding parental (P) plants were SSYY and ssyy.

Mendel's F_1 progeny are *dihybrids*—the product of two plants that differ from one another in two traits. As in his monohybrid crosses, all the F_1 progeny had smooth, yellow seeds—the dominant phenotype. Mendel then allowed the F_1 dihybrids to self-fertilize. The seeds from these self-crosses grew into 315 plants having smooth, yellow seeds; 101 plants having wrinkled, yellow seeds; 108 plants having smooth, green seeds; and 32 plants having wrinkled, green seeds—an approximate ratio of 9:3:3:1. Other dihybrid crosses also produced offspring having the same approximate ratio.

Analyzing the Results of Mendel's Dihybrid Crosses

Mendel reasoned that if the alleles for seed color and seed shape segregated into gametes independently of one another and were therefore inherited independently of one another, then the outcome for each trait would exhibit the 3:1 ratio of a monohybrid cross. Looking at Mendel's results, 315 + 108 (423) plants had smooth seeds and 101 + 32 (133) plants had wrinkled seeds. Put simply, *three times* as many plants had smooth seeds as had wrinkled seeds—a 3:1 ratio. Study these figures for the trait "seed color." How many plants produced yellow seeds? How many produced green seeds? What is the approximate ratio of plants having yellow seeds to plants having green seeds?

The contrasting alleles for the genes of seed shape and seed color assort independently from one another during gamete formation. This concept is referred to as **Mendel's law of independent assortment.** Put in today's terms with today's understandings: the distribution of alleles for one trait into the gametes does not affect the distribution of alleles for other traits.

Even though you may understand the concept of the independent assortment of alleles, you may not understand how two 3:1 ratios combine to produce a 9:3:3:1 ratio. These ratios and the relationship between these ratios are governed by the laws of *probability,* or chance. Each ratio is not simply a statement of the comparative numbers of plants Mendel found but is also a predictive statement of events that could occur in the future under the same conditions.

> *Mendel's second law of inheritance, the law of independent assortment, states that the distribution of alleles for one trait into the gametes does not affect the distribution of alleles for other traits.*

Analysis of a Dihybrid Cross Using Probability Theory

Figure 19-9 shows the self-cross of hybrid F_1 plants having smooth seeds. The genotype of these plants is Ss. The male gametes S and s are shown along one side of the Punnett square, and the female gametes S and s are shown along the other. These four gametes can combine in four ways. What is the probability that this cross will produce a plant having wrinkled seeds? Because there are only four possible combinations of gametes and the alleles segregate randomly as the gametes are formed, there is only a one-in-four possibility that the ss gametes will combine. Likewise, there is a three-in-four probability that gametes can combine in such a way that they will produce plants having smooth seeds: SS, Ss, and sS. Although Ss and sS are the same genotypes, the reversed position of the alleles represents the dominant and recessive alleles as contributed by each of two parent plants, hence the 3:1 phenotypic ratio. The alleles for the trait seed color, when looked at independently of seed shape, segregate in the same manner to produce plants having yellow seeds and plants having green seeds in a 3:1 ratio, respectively (see Figure 19-9).

To illustrate further, you can simulate the combination of the gametes S and s of two parents by per-

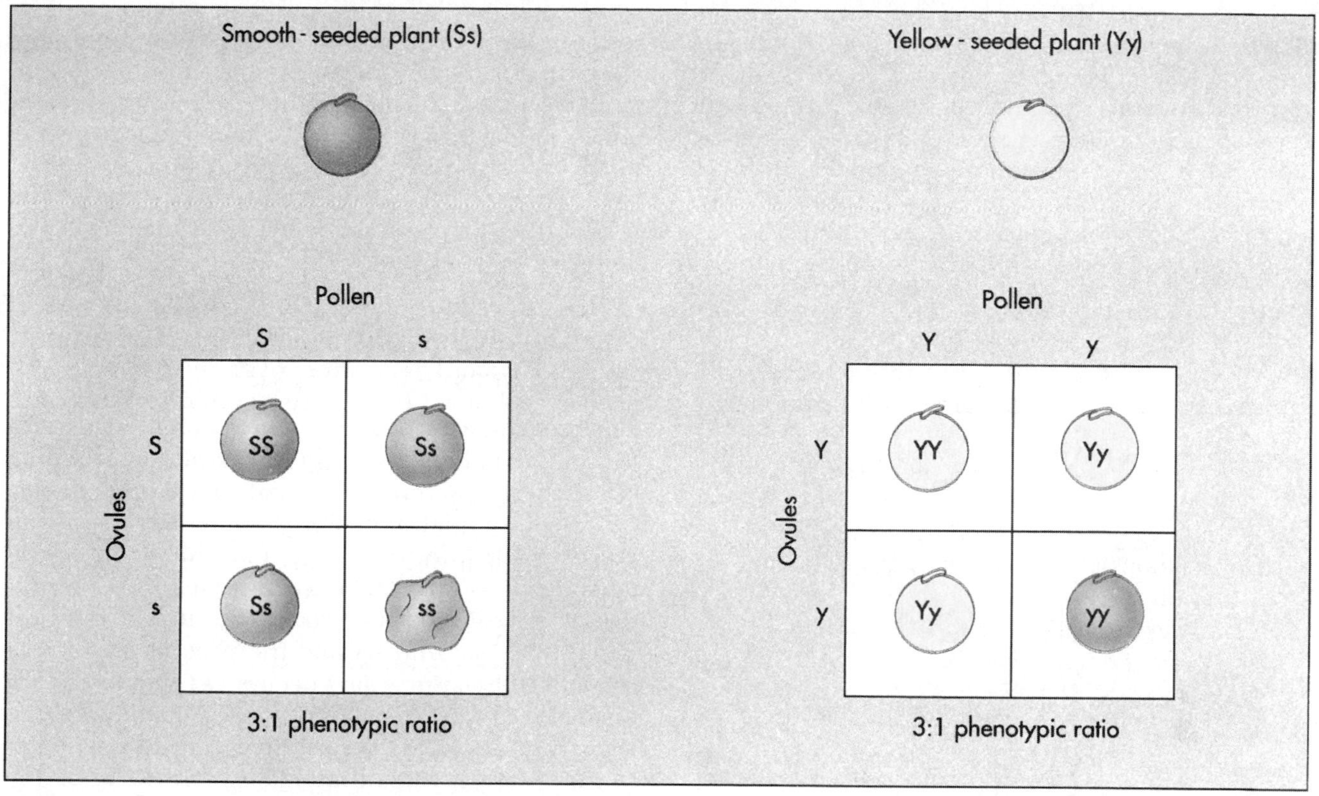

FIGURE 19-9 Mendel's law of independent assortment. As this diagram demonstrates, the contrasting alleles for seed shape (S and s) and color (Y and y) assort independently from one another during gamete formation.

forming a simple activity. Take two pennies and tape the letter "S" on one side of each and "s" on the other. Flip the two pennies 100 times and record your results. The probability of flipping an S on one of the pennies is one out of two, or $^1/_2$. The chance of flipping an S on the other penny is also $^1/_2$. The probability that an event will occur at the same time as another independent event is simply the product of their individual probabilities. Therefore the probability of flipping SS with the two pennies is $^1/_2 \times {}^1/_2$, or $^1/_4$. (Did you flip SS approximately 25 times out of 100?) Likewise, the probability of flipping ss is $^1/_4$, flipping Ss is $^1/_4$, and flipping sS is $^1/_4$. (Do your data match?) If the pennies were in fact gametes, $^3/_4$ would produce plants having smooth seeds and $^1/_4$ would produce plants having wrinkled seeds—a 3:1 ratio.

Now look at these traits as they occur together in a dihybrid cross. The probability that a plant with wrinkled, green seeds will appear in the F_2 generation is equal to the probability of observing a plant with wrinkled seeds ($^1/_4$) times the probability of observing a plant with green seeds ($^1/_4$), or $^1/_{16}$. The probability that a plant with smooth, green seeds will appear in the F_2 is equal to the probability that the F_1 parents will produce a plant with smooth seeds ($^3/_4$) times the probability that they will produce a plant with green seeds ($^1/_4$), or $^3/_{16}$. Can you figure out the probability that a plant with wrinkled, yellow seeds will appear in the F_2 or that a plant with smooth, yellow seeds will appear?

Analysis of a Dihybrid Cross Using a Punnett Square

Figure 19-10 portrays the self-cross of hybrid F_1 plants having smooth seeds and yellow seeds, showing how the gametes segregate to produce the F_2 plants. The eight gametes of the two parents can combine in 16 different ways. What is the probability that this cross will produce a plant having wrinkled, green seeds? Because there are only 16 possi-

Mendel's law of independent assortment the concept that the distribution of alleles for one trait into the gametes does not affect the distribution of alleles for other traits.

ble combinations of gametes and if you assume that the alleles segregate randomly and assort independently as the gametes are formed, there is only a 1 in 16 possibility that an sy egg will combine with an sy sperm (producing ssyy offspring). Likewise, there are only 3 combinations of gametes out of 16 possible combinations that will produce plants having wrinkled, yellow seeds: ssYY, ssYy, and ssyY. Following the same reasoning and using the Punnett square, notice that there are 3 possible combinations of gametes that will form plants producing smooth, green seeds and 9 possible combinations of gametes that will form plants producing smooth, yellow seeds. So whether you calculate the expected offspring in a dihybrid cross by using a Punnett square or whether you do this analysis by using the probability theory, the ratio of the F_2 progeny is predicted to be 9:3:3:1.

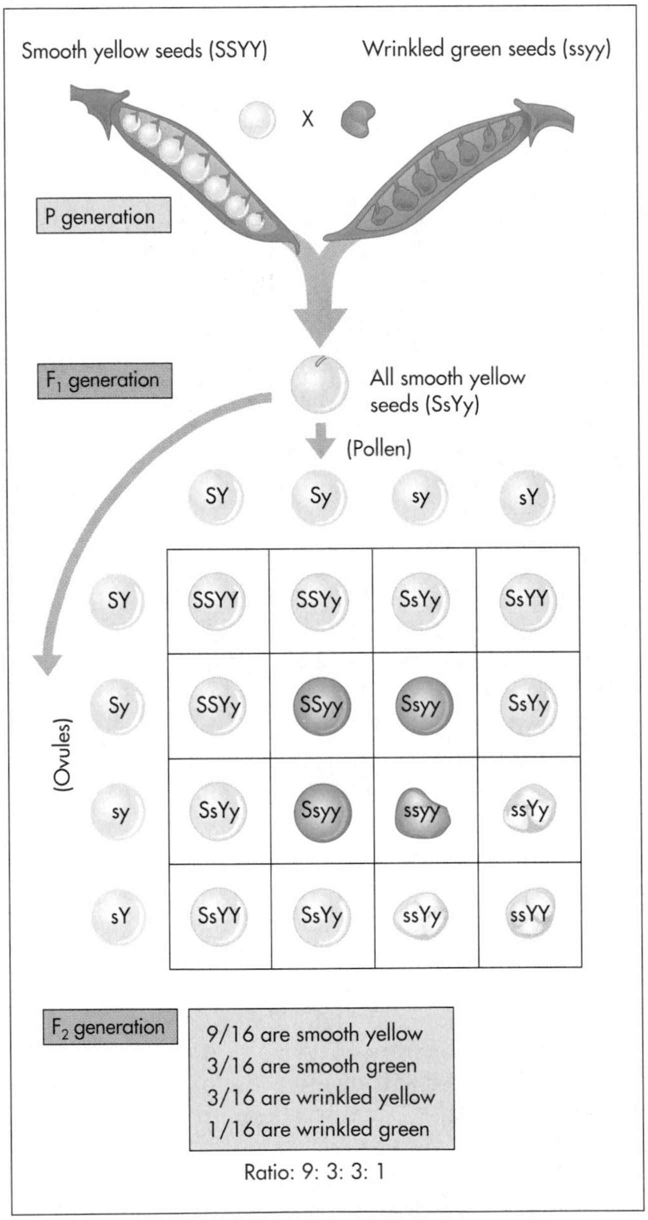

FIGURE 19-10 A Punnett square showing the results of Mendel's self-cross of dihybrid smooth yellow-seeded plants. The approximate ratio of the four possible combinations of phenotypes is predicted to be 9:3:3:1, the ratio that Mendel found.

The Connection Between Mendel's Factors and Chromosomes

During the years Mendel was experimenting with pea plants to determine the principles of heredity, other scientists were studying the structure of cells. Although Mendel published his theories of inheritance in 1866, few scientists read his work or understood its significance. In the late 1870s, scientists had described the process of nuclear division now known as mitosis and the process of nuclear reduction-division that occurs as gametes are formed now known as meiosis (see Chapter 18). In fact, in 1888, scientists gave the name *chromosomes* to the discrete, threadlike bodies that form as the nuclear material condenses during these processes of cell division. But they still had no idea of the link between Mendel's *factors* of inheritance and the newly named chromosomes—both important keys to unlocking the mysteries of heredity.

In 1900, three biologists, Carl Correns, Hugo de Vries, and Eric von Tschermak, independently worked out Mendel's principles of heredity. However, they knew nothing of Mendel's work until they searched the literature before publishing their results. The rediscovery of Mendel's work—his hypotheses supported by the independent work of others—helped scientists begin to make connections between Mendel's ideas and chromosomes.

By the late 1800s, scientists had observed the process of fertilization and knew that sexual reproduction required the union of an egg and a sperm. However, they did not know how each contributed to the development of a new individual. Many scientists hypothesized that the sperm merely stimulated the egg to develop. An American graduate student, Walter Sutton, suggested that if Mendel's hypotheses were correct, then each gamete must make equal hereditary contributions. Sutton also suggested that because sperm contain little cytoplasm, the hereditary material must reside within the nuclei of the gametes. He noted that chromosomes are in pairs and segregate during meiosis, as did Mendel's factors. In fact, the behavior of chromosomes during the meiotic process paralleled the behavior of the hereditary factors. Using this line of reasoning, Sutton suggested that Mendel's factors were located on the chromosomes. However, Sutton had no experimental evidence to support his hypothesis. Years later, various scientists worked on this problem, but the most conclusive evidence to uphold Sutton's chromosomal theory of inheritance was provided by a single, small fly.

Sex Linkage

In 1910, Thomas Hunt Morgan, studying the fruit fly *Drosophila melanogaster,* detected a male fly that differed strikingly from normal flies of the same species. This fly had white eyes (Figure 19-11) in-

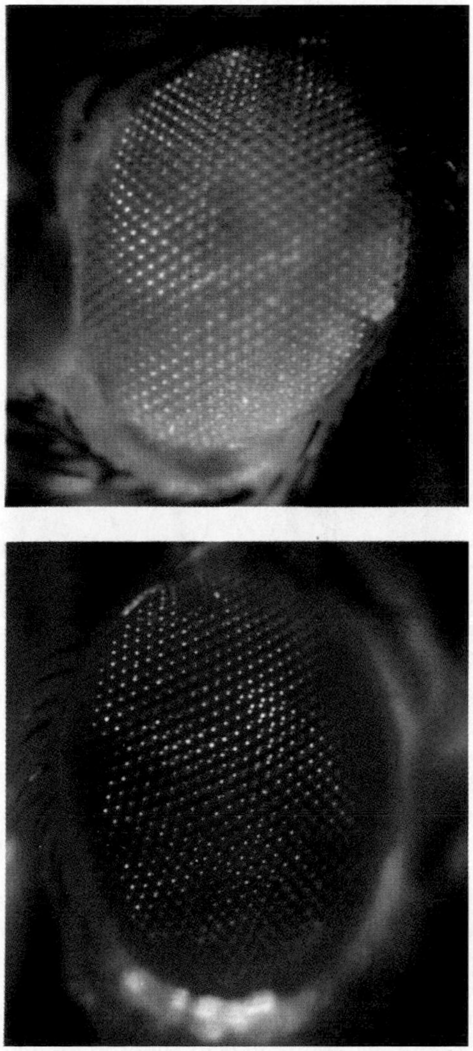

FIGURE 19-11 Red-eyed and white-eyed *Drosophila melanogaster.* The white-eyed defect in eye color is hereditary, the result of a mutation in a gene located on the sex-determining X chromosome. It was by studying this mutation that Morgan first demonstrated that genes are on chromosomes.

JUST WONDERING. . . .

Why do scientists study tiny organisms like fruit flies? Why don't they choose something larger?

Drosophila melanogaster, known as a fruit fly to most people, is an excellent organism for geneticists to study for many reasons. First, fruit flies are easy to breed and their life cycles are short. A single female can lay several hundred eggs, which develop into adults within 12 days. Therefore, a geneticist can study multiple generations of flies within a short period of time. Second, fruit flies are easy to maintain and take up little room in the laboratory. Populations of flies can be kept in small containers with easily prepared media. (Thomas Hunt Morgan, one of the first to study fruit-fly genetics, housed fruit flies in half-pint bottles.) Third, fruit flies exhibit variations in certain inherited traits such as eye color and wing formation that are easy to see under a dissecting microscope. The geneticist can easily anesthetize the population, study the flies, and place them back in the bottle before they revive. Fourth, fruit flies have only four pairs of chromosomes: three pairs of autosomes (body chromosomes) and one pair of sex chromosomes. Experiments can often be simpler using organisms with few chromosomes rather than organisms with a large number of chromosomes.

Underlying the use of various organisms for genetic studies is the assumption that genetic principles are universal. Thus, Mendel and Morgan were able to articulate laws of genetics that apply to all living things, even though they studied only fruit flies and peas.

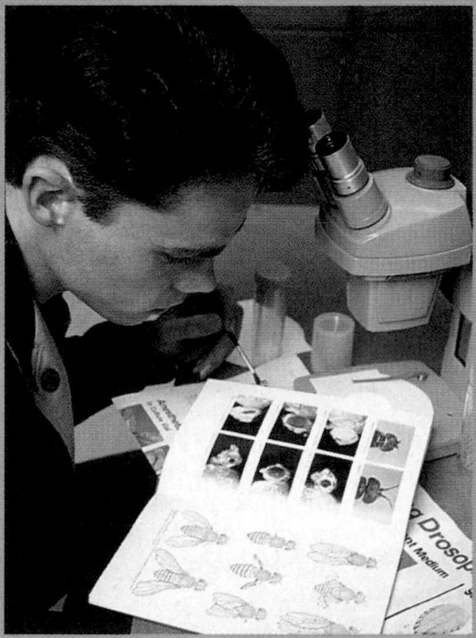

stead of the normal red eyes. Morgan quickly designed experiments to determine whether this new trait was inherited in a Mendelian fashion.

Morgan first crossed the white-eyed male fly to a normal female to see whether red or white eyes were dominant. All F_1 progeny had red eyes, and Morgan therefore concluded that red eye color was dominant over white. Following the experimental procedure that Mendel had established long ago, Morgan then crossed flies from the F_1 generation with each other. Eye color did indeed segregate among the F_2 progeny as predicted by Mendel's theory. Of 4252 F_2 progeny that Morgan examined, 782 had white eyes—an imperfect 3:1 ratio but one that nevertheless provided clear evidence of segregation. Something was strange about Morgan's result, however—something totally unpredicted by Mendel's theory: all of the white-eyed F_2 flies were males!

How could this strange result be explained? The solution to this puzzle involves gender. In *Drosophila* (as in most animals), the gender of the fly is determined by specific chromosomes called X and Y chromosomes. Female flies have four pairs of chromosomes, with one of those pairs being two X chromosomes. Male flies, on the other hand, have three pairs plus an X and a Y chromosome. To explain his results, Morgan deduced that the white-eyed trait is located on the X chromosome but is absent from the Y chromosome. (Scientists now know that the Y chromosome carries relatively few functional genes.) Because the white-eye trait is recessive to the red-eye trait, Morgan's result was a natural consequence of the Mendelian segregation of alleles (Figure 19-12). Morgan's experiment is one of the most important in the history of genetics because it presented the first clear evidence upholding Sutton's theory that the factors determining Mendelian traits are located on the chromosomes. When Mendel observed the segregation of alternative traits in pea plants, he was observing the outward reflection of the meiotic segregation of homologous chromosomes.

From his observation of chromosomes during meiosis and his knowledge of Mendel's work, Walter Sutton hypothesized that the hereditary factors were on the chromosomes. Thomas Hunt Morgan provided experimental evidence to uphold this hypothesis, also revealing that certain traits may be located on the chromosomes that determine the gender of an organism. Such traits are said to be sex-linked.

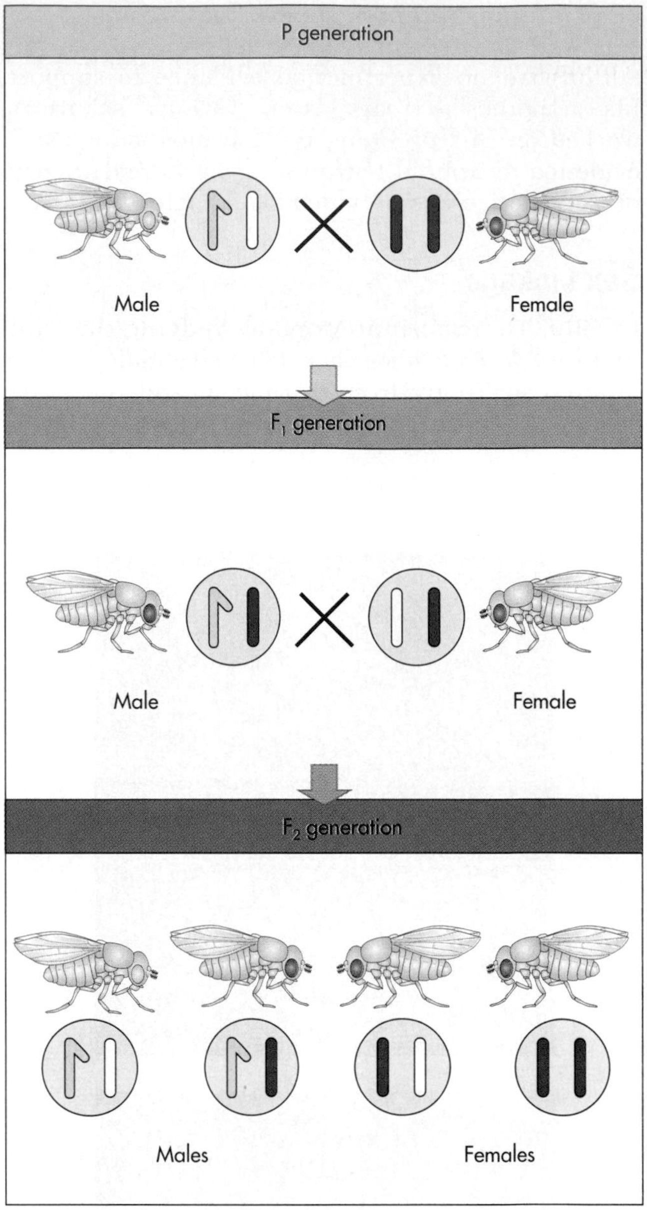

FIGURE 19-12 Morgan's experiment demonstrating the chromosomal basis of sex-linkage in *Drosophila melanogaster*. White-eyed mutant male flies were crossed with normal red-eyed females. The F_1 generation flies all exhibited red eyes. The male flies carried the normal red-eyed allele along with a Y chromosome that contains no allele for eye color, whereas the female flies carried one red-eyed allele and one white-eyed allele. Therefore in the F_2 generation, all female progeny were red eyed, being either homozygous or heterozygous for this dominant trait. The F_2 male flies, however, exhibited the trait inherited on their single X chromosome. Because half of them bear the white allele (the F_1 mother is heterozygous), half the male progeny will exhibit white eyes. This is exactly the result that Morgan observed: all of the white-eyed generation flies were male.

Solving the Mysteries of Inheritance

Throughout human history, an understanding of the nature of heredity was highly speculative until the work of Mendel, Sutton, and Morgan. These early geneticists took the first real steps toward solving the puzzles of inheritance and laid the foundation for one of the great scicntific advancements of the twentieth century: an understanding of the nature of genetic material and how it is transmitted from generation to generation. This basic outline of heredity led to a long chain of questions and was to be modified and refined as scientists provided additional experimental data and evidence. In one line of questioning described in Chapter 18, scientists such as Beadle and Tatum probed the structure and function of genes at the molecular level. Another line of questioning regards inheritance in humans and is the topic of Chapter 20.

Summary

- Scientists and philosophers hypothesized about the nature of inheritance over many centuries. Some suggested that parents gave off particles from their bodies that traveled to their sex organs during reproduction. Many assumed that the male and female contribution blended in the offspring. Others held that either sperm or eggs carried tiny preformed beings within them. Not until the mid-1800s did scientists realize that new life arose from old life in the form of new cells arising from old cells.
- In the mid-1800s, Gregor Mendel, an Austrian monk trained in botany and mathematics, studied patterns of inheritance using garden pea plants. Using true-breeding varieties of pea plants that exhibited alternative forms of seven different traits, Mendel artificially cross-fertilized them. The purpose of his experiments was to observe the offspring from the crossing of each pair of plants and to look for patterns in the transmission of traits.
- Mendel found that alternative traits segregate in crosses and may mask each other's presence. From counting progeny types, Mendel learned that the alternatives that were masked in hybrids appeared only 25% of the time when the hybrids were self-crossed. This finding, which led directly to Mendel's model of heredity, is usually referred to as the *Mendelian ratio* of 3:1 dominant to recessive traits.
- Mendel deduced from the 3:1 ratio that traits are specified by discrete factors that do not blend in the offspring. Today, scientists refer to Mendel's factors as *genes* and to alternative forms of his factors as *alleles.*
- Because Mendel observed that traits did not blend in the offspring and because the results he obtained from experiments on various traits showed similar results, Mendel hypothesized that alleles separate from one another during gamete formation. This concept is called *Mendel's law of segregation.*
- Working with pea plants that differed from one another in two characteristics, Mendel discovered that alleles of different genes assort independently during gamete formation. This concept is referred to as *Mendel's law of independent assortment.*
- The first clear evidence that genes reside on chromosomes was provided by Thomas Hunt Morgan. Morgan demonstrated that the segregation of the white-eye trait in *Drosophila melanogaster* was associated with the segregation of the X chromosome, the one responsible for sex determination.

Knowledge and Comprehension Questions

1. Genetically, how do offspring of sexual reproduction differ from those produced by asexual reproduction? Compare the amount of variation introduced within a species that reproduces sexually with a species that reproduces asexually.
2. Distinguish between self-fertilization and cross-fertilization in plants.
3. If you are told that an individual is heterozygous for a particular trait (brown eyes, for example), what information do you know? Will one of the alleles be dominant over the other in all cases?
4. What is Mendel's law of segregation? When does this segregation occur?
5. Assume that "L" represents the dominant trait of having long leaves, and "l" represents the recessive short-leaved trait in a plant. In the parental generation, you cross a homozygous long-leaved plant with a homozygous short-leaved plant. Draw a Punnett square illustrating this cross, and give the genotypes and phenotypes of the F_1 generation.
6. Using a Punnett square, show the genotypes and phenotypes of the F_2 generation if the F_1 plants in question 5 are self-fertilized.
7. In question 6, what is the probability that the F_2 plants will have short leaves?
8. What was Mendel testing when he used a testcross? What procedure did he use, and what was the outcome?
9. What is Mendel's law of independent assortment?
10. In a dihybrid cross between organisms that are heterozygous for both traits, what is the probability that their offspring will exhibit the phenotype for (a) both recessive traits? (b) both dominant traits?
11. How did the work of Walter Sutton and Thomas Hunt Morgan change the way scientists viewed the role of sperm in heredity and reproduction?

Critical Thinking

1. Was Mendel, in your opinion, able to successfully assess the phenotype of his plant crosses? Was he able to assess genotype with complete accuracy? What may affect the assessment of a genotype?
2. Can you think of a distinctive human trait that is *not* inherited? How would you test this hypothesis?

Genetics Problems

1. Among Hereford cattle there is a dominant allele called *polled;* the individuals that have this allele lack horns. After college, you become a cattle baron and stock your spread entirely with polled cattle. You have many cows and few bulls. You personally make sure that each cow has no horns. Among the calves that year, however, some grow horns. Angrily you dispose of them and make certain that no horned adult has gotten into your pasture. The next year, however, more horned calves are born. What is the source of your problem? What should you do to rectify it?
2. Many animals and plants bear recessive alleles for albinism, a condition in which homozygous individuals completely lack any pigments. An albino plant lacks chlorophyll and is white. An albino person lacks melanin. If two normally pigmented persons heterozygous for the same albinism allele have children, what proportion of their children would be expected to be albino?
3. Your uncle dies and leaves you his race horse, Dingleberry. To obtain some money from your inheritance, you decide to put the horse out to stud. In looking over the stud book, however, you discover that Dingleberry's grandfather exhibited a rare clinical disorder that leads to brittle bones. The disorder is hereditary and results from homozygosity for a recessive allele. If Dingleberry is heterozygous for the allele, it will not be possible to use him for stud because the genetic defect may be passed on. How would you go about determining whether Dingleberry carries this allele?
4. In *Drosophila,* the allele for dumpy wings (d) is recessive to the normal long-wing allele (D). The allele for white eye (w) is recessive to the normal red-eye allele (W). In a cross of DDWw × Ddww, what proportion of the offspring are expected to be "normal" (long wing, red eye)? What proportion "dumpy, white"?
5. Your instructor presents you with a *Drosophila* named Oscar. Oscar has red eyes, the same color that normal flies possess. You add Oscar to your fly collection, which also contains Heidi and Siegfried, flies with white eyes, and Dominique and Ronald, which are from a long line of red-eyed flies. Your previous work has shown that the white-eyes trait exhibited by Heidi and Siegfried is caused by their being homozygous for a recessive allele. How would you determine whether Oscar was heterozygous for this allele?
6. In some families, children are born who exhibit recessive traits (and therefore must be homozygous for the recessive allele specifying the trait), even though one or both of the parents do not exhibit the trait. What can account for this occurrence?

CHAPTER 20

HUMAN GENETICS

ABOUT 100 YEARS AGO, this is what family photos looked like. But this family is not just any family—it is a royal family. Queen Victoria of England, matriarch of this family, is standing in the center surrounded by children and is wearing a crown. At the time this photo was taken, Queen Victoria did not know that she would have an enormous effect on the lives of some of the family members surrounding her and on the lives of future members of the royal families of both Russia and Spain.

Queen Victoria affected many of her descendants because of a single allele—a mutant, or changed, gene that coded for a disorder known as *hemophilia.* Hemophilia is a hereditary condition in which the blood clots slowly or not at all. It is a recessive disorder, expressed only when an individual does not have a normal blood-clotting allele that masks the mutant's appearance. In addition to being recessive, the allele for this type of hemophilia is sex-linked: it is located on the X chromosome, one of the chromosomes that determines gender. This mutant gene probably arose in one of the sex cells from which Queen Victoria developed because the disease was not manifested in earlier generations.

Genetically normal females have two X chromosomes; normal males have one X and one Y chromosome. Although the X and Y chromosomes act as pairing mates during meiosis, only a portion of these two chromosomes have "matching" genes. A male who inherits an X chromosome with a mutant clotting allele will develop hemophilia because his Y chromosome does not have a corresponding allele to mask the allele on the X chromosome. A woman who inherits a mutant allele on one X chromosome and a normal allele on the other will not have the disorder but can pass the mutant allele along to her offspring. Such women are called *carriers*.

Of Queen Victoria's four daughters who lived to bear children, two daughters—Alice and Beatrice—were carriers of hemophilia. Two of Alice's daughters, Princess Irene of Prussia and Czarina Alexandra of Russia, were also carriers of hemophilia.

KEY CONCEPTS

- Genetic disorders can occur when a person inherits a fewer or greater number of chromosomes than is normal, chromosomes with pieces that have been added or removed, or chromosomes having changes in the structure of their DNA.
- Geneticists use pictures of chromosomes, arranged in pairs according to size, and pedigrees, diagrams of genetic relationships among family members over several generations, to study the inheritance of genetic disorders.
- Inherited traits do not always exhibit clear-cut dominance and recessiveness; certain traits exhibit incomplete dominance or codominance or may be coded by multiple alleles.
- Although most genetic disorders cannot yet be cured, scientists are making progress in developing gene therapies that may have a drastic impact on a physician's future ability to treat genetic disorders before birth.

OUTLINE

The Use of Karyotypes to Study Inheritance Patterns in Humans

In addition to a pair of **sex chromosomes,** each human body cell also contains 22 pairs of **autosomes.** Unlike sex chromosomes, autosomes (or "body" chromosomes) are the same in both sexes. The 23 pairs of human chromosomes are shown in Figure 20-1. Arranged in this manner according to size, shape, and other characteristics, the chromosome pairs make up a *karyotype.*

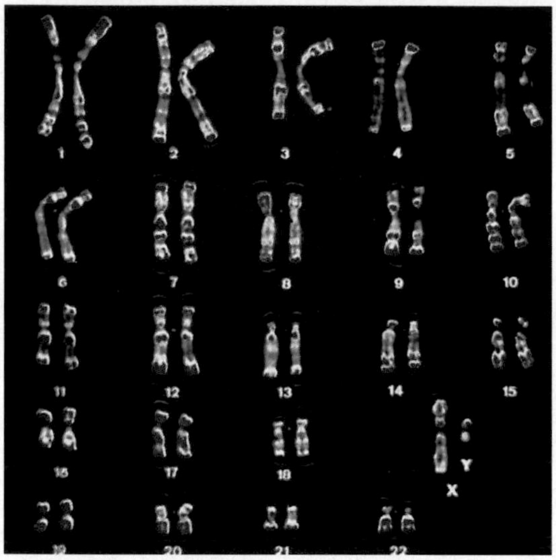

FIGURE 20-1 A normal karyotype of a human male. Humans have 23 pairs of chromosomes, including the two sex chromosomes. Notice the small Y chromosome in pair 23 that is characteristic of males.

Looking at a karyotype can often help researchers see genetic disorders if they are caused by the loss of all or part of a chromosome or by the addition of extra chromosomes or chromosome fragments. Changes in single genes *cannot* be seen. The differences between alleles that code for alternative forms of a trait lie in the chemical structure of the DNA, so they are invisible in a karyotype. Permanent changes in the genetic material, whether they affect single genes, pieces of chromosomes, whole chromosomes, or entire sets of chromosomes, are called **mutations.**

> *Each human cell contains 23 pairs of chromosomes. A karyotype is a picture of these chromosome pairs and shows whether a person has inherited a fewer or greater number of chromosomes than is normal or has lost or gained pieces of chromosomes.*

The Inheritance of Abnormal Numbers of Autosomes

Look at the karyotype in Figure 20-2, *A,* and compare it with that in Figure 20-1. What differences do you see? Carefully examine chromosome 21. One karyotype contains an extra copy of this chromosome, a situation called *trisomy 21.* The developmental defect produced by trisomy 21 was first described in 1866 by J. Langdon Down. For this rea-

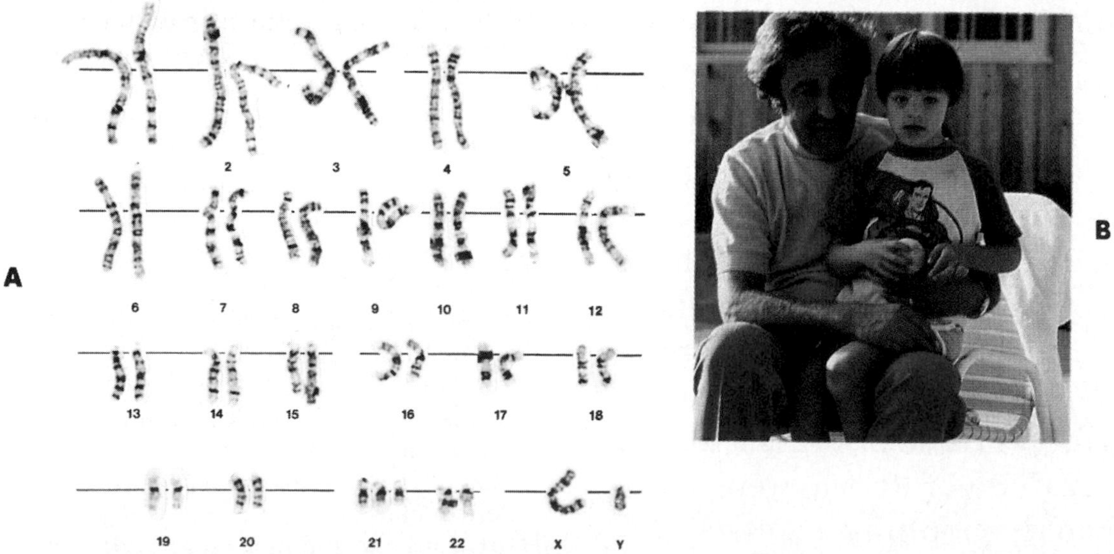

FIGURE 20-2 Down syndrome. A, This karyotype is of a male with Down syndrome. Notice the three number 21 chromosomes. **B,** This child was born with Down syndrome, but with early intervention and schooling, he may be able to learn to read and write and to find employment as an adult.

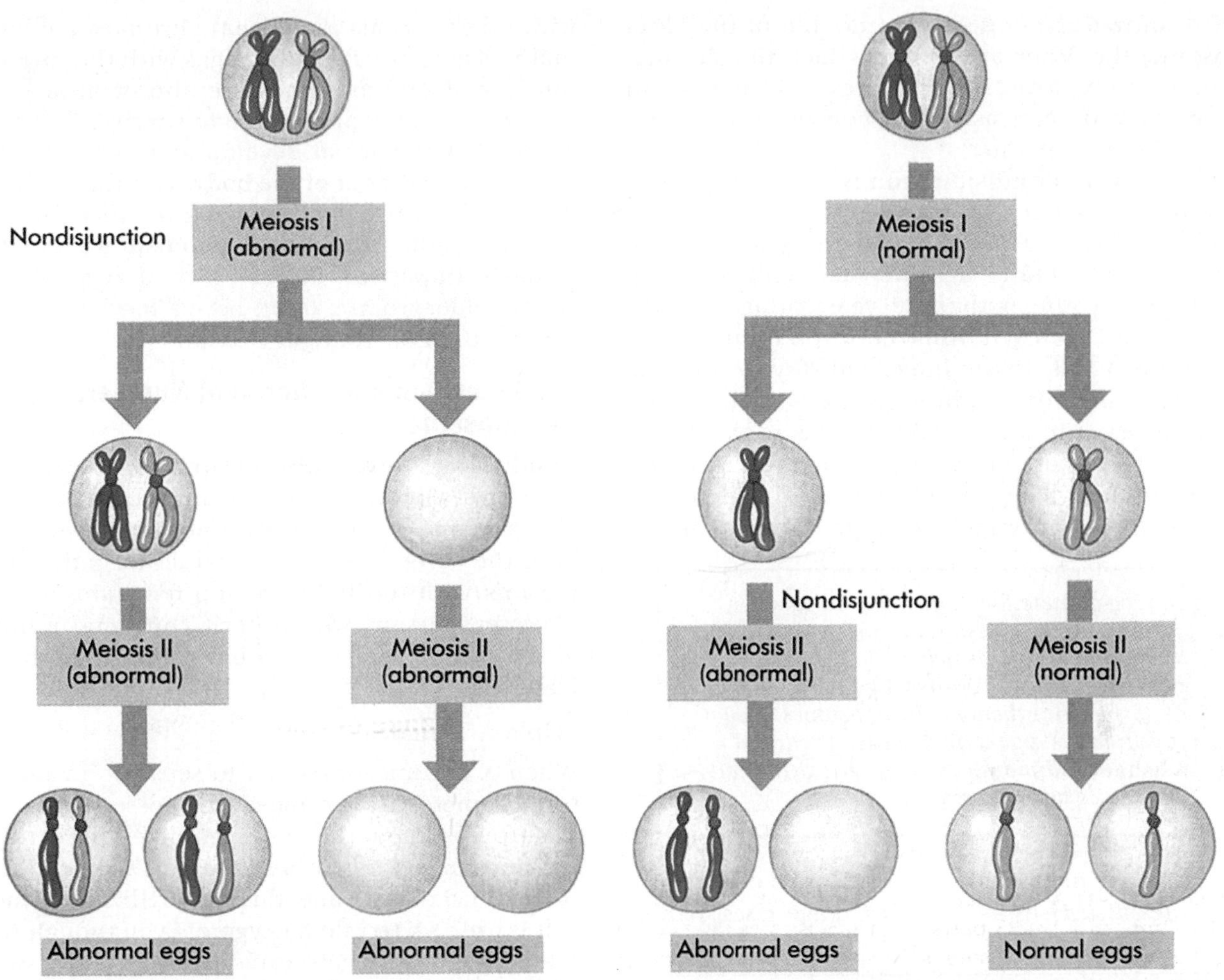

FIGURE 20-3 Nondisjunction. Nondisjunction results from an error in meiosis in which homologous chromosomes fail to separate during meiosis I (left). Nondisjunction can also occur during meiosis II when two sister chromatids fail to separate (right).

son, it is called *Down syndrome*. In these individuals the maturation of the skeletal system is delayed, so persons having Down syndrome are generally short and have poor muscle tone. In addition, they are mentally retarded.

Down syndrome is only one genetic disorder caused by the inheritance of an abnormal number of autosomal or sex chromosomes. How does such a situation arise? In humans, it comes about almost exclusively as a result of errors during meiosis. Meiosis is the process of nuclear division in which the number of chromosomes in cells is halved during gamete formation (see Chapter 18).

Early in meiosis, pairs of chromosomes called *homologues* (the pairs of chromosomes shown in the karyotype) line up side by side in a process called *synapsis*. Because, at this time, each chromosome is made up of two sister chromatids, the paired homologous chromosomes together have four chromatids. Gametes can gain or lose chromosomes at this point in the meiotic process if two homologous chromosomes fail to separate, or disjoin. If **nondisjunction** occurs (Figure 20-3), two of

sex chromosomes (**kro** muh **somes**) chromosomes that determine the gender of an individual as well as certain other characteristics.

autosomes (**aw** tuh **somes**) chromosomes that carry the majority of an individual's genetic information but do not determine gender.

mutations (myoo **tay** shuns) permanent changes in the genetic material that alter the original expression of a gene or genes. Mutations can affect single genes, pieces of chromosomes, whole chromosomes, or entire sets of chromosomes.

nondisjunction (non dis **jungk** shun) the failure of homologous chromosomes to separate after synapsis during meiosis, resulting in gametes with abnormal numbers of chromosomes.

the resulting gametes carry a double of the chromosome; the other two gametes lack the chromosome entirely. Nondisjunction can occur later in the meiotic process as well, when two sister chromatids fail to separate.

The cause of nondisjunction is not known. However, the occurrence of nondisjunction of chromosome 21, which results in Down syndrome, increases as a woman's age increases (Figure 20-4). In mothers younger than 20 years of age, the occurrence of Down syndrome children is only about 1 per 1700 births. In mothers 20 to 30 years old, the incidence is only slightly greater—about 1 per 1400. However, in mothers 30 to 35 years old, the incidence almost doubles to 1 per 750. And in mothers older than 45 years of age, the incidence of Down syndrome babies is as high as 1 in 16 births.

> *During gamete formation, homologous chromosomes occasionally fail to separate after synapsis. This occurrence, called* nondisjunction, *results in gametes with abnormal numbers of chromosomes. Down syndrome is a genetic disorder produced when an individual receives three (instead of two) 21 chromosomes.*

The incidence of nondisjunction of other chromosomes also rises as a woman's age rises. Babies with other serious autosomal chromosome abnormalities are rare. Fertilized eggs with the improper number of chromosomes are almost always inviable; that is, they are unable to survive. These eggs do not begin normal development and implantation but are cast out of the body with the menstrual flow—a process called *spontaneous abortion.* This increase in the incidence of nondisjunction and its negative impact on the viability of zygotes is one reason older women often have a harder time conceiving than do younger women.

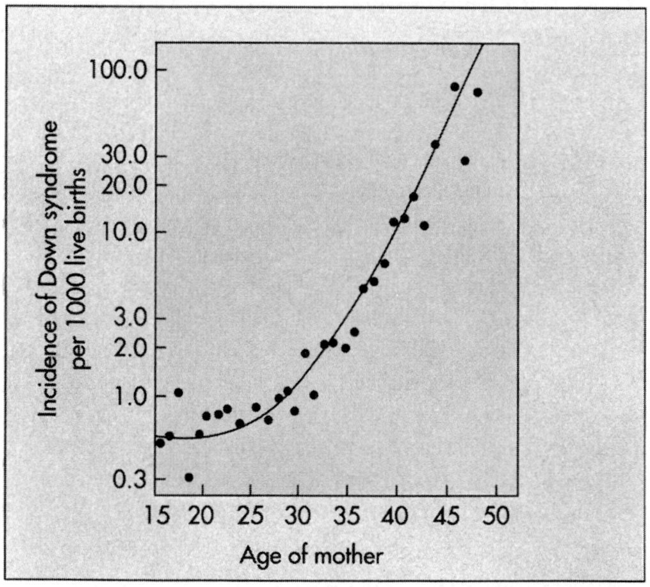

FIGURE 20-4 Incidence of Down syndrome vs. maternal age. The occurrence of this nondisjunction increases sharply as a woman grows older; thus the Y axis uses an exponential scale. The 10-fold increase from 1 to 10 is graphed using the same scale as the 10-fold increase from 10 to 100.

The Inheritance of Abnormal Numbers of Sex Chromosomes

Nondisjunction can also occur with the sex chromosomes. However, persons inheriting an extra X chromosome or inheriting one X too few do not have the severe developmental abnormalities that persons with too many or too few autosomes do. However, persons who inherit abnormal numbers of sex chromosomes often have abnormal physical features.

Triple X Females

When X chromosomes fail to separate in meiosis, some gametes are produced that possess both of the X chromosomes; the other gametes have no sex chromosome and are designated *O.* If an XX gamete joins an X gamete during fertilization, the result is an XXX (triple X) zygote. Even though triple X females usually have underdeveloped breasts and genital organs, they can often bear children. In addition, a small number of XXX people have lower-than-average intelligence. Although rare, a few individuals have been discovered to have tetra X (XXXX) and penta X (XXXXX) genotypes. Individuals having these genotypes are similar phenotypically to triple X individuals but are usually mentally retarded.

Klinefelter Syndrome

If the XX gamete joins a Y gamete, the result is quite serious. The XXY zygote develops into a sterile male, who has, in addition to male genitalia and characteristics, some female characteristics, such as breasts (Figure 20-5) and a high-pitched voice. In some cases, XXY individuals have lower-than-average intelligence. This condition, called *Klinefelter syndrome,* occurs in about 1 out of every 600 male births.

Turner Syndrome

If an O gamete (no X) from the mother fuses with a Y gamete, the resulting OY zygote is nonviable and fails to develop further. If, on the other hand, an O gamete from either the mother or the father

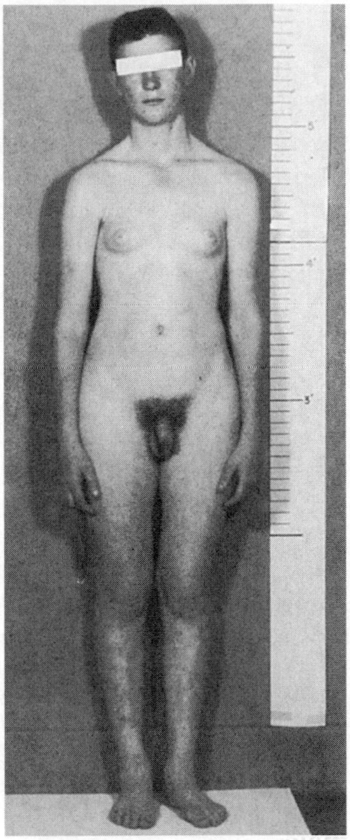

FIGURE 20-5 Klinefelter syndrome. A male with Klinefelter syndrome (XXY) exhibits some female characteristics, such as enlarged breasts and a high-pitched voice.

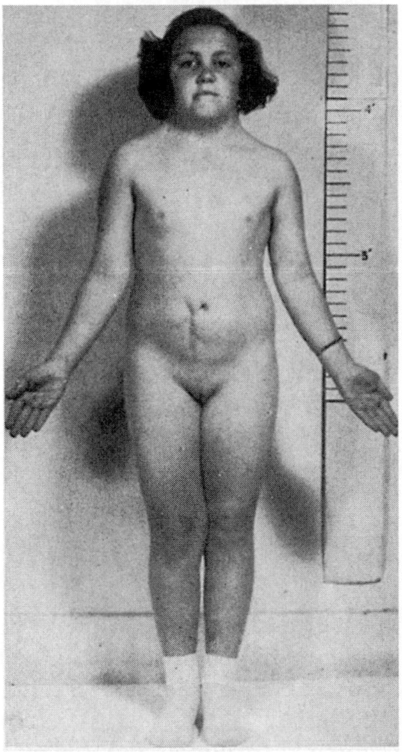

FIGURE 20-6 Turner syndrome. A female with Turner syndrome (XO) exhibits short stature, webbed neck, low-set ears, broad chest, and immature sex organs.

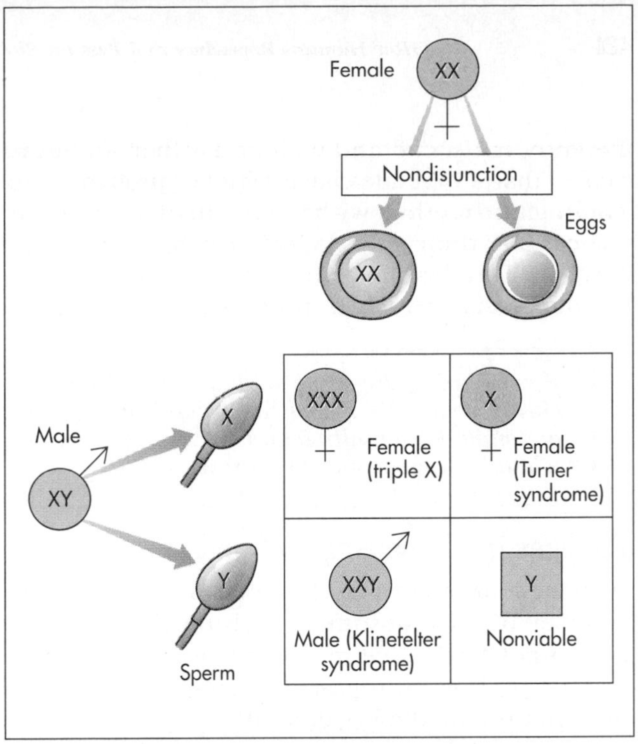

FIGURE 20-7 How nondisjunction can result in abnormalities in the number of sex chromosomes. Both Klinefelter syndrome and Turner syndrome result from nondisjunction of either the male or female gamete. This diagram shows how these genetic disorders occur when nondisjunction takes place in female gametes. It also shows how nondisjunction in female gametes can result in a triple X female.

fuses with an X gamete to form an XO zygote, the result is a sterile female of short stature, a "webbed" neck, low-set ears, a broad chest, and immature sex organs that do not undergo puberty changes (Figure 20-6). The mental abilities of an XO individual are slightly below normal. This condition, called *Turner syndrome,* occurs roughly once in every 3000 female births. The ways in which nondisjunction can result in abnormal numbers of sex chromosomes are shown in Figure 20-7.

XYY Males

The Y chromosome occasionally fails to separate from its sister chromatid in meiosis II (see Chapter 18). Failure of the Y chromosome to separate leads to the formation of YY gametes and viable XYY zygotes that develop into males who are unusually tall but have normal fertility. The frequency of XYY among newborn males is about 1 per 1000. The XYY syndrome has some interesting history associated with it. In the 1960s, the frequency of XYY males in penal and mental institutions was reported to be approximately 2% (that is, 20 per 1000)—20 times higher than in the general population. This observation led to the suggestion that XYY males are inherently antisocial and violent. Further studies revealed that XYY males have a higher probability of coming into conflict with the law than XY (normal) males, but their crimes are usually nonviolent. Most XYY males, however, lead normal lives and cannot be distinguished from other males.

> *Persons inheriting abnormal numbers of sex chromosomes often have abnormal features and may be mentally retarded. Zygotes without an X chromosome are not viable.*

Changes in Chromosome Structure

Although a person may have inherited the proper number of chromosomes, a chromosome that is structurally defective may occur if a chromosome breaks and the cell repairs the break incorrectly or does not repair the break at all. These broken and misrepaired chromosomes can then be passed on in the gametes of the parents to produce disorders in the offspring.

Chromosomes may break naturally, or breaks may be caused by outside agents, such as ionizing radiation and chemicals. **Ionizing radiation,** such as x-rays and nuclear radiation, is a form of energy known as *electromagnetic energy.* Sunlight is also a type of electromagnetic energy. Ionizing radiation, however, has a higher level of energy than ordinary light does. When ionizing radiation reaches a cell, it transfers energy to electrons in the outer shells of the atoms it encounters, raising them to a still higher energy level and ejecting them from their shells. The result is the breaking of covalent bonds and the formation of charged fragments of molecules with unpaired electrons. The charged molecular fragments, or *free radicals,* are highly reactive. They may interact with DNA, producing chromosomal breaks or changes in the nucleotide structure.

When chromosomes break and are improperly repaired by the cell, some chromosomal information may be added, lost, or moved from one location on the chromosome to another. Chromosomal rearrangement changes the way that the genetic message is organized, interpreted, and expressed. Newly added chromosomal information can be caused by a *duplication* of a section of a chromosome. Seen in a karyotype, a chromosome having a duplication appears longer than its homologue. A *translocation* could also produce an abnormally long chromosome. In this situation, a section of a chromosome breaks off and then reattaches to another chromosome. The chromosome losing a section would then appear shorter because of this *deletion* of information. As gametes form, the progeny could receive one or both of these defective chromosomes. If a chromosomal *inversion* occurs, the broken piece of chromosome reattaches to the same chromosome but in a reversed direction (Figure 20-8).

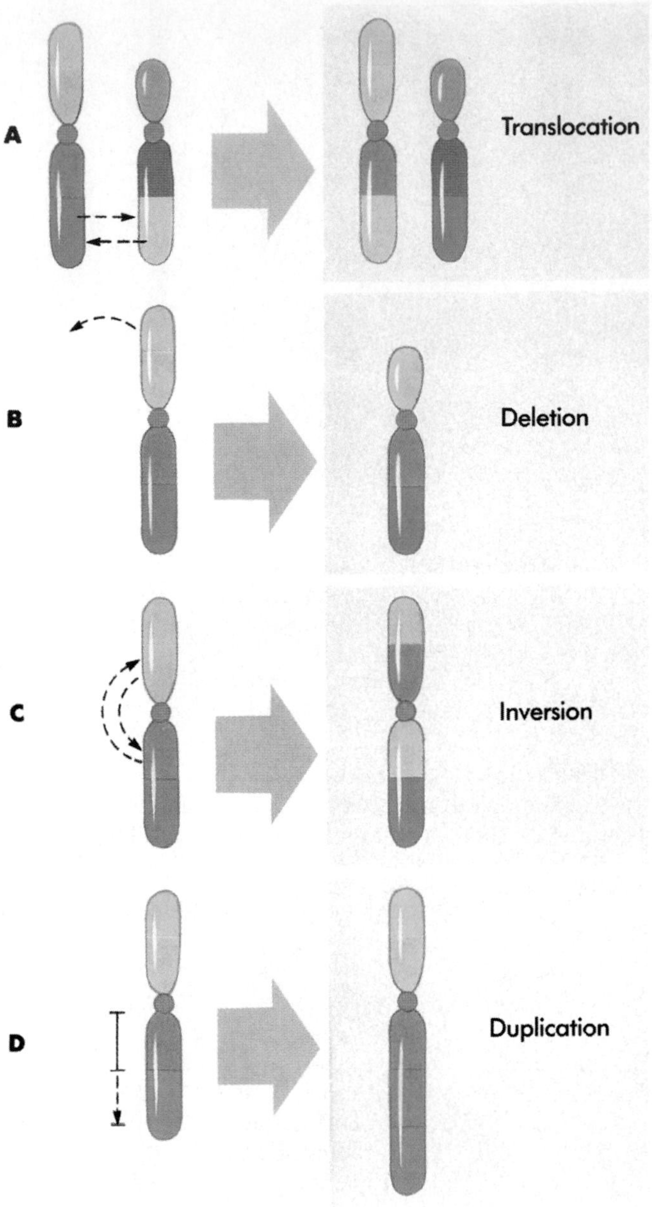

FIGURE 20-8 Types of chromosomal rearrangement. **A,** Translocation occurs when a section of a chromosome breaks off and attaches to another chromosome. **B,** A deletion occurs when a chromosome loses a section. **C,** Inversion occurs when a broken piece of a chromosome reattaches to the same chromosome but in a reversed direction. **D,** A duplication occurs when chromosomal information is repeated within the chromosome.

A well-known disorder associated with a chromosomal deletion in humans is the *cri du chat* syndrome. Described in 1963 by a French geneticist, *cri du chat* means "cat cry" and describes the catlike cry made by cri du chat babies. Other symptoms of

TABLE 20-1

Summary of Chromosomal Aberrations

CONDITION	CHARACTERISTICS	ABERRATION
Trisomy 13	Multiple defects, including severe mental retardation and deafness Death before 6 months of age for 90% of those who survive birth	3 13 autosomes
Trisomy 18	Facial deformities Heart defects Death before 1 year of age for 90% of those who survive birth	3 18 autosomes
Down syndrome (trisomy 21)	Developmental delay of skeletal system Mental retardation	3 21 autosomes
Trisomy 22	Similar features to Down syndrome More severe skeletal deformities	3 22 autosomes
Cri du chat	Moon face Severe mental retardation	Deletion on chromosome 5
Triple X female	Underdeveloped female characteristics	XXX XXXX XXXXX
Turner syndrome	Sterile female Webbed neck Broad chest	XO
Klinefelter syndrome	Sterile male Male and female characteristics	XXY
XYY male	Fertile male Usually quite tall	XYY

this disorder include severe mental retardation and a round "moon face." (Table 20-1 summarizes the chromosomal aberrations discussed.)

Gene Mutations

Sometimes changes take place in a single allele rather than entire sections of chromosomes. A change in the genetic message of a chromosome caused by alterations of molecules within the structure of the chromosomal DNA is called a **point mutation** or **gene mutation.** Through mutation, a new allele of a gene is produced. Mutations may occur spontaneously (although their occurrence is rare) or may be caused by ionizing radiation, ultraviolet radiation, or chemicals.

Ultraviolet (UV) radiation, the component of sunlight that leads to suntan (and sunburn), is much lower in energy than x-rays but still higher in energy than ordinary light. Certain molecules within the structure of chromosomes absorb UV radiation, developing chemical bonds among them that are not normally present. These unusual bonds produce a kink in the molecular structure of the DNA of chromosomes. Normally, a chromosome is able to repair itself by removing the affected molecules and synthesizing new, undamaged molecules. Sometimes, however, mistakes can take place in some part of the repair process and a change may occur in a gene.

Sometimes chemicals damage DNA directly. LSD (the hallucinogenic compound lysergic acid), marijuana (leaves of the *Cannabis sativa* plant), cyclamates (compounds widely used as non-nutritive sugar substitutes), and certain pesticides (compounds used to kill insects on crops and other plants) are a few examples of chemicals known to damage DNA. In general, chemicals alter the molecular structure of nucleotides, resulting in a mispairing of bases. As with UV radiation, mistakes can take place as the cell works to repair the DNA and changes may then occur in a gene.

> *Three major sources of environmental damage to chromosomes are (1) high-energy radiation, such as x-rays; (2) low-energy radiation, such as UV light; and (3) chemicals, such as LSD, marijuana, cyclamates, and certain pesticides.*

Mutations and the Molecular Biology of Cancer

Mutations in the DNA of body cells play a key role in the development of cancer. Although cancer is often thought of as a single disease, more than 200 types exist. All cancers, however, have four characteristics: (1) uncontrolled cell growth, (2) loss of cell differentiation (specialization), (3) invasion of normal tissues, and (4) **metastasis,** or spread, to multiple sites. Uncontrolled cancer eventually causes death be-

ionizing radiation (eye uh nye zing ray dee ay shun) a form of electromagnetic energy that can cause chromosomes to break or can cause changes in the nucleotide structure of DNA. X-rays and nuclear radiation are kinds of ionizing radiation.

point mutation (gene mutation) a change in the genetic message of a chromosome caused by alterations of molecules within the structure of the chromosomal DNA.

metastasis (muh tas tuh sis) one of the characteristics of cancer cells: the ability to spread to multiple sites throughout the body.

cause the cancer cells continually increase in number while spreading to vital areas of the body, occupying the space in which normal cells would reside and carry out normal body functions.

Cancer cells arise through a series of stepwise, progressive mutations of the DNA in normal body cells. The first set of mutations may involve protooncogenes and/or tumor-suppressor genes. *Protooncogenes* are latent (dormant) forms of cancer-causing genes, or **oncogenes,** that are present in all people. (The word oncogenes is derived from the Greek word *onkos* meaning "mass" or "tumor.") Protooncogenes become oncogenes when some part of their DNA undergoes mutation. Oncogenes can be thought of as "on" switches in the development of cancer, signaling cells to speed up their growth and decrease their levels of differentiation.

Tumor-suppressor genes are a separate class of genes that can be thought of as "off" switches in the development of cancer, signaling cells to slow their growth and increase their levels of differentiation. The expression of tumor-suppressor genes is required for the normal functioning of a cell; they appear to allow normal cells to differentiate into mature cell types with reduced or no growth potential. These genes are inactivated when they undergo mutation, thereby allowing cancerous growth.

Mutations in protooncogenes and in tumor-suppressor genes may be inherited or caused by viruses, chemicals, or radiation. All four of these mutagenic agents are referred to as *initiators,* and the process of protooncogene or tumor-suppressor gene mutation is called *initiation.* Unfortunately, initiation may occur after only brief exposure to an initiator. Initiation does not directly result in cancer but in a mutated cell (precancerous cell) that may or may not look abnormal and that gives rise to other initiated cells when it divides.

> *Cancer cells arise through a series of progressive mutations in the DNA of normal body cells. These initial mutations may be inherited or may be caused by certain viruses, chemicals, or types of radiation; they activate protooncogenes (dormant cancer-causing genes) and/or inactivate tumor-suppressor genes.*

For cancer to occur, initiated cells must undergo *promotion.* Promotion is a process by which the DNA of initiated cells is damaged further, eventually stimulating these cells to grow and divide. It is a gradual process and happens over a long time as opposed to the short-term nature of initiation. Research suggests that if the promoter (the substance causing promotion) is withdrawn in the early stages, cancer development can be reversed. For example, if a smoker stops smoking, that person's risk of lung cancer returns eventually to that of a nonsmoker (depending on how long and how much smoking has occurred).

The same agent that caused initiation may cause promotion in the same cell or it may be caused by another initiator. The mode of action of promoters is unclear, but evidence suggests that they, like initiators, mutate genes or damage chromosomes. Those agents that can both initiate and promote cancer are called complete **carcinogens,** or cancer-causing substances. Most substances linked with the development of cancer are complete carcinogens. A few, however, act only as an initiator or as a promoter. Heredity, for example, acts only as an initiator. Conversely, asbestos acts only as a promoter. Asbestos promotes cells initiated by other agents such as cigarette smoke and air pollution.

As promotion proceeds, damage to the DNA accumulates and the expression of oncogenes begins. That is, the oncogenes begin to be transcribed, producing polypeptides. These polypeptides affect specific cells in specific ways, causing cells to grow and divide when normal cells would not and causing a variety of other changes such as modifications in cells' shapes and structures, cell-to-cell interactions, membrane properties, cytoskeletal structure, protein secretion, and gene expression. The mechanisms by which a particular polypeptide interacts with a cell to produce one or more of these effects vary also. One way in which oncogene-transcribed polypeptides affect cell growth, for example, is by mimicking growth factors. This type of polypeptide binds to receptor sites on the surfaces of certain cells, activating specific enzymes within them. The activated enzymes cause these cells to grow and divide when they would normally *not* be growing and dividing.

When damage to the DNA of a cell is not drastic, most of the normal components of the cell are produced and it still responds to normal growth-inhibiting factors. Sometimes initiated cells or cells in the early stages of promotion grow and divide abnormally, forming a benign (noncancerous) tumor. **Benign tumors** are growths or masses of cells that are made up of partially transformed cells, are confined to one location, and are encapsulated, shielding them from surrounding tissues. Such tumors are not life threatening. Some benign growths, although not life threatening at the time, exhibit growth patterns that are characteristic of the development of cancer cells. These cells are said to exhibit *dysplasia* (Figure 20-9).

A
B

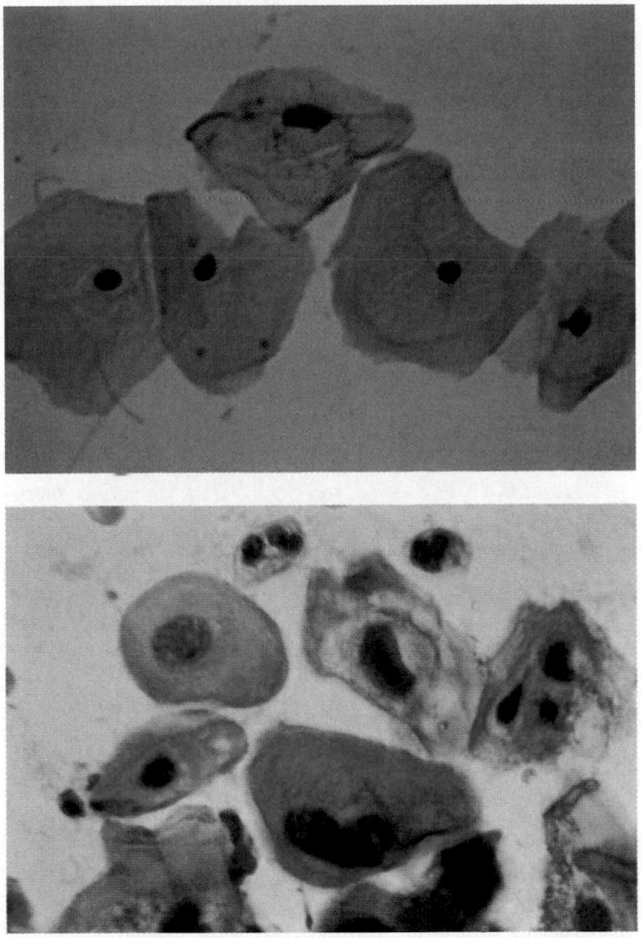

FIGURE 20-9 A comparison of normal and dysplastic cells. A, Normal cells. These cells are all approximately the same size and have nuclei that look similar. **B,** Dysplastic cells. These cells (stained differently from A) are irregular in size and the appearance of their nuclei. Both white and red blood cells can be seen at the top of this photo.

For cancer to occur, initiated (DNA-mutated) cells must undergo promotion, a process by which the DNA is damaged further. As promotion proceeds, damage to the DNA accumulates and the expression of oncogenes begins. The polypeptides produced cause cells to grow and divide when normal cells would not, as well as a variety of other cellular changes.

As promoters continue to damage the DNA, partially transformed cells reach a point where they irreversibly become cancer cells. This point marks the beginning of the third stage in the development of cancer called *progression.* During progression, the transformed cells usually become less differentiated than benign cells and increase their rate of growth and division without regard to the body's needs. In addition, these cancer cells have the ability to (1) *invade and kill* other tissues and (2) *metastasize,* or move to other areas of the body. Tumors with these properties are **malignant.** Interestingly, the word *malignant* is derived from two Latin words meaning "of an evil nature," whereas the word *benign* means "kind-hearted."

The spread of cancer during progression is a multistage process as is the development of the cancer cells themselves. At first the cancer cells proliferate, forming tumors. These cancers are referred to as *in situ,* meaning "in place," and are small, localized tumors that have not invaded the surrounding normal tissue. During the next stage of cancer progression, cancer cells invade the surrounding tissue by secreting chemicals that break down the intercellular matrix—the substances that hold cells together. Other secretions cause the cells to break apart. Figure 20-10 shows these stages in the development of a carcinoma, a cancer of the epithelial tissues that tends to invade surrounding tissue. Skin, breast, lung, and colon cancers are examples of carcinomas. Once cancer cells invade surrounding tissues, they can travel to other parts of the body by entering the blood and/or lymphatic vessels.

As DNA is damaged more and more, partially transformed cells irreversibly become cancer cells. Cancer cells are malignant cells, meaning they can invade and kill other tissues and metastasize, or move to other areas of the body.

Researchers suggest that 90% of all cancers are environmentally induced; that is, they are not inherited but result from external factors. Table 20-2 lists factors that are initiators and/or promoters of cancer. Notice that many of these factors are implicated in other diseases or conditions that endanger health. Avoiding these factors may increase

oncogenes (**on** ko **jeens**) cancer-causing genes.
carcinogens (kar **sin** uh jens) cancer-causing substances.
benign tumors (buh **nine**) growths or masses of cells that are made up of partially transformed cells, are confined to one location, and are encapsulated, shielding them from surrounding tissues.
malignant (muh **lig** nunt) a term describing cancerous tumors that have the ability to invade and kill other tissues and move to other areas of the body.

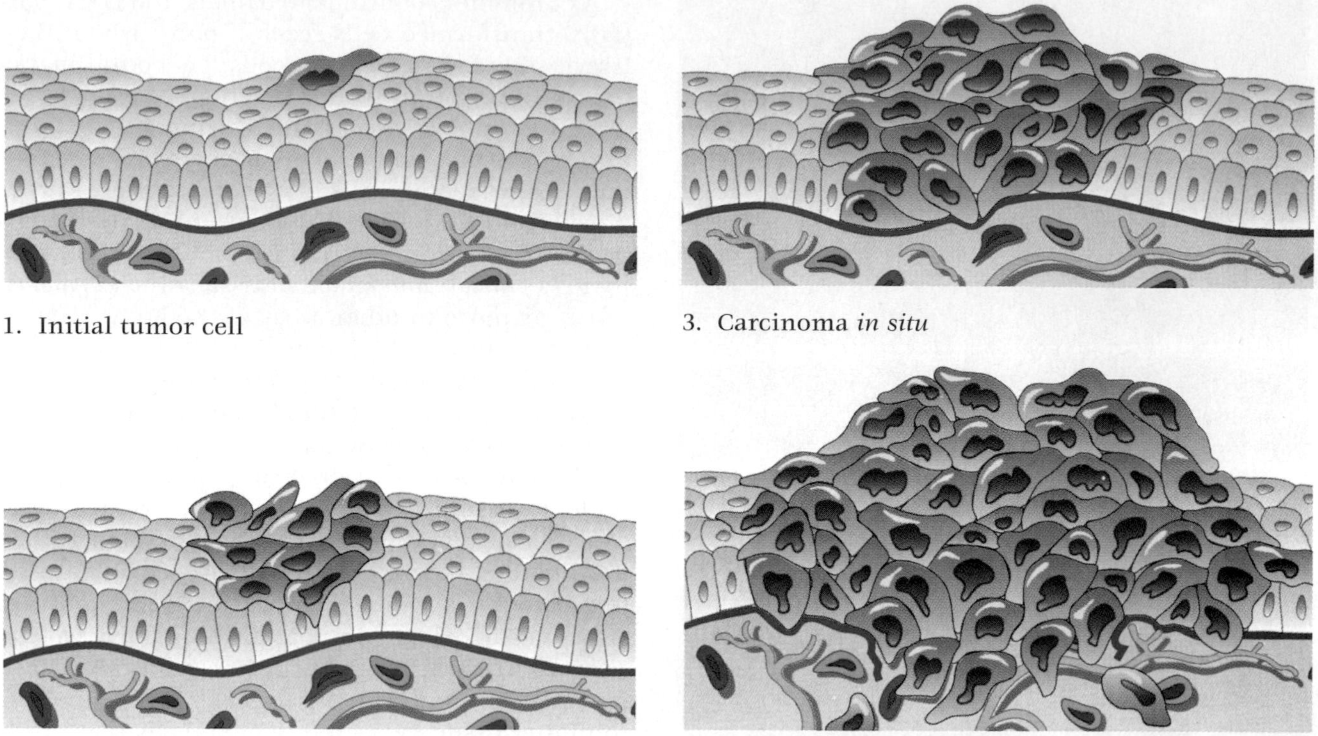

1. Initial tumor cell
2. Epithelial dysplasia
3. Carcinoma *in situ*
4. Invasive carcinoma

FIGURE 20-10 How cancer cells multiply and spread. In (1) a partially transformed tumor cell is shown in the epidermis of the skin. This cell multiplies (2), forming a mass of dysplastic cells. As the DNA of these cells is further damaged, they irreversibly become cancer cells. Their growth rate increases, and they form a cancerous tumor that is localized (3). These cells then invade the underlying dermis (4), entering blood vessels and traveling throughout the body.

TABLE 20-2

Major Factors That Increase the Risk of Cancer

FACTOR	EXAMPLES OF IMPLICATED CANCERS	COMMENTS
Heredity	Retinoblastoma (childhood eye cancer) Osteosarcoma (childhood bone cancer)	Most cancers are not caused by heredity alone. Persons having family histories of certain cancers should follow physicians' recommendations.
Tumor viruses	Liver cancer Adult T cell leukemia/lymphoma Cervical cancer	Five viruses are initiators of certain cancers. (See Table 25-1.)
Cigarette smoke	Lung cancer Cancers of the oral cavity and throat Cancers of the kidney and bladder	Cigarette smoking is responsible for approximately one third of all cancers. Nonsmokers have an increased risk of smoking-related cancers if they regularly breathe in sidestream smoke.
Industrial hazards	Lung cancer	Certain fibers, such as asbestos, chemicals such as benzene and arsenic, and wood and coal dust are prominent industrial hazards.
Ultraviolet radiation from the sun	Skin cancers	Those at greatest risk are fair-skinned persons who burn easily. However, everyone is at risk and should wear sunscreens and protective clothing when in the sun for extended periods of time.

TABLE 20-2

Major Factors That Increase the Risk of Cancer—cont'd

FACTOR	EXAMPLES OF IMPLICATED CANCERS	COMMENTS
Ionizing radiation	Related to location and type of exposure	Eliminate unnecessary medical x-rays to lower cancer risk. Infants and children are particularly susceptible to the damaging effects of ionizing radiation. Check your home to detect high levels of radon gas.
Hormones (estrogen and possibly testosterone)	Breast, endometrial, ovarian, and prostate cancers	When estrogen is used with progesterone in birth control pills and hormone-replacement therapy, the risk of cancer from this hormone is lowered and may become *less* than normal. The role of testosterone in prostate cancer is unclear.
Diet	Colon, breast, and prostate cancers (fat) Stomach and esophageal cancers (nitrites)	The primary dietary factor that increases the risk of cancer is the consumption of fat, particularly polyunsaturated fats. Nitrites found in salt-cured, salt-pickled, and smoked foods also increase the risk of cancer.

your chances for a longer, healthier life. Although your heredity is not a factor you can control, knowledge about your hereditary background can help you control other factors important to your health.

One way to decrease your risk of cancer is to determine which of the previously described cancer risks applies to you and to change your behavior accordingly. Another way to decrease your cancer risk is to follow the recommendations of the American Cancer Society (ACS) regarding diagnostic tests. Your local chapter of the ACS will send these recommendations to you on request. In addition, follow the general dietary recommendations listed in Table 20-3 and look for the danger signs of cancer listed in Table 20-4. If you notice any of these signs, see your doctor immediately. Otherwise, the ACS recommends a cancer-related checkup by a physician every 3 years for persons aged 20 to 39 years and annually for those 40 years of age and older. Persons at risk for particular cancers may need to see their physician more often.

TABLE 20-3

General Dietary Recommendations to Reduce the Risk of Cancer*

1. Avoid obesity
2. Reduce fat intake to 30% of total kilocalories
3. Eat more high-fiber foods, such as fruits, vegetables, and whole-grain cereals
4. Include foods rich in vitamins A, E, and C, as well as carotenes, in the daily diet
5. If alcohol is consumed, do not drink excessively
6. Use moderation when consuming salt-cured, smoked, and nitrite-cured foods

*The National Cancer Institute (U.S.) endorses all the above but warns not to exceed 35 grams of dietary fiber intake per day.
The American Cancer Society endorses all the above but sets no percentage for fat intake and adds a recommendation to include cruciferous vegetables (cabbage, broccoli, and brussels sprouts). These may decrease carcinogen activation.
The Canadian Dietetic Association generally endorses all of the above, but the specific language differs.
From Wardlaw, Insel: *Perspectives in Nutrition,* ed 2, St Louis, 1993, Mosby.

TABLE 20-4

The Warning Signs of Cancer

Sores that do not heal
Unusual bleeding
Changes in a wart or mole
A lump or thickening in any tissue
Persistent hoarseness or cough
Chronic indigestion
A change in bowel or bladder functions

HOW *Science* WORKS

Discoveries

The Genetics of Cancer

Many people are confused when they hear scientists refer to cancer as a genetic disease. We tend to think of genetic diseases as those inherited from our parents. However, cancers are not directly inherited, although we can inherit our individual tendencies to develop various types of cancer. Although the popular press may report the discovery of, for example, the "colon cancer gene," the common forms of cancer, such as colon and breast cancer, require the interaction of multiple and varied genes. Because the term *cancer* includes such a variety of different diseases, there is a tremendous diversity of genetic elements involved in the process of carcinogenesis, which is basically the result of accumulating specific types of genetic changes over a period of time. Some individuals inherit one or more of these changes, which increases the possibility that the required amount of additional genetic damage will occur during that individual's lifetime, leading to cancer. Although only 5% of all cancer cases appear to involve a strong hereditary predisposition, understanding the mechanisms involved in cancer genetics can be helpful in understanding the development of all cancers.

Physicians and others have long been aware that people are more likely to develop the same type of cancer that has been diagnosed in close family members, such as siblings and parents. In 1985, it was shown that children who inherited a specific genetic mutation were strongly predisposed to develop retinoblastoma, a rare type of eye cancer. Since then, many "susceptibility genes" (mutated genes that contribute to development of specific cancer types) have been studied. These forms of cancers have been documented to occur in certain families (see list above for a partial list), although it must be emphasized that most cases of these cancers do not have a clear genetic pattern. For example, only about 5% of patients with breast cancer have a strong family history of breast cancer (discussed later).

Examples of Cancers with Inherited Susceptibility Genes

Genes
Breast cancer
Colon cancer
Kidney cancer
Liver cancer
Lung cancer
Melanoma
Neuroblastoma
Osteosarcoma
Retinoblastoma

Identifying and studying cancer susceptibility genes offers a number of important potential benefits. It will eventually be possible to screen individuals and predict their likelihood of developing specific cancers, which could lead to life-style modifications and ultimately prevention of the cancer. Individuals who know that they are predisposed to certain cancers should have regular medical checkups and cancer tests, leading to early diagnosis if cancer does develop. Ultimately, scientists hope to be able to counteract cancer predisposition genes, either by reversing the biochemical effects of these genes or by modifying or replacing the genes themselves. However, we will need a great deal of knowledge about these genes before we can reach these goals. In the meantime, research continues on the genetics of many specific cancers.

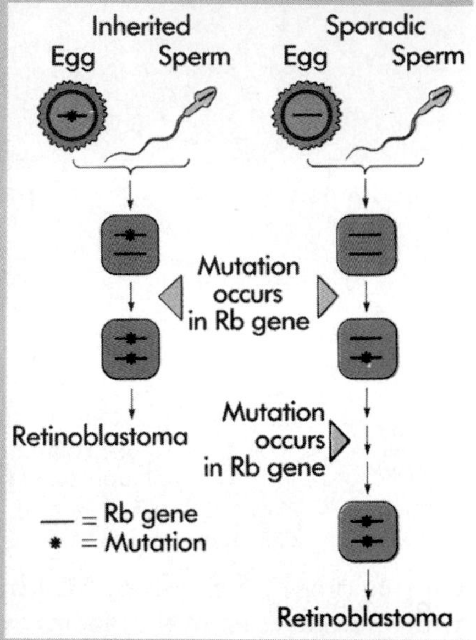

FIGURE 20-A The two-hit model for development of retinoblastoma.

Retinoblastoma

Retinoblastoma is a widely studied model for human cancer genetics, in part because it has the most simple genetic mechanism yet identified. This rare form of childhood eye cancer occurs in two main disease patterns. In one type, infants and very young children develop multiple tumors, frequently in both eyes. In these cases, there is almost always a family history of retinoblastoma. In the second pattern, older children with no family history of retinoblastoma develop a single tumor in one eye. Dr. Alfred Knudson proposed, in 1971, that in all cases, retinoblastoma is the result of two specific occurrences of genetic damage within a single eye cell and that the first type of patient

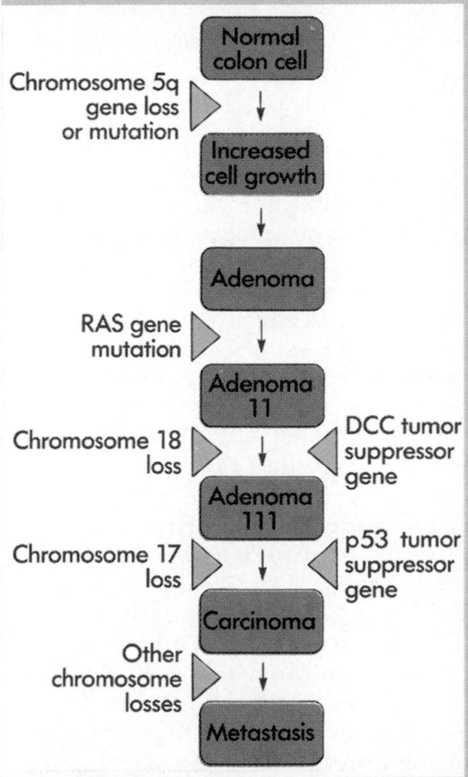

FIGURE 20-B Stages in the evolution of colon cancer.

Genes currently known to be involved in color cancer

GENE	% OF TUMORS WITH MUTATIONS	TYPE OF GENE	GENE ACTIVITY
FCC	~15%	?	Aids accurate DNA replication
ras	~50%	Oncogene	Intracellular signals
Cyclins	4%	Oncogene	Regulates cell growth cycle
neu/Her2	2%	Oncogene	Receptor for cell growth factors
myc	2%	Oncogene	Regulates gene activity
APC	>70%	Tumor suppressor	Unknown
DCC	>70%	Tumor suppressor	Cell adhesion
p53	>70%	Tumor suppressor	Regulates gene activity

Adapted from "New Colon Cancer Gene Discovered," *Science* 260:752, 1993.

inherits one of these genetic events in all of their cells (see Figure 20-A). In these patients with "inherited" retinoblastoma, one mutation is already present, so that only one additional mutation needs to occur for a cell to become cancerous. In the second type, sporadic retinoblastoma, both mutations must occur in the same eye cell for cancer to develop. This accounts both for the later age of onset and for the fact that usually only one tumor develops. In the 1980s, scientists discovered the site of the genetic damage responsible for retinoblastoma. Normally, the Rb tumor–suppressor gene apparently prevents tumor development. When both copies of this gene in an eye cell are damaged, retinoblastoma results.

Colon cancer

Unlike retinoblastoma, colon cancer is one of the most common forms of cancer. A significantly more complex series of genetic changes occurs during the development of colon cancer than has been found in retinoblastoma. Study of this disease has led to the recent exciting discovery of a new class of cancer gene and the hope for a predictive test to identify individuals at increased risk. An estimated 55,000 Americans die of colon cancer each year. Most of these individuals are diagnosed at a relatively late stage of the disease, after cancer has spread to other organ systems. Although colon cancer is highly curable in its early stages, some people avoid medical checkups for colon cancer because of embarrassment or fear of discomfort. Many lives could be saved if a reliable noninvasive test were available for this disease. Cancers of the colon show a distinct set of cellular changes as normal tissue begins to overproliferate, first forming benign tumors (adenomas) that become malignant and eventually spread throughout the body (Figure 20-B). These cellular changes are paralleled by genetic changes in the tumor cells. In contrast to retinoblastoma, current understanding suggests that at least five genes must be altered during the development of a colon cancer. These target genes include both oncogenes and tumor–suppressor genes, as shown in the table above. Activation of the oncogenes is found in many types of cancer, as is the inactivation of the p53 tumor–suppressor gene. Inactivation of the adenomatous polyposis coli (APC) tumor–suppressor gene, however, is characteristic of colon cancers, where it appears to be an early event in carcinogenesis. This gene was originally identified as the heritable element in a condition called *familial adenomatous polyposis* (*FAP*). Individuals with FAP develop large numbers of polyps (noncancerous growths) in the colon throughout their lifetimes. If these polyps are not removed, some will inevitably progress to become malignant tumors. FAP results from inherited alterations in the APC gene, although, as shown in Figure 20-B, many additional changes are needed for the progression of polyps to colon cancer in these patients.

Although FAP is the inherited precursor of approximately 1% of all colon cancers, some families have strong histories of colon cancer without the presence of FAP. Scientists have recently identified another inherited gene, FCC (for "familial colon cancer"), a susceptibility gene that occurs in about 1 out of every 200 people in the United States and that may ac-

Discoveries

HOW *Science* WORKS

The Genetics of Cancer—cont'd

count for as many as 15% of all colon cancer cases. Of the individuals who carry this defective gene, scientists estimate that 90% to 100% will develop some form of cancer. Most will have colon cancer, but many women with this gene will develop uterine cancers. Many other cancer types are also associated with this defect.

From a scientific standpoint, the FCC gene is an exciting discovery because it presents for study a new mechanism by which a gene can cause cancer. Instead of acting as either an oncogene or a tumor–suppressor gene, the FCC gene somehow seems to encourage the occurrence of accumulation of multiple mutations in affected cells. Because colon and other cancers require a series of genetic events for progression, any cellular process that increases genetic damage would certainly speed up the carcinogenic process. It is possible that the activity of this gene could be responsible for many of the other genetic changes seen in cancer cells with this defect.

As well as providing a new way for scientists to understand cancer, the discovery of this common cancer gene may have great public impact. A predictive test for FCC will be developed as soon as the gene itself is isolated and characterized, with the potential of identifying individuals at high risk for developing colon and other types of cancer. Because early detection is the key to curing colon cancer, this test, with appropriate follow-up, is likely to save many lives.

Breast cancer

At least 10% of all women in the United States will develop breast cancer at some point in their lives. For some women, the risk is significantly higher because of genetic factors. Breast cancer is so common that many families include at least one member with this disease. In the vast majority of these cases, recent evidence indicates that other female family members have only a slightly increased risk of developing breast cancer. However, in some families the risk is clearly higher. A woman is considered to have a significant family history and therefore a higher risk of developing breast cancer if she has more than one sister with breast cancer, if her mother or sister developed breast cancer before 50 years of age, or if any family member has had cancer in both breasts. Many of these families still have only moderate increases in breast cancer risk, and it is clear from the different patterns that several inherited factors may be responsible for familial breast cancer.

About 5% of all breast cancer cases occur in families with extremely high-risk patterns. This group includes all of the family characteristics previously mentioned, with women developing the disease at very young ages (teens to twenties) as well as significant numbers of males who also develop breast cancer. The mortality rate from breast cancer is so high in these families that some female members will choose to have both breasts removed surgically before cancer can develop. Progress has recently been made in discovering the gene responsible for this severe pattern of predisposition. A susceptibility gene, called *BRCA1,* has been located on chromosome 17. It is estimated that 1 out of every 200 women in the United States may carry this genetic defect, which confers an 80% to 90% risk of developing breast cancer. The function of this gene is not known at present, but it appears to act as a tumor–suppressor gene. A screening test for this marker is already in use for members of some high-risk families. This screening test reassures those individuals who have not inherited this gene and allows those whose test result is positive for this gene to make more informed choices about preventive measures.

The future of cancer genetics

Cancer is such a complex genetic problem that there will undoubtedly be new and sometimes unexpected discoveries about the carcinogenic process for many years. Scientists are constantly identifying new oncogenes and tumor–suppressor genes, and each discovery brings us closer to an understanding of how cancer develops. In the past 4 years, inherited genes that predispose individuals to develop two of the most common causes of cancer death in the United States have been identified. (Only lung cancer kills more Americans than do colon and breast cancer.) The information gained from these discoveries promises not only a means of identifying high-risk individuals but also the hope of producing preventive strategies for these cancers.

Cynthia J. Moore, Ph.D.
Washington University, St. Louis, Mo.

The Use of Pedigrees to Study Inheritance Patterns in Humans

Karyotypes are useful in studying diseases or disorders caused by an abnormal number of chromosomes or the addition or deletion of chromosome pieces. However, most inherited disorders, such as the type of hemophilia discussed in the chapter opener, are caused by point mutations, which cannot be seen in a karyotype.

To study how this and other human traits are inherited, scientists study family histories—the records of relevant genetic features—to draw pedigrees. **Pedigrees** are diagrams of genetic relationships among family members over several generations. By studying which relatives exhibit a trait, it is often possible to say if the gene producing the trait is sex linked (occurs on one of the sex-determining chromosomes) or autosomal (occurs on a non–sex-determining chromosome). Pedigre analysis also helps a geneticist determine whether the trait is a dominant or a recessive characteristic. In many cases, it is also possible to infer which individuals are homozygous or heterozygous for the allele carrying the trait.

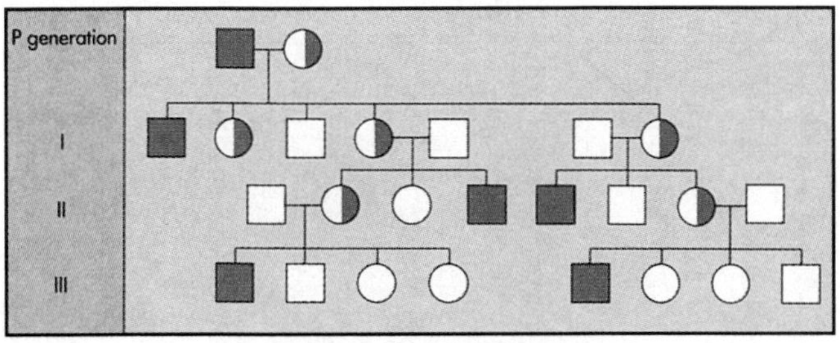

FIGURE 20-11 A pedigree showing red-green color blindness. *Squares* indicate males, *circles* indicate females. From this pedigree, it can be determined that the carriers of the color blindness gene are the female in the P generation, the three females in generation I, and the first and third females in generation II. The status of the other females is unknown.

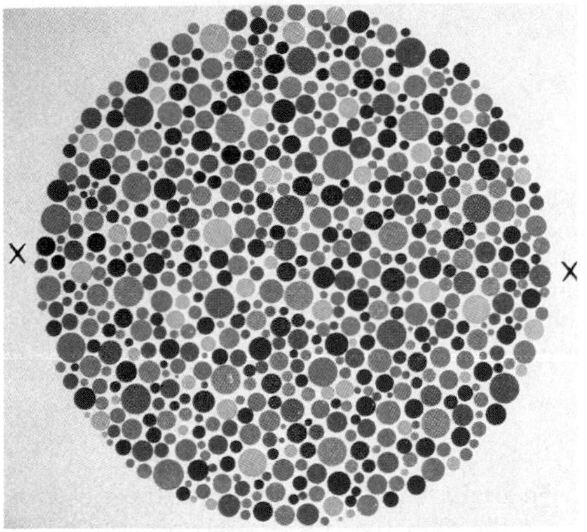

FIGURE 20-12 A test for red-green color blindness. About 8% of Caucasian males and 0.5% of Caucasian females have some loss of color vision, which is almost always caused by an X-linked recessive allele. This recessive allele causes a defect in one of three groups of color-sensitive cells in the eye. The most common form of color blindness is dichromatic vision, in which affected individuals cannot distinguish red from green or yellow from blue. The pattern tests for an inability to distinguish red from green. Those with normal vision will see a path of green dots between the points marked *X*. Those with red-green color blindness will not be able to distinguish the green dots from red and therefore will not see the green path.

Figure 20-11 is a pedigree that shows the history of red-green color blindness in a family. (A test for color blindness and a description of this disorder are provided in Figure 20-12.) The circles in Figure 20-11 represent females, and the squares represent males. Solid-colored symbols stand for individuals who exhibit the trait being studied. In this case, those individuals would be color blind (Figure 20-12). Parents are represented by horizontal lines connecting a circle and a square. Vertical lines coming from two parents indicate their children, arranged along a horizontal line in order of their birth.

First, to determine whether a trait is sex linked or autosomal, notice whether the trait is expressed more frequently in males than in females. If so, the trait is most likely a sex-linked trait. In addition, the trait would then likely be recessive because few dominant sex-linked traits are known. An autosomal trait (whether dominant or recessive) is expressed equally in both males and females. A sex-linked trait, however, is always expressed in a male, because the relatively inert Y chromosome lacks alleles that correspond to alleles on the X chromosome; one deleterious gene on a male's X chromosome results in its phenotypic expression. Is red-green color blindness a sex-linked trait?

pedigrees (ped uh greez) diagrams of genetic relationships among family members over several generations.

Second, to determine whether a trait is dominant or recessive, notice whether each person expressing the trait has a parent who expressed the trait. In this case, if color blindness is dominant to normal color vision, then each color-blind person will have a color-blind parent. However, if the trait is recessive, a person expressing the trait can have parents who do not express the trait. Both parents may be heterozygous (may carry the trait). If each parent passes on the recessive gene to a child, the child is then homozygous recessive and the trait will be expressed. Is color blindness a dominant or a recessive characteristic?

Next, determine which individuals are carriers of the color blindness gene. By now you probably realize that color blindness is a sex-linked recessive trait. A color blind male always contributes an X chromosome with the defective allele to his daughters; it is the only X he can contribute. His sons, however, get a Y chromosome from him and an unaffected X chromosome from their (normal) mother. A color-blind female can contribute only a chromosome with the defective allele to her children because both her X chromosomes are affected. Women who are carriers of the trait, however, may contribute either a normal or a defective X. The Punnett square in Figure 20-13 illustrates the contributions of genes by a normal mother and a color-blind father (Figure 20-13, *A*) and a color-blind mother and normal father (Figure 20-13, *B*). Can you determine which individuals are carriers of color blindness in the pedigree?

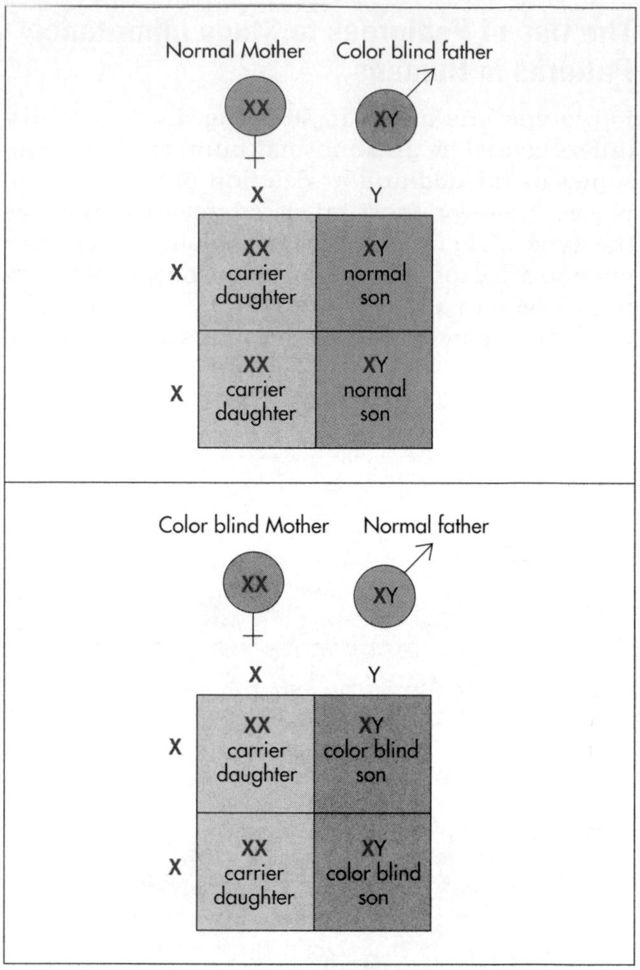

FIGURE 20-13 Punnett squares depicting the outcomes for the sex-linked disorder color blindness. **A** shows a Punnett square for a normal mother and color-blind father. **B** shows a Punnett square for a color-blind mother and normal father.

> *Pedigrees, diagrams of genetic relationships among families over several generations, are useful in determining whether traits are sex linked or autosomal and dominant or recessive. Carriers of recessive genes can also be identified.*

Queen Victoria's family pedigree is shown in Figure 20-14. Persons carrying a mutant allele but not having the disorder are shown by partially colored symbols. Using the strategy outlined in the preceding paragraphs, analyze the pedigree to determine whether hemophilia is sex linked or autosomal and whether it is a dominant or recessive trait. Look back to the chapter opener to confirm your results.

As shown in the royal pedigree, in the six generations since Queen Victoria, 10 of her male descendants have had hemophilia. The British royal family escaped the disorder because Queen Victoria's son King Edward VII did not inherit the defective allele. Three of Victoria's nine children did receive the defective allele, however, and carried it by marriage into many of the royal families of Europe. It is still being transmitted to future generations among these family lines, except in Russia, where the five children of Alexandra, Victoria's granddaughter, were killed soon after the Russian revolution (Figure 20-15).

Dominant and Recessive Genetic Disorders

Most genetic disorders are recessive; these mutant genes are able to persist in the population among carriers. Unfortunately, persons receiving a mutant allele from each of their parents are affected

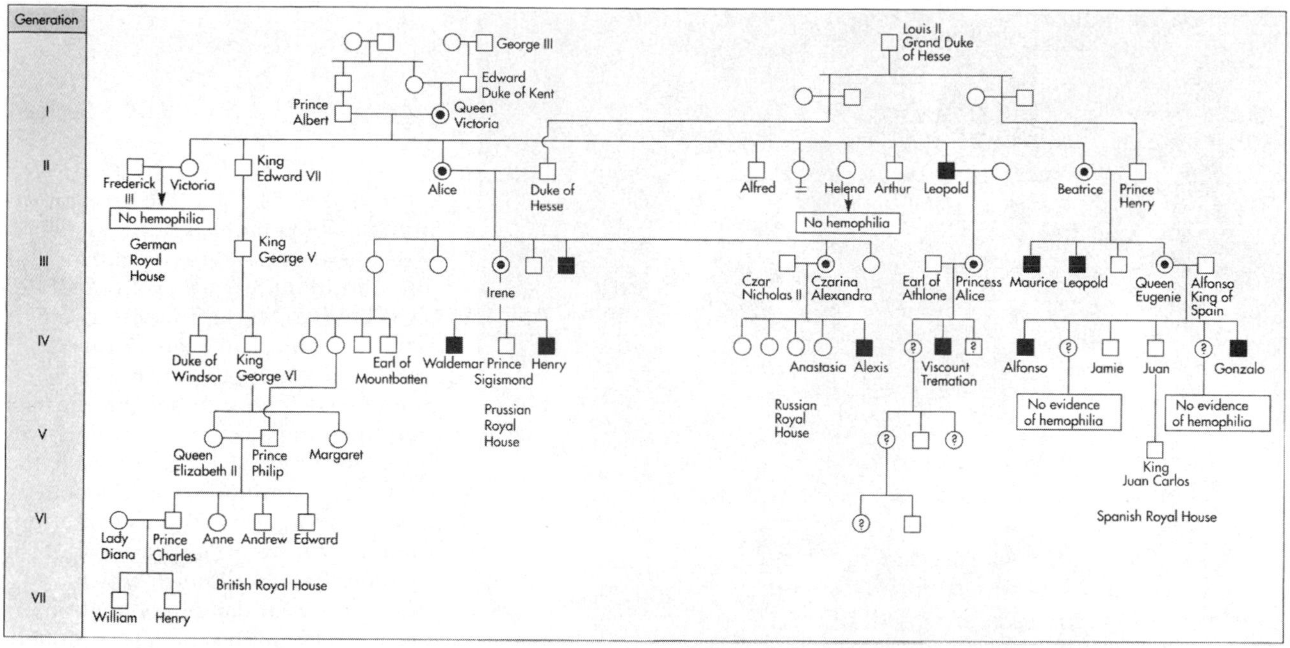

FIGURE 20-14 The royal hemophilia pedigree. From Queen Victoria's daughter, Alice, the disorder was introduced into the Prussian and Russian royal houses, and from her daughter Beatrice, it was introduced into the Spanish royal house. Victoria's son Leopold, himself a victim, also transmitted the disorder in a third line of descent.

with the disease and, in the case of some diseases, may die before reaching adulthood. Likewise, persons having certain dominant disorders may die before reaching reproductive age; therefore lethal dominant disorders are less likely to persist in the population.

Huntington's disease, the disorder that killed folk singer and songwriter Woody Guthrie, is caused by a mutant dominant allele that does not show up until individuals are older than 30 years of age—after they may have had children. This disorder causes progressive deterioration of brain cells. Other dominant genetic disorders include the following:

- *Marfan's syndrome,* which results in skeletal, eye, and cardiovascular defects
- *Polydactyly,* which results in extra fingers or toes
- *Achondroplasia,* which results in a form of dwarfism
- *Hypercholesterolemia,* which results in high blood cholesterol levels and a higher likelihood of developing coronary artery disease

Recessive genetic disorders are often seen primarily within specific populations or races of people unless many members of the population marry and have children outside of their population or race. Among the Caucasian population, for example, the most common fatal genetic disorder is *cystic fibrosis.* Affected individuals secrete a thick mucus that clogs the airways of their lungs and the passages of their pancreas and liver. Among Caucasians, about 1 in 20 individuals has a copy of the defective gene but shows no symptoms. Approximately 1 in 1800 has two copies of the gene and therefore has the disease. These individuals inevitably die of complications that result from their disease. It was learned only recently that the cause of cystic fibrosis is a defect in the way cells regulate the transport of chloride ions across their membranes.

Sickle cell anemia is a recessive disorder most common among African blacks and their descendants. In the United States, for example, about 9% of African-Americans are heterozygous for this allele; about 0.2% are homozygous and therefore have sickle cell anemia. In some groups of people in Africa, up to 45% of the individuals are heterozygous for this allele. Heterozygous carriers of the sickle cell gene do not have the disease but can pass the gene along to their children.

Individuals afflicted with sickle cell anemia are unable to transport oxygen to their tissues properly because the molecules within red blood cells that carry oxygen—molecules of the protein hemoglobin—are defective. Red blood cells that contain large proportions of such defective molecules become sickle shaped and stiff; normal red blood cells are disk shaped and much more flexible (Fig-

FIGURE 20-15 The last Russian royal family. Shown here are Czar Nicholas II of Russia, the last Russian czar, and his wife Alexandra, with their five children, Olga, Tatiana, Maria, Anastasia, and Alexis. Alexandra, Queen Victoria's granddaughter, was a carrier of hemophilia and passed it on to her son Alexis. Because the kind of hemophilia they had is caused by a recessive mutant allele located on the X chromosome, it is not expressed in the heterozygous condition in women because women have two X chromosomes. It is not known which of Alexandra's four daughters might have received the hemophilia allele from her; none of the four daughters lived to bear children.

ure 20-16). As a result of their stiffness and irregular shape, the sickle-shaped red blood cells are unable to move easily through capillaries. Therefore they tend to accumulate in blood vessels, reducing the blood supply to the organs they serve and causing pain, tissue destruction, and an early death.

Interestingly, the gene for sickle cell anemia is most prevalent in the regions of Africa where malaria is prevalent. Malaria is a disease caused by microorganisms that live in a person's red blood cells. These microbes are injected into a person's bloodstream by the bite of a female *Anopheles* mosquito. A long-lasting disease, malaria affects the physical and mental development of its victims, causing damage to many body organs. Scientists have discovered that the defective hemoglobin molecules of the person with sickle cell anemia produce conditions that are unfavorable to the growth of the malaria organism. However, these persons eventually die of their anemia. But persons heterozygous for the sickle cell gene, although not afflicted with the disease, are more resistant to malaria. Heterozygotes, then, have a survival advantage with respect to malaria and sickle cell anemia (although their offspring may be afflicted with sickle cell anemia if they reproduce with persons also carrying the defective gene).

Most genetic disorders are recessive because mutant genes are able to persist in the population among carriers. Some more common recessive genetic disorders among humans are color blindness, hemophilia, cystic fibrosis, and sickle cell anemia.

Tay-Sachs disease is an incurable, fatal recessive hereditary disorder. Although rare in most human populations, Tay-Sachs has a high incidence among Jews of Eastern and Central Europe and among American Jews (90% of whom are descendants of Eastern and Central European ancestors). In fact, geneticists estimate that 1 in 28 individuals in these Jewish populations carries the allele for this disease and that approximately 1 in 3600 infants within this population is born with this genetic disorder. Affected children appear normal at birth but begin to show signs of mental deterioration at about 8 months of age. As the brain begins to deteriorate, affected children become blind; they usually die by the age of 5 years.

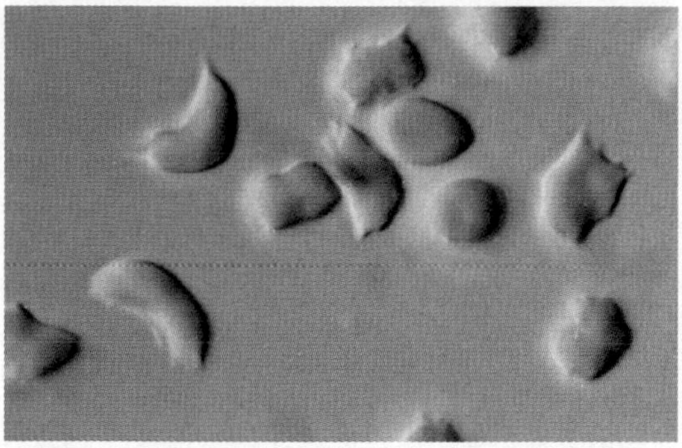

FIGURE 20-16 Sickle-shaped red blood cells. In individuals who are homozygous for the recessive sickle cell trait, many of their red blood cells are sickle-shaped and stiff. These irregularities are caused by defective hemoglobin molecules within the red blood cells. Because of their irregular shape, sickle-shaped red blood cells are unable to move easily through capillaries and become lodged in these blood vessels.

Table 20-5 summarizes the dominant and recessive disorders discussed. Many common human traits are also coded by either dominant or recessive genes. Table 20-6 lists some of these characteristics.

Incomplete Dominance and Codominance

Alleles do not always exhibit clear-cut dominance or recessiveness. The pea plants that Gregor Mendel worked with exhibited complete dominance, and for this reason his work disputed the concept of blended inheritance (see Chapter 19). However, since Mendel's time, researchers have discovered many cases of **incomplete dominance** in which alternative alleles are *not* dominant over or recessive to other alleles governing a particular trait. Instead, heterozygotes are phenotypic "intermediates."

Incomplete dominance is found in various plant and animal traits. For example, a cross between snapdragons having red flowers and those having white flowers yields plants having pink flowers. Crossing black Andalusian chickens with white Andalusian chickens yields an intermediate hybrid that is slate blue. In humans, a curly haired Caucasian and a straight-haired Caucasian will have children with wavy hair. And persons with wavy hair will have children that are either straight haired, wavy haired, or curly haired. Figure 20-17 shows a cross between two wavy-haired people. Notice that an uppercase H denotes the "hair" allele. No lowercase letters are used because neither allele (curly or straight) is recessive. Instead, one allele is designated as *H* and the other as *H'*.

A slightly different situation occurs with alleles that are codominant. **Codominant** alleles are *both* dominant; both characteristics are exhibited in the phenotype. In humans the alleles that code the A,

TABLE 20-5

Dominant and Recessive Disorders

NAME	CHARACTERISTICS
DOMINANT DISORDERS	
Huntington's disease	Progressive deterioration of brain cells
Marfan's syndrome	Skeletal, eye, and cardiovascular defects
Polydactyly	Extra fingers or toes
Achondroplasia	Dwarfism
Hypercholesterolemia	High blood cholesterol levels
RECESSIVE DISORDERS	
Albinism	Lack of pigment in skin, hair, and eyes
Cystic fibrosis	Production of thick mucus that clogs airways
Sickle cell anemia	Defective hemoglobin in red blood cells, effect on oxygen transport Misshapen red blood cells, tendency to form clots in vessels
Tay-Sachs disease	Brain deterioration beginning at 8 months of age
SEX-LINKED RECESSIVE DISORDERS	
Color blindness	Inability to discern certain colors
Hemophilia	Improper clotting of blood
Duchenne muscular dystrophy	Degeneration of muscle
Fragile X syndrome	Easy breakage of tip of X chromosome Long face, squared forehead, large ears Mental retardation

incomplete dominance a term referring to traits in which alternative alleles are neither dominant over nor recessive to other alleles governing a particular trait; heterozygotes for incomplete dominance traits are phenotypic "intermediates."

codominant (**ko dom** uh nunt) a term referring to traits in which the alternative forms of an allele are both dominant, and both characteristics are exhibited in the phenotype.

B, and AB blood types are a good example of codominance. For example, a person having allele A and allele B has blood type AB. In addition to being coded by codominant alleles, blood types are coded by more than two alleles. So far, genes that consist of only two—a pair of—alleles have been discussed. But a gene may be represented by more than two alleles within the population. Some genes, such as the ABO blood type genes, consist of a system of alleles, or **multiple alleles.** In this system of multiple alleles, the alleles A and B exhibit codominance and the O allele is recessive.

> *In traits exhibiting incomplete dominance, alternative forms of an allele are neither dominant nor recessive; heterozygotes are phenotypic intermediates. In traits exhibiting codominance, alternative forms of an allele are both dominant; heterozygotes exhibit both phenotypes.*

Multiple Alleles

As mentioned, human ABO blood types are coded by multiple alleles. In fact, four phenotypes (A, B, AB, and O blood types) are determined by the presence of two out of three possible alleles (A, B, or O)

TABLE 20-6

Common Dominant and Recessive Human Traits

DOMINANT TRAITS
Widow's peak
Hair on middle segment of fingers
Short fingers
Inability to straighten little finger
Brown eyes
A or B blood factor

RECESSIVE TRAITS
Common baldness
Blue or gray eyes
O blood factor
Attached ear lobes

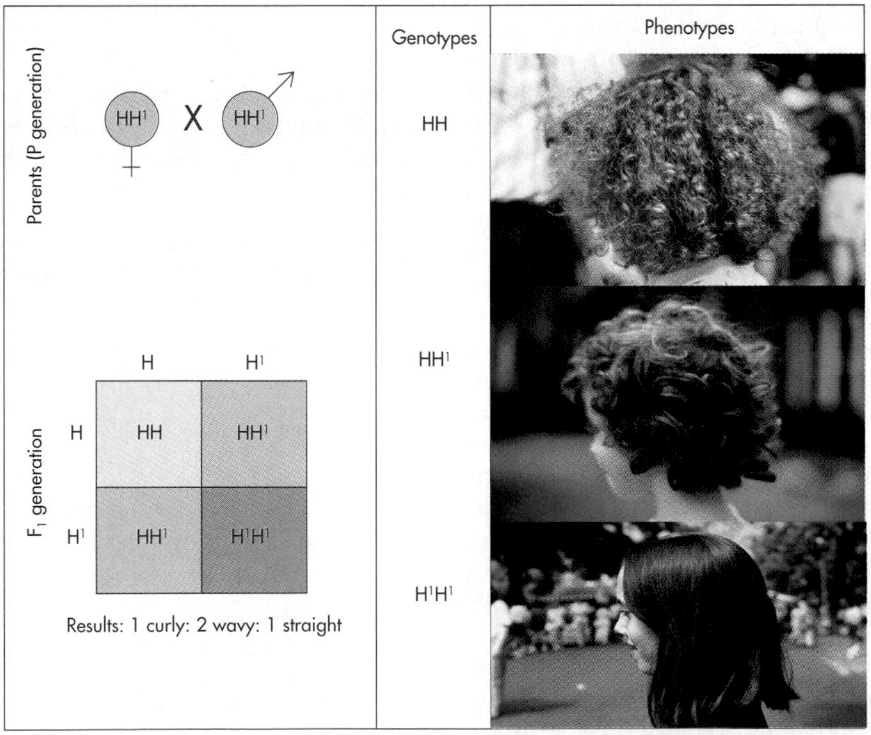

FIGURE 20-17 Incomplete dominance. Sometimes, alternative alleles are not dominant or recessive. For example, neither straight nor curly hair is dominant in Caucasians. The predicted ratio of the offspring of two wavy-haired people will be one curly-haired individual, one straight-haired individual, and two wavy-haired individuals. Wavy hair is the phenotypic intermediate.

JUST WONDERING . . . .

How can you tell who the father of a baby is just by doing blood tests?

No test can *prove* that a particular man is the father of a child, but the paternity tests being performed today can yield a probability of paternity in excess of 99%. Paternity testing involves analyzing the blood of a child, the child's mother, and the suspected father to determine which blood antigens they possess and then using the principles of genetics to determine whether the child could have received its complement of antigens from the hypothetical parents.

There are several types of antigens that are analyzed in paternity testing. The most familiar are the antigens associated with the ABO system. The presence or absence of these antigens is governed by alleles inherited from one's parents (see "Multiple Alleles"). The alleles for each blood type are shown in Figure 20-C.

How can you rule out a suspected father using ABO blood groups? Say a child's mother has type A blood with the genotype shown in the Punnett square, and the suspected father has type AB blood. The child is type O. Is the suspect the father? Examine the Punnett square in Figure 20-D. The blood types possible from an A mother (who must be heterozygous [$I^A i$] to have an O child) and an AB father are types A, B, and AB. An O child could not be born to such parents, and thus the suspect is not the father.

Paternity testing, however, is based on more than ABO blood groups. DNA analysis is now used in conjunction with antigen analysis to test paternity. It is this combination of tests that contributes to the high accuracy of the results.

C

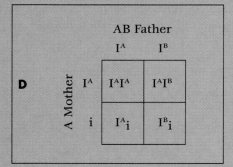

Type	Possible alleles
A	I^AI^A, I^Ai
B	I^BI^B, I^Bi
AB	I^AI^B
O	i i

D

A Mother \ AB Father	I^A	I^B
I^A	I^AI^A	I^AI^B
i	I^Ai	I^Bi

FIGURE 20-C, D

in an individual. The A and B alleles each code for the production of different enzymes. These enzymes add certain sugar molecules to lipids on the surface of red blood cells. These sugars act as recognition markers for the immune system and are called *cell surface antigens.* The enzyme produced by allele A results in the addition of one type of sugar; allele B adds a different sugar. Allele O adds no sugar. Different combinations of the three possible alleles occur in different individuals, with each individual having *one pair* of alleles. Therefore a person having an AA genotype produces only the A sugar. A person having an AB genotype produces both A and B sugars. (The A and B alleles are codominant and both are expressed.) A person having an AO phenotype will produce the A sugar (so is said to have type A blood). The A allele is expressed because it is dominant to the O allele.

Some genes consist of a system of alleles, or multiple alleles, which are usually codominant to one another. The ABO system of human blood types is an example of codominant multiple alleles.

multiple alleles (uh leels) a system of more than two alleles that govern certain traits.

Genetic Counseling

Although most genetic disorders cannot yet be cured, research scientists and physicians are continually learning a great deal about them. Exciting methods of genetic therapy are being studied. Geneticists can identify couples at risk of having children with genetic defects and can help them have healthy children. This process is called **genetic counseling.**

By analyzing family pedigrees and applying the laws of statistical probability, genetic counselors can often determine the probability that a person is a carrier of a recessive disorder. When a couple is expecting a child and both parents have a significant probability of being carriers of a serious recessive genetic disorder, the pregnancy is said to be a high-risk pregnancy. In such a pregnancy, there is a significant probability that the child will exhibit the clinical disorder. Another class of high-risk pregnancies occurs with mothers who are older than 35 years of age. Remember, the frequency of birth of infants with Down syndrome increases dramatically among the pregnancies of older women (see Figure 20-4).

When a pregnancy is diagnosed as being high risk, many women select to undergo *amniocentesis,* a procedure that permits the prenatal diagnosis of many genetic disorders (Figure 20-18). In the fourth month of pregnancy, a sterile hypodermic needle is used to obtain a small sample of amniotic fluid from the mother. The amniotic fluid, which bathes the fetus, contains free-floating cells derived from the fetus. Once removed, these cells can be grown as tissue cultures in the laboratory. Studying these tissue cultures, genetic counselors can test for many of the most common genetic disorders.

During amniocentesis, the position of the needle in relationship to the fetus is observed by means of a technique called ultrasound. The term *ultrasound* refers to high-frequency sound waves. Pulses of these waves are sent into the body and are reflected back in various patterns depending on the tissues or fluids the waves hit. These patterns of sound-wave reflections are then mapped to produce

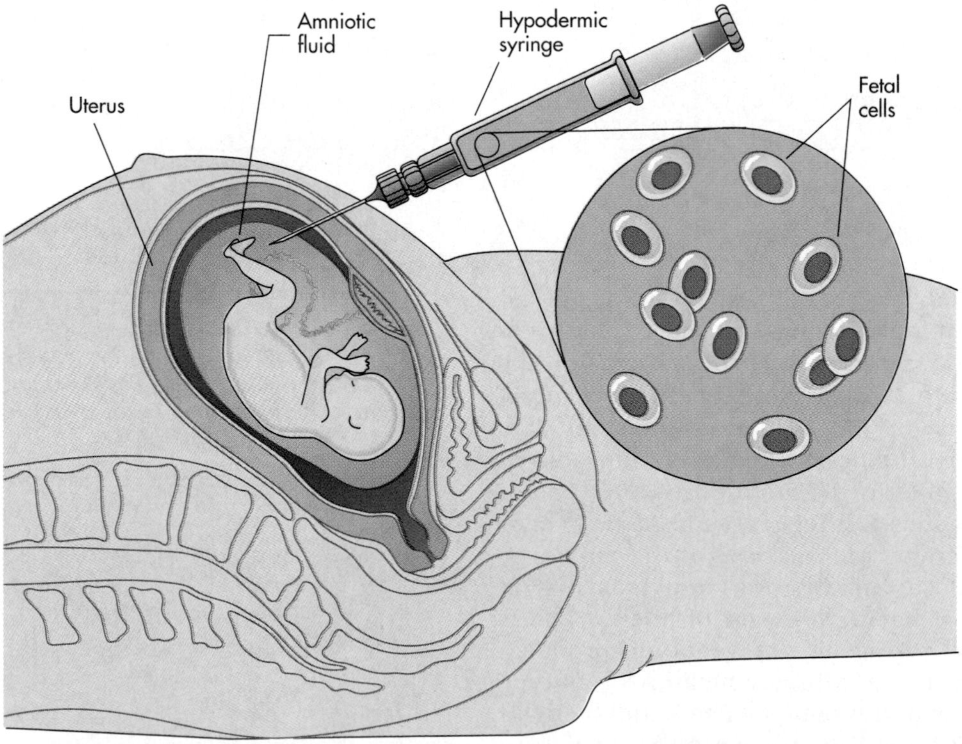

FIGURE 20-18 Amniocentesis. A needle is inserted into the amniotic cavity, and a sample of amniotic fluid, containing some free cells derived from the fetus, is drawn into a syringe. The fetal cells are then grown in tissue culture so that their karyotype and many of their metabolic functions can be examined.

a picture of inner tissues as in Figure 20-19. Because ultrasound allows the position of the fetus to be determined, the person performing the amniocentesis can avoid damaging the fetus. Ultrasound also allows the fetus to be examined for the presence of major abnormalities.

At this time, most of the problems that can be diagnosed by pedigree analysis, amniocentesis, chorionic villus sampling and ultrasound techniques cannot be treated. Sometimes the only options available to a couple are to continue the pregnancy and deal with the problems after birth or to have a therapeutic abortion. However, continual progress in developing gene therapies may, in the future, have a drastic impact on a physician's ability to treat genetic disorders before birth.

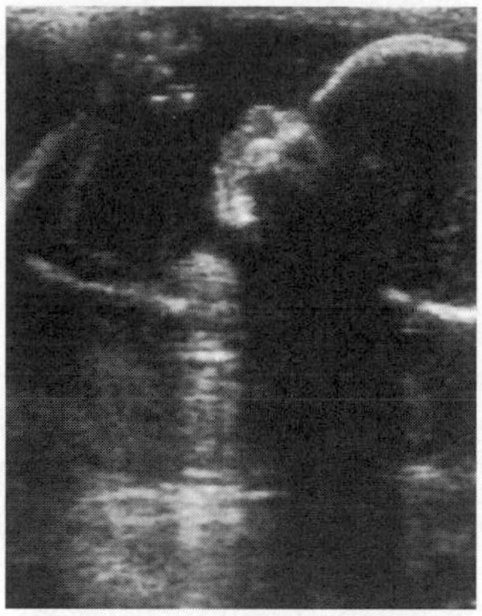

FIGURE 20-19 Ultrasound. Before amniocentesis, the fetus is observed with ultrasound to determine position. Imaging the fetus using ultrasound before amniocentesis is important because the amniocentesis needle must be inserted precisely without damaging the fetus.

> **genetic counseling** the process in which a geneticist discusses a couple's genetic history early in a pregnancy or before conception to determine if their offspring may be at risk for a variety of genetic disorders.

Summary

- Human cells contain 46 chromosomes: 44 autosomes and 2 sex chromosomes. The autosomes form 22 pairs of homologous (matched) chromosomes. In females, the two sex chromosomes are similar in size and appearance and are designated XX. In males, the two sex chromosomes are not similar in size and appearance. The smaller of the two is designated Y and the larger X. The genes that determine male characteristics are located on the Y chromosome.
- Researchers often use karyotypes to study the inheritance of genetic disorders caused by the loss of all or part of a chromosome or by the addition of extra chromosomes or chromosome fragments. A karyotype is a picture of chromosomes paired with their homologues and arranged according to size.
- In humans the inheritance of one autosome too few usually results in a nonviable zygote. The inheritance of one autosome too many, or a trisomy, also results in a nonviable zygote with the exception of chromosome 21. Individuals with an extra copy of this chromosome are retarded and have Down syndrome. Down syndrome is much more frequent among pregnant mothers older than 35 years of age, and it occurs when two sister chromatids fail to separate, or disjoin, during the meiotic process.
- Nondisjunction can occur with the sex chromosomes. Persons who inherit abnormal numbers of sex chromosomes often have abnormal features and may be mentally retarded. Examples of genetic disorders caused by abnormal numbers of sex chromosomes are poly-X females, XXY males (Klinefelter syndrome), XO females (Turner syndrome), and XYY males.
- Persons may inherit chromosomes that are structurally defective due to chromosomal breaks and misrepairs or changes taking place in the molecular structure of the chromosomal

DNA. The three major sources of environmental damage to chromosomes are (a) high-energy radiation, such as x-rays; (b) low-energy radiation, such as UV light; and (c) chemicals, such as LSD, marijuana, cyclamates, and certain pesticides.

- Cancer is not a single disease but is a cluster of more than 200 diseases. All cancers have four characteristics: (a) uncontrolled cell growth, (b) loss of cell differentiation (specialization), (c) invasion of normal tissues, and (d) metastasis, or spread, to multiple sites.
- Cancer cells arise through a series of stepwise, progressive mutations, or changes, to the DNA of normal body cells. The first set of changes involves protooncogenes (dormant forms of cancer-causing genes) and/or tumor–suppressor genes (genes that can be thought of as "off" switches in the development of cancer). These first mutations result in precancerous cells, which are termed *initiated* or *transformed cells.*
- For cancer to occur, initiated cells must undergo promotion, a process by which the DNA of initiated cells is damaged further, eventually stimulating these cells to grow and divide. Research suggests that if the promoter (the substance causing promotion) is withdrawn in the early stages, cancer development can be reversed.
- Patterns of inheritance observed in family histories, or pedigrees, can be used to determine the mode of inheritance of a particular trait. By such analysis, it can often be determined whether a trait is associated with a dominant or a recessive allele and whether the gene determining the trait is sex linked.
- Some genetic disorders are relatively common in human populations, whereas others are rare. Many of the most common genetic disorders are associated with recessive alleles, the functioning of which may lead to the production of defective versions of enzymes that normally perform critical functions. Because such traits are determined by recessive alleles and therefore expressed only in homozygotes, the alleles are not eliminated from the human population. Dominant alleles that lead to severe genetic disorders are less common; in some of the more frequent ones, the expression of the alleles does not occur until after the individuals have reached their reproductive years.
- Alleles do not always exhibit clear-cut dominance or recessiveness but may instead exhibit incomplete dominance or codominance. In traits exhibiting incomplete dominance, alternative forms of a trait are neither dominant nor recessive; heterozygotes are phenotypic intermediates. In traits exhibiting codominance, alternative forms of an allele are both dominant; heterozygotes exhibit both phenotypes. Some genes consist of a system of alleles, or multiple alleles, which are usually codominant to one another. The ABO system of human blood types is an example of multiple alleles with two of the alleles (A and B) codominant to one another and one allele (O) recessive to the other two.
- Although most genetic disorders cannot yet be cured, research scientists are making progress in developing gene therapies that may, in the future, have an important impact on a physician's ability to treat genetic disorders before birth.

Knowledge and Comprehension Questions

1. What is Down syndrome and what causes it? Give another term for this condition, and summarize its symptoms.
2. What happens when human offspring inherit abnormal numbers of sex chromosomes? Summarize the four examples discussed in the chapter, and give the genotype of each.
3. Which sex chromosome is necessary for the survival of the developing zygote?
4. What are the three major sources of damage to chromosomes?
5. Describe how heavy use of drugs could affect a woman's ova or a man's sperm.
6. What common factors must exist for a disease to be labeled as a form of cancer?
7. In what manner does mutation affect protooncogenes? In what manner does mutation affect tumor-suppressor genes? What are the clinical results?
8. Lynn's doctor informs her that the pathology report on her cyst indicates a benign growth. What does this mean?
9. What would you conclude about the inheritance pattern of each human trait in each of the following situations?
   a. The trait is expressed more frequently in males than in females.
   b. Offspring who exhibit this trait have at least one parent who exhibits the same trait.
   c. Offspring can exhibit this trait even though the parents do not.

d. The trait is expressed equally in both males and females.

10. Why do most genetic disorders in humans result from recessive genes? Name several examples.
11. Distinguish between incomplete dominance and codominance. Describe the phenotype of a heterozygote in each case.
12. A woman whose blood type is AB marries a man with the same blood type. Draw a Punnett square to illustrate the possible genotypes of their children. What blood type will each genotype have?
13. If a couple want to have a child but suspect that they may be at risk for a genetic disorder, what can they do? If a pregnancy turns out to be a high risk, what options are available?

Critical Thinking

1. The extra chromosome 21 that is found in persons with Down syndrome is the cause of multiple developmental defects. What might this tell you about the interaction of genes on a particular chromosome?
2. Design a family medical tree identifying known cancers in your medical history. Go back as many generations as possible and include aunts, uncles, and cousins. What cancers emerge as hereditary or potentially a health trend in your family?
3. What factors in your life and environment predispose you toward cancer (excluding heredity)? How many of these can you personally restrict or eliminate? Have you done so?

Human Genetics Problems

1. George has the same type of hemophilia as did Queen Victoria and some of her descendants. He marries his mother's sister's daughter Patricia. His maternal grandfather also had hemophilia. George and Patricia have five children: two daughters are normal, and two sons and one daughter develop hemophilia. Draw the pedigree.
2. A couple with a newborn baby are troubled that the child does not appear to resemble either of them. Suspecting that a mix-up occurred at the hospital, they check the blood type of the infant. It is type O. Because the father is type A, and the mother is type B, they conclude that a mistake must have been made. Are they correct?
3. How many chromosomes would you expect to find in the karyotype of a person with Turner syndrome?
4. A woman is married for the second time. Her first husband was ABO blood type A, and her child by that marriage was type O. Her new husband is type B, and their child is type AB. What is the woman's ABO genotype and blood type?
5. Total color blindness is a rare hereditary disorder among humans in which no color is seen, only shades of gray. It occurs in individuals homozygous for a recessive allele and is not sex linked. A non–color-blind man whose father is totally color-blind intends to marry a non–color blind woman whose mother was totally color-blind. What are the chances that they will produce offspring who are totally color-blind?
6. This pedigree is of a rare trait in which children have extra fingers and toes. Which one of the following patterns of inheritance is consistent with this pedigree?
   a. Autosomal recessive
   b. Autosomal dominant
   c. Sex-linked recessive
   d. Sex-linked dominant
   e. Y-linkage

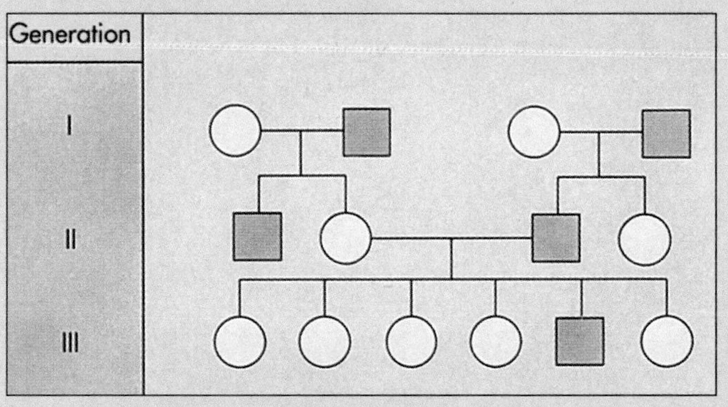

CHAPTER 21

SEX AND REPRODUCTION

THEIR NUCLEI GLOWING with a special fluorescent dye, these sperm are all moving in the same direction, propelled by their whiplike flagella. Only one of the hundreds of millions introduced into the vagina will be the first to encounter the female gamete—a developing egg. Enzymes in the sperm head will dissolve the jellylike covering of the egg and pierce its membrane. Once the union of egg and sperm takes place, changes on the surface of the egg block the entry of additional sperm, ensuring that the fertilized egg will have only one complete set of hereditary material. Having more than one complete set could be fatal to a new embryo's existence. But with a set of genes from each of its parents and with conditions right for survival, a single human embryo has the potential to develop into the complex network of 100 trillion cells that makes up the human body.

Reproduction

Each male sex cell, or **spermatozoon (sperm),** is a highly motile cell that can penetrate the membrane of a secondary oocyte (potential egg). After penetration, a sperm joins its half of the hereditary message—the "blueprints" for the development of a new individual—with the hereditary material of the oocyte. This process, whereby a male and a female sex cell combine to form the first cell of a new individual, is called *sexual reproduction.* The organs of the human male and female reproductive systems are the organs of sexual reproduction. The male and female reproductive organs are called **gonads** and produce the sex cells, or **gametes** (eggs and sperm). Other organs and ducts store, transport, and receive the sex cells. Accessory glands help nourish the sex cells.

The Male Reproductive System

The function of the male reproductive system is to produce sperm cells and transport them to the female reproductive tract. Sperm are produced in the male gonads, the **testes.**

The Production of Sperm: The Testes

The testes, or testicles (Figure 21-1), are located outside the lower pelvic area of the male, housed within a sac of skin called the **scrotum** (Figure 21-2). The placement of this organ allows the sperm to successfully complete their development at a temperature slightly lower than the 37° C (98.6° F) internal temperature of the human body. Tight jeans or other clothing that presses the testes close to the body can have a detrimental effect on sperm development, lowering the number of viable sperm that a man produces.

Each testis is packed with approximately 750 feet of tightly coiled tubes called **seminiferous tubules** (Figure 21-3, *A*). Male gametes, or sperm cells, develop within the tissue lining these tubules during a process of meiosis (see Chapter 18) and development called *spermatogenesis* (literally, "the making of sperm"). Hundreds of millions of sperm are produced by this process each day! Spermatogenesis usually begins during the teenage years when a boy reaches sexual maturity, or *puberty,* and continues

KEY CONCEPTS

- Sexual reproduction in humans, as well as in most sexually reproducing organisms, is a process in which a male and female sex cell unite to form the first cell of a new individual.
- The male reproductive system produces male sex cells called *sperm* and deposits them in the female reproductive tract during sexual intercourse.
- The female reproductive system receives the male sex cells, produces female sex cells called *eggs,* and provides an environment for a fertilized egg to grow and develop into a new individual.
- The human sexual response promotes sexual reproduction; techniques to avoid conception during sexual activity vary in their methods and success rates.

OUTLINE

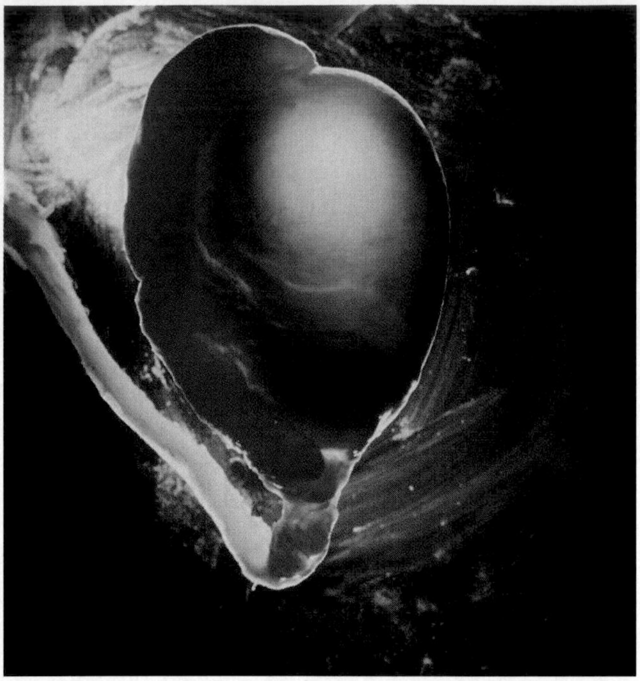

FIGURE 21-1 Human testis. The testis is the ovoid structure in the center of the photograph; sperm are formed within it. Cupped alongside of the testis is the epididymis, a highly coiled passageway within which sperm complete their maturation. Extending away from the epididymis is a long tube, the vas deferens.

throughout his life. Sperm production is triggered by follicle-stimulating hormone (FSH).

Luteinizing hormone (LH) regulates the testes' secretion of testosterone, a sex hormone responsible for the development and maintenance of male secondary sexual characteristics. These characteristics, such as a deepening of the voice and the growth of facial hair, begin to develop at puberty and are signs that sexual maturation is taking place.

Both FSH and LH are secreted by the pituitary, a tiny gland that hangs from the underside of the brain (see Chapter 14). The secretion of these hormones is regulated by means of a negative feedback loop involving the hypothalamus, a tiny portion of the brain that controls the pituitary. A high level of testosterone in the blood inhibits the hypothalamus from triggering the pituitary to release FSH and LH. Without stimulation by these hormones, the testes produce fewer sperm and less testosterone. When the testosterone level drops too low, the hypothalamus once again triggers the release of FSH and LH from the pituitary, causing the testes to step up their production of sperm and testosterone.

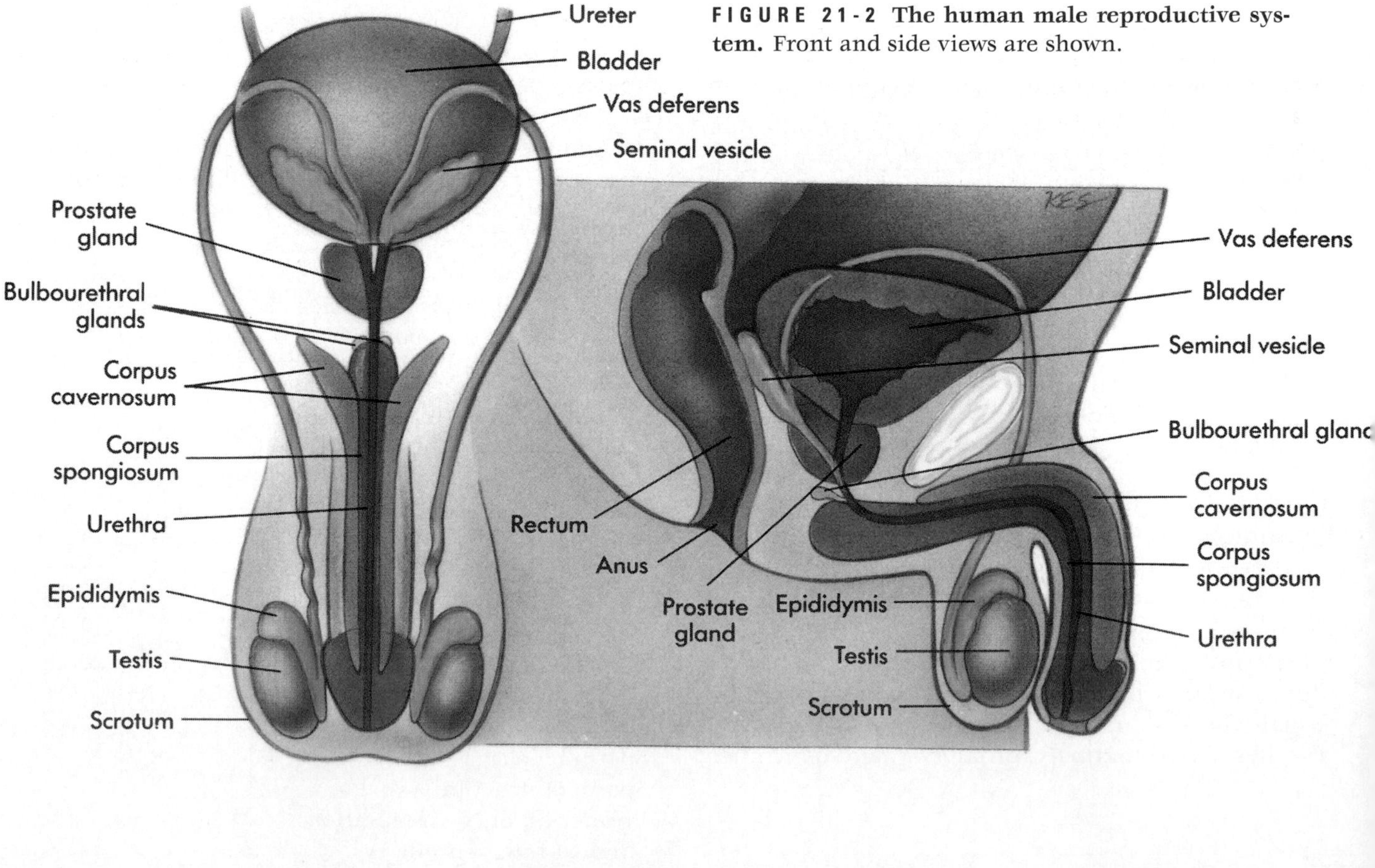

FIGURE 21-2 The human male reproductive system. Front and side views are shown.

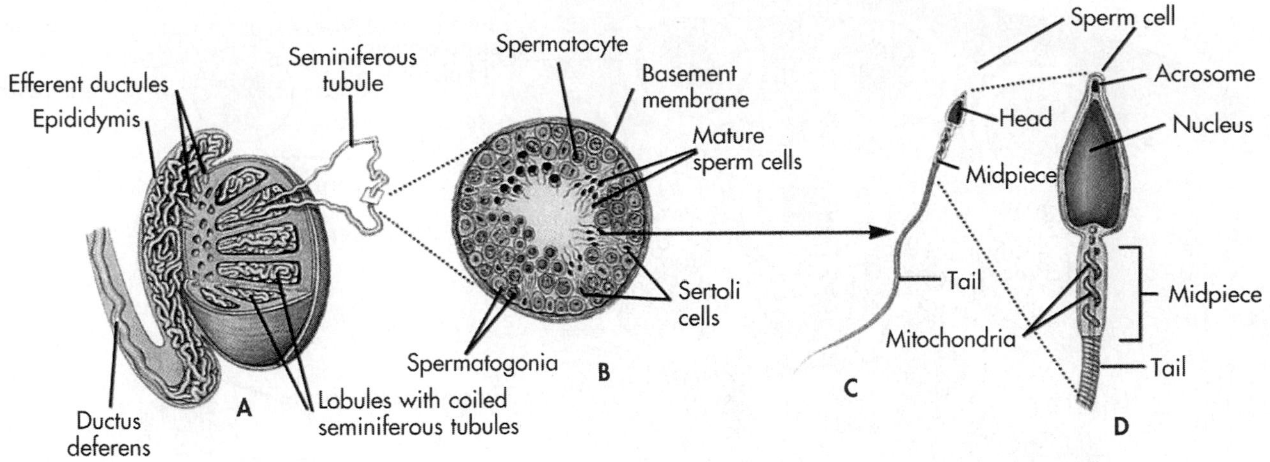

FIGURE 21-3 The interior of the testis, site of spermatogenesis. A, Seminiferous tubules within the testis is where the process of spermatogenesis occurs. **B,** Cross-section of the seminiferous tubule in which diploid cells (spermatogonia) develop into haploid cells (sperm), passing through the spermatocyte and spermatid stages (see Figure 21-4). Each sperm cell (**C**) has a long tail attached to a head. **D,** An enlarged view of the head of a sperm showing the haploid nucleus, the acrosome that contains enzymes that aid in penetrating the egg, and mitochondria, which provide the energy necessary to propel the sperm.

> *Sperm production, or spermatogenesis, takes place within the coiled tubules of the testes and is triggered by the hormone FSH. The development of functionally mature sperm also requires LH and testosterone.*

Packed within the walls of the seminiferous tubules are cells called *spermatogonia,* the cells that give rise to sperm (Figure 21-3, *B*). These cells have a full complement (diploid amount) of hereditary material and constantly produce new cells by the process of mitosis (see Chapter 18). Some of the new (daughter) cells produced by mitotic cell division move inward toward the interior of the tubule and undergo meiosis (Figure 21-4). The reduction-division of meiotic cell division produces four cells from each original cell, each one having *half* the normal amount (or haploid amount) of hereditary material. These haploid cells are now called *spermatids.* Each spermatid contains one member from each pair of the 23 chromosome pairs present in the cells of the human body. Included in these 23 chromosomes is either an X sex chromosome or a Y sex chromosome, which will determine the gender of the new individual.

The spermatids then undergo a process of development, producing sperm cells, or spermatozoa. Spermatozoa are relatively simple cells (see Figure 21-3, *C* and *D*). Each has an anterior portion, or head, that consists primarily of a cell membrane encasing hereditary material. A vesicle called an *acrosome,* derived from the Golgi body, is located at the leading tip of the sperm cell. It contains enzymes that aid in the penetration of the protective layers surrounding the egg. In addition, each spermatozoon has a whiplike tail, or flagellum, that propels the cell and mitochondria, which produce adenosine triphosphate (ATP) from which a sperm derives energy that powers its flagellum. Sperm development takes about 2 months from spermatogonia to mature spermatozoa. On average, 300 million mature sperm are produced per day.

spermatozoon (sperm) (spur **mat** uh **zoo** un) a male sex cell.
gonads (**go** nads) the male and female reproductive organs that produce the sex cells, or gametes.
gametes (**gam** eets *or* guh **meets**) sex cells; the female gamete is the egg, and the male gamete is the sperm.
testes (**tes** teez) the male gonads, where sperm production occurs.
scrotum (**skro** tum) a sac of skin, located outside the lower pelvic area of the male, which houses the testicles, or testes.
seminiferous tubules (sem uh **nif** uhr us **too** byools) tightly coiled tubes within each testis, where sperm cells develop.

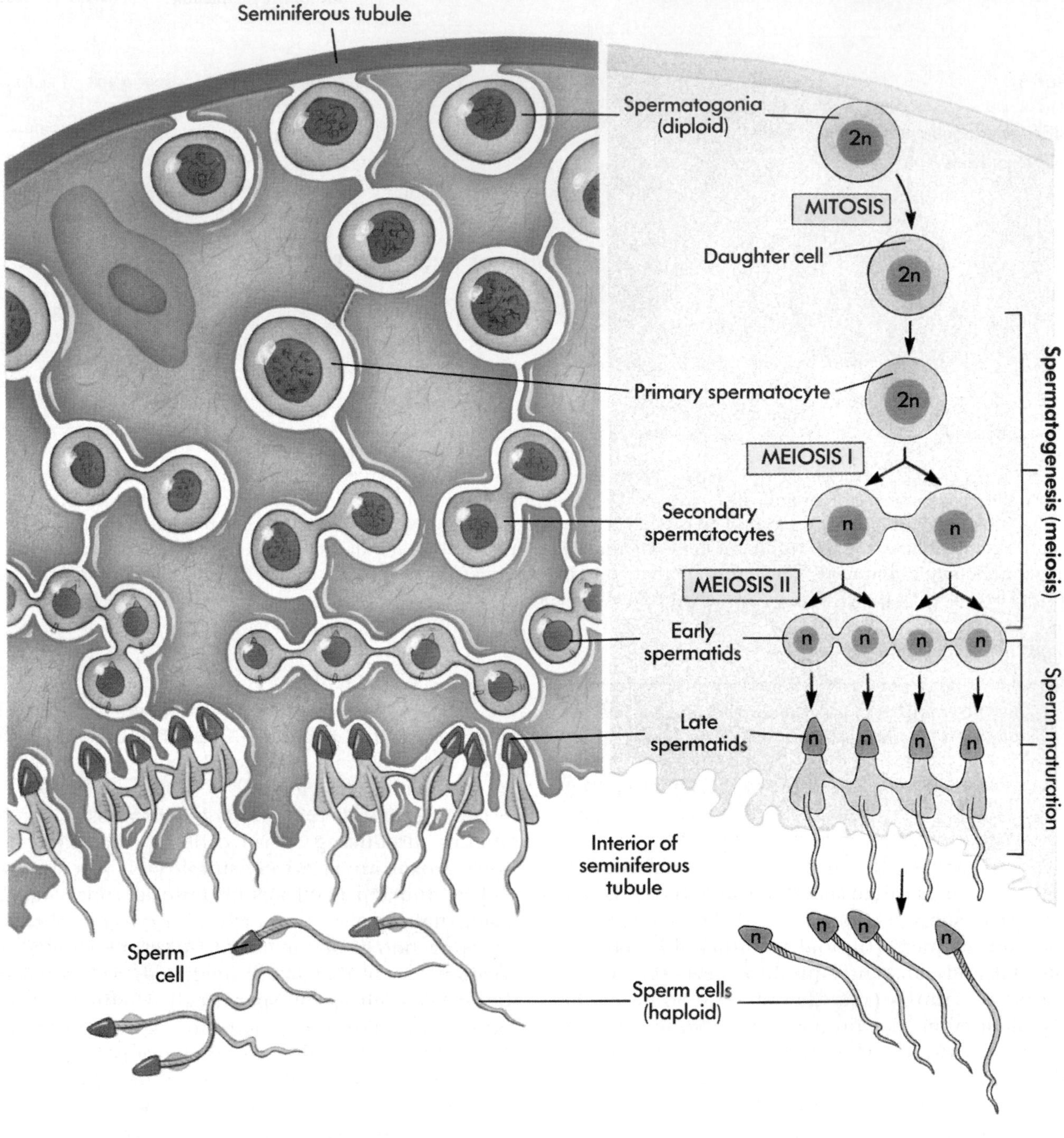

FIGURE 21-4 Spermatogenesis. Spermatogonia undergo mitosis to produce more spermatogonia. Some of the spermatogonia, however, move to the interior of the seminiferous tubule and undergo meiosis. Meiotic cell division produces four haploid spermatids that will eventually develop into spermatozoa, or sperm.

In the testes, diploid cells called spermatogonia give rise to haploid cells called spermatids. These haploid cells contain half the normal amount of hereditary material of body cells. The spermatids then develop into sperm cells, or spermatozoa, that have a head containing the hereditary material, a vesicle containing enzymes to penetrate an egg, and a whiplike tail.

Maturation and Storage: The Epididymis and Vas Deferens

After formation within the testes, the sperm moves through its tubules to a nearby long, coiled tube called the **epididymis** (see Figure 21-3, *A*) located on the back side of the testes. Here the sperm undergo further development and may be stored for about 18 hours to 10 days, becoming capable of propelling themselves and penetrating an oocyte. The sperm then move toward a connecting tube, the **vas**

deferens, a long tube that ascends from the epididymis into the pelvic cavity, looping over the side of the urinary bladder (see Figure 21-2). Stored in the epididymis and the very beginning of the vas deferens, sperm retain their ability to fertilize an oocyte for a short while. If they are not ejaculated, "old" sperm are reabsorbed by the body to make room for "new" sperm.

> *Spermatozoa mature in the epididymis, a long, coiled tube that sits on the back side of the testes and leads to the long ascending vas deferens.*

Nourishment of the Sperm: The Accessory Glands

When a male ejaculates (see p. 450), the sperm are propelled through the vas deferens to the **urethra,** where the reproductive and urinary tracts join. On the way, accessory glands add fluid to the sperm, producing an ejaculate called *semen.* The semen leaves the body through the urethral opening at the tip of the penis.

The first accessory glands to add fluid to the traveling sperm are the **seminal vesicles.** This pair of glands has short ducts that empty a thick, clear fluid into each vas deferens just before they join and flow into the urethra. The fluid secreted by the seminal vesicles is primarily composed of the sugar fructose, which serves as a source of energy for the sperm.

Near its beginning, the urethra is surrounded by a gland called the **prostate.** About the size of a chestnut, the prostate adds a milky alkaline fluid to the semen. This fluid neutralizes the acidity of any traces of urine left in the urethra and also neutralizes the acidity of the female vagina. Sperm are unable to swim in an acid environment, so the secretion of the prostate ensures their motility.

Just beneath the prostate lies a set of tiny, round accessory glands called the **bulbourethral glands.** The purpose of these glands is somewhat of a mystery. Like the prostate, the bulbourethral glands secrete an alkaline fluid into the semen. However, they secrete only a drop or two, which precedes the ejaculate. This secretion does carry sperm to the outside before ejaculation, making the withdrawal method of birth control highly unreliable (see p. 458). Scientists think that the purpose of this secretion may be to neutralize the acidity of the urethra just before ejaculation as a further protection for sperm.

> *During ejaculation, sperm are propelled through the vas deferens to the urethra and out the penis. Fluid is added to the sperm from various accessory glands along the way. This fluid contains substances that nourish the sperm and neutralize the acidity of the vagina.*

The Penis

The scrotum (which encloses the testes and epididymides) and the penis are the two male external sexual organs, or **external genitals.** The **penis** is a cylindrical organ that transfers sperm from the male reproductive tract to the female reproductive tract. As shown in the cross-sectional view of the penis in Figure 21-5, this organ is composed of three cylinders of spongy erectile tissue. Two veins run along the top surface of the penis. Beneath these veins, two of the cylinders of erectile tissue sit side by side. And beneath them lies a third cylinder, surrounding the urethra.

The tissue that makes up the three erectile cylinders is riddled with spaces between its cells. These spaces normally contain a small amount of blood and are lined by a layer of flattened cells similar to the inner lining of blood vessels. During sexual stimulation, nerve impulses from the central nervous system cause the arterioles leading into this tissue to widen, causing additional blood to collect within the spaces. Pressure from the increased vol-

epididymis (**ep** ih **did** uh mis) a long, coiled tube that sits on the back side of the testes and in which sperm mature.

vas deferens (vas **def** uh renz) paired tubes that ascend from the epididymides into the pelvic cavity, looping over the side of the urinary bladder and eventually joining at the urethra.

urethra (yoo **ree** thruh) a muscular tube that brings urine from the bladder to the outside; in men, the urethra also carries semen to the outside of the body during ejaculation.

seminal vesicles (**sem** un uhl **ves** uh kuls) two accessory glands that secrete a thick, clear fluid, primarily composed of fructose, that forms a part of the semen.

prostate (**pross** tate) a gland that surrounds the male urethra and adds a milky alkaline fluid to the semen.

bulbourethral glands (**bull** bo yoo **ree** thrul) a set of tiny glands lying beneath the prostate that secrete an alkaline fluid into the semen.

external genitals (**jen** uh tuls) sexual organs, such as the penis, that are located on the outside of the body.

penis (**pee** nis) a cylindrical organ that transfers sperm from the male reproductive tract to the female reproductive tract; it is also the male urinary organ.

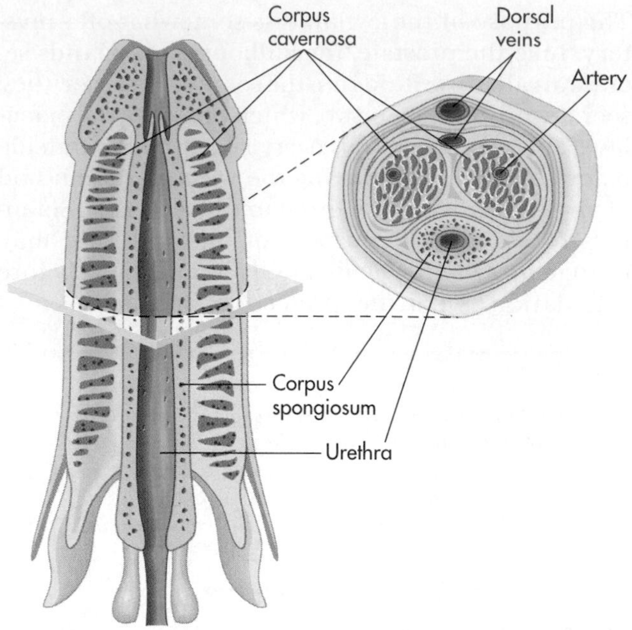

FIGURE 21-5 A penis in cross-section and longitudinal section. The three cylinders of spongy tissue that make up the penis are the corpus spongiosum (middle) and the corpora cavernosa (one cylinder on each side).

ume of blood within the spaces causes the erectile tissue to become distended and compresses the veins that normally drain blood from the penis, causing it to become erect and rigid. Continued stimulation by the central nervous system is required for this erection to be maintained.

Erection can be achieved without any physical stimulation of the penis. However, physical stimulation usually is required for the ejaculation of semen, which corresponds to the culmination of sexual excitement, or orgasm. Prolonged stimulation of the penis, as by repeated thrusts into the female's vagina, leads first to the mobilization of the sperm. In this process, muscles encircling the vas deferens contract, moving the sperm into the urethra. Then ejaculation takes place: muscles at the base of the penis and within the walls of the urethra contract repeatedly, ejecting approximately a teaspoon of semen containing about 200 million sperm out of the penis.

The Female Reproductive System

The part of the female reproductive tract that receives the sperm is called the **vagina,** a muscular tube 9 to 10 centimeters ($3^1/_2$ inches) long (Figure 21-6). But the female reproductive tract does more

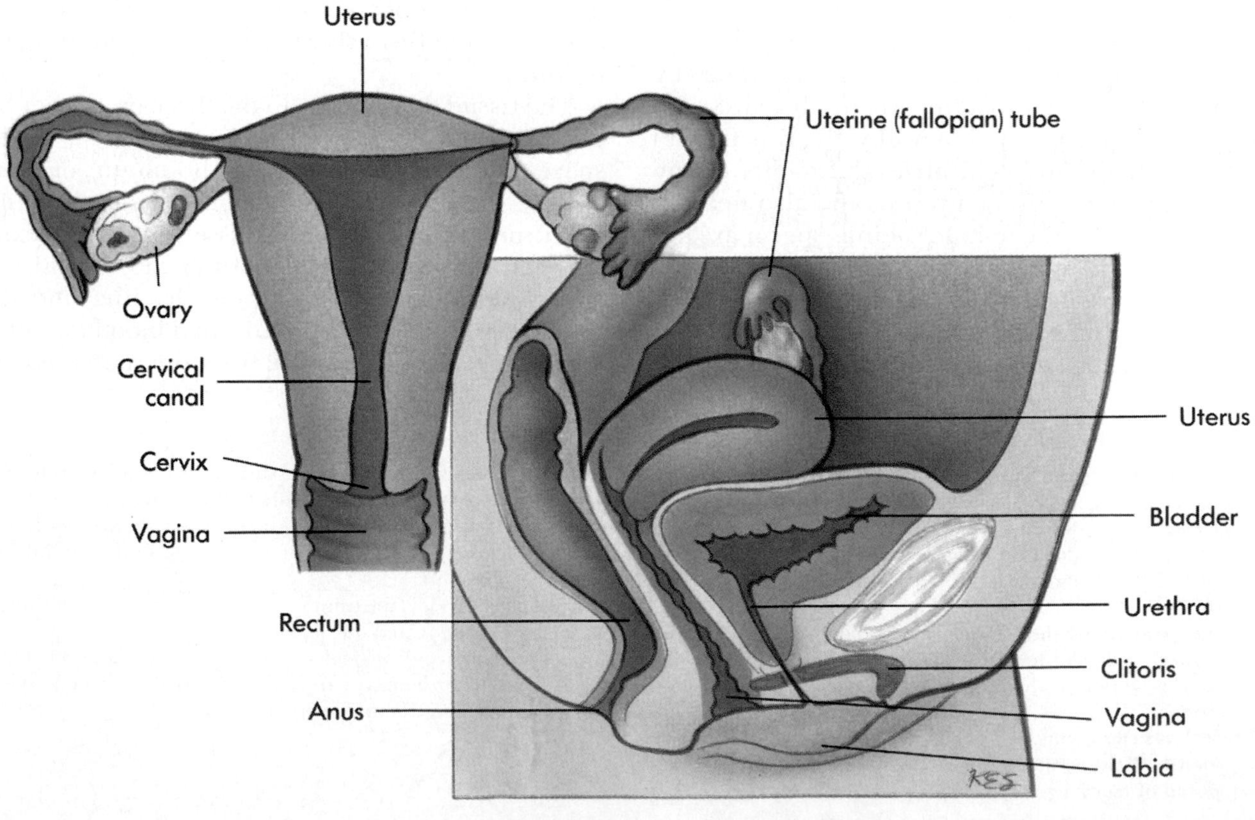

FIGURE 21-6 The human female reproductive system. Front and side views are shown.

than receive sperm. It usually produces one secondary oocyte (potential egg) each month and transports it to where it can be fertilized and the resultant zygote nourished as it develops into an embryo and then a fetus (see Chapter 22).

The Production of Eggs: The Ovaries

Secondary oocytes are produced in the female gonads, or **ovaries** (see Figure 21-6). These two almond-shaped organs are about 3 centimeters long, slightly larger than the distance from the last knuckle on your thumb to the thumb's tip. They are located in the lower portion of the abdominal cavity in an area called the *pelvic cavity.* The ovaries contain *primary oocytes,* cells that have the potential to develop into eggs. They are surrounded by supporting cells called *follicular cells.* Together the follicular cells and the oocyte are called a *follicle,* a word derived from a Latin word meaning "bag" or "sac."

The process of meiosis and development that produces mature female sex cells, or eggs, is called *oogenesis* (literally, "the making of eggs"). Unlike males, whose spermatogonia are constantly dividing to produce new spermatocytes, females have all of the primary oocytes at birth that they will ever produce—about 1 to 2 million of them! (By the age of puberty, more than half these oocytes have degenerated, leaving about 400,000 viable oocytes.) These primary oocytes were produced during fetal development from diploid cells called *oogonia.* The oocytes in the ovaries of a female began the process of meiosis before birth, but the process stopped during the first meiotic division (Figure 21-7). In other words, a female's primary oocytes are in a developmental holding pattern during her lifetime. As a woman ages, her oocytes age. The oocytes are continually exposed to chromosomal mutation, or change, throughout life. After 35 years, the odds of a harmful mutation having occurred are appreciable. This is one reason that developmental abnormalities occur with increasing frequency in pregnancies of women who are over 35 years old.

With the onset of puberty, one oocyte completes the first meiotic division approximately every 28 days. As a result of this division, two cells of unequal size are produced. One is called the *secondary oocyte* and contains most of the cytoplasm of the primary oocyte. The other cell, called a *polar body,* contains little cytoplasm and soon dies. The secondary oocyte breaks away from the ovary during **ovulation** and continues with the second meiotic division only if it is fertilized by a sperm. After the completion of the second meiotic division, the second polar body dies and the egg, or *ovum,* is mature. The haploid nuclei of the sperm and the egg fuse, producing a **zygote,** the first cell of a new individual.

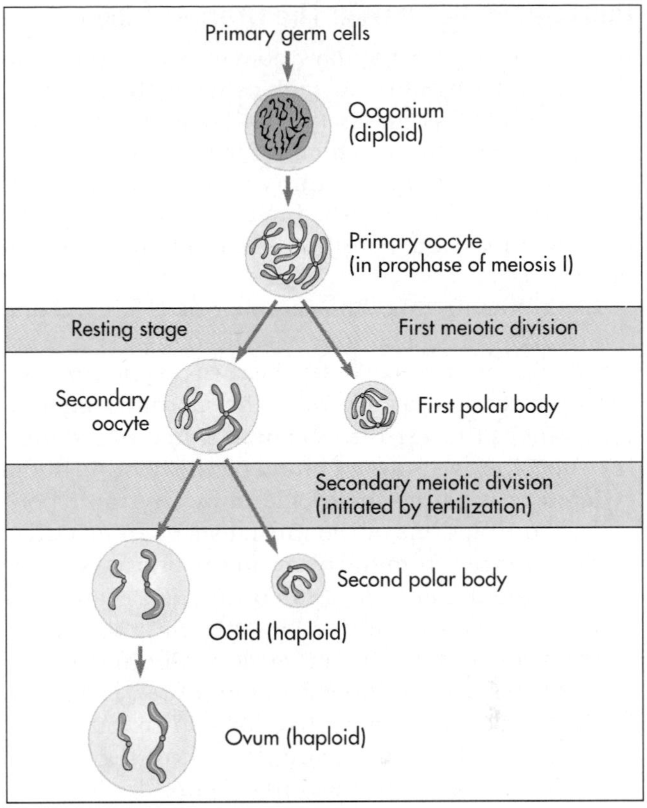

FIGURE 21-7 Oogenesis. Notice that the oocytes have already completed part of the first meiotic division and are in a holding pattern until puberty. One oocyte completes the first meiotic division approximately every 28 days.

The ovaries are made up of primary oocytes—cells that have the potential to develop into eggs—surrounded by supporting cells called follicular cells. At birth, females have all of the primary oocytes that they will ever produce. With the onset of puberty, one primary oocyte matures into a secondary oocyte capable of breaking away from the ovary each month during a process called ovulation.

vagina (vuh **jine** uh) an organ in the body of a female whose muscular, tubelike passageway to the exterior has three functions: it accepts the penis during intercourse; it is the lower portion of the birth canal; and it provides an exit for the menstrual flow.

ovaries (**oh** vah reez) the female gonads, in which secondary oocytes develop.

ovulation (**ov** yuh **lay** shun) the monthly process by which a secondary oocyte (potential egg) is expelled from the ovary.

zygote (zye gote) a cell produced by the fusion of the haploid nuclei of the sperm and egg; the fertilized ovum.

Passage to the Uterus: The Uterine Tubes

At ovulation the secondary oocyte is released from the ovary. Lying close to each ovary is the entrance to a tubelike passageway that will take the egg to the **uterus,** the organ in which a fertilized egg can develop. Each passageway from an ovary to the uterus is called an *oviduct,* or **uterine tube.** In humans, the uterine tubes are commonly called the *fallopian tubes.*

Each uterine tube has an expanded, fringed end near each ovary. The tube and its fringed ends are lined with ciliated cells that beat in a synchronized manner, creating a current in the surrounding fluid that sweeps the egg into the tube. Once in the uterine tube, the egg is swept along by the beating of the ciliated cells lining the tube and by rhythmic peristaltic contractions of the muscles within its walls.

The journey from the ovary to the uterus is a slow one, taking about 3 days to complete. Because unfertilized oocytes can live for only 24 hours, a secondary oocyte must be fertilized while in the uterine tube, or it is no longer viable (able to be fertilized). In addition, sperm live for approximately 48 to 72 hours; they cannot simply wait in the uterine tubes for an egg to come along. Therefore fertilization can take place only if sexual intercourse occurs within 3 days before to 1 day after ovulation. (However, knowing exactly when ovulation will occur is usually a difficult task unless specific testing procedures are used, see pp. 457-458.) If fertilization occurs, the egg completes its development, its hereditary material joins with that of the sperm, and development of the zygote begins as it moves along the uterine tube. By the time the new embryo reaches the uterus, it is a tiny ball of cells. After a few days, it implants in the lining of the uterus to receive nourishment. (Further development of the embryo is discussed in the next chapter.)

> *The secondary oocyte journeys to the uterus, or womb, by means of a ciliated muscular tube called a uterine tube. One uterine tube extends from near each ovary to the uterus. If fertilized while in the tube, an oocyte will complete its maturation, join its hereditary material with that of a sperm, and begin its development as an embryo. A few days after reaching the uterus, the embryo will implant in the uterine lining to continue development.*

The Site of Prenatal Development: The Uterus

As shown in Figure 21-6, the uterine tubes lead directly to the uterus. The uterus (from a Latin word meaning "womb") sits above the urinary bladder and in front of the rectum and connects directly with the vagina. Its walls are made up of thick layers of muscle that contract during the birth process and during menstruation (see p. 453). The interior of the uterus is hollow and provides a cavity for the development of a fertilized egg. Amazingly, the uterus of a woman who has never been pregnant is only 7.5 centimeters (3 inches) long and 5 centimeters (2 inches) wide; yet it can stretch to accommodate the size of a newborn!

The inner lining of the uterus is called the **endometrium.** This lining has two layers: one functional, transient layer that is in contact with the uterine cavity and an underlying permanent layer. The functional layer grows and thickens each month, readying itself for the implantation of an embryo. If an embryo is not present, the functional layer is shed. This monthly development and shedding of the functional layer of the endometrium (called the *menstrual cycle*) and the monthly maturation of an egg and its release (called the *ovarian cycle*) are both governed by levels of various female sex hormones. Together, these two cycles are commonly referred to as the *reproductive cycle.*

The Reproductive Cycle

Each month, a small number of primary oocytes (20 to 25) continue their development within an ovary. This initial development is triggered by follicle-stimulating hormone (FSH) released by the pituitary gland in the brain (see Chapter 14). After they are stimulated, these follicles start producing very low levels of a group of hormones called **estrogens.** Estrogens develop and maintain the female reproductive structures such as the ovarian follicles, as well as the lining of the uterus and the breasts. Therefore as the follicles develop in the ovaries and secrete estrogens, the lining of the uterus begins to thicken, preparing for the implantation of a fertilized egg. As the developing follicles mature, a membrane forms around each potential egg, fluid builds up within each follicle (Figure 21-8), and each follicle's cells produce increasing amounts of estrogens. During the second week of the reproductive cycle, most of the developing follicles die; usually only one follicle continues to mature. The hypothalamus senses the rise in blood level of estrogens and triggers the pituitary to secrete a surge of luteinizing hormone, or LH. A smaller surge of FSH occurs also. The surge in LH triggers ovulation. During ovulation the follicle breaks open and releases its immature egg into the abdominal cavity, close to the opening of the uterine tube. This rise in LH can be detected by home test kits, informing a woman 1 day before ovulation occurs. A couple can then use this information to help them conceive a child.

Ovulation usually (but not always) occurs at the midpoint of the reproductive (menstrual) cycle—

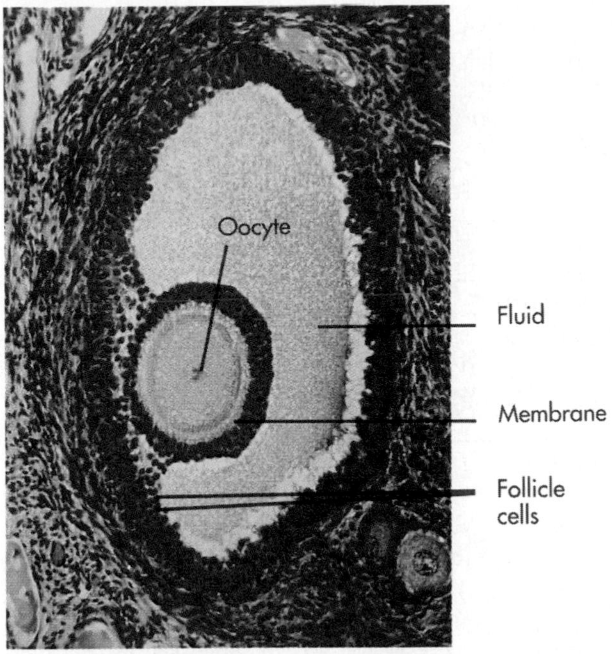

FIGURE 21-8 A mature egg within an ovarian follicle of a cat. The follicle produces estrogens, which develop and maintain other follicles within the ovary.

around day 14 in a 28-day cycle (Figure 21-9). After ovulation the movement of the ciliated cells at the fringes of the nearby uterine tube sweeps the egg into the tube. The egg then begins its 3- to 4-day journey to the uterus. The ruptured follicle collapses, and under the continued influence of LH, it begins to enlarge, forming a yellowish structure called the *corpus luteum,* or "yellow body." The corpus luteum plays an important role in the preparation of the endometrium for the implantation of the fertilized egg by secreting increasing quantities of estrogens and progesterone. The progesterone steps up the thickening of the lining of the uterus and dramatically increases its blood supply.

The rise in progesterone and estrogens also acts as a negative feedback mechanism to the hypothalamus, ultimately inhibiting the secretion of FSH and LH from the pituitary. If the immature egg is fertilized while journeying down the uterine tube, it completes a second meiotic division, joins its hereditary material with that of the sperm, and begins dividing by mitosis (see Chapters 18 and 22). A few days after reaching the uterus, the developing embryo nestles into the thickened uterine lining to continue its development. The placenta, a flat disk of tissue that provides an exchange of nutrients and wastes between the embryo and the mother, begins to form. Its embryonic tissues secrete a hormone called *human chorionic gonadotropin (HCG).* This hormone maintains the corpus luteum so that it will continue to secrete progesterone and estrogen. Progesterone is needed for the maintenance of the uterine lining and the continued attachment of the fetus to this lining. Secretion of HCG occurs about 8 days after fertilization. Because this hormone is excreted in the urine, its detection is the basis for both laboratory and home pregnancy tests.

If the egg is not fertilized while in the uterine tube, it dies and cannot implant in the uterus and develop. Therefore no placenta develops and no HCG is produced. The corpus luteum begins degenerating on approximately day 24 of the cycle. As it degenerates, it stops producing estrogens and progesterone, hormones necessary to maintain the buildup of the lining of the uterus. As the blood level of these hormones falls, the endometrial lining sloughs off in a process known as **menstruation,** or menses. The lining degenerates over a period of approximately 5 days, causing a somewhat steady flow of blood, tissue, and mucus from the uterus and out through the vagina, usually heaviest during the first 3 days of menses. The process of menstruation is traditionally considered to mark the *beginning* of the reproductive cycle.

> *The reproductive cycle of females occurs approximately every 28 days. This cycle includes the maturation of the primary oocyte, its release from the ovary, its journey to the uterus, and the development of the former follicular cells into the corpus luteum. In addition, it includes the thickening of the endometrial lining of the uterus in preparation for the implantation of a new embryo. The hormones FSH, LH, estrogen, and progesterone orchestrate these events.*

uterus (**yoo** tur us) the female organ in which a fertilized egg can develop; the womb.

uterine tube (**yoo** tur in) the passageway from an ovary to the uterus; commonly called a fallopian tube.

endometrium (en do **mee** tree um) the inner lining of the uterus. This lining has two layers: one functional, transient layer that is in contact with the uterine cavity, and an underlying permanent layer. The functional layer is shed each month an embryo is not present in the uterus.

estrogens (**es** truh jens) various hormones that develop and maintain the female reproductive structures, such as the ovarian follicles, the lining of the uterus, and the breasts.

menstruation (**men** stroo ay shun) the monthly sloughing of the blood-enriched lining of the uterus when pregnancy does not occur; the lining degenerates and causes a flow of blood, tissue, and mucus from the uterus out through the vagina.

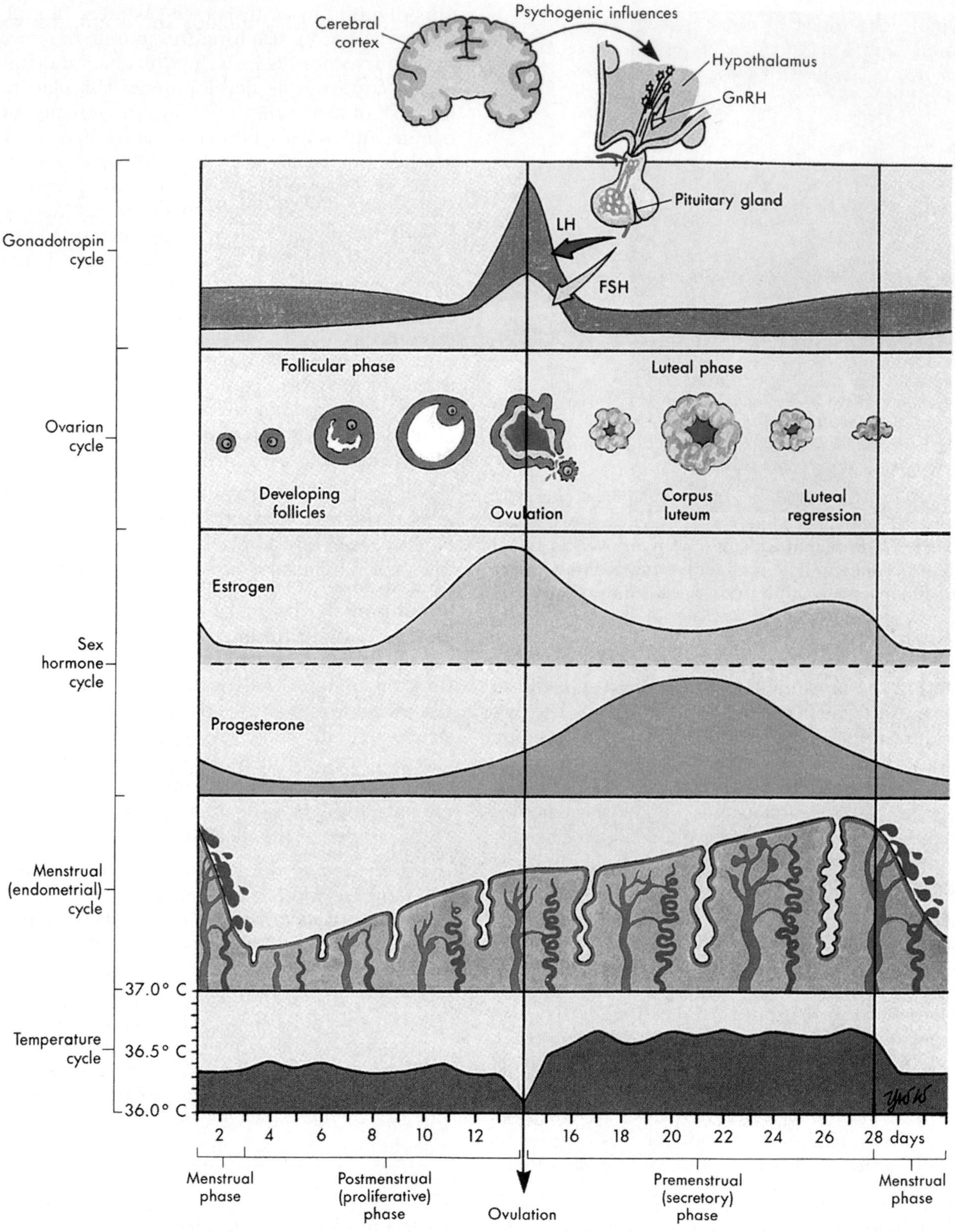

FIGURE 21-9 The human menstrual cycle. The growth and thickening of the uterine lining is governed by increasing levels of estrogens and progesterone; menstruation, the sloughing off of the blood-rich tissue, is initiated by decreasing levels of these hormones.

Before the onset of a menstrual period, many women experience symptoms that are collectively referred to as *premenstrual tension syndrome (PMS)*. Premenstrual symptoms are thought to be related to the cutoff of progesterone and the retention of water. Symptoms vary from woman to woman and include breast swelling and tenderness, bloating, weight gain, headaches, fatigue, and backache. Physicians suggest that a diet high in protein and low in salt, alcohol, and carbohydrates can reduce water retention during this time and lessen the symptoms of PMS.

The reproductive cycle of women generally begins during adolescence and ends between the ages of 50 and 55 years. This cessation of menses is termed **menopause.** As a woman ages, changes take place in the reproductive cycle that lead to menopause. These changes usually begin between the ages of 40 and 50 years. The menstrual cycles may become irregular, varying in length. Generally they become less frequent. These changes result from the failure of the ovaries to respond to stimulation by FSH and LH secreted by the pituitary. As a result of these changes, follicles do not develop; the ovaries therefore produce less estrogen. The drop in estrogen causes the two major symptoms associated with menopause: flushes, the reddening of the face and a feeling of warmth, and flashes, short periods of intense flushing. Although some people think that depression often accompanies menopause, this link does not exist. Recent psychiatric studies show that depression occurs no more often during this time of a woman's life than at any other.

The Vagina

The uterus is shaped somewhat like an upside-down pear: wider at its top and narrower at its bottom. The narrower portion, called the **cervix** (or neck), opens into the vagina. The vagina, a muscular, tubelike passageway to the exterior of the body, has three functions: it accepts the penis during sexual intercourse and is the place where sperm are deposited, it is the lower portion of the birth canal during childbirth, and it provides a passageway for the exit of the menstrual flow. Although only about 9 centimeters ($3^1/_2$ inches) in length, the vagina can stretch considerably during intercourse and childbirth because its walls are composed of elasticlike smooth muscle tissue. The mucous membrane that lines the inside of the vagina produces a viscous fluid that keeps the vagina moist and provides lubrication during intercourse.

Bacteria that normally reside in the vagina use the large amounts of glycogen in the vaginal secretion as a food source. As a byproduct of the digestion of this glycogen, these normal vaginal bacteria produce acid, which is harmful to sperm. However, the ejaculate contains substances that neutralize this acidity (see p. 449). These bacteria have a positive effect in that they discourage the growth of unwanted bacteria and yeasts. Often, when a woman takes antibiotics to kill bacteria infecting another part of her body, the antibiotic also kills the bacteria residing in the vagina. When these bacteria die, yeast cells or other bacteria can begin to grow. A vaginal infection can be a troublesome side effect of antibiotic therapy in women because it can produce itching, discomfort, and vaginal discharges.

The External Genitals

The external genitals of a female are collectively called the **vulva.** Interestingly, parts of the vulva are homologous to the external genitals of the male: the organs have a similar structure and anatomical position, and they originated from the same tissues during embryonic development. The most anterior structure is the **mons pubis** (Figure 21-10). The word *mons* comes from a Latin word meaning "prominence" and refers to the mound of fatty tissue that lies over the place of attachment of the two pubic bones. Two longitudinal folds of skin called the **labia majora** (*labia* means "lips") run posteriorly from the mons. These folds are homologous to the scrotum in the male. Covered by the labia majora are additional folds of skin called the **labia minora.** Both sets of skin folds protect the vaginal and urethral openings beneath. Figure 21-10 shows that the urethra opens to the outside slightly in front of the vagina and is not connected to the reproductive system as in males. In addition to the vaginal and

menopause (**men** uh pawz) the permanent cessation of menstrual activity in a woman, usually between the ages of 50 and 55; the end of the menses.

cervix (**sur** viks) the narrower, bottom part of the uterus that opens into the vagina.

vulva (**vul** vuh) the collective term for the external genitals of a female.

mons pubis (**monz pyoo** bis) the mound of fatty tissue that lies over the place of attachment of the two pubic bones in females.

labia majora (lay bee uh muh **jore** uh) two longitudinal folds of skin that run posteriorly from the mons in the exterior genitals of the female.

labia minora (lay bee uh mu **nore** uh) folds of skin covered by the labia majora in the exterior genitals of the female.

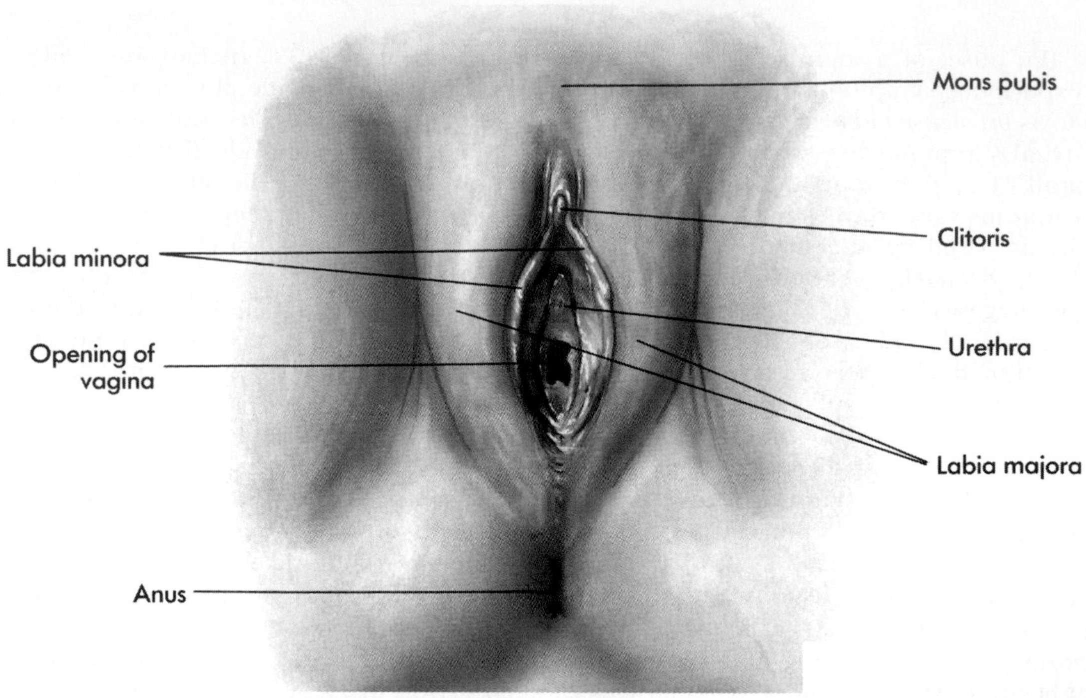

FIGURE 21-10 The external genitals of the female. The vagina can stretch considerably to accommodate a child's head during childbirth because its walls are made of elastic muscle tissue.

urethral openings, the labia minora also cover the openings of several tiny ducts that produce a mucuslike secretion during sexual stimulation. This secretion helps lubricate the vagina and penis during sexual intercourse. Slightly anterior to the convergence of the labia minora is the **clitoris,** a small mass of erectile and nervous tissue that responds to sexual stimulation. The clitoris is homologous to the tip of the penis in males.

The Mammary Glands

The **mammary glands,** or milk-producing glands, lie over the chest muscles. Each mammary gland is made up of 15 to 20 lobes that are separated by fat. The amount of fat determines the size of the breast and is not an indicator of a woman's ability to produce milk. Ducts extend from the glands to the nipple of the breast.

The mammary glands are secondary sexual characteristics that begin to develop in adolescent girls as the hormonal changes of puberty take place. These changes are initiated as the ovary begins to produce estrogens, and they accelerate as the menstrual cycle begins and as the ovaries produce higher blood levels of estrogens.

The function of the mammary glands is lactation, the production and secretion of milk. Prolactin, a hormone secreted by the anterior pituitary during pregnancy, is the hormone primarily responsible for stimulating the production of milk. As an infant nurses, sucking sends nervous impulses to the hypothalamus. These messages stimulate the continued production of prolactin and therefore of milk. In addition, the nervous impulses trigger the hypothalamus to produce oxytocin, a hormone released by the posterior pituitary that causes milk to be ejected into the ducts of the mammary glands. This process is called *milk letdown.*

The Sexual Response

The sexual act is referred to by a variety of names, including *intercourse, copulation,* and *coitus,* as well as a host of more informal ones. The physiological events associated with sexual intercourse can be described as having four phases, although the division of the sexual response into phases is somewhat arbitrary. These phases are excitement, plateau, orgasm, and resolution (Figure 21-11).

The *excitement phase* refers to the sexual activity that precedes intercourse. This phase therefore varies in length depending on the needs and desires of the sexual partners. During excitement the heartbeat, blood pressure, and rate of breathing increase. In addition, the blood vessels of the face, breasts, and genitals widen, producing a reddening of the skin in these areas called the *sex flush.* The nipples commonly harden and become more sensitive. In the genital area, increased circulation leads to an erection in the male and a swelling and parting of the labia in females. The vaginal walls become moister and the muscles of the vagina relax.

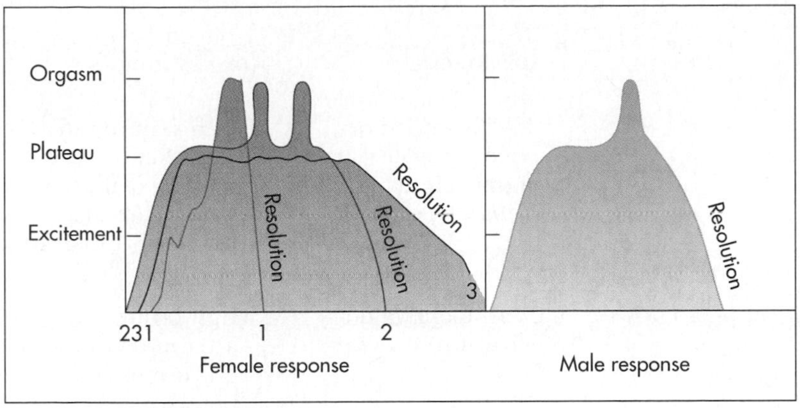

FIGURE 21-11 The human orgasmic response. Among females, the response is highly variable. It may be typified by one of the three patterns illustrated here. Among males, the response is not as variable.

The act of sexual intercourse signals the beginning of the *plateau phase,* a period of intensifying physiological changes initiated in the excitement phase. During sexual intercourse the penetration of the vagina by the thrusting penis results in the repeated stimulation of nerve endings both in the tip of the penis and in the clitoris. The clitoris, which is now swollen, becomes very sensitive and withdraws up into a sheath, or hood. Once it has withdrawn, the stimulation of the clitoris is indirect, with the thrusting movements of the penis rubbing the clitoral hood against the clitoris. This type of stimulation of the clitoris is insufficient for many women, however, who require a more direct stimulation of this organ.

The climax of intercourse (*orgasm*) is reached when the stimulation is sufficient to initiate a series of reflexive muscular contractions. The nerve signals producing these contractions are associated with other nervous activity within the central nervous system, activity that you experience as sexual pleasure. In females the contractions are initiated by impulses in the hypothalamus that cause the pituitary to release large amounts of the hormone oxytocin. This hormone, in turn, causes the muscles in the uterus and around the vaginal opening to contract and the cervix to be pulled upward. There may be one or several intense peaks of contractions (orgasm), or the peaks may be more numerous but less intense (see Figure 21-11).

Analogous contractions occur in the male, initiated by nerve signals from the brain. These signals first cause emission, in which rhythmic peristaltic contractions of the vas deferens and the prostate gland cause the sperm and seminal fluid to move to the base of the penis. Shortly after, nerve signals from the brain induce contractions of the muscles at the base of the penis, resulting in the ejaculation of the semen from the penis.

The *resolution phase* begins after orgasm. The bodies of both men and women return to their normal physiological states after several minutes. After ejaculation, males rapidly lose their erection and enter a period lasting 20 minutes or longer in which sexual arousal is difficult to achieve and ejaculation is almost impossible. By contrast, women can be aroused again almost immediately.

Contraception and Birth Control

In most vertebrates, sexual intercourse is associated solely with reproduction. In humans, however, sexual behavior serves a second important function—the reinforcement of pair bonding, the emotional relationship between two individuals. Couples who wish to limit the pregnancies resulting from sexual intercourse may use one of many methods of family planning, or birth control (Table 21-1). Several different approaches are commonly taken to achieve birth control. These methods differ from one another in their effectiveness and in their acceptability to different couples. Some of them are shown in Figure 21-12. In addition, some methods of birth control protect against sexually transmitted diseases.

Abstinence

The most reliable way to avoid pregnancy and sexually transmitted diseases is abstinence. A variant of this approach (for birth control purposes only) is to avoid sexual relations for 3 days preceding and the day following ovulation because this is the only time fertilization can occur. This approach, called the **rhythm method,** is satisfactory in principle but difficult in application because ovulation is not easy to predict and may occur unexpectedly. The effectiveness of the rhythm method is low. However, a

clitoris (klit uh ris) a small mass of erectile and nervous tissue in the female genitalia that responds to sexual stimulation.

mammary glands milk-producing glands that lie over the chest muscles in females.

rhythm method a form of birth control that involves avoiding sexual relations for 3 days preceding and 1 day following ovulation.

TABLE 21-1

Methods of Birth Control*

DEVICE	ACTION	FAILURE RATES (%)	ADVANTAGES	DISADVANTAGES
Rhythm method	Avoidance of sexual intercourse during woman's fertile period	13-21	Contraceptive devices unnecessary; agreeable to some religious groups	High failure rate due to difficulty of predicting ovulation; no protection against sexually transmitted diseases
Withdrawal	Removal of penis from vagina before ejaculation	9-25	Contraceptive devices unnecessary	High failure rate attributed to the secretion of sperm before ejaculation; self-control necessary
Condom	Thin rubber sheath for penis that collects semen	3-15	Easy to use, effective, and inexpensive; the most reliable protection against some sexually transmitted diseases with the exception of abstinence	Possible decrease in spontaneity; possible deterioration on the shelf
Diaphragm	Soft rubber cup that covers entrance to uterus, prevents sperm from reaching egg, and holds spermicide	4-25	No dangerous side effects; reliable if used properly; some protection against cervical cancer	Careful fitting required; some inconvenience associated with insertion and removal; possible dislodgment during sex
Cervical cap	Miniature diaphragm that covers cervix closely, prevents sperm from reaching egg, and holds spermicide	Probably comparable to diaphragm	No dangerous side effects; fairly effective; able to remain in place longer than diaphragm	Problems with fitting and insertion; limited number of sizes; not available in the United States

*Approximate effectiveness of these reversible methods of birth control is measured in pregnancies per 100 actual users per year.

woman can use techniques that involve taking her basal body temperature or analyzing her cervical mucus to help determine when ovulation takes place. Home test kits are also available that detect the time of ovulation. (These methods can be used to help a couple conceive a child, as well as to avoid conception.) As an added birth control measure, some couples using the rhythm method of birth control conservatively avoid sexual intercourse many days before and a few days after ovulation. The failure rate of the rhythm method is estimated to be 13% to 21% (13 to 21 pregnancies per 100 women practicing the rhythm method, per year).

Another variant of this approach is to withdraw the penis before ejaculation. This method is as unreliable as the rhythm method, with a failure rate estimated between 9% and 25%. The reason for this high failure rate is that the penis can secrete prematurely released sperm within its lubricating fluid and a second sexual act may transfer sperm ejaculated earlier.

Sperm Blockage

If sperm do not reach the uterine tubes, fertilization cannot take place. One way to prevent sperm from reaching the uterine tubes is for the male partner to use a **condom.** A condom is a sheath for the penis, constructed of thin rubber or other natural or synthetic materials. In principle, this method is easy to use and should be highly successful. However, in

TABLE 21-1

Methods of Birth Control—cont'd

DEVICE	ACTION	FAILURE RATES (%)	ADVANTAGES	DISADVANTAGES
Foams, creams, jellies, vaginal suppositories	Chemical spermicides inserted in vagina before intercourse that also prevent sperm from entering uterus	10-25	Possible use by anyone who is not allergic, no known side effects	Relatively unreliable, sometimes messy; necessary to use 5 to 10 minutes before each act of intercourse
Oral contraceptives (birth control pills)	Hormones, either in combination or progestin only, that primarily prevent release of egg	1-5, depending on type	Convenient and highly effective; significant noncontraceptive health benefits, such as protection against ovarian and endometrial cancers	Necessary to take regularly; possible minor side effects, which new formulations have reduced; not for women with cardiovascular risks, mostly those over 35 who smoke
Implant (Norplant)	Capsules surgically implanted under skin that slowly release a hormone that blocks release of eggs	0.3	Very safe, convenient, and effective; very long lasting (5 years); possible nonreproductive health benefits like those of oral contraceptives	Irregular or absent periods, necessity of minor surgical procedure to insert and remove
Intrauterine devices	Small plastic devices placed in the uterus that prevent fertilization or implantation	1-5	Convenient, highly effective, infrequent replacement	Possible excess menstrual bleeding and pain; danger of perforation, infection, and expulsion; not recommended for those who are childless or not monogamous; risk of pelvic inflammatory disease or infertility; dangerous in pregnancy

practice the failure rate of condoms is from 3% to 15%, primarily due to improper use or failure to use a condom consistently. Nevertheless, it is the most commonly used form of birth control in the United States. The use of latex condoms is also an excellent way to reduce the risk of contracting sexually transmitted diseases, including AIDS.

The **female condom** was recently introduced but is not a reliable method of birth control: it has a 25% failure rate. It is shaped somewhat like a male condom but has a ring at both ends. One ring fits over the cervix and the other hangs outside the vagina.

A second way to prevent the entry of sperm into the uterine tubes is to place a cover over the cervix. The cover may be a relatively tight-fitting *cervical cap* (not available in the United States at this time), which is worn for days at a time, or a rubber dome called a **diaphragm,** which is inserted immediately before intercourse. In addition, the vaginal side of a diaphragm is covered with a spermicide before in-

condom (kon dum) a sheath for the penis, constructed of thin rubber or other natural or synthetic materials, that is designed to prevent sperm from reaching the uterine tubes. The latex condom helps reduce the risk of contracting sexually transmitted diseases.

female condom a birth control device that is shaped somewhat like a male condom, but has a ring at each end. One ring fits over the cervix and the other hangs outside the vagina.

diaphragm a rubber dome that is inserted immediately before intercourse to cover the cervix and prevent the entry of sperm into the uterine tubes.

JUST WONDERING . . . .

Are any new birth control products being developed for men?

Unfortunately, new contraceptives for men have been difficult to develop and are probably at least a decade away. The most promising line of research is the development of chemical contraceptives for men. Scientists are testing a substance that blocks sperm production by mimicking an important "releasing hormone," so named because it affects the release of other hormones from the anterior pituitary. This releasing hormone, gonadotropic-releasing hormone (GnRH), controls secretion of follicle-stimulating hormone (FSH) and luteinizing hormone (LH) in both men and women. In men, FSH triggers sperm production and LH controls the release of testosterone from the testes. Researchers have found that injecting men with a modified, inactive form of GnRH stops sperm production. This inactive GnRH binds to receptors on the membranes of cells in the anterior pituitary, preventing secretion of both FSH and LH.

A major side effect of this method was discovered during testing, however. When the modified form of GnRH binds to receptors in the anterior pituitary, it prevents secretion of LH. When LH secretion is blocked, testosterone levels fall, causing impotence. This effect can be avoided by accompanying injections of the modified GnRH with injections of testosterone. However, testosterone also stimulates the development of sperm. At this stage of development, men must have daily injections of these chemicals—a costly and inconvenient process. If scientists can develop an oral form of the two substances, a male birth control pill could become a reality.

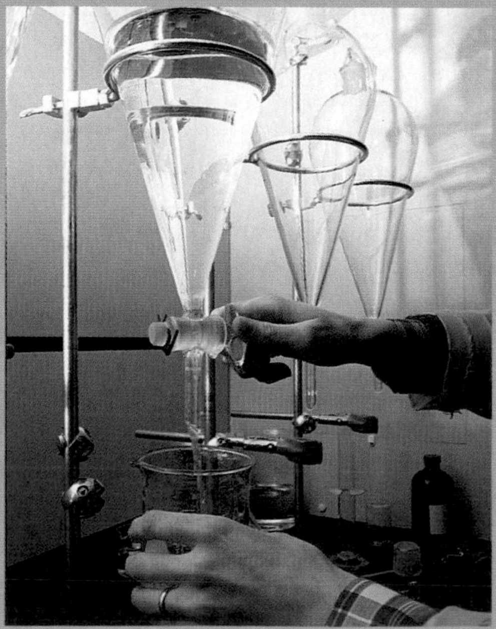

A

B

C

D

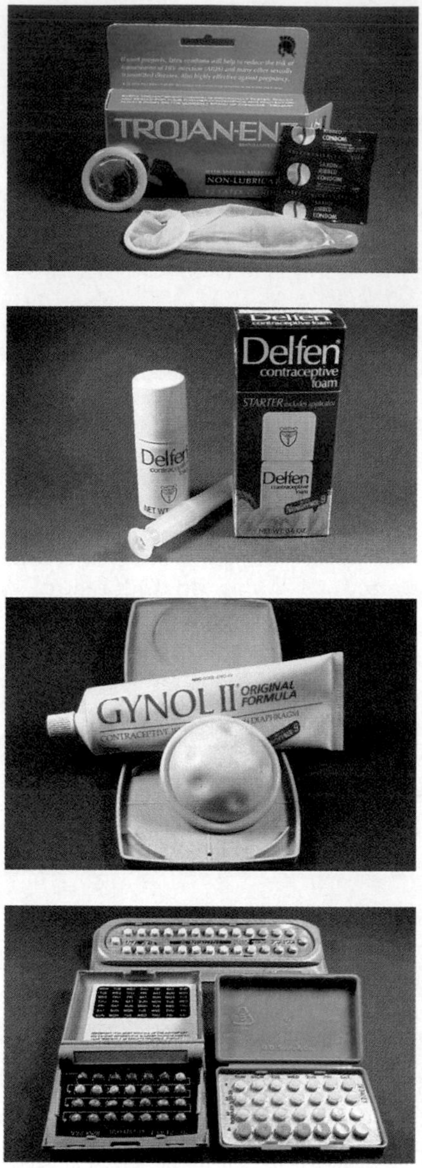

FIGURE 21-12 Four common methods of birth control. A, Condom. **B,** Foams. **C,** Diaphragm and spermicide. **D,** Oral contraceptives.

sertion and is then kept in place for a minimum of 6 hours after intercourse. Because the dimensions of individual cervices vary, a cervical cap or diaphragm must be fitted by a physician. Failure rates average from 4% to 25% for diaphragms and are somewhat lower for cervical caps. These high failure rates are due, in part, to incorrect use of these birth control devices. Neither form of birth control is effective against sexually transmitted diseases.

Sperm Destruction

The spermicidal (sperm-killing) sponge is inserted and used much like a diaphragm, but it is composed of a spongelike material impregnated with a spermicide. The sponge provides somewhat of a barrier to the uterus but also absorbs the sperm into the sponge to be killed there. (As of this writing, contraceptive sponges are no longer available due to contamination problems during their manufacture. It is uncertain when or if they will become available again.) Sperm can also be destroyed in the vagina by the use of spermicidal jellies, creams, or foams applied immediately before intercourse. The failure rate of these methods varies widely from 10% to 25%, and they are not useful in the prevention of sexually transmitted diseases.

Prevention of Egg Maturation

Since about 1960, a widespread form of birth control has been the use of **birth control pills** by women. These pills contain estrogen and progesterone, either taken together in the same pill or in separate pills taken sequentially. In the normal reproductive cycle of a female, these hormones act to shut down the production of the pituitary hormones FSH and LH. The artificial maintenance of high levels of estrogen and progesterone in a woman's bloodstream causes the body to act as if ovulation had already occurred when in fact it has not. The ovarian follicles do not mature in the absence of FSH, and ovulation does not occur in the absence of LH. For these reasons, birth control pills provide a very effective means of birth control, with a failure rate of 0% to 5%. They play no role in the prevention of sexually transmitted diseases, but research shows that the synthetic progesterone (progestin) in the pill helps reduce a woman's risk of cancers of the uterus and ovaries. Scientists are concerned, however, that the estrogen in the pill may increase the risk of breast cancer. Medical researchers are now experimenting with a pill that includes LH-releasing hormone. This pill is currently undergoing clinical trials. Another, more highly experimental pill substitutes melatonin for the estrogen but is not yet approved for testing.

In 1990 a product that works in a manner similar to the pill but that prevents pregnancy for up to 5 years was introduced in the United States. This product is called *Norplant* (Figure 21-13). In a 15-minute operation in the physician's office, the six matchstick-sized tubes are surgically implanted under the skin of the upper arm. These tubes contain the synthetic hormone progestin, the active ingredient in most birth control pills. Norplant prevents pregnancy by inhibiting ovulation as the pill does, but it also thickens the cervical mucus so that sperm cannot move through the female reproductive tract. Although 99.8% reliable, Norplant causes irregular menstrual bleeding in 75% of its users. Scientists are researching ways to reduce the side effects of these implants and make them smaller.

FIGURE 21-13 The contraceptive implant. The tubes shown here are implanted in a woman's upper arm in a simple surgical procedure. Once implanted, the tubes release a constant, low-level flow of progestin that prevents ovulation. The implant is effective for about 5 years.

Some researchers hope to develop implants that will degrade within the body (at the appropriate time) so that they will not need to be removed.

Surgical Intervention

A completely effective means of birth control is the surgical removal of a portion of the tube through which gametes move from one reproductive structure to another (Figure 21-14). In males such an operation involves the removal of a portion of the vas deferens, the tube through which sperm travel to the penis. This procedure, called a **vasectomy,** can be carried out in a physician's office. Reversing a vasectomy is difficult but can sometimes be done. In females the comparable operation involves the removal of a section of each of the two uterine tubes through which the oocyte travels to the uterus. Because these tubes are located within the abdomen, the operation, called a **tubal ligation,** is more difficult to perform than a vasectomy and is even more difficult to reverse.

birth control pills drugs that contain estrogen and progesterone, which shut down the production of the pituitary hormones FSH and LH, preventing the maturation of the secondary oocyte and ovulation.

vasectomy (va **sek** tuh mee) in males, an operation that involves the removal of a portion of the vas deferens, the tube through which sperm travels to the penis, in order to bring about sterilization.

tubal ligation (**too** bul lye **gay** shun) in females, an operation that involves the removal of a section of each of the two uterine tubes, through which the oocyte travels to the uterus, in order to bring about sterilization.

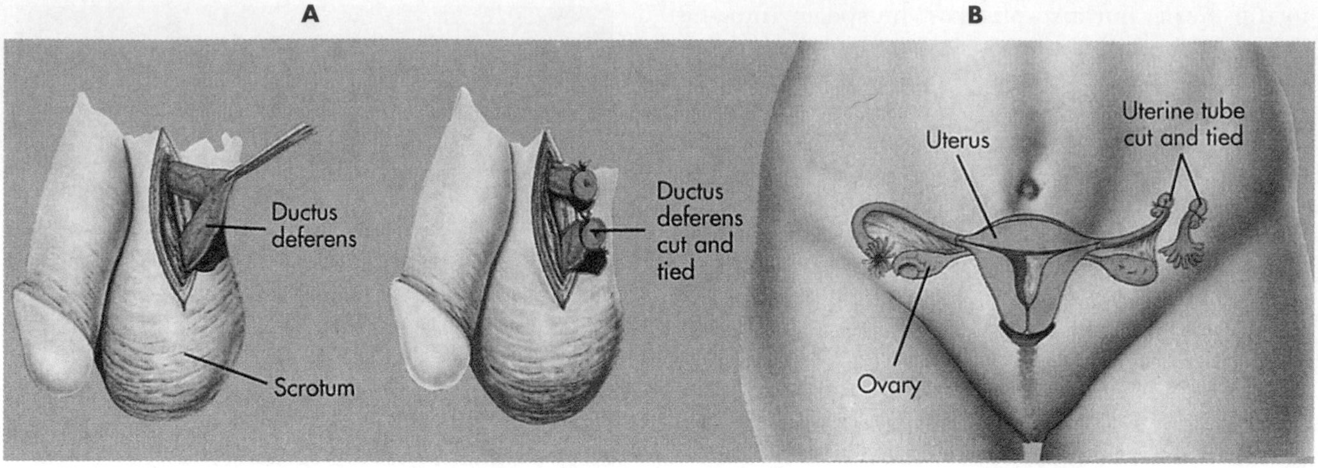

FIGURE 21-14 Surgical means of birth control. **A,** Vasectomy. **B,** Tubal ligation.

> *Among methods of birth control, the rhythm method and withdrawal are not highly reliable. Condoms and diaphragms are effective when used correctly, but mistakes are common. Birth control pills and implanted hormonal tubes are very effective. Vasectomies and tubal ligations are completely effective although usually permanent.*

Sexually Transmitted Diseases

Sexually transmitted diseases (STDs) are caused by the transmission of certain agents of infection from one person to another during sexual activity. The ideas that STDs are caused by poor personal hygiene and are transmitted only by persons of low socioeconomic status are *misconceptions.* Statistics show that the incidence of STDs crosses all boundaries of race, gender, ethnicity, social class, and economic status. Although the incidence of STDs crosses boundaries of age too, young people are hardest hit. Today, 12 million new sexually transmitted infections occur each year, with two thirds of these cases in persons younger than 25 years. Currently, one out of every four Americans has an STD.

All STDs are communicable, or *contagious,* meaning that they can be spread from one person to another. The most common causative agents of STDs are viruses and bacteria. They cause infection by entering the tissues, multiplying, and causing damage. The most common STDs in the United States are gonorrhea and chlamydial infection, which are both caused by bacteria, and genital herpes, which is caused by a virus. Taken together, these three diseases make up about 10 million of the 12 million annual cases of STDs. (A discussion of STDs caused by bacteria and viruses can be found in Chapter 25.)

Other organisms such as certain *fungi, protozoans, mites,* and *lice* can be transferred from person to person by sexual contact, but these are regularly transferred in other ways as well. (STDs caused by these organisms are discussed in Chapter 26.) Mites and lice, members of a diverse phylum of animals called *arthropods,* cause *infestation.* They live and/or feed on the skin and underlying tissues. (These organisms are discussed in Chapter 29.)

The only way to totally protect yourself against contracting an STD is to abstain from sexual activity. If you do engage in sexual activity, your risk of contracting an STD increases as your number of partners increases. A monogamous relationship is the safest situation in terms of disease transmission.

How else can you protect yourself from STDs? The best protection is the proper and consistent use of latex condoms. The condom puts a barrier between yourself and any disease-producing organisms, but *condoms are no guarantee against contracting disease.* The use of the spermicide nonoxynol-9 in addition to a condom can provide further protection because it inactivates viruses. In addition, observe your partner, if possible, for the presence of any of the signs or symptoms of disease summarized in Table 21-2. However, absence of symptoms does not mean that a person is free of disease. You may choose to ask your partner whether he or she is currently infected with an STD so that you can make an informed choice regarding your behavior. Last, if you observe any of the signs or symptoms of an STD in yourself, see your physician immediately.

TABLE 21-2

Sexually Transmitted Diseases and Infestations

DISEASE	CAUSATIVE AGENT	TYPE OF AGENT	BRIEF DESCRIPTION OF DISEASE	TREATMENT
Genital herpes	Herpes simplex virus (HSV)	Virus	Disease causes painful sores in genital area. Virus causes latent infection so recurrences common.	Acyclovir and other antiviral drugs (no cure)
Genital warts	Human papillomavirus (HPV)	Virus	Disease causes soft, pink, flat or raised growths on external genitals, rectum, vagina, cervix.	Electrocauterization, cryotherapy, surgery (no cure)
Acquired immunodeficiency syndrome (AIDS)	Human immunodeficiency virus (HIV)	Virus	Virus attacks and destroys T cells, a key component in the body's immune system.	Various antiviral drugs (no cure)
Syphilis	*Treponema pallidum*	Bacterium	Disease progresses through stages of localized infection to widespread infection. If untreated, it can result in death.	Penicillin
Gonorrhea	*Neisseria gonorrhoeae*	Bacterium	Primary infection is of urethra (in men and women) and vagina and cervix (in women). Other urogenital structures may become infected. Disease often causes puslike discharge. Disease can result in sterility and/or damage to other organs.	Penicillin and other antibiotics
Chlamydial infection (nongonococcal urethritis)	*Chlamydia trachomatis*	Bacterium	Infection is similar to gonorrhea but usually has milder symptoms.	Tetracycline and sulfa drugs
Candidiasis (yeast infection)	*Candida albicans*	Yeast	Vaginal infection results in raised gray or white patches on the vaginal walls, a thick whitish discharge, and itching. Penile infection results in the growth of small, raised yeast colonies on the penis.	Topical antifungal drugs
Trichomoniasis	*Trichomonas vaginalis*	Protozoan	Infection in women results in vaginal itching, burning, and a profuse discharge that may be bloody or frothy. Infection in men results in a slight discharge from the urethra, painful urination, and increased urination.	Antiparasitic drugs
Scabies	*Sarcoptes scabiei*	Itch mite	Common sites of infestation are the base of the fingers, the wrists, the armpits, the skin around nipples, and the skin around belt line. Female mites bore into the top layers of skin of an infected person to lay their eggs. Symptoms are dark, wavy lines in the skin and itching.	Topical antiparasitic medications
Pubic lice (crabs)	*Phthirius pubis*	Pubic louse	Females lay eggs in pubic hairs. Disease causes intense itching.	Topical antiparasitic medications

Summary

- Sexual reproduction is the process whereby a male and a female sex cell combine to form the first cell of a new individual. The male and female reproductive systems are made up of organs and ducts that produce, store, transport, and receive the sex cells. Accessory glands help nourish the sex cells.
- The male reproductive system produces male sex cells called *sperm* and transports them to the female reproductive tract. Sperm cells develop in the tightly packed tubules of the testes, which are covered with a pouch of skin called the *scrotum* and hang from the lower pelvic area of the male. Sperm production is triggered by follicle-stimulating hormone (FSH).
- During the development of sperm, diploid cells called *spermatogonia* give rise to haploid cells called *spermatids.* These haploid cells contain half the normal amount of hereditary material of body cells. The spermatids then develop into sperm cells, or spermatozoa, which have a head containing the hereditary material, a vesicle containing enzymes to penetrate the oocyte membrane, and a whiplike tail. They mature in the epididymis, a long, coiled tube that sits on the back side of the testes.
- The accessory glands of the male reproductive system secrete a fluid that combines with the sperm during ejaculation. This alkaline fluid nourishes the sperm and neutralizes the acidity of the female vagina. During ejaculation, sperm are propelled through the vas deferens to the urethra and out the penis.
- The female reproductive tract produces one potential egg (secondary oocyte) each month. Each potential egg develops in the ovaries, travels to the uterus via one of the two uterine tubes, and implants in the inner lining of the uterus if fertilized while journeying down the tube.
- The reproductive cycle of the female encompasses the events of the maturation of each potential egg, or primary oocyte; its release from the ovary as a secondary oocyte; its journey to the uterus; and the "healing" of its follicle. Certain primary oocytes in the ovaries continue the development they began before birth when stimulated by the hormone FSH. Their surrounding cells start producing estrogens, which promote the thickening of the endometrial lining of the uterus. Of the developing oocytes, one matures into a secondary oocyte that bursts from the ovary as a surge of luteinizing hormone (LH) occurs midcycle. The ruptured follicular cells that surrounded the oocyte secrete estrogens and progesterone. If fertilization of the oocyte does not take place, the progesterone and estrogens maintain the endometrial lining until diminished levels of these hormones cause it to degenerate. A sloughing of the endometrial lining follows, which is referred to as *menstruation.*
- The human sexual response has four physiological periods: excitement, plateau, orgasm, and resolution. Orgasm in women is variable and may be prolonged. Orgasm in men is more uniform and abrupt; it coincides with the ejaculation of sperm.
- Humans practice a variety of birth control procedures; men using condoms and women using birth control pills are the most common.
- Sexually transmitted diseases (STDs) are caused by the transmission of certain agents of infection from one person to another during sexual activity. The most common STDs are caused by bacteria and viruses. Other organisms such as certain fungi, protozoans, mites, and lice can be transferred from person to person by sexual contact but are regularly transferred in other ways as well.
- The best protection against the transmission of STDs, other than abstention from sexual activity, is the consistent and proper use of latex condoms.

Knowledge and Comprehension Questions

1. What is sexual reproduction?
2. Discuss the roles of the hormones involved in the development of male gametes and secondary sexual characteristics.
3. Relate the structure of a spermatozoon to its function.
4. What are the seminal vesicles, prostate gland, and bulbourethral glands? Describe the contribution made by each during ejaculation.
5. Two monthly cycles are involved in a woman's reproductive cycle. What are they, and what happens in each one?
6. Summarize the reproductive cycle of a human female when fertilization does not occur. What hormones are involved?
7. Summarize what happens during the reproductive cycle of the human female if fertilization occurs. What hormone is secreted?
8. John and Mary have had difficulty conceiving a child, so their doctor recommends that they use in an in-home ovulation test to determine the day of Mary's ovulation. Given this information, during what window of time could fertilization take place?
9. Which methods of birth control can help prevent the spread of sexually transmitted diseases?
10. Explain how birth control pills prevent conception. What hormones are involved?
11. Compare the effectiveness of the various birth control methods discussed in the chapter. Which are most effective? Least effective?
12. Are high standards of personal hygiene sufficient protection against sexually transmitted diseases?
13. Are condoms a completely adequate barrier to sexually transmitted diseases? Is there still a risk factor present if condoms are used correctly?

Critical Thinking

1. Female gametes begin their development prenatally, so at age 20, a woman's gametes are all 20 years old. A male's gametes develop all during his life, so at age 20, a man's gametes are only a few days old. What are some implications inherent in this difference?
2. Can you assess why a cervical cap or diaphragm is ineffective in preventing the transfer of sexually transmitted diseases?

CHAPTER 22

DEVELOPMENT BEFORE BIRTH

THE IMAGE OF A MOTHER nursing, rocking, and cuddling her baby is a warm and loving picture. But scientists have only recently validated that not only are these activities emotionally good for moms and babies but they also help babies grow properly. Cuddling and caressing an infant stimulate respiration, blood flow, and growth rate. The baby is healthier and cries less. For these reasons, volunteers at Duke University Medical Center cuddle hospital-bound premature infants whose parents must work and are unable to be with them continually. Interestingly, gentle rocking of an infant stimulates the baby's cerebellum—the part of the brain that controls coordination. Rocking hastens the maturation of this control center. And while a baby is being cuddled, rocked, and nursed by its mother, her breast milk is providing nourishment and antibodies that will help stave off disease until the baby's own defenses are produced. In fact, this peaceful nurturing provides the baby with medical benefits that even the most high-tech pediatric unit would find hard to match.

KEY CONCEPTS

- Life before birth, or prenatal development, lasts approximately $8^1/_2$ months in humans and consists of the growth and changes that take place from conception until birth.
- Conception takes place in the uterine tube when a sperm penetrates an ovum and the nuclear material of both fuse to form the first cell of a new individual.
- Most of the body systems of the humans develop during the first 9 weeks of pregnancy; the ensuing 29 weeks is primarily a time of growth and refinement of these organ systems and of the body form of the developing individual.
- Birth takes place 38 weeks from conception. Hormonal changes in the fetus and the mother trigger waves of uterine contractions that push the fetus through the birth canal.

OUTLINE

Development

What happens after the birth of an infant is crucial to its survival. However, what happens before birth may be even more crucial. In fact, if an infant does not develop properly before birth, normal growth after birth becomes difficult and sometimes impossible. This chapter focuses on life before birth, or **prenatal development,** the gradual growth and progressive changes in a developing human from conception until the time the fetus leaves the mother's womb. This time of development is also called *gestation* and lasts approximately $8^1/_2$ months. (The 9-month pregnancy calculation is computed from the first day of a woman's last menstrual period, but conception usually takes place in the middle of her cycle.) The gestation period is commonly referred to as *pregnancy*.

Fertilization

The union of a male gamete (sperm) and a female gamete (egg) is called **fertilization,** or conception. In the human, fertilization usually takes place in the uterine tube of the female. (See Chapter 21 for a discussion of gamete formation and the events leading up to fertilization.) Therefore development begins in the uterine tube.

Figure 22-1 shows that a human egg, or **ovum,** is a relatively large cell. In fact, ova are among the largest cells within an animal's body. Making up most of the substance of an ovum is the *yolk,* the nutrient material that the developing individual lives on until nutrients can be derived from the mother. In humans, the yolk provides nutrition for a week or so. In the case of birds, however, the young develop totally within the egg. Therefore the yolk of these eggs must be substantial enough to maintain the animal throughout its entire course of development. The yolk of an ostrich egg, for ex-

A

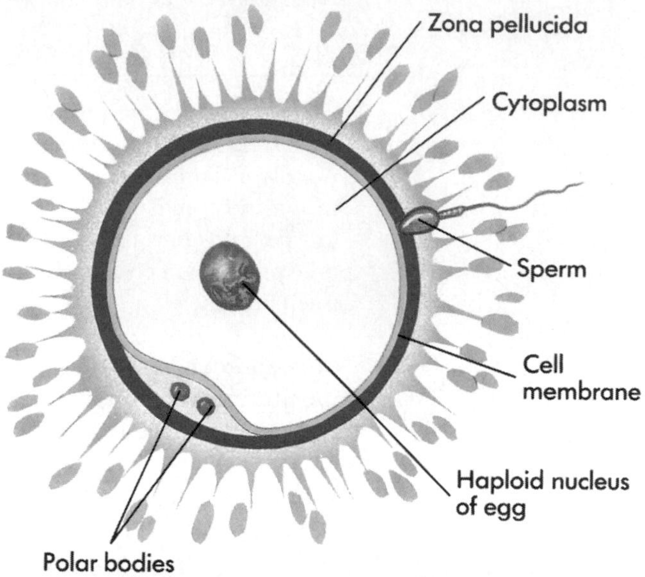

B

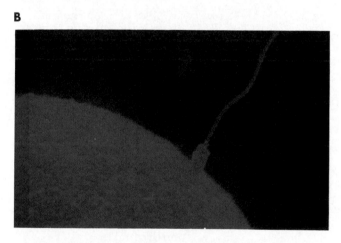

FIGURE 22-1 **The egg. A,** A human egg is surrounded by a membrane, is nourished by yolk, and contains a haploid nucleus. **B,** The membrane of this sea urchin egg has been penetrated by a sperm cell. Once the sperm enters the egg and its nucleus reaches the egg nucleus, the fused cell begins to divide.

ample, is so large that it is approximately the size of a baseball!

As mentioned in Chapter 21, the secondary oocyte breaks away from the ovary during ovulation in placental mammals (which includes people) and is swept into the uterine tube. This cell is not yet a mature ovum, or egg. Only after penetration of the oocyte by a sperm will the second meiotic division be completed, yielding the mature ovum and a second polar body. And, although fertilization may sound like an easy job for sperm, only 50 to 100 sperm make it to the egg from an ejaculate that contains approximately 200 to 300 million!

Many things can go wrong from the time the sperm are manufactured in the testes, are stored in the epididymis, and travel out of the man's body and into the vagina of a woman. For example, up to 20% of sperm are deformed in some way. A common abnormality is having two flagella instead of one. Sperm with this abnormality do not have normal motility and cannot make the lengthy journey from the vagina, through the constricted cervix, up through the uterus, and along the uterine tube.

Although the acid environment of the female vagina is treacherous territory for sperm and although they constantly swim upstream against the downward currents of female secretions, they are also aided by the female body. During the time of ovulation, a female produces strands of a special protein called *mucin.* This substance becomes a part of the cervical mucus her body produces and provides threadlike highways along which the sperm can travel. The sperm make their way up these strands of mucin to the uterus. As they move along, enzyme inhibitors located at the tips of their heads are slowly worn away, gradually uncovering enzymes that are capable of penetrating the egg's outer protective layers and its membrane. Figure 22-2, *A,* pictures a sperm that has penetrated an egg's jellylike covering, the *zona pellucida.*

The moment of penetration of a human oocyte by a sperm is an event only recently captured on film. On entry of the cytoplasm of the oocyte, the sperm sheds its tail and its head swells. Immediate changes in the surface of the egg allow no other sperm to penetrate, and oocyte meiosis is completed. Finally, the sperm and ovum nuclei, clearly seen as separate from one another in Figure 22-3, *A,* fuse to form the nucleus of the first cell of a new individual (Figure 22-3, *B*). This new cell, which contains intermingling genetic material from both the mother and the father, is called a **zygote.**

> *The penetration of the secondary oocyte by the sperm is called fertilization. This penetration causes changes in the surface of the egg that allow no other sperm to enter and triggers the completion of meiosis. After the sperm sheds its tail, the nuclei of the egg and sperm fuse to form a zygote, the first cell of a new individual.*

With the union of the chromosomal material of both gametes, the zygote begins to divide by mitotic cell division. As division proceeds during the next 2 weeks, the developing cell mass is referred to as a **preembryo.**

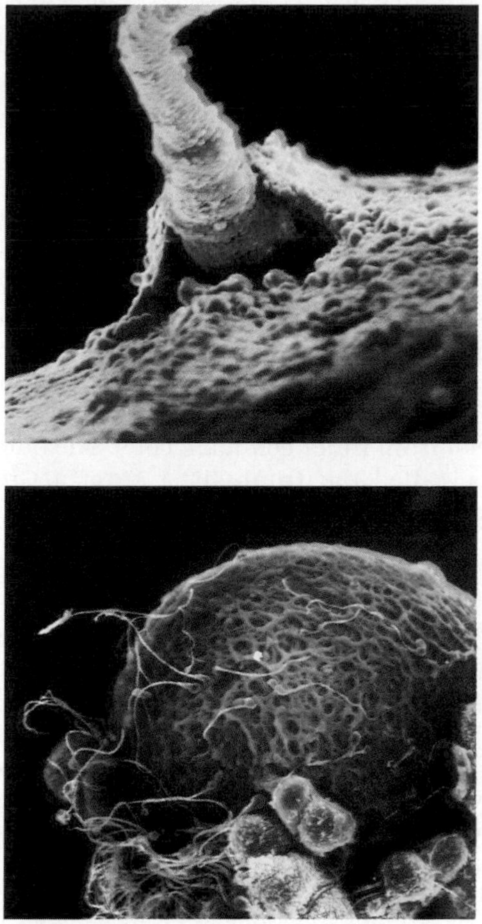

A

B

FIGURE 22-2 Fertilization. The moment of penetration of a human egg by a sperm cell **(A)** initiates changes in the surface of the egg that prevent other sperm from entering **(B)**.

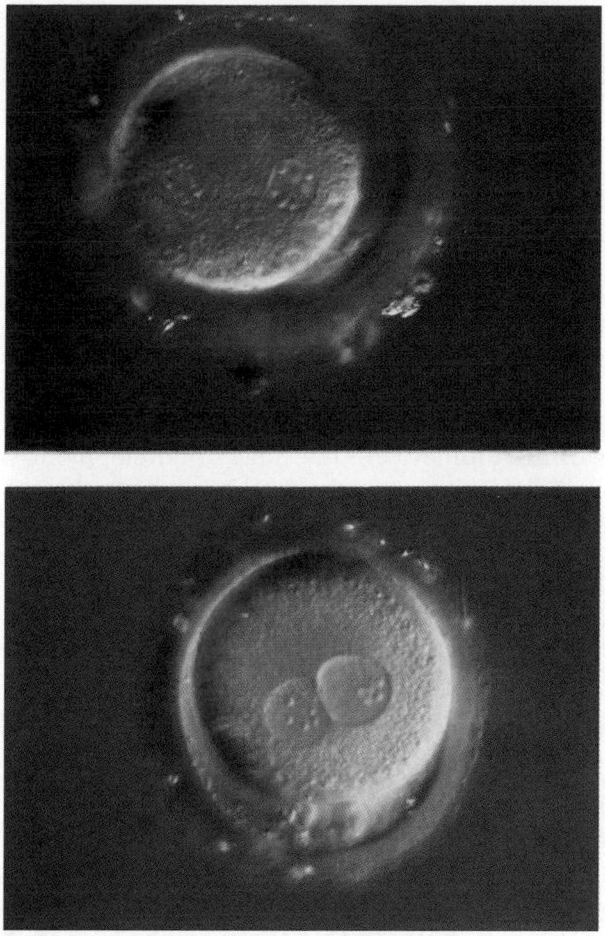

A

B

FIGURE 22-3 The fusion of nuclei. A, The nuclei of the sperm and ovum are clearly separate. **B,** The nuclei have merged to form the nucleus of the first cell of a new individual.

The First and Second Weeks of Development: The Preembryo

Within 30 hours, the one-celled zygote begins to divide rapidly: one cell into two, two into four, four into eight, and so forth, producing a cluster of cells. This process of cell division occurs without cell growth and is called *cleavage.* Occasionally, a single, fertilized ovum splits into two cell clusters during cleavage, and each cluster continues developing on its own. Because both cell clusters arose from the same cell, they have identical genetic information and will result in the development of identical twins. Identical twins are always the same sex. Fraternal twins arise from the release of two secondary oocytes and the fertilization of both. Fraternal twins are genetically different from one another because they are two separate oocytes fertilized by two different sperm cells. Therefore they may be different sexes.

prenatal development life before birth; the gradual growth and progressive changes in a developing human from conception until the time the fetus leaves the mother's womb.

fertilization the union of a male gamete (sperm) and a female gamete (egg).

ovum (**oh** vum) a mature human egg cell.

zygote (**zye** gote) a new cell that contains intermingling genetic material from both the sperm and egg cells; the fertilized ovum.

preembryo a term referring to the developing cell mass formed by the zygote as it begins to divide by mitotic cell division during the first 2 weeks of development.

As the cell cluster (preembryo) divides, it journeys along the remaining two thirds of the uterine tube. This trip takes approximately 3 days. Occasionally, a preembryo gets caught in the folded inner lining of the uterine tube and implants there, creating an *ectopic pregnancy.* The possibility of this situation occurring is increased if the tubes are scarred from previous infections. Acute pelvic pain usually signals this problem, which requires immediate medical attention. Normally, however, the preembryo reaches the uterus, or womb. Still the same size as a newly fertilized ovum, the preembryo now consists of about 16 densely clustered cells and is called a **morula,** from a Latin word meaning "mulberry." This stage of development is common to all vertebrates and even invertebrates. In fact, the processes of development in all vertebrates are very similar to the course of human development before birth.

> *Cleavage, or cell division of the zygote without cell growth, is the first stage in the development of humans as well as other multicellular animals. It results in the formation of a mass of cells known as a morula approximately 3 days after fertilization.*

The preembryo floats free in the uterus as its cells continue to divide. After 2 days in the uterus, the morula has developed into the **blastocyst,** a stage of development in which the preembryo is a hollow ball of cells (Figure 22-4). The center is filled with fluid from the uterine cavity. Like the term *morula,* the term *blastocyst* is descriptive, derived from two Greek words meaning "germ (germinal) sac."

One portion of the blastocyst contains a concentrated mass of cells destined to differentiate into the various body tissues of the new individual. It is referred to as the *inner cell mass.* Interestingly, each cell in this mass has the ability to develop into a complete individual. In fact, scientists have been able to produce test-tube mice using transplanted nuclei from inner cell mass cells. The outer ring of cells, called the **trophoblast** from a Greek word meaning "nutrition," will give rise to most of the extraembryonic membranes, including much of the placenta, an organ that helps maintain the developing embryo.

The general term **blastula** is used to describe the saclike blastocyst of mammals and, in other animals, the stage that develops a similar fluid-filled cavity. This stage of development is significant for all vertebrates; it is the first time that cells begin to move, or migrate, to shape the new individual in a process called **morphogenesis.** However, the major morphogenetic events occur during the third to eighth weeks.

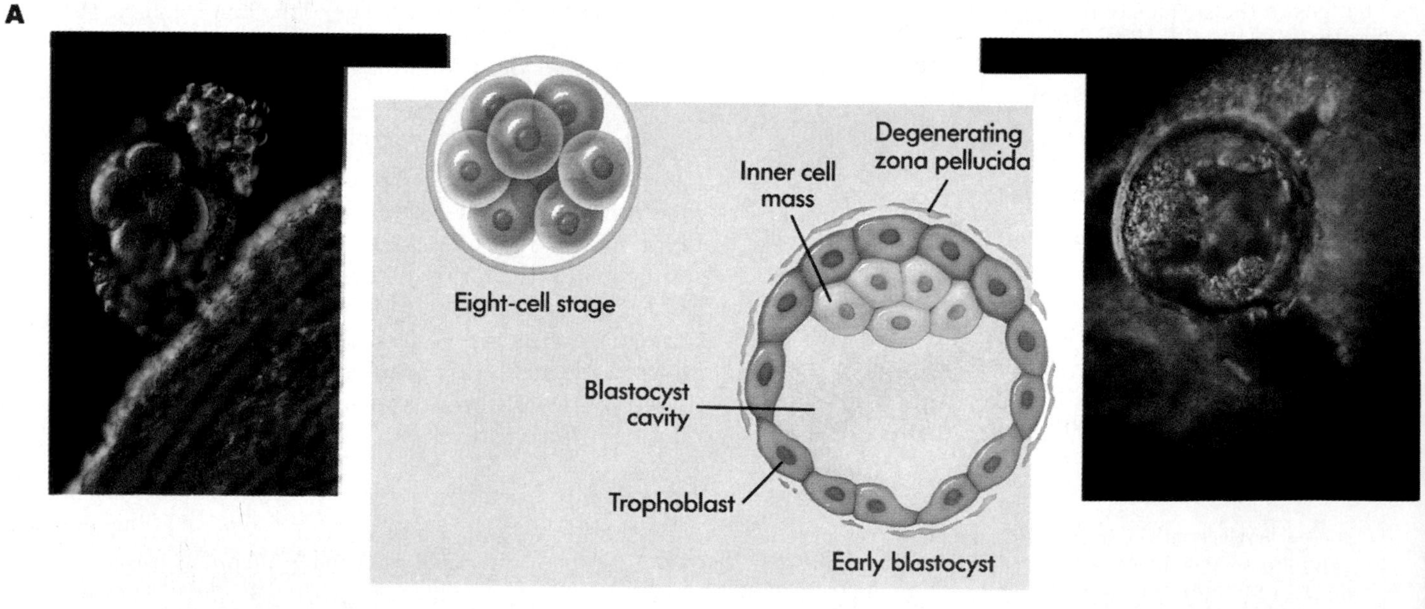

FIGURE 22-4 Development of the preembryo. A, View of the preembryo at the eight-cell stage. At this stage, the preembryo is floating free in the uterus as its cells continue to divide. **B,** After 2 days in the uterus, the ball of cells has developed into a blastocyst. The inner cell mass contains cells that will develop into the different body tissues of the new individual.

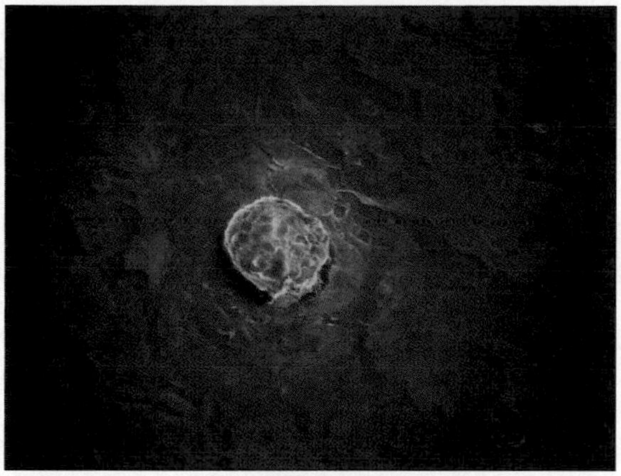

FIGURE 22-5 Implantation. About 7 to 8 days after fertilization, the blastocyst imbeds itself into the wall of the uterus.

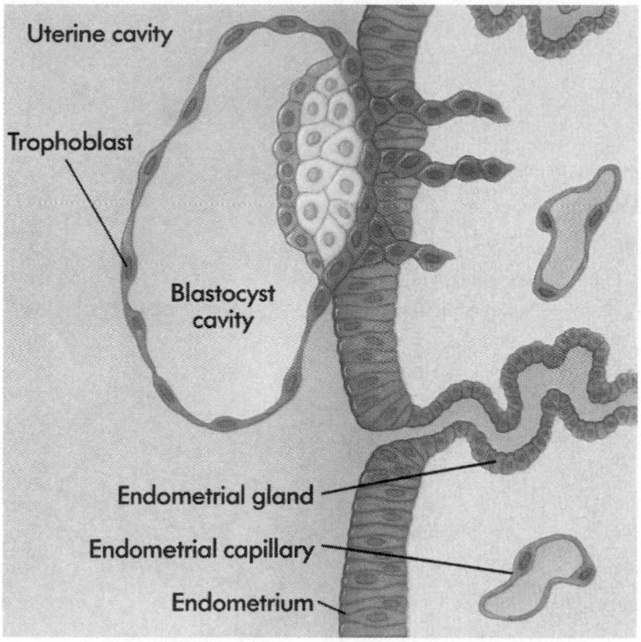

FIGURE 22-6 A detailed view of implantation. Special fingerlike projections produced by the trophoblast anchor the blastocyst to the wall of the uterus, or endometrium.

Cell migration, which begins to shape the preembryo, is the process that begins approximately 4 days after fertilization. Cell migration helps shape the developing individual in a process called morphogenesis. At this early stage, cell movement results in the formation of a hollow ball of cells known as a blastocyst in humans and in other mammals a stage more generally called the blastula.

Approximately 1 week after fertilization, the blastocyst secretes enzymes that digest a microscopic portion of the uterus. It then nestles into this site, nourished by the digested uterine cells, in a process called **implantation** (Figure 22-5). Barely visible to the naked eye, the blastocyst most often attaches to the posterior wall of the uterus (Figure 22-6). Figure 22-7 summarizes the events that take place from fertilization to implantation.

By this time, a woman is nearing the end of her menstrual cycle (see Chapter 21), and the blastocyst is in danger of being swept away during menstruation. However, the blastocyst secretes a hormone called *human chorionic gonadotropin* (*HCG*), which maintains the corpus luteum. The corpus luteum is a structure derived from the ruptured follicle from which the egg was cast out of the ovary. Sustained by HCG, the corpus luteum continues to secrete progesterone and estrogen, hormones that maintain the uterine lining and allow the development of the preembryo to proceed. After the first 3 months of development, the placenta (see p. 474) begins to secrete the estrogens and progesterone that maintain the pregnancy.

Approximately 1 week after fertilization, the blastocyst implants in the lining of the uterus and secretes human chorionic gonadotropin (HCG). HCG acts on the corpus luteum in the ovary. The corpus luteum responds by continuing to produce estrogens and progesterone, hormones that sustain the implanted blastocyst as the placenta develops.

morula (**more** yuh luh) a stage of development in which the preembryo consists of about 16 densely clustered cells and is still the same size as a newly fertilized ovum.

blastocyst (**blas** tuh sist) a stage of development in which the preembryo is a hollow ball of cells; its center is filled with fluid.

trophoblast (**trof** uh blast) the outer ring of cells of the blastocyst that will give rise to most of the extraembryonic membranes as the embryo develops.

blastula (**blas** chuh luh) the general term used to describe the saclike blastocyst of mammals and, in other animals, the embryonic stage that develops a similar fluid-filled cavity.

morphogenesis (**more** fuh **jen** uh sis) the early stage of development in a vertebrate when cells begin to move, or migrate, thus shaping the new individual.

implantation (**im** plan **tay** shun) the embedding of the developing blastocyst into the posterior wall of the uterus approximately 1 week after fertilization.

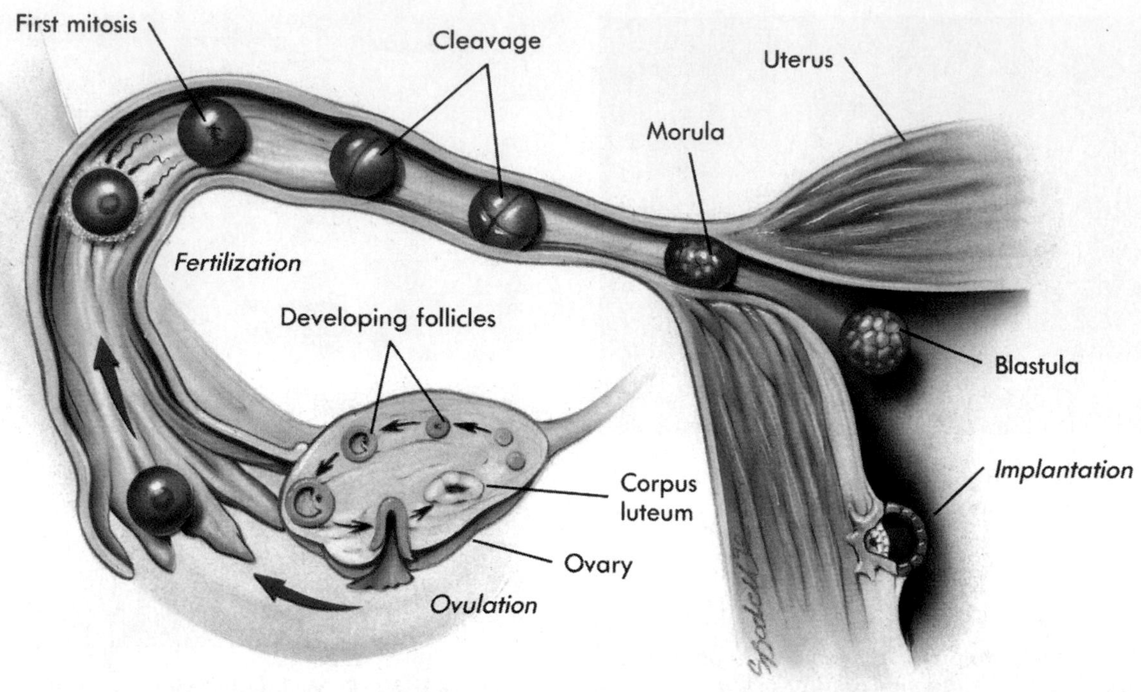

FIGURE 22-7 From fertilization to implantation. The unfertilized oocyte is fertilized by a sperm in the uterine (fallopian) tube. The fertilized egg (zygote) continues to divide as it makes its way through the tube to the uterus. About 7 to 8 days after fertilization, the fertilized egg, now a blastocyst, implants itself in the wall of the uterus.

During its second week of development, the preembryo completes its implantation within the uterine wall, two of its three **primary germ layers** develop, and the extraembryonic membranes begin to form. The primary germ layers are three layers of cells that develop from the inner cell mass of the blastocyst and from which all the organs and tissues of the body develop. These three layers are called the **ectoderm** ("outside skin"), **endoderm** ("inside skin"), and **mesoderm** ("middle skin"). The ectoderm forms the outer layer of skin, the nervous system, and portions of the sense organs. The endoderm gives rise to the lining of the digestive tract, the digestive organs, the respiratory tract, and the lungs; the urinary bladder; and the urethra. The mesoderm differentiates into the skeleton, muscles, blood, reproductive organs, connective tissue, and the innermost layer of the skin (Figure 22-8). Figure 22-9 diagrams the completely implanted preembryo buried within the uterine lining, visualizing two of the germ layers and the beginnings of the third. At the beginning of the second week only the endoderm and the ectoderm have formed.

Early Development of the Extraembryonic Membranes

In Figure 22-9, the extraembryonic membranes have begun to form even at this early stage of development. The **extraembryonic membranes** all play some role in the life support of the preembryo, the embryo, and then the fetus. (The term **fetus** is used to describe the stage of development after 8 weeks.) Called *extraembryonic membranes* because they are not a part of the body of the embryo, these structures provide nourishment and protection. They form from the trophoblast, the ring of cells surrounding the inner cell mass. At the same time the preembryo is forming from the inner cell mass cells. The further development of these membranes continues into the fetal period.

The **amnion** is a thin, protective membrane that grows down around the embryo during the third and fourth weeks, fully enclosing the embryo in a membranous sac. The amniotic sac can be thought of as a shock absorber for the embryo. This thin membrane encloses a cavity that is filled with a fluid—*amniotic fluid*—in which the fetus floats and moves. This fluid also helps keep the temperature of the embryonic and fetal environment constant. The amniotic cavity is first seen at about 8 days as a slitlike space that appears between the inner cell mass and the underlying trophoblast cells that are invading the uterine wall. It can be seen as a better-developed structure in Figure 22-9, at the end of 2 weeks of development.

Notice also in Figure 22-9 that a structure called the **chorion** is beginning to develop as tissue outside of and including the trophoblast cells that ring

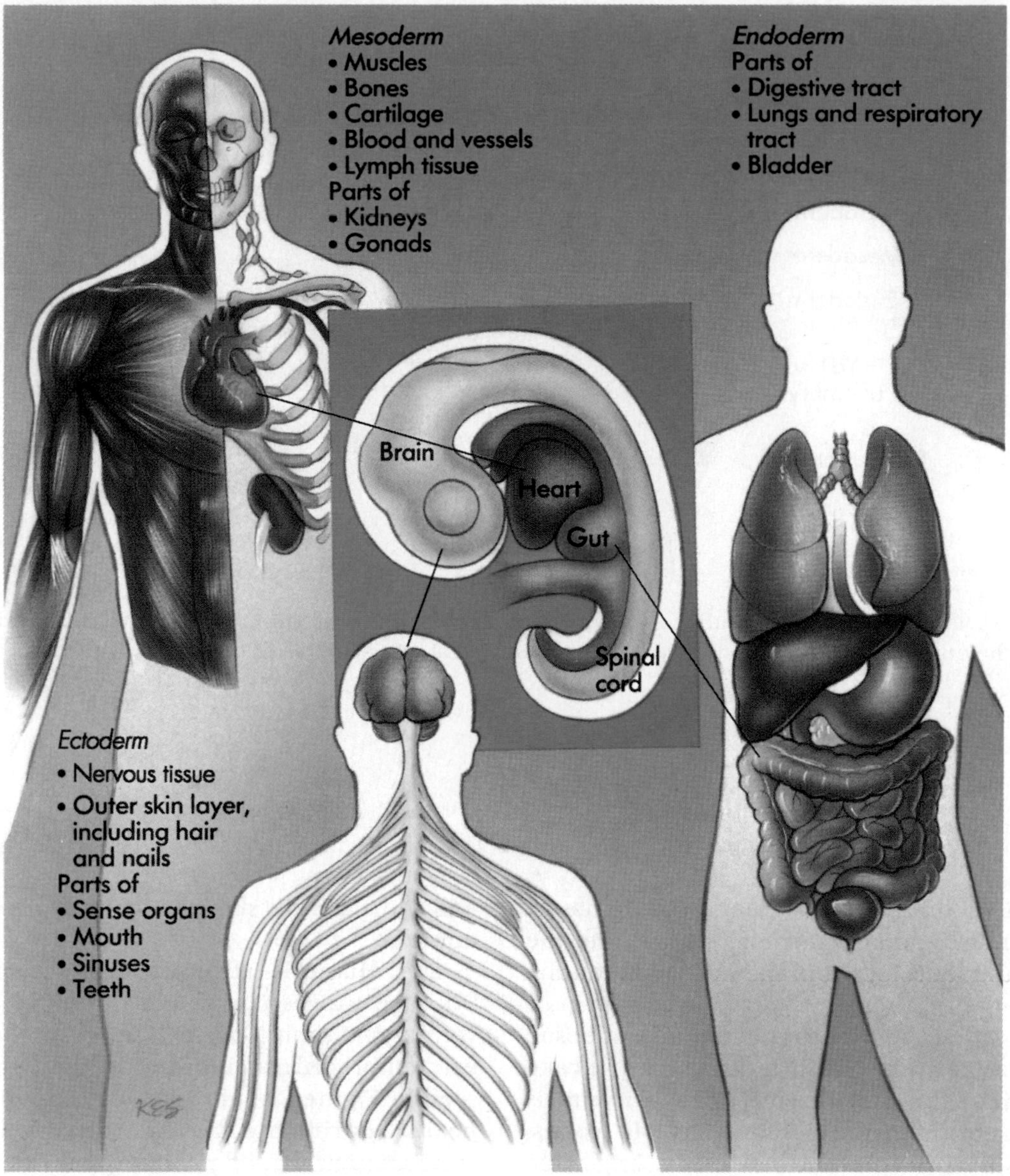

FIGURE 22-8 Structures produced by the three primary germ layers. The diagram shows the body systems and tissues into which the germ layers develop.

primary germ layers three layers of cells that develop from the inner cell mass of the blastocyst and from which all the organs and tissues of the body develop.

ectoderm (**ek** toe durm) the primary germ layer that gives rise to the outer layer of skin, the nervous system, and portions of the sense organs.

endoderm (**en** doe durm) the primary germ layer that gives rise to the digestive tract lining, the digestive organs, the respiratory tract, the lungs, the urinary bladder, and the urethra.

mesoderm (**mez** oh durm) the primary germ layer that gives rise to the skeleton, muscles, blood, reproductive organs, connective tissue, and the innermost layer of the skin.

extraembryonic membranes (ek struh **em** bree **on** ik) structures that form from the trophoblast and provide nourishment and protection for the preembryo, the embryo, and then the fetus.

fetus (**fee** tus) the term used to describe the stage of prenatal development after 8 weeks and until birth.

amnion (**am** nee on) a thin, protective membrane that grows down around the embryo during the third and fourth weeks. Fluid fills the cavity between the amnion and the embryo or fetus.

chorion (**kore** ee on) an extraembryonic membrane that facilitates the transfer of nutrients, gases, and wastes between the embryo and the mother's body. It is a primary part of the placenta.

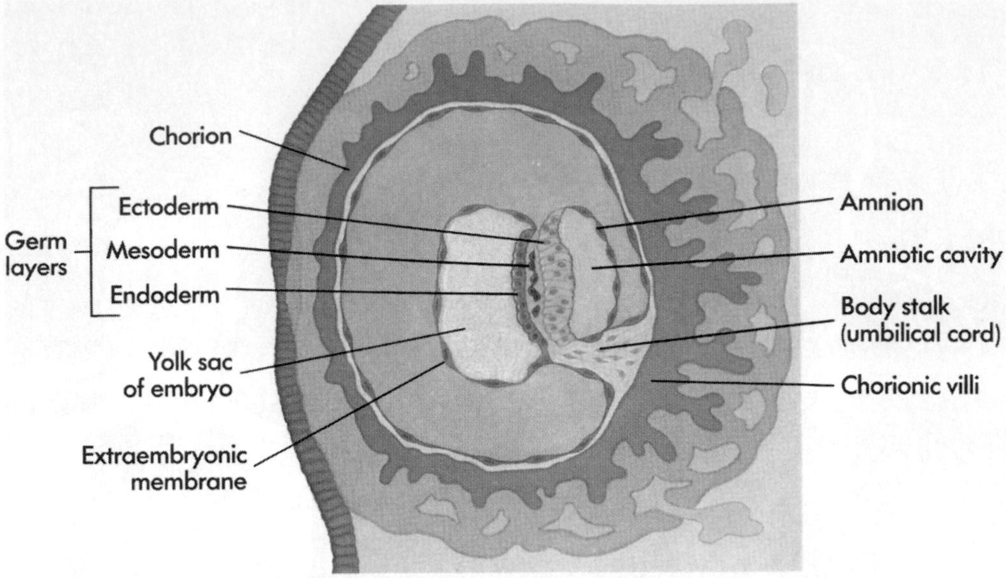

FIGURE 22-9 The implanted preembryo showing the three primary germ layers during the third week of development. The endoderm and ectoderm have already formed at this stage, with the mesoderm just beginning its development.

the developing embryo and extraembryonic tissue. The chorion is highly specialized to facilitate the transfer of nutrients, gases, and wastes between the embryo and the mother's body. It is a primary part of an organ called the **placenta,** a flat disk of tissue about the size (at birth) of a large, thick pancake that grows into the uterine wall. The placenta is made up of both chorionic and maternal tissues. Digging deeply into the uterine lining, fingerlike extensions of the chorion come into close contact with the blood-filled uterine tissues at the placenta. When the embryo reaches a stage of development in which the heart begins to beat, oxygen-poor blood filled with wastes is sent from the embryo's body through the *umbilical arteries* to the placenta.

The two umbilical arteries and a single vein are embedded in the connective tissue of the **umbilical cord,** the developing embryo's lifeline to the mother. This "highway" joins the circulatory system of the embryo with the placenta. At the placenta, embryonic wastes are exchanged for nutrients and oxygen through a thin layer of cells that separates the embryo's blood from the mother's blood. The embryonic blood and maternal blood do not mix. The fetal blood then travels through the umbilical vein back to the embryo. The umbilical cord develops from the body stalk, the yolk sac, and the allantois during the fourth week of development.

The **allantois** (from a Greek word meaning "sausage") gives rise to the umbilical arteries and vein as the umbilical cord develops. Although this extraembryonic membrane is small and cannot be seen in Figure 22-9, it is shown in Figure 22-10. Appearing during the third week of development as a tiny, sausage-shaped pouching on the yolk sac, the allantois is initially responsible for the formation of the embryo's blood cells; it later develops into the umbilical blood vessels. Notice that the **yolk sac,** a structure established during the end of the second week, also becomes a part of the umbilical cord. Before becoming a nonfunctional part of the cord, the yolk sac produces blood for the embryo until its liver becomes functional during the sixth week of development. In addition, part of the yolk sac becomes the lining of the developing digestive tract.

During the second week of development, the preembryo completes implantation within the uterine wall, two of the three layers of cells develop from which organs and tissues will arise, and the extraembryonic membranes—the amnion, chorion, yolk sac, and allantois—begin to form from the trophoblast.

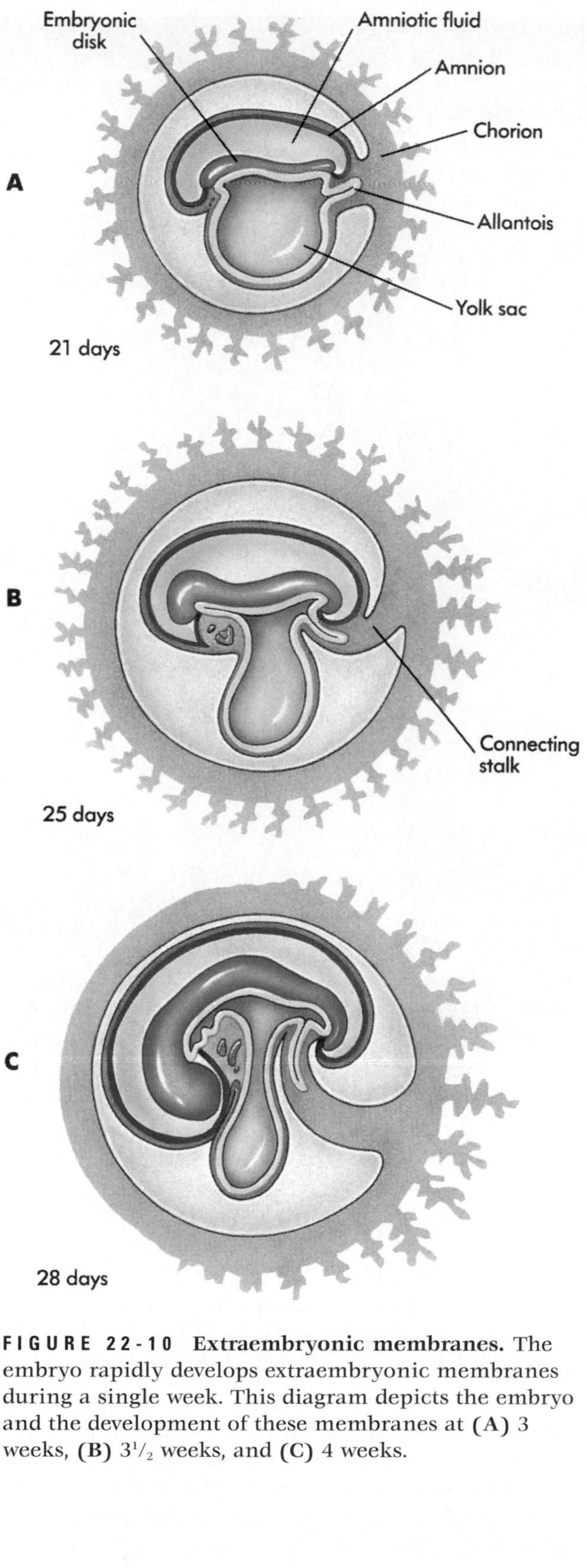

FIGURE 22-10 Extraembryonic membranes. The embryo rapidly develops extraembryonic membranes during a single week. This diagram depicts the embryo and the development of these membranes at **(A)** 3 weeks, **(B)** $3^1/_2$ weeks, and **(C)** 4 weeks.

Development from the Third to Eighth Weeks: The Embryo

During this crucial time period, the developing individual is termed an **embryo.** Throughout the next 6 weeks, the embryo takes on a human shape by means of the ongoing process called *morphogenesis* mentioned earlier. During morphogenesis, cells move to shape the developing embryo. In addition, the organs are established but need further development during the third through eighth weeks in a process called *organogenesis.* Because of the incredible array of developmental processes proceeding simultaneously during this period, the embryo is particularly sensitive to *teratogens,* certain agents such as alcohol that can induce malformations in the rapidly developing tissues and organs (see "Just Wondering").

The Third Week

At the end of the second week and continuing into the third week, various cell groups of the inner cell mass begin to divide, move, and differentiate, changing the two-layered preembryo into a three-layered embryo. This process is called **gastrulation.** This word is derived from a Greek word *gastros* meaning "belly." In fact, the prefix *gastr-* is found in many words denoting parts of the human body, such as gastric, referring to the stomach. This term is descriptive of the fact that at this stage in many animals, a primitive gut is formed by the invagination (infolding) of the blastula, but the intestines develop differently in humans and many animals.

placenta (pluh **sen** tuh) a flat disk of tissue that grows into the uterine wall, made up of both chorionic and maternal tissues, through which the embryo and fetus are supplied with food, water, and oxygen and through which wastes are removed.

umbilical cord (um **bil** uh kul) the developing embryo's lifeline to the mother; it joins the circulatory system of the embryo with the placenta.

allantois (ah **lan** toe us) an extraembryonic membrane that gives rise to the umbilical arteries and vein as the umbilical cord develops.

yolk sac a membranous sac that surrounds the food yolk and produces blood for the embryo until its liver becomes functional. In addition, part of the yolk sac becomes the lining of the developing digestive tract.

embryo (**em** bree oh) the early stage of development in humans, from the third to eighth weeks.

gastrulation (gas truh **lay** shun) during prenatal development, the process by which various cell groups of the inner cell mass migrate, divide, and differentiate resulting in a three-layer embryo.

JUST WONDERING. . . .

Can I drink a glass of wine occasionally during my pregnancy or will it hurt my baby?

Unfortunately, scientists have not yet determined how much alcohol, if any, a woman can drink during pregnancy and not risk damage to her fetus. Complicating the matter, researchers are unsure when the fetus is most vulnerable to the effects of alcohol. So having two drinks on day 39 may not harm the fetus, for example, but having two drinks on day 40 may cause fetal brain damage.

When a pregnant woman drinks, the alcohol she consumes crosses the placenta and intoxicates the fetus. A recent study has demonstrated that women have less of a stomach enzyme to neutralize alcohol than men do, so women become intoxicated faster because of higher levels of alcohol in their bloodstream. These high blood alcohol levels also mean that more alcohol is passed along to the fetus.

The damage that is inflicted on the fetus can be severe, or it can be more subtle. There seems to be a rough correlation between the amount of alcohol consumed and the severity of the birth defects that result. The most serious effects of drinking on the developing fetus are seen in babies born to chronic alcoholics who drank heavily during pregnancy. These effects are called *fetal alcohol syndrome* and are characterized by certain psychological, behavioral, mental, and physical abnormalities. These children require special education and sometimes physical therapy to help them contend with their disabilities. Women who drink less than alcoholics but drink consistently throughout their pregnancies often give birth to children with *fetal alcohol effects*. These children have learning difficulties, show poor judgment, are impulsive, are unable to learn from their mistakes, and are undisciplined. These children often have serious difficulties in school.

Experts agree that the surest way to protect the unborn child from alcohol-related effects is not to drink at all during pregnancy. Some researchers believe that a woman even contemplating pregnancy should not drink. This position makes sense because many women do not know they are pregnant until weeks after conception, and unwitting alcohol consumption may have already caused damage to the fetus. Given the uncontrollable factors influencing the birth of a healthy baby (such as heredity), it makes sense to control the factors we can. Not drinking is one thing a pregnant woman can do to contribute to the health of her baby.

At the beginning of the third week, the embryo is an elongated mass of cells barely one tenth of an inch long. A streak (called the *primitive streak*) runs down the midline of what will be the back side of the embryo. Cells at the streak migrate inward, producing the mesoderm that develops during the gastrulation period. Cells at the head end of the streak grow forward to form the beginnings of the **notochord.** The notochord is a structure that forms the midline axis along which the vertebral column (backbone) develops. An embryonic notochord forms in all vertebrate animals. However, in humans and in most other vertebrates, it degenerates and disappears long before birth. Gastrulation ends with the completion of the notochord midway through the third week. By the end of gastrulation, a layer of ectoderm covers the notochord tissue. Figure 22-11 diagrams the events of gastrulation.

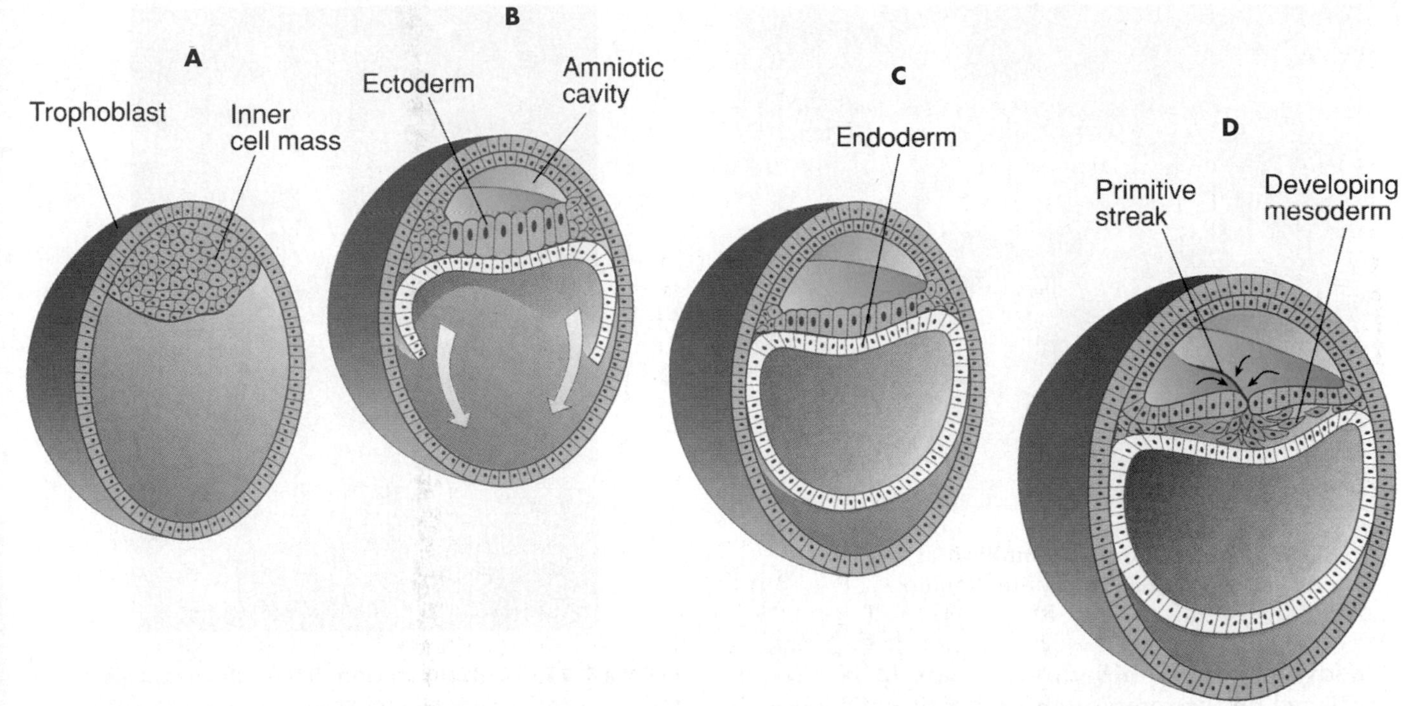

FIGURE 22-11 Human gastrulation. The amniotic cavity forms within the inner cell mass **(A)**, and in its base, layers of ectoderm and endoderm differentiate **(B** and **C)**. A primitive streak develops, through which cells destined to become mesoderm migrate into the interior **(D)**.

Gastrulation is a process by which cell groups of the inner cell mass migrate, divide, and differentiate into three primary germ layers from which all the organs and tissues of the body will develop. This process in humans as well as all vertebrates results in a developmental stage called the gastrula.

In the photo of the embryo at 4½ weeks (Figure 22-12), its head can be seen to have the beginnings of eyes, a sign that **neurulation** has taken place. Neurulation is the development of a hollow nerve cord, which later develops into the brain, spinal cord, and related structures such as the eyes.

Neurulation begins in the third week with the folding of the ectoderm lying above the notochord, forming an indentation along the back of the embryo. This indentation is called the *neural groove.* On either side of the groove are areas of tissue called *neural folds.* In Figure 22-13 the neural folds of the 3-week embryo have come together at one spot and have fused. This spot is destined to be the neck region of the developing individual. The tissue above the fused region, looking somewhat like a pair of lips, will develop into the brain. The less broad area of tissue below the fused region will develop into the spinal cord. Eventually the edges of the groove will fuse along its length, forming a neural tube, the precursor to these structures.

During neurulation, the mass of ectodermal cells lying over the notochord curls up, forming a groove. The edges of the groove move together, eventually forming a tube that will develop into the brain and spinal cord—the central nervous system.

Neurulation results in an embryo called a *neurula* and signals the developmental process of **tissue differentiation.** During tissue differentiation, groups of cells become distinguished from other groups of cells by the jobs they will perform in the

notochord (**no** toe kord) a structure that forms the midline axis along which the vertebral column (backbone) develops in all vertebrate animals.

neurulation (**noor** oo **lay** shun) the development of a hollow nerve cord.

tissue differentiation a prenatal developmental process in which groups of cells become distinguished from other groups of cells by the jobs they will perform in the body.

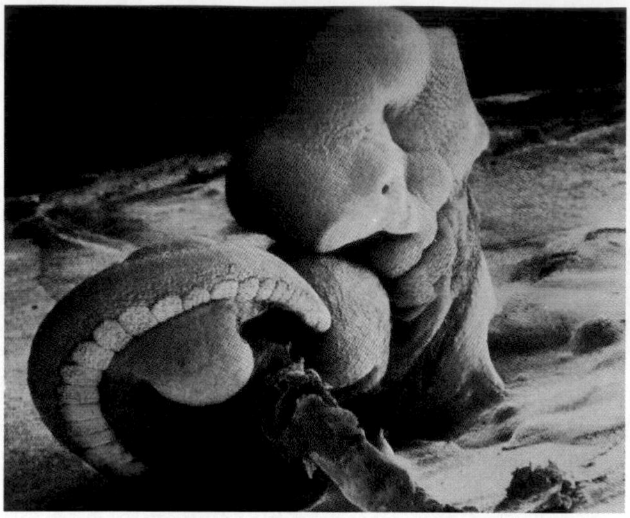

FIGURE 22-12 A human embryo at $4^1/_2$ weeks. Note that the eyes have begun to develop.

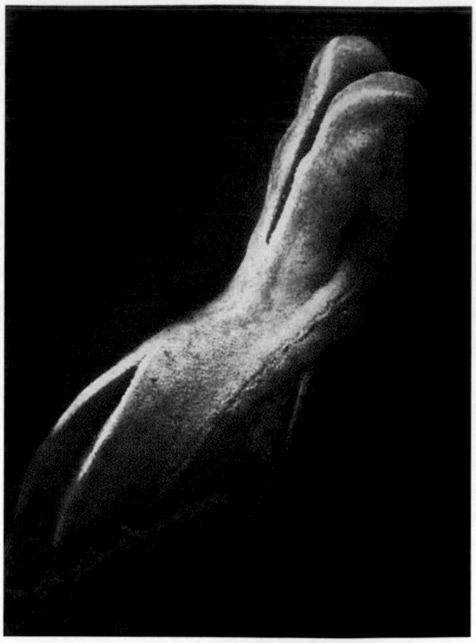

FIGURE 22-13 Neurulation. The ectoderm folds inward to form the neural groove and, at 3 weeks gestation, fuses. The spot at which the fusion occurs will become the neck, and the tissue above this spot will become the brain. The area below the fused region will become the spinal cord.

body. This differentiation appears to be accomplished by the interaction of cytoplasmic stimulators and inhibitors of certain cells with various genes of other cells. Put simply, some cells act like on and off switches for the genes of neighboring cells. This process is **induction.**

> *Important processes of preembryonic and embryonic development are cleavage, cell migration, morphogenesis, and tissue differentiation. The four stages of development of the preembryo and the embryo are the morula, the blastula, the gastrula, and the neurula.*

The embryo's heart also begins its development during the third week as a pair of microscopic tubes. In fact, the cardiovascular system is the very first system to become functional in the embryo. At the end of the third week of development, the heart tubes have fused and have linked with blood vessels in the embryo, body stalk, chorion, and yolk sac to provide a primitive circulation of blood (Figure 22-14), and the heart actually begins to beat! At the same time, fingerlike structures called *villi* begin to protrude from the chorion. These blood-filled projections dig into the uterine tissues of the mother, increasing the surface area over which gases, nutrients, and wastes can be exchanged. At 8 to 10 weeks of development, doctors can take some of this tissue to detect whether genetic abnormalities exist in the embryo. Termed *chorionic villus sampling* (*CVS*), this procedure is performed by inserting a suction tube through the vagina, into the uterus, and to the chorionic villi. Because chorion cells and fetal cells contain identical genetic information, doctors can use these cells for genetic studies.

Colorized with a blue-gray tint, paired segments of tissue are also prominent in Figure 22-12. These chunks of mesoderm are called *somites* (from a Greek word meaning "a body"). They will give rise to most of the axial skeleton (see Chapter 15) with its associated skeletal muscles and most of the dermis of the body—tissues that underlie the epidermis of the skin. The first somites appear during the third week of development and are added as the embryo grows. Eventually, 42 to 44 pairs develop by the middle of the fifth week.

The Fourth Week

During the fourth week of development, the embryo begins to curl (see Figure 22-12). The curling occurs as folds are formed at the head and tail end of the embryo. As the embryo curls, the yolk sac becomes squeezed into a narrow stem that fuses with the body stalk connecting the embryo to the placenta (see Figure 22-10). As this fused body stalk lengthens and as its surface is tightly covered by the growing amnion membrane, it is properly called the *umbilical cord.* Blood cells continue to be

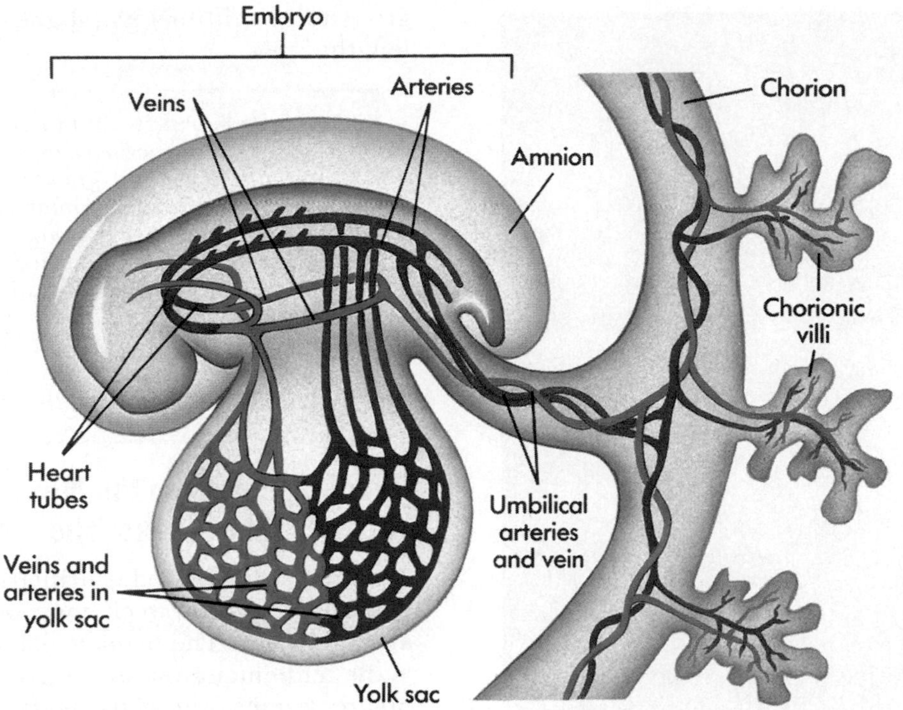

FIGURE 22-14 Cardiovascular system of a 3-week-old embryo. At the end of the third week, the heart begins to beat. Villi extending out from the chorion dig into the uterine wall and provide a surface for the exchange of gases, nutrients, and wastes.

produced by the yolk sac until the liver completely takes over this job in the sixth week. The umbilical arteries and vein, already functional, arose from the pouch of the allantois in the body stalk and have been transporting blood between the embryo and the placenta since the third week.

Four visible sets of swellings called the *branchial arches* develop on either side of the head end of the embryo during the fourth week. The word *branchial* is derived from a Greek word meaning "gill" and better describes similar structures that develop into gill supports in fish. In fact, the head end of the human embryo at this stage somewhat resembles a fish embryo.

The presence of gill arches and a tail (see Figure 22-12) in the human embryo indicates a relationship between humans and the lower vertebrates (fish and amphibians). In the nineteenth century, the German scientist Ernst Haeckel described this link as "ontogeny recapitulates phylogeny." What Haeckel meant was that the embryological development of an individual organism (ontogeny) repeats (recapitulates) the evolutionary history of its ancestors (phylogeny). This statement, however, is untrue. Embryonic stages of particular vertebrates are not a replay of the succession of its adult ancestors. Rather, the embryonic stages of an individual organism often reflect embryonic stages of its ancestors. For example, gill arches form during the embryological development of humans. These arches develop into structures such as the middle ear, eustachian tube, tonsils, thymus, and parathyroids.

The embryo's heart begins development during the third week. Somites—chunks of mesoderm—also become prominent at this time. They will give rise to most of the axial skeleton. During the fourth week, the gill arches appear, which will later develop into such structures as the middle ear and the tonsils.

The Fifth through Eighth Weeks

During the fifth week, the embryo doubles in length from 4 millimeters ($^3/_{16}$ inch) to about 8 millimeters ($^3/_8$ inch). A nose begins to take shape as tiny pits. Although *limb buds* were first visible during the fourth week, first the arms and then the

induction the process by which some cells act like off and on switches for the genes of neighboring cells.

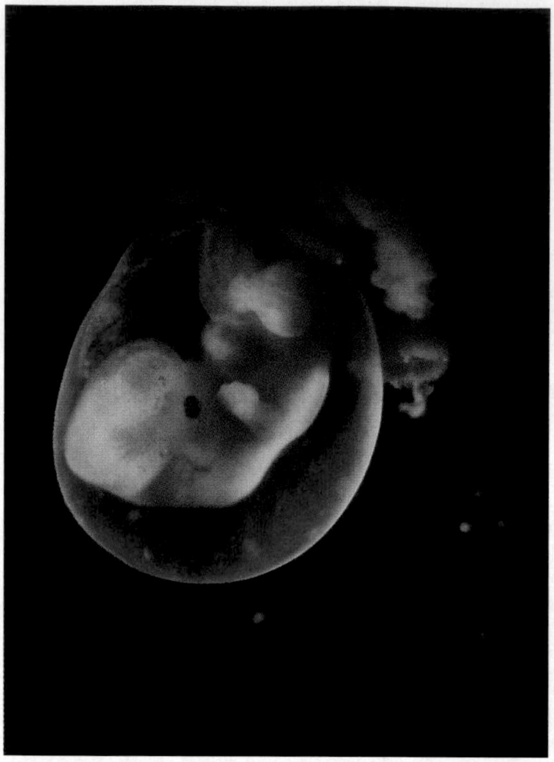

FIGURE 22-15 Embryo at 6 weeks of age. At this stage, the arms and legs are beginning to form. No bones have formed in the skull, so the brain can be clearly seen above the eye.

legs, at this stage of development the limbs look like microscopic flippers. The brain grows rapidly this week, resulting in an embryo with a large head in proportion to the rest of the developing body.

Seen floating within the fluid-filled amniotic sac in Figure 22-15, the 6-week-old embryo now has arm buds with distinct wrists. Fingers are just beginning to form as are the ears. The retina of the eyes is now darkly pigmented. Although the trunk of the body straightens out somewhat as the liver and digestive system grow, the neck area remains bent, forcing the head to rest on the chest above the red protruding heart, which now beats 150 times per minute. At the end of 6 weeks, the embryo is little more than half an inch long.

During the seventh week of development, eyelids begin to partially cover the eyes, and the ears develop more fully. The face begins to take on a human appearance. Each arm develops an elbow. Individual fingers can be distinguished, but they are connected with webs of skin that disappear during the eighth week. The legs, slower in their development than the arms, develop ankles and the suggestion of toes. During the eighth week—the last week as an embryo—the developing individual grows to 30 millimeters (slightly over an inch) in length.

From the fourth to eighth weeks of development, dramatic changes take place in the embryo. The embryo grows from a length of 2 millimeters to approximately 30 millimeters (slightly longer than an inch). A primitive circulation is established within the embryo, and a maternal-fetal exchange of nutrients, gases, and wastes begins. The central nervous system and its associated structures begin to develop. The body form is established, including the appendages.

Development from the Ninth to Thirty-Eighth Weeks: The Fetus

The embryonic period is primarily one of development. By the ninth week, most of the body systems are functional. The primary job of the fetal period is the refinement, maturation, and growth of these organ systems and of the body form of the developing individuals.

The Third Month

The **first trimester** ends with the completion of the third month of pregnancy. During the third month, the fetus grows rapidly, tripling its length to approximately 85 millimeters ($3^1/_2$ inches). Fine hair called *lanugo,* meaning "down," appears over its body, but this downy coat is lost before birth. Changes in the shape of the face cause the eyes to face forward and appear closer together, although the forehead remains prominent (Figure 22-16). The eyes also become well developed during this month. The eyelids have continued their development and now fuse shut. They will not open until the fetus is 7 months old. A suggestion of an ear can be seen on the side of the head in line with the jaw bone in the 12-week-old fetus in Figure 22-17.

During the embryonic period, male babies cannot be distinguished from female babies in their outward appearance. During early fetal development, outward differences between the sexes begin to take shape. The differentiation of male or female sex organs depends, of course, on the genetic makeup of the fetus but also on a complex interplay between hormones and tissues in the developing fetus. This interplay of factors results in the development of strikingly different reproductive systems by the end of the third month, with the penis in the male and the clitoris in the female arising from the same embryonic tissues. Likewise, the scrotum in the male is homologous to the labia majora in the female.

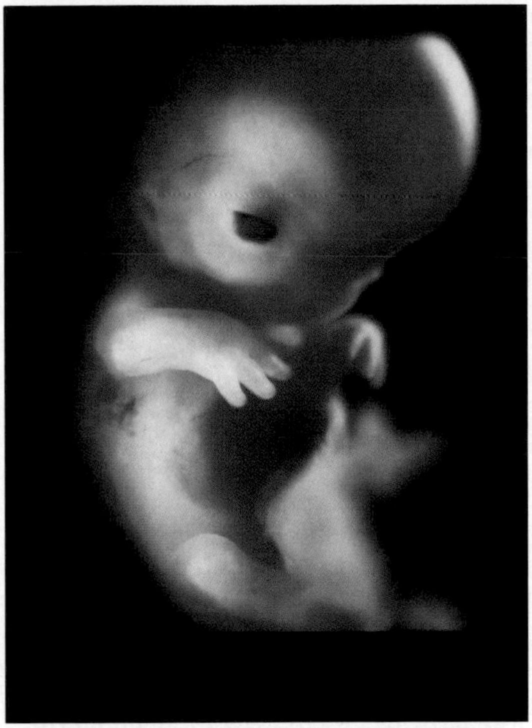

FIGURE 22-16 Fetus at 11 weeks of age. Fingers have begun to develop at this stage. The eyes are well formed and are covered with eyelids.

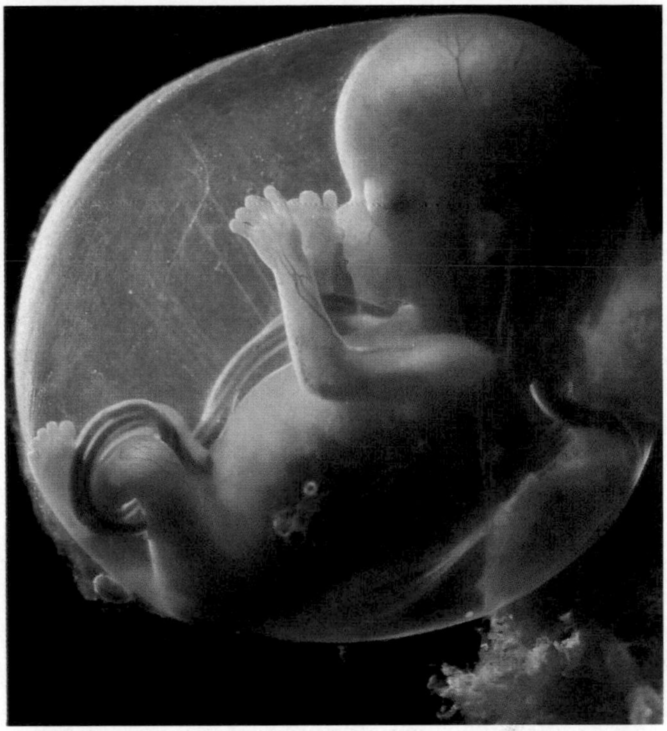

FIGURE 22-17 Fetus at 12 weeks of age. Ears can be seen, as well as the fingers and toes.

During the third month the fetus grows to $3^1/_3$ inches in length. Eyes become well developed, and ears begin to form.

The Fourth through Sixth Months

During the **second trimester,** the fetus grows to about 0.6 kilogram (about $1^1/_2$ pounds) and 0.3 meter (1 foot) long. The 4-month-old fetus in Figure 22-18 looks quite human. Halfway through the fourth month, the fetus can bring the hands together and suck the thumb. By 15 weeks, the sensory organs are almost completely developed, and by 16 weeks the fetus is actively turning inside the mother. Many of the bones are forming, replacing areas of cartilage. The process of bone formation, termed **ossification,** began at approximately 8 weeks of development and will continue beyond birth to the age of 18 or 19 years. By the end of the fifth month, the heartbeat of the fetus can be heard through a stethoscope. Although the body systems have been rapidly continuing their development, the fetus is still unable to survive outside of the mother's womb. At the end of the sixth month, the head is no longer quite as large compared with the rest of the body, the eyelids have separated, and the eyelashes have formed. The fetus is capable of independent survival at the sixth month but only with special medical intervention.

During the fourth through sixth months, the fetus grows to 30 centimeters (12 inches) in length and begins to look human. The sensory organs become well developed, and the bones begin to form, replacing cartilage.

first trimester the first 3 months of pregnancy.
second trimester the second 3-month period of pregnancy.
ossification (os uh fih **kay** shun) the process of bone formation, which in humans begins approximately at the eighth week of development, and continues beyond birth to the age of 18 or 19 years.

FIGURE 22-18 Fetus at 4 months of age. The fetus at this point has a human appearance and can even bring the hands together and suck the thumb.

The Seventh through Ninth Months

The **third trimester** is predominantly a period of growth rather than one of development. In the seventh, eighth, and ninth months of pregnancy, the weight of the fetus doubles several times because of growth but also because of fat that is laid down under its skin.

One important developmental change, however, is that most of the major nerve tracts in the brain, as well as many new brain cells, are formed during this period. The mother's bloodstream fuels all of this growth by the nutrients it provides. Within the placenta these nutrients pass into the fetal blood supply (Figure 22-19). If the fetus is malnourished because the mother is malnourished, this growth can be adversely affected. The result is a severely retarded infant. Retardation resulting from fetal malnourishment is a serious problem in many underdeveloped countries where poverty is common.

By the end of the third trimester, the neurological growth of the fetus is far from complete and, in fact, continues long after birth. By this time, however, the fetus is able to exist independently.

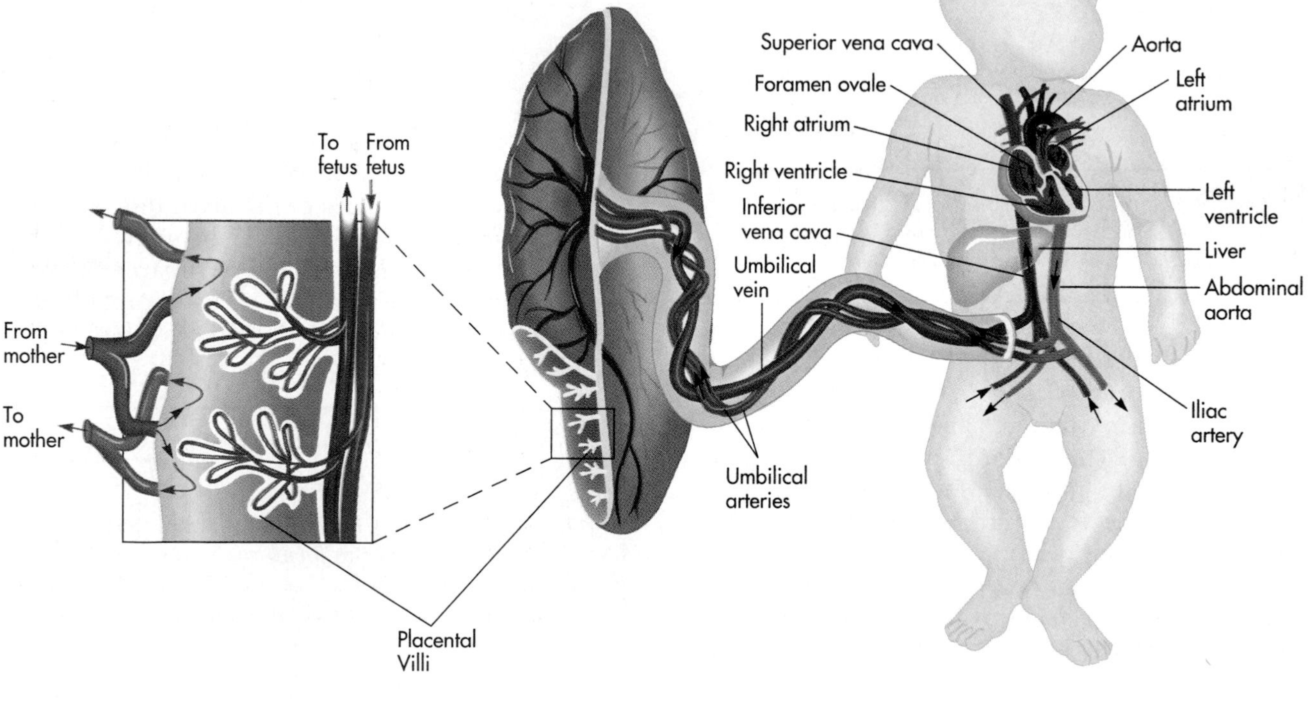

FIGURE 22-19 Placental-fetal circulation. During its development, the fetus receives all its nutrients from its mother across the umbilical cord. Wastes are removed from the fetus through the umbilical cord to be excreted by the mother.

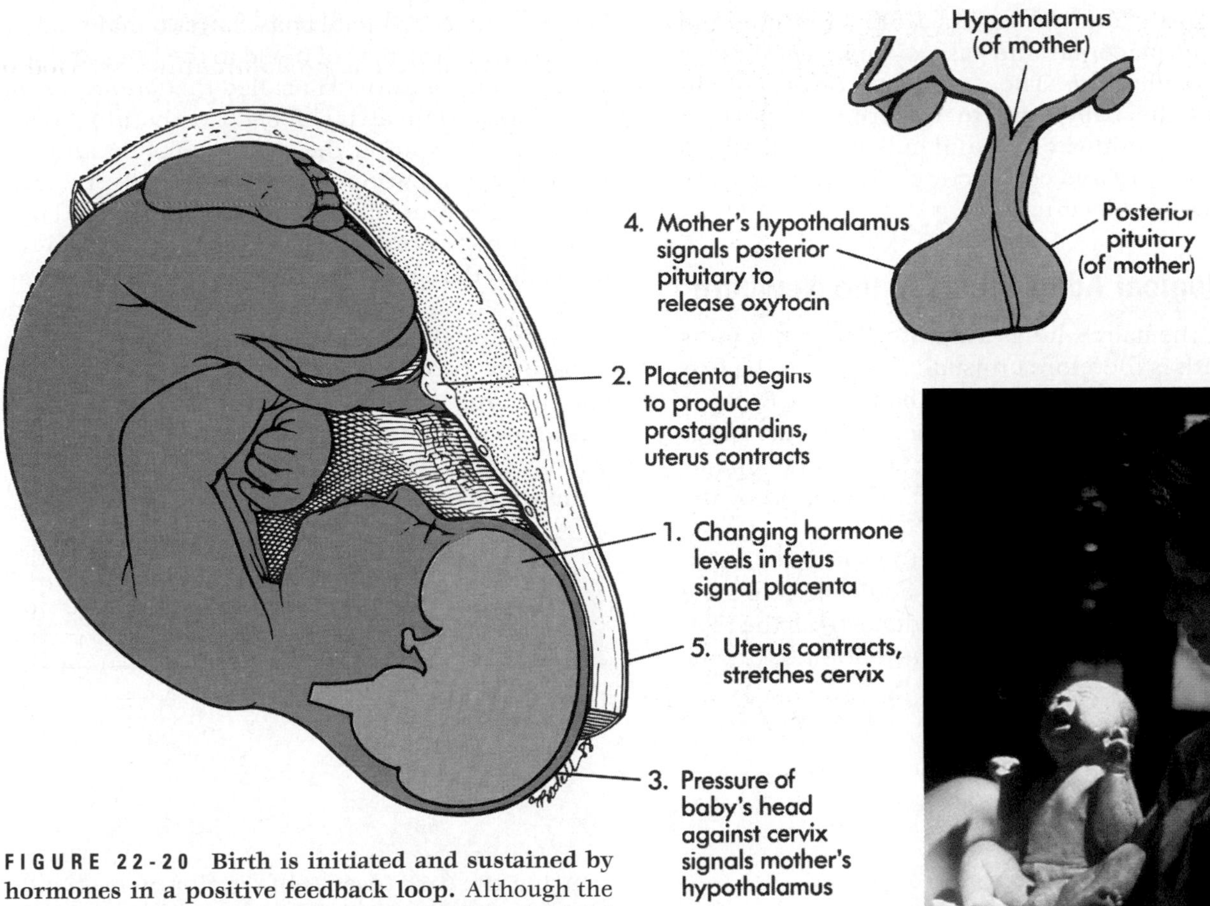

FIGURE 22-20 Birth is initiated and sustained by hormones in a positive feedback loop. Although the exact mechanism is not known, it is thought that birth is triggered by the following mechanism: changing hormone levels in the fetus *(1)* signal the placenta to produce prostaglandins, causing the uterus to contract *(2)*. In addition, the fetus' head pushing against the cervix *(3)* signals the hypothalamus to release oxytocin from the pituitary *(4)*, which causes the uterus to contract even harder *(5)*. This feedback loop, in which the pressure of the fetus' head causes the release of oxytocin, continues until the baby is born.

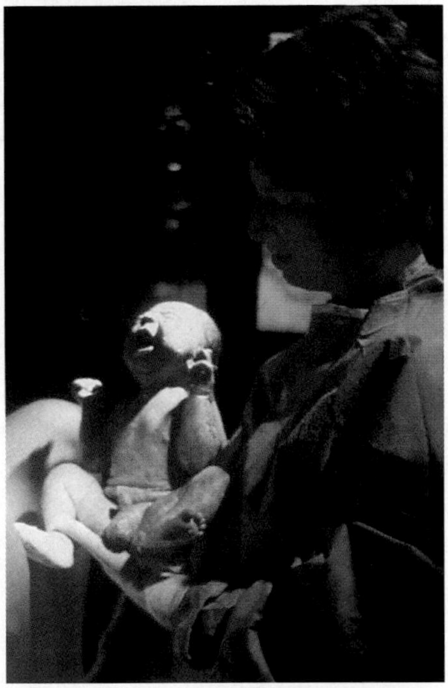

> *The critical stages of human development take place quite early. All the major organs of the body have been established by the end of the third month. The following 6 months are essentially a period of growth.*

Birth

Birth takes place at the end of the third trimester, 38 weeks from conception (40 weeks from the start of the last menstrual period). Although the exact mechanism of the onset of **labor,** the sequence of events that leads to birth, is not well understood, scientists know that changing hormone levels in the developing fetus initiate this process. These hormones induce placental cells of the mother to manufacture *prostaglandins* (see p. 360), hormonelike substances that cause the smooth muscle of the uterine wall to contract. In addition, the pressure of the fetus' head against the cervix sends nerve impulses to the mother's brain that trigger the hypothalamus to release the hormone oxytocin from her pituitary. Working together, oxytocin and prostaglandins stimulate waves of contractions in the walls of the uterus, forcing the fetus downward (Figure 22-20). Initially, only a few contractions occur each hour, later increasing to one every 2 to 3 minutes. Eventually, strong contractions, aided by the mother's pushing, expel the fetus through the vagina, or **birth canal.** The fetus is now a newborn.

third trimester the third 3-month period of pregnancy.
labor the sequence of events that leads to birth.
birth canal the vagina, through which the fetus passes during birth.

After birth, uterine contractions continue and expel the placenta and associated membranes called the *afterbirth*. The umbilical cord is still attached to the baby, and to free the newborn, a physician or midwife ties and cuts the cord; blood clots in the cord and contraction of its muscles prevent excessive bleeding.

Physiological Adjustments of the Newborn

At birth, the baby's lungs are not filled with air. Its first breath is therefore unusually deep. For the first time, the lungs are inflated; the baby cries. Because the baby is now obtaining oxygen from the lungs rather than from the placenta, several major changes must take place in the circulation of the blood.

Until birth, the placenta was the source of nutrients and oxygen for the fetus; in addition, it was the site for the removal of waste products from the fetal circulation. The lungs were not functional as organs of gas exchange. The fetal body had two major adaptations to limit the flow of blood to the lungs. First, a hole between the two atria called the *foramen ovale* shunted most right atrial blood directly into the left atrium, thus avoiding the right ventricle and the pulmonary circulation (see Figure 22-19). In addition, the right ventricular blood that is pumped into the pulmonary artery was mostly shunted into the aorta rather than through the lungs by the *ductus arteriosus,* a direct connection between the pulmonary artery and the aorta. At birth, the foramen ovale is closed by two flaps of heart tissue that fold together and fuse. The ductus arteriosus is shut off by contractions of muscles in its walls. Complete closure may take several months. The umbilical arteries and vein also must close off.

> *At birth, the circulation of the newborn changes as the lungs rather than the placenta become the organ of gas exchange.*

Summary

- Human development before birth, or gestation, lasts approximately $8^1/_2$ months. Conception takes place when a sperm, or male sex cell, penetrates a secondary oocyte, or female sex cell. This penetration, or fertilization, usually takes place in the uterine tube.
- The fertilized egg, or zygote, begins to divide after the genetic material of the egg and sperm unite. This process of cleavage is the first stage in the development of humans, as well as other multicellular animals. Cleavage results in a ball of cells, or morula.
- After 2 days in the uterus, the morula has developed into the blastocyst, a stage of development in which the embryo is a hollow ball of cells. This stage of development is significant; it is the first time that cells begin to move, or migrate, to shape the new individual. This process of cell movement to shape the developing individual is morphogenesis.
- At approximately 1 week after fertilization, the blastocyst nestles into the lining of the uterine wall, which is the process of implantation. It also secretes a hormone that acts on the corpus luteum in the ovary, stimulating the body to produce estrogens and progesterone to maintain the uterine lining rather than having it slough off during a menstrual period.
- During the first 2 weeks of development, the developing mass of cells is generally referred to as a *preembryo*; from 3 to 8 weeks, an *embryo*; and from 9 to 38 weeks, a *fetus*. Structures also develop that are not a part of the growing individual but help sustain the preembryo, embryo, and fetus through development. These extraembryonic membranes are the amnion, chorion, yolk sac, and allantois.
- At the end of the second week and continuing into the third week, various cell groups of the blastula begin to divide, move, and differentiate in a process called *gastrulation*. During this time a three-layered embryo is formed. Each of these three layers will give rise to specific tissues and organs of the developing individual.
- Neurulation, the development of a hollow nerve cord that becomes the central nervous system, begins in the third week. However, the development of the nervous system continues even after birth.
- Dramatic changes take place in the embryo from the fourth to the eighth weeks of development. Limb buds become visible in the fourth week and are quite well developed by the seventh week. A primitive circulation is established, and a maternal-fetal exchange of nutrients, gases, and wastes begins. By the eighth

week of development, the embryo is about an inch long, and most of the body systems are functional.

- The fetal period, from 9 to 38 weeks, is a time of refinement, maturation, and growth.
- Changing hormone levels in the fetus at about the thirty-eighth week trigger the onset of labor, the sequence of events that leads to birth. Waves of contractions in the walls of the uterus force the fetus downward and out the birth canal.
- At birth, the circulation of the newborn changes as the lungs rather than the placenta become the organ of gas exchange.

Knowledge and Comprehension Questions

1. At what point in the prenatal development of a human does the zygote exhibit a new genetic makeup different from that of either parent?
2. What happens to a sperm and a secondary oocyte immediately after fertilization occurs?
3. Does the development of the zygote during the first 2 weeks after fertilization involve meiotic or mitotic cell division?
4. Some birth control methods prevent implantation of the blastocyst in the uterine wall. What happens to the blastocyst if these methods are successful?
5. Identify the three primary germ layers of the developing preembryo, and list the structures into which they will develop.
6. What are the extraembryonic membranes? State the function of each.
7. Summarize the events that occur during gastrulation.
8. Define neurulation, and explain its significance.
9. What is a notochord, and what role does it play in humans?
10. Maternal nutrition is a key element in normal development of the fetus. Which fetal systems are affected by maternal nutritional habits during the first 3 weeks of development?
11. Summarize the changes that occur in the embryo between the fourth and eighth weeks of development.
12. Summarize the changes that occur in the embryo/fetus during the first, second, and third trimesters. When do most body systems become functional?
13. Explain why good nutrition is so important during pregnancy.
14. Describe the physiological adjustments that a newborn must undergo to survive.

Critical Thinking

1. What changes occur in a fertilized egg that prevent the penetration of a second sperm? What might be the consequences of a second sperm gaining entry?
2. What do you think might be the consequences if the foramen ovale did not close after birth? What symptoms of this condition might be seen in the newborn?

PART FIVE

EVOLUTION: HOW ORGANISMS CHANGE OVER TIME

CHAPTER 23

THE SCIENTIFIC EVIDENCE OF EVOLUTION

BROUGHT BACK TO LIFE with computer chips and fleshlike polymers, the eighteenth-century English naturalist Charles Darwin speaks to visitors from his library at the St. Louis Zoo in St. Louis, Missouri. Darwin's library is located in The Hall of Animals, one of the indoor exhibit areas at The Living World, the Zoo's educational complex. Darwin explains:

> A finch, not very different from this one, was one of the many things that led my mind to an idea of which you may have heard—that the great diversity of creatures on Earth, the wealth of form and color, is the result of evolution by natural selection.

Darwin then continues to explain the experiences that affected his thinking and led to the development of his theory:

> As a young man, I sailed around the world on Her Majesty's ship *Beagle,* and on the little Galapagos Islands—six hundred miles at sea—I saw finches. Some had big bills with which they cracked seeds like sparrows. Others darted around catching insects as a war-

bler might. There was even one that used a twig to dig for grubs, trying to be a woodpecker. And I noticed a peculiar thing . . . the birds that I saw on the Galapagos looked like finches I had seen before, earlier in my voyage in South America. This was a clue. An ancestor of these Galapagos finches must have come from South America to the islands, and these finches are its descendants. But how could Galapagos finches be so different from one another if they all have the same ancestor? Somehow, they must have changed. And that is one of the many pieces of evidence that led me to the heart of it . . . descent with modification . . . evolution.

Visitors at the St. Louis Zoo have a chance to better understand Darwin's statements. Likewise, this chapter describes the history of evolutionary thought and explains the scientific evidence upholding the theory of evolution. Then you, too, can better understand Darwin's statements and your own evolutionary past.

KEY CONCEPTS

- One of the central theories of biology is the theory of evolution, which states that living things change over time.
- Charles Darwin, a nineteenth-century English naturalist, developed this theory based on observations of plants and animals of oceanic islands, geological evidence that the Earth has changed over time, the results of artificial breeding, and Malthus' ideas on population growth and decline.
- Darwin proposed that evolution occurs as a result of natural selection.
- Although proposed more than 100 years ago, the theory of evolution is still upheld and is further supported by the fossil record, the molecular record, progressive changes in homologous structures, and the presence of vestigial structures.

The Development of Darwin's Theory of Evolution

The story of Darwin and his theory begins in the early 1800s, when he was a medical student at Edinburgh. He then attended Cambridge University. In 1831, on the recommendation of one of his professors, he was selected to join a 5-year voyage (from 1831 to 1836) around the coasts of South America on H.M.S. *Beagle* (Figure 23-1). Darwin had the chance to study plants and animals on continents, islands, and seas distant from his native England. He was able to experience firsthand the remarkable diversity of living things on the Galapagos Islands off the west coast of South America. Such an opportunity clearly played an important role in the development of his thought about the nature of life on Earth.

When the *Beagle* set sail, Darwin was fully convinced that species were unchanging. (At this time in his career, Darwin's concept of species was remarkably close to the following modern definition.) A **species** is a population of organisms that interbreeds freely in the wild and does not interbreed with other populations. In other words, species are defined by their reproductive isolation from one another. (Organisms that do not repro-

OUTLINE

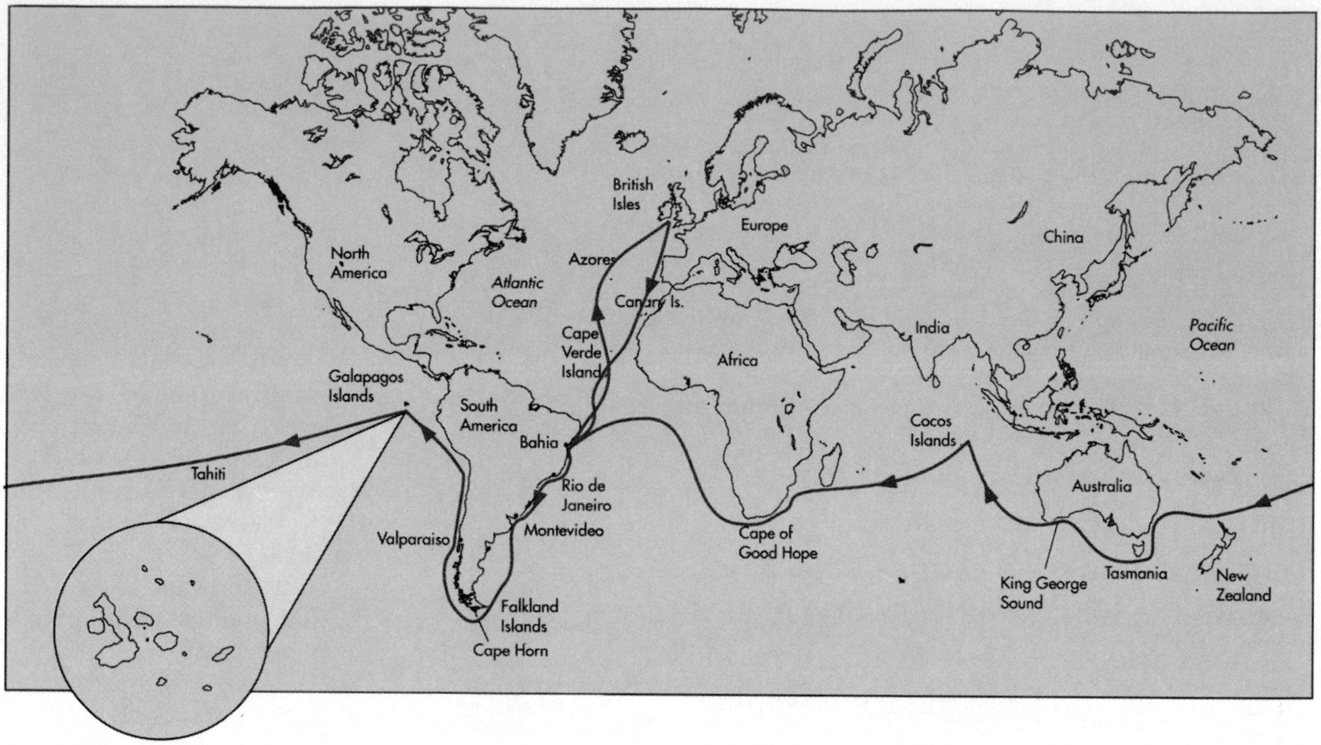

FIGURE 23-1 **Voyage of the H.M.S. *Beagle*.** Most of Darwin's time was spent exploring the coasts and coastal islands of South America, such as the Galapagos Islands. Darwin's studies of the animals of the Galapagos Islands played a key role in his development of the theory of evolution by natural selection.

duce sexually are designated as species by means of their morphological and biochemical characteristics.) Darwin wrote that it was not until 2 or 3 years after his return that he began to consider seriously the possibility that species could change. A couple of years after the voyage, Darwin began to formulate a theory integrating his observations of the trip and his understanding of geology, population biology, and the fossil record. Beginning in 1838, Darwin began to write his explanation of the diversity of life on Earth and the ways in which living things are related to one another.

Darwin's Observations

During his 5 years on the ship, Darwin observed many phenomena that were of central importance to the development of his theory of evolution. While in southern South America, for example, Darwin observed fossils (preserved remains or impressions) of extinct armadillos that were similar to armadillos still living in that area (Figure 23-2). He found it interesting that such similar yet distinct living and fossil organisms were found in this same small geographical area. This observation suggested to Darwin that the fossilized armadillos were related to the present-day armadillos—that they were "distant" relatives.

Another observation made by Darwin was that geographical areas having similar climates, such as Australia, South Africa, and Chile, are each populated by *different* species of plants and animals. These differences suggested to Darwin that factors other than or in addition to climate must play a role in plant and animal diversity. Otherwise, all lands having the same climate would have the same species of animals and plants. However, he noted that these organisms are often similar to one another, "shaped" by environmental similarities.

Darwin was also struck by the fact that the relatively young Galapagos Islands (formed by undersea volcanoes) were home to a profusion of living organisms resembling plants and animals that lived on the nearby coast of South America. Notice, for example, the similarity of the two birds in Figure 23-3. The bird on the left is a medium ground finch and is found on the Galapagos. The blue-black grassquit, shown on the right, is found in grasslands along the Pacific Coast from Mexico to Chile. These observations suggested to Darwin that the Galapagos organisms were related to ancestors who long ago flew, swam, or "hitchhiked" (were transported by other organisms) to the islands from the mainland.

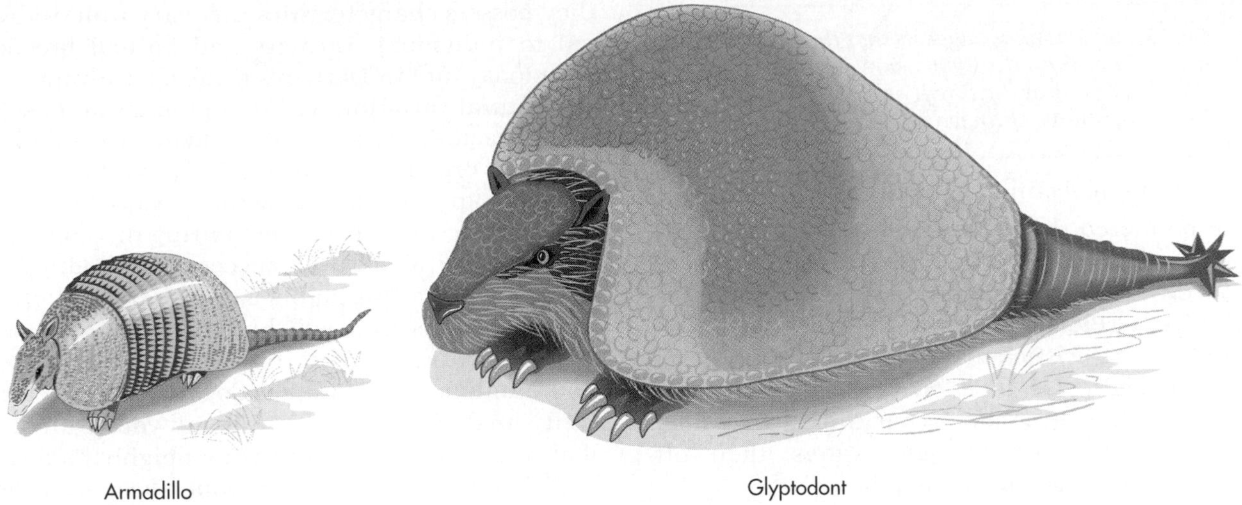

FIGURE 23-2 Distant relatives. Reconstruction of a glyptodont, a 2-ton fossil of a South American armadillo, compared with a modern armadillo, which averages about 10 pounds.

Darwin made observations during his 5-year voyage that suggested to him the following ideas:

1. *Organisms of the past and present are related to one another.*
2. *Factors other than or in addition to climate play a role in the development of plant and animal diversity.*
3. *Members of the same species often change slightly in appearance after becoming geographically isolated from one another.*
4. *Organisms living in oceanic islands often resembled organisms found living on a close mainland.*

Factors That Influenced Darwin's Thinking

As Darwin studied the data he collected during his voyage, he reflected on their significance in the context of what was known about geology, the breeding of domesticated animals, and population biology.

The Influence of Geology

In the late eighteenth and early nineteenth centuries, scientists studying the rock layers of the Earth noticed two things. First, scientists saw evidence that the Earth had changed over time, acted on by natural forces such as the winds, rain, heat, cold, and volcanic eruptions. Geologists began to hypothesize that the Earth was much colder than the 6000 to 10,000 years many had originally thought it to be. Second, they noticed that the fossils found within the Earth's rock layers were similar to but different in many ways from living organisms—an observation Darwin himself had made on his voyage. Not only had the Earth changed, thought scientists, but evidence existed that the organisms living on Earth had changed also.

FIGURE 23-3 Evidence from the finches. A, One of Darwin's finches, the medium ground finch. **B,** The blue-black grassquit, which is found in grasslands along the Pacific coast from Mexico to Chile. This bird may have a common ancestor with Darwin's finches.

species (**spee** sheez *or* **spee** seez) a population of organisms that interbreeds freely in the wild and does not interbreed with other populations.

> *Geological evidence suggests that the Earth is much older than the 6000 years originally thought and that the Earth and organisms living on it have changed over time.*

The Results of Artificial Breeding

As he pondered these ideas that the Earth and its organisms may have changed over time, Darwin reflected on the results of a process called *artificial selection.* In artificial selection a breeder selects for desired characteristics, such as those of the pigeons shown in Figure 23-4. At one time these pigeons came from the same stock, but through artificial breeding over successive generations, their offspring have changed dramatically.

Artificial selection is based on the *natural variation* all organisms exhibit. For example, looking back at p. 400, you can see that although individuals within this population of flamingos are similar, they possess characteristics that vary from individual to individual. Farmers and animal breeders, both today and in Darwin's time, take advantage of the natural variation within a population to select for characteristics they find valuable or useful. By choosing organisms that naturally exhibit a particular trait and then breeding that organism with another of the same species exhibiting the same trait, breeders are able (over successive breedings) to produce animals or plants having a desired, inherited trait. This trait will breed true in successive generations when these organisms are bred with one another. For example, dogs have been artificially bred for centuries. Although your collie may look much different from your neighbor's terrier, both animals belong to the same species but have been artificially bred to retain traits that are characteristic of their breeds (Figure 23-5). Even the

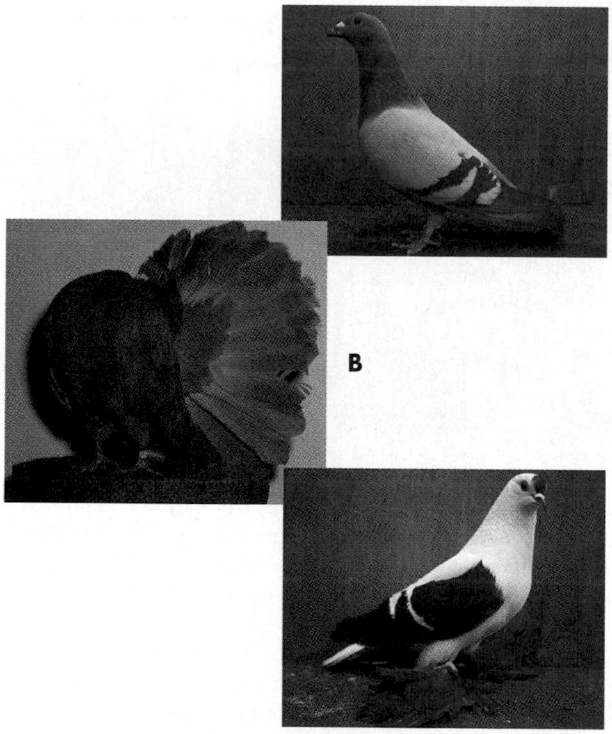

FIGURE 23-4 Artificial selection: another clue in the natural selection puzzle. The differences that have been obtained by artificial selection of the wild European rock pigeon **(A)** and domestic races such as the red fantail **(B)** and the fairy swallow **(C)** are so great that these birds probably would, if wild, be classified in entirely different major groups. In a way similar to that in which these races were derived, widely different species have originated in nature by means of natural selection.

FIGURE 23-5 Artificial selection leads to different breeds of the same species. All dogs belong to the species *Canis familiaris,* but through artificial selection, breeders have been able to choose which traits each breed should retain. Notice that the collie in **A** has long, straight hair and a pointed snout in contrast to the short, curly hair and less pointed snout of the terrier in **B**. How many other differences can you note?

turkey you eat on Thanksgiving has been artificially bred for large cavities for stuffing.

> *Breeders of plants and animals are able to alter the characteristics of organisms by selecting those with desired, inheritable traits and breeding them. After successive breedings, these inherited traits will consistently appear in offspring. Darwin hypothesized that a similar type of selection might take place in nature and result in changes within populations of organisms over time.*

The Study of Populations

Pondering his observations, Darwin began to study Thomas Malthus' *Essay on the Principles of Population.* Malthus, an economist who lived from 1766 to 1834, pointed out that populations of plants and animals (including humans) tend to increase exponentially. In an exponential progression, a population (for example) increases as its number is multiplied by a constant factor. In the exponential progression 2, 4, 8, 16, and so forth, each number is two times the preceding one. (These numbers can also be expressed as 2^1, 2^2, 2^3, and 2^4. The number, or power, to which 2 is raised is termed the *exponent,* from which the term *exponential growth* is derived.) Figure 23-6, *A,* shows how the numbers in a exponential progression increase quickly! Malthus suggested that although populations grow exponentially, food supplies increase only arithmetically. An arithmetic progression, in contrast, is one in which the elements increase by a constant difference, as the progression 2, 6, 10, 14, and so forth. In this progression, each number is 4 greater than the preceding one. Figure 23-6, *B,* shows graphically how the numbers in an arithmetic progression increase.

Although Malthus suggested that populations grow at a exponential rate, he realized that factors existed to limit this astounding growth. If populations grew unchecked, organisms would cover the entire surface of the Earth within a surprisingly short time. But the world is not covered in ants, spiders, or poison ivy. Instead, populations of organisms vary in number within a certain limited range. Space and food are limiting factors of population growth; death limits infinite population growth. Malthus noted that in human populations, death was caused by famine, disease, and war. Sparked by Malthus' ideas, Darwin saw that in nature, although every organism has the potential to produce many offspring, thereby contributing to a exponential growth rate of its population, only a limited number of organisms actually survive to reproductive age. Darwin realized that factors similar

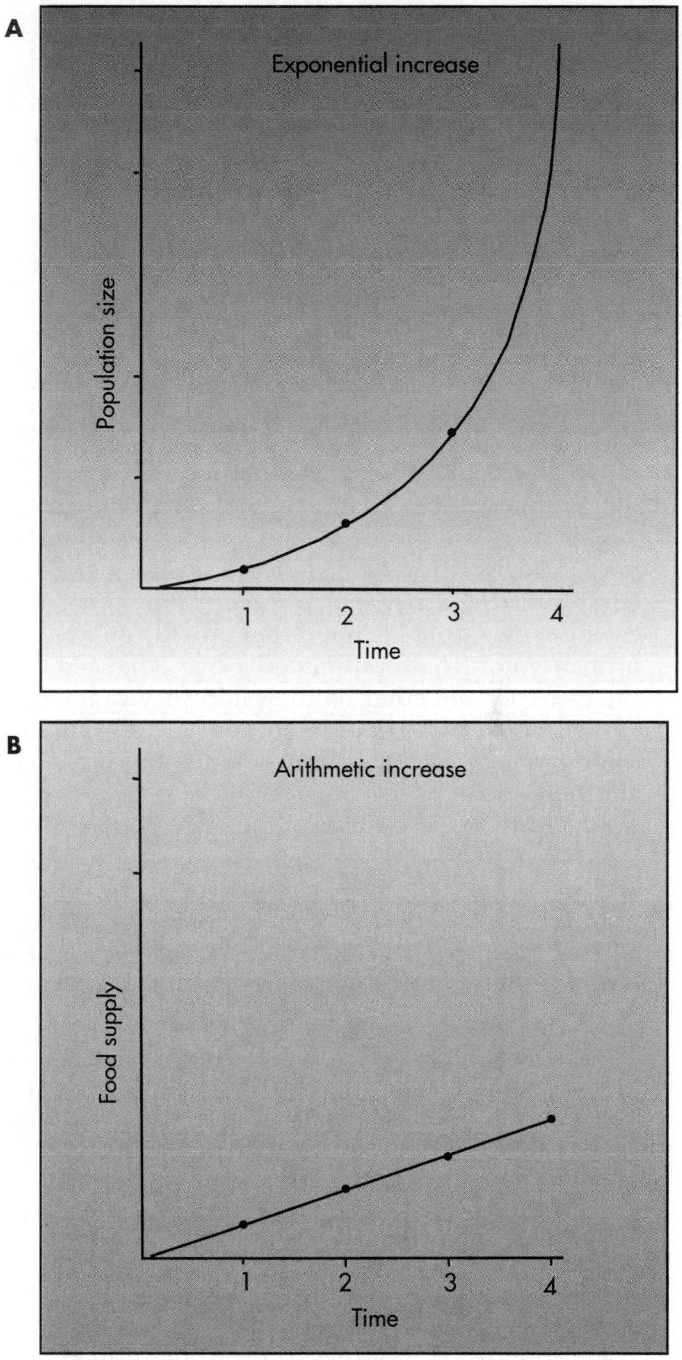

FIGURE 23-6 Types of mathematical progressions. **A,** In an exponential progression, a population increases as its number is multiplied by a constant factor. The numbers in an exponential progression increase rapidly. **B,** In an arithmetic progression, the numbers increase by a constant difference.

to those limiting human populations must also act to limit plant and animal populations in nature.

> *A key contribution to Darwin's thinking was Malthus' concept of exponential population growth. Real populations do not expand at this rate, and this implies that nature acts to limit population numbers.*

Natural Selection: A Mechanism of Evolution

Darwin realized that environmental factors could influence which organisms in a population lived and which ones died. Berry-eating birds, for example, with variations in the structure of their beaks that allowed them to crush seeds would survive longer than other birds of their species who did not have seed-crushing beaks—if the bushes bore few berries during a particular season. The non-seed-eating birds would die out rather quickly and would probably not live to reproductive age. The seed-eating birds, on the other hand, would survive the bad berry season and would likely reproduce. Many of their progeny would have seed-crushing beaks. If another season of few berries followed, these birds would have a survival advantage over other birds that were unable to live on an alternative food source.

Darwin made associations between the process of artificial breeding and reproduction within natural populations. His ideas were expressed in his autobiography:

> I soon perceived that selection was the keystone of man's success in making useful races of animals and plants. But how selection could be applied to organisms living in a state of nature remained for some time a mystery to me.
>
> In September 1838, that is, fifteen months after I had begun my systematic enquiry, I happened to read for amusement "Malthus on Population" and being well prepared to appreciate the struggle for existence which everywhere goes on from long-continued observation of the habits of animals and plants, it at once struck me that under these circumstances favourable variations would tend to be preserved, and unfavourable ones to be destroyed. The result of this would be the formation of new species. Here then I had at last got a theory by which to work. . . .

Darwin was saying that those individuals that possess physical, behavioral, or other attributes well-suited to their environment are more likely to survive than those that possess physical, behavioral, or other attributes less suited to their environment. The survivors have the opportunity to pass on their favorable characteristics to their offspring. These characteristics are naturally occurring inheritable traits found within populations and are called **adaptations.** (Populations are individuals of a particular species inhabiting a locale or region.)

Notice that the term *adaptation* is used differently than in its everyday sense. Here, it refers to naturally occurring inheritable traits present in a population of organisms rather than noninheritable traits in individuals. Adaptive traits are inherited characteristics that confer a reproductive advantage to the portion of the population possessing them. In its everyday sense, an *adaptation* refers to something a single individual does to change how it responds to the environment. For example, you may adapt to getting up early for an 8:00 AM class. But this is not an inherited trait in the entire population that confers a reproductive advantage!

As adaptive, or reproductively advantageous, traits are passed on from surviving individuals to their offspring, the individuals carrying these traits will increase in numbers within the population, and *the nature of the population as a whole will gradually change.* Darwin called this process, in which organisms having adaptive traits survive in greater numbers than those without such traits, **natural selection.** Change in populations of organisms therefore occurs over time because of natural selection: the environment imposes conditions that determine the results of the selection and thus the direction of change. The driving force of change—natural selection—is often referred to as *survival of the fittest.* Again, the term *fittest* does not have the everyday meaning of the healthiest, strongest, or most intelligent. You may be fit if you work out at the local health club regularly. But fitness in the context of natural selection refers to reproductive fitness—the ability of an organism to survive to reproductive age in a particular environment and produce viable offspring.

Natural selection provides a simple and direct explanation of biological diversity—why animals are different in different places. Environments differ; thus organisms are "favored" by natural selection differently in different places. The nature of a population gradually changes as more individuals are born that possess the "selected" traits. **Evolution** by means of natural selection is this process of change over time by which existing populations of organisms develop from ancestral forms through modification of their characteristics.

An Example of Natural Selection at Work: Darwin's Finches

Interestingly, the results of evolution by natural selection can actually be seen if the process takes place relatively quickly (over a period of years to several thousand years), resulting in the existence of groups of closely related species from an original ancestral species. The results that can be seen are clusters of these closely related species found living near one another. Such clusters of species are often found on a group of islands, in a series of lakes, or in other environments that are close to but separated from one another. Organisms living in such sharply discontinuous habitats are said to be *geographically isolated* from one another.

The Galapagos Islands are a particularly striking example of sharply discontinuous habitats, providing a "natural laboratory" to view the results of natural selection. The islands are all relatively young in geological terms (several million years) and have never been connected with the adjacent mainland of South America or with any other area. Made up of 13 major islands (and some very tiny islands), the Galapagos are separated from one another by distances of up to 100 miles and are 600 miles from the South American mainland (see Figure 23-1). As a group, they exhibit diverse habitats. For example, the lowlands of the Galapagos are covered with thorn scrub. At higher elevations, attained only on the larger islands, there are moist, dense forests.

Formed by undersea volcanoes, the Galapagos Islands were uninhabited when they appeared above the surface of the water. The ancestors of all the organisms found on the Galapagos today reached these islands by crossing the sea by water or wind or on the bodies of other organisms. Only eight species of land birds reached the islands. One of these species was the finch, which fascinated Darwin. Hypothetically, the ancestor of Darwin's finches reached these islands earlier than any of the other birds. If so, all the types of habitats where birds occur on the mainland were unoccupied on the Galapagos—and the ancestral finches were able to take advantage of them all!

As the finches moved into these vacant habitats, the ones best suited to each particular habitat were selected for by nature. In other words, those birds possessing naturally occurring variations in their characteristics that were beneficial to survival lived to reproduce. Their offspring also possessed these inheritable traits. Over time, the population of finches occupying each habitat changed, and the ancestral finches split into a series of diverse populations. This phenomenon, by which a population of a species changes as it is dispersed within a series of different habitats within a region, is referred to as **adaptive radiation.** Some of these populations became so changed from the others that interbreeding was no longer possible: new species of finches were formed. This process, by which new species are formed during the process of evolution, is termed **speciation.**

The evolution of Darwin's finches on the Galapagos Islands provides one of the classic examples of speciation. The descendants of the original finches that reached the Galapagos Islands now occupy many different kinds of habitats on the islands (Figure 23-7) and are found nowhere else in the world. Among the 13 species of Darwin's finches that inhabit the Galapagos, there are three main groups: ground finches, tree finches, and warbler finches. The ground finches feed on seeds of different sizes. The size of their bills is related to the size of the seeds on which the birds feed. The tree finches, as their name suggests, eat insects, buds, or fruit found in the trees. Again, the size and shape of their bills are related to their food. The most unusual member of this group is the woodpecker finch. This bird carries around a twig or a cactus spine, which it uses to probe for insects in deep crevices. It is an extraordinary example of a bird that uses a tool. And last, the warbler finches, named for their beautiful singing, search continually with their slender beaks over leaves and branches for insects.

The evolution of Darwin's finches illustrates the same kinds of processes by which species are originating continuously in all groups of organisms. Isolated populations subjected to unique combinations of selective pressures (the conditions imposed by nature) diverge biologically from one another and may ultimately become so different that they are distinct species.

adaptations naturally occurring inheritable traits present in a population of organisms that confer reproductive advantages on organisms that possesses them.

natural selection the process in which organisms with adaptive traits survive in greater numbers than organisms without such traits.

evolution (ev uh **loo** shun) the process of change over time by which existing populations of organisms develop from ancestral forms through modification of their characteristics.

adaptive radiation the phenomenon by which a population of a species changes as it is dispersed within a series of different habitats within a region.

speciation (**spee** shee **ay** shun) the process by which new species are formed during the process of evolution.

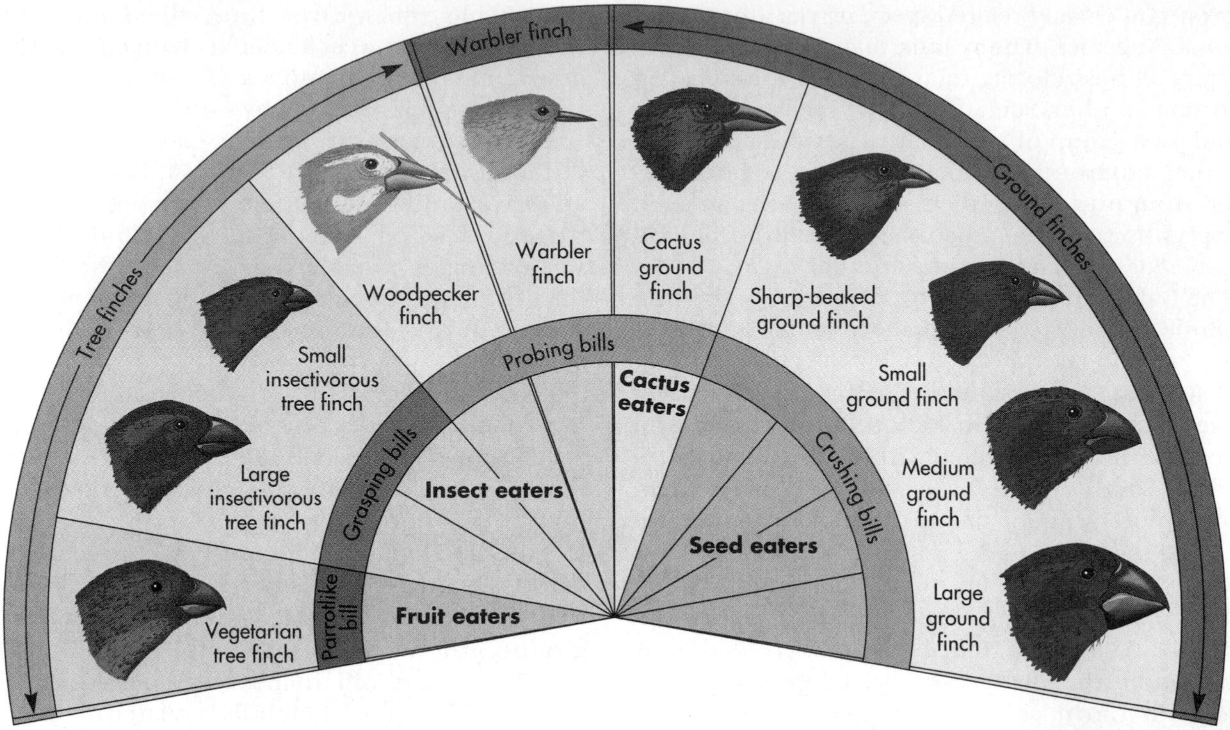

FIGURE 23-7 Darwin's finches. Ten species of Darwin's finches from Indefatigable Island, one of the Galapagos Islands, showing differences in bills and feeding habits. The bills of several of these species resemble those of different, distinct families of birds on the mainland. All these birds are thought to have been derived from a single common ancestor.

The Publication of Darwin's Theory

Darwin drafted the overall argument for evolution by natural selection in 1842 and continued to refine it for many years. The stimulus that finally brought it into print was an essay that he received in 1858. A young English naturalist named Alfred Russel Wallace (1823-1913) sent the essay to Darwin from Malaysia; it concisely set forth the theory of evolution by means of natural selection (Figure 23-8). Like Darwin, Wallace had been influenced greatly in his development of this theory by reading Malthus' 1798 essay. After receiving Wallace's essay, Darwin arranged for a joint presentation of their ideas at a seminar in London. Darwin then proceeded to complete his own book, which he had been working on for some time, and submitted it for publication in what he considered an abbreviated version.

Darwin's book, *On the Origin of Species by Means of Natural Selection,* appeared in November 1859 and caused an immediate sensation. Some called the book "glorious" and were in complete agreement with Darwin's theories. Others criticized the book, admiring some parts and asserting that other parts were "totally false." Still others attacked Darwin on religious grounds. Many of the clergy, however,

FIGURE 23-8 Alfred Russel Wallace. Wallace arrived at the theory of natural selection independently of Darwin, but the ideas of the two were presented together at a seminar in 1859.

openly agreed with Darwin, saying that the theory of evolution did not deny the existence of God.

> *Darwin published his argument for the theory of evolution in 1859, in a book entitled* On the Origin of Species by Means of Natural Selection. *His ideas were presented at a scholarly meeting along with those of another scientist, Alfred Russel Wallace, who had independently developed a theory of evolution. These ideas were hotly debated at that time; the mechanism of evolution—natural selection—was not well accepted or understood.*

At the end of June 1860, a debate was held at a meeting of the British Association for the Advancement of Science. The debate lasted many days and attracted huge crowds of people. The debate was heated, but the outcome pleased Darwin: people read his book and gave his arguments serious consideration.

Within 20 years or so after the publication of *Origin,* the concept that species have changed over time was well accepted. However, scientists continually disagreed concerning the definition of the word "species." In fact, Darwin's own concept of species had changed since the 1830s; he eventually stopped trying to define the term, explaining that species continue to evolve, therefore they cannot be defined.

Another hot debate within the scientific community regarded the mechanism of evolutionary change—natural selection. At that time, no one had any concept of genes or of how heredity works, and so it was impossible for Darwin to explain completely how evolution occurs. Gregor Mendel (see Chapter 19) had begun his ground-breaking work in the study of inheritance but had not yet published it. In fact, the science of genetics was not established until the beginning of the twentieth century, 40 years after the beginning of the publication of Darwin's book. An understanding of the laws of inheritance and the mechanism by which inheritable traits are passed on from one generation to the next helped scientists understand the process of natural selection.

Testing the Theory

More than a century has elapsed since Charles Darwin's death in 1882. During this period the evidence supporting his theory has grown progressively stronger. In fact, evolution is a theory that has been upheld countless times as it is tested and retested. The phrase *the theory of evolution* does *not* suggest that evolution by means of natural selection is a highly tentative concept. That suggestion applies only to the everyday use of the word *theory;* it does not apply to its scientific use. Scientists are continuously learning more, however, about the intricacies of the mechanism of natural selection and the history of the evolution of life on Earth.

> *The concept of evolution by means of natural selection is established as valid within the scientific community today, and scientists are continually developing their understanding of natural selection.*

What is the scientific evidence that upholds the theory of evolution? Scientists find evidence in the fossil record, using widely accepted techniques to assess the age of the rocks of which the fossils are formed or in which the fossil remains are found, while gathering a picture of the history of the Earth and its organisms. In addition, the tools of comparative anatomy help researchers understand relationships among organisms alive today. Significant scientific advances, such as those in genetics and molecular biology, have given scientists tools Darwin did not have; consequently, scientists today understand more fully than Darwin ever could how organisms change over time.

The Fossil Record

A **fossil** is any record of a dead organism. Fossils may be nearly complete impressions of organisms or merely burrows, tracks, molecules, or other traces of their existence. Unfortunately, only a minute fraction of the organisms living at any one time are preserved as fossils.

Most fossils are preserved in *sedimentary* rocks. Sedimentary rocks are made up of particles of other rocks, cast off as they weather and disintegrate. Running water, such as a river or stream, picks up these pieces of rock and carries them to lakes or oceans where they are deposited as *sediment,* better known as mud, sand, or gravel. Over time, the sediment hardens into rock. But while some sediment is hardening, other sediment is still being deposited, creating layers of rock formed one on top of the other. Therefore most sedimentary rock has a stratified appearance, such as that seen in the Grand Canyon (Figure 23-9).

During the formation of sedimentary rock, dead organisms are sometimes washed along with the

> **fossil** any record of a dead organism.

FIGURE 23-9 The Grand Canyon. In this photo, the layers of sedimentary rock can be clearly seen.

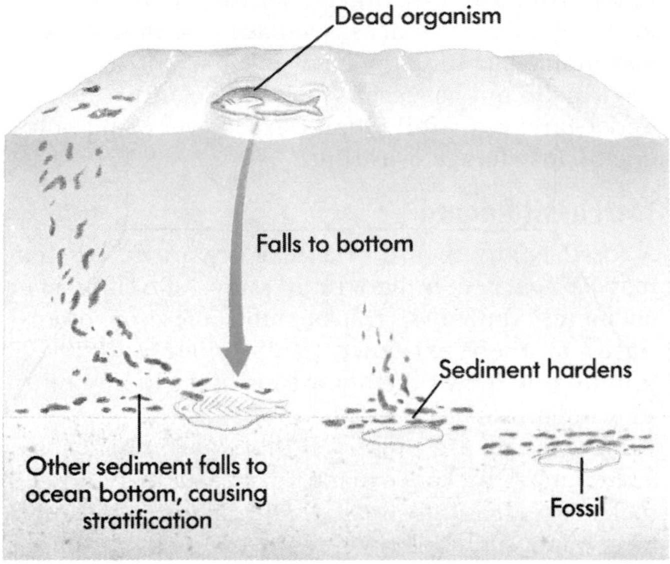

FIGURE 23-10 The process of fossil formation. Fossils are formed when dead organisms are trapped in hardening sediment.

mud or sand and eventually reach the bottom of a pond or lake. Dead marine organisms fall to the bottom of the ocean. As the sediments harden into rock, they harden around the bodies of these dead organisms (Figure 23-10). The hard parts of these organisms, such as their skeletons, may become preserved or may be broken down and replaced with other minerals.

Fossils of organisms having hard parts are the type most often found (Figure 23-11) rather than fossils formed from soft body parts, which usually decay quickly and leave no trace of their existence. Sometimes, however, soft-bodied animals are preserved in exceptionally fine-grained muds, in conditions in which the supply of oxygen was poor while the muds were being deposited, thus slowing the decomposition of the organism. Eventually, the soft parts of an organism decay completely, leaving behind a *mold,* or impression, of its body. Molds may become filled with minerals, such as lime or silica found in underground water, forming *casts,* which resemble the original organism or body part (Figure 23-12). In general, however, fossils of soft-bodied organisms such as worms are rare. Even though soft-bodied animals undoubtedly evolved before their hard-bodied counterparts, there is comparatively little evidence of their history in the fossil record.

Fossils provide an actual record of organisms that once lived, an accurate understanding of where and when they lived, and some appreciation of the environment in which they lived. Limestone that contains corals, for example, would have been deposited when the location was an ocean. Oak leaves found in sandstone suggest that a location was once a continent. In this way, fossils and the rock in which they are embedded provide information about the history of an area, which gives scientists clues to the location of the continents, ponds, lakes, and oceans and how their positions have changed over time.

Scientists can determine the age of fossils and use this information to establish the broad patterns

FIGURE 23-11 ***Archaeopteryx*, a prehistoric ancestor of modern birds.** A well-preserved fossil of this bird, about 150 million years old, was discovered within 2 years of the publication of *On the Origin of Species*. This particular specimen is an excellent example of a fossil formed from the hard parts of the organism. The skeleton can be clearly seen.

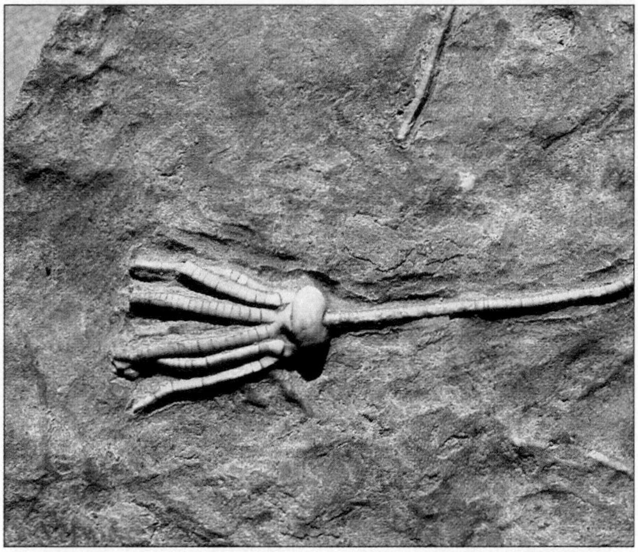

FIGURE 23-12 **A fossil "mold."** Shown here is *Mawsonites spriggi*, a fossil jellyfish from South Australia. This fossil illustrates the remarkable preservation of soft-bodied fossils in fine-grained sedimentary rocks.

of the progression of life on Earth (Figure 23-13). One of the ways of determining the age of particular fossils is to compare the sequences in which they appear in different layers, or strata, of sedimentary rock. Sedimentation deposits new layers mostly on older ones; thus the fossils in upper layers mostly represent younger species than the fossils in lower layers. Fossils found in the same strata are assumed to be of the same age. Such correlations, in fact, were known at the time of Darwin and were used by him to formulate the theory of evolution.

Fossils, impressions of organisms that once lived, provide a record of the past.

The Age of the Earth: Fossil Dating

Direct methods of dating fossils first became available in the late 1940s. This process depends on naturally occurring isotopes of certain elements that are found in rock. Isotopes are atoms of an element that have the same number of protons but different numbers of neutrons in their nuclei (see Chapter 2). They therefore differ from one another in their atomic numbers. *Radioactive isotopes* are unstable; their nuclei decay, or break apart, at a steady rate, producing other isotopes and emitting energy. After decay some radioactive isotopes may give rise to elements. Many different isotopes are used in radioactive dating. Some methods give scientists information about the age of rocks; others measure the length of time since the death of an organism.

One of the most widely used methods of dating, the carbon-14 method, estimates the relative amount of the different isotopes of carbon present in a fossil (or other organic material). Most carbon atoms have an atomic weight of 12 (6 protons and 6 neutrons); the symbol of this particular isotope of carbon is ^{12}C. A fixed proportion of the atoms in a given sample of carbon, however, consists of carbon with an atomic weight of 14 (^{14}C), an isotope that has two more neutrons than ^{12}C. Interestingly, ^{14}C is produced from ^{14}N (the chemical symbol for nitrogen-14) as these atoms are bombarded by cosmic rays—high-energy particles from space. The cosmic rays (usually protons) bump a proton from

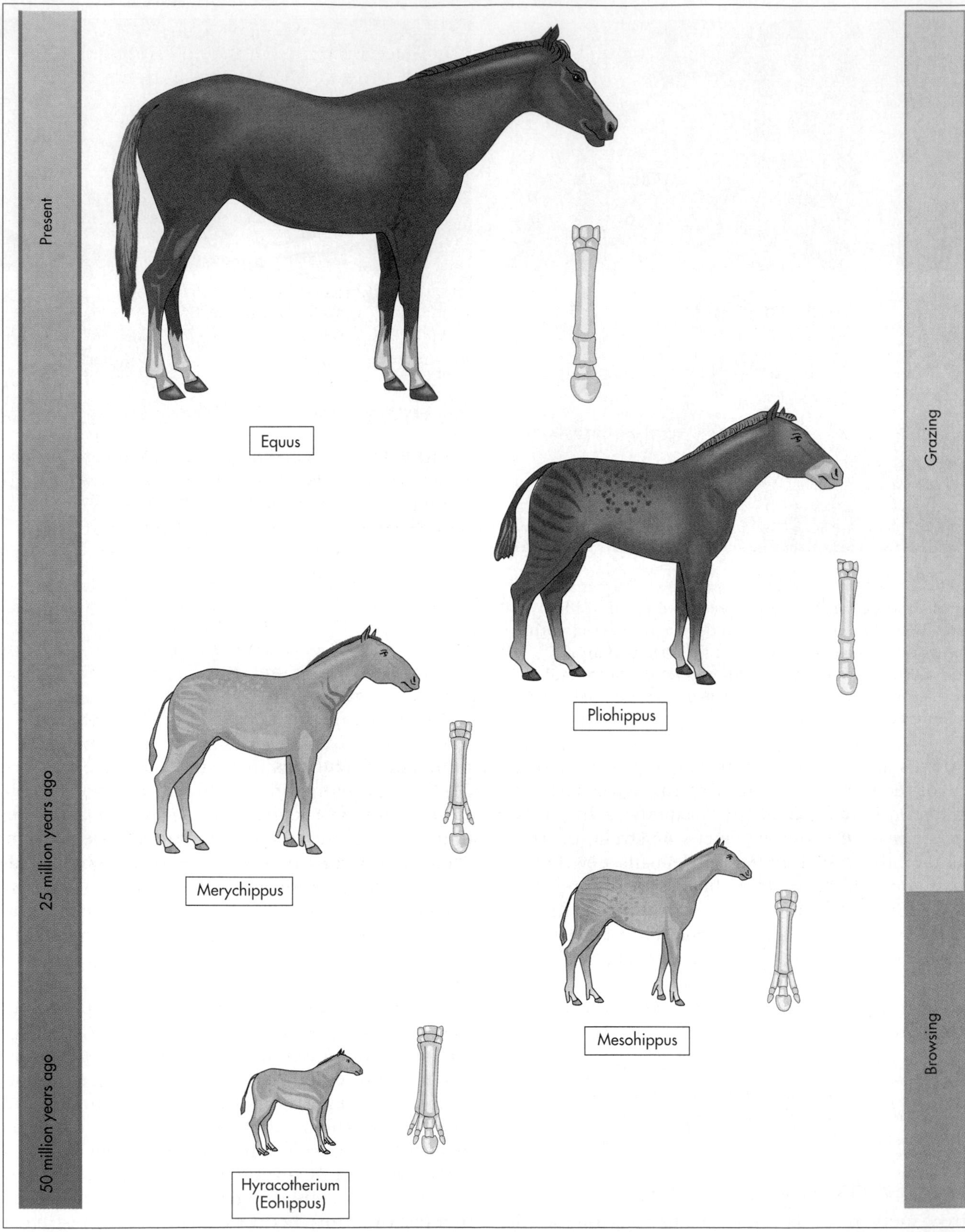

FIGURE 23-13 Reconstruction of the evolution of horses based on fossils. Animals called *hyracotheres,* which included the earliest member of the evolutionary line that is illustrated, *Hyracotherium,* gave rise to several groups of mammals, including tapirs, rhinoceroses, and horses. Horses provide an excellent example of the way that abundant fossil evidence has allowed the evolution of a particular vertebrate group to be reconstructed.

JUST WONDERING. . . .

How is the age of fossils determined and how accurate is it?

There are two methods of geological dating: *relative dating* and *absolute dating*. Relative dating is briefly described on pages 498-499. This method of observing the layers of the Earth's crust results in a time scale with no dates but tells the age of rocks relative to one another on the basis of the unique sets of fossils embedded within them. The Earth's crust is seen as a calendar of sorts, from which scientists derived the five main divisions, or "eras," associated with the five major rock strata. Most have a number of subdivisions called *periods*, and some are further subdivided into epochs (see Table 24-1).

Well before the theory of evolution was developed, fossil-bearing rocks were used to determine the stages of the history of the world. As new specimens are found, they are matched with known species. This new information is the evidence by which scientists continually revise (usually simply extending) the geologic ranges of fossil species.

The only evidence as to the "accuracy" of this system is from the mineral and petroleum industries. Geologists in these industries use fossils to determine where to drill. If, for example, a petroleum company finds oil in a particular rock stratum in one part of the world, they look for oil in that same stratum in another location. This method is much more successful in finding oil than is random drilling.

Absolute dating, the process by which radioactive isotopes are used to determine the age of fossils and the rocks within which they are embedded, is described in detail on pages 499 and 501. This method (also called *radiometric dating*) does have sources of error but has yielded a great deal of consistent data. In addition, the relative ages of the rocks correlates well with the absolute dates determined by the use of isotopes. Therefore, radiometric dating is a reliable way to determine the exact ages of the rock strata.

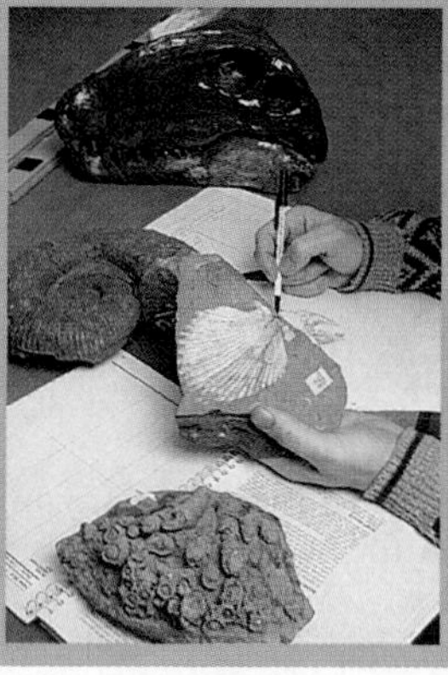

the nucleus of a nitrogen atom, leaving it with 6 protons (the atomic number of carbon) and 7 neutrons. As a result of the collision, the atom also captures a neutron and becomes an atom of ^{14}C. This newly created ^{14}C reacts with oxygen, becoming carbon dioxide, a gas commonly found in the air. Plants use this carbon during photosynthesis, incorporating it into the sugars and starches they make. By eating plants, animals incorporate ^{14}C into their bodies as well.

The carbon that is incorporated into the bodies of living organisms consists of the same fixed proportion of ^{14}C and ^{12}C that occurs in the atmosphere. After an organism dies, however, and is no longer incorporating carbon, the ^{14}C in it gradually decays back to nitrogen by emitting a beta particle. (A beta particle is an electron discharged from the nucleus when a neutron splits into a proton and an electron.) It takes 5730 years for half of the ^{14}C present in a sample to be converted by this process; this length of time is called the *half-life* of the ^{14}C isotope. By measuring the amount of ^{14}C in a fossil, scientists can estimate the proportion of the ^{14}C to all other carbon that is still present and compare that with the ratio of these isotopes as they occur in the atmosphere. In this way, scientists can then estimate the length of time over which the ^{14}C has been decaying, which is the same as the length of time since the organism died.

For fossils older than 50,000 years, the amount of ^{14}C remaining is so small that it is not possible to measure it precisely enough to provide accurate estimates of age. These fossils may be dated by using the isotope thorium-230, which has a half-life of 75,000 years and decays from uranium-238. These techniques have been most useful to scientists who study deep-sea sediments too old to be dated with ^{14}C.

With the use of radioactive dating methods, knowledge of the ages of various rocks has become more precise. The oldest rocks on Earth that have been dated include rocks from South Africa, southwestern Greenland, and Minnesota that are approximately 3.9 billion years old. Meteorites have been dated at about 4.6 billion years. Recently, rocks brought back to Earth from the moon have been dated from 3.3 to 4.6 billion years old. These pieces of evidence suggest that the Earth and the moon, most likely formed from the same processes at the same time, are about 4.6 billion years old.

What significance does the age of the Earth hold for the theory of evolution? The "accumulation" of adaptations and the development of new species usually take thousands and probably millions of years. Until the time of Darwin, most held the belief that the Earth was approximately 6000 to

10,000 years old. This time frame would not allow enough time for the process of evolution to take place. In fact, Sir Isaac Newton (1642-1727), an English physicist, calculated that it would take 50,000 years just for the Earth, after its formation, to cool to a temperature that would sustain life. (See Chapter 24 for a discussion of the formation of the Earth.) Although scientists such as William Thomson, Lord Kelvin (1824-1907), used nineteenth century techniques to date the Earth at about 100 million years, this time span still seemed too short to allow for evolution. Not until the present techniques of radioactive dating were developed could scientists begin to solve the "time problem" of evolution and accurately measure the geological age of the Earth and its fossils.

> *By studying the comparative amounts of certain radioactive isotopes found within fossils, scientists can determine the age of fossils and use this information to establish patterns of life's progression. Studies based on radioactive dating methods provide evidence that the Earth is 4.6 billion years old. This time is sufficient for organisms to have developed and evolved from "original" forms.*

Comparative Anatomy

Comparative studies of animal anatomy provide strong evidence for evolution. After Darwin proposed his theory, scientists began looking for evolutionary relationships in the anatomical structures of organisms. If derived from the same ancestor, organisms should possess similar structures, with modifications reflecting adaptations to their environments. Such relationships have been shown most clearly in vertebrate animals.

Within the subphylum Vertebrata, the classes of organisms, such as birds, mammals, and amphibians, have the same basic anatomical plan of groups of bones (as well as nerves, muscles, and other organs and systems), but these bones are put to different uses among the classes. (See Chapter 1 for a discussion of classification.) For example, the forelimbs seen in Figure 23-14 are all constructed from the same basic array of bones, modified in one way in the wing of a bat, in another way in the fin of a porpoise, and in yet another way in the leg of a horse. The bones are said to be **homologous** in the different vertebrates—that is, of the same evolutionary origin, now differing in structure and function. Although these vertebrates deviated from one another in their evolution, they all use the same bones in the same relative positions to do different jobs.

In some cases, homologous structures exist among related organisms but are no longer useful. These structures have diminished in size over time. Figure 23-15 shows tiny leg bones of the python that no longer serve a purpose. Such organs or structures that are present in an organism in a diminished size but are no longer useful are called *vestigial organs.* Humans have a tiny pouch called the appendix that hangs like a little worm at the junction of the small and large intestines. It serves no useful purpose in humans, but it helps in digestion of

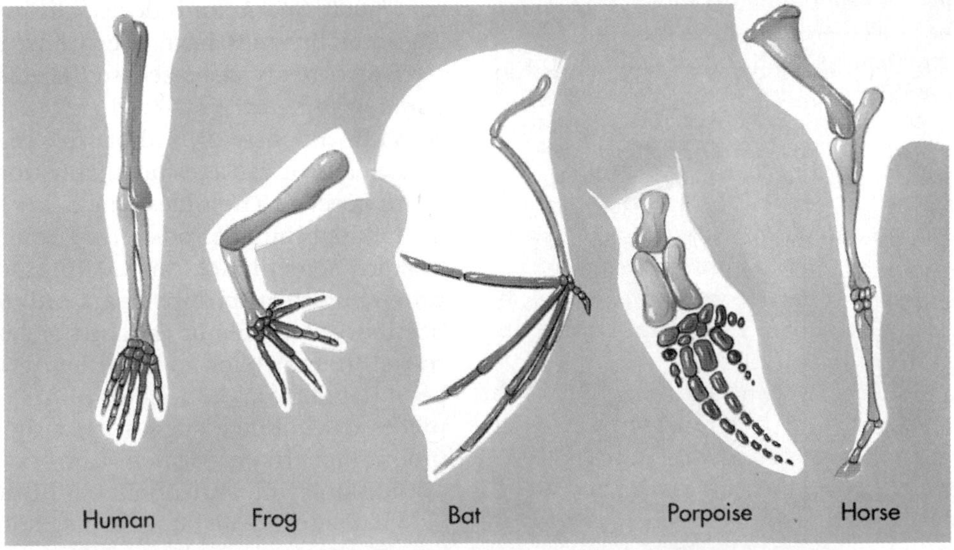

FIGURE 23-14 Homology among vertebrate limbs. Homologies among the forelimbs of four mammals and a frog show the ways that the proportions of the bones have changed in relation to the particular way of life of the organism and that the forelimb of each animal has the same basic bone structure.

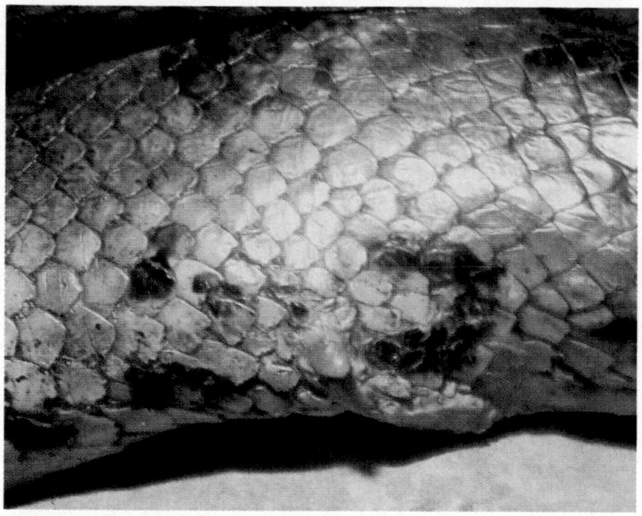

FIGURE 23-15 A vestigial organ in a python. Pythons possess tiny leg bones that serve no purpose in locomotion. Humans also possess a vestigial organ—the appendix—that is not needed for digestion or any other purpose.

plant material in organisms that are evolutionarily related to humans. And if you can wiggle your ears, you are using external ear muscles, inherited from a distant ancestor. They are also no longer useful except to amuse your friends.

> *Comparative studies of animal anatomy show that many organisms have groups of bones, nerves, muscles, and organs with the same anatomical plan but with different functions. These homologous structures provide evidence of evolutionary relatedness.*

In contrast to homologous structures, similar structures often evolve within organisms that have developed from different ancestors. Such body parts or organs, called **analogous** structures, have a similar form and function but have different evolutionary origins. To show that structures are indeed analogous, comparative anatomists do detailed dissections and study the embryological development of organisms. Analogous structures arise developmentally from different tissues.

The eyes of vertebrates and octopuses, which evolved independently but are similar in design, are analogous structures, as well as the wings of birds and insects. Plants also show analogous structures. For example, three different families of flowering plants—the cacti, euphorbia, and milkweeds—have all developed thick, barrellike fleshy stems that store water as an adaptation to a desert environment. In fact, these plants look so much

A

B

FIGURE 23-16 An example of convergent evolution. A, North American cactus. **B,** *Euphorbiaceae* found in Africa.

like one another the casual observer might think they were all cacti. However, they evolved independently of one another in different parts of the world (southwest North America, Africa, and the Mediterranean, respectively). Figure 23-16, *A* and *B* show the striking similarity between the cacti and euphorbia.

The presence of analogous structures shows how in similar habitats, natural selection can lead to similar but not identical anatomical structures. Change over time among different species of organisms having different ancestors that results in similar structures and adaptations is called **convergent evolution.**

homologous (hoe **mol** eh gus) of the same evolutionary origin, now differing in structure and function.
analogous (uh **nal** eh gus) of differing evolutionary origins, now similar in form and function.
convergent evolution change over time among different species of organisms having different ancestors that results in similar structures and adaptations.

Comparative Embryology

Embryologists, scientists who study the development of organisms from conception to birth, noticed as early as the nineteenth century (around the time of Darwin) that various groups of organisms, although different as adults, possessed early developmental stages that were quite similar. For example, the embryological development of vertebrate animals is similar in that all vertebrate embryos have a similar number of gill arches, seen as pouches below the head. Only fish, however, actually develop gills. Likewise, the embryological development of the backbone of vertebrates is similar, but some organisms develop a tail, and others such as humans do not (Figure 23-17). "Tailbones"—the fused coccyx bones at the end of the spine—are actually vestigial structures.

These similar developmental forms tell scientists that similar genes are at work during the early developmental stages of related organisms. The genes active during development have been passed on to distantly related organisms from a common ancestor. Over time, new instructions are added to the old, but both are expressed at different times, resulting in similar embryos that develop into organisms quite different from one another.

Comparative embryological studies show that many organisms have early developmental stages that are quite similar. These similar developmental forms provide evidence of evolutionary relatedness.

Molecular Biology

Today, biochemical tools provide additional evidence for evolution and give scientists new insights into the evolutionary relationships among organisms. Molecular biologists study the progressive evolution of organisms by looking at their hereditary material, DNA. According to evolutionary theory, every evolutionary change involves the formation of new alleles from the old by mutation; favorable new alleles persist because of natural selection. In this way, a series of evolutionary changes in a species involves a progressive accumulation of genetic change in its DNA. Organisms that are more distantly related will therefore have accumulated a greater number of changes in their DNA than organisms that more recently evolved from a common ancestor.

Within the last 30 years, molecular biologists have learned to study DNA by "reading" genes, much

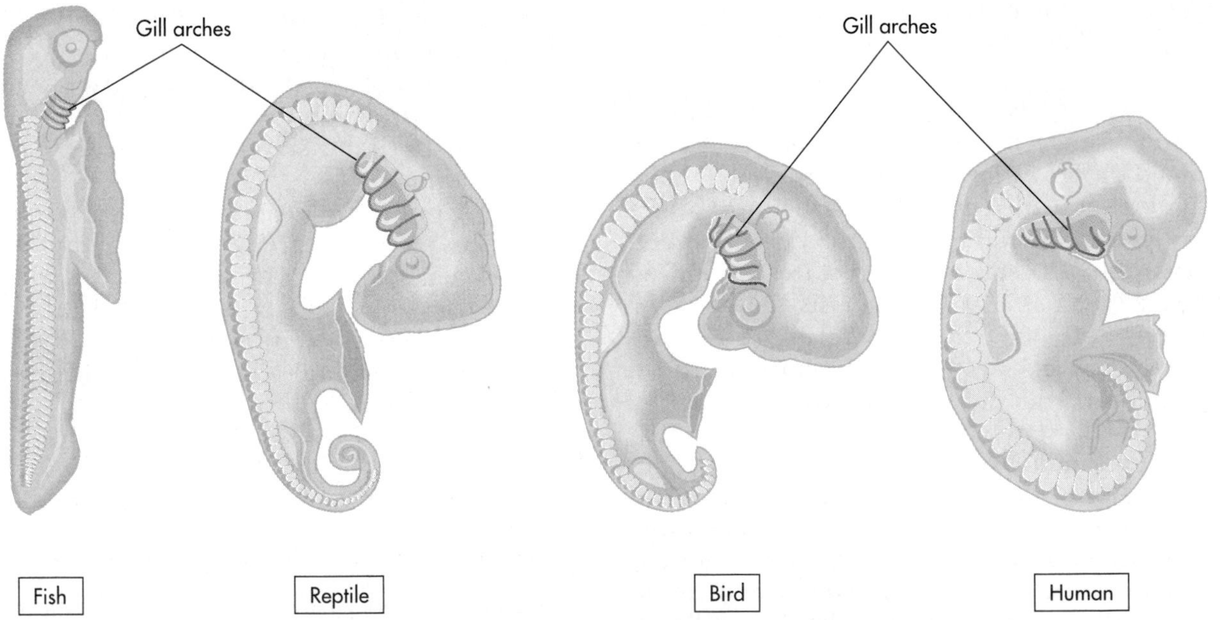

FIGURE 23-17 Embryos show our evolutionary history. The embryos of various groups of vertebrate animals show the primitive features that all vertebrate animals share early in development, such as gill arches and a tail.

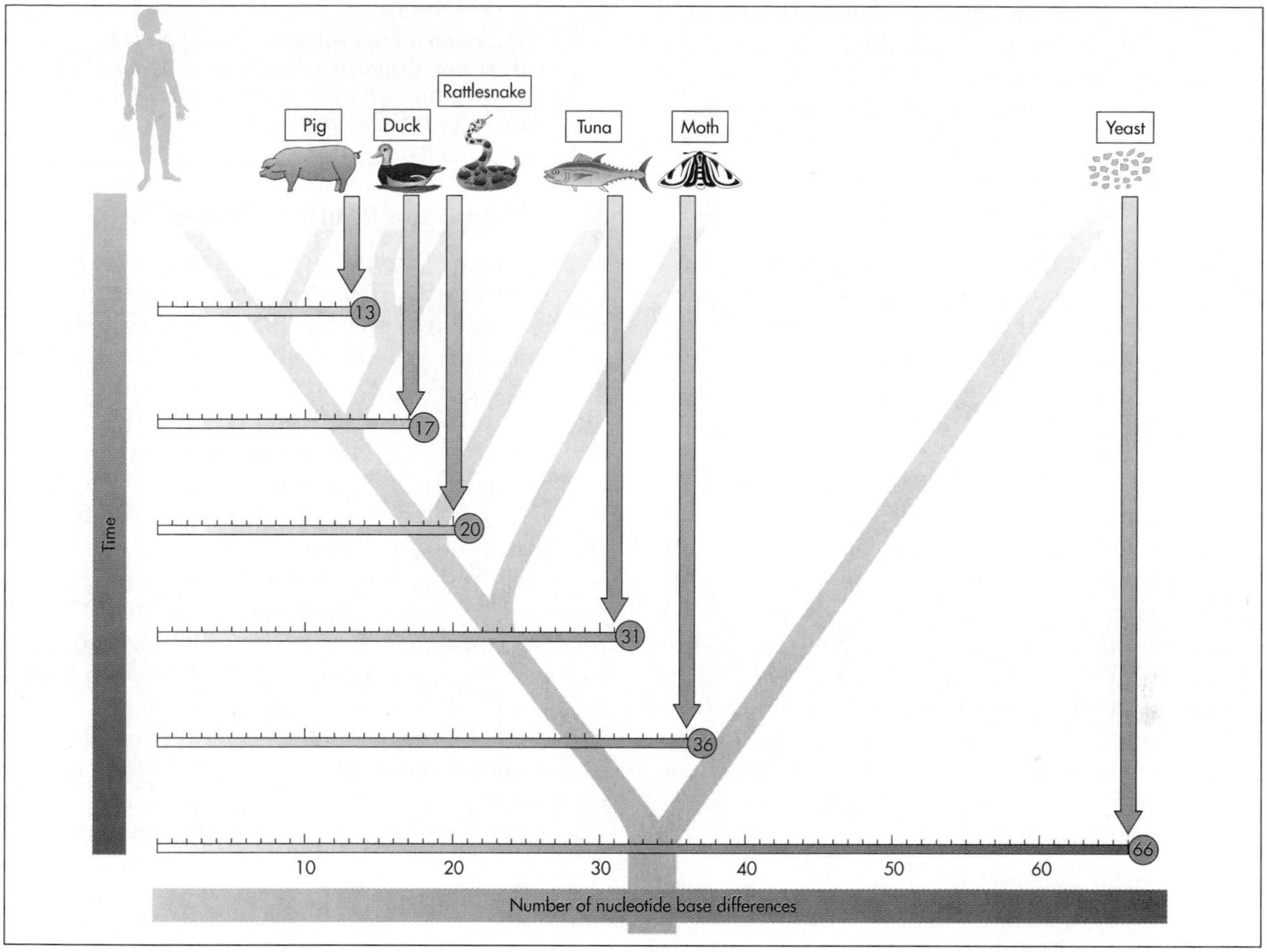

FIGURE 23-18 An evolutionary tree constructed using information from molecular biology. All of the organisms shown in this illustration have an enzyme called cytochrome C oxidase, which helps transfer electrons in the electron transfer system (see Chapter 5). The genes that code for this enzyme in these organisms are not the same, due to changes over time as organisms diverged from one another. This evolutionary tree shows the nucleotide base differences between humans and other organisms. Comparing these differences can provide information about the evolutionary relationships of these organisms. The fewer differences in base sequences between two organisms, the more recently they diverged from a common relative.

as you read this page. They have learned to recognize the order of the nitrogenous bases—the "letters"—of the long DNA molecules. By comparing the sequences of bases in the DNA of different groups of animals or plants, scientists can show the degree of relatedness among groups of organisms.

Tracing the Lineage of Organisms: Evolutionary Trees

As you can see, various lines of inquiry help scientists determine the evolutionary relationships among organisms. In the mid-1800s, Ernst Haeckel, an artist and a scientist, came up with the idea of pictorially representing the taxonomic groups of organisms (see Chapter 1) as trees that depicted these groups' evolutionary relationships. Scientists still use this technique today, but today's evolutionary trees are structured more like graphs and less like the actual trees of Haeckel's time.

An **evolutionary tree** depicts the pattern of relationships among major groups of organisms. Most evolutionary trees place information about the pattern of relationships among organisms on the horizontal axis and information about time on the vertical axis. However, scientists who develop evolutionary trees use a wide variety of data to construct trees. Some scientists may use only a single set of data and a particular type of data. For example, the evolutionary tree in Figure 23-18 derives some

evolutionary tree a representation of the pattern of relationships among major groups of organisms.

information from molecular biology. Organisms are positioned on this tree on the basis of the differences in the nucleotide base sequence of the gene that codes for an enzyme called *cytochrome C oxidase.* The theory that underlies the development of the tree is that these differences exist because of mutations in the bases that occurred sometime in the past. The higher the number of differences among the bases, the greater the number of mutations and therefore the longer ago in time these organisms had a common ancestor (the more distantly they are related). The fewer the number of nucleotide base differences, the more recently these organisms had a common ancestor (the more closely they are related).

Interestingly, evolutionary trees constructed from analyses of molecular differences are similar to those built by use of anatomical studies. For example, both approaches show whales, dolphins, and porpoises clustering together, as do the primates and the hoofed animals. (Note the evolutionary closeness of humans in Figure 23-18.) By the study and interpretation of the evidence from the fossil record, comparative anatomy, and genetic studies, it is often possible for scientists to estimate the rates at which evolution is occurring in different groups of organisms.

> *The pattern of change seen in the molecular record supports other scientific evidence for evolution and provides strong, direct evidence for change over time.*

The Mode and Tempo of Evolution

In 1972, a new perspective was brought to the scientific community regarding the manner and pace of evolutionary change. This theory, developed by American paleontologists Niles Eldredge and Stephen Jay Gould, is called **punctuated equilibrium.** This theory states that new species arise suddenly and rapidly as small subpopulations of a species split from the populations of which they were a part. This theory differs from the theory of **phyletic gradualism** proposed by Darwin, which asserts that new species develop slowly and gradually as an entire species changes over time. (The term *phyletic gradualism* is a modern phrase used to refer to these ideas of Darwin.) Table 23-1 compares the main ideas embodied in both theories.

The *gradualist* idea of slow change means that an entire species changes little by little over time until the species has transformed to a new species. The *punctuationalist* idea of rapid change means

TABLE 23-1

A Comparison of the Tenets of Phyletic Gradualism and Punctuated Equilibrium

NAME: PHYLETIC GRADUALISM	NAME: PUNCTUATED EQUILIBRIUM
PRINCIPAL PROPONENT: DARWIN	PRINCIPAL PROPONENTS: ELDREDGE AND GOULD
• New species develop gradually and slowly with little evidence of stasis (no significant change).	• New species develop rapidly and then experience long periods of stasis.
• The fossil record should contain numerous transitional forms within the lineage of any one type of organism.	• The fossil record should contain few transitional forms with the maintenance of given forms for long periods of time.
• New species arise via the transformation of an ancestral population.	• New species arise as lineages are split.
• The entire ancestral form usually transforms into the new species.	• A small subpopulation of the ancestral form gives rise to the new species.
• Speciation usually involves the entire geographical range of the species (sympatry).	• The subpopulation is in an isolated area at the periphery of the range (allopatry).

From Alters BJ et al: *The American Biology Teacher, 56* (6), 334-339, 1994.

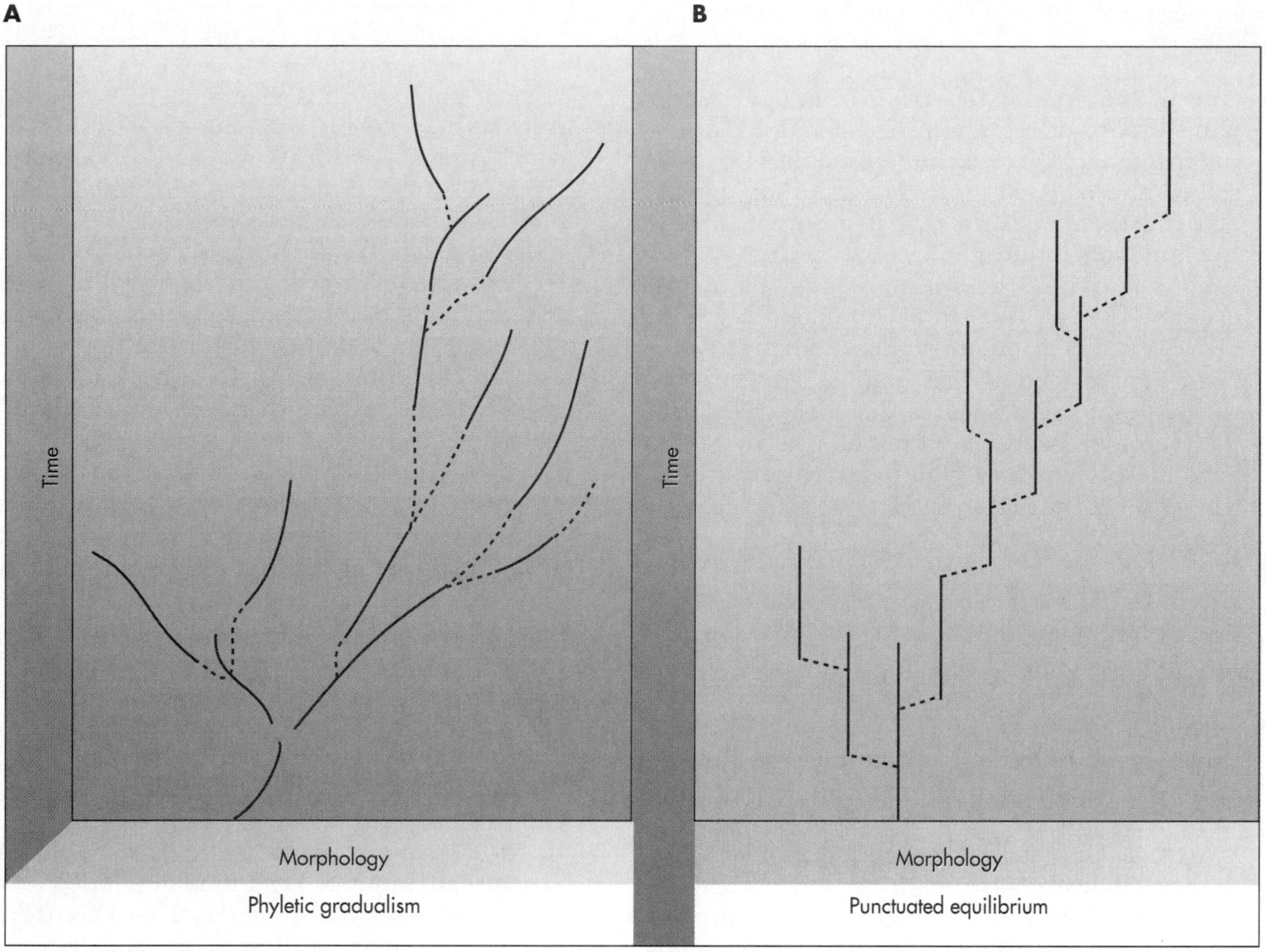

FIGURE 23-19 Phyletic gradualism compared to punctuated equilibrium. A, Phyletic gradualism: time versus morphology. **B,** Punctuated equilibrium: time versus morphology. In both representations, time progresses up the Y axis. Morphology, however, does *not* progress to the right along the X axis. This axis simply denotes morphological change, whether to the left or the right.

that the period of time during which speciation occurs is short with respect to the period of time of *stasis,* small variations that do not lead to speciation. The term *rapid* does not imply that speciation takes any particular (short) length of time. That time may vary from species to species and has been estimated to fall between 50,000 and 500,000 years—incredibly short in geological terms.

The evolutionary trees shown in Figure 23-19 illustrate these ideas. In Figure 23-19, *A,* the dots depict apparent gaps in the fossil record from a gradualist point of view: gradual change of a species but no speciation event. *Apparent gaps* are areas in which few if any intermediate forms in the fossil record are found. (Speciation is depicted in the gradualist model by one line splitting from another.) In Figure 23-19, *B,* the dots show the same apparent gaps from a punctuationalist point of view: a speciation event occurring within a short period of time. Nondirectional variation is implied in both models.

punctuated equilibrium (**pungk** choo ay ted **ee** kwuh **lib** ree um) a theory that states that new species arise suddenly and rapidly as small subpopulations of a species split from the populations of which they were a part.

phyletic gradualism (fye **let** ik **gradj** oo uh **liz** um) a theory that states that new species develop slowly and gradually as an entire species changes over time.

Summary

- One of the central theories of biology is Darwin's theory of evolution, which states that living things change over time by means of natural selection. Proposed more than a hundred years ago, this theory has stood up to a century of testing and questioning.
- While studying the animals and plants of oceanic islands, Darwin accumulated a wealth of evidence that organisms have changed over time. Other factors that influenced Darwin's thinking were geological evidence that the Earth had changed over time, the changes in organisms that farmers were able to attain by using artificial breeding methods, and Malthus' ideas on population dynamics.
- Darwin proposed that evolution occurs as a result of natural selection: some individuals have traits that make them better-suited to a particular environment, allowing more of them to survive to reproductive age and produce more offspring than other individuals lacking these traits. These characteristics, or adaptations, are naturally occurring inherited traits found within populations.
- Fitness is a measure of the tendency of some organisms to leave more offspring than competing members of the same population. Therefore genetic traits possessed by the fitter individuals will appear in greater proportions among members of succeeding generations. As a result of this process, the traits allowing greater reproduction will increase in frequency over time. The environment imposes conditions that determine the direction of selection and thus the direction of change.
- Darwin published his theory in 1859 in a book titled *On the Origin of Species.* A wealth of evidence since Darwin's time has supported his proposals that evolution occurs and that its mechanism is natural selection. By the 1860s, natural selection was widely accepted as the correct explanation for the process of evolution, but the mechanism of natural selection was not understood. The field of evolution did not progress much further until the 1920s because of the lack of a suitable explanation of how hereditary traits are transmitted.
- Two direct lines of evidence uphold the theory of evolution: (a) the fossil record, which exhibits a record of progressive change, whether punctuated and/or gradualistic, correlated with age, and (b) the molecular record, which exhibits a record of accumulated changes, the amount of change correlated with age as determined in the fossil record.
- Several indirect lines of evidence uphold the theory of evolution, including progressive changes in homologous structures, the existence of vestigial structures, and changes in DNA sequences.
- In 1972, American paleontologists Niles Eldredge and Stephen Jay Gould developed a theory called *punctuated equilibrium.* This theory states that new species arise suddenly and rapidly as small subpopulations of a species split from the populations of which they were a part. This theory differs from the theory of *phyletic gradualism,* which asserts that new species develop slowly and gradually as an entire species changes over time.

Knowledge and Comprehension Questions

1. Summarize four of Darwin's conclusions that were inspired by his observations during his voyage.
2. What three factors, in addition to his voyage on the H.M.S. *Beagle,* influenced Darwin's thinking on evolution?
3. A dachshund and a Siberian husky are both dogs, but they look very different from each other. By what process did the two breeds come to look so different? What observable genetic principle is this process based on?
4. Distinguish between an exponential progression and an arithmetic progression. Which increases more quickly? Relate these concepts to the development of Darwin's theory of evolution.
5. What are adaptations? Explain their significance.
6. Explain the phrase *survival of the fittest.* What does "fit" mean in this context?
7. Distinguish between adaptive radiation and speciation.
8. Fill in the blanks: A ________ is a record of a dead organism, often preserved in ________ rocks. It can be dated with use of naturally occurring ________ of certain elements found in rocks.
9. Explain how scientists can use the ^{14}C method to date a fossil.
10. How old is the Earth according to radioactive dating methods? Why is this significant?
11. What are vestigial organs? Give an example from the human body.
12. Distinguish between homologous and analogous structures.
13. Explain how studies of comparative anatomy and comparative embryology support the theory of evolution.
14. What is an evolutionary tree? How does it support the concept of evolution?

Critical Thinking

1. Most species of bears are black, brown, or gray. Why are polar bears white?
2. If a population were composed of only a few organisms, what effects might natural selection have on its continued survival? Would a larger population have a better chance of survival? Why or why not?

CHAPTER 24

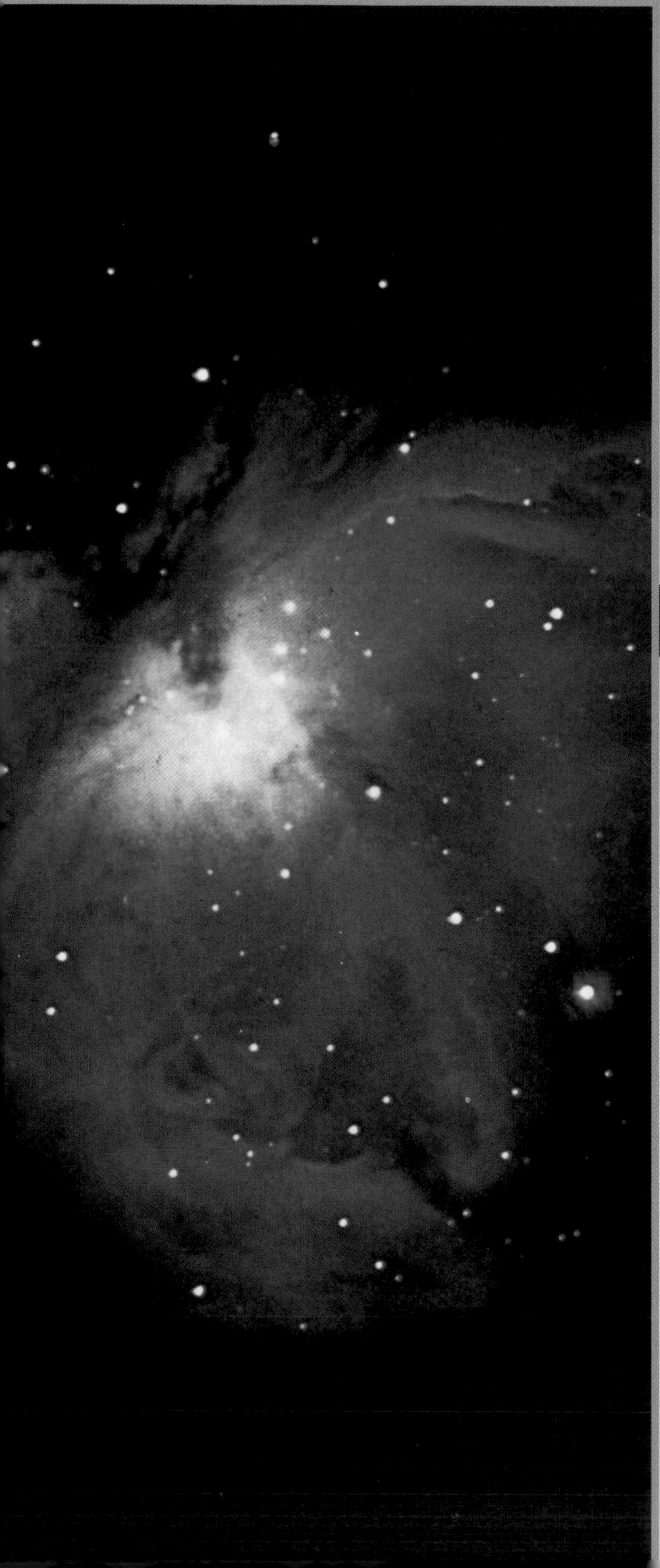

THE EVOLUTION OF THE FIVE KINGDOMS OF LIFE

VAST AREAS of gas and dust particles swirl in the Orion nebula M42. Orion is a specific constellation, or group of stars, that can be seen in the night sky of the northern hemisphere. Surrounding some of its stars are concentrations of gases and "stardust" called *nebulae,* a name derived from a Latin word meaning "cloud." In fact, nebulae are found in many sections of the Milky Way galaxy and are part of a cosmic life cycle—the birth, life, and death of stars.

Nebulae are "born" from the dust and gases hurled from unstable stars as they explode. Over time, this material condenses, giving birth to second- and later-generation stars. Then, in turn, these stars become unstable and explode, spewing dust and gases into space. And so the nebula life cycle continues.

KEY CONCEPTS

- Approximately 15 billion years ago, the universe condensed and then exploded in the "big bang"; Earth and the other planets of the solar system were formed approximately 4.6 billion years ago.
- The long-held but highly controversial "primordial soup" theory states that life arose in the sea as elements and simple compounds reacted to form organic molecules about 3.5 billion years ago.
- The evolution of oxygen-producing bacteria was a key event, making possible an "explosion of life" during the Cambrian period, approximately 530 million years ago.
- Vertebrates evolved from the chordates and include a class of organisms called mammals, animals that are warm blooded and have hair and whose females secrete milk from mammary glands to feed their young.

OUTLINE

The American astronomer, Edwin Hubble (1889-1953) observed that the galaxies in the universe are moving away from one another. Put simply, the universe is expanding. This movement may mean that at one time all the matter of the universe was in one spot, a theory that leads scientists to further theorize that the Earth (and the solar system) originated in the same way as nebulae do. Approximately 15 billion years ago, scientists theorize, *all* the material of the universe was condensed into a single small space. It then exploded in an event called the big bang. The universe—as it is known today—did not exist before that time. The gases and dust from the big bang produced an early generation of stars. Then, over billions of years, these stars exploded and their "space debris" formed other stars and planets. The solar system formed in this way some 4.6 billion years ago. For almost a billion years or so the molten Earth cooled, eventually forming a hardened, outer crust. Approximately 3.8 billion years ago, the oldest rocks on Earth formed. After that time, some 3.5 billion years ago, life began.

Theories about the Origin of Simple Organic Molecules

Primitive Earth was an incubator for life, but scientists know very little about that incubator. In trying to understand what the Earth was like 3.8 billion years ago, scientists often use information from events that are observable today as starting points to build hypotheses. For example, scientists have determined that the dust and gases ejected from unstable stars contain mostly hydrogen as well as helium and varying amounts of other elements such as nitrogen, sodium, sulfur, and carbon. Some of these elements combine to form compounds such as hydrogen sulfide, methane, water, and ammonia. As the Earth coalesced, or came together into one mass, it may have contained some of these gases and vapors produced by the planet as it cooled. As the water vapor in this mix condensed, it could have caused millions of years of torrential rains extensive enough to form oceans. Other scientists suggest that water and gas deep within the Earth were vented to the surface by volcanoes.

Scientists agree that the environment of primitive Earth was harsh and violent, bathed in ultraviolet radiation from the sun and subjected to violent electrical storms and constant volcanic eruptions. Under these conditions, many scientists hypothesize that the elements and simple compounds of the primitive atmosphere reacted with one another. New and more complex molecules were formed, capturing the surrounding energy within their bonds. In 1953, Stanley Miller and Harold Urey, working at the University of Chicago, designed an experiment that modeled an environment that included methane (CH_4), ammonia (NH_3), water vapor (H_2O), and hydrogen gas (H_2). They wondered if organic molecules—the molecules that make up the structure of all living things—could have spontaneously formed under such conditions.

Using an apparatus similar to that in Figure 24-1, Miller and Urey filled a glass chamber with the four gases. Electrodes within the chamber shot sparks of electricity through the mixture while condensers cooled it. Miller and Urey wondered if complex molecules formed in their "atmosphere," dissolved in the water vapor, and fell into their "ocean." Within a week, they had their answer. A total of 15% of the carbon that was originally present as methane gas had been converted into other, more complex compounds of carbon!

Miller and Urey, joined later by many other scientists in laboratories around the world, performed experiment after experiment using various combinations of simple compounds that might have made up the primitive atmosphere. In addition, scientists experimented with a variety of energy sources, such as ultraviolet light, heat, and radioactivity. The outcome was always the same: more complex organic molecules formed from simpler ones. These newly formed molecules included amino acids—the building blocks of protein. Sugars, including glucose, ribose, and deoxyribose, were also shown to form under the conditions of the primitive atmosphere. As scientists experimented further, they discovered that many of the molecules important to life—and important to the structure and function of cells—could be synthesized under certain primordial conditions. Not

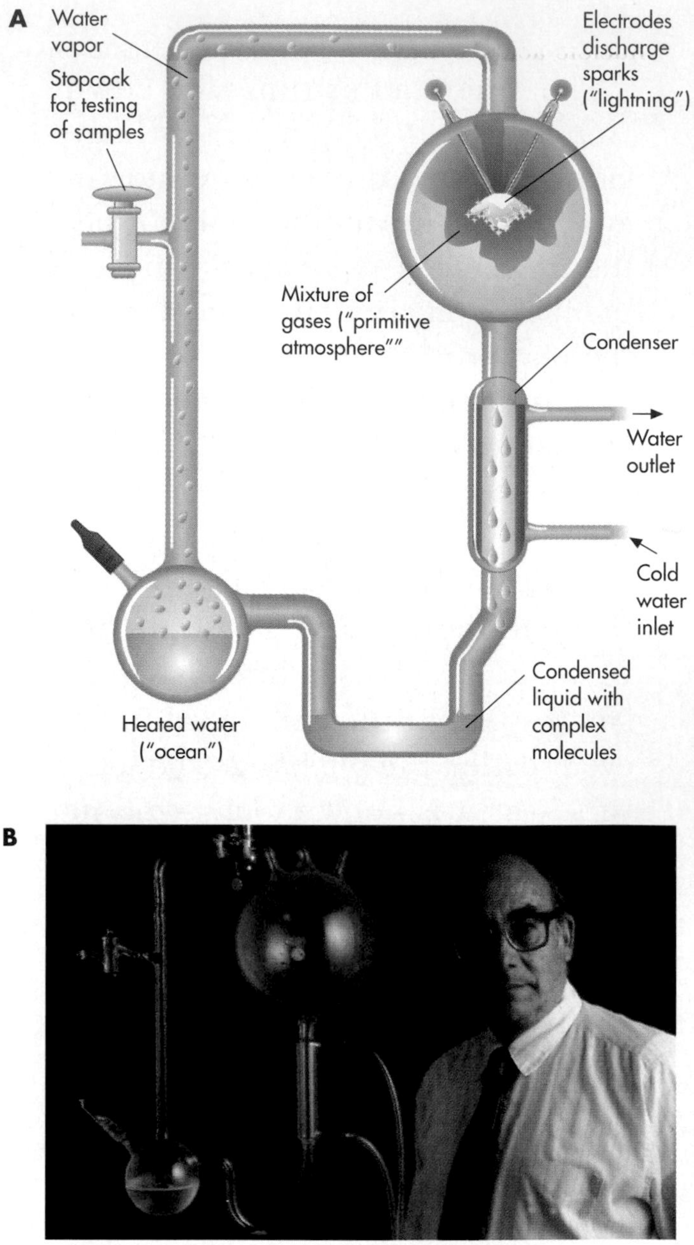

FIGURE 24-1 The Miller-Urey experiment. **A**, Miller and Urey's apparatus consisted of a closed tube connecting two chambers. The upper chamber contained a mixture of gases thought to resemble the Earth's atmosphere. Any complex molecules formed in the atmosphere chamber would be carried and dissolved in these droplets to the lower "ocean" chamber, from which samples were withdrawn for analysis. **B**, Dr. Stanley Miller standing in front of the apparatus in 1991.

only were ribose and deoxyribose, components of nucleic acids, synthesized in their methane environment, but so were purine and pyrimidine bases. Fatty acids, used in membranes and storage tissues of living organisms, were also made. From this array of experiments, scientists developed the primordial soup theory, which states that life arose in the primitive seas or in smaller lakes as complex organic molecules formed from the simple molecules in the ancient atmosphere.

During the past decade, some scientists have begun to doubt the primordial soup theory as described by the experiments of Miller, Urey, and others. Using computerized reconstructions of the atmosphere, James C. G. Walker of the University of Michigan at Ann Arbor suggests that the major components of the primitive atmosphere were carbon dioxide and nitrogen gases spewed forth from volcanoes. But these gases could not have formed complex organic molecules—the precursors to life—in the seas or tidal pools of the Earth's surface.

Recently, some scientists have shifted from adherence to the primordial soup theory to thinking that life might have arisen in hydrothermal vents, spots deep in the oceans where hot gases and sulfur compounds shoot from cracks in the Earth's crust (Figure 24-2). Some scientists hypothesize that the vents could have supplied the energy and nutrients needed for life to arise and be sustained. In fact, the hydrothermal vents in existence today support extensive communities of living organisms, such as tube worms, clams, and bacteria.

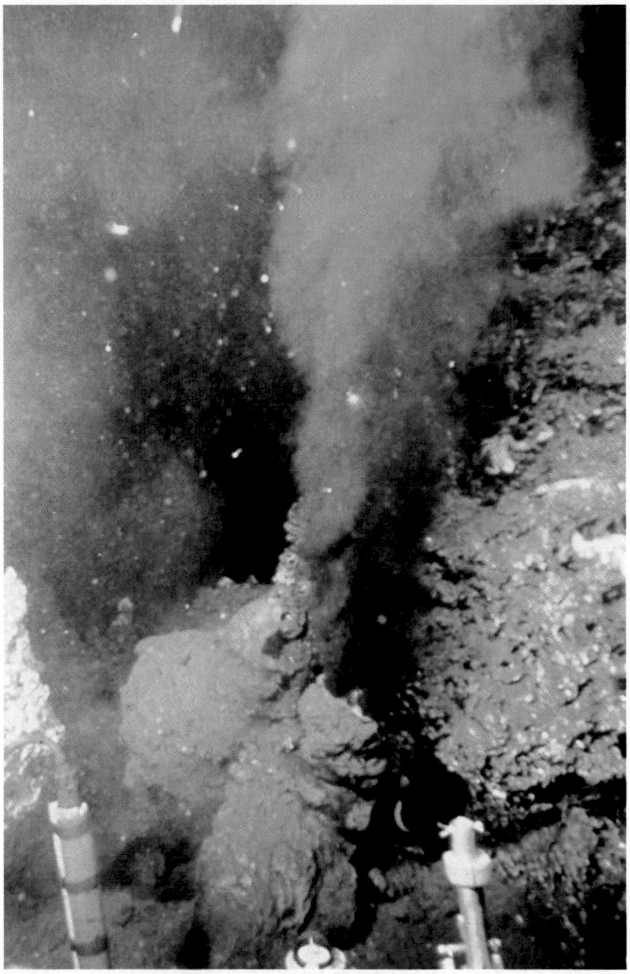

FIGURE 24-2 Hydrothermal vents. Located under deep oceans, hydrothermal vents are cracks in the Earth's crust that release hot gases and sulfur compounds. Some researchers think that these vents could have provided enough energy to start and sustain life.

The long-held primordial soup theory states that life arose in the sea as elements and simple compounds of the primitive atmosphere reacted with one another to form simple organic molecules such as amino acids and sugars. Recently, scientists have begun to reexamine the premises of this theory and have proposed alternative hypotheses.

Theories about the Origin of Complex Molecules

Although scientists can replicate various plausible conditions of the primitive Earth, they cannot duplicate the millions of years that passed. Although scientists can observe the "birth" of simple organic molecules, they have not generated life from nonlife. But at some point in the history of the Earth, organisms appeared. Scientists are still hard at work trying to unravel this mystery of life's beginnings. The work of many scientists over many years has led to various lines of reasoning regarding the evolution of the precursors to life and life itself. However, most of these hypotheses have been based on the existence of a primordial soup.

The first step toward the evolution of life must have been the synthesis of even more complex organic molecules than the ones previously described. However, scientists know that complex molecules such as *polymers* (for example, polysaccharides and proteins) do not spontaneously develop from a mixture of their simpler building blocks, or *monomers* (in this case, sugars and amino acids). Most synthesis reactions are dehydrations and depend on the removal of water and the input of energy to chemically link one molecule to another. This process, called *dehydration synthesis* (see Chapter 2), could have taken place if condensing agents were present with the monomers in a primordial soup. Condensing agents are molecules that combine with water and release energy. If condensing agents were not present, heat and evaporation could also promote dehydration synthesis and the synthesis of polymers.

In the 1950s, the American scientist Sidney Fox and his co-workers developed a technique in which they used heat to produce polymers, which Fox called *proteinoids,* from dry mixtures of amino acids. Perhaps such complex molecules formed over 3.5 billion years ago as pools and puddles of a primordial soup evaporated under the heat of the sun or were heated near areas of volcanic activity. Interestingly, the proteinoid molecules synthesized by Fox act like enzymes because they are able to increase the rate of various organic reactions. Metal ions and clays (composed chiefly of minerals) could also have served as early enzymes. When combined into sequences, enzymatic reactions can be thought of as the beginning of metabolic systems. To be considered "life" as scientists define it, metabolic systems must be organized within a cellular structure, be able to "carry" information about themselves, and be able to pass this information on by the process of replication.

> *The first step toward the evolution of life is the synthesis of highly complex organic molecules. Such complex molecules might have formed as pools of a primordial soup evaporated under the heat of the sun, promoting the linking of small molecules. Another theory suggests that metal ions and clays might have served as early enzymes promoting such reactions.*

Recently, scientists have begun to suspect that self-replicating systems of RNA molecules may have started the process of evolution. RNA molecules have the ability, as do all polynucleotides, of acting as a *template,* or guide, for the synthesis of a second polynucleotide based on the complementarity of its bases. As mentioned previously, the building blocks of nucleotides were probably present in a primordial soup. Theoretical work by Eigen and his coworkers suggests that short strands of self-replicating RNA eventually assembled enzymes from the amino acids present in the primordial soup.

For replicating DNA to be considered life, however, it must have become enclosed by some sort of a membrane forming the first cells. Chemical reactions could then take place in a closed environment—not the open environment of a primordial soup. Within the boundaries of a membrane, enzymes could be organized to carry on life functions. Chemicals could be selectively allowed into or kept out of the cell's interior. Wastes could be eliminated from the cell, and hereditary material could be passed on from cell to cell. How much time passed before biochemical evolution became biological evolution? How did this evolution take place? There is no scientific consensus on these points. Scientists do agree, however, that single-celled life existed in the shallow seas of the primitive Earth.

> *After the synthesis of complex organic molecules had taken place, self-replicating systems of RNA molecules may have begun the process of evolution. To be considered life, however, this "early" genetic material must have been organized within a cellular structure.*

JUST WONDERING . . .

How does science reconcile the differences between religious beliefs and scientific theories such as evolution?

It is not within the scientific domain to reconcile science with other disciplines such as social studies, English, or theology. What you may really be asking is how individual scientists reconcile such differences.

With respect to evolution, this theory fits easily into the belief systems of the majority of the world's religions. How the theory fits varies from religion to religion. For those scientists embracing religious beliefs in opposition to the theory of evolution, many hold their religion separate from their scientific work and do not try to reconcile the two. A small minority of scientists find the differences irreconcilable.

The History of Life

When discussing the history of life on Earth, scientists divide the time from the formation of the Earth until the present day into five major time periods or **eras** shown in the table of geological time (Table 24-1). The eras are subdivided into shorter time units called **periods.** In addition, the periods of the Cenozoic era are subdivided into **epochs.** However, each time unit (such as an era or a period) does not stand for a consistent length of time because early geologists defined and named these time units as they discovered them and because the distinctive events that mark the beginning and end of a time unit occurred over varying amounts of time. A 1-year geological "calendar" is shown on p. 517, which relates the geological time scale to a single year, making it easier to understand how long each geological time unit is in relation to the others.

The Archean Era: Oxygen-Producing Cells Appear

Scientists speculate that the first cells to evolve fed on organic materials in the environment. Because there would have been a very limited amount of suitable organic food available, a type of nutrition must have soon evolved in which cells were able to capture energy from inorganic chemicals. The evolution of a pigment system that could capture the energy from sunlight and store it in chemical bonds most likely evolved after that. Then, oxygen-producing bacteria evolved. Probably similar to present-day *cyanobacteria* (formerly called "blue-green algae"), these single-celled organisms became key figures in the evolution of life as it is known today. Their photosynthesis gradually oxygenated the atmosphere and the oceans around 2 billion years ago as evidenced by Archean and Proterozoic sediments. Scientists have also discovered fossils of the cyanobacteria.

Although fossils of single-celled organisms are usually difficult to find, the cyanobacteria sometimes grew in "piles," creating fossilized columns of organisms and the sediments that collected around them. These ancient columns of cyanobacteria are called *stromatolites* and date back 3.5 billion years. In Figure 24-3, *A*, the tops of stromatolites are visible above the surface of the water in which they form. Like icebergs, much of the stromatolite is hidden under water, as shown in Figure 24-3, *B*. Actively developing stromatolite formations can be observed even today in the warm, shallow waters of places such as the Gulf of California, western Australia, and San Salvador, Bahamas. Interestingly,

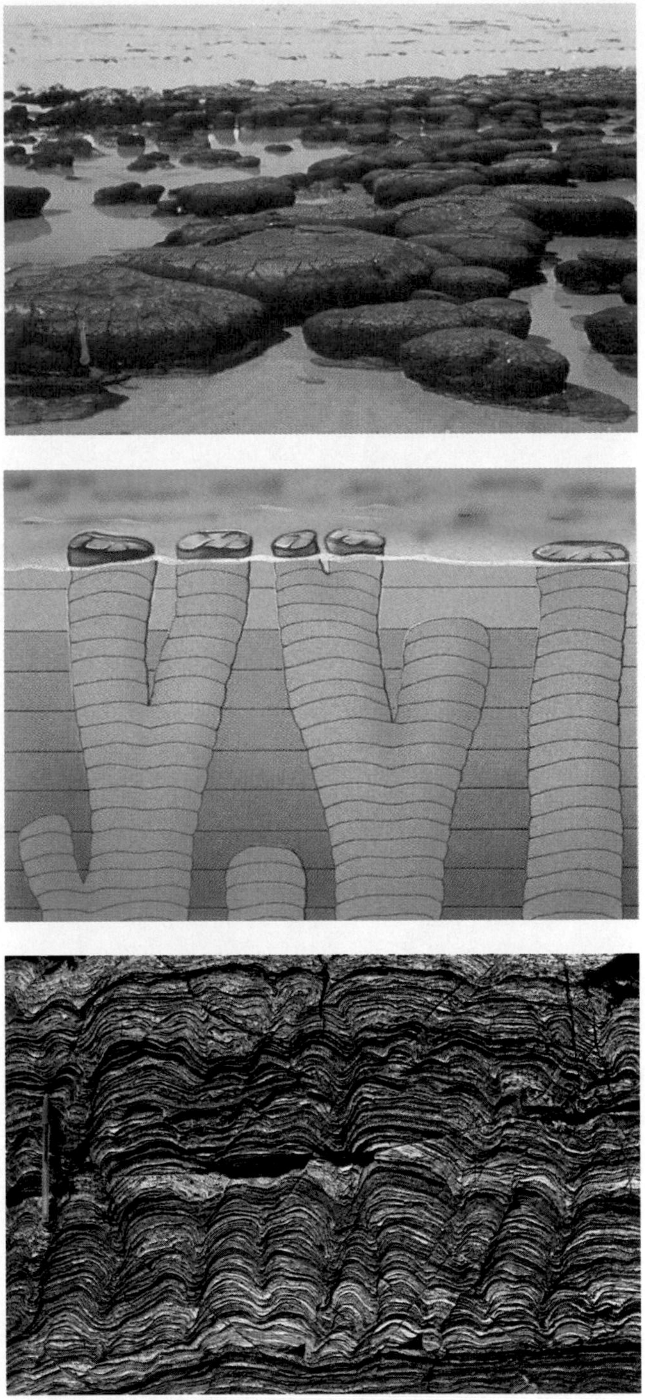

FIGURE 24-3 Stromatolites. A, These stromatolites are located in Shark Bay, Western Australia. The largest structures are about 1.5 meters (approximately $4\frac{1}{2}$ feet) across. These stromatolites formed about 4000 years ago. **B,** This diagram shows that much of the stromatolite formation is underwater. **C,** The internal, laminated structure of a fossilized stromatolite in Great Slave Lake, Canada.

TABLE 24-1

Table of Geological Time

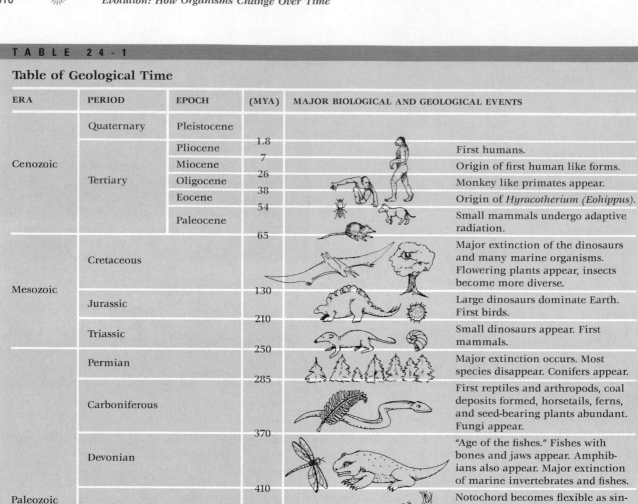

ERA	PERIOD	EPOCH	(MYA)	MAJOR BIOLOGICAL AND GEOLOGICAL EVENTS
Cenozoic	Quaternary	Pleistocene		
	Tertiary	Pliocene	1.8	First humans.
		Miocene	7	Origin of first human like forms.
		Oligocene	26	Monkey like primates appear.
		Eocene	38	Origin of *Hyracotherium (Eohippus).*
		Paleocene	54	Small mammals undergo adaptive radiation.
Mesozoic	Cretaceous		65	Major extinction of the dinosaurs and many marine organisms. Flowering plants appear, insects become more diverse.
	Jurassic		130	Large dinosaurs dominate Earth. First birds.
	Triassic		210	Small dinosaurs appear. First mammals.
Paleozoic	Permian		250	Major extinction occurs. Most species disappear. Conifers appear.
	Carboniferous		285	First reptiles and arthropods, coal deposits formed, horsetails, ferns, and seed-bearing plants abundant. Fungi appear.
	Devonian		370	"Age of the fishes." Fishes with bones and jaws appear. Amphibians also appear. Major extinction of marine invertebrates and fishes.
	Silurian		410	Notochord becomes flexible as single rod is replaced with separate pieces, as seen in the Ostracoderms (armored fish without bones, jaws, or teeth). Plants invade the land.
	Ordovician		430	First vertebrates appear. Major extinction of marine species.
	Cambrian		505	Major extinction of the trilobites. Origin of the main invertebrate phyla.
Proterozoic			544	Multicellular eukaryotic animals appear. First eukaryotic cells appear. Oxygen-producing bacteria present; atmosphere and oceans oxygenated.
Archean			2500	Earth is born. Chemical evolution resulting in formation of first cells. Stromatolites formed. First rock formed.

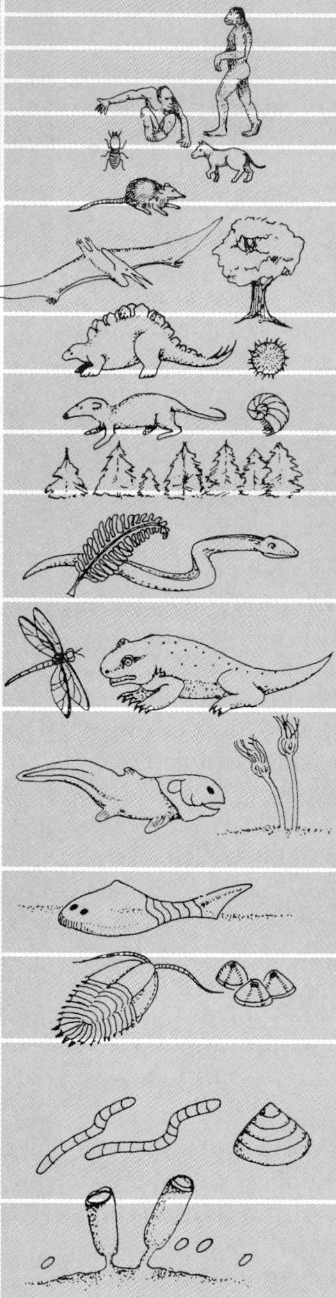

the shapes and sizes of the bacteria within present-day stromatolites look very much like the bacteria found within fossil stromatolites (Figure 24-4). In the absence of competing organisms, cyanobacteria produced stromatolites abundantly in all fresh water and marine communities until about 1.6 billion years ago.

> *Scientists speculate that the first cells to evolve fed on organic materials in the environment. Because of limited organic material, the evolution of a pigment system that could capture the energy from sunlight and store it in chemical bonds most likely evolved relatively quickly. Then, oxygen-producing bacteria evolved, gradually oxygenating the atmosphere.*

TABLE 24-1

Table of Geological Time

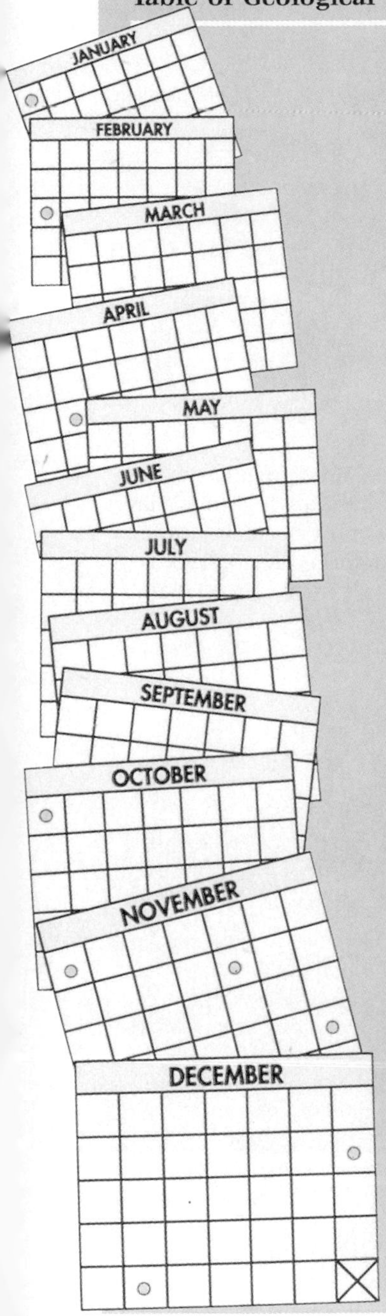

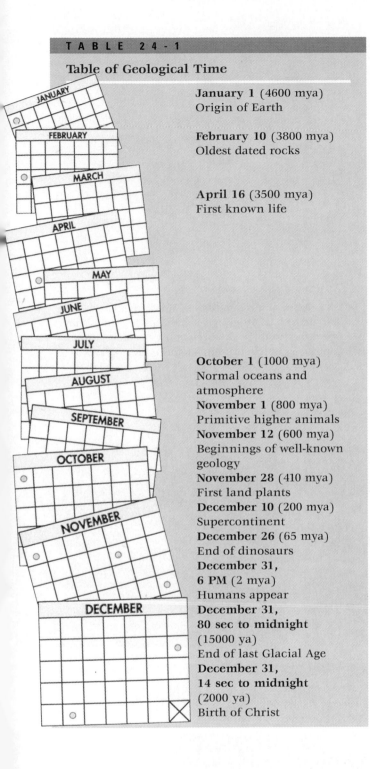

January 1 (4600 mya)	Origin of Earth
February 10 (3800 mya)	Oldest dated rocks
April 16 (3500 mya)	First known life
October 1 (1000 mya)	Normal oceans and atmosphere
November 1 (800 mya)	Primitive higher animals
November 12 (600 mya)	Beginnings of well-known geology
November 28 (410 mya)	First land plants
December 10 (200 mya)	Supercontinent
December 26 (65 mya)	End of dinosaurs
December 31, 6 PM (2 mya)	Humans appear
December 31, 80 sec to midnight (15000 ya)	End of last Glacial Age
December 31, 14 sec to midnight (2000 ya)	Birth of Christ

The Proterozoic Era: The First Eukaryotes Appear

The beginning of Proterozoic time, which extends from 2.5 billion to 544 million years ago, is marked by the formation of a stable oxygen-containing atmosphere. The term *Proterozoic* is derived from two Greek words meaning "prior" (*proteros*) "life" (*zoe*) and is descriptive of the fact that, until a few decades ago, scientists did not find evidence of life during this time (or during Archean time) because they were not looking for microfossils.

During the Proterozoic era, the first eukaryotic organisms appeared on Earth. A fossil stromatolite found in California dating back 1.4 to 1.2 billion years ago contains larger and more complex microfossils than the fossil cyanobacteria previously found. Scientists think these microfossils may have been eukaryotes, organisms separate from the cyanobacteria that may have lived in harmony with stromatolites or have fossilized with them. Other fossils in rocks of the same time add additional evidence to the theory that eukaryotes appeared around 1.4 billion years ago.

How did the eukaryotes arise? The most widely accepted hypothesis at this time is called the **en-**

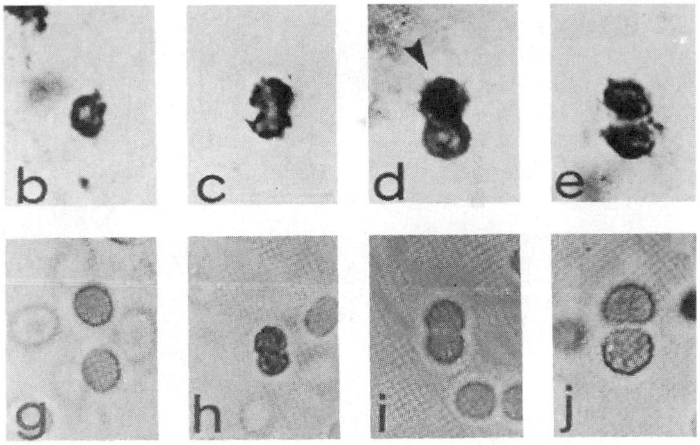

FIGURE 24-4 Fossilized bacteria. The fossilized bacteria in **B** through **E** were found in stromatolites located in South Africa. These stromatolites are about 3.4 billion years old. The bacteria in **G** through **J** show living, similar bacteria. The fossilized bacteria bear a striking resemblance to the living bacteria.

eras (ear uhs *or* air uhs) geological time periods. Scientists divide the time from the formation of the Earth until the present day into five eras.

periods shorter time units into which eras are subdivided.

epochs (ep uks *or* ee poks) subdivisions of geological periods.

endosymbiotic theory (en do **sim** bye **ot** ik) the most widely accepted theory regarding how eukaryotes arose. According to this theory, bacteria became attached to or were engulfed by host prokaryotic cells. Mitochondria originated from aerobic bacteria and chloroplasts originated from anaerobic bacteria.

dosymbiotic theory. (*Symbiosis* means "living together," and *endo-* means "within.") According to the endosymbiotic theory, bacteria became attached to or were engulfed by preeukaryotic (host prokaryotic) cells. The bacteria carried out chemical reactions necessary for living in an atmosphere that was increasing in its amounts of oxygen, which benefited their host cells. The bacteria, or endosymbionts, benefited by living within the protective environments of their hosts. The cells that engulfed the bacteria had a reproductive advantage over those cells not associated with bacteria and therefore flourished.

The cell membranes of the preeukaryotes are thought to have pouched inward during these symbiotic relationships, loosely surrounding the bacteria. The bacteria are thought to be early mitochondria. Symbiotic events similar to those postulated for the origin of mitochondria also seem to have been involved in the origin of chloroplasts, which are thought to be derived from symbiotic photo-

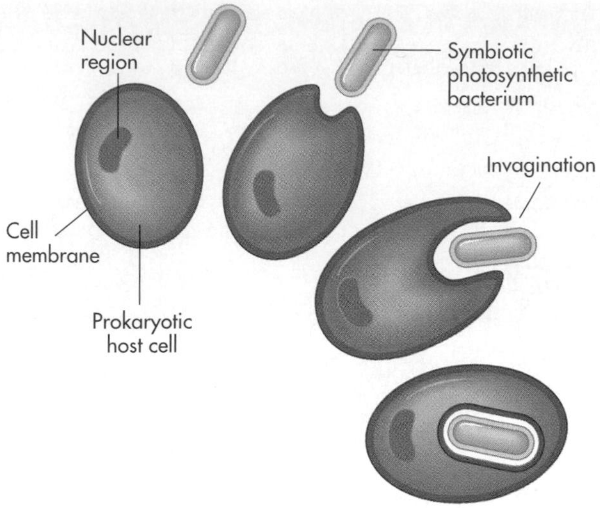

FIGURE 24-5 Endosymbiosis. How preeukaryotes are thought to have engulfed bacteria, which may have become mitochondria (heterotrophic bacteria) and chloroplasts (photosynthetic bacteria) as eukaryotic cells evolved.

FIGURE 24-6 Cambrian period fossils from the Burgess Shale, British Columbia, Canada, about 530 million years old. A, a model of Sidneyia inexpectans and **B,** a fossil of this organism. This creature was an arthropod. **C,** *Hallucigenia sparsa.* This specimen is 12.5 millimeters (approximately 1/2 inch) long. This animal seems to have been supported on seven pairs of spines; its trunk bore seven long tentacles and an additional group of tentacles near the end. *H. sparsa* may have been a scavenger on the bottom of the sea. **D,** *Wiwaxia corrugata.* The body of this animal was covered with scales and also bore spines. This specimen is 30.5 millimeters (approximately 1 1/4 inches) across, excluding the spines. It may have been a distant relative of the mollusks. **E,** *Burgessochaeta setigera,* a segmented worm. The photograph shows the front end of the worm, which had a pair of tentacles. Footlike structures are attached to the worm in pairs along the sides of the body. The specimen is 16.5 millimeters (approximately 5/8 inch) long.

synthetic bacteria (Figure 24-5). The double membrane of these organelles is thought to be derived from the bacterium's plasma membrane and the surrounding cell membranes of the preeukaryotes that engulfed them. In addition, both mitochondria and chloroplasts are similar in size to bacteria, have their own DNA, have ribosomes similar to those of bacteria, and produce a limited portion of their own enzymes and proteins.

> *Scientists theorize, based on evidence in the fossil record, that the first eukaryotes, larger and more complex than the cyanobacteria, appeared around 1.4 billion years ago. The chloroplasts and mitochondria of eukaryotic cells are thought to be derived from bacteria (referred to as endosymbionts) that came to live within preeukaryotic cells.*

The Paleozoic Era: The Occupation of the Land

The Paleozoic era (544 to 250 million years ago) is marked by an abundance of easily visible, multicellular fossils. In fact, the earliest such fossils are found in rocks of southern Australia and are approximately 630 million years old. For many years such fossils were the oldest known, which led scientists to name the era in which these fossils became plentiful as the *Paleozoic,* meaning "old" *(paleos)* "life" *(zoos).* However, of the roughly 250,000 different kinds of fossils that have been identified, described, and named, only a few dozen are more than 630 million years old.

The Paleozoic era is divided into six shorter time spans called *periods.* (These periods are shown in Table 24-1.) The Cambrian period, which ended roughly 505 million years ago, is the oldest period within the Paleozoic era. It represents an important point in the evolution of life; all of the main phyla and divisions of organisms that exist today (except for the chordates and land plants) evolved by the end of the Cambrian period. Because so many new kinds of organisms appeared in such a relatively short time span, paleontologists (scientists who study fossil life) speak of a Cambrian "explosion" of living forms and often refer to geological strata older than Cambrian time as Precambrian.

The evolution of Cambrian organisms took place in the sea. Figure 24-6 shows fossils of the organisms that lived on the sea floor during the Cambrian period—quite unusual by today's standards! These fossils, together with well over 100 other species, have been found in the Burgess Shale, which are geological strata formed from fine-grained mud. During past movements of the Earth's crust, these geological strata (which were originally underwater) were

A

B

FIGURE 24-7 Trilobites and horseshoe crabs. The trilobites flourished in the seas of the Cambrian period, 505 to 544 million years ago. The example in **A** was found in the Burgess Shale. Trilobites appear to be directly related to modern-day horseshoe crabs **(B)**.

uplifted and are now part of the Rocky Mountains of British Columbia, Canada.

Many kinds of multicellular organisms that thrived in the early Paleozoic era have no relatives living today, while others ultimately led to the contemporary phyla of organisms. The trilobites, for example (Figure 24-7, *A*), appear to be derived from the same evolutionary line that gave rise to one living group of arthropods, the horseshoe crabs (Figure 24-7, *B*).

> *During the Cambrian period of the early Paleozoic era, an array of species evolved in the sea. Some of these species still exist today.*

As the Paleozoic era continued with the Ordovician period (505 to 430 million years ago), wormlike aquatic animals similar to lancelets and lampreys (Figure 24-8, *B*) began to evolve from ancient flatworms. Characteristic of these organisms was the stiff, internal rod that ran down the back, parallel to the central nerve. This stiffening rod is called a *notochord* (meaning "back cord"); these organisms are therefore called **chordates.** One of the earliest known chordate fossils is shown in Figure 24-8, *A*.

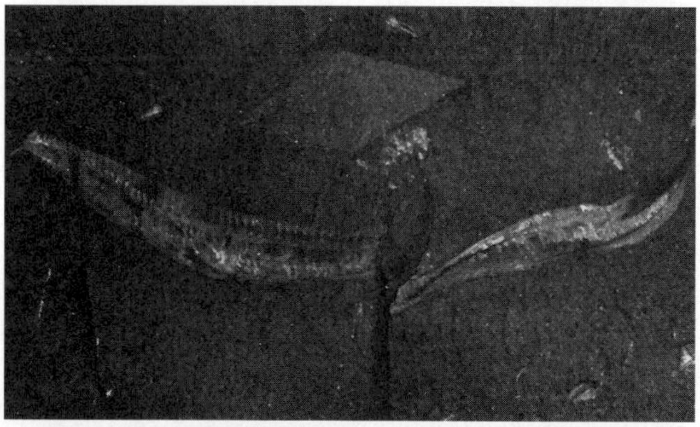

A

B

FIGURE 24-8 The earliest known chordate. A, *Pikaia gracilens* is a small fishlike animal, and it is one of the first organisms with a notochord. This particular fossil was found in the Burgess Shale. **B,** A lancelet. Note the resemblance to *Pikaia.*

The approximately 42,500 species of chordates that exist today include fishes, amphibians, reptiles, birds, and mammals. (**Mammals** are warm-blooded vertebrates that have hair and whose females secrete milk from mammary glands to feed their young.) Chordates are distinguished by three principal features: (1) a single hollow *nerve cord* located along the back; (2) a rod-shaped *notochord,* which forms between the nerve cord and the developing gut (stomach and intestines); and (3) *pharyngeal (gill) arches* and *slits,* which are located at the throat (pharynx). (These three chordate features are shown in Figure 24-9 as they appear in the embryo.) All the features of chordates are evident in their embryos, even if they are not present in the adult form of the organism.

Each of the three chordate characteristics played an important role in the evolution of the chordates and the vertebrates. The dorsal nerve root increased the animals' responsiveness to the environment. In the more advanced vertebrates, it became differentiated into the brain and spinal cord. The notochord was the starting point for the development of an internal stabilizing framework, the backbone and skeleton, which provided support for locomotion. The bony structures that support the pharyngeal arches evolved into jaws with teeth, al-

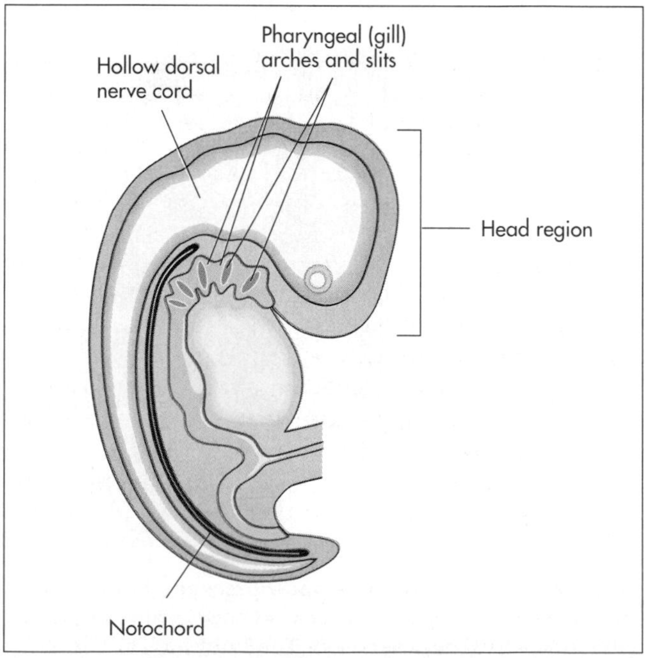

FIGURE 24-9 Chordate features. Embryos reveal the three principal features of the chordates: a nerve cord, a notochord, and pharyngeal (gill) arches and slits.

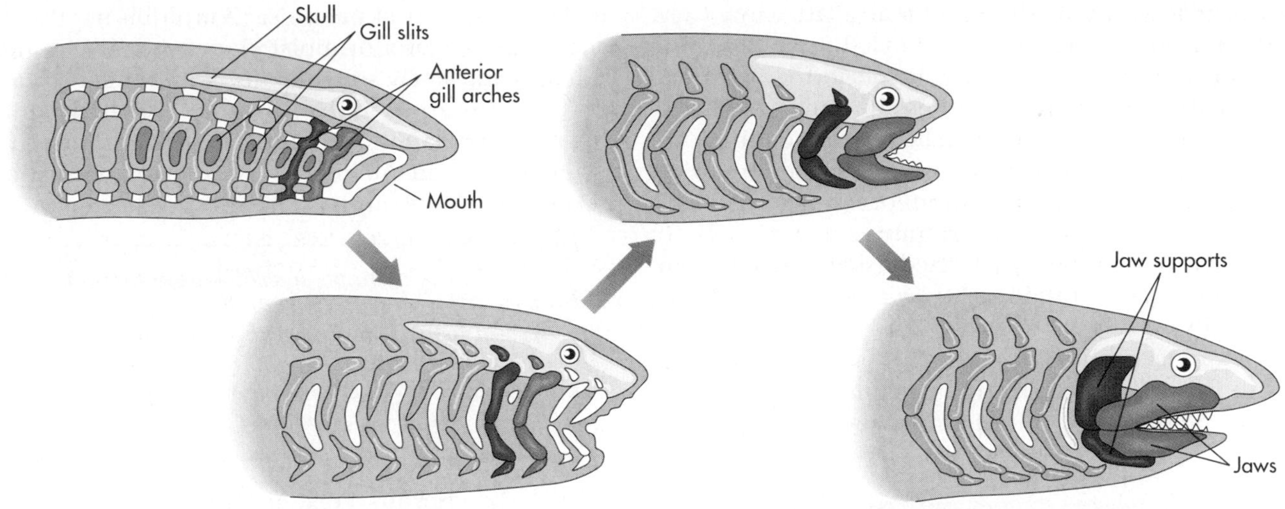

FIGURE 24-10 Evolution of the jaw. Jaws evolved from the anterior gill arches of the jawless fishes.

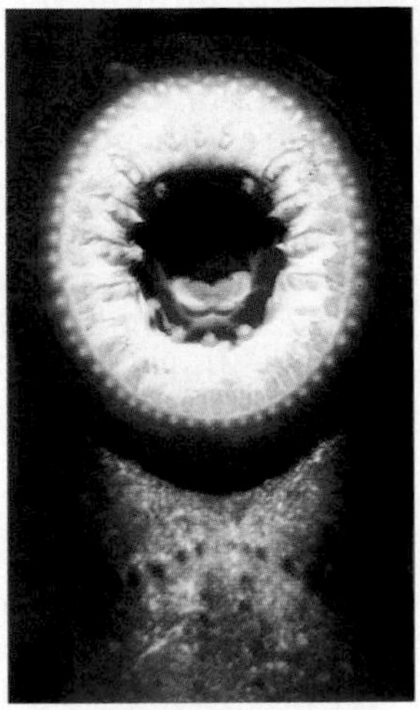

FIGURE 24-11 A lamprey, member of class Agnatha. Lampreys attach themselves to fish by means of their suckerlike mouths, rasp a hole in the body cavity, and suck out blood and other fluids from within. Lampreys do not have a distinct jaw.

lowing these organisms to feed differently than their ancestors (Figure 24-10).

With the exception of two groups of relatively small marine animals, the tunicates and lancelets (see Figure 24-8, *B*), all chordates are **vertebrates.** Vertebrates (a subphylum of the chordates) differ from these two chordate groups in that the adult organisms have a *vertebral column,* or backbone, that develops around and replaces the embryological notochord. In addition, most vertebrates have a distinct head and a bony skeleton, although the living members of two classes of fishes, *Agnatha* (lampreys and hagfishes) (Figure 24-11) and *Chondrichthyes* (sharks, skates, and rays), have a cartilaginous skeleton.

The Devonian Period: Evolution of the Fishes

The first vertebrates to evolve were jawless fishes, members of the class *Agnatha,* about 500 million years ago (see Figure 24-11). Although traces of agnathan fossils are found in Cambrian strata, most date back to the Ordovician and Silurian periods, about 410 to 505 million years ago. Jaws first developed among vertebrates that lived about 410 million years ago, toward the end of the Silurian and beginning of the Devonian periods. The first jawed fishes that evolved were the placoderms, ancestors of today's bony fishes (such as salmon, trout, cod, and tuna) and the cartilaginous fishes (such as sharks, skates, and rays). They were also one of the first groups of fishes to have paired fins. During the Devonian period, fishes having a bony skeleton appeared: the ray-finned fishes and the fleshy-finned fishes. The ray-finned fishes are the ancestors of the wide variety of bony fishes that exists today.

In the late Devonian period, the lungfishes and the lobe-finned fishes appear in the fossil record. Fleshy-finned fishes are ancestors to both. Only a few species of lungfishes exist today (Figure 24-12). The lungs of these fishes are supplementary to

chordates (**kor** dates) organisms distinguished by three principal features: a nerve cord, a notochord, and pharyngeal slits; includes fishes, amphibians, reptiles, birds, mammals, and humans. Chordates evolved during the Ordovician period of the Paleozoic era.

mammals (**mamm** uhls) warm-blooded vertebrates that have hair and whose females secrete milk from mammary glands to feed their young.

vertebrates (**ver** tuh bruts *or* **ver** tuh braytes) a subphylum of chordates; vertebrates have a bony (vertebral) column surrounding a dorsal nerve cord.

their gills but allow them to breathe air when they are buried in the mud of dried-up lakes and streams. Most of the Devonian period lobe-finned fishes also had primitive lungs similar to the lungfishes. Interestingly, these fishes were thought to be extinct until the discovery of a living member of this group in 1938. Since then, other specimens of this same species have been found.

One group of the lobe-finned fishes is believed to be the ancestor of the land-living *tetrapods,* or four-limbed vertebrates. The first land vertebrates were *amphibians,* animals able to live both on land and in the water. Although a mature amphibian spends time on land, it must return to the water frequently to keep its thin skin moist. Amphibians also lay their eggs in water or moist places and live in water during their early stages of development. The early amphibians arose in the Devonian and had fishlike bodies, short stubby legs, and lungs (Figure 24-13). They flourished during the next period of the Paleozoic era: the swampy Carboniferous period.

FIGURE 24-12 A lungfish. Lungfishes have supplemental lungs that allow them to breathe air when they are buried in mud.

> *The first vertebrates to evolve were the jawless fishes, about 500 million years ago. Jawed fishes evolved toward the end of the Silurian period, about 410 million years ago. Amphibians arose from lobe-finned fishes during the Devonian period.*

The Carboniferous Period: Land Plants and Fungi Arise

At the interface of the Silurian and Devonian periods, approximately 410 million years ago, the first terrestrial plants evolved. The ancestors of these plants are believed to be the green algae, because their chloroplasts are biochemically similar to those of the plants. The green algae are an extremely varied group of more than 7000 protist species, found growing in the water as well as on tree trunks or in the soil. Many of these algae are microscopic and unicellular. However, the plants that appeared during the Silurian were multicellular and had developed mechanisms for transporting water within their bodies as well as for conserving it. Within the next 100 million years, plants became abundant and diverse on the land and eventually formed extensive forests.

The Carboniferous period (370 to 285 million years ago) is named for the great coal deposits formed during this time. Much of the land was low

FIGURE 24-13 Reconstruction of *Ichthyostega,* one of the early amphibians. *Ichthyostega* had efficient limbs for crawling on land, an improved olfactory sense associated with a lengthened snout, and a relatively advanced ear structure for picking up airborne sounds. Despite these features, *Ichthyostega,* which lived about 350 million years ago, was still quite fishlike in overall appearance.

Processes and Methodologies

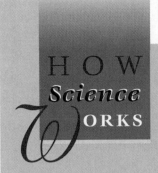

HOW *Science* WORKS

The Merging of Sciences: Using the Theory of Plate Tectonics to Understand Evolution

During the Mesozoic era, the continents did not exist as they do today. Instead, they were all joined in one giant continent scientists called *Pangea,* meaning "all Earth." Pangea began to break up into smaller pieces during the Mesozoic era, with this movement continuing during the following era, the Cenozoic. These changes greatly affected evolution.

After the Earth coalesced approximately 4.5 billion years ago, its surface was hot and violent for about 600 to 700 million years. At first, these conditions were too inhospitable for living things or even for biochemical evolution to take place. Scientists hypothesize that geological activity eventually resulted in the formation of fragments of continents. Some geologists think that these land masses were formed as melted rock material rose to the surface of the Earth, pushed upward by other heavier, sinking, melted metals such as iron and nickel. Others hold that the highly active volcanoes prevalent at that time spewed so much lava into the seas that it eventually accumulated, forming land masses.

Approximately 200 million years ago, small land masses united to form the single, large "supercontinent" Pangea. Its northern half was called *Laurasia* and consisted of the present-day North America, Europe, and Asia. Its southern half was called *Gondwana* and was a combination of present-day South America, Africa, India, Antarctica, and Australia. Pangea remained as a supercontinent for approximately 100 million years. However, forces were at work beneath it that eventually divided it—once again—into smaller land masses.

Scientists now have some insight into the movement of land masses over time, such as those that resulted in the formation of Pangea and its subsequent division. No one knows exactly the speed of the movement long ago; the current rate of movement is 1 to 15 cm per year—incredibly slow.

Today, the outer shell of the Earth is made up of six large "plates" and several smaller ones (Figure 24-A). According to the *plate tectonics theory,* these plates are rigid pieces of the Earth's crust, which are from 75 to 150 kilometers (approximately 50 to 100 miles) thick. The rocks beneath these plates are less rigid than the plates they support. The surface plates move, or drift, over this underlying semiliquid rock. Scientists think that this "liquid" rock flows slowly, rising as it heats, and sinking as it cools (Figure 24-B). At the mid-oceanic ridges—long, narrow mountain ranges under the sea—the plates move away from one another in a process called *sea floor spreading.* Molten rock wells up from deep within the Earth, pushing the plates apart and filling in the space between. At other places the plates move toward each other. As they meet,

24-A

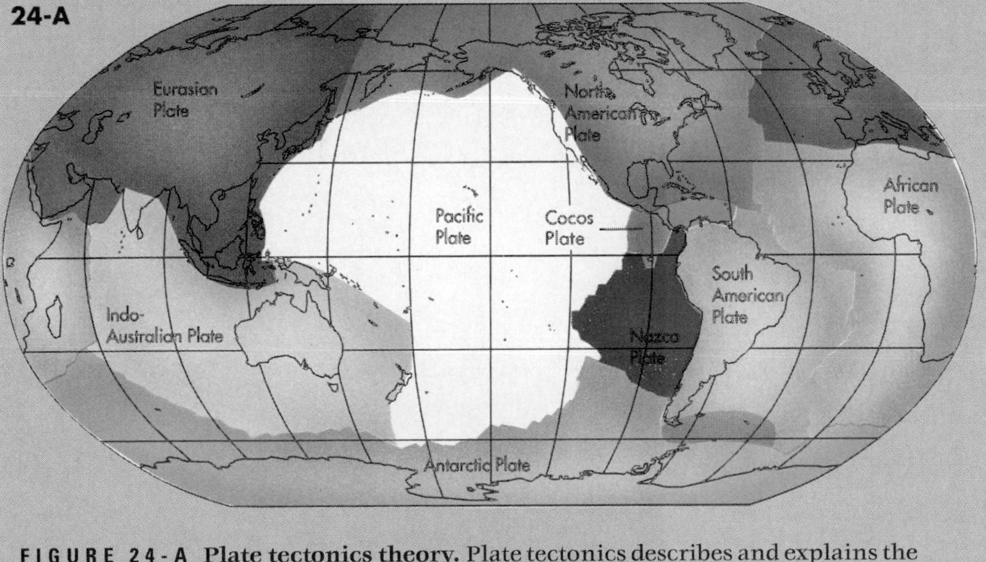

FIGURE 24-A Plate tectonics theory. Plate tectonics describes and explains the movement of the continents. According to this theory, Earth is made up of six large plates and several smaller plates. The rocks beneath these plates are semiliquid. The plates drift over this semiliquid rock, occasionally colliding. When this occurs, one plate is ususally pushed below the other, a process called *subduction.* When plates move apart, the semiliquid rock wells up in the space created, a process called *sea floor spreading.*

Processes and Methodologies—cont'd

HOW Science WORKS

one plate may sink below the other, or they may collide, pushing up mountain ranges in the process. The Himalayas, for example, have been thrust up to the highest ele-vations on Earth as a result of the grinding, prolonged collision of the Indian subcontinent with Asia. Other types of violent geological activity were also associated with the movements of the crustal plates. The earthquakes that destroyed San Francisco in 1906 and the one that occurred in the same region in 1989 resulted from two plates sliding past each other.

Figure 24-C summarizes the movements of the continents as they gradually moved apart from their positions as parts of Pangea 200 million years ago. The South Atlantic Ocean opened about 125 to 130 million years ago as Africa and South America moved apart from one another. South America then moved slowly toward North America and, approximately 3.6 to 3.1 million years ago, became connected to it by the Isthmus of Panama. Changes occurred in the positions of all the continents, which played a major role in the distribution of organisms that are seen today.

As land masses move away from one another, populations inhabiting those lands become separated from one another. Geographically isolated, populations of the same species often undergo speciation as differing pressures of natural selection act on them. Similarly, as land masses collide, species are thrust together and must compete with one another to survive. Shorelines disappear and with them go a variety of habitats and species. And as continents move, ocean currents change, causing climatic changes and yet new selection pressures.

24-B

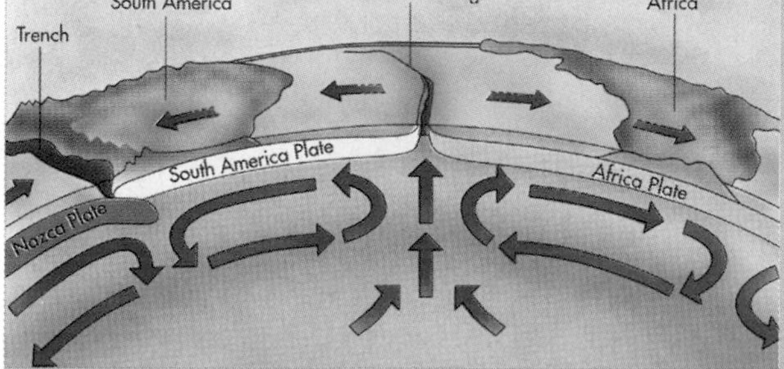

FIGURE 24-B Why do the plates move? The semiliquid rock on which the plates rest rises as it heats and sinks as it cools, causing movement.

24-C

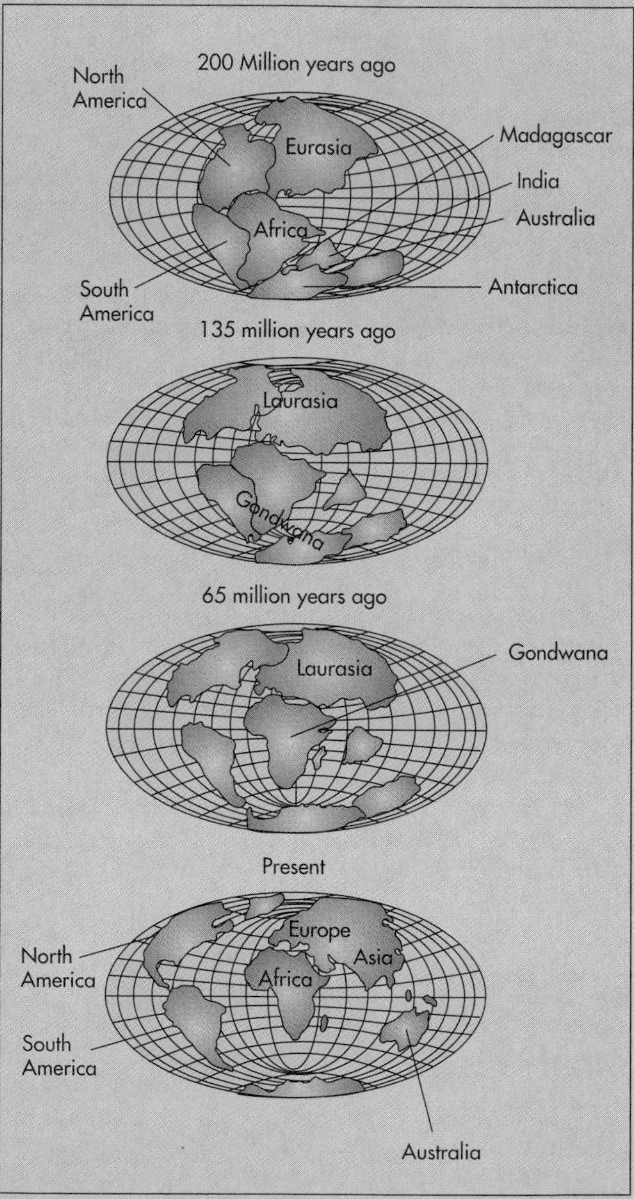

FIGURE 24-C Movement of the continents. The changes that occurred in the position of all the continents as a result of shifting plates have played a major role in the distribution of organisms seen today.

and swampy due to a worldwide moist, warm climate—conditions that contributed to the fossil preservation of the still-prevalent forests. These coal deposits provide a relatively complete record of the horsetail plants, ferns, and primitive seed-bearing plants of the Carboniferous period. The conifers—a group of seed-bearing plants that is represented today by pines, spruces, firs, and similar trees and shrubs—also originated then.

The fungi, a kingdom of organisms that live on dead and decaying matter, also invaded the land in the late Carboniferous period. Scientists think fungi were successful because of their chitin-rich cell walls. Chitin is a stiff, hard substance (also found in the outer skeletons of insects and some other arthropods) that helps fungi remain "drought resistant." Fungi are also thought to be successful land organisms because of their roles (along with certain bacteria) as decomposers.

The Mesozoic Era: The Age of Reptiles

The Mesozoic era, or "middle life" (250 to 65 million years ago), was a time of adaptive radiation of the terrestrial plants and animals that had been established during the mid-Paleozoic era but that had not been wiped out by the mass extinctions of the Permian period (see "How Science Works"). The radiation of these organisms led to the establishment of the major groups of organisms living today. The amphibians, the dominant land vertebrates for about 100 million years, were gradually replaced by the reptiles as the dominant organisms as the climate became drier.

In contrast to amphibians, reptiles have water-resistant skin and more efficient lungs. These adaptations to dry climates gave reptiles a survival advantage over the amphibians. In addition, reptiles developed a reproductive advantage—the amniotic egg. An *amniotic egg* has a thick shell that encloses the developing embryo and a nutrient source within a water sac (see Chapter 30). This type of egg protects the embryo from drying out, nourishes it, and enables it to develop away from water. Birds too have amniotic eggs. During the course of the Mesozoic, one group of reptiles evolved into mammallike reptiles, precursors to mammals; another group led to the evolution of the birds; and yet another developed into the dinosaurs. Reptiles also invaded the sea and the air, but many of these kinds of reptiles eventually died out.

The Mesozoic is divided into three periods: the Triassic, the Jurassic, and the Cretaceous. These names refer to geological events or formations of the eras and do not chronicle the evolution of life during this time. During the Triassic period, small dinosaurs and primitive mammals appeared. During the Jurassic period, large dinosaurs dominated the Earth, and birds first appeared. The earliest known bird, *Protoavis* ("first bird") (Figure 24-14, *B*), was reported in 1986 by Sankar Chaterjee of Texas Tech University. The fossil bones of this organism were found in rocks dated to be 225 million years old, very early in the Mesozoic era. Until this find, the oldest known fossil bird was *Archaeopteryx*, estimated to be about 150 million years old (Figure 24-14, *A*). There is no consensus regarding whether *Archaeopteryx* arose from the same evolutionary line as *Protoavis* or independently from another dinosaur. In addition, there is no consensus among paleontologists whether *Protoavis* is actually a bird. Many scientists question the validity of Chaterjee's "find."

Archaeopteryx had feathers but apparently not enough of them to fly very effectively. Feathers, thought to have evolved from reptilian scales, probably helped early birds glide from tree to tree and provided them with insulation. Ultimately, feathers made possible the subsequent evolution of birds that could fly well. This ability allowed birds to inhabit unoccupied habitats and resulted in the evolution of a large and diverse class of organisms. Today, there are about 9000 different species of birds.

During the Mesozoic era, the reptiles, which had evolved earlier from amphibians, became dominant and in turn gave rise to mammals (about 200 million years ago) and birds (at least 150 million years ago).

The Cretaceous Period: Flowering Plants Appear

During the early Cretaceous period, about 140 million years ago, flowering plants (angiosperms) began to appear, becoming the dominant form of plant life about 100 million years ago. Seed-bearing plants with fernlike leaves, similar to the living cycads, were abundant at that time. Today, with about 240,000 species of flowering plants, the angiosperms still dominate the plant kingdom, greatly outnumbering all other kinds of plants.

Until recently, most scientists thought that insects and flowering plants *coevolved*. That is, as flowering plants changed over time, so did insects because their feeding habits were linked to the characteristics of the flowers. As the flowering plants became more diverse and plentiful, so did the insects. However, recent fossil evidence analyzed by American paleontologists Conrad Labandeira and Jack Sepkoski suggests that insects (which arose in the Devonian period) experienced a major radiation in diversity during the Carboniferous period.

Then, during the mass extinctions of the Permian period, about 30% of all insect orders died out. The adaptive radiation of the insects began again in the Triassic period, *before* the diversification of the angiosperms, and it continues today. In fact, Labandeira's and Sepkoski's research suggests that the diversification of the insects actually *slowed down* as angiosperm diversity speeded up.

The End of the Mesozoic Era: Mass Extinctions and Continued Change

Around 65 million years ago, as the Cretaceous period (and the Mesozoic era) ended, sudden shifts occurred in the kinds of marine organisms that existed. Some became extinct and others began to flourish. Scientists can infer these changes in the populations of marine organisms living at that time by studying marine fossils exposed in certain European geological strata. For example, many of the larger plankton (free-drifting aquatic organisms) disappeared about 65 million years ago, and a fewer number of smaller ones took their place. The same rapid changes occurred in at least some nonplanktonic marine animal groups, such as the bivalve mollusks (clams and their relatives). The ammonites, a large and diverse group related to octopuses but with shells, abruptly disappeared. And, of course, a major extinction occurred on land: the dinosaurs disappeared, although at a pace much slower than that of the large plankton just mentioned.

B

A

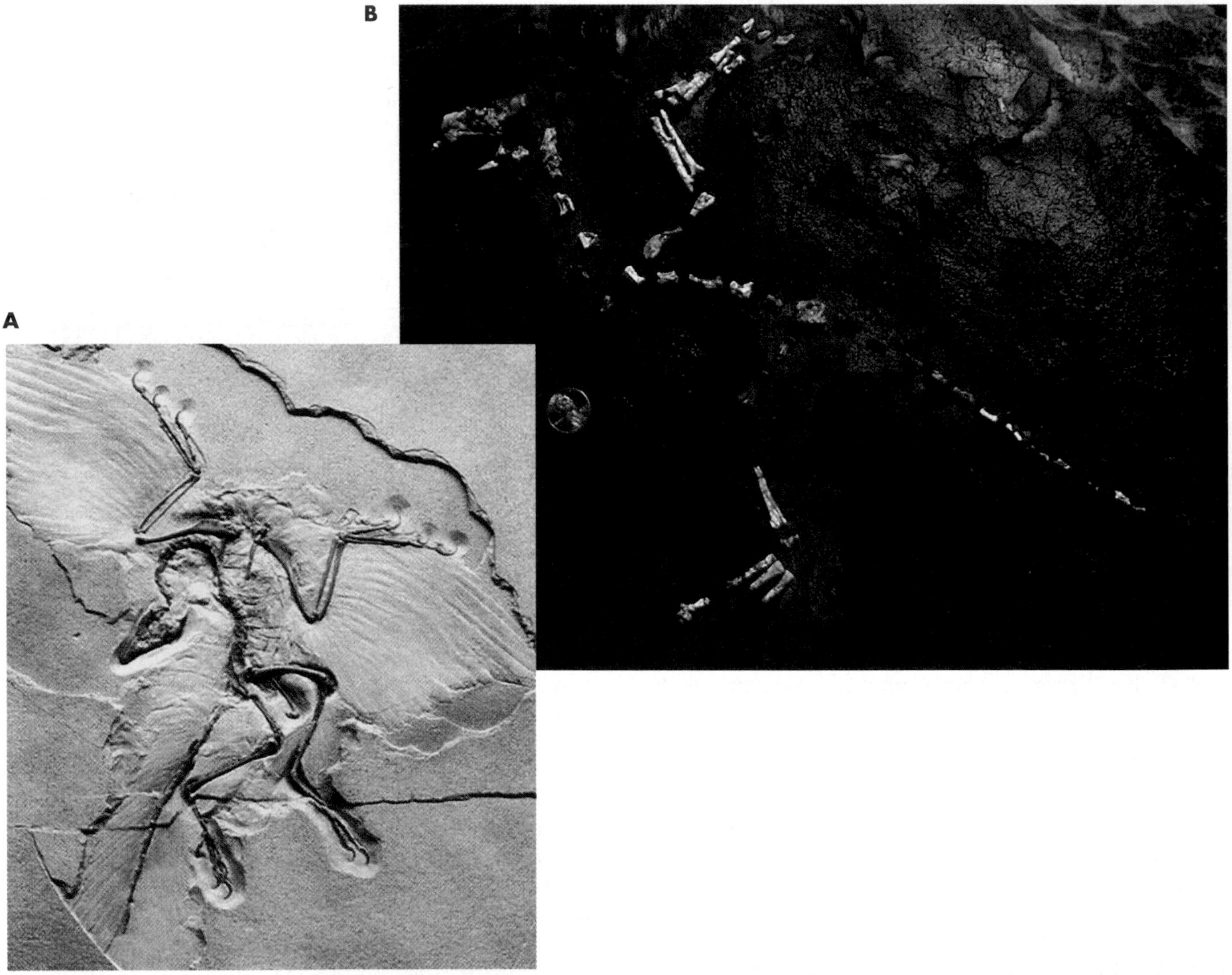

FIGURE 24-14 Links to the birds. **A**, About the size of the crow, *Archaeopteryx* lived in the forests of central Europe 150 million years ago. The teeth and long, jointed tail are features not found in any modern birds. Discovered in 1862, *Archaeopteryx* was cited by Darwin in later editions of *On the Origin of Species* in the support of his theory of evolution. **B**, *Protoavis,* a 225 million-year-old fossil from west Texas, may be an ancestor of the birds. If so (this claim is highly disputed) it will extend the history of birds back 75 million years.

Processes and Methodologies

HOW *Science* WORKS

Mass Extinctions: The Development of Hypotheses to Explain These Events

One of the most prominent features of the history of life on Earth has been periodic *mass extinctions,* the global death of whole groups of organisms. There have been five such events. Four of these events occurred during the Paleozoic era, the first occurring near the end of the Cambrian period (about 500 million years ago). At that time, most of the existing families of trilobites became extinct. Additional mass extinctions occurred about 440 million years ago and about 360 million years ago. The fourth and most drastic mass extinction happened at the close of the Permian period. It is estimated that approximately 96% of all species of marine animals living at that time died out! All of the trilobites and many other groups of organisms as well disappeared forever. The fifth major extinction event occurred at the close of the Mesozoic era, 65 million years ago. This was the famous event when dinosaurs became extinct.

Major episodes of extinction produce conditions that can lead to the rapid evolution of those relatively few plants, animals, and microorganisms that survive. Habitats, or places to live, that were formerly occupied suddenly open up, providing the "raw material" for *adaptive radiation,* a phenomenon by which populations of a species change as they disperse within a series of different habitats within a region (see Chapter 23). In fact, many paleontologists suggest that extinction episodes have been key events in the pattern of evolution.

Scientists have long discussed hypotheses regarding the extinction event of the late Cretaceous. Some propose that the climate cooled. Others have suggested that the sea level changed or that the vegetation cover of the Earth changed. These reasons, however, would not account for the change in the plankton community.

In 1980, a group of scientists headed by Luis Alvarez of the University of California, Berkeley, presented a dramatic hypothesis about the reasons for these changes. Alvarez and his associates observed that the usually rare element iridium was abundant in a thin layer in the geological strata that marked the end of the Cretaceous period. Alvarez and his colleagues proposed that if a large, iridium-rich meteorite or asteroid had struck the surface of the Earth, a dense particulate cloud would have been thrown up. This cloud would have darkened the Earth for a time, greatly slowing or temporarily halting photosynthesis and driving many kinds of organisms (particularly photosynthetic plankton alive at that time) to extinction. By disrupting and killing plant life, other organisms dependent on the plants would die out as well. Then, as its particles settled, the iridium in the cloud would have been incorporated in the layers of sedimentary rock that were being deposited at that time. Although the "meterorite hypothesis" is the most plausible one put forth to date, scientists have not reached consensus regarding this hypothesis.

The Cenozoic Era: The Age of Mammals

The mass extinctions of the Cretaceous marked the end of the Mesozoic era and heralded in the Cenozoic era about 65 million years ago. Extending to the present, the Cenozoic era has two periods—the Tertiary and the Quaternary—which are subdivided into many smaller time units called *epochs.* The epochs of the Tertiary period are listed in Table 24-1. Notice also that most of the "year" has passed on the geological calendar. Although 65 million years may seem like an incredibly long time, in terms of the history of life, it is short. During this very short time, life as it is known has evolved.

At the beginning of the Cretaceous during the Paleocene epoch, the small mammals that survived the extinctions of the Mesozoic underwent adaptive radiation, quickly filling the habitats vacated by the dinosaurs. Organisms having *both* mammalian and reptilian characteristics, called *transitional forms,* first appear in the fossil record approximately 245 million years ago. Then, about 200 million years ago in the early Mesozoic era, the first known mammals appear. These early mammals resemble today's shrews (Figure 24-15). They were small, fed on insects, and were probably *nocturnal,* or active at night. About 65 million years ago, following the

FIGURE 24-15 A shrew. Shrews are small nocturnal animals that feed on insects. They resemble the first known mammals that appeared about 200 million years ago.

extinction of the dinosaurs, mammals became abundant.

Various natural selection pressures resulted in the emergence of a number of significant anatomical and physiological changes in the mammals as they evolved from their reptilian ancestors. These changes resulted in a class of organisms that not only survived but flourished in a wide variety of habitats. Changes occurred in the reptilian arrangement of limbs as the mammals evolved, raising them high off the ground and allowing them to walk quickly and to run. A hinge developed between the lower jaw and the skull, and the teeth became differentiated, allowing mammals to eat a wide variety of food. The reptilian heart, having two ventricles with an incomplete separation, developed a complete wall in the mammals, preventing the mixing of oxygenated and deoxygenated blood. This change was significant in that mammals had also developed *warm bloodedness:* a constant internal body temperature. Warm-blooded animals are also called *endotherms,* meaning "within" *(endo-)* "temperature" *(-therm).* Their ancestors were *cold blooded,* having internal body temperatures that followed the temperature of their environments within certain limits. Cold-blooded animals are also called *ectotherms,* meaning "outside" (*ecto-*) "temperature"(*-therm*). Endotherms have a higher rate of metabolism than ectotherms and need more oxygen for the increase in the rate of cellular respiration.

Important changes also occurred in reproduction as the mammals evolved from the reptiles. The most "reptilian" type of reproduction is seen in the *monotremes,* one of the three subclasses of mammals living today. Monotremes lay eggs with leathery shells and incubate these eggs in a nest. The underdeveloped young that hatch from these eggs feed on their mother's milk until they mature. The ability of a mother to feed her young with milk she produces in mammary glands is another characteristic that has been advantageous to the adaptive radiation of the mammals. Present-day examples of monotremes are platypuses and spiny anteaters of Australia. These organisms are thought to have arisen on Gondwana, which is today's South America, Africa, and Australia (see the boxed essay on p. 523), and were then isolated as the continents drifted apart.

The *marsupials,* a second subclass of mammals, do not lay eggs but give birth to immature young. These blind, embryonic-looking creatures crawl to the mother's pouch and nurse until they are mature enough to venture out on their own. Like the monotremes, marsupials were present in Gondwana; today most marsupials, such as kangaroos, wombats, and koalas, are found in Australia. Interestingly, the marsupials of Australia resemble the placental mammals that are present on the other continents. Figure 24-16 compares individual members of these two sets of mammals. The members of each pair have similar habitats and find their food in similar ways. These characteristics, along with their strikingly similar anatomy, suggest that the evolution of these two subclasses of mammals is a product of **convergent evolution:** the development of similar structures having similar functions in different species as the result of the same kinds of selection pressures.

> *Mammals evolved from the reptiles approximately 200 million years ago. Changes in the structure and placement of their limbs, the structure of their heart, and reproductive strategies resulted in their ability to survive in a variety of climates and to eventually become abundant.*

In *placental mammals,* the young develop to maturity within the mother. They are named for an organ formed during the course of their embryonic development, the placenta. The placenta is located within the walls of the uterus, or womb. Composed of both maternal and fetal tissues, the placenta is connected to the fetus by the umbilical cord. At the placenta, fetal wastes pass into the bloodstream of the mother, and oxygen and nutrients in the mother's bloodstream pass into the bloodstream of

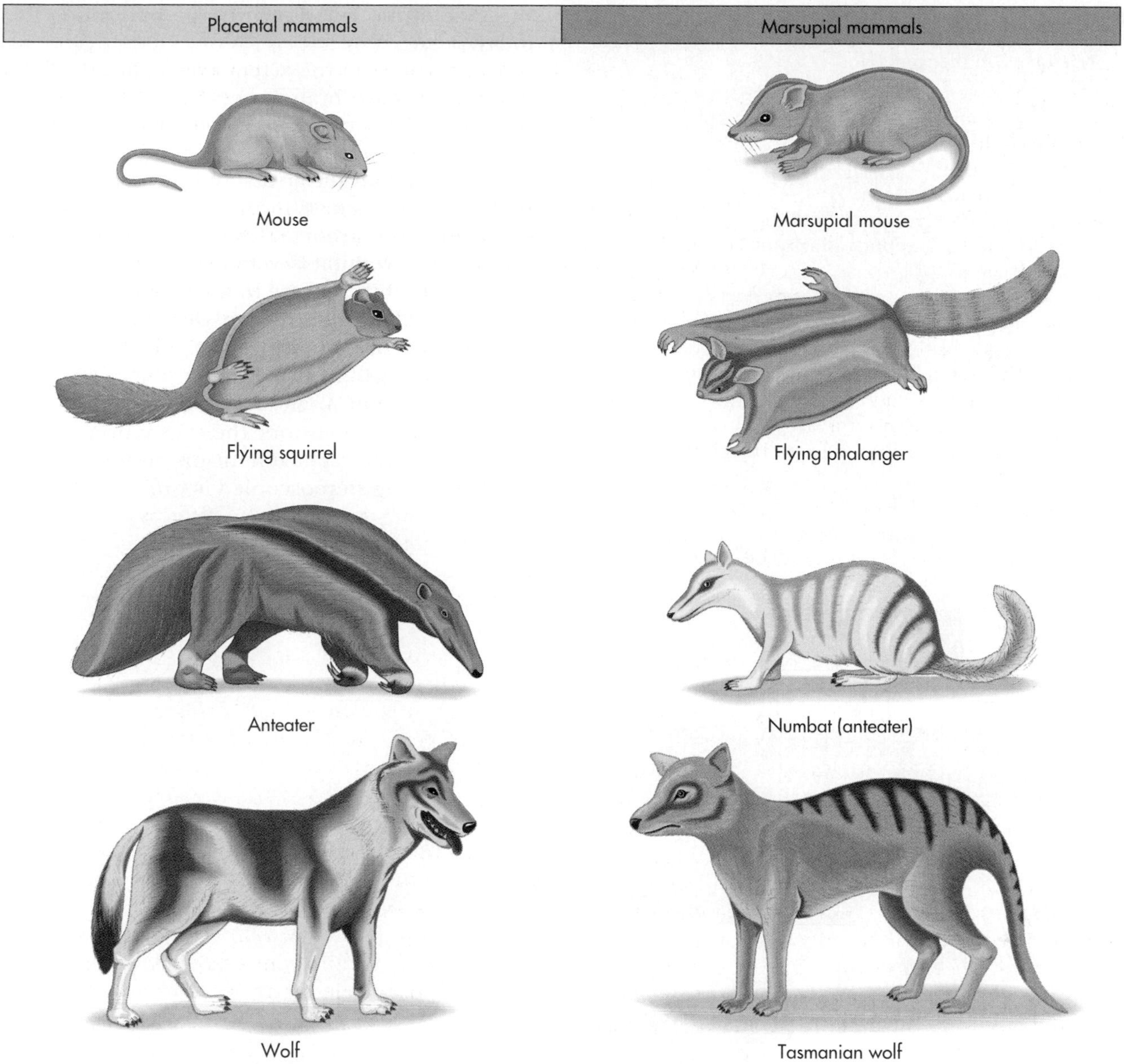

FIGURE 24-16 Convergent evolution. This illustration demonstrates the convergent evolution of marsupials in Australia and placental mammals in the rest of the world.

the fetus, a mechanism that allows the fetus to develop within the mother until it reaches a certain age of maturity.

Evolution of the Primates

There are 14 orders of placental mammals (see Chapter 1 for a discussion of classification), which include many animals familiar to you, such as dogs, cats, horses, whales, squirrels, rabbits, bats, and a variety of others. The order of placental mammals that humans belong to is the **primates.** Primates are mammals that have characteristics reflecting an arboreal, or tree-dwelling, lifestyle. Among these characteristics are hands and feet that are able to grasp objects (such as tree branches), flexible limbs, and a flexible spine. Table 24-2 lists primate characteristics. (Figure 24-17 diagrammatically represents

convergent evolution (kun **vur** junt ev uh **loo** shun) the development of similar structures having similar functions in different species as the result of the same kinds of selection pressures.
primates placental mammals that have characteristics reflecting an arboreal, or tree-dwelling, lifestyle.

TABLE 24-2

Primate Characteristics

- Ability to spread toes and fingers apart
- Opposable thumb (thumb can touch the tip of each finger)
- Nails instead of claws
- Omnivorous diet (teeth and digestive tract adapted to eating both plant and animal food)
- A semi-erect to an erect posture
- Binocular vision (overlap of visual fields of both eyes) resulting in depth perception
- Well-developed eye-hand coordination
- Bony sockets protecting eyes
- Flattened face without a snout
- A complex brain that is large in relation to body size

the taxonomic relationships of mammals living today.)

Some time in the Cretaceous period, the primates may have branched off from the shrewlike mammals. As thousands of years passed, primates' ancestors probably moved into the trees, eating insects that fed on the fruits and flowers growing on tree branches. Selection pressures must have been great for these arboreal creatures. Those that survived in this habitat developed excellent *depth perception,* the ability to see in more than one plane as they moved from tree to tree. Depth perception is a function of *stereoscopic vision,* vision created by two eyes focusing on the same object but from a slightly different angle. As primates evolved from their shrewlike ancestors, the eyes moved closer together from their placement on either side of the head, allowing stereoscopic vision.

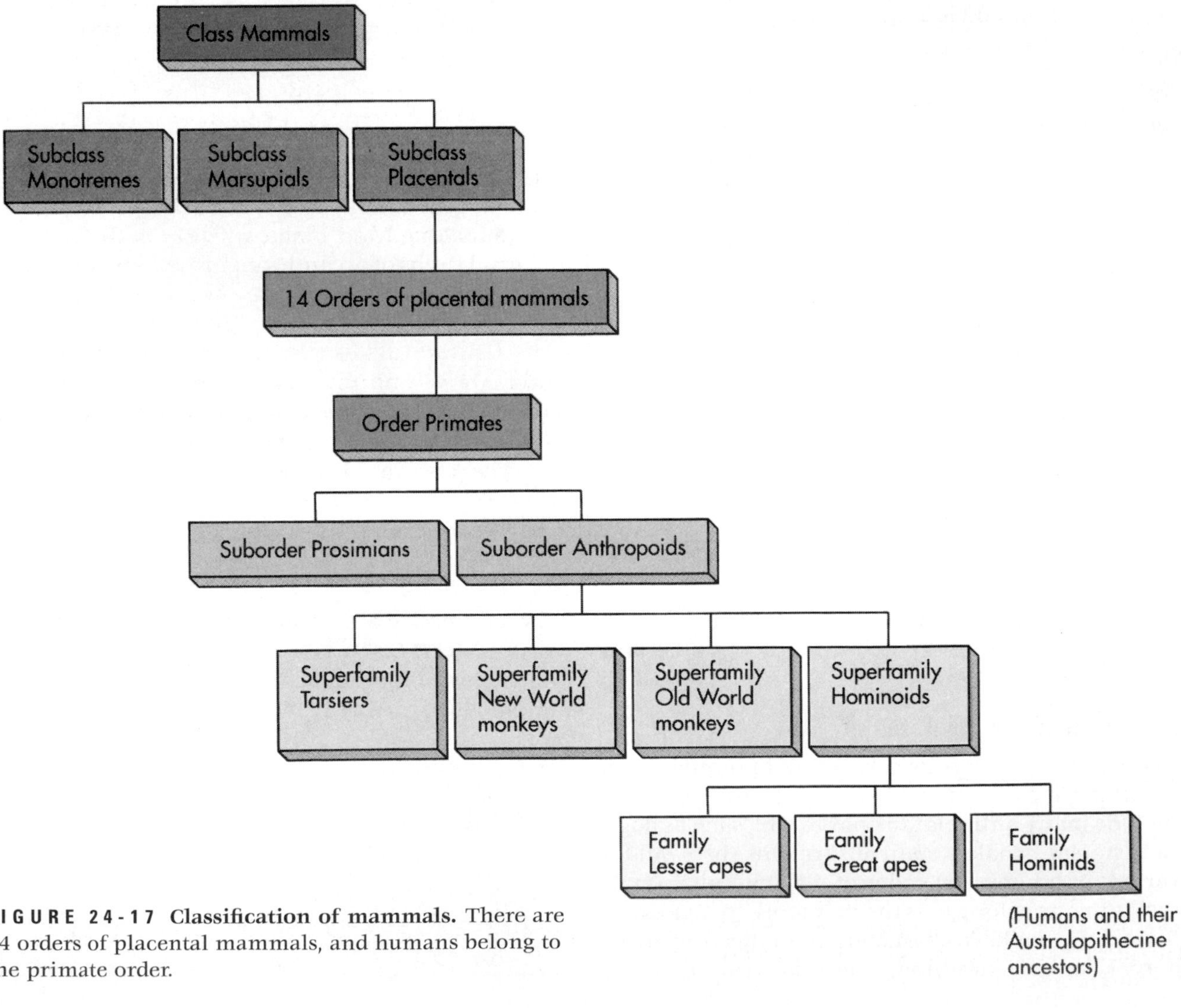

FIGURE 24-17 Classification of mammals. There are 14 orders of placental mammals, and humans belong to the primate order.

The primates also developed long limbs with flexible hands and feet adapted to grasping and swinging from branch to branch. Primates have two bones in the lower part of a limb that enable the wrists and ankles to rotate. In addition, the hands and feet of primates have digits that can be spread apart from one another, helping primates balance themselves when walking or running or helping them grasp objects. An *opposable thumb,* one that can touch the tip of each finger, helps in grasping. Most primates also developed flattened nails at the end of the digits, replacing the claws of their mammalian relatives.

The primates, an order of mammals that humans belong to, reflect an arboreal heritage. The earliest primates arose about 75 million years ago and lived on the ground, feeding on plants and insects. Over time, the primates moved into the trees, developing excellent depth perception and flexible, grasping hands.

The Prosimians

The primates are divided into two suborders: the **prosimians** and the **anthropoids** (the suborder that includes humans). The *prosimians* (meaning "before ape") are small animals such as lemurs, indris, aye-ayes, and lorises; they range in size from less than a pound to approximately 14 pounds—about the size of a cat or a small dog. Most are nocturnal and have large ears and eyes to help them see and hear at night. Prosimians have elongated snouts, reflecting their highly developed sense of smell. These characteristics are clearly seen in the prosimian pictured in Figure 24-18, in addition to its elongated rear limbs, which help it leap from tree to tree in its tropical rain forest habitat. By the end of the Eocene epoch (38 million years ago), prosimians were abundant in North America and Eurasia and were probably also present in Africa. However, their descendants now live only in the tropics of Asia, in tropical Africa, and on the island of Madagascar.

FIGURE 24-18 Prosimian. Prosimians are nocturnal and have large ears and eyes and an elongated snout that aids in their extraordinary sense of smell. The prosimian shown in the picture is a ringtail lemur, *Lemur catta.* All living lemurs are restricted to the island of Madagascar, and all are in danger of extinction as the rain forests are destroyed.

The Anthropoids

The *anthropoids* (meaning "humanlike") include the monkeys, apes, gorillas, chimpanzees, and humans. These primates differ from the prosimians in the structure of their teeth, brain, skull, and limbs. The prosimians have pointed molars and horizontal lower front teeth. These horizontal teeth are used to comb the coat or to get at food. The anthropoids have more rounded molars and no horizontal lower front teeth. The brain of the prosimian is much smaller in relation to body size than the brain of an anthropoid. The face of an anthropoid is somewhat flat. In addition, the eyes of an anthropoid are closer together than those of a prosimian. Lastly, a prosimian's front limbs are short in relation to its long hind limbs, whereas both the front and hind limbs of an anthropoid are long. Compare these features in the prosimians (see Figure 24-18) and the anthropoids (Figures 24-19 and 24-20).

In addition to these structural differences, prosimians and anthropoids also exhibit behavioral differences. Anthropoids are *diurnal,* that is, active during the day, whereas the prosimians are nocturnal. The anthropoids have evolved color vision, probably in relation to their diurnal existence. Also, the anthropoids live in groups in which complex social interaction occurs. In addition, they tend to care for their young for prolonged periods.

The **hominoids** (also meaning "humanlike") are one of four superfamilies of the anthropoid subor-

prosimians (pro **sim** ee uns) a suborder of primates that includes lemurs, indris, aye-ayes, and lorises; they are small animals, mostly nocturnal, with large ears and eyes, and elongated snouts and rear limbs.

anthropoids (**an** thruh poyds) a suborder of primates that includes the monkeys, apes, gorillas, chimpanzees, and humans; they differ from prosimians in the structure of the teeth, brain, skull, and limbs.

hominoids (**hom** uh noyds) one of four superfamilies of the anthropoid suborder. This superfamily includes apes and humans.

A

B

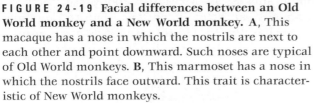

FIGURE 24-19 Facial differences between an Old World monkey and a New World monkey. A, This macaque has a nose in which the nostrils are next to each other and point downward. Such noses are typical of Old World monkeys. **B**, This marmoset has a nose in which the nostrils face outward. This trait is characteristic of New World monkeys.

der. This superfamily includes apes and humans. The other three anthropoid superfamilies include (1) tarsiers (formerly classified with the prosimians [Figure 24-21]), (2) New World monkeys, and (3) Old World monkeys. In general, New World monkeys are found in South and Central America, and they have flat noses with nostrils that face outward (Figure 24-19, *B*). Old World monkeys have noses similar to humans with nostrils next to each other that point downward (Figure 24-19, *A*); these monkeys are found in Africa, southern Asia, Japan, and Indonesia.

The hominoid superfamily is divided into three families: the lesser apes, the great apes, and the hominids, or humans. A variety of characteristics put the apes in a different superfamily than that of the monkeys: most apes are bigger than monkeys, have larger brains than monkeys, and lack tails.

The *lesser apes* include the gibbons, which are the smallest hominoids and closest in size to monkeys. They are about the size of a small dog, weighing 4 to 8 kilograms (9 to 18 pounds). Like all monkeys and apes, they live in tropical rain forests. Here, they leap and swing from tree to tree with their long arms.

The *great apes* include the orangutans, gorillas, and chimpanzees. These apes are much larger then the gibbons (see Figure 24-20). The orangutans are about the size of humans, weighing between 50 and 100 kilograms (110 to 220 pounds). These apes exhibit sexual dimorphism; the females weigh about half that of the males. Like the gibbons, the orangutans have long arms, but they walk—they do not swing—from branch to branch of the rain forest trees. The gorillas are the largest of the apes; they weigh about 160 kilograms, or 350 pounds. The females are smaller, ranging in weight from 165 to 240 pounds. Gorillas spend most of their time on the ground. When alarmed, male gorillas beat on their chests in a behavioral display. However, they are not the fierce animals humans think they are. In fact, gorillas are quite peaceable. The smallest of the great apes are the chimpanzees—humans' closest relatives. These animals weigh between 40 and 50 kilograms, or 90 to 110 pounds; females are slightly lighter. Like the other apes (except the gorillas), they spend their time in the trees.

A

B

C

D

FIGURE 24-20 The apes. A, Gorilla, *Gorilla gorilla*. **B**, Mueller gibbon, *Hylobates muelleri*. **C**, Chimpanzee, *Pan troglodytes*. **D**, Orangutan, *Pongo pygmaeus*.

FIGURE 24-21 Anthropoid. Tarsier, *Tarsius syrichta*, tropical Asia. Note the large eyes of the tarsier, an adaptation to nocturnal living.

The **hominids,** the family that includes humans of today, are the most intelligent of the hominoids. They are distinguished from the other families of hominoids in that they are *bipedal;* that is, they walk upright on two legs. In addition, hominids communicate by language and exhibit culture—a way of life that is passed on from one generation to another. Humans are the only living hominids.

Scientists have found jaw fragments that suggest that the ancestors of monkeys, apes, and humans began their evolution approximately 50 million years ago. The tarsioids may have begun their evolution even earlier—about 54 million years ago. Scientists do not know what adaptive pressures resulted in the appearance of these early anthropoid primates, nor do they know which prosimian was their ancestor.

Although the only fossil evidence of the emergence of the anthropoids consists of pieces of jaw, biochemical studies complement the knowledge gained from the fossil record. Together, these techniques tell scientists a great deal about the evolution of the anthropoids and the relationships among humans, apes, and monkeys. The evolutionary tree of the anthropoids is shown in Figure

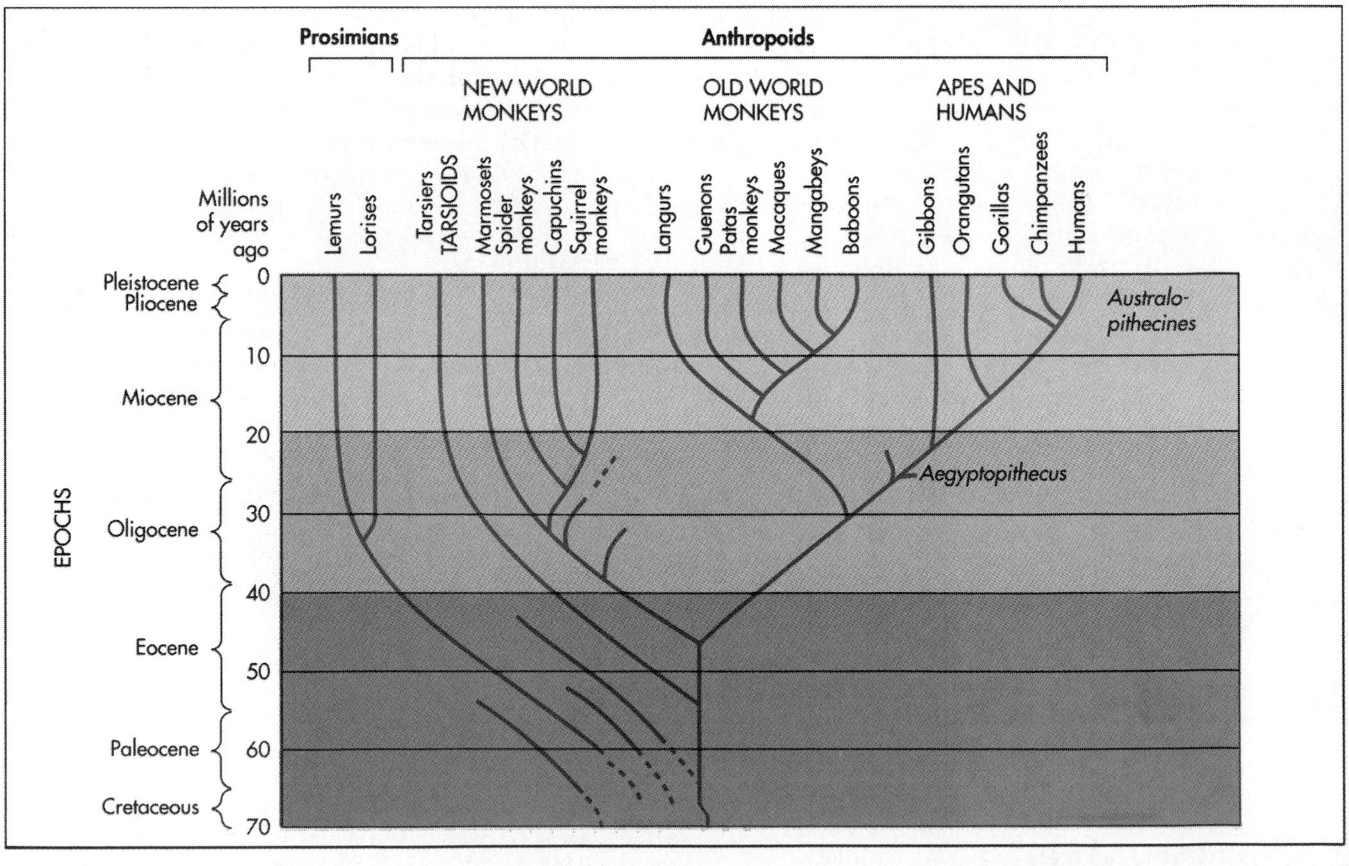

FIGURE 24-22 The primate family tree. This tree is based on DNA comparisons between living species. The dating of the branches is based on the ages of fossils.

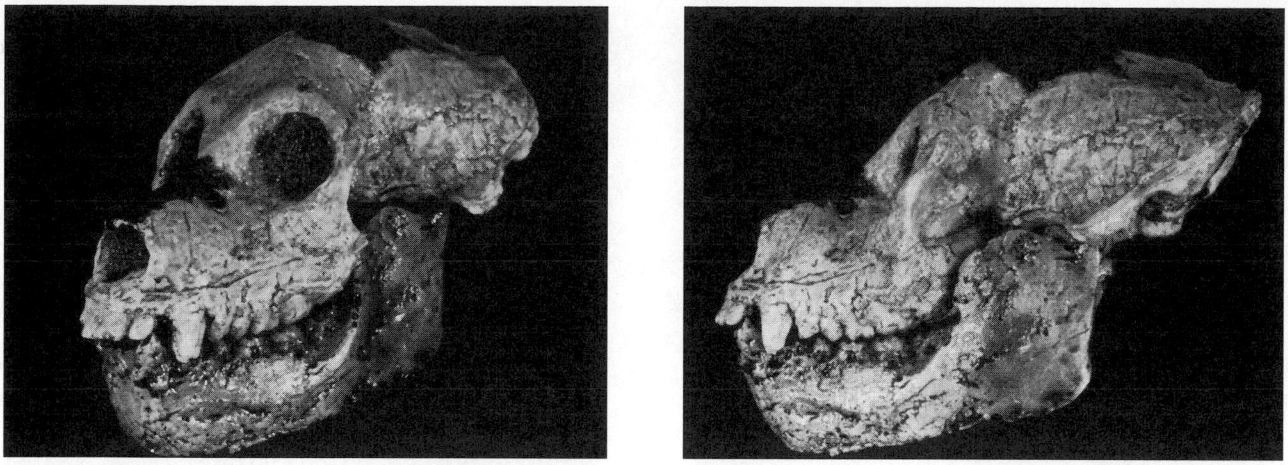

FIGURE 24-23 ***Aegyptopithecus zeuxis.*** This primate fossil dates to the late Oligocene era and is thought to be the ancestor of the hominoids that lived in Africa during the early Miocene era.

24-22. (See Chapter 23 for a discussion of evolutionary trees.) Fossils and biochemical studies show that the New World monkeys branched from the line leading to the Old World monkeys and the hominoids about 45 million years ago in the mid-Eocene epoch. In addition, scientists have discovered many fossils in North Africa that date from 25 to 30 million years ago in the Oligocene epoch. One of the earliest of these fossils, which scientists have named *Parapithecus,* probably led to the line of Old World monkeys, splitting from the hominoids approximately 30 million years ago. The others, scientists think, are probably members of the evolutionary line that leads to the hominoids.

The most well-known of the Oligocene fossils found in North Africa is called *Aegyptopithecus.* This fossil is from the late Oligocene epoch and is thought to be the ancestor to the early Miocene hominoids of Africa. Looking at the partially restored skull in Figure 24-23, you can see that *Aegyptopithecus* had some prosimian characteristics. It had a pronounced snout, leading scientists to believe that its sense of smell was still highly specialized, much like that of the prosimians. In addition, like the prosimians, *Aegyptopithecus* lived singly rather than in social groups.

> *Fossil evidence and biochemical studies show that the anthropoids, a suborder of primates that includes humans, monkeys, and apes, first appeared about 50 million years ago. The New World monkeys branched from the evolutionary line leading to the Old World monkeys and the hominoids about 45 million years ago. The Old World monkeys split from the hominoids about 30 million years ago.*

Fossil evidence is still being accumulated that will help tell the story of hominoid evolution. From mid-Miocene times on, the hominoid fossil record is quite extensive but consists primarily of skull and teeth fragments. Investigators have calculated that the evolutionary line leading to gibbons diverged from the line leading to the other apes about 18 to 22 million years ago, the line leading to orangutans split off roughly 13 to 16 million years ago, the line leading to gorillas diverged 8 to 10 million years ago, and the split between hominids and chimps occurred approximately 5 to 8 million years ago. This last statement suggests that chimpanzees and gorillas are humans' closest relatives, with a common relative alive 5 to 8 million years ago.

The Hominids

The First Hominids: Australopithecus

The two critical steps in the evolution of humans were the evolution of bipedalism and the enlargement of the brain. For many years, scientists have hypothesized how and when bipedalism arose. The current theory is that bipedalism arose as a *preadaptation* because of the way humans' arboreal, apelike ancestors moved through the trees. In other words, the skeletons and muscles of human ancestors were structured in a way that allowed these hominoids to walk bipedally even though they lived in the trees. Interestingly, these structural adaptations developed as a part of their arboreal life and the types of locomotion they exhibited in the trees.

> **hominids** (**hom** uh nids) the family of hominoids consisting of humans; the only living hominid is *Homo sapiens sapiens.*

FIGURE 24-24 Brachiation and bipedalism. Brachiators, such as gibbons and siamangs, locomote by hanging from branches with their arms and reaching from hold to hold. Brachiation has preadapted these animals to bipedal walking, which they do on broad branches (as the siamang in the illustration is doing) or on the ground.

Figure 24-24 shows the movements of gibbons and other related brachiators—those hominoids that move through trees by hanging from the branches with their arms, "walking" themselves along. As the weather in Africa cooled somewhat and became seasonal, patches of savannah grasslands began to invade former areas of tropical forest. The hominoids best able to walk efficiently on the ground survived, and bipedalism evolved.

Although hominids may have first appeared as long ago as 5 million years, the oldest undisputed evidence of the hominids is 3.6 to 3.8 million years old. This find was made relatively recently, in 1976, by Mary Leakey and an international team of scientists who found the fossil footprints shown in Figure 24-25. Anthropologists think that the individuals who made the footprints are human ancestors or are very closely related to human ancestors. Fossil bones discovered a few years earlier, at a site in Ethiopia about 2000 kilometers (1250 miles) north of the footprints, support this hypothesis. At 3.5 million years old, these hominid bones are somewhat younger than the footprints but are the oldest hominid bones ever found. These fossils were named *Australopithecus afarensis,* meaning "southern ape of Afar," by Donald Johanson and an international team of scientists who found them in the Afar region of Ethiopia. In 1974, this team also found and pieced together one of the most complete fossil skeletons of *A. afarensis,* which has since become famous (Figure 24-26). This "first" hominid was named Lucy after the Beatles song "Lucy in the Sky with Diamonds," which was playing on their tape machine at the time.

Although the australopithecines are considered to be the first hominids, they are not humans (members of the genus *Homo*). Their brains were still small in comparison to present-day human brains, and they had long, monkeylike arms. In addition, their faces were apelike as shown in the photo of a reconstruction of Lucy.

A. afarensis evolved into two (and possibly more) lineages, including the species *A. africanus, A. robustus,* and *A. boisei.* These australopithecines lived on the ground in the open savannah of eastern and southern Africa. Their diets consisted primarily of plants. At night they probably slept in the few trees that existed in these grasslands, much

FIGURE 24-25 Fossil footprints. These fossil footprints, found in the Afar region of Ethiopia, have been preserved in volcanic ash for over $3^1/_2$ million years. They were made by our nonhuman ancestors, *Australopithecus afarensis,* and give scientists important clues to our heritage. The differing sizes of the footprints may reflect our ancestors' sexual dimorphism, or size difference with gender. Also, the footprints show that these organisms walked erect on two legs instead of four, a characteristic called bipedalism. Bipedalism is only one of the many evolutionary changes that led to the appearance of modern humans.

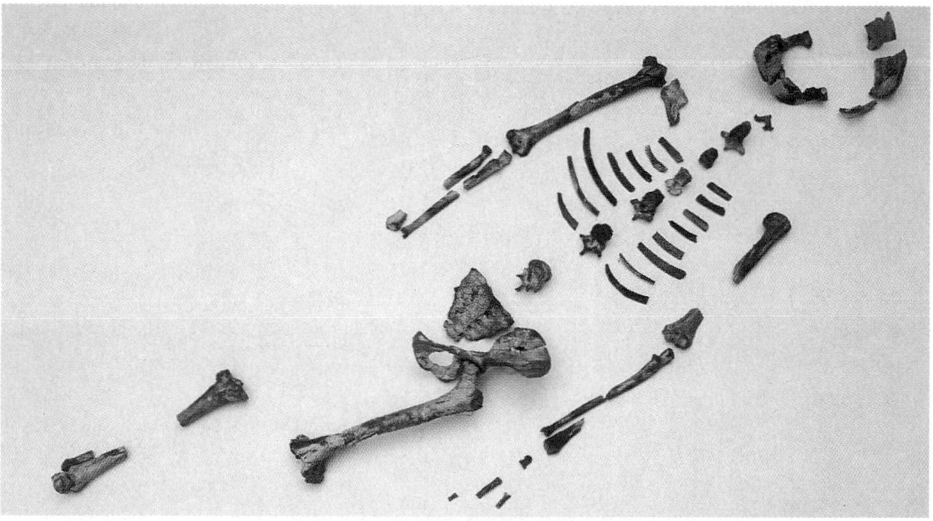

FIGURE 24-26 Lucy, from Ethiopia, is the most complete skeleton of *Australopithecus* discovered so far. The reconstruction was made by a careful study of muscle attachments to the skull.

like the savannah baboons do today to protect themselves from predators.

> *The oldest evidence of the first appearance of the hominids is 3.6 to 3.8 million-year-old footprints. The oldest hominid fossil skeleton is 3.5 million years old and is not classified as human. This hominid is of the species* Australopithecus afarensis *and is nicknamed Lucy.*

The evolutionary relationships among these australopithecines is not clear; a family tree that is widely accepted at this time is shown in Figure 24-27. Notice from the diagram that no australopithecines are alive today; the last ones disappeared about 1 million years ago.

The First Humans: Homo habilis

Climatic changes during the Pleistocene epoch (1.8 million to 12,000 years ago) may have contributed to the disappearance of the australopithecines and the evolution of a new, more intelligent genus of hominids: the human (genus *Homo*). During the Pleistocene epoch, the Earth cooled, repeatedly undergoing *ice ages* during which vast regions of the Earth (from the poles to a latitude of about 40 degrees) were covered by massive glaciers. For example, ice covered land from the north pole southward over Canada and parts of the northern United States. Organisms such as the australopithecines had a low level of intelligence and no innate cold weather survival behaviors, and they died. Apparently, however, a species of hominid more intelligent than the australopithecines was evolving from

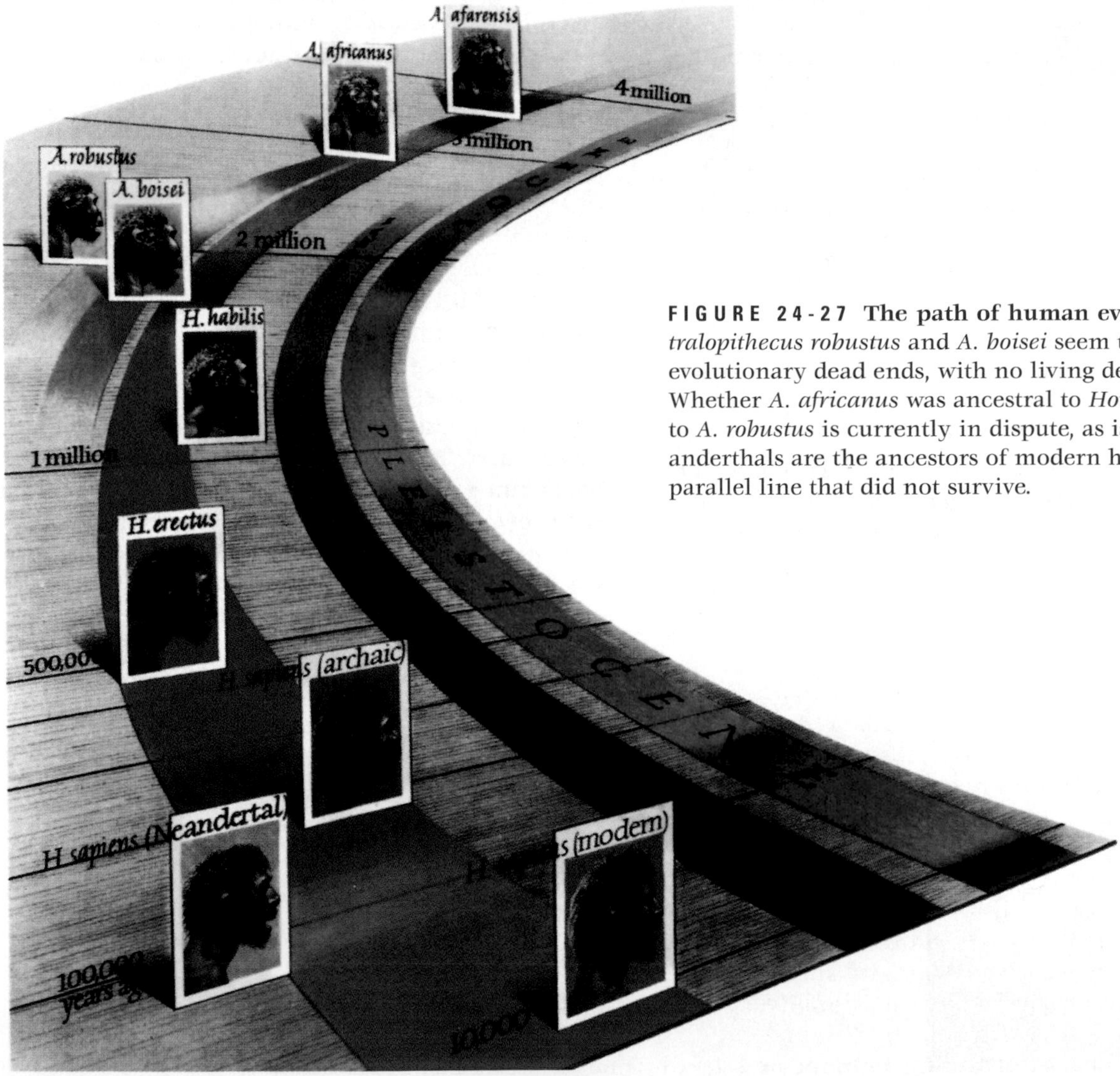

FIGURE 24-27 The path of human evolution. *Australopithecus robustus* and *A. boisei* seem to represent evolutionary dead ends, with no living descendants. Whether *A. africanus* was ancestral to *Homo habilis* or to *A. robustus* is currently in dispute, as is whether Neanderthals are the ancestors of modern humans on a parallel line that did not survive.

A. afarensis. These hominids protected themselves from the cold by building shelters and wearing clothes. They are thought to have been the first humans and were given a name to emphasize their intelligence—***Homo habilis,*** or "skillful human." Scientists have fossil bones and tools of *H. habilis* dating back about 2 million years. The first *H. habilis* fossils were discovered in 1964 by Louis Leakey, Philip Tobias, and John Napier in the Olduvai Gorge of eastern Africa.

Judging from the structure of the hands, *H. habilis* regularly climbed trees as did their australopithecine ancestors, although the *H. habilis* people spent much of their time on the ground and walked erect on two legs. Skeletons found in 1987 indicate that the *H. habilis* people were small in stature like the australopithecines, but fossil skulls and teeth reveal that the diet of *H. habilis* was more diverse than that of the australopithecines, including meat as well as plants. The tools found with *H. habilis* were made from stones fashioned into implements for chopping, cutting, and pounding food. The use of stone tools by these early humans marks the beginning of the *Stone Age,* a time that spans approximately 2 million to 35,000 years ago.

> *The first human fossils date back about 2 million years and are of the extinct species* H. habilis, *meaning "skillful human." This species is considered human because by making tools and clothing, they exhibited an intelligence far greater than their ancestors.*

Human Evolution Continues: Homo erectus

All of the early evolution of the genus *Homo* seems to have taken place in Africa. There, fossils belonging to the second, also extinct species of *Homo*—***Homo erectus***—are widespread and abundant from 1.6 million to about 300,000 years ago. By 1 million years ago, however, *H. erectus* had migrated into Asia and Europe.

The *H. erectus* people were about the size of modern-day humans, were fully adapted to upright walking, and had brains that were roughly twice as large as those of their ancestors. However, they still retained prominent brow ridges, rounded jaws, and large teeth. The tools of *H. erectus* were much more sophisticated than those of *H. habilis* and were used for hunting, skinning, and butchering animals. These people were hunter-gatherers, which means that they collected plants, small animals, and insects for food while occasionally hunting large mammals. Researchers have found the first evidence of the use of fire by humans at 1.4 million-year-old campsites of *H. erectus* in the Rift Valley of Kenya, Africa. Fire is characteristically associated with populations of this species from that time onward. All of these activities (tool making, hunting, using fire, and building shelters) are signs of *culture,* or a way of life that depends on intelligence and the ability to communicate knowledge of the culture to succeeding generations. Inherent in the concept of culture, then, is concept of language. The ability to communicate enhances the ability of a species to survive, especially in harsh conditions such as an ice age, by sharing survival tactics and warning each other of danger. Anthropologists think that the development of language was probably one of the most important factors in the appearance of *Homo sapiens.*

> *Fossils of a second extinct species of human,* H. erectus, *date back to 1.6 million years ago. These humans made sophisticated tools, built shelters, used fire, and probably communicated with language.*

Modern Humans: Homo sapiens

The earliest fossils of ***Homo sapiens*** (meaning "wise humans") are about 200,000 years old. They most likely evolved from the *H. erectus* species in Africa, although some scientists contend they evolved simultaneously from *H. erectus* populations in Asia, Europe, and Africa. The oldest *H. sapiens* are not considered to be anatomically modern, that is, to have the same anatomical features of today's humans. This species is therefore referred to as an early or archaic form (see Figure 24-27). In general, these early *H. sapiens* had larger brains, flatter heads, more sloping foreheads, and more protruding brow ridges and faces than today's humans.

The fossil record shows gradual change of the species *H. sapiens,* with the early form evolving

Homo habilis (**hoe** moe **hab** uh lus) an extinct species of hominids whose fossil record dates back about 2 million years; members of this species are thought to have been the first humans and have been given a name that emphasizes their intelligence (*Homo habilis* means "skillful human").

Homo erectus (**hoe** moe ih **rek** tus) a second extinct species of hominids whose fossil record dates back 1.6 million years; members of this species were fully adapted to upright walking, made sophisticated tools, and probably communicated with language.

Homo sapiens (**hoe** moe **say** pee unz) an early or archaic form of modern humans (*Homo sapiens sapiens*) that are extinct. Their fossil record dates back about 200,000 years.

over a 75,000-year span to a subspecies of *H. sapiens* called **Neanderthal.** This subspecies was named after the Neander Valley in Germany where their fossils were first found. The Neanderthals lived from about 125,000 to 35,000 years ago in Europe and the Middle East.

Compared with modern humans, the Neanderthal people were powerfully built, short, and stocky. Their skulls were massive, with protruding faces, projecting noses, and rather heavy bony ridges over the brows. Their brains were even larger than those of modern humans, a fact that may have been related to their heavy, large bodies. The Neanderthals made diverse tools, including scrapers, borers, spear heads, and hand axes. Some of these tools were used for scraping hides, which they used for clothing. They lived in hutlike structures or in caves. Neanderthals took care of their injured and sick and commonly buried their dead, often placing food and weapons and perhaps even flowers with the bodies. Such attention to the dead strongly suggests that they believed in a life after death. For the first time, the kinds of thought processes characteristic of modern *H. sapiens,* including symbolic thought, are evident in these acts.

Approximately 10,000 years before the Neanderthal subspecies died out, the "modern" subspecies of *H. sapiens* made their appearance. This modern subspecies (our subspecies) is called ***Homo sapiens sapiens.*** The early members of this subspecies are the Cro-Magnons and are named after a cave in southwestern France where scientists found some of their fossils.

FIGURE 24-28 Cave painting. Cave paintings, almost always showing animals and sometimes hunters, were made by Cro-Magnon people, our immediate ancestors. These paintings are found primarily in Europe and were made for about 20,000 years, until 8000 to 10,000 years ago.

The **Cro-Magnons** had a stocky build, much like the Neanderthals, but their heads, brow ridges, teeth, jaws, and faces were much smaller than the Neanderthals and were more similar to today's humans. However, just as modern humans show variation among races, so too did the Cro-Magnons. The Cro-Magnons used sophisticated tools that were

FIGURE 24-29 The beginnings of civilization. This photo shows the remains of a house in Jericho, which dates to about 7000 B.C. The ruins of Jericho also contain the remnants of city walls and towers, demonstrating that by about 9000 years ago, many of our ancestors had moved away from the hunting-gathering lifestyle into an agricultural lifestyle.

made not only from stone but from bone, ivory, and antler—materials that were not used by earlier peoples. Hunting was an important activity for the Cro-Magnons, evidenced by the abundance of animal bones found with human bones and elaborate cave paintings of animals and hunt scenes (Figure 24-28). The paintings appear to have been part of a ritual to ensure the success of the hunt.

The subspecies of *H. sapiens* that preceded us showed a gradual development of culture and society, which was the foundation for the development of "modern" culture and society. About 10,000 years ago, the last ice age came to a close, the global climate began to warm, and various groups of *H. sapiens sapiens* began to cultivate crops and breed animals for food. Archeologists have uncovered the remains of small, ancient cities, such as those of Jericho shown in Figure 24-29, which give evidence that by 9000 years ago humans had developed complex social structures. By 5000 years ago the first large cities and great civilizations appeared, such as those in Egypt (3100-1090 BC) and Mesopotamia (3100-1200 BC). The final break occurred with the hunter-gatherer way of life.

Neanderthal (nee **an** dur thol *or* nee **an** dur tal) a subspecies of *Homo sapiens* named after the Neander Valley in Germany, where their fossils were first found. Neanderthals were short and powerfully built, with large brains; they made diverse tools, took care of the sick and injured, and buried their dead.

Homo sapiens sapiens modern humans; the subspecies of hominids who made their appearance 10,000 years before the Neanderthal subspecies died out, and whose early members are called Cro-Magnons.

Cro-Magnons (**kro-mag** nuns) early members of *H. sapiens sapiens* whose anatomical features were similar to modern humans; they made sophisticated tools, hunted, and made elaborate cave paintings.

Summary

- Approximately 15 billion years ago, scientists theorize, all the material of the universe condensed into a single small space and then exploded in an event called the "big bang," forming an early generation of stars. Then, approximately 4.6 billion years ago, these stars exploded, forming the Earth and the other planets of the solar system.
- Somehow, in the 300 million years between the time when the Earth cooled and the first bacteria appeared, biochemical evolution became biological evolution. The long-held primordial soup theory states that life arose in the sea as elements and simple compounds of the primitive atmosphere reacted with one another to form simple organic molecules such as amino acids and sugars. Recently, scientists have begun to reexamine the premises of this theory and have proposed alternative hypotheses. After the synthesis of complex organic molecules had taken place, self-replicating systems of RNA molecules may have begun the process of biological evolution. To be considered life, however, this "early" genetic material must have been organized within a cellular structure.
- Scientists speculate that the first cells to evolve fed on organic materials in the environment. Because of limited organic material, the evolution of a pigment system that could capture the energy from sunlight and store it in chemical bonds most likely evolved relatively quickly. Then, oxygen-producing bacteria evolved, gradually oxygenating the atmosphere.
- The first unicellular eukaryotes appeared about 1.4 billion years ago during the Proterozoic era. Eukaryotic cells likely arose as "preeukaryotic" cells became hosts to endosymbiotic prokaryotes. Multicellular organisms first appeared about 630 million years ago.
- Many of the phyla of organisms in existence today, except the chordates and the land plants, appear to have evolved during the Cambrian period (544 to 505 million years ago) of the Paleozoic era. These organisms evolved in the sea. Chordates and then vertebrates evolved later in the Paleozoic era, with the first vertebrates appearing about 430 million years ago.
- Terrestrial plants appeared about 410 million years ago and flourished during the Carboniferous period (370 to 285 million years ago). Fishes became abundant and diverse during the Devonian period (410 to 370 million years ago). The amphibians, organisms that spend part of their time on land and part in the water, evolved from the fishes, appearing about 360 million years ago. Then, about 300 million years ago, the amphibians gave rise to the first reptiles.
- A major extinction event occurred during the Permian period, wiping out many of the species living at that time. The Mesozoic era (250 to 65 million years ago) was a time of the adaptive radiation of the organisms surviving the Permian extinction, dominated by the evolution of the reptiles. The reptiles gave rise to the mammals (about 200 million years ago) and the birds (at least 150 million years ago).
- Mammals are warm blooded, or able to maintain a constant internal body temperature. Other than a few organisms such as the birds, all other living animals are cold blooded; their body temperatures vary with the temperature of the environment. Changes in the structure and placement of the mammalian limbs, the structure of the heart, and reproductive strategies resulted in mammals' ability to survive in a variety of climates and to eventually become abundant.
- Primates, one of the 14 orders of mammals, first appeared 75 million years ago. Primates have large brains in proportion to their bodies, binocular vision, and five digits, including an opposable thumb; they exhibit complex social interactions.
- The primates are divided into two suborders: the

prosimians (small animals such as lemurs, indris, and aye-ayes) and the anthropoids (a group that includes monkeys, apes, and humans).

- The hominoids (meaning "humanlike") are one of four superfamilies of the anthropoid suborder. This superfamily includes apes and humans. The other three anthropoid superfamilies include tarsiers, New World monkeys, and Old World monkeys. The New World monkeys branched from the evolutionary line leading to the Old World monkeys and the hominoids about 45 million years ago. The Old World monkeys split from the hominoids approximately 30 million years ago. Ancestors to the apes gave rise to the gibbons, orangutans, chimpanzees, gorillas, and hominids.
- The two critical steps in the evolution of humans were the evolution of bipedalism (walking on two feet) and the enlargement of the brain. The earliest hominids and the direct ancestors of humans belong to the genus *Australopithecus.* They appeared in Africa about 5 million years ago.
- The genus *Australopithecus* gave rise to humans belonging to the genus *Homo.* The first species of this genus, *H. habilis,* appeared in Africa about 2 million years ago. Now extinct, the people of this species are considered human because they exhibited an intelligence far greater than their ancestors by making tools and clothing.
- The second species of *Homo, H. erectus,* appeared in Africa approximately 1.6 million years ago. These people used fire, built shelters, fashioned sophisticated tools, and exhibited culture. *H. sapiens* probably evolved from *H. erectus* about 200,000 years ago.

Knowledge and Comprehension Questions

1. What is the primordial soup theory, and what does it attempt to explain? Describe another theory that explains the same event.
2. Explain the importance of a cell membrane to the evolution of early cells.
3. Create a table that lists the major eras and periods of geological time.
4. What are cyanobacteria, and when did they first appear? Explain their evolutionary significance.
5. When was the Cambrian period, and why was it important?
6. Describe the major events of the Carboniferous period. What were the living conditions on Earth during this time?
7. Summarize the events of the Cenozoic era. When did it begin?
8. Describe the three features that distinguish chordates from other organisms.
9. What do monotremes, placental mammals, and marsupials have in common? How do they differ?
10. What are primates? What two characteristics have helped them to be successful?
11. Distinguish between the prosimians and the anthropoids. Give an example of each.
12. Which statement is true? (Or are they both false?) Explain your answer:
    a. All hominids are hominoids.
    b. All hominoids are hominids.
13. How did *Homo sapiens sapiens* differ from earlier *Homo sapiens?*

Critical Thinking

1. Many movies have shown early humans battling ferocious dinosaurs. Is this correct? Explain.

PART SIX

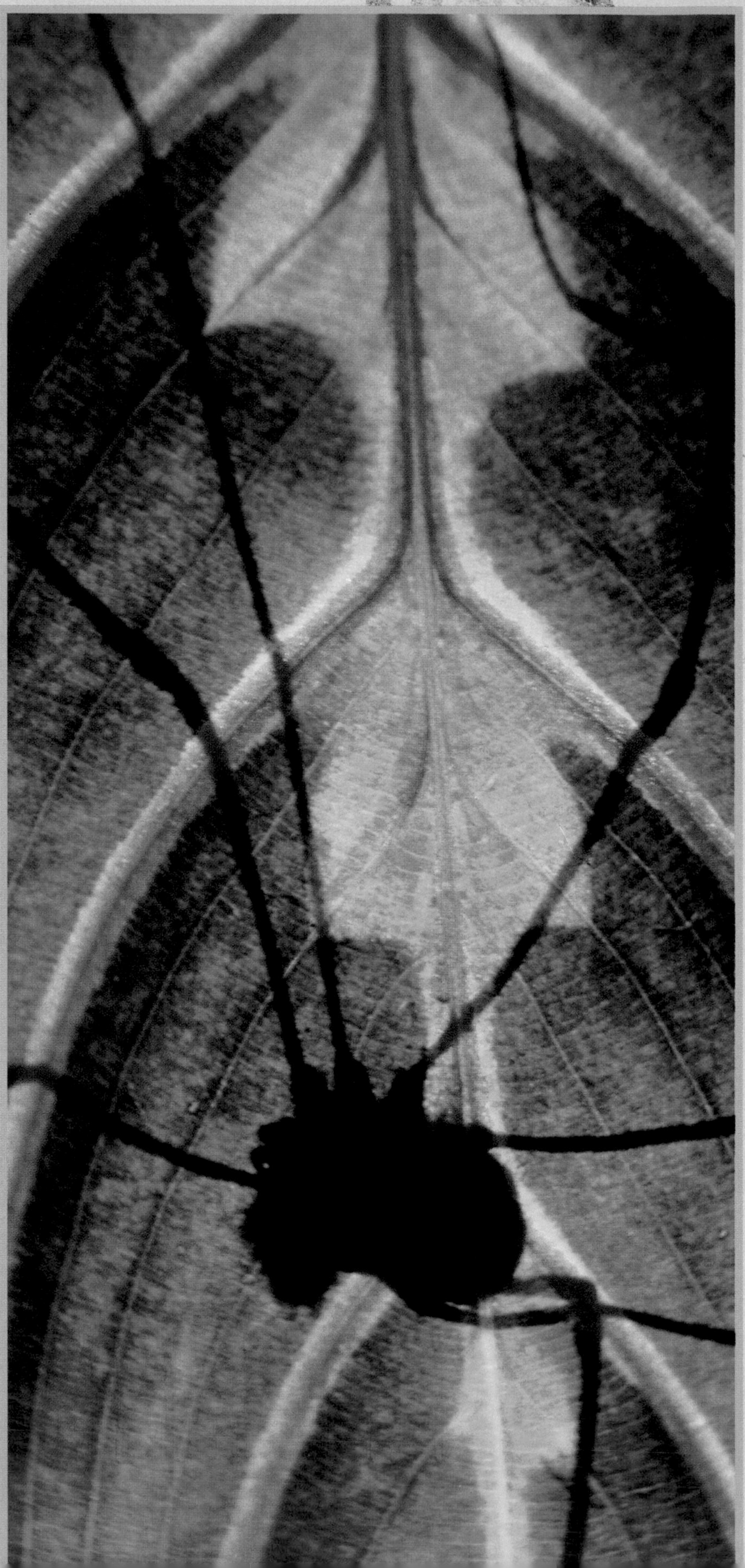

THE UNITY AND DIVERSITY OF ORGANISMS

CHAPTER 25

VIRUSES AND BACTERIA

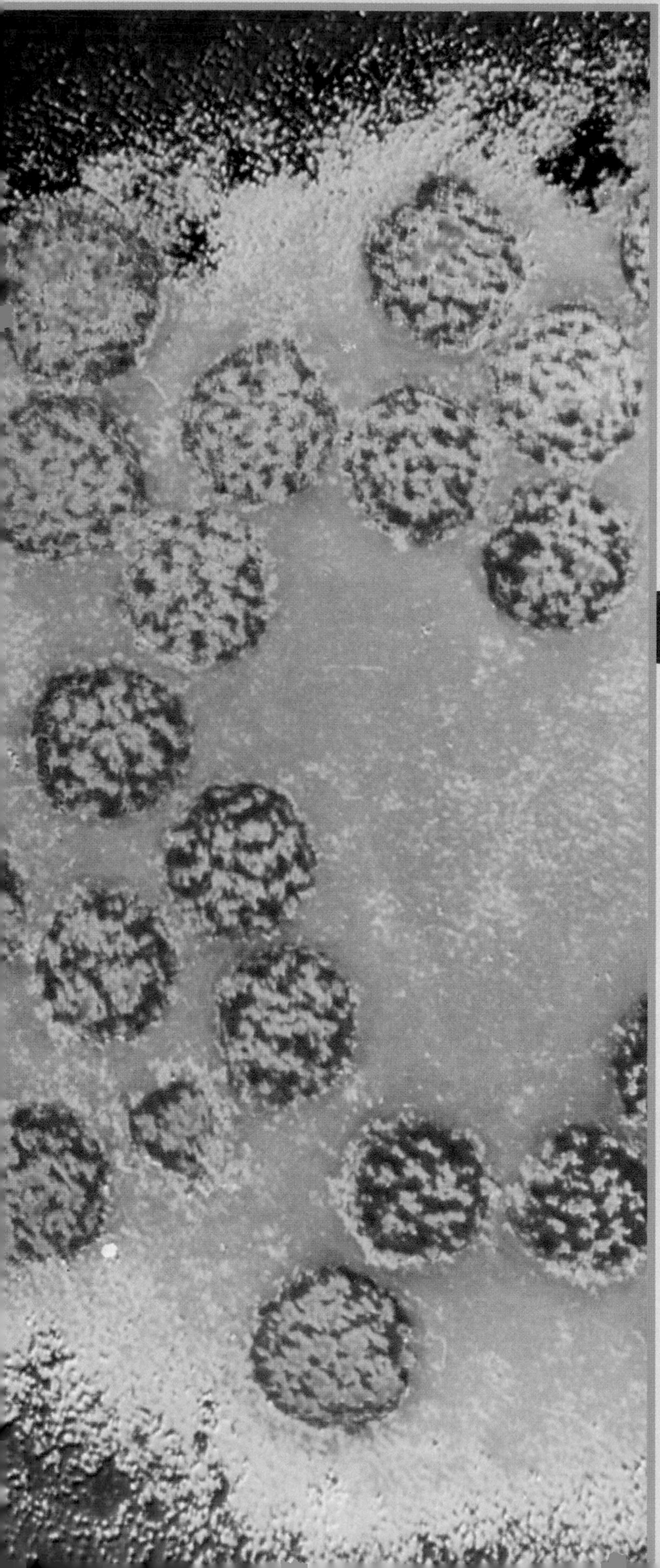

THE STATISTICS ARE ASTOUNDING! A recent study revealed that nearly one out of every four college women tested in Seattle, Washington, is harboring a sexually transmitted virus that is strongly associated with cancer of the cervix. Computer color-enhanced, these deadly virus particles (formally known as *human papillomavirus 16*) are shown in the photo as green spheres speckled with red. The human cell they are destroying is yellow.

The human papillomavirus (HPV) causes a sexually transmitted disease commonly known as *genital warts.* HPV comes in 60 different varieties. Infection with any one of the 60 types can lead to an outbreak of warts, although some are more likely to cause warts than others. Ironically, the types that most often result in warts are the least dangerous. Infection with HPV types such as HPV 16 rarely cause warts on infection but are those most likely to be associated with cancer and the least likely to be detected.

Two other viral diseases transmitted by sexual contact are genital herpes and acquired immunodeficiency syndrome (AIDS). Unlike sexually transmitted diseases (STDs) spread by bacteria, viral STDs cannot be cured. The drugs available to treat these diseases only help ease the symptoms but cannot destroy the viruses. Why are these diseases different in that regard? What makes a virus so hard to control?

KEY CONCEPTS

- Viruses are noncellular infectious agents capable of causing certain diseases and cancers.
- Three sexually transmitted diseases caused by viruses are genital herpes, genital warts, and acquired immunodeficiency syndrome.
- Bacteria have a prokaryotic cell structure, occur in many habitats, play key ecological roles, are important in producing food, and can cause disease.
- The sexually transmitted diseases caused by bacteria include syphilis, gonorrhea, and nongonococcal urethritis (NGU).

OUTLINE

Viruses

Viruses are *infectious agents.* They enter living organisms and cause disease. But although viruses invade living things and cause cells to make more viruses, the viruses themselves are not living! They do not have a cellular structure, which is the basis of all life. They are nonliving *obligate parasites,* which means that viruses cannot reproduce outside of a living system. They must exist in association with and at the expense of other organisms. Unfortunately, that "other organism" may be you!

The Discovery of Viruses

At the end of the nineteenth century, several groups of European scientists working independently first realized that viruses existed. As they filtered fluids derived from plants with tobacco mosaic disease and cattle with hoof-and-mouth disease, the scientists discovered that the infectious agents passed right through the fine-pored filters they used, which were designed to hold back bacteria. They concluded that the infectious agents associated with these diseases were *not* bacteria—they were too small. As they studied the filtrate containing these mysterious agents, the scientists also discovered that the disease-causing agents could multiply only within living cells. These infection agents, they hypothesized, must lack some of the critical "machinery" cells use to reproduce.

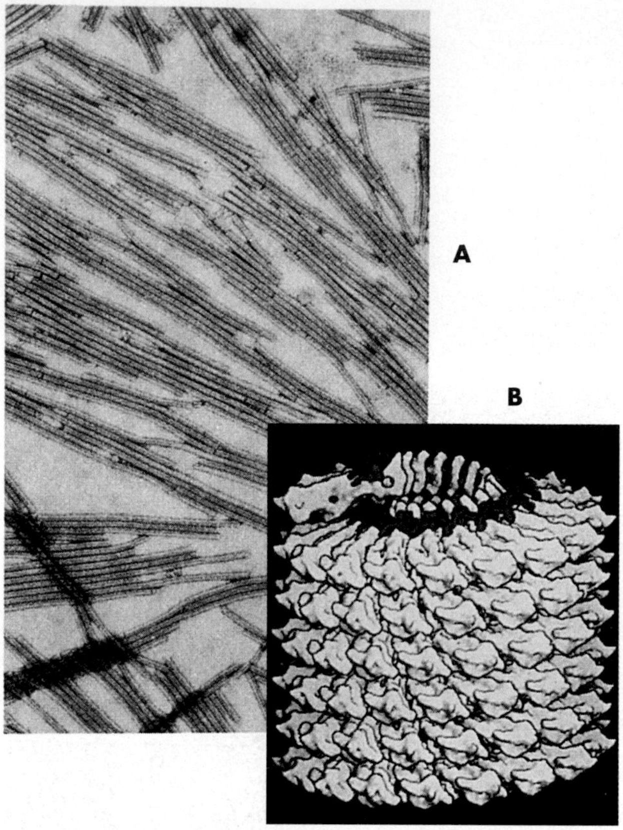

FIGURE 25-1 Tobacco mosaic virus. A, An electron micrograph of purified tobacco mosaic virus. **B,** Computer-generated model of a portion of tobacco mosaic virus. An entire virus consists of 2130 identical protein molecules—the yellow knobs—which form a cylindrical coat around a single strand of RNA *(red)*.

For many years after their discovery, viruses were regarded as very primitive forms of life, perhaps the ancestors of bacteria. Today, scientists know that this view is incorrect—viruses are not living organisms. The true nature of viruses became evident in the 1930s after the groundbreaking work of an American scientist, Wendell Stanley. Stanley prepared an extract of tobacco mosaic virus (TMV), purified it, and studied its chemical composition. His conclusion: TMV was a protein—and he was partially right. Scientists later discovered that TMV also contains ribonucleic acid (RNA). In the late 1930s, with the development of the electron microscope, scientists were able to see the virus that Stanley purified (Figure 25-1, *A*).

The Structure of Viruses

Viruses infect primarily plants, animals, and bacteria. A specific virus can infect only a certain species. So you cannot be infected by a bacterial virus (bacterial viruses can only infect bacteria), nor can your dog catch your cold. Some viruses, however, can infect more than one species; for example, the human immunodeficiency virus (HIV virus that causes AIDS) is thought to have been introduced to humans from African monkeys.

Each virus has its own unique shape (Figure 25-2), but all contain the same basic parts: a nucleic acid **core** (either DNA or RNA) and a protein "overcoat" called a **capsid.** The structure of the TMV is shown in Figure 25-1, *B*, and illustrates one way that a virus is put together. This virus is helical, with its single strand of RNA coiled like a spring, surrounded by a spiraling capsid of protein molecules. Many viruses have another chemical layer over the capsid called the **envelope,** which is rich in proteins, lipids, and carbohydrate molecules. Figure 25-3, *A*, is an electron micrograph of a typical enveloped virus, the causative agent of herpes. Its structure is illustrated in Figure 25-3, *B*.

It is hard to conceptualize how small viruses are, but Figure 25-2 helps by showing the size of a few viruses relative to the size of a bacterium and a human red blood cell. Some viruses, such as the poliovirus, are as small as the width of the plasma membrane on a human cell. The largest viruses are barely visible with a light microscope. Most viruses can be seen only by using an electron microscope.

> *Viruses are made up of a nucleic acid core surrounded by a protein covering called a capsid. Some viruses also have an additional covering, or envelope.*

Viral Replication

Viruses cannot multiply on their own. They must enter a cell and use the cell's enzymes and ribosomes to make more viruses. This process of viral multiplication within cells is called *replication.* Various patterns of viral replication exist. Some viruses enter a cell, replicate, and then cause the cell to burst, releasing new viruses. This pattern of viral replication is called the **lytic cycle.** Other types of viruses enter into a long-term relationship with the cells they infect, their nucleic acid replicating as the cells multiply. This pattern of viral replication is called the **lysogenic cycle.**

The Lytic Cycle

The process of viral replication has been studied most extensively in bacteria because bacteria are easier to grow in the laboratory and to infect with viruses than are plant or animal cells. Many bacte-

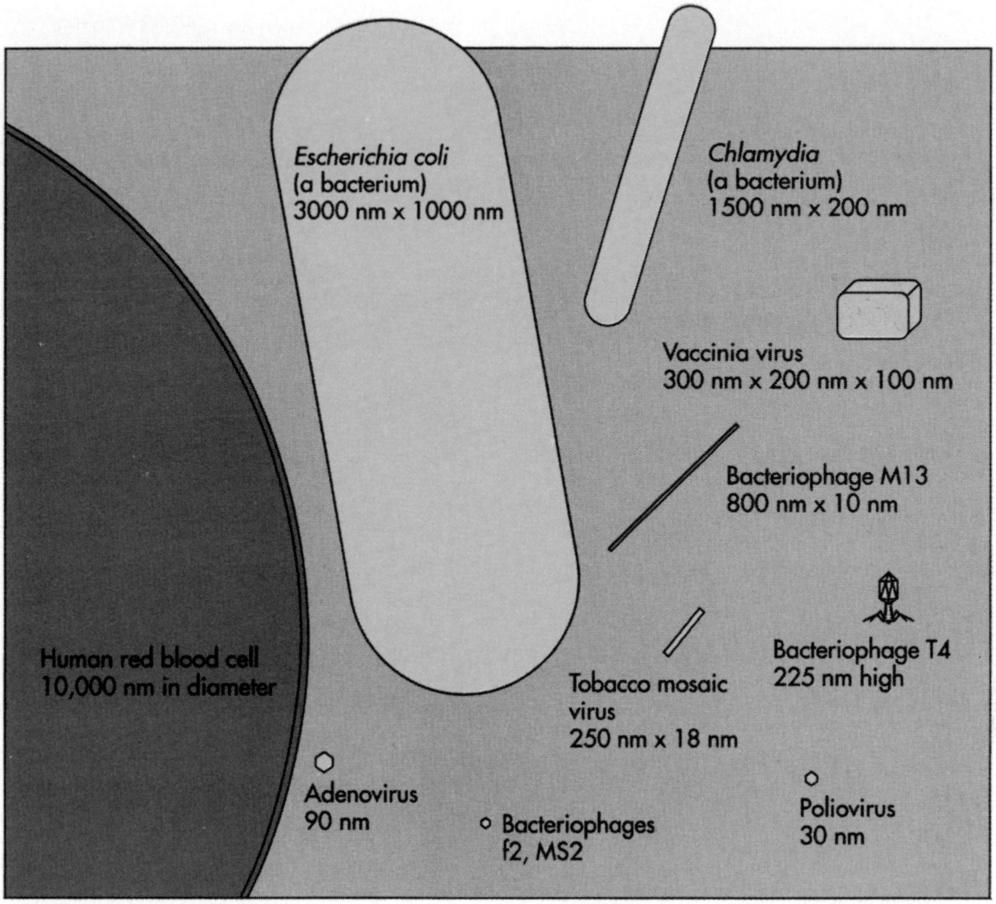

FIGURE 25-2 The shapes and sizes of viruses. The size of various viruses are shown here in relation to a bacterium and a human red blood cell. Dimensions are given in nanometers.

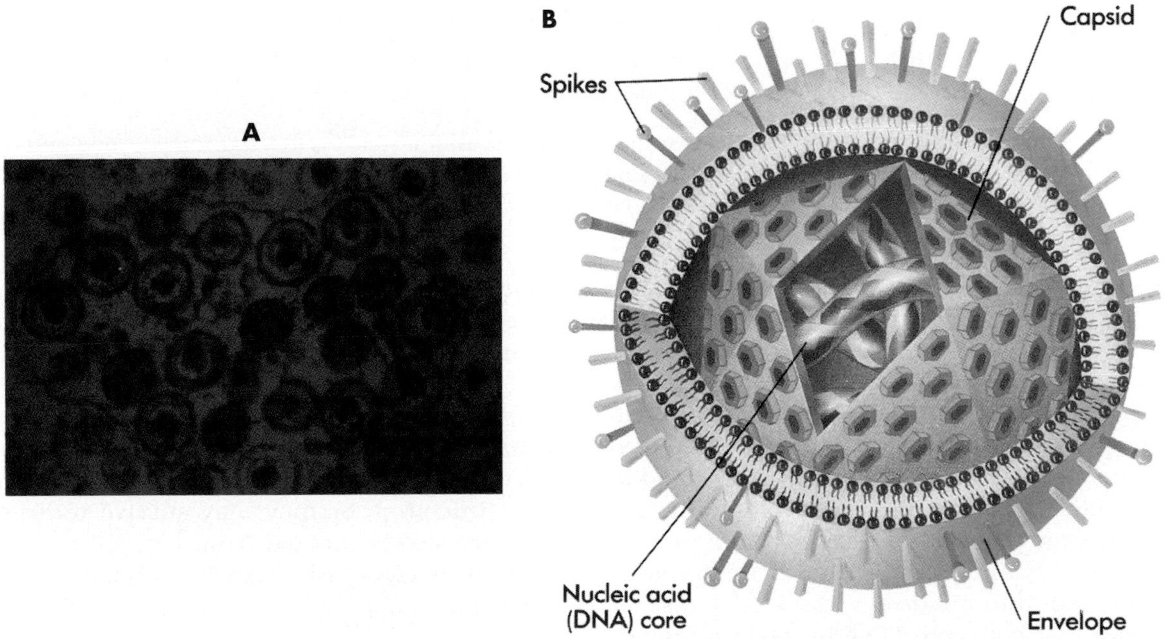

FIGURE 25-3 Structure of a typical virus. A, Electron micrograph of the herpesvirus. **B,** Structure of the herpesvirus.

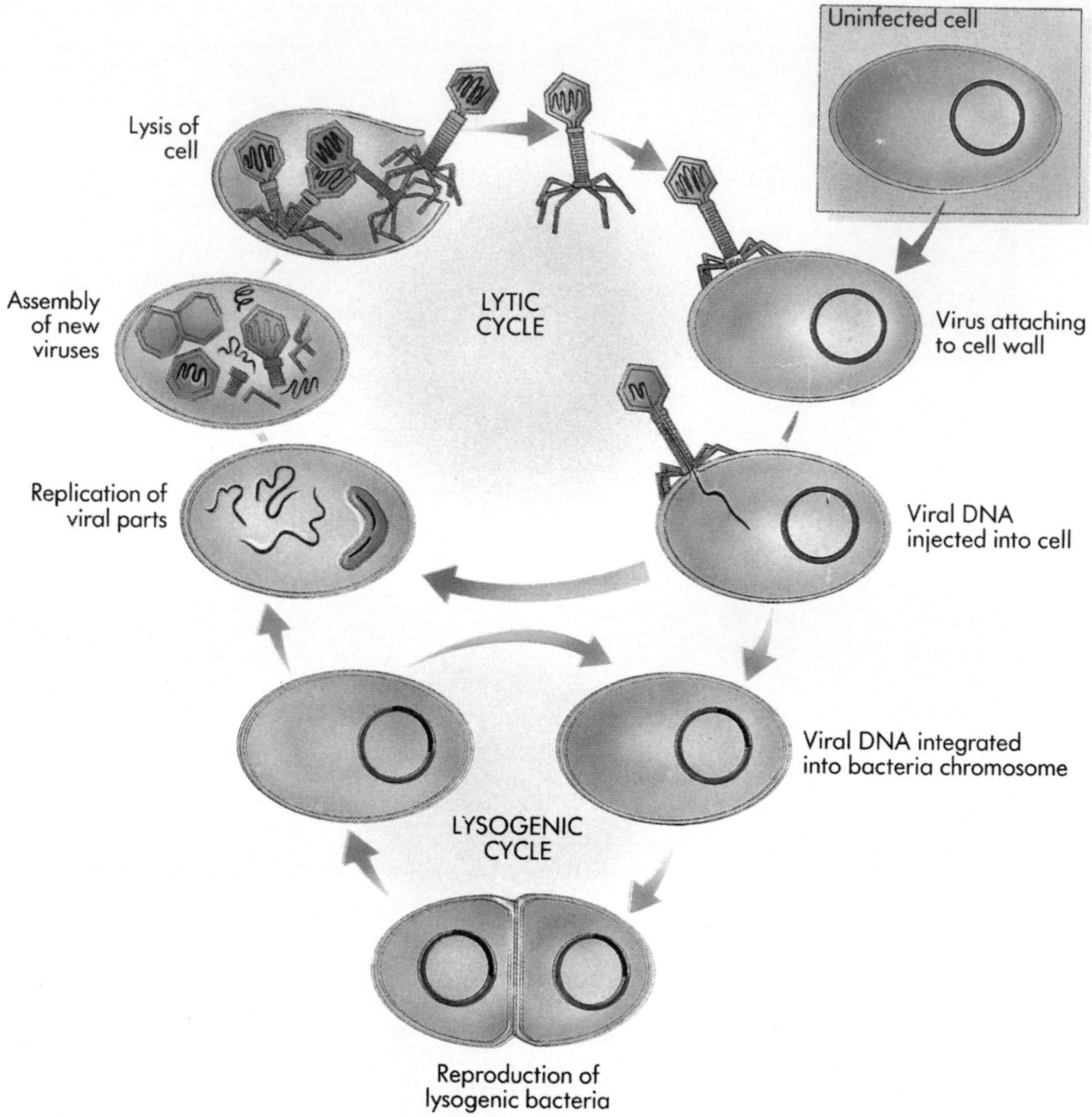

FIGURE 25-4 Viral replication: the lytic and lysogenic cycles. In the lytic cycle, viral nucleic acid enters a cell and causes it to burst, releasing new virus. In the lysogenic cycle the virus is integrated into the host cell, and the virus' nucleic acid is replicated as the host cell multiplies.

rial viruses (usually called *bacteriophages* or simply *phages*) follow a lytic cycle pattern of replication. As shown in Figure 25-4, a bacteriophage first attaches to a receptor site on a bacterium. The virus then injects its nucleic acid into the host cell while its protein capsid is left outside the cell. Next, the viral genes take over cellular processes and direct the bacterium to produce viral "parts" that will be used to assemble whole viruses. After their manufacture, these strands of nucleic acid and proteins are assembled into mature viruses that *lyse,* or break open, the host cell. Each bacterium releases many virus particles. Each "new" virus is capable of infecting another bacterial cell. Some animal viruses infect cells in a manner similar to bacterial virus infection, but they enter animal cells by endocytosis and must be uncoated before they can cause the cell to manufacture viruses. These cells may die as new virus particles are released in a lytic infection, or they may survive if virus particles are slowly budded from the cell by a process similar to exocytosis. This "slow budding" causes a persistent infection.

During a lytic pattern of viral infection, a bacterial virus injects its nucleic acid into a host cell, whereas an animal virus enters by endocytosis. The viral nucleic acid directs the cell to produce "new" viral nucleic acid and protein coats. After assembly of these parts, the bacterial virus particles cause the cell to burst open, releasing them. Animal viruses often leave the cell by slow budding.

Plant cells are somewhat protected from viral infection by their rigid cell walls and protective outer, waxy cuticles. Viruses can enter plants only if they are damaged or if other organisms, such as sucking insects or fungi, assist them.

The Lysogenic Cycle

Some viruses, instead of killing host cells, integrate their genetic material with that of the host. Then, each time the host cell reproduces, the viral nucleic acid is replicated as if it were a part of the cell's genetic makeup (see Figure 25-4). In this way, the virus is passed on from cell to cell. Infection of this sort is called a *latent infection.* These integrated latent genes may not cause any change in the host for a long time. Then, triggered by an appropriate stimulus, the virus may enter a lytic cycle and produce symptoms.

The herpes simplex virus causes latent infections of the skin. Herpes nucleic acid remains in nerve tissue (sensory ganglia) without damaging the host until a cold, a fever, or other factor such as ultraviolet radiation from the sun acts as a trigger, and the cycle of cell damage begins. This "damage" manifests as cold sores or fever blisters. The herpes zoster virus (the chickenpox virus) can also act in the same way. This virus may remain latent in the nerve tissue of a person having had chickenpox, only to be triggered at a later time to cause the painful nerve disorder *shingles.*

During a lysogenic pattern of viral infection, a virus integrates its genetic material with that of a host and is replicated each time the host cell reproduces.

The Classification of Viruses

Because viruses are not living things, they are not included in the five kingdoms of life. Scientists have, however, devised a classification scheme for viruses that is based on the host they infect. Viruses are first grouped according to whether they infect plants, animals, or bacteria. Further classification usually focuses on differences in morphology (shape and structure), type of nucleic acid, and manner of replication.

Viruses and Cancer

Five viruses are initiators (see p. 552) of cancer: hepatitis B virus, human T-cell lymphotropic/leukemia virus, human papillomavirus, human cytomegalovirus, and the Epstein-Barr virus (Table 25-1).

The *hepatitis B virus* (HBV) is transmitted in blood or blood products (such as blood serum or plasma) and by contaminated needles or syringes. Therefore, the persons most likely to contract this virus are health care workers (although their precautions are strict), intravenous (IV) drug users who share contaminated needles, and persons having sexual contact with infected persons. (Hepatitis is not usually classified as a sexually transmitted disease, however, because sexual contact is not the primary route of transmission.) In addition, an infected pregnant woman can pass this virus to her developing fetus.

HBV causes a serious infection of the liver. Persons with prolonged hepatitis B liver disease, especially persons who develop cirrhosis, are at risk for developing liver cancer. (Cirrhosis is a condition in which liver tissue is destroyed and is replaced by scar tissue.) Liver cancer as a result of hepatitis B

viruses (vye russ es) nonliving infectious agents that enter living organisms and cause disease.
core viral nucleic acid.
capsid (**kap** sid) a protein "overcoat" that covers the nucleic acid core of a virus.
envelope a chemical layer over the capsid of many viruses that is rich in proteins, lipids, and carbohydrate molecules.
lytic cycle (**lit** ik) a pattern of viral replication in which a virus enters a cell, replicates, and then causes the cell to burst, releasing new viruses.
lysogenic cycle (lye suh **jen** ik) a pattern of viral replication in which a virus integrates its genetic material with that of a host and is replicated each time the host cell replicates.

TABLE 25-1

Tumor Viruses

VIRUS	CANCER TYPE
Hepatitis B virus (HBV) 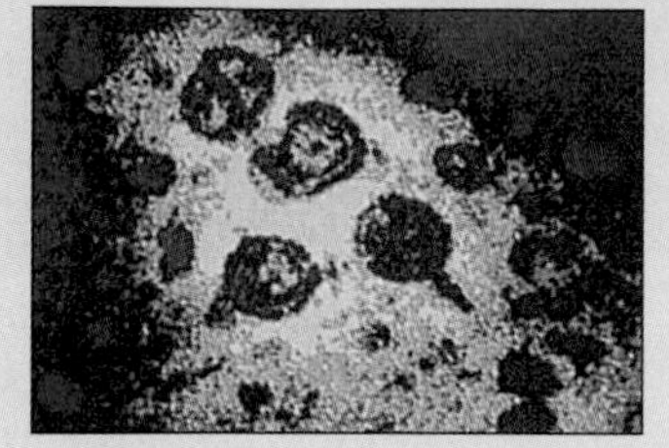	Primary liver cancer
Human T cell lymphotropic/leukemia virus (HTLV) 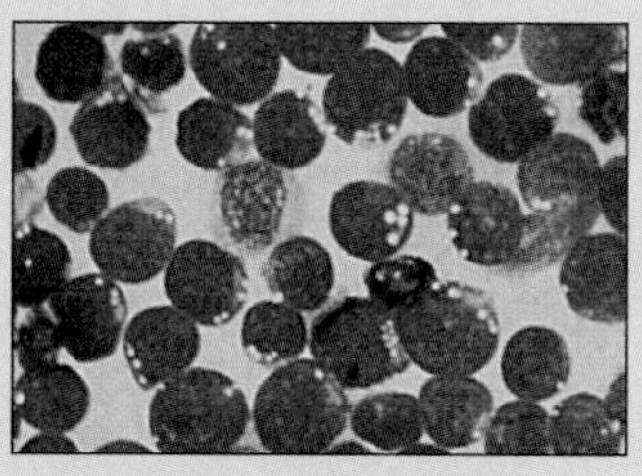	Leukemias and lymphomas
Cytomegalovirus (CMV) 	Kaposi's sarcoma
Human papillomavirus (HPV) 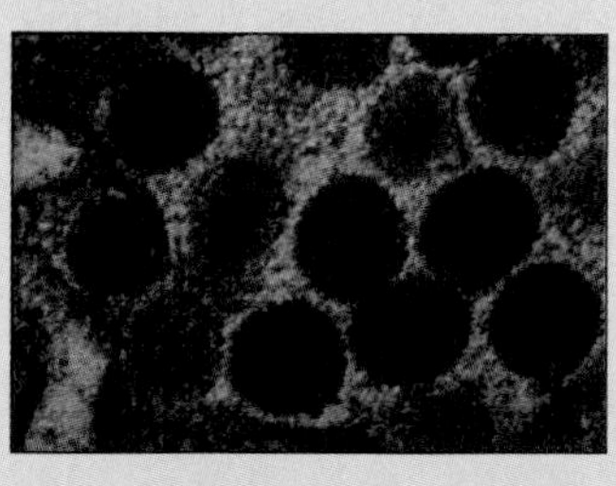	Vaginal/vulval/cervical cancer Penile cancer
Epstein-Barr virus (EBV) 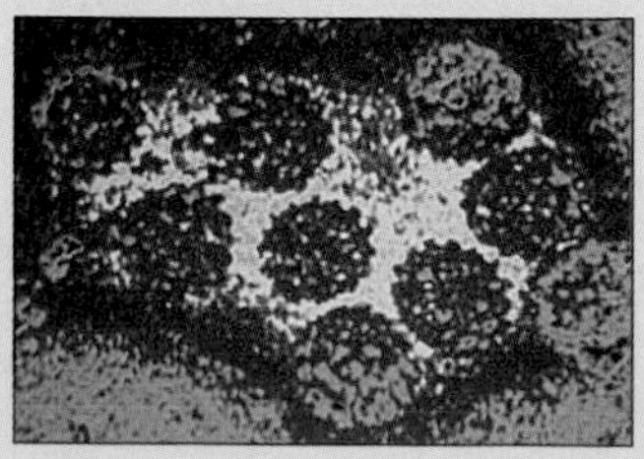	Burkitt's lymphoma Nasopharyngeal cancer

JUST WONDERING. . . .

How come scientists haven't found a cure for the common cold?

Although we often refer to the causative agent of colds as "*the* cold virus," there are actually more than 200 viruses that cause colds. Rhinoviruses and coronaviruses top the list, but within these viral groups alone there are well over 100 cold-causing viral types. Developing a vaccine against 200 agents of infection is not practical.

Antibiotics, those "magic bullets" against bacterial diseases, do not work against viruses. Antibiotics attack bacterial structures and metabolic processes, which is the reason they cause little harm to human cells. Viruses are much different from bacteria; they too are not affected by these infection-fighting drugs.

One line of investigation that has been suggested is to develop an antibody to be injected into humans that would block the receptor sites to which viruses adhere. Although more than 200 viruses cause colds, most of them use the same two receptor sites on cells of the upper respiratory tract mucosa. As yet, this cold-fighting antibody has not been developed.

There is good news, however: you develop immunity against colds as you age. Each time you contract a cold, you develop an immunity to the particular cold virus that infected you. Therefore, by the time you turn 60 years of age, you should be catching only one cold per year, on average, compared with the three or four you probably contracted when you were a child.

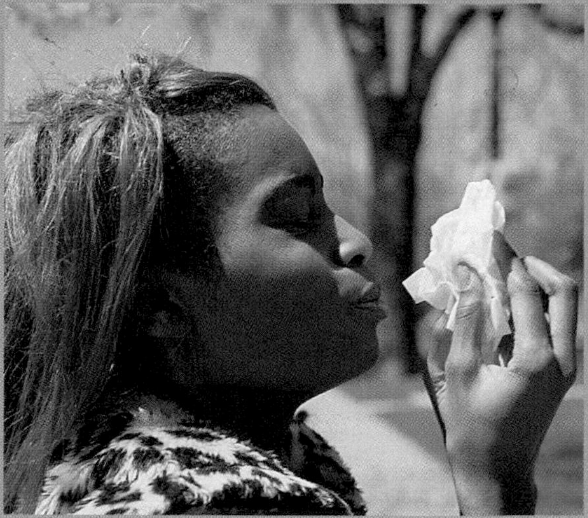

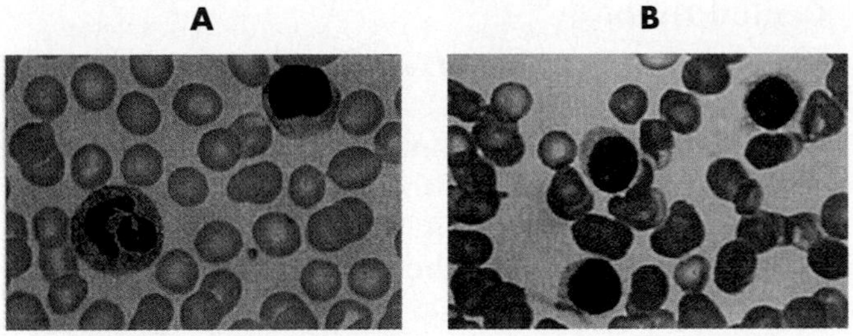

FIGURE 25-5 A normal blood smear contrasted with a hairy cell leukemia blood smear. A, This micrograph of a normal blood smear shows one neutrophil and one monocyte, both types of white blood cells, among red blood cells. **B,** This micrograph of a hairy cell leukemia blood smear shows three abnormal white blood cells among red blood cells.

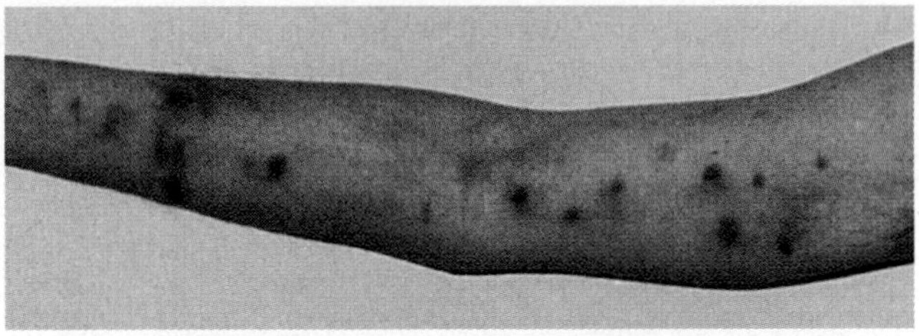

FIGURE 25-6 Kaposi's sarcoma. This photograph shows the multiple lesions of the skin cancer, Kaposi's sarcoma, on the arm of a patient with AIDS.

infection is relatively rare in the United States but is quite common in developing countries of Africa and Asia. This cancer is usually fatal; most patients die within 6 months to a year after diagnosis.

The *human T-cell lymphotropic/leukemia virus (HTLV)* causes adult T-cell leukemia/lymphoma (ATLL). This disease is rare in the United States but is common in Japan and the Caribbean. The cancers caused by HTLV are T-cell leukemias or lymphomas. *Leukemia* is a disease of the red bone marrow (the substance that produces most of the body's blood cells), which results in the manufacture of a greater than normal number of white blood cells that are immature, abnormal, and unable to perform their infection-fighting roles (Figure 25-5). *Lymphoma* is a malignant condition of the lymphoid tissue—the fluid or tissues of the lymphatic system (see Chapter 10). T lymphocytes, or T cells, are a type of white blood cell that develops in the bone marrow but matures in the thymus, an organ of the lymphatic system. T cells are integral in the immune response—your defense against disease (see Chapter 11).

ATLL spreads quickly in its victims and often results in an enlarged liver, spleen, and lymph nodes. Researchers are still uncertain how the HTLV is transmitted, but they think possible routes may involve sexual activity, the sharing of needles and syringes among IV drug users, and transfusion with contaminated blood. These pathways of infection are the same as those of HBV and the human immunodeficiency virus (HIV), which causes AIDS.

Although HIV is a "cousin" to HTLV (it was previously known as "HTLV-III"), the AIDS virus is not a tumor virus itself; however, it causes a breakdown of immunity that leaves its victims susceptible to other infections and cancers. The most common of the "AIDS cancers" is Kaposi's sarcoma, or skin cancer characterized by flat or raised red or purplish lesions (Figure 25-6). The cytomegalovirus (CMV) is thought to be the initiator of this cancer.

More than 60 types of *human papillomaviruses (HPV)* have been isolated in humans. Some of these viruses initiate only benign tumors, such as warts of the hands and feet. (Warts on the soles of the feet are called plantar warts.) Other human papillomaviruses cause a sexually transmitted disease commonly known as "venereal warts"; at least two types of these papillomaviruses (referred to as HPV-16 and HPV-18) are strongly linked to the development of cancer, particularly cervical carcinoma. (The cervix is the tissue surrounding the opening to the uterus, or womb.) Although venereal warts may appear on the cervix, within the vagina, or on the labial tissues in women, researchers think that the cervix has an area of tissue that is particularly vulnerable to viral infection. Researchers also suggest that the development of cancers of the cervix linked to HPV infection may be due to an interaction of factors, and they speculate that infection with other viruses, as well as smoking, may act as tumor initiators, with HPV types 16 or 18 functioning as promoters.

Women contract the papillomavirus during sexual intercourse with an infected partner. The risk of infection with the virus and the development of cervical cancer rise as a woman's number of sexual partners rises. Also, women who become sexually active at an early age are at a greater risk than women who become sexually active at an older age. Scientists are unsure whether the tissues of a young woman are simply more vulnerable or whether women who are sexually active at an early age are more likely to have a greater number of partners than other women.

Cancers caused by the *Epstein-Barr virus (EBV)* are rare in the United States. Americans are most familiar with the noncancerous condition it causes: mononucleosis. The virus is found in saliva because it is shed from cells lining the nose and

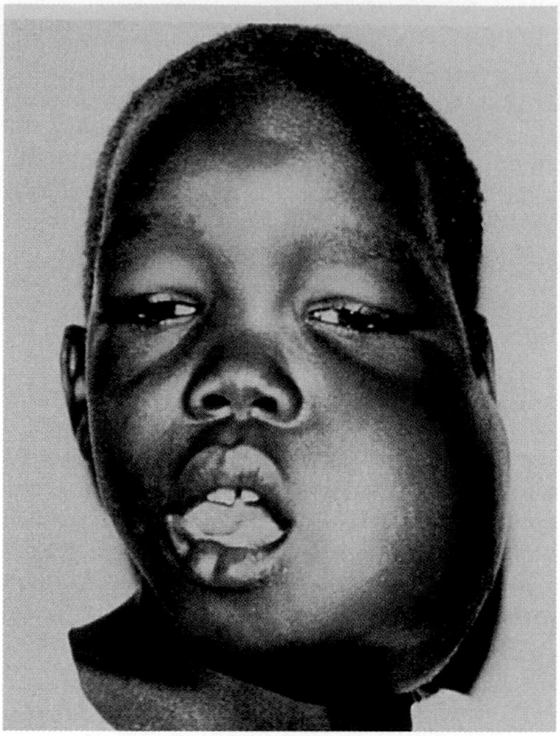

FIGURE 25-7 Burkitt's lymphoma. Burkitt's lymphoma is a cancer of the lymphatic system and is usually manifested as a large mass in the jaw or abdomen.

throat. For this reason, mononucleosis is often called the "kissing disease," but the virus can also be transmitted on contaminated cups, glasses, eating utensils, and similar objects.

The cancer-causing effects of the Epstein-Barr virus are seen in people living in Africa and China. In fact, cancers of the nose and upper throat linked to this virus are one of the leading causes of cancer deaths in China. In Africa, the more common cancer in which the EBV is implicated is Burkitt's lymphoma, a cancer of the lymph system (Figure 25-7).

Viral Sexually Transmitted Diseases

Three sexually transmitted diseases caused by viruses are *genital herpes, genital warts,* and *acquired immunodeficiency syndrome.*

Acquired Immunodeficiency Syndrome

Acquired immunodeficiency syndrome (AIDS) is discussed extensively in Chapter 11. The causative agent of AIDS, *human immunodeficiency virus (HIV),* is present in the blood, semen, and vaginal secretions of infected individuals. Therefore, the virus is spread by sexual contact as well as by the sharing of needles and syringes contaminated with blood from an infected person. In addition, HIV can be transmitted by mothers to their babies before or during delivery.

Genital Herpes

Genital herpes is an STD caused most often by the *herpes simplex virus (HSV) type 2.* However, HSV-1, which is associated primarily with the development of cold sores and fever blisters, is found to be the causative agent of genital herpes approximately 10% to 15% of the time. These conditions result from transmission through oral sex.

Genital herpes develops within a few days after sexual contact with an infected person. The first symptoms are an itching or throbbing in the genital area. Blisterlike, painful, itchy lesions and swollen lymph nodes in the groin area develop soon after. In women, lesions develop in the vagina and on the cervix, vulva, and thighs. In men, lesions develop primarily on the penis (Figure 25-8). The virus also causes an inflammation of the urethra in men, resulting in a watery discharge from the urethral opening.

The herpesviruses cause latent infections. In genital herpes, the virus particles travel via sensory nerve fibers to ganglia in the sacral region of the spinal cord after the initial infection appears to subside. There, the virus remains dormant or slowly replicates. As is typical with a latent infection, the virus is reactivated by one of a variety of stimuli such as stress, fever, and menstruation. Once reactivated, the virus moves along the nerve fibers back to the genital area where it replicates, causing lesions once again. This cycle of dormancy followed by a recurrence of lesions continues, but successive recurrences usually become milder. Eventually the recurrences may cease, but the viruses remain latent in the ganglia and can be reactivated by severe emotional or physical stress even a long time later. Along with the pain and suffering this disease causes both genders, women face additional problems. Women infected with HSV have a higher rate of miscarriage than do uninfected women. Also, pregnant women infected with HSV must have their babies delivered by cesarean section because the baby becomes infected as it travels through the birth canal.

A person with genital herpes (or oral herpes) can pass the virus on to another person any time viruses are being shed, or cast off. Unfortunately, it is impossible to know when a person is *not* shedding virus. Scientists know that shedding occurs when lesions are present and usually starts a few days before lesions appear, but shedding may *never* cease. In addition, there is no way of knowing if a person is infected unless lesions are present and observable or unless the person says he or she is infected. Unfortunately, there is no way to completely protect yourself from infection if you have sex with an infected person. Using latex condoms only reduces the chance of transmission of the virus.

A
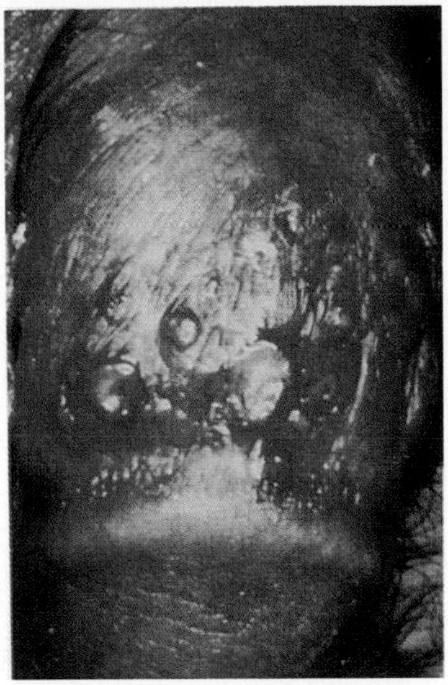
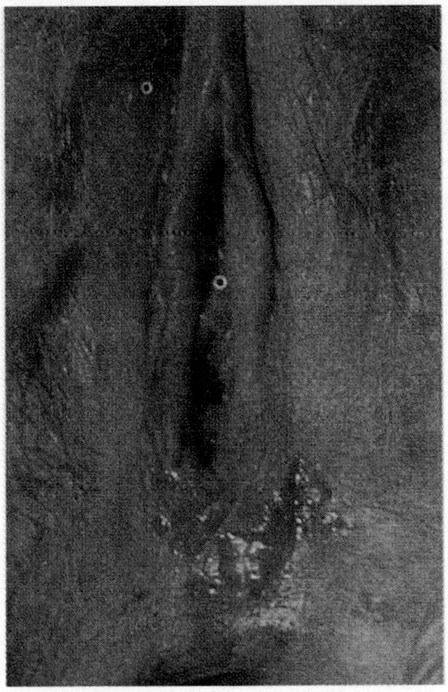
B

FIGURE 25-8 A genital herpes lesion on the penis (**A**) and on the female genitalia (**B**).

Genital herpes is an incurable disease. The antiviral drug acyclovir, which interferes with the replication of HSV, is usually used to treat the symptoms. The topical form is used directly on lesions and helps reduce the pain and itching they cause and the length of time they are present. Oral acyclovir is also available. Both forms of the drug decrease the duration of viral shedding and reduce the time required for healing.

Genital Warts

Genital warts are soft, pink, flat, or raised growths that appear singly or in clusters on the external genitals and rectum (Figure 25-9, *A*). In women, these irritating and itchy growths can also appear on the vaginal walls and on the cervix. In men, they can in-

A
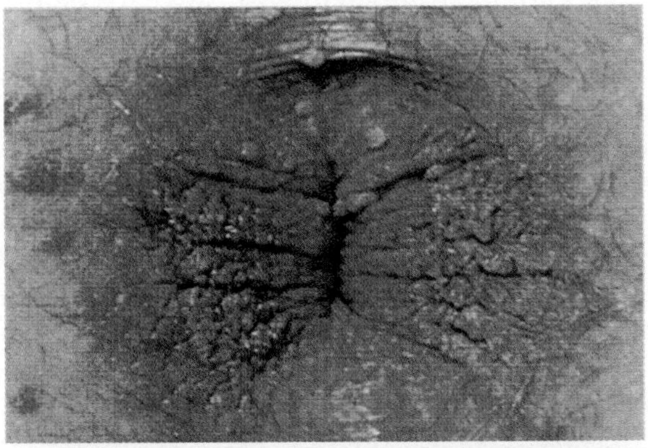

B
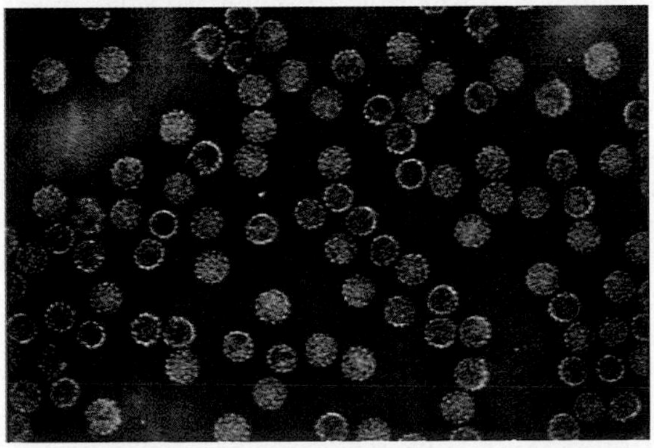

FIGURE 25-9 Genital warts. A, Perianal genital warts. **B,** An electron micrograph of the causative agent of genital warts, the human papillomavirus, or HPV.

acquired immunodeficiency syndrome (AIDS) (uh **kwy** urd ih **myoon** dih **fish** un see **sin** drome) a disease transmitted by the exchange of body fluids, such as blood or semen, containing the human immunodeficiency virus (HIV). This virus attacks and destroys T cells, a key component of the body's immune system.

genital herpes (jen ih tul **hur** peez) a sexually transmitted disease, caused most often by the herpes simplex virus type 2 (HSV-2), that produces blisterlike sores on the genitals.

genital warts a sexually transmitted disease caused by the human papillomavirus (HPV). The warts are soft, pink, flat, or raised growths that appear singly or in clusters on the external genitals and rectum.

vade the urethra. Caused by the human papillomavirus (HPV) shown in Figure 25-9, *B,* genital warts are becoming an increasingly prevalent sexually transmitted disease in the United States. Although the warty growths, or papillomas, caused by HPV are usually benign (noncancerous), genital warts that persist for many years can transform into malignant growths. In addition, HPV types 16 and 18 are tumor viruses that can initiate cancers of the cervix (see p. 546). Women infected with these virus types are at a higher risk for cervical cancer than are uninfected women; they should have a Papanicolaou (Pap) smear to detect this type of cancer once every 6 months.

At this time, genital warts are treated with electrocauterization (burning the infected tissue by means of electric current) or cryotherapy (freezing the infected tissue with liquid carbon dioxide and then removing it surgically). Sometimes, surgery alone is used. This virus can also infect babies during delivery as does the herpes simplex virus and most sexually transmitted disease organisms.

Bacteria

Bacteria and viruses both have "reputations" as being agents of disease. But the role of bacteria in the world of living things is much broader than that of a pathogen (disease-causing agent of infection). Bacteria make life on Earth possible because they perform integral functions as decomposers of organic material and are natural recyclers of nitrogen and other organic compounds in ecosystems (see Chapter 36). They are also used to produce certain foods, such as yogurt, sauerkraut, dill pickles, and olives.

Bacteria are the oldest, most abundant, and simplest organisms. Bacteria were abundant for well over 2 billion years before eukaryotes appeared in the world. They were largely responsible for creating the properties of the atmosphere and the soil during the long ages in which they were the only form of life on Earth (see Chapter 24). Bacteria are present on and in virtually everything you eat and you touch. Bacteria are also the only organisms with a prokaryotic cellular organization. (This type of cellular organization is described in Chapter 3.) Their organization differs from eukaryotic cellular organization in primarily two ways: the prokaryotic cell (1) has no membrane-bounded nucleus and (2) contains no membrane-bounded organelles that compartmentalize the cell. This structural uniqueness places bacteria in a kingdom all their own: *Monera,* meaning "alone."

Bacteria have a cell structure different from other organisms: the cytoplasm contains no internal compartments or organelles; the hereditary material is not enclosed by a membrane to form a nucleus; and the cell is bounded by a membrane encased within a cell wall. For these reasons, bacteria are classified as a separate kingdom of organisms, Monera.

Bacterial Reproduction

Reproduction among bacteria is asexual; one cell divides into two with no exchange of genetic material among cells. This process is called **binary fission.** Before fission, or division of the cell, the genetic material replicates and divides.

Bacterial DNA exists as a single, circular molecule that is attached at one point to the interior surface of the cell membrane. As eukaryotic cells do, bacteria make a copy of their genetic material before cell division. The bacterium also grows in size and manufactures sufficient ribosomes, membranes, and macromolecules for two cells before dividing. When the cell reaches an appropriate size and the synthesis of cellular components is complete, binary fission begins.

The first step of binary fission is the formation of a new cell membrane and cell wall between the attachment sites of the two DNA molecules (Figure 25-10). As the new membrane and wall are added, the cell is progressively constricted in two. Eventually the invaginating membrane and wall reach all the way into the cell center, forming two cells from one.

Most bacteria reproduce every 1 to 3 hours. Some bacteria take a great deal longer. However, bacteria having conditions favoring their growth could produce a population of billions in little more than a day! This type of growth is referred to as *exponential growth,* a period of rapid doubling of cells. However, the population cannot grow unchecked indefinitely. Many cells compete for food and other growth factors they need to survive. With resources limited, the entire population cannot be maintained, and cells begin to die. Wastes also accumulate, poisoning some of the cells. The growth of the population "levels out." Eventually, if the growth requirements of the bacterial population are no longer met, the population may begin to die as rapidly as it once grew.

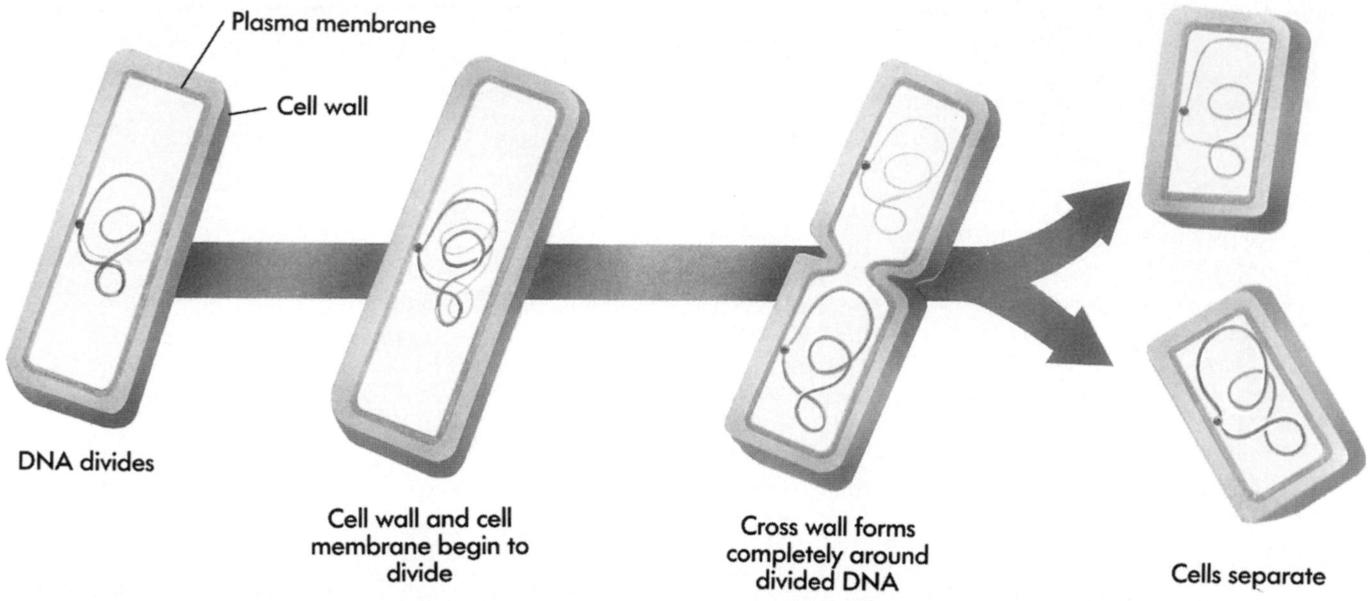

FIGURE 25-10 Binary fission in bacteria. During binary fission, one bacterium divides into two bacteria.

> *Bacteria reproduce asexually by binary fission. First they produce sufficient cell parts for two cells and replicate their single, circular molecule of DNA. Then the cell wall and membrane grow inward, between the attachment sites of the two DNA molecules, and literally split the cell in two.*

Bacterial Diversity and Classification

Bacteria occur in a wide range of habitats and play key ecological roles in each of them. Some thrive in hot springs in which the water temperature can be as high as 78°C (172°F). Others live more than a quarter of a mile beneath the surface of the ice in Antarctica. Still other bacteria, capable of dividing only under high pressures, exist around deep-sea vents formed by undersea volcanoes. Bacteria live practically everywhere, even in ground water where they were once thought to be absent. They are able to play many ecological roles because they make up a kingdom of organisms that is extremely diverse in its physiology.

This diverse kingdom of organisms is classified according to criteria listed in Table 25-2. These organisms are not grouped into phyla but into four divisions that reflect differences in the chemistry of their cell walls. Each division is further split into sections rather than classes of organisms. Classification from this point on is similar to the classification of the eukaryotes in that further subdivisions use the titles of order, family, and genus. Within this classification scheme, most of the bacteria are considered *eubacteria,* or "true" bacteria. But one section of organisms is different from all the rest. These are the *archaebacteria,* or "ancient bacteria." Many taxonomists consider archaebacteria to be a separate, sixth kingdom.

> *The kingdom of bacteria, Monera, is divided into four divisions, which are further subdivided into sections, orders, families, genera, and species.*

TABLE 25-2

Criteria for Classifying Bacteria

- Shape of individual cells
- Arrangements of groups of cells
- Presence of flagella
- Staining characteristics
- Nutritional characteristics
- Temperature, pH, and oxygen requirements
- Biochemical nature of cellular components, such as RNA and ribosomes
- Genetic characteristics, such as percentages of DNA bases

binary fission (bye nuh ree **fizh** un) a type of asexual reproduction in which one cell divides into two with no exchange of genetic material among cells; bacteria reproduce in this way.

HOW *Science* WORKS

Discoveries

Ulcers: An "Old" Malady With a New Treatment

Everyone has seen the portrayal of the hard-hitting executive who barks orders, has no time for relaxation, and is the owner of a "bleeding" ulcer. But ulcers are not just confined to business persons who experience continual stress. People from all walks of life have ulcers, including children and college students. In addition, the incidence of ulcers among women is on the rise.

An ulcer is a sore, usually about $^1/_2$ to 1 inch wide, that develops in the lining of the esophagus, stomach, or duodenum (Figure 25-A). Stomach ulcers develop because the protective mucosal barrier of the stomach is breached, and powerful hydrochloric acid invades the cells of the stomach wall. These cells are usually protected by a thick mucous covering. In addition, the surface cells, which are closest to the acid, are replaced with sturdy, new cells every 3 days. This barrier can be violated, however, if there is an oversecretion of hydrochloric acid or a breakdown in the barrier. Once the stomach wall contacts the acid, it releases histamine in an inflammatory response. Histamine, unfortunately, triggers the release of more acid, and a vicious cycle begins.

As the stomach contents empty into the duodenum (the first segment of the small intestine), they are neutralized by sodium bicarbonate secreted by the pancreas. However, the pancreas may not secrete enough sodium bicarbonate to neutralize excess acid. This acid can irritate the duodenum and cause ulcers there. Duodenal ulcers are more likely than stomach ulcers to perforate (punch through) the lining of the tract, causing the serious consequences of bleeding and potential infection of the surrounding organs by the contents of the stomach.

The classic symptom of an ulcer is a boring pain in the stomach several hours after eating, as stomach acid continues to be secreted while much of the meal has left the stomach. The pain may be relieved when the affected individual eats again.

The factors contributing to ulcers are numerous. Smoking seems to be linked to ulcers because nicotine inhibits the alkaline secretions of the pancreas. In fact, experts think that the increase in smoking among women is a prime factor in their increased incidence of ulcers. Prolonged aspirin and alcohol intake has also been implicated in the incidence of ulcers because they break down the mucosal barrier. Of course, stress and emotional tension have always been linked to ulcers; these responses appear to result in increased stomach acid secretion.

Recently, researchers have discovered that a bacterium called *Helicobacter pylori* is present in the stomach of approximately 90% of all patients with ulcers. This bacterium is extremely common and enters the body through human contact and possibly through contaminated drinking water. Researchers estimate that as many as 50% of all Americans harbor this bacterium in their stomachs. The good news is that once diagnosed, the patient can be treated with antibiotics, which kill the bacteria and cure the ulcers.

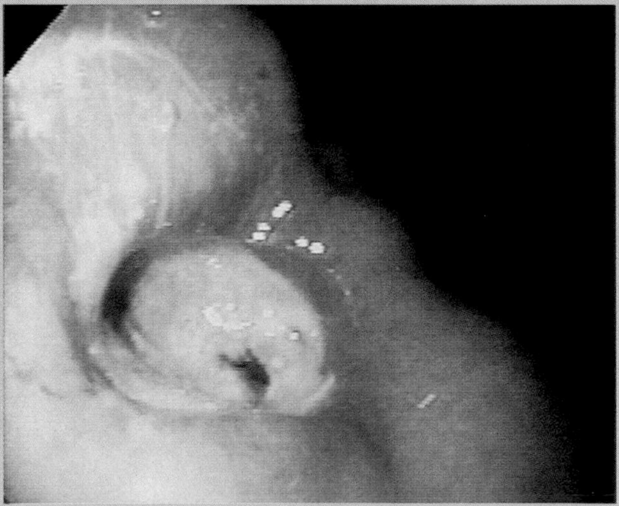

You can do a lot to prevent ulcers. First, don't smoke, and drink moderately or not at all. Watch your intake of aspirin, and if it is a problem for you, learn to deal more effectively with stress. By protecting your digestive tract with common sense, you can avoid the pain of ulcers.

Archaebacteria

The cell walls and the cell membranes of the archaebacteria are chemically different from those of the eubacteria, as are certain of their physiological processes. These bacteria have such unusual chemical processes that they live in quite unusual places!

One group of archaebacteria produces methane (also known as "marsh gas") from carbon dioxide and hydrogen. These bacteria can be found in places such as the gut of cattle and the depths of landfills. Unfortunately, the methane they produce adds to the blanket of greenhouse gases surrounding the Earth and therefore to the problem of global warming (see Chapter 38). Another group of archaebacteria live only in areas having high concentrations of salt, such as in salt marshes in the intertidal zone, where fresh water meets seawater in stagnant, concentrated salt pools. A third group of archaebacteria was already mentioned: those that live in the incredible heat and pressure of the deep-sea vents.

The archaebacteria are a taxonomic section of bacteria that are chemically different in certain structures and metabolic processes from all other bacteria.

Eubacteria

The eubacteria make up all the rest of the bacteria—quite a diverse collection. However, these thousands of species can be placed into one of three groups according to their mode of nutrition.

One of these groups is the photoautotrophic (photosynthetic) bacteria. Like plants, photosynthetic bacteria use energy from the sun to produce "food" in the form of carbohydrates. They are therefore called **photoautotrophs,** which means "light" (*photo*) "self" (*auto*) "feeders" (*trophs*). Photosynthesis takes place in green bacteria, purple bacteria, and cyanobacteria (formerly known as "blue-green algae"). These bacteria contain chlorophyll (chemically different from the chlorophyll in plants), which is found within a system of membranes that ring the interior periphery of the cell. In the green and purple bacteria, photosynthesis does not take place exactly as it does in plants, and oxygen is not a by-product. In fact, oxygen tends to break down their chlorophyll, so these species usually live in polluted water that contains little oxygen. The cyanobacteria, however, release oxygen during photosynthesis as plants do and can be found living near the surface of lakes and ponds (Figure 25-11). The cyanobacteria were among the first cells to evolve and probably oxygenated Earth's primitive environment.

A second group of eubacteria are the chemoautotrophic bacteria. "Chemical self-feeders," or **chemoautotrophs,** derive the energy they need from inorganic molecules such as ammonia (NH_3), methane (CH_4), and hydrogen sulfide (H_2S) gases. With this energy and carbon dioxide (CO_2) as a carbon source, they can manufacture all their own carbohydrates, fats, proteins, nucleic acids, and other growth factors. The archaebacteria that live near deep-sea vents are chemoautotrophs, as are eubacteria called *nitrifying bacteria.* Nitrifying bacteria play an important role in the cycling of nitrogen between organisms and the environment. These bacteria live in nodules (spherical swellings) in the roots of legumes, such as beans, peas, and clover (Figure 25-12). They convert ammonia to nitrates, a form of nitrogen used by plants. Other species of chemoautotrophs play a key role in the sulfur cycle, using H_2S or elemental sulfur for energy and converting it in the process to sulfates, which are other plant nutrients.

The third group of eubacteria are heterotrophic bacteria. Most bacteria are "other feeders" or **heterotrophs,** obtaining their energy from organic

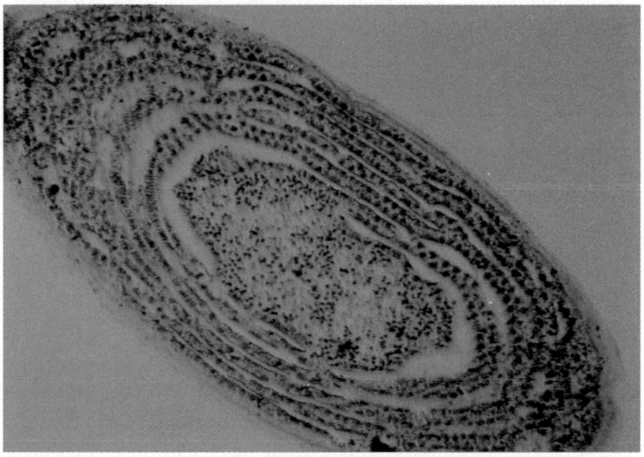

FIGURE 25-11 A cyanobacterium. The outer regions of the cell are filled with photosynthetic membranes. The dark spots between the membranes are storage areas for the carbohydrates produced by photosynthesis.

photoautotrophs (**fote** oh **awe** toe trofes) organisms that make their own food by photosynthesis, using the energy of the sun.

chemoautotrophs (**kee** mo **awe** toe trofes) organisms that make their own food by deriving energy from inorganic molecules.

heterotrophs (**het** ur oh **trofes**) organisms that cannot produce their own food; they obtain their energy by breaking down organic material.

FIGURE 25-12 Soybean root nodules. The nodules on the roots of this soybean plant contain nitrogen-fixing bacteria. These bacteria convert atmospheric nitrogen (N_2) to ammonia (NH_3), which can then be used as a nitrogen source by the plants.

FIGURE 25-13 Crown gall disease on a tobacco plant. This disease is caused by bacteria of the genus *Agrobacterium.*

material that enters these cells by diffusion and active transport. Humans are heterotrophs too, eating plants and animals—organic material that once lived. Heterotrophic bacteria are considered **decomposers** and play a key role in the carbon cycle. They break down large organic compounds, such as proteins and carbohydrates, into small compounds, such as CO_2, which is released into the atmosphere to be recycled as it is "fixed" by plants during the process of photosynthesis.

Bacteria as Disease Producers

Most plant diseases are not caused by bacteria but by fungi (see Chapter 26). However, a few genera of bacteria do infect plants, primarily causing types of plant rot and wilt. Wilt occurs when bacteria block water from moving up the xylem in the plant. In addition, some genera of bacteria cause tumorlike growths called *galls* in plants. Figure 25-13 is a photograph of a plant with crown gall disease caused by bacteria of the genus *Agrobacterium.* Interestingly, scientists have learned ways to use these disease-causing bacteria in productive ways in the genetic engineering of plants.

In contrast to plant diseases, many human diseases are caused by bacteria, including the diseases listed in Table 25-3 with which you may be familiar. The bacteria that cause disease are all heterotrophic. Most disease-producing bacteria use their hosts for food, but some poison their hosts. To cause disease, bacteria or the poisons they produce must first get into the body. Usually this happens if you eat contaminated food or drink contaminated water. You may inhale bacteria present in the air after an infected person coughs or sneezes, touch a contaminated object, or have sexual intercourse with an infected partner. Sometimes bacteria enter the body through broken skin as the result of an injury or injection with a contaminated needle.

After entering the body, bacteria then attach to body cells and cause various types of tissue damage. The chemicals bacteria produce digest the tissues so that the breakdown products can be taken into the bacteria and metabolized. In addition, certain bacteria, such as the bacteria that cause staphylococcal food poisoning, gas gangrene, tetanus, or cholera, produce *toxins,* or poisons. These toxins can cause powerful effects, such as high fevers, violent muscle spasms, vomiting, diarrhea, heart damage, and respiratory failure. All the while, your body combats these cells using its nonspecific and specific defense systems (see Chapter 11). The ability of your body to wage this war affects the degree and length of the illness and its eventual outcome.

Bacterial Sexually Transmitted Diseases

The sexually transmitted diseases caused by bacteria include *syphilis, gonorrhea,* and *chlamydial infection.* The bacteria that cause these diseases enter the body via the mucous membranes of the genitals. Persons harboring these organisms transmit them to other individuals during sexual acts. After entering the body, bacteria attach to body cells and cause various types of tissue damage.

TABLE 25-3

Some Diseases Caused by Bacteria

RESPIRATORY DISEASES
Bacterial pneumonia
Bacterial meningitis
Strep throat
Tuberculosis
Legionnaire's disease
GASTROINTESTINAL DISEASES
Cholera
Dysentery
Typhoid fever
GASTROINTESTINAL POISONING
Staph food poisoning
Botulism
SEXUALLY TRANSMITTED DISEASES
Syphilis
Gonorrhea
Chancroid
SKIN DISEASES
Staph infections
Leprosy
Yaws

Syphilis

Syphilis is caused by a spiral bacterium, *Treponema pallidum* (Figure 25-14). Although it is curable if treated with penicillin, medical researchers estimate that approximately 400,000 persons in the United States have this disease. If left untreated, syphilis can result in death.

The disease progresses through a variety of stages. Approximately 3 weeks after sexual contact with an infected partner, the newly infected person usually develops a lesion on either the cervix (in women) or on the penis (in men) where the organism entered the body. This lesion, which is called a *hard chancre,* is the hallmark of the *primary stage* of syphilis. At first the lesion is painless. It is raised above the skin and has a hard base. However, it soon develops into a painful, ulcerated sore. During this time, the bacterium enters the lymphatic system, and the lymph nodes closest to the lesion enlarge. Once in the lymph, the organisms quickly travel to the bloodstream and throughout the body, infecting other tissues. Within 4 to 12 weeks the sore heals and the primary phase is over.

The *secondary stage* of syphilis begins from 6 weeks to several months after infection. Because the organisms are now dispersed throughout the body, lesions of the skin and mucous membranes develop away from the site of the original infection—usually on the trunk, arms, and legs. Sometimes the lesions appear as discolored, flat spots on the skin. Other times they are small, solid, raised bumps. A person with syphilis is highly contagious during the primary stage and through the secondary stage until these secondary lesions heal.

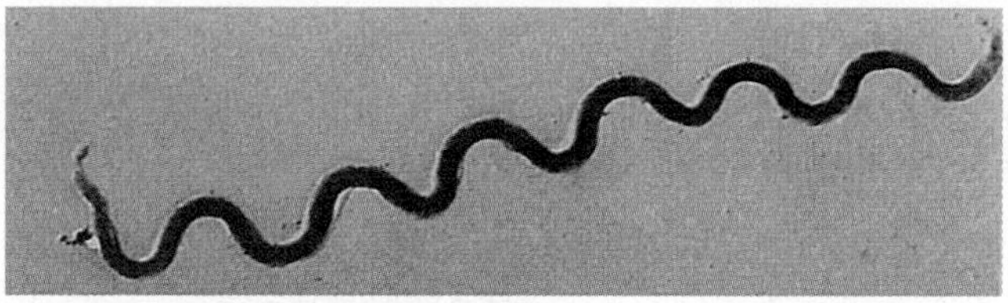

FIGURE 25-14 Causative agent of syphilis. This photomicrograph shows the spiral bacterium *Treponema pallidum,* which is the cause of syphilis.

During the next stage, an infected person has no symptoms and may think the disease has been successfully battled. Unfortunately, the disease is only in a *latent stage.* This stage may last a lifetime but, more often, leads to the final, or *tertiary,* stage. Once again, lesions develop, but the disease is no longer communicable. The most typical type of tertiary lesions are *gummas.* These lesions are tumorlike masses that can invade the skin, tissues beneath the skin, mucous membranes, bones, and internal organs. When gummas develop in the cardiovascular and central nervous systems, paralysis and death often result. Figure 25-15 shows lesions typical of the primary, secondary, and tertiary stages of syphilis.

Gonorrhea

Gonorrhea is caused by *Neisseria gonorrhoeae,* a paired, coffee bean–shaped bacterium (Figure 25-16, *A*). The word *gonorrhea* is derived from the two Greek words meaning "a flow of semen." This name was given to the disease in 130 AD by the Greek physician Galen when he mistook the genital discharge caused by gonorrhea for seminal fluid. This discharge, a fluid composed of mucus intermingled with pus, is a result of infection of tissues of the urogenital system by *N. gonorrhoeae.* Figure 25-16,

decomposers (**dee** kum **poe** zurs) organisms such as bacteria and fungi that obtain their energy by breaking down organic material in dead organisms and contribute to the recycling of nutrients to the environment.

syphilis (**sif** uh lis) a sexually transmitted disease, caused by the bacterium *Treponema pallidum,* that produces three stages of infection, from localized to widespread.

gonorrhea (**gon** uh **ree** uh) a sexually transmitted disease, caused by the bacterium *Neisseria gonorrhoeae,* that causes a primary infection and inflammation of the urethra (in men and women) and the vagina and cervix (in women).

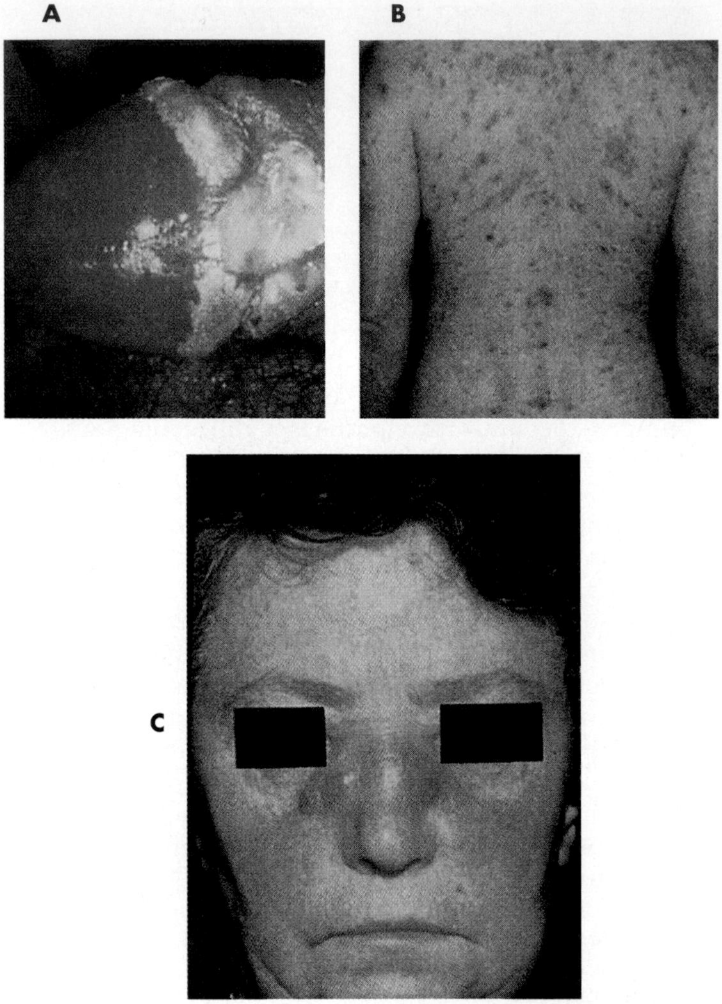

FIGURE 25-15 **The three stages of syphilis.** **A,** Genital lesions of the primary stage. **B,** Skin lesions of the secondary stage. **C,** Gummata of the nasal bones in the tertiary stage.

B, shows this typical symptom of gonorrheal infection in the male.

Men who become infected usually develop an inflammation of the urethra, with initial symptoms of discomfort and pain. Within 2 to 5 days, a discharge like that shown in Figure 25-16, *B,* becomes evident. The infection then passes up the urethra. Although urination becomes more difficult at this stage of the disease, the urge to urinate becomes more frequent. Occasionally these symptoms are accompanied by a headache and fever. If the infection remains untreated, other urogenital structures may become infected. Infection of the epididymis (ducts within which sperm are stored) or vas deferens (ducts that conduct sperm away from the testes) is serious; scar tissue may develop and block these passageways, resulting in sterility.

Women who become infected can develop an inflammation of the urethra, vagina (and nearby glands), cervix, and rectum. The initial symptoms of infection are abdominal or pelvic pain, vaginal discharge, and painful or difficult urination. If the infection remains untreated, a chronic (persistent) infection develops, with symptoms such as tenderness of the lower abdomen, backaches, inflammation of the urethra or other urogenital structures, and profuse menstrual bleeding. Both women and men may have no symptoms and so may not seek treatment, which puts them at risk for chronic infection and places their partners at risk for contracting this disease.

Chronic infections in women can lead to pelvic inflammatory disease (PID) if the bacteria migrate up through the uterus and to the uterine (fallopian) tubes. PID is also caused by *Chlamydia trachomatis* and a variety of other bacteria, but a large portion of the cases are caused by *N. gonorrhoeae.* PID can cause sterility if scar tissue develops and blocks the uterine tubes, which are passageways that lead from the ovaries to the uterus. If the tubes become constricted rather than blocked, sperm may be able to reach and fertilize an egg, but the developing embryo may become stuck in the tube as it travels to the uterus. An ectopic pregnancy results, in which the egg embeds in the wall of the tube. Such a pregnancy is fatal to the developing fetus and may cause the tube to rupture. In addition, pregnant women with gonorrhea can infect the eyes of their newborns as they pass through the birth canal; blindness can result. Babies' eyes are treated with a solution of silver nitrate or the antibiotic erythromycin to avoid such infection.

In addition to causing infection of the urogenital tract, *N. gonorrhoeae* can cause infections of the joints, skin, and blood if it invades the bloodstream. It commonly infects the rectum because of its proximity to the site of initial infection. In addition, the organism can infect the throat as a result of oral-genital contact. Although gonorrhea is most often transmitted by sexual contact, the bacteria can survive for several hours on objects such as bed linens. Therefore, unsanitary conditions can sometimes be a route of transmission. Most infections respond to penicillin; other antibiotics are sometimes used as well.

Chlamydial Infection

Chlamydial sexually transmitted infection is caused by an unusual bacterium, *Chlamydia trachomatis.* These organisms, among the smallest of the bacteria, grow only within other cells and obtain adenosine triphosphate (ATP) (see Chapter 4) from their host cells. For this reason, scientists call the chlamydiae "energy parasites." Another unusual feature of the chlamydiae is that they exhibit two different forms depending on whether they are inside or outside host

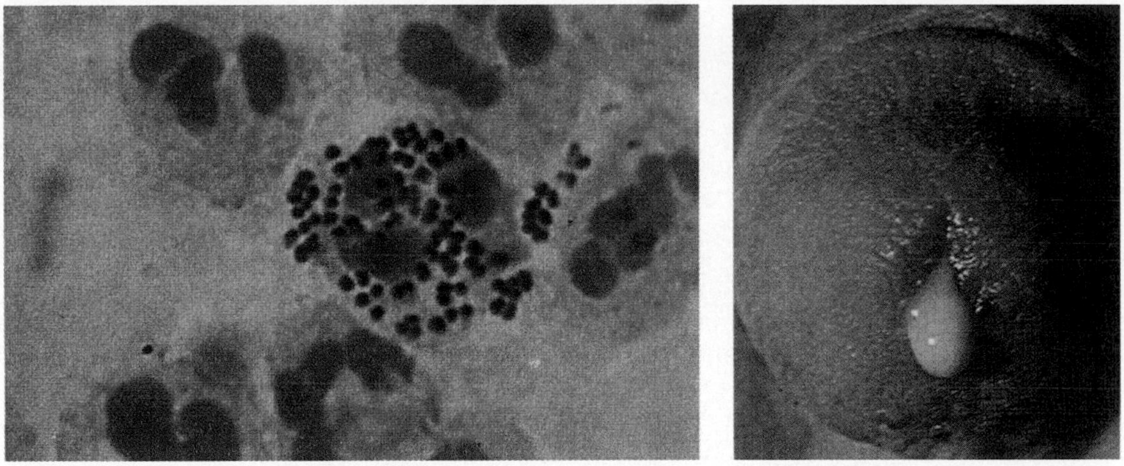

FIGURE 25-16 Gonorrhea. A, *Neisseria gonorrhoeae,* the causative agent of gonorrhea, appears as pairs of coffee bean–shaped cells. **B,** The mucuslike discharge of gonorrhea from the urethra of a male.

cells. The chlamydiae capable of infecting cells are metabolically inactive and small and have dense bodies with rigid cell walls. These infective forms of chlamydiae are called *elementary bodies.* They attach to host cells and are taken into these cells by phagocytosis. Once inside their hosts, the elementary bodies transform into *reticulate bodies* (Figure 25-17): relatively large, metabolically active cells having flexible cell walls. The reticulate bodies then multiply, condense to form elementary bodies, and cause the host cell to burst, releasing them. Each new elementary body is capable of infecting a new host cell.

Chlamydia trachomatis is the primary causative agent of **nongonococcal urethritis (NGU),** a disease similar to gonorrhea but caused by organisms other than *N. gonorrhoeae.* In addition, one highly invasive strain, or group, of *C. trachomatis* causes *lymphogranuloma venereum,* a sexually transmitted disease that is rare in the United States. However, chlamydial NGU infections are common; this realization is relatively recent due to the development of new diagnostic techniques to detect this organism.

After infection with *Chlamydia trachomatis,* mild, gonorrhealike symptoms appear within 1 to 3 weeks. A discharge is present, but it is more watery than the discharge of gonorrhea. Unfortunately, many persons with chlamydial NGU have no symptoms, so do not seek treatment and transmit the organism to other partners. As with gonorrhea, chronic infection with *Chlamydia* can lead to PID and sterility in women and to inflammation of the epididymis and sterility in men. Infants passing through the birth canal can also become infected. Erythromycin is used to treat the eyes of infants. Tetracycline and sulfa drugs (antibacterial chemical compounds containing the element sulfur) are used to treat chlamydial NGU.

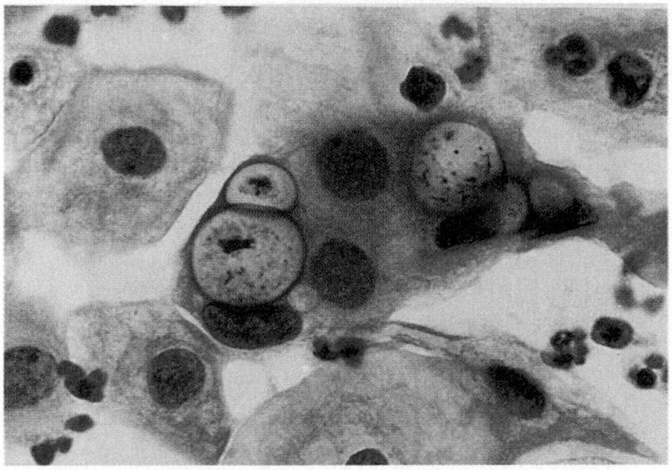

FIGURE 25-17 Chlamydial reticulate and elementary bodies in a host cell. This micrograph shows a Pap smear of cells taken from the cervix. The cytoplasm of the cells is stained blue, with nuclei stained dark pink. Normal cells surround a cell infected with *Chlamydia.* The infected cell has a thin ring of cytoplasm surrounding the chlamydial cells, which appear as various shades of pink and take up the rest of the cell. The reticulate bodies are dark pink cells without red centers. These cells are the metabolically active, reproductive form of *Chlamydia.* The light pink cells with red centers are in the process of condensing to form elementary bodies: the infective, dormant form of *Chlamydia.*

nongonococcal urethritis (NGU) a sexually transmitted disease, caused by the bacterium *Chlamydia trachomatis,* that has gonorrhealike symptoms.

Summary

- Viruses were discovered at the end of the nineteenth century. For many years after their discovery, viruses were regarded as primitive forms of life. Today, scientists know that viruses are not living organisms because they are not cells. Viruses are protein-coated nucleic acids that replicate (multiply) within living cells.
- Viruses primarily infect plants, animals, and bacteria, replicating within their cells. Various patterns of viral replication exist. In the lytic cycle a virus enters a cell and causes it to produce viral nucleic acid and protein coats. After these viral parts are assembled, the new virus particles may burst from the host cell or may leave the host cell by budding. In the lysogenic cycle, viruses enter into a long-term relationship with the cells they infect, their nucleic acid replicating as the cells multiply.
- Some viruses can seriously disrupt the normal functioning of the cells they infect, transforming them into rapidly growing, invasive cells. These cells grow out of control, forming cancerous tumors that destroy body tissues.
- Three sexually transmitted diseases caused by viruses are genital herpes, genital warts, and acquired immunodeficiency syndrome (AIDS).
- Bacteria (kingdom Monera) have a prokaryotic cell structure. They differ from eukaryotic cells in many ways but primarily in that they have no membrane-bounded nucleus or membrane-bounded cellular organelles.
- Bacteria reproduce asexually by binary fission, a splitting in two, after replication of the genetic material takes place. The numbers of bacteria within a population increase rapidly when growth conditions are favorable. These growth conditions vary among bacteria.
- Most bacteria are considered eubacteria, meaning "true bacteria." These thousands of genera of bacteria can be placed into one of three groups according to their mode of nutrition: photoautotrophs, which make their own food by photosynthesis using the energy of the sun; chemoautotrophs, which make their own food by deriving energy from inorganic molecules; and heterotrophs, which obtain energy from organic material.
- The archaebacteria differ from the eubacteria in their structural and physiological chemistry. Archaebacteria inhabit harsh environments, such as landfills, salt marshes, and deep-sea vents.
- Bacteria play diverse ecological roles and are extremely important as decomposers. Bacteria are also important in the manufacture of certain foods. A few genera of bacteria cause diseases in plants, and many genera cause diseases in animals, including humans.
- The sexually transmitted diseases caused by bacteria include syphilis, gonorrhea, and chlamydial infection. The bacteria that cause these diseases enter the body via the mucous membranes of the genitals, usually during sexual activity. After entering the body, bacteria attach to body cells and cause various types of tissue damage. The chemicals bacteria produce digest the tissues so that the breakdown products can be taken into the bacteria and metabolized.

Knowledge and Comprehension Questions

1. What are viruses? What do scientists mean when they say that viruses are not alive?
2. Diagram the structure of a generalized virus. Label the core, capsid, and envelope.
3. Distinguish between the lytic cycle and the lysogenic cycle.
4. One of your friends usually develops a cold sore on his mouth when he gets a cold. Why? What pattern of viral replication is involved?
5. Why are bacteria classified into a kingdom that is unique from all other organisms?
6. Although many bacteria can cause dangerous diseases, in general, bacteria make life on earth possible. Why?
7. Are high standards of personal hygiene sufficient protection against sexually transmitted diseases?
8. Are genital warts cancerous?
9. Many scientists think that bacteria are the most primitive biological organisms and probably evolved long before eukaryotes. Can you support this theory?
10. What are gummas? What damage can they cause to internal organs and systems?
11. Describe how gonorrhea may cause sterility in both men and women.
12. Explain how persons with chlamydial NGU, gonorrhea, or syphilis may unknowingly pass on these sexually transmitted diseases.
13. Are condoms a completely adequate barrier to sexually transmitted diseases? Is there still a risk factor if condoms are used correctly?

Critical Thinking

1. Your doctor prescribed an antibiotic for you when she diagnosed your strep throat. Antibiotics kill or inhibit the growth of bacteria. The instructions were to take it for 10 days, but because you felt much better after 4 days, you stopped taking the medication. Why do you think the doctor told you to take the antibiotic for 10 days, and what could happen by stopping the drug prematurely?

CHAPTER 26

PROTISTS AND FUNGI

"PLEASE DON'T DRINK THE WATER!" may seem like an unnecessary warning if you were camping along this pristine-looking mountain stream. But after an invigorating hiking trip, you could return home with nausea, cramps, bloating, and diarrhea—all symptoms of giardiasis, or "hiker's diarrhea." The culprit of this uncomfortable ailment is a single-celled organism no larger than a red blood cell: *Giardia lamblia.*

The organism *G. lamblia* is found throughout the world, including all parts of the United States and Canada. It occurs in water, including the clear water of mountain streams and the water supplies of some cities. In addition to humans, it infects at least 40 species of wild and domesticated animals. These animals can transmit *G. lamblia* to humans by contaminating water with their feces. Flagella protrude from one end of *G. lamblia,* allowing it to move along the intestinal wall of its host. When feeding, it attaches to the wall, sucking blood for nutrition and leaving the suction cup–shaped marks shown in Figure 26-1. This motile form of the protist exists only while inside the body of its victim. Dormant, football-shaped cysts are expelled in the feces, which can survive for long periods of time outside their hosts—especially in the cool water of mountain streams. When ingested by other hosts, the cysts develop into their motile, feeding form.

What should you do to prevent infection by *G. lamblia* when hiking or camping? First, *never* drink untreated water, no matter how clean it looks. Because *G. lamblia* is resistant to usual water treatment agents, such as chlorine and iodine, you should boil the water you drink for at least a minute. Better still, *do not* drink the water—bring your own bottled water to be safe.

KEY CONCEPTS

- Protists make up an extremely diverse kingdom of eukaryotic organisms that are primarily single celled but that include certain phyla containing multicellular forms.
- Protists can be grouped according to their mode of nutrition into those that are animallike heterotrophs, plantlike photosynthetic autotrophs, and funguslike saprophytes, although taxonomists differ in the classification of many forms.
- Fungi make up a kingdom of multicellular eukaryotic organisms that are primarily saprophytic; that is, they feed on dead or decaying organic material.
- Lichens are symbiotic associations between fungi and algae that allow both to survive in extremely harsh environments.

OUTLINE

Protists
Animallike protists: Protozoans
Plantlike protists: Algae
Funguslike protists
Fungi
Zygote-forming fungi
Sac fungi
Club fungi
Imperfect fungi
Lichens

Protists

The **protists** (kingdom Protista) are a varied group of eukaryotic organisms. Many are single celled, although some phyla of protists include multicellular or colonial forms (single cells that live together as a unit). Within this kingdom are animallike, plantlike, and funguslike organisms (Figure 26-2)—quite a diverse array! The animallike protists are called **protozoans** and are considered animallike because they are heterotrophs: they take in and use organic matter for energy. *G. lamblia,* for example, is a protozoan. The plantlike protists are the **algae** (including *diatoms*). These organisms are plantlike because they are eukaryotic photosynthetic autotrophs: they manufacture their own food using energy from sunlight. The funguslike protists consist of two phyla of **slime molds** and one of **water molds.** They secrete enzymes onto food sources to predigest the food before they absorb it as the fungi do.

Although many protists are single celled, these organisms are incredibly different from prokaryotic single-celled organisms, the bacteria. The single-celled protists are much larger than bacteria, having approximately 1000 times the volume, and they contain typical eukaryotic cellular organelles. This compartmentalization of eukaryotic cells by membrane-bounded organelles increases their organization to a level much more complex than that of bacteria and allows single-celled eukaryotic organisms to carry out cell functions that support their larger cell volume.

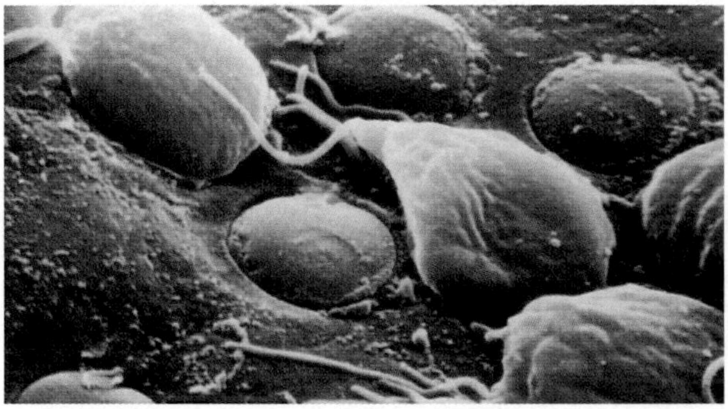

FIGURE 26-1 ***Giardia lamblia.*** These single-celled organisms live in water. When humans drink water infected with *Giardia* organisms, they can experience nausea, cramps, bloating, vomiting, and diarrhea. It is important not to drink *any* untreated water, even though it may look safe.

Most of the protist phyla are unicellular eukaryotes, although some phyla contain multicellular forms. Protists can be grouped as animallike (protozoans), plantlike (algae), or funguslike (slime molds and water molds) according to their mode of nutrition.

Animallike Protists: Protozoans

The animallike protists, or *protozoans,* obtain their food in diverse ways. These heterotrophic characteristics are linked to the ways that these organisms move and provide a means by which they can be grouped. There are four protozoan phyla: the amebas, flagellates, ciliates, and sporozoans.

Amebas

Amebas appear as soft, shapeless masses of cytoplasm. Within the cytoplasm lie a nucleus and other eukaryotic organelles. The ameba's cytoplasm continually flows, pushing out certain parts of the cell while retracting others. These cytoplasmic extensions are called *pseudopods* (from the

A

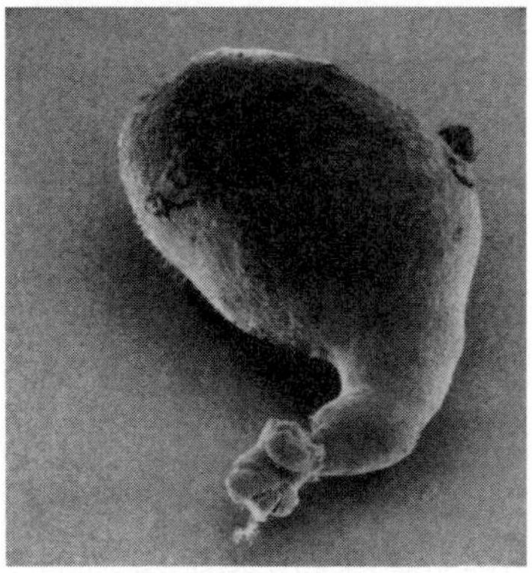

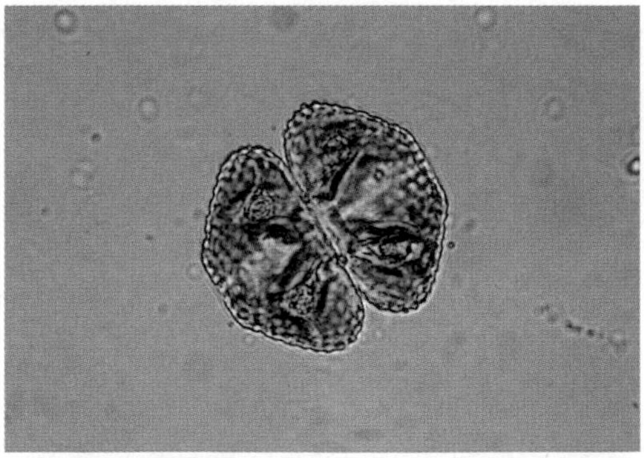

B

C

FIGURE 26-2 Single-celled protists. A, Animallike protist *Pelomyxa palustris,* a unique, single-celled organism that lacks mitochondria and does not divide mitotically. **B,** Plantlike protist, green alga from the genus *Cosmalium.* **C,** Funguslike protist, the plasmodial slime mold. The individual cells cannot be distinguished in such a plasmodium.

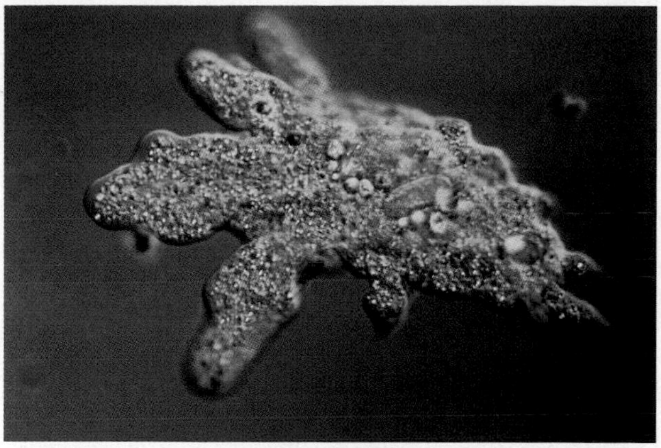

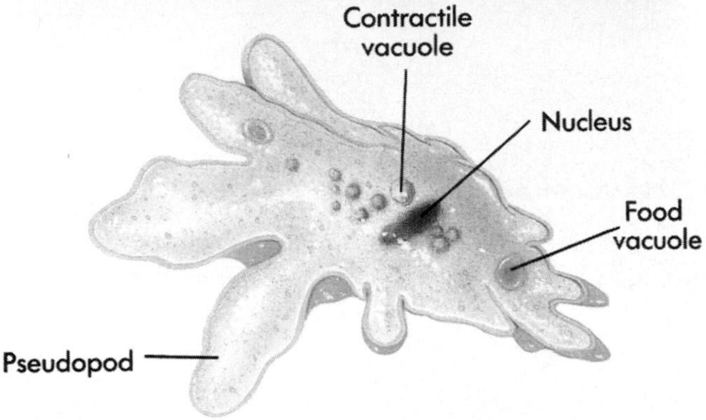

FIGURE 26-3 Structure of an ameba. The pseudopods, literally "false feet," allow the ameba to move and to trap food particles

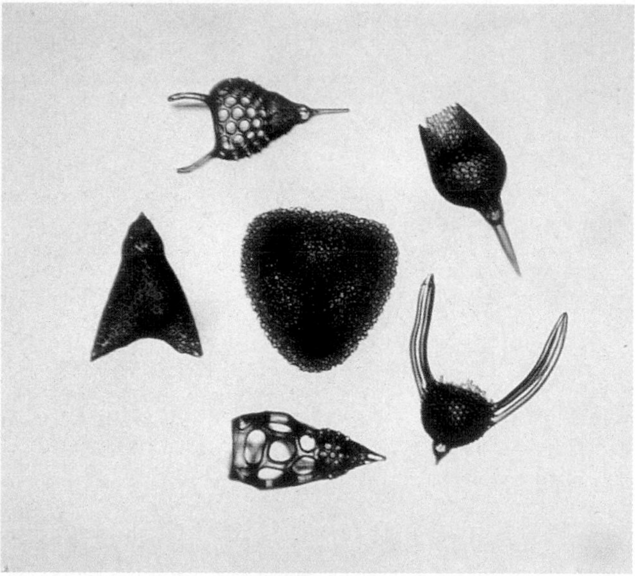

FIGURE 26-4 Radiolarian shells. The shells of radiolarians resemble delicate glass sculptures.

Greek meaning "false feet") and are a means of both locomotion and food procurement. In fact, these cell extensions give the amebas their phylum name, *Rhizopoda,* which means "rootlike feet." Shown in Figure 26-3, the pseudopods simply stream around the ameba's prey, engulfing it within a *food vacuole.* Enzymes digest the contents of the vacuole, which are then absorbed into the cytoplasm to be further broken down for energy and biosynthesis. A second type of vacuole, the *contractile vacuole,* pumps excess liquid from the ameba. Amebas reproduce by binary fission: the nucleus reproduces by mitosis and then the cell splits in two.

Amebas are abundant throughout the world in fresh water and saltwater, as well as in the soil. Many species are parasites of animals, including humans, and can cause diseases such as amebic dysentery, an infection of the digestive system that produces a diarrhea containing blood and mucus. Although amebic dysentery is a disease associated with poor sanitation and is found primarily in the tropics, medical researchers estimate that about 2 million Americans are infected with the causative agent, *Entamoeba histolytica.*

Certain groups of ameba secrete shells that cover and protect their cells. The *radiolarians* secrete shells made of silica that are glasslike and delicate (Figure 26-4). The *foraminifera* (or forams) secrete beautifully sculpted shells made of calcium carbonate ($CaCO_3$) (Figure 26-5, *A*). The name *fo-*

protists members of the kingdom Protista, which includes unicellular eukaroytes as well as some multicellular forms.

protozoans (**pro** tuh **zoe** uns) animallike protists that are heterotrophs: they take in and use organic matter for energy.

algae (**al** jee) plantlike protists that are photosynthetic autotrophs. Algae contain chlorophyll and manufacture their own food using energy from sunlight.

slime molds protists that are funguslike in one phase of their life cycle and amebalike in another phase of their life cycle.

water molds funguslike protists that predigest and absorb food as fungi do. Water molds thrive in moist places and aquatic environments and parasitize plants and animals.

amebas (uh **mee** buhs) a protozoan phylum whose members have changing shapes brought about by cytoplasmic streaming, which forms cell extensions called pseudopodia.

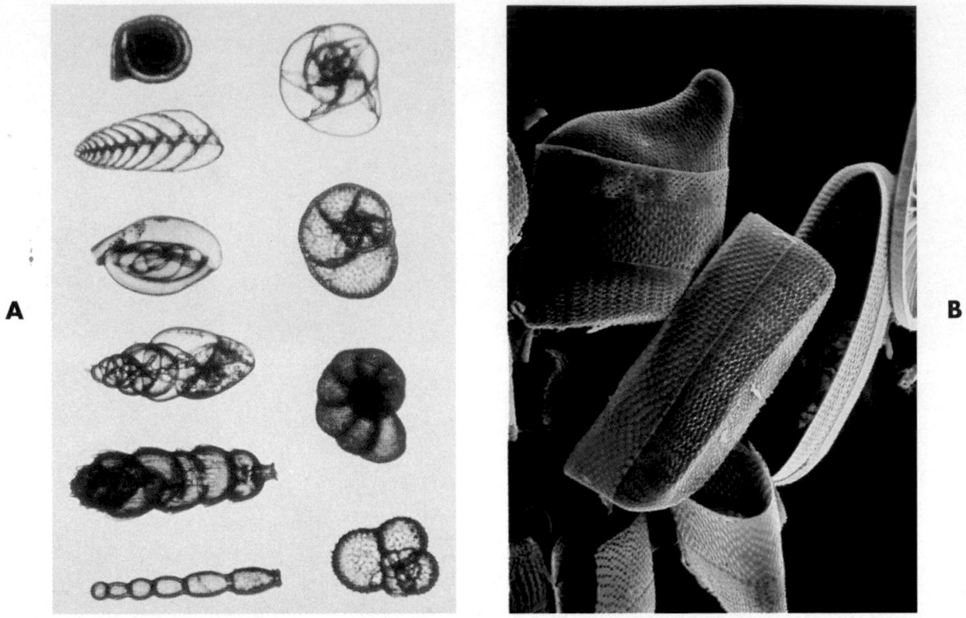

FIGURE 26-5 **Foraminiferan shells. A,** These beautiful foraminiferan shells are made out of calcium carbonate. **B,** The shells of foraminifera are punctured with tiny holes through which the foraminiferan pseudopods poke.

raminifera means "hole bearers" and refers to the microscopic holes in their shells through which their pseudopods protrude (Figure 26-5, *B*). Food particles stick to these cellular extensions and are then absorbed into the cell.

Forams are abundant in the sea—so abundant, in fact, that their shells litter the sea floor. When studying geological strata, scientists often use the forams as indicators of geological age by noting the types of forams present in ancient rock. Interestingly, the white cliffs of Dover (Figure 26-6) are actually masses of foram shells, uplifted millions of years ago with the sea floor in an ancient geological event.

Amebas are protozoans that have changing shapes brought about by cytoplasmic streaming, which forms cell extensions called pseudopodia.

Flagellates

Flagellates are an interesting group because they are so diverse; a few representative genera of flagellates are shown in Figure 26-7. Although they all have at least one *flagellum* (a long, whiplike organelle of motility), some members of this phylum have many flagella, and some have thousands! Their phylum name is *Zoomastigophora,* which means "animal whip."

FIGURE 26-6 **The white cliffs of Dover.** The picturesque white cliffs are actually composed of masses of foram shells, which were uplifted from the sea floor millions of years ago.

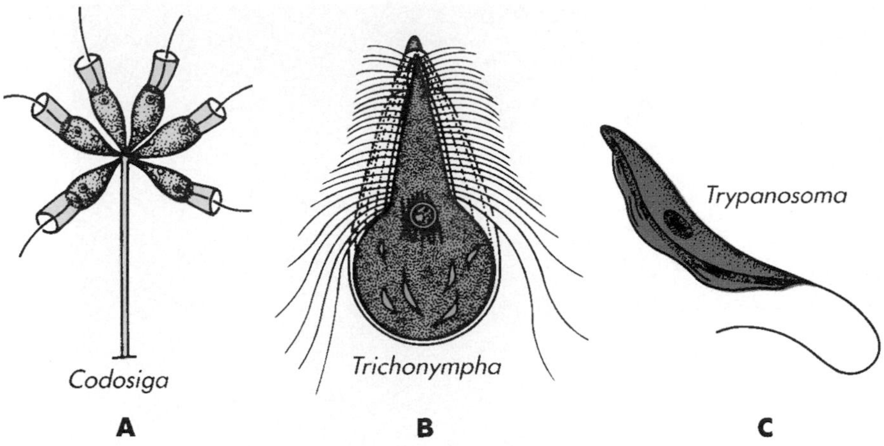

FIGURE 26-7 The flagellates. A, *Codosiga,* a colonial flagellate that remains attached to its substrate. **B,** *Trichonympha,* one of the flagellates that inhabits the gut of termites and woodfeeding cockroaches. *Trichonympha* ingests wood cellulose, which it finds in abundance in the digestive tracts of its hosts. **C,** *Trypanosoma* causes sleeping sickness in humans. It has a single flagellum.

All flagellates have a relatively simple cell structure. They do not have cell walls or protective outer shells as some of the amebas or ciliates do. They also have no complex internal digestive system of organelles as the ciliates do. A flagellate simply absorbs food through its cell membrane, sometimes using its flagella to ensnare food particles.

The flagellates are generally found in lakes, ponds, or moist soil where they can absorb nutrients from their surroundings. The colonial flagellate *Codosiga* shown in Figure 26-7, *A,* for example, consists of groups of cells that are often found anchored to the bottom of a lake or pond by a cellular stalk. The flagella create currents in the water that draw food toward the cells. The flagellate *Trichonympha,* shown in Figure 26-7, *B,* lives a protected life in the gut of termites, digesting wood particles the termite eats. Many flagellates are found living within other organisms; some of these relationships are not harmful to the hosts but other relationships are. Figure 26-7, *C,* shows the flagellate *Trypanosoma* that lives in a parasitic relationship with certain mammals, including humans, causing the disease sleeping sickness.

> *Flagellates are protozoans characterized by fine, long, hairlike cellular extensions called flagella.*

One class of flagellates, the **euglenoids,** generally have chloroplasts and make their own food by photosynthesis. Each euglenoid has two flagella, a short one and a long one, that are located on the anterior end of the cell. They move by whipping the long flagellum (Figure 26-8). The euglenoids reproduce asexually by transverse fission, a process in which the parent cell divides across its short axis (see Figure 26-12, *A*). No sexual reproduction is known among this group, which is named after its most well-known member: *Euglena.*

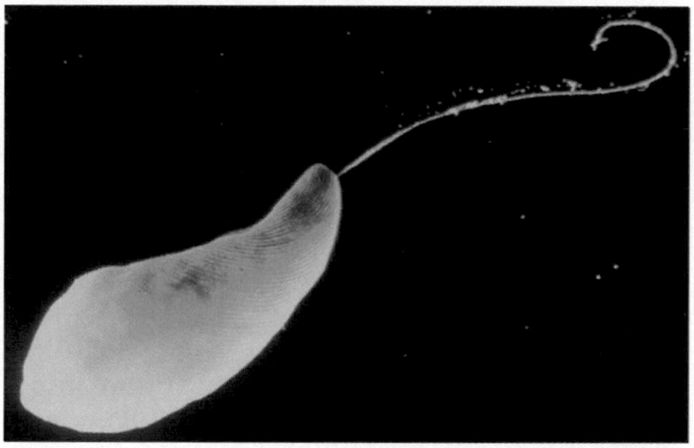

FIGURE 26-8 ***Euglena,* a euglenoid.** Euglenoids make their own food by photosynthesis and swim by means of flagella.

Euglena lives in ponds and lakes and can withstand stagnant water. It has a hard yet flexible covering beneath its plasma membrane called a *pellicle,* with ridges spiraling around its body. These ridges can be seen clearly in Figure 26-8. Two organelles, an *eyespot* and a *photoreceptor,* help *Euglena* stay near the light. (The name *Euglena,* in fact, means "true eye.") The photoreceptor, located

flagellates (flaj uh lates) a protozoan phylum whose members are characterized by fine, long, hairlike cellular extensions called flagella.
euglenoids (yoo glee noyds) members of a class of flagellates that have chloroplasts and make their own food by photosynthesis.

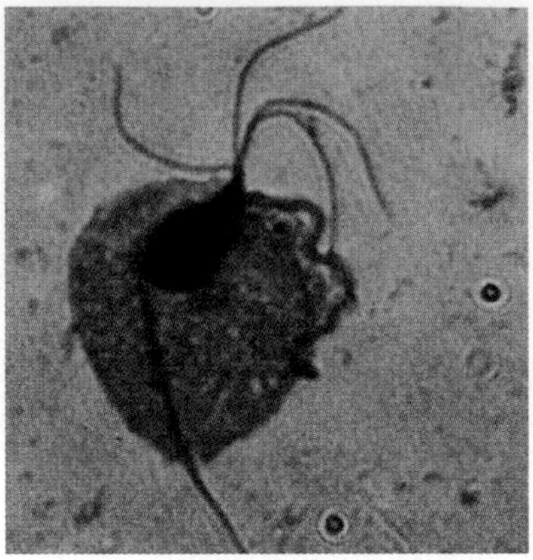

FIGURE 26-9 The causative agent of trichomoniasis. The flagella of the protozoan *Trichomonas vaginalis* can be clearly seen in this photomicrograph.

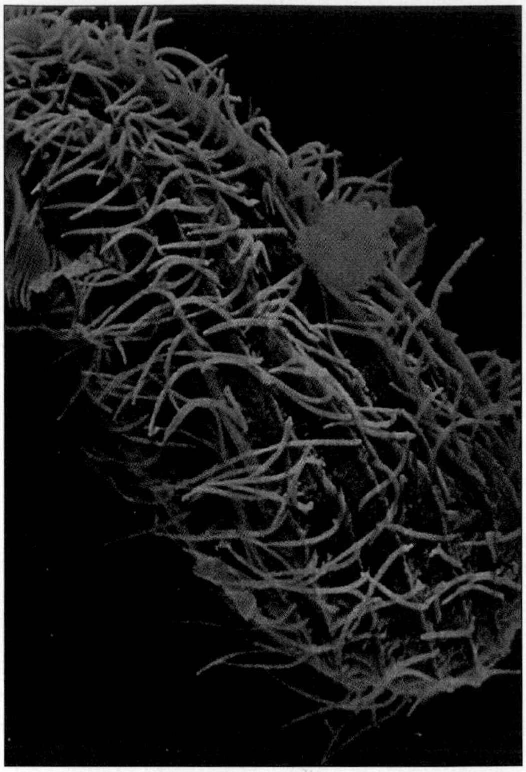

FIGURE 26-10 The ciliates. Cilia can be seen covering the paramecium shown in this photo. Note the concentration of cilia at the gullet, visible in the upper left corner.

near the base of its longer flagellum, is shaded by the nearby eyespot. As light filters through the pigment of the eyespot, the receptor senses the direction and intensity of the light source. Information from the receptor assists the movement of *Euglena* toward the light, a behavior known as **positive phototaxis.**

Interestingly, *Euglena* can survive without light—the chloroplasts simply become small and nonfunctional, and the organism begins to absorb food like a heterotroph. *Euglena* is a good example of an organism that is difficult to classify because of its ability to change its mode of nutrition. In fact, some euglenoids never have chloroplasts and live in a totally heterotrophic existence.

> *Euglenoids are flagellates, most of which have chloroplasts.*

One species of flagellate, *Trichomonas vaginalis* (Figure 26-9), causes genital infections in both women and men (despite its name). *T. vaginalis* feeds on bacteria and cell secretions in the urogenital structures it infects. In women, it infects the vagina, cervix, and vulva. However, it can cause infection only when the pH level of the vagina is elevated because it cannot survive in the vagina's normally acidic environment. (The pH of the vagina may become elevated when a woman is taking antibiotics. These drugs sometimes kill the resident vaginal bacteria that secrete acids.)

A *Trichomonas* infection in females is characterized by itching, burning, and a profuse discharge that may be bloody or frothy. This organism can survive on objects for some time; thus an infection can be contracted from a toilet seat or garments that are contaminated with the organism. Transmission is usually by sexual intercourse, however.

In males, *Trichomonas* usually causes an inflammation of the urethra, but the epididymis, prostate gland, and seminal vesicles can also become infected. An infected man may have no symptoms or may experience a slight discharge from the urethra, painful urination, or increased urination. Infection with *Trichomonas vaginalis* is treated with antiparasitic drugs.

Ciliates

Ciliates (phylum Ciliophora) get their name from a Latin word meaning "eyelash"—a name that is descriptive of the fact that all or parts of these cells are covered with hairlike extensions called *cilia* (Figure 26-10). These cilia beat in unison, moving the cell about or creating currents that move food particles toward the gullet of the cell.

Ciliates possess a wide array of cellular organelles that perform functions similar to the organs of multicellular organisms. An example of this interesting cellular organization is shown in the diagram of *Paramecium* in Figure 26-11. The paramecium is a ciliate that is classically used as one example of this group. Cilia protrude through

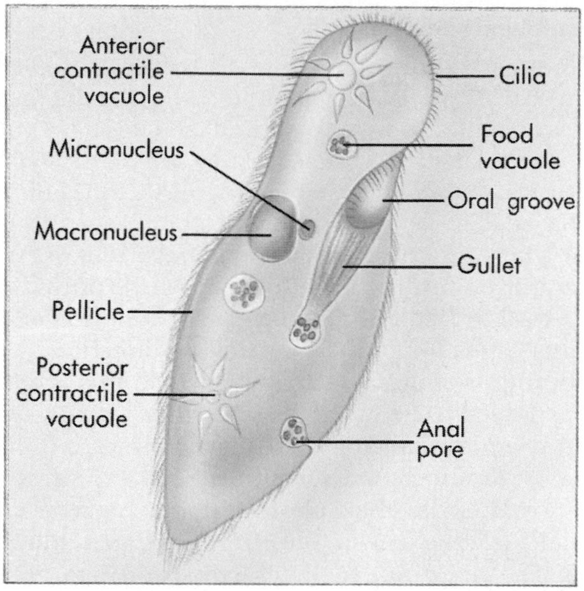

FIGURE 26-11 ***Paramecium.*** This diagram shows the location of the contractile vacuoles, the pellicle, micronucleus, macronucleus, cilia, oral groove, gullet, food vacuole, and anal pore.

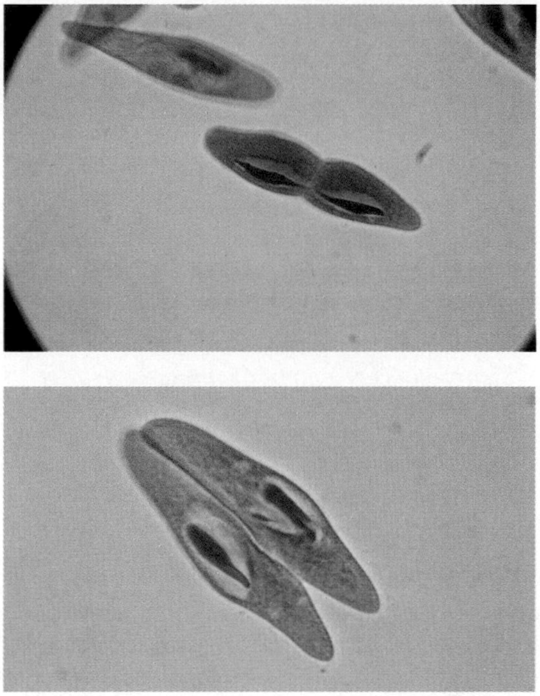

FIGURE 26-12 Reproduction and the exchange of genetic material in *Paramecium*. A, Transverse fission. When a mature *Paramecium* divides, two complete individuals result. **B,** Conjugation. In this process, two Paramecia exchange a portion of their genetic material, which promotes genetic variability, but they do not reproduce.

holes in the paramecium's outer covering, or *pellicle.* The micronuclei (there may be several) function in sexual reproduction, whereas the single macronucleus controls cell metabolism and growth. The beating cilia of the paramecium sweep food into its *gullet.* From the gullet, food passes into a *food vacuole* where enzymes and hydrochloric acid aid in digestion. After absorption of the digested material is complete, the vacuole empties its waste contents into the *anal pore,* located in a special region of the pellicle. The waste then leaves the cell by a process similar to exocytosis. The contractile vacuoles expel excess water from the cell.

Paramecia reproduce asexually by *transverse fission* (Figure 26-12, *A*). In addition, paramecia undergo a type of sexual reproduction shown in Figure 26-12, *B,* called *conjugation.* However, conjugation is not really a reproductive process. Instead, two cells exchange a portion of their genetic material. Although conjugation does not produce offspring cells, it does promote genetic variability among cells that normally produce clones of identical cells when they reproduce. Genetic variability enhances the ability of the population to survive. Some algae, fungi, and bacteria also exchange genetic material in similar processes, which are also termed conjugation.

Most ciliates live in fresh water or salt water and do not infect other organisms. However, one species called *Balantidium coli* inhabits the intestinal tracts of pigs and rats. People who come in contact with this protozoan on farms or in slaughterhouses can become infected. The ciliates embed themselves into the lining of the intestines, producing sores and causing dysentery, similar to amebic dysentery. Occasionally, epidemics of balantidiasis occur in areas having poor sanitation.

> *Ciliates are protozoans characterized by fine, short, hairlike cellular extensions called cilia.*

Sporozoans

All **sporozoans** (phylum Sporozoa) are nonmotile, spore-forming parasites of vertebrates, including

positive phototaxsis (foe toe **tak** sis) an innate movement of an organism toward a source of light.

ciliates (**sil** ee uts *or* **sil** ee **ates**) a protozoan phylum whose members are characterized by fine, short, hairlike extensions called cilia.

sporozoans (**spor** uh **zoh** uns) a protozoan phylum whose members are nonmotile, spore-forming parasites of vertebrates, including humans.

humans. Sporozoans have complex life cycles that involve both asexual and sexual phases. Their spores are small, infective bodies that are transmitted from host to host by various species of insects.

The sporozoan used classically to represent this phylum of protists is *Plasmodium,* the causative agent of malaria. Approximately 1 million people die of this disease each year; therefore it is considered one of the most serious diseases in the world.

> *Sporozoans are nonmotile protozoans that are parasites of vertebrates, including humans. They undergo complex life cycles in which they are passed from host to host by various species of insects.*

Plantlike Protists: Algae

Algae are widely distributed in the oceans and lakes of the world, floating on or near the surface of the water no lower than the sun's rays can reach. They are eukaryotic organisms that contain chlorophyll and carry out photosynthesis, so they can be thought of as plantlike. However, algae lack true roots, stems, leaves, and vascular tissue (an internal water-carrying system). In addition, many contain other pigments that mask the green chlorophyll. Different pigments are associated with different groups of algae; this characteristic is used in the classification of these organisms.

Some systems of classification place the unicellular algae in the protist kingdom and the multicellular algae in the plant kingdom. However, current thinking favors placing all the algae with the protists, broadening the scope of this kingdom. Formerly, the protist kingdom included only the protozoans and the unicellular algae. It now includes multicellular algae and the funguslike protists, each of which lacks some of the characteristics of either plants or fungi. These changes reflect the ongoing debate among scientists regarding the evolutionary relationships among organisms and how organisms should be classified to reflect these relationships. Classifying all algae as protists suggests closer evolutionary relationships among the various phyla of algae and other protists than between the algae and the plants. The green algae, thought to be the ancestors to the land plants, are sometimes still placed in the plant kingdom.

There are three phyla of multicellular algae: the **brown algae,** the **green algae** (which includes many unicellular forms), and the **red algae.** These are the phyla that were formerly classified with the plants. There are two phyla of unicellular algae: the dinoflagellates and the **golden algae.**

Dinoflagellates

Although the **dinoflagellates** (phylum Dinoflagellata) have flagella, they look nothing like the flagellates or euglenoids. Many dinoflagellates have outer coverings of stiff cellulose plates, which give them very unusual appearances (Figure 26-13). Their flagella beat in two grooves, one encircling the cell like a belt and the other perpendicular to it. As they beat, the encircling flagellum causes the dinoflagellate to spin like a top; the perpendicular flagellum causes movement in a particular direction. This spinning is a characteristic for which this phylum was named. The word *dinoflagellate* comes from Latin words meaning "whirling swimmer."

Most dinoflagellates live in the sea and carry on photosynthesis. Their photosynthetic pigments are usually golden brown, but some are green, blue, or red. The red dinoflagellates are also called *fire algae.* In coastal areas, these organisms often experience population explosions or "blooms," causing the water to take on a reddish hue referred to as a *red tide.* Red tides destroy other living things because many species of dinoflagellates produce powerful toxins. These poisons kill fishes, birds, and marine mammals (Figure 26-14). In addition,

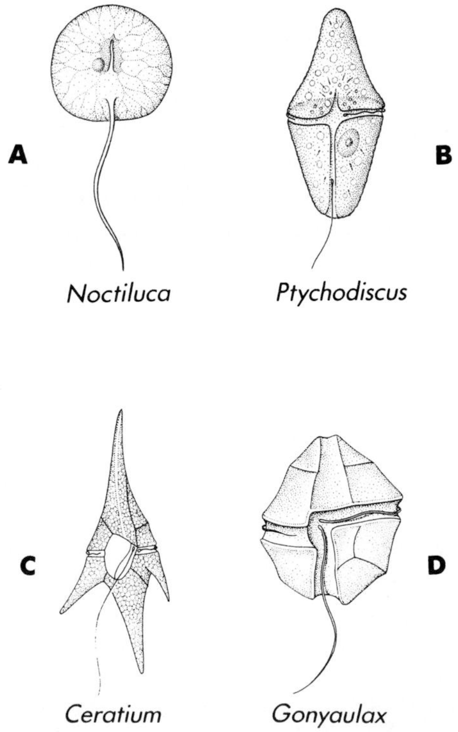

FIGURE 26-13 The dinoflagellates. A, *Noctiluca* lacks the heavy cellulose armor characteristic of most dinoflagellates. **B,** *Ptychodiscus.* **C,** *Ceratium.* **D,** *Gonyaulax.*

FIGURE 26-14 Red tide. Red tides are caused by a population explosion of certain dinoflagellates. The release of toxins from these dinoflagellates can poison marine life.

shellfish strain these dinoflagellates from the water and store them in their bodies. Although the shellfish are not harmed, they are poisonous to humans and other animals that eat them.

Dinoflagellates reproduce primarily by longitudinal cell division, but sexual reproduction has also been shown to occur in more than 10 genera of dinoflagellates.

> *Dinoflagellates, a type of flagellated unicellular algae, are characterized by stiff outer coverings. Their flagella beat in two grooves, one encircling the cell like a belt and the other perpendicular to it.*

Golden Algae

The three types of organisms found in phylum Chrysophyta are the **yellow-green algae,** the **golden-brown algae,** and the **diatoms.** These organisms are named after the gold-green color of the photosynthetic pigments in their chloroplasts. (The prefix *chryso-* means "color of gold.") An unusual characteristic of these organisms is that they store food as oil. Some forms smell "fishy," giving unpleasant odors to the freshwater lakes they inhabit.

The most well-known members of this group, abundant in both the ocean and in fresh water, are the diatoms—microscopic aquatic pillboxes. Diatoms are made up of top and bottom shells that fit together snugly (Figure 26-15, *A*). These organisms reproduce asexually by separating their top from their bottom, each half then regenerating another top or bottom shell within itself. The shells are composed of silica, so like the radiolarians, the diatoms look somewhat glasslike. However, the shells of the diatoms are so characteristically striking and intricate (Figure 26-15, *B*) that it would be hard to confuse them with any other group of protists. Interestingly, the shells of fossil diatoms often form very thick deposits on the sea floor, which are sometimes mined commercially. The resulting "diatomaceous earth" is used in water filters, as an abrasive, and to add the sparkling quality to products such as the paint used on roads and frosted fingernail polish.

brown algae one of the three phyla of multicellular algae that were formerly classified with the plants; brown algae are found predominantly on northern, rocky shores, and can grow to enormous sizes.

green algae one of the three phyla of multicellular algae; green algae include both unicellular and multicellular forms. Most forms of green algae are aquatic but some species live in moist places on land.

red algae one of the three phyla of multicellular algae; red algae play an important role in the formation of coral reefs and produce gluelike substances that make them commercially useful.

golden algae a phyla of unicellular algae that have gold-green photosynthetic pigments and store food as oil.

dinoflagellates (**dye** no **flaj** uh luts *or* **dye** no **flaj** uh lates) a type of flagellated unicellular algae that is characterized by stiff outer coverings. Their flagella beat in two grooves, one encircling the cell like a belt and the other perpendicular to it.

yellow-green algae a type of golden algae.

golden-brown algae a type of golden algae.

diatoms (**die** uh toms) the most well-known type of golden algae. These organisms look like microscopic pillboxes because they are made up of top and bottom shells that fit together.

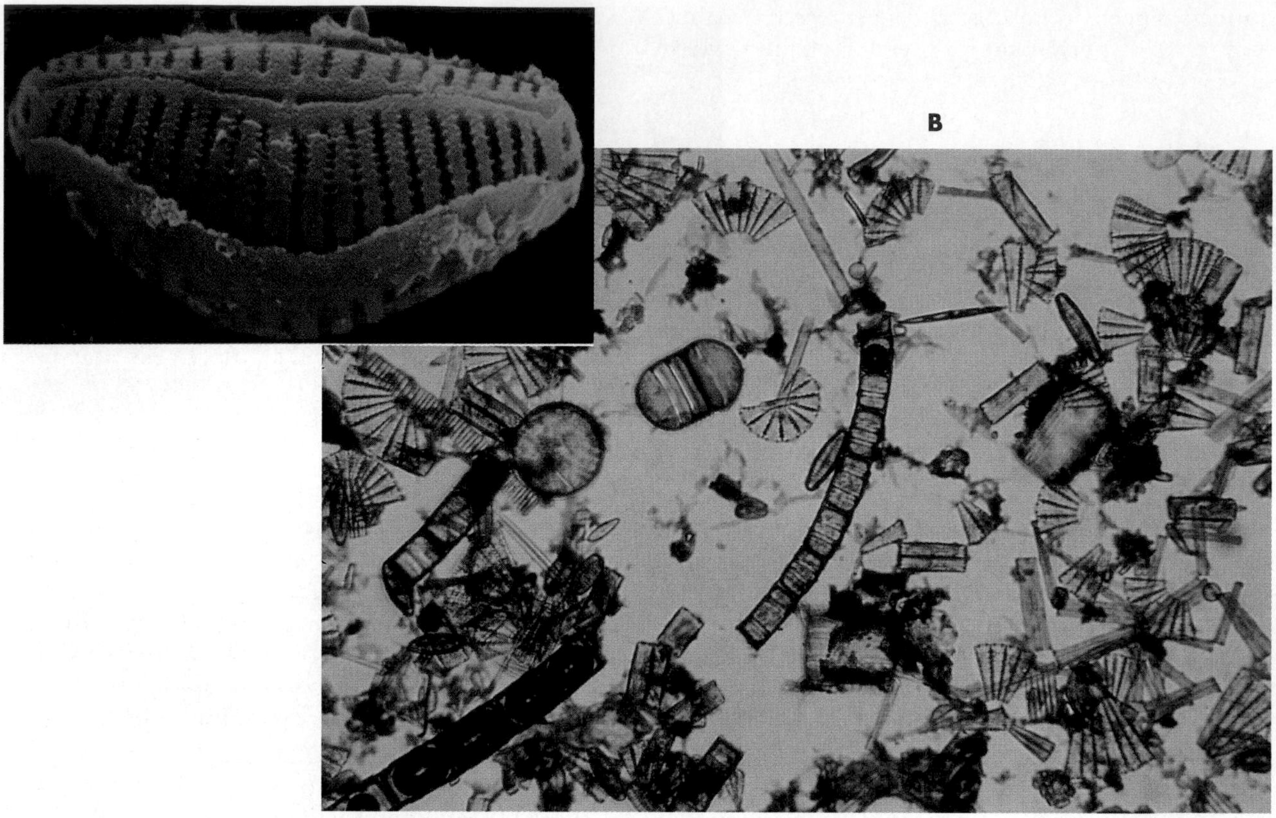

FIGURE 26-15 Diatoms. A, Diatoms are composed of a top and bottom shell that fit together. **B,** Several different types of diatoms.

> *The golden algae have gold-green photosynthetic pigments and store food as oil. They are often represented by diatoms, organisms that look like microscopic pillboxes.*

Brown Algae

The brown algae (phylum Phaeophyta) are the dominant algae of the rocky, northern shores of the world. The types of brown algae that grow attached to rocks at the shoreline are known as *rockweed* (Figure 26-16, *A*). Their puffy air bladders keep the plant afloat during high tide. One type of rockweed is also called *sargasso weed* and gave the Sargasso Sea its name. The Sargasso Sea is an area of ocean in the mid-Atlantic, east of Bermuda, with unusual water and current patterns that cause it to be quite calm. This calmness allows floating species of the Sargasso weed to proliferate and dominate the area.

The large brown alga with enormous leaflike structures is kelp (Figure 26-16, *B*). These algae are an important source of food for fish and invertebrates as well as some marine mammals and birds that live among these seaweeds. Some genera of the kelps are among the longest organisms in the world (rivaling the height of the giant sequoia trees), reaching lengths of up to 100 meters (328 feet)! These algae usually have a structure in which a rootlike portion, descriptively termed a *holdfast,* anchors the seaweed to the ocean floor or to rocks. A stemlike *stipe* carries the leaflike *blades* of the seaweed, which float on or in the water, capturing the sun's rays (Figure 26-16, *C*).

Most brown algae have life cycles that parallel the generalized life cycle of plants (see Chapter 27 and Figure 27-1). The other multicellular algae (brown algae and red algae) also have plantlike life cycles. However, the life cycles of the red algae are quite complex.

> *Brown algae are large, multicellular algae found predominantly on northern, rocky shores. Some grow to enormous sizes.*

Green Algae

The green algae, or Chlorophyta (literally, "green plants"), are an extremely varied phylum of protists. In fact, more than 7000 species exist! Of these species, most are aquatic (as are other algae), but some are *semiterrestrial.* Semiterrestrial algae live in moist places on land, such as on tree trunks, on snow, or in the soil. These algae are primarily unicellular microscopic forms, but some are multicellular. Green algae show many similarities to land plants: they store food as starch, they have a similar chloroplast structure, many genera have cell walls composed of cellulose, and their chloroplasts contain chlorophyll *a* and *b*. For these reasons, scientists think the green algae were ancestors to the plant kingdom.

Well-known among the unicellular green algae is the genus *Chlamydomonas* (Figure 26-17, *A*). Individuals are microscopic, green, and rounded and have two flagella at their anterior ends. They are aquatic and move rapidly in the water as a result of the beating of their flagella in opposite directions. They have eyespots that help direct the algae to the sunlight. The life cycle of *Chlamydomonas* is very simple. This haploid organism reproduces asexually by cell division and sexually by the functioning of some of its cells as gametes. After gametes fuse, the diploid zygote divides by meiosis, restoring the haploid state.

Some genera of green algae live together in groups. *Volvox* is one of the most familiar of these colonial green algae. Each colony is a hollow sphere made up of a single layer of individual, biflagellated cells (Figure 26-17, *B*). The flagella of all of the cells beat in such a way as to rotate the colony in a clockwise direction as it moves through the water. In some species of *Volvox* there is a division of labor among the different types of cells, making them truly multicellular organisms.

The multicellular forms of green algae grow in either fresh water or salt water, such as the sea let-

A

B

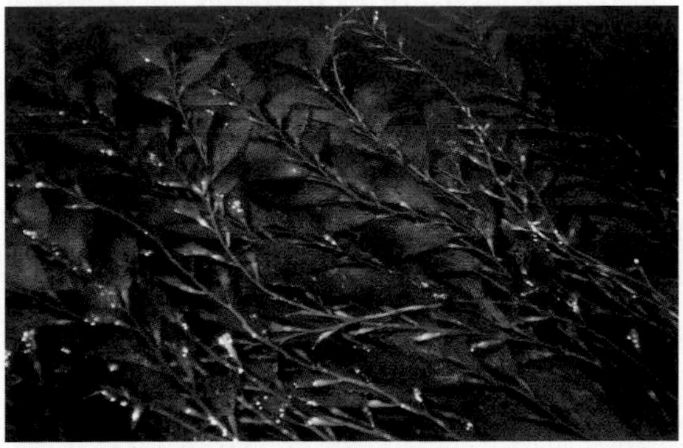

C

FIGURE 26-16 The brown algae. A, Rockweed. The air bladders attached to the algae help keep the plant afloat. **B,** Kelp. Kelp have large blades (leaflike structures) and are an important food source for marine fishes, birds, mammals, and invertebrates. **C,** A kelp forest. The blades of kelp capture sunlight, which they use for photosynthesis. Kelp forests can be quite large.

A

B

C

D

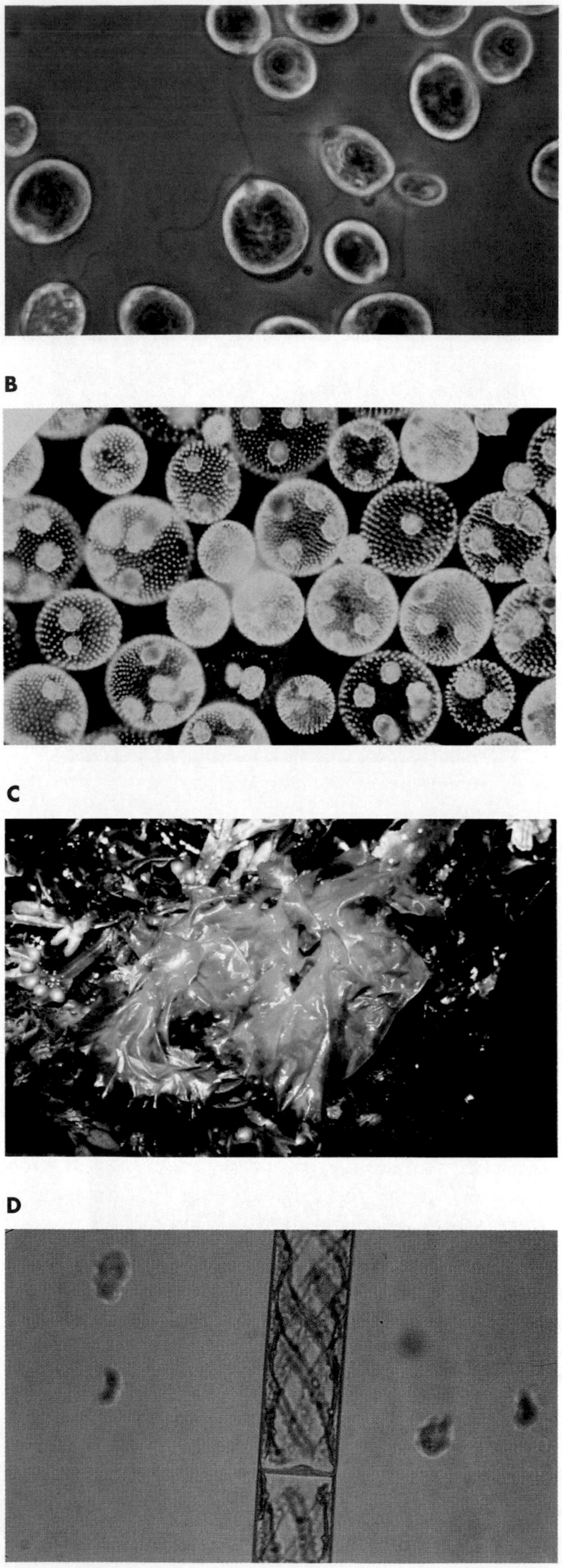

tuce *Ulva* shown in Figure 26-17, *C*. Sea lettuce is extremely plentiful in the ocean and is often found clinging to rocks or pilings. Interestingly, this alga consists of sheets of tissue only two cells thick. Another familiar multicellular green alga is the filamentous *Spirogyra*. This alga is interesting because it has spiral chloroplasts (Figure 26-17, *D*).

> *Green algae include both unicellular and multicellular forms. Most forms are aquatic, as are other algae, but some species live in moist places on land.*

Red Algae

Almost all red algae, or Rhodophyta ("red plants"), are multicellular, and most of their species are marine. Their color comes from the types and amount of photosynthetic pigments present in their chloroplasts. Many species have a predominance of red pigments in addition to chlorophyll and so are red (as their name suggests). Some red algae have a predominance of other pigments so that they look green, purple, or greenish black (Figure 26-18). The red pigment, however, is especially efficient in absorbing the green, violet, and blue light that penetrates into the deepest water. (It *reflects* red light.) This enables some of the red algae to grow at greater depths than other algae and inhabit areas in which most algae cannot exist.

The red algae produce substances that make them interesting both ecologically and economically. The coralline algae, for example, deposit calcium carbonate (limestone) in their cell walls. Along with coral animals, these red algae play a major role in the formation of coral reefs. Also, all red algae have gluelike substances in their cell walls: agar and carrageenan. Agar is used to make gelatin capsules, the material for making dental impressions and a base for cosmetics. It is also a main ingredient in the laboratory media on which bacteria, fungi, and other organisms are often grown. Carrageenan is used mainly as a stabilizer and thickener in dairy products, such as creamed soups, ice cream, puddings, and whipped cream, and as a stabilizer in paints and cosmetics. Some of the red algae are used as food in certain parts of the world, such as in Japan.

FIGURE 26-17 The green algae. A, *Chlamydomonas* is a unicellular green alga that has two flagella and an eyespot sensitive to sunlight. **B,** *Volvox* is a colonial green alga. Each colony is made up of individual cells. **C,** *Ulva,* or sea lettuce. **D,** *Spirogyra,* a filamentous green alga.

A

B

C

FIGURE 26-18 The red algae. Red algae do not always appear red. Some algae show different colors depending on their daily exposure to light. **A,** *Ahnfeltia plicata* growing on rocks. This type of alga is edible and is considered a delicacy in many Asian cultures. **B,** *Bossiella* is a coralline red alga, which means that it secretes the hard substance calcium carbonate. **C,** *Ahnfeltia concinna,* a common red alga.

> *The red algae play a major role in the formation of coral reefs and produce gluelike substances that make them commercially useful.*

Funguslike Protists

The slime molds make up two unique and interesting phyla of protists. They are called molds because they give rise to moldlike (funguslike) spore-bearing stalks during one stage of their life (see p. 584). However, slime molds look very much like amebas at other times in their life cycle, forming visible "slimy" masses. In fact, scientists now think slime molds may be most closely related to the amebas.

Slime molds are heterotrophs and feed on bacteria, which they find in damp places rich in nutrients, such as rotting vegetation (especially rotting logs), damp soil, moist animal feces, and water. The slime molds are divided into two phyla: the **cellular slime molds** and the **plasmodial (acellular) slime molds.**

> *Slime molds are organisms that are funguslike in one phase of their life cycle and amebalike in another phase of their life cycle.*

Cellular Slime Molds

Most of their lives, the cellular slime molds look and behave like amebas, moving along and capturing bacteria by means of pseudopods (Figure 26-19, *A*). At a certain phase of their life cycle, which is often triggered by a lack of food, the individual organisms move toward one another (Figure 26-19, *B*) and form a large, moving mass called a *slug* (Figure 26-19, *C*). (Their phylum name, Acrasiomycota, comes from the name of the chemical attractant, acrasin, the cellular form produces, which causes the cells to aggregate.) The slug eventually transforms into a fruiting body (Figure 26-19, *D* to *F*).

cellular slime molds a phylum of funguslike protists that look and behave like amebas during much of their life cycle but can group together and form dormant, cystlike spores when food is scarce. These spores then revert to the amebalike form when conditions are favorable again.

plasmodial (acellular) slime molds (plaz **moe** dee uhl) a phylum of funguslike protists that have an ameboid stage to their life cycle; they spend much of their time as a nonwalled, multinucleated mass of cytoplasm called a plasmodium, but can form spores when food or moisture is in short supply.

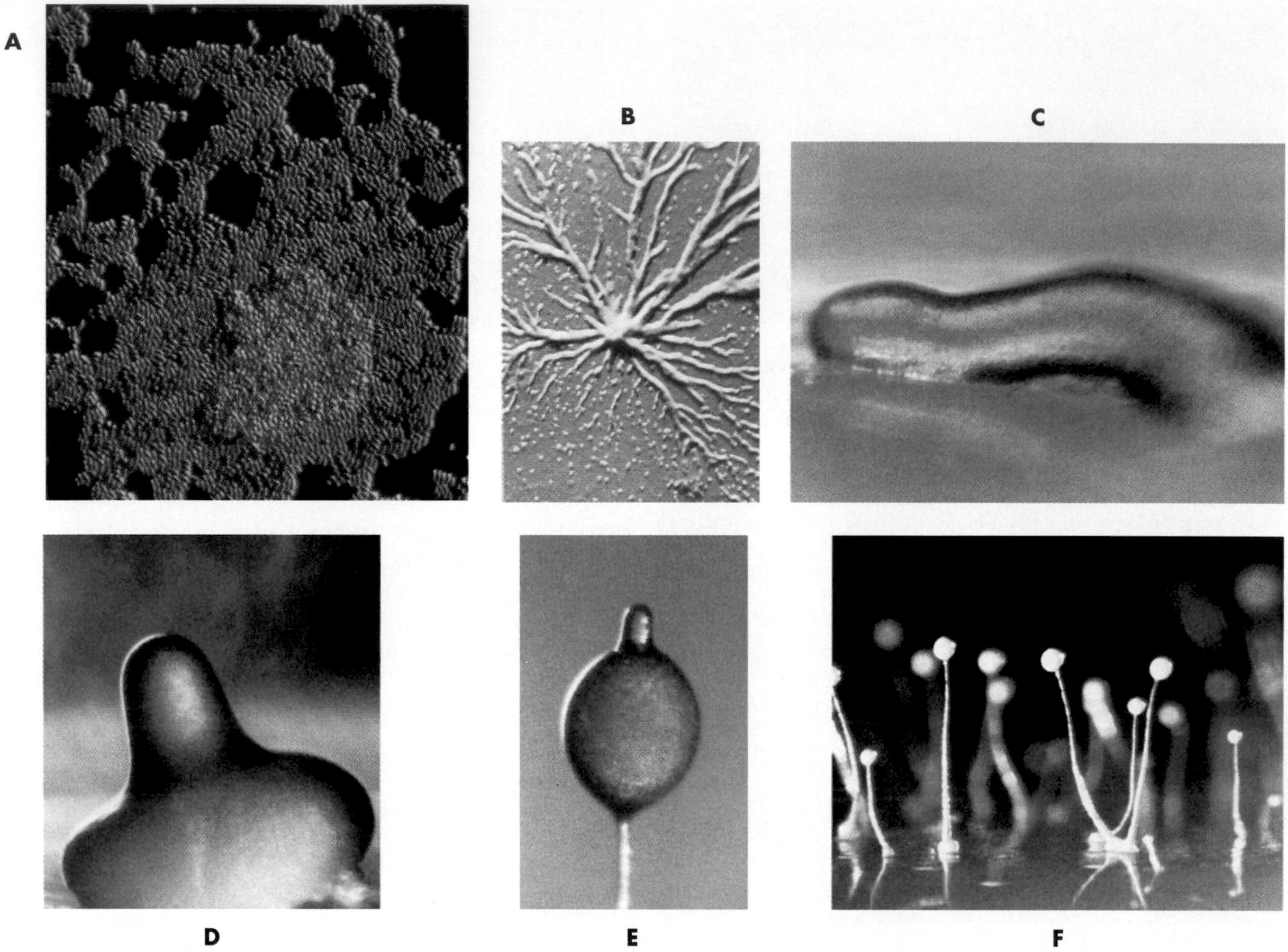

FIGURE 26-19 Development in a cellular slime mold. **A,** The ameba stage, in which the mold looks and behaves like an ameba. **B,** The amebas aggregate and move toward a fixed center. **C,** They form a slug that migrates toward light. **D,** The slug stops moving and begins to form into a fruiting body. **E,** The head of a fruiting body. **F,** Many fruiting bodies together.

The tips of the fruiting bodies contain dormant, cystlike forms of the amebalike cells and are called *spores*. Some of these spores fuse and undergo a type of sexual reproduction before being released; others do not. Each spore becomes a new "ameba" if it falls onto a suitably moist habitat. The amebas begin to feed and continue the life cycle.

Plasmodial (Acellular) Slime Molds

The plasmodial slime molds also have an ameboid stage to their life cycle, but these "amebas" are quite unusual. These bizarre organisms stream along as a **plasmodium**—a nonwalled, multinucleate mass of cytoplasm—which resembles a moving mass of slime (Figure 26-20). Their phylum name, Myxomycota, literally means "mucous mold." The plas-

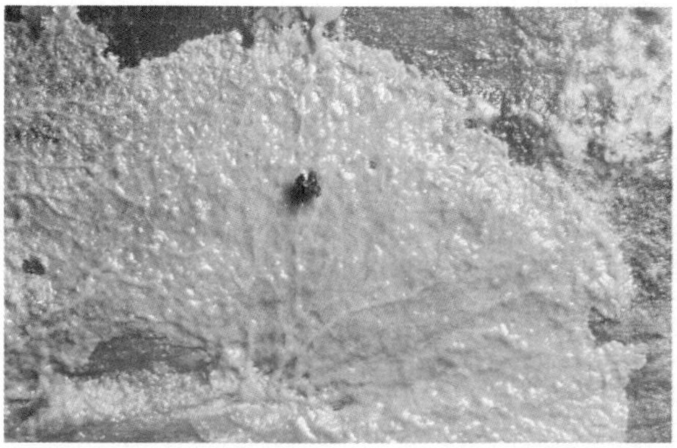

FIGURE 26-20 The plasmodial slime mold. Plasmodial slime molds move about as a plasmodium, a multinucleated mass of cytoplasm.

A
B
C

FIGURE 26-21 Spore cases of three types of plasmodial slime molds. A, *Arcyria*. B, *Lycogala*. C, *Physarium*.

modia engulf and digest bacteria, yeasts, and other small particles of organic matter as they move along. At this stage of its life cycle, a plasmodium may reproduce asexually; the nuclei undergo mitosis simultaneously, and the entire mass grows larger.

When food or moisture is in short supply, the plasmodium moves to a new area and forms spores. These spores are held in spore cases, which have a characteristic look for each genera of acellular slime mold. Figure 26-21 shows three different types of spore cases. The spores are resistant to unfavorable environmental influences and may last for years if they remain dry. Meiosis occurs in the spores. When conditions are favorable, the spore cases open and release spores that germinate into flagellated, haploid cells called *swarm cells.* The swarm cells can divide, producing more swarm cells, or can act as gametes. Gametes can fuse and form a new plasmodium by repeated mitotic divisions.

Water Molds

Taxonomists disagree as to whether the water molds should be considered protists or fungi. A primary issue is that water molds have flagellated spores, which are not characteristic of fungi. In this textbook, these organisms are classified with the funguslike protists.

If you have an aquarium and have seen white fuzz on any of your fish, you have been introduced to the water molds. They live not only in fresh water and salt water and on aquatic animals but also in moist soil and on plants. Some of the plant diseases caused by this group are late blight and downy mildew. Although this group gets the name *water molds* because many species thrive in moisture, they are sometimes called *egg fungi* because of the large egg cells present during their sexual reproduction.

> *Water molds thrive in moist places and aquatic environments, parasitizing plants and animals. During sexual reproduction, they produce large egg cells.*

Fungi

Fungi are a separate kingdom of mostly multicellular eukaryotic organisms that are **saprophytic;**

plasmodium a nonwalled, multinucleate mass of cytoplasm that resembles a moving mass of slime. Plasmodial slime molds spend much of their life cycle in this form.

saprophytic (**sap** roe **fit** ik) a term describing an organism that feeds on dead or decaying organic matter; most fungi are saprophytic.

that is, they feed on dead or decaying organic material. To do this, fungi secrete enzymes onto a food source to break it down and then absorb the breakdown products. Some fungi are **parasites** and feed off living organisms in the same way (as happens in ringworm and athlete's foot). Most fungi are multicellular. Unlike plants, fungi have no chloroplasts and do not produce their own food by photosynthesis. They are composed of slender filaments that may form cottony masses or that may be packed together to form complex structures, such as mushrooms.

The slender filaments of fungi are barely visible to the naked eye. Termed *hyphae* (sing. hypha), they may be divided into cells by cross walls called *septa* (sing. septum) or may have no septa at all. The septa rarely form a complete barrier, however; thus cytoplasm streams freely throughout either type of hypha. Because of this streaming, proteins made throughout the hyphae may be carried to their actively growing tips. As a result, the growth of fungal hyphae may be very rapid when food and water are abundant and the temperature is optimum.

A mass of hyphae is called a *mycelium* (pl. mycelia). Part of the mycelial mass grows above the food source (substrate) and bears reproductive structures (Figure 26-22). The rest grows into the substrate. The part of the mycelium embedded in the food source secretes enzymes that digest the food. For this reason, many fungi are harmful because their mycelia grow into and decay, rot, and spoil foods (and sometimes other organic substances, such as leathers). In addition, some fungi cause serious diseases of plants and animals, including humans.

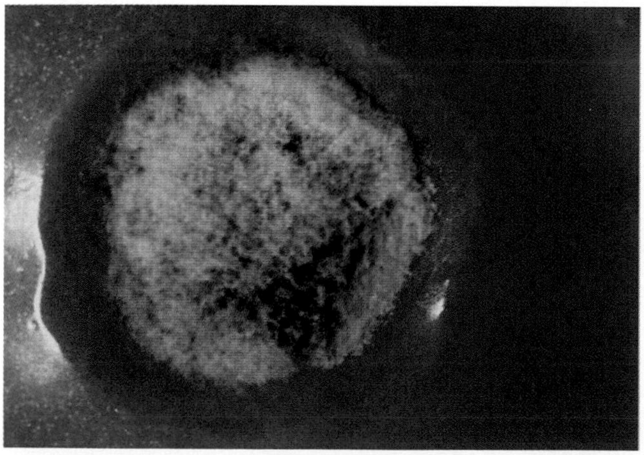

FIGURE 26-22 The structure of mold. A cottony mold growing on a tomato.

In their roles as decomposers, fungi may seem troublesome to humans, but they are essential to the cycling of materials in ecosystems (see Chapter 38). In addition, many antibiotics (drugs that act against bacteria) are produced by fungi. Yeasts, which are single-celled fungi, are also useful in the production of foods such as bread, beer, wine, cheese, and soy sauce. These foods all depend on the biochemical activities of yeasts, in which they produce certain acids, alcohols, and gases as byproducts of their metabolism of sugar. Yeasts can also cause infections in humans, such as vaginal and oral yeast infections.

Most fungi reproduce both asexually and sexually. (Figure 26-23 shows a generalized life cycle for many fungi.) During sexual reproduction, hyphae of two genetically different mating types (called + and − strains) come together, and their haploid nuclei fuse and produce a diploid zygote. This zygote germinates into diploid hyphae, which bear **spores.** Spores are reproductive bodies formed by cell division (mitosis or meiosis) in the parent organism. These spores are formed by meiosis in a diploid parent and by mitosis in a haploid parent; the spores are always haploid. They germinate into haploid hyphae after being released and finding appropriate growth conditions. The cycle comes full circle when their hyphae fuse with those of a genetically different mating type, forming zygotes. Some fungi may reproduce asexually by budding, by growing new hyphae from fragments of parent hyphae, or by producing spores by mitosis.

Fungi are classified into divisions rather than into phyla as plants are. Three divisions of fungi include those organisms that have a distinct sexual phase of reproduction: the **zygote-forming fungi,** the **sac fungi,** and the **club fungi.** A fourth group is a "catch-all" division in which organisms are placed because a sexual stage of reproduction has never been observed. This division is termed the **imperfect fungi.**

Fungi are eukaryotic organisms that feed on dead or decaying organic material or parasitize living organisms. Although some fungi cause disease in plants and animals, some are important in the production of food, and most are ecologically important decomposers.

Zygote-Forming Fungi

You may have seen black mold *(Rhizopus)* growing on bread or other food (Figure 26-24, *A*), but you may not have seen this mold under a microscope

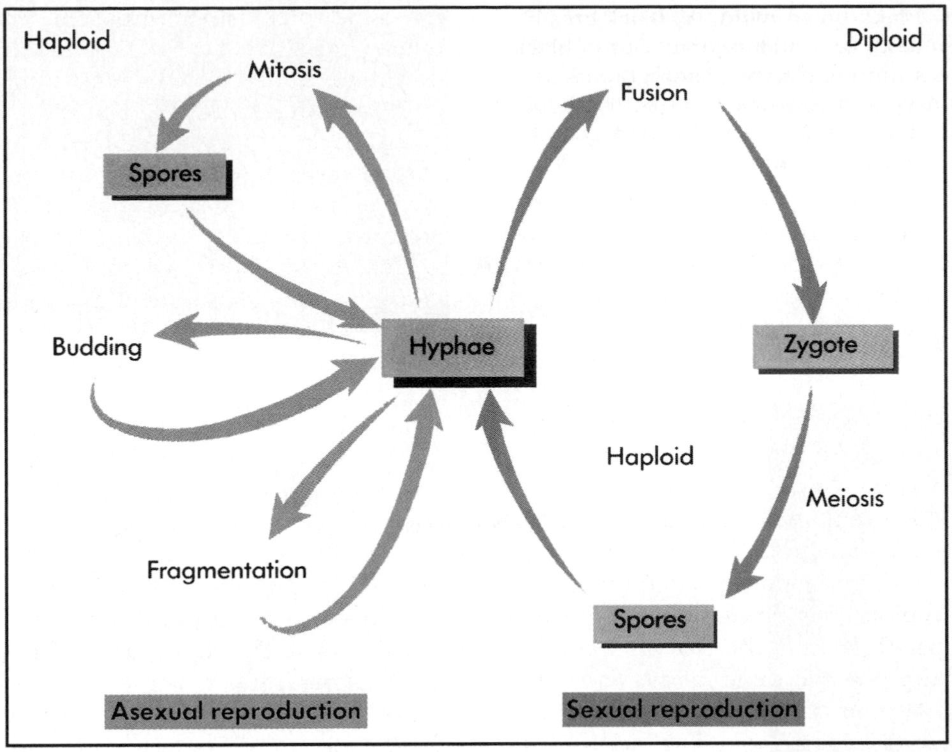

FIGURE 26-23 A generalized life cycle for fungi. Fungi alternate between sexual and asexual reproductive stages.

(Figure 26-24, *B*). Members of this group are found on decaying food and other organic material and are characterized by their formation of sexual spores called **zygospores.** For this reason, this group of fungi is called the *zygote-forming fungi* (division Zygomycota). The life cycle of these fungi (Figure 26-24, *C*) parallels the generalized life cycle shown in Figure 26-23 quite closely.

Zygote-forming fungi, such as black bread mold, live on decaying organic material and are characterized by their formation of sexual spores called zygospores.

Sac Fungi

The sac fungi are the largest division of fungi, having at least 30,000 named species. They live in a wide variety of places, such as in the soil, in salt water and fresh water, on dead plants and animals, and on animal feces. They are called *sac fungi* because their sexual spores, which occur in groups of eight, are enclosed in saclike structures called **asci** (sing. ascus). Their life cycles are similar to the club fungi, which will be discussed shortly.

Among the sac fungi (division Ascomycota) are such familiar and economically important fungi as yeasts, cup fungi (Figure 26-25), and truffles. Unfortunately, this class of fungi also includes many

parasites organisms that feed on other, living organisms; some fungi are parasitic.

spores reproductive bodies formed by cell division (mitosis or meiosis) in the parent organism. Spores are formed by meiosis in a diploid parent and mitosis in a haploid parent; the spores are always haploid.

zygote-forming fungi one of the three divisions of fungi. Zygote-forming fungi have a distinct sexual phase of reproduction that is characterized by the formation of sexual spores called zygospores.

sac fungi one of the three divisions of fungi. Sac fungi live in both aquatic and terrestrial environments and are characterized by sexual spores borne in saclike structures. Familiar examples of this group are cup fungi and yeasts.

club fungi one of the three divisions of fungi. Club fungi have club-shaped structures from which unenclosed spores are produced. Mushrooms, puffballs, and shelf fungi are all types of club fungi.

imperfect fungi one of the three divisions of fungi. Imperfect fungi have no sexual stage of reproduction and reproduce asexually by spores. Some members of this division are sources of antibiotics or are important in food production, but may cause disease in humans.

zygospores (zye go **sporz**) sexual spores formed by the zygote-forming fungi.

asci (**as** kye) saclike structures that enclose the sexual spores of sac fungi.

FIGURE 26-24 Black bread mold. A, Black bread mold growing on food. **B,** A microscopic view of black bread mold. This scanning electron micrograph has been color enhanced to bring out contrasts. The hyphae (*green strands*) and the sporangia (*yellow balls*) can be clearly seen. **C,** The life cycle of black bread mold.

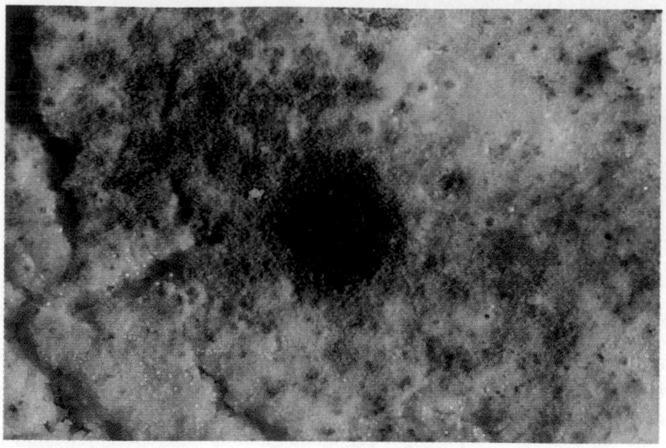

A

B

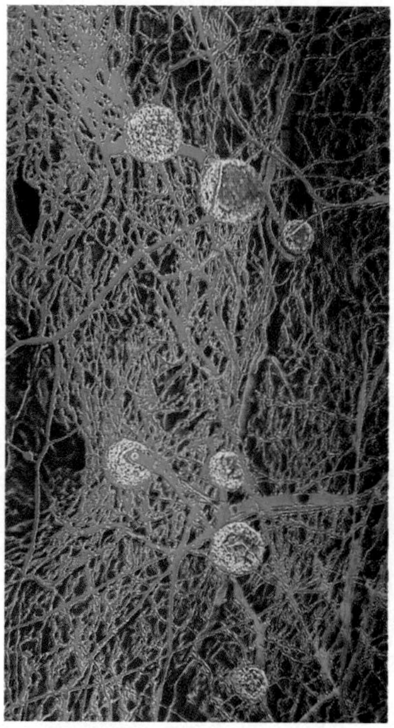

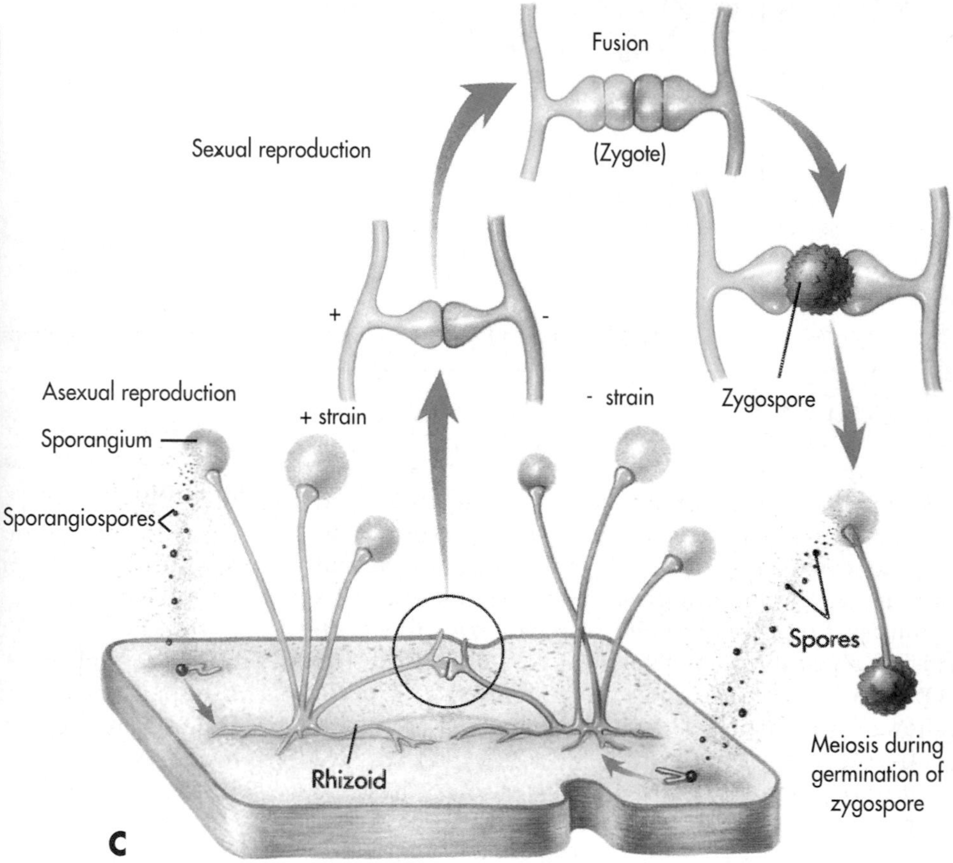

C

of the most serious plant pathogens, including the causative agent of Dutch elm disease *(Ceratocystis ulmi),* which has killed millions of elms in North America and Europe (Figure 26-26). The spores of the mold causing this disease are carried from tree to tree by a bark beetle.

Although single cells, the yeasts are also sac fungi and are one of the most economically important of the class because of their use in the production of various foods. Most of the reproduction of the yeasts is asexual and takes place by binary fission or by budding (the formation of a smaller cell from a larger one). Sometimes, however, whole yeast cells may fuse, forming sacs with zygotes. These cells divide by meiosis and then mitosis, forming eight spores within each sac. When the spores are released, each functions as a new cell.

> *The sac fungi live in both aquatic and terrestrial environments and are characterized by sexual spores borne in saclike structures. Familiar examples of this group are cup fungi and yeasts.*

One species of yeast, *Candida albicans,* causes superficial infections of the mucous membranes of the vagina. (Other species may cause such infections, but *C. albicans* is most often the cause.) *Candida* organisms are normally found in small num-

A

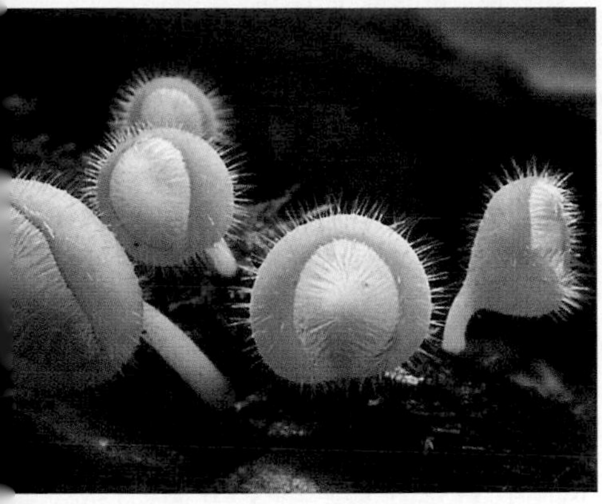

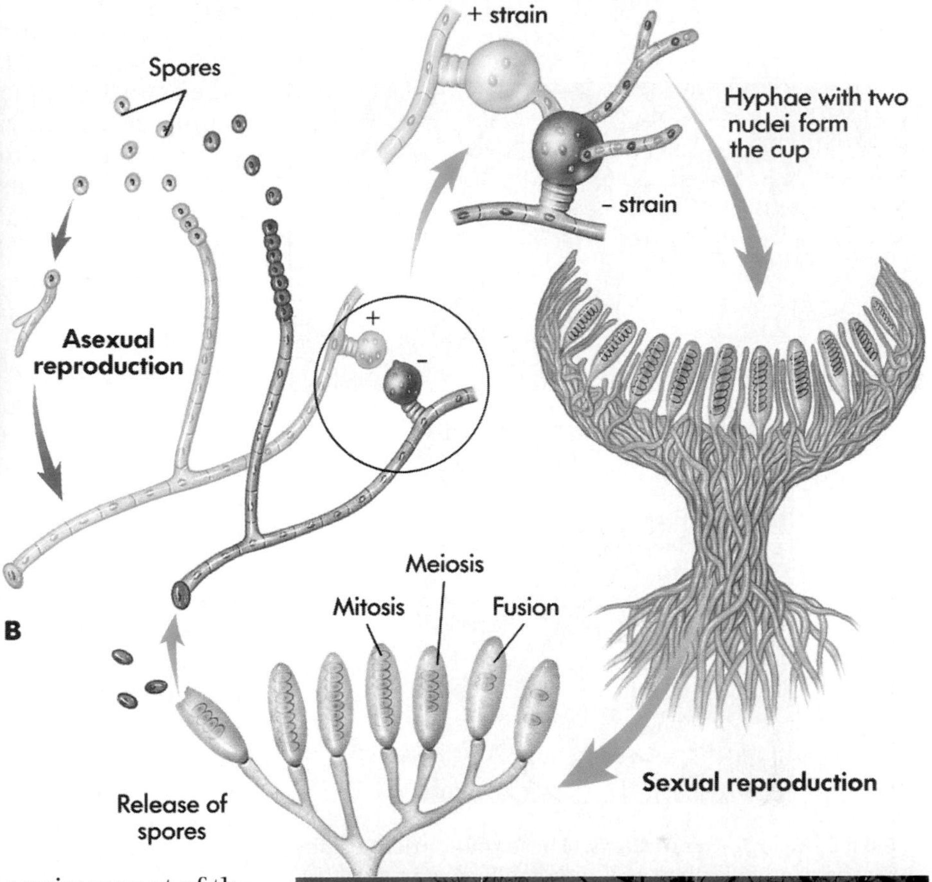

FIGURE 26-25 Sac fungi (division Ascomycota). A, A cup fungus, *Cookeina tricholoma,* from the rain forest of Costa Rica. **B,** The life cycle of a cup fungus.

bers in the vagina, but the acid environment of the vagina and the proliferation of normal vaginal bacteria limit its growth. When the environment of the vagina changes, this organism often flourishes, producing raised gray or white patches on the vaginal walls, a scanty but thick whitish discharge, and itching. Situations that change the environment of the vagina are pregnancy and the use of birth control pills, which often change the vaginal pH, and taking of broad-spectrum antibiotics such as tetracycline, which kills the normal vaginal bacteria. In addition, patients with AIDS or uncontrolled diabetes often develop yeast infections. Candidiasis can also occur in other areas of the body, such as the mouth, hands, feet, skin, and nails. Figure 26-27, *A,* is a photomicrograph that shows the elongated structures termed *pseudohyphae* that *Candida* produces as it grows.

Although candidiasis is not considered a sexually transmitted disease because it is usually contracted without sexual contact, a vaginal yeast infection can spread to a partner during sexual activity. The infection in the male is characterized by the growth of small, elevated yeast colonies on the

FIGURE 26-26 Dutch elm disease. Millions of elms in North America and Europe have been killed by Dutch elm disease, which is caused by an ascomycete, *Ceratocystis ulmi.* The spread of Dutch elm disease can be controlled by a combination of sanitation (removing dead and dying trees promptly), killing beetles in traps baited with chemicals, and treating the trees in advance with fungicides and bacteria that inhibit the growth of the fungus.

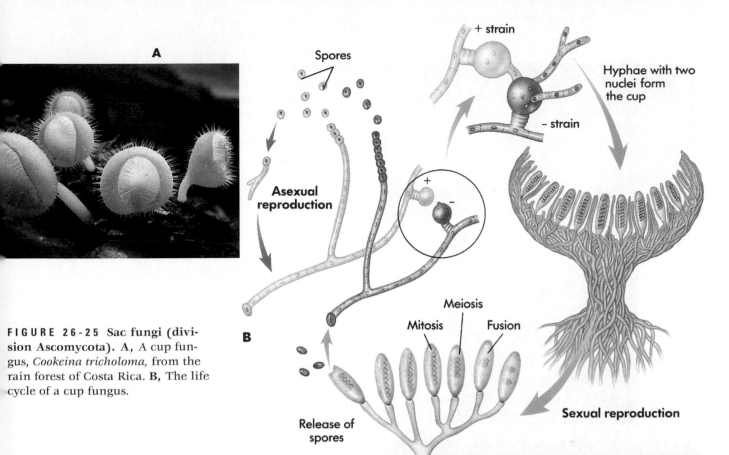

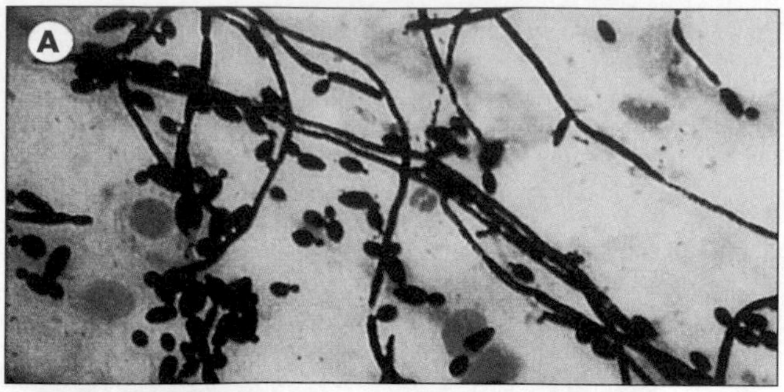

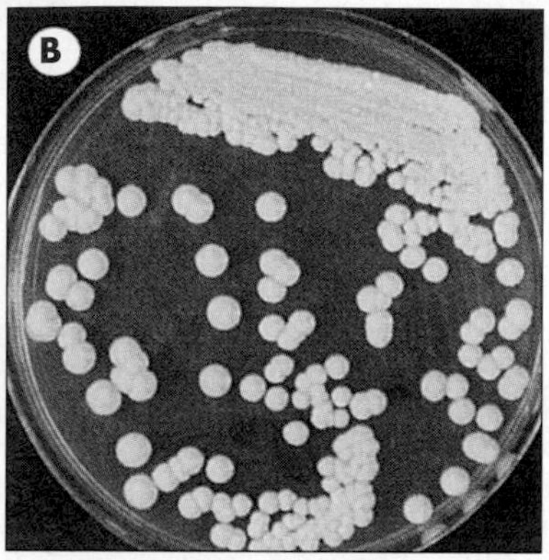

FIGURE 26-27 Candidiasis. These photomicrographs show the pseudohyphae of *Candida albicans* in a vaginal smear (**A**) and the colonial form (**B**).

penis, similar to those shown in Figure 26-27, *B*. In addition, like most sexually transmitted diseases, a vaginal yeast infection can be transmitted to a newborn as it passes through the birth canal. A number of antifungal drugs are available that can be applied locally to treat infections caused by *Candida* organisms.

Club Fungi

Club fungi include the mushrooms, toadstools, puffballs, jelly fungi, and shelf fungi (Figure 26-28). Some species of club fungi are commonly cultivated; for example, the button mushroom—commonly served in salad bars—is grown in over 70 countries, producing a crop with a value of over $15 billion. Other kinds of club fungi are represented by the rusts and smuts, which are devastating plant pathogens. Wheat rust, for example, causes huge economic losses to wheat wherever it is grown.

The club fungi are so-named because they have club-shaped structures from which unenclosed spores are produced. These structures are called **basidia** (sing. basidium) and give the division name of Basidiomycota. In mushrooms, basidia line the gills found under the cap (Figure 26-29). In rusts and smuts, basidia arise from hyphae at the surface of the plant.

As shown in Figure 26-29, a mushroom is formed when pairs of hyphae (often of two different mating strains) fuse and intermingle their cytoplasm and haploid nuclei. The fused hyphae develop into the mushroom, including the gills on the underside of its cap. The basidia develop as single cells on the free edges of the gills. Within the cells that develop into basidia, the nuclei present from the fusion of hyphae now fuse to form zygotes. Each zygote undergoes meiosis, producing four haploid spores in each basidia. The zygotes within some basidia undergo mitosis, producing basidia having eight haploid spores. Turgor pressure (water buildup within the basidia) bursts the basidia and hurls the spores from the mushroom—at an average rate of 40 million per hour! These spores germinate to produce hyphae, which fuse to begin a sexual cycle of reproduction once again. Asexual reproduction is rare in this class of fungi.

The club fungi, named because they have club-shaped structures from which unenclosed spores are produced, include the mushrooms, puffballs, and shelf fungi.

FIGURE 26-28 Club fungi (division Basidiomycota). **A,** *Tremella,* witches' butter. **B,** *Cyathus,* bird's nest fungi. **C,** *Gymnosporangium,* apple-cedar rust.

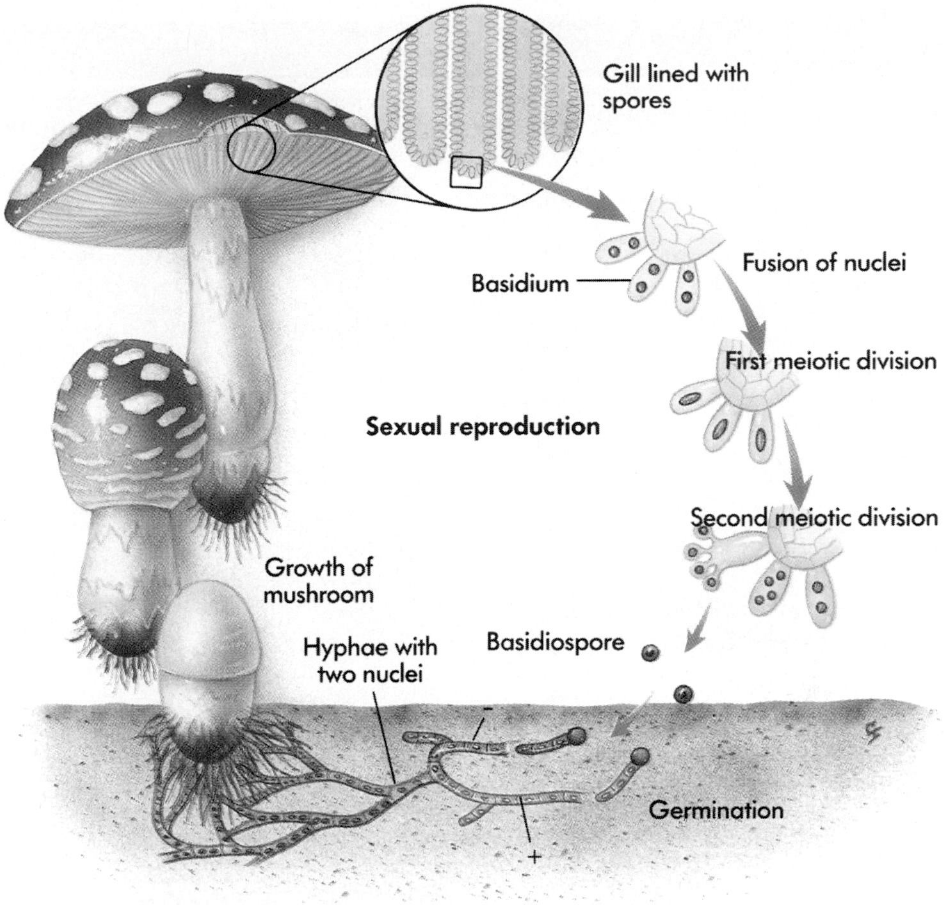

FIGURE 26-29 Life cycle of a common club fungus—a mushroom. This class of fungi rarely undergoes an asexual stage of reproduction.

Imperfect Fungi

Imperfect fungi include members whose hyphal structure and asexual reproduction suggest that they are zygote-forming fungi, sac fungi, or club fungi. However, the fungi in this group have lost the ability to reproduce sexually, or their sexual stages of reproduction have not been observed. Therefore these genera cannot be placed in their appropriate division because placement is based on features related to sexual reproduction. They are therefore referred to as *imperfect*—the Fungi imperfecti. Although these fungi do not reproduce sexually, hyphae of different mating types often fuse, providing some genetic recombination. This characteristic is important to the survival of each species, especially in the plant pathogens, because mutation and recombination produce new strains that can still infect plants bred for resistance to these fungal diseases.

The imperfect fungi reproduce asexually by spores. An enormous range of diversity occurs in the structure of the spore cases found at the tips of their hyphae. These variations represent adaptations of the fungi to dispersal of their spores. For example, spores distributed by insects usually have sticky, slimy spore cases with odors attractive to insects. Those dispersed by the wind have dry spore cases.

Among the economically important genera of imperfect fungi are *Penicillium* and *Aspergillus* (Figure 26-30). Some species of *Penicillium* are sources of the well-known antibiotic penicillin, and other species of the genus give the characteristic flavors and aromas to cheeses such as Roquefort and Camembert. Species of *Aspergillus* are used for fermenting soy sauce and soy paste, processes in which certain bacteria and yeasts also play important roles.

Most of the fungi that cause diseases in humans are members of the imperfect fungi. Some of these fungi cause infections of the skin, such as athlete's foot (Figure 26-31) and other forms of ringworm. The term *ringworm* refers to any fungal infection of the skin, including the scalp and nails. There is also a small group of fungi that causes diseases of various organs, including the brain. Some of these diseases are spread by birds, bats, and contaminated

basidia (buh **sid** ee uh) in club fungi, club-shaped structures from which unenclosed sexual spores are produced.

JUST WONDERING. . . .

I love to eat mushrooms and often see them growing in my yard or when I go hiking in the woods. How do I tell which are safe to eat?

You should never eat mushrooms unless you are absolutely sure they are safe. For most people, that means eating only the ones they buy at the grocery store.

The mushrooms most often sold in grocery stores are *Agaricus bisporus,* a fungus closely related to the common field mushroom. Other popular edible mushrooms are shiitake mushrooms, *Lentinus edodes.* Together these two species make up about 86% of the world crop of mushrooms.

Within other genera of mushrooms such as *Amanita,* for example, there are both poisonous and edible species. For an untrained person, these closely related organisms are difficult to tell apart; even mushrooms from different genera can be difficult to identify with certainty. Because the stakes are so high (one bite of a highly poisonous *Amanita* could kill you), it makes no sense to take a chance eating wild mushrooms. In addition, certain mushrooms produce powerful hallucinogenic drugs; ingesting these drugs can be extremely dangerous. Therefore, unless you are a trained mushroom expert—stick with the grocery store varieties!

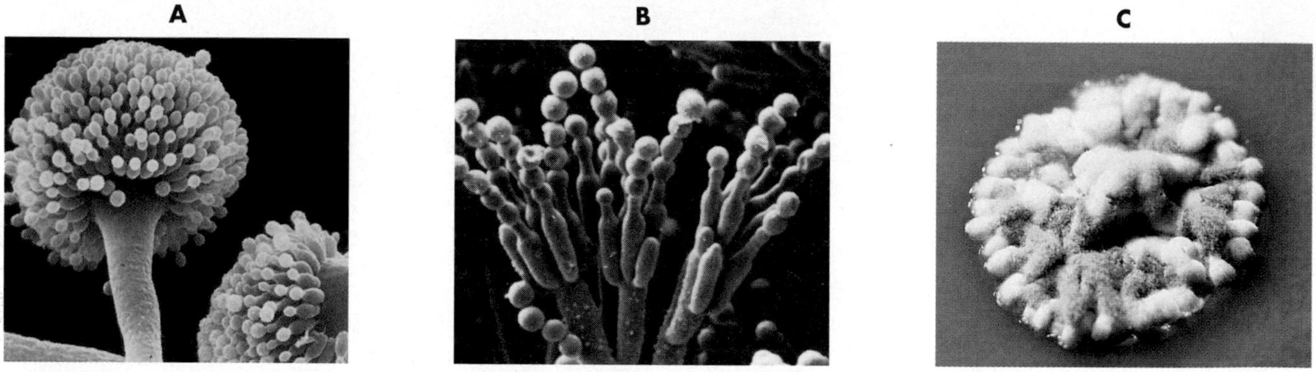

FIGURE 26-30 Imperfect fungi. A, *Aspergillis.* **B,** *Penicillium* growing on an orange. **C,** *Penicillium* growing in a petri dish.

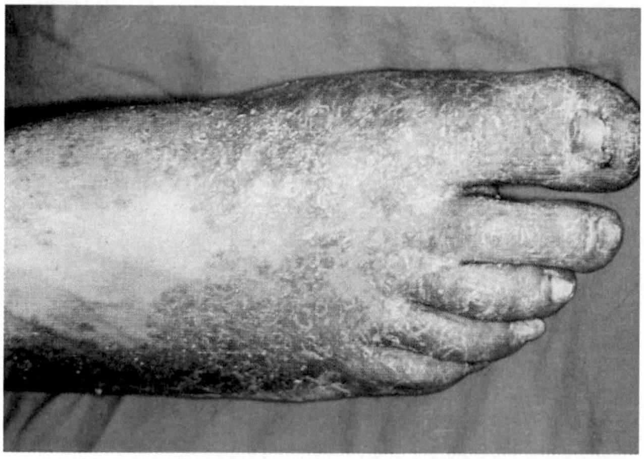

FIGURE 26-31 Athlete's foot. This itchy, painful condition is caused by a member of the imperfect fungi, *Tinea pedis.*

soil. Such diseases are usually contracted by breathing air heavily contaminated with spores of the causative agent. Other fungal diseases occur only in persons weakened by other diseases, making them more susceptible to infection by pathogenic fungi.

> *The imperfect fungi have no sexual stage of reproduction and reproduce asexually by spores. Some members of this division are sources of antibiotics or are important in food production, but many cause disease in humans.*

Lichens

Lichens (Figure 26-32) are associations between fungi and photosynthetic partners and are classified with the fungi. They provide an example of

FIGURE 26-32 Three types of lichens. A, Crustose (encrusting) lichens growing on a rock. **B,** A fruticose (shrubby) lichen. Fruticose lichens predominate in deserts because they are more efficient in capturing water from moist air than either of the other two types. **C,** A foliose (leafy) lichen growing on the bark of a tree.

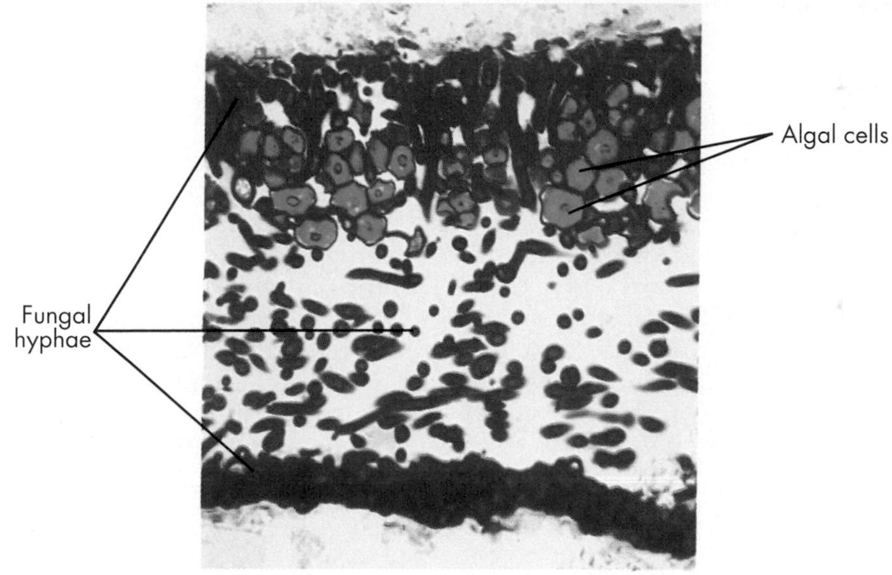

FIGURE 26-33 Section of a lichen. The fungal hyphae are more densely packed into a protective layer on the top and bottom layers of the lichen. The cells (*blue*) near the surface of the lichen are a green alga. This alga supplies carbohydrates to the fungus.

mutualism, a living arrangement in which both partners benefit. Most of the visible body of a lichen consists of fungus, but cyanobacteria or green algae (or sometimes both) live within the fungal tissues (Figure 26-33). The photosynthetic organism provides nutrients for itself and the fungus. Specialized fungal hyphae penetrate or envelop the photosynthetic cells and transfer water and minerals to them, although this part of the mutualistic relationship has been questioned recently. Researchers now suspect that the fungus acts more like a parasite than a mutualistic partner.

Lichens are able to invade the harshest of habitats at the tops of mountains, in the farthest northern and southern latitudes, and on dry, bare rock faces in the desert. In such harsh, exposed areas, lichens are often the first colonists, breaking down the rocks and setting the stage for the growth of other plants. These amazing algae-fungi partnerships are able to dry or freeze and then recover quickly and resume their normal metabolic activities. The growth of lichens may be extremely slow in harsh environments—so slow, in fact, that many small lichens appear to be thousands of years old and are among the oldest living things on Earth.

Summary

- The protists (kingdom Protista) are a varied group of eukaryotic organisms. Many are single celled, although some phyla of protists include multicellular or colonial forms. Within this kingdom of organisms are animallike cells, which take in organic matter for nutrition; plantlike cells, which manufacture their own food; and funguslike cells, which live on dead or decaying organic matter.
- Animallike protists, the heterotrophs, are usually referred to as *protozoans*. The protozoans are classified according to the means by which they move and feed. The amebas move and eat by means of cell extensions, or pseudopods. The flagellates move by means of long, whiplike cellular extensions, their flagella. The ciliates move and sweep food toward themselves by means of short, hairlike cellular processes, their cilia. And the sporozoans are nonmotile parasites of vertebrate animals, carried from one host to another by insects.
- The algae contain chlorophyll, carry out photosynthesis, but lack true roots, stems, leaves, and vascular tissue. There are five phyla of plantlike protists: the dinoflagellates, the golden algae, the brown algae, the green algae, and the red algae. The dinoflagellates and the golden algae are unicellular organisms, whereas brown algae, green algae, and red algae are multicellular. However, the green algae has certain unicellular species.
- The slime molds have unique life cycles that have funguslike spore-bearing stages and "slimy" amebalike stages. The water molds grow in fresh water and salt water, on aquatic animals, in moist soil, and on plants. These protists produce large egg cells during sexual reproduction.
- Fungi are multicellular, eukaryotic organisms, most of which feed on dead or decaying organic material. They are important decomposers that are essential to the cycling of materials in ecosystems, are important in the production of certain foods, and also cause certain diseases. Most fungi are composed of slender microscopic filaments, or hyphae, that form cottony masses or compact plantlike structures.
- Three divisions of fungi have sexual stages of reproduction. These are the zygote-forming fungi (such as black bread mold), the sac fungi (such as cup fungi and yeasts), and the club fungi (such as mushrooms, puffballs, and shelf fungi).
- The imperfect fungi are a fourth division of fungi. Members of this division have no known sexual stages of reproduction. The imperfect fungi include organisms that produce antibiotics and are used in the production of certain foods but that also cause certain diseases in humans.
- Lichens are associations of fungi with green algae and/or cyanobacteria. The photosynthetic organism provides nutrients for itself and the fungus. Researchers now question whether the fungus helps the photosynthetic organism; it has been thought that fungal hyphae penetrated or enveloped the photosynthetic cells and transferred water and minerals to them. Lichens are able to exist in harsh environments and live for long periods of time. Some lichens may be among the oldest organisms on Earth.

Knowledge and Comprehension Questions

1. Explain how the following organisms are related: protists, protozoa, algae, diatoms, and slime molds.
2. Describe an ameba. How does it move and take in food?
3. What do flagellates and ciliates have in common? How do they differ?
4. What are sporozoans? Briefly summarize a sporozoan's life cycle.
5. What do euglenoids, dinoflagellates, and golden algae have in common? Describe each one.
6. Identify the phylum or phyla to which the following belong: yellow-green algae, diatoms, and golden-brown algae.
7. Fill in the blanks: The slime molds are protists that are __________like in one phase of their life cycle, and __________like in another phase of their life cycle.
8. Fill in the blanks: A fungus that feeds on a fallen tree in a forest is a(n) __________. A fungus that invades the skin between your toes, creating the uncomfortable condition known as athlete's foot, is a(n) __________.
9. Fungi are both helpful and harmful to humans. Give some examples of each.
10. Match each term with the most appropriate statement:
    Club fungi
    Zygote-forming fungi
    Sac fungi
    Water molds
    Imperfect fungi
    a. Their sexual spores are borne in saclike structures.
    b. No sexual stage of reproduction has been observed in these organisms.
    c. Their sexual spores are produced by basidia.
    d. Their sexual spores are called *zygospores*.
    e. During sexual reproduction, they produce large egg cells.
11. What are lichens, and where are they found?
12. Fill in the blanks: Amebas that secrete delicate shells made of silica are called __________; amebas that secrete shells made of calcium carbonate or limestone are called __________.

Critical Thinking

1. Some single-celled protists like *Volvox* form spherical colonies of hundreds of cells. Within the colony, some of the cells become specialized for sexual reproduction. Would such a colony be considered a multicellular organism or a colonial organism? Why?
2. If the protist *Euglena* is "plantlike" because it photosynthesizes in the presence of light, and it is "animallike" because it ingests food in the absence of light, in which protist group does it belong? What does this quandary say about the usefulness of such an approach to classifying protists?
3. Many freshwater lakes, when polluted by high-phosphate detergents, become overgrown with mats of photosynthetic green algae. If the pollution continues, all other life in the lake soon dies. Why? Why can you not avoid this lethal result by simply poisoning the algae?

CHAPTER 27

PLANTS: REPRODUCTIVE PATTERNS AND DIVERSITY

ALTHOUGH THIS DOES NOT look like a family photo of two generations of individuals—it is! The leaflike greenery are plants of one generation, and the stalks are of another generation. But interestingly, both are the same plant—the hairy-cap moss *Polytrichum*.

The green plants are the gametophyte generation. Gametophytes do as their name suggests: they produce gametes, or sex cells. These male and female gametes fuse during fertilization to form zygotes.

The generation of plants that develops from the zygotes, the stalklike plants called sporophytes, looks very different from its gametophyte parents. The sporophyte generation is named because it produces spores. And the gametophyte generation is in turn produced from these spores, resulting in a cycling of two very different generations of the same plant.

Characteristics of Plants

Because plants have the ability to reproduce sexually, a characteristic they have in common with most living things, other characteristics differentiate them from other organisms and place them in their own kingdom. Plants are multicellular, eukaryotic, photosynthetic autotrophs. That is, they produce their own food (which most plants store as starch) by using energy from the sun and carbon dioxide from the atmosphere (a process described in Chapter 6). However, not all organisms fitting this description are plants. The multicellular algae, at one time classified as plants and now classified by most systematists (taxonomists) as protists, also fit this description. Today, plants are often defined as multicellular, photosynthetic eukaryotes that *live on land,* although some botanists disagree with this definition. The few species of plants that live in the water evolved on land.

Another characteristic that differentiates the multicellular algae from the plants is their type of chlorophyll, the pigment that absorbs energy from the sun during the process of photosynthesis. Plants have the pigments chlorophyll *a* and chlorophyll *b*. The red and brown algae do not have chlorophyll *b* and so are thought not to have evolved from the same ancestors as the plants. The green algae do have chlorophyll *a* and *b* but, because they have many single-celled forms, are also classified as protists. The green algae are thought by most scientists to be the evolutionary predecessors of the land plants.

All plants can be placed into one of two groups: the nonvascular plants and the vascular plants. Most plants (more than 80% of all living plant species) are **vascular plants,** those having specialized tissues to transport fluids. There are three major groups of vascular plants: seedless vascular plants (such as ferns), vascular plants with naked seeds (such as pine trees), and vascular plants with protected seeds (such as flowering plants). Plants lacking these specialized transport tissues are called **nonvascular plants.** There is one major group of nonvascular plants: the bryophytes (such as the "true" mosses).

Traditionally, plants are taxonomically separated into **divisions** rather than into phyla, but these categories are basically equivalent to one another. In this classification scheme, the plant kingdom has 10 divisions. Table 27-1 lists these divisions, including examples. It shows how the divisions fit into the groupings of plants previously described.

KEY CONCEPTS

- Plants are multicellular, eukaryotic organisms that typically live on land and make their own food by the process of photosynthesis.
- Plants, unlike animals, have two different multicellular phases that occur alternately within their sexual life cycles.
- The differences between the life cycles of the lower, nonvascular plants and the higher, vascular plants reflect an adaptation of the vascular plants for life on land.
- Plants are also able to reproduce asexually by vegetative propagation, a process in which a new plant develops from a portion of a parent plant.

OUTLINE

TABLE 27-1

The ten divisions of plants

GROUP	DIVISION	EXAMPLES
Nonvascular plants	Bryophyta	"True" mosses, liverworts, hornworts
Seedless vascular plants	Psilophyta	Whisk ferns
	Lycophyta	Club mosses
	Sphenophyta	Horsetails
	Pterophyta	Ferns
Vascular plants with naked seeds	Coniferophyta	Conifers
	Cycadophyta	Cycads
	Ginkgophyta	Ginkgos
	Gnetophyta	Gnetae
Vascular plants with protected seeds	Anthophyta	Flowering plants
	Class Monocotyledons	Grasses, irises
	Class Dicotyledons	Flowering trees, shrubs, roses

The General Pattern of Reproduction in Plants

It is necessary to take a broader look at the patterns of reproduction of all living things to investigate patterns of reproduction in specific divisions of plants. The series of events that take place from one stage during the life span of an organism, through a reproductive phase, and until a stage similar to the original is reached in the next generation is called a **life cycle.** For example, a bacterium's life cycle can be described as beginning when it (along with another daughter cell) splits from a mother cell during binary fission. This life cycle is complete when the bacterium itself splits into two new cells. This is an example of an asexual life cycle.

Animals, plants, some protists, and some fungi have sexual life cycles, which are characterized by the alternation of meiosis and fertilization. For example, your life cycle could be described as all of the events of growth and development that began at fertilization and continued after your birth through the time of your sexual maturation. This is the time you began producing sex cells by the process of meiosis. The cycle would begin again, or have come "full circle," when you have a child—the next generation.

Figure 27-1, *A,* diagrams this type of sexual life cycle, which is common to most animals. Notice in the diagram that during the animal life cycle, the multicellular organism has a double set of hereditary material; it is diploid. (One set was contributed by a father and one set was contributed by a mother.) This diploid, multicellular phase is the dominant phase of the animal life cycle. The gametes represent the haploid phase of the life cycle; they are individual, haploid cells.

Animal life cycles are characterized by a dominant, multicellular diploid phase and a unicellular haploid phase. The multicellular diploid organisms produce haploid gametes by meiosis. These gametes fuse during fertilization, forming the first cell of a new diploid multicellular organism.

In some protists and many fungi, the haploid phase of the life cycle is multicellular—the phase that is dominant and noticeable. Gametes are produced from these organisms by mitosis rather than by meiosis. Only the zygotes are diploid (see Chapter 26). This phase may go unnoticed to the casual observer.

Plants and some species of algae have life cycles that differ from the two sexual life cycles already described. They have both a multicellular haploid phase and a multicellular diploid phase. Because both phases of the life cycle are multicellular, this type of life cycle is called **alternation of generations** (Figure 27-1, *B*). The alternating generations of plants are called the **sporophyte (spore-plant) generation** and the **gametophyte (gamete-plant) generation**. However, the gametophyte (haploid) generation often dominates the life cycles of the nonvascular plants, whereas the sporophyte (diploid) generation dominates the life cycles of the vascular plants.

As previously mentioned, gametophytes form gametes, or sex cells. Because they are **haploid** individuals, having *half* the usual number of chromosomes for that species, they form gametes (which are haploid) by the process of mitotic cell division. These gametes fuse during fertilization, a process that depends on the presence of water in which sperm swim to the egg. The plants produced from fertilized eggs are **diploid** plants, having a full complement of genetic material. These diploid

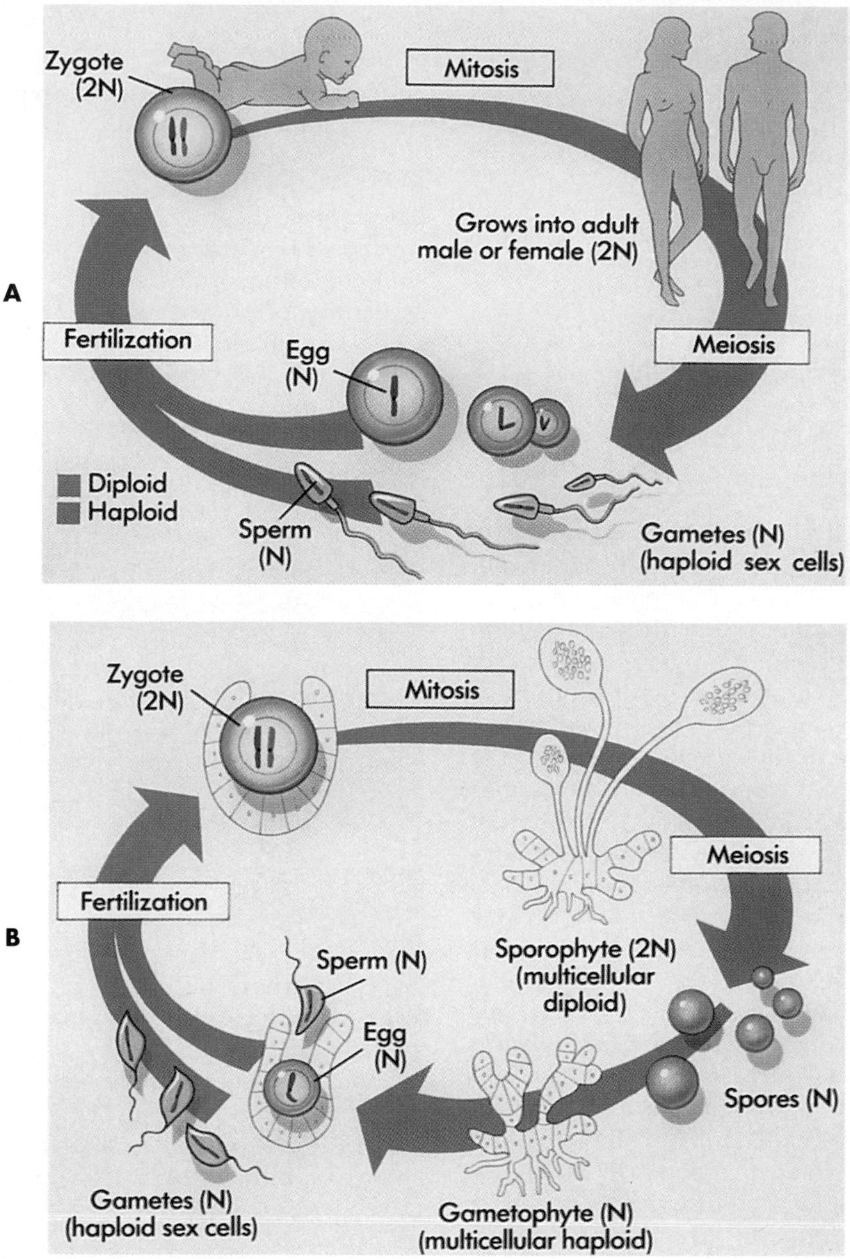

FIGURE 27-1 Sexual life cycles. A, The animal sexual life cycle. The organism that results from the union of haploid gametes is diploid. **B,** A generalized plant life cycle showing the alternation of generations. The sporophyte generation (diploid) alternates with the gametophyte generation (haploid). Haploid spores are produced by the sporophyte generation by meiosis. When dispersed, these haploid spores grow into gametophytes.

vascular plants (**vas** kyuh ler) plants that have systems of specialized vessels that transport water and nutrients; more than 80% of all living plant species are vascular plants.

nonvascular plants (**non vas** kyuh ler) plants that lack specialized transport tissues.

divisions major taxonomic categories corresponding to phyla, used in botany.

life cycle the series of events that take place from one stage during the life span of an organism, through a reproductive phase, and until a stage similar to the original is reached in the next generation.

alternation of generations a type of life cycle that has both a multicellular haploid phase and a multicellular diploid phase.

sporophyte (spore-plant) generation (**spor** uh fite) the diploid phase of a plant life cycle. Sporophytes produce haploid spores by meiosis; this generation tends to dominate the life cycles of vascular plants.

gametophyte (gamete-plant) generation (guh **mee** toe fite) the haploid phase of a plant life cycle. Gametophytes produce haploid gametes by mitosis; this phase tends to dominate the life cycles of nonvascular plants.

haploid (**hap** loyd) having half the usual number of chromosomes for that species.

diploid (**dip** loyd) having a full complement of genetic material for that species.

plants are the sporophytes. Because sporophytes are diploid, they use the process of meiosis to produce haploid spores. When dispersed, spores grow into gametophyte plants.

> *Plant life cycles are marked by an alternation of generations of diploid sporophytes with haploid gametophytes. As a result of meiosis, sporophytes produce spores, which grow into gametophytes. Gametophytes produce gametes as a result of mitosis. These gametes fuse during fertilization and grow into sporophytes.*

Looking at general characteristics of the bryophytes and the vascular plants and their patterns of reproduction, you will notice that each group has its own variation of a life cycle within the general pattern of a dominant gametophyte generation in the nonvascular plants and a dominant sporophyte generation in the vascular plants. This difference reflects an adaptation of the vascular plants for life on land as they evolved from earlier, nonvascular forms. These adaptations included the development of spores with protective walls able to tolerate dry conditions, efficient water and food-conducting systems, and gametophyte generations that became protected by and nutritionally dependent on the sporophyte generation.

Plants produce gametes and spores in specialized structures. In the bryophytes and several divisions of vascular plants, eggs are formed in structures called **archegonia** (sing. **archegonium**). Sperm are produced in structures called **antheridia** (sing. **antheridium**). In the more specialized vascular plants, except in a few species of seed-bearing plants, these specialized gamete-producing structures have been reduced in size during the course of evolution as the sporophyte generation began to dominate the life cycle. Eggs and sperm differentiate from a small number of haploid cells within the sporophyte. These haploid cells are actually the gametophyte generation.

Patterns of Reproduction in Nonvascular Plants

The Bryophytes

Bryophytes are a single division of small, low-growing plants that are commonly found in moist places. The three classes of bryophytes are mosses (such as the hairy-cap moss shown on p. 592), liverworts (Figure 27-2, *A*), and hornworts (Figure 27-2, *B*). In all three, notice that the gametophyte generation is dominant, whereas the sporophyte generation grows out of the gametophyte and is dependent on it for nutrition.

Mosses

The largest class of bryophytes and probably the one most familiar to you is the mosses. The gametophytes of most mosses have small, simple leaflike structures often arranged in a spiral around stemlike structures. These bryophyte structures are

A

B

FIGURE 27-2 Bryophytes. A, A liverwort, *Marchantia.* The sporophytes are carried within the tissues of the umbrella-shaped structures that arise from the surface of the flat, green, creeping gametophyte. **B,** *Antheros,* a hornwort. This photo shows the long, slender sporophyte with the gametophyte below.

FIGURE 27-3 Peat mosses, *Sphagnum*. A, A peat bog being drained by the construction of a ditch. In many parts of the world, accumulations of peat, which may be many meters thick, are used as fuel. **B,** The glistening black, round objects are the spore cases, which contain spores. The spore cases of *Sphagnum* have a lid that blows off explosively, releasing the spores.

different from the stems and leaves of vascular plants; these differences are described in Chapter 28. Many small, carpetlike plants are mistakenly called mosses. For example, *Spanish moss* is actually a flowering plant, a relative of the pineapple. Even eliminating these impostors, however, there are still some 10,000 species of true mosses, and they are found almost everywhere on Earth.

One kind of moss that is most important economically is *Sphagnum* (Figure 27-3). This moss grows in boggy places (low-lying, wet, spongy ground), forming dense and deep masses that are often dried and sold as peat moss. Peat moss is used in gardening as a mulch; it is layered around trees or plants to protect the roots from temperature fluctuations and to retain moisture, control weeds, and enrich the soil. It functions well as a mulch and a soil additive because its tissues have special water storage cells. These cells allow the peat to absorb and retain up to 90% of its dry weight in water, whether or not the moss is alive.

Figure 27-4 shows the life cycle of the hairy-cap moss. Flask-shaped archegonia are found among the top "leaves" of the female gametophytes. Each archegonium produces one egg. Antheridia are found in a similar place in the male gametophytes. Each antheridium produces many sperm. The flagellated sperm swim through drops of water from rain, dew, or other sources into the neck of the archegonium and then to the egg. After fertilization takes place, the zygote develops into a young sporophyte within the archegonium. It then grows out of the archegonium and differentiates into a slender stalk. This stalk is initially green, but its chlorophyll disintegrates as it matures, leaving the stalk yellow or brown. At this stage the sporophyte stalk derives its nourishment from the gametophyte. Each sporophyte stalk bears a spore capsule, or *sporangium,* near its tip. Haploid spores are produced by meiosis within this capsule. When the top of the spore capsule pops off, the spores are freed. Under the proper conditions, these spores germinate into threadlike filaments; the characteristic leafy gametophytes arise from buds that form on these filaments.

Liverworts

Liverworts were given their name in medieval times when people believed that plants resembling particular body parts were good for treating disease of those organs. Some liverworts are shaped like a liver and were thought to be useful in treating liver ailments. The ending *-wort* simply means "herb."

A well-known example of a liverwort is *Marchantia,* shown in Figure 27-2, *A.* It has green, leafy gametophytes that grow close to the ground. Antheridia and archegonia develop within the umbrellalike

archegonia (singular, archegonium) (**ar** kih **go** nee uh / **ar** kih **go** nee um) specialized structures in bryophytes and several divisions of vascular plants in which eggs are formed.

antheridia (singular, antheridium) (**an** thuh **rid** ee uh / **an** thuh **rid** ee um) structures in several divisions of vascular plants in which sperm are produced.

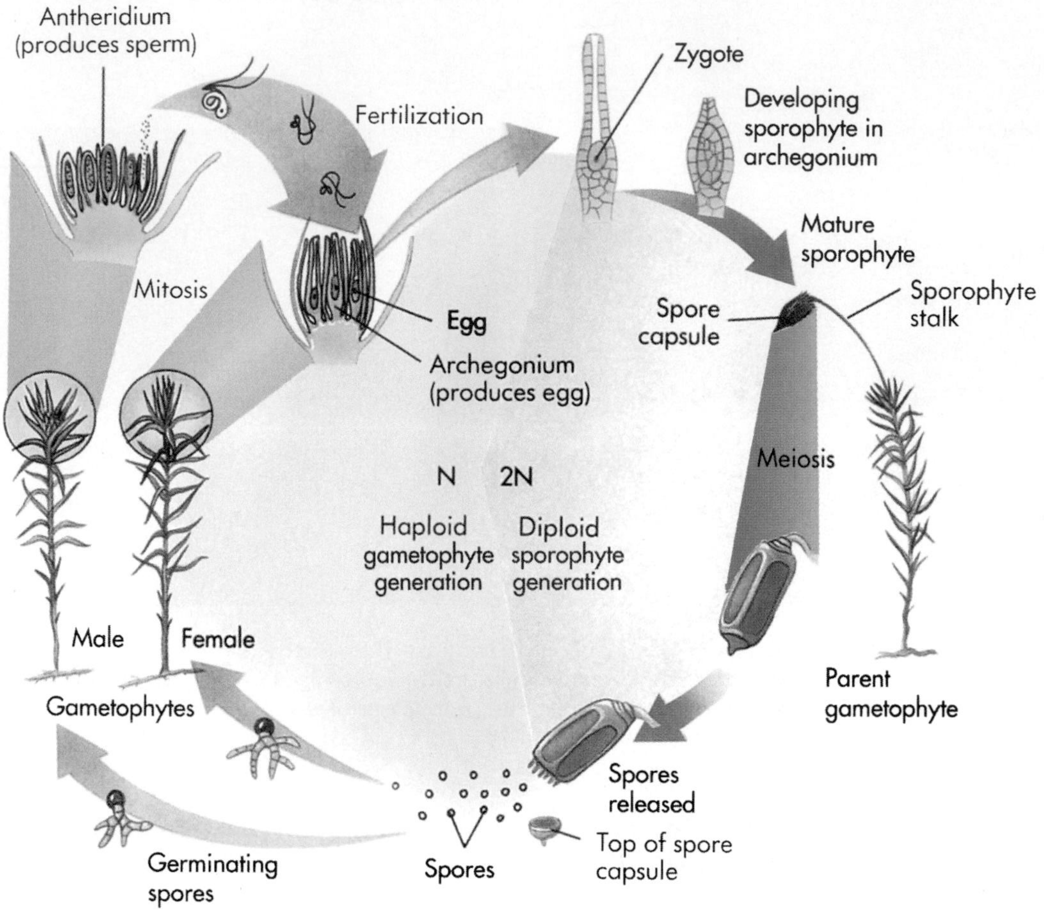

FIGURE 27-4 Life cycle of a moss. Sperm from the antheridium and an egg from an archegonium fuse to form a zygote. This zygote develops into a young sporophyte, which eventually grows out of the archegonium and into a slender stalk. At the end of the stalk is a spore case. The spore case breaks open, releasing spores, which germinate and produce gametophytes. The dominant gametophyte generation is shown in gold, and the sporophyte generation is shown in blue.

portion of the stalks that grow up from the leaflike gametophyte. The sporophytes develop encased within these tissues; spores are freed from these structures.

Hornworts

The *hornwort* is so named because it has elongated sporophytes that protrude like horns from the surface of the creeping gametophytes (Figure 27-2, *B*). These tall sporophytes are made up of a foot and a long, cylindrical sporangium (spore case). Its life cycle parallels quite closely those of the mosses and liverworts.

> *The life cycles of all three classes of bryophytes—the mosses, liverworts, and hornworts—are somewhat uniform, all having two distinct phases. The sporophyte generation grows out of or is embedded in the tissues of the gametophyte generation (the dominant generation) and depends on the gametophyte generation for nutrition.*

Patterns of Reproduction in Vascular Plants

Seedless Vascular Plants

As listed in Table 27-1, the members of four divisions of vascular plants do not form seeds. **Seeds** are structures from which new sporophyte plants grow; they protect the embryonic plant from drying out or being eaten when it is at its most vulnerable stage. Seeds also contain stored food for the new plant. The seedless plants overcome these problems in interesting ways. Figure 27-5 diagrams the life cycle of a fern—a familiar member of the **seedless vascular plants.** Its life cycle is representative of this group.

When a fern plant is mature, it produces spores by meiosis. Each cluster of spore cases, or *sporangia,* looks like a dot on the underside of the fern leaf, or *frond.* (These clusters of spore cases are also called *sori* [sing. *sorus*] in the fern.) Because it produces sporangia and spores, the fern is the sporophyte generation—the dominating form in the life cycle of the seedless vascular plants. After its spores are dispersed, those that settle in a moist environment will germinate into haploid plants that look very *un-*

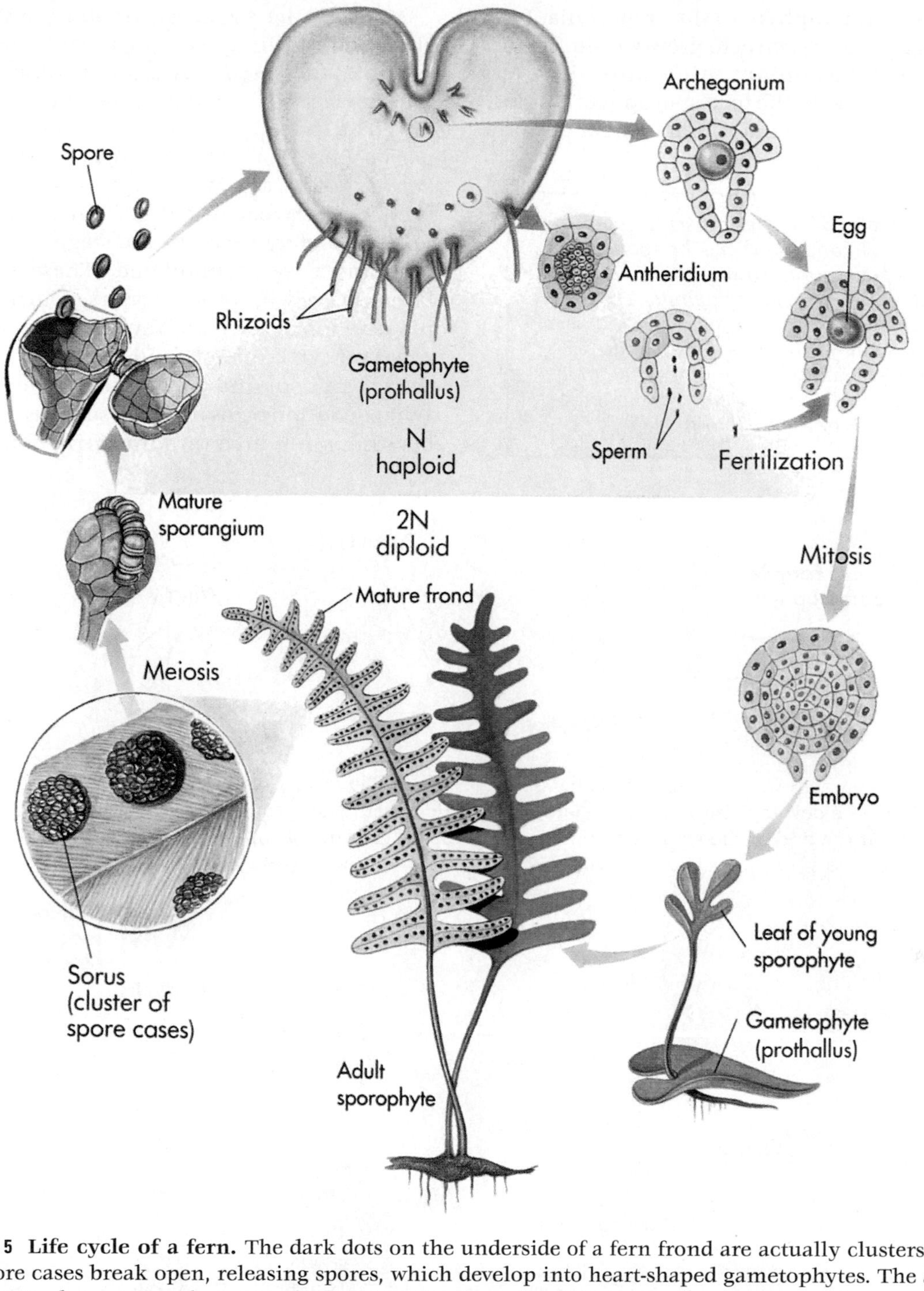

FIGURE 27-5 Life cycle of a fern. The dark dots on the underside of a fern frond are actually clusters of spore cases. The spore cases break open, releasing spores, which develop into heart-shaped gametophytes. The archegonia and antheridia produce eggs and sperm, which fuse to form zygotes. Zygotes develop into sporophytes, which become much larger than gametophytes.

like ferns. Each plant is a small, ground-hugging, heart-shaped gametophyte called a *prothallus* (pl. prothalli), which is anchored to the ground by filaments of cells called *rhizoids.* Their antheridia and archegonia are protected somewhat by being located on the underside of the plant. Sperm, when released, swim through moisture collected on the un-

seeds structures from which new sporophyte plants grow; seeds protect the embryonic plant from drying out or being eaten when it is at its most vulnerable stage.

seedless vascular plants plants that reproduce by means of spores rather than seeds; a familiar member of this group is the fern.

derside of the gametophyte to the archegonia. Each fertilized egg (zygote) starts to grow within the protection of the archegonium. After this initial protected phase of growth, the fern sporophyte is able to grow on its own and becomes much larger than the gametophyte.

> *The life cycles of seedless vascular plants are similar to one another. The sporophyte (diploid) generation is dominant and lives separately from the gametophyte. The gametophytes produce motile sperm that need water to swim to the eggs. The fertilized eggs produce young sporophytes that grow protected within gametophyte tissues but eventually become free living.*

Many of the seedless vascular plants that lived about 270 million years ago were converted long ago to a fuel used today—coal. Club mosses that grew on trees, horsetails, ferns, and tree ferns made up great swamp forests during the Carboniferous period (see Chapter 24). Areas of New York State, Pennsylvania, and West Virginia, for example, were lying near the equator at that time. Dead plants did not completely decay in the stagnant, swampy waters, and they accumulated. These swamps were later covered by ocean waters. Marine sediments piled on top of the plant remains. Pressure and heat acted on the layers of dead plant material beneath the ocean floor and converted the remains to coal. When you burn fossil fuels such as coal, you are burning a resource that was formed under special

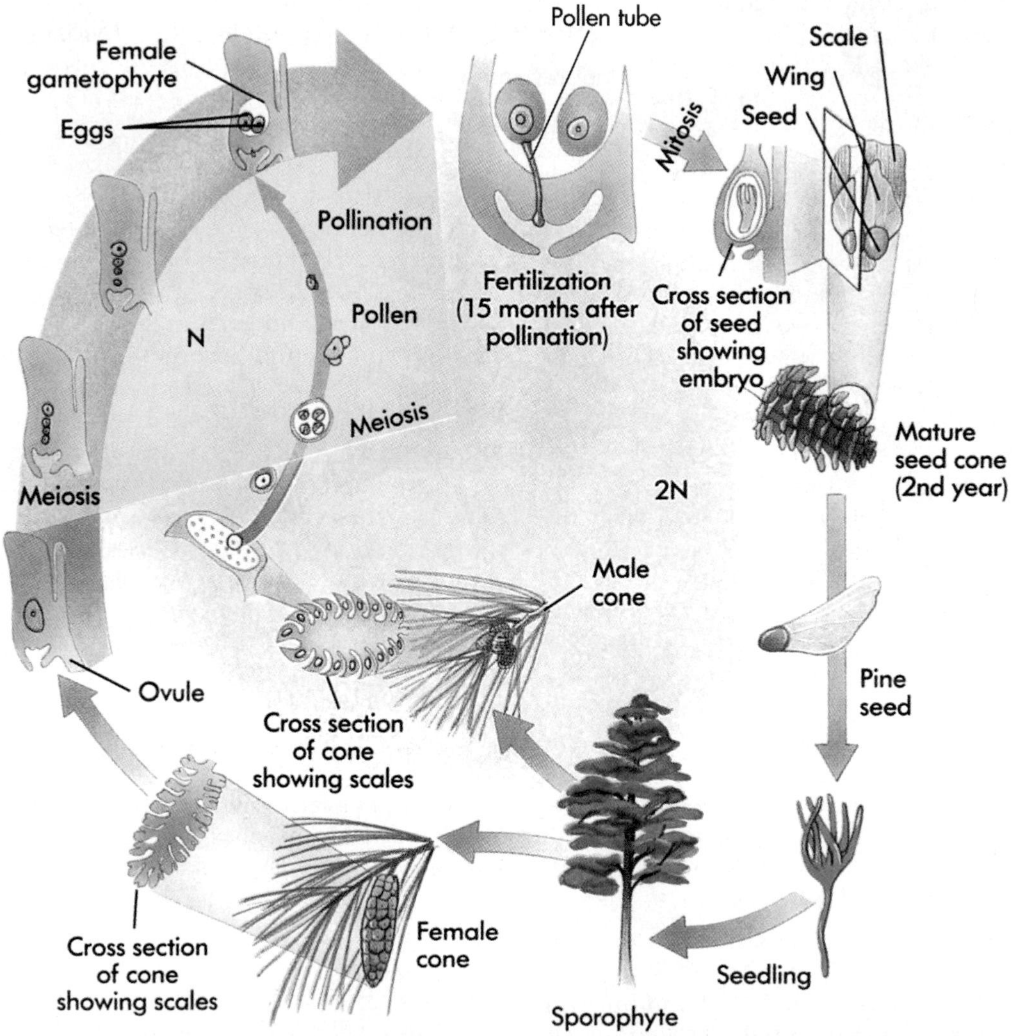

FIGURE 27-6 Life cycle of a pine. Start at the sporophyte, which is a full-grown pine tree. The gametophyte generation of a pine is so small that it is not visible to the naked eye! The male gametophyte is the pollen. The female gametophyte (archegonium) is multicellular and contains two eggs, of which only one will develop into an embryo. The female gametophyte is depicted as the white area around the eggs and is multicellular tissue.

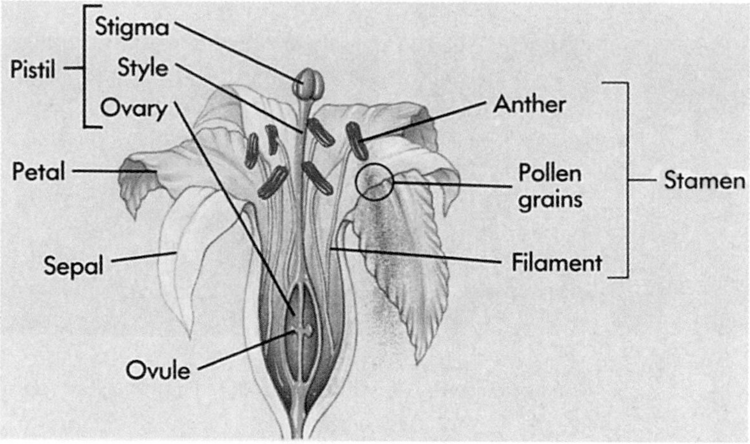

FIGURE 27-7 An angiosperm flower. The main parts of a flower, which contain both male and female structures.

conditions that have not been repeated in the last 270 million years. Coal is therefore called a *nonrenewable resource.* Burning coal adds carbon dioxide to the atmosphere that had been locked in this fossil fuel for millions of years. This gas adds to the amount of carbon dioxide in the atmosphere and contributes to the problems of global warming and acid rain (see Chapter 38).

Vascular Plants With Naked Seeds

Four divisions of vascular plants belong in the category of vascular plants with naked seeds: the conifers, cycads, ginkgo, and gnetophytes. This group is called the **gymnosperms,** a name derived from Greek words meaning "naked seed." The term *naked* refers to the seeds of gymnosperms, which are not completely enclosed by the tissues of the parent at the time it is pollinated.

The most familiar division of gymnosperm is the conifers, or cone-bearing trees. The conifers include pines, spruces, firs, redwoods, and cedars. Figure 27-6 diagrams the life cycle of a pine. The pine tree is the sporophyte (diploid) generation. Interestingly, the gametophyte generation is so reduced that it is not visible to the naked eye and is comparatively short-lived.

Pines and most other conifers bear both male cones and female cones on the same tree. You can tell them apart because the female cones are the larger of the two. These cones produce spores that undergo meiosis, producing the male and female gametophytes. Male gametophytes are pollen grains, each consisting of four cells. The male gametophytes produce sperm, which remain located in the pollen grains. Each of the multicellular female gametophytes produces two to three eggs. The eggs develop within protective structures called *ovules* within the scales of the female pine cones. In the spring, the male cones release their pollen, which is blown about by the wind. (Those allergic to pine pollen know this fact quite well!) As some of this pollen passes by female cones, it gets trapped there by a sticky fluid produced by the now open female cones. As this fluid evaporates, the pollen is drawn further into the cone. When the pollen comes into contact with the outer portion of the ovule, it germinates and forms a pollen tube that slowly makes its way into the ovule—to the egg. After 15 months the tube reaches its destination and discharges its sperm. Fertilization takes place, producing a zygote. The development of the zygote into an embryo takes place within the ovule, which matures into a seed. Eventually the seed falls from the cone and germinates, and the embryo resumes growing and becomes a new pine tree.

Vascular Plants With Protected Seeds

Flowers are the organs of sexual reproduction in the vascular plants with protected seeds—the **angiosperms**. These plants bear seeds in a *fruit*. The flower is structured to promote sexual reproduction—the union of gametes that ultimately develops into an embryo within the seed.

Developmentally, flower parts are actually modified leaves. The outermost whorl, or ring, of modified leaves are the *sepals* (Figure 27-7). Sepals enclose and protect the growing flower bud. In flowers such as roses, the sepals remain small, green, and somewhat leaflike. In other flowers, such as tulips, the sepals become colored and look like the *petals,* the next whorl of flower parts.

gymnosperms (jim no spurms) vascular plants with naked seeds. Four divisions of plants fall into this category: the conifers, cycads, ginkgo, and gnetophytes.

angiosperms (an jee oh spurms) vascular plants with protected seeds, and flowers structured to promote sexual reproduction.

A B

FIGURE 27-8 Different modes of pollination. A, Pollination by a bumblebee. As this bumblebee, *Bombus,* collects nectar from the flame azalea, the stigma contacts its back and picks up any pollen that the bee might have acquired there during a visit to a previous flower. **B,** Pollination by a butterfly. This copper butterfly, *Lycaena gorgon,* is probing a flower with its proboscis, a coiled, tonguelike organ, to extract its nectar. **C,** Pollination by a hummingbird. A long-tailed hermit hummingbird extracting nectar from the flowers of *Helinconia imbricata* in the forests of Costa Rica. Notice the pollen grains on the bird's beak.

Flower petals are frequently prominent and colorful and attract pollinating animals, especially insects. Although the sepals and petals are the dominant outward features of flowers, they are not the organs of sexual reproduction. Many flowers either do not have sepals or petals or have inconspicuous sepals or petals—particularly flowers that are pollinated by the wind. The sex organs of the flower are the innermost modified leaves, the male *stamens,* and, at the center of the flower, the female *pistil.*

Each stamen consists of an *anther,* a compartmentalized structure where haploid *pollen grains* are produced. Each pollen grain is a male gametophyte enclosed within a protective outer covering. This gametophyte produces sperm by mitosis, which remain enclosed within the pollen grain. A long, thin filament bears and supports the anther, exposing the pollen to wind or pollinating animals.

The pistil consists of three parts. At its tip is a sticky surface called the *stigma* to which pollen grains can adhere. At the base of the pistil is the *ovary,* a chamber that completely encloses and protects the ovules. Within each ovule, a single mother cell develops and then divides meiotically, producing four cells. One of these cells develops into a female gametophyte. When mature, the female gametophyte is called an *embryo sac.* Within this sac are typically eight cells, one of which is the egg. The ovary will become a seed when its eggs are fertilized by the male gamete.

The *style* is a narrow stalk arising from the top of the ovary that bears the stigma. The style may be either long or short to facilitate the best exposure of the stigma to the method of pollination that characterizes the plant (Figure 27-8). Pollination in the flowering plants is the transfer of pollen from the anther to the stigma.

Figure 27-9 diagrams the life cycle of an angiosperm showing an ovary containing only one ovule. After a pollen grain lands on the stigma, it produces a long pollen tube that grows from the pollen grain down the style and penetrates the ovary, entering an ovule. One of the two haploid nuclei (sperm) in the pollen tube fertilizes the haploid egg nucleus in the ovule, producing a diploid zygote. The zygote will become a new plant embryo. The other sperm nucleus fuses with two other nuclei of the embryo sac, producing a *triploid* endosperm nucleus. The endosperm nucleus develops into tissue that will feed the embryo as it grows into a plant. The ovule becomes the seed within which the embryo develops, and the ovary ripens into a fruit. The fruit is a food that attracts animals, which play a role in seed dispersal.

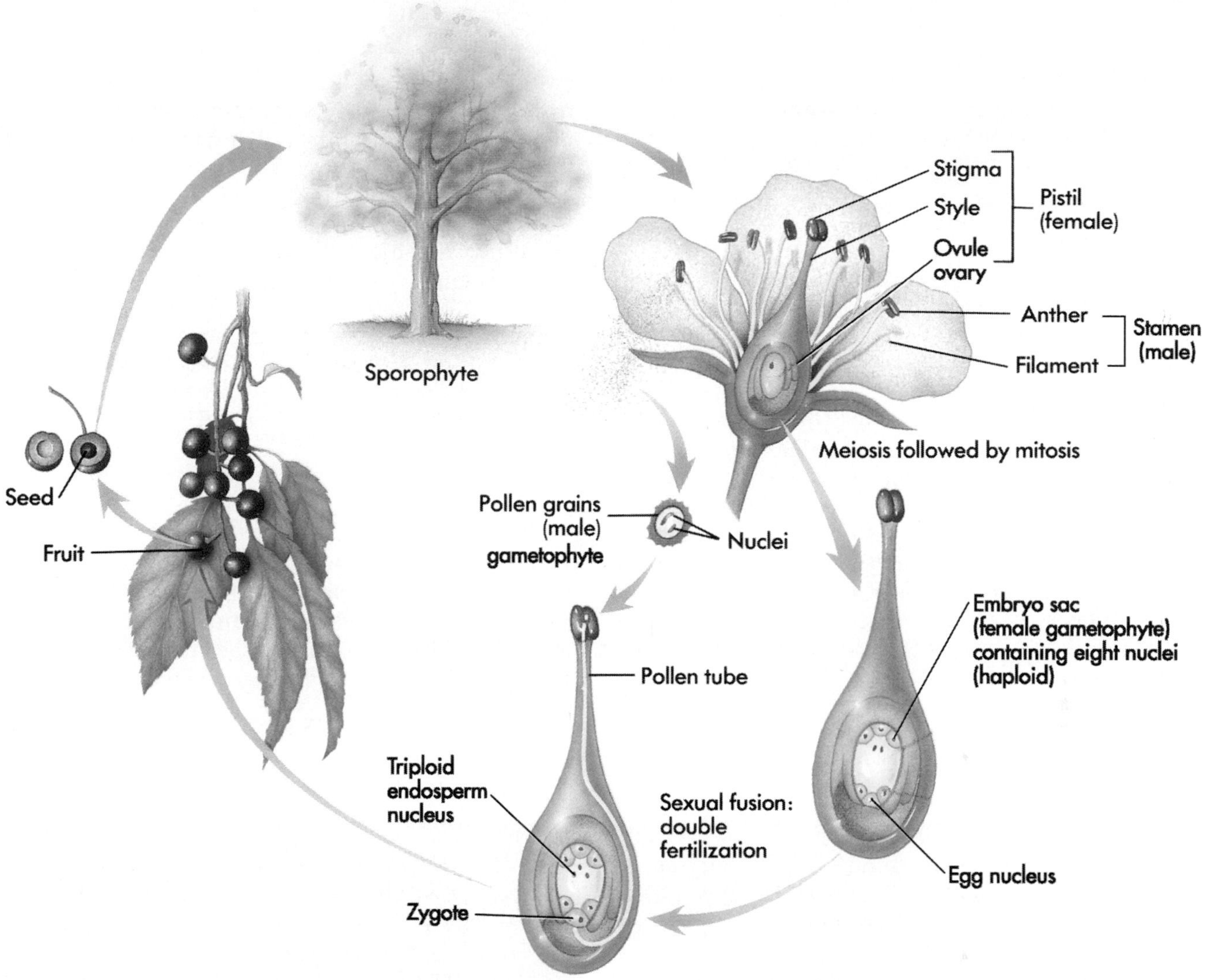

FIGURE 27-9 Angiosperm life cycle. The embryo sac is fertilized twice by sperm in the process of double fertilization. The embryo and endosperm that result develop inside a seed, which forms inside a fruit.

Sexual reproduction in gymnosperms and angiosperms produces seeds by the union of male gametes released from pollen grains and female gametes within ovules. Germinating pollen tubes convey sperm to eggs. The seeds, formed within the ovary after the union of sperm and eggs, protect and nourish the young sporophytes as they begin to develop into new plants.

Mechanisms of Pollination

Pollination takes place in many ways, such as by insects, animals, and the wind. Insects and other animals often visit the flowers of angiosperms for a liquid called *nectar.* Nectar is a food rich in sugars, amino acids, and other substances.

In certain angiosperms and in most gymnosperms, pollen is blown about by the wind and reaches the stigmas passively. However, compared with insects or other animals, wind does not carry pollen far or precisely. Therefore plants pollinated by the wind usually grow close together. These plants typically produce large quantities of pollen.

In some angiosperms, the pollen does not reach other individuals at all. Instead, it is shed directly onto the stigma of the same flower, sometimes in bud. This process is termed *self-pollination.*

For plants to be effectively pollinated by animals, a particular insect or other animal must visit many plants of the same species. Flowers have evolved various colors and forms that attract certain pollinators, thereby promoting effective pollination. Yellow flowers, for example, are particularly attractive to bees, whereas red flowers attract birds but not insects. Insects in turn have evolved a number of special traits that enable them to obtain food efficiently from the flowers of the plants they visit.

Plants can be pollinated by animals, insects, or the wind.

Fruits and Their Significance in Sexual Reproduction

Parallel to the evolution of the angiosperms' flowers and nearly as spectacular has been the evolution of their fruits. Fruits have evolved a diverse array of shapes, textures, and tastes and exhibit many differing modes of dispersal.

Fruits that have fleshy coverings—often black, bright blue, or red—are normally dispersed by birds and other vertebrates. Just as red flowers attract birds, the red fruits signal an abundant food supply. By feeding on these fruits, birds and other animals carry seeds from place to place and thus transfer the plants from one suitable habitat to another (Figure 27-10, *B*). Other fruits, such as snakeroot, beggar ticks, and burdock (Figure 27-10, *A*), have evolved hooked spines and are often spread from place to place because they stick to the fur of mammals or the clothes of humans. Others, like the coconut and those that occur on or near beaches, are regularly spread by water (Figure 27-10, *C*). Some fruits have wings and are blown about by the wind. The dandelion is a familiar example (Figure 27-10, *D*).

Seeds are dispersed by sticking to or being eaten by animals, by being blown by the wind, or by floating to new environments across bodies of water.

Seed Formation and Germination

The long chain of events between fertilization and maturity is **development.** During development, cells become progressively more specialized, or differentiated. Development in seed plants results first

A

B

FIGURE 27-10 How seeds are dispersed. A, Animal dispersing seeds. This English setter is covered with hound's tongue weed seeds that have stuck to his coat. **B,** Bird dispersing seeds. This cedar waxwing feeds berries to the waiting young in the nest. Birds that eat fruits digest them rapidly so that much of the seed is left intact. What is excreted from the bird can grow into a mature plant. **C,** The seeds of a coconut, *Cocos nucifera.* One of the most useful plants for humans in the tropics, coconuts have become established even on the most distant islands by drifting in the waves. **D,** The seeds of a dandelion, *Pyropappus caroliniana,* are dispersed by the wind. The "parachutes" disperse the fruits of dandelions widely in the wind, much to the gardener's despair.

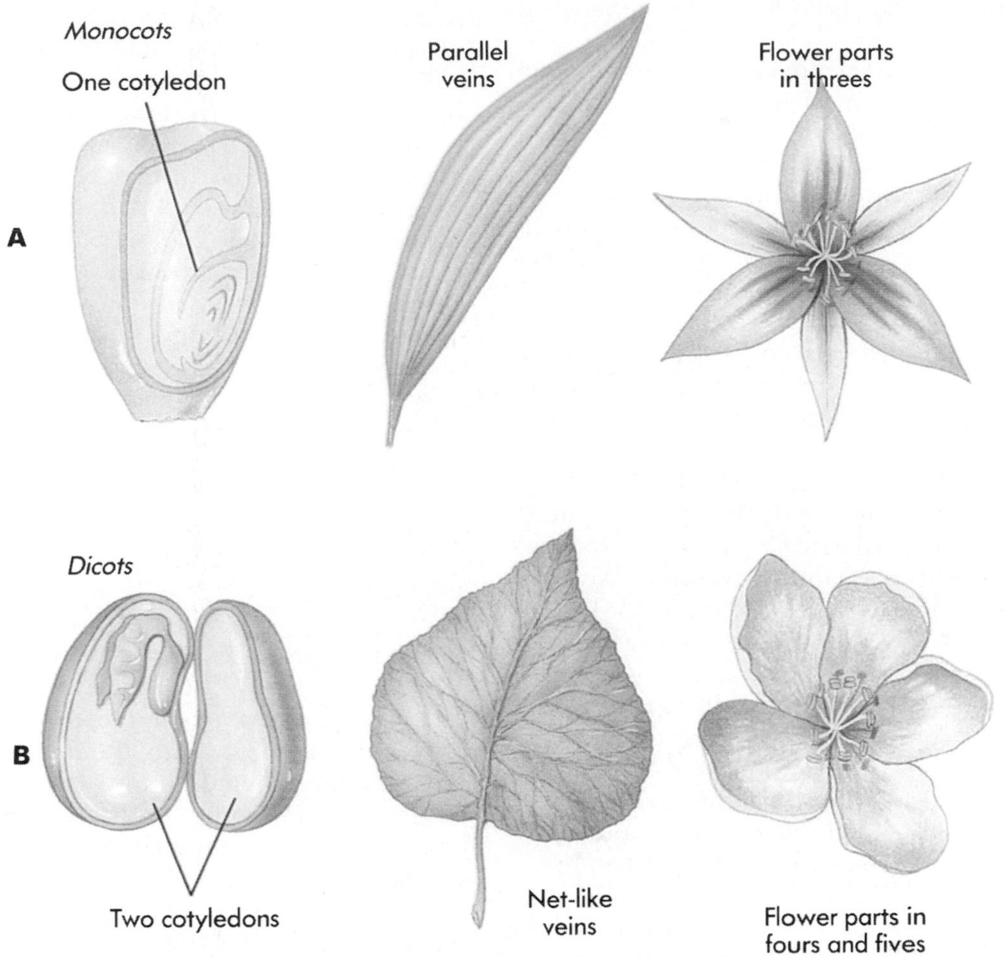

FIGURE 27-11 Monocots and dicots. A, In monocots, most of the food for the embryo is stored in endosperm. Other characteristics of monocots are the parallel veins in their leaves and the occurrence of their flower parts in threes. **B,** Dicots store their food in cotyledons, or seed leaves. They have netlike veins in their leaves, and their flower parts occur in fours or fives.

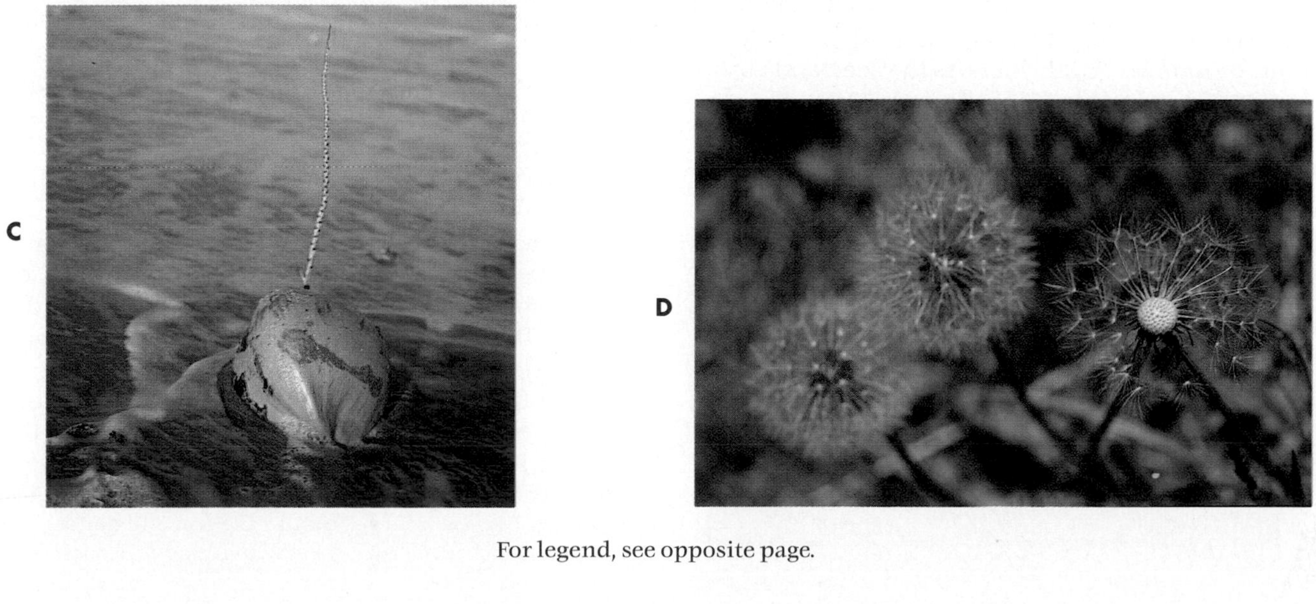

For legend, see opposite page.

in the production of an embryo, which remains dormant within a seed until the seed germinates, or sprouts.

Seed **germination** depends on a variety of environmental factors, the most important of which is water. However, the availability of oxygen (for aerobic respiration in the germinating seed), a suitable temperature, and sometimes the presence of light are also necessary. The first step in germination of a seed occurs when it imbibes, or takes up, water. Once this has taken place, metabolism within the embryo resumes.

Germination and early seedling growth require the mobilization of food storage reserves within the seed. A major portion of almost every seed consists of food reserves. Angiosperms fall into two groups regarding the placement of stored food in their seeds: the **monocots** and the **dicots.** In dicots, most of the stored food is in the *cotyledons,* or seed leaves. In monocots, most of the food is stored in extraembryonic tissue called *endosperm* (Figure 27-11). (Although dicots store some food in endosperm, the amount of endosperm varies among species of plants.) Its single cotyledon absorbs food from the endosperm and shuttles it to the embryo. Monocots and dicots also differ from one another in a number of features. Monocots usually have parallel veins (fluid-carrying tissues) in their leaves, and their flower parts are often in threes. Among the monocots are the lilies, grasses, cattails, orchids, and irises (Figure 27-12, *A* to *C*). Dicots usually have netlike veins in their leaves, and their flower parts are in fours or fives. The dicots include the great majority of familiar plants: almost all kinds of flowering trees and shrubs and most garden plants, such as snapdragons, chrysanthemums, roses, and sunflowers (Figure 27-12, *D* to *F*).

Usually the first portion of the embryo to emerge from the germinating seed is the young root, or *radicle,* which anchors the seed and absorbs water and minerals from the soil. Then, the shoot of the young seedling elongates and emerges from the ground.

As the shoot emerges from the soil, the first true leaves are protected by a straight sheath or a curved stem. Multiplication of the cells in the tips of the stem and roots, along with their elongation and differentiation, initiates and continues the growth of the young seedling.

> *The embryo of a plant remains dormant within the seed until the seed germinates. Germination begins when the seed takes up water and begins to sprout.*

Types of Vegetative Propagation in Plants

Vegetative propagation is an asexual reproductive process in which a new plant develops from a portion of a parent plant. Some plants, such as irises and grasses, produce new plants along underground stems called *rhizomes.* Other plants, such as strawberries, have horizontal stems that grow above the ground called *runners* or *stolons.* These plants produce new roots and shoots at nodes (places where one or more leaves are attached) along these stems. New plants can also arise vegetatively from specialized underground storage stems called *tubers.* A white potato, for example, is a tuber and can grow a new plant from each of its eyes (which are nodes containing lateral buds).

In some plants, new shoots can arise from roots and grow up through the surface of the soil. For example, a group of aspen trees often consists of a single individual that has given rise to a colony of genetically identical trees by producing new shoots from its horizontal roots (Figure 27-13).

In a few species of plants, even the leaves are reproductive. The plant in Figure 27-14 is commonly called the *maternity plant* because small plants arise in the notches along the margins of the leaf. When mature, they drop to the soil and take root. Gardeners commonly propagate African violets from leaf cuttings and many other plants from stem cuttings.

A major breakthrough in the asexual propagation of plants has been the development of **cell culture techniques** (Figure 27-15). Using these techniques, scientists are able to remove individual cells from a parent plant and grow these cells into new individuals. These techniques are successful, however, only because individual plant cells (unlike animal cells) have the inherent capability to direct the growth and development of a new plant.

Using cell culture techniques, botanists are able to produce virtually unlimited numbers of genetically identical offspring. These techniques have been particularly useful in propagating plants that are slow to multiply on their own, such as coconut palms and redwoods, and in cultivating varieties of individual plants with special characteristics, such as large flowers. Award-winning varieties of orchids, for example, are often produced in this way. Cell culture is also used for the commercial production of tremendous numbers of plants in a short period of time—such as the chrysanthemums you may buy at the grocery store. The timber industry uses cell culture techniques in developing rapidly growing conifers such as the Douglas fir.

FIGURE 27-12 Examples of monocots and dicots. Monocots include the cattail (**A**), the crested dwarf iris (**B**), and the pink lady's slipper (**C**). All of these plants show the parallel leaf veination typical of monocots. Dicots include the rose (**D**), the sea grape (**E**), and the bloodroot (**F**). These examples show the netlike veination typical of dicots.

FIGURE 27-13 Clones of aspen. The contrast between the golden quaking aspens (*Populus tremuloides*) and the dark green Engelmann spruces (*Picea engelmannii*) evident in this autumn scene near Durango, Colorado, makes it possible to see their clones, large colonies produced by individuals' roots spreading underground and sending up new shoots periodically.

FIGURE 27-14 *Kalanchoe daigremontiana*. The small plants growing in the notches along the leaf margins will drop to the soil and take root. For this reason, *Kalanchoe daigremontiana* is called the "maternity plant."

FIGURE 27-15 A plant growing in a cell culture. The small shoots in the petri dish are growing from a mass of undifferentiated callus tissue and are genetic clones of carrots. Researchers use clones and tissue cultures to find heartier and more insect-resistant varieties of plants.

JUST WONDERING. . . .

I find the Venus flytrap intriguing. But how can a plant eat an animal?

The Venus flytrap (*Dionaea muscipula*) is one of a few carnivorous (animal-eating) plants. Carnivorous plants often grow in soils depleted of nutrients, especially nitrogen. Digesting insects is the way in which these plants obtain nutrients they are unable to get from the soil. (Like other plants, however, carnivorous plants make their own carbohydrates during the process of photosynthesis.)

All carnivorous plants must first trap their prey and then digest it to absorb nutrients. The Venus flytrap ensnares prey by means of its touch response, which works as follows. On the inner sides of each pair of cusplike modified leaves of the plant are sensitive hairs. Attracted by nectar on the leaves' surfaces, insects touch the hairs as they walk onto the leaves. If two hairs are touched in succession or if one hair is touched twice (the plant's way of distinguishing between a living and a nonliving stimulus) the leaves close on the insect, imprisoning it. The leaf closure results from biochemical changes that occur within the epidermal cells of the leaves when the sensitive hairs are stimulated. These biochemical changes result in the expansion of the outer epidermal cells of each leaf, whereas the inner epidermal cells do not change. As the leaves change shape because of the cellular changes, they come together.

The Venus flytrap produces digestive enzymes, storing them in vacuoles within the cells of the leaves. As the insect becomes enclosed by the leaves, it gets pressed against the inner surfaces of the leaves. This stimulus results in the discharge of the digestive enzymes by the vacuoles onto the trapped prey. As the organism is digested, the nutrients are absorbed by the leaf cells. Seems like quite a gory story for the plant world, doesn't it?

development the events between fertilization and maturity in an organism's life cycle.

germination the sprouting of a seed, which begins when it receives water and has appropriate environmental conditions.

monocots one of two groups of angiosperms that differ in the placement of stored food in their seeds; monocots store most of their extra food in extraembryonic tissue called endosperm.

dicots one of two groups of angiosperms that differ in the placement of stored food in their seeds; dicots store food in their cotyledons, or seed leaves.

vegetative propagation an asexual reproductive process in which a new plant develops from a portion of a parent plant.

cell culture techniques methods that enable scientists to remove cells from a parent plant and grow them into new plants.

hormones (hore mones) chemical substances produced in small, often minute, quantities in one part of an organism and then transported to another part of the organism, where they bring about physiological responses.

The use of cell culture techniques in agriculture will revolutionize the way in which farmers produce some crops. In the future, farmers will be able to grow certain crop plants that have been mass produced by cell culture and are genetically superior in some way, such as disease resistance, high yield per plant, or desirable taste. However, variations (mutations) arise frequently in cell culture and limit the usefulness of cell culture techniques in the mass production of many crops.

The use of genetic engineering techniques is another way in which scientists may revolutionize worldwide agriculture. Scientists are able to insert genes that express favorable characteristics into the chromosomes of culture cells, producing plants with desirable characteristics. Genetic engineering and its applications are described in Chapter 31.

Many plants can reproduce vegetatively by growing new plants from their roots, stem, or leaves. Scientists, using cell culture techniques, can also propagate many plants from a single cell of the parent. Use of these techniques, along with the genetic engineering of plants, will revolutionize the way that farmers produce some crops.

Regulating Plant Growth: Plant Hormones

As a plant grows, it is influenced by environmental factors such as the amount of water and light it receives. However, a plant's growth, differentiation, maturation, flowering, and many other activities are also regulated by chemicals called **hormones.** Hormones are chemical substances produced in

FIGURE 27-16 The effect of the plant hormone gibberellin. This cabbage plant has been treated with gibberellin, which promotes growth through cell elongation.

FIGURE 27-17 Phototropism. Plants grow toward the light because auxin, a plant hormone that controls plant growth, is redistributed in a plant that is darker on one side, resulting in a higher concentration of this hormone on the darker side of the stem. Thus the side of the plant away from the light will grow faster, causing the plant to bend toward the light. The stems of this blood sorrel are oriented toward the window, which lets in plenty of sunlight.

small, often minute quantities in one part of an organism and then transported to another part of the organism where they bring about physiological responses.

There are at least five major kinds of hormones in plants. *Auxins, gibberellins,* and *cytokinins* promote and regulate growth. Cytokinins stimulate cell division, and auxins and gibberellins promote growth through cell elongation (Figure 27-16). The differences in concentration of auxin from one side of the stem or root to another, for example, control the bending of plants toward or away from light (phototropism) (Figure 27-17) and toward or away from gravity (gravitropism). In contrast, *abscisic acid* is a growth inhibitor that induces and maintains dormancy or otherwise opposes the three growth-promoting hormones. *Ethylene* is released by plants as a gas and effects the ripening of fruit and leaf drop in nearby plants.

Other chemicals, such as special plant photoreceptor pigments called *phytochromes,* also affect a plant's response to its environment. Phytochromes change form in response to the length of the day and thereby stimulate or inhibit flowering. The study of plant hormones and other regulatory chemicals, especially how they produce their effects, is an active and important field of research today.

Summary

- Plants are multicellular, eukaryotic, photosynthetic autotrophs that live on land: the bryophytes (nonvascular plants) and the vascular plants. Current classification schemes usually classify the multicellular algae, once classified as plants, with the protists.
- Animals, plants, some protists, and some fungi have sexual life cycles that are characterized by the alternation of the processes of meiosis and fertilization. Plants and some species of algae have life cycles that have both a multicellular haploid phase and a multicellular diploid phase. Because both phases of the life cycle are multicellular, this type of life cycle is called *alternation of generations.*
- In plant life cycles, multicellular haploid plants produce gametes by mitosis and are gametophytes. These gametes fuse during fertilization and grow into the multicellular, diploid spore-producing plants, or sporophytes. These individuals form haploid spores by meiosis, which grow into gametophytes.
- In general, the gametophyte generation often dominates the life cycles of the nonvascular plants, whereas the sporophyte generation dominates the life cycles of vascular plants. This difference reflects an adaptation of the vascular plants for life on land.
- The bryophytes include three divisions of nonvascular plants: the mosses, liverworts, and hornworts. The life cycles of most have two distinct phases, with the gametophyte phase dominating. The sporophyte plants live on and derive nutrients from the gametophyte plants.
- The sporophyte phase of the life cycle is the dominant phase of the life cycle of vascular plants. The seedless vascular plants produce motile sperm that swim through moisture on the gametophyte to fertilize the egg. The sporophytes that develop are protected initially by the gametophyte tissues. The gametophytes in seed-bearing plants consist of only a few cells. These cells produce eggs and sperm. After fertilization the embryonic sporophyte is protected and nourished within a seed.
- The organs of sexual reproduction in the flowering plants, or angiosperms, are in the flower. Pollen grains are male gamete-producing cells. The female gamete-producing cells are in the ovules in the ovary.
- The flowers of angiosperms make possible the precise transfer of pollen from the anther to the stigma, an event that precedes fertilization and formation of a seed. Pollen is transferred by insects, particularly bees; birds and other animals; and the wind.
- The seeds of angiosperms remain within the ovary, which develops into a fruit. Many fruits are fleshy and often sweet; animals that consume fruit may carry the seeds for long distances before excreting them as waste. The seeds can then germinate, or sprout, in the new location.
- Seeds germinate only when they receive water and appropriate environmental cues. In the germination of seeds, mobilization of the food reserves stored in the cotyledons and in the endosperm is critical.
- The change of a zygote into a mature individual, initiated immediately after fertilization, is development. Development is a process of progressive specialization that results in differentiation, the production of highly individual tissues and structures.
- Plants can also reproduce asexually by vegetative propagation; a new plant grows from a portion of another plant.

Knowledge and Comprehension Questions

1. Fill in the blanks: __________ plants contain specialized tissues within them to transport fluids. __________ plants lack these specialized tissues.
2. Fill in the blanks: __________ produce spores, which grow into __________. These produce gametes, which fuse during fertilization and grow into __________.
3. Name the three classes of bryophytes, and summarize their life cycles.
4. What are seeds? Explain their function(s).
5. Summarize the life cycle of seedless vascular plants. Give an example of this type of plant.
6. Match each of the following with the most appropriate term:
   Gymnosperms
   Angiosperms
   Bryophytes
   Ferns
   a. Seedless vascular plants
   b. Nonvascular plants
   c. Vascular plants with naked seeds
   d. Vascular plants with protected seeds
7. Draw a generalized diagram of a flower. Label the following: sepals, petals, stamens, pistil, anther, pollen grains, stigma, ovary, ovules, and style.
8. Summarize the commonalities of sexual reproduction in the angiosperms and the gymnosperms.
9. Summarize the significance of fruits in sexual reproduction.
10. Place the following events in the correct sequence:
    a. The radicle emerges from the seed.
    b. Fertilization occurs.
    c. The shoot emerges from the ground.
    d. An embryo develops.
    e. The seed germinates.
11. What do monocots and dicots have in common? How do they differ?
12. What is vegetative propagation? Give an example.
13. What are hormones? Summarize the function(s) of the five major kinds of plant hormones

Critical Thinking

1. You probably know what a dandelion looks like when it goes to seed, a delicate white sphere at the tip of a slender stalk. Blowing on these dandelion "flowers" is something almost all children do. Dandelions reproduced asexually; they never undergo sexual reproduction. Because flowers attract insects and aid in sexual reproduction, what do you think is the reproductive advantage of this "flower" to dandelions?

CHAPTER 28

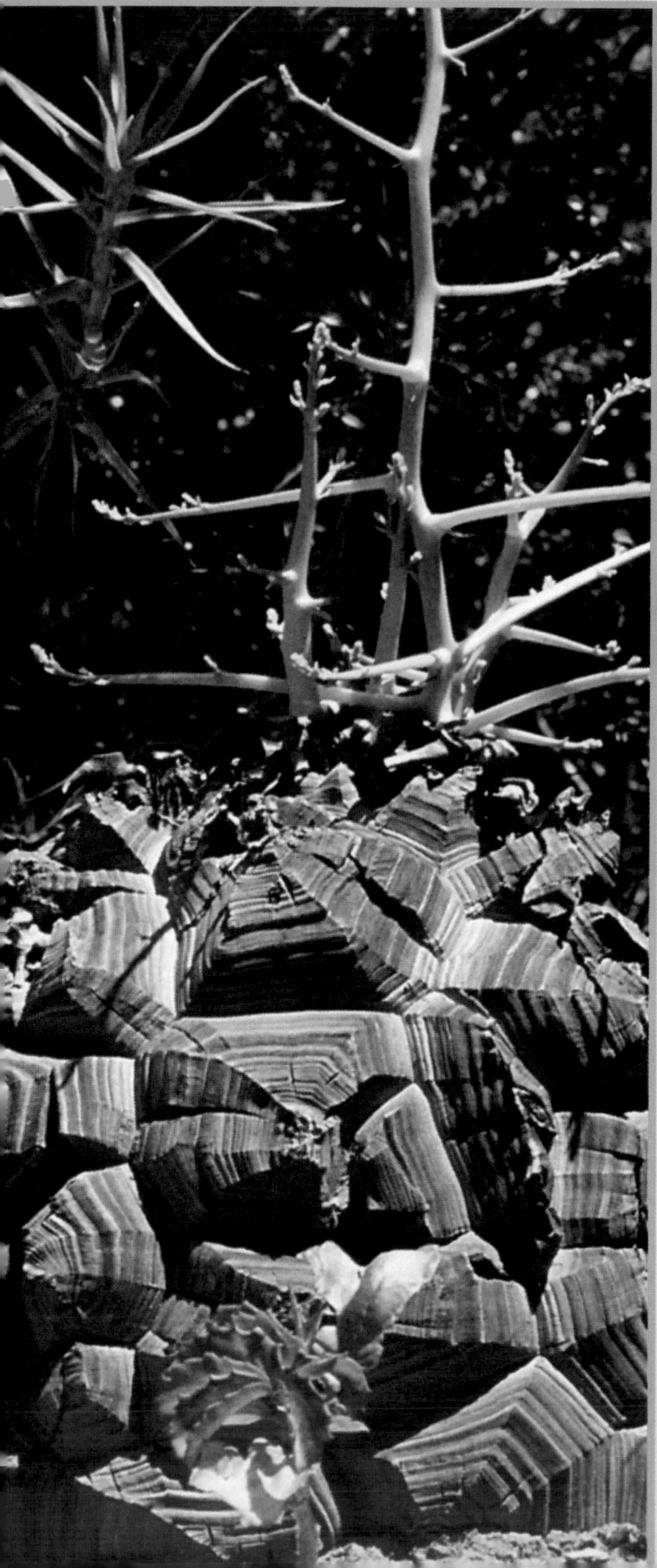

PLANTS: PATTERNS OF STRUCTURE AND FUNCTION

ALTHOUGH THE PHOTO looks like the discarded shell of a huge turtle, it is actually a plant. In fact, one nickname for this plant is the "turtleback." The turtleback's "shell" is really a modified stem that stores water for the plant. This helps the turtleback survive the dry season in its tropical environment. During the rainy season, a system of green branches grows out its top, manufacturing food and bearing leaves, flowers, and fruits. Through the years these branches die, covering the turtleback with an intertwining mass.

Turtlebacks commonly grow in Africa. The native people, however, call this plant *Hottentot's bread* because its fleshy storage organ can be eaten and has a texture similar to a giant turnip. This organ commonly grows to 1 meter (3 feet) in diameter.

Some grow slightly more than 2 meters tall (approximately 7 feet) and weigh up to 318 kilograms (700 pounds)—rather large for the produce section of the grocery store!

The Organization of Vascular Plants

For any plant to grow as tall as the turtleback, it must have a means to transport substances to and from its various parts. Most plants (more than 80% of all living species) have a system of specialized tissues that performs this function. Plants with these tissues are called **vascular plants** (see Chapter 27 for discussion of their reproductive cycles). Despite their reproductive diversity, however, vascular plants all have the same basic architecture.

A vascular plant is organized along a vertical axis (Figure 28-1). The part below the ground is called the **root.** The part above the ground is called the **shoot.** The root penetrates the soil and absorbs water and various ions crucial for plant nutrition. It also anchors the plant. The shoot consists of a **stem** and **leaves.** The stem serves as a framework for the positioning of the leaves, where most photosynthesis takes place. The arrangement, size, and other characteristics of the leaves are critically important in the plant's production of food. Flowers and ultimately fruits and seeds are also formed on the shoot.

Vascular plants are those having a system of tissues that transports water and nutrients. Vascular plants have the same basic architecture: they are made up of underground roots and aboveground shoots. The shoot consists of a stem and leaves.

Tissues of Vascular Plants

The organs of a vascular plant—the leaves, roots, and stem—are made up of different mixtures of tissues, just as your legs are composed of different tissues such as bone and muscle. A tissue is a group of cells that work together to carry out a specialized function. Vascular plants have three types of differentiated tissues, groups of cells having specific structures to perform specific functions: **vascular tissue, ground tissue,** and **dermal tissue** (see Figure 28-1).

The word *vascular* comes from a Latin word meaning "vessel." Thus vascular tissue forms the

KEY CONCEPTS

- Vascular plants are organized along a vertical axis with underground roots and aboveground shoots made up of stems and leaves.
- Vascular plants have specialized tissues that make up the structure of roots, stems, and leaves, enabling them to absorb nutrients, transport fluids, store food, carry on photosynthesis, and produce new cells for plant growth.
- Water and dissolved inorganic nutrients move upward from root to stem and leaves by transpirational pull; water and dissolved sugar are transported throughout the plant by mass flow.
- The nonvascular plants have various levels of organization within their groups; all have a much simpler organization than do the vascular plants.

OUTLINE

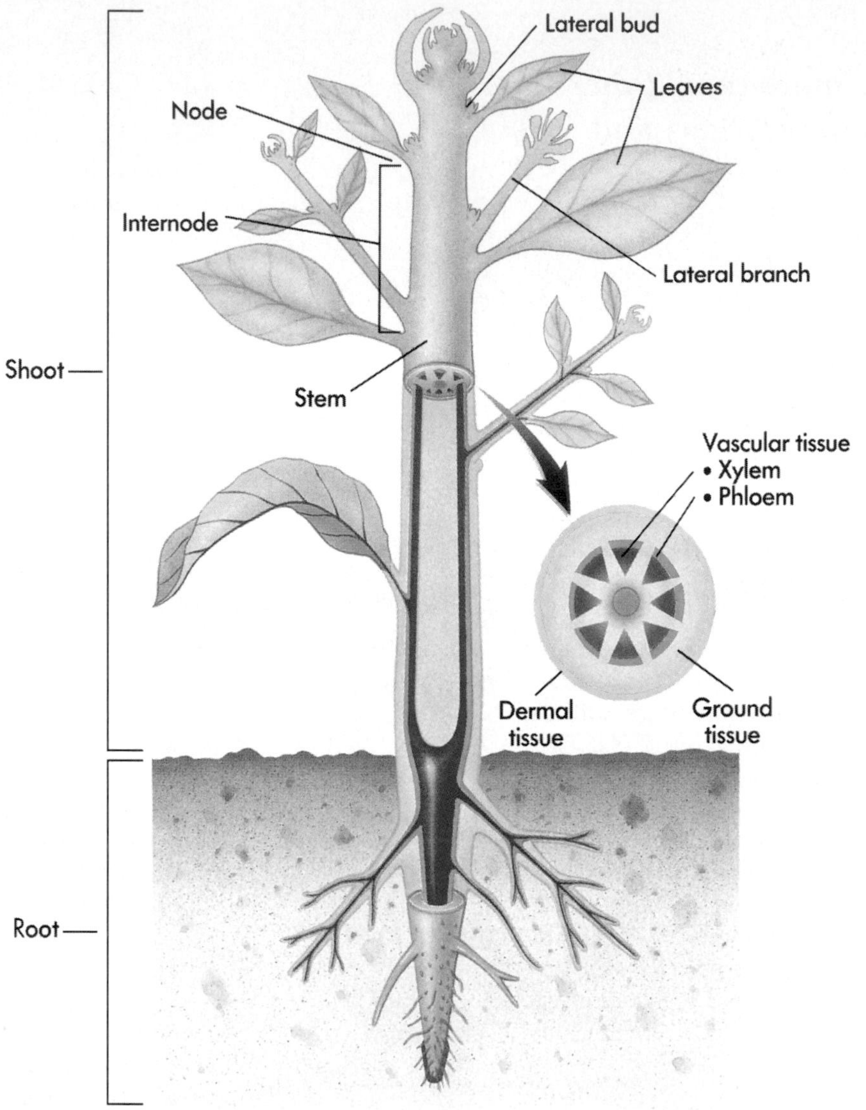

FIGURE 28-1 Structure of a typical plant. The root anchors the plant while the shoot supports the leaves, structures in which photosynthesis takes place. The vascular tissue is the circulatory system of plants and transports water and nutrients.

"circulatory system" of a plant, conducting water and dissolved inorganic nutrients (see Table 28-1) up the plant and the products of photosynthesis throughout the plant. Ground tissue stores the carbohydrates the plant produces. It is the tissue in which the vascular tissue is embedded and forms the substance of the plant. Dermal tissue covers the plant, protecting it—like skin—from water loss and injury to its internal structures. Plants also contain **meristematic tissue,** or growth tissue, an undifferentiated tissue in which cell division occurs. This tissue, often just called *meristem,* is considered undifferentiated because the cells it produces will eventually become one of the other three cell types.

> *The three types of differentiated tissues in vascular plants are (1) dermal tissue, which protects the plant; (2) vascular tissue, which conducts water, inorganic nutrients, and carbohydrates throughout the plant; and (3) ground tissue, which stores food the plant manufactures. In addition, an undifferentiated type of tissue called meristem produces new plant cells during growth.*

Fluid Movement: Vascular Tissue

There are two principal vascular tissues: **xylem,** which conducts water and dissolved inorganic nu-

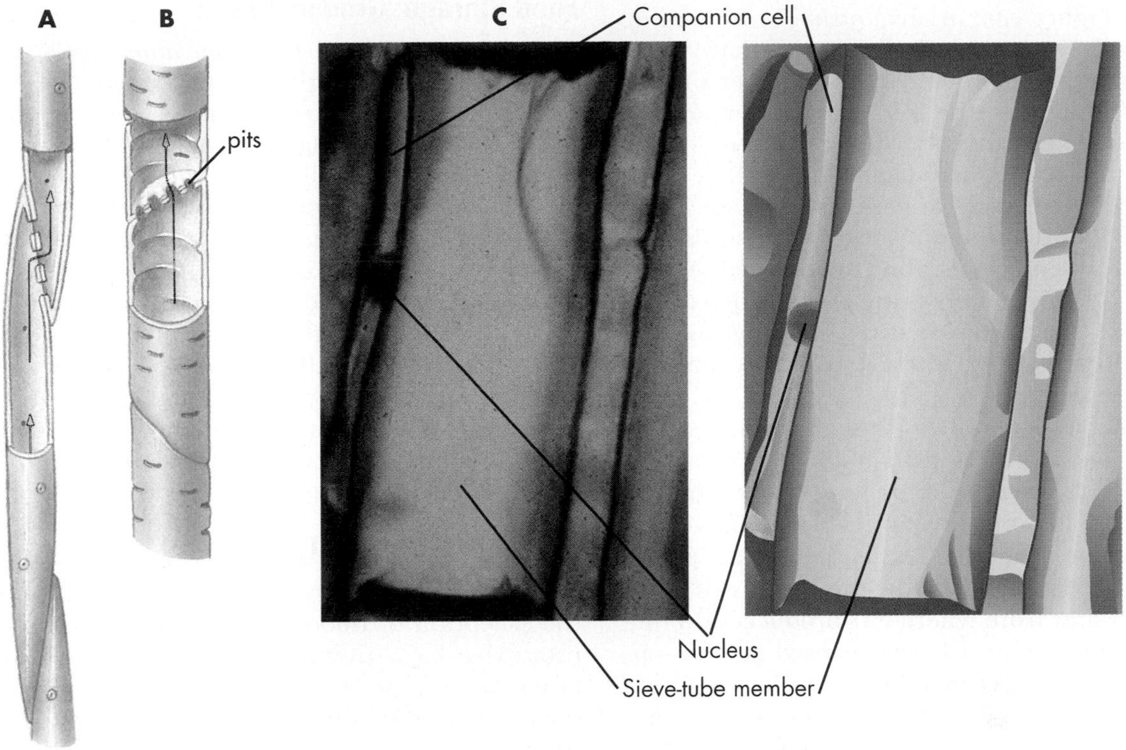

FIGURE 28-2 Vascular tissues of a plant. **A** and **B** show two different types of xylem cells. **C** shows two types of phloem cells. **A,** Tracheid. In tracheids water passes from cell to cell through pits, which are covered with a layer of cell wall. **B,** Vessel elements. In vessel elements, water passes from cell to cell through perforations, which are not covered by tissue. **C,** Sieve tube with companion cell. Individual cells or sieve-tube members are stacked end-to-end, forming sieve tubes. Sugars produced during photosynthesis move from cell to cell through pits. Companion cells secrete substances into and remove substances from the sieve-tube members.

trients, and **phloem,** which conducts carbohydrates the plant uses as food along with other needed substances. Each vascular tissue is made up of different, specialized conducting cells.

Figure 28-2, *A* and *B,* shows two different types of xylem cells. Both types of cells conduct water and dissolved inorganic nutrients only after they die and lose their cytoplasm, becoming hollow and thick walled. Stacked end to end, they form pipelines that extend throughout the plant. The cells

vascular plants plants that have systems of specialized tissues that transport water and nutrients; they are made up of underground roots and aboveground shoots.

root the part of a vascular plant that exists below the ground. The root penetrates the soil and absorbs water and various ions that are crucial for plant nutrition. It also anchors the plant.

shoot the part of a vascular plant that exists above the ground; the stem, leaves, and other plant structures are formed at the shoot.

stem a part of a plant's structure that serves as a framework for the positioning of its leaves.

leaves the parts of a plant's structure where most photosynthesis takes place.

vascular tissue the tissue that forms the "circulatory system" of a vascular plant. Vascular tissue conducts water and dissolved inorganic nutrients up the plant and carries the products of photosynthesis throughout the plant.

ground tissue plant tissue that stores the carbohydrates the plant produces. It forms the substance of the plant and is the tissue in which vascular tissue is embedded.

dermal tissue the outer protective covering of virtually all plants. Dermal tissue acts as a plant's skin and protects it from water loss and injury to its internal structures.

meristematic tissue (**mer** uh stuh **mat** ik) an undifferentiated type of tissue in a vascular plant in which cell division occurs during growth.

xylem (**zy** lem) a type of vascular tissue that conducts water and dissolved inorganic nutrients.

phloem (**flo** em) a type of vascular tissue that conducts carbohydrates the plant uses as food, along with other needed substances.

pictured in Figure 28-2, *A*, are *tracheids*, which are stacks of cells with tapering ends that have connections between them called *pits*. A layer of cell wall material covers the pits, but water moves through them nevertheless. Figure 28-2, *B*, shows *vessel elements*, which are stacks of cells that have connections between them called *perforations*. Because these openings are not covered by tissues, water moves through perforations unimpeded. Along the lengths of both cell types are *pores*, which allow water to move laterally to cells surrounding the xylem pipelines. They work much like soaking hoses used in gardens.

Figure 28-2, *C*, shows two types of phloem cells: a *sieve-tube member* and *companion cell*. The conducting cells (sieve-tube members) are alive and contain cytoplasm but are not typical cells. They lack nuclei, for example. The ends of these elongated cells have pits that allow the easy passage of sugar-filled water from where it is produced—in the leaves, for example—to where it is used rapidly—at the reproductive structures, for example. Companion cells, which contain all of the organelles commonly found in plant cells (including nuclei), secrete substances into and remove substances from the sieve-tube members. These substances include sugars produced during photosynthesis.

A

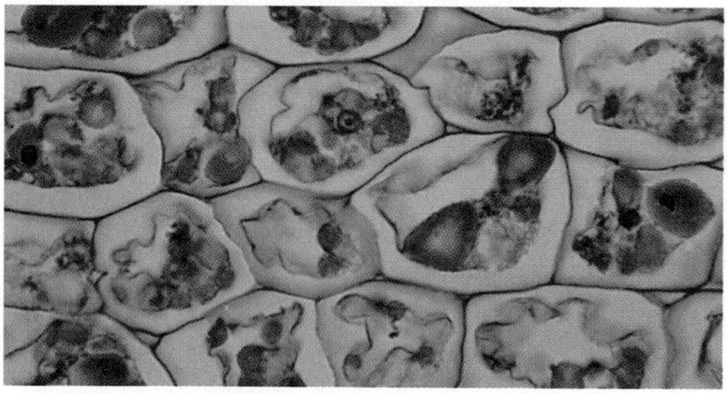

B

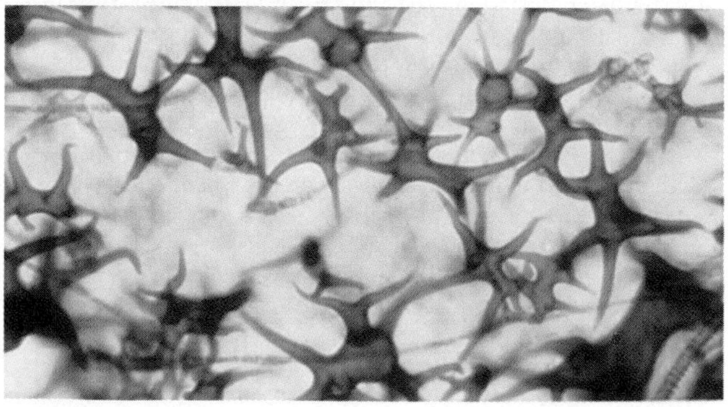

FIGURE 28-3 Ground tissue. A, Parenchyma cells. Parenchyma cells store starch and contain chloroplasts. **B,** Sclerenchyma cells. These cells are hollow, with strong walls. Sclerenchyma cells help support the plant.

Food Storage: Ground Tissue

Ground tissue contains *parenchyma cells*, which function in photosynthesis and storage. These thin-walled cells contain large vacuoles and may also be packed with chloroplasts. They are the most common of all plant cell types and form masses in leaves, stems, and roots. Fleshy storage roots, such as carrots or sweet potatoes, contain a predominance of storage parenchyma. The flesh of most fruits is also made up of these cells. *Sclerenchyma* cells, also found in the ground tissue, are hollow cells with strong walls. These cells help support and strengthen the ground tissue. The gritty texture of a pear is due to schlerenchyma cells that are dispersed among the softer parenchyma cells. Both types are shown in Figure 28-3.

Protection: Dermal Tissue

Dermal tissue covers the outside of a plant, with the exception of woody shrubs and trees that have protective bark in its place. Bark is made up of other tissue types (see p. 625). *Epidermal cells* are the most abundant type of cell found in the dermal tissue. These cells are often covered with a thick, waxy layer called a *cuticle* that protects the plant and provides an effective barrier against water loss (Figure 28-4, *A*). Other types of cells found in the dermis are guard cells, which surround openings in the leaves through which gases and water vapor enter and leave (Figure 28-4, *B*), and trichomes (Figure 28-4, *C*), which are outgrowths of the epidermis (much like the hairs on your body) that have various functions. For example, on "air plants," trichomes help provide a large surface area for the absorption of water and inorganic nutrients. On leaves or fruits, fuzzlike trichomes reflect sunlight, which helps control water loss. In some desert plants, white multicellular trichomes reflect enough sunlight to reduce temperatures in internal tissues. Some trichomes defend a plant against insects or larger animals with their sharpness or by secreting chemicals.

Growth: Meristematic Tissue

Plants contain meristems, areas of undifferentiated cells that are centers of plant growth. Every time one of these cells divides, one of the two resulting cells remains in the meristem. In this way, meristem cells remain "forever young," capable of repeated cell division. The other cell goes on to differentiate into one of the three kinds of plant tissue, ultimately becoming part of the plant body.

Plants can grow only in relationship to where their meristematic tissue is located. Located at the tips of the roots and the tips of the shoots, these tis-

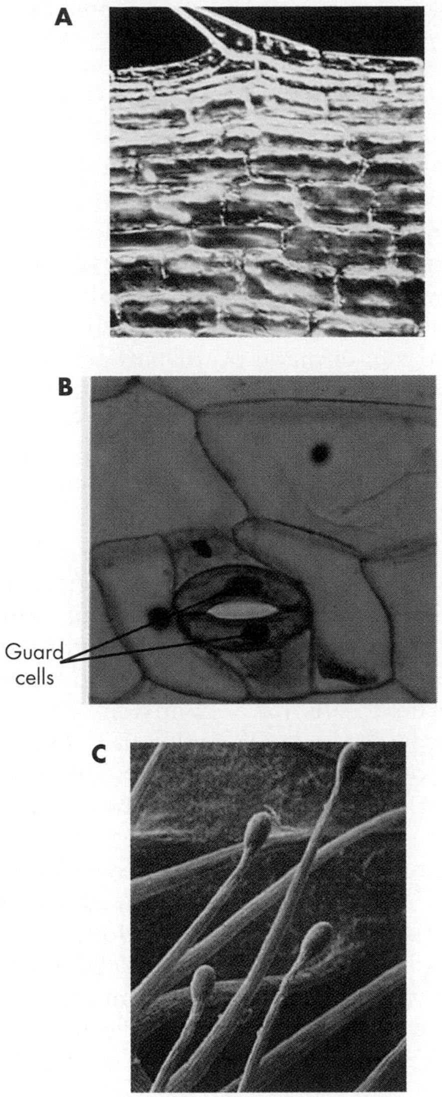

FIGURE 28-4 Dermal tissue. A, Epidermal cells. These cells are covered with a waxy substance and thus protect the plant from water loss. **B,** Guard cells. These cells regulate the movement of water vapor and gases into and out of the plant. **C,** Trichomes. Trichomes perform various functions: they increase surface area, defend the plant against insects and other predators, and reflect sunlight away from the plant, reducing water loss.

sues are called *apical meristems* (*apex* meaning "tip"). Apical meristems allow plants to grow taller and their roots to grow deeper into the ground. This type of plant growth, which occurs mainly at the tips of the roots and shoots, is called **primary growth.**

Some plants not only grow taller but grow thicker as well. This type of growth is called **secondary growth** and occurs in all woody trees and shrubs such as pines, oaks, and rhododendrons. These plants have a cylinder of meristematic tissue along the length of their stems and branches, which is called *lateral meristem*. Herbaceous (nonwoody) plants, such as tulips, have only primary growth. However, their stems do grow somewhat thicker as they develop. One reason for this growth is that plant development involves both cell division *and* cell growth; as cells enlarge, plant parts enlarge. In addition, some cell division occurs in the cells differentiating from meristematic tissue.

Organs of Vascular Plants

Roots, stems, and leaves are the organs of vascular plants. Together, the stems and leaves make up the shoot. The roots and shoots of all vascular plants share the same basic architecture, but there are differences. This chapter discusses the structure of the plants that dominate the plant world today: the flowering plants, or **angiosperms.** The principal differences between the structures of the roots, stems, and leaves in the seedless vascular plants and those with naked seeds lie primarily in the relative distribution of the vascular and ground tissue systems.

Roots

The function of a root system is to anchor a plant in the soil and to absorb water and inorganic nutrients. (The essential inorganic nutrients that plants use are listed in Table 28-1.) During the process of photosynthesis, plants use the water they absorb along with carbon dioxide they capture from the air to produce carbohydrates (see Chapter 6). But primarily, water absorbed at the roots replaces the water released by the plant into the air in a process called *transpiration* (see p. 627). Leaves are the principal organs of transpiration. But why do plants need to absorb inorganic nutrients? The answer is simple. Although plants manufacture carbohydrates during photosynthesis, these sugars are not the only substances that plants need to live. Plants need nucleic acids, proteins, fats, and vitamins. These substances are formed from the carbohydrates plants manufacture and from the inor-

primary growth a type of plant growth that occurs mainly at the tips of the roots and shoots. During primary growth the plant grows taller and its roots grow deeper into the ground.

secondary growth a type of plant growth in which a plant grows wider (thicker).

angiosperms (an jee oh spurms) vascular plants with protected seeds and flowers that act as their organs of sexual reproduction.

TABLE 28-1

Inorganic nutrients important to plants

NUTRIENT	RELATIVE ABUNDANCE IN PLANT TISSUE (PPM)
MACRONUTRIENTS (nutrients required in greater concentration)	
Hydrogen	60,000,000
Carbon	35,000,000
Oxygen	30,000,000
Nitrogen	1,000,000
Potassium	250,000
Calcium	125,000
Magnesium	80,000
Phosphorus	60,000
Sulfur	30,000
MICRONUTRIENTS (nutrients required in minute quantities)	
Chlorine	3000
Iron	2000
Boron	2000
Manganese	1000
Zinc	300
Copper	100
Molybdenum	1

ppm, Parts per million. Parts per million equals units of an element by weight per million units of oven-dried plant material.

ganic nutrients that plants take in from the soil and concentrate. Many plants therefore are an important source of inorganic nutrients in the human diet. Broccoli and cabbage, for example, are excellent sources of calcium. Bananas provide you with potassium.

> *Roots, the part of a plant usually found below ground, absorb water and minerals as well as anchor the plant.*

Structurally, the roots of dicots (plants having netlike veins in their leaves; see Chapter 27) have a central column of xylem with radiating arms. Between these arms are strands of phloem (Figure 28-5). Ringing this column of vascular tissue (often called the *vascular cylinder*) and forming its outer boundary is another cylinder of cells called the *pericycle.* This tissue is made up of parenchyma cells able to undergo cell division to produce branch roots (roots that arise from other older roots). Surrounding the pericycle is a mass of parenchyma called the *cortex.* These cells store food for the growth and metabolism of the root cells. The innermost layer of the cortex is called the *endodermis,* which consists of specialized cells that regulate the flow of water between the vascular tissues and the outer portion of the root. The outer layer of the root is the *epidermis,* which absorbs water and inorganic nutrients. These cells have extensions called *root hairs* that provide the epidermal cells with a larger surface area over which absorption can take place—like the function of microvilli in the intestinal walls of animals.

Monocot roots (plants having parallel veins in their leaves; see Chapter 27) are similar to dicot roots with one important exception: monocot roots often have centrally located parenchyma (storage) tissue called *pith.* The xylem and phloem are arranged in rings around the pith (Figure 28-6).

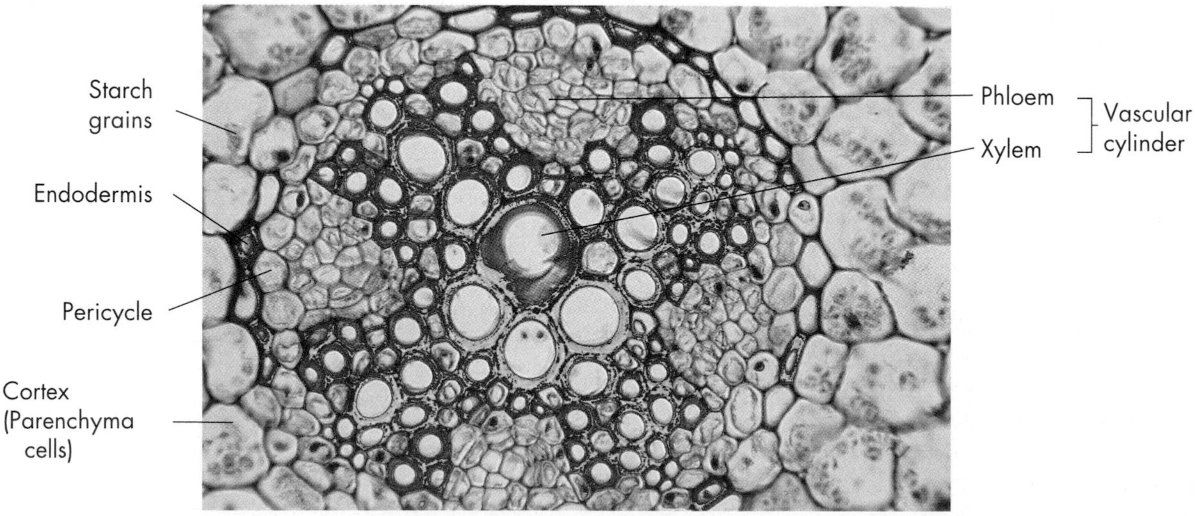

FIGURE 28-5 Dicot root. This typical dicot root from a buttercup has a central column of xylem with radiating arms. Phloem tissue (*green*) surrounds the xylem (*red*).

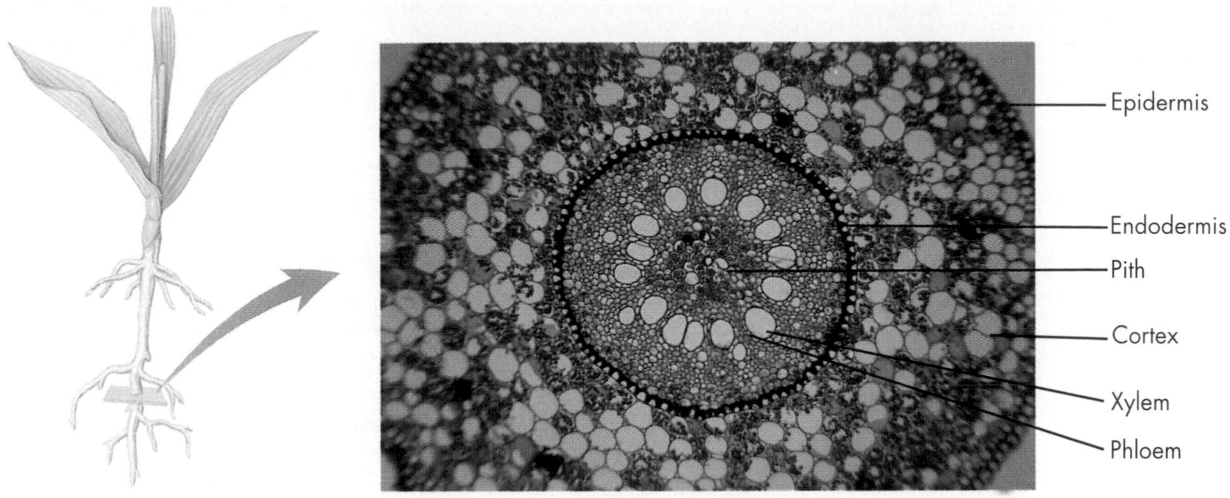

FIGURE 28-6 Monocot root. The important difference between a dicot and monocot root is that the monocot root has a centrally located storage tissue, or pith. The xylem and phloem are arranged in rings around the pith.

> *Most dicot roots have a central column of xylem with radiating arms and strands of phloem between these arms. Surrounding this vascular tissue is a layer of cells called the pericycle that is capable of cell division. Parenchyma (storage) cells of the cortex surround the pericycle. The entire root is covered with a protective epidermis. Monocot roots are structured similarly but often have an additional storage tissue called pith, which is located in the center of the root, and they contain more xylem elements.*

The end of a root is tipped with apical meristem. These growth cells divide and produce cells inwardly, back toward the body of the plant, and outwardly. Outward cell division results in the formation of a thimblelike mass of relatively unorganized cells, the *root cap,* which covers and protects the root's apical meristem as it grows through the soil. Just behind its tip, the root cells elongate. Velvety root hairs, the tiny projections from the epidermis, form above this area of elongation (Figure 28-7). Here too the cells differentiate to produce specialized cell types.

If you have ever pulled a plant up by its roots, you will have noticed that the roots of one type of plant may look different from those of another type of plant. These differences in appearance are linked to differences in function. Many dicots, such as dandelions, for example, have a single, large root called a *taproot* (Figure 28-8, *A*). Taproots grow deep into the soil, firmly anchoring the plant. Some taproots, like carrots and radishes, are fleshy because they are modified for food storage. The plant draws on these food reserves when it flowers or produces fruit; thus taproot crops are harvested before that time. Plants with taproots also have extensive secondary root systems, which are much smaller in diameter than the taproot and so are lost and therefore not seen when the taproot is pulled up. The secondary root system is the major water and nutrient absorption organ of the plant.

The taproot that develops in monocots, on the other hand, often dies during the early growth of the plant, and new roots develop from the lower part of the stem. These roots are called *adventitious roots,* roots that develop from an aboveground structure. Often, adventitious roots help anchor a plant, such as "prop" roots in corn (Figure 28-8, *B*). Certain dicots, such as ivy plants, also develop adventitious roots. The adventitious roots of ivy plants help them cling to walls.

Have you ever pulled up a clump of grass and looked at its roots? Grass has *fibrous roots,* a type of root system that has no predominant root. Most monocots have fibrous roots also, such as the daffodil shown in Figure 28-8, C. They work well to anchor the plant in the ground and absorb nutrients efficiently because of their large surface area. They also help prevent soil erosion by holding soil particles together.

Shoots

Plant shoots grow aboveground and are made up of stems and leaves. As with roots, the growing end of a shoot is tipped with apical meristematic tissue.

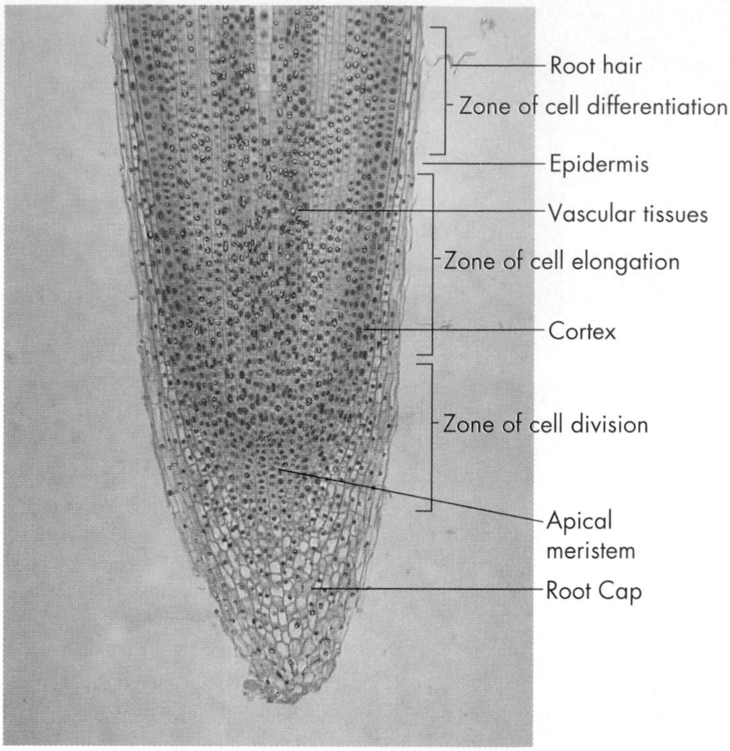

FIGURE 28-7 **A root tip.** Section of an onion root tip showing the differentiation of epidermis, cortex, and column of vascular tissues.

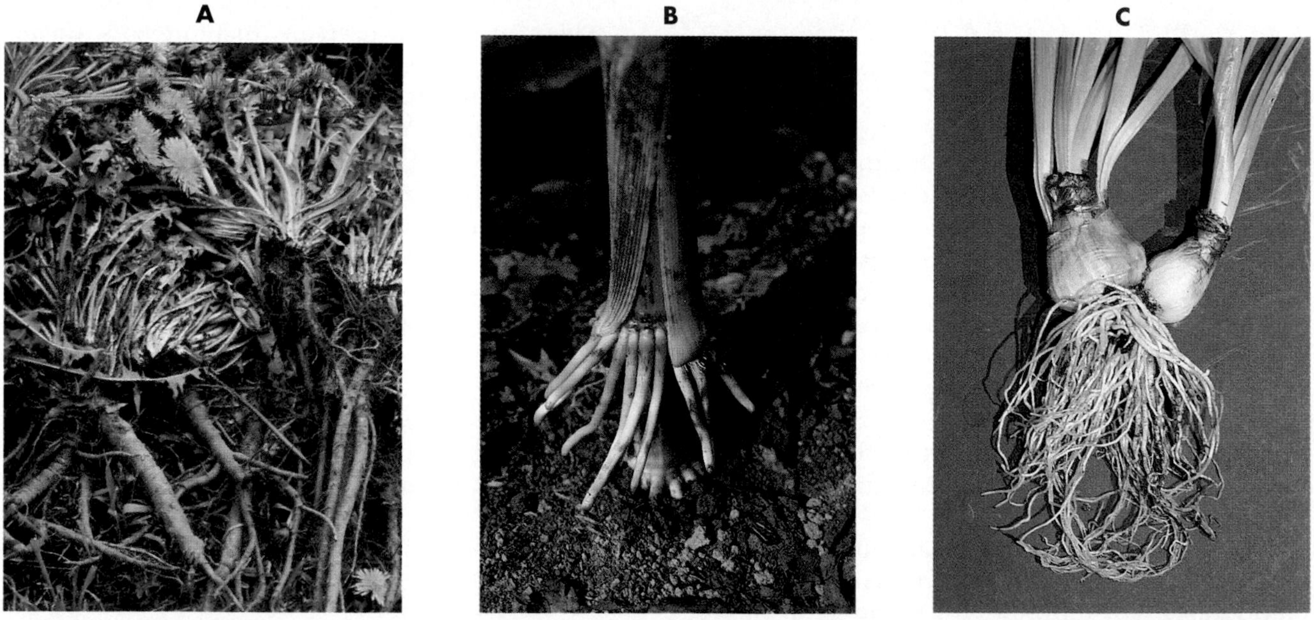

FIGURE 28-8 **Types of roots. A,** Taproots in a dandelion, *Taraxacum officinale.* Even a small section of these taproots can regenerate a new plant, which is one reason why dandelions are so difficult to eliminate from lawns and gardens. **B,** Prop roots in corn, *Zea mays,* are adventitious—they arise from stem tissue and take over the function of the main root. **C,** Fibrous roots in a daffodil. These roots have no central, predominant root. Fibrous roots are especially efficient in anchoring the plant and absorbing nutrients.

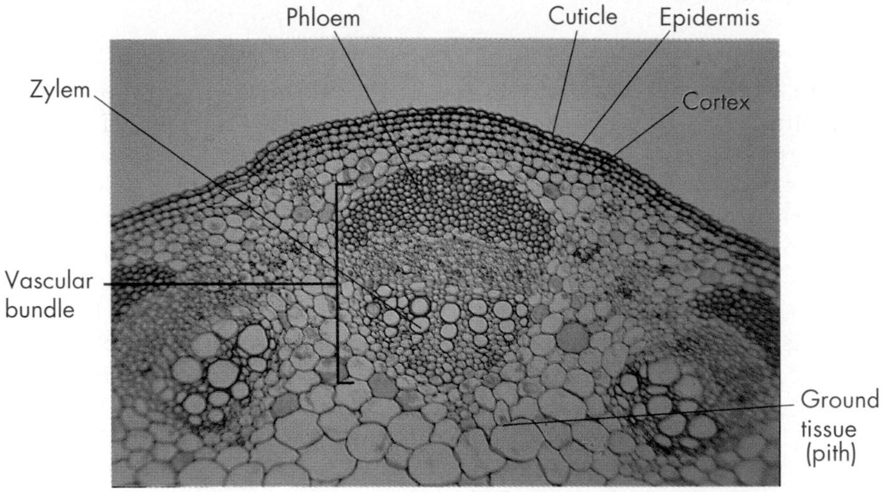

A

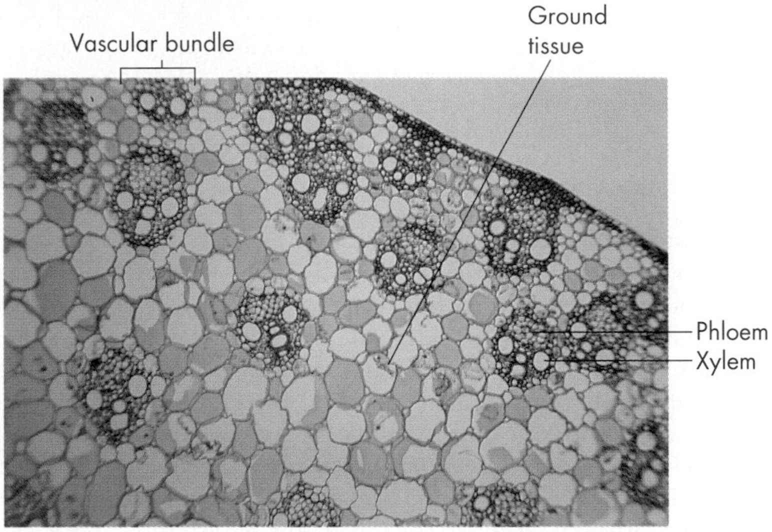

B

FIGURE 28-9 **Stems. A,** A dicot stem from the common sunflower, in which the vascular bundles are arranged around the outside of the stem. **B,** A monocot stem from corn, showing the scattered vascular bundles characteristic of monocots.

Young leaves cluster around the apical meristem, unfolding and growing as the stem itself elongates (see Figure 28-1).

Leaves form on the stem at locations called **nodes.** The portions of the stem between the nodes are called the **internodes.** As the leaves grow, tiny undeveloped side shoots called **lateral buds** develop at the angles between the leaves and the stem. Given the proper environmental conditions, these buds, which contain their own embryonic leaves, may elongate and form lateral branches.

Stems

One purpose of stems is to support the parts of plants that carry out photosynthesis. This process takes place primarily in the leaves, which are arranged on the stem so that light will fall on them. In addition, stems conduct water and inorganic nutrients from the roots to all plant parts and bring the products of photosynthesis to where they are needed or stored.

The stem "transportation system" is made up of strands of xylem and phloem tissue that are positioned next to each other, forming cylinders of tissue called *vascular bundles* (often referred to as the *veins* of leaves) (Figure 28-9). The xylem tissue characteristically forms the part of each bundle closer to the interior of the stem. The phloem lies closer to the epidermis. In herbaceous dicots (those with soft stems rather than woody, treelike stems), the vascular bundles are arranged in a ring near the periphery of the stem (Figure 28-9, *A*). In monocots, the vascular bundles are scattered throughout the stem (Figure 28-9, *B*).

nodes locations on the stem of a plant where leaves form.
internodes the portions of a plant's stem that lie between the nodes.
lateral buds tiny undeveloped side shoots that develop at the angles between a plant's leaves and its stem.

Because of the arrangement of their vascular bundles, dicot stems have a mass of ground tissue in the center of the stem called the *pith* and a ring of ground tissue between the epidermis and the vascular bundles called *cortex.* Monocot stems also have ground tissue, but because it surrounds scattered vascular bundles, ground tissue does not form areas of pith or cortex. The epidermis of both monocot and herbaceous dicot stems is covered with a protective, waxy coating called the *cuticle.*

Stems support the photosynthetic structures—the leaves—and transport water and dissolved substances throughout plants. Herbaceous dicot stems are characterized by an inner cylinder of ground tissue called pith surrounded by a ring of vascular bundles. Encircling the ring of vascular bundles is additional ground tissue called cortex. Monocot stems are characterized by scattered vascular bundles embedded in ground tissue.

Dicots and gymnosperms (such as pine trees) with woody stems (or trunks), such as flowering trees, have lateral meristems called *cambia.* One type of cambium in woody stems is called *vascular cambium.* As the name suggests, this growth tissue lies between the vascular tissue—the xylem and phloem—connecting the bundles to form a ring (Figure 28-10). As the cambial cells divide during secondary growth, one of the resulting daughter cells remains as a cambial cell, and the other differentiates into either a xylem or a phloem cell. This new xylem and phloem is called *secondary xylem* and *secondary phloem* (Figure 28-11).

The wood of trees is actually accumulated secondary xylem. The wood of dicot trees (such as cherry, hickory, oak, and walnut) is commonly referred to as *hardwood,* whereas the wood of conifers (such as fir, cedar, pine, and spruce) is called *softwood.* However, these names are not accurate descriptions of each group. Each contains trees having woods of varying hardness. The hardness of wood relates to its density, which depends on its proportion of wall substance to the space bounded by the cell wall. The denser a wood, the more wall substance it has in relation to the space it bounds and the stronger it is. When used for building, denser woods are harder to nail and machine, but they generally shrink and swell less than less dense, softer woods. Denser woods are also better fuel woods.

When growth conditions are favorable, as in the spring and early summer in most temperate regions, the cambium divides most actively, producing large, relatively thin-walled cells. During the rest of the year, the cambium divides more slowly, producing small, thick-walled cells. This pattern of growth results in the formation of rings in the wood. These rings are called *annual rings* and can be used to calculate the age of a tree (Figure 28-12).

Other lateral meristem tissue called *cork cambium* lies just under the epidermis. The outermost cells of this growth tissue produce densely packed cork cells. Cork cells have thick cell walls that contain fatty substances, making these cells waterproof and resistant to decay. When mature, the cork cells lose their cytoplasm and become hardened in a manner similar to the epidermal cells on your skin. The innermost cells of the cork cambium produce a dense layer of parenchyma cells. The cork, the cork cambium, and the parenchyma cells make up the outer protective covering of the plant called the *bark.*

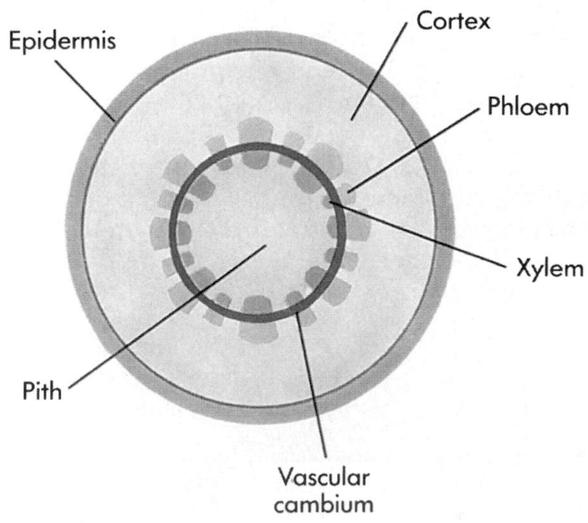

FIGURE 28-10 The vascular cambium in a stem. The vascular cambium is the growth tissue in woody stems and lies between the xylem (*green*) and the phloem (*blue*).

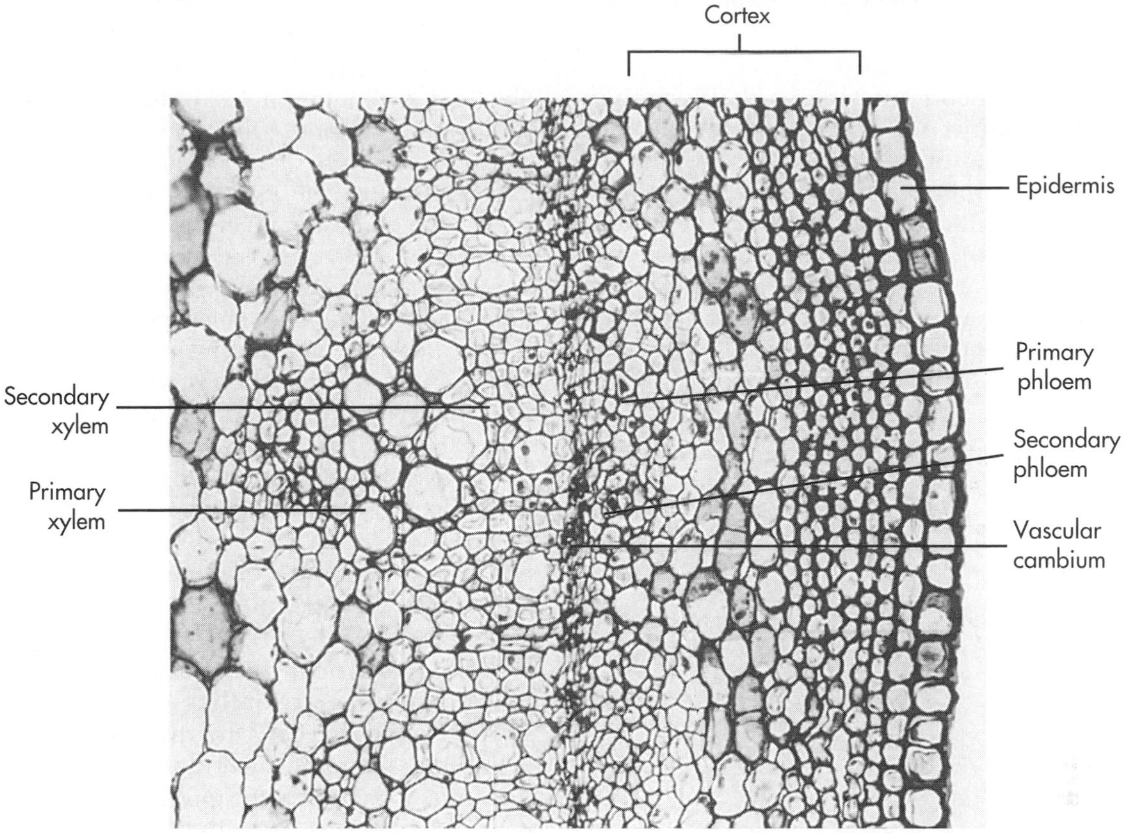

FIGURE 28-11 Secondary growth in woody stems. A, A cross-section of a woody stem. This older stem has an outer layer of cork and underlying cork cambium. Its epidermal cells have died. **B,** A photomicrograph of a cross-section of a woody stem. This younger stem has an epidermis as its outer layer. The cork and cork cambium have not yet formed. Secondary growth can be observed on both stems. During secondary growth, cells of the vascular cambium divide, each forming two "daughter" cells from one original "mother" cell. One of the resulting daughter cells remains a cambial cell, retaining the ability to divide again. The other daughter cell becomes either a new xylem cell or a new phloem cell. "New" xylem and phloem is called secondary xylem and secondary phloem. Note these areas of secondary xylem and phloem on either side of the vascular cambium in both **A** and **B.**

FIGURE 28-12 The structure of a tree trunk. The yearly growth of the vascular cambium produces growth rings.

Woody stems have lateral meristem tissue that provides for secondary growth, which increases the diameter of the stem. Other lateral meristem tissue produces a dense layer of cells called bark that protects the plant.

Leaves

Leaves, outgrowths of the shoot apex, are the light-capturing photosynthetic organs of most plants. Some exceptions exist; a major exception is found in most cacti, in which the stems are green and have largely taken over the function of photosynthesis for the plants.

Most leaves have a flattened portion, the **blade,** and a slender stalk, the **petiole** (Figure 28-13). Veins, consisting of both xylem and phloem, run through the leaves. In monocots, veins are usually parallel, and in dicots, they are usually netlike (Figure 28-14). Many conifers (vascular plants with naked seeds) have needlelike leaves suited for growth under dry and cold conditions. These modified leaves have thick, waxy coverings beneath, which are compactly arranged, thick-walled cells.

Microscopically, a cross-section of a typical leaf looks somewhat like a sandwich: parenchyma cells in the middle, bounded by epidermis. The vascular bundles, or veins, run through the parenchyma. The leaf parenchyma is appropriately called the *mesophyll,* or "middle leaf" (Figure 28-15).

The mesophyll of most dicot leaves is divided into two layers: the *palisade layer* and the *spongy* layer. The palisade layer lies beneath the upper epidermis of the leaf and consists of one or more layers of closely packed, columnlike cells. These cells and the cells of the spongy layer contain chloroplasts, the organelles in which photosynthesis takes place. The spongy layer lies beneath the palisade layer and is made up of irregularly shaped cells. These cells are not closely packed like those in the palisade layer but instead have many air spaces between them. These spaces are connected, directly or indirectly, with openings to the outside called **stomata** (sing. **stoma**). Each stoma is bracketed by two **guard cells** that regulate its opening and closing (Figure 28-16).

The stomata open and close because of changes in the water pressure of their guard cells. As Figure 28-16 shows, guard cells are thicker on the side next to the stomatal opening and thinner on their other sides and ends. When the guard cells are plump and swollen with water, the cells bow. This change in the shape of the guard cells opens the stomata. When photosynthesis is taking place, water enters the guard cells, opening the stomata because the guard cells actively transport potassium ions to their interior. The water follows the potassium ions by osmosis because of the osmotic gradient the ions create. The oxygen produced by photosynthesis diffuses

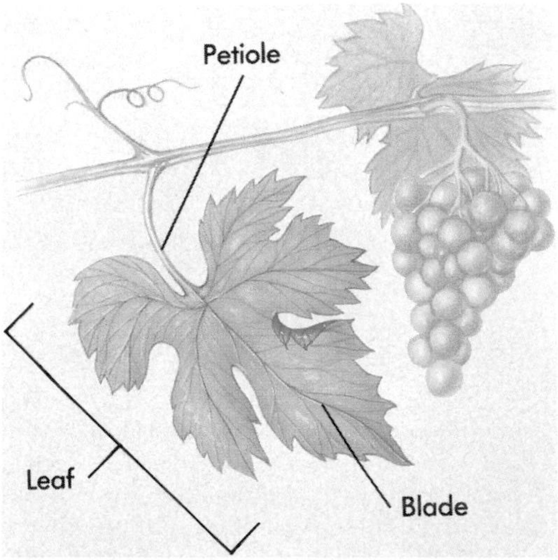

FIGURE 28-13 Structure of a leaf. A leaf consists of a petiole and blade.

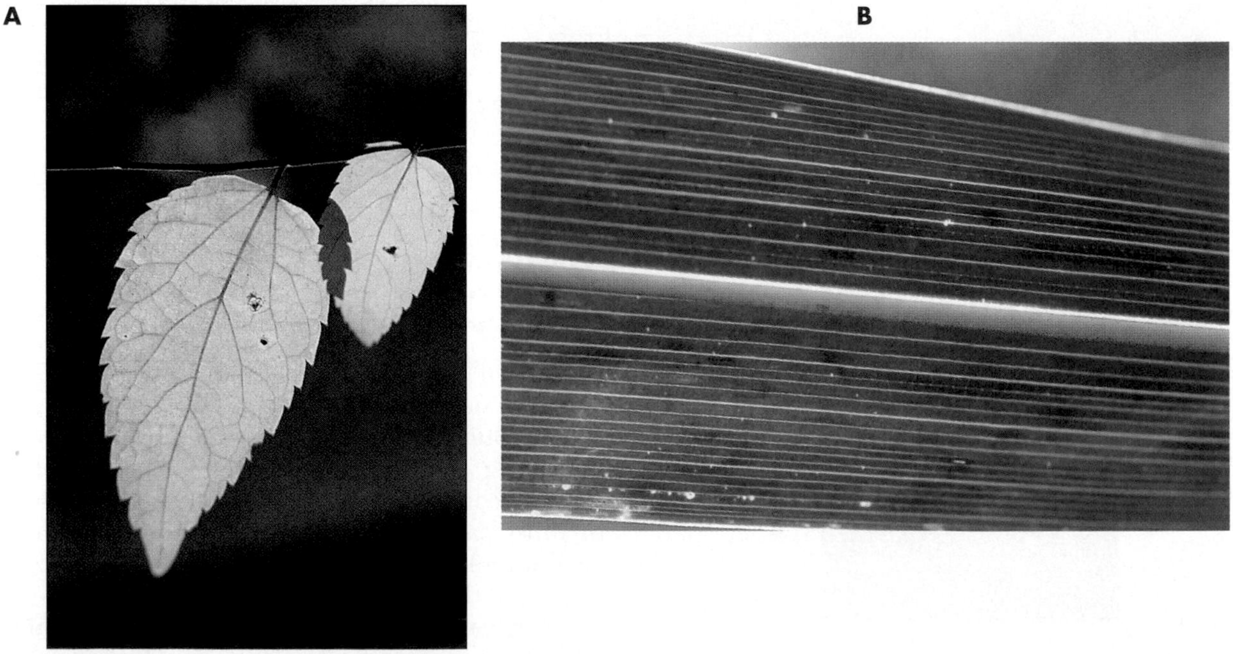

FIGURE 28-14 Dicot versus monocot leaves. A, The leaves of dicots have net veination; those of monocots, like this palm (**B**), have parallel veination.

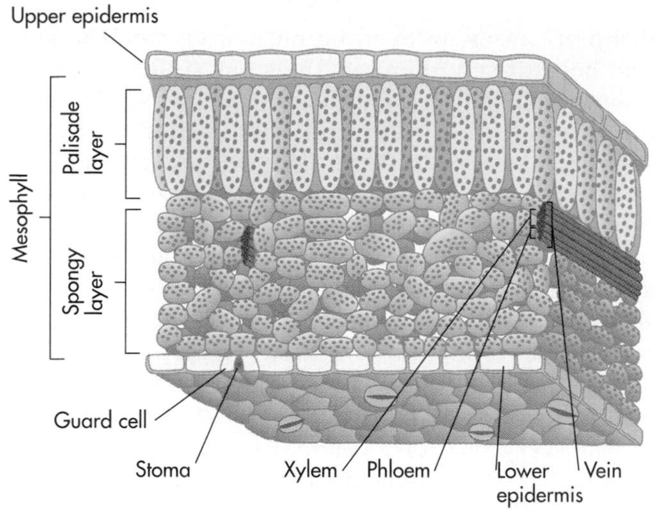

FIGURE 28-15 Internal structure of a leaf. This diagram shows the internal structure of a typical leaf.

into the atmosphere through the stomatal openings, whereas the carbon dioxide needed for photosynthesis to take place diffuses in. In addition, water leaves the leaf through these openings in the form of water vapor, a process called **transpiration.**

> *Leaves, the light-capturing organs of most vascular plants, are made up of parenchyma cells bounded by epidermis. The parenchyma cells contain chloroplasts, the organelles of photosynthesis.*

Movement of Water and Dissolved Substances in Vascular Plants

Did you ever wonder how trees manage to move water to their uppermost leaves? Osmosis alone cannot account for this amazing feat. In fact, many

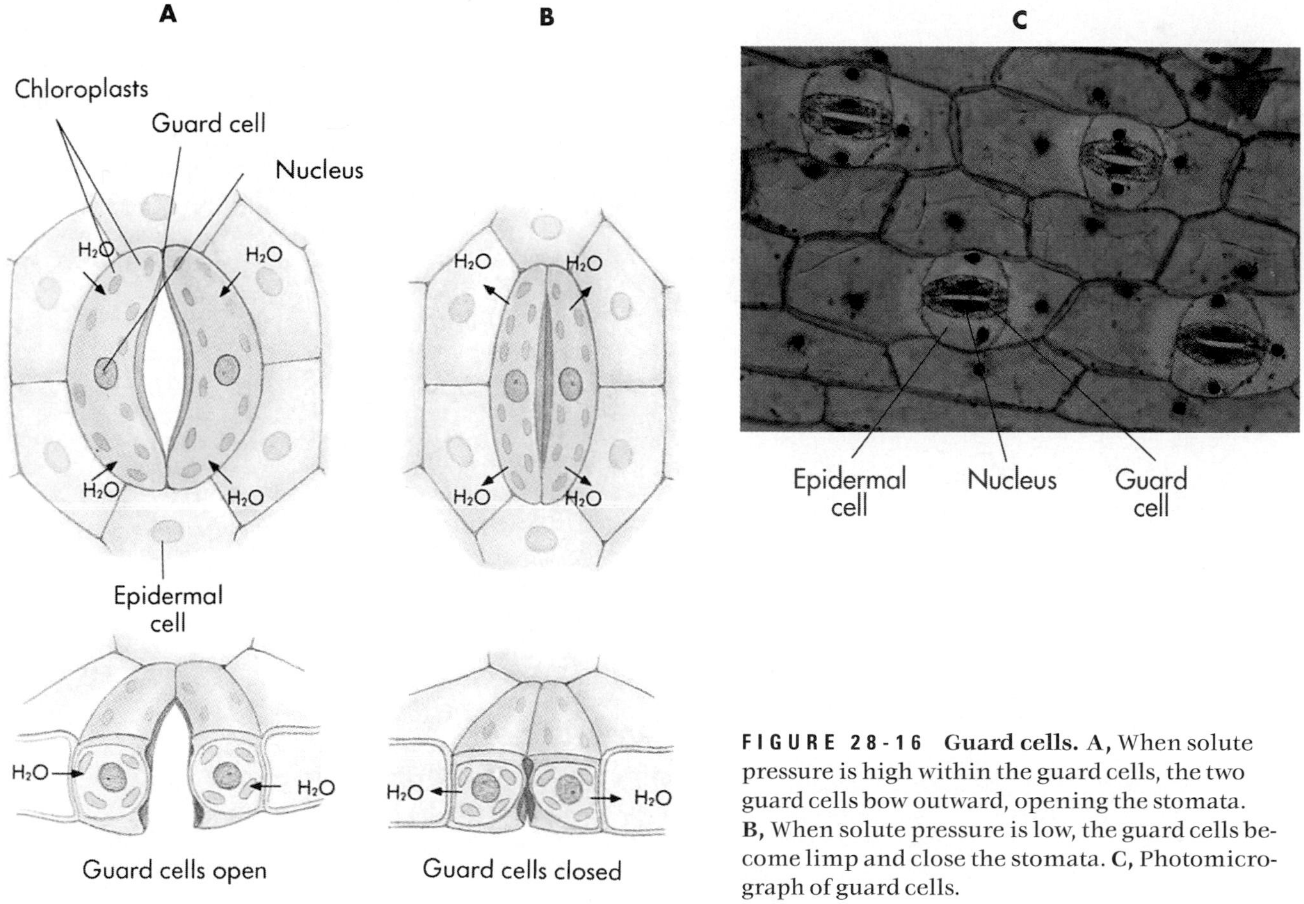

FIGURE 28-16 Guard cells. A, When solute pressure is high within the guard cells, the two guard cells bow outward, opening the stomata. **B,** When solute pressure is low, the guard cells become limp and close the stomata. **C,** Photomicrograph of guard cells.

blade the flattened portion of a leaf.
petiole (**pet** ee ole) the slender stalk of a leaf.
stomata (singular, stoma) (**stow** muh tuh/**stow** muh) openings in the epidermis of a leaf bordered by guard cells.
guard cells a pair of cells that brackets a stoma and regulates its opening and closing.
transpiration the process by which water vapor leaves a leaf through the stomata.

forces are at work. To understand these forces and their interactions, try a little experiment. Fill a glass jar with water—a jar that apple or cranberry juice comes in, for example. Now fill a dishpan about three-fourths full with water. Put a piece of cardboard over the mouth of the jar, quickly turn it upside down as you place the mouth of the jar under water, and then remove the cardboard. What happens? Unless some water escaped from the jar before the mouth was submerged, the water stayed in the jar well above the level of the water in the dishpan. But why? The answer to this question is one of the reasons plants can move water up their vascular system from the roots to the leaves. Interestingly, the answer is air pressure.

The force of gravity pulls on the air that is over the water in the dishpan, creating a force called *air pressure.* The air pushing down on the water in the dishpan causes water to be pushed up into the jar and helps it stay there. At the same time, however, gravity pulls down the water within the jar. The interaction of these two forces (the push up and the pull down) are both dependent on gravity and determine the level of water in the jar. At sea level, air pressure overcomes the pull of gravity on a column of water in a microscopically thin tube (such as xylem in plants), pushing it to a height of about 10.4 meters (about 31 feet or $3^1/_2$ stories)! Two other forces, adhesion and cohesion, help prevent the collapse of this column. The term *adhesion* means that water molecules adhere, or stick to, the walls of the very narrow xylem of plants. *Cohesion* means that water molecules tend to stick to one another. (See Chapter 2 for a more complete discussion of the properties of water molecules.) These two forces link water molecules together and to the

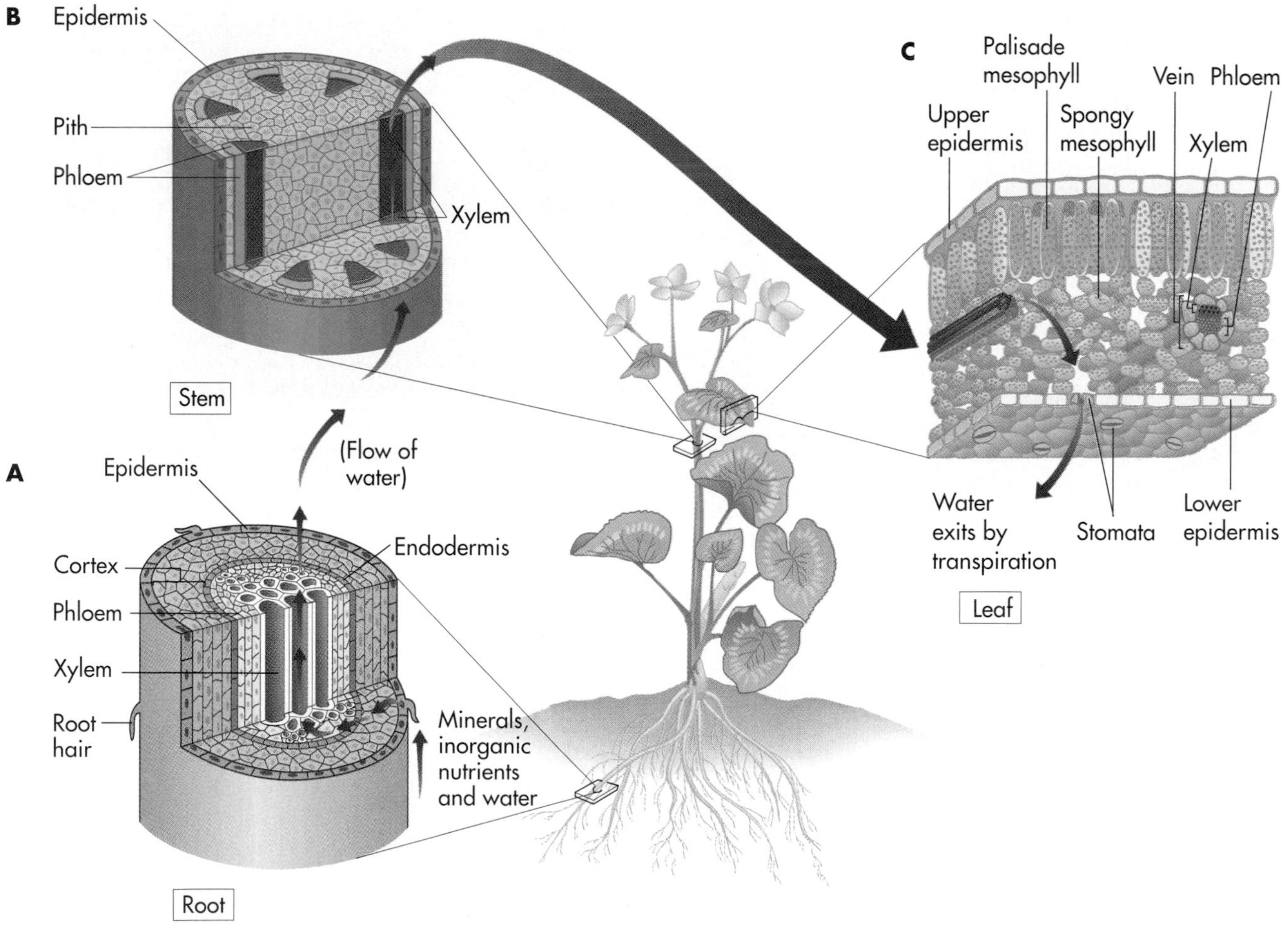

FIGURE 28-17 Water movement through a plant (transpiration). Water is pulled up through the plant by a combination of forces. **A,** Minerals and other inorganic nutrients enter root hairs by active transport. Water enters root hairs by osmosis. **B,** These substances travel up the plant through the xylem. **C,** Water vapor passes out of the plant through the stomata.

sides of the xylem with weak chemical bonds called *hydrogen bonds*.

Where does the water come from that is in the xylem of plants? Most of the water absorbed by a plant comes in through its root hairs. Inorganic nutrients also pass into the cells of the root hairs by means of special cellular "pumps." In this way, root cells maintain a higher concentration of dissolved inorganic nutrients than the concentration of inorganic nutrients in the water of the soil. Therefore water tends to steadily move into the root hair cells from an area of higher concentration (in the soil) to an area of lower concentration (in the root hairs) by osmosis, developing a force called *root pressure*. Once inside the roots, the water and dissolved inorganic nutrients pass inward to the conducting elements of the xylem.

Almost all of the time, however, the water in xylem is pulled upward by forces associated with transpiration. Transpiration is a process by which water vapor passes out of a plant through the stomatal openings in its leaves. The water moves into the leaves after traveling up the plant through the xylem. When it reaches the end of the xylem, it diffuses into the photosynthetic mesophyll cells. Many air spaces surround the mesophyll cells, and water molecules easily pass through the membranes of the cells and into the spaces. Notice the close placement of the xylem, the mesophyll cells, and the air spaces in the cross-sectional view of the leaf in Figure 28-15. From the air spaces, the water vapor passes through open stomata. The transpiration of water creates a concentration gradient of water molecules, which causes the water molecules in areas of higher concentration (the xylem and mesophyll cells) to move to areas of lower concentration (the air spaces), creating a tension, or pull, on the column of water. Figure 28-17 shows how transpiration moves water from the roots to the leaves of a plant.

Water rises in a vascular plant beyond the point at which it would be supported by air pressure because transpiration from its leaves produces a force that pulls up on the entire water column all the way down to the roots. The forces of cohesion and adhesion work to maintain an unbroken column of water.

Fluid is also transported in plants by the phloem. As mentioned previously, these tissues transport the products of photosynthesis—sugars—dissolved in water. This sugary solution is commonly called *sap*. Have you ever seen the collection of sap from maple trees? If so, you realize that plants can rapidly transport large volumes of fluid very quickly. Phloem sap contains 10% to 25% sucrose in addition to inorganic nutrients, amino acids, and plant hormones and may travel as fast as 1 meter per hour!

The forces of diffusion and osmosis alone cannot account for this rapid movement of phloem sap. Instead, a pressure-flow, or mass-flow, system performs this function. The *mass-flow* system is shown in Figure 28-18 and works in the following way. Sucrose is produced at a *source*, such as a photosynthesizing leaf, and is actively transported into sieve tube members by companion cells. As the concentration of sucrose increases in the phloem, water follows by osmosis. In the roots below or at some other *sink* where sucrose is needed, companion cells actively transport sucrose out of the phloem. Water again follows by osmosis. The high hydrostatic (water) pressure in the phloem near the source and the low pressure near the sink cause the rapid flow of the sap. After the sap reaches its destination in the plant, the water can be recycled by moving back to the source through the xylem.

The Organization of Nonvascular Plants

The nonvascular plants, the bryophytes, are made up of three groups of plants: the liverworts, the hornworts, and the mosses (see Chapter 27). These plants are not organized the same as the vascular plants. Only certain genera have specialized vascularlike tissue. None have true roots, stems, or leaves. Bryophytes range from those forms that look like filamentous algae to those that look somewhat like certain vascular plants. As a consequence of having less sophisticated transport systems than the vascular plants do, the bryophytes do not grow very tall. Some look as though they are creeping over their substrate, or food source.

Some bryophytes have distinct stems and leaflike structures, such as most mosses (see Chapter 27), whereas other bryophytes do not. Bryophyte stems and leaflike structures look outwardly different from the stems and leaves of vascular plants and are anatomically similar but simpler. These organs have an outer layer of epidermis made up of protective cells and growth cells. The cortex is made of

rhizoids (rye zoyds) in a bryophyte, slender, rootlike projections that anchor the plant to its substrate; unlike roots, however, they consist of only a few cells and do not play a major role in the absorption of water or minerals.

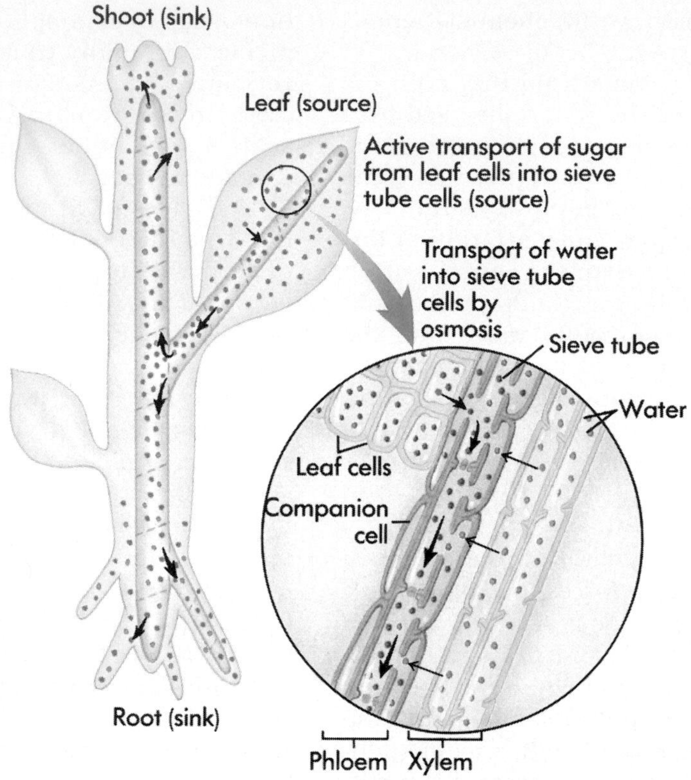

FIGURE 28-18 **Mass flow.** Sucrose from the source is transported into sieve-tube members by companion cells, and water follows by osmosis. At a sink—a place where sucrose is needed—sucrose is actively transported from the phloem, and water again follows by osmosis. The high water pressure in the phloem near the source and the low pressure near the sink cause this flow of sucrose, which can be very rapid.

parenchyma cells, much like those found in vascular plants. In addition, these stems and leaflike structures have a central area of water-conducting tissue.

Bryophytes have no roots, but some have slender, usually colorless projections called **rhizoids** that anchor these simple plants to their substrate. Unlike roots, however, rhizoids consist of only a few cells and do not play a major role in the absorption of water or inorganic nutrients. These substances often enter a bryophyte directly through its stems or leaves.

In general, the bryophytes have no specialized vascular tissues. The sporophytes of many moss species and the gametophytes of some moss species do have a central strand of somewhat specialized water-conducting tissue in their stems, and food-conducting tissue has been identified in a few genera. Even when such tissues are present, however, their structures are much less complex than those found in the vascular plants.

The bryophytes—liverworts, hornworts, and mosses—are primarily low-growing plants. Some have stems and leaflike structures, but these structures are anatomically much simpler than those of vascular plants. Although most bryophytes have no specialized vascular tissues, a few species have somewhat specialized water-conducting tissues. Food-conducting tissues are rare.

JUST WONDERING. . . .

I heard that one reason we need to save the tropical rain forests is so that we won't lose plants that could cure diseases. How do plants cure diseases? And how do scientists find these plants?

Some plants produce chemicals that affect animals. These chemicals can be poisons, such as the blowgun poison curare, for example, or the recently discovered breast cancer drug taxol. You may have used a medicine derived from plants yourself, because approximately 25% of all prescription drugs dispensed each year in North America were originally derived from flowering plants and ferns alone! (Many are now synthesized in the laboratory.) Interestingly, the science of botany (the study of plants) was considered a branch of medicine until the early to mid-1800s.

The only way that scientists can discover whether plants produce biologically active chemicals is to test them in the laboratory. Testing all the plants in the rain forest (or in any area) for their medicinal benefits is, of course, impossible. Therefore, researchers have devised various methods to hunt for therapeutic plants.

One method drug hunters use is to study the therapeutic plant uses of those people indigenous to a particular area. They often choose cultures that have populated a particular area for many generations and pass their "folk medicine" knowledge from generation to generation. Scientists also choose cultures living in areas exhibiting diversity in plant life. Researchers then isolate and study the biologically active chemicals they extract from these plants and determine whether they are active against the type of condition for which they are used by folk healers. Many useful drugs have been developed this way, such as reserpine, which is derived from the Indian snakeroot *(Rauwolfia serpentina)* and is used to treat hypertension, and two cancer drugs, which are both extracted from rosy periwinkle *(Catharanthus roseus)*.

Another approach researchers use is to study plants that are closely related to those shown to have medicinal value. Evolutionarily, plants that produce chemicals that help them survive are more reproductively fit than other plants in their species. These plants would flourish and carry this adaptive trait from generation to generation. As species diverge from one another, traits conferring a survival advantage tend to be retained in a population. Therefore, it makes evolutionary sense that closely related plants might have similar biochemical traits.

Yet another approach is to study plants in their natural surroundings and to choose those plants that remain untouched by predators such as insects. These plants are likely candidates to produce chemicals that affect animals; some may produce chemicals that have desirable and useful druglike characteristics.

In their laboratories, scientists, using particular solvents, now run automated tests to screen the chemicals they extract from plants. Plant extracts can be quickly screened to see whether they are active against certain types of cancer cells or whether they affect enzymes active in particular diseases. After initial screenings, drug companies often take over further analyses to decide whether the drug holds promise as an addition to their pharmaceutical line.

It takes many years for a plant with suspected healing powers to reach your local pharmacy as a prescription drug. And many never make it. However, the drugs that prove to be useful are extremely important additions to our arsenal against disease. Thus, saving the rain forest, where about two thirds of the world's plant species are found, not only will preserve biological diversity but will preserve plants that one day may cure many of our most devastating illnesses.

Summary

- The body of a vascular plant has two parts, a root and a shoot. These parts are made up of three principal tissues: vascular tissue, ground tissue, and dermal tissue. Vascular tissue conducts water and dissolved inorganic nutrients up the plant, ground tissue stores the carbohydrates the plant produces and forms the substance of the plant, and dermal tissue covers the plant, protecting it. In addition, plants contain meristematic tissue, an undifferentiated tissue in which cell division occurs.
- Vascular tissue is of two types: xylem and phloem. The xylem conducts water and dissolved inorganic nutrients from the roots through the stem and to the leaves. Phloem conducts water and dissolved sugars (sap).
- Plants grow in length by cell elongation and apical meristems, zones of active cell division at the ends of the roots and the shoots. This type of growth is primary growth. Secondary growth in both stems and roots takes place in woody trees and shrubs by means of lateral meristems along the length of their stems and branches.
- Roots, stems, and leaves are the organs of vascular plants. The root system anchors a plant in the ground and absorbs water and inorganic nutrients from the soil. Stems support the leaves—the primary organs of photosynthesis. In addition, stems conduct water and inorganic nutrients from the roots to all plant parts and bring the products of photosynthesis to where they are needed or stored.
- Together, the xylem and phloem are called *vascular bundles.* They form different, distinct organizational patterns in the roots, stems, and leaves of monocots (plants with parallel vascular bundles, or veins, in their leaves) and dicots (plants with a weblike pattern of veins in their leaves).
- Dicots with woody stems have lateral meristem tissue, or cambium. The vascular cambium lies between the xylem and phloem; its dividing cells form xylem toward the interior (secondary xylem) and phloem (secondary phloem) toward the exterior. As a result, the diameter of a plant increases.
- Wood is accumulated secondary xylem; it often displays rings because it exhibits different rates of growth during different seasons.
- Leaves, the photosynthetic organs of most vascular plants, are made up of specialized ground tissue cells, or parenchyma, bounded by epidermis. Vascular bundles run through the parenchyma, and its cells contain chloroplasts, the organelles in which photosynthesis takes place. Openings in the epidermis—stomata—allow the carbon dioxide needed for photosynthesis to enter and allow the oxygen produced by photosynthesis to escape. Water vapor also evaporates from the plant through the stomata.
- Water flows through plants in a continuous column, driven mainly by the evaporation of water vapor from the stomata. The force of air pressure helps maintain the height of the column of water in the xylem. In addition, the cohesion of water molecules and their adhesion to the walls of the narrow xylem through which they pass are important factors in maintaining the flow of water to the tops of plants.
- Sucrose moves from where it is produced in a plant to where it is used by the process of mass flow. First, it is actively transported into phloem cells where it is produced and is actively transported out of the phloem cells where it is used. These active transport processes produce a sugar gradient and a water pressure gradient, which cause their movement.
- The bryophytes are not organized the same as the vascular plants. Only large, multicellular forms of these plants have tissue differentiation and specialized transport cells. However, the organization and transport systems of these genera are much less complex than those of the vascular plants.

Knowledge and Comprehension Questions

1. What are vascular plants?
2. Fill in the blanks: In a vascular plant, the part below ground is called the ______. The part above ground, called the ______, consists of ______ (structures in which most photosynthesis takes place) and a(n) ______ (which serves as a supporting framework).
3. Match each type of tissue to its function:
   a. Stores food manufactured by the plant
   b. Protects the plant
   c. Produces new plant cells during growth
   d. Conducts water, inorganic nutrients, carbohydrates, and other substances throughout the plant
   (1) Meristem
   (2) Dermal tissue
   (3) Ground tissue
   (4) Vascular tissue
4. What do xylem and phloem have in common? How do they differ?
5. Identify and give a function for each of the following: (a) parenchyma cells, (b) epidermal cells, (c) stomata, (d) root cap.
6. Distinguish between primary and secondary growth.
7. Parents are always telling their children to eat lots of vegetables. Using your knowledge of plant physiology from this chapter, give one benefit of eating vegetables.
8. Draw two diagrams, one showing the root of a "typical" monocot, the other the root of a "typical" dicot. Label the pericycle, xylem, cortex, endodermis, epidermis, and pith.
9. What type of root system would you expect to find in a dandelion, an ivy plant, and a clump of grass? What are the advantages of each type of root?
10. How can annual rings help you estimate the age of a tree? What type of tissue is involved? In which type of growth does this result?
11. Describe the process by which water rises in a vascular plant. What forces are involved?
12. Diagram the movement of phloem sap in a mass-flow system. Label the source and sink, and show the direction of flow.
13. In general, how does the appearance of nonvascular plants differ from that of vascular plants? Why?

Critical Thinking

1. In tropical climates, many tall plants shut their stomata during the hot days and open them at night. If their stomata are closed during the day, why doesn't the water within the plant fall down the stem?
2. The roots of many plants have permanent mutualistic associations with fungi. What might be the advantage of this association to the plant?

CHAPTER 29

INVERTEBRATES: PATTERNS OF STRUCTURE, FUNCTION, AND REPRODUCTION

CORAL REEFS ARE one of the most beautiful, exotic, and colorful living landscapes. They are only partially made up of corals, which are animals that grow on rocks in shallow, tropical waters (from 30 degrees north to 30 degrees south of the equator). In fact, the organisms that make up a coral reef are quite diverse. Many organisms such as red algae help build and strengthen coral reefs. Even the corals themselves show incredible diversity.

The term *coral* is a general term that refers to a variety of invertebrate animals (those without a backbone) of the phylum Cnidaria. Some corals are hard corals—those that have a hard outer shell. Others are soft corals—those that lay down fragments of shell-like material within their tissues. The tissues of corals contain pigments that produce a wide range of hues, as do the microscopic algae that live within the coral animals. Corals therefore exist in many colors: brown, blue, green, yellow, pink, purple, red, and even black. Later in this chapter you will learn about the patterns of structure and function that corals and other invertebrate animals have in common despite their incredible diversity. But first you will look at the "bigger picture" of the animal kingdom to discover its unifying themes.

Characteristics of Animals

How are animals different from the other kingdoms of living things? Like the plants, fungi, and protists, animals are eukaryotic organisms, having a distinct nucleus and a cellular structure different from the prokaryotic structure of bacteria (see Chapter 25). Also like the plants, most fungi, and some protists, animals are multicellular. No single-celled animals exist. Only their gametes are single celled, but these cells are not independently living organisms. As soon as fertilization takes place, the development of a new multicellular individual begins. (See Figure 27-1, *A*, which diagrams the sexual life cycle of animals.)

Animals are heterotrophs, unable to make their own food. Therefore animals must eat plants, other organisms, or organic matter for food. Some simple animals, such as the sponges, take organic matter directly into their cells. Most animals digest food within a body cavity. The resulting molecules are then taken into the body cells to be broken down further by the chemical reactions of cellular respiration (see Chapter 5). The end product of cellular respiration is energy, which is used to drive the activities of life, including growth, maintenance, reproduction, and response to the external environment. As part of this response, most animals are capable of movement to capture food or to protect themselves from injury.

The cells of animals are organized into tissues, which are groups of cells combined into structural and functional units (see Chapter 7). In most animals, the tissues are organized into organs,

KEY CONCEPTS

- Animals are a diverse group of multicellular, heterotrophic, eukaryotic organisms that most likely evolved from protist ancestors.
- Other than the sponges, all animals exhibit either radial symmetry or bilateral symmetry.
- Most bilaterally symmetrical organisms (except roundworms and flatworms) have coeloms, which are lined enclosures that contain body organs suspended by thin sheets of connective tissue.
- The echinoderms have embryological developmental patterns suggesting that they are more closely related to the vertebrates than to the other invertebrates.

OUTLINE

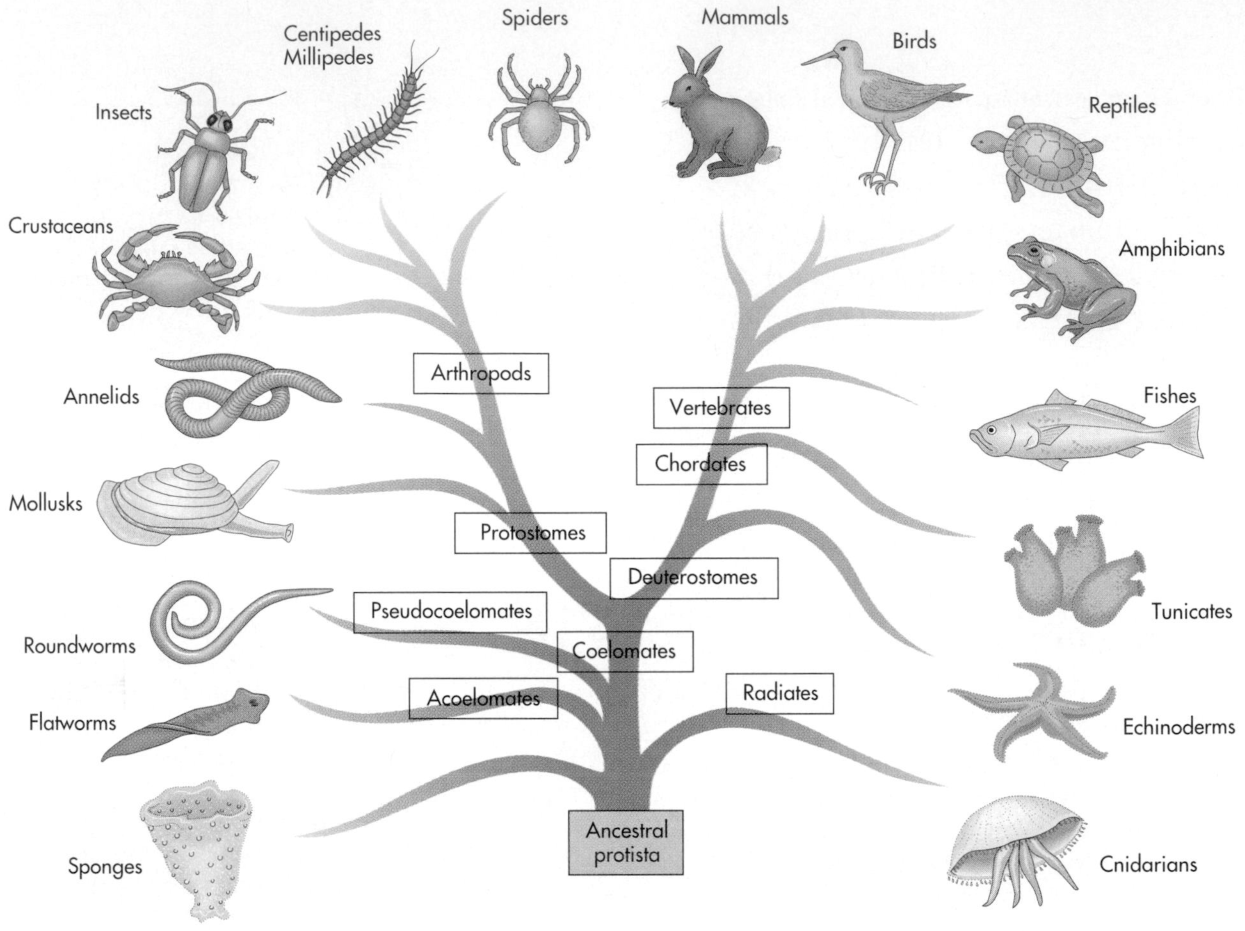

FIGURE 29-1 The animal ancestral tree. All animals on the lower branches of the tree are phyla. The animals on the upper right branches are classes of phylum Chordata, subphylum Vertebrata (as shown by the boxed labels). The animals on the upper left branches of the tree are the subphyla of phylum Arthropoda (as shown by the boxed labels).

complex structures made up of two or more kinds of tissues. Organs that work together to perform a function are organ systems.

Animals are extraordinarily diverse in their forms and how they function. This diverse group is often divided into two subgroups: the invertebrates and the vertebrates. The invertebrates are animals without a backbone, a series of bones that surrounds and protects a dorsal (back) nerve tube and that is used as a lever system in movement. In some invertebrates, the dorsal nerve, as well as the backbone, does not exist. Examples of invertebrates are spiders, sponges, jellyfish, snails, and worms. The vertebrates (a group that includes you) have both a backbone and a dorsal nerve tube. Examples of vertebrates are bears, fish, dogs, cats, birds, and frogs. Interestingly, although the vertebrates are usually larger and more commonly known, the invertebrates make up more than 95% of all animal species.

Most zoologists think that the animals arose from protist ancestors (see Chapter 24). (The evolutionary relationships among the animals and their ancestors are shown in Figure 29-1.) Each animal on the lower branches of the tree represents a present-day phylum in the animal kingdom. The animals on the upper right branches—the mammals, birds, reptiles, amphibians, and fishes—are all classes of animals in the phylum Chordata, subphylum Vertebrata. The chordate phylum includes three subphyla: the lancelets, the tunicates, and the vertebrates. Lancelets and tunicates (sea squirts) are both groups of marine animals that have a notochord instead of a bony vertebral column. In addition, the lancelets and tunicates lack a brain, an organ common to all vertebrates. The animals on the upper left branches in Figure 29-1—the crustaceans, insects (with the centipedes and millipedes), and spiders—are the three subphyla of the phylum Arthropoda. Arthropods are animals that have a hard outer shell, jointed legs, and a segmented body.

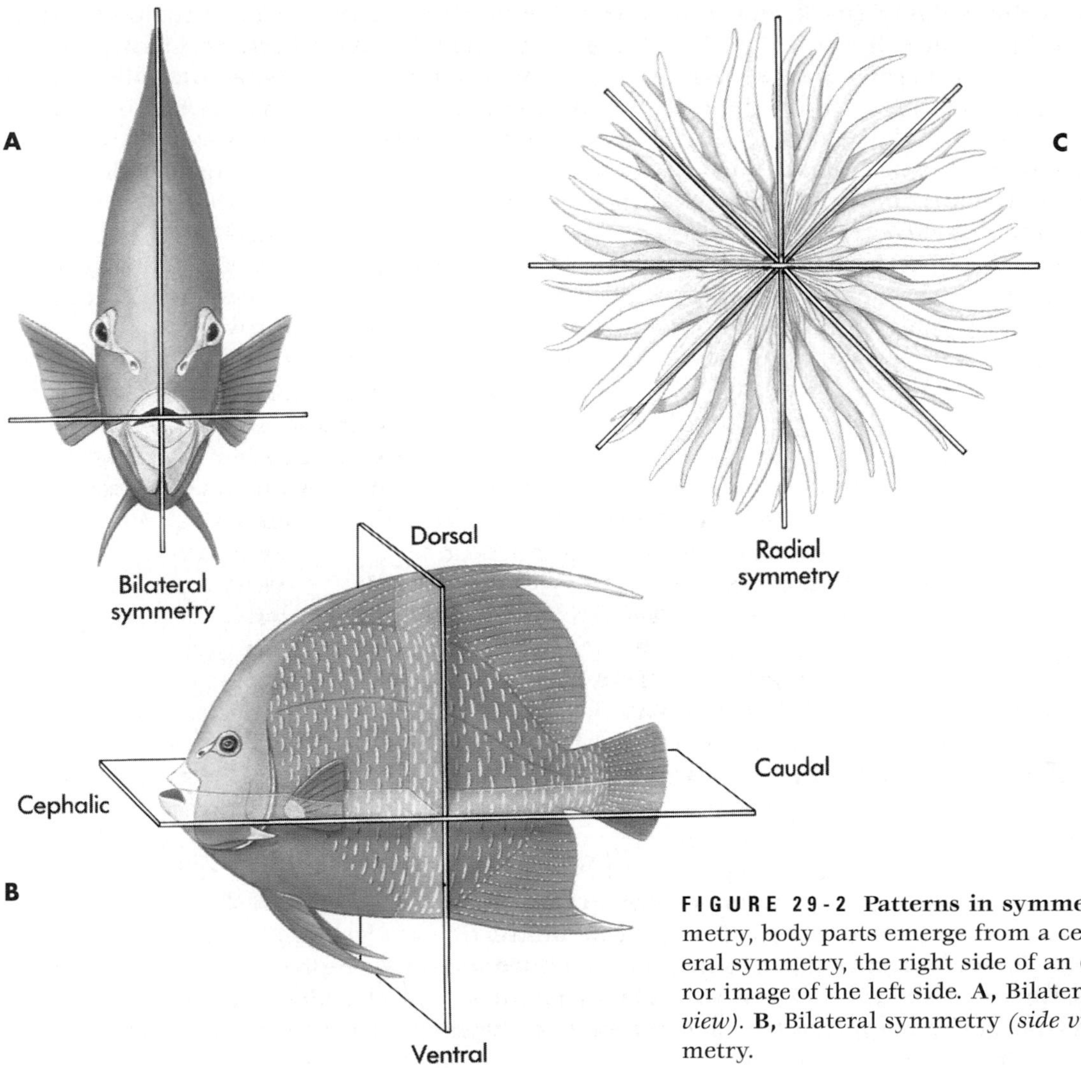

FIGURE 29-2 Patterns in symmetry. In radial symmetry, body parts emerge from a central point. In bilateral symmetry, the right side of an organism is a mirror image of the left side. **A,** Bilateral symmetry *(front view)*. **B,** Bilateral symmetry *(side view)*. **C,** Radial symmetry.

> *Animals are a diverse group of eukaryotic, multicellular, heterotrophic organisms. This diverse group is often divided into two subgroups: the invertebrates and the vertebrates. The invertebrates are animals without backbones; the vertebrates are animals with a backbone that surrounds and protects a dorsal nerve cord. Most zoologists think that the animals evolved from protist ancestors.*

Figure 29-1 shows that the sponges, cnidarians (jellyfish, corals), and flatworms are the present-day phyla of animals most closely related to the evolutionary ancestors of the animals. Likewise, animals higher on the tree represent present-day phyla, subphyla, or classes of organisms that are less closely related to the ancestors of the animals. Each branch of this phylogenetic tree represents an evolutionary pathway that diverged from an ancestral pathway.

Patterns in Symmetry

The word *symmetry* refers to the distribution of the parts of an object or living thing. All animals (except the sponges) exhibit either **radial symmetry** or **bilateral symmetry.** Sponges are asymmetrical, or without symmetry.

Radial symmetry means that body parts emerge, or radiate, from a central point, much like spokes on a wheel (Figure 29-2). As noted in Figure 29-1, the cnidarians and their relatives are radially symmetrical. These organisms have a top and a bot-

radial symmetry having parts that emerge, or radiate, from a central point, much like spokes on a wheel.
bilateral symmetry having two sides in which the right side is a mirror image of the left side.

tom, but they have neither a dorsal (back) side nor a ventral (belly) side (see Figure 29-6).

Animals exhibiting radial symmetry are aquatic organisms. Many are sessile; that is, they are anchored in one place. Their radially symmetrical body allows them to interact with the watery environment in all directions. Other organisms such as starfish (phylum Echinodermata) are also radially symmetrical but are not grouped with the cnidarians because their embryonic development and internal anatomy suggest that they are related to organisms having bilateral symmetry. In addition, their larvae are bilaterally symmetrical.

Bilateral symmetry means that the right side of an object or an organism is a mirror image of the left side. Animals with bilateral symmetry have a dorsal and a ventral side and a cephalic (head) end and a caudal (tail) end. All phyla of animals other than the sponges, jellyfish, and echinoderms exhibit bilateral symmetry.

Patterns in Body Cavity Structure

A **coelom,** or body cavity, is a fluid-filled enclosure lined with connective tissue within most bilaterally symmetrical organisms and the echinoderms. It functions as a transport system and a waste disposal system. In organisms lacking a skeletal system, the coelom acts as a hydroskeleton. Organs are located within this body cavity. For example, you have an abdominal cavity lined with a thin, nearly transparent sheet of connective tissue (peritoneum). Suspended by thin sheets of connective tissue arising from the peritoneum, the stomach and intestines hang in this coelom. Most bilaterally symmetrical organisms have coeloms, except organisms such as roundworms and flatworms.

Organisms having a coelom are called **coelomates.** Most of the organisms in Figure 29-1 are coelomates. However, the roundworms are labeled **pseudocoelomates.** These organisms (along with tiny aquatic organisms called *rotifers*) have a so-called false coelom: a fluid-filled cavity that houses the organs but that is not completely lined by tissue. In addition, the organs are not suspended within the cavity by thin sheets of connective tissue.

The **acoelomates,** the flatworms, have no body cavity. Embedded within the other tissues of the body, their organs are not protected from the movements and compressions of other body tissues.

The three patterns of body cavity structure are shown in Figure 29-3.

Coelomates have lined body cavities in which organs are suspended by thin sheets of connective tissue. Pseudocoelomates have body cavities that are not lined, nor are the organs within them suspended by tissues. Acoelomates have no body cavity surrounding the gut.

Patterns in Embryological Development

Two main branches arise from the trunk of the animal family tree in Figure 29-1. These two branches of coelomate animals represent two distinct evolutionary lines. One includes the mollusks (clams), annelids (segmented worms), and arthropods (lobsters, insects, and spiders). These organisms are

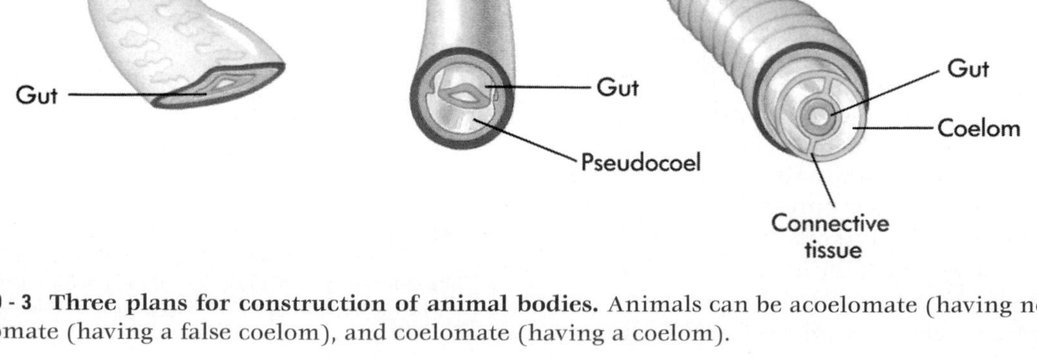

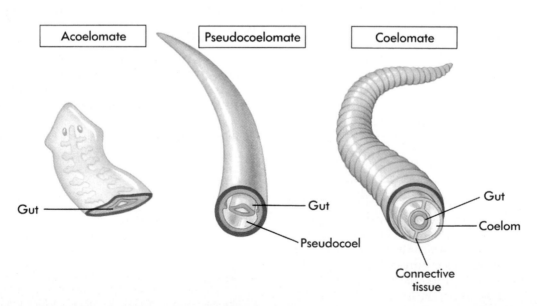

FIGURE 29-3 Three plans for construction of animal bodies. Animals can be acoelomate (having no coelom), pseudocoelomate (having a false coelom), and coelomate (having a coelom).

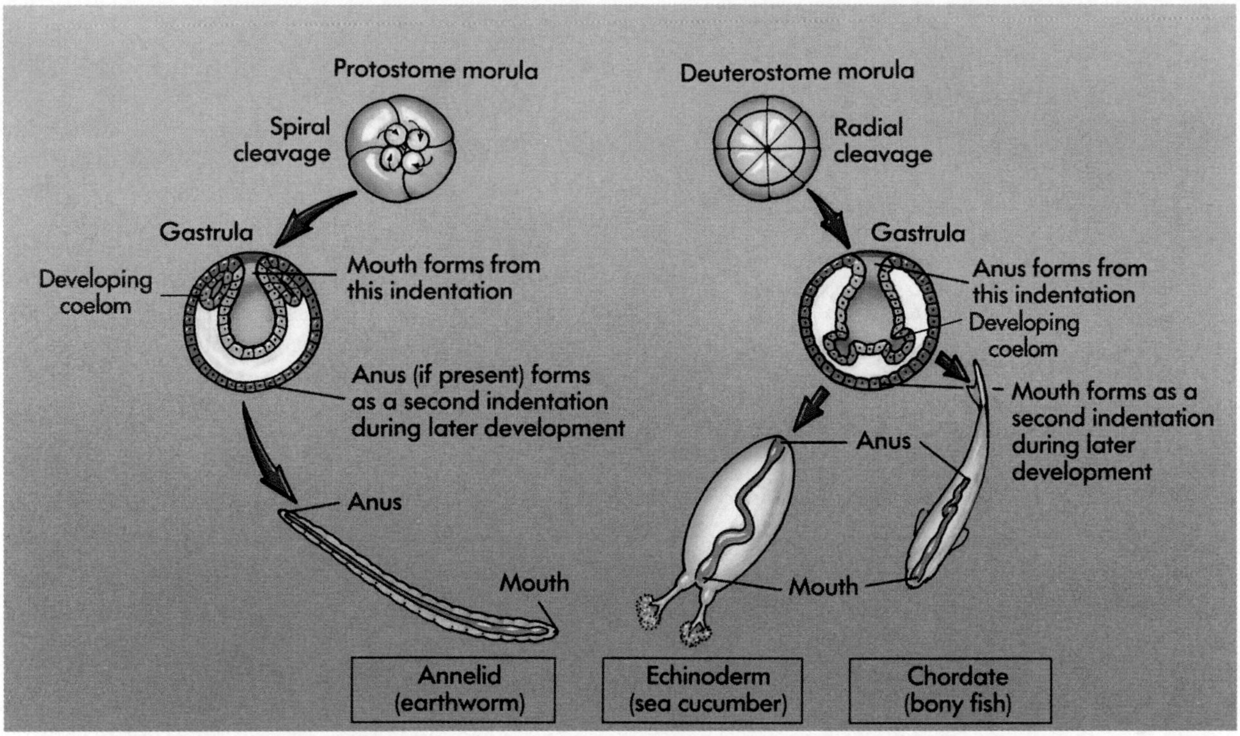

FIGURE 29-4 Protostomes and deuterostomes. The difference between protostomes and deuterostomes arises during embryonic development. The mouth of the protostome develops from the first indentation of the gastrula, and the anus, if present, forms as a second indentation later. The mouth of a deuterostome develops from a second indentation. The deuterostome anus develops from the first indentation.

called *protostomes,* meaning "first" *(proto)* "mouth" *(stome).* This strange name refers to events in the embryological development of these organisms.

During early embryological development (see Chapter 22), all animals consist of a solid ball of cells usually called a *morula.* As development proceeds, the cells secrete fluid that fills the interior of the ball and pushes the cells to the edges of the sphere. This stage is called the *blastula.* This fluid-filled ball of cells then forms an indentation, assuming the shape of a blown-up balloon with a fist pushing in one side. This stage of development, the *gastrula,* gives rise to a two-layered embryo and begins the formation of the gut. In the **protostomes,** this first indentation becomes the mouth of the organism. In another group of organisms called **deuterostomes,** which includes the echinoderms (starfish) and chordates, the first indentation becomes the anus and a second one becomes the mouth. The term *deuterostome* means "second mouth." Figure 29-4 shows these differences.

The protostomes and the deuterostomes also show different cleavage patterns during embryological development. During the morula stage of development, the cells of the protostomes divide in a way that forms a spiral pattern. The cells of the deuterostomes cleave in a radial pattern. These patterns of embryological development (along with other developmental similarities) suggest close evolutionary relationships among the organisms within each group. In addition, because the protostome developmental pattern occurs in all acoelomates, scien-

coelom (**see** lum) a body cavity, found within most bilaterally symmetrical organisms and the echinoderms, that is a fluid-filled enclosure lined with connective tissue.

coelomates (**see** lum aytes) organisms that have a coelom.

pseudocoelomates (soo doe **see** lum aytes) organisms that have a so-called "false" coelom: a fluid-filled cavity that houses the organs but that is not completely lined with tissue.

acoelomates (ay **see** lum aytes) organisms that have no body cavity; the flatworms.

protostomes (**pro** toe stowmz) one of two distinct evolutionary lines of coelomates that includes the mollusks, annelids, and arthropods. Their embryological development is characterized by the mouth developing from the first indentation of the gastrula and a spiral pattern of cleavage.

deuterostomes (**doot** uh row stowmz) one of two distinct evolutionary lines of coelomates that includes the echinoderms and chordates. Their embryological development is characterized by the anus developing from the first indentation of the gastrula and a radial pattern of cleavage.

tists think that it was the pattern of development in the ancestors of modern animals.

> *The protostomes (mollusks, annelids, and arthropods) differ from the deuterostomes (echinoderms and chordates) with regard to certain embryological events. These differences suggest close evolutionary relationships among the organisms within each group.*

Invertebrates: Diversity in Symmetry and Coelom

Asymmetry: Sponges

Sponges (phylum Porifera) are aquatic organisms; most species live in the ocean rather than in fresh water. These sessile creatures are considered simpler than other animals in their organization because they have no tissues, no organs, and no coelom. The asymmetrical bodies of sponges consist of little more than masses of cells embedded in a gelatinous material, or matrix (Figure 29-5, *A*).

A

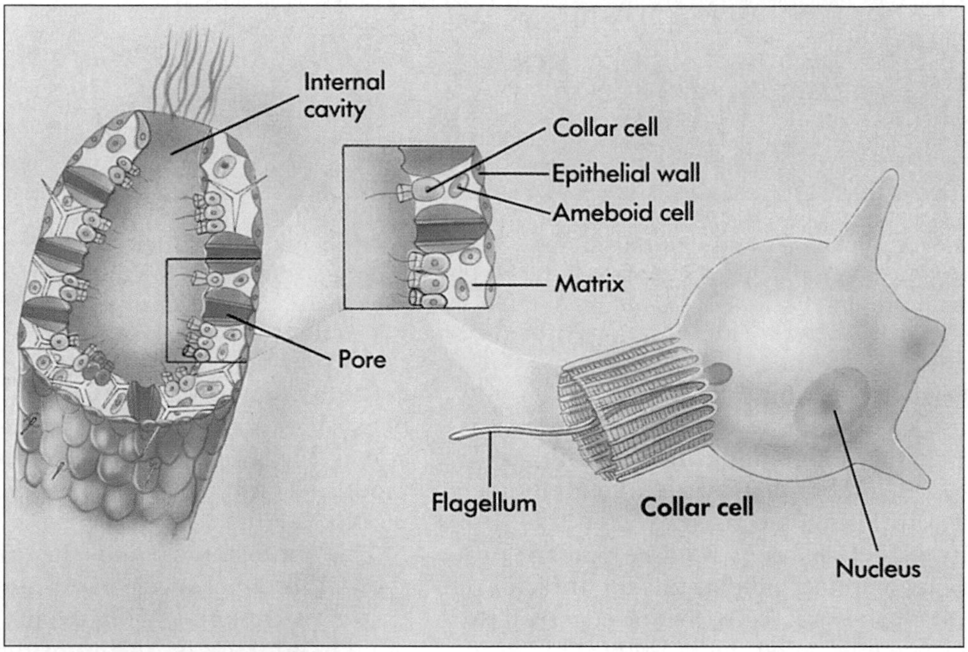

B

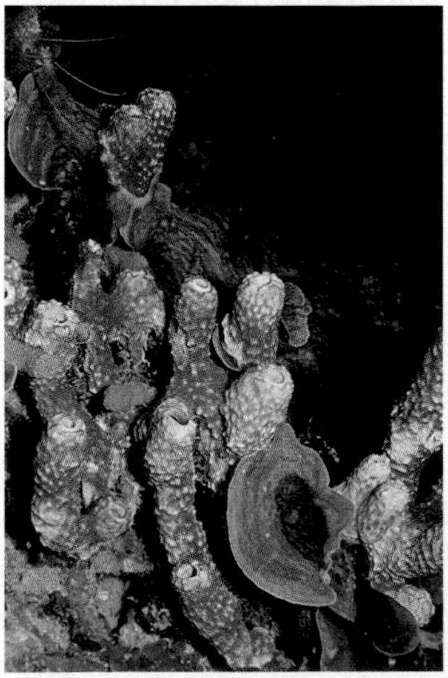

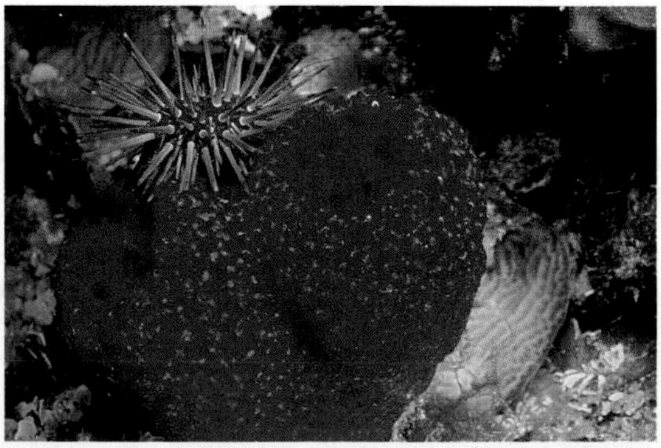

C

FIGURE 29-5 The sponge (phylum Porifera). A, Diagram of a sponge, with detail of a collar cell. **B,** Yellow tube sponges. **C,** A red vase sponge.

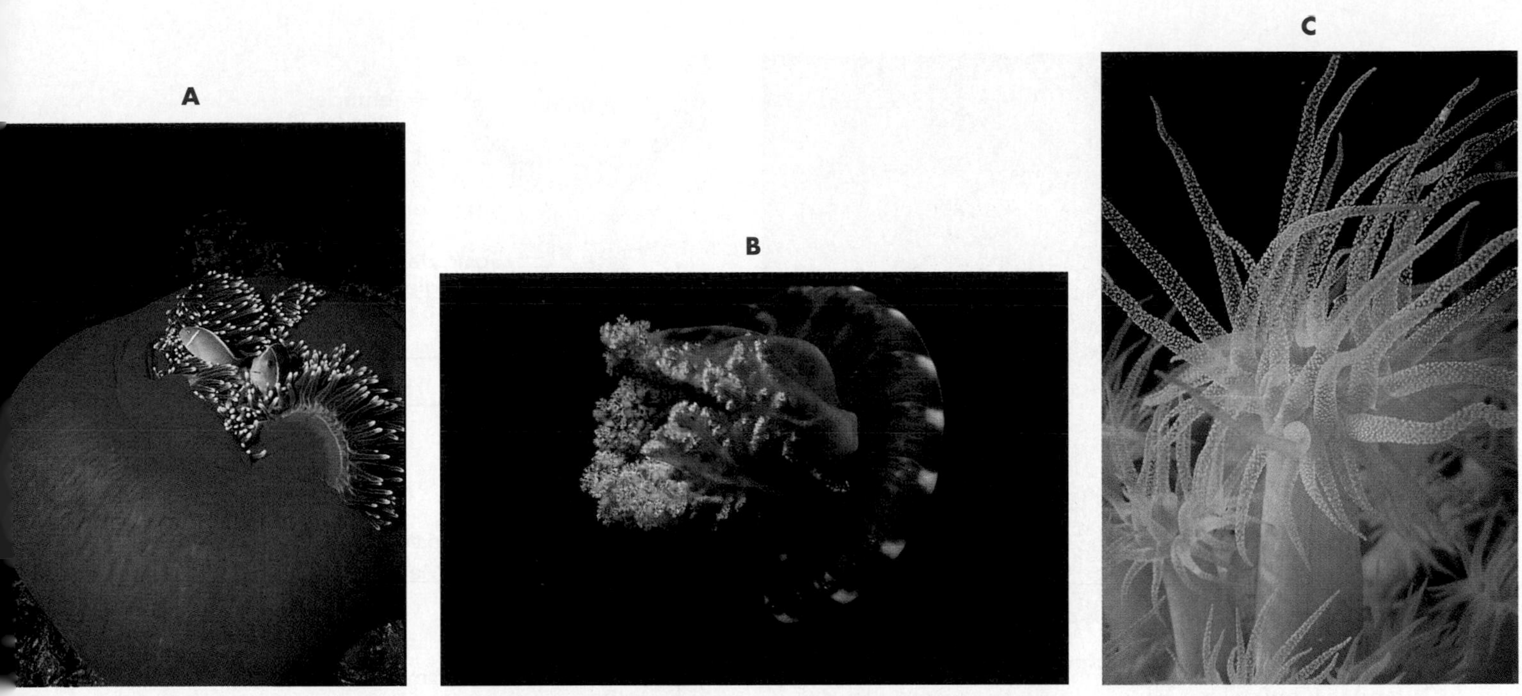

FIGURE 29-6 Representatives of three classes of cnidarians (phylum Cnidaria). A, Sea anemone (those are clown fish swimming among the tentacles). **B,** Jellyfish. **C,** Orange cup coral.

The body of a sponge is shaped like a sac or vase (Figure 29-5, *B* and *C*). The body wall is covered on the outside by a layer of flattened cells called the *epithelial wall.* Lining the inside cavity of the sponge are specialized, flagellated cells called *collar cells.* The matrix makes up the substance of the sponge, sandwiched between the outer epithelial layer and the inner layer of collar cells. Within the matrix are ameboid-type cells, needlelike crystals of calcium carbonate or silica, and tough protein fibers. Pores, channellike openings that span the matrix, are dispersed throughout the sponge. These pores are integral to the movement of water, dissolved substances, and particulate matter to the interior of the sponge. They give the sponges their phylum name Porifera, which means "pore bearers."

As they beat, the flagella of the collar cells create a current of water that flows from the outside of the sponge, through pores in the matrix, to the internal cavity of the sponge, and then out again through the large opening at the top of the sponge. The circulation of water in this way brings the nutrients in the water to the collar cells.

Sponges frequently reproduce asexually by fragmentation; groups of cells become separated from the body of the sponge and develop into new individuals. In addition, sponges may develop branches that grow over the rocks on the sea floor, much like a plant develops underground runners. Colonies of sponges grow along these branches.

Sponges reproduce sexually and asexually. Most species of sponges produce both female sex cells (eggs) and male sex cells (sperm), which arise from cells in the matrix. Both types of sex cells are produced within the same organism. Such individuals are called **hermaphrodites,** after the Greek male god Hermes and the female goddess Aphrodite. The sperm are released into the cavity of the sponge and are carried out of the sponge with water currents and into neighboring sponges through their pores. Fertilization occurs in the gelatinous matrix where the eggs are held. There, the fertilized eggs develop into flagellated, free-swimming larvae that are released into the sponge's cavity. After the larvae leave the interior of the sponge, they settle on rocks and develop into adults.

> *Sponges (phylum Porifera) are aquatic, asymmetrical, acoelomate organisms shaped somewhat like a vase. They are considered the simplest of animals because they have no tissues or organs.*

Radial Symmetry: Hydra and Jellyfish

Animals other than the sponges have a definite shape and symmetry. Only two phyla are classified as radially symmetrical: the Cnidaria, which includes jellyfish, hydra, sea anemones, and corals (Figure 29-6), and Ctenophora, a minor phylum that includes the comb jellies (Figure 29-7).

> **hermaphrodites** (hur **maf** rah dytes) an animal or plant having both male and female reproductive organs.

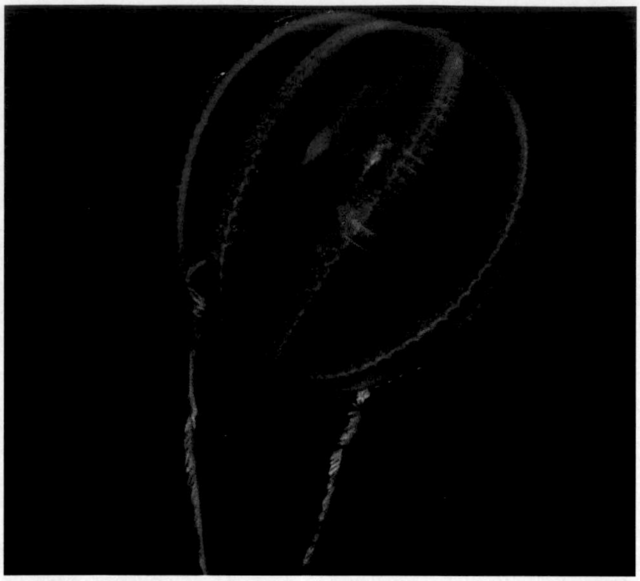

FIGURE 29-7 A comb jelly. Note the comblike plates and two tentacles.

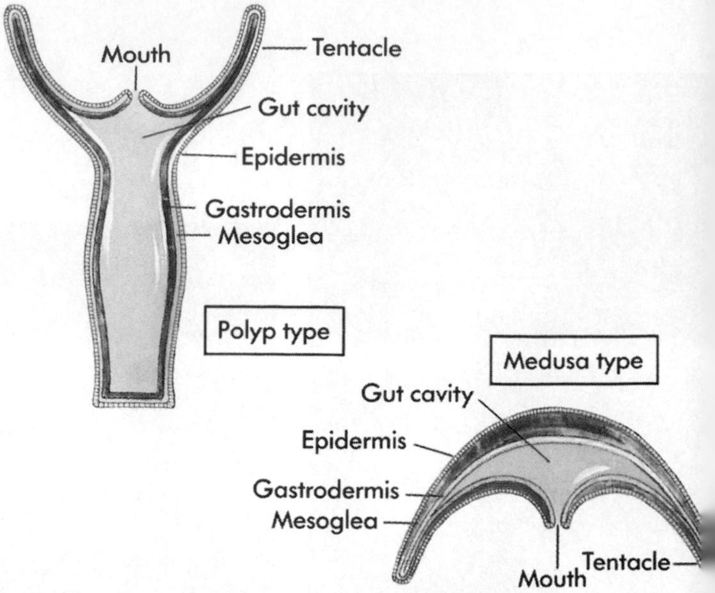

FIGURE 29-8 Body forms of cnidarians: the medusa and the polyp. These two phases alternate in the life cycles of many cnidarians, but a number—including the corals and sea anemones—exist only as polyps. Other cnidarians exist only as medusae.

There are two basic body plans exhibited by the cnidarians: **polyps** and **medusae** (Figures 29-6 and 29-8). Polyps are aquatic, cylindrical animals with a mouth at one end that is ringed with tentacles. (The name *polyp* actually means "many feet.") Polyps such as the sea anemones and corals live attached to rocks. Like the corals, many polyps build up a hard outer shell, an internal skeleton, or both. Some polyps are free floating, such as the freshwater *Hydra.* In contrast, most medusae are free floating and are often umbrella shaped. Commonly known as *jellyfish,* medusae have a thick, gelatinous interior. The mouth of a medusa is usually located on the underside of its umbrella shape, with its tentacles hanging down around the umbrella's edge.

Structurally, epithelial tissue covers the outside of cnidarians, and an inner tissue layer, the gastrodermis, lines the gut cavity (see Figure 29-8). The mesoglea (literally, "middle glue") lies between. This layer is quite thick within medusae and gives them their jellylike appearance. A network of nerve cells extends through cnidarians, but they have no brainlike controlling center.

Some cnidarians such as the hydra, sea anemones, and corals occur only as polyps. Simple polyps such as hydra usually reproduce asexually by budding (see Figure 18-21). However, some of their tissue is organized into primitive ovaries that produce eggs and testes that produce sperm, as do the more complex polyps such as sea anemones and corals. (Some species of these organisms *never* reproduce by budding.) Some hydra, like sponges, are hermaphrodites. Others exist as males and females. Eggs remain attached to the hydra but exposed to the water. Sperm are discharged from the testes and swim to the egg. After fertilization, developing hydras grow while attached to the parent.

Some cnidarians exist only as medusae. Medusae reproduce sexually. Ovaries hang from the underside of female medusae, and testes hang from the males. Eggs and sperm are shed into the water, where fertilization takes place. The fertilized egg develops into a larva that never settles down to become a sessile polyp but develops directly into a medusa. However, in most species, medusae have a life cycle in which the larvae develop into polyps. Some of these polyps produce medusae. This type of alternating life cycle is shown in Figure 29-9.

The tentacles of a cnidarian help it capture prey—other animals such as small fishes, shrimp, and aquatic worms. The tentacles bear stinging cells called *cnidocytes,* which give the phylum its name. You can see these cells as tiny dots in the tentacles of the yellow cup coral shown in Figure 29-6, *C.* If you have ever been stung by a jellyfish, you know how powerful its sting can be. These stinging cells work much like harpoons. Powered by water pressure, threadlike stingers are jettisoned out of the cells, spearing and immobilizing the prey. The tentacles then draw the prey back to the mouth.

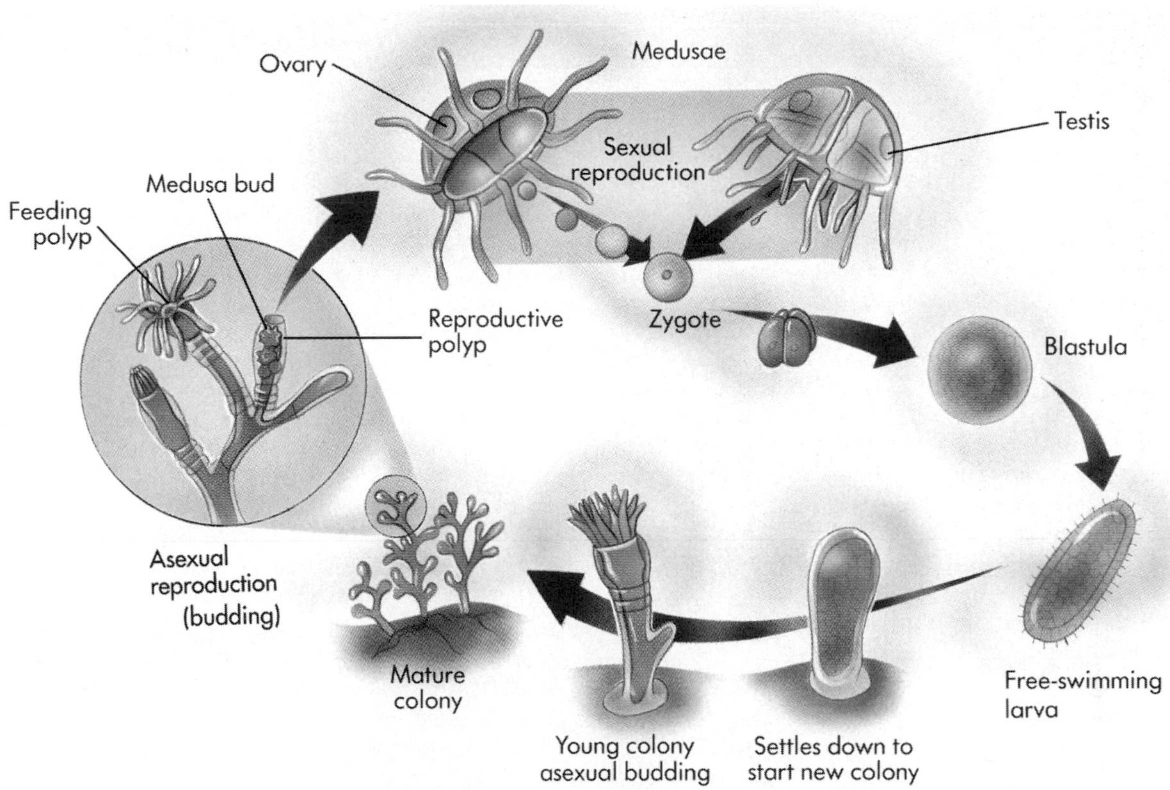

FIGURE 29-9 The life cycle of *Obelia*, a marine cnidarian. This life cycle demonstrates the alternation of medusa and polyp forms found in cnidarians.

> *Cnidarians have two layers of tissues and a nerve net that coordinates cell activities. They exist either as polyps (corals and sea anemones), which are cylindrically shaped animals that anchor to rocks, or as medusae (jellyfish), which are free-floating, umbrella-shaped animals. In some cnidarians, these two forms alternate during the life cycle of the organism.*

Bilateral Symmetry: Variation on a Theme in Many Phyla

The remaining phyla of animals are all bilaterally symmetrical or (in the case of the echinoderms) are thought to have evolved from bilaterally symmetrical forms. In addition, they all develop embryologically from three layers of tissue: an inner layer or **endoderm,** an outer layer or **ectoderm,** and a middle layer or **mesoderm.** Although cnidarians have a middle layer between their two tissue layers, it consists of a jellylike material with only widely dispersed cells.

Platyhelminthes: Flatworms

The bilaterally symmetrical animals with the simplest body plan are the flatworms (phylum Platyhelminthes). This phylum name comes from Greek words meaning "flat" *(platys)* "worm" *(helminthos)* and describes their flattened ribbon or leaflike shapes. Although simple in structure, the flatworms have organs and some organ systems, but because they have no coelom, the organs are embedded within the body tissues (see Figure 29-3). Flatworms are also the simplest animals to have a dis-

polyps (**pol** ups) aquatic, cylindrical animals with a mouth at one end that is ringed with tentacles.

medusae (meh **doo** see *or* meh **doo** zee) free-floating and often umbrella-shaped aquatic animals with the mouth usually located on the underside of the umbrella shape and tentacles hanging down around the umbrella's edge.

endoderm (**en** doe durm) the inner layer of cells formed during the early development of the embryos of all bilaterally symmetrical animals.

ectoderm (**ek** toe durm) the outer layer of cells formed during the early development of the embryos of all bilaterally symmetrical animals.

mesoderm (**mez** oh durm) the middle layer of cells formed during the early development of the embryos of all bilaterally symmetrical animals.

JUST WONDERING....

How do people get tapeworms? Are they common in the United States?

The most common tapeworm that affects humans is the beef tapeworm *(Taenia saginata)*. The life cycle of this tapeworm begins when cattle (the secondary hosts) graze in areas in which tapeworm eggs contaminate the soil. After the eggs are eaten, they hatch within the intestine, and the larvae burrow through the intestinal wall until they reach muscle. Here, the larvae encyst—they become encapsulated and quiescent.

The person who eats raw or undercooked infected beef ingests the cysts along with the meat. Digestive enzymes break down the capsule surrounding the larvae, and they attach themselves to the intestinal wall. Feeding on the digested food of the host, the tapeworms mature within a few weeks. As adults, they consist of long chains of segments called *proglottids*. Each proglottid has both male and female sex organs and can produce up to 100,000 eggs, which are shed with proglottids in the feces. If the feces contaminate areas where cattle graze and they are ingested, the life cycle begins once again.

Parasites such as tapeworms are more prevalent in countries having poor sanitation systems (or no sanitation systems at all) than in countries with effective sewage treatment. In the United States, water is treated and purified in sewage treatment plants before it is returned to rivers, streams, or the ocean. In addition, beef is inspected before it can be sold. However, infection is possible. To protect yourself, do not eat beef that is raw or undercooked.

Two other tapeworms, pork tapeworms and fish tapeworms, can also infect humans. Their life cycles are similar to that of the beef tapeworm; humans are their primary hosts. Therefore, avoid eating undercooked pork, raw fish (sushi), or undercooked fish.

Infection with tapeworms is serious. The organisms can live for years within the body and absorb nutrients essential for proper nutrition. Developing malnutrition due to a tapeworm infestation is common. In addition, long tapeworms can block the movement of materials through the intestine. Tapeworm infections can be treated with certain medications, but treatment is difficult if the worms invade tissues beyond the intestines, as sometimes happens. Following the precautions mentioned here will help you avoid sharing your body with these unpleasant creatures!

tinct head, a characteristic common to many of the bilaterally symmetrical animals.

There are three classes of flatworms: the turbellarians, the flukes, and the tapeworms (Figure 29-10). The turbellarians are free living and found in fresh water, salt water, or damp soil. The flukes and tapeworms are parasites and live on or in other animals, deriving nutrition from their hosts.

Free-living flatworms move from place to place, feeding on a variety of small animals and bits of organic debris. They move by means of ciliated epithelial cells that are concentrated on their ventral surfaces. In fact, the name *turbellarian* comes from a Latin word meaning "to bustle or stir" and refers to the water turbulence created by their movement. Sensory pits or tentacles along the sides of their heads detect food, chemicals, and movements of the fluid in which they are moving. They also have eyespots on their heads, which contain light-sensitive cells that enable the worms to distinguish light from dark. These organs are part of the flatworm nervous system and connect to a ladderlike paired nerve cord that extends down the length of the animal. Tiny swellings at the cephalic, or head, end of the organism are considered a primitive brain.

The flatworm has a digestive system consisting of

FIGURE 29-10 Flatworms (phylum Platyhelminthes). A, A marine, free-living turbellarian. **B,** The human liver fluke, *Clonorchis sinensis.*

a digestive sac, or gut, open only at one end (Figure 29-11). Muscular contractions in the upper end of the gut of flatworms cause a strong sucking force by which the flatworms ingest their food and tear it into small bits. The cells making up the gut wall engulf these particles; most digestion takes place within these cells. Wastes from within the cells diffuse into the digestive tract and are expelled through the mouth. In addition, the flatworm excretes excess water and some wastes by means of a primitive excretory system: a network of fine tubules that runs along the length of the worm. Specialized bulblike cells, or flame cells, are located along these tubules (see Figure 29-11). As cilia within them beat (looking like a flickering flame), they move the water and wastes into the tubules and out excretory pores.

Reproduction in flatworms is much more complicated than in sponges or cnidarians. Although most flatworms are hermaphroditic, a characteristic exhibited by the sponges and cnidarians, the organs of reproduction are better developed. When flatworms mate, each partner deposits sperm in a copulatory sac of the other. The sperm travel along special tubes to reach the egg. In free-living flatworms, the fertilized eggs are laid in cocoons and hatch into miniature adults. In some parasitic flatworms, there is a complex succession of distinct larval forms. Flatworms are also capable of asexual reproduction. In some genera, when a single individual is divided into two or more parts, each part can regenerate an entire new flatworm.

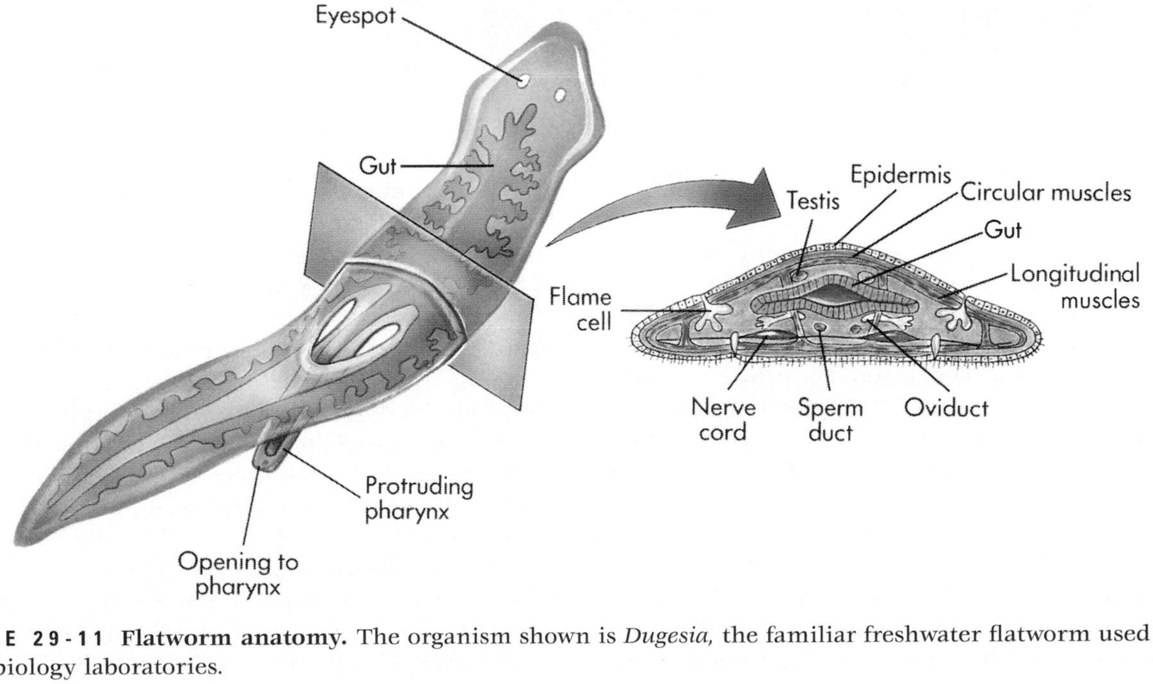

FIGURE 29-11 Flatworm anatomy. The organism shown is *Dugesia,* the familiar freshwater flatworm used in many biology laboratories.

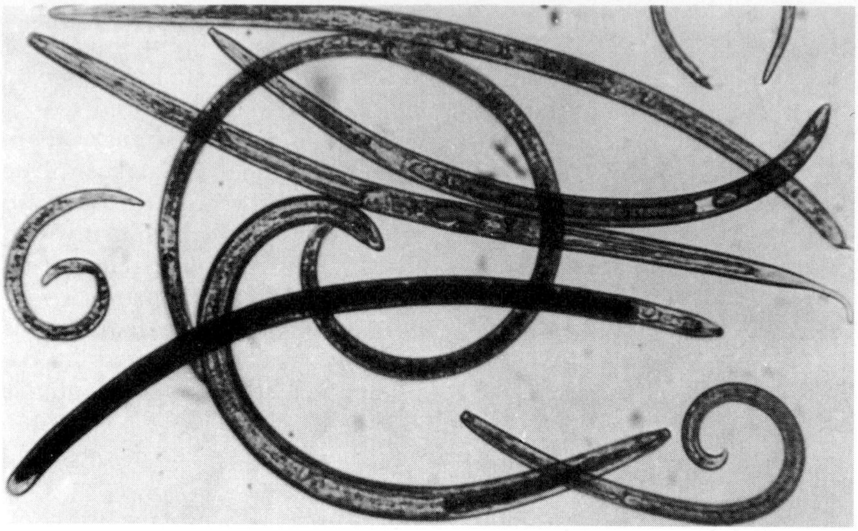

FIGURE 29-12 Living dirt. One square meter of ordinary lawn soil teems with 2 to 4 million roundworms. Although most are similar in form, they range in length from about 0.2 millimeters (about the width of a human hair) to about 6 millimeters (the diameter of the head of a tack).

The flukes and the tapeworms are two classes of flatworms that live within the bodies of other animals. The adult form of both classes parasitizes humans. The life cycles of these organisms are discussed in the boxed essay.

> *The acoelomates, typified by the flatworms, are the most primitive bilaterally symmetrical animals. Although simple in structure, the flatworms have organs and some organ systems, including a primitive brain. Because flatworms do not have a coelom, their organs are embedded within the body tissues.*

Seven phyla have a pseudocoelomate body plan (see Figure 29-3)—a body cavity with no lining. Only one of the seven, the roundworms, includes a large number of species.

Nematodes: Roundworms

Roundworms are classified in the phylum Nematoda, a word that comes from a Greek word meaning "thread." These cylindrical worms are diverse in size, but some of them are so small and slender that they look like fine threads. In fact, they can be so microscopically small that a spadeful of fertile soil may contain millions of these worms (Figure 29-12). Some species are abundant in fresh water or salt water.

Many members of this phylum are parasites of vertebrates. About 50 species of roundworms parasitize humans, causing problems such as blockage of the lymphatic vessels (Figure 29-13) or intestines and infections of the muscles or lungs. These worms are transmitted to humans as larvae by the bite of an infected mosquito. Roundworms also parasitize invertebrates and plants. For this reason, some nematodes are being investigated as agents of biological control of insects and other agricultural pests.

Roundworms are covered by a flexible, tough, transparent multilayered tissue (cuticle) that is shed as they grow. A layer of muscle lies beneath this epidermal layer and extends lengthwise (Figure 29-14). These longitudinal muscles pull against both the cuticle and the firm, fluid-filled pseudocoelom, similar to how your muscles pull against your bones. All this effort gets them nowhere in clear water, but they can move in muddy water or soil, which provides surfaces against which the worm's body pushes. The movements do, however, push on the fluid-filled pseudocoelom and aid in the distribution of food and oxygen throughout the worm.

The roundworm digestive system has two openings—a mouth and anus. Most roundworms have raised, hairlike sensory organs near their mouths. The mouth itself often has piercing organs, or stylets. Food passes through the mouth as a result of the sucking action of a muscular pharynx. After passing through these organs, food continues through the digestive tract where it is broken down and then absorbed. The roundworms that parasitize animals take in digested food of the host; the cells lining the digestive system simply absorb these nutrients.

The roundworms also contain primitive excretory and nervous systems. The nervous system consists of a ring of tissue surrounding the pharynx and a solid dorsal and a ventral nerve cord (unlike

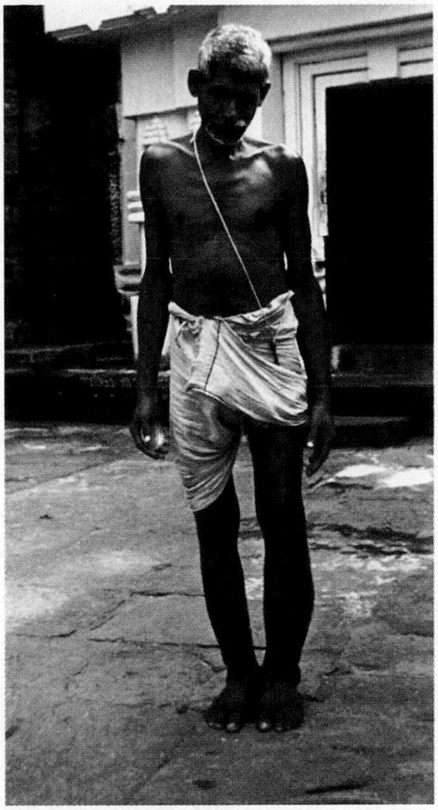

FIGURE 29-13 Elephantiasis. This condition is caused by roundworms that live in the lymphatic passages and block the flow of lymph. As a result, fluids cannot drain, and swelling occurs.

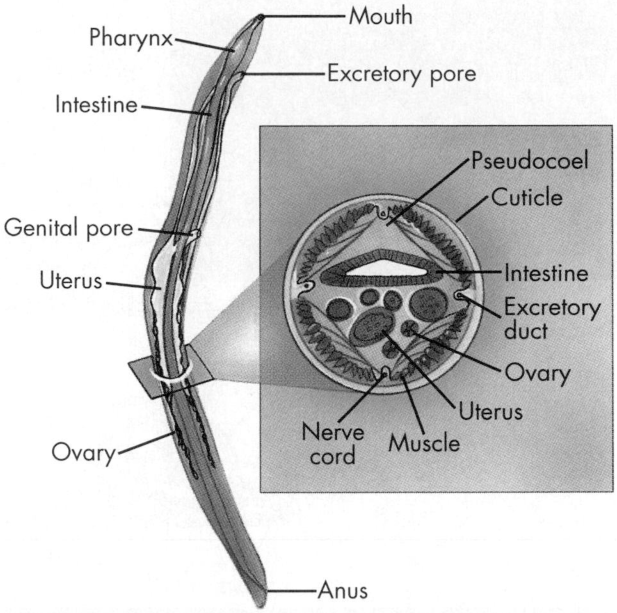

FIGURE 29-14 Anatomy of *Ascaris*, a parasitic roundworm of humans. A layer of muscle extending lengthwise along the body lies underneath the epidermal layer, or cuticle. The body organs are located within the pseudocoelom.

the hollow nerve tube of a chordate [see p. 659]). The excretory system consists of two lateral canals that unite near the anterior end to form a single tube ending in an excretory pore.

> *The pseudocoelomates are typified by the roundworms. Muscles attached to their thick, outer cuticle push against the fluid-filled pseudocoelom, resulting in a whiplike movement. Many of the roundworms are parasites of invertebrates, vertebrates (including humans), and plants.*

The rest of the invertebrate animals have a "true" coelom. The next three phyla of coelomates—the mollusks (clams, snails, and octopuses), the annelids (earthworms and leeches), and the arthropods (lobsters, insects, and spiders)—are all protostomes. The acoelomate animals already discussed are not protostomes but exhibit similar developmental patterns to the protostomes. These similarities suggest an evolutionary closeness among these groups.

Mollusks: Clams, Snails, and Octopuses

The mollusks are a large phylum of invertebrate animals having a muscular foot and a soft body covered by a mantle, and they are usually covered with a hard shell. The phylum name Mollusca comes from a Latin word meaning "soft bodied." The shelled mollusks include the snails, clams, scallops, and oysters. Unshelled mollusks are represented by the octopuses, squids, and slugs.

Mollusks are widespread and often abundant in marine and freshwater environments, and some, such as certain snails and the slugs, live on land. They range in size from being near microscopic to having the huge proportions of giant squid. Large giant squid measure approximately 21 meters long (almost 70 feet) and weigh approximately 250 kilograms (550 pounds). The largest giant squid ever documented, however, is over 140 feet long!

Mollusks exhibit four body plans: gastropod, cephalopod, bivalve, and chiton (Figure 29-15). The name of each group describes its prominent features. Although each group (with the exception of the bivalves) has a head end, the cephalopods (literally, "head-foot") have the most well-differentiated head and the most well-developed nervous system. The bivalves ("two-shelled") gastropods have the least well-developed nervous system of these groups. All four groups have a *visceral mass,* or group of organs, consisting of the digestive, excre-

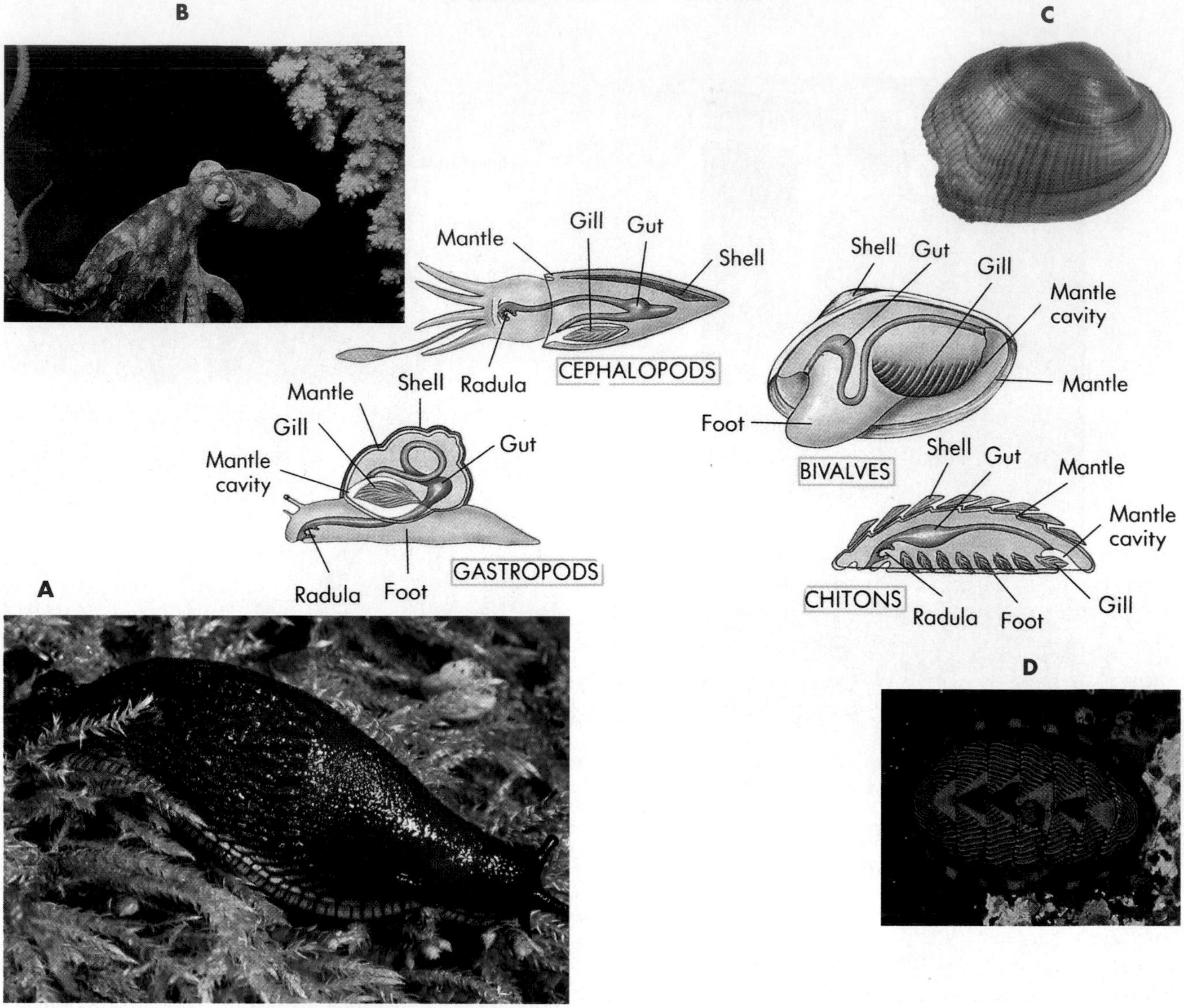

FIGURE 29-15 Body plans among the mollusks. The name of each group describes its prominent features. **A,** Gastropods. The example shown in the photo is a slug. **B,** Cephalopods. The example shown in the photo is an octopus. **C,** Bivalves. The example shown in the photo is a mussel. **D,** A chiton (shown in both illustration and photo).

tory, and reproductive organs. The visceral mass is covered with a soft epithelium called the *mantle,* which arises from the dorsal body wall and encloses a cavity between itself and the visceral mass. This cavity is *not* the coelom; the coelom surrounds the heart only. The mollusk's gills, the organs of respiration, lie within the mantle cavity. *Gills* are a system of filamentous projections of the mantle tissue that is rich in blood vessels. These projections greatly increase the surface area available for gas exchange. In land-dwelling mollusks, a network of blood vessels within the mantle cavity serves as a primitive lung.

Mollusks exhibit both open circulatory systems (in clams, for example) and closed circulatory systems (in squid, for example). Both types of circulatory system have a heart to pump blood. In a **closed circulatory system,** blood is enclosed within vessels as it travels throughout the body of the organism. In an **open circulatory system,** blood flows in vessels leading to and from the heart but through irregular channels called *blood sinuses* in many parts of the body.

Most coelomate animals have a closed circulatory system (with the notable exception of the insects); their blood vessels are intimately associated with the excretory organs, making the direct exchange of materials between these two systems possible. Mollusks were one of the earliest evolutionary lines to develop an efficient excretory system. Wastes are removed from the mollusk by tubular structures called *nephridia.* In mollusks with open circulatory systems, wastes move from the coelom into the nephridia and are discharged into the

A 

B 

C 

FIGURE 29-16 Representative annelids. A, Shiny bristleworm. **B,** Earthworms mating. **C,** A freshwater leech, with young leeches visible in the brood pouch.

mantle cavity. From there they are expelled by the continuous pumping of the gills. In animals with closed circulatory systems, such as annelids, some mollusks, and the vertebrates, the coiled tubule of a nephridium is surrounded by a network of capillaries. Wastes move from the circulatory system to the nephridium for removal from the body. All coelomates (except for arthropods and chordates) have basically similar excretory systems.

> *Mollusks, widespread in marine and freshwater environments, exhibit four body plans as represented by snails, clams, squid, and chitons. These animals use gills for respiration in water; terrestrial species have adaptations for breathing on land. They exhibit both open and closed circulatory systems and well-developed excretory systems.*

The other two major phyla of protostomes, the annelids and arthropods, have segmented bodies, whereas the mollusks do not. Segmentation underlies the organization of all of the more complex phyla of animals: the annelids (earthworms and leeches), arthropods (lobsters, insects, and spiders), echinoderms (sea urchins and starfish), and chordates (tunicates, fishes, amphibians, reptiles, birds, and mammals). In some adult arthropods the segments are fused; their segmentation is apparent only in their embryological development. So, too, embryological development reveals segmentation in the chordates. Segmentation in this phylum is exhibited only by the vertebrates in the repeating units of their backbones and as muscle blocks, such as the abdominal muscles of humans. Because segmentation in animals is different among phyla, scientists think it arose independently in more than one line of evolution.

closed circulatory system a body system in which blood is enclosed within vessels as it travels throughout an organism.

open circulatory system a body system in which blood flows in vessels leading to and from the heart but through irregular channels called blood sinuses in many parts of an organism.

Annelids: Earthworms and Leeches

The annelids (phylum Annelida) are worms characterized by a soft, elongated body composed of a series of ringlike segments. In fact, the word *annelid* means "tiny rings." Annelids are abundant in the soil and in both marine and freshwater environments throughout the world. Internally, their segments are divided from one another by partitions called *septa.* Digestive, excretory, and neural structures are repeated in each segment.

There are three classes of annelids: marine worms, freshwater and terrestrial worms, and leeches (Figure 29-16). Exhibiting unusual forms and sometimes iridescent colors, the marine worms live in burrows, under rocks, inside shells, and in tubes of hardened mucus they manufacture. Leeches occur mostly in fresh water, although a few are marine and some tropical leeches are found in terrestrial habitats. Most leeches are predators or scavengers, and some suck blood from mammals, including humans. The best-known leech is the medicinal leech, which was used for centuries to remove what was thought to be excess blood responsible for certain illnesses. The medicinal leech is now used as a source of anticoagulant in research that focuses on blood clotting. This animal is also used by some physicians to remove excess blood after surgery and to help restore circulation to severed body parts (such as fingers) after reattachment.

The earthworm exhibits the generalized body plan of this phylum: a tube within a tube (Figure 29-17). The digestive tract, a straight tube running from mouth to anus, is suspended within the coelom. An earthworm sucks in organic material by contracting its strong pharynx. It grinds this material in its muscular gizzard, aided by the presence of soil particles it takes in with its food.

The anterior segments of an earthworm contain a well-developed *cerebral ganglion,* or brain, and a few muscular blood vessels that act like hearts, pumping the blood through the closed circulatory system. Sensory organs are also concentrated near the anterior end of the worm. Some of these organs are sensitive to light, and elaborate eyes with lenses and retinas have evolved in certain members of the phylum. Separate nerve centers, or ganglia, are located in each segment and are connected by nerve cords. Each segment also contains both circular and longitudinal muscles, which annelids use to crawl, burrow, and swim. *Setae,* or bristles, help anchor the worms during locomotion or when they are in their burrows.

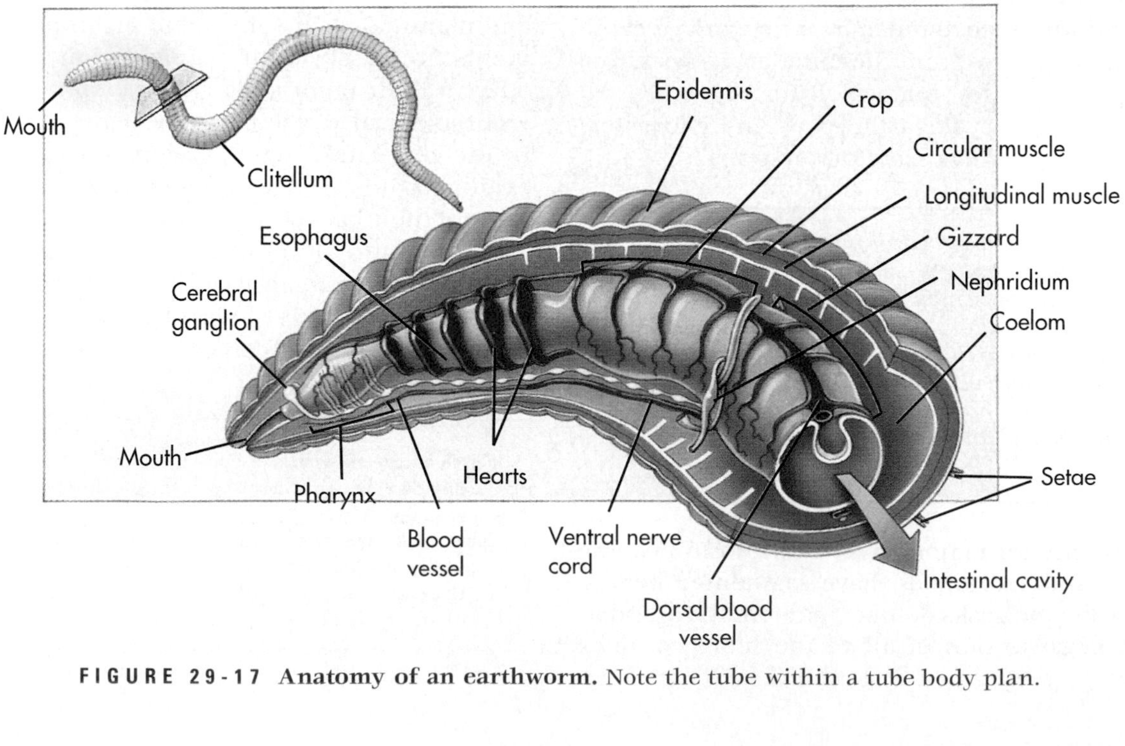

FIGURE 29-17 Anatomy of an earthworm. Note the tube within a tube body plan.

A

B

C

FIGURE 29-18 Representatives of the phylum Arthropoda. A, American spiny lobster. **B,** Praying mantis. **C,** The arrowhead spider.

The annelids are characterized by serial segmentation and a tube within a tube body plan. The body is composed of numerous similar segments, each with its own circulatory, excretory, muscular, and neural structures.

Reproduction differs among the annelid classes. In the marine worms, the sexes are usually separate, and fertilization is often external, occurring in the water and away from both parents. The earthworms and leeches, on the other hand, are hermaphroditic. When they mate, their anterior ends point in opposite directions, and their ventral surfaces touch. The *clitellum,* a thickened band on an earthworm's body, secretes a mucus that holds the worms together as they exchange sperm. Ultimately, the worms release the fertilized eggs into cocoons also formed by mucous secretions of the clitellum.

Crustaceans and Arachnids: Lobsters, Insects, and Spiders

Lobsters (crustaceans), insects, and spiders (arachnids) are representatives of the diverse phylum Arthropoda (Figure 29-18). The name *arthropod* comes from the two Greek words "arthros" (jointed) and "podes" (feet) and describes the characteristic jointed appendages of all arthropods. The nature of the appendages differs greatly in different subgroups; appendages may take the form of antennae, mouthparts of various kinds, or legs.

The arthropods have a rigid external skeleton, or **exoskeleton,** which varies greatly in toughness and thickness among arthropods (Figure 29-19). The exoskeleton provides places for muscle attachment, protects the animal from predators and injury, and most important, protects arthropods from water loss. As an individual outgrows its exoskeleton, that exoskeleton splits open and is shed. A new soft exoskeleton lies underneath, which subsequently hardens. The animal then grows into its new "shell."

All arthropods can be placed into one of two groups: those with jaws and those without jaws. The crustaceans, insects, centipedes, millipedes, and a few other small groups of arthropods have jaws, or *mandibles.* These jaws are formed by the modification of one of the pairs of anterior appendages (but *not* the first pair). The appendages nearest the anterior end are sensory antennae (Figure 29-20, *B*).

exoskeleton (ek so **skel** uh tun) a rigid structure, located on the outside of an organism, that maintains its shape and supports structures of the body.

FIGURE 29-19 Exoskeletons. Some arthropods have a tough exoskeleton, like this South American scarab beetle (**A**); others have a fragile exoskeleton, like the green darner dragonfly (**B**).

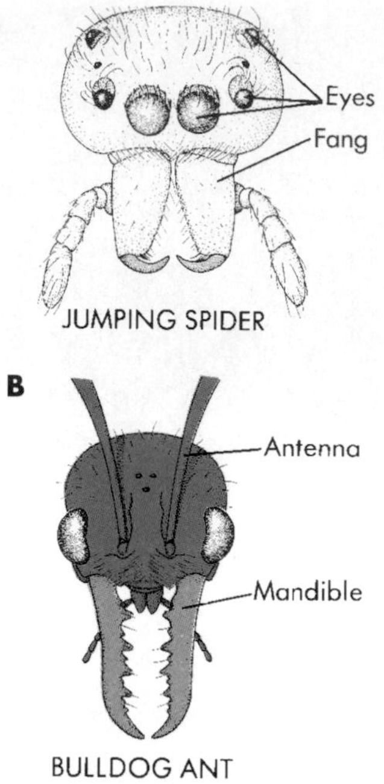

FIGURE 29-20 Arthropod mouthparts. A, The jumping spider lacks jaws but does have fangs with which it catches prey. **B,** The bulldog ant has jaws (mandibles).

FIGURE 29-21 The compound eye. This compound eye is found on a robberfly.

The remaining arthropods, which include the spiders, mites, scorpions, and a few other groups, lack mandibles. Their mouthparts usually take the form of fangs, which evolved from the appendages nearest the anterior end of the animal (Figure 29-20, *A*).

An important structure of many arthropods such as bees, flies, moths, and grasshoppers is the *compound eye* (Figure 29-21). Compound eyes are composed of many independent visual units, each containing a lens. *Simple eyes,* composed of a single visual unit having one lens, are found in many arthropods with compound eyes and function in distinguishing light and darkness. In some flying insects, such as locusts and dragonflies, simple eyes function as horizon detectors and help stabilize the insects during flight.

In the course of arthropod evolution, the coelom has become greatly reduced, consisting only of the cavities that house the reproductive organs and some glands. Figure 29-22 illustrates the major structural features of a grasshopper as a representative of the arthropods. Like the annelids, the arthropods have a tubular gut that extends from the mouth to the anus. The circulatory system of arthropods is open; their blood flows through cavities between the organs. One longitudinal dorsal vessel functions as a heart, helping move the blood along.

Most aquatic arthropods breathe by means of gills. Their feathery-looking structure provides a large surface area over which gas exchange takes place between the surrounding water and the animal's blood. The respiratory systems of terrestrial arthropods generally have internal surfaces over which gas exchange takes place. The respiratory systems of insects, for example, consist of small,

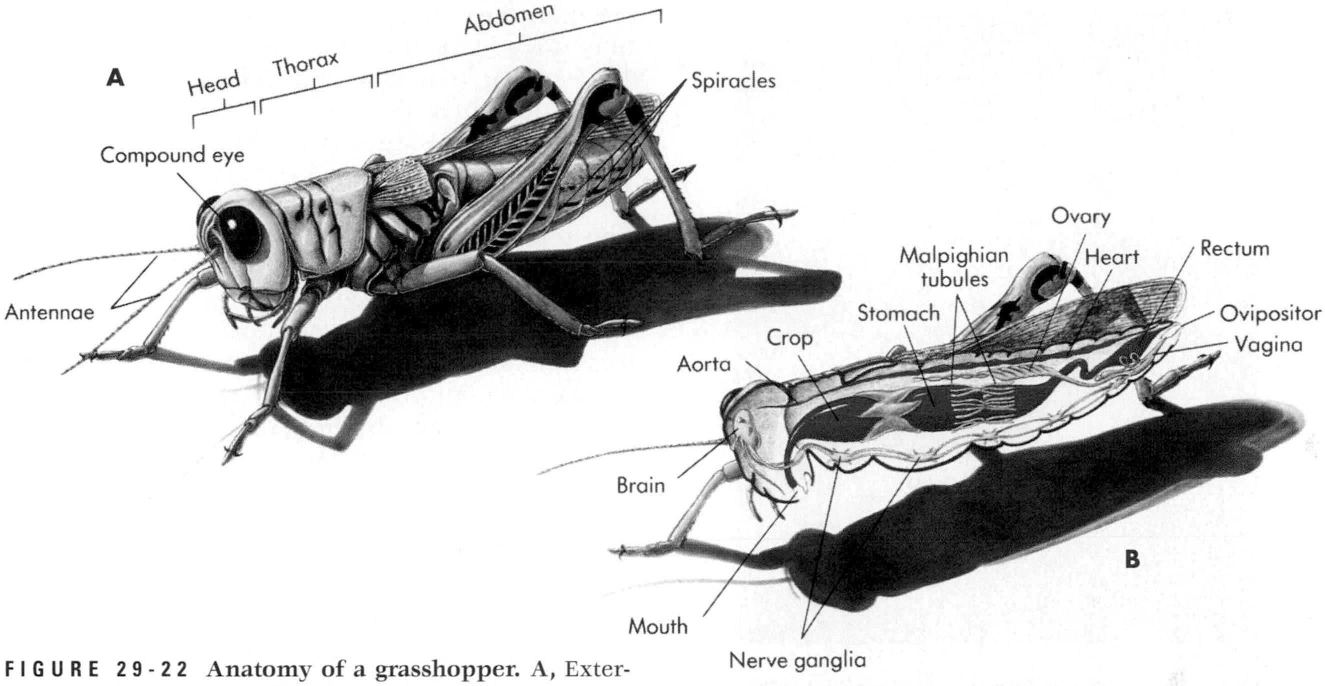

FIGURE 29-22 Anatomy of a grasshopper. A, External anatomy. **B,** Internal anatomy.

branched air ducts called *tracheae.* These tracheae, which ultimately branch into very small *tracheoles,* are a series of tubes that transmits oxygen throughout the body. The tracheoles are in direct contact with the individual cells, and oxygen diffuses from them to other cells directly across the cell membranes. Air passes into the trachea by way of specialized openings called *spiracles,* which in most insects, can be closed and opened by valves.

Although there are various kinds of excretory systems in different groups of arthropods, a unique excretory system evolved in terrestrial arthropods in relation to their open circulatory system. The principal structural element is the *malpighian tubules,* which are slender projections from the digestive tract. Fluid is passed through the walls of the malpighian tubules to and from the blood in which the tubules are bathed. The nitrogenous wastes in it are precipitated and then emptied into the hindgut (the posterior part of the digestive tract) and eliminated. Most of the water and salts in the fluid is reabsorbed by the hindgut, thus conserving water.

Arthropods are a diverse phylum of organisms having jointed appendages and rigid exoskeletons.

Two members of the phylum Arthropoda, *Sarcoptes scabiei* (Figure 29-23) and *Phthirus pubis* (Figure 29-24), cause two common contagious parasitic skin infestations of humans: scabies and pubic lice (commonly known as *crabs*). Both of these organisms are transmitted by close physical contact and are often spread by sexual contact. Although they do not cause serious disease, both mites (organisms closely related to spiders) and lice (organisms closely related to fleas) can be vectors, or carriers, of other diseases. Both respond promptly to treatment with antiparasitic medications applied to the skin.

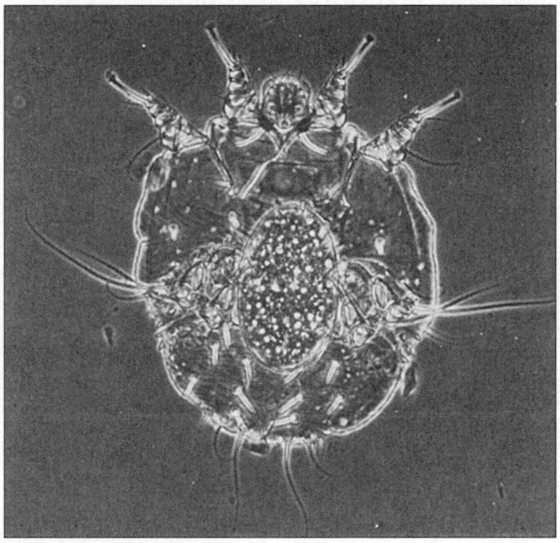

FIGURE 29-23 The itch mite, *Sarcoptes scabiei.*

During an infestation of scabies, female itch mites bore into the top layers of their host's skin to

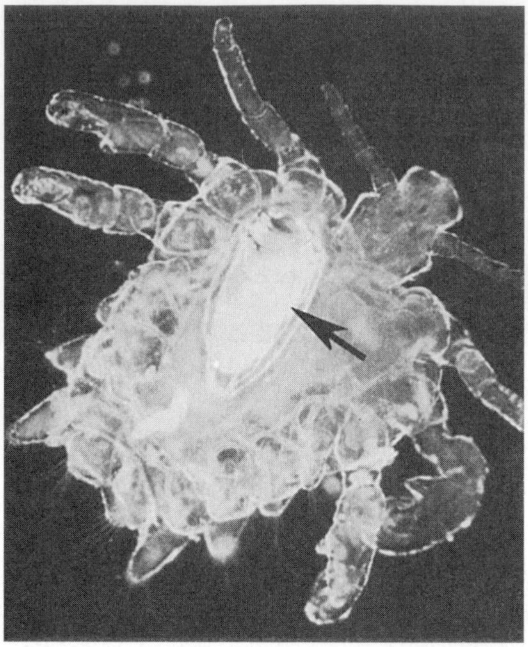

FIGURE 29-24 The pubic louse, *Phthirius pubis*. The arrow shows a developing egg.

lay their eggs. Burrows are formed that look like fine, wavy, dark lines on the surface of the skin. The eggs hatch in a few days, increasing the level of infestation. Itching becomes intense, but scratching abrades the skin surface, which creates an environment in which a secondary bacterial infection can take hold. The common sites of infestation on the body are the base of the fingers, the wrists, the armpits, the skin around nipples, and the skin around the belt line.

The pubic louse is the most common louse to infect humans. It lives in the hairs of the anal and genital area, and the females lay their eggs there, attaching the eggs to the hairs. Infestation causes intense itching.

Echinoderms: Sea Urchins and Starfish

Sea urchins and starfish are representatives of the phylum Echinodermata. The term *echinoderm* means "spine skin," an appropriate name for many members of this phylum (Figure 29-25). The five living classes of echinoderms are shown: the sea lilies, sea stars (starfish), brittle stars, sea urchins

A

B

C

D

FIGURE 29-25 Representatives of the phylum Echinodermata. A, Starfish off the island of Belize (class Asteroidea). **B,** Sea cucumber (class Holothuroidea). **C,** Sea lillies (class Crinoidea). **D,** A brittle star (class Ophiuroidea) on a red vase sponge. A sea urchin (class Echinodea) is attached to the left side of the sponge. **E,** Sand dollars also belong to the class Echinodea.

and sand dollars, and sea cucumbers. These marine animals live on the sea floor, with the exception of a few swimming sea cucumbers.

The echinoderms are different from the other invertebrates in that they are deuterostomes.The other invertebrates are protostomes. The embryological differences that distance them evolutionarily from the other invertebrates connect them with the chordates. It is thought that the echinoderms and the chordates, along with two smaller phyla not mentioned here, evolved from a common ancestor.

Echinoderms are bilaterally symmetrical as larvae but radially symmetrical as adults. They are closely related to and grouped with the bilaterally symmetrical animals, however. Adult echinoderms have a five-part body plan corresponding to the arms of a sea star or the design on the "shell" of a sand dollar. Five *radial canals,* the positions of which are determined early in the development of the embryo, extend into each of the five parts of the body (Figure 29-26). This water vascular system is used for locomotion and is unique to echinoderms.

As adults, these animals have no head or brain. Their nervous systems consist of central *nerve rings* from which branches arise. The animals are capable of complex response patterns, but there is no centralization of function.

There is no well-organized circulatory system in echinoderms. Food from the digestive tract is distributed to all the cells of the body in the fluid that lies within the coelom. In many echinoderms, respiration takes place by means of skin gills, which are small, fingerlike projections that occur near the spines. Waste removal also takes place through these skin gills. The digestive system is simple but usually complete, consisting of a mouth, gut, and anus.

Although the echinoderms are well-known for regenerating lost parts, most reproduction is sexual and external. The sexes in most echinoderms are separate, although usually little external difference exists between them. The fertilized eggs of echinoderms usually develop into free-swimming bilaterally symmetrical larvae that look quite different from the adults. They eventually change through a series of stages, or metamorphose, to adult forms.

> *The echinoderms, or spiny-skinned animals, are marine organisms represented by sea urchins and starfish. They are closely related to the chordates.*

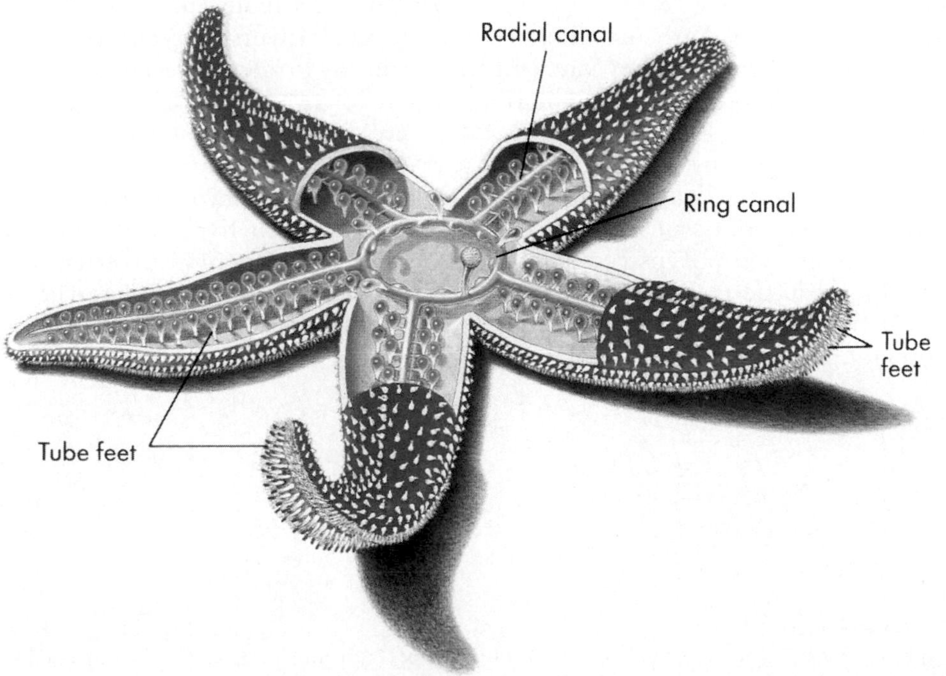

FIGURE 29-26 Anatomy of an echinoderm. Notice the water vascular system of a sea star *(blue)*.

Summary

- Animals are a diverse group of multicellular, eukaryotic, heterotrophic organisms. The cells of all animals (with the exception of sponges) are organized into tissues, and in most animals the tissues are organized into organs. Most zoologists think that the animals arose from protist ancestors.
- Other than the sponges, all animals exhibit either radial symmetry (the jellyfish and hydra) or bilateral symmetry (all other animals). Radial symmetry means that body parts emerge from a central point. Bilateral symmetry means that the right side of an organism is a mirror image of the left side.
- A body cavity, or coelom, is a lined enclosure that contains body organs that hang suspended by thin sheets of connective tissue. Except for roundworms and flatworms, most bilaterally symmetrical organisms (including the echinoderms) have coeloms. Roundworms have pseudocoeloms, or false body cavities, which are not lined; roundworms' organs are not suspended by thin sheets of connective tissue. Flatworms have no coeloms surrounding their gut cavity.
- Two main branches of coelomate animals, representing two distinct evolutionary lines, are the protostomes and the deuterostomes. The protostomes (mollusks, annelids, and arthropods) differ with regard to certain embryological events from the deuterostomes (echinoderms and chordates).
- Sponges (phylum Porifera) are acoelomate, asymmetrical, aquatic animals. They are considered simple animals because they have no tissues or organs.
- The hydra and jellyfish (phylum Cnidaria) are acoelomate, radially symmetrical, aquatic animals. They have two layers of tissues and exist either as cylindrically shaped polyps (corals and sea anemones) or medusae (jellyfish). In some cnidarians, these two forms alternate during the life cycle of the organism.
- All animals other than the sponges and the cnidarians are bilaterally symmetrical or, in the case of the echinoderms, evolved from bilaterally symmetrical ancestors. In addition, they all develop embryologically from three tissue layers: endoderm, ectoderm, and mesoderm.
- The flatworms (phylum Platyhelminthes) are ribbonlike worms that live in the soil or water or within other organisms. Having no coelom, they are considered to have the simplest body plan of the bilaterally symmetrical animals. However, the flatworms have organs and some organ systems, including a primitive brain.
- The roundworms (phylum Nematoda) are pseudocoelomate, cylindrical worms that live in the soil, in water, or within other organisms. They have simple digestive, excretory, and nervous systems.
- The mollusks (phylum Mollusca) are a large phylum of coelomates that exhibit four body plans. All four groups have a visceral mass, or group of organs, consisting of the digestive, excretory, and reproductive organs. They exhibit both closed and open circulatory systems.
- The annelids (phylum Annelida) are worms characterized by a soft, elongated body composed of a series of ringlike segments, each with its own circulatory, excretory, muscular, and neural structures. They have a tube within a tube body plan.
- The arthropods (phylum Arthropoda) are an extremely diverse phylum that includes organisms such as lobsters, insects, and spiders. Arthropods are characterized by a rigid external skeleton and jointed appendages.
- The phylum Echinodermata, represented by sea urchins and starfish, are spiny-skinned marine animals that live on the sea floor. They are the only deuterostome invertebrates, a characteristic that shows relatedness with the chordates. Although adult echinoderms are radially symmetrical, their larvae are bilaterally symmetrical.

Knowledge and Comprehension Questions

1. What do you and jellyfish have in common? What is an important taxonomic difference between the two of you?
2. "Animals are a diverse group of eukaryotic, multicellular, heterotrophic organisms." Explain this statement.
3. Fill in the blanks: If an organism's body parts all emerge from a central point, it exhibits ____________. If an organism has two sides that are a mirror image of each other, it shows ____________.
4. Distinguish among coelomates, pseudocoelomates, and acoelomates. Give an example of each.
5. Which organisms are considered the "simplest animals"? Why?
6. What two basic body plans are shown by cnidarians? Summarize the differences between these two plans.
7. Which organisms are the most primitive bilaterally symmetrical animals? Briefly describe these animals' structure.
8. People can become very ill if infested by roundworms. Describe these organisms and why they can be a health risk to humans.
9. Distinguish between open and closed circulatory systems. Which do you have?
10. Summarize the characteristics of mollusks.
11. If you have ever tried to swat a fly, you know that flies are quick to respond to even the slightest movements. What feature of a fly's body allows it to perceive motion so rapidly?
12. Briefly summarize the characteristics of arthropods.
13. In casual conversation, you refer to a spider on the wall as an insect. A friend who has studied biology informs you that spiders are not insects. Explain what she means.
14. How do echinoderms differ from all other invertebrates? What is the significance of this fact?

Critical Thinking

1. Some invertebrates, such as most flatworms, are hermaphroditic. Wouldn't this characteristic result in a decrease in genetic diversity in populations of these worms?

CHAPTER 30

CHORDATES AND VERTEBRATES: PATTERNS OF STRUCTURE, FUNCTION, AND REPRODUCTION

ALTHOUGH BATS OFTEN conjure up images of blood-sucking vampires, these bats would be totally disinterested in your neck. Nicknamed "little red flying foxes," they find the eucalyptus blossoms of their Australian homeland a delicacy. When this favorite food is in short supply, they often raid orchards. Although these bats can become pests, you can leave the vampire-fighting garlic and pointed stakes at home if you venture on a trip "down under." Little red flying foxes only bother the fruit growers.

Interestingly, bats are mammals—the same class of animals to which humans belong. Bats are more like you than they are like birds! As this chapter progresses, you will come to understand those similarities as you explore the patterns of structure, function, and reproduction of the chordates (including the vertebrates).

Chordates

The chordates are a phylum of animals that include three subphyla: tunicates, lancelets, and vertebrates. Chordates are characterized by three principal features (see Chapter 24): (1) a single, hollow **nerve cord** located along the back, (2) a rod-shaped **notochord,** which forms between the nerve cord and the gut (stomach and intestines) during development, and (3) **pharyngeal (gill) arches,** which are located at the throat (pharynx). (These three chordate features are shown in Figure 24-9 as they appear in the embryo, because they are present in the embryos of all chordates.) In addition, lancelets exhibit these characteristics as adults, and tunicates exhibit them as larvae. In vertebrates, the nerve cord differentiates into a brain and spinal cord. The notochord serves as a core around which the vertebral column develops, encasing the nerve cord and protecting it. The pharyngeal arches develop into the gill structures of the fishes and into ear, jaw, and throat structures of the terrestrial vertebrates. The presence of the gill arches in all vertebrate embryos provides a clue to the aquatic ancestry of the subphylum Vertebrata.

Along with having these three traits, chordates have many other characteristics in common. All have a true coelom and bilateral symmetry. The embryos of chordates exhibit segmentation. Figure 30-1 shows segments of tissue called *somites* in the human embryo, which develop into the skeletal muscles. Most chordates have an internal skeleton to which their muscles are attached and work against, providing movement. (Larval tunicates and adult lancelets do not have internal skeletons; their muscles are attached to their notochords.) Finally, chordates have tails that extend beyond the anus, at least during embryonic development. Nearly all other animals have a terminal anus.

> *Chordates are characterized by a single, hollow nerve cord located along the back, a rod-shaped notochord, and gill arches located at the throat. The vertebrates are an important subphylum of the chordates.*

Tunicates

The **tunicates** comprise a group of about 2500 species of marine animals, most of which look like

KEY CONCEPTS

- Chordates are organisms that, at some time during their development, have a single, hollow nerve cord along the back, a rod-shaped notochord between the nerve cord and the gut, and gill arches located at the throat.
- There are three subphyla of the phylum chordates: the tunicates, saclike sessile marine organisms; the lancelets, tiny scaleless fishlike organisms, and the vertebrates.
- A vertebrate has a vertebral column surrounding the dorsal nerve cord, a distinct head with a skull that encases the brain, a closed circulatory system with a heart to pump the blood, and many other complex organs and organ systems.
- There are seven classes of living vertebrates: three classes are fishes, and four classes are land-dwelling, four-footed animals.

OUTLINE

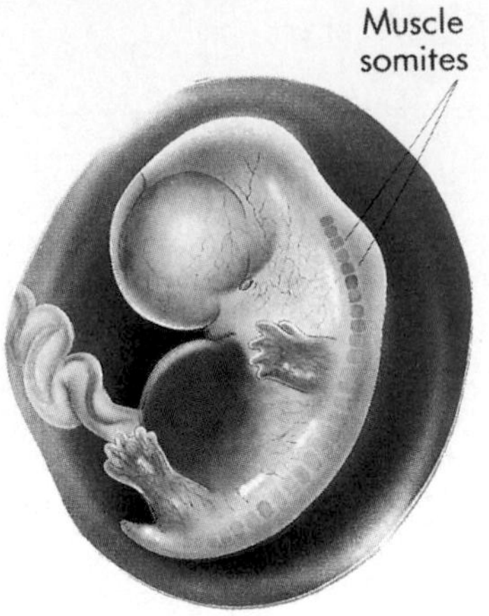

FIGURE 30-1 A chordate embryo. All chordate embryos have segments of tissue called somites. The somites eventually develop into the skeletal muscles.

living sacs attached to the floor of the ocean (Figure 30-2, *A* and *B*). As shown in Figure 30-2, *C,* the tunicates are not much more than a large pharynx covered with a protective *tunic.* The tunic is a tough outer "skin" composed mainly of cellulose, a substance found in the cell walls of plants and algae but rarely found in animals. Some colonial tunicates live in masses on the ocean floor and have a common sac and a common opening to the outside (Figure 30-3). Colonial tunicates reproduce asexually by budding. Individual tunicates are hermaphrodites, with each organism having both male and female sex organs.

The pharynx of a tunicate is lined with numerous cilia. As these cilia beat, they draw a stream of water through the incurrent siphon into the pharynx, which is lined with a sticky mucus. Food particles are trapped within the pharynx, and the fil-

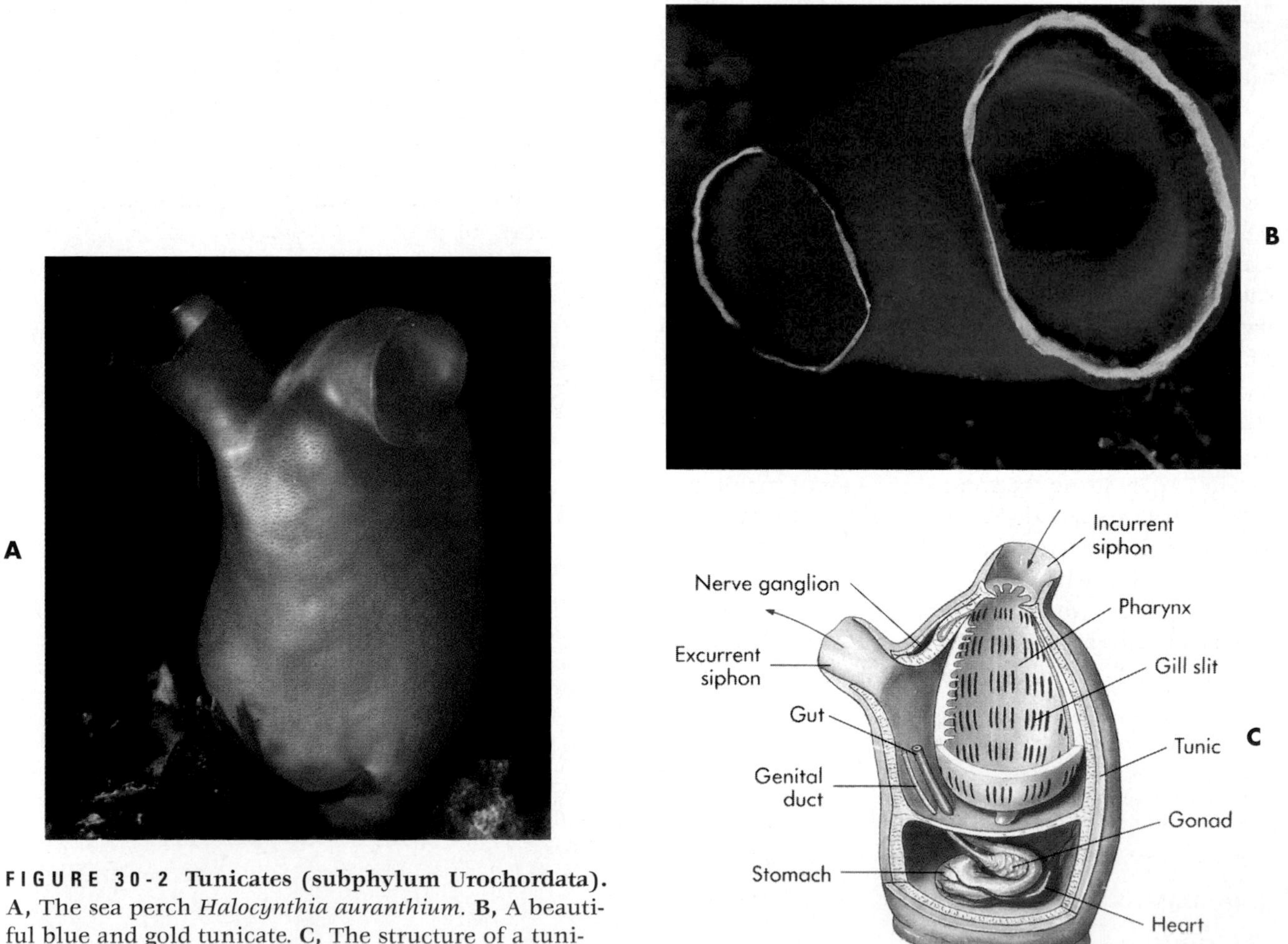

FIGURE 30-2 Tunicates (subphylum Urochordata). **A,** The sea perch *Halocynthia auranthium.* **B,** A beautiful blue and gold tunicate. **C,** The structure of a tunicate.

FIGURE 30-3 Colonial tunicates. This colony was found at a depth of about 2 meters. The colony on the right is in the process of dividing.

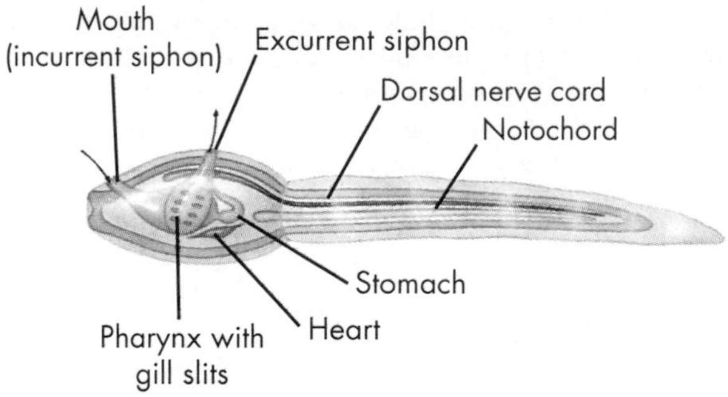

FIGURE 30-4 The structure of a larval tunicate. The larvae of tunicates resemble tadpoles and contain notochords and nerve cords.

tered water flows out of the animal through the excurrent siphon. Because 90% of all species of tunicate have this structure and forcefully squirt water out their excurrent siphons when disturbed, they are also called *sea squirts.*

As adults, tunicates lack a notochord and a nerve cord. The gill slits (which develop from the gill arches) are the only clue that adult tunicates are chordates. Only the larvae, which look like tadpoles (Figure 30-4), have notochords and nerve cords. The notochords are in their tails, and their nerve cords run dorsal to their notochords almost the entire length of their bodies. The subphylum name Urochordata comes from the placement of the larval notochord and literally means "tail chordate." The larvae remain free swimming for no more than a few days. Then they settle to the bottom and become sessile, attaching themselves to a suitable substrate by means of a sucker. As they mature, they adjust to a filter-feeding existence. Some tunicates that live in the world's warmer regions retain the ability to swim and never develop into a sessile form.

> *Adult tunicates are saclike, sessile, marine filter feeders. Their only chordate characteristic is their gill slits. Their larvae, however, have notochords and nerve cords in the tails of their tadpolelike bodies.*

Lancelets

The **lancelets** are tiny, scaleless, fishlike marine chordates that are just a few centimeters long and pointed at both ends (Figure 30-5). They look very much like tiny surgical blades called *lancets,* from which they get their name. You may have had a few drops of blood taken in the doctor's office from a "fingerstick" done with a lancet.

Lancelets have a segmented appearance because of blocks of muscle tissue that are easily seen through their thin, unpigmented skin. Although they have pigmented light receptors, lancelets have no real head, eyes, nose, or ears. However, unlike the tunicates, the lancelet's notochord runs the entire length of its dorsal nerve cord, so these organisms are called *cephalochordates,* or *head chordates.* The lancelet retains its notochord throughout its lifespan.

The 23 species of lancelets live in the shallow waters of oceans all over the world. They spend

nerve cord a single, hollow cord along the back that carries sensory and motor impulses and that is a principal feature of chordates; in vertebrates, the nerve cord differentiates into a brain and spinal cord.

notochord (**no** toe kord) a rod-shaped structure that forms between the nerve cord and the gut (stomach and intestines) during the development of all chordates.

pharyngeal (gill) arches (fuh **rin** jee uhl *or* **far** in **jee** uhl) a principal feature of the embryos of all chordates. Pharyngeal arches develop into the gill structures of fish, and the ear, jaw, nose, and throat structures of terrestrial vertebrates.

tunicates (**too** nih kits *or* **too** nih **kates**) a subphylum of the chordates. A group of about 2500 species of marine animals, most of which look like living sacs attached to the floor of the ocean. The only chordate characteristic adult tunicates exhibit is their gill slits; however, their larvae have notochords and nerve cords.

lancelets (**lans** lets) a subphylum of the chordates. Tiny, scaleless, fishlike marine organisms that are just a few centimeters long and pointed at both ends. Their adult forms exhibit chordate characteristics.

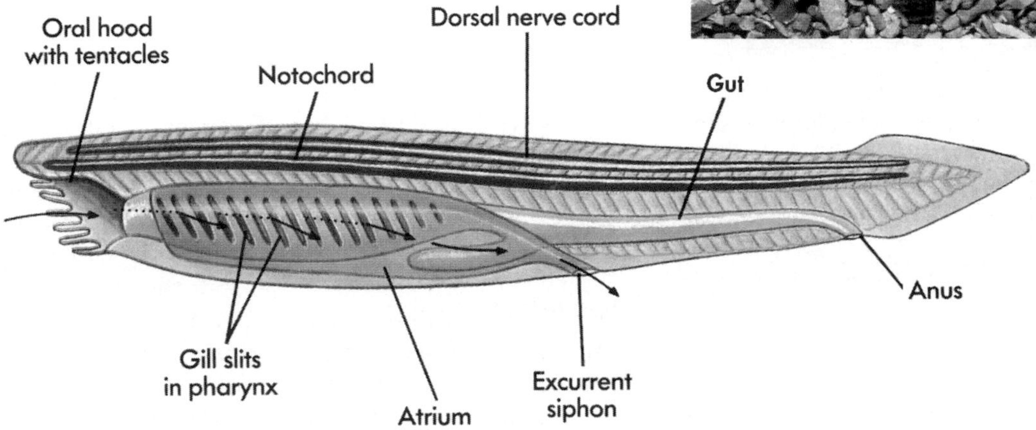

FIGURE 30-5 Lancelets (subphylum Cephalochordata). A, Two lancelets partly buried in shell gravel, with their heads protruding. The muscle segments are clearly visible. **B,** The structure of a lancelet, showing the path of water flow through the animal.

most of their time partly buried in the sandy or muddy bottom with only their anterior ends protruding, feeding on microscopic plankton (floating plants and animals). In a manner similar to the tunicates, lancelets filter plankton from the water. Cilia line the anterior ends of their alimentary canals, and these beating cilia create an incoming current of water. The filtered water exits at an excurrent siphon. An oral hood projects beyond the mouth, or incurrent siphon, and bears sensory tentacles. The males and females are separate, but no obvious external differences exist between them.

The lancelets are scaleless, fishlike marine chordates. The adult forms exhibit chordate characteristics.

Vertebrates

The **vertebrates** differ from the other chordates in that most have vertebral columns in place of notochords. A vertebral column, or backbone, is a stack of bones, each with a hole in its center, that forms a cylinder surrounding and protecting the dorsal nerve cord. (Each bone in the column is a vertebra.) One class of vertebrates that does not have a vertebral column are the jawless fishes. Present-day jawless fishes have notochords, but their ancestors had bony skeletons and vertebral columns. Another class, the cartilaginous fishes have skeletons and vertebral columns composed of cartilage (a tough yet elastic type of connective tissue) rather than bone.

In addition to having vertebral columns, vertebrates have distinct heads with skulls that encase their brains. They have closed circulatory systems (their blood flows within vessels) and a heart to pump the blood. Most vertebrates also have livers, kidneys, and endocrine glands. (Endocrine glands are ductless glands that secrete hormones, which play a critical role in controlling the functions of the vertebrate body.)

The human body plan is representative of the vertebrate body plan. Even though vertebrates have many similarities, they are a diverse group, consisting of animals adapted to life in the sea, on land, and in the air. There are seven classes of living vertebrates: three classes are fishes, and four classes are land-dwelling *tetrapods,* or four-footed animals.

Vertebrates comprise a subphylum of chordates characterized by a vertebral column surrounding a dorsal nerve cord or by descent from ancestors with these features.

Jawless Fishes

Other than the lamprey eels and hagfishes, the major groups of the **jawless fishes,** or agnatha (*a* meaning "not," *gnatha* meaning "jaws"), have been extinct for hundreds of millions of years. Only about 20 to 30 species of each of these two groups are alive today. Both groups are long, tubelike aquatic animals that usually live in the sea or in brackish (somewhat salty) water where the fresh water of a river meets the ocean. In addition to their lack of jaws, they have no paired fins to help them swim, and they have no scales. They do have notochords, however, and portions of cartilaginous skeletons that are remnants from their extinct ancestors.

Lampreys parasitize other fishes. In fact, they are the only parasitic vertebrates. They have round mouths that function like suction cups (see Figure 24-11), which they attach to their prey—the bony fishes. When a lamprey attaches to a fish, it uses its spine-covered tongue like a grater, rasping a hole through the skin of the fish and then sucking out its body fluids. Sometimes lampreys can be so abundant that they are a serious threat to commercial fisheries, preying on salmon, trout, and other commercially valuable fishes. Entering the Great Lakes from the sea, they have become important pests there; millions of dollars are spent annually on their control. The hagfishes, although similar in size and shape to lampreys, are not parasitic. They are scavengers, often feeding on the insides of dead or dying fishes or large invertebrates.

Lampreys and hagfishes are tubular, scaleless, jawless organisms that lack paired fins and that live in the sea or in brackish water. Lampreys parasitize bony fishes and pose a serious threat to some commercial fisheries. Hagfishes are scavengers.

To reproduce, jawless fishes *spawn* as do most fishes, amphibians, and shellfish. During spawning, the males and females deposit eggs and sperm directly into the water. Fertilization that takes place outside the body of the female is called **external fertilization.** Lampreys swim upstream to spawn as salmon do and create nests in which they deposit eggs and sperm. The fertilized eggs develop into larvae that feed on plankton. Over a period of years, the larvae mature and metamorphose (change) into parasitic adults. In contrast, the eggs of hagfishes do not develop into larvae. Completely formed hagfishes hatch directly from fertilized eggs.

Most fishes, amphibians, and shellfish have an external type of fertilization. Eggs are fertilized by sperm outside the body of the organism.

Cartilaginous Fishes

The Chondrichthyes (*chondri* meaning "cartilage," *ichthyes* meaning "fishes") include the sharks, skates, and rays (Figure 30-6). Hundreds of extinct species of **cartilaginous fishes** are known from the fossil record, but less than 800 species exist today.

The skin of sharks (as well as the skates and rays) is covered with small, pointed, toothlike scales called *denticles,* which give the skin a sandpaper texture. Sharks have streamlined bodies and two pairs of fins: pectoral fins just behind the gills and pelvic fins just in front of the anal region. The dorsal (back) fins provide stability, and motions of the other fins (including the asymmetrical tail fin), as well as sinuous motions of the whole body, give the shark lift and propel it through the water.

Many sharks are predators and eat large fishes and marine mammals. Their sharp, triangular teeth saw and rip off pieces of flesh as the shark thrashes its head from side to side. Some sharks feed on plankton rather than prey on other animals. These sharks swim with their mouths open, and the plankton is strained from the water by specialized denticles on the inner surfaces of the gill arches.

Skates and rays are generally smaller than sharks and have flattened bodies with enlarged pectoral fins (see Figure 30-6, *B* and *C*) that undulate when these fishes move. Their tails, which are not a principal means of locomotion as in sharks, are thin and whiplike. They are sometimes armed with poisonous spines that are used as defense mechanisms rather than for predation. These animals have a mouth on their underside and feed mainly on invertebrates on or near the ocean floor.

vertebrates (**ver** tuh bruts *or* **ver** tuh braytes) a subphylum of the chordates, characterized by a vertebral column surrounding a dorsal nerve cord.

jawless fishes tubular, scaleless, jawless vertebrates that live in the sea or in brackish water. They have no paired fins.

external fertilization the union of a male gamete (sperm) and a female gamete (egg) outside the body of the female.

cartilaginous fishes (kar tuh **laj** uh nus) a class of vertebrates whose members have skeletons made of cartilage rather than bone and that includes the sharks, skates, and rays.

A

B

C

FIGURE 30-6 Cartilaginous fishes (class Chondrichthyes). Members of this class spend most of their time in graceful motion. **A,** Blue shark. **B,** Diamond sting ray. **C,** Manta ray.

> *The cartilaginous fishes include the sharks, skates, and rays. The skin is covered with small, pointed, toothlike scales that result in a sandpaper texture. Many sharks are predators of large fishes and marine mammals; skates and rays feed on invertebrates on the ocean floor.*

The sensory systems of sharks, skates, and rays are quite sophisticated and diverse. Sharks have a lateral line system as all fishes and amphibians do. A **lateral line system** is a complex system of *mechanoreceptors* that lies in a single row along the sides of the body and in patterns on the head. These receptors can detect mechanical stimuli such as sound, pressure, and movement. Sharks often detect prey by means of their lateral line systems, but, using their electroreceptors, they can also detect electrical fields emitted by other fishes. Sharks have these receptors on their heads, and rays have them on their pectoral fins. Scientists think these fishes may use their electroreceptors for navigation as well. Chemoreception is another important sense in the Chondrichthyes. In fact, sharks have been described as "swimming noses" because of their acute sense of smell. Vision is also important to the feeding behavior of sharks; they have well-developed mechanisms for vision at low-light intensities.

> *All fishes and amphibians have a complex system of mechanoreceptors called a lateral line system. These receptors, which lie in a row on the sides of the body and on the head, detect mechanical stimuli such as sound, pressure, and movement.*

Another important characteristic of fishes is their ability to regulate their buoyancy, or ability to float at various depths in the water. Bony fishes have gas-filled *swim bladders* that regulate their buoyancy. Cartilaginous fishes do not have swim bladders but can adjust the size and oil content of their livers. Because oil is less dense than water and the liver has a high oil content, adjusting the oil content and size of the liver can regulate buoyancy to a certain degree. Most species of sharks, however, swim continually to keep from sinking and to keep water flowing over their gills.

Aquatic organisms show various adaptations regarding **osmoregulation,** the control of water movement into and out of their bodies. Marine organisms tend to lose water to their surroundings because their body fluids usually have a solute concentration (concentration of dissolved substances) lower than that of seawater. The cartilaginous fishes maintain solute concentrations close to that of seawater, however, because they change potentially toxic nitrogen-containing wastes (such as ammonia) into a less toxic compound called *urea* and retain it in their bodies rather than excreting it (Figure 30-7). (In fact, shark meat must be soaked in freshwater before it is eaten to remove most of the urea.) With similar solute concentrations both inside and outside the fish, water movement remains relatively equal in both directions.

Although most aquatic animals reproduce by means of external fertilization, the cartilaginous fishes have developed a method of internal fertilization. These fishes have *pelvic claspers,* which are rodlike projections between the pelvic fins of the male fish. During *copulation,* or coupling of the male and female animals, the male inserts his clasper into the female's **cloaca.** The cloaca is the terminal part of the gut into which ducts from the kidney and reproductive systems open. The fish either swim side-by-side, or the male wraps around

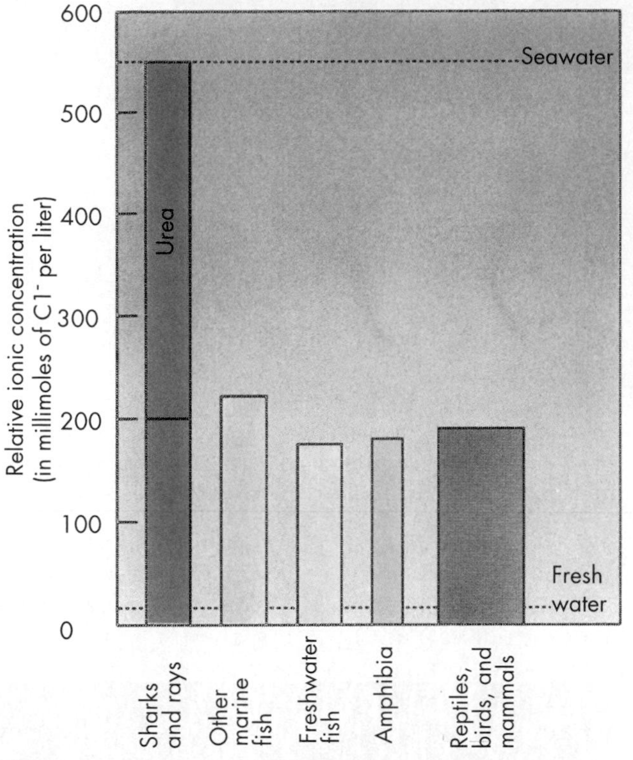

FIGURE 30-7 Ion concentrations for different classes of vertebrates. Ion concentrations in the bodies of different classes of vertebrates are roughly similar, with the exceptions of sharks and rays. Sharks and rays keep their ion concentrations close to that of seawater by adding urea to the bloodstream.

the female or holds onto her with his jaws. He then ejaculates into a groove in the clasper, which directs the sperm into the female's cloaca. The sperm then swim up the female's reproductive tract.

In addition to having a mechanism of internal fertilization, the cartilaginous fishes are ovoviviparous. **Ovoviviparous** organisms retain fertilized eggs within their oviducts until the young hatch (*ovum* meaning "egg," *vivus* meaning "alive," *pario* meaning "to bring forth"). However, although the young are born alive, they do not receive nutrition from the mother but from the yolk of the egg. The mother acts like an internal "nest" for the eggs until the young hatch. Certain fishes, some reptiles, and many insects are ovoviviparous. In contrast, **oviparous** organisms lay their eggs, and the young hatch from the eggs outside of the mother. Many skates and rays, for example, release eggs within protective egg cases. Birds and most reptiles, along with some of the cartilaginous fishes, are oviparous. **Viviparous** organisms such as mammals bear live young as do ovoviviparous organisms, but the developing embryos primarily derive nourishment from the mother and not the egg. A few sharks, such as hammerhead sharks, blue sharks, and lemon sharks, are viviparous fishes (Figure 30-8).

> *Most cartilaginous fishes fertilize their eggs internally and are ovoviviparous, retaining fertilized eggs within their oviducts until the young hatch.*

Bony Fishes

The vast majority of known species of fishes belong to the class Osteichthyes (*oste* meaning "bone"). **Bony fishes** live in both salt water and fresh water, with many species spending a portion of their lives in fresh water and another portion in the sea.

The Osteichthyes get their name from their bony internal skeletons. However, the skin of Osteichthyes is also covered with thin, overlapping bony scales (Figure 30-9). Sometimes these scales have spiny edges, which provide some protection for the animal. Bony fishes also have a protective flap that extends posteriorly from the head and protects the gills. This flap is called the **operculum.** Along with protecting the gills, the movement of the opercu-

lateral line system a complex system of mechanoreceptors possessed by all fishes and amphibians that detects mechanical stimuli such as sound, pressure, and movement.

osmoregulation (**oz** mo reg you **lay** shun *or* **os** mo reg you **lay** shun) the control of water movement within an organism.

cloaca (kloe **ay** kuh) the terminal part of the gut of most vertebrates into which ducts from the kidney and reproductive systems open.

ovoviviparous (**oh** vo vye **vip** uh rus) a term that describes an organism that retains fertilized eggs within the oviducts until the young hatch.

oviparous (oh **vip** uh rus) a term that describes an organism that lays eggs in which the embryo develops after egg laying.

viviparous (vye **vip** uh rus) a term that describes an organism that gives birth to live offspring.

bony fishes a class of vertebrates whose members have skeletons made of bone.

operculum (oh **pur** kyuh lum) a flap in a bony fish that extends posteriorly from the head over the gills, protects the gills, and enhances water flow over the gills.

FIGURE 30-8 A viviparous fish giving birth. Viviparous fishes such as this lemon shark carry live young within their bodies. The young complete their development inside their mother's body and are then released as small but competent adults.

A

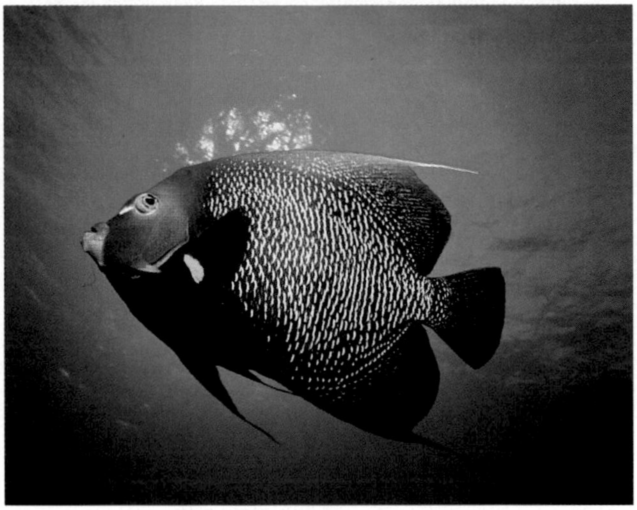

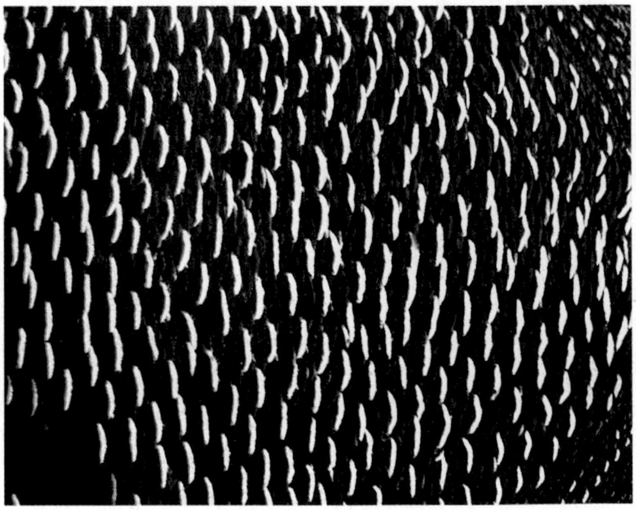

B

FIGURE 30-9 Bony scales. A, French anglefish. **B,** Scales of the French anglefish.

lum enhances the flow of water over the gills, bringing more oxygen in contact with the gas-exchanging surfaces and allowing a fish to breathe while stationary. Unlike most of the cartilaginous fishes, bony fishes can remain motionless at various depths because of their ability to more finely regulate their buoyancy with their swim bladders.

The swim bladders of most bony fishes, the *ray-finned fishes* (the largest subclass of bony fishes), evolved from lunglike sacs of their ancestors. These sacs aided in respiration. In the ray-finned fishes, gas exchange between the swim bladder and the blood regulates the density of a fish and allows it to remain suspended in water without sinking to the bottom. The ray-finned fishes are the most familiar fishes and include perch, cod, trout, tuna, herring, and salmon. Another subclass of bony fishes continues to use lungs to aid the gills in breathing: the *lobe-finned fishes.* Only four genera of lobe-finned fishes exist today; all are found in the southern hemisphere. Ancestors of this subclass of fishes gave rise to land-dwelling tetrapods.

Bony fishes live in both saltwater and freshwater and get their name from their bony internal skeletons. By means of gas exchange between their swim bladders and their blood, bony fishes can regulate their buoyancy better than the cartilaginous fishes do. Cartilaginous fishes regulate buoyancy by controlling the amount of oil in their livers.

All chordates have closed circulatory systems with a system of blood vessels and a pump to push the blood through these vessels. The cartilaginous fishes and the bony fishes have tubelike hearts with four chambers, one right after the other (see Figure 30-13, *A*), but these four chambers really function as a two-chambered heart. The first two chambers (*sinus venosus* and *atrium*) collect blood from the organs. The second two (*ventricle* and *conus arteriosus*) pump blood to the gills. From the gills the blood moves to the rest of the body. Because of the great resistance in the narrow passageways of the capillaries at the gills, the movement of blood to the rest of the body is sluggish.

Although the cartilaginous fishes maintain a solute concentration of their body fluids near that of seawater, bony fishes do not. Only some bony fishes live in the sea; many groups live in fresh water. Marine and freshwater bony fishes have opposite situations regarding osmoregulation. The body fluids of marine fishes are *hypoosmotic* with respect to seawater. That is, their body fluids contain a lower concentration of dissolved substances than seawater does (see Figure 30-7). Consequently, water tends to leave these fishes (at the gill epithelium) by osmosis. To regulate water balance, marine fishes drink seawater and then excrete the salt by means of active transport at the gill epithelium. The fish kidney is not able to get rid of excess salt in the urine because it has no loop of Henle (see Chapter 12). This part of the kidney nephron works to concentrate urine and enable an organism to produce urine with a high solute concentration. Only birds and mammals have loops of Henle.

FIGURE 30-10 Amphibians (class Amphibia). An example of an amphibian that has a tail is the salamander **(A).** Examples of amphibians without tails are frogs **(B)** and toads **(C).**

In contrast to marine fishes, the body fluids of freshwater fishes are *hyperosmotic* with respect to their freshwater environment. That is, their body fluids contain a higher concentration of dissolved substances than fresh water does. Consequently, water tends to enter freshwater fishes by osmosis. To regulate water balance, freshwater fishes do not drink water and excrete large amounts of very dilute urine. They reclaim some of the ions they lose in their urine by the uptake of sodium and chlorine ions by the gills and in the food they eat.

Because they tend to lose water by osmosis, marine fishes drink seawater and excrete salt at the gill epithelium. Because they tend to take on water, freshwater fishes do not drink water and excrete large amounts of very dilute urine.

Most Osteichthyes are oviparous, fertilizing their eggs externally. These eggs are often food for other marine organisms, however, and must survive other risks such as drying out. Most fishes (as well as other organisms that externally fertilize their eggs) lay large numbers of eggs, with some surviving these dangers. Many species of fishes build nests for their eggs and watch over them, which also enhances the chances of survival.

Amphibians

The word *amphibian* means "two lives" (*amphi* meaning "both," *bios* meaning "life") and refers to both the aquatic and terrestrial existence of this class of animals (Figure 30-10). **Amphibians** (unlike reptiles, birds, and mammals) depend on water during their early stages of development. Many amphibians live in moist places like swamps and in tropical areas even when they are mature, which lessens the constant loss of water through their thin skins. The two most familiar orders of amphibians are those that have tails—the salamanders, mudpuppies, and newts—and those that do not have tails—the frogs and toads.

Most frogs and toads fertilize their eggs externally. The male grasps the female and sheds sperm over the eggs as they are expelled from the female. Most salamanders use internal fertilization but are still oviparous (lay their eggs). Because amphibian eggs have no shells or membranes to keep them from drying out, amphibians lay their eggs directly in water or in moist places. Figure 30-11, *A,* shows the "nests" of foam that tropical tree frogs create to incubate their eggs. When the tadpoles develop, they drop from the tree branches into the water (Figure 30-11, *B*). Some amphibians protect their eggs by incubating them in their mouths, on their backs, or even in their stomachs! A few amphibian species are ovoviviparous and incubate the eggs within their reproductive organs until they hatch, and a few are viviparous.

The young of frogs and toads undergo *metamorphosis,* or change, during development from a larval to an adult form. The larvae are immature forms that do not look like the adult. The larvae of frogs and toads are tadpoles, which usually live in the water and have internal gills and a lateral line

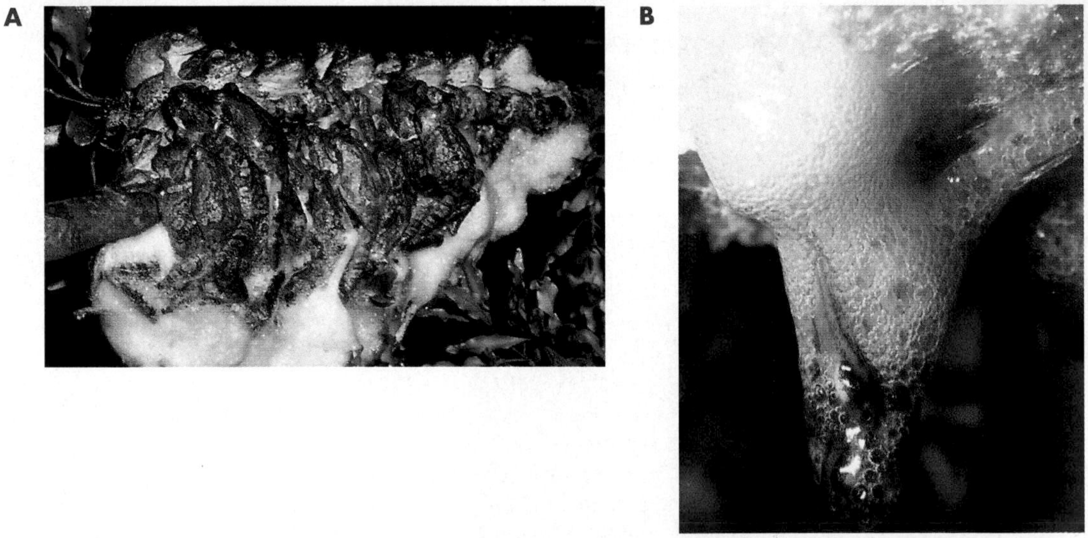

FIGURE 30-11 An amphibian egg-protection strategy. To keep their eggs from drying out, tropical tree frogs lay their eggs in nests of foam **(A)**. When the tadpoles hatch, they drop from their foam nest into a pond of water below **(B)**.

A B C D

FIGURE 30-12 Reptiles (class Reptilia). A, American alligator. Alligators, like birds, are related to the dinosaurs. **B,** Box turtle. **C,** An Australian bearded lizard. Although this species of lizard has legs, some species lack legs. Snakes evolved from one line of legless lizard. **D,** Coral snake.

system like that of fishes. They feed on minute algae. These fishlike forms develop into carnivorous adults having legs and lungs; their gills and lateral line system disappear. The lungs of the adults are inefficient, however. Much of the gas exchange takes place across the skin and on the surfaces of the mouth. The skin of amphibians must therefore remain damp to allow gases to diffuse in and out.

The adults of certain salamanders (such as mudpuppies) live permanently in the water and retain gills and other larval features as adults. Other salamanders are terrestrial but return to water to breed. They, like frogs and toads, usually live in moist places such as under stones or logs or among the leaves of certain tropical plants.

Amphibians live both aquatic and terrestrial existences. Because they fertilize their eggs externally and their eggs have no shells or membranes, amphibians lay their eggs in water or in moist places. The young of frogs and toads change from a larval form to an adult form during their development.

amphibians (am **fib** ee uns) a class of vertebrates capable of living on land and in the water. Amphibians depend on water during their early stages of development.

Along with having lungs rather than gills (in most cases), amphibians have a pattern of blood circulation different from that of the fishes. After the blood is pumped through the fine network of capillaries in the amphibian lungs, it does not flow directly to the body as it does in fishes. Instead, it returns to the heart. It is then pumped out to the body at a much higher pressure than if it were not returned to the heart. However, the oxygenated blood that returns to the heart from the lungs mixes with the deoxygenated blood returning to the heart from the rest of the body. Consequently, the heart pumps out a mixture of oxygenated and deoxygenated blood rather than fully oxygenated blood. Figure 30-13, *B,* shows this pathway of blood flow. The blood from the lungs and the blood from the body enter the right and left atria of the heart, respectively. The blood in both these chambers flows into the single ventricle of the three-chambered amphibian heart and is pumped through two large vessels to both the lungs and the body.

Amphibians have a three-chambered heart rather than the two-chambered heart of fishes. However, oxygenated and deoxygenated blood mix in the heart.

Reptiles

The three major orders of **reptiles** are the crocodiles and alligators, the turtles and tortoises, and the lizards and snakes (Figure 30-12). Reptiles have dry skins covered with scales that help retard water loss. As a result, reptiles can live in a wider variety of environments on land than amphibians can, but the crocodiles, alligators, and turtles are aquatic organisms.

The hearts of reptiles differ from amphibian hearts in that a partition called a *septum* subdivides the ventricle, the pumping chamber of the heart. The septum reduces the mixing of oxygenated and deoxygenated blood in the heart. In most crocodiles, the separation is complete. Figure 30-13, *C,* shows that the reptilian heart closely resembles the four-chambered heart of birds and mammals shown in Figure 30-13, *D.* It is still considered to be a three-chambered heart, however.

One of the most critical adaptations of reptiles to life on land is the evolution of the shelled **amniotic egg.** Amniotic eggs are also characteristic of birds and egg-laying mammals (monotremes). The amniotic egg protects the embryo from drying out, nourishes it, and enables it to develop outside of water (Figure 30-14). This type of egg contains a yolk and albumin (egg white). The yolk is the primary food supply for the embryo, and the albumin provides additional nutrients and water. The embryo's nitrogenous wastes are excreted into the allantois, a sac that grows out of the embryonic gut. Blood vessels grow out of the embryo through the sac surrounding the yolk and through the allantois to the egg's surface, where gas exchange takes place. The amnion surrounds the developing embryo, enclosing a liquid-filled space within which the embryo develops and is protected. Lying just within the shell, a membrane called the *chorion* surrounds the embryo, amnion, yolk sac, and allantois and, along with the shell, controls the movement of gases into and out of the egg. In most reptiles, the eggshell is leathery, unlike the hard shell of bird eggs. Because of this difference, reptile eggs are somewhat permeable to water, whereas bird eggs are not.

Reptiles, like amphibians and fishes (but unlike birds and mammals), are **ectothermic** (*ectos* meaning "outside," *thermos* meaning "heat"). Ectothermic animals regulate body temperature by taking in heat from the environment. Even though ectothermic animals are often called "cold blooded" (a misleading term), they often maintain body temperatures much warmer than their surroundings. Did you ever wonder why fishes do not freeze in wa-

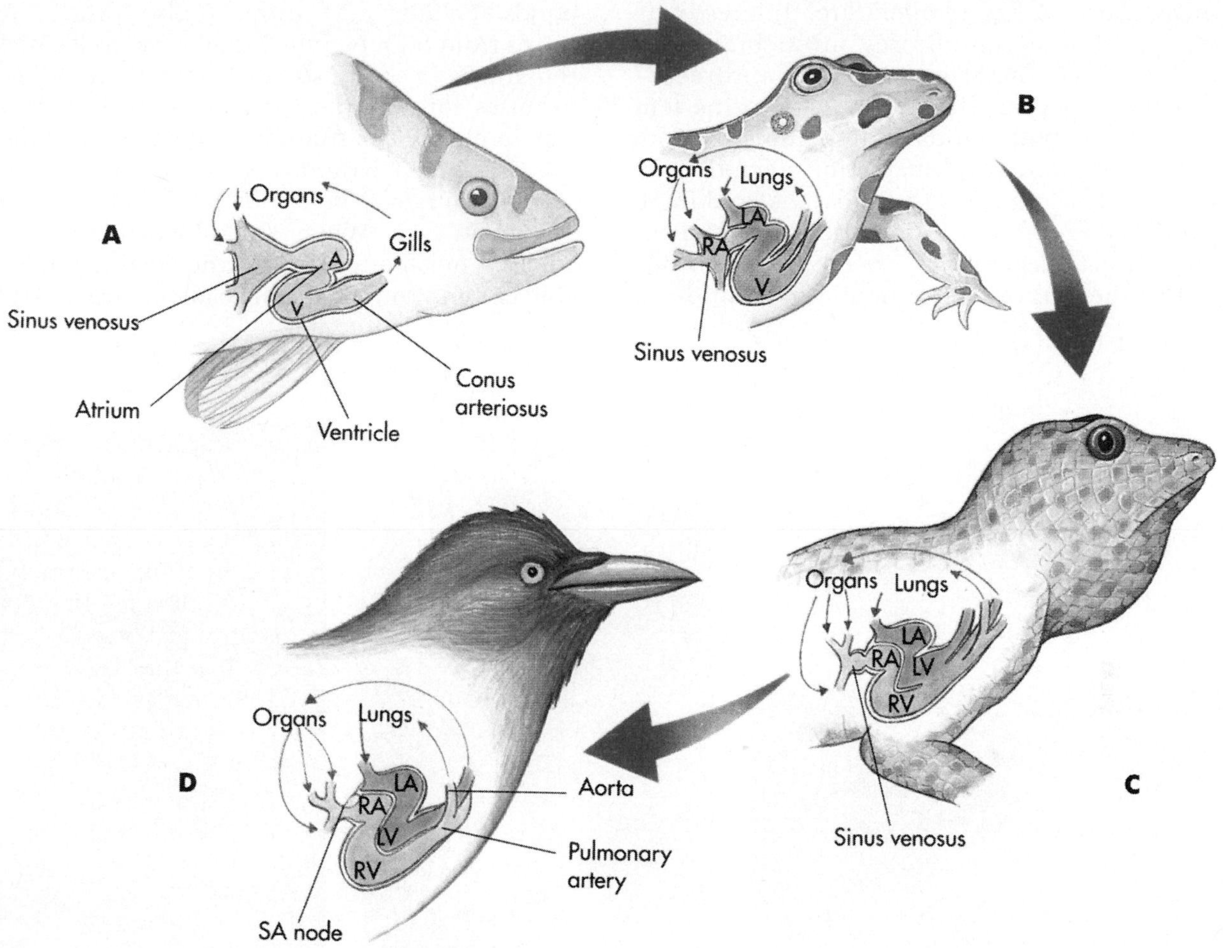

FIGURE 30-13 Evolution of the vertebrate heart. **A,** Fishes (two chambers). **B,** Amphibians (three chambers). **C,** Reptiles (three chambers—note that the ventricles open to one another). **D,** Birds and mammals (four chambers).

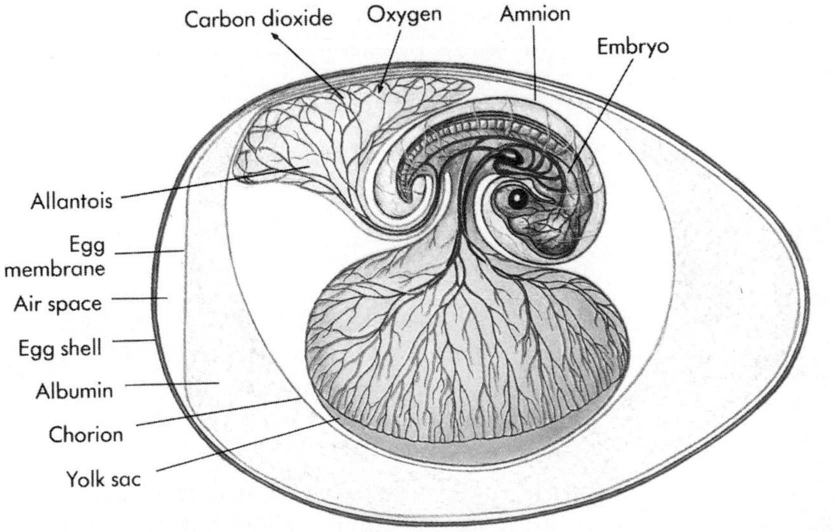

FIGURE 30-14 The amniotic egg. The amniotic egg is an important adaptation that allows reptiles (and birds and monotremes) to live in a wide variety of terrestrial habitats.

ter near the freezing temperature? Interestingly, fishes (as well as certain lizards, invertebrates, and plants) produce their own internal antifreeze—chemical compounds that lower the freezing temperature of the body fluids of an organism. Along with such physiological adaptations, ectothermic animals protect themselves against the cold in behavioral ways. For example, frogs help protect themselves against freezing by spending the winter buried in the soil or in the mud at the bottom of ponds. Ectothermic animals also protect themselves from high heat by burrowing under rocks or remaining in shady, somewhat cooler areas. Desert tortoises, for example, construct shallow burrows to stay in during the summer and deeper burrows for hibernation in winter. Reptiles often bask in the sun, which raises their body temperature and their metabolic rate. When cold-blooded animals are cold, the metabolic rate slows down and they are unable to hunt for food or move about very quickly.

FIGURE 30-15 Birds (class Aves). The birds are a large and successful group of about 9000 species, more than any other class of vertebrates except the bony fishes. **A,** Australian pelicans. **B,** Gold and blue macaw. **C,** Egret. **D,** A pair of wood ducks, showing the marked difference between the coloration of males and females. **E,** Tufted puffin, one of many unique groups of sea birds.

Reptiles have dry skins covered with scales that help retard water loss. Another critical adaptation to their life on land (although some reptiles are aquatic) is the development of the amniotic egg, which protects the embryo from drying out, nourishes it, and enables it to develop outside of water. Amniotic eggs are also characteristic of birds and egg-laying mammals.

Birds

There are approximately 9000 species of birds living today (Figure 30-15). In birds, the wings are homologous to the forearms of other vertebrates. That is, they are derived from the same evolutionary origin but have been modified in the course of evolution. Birds have reptilianlike scales on their legs and lay amniotic eggs as reptiles do. They have hard, horny extensions of the mouth called *beaks* that tear, chisel, or crush their food. They also have digestive organs called *gizzards,* often filled with grit, that grind food. Beaks are not limited to birds. Many reptiles have beaks, including turtles, and so do some fishes, including the parrot fish, which uses its beak to rip away fragments of coral reef. Birds, however, are the only animals that have feathers.

Feathers are flexible, light, waterproof epidermal structures. Several types of feathers form the body covering of birds, including contour feathers and down feathers. Contour feathers are flat (except for a fluffy, downy portion at the base) and are held together by tiny barbules (Figure 30-16). These feathers provide a streamlined surface for flight, and some are modified to reduce drag on the wings or act like individual propeller blades. In addition, birds can alter the area and shape of their wings by altering the positions of their feathers. Feathers also provide birds with waterproof coats and play an important role in insulating birds against temperature changes.

Along with feathers, birds have light, hollow bones adapted to flight. Birds also have highly efficient lungs that supply the large amounts of oxygen necessary to sustain muscle contraction during prolonged flight. Unlike fishes, amphibians, and reptiles (except for the crocodiles), birds and mammals have hearts that act as a double pump. The sides of the heart are completely separated with a septum (see Figure 30-13). The right side of the heart pumps blood to the lungs. The left side of the heart pumps blood to the body. Oxygenated blood and deoxygenated blood do not mix. As you compare the pathway of blood in the hearts of fishes, amphibians, reptiles, and birds in Figure 30-13, note that the four-chambered heart of birds and

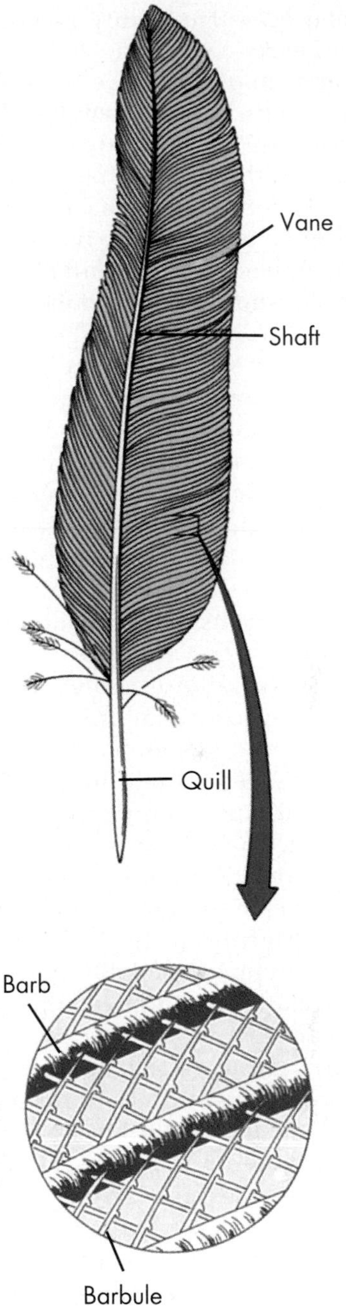

FIGURE 30-16 A feather. The enlarged view shows the way in which the vanes are linked together by the microscopic barbules.

reptiles a class of vertebrates characterized by dry skins covered with scales that help retard water loss.

amniotic egg (**am** nee **ot** ik) an egg, characteristically produced by birds, reptiles, and monotremes, that protects the embryo from drying out, nourishes it, and enables it to develop outside of water.

ectothermic (**ek** toe **thur** mik) a term referring to animals like reptiles, amphibians, and fishes that regulate body temperature by taking in heat from the environment and by their behavior. Their internal body temperature fluctuates.

mammals evolved from only two of the chambers of the fish heart.

Birds and mammals are **endothermic** (*endo* meaning "within"); they regulate body temperature internally. The evolution of the four-chambered heart with separate pathways to the lungs and the body is thought to have been important in the evolution of endothermy in the birds and mammals. More efficient circulation is necessary to support the great increase in metabolic rate that is required to generate body heat internally. In addition, blood is the carrier of heat in the body, and an efficient circulatory system is required to distribute heat evenly throughout the body. Although endothermic animals are sometimes called "warm blooded," body temperature may often be cooler than that of the surroundings. Because endotherms maintain high internal body temperatures (37° C [98.6° F] in humans, for example), their internal temperatures are usually higher than that of the environment.

Endotherms maintain a constant high body temperature by adjusting heat production to equal heat loss from their bodies under various environmental conditions. The high metabolic rate of endotherms and the energy released during these chemical reactions produce much of this body heat. Increasing the action of skeletal muscles increases the metabolic rate and the amount of heat produced. Shivering in the cold, for example, is an action that produces body heat. But because endotherms usually live in environments cooler than body temperature, restricting heat loss is usually the concern. Animals that live in extremely cold temperatures, such as arctic birds and polar bears, are well-insulated with either feathers or hair that traps air. Raising or lowering the feathers or hair adjusts the insulating capacity. Getting "goosebumps" when you are cold is your body's reaction to raise your hairs and increase your insulation. Although humans no longer have substantial body hair, this mechanism is important in reducing heat loss in other endothermic vertebrates.

Birds are winged vertebrates that are covered with feathers and have scales on their legs. They have beaks, which are horny extensions of the mouth. Their light, hollow bones are an adaptation to flight. They have a four-chambered heart as mammals do. Unlike the reptiles, amphibians, and fishes, birds can regulate body temperature internally.

Mammals

There are about 4500 species of living **mammals,** including humans. Mammals are endothermic vertebrates that have hair and whose females secrete milk from mammary glands to feed their young. Mammals, like birds and most crocodiles, have a four-chambered heart with circulation to the lungs and separate circulation to the body. The locomotion of mammals is advanced over that of the reptiles, which in turn is advanced over that of the amphibians. The legs of mammals are positioned much farther under the body than those of reptiles and are suspended from limb girdles, which permit greater leg mobility.

The evolution of specialized teeth in mammals represents a major evolutionary advance. In fishes, amphibians, and reptiles, all teeth are essentially the same size and shape. In mammals, evolutionary specialization has resulted in incisors, which are chisellike teeth used for cutting; canines, which are used for gripping and tearing; and molars, which are used for crushing and breaking (see Figure 8-2, *A*).

There are three subclasses of mammals: monotremes, marsupials, and placental mammals. The only **monotremes** that exist are the duckbilled platypus and two genera of spiny anteaters (Figure 30-17). Monotremes lay eggs with leathery shells similar to those of reptiles. The platypus generally lays one egg and incubates it in a nest. The spiny anteater generally lays two eggs and incubates them in a pouch. When the young hatch, they feed on milk produced by specialized sweat glands of the mother. **Marsupials** are mammals in which the young are born early in their development and are retained in a pouch. After birth, the embryos crawl to the pouch and nurse there until they mature. The kangaroo and koala are familiar examples of marsupials (Figure 30-18). In **placental mammals,** the young develop to maturity within the mother. They are named for the first organ to form during the course of their embryonic development, the placenta. (Human development as an example of the development of a placental mammal is described in detail in Chapter 22.)

Placental mammals are extraordinarily diverse (Figure 30-19). There are 14 orders of mammals (see Figure 24-17), of which one is the primates, the order that includes monkeys, apes, and humans. The bats in the opening photograph (as well as all bats) belong to the order Chiroptera, a name that means the forelimbs are modified to form wings (*chiro* meaning "hand," *pteron* meaning "wing"). So although humans and bats are quite different, they have remarkable similarities in their

A

B

FIGURE 30-17 Monotremes (class Mammalia). A, Duckbilled platypus at the edge of a stream in Australia. **B,** Echidna.

circulatory, thermoregulatory, osmoregulatory, and reproductive patterns.

> *Mammals are endothermic vertebrates that have hair and whose females secrete milk from mammary glands to feed their young. There are three subclasses of mammals: monotremes, marsupials, and placentals.*

A

B

FIGURE 30-18 Marsupials (class Mammalia). A, Kangaroo with young in its pouch. **B,** Koala in a eucalyptus tree.

endothermic (**en** doe **thur** mik) a term referring to organisms such as birds and mammals that maintain a stable internal body temperature by means of internal regulatory mechanisms.

mammals endothermic vertebrates that have hair and whose females secrete milk from their mammary glands to feed their young.

monotremes (**mon** oh treems) a subclass of mammals that lays eggs with leathery shells similar to those of reptiles.

marsupials (mar **soo** pee uhls) a subclass of mammals that gives birth to immature young that are carried in a pouch.

placental mammals (pluh **sent** uhl) a subclass of mammals that nourishes their developing young within the body of the mother.

FIGURE 30-19 Placental mammals (class mammalia). A, Addax from northern Africa. **B,** Bottle-nosed dolphin. **C,** Grey timber wolf. **D,** Greater horseshoe bat in flight. **E,** Orca, a carnivorous whale.

JUST WONDERING. . . .

Why do scientists use animals in research? Do they have any guidelines they must follow regarding how they treat the animals?

Scientists use animals in research (often referred to as "whole-body" research) in studies that could not be conducted with use of tissue cultures or other research methods alone, such as studying the regulation of bodily processes or the functioning of systems. Computer modeling strategies cannot be used unless appropriate animal data are first collected. (Scientists cannot ask a computer to answer a question for which it has no data.)

Scientists' choices of which animals to use are also important. If a particular line of animal research has applications for humans, vertebrates are generally used because they best model human systems. Rodents (such as rats and mice) are used in 90% of all such animal research.

Two governmental agencies provide regulations that scientists must follow in animal research: the U.S. Department of Agriculture (USDA) and the Department of Health and Human Services (DHHS). If scientists do not follow these regulations, they become ineligible for any type of federal funding for their research.

USDA representatives (veterinarians and occasionally animal care technicians) make unannounced semiannual visits to facilities using warm-blooded animals (other than rodents) in biomedical research, teaching, or exhibits. In addition, they require an annual review of all animal activities. Protocols require that researchers document their training in animal welfare and sign written statements that any proposed research is not unnecessarily duplicative. If the proposed activity has the possibility of causing "unrelieved pain or distress," the researcher must provide a literature search to support the argument that there is no alternative approach that might have less potential for causing distress. (Only 6% of animal experimentation involves pain. The goal of such experiments is often to develop safer and more effective pain relievers or anesthetics.) The USDA also requires that any research, teaching, or exhibits involving warm-blooded animals be reviewed and approved by an Institutional Animal Care and Use Committee (IACUC). This committee must be composed of a veterinarian, a chairperson, a nonscientist, and a member with no affiliation to the institution. The last two criteria can be fulfilled by a single person.

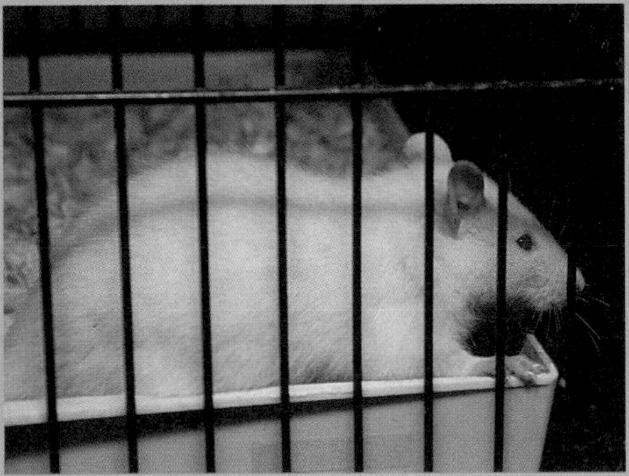

In addition to the USDA regulations, the Public Health Service (PHS), a branch of the DHHS that funds most biomedical research, requires that all research funded by them that involves animals (*including* rats and mice) must also have a IACUC (with the same criteria as the USDA committee), but it must have five committee members. The IACUC must ensure that animals are being cared for appropriately, must meet and review research protocols for appropriateness of animal use, and must visit the animal holding facilities in their institutions (announced) semiannually.

The American Association for the Accreditation of Laboratory Animal Care (AAALAC) also conducts extremely rigorous reviews of animal facilities (including those using rodents). If the facilities and their animal care meet their strict standards, they receive the AAALAC seal of approval. Approval from this agency, which is highly regarded by scientists and industry, is frequently sought.

Animal research has been essential for the development of many therapies for treating heart problems, cancer, and other fatal and serious diseases, as well as for the development of anesthetics used in surgery. When a loved one is saved from cancer, when more information is gathered about dreaded diseases, or an operation is performed painlessly, chances are animal research played some role in this ability to combat death and reduce suffering in humans.

Summary

- The chordates are characterized by three main features: (1) a single, hollow nerve cord located along the back; (2) a rod-shaped notochord, which forms between the nerve cord and the gut (stomach and intestines) during development; and (3) pharyngeal (gill) arches, which are located at the throat (pharynx).
- There are three subphyla of chordates: the tunicates (subphylum Urochordata), the lancelets (subphylum Cephalochordata), and the vertebrates (subphylum Vertebrata). The tunicates are sessile, saclike marine organisms that filter food from the surrounding water. Although the adults have gill slits, only their larvae have notochords and nerve cords. The lancelets are tiny, scaleless, fishlike marine chordates that are just a few centimeters long.
- Vertebrates are a subphylum of chordates characterized by a vertebral column surrounding a dorsal nerve cord. Vertebrates have distinct heads with skulls that encase their brains, closed circulatory systems, and a heart to pump the blood. Most vertebrates also have livers, kidneys, and endocrine glands. In spite of these similarities, the vertebrates are an extremely diverse subphylum of organisms. Three classes of vertebrates are fish; four are tetrapods—animals with four limbs.
- The jawless fishes—lampreys and hagfishes—are long, tubelike animals that live in the sea or brackish water. They lack paired fins and scales. Lampreys are parasites of bony fishes, and hagfishes are scavengers.
- The cartilaginous fishes—sharks, skates, and rays—are covered with toothlike scales. They have sophisticated and diverse sensory systems. They fertilize their eggs internally, and most are ovoviviparous, retaining fertilized eggs within their oviducts until the young hatch.
- The vast majority of fishes are the bony fishes. Along with having bony internal skeletons, these fishes have thin, bony, platelike scales. Both the bony fishes and the cartilaginous fishes have two-chambered hearts that pump blood to the gills. From there, the blood moves sluggishly around the body.
- The amphibians live both in water and on land. Because their eggs have no shells or membranes to keep them from drying out, amphibians lay their eggs directly in water or moist places. The young of frogs and toads undergo change from larval to adult forms during development. The adults of certain salamanders live permanently in the water and retain gills and other larval features as adults. Although amphibians have a three-chambered heart, oxygenated and deoxygenated blood mix in the heart.
- Reptiles are better adapted to life on land than the amphibians because of their dry, scaly skin that retards water loss and their shelled (amniotic) egg. The amniotic egg retains a watery environment within the egg while protecting and nourishing the developing embryo.
- Fishes, amphibians, and reptiles are ectothermic; that is, they regulate body temperature by taking in heat from the environment. Ectothermic animals protect themselves from the cold and high heat in behavioral ways.
- Birds are winged vertebrates that are covered with feathers and are adapted to flight. They lay amniotic eggs like the reptiles but have a four-chambered heart like the mammals. Birds, like mammals and unlike reptiles, amphibians, and fishes, are endothermic; that is, they regulate body temperature internally.
- Mammals are endothermic vertebrates that have hair and whose females secrete milk from mammary glands to feed their young.

Knowledge and Comprehension Questions

1. Fill in the blanks: The three major features that are characteristic of all vertebrates are: a(n) ______, a(n) ______, and ______.
2. Into what structure(s) does each of the features in question 1 develop (or serve as templates)?
3. Match each term to the most appropriate description:

   Descriptions

   a. Scaleless, fishlike marine chordates that retain a notochord throughout life
   b. Tubular, scaleless organisms without paired fins; most members of this class have been extinct for millions of years
   c. Saclike, sessile, marine filter feeders

   Terms

   (1) Jawless fishes
   (2) Tunicates
   (3) Lancelets
4. List four distinguishing characteristics of vertebrates.
5. What is a lateral line system? Which animals have one, and why is it important?
6. What is osmoregulation, and what is its significance?
7. Distinguish among ovoviviparous, oviparous, and viviparous.
8. What does the term *amphibian* mean?
9. Distinguish between ectothermic and endothermic. Give an example of an ectotherm and an endotherm.
10. Describe two important adaptations of reptiles to life on land.
11. What are feathers, and what functions do they perform for birds?
12. What characteristics differentiate mammals from other vertebrates?

Critical Thinking

1. How are the characteristics of the three subphyla of chordates related to the way of life of each group?
2. What limits the ability of amphibians to occupy the full range of terrestrial habitats and allows other terrestrial vertebrates to live in them successfully?

CHAPTER 31

BIOTECHNOLOGY AND GENETIC ENGINEERING

THE BUSINESS OF MANUFACTURING has taken on a new look that is far different from the stereotyped notion of the assembly line in a noisy factory. This laboratory technician is putting bacteria to work to create products for human use. The bacteria are growing in the small, table-top fermentors lining the laboratory bench. Working in an environmentally controlled room, the technician is engaged in a scale-up experiment—figuring out how to make a procedure work on a large scale that has been perfected on a small scale. These fermentors are only an intermediate step. Eventually, the bacteria used in this process will be grown in huge vats like the one shown in Figure 31-7.

To turn bacteria into living "factories," scientists isolate individual genes (portions of the hereditary material) and transfer them from one kind of organism (a human, for example) to another such as a bacterium. The bacteria are then propagated in vast quantities, producing the substance they are now genetically programmed to manufacture.

This newly found ability to isolate individual genes and transfer them from one kind of organism to another has revolutionized scientists' ability not only to create new products for therapeutic use but also to improve the characteristics of plants and animals. Gene manipulation offers enormous potential for use in agriculture and medicine of the future, and it has already produced many important applications. Genetic engineering is also an important tool that helps scientists learn about gene structure, function, and regulation. Thus these powerful techniques are used not only to produce new products but also to gain a better understanding of the molecular basis of life.

A Short History of Biotechnology

Biotechnology is the use of scientific and engineering principles to manipulate organisms, producing one or more of the following:

1. Organisms with specific biochemical, morphological, and/or growth characteristics
2. Organisms that produce useful products
3. Information about an organism or tissue that would otherwise not be known

Although *biotechnology* is a relatively new word, classical biotechnology is not new. Molecular biotechnology is the "new" biotechnology.

KEY CONCEPTS

- Selection, mutation, and hybridization, the basis of classical biotechnology, have been used for thousands of years to produce organisms with desired characteristics and to produce foods and other products for human use.
- Some bacteria can transfer genetic material, naturally recombining genes.
- In recent decades, scientists have learned how to intervene in and direct these natural mechanisms of gene transfer among organisms.
- Scientists use genetic engineering to improve the characteristics of plants and animals, to produce medically important products, and to treat genetic disorders in humans.

OUTLINE

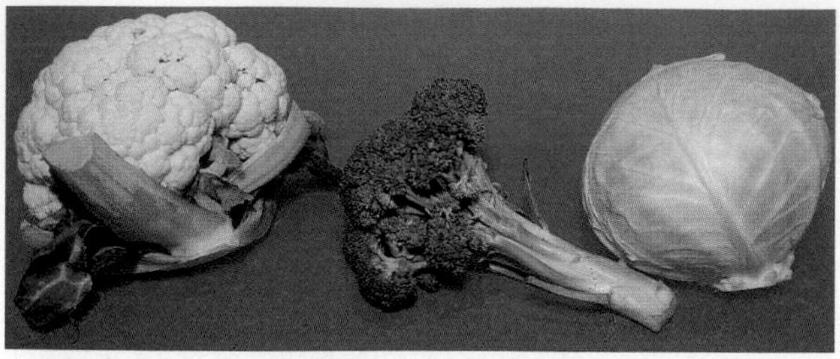

FIGURE 31-1 Artificial selection. The cauliflower, broccoli, and cabbage shown here have all been bred from one species of wild mustard. By selecting certain characteristics in each plant they wished to cultivate, breeders were able to develop plants that are quite different from one another.

Classical Biotechnology

As far back as 10,000 years ago, humans selected plants and animals with specific characteristics to propagate from the wild. In a simple way, this process is considered biotechnological: humans were producing organisms with specific characteristics by selection. For more than 8000 years, bacteria and yeasts have been used to produce products such as beer, vinegar, yogurt, and cheese, although the processes involved were not understood at the time. Loaves of yeast breads have even been found in Egyptian pyramids built 6000 years ago. And more than 3000 years ago, the Chinese and the Central American Indians used products produced by molds and fungi to treat infections. Today we know that certain molds and bacteria are natural sources of bacteria-fighting antibiotics.

Even before the late 1800s, when Gregor Mendel and others gathered evidence regarding the nature of the variability among organisms, plant and animal breeders selectively bred organisms to develop hybrids having certain desired characteristics (Figure 31-1). (*Hybrids* are the offspring produced by crossing two genetically dissimilar varieties of a species.) At the turn of the century, scientists began developing hybrids using the scientific principles elucidated by Mendel: the selection and recombination of *hereditary factors* now known as genes. (A gene is a sequence of nucleotides that codes for the amino acid sequence of a particular polypeptide—a portion of an enzyme or other protein.) In breeding plants, scientists also induced mutations to obtain plants with new genetic combinations. These techniques of selection, mutation, and hybridization are the basis of **classical biotechnology.**

Using carefully planned and controlled breeding programs, scientists found that they could produce hybrids that were stronger or improved in certain ways from their parents—a feature termed *hybrid vigor.* During the 1930s, a worldwide effort was initiated to increase food production in developing countries. By developing hybrid varieties of food crops using the techniques of selection and recombination, agricultural researchers began a *green revolution,* dramatically increasing the yields of food crops. For example, hybrid corn developed in the United States at that time helped double crop yields. New dwarf varieties of wheat introduced to farmers in Mexico in 1960 resulted in a 300% increase in wheat yields there.

At about the same time that Mendel was conducting genetic research with pea plants in the mid-1800s, strides were also being made in classical biotechnology in the field of microbiology. For example, processes of fermentation that had been used without being fully understood for thousands of years were slowly explained during the 1800s. In 1837–1838, researchers concluded that yeasts are alive. Almost 30 years later, Louis Pasteur confirmed that certain bacteria and yeasts formed molecules such as acetic acid, lactic acid, butyric acid, alcohol, and carbon dioxide. He determined why and how wine often turned to vinegar, and he developed a process to kill the microorganisms causing this fermentative change. This process, termed *pasteurization,* is still used to help preserve wine and other foods such as milk.

Knowledge of fermentative processes was also put to work during World War I. Because of blockades during the war, the Germans were unable to obtain the vegetable oils from which they extracted the glycerol necessary for manufacturing explosives. As a result, the Germans devised a biological way to produce glycerol using the fermentative abilities of yeast. The British, likewise, developed biological methods to produce acetone and butanol using the bacterium *Clostridium acetobutylicum.* (Acetone was used in manufacturing ammunition and butanol in producing artificial rubber.)

As strides were being made in the use of microorganisms to synthesize products for humans and in the understanding of those processes, the use of classical biotechnology to fight disease progressed rapidly. In the late 1800s, Robert Koch developed the germ theory of disease, which explains that microbes caused infection. He discovered both the bacterium that causes anthrax, a disease of cattle that sometimes affects humans, and the bacterium that causes tuberculosis. During this time, Koch also developed pure culture methods. Using these methods, scientists could isolate and work with a specific organism, free from contamination by other organisms.

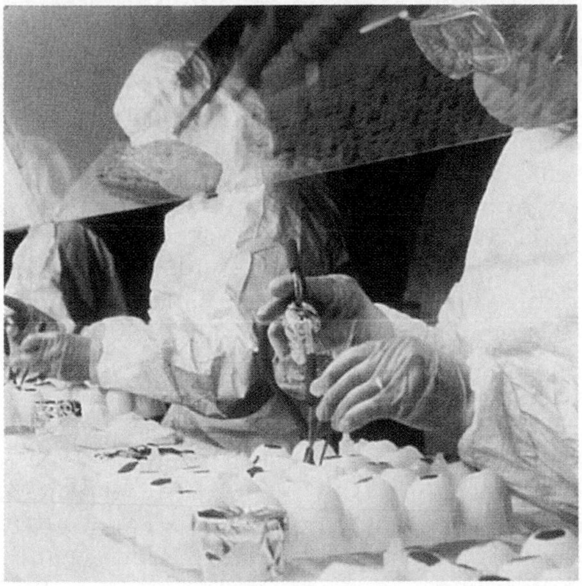

FIGURE 31-2 Growing viruses in eggs. The shells have been broken off the tops of the raw eggs, and laboratory technicians are inoculating the eggs with viruses, which will be used to develop influenza vaccine.

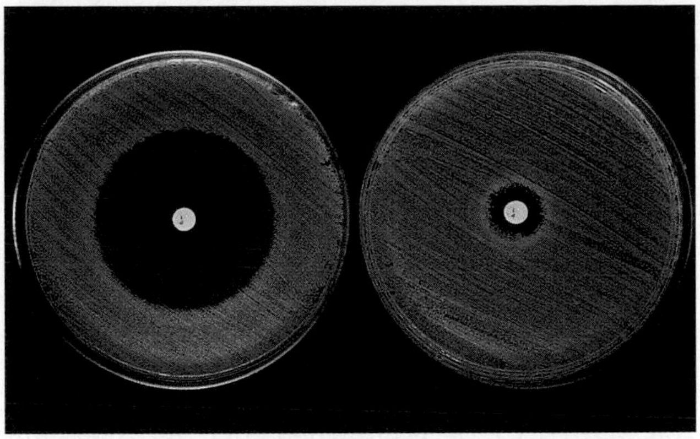

FIGURE 31-3 Penicillin resistance (*right*) and sensitivity (*left*) in culture of bacteria *Staphylococcus aureus*. Bacteria were spread on the nutrient surface of each of these two petri plates. A disk containing the antibiotic penicillin was then placed in the center of each plate. The plates were incubated, and the bacteria grew. The bacteria in the plate on the right carry a gene that confers resistance to penicillin. Therefore, these bacteria grew quite close to the disk. However, the bacteria growing in the plate to the left do not carry this resistance gene. The antibiotic inhibited the bacteria from growing in a large area surrounding the disk, which is called a zone of inhibition.

Microorganisms were also used to produce various disease-fighting products such as vaccines. Although Edward Jenner did not understand the role of microorganisms in disease, he is credited with developing the first successful vaccine against smallpox in 1796. Without understanding the microbiology and immunology behind his work, Jenner injected volunteers with the relatively harmless *Vaccinia* virus, which conferred resistance against the deadly smallpox virus. Almost 100 years later, having a greater understanding than Jenner had of the disease process, Louis Pasteur and others developed the first cholera, diphtheria, and tetanus vaccines. By the early 1900s, mammalian cell culture techniques had been developed. Scientists were then able to replicate viruses within these cultures (Figure 31-2), harvest and attenuate (weaken) them, and use them in vaccine preparations. Thus, microbes were themselves being used as useful products in fighting human disease.

An important step in the development of disease-fighting products occurred quite by accident in 1927, when Alexander Fleming discovered antibiotics. He noticed that a mold, which had contaminated his bacterial cultures, inhibited the growth of surrounding bacteria (Figure 31-3). That mold was *Penicillium notatum,* which naturally produces the antibacterial product we now call penicillin. By 1940, the first purified preparations of penicillin were available. This antibiotic played a major role in preventing the death of soldiers from infection during World War II. An intensive search for new antibiotics began as pharmaceutical companies began screening bacteria and molds for their antibiotic potential.

> *For thousands of years, even before they had any understanding of genetics, humans have selected plants and animals with specific characteristics to propagate from the wild, and they have selectively bred organisms to develop individuals having certain desired characteristics. In addition, without knowledge of microbial biochemistry, humans have developed various products by using microbes.*

Molecular Biotechnology

During the 1950s, a revolution took place that set the stage for the birth of molecular biotechnology:

biotechnology (bye oh tek **nol** uh jee) the use of scientific and engineering principles to manipulate organisms.
classical biotechnology the use of scientific and engineering principles to manipulate organisms, relying on the traditional techniques of selection, mutation, and hybridization.

the discovery of the molecular structure of the hereditary material deoxyribonucleic acid (DNA). This revolution involved more than 50 years of work by a succession of scientists building on the work of Gregor Mendel. Mendel's work suggested that traits are inherited as discrete packets of information. As the years progressed, experimental data showed that this information is stored within the nucleus of the cell in the chromosomes (see "Just Wondering," Chapter 3). Scientists were then able to determine that the chromosomes are composed of DNA and protein. In 1952, Alfred Hershey and Martha Chase established that the DNA, not the protein, *is* the hereditary material. After Rosalind Franklin and Maurice Wilkins demonstrated that DNA molecules are helically shaped, James Watson and Francis Crick deduced the structure of the DNA molecule, for which they were awarded the Nobel prize. (These scientific discoveries are described in more detail in Chapter 18.)

Locating and characterizing the hereditary material were the first steps in what might be termed a *gene revolution*. Now, genes and genetic diversity could be studied at the molecular level. Scientists then began to manipulate genes in bacteria, intervening in and directing their natural methods of genetic recombination.

Natural Gene Transfer among Bacteria

Certain bacteria can naturally transfer genetic material from one cell to another. Genetic material is exchanged between bacteria by three methods: transformation, transduction, and conjugation (Figure 31-4).

In **transformation,** pieces of genetic material from a donor bacterial cell that has lysed (ruptured) is released into the surrounding medium. A certain type of bacterial cell, called a *competent cell,* is then able to "take up" this DNA and incorporate pieces into its genome, thereby becoming transformed, or changed. (The **genome** of an organism is its total complement of genetic material.) Not all bacteria are able to be transformed; the ability to take up DNA fragments is an inherited characteristic. In addition, genetic engineers can induce certain bacteria to take up DNA by cultivating them in the presence of certain chemicals.

During **transduction,** DNA from a donor bacterium is transferred to a recipient bacterium by a virus. For this transfer to occur, the virus must first combine its genetic material with that of the bacterium it has infected. Only certain lysogenic viruses (viruses that infect a cell but do not immediately replicate) or damaged viruses are capable of incorporating bacterial genes with their genome.

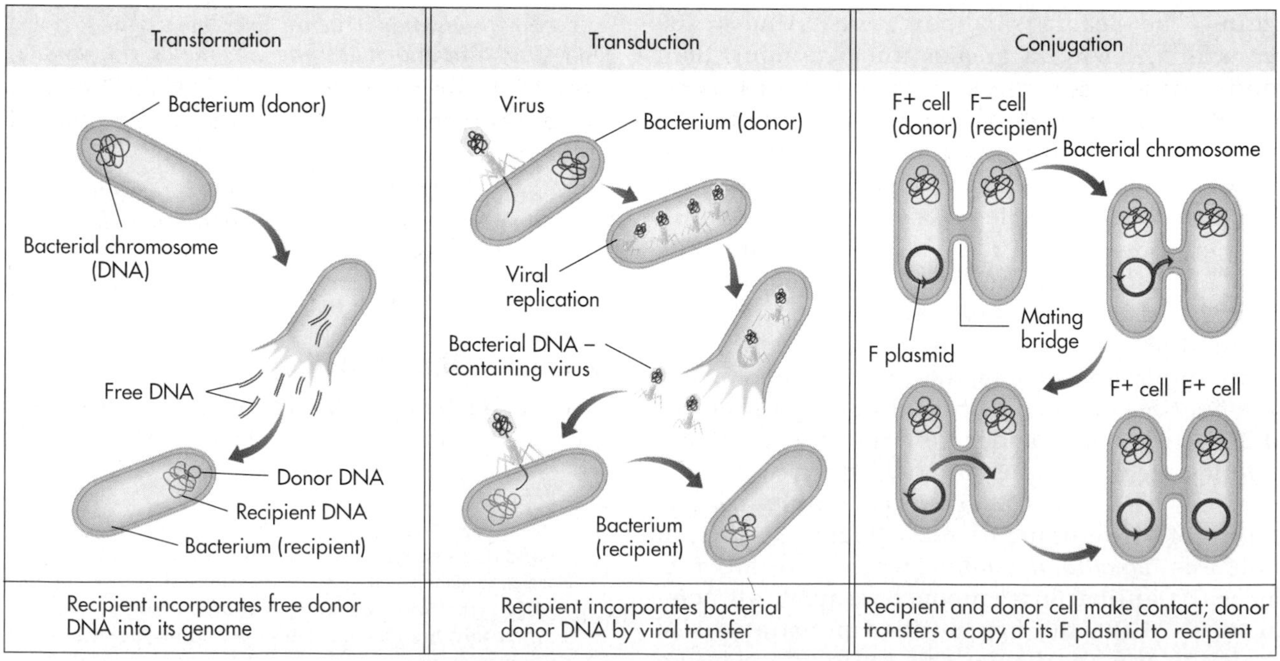

FIGURE 31-4 Natural gene transfer in bacteria: transformation, transduction, and conjugation.

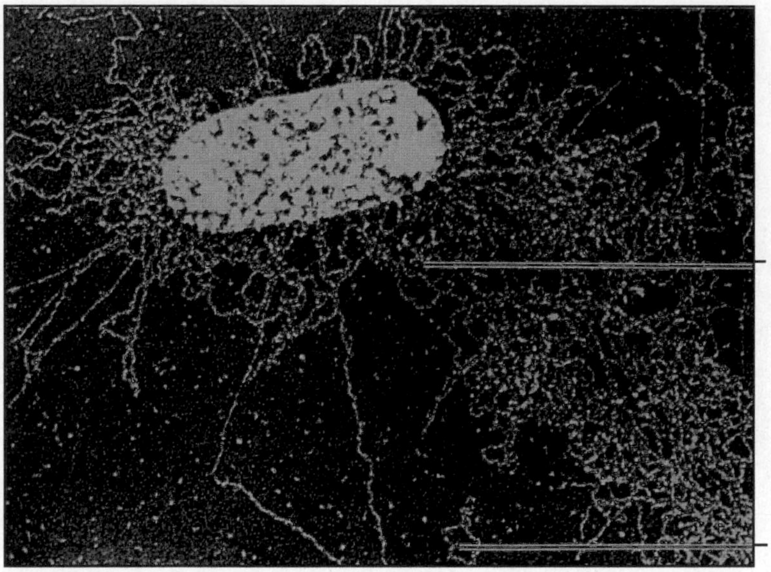

FIGURE 31-5 Bacterium with DNA extruded and plasmid visible. Notice how small the plasmid is with respect to the total chromosomal DNA.

It is also true that not all bacteria are capable of being infected by a virus carrying bacterial genes. But this process does occur in certain viral-bacterial interactions and results in genes from one bacterium being transferred to another bacterium.

During **conjugation,** a donor and a recipient bacterium make contact, and the DNA from the donor is transferred to the recipient cell. This transfer takes place in bacteria having extra chromosomal pieces of DNA called **plasmids.** These plasmids have genes that code for the ability to transfer chromosomes. Plasmids are fragments of DNA separate from the main circular DNA molecule of bacteria. They replicate independently of the main chromosome and make up about 5% of the DNA in many bacteria. (Figure 31-5 shows the size of a plasmid relative to the chromosomal DNA.) Some plasmids have several special genes that promote the transfer of the plasmid to other cells. These genes are referred to as a fertility factor, and such plasmids are called *F plasmids.* Fertility genes code for proteins that form a tube called an *F pilus* (or *sex pilus*) on the surface of the bacterial cell (Figure 31-6). When the F pilus of one cell makes contact with the surface of another cell that lacks F pili (and therefore does not contain the F plasmid), the replication and then transfer of the plasmid occurs (see Figure 31-4). If the F plasmid has been incorporated into the larger bacterial chromosome, as sometimes happens, the cell may copy and transfer the entire newly formed bacterial chromosome to the recipient cell.

transformation one of the ways in which bacteria transfer genetic material. Transformation occurs as DNA from a lysed donor cell is taken up from the surrounding medium by a competent cell.

genome (**gee** nome) an organism's total complement of genetic material.

transduction (tranz **duk** shun) one of the ways in which bacteria transfer genetic material. Transduction occurs as DNA from a donor bacterium is transferred to a recipient bacterium by a virus.

conjugation (**con** juh **gay** shun) One of the ways in which bacteria transfer genetic material. Conjugation occurs as a donor and a recipient bacterium make contact, and the DNA from the donor is transferred to the recipient cell.

plasmids (**plaz** mids) extrachromosomal pieces of DNA that replicate independently of the main chromosome in bacteria.

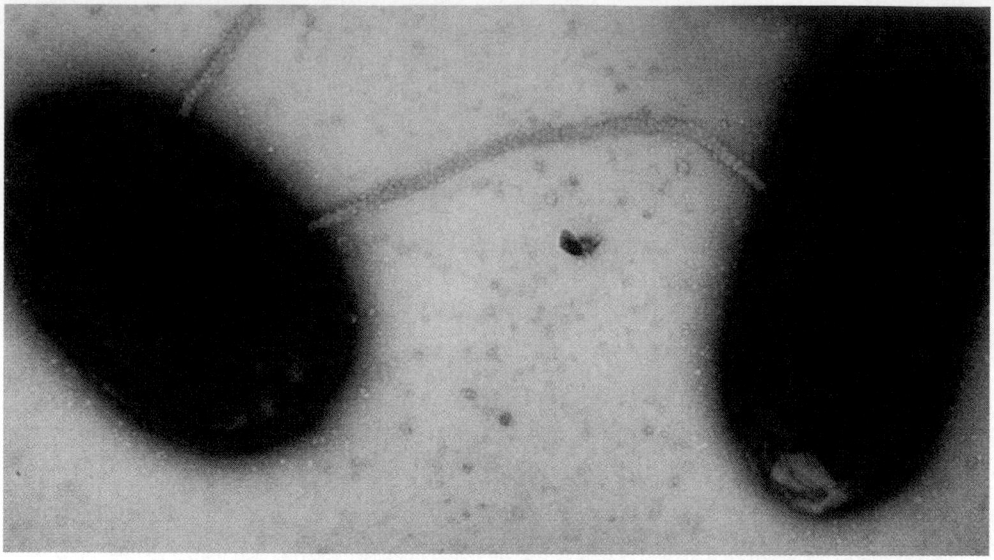

FIGURE 31-6 Genetic recombination between bacteria. In this colorized electron micrograph, the mating *E. coli* bacteria are seen as purple, and the F pilus (sex pilus) joining them is blue.

Certain bacteria can transfer genetic material from one cell to another. In one method called transformation, free pieces of DNA move from a donor cell to a recipient cell. During transduction, a second method of natural gene transfer, pieces of DNA from a donor cell are transferred to a recipient cell by a virus. During conjugation, a third method, a donor and a recipient cell make contact, and DNA from the donor is transferred to the recipient cell.

Human-Engineered Gene Transfer Using Eukaryotes

Although scientists were able to manipulate the natural mechanisms of bacterial (prokaryotic) gene transfer in the 1950s and 1960s, it was not until the mid-1970s that scientists were able to combine DNA from different species of organisms and manipulate genes in eukaryotes. These techniques of molecular biology that involve the manipulation of genes themselves (not just the organism) are called **genetic engineering,** or **recombinant DNA technology.** In the 1980s, molecular biotechnology became a major area of growth in business. The first commercial application of recombinant DNA technology was the production of proteins for therapeutic use.

The first techniques of molecular biotechnology, the artificial manipulation of the genes of organisms, were developed in the 1970s. These techniques differ from the techniques of classical biotechnology in that the genes, not just the organisms, are used and manipulated.

Applications and Methods of Molecular Biotechnology

The genetic engineer uses a variety of complex technological methods and laboratory equipment to study and manipulate genes. As scientists learn more about the process of genetic engineering, they develop new methods and equipment to advance their studies. The field of DNA technology is constantly expanding and changing. The following sections explain some of the basic methods employed today in the context of current applications.

The Production of Proteins for Therapeutic Use

One of the first applications of recombinant DNA technology was the insertion of human genes into bacteria and the manipulation of these organisms to produce human proteins for therapeutic use. The first two genetically engineered proteins were developed in 1979 and became available to the public after the completion of clinical trials and with

FIGURE 31-7 Large fermentors are used for the commercial production of genetically engineered bacterial products. Large tanks such as these hold hundreds (and some hold thousands) of gallons of bacterial culture.

the approval of the Food and Drug Administration a few years later. These proteins were human insulin, used in the treatment of diabetes mellitus, and human growth hormone, given to children who do not produce enough of this hormone for normal growth. Genetically engineered bacteria grown in large numbers, such as shown in Figure 31-7, produce substantial quantities of these hormones, which can be purified and used in patients.

Finding the Gene of Interest

The first step in producing human proteins using the recombinant DNA technology may be the hardest of all: the identification and characterization of the desired gene. In other words, before scientists can insert a gene from one organism into another, they must find the gene (such as the human insulin gene) among all the genes in the genome. There are an estimated 100,000 genes in the human genome, so finding one gene among them is not an easy task. In addition to finding a specific gene, genetic engineers need copies of the gene to insert into microbes. These microbes are usually bacteria or yeast, and they will produce the protein desired.

Scientists use various methods to find genes of interest. The methods a researcher uses in any particular situation depends on the gene itself, the amount of information known about the gene, and the research goals. One method of finding a gene is **shotgun cloning.** This method is a somewhat long and complex process, and it is used when researchers know very little about the gene they are trying to find. In brief, shotgun cloning involves cutting the DNA of the entire genome into pieces, isolating and purifying the genomic DNA, and inserting these pieces or fragments into bacteria or yeast. These transformed organisms then reproduce and make copies, or **clones,** of the DNA fragments. The term *shotgun* means that no one gene is targeted for cloning—all the genes are cloned. The result is a complete **gene library:** a collection of clones of DNA fragments, which together represent the entire genome of an organism. The library is then screened to find the desired gene. Each step of this process is described in more detail as follows.

genetic engineering (recombinant DNA technology) techniques of molecular biology that involve the manipulation of genes.

shotgun cloning the process of synthesizing DNA by cutting the DNA of an entire genome into pieces, isolating and purifying the genomic DNA, and then making copies of this DNA using yeast or bacterial cells.

clones copies.

gene library a collection of clones of DNA fragments, which together represents the entire genome of an organism.

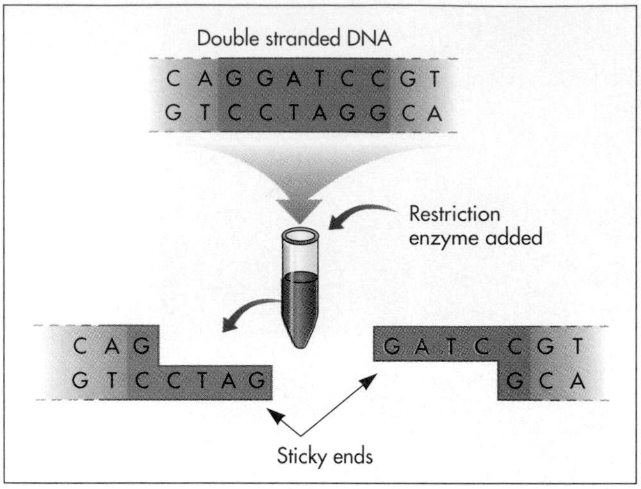

FIGURE 31-8 Restriction enzymes act like chemical scissors. Restriction enzymes recognize short sequences of nucleotides in DNA and cut them at a certain point within the sequence. The restriction enzyme in this illustration recognizes the base (nucleotide) sequence GGATCC and cuts the sequence between the two guanine (G) bases. Because each strand of DNA in double-stranded DNA is "read" in a direction opposite to the other, the cut piece of DNA has ends in which one strand is longer than the other. These ends are called sticky ends.

Cutting DNA into Gene Fragments

The first step in cutting the genomic DNA into pieces is to extract the DNA from human cells that have been grown in tissue culture. The freed DNA is then clipped into pieces at the molecular level using the restriction enzymes. **Restriction enzymes** act like chemical scissors, cutting long, intact DNA strands into smaller pieces of various sizes, which are more easily studied than the intact DNA. Restriction enzymes recognize certain nucleotide (base) sequences in DNA molecules and break the bonds between the nucleotides (Figure 31-8). These linkage points are called **recognition sites.** In nature, restriction enzymes are produced by bacteria to protect themselves from bacteriophage (viral) invasion. They protect the bacteria by cutting viral DNA into pieces before the virus can infect the cell. Discovered in the late 1960s, these naturally occurring bacterial enzymes have now become one of the many research tools of the genetic engineer.

Hundreds of restriction enzymes exist and are now available for use by scientists. Different restriction enzymes identify different recognition sites, allowing scientists to cleave the DNA in a variety of ways. Carefully using these enzymes, scientists can produce a range of sizes of fragmented DNA, called **restriction fragments.**

Cloning Genes

The next step in shotgun cloning is to insert the restriction fragments (each of which may contain one or more genes) into bacteria or yeast cells using either plasmids or viruses. Plasmids can be extracted from bacteria or yeasts, induced to incorporate restriction fragments into their genomes, and reinserted into the cell. Scientists can also incorporate restriction fragments into viral genomes and use the viruses to infect bacteria or yeasts. Plasmids and viruses used to insert restriction fragments into cells are called *cloning vectors.*

To incorporate restriction fragments into cloning vectors (such as plasmids, for example), the genetic engineer must first cut the DNA of both using the same restriction enzyme. This results in cut pieces of DNA that will bond with one another for the following reasons. Nucleotides of double-stranded DNA are complementary to one another and are "read" in opposite directions. Therefore a restriction enzyme recognizes a base sequence (such as GGATCC) in different places on each strand. It then cuts the strand at a specific point in the sequence. In Figure 31-8, you can see the result: the cut piece of DNA has ends in which one strand is longer than the other (in most cases). These ends are called *sticky ends* because their bases are complementary. They will bond with another piece of DNA cut with the same restriction enzyme (Figure 31-9). Therefore fragments of human DNA and bacterial plasmid DNA cleaved by the same restriction enzyme have the same complementary nucleotide sequences at their ends and can be joined to one another. A sealing enzyme called a **ligase** helps re-form the bonds.

The next step is to insert these "hybrid" plasmids into bacteria (or yeasts) by transformation. When these plasmids are mixed with specially prepared bacteria under the proper conditions needed for transformation, the bacteria will act as competent cells and take up the plasmids from the mixture. (This procedure of genetically engineering bacteria is summarized in Figure 31-10.)

Inserting fragments of foreign DNA into bacterial or yeast cells using plasmids is a technique that researchers employ frequently. Occasionally certain viruses are used to insert DNA fragments into bacteria, plant cells, or animal cells by introducing the DNA into the genome of a virus instead of into a bacterial plasmid. The virus then carries the DNA fragment into a cell that it is capable of infecting and begins replicating.

Once the restriction fragments have been inserted into the bacteria or yeasts, the genetic engineer then cultures the organisms. This results in

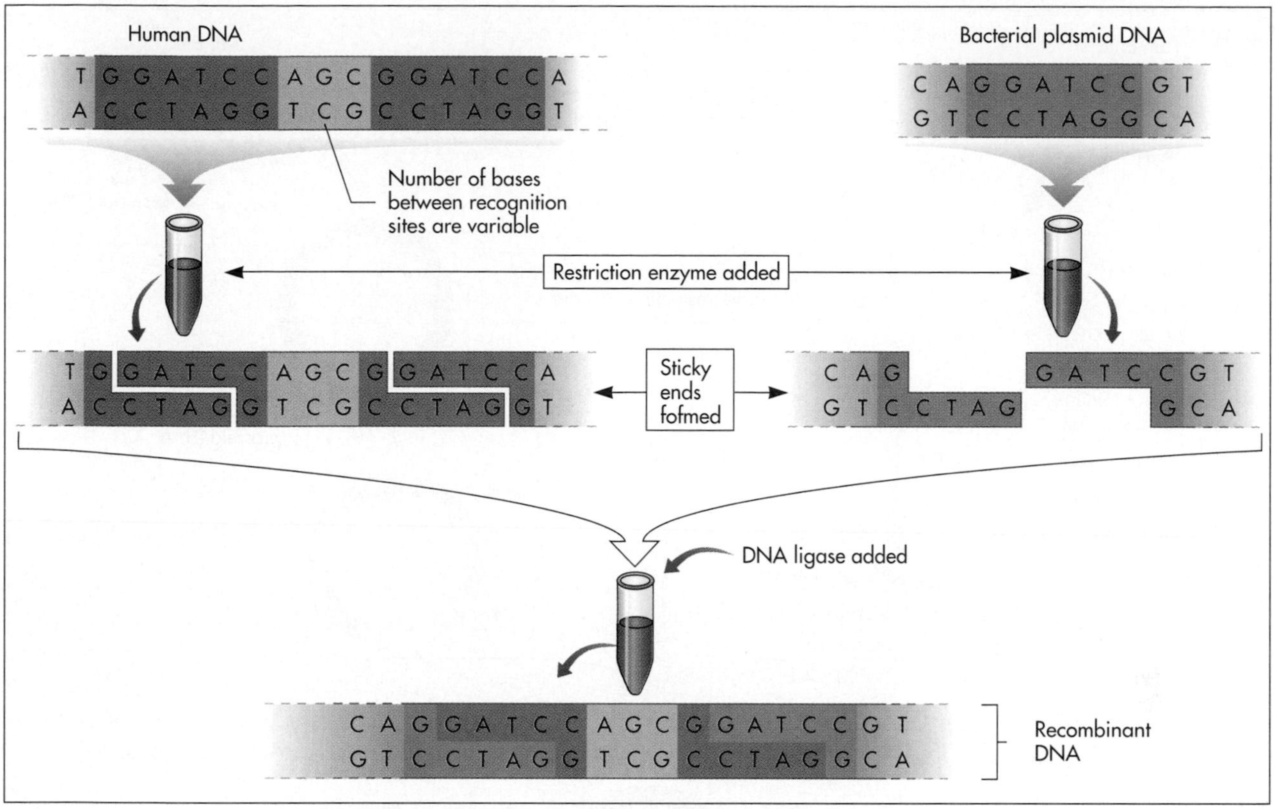

FIGURE 31-9 The cutting and splicing of DNA from two different organisms. When human DNA (*left*) and bacterial plasmid DNA (*right*) are cut with the same restriction enzyme, the restriction fragments formed can be recombined with the use of a sealing enzyme called DNA ligase. Here, a restriction enzyme cuts the human DNA at two restriction sites. The piece cut from the human DNA is then inserted into the "opened" bacterial plasmid DNA, which was cut at one site by the same restriction enzyme.

large populations of the organisms that contain the inserted DNA. Both bacteria and yeasts reproduce asexually—one cell splitting into two, two into four, and so forth. If these cells are grown on a semisolid culture medium in a small, flat dish (a Petri dish) they grow into discrete "piles" of cells called *colonies,* which are visible to the naked eye (Figure 31-11). All the cells in the colony are genetically identical to one another and are considered clones because each colony grew from an individual cell.

Scientists may use approaches other than the shotgun technique to clone genes. A process called **complementary DNA (cDNA) cloning** can be performed if the messenger RNA (mRNA) transcribed from the gene of interest is available. When a protein is synthesized by a particular tissue in the body, its mRNA is usually found in abundance there. For example, human insulin is manufactured in the pancreas; these cells are rich in insulin-specific mRNA. Scientists can extract this mRNA from pancreatic cells and manufacture molecules of DNA from mRNA using the enzyme reverse transcriptase. This DNA can then be inserted into bacterial or yeast cells and cloned.

Complementary DNA cloning also avoids a problem inherent in the shotgun cloning technique: the removal of *introns,* noncoding regions of DNA found in eukaryotes. As described in Chapter 18, eukaryotic cells cut introns out of mRNA and then splice together the exons, or coding portion of the nucleotide chain, before genes can be trans-

restriction enzymes proteins that recognize certain nucleotide (base) sequences in a DNA strand and break the bonds between the nucleotides at those points.

recognition sites certain nucleotide sequences in DNA molecules that are acted upon by restriction enzymes.

restriction fragments the pieces of DNA that have been cut from larger pieces of DNA by restriction enzymes.

ligase (lye gase) a sealing enzyme that helps re-form the bonds between pieces of DNA that have been cut with the same restriction enzymes.

complementary DNA (cDNA) cloning the process of synthesizing DNA from mRNA and then making copies of this DNA using yeast or bacterial cells.

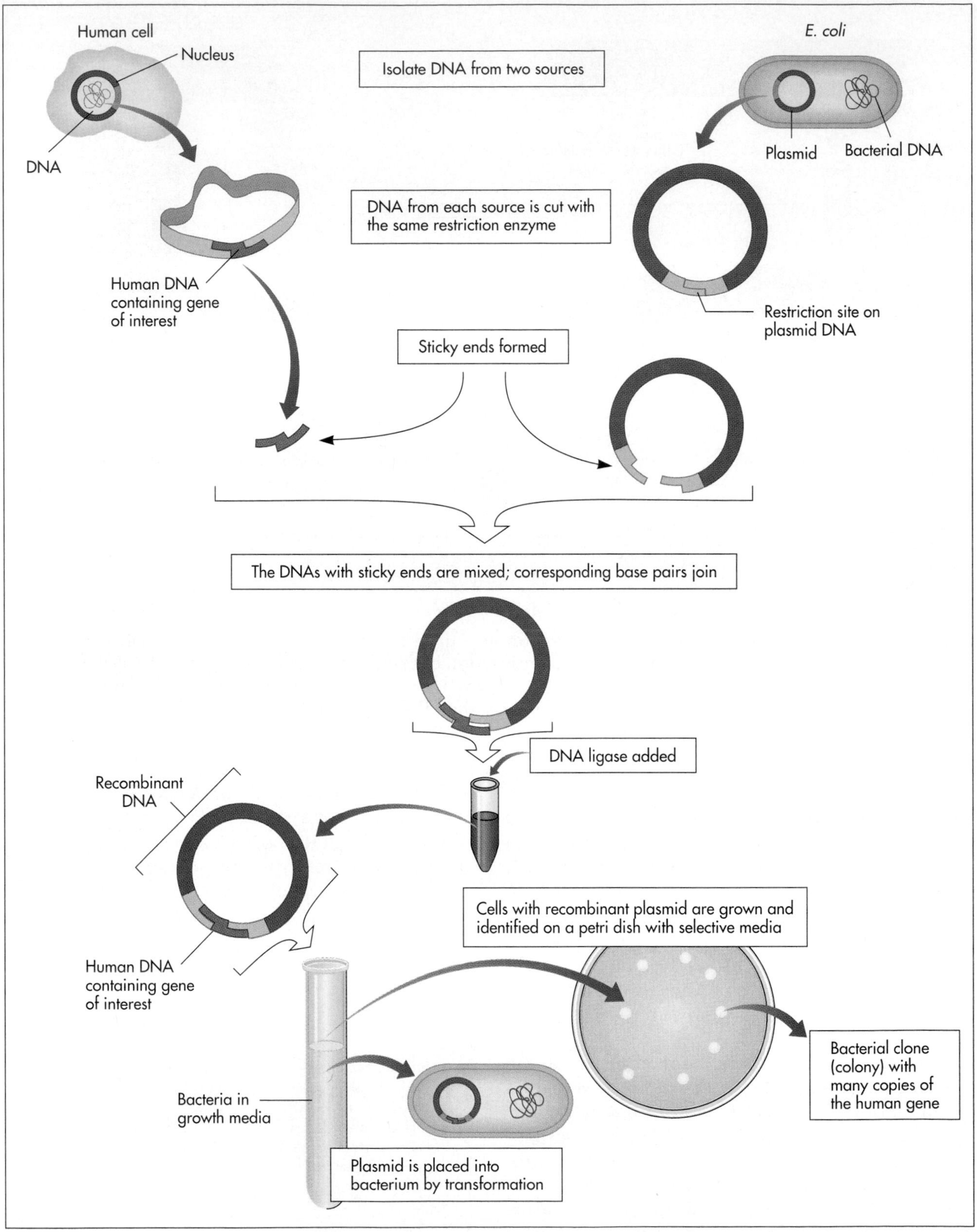

FIGURE 31-10 Cloning a human gene in a bacterium. Both human and bacterial (plasmid) DNA are cut using the same restriction enzyme. The human DNA is inserted into plasmids, and bacteria take up the plasmids by transformation. The bacteria are then cultured and plated on selective media to identify colonies made up of recombinant cells.

FIGURE 31-11 Bacterial colonies growing on culture medium in a Petri dish. Several discrete piles of bacterial cells called colonies can be seen in this dish. Each colony is composed of millions of bacterial cells that arose from a single cell. Each cell in a colony is genetically identical to all others in the colony. However, the cells among colonies may differ from one another. This photo is of a streak plate, in which a laboratory technician has used a sterile inoculating loop to "drag" bacteria across the nutrient surface. In some areas, so many bacteria were deposited on the medium that confluent areas of growth appear. In other areas, single cells were deposited in areas separate from other cells, so individual colonies are visible.

lated into proteins. During shotgun cloning, genes having introns will not be correctly expressed in a bacterial cloning host. Yeasts can remove introns, but using them for shotgun cloning is less efficient than using bacteria. In cDNA cloning, the mRNA used as the template for reverse transcriptase has no introns. Consequently, the newly synthesized DNA can readily be cloned and expressed in bacterial systems.

Gene synthesis cloning, another approach to cloning genes, can be performed if the nucleotide sequence of the gene of interest is known and the genetic engineer is able to synthesize the gene. (DNA nucleotide sequencing techniques are described in "How Science Works.") The laboratory-made gene can be inserted into bacterial or yeast cells (as described previously) and cloned. This method of cloning also avoids the intron problem.

To use microbes to produce proteins for human use, scientists first find the gene in the human genome and isolate it. One method used when the researcher knows the least about the gene is shotgun cloning. This method involves cutting the DNA of the entire human genome into pieces with restriction enzymes, inserting these pieces or fragments into bacteria or yeast using cloning vectors, and allowing the organisms to reproduce, thereby making copies, or clones, of the DNA fragments. Two other methods used under different circumstances are complementary DNA cloning and gene synthesis cloning.

Producing proteins by means of recombinant DNA technology is not as simple as inserting genes into microorganisms and then having these microorganisms synthesize functional proteins. There are many problems intrinsic to the cloning processes described thusfar; the intron problem is only one. Scientists solve this problem in various ways depending on the protein being synthesized and the procedures being used. Another problem involves the modification of the newly synthesized protein so that it can be secreted from the cells producing it. Many other challenges exist for genetic researchers in the development of new and novel approaches to meeting their goals.

Screening Clones

Screening clones to find a desired gene is necessary to select the proper clones to make large amounts of the desired gene or gene product. With shotgun cloning, the entire genome has been cloned, and consequently, screening for particular genes can be a lengthy procedure. Complementary DNA cloning results in a small variety of clones, so screening is a relatively easy process. Screening is not necessary for genes that are directly synthesized because only the synthesized genes are available for cloning.

Most screening techniques involve DNA hybridization and the use of *molecular probes.* The process is illustrated in Fig. 31-12. A *replica plate* is made of the original master plate that contains the genetically engineered bacterial clones (see Fig. 31-12, *1*

gene synthesis cloning the process of synthesizing DNA in the laboratory based on knowledge of its nucleotide sequence and then making copies of this DNA using yeast or bacterial cells.

probe a molecule that binds to a specific gene or nucleotide sequence.

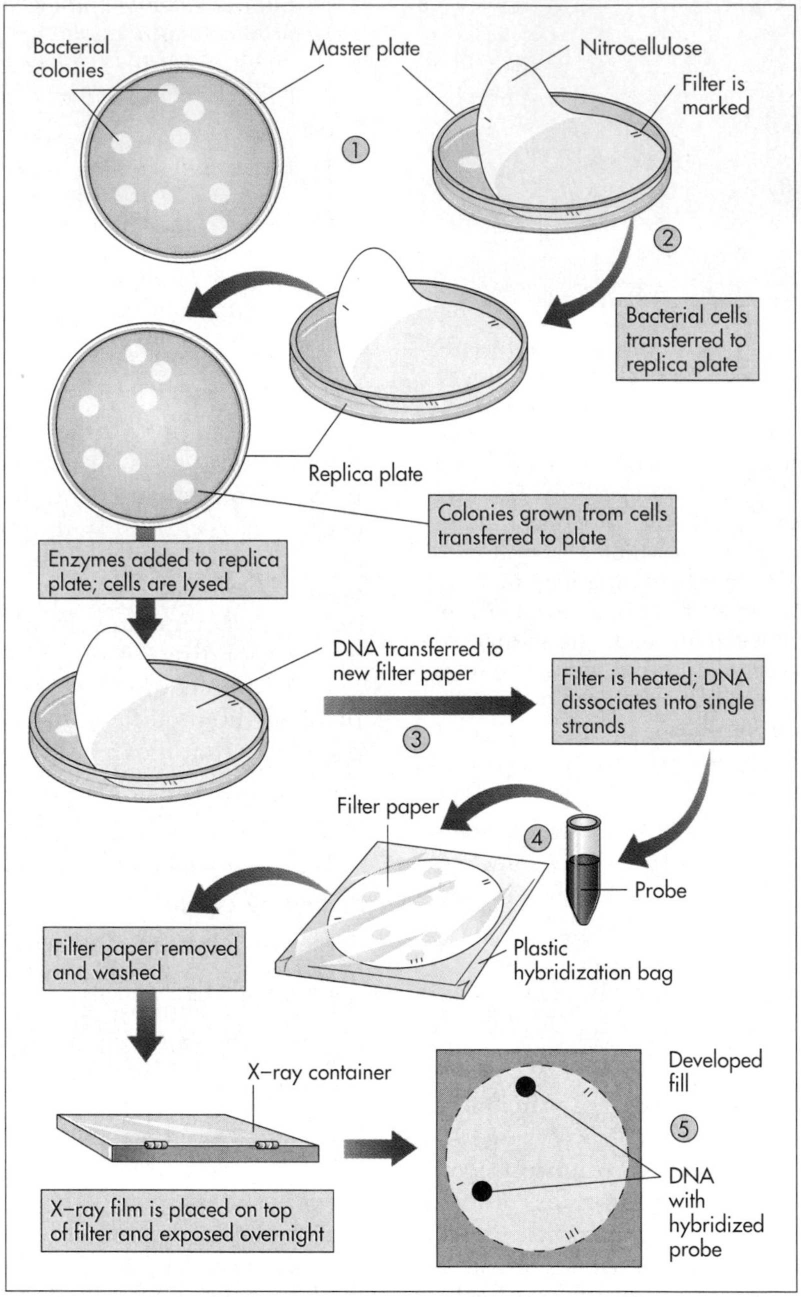

FIGURE 31-12 Screening cloned genes using molecular probes. A replica plate is first made of the genetically engineered bacterial colonies. A nitrocellulose filter is used to transfer cells from colonies in a master plate to a fresh, sterile plate (1 & 2). (Needle marks are poked through the filter when on the master plate to orient all subsequent filters and x-rays.) After the cells are grown on the replica plate, enzymes are added to break open the cells, causing them to release their DNA. The DNA is then transferred to fresh filter paper and it is heated, dissociating the double-stranded DNA into single strands (3). A radioactive molecular probe (4), which will bind to the gene of interest, is then added to the filter paper. The DNA and the probe are allowed to hybridize (bind). After washing the filter paper of excess probe, it is placed on X-ray film overnight, which visualizes the hybridized DNA and probe—and therefore the cells containing the gene of interest.

and 2). This replica is made by touching the surface of the master plate with a circular piece of sterile nitrocellulose (a thin, filter paper type material). The nitrocellulose, now containing cells from the master plate, is then touched to the nutritive surface of a sterile plate, transferring cells from the colonies of the master plate onto the replica plate. The original plate is saved for comparison purposes and as a source of the cloned cells. During incubation of the replica, (see Fig. 31-12, *2*) colonies of cells grow in the same positions as on the master plate. Then the replica is treated with enzymes that lyse (break open) the cells, releasing the cell contents (including the DNA) (see Fig. 31-12, *3*). A special filter is placed on the surface of the replica plate, and the DNA (and other molecules from within the cells) is transferred to the filter. The filter is heated, which causes the doubled-stranded DNA to dissociate into single strands (see Fig. 31-12, *4*). A radioactively labeled probe is then added to the filter.

A **probe** is a molecule that binds to a specific gene, or nucleotide sequence. Probes can be (a) molecules of mRNA purified from tissue or cells that produce the protein, (b) complementary sequences of DNA or RNA that have been synthesized using knowledge of the codons (triplet-nucleotide sequences on mRNA) for the amino acid sequence of the protein, or (c) antibodies specific for the desired protein. After the probe has been added to the filter that contains single-stranded DNA fragments or protein, the probe will bind with the complementary DNA sequences (a process called *hybridization*) or the antibodies will bind with the desired protein. Molecules of the probe that did not bind, then, are washed from the filter, leaving the larger complexes of bound probe and target in the filter. These filters are then placed under x-ray film in sealed containers overnight. This technique is called *autoradiography*. The radioactivity exposes the x-ray film to create a photographiclike image, and the colonies containing DNA with the hybridized probe are visualized, pointing out the clones containing the searched-for gene (see Figure 31-12, *5*).

> *Most screening techniques used to find a gene of interest within a gene library involve DNA hybridization and the use of molecular probes, molecules that bind to specific genes, or nucleotide sequences.*

Making Copies of a Gene: The Polymerase Chain Reaction

After a gene has been isolated, many copies of it are needed to make enough recombinant cells as an inoculum for large-scale culturing. Developed in 1983 by American scientist Kary B. Mullis, the polymerase chain reaction (PCR) is a method used to make unlimited copies of genes. This is an extremely important technique in genetic engineering today. Its applications vary widely because the DNA can be in a mixture with other molecules and can also come from anywhere, such as a cloned cell, a blood stain, or even an organism that has been dead for hundreds of years! Not only has this technique revolutionized genetic research, it has also enabled the analysis of DNA in extremely small blood stains, semen samples, or pieces of hair at crime scenes that were previously impossible to analyze. Scientists are even able to amplify small amounts of DNA found in the preserved remains of long-extinct organisms to compare their relatedness with other extinct or present-day organisms.

The polymerase chain reaction is illustrated in Figure 31-13 and proceeds the following way. First, the DNA is heated to separate the double DNA strands into single-stranded DNA (see Figure 31-13, *1*). A probe a few nucleotides in length (referred to here as a *primer*) is then hybridized with a portion of the DNA at the place where the "copying" is to begin (see Figure 31-13, *2*). The enzyme DNA polymerase and free nucleotides (adenine [A], cytosine [C], guanine [G], and thymine [T]) are added to the solution (see Figure 31-13, *3*). DNA polymerase, "primed" by the hybridized probe, extends the primer, adding nucleotides in sequence from the site of the primer to the end of the fragment. Each sDNA fragment is double stranded once again and is an exact replica of one another and the original double strand. This process is repeated over and over again: one DNA strand is separated into two single strands; each strand is hybridized with a nucleotide primer; and DNA polymerase adds free nucleotides to complete the strands, forming two new, double-stranded molecules from one original. Two strands are then used to produce four, the four produce eight, and the numbers quickly escalate. In an afternoon, one original DNA molecule can be used to make a billion copies!

Today, PCR can be performed by machine (Figure 31-14). In fact, researchers at Lawrence Livermore National Laboratory are currently developing a miniature PCR machine. This technology would have many applications in medical diagnostics and in the military. For example, researchers think this portable device could be used by soldiers in the field to detect low levels of microorganisms contaminating water or supplies, or the spores of organisms used in biological warfare.

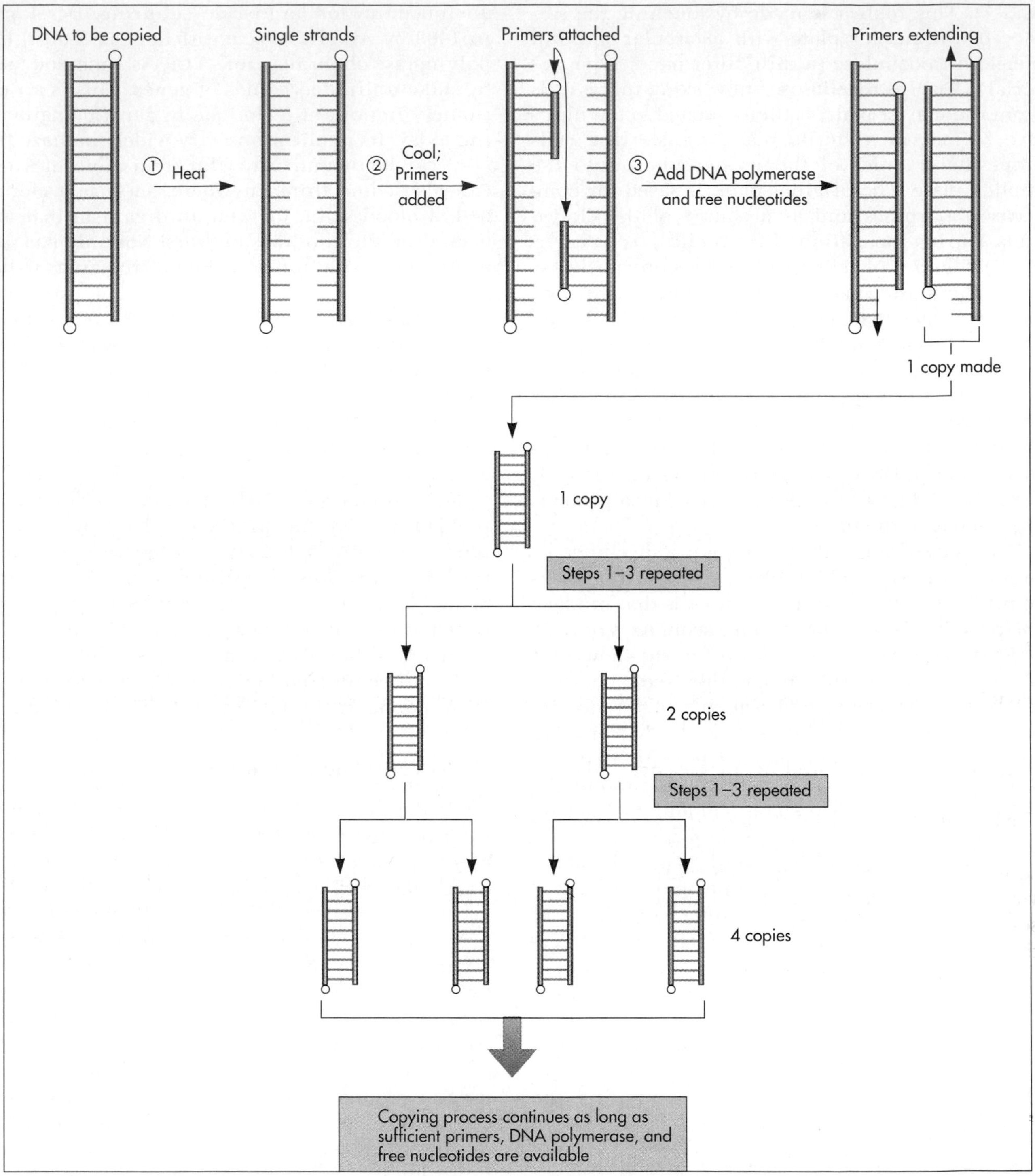

FIGURE 31-13 The polymerase chain reaction (PCR). The double-stranded DNA fragment to be copied is heated (1), which separates the strands. Primers then bind to spots on the DNA where the copying is to begin (2). DNA polymerase and free nucleotides are added to the mixture (3), and the DNA polymerase extends the DNA. The result is two strands of (double-stranded) DNA identical to one another and identical to the original strand. Repeating these steps with these two strands yields four strands from one original. This process can be continued as long as the researcher desires, generating an unlimited number of copies of DNA from one original double-stranded molecule.

FIGURE 31-14 The polymerase chain reaction (PCR) can be carried out by machine. The PCR machine is a thermal cycler—it goes through cycles of heating and cooling to carry out this chain of reactions. The laboratory technician simply loads the DNA to be copied, primer, free nucleotides, and DNA polymerase in the pink tubes visible in the machine. A special type of DNA polymerase that is resistant to destruction by high temperature is used. The reactions will take place in a cyclic fashion as long as sufficient reactants are available.

In our example of finding, extracting, and copying a human gene for genetic engineering, PCR is used to produce as many copies of the gene as the genetic engineer needs. These genes are then inserted into bacteria or yeasts by means of cloning vectors. Populations of cells are grown on a large scale, from which the protein is extracted. As mentioned previously, technical problems are often encountered in the identification, characterization, isolation, and use of human genes to produce genetically engineered microbes for manufacturing human proteins. However, these problems can usually be overcome, as in the case of the human insulin gene. The product produced by genetic engineering is called Humulin and is used to treat more than half the new cases of diabetes in the United States. In addition to avoiding the unwanted allergic reactions sometimes caused by using insulin extracted from cattle or pig pancreas, Humulin is less expensive than animal preparations.

The polymerase chain reaction is used to produce multiple copies of a gene and has wide application in molecular biotechnology.

Genetically Engineered Proteins in Use Today

Along with producing human insulin and human growth hormone, gene technology has produced proteins useful in treating a variety of human disorders or diseases. For example, the three primary types of human interferon (alpha, beta, and gamma) have been produced by genetic engineering. Interferons are proteins that interfere with the ability of a virus to invade a cell. These proteins also modulate the activity of the immune system and, in doing so, help the body stave off invasion from other disease-causing agents. Although interferons have been used to treat certain viral diseases and rheumatoid arthritis, the most successful use of genetically engineered interferon has been the use of alpha-interferon to treat hairy-cell leukemia, a rare form of cancer. An amazing 75% to 90% of new patients treated thus far have experienced remission of their disease. A variety of other medically important proteins have also been genetically engineered, such as interleukin-2, which is used to treat certain deficiencies of the human immune system, and tissue-type plasminogen activator, a clot dissolver used in the treatment of heart attacks.

JUST WONDERING. . . .

Do scientists have to follow ethical guidelines for genetic engineering?

There has been considerable discussion in the scientific community and among nonscientists about the concerns that underlie your question—the potential danger of inadvertently creating undesirable or potentially dangerous organisms in the course of a recombinant DNA experiment. For example, what if someone fragmented the DNA of a cancer cell and then incorporated these DNA fragments at random into viruses that are propagated within bacterial cells? Might there not be a danger that one of the resulting bacteria or viruses could be capable of infecting humans and causing a disease—even cancer?

Even though most recombinant DNA experiments are not dangerous, such concerns are taken seriously. It was the scientists themselves who first realized the possible dangers associated with genetic engineering technology. These scientists, along with government agencies such as the National Institutes of Health, drew up formal guidelines that govern all such experimentation in this country. Many other countries have done the same. Many scientific organizations have ethics committees to deal with such issues also, especially as they relate to gene therapy in humans.

Scientists and governmental regulatory agencies not only monitor experiments but study products to make sure that these products do not cause increased health risks to consumers or pose environmental hazards. Many genetically engineered products, especially pharmaceutical and agricultural products, are on the market today; these products have been shown to be safe and effective.

Although genetic engineering may sound rather ominous to some, it is important to remember that many of the ways that scientists manipulate genes occur naturally. These processes are occurring right now—*without* the help of scientists. In addition, scientists have been manipulating genes for centuries using classical biotechnological techniques—techniques with which most people are comfortable. Molecular biotechnology is simply more specific, faster, and versatile.

The Development of Genetically Engineered Vaccines

A vaccine is a substance that is either injected into the body or taken orally to stimulate the immune response. Vaccines are often produced by culturing the disease-causing agent and then killing (inactivating) it or attenuating (weakening) it. The techniques of genetic engineering have been applied to the development of vaccines, producing a successful preparation of hepatitis B vaccine, which helps protect against the leading cause of liver cancer. But traditional methods of producing vaccines do not always work. Some organisms, such as the malaria parasite, are difficult to culture. Some vaccine preparations do not confer resistance well enough, whereas others produce unwanted—and sometimes dangerous—side effects. However, scientists are currently working on a variety of new genetically engineered vaccines, including preparations to provide immunity against cholera, malaria, ear infections, and of course, human immunodeficiency virus (HIV) infection.

In one approach, scientists use the gene-splicing techniques described earlier in this chapter to insert one or more of a pathogen's genes into a nonpathogenic organism. (A pathogen is a disease-causing agent of infection.) Such a technique can produce, for example, a noninfective virus whose coat contains proteins of the pathogen as a result of the inserted genes being expressed. When injected into humans, these viruses do not cause disease, but they do stimulate the immune system to produce antibodies specific for the infective form of the virus.

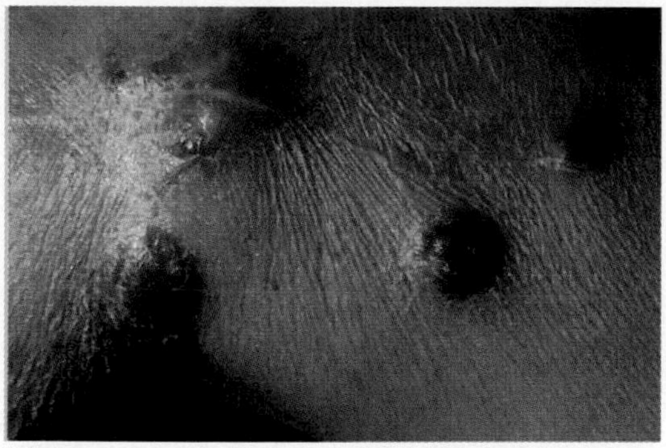

FIGURE 31-15 Malignant melanoma. This form of skin cancer is one of the fastest-spreading cancers known and can be caused by excessive exposure to the sun.

Gene Therapy

Recombinant DNA technology entered a new phase of application on September 14, 1990, when a 4-year-old girl became the first person to undergo **gene therapy.** Gene therapy is the treatment of a genetic disorder by the insertion of "normal" genes into the cells of a patient. The young girl who was treated suffered from the rare genetic disorder severe combined immunodeficiency (SCID). Key immune system cells called *T cells* (see Chapter 11) were not working because they lacked the enzyme ADA (adenosine deaminase). Without these cells, this girl had no defense against infection.

To treat the girl, doctors removed some of her white blood cells and cultured her defective cells. They then added viruses that contained working copies of the ADA gene to the cell culture. When the viruses infected the blood cells, they inserted these ADA genes into the cells. Doctors grew these altered blood cells in the laboratory until they numbered in the billions and then injected them into the child's blood. Although the genetically repaired cells produced ADA, they did not multiply and populate the girl's immune system. Therefore, the therapy had short-term effects and had to be repeated every few months. In a second approach in May 1993, bone marrow stem cells (which give rise to the affected cells) were flushed from the girl's bone marrow into her bloodstream. Doctors then drew her blood, isolated the stem cells, genetically repaired them as they had previously repaired fully mature white cells, and reinjected them into her body. To date, several more children with this disease have also started treatment. Studies suggest that thusfar, the children's immune systems have only small numbers of stem cells and T cells that carry the "normal" gene. The children are still being given the missing enzyme in drug form. Researchers are hopeful, however, that with time sufficient "healthy" T cells will develop from the stem cells.

Medical researchers have also been using gene therapy in another way. In an effort to strengthen the body's defenses against certain forms of cancer, researchers have been splicing genes that code for cancer-fighting proteins into patients' white blood cells. The white blood cells that are the target for this gene therapy are called *tumor-infiltrating lymphocytes* (TILs). These cells are part of the immune system's cancer surveillance system. They normally seek out and attack cancerous tumors, but they are not strong enough to control certain types of tumors such as those of malignant melanoma, a particularly deadly form of cancer (Figure 31-15).

In the fall of 1991, doctors injected genetically engineered cells into the thigh of a melanoma patient in an attempt to use gene therapy to help his immune system destroy the cancer. Researchers first removed TIL cells from the patient and inserted a gene that codes for the protein *tumor necrosis factor* (TNF). This protein kills tumor cells by preventing them from establishing a blood supply. The engineered TIL cells were then returned to the patient's bloodstream to seek out and invade the malignant melanoma tumors. As each genetically altered TIL cell finds and enters a tumor, it is able to attack the tumor with the TNF. The engineered TIL cell, in effect, becomes a factory that makes the tumor-killing protein inside the tumor itself.

This research is still ongoing; although this patient had a substantial regression of his cancer, results with other patients have been variable. Other related therapies for patients with advanced melanoma are currently being developed.

> *Gene therapy is the treatment of a genetic disorder by the insertion of "normal" genes into a patient's cells. It is currently being used to treat a rare genetic disorder called severe combined immunodeficiency (SCID). Researchers have also used a variation of gene therapy to fight certain cancers by splicing genes that code for cancer-fighting proteins into patients' white blood cells.*

The Human Genome Project and Gene Therapy

SCID is only one of more than 4000 diseases such as cystic fibrosis, sickle cell anemia, and achon-

> **gene therapy** the treatment of a genetic disorder by the insertion of "normal" genes into the cells of a patient.

droplasia (a form of dwarfism) known to result from an abnormality in a gene. The underlying idea of transferring a "normal" copy of a gene into an individual who has an "error" copy requires that the location of the gene is known and that the gene has been isolated. If gene transfer therapy is going to be applied to a wide range of genetic defects, scientists know that they need a detailed map of the human chromosomes.

Launched in 1990, the **Human Genome Project (HGP)** is a worldwide effort to map the positions of all the genes and to sequence the DNA base pairs of the entire human genome. Scientists from Paris to California are working on developing three increasingly detailed maps of the DNA in cells: (1) *a genetic linkage map,* which shows the distances between genetic markers (identified reference points such as genes for particular diseases) on the chromosomes; (2) *a physical map,* which shows the number of nucleotides between the markers; and (3) *an ultimate map,* which shows the sequence of nucleotides in a chromosome and describes its genes and the proteins they make. Work on the genetic linkage and physical maps is just about completed; sequencing the entire genome will take much longer. This project is significant not only because it will give scientists a clearer picture of the genes that cause various diseases but also, with this information, scientists will be able to study the relationship between the structure of genes and the proteins they produce.

Targeted Gene Replacement and Gene Therapy

Gene therapy is being helped by recent research in targeted gene replacement. This research is currently being conducted at the University of Utah by Mario Capecchi and his associates. The process of gene targeting involves changing the nucleotide sequence of a particular gene, which changes the function of the gene, and then observing the resultant changes in the anatomy, physiology, or behavior of the organism. Pooling these data with data from the Human Genome Project will give scientists great insight into the connection between the structure and function of genes.

Targeted gene replacement research is currently being conducted on mice, whose genome is surprisingly similar to the human genome. In fact, approximately 99% of the genes in mice and in humans are the same and serve the same functions. Therefore, scientists are hopeful that this research will provide data regarding how specific parts of the body (such as the brain) operate; how mutations in cells cause diseases such as cancer; how genes affect the development of cells, tissues, and organs; and as previously mentioned, how inherited defects in the genome result in inherited disorders.

Food Biotechnology

Biotechnology is being used in a variety of ways in the food industry. It focuses on increasing the yields of various foods and creating products with superior qualities. The most widespread use of biotechnology in the food industry is in agriculture. Through the use of gene splicing, scientists are working to improve crops and forest trees by making them more resistant to disease, frost, and herbicides (chemicals that kill weeds). Scientists expect that plant improvement through the use of biotechnology will be an important part of increasing food production to supply the burgeoning world population, which is predicted to reach 10.7 billion by the year 2030.

In early 1994, the U.S. Food and Drug Administration approved the first genetically engineered food: a tomato having a gene that allows it to ripen longer on the vine yet reach the supermarket without softening. Later in 1994, many genetically altered foods were approved, such as a squash that resists viruses, cotton and soybean plants that resist certain herbicides, and a potato that produces a pesticide that kills Colorado potato beetles. Such genetically altered plants are called **transgenic plants.**

Since 1980, a plasmid of the bacterium *Agrobacterium tumefaciens,* which causes a tumorlike disease called *crown gall* in plants, has been the main vehicle used to introduce foreign genes into broadleaf plants such as tomatoes, tobacco, and soybeans. But to use the plasmid to genetically engineer plants, scientists first had to remove *A. tumefaciens'* disease-causing genes—a technique known as disarming—while leaving its natural ability to transfer DNA intact.

After *A. tumefaciens* has been disarmed, it can be used as a vector to shuttle desired genes into plant cells. A part of its plasmid integrates into the plant DNA, carrying whichever genes the genetic engineer has inserted into the plasmid genome. (Figure 31-16 diagrams the steps of this process.) New plants can be micropropagated, or grown from these transformed cells in tissue culture. The cells develop into plantlets, which can be grown in conventional ways.

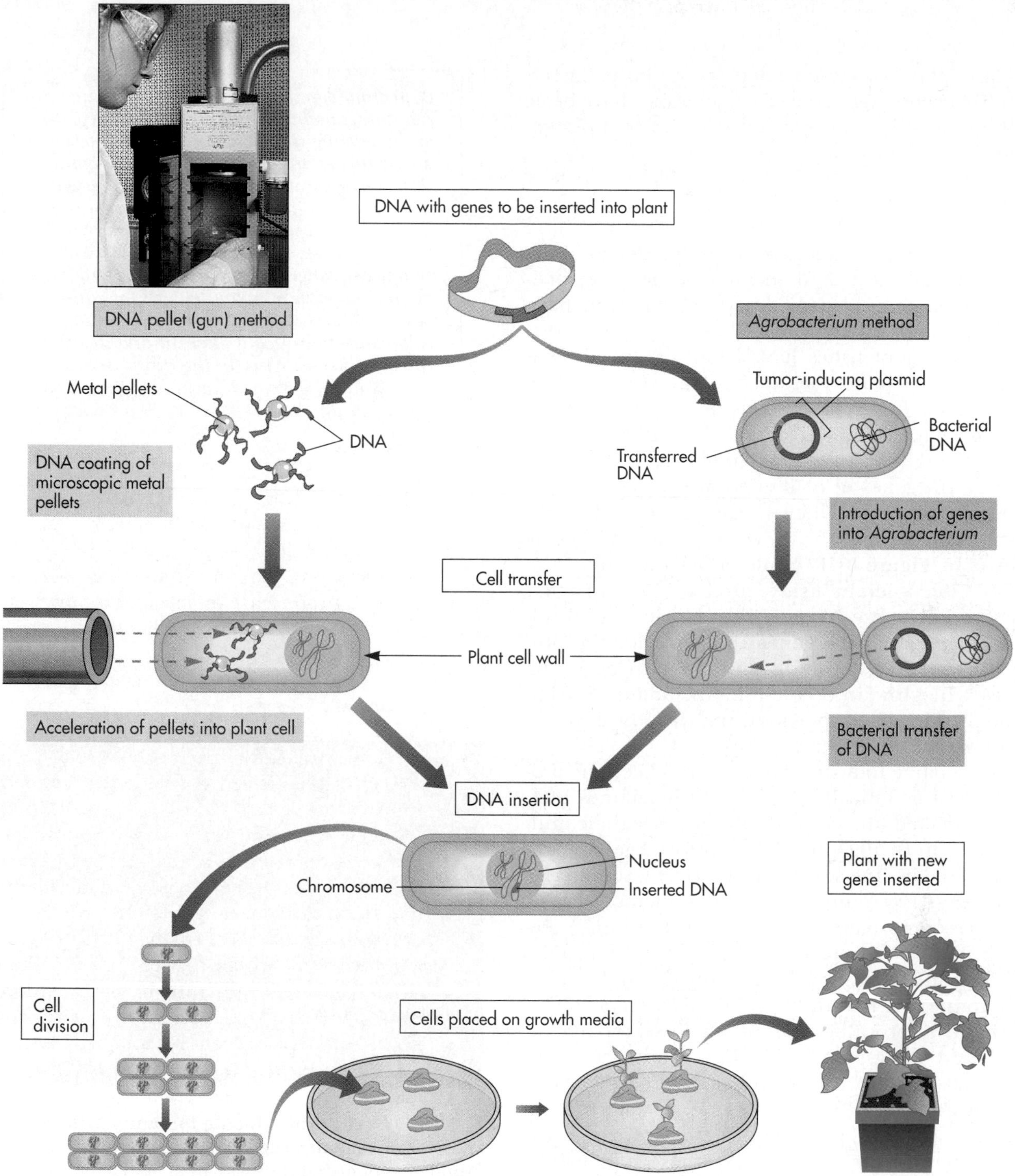

FIGURE 31-16 Genetic engineering of plants. This diagram shows two ways to insert a gene of interest into a plant. First, the gene is cut from the DNA of the donor organism by using restriction enzymes. Then, metal pellets can be coated with the DNA (*left*) and "shot" into plant cells. The DNA is taken up by the nucleus of the cell, and the genetically engineered cells are then cloned and propagated into new plants. Alternatively, the DNA can be inserted into bacterial plasmids of the tumor-inducing bacterium *Agrobacterium* (*right*). This bacterium has the ability to transfer plasmid DNA to plant cells. The DNA is then taken up by the nucleus of the cell, and the genetically engineered cells are cloned and propagated into new plants as with the "metal pellet" method.

The Human Genome Project (HGP) a worldwide scientific project to decipher the DNA code of all 46 human chromosomes.

transgenic plants (tranz **gee** nik) plants that are genetically altered using the techniques of genetic engineering.

Many plant species are not natural hosts for the *Agrobacterium* organisms; thus scientists have been looking for other ways to insert genes into plants. One commonly used method shoots microscopic metal pellets coated with DNA into plant cells (see Figure 31-16). Then the cells are cultured and propagated.

Transgenic animals have been produced since the 1980s and are used in targeted gene replacement research as already described, but none have been approved for commercial use. So far, 13 transgenic species of fishes have been developed. One major goal of marine biotechnological research is to produce fishes (the focus has been Atlantic salmon) with an enhanced tolerance to cold water. Genetic engineers are experimenting with injecting certain genes of cold-tolerant species of fishes into salmon eggs, which are large and relatively easy to microinject with DNA (A mouse egg is shown in Figure 31-17 undergoing the same process.). Cold-tolerant fishes produce proteins that act like antifreeze in their blood; scientists think that by injecting the genes that code for the production of these proteins into salmon eggs, the fishes will grow into cold-tolerant adults. Marine biologists are also experimenting with the development of disease-resistant fishes.

In the dairy industry, the focus has been the production of genetically engineered hormones that, when injected into cows, would increase their milk production. In 1993, the FDA approved genetically engineered bovine somatotropin (bST), sometimes called rbST (recombinant bovine somatotropin) or rbGH (recombinant bovine growth hormone). Recombinant bST is made by a process similar to the production of synthetic human insulin and is identical to the one naturally produced by cows. Before giving its approval for commercial use, the federal Food and Drug Administration conducted more than 120 studies on bST and concluded that milk and meat from bST-treated cows are safe to consume. In addition, pasteurization destroys approximately 90% of the bST in milk. These conclusions have been affirmed by a variety of regulatory agencies, such as the National Institutes of Health in the United States as well as those in Canada and abroad.

Biotechnology, which is being used in a variety of ways in the food industry, focuses on increasing the yields of various foods and creating products with superior qualities. The most widespread use is in agriculture; scientists are working to improve crops and forest trees by making them more resistant to disease, frost, and herbicides. Many genetically engineered agricultural products are on the market today. Regarding animals, research is being conducted to develop varieties of cold-tolerant and disease-resistant fishes. Also, in the dairy industry, genetically engineered hormones have been produced that increase the milk production of cows.

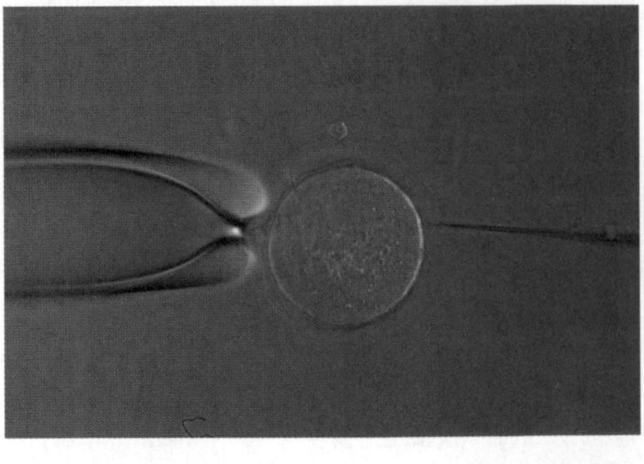

FIGURE 31-17 Introducing foreign genes into the nucleus of a mouse egg. The blunt tip of the glass micropipet supporting the egg with a slight suction can be seen on the left. The egg is in the center, and the microneedle delivering the DNA is on the right. After this process is completed, the egg will be placed in the oviducts of a foster mother so it can develop.

HOW *Science* WORKS

Tools of Scientists

Gel Electrophoresis

If you were a genetic engineer, much of your laboratory work would involve separating, analyzing, and purifying proteins and other macromolecules such as DNA restriction fragments. What methods would you use? One standard laboratory technique widely used by molecular biologists is *gel electrophoresis*—a technique developed more than 100 years ago! Gel electrophoresis, a tool in molecular biology with wide application in genetic engineering today, is one of many technologies—such as the microscope—that was developed long ago and has been refined and adapted to help answer new questions in new contexts over time. It is a technology that has never lost its usefulness. In fact,

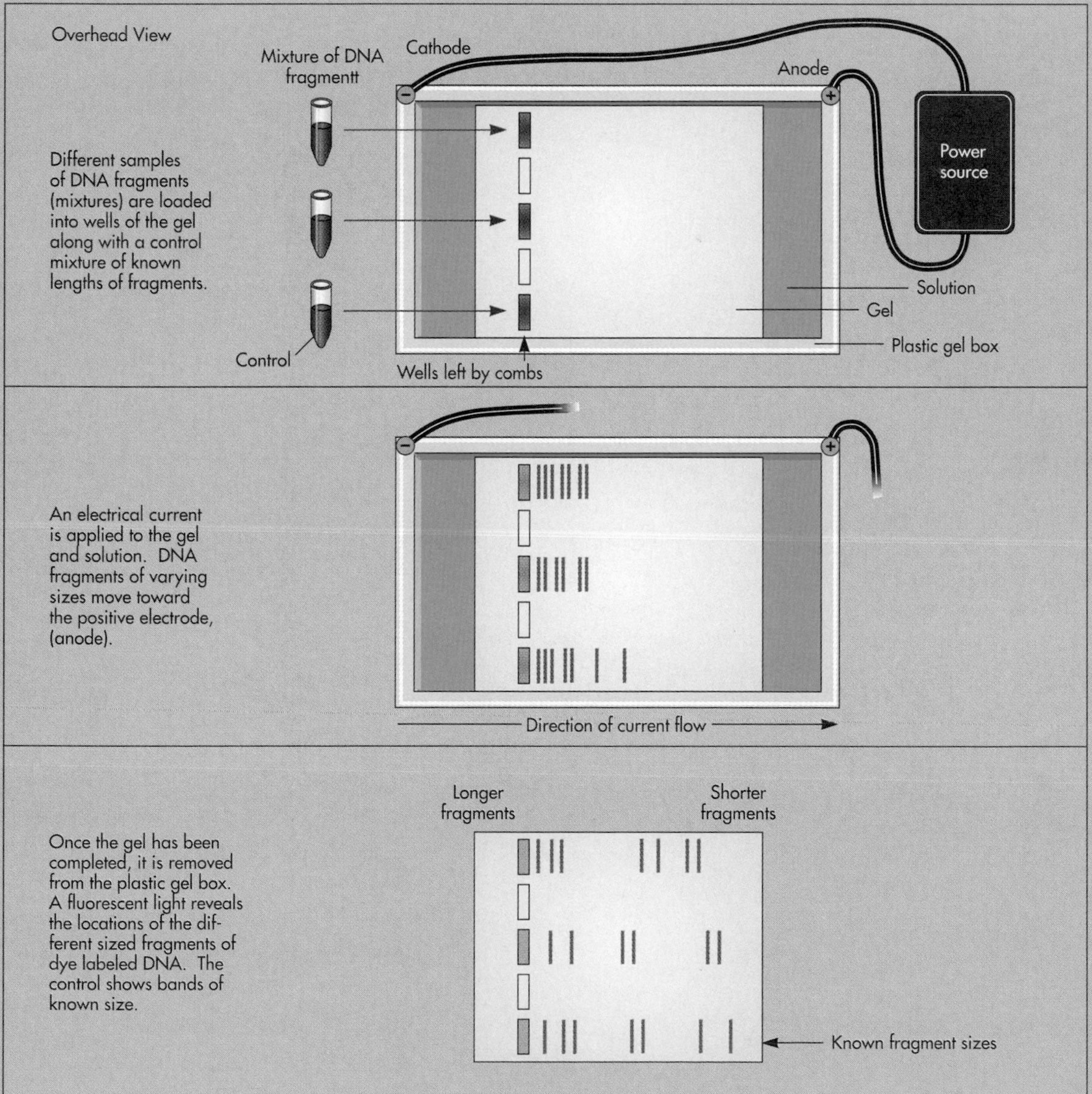
Overhead View
Mixture of DNA fragmentt
Cathode
Anode
Power source
Different samples of DNA fragments (mixtures) are loaded into wells of the gel along with a control mixture of known lengths of fragments.
Solution
Gel
Plastic gel box
Control
Wells left by combs
An electrical current is applied to the gel and solution. DNA fragments of varying sizes move toward the positive electrode, (anode).
Direction of current flow
Longer fragments
Shorter fragments
Once the gel has been completed, it is removed from the plastic gel box. A fluorescent light reveals the locations of the different sized fragments of dye labeled DNA. The control shows bands of known size.
Known fragment sizes

Tools of Scientists

HOW Science WORKS

Gel Electrophoresis—cont'd

Today, gel electrophoresis helps accomplish such procedures as analyzing DNA fragments from crime scenes and comparing the DNA of organisms to determine their evolutionary relatedness—tasks never thought of in the nineteenth century.

The word *electrophoresis* refers to a phenomenon in which molecules with net charges migrate, or move, in an electric field. Electrodes are placed at either end of a thin slab of a gel-like material that feels like firm Jell-O to create an electric field. The gel is formed by pouring the hot liquid gel medium into a glass or plastic container. Notches called *wells* are formed with "combs" at one end of the gel as it hardens. The samples to be electrophoresed are pipetted into the wells once the gel has cooled, the combs have been removed, and a solution (through which the current will run) has been poured over the gel (Figure 31-A). (A pipette is a mechanical measuring device that looks somewhat like a pointed glass straw.) A negatively charged electrode is placed at the end of the gel containing the wells, and a positively charged electrode is placed at the opposite end. When an electric current is applied, the molecules move from the positive end to the negative end of the gel but at different rates depending on their net charges and their sizes.

Macromolecules often have portions of their structures that are positively or negatively charged. These charges do not usually balance each other exactly; the remaining charge is the net charge. The more negatively charged a particle is, the more quickly it will move toward the positive pole and vice versa. Another factor that affects the movement of molecules through the gel is the size and shape of the molecules. The gel acts like a strainer, having pore-like spaces through which the molecules move. Molecules that are small move through the pores easily and quickly, whereas larger molecules move more slowly as they "squeeze" through the pores (Figure 31-B). Scientists can control the movement of the molecules by manipulating the concentration of the gel ingredients, thereby controlling the pore size, and by manipulating the strength of the electric current and the manner in which it is applied (pulsing or not pulsing the current, for example). The result is that the molecules within the samples placed in each will move through the gel at differing

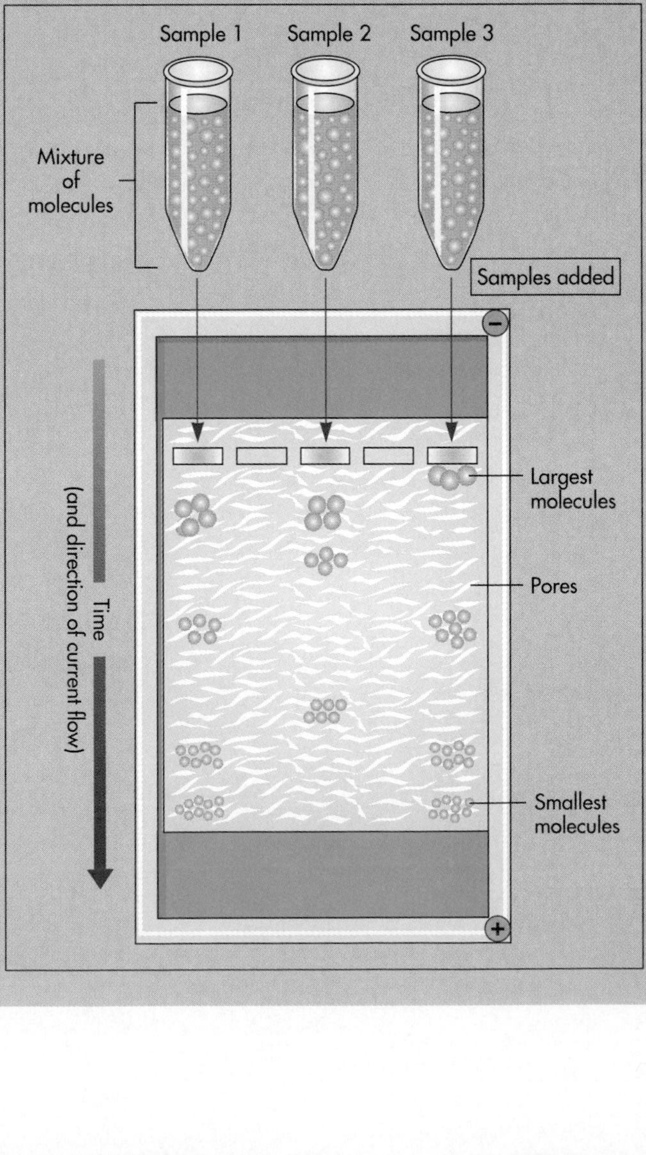

31-B

Tools of Scientists

Gel Electrophoresis—cont'd

rates, creating a series of lines, or bands (see Figure 31-A). These bands are visualized by staining the gel with a dye.

Gel electrophoresis has become an important tool for a variety of researchers and scientists, but the most visible use to the general public is in forensics. Because human genomes reflect the variability that exists within the human species, any material containing DNA found at a crime scene can be used to determine whether blood, semen, or hair, for example, came from the victim, the suspect; or another (unknown) individual.

A sample of restriction fragments of an individual will show differences in the length of one or more of the fragments when compared with the same restriction fragments of another individual, reflecting slight variations within their genetic codes. These differences occur in various locations in the human genome and result in differences in the length of a variety of restriction fragments. Particular restriction fragments from a single individual will always produce the same banding pattern (or *DNA fingerprint*) on gel electrophoresis under the same conditions. These same restriction fragments from another individual will show differences in their banding patterns; the fingerprint will be different. These techniques and related techniques have been used increasingly in recent years to free falsely accused rapists in prisons across the United States and to provide evidence in a variety of murder trials. It seems amazing that a nineteenth-century technology is helping solve twentieth-century crimes in a very high-tech way.

In the future, electrophoresis technology will surely become even more high-tech. In 1994, Dr. Jed Harrison and associates at the University of Alberta in Edmonton, Canada, reported their progress in making a micro-sized electrophoresis apparatus. They cut channels only microns deep in a centimeter-square piece of glass—that is, less than one-fourth the size of a postage stamp! After attaching microelectrodes to either side of this glass chip, they introduced amino acids tagged with fluorescent dyes at one end of the channels and separated them into discrete bands in about 15 seconds, about 60 times faster than the speediest electrophoresis equipment in use today. Electrophoresis on a chip may be the next modification in this enduring technology.

Tools of Scientists

HOW *Science* WORKS

DNA Sequencing

In order to study gene organization, regulation, and function; clone and manipulate genes in the laboratory; prepare probes and primers; and for a variety of other uses scientists must first identify genes and then determine their structures. The primary structure of a gene is determined by the sequence of its nucleotide bases.

One popular method used to determine the structures of genes, a process called *DNA sequencing,* is the controlled interruption of replication. This method is usually referred to by its developer's name—the Sanger method. This method was the starting point for Kary Mullis when he developed the Polymerase Chain Reaction.) Variations of this method allow one to complete single- or double-stranded DNA sequencing. To carry out single-stranded DNA sequencing using the Sanger method, DNA fragments are first heated and treated with chemicals, which unwind and dissociate the double-stranded molecules into complementary single strands. A preparation is made of one of these complementary DNA strands; each piece is referred to as a *template*. The template is the piece of DNA with the unknown sequence of bases, which will be determined by the sequencing process.

As shown in Figure 31-C, the template is mixed with a small amount of a short piece of nucleic acid called a *primer* that is complementary to a short nucleotide base sequence on the DNA template. In our example, the primer is radioactively "labeled." This radioactive labeling is the means by which reaction products can be detected during the last step of the sequencing procedure. In addition, individual nucleotides (called deoxynucleotides and shown as dATP, dCTP, dGTP, and dTTP) and the enzyme DNA polymerase are added to the mix.

A small amount of this mixture is then added to four test

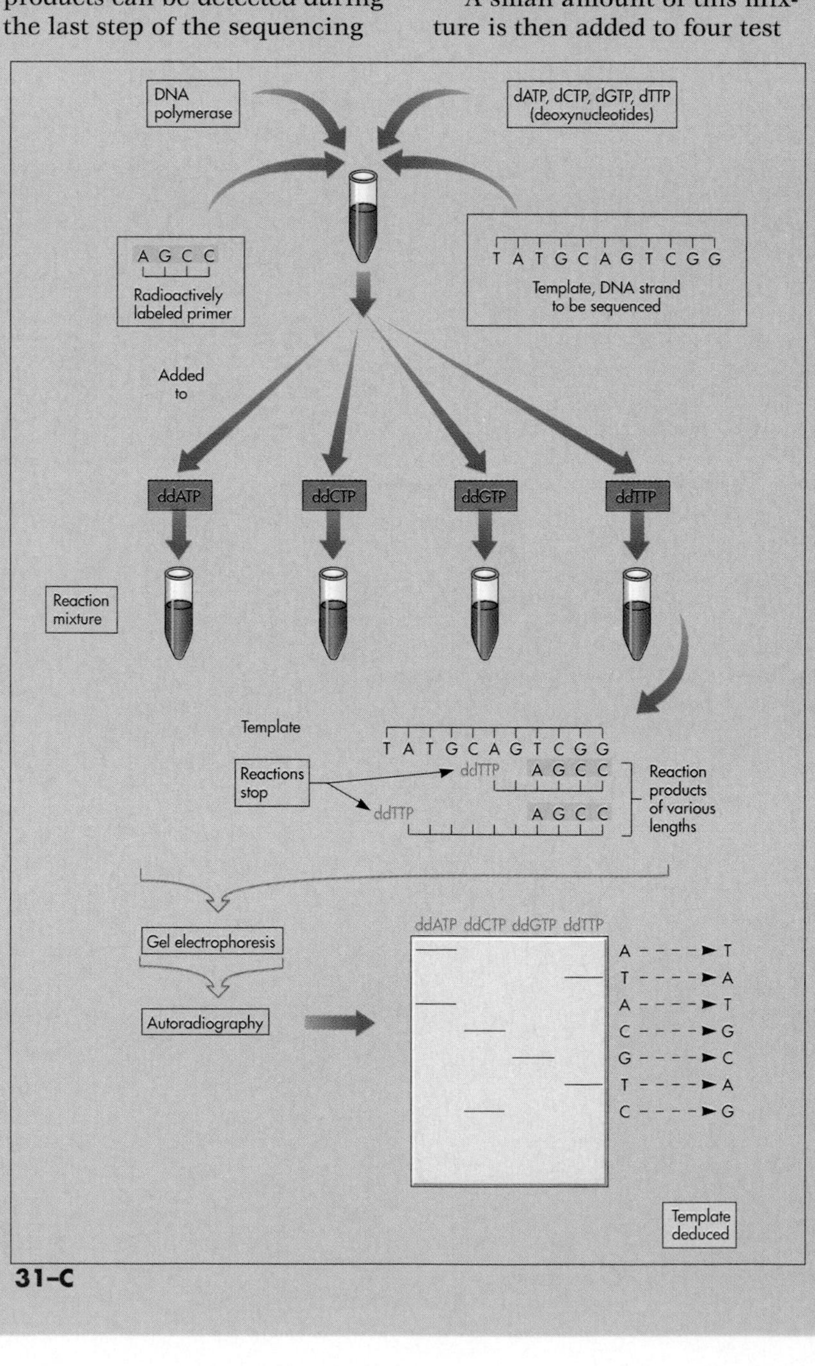

31-C

Tools of Scientists

DNA Sequencing—cont'd

tubes, each of which contains an *analog* of one of the nucleotides (shown as ddATP, ddCTP, ddGTP, and ddTTP). An analog is something that is similar to something else; an analog nucleotide is the same as other nucleotides except where it bonds with the next nucleotide in sequence, resulting in its inability to bond at that site. Each test tube, therefore, contains a solution of only one of the four possible modified nucleotides.

DNA polymerase sequentially adds nucleotides to templates hybridized with primers, extending the primers. However, where ever the analog bonds to the DNA template, the extension of the primer stops. As a result, four sets of DNA fragments are produced (one set for each analog) that each have hybridized strands of differing lengths. The strands in each test tube, however, no matter how long or short, always end with the nucleotide analog present in that tube.

The four samples are then electrophoresed (see Boxed Essay), and the base sequence of the new DNA is read from the autoradiogram (see p. 693) of the four lanes as is illustrated. The template sequence is deduced from this sequence on the basis of the complementarity of the nucleotide bases of DNA.

DNA sequencing is automated today, using a variation of the Sanger method. Instead of radioactively-labelling the primers, scientists use fluorescent dyes of different colors to label the primers. The primers added to each of the four test tubes are each labeled with a different color, such as using blue-emitting primers in the tube in which the adenine analogs are placed, green-emitting primers in the tube in which the thymine analogs are placed, yellow in the guanine tube, and red in cytosine tube. After the reactions are completed, the solutions are mixed and electrophoresed together. The sequence of the colors, which is determined by measuring the wavelengths of the bands, shows the sequence of the bases. Figure 31-D shows and example using a sequence different from the one shown in 31-C.

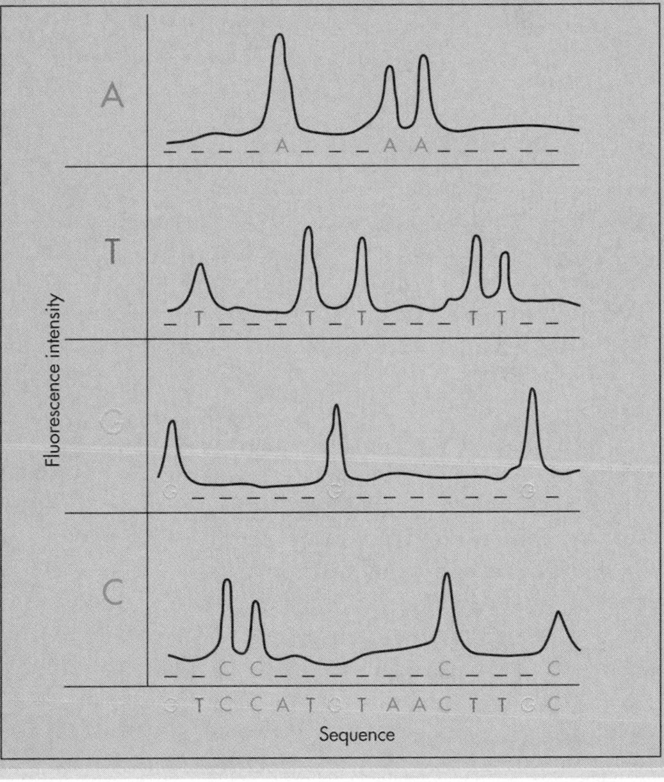

31-D

Summary

- Biotechnology is the manipulation of organisms to yield organisms with specific characteristics, organisms that produce useful products, and information about an organism or tissue that would otherwise not be known. It has been practiced for more than 10,000 years. Classical biotechnology uses the techniques of selection, mutation, and hybridization. The advent of molecular biotechnology, which involves the manipulation of the genes themselves, began with the discovery of the structure of the genetic material DNA in the 1950s.
- Certain bacteria can naturally transfer genetic material from one cell to another by transformation, transduction, and conjugation. Scientists were able to manipulate these mechanisms of gene transfer in the 1950s and 1960s and, by the 1970s, discovered how to combine DNA from different species of organisms and manipulate genes from eukaryotes. Techniques of gene manipulation are called *genetic engineering* or *recombinant DNA technology.*
- One of the first applications of recombinant DNA technology was the insertion of human genes into bacteria, causing these organisms to produce human proteins for therapeutic use. The genes that code for the desired protein must first be identified within the human genome to produce human proteins by means of recombinant DNA technology. Scientists use various methods to find genes in which they are interested.
- One method of finding a gene is shotgun cloning, which is used when researchers know very little about the gene they are trying to find and which results in a complete gene library: a collection of copies of DNA fragments that represent the entire genome of an organism. Shotgun cloning involves cutting the DNA of the entire genome into pieces with restriction enzymes, inserting these pieces or fragments into bacteria or yeast with plasmids or viruses, and allowing the organisms to reproduce, making copies, or clones, of the DNA fragments. Two other methods, used under different circumstances, are complementary DNA cloning and gene synthesis cloning.
- Most screening techniques used to find a gene of interest within a gene library involve DNA hydridization of molecular probes, molecules that bind to specific genes, or nucleotide sequences.
- After a gene has been isolated, many copies of it are needed to make enough recombinant cells as an inoculum for large-scale culturing. The polymerase chain reaction is a process used to produce multiple copies of a gene. This technique has wide application in molecular biotechnology.
- Many products have been developed by genetic engineering to treat a variety of human disorders or diseases. These products include human insulin, used to treat diabetes; human growth hormone, used to treat children who do not produce enough of this hormone for normal growth; the human interferons, proteins that interfere with the ability of a virus to invade a cell; and hepatitis B vaccine, which helps protect against the leading cause of liver cancer.
- Genetic engineering has also been used to treat diseases by means of a technique called *gene therapy.* Gene therapy is the treatment of a genetic disorder by the insertion of "normal" genes into the cells of a patient to replace nonfunctional or malfunctional genes. It is currently being used to successfully treat a rare genetic disorder called *severe combined immunodeficiency* as well as certain types of cancer.
- The Human Genome Project (HGP) is a worldwide effort to sequence the DNA of the entire human genome. This project will help identify genes that cause various diseases. Along with information from targeted gene replacement research, the HGP will help elucidate relationships between the structure of genes and the proteins they produce. The process of gene targeting involves changing the nucleotide sequence of a particular gene, which changes the function of the gene, and then observing the resultant changes in the anatomy, physiology, or behavior of the organism.
- Genetic engineering is being used in a variety of ways in the food industry. There, it focuses on increasing the yields of various foods and creating products with superior qualities. In agriculture, scientists are improving crops and forest trees by making them more resistant to disease, frost, and herbicides. Regarding animals, research is being conducted to develop varieties of cold-tolerant and disease-resistant fishes. In the dairy industry, genetically engineered hormones have been produced that increase the milk production of cows.

Knowledge and Comprehension Questions

1. Define the term *biotechnology*. Distinguish between biotechnology and genetic engineering.
2. What are the differences between classical biotechnology and molecular biotechnology? Are both being practiced today? Support your answer with evidence.
3. Name and describe the three methods of natural gene transfer among bacteria.
4. Describe one way in which scientists insert human genes into bacteria using the natural ability of certain bacteria to transfer genetic material.
5. Define the term *gene cloning*. Name three types of gene cloning and briefly characterize each.
6. What is the role of restriction enzymes in genetic engineering?
7. What is a molecular probe? What is its function in screening clones?
8. Why is the polymerase chain reaction an important technique in genetic engineering today?
9. List two genetically engineered proteins in use today and describe their use.
10. How does the process of gene therapy treat disease? Is this technique in widespread use today? Support your answer with evidence.
11. What are the main goals of the Human Genome Project and the purpose for achieving these goals? How will research in targeted gene replacement help scientists realize their purpose?
12. Describe an application of genetic engineering in agriculture.

Critical Thinking

1. If scientists can engineer bacteria to produce human proteins, why can't they genetically engineer bacteria to synthesize gold or other precious metals?
2. What advantages do you think exist for a person using genetically engineered human insulin versus insulin extracted and purified from an animal source? Discuss a recipient's possible reaction to human (bacterial) insulin versus animal-derived insulin.

PART SEVEN

HOW ORGANISMS INTERACT WITH EACH OTHER AND WITH THE ENVIRONMENT

CHAPTER 32

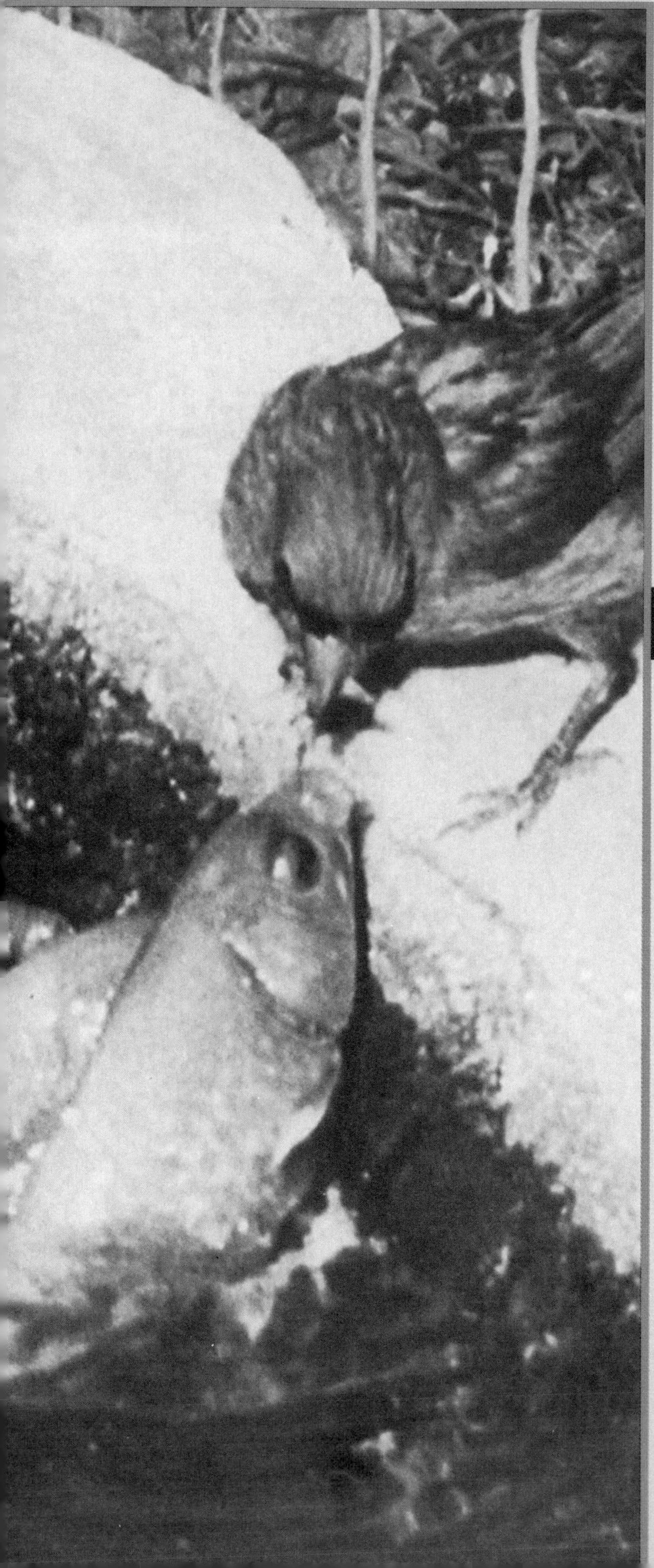

*I*NNATE BEHAVIOR AND LEARNING IN ANIMALS

A CASE OF MISTAKEN IDENTITY? It would certainly seem to be so. Scientists, however, would describe this "identity crisis" as a case of misfiring. Misfiring refers to a behavior that is normally appropriate being exhibited in an inappropriate circumstance.

Female birds like this cardinal are stimulated to feed their young when signaled by their tiny, gaping mouths. Such a signal is called a *releaser.* Releasers are structures or behavior patterns exhibited by one member of a species that trigger a response—a series of precise physical movements—in another member of that species. The behaviors triggered by releasers are also influenced by hormones or other internal stimuli. This cardinal recently lost her nest of young but is still hormonally "driven" to feed them. Therefore the tiny, wide-open mouth triggered a response in the cardinal—a response that will be quite a surprise for the fish!

The Study of Animal Behavior

Animals (including humans) continually exhibit a wide range of behaviors. **Behaviors** include the patterns of movement, sounds (vocalizations), and body positions (postures) exhibited by an animal. In addition, behaviors include any type of change in an animal, such as a change in coloration or the releasing of a scent, that can trigger certain behaviors in another animal.

Some behaviors are simple, automatic responses to environmental stimuli. A bacterium "behaves" when it moves toward higher concentrations of sugar. You behave when you slam on the brakes to avoid a car accident. Other types of behaviors, such as the eating behavior of the sea otter, are quite complex (Figure 32-1). Complex behaviors are limited to multicellular animals with a neural network that can sense stimuli, process these stimuli in a central nervous system (brain), and send out appropriate motor impulses. You exhibit complex behavior when you walk and talk, drive a car, interact with your family, and do your job.

Before the late 1950s, the study of animal behavior was dominated by **ethology.** Using a physiological perspective, ethologists observed and interpreted the behavior of animals in their natural environments. They broke down behavior patterns into recognizable units, named these patterns, and categorized them. They often focused on the activity of the nervous system and also sought to understand the connection between an animal's behavior and its genetic makeup. They were particularly interested in the types of motor movements and responses to stimuli that typically occurred within closely related species as they attempted to uncover the biological significance of animal behavior—the importance of a behavior for a particular species in its natural environment.

One of the most famous ethologists is the Austrian scientist Konrad Z. Lorenz (see Figure 32-7), referred to by some as the father of modern ethology. In 1973, Lorenz, Dutch ethologist Nikolaas Tinbergen, and Austrian zoologist Karl von Frisch won the Nobel prize for their contributions to the study of animal behavior. Lorenz based his work on the premise that animal behaviors are evolutionary adaptations, as are physical adaptations. He re-

KEY CONCEPTS

- Ethology, the study of animal behavior in the natural environment, examines the biological basis of the patterns of movement, sounds, and body positions of animals. This science provided a foundation for the study of animal behavior, which is now conducted by scientists from multiple disciplines.
- All behaviors depend on nerve impulses, hormones, and other physiological mechanisms; behaviors are therefore directed in part by the genes that control the development of these systems and mechanisms.
- Certain animal behaviors are unchangeable and are performed correctly the first time they are attempted; these innate, or instinctive, behaviors are directed by nerve pathways developed before birth.
- Animals that can change their behaviors based on experience are capable of learning, a process that helps animals adapt to changes in the environment and increases their reproductive fitness.

OUTLINE

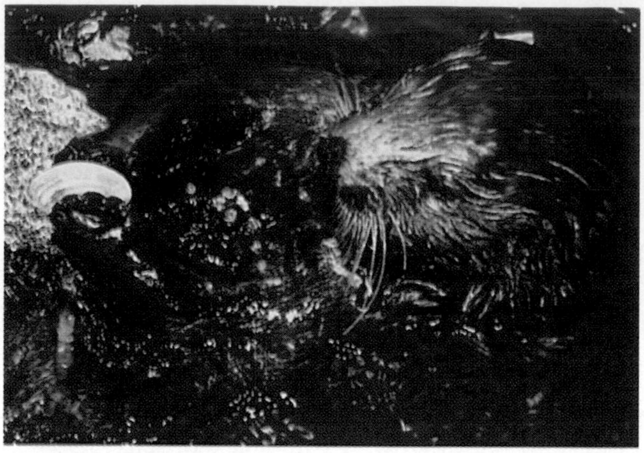

FIGURE 32-1 Complex behavior in a sea otter. This sea otter is having dinner while swimming on its back. It is using the rock as a hard surface against which to hit the clam and break it open. Often a sea otter will keep a favorite rock for a long time, suggesting that it has a clear idea of what it is going to use the rock for. The sea otter may learn this pattern of eating behavior from others while young, but the capacity to use tools and consciously foresee their future use certainly depends on inherited abilities.

ferred to behaviors as being part of an animal's "equipment for survival." This concept is still a fundamental premise in the study of animal behavior.

> *Before the late 1950s, ethology dominated the study of animal behavior. Ethologists study the behaviors of animals in their natural environments, focusing on the patterns of movement, sounds, and body positions exhibited by animals.*

Another group of animal behavior researchers prominent before the late 1950s was the *behaviorists.* These psychologists focused on behaviors themselves, studying them in the laboratory *without* focusing on the mental (cognitive) events that took place during the behaviors. In the late 1950s and the early 1960s, the fields of ethology and behavioral psychology merged to form the discipline of **animal behavior.** This approach includes many features derived from both the behaviorists and the ethologists.

The field of animal behavior expanded in the late 1970s as the science of *behavioral ecology* emerged. This biological discipline has an evolutionary focus. The central assumption of behavioral ecology is that an animal will behave in ways that will benefit it in the short term and that will maximize its Darwinian fitness, or ability to achieve reproductive success.

Today, the field of animal behavior is composed of researchers from a variety of disciplines: psychology, sociobiology (see Chapter 33), neurobiology, and behavioral ecology. Researchers from these fields study animal behavior from distinct perspectives, using differing methodologies and basing their work on differing underlying assumptions and philosophies. Many biologists suggest that behavioral ecology is the dominant approach to the study of animal behavior today. Other scientists disagree. Whatever the case, a main difference between the work of animal behaviorists today and the ethologists of the past is explanation. Ethologists described, characterized, and cataloged behavior; animal behaviorists today pose hypotheses and test their predictions in an effort to determine *how* behavioral mechanisms work and *why* they evolved.

"How" and "why" questions are two types of causal questions. "How" questions refer to the *immediate cause*—the mechanisms underlying the particular behavior. "Why" questions refer to the *ultimate cause.* Behavioral ecologists, for example, frame "why" questions in evolutionary terms, asking why natural selection favored a particular behavior and not another.

> *The study of animal behavior today is conducted by researchers from a variety of disciplines. Behavioral ecology, a prominent biological approach to the study of animal behavior, has an evolutionary focus.*

The Link between Genetics and Behavior

All behaviors depend on nerve impulses, hormones, and other physiological mechanisms such as sensory receptors. Therefore genes play a role in the development of behaviors because they direct the development of the nervous system. In addition, automatic responses depend on specific nerve pathways within the central nervous system of an organism. These pathways are neural programs and are genetically determined.

Margaret Bastock of Oxford University conducted classic experiments in the 1950s to show that certain behavioral traits are under the control of single genes. She found that male fruit flies *(Drosophila melanogaster)* having a mutant sex-linked gene termed "yellow" displayed a courtship pattern with less wing vibration than that of males having the "normal" gene. This altered courtship pattern resulted in the mutant males being less successful than normal males in mating normal females.

JUST WONDERING. . . .

Why do people yawn?

Your question could be broadened to ask why most animals yawn—there is nothing particularly human about yawning. Most carnivores yawn. Few herbivores seem to yawn, although there are exceptions—hippos have enormous yawns. Frogs yawn. Even fish appear to yawn. Whatever is going on, it must be a basic behavior, like eating or sleeping, that is exhibited in many animals.

It used to be hypothesized that a yawn was a silent scream for oxygen that usually occurs when people are tired or bored—a type of deep breath to increase oxygen in the blood or to get rid of excess carbon dioxide. Not so. When a group of freshman psychology students at the University of Maryland inhaled air containing different mixtures of oxygen and carbon dioxide and counted their yawns, only breathing rates went up or down to compensate for changing levels of oxygen and carbon dioxide. Yawning rates did not change.

Some researchers think that yawning is the body's way of promoting arousal in situations where you have to stay awake. Most humans yawn when stimulation is lacking. When a team of yawn counters observed people engaged in various activities, they found data to support this "lack of stimulation" idea. For example, they found that people riding subway cars yawned far more often when the cars were empty. The arousal hypothesis helps explain why people driving late at night on the highway yawn a lot and why very few people yawn when they are actually in bed—they don't need to stimulate themselves with a yawn because it's okay to go to sleep.

Whether this hypothesis is plausible or not, one thing is certain: yawning is highly contagious. Seeing another person yawn releases a powerful urge for you to yawn yourself. This trait seems to be restricted to humans; no other animal responds in this way. Although yawning is not a hotbed of research, there is still much to learn about this behavior, because yawning seems associated with many diseases in ways not yet understood—brain lesions and epilepsy often lead to excessive yawning, but schizophrenics yawn very little.

Genetically determined neural programs are part of the nervous system at the time of birth or develop at an appropriate point in maturation, resulting in **innate behaviors.** Interestingly, these instinctive or inborn behaviors are performed in a reasonably complete form the first time they are exhibited. A human newborn, for example, will turn to suckle when touched on the cheek near the mouth. Innate behaviors are important to the survival of an animal because they help it stay alive in

behaviors the patterns of movement, sounds (vocalizations), and body positions (postures) exhibited by an animal. Behaviors also include any type of change in an animal, such as a change in coloration or the releasing of a scent, that can trigger certain behaviors in another animal.

ethology (ee **thol** uh jee) the study of animal behavior in the natural environment. Ethology examines the biological basis of the patterns of movement, sounds, and body positions of animals.

animal behavior a scientific discipline that was formed when the fields of ethology and behavioral psychology merged. The field of animal behavior applies both the physiological perspective of the ethologist and the psychological perspective of the behaviorist to animal study. Today, the field of animal behavior is composed of researchers from a variety of disciplines.

innate behaviors behaviors resulting from genetically determined neural programs that are part of the nervous system at the time of birth or develop at an appropriate point in maturation.

certain situations and provide adaptive advantages that contribute to its fitness, or ability to achieve reproductive success.

Ethologists such as Lorenz and animal behaviorists of today would argue about the role of environment and learning on innate behaviors. **Learning** is an alteration in behavior based on experience. Today, the rigid distinction between innate and learned behavior that was once held by ethologists no longer exists. Evidence suggests that many aspects of behavior are influenced by both genetic factors *and* the experience of the individual. Innate behaviors are now thought to be those that occur without *obvious* environmental influence.

> *The development of the nervous system and certain automatic nervous responses that are "preprogrammed" within the nervous system are directed by genes and occur without obvious environmental influence.*

Innate Behaviors: Coordination and Orientation

To survive, animals must respond to the environment. To do this, they must coordinate their movements in ways that result in effective responses. Such behaviors are called *coordination behaviors.* In addition, they must orient their movements in relation to external stimuli. These types of behaviors are termed *orientation behaviors.* Certain types of responses are characteristic of a species and thus are considered innate. A reflex is the simplest type of innate reaction to a stimulus and is an example of a coordination behavior involving various muscles. Kineses and taxes are simple types of orientation behaviors.

Reflexes

A **reflex** is an automatic response to nerve stimulation. The knee jerk is one of the simplest types of reflexes in the human body. If the tendon just below the kneecap is struck lightly, the lower leg automatically "kicks" or extends. This behavior occurs as the tendon is stretched and pulls on the muscles of the upper leg that are attached to it. Stretch receptors in these muscles immediately send an impulse along sensory nerve fibers to the spinal cord. At the spinal cord these fibers synapse directly with motor neurons that extend back to upper leg muscles. The upper leg muscles are stimulated to contract and jerk the leg upward while opposing muscles simultaneously relax (see Figure 14-13). These reflexes play an important role in maintaining posture. More complex reflexes involve a relay of information from a sensory neuron through one or more interneurons to a motor neuron. The outcomes of these reflexes are modulated by the interneurons.

In complex organisms, reflexes play a role in survival, such as when you jerk your hand away from a hot stove before you consciously realize that your hand hurts. In animals with extremely simple nervous systems such as the cnidarians (hydra, jellyfish, sea anemones, and coral, see pp. 641-642), most behaviors are the result of reflexes, although some simple learning can take place. In these animals a stimulus is detected by sensory neurons, and the impulse is passed on to other neurons in the animal's nerve net, eventually reaching the body muscles and causing them to contract. There is no associative activity in which other neurons can influence the outcome, no control of complex actions, and little coordination. The nerve net of the cnidarians possesses only the barest essentials of nervous reaction.

Kineses

A **kinesis** (pl. kineses) is the change in the speed of the random movements of an animal with respect to changes in certain environmental stimuli. Put simply, movement slows down in an environment favorable to the animal's survival and speeds up in an unfavorable one. Have you ever picked up a rotting log or a clump of damp leaves in a wooded area? The pillbugs that you may have seen living under the leaves or the log (Figure 32-2) stay in this favorable environment because of their low levels of activity and then scurry when the log is rolled over.

FIGURE 32-2 Kinesis. These pillbugs scurry when the log they are living under was moved. This change in the level of activity that occurs with a change in the environmental stimulus is called kinesis.

Taxes

Kineses are nondirected types of movements. In other words, an animal is not attracted by a favorable environment; it tends to "blunder" there as it moves about quickly and then stays there as it moves about slowly. A **taxis** (pl. taxes), however, is a directed movement toward or away from a stimulus, such as light, chemicals, or heat. Animals having preprogrammed taxes also have receptors that sense the particular stimuli to which the animal can orient. Female mosquitoes and ticks, for example, have sensory receptors that detect warmth, moisture, and certain chemicals emitted by mammals. Sensing these stimuli helps the insects orient to their victims. In fact, some mosquito repellents work by "blocking" the insect's receptors so that it cannot sense and then locate its victim. The crowding of flying insects about outdoor lights is another familiar example of a taxis called *phototaxis*. Other insects, such as the common cockroach, avoid light (are negatively phototactic). Fish swim upright by orienting to gravity and light. If the gravity-detecting organ in the inner ear is removed or if a light is shone into the sides or bottom of a fish tank, the fish become disoriented and do not swim upright (Figure 32-3). Certain species of fishes such as trout and salmon also automatically orient against a current and therefore face and swim upstream

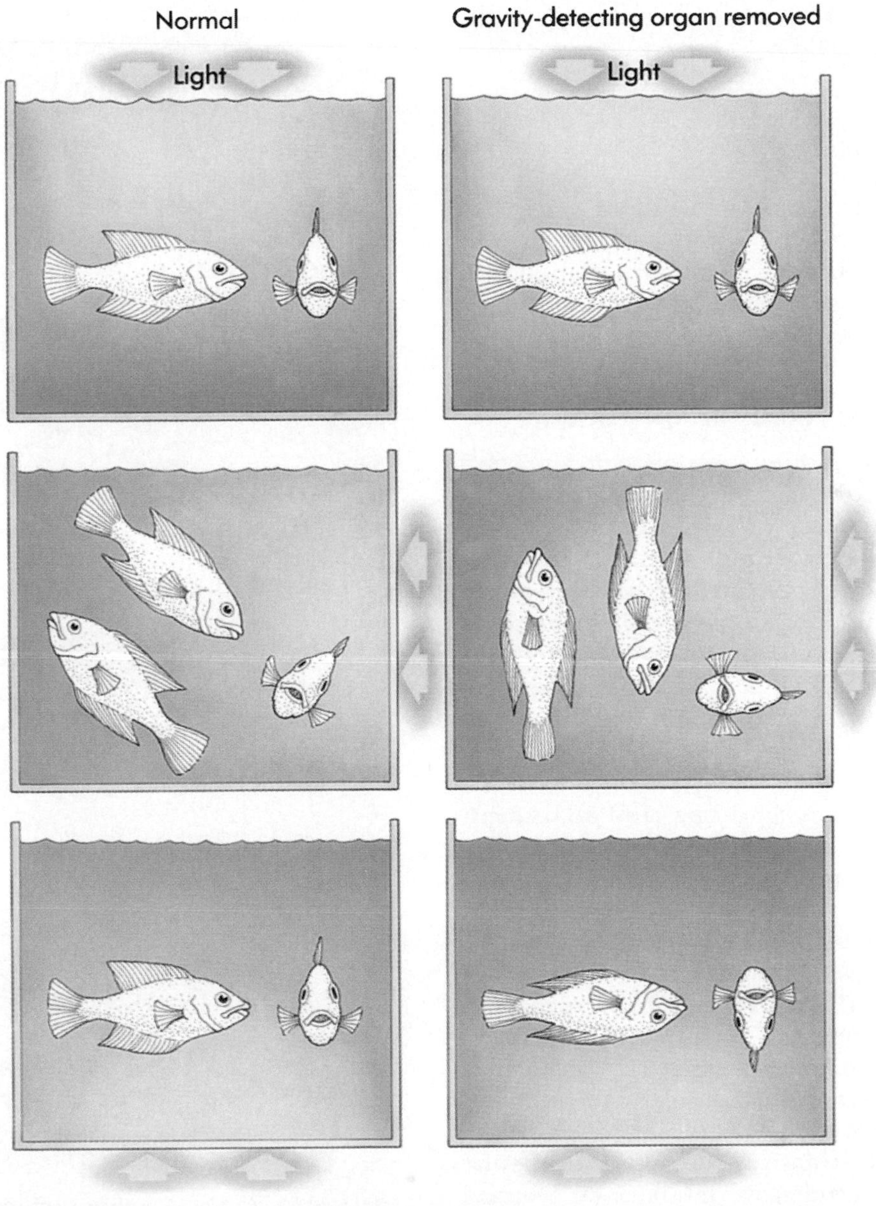

FIGURE 32-3 Phototaxis and gravitaxis in fish. The normal fish in the left column orient to both light and gravity. The fish in the right column, which have had their gravity-detecting organ removed, orient to light only and have difficulty staying upright. The lower left diagram points out that orientation to gravity overrides orientation to light.

FIGURE 32-4 Taxis in a salmon. Salmon automatically orient themselves against a current and therefore always swim upstream.

(Figure 32-4). Even organisms without a nervous system such as protozoans and bacteria exhibit taxes. These organisms have cellular organelles or inclusions that act as receptors or that react to specific environmental stimuli.

Innate Behaviors: Fixed Action Patterns

More complex than other innate behaviors are **fixed action patterns.** Fixed action patterns are sequences of innate behaviors in which the actions follow an unchanging order of muscular movements, such as the mother cardinal popping an insect into the mouth of her young. This sequence is a recognizable "unit" of behavior. When a releaser triggers the behavior, the sequence of activity begins and is carried through to completion. Fixed action patterns are often seen in body maintenance behaviors (such as a cat washing its face), courtship behavior, nest building, and attainment of food.

The classical example used to illustrate a fixed action pattern behavior is the retrieval of an egg that rolls out of the nest by the graylag goose (Figure 32-5). If a goose notices that an egg has been knocked out of the nest, it will extend its neck toward the egg, reach up, and roll the egg back into the nest with its bill. Because this behavior seems so logical, it is tempting to believe that the goose saw the problem and figured out what to do. But, in fact, the entire behavior is totally instinctive. During experimentation, ethologists have discovered that any rounded object, regardless of size or color, acts as a releaser and triggers the response. Beer bottles, for example, are an effective releaser. And, as a fixed action pattern behavior, the goose completes the action even if the egg rolls away from its retrieving bill. Instead of stopping in the middle of the behavior, the goose will continually repeat the entire behavior until the egg is brought back to the nest.

Coordination behaviors such as reflexes and orientation behaviors such as kineses and taxes are simple types of innate behaviors. Fixed action patterns—sequences of innate behaviors in which the actions follow an unchanging order of muscular movement—are more complex than other innate behaviors.

FIGURE 32-5 Fixed action pattern in a graylag goose. The egg retrieval movement in the graylag goose is an entirely programmed behavior, completely instinctive.

Learning

Innate behaviors are certainly important to the survival of an animal. An animal with protective coloration, for example, does not have time to learn to "freeze" when it detects a predator; its survival depends on its instinct to do so (Figure 32-6). Many social behaviors, such as certain mating or food-attainment behaviors, also depend on each individual's performance of instinctive behaviors. In addition, organisms with simple nervous systems, such as the cnidarians, have an extremely limited capacity for learning and must rely primarily on hard-wired innate behavior patterns for their survival.

Although important in many respects to the survival and fitness of animals, innate behaviors can become a liability if environmental conditions change and an animal's behavior cannot change to adapt to new conditions. For this reason, behavior patterns that can change in response to experience have adaptive advantages over the set programs of instinct. In most animals, only some behaviors are innate and immutable; many behaviors can be changed or modified by an individual's experiences during the process of learning. And, as mentioned previously, the line between learned behaviors and innate behaviors is blurring as scientists learn more about the nature of animal behavior.

> *Behaviors based on experience are learned behaviors. Learning helps an animal change its behavior to adapt to changes in environmental conditions.*

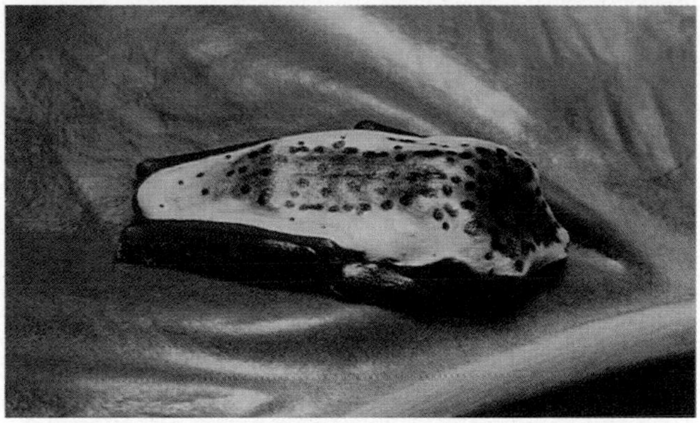

FIGURE 32-6 Protection against predation is instinctive. This tiny tropical frog has markings that resemble a bird dropping. The effectiveness of this protective coloration depends on the frog remaining completely still. This stillness is *not* learned but is an instinctive behavior on which the animal depends for survival.

Learned behaviors can help an animal become better suited to a particular environment or set of conditions, and these behaviors can be grouped into five categories: imprinting, habituation, trial-and-error learning, classical conditioning, and insight. As you read about these types of learning, notice that the first four categories reflect learning as automatic and machinelike. Learning that takes place in a stimulus/response fashion, possibly reinforced by some type of reward that may or may not be readily apparent, reflects a school of thought called **behaviorism.** Behaviorism, which flourished from 1900 to 1960, is the theoretical basis for many of the rote, drill-and-practice type approaches to learning in schools during that time.

More recently, another school of thought regarding how we learn has emerged. **Cognitivism,** the prominent view in psychology today, suggests that individuals acquire and then store information in memory. Learning takes place as new information builds and merges with the old, leading to new types of behavior. Cognitivism is the theoretical basis for the constructivist approach to learning, which has gained wide acceptance in education.

Interestingly, neuropsychological experiments show that we learn *both* ways. Each approach to learning uses different mechanisms and neural systems in the brain. Recent studies have confirmed that the mammalian brain (which includes your brain) has both systems.

Imprinting

Innate behavior and learning interact very closely in a time-dependent form of learning known as

learning an alteration in behavior based on experience.
reflex an automatic response to nerve stimulation.
kinesis (kih **nee** sis) the change in the speed of the random, nondirected movements of an animal with respect to changes in certain environmental stimuli.
taxis (**tak** sis) a directed movement by an animal toward or away from a stimulus, such as light, chemicals, or heat.
fixed action patterns sequences of innate behaviors in which the actions follow an unchanging order of muscular movements.
behaviorism a school of thought that suggests that learning takes place in a stimulus/response fashion, possibly reinforced by some type of reward that may or may not be readily apparent.
cognitivism (**kog** nih tiv izm) a school of thought that suggests that individuals acquire and then store information in memory; learning takes place as new information builds and merges with the old, leading to changes in behavior.

imprinting. Imprinting is a rapid and irreversible type of learning that takes place during an early developmental stage of some animals. Various types of imprinting exist. One type is object imprinting and has been observed in birds such as ducks, geese, and chickens. During a short time early in the bird's life (optimally 13 to 16 hours after hatching in mallard ducks, for example), the young animal forms a learned attachment to a moving object. Usually this object is its mother. However, animals can imprint on various objects regardless of size or color, such as balloons, clocks, and people. From an evolutionary viewpoint, parent-offspring imprinting enhances reproductive fitness by allowing parents and offspring to "recognize" one another, enabling parents to care for their offspring. Researchers have found that imprinting tends to occur in species that have a social organization in which attachment to parents, to the family group, or to a member of the opposite gender is important.

Konrad Lorenz performed classical experiments regarding object imprinting during the 1930s. In one of his most famous experiments, Lorenz divided a clutch of graylag goose eggs in half. He left one half of the eggs with the mother goose and put the other half in an incubator. The half that hatched with their mother displayed normal behavior, following her as she moved. As adults, these geese also exhibited normal behavior and mated with other graylag geese. The other half of the clutch, however, hatched in the incubator and then spent time with Lorenz. These goslings did not behave in the same manner as their siblings; they followed Lorenz around as if he were their mother (Figure 32-7). After their initial time with him, Lorenz introduced the geese to their mother, but they still preferred to follow Lorenz. As adults, these geese tried to "court" adult humans!

FIGURE 32-7 Imprinting is a time-dependent form of learning. The eager goslings following Konrad Lorenz think he is their mother. He is the first animal they saw when they hatched, and they have become imprinted with his image. The goslings will always recognize Lorenz as their parent.

Other types of imprinting also exist. Many migrating birds and fishes learn to recognize their birthplace by locality imprinting. Pacific salmon, for example, are imprinted with the odor of the stream or lake in which they were born. Amazingly, 2 to 5 years later when they return from the sea to spawn, they are able to find their birthplace by its odor. Such long-range, two-way movements are called **migrations.** The physiological and behavioral mechanisms that interact to help migrating animals navigate have interested and puzzled biologists and naturalists for centuries.

In many animals, migrations occur once a year and result from interactions among various environmental factors (such as day length) with animals' physiological and hormonal changes. Ducks and geese, for example, migrate down flyways from Canada across the United States each fall and return each spring. Caribou migrate across the top of North America each year (Figure 32-8). Monarch butterflies migrate from the eastern United States to Mexico and back, a journey of more than 3000 kilometers (approximately 2000 miles), which takes from two to five generations of butterflies to complete. Perhaps the longest migration is that of the golden plover, a bird that flies from Arctic breeding grounds to wintering areas in southeastern South America, a distance of approximately 13,000 kilometers (about 8000 miles).

Migrating animals have the ability to orient themselves in relation to an environmental cue, such as the sun, the stars, or the Earth's magnetic field. In addition, they can navigate, or set a course and follow it. Day migrators such as some birds, ants, and bees use the sun's position to chart a course. Night migrators, a group that includes birds such as the indigo bunting, use the stars to chart a course. All these animals also have a biological clock, some sort of internal "timepiece" (of which scientists have limited understanding) that inter-

FIGURE 32-8 Caribou migration. In caribou the seasonal migration across northern Canada may cover 1500 kilometers (approximately 1000 miles), although hunting and development are disrupting the long-used migration routes.

acts with information from the environment to help migrating animals find their way. It helps animals compensate for the daily movements of the sun.

> *Imprinting is a rapid and irreversible type of learning that takes place during an early developmental stage of some animals. Imprinting helps animals recognize kin or recognize their "home" or birthplace. The last type, locality imprinting, plays an important role in the migrations of some animals.*

Habituation

Habituation is the ability of animals to "get used to" certain types of stimuli. People get used to stimuli all the time. A single shotgun blast would startle most people, but this response would fade at the end of a day in a firing range. Not only would you get used to the noise level (and wear ear protection), but you would perceive the gunshots as nonthreatening. Likewise, you quickly get used to the feel of your clothing after dressing in the morning. Animals also stop responding to stimuli that they learn are neither harmful nor helpful. Learning to ignore

FIGURE 32-9 Habituation in gull chicks. At first, a chick will crouch at any object that flies overhead. Gradually, the chick stops crouching at the familiar objects that fly overhead—the chick has gotten used to them. However, the chick will crouch when an unfamiliar flying object poses a potential threat.

unimportant stimuli is a critical ability in an animal confronting a barrage of stimuli in a complex environment and can help an animal conserve its energy. Hydra, for example, stop contracting if they are disturbed too often by water currents. Sea anemones stop withdrawing if they are touched repeatedly. Young black-headed gull chicks stop crouching as familiar-shaped birds fly overhead. However, they still crouch for the occasional unfamiliar shape that may pose a threat (Figure 32-9).

Trial-and-Error Learning (Operant Conditioning)

More complex than imprinting or habituation, **operant conditioning** is a form of learning in which an animal associates something that it does with a reward or punishment. (The word *operant* means "having the power to produce an effect" and is a form of the verb "to operate.") In operant conditioning an animal must make the proper associa-

imprinting a rapid and irreversible type of stimulus/response learning that takes place during an early developmental stage of some animals.

migrations long-range, two-way movements by animals, often occurring yearly with the change of seasons.

habituation (huh **bich** yoo **ay** shun) the ability of animals to "get used to" certain types of stimuli that they perceive as nonthreatening.

operant conditioning (**op** uh runt) a form of learning in which an animal associates something that it does with a reward or punishment.

tion between its response (such as pressing a lever) and a reward (the appearance of a food pellet) before it receives this reinforcing stimulus. It may also learn to avoid a behavior when the stimulus is negative. Animal trainers use the techniques of operant conditioning.

The American psychologist B.F. Skinner studied such conditioning in rats by placing each in a specially designed box (today called a *Skinner box*) fitted with levers and other experimental devices (Figure 32-10). Once inside, the rat would explore the box feverishly, running this way and that. Occasionally, it would accidentally press a lever, and a pellet of food would appear. At first, a rat would ignore the lever and continue to move about, but soon it learned to press the lever to obtain food.

This sort of trial-and-error learning is of major importance to most vertebrates in nature. The toad in Figure 32-11 learned quickly, for example, not to eat a bumblebee. The frog did not use reasoning to determine that the bumblebee was not good to eat—it merely became conditioned by experience to avoid a response that caused it pain.

Behavioral psychologists used to think that animals could be conditioned in the laboratory to perform *any* learnable behavior in response to *any* stimulus by operant conditioning. However, through experimentation, researchers have discovered that animals tend to learn only in ways that are compatible with their neural programs. Put simply, operant conditioning works only for stimuli and responses that have meaning for animals in nature. Rats can be conditioned, for example, to press levers with their paws to obtain food because in nature they obtain food with their paws. They cannot, however, be conditioned to obtain food by jumping, an unnatural food-attaining behavior for a rat.

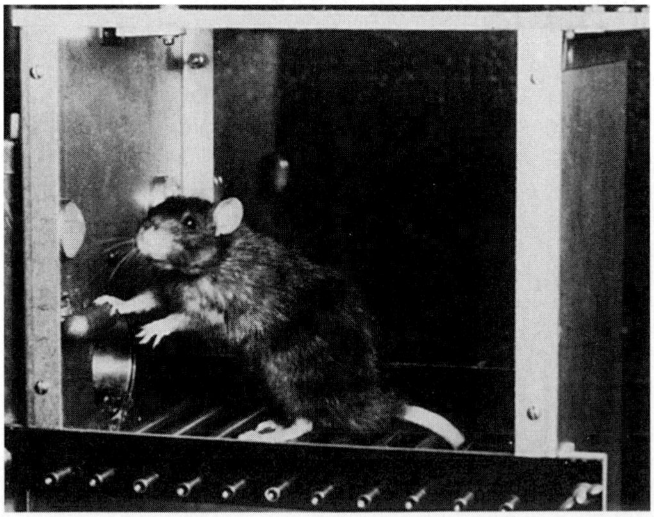

FIGURE 32-10 Operant conditioning. This rat in a Skinner box rapidly learns that pressing the lever results in the appearance of a food pellet. This kind of learning is based on trial and error with a reward for success.

FIGURE 32-11 A toad is taught a lesson through trial and error. A toad gobbles a bumblebee, but as it does, it gets a violent sting on its tongue and spits out the bee.

> *Some animals can learn by trial-and-error; by associating a specific behavior with a reward or punishment, they can learn to apply (or avoid) certain behaviors to affect outcomes in particular situations. Using the trial-and-error method, scientists can condition animals to perform specific behaviors in the laboratory. This type of learning is operant conditioning.*

Classical Conditioning

Classical conditioning is a form of learning in which an animal is taught to associate a new stimulus with a natural stimulus that normally evokes a response in the animal. Repeatedly presenting an animal with the new stimulus in association with the natural stimulus can cause the animal's

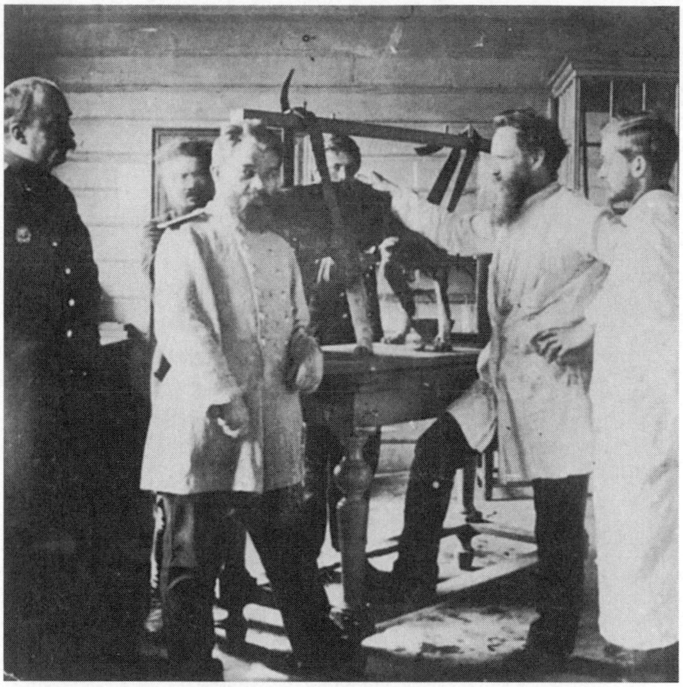

FIGURE 32-12 Classical conditioning. The Russian psychologist Ivan Pavlov (second from right) in his laboratory. In the background a dog is suspended in a harness that Pavlov used to condition salivation in response to light.

brain to form an association between the two stimuli. Eventually, the animal will respond to the new stimulus alone; it will act as a substitute for the natural stimulus. The connection between the new stimulus and the natural stimulus is the result of a learning process but must be reinforced periodically with the presence of the natural stimulus or the animal will stop responding to the substitute.

In his famous study of classical conditioning, the Russian psychologist Ivan Pavlov worked with dogs to condition them to salivate in response to a stimulus normally unrelated to salivation. A natural stimulus to trigger the response of salivation in a dog is meat. In his experiments, Pavlov also presented the dog with a second, unrelated stimulus. He shone a light on the dog at the same time that meat powder was blown into its mouth. As expected, the dog salivated. After repeated trials, the dog eventually salivated in response to the light alone. The dog had learned to associate the unrelated light stimulus with the meat stimulus (Figure 32-12).

The natural stimulus-response connection in an animal is an inborn reflex, or an *unconditioned* response. Such innate responses are important to animals; many of them are protective, such as blinking, sneezing, vomiting, and coughing. Even the knee-jerk reflex mentioned earlier is part of the mechanism by which humans (and other vertebrates) maintain their posture. The work of researchers such as Pavlov contributed to the understanding that animal behavior depends on innate neural circuitry but that these neural programs can most often be modified and directed by the processes of learning.

Insight

Best developed in primates such as chimpanzees and humans, **insight,** or **reasoning,** is the most complex form of learning. An animal capable of insight can recognize a problem and solve it mentally before ever trying out a solution. Therefore the animal is able to perform a correct or appropriate behavior the first time it tries, without having been exposed to the specific situation.

German psychologist Wolfgang Kohler was the first to describe learning by insight, performing extensive experiments on chimpanzees in the 1920s.

classical conditioning a form of learning in which an animal is taught to associate a new stimulus with a natural stimulus that normally evokes a response in the animal.

insight (reasoning) The capability of recognizing a problem and solving it mentally before ever trying out a solution.

FIGURE 32-13 Insight is the most complex form of learning. A, A chimpanzee is able to see a problem and develop a solution, even before taking any action. To reach the bananas, the chimpanzee sees a solution in stacking the boxes and fitting the poles together. **B,** A dog, on the other hand, lacks insight and cannot develop a solution to a problem. A dog will eventually learn by trial and error to go around the tree, unwinding his leash to reach the food.

Kohler showed that animals must perceive relationships and manipulate concepts in its mind to solve a problem on the first try. In his classical experiments, he placed chimpanzees in a room with a few crates, poles that could be joined together, and a banana hung high above the grasp of the animals. The chimpanzees were able to stack the crates and join the poles appropriately to retrieve the banana (Figure 32-13, *A*). Unlike chimpanzees, most animals are unable to use insight to solve problems. Your dog or cat, for example, must use trial-and-error to "try out" solutions to problems. Your dog has no insight into the problem that having its leash wrapped around a tree is keeping it from reaching its food (Figure 32-13, *B*). The dog can merely continue walking in various directions and ultimately free itself by chance. This experience, however, may help the dog perform the appropriate behavior in the future, having learned by trial-and-error.

Summary

- The study of animal behavior encompasses patterns of change in animals, including movement, sounds, and body positions. Ethologists, the prominent animal behavior researchers before the late 1950s, study the behavior of animals in their natural environments, observing and interpreting their behavior in the context of physiology while attempting to uncover the biological significance of the behavior. Today, the field of animal behavior is composed of researchers from a variety of disciplines: psychology, sociobiology, neurobiology, and behavioral ecology.
- Some patterns of behavior are inborn, or innate, resulting from neural pathways developed before birth. These nervous system responses are directed by genes and occur without obvious environmental influence. Such instinctive behaviors are important to animals, providing them with fixed patterns of survival, such as those used to elude predators, to mate, and to care for young.
- Certain animals slow down in suitable environments and speed up in unsuitable ones by means of kineses. Some move toward or away from certain stimuli by means of taxes. Both kineses and taxes are simple types of innate orientation behaviors.
- During innate coordination behaviors called *reflexes,* animals automatically respond to nerve stimulation. Fixed action patterns are more complex innate behaviors in which organisms perform sequences of movements.
- Animals can learn, or change their behaviors, on the basis of experience. Behaviors are changed within certain limits by means of various types of learning: imprinting, habituation, trial-and-error, and conditioning.
- Certain animals learn to recognize their kin or birthplace by means of imprinting, a rapid and irreversible type of learning that takes place during an early developmental stage. Habituation helps animals adapt to certain types of stimuli. Some animals learn to associate a behavior with a reward or punishment by means of trial-and-error learning. During classical conditioning, animals learn to associate a new stimulus with a natural stimulus that normally evokes a response in the animal.
- Insight, or reasoning, is the most complex form of learning. An animal capable of insight can recognize a problem and solve it mentally by using experience, thereby performing an appropriate behavior the first time it tries. Insight is best developed in primates such as chimpanzees and humans.

Knowledge and Comprehension Questions

1. What are behaviors?
2. Explain this statement: Behaviors are part of an animal's equipment for survival.
3. Fill in the blanks: Behaviors that are performed in a fairly complete form the first time they are exhibited are ______ behaviors. Behaviors that are based on experience are ______ behaviors.
4. Fill in the blanks: Two types of orientation behaviors are ______ and ______. A type of coordination behavior is ______. A more complex type of innate behavior involving sequences of behaviors is a ______.
5. Define and give an example of each of the categories you listed for question 4.
6. Explain the importance of innate and learned behaviors. How do they complement each other?
7. Fill in the blanks: Five types (categories) of learning are: ________, ________, ________, ________, and ________.
8. Identify the specific type of behavior involved in each of the following situations:
   a. Fireflies are attracted to the flashing luminescence of other fireflies.
   b. While studying your biology text, you hear a screen door slam. It startles you at first, but then you barely notice it when you realize the wind is occasionally opening and closing it.
   c. You praise your puppy every time it sits down when you say "sit." Soon you have trained it to sit on command.
9. Distinguish between operant conditioning and classical conditioning.
10. What is insight? Explain its significance.
11. Identify the specific type of behavior involved in each of the following situations:
    a. Every year, purple martins fly to Brazil and return to the United States around April.
    b. When a bright light flashes near your eyes, you automatically blink.
    c. Your cat frequently grooms itself by licking its fur and rubbing its paws over its face.

Critical Thinking

1. Pacific salmon are born at the headwaters of rivers, then swim downstream hundreds of miles to the sea where they spend their adult lives. Years later, when it is time to spawn (that is, to lay and fertilize the eggs that will be the next generation), the adults swim up the same rivers to the precise location where they were born. Using information from this chapter, propose a hypothesis regarding how the fish know which way to go when they come to a fork in the river.

CHAPTER 33

SOCIAL BEHAVIOR IN ANIMALS

THESE KISSING FISH only appear to be kissing. In fact, their behavior is one of aggression, not affection. Engaged in a territorial "war," each wants to win the rights to its own "space" within the water. Neither fish will die during this bloodless battle, but one will back down, allowing the other to lay claim to the win.

Their fight begins as they encounter one another at a territorial boundary. First, each tries to intimidate the other, puffing up its body with air to appear as big and threatening as possible. Then they swim side by side, pushing currents of water past one another with their fins. These currents provide clues to each contender regarding the size of the opponent. If neither retreats, the tug-of-war begins, mouths locked in battle. Soon, one fish emerges the victor. The loser signals submission, and the "tournament" is over.

Communication via Social Behaviors

The aggressive behaviors exhibited by the kissing gouramis are types of **social behaviors,** which help members of the same species communicate and interact with one another, each responding to stimuli from others. The advantages of social behaviors are numerous, and these advantages differ from species to species. In general, however, all species that reproduce sexually exhibit interactive patterns of behavior for the purposes of reproduction, the care of offspring, and the defense of a territory. In addition, some animals use social behaviors to hunt for and share food and to warn and defend against predators.

The biology of social behavior is called **sociobiology.** This science applies the knowledge of evolutionary biology to social behavior. Its purpose is to develop general laws of the biology and evolution of social behavior.

> *Social behaviors are interactive patterns of animal behavior that help members of the same species communicate and interact with one another. The biology of social behavior is sociobiology.*

Competitive Behaviors

When two or more individuals strive to obtain the same needed resource, such as food, water, nesting sites, or mates, they are exhibiting **competitive behavior.** This type of behavior occurs when resources are scarce.

Threat Displays

How do animals compete? Many first engage in **threat** or **intimidation displays,** a form of aggressive behavior. The purpose of these displays is to do as the name suggests: scare other animals away or cause them to "back down" before fighting takes place. Exact forms of this behavior vary widely among species but usually involve such behaviors as showing fangs or claws, making noises such as growls or roars, changing body color to one that is a releaser of aggression in an opponent, and making the body appear larger by standing upright, making the fur or hair stand on end, or inflating a

KEY CONCEPTS

- Sociobiology, the biology of social behavior, applies the knowledge of evolution to the study of animal behavior.
- Social behaviors—those that help members of the same species communicate and interact with one another—are valuable mechanisms that ultimately aid the reproductive fitness of individuals.
- Many species of animals live in social groups; some insects, in fact, live in societies having a division of labor.
- Many researchers question the scientific validity of applying sociobiology to human behavior and propose that the application may lead to racism, sexism, or other types of group stereotyping.

OUTLINE

FIGURE 33-1 Defense by deception. This common toad, confronted by a snake, inflates its body and sways from side to side in an attempt to make itself appear larger.

throat sac (Figure 33-1). Threat displays are important social signals and communicate the intent to fight. Interestingly, some of the movements and body postures (body language) of threat displays that repel competitors also attract members of the opposite gender. Biologically, this makes sense because the competition may be over a mate.

Submissive Behavior

Animals use other behaviors to avoid fighting. One type of behavior is called **submissive behavior** and is usually a behavior opposite to a threat display. The behavior might include making the body appear smaller, "putting away weapons" of fangs or claws, turning a vulnerable part of the body to an opponent, or removing body colors that are releasers of aggression. As Figure 33-2 shows, an animal may, in fact, possess an array of color "signals." If an animal is losing a fight, it might also display submissive behaviors to stop the fight. Contrary to popular belief, animals of the same species rarely fight to the death. One reason is that most animals do not have the means to do so—they do not have sufficiently dangerous fangs, claws, or horns, for example. Species with dangerous weapons usually have defenses against those weapons too, such as a strong hide, long hair, or a thick layer of body fat. Scientists hypothesize that aggression within a species is meant to chase off rather than kill the rival.

Territorial Behavior

A territory is an area that an animal marks off as its own, defending it against the same-gender members of its species. These behaviors are called **territorial behaviors.** Members of the opposite gender are sometimes allowed into the territory, often for mating purposes.

Territoriality is very common in all classes of vertebrates and is even found in some invertebrates such as crickets, wasps, and praying mantises. Some animals such as dragonflies mark their territories by conspicuously patrolling their borders, resting and then moving from prominent landmarks. As an animal demonstrates the borders of its territory (Figure 33-3), it may exhibit specific movements or body postures as signals. Some animals, such as song birds, monkeys, apes, frogs, and lizards, mark their territory by producing sounds that announce their ownership. Animals with a well-developed sense of smell, such as wolves, hippopotamuses, rhinoceroses, some rodents, and even domestic dogs, mark their territories with substances that have an odor such as urine. Many species, in fact, possess special scent glands that secrete substances just for this purpose. The odor marking helps keep out members of the same gender and helps owners of large territories orient themselves to its borders.

Territorial behavior has several adaptively important consequences. In many mammalian societies, territoriality often has an important influence on the establishment of new populations, with the dominant male defeating the lesser males and leaving them to found their own populations. Individuals surviving in the surrounding marginal areas repopulate any vacant territories.

Although territoriality evolved as a result of the advantages to individuals, this system has an effect on the resources available to various members within a population. When resources such as nesting sites and food are limited (Figure 33-4), each member of a population is in danger of the resources being spread too thin, so no member gets an adequate amount of what it needs. In addition, breeding pairs of animals will not reproduce unless they have adequate resources; territorial be-

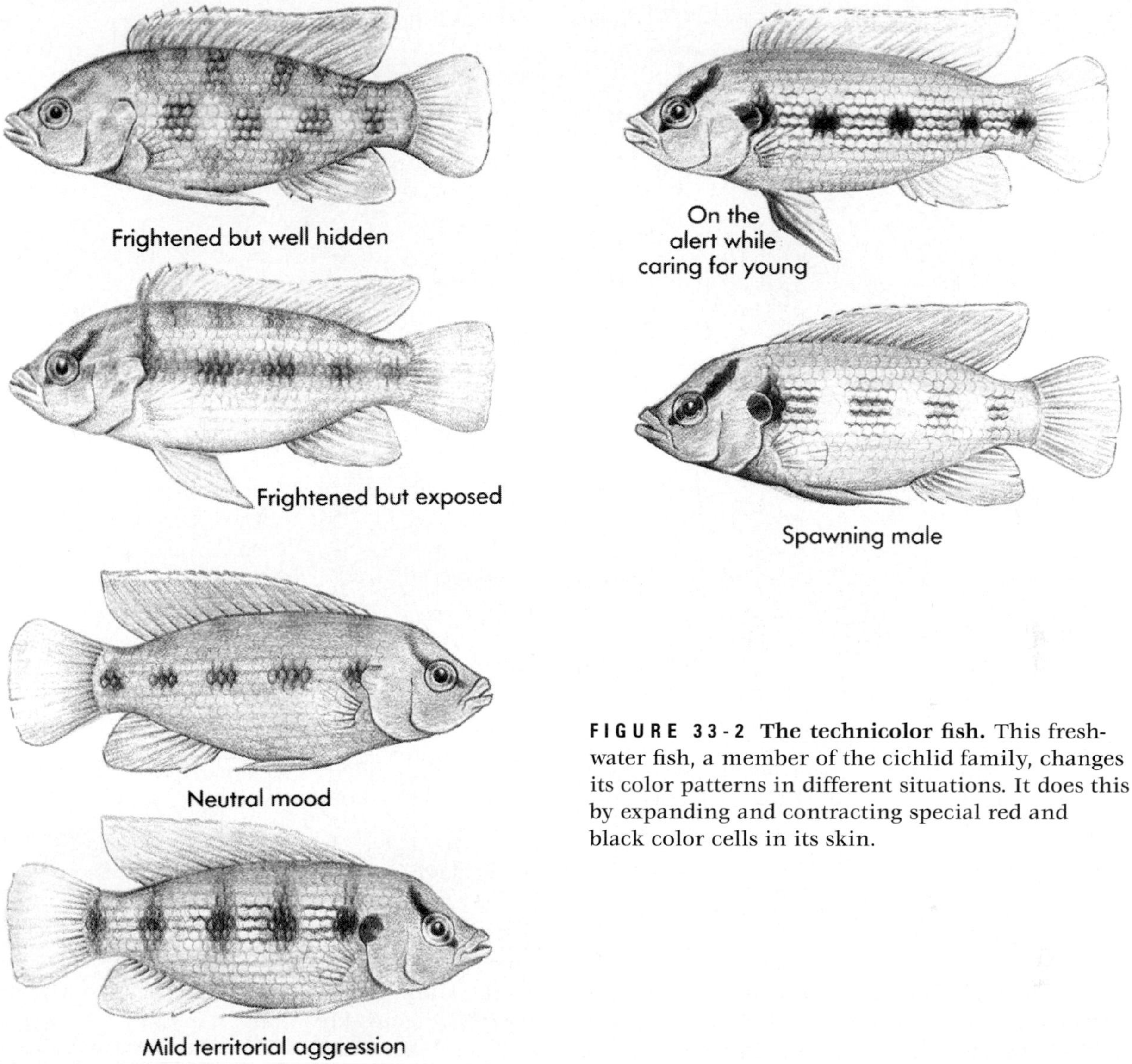

FIGURE 33-2 The technicolor fish. This freshwater fish, a member of the cichlid family, changes its color patterns in different situations. It does this by expanding and contracting special red and black color cells in its skin.

havior ensures that at least some members of the population will have their own space, food, and nesting sites so that they will survive and reproduce. Territorial behaviors also enhance reproductive capabilities by placing the peak time of competition and aggressive behavior at the time of the marking of the territory, *before* the time of reproduction and raising of young.

social behaviors activities of animals that help members of the same species communicate and interact with one another, each responding to stimuli from others.

sociobiology (so see oh bye **ol** uh gee *or* so shee oh bye **ol** uh gee) the biology of social behavior. This science applies the knowledge of evolutionary biology to the study of animal behavior.

competitive behavior a type of behavior that occurs when two or more animals strive to obtain the same needed resource, such as food, water, nesting sites, or mates.

threat displays/intimidation displays a form of aggressive behavior that animals may exhibit during a competitive situation. The purpose of these displays is to scare other animals away or cause them to "back down" before fighting takes place.

submissive behavior behavior that animals often exhibit in response to a threat display in order to avoid fighting. Submissive behavior may include making the body appear smaller or withdrawing fangs or claws.

territorial behaviors behaviors an animal may exhibit that involve marking off an area as its own, and defending it against the same sex members of its species.

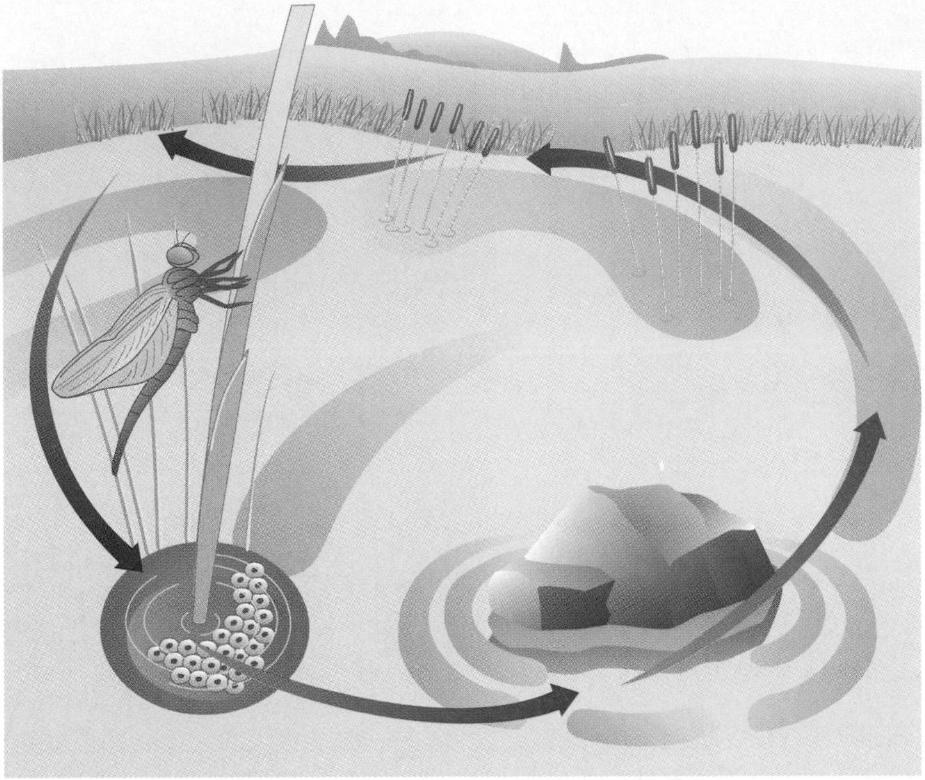

FIGURE 33-3 Territory of a male Demoiselle dragonfly. The male dragonfly sits on a perch above the egg deposit site. He periodically flies around his territory *(red arrows)* or circles within the territory.

Individuals within the same species compete with each other for various resources. Competition among animals includes a repertoire of behaviors such as threat displays, submissive behavior, and territorial behavior. Such competitive behaviors evolved as a result of the advantages to individuals, but they have an effect on the resources available to various members within a population.

Reproductive Behaviors

Sexual reproduction—the union of a male sex cell (sperm) and a female sex cell (egg) to form the first cell of a new individual (zygote)—takes place in most animals. Even some animals that can reproduce asexually, such as sponges that can bud off bits of themselves that grow into new individuals, reproduce sexually at certain times. Sexual reproduction is biologically significant because it provides genetic variability among organisms of the same species. Put simply, the combining of genes

FIGURE 33-4 The emperor penguins in Antarctica nest in rookery sites. There are only a limited number of such sites available, which may reflect the limited amount of resources available to the breeding penguins.

JUST WONDERING . . .

Are animals other than humans capable of expressing their sexuality?

First, we need to define the term *sexuality*. Sexuality refers to the gender (sex) of an organism and its sexual behavior. With organisms other than humans, these aspects of sexuality are simply biological: does the animal have male or female reproductive organs (or both), and what courting and mating behaviors does it exhibit? Human sexuality is based on these biological aspects, but the biological dimension of human sexuality is influenced by its social (cultural), psychological, and moral dimensions. The expression of sexuality in other animals does not include these dimensions. However, their sexuality (especially in vertebrates) is expressed in a variety of chemical, anatomical, and behavioral ways, many of which are mentioned in this chapter.

Scientists have developed some interesting insights into sexuality in the past few decades with research into the nature of sexuality of mammals. Scientist know that a mammal's gender is hereditary—whether it is male or female depends on the complement of DNA inherited from its parents. As the organism develops, its sex organs produce hormones that shape its masculine and feminine features. Accumulating evidence suggests that this hormonal environment during fetal development is crucial to adult sexuality because sex hormones act directly on the developing circuitry of the brain as well as on other tissues. The neural circuits that are developmentally influenced by sex hormones seem to provide the impetus for behavioral differences between males and females. The controversial research into homosexuality in humans supports this hypothesis, suggesting a link between sexual behavior and brain structure in humans. Sexuality in mammals, then, appears to depend on subtle hormonal controls that influence the development of both secondary sexual characteristics and the brain, which underlies sexual behavior.

The factors that influence the sexuality of mammals differ in certain respects from factors that influence sexuality in other vertebrates. For example, the incubation temperatures of eggs, rather than genetic inheritance, determines whether certain reptiles are male or female. In leopard geckos, incubation temperature has even more far-reaching effects than gender determination. Among females and males, the sexual behaviors of adults are also influenced by incubation temperature: whether they were incubated at the high end of the temperature range for males or the low end determines their levels of aggressiveness and submissiveness. This idea also holds true for females.

The maleness or femaleness of certain other vertebrates that do not inherit their sexual status, such as anemone fish and sea bass, is determined by their social environment. These animals have both male and female organs; they are *hermaphrodites.* Anemone fish and sea bass exhibit maleness and femaleness at separate times during their lives. A change from one gender to the other is precipitated by a change in their social environments, such as the disappearance of a dominant male or female. Other species of hermaphroditic fish exhibit both maleness and femaleness at the same time; they simultaneously have both male and female sex organs.

Single-gender species also exist. Certain species of whiptail lizards, for example, are only female; males do not exist. They reproduce by a method called *parthenogenesis,* laying unfertilized eggs that develop into new individuals. (Many insects reproduce by parthenogenesis also.) Interestingly, these lizards exhibit male and female reproductive behaviors that are similar to sexually reproducing species of whiptail lizards. These behaviors, however, are not futile; lizards engaging in sexual behaviors lay more eggs than lizards that do not.

Thus, the simple answer to your question is "yes," animals other than humans express their sexuality. But scientists are still in the process of discovering the underlying mechanisms by which this occurs and the connections regarding sexuality among animals.

from two parents produces offspring that are similar but different from their parents. In addition, because the genes of each parent become assorted independently during the meiotic process that produces sex cells (see Chapters 18 and 21), the offspring are different not only from their parents but from one another. Differences among organisms within a species are the "raw materials" needed for adaptation to environmental change. The ability of some organisms to survive and reproduce in an environment in which others of their species cannot is, in turn, the basis of natural selection.

Genetic variability does not exist (with the exception of mutation) in organisms that reproduce asexually; these organisms are clones of one parent. That is, organisms produced by means of asexual reproduction develop from one or a few cells of a single parent (Figure 33-5) and are therefore genetically identical to that parent. Because organisms within a population of clones do not vary from one another, the entire population can be at risk if environmental conditions change.

For reproduction to take place, animals of most species must communicate and cooperate with one another. Individuals within many species of animals form pair bonds for the purposes of reproduction and parenting. An important outcome of these processes is the retention in the population of the genes responsible for the various facets of social behavior. For natural selection to favor a particular gene, the advantages of the particular social trait it encodes must outweigh the disadvantages—for the individual or for the individual and its close relatives (*kin*).

FIGURE 33-5 Asexual reproduction. Offspring of sea anemones can be produced by budding. A bud grows from a few cells of a single parent and, when developed, breaks off and becomes a completely separate individual. Budding results in genetically identical parents and offspring. At the sea anemone's midsection, a new individual has formed by budding and can be seen as a large, rounded mass. It looks identical to the parent and is, therefore, hard to see from this angle.

Behaviors that promote successful sexual reproduction, then, are highly adaptive behaviors. Organisms that reproduce sexually must have patterns of male-female interactions that lead to fertilization—the penetration of an egg by a sperm. Fertilization can be preceded by copulation, the joining of male and female reproductive organs. However, some organisms such as many fish and frogs do not copulate; instead the male releases sperm over the eggs after they have been deposited by the female. The term **mating** refers to male-female behaviors that result in fertilization, regardless of whether copulation occurs. The term **courtship** refers to the behavior patterns that lead to mating.

For courtship to take place, males and females must first find each other (see "How Science Works"). In some species, males and females live together in pairs or groups, but many animal species do not. Most often the job of attracting a mate goes to the male of the species. Notable exceptions, however, are often found within insect populations.

Males often attract a mate by marking a territory and defending it against intrusion by other males. As mentioned previously, aggressive behaviors against other males are often the same behaviors that attract a female. In addition, males can attract females by displaying body colors or markings like the peacock shown in Figure 33-6.

FIGURE 33-6 A peacock. Males of certain species, such as the peacock, use fantastic displays of color or markings to attract females, or peahens.

HOW *Science* WORKS

Discoveries

Songs, Sex, and . . . . Elephants?

How do male and female elephants find each other to reproduce? Consider the obstacles: male and female elephants live apart and have no fixed breeding season. Females spend 2 years in gestation (the time from fertilization until birth) and another 2 years nursing their young. During these periods, female elephants will not mate. The period of estrus, in which the female elephant is receptive to mating, occurs only once every 4 years—and lasts only a few days! It seems a miracle that elephants reproduce at all.

But this seeming miracle is understandable in light of the knowledge scientists uncovered about elephant reproductive behavior and communication. Researchers investigated the possibility that a female elephant in estrus sends out a signal that informs males for miles around of her condition. Using specialized microphones, the researchers recorded a sequence of intense, low-frequency calls emitted from an elephant in estrus. The estrus call is the same in all female elephants and is a song. The low frequency of the song places it in the infrasound frequency range—the same range that includes the sounds of thunderstorms and earthquakes. Humans can barely detect infrasound—you "feel" more than hear the low, dull throbbing associated with these low-frequency sounds. The female elephant sings her song for only half an hour, and within a day she is surrounded by male elephants, some of whom have heard the song from several miles away.

Infrasound is used by elephants for purposes other than mating as well. Warnings are issued by infrasound, and the whereabouts of groups of elephants are transmitted to other groups by infrasound. Observers have noted for years that entire groups of elephants will suddenly freeze, raising and spreading their ears. Researchers surmise that these elephants are listening to faint, distant calls and are remaining as silent as possible to do so. Moments later, the elephants will come out of their trance and rush off together to follow the signal.

The elephants' "talk" is a factor that enhances elephants' survival by communicating useful information and facilitating the reproductive process. Researchers are currently looking into the possibility that elephants use their infrasound calls to maintain their elaborate hierarchical society over distances of several miles.

Many birds sing songs to attract a mate, and certain frogs and many insects use other sounds or mating calls for this purpose. Many mammals and insects produce odors that are attractive to females. In some species, the males congregate and perform various dances or songs as a group to attract their mates (Figure 33-7).

After the formation of a male-female pair, a **courtship ritual,** unique to each species, leads to mating. Courtship behaviors usually consist of a series of fixed action patterns of movement (see Chapter 32), each triggered by some action of the partner that acts as a releaser and triggers in turn a preprogrammed movement pattern by the potential mate (Figure 33-8). For example, cranes dance around one another, peacocks strut with their tail feathers extended in an elaborate fan, and hummingbirds participate in a courtship flight (Figure 33-9). In addition, some species, such as various types of geese, exhibit postcopulatory behaviors. Both birds rise out of the water, wings extended, facing one another. Such behaviors appear to strengthen the pair bond, a behavior that is necessary when both parents care for the young.

> *Reproductive behaviors within a sexually reproducing species serve to bring males and females together in a manner that leads to the fertilization of eggs by sperm.*

Parenting Behaviors and Altruism

For the offspring to survive, grow, and eventually reproduce, parents must either make preparations for the care of their young or care for them themselves. Parenting behaviors are a type of **altruistic behavior.** Altruistic behaviors are those that benefit one at the cost of another. In animal behavior, this term is different from the everyday use of the word. When we speak of altruistic behaviors of humans, we imply that the person performing the altruistic act was knowingly and willingly risking or giving up something for another. For example, good samaritans risk their lives to save another are performing an altruistic act. With animals other than humans, however, altruism does not carry this meaning of performing the act with knowledge of its altruistic nature. In addition, its focus is evolutionary: an altruistic act is one that increases the individual fitness of the recipient while decreasing the individual fitness of the donor. Natural selection favors genes that promote altruistic behavior among closely related individuals (kin) such as parents and offspring.

Parenting behaviors are found in almost all animal groups, but the participation of the parents varies. In many species of birds and fishes, both parents care for the young. In these species, a division of labor usually exists, with each parent carrying out a specific role. Because female mammals

FIGURE 33-7 Courtship dance in male prairie chickens. Male prairie chickens perform a courtship dance to attract females. They produce sounds as part of their behaviors, puffing up the orange air sacs on the sides of their necks. In this photograph, the males are preparing to begin the dance by encircling the female. When they carry out the dance, they will posture and twirl in unison.

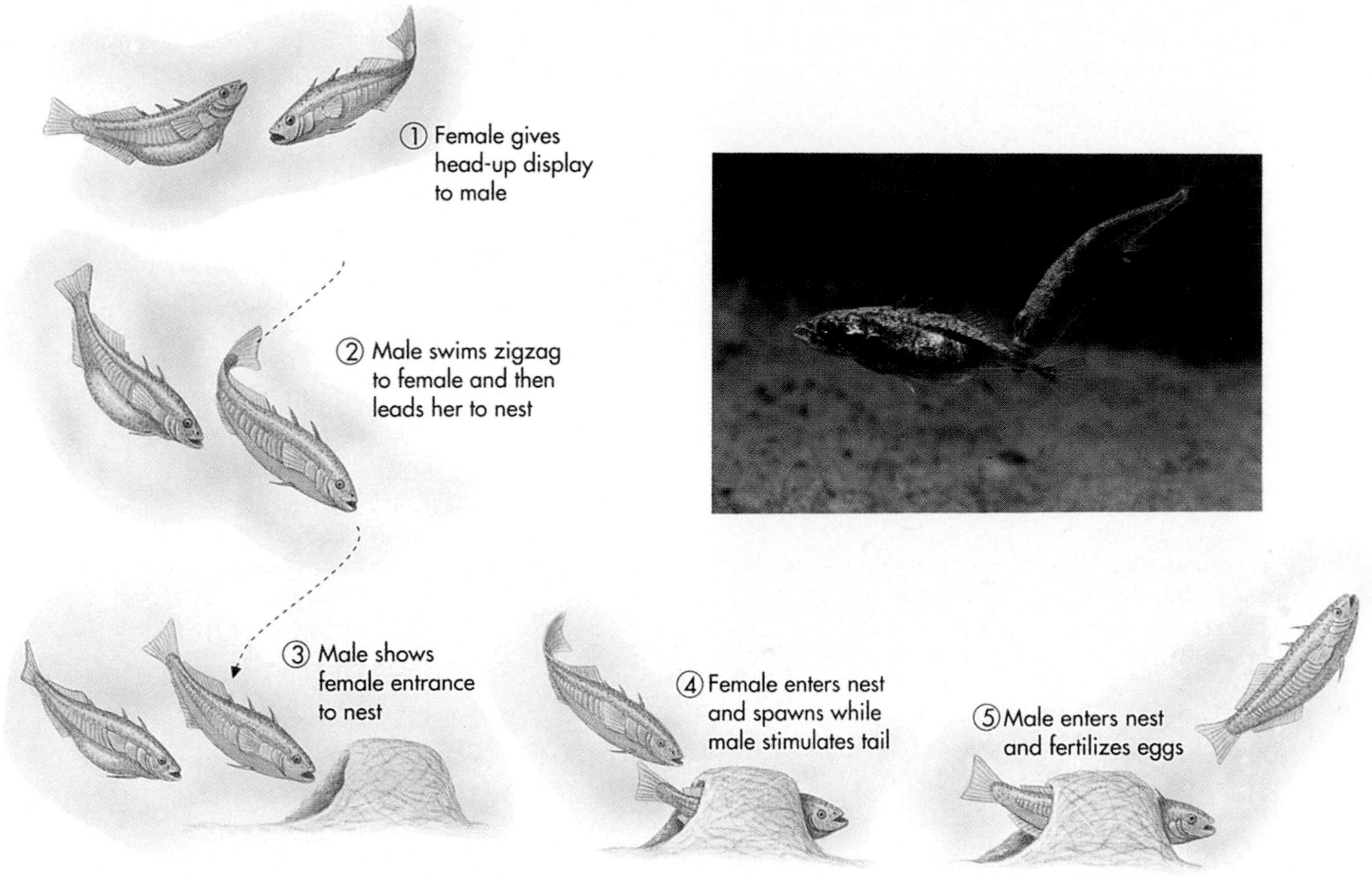
① Female gives head-up display to male
② Male swims zigzag to female and then leads her to nest
③ Male shows female entrance to nest
④ Female enters nest and spawns while male stimulates tail
⑤ Male enters nest and fertilizes eggs

FIGURE 33-8 Stickleback courtship. Each set of movements in the courtship ritual is a fixed action pattern that triggers the next set of movements by the partner.

produce milk to nourish the young, most often the mother assumes the parenting role in this class of animals. In only a few species, such as certain fishes, sea horses, and birds, does the male alone take care of the offspring.

To prepare for their young, animals build protective structures such as nests or cocoons. Usually a food supply is gathered, and the eggs are deposited in areas that are protected and that may also lie near a ready food source. Marsupials—mammals in which the young are born developmentally early—keep their young safe and nourish them as they complete their development by storing them in a pouch, conveniently situated at the mammary glands (Figure 33-10). Some animals simply carry their young to a safe place when they sense danger.

Some species exhibit quite interesting and unique parenting behaviors. Very early in his studies of behavior, Nikolaas Tinbergen noted that certain birds remove the shell fragments of broken eggs from their nests. Any shell fragment that Tin-

mating male-female behaviors that result in fertilization, regardless of whether copulation occurs.
courtship behavior patterns that lead to mating.
courtship ritual patterns of behavior, usually consisting of a series of fixed action patterns of movement, designed to lead to mating. Courtship rituals are unique to each species and occur after the formation of a male-female pair.
altruistic behavior (al troo **iss** tik) a kind of behavior that benefits one at the cost of another; parenting behaviors are a type of altruistic behavior.

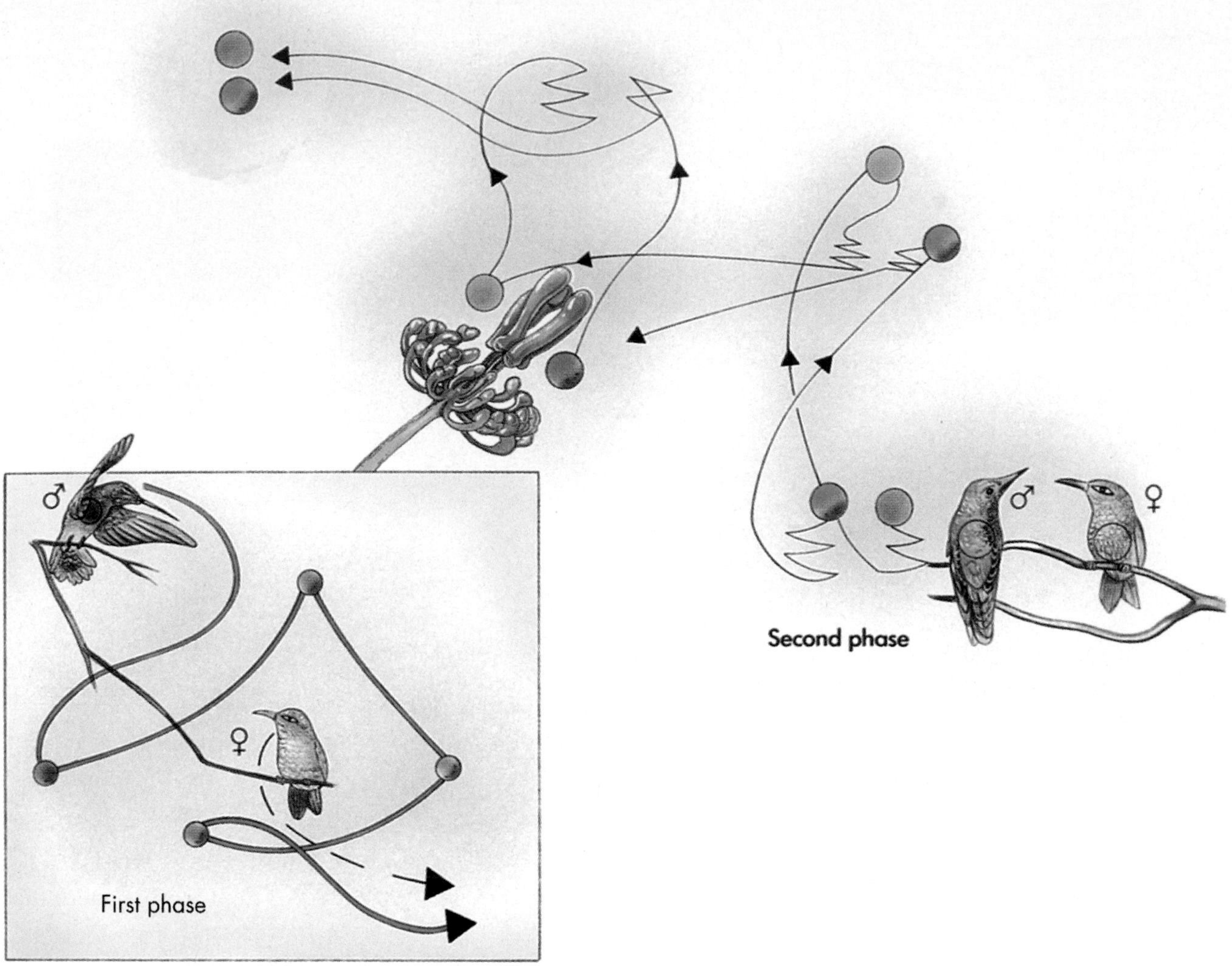

FIGURE 33-9 Courtship flight in hummingbirds. In the first phase of the courtship flight, the male flies around the female in an undulating pattern (box). In the second phase, both partners fly around each other and may mirror their partner's movements (above).

FIGURE 33-10 A marsupial with young. The kangaroo protects its young in its pouch. The young are nourished by mammary glands located in the pouch.

FIGURE 33-11 Defensive formation of a herd of musk-oxen. Like the pioneers who circled their wagons against danger, these musk-oxen form a defensive ring with horns pointing outward. The young musk-oxen are inside the ring.

bergen added was removed quickly. The adaptive advantage of this parenting behavior, Tinbergen discovered, was that the odor from shell fragments attracted predators and their prompt removal helped protect the other eggs in the nest.

> *Parenting behaviors lead to the reproductive fitness of offspring by helping individuals survive and grow to reproductive age. These acts are altruistic; they increase the individual fitness of the recipient while decreasing the individual fitness of the donor.*

Group Behaviors

In addition to pair-bond relationships, the individuals of many species form temporary or permanent associations with many other members of their species, forming social groups. Animals within social groups work together for common purposes. Individuals within a group, for example, have more protection against predators than do single individuals or pairs. They can warn each other of danger by using calls or chemical signals or cluster together when a predator appears, forcing the predator to either separate individuals from the group or fight the entire group. In addition, some animals such as musk oxen and bison place their young within the center of the group (Figure 33-11), forming a circle around them much like the early American settlers did with their covered wagons.

Many animals living within social groups search for food together and, scientists have observed, are more successful as group hunters than as single individuals. Food-gathering strategies become group strategies; often animals such as those that prey on fish will drive the prey to one location, encircle them, and feast. The eastern white pelicans shown in Figure 33-12 hunt for fish in this way.

In groups, animals can also ration the "work load" so that each animal does not have to perform an array of daily tasks but may focus on one task (Figure 33-13). Many species of insects, particularly bees and ants, are organized into social

A

B

FIGURE 33-12 Hunting behaviors of pelicans. Eastern white pelicans hunt in groups (A) and then circle their prey (B).

FIGURE 33-13 Ant society. This worker ant, a member of the genus *Polyergus,* is responsible for capturing slaves for the nest. The slave, a member of the genus *Formica,* is in its pupa stage. When it emerges as an adult, it will be a nonreproductive worker slave.

FIGURE 33-14 A bee colony. The queen bee has a red spot on her back painted on by a researcher.

groups having a division of labor. These social groups are called **insect societies.**

Insect Societies

In insect societies, individuals are organized into highly integrated groups in which each member of the society performs one special task or a series of tasks that contributes to the survival of the group. The role an individual plays depends on its body structure, which in turn depends on its gender or age. In some animal societies the same animal performs different roles during the course of its life. In addition, a common feature among social insects is a queen that outlives other members of the society. The role of the queen is to produce offspring and to promote cooperation among members of the group.

A honeybee colony, for example, is made up of three different types of bees called *castes:* the queen, the workers, and the drones. The queen is the focus of the colony; she lays the eggs from which all the other bees develop (Figure 33-14).

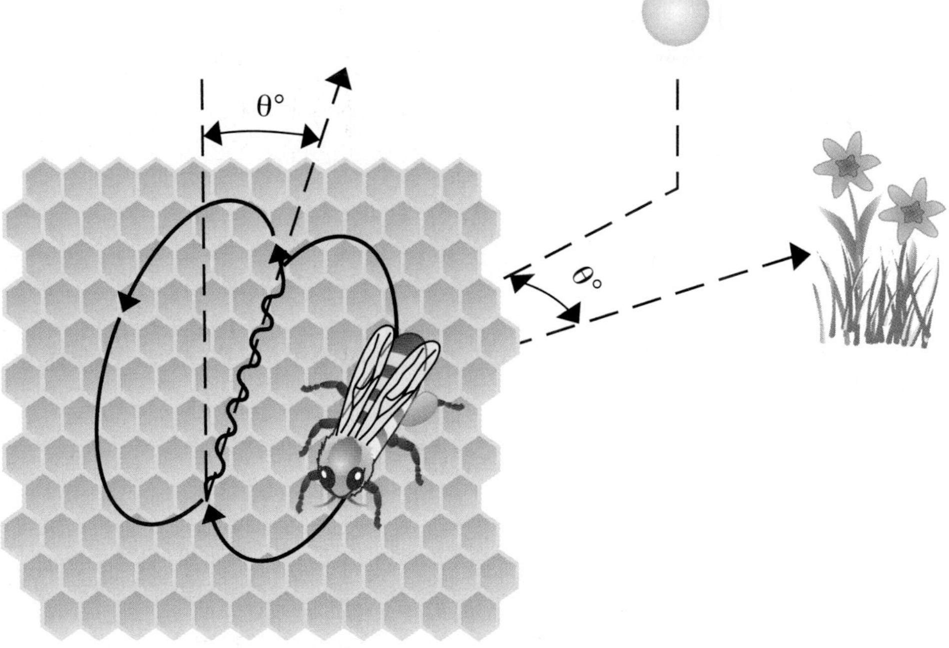

FIGURE 33-15 The waggle dance. This dance is performed by a forager bee who, on spotting a food supply (a flower), goes back to the hive and communicates to the other bees through dance the exact location of the flower. The dance tells how far away the flower is and also the angle of the flower to the sun. The part of the dance that yields both kinds of information is the straight run, *(zigzag line between two circles).* The forager shakes its abdomen back and forth and buzzes with its wings while moving in the straight run. Then it makes a semicircle and returns to the starting point of the straight run. After completing this run, it circles in the opposite direction to land again at the beginning of the straight run and so on. How fast the bee performs the straight run determines how far away the flower is—a distance of 500 meters equals a straight run duration of about 1 second. The direction of the flower is indicated by the angle that the straight run deviates from the vertical. This angle equals the angle of the flower to the sun.

The drones are the male bees that fertilize the eggs of the queen. Interestingly, they develop from unfertilized eggs and so have no male parent themselves—and only one set of chromosomes! The workers are females that develop from fertilized eggs. Genetically, a queen and a worker are no different; any fertilized egg can develop into either a queen or a worker depending on how the egg is housed and fed.

In one hive, there may be up to 50,000 workers, 5000 drones, and of course 1 queen. The workers perform all the tasks of the hive except for mating and egg laying. In the winter months, the average length of the worker bee's life is 6 months. In the summer, workers live only 38 days! During the first 20 days of a worker's life, she performs various duties within the hive sequentially as she matures: feeding larvae, producing wax, building the honeycomb, passing out food, and guarding the hive. Although a worker may go on short play flights around the hive during this time, she does not venture very far. It is only after the first 20 days of her life that the worker takes long flights away from the hive to forage for food.

An interesting means of communication among the workers has evolved that enables them to communicate the location of flowers at which they can collect pollen and nectar. This communication is a form of body language called the *waggle dance*. The manner in which a worker shakes, or waggles, her abdomen communicates the distance to the flowers. The angle at which the worker dances indicates the direction her co-workers must fly (Figure 33-15).

The drones play almost no role in the hive other than to mate with the queen. They help keep the larvae warm by congregating near them in the morning while the temperature is still cool. Drones also help distribute food. At the time of mating, many drones follow the queen out of the hive, following her pheromone trail, and mate in flight. After mating, the drones die. Drones that do not mate have no better fate, because the workers pinch them, sting them, and then throw them out of the hive as the winter approaches.

The queen bee controls the worker bees with a chemical she produces known as *queen substance*. This substance is a **pheromone,** a chemical produced by one individual that alters the physiology or behavior of other individuals of the same species. The bees ingest the queen substance by licking the queen and then pass the substance around from one to the other as they pass food. Queen substance renders the worker bees sterile and also inhibits the workers from making queen

FIGURE 33-16 A swarm resting on a horse chestnut branch. These bees are in the process of establishing a new hive.

cells, compartments in which fertilized eggs can develop into queens. If the hive becomes too congested, however, the queen may begin producing less queen substance, resulting in the removal of the inhibition to produce queen cells. Workers make a half dozen or more new queen cells in which replacement queens begin to develop. The old queen and a swarm of females and male drones leave to establish a new hive (Figure 33-16). The first new queen to emerge may kill the other candidate queens and assume rule or may create another swarm and leave to establish another hive.

insect societies social groups that are formed by many species of insects, particularly bees and ants, and are characterized by a division of labor.

pheromone (fare uh moan) a chemical produced by one individual that alters the physiology or behavior of other individuals of the same species.

The lifestyles of social insects can be so unusual as to be bizarre, none more so than the little reddish ants—the leafcutters (Figure 33-17). Leafcutters live in the tropics, organized into colonies of up to several million individuals. These ants are farmers, growing crops of fungi beneath the ground. Their moundlike nests look like tiny underground cities covering more than 100 square yards, with hundreds of entrances and chambers as deep as 16 feet underground. Long lines of leafcutters march daily from the mound to a tree or bush, hack its leaves into small pieces, and carry the pieces back to the mound. Each ant finds it way by following a trail of secretions left by those who came before it. Although these ants are nearly blind, they follow the scent by holding their antennae close to the ground. At the underground site, worker ants chew the leaf fragments into a mulch that they spread like a carpet in underground chambers. They wet the leaf mulch with their saliva and fertilize it with their feces. Soon a luxuriant lawn of fungi is growing, which serves as the sole food for all the ants, no matter what their age. Nurse ants even feed this fungus to the larvae, carrying them around to browse on choice spots.

The complex caste system of bees and ants apparently evolved a long time ago. Ant fossils 80 million years old exhibit three castes (males, queens, and workers), indicating that their complex social system had already evolved.

FIGURE 33-17 Leafcutter ants. These leafcutter ants are carrying their day's harvest back to the nest. The ants do not live on the leaf material—they use it as a substrate on which to grow the fungi that they cultivate in underground farms.

> *Many animals not only interact with other members of their species but form social groups and aid one another in various ways. Many species of insects, particularly bees and ants, form highly organized social groups exhibiting a division of labor.*

Rank Order in Vertebrate Groups

A social hierarchy, or **rank order,** exists in groups of fishes, reptiles, birds, and mammals. In chickens, the rank order is often called a **peck order** because these animals establish the order by pecking at one another. In many species of animals, the rank order is linear, with the highest-ranking individual dominant over all the others and the lowest-ranking individual submissive to all others. Some rank orders show certain complexities within the ranking, such as a rank order in which animal "A," for example, may be dominant over animal "B" and "B" dominant over "C," but "C" may be dominant over "A." In some rankings, the sex of the animal may play no role. In other rankings, males and females are ranked separately, and in still other vertebrate groups, the female takes on the rank of the male. Such rankings help reduce aggression and fighting within social groups, focusing aggressive behavior on the time that the ranking is developed. This behavior is only one of a repertoire of behaviors, including territoriality and submissive behaviors, that helps contribute to the stability of the social relationships among animals and within groups.

Human Behavior

Humans are unique animals. Because of this uniqueness, many scientists disagree with applying the principles of sociobiology to human behavior. Many scientists question the scientific validity of sociobiological research on human activity. Others question the social and political implications of human sociobiology, proposing that such endeavors might lead to yet another form of racism, sexism, or other types of group stereotyping. In addition, the study of human sociobiology has inherent limitations: scientists cannot experimentally manipulate the genes or vary the environments of humans for scientific studies. It is therefore extremely difficult to learn whether observed behaviors are a product of heredity or a person's environment.

This "nature or nurture" dichotomy continually raises questions that scientists have a difficult time researching. Interesting observations have been made, however, in the study of identical twins raised by different families in different environments. From these studies and others, most biologists hold that much of the behavioral *capacity* of humans is genetically determined but that the neural circuits specified by genes can be shaped and molded—within certain limits—by learning.

rank order a social hierarchy that exists in many groups of fishes, reptiles, birds, and mammals.

peck order in chickens, another term for the rank order chickens establish by pecking at one another.

Summary

- Social behaviors help members of the same species communicate and interact with one another. Scientists who study the social behaviors of animals are sociobiologists. They seek to determine general laws of the biology and evolution of behavior.
- Animals of the same species naturally compete for resources they need to survive. Within their repertoire of behaviors are aggressive behaviors that threaten and intimidate but do not kill rivals. Other behaviors also avoid fights to the death, such as submissive behaviors—signals of "backing down."
- Animals also use territorial behavior to limit aggression within their species. Territoriality is very common in all classes of vertebrates and in some invertebrates. A territory is an area that an animal marks off as its own, defending it against same-sex members of its species.
- Reproductive behaviors include patterns of male-female interactions that lead to successful fertilization—the penetration of an egg by a sperm. These behaviors include methods by which mates attract each other and then interact with one another in types of courtships that lead to mating and fertilization.
- Parenting behaviors include preparations for the care of young and their direct care to aid their survival and ultimate reproduction. Parenting behaviors are an example of altruistic behaviors: those that are advantageous to the recipient, increasing its individual fitness, while disadvantageous to the donor, decreasing its individual fitness. Natural selection favors genes that promote altruistic behavior among closely related individuals (kin) such as parents and offspring.
- Individuals within species form temporary or permanent associations within social groups. Animals within social groups work together for a variety of common purposes. Some species of insects, particularly various species of bees and ants, are organized into social groups having a division of labor.
- A social hierarchy, or rank order, exists in groups of fishes, reptiles, birds, and mammals. Behaviors relating to rank order help contribute to the stability of the social relationships among animals and within groups.
- Some scientists question the scientific validity of applying sociobiological principles to human behavior.

Knowledge and Comprehension Questions

1. Distinguish between behaviors and social behaviors.
2. Explain the functions of social behaviors.
3. What is competitive behavior?
4. Identify the type of behavior involved in each of the following situations:
   a. The hair on a dog's back stands on end as the dog growls at an approaching stranger.
   b. When you scold the dog for growling, it rolls over on its back and exposes its belly.
5. Define the term *territorial behavior.* Why is it important?
6. What is sexual reproduction? Explain the advantages that it can offer to a species.
7. Distinguish between mating and courtship.
8. Identify the type of behavior shown in each of the following situations:
   a. A robin builds its nest.
   b. A male peacock extends its tail feathers into a colorful fan.
   c. A cat chases another cat away from its food bowl.
9. Why do some animals form social groups? Why are genes that encode these behaviors retained in the gene pool (the total of all the genes of the breeding individuals in a population at a particular time)?
10. Summarize the social structure of a honeybee colony.
11. Explain the term *rank order.* What is its significance, and in what animal(s) does it appear?
12. Do you think the principles of sociobiology can be applied to human behavior? Explain your answer.

Critical Thinking

1. Very few mammals exhibit the complex societies seen among bees and ants—but a few do. Naked mole rats of the Middle East, for example, maintain large colonies with queens and special worker castes, organized remarkably like the colonies of bees. Why do you think such societies are so much rarer among mammals?

CHAPTER 34

Population Ecology

Looking like homes for oversized mud wasps, these cliff swallow nests hang precariously from the face of a rock outcropping. Any rough, vertical surface will serve as a nesting site for these birds. In fact, cliff swallows have recently discovered that the sides of bridges work well for this purpose ... and come complete with protective overhangs! As a result, cliff swallows, once found mainly in the western part of the United States, can now be found inhabiting the prairie, nesting on the sides of bridges that span the major prairie rivers.

The cliff swallows that inhabit these nests are a population of organisms. A **population** consists of the individuals of a given species that occur together at one place and at one time. This flexible definition allows the use of this term in many contexts, such as the world's human population, the population of protozoans in the gut of an individual termite, and the population of blood-sucking bugs living in the feathers of a cliff swallow.

An Introduction to Ecology

Population ecologists study how populations grow and interact. The science of ecology is much broader, however, and includes the study of interactions between organisms and the environment. Underlying questions to this study are "What factors determine the distribution patterns of organisms?" and "What factors control their numbers in the locations in which they are distributed?"

Scientists usually classify the study of ecological interactions into four levels: populations, communities, ecosystems, and the biosphere. This chapter discusses the first level of this hierarchy: populations.

Population Growth

Most populations will grow rapidly if optimal conditions for growth and reproduction of its individuals exist. But why, then, is the Earth not completely covered in bacteria, cockroaches, or houseflies? Why do some populations change from season to season or year to year?

To answer these questions, you need to understand how the size of a population is determined. The size of a population at any given time is the result of additions to the population from births and from **immigration,** the movement of organisms into a population, and deletions from the population from deaths and **emigration,** movement of organisms out of a population. Put simply:

$$(\text{Births} + \text{immigrants}) - (\text{Deaths} + \text{emigrants}) = \text{Population change.}$$

These statistics (births and deaths) are often expressed as a *rate:* numbers of individuals per thousand per year. For example, the population of the United States at the beginning of 1992 was approximately 254 million people. During 1992 the following occurred:

- 4,087,000 live births: The birth rate was 4,087,000 per 254,000,000 people, or 16.1 births per 1000.
- 2,166,000 deaths: The death rate was 2,166,000 per 254,000,000 people, or 8.5 deaths per 1000.

KEY CONCEPTS

- Under ideal conditions, populations grow at an exponential rate, leveling off at the carrying capacity of the environment.
- Very small populations are less able to survive than large populations; in addition, inbreeding in small populations leads to a loss of genetic diversity and increases the probability of the extinction of that species.
- Large populations face threats to survival such as competition for resources—for example, food, light, and shelter—and predation by other organisms.
- In 1992, the global human population of more than 5.3 billion people was growing at approximately 1.8% per year; at that rate, the human population will reach well over 10 billion people by 2029.

OUTLINE

- 973,977 (legal) immigrations: The immigration rate was 973,977 per 254,000,000 people, or 3.8 legal immigrants per 1000.
- 6977 emigrations: The emigration rate was 6977 per 254,000,000 people, or 0.03 emigrants per 1000.

The population change in the United States in 1992 can be calculated as follows:

(16.1 births/1000/year + 3.8 legal immigrants/1000/year) −
(8.5 deaths/1000/year + 0.03 emigrants/1000/year) =
11.37 people/1000/year.

This figure can also be expressed as a population change of 1.137% (rounded off to 1.1%)—an increase of approximately 1%.

The size of a population is the result of additions to the base population due to births and immigrations, and deletions from the population due to deaths and emigrations.

In natural populations of plants and animals, immigration and emigration are often minimal. Therefore, a determination of the **growth rate** of a population does not include these two factors. Growth rate (r) is determined by subtracting the death rate (d) from the birth rate (b):

$$r = b - d$$

Time (hours)	Number of bacteria
10	1,048,576
$9\frac{1}{2}$	524,288
9	262,144
$8\frac{1}{2}$	131,072
8	65,536
$7\frac{1}{2}$	32,768
7	16,384
$6\frac{1}{2}$	8,192
6	4,096
$5\frac{1}{2}$	2,048
5	1,024
$4\frac{1}{2}$	512
4	256
$3\frac{1}{2}$	128
3	64
$2\frac{1}{2}$	32
2	16
$1\frac{1}{2}$	8
1	4
$\frac{1}{2}$	2
0	1

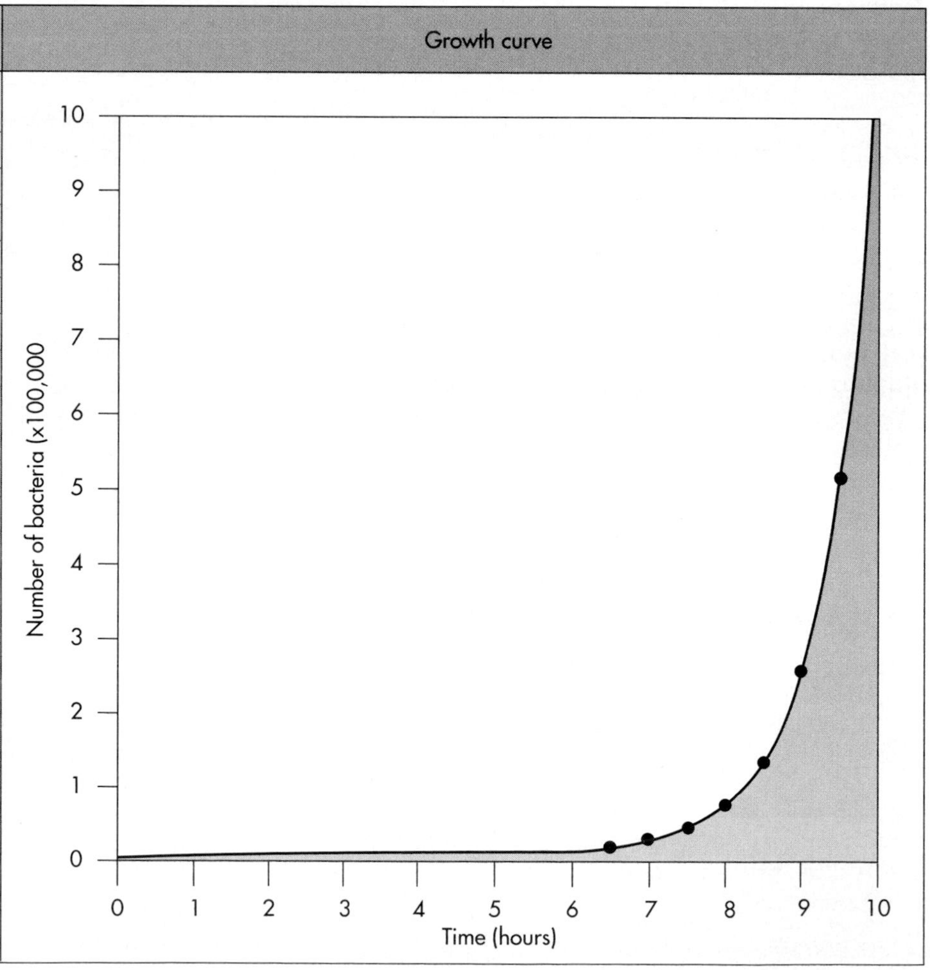

FIGURE 34-1 Exponential growth in a population of bacteria. This period of rapid exponential growth can only be sustained as long as resources are abundant.

Using the figures from our previous example:

$$r = 16.1 \text{ births/1000/year} - 8.5 \text{ deaths/1000/year}$$
$$r = 7.6/1000 \text{ or } 0.0076$$

To figure out the number of individuals added to a population of a specific size (N) in a given time *without* regard to immigration and emigration, r is multiplied by N:

$$\text{Population growth} = rN$$

Therefore the population growth in the United States in 1992 solely from births and deaths was 0.0076 × 254,000,000 people = 1,930,040 people.

> *The determination of the growth rate of a population does not include the factors of immigration and emigration. It is determined by subtracting the death rate from the birth rate.*

Exponential Growth

Although the *rate* of increase in population size *may stay the same,* the *actual increase* in the number of individuals *grows.* This sort of growth pattern is similar to the growth pattern of money in the bank as interest is earned and compounded. If you put $1000 in the bank at 8% per year, the first year you will earn $80. The second year you will earn 8% interest on $1080, or $86.40. Although your interest rate has stayed the same, the amount of money you earn grows as your money grows. Actually, in a bank your earnings would grow even more quickly because interest would be posted and compounded more often than once a year.

Figure 34-1 illustrates this principle with a population of bacteria in which each individual divides into two every half hour. The rate of increase remains constant, but the actual increase in the number of individuals accelerates rapidly as the size of the population grows. This type of mathematical progression found in the growth pattern of bacteria is termed **exponential growth.** (For example, two cells split to form 2^2 or 4, 4 becomes 2^3 or 8, and so on. The number, or power, to which 2 is raised is called an *exponent.*) Exponential growth refers to the rapid growth in numbers of a population of any species of organism, even though most organisms reproduce sexually and one organism does not split into two "new" organisms.

A period of exponential growth can occur only as long as the conditions for growth are ideal. In nature, exponential growth often takes place when an organism begins to grow in a new location having abundant resources. Such a situation occurred when the prickly pear cactus was introduced into Australia from Latin America. The species flourished, overrunning the ranges. In fact, the cactus became so abundant that cattle were unable to graze (Figure 34-2, *A*). Scientists regulated the population by introducing a cactus-eating moth to the area. The larvae of the moth fed on the pads of the cactus and rapidly destroyed the plants. Within relatively few years, the moth had reduced the population; the prickly pear cactus became rare in many regions where it was formerly abundant (Figure 34-2, *B*).

Carrying Capacity

No matter how rapidly a population may grow under ideal conditions, however, it cannot grow at an exponential rate indefinitely. As a population grows, each individual takes up space, uses resources such as food and water, and produces wastes. Eventually, shortages of important growth factors will limit the size of the population. In some populations such as bacteria, a buildup of toxic wastes may also limit population growth. Ultimately, a population stabilizes at a certain size, called the **carrying capacity** of the particular place where it lives. The carrying capacity is the number of individuals within a population that can be supported within a particular environment for an indefinite period. A population actually rises and

population the individuals of a given species that occur together at one place and at one time.

population ecologists (ih **kol** uh jists *or* ee **kol** uh jists) scientists who study how populations grow and interact.

immigration (im uh **gray** shun) the movement of organisms into a population.

emigration (**em** uh **gray** shun) the movement of organisms out of a population.

growth rate the number of individuals added to a population during a given time. The growth rate of a population is determined by subtracting its death rate from its birth rate.

exponential growth (ek spo **nen** shul) a type of growth pattern in which a population (for example) increases as its number is multiplied by a constant factor.

carrying capacity the number of individuals within a population that can be supported within a particular environment for an indefinite period.

FIGURE 34-2 A cactus takes over Australia. After an initial period in which prickly pear cacti, introduced from Latin America, choked many of the pastures of Australia with their rampant growth, they were controlled by the introduction of a cactus-feeding moth from the areas where the cacti were native. **A,** An infestation of prickly pear cacti in scrub in Queensland, Australia, in October 1926. **B,** The same view in October 1929, after the introduction of the cactus-feeding moth.

falls in numbers at the level of the carrying capacity but tends to be maintained at an average number of individuals (Figure 34-3). The exponential growth of a population and its subsequent stabilization at the level of the carrying capacity is represented by an S-shaped **sigmoid growth curve** (after the Greek letter *sigma*).

Under ideal conditions, populations grow at an exponential rate and show some stability in size at the carrying capacity of that place for that species.

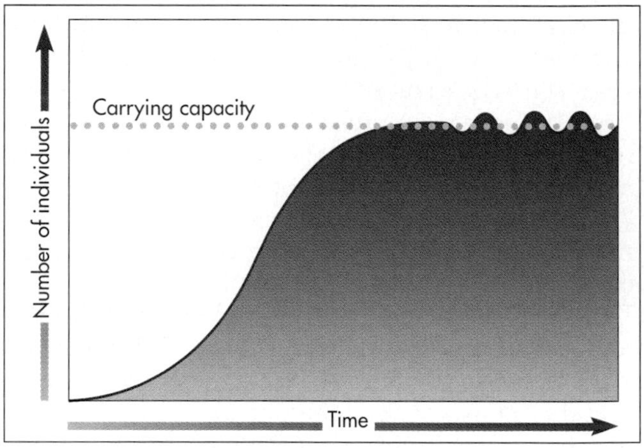

FIGURE 34-3 The sigmoid growth curve. The sigmoid growth curve, characteristic of populations, begins with a slow period of growth quickly followed by a period of exponential growth. When the population approaches its environmental limits, the growth begins to slow down, and it finally stabilizes, fluctuating around the maximum number of individuals that the environment will hold. This maximum number is the carrying capacity.

Population Size

The size of a population has a direct bearing on its ability to survive. Very small populations are less able to survive than large populations and are more likely to become extinct. Random events or natural disturbances can wipe out a small population, whereas a large population—simply due to its larger numbers and wider geographical distribution—is more likely to have survivors. Inbreeding—reproduction between closely related individuals—is also a negative factor in the survival of small populations. Inbreeding tends to produce many homozygous offspring (see Chapter 19), which results in the expression of many recessive deleterious traits that are usually masked by dominant genes. In addition, inbreeding reduces the level of variability in the gene pool (the genes of all breeding individuals) of the population, detracting from the population's ability to adjust to changing conditions. Loss of genetic diversity therefore increases the probability of extinction of that species.

Population Density and Dispersion

In addition to a population's size, its **density**—the number of organisms per unit of area—influences its survival. For example, if the individuals of a population are spaced far from one another, they may rarely come into contact. Sexually reproducing animals cannot produce offspring if they do not mate. Therefore the future of such a population may be limited even if the absolute numbers of individuals over a wide area are relatively high.

A factor related to population density is **dispersion,** the way in which the individuals of a population are arranged. In nature, organisms within a population may be distributed in one of three different patterns: uniform, random, and clumped (Figure 34-4). Each of these patterns reflects the interactions between a given population and its environment, including the other species that are present.

Uniform, or evenly spaced, distributions are rare in nature and generally are indications of competition or interference. For example, populations of plants exhibiting *allelopathy,* the secretion of toxic chemicals that harm other plants, often show a uniform distribution. The creosote bush, often the dominant vegetation covering wide areas of deserts of Mexico and the southwestern United States, grows well spaced and evenly dispersed. This uniform pattern of distribution is probably due to chemicals secreted by the bush that retard the establishment of other individuals near established ones.

Random distributions occur if individuals within a population do not influence each other's growth and if environmental conditions are uniform—that is, if the resources necessary for growth are distributed equally throughout the area. Random distributions are often seen in plants as the result of certain types of seed dispersal, such as scattering by the wind. Because environmental conditions are rarely "purely" uniform, most ecologic patterns are clumped to some degree.

Clumped distributions are by far the most frequent in nature. Organisms that show a clumped distribution are close to one another but far from others within the population. Clumping occurs as a result of the interactions among animals, plants, microorganisms, and unevenly distributed resources in an environment. Organisms are found grouped in areas of the environment that have resources they need. Furthermore, animals often congregate for a variety of other reasons, such as for hunting, mating, and caring for their young.

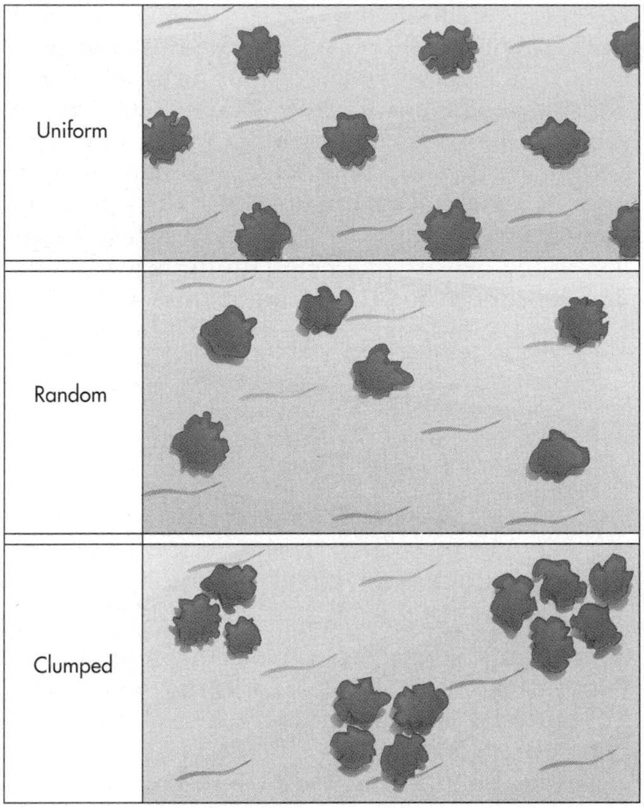

FIGURE 34-4 Distribution patterns in populations. Uniform distribution patterns are rare in nature but are sometimes caused by allelopathy, in which plants secrete toxic chemicals that discourage the growth of other plants near them. Random patterns in plants are often caused by random seed dispersal, such as by wind. Clumped patterns, the most common type, are often the result of an environment in which resources are unevenly distributed.

Regulation of Population Size

As a population grows and its density increases, competition among organisms for resources such as food, shelter, light, and mating sites increases and toxic waste products accumulate. Factors that result from the growth of a population *regulate its subsequent growth* and are "nature's way" of keep-

sigmoid growth curve the exponential growth of a population and its subsequent stabilization at the level of its environment's carrying capacity.
density the number of organisms or individuals in a population per unit of area.
dispersion the way in which the individuals of a population are arranged within their environment.

ing the population size of every species in check. Such factors increase in effectiveness as population density increases and are appropriately termed **density-dependent limiting factors.** Other factors such as the weather, availability of soil nutrients, and physical disruptions of an area (such as volcanoes or earthquakes) can also limit the growth of a population. Because these factors operate regardless of the density of a population, they are called **density-independent limiting factors.**

> *Density-dependent limiting factors come into play when a population size increases in a given area; density-independent limiting factors operate regardless of population size.*

Density-Independent Limiting Factors

A variety of environmental conditions can limit populations. For example, freak snowstorms in the Rocky Mountains of Colorado in the summer can kill butterfly populations there. The size of insect populations that feed on pollen and flower tissues varies seasonally with the blooming of flowering plants. Humans, too, can affect the sizes of populations. Poachers have killed so many African elephants for their ivory, for example, that the species may become extinct.

Density-Dependent Limiting Factors

Individuals within a species and individuals of differing species compete for the same limited resources. (Chapter 35 discusses interspecific competition in detail and points out that competition among different species of organisms is often greatest between those that obtain their food in similar ways.) In fact, if two species are competing with one another for the same limited resource in a specific location, the species able to use that resource most efficiently will eventually eliminate the other species in that location. This concept is called the principle of **competitive exclusion.** In fact, competition among organisms of the same and differing species was described by Charles Darwin as resulting in natural selection and survival of the fittest or most well-adapted organisms. Competition, therefore, not only limits the sizes of populations but is one of the driving forces of evolutionary change. (However, the fittest members of populations are often the ones that successfully avoid competition.)

Predation is another factor that limits the size of populations and works most effectively as the density of a population increases. Predators are organisms of one species that kill and eat organisms of another—their prey. Predators include animals that feed on plants (such as cows grazing on grass) and plants that feed on insects (such as the Venus flytrap; see p. 609). The intricate interactions between predators and prey are an essential factor in the maintenance of diverse species living in the same area. By controlling the levels of some species, the predators make the continued existence of other species in that same community possible. In other words, by keeping the numbers of individuals of some of the competing species low, the predators prevent or greatly reduce competitive exclusion. In fact, a given predator may very often feed on two or more different kinds of plants or animals, switching from one to the other as their relative abundance changes. Similarly, a given prey species may be a primary source of food for an increasing number of predator species as it becomes more abundant, a factor that will limit the size of its population automatically.

Parasitism also limits the size of populations by weakening or killing host organisms. Parasites live on or in larger species of organisms and derive nourishment from them. As a population increases in density, parasites such as bacteria, viruses, and a variety of invertebrates can more easily move from one organism to another, infecting an increasing proportion of a population. Once again, this limiting factor to population size acts in negative feedback fashion, becoming more effective as the density of the population increases.

> *Three density-dependent limiting factors are competition, predation, and parasitism.*

Mortality and Survivorship

A population's growth rate depends not only on the availability of needed resources and on the ability of its individuals to survive and compete effectively for those resources but also on the ages of the organisms in it. Interestingly, when a population lives in a constant environment for a few generations, its **age distribution**—the proportion of individuals in the different age categories—becomes stable. This distribution, however, differs greatly from species to species and even to some extent within a given species from place to place.

Scientists express the mortality characteristics of a population by means of a survivorship curve. **Mortality** is the death rate. **Survivorship** refers to the proportion of an original population that survives to a certain age. The curve is developed by graphing the number of individuals within a population that

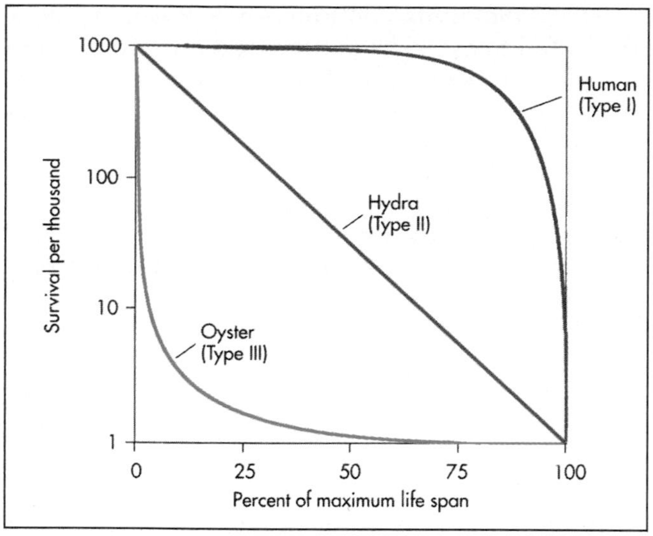

FIGURE 34-5 Survivorship curves. The shapes of the respective curves are determined by the percentages of individuals in populations that are alive at different ages.

survive through various stages of the life span. Samples of survivorship curves that represent certain types of populations are shown in Figure 34-5.

In the hydra, individuals are equally likely to die at any age, as indicated by the straight survivorship curve (type II) shown in Figure 34-5. This type of survivorship curve is characteristic of organisms that reproduce asexually, such as hydra, bacteria, and asexually reproducing protists.

Oysters, on the other hand, produce vast numbers of offspring, but few of these offspring live to reproduce. The death rate of organisms that survive and reach reproductive age is extremely low, however (type III survivorship curve). This type of survivorship curve is characteristic of organisms producing offspring that must survive on their own and therefore die in large numbers when young because of predation or their inability to acquire the resources they need.

The survivorship curve for humans and other large vertebrates is much different from that for hydra and oysters. Humans, for example, produce few offspring but protect and nurture them; therefore most humans (except in areas of great poverty, hunger, and disease) survive past their reproductive years (type I survivorship curve).

Many animal and protist populations have survivorship curves that lie somewhere between those characteristic of type II and type III. Many plant populations, with high mortality at the seed and seedling stages, have survivorship curves close to type III. Humans have probably approached type I more and more closely through the years, with the life span being extended because of better health care and new medical technology.

When a population lives in a constant environment for a few generations, the proportion of individuals in various age categories becomes stable. Characteristic survivorship curves occur among species and are closely linked in animal populations to parental care for offspring.

Demography

Demography is the statistical study of human populations. The term comes from two Greek words: *demos,* "the people" (the same root in the word *democracy*), and *graphos,* "to write." It therefore means the description of peoples and the characteristics of populations. Demographers predict the ways in which the sizes of populations will alter in the future, taking into account the age distribution of the population and its changing size through time.

density-dependent limiting factors environmental factors that result from the growth of a population but act to limit its subsequent growth. Density-dependent limiting factors increase in effectiveness as population density increases.

density-independent limiting factors environmental factors that operate to limit a population's growth, regardless of its density.

competitive exclusion a principle that states than when two species are competing with one another for the same limited resource in a specific location, the species able to use that resource most efficiently will eventually eliminate the other species in that location.

predation an interaction among populations in which an organism of one species kills and eats an organism of another.

parasitism (**pare** uh suh tiz um) an interaction in which an organism of one species (the parasite) lives in or on another (the host).

age distribution the proportion of individuals in the different age categories of a population.

mortality the death rate of a population.

survivorship the proportion of an original population that lives to a certain age.

demography (dih **mog** ruh fee) the statistical study of human populations.

A population whose size remains the same through time is called a **stable population.** In such a population, births plus immigration exactly balance deaths plus emigration. In addition, the number of females of each age group within the population is similar. If this were not the case, the population would not remain stable. For example, if there were many more females entering their reproductive years than older females leaving the population, the population would grow.

The age distribution of males and females in human populations of Kenya, the United States, and Austria is shown in Figure 34-6. A **population pyramid** is a bar graph that shows the composition of a population by age and gender. Males are conventionally enumerated to the left of the vertical age axis and females to the right. By using population pyramids, scientists can predict the future size of a population. First, the number of females in each age group is multiplied by the average number of female babies that women in that age group bear. These numbers are added for each age group to see whether the new number will exceed, equal, or be less than the number of females in the population being studied. By such means the future growth trends of the human population as a whole and of individual countries and regions can be determined.

The population pyramids in Figure 34-6 show the differences in the pattern of a rapidly growing population (Kenya), a slowly growing population (the United States), and a country experiencing negative growth (Austria). The population pyramid of Kenya is characteristic of developing countries, those that have not yet become industrialized, such as countries in Africa, Asia, and Latin America. Each of these countries has a population pyramid with a broad base, reflecting the large numbers of individuals yet to enter their reproductive years. In Kenya, for example, 50% of the population is younger than 15 years of age. These children will reach reproductive age in the near future, leading to a population explosion in Kenya over the next decade.

In the United States, birth rates are higher than death rates at present, producing a growth rate of approximately 0.7%. The high birth rate is not due to couples having large families but to the large size of the "baby boom" generation that is at the peak of its reproductive years. Individuals in this age group were born within the 20 years or so after World War II. The large number of women in this group causes the births to still outnumber the deaths. Austria and the United States are both experiencing a decline in fertility and mortality. However, Austria's population does not include as high a percentage of women

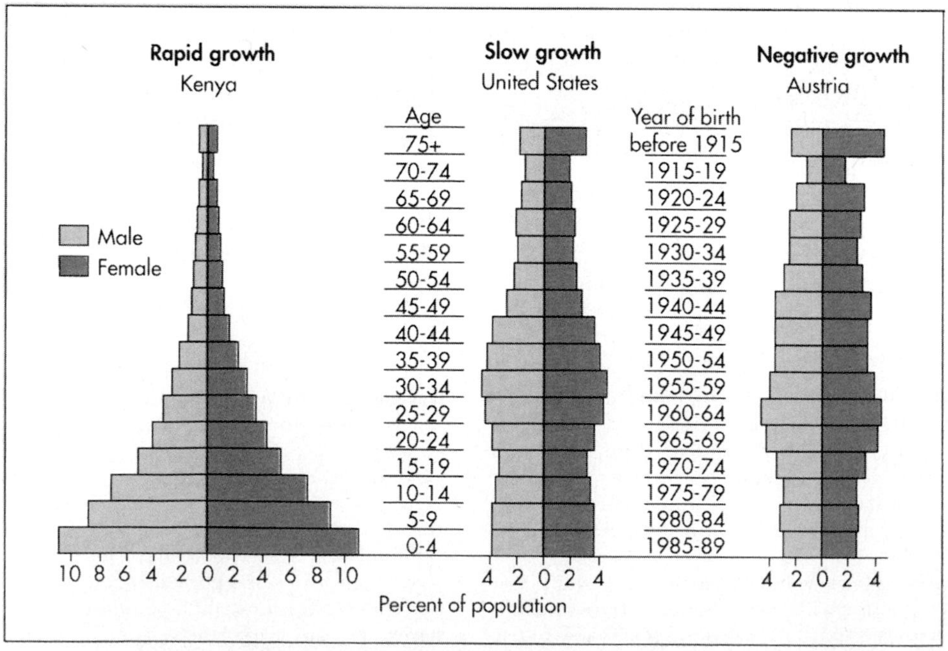

FIGURE 34-6 Three patterns of population change. Kenya shows a rapid growth pattern—its pyramid has a broad base because of high fertility. The United States and Austria both show declining fertility and mortality, but the United States' "baby boom" has resulted in higher fertility rates than Austria, whose population is declining.

FIGURE 34-7 The development of agriculture was a key step in the growth of human populations. By producing abundant supplies of food, agriculture made the growth of cities and the future development of human culture possible.

in their childbearing years as the United States, so deaths are outnumbering births.

Developing countries—those that have not yet become industrialized—have proportionately young populations and are experiencing rapid growth. Developed countries have populations with similar proportions of their populations in each age group and are growing very slowly or not at all.

The Human Population Explosion

Although some countries have populations that are no longer growing, such as Denmark, West Germany, Hungary, and Italy, and some countries such as Austria are declining in numbers, the population of the world as a whole is growing at the rate of 1.8% a year. This growth rate may sound low, but with the world population numbering more than 5.3 billion, it adds over 95 *million* people to the population each year. Scientists estimate that the world population will double between 1990 and 2029!

How did the human population reach its present-day size? With the development of agriculture 11,000 years ago (Figure 34-7), human populations began to grow steadily (Figure 34-8). Villages and

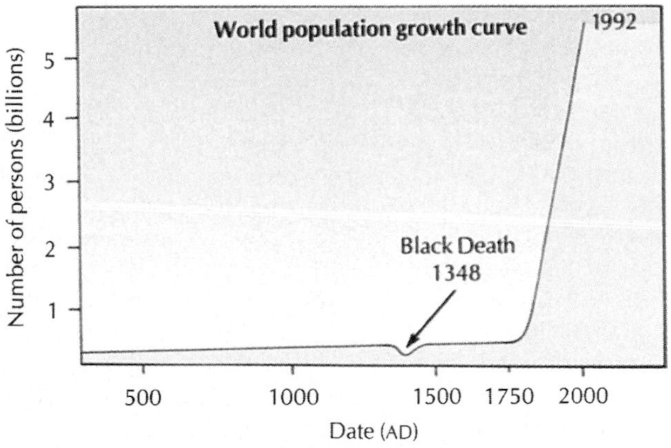

FIGURE 34-8 World population growth through history. Except for the blip caused by the Black Death (25 to 35 million people died in Europe alone), the human population has grown steadily since the Stone Age.

stable population a population whose size remains the same through time.
population pyramid a bar graph that shows the composition of a population by age and sex.

HOW *Science* WORKS

Applications

Managing Urban Explosions

Cities are a relatively new phenomenon. Villages and towns were first organized only about 9000 years ago, long after the development of agriculture. The first great cities of Mesopotamia date soon after that, but relative to a modern metropolis, the cities of ancient times were small. Only in the last thousand years have people lived in cities containing hundreds of thousands of people. Until the Industrial Revolution, only one person in five lived in a town of more than 10,000 people. In the past hundred years, however, there has been a mass influx of people into cities.

In 1920 the world's urban population was 360 million. By the year 2000, it will be near 3 billion—more than 50% of the world's population will be urbanized. Two thirds of all people will be in cities of 100,000 or more, with over 20% in cities with more than a million people.

This massive movement of people, along with increases in population, is alarming because it is difficult to manage so many people crowded together. Tokyo-Yokohama, Japan, is the world's largest and most densely populated urban region today. This megacity is plagued by smog and traffic problems as are other big cities, all worsened by the incredible congestion of more than 27 million inhabitants (Figure 34-A). The prospects of supplying adequate food, water, and sanitation to people in a city whose population will increase to nearly 30 million by the end of the century are almost unimaginable.

Tokyo-Yokohama's astonishing urban growth is not unique. Only 7 cities had a population larger than 5 million in 1950, but by 2000, there will be 43 such megacities. In 1992 there were 16 cities with populations greater than 10 million. By 2000 there will be 19 such cities—and all but 4 of these will be in the northern hemisphere. Up near the top of the list in population is New York City, with almost 14 million inhabitants—and Mexico City, Mexico; San Paulo, Brazil; and Seoul, South Korea, are bigger than New York! These four cities alone contain more than 70 million people—half as many as the entire human population in the year 1 AD.

Management of cities will be a priority for governments in the twenty-first century. Scientists, engineers, and government officials must work together to develop environmentally sound ways of disposing of waste, providing livable housing, and supplying safe food and water to the multitudes living in large urban areas, and these are just *some* of the concerns that will be at the forefront of city management.

towns were first organized about 5000 years ago, and human effects on the environment began to intensify. In these centers of civilization, however, the specialization of professions such as metallurgy became possible; technology advanced. By 1660, the world population totaled approximately 500 million people. The Renaissance in Europe, with its renewed interest in science, ultimately led to the establishment of industry in the seventeenth century and to the Industrial Revolution of the late eighteenth and early nineteenth centuries. By the mid-nineteenth century, Louis Pasteur put forth the germ theory of disease, the understanding that microbes caused infection. With this understanding came new medical technology and discoveries.

In the 1920s, Alexander Fleming accidentally discovered penicillin and opened the door to antibiotic therapy—medicine's "magic bullets" against bacterial infection. These medical advancements decreased the death rate by increasing the number of individuals surviving infection.

The advent of the Industrial Revolution also heralded new farming and transportation technology, which helped provide better nutrition for many people, especially those in industrialized countries. With better nutrition and increased medical understanding and technology, the death rate fell steadily and dramatically from the mid-nineteenth century on (Figure 34-9, *A*). In developing countries, international foreign aid imported this new technology along with food aid after World War II. The mortality rate plunged in a matter of years (Figure 34-9, *B*). However, birth rates remained largely unchanged, and as a result, the world's population-growth rate soared. In fact, at their present rates of growth, the population of tropical South America will double in 31 years and that of tropical Africa will double in only 24 years. By the year 2000, about 60% of the people in the world will be living in countries that are at least partly tropical or subtropical, 20% will be living in China, and the remaining 20% will be located in the developed countries of Europe, the Commonwealth of Independent States (formerly the Soviet Union), Japan, the United States, Canada, Australia, and New Zealand (Figure 34-10). Putting it another way, for every person living in an industrialized country such as the United States in 1950, there were two people living elsewhere; by 2020, just 70 years later, there will be five.

> *The world population rose sharply and dramatically after the Industrial Revolution because of new technology in agriculture, transportation, industry, and medicine.*

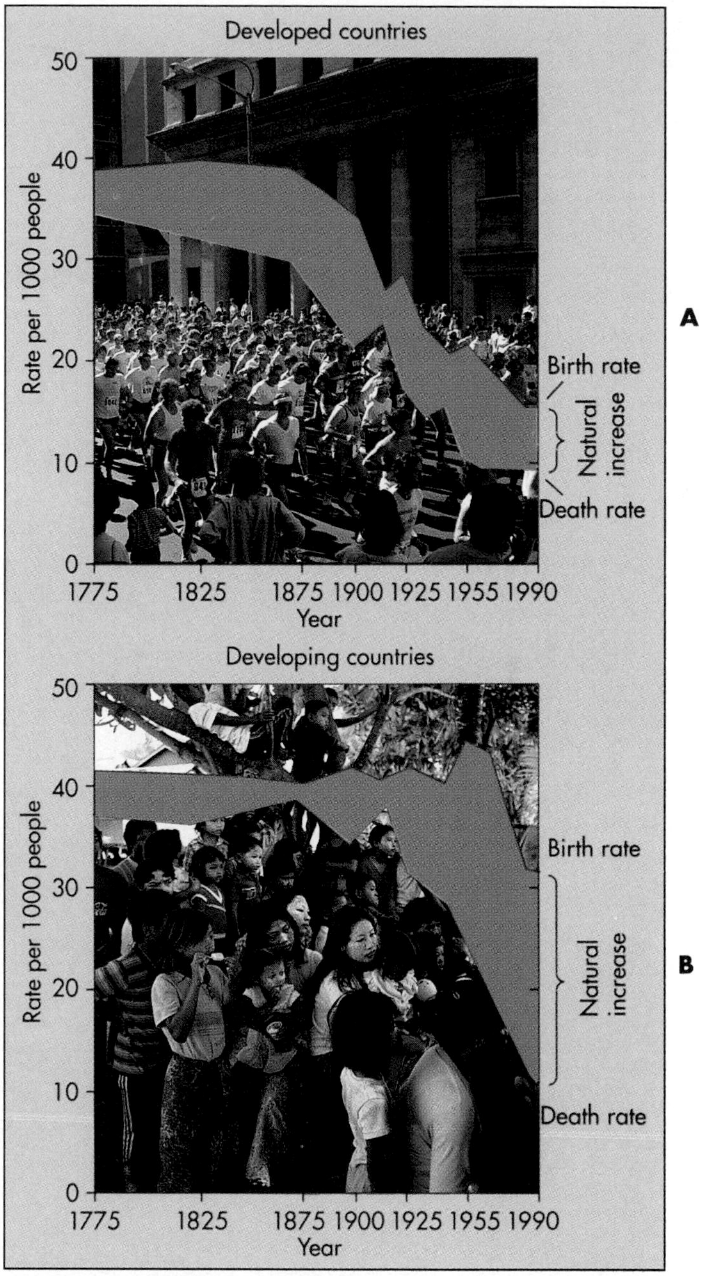

FIGURE 34-9 Population growth from 1775 to 1990. A, In developed countries, the Industrial Revolution caused mortality (death) to drop. Birth rates also began to drop at the turn of the century. **B,** In developing countries, mortality began to decline after World War II, but birth rates remained high. Taken together, these changes caused human population growth rates to soar, especially in developing countries.

Of the estimated 2.8 billion people living in the tropics in the late 1980s, the World Bank estimated that about 1.2 billion people were living in absolute poverty. (The World Bank is an international bank that provides loans and technical assistance for economic development projects in developing member countries.) These people cannot reasonably ex-

JUST WONDERING . . . .

I see photos of starving people all the time on the news. Aside from the politics involved, is the world running out of enough food to feed everybody?

Scientists estimate that the world population of more than 5.3 billion people will reach well over 10 billion by 2029. Your question raises an issue of grave concern as the world faces such astonishing growth. Unfortunately, there are no easy answers, and scientists themselves are polarized on this issue. The only thing both sides can agree on is that the demand for food will grow enormously as the population swells.

One group within the scientific community contends that new technologies, if managed properly, will allow humans to expand the world food production and feed all of its people. Recent research assessing the climate and soil conditions in 93 developing countries suggests that 3 times as much land as is currently cultivated could be put to agricultural use. In areas without additional land to farm, the number of crops grown each year on land presently in use could be increased—an agricultural practice called *multicropping.* In addition, the use of high-yield crop varieties, fertilizer, and irrigation in areas where they are not currently employed could increase crop yields.

An opposing view is held by many scientists who suggest that intensifying agricultural practices as suggested will cause serious ecological damage to our world, such as extensive deforestation, loss of species diversity, erosion of the soil, and the pollution of aquifers, streams, and rivers from pesticides and fertilizers. In addition, they assert that our natural resources will be unable to support this future demand, suggesting that crop yields would have to rise by 112% to feed all the people of the developing (not yet industrialized) nations in 2050. To also raise the standards of their presently inadequate diets would translate into each acre of land increasing its yield more than 6 times. Many scientists think that these goals are impossible to achieve unless new technologies are developed.

Unfortunately, in countries torn apart by war, food production is only one of the many serious problems facing its citizens. War-torn countries and those with inadequate natural resources and technologies will have to rely on food aid, placing even more pressure on the resources available in the rest of the world. There are no easy answers to your question, and feeding the people of the world will continue to be a critical issue. One hopes researchers in biotechnology, agriculture, and related areas will make contributions that will help solve this problem for future generations. However, even if we solve our immediate problems of food production but the world population continues to grow at the same rate it does today, what will happen beyond 2050? Will we reach a final limit?

pect to be able to consistently provide adequate food for themselves and their children. Even though some experts estimate that enough food is produced in the world to provide an adequate diet for everyone in it (see "Just Wondering . . ."), the distribution is so unequal that large numbers of people live in hunger. The United Nations International Children's Emergency Fund (UNICEF) estimates that in the developing world, about 14 million children younger than 5 years of age—40,000 per day—die each year, mainly of malnutrition and the complications associated with it.

The size of human populations, like those of other organisms, is or will be controlled by the environment. Early in its history, human populations were regulated by both density-dependent and den-

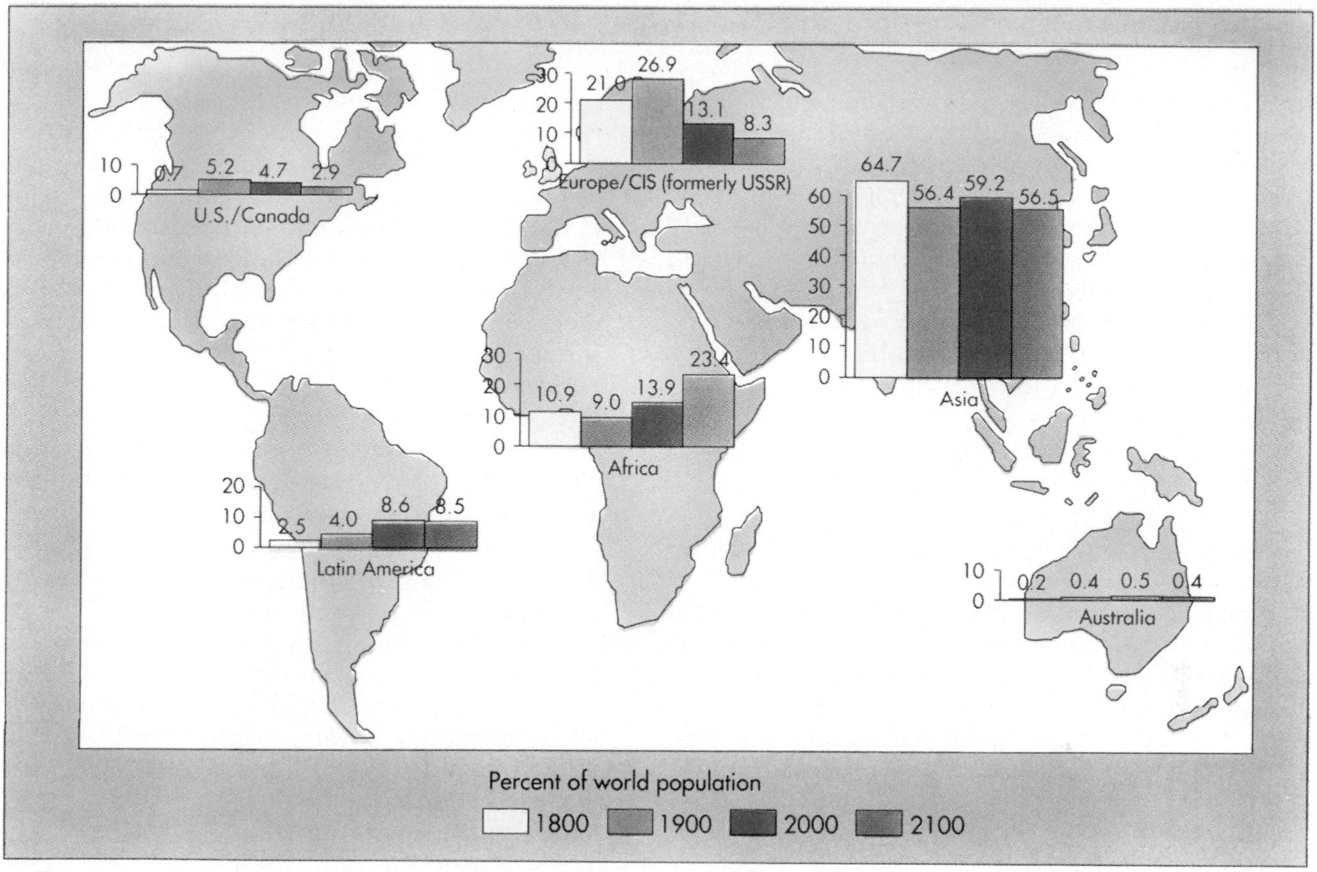

FIGURE 34-10 World population distribution by region, 1880 to 2100. If current trends continue, Asia will have 57% of the total world population in 2100, Africa nearly a quarter, and Europe's share will drop to less than 10%.

sity-independent limiting factors, including food supply, disease, and predators; there was also ample room on Earth for migration to new areas to relieve overcrowding in specific regions. In the past century, however, humans have been able to expand the carrying capacity of the Earth because of their ability to develop technological innovations. Gradually, changes in technology have given humans more control over their food supply and enabled them to develop superior weapons to ward off predators as well as the means to cure diseases. Improvements in transportation and housing have increased the efficiency of migration. At the same time, improvements in shelter and storage capabilities have made humans less vulnerable to climatic uncertainties.

As a result of the ability to manipulate these factors, the human population has been able to grow explosively to its present level of more than 5.3 billion people. Both the current human population level and the projected rate of growth have potential consequences for the future that are extremely grave. Pressures are placed on the land, water, forests, and other natural resources. Industrialization, although raising the standard of living by increasing the availability of goods and services, adds to air and water pollution. In the developed countries, nations of consumers have developed into "throwaway" societies, adding billions of tons of solid waste to landfills every year.

The most effective means of dealing with the population explosion has been the support of governments to encourage small families, the establishment of family planning clinics, improvement in education, and socioeconomic development. The developed countries of Western Europe and North America, Japan, and Australia have very low birth rates at this time. China, Indonesia, Thailand, South Korea, Hong Kong, and Singapore have had considerable success with lowering their birth rates. Countries such as Mexico and India have had some success in reducing their birth rates but are still striving toward this goal. The countries in sub-Sahara Africa have the highest birth rate of any countries in the world—some as high as 4.1%. Family planning programs are now being implemented across the continent.

Summary

- Populations consist of the individuals of a given species that occur together at one place and at one time. They may be dispersed in an evenly spaced, clumped, or random manner. Clumped patterns are the most frequent, with random distributions being rare (most showing some degree of clumping).
- The rate of growth of any population is the difference between the birth rate and the death rate per individual per unit of time. The actual rate may also be affected by emigration from the population and immigration into it.
- Most populations exhibit a sigmoid growth curve, which implies a relatively slow growth, a rapid increase, and then a leveling off when the carrying capacity of the species' environment is reached.
- Survivorship curves are used to describe the characteristics of growth in different kinds of populations. Type I populations are those in which a large proportion of the individuals approach their physiologically determined limits of age. Type II populations have a constant mortality throughout their lives. Type III populations have very high mortality in their early stages of growth, but an individual surviving beyond that point is likely to live a very long time.
- Each population grows in size until it eventually reaches the limits of its environment to support it; resources are always limiting. Some of the limits to the growth of a population are related to the density of that population, but others are not. Competition within the populations and between populations of any two species limits their coexistence. Predation and other forms of interaction among populations also play an important role in limiting population size.
- Agriculture was developed in several centers about 11,000 years ago when about 5 million people lived throughout the world. By 1990, the global population was more than 5.3 billion people and was growing at a rate of 1.8% per year, a rate that will double the population in approximately 39 years.
- In the 1980s, the World Bank estimated that about 1.2 billion of the more than 2.75 billion people living in the tropics and subtropics existed in absolute poverty. Some 14 million children younger than 5 years of age were starving to death each year. With the populations of tropical countries growing from 1.8% to 4.1% per year, the task of feeding these people will be extraordinarily difficult.

Knowledge and Comprehension Questions

1. What is ecology? State one question that underlies ecological inquiry.
2. In the early twentieth century, the United States experienced a large influx of immigrants from many European nations. Does immigration affect the size or growth rate of the population in the United States?
3. How is the work of population ecologists similar to that of demographers? How does it differ?
4. Which is more likely to survive, a small population or a large one? Why?
5. Distinguish between density and dispersion. How does each affect a population's chances for survival?
6. Identify the three patterns of population dispersion found in nature. Into which pattern do human populations fall?
7. Distinguish between density-dependent and density-independent limiting factors. Give an example of each.
8. What do predation and parasitism have in common? How do they differ?
9. Draw type I, type II, and type III survivorship curves. Summarize the types of organisms that are characteristic of each curve, and give an example of each.
10. Compare the typical population pyramids of a developing country and industrialized country.
11. Fill in the blanks: The population of the world as a whole is growing at an annual rate of __________%, adding more than __________ people to the population each year. Scientists estimate the world population will __________ between 1990 and 2029.

Critical Thinking

1. Both the current human population level and the projected rate of growth worldwide have potential consequences for the future that are extremely grave. Explain why.
2. Suppose that you were given the political power to deal with the world's population explosion. What steps would you take?

CHAPTER 35

INTERACTIONS WITHIN COMMUNITIES OF ORGANISMS

THIS BEAUTIFUL CORAL REEF is home to a variety of organisms that each contribute to the array of colors evident here. The reef provides a shallow-water environment favorable to many organisms. Nutrients are abundant as are surfaces for attachment and hiding places into which animals can burrow. Although a number of species characteristic of coral reef populations are visible in the photograph, more than 3000 species may coexist in a large reef. Its populations interact in a variety of ways. The coral reef organisms, primarily various species of soft corals such as the colorful flowery stalks in the photograph, compete with one another and with other organisms for space and food. Some organisms attach to others as does the sponge growing on the green soft coral in the upper left of the photograph. The sponge is actually boring into the coral and will eventually destroy it. Schools of fish swim within and around the reef, eating algae and plankton that drift in the surrounding water. Together, the interacting populations of a coral reef form a vibrant, colorful marine community.

KEY CONCEPTS

- A community is a grouping of populations of different species living together in a particular area at a particular time.
- Each organism plays a special role within a community; this role is its niche.
- Members of the various species within a community interact by competing with one another for resources, killing one another for food, and living in close associations that both benefit and harm one another; these interactions affect the niche that a species can occupy in that community.
- Communities, like species, change over time in the dynamic process of succession.

OUTLINE

Ecosystems

Scientists have long known that the nonliving or **abiotic factors** within the environment—such as air, water, and even rocks—affect an organism's survival, as do the living or **biotic factors**—such as surrounding plants, animals, and microorganisms. All the biotic and abiotic factors together within a certain area are called an **ecosystem.** Therefore the rock is as important to the makeup of the ecosystem as the barnacles.

Within an ecosystem, each living thing has a home, an actual area in which it resides. This space, including the factors within it, is an organism's **habitat.** Organisms not only reside in their habitats, they interact with the biotic and abiotic factors within it and use them to survive. Each organism also plays a special role within an ecosystem; this role is called a **niche.** The term *niche* refers to the organism's use of the biotic and abiotic resources in its environment. Thus it may be described with reference to space, food, temperature, appropriate conditions for mating, and requirements for moisture, for example. A full portrait of an organism's niche also includes the organism's behavior and the ways in which this behavior changes at different seasons and at different times of the day. These concepts are important to the understanding of the concept of communities.

Communities

The interactions among organisms within ecosystems are varied. Individuals of the same species make up a **population** of organisms. (The individuals within a population interact with one another in ways described in Chapters 33 and 34.) Populations also interact with one another, forming communities. A **community** is a grouping of populations of different species living together in a particular area at a particular time. Thus, an ecosystem can be thought of as a community of organisms, along with the abiotic factors with which the community interacts.

The magnificent redwood forest that extends along the coast of central and northern California and into the southwestern corner of Oregon is an example of a community. Within it, the most obvious organisms are redwood trees (Figure 35-1). However, populations of other organisms, such as the sword fern and the ground beetle shown in Figure 35-2, make up this community too. The coexistence of these various populations is made possible in part because of the special conditions that are created by the redwoods: shade, water (dripping from the branches), and relatively cool temperatures. For this reason and because the redwoods visually dominate the area, this distinctive group of populations is the redwood community.

FIGURE 35-1 The redwood community. Although a community, the redwood forests in coastal California and southwestern Oregon are dominated by the redwoods.

A

B

FIGURE 35-2 Plants and other organisms that live in the redwood community. A, The redwood sorrel. **B,** A ground beetle is feeding on a slug on a leaf of sword fern.

> *Individuals of a species compose a population; many populations make up a community. These populations of living things interact with one another in many complex ways.*

Types of Interactions Within Communities

Interactions among populations can be grouped into the following five categories, some of which are illustrated in Figure 35-3:

1. *Competition.* Organisms of different species that live near one another strive to obtain the same limited resources.
2. *Predation.* An organism of one species kills and eats an organism of another. (Animals that kill and eat members of their own species are cannibals.)
3. *Commensalism.* An organism of one species benefits from its interactions with another, whereas the other species neither benefits nor is harmed.
4. *Mutualism.* Two species live together in close association, both benefiting from the relationship.
5. *Parasitism.* An organism of one species (the parasite) lives in or on another (the host). The parasite benefits from this relationship, whereas the host is usually harmed. Parasitism is sometimes considered a form of predation. However, unlike a true predator, the successful parasite does not kill its host.

Scientists study the interactions among organisms and between organisms and their environments in the laboratory and in nature. This specialized field of biology is called **ecology;** the scientists who work within this field are ecologists. Ecologists study the physical and biological vari-

FIGURE 35-3 Living—and dying—together. Organisms that live within a community interact in a number of ways. **A,** Competition. The kudzu vine in the southern United States competes with other plants in the area for resources. However, the kudzu vine seems to be winning in many locations. **B,** Predation. This spider has successfully caught its prey. **C,** Mutualism. This moray eel shelters a smaller fish in its mouth. The eel benefits from a thorough teeth cleaning, and the smaller fish receives an undisturbed meal.

ables governing the distribution and growth of living things. Ecologists also study the theoretical bases of these interactions, and some use computers to develop mathematical models of ecological systems. The knowledge gained by ecologists is essential to the basic understanding of the world and provides a foundation to find solutions to the many environmental problems created by humans.

Competition

When two kinds of organisms use the same limited resource for survival, they compete for that resource. Complex animals such as vertebrates compete by using innate behaviors such as threat displays and territorial behavior (see Chapter 33). Lower forms of animals and plants do not exhibit complex behaviors; they compete with one another

abiotic factors (aye bye **ot** ik) nonliving factors within the environment, such as air, water, and rocks.

biotic factors (bye **ot** ik) living factors within the environment, such as plants, animals, and microorganisms.

ecosystem (eh koe **sis** tem or ee koe **sis** tem) all the biotic and abiotic factors within a certain area. A community consisting of plants, animals, and microorganisms that interact with one another and with their environments and are interdependent on one another for survival.

habitat a place where an organism lives or grows.

niche the role each organism plays within an ecosystem.

population a group of organisms that consists of individuals of the same species.

community a grouping of populations of different species living together in a particular area at a particular time.

ecology (eh **kol** uh gee *or* ee **kol** uh gee) the study of the interactions among organisms and between organisms and their environments.

simply by their adaptive fitness—by reproducing more and better-adapted offspring that crowd out opponents.

Competition among different species of organisms is often greatest between organisms that obtain their food in similar ways. Thus green plants compete mainly with other green plants, meat-eating animals with other meat-eating animals, and so forth. Competition within a genus or between individuals of the same species occurs as well.

> *Organisms of different species compete with one another when they need the same limited resource for survival. This competition is greatest among organisms that obtain their food in similar ways.*

Competition among organisms has been observed by scientists for a long time. More than 50 years ago, the Russian scientist G.F. Gause formulated the principle of **competitive exclusion,** which was based on his experimental work. This principle states that if two species are competing with one another for the same limited resource in a specific location, the species able to use that resource most efficiently will eventually eliminate the other species in that location. A modification of this principle is the concept of *coexistence with niche subdivision.* In this situation, two species still exist in the same geographical area, and one or both species occupies a more restricted niche as the result of competition with the other. This concept is illustrated in the next section.

The Study of Competition in the Laboratory

Scientists sometimes study competition between species in the laboratory so that they can control the environmental conditions. John Harper and his colleagues at the University College of North Wales, Australia, for example, performed competition experiments with two species of clover: white clover and strawberry clover. Each species was sown with the other at one of two densities: 36 or 64 plants per square foot. Various plots of the two species were planted, using all the possible combinations of the two densities of plants. One plot contained 36 white clover plants and 36 strawberry clover plants per square foot. Another contained 64 white and 36 strawberry per square foot and so forth. The white clover initially formed a dense canopy of leaves in each experimental plot. However, the slower-growing strawberry clover, whose leaf stalks are taller, eventually produced leaves that grew above the white clover leaves. In competing more effectively for light, the strawberry clover overcame the white clover, causing it to die out. The outcome was the same, regardless of the initial densities at which the seeds of the plants were sown.

> *In a competitive situation and under controlled laboratory conditions, the species able to use a particular resource more efficiently will be the species to survive.*

The Study of Competition in Nature

As well as competing for sunlight, plants compete for soil nutrients. The roots of one species, for example, may outcompete another species by using up minerals in the soil essential to both species. In addition, one species may secrete poisonous substances that depress the growth of other species. Sage plants, for example, inhibit the establishment of other plant species nearby, producing bare zones around populations of these plants (Figure 35-4). These interactions are poorly understood, but experimental studies are beginning to resolve their complexity.

Another interesting view of competition in nature is demonstrated by the acorn barnacles. Highly adapted to their environment, acorn barnacles are typically found in the intertidal zone of rocky shores—the narrow strip of land exposed during low tide and covered during high tide. When submerged, an acorn barnacle feeds by extending appendages from the hole in its shell. Spread out, these appendages act like a net, sweeping the water and collecting food that it then brings into its shell and eats. When exposed to the air, an acorn barnacle pulls in its feeding appendages and shuts down, actually us-

FIGURE 35-4 Competition among plants. The bare zones around these colonies of sage are plainly visible in this aerial photograph taken in the mountains above Santa Barbara, California. The plants secrete chemicals that suppress the growth of other plants nearby.

FIGURE 35-5 ***Chthamalus* and *Balanus* barnacles growing together on a rock.** Competing species belonging to these two genera were studied by J.H. Connell along the coast of Scotland. *Balanus* are the larger organisms of the two genera.

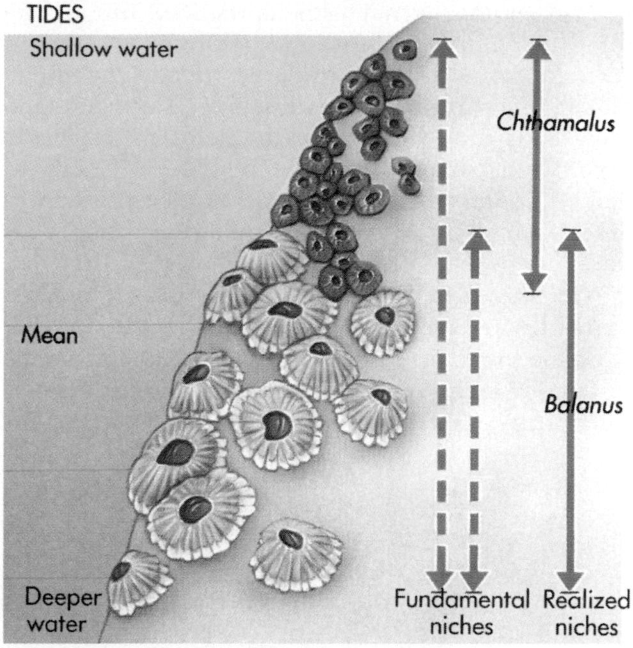

FIGURE 35-6 **Competition can limit niche use.** The distribution of *Chthamalus* and *Balanus* barnacles with respect to different water levels is shown. In theory, *Chthamalus* barnacles can live in both deep and shallow zones, whereas *Balanus* organisms can live only in the deeper zone. These are the barnacles' fundamental niches. In reality, however, Balnus organisms outcompete *Chthamalus* barnacles in the deeper zone because *Balanus* barnacles use the resources of the deeper zone more efficiently. These are the barnacles' realized niches.

ing much less oxygen than when underwater. Interestingly, barnacles of the genus *Balanus* have been kept out of the water as long as 6 weeks without detectable ill effects. However, a relative of *Balanus* organisms, barnacles of the genus *Chthamalus* (pronounced with the first two letters silent), have been kept out water for 3 years, being submerged only 1 or 2 days a month—and they survived!

Although both organisms have adaptations that make them well suited to the intertidal environment, their differences play an important role in determining where each genus lives. Of the two, *Chthamalus* barnacles live in shallower water, where they are often exposed to air as the tide rolls in and out. *Balanus* barnacles live deeper in the intertidal zone and are covered by water most of the time (Figure 35-5).

In studying these two genera of barnacles, J.H. Connell of the University of California, Santa Barbara, found that in this deeper zone, *Balanus* barnacles could always outcompete *Chthamalus* barnacles. *Balanus* organisms would crowd *Chthamalus* barnacles off the rocks, replacing it even where it had begun to grow. When Connell removed *Balanus* barnacles from the area, however, *Chthamalus* organisms were easily able to occupy the deeper zone, indicating that no physiological or other general obstacles prevented it from becoming established there. *Balanus* barnacles, however, must use the resources of the deeper zone more efficiently than *Chthamalus* organisms do, even though *Chthamalus* barnacles are able to survive there in the absence of its competitor. In contrast, *Balanus* barnacles cannot survive in the shallow water where *Chthamalus* organisms normally occur. It evidently does not have the special physiological and morphological adaptations that allow *Chthamalus* barnacles to occupy this zone.

Along with illustrating the principle of competitive exclusion, these experiments with *Balanus* and *Chthamalus* barnacles illustrate that the role an organism plays in an ecosystem—its niche—can vary depending on the biotic and abiotic factors in the ecosystem. In this example, the niche occupied

competitive exclusion a principle that states that if two species are competing with one another for the same limited resource in a specific location, the species able to use that resource most efficiently will eventually eliminate the other species in that location.

by *Chthamalus* barnacles is its **realized niche**—the role it *actually plays* in the ecosystem. It is distinguished from its **fundamental niche**—the niche that it *might* occupy if competitors were not present. Thus the fundamental niche of the barnacle *Chthamalus* in Connell's experiments included that of *Balanus* barnacles, but its realized niche was much narrower because *Chthamalus* organisms were outcompeted by *Balanus* organisms. However, the realized and fundamental niches of *Balanus* barnacles are the same (Figure 35-6).

Gause's principle of competitive exclusion can be restated in terms of niches as follows: no two species can occupy exactly the same niche indefinitely. Certainly, species can and do coexist while competing for the same resources. Nevertheless, Gause's theory predicts that when two species do coexist on a long-term basis, one or more features of their niches will always differ; otherwise the extinction of one species will inevitably result. The factors that are important in defining a niche are often difficult to determine, however; thus Gause's theory can sometimes be difficult to apply or investigate.

> *The role an organism plays in an ecosystem can vary depending on the biotic and abiotic factors in the ecosystem.*

A

B

C

FIGURE 35-7 Different forms of predation. A, A cheetah eats its kill, an impala, while vultures wait patiently in the background. **B,** A gazelle feeds on the grasses of the African plain. **C,** This predatory plant, the Venus flytrap, inhabits the low, boggy ground of North and South Carolina. The Venus flytrap traps insects within its snapping leaves.

Predation

A **predator** is an organism that kills and eats other organisms, or **prey.** Predation includes one kind of animal capturing and eating another, an animal feeding on plants, and even a plant, such as the Venus flytrap shown in Figure 35-7, capturing and eating insects. In a broad sense, parasitism is also considered a form of predation. Parasitism is a close association between two organisms in which the parasite (predator) is much smaller than the prey but feeds on the prey, harming it and benefiting the parasite.

How do predator populations affect prey populations? When experimental populations are set up under very simple conditions in the laboratory, the predator often exterminates its prey and then becomes extinct itself because it has nothing to eat. This fact was illustrated nicely in experiments performed by Gause. In his experiments, Gause used populations of the two protozoans shown in Figure 35-8: *Didinium,* and its prey *Paramecium.* As shown in Figure 35-9, *A,* when *Didinium* protozoans are introduced into a growing population of *Paramecium* protozoans, the population instantly begins to decline and quickly dies out. The *Didinium* population lives on for a short while, then dies out itself.

If refuges are provided for the prey, however, its population can be driven to low levels but can recover. In another of Gause's experiments, he provided sediment in the bottom of the test tubes in which he was growing *Didinium* and *Paramecium* protozoans. Interestingly, as *Didinium* began to prey on *Paramecium,* only those organisms in the clear fluid of the test tubes were killed. Those in the sediment were not eaten. Eventually, *Didinium* protozoans died from lack of food; meanwhile, the *Paramecium* prey multiplied (Figure 35-9, *B*) and overtook the culture!

In another series of experiments, Gause discovered that when he introduced new prey at successive intervals (Figure 35-9, *C*), the decline and rise in the numbers of the predator-prey populations followed a cyclical pattern. As the number of prey increased, the number of predators increased. As the numbers of the prey were lowered by predation, the large predator population did not have enough food to eat; some died and the predator population declined. As this decline occurred, the prey recovered (aided by the addition of new organisms) and again became abundant, starting the cycle once again.

At one time, scientists thought that predator-prey populations always cycled in this manner. However, they have come to realize that in nature, conditions for survival are complex and do not always lead to such a cycling of populations. From experiments such as those described, scientists know that predators cannot survive when the prey population is low. Immigration of prey (movement of new prey into the community) may be necessary to sustain the predator population. From other experiments, scientists learned that changes in predator-prey populations also depend on how prey are dispersed in an area and the manner in which the predator searches for the prey. Factors other than the relationship between a single predator population and a single prey population also influence the survival and abundance of both predator and prey. For example, adverse weather conditions may result in the death of the predator and/or prey species, the prey may be eaten by more than one predator, or fluctuations may occur in the food source of the prey, limiting the survival of this population.

The effect of predator populations on prey populations is difficult to predict because their complex interactions depend on their interactions with other organisms in the community, movement of new organisms into and out of the community, and the abiotic factors that influence their survival.

The intricate interactions between predators and prey often affect the populations of other organisms in a community. By controlling the levels of some species, for example, predators help species survive that may compete with their prey. In other words, predators sometimes prevent or greatly reduce competitive exclusion by limiting the population of one of the competing species. Such interactions among organisms involving predator-prey relationships are key factors in determining the balance among populations of organisms in natural communities.

realized niche the role an organism actually plays in the ecosystem.

fundamental niche the role that an organism might play in an ecosystem if competitors were not present.

predator an organism of one species that kills and eats organisms of another.

prey an organism killed and eaten by a predator.

A

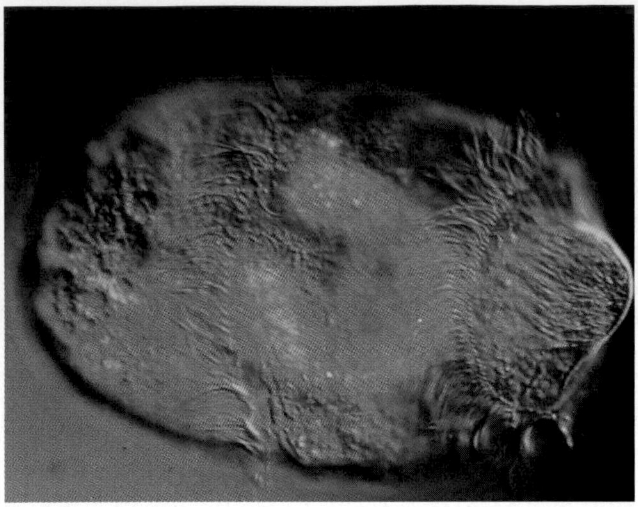

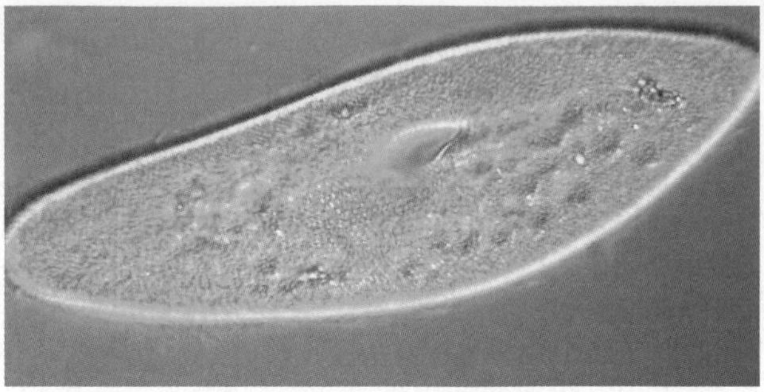

B

FIGURE 35-8 **Gause's experimental subjects. A,** *Didinium.* **B,** *Didinium's* prey, *Paramecium.*

A

B

C

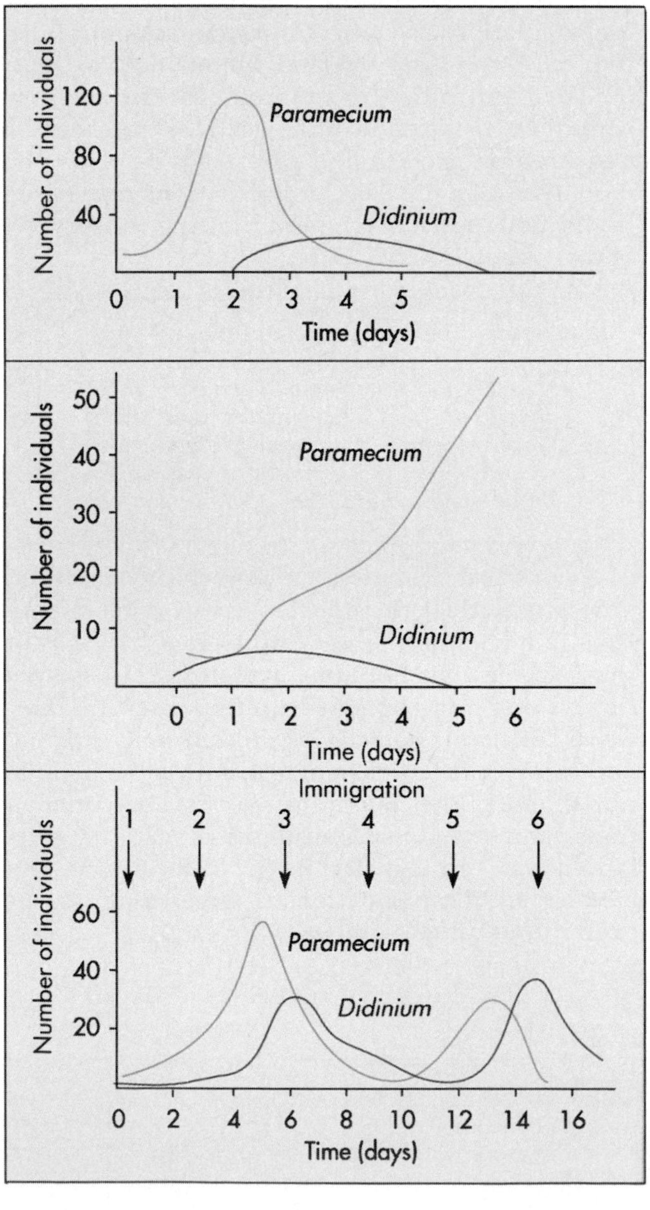

FIGURE 35-9 **Outcome of Gause's experiments with *Paramecium* and *Didinium*. A,** Results obtained when *Didinium* is introduced into a population of *Paramecium* protozoans. **B,** Results of the same experiment, only this time sediment was added to the test tubes, providing a place for *Paramecium* protozoans to hide. **C,** Results from a third experiment in which new prey were introduced at intervals. As the number of prey increased, so did the number of predators.

Plant-Herbivore Coevolution

Plants, animals, protists, fungi, and bacteria that live together in communities have changed and adjusted to one another continually over millions of years. Such interactions, which involve the long-term, mutual evolutionary adjustment of the characteristics of the members of biological communities in relation to one another, are examples of *coevolution.*

Plants and plant-eating predators called *herbivores* are a group of organisms that change and adjust to one another over time. Natural selection favors plants that have developed some means of protection against herbivores. In the dynamic equation of coevolution, however, natural selection also favors adaptations that enable animals to prey on plants in spite of their protective mechanisms. To avoid being eaten, for example, some plants have developed hard parts that are difficult to eat or are unpalatable. In fact, certain grasses defend themselves by incorporating silica (a component of glass) in their structure (Figure 35-10). If enough silica is present, the plants may simply be too tough to eat. Some groups of herbivores, however, have developed strong, grinding teeth and powerful jaws. In addition, herbivores such as cattle have developed adaptations of their digestive systems. One such adapta-

JUST WONDERING. . . .

Why were killer bees given that name? Is their sting so serious that you could die of it?

African honeybees were dubbed "killer bees" by the news media, not scientists. The sting of a so-called killer bee will not kill a person (unless he or she is allergic to bee stings); one bee dispenses no more venom than any other bee. However, there are aspects of the behavior of African honeybees that make them potentially life threatening.

When a bee stings a victim, glands attached to the stinger release alarm pheromone. Pheromones are chemicals produced by one individual that alter the physiology or behavior of other individuals of the same species. African honeybees release more alarm pheromone when they sting than do other bees, and they are more sensitive to this chemical trigger than are other bees. Instead of a dozen bees pursuing a victim for 100 yards as a reaction to alarm pheromone, an entire colony of "killer bees" may pursue a victim for a mile. (A small swarm may contain 30,000 bees.) Luckily, however, people can outrun African honeybees. The problem arises if the victim is an animal that is tied and cannot flee or a human who falls while fleeing, allowing the bees to catch up and inflict numerous stings. One death occurred in 1986 when a University of Miami graduate student, on a field trip in Costa Rica stepped in a crack inside a cave, which disturbed a killer bee colony. The student caught his foot in the crack and was unable to run. He died of 8000 bee stings.

Anita Collins, a U.S. Department of Agriculture research leader specializing in honeybees, thinks that African honeybees are a minimal threat to the public. She is more concerned that African bees will upset the pattern of crop pollination established by the European honeybees now in the United States. Steps are currently being undertaken to inject European honeybee queens with semen from African honeybees to produce a hybrid that will be as high a honey producer as the African variety and, it is hoped, that will be less sensitive to alarm pheromone. If scientists are successful in their breeding goals, interactions within the bee community and between the bee community and human populations may become more productive and less threatening!

tion allows them to store the grass they have eaten in a digestive pouch called a *rumen.* Bacteria that live in the rumen attack the grass chemically, aiding in the digestive process. This stored food is then regurgitated and rechewed at a later time, providing a better breakdown of the cell walls within the grass.

Some plants have developed chemical defenses against herbivores. The best known plant groups with toxic effects are the poison ivy, poison oak, and poison sumac plants. All contain the contact poison urushiol. Castor bean seeds are also toxic to a wide variety of animals, producing a protein that attaches to ribosomes, blocking protein synthesis. Other plants produce toxins that inhibit the growth of bacteria, fungi, and roundworms. Still others produce chemicals having odors that act as a warning or as a repellent to a predator. Today, using the techniques of genetic engineering, scientists have been able to grow plants that chemically repel certain predators, thereby reducing the need for artificial pesticides (Figure 35-11).

FIGURE 35-10 Plants can defend themselves. These zebras grazing on the East African savanna dislike eating grasses with leaves reinforced with silica. The more silica in the leaf cells, the less likely zebras and other grazing animals are to eat that particular kind of grass.

FIGURE 35-11 Defense through genetic engineering. Recently, scientists have developed plants that, through genetic engineering, chemically repel predators. The tobacco plant on the right is a nonengineered plant and shows the effects of insect predation. The tobacco plant on the left, however, has been engineered to produce an insect toxin that deters insects and protects the plant from insects.

However, associated with each family or other group of plants naturally protected by a particular kind of chemical compound are certain groups of herbivores that are adapted to feed on these plants, often as their exclusive food source. For example, the larvae of cabbage butterflies feed almost exclusively on plants of the mustard and caper families, which are characterized by the presence of protective chemicals—the mustard oils. Although these plants are protected against most potential herbivores, the cabbage butterfly caterpillars have developed the ability to break down the mustard oils, rendering them harmless. In a similar example of coevolution, the larvae of monarch butterflies are able to harmlessly feed on the toxic plants of the milkweed and dogbane families (Figure 35-12).

Over time, plants have developed various morphological and chemical adaptations that help protect them against plant eaters, or herbivores. In turn, however, herbivores have changed and adjusted to the plants and their adaptations. Such interactions, involving long-term mutual evolutionary adjustment, characterize coevolution.

Protective Coloration

Some groups of animals that feed on toxic plants receive an extra benefit—one of great ecological importance. When the caterpillars of monarch butterflies feed on plants of the milkweed family, for example, they do not break down the chemicals that protect these plants from most herbivores. Instead, they store them in fat within their bodies. As a result, the caterpillars and all developmental stages of the monarch butterfly are protected against predators by this "plant" poison. A bird that eats a monarch butterfly quickly regurgitates it. Although this is no help to the eaten insect, the bird will soon learn not to eat another butterfly with the bright orange and black pattern that characterizes the adult monarch. Such conspicuous coloration, which "advertises" an insect's toxicity, is called *warning coloration.* Warning coloration is characteristic of animals that have effective defense sys-

FIGURE 35-12 Monarch butterflies make themselves poisonous. All stages of the life cycle of the monarch butterfly are protected from predators by the poisonous chemicals that occur in the milkweeds and dogbanes on which they feed as larvae. Both caterpillars and adult butterflies advertise their poisonous nature with warning coloration.

tems, such as poisons, stings, or bites. Other examples of animals that exhibit warning coloration are shown in Figure 35-13.

During the course of their evolution, many unprotected species have come to resemble distasteful ones that exhibit warning coloration. Provided that the unprotected animals are present in low numbers relative to the species they resemble, they too will be avoided by predators. If the unprotected animals are too numerous, of course, many of them will be eaten by predators that have not yet learned to avoid individuals with a particular set of characteristics. Such a pattern of resemblance is called *Batesian mimicry,* after the British naturalist H.W. Bates, who first described this concept in the 1860s. Many of the best-known examples of Batesian mimicry occur among butterflies and moths (Figure 35-14, *A* and *B*).

Another kind of mimicry, *Müllerian mimicry,* was named for the German biologist Fritz Müller, a contemporary of Bates. Interestingly, in Müllerian mimicry, the protective colorations of different animal species come to resemble one another as in Batesian mimicry. However, unlike Batesian mimicry, the organism and its mimic *do* possess similar defenses (Figure 35-14, *C*).

Some organisms are colored so as to blend in with their surroundings—a protective coloration called *camouflage.* Both cabbage caterpillars and cabbage butterflies have evolved a green coloration, allowing them to hide while feeding (Figure 35-15). Unlike monarchs, these insects do not store the

A B C D

FIGURE 35-13 Warning coloration. The coloring of all these animals is meant to warn other animals to stay away. **A,** The skunk ejects a foul-smelling liquid when threatened. **B,** The poisonous gila monster is a member of the only genus of poisonous lizards in the world. **C,** The red and black African grasshopper feeds on highly poisonous *Euphorbia* plants. **D,** This tropical frog is so poisonous that Indians in western Columbia use frog venom to poison their blow darts.

A
B
C

FIGURE 35-14 Mimicry. A, The viceroy butterfly is a Müllerian mimic of the poisonous monarch. **B,** Larvae of the viceroy feed on willows or other nontoxic plants. **C,** The red-spotted purple is a Batesian mimic of another poisonous butterfly, the pipevine swallowtail, which obtains its toxic chemicals from plants of the pipevine family on which its larvae feed.

A
B

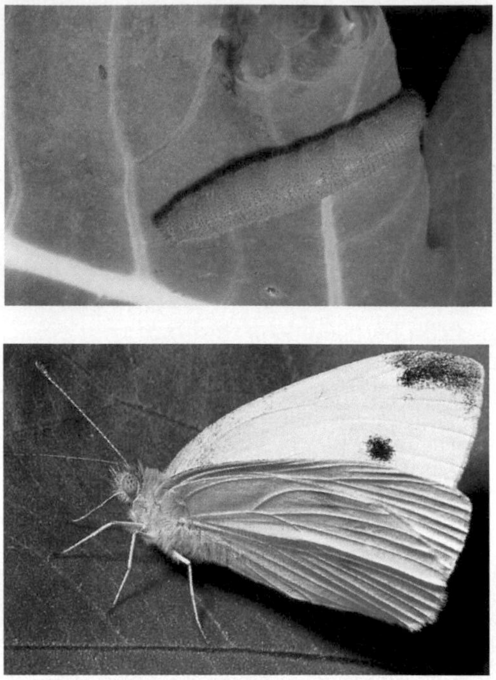

FIGURE 35-15 Insect herbivores are well suited to their hosts. The green caterpillars of the cabbage butterfly are camouflaged on the leaves of cabbage and other plants on which they feed. These caterpillars are able to break down the toxic mustard oils that prevent most insects from eating cabbage.

toxic chemical in the plants they eat for use against predators. Instead, the cabbage butterflies break down the toxin. Similarly, insects that eat plants lacking specific chemical defenses are seldom brightly colored. Figure 35-16 shows examples of other animals who are camouflaged from their predators.

> *Some organisms within communities exhibit conspicuous coloration, or warning coloration, that advertises their ability to poison, sting, or bite. Organisms without specific defenses are often colored, or camouflaged, so as to blend in with their surroundings.*

Commensalism

Many examples of the "one-sided" relationship of commensalism exist in nature. Often, the individuals deriving benefit are physically attached to the other species in the relationship. For example, plants called *epiphytes* grow on the branches of other plants. The epiphytes derive their nourishment from the air and the rain—not from the plants to which they attach for support (Figure 35-17). Similarly, various marine animals such as barnacles grow on other, often actively moving sea ani-

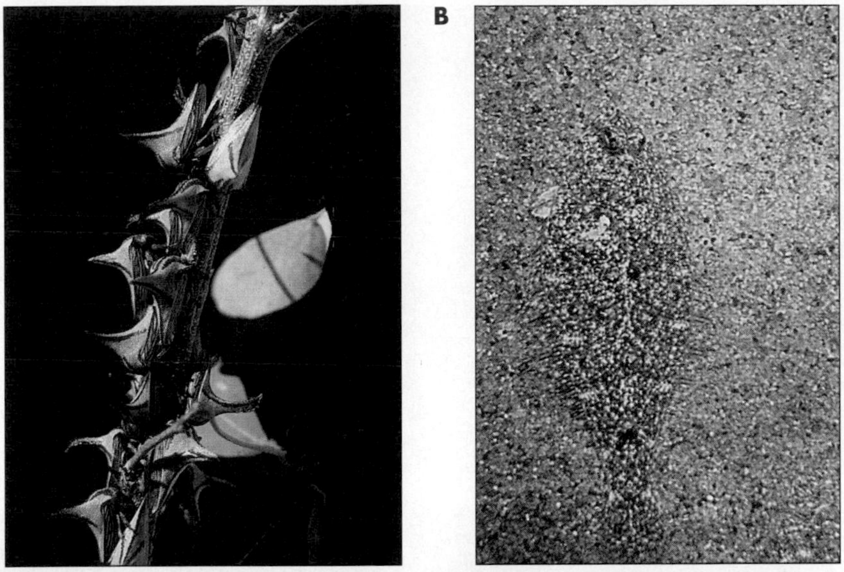

FIGURE 35-16 A few striking examples of camouflage. A, Leaf hoppers disguised as thorns. **B,** Protective coloration in a flatfish on the sea floor. **C,** A tropical hawkmoth at rest, blending in perfectly with the moss-covered and lichen-covered tree bark. **D,** Orange dog caterpillar camouflaged as feces. **E,** An inchworm caterpillar, which closely resembles a twig.

FIGURE 35-17 Commensalism: epiphytes and trees. The growths attached to the tree limb are epiphytes. These plants derive their nourishment from the rain and air and use the tree limb for support only.

mals (Figure 35-18). These "hitchhikers" gain more protection from predation than if they were fixed in one place, and they continually reach new sources of food. They do not, however, harm the organisms to which they are attached.

Possibly one of the best-known examples of commensalism involves the relationship between certain small tropical fish—the clownfish—and sea anemones, marine animals that have stinging tentacles. The fish have developed an adaptation that allows them to live among the deadly tentacles of the anemones (Figure 35-19). These tentacles quickly paralyze other species of fishes, protecting the clownfish against predators.

Mutualism

When both species benefit in a close relationship, it becomes one of mutualism. A particularly striking example of mutualism involves one genus of stinging ants and a Latin American plant of the genus *Acacia*. The modified leaves of acacia plants appear as paired, hollowed thorns. These thorns provide a home for the ants, protecting them and their larvae.

A

B

FIGURE 35-18 Commensalism: barnacles and whales. A, This breaching whale displays the barnacles growing on its skin. **B,** A close-up of a gray whale's skin reveals hitchhikers—lice and barnacles. The lice are actually parasites, whereas the barnacles cause no harm to the whale.

FIGURE 35-19 Commensalism: clownfish and sea anemones. Two clownfish peer out from the tentacles of a large red sea anemone off the coast of Australia.

In addition, the ants eat nectar the plants produce. In turn, the ants attack any herbivore that lands on the branches or leaves of an acacia and clear away vegetation that comes in contact with their host shrub, increasing the plant's ability to survive.

Many other interesting examples of mutualism exist in nature. Certain birds, for example, spend most of their time clinging to grazing animals (such as cattle), picking insects from their hides. In fact, the birds carry out their entire life cycles in close association with the cattle. The birds are provided with food, and the cattle benefit by having their parasites removed. In another similar mutualistic relationship, ants use the tiny insects aphids, or greenflies, as a provider of food. The aphids suck fluids from the phloem of plants, extracting a certain amount of sucrose and other nutrients. However, many of these nutrients are not absorbed within the digestive tract of the aphid. A substantial portion runs out (somewhat altered) through the anus. The ants use this nutritional excrement as a food source and, in turn, actually carry the aphids to new plants so that they can continue eating!

Parasitism

Parasites include viruses, many bacteria, fungi, and an array of invertebrates. A different species of organism, usually larger than the parasite itself, is "home" for a parasite. During the intimate relationship between parasite and host, the parasite derives nourishment and the host is harmed.

FIGURE 35-20 Parasitism: hookworms and humans. These parasites of humans live in the intestine, feeding on the blood of the host.

FIGURE 35-21 Parasitism: dodder plants. Dodder has lost its chlorophyll, along with its leaves, in the course of evolution. Since it is unable to make its own food, it obtains food from host plants on which it grows.

Many instances of parasitism are well known. Intestinal hookworms, for example, are parasites (Figure 35-20). A person is infected when walking barefoot in soil containing hookworm larvae. These larvae are able to penetrate the skin, entering the bloodstream. The blood carries the larvae to the lungs. From there, they are able to migrate up the windpipe to the esophagus. The larvae are then swallowed and reach the intestines. After growing into adult worms, they attach to the inner lining of the intestines. They remain attached there, feeding on the blood of the host.

Some parasites do not live within an organism as hookworms do but attach to the outer surface of a plant or an animal. The attachment may be fleeting, as with the bite of a mosquito, or may take place over a longer period, such as the burrowing of mites.

Many fungi and some flowering plants are parasitic. The dodder plant, for example, has lost its chlorophyll and leaves in the course of its evolution and is unable to manufacture food. Instead, it obtains food from the host plants on which it grows (Figure 35-21).

The more closely the life of a parasite is linked with that of its host, the more its morphology and behavior are likely to have been modified during the course of its evolution. The human flea, for example, is flattened from side to side and slips easily through hair. The ancestors of this species of flea were brightly colored, large, winged insects. The structural and behavioral modifications of the human flea have come about in relation to a parasitic way of life.

Changes in Communities over Time: Succession

Communities, like species, change over time. Even when the climate of a given area remains stable year after year, the composition of the species making up a community, as well as the interactions within the community, shows a dynamic process of change known as **succession.** During the process of succession, a sequence of communities replaces one another. This process is familiar to anyone who has seen a vacant lot or cleared woods slowly become occupied with plants and animals or has seen a pond become filled with vegetation (Figure 35-22).

Primary succession takes place in areas not previously supporting organisms. Primary succession occurs in lakes formed from the retreat of glaciers, for example, or on volcanic islands that may rise above the sea. In the latter case, succession may begin as lichens take hold on bare rock. Lichens are made up of an alga and a fungus living together in a symbiotic relationship. As lichens grow, they produce acids that can break down rock, forming small pockets of soil. When enough soil has accumulated, mosses may begin to grow. This first community, a pioneer community, consists of plants

succession the dynamic process of change during which a sequence of communities replaces one another in an orderly and predictable way.
primary succession succession that takes place in areas not previously supporting organisms.

A B C

FIGURE 35-22 Succession of a pond. This pond in central Maine is beginning to fill with aquatic vegetation **(A)**. As the process of succession proceeds, the pond is slowly filled **(B)**. Gradually, the area where the pond once existed will become indistinguishable from the surrounding vegetation **(C)**.

FIGURE 35-23 Climax community. A stage in succession toward a climax community on the west side of the San Francisco Peaks in northern Arizona. Here, Ponderosa pine *(Pinus ponderosa)* is replacing aspen *(Populus tremuloides)*.

A
B
C

FIGURE 35-24 Secondary succession. Mount St. Helens in the state of Washington erupted violently on May 18, 1980. The lateral blast devastated more than 600 square kilometers of forest and recreation lands within 15 minutes. **A** shows an area near Lang Ridge 4 months after the blast; 4 years later **(B)**, succession was underway at the same spot, with shrubs, blueberries, and dogwoods following the first plants that became established immediately after the blast. In 1989, more plants have become established and stable communities have formed **(C)**.

that are able to grow under harsh conditions. The pioneer community paves the way for the growth and development of vegetation native to that climate. Over many thousands of years or even longer, the rocks may be completely broken down, and vegetation may cover a once-rocky area. As plants take hold, the area becomes able to sustain other forms of life. And as the plant community changes, so too do the other living things. Eventually, the mix of plants and animals becomes somewhat stable, forming what is termed a **climax community** (Figure 35-23). However, with an increasing realization that (1) climates may change, (2) the process of succession is often very slow, and (3) the nature of a region's vegetation is determined to a great extent by human activities, ecologists do not consider the concept of a climax community as useful as they once did.

Succession occurs not only within terrestrial communities but in aquatic communities as well. A lake poor in nutrients, for example, may gradually become rich in nutrients as organic materials accumulate (see Figure 35-22). Plants growing along the edges of the lake, such as cattails and rushes, and those growing submerged, such as pondweeds, may contribute to the formation of a rich organic soil as they die and are decomposed by bacteria. As this process of soil formation continues, the pond may become filled in with terrestrial vegetation. Eventually, the area where the pond once stood may become an indistinguishable part of the surrounding vegetation.

Secondary succession occurs in areas that have been disturbed and that were originally occupied by organisms. Humans are often responsible for initiating secondary succession throughout portions of the world that they inhabit. Abandoned farm fields, for example, undergo secondary succession as they revert to forest. Secondary succession may also take place after natural disasters such as a forest fire or a volcanic eruption producing ash (rather than lava flows) (Figure 35-24).

Communities change over time by means of the dynamic process of succession. During this process, a sequence of communities replaces one another.

climax community a community in which the mix of plants and animals becomes stable; the last stage of succession.

secondary succession succession that takes place in areas that have been disturbed and that were originally occupied by organisms.

Summary

- All of the living (biotic) factors and nonliving (abiotic) factors within a certain area are an ecosystem. Within an ecosystem, each living thing has a habitat, an area in which it lives that is characterized by its physical (abiotic) properties. In addition, each organism plays a special role within an ecosystem, which is its niche.
- Of the biotic factors within an ecosystem, groups of organisms of the same species are populations. Various populations of organisms living together in a particular area at a particular time make up a community. Thus an ecosystem can be thought of as a community of organisms, along with the abiotic factors with which the community interacts.
- The interactions within communities fall into five categories: competition, predation, commensalism, mutualism, and parasitism. Competition involves organisms of different species striving to obtain the same needed resource. During predation, one species (the prey) becomes a resource, being killed and eaten by another species (the predator). Commensalism is a one-sided relationship: one species in a relationship between organisms of different species benefits, whereas the other species neither benefits nor is harmed. In mutualism, both species benefit. In a parasitic relationship, one species (the parasite) benefits but the other (the host) is harmed.
- Organisms living together in communities continually change in relationship to one another. Such interactions, which involve the long-term, mutual evolutionary adjustment of the characteristics of the members of biological communities in relation to one another, are forms of coevolution.
- Communities, like species, change over time. This dynamic process of change, during which a sequence of communities replaces one another, is succession. Primary succession takes place in areas that are originally bare, such as rocks or open water. Secondary succession takes place in areas where the communities of organisms that existed initially have been disturbed.

Knowledge and Comprehension Questions

1. Distinguish among population, community, and ecosystem.
2. Fill in the blanks: Within an ecosystem, each organism has a(n) __________, an area in which it resides. Each organism plays a role within the ecosystem, which is its __________.
3. Interactions within communities can be grouped into five categories. Name and describe each category.
4. True or false? (If it is false, change the statement so that it is correct.) Competition among different species is often greatest between organisms that obtain their food in different ways.
5. State the principle of competitive exclusion in your own words.
6. Explain the terms *realized niche* and *fundamental niche.*
7. Fill in the blanks: A(n) __________ is an organism that kills and eats another organism called its __________. A(n) __________ feeds on another organism but does not necessarily kill it.
8. Summarize the pattern of the predator-prey relationship as shown in Gause's experiments.
9. What other factors, in addition to those revealed in Gause's work, can affect the balance between predators and prey?
10. What is coevolution? Give an example of coevolution involving a herbivore.
11. What type of adaptation is shown in each of the following:
    a. The bright yellow and black stripes of a bee's body
    b. The unobtrusive color of some lizards that make them difficult to see against surrounding rocks
12. What is primary succession? Summarize the process.
13. Fill in the blank: Forests that have been cut down for logging can often no longer support the same animals and birds that lived in that ecosystem previously. Over time, the changes that take place in the communities of organisms in this logged area are examples of __________.

Critical Thinking

1. Give a hypothetical example in which a seemingly commensal relationship could become mutualistic.
2. How could you determine whether a resemblance between two butterfly species is mimicry or not?

CHAPTER 36

Ecosystems

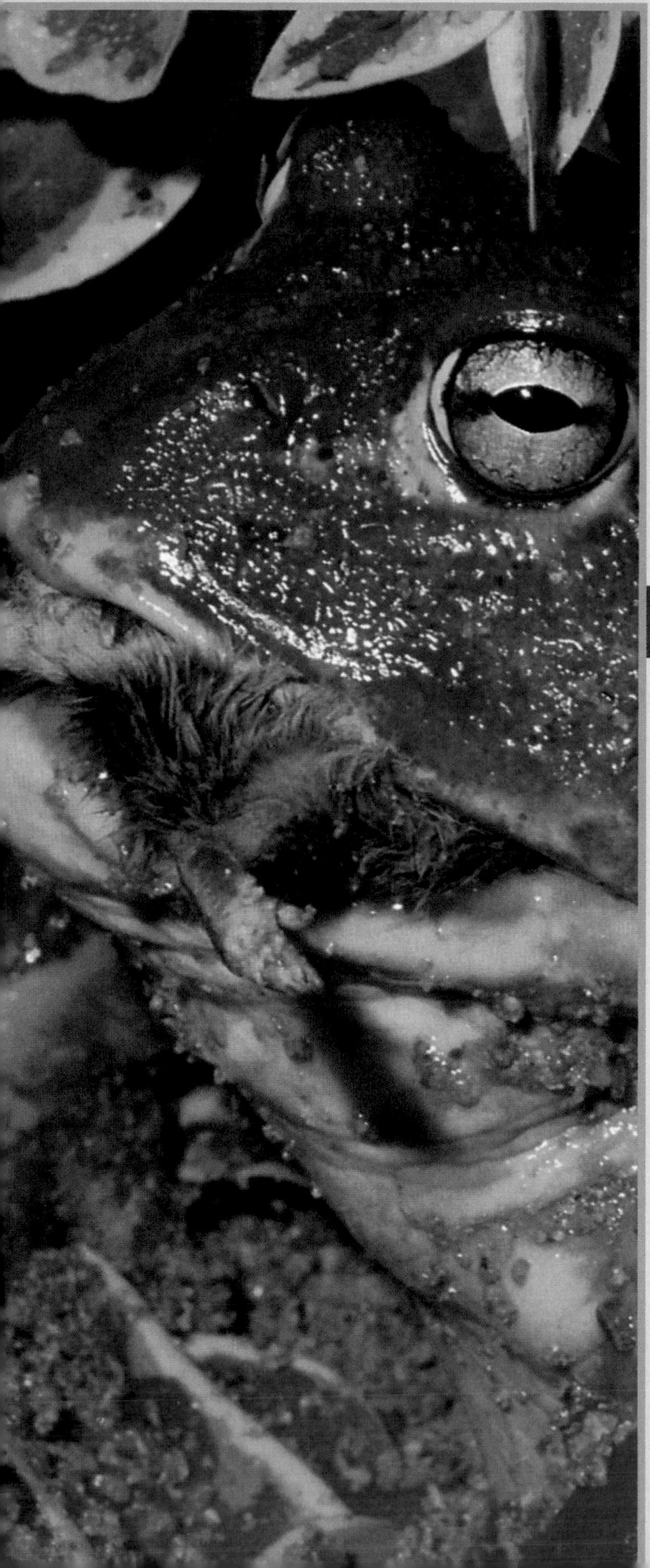

A WHOLE, RAW MOUSE might not be your idea of a gourmet meal, but to this bullfrog it is a five-star dinner. Located in a rural village in South Africa, this "frog restaurant" is part of a food chain, a series of organisms that feed on one another. The bullfrog crushed the mouse with its powerful jaws. Its lower jaw has bony projections that help pull the prey into the frog's cavernous mouth and will soon help push it down into the stomach. Minutes before this photograph was taken, it was the mouse that was feasting—on the seeds of nearby plants. And, unknown to the frog, a snake lies in wait behind the next branch, ready to add yet another link to this chain of gourmet relationships among the plants and animals of this community.

Populations, Communities, and Ecosystems

Individuals of a species, such as bullfrogs and mice, are each part of an individual population of organisms. Together, interacting populations are communities (see Chapters 34 and 35). The living organisms in a community interact not only with each other but with the nonliving substances in their environment, such as the soil, water, and air, to form an ecological system, or ecosystem.

An **ecosystem** is a community consisting of plants, animals, and microorganisms that interact with one another and with their environments and that are interdependent on one another for survival. The living, or **biotic,** components of an ecosystem are made up of two types of organisms: those that can make their own food, or **producers,** and those that eat other organisms for food, or **consumers.** Many consumers kill and eat their food. A special group of consumers, called **decomposers,** obtains nourishment from dead matter such as fallen leaves or the bodies of dead animals. You know the decomposers as bacteria and fungi (although some species of bacteria can manufacture their own food and are therefore producers).

Figure 36-1 is a diagram of an ecosystem. Notice that the **abiotic,** or nonliving, components of the environment contribute substances needed for the ecosystem to function. In addition, notice that the exchanges of nutrients and other chemical substances among the organisms within the ecosystem form a cyclical pattern. Energy, however, does not cycle within the ecosystem but flows *through* the ecosystem, first captured from the sun by the producers, then used by herbivores that eat the producers, and ultimately used by all the consumers living in the ecosystem. Ecosystems are therefore systems in which there is a regulated transfer of energy and an orderly, controlled cycling of nutrients. The individual organisms and populations of organisms in an ecosystem act as parts of an integrated whole, adjust over time to their roles in the ecosystem, and relate to one another in complex ways that are only partly understood.

> *An ecosystem is made up of communities of organisms living within a defined area and the nonliving environmental factors with which they interact. The organisms of an ecosystem—the producers, consumers, and decomposers—each play a specific role within it, contributing to the flow of energy and the cycling of nutrients.*

KEY CONCEPTS

- Ecosystems are communities of organisms and the nonliving environment with which they interact.
- Energy flows through ecosystems, initially captured from the sun by green plants and then passed from organism to organism as they feed on one another.
- The elements essential to life, primarily hydrogen, carbon, oxygen, and nitrogen, are cycled from the atmosphere, through living things, and back to the atmosphere once again.
- Other elements such as phosphorus, potassium, calcium, and sodium are held in the soil in small amounts, incorporated in the tissues of living things, and cycled to the soil once again.

OUTLINE

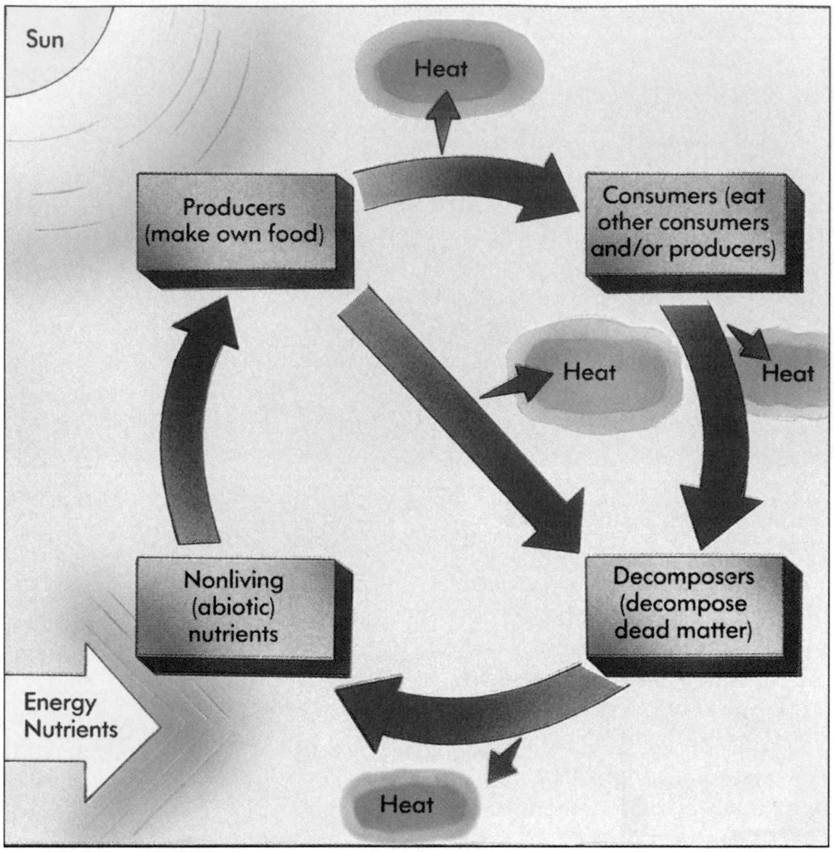

FIGURE 36-1 An ecosystem. Nutrients flow within an ecosystem in a cyclic pattern, whereas energy flows through an ecosystem.

Where does one ecosystem begin and another end? Some ecosystems have clearly recognizable boundaries, such as that of a pond or a puddle or those found within the coastal ranges of California (Figure 36-2). Sometimes humans produce artificial ecosystems with human-made boundaries, such as the glass walls of an aquarium or the fencing surrounding a cultivated field. But the boundaries of many natural ecosystems blend with one another, sometimes almost imperceptibly. Ecosystems also change over time and with climate changes, slowly becoming modified into new ecosystems whose characteristics differ increasingly from those that preceded them.

The Flow of Energy through Ecosystems

The energy that flows through an ecosystem comes from the sun. Green plants, the primary producers of terrestrial ecosystems, are able to capture some of the sun's radiant energy that falls on their leaves and convert it to chemical energy during the process of photosynthesis (see Chapter 6). Producers, then, are the key to life on Earth, because no other organisms can capture this energy for use in living systems. **Primary consumers,** or *herbivores*, feed directly on the green plants, incorporating some of this energy into molecules that make up their bodies and using the rest to perform the activities of life. **Secondary consumers** are meat eaters, or *carnivores*, that feed in turn on the herbivores. And so the chain continues, with one living thing feeding on another, passing energy along that was once captured from the sun.

The refuse or waste material of an ecosystem is known as **detritus.** Organisms that are decomposers break down the organic materials of detritus into inorganic nutrients that can be reused by plants (Figure 36-3). Some of the energy still held in the tissues of once-living things is used by the decomposers, but they are the last link in this transfer of energy among organisms.

FIGURE 36-2 Two distinct ecosystems. In the coastal ranges of California, the boundary between the evergreen shrub ecosystem known as chaparral and the grassland ecosystem is often sharp, as shown in this photograph taken along the western edge of the Santa Clara Valley near Morgan Hill. These two ecosystems also include characteristic sets of nonliving factors.

The producers of an ecosystem capture energy from the sun and convert it to chemical energy usable by themselves and consumers. Consumers feed on the producers and other consumers in the ecosystem, passing energy along that was once captured from the sun. Decomposers break down the organic molecules of dead organisms, serving as the last link in the flow of energy through an ecosystem and contributing to the recycling of nutrients within the environment.

Food Chains and Webs

All the feeding levels previously described and additional levels such as tertiary consumers are represented in any fairly complicated ecosystem. These feeding levels are called **trophic levels,** from the Greek word *trophos,* which means "feeder." (In fact, this Greek word is the root for the words **heterotroph,** or "other feeder"—another word for consumer, and for **autotroph,** or "self-feeder"—another word for producer.) Organisms from each of these levels, feeding on one another, make up a series of

ecosystem (ee ko **sis** tem or **eh** ko **sis** tem) a community consisting of plants, animals, and microorganisms that interact with one another and with their environments and are interdependent on one another for survival. All the biotic and abiotic factors within a certain area.

biotic (bye **ot** ik) living factors within the environment such as plants, animals, and microorganisms.

producers organisms that can make their own food.

consumers organisms that eat other organisms for food.

decomposers (dee kum **poe** zurs) a special group of consumers that obtains nourishment from dead matter such as fallen leaves or the bodies of dead animals; most varieties of bacteria and fungi are decomposers.

abiotic (aye bye **ot** ik) nonliving factors within the environment such as air, water, and rocks.

primary consumers organisms that feed directly on green plants.

secondary consumers meat eaters that feed on the primary consumers (herbivores).

detritus (dih **trite** us) the refuse or waste material of an ecosystem.

trophic levels (**trow** fik) the different feeding levels within an ecosystem.

heterotroph (**het** uhr uh **trofe**) an organism that cannot produce its own food; a consumer in an ecosystem or a food chain.

autotroph (**aw** tuh **trofe**) a self-feeder; an organism that produces its own food by photosynthesis.

FIGURE 36-3 A decomposer doing its job. The shelf fungi (division Basidiomycota) growing on this hardwood stump are decomposing it, converting the organic materials contained within it to nutrients that can be reused by plants.

organisms called a **food chain.** An example of a food chain can be seen in a pond ecosystem in which water fleas (primary consumers) feed on green algae (producers). Sunfish (secondary consumers) eat the water fleas but in turn are eaten by green heron (tertiary consumers) (Figure 36-4). The length and complexity of food chains vary greatly.

In reality, it is rare for any species of organism to feed on only one other species. Organisms feed on many different species and types of organisms and are, in turn, food for two or more other kinds. Shown in a diagram such as Figure 36-5, these relationships appear as a series of branching and overlapping lines rather than as one straight line. The organisms in an ecosystem that have such interconnected and interwoven feeding relationships make up a **food web.** Figure 36-5 shows a food web in a salt marsh ecosystem.

> *A linear relationship among organisms that feed on one another is a food chain. Food chains that interweave are food webs.*

Food Pyramids

In any ecosystem the number of organisms (or their total mass [biomass]) and the amount of en-

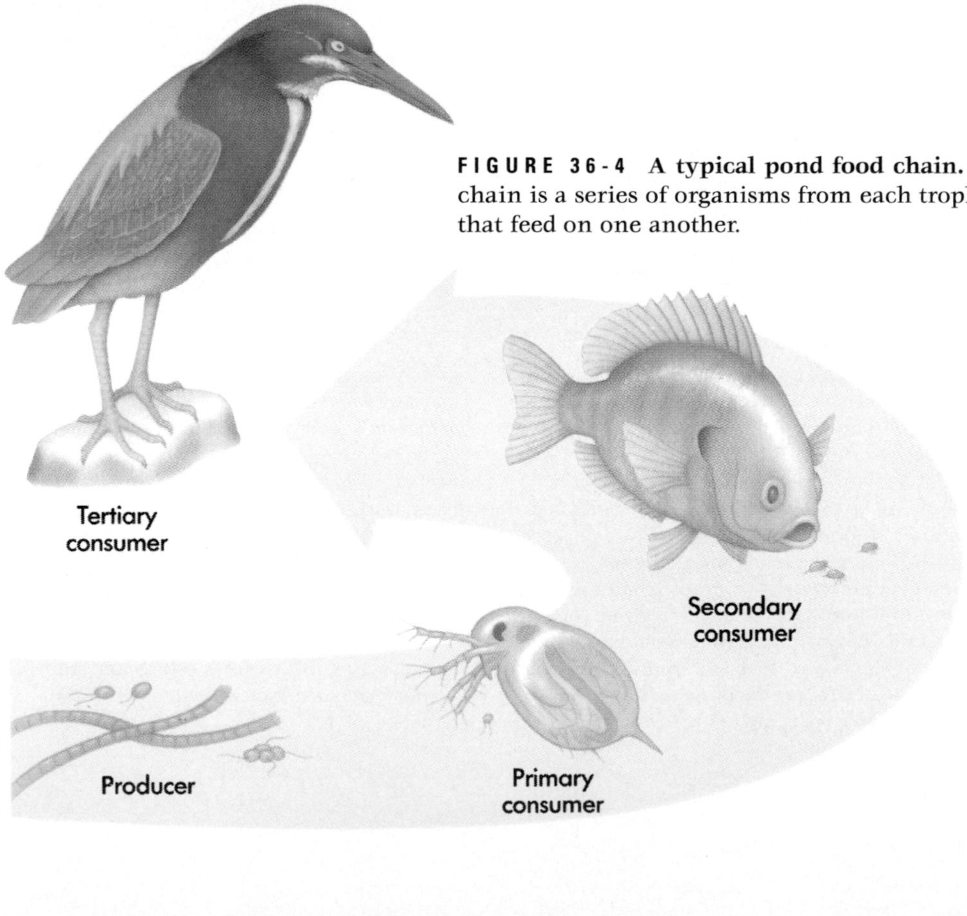

FIGURE 36-4 A typical pond food chain. A food chain is a series of organisms from each trophic level that feed on one another.

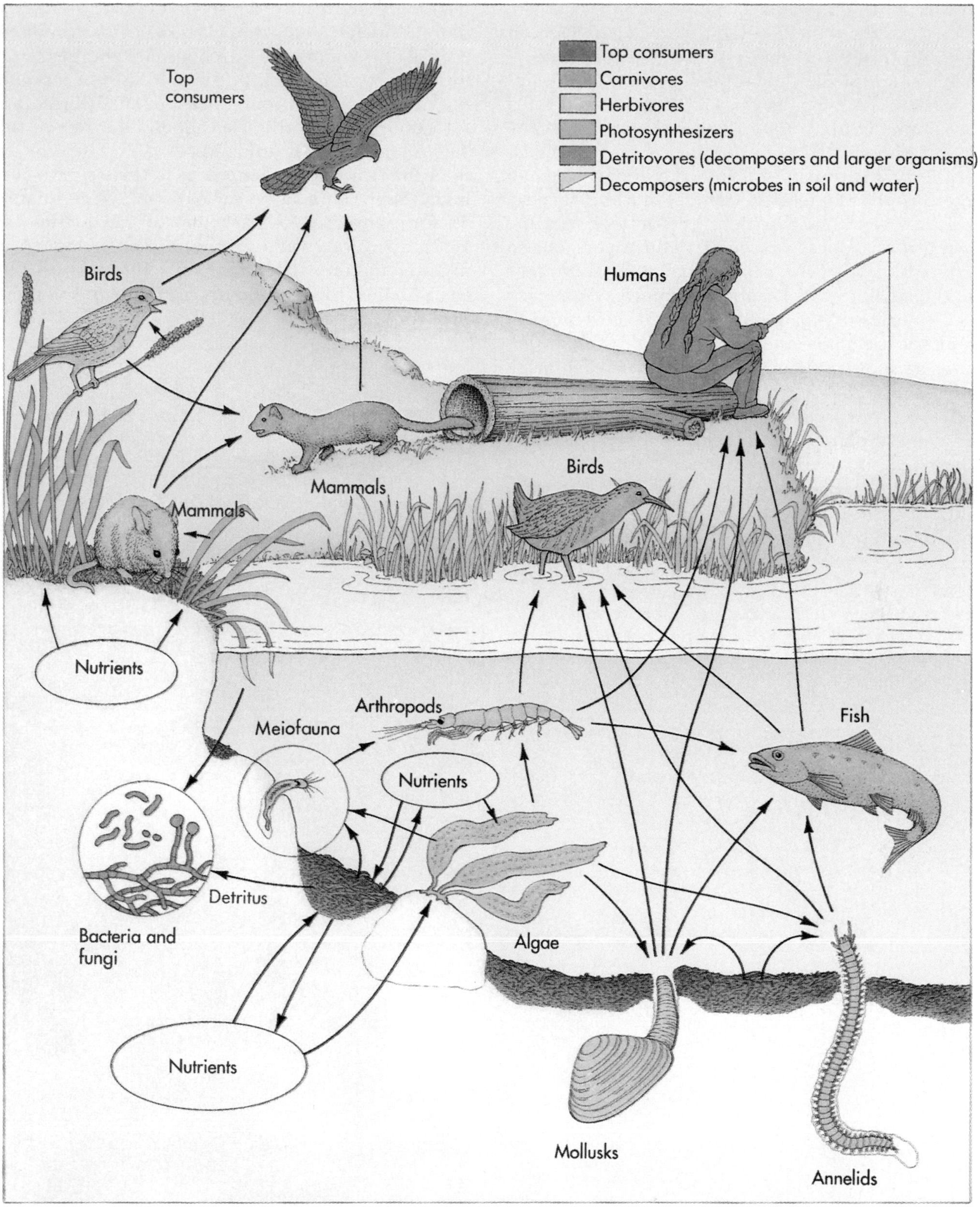

FIGURE 36-5 The food web in a salt marsh. A food web is a group of interwoven food chains within an ecosystem. In this diagram, each color represents a trophic level. Each trophic level feeds on, or gains energy from, the layer below. (The term *meiofauna* refers to a group of animals that live in the spaces between grains of sand.)

ergy making up each successive trophic level are often less than the level that preceded. Lamont Cole of Cornell University illustrated this concept in his study of the flow of energy in a freshwater ecosystem in Cayuga Lake, New York. He calculated that approximately 150 calories of each 1000 calories of energy "fixed" by producers during photosynthesis was transferred into the bodies of small heterotrophs that feed on these plants and bacteria (Figure 36-6). Smelt, which are tiny fish, eat the heterotrophs; these secondary consumers obtain about 30 calories of each original 1000. If humans eat the smelt, they gain about 6 calories from each 1000 calories that originally entered the system. If trout eat the smelt and humans eat the trout, humans gain only about 1.2 calories from each original 1000.

These types of calculations show that, on average, only 10% of plants' accumulated energy is actually converted into the bodies of the organisms that consume them. What happens to the rest? A certain amount of the energy that is ingested by organisms goes toward heat production. (It is actually "lost" as heat, consistent with the second law of thermodynamics, which states that no transformation of energy is 100% efficient.) A great deal of energy is used for digestion and work, and usually 40% or less goes toward growth and reproduction. An invertebrate, for example, typically uses about a quarter of this 40% for growth. In other words, about 10% of the food that an invertebrate eats is turned into new body tissue. This figure varies from approximately 5% for carnivores to nearly 20% for herbivores, but 10% is an average value for the amount of energy (or organic matter) that organisms incorporate into their bodies from the energy available in the previous trophic level.

A plant captures some of the sun's energy that falls on its green plants. The successive members of a food chain process about 10% of the energy available in the organisms on which they feed into their own bodies. The rest is released as heat (and is lost from the food chain) or is used during various metabolic activities.

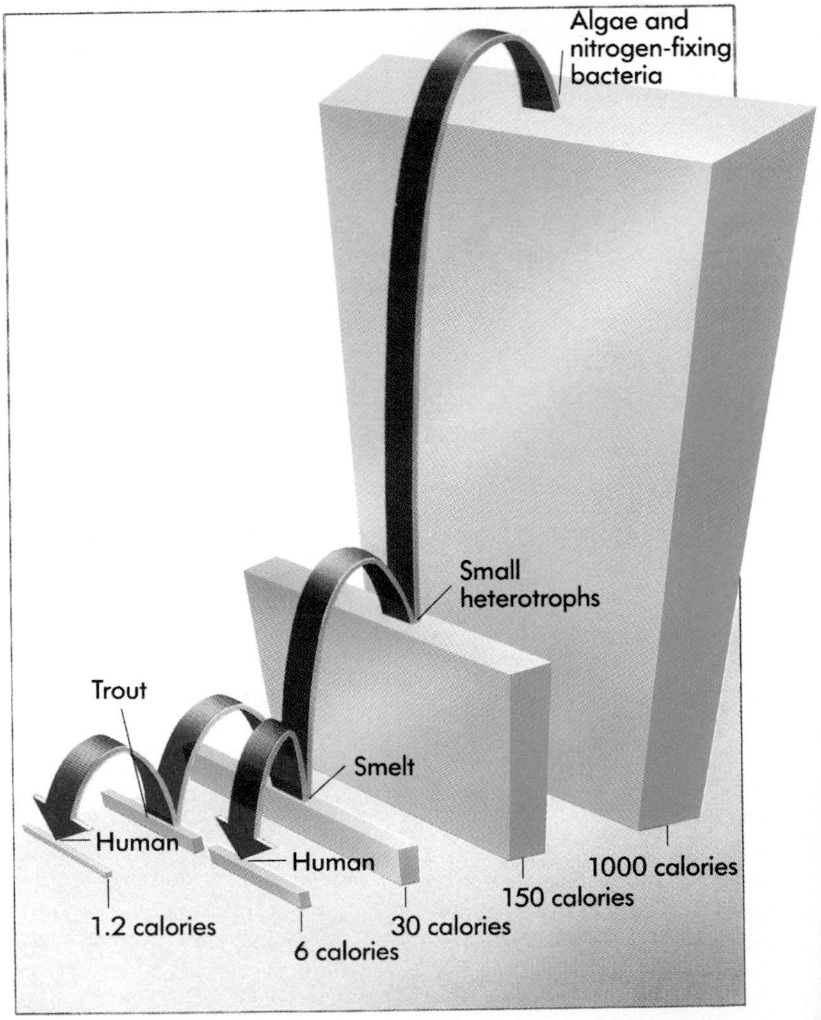

FIGURE 36-6 The flow of energy in Lake Cayuga. The experiments in Lake Cayuga demonstrated that the number of organisms and the amount of energy making up each successive trophic level is smaller than the preceding level.

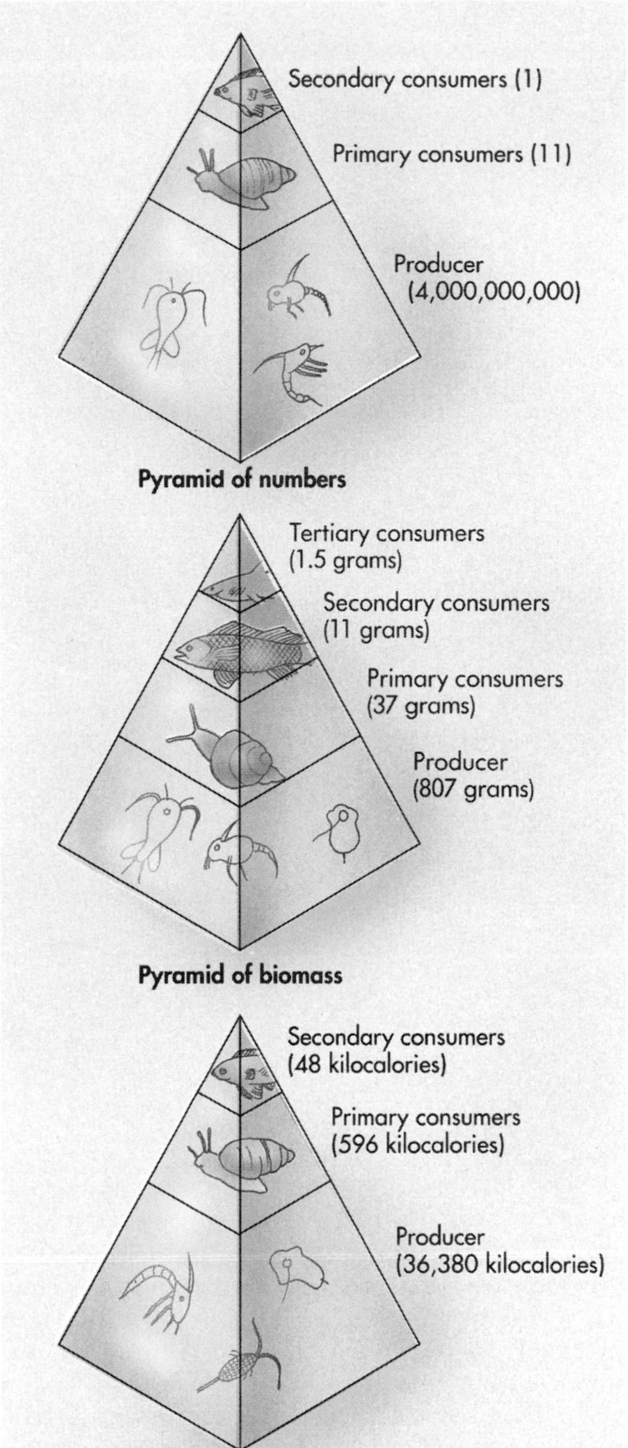

FIGURE 36-7 Pyramids of numbers, biomass, and energy for an aquatic ecosystem. The pyramid of numbers shows the number of organisms at each trophic level in the ecosystem. The pyramid of biomass depicts the total weight of organisms supported at each level. The pyramid of energy depicts the amounts of energy available at each level.

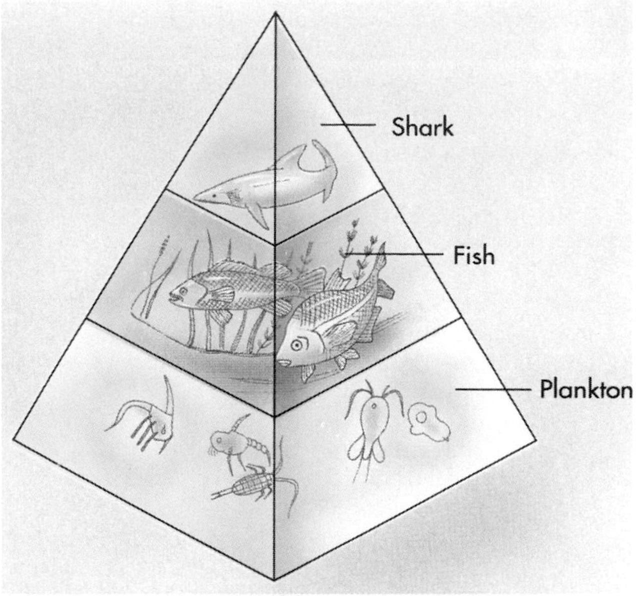

FIGURE 36-8 Food pyramid. All food pyramids are basically the same—each successive trophic level is smaller than the preceding level (see Figure 36-7).

If shown diagrammatically, the relationships already described between trophic levels appear as pyramids. Diagrams that depict the energy flow through an ecosystem are called *pyramids of energy* (Figure 36-7). Those that depict the total weight of organisms supported at each trophic level in an ecosystem are referred to as *pyramids of biomass* (see Figure 36-7). *Pyramids of number* depict, as the name suggests, the total number of organisms at each feeding level (see Figure 36-7). Figure 36-8 shows a food pyramid that incorporates all three concepts.

Occasionally, pyramids of biomass and/or pyramids of number can be inverted. For example, in an ocean community, the photosynthesizing organisms (usually monerans [bacteria] and protists) reproduce so rapidly that a small but constant biomass can feed a much larger biomass of herbivores. This relationship would be depicted by an inverted pyramid of biomass. In another example, within a forest community, a single tree can support many herbivorous insects, resulting in an inverted pyramid of numbers.

The relationships among organisms at the various trophic levels of a food pyramid make it clear

food chain a series of organisms, from each trophic level, that feed on one another.

food web the interwoven and interconnected feeding relationships of an ecosystem; interwoven food chains.

JUST WONDERING . . . .

What happens if one species in a food chain dies off? Do all the other species in the chain die too?

A food chain is part of an intricate food web, and the organisms that make up the chain enter into a variety of relationships (in addition to predator-prey relationships) with other organisms within an ecosystem. So as species die off, new species move in and other species move out of the ecosystem as relationships among organisms change. Species that have the greatest impact on ecosystems are *keystone species.*

Keystone species are those that influence the structure and function of an ecosystem to a much greater degree than would be expected simply from the size of their populations. When a keystone is removed from an ecosystem (and therefore from a food chain and web), changes are certain to take place, and those changes depend on the role of the keystone species. For example, in the forests of Peru (located on the Pacific coast of South America), only a dozen or so species of fig and palm trees support an entire community of fruit-eating birds and mammals during the time of year when fruits are least available. Loss of these few tree species would probably result in the loss of most of the fruit-eating animal species even if hundreds of other tree species remained. Conversely, if a keystone predator species dies, the populations of the species it kept in check through predation may soar. These prey may now overtake their competitors as predators of other species, or they may become prey for species that move into the area.

Thus, the loss of a keystone species can have a great impact on a food chain and the ecosystem of which it is an integral part. Scientists have also discovered that the number of species lost is critical to ecosystems. As more species are lost, the more the functioning of an ecosystem is degraded. In other words, diversity contributes to the health of an ecosystem—it survives better.

Therefore, the effects of the loss of a single species in a food chain vary, and they depend on whether the species was a keystone species and on the other adjustments made among species as relationships change in response to the loss. Because of the importance of keystone species, however, scientists are anxious to identify them, because their extinction *could* result in the extinction of many other species in food chains, food webs, communities, and ecosystems.

that herbivores have more food available to them than do carnivores. In other words, the lower a population eats in the food chain, the higher the number of individuals that can be fed. Such considerations are increasingly important as humans work to maximize the food available for a hungry and increasingly overcrowded world.

The Cycling of Chemicals within Ecosystems

Although energy flows through ecosystems and most is lost at each successive level in food pyramids, the matter making up the organisms at each level is not lost. All of these substances are recycled and are used only temporarily by living things. Hydrogen, carbon, nitrogen, and oxygen—the principal elements that make up all living things—are primarily held in the atmosphere in molecules of water, carbon dioxide, nitrogen gas, and oxygen gas. Other recycled substances necessary for life such as phosphorus, potassium, sulfur, magnesium, calcium, sodium, iron, and cobalt are held in rocks and, after weathering, enter the soil. The atmosphere and rocks are therefore referred to as the *reservoirs* of inorganic substances that cycle within ecosystems.

The cycling of materials in ecosystems is usually described as beginning at the reservoirs. Living things incorporate substances into their bodies

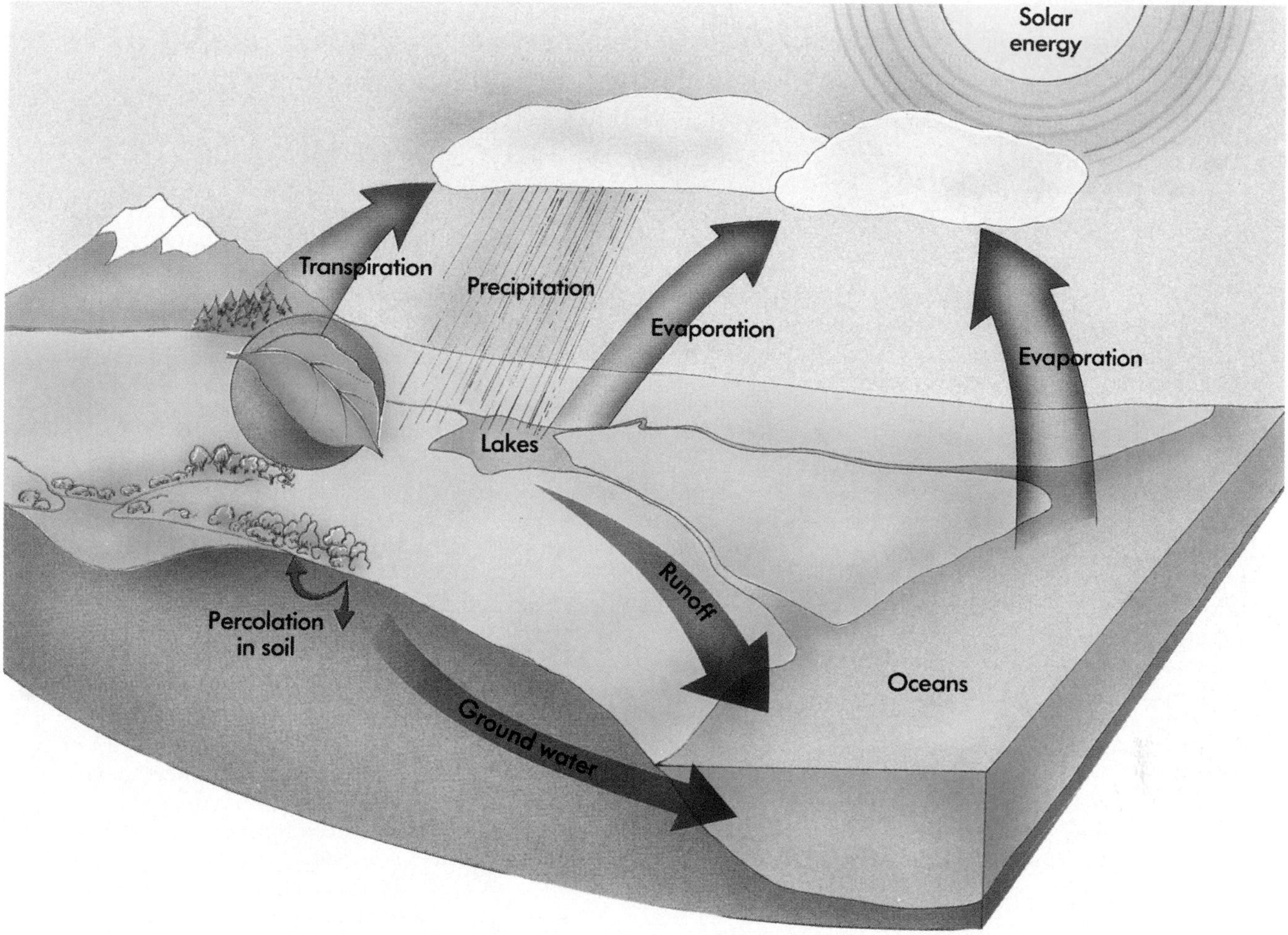

FIGURE 36-9 The water cycle. In terrestrial ecosystems, 90% of the water that reaches the atmosphere comes from plant transpiration. In general, however, oceans contribute the most water to the Earth's atmosphere.

from their reservoirs or from other living things, passing these materials along the food chain. Ultimately these substances, with the help of decomposers, move from the living world back to the nonliving world, becoming part of the soil or the atmosphere once again.

> *Within ecosystems, matter cycles from its reservoir in the environment, to the bodies of living organisms, and back to the environment.*

The Water Cycle

Heated by the sun, water evaporates into the atmosphere from the surfaces of oceans, lakes, and streams (Figure 36-9). In terrestrial (land-based) ecosystems, as much as 90% of the water that reaches the atmosphere comes from plants as they release water vapor into the air during the process of transpiration (see Chapter 28). Because oceans cover three fourths of the Earth's surface, these bodies contribute most of the water to the atmosphere worldwide.

Atmospheric water condenses in clouds and eventually falls back to the Earth as precipitation. Most of it falls directly into the oceans. But some falls onto the land, flowing into surface bodies of water or trickling through layers of soil and rock to form subsurface bodies of fresh water called *groundwater.* Plants take up water as it trickles through the soil, almost in a continuous stream. Crop plants, for example, use about 1000 kilograms of water just to produce 1 kilogram of biomass. Animals obtain water directly from surface water or from the plants or other animals they eat. In the United States, groundwater provides about a quarter of the water used by humans for all purposes

HOW *Science* WORKS

The Hubbard Brook Experiment: What Have We Learned about the Cycling of Nutrients?

In 1963, scientists at the Yale School of Forestry and Environmental Studies began an interesting experiment designed to show the effects of the cycling of nutrients (chemicals) within ecosystems. (Their work is still ongoing today and is called *long-term ecological research.*) Their studies have yielded much of the information that is now known about nutrient cycles, and their ingenious experimental design has provided the basis for the development of the experimental methods used in the study of other ecosystems.

Hubbard Brook is the central stream located in a temperate deciduous forest in New Hampshire. For measurement of the flow of water and nutrients within the Hubbard Brook ecosystem, concrete weirs with V-shaped notches were built across six tributary streams that were selected for study. All of the water that flowed out of the six valleys (the areas surrounding the six streams) had to pass through the notches. The precipitation that fell in the six streams was measured, and the amounts of nutrients present in the water flowing in the six streams were also determined. By these methods the scientists demonstrated that the undisturbed forests in this area were very efficient in retaining nutrients. The small amounts of nutrients that fell from the atmosphere with the rain and snow were approximately equal to the amounts of nutrients that ran out of the six valleys. For example, there was only a small net loss of calcium—about 0.3% of the total calcium in the system per year—and small net gains of nitrogen and potassium.

Then the researchers disturbed the ecosystem. In 1965, the investigators felled all of the trees and shrubs in one of the six valleys and then prevented their regrowth by spraying the area with herbicides. The effects of these activities were dramatic. The amount of water running out of the valley increased by 40%, indicating that water normally taken up by the trees and shrubs—and evaporated into the atmosphere from their leaves—was now running off. The amounts of nutrients running out of the system also increased. The loss of calcium was 10 times higher than it had been previously. The change in the status of nitrogen was particularly striking. The undisturbed ecosystem in this valley had been accumulating nitrogen at a rate of 2 kilograms per hectare per year, but the cut-down ecosystem lost it at a rate of about 120 kilograms per hectare per year! The nitrate level of the water rapidly increased to a level exceeding that judged safe for human consumption, and the stream that drained the area generated massive blooms of cyanobacteria and algae.

The Hubbard Brook experiment demonstrated that nutrient cycling depends, among other things, on the vegetation present in the ecosystem. The fertility of the deforested valley decreased rapidly, and at the same time the danger of flooding greatly increased. The Hubbard Brook experiment is particularly instructive in the 1990s because large areas of tropical rain forest are being destroyed to make way for cropland. Many of the insights gleaned from the Hubbard Brook experiment can help scientists understand some of the consequences of rain forest destruction that may not have been readily apparent otherwise.

and provides about half of the population with drinking water.

> *Water in the atmosphere condenses in clouds and falls back to the Earth as precipitation. Plants take up water from the soil, and animals obtain water from surface water or from the plants or other animals they eat. Water returns to the atmosphere through the evaporation of surface water and transpiration by plants.*

Unfortunately, about 2% of the groundwater in the United States is polluted, and the situation is worsening. Pesticides are one source, being carried to *aquifers,* underground reservoirs in which the groundwater lies within porous rock as rain washes the chemicals from the surfaces of leaves and the topsoil. Chemical wastes, stored in surface pits, ponds, and lagoons, are another key source of groundwater pollution. Scientists have no technology that will remove pollutants from underground aquifers.

The Carbon Cycle

The carbon cycle is based on carbon dioxide (CO_2), which makes up about 0.03% of the atmosphere and is found dissolved in the oceans (Figure 36-10). Terrestrial as well as marine producers use CO_2—along with energy from the sun—to build carbon compounds such as glucose during the process of photosynthesis (see Chapter 6). The producers and the consumers that eat them break down these carbon compounds during cellular respiration and use the energy locked in their chemical bonds to carry on the metabolic processes of life. Consumers use some of the carbon atoms and compounds from the food they eat to produce needed substances. However, most of this carbon is waste and is released to the atmosphere (or to the oceans) as CO_2.

Intimately linked to the cycling of carbon is the cycling of oxygen (O_2). As plants use CO_2 during the process of photosynthesis, they produce O_2 as a by-product. This O_2 is released into the atmosphere and becomes available to organisms for the process of cellular respiration (Figure 36-11).

Some aquatic organisms such as mollusks use the CO_2 dissolved in water and combine it with calcium to form their calcium carbonate ($CaCO_3$) shells. When these organisms die, their shells collect on the sea floor. Years of exposure to water slowly dissolves the $CaCO_3$, releasing the CO_2 and making its carbon once again available to aquatic producers to use in the process of photosynthesis.

When organisms die, decomposers break down

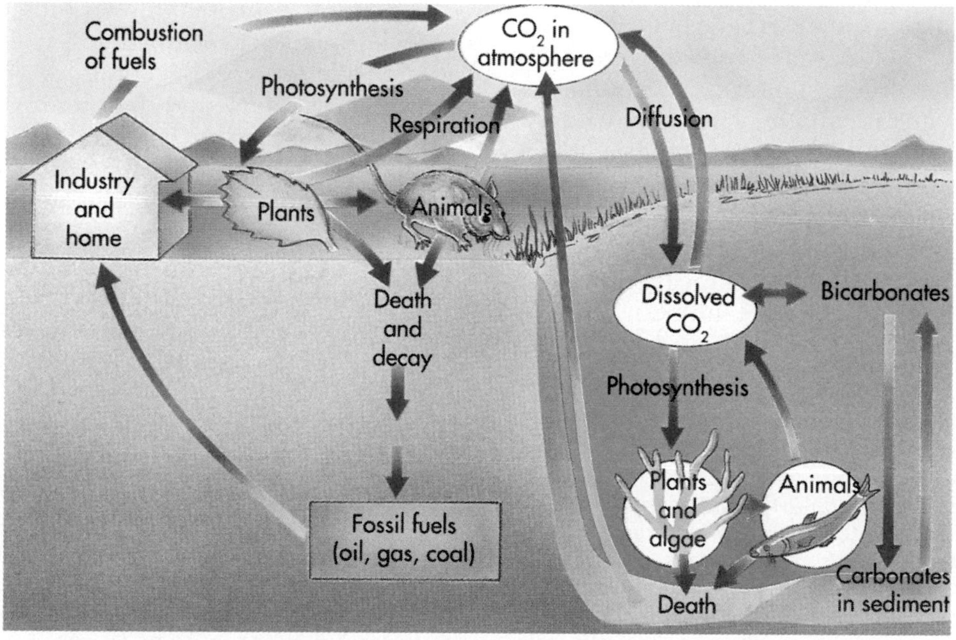

FIGURE 36-10 The carbon cycle. Carbon dioxide is found in the oceans and in the atmosphere. Producers and some consumers incorporate this carbon into substances necessary for life.

FIGURE 36-11 How carbon dioxide is linked to oxygen cycling. Plants use carbon dioxide in photosynthesis and give off oxygen. This oxygen is available to animals for the process of cellular respiration. Animals breathe out carbon dioxide; some carbon dioxide is liberated from the decomposition of dead organisms.

the carbon compounds making up their bodies. Certain carbon-containing compounds, such as the cellulose found in the cell walls of plants, are more resistant to breakdown than others, but certain bacteria, fungi, and protozoans are able to accomplish this feat. Some cellulose, however, accumulates as undecomposed organic matter. Over time and with heat and pressure, this undercomposed matter results in the formation of fossil fuels such as oil and coal. When these fuels are burned, as when wood is burned, the CO_2 is returned to the atmosphere. The release of this carbon as CO_2, a process that is proceeding rapidly as a result of human activities, may change global climates.

Carbon is held in the atmosphere in the form of carbon dioxide. This gas is taken in by photosynthetic organisms and is used to build the carbon compounds that plants manufacture during the process of photosynthesis. Both producers and consumers use these compounds as an energy source, metabolizing them during cellular respiration and releasing carbon dioxide to the atmosphere.

The Nitrogen Cycle

Although nitrogen gas (N_2) makes up 78% of the Earth's atmosphere, only a minute amount is incorporated into chemical compounds in the soil, oceans, and bodies of organisms. However, this N_2 is an essential part of the proteins within living things. Relatively few kinds of organisms—only a few genera of bacteria—can convert N_2 into a form that can be used for biological processes, playing a crucial role in the cycling of nitrogen (Figure 36-12). These bacteria are called *nitrogen-fixing bacteria,* and they convert N_2 to ammonia (NH_3). Living things depend on this process of nitrogen fixation. Without it, they would ultimately be unable to continue to synthesize proteins, nucleic acids, and other necessary nitrogen-containing compounds.

Certain of the nitrogen-fixing bacteria are free living in the soil. Others form mutualistic relationships (see Chapter 35) with plants by living within swellings, or nodules, of plant roots. Some of these plants are legumes—plants such as soybeans, alfalfa, and clover. Plants having mutualistic associations with nitrogen-fixing bacteria can grow in soils having such low amounts of available nitrogen that they are unsuitable for most other plants. Growth of a leguminous crop can enrich the nitrate level of poor soil enough to benefit the next year's nonleguminous crop. This is the basis for crop rotation in which, for example, a field may be planted with soybeans (a legume) and corn (a nonlegume) in alternating years.

The roots of a few other kinds of plants form associations with nitrogen-fixing bacteria of the group actinomycetes. Some of the plants involved are alders and mountain lilac. In addition, nitrogen-fixing bacteria of the genus *Anabaena* contribute large amounts of nitrogen to the rice paddies of China and Southeast Asia (Figure 36-13).

Other bacteria—certain decomposers—play another key role in the nitrogen cycle, producing NH_3 from amino acids that make up the proteins and wastes of dead organisms. Still other bacteria convert NH_3 to nitrates (NO_3^-). NO_3^- is also produced by lightning, which causes N_2 to react with O_2 in the atmosphere. Humans add NO_3^- to the soil by spreading chemical fertilizers. Plants are able to use the nitrogen within molecules of NH_3 and NO_3^- to build their own proteins, nucleic acids, and vitamins. The nitrogen cycle comes full circle as nitrogen is continuously returned to the environment by bacteria that break down NO_3^-, liberating N_2 to the atmosphere.

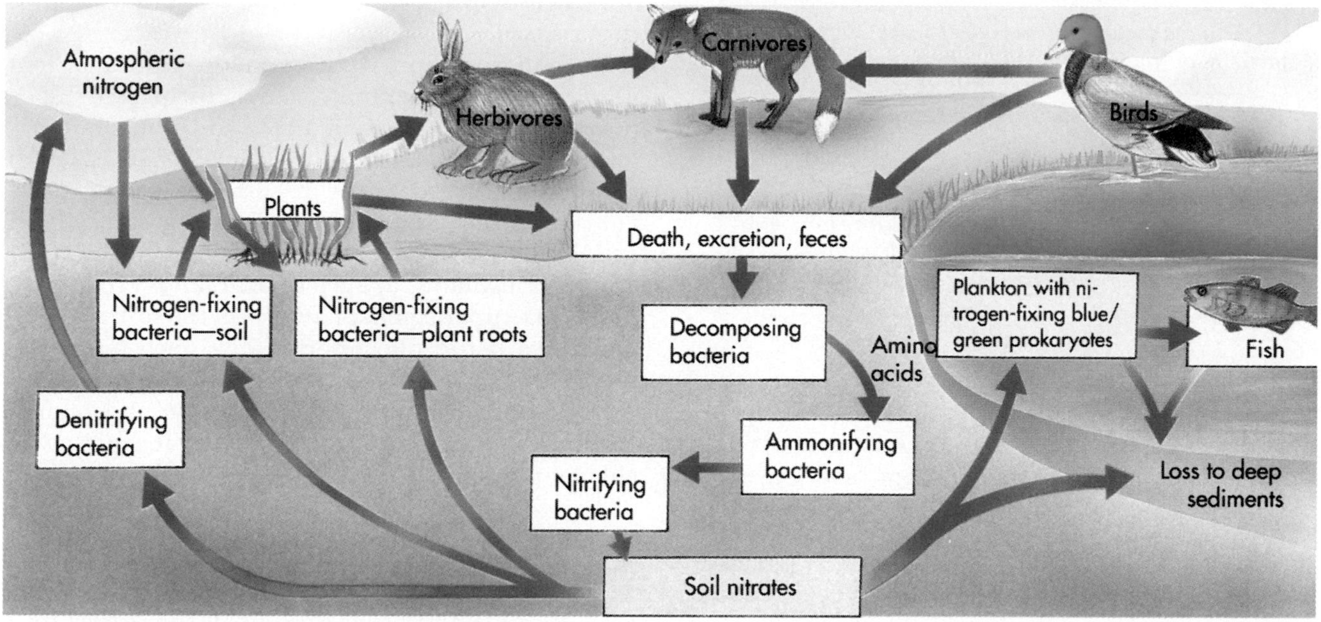

FIGURE 36-12 The nitrogen cycle. Bacteria that can convert nitrogen into a usable form (nitrogen-fixing bacteria) are extremely important in the nitrogen cycle.

FIGURE 36-13 The use of nitrogen-fixing bacteria in agriculture. The nitrogen-fixing bacteria *Anabaena azollae* (**A**) lives in the spaces between the leaves of the floating water fern *Azolla* (**B**), which is deliberately introduced into the rice paddies of the warmer parts of Asia. Rice, here cultivated in Bali (**C**), is the major food for well over one fourth of the human race.

Although nitrogen gas constitutes about 78% of the Earth's atmosphere, it becomes available to organisms only through the metabolic activities of a few genera of bacteria, some of which are free living and others of which live mutualistically on the roots of legumes and some other plants. This bound nitrogen is released back to the atmosphere by other microorganisms capable of breaking down certain nitrogen compounds.

The Phosphorus Cycle

The reservoir of the nutrients phosphorus, potassium, sulfur, magnesium, calcium, sodium, iron, and cobalt is in rocks and minerals rather than in the atmosphere.

Phosphorus, more than any of the other required plant nutrients except nitrogen, is apt to be so scarce that it limits plant growth. In the soil, phosphorus is found as relatively insoluble phosphate (PO_4^{-3}) compounds, present only in certain kinds of rocks. For this reason, phosphates exist in the soil only in small amounts. Therefore humans add millions of tons of phosphates to agricultural lands every year. Plants take up the phosphates from the soil, and animals obtain the phosphorus they need by eating plants or other plant-eating animals. When these organisms die, decomposers release the phosphorus incorporated in their tissues, making it again available for plant use (Figure 36-14).

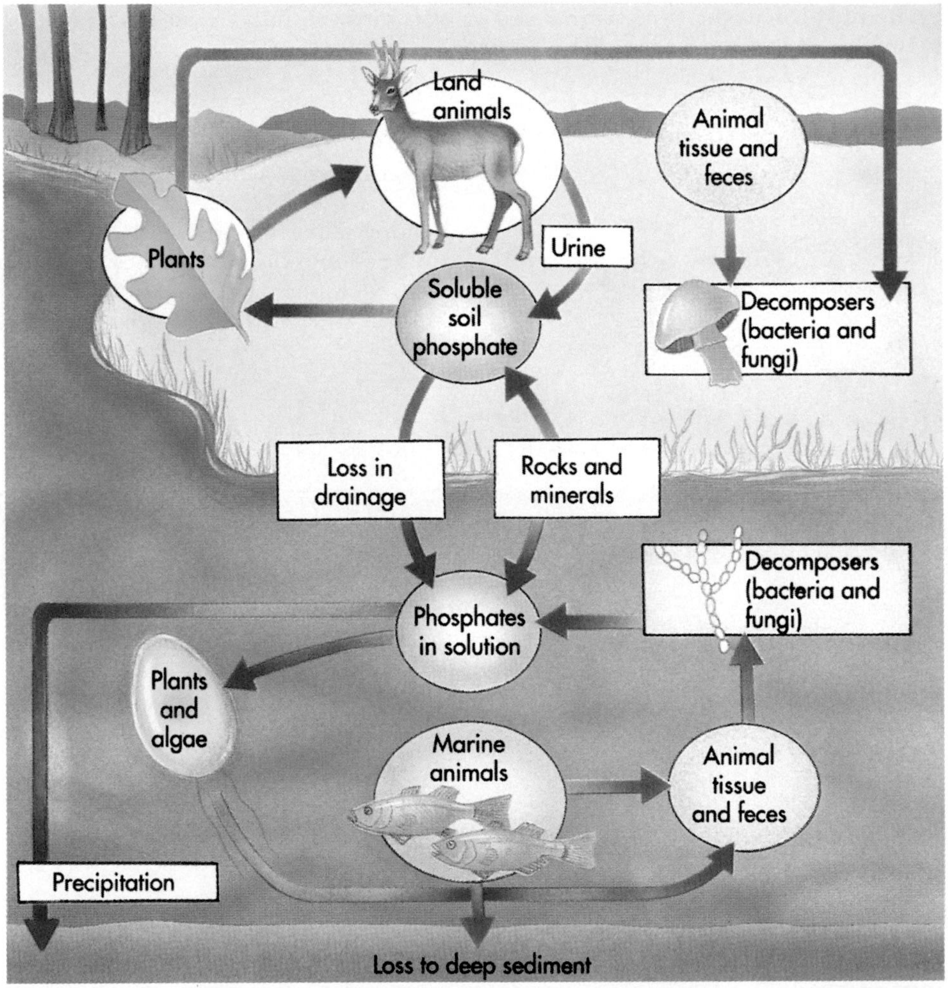

FIGURE 36-14 The phosphorus cycle. Phosphorus is found in only very small amounts in the soil; to make up for this lack of phosphorus, humans add phosphates to the soil. Plants take up phosphate from the soil, and animals obtain the necessary levels of phosphate by eating plants.

Phosphorus is held in the soil in the form of relatively insoluble phosphate compounds, found in most soils in small amounts. Therefore humans add phosphate compounds to the soil for plants to take up, and animals obtain the phosphates they need by eating plants or plant-eating animals. When organisms die, decomposers release phosphorus back to the soil.

Humans unnecessarily alter the phosphorus cycle by adding up to four times as much phosphate as a crop requires each year. Much of this phosphate runs off the land or is eroded into rivers, streams, and the oceans. In addition, phosphates in sewage and waste water from homes and industry find their way into surface water. Rivers and streams become *eutrophic,* or "well fed," causing algae and other aquatic plants to overgrow. As they die, the decomposers feed on them, using up much of the oxygen in the water and choking out other forms of life.

Rivers and streams carry phosphates to the oceans. Some of these phosphates become incorporated into the bodies of fishes and other marine animals. The seabirds that eat these animals deposit enormous amounts of guano (feces) rich in phosphorus along certain coasts (these deposits have traditionally been used for fertilizer) (Figure 36-15). Phosphates not incorporated into the bodies of animals precipitate out of the water and become part of the bottom sediment. These phosphates become available again only if the sea floor rises up during eras of climatic change, such as along the Pacific coast of North and South America.

FIGURE 36-15 A guano coast. Marine animals incorporate phosphates from runoff into their bodies. The seabirds that feast on these marine animals deposit feces rich in phosphorus on certain coasts.

Summary

- Ecosystems are communities of organisms and the nonliving factors of their environments through which energy flows and within which nutrients cycle.
- Through photosynthesis, plants growing under favorable circumstances capture and lock up some of the sun's energy that falls on their green parts. They may then be eaten by herbivores (primary consumers), which in turn may be eaten by secondary consumers (carnivores). The remains of all organisms are broken down by the decomposers. This sequence of organisms, one feeding on another, constitutes a food chain.
- Each link in a food chain is a trophic, or feeding, level. Each trophic level includes organisms that can transfer about 10% of the energy that exists at each level to the next level. The rest is lost as heat or is used for various metabolic activities.
- Carbon dioxide, nitrogen gas, oxygen gas, and water are the atmospheric reservoirs of the carbon, nitrogen, oxygen, and hydrogen used in biological processes. All of the other elements that organisms incorporate into their bodies come from the Earth's rocks.
- Water in the atmosphere condenses in clouds and falls to the Earth as precipitation. Plants take up water from the soil, and animals obtain water from surface water or from the plants or other animals they eat. Water returns to the atmosphere through the evaporation of surface water and transpiration by plants.
- Carbon is held in the atmosphere in the form of carbon dioxide. This gas is taken in by photosynthetic organisms and is used to build the carbon compounds that plants manufacture during the process of photosynthesis. Both producers and consumers use these compounds as an energy source, metabolizing them during cellular respiration and releasing carbon dioxide to the atmosphere.
- Atmospheric nitrogen is converted to ammonia by several genera of symbiotic and free-living bacteria. The ammonia, in turn, is assimilated into amino groups in proteins of cells or is converted to nitrites and then to nitrates by other bacteria. Nitrates are incorporated into the bodies of plants and are converted back into ammonium ions, which are used in the manufacture of many kinds of molecules in the bodies of living organisms. The breakdown of these molecules either converts them to recyclable forms or results in the release of atmospheric nitrogen.
- Phosphorus is a key component of many biological molecules; it weathers out of soils and is transported to the world's oceans where it tends to be lost. Phosphorus is relatively scarce in rocks; this scarcity often limits or excludes the growth of certain kinds of plants. Therefore humans add phosphate compounds to the soil for plants to take up, and animals obtain the phosphates they need by eating plants or plant-eating animals. When organisms die, decomposers release phosphorus back to the soil.

Knowledge and Comprehension Questions

1. What is an ecosystem? Briefly describe three of the living groups it contains with respect to mode of nutrition.
2. To which of the groups in question 1 do you belong?
3. Draw a diagram that illustrates the relationships between the biotic and abiotic components of a generalized ecosystem.
4. Fill in the blanks: _________ feed directly on green plants. _________ feed on herbivores. Other organisms known as _________ live on the refuse or waste material of an ecosystem, called __________.
5. Distinguish among trophic level, food chain, and food web.
6. Summarize what happens to the sun's energy as it travels through a food chain.
7. Explain how chemicals are cycled within an ecosystem.
8. Describe how water is cycled within ecosystems.
9. What is groundwater, and how can it become polluted? Explain why this is a serious problem.
10. Summarize the carbon cycle.
11. Why can crop rotation make soil more fertile? To what chemical cycle does this relate?
12. Many farmers fertilize their crops heavily with phosphates, believing that this will improve the soil. Explain how this practice can affect an ecosystem

Critical Thinking

1. Imagine you had a hollow glass sphere the size of a basketball and you wanted to create within it a self-sustaining stable ecosystem that needed only sunlight and moderate temperature to persist indefinitely. What would you put in this ecosphere?
2. Most terrestrial food chains have only three or rarely four links. Why do you think this is so? Why not forest food chains with 10 links?
3. Extensive cutting and burning of tropical rain forests often result in a drastic and permanent lowering of rainfall in the cleared area (an effect noted by Alexander von Humboldt more than 100 years ago). Why?

CHAPTER 37

BIOMES AND LIFE ZONES OF THE WORLD

DOES YOUR BACKYARD look like the photo? Probably not, unless you happen to live near the equator. In this steamy climate, the vegetation is lush and the trees are tall, "topping out" at about 160 feet! Very little light penetrates this screen of foliage and finds its way to the forest floor beneath. Vines and other plants cling high up on the trunks of trees, competing for the light that filters through their leaves. Animal and plant species abound; so many species inhabit the tropical rain forest, in fact, that thousands remain unknown, unclassified, and unstudied.

Interestingly, not all regions of the Earth lying near the equator are tropical rain forests. Latitude is only one factor having an impact on the growth and distribution of living things. This chapter describes other factors, explaining why organisms live where they do . . . and why your backyard is probably not a tropical rain forest.

Biomes and Climate

Biomes are ecosystems of plants and animals that occur over wide areas of land within specific climatic regions and are easily recognized by their overall appearance. Each biome is similar in its structure and appearance wherever it occurs on Earth and differs significantly from other biomes. Biomes are sometimes named by the climax vegetation (stable plant communities) of the region (see Chapter 35), such as the tropical rain forest.

The characteristics of biomes are a direct result of their temperature and rainfall patterns. These patterns result from the interaction of the features of the Earth itself (such as the presence of mountains and valleys) with two physical factors:

1. The amounts of *heat from the sun* that reach different parts of the Earth and the seasonal variations in that heat
2. *Global atmospheric circulation* and the resulting patterns of oceanic circulation

Together these factors determine the local climate, including the amounts and distribution of precipitation.

> *Biomes are the largest recognizable terrestrial ecosystems, which occur over wide areas and within specific climatic regions. Temperature and rainfall patterns are the primary determinants of the characteristics of these huge ecosystems.*

The Sun and Its Effects on Climate

Because the Earth is a sphere, some parts receive more energy from the sun than others do. The tropics are warmer than the temperate regions because the sun's rays arrive almost perpendicular to regions near the equator, whereas near the poles the rays' angle of incidence spreads them over a much greater area (Figure 37-1). Therefore the greater the latitude (distance from the equator), the colder the climate.

The northern and southern hemispheres also experience a change of seasons as well as the gradations in temperature that vary with latitude. Seasons occur because the Earth is tilted on its axis;

KEY CONCEPTS

- Biomes are the largest recognizable terrestrial ecosystems, which occur over wide areas and within specific climatic regions.
- Seven categories of biomes, listed by their distance from the equator, are tropical rain forests, savannas, deserts, temperate grasslands, temperate deciduous forests, taiga, and tundra.
- The fresh water of ponds, lakes, streams, and rivers also provides various habitats for organisms to live; estuaries, places where fresh water meets salt water as rivers open into oceans, provide nutrient-rich environments intermediate between fresh water and salt water.
- Marine organisms occur most plentifully in the intertidal zone and the photic zone of the neritic and open sea; here they have abundant light and oxygen.

OUTLINE

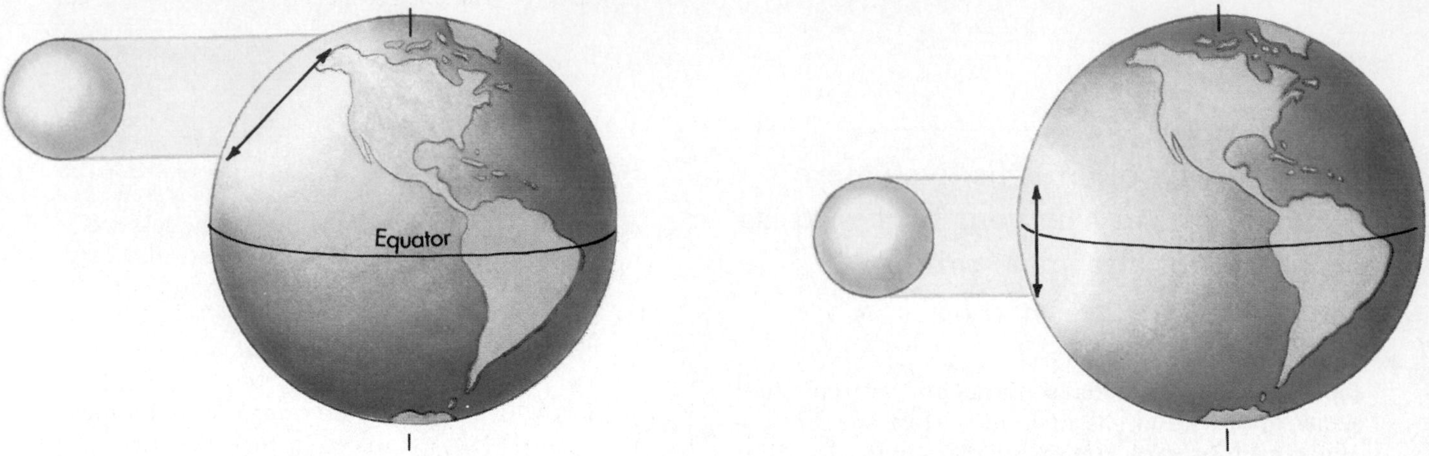

FIGURE 37-1 How the angle of solar energy striking the Earth affects climate. The angle of a beam of solar energy striking the Earth at the equator is perpendicular to the surface of the Earth (*right*). The angle of this same beam striking an area at the middle latitude is larger, and the solar energy is more spread out over a larger area (*left*). In general, the greater the latitude, the colder the climate.

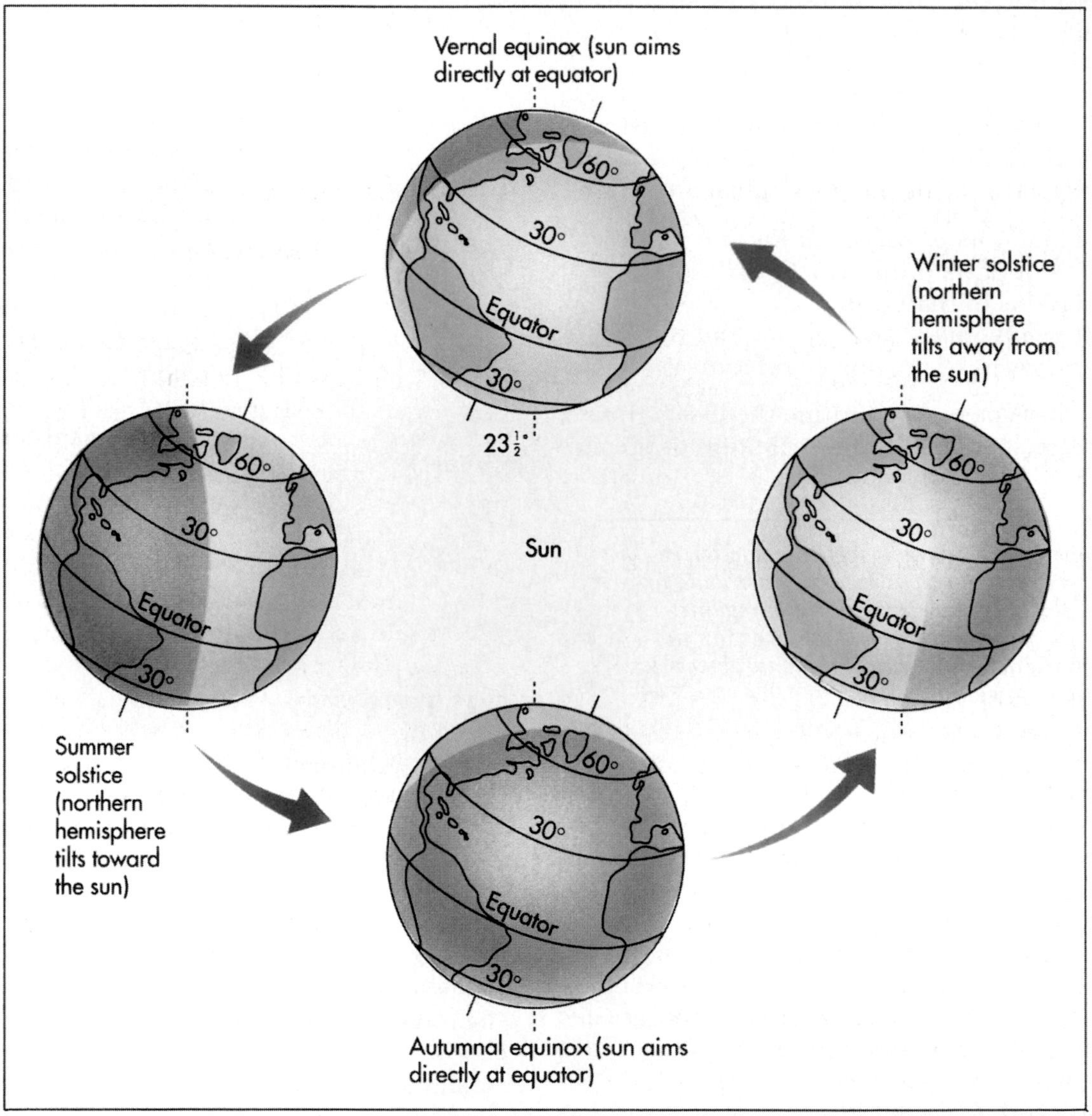

FIGURE 37-2 The rotation of the Earth around the sun has a profound effect on climate. In the northern and southern hemispheres, temperatures change in an annual cycle because the Earth is slightly tilted on its axis in relation to its pathway around the sun.

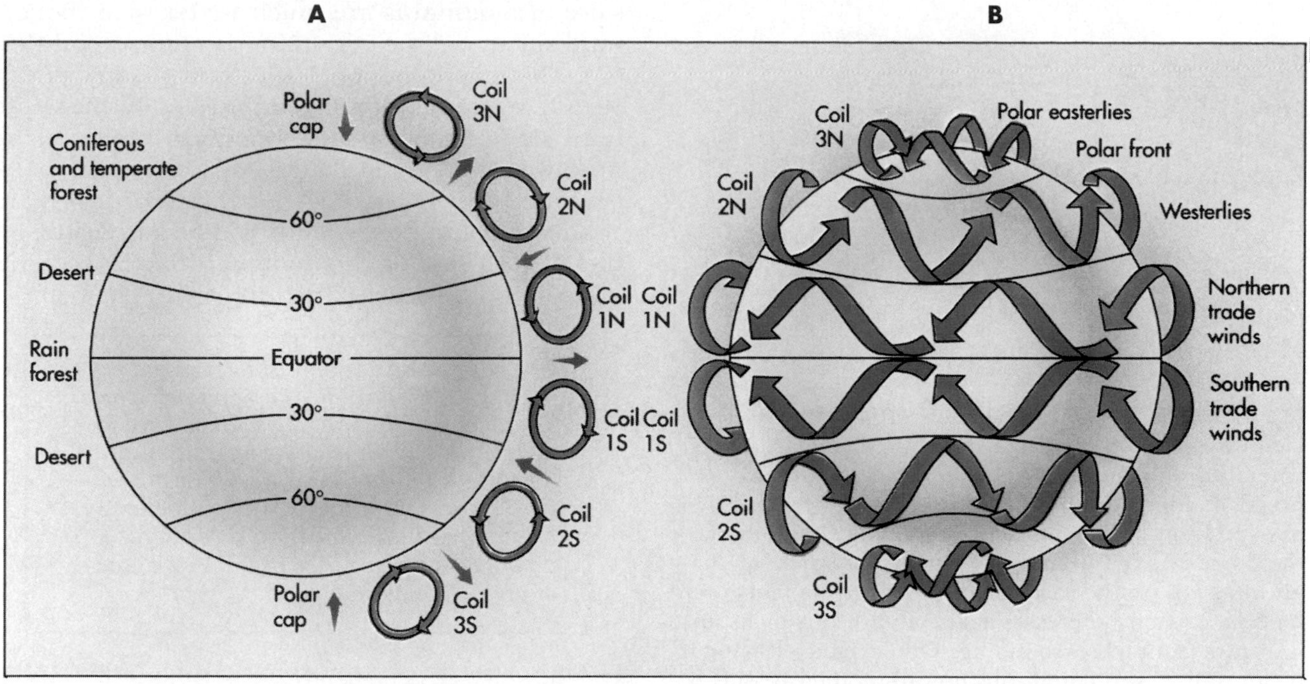

FIGURE 37-3 How the Earth's atmospheric circulation influences climate. A, Air surrounding the rotating Earth is broken into six coils of rising and falling air. These masses of air are associated with certain climates where air masses rise and fall. **B,** The wind patterns around the Earth are set up by the six moving air masses.

thus the northern and southern hemispheres receive unequal amounts of sunlight at various times during the Earth's year-long journey around the sun. One of the poles is closer to the sun than the other, except during the spring and autumn equinoxes (Figure 37-2).

Atmospheric Circulation and Its Effects on Climate

Near the equator, warm air rises and flows toward the poles. If the Earth were a stationary sphere, this air would reach the poles, cool, fall to the ground, and flow back toward the equator. The Earth is not stationary, however; as it spins, its rotary movement breaks the air into six "coils" of rising and falling air that surround the Earth (Figure 37-3, *A*). With increasing latitude, each air mass is cooler than the one before but warmer than the next. Therefore the air in each mass rises at the region of its lowest latitude, moves toward the poles, sinks to the ground at the region of its highest latitude, and flows back toward its lowest latitude. As it warms, it rises again, completing the cycle.

These moving masses of air that encircle the Earth set up the prevailing wind patterns (Figure 37-3, *B*) and also have an effect on precipitation. The moisture-holding capacity of air increases when it is warmed and decreases when it is cooled. Therefore precipitation is relatively high near 60 degrees north and south latitude and at the equator, where it is rising and being cooled. Conversely, precipitation is generally low near 30 degrees north and south latitude, where air is falling and being warmed (see Figure 37-3, *A*).

Partly as a result of these factors, all the great deserts of the world lie near 30 degrees north or 30 degrees south latitude, and some of the great temperate forests are near 60 degrees north or south latitude. But other factors also come into play. Some major deserts are formed in the interiors of the large continents; these areas have limited precipitation because of their distance from the sea, the ultimate source of most precipitation. Other deserts sometimes occur because mountain ranges intercept the moisture-laden winds from the sea (Figure 37-4).

As the air travels up a mountain (its windward side), it is cooled, and precipitation forms. As the air descends the other side of the mountain (its leeward side), it is warmed; its moisture-holding capacity increases. For these reasons, the windward

biomes (bye omes) ecosystems of plants and animals that occur over wide areas of land within specific climatic regions easily recognized by their overall appearance.

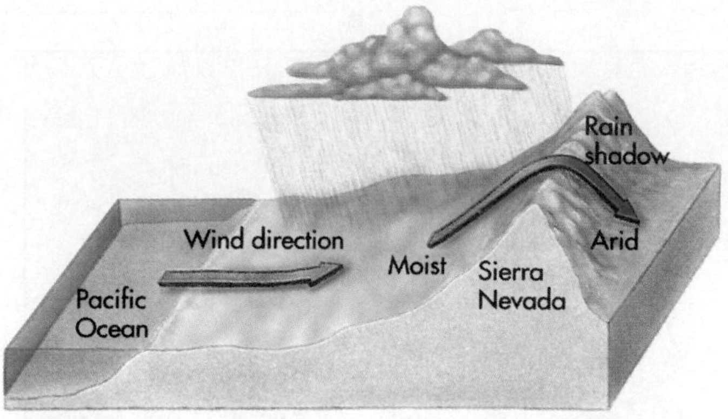

FIGURE 37-4 The rain shadow effect. Moisture-laden winds from the Pacific Ocean rise and are cooled when they encounter the Sierra Nevada mountain range. As their moisture-holding capacity decreases, precipitation occurs, making the middle elevation of the range one of the snowiest regions on Earth. As the air descends on the east side of the range, its moisture-holding capacity increases again, and the air picks up moisture from its surroundings rather than releasing it. As a result, desert conditions prevail on the east side of the mountains.

sides of mountains are much wetter than the leeward sides, and the vegetation is often very different. This phenomenon is the rain shadow effect. Seattle, Washington, for example, lies on the windward side of the Cascade Mountain range in the northwestern United States. This city receives 99 centimeters (39 inches) of rainfall per year. Yakima, Washington, slightly south from Seattle on the leeward side of this mountain range, receives only 20 centimeters (8 inches) of rain per year.

> *The climate of a region is determined primarily by its latitude and wind patterns. These factors, interacting with surface features of the Earth such as mountains and distance from the ocean, result in the particular rainfall patterns. The temperature, rainfall, and altitude of an area, in turn, provide conditions that result in the growth of vegetation characteristic of that area.*

The patterns of atmospheric circulation influence patterns of circulation in the ocean, modified by the location of the land masses around and against which the ocean currents must flow. Oceanic circulation is dominated by huge surface gyrals (Figure 37-5), which move around the subtropical zones of high pressure between approxi-

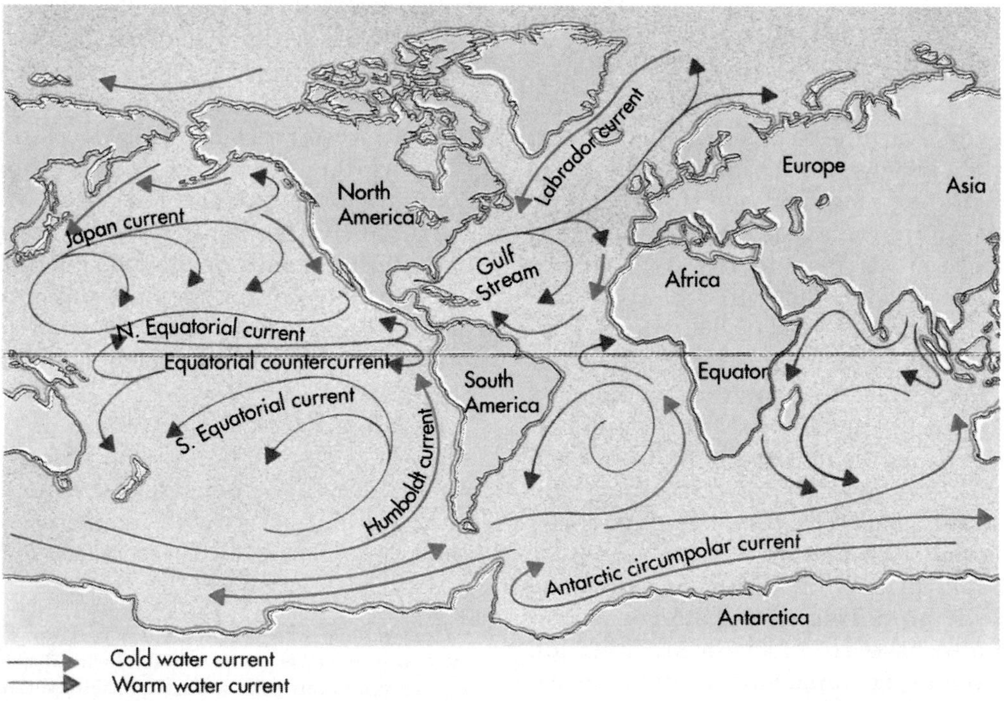

FIGURE 37-5 Ocean circulation. The circulation in the oceans moves in great surface spiral patterns, or gyrals, and affects the climate on adjacent lands.

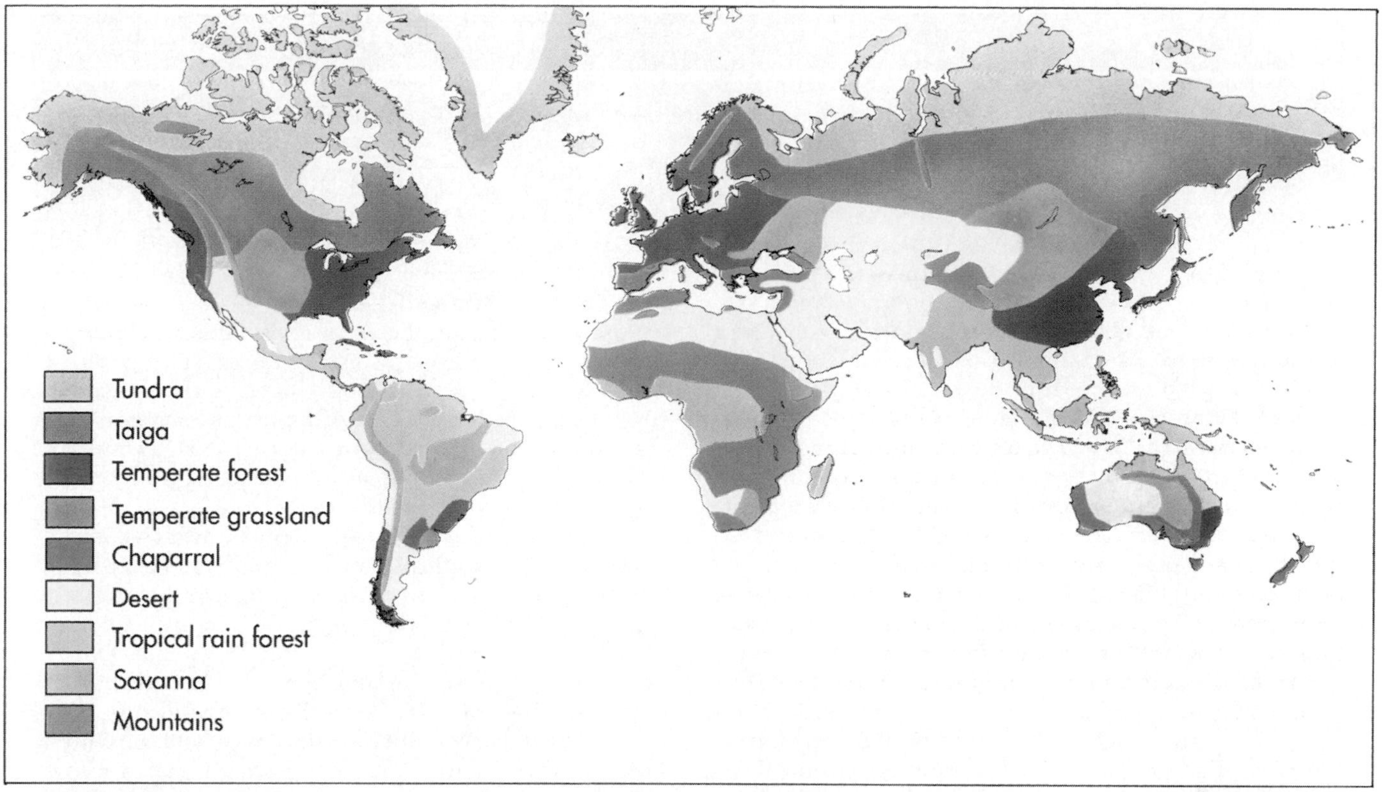

FIGURE 37-6 Distribution of biomes. Mountain ranges and chaparrals are also shown on this map. Chaparrals are temperate shrublands that border grasslands and deserts in certain parts of the world.

mately 30 degrees north and 30 degrees south latitude. These gyrals move clockwise in the northern hemisphere and counterclockwise in the southern hemisphere. They profoundly affect life not only in the oceans but also on coastal lands because they redistribute heat. For example, the gulf stream in the North Atlantic swings away from North America near Cape Hatteras, North Carolina, and reaches Europe near the southern British Isles. Because of the gulf stream, western Europe is much warmer and thus more temperate than eastern North America at similar latitudes. As a general principle, the western sides of continents in the temperate zones of the northern hemisphere are warmer than their eastern sides; the opposite is true in the southern hemisphere.

Life on Land: The Biomes of the World

Biomes are often classified in seven categories: (1) tropical rain forests, (2) savannas, (3) deserts, (4) temperate grasslands, (5) temperate deciduous forests, (6) taiga, and (7) tundra (Figure 37-6). This list is arranged by distance from the equator, but the biomes do not encircle the Earth in neat bands. Their distribution is greatly affected by the climatic effects caused by the presence of mountains, the irregular outlines of the continents, and the

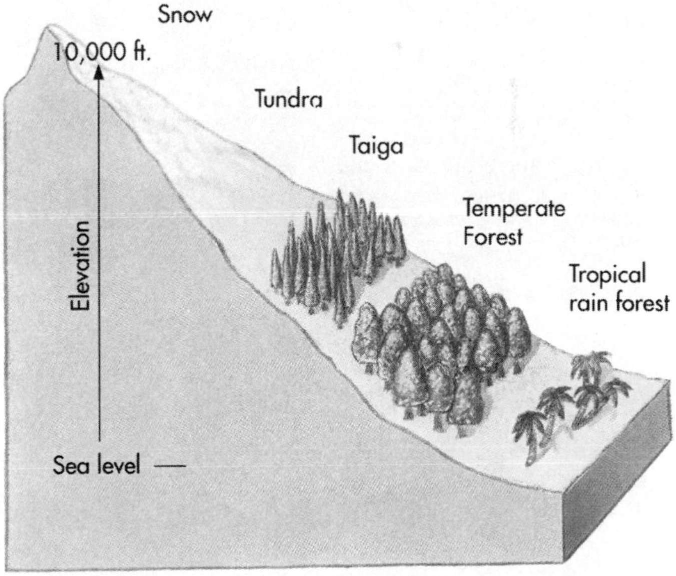

FIGURE 37-7 Elevation and biomes. Biomes that normally occur far north and far south of the equator at sea level occur also in the tropics (and at other latitudes) but at high mountain elevations. Thus on a tall mountain in southern Mexico or Guatemala, you might see a sequence of biomes like the one illustrated here.

JUST WONDERING. . . .

I often hear references to "El Niño" during the weather reports on TV. What is "El Niño," and how does it affect the weather?

El Niño is an oceanic phenomenon. The story of El Niño starts with the Humbolt current, which sweeps up the west coast of South America from the south (see Figure 37-5). Simultaneously, westerly winds push the warm, nutrient-poor surface water away from the coast, resulting in an upwelling of the Humbolt water. An upwelling brings up cool, nutrient-rich waters from the depths, replacing the surface water as it is blown away.

Once every 3 to 7 years the westerly winds diminish. The westerlies arise from the pressure gradient formed by an area of high pressure that normally sits over the eastern Pacific Ocean and an area of low pressure that dominates the western Pacific Ocean. The westerly winds diminish because of the southern oscillation: the low-pressure system expands while the high-pressure system shrinks. The pressure gradient changes; therefore the winds diminish or may even reverse direction. Therefore the upwelling of the Humbolt current diminishes, and warm water flows southerly, down the coast to southern Peru and northern Chile. This warm water flow has been named "El Niño" (meaning "the child") by local fisherman because the warm current occurs around the Christmas season; their reference is to the Christ child. Fishermen are familiar with this event because the warm, nutrient-poor water causes massive fish kills.

Together, these events are often referred to as *El Niño southern oscillation (ENSO)*. These events can be devastating to South American fisheries, especially in Chile and Peru. An ENSO event also influences weather patterns throughout the Pacific Ocean and as far away as India and Africa. It may cause such occurrences as abnormal warmth in Alaska and western Canada, as well as more Pacific lows to hit the California coast instead of going farther north. The severe ENSO of 1986 to 1988 correlated with unusual drought in the states of Washington and Oregon in the summer of 1987. In contrast, the ENSO event of 1991 to 1992 brought heavy flooding to southern California and parts of Texas.

Not all the effects of an ENSO are negative and catastrophic. The moist, rainy conditions that accompany an ENSO in some regions are favorable for crops. The Peruvian rice crop, for example, benefits from the ENSO rains. Land birds, such as Darwin's finches on the Galapagos Islands, breed abundantly during an ENSO, and their populations increased remarkably during the ENSO of 1982 and 1983.

Scientists still do not know what causes an ENSO event. Because the cause is still unknown, an ENSO cannot be predicted. Nevertheless, scientists have used the knowledge they have gained to recognize ENSO events in their early stages. Further study of the ENSO may hold the key to understanding the complex interplay of atmospheric and water currents around the globe and the effects of El Niño and the southern oscillation on weather patterns.

temperature of the surrounding sea. In addition, the climate (and vegetation) changes with the elevation of land similar to changes with increasing latitude (Figure 37-7). Thus tundralike vegetation occurs near the top of a tall mountain in the tropics, as well as near the North and South Poles.

Tropical Rain Forests

Tropical rain forests occur in Central America; in parts of South America, particularly in and around the Amazon Basin; in Africa, particularly in central and west Africa; and in southeast Asia. As the name suggests, the tropical rain forests occur in regions of high temperature and rainfall, generally 200 to 450 centimeters (80 to 175 inches) per year, with little difference in its distribution from season to season. The temperature averages 25° C (77° F). As a comparison, Houston, Texas—one of the hottest and wettest cities in the United States—receives an average of 121 centimeters (48 inches) of rainfall per year; the average temperature is 20.5° C (69° F). New Orleans, Louisiana—another hot, wet city—receives an average of 145 centimeters (57 inches) of rainfall per year; the average temperature is 20° C (68° F). These cities almost seem dry and cool in comparison to the tropical forest!

You may have the idea that the tropical rain forest is thick with lush vegetation and creeping vines, creating a network too dense to penetrate without a machete. These forests *are* thick and lush—but not at the forest floor. Little can grow on the ground far beneath the canopy of trees whose branches and leaves form an overlapping roof to the forest. In fact, only 2% of the light shining on the forest canopy reaches its floor! Plants that do grow there have large, dark green leaves adapted to conducting photosynthesis at low light levels. Other types of vegetation have interesting adaptations that enable them to compete successfully with the large trees for sunlight. Vines, for example, have their roots anchored in the soil but climb up the trees with their long stems. They reach the canopy where leaves grow to capture sunlight. Other interesting plants are the epiphytes, or "air plants." *Epiphytes* grow on the trees or other plants for support but draw their nourishment mostly from rain water. Some epiphytes catch moisture with modified leaves or flower parts; others have roots that hang free in the air and absorb water (with its dissolved minerals) from the rain (Figure 37-8).

The giant trees of the tropical forest support a rich and diverse community of animals on their

A

B

FIGURE 37-8 Tropical rain forests: lush equatorial forests. A, Epiphytes, plants that anchor themselves to trees but derive nutrients and moisture from the air and rain. **B,** Blooming only for 24 hours, the flower of a passion vine is a specialist in rain forest competition. Sweet spots reward vigilant guard ants, and poisonous leaves deter unwanted intruders.

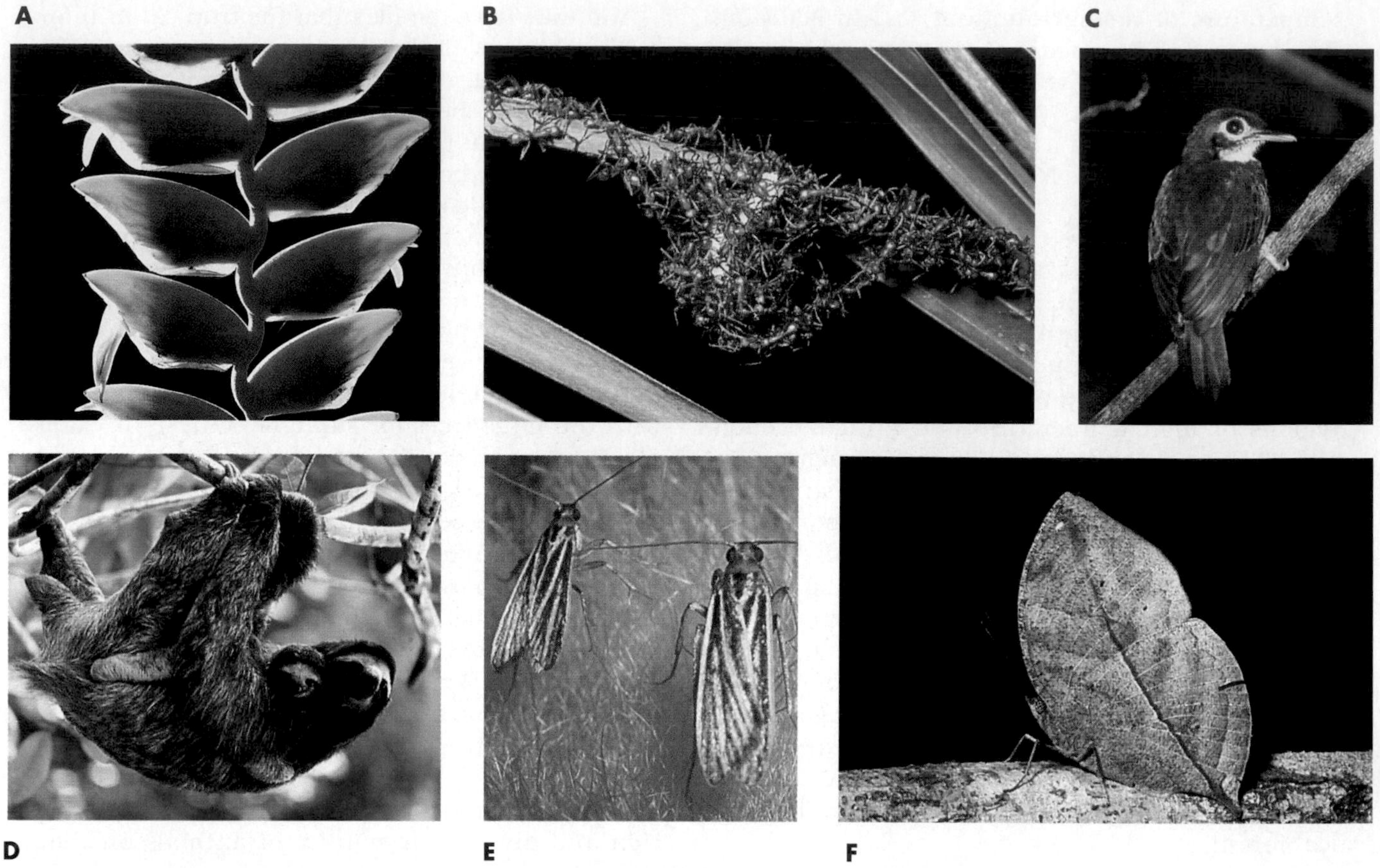

FIGURE 37-9 Ecological specializations in the tropical rain forest. A, The native South American flower, *Heliconia,* has colors that stand out like beacons. These flowers are visited and pollinated by the hummingbirds that they attract. **B,** Army ants live as huge, mobile, foraging communities in the tropical rain forest. Here a small column is transporting a wasp larva to be used as food. **C,** Some species of birds, such as the bicolored antbird, fly above the columns of army ants and feed on the insects that they flush from the foliage. **D,** A mother and baby three-toed sloth hang from a tree limb. Sloths reside in the canopy of the tropical rain forest, as shown here in Panama. **E,** Sloth moths carry out their entire life cycle in the fur of the three-toed sloth, where their larvae feed on the green algae that grow luxuriantly there. **F,** A dead leaf butterfly in the forests of Sumatra is superbly camouflaged.

branches. Figure 37-9 provides examples of just a few of these interesting organisms. The roots of the trees are interesting also, spreading out from thickened trunks into a thin layer of soil, often no more than a few centimeters deep. These roots transfer the nutrients from fallen leaves and other organic debris quickly and efficiently back to the trees after bacteria and fungi break down the debris. Very few nutrients remain in the soil. Therefore when humans cut down and then burn these trees to clear the land for agriculture, they are "burning away" the nutrients held in the trees, as well as breaking down organic matter to carbon dioxide. The small amount of ash that is left provides few nutrients for the crops farmers try to grow. In 2 to 3 years of farming, these few remaining nutrients are depleted from the soil, and the land remains barren (Figure 37-10).

The tropical rain forest, a biome that occurs in regions of high temperature and rainfall, is characterized by tall trees that support a variety of plant and animal life on their branches.

Savannas

Not all areas near the equator are wet; some areas experience prolonged dry seasons or have a lower annual rainfall than the tropical forests do (generally, about 90 to 150 centimeters per year, or 35 to 60 inches). The heat, periodic dryness, and poor soils cannot support a forest but have led to the evolution of savannas: open grasslands with scattered shrubs and trees. These areas, situated between the tropical forests and deserts, cover much of central and southern Africa, western India, northern Aus-

FIGURE 37-10 The destruction of the rain forest. To clear land for agriculture, farmers first cut down the trees and then burn the stumps in the technique of slash-and-burn agriculture. Here, a rain forest is being destroyed in the Amazon.

tralia, large areas of northern and east-central South America, and some of Malaysia.

The vegetation of the savanna supports large grazing herbivores such as buffalo, wildebeests, and zebra (Figure 37-11); these animals, in turn, are food for carnivores such as lions. The savanna also supports a large number of plant-eating invertebrates, such as mites, grasshoppers, ants, beetles, and termites. The termites, in fact, are one of the most important soil organisms, breaking down dried twigs, leaves, and grass to usable nutrients. In addition, their huge, complex mounds (Figure 37-12) provide passageways for rain water to deeply penetrate the ground rather than just running off or evaporating from the surface.

Like tropical forests, savannas are found near the equator but in areas having less annual rainfall. This climate supports grasslands with only scattered trees and shrubs.

Deserts

Deserts are biomes that have 25 centimeters (10 inches) or less of precipitation annually. For this reason, the vegetation in deserts is characteristically sparse. The higher the annual rainfall a desert has, however, the greater the amount of vegetation it will be able to support. In fact, ecologists classify deserts based on their annual rainfall: semideserts receive about 25 centimeters (10 inches) per year (Phoenix, Arizona, and San Diego, California, for example), true deserts receive less than 12 centimeters (4.7 inches) per year (Las Vegas, Nevada), and extreme deserts average below 7 centimeters (2.8 inches) per year (Namib Desert in southwestern Africa). The photos of these three types of deserts shown in Figure 37-13 point out the differences in their patterns of vegetation.

Major deserts occur around 20 to 30 degrees north and south latitude, where the warm air that rose from the equator falls. As previously mentioned, the air at the equator rises, cooling and releasing its moisture, which falls on the tropical forests. The dry air then falls over desert regions, resulting in little precipitation. Deserts also occur in the interiors of continents far from the moist sea air, especially in Africa (the Sahara Desert), Eurasia, and Australia. Some deserts, such as the Baja region of California, are near the ocean yet are dry; the winds blow from the north, carrying little moisture because they are cool. High pressure areas off the West Coast of the United States also

FIGURE 37-11 **Savannas: dry, tropical grasslands.** These zebras are grazing on a savanna in Tanzania.

FIGURE 37-12 **Termite mound in Colombia, South America.** Termite mounds allow rain water to penetrate the ground, thwarting run-off and evaporation.

A
B
C

FIGURE 37-13 Deserts: arid and hot. The semi-desert in **A** is in South Mountain Park in Phoenix, Arizona. The true desert in **B** is the Great Basin desert near Baker, Nevada. The extreme desert in **C** is the Namib desert in Namibia.

deflect storms moving down from the north. In addition, some deserts form on the leeward side of mountain ranges, such as in the Great Basin of Nevada and Utah in the United States.

Because desert vegetation is sparse and the skies are usually clear, deserts radiate heat rapidly at night. This situation results in substantial daily changes in temperature, sometimes more than 30 degrees centigrade (approximately 55 degrees Fahrenheit) between day and night. Although both hot deserts (the Sahara, for example) and cool deserts (the Great Basin of North America) exist, summer daytime temperatures in all deserts are extremely high, frequently exceeding 40° C (104° F). In fact, temperatures of 58° C (136.4° F) have been recorded both in Libya and in San Luis Potosi, Mexico—the highest that have been recorded on Earth.

Plants have developed a wide variety of adaptations to survive in this difficult environment. Annual plants are often abundant in deserts and simply bypass the unfavorable dry season in the form of seeds. After sufficient rainfall, many germinate and grow rapidly, sometimes forming spectacular natural displays. Characteristic of deserts, of course, are the many species of succulent plants, those with tissues adapted to store water, such as cacti. The trees and shrubs that live in deserts often have deep roots that reach sources of water far below the surface of the ground. The woody plants that grow in deserts may be either deciduous, losing their leaves during the hot, dry seasons of the year, or evergreen, with hard, reduced leaves. The creosote bush of the deserts of North and South America is an example of an evergreen desert shrub. Near the coasts in areas where there are cold waters offshore, deserts may be foggy, and the water that the plants obtain from the fog may allow them to grow quite luxuriantly.

Desert animals, too, have fascinating adaptations that enable them to cope with the limited water of the deserts (Figure 37-14). Many limit their activity to a relatively short period of the year when water is available or even plentiful; they resemble annual plants in this respect. Many desert vertebrates live in deep, cool, and sometimes moist burrows. Organisms that are active for much of the year emerge from their burrows only at night, when temperatures are relatively cool. Other organisms, such as camels, can drink large quantities of water when it is available and store it for use when water is unavailable. A few animals simply migrate to or through the desert and exploit food that may be abundant seasonally; when the food disappears, the animals move on to more favorable areas.

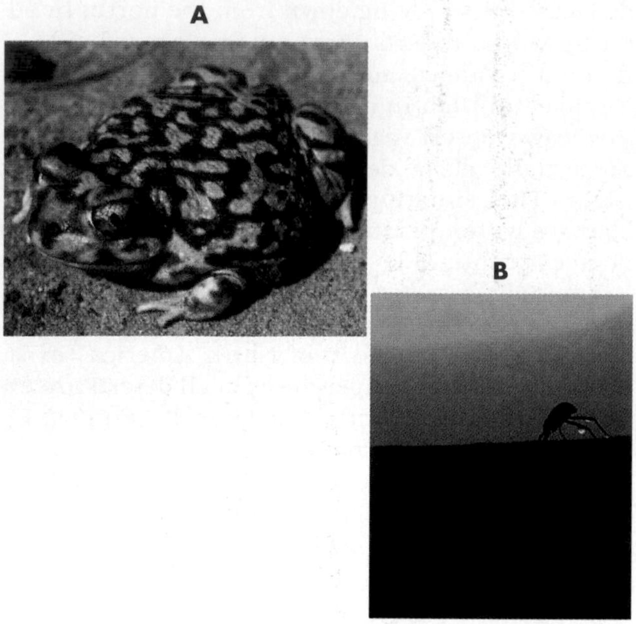

FIGURE 37-14 How two desert animals conserve water. A, Spadefoot toads, which live in the deserts of North America, can burrow nearly a meter below the surface and remain there for as long as 9 months out of each year. Under such circumstances, their metabolic rate is greatly reduced, and they depend largely on their fat reserves. When moist, cool conditions return to the desert, they emerge and breed rapidly. The young toads mature quickly and burrow back underground, using the horny projections on their feet that gave them their name. **B,** On the very dry sand dunes of the Namib desert in southwestern Africa, the beetle *Onymacris unguicularis* collects fog water by holding up its abdomen at the crest of a dune, thus gathering condensed water on its body.

> *Deserts occur around 20 to 30 degrees north and south latitude and in other areas that have 25 centimeters (10 inches) or less of precipitation annually. Desert life is somewhat sparse but exhibits fascinating adaptations to life in a dry environment.*

Temperate Grasslands

Temperate grasslands have various names in different parts of the world: the prairies of North America, steppes of Russia, pusztas of Hungary, veld of South Africa, and pampas of South America. All temperate grasslands have 25 to 75 centimeters (10 to 30 inches) of rainfall annually, much less than that of savannas but more than that of deserts. Temperate grasslands also occur at higher latitudes than savannas but are often found bordering deserts as savannas do.

FIGURE 37-15 Temperate grasslands: seas of grass. Among the other names for these grasslands are prairie, steppe, veld, and pampa.

Temperate grasslands are characterized by large quantities of perennial grasses; the rainfall is insufficient to support forests or shrublands. Grasslands are often populated by burrowing rodents, such as prairie dogs and other small mammals, and herds of grazing mammals, such as the North American bison (Figure 37-15). The grazing of herbivores contributes to the maintenance of this biome by preventing woody vegetation from becoming established.

The soil in temperate grasslands is rich; in fact, much of the temperate grasslands are farmed for this reason. Grasslands are often highly productive when they are converted to agriculture, and many of the rich agricultural lands in the United States and southern Canada were originally occupied by prairies.

> *Temperate grasslands experience a greater amount of rainfall than deserts but a lesser amount than savannas. They occur at higher latitudes than savannas but, like savannas, are characterized by perennial grasses and herds of grazing mammals.*

FIGURE 37-16 Temperate deciduous forests: rich hardwood forests. The leaves of the trees in temperate deciduous forests often change color in the autumn before they fall from the trees as winter approaches.

FIGURE 37-17 The floor of the temperate deciduous forest is rich in plant growth. Plants such as herbs, ferns, and mosses inhabit the forest floor.

Temperate Deciduous Forests

The climate in areas of the northern hemisphere such as the eastern United States and Canada and an extensive region in Eurasia supports the growth of trees that lose their leaves during the winter. Such trees are called *deciduous* (from a Latin word meaning "to fall"), dropping their leaves and remaining dormant throughout the winter. These vast areas of trees are therefore temperate deciduous forests (Figure 37-16) and thrive in climates where summers are warm, winters are cold, and the precipitation is moderate, generally from 75 to 150 centimeters (30 to 60 inches) annually. Precipitation is well distributed throughout the year, but water is generally unavailable during the winter because it is frozen.

The temperate forest differs from the tropical forest in that more vegetation grows near the forest floor than in the tropical forest. Although the temperate forest does have an upper canopy of dominant trees such as beech, oak, birch, and maple, there is a lower tree canopy and a layer of shrubs beneath. On the ground, herbs, ferns, and mosses abound (Figure 37-17). In addition, animal life in the temperate forest is abundant on the ground as well as in the trees; in the tropical forest, animal life is primarily arboreal, with the exception of large mammals, a few bird species, and soil invertebrates.

In areas having less than 75 centimeters (30 inches) of precipitation annually, temperate deciduous forests are replaced by grassland, as in the prairies of North America and the steppes of Eurasia. Where conditions are more limiting—restricted, for example, by intense cold—these forests may be replaced by coniferous forests.

> *Temperate deciduous forests occur in areas having warm summers, cold winters, and moderate amounts of precipitation. The trees of this forest lose their leaves and remain dormant throughout the winter.*

Taiga

The northern coniferous forest is called *taiga*. The cone-bearing trees of this forest are primarily spruce, hemlock, and fir and extend across vast areas of Eurasia and North America (Figure 37-18). The taiga is characterized by long, cold winters with little precipitation; most of the precipitation falls in the summers. Because of the latitude where taiga occurs, the days are short in winter (as little as 6 hours) and correspondingly long in summer.

FIGURE 37-18 Taiga: great evergreen forests of the north. The taiga located in Alaska is dominated by spruces and alders.

The light, warmth, and rainfall of the summer allows plants to grow rapidly, and crops often attain a large size in a surprisingly short time.

The trees of the taiga occur in dense stands of one or a few species of cone-bearing trees. Alders, a common species, harbor nitrogen-fixing bacteria in nodules on their roots; for this reason, they are able to colonize the infertile soils of the taiga. Marshes, lakes, and ponds also characterize the taiga; they are often fringed by willows or birches. Many large mammals can also be found there, such as the elk (Figure 37-19). Other herbivores, including moose and deer, are stalked by carnivores such as wolves, bear, lynx, and wolverines.

To the south, taiga grades into temperate forests or grasslands, depending on the amount of precipitation. Coniferous forests also occur in the mountains to the south, but these are often richer and more diverse in species than those of the taiga. Northward, the taiga gives way to open tundra.

> *The taiga, or northern coniferous forest, consists of evergreen, cone-bearing trees. The climate of this biome is characterized by long, cold winters with little precipitation.*

Tundra

Farthest north in Eurasia, North America, and their associated islands—between the taiga and the permanent ice—is the open, often boggy community known as the *tundra* (Figure 37-20). Dotted with lakes and streams, this enormous biome encircles the top of the world, covering one fifth of the Earth's land surface. (A well-developed tundra does not occur in the Antarctic because there is no land at the right latitude.) The tundra is amazingly uniform in appearance, dominated by scattered patches of grasses and sedges (grasslike plants), heathers, and lichens. Some small trees do grow but are primarily confined to the margins of streams and lakes.

Annual precipitation in the tundra is very low, similar to desertlike precipitation of less than 25 centimeters (10 inches) annually. In addition, the precipitation that falls remains unavailable to plants for most of the year because it freezes. During the brief Arctic summers, some of the ice melts. The permafrost, or permanent ice, found about a meter down from the surface never melts, however, and is impenetrable to both water and roots. When the surface ice melts in the summer, it has nowhere to go and forms puddles on the land. In contrast, the alpine tundra found at high eleva-

FIGURE 37-19 Bull elk crossing a mountain stream. Elk, which are herbivores, are hunted by carnivores such as wolves, bears, lynx, and humans.

tions in temperate or tropical regions does not have this layer of permafrost.

The tundra teems with life during its short summers. As in the taiga, perennial herbs grow rapidly then, along with various grasses and sedges. Large grazing mammals, including musk oxen, caribou, and reindeer, migrate from the taiga. Many species of birds and waterfowl nest in the tundra in the summer and then return to warmer climates for the winter. Populations of lemmings, small rodents that breed throughout the year beneath the snow, rise rapidly and then crash on a 3- to 4-year cycle, influencing the populations of the carnivores that prey on them, such as snowy owls and arctic foxes.

The tundra encircles the top of the world. This biome is characterized by desertlike levels of precipitation, extremely long and cold winters, and short, warmer summers.

Life in Fresh Water

Only 2% of the Earth is covered by fresh water, found standing in lakes and ponds or moving in rivers and streams. Freshwater ecosystems lie near and are intertwined with terrestrial ecosystems.

FIGURE 37-20 Tundra: cold boggy plains of the north. Mount McKinley National Park in Alaska.

For example, some organisms such as amphibians may move from one ecosystem to another. In addition, organic and inorganic material continuously enters bodies of fresh water from terrestrial communities. Often, the wet, spongy land of marshes and swamps provides habitats intermediate between the two.

Ponds and lakes have three life zones, or regions, in which organisms live: the shore zone, the open-water zone, and the deep zone. The **shore zone** is the shallow water near edges of a lake or pond in which plants with roots, such as cattails and water lilies, may grow. Consumers such as frogs, snails, dragonflies, and tiny shrimplike organisms live among these producers. The **open-water zone** is the main body of water through which light penetrates. Floating and drifting algae and plantlike organisms, or phytoplankton, grow here. Floating "animals" such as protozoans (zooplankton) feed on the phytoplankton in this aquatic food chain. They, in turn, are eaten by small fish, which are eaten by larger fish. The **deep zone,** the water into which light does not penetrate, is devoid of producers. This dark zone is inhabited mainly by decomposers and other organisms such as clams that feed on the organic material that filters down to them (Figure 37-21). Ponds differ from lakes in that they are smaller and shallower. Therefore light usually reaches to the bottom of all levels of a pond; it has no deep zone.

Rivers and streams differ from ponds and lakes primarily in that their water flows rather than remains stationary. The nature of this ecosystem is therefore different from that of a pond or lake. One difference is that the level of dissolved oxygen is usually much higher in a river or stream than in a standing body of water because moving water mixes with the air as it churns and bubbles along. A high level of dissolved oxygen allows an abundance of fish and invertebrates to survive. In addition, only a few types of producers inhabit rivers and streams: various species of algae that grow on rocks and a few types of rooted plants such as water moss.

A river or stream is characterized as an open ecosystem; that is, it derives most of its organic material from sources other than itself. Detritus (debris or decomposing material) flows from upstream or enters from the land. Leaves and woody material drop into the stream from vegetation bordering its banks. Rain water washes organic material from overhanging leaves. In addition, water

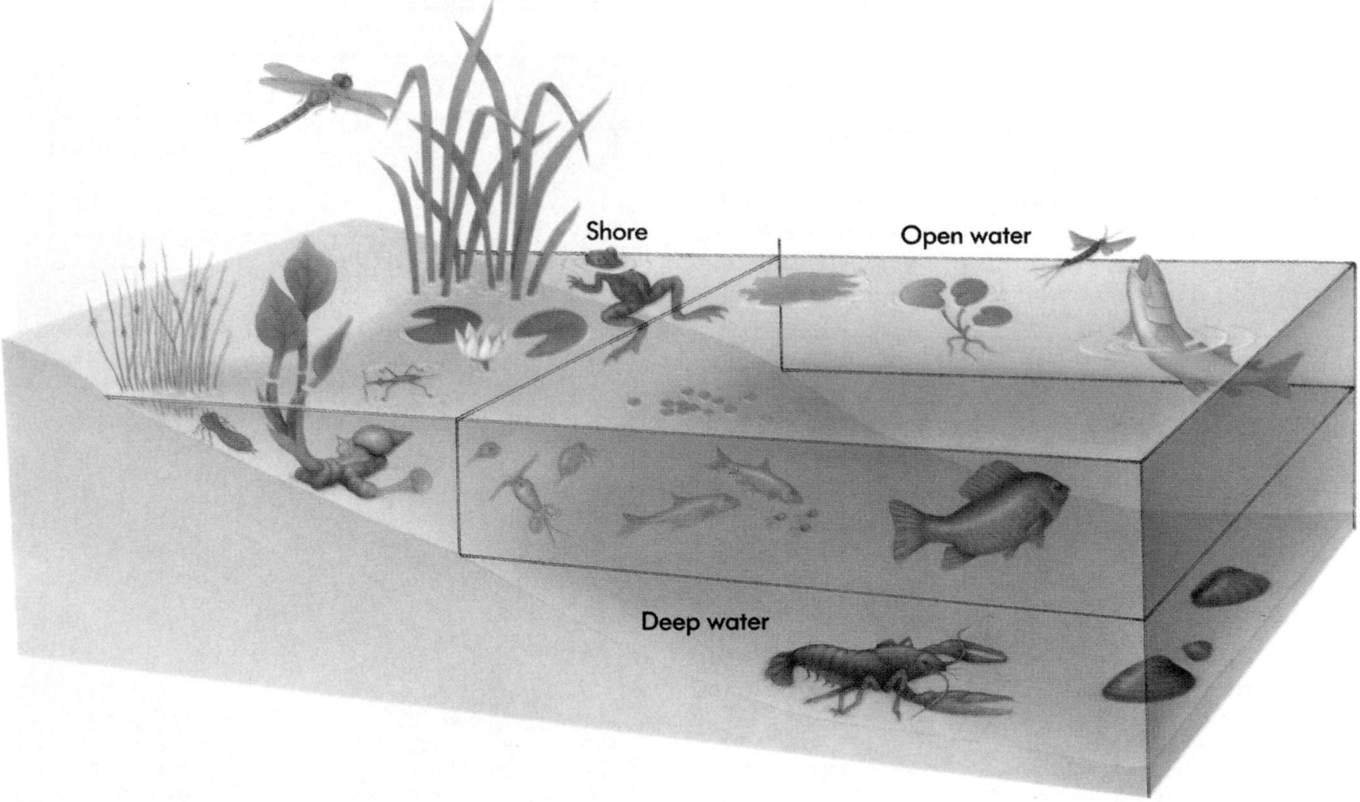

FIGURE 37-21 The three life zones of a lake. Organisms live in the shore zone, the open-water zone, and the deep-water zone.

seeps into a river or stream from below the surface of adjoining land, carrying with it organic materials and, in some cases, fertilizers and other chemicals. These nutrients feed the producers and small consumers. As commonly occurs in pond and lake food chains, large fish feed on smaller fish that feed on tiny invertebrates. The river/stream ecosystem is largely heterotrophic and is strongly tied to terrestrial ecosystems that surround it.

Freshwater ecosystems lie near and are intertwined with terrestrial ecosystems. From them, organic and inorganic material continuously enters freshwater ecosystems.

As the water from a stream or river flows into a lake or the sea, the velocity of the water decreases. As the water slows, sediment carried along by faster water now sinks to the bottom. These deposits form fan-shaped areas called **deltas.** Accumulated sediment breaks the path of the water into many small channels that course through the delta. Occasionally, the path of the water is actually stopped as sediment accumulates. For this reason a delta created by a large river, such as the Mississippi River delta, consists of a great deal of swampy, marshy land.

Estuaries: Life between Rivers and Oceans

As rivers and streams flow into the sea, an environment called an **estuary** is created where fresh water joins salt water. In the shallow water of the estuary, rooted grasses often grow. Other producers of the estuary are various types of algae and phytoplankton. Consumers are primarily mollusks, crustaceans, fish, and various zooplankton. All organisms inhabiting estuaries, however, have adaptations that allow them to survive in an area of moving water and changing salinity. (*Salinity* refers to the concentration of dissolved salts in the water.) For example, many estuarine organisms are bottom dwellers and attach themselves to bottom material or burrow in the mud. Each species of organism is found living in a region of salinity optimal for its survival.

Oysters are one of the most important bottom dwellers in estuaries, providing a habitat for many other species of organisms. These mollusks either bury themselves in the mud, forming oyster beds, or cement themselves in clusters to the partially buried shells of dead oysters, forming oyster reefs. Many other invertebrates, such as sponges and barnacles, attach themselves to the oysters and feed on plankton. Other species such as crabs, snails, and worms live on, beneath, and between the oysters, feeding on the oysters themselves or on detritus (dead, decaying material) trapped in the oyster reef. One researcher, in fact, has documented more then 300 species of organisms living in association with a single oyster reef!

The motile organisms of the estuary are primarily crustaceans such as crabs, lobsters, and shrimp and various species of fish. Fish exhibit interesting reproductive adaptations to the varying salinities of the estuary. Some species, such as the striped bass, spawn upstream from the estuary where the salinity of the water is low. The larvae and young fish move downstream through increasing concentrations of salt as they develop, moving into the ocean in adulthood. In a similar manner, shad spawn upstream in fresh water, and the young spend their first summer in the estuary before swimming to the sea.

Estuaries are places where the fresh water of rivers and streams meets the salt water of oceans. These ecosystems are made up of plentiful communities of organisms exhibiting behaviors and growth characteristics adapted to changing salinity.

Nutrients are more abundant in estuaries than in the open ocean because estuaries are close to terrestrial ecosystems and derive much of their nutrients from them as rivers and streams do. Unfortunately, estuaries are also easily polluted from these sources. In Chesapeake Bay (Figure 37-22), for example, complex systems of rivers enter the Atlantic Ocean, forming one of the most biologically productive bodies of water in the world. In the 1960s, the bay yielded an annual average of about

shore zone the shallow water near the edges of a lake or pond in which plants with roots, such as cattails and water lilies, may grow.

open-water zone the main body of pond or lake water through which light penetrates.

deep zone the area of pond or lake water into which light does not penetrate.

deltas fan-shaped areas of accumulated sediment deposited by a stream or river as it enters an open body of water such as a lake or the ocean.

estuary (ess choo er ree) a place where the fresh water of rivers and streams meets the salt water of oceans.

FIGURE 37-22 For legend, see opposite page.

FIGURE 37-22 Chesapeake Bay. Chesapeake Bay has more than 11,300 kilometers (7020 miles) of shoreline and drains more than 166,000 square kilometers (64,100 square miles) in one of the most densely populated and heavily industrialized areas in North America. The body of open water is about 320 kilometers (199 miles) long and, at some points, nearly 50 kilometers (31 miles) wide. **A,** The large metropolitan areas (noted on the map) and shipping facilities make the bay one of the busiest natural harbors anywhere. **B,** Uncontrolled erosion from certain agricultural practices, pesticides, and increases in nutrients block the light needed for photosynthesis and upset the delicate ecological balance on which the productivity of the bay depends. This material can be seen in some of the rivers that empty into the bay in this aerial view. The states that border the bay are cooperating, with the assistance of the Environmental Protection Agency, to try to bring the bay back to its former productivity. **C,** One of the most biologically productive bodies of water in the world, the bay yielded an annual average of about 275,000 kilograms (600,000 pounds) of fish in the 1960s but only a tenth as much in the 1980s. The human population grew 50% during the same period. **D,** Oil transport and commercial shipping is expected to double by the year 2020. This grebe is coated from a recent oil spill off the mouth of the Potomac River.

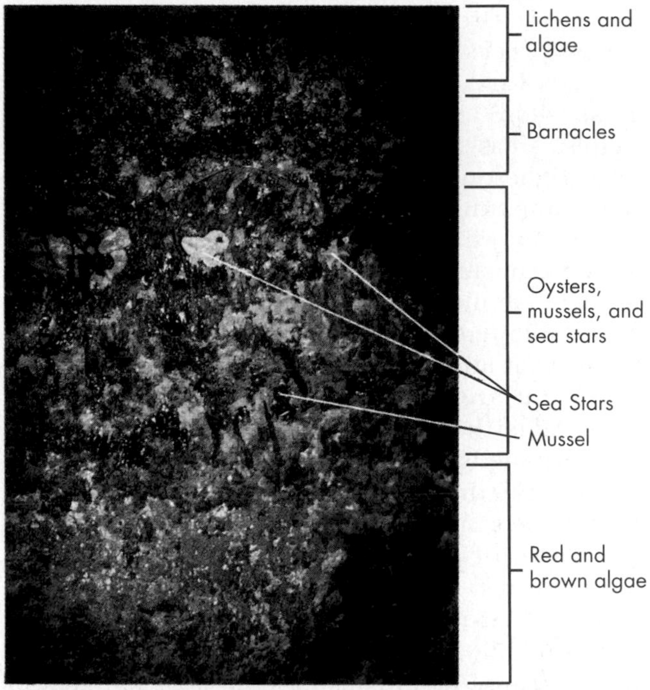

FIGURE 37-23 A rocky shore in Pacific Grove in Monterey Bay, California. The different areas on the rock all receive different amounts of moisture, providing a variety of habitats for a variety of organisms.

275,000 kilograms (600,000 pounds) of fish. As the human population in this area increased, however, along with oil transport and commercial shipping, pollution also increased. More than 290 oil spills were reported in the bay in 1983 alone. In addition, uncontrolled erosion of the land into the water and leaching of pesticides has caused a 90% decrease in the yield of fish. Maryland and Virginia, the states that border the bay, along with the Environmental Protection Agency, are working to bring the Chesapeake Bay estuary back to its former productivity.

Life in the Oceans

Although only 2% of the Earth is covered by fresh water, nearly three quarters of the Earth's surface is covered by ocean. These seas have an average depth of 4 kilometers (approximately $2^1/_2$ miles), and they are, for the most part, cold and dark. The concentration of oxygen, as well as the availability of light and food, is a factor that limits life in the ocean. Although cold water is able to "hold" more oxygen than warm water, the warmer sea water near the surface of the ocean mixes with the oxygen in the atmosphere. Therefore oxygen is present in its highest concentrations in the upper 200 meters (650 feet) or so of the sea. Light is most abundant in the top 100 meters (325 feet).

The marine environment provides a variety of habitats, but it can be divided into three major life zones:

1. The *intertidal zone,* the area between the highest tides and the lowest tides
2. The *neritic zone,* the area of shallow waters along the coasts of the continents, which extends from the low tide mark to waters down to 200 meters deep
3. The *open-sea zone,* composing the remainder of the ocean

The Intertidal Zone

The wind-swept shoreline is a harsh place for organisms to live. As the tide rolls in and out, environmental conditions change from hour to hour: wet to dry, sun protected to sun parched, and wave battered to calm. Nevertheless, life abounds in the intertidal zone, exhibiting interesting adaptations and characteristics necessary for survival.

Figure 37-23 shows the rocky shore at low tide at

the Pacific Grove Monterey Bay in California. The exposed rocks are teeming with life, but these organisms vary along a continuum from the driest areas (those least often covered with water) to the wettest areas (those most often covered with water). Highest up on the rocks, in areas that the high tide sometimes does not reach, grow certain lichens and algae. Somewhat lower on the rocks grow barnacles. Then oysters, blue mussels, and limpets (mollusks that have conical shells) take over, followed by brown algae and red algae. "Forests" of large brown algae, or kelp, take over in areas that are exposed for only short periods of time. All of these organisms have adaptations such as hard shells or gelatinous coverings that keep them from drying out and are either anchored within the sand or stick to the surfaces of rocks so that they will not be washed away (Figure 37-24).

In contrast to the rocky shore, the sandy shore (Figure 37-25) looks as though no life is present. However, life is plentiful beneath the sand and mud. Because organisms have no large surfaces to which they can attach, they are adapted to burrowing under the sand during low tide. Copepods, tiny "micro" crustaceans, are predominant organisms. In addition, worms, crabs, and mollusks such as clams burrow to safety when the tides roll out. In areas near the low tide mark, sea anemones, sea urchins, and sea stars make their home.

The intertidal zone has plentiful light and is home to a variety of producers. Along with the algae, phytoplankton float in the water and are used for food by the zooplankton and many other consumers. In addition, the heterotrophs of the intertidal zone have the waves to thank for bringing fresh organic material to them and for washing away their wastes.

The Neritic Zone

Surrounding the continents of the world is a shelf of land that extends out from the intertidal zone usually 50 to 100 kilometers (30 to 60 miles), sloping to a depth of about 200 meters (approximately 650 feet) beneath the sea. This margin of land is called the **continental shelf.** The waters lying above it make up the neritic zone, which is derived from a Greek root referring to the sea. Because light reaches the waters of most of this zone, it supports an abundant array of plant and therefore animal life.

One outstanding community in the neritic zone is the **coral reef.** The term *reef* refers to a mass of rocks in the ocean lying at or near the surface of the water. Coral reefs are built by marine animals, or corals (phylum Cnidaria), that secrete calcium carbonate, a hard, shelllike substance. With the help of algae that reside in their bodies, the coral build on already shallow portions of the continental shelf or on submerged volcanoes in the ocean. Complex and fascinating ecosystems, coral reefs provide habitats for a variety of invertebrates and fishes (Figure 37-26). Along with the tropical rain forests, coral reefs are the most highly productive ecosystems in terms of biomass (see Chapter 36).

FIGURE 37-24 Rocky shore organisms. A rocky shore in Maine shows a few of the organisms that make this shore their home. The pink sea star clings to the rocks, as do barnacles, which use special filaments to anchor themselves.

FIGURE 37-25 A sandy shore. In contrast to the rocky shore, there seems to be no life present on this sandy shore in Hawaii. However, microscopic animals make their homes in the spaces between grains of sand, and larger animals burrow beneath the sand during low tide.

A

C

B

FIGURE 37-26 The coral reef: a delicate biome in danger. The living corals in coral reefs depend on constant conditions for survival. Increasingly, the conditions in which coral reefs live are being disrupted by pollution and human carelessness. Preventative measures and a sensitivity to the precarious balance necessary for the coral reefs' survival will be instrumental in saving the reefs from destruction. **A,** Coral reef in Fiji, showing hard and soft corals. **B,** Coral reef in Borneo, showing the brilliant colors of the different coral formations. **C,** Scallop attached to a limestone reef in the Indo-Pacific.

The Open-Sea Zone

Beyond the continental shelf lies the great expanse of the open ocean. This open-sea zone is often referred to as the **pelagic zone,** a term derived from another Greek word meaning "ocean" (Figure 37-27). Within this huge ecosystem exists many diverse forms of life—some with which you are familiar, such as the floating plankton and various species of fishes. But other forms of life are unfamiliar—even bizarre—such as the anglerfish shown in Figure 37-28.

Organisms live in the vast expanse of the ocean in relationship to available light and food. Temperature, salinity, and water pressure also play roles in creating the various habitats of the ocean. Light is available to organisms from the water's surface only to an approximate depth of 200 meters (650 feet). In fact, this area of the open ocean is called the **photic zone** for this reason. (*Photic* means "light".) Phytoplankton thrive in this well-lighted layer of the ocean (especially within the better-lighted upper 100 meters [325 feet]), drifting freely with the ocean currents and serving as the base of oceanic food webs. Zooplankton float with the

continental shelf the margin of land that extends out from the intertidal zone usually 50 to 100 kilometers (30 to 60 miles) and slopes to a depth of about 200 meters (approximately 650 feet) beneath the sea.

coral reef (**kor** ul) a community of organisms that live on masses of rocks in the ocean lying at or near the surface of the water. Coral reefs are partially made up of marine animals called corals (phylum *Cnidaria*) that secrete calcium carbonate, a hard shelllike substance. Other marine organisms also grow on the rocks and make up part of the coral reef.

pelagic zone (puh **laj** ik) the open ocean.

photic zone (**foe** tik) an area of the open ocean where light is available to organisms from the water's surface to an approximate depth of 200 meters (650 feet).

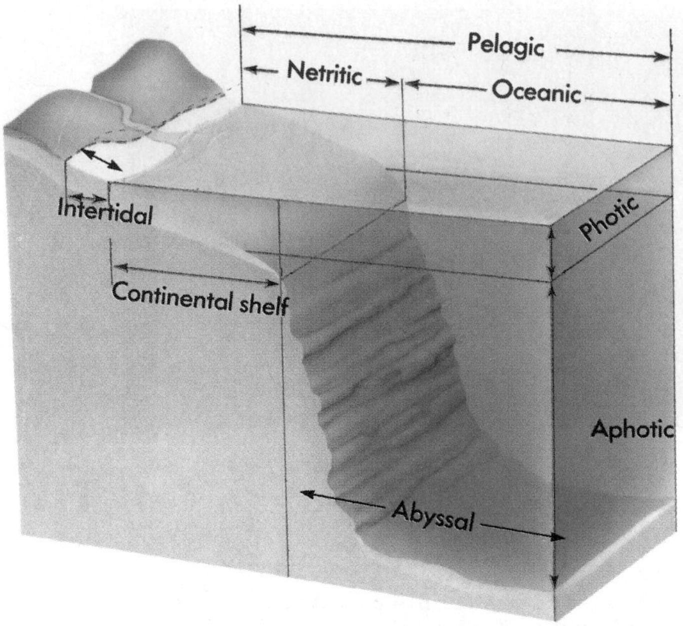

FIGURE 37-27 Zones of the ocean. The pelagic zone encompasses the oceanic and neritic zones. Nearer to shore is the continental shelf and the intertidal zone. The photic zone and aphotic zones are measured by the amount of light that penetrates the water.

phytoplankton and are first-order consumers in many photic food webs. Other typical heterotrophs of this zone are most air-breathing mammals such as whales, porpoises, dolphins, seals, and sea lions and fishes such as herring, tuna, and sharks. These fishes and mammals of the sea, which are called *nekton,* feed on the plankton and on one another. Together, the organisms that make up the plankton and the nekton provide all of the food for those that live below.

Little light penetrates the ocean from 200 to 1000 meters (650 to 3250 feet). Therefore no photosynthetic organisms live in this zone known as the **mesopelagic zone,** or middle ocean. At this depth, temperatures remain somewhat steady throughout the year but are cooler than the water above; water pressure increases steadily with depth. Under these conditions, many bizarre organisms have evolved, such as those that exhibit bioluminescence (see Figure 1-2, C), which they use to communicate with one another or to attract prey. The organisms common to middle-ocean life are fish with descriptive names such as swordfish, lanternfish, and hatchetfish; certain sharks and whales; and cephalopods such as octopi and squid.

Peculiar creatures also live in the ocean depths of 1000 or more meters—a mile or so (and more) beneath the surface. This region of the ocean is called the **abyssal zone,** meaning "bottomless," because it is so deep that it seems to be bottomless. The water at this tremendous depth contains high concentrations of salt, is under immense pressure from the water above, and is very cold. The organisms living at this level cannot make trips into the photic zone to capture food as some mesopelagic organisms do but must feed on material that settles from above.

Organisms collectively referred to as *benthos,* or bottom dwellers, live on the ocean floor. Sea cucumbers and sea urchins crawl around eating detritus. Various species of clams and worms burrow in the mud, feeding on a similar array of decaying organic material. Bacteria are also rather common in the deeper layers of the sea, playing important roles as decomposers as they do on land and in freshwater habitats. Such organisms (sea cucumbers, sea urchins, clams, worms, and bacteria) are found in other ocean zones as well; however, different species inhabit different ocean zones.

Of the three main life zones in the ocean, the intertidal zone, the neritic zone, and the photic area of the open-sea zone support the most life because of the presence of light and the availability of oxygen. Organisms that live in waters beneath the photic zone must feed on organic material and detritus that falls from above and must be adapted to increased water pressure and salinity.

mesopelagic zone (**mez** o puh **laj** ik) a term referring to the middle ocean zone; little light penetrates this ocean zone from 200 to 1000 meters (650 to 3250 feet).

abyssal zone (uh **biss** uhl) the deepest region of the ocean. The abyssal zone includes ocean depths of 1000 or more meters.

A

B

C

D

E

FIGURE 37-28 The open ocean. A, What mysteries lie beneath the ocean's deepest waters? **B,** Flying fish in the eastern Pacific ocean. **C,** School of yellowfin tuna swimming around a coral reef. **D,** A deep-sea anglerfish showing the lure used by the fish to attract prey. The lure is actually part of the fish, but to prey, the lure looks like a tempting bit of food. When the prey comes close to investigate, the anglerfish quickly eats the prey. **E,** Hydrothermal vent community located in the deep waters off the Galapagos Islands.

Summary

- Biomes are the largest recognizable terrestrial ecosystems, which occur over wide areas and within specific climatic regions. Their characteristics are a direct result of temperature and rainfall patterns.
- The climate of a region is determined primarily by its latitude and wind patterns. These factors interact with the surface features of the Earth, resulting in particular rainfall patterns. The temperature, rainfall, and altitude of an area, in turn, provide conditions that result in the growth of vegetation characteristic of that area.
- Seven categories of biomes, arranged by distance from the equator, are (1) tropical rain forest, (2) savannas, (3) deserts, (4) temperate grasslands, (5) temperate deciduous forests, (6) taiga, and (7) tundra.
- The tropical rain forests and savannas lie nearest the equator. The tropical rain forest receives an enormous amount of rain year round, whereas the savanna experiences less rain and prolonged dry spells. These differences promote the growth of tall trees and other lush vegetation in the tropical rain forest and open grasslands, with scattered trees and shrubs in the savanna.
- Deserts are extremely dry biomes. Hot deserts are hot year round, whereas cool deserts are hot only in the summer. Deserts are of great biological interest due to the extreme behavioral, morphological, and physiological adaptations of the plants and animals that live there.
- Temperate grasslands receive less rainfall than savannas but more than deserts. The soil in these grasslands is rich, so they are well suited to agriculture.
- The temperate deciduous forests receive moderate precipitation that is well distributed throughout the year. The climate of these forests differs from tropical forests in that they receive less rainfall, are found at a higher and cooler latitude, and experience cold winters. The trees are therefore deciduous, losing their leaves and remaining dormant throughout the winter.
- The taiga is the coniferous forest of the north. It consists primarily of cone-bearing evergreen trees, which are able to survive long, cold winters and low levels of precipitation. Even farther north is the tundra, which covers about 20% of the Earth's land surface and consists largely of open grassland, often boggy in summer, which lies over a layer of permafrost.
- Freshwater ecosystems make up only about 2% of the Earth's surface; most of them are ponds and lakes. Ponds and lakes have a shore zone and an open-water zone. Lakes have a deep zone. Rivers and streams differ from these bodies of water because they contain moving water that mixes with the air to provide high levels of oxygen for its fish and invertebrate inhabitants.
- Estuaries are places where fresh water meets salt water as rivers empty into the ocean. These ecosystems receive nutrients from the surrounding land and support a large number of organisms.
- The marine environment consists of three major life zones: the intertidal zone, between the highest tides and the lowest tides; the neritic zone, the area of shallow water that lies over the continental shelf; and the open-sea zone. Within these zones, the ocean supports the most life in areas that have light and sufficient quantities of dissolved oxygen.

Knowledge and Comprehension Questions

1. What is a biome?
2. January is a winter month in the United States but part of summer in Australia. Why?
3. Summarize how atmospheric circulation affects climates around the world.
4. What factors determine the climate of a region?
5. Identify the terrestrial biome associated with each of the following:
   a. Animals in these areas live mostly in trees.
   b. Dominant trees such as oak and beech allow enough light so that herbs, ferns, and mosses grow on the ground.
   c. Because of permafrost, when surface ice melts, the ground cannot absorb the water.
   d. Although located near the equator, these regions are mostly open grasslands.
   e. These coniferous forests receive most of their precipitation during the summer.
   f. Herds of North American bison once roamed these areas, grazing on large expanses of perennial grasses.
   g. These regions radiate heat rapidly at night, causing wide temperature differences between day and night.
6. What factors limit rainfall in a desert?
7. Distinguish between taiga and tundra.
8. Describe the three life zones found in fresh water. What types of organisms live in each?
9. What is an estuary? Summarize how its inhabitants are adapted to its conditions.
10. Why are estuaries and freshwater regions so vulnerable to human pollution?
11. Identify and briefly describe the three major life zones of the ocean.
12. The ocean is divided into three zones based on depth. Describe what types of organisms are found in each zone.

Critical Thinking

1. Much of the world's most productive agriculture is carried out on soil of temperate grasslands. Agriculture in the tropics is far less productive. Why do you think temperate grassland soil is so much richer than soil in a tropical rain forest?

CHAPTER 38

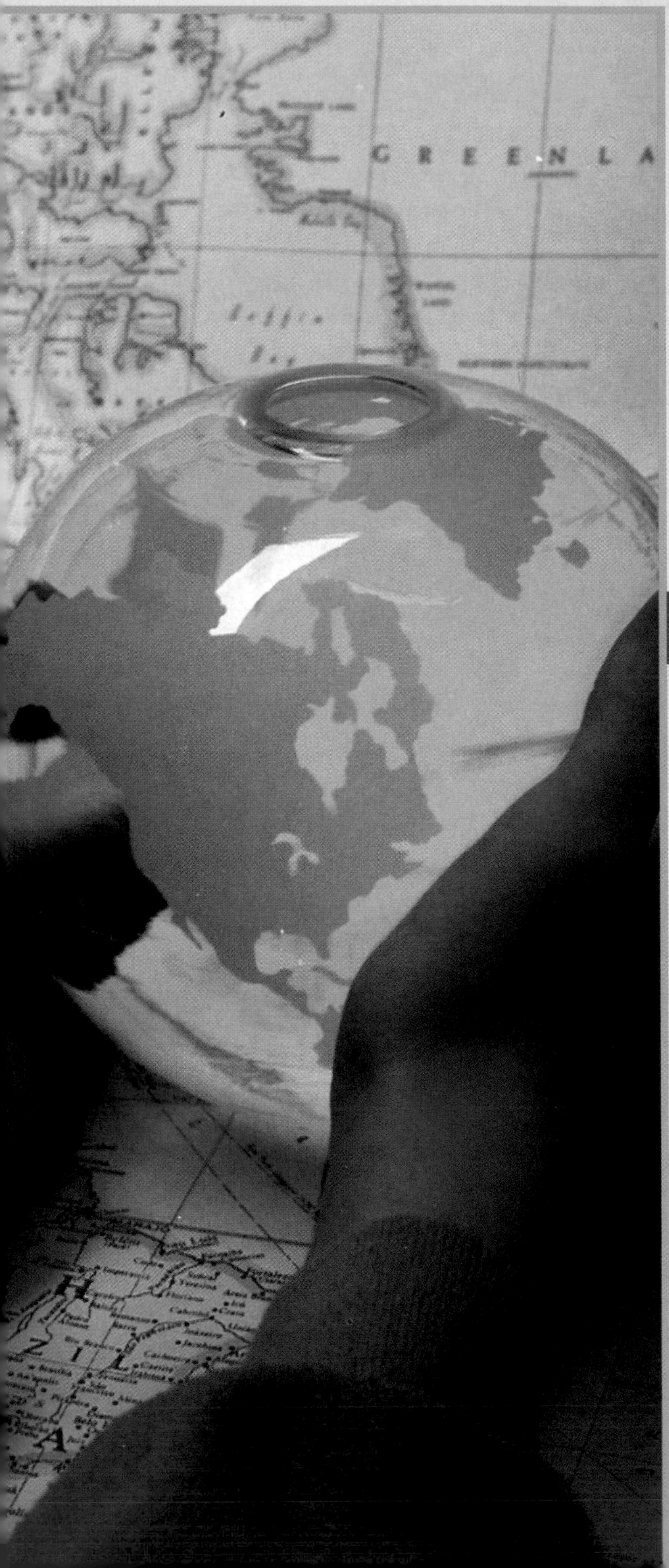

THE BIOSPHERE: TODAY AND TOMORROW

JUST IMAGINE that you could hold the world in your hands and make a difference in its future. Well, imagining isn't even necessary. All people do hold the world in their hands in a figurative way. Everyone's actions have a direct impact on the Earth and on the quality of life that you and others will experience for generations to come.

However, you must ask yourself some crucial questions. How *are* you affecting the Earth? What environmental problems do people face today? And what can you do to deal with those problems to be sure that the quality of life on Earth will be enhanced for yourself and your children? The answers to these questions are the focus of this chapter and will help you understand more about how you shape this fragile planet on which you live.

The Biosphere

Life on Earth is confined to a region called the **biosphere,** the global ecosystem in which all other ecosystems exist. The biosphere extends from approximately 9000 meters (30,000 feet) above sea level to about 11,000 meters (36,000 feet) below sea level. You can think of it as extending from the tops of the highest mountains (such as Mount Everest or some of the Himalayas) to the depths of the deepest oceans (such as the Mariana Trench of the Pacific Ocean)—the part of our Earth in which the land, air, and water come together to help sustain life (Figure 38-1).

The biosphere is often spoken of as the **environment.** This general term refers to everything around you—not only the land, air, and water but other living things as well. You can speak, for example, of the environment of an ant, a water lily, or all the peoples of the world. Your particular environment can change during the day from a home environment, to a classroom environment, and then to an office environment. The environment can include a great deal—or very little—of the total biosphere and its living things.

The Land

Humans interact with the land and its inhabitants in myriad ways. Its natural resources are used. Wastes are produced that are filling and in some cases polluting the land. Forests are cut down, stripping many species of their habitats and therefore their life. And people continue to procreate at a rate that will result in the doubling of the world's population by the year 2030. What can people do to stop using the land in destructive ways? What can people do to help heal the Earth?

Diminishing Natural Resources

The land gives humans many things: fossil fuels, timber, food, and minerals. Water, too, is an important natural resource. Some of these resources—fossil fuels and minerals—are finite or **nonrenewable resources;** they formed at a rate much slower than their consumption. Coal and oil are examples of nonrenewable resources. Forests *can* be used and replaced *if managed wisely.* However, when whole sections of forests are cut and burned and their soil depleted, they cannot be replaced . . . nor can the wildlife that once lived there.

Renewable resources are those produced by natural systems that replace themselves quickly

KEY CONCEPTS

- The biosphere extends from the tops of the mountains to the depths of the seas; it is all the parts of the Earth where biological activity occurs.
- Human beings need to work toward a sustainable society that uses nonfinite and renewable sources of fuel, recycles resources to the fullest extent, and manages forests to meet global needs while not compromising the ability of future generations to survive.
- Water is being polluted by multiple sources, causing physical or chemical changes that harm living and nonliving things; prevention of this contamination is essential to ensure a safe water supply.
- A primary cause of air pollution is the combustion of fossil fuels; energy conservation measures and new energy technology are needed.

OUTLINE

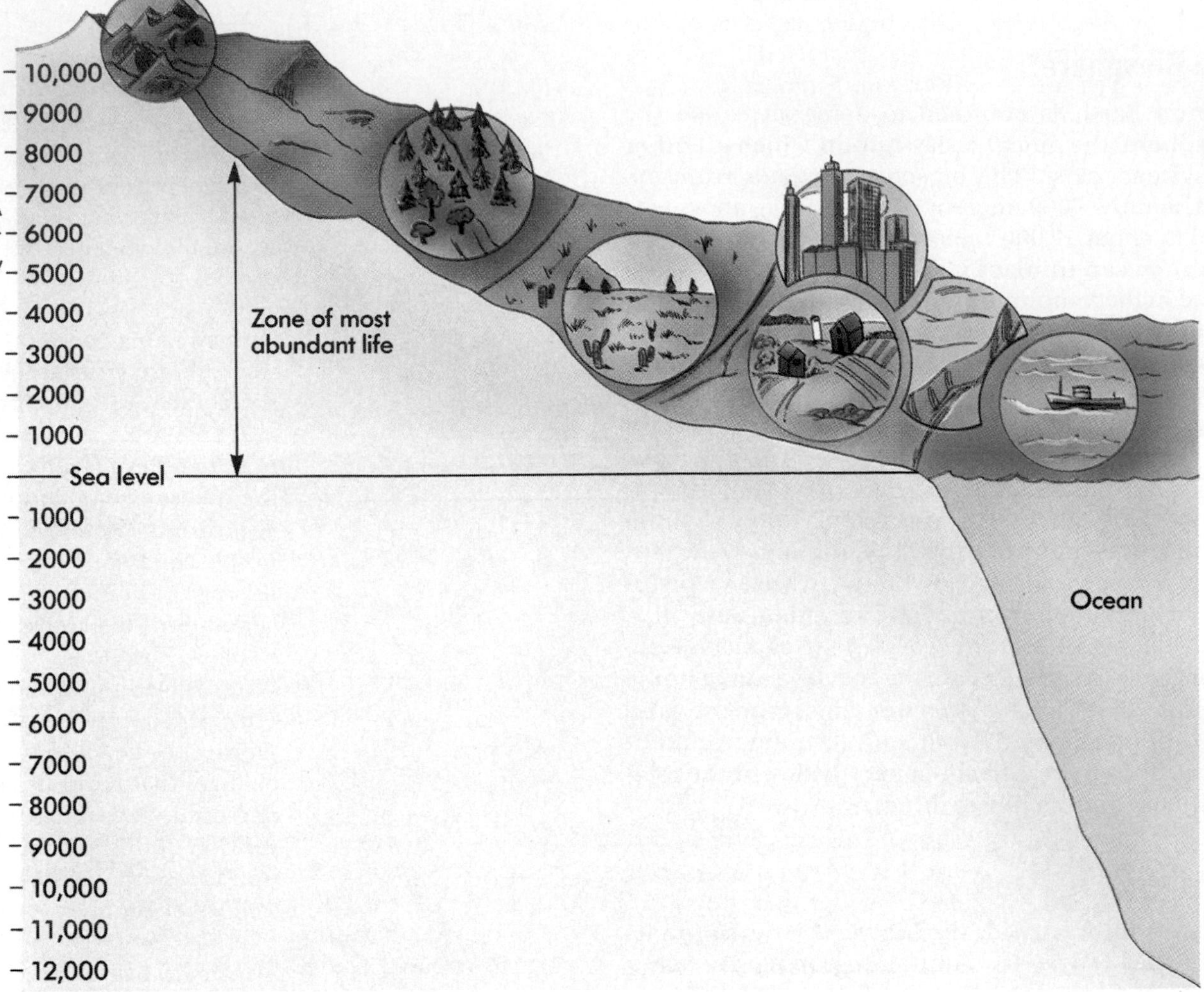

FIGURE 38-1 The biosphere. The biosphere is that part of the Earth where life can be found, from the highest mountain peaks to the depths of the ocean.

enough to keep pace with consumption. Examples of renewable resources are wind power and solar power.

Nonrenewable Resources

Many of the fuels used to heat homes and run cars are **fossil fuels.** These substances—coal, oil, and natural gas—are formed over time (acted on by heat and pressure) from the undecomposed carbon compounds of organisms that died millions of years ago. Environmental scientists suggest that instead of depending on these fuels for energy—fuels that are finite—people should work toward a *sustainable society* that uses nonfinite and renewable sources of fuel (see "How Science Works"). In a sustainable society, the needs of the society are satisfied without compromising the ability of future generations to survive and without diminishing the available natural resources.

Nuclear power, once considered a viable alternative energy source, will probably contribute only 6% to 8% of the world's energy by the year 2000. Although nuclear power is dependent on uranium, a finite but abundant natural resource, its problems lie primarily with the cost of building nuclear power plants, disposal of highly radioactive wastes, and the public fear regarding safety. The waste disposal problem would be virtually eliminated if nuclear fusion reactors can be developed to replace today's reactors, all of which use nuclear fission (splitting) reactions. Nuclear power could become a primary source for generating electricity. However, this method of energy production has yet to be shown to be feasible because of the very high reaction temperatures (100 million to 1 billion degrees Celsius). Even if successful, this energy source would probably not be commercially available before the middle of the next century.

Bioenergy refers to the use of living plants to produce energy. The most obvious type of bioenergy, the burning of wood, was first used by our ancestors more than 1 million years ago. In fact, until the Industrial Revolution of the 1800s, wood, not coal, supplied most of the world's energy. Today, wood supplies 12% of the world's energy, primarily in Latin America, Asia, India, and Africa. Unfortunately, however, the world is experiencing a fuel wood crisis—demand is exceeding the supply. The reasons for this crisis are complex but include the high world population and therefore high demand for wood, the degradation of woodlands without proper reforestation (replanting) techniques, and the cutting and burning of huge areas of tropical rain forests.

Renewable Resources

The primary renewable and nonfinite sources of power that are available are the sun, the wind, moving water, geothermal energy, and bioenergy. **Solar power,** the use of the sun for heating, is not a new technology; panels mounted on rooftops, for example, heat water that can be used for bathing, cooking, and space heating (Figure 38-2, *A*). In addition, homes and businesses can be built using the strategies of passive solar architecture in which windows are located such that the rays of the winter sun, low in the sky, enter the home or business, heating it. In the summer, an overhang protects the windows from the sun, which is high in the summer sky. The technology is also available to use solar power to produce electricity by converting the sun's energy to electricity after it has been captured within a fluid such as oil or by the use of photovoltaic solar cells that can convert sunlight directly into electricity.

Water has been used as an energy source for decades (Figure 38-2, *B*) and is currently supplying 20% of the world's electricity. A new form of hydropower called *wave power* is now being tested. Wave power uses the vertical motion of sea waves to produce electricity. In addition, *tidal power,* the use of the movement of the tides of the oceans to generate electricity, is currently under study in France and England (Figure 38-2, *C*). The wind, too, has been used not for decades but for centuries as a source of energy. Today, windmills are being used in developing countries (those not yet industrialized) for pumping water to livestock and to irrigate the land. In developed countries, however, windmills are being used in an entirely new way—to generate electricity. This "new breed" of windmill has rigid blades fashioned from lightweight materials (Figure 38-2, *D*) and looks unlike the picturesque windmills of Holland and Cape Cod, Massachusetts. At present, the United States and China are the two countries that lead in the use of **wind power** to generate electricity. In fact, some scientists predict that the United States will be generating 10% to 20% of its electricity from wind power by the year 2030.

Geothermal energy refers to the use of heat deep within the Earth. In some places, reservoirs of hot water exist that can be extracted from the Earth by drilling procedures much like those used to tap into the Earth's oil and natural gas reserves. Alternatively, dry, hot rock can be drilled and water flushed through it. After the water is heated within the Earth, it can be used directly for heating purposes or as part of a process to produce electricity. Currently, this technology is being used and further developed in the United States, England, Italy, New Zealand, and Japan.

biosphere (**bye** oh sfear) the global ecosystem of life on Earth that extends from the tops of the tallest mountains to the depths of the deepest seas.

environment a general term for the biosphere, encompassing the land, air, water, and every living thing on Earth.

nonrenewable resources resources that are formed at a rate much slower than their consumption; they are thus finite in supply. Fossil fuels and minerals are nonrenewable resources.

renewable resources resources that are produced by natural systems that replace themselves quickly enough to keep pace with consumption; wind power and solar power are renewable resources.

fossil fuels substances such as coal, oil, and natural gas that are formed over time (acted on by heat and pressure) from the undecomposed carbon compounds of organisms that died millions of years ago.

nuclear power (**noo** klee ur) an energy source that derives its power from the splitting apart of the nuclei of large atoms (nuclear fission) or from the combining of the nuclei of certain small atoms (nuclear fusion). All nuclear reactors in use today use nuclear fission reactions to produce energy.

bioenergy (**bye** oh **en** ur jee) the use of living plants to produce energy; the burning of wood to produce energy is an example of bioenergy.

solar power the use of the sun for heating or electricity.

wind power the use of the wind to generate electricity, usually by means of windmills.

geothermal energy (jee oh **thur** mul) the use of heat from deep within the Earth for heating purposes or as part of a process to produce electricity.

A

B

C

D

FIGURE 38-2 Alternative energy sources. To decrease reliance on fossil fuels, many individuals and nations are experimenting with alternative energy sources. **A,** Solar power. These panels collect the energy from sunlight and use this energy to heat homes and water. **B,** Hoover Dam in Arizona. Water power has been used for many centuries. **C,** Tidal power plant in the Bay of Fundy. Using the ocean's tides to generate power is still in the experimental stages. **D,** Modern windmills in southern California.

In developing countries, the use of *biogas machines* is helping ease the shortage of fuel wood already described. These stoves use microorganisms to decompose animal or human excrement or other organic wastes in a closed container. This process yields a methane-rich gas that can be used to fuel stoves, light lamps, and produce electricity. The use of animal waste to produce fuels such as methane, however, has severe environmental effects by depriving the soil of natural fertilizers.

Environmental scientists suggest that humans use renewable and nonfinite energy sources such as solar, wind, and water power, as well as tap the Earth's geothermal energy. The use of such fuels rather than the extensive use of fossil fuels will help satisfy people's needs without diminishing the available natural resources.

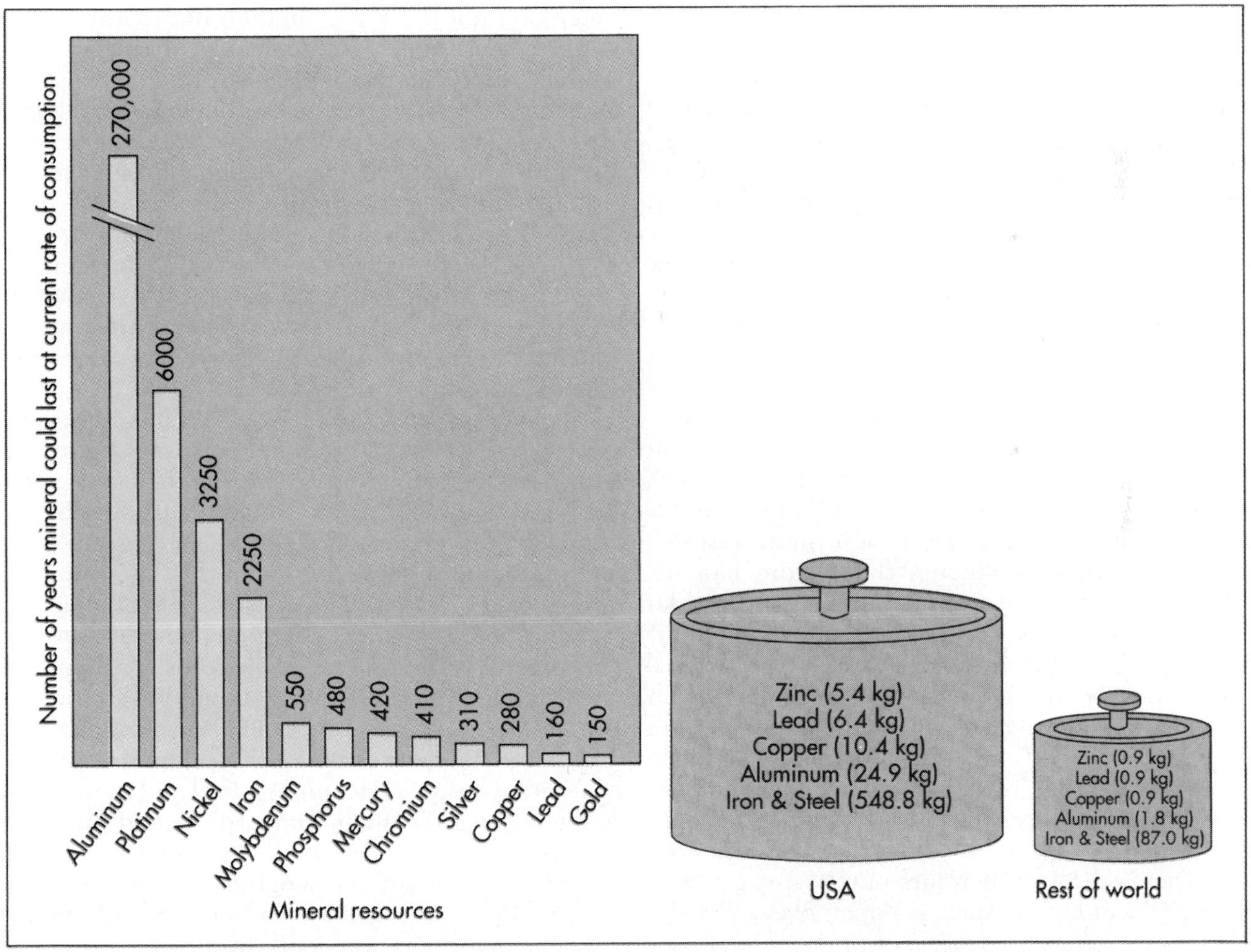

FIGURE 38-3 Consumption of mineral resources. According to *The Earth Report,* modern industrialized societies require large amounts of minerals to sustain their high standard of living. The graph on the left shows a number of major minerals and the years they could last at current rates of consumption. The per capita consumption of major minerals by the United States compared with the rest of the world is shown on the right.

A newer form of bioenergy is the use of plants such as corn and sugar cane to produce carbohydrates that are fermented, producing liquid fuels like ethanol. This technology, however, produces a product that appears to be an expensive alternative to fossil fuels. In addition, many cars do not function well on gasoline with ethanol additives. Critics say that food crops should be used to feed people, not produce energy (see "Just Wondering," Chapter 34).

Mineral Resources

Minerals are inorganic substances that occur naturally within the Earth's crust. Zinc, lead, copper, aluminum, and iron are the minerals that humans use in the greatest quantities. However, other minerals are also mined, such as gold, silver, copper, and mercury. These minerals are present in the Earth in fixed amounts; once they are used, they are gone. Figure 38-3 shows the number of years the world's reserves of various minerals will last based on the rate of current consumption.

What can be done to avert a mineral shortage in years to come? First of all, people can cut down on their consumption. An American uses, on average, six times more zinc, lead, and iron than does a person living outside the United States. In addition, an American uses 11 times more copper and 14 times more aluminum! Researchers agree that increased use of plastics and technology such as microelectronics will lower the demand for certain minerals. Plastics, however, are derived from petroleum products, so their manufacture increases consumption of fossil fuels.

Recycling is another way to help avert a mineral shortage in the coming years. Aluminum soda cans are produced from the claylike mineral bauxite. Producing cans from recycled aluminum uses 95% less energy than producing them from bauxite, saves mineral resources, and saves waste disposal costs and problems. Plastics are not easily recyclable and are not "truly" recyclable. That is, plastics cannot be refashioned into the product from which they were claimed; plastic milk jugs cannot be made into plastic milk jugs but must be made into park benches or parking lot curbs. Grocery store plastic bags are the closest truly recycled plastic products, being made from used plastic bags and other additives. Therefore purchasing goods in recyclable containers such as paper, glass, and aluminum *and recycling those containers* is a wise choice. Scrap metal and old automobiles are also recyclable, as are automobile tires and batteries.

Reuse is a third way to avert a mineral shortage, and it avoids the consumption of energy necessary to produce recycled items. Obviously, some items cannot be reused, but some can be put to the same or a different use: a plastic bread bag can be used to carry your lunch to work, glass juice bottles can be used as a container for a variety of liquids or other items, and paper printed on one side only can be cut up and used as notepaper. Many of the items we use regularly can be put to creative reuse.

> *Recycling and reusing paper, glass, and various metals will help preserve mineral resources as well as other natural resources.*

Deforestation

Paper makes up 40% of the garbage in the typical American household. Recycling just your daily newspaper will save four trees every year. Buying products made from recycled paper will save even more. But recycling is not enough. The forests of the world are in severe crisis and will be lost if steps are not taken to stop their destruction.

The most severe crisis is that of **tropical rain forest deforestation.** Tropical rain forests (see Chapter 37) are located in Central and South America, tropical Asia, and central Africa, forming a belt around the equatorial "waist" of the Earth. Although this belt of forest covers only 2% of the Earth's surface, it is home to more than *half* the world's species of plants, animals, and insects. These organisms contribute 25% of medicines, along with fuel wood, charcoal, oils, and nuts. In addition, the tropical rain forests play an important role in the world climate.

Population and poverty are both high in rain forest countries. People with few resources move from towns and cities to the rain forest and cut the trees for sale as lumber or burn them to clear a patch of land to grow crops and raise cattle to sustain themselves and their families (Figure 38-4). Commercial ranchers also cut and burn the forests to make way for pastureland to feed beef-producing cattle. Unfortunately, the soil of the rain forests is poor, with few nutrients, and does not support crops. Before it is cut, the forest sustains itself because of mutualistic relationships between the trees and microorganisms that quickly decompose dead and dying material on the forest floor. These "processed" nutrients are quickly reabsorbed by the tree roots. Few nutrients stay in the soil; most of the nutrients are in the vegetation. Cutting these trees down and burning them release the nutrients from the trees. Crops grow poorly on this land after it is stripped. After a year or two, crops will not grow at all. The people move on, cutting yet another portion of the forest.

Commercial logging also takes its toll on the tropical rain forests. Many of the trees are cut to

FIGURE 38-4 Deforestation: a desperate effort to stave off poverty results in an environmental disaster. These farmers live near the Andasibe reserve in Madagascar, an island where the per capita income is less than $250 a year. Clearing the rain forest allows these impoverished farmers to plant rice and graze cattle. But the environmental price may be too steep. In Madagascar alone, 80% of the rain forest has been destroyed.

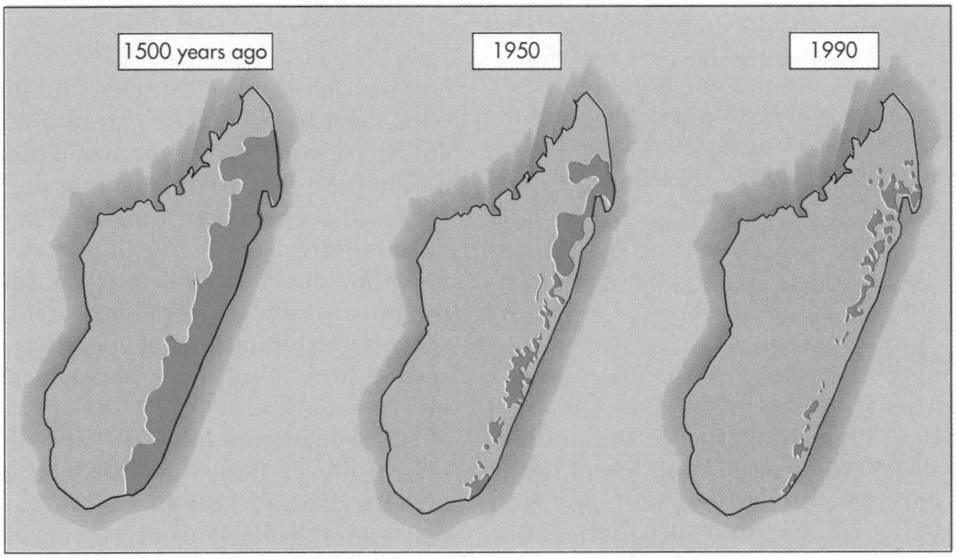

FIGURE 38-5 Destruction of the rain forests in eastern Madagascar. These maps show the progressive deforestation that has taken place in Madagascar. Areas of rain forest are shown in green.

supply fuel wood, paper, wood panels such as plywood, and charcoal and to supply furniture manufacturers with mahogany and other woods demanded by consumers around the world. At this time, the tropical forest is being slashed and burned at a devastating rate. By 1950, two thirds of the forests of Central America had been cleared. Madagascar, an island off the southwestern coast of Africa, had lost about half its rain forest by 1950, but as Figure 38-5 shows, has lost half again since then! The United Nations Food and Agriculture Organization estimates that 100,000 square kilometers (62,000 square miles) of rain forest are now being lost each year worldwide—about the size of New England. At the present rate, scientists estimate, nearly all the tropical rain forests will be gone—including their rich diversity of animal life—within the next 50 years.

In addition to losing a rich natural resource, many scientists agree that the burning of the tropical forests is adding tremendous quantities of carbon dioxide (CO_2) to the air. CO_2 acts like the glass in a greenhouse (or the windows in your car), allowing heat to enter the Earth's atmosphere but

tropical rain forest deforestation the loss of the plants and animals of the tropical rain forest (a biome that occurs in regions of high temperature and rainfall) through cut-and-burn agriculture and logging.

preventing it from leaving. This selective energy absorption by CO_2 in the atmosphere is called the **greenhouse effect.** Rising levels of atmospheric CO_2 may therefore result in a worldwide temperature increase, a situation called **global warming.** Global warming could melt the arctic and antarctic ice packs, causing a rise in the sea level and the flooding of one fifth of the world's land area. In addition, a rise in global temperatures could affect rain patterns and agricultural lands.

To help stop rain forest destruction, individuals can join and support organizations involved in rain forest conservation, as well as avoid buying furniture constructed from tropical hardwoods. Research programs are currently being conducted that are exploring the concept of conserving forest "fragments" as reserves that will provide suitable habitats for species (Figure 38-6).

FIGURE 38-6 A rain forest reserve. In 1979, the World Wildlife Fund established a research program—the Biological Dynamics of Forest Fragments Project. The program isolates "islands" of rain forests out of the larger forest during conversion to pastureland. Scientists can study these islands to determine what their optimal size and shape should be if they are to provide a suitable habitat for various species.

The world's forests are being destroyed at an alarming rate, resulting in the loss not only of this precious resource but of the habitats of thousands of species of organisms. In addition, the burning of the forests adds tremendous quantities of carbon dioxide to the air, a situation that may result in a worldwide temperature increase, or global warming.

Species Extinction

Although scientists have classified approximately 1,700,000 species of organisms, they estimate that approximately 40,000,000 exist. But many of these species are becoming extinct—dying out—and will never be seen again. An estimated 1000 extinctions are taking place each year, a number that translates into more than two species per day. Although such mass extinctions have occurred in the past, as with the dying out of the dinosaurs, these former extinctions were caused by climatic and geophysical factors. The extinctions today are caused by human activity.

One way in which humans destroy species is by destroying their habitats. Widespread habitat destruction is taking place as the tropical rain forests are cut down. This one factor alone will cause the extinction of one third of the world's species.

Humans are also destroying coral reefs at an alarming rate. Because the base of the coral reef food chain is algae, the reef ecosystem depends on sunlight for its existence. In areas where soil erosion muddies the water, the reef dies from insufficient sunlight. In some cases the water may become polluted with fertilizer runoff or with sewage, which causes the algae to overgrow, smothering the corals. In some cases, humans use dynamite to kill and harvest fish, a practice that obviously destroys the coral reef. In addition, coral is often harvested to sell to tourists.

Although habitat destruction and pollution are the most serious threats to the existence of certain species, exploitation of commercially valuable species threatens them as well. The Convention on International Trade in Endangered Species (CITES) regulates trade in live wildlife and products. However, illegal trade still takes place because of smuggling and the inability of law-enforcement inspectors to check all shipments of goods. Certain types of alligator, crocodile, sea turtle, snake, and lizard trade are illegal and endanger the existence of various species. Buying other products such as coral and ivory endangers the existence of coral reefs and African elephants. Even certain plants such as cacti and succulents should be purchased only if cultivated in greenhouses; they should not be taken from the wild. Organizations such as the World Wildlife Fund are invaluable sources of information regarding which products should be avoided by consumers so that they can help stop trade that threatens certain species.

Humans are destroying species by destroying their habitats, polluting their habitats, and illegally trading endangered species.

The extinction of many species leads to a reduction in **biological diversity,** a loss of richness of species. This loss is certainly regrettable; each species is not only the result of millions of years of evolution and can *never* be reproduced once it is gone but also is an intricate part of the interwoven relationships among organisms within the ecosystems of the world. The extinction of just one species can affect ecosystems in many unforeseen ways. In addition, many people feel that humans have no right to influence the world in such a devastating way. And looking at the issue from a utilitarian standpoint, humans rely on the other species of the world as sources of food, medicine, and other substances.

Recently, for example, the genetic diversity in crop plants has declined as scientists have used selective breeding techniques to produce plants with specific characteristics, such as resistance to particular diseases or pests, hardiness, and other characteristics considered important. However, when only a few species or varieties of a species are cultivated or survive, the genetic diversity of the organism declines. Populations of species that have little diversity are more vulnerable to being wiped out by new diseases or climatic changes.

In the 1960s, the United Nations' Food and Agriculture Organization (FAO) made recommendations that have led to the establishment of "banks" that store plant seeds and genetic material, or germ plasm (Figure 38-7). Zoos around the world are making similar efforts to preserve genetic diversity among animals by establishing and carrying out sophisticated breeding programs to increase genetic diversity among endangered species (Figure 38-8).

FIGURE 38-7 A germ plasm bank at the United States Department of Agriculture. Researchers are in a cold storage room that contains seeds from around the world.

Solid Waste

The average American produces about 19 pounds of garbage and trash—referred to as **solid waste**—per week. If that does not sound like very much, multiply that number by 52, and you will realize that each person produces approximately a *half ton* of solid waste per year. In 1992, the population of the United States was over 250 million—a population that produced approximately 125 million tons of waste per year, not including the waste produced by schools, stores, and manufacturing. Americans, often described as having a "disposable society," are beginning to realize that much of the wastes produced should be viewed as resources that must be reclaimed rather than as substances to throw away and clog the landscapes.

What happens to your trash when you put it out for collection? The burial site for your throwaways is the **sanitary landfill,** an enormous depression in the ground where trash and garbage is dumped, compacted, and then covered with dirt (Figure 38-9). In 1983, the Federal Resource Conservation and Recovery Act forced the closing of all *open dumps* or required them to be converted to landfill sites. At an open dump, solid waste is heaped on the ground, periodically burned, and left uncovered. Landfills are considered superior to dumps because landfill wastes are covered, reducing the number of flying insects and rodents that are attracted to the site and reducing the odor produced by open, rotting organic material. In addition, wastes are not burned at landfills, decreasing the

greenhouse effect The blocking of outward heat radiation from the Earth by carbon dioxide in the atmosphere.

global warming a worldwide temperature increase that could result from the greenhouse effect.

biological diversity the richness and variety of species on Earth.

solid waste garbage and trash.

sanitary landfill an enormous depression in the ground where trash and garbage are dumped, compacted, and then covered with dirt.

FIGURE 38-8 Zoos are now responsible for preserving genetic diversity in endangered species. These zookeepers are responsible for keeping these elephants healthy for possible breeding. The efforts of zoos around the world will be instrumental in saving some endangered species from extinction.

FIGURE 38-9 Sanitary landfill, New York City. Sanitary landfills around the country are filling up rapidly. Communities must quickly find new ways to dispose of their trash. One of the most popular and efficient methods is recycling.

problem of air pollution. Further, when the capacity of a landfill site is reached, it may be used as a building site or recreational area. Examples of landfill reuse are Mount Trashmore recreational complex in Evanston, Illinois, and Mile High Stadium in Denver, Colorado.

Problems do exist with landfills, however. First of all, space is running out! In just one year, the population of New York City alone will produce enough trash to cover more than 700 acres of land 10 feet deep. Another problem with landfills is that liquid waste can trickle down through a landfill, reaching and contaminating groundwater below. Liquids leaching from landfills can also pollute nearby streams, lakes, and wells (therefore *never* put batteries, paint solvents, drain cleaners, and pesticides in with your trash). In addition, as the organic material compacted in landfills is decomposed in the absence of oxygen, methane gas is produced. This highly explosive gas rises from landfills and can seep into buildings constructed on or near reclaimed sites.

To reduce solid waste and landfill problems, everyone should recycle paper, glass, aluminum cans, and even clothing. An important part of recycling is buying recycled goods so that a market is maintained for them. In addition, people should purchase products in recyclable containers, avoid purchasing "overpackaged" products, reuse products when feasible, and compost yard waste. Composting involves piling (and periodically turning to aerate) grass clippings, wood shavings, and similar yard wastes. Bacteria will degrade these substances, and they can be used to fertilize flower

JUST WONDERING. . . .

Did scientists learn anything from Biosphere 2 that will help us manage the planet better?

Scientists are still learning from Biosphere 2 and will continue to learn—if interest and funding continue—for about 100 years. The first set of eight "bionauts" lived in the 3-acre, glass-enclosed, self-sustaining ecosystem in the Arizona desert for 2 years (fall 1991 to fall 1993). Another crew entered in 1994 for a 10-month mission, but this time outside scientists were welcome to join the bionauts for stays from a few days to a few months.

Named after Biosphere 1 (the Earth), Biosphere 2 comprises various biomes such as an ocean, a rain forest, and a desert. Providing human life support and self-supporting ecosystems, Biosphere 2 has a generalized goal: to develop new insights into the workings of our planet. Biosphere's first crew set the stage for crews and experiments to come. Their accomplishments included recycling all their waste and water, producing 80% of their food, and maintaining the various biomes and ecosystems.

Many scientists criticize the lack of "hard science" undertaken during those first 2 years, whereas others suggest that the first bionauts did all they could while facing the challenge of surviving and making Biosphere 2 work as the largest, closed life support system ever built. In addition to those awesome tasks, the bionauts were plagued by problems: 2 years of unusually cloudy weather, a low amount of oxygen and high amount of carbon dioxide, and pest invasions of cockroaches and black ants that were accidentally introduced into the enclosure.

Thus, the first years of the Biosphere 2 project may not have taught us many lessons about Biosphere 1, but the experience left no doubt about how difficult it is to build a substitute that works. We will all have to look to future work in Biosphere 2 for advancements in knowledge of ecosystem function and the development of strategies to restore dying ecosystems. Perhaps not-yet-formulated research questions will be asked and then answered, because a strength of the long-term design of Biosphere 2 is the wide scope it offers for potential investigations.

beds and vegetable gardens. By composting, you will be eliminating the second largest waste by volume in landfills.

In the future, careful planning will be necessary to ensure that landfill sites are located away from streams, lakes, and wells and that they have proper drainage and venting systems. Most likely, landfills will be only a portion of a solid waste disposal system that incorporates recycling, safe incineration, and waste-to-energy reclamation.

> *Solid waste is disposed of in sanitary landfills, depressions in the ground where trash and garbage is dumped, compacted, and covered with dirt. Solid waste management could be improved with increased and better-organized recycling efforts, safe incineration, better-engineered landfill sites, and waste-to-energy reclamation.*

FIGURE 38-10 A toxic chemical dump in northern New Jersey. Toxic chemical dumps are serious threats to groundwater and surface water. Pollution occurs when the drums rust through and release their contents, which then enter the surface water and may eventually percolate down to the groundwater.

The Water

The water, or hydrosphere, of this planet lies mainly in the oceans but is also found in freshwater lakes and ponds, in the atmosphere as water vapor, and as subsurface reservoirs called *groundwater.* Heated by the sun, water continually cycles from the land to the air, condenses, and falls back to the Earth (see Chapter 36). As this cycling of water repeats continually, contaminants may mix with the water, *polluting* it—causing physical or chemical changes in the water that harm living and nonliving things.

Surface Water Pollution

Surface water can be polluted by factories, power plants, and sewage treatment plants that dump waste chemicals, heated water, or human sewage into a lake, stream, or river. These sources of pollutants are called *point sources* because they enter the water at one or a few distinct places. Other types of pollutants may enter surface water at a variety of places and are called *nonpoint sources.* Examples of nonpoint sources of pollution are (1) sediments in land runoff caused by erosion from poor agricultural practices (a major type of water pollution); (2) metals and acids draining from mines; (3) poisons leaching from hazardous waste dumps (Figure 38-10); and (4) pesticides, herbicides, and fertilizers washing into surface waters after a rain. These pollutants affect aquatic organisms and also affect terrestrial organisms that drink the water. The manner in which a pollutant affects living things depends on its type: nutrient, infectious agent, toxin (poison), sediment, or thermal pollutant.

Organic nutrients are sometimes discharged into rivers or streams by sewage treatment plants, paper mills, and meat-packing plants. These "organics" are food for bacteria. If high amounts of organic nutrients are available to bacteria, their populations will grow exponentially. As they grow and reproduce, they use oxygen—oxygen that fish need. Therefore as the bacterial populations rise, the only organisms that survive are those that can live on little oxygen. So-called "trash fish" such as carp can outsurvive other species such as trout and bass, but if oxygen levels become extremely low, all the fish die, survived only by various worms and insects.

The accumulation of *inorganic nutrients* in a lake is called **eutrophication,** meaning "good feeding." Certain inorganics such as nitrogen and phosphorus, which come from croplands or laundry detergents, stimulate plant growth. Although heavy plant growth makes swimming, fishing, or boating difficult, it does not cause most of its problems until the autumn (in most regions of the United States), when the plants die. At that time, bacteria decompose the dead plant material, and problems similar to those of organic nutrient pollution arise. In addition, the decomposed materials begin to fill the bottom of the lake. Eventually, the lake may become transformed into a marsh and then into a terrestrial community by an accelerated process of succession (see Chapter 35). *Sediment* that flows into lakes from erosion of the land caused by certain agricultural practices, mining, and road construction also fills in lakes and hastens natural succession.

An example of eutrophication that began leading to the successional "death" of a lake occurred in Lake Washington near Seattle. The problem stemmed from local communities dumping their wastes into the lake. In 1968 the dumping was halted, and Lake Washington is now fully recovered.

Surface waters are rarely polluted with *infectious agents*, or disease-causing microbes, in the United States. However, in areas such as Africa, Asia, and Latin America, waterborne diseases are common. Surface waters become polluted from untreated human wastes and from animal wastes, causing diseases such as hepatitis, polio, amebic dysentery, and cholera.

An array of *toxic substances* pollutes surface waters worldwide. Toxic substances include both organic compounds such as PCBs (polychlorinated biphenyls) and phenols and inorganic substances such as metals, acids, and salts. These toxic, or poisonous, substances come from a diverse array of sources such as industrial discharge, mining, air pollution, soil erosion, old lead pipes, and many natural sources. The effects on humans from drinking these substances in water range from numbness, deafness, vision problems, and digestive problems to the development of cancers.

Unfortunately, most toxic pollutants do not *degrade,* or break down, and are therefore present in bottom sediments of surface waters for decades or more. In fact, some organisms accumulate certain chemicals (often deadly ones) within their bodies, a process called **biological concentration.** Oysters, for example, accumulate heavy metals such as mercury; thus these organisms might be highly toxic when living in waters with relatively low concentrations of this metal. Also, as organisms higher on the food chain eat organisms lower on the food chain, toxins accumulate in the predators in a high concentration, a concept called **biological magnification.** As discussed in Chapter 36, the relationships between trophic levels of food chains are pyramidal, so a smaller number of organisms eat a larger number of organisms as you progress "up" the chain. (Figure 38-11 illustrates this concept.) Scientists estimate that the concentration of a toxin in polluted water may be magnified from 75,000 to 150,000 times in humans who consume tainted fish.

The electric power industry and various other industries such as steel mills, refineries, and paper mills use river water for cooling purposes, discharging the subsequently heated water back into the river (Figure 38-12). Small levels of **thermal pollution** do not cause serious problems in aquatic ecosystems, but sudden, large temperature changes kill heat-intolerant plants and animals. Interestingly, ecosystems adjust to artificially heated waters and are damaged if the heat source is shut down, as when a power plant closes.

> *Surface water can be polluted by waste chemicals, heated water, or human sewage put into the water by factories, power plants, and sewage treatment plants or by silt or chemicals that can leach into the water from surrounding sites. These substances all affect the aquatic ecosystem in different ways.*

Groundwater Pollution

The surface water and rainwater that trickles through the soil to underground reservoirs is groundwater. This water can be contaminated with some of the same substances as surface water. The primary groundwater contaminants are toxic chemicals that seep into the ground from hazardous waste dump sites and chemicals such as pesticides used in agriculture. Although groundwater does get filtered and "cleansed" of some substances as it trickles through the soil, toxic chemicals consist of molecules too small to be filtered in this manner. At this time, prevention of contamination is the cheapest and most feasible way to end groundwater contamination. Pumping contaminated water from underground sources to the surface, purifying it, and returning it to the ground is extremely costly.

Acid Rain

In recent years, the water in the atmosphere has become polluted with sulfur dioxide and nitrogen dioxide, two chemical compounds that form acids

eutrophication (yoo **trowf** uh **kay** shun) the accumulation of inorganic nutrients in a lake, which stimulate plant growth. When the plants die, the decomposition process takes oxygen from the water, leading to the death of other organisms.

biological concentration a process by which some organisms accumulate certain harmful or deadly chemicals within their bodies that are present in their environments or in the food they eat.

biological magnification a process by which toxins accumulate in organisms in high concentrations when they consume tainted organisms lower on the food chain. This effect increases with progression up the food chain.

thermal pollution a change in water temperature that occurs when industries release heated water back into rivers after using the water for cooling purposes.

acid rain precipitation having a low ph (usually between 3.5 and 5.5) produced when sulfur dioxide and nitrogen dioxide pollutants combine with water in the atmosphere.

FIGURE 38-11 Biological magnification. The amount of DDT in an organism increases as you go up the food chain. This illustration depicts an 8 million-fold amplification of the DDT.

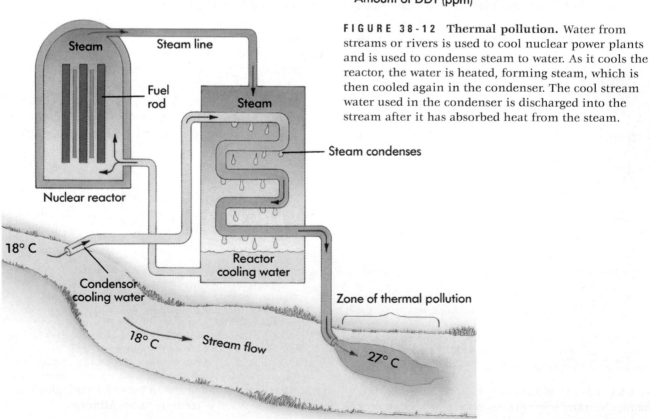

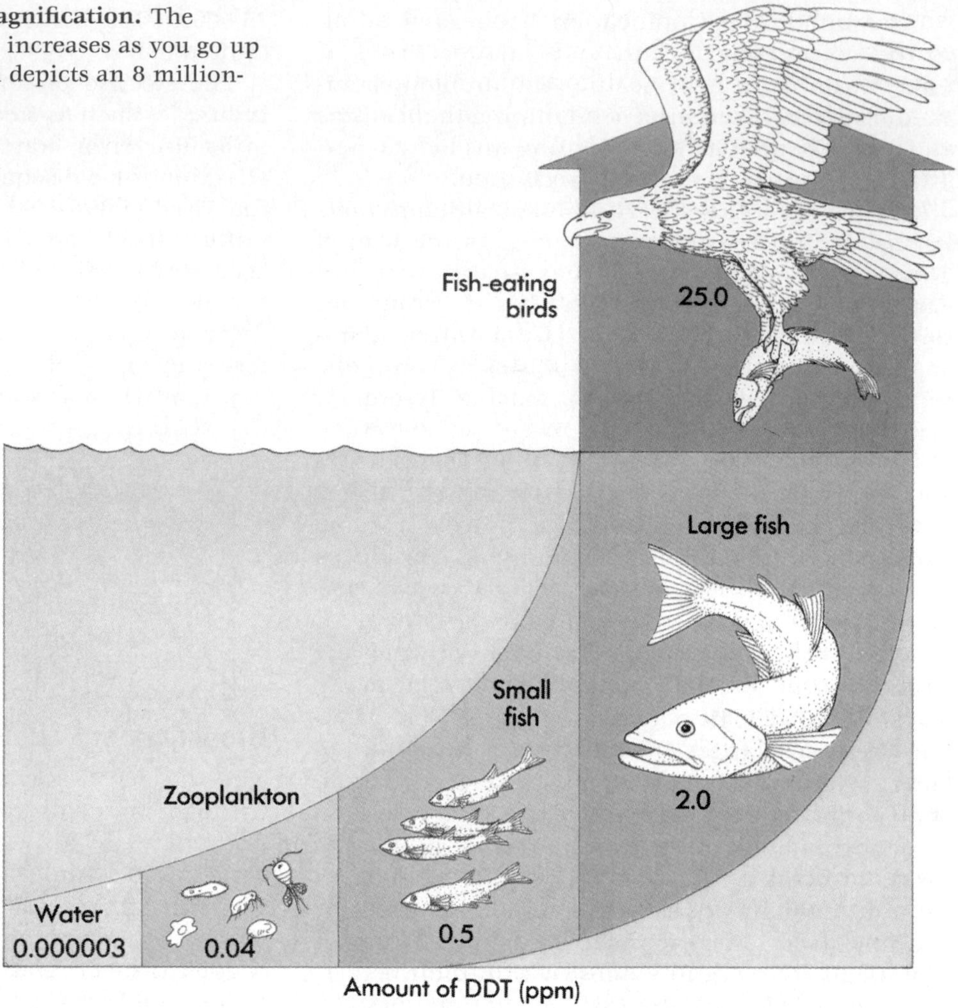

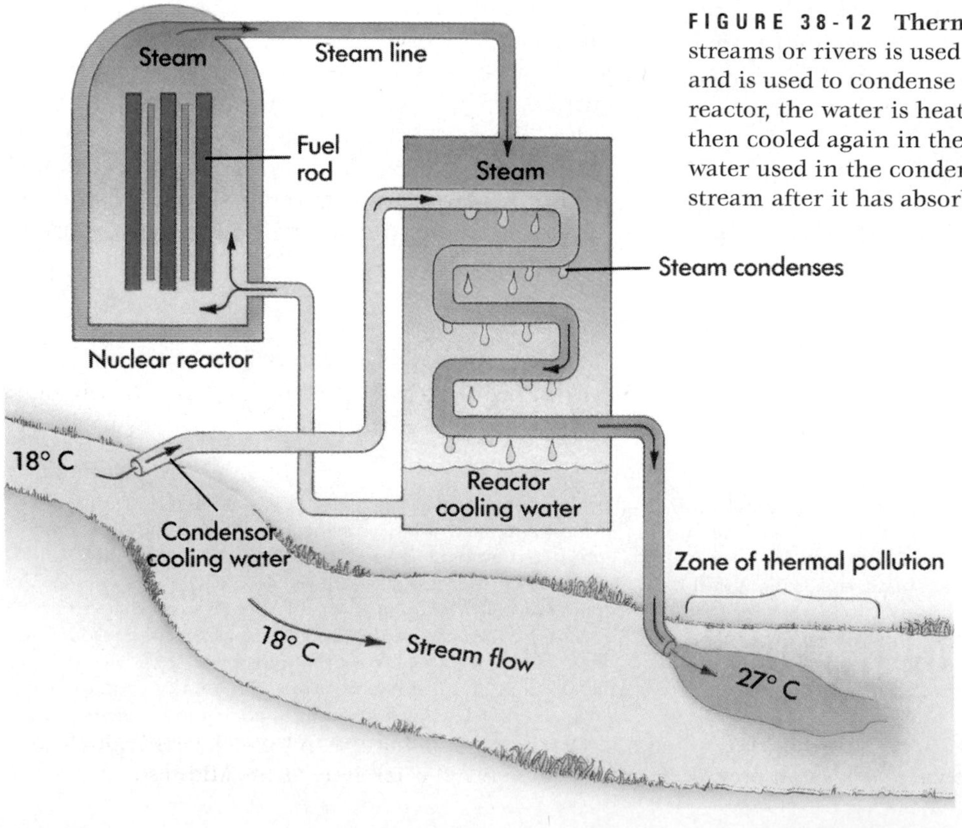

FIGURE 38-12 Thermal pollution. Water from streams or rivers is used to cool nuclear power plants and is used to condense steam to water. As it cools the reactor, the water is heated, forming steam, which is then cooled again in the condenser. The cool stream water used in the condenser is discharged into the stream after it has absorbed heat from the steam.

when combined with water. As this water vapor condenses and falls to the ground, it is commonly referred to as **acid rain,** although acid precipitation also falls as snow or as dry "micro" particles, mixing with water when it reaches surfaces on the ground. "Normal" rain has a pH of approximately 5.7, primarily because of dissolved carbon dioxide (see Chapter 2 for a discussion of pH). The pH of acid rain is lower than 5.7 and usually falls between 3.5 and 5.5. However, rainfall samples taken in the eastern United States have measured as low as 1.5—a pH lower than that of lemon juice and approaching that of battery acid!

Acid rain results in many devastating effects on the environment. As it mixes with surface water, it acidifies lakes and streams, killing fish and other aquatic life. It seeps into groundwater, causing heavy metals to leach out of the soil. The result is that these heavy metals enter the groundwater and surface water, posing health problems for humans as well as fish. Acid rain also eats away stone buildings, monuments, and also metal and painted surfaces (Figure 38-13).

The effect of acid rain falling on plant life has been hypothesized but undocumented until recently. Botanists realized that acid rain has not only leached many of the minerals essential to plant growth from the soil but also has liberated toxic minerals such as aluminum. Recent research suggests that acid rain also kills or damages microorganisms living in symbiotic associations with forest trees, helping them extract water and needed minerals from the soil. Without these organisms, the trees die (Figure 38-14).

Although sulfur and nitrous oxides are produced naturally during volcanic eruptions and forest fires, humans produce more than half these chemicals from the burning of coal by electricity-generating plants, industrial boilers, and large smelters that obtain metals from ores. In addition, nitrogen oxides are emitted by cars and trucks. The situation also becomes complicated because countries pol-

FIGURE 38-13 Acid rain damage. This statue has been eroded by acid precipitation.

FIGURE 38-14 Effects of acid rain. These balsam fir trees in North Carolina have been killed by acid rain. The chemicals in the acid rain, which are carried on prevailing winds, come from as far away as the Midwest.

lute their own air as well as the air of other countries. Emissions produced in the Midwest and eastern portions of the United States affect not only those areas, for example, but are carried by the wind into Canada. Emissions produced in England move into the Scandinavian countries.

> *Sulfur dioxide and nitrogen dioxide, gases emitted primarily by coal-fired power plants, combine with water in the atmosphere to produce acid rain, a precipitation that harms both living and nonliving things.*

Solving the problem of acid rain and curtailing these emissions is not easy or inexpensive. One technique used to reduce sulfur dioxide is to put *scrubbers* on coal-burning power plants. This technology can remove up to 95% of the sulfur dioxide emissions produced. Unfortunately, the United States lags behind most sulfur-emitting countries of the world in implementing this technology. A solution to this problem depends on international agreements to reduce emissions, to employ energy-conservation measures, and to increase the use of public transportation.

The Atmosphere

The Earth's atmosphere actually extends much higher than the portion within the biosphere, a part of the atmosphere more technically called the *troposphere.* The troposphere extends approximately 11 kilometers (36,000 feet) into the atmosphere but slopes downward toward the poles and upward toward the equator. The word *troposphere* literally means "turning over," a name extremely descriptive of the atmosphere. As the Earth is heated by the sun, air rises from its surface, cooling as it ascends. Cooler air then falls, resulting in a constant turnover of the air, aided by the prevailing winds.

Air Pollution

About 99% of the clouds, dust, and other substances in the atmosphere are located in the troposphere. Some of these "other substances" are nitrogen and sulfur oxide pollutants. Other major air pollutants are carbon monoxide, hydrocarbons, and tiny particles, or particulates. As Figure 38-15 shows, sulfur oxides and particulates are produced primarily by the burning of coal in electricity-generating plants. Carbon monoxide and hydrocarbons are emitted primarily by cars, buses, and trucks; all these sources spew out nitrogen oxides.

The type of air pollution in a city depends not only on which of the pollutants are in the air but on the climate of the city. *Gray-air cities,* such as New York, Philadelphia, and Pittsburgh, are located in relatively cold but moist climates and have an abundance of sulfur oxides and particulates in the air. The haze, or **smog,** that can be seen in the air is the result of the burning of fossil fuels in power plants, other industries, and homes. *Brown-air cities,* such as Los Angeles, Denver, and Albuquerque, have an abundance of hydrocarbons and nitrogen oxides in the air. In these sun-drenched cities the hydrocarbons and nitrogen oxides undergo photochemical reactions that produce "new" pollutants called *secondary pollutants.* The principal secondary pollutant formed is **ozone** (**O_3**), a chemical that is extremely irritating to the eyes and the upper respiratory tract. Smog caused by pollutants reacting in the presence of sunlight is called **photochemical smog** (Figure 38-16).

The solutions to cutting down on smog are the same solutions to problems of acid rain formation and fossil fuel depletion: the use of coal-fired power plant scrubbers, energy conservation, and the recycling of resources leading to a reduction in energy use for manufacturing.

> *The primary type of air pollution in the first 11 kilometers of the atmosphere consists of smoke and fog, or smog. Smog is caused by sulfur oxides, nitrogen oxides, hydrocarbons, and particulates in the air as a result of the combustion of fossil fuels.*

Ozone Depletion

Ironically, humans are producing ozone in the troposphere that is polluting the environment but destroying it in the *stratosphere* where it is needed. The stratosphere is the layer of the atmosphere directly above the biosphere. It contains a layer of ozone that is formed when sunlight reacts with oxygen. Although ozone is harmful in air that is breathed, it is helpful in the stratosphere, acting as a shield against the sun's powerful ultraviolet (UV) rays. Excess exposure to UV rays can cause serious burns, cataracts (an opacity of the lens of the eye), and skin cancers and can harm or kill bacteria and plants.

Scientists have measured "holes" in the ozone layer where it is thinnest, over the polar ice caps. The chemical chlorofluorocarbon (CFC) is the main culprit. The spray-can propellant and refrigerant used in air conditioners, Freon, is a CFC. In addition, CFCs are used in the manufacture of Sty-

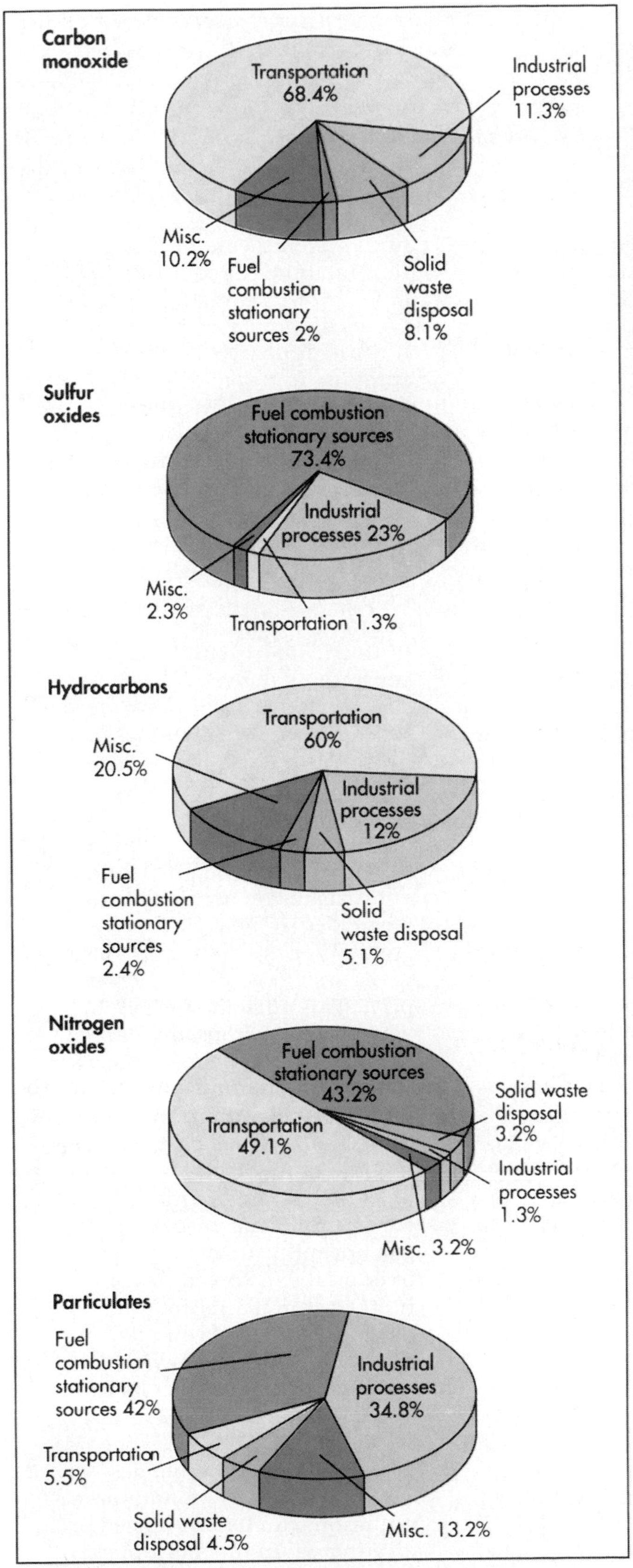

FIGURE 38-15 Major air pollutants and their sources. Transportation and fuel combustion at stationary sources are the main contributors to pollution.

FIGURE 38-16 Air pollution. A, "Brown air" in Los Angeles. Sunlight reacts with the chemicals spewed from automobiles to create ozone, which in the upper atmosphere protects you from the sun's harmful rays but in the lower atmosphere is poisonous. **B,** "Gray air" in New York City is caused by fossil fuel pollutants from power plants.

rofoam and foam insulation. To protect the ozone layer, people should have their air conditioners serviced by persons who use equipment that does not allow the escape of Freon. In addition, they should not use aerosol products containing CFC and should avoid the use of Styrofoam products.

smog a type of air pollution that results from the burning of fossil fuels.
ozone (O_3) (oh zone) a chemical air pollutant formed as a secondary pollutant when hydrocarbons and nitrogen oxides undergo photochemical reactions; it is extremely irritating to the eyes and the respiratory tract.
photochemical smog a type of air pollution caused by reactions between hydrocarbons and nitrogen oxides taking place in the presence of sunlight.

Applications

HOW *Science* WORKS

Can Technology Save Us from Ourselves?

As the human population grows, the amount of resources available to sustain our population shrinks. In addition, humans create waste and pollute the environment. One result of these increases in population, pollution, and wastes and the decrease in natural resources is that global ecological stability becomes difficult to maintain. Are we on the brink of disaster? Some think the picture may not be as gloomy as it looks. Humans have made astonishing technological advances in a variety of areas while their numbers were "exploding." Can technology be an answer to our problems? Can technology help avert an ecological crisis resulting from the human population explosion and all its ramifications?

In recent years, advances in gene technology have allowed scientists to genetically alter bacteria, producing new strains that can break down hazardous pollutants such as oil from coastal oil spills. Other bacteria have been "gene spliced" to consume potentially harmful medical refuse such as waste blood. Although solutions like these are important in that they alleviate serious ecological damage that would otherwise occur, they respond to immediate problems only. To ensure a bright ecological future for all organisms living on the Earth, scientists must develop structured strategies that allow for economic development and ensure *sustainability* of the environment. To approach sustainability, we must meet human needs without continuing to degrade our environment and diminish our resource base, while concurrently using technologies that help our human ecosystem imitate biological ecosystems.

In biological ecosystems, plants produce their own food using carbon dioxide from the atmosphere, nutrients from the soil, and energy from the sun. This food allows plants to grow but also provides food for plant-eating organisms, the herbivores. Herbivore populations grow and, in turn, feed meat-eating organisms, the carnivores. Wastes from these organisms and their remains cycle through yet another group of organisms, the decomposers (fungi and bacteria), serving eventually as nutrients for plants.

Learning from this process of cycling by-products and degraded materials in natural systems, individuals and organizations are finding ways to reduce waste products or reclaim them. For example, one industrial manufacturing leader recovers hydrochloric acid, a hazardous waste product of their manufacturing process, which was formerly discarded. Now this company either reuses the acid or sells it to others who use it. Another company, which disposed of its used (and hazardous) alumina catalysts in the past, now sells them to cement makers for use in their manufacturing processes. And fly ash, previously a problematic waste product for energy companies, is being used to build roads.

Similarly in agriculture, with the help of new technologies and the resurgence of ancient methods, wastes are being minimized or reclaimed, providing more ecologically sound agricultural systems. One new irrigation method, called *trickle irrigation,* involves carrying water to plants by tubing and delivering it through nozzles rather than spraying it into the air to fall on the plants. In standard irrigation systems, nearly two thirds of the water sprayed into the air is lost by evaporation; trickle irrigation decreases this amount dramatically. In addition, as water evaporates through standard methods, the salts dissolved in the water become concentrated. With trickle irrigation, salt concentrations increase only slightly and the soil suffers less salt buildup. An additional technology also helps trickle irrigation systems trap runoff water and reuse it, thereby further reducing water use and reducing the pollution of nearby lakes and streams by herbicides and pesticides in the runoff.

Another agricultural management system is multiple cropping. This ancient method, in which several crops are planted together, results in a decrease in the growth of crop pests such as weeds, insects, and fungi because the multiple-crop environment is more diverse than that of a single-crop environment. For example, if a particular fungus normally attacks one type of crop, it can spread less easily if other crop plants are growing between the fungus and its host. This type of diverse environment also allows predators from one crop to inhibit pests from another crop. Thus, even the use of pesticides and herbicides, sources of serious groundwater pollution, can be minimized or eliminated. In one multiple cropping system in the tropics, family gardens are created by collecting soil from unprofitable

Applications—cont'd

swamp land and building raised gardens on which several crops are grown. While the garden is established in such a way as to recycle nutrients from the garden, it also makes previously undervalued swamp forests more valuable, decreasing the possibility that these forests will be destroyed.

In our households, many Americans have focused on reclaiming and recycling materials to prevent the by-products of our daily lives from being discarded as waste products. Aluminum recycling not only maintains the use of the aluminum involved but also conserves nearly 20 times the energy it would cost to manufacture the same amount of aluminum from ore. Similarly, recycling plastic bottles and jugs into other plastic products such as polyester fiber and synthetic lumber substantially reduces the cost of fuel to create the products from "scratch" and conserves natural resources.

A sustainable society is indeed a future possibility if we can integrate the ecosystem concept into our farming, industrial, business, and home activities, recycling by-products and spent products repeatedly through our human ecosystem. There are two major challenges to actualizing such a system, however. The first challenge is developing more technologies that can be based on the use of by-products and reused products. The second challenge is more difficult and is where the analogy of biological ecosystems to human-centered ecosystems breaks down. The source of energy for natural ecosystems is almost entirely from solar energy, whereas the source of energy for human-centered ecosystems arises mostly from fossil fuels such as coal and oil—fuels that will someday be depleted. Until this problem is rectified, the maintenance of a sustainable human ecosystem will not be possible. And even then—as we approach the "biological model" of a human ecosystem—can the Earth support a continually growing human population? Although recycling and reuse technologies, as well as the use of resources that are infinite such as solar power, may help increase our carrying capacity, we cannot raise the carrying capacity of the Earth past the levels of the amount of food we can produce and the technologies we can provide amid the space that is necessary for humans to enjoy civilized, healthy lives

Overpopulation and Environmental Problems

Some scientists hold that the enormous world population is the key to the problems of pollution and diminishing natural resources. Others state that *technology,* the application of science to industrial use, is the culprit. Most, however, would agree that neither factor alone is the cause of these complex problems. Many factors interact to affect population size and the "health" of the environment.

The size of the human population influences the environment because it puts a demand on resources. However, the social, economic, and technological development of a country affects the demand its population places on resources. A person living in a rural area of Kenya, for example, does not use the same amount and kind of environmental resources as a person living in a large city in the United States. The lifestyle and per capita consumption of a population make a big difference on the impact of that population on the environment.

Other factors regarding impact of a population on the environment is how a population uses its resources and the effects of that use. Using resources and disposing of wastes in certain ways pollute the environment. The pollution of one aspect of the environment (such as the air) can lead to pollution of other aspects of the environment (such as the water through acid rain). Pollution, in turn, can limit population size by increasing the death rate. However, living with pollution can lead to attitudinal changes among members of a population, which may result in the development of laws to better manage the use of resources and to curb the pollution resulting from their use.

The impact of the human population on the environment is extremely complex and has multiple causes and effects. Many factors, such as population size, per capita consumption, technology, and politics, interact in complex ways, resulting not only in the problems faced today but in solutions for the future. Through research, environmental scientists will help everyone understand how each person can live on this planet without harming it and everyone's future existence.

Summary

- All life on Earth exists in the biosphere. The biosphere is the interface of the land, air, and water, extending from approximately 9 kilometers (30,000 feet) above sea level to 11 kilometers (36,000 feet) below.
- Many of the natural resources of the Earth are finite and nonrenewable; that is, they cannot be replaced. Coal, oil, and natural gas are fossil fuels, nonrenewable resources formed over time from the remains of organisms that lived long ago. To curb the use of these resources, scientists are testing and refining alternative renewable energy sources, such as solar, wind, water, and geothermal power.
- The most severe crisis of natural resource destruction is occurring today in the tropical rain forests. Approximately 100,000 square kilometers (62,000 square miles) of rain forest are being lost each year because of slashing and burning for agriculture and logging and to create pastures for cattle. Scientists estimate that at the present rate, nearly all the tropical rain forest will be gone within the next 50 years.
- The atmospheric rise in carbon dioxide levels, which is caused by the burning of the tropical forests and the combustion of fossil fuels, is acting as a barrier against the escape of heat from the surface of the Earth. This greenhouse effect could affect global temperatures, rainfall patterns, and agricultural lands.
- Species are dying out at a rate of 1000 extinctions per year. A major cause of species extinction is habitat loss, which occurs when the tropical rain forests are cut. The extinction of many species leads to a reduction in biological diversity, a loss of species richness. In addition, the loss of species affects ecosystems and diminishes future sources of food, medicine, and other substances.
- The populations of the world are literally drowning in their trash and garbage, running out of room in which to put it. Solutions to this problem lie in recycling, reusing, lowering consumption, composting yard waste, and implementing landfill technology that incorporates recycling, safe incineration, and energy reclamation.
- Surface water is contaminated by factories, power plants, and sewage treatment plants that dump waste chemicals, heated water, and human sewage into lakes, streams, and rivers. Other sources of surface water pollution are sediment runoff during erosion and the leaching of chemicals from mines, hazardous waste dumps, and croplands. These pollutants affect aquatic life in different ways and can become concentrated in their bodies.
- Underground reservoirs of water can be contaminated by the same sources as surface water when the contaminants trickle through the soil; however, the primary groundwater contaminants are toxic chemicals. At this time, prevention of contamination is the cheapest and most feasible way to end groundwater contamination.
- The atmosphere has become polluted with sulfur and nitrogen oxides, carbon monoxide, hydrocarbons, and particulates. The primary source of these pollutants is the combustion of fossil fuels in automobiles and electricity-generating plants. As the sulfur and nitrogen oxides mix with water vapor in the atmosphere, they form acid rain, which harms both living and nonliving things. As the nitrogen oxides and hydrocarbons react with sunlight, they form photochemical smog, an upper respiratory irritant that is a health hazard.
- The environmental problems of this and the next century are the result of many interacting factors. Through research, environmental scientists will help people understand how they can live on this planet without harming it and their future existence.

Knowledge and Comprehension Questions

1. What is the biosphere?
2. Many scientists think people should work toward creating a "sustainable society." What does this mean?
3. Name several energy sources that could be used to help create a sustainable society.
4. What can be done to help prevent a mineral shortage in the future? Why are preventive measures important?
5. Why is the destruction of tropical rain forests a serious problem for everyone—not just the people living in tropical countries?
6. What is meant by global warming? Why is it dangerous?
7. Define *biological diversity*. How are humans affecting the biological diversity of the world's species? Why is this serious?
8. How can the amount of solid waste thrown away be reduced and waste management improved?
9. Distinguish between point and nonpoint sources of surface water pollution. Give an example of each.
10. Describe the different effects of pollution by organic nutrients, inorganic nutrients, infectious agents, and toxic substances in surface water.
11. How is acid rain produced? What are its effects?
12. Explain the terms *gray-air cities* and *brown-air cities*. What pollutants are involved?
13. Ozone is a dangerous pollutant in air that is breathed. Why should people be concerned that the ozone layer in the stratosphere is being depleted?

Critical Thinking

1. Suppose the president of the United States asked you to put together a plan of action for addressing environmental problems. What steps would you recommend?
2. Imagine for a moment that no serious effective program is mounted in the next decades to address the Earth's environmental woes. Outline a scenario of the likely consequences—what would such a world be like in 20 years?

APPENDIX A

ANSWERS

CHAPTER 1

Knowledge and Comprehension Questions

1. The study of biology includes a wide variety of topics, for example, studying hormonal changes in one organism, researching changes in populations over time, and learning how your body protects itself against disease. You may have suggested other topics, but they should focus on the study of life.
2. Scientists can only disprove hypotheses or support them with evidence. Future evidence may disprove a hypothesis that is supported with present evidence.
3. A scientific theory is a synthesis of hypotheses that are supported by so much evidence that they are commonly accepted by the scientific community, although conflicting theories occasionally exist. The everyday use of the word *theory* simply means a hunch or an idea. A scientific theory is a powerful explanation supported by scientific evidence, whereas an "everyday theory" is not usually powerful nor well-supported with evidence.
4. (1) Living things display both diversity and unity. (2) Living things are composed of cells and are hierarchically organized. (3) Living things in teract with each other and with their environments. (4) Living things transform energy and maintain a steady internal environment. (5) Living things exhibit forms that fit their functions. (6) Living things reproduce and pass on biological information to their offspring. (7) Living things change over time, or evolve.
5. The levels are (1) the cell (microscopic mass of protoplasm), (2) tissue (group of similar cells that work together to perform a function), (3) organ (group of tissues forming a structural and functional unit), and (4) organ system (group of organs that function together to carry out the principal activities of the organism).
6. The four levels are (1) population (individuals of a given species that occur together at one place and time), (2) community (populations of different species that interact with each other), (3) ecosystem (community of plants, animals, and microorganisms that interact with each other and with their environment), and (4) biosphere (the part of the Earth where biological activity takes place).
7. Unlike all other organisms, people use many nonrenewable resources and fill the biosphere with wastes. In addition, people are overpopulating the land and destroying habitats, killing off various species, and using up resources that cannot be renewed.
8. Ultimately, all energy comes from the sun. Solar energy is captured by producers, which use photosynthesis to convert it into chemical energy. Other organisms (consumers) feed on the producers and on each other, passing the energy along. When you eat plants or animals, you are harvesting their stored energy.
9. DNA is the code of life—instructions that are capable of becoming translated into an organism. Using only four different nucleotides, DNA codes for all the structural and functional components of an organism.
10. The theory of evolution states that organisms alive today are descendants of organisms that lived long ago and that organisms have changed and diverged from one another over time. Fossils are the preserved remains of earlier cells and organisms; they provide a record of this change.
11. (1) Monera (bacteria), (2) Protista (algae), (3) Plantae (rosebushes), (4) Fungi (mushrooms), (5) Animalia (you). The organisms on your list may vary from these; check Appendix B for further information.
12. Kingdom, phylum, class, order, family, genus, species.

Critical Thinking

1. Your hypothesis could be the following: If the students walk 1 hour a day, 4 days a week, in addition to their usual activity, then they will each lose at least 5 pounds during the semester. The independent variable is the walking; the dependent variable is the change in their weights.
2. The control is the standard against which the subjects who go through the experiment's treatment would be compared. In this case, it would be a group

of students who do not take the hour-long walks that the other students take.

3. a. True. Classes are more inclusive than orders. That is, a single class contains many orders of organisms. Organisms of the same order are, by definition, of the same class, phylum, and kingdom.
   b. False. The two organisms might be related closely enough to belong to the same genus, but this cannot be assumed. Many genera are included in a single order.
   c. False. Binomial nomenclature describes only the genus and species of an organism. These organisms may not belong to the same genus.
4. There are many characteristics of life; this textbook describes them as seven unifying themes. The scenario described in the question shows that computer-driven machines ostensibly have many of the characteristics of life; however, they do not carry out these activities in the same way as living things. For example, their "evolution" is driven by the marketplace and advances in technology, and is engineered by humans. Organismal evolution is driven by environmental selection pressures. Although humans may intervene to change environments, humans do not engineer organic change in the way they engineer new models of machines. They reproduce by manufacturing processes in factories, not by means of natural processes involving hereditary material. In addition, machines are not composed of cells, which are the smallest functional unit of living things. To be considered living, a machine would have to be composed of cells and carry on the basic processes of life in a naturalistic manner.

CHAPTER 2

Knowledge and Comprehension Questions

1. a. See Figure 2-1. Your atom would be hydrogen and would have one proton, one electron, and have from zero to two neutrons.
   b. The atom would still have one proton and one electron, but it would have a different number of neutrons from the atom in *a* within the range mentioned.
2. Three factors that influence how an atom interacts with other atoms are (1) the tendency of electrons to occur in pairs, (2) the tendency of atoms to balance positive and negative charges, and (3) the tendency of the outer shell (energy level) of electrons to be full. The third point is known as the octet rule because an atom with an unfilled outer shell tends to interact with other atoms to fill this outer shell. For many atoms, eight (an octet of) electrons fill the outer shell.
3. The atoms in nitrogen gas are bound together by covalent bonds as are the hydrogen and oxygen atoms of water molecules. Both involve the sharing of electrons between molecules. However, nitrogen gas molecules have triple bonds, which are stronger bonds than the single covalent bonds in water.
4. Water is unusual because it is the only common molecule on Earth that exists as a liquid at the Earth's surface. This liquid enables other molecules dissolved or suspended in it to move and interact. A second important characteristic is its ability to form hydrogen bonds with itself and other molecules because it is a polar molecule. Water molecules are strongly attracted to ions and other polar molecules, giving water another important trait: it is an excellent solvent. Chemical interactions readily take place in water because so many molecules are water soluble.
5. Organic molecules tend to be large, covalently bonded, and carbon based. Inorganic molecules tend to be small, do not usually contain carbon, and interact by means of ionic bonding. If you wanted to study living things—such as humans—you would primarily learn about organic molecules.
6. This environment is slightly basic. Like all basic environments, the intestinal environment would have a relatively low level of free H^1ions.
7. Water is a polar molecule. It quickly and easily forms hydrogen bonds, which, although not very strong, allow water to react with many other molecules. Carbon lends itself to being the basis of living material because of its ability to form four bonds with other atoms and therefore interact with other molecules in myriad ways.
8. Monosaccharides, disaccharides, and polysaccharides are all carbohydrates. Most organisms use carbohydrates as a primary fuel. Monosaccharides are among the least complex carbohydrates. Many organisms link mono-saccharides to form disaccharides that are less readily broken down as they are transported within the organism. To store the energy from carbohydrates, organisms convert monosaccharides and disaccharides into polysaccharides, long insoluble polymers of sugars.
9. Plants and animals must store energy as insoluble polysaccharide compounds, and glucose is a very soluble sugar in either blood or water. Plants generally store sugars as starch, whereas animals store it as glycogen.
10. Both DNA and RNA are nucleic acids composed of a five-carbon sugar, a phosphate group, and a nitrogenous base. Human cells store and use hereditary information by means of these nucleic acids. They vary in that RNA contains the base uracil and is responsible for directing protein synthesis, whereas DNA contains thymine and stores the hereditary information of the cell.
11. The three classes of macromolecules taken in as food energy by humans are carbohydrates, fats, and proteins. All three are made up of smaller units that, when disassembled, release energy. Because these molecules were all put together by dehydration synthesis, breaking them apart (digesting them) entails hydrolysis, an opposite process.

Critical Thinking

1. Oil is a nonpolar molecule, which means that it does not form hydrogen bonds with water and therefore does not dissolve. Answers will vary as to why an oil spill should be cleaned up. But one reason based on its nonpolarity is that the oil will form a film on the top of the water, affecting the amount of light entering the water. Lowered light levels would affect aquatic plant life. In addition, organisms that come to the surface of the water would get coated with this film of oil, which would probably affect their ability to survive.
2. The macromolecules present in this food are lipids, carbohydrates, and proteins. This food could be a part of a heart-healthy diet because it supplies only 17% of its calories from fat, well under the 25% to 30% recommendation. In addition, the amount of saturated fats as compared with the total fat is low—30%.
3. In living systems, proteins are used as enzymes and as structural components. Because DNA codes for the sequence of amino acids in a protein, an error in DNA can result in an error in the sequence of amino acids in a protein. This sequence is called the primary structure of a protein, and it affects the formation of the secondary and tertiary levels of structure. If a protein does not have the correct sequence of amino acids and is not structurally configured correctly, it cannot perform its job.

CHAPTER 3

Knowledge and Comprehension Questions

1. The basic principles of the cell theory are: (a) All living things are made up of one or more cells; (b) The smallest living unit of structure and function of all organisms is the cell; and (c) All cells arise from preexisting cells.
2. The human body is composed of millions of cells that interact and work together, allowing the organism to survive. Cells are specialized for different tasks, yet all cells take in nutrients, maintain homeostasis, react to stimuli, and are composed of smaller units called organelles.
3. Large, complex cells that are active often have many nuclei. A single nucleus could not control the activities of such a large cell.
4. The cell would have produced the protein along ribosomes located on the rough endoplasmic reticula. The protein would then enter the inner space of the ER, and eventually be encased in a vesicle at the smooth ER. This vesicle would travel to the Golgi complex and fuse with its membrane. Within the Golgi complex it might be modified. The "finished" protein would be encased in a vesicle that would "bud" from the Golgi, pass to the plasma membrane, and leave the cell via exocytosis.
5. Cilia in the respiratory tract of humans sweep invading bacteria and particles up the trachea and away from the lungs. If ciliary effectiveness is reduced because of smoking, invading organisms or particles could pass down the windpipe and enter the lungs, causing damage to the delicate lung tissue that might result in disease.
6. If the lysosomes stopped working, your cells would soon fill up with old cell parts and foreign substances such as bacteria. Your cells would probably die as these substances accumulated.
7. See Table 3-1.
8. See Table 3-2 for a summary of the differences. For example, the cell with chloroplasts would be the plant, the cell with mitochondria but no chloroplasts would be the animal, and the one with neither mitochondria nor chloroplasts would be the bacterium.
9. No. The ancient bacteria that were endosymbionts to other cells and developed into chloroplasts or mitochondria appeared millions of years ago. Today mitochondria and chloroplasts are functioning organelles that are part of the organism itself, as are other cell parts.
10. Both diffusion and facilitated diffusion occur along a concentration gradient and do not require energy. However, unlike diffusion, facilitated diffusion involves specific transport proteins that assist the movement of molecules across the membrane.
11. See Figure 3-26. White blood cells use endocytosis to engulf bacteria and other foreign substances.

Critical Thinking

1. Nutrients will enter the cell in different ways, depending on the nutrient. Water (because it is a small molecule) and the lipid-soluble vitamins (A, D, and E) are the only nutrients that can pass directly through the plasma membrane. The other nutrients, fatty acids and glycerol (lipids), amino acids (proteins), sugars (carbohydrates), and minerals must move through the membrane by means of carrier molecules or endocytosis.
2. a. Facilitated diffusion (passive).
   b. Osmosis (passive).
   c. Diffusion (passive).
   d. Active transport (active).
3. The solute is the instant coffee crystals, the solvent is water, and the solution is the liquid coffee.
4. Blood is made up of a solution of water with various dissolved substances and cells; the fluid portion of the blood is isotonic with respect to the red and white blood cell contents. Plain water is hypotonic to blood and to blood cells. The dehydration patient would run the risk of the blood cells absorbing water until they eventually lyse if water was put into the blood, causing the fluid portion of the blood to become hypotonic to the blood cells.

CHAPTER 4

Knowledge and Comprehension Questions

1. Free energy of activation destabilizes existing chemical bonds in the substrates of a reaction, allowing the reaction to proceed.
2. The reactions involved in building complex molecules, such as blood cells, from simpler substrates would be anabolic reactions. Energy is used to build the chemical bonds, and the product contains more energy than the substrates. The process of breaking down monosaccharides, such as glucose, is a catabolic reaction, which releases energy.
3. Both glycogen and lipid storage are examples of potential energy, as the energy within the chemical bonds of these substances is capable of performing work but is presently stored.
4. The first law of thermodynamics states that energy cannot be created or destroyed; it can only be changed from one form to another. The second law states that disorder in the universe constantly increases and energy is continually lost (as heat) to this disorder.
5. Entropy is the energy lost to disorder. Your friend is restating the second law of thermodynamics, which holds that disorder in the universe is constantly increasing.
6. The Earth constantly receives energy from the sun. Photosynthetic organisms capture this energy and convert it to chemical energy, storing it in the bonds of the carbohydrates they manufacture. When photosynthesizers are eaten by other organisms, this energy becomes available to them as these carbohydrates are broken down. The photosynthetic organisms are key to capturing the energy from the sun for all the organisms that eat plants or eat the organisms that eat plants.
7. Metabolic pathways are the chains of reactions within your body that move, store, and free energy. By means of metabolic pathways, people obtain energy from food, repair damaged tissues, and in general, avoid increasing entropy.
8. A catalyst increases the rate and often the probability of a chemical reaction while remaining unchanged itself. It may reduce the free energy of activation needed by placing stress on the bonds of a substrate and bringing substrates together. The cells of your body require such a high number of enzymes because there is a high degree of specificity in enzymatic reactions; only specific enzymes catalyze specific reactions.
9. Both activators and inhibitors bind to specific sites on enzymes and affect enzymatic activity. However, activators stimulate catalysis whereas inhibitors restrict it. Inhibitors play a key role in negative feedback loops. As end products are produced in metabolic pathways, they may serve as inhibitors by binding to the first enzyme of the metabolic pathway and causing enzymatic action to stop at that point, thus shutting down the pathway.
10. Cofactors are special nonprotein molecules that help enzymes catalyze chemical reactions; one example is a zinc ion that helps digestive enzymes break down proteins in food. Coenzymes are cofactors that are nonprotein organic molecules. Many vitamins are synthesized into coenzymes in your body.
11. Your body can store energy for later use by converting it to fat (long-term storage), glycogen (medium-term storage), or ATP (short-term storage). This is important because otherwise you would have energy only when it became available during exergonic reactions. The energy not captured in coupled reactions at that time would be lost as heat.

Critical Thinking

1. ATP is a high-energy compound that can be broken down easily to release energy for use by cells. It is called "energy currency" because cells can save energy released in exergonic reactions by storing it as ATP; cells can also spend ATP to provide necessary energy for endergonic reactions. The speed of the ATP recycling process allows the cell to fuel a large variety of cell processes as quickly as possible. The process also has endergonic and exergonic aspects so that it can make full use of the cell's nutrient supply as it varies.
2. Most enzymes function best within a narrow range of pH. Water and ions enter the cells of the plants from their environment, possibly changing the internal environment of the plants. If the plants' internal environment becomes too acidic, enzymes could be affected, causing chemical reactions to stop, which ultimately would affect the metabolic pathways of organisms.
3. Enzymes are proteins. The directions for making proteins are encoded in the DNA of an organism. A feedback system might trigger particular enzymes to begin transcribing those portions of the DNA that code for the particular enzymes needed.
4. The energy that is released when ATP is broken down to ADP and inorganic phosphate is used to do cell work or synthesize molecules. Some of this energy may be lost as heat. It is no longer available to be stored once again in ATP. "New" energy stored in the chemical bonds of food is used to keep the ADP-ATP cycle functioning in living systems.

CHAPTER 5

Knowledge and Comprehension Questions

1. ATP is the energy currency of cells; it serves as a short-term storage form of energy. The splitting of ATP into ADP and inorganic phosphate is coupled toendergonic reactions; in this way ATP provides the energy needed to drive these life-sustaining reactions.
2. Literally, this formula "says" glucose plus carbon dioxide yields carbon

dioxide, water, and energy. It is the overall formula for cellular respiration, the process by which living things make the ATP that powers cellular operations. During respiration, fuel molecules are broken down in the presence of oxygen to release energy.

3. This simplified formula summarizes the process of cellular respiration. It shows that the net effect is to break down one molecule of glucose in the presence of six oxygen molecules to yield six molecules of carbon dioxide and six molecules of water.
4. No. Oxidation describes the process in which an atom or molecules gives up an electron. The atom or molecule gaining the electron is "reduced." The term *oxidation* is used because in biological systems oxygen is most often the electron acceptor.
5. Both NAD[1] and FAD are electron carriers in the process of cellular respiration, playing key roles in the ultimate synthesis of ATP. They are different, however, in that NAD[1] can accept 2 electrons and 1 proton to be reduced to NADH, whereas FAD can accept 2 electrons and 2 protons to be reduced to $FADH_2$.
6. Your body metabolizes foods by breaking down complex molecules into simple ones. The digestive system breaks down the proteins (hamburger meat) into amino acids, the carbohydrates (bun and fries) into simple sugars, and the lipids (within hamburger meat and fried potatoes) into fatty acids and glyceralol The carbohydrates (now glucose) are further broken down to release energy within the cells by means of glycolysis, the Krebs cycle, and the electron transport chain. The amino acids, fatty acids, and glycerol enter the Krebs cycle without undergoing glycolysis.
7. Three main events of glycolysis are that (1) glucose is converted to pyruvate, (2) ADP 1 P_i is converted to ATP, and (3) NAD[1] is converted to NADH.
8. Glycolysis is inefficient because it captures only about 2% of the available chemical energy of glucose. However, cells have retained the enzymatic pathways to break down glucose in this way; glycolysis was probably one of the first biochemical processes to evolve. As cells evolved, they retained glycolysis as the first step in catabolic metabolism but added additional steps that were more efficient.
9. The Krebs cycle consists of nine reactions. The first three reactions are preparation reactions in which acetyl-CoA joins a four-carbon molecule from the previous cycle to form citric acid,and the chemical groups are rearranged. The last six reactions are energy-extraction reactions. At every turn of the cycle, acetyl-CoA enters and is oxidized to CO_2 and H_2O, and the hydrogen ions and electrons are donated to electron carriers.
10. See Figure 5-11.
11. Oxygen is the electron acceptor (thus it is reduced) for both the electrons and protons of the electron transport chain. Water is formed as a result of this reaction. It is an important end product of cellular respiration; 1 H_2O molecule is produced for each NAD or $FADH_2$ molecule that enters the electron transport chain.
12.

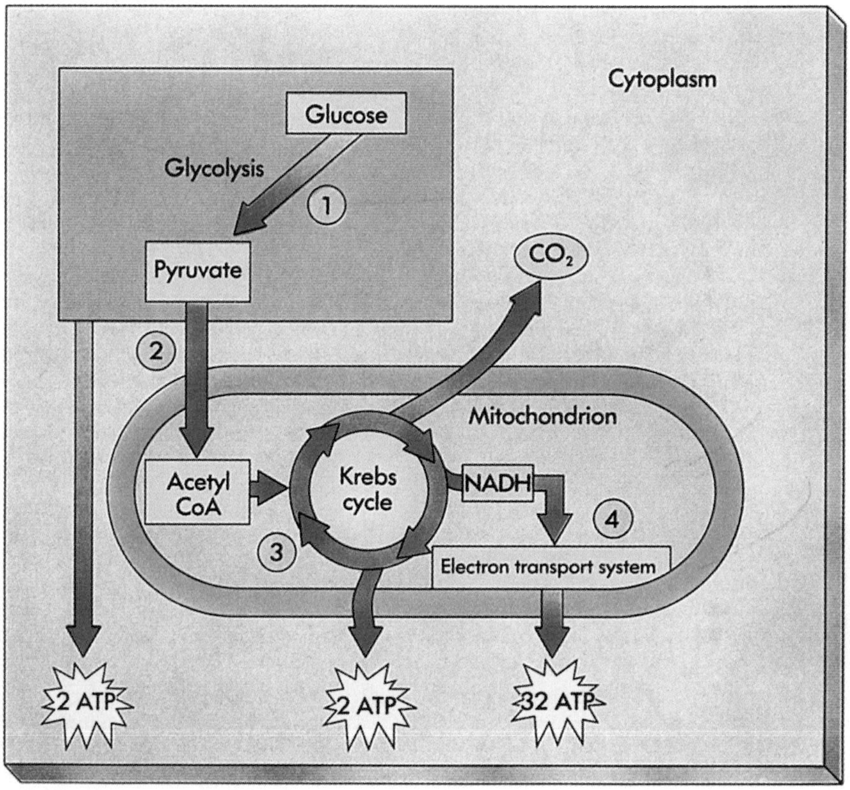

Critical Thinking

1. True. The breakdown of glucose during cellular respiration is exergonic in that energy is released from the glucose being cleaved. The energy is then used endergonically to produce molecules of ATP.
2. Spraying water into the air aerates (adds oxygen to) the water. Water aeration is done so that any organic particles in the water can be completely broken down by bacteria, which helps keep the water clean. Otherwise, an aerobic processes begin and fermentative by-products build up in the water. Bacteria multiply and the water begins to smell. Sewage treatment plants likewise aerate water in a variety of ways so that bacteria can completely break down the organic materials.
3. Answers may vary, but this chapter shows how the mitochondria are the place where a more complete breakdown of food molecules occurs. It seems likely that a symbiotic relationship with bacteria could result millions of years later in eukaryotic cells having bacteria-like organelles. The structure of the mitochondria is discussed more fully in this chapter in terms of its function—the inner membrane (possibly the original bacterial membrane) is the site of biochemical activity. The outer membrane (which is relatively inactive in cellular respiration) could be derived from the plasma membrane of the cell, and not be bacterial in origin.

CHAPTER 6

Knowledge and Comprehension Questions

1. Autotrophs are organisms that produce their own food by photosynthesis or by using chemical energy rather than light energy. Heterotrophs are organisms that cannot produce their own food; they depend on the autotrophs and the sun for survival. Humans are heterotrophs.
2. The visible light in sunlight consists of many wavelengths. The prism separated the various wavelengths so that each color became visible as the light shone on the floor.
3. Electromagnetic energy; radiation; waves; photons.
4. A blue sweater is dyed with pigment; a pigment is a molecule that absorbs some wavelengths of light and reflects others. The sweater appears blue because the pigment it is dyed with absorbs variable amounts of different wavelengths, including blue wavelengths, but reflects light enriched in blue wavelengths. Those wavelengths bounce off the sweater and are reflected back to your eyes.
5. The equation describes the process of photosynthesis: carbon dioxide and water, in the presence of light, produce glucose and oxygen. By means of photosynthesis, energy from the sun is captured and stored in molecules of food. This food ultimately builds and fuels the bodies of virtually all living things.
6. Sugar is produced during the Calvin cycle.
7. Photosynthesis involves two series of reactions: (a) synthesis of ATP and NADPH and (b) the use of ATP and NADPH to produce sucrose and other sugars. The reactions that produce ATP and NADPH are called *light-dependent reactions* because they occur only in the presence of light. During the light-independent reactions (which do not need light to drive the reactions), ATP and NADPH are used to produce glucose from carbon dioxide.
8. A photocenter is an array of pigment molecules within the thylakoid membranes of a chloroplast. It channels photon energy to chlorophyll *a*, a special molecule of chlorophyll that participates in photosynthesis. Two kinds of chlorophyll *a* act as reaction centers in the two photosystems of photosynthesis, absorbing and then ejecting energized electrons that provide the energy (originally from the sun) for the reactions of photosynthesis.
9. In photosystem I, the first to evolve in photosynthetic organisms, energy is transferred to a molecule of chlorophyll *a* called P700. In photosystem II a different chlorophyll *a* molecule, P680, acts as a reaction center. Both photosystems absorb light at the same time and are part of the mechanism of light-dependent reactions of photosynthesis. Energized electrons ejected from photosystem II are passed down an electron transport chain, triggering ATP production. Their places in photosystem II are taken by electrons from the breakdown of water, a process that releases oxygen as a by-product. Energized electrons from photosystem I, along with hydrogen ions, reduce $NADP^+$ to NADPH.
10. During the light-independent reactions (a) carbon is fixed with the attachment of CO_2 to an organic molecule, (b) sucrose and other sugars are produced from molecules of CO_2, and (c) the organic molecule that attaches to CO_2 reforms to begin the cycle again.
11. Such a dust cloud would interfere with photosynthesis by blocking the light from the sun. All life on Earth ultimately depends on the process of carbon fixation (photosynthesis) powered by the sun's light, and on the by-product, oxygen. Eventually all organisms would die if photosynthesis did not take place.

Critical Thinking

1. The extra plant mass came from the process of carbon fixation (photosynthesis), in which carbon dioxide and energy from the sun were used to manufacture sucrose and other carbohydrates for the plant. This plant had made more organic material than it needed as fuel, and it was used to build new plant parts for growth, as well as to provide starch for storage. In addition, as the plant grows, more water is taken up by the plant and held in its tissues, which would be reflected in its weight.
2. The roots do not manufacture ATP via photosynthesis but via cellular respiration. Sucrose would be transported to the roots from the photosynthetic parts of the plant, where it would be broken down during respiration, yielding ATP.
3. The absorption spectrum for chlorophyll *a* depicted in Figure 6-4 shows that chlorophyll *a* absorbs many violet and red wavelengths. *These* are the wavelengths effective for photosynthesis because they are the wavelengths that are absorbed. The others are reflected. Other wavelengths effective for photosynthesis are absorbed by other pigments; chlorophyll *b* absorbs many of the violet and red wavelengths chlorophyll *a* does not. Although it is not shown in the figure, the carotenoids absorb wavelengths in the green and blue range (as well as violet), increasing the wavelengths effective for photosynthesis.

CHAPTER 7

Knowledge and Comprehension Questions

1. (Examples will vary.) A cell is the smallest structure in an organism that is capable of performing all the functions necessary for life. An example is a cardiac muscle cell. The cells of the body are organized into tissues (groups of similar cells that work together to perform a function, such as cardiac muscle tissue). Several different tissues group together to form a structural and functional unit—an organ (the heart.) An organ system is a group of organs that works together to carry out the body's principal activities (such as the circulatory system).
2. (Examples will vary.) The four types of tissue in the body are (1) epithelial (lining of the body cavities), (2) connective (tendons), (3) muscle (skeletal muscles such as leg muscles), and (4) nervous (the brain).
3. (Examples will vary.) See Table 6-2. Squamous (flattened) cells are found in the air sacs of the lungs, cuboidal cells line ducts in glands, and columnar cells line much of the digestive tract.
4. Lymphocytes, erythrocytes, and fibroblasts are all cells found in connective tissue, but they have different functions. In addition, lymphocytes and erythrocytes are found in the blood, but fibroblasts are found in other types of connective tissue. Lymphocytes defend the body against infection, erythrocytes pick up and deliver gases in the blood, and fibroblasts produce fibers that are found in various types of connective tissue.
5. Oxygen delivery by means of red blood cells is necessary for cells to use the aerobic process of cellular respiration. This process is the primary means of ATP (hence, energy) production in our bodies.
6. Yes, fat cells have their uses. In addition to storing fuel, fat tissue helps shape and pad the body and insulate it against heat loss.
7. Skeletal muscle; cardiac muscle.
8. Neurons are specialized to carry electrical impulses that are transmitted almost instantly. This system of communication is far quicker than hormone release and response.
9. See Table 7-1.
10. Homeostasis is the maintenance of a stable environment inside your body despite varying conditions in the external environment. The molecules, cells, tissues, organs, and organ systems in your body must all work together to maintain this internal equilibrium.

Critical Thinking

1. Like most living tissues, bone is constantly renewing itself. Blood vessels within the bone bring calcium and other substances necessary to heal a break to the area. In addition, the bone cells are able to lay down new matrix to help repair a break.
2. This is part of the body's inflammation response to injury. Mast cells produce histamine, which dilates blood vessels and increases blood flow to the area. This makes the area turn red and feel warm.
3. The connective tissue matrix may provide a nutrient pool for connective tissue cells, allow for waste removal, and provide an environment enhancing cell mobility. Proteins secreted into the matrix may provide a fibrous, stronger structure to matrix, such as that found in bone and cartilage.
4. (Answers may vary, for the chapter gave only the basics on which to base an answer.) Nervous tissue is specialized for conducting electrical impulses. The brain is specialized, in addition, for interpreting electrical impulses. This suggests that neurons are connected to one another in particular pathways, unlike other tissues such as skin. Therefore a reasonable suggestion might be that nervous tissue in the brain and spinal cord is incapable of regeneration based on properties of its specialization regarding carrying messages and, more importantly, interpretation (in the brain). Other tissues do not have such a specialization. These pathways, once severed, may be too complex to "hook up" to one another again.

CHAPTER 8

Knowledge and Comprehension Questions

1. The six classes of nutrients are carbohydrates, fats, proteins, vitamins, minerals, and water. The first three are organic compounds that your body uses as a source of energy; they are also used as building blocks for growth and repair and to produce other important substances. Vitamins, minerals, and water help body processes take place. Some minerals are also part of body structures.
2. Three types of enzymes help to digest the energy nutrients. Proteases break down proteins into peptides and amino acids. Amylases break down carbohydrates (starches and glycogen) to sugars. Lipases break down the triglycerides in lipids to fatty acids and glycerol.
3. The process of digestion begins in the mouth, as the saliva lubricates food and the salivary amylase breaks down starches into maltose.
4. You would bite off the mouthful with your incisors, your canines would help you tear the food away, and your premolars and molars would allow you to

crush and grind the food thoroughly.
5. Peristalsis is a series of involuntary muscular contractions. No, the esophagus transports the food from mouth to stomach, but no digestive processes occur here (except for any continuing digestion by salivary amylase mixed with the food).
6. The hydrochloric acid in the stomach converts pepsinogen to the active form of pepsin, softens connective tissue in food, denatures large proteins, and kills bacteria. The stomach wall is protected by thick mucus.
7. The duodenum is the first part of the small intestine; digestion is completed there. Starch and glycogen are broken down to disaccharides and then to monosaccharides. Proteins and polypeptides are broken down to shorter peptides and then to amino acids. Triglycerides are digested to fatty acids and glycerol.
8. These are hormones that help regulate digestion. Gastrin controls the release of hydrochloric acid in the stomach. Secretin stimulates the release of sodium bicarbonate to neutralize acid in the chyme and increases bile secretion in the liver. Cholecystokinin stimulates the gallbladder to release bile into the small intestine and stimulates the pancreas to release digestive enzymes.
9. Increasing the internal surface area of the small intestine helps absorb more nutrients from the same amount of food. Three features accomplish this: inner folds, projections (villi) on the folds, and additional projections on the projections (microvilli).
10. Glucose is broken down by glycolysis, the Krebs cycle, and the electron transport chain. Amino acids and triglycerides are converted to substances that can be metabolized by two of these pathways. Fatty acids are metabolized by another pathway.

Critical Thinking

1. Meat and animal products provide protein to the diet and some fat. Because we Americans eat too much fat, eating a vegetarian diet would most likely lower fat intake, a positive dietary consequence. To replace the protein, a vegetarian could eat legumes (dried peas, beans, peanuts, or soy-based food), grains, nuts, and seeds. To take in the essential amino acids, legumes should be combined with any grain, nut, or seed supplemented with small amounts of milk, cheese, yogurt, or eggs. Persons who are not strict vegetarians might also consider eating small amounts of fish or poultry.
2. The shape of the teeth on the skull would indicate the type of food eaten by the animal. Herbivores (plant-eaters) have flat, grinding teeth that look like human molars.
3. The liver and pancreas are not organs in which the digestive products are altered and pass through and so are called *accessory organs.* They are not a stop in the digestive tract itself.
4. Sodium bicarbonate neutralizes the acid in chyme. Thus it can help reduce the uncomfortable symptoms of acid indigestion.
5. Answers will vary. See Figure 8-13.

CHAPTER 9

Knowledge and Comprehension Questions

1. Respiration is the uptake of oxygen and the release of carbon dioxide by the body. Cellular respiration is the process at the cellular level that uses the oxygen you breathe in and produces the carbon dioxide you breathe out. Internal respiration is the exchange of oxygen and carbon dioxide between the blood and tissue fluid; the exchange of the two gases between the blood and alveoli is external respiration.
2. Your vocal cords produced the sound for speech as air rushed by and made them vibrate. Your lungs served as a power supply and volume control for your voice, and your lips and tongue formed the sounds into words.
3. As the volume of the thoracic cavity increases, the air pressure within it decreases. Outside air is pulled in, equalizing the pressures inside and outside the thoracic cavity. As the volume of the thoracic cavity decreases, the air pressure inside it increases and forces air out, equalizing the pressures.
4. At the alveoli, carbon dioxide diffuses out of the blood into the alveoli because it moves down the pressure gradient (its pressure in the blood is greater than its pressure within the alveoli.) Meanwhile, oxygen diffuses from the alveoli into the blood because its pressure within the lungs is greater than in the blood.
5. Hemoglobin is known as the oxygen-carrying molecule, but it also carries almost 25% of circulating carbon dioxide in the bloodstream.
6. The hemoglobin in the deoxygenated blood of your veins absorbs oxygen in the air as soon as it leaves the body. Thus, wounds appear to bleed bright red blood even when a vein is involved.
7. At the capillaries, oxygen moves down a pressure gradient to diffuse from the blood into the tissue fluid. Meanwhile, carbon dioxide moves down its own pressure gradient and diffuses from the tissue fluid into the blood.
8. You apply the Heimlich maneuver when a victim cannot cough or talk and is unable to breathe.
9. At high altitudes the pressure of the oxygen molecules in the air is lower than at sea level. This means that the pressure gradient is lower at the alveoli and less oxygen diffuses into the blood. This can cause you to feel dizzy and short of breath.

Critical Thinking

1. The sound of the hiccup is produced as the sudden inhaled air of the spasm causes the glottis at the back of the throat to close suddenly. The opening of the vocal cords and quick closing produce a loud gasp.
2. The circulatory system provides a transport system that distributes gases throughout the body. Without it you would not survive because it would take too long for oxygen to diffuse from the lungs to the rest of the body.
3. One of the primary symptoms of respiratory diseases is that patients tire easily. This is because their decreased lung capacity does not allow for sufficient reoxygenation of blood and produces an overall effect of less oxygen reaching the brain and other organ systems.

CHAPTER 10

Knowledge and Comprehension Questions

1. The circulatory system has four main functions: (a) nutrient and waste transport, (b) oxygen and carbon dioxide transport, (c) temperature maintenance, and (d) hormone circulation.
2. The blood carries sugars, amino acids, and fatty acids to the liver, where some of the molecules are converted to glucose and released into the bloodstream. Excess energy molecules are stored in the liver for later use. Essential amino acids and vitamins pass through the liver into the bloodstream. The cells release their metabolic waste products into the blood, which carries them to the kidneys.
3. A regulatory center in your brain constantly monitors your body temperature. If your temperature gets too high, signals from this center dilate your surface blood vessels and increase blood flow to the surface of your skin. This increases heat loss and lowers your body temperature.
4. The circulatory system has three components: the heart, the blood vessels, and the blood. The term *cardiovascular system* refers to the "plumbing" (heart and blood vessels) of the circulatory system.
5. As blood flows through tissue, from capillaries to venules, it loses gases (primarily oxygen) and nutrients to the tissues. The blood in the venules is deoxygenated and carries waste products from the tissues.
6. When people are scared (or cold), the walls of the arterioles can contract and the blood flow decreases, routing more blood to other areas of the body. This can cause light-skinned people to "turn pale."
7. The entry to a capillary is guarded by a ring of muscle, which can constrict to block blood flow into that vessel. This is an important means of limiting heat loss. It also permits adjustments in blood flow, allowing more blood to go to areas where it is needed, such as to the intestines after a meal.
8. The walls of arteries have more elastic tissue and smooth muscle than those of veins; this helps arteries accommodate the pulses and high pressures of blood pumped from the heart. The lumen of veins is larger because blood flows through them at lower pressures; a larger diameter offers less resistance. The walls of capillaries are only one cell thick, which permits the exchange of substances between the cells and the blood.
9. The heart has left and right sides, which both serve as pumps yet are not directly connected with one another. Oxygenated blood from the lungs enters and is pumped from the left side, while deoxygenated systemic blood passes through the right side into the lungs. These two types of blood must be separated—hence the two-pump system.
10. The heart is composed of cardiac muscle. Like skeletal muscle, cardiac muscle may be voluntarily strengthened by means of aerobic exercise.
11. The heart valves prevent backflow during the strong contractions of the heart. They shut in response to backflow, allowing pressure to push blood in one direction only.
12. The SA node is the pacemaker of the heart; this cluster of cells initiates the excitatory impulse that causes the atria to contract and the impulse to be passed along to other cardiac cells. The AV node conducts the impulse from the SA node to other cardiac cells, initiating the contraction of the ventricles.
13. In general, normal blood pressure is 110 to 130 mm Hg systolic and 70 to 90 mm Hg diastolic. Answers to the rest of the question will vary.
14. A high-fat diet can encourage the development of fatty deposits (plaques) on the inner walls of arteries, making the vessels narrower and less elastic. This can lead to atherosclerosis, heart attacks, and strokes.

Critical Thinking

1. The student should list family members who have (or have had) heart disease. The student should also list lifestyle factors that may contribute to heart disease such as whether or not they consume a high-fat diet, if they are overweight, if they smoke cigarettes, and if they exercise regularly. Obviously, one cannot change one's genetics. The other factors are behavioral and can be changed.
2. The left ventricle of the heart is the "workhorse" portion of the heart. It has the thickest musculature, pumping blood out to the body. The right ventricle has a less ominous job, pumping blood to the nearby lungs. The atria simply squeeze blood to the ventricles. Therefore, replacing the pumping action of the left ventricle is the most important job in heart-replacement technology.

CHAPTER 11

Knowledge and Comprehension Questions

1. Nonspecific defenses act against any foreign invader, whereas specific defenses work against particular types of microbes. Nonspecific defenses are the first line of defense because they are the first barriers that a microorganism must bypass to enter your body.
2. The nonspecific defenses include (1) the skin (which if unbroken provides a barrier against invasion), (2) mucous membranes (mucus can trap particles and move them away from delicate areas), (3) chemicals that kill bacteria (many membranes are washed by fluids deadly to bacteria or have an acidic environment), and (4) the inflammatory process (a series of events that removes the irritation, repairs the damage, and protects the body from further damage).
3. In this case, the immune response (specific resistance) is not involved. Nonspecific resistance includes cases where foreign matter is trapped by mucous membranes and expelled from the body, probably via tears or eye watering.
4. The immune system differs from other body systems because it is not a system of organs and it lacks a single "controlling" organ. Instead, it consists of cells scattered throughout the body that have a common function: reacting to specific foreign molecules.
5. The transplanted cell would be identified as "non-self"; therefore it would act as an antigen and probably produce an immune response in the new environment.
6. The five principal types of T cells are (a) helper T cells (which initiate the immune response), (b) cytotoxic T cells (which break apart infected and foreign cells), (c) inducer T cells (which oversee the development of T cells in the thymus), (d) suppressor T cells (which limit the immune response), and (e) memory T cells (which respond quickly and vigorously if the same foreign antigen reappears.
7. During the cell-mediated immune response, cytotoxic T cells recognize and destroy infected body cells in addition to destroying transplanted and cancer cells. Helper T cells initiate the response, activating cytotoxic T cells, macrophages, inducer T cells, and finally, suppressor T cells.
8. During the antibody-mediated response, B cells recognize foreign antigens and, if activated by helper T cells, produce antibodies. The antibodies bind to the antigens and mark them for destruction.
9. Antibodies have highly specific binding sites that fit only one specific antigen with the exact amino acid composition necessary for complete binding.
10. Injection with antigens causes an antibody (B cell) response, with the production of memory cells. A booster shot induces these memory cells to differentiate into antibody-producing cells and form still more memory cells.
11. The human immunodeficiency virus (HIV), which is responsible for AIDS, is dangerous because it destroys helper T cells, thus destroying the immune system's ability to mount a defense against any infection.
12. An allergic reaction occurs when the immune system mounts a defense against a harmless antigen. When class E antibodies are involved and bind to the antigens, they cause a strong inflammatory response that can dilate blood vessels and lead to symptoms ranging from uncomfortable to life-threatening.

Critical Thinking

1. These phagocytes serve to sound the alarm for an immune response if the body is being invaded by foreign microbes. They offer a quick response by engulfing the invader, stimulating the maturation of large numbers of monocytes into macrophages, and inducing an inflammation or fever to resist microbial growth.
2. Memory T cells were developed by your immune system as part of your immune response. These cells give you an accelerated response during subsequent exposures to measles, promptly defending you against infection.
3. This is a case of passive immunity, in which an organism is protected not by its own cell-mediated immune response but by short-term protection from another source (in this case, the maternal immune system). This protection is short-term because these cells are not the baby's own. Eventually they die; only the baby's immune cells are propagated by the baby's body.

CHAPTER 12

Knowledge and Comprehension Questions

1. Excretion is the removal of metabolic wastes and excess water from the body. Without excretion, your body would soon become poisoned by the build-up of metabolic waste products. The process also helps maintain the balance of water and ions that is necessary for life.
2. The primary metabolic waste products are carbon dioxide, water, salts, and nitrogen-containing molecules.
3. Urea, urine, and uric acid are all forms of metabolic waste products (nitrogenous wastes). Urea is the primary excretion product from the deamination of amino acids and is a compound formed from ammonia and carbon dioxide. Uric acid forms from the

breakdown of the nucleic acids you eat and from the metabolism of your own nucleic acids and ATP. Urine is the excretion product consisting of water, urea, creatinine, uric acid, and other substances.
4. The kidneys help maintain the fluid balance and isotonicity of the blood plasma.
5. The kidneys control the ion concentration and pH of the blood, as well as ion balances necessary for the electrical activity in muscle responsible for muscle movement.
6. See Figures 12-4 and 12-5.
7. Kidneys help conserve water by reabsorbing it from the filtrate. By changing the concentrations of salts and urea in the filtrate, the kidneys create an osmotic gradient that results in water moving from the filtrate into the surrounding tissue fluid.
8. Urine can reveal the presence of certain substances in the body because the kidney detoxifies the blood. When these substances, such as drugs, are removed from the blood, they become part of the filtrate.
9. Urine is formed when the blood that flows through the kidneys is filtered, removing most of its water and all but its largest molecules and cells. The kidneys then selectively reabsorb certain substances.
10. Homeostasis is the maintenance of constant physiological conditions in the body. ADH and aldosterone are two hormones that help regulate the water and ion balance necessary for homeostasis. ADH regulates that amount of water reabsorbed at the collecting ducts; aldosterone promotes the retention of sodium (and therefore water), while promoting the excretion of potassium.
11. Two problems with kidney function are kidney stones and renal failure. Kidney stones are crystals of certain salts that can block urine flow. Renal failure is a reduction in the rate of filtration of blood in the glomerulus.
12. The process of donor rejection involves the immune system of the recipient.

Critical Thinking

1. Carbon dioxide is a waste product of cellular respiration and is used by the liver to form urea. Other excess carbon dioxide is excreted by the lungs.
2. Neither. You need to conserve water. Alcohol acts as a diuretic; drinking the whiskey will cause they body to excrete water. Salt will be re-absorbed to its threshold; remember, the kidney's main job is regulating salt and water balance. Drinking salt water will cause the body to excrete water with the excess salt that is taken in, because water follows salt osmotically at the collecting ducts.

CHAPTER 13

Knowledge and Comprehension Questions

1. Neurons and hormones are both forms of communication that integrate and coordinate body functions. They differ in that neurons transmit rapid signals that report information or initiate quick responses in specific tissues. Hormones are chemical messengers that trigger widespread prolonged responses, often in a variety of tissues.
2. A sensory neuron in your hand would transmit the sensation of heat immediately when the coffee contacted you. Usually in such a case an interneuron in your spinal cord would stimulate a motor neuron, causing your hand to move almost instantaneously.
3. *Resting potential* refers to the difference in electrical charge along the membrane of the resting neuron. The action of sodium-potassium transmembrane pumps and ion-specific membrane channels separates positive and negative ions along the inside and outside of the membrane, which creates the resting potential. The difference in electrical charges is the basis for the transmission of nerve impulses.
4. A nerve impulse travels because nearby transmembrane proteins respond to the electrical changes that accompany depolarization of the nerve cell membrane. The adjacent section of membrane depolarizes, followed by another section, leading to a wave of depolarization.
5. Saltatory conduction occurs along myelinated neurons when impulses jump to unmyelinated areas and skip over myelinated portions. Impulses travel much more quickly by this method than by continuous waves of depolarization that occur along continuous portions of unmyelinated membrane.
6. Synapses between neurons and skeletal muscle cells are neuromuscular junctions. At a neuromuscular junction, acetylcholine released from an axon tip depolarizes the muscle cell membrane, releasing calcium ions that trigger muscle contraction.
7. The enzyme acety cholinesterase breaks down acetylcholine. If the neurotransmitter were not broken down, it would continue to signal the muscles to contract.
8. The postsynaptic neuron integrates the information it receives from presynaptic neurons. The summed effect of excitatory and inhibitory signals either facilitates or inhibits depolarization.
9. All drugs, whether stimulatory or inhibitory, act by interfering with the normal activity of neurotransmitters.
10. Although a few alcoholic drinks or sleeping pills (barbiturates) are not themselves dangerous, the combination may be deadly. Barbiturates and alcohol enhance the sedative activities of each other, leading to a potentially dangerous combination.
11. These stimulants actually block the breakdown and recycling of neurotransmitters, causing a depletion of the former. At this point, the drug abuser needs consistently higher doses of stimulant (tolerance has developed).

Critical Thinking

1. The nerve impulse is not a chemical reaction because there are no changes in molecules in which one substance is changed to another. It consists of the movement of charged particles called *ions,* along the length of the neuron membrane. Therefore, the nerve impulse is actually an electrical-chemical event.
2. Psychoactive drugs, by affecting neurotransmitter transmission in certain parts of the brain, can produce emotional and physical states that are, at least temporarily, extremely exciting or pleasant to the drug abuser. This effect often outweighs the person's intellectual considerations of the drug's potential for harm if abused. Drug addiction may occur quickly, and withdrawal symptoms make it very difficult to stop the drug use without professional help.
3. Drug tolerance develops over time, not only causing an addict to require increased drug doses for the desired effect but also increasing the drug's side effects and swifter degradation in the body. These higher doses may become financially unmanageable and lead to criminal behavior or cause side effects that are antisocial and harder for an addict to hide. Heavy drug use also may interfere with judgment and assessment of situations and interactions between people.
4. Drug addiction, as defined in the text, is the compulsive urge to continue drug use, a tendency to increase drug dosage, and either physical or psychological (or both) dependence on the drug. Marijuana abuse meets all of these criteria for drug addiction.

CHAPTER 14

Knowledge and Comprehension Questions

1. The nervous system is a complex network of neurons that gathers information about the internal and external environment and processes and responds to this information. The central nervous system (brain and spinal cord) is the site of information processing. The peripheral nervous system (nerves) shuttles messages to and from the central nervous system.
2. The somatic nervous system consists of motor neurons that control voluntary responses (for example, moving your leg to take a step). The autonomic nervous system consists of motor neurons that control involuntary activities (for example, digesting a meal).
3. Integration would occur in the spinal cord for such a simple reflex.
4. Integration would occur, ultimately, in the cerebrum, but the sensory input travels to the cerebrum through the limbic system.
5. The brainstem (midbrain, pons, medulla) contains tracts of nerve fibers that carry messages to and from the spinal cord. Nuclei located there control important body reflexes.
6. Sensory receptors change stimuli into nerve impulses. The nerve fibers of sensory neurons carry this information to the central nervous system, where interneurons interpret them and direct a response (in this case, turning your head). The axons of motor neurons conduct these impulses to the appropriate muscles.
7. How you perceive stimuli depends on which receptors are stimulated. You registered this stimulus as a sound because it stimulated your auditory nerve.
8. These systems act in opposition to each other to maintain homeostasis and help you respond to environmental changes. The sympathetic nervous system generally mobilizes the body for greater activity (faster heart rate, increased respiration), whereas the parasympathetic nervous system stimulates normal body functions such as digestion.

Critical Thinking

1. The areas of the hands and face are associated with the primary senses (touch, taste, sight) and, as such, require more of the cortex area to interpret and react to increased input from sensory and motor neurons.
2. Infants and toddlers are quickly developing nervous system connections and expanding motor skills. The myelin sheath of many nerve fibers, particularly in the cerebral white matter and motor neurons, is critical for nerve signal transmission and normal development.
3. The limbic systems of both a lion and a human respond to similar stimuli (pain, sexual drive, hunger, thirst) and would be somewhat similar, whereas their associative cortexes would be quite dissimilar in function, because primates (such as humans) have a larger portion of the cerebrum devoted to associative activities than do nonprimate mammals (such as lions).
4. As the spinal cord extends down the back from the brain, many sensory and nerve tracts emerge from it along its length. The higher the point of an injury on the spinal cord that severs it from the brain, the greater the number of sensory and motor nerve tracts that are no longer connected to the brain because of the injury.

CHAPTER 15

Knowledge and Comprehension Questions

1. Pain is, for the most part, a subjective experience. Although a doctor may assess the potential for pain in a given situation, individuals' responses and sensitivities vary widely. The differences between the subjective assessment of pain by an individual not experiencing it and the actual pain experienced by an individual makes pain a very difficult medical management problem.
2. For you to sense a stimulus, it must be of sufficient magnitude to open ion channels within the membrane of the receptor cell. This depolarizes the membrane, creating a generator potential that leads to an action potential (nerve impulse) in the sensory neurons with which the receptor synapses. Nerve fibers conduct the impulse to the central nervous system.
3. A roller coaster ride would most probably include an extreme rise in proprioception stimuli as the person's spatial orientation changes quickly.
4. Satiation is a sensation involving the body's information about the internal environment, particularly those receptors in the digestive system.
5. Nasal congestion would most probably affect the special senses of both smell and taste by reducing sensory input in these areas.
6. Olfactory receptors detect different smells because of specific binding of airborne gases with the receptor chemicals located in the cilia of the nasal epithelium.
7. On its way to the olfactory area of the cerebral cortex, the nerve impulse travels through the limbic system, the area of the brain that is responsible for many of your drives and emotions. Thus, certain odors become linked in your memory with emotions and events.
8. Taste buds are taste receptors concentrated on the tongue. They detect chemicals in food and register an overall taste that consists of different combinations of sweet, salty, sour, and bitter. The sense of taste interacts with the sense of smell to produce a taste sensation.
9. Color vision is sensed by cone cells. There are three types of cones, each of which absorbs a specific wavelength of light. The color you perceive depends on how strongly each group of cones is stimulated by a light source.
10. You see objects because light enters the eye through the pupil and the lens focuses it on the fovea, which is rich in cone cells. These cells, along with surrounding rods and cones, initiate nervous impulses that travel to the cerebrum via the optic nerve.
11. See Figure 15-9.
12. Rod cells, which can detect low light levels, are the primary cell type stimulated in a dark room, but in a well-lit setting the cone cells (which sense color) would be equally activated.
13. You have depth perception because of the positioning of your two eyes on either side of the head, which sends slightly different images to the brain. Covering one eye makes it more difficult to judge distance.
14. Your outer ear funnels sound waves toward the eardrum, which changes these waves into mechanical energy. This energy is then transmitted to the bones of the middle ear, which increases its force. Receptor cells in the inner ear change the mechanical energy into nerve impulses.
15. You can detect movements of your head because of the otoliths embedded in the jellylike layer that covers the cilia in your inner ear. When you move your head, the otoliths slide and bend the underlying cilia. This generates signals to the brain, which interprets the type and degree of movement.

Critical Thinking

1. The birds oriented in some way to the direction of migration, but that orientation was superseded by an orientation to a magnet. Because the birds were placed in a cage lined with carbon paper, they would be unable to see movements of the sun or moon for navigational purposes. Possibly these birds have a sensory system in which they can detect and orient to the Earth's magnetic field, which is their means of navigation. However, placing a magnet near them would provide a stronger stimulus than that of the Earth's magnetic field, so the birds would orient to it.
2. The inability to feel pain is a disadvantage to a person's well-being. The reason we avoid many injuries is that we detect potentially injurious stresses we place on our bodies because of the pain it causes, and then we stop the activity. For example, a child who jumped from a staircase and slightly twisted an ankle would not continue to walk on it and would probably not jump from that height again. However, the child who felt no pain would continue to walk and jump, injuring the ankle further, and might not be deterred from this behavior in the future.

CHAPTER 16

Knowledge and Comprehension Questions

1. Your movement resulted from the contraction of your skeletal muscles, which are anchored to bones. The muscles use bones like levers to direct force against an object. When you raised your hand, your skin stretched to accommodate the change in position.
2. The role of skin as a sensory surface is of great use to small children. As we age, we tend to rely less heavily on such input and more on visual perception.

3. Bone is a type of connective tissue consisting of living cells that secrete collagen fibers into the surrounding matrix. The bones of the skeletal system support the body and permit movement by serving as points of attachment and acting as levers against which muscles can pull. They also protect delicate internal structures, store important minerals, and produce red and white blood cells.
4. Bone marrow transplants entail the removal and transfer of red bone marrow from spongy bone in the end of long bones.
5. Spongy bone makes up most of the tissue of the smaller bones of the axial skeleton.
6. Excess calcium taken in by the body and not needed at the moment in cellular activities is stored in bone tissue until a later time.
7. A "slipped disk" refers to one of the intervertebral disks of fibrocartilage that separate the vertebrae from each other (except in the sacrum and coccyx). These disks act as shock absorbers, provide the means of attachment between vertebra, and allow the vertebral column to move.
8. These are all types of synovial (freely movable) joints. They differ in the type of movement they allow. A hinge joint allows movement in one plane only, a ball-and-socket joint allows rotation, and a pivot joint permits side-to-side movement.
9. Tendons connect muscle to bone. Ligaments connect muscles together.
10. Osteoarthritis generally strikes most severely in synovial joint tissue.
11. Actin is a protein that makes up the thin myofilaments of muscle; the protein myosin makes up the thick myofilaments. Myofilaments are the microfilaments of muscle cells. Muscle cells contract when the actin and myosin filaments slide past each other. Changes in the shape of the ends of the myosin molecules (located between adjacent actin filaments) cause the myosin molecule to move along the actin, causing the myofilament to contract.
12. When depolarization from a nerve impulse reaches a neuromuscular junction, it triggers the release of the neurotransmitter acetylcholine. The acetylcholine crosses over to the muscle fiber membrane and opens the ion channels of that membrane, depolarizing it. This sets off a series of reactions that release calcium ions; the calcium ions initiate the chemical reactions of contraction.

Critical Thinking

1. Answers may vary. People generally exhibit less elasticity, wrinkling, and stretching or sagging skin with age as well as lower touch sensation. Bones often become brittle, causing more fractures and breaking more easily. Joints often stiffen, and osteoarthritis is common.
2. Bone functions as one type of connective tissue that supports the skin, giving the body some of its shape, and strength, as well as providing a place of attachment for muscles. Certain minerals, such as calcium, give hardness to bone but also provide a storage place for them. In addition, the bone marrow is located within spaces of bones. Red bone marrow, located within spongy bone, produces most of the body's blood cells. Artificial bones can support the skin, helping to impart shape to the body. Artificial joints can allow for movement where bones meet bones. However, artificial bones cannot be a storehouse for minerals, nor can it provide bone marrow.

CHAPTER 17

Knowledge and Comprehension Questions

1. Hormones are chemical messages secreted by cells that affect other cells. They are an important method the body uses to integrate the functioning of various tissues, organs, and organ systems.
2. The nervous system sends messages to glands and muscles, regulating glandular secretion and muscular contraction. Endocrine hormones carry messages to virtually any type of cell in the body.
3. Peptide hormones are not lipid soluble and so do not pass through lipid bilayers of cell membranes. The binding of peptide hormones to a cell membrane receptor triggers the production of a second messenger within the cell. Cyclic AMP is this second messenger, and it stimulates the activity of enzymes within the cell that cause the cell to alter its functioning.
4. The sexual and physiological changes in puberty are primarily triggered by high steroid hormone levels, which control physiological processes such as growth and development.
5. A feedback loop controls hormone production by initially stimulating a gland to produce the hormone. After the hormone has exerted its effect on the target cell, the body feeds back information to the endocrine gland. In a positive feedback loop, the feedback causes the gland to produce more hormone; in a negative feedback loop, it causes the gland to slow down or stop hormone production.
6. ADH levels would be monitored and controlled by the hypothalamus. This structure produces "releasing hormones" that trigger ADH production and release from the posterior pituitary.
7. Tropic hormones stimulate other endocrine glands. The four tropic hormones are—(1) Follicle-stimulating hormone (FSH): In women, FSH triggers the maturation of eggs in the ovaries and stimulates the secretion of estrogens. In men, it triggers the production of sperm. (2) Luteinizing hormone (LH): In women, LH stimulates the release of an egg from the ovary and fosters the development of progesterone. In men, it stimulates the production of testosterone. (3) Adreno-

corticotropic hormone (ACTH): ACTH stimulates the adrenal cortex to produce steroid hormones. (4) Thyroid-stimulating hormone (TSH): TSH triggers the thyroid gland to produce the thyroid hormones.
8. No. Alcohol suppresses the release of ADH (antidiuretic hormone), which means that it encourages more water to leave your body in your urine. If you are already hot and thirsty, this will only dehydrate you further.
9. Parathyroid hormone (PTH) and calcitonin (CT) work antagonistically to maintain appropriate calcium levels. If the level becomes too low, PTH stimulates osteoclasts to liberate calcium from the bones and stimulates the kidneys and intestines to reabsorb more calcium. When levels grow too high, more CT is secreted, which inhibits the release of calcium from bones and speeds up its absorption.
10. Over a prolonged period of stress, the body reacts in three stages: (1) alarm reaction (quickened metabolism triggered by adrenaline and noradrenaline); (2) resistance (glucose production and rise in blood pressure); hormones involved are ACTH, GH, TSH, mineralocorticoids and glucocorticoids; and (3) exhaustion (loss of potassium and glucose and organs become weak and may stop functioning).
11. The pancreatic islets of Langerhans secrete two hormones that act antagonistically to one another to regulate glucose levels. Glucagon raises the glucose level by stimulating the liver to convert glycogen and other nutrients into glucose, whereas insulin decreases glucose levels in the blood by helping cells transport it across their membranes.

Critical Thinking

1. People living far from the ocean faced little chance of receiving their iodine from seafood, a primary source of iodine. Until this century, iodized salt was not freely available. This lack of dietary iodine caused goiter.
2. Chronic stress leads to long-term stimulation by the autonomic nervous system which heightens metabolism, raises blood pressure, and speeds up internal chemical reactions. Over time, this can physically stress the body and lead to health problems.

CHAPTER 18

Knowledge and Comprehension Questions

1. The chemical instructions that determine our specific personal characteristics are located within chromosomes. The DNA within our chromosomes directs the millions of complex chemical reactions that govern our growth and development.
2. The two types of nucleic acid are DNA (deoxyribonucleic acid) and RNA (ribonucleic acid). Both consist of nucleotides that are made up of three molecular parts: a sugar, a phosphate group, and a base. The sugar in RNA is ribose, whereas the sugar in DNA is deoxyribose. DNA contains the bases adenine, thymine, cytosine, and guanine; RNA contains uracil instead of thymine.
3. *Double helix* refers to the DNA molecule, which is shaped like a double-stranded helical ladder. The bases of each strand together form rungs of uniform length, and alternating sugar-phosphate units form the ladder uprights.
4. Chargaff's experiments showed that the proportion of bases varied in the DNA of different types of organisms. This suggested that DNA has the ability to be used as a molecular code, because its base composition varies as its code varies from organism to organism.
5. The type of RNA found in ribosomes is ribosomal RNA (rRNA). Transfer RNA (tRNA) is in the cytoplasm; during polypeptide synthesis, tRNA molecules transport amino acids to the ribosomes and position each amino acid at the correct place on the elongating polypeptide chain. Messenger RNA (mRNA) brings information from the DNA in the nucleus to the ribosomes in the cytoplasm to direct which polypeptide is assembled.
6. Sexual reproduction, whereby a new human is formed by the fusion of egg and sperm meiotically produced, leads to tremendous variation. The alternative, mitosis, produces identical daughter cells and is responsible for human growth during the life cycle.
7. The correct sequence is *c* (prophase), *a* (metaphase), *d* (anaphase), *b* (telophase). This is the process of mitosis.
8. The correct sequence is *c* (prophase I), *d* (metaphase I), *a* (anaphase I), *b* (telophase I), *e* (interphase). This is the process of meiosis I.
9. Mitosis produces two daughter cells from one parent cell; the daughter cells are identical with each other and with the parent. Meiosis produces four sex cells, each of which contains half the amount of the parent cell's hereditary material.
10. Genetic recombination during meiosis generates variability in the hereditary material of the offspring. It is the principal factor that has made the evolution of eukaryotic organisms possible.
11. Meiosis II produces four haploid cells that may function as gametes (animals) or spores (plants). The mother cells are diploid and do not function as gametes or spores.

Critical Thinking

1. There are many answers, and student responses may vary. Possible answers include the following: a nonviable DNA strand is produced; an error causes the chemical message to vary and results in genetic mutation of some sort; and the change does not directly affect instruction because of repetition of the instruction elsewhere on the strand.
2. There may be only one gene coding for a particular enzyme, and many enzymes are individually essential to the viability of an organism. A "backup" system of repetition may ensure enzyme production despite a single mutation or alteration in genetic code.

CHAPTER 19

Knowledge and Comprehension Questions

1. Offspring of sexual reproduction inherit characteristics from both parents; those of asexual reproduction are exact duplicates (with the exception of mutations) of one parent. Sexual reproduction introduces variation within a species; asexual reproduction does not.
2. Self-fertilization occurs when the male gametes of one plant fertilize the female gametes of the same organism. Cross-fertilization occurs when the male gametes of one plant fertilize the female gametes of another.
3. An individual who is heterozygous for a particular trait has two different alleles for that trait. Often alleles express dominance or recessiveness in relation to one another, but there are also many instances where incomplete dominance or codominance occurs.
4. Mendel's law of segregation states that each gamete receives only one of an organism's pair of alleles. Random segregation of the pair members occurs during meiosis.
5. All of the F_1 offspring would have an Ll genotype and the phenotype of long leaves:

	L	L
l	Ll	Ll
l	Ll	Ll

6. The F_2 genotype would be 1LL:2Ll:1ll

	L	l
L	LL	Ll
l	Ll	ll

the phenotype would be 3 long-leaved: 1 short-leaved

	L	l
L	LL	Ll
l	Ll	ll

7. The probability is $^1/_4$.
8. Mendel used a testcross to determine whether a phenotypically dominant plant was homozygous or heterozygous for the dominant trait. He crossed the plant with a homozygous recessive plant. When the test plant was homozygous, the offspring were hybrids and phenotypically dominant. When it was heterozygous, half the progeny were heterozygous and looked like the test plant, and half were homozygous recessive and resembled the recessive parent.
9. Mendel's law of independent assortment states that the distribution of alleles for one trait into the gametes does not affect the distribution of alleles for other traits, unless they are on the same chromosome.
10. a, $^1/_{16}$; b, $^9/_{16}$.
11. Sutton and Morgan provided a framework that explained the equal genetic role of both egg and sperm in the hereditary material of offspring. Sutton suggested the presence of hereditary material within the nuclei of the gametes, whereas Morgan gave the first clear evidence that genes reside on chromosomes.

Critical Thinking

1. Phenotype assessment, because it involves the outward appearance of an individual, is relatively easy and accurate. However, genotype determination is far more complex, especially in the era in which Mendel lived. Genotype may be complicated by cases of incomplete dominance, more than one gene coding of the same trait, genetic linkage, and other complex relationships between individual genes or chromosomes.
2. Answers will vary regarding distinctive human traits that are not inherited. All physical traits are inherited (other than those caused by environmental influences in utero or later in life). Students may mention traits that are behavioral and suggest tracing these traits in their own families for evidence of inheritance. (Scientists are still probing this question; thus student results are merely speculative.)

Genetics Problems

1. Somewhere in your herd you have cows and bulls that are not homozygous for the dominant gene "polled." Because you have many cows and probably only one or some small number of bulls, it would make sense to concentrate on the bulls. If you have only homozygous "polled" bulls, you could never produce a horned offspring regardless of the genotype of the mother. The most efficient thing to do would be to keep track of the matings and the phenotype of the offspring resulting from these matings and prevent any bull found to produce horned offspring from mating again.

2. Albinism, *a,* is a recessive gene. If heterozygotes mated, you would have the following:

	A	a
A	AA	Aa
a	Aa	aa

One fourth would be expected to be albinos.
3. The best thing to do would be to mate Dingleberry to several dames homozygous for the recessive gene that causes the brittle bones. Half o f the offspring would be expected to have brittle bones if Dingleberry were a heterozygous carrier of the disease gene. Although you could never be 100% certain Dingleberry was not a carrier, you could reduce the probability to a reasonable level.
4. Your mating of DDWww and Ddww individuals would look like the following:

	Dw	Dw	dw	dw
DW	DDWw	DDWw	DdW	DdWw
Dw	DDww	DDww	Ddww	Ddww
DW	DDWw	DDWw	DdWw	DdWw
Dw	DDww	DDww	Ddww	Ddww

Long-wing, red-eyed individuals would result from 8 of the possible 16 combinations, and dumpy, white-eyed individuals would never be produced.
5. Breed Oscar to Heidi. If half of the offspring are white eyed, Oscar is a heterozygote.
6. Both parents carry at least one of the recessive genes. Because it is recessive, the trait is not manifested until they produce an offspring who is homozygous.

CHAPTER 20

Knowledge and Comprehension Questions

1. Down syndrome (trisomy 21) is a genetic disorder caused by the presence of three 21 chromosomes. Individuals who inherit this condition show delayed maturation of the skeletal system and are often mentally retarded.
2. People who inherit abnormal numbers of sex chromosomes often have abnormal features and may be mentally retarded. Examples are (a) triple X females (XXX zygote), underdeveloped females who may have lower-than-average intelligence, (b) Klinefelter syndrome (XXY zygote), sterile males with some female characteristics, (c) Turner syndrome (XO zygote), sterile females with immature sex organs, and (d) XYY males, fertile males of normal appearance.
3. An X chromosome is necessary for zygote viability.
4. The three major sources of damage to chromosomes are high-energy radiation (such as x-rays), low-energy radiation (such as UV light), and chemicals (such as certain legal and illegal drugs).
5. Heavy use of drugs could affect ova and sperm by damaging the DNA. In general, chemicals add or delete molecules from the structure of DNA. This can result in chromosomal breaks or changes as the cell works to repair the damage.
6. All cancers exhibit uncontrolled cell growth, loss of cell differentiation, invasion of normal tissues, and metastasis to other sites.
7. Mutation of the DNA of proto-oncogenes causes them to become active oncogenes. These act as the "on" switches in the development of cancer. Mutation of the DNA of tumor-suppressor cells causes them to be inactivated and they cannot stop the development of cancerous tissue. As a result of these mutations, cancer will develop in a patient.
8. If a cyst or tumor is benign, it indicates that the growth of this cell mass is confined and encapsulated and poses no serious health threat.
9. a, Sex-linked; b, dominant; c, recessive; d, autosomal.
10. Most human genetic disorders are recessive because those genes are able to persist in the population among carriers; people carrying lethal dominant disorders are more likely to die before reproducing. Recessive disorders include cystic fibrosis, sickle cell anemia, and Tay-Sachs disease.
11. In incomplete dominance, alternative forms of an allele are neither dominant nor recessive; heterozygotes are phenotypic intermediates. In codominance, alternative forms of an allele are both dominant; thus heterozygotes exhibit both phenotypes.
12. Their children's genotypes would be 1:2:1 (AA [type A], AB [type AB], BB [type B]).

	A	B
A	AA	AB
B	AB	BB

13. Couples who suspect that they may be at risk for genetic disorders can undergo genetic counseling to determine the probability of this risk. When a pregnancy is diagnosed as high risk, a woman can undergo amniocentesis (analysis of a sample of amniotic fluid) to test for many common genetic disorders.

Critical Thinking

1. The fact that a genetic abnormality such as trisomy 21 can exist shows that there are often groups of genes located near each other on a chromosome that may be transferred or altered as a group and cause multiple alterations that occur as a discrete group.
2. Answers will vary.
3. Answers will vary.

Human Genetics Problems

1.

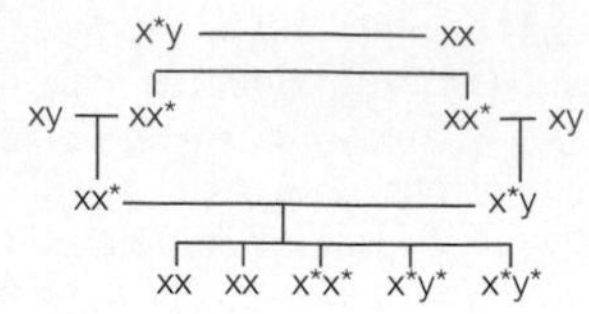

2. No. A type O child is possible.
3. 45 (44 autosomes 1 one X)
4. I^AI^O (blood type A)
5.

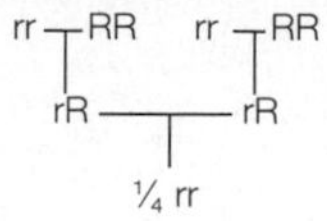

6. e—Y-linkage

CHAPTER 21

Knowledge and Comprehension Questions

1. Sexual reproduction is the process in which a male and a female sex cell combine to form the first cell of a new individual.
2. Follicle-stimulating hormone (FSH) triggers sperm production. Luteinizing hormone (LH) regulates the testes' secretion of testosterone, a hormone responsible for the development of male secondary sexual characteristics.
3. A spermatozoon has a head that contains the hereditary material. Located at its leading tip is an acrosome that contains enzymes helping the sperm penetrate an egg's membrane. The sperm also has a flagellum that propels it and mitochondria that produce the ATP from which sperm derives the energy to power the flagellum.
4. These are accessory glands that add fluid to the sperm to produce semen. The seminal vesicles supply a fluid containing fructose, which serves as a source of energy for the sperm. The prostate gland adds an alkaline fluid that neutralizes the acidity of any urine in the urethra and the acidity of the female vagina. The bulbourethral glands also contribute an alkaline fluid.
5. The reproductive cycle of a woman consists of both her menstrual cycle (monthly development and shedding of the functional layer of the endometrium) and her ovarian cycle (monthly maturation of her egg and its release).
6. The reproductive cycle of females occurs roughly every 28 days. The primary oocyte matures, is released from the ovary during ovulation, and journeys through the uterine tube to the uterus. The endometrial lining of the uterus has thickened to prepare for implantation; if fertilization does not occur, it sloughs off during menstruation. The hormones FSH, LH, estrogen, and progesterone orchestrate these events.
7. Fertilization occurs in the uterine tube. After the egg is fertilized, it completes a second meiotic division, joins its hereditary material with that of the sperm, begins dividing by mitosis, and implants itself on the thickened uterine wall. The placenta begins to form and human chorionic gonadotropin (HCG) is secreted.
8. Once the date of ovulation is determined, a couple must have intercourse within 24 hours to fertilize the egg. The egg is viable only up to 1 day after ovulation.
9. The only methods of birth control that may prevent the transmission of sexually transmitted diseases are abstinence and condoms. Nonoxyl–9, a spermacide, may inactivate some viruses
10. Birth control pills contain estrogen and progesterone, which shut down the production of FSH and LH. By maintaining high levels of estrogen and progesterone, the pills cause the body to act as if ovulation has already occurred; the ovarian follicles do not mature, and ovulation does not occur.
11. The most effective birth control methods are vasectomies and tubal ligation. Birth control pills are very effective. Condoms and diaphragms are effective when used correctly, but mistakes are common. Least reliable are the rhythm method and withdrawal, which have high failure rates.
12. No. Sexually transmitted diseases are not caused by poor personal hygiene or socioeconomic class. They are communicable illnesses transmitted between persons during sexual contact.
13. The only completely effective protection against sexually transmitted diseases is abstinence. There is decreased risk if latex condoms are used correctly along with the spermicide nonoxynol-9.

Critical Thinking

1. At birth, females have all of the oocytes that they will ever produce. As a woman ages, so do her oocytes, and the odds of a harmful mutation increase appreciably after age 35. Because a man's gametes are (at the most) a few days old, this risk of mutation due to aging of the gametes does not exist in men. Therefore, women older than 35 years of age are more likely to bear children with genetic abnormalities than are younger women (in general), but this is not the case with men.
2. When a cervical cap or diaphragm is used, there is still fluid-to-fluid contact between sexual partners as the penis contacts the vaginal walls. This contact permits the transfer of sexually transmitted diseases.

CHAPTER 22

Knowledge and Comprehension Questions

1. At the instant of union and fertilization between a male and female gamete, the zygote formed has a unique genetic composition.
2. After fertilization, the oocyte's surface changes so that no other sperm can penetrate, and oocyte meiosis is completed. The sperm sheds its tail, and the sperm and egg nuclei fuse.
3. This period consists of mitotic cell division.
4. If the blastocyst does not successfully implant in the uterine wall, it will be swept away in the consequent menstruation. If implanted, the blastocyst secretes human chorionic gonadotropin, which maintains the corpus luteum and prevents menstruation.
5. The three primary germ layers are the (a) ectoderm (outer layer of skin, the nervous system, and portions of sense organs), (b) endoderm (lining of the digestive tract and digestive organs, respiratory tract and lungs, and urinary bladder and urethra), and (c) mesoderm (skeleton, muscles, blood, reproductive organs, connective tissues, and innermost layer of skin).
6. The extraembryonic membranes play a role in the life support of the preembryo/embryo/fetus. They are the (a) amnion (cushions the embryo in amniotic fluid and keeps the temperature constant), (b) chorion (facilitates the exchange of nutrients, gases, and wastes between the embryo and mother), (c) yolk sac (produces blood for the embryo before its liver is functional and becomes the lining of the digestive tract), and (d) allantois (responsible for formation of the embryo's blood cells and vessels).
7. During gastrulation, groups of inner mass cells differentiate into the three primary germ layers from which all the organs and tissues will develop.
8. Neurulation is the development of a hollow nerve cord. During neurulation, cells lying over the notochord curl upward to form a tube that will develop into the central nervous system.
9. The notochord is a structure that forms the midline axis of an embryo. In humans and in most other vertebrates, it develops into the vertebral column.
10. Along with general systemic development, neural development is a key aspect of the first 3 weeks of fetal development, and the fate of the central nervous system is dependent on good maternal nutrition.
11. Between the fourth and eighth weeks, the embryo grows much longer, establishes a primitive circulation, and begins to exchange gases, nutrients, and wastes with the mother. The central nervous system and body form begin to develop.
12. The first trimester is primarily a time of development; by the ninth week, most body systems are functional. During the second trimester the fetus grows, ossification is well underway, and the fetus has a heartbeat. The third trimester is primarily a period of growth; by birth the fetus is able to exist on its own.
13. During pregnancy the fetus obtains all of its nutrients from the mother. Poor nutrition in the mother can damage the child, possibly resulting in retardation and stunted growth.
14. At birth the circulation of the newborn changes as the lungs, rather than the placenta, become the organ of gas exchange. The hole between the two atria close, and the ductus arteriosis (the direct connection between the pulmonary artery and the aorta) is shut off.

Critical Thinking

1. Cell surface changes occur on the egg that prevent entry of a second sperm. If a second sperm gained entry, the zygote would not be viable because it would have $1^1/_2$ times the normal amount of hereditary material.
2. If the foramen ovale did not close, some of the deoxygenated blood from the right side of the heart would flow directly to the left side of the heart without going to the lungs to be oxygenated. This condition results in blood with a deep red color rather than the bright red of oxygenated blood. Such babies are termed "blue babies," because the blood has a bluish cast as seen through the skin. This abnormality must be repaired by surgery.

CHAPTER 23

Knowledge and Comprehension Questions

1. Darwin's observations led him to conclude that: (a) organisms of the past and present are related, (b) factors other than or in addition to climate are involved in the development of plant and animal diversity, (c) members of the same species often change slightly in appearance after becoming geographically isolated from each other, and (d) organisms living on oceanic islands often resembled those living on a nearby mainland.
2. Three other factors that influenced Darwin were (a) evidence of geological layers and fossils that suggested the Earth was much older than traditionally thought, (b) the observed changes in species that occur with artificial selection, and (c) Malthus' writings on geometric population growth.
3. The two dogs look so different today because of artificial selection; people who bred them selected for desired characteristics, so that over time the breeds changed. Artificial selection is based on the principle that all organisms within a sexually-reproducing species exhibit variation.
4. An exponential progression increases more rapidly because it involves multiplying a number by a constant factor; an arithmetic progression is one in which the elements increase by a constant difference. Malthus theorized that populations grow exponentially, but certain factors limit this growth. Darwin

realized that nature acts to limit population numbers.

5. Adaptations are naturally occurring inheritable traits found within populations that confer a reproductive advantage on organisms that possess them. As adaptive (advantageous) traits are passed on from surviving individuals to their offspring, the individuals carrying those traits will increase, and the nature of the general population will gradually change.
6. *Survival of the fittest* refers to the idea that natural selection tends to favor those organisms that are most fit to survive to reproductive age in a particular environment and to produce offspring.
7. In adaptive radiation, the population of a species changes as it disperses into different habitats. Eventually some of these populations change with the result that interbreeding is no longer possible. This is speciation, the formation of new species through the process of evolution.
8. Fossil; sedimentary; radioactive isotopes.
9. The carbon-14 method can help establish a date by estimating the relative amounts of the different isotopes of carbon present in a fossil. The half-life of carbon-14 is 5730 years (the amount of time for half of the ^{14}C to decay into nitrogen). Scientists can estimate the length of time that the carbon-14 has been decaying, which means the time that has elapsed since the organism died.
10. According to radioactive dating, the Earth is 4.6 billion years old. This is significant because it allows sufficient time for evolution to occur.
11. Vestigial organs are structures that are present in an organism of today but are no longer useful. A human example is the appendix.
12. Homologous structures share the same evolutionary origin but now differ in structure and function. Analogous structures have a similar form and function but different evolutionary origins.
13. Comparative studies of anatomy show that many organisms have groups of bones, nerves, muscles, and organs with the same anatomical plan but with different functions. These homologous structures imply evolutionary relatedness. Studies in comparative embryology show that many organisms have early developmental stages that are similar. This again implies evolutionary relatedness.
14. An evolutionary tree diagrams the degree of relatedness among groups of organisms. Evolutionary trees support evolution by showing the same evolutionary relationships as revealed by anatomical studies.

Critical Thinking

1. Polar bears live in the Arctic where there is snow year round. Those bears that are white survive in greater numbers than those that have discolorations, spots, or are more darkly colored. The white color serves as a camouflage. Over the years, natural selection "weeds out" un-white bears.
2. A population comprised of only a few organisms has limited genetic variation and mating possibilities. Should conditions for survival change and should some of these organisms die due to their lack of characteristics that could accommodate that change, the population would become smaller over time and could eventually die out. Larger populations generally have wider genetic variation and are better able to withstand changes that affect the survival of the population.

CHAPTER 24

Knowledge and Comprehension Questions

1. The primordial soup theory attempts to explain how life began on Earth. It states that life arose in the sea as elements and simple compounds in the atmosphere interacted to form simple organic molecules. Another theory holds that life might have arisen in hydrothermal vents in the oceans.
2. The formation of some sort of membrane would have allowed chemical reactions to occur in a closed environment. Enzymes could be organized to carry on life functions, chemicals could be selectively absorbed, wastes could be eliminated, and hereditary material could be passed on to future cells.
3. See Table 24-1.
4. Cyanobacteria are single-celled organisms that are similar to the first oxygen-producing bacteria. These evolved during the Archean era and played an important evolutionary role by gradually oxygenating the atmosphere and oceans.
5. The Cambrian period was the oldest period in the Paleozoic era (roughly 590 to 500 million years ago). All of the main phyla and divisions of organisms that exist today (except chordates and land plants) evolved by the end of the Cambrian period.
6. During the Carboniferous period, reptiles developed from amphibians and anthropods moved from the sea onto land. Insects, one arthropod group, evolved water-conserving characteristics such as a cuticle. Fungi evolved during the late Carboniferous period. Much of the land was low and swampy with extensive forests; the worldwide climate was warm and moist.
7. The Cenozoic era, which began 65 million years ago, saw the rapid evolution and growth of mammals, including primates.
8. Chordates are distinguished by three features: (1) a single, hollow nerve cord along the back, (2) a rod-shaped notochord located between the nerve cord and the developing gut, and (3) pharyngeal arches.
9. These are all subclasses of mammals that are living today. Monotremes lay

eggs with leathery shells and incubate them in a nest. Marsupials bear immature young and nurse them in a pouch until they are old enough to be on their own. In placental mammals, the young develop to maturity inside the mother.

10. Primates are the order of mammals with characteristics reflecting an arboreal life-style. They have developed two especially helpful characteristics: depth perception resulting from stereoscopic vision and flexible, grasping hands with opposable thumbs.
11. These are both suborders of the primates. Prosimians (lower primates such as lemurs) are small animals, usually nocturnal, with a well-developed sense of smell. Anthropoids (higher primates such as apes and humans) have larger brains, flatter faces, eyes that are closer together, and relatively long front and hind limbs. Anthropoids are also diurnal and possess color vision.
12. Statement (a) is correct; statement (b) is false. All hominids are a family within the hominoid superfamily.
13. More modern *Homo sapiens* had smaller heads, brow ridges, teeth, jaws, and faces than earlier species did. They also fashioned sophisticated tools from stone and from other substances such as bone and ivory. Over time, groups of *Home sapiens sapiens* moved away from the hunter-gatherer life-style to develop agriculture, complex social structures, and civilization.

Critical Thinking

1. This portrayal is inaccurate. Dinosaurs died out at the end of the Cretaceous period (about 65 million years ago), long before the first primates appeared.

CHAPTER 25

Knowledge and Comprehension Questions

1. Viruses are infectious agents that lack a cellular structure, so they are not living. They are nonliving obligate parasites that must exist in association with and at the expense of other organisms.
2. See Figure 25-3.
3. Both cycles are patterns of viral replication. In the lytic cycle, viruses enter a cell, replicate, and cause the cell to burst and release new viruses. In the lysogenic cycle, viruses enter into a long-term relationship with the host cells, their nucleic acid replicating as the cells multiply.
4. A cold sore results from the lysogenic cycle of cell damage caused by the herpes simplex virus. The virus remains latent in nervous tissues until something triggers it, such as a cold.
5. Bacteria are classified into their own kingdom because their cell structure is different from that of all other organisms. (They do not contain membrane-bounded organelles, they lack a nucleus, and the bacterial cell is bounded by a membrane encased within a cell wall.)
6. Bacteria perform many vital functions, such as decomposing organic materials and recycling inorganic compounds. They were also largely responsible for creating the properties of the atmosphere and soil that are present today.
7. No. Sexually transmitted diseases are not caused by poor personal hygiene or socioeconomic class. They are communicable illnesses transmitted between persons during sexual contact.
8. No. The human papillomavirus that causes genital warts has various types; types 16 and 18 are tumor viruses that may cause cervical cancer in women. These women should have Pap smears every 6 months.
9. The extreme simplicity of bacterial design—small size, lack of discrete organelles, lack of a membrane-bound nucleus—seems to support the theory that they evolved before eukaryotes.
10. Gummas are tumorlike masses exhibited during the tertiary stage of syphilitic infection. These lesions are not communicable but may develop in the cardiovascular or central nervous systems and cause paralysis or death.
11. In men, gonorrheal infection, if untreated, may spread to other urogenital structures from the urethra. Infection of the epididymis or vas deferens may result in scar tissue, causing male sterility. Chronic gonorrheal infections in women may lead to pelvic inflammatory disease (PID), and if uterine tubes are involved, scar tissue may develop, causing sterility.
12. These diseases may be acquired without the knowledge of the person or partner. Occasionally, patients do not have extreme or even normal symptoms of infection and may unwittingly pass on these diseases.
13. The only completely effective protection against sexually transmitted diseases is abstinence. There is decreased risk if latex condoms are used correctly along with the spermicide nonoxynol 9.

Critical Thinking

1. Apparently, it takes 10 days for the antibiotic your physician prescribed to kill the bacteria causing your infection, or inhibit their growth so that your body can mount its defenses and destroy these disease-causing cells. Some cells may still be living and able to reproduce after only 4 days of the antibiotic. These cells could begin to reproduce; when they enter the exponential phase of growth, they would trigger a strep infection once again.

CHAPTER 26

Knowledge and Comprehension Questions

1. Protists are single-celled eukaryotic organisms; this kingdom contains organisms that are animallike (protozoa), plantlike (algae and diatoms), and funguslike (slime molds).
2. An ameba is a protozoan that has a changing bloblike shape. It moves and obtains food by means of pseudopods (cytoplasmic extensions). Its pseudopods engulf organic matter, which the ameba then digests.
3. Both flagellates and ciliates are protozoans, and both have hairlike cellular processes that help them move and obtain food. However, flagellates have flagella (long hairlike processes), whereas ciliates have cilia (short hairlike extensions). Ciliates have a more complex internal organization than flagellates do.
4. Sporozoans are nonmotile protozoans that parasitize animals. They undergo complex life cycles in which they are passed from host to host by means of a vector.
5. Euglenoids, dinoflagellates, and golden algae are all phyla of plantlike protists. Euglenoids are flagellated protists with chloroplasts. Dinoflagellates have stiff outer coverings, and their flagella beat in two grooves. The golden algae have gold-green photosynthetic pigments and store food as oil.
6. All three organisms belong to the phylum of golden algae.
7. Fungus; ameba.
8. Saprophytic; parasite.
9. All fungi could be described as helpful, because as decomposers they are crucial to the cycling of materials in the environment. More specifically, helpful fungi include those that produce antibiotic drugs. Yeasts are fungi that are crucial in the production of bread, cheese, and beer. Other fungi, however, cause diseases in plants and animals, such as Dutch elm disease, wheat rust, and yeast infections.
10. Club fungi (c), zygote-forming fungi (d), sac fungi (a), water molds (e), imperfect fungi (b).
11. Lichens are associations between fungi and photosynthetic partners. Lichens are able to tolerate harsh living conditions; thus they are found in a wide range of habitats (such as deserts, extreme northern and southern latitudes, and mountaintops).
12. Radiolarians; foraminifera.

Critical Thinking

1. It is difficult to give a definitive answer to this question. As groups of cells within colonial organisms specialize, taking on particular jobs, many would suggest that they should be considered multicellular organisms. However, in this case, only one group of cells is specialized; if the cells could exist independently, the group might still be considered as a colonial organism.
2. *Euglena* is difficult to place using this frame of reference. Clearly, classifying protists on the basis of such criteria is difficult and, in some cases, may be impossible. Protists are a varied group; classifying these organisms is the substance of debates among scientists who still ponder the evolutionary relationship among members of this group. Using a simplistic approach such as "plantlike" and "animallike" may not appropriately reflect evolutionary closeness or distance among organisms, and it may be an arbitrary way to group organisms.
3. Eventually the exponential growth of the algae will slow as nutrients are used and the surface of the water becomes covered. Some algae will then begin to die. As the algae die, bacteria and other decomposers feed on it. As they do, they use oxygen dissolved in the water. As the oxygen level decreases, other organisms dependent on the oxygen begin to die also, providing yet more food for decomposers. Poisoning the algae would have a devastating effect on the lake, because a great deal of food would become available to decomposers. Unless oxygen is returned to the system, as happens in a rapidly flowing stream splashing over rocks, the lake will die.

CHAPTER 27

Knowledge and Comprehension Questions

1. Vascular; Nonvascular.
2. Sporophytes; gametophytes; sporophytes.
3. The three classes are the mosses, liverworts, and hornworts. All three share a fairly uniform life cycle that involves two phases. The sporophyte grows out of or is embedded in the tissues of the gametophyte and depends on it for nutrition. It produces spores by meiosis. These spores germinate into male and female gametophytes, which produce sperm and eggs, respectively. Fertilization results in a zygote capable of developing into a mature sporophyte.
4. Seeds are structures from which sporophyte plants grow. They protect the embryonic sporophyte plant from drying out or being eaten, and they contain stored food for the plant.
5. In seedless vascular plants, the sporophyte generation is dominant and lives separately from the gametophyte. The gametophytes produce motile sperm that swim to the eggs. The fertilized eggs produce sporophytes that at first grow within gametophyte tissues but eventually become free-living. A fern is an example of this type of plant.
6. Gymnosperms (c), angiosperms (d), bryophytes (b), ferns (a).
7. See Figure 27-7.
8. Sexual reproduction in gymnosperms

and angiosperms produces seeds by the union of male and female gametes. Germinating pollen tubes convey the sperm to the eggs. The seeds protect and nourish the young sporophytes as they develop into new plants, the dominant generation. The sporophytes produce the male and female gametophytes.

9. Fruits have evolved a range of shapes, textures, and tastes that help disperse their seeds effectively. Fruits with brightly colored flesh are often eaten by animals and thus dispersed to different habitats. Other fruits have evolved hooked spines that stick to animals' fur; others have wings that are blown by the wind.
10. b, d, e, a, c.
11. Monocots and dicots are both angiosperms. They differ in the placement of the stored food in their seeds. In dicots, most of the stored food is in the cotyledons (seed leaves), whereas in monocots, most of it is stored in endosperm. In addition, monocots usually have parallel veins in their leaves, and dicots have netlike veins. Monocots also have flower parts in threes, whereas dicots have flower parts in fours and fives.
12. Vegetative propagation is an asexual reproductive process in which a new plant develops from a portion of a parent plant. An example is the white potato, a tuber which can grow new plants from its "eyes."
13. Hormones are chemical substances produced in one part of an organism and then transported to another part of the organism, where they bring about physiological responses. The five kinds of plant hormones are auxins, gibberellins, cytokinins (promote and regulate growth), abscisic acid (inhibits growth), and ethylene (affects the ripening of fruit).

Critical Thinking

1. The reproductive advantage of the dandelion flower is the adaptation to dispersal of its seeds by the wind. (Actually the dandelion "flower" is not a flower at all but is many fruits.)

CHAPTER 28

Knowledge and Comprehension Questions

1. Vascular plants are plants having a system of vessels that transport water and nutrients.
2. Root; shoot; leaves; stem.
3. (1) c, (2) b, (3) a, (4) d.
4. Xylem and phloem are both types of vascular tissue, but they have different functions and contain different types of conducting cells. Xylem conducts water and dissolved inorganic nutrients, whereas phloem conducts carbohydrates and other substances.
5. a. Cells in ground tissue that function in photosynthesis and storage.
   b. The most abundant type of cell in dermal tissue, often covered with a thick waxy layer that protects the plant and retards water loss.
   c. Openings that connect air spaces inside dicot leaves with the exterior; they are opened and closed by the change in shape of surrounding guard cells. Water vapor and gases move into and out of the leaves through these openings.
   d. Thimblelike mass of unorganized cells that covers and protects a root's apical meristem as it grows through soil.
6. Primary growth occurs mainly at the tips of roots and shoots, making plants taller; secondary growth occurs in the lateral meristem and makes plants larger in diameter.
7. Vegetables are a good source of inorganic nutrients (we call them minerals in our diets). Plants take in inorganic nutrients from the soil and form proteins, fats, and vitamins.
8. See Figures 28-5 and 28-6.
9. A dandelion has a taproot, a single major root that can grow deep as a firm anchor. Some taproots also store food for the plant. An ivy plant has adventitious roots that develop from the lower part of the stem and help anchor the plant. Grass has fibrous roots, a root system without one major root; this type of system anchors the plant and absorbs nutrients over a large surface area.
10. Annual rings provide a clue to a tree's age by illustrating the annual pattern of rapid and slow growth in the cambium (lateral meristem tissue). This results in secondary growth.
11. Water rises beyond the point at which it would be supported by air pressure because evaporation from the plant's leaves produces a force that pulls upward on the entire column of water. The forces involved include air pressure, adhesion, cohesion, and, primarily, transpiration.
12. See Figure 28-19.
13. In general, nonvascular plants grow closer to the ground than vascular plants because they lack the sophisticated transport systems of vascular plants.

Critical Thinking

1. Air pressure still pushes on water in the ground, pushing water up the thin xylem tubes of the plant. In addition, the water molecules adhere to the walls of the xylem tubes and stick to one another (cohesion), which helps maintain the column.
2. The fungi, because they are at the roots, probably function in making inorganic nutrients in the soil available to the plants.

CHAPTER 29

Knowledge and Comprehension Questions

1. Both you and jellyfish are animals. An important difference is that jellyfish are invertebrates (they lack a backbone), whereas you are a vertebrate (with a backbone and dorsal nerve cord).
2. Eukaryotic organisms have a distinct nucleus and a cellular structure different from that of bacteria. Animals are also multicellular (having more than one cell), and as heterotrophs, they are unable to make their own food. Therefore they must eat other organic matter for food. Beyond these basic similarities, however, animals are a diverse group.
3. Radial symmetry; bilateral symmetry.
4. Coelomates (such as humans) have lined body cavities in which organs are suspended by thin sheets of connective tissue. Pseudocoelomates (such as roundworms) have body cavities, but the cavities are not fully lined and their organs are not suspended by connective tissue. Acoelomates (such as flatworms) have no body cavity surrounding the gut.
5. Sponges are called the simplest animals because they lack tissues and organs.
6. Cnidarians exist as either polyps (cylindrically shaped animals that anchor to rocks) or as medusae (free-floating, umbrella-shaped animals).
7. The acoelomates are the most primitive bilaterally symmetrical animals. They have organs and some organ systems, including a primitive brain. Their organs are embedded within body tissues.
8. Roundworms can be dangerous if they parasitize humans, absorbing digested food from their hosts. Roundworms are cylindrical pseudocoelomates encased in a flexible outer cuticle; they have primitive excretory and nervous systems. They move by using muscles attached to the cuticle that push against the pseudocoelom.
9. In a closed circulatory system, blood is enclosed within vessels as it travels through the body. In an open system, blood flows through irregular channels (blood sinuses) in many parts of the body. Since you have blood vessels, you have a closed system.
10. Mollusks are coelomates with a muscular foot and soft body, usually covered by a shell. They use gills for respiration in water, whereas terrestrial species have adaptations to breathing on land. They show both open and closed circulatory systems and well-developed excretory systems.
11. Like many arthropods, flies have compound eyes that contain many independent visual units, each with its own lens. This feature helps them detect motion quickly.
12. Arthropods are a diverse phylum of organisms having jointed appendages and rigid exoskeletons.
13. Insects belong to a group of arthropods that have jaws (mandibles). Spiders belong to a different group of arthropods that lack mandibles.
14. Echinoderms are deuterostomes, whereas the other invertebrates are protostomes. These embryological differences make echinoderms more similar to the chordates, which implies that echinoderms and chordates evolved from a common ancestor.

Critical Thinking

1. Generally, no. Most hermaphroditic invertebrates such as flatworms and earthworms still reproduce sexually although each organism has both male and female reproductive glands. Sponges produce both sperm and eggs, but these gametes float through the water and fertilization generally takes place between gametes of different organisms.

CHAPTER 30

Knowledge and Comprehension Questions

1. Nerve cord; notochord; pharyngeal (gill) slits.
2. The nerve cord differentiates into a brain and spinal cord; the notochord serves as a core for the development of the vertebral column, which encloses the nerve cord; and the pharyngeal arches develop into gills (in fishes) or into ear, jaw, and throat structures (terrestrial vertebrates).
3. a (3), b (1), c (2).
4. Vertebrates have (a) a vertebral column, (b) a distinct head with a skull enclosing their brain, (c) a closed circulatory system, and (d) a heart that pumps their blood.
5. A lateral line system (found in all fishes and amphibians) is a system of mechanoreceptors that detects sound, pressure, and movement.
6. Osmoregulation is the control of water movement into and out of organisms' bodies. Maintaining the proper balance of water and solutes is vital to life. Terrestrial organisms must conserve water, whereas aquatic organisms must have mechanisms to maintain an appropriate water balance living in either a freshwater or saltwater environment. Freshwater fishes tend to take on water whereas marine fishes tend to lose water.
7. Ovoviviparous organisms (some fish, reptiles; many insects) retain fertilized eggs within their oviducts until the young hatch; the young receive nourishment from the egg. Oviparous animals (birds) lay eggs, and their young hatch outside the mother. Viviparous

organisms (humans) bear their young alive, and the young are nourished by the mother, not the egg.

8. *Amphibian* means "two lives." This refers to the fact that amphibians live both aquatic and terrestrial existences.
9. Ectothermic animals (fish, amphibians, reptiles) regulate body temperature by taking in heat from the environment. Endothermic animals (birds, mammals) regulate their body temperatures internally.
10. Reptiles have dry skins covered with scales, retarding water loss. They also lay amniotic eggs, which contain nutrients and water for the embryos and protect the embryos from drying out.
11. Feathers are flexible, light, waterproof epidermal structures; birds are the only animals that have them. Feathers are important to flight; they also give birds waterproof coverings and insulate them against temperaturechanges.
12. Mammals are endothermic vertebrates that have hair and whose females secrete milk from mammary glands to feed their young.

Critical Thinking

1. The three subphyla of chordates are tunicates, lancelets, and vertebrates. The tunicates are marine saclike forms with no notochord in the adult. The structure of these bloblike creatures fits their life-style as sessile filter-feeders. Their structure allows them to draw in organisms and allows filtered water to exit. The lancelets are tiny, scaleless fishlike marine chordates. Their knifelike shape allows them to bury the posterior parts of their bodies in the sand while their anterior ends stick out and feed on plankton. Vertebrates each have a vertebral column, distinct head, and a closed circulatory system; they are much more complex organisms than the other two subphyla of chordates. These characteristics allow for the range of life-styles of the vertebrates, from those that swim in the oceans, fly through the air, walk on four legs on land, or walk upright. The vertebral column provides support and acts as a lever system for movement.
2. The skin of amphibians limits their ability to remain on land for long periods of time because they dry out unless their skin is kept moist. Likewise, other terrestrial organisms are able to live successfully on land because of their skins (and their osmoregulatory mechanisms), all of which are designed to conserve water.

CHAPTER 31

Knowledge and Comprehension Questions

1. Biotechnology is the manipulation of organisms to yield organisms with specific characteristics, organisms that produce particular products, and information about an organism or tissue that would otherwise not be known. Genetic engineering refers to techniques of molecular biotechnology (as opposed to classical biotechnology) in which genes are manipulated, not just organisms.
2. Classical biotechnology employs the techniques of selection, mutation, and hybridization, manipulating the genetics of organisms at the organism level. Molecular biotechnology employs the techniques of genetic engineering, or recombinant DNA technology, in which genes of organisms are manipulated. Yes, both are being practiced today. Dog breeders, for example, still practice classical biotechnology in breeding programs, whereas molecular biologists continually break new ground in the applications of molecular biotechnology. (Evidence will vary among students.)
3. Three methods of natural gene transfer among bacteria are transformation, transduction, and conjugation. In transformation, free pieces of DNA move from a donor cell to a recipient cell. During transduction, DNA from a donor cell is transferred to a recipient cell by a virus. During conjugation, a donor and a recipient cell make contact, and the DNA from the donor is injected into the recipient cell.
4. After identifying and isolating a desired human gene, scientists can insert the gene or genes into bacteria using either plasmids or viruses. Plasmids can be extracted from bacteria or yeasts, induced to incorporate genes (restriction fragments) into their genomes, and reinserted into the cell, much like a cell would take up DNA fragments during the natural process of transformation. Scientists can also incorporate restriction fragments into viral genomes and use the viruses to infect bacteria similar to the natural process of transduction.
5. Gene cloning is the process of making copies of genes. In shotgun cloning, no one gene is targeted for cloning—all the genes of a genome are cloned from restriction fragments. In complementary DNA cloning, genes are "manufactured" by using the messenger RNA of the gene and the enzyme reverse transcriptase. Genes can also be synthesized in the laboratory when the nucleotide sequence of the gene is known; this process is called *gene synthesis cloning*.
6. Restriction enzymes play an important role in genetic engineering because they are tools that scientists use to cut long, intact DNA strands into fragments that contain one or a few genes. These small pieces are more easily studied than long DNA strands.
7. Molecular probes are molecules that bind to specific genes or nucleotide sequences. Probes help scientists identify searched-for genes.
8. The polymerase chain reaction (PCR) makes many copies of a gene or segment of DNA. This is necessary so that (a) there are sufficient copies of a gene available for large-scale culturing or (b) extremely small amounts of DNA can be amplified for analysis.
9. Answers may vary, but two genetically engineered proteins in use today are human insulin, used to treat diabetes, and human growth hormone, used to treat growth disorders in children.
10. Gene therapy is the treatment of a genetic disorder by the insertion of "normal" genes into a patient's cells. This technique is not in widespread use today; only a few persons have under-

gone gene therapy, with the first case occurring in the fall of 1990.

11. The goals of the Human Genome Project (HGP), a worldwide effort to sequence the DNA of the entire human genome, are to develop three increasingly detailed maps of the DNA in human cells. These maps will show the number of nucleotides and the distances between reference points on the DNA, as well as the sequence the nucleotides in each chromosome. The purpose is to give scientists a clearer picture of the genes that cause various diseases so that scientists will be able to study the relationship between the structure of genes and the proteins they produce. Pooling the data of targeted gene replacement research with HGP research will give scientists insight into the connection between the structure and function of genes. The process of gene targeting involves changing the nucleotide sequence of a particular gene, which changes the function of the gene, and then observing the resultant changes in the anatomy, physiology, or behavior of the organism.
12. Answers will vary. One application of genetic engineering in agriculture is the improvement of crops and forest trees by making them more resistant to disease, frost, and herbicides.

Critical Thinking

1. Scientists must insert human genes into the bacterial genome so that bacteria can express those genes, producing human protein. Gold and other precious metals are not produced by living things and thus are not coded for by DNA. Although it sounds like a great idea to have bacteria produce gold, it won't work.
2. Humans produce antibodies against foreign proteins. Although genetically engineered insulin is synthesized by bacteria, it is still human (not foreign) protein because it is coded for by human DNA. Animal-derived insulin, on the other hand, is coded for and synthesized by an animal other than a human. The human body is likely to react by producing antibodies against such protein.

CHAPTER 32

Knowledge and Comprehension Questions

1. Behaviors include the patterns of movement, sounds, and body positions exhibited by animals. They also include any type of change in an animal that can trigger behaviors in other animals.
2. This statement expresses a fundamental idea in the study of animal behavior—that animal behaviors are evolutionary adaptations that make an organism more fit to survive in and adjust to its environment.
3. Innate (inborn); learned.
4. Kineses; taxes; reflexes; fixed action patterns.
5. A kinesis is a change in the speed of an animals's random movements in response to environmental stimuli (for example, insects moving around underneath a fallen log). A taxis is a directed movement toward or away from a stimulus (for example, moths being attracted to a light at night). A reflex is an automatic response to nerve stimulation (for example, knee jerk reflex in humans). A fixed action patterns is a sequence of innate behaviors in which the actions follow an unchanging order of muscular movements (for example, a mother bird stuffing an insect into the gaping beak of her offspring).
6. Innate behaviors protect an animal from environmental hazards without it having learned to do so; many social behaviors also depend on innate behaviors. Organisms with simple nervous systems rely primarily on innate behaviors for survival. Learned behaviors, however, allow an animal to adjust its behavior on the basis of its experiences, helping it to adapt better to its environment or to environmental changes.
7. Imprinting; habituation; trial-and-error learning; classical conditioning; insight.
8. Type of behavior:
   a. Taxis (phototaxis).
   b. Habituation.
   c. Trial-and-error conditioning (operant conditioning).
9. Operant conditioning is a form of learning in which an animal associates something that it does with a reward or punishment. In classical conditioning, an animal learns to associate a new stimulus with a natural stimulus that normally evokes a particular response. Eventually the animal will respond to the new stimulus alone.
10. Insight (reasoning) is the most complex form of learning in which an animal recognizes a problem and solves it mentally before trying out the solution. Insight allows an animal to perform a correct behavior the first time it tries; it helps the animal adjust and respond to new situations and perhaps determine better ways to deal with its environment.
11. Type of behavior:
    a. Locality imprinting (migration).
    b. Reflex.
    c. Fixed action pattern.

Critical Thinking

1. Answers will vary, but hypotheses should focus on imprinting and other means of animal migration mentioned in the chapter. Research shows, however, that Pacific salmon become imprinted with the smells of their native stream. Because they can discriminate between the waters of two rivers coming together at a fork, they know which way to swim.

CHAPTER 33

Knowledge and Comprehension Questions

1. Behaviors are the patterns of movement, vocalizations, and body positions exhibited by animals. Social behaviors are behaviors that help members of the same species communicate and interact.
2. Social behaviors help sexually reproducing species reproduce, care for offspring, and defend territories. Other behaviors may help animals to hunt for and share food and to warn and defend against predators.
3. Competitive behavior results when two or more individuals are striving to obtain the same resource (such as food, territories, or mates).
4. Type of behavior
   a. Threat display.
   b. Submissive behavior.
5. An animal exhibiting territorial behavior marks off an area as its own and defends it against same-sex members of its species. Although territoriality evolved as a result of the advantages to individuals, this system has an effect on the resources available to various members within a population. When resources such as nesting sites and food are limited, each member of a population is in danger of the resources being spread too thin, so that no member gets an adequate amount of what it needs. In addition, breeding pairs of animals will not reproduce unless they have adequate resources; territorial behavior ensures that at least some members of the population will have their own space, food, and nesting sites so that they will survive and reproduce. Territorial behaviors also enhance reproductive capabilities by placing the peak time of competition and aggressive behavior at the time of the marking of the territory, *before* the time of reproduction and raising of young.
6. Sexual reproduction involves the union of male and female sex cells to form the first cells of new individuals. It provides genetic variability to a species, giving it the raw materials needed to adapt to environmental change.
7. Mating includes male-female behaviors that result in fertilization; courtship includes the behavior patterns that lead to mating.
8. Type of behavior
   a. Parenting behavior.
   b. Courtship ritual (reproductive behavior).
   c. Territorial behavior.
9. Animals form social groups to work together toward common goals: to protect against predators, hunt for food, and create a division of labor. The genes encoding these behaviors are retained in the gene pool if the advantages for the individual outweigh the disadvantages.
10. A honeybee colony consists of three castes: (a) the queen, which lays the eggs, (b) the drones, male bees that fertilize the queen's eggs, and (c) the workers, female bees that develop from fertilized eggs and do everything around the hive except lay eggs and fertilize them. (This includes feeding larvae, guarding the hive, and foraging for food).
11. A rank order is the social hierarchy that appears in groups of fishes, reptiles, birds, and mammals. This hierarchy helps reduce aggression and fighting within social groups, focusing these behaviors to a short time when the rank order develops.
12. Answers will vary; human behavior appears to result from a combination of hereditary and environmental factors.

Critical Thinking

1. Answers will vary. However, the evolutionary basis of group social behaviors is that the advantages of a particular social trait must outweigh the disadvantages for the individual and its kin. Individuals within colonies of bees, for example, are closely related. Mammals do not produce large numbers of offspring as occurs in bee and ant colonies; large groups of mammals are generally not closely related. Research has shown, however, that naked mole-rats live within colonies and are closely related—an unusual situation for mammals.

CHAPTER 34

Knowledge and Comprehension Questions

1. Ecology is the study of interactions between organisms and the environment. Underlying questions to this study are "What factors determine the distribution patterns of organisms?" and "What factors control their

numbers in the locations in which they are distributed?

2. The population size is affected by immigration. The growth rate does not include factors of emigration and immigration.
3. Demographers study human populations, describing peoples and the characteristics of populations. They predict the ways in which the sizes of populations will alter in the future, taking into account the age distribution of the population and its changing size through time. Population ecologists study how populations of any species grow and interact, investigating the factors that determine the distribution patterns of organisms and the factors that control their numbers in the locations in which they are distributed. Although both study populations, demographers are concerned only with human populations.
4. A large population is more likely to survive because it is less vulnerable to random events and natural disasters and because its size reflects greater genetic diversity than does a small population.
5. Density refers to the number of organisms per unit of area; dispersion refers to the way in which the individuals of a population are arranged. If individuals are too far apart, they may not be able to reproduce. If population becomes too dense, however, factors can arise that limit the population's size (disease, predation, starvation).
6. The three patterns are (a) uniform (evenly spaced), (b) random, and (c) clumped (organisms are grouped in areas of the environment that have the necessary resources). Human populations are distributed in the clumped pattern.
7. Density-dependent limiting factors (such as competition) come into play when a population size is large, whereas density-independent factors (such as weather) operate regardless of population size.
8. Both predation and parasitism are density-dependent limiting factors. Predators are organisms that kill and eat organisms of another species; parasites live on or in other species and derive nourishment from them but do not necessarily kill them.
9. See Figure 34-7. Type I is typical of organisms that tend to survive past their reproductive years (such as humans). Type II characterizes organisms that reproduce asexually and are equally likely to die at any age (such as the hydra). Type III is characteristic of organisms that produce offspring that must survive on their own and therefore die in large numbers when young (oysters).
10. Developing countries tend to have a population pyramid with a broad base, reflecting a rapidly growing population with large numbers of individuals entering their reproductive years. Industrialized countries tend to have populations with similar proportions of their populations in each age-group and that are growing slowly or not at all.
11. 1.8%; 95 million; double.

Critical Thinking

1. People are already putting pressure on our natural resources (land, water, forests, atmosphere). Increasing industrialization leads to rising levels of pollution and solid waste. Even though some experts suggest that there is enough food produced to feed the world's population, many people live—and die—in poverty and hunger.
2. Answers will vary.

CHAPTER 35

Knowledge and Comprehension Questions

1. A population consists of individuals of the same species. Populations that interact with one another form a community. An ecosystem is a community of organisms along with the abiotic factors with which the community interacts.
2. Habitat; niche.
3. Interactions can be grouped into (a) competition (organisms of different species living near each other strive to obtain the same limited resources), (b) predation (organisms of one species kill and eat organisms of another species), (c) commensalism (an organism of one species benefits from its interaction with another while the other species neither benefits nor is harmed), (d) mutualism (two species live together in close association, both benefiting from the relationship), and (e) parasitism (an organism of one species [the parasite] lives in or on another [the host]. The parasite benefits from this relationship while the host is usually harmed.)
4. False. Competition is often greatest between organisms that obtain their food in similar ways.
5. The principle of competitive exclusion states that if two species are competing for the same limited resource in a location, the species that can use the resource most efficiently will eventually eliminate the other species in that area.
6. An organism's realized niche is the role it actually plays in the ecosystem; its fundamental niche is the role it might play if competitors were not present.
7. Predator; prey; parasite.
8. Gause's experiments showed that predator-prey relationships follow a cyclical pattern: as the number of prey increases, the number of predators increases. When the prey decrease because of predation, the predator population also declines until an upsurge in the prey population starts the cycle again.
9. The balance between predator and prey populations is complex because interactions depend on many factors: interactions with other organisms, movement of new organisms into or out of the community, and abiotic factors in the ecosystem.
10. Coevolution is the long-term, mutual evolutionary adjustment of characteristics of members of biological communities is relation to one another. Cattle are one example: their digestive systems have adapted (grinding teeth, strong jaws, a rumen) to allow them to digest silica in certain grasses.
11. Type of adaptation
    a. Warning coloration.
    b. Camouflage.

12. Primary succession takes place in areas originally bare. It could begin, for instance, with lichens growing on bare rock, producing acids that break down the rock to form soil. This gives rise to a pioneer community, which in turn leads to more vegetation, eventually making the area able to support other forms of life. At some point the mix of plants and animals becomes somewhat stable, forming a climax community.
13. Secondary succession.

Critical Thinking

1. Answers will vary. However, students must show how a one-sided relationship (commensalism) between two organisms of different species really benefits both rather than only one.
2. To determine whether the resemblance between two butterfly species is mimicry or not, one could determine if the supposed mimic existed in small numbers in the community with respect to the organism it was mimicking. If so, it is probably a mimic. One could also test the butterfly for the presence of toxin. If toxin was present, it could exist in the community in large numbers also and be a mimic.

CHAPTER 36

Knowledge and Comprehension Questions

1. An ecosystem is a community of plants, animals, and microorganisms that interact with one another and their environment and that are interdependent. The living (biotic) groups it contains include producers (organisms that can make their own food), consumers (organisms that eat other organisms for food), and decomposers (consumers that eat dead matter).
2. Humans are consumers.
3. See Figure 36-1.
4. Primary consumers; secondary consumers; decomposers; detritus.
5. A trophic level is a feeding level within a food chain. A food chain is a linear relationship among organisms that feed one on another. Food webs are food chains that interweave with one another.
6. A green plant captures some of the sun's energy. The successive members of a food chain, in turn, incorporate into their own bodies about 10% of the energy in the organisms on which they feed. The rest is released as heat or is used to power metabolic activities.
7. Chemicals are stored in the atmosphere and in rocks (reservoirs). These substances enter the soil, are incorporated into the bodies of living organisms, and are passed along the food chain. Ultimately, the chemicals, with help from decomposers, return to the nonliving reservoirs.
8. Water in the atmosphere falls to the Earth as precipitation. Plants take up water from the soil, and animals obtain water from drinking it or from eating plants or other animals. Water returns to the atmosphere through the evaporation of surface water and transpiration by plants.
9. Groundwater includes subsurface bodies of fresh water. It can be polluted by pesticides that are washed from plants and topsoil by the rain and by chemical wastes that are dumped in surface water. It is serious because at the present time there is no technology to remove these pollutants from groundwater; already about 2% of groundwater in the United States is polluted, and the situation is getting worse.
10. Carbon is held in the atmosphere as carbon dioxide. Photosynthetic organisms take in this gas and use it to build carbon compounds. Both producers and consumers use carbon compounds as an energy source, metabolizing them and releasing carbon dioxide to the atmosphere again.
11. Crop rotation can improve soil by alternating a nonleguminous crop (which consumes nitrogen from the soil) with a leguminous crop (which can enrich the nitrate level of depleted soil by forming mutualistic associations with nitrogen-fixing bacteria). This relates to the nitrogen cycle.
12. Adding more phosphates than crops need causes the extra phosphates to wash off or erode into water. Bodies of water can become eutrophic; algae and other aquatic plants overgrow and decomposers feed on them, using much of the water's oxygen and choking out other organisms.

Critical Thinking

1. Student answers will vary. However, the ecosphere must include producers to serve as food for the consumers. Decomposers are necessary to break down dead material, making the break-down products available for re-use. The photosynthesizing organisms (producers) will generate oxygen for the nonphotosynthesizing organisms (consumers). Both producers and consumers will generate carbon dioxide through cell respiration, which is needed by the producers. A moderate amount of water must be present in a terrestrial environment to be continually cycled throughout the system. The specific organisms would, of course, need to be carefully chosen for their ability to survive in such a system. An aquatic self-sustaining closed system might consist of snails and elodea plants.
2. Producers ultimately support all the other organisms in a food chain. Organisms highest in the food chain have the least food available to them; the total number of organisms able to be fed at each succeeding level of the food chain

diminishes because energy is lost as heat and is used for work at each level. Only 10% of the energy from food is incorporated into the bodies of the organisms at each level. Therefore, there is a limit to the number of links in the chain and how far removed an organism can be from producers.

3. Plants transpire, or release water vapor into the atmosphere. This water vapor is a critical part of the water cycle. As the rain forest is cut down, the plants are lost and less water is available in the atmosphere to return to the ground as precipitation.

CHAPTER 37

Knowledge and Comprehension Questions

1. A biome is an ecosystem of plants and animals that occurs over wide areas within specific climatic regions.
2. Australia is in the southern hemisphere, whereas the United States is in the northern hemisphere. Thus when the Earth tilts so that the United States is tilted away from the sun, people in the United States experience winter but it is summer in Australia (which is tilted toward the sun). When the Earth tilts so that the United States is tilted toward the sun, it is summer in America but winter in Australia.
3. Near the equator, warm air rises and flows toward the poles; the Earth's rotation breaks the air into six coils of rising and falling air that surround the Earth. The air in each coil rises at the region of its lowest latitude, moves toward the poles, sinks at the region of its highest latitude, and flows back to the equator. These air masses affect wind patterns and precipitation worldwide.
4. The climate of an area is determined by its latitude, wind patterns, and surface features of the Earth.
5. Terrestrial biome:
   a. Tropical rain forests
   b. Temperate deciduous forests
   c. Arctic tundra
   d. Savannas
   e. Taiga
   f. Temperate grasslands
   g. Desert
6. Desert rainfall is limited by latitude because the air, which has released its moisture over the rain forests, is dry over desert latitudes. Deserts can also occur in the interior of continents, far from the ocean, and on the leeward side of mountains that block rainfall.
7. The taiga is the northern coniferous forest with long, cold winters and little precipitation. The tundra lies farther north and is open, often boggy, with desertlike levels of precipitation.
8. The three freshwater life zones are (a) shore zone (shallow water in which plants with roots may grow; some consumers live here), (b) open-water zone (the main body of water through which light penetrates; floating plants, microscopic floating animals, and fish live here), and (c) deep zone (light does not penetrate, so there are no producers; inhabited mainly by decomposers or organisms that feed on organic material that filters down).
9. An estuary is an environment where fresh water and salt water meet. All organisms living here are adapted to living in an area of moving water and changing salt concentrations. Some are bottom dwellers that attach to bottom material or burrow into the mud; others spawn in less salty water.
10. Estuaries are ecosystems where freshwater rivers and streams meet the ocean. They are vulnerable to pollutants that flow from upstream, as well as to pesticides that leach from the land and soil.
11. The three major life zones of the ocean are (a) intertidal (area between high and low tides), (b) neritic (shallow waters along the coasts of the continents), and (c) open sea (the rest of the ocean).
12. Types of organisms
    a. Photic zone (well-lit layer of the ocean, which contains phytoplankton, zooplankton, fish, and air-breathing aquatic mammals (nekton)
    b. Mesopelagic zone (organisms exhibiting bioluminesence, some fishes [swordfish, certain sharks], whales, and cephalopods)
    c. Abyssal zone (benthos or bottom dwellers that eat detritus)

Critical Thinking

1. Tropical rain forests make poor farmland because most nutrients are found in the vegetation, not the soil. Root systems are poor and shallow; the soil itself is thin, often only a few centimeters deep. Cutting down and burning the trees remove the nutrients and break down the organic matter to carbon dioxide. In 2 to 3 years the land becomes barren. Conversely, in grasslands, root systems are extensive and deep. Much of the nutrients are held in the root systems and in the soil rather than in the above-ground vegetation.

CHAPTER 38

Knowledge and Comprehension Questions

1. The biosphere is the global ecosystem in which all other ecosystems on the Earth exist.
2. A sustainable society uses nonfinite, renewable sources of energy. Society's needs are satisfied without compromising future generations and natural resources.
3. People could use solar, wind, and water power, as well as the Earth's geothermal energy.
4. People can reuse and recycle paper, glass, and metal products to help preserve mineral resources. This is important because mineral supplies are finite; once used up, they are gone forever.
5. Tropical rain forests are home to more than half the world's species of plants, animals, and insects from which people get many medicines as well as wood, charcoal, oils and nuts. Burning the forests adds carbon dioxide to the air, which may raise temperatures worldwide. Destroying the forests also destroys the habitat for many species and leads to extinctions.
6. Global warming refers to a worldwide increase in temperature. Even a small increase in temperature could melt ice at the poles, raising the sea level and flooding one fifth of the world's land area. Higher temperatures also affect rain patterns and agriculture.
7. Biological diversity refers to the richness of species. Humans are reducing the world's biological diversity through habitat destruction, pollution, exploitation of commercially valuable species, and selective breeding. Reducing diversity makes species more vulnerable to being wiped out by disease or environmental changes.
8. Humans can improve solid waste management by increasing and improving recycling, using safe incineration, engineering better landfill sites, and practicing waste-to-energy reclamation.
9. Point sources are sources of pollutants that enter the water at one or a few distinct places (industrial waste dumped by a specific factory). Nonpoint sources refer to pollutants that enter water at a variety of places (metals and acids draining from mines).
10. Organic nutrients are food for bacteria; as the bacteria multiply, they use up oxygen in the water, killing the fish. Inorganic nutrients can stimulate plant growth in the water. Bacteria decompose the plants after they die; again, when the bacteria multiply, they consume the water's oxygen. The decomposed materials also begin to fill in the body of water, eventually turning it into a terrestrial community. Pollution with infectious agents can cause serious diseases. Toxic substances in water can poison the organisms that take in this water; sometimes organisms accumulate these substances in their bodies.
11. Acid rain results from gases such as sulfur dioxide and nitrogen dioxide that combine with water in the atmosphere. When this acidic rain falls, it kills aquatic life, pollutes groundwater, damages plants, and eats away at stone, metal, and painted surfaces.
12. Gray-air cities are those with relatively cold, moist climates. They develop a layer of smog because of the burning of fossil fuels and the abundance of sulfur oxides and particulates in the air. The air over brown air cities contains hydrocarbons and nitrogen oxides that react with the sunny climate to produce secondary pollutants such as ozone.
13. Ozone in the stratosphere helps shield you from the sun's ultraviolet rays. Depletion of this layer can damage bacteria and plants and cause burns and skin cancers.

Critical Thinking

1. Answers will vary.
2. Answers will vary.

APPENDIX B

CLASSIFICATION OF ORGANISMS

The purpose of the classification scheme presented here is for student reference. It was compiled from a variety of sources in microbiology, botany, and zoology, taking into account the most widely accepted schemes at this time. It is not intended to include all of the phyla of organisms currently identified by taxonomists, but it includes all of the phyla and divisions described and referred to in *Biology: Understanding Life*. Because this textbook describes humans in its extensive physiology section and includes a chapter on chordate and human evolution, the classification of the chordates is listed in more detail than other phyla.

Kingdom Monera

Prokaryotic, single-celled organisms that have neither membrane-bounded nuclei nor membrane-bounded organelles.

Subkingdom Archaebacteria: Methanogens (methane-producing bacteria), halophiles (bacteria that live in salt marshes), thermoacidophiles (bacteria that live in hot springs and deep-sea vents). A separate kingdom has been proposed for this subkingdom.

Subkingdom Eubacteria: All other bacteria, including nitrogen-fixing bacteria and cyanobacteria.

Kingdom Protista

A varied group of eukaryotic organisms. Many are single celled, although some phyla include multicellular or colonial forms. Some forms are photosynthetic.

Phylum Rhizopoda: Amebas
Phylum Zoomastigophora: Flagellates
Phylum Ciliophora: Ciliates
Phylum Sporozoa: Sporozoans
Phylum Dinoflagellata: Dinoflagellates
Phylum Chrysophyta: Diatoms, yellow-green algae, golden-brown algae
Phylum Phaeophyta: Brown algae
Phylum Chlorophyta: Green algae
Phylum Rhodophyta: Red algae
Phylum Acrasiomycota: Cellular slime molds

Phylum Myxomycota: Plasmodial slime molds
Phylum Oomycota: Water molds, white rusts, downy mildews

Kingdom Fungi

Mostly multicellular eukaryotic organisms that are saprophytes, feeding on dead or decaying organic material. Some fungi are parasitic, feeding on living organisms.

Division Zygomycota: Zygote-forming fungi (black bread mold)
Division Ascomycota: Sac fungi (yeasts, cup fungi)
Division Basidiomycota: Club fungi (mushrooms, puffballs, shelf fungi)
Division Fungi Imperfecti: Imperfect fungi (fungi that have no sexual stages of reproduction, such as *Penicillium* and *Aspergillus*)
Lichens: Associations between fungi and green algae and/or cyanobacteria

Kingdom Plantae

Multicellular, eukaryotic organisms that evolved on land and that perform photosynthesis, producing their own food by using energy from the sun and carbon dioxide from the atmosphere.

Division Bryophyta: Mosses, hornworts, liverworts
Division Psilophyta: Whisk ferns
Division Lycophyta: Club mosses
Division Sphenophyta: Horsetails
Division Pterophyta: Ferns
Division Coniferophyta: Conifers
Division Cycadophyta: Cycads
Division Ginkgophyta: Ginkgos
Division Gnetophyta: Gnetae
Division Anthophyta: Flowering plants (angiosperms)
- Class Monocotyledons: Grasses, irises
- Class Dicotyledons: Flowering trees, shrubs, roses

Kingdom Animalia

Multicellular, eukaryotic organisms that are heterotrophic, eating other organisms for food.

Phylum Porifera: Sponges
Phylum Cnidaria
- Class Hydrozoa: Hydra
- Class Scyphozoa: Jellyfish
- Class Anthozoa: Corals, sea anemones

Phylum Ctenophora: Comb jellies, sea walnuts
Phylum Platyhelminthes
- Class Turbellaria: Free-living flatworms
- Class Trematoda: Flukes
- Class Cestoda: Tapeworms

Phylum Nematoda: Roundworms
Phylum Mollusca
- Class Polyplacophora: Chitons
- Class Gastropoda: Snails and slugs
- Class Bivalvia: Bivalves
- Class Cephalopoda: Octopuses, squids, nautilus

Phylum Annelida
- Class Polychaeta: Marine worms

Class Oligochaeta: Earthworms and fresh water worms
Class Hirundinea: Leeches
Phylum Arthropoda
Subphylum Chelicerata
Class Arachnida: Spiders, mites, ticks
Class Merostomata: Horseshoe crabs
Class Pycnogonida: Sea spiders
Subphylum Crustacea
Class Crustacea: Lobsters, Crayfish, shrimps, crabs
Subphylum Uniramia
Class Chilopoda: Centipedes
Class Diplopoda: Millipedes
Class Insecta: Insects
Phylum Echinodermata
Class Crinoidea: Sea lilies
Class Asteroidea: Sea stars (starfish)
Class Ophiuroidea: Brittle stars
Class Echinoidea: Sea urchins and sand dollars
Class Holothuroidea: Sea cucumbers
Phylum Chordata: Chordates
Subphylum Urochordata: Tunicates (sea squirts)
Subphylum Cephalochordata: Lancelets
Subphylum Vertebrata: Vertebrates
Class Agnatha: Jawless fish (lamprey, eels, hagfishes)
Class Chondrichthyes: Cartilaginous fishes (sharks, skates, rays)
Class Osteichthyes: Bony fishes (perch, cod, and trout)
Class Amphibia: Salamanders, frogs, toads
Class Reptilia: Reptiles (lizards, snakes, turtles, crocodiles)
Class Aves: Birds
Class Mammalia: Mammals
Subclass Prototheria: Egg-laying mammals (duck-billed platypus, spiny anteater)
Subclass Metatheria: Pouched mammals or marsupials (oppossums, kangaroos, wombats)
Subclass Eutheria: Placental mammals
Order Edentata: Anteaters, armadillos, sloths
Order Lagomorpha: Rabbits, hares, pikas
Order Rodentia: Squirrels, rats, woodchucks
Order Insectivora: Hedgehogs, tenrecs, moles, shrews
Order Carnivora: Dogs, wolves, cats, bears, weasels
Order Pinnipeda: Sea lions, seals, walruses
Order Cetacea: Whales, dolphins, porpoises
Order Proboscidea: Elephants
Order Perissodactyla: Horses, asses, zebras, tapirs, rhinoceroses
Order Artidactyla: Swine, camels, deer, hippopotamuses, antelopes, cattle, sheep, goats
Order Tubulidentia: Aardvarks
Order Scandentia: Tree shrews
Order Chiroptera: Bats
Order Primates: Lemurs, monkeys, humans
Suborder Strepsirhini: Prosimians (lemurs, aye-ayes)
Suborder Haplorhini: Anthropoids
Superfamily Tarsioidea: Tarsiers
Superfamily Ceboidea: New World monkeys
Superfamily Cercopithecoidea: Old World monkeys

Superfamily Hominoidea
- Family Hylobatidae: Gibbons, orangutans, chimpanzees
- Family Pongidae: Gorillas
- Family Hominidae: Humans

APPENDIX C

PERIODIC TABLE OF THE ELEMENTS

- Metals (main-group)
- Metals (transition)
- Metals (inner transition)
- Nonmetals
- Metalloids

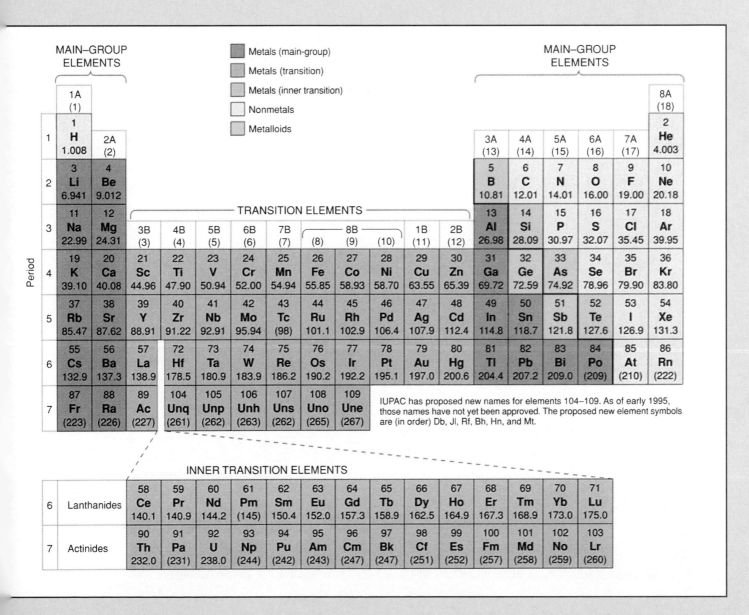

MAIN–GROUP ELEMENTS | TRANSITION ELEMENTS | MAIN–GROUP ELEMENTS

Period	1A (1)	2A (2)	3B (3)	4B (4)	5B (5)	6B (6)	7B (7)	8B (8)	8B (9)	8B (10)	1B (11)	2B (12)	3A (13)	4A (14)	5A (15)	6A (16)	7A (17)	8A (18)
1	1 **H** 1.008																	2 **He** 4.003
2	3 **Li** 6.941	4 **Be** 9.012											5 **B** 10.81	6 **C** 12.01	7 **N** 14.01	8 **O** 16.00	9 **F** 19.00	10 **Ne** 20.18
3	11 **Na** 22.99	12 **Mg** 24.31											13 **Al** 26.98	14 **Si** 28.09	15 **P** 30.97	16 **S** 32.07	17 **Cl** 35.45	18 **Ar** 39.95
4	19 **K** 39.10	20 **Ca** 40.08	21 **Sc** 44.96	22 **Ti** 47.90	23 **V** 50.94	24 **Cr** 52.00	25 **Mn** 54.94	26 **Fe** 55.85	27 **Co** 58.93	28 **Ni** 58.70	29 **Cu** 63.55	30 **Zn** 65.39	31 **Ga** 69.72	32 **Ge** 72.59	33 **As** 74.92	34 **Se** 78.96	35 **Br** 79.90	36 **Kr** 83.80
5	37 **Rb** 85.47	38 **Sr** 87.62	39 **Y** 88.91	40 **Zr** 91.22	41 **Nb** 92.91	42 **Mo** 95.94	43 **Tc** (98)	44 **Ru** 101.1	45 **Rh** 102.9	46 **Pd** 106.4	47 **Ag** 107.9	48 **Cd** 112.4	49 **In** 114.8	50 **Sn** 118.7	51 **Sb** 121.8	52 **Te** 127.6	53 **I** 126.9	54 **Xe** 131.3
6	55 **Cs** 132.9	56 **Ba** 137.3	57 **La** 138.9	72 **Hf** 178.5	73 **Ta** 180.9	74 **W** 183.9	75 **Re** 186.2	76 **Os** 190.2	77 **Ir** 192.2	78 **Pt** 195.1	79 **Au** 197.0	80 **Hg** 200.6	81 **Tl** 204.4	82 **Pb** 207.2	83 **Bi** 209.0	84 **Po** (209)	85 **At** (210)	86 **Rn** (222)
7	87 **Fr** (223)	88 **Ra** (226)	89 **Ac** (227)	104 **Unq** (261)	105 **Unp** (262)	106 **Unh** (263)	107 **Uns** (262)	108 **Uno** (265)	109 **Une** (267)									

IUPAC has proposed new names for elements 104–109. As of early 1995, those names have not yet been approved. The proposed new element symbols are (in order) Db, Jl, Rf, Bh, Hn, and Mt.

INNER TRANSITION ELEMENTS

Period	Series														
6	Lanthanides	58 **Ce** 140.1	59 **Pr** 140.9	60 **Nd** 144.2	61 **Pm** (145)	62 **Sm** 150.4	63 **Eu** 152.0	64 **Gd** 157.3	65 **Tb** 158.9	66 **Dy** 162.5	67 **Ho** 164.9	68 **Er** 167.3	69 **Tm** 168.9	70 **Yb** 173.0	71 **Lu** 175.0
7	Actinides	90 **Th** 232.0	91 **Pa** (231)	92 **U** 238.0	93 **Np** (244)	94 **Pu** (242)	95 **Am** (243)	96 **Cm** (247)	97 **Bk** (247)	98 **Cf** (251)	99 **Es** (252)	100 **Fm** (257)	101 **Md** (258)	102 **No** (259)	103 **Lr** (260)

APPENDIX D

Units of Measurement

UNIT	METRIC EQUIVALENT	SYMBOL	U.S. EQUIVALENT
Measures of length			
1 kilometer	= 1000 meters	km	0.62137 mile
1 meter	= 10 decimeters or 100 centimeters	m	39.37 inches
1 decimeter	= 10 centimeters	dm	3.937 inches
1 centimeter	= 10 millimeters	cm	0.3937 inch
1 millimeter	= 1000 micrometers	mm	No U.S. equivalent
1 micrometer	= 1/1000 millimeter or 1000 namometers	μ	No U.S. equivalent
1 nanometer	= 10 angstroms or 1000 picometers	nm	No U.S. equivalent
1 angstrom	= 1/10,000,000 millimeter	Å	No U.S. equivalent
1 picometer	= 1/1,000,000,000 millimeter	pm	No U.S. equivalent
Measures of volume			
1 cubic meter	= 1000 cubic decimeters	m^3	1.308 cubic yards
1 cubic decimeter	= 1000 cubic centimeters	dm^3	0.03531 cubic feet
1 cubic centimeter	= 1000 cubic millimeters or 1 milliliter	cm^3(cc)	0.06102 cubic inch
Measures of capacity			
1 kiloliter	= 1000 liters	kl	264.18 gallons
1 liter	= 10 deciliters	L	1.0567 quarts
1 deciliter	= 100 milliliters	dl	0.4227 cup
1 milliliter	= volume of 1 gram of water at standard temperature and pressure	ml	0.3381 ounces
Measures of mass			
1 kilogram	= 1000 grams	kg	2.2046 pounds
1 gram	= 100 centrigrams or 1000 milligrams	g	0.0353 ounces
1 centigram	= 10 milligrams	cg	0.1543 grain
1 milligram	= 1/1000 gram	mg	
1 microgram	= 1/1,000,000 gram	μg	
1 nanogram	= 1/1,000,000,000 gram	ng	
1 picogram	= 1/1,000,000,000,000 gram	pg	

Note that a micrometer was formerly called a micron (μ), and a nanometer was formerly called a millimicron (mμ).

COMPARATIVE TEMPERATURE SCALES

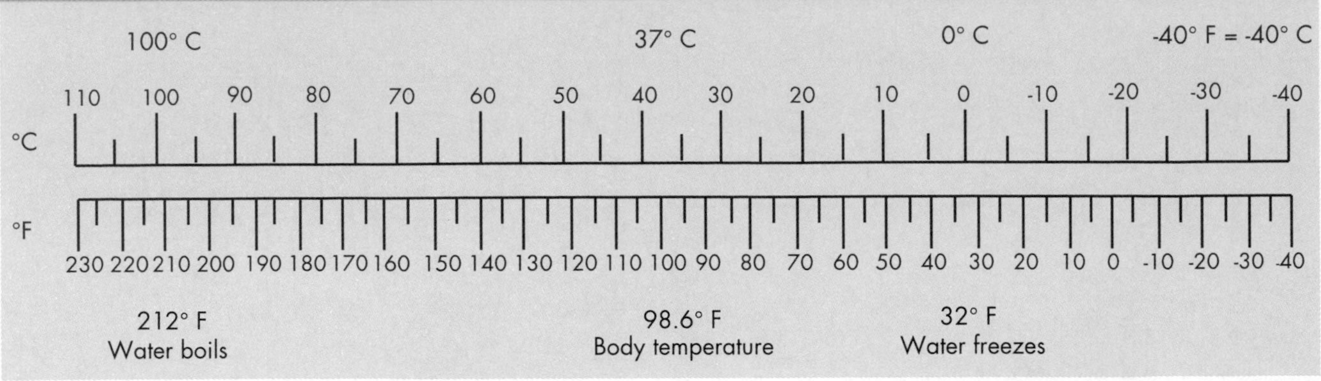

Temperature conversions:
Fahrenheit (°F) to Celsius (°C) = $^5/_9$ (°F − 32)
Celsius (°C) to Fahrenheit (°F) = $^9/_5$ (°C + 32)

GLOSSARY

abiotic Nonliving factors within the environment such as air, water, and rocks. *Ch. 35*

abscisic acid (L. ab, away, off + scisso, dividing) A hormonal growth inhibitor that induces and maintains dormancy in plants. *Ch. 30*

abyssal zone The deepest region of the ocean. The abyssal zone includes ocean depths of 1000 or more meters. *Ch. 37*

acetyl CoA A compound combining the two-carbon acetyl fragment removed during the oxidation of pyruvate with the carrier molecule coenzyme A. (CoA). *Ch. 5*

acetylcholine The neurotransmitter found at neuromuscular junctions that depolarizes the muscle cell membrane, releasing calcium ions that trigger muscle contraction. *Ch. 13*

acetylcholinesterase The enzyme that stops the action of acetylcholine; it is one of the fastest-acting enzymes in the blood. *Ch. 13*

acid Any substance that dissociates to form H^+ ions when it is dissolved in water. *Ch. 2*

acid rain Precipitation with a pH of 5.6 or less, caused primarily by coal-fired plants' emission of sulfur dioxide and nitrogen dioxide, which then combines with water in the atmosphere; results in many devastating environmental effects. *Ch. 38*

acoelomates Organisms that have no body cavity; the flatworms. *Ch. 29*

acquired immunodeficiency syndrome (AIDS) A disease resulting from infection with the human immunodeficiency virus (HIV). This virus attacks and destroys T cells, a key component of the body's immune system. *Ch. 25*

acrosome (Gr. akron, extremity + soma, body) A vesicle located at the tip of a sperm cell that contains enzymes, that aid in the penetration of the protective layers surrounding the egg. *Ch. 21*

ACTH *See* adrenocorticotropic hormone.

actin (Gr. actis, ray) A protein that makes up the thin myofilaments in a muscle fiber; provides support and helps determine cell shape and movement. *Ch. 16*

action potential The rapid change in a membrane's electrical potential caused by the depolarization of a neuron to a certain threshold. *Ch. 13*

active immunity A type of immunity conferred by vaccination, which causes the body to build up antibodies against a particular disease without getting the disease. *Ch. 11*

activator A chemical that binds to an enzyme and changes its shape so that catalysis can occur. *Ch. 4*

active site The grooved or furrowed location on the surface of an enzyme where catalysis occurs. *Ch. 4*

active transport The movement of a molecule across a membrane against the concentration gradient with the expenditure of chemical energy. This process requires the use of a transport protein specific to the molecule(s) being transported. *Ch. 3*

adaptation (L. adaptare, to fit) A naturally occurring inherited trait found within a population that makes individuals carrying this trait better suited to a particular environment than those not carrying the trait. *Ch. 23*

adaptive radiation The phenomenon by which a population of a species changes as it is dispersed within a series of different habitats within a region. *Ch. 23*

adenine An organic compound that is one of the two purine bases of RNA and DNA. *Ch. 2*

adenosine diphosphate (ADP) The molecule remaining after the removal of a phosphate group from adenosine triphosphate. *Ch. 4*

adenosine triphosphate (ATP) A molecule composed of three subunits: ribose, adenine, and a triphosphate group; ATP captures energy in its high-energy bonds and can be used to fuel a variety of cell processes. *Ch. 4*

ADP *See* adenosine diphosphate.

adrenal cortex (L. near, + ren, kidney; L. rind) The outer, yellowish portion of each adrenal gland that secretes a group of hormones known as corticosteroids in response to ACTH. *Ch. 17*

adrenal gland (L. near, + ren, kidney) Either of two triangular glands, named for their position in the body, having two parts with two different functions: the adrenal cortex and the adrenal medulla. *Ch. 17*

adrenal medulla (L. near, + ren, kidney; L. marrow) The inner, reddish portion of each adrenal gland surrounded by the cortex that secretes the hormone adrenaline and noradrenaline. *Ch. 17*

adrenaline *See* epinephrine. *Ch. 17*

adrenocorticotropic hormone (ACTH) (L. near, + ren, kidney + cortex, bark + Gr. tropikos, turning) A hormone secreted by the anterior pituitary that triggers the adrenal cortex to produce certain steroid hormones. *Ch. 17*

aerobic (Gr. aer, air + bios, life) Oxygen dependent. *Ch. 5*

aerobic respiration A series of chemical reactions in which glucose is broken down in the presence of oxygen. These reactions are categorized into three groups: glycolysis, the Krebs cycle, and the electron transport chain. *Ch. 5*

afferent neuron *See* sensory neuron.*Ch. 14*

age distribution The proportion of individuals in the different age categories of a population. *Ch. 34*

agranulocyte (Gr. a-, not + L. granulum, granule + Gr. kytos, cell) One of the two major groups of leukocytes; they have neither cytoplasmic granules nor lobed nuclei. *Ch. 10*

albumin (L. albumen, white of egg) A protein found in blood that elevates its solute concentration to match that of the tissues so that the movement of water molecules between the blood and the tissues is regulated. *Ch. 3*

aldosterone A mineralocorticoid hormone produced by the adrenal cortex that regulates the level of sodium and potassium ions in the blood, thereby promoting the conservation of sodium and water and the excretion of potassium. *Ch. 12*

alga, pl. algae (L. seaweed) A plantlike, photosynthetic protist that contains chlorophyll. *Ch. 26*

allantois (Gr. allas, sausage + eidos, form) An extraembryonic membrane that gives rise to the umbilical arteries and vein as the umbilical cord develops. *Ch. 22*

allele (Gr. allelon, of one another) Each member of a factor pair containing information for an alternative form of a trait that occupy corresponding positions on paired chromosomes. *Ch. 19*

allergen (Gr. allos, other + ergon, work) Any substance that causes manifestations of allergy, a hypersensitive immune response. *Ch. 11*

allergic reactions An immune response that results from the immune system mounting a major defense against a harmless antigen. *Ch. 11*

all-or-nothing Refers to the nerve impulse response; the amount of stimulation applied to the receptor or neuron must always be sufficient to open enough sodium channels to generate an action potential; otherwise, the cell membrane will simply return to the resting potential. *Ch. 13*

alternation of generations A type of life cycle that has both a multicellular haploid phase and a multicellular diploid phase. *Ch. 27*

altruistic behavior A kind of behavior that benefits one at the cost of another; parenting behaviors are a type of altruistic behavior. Altruistic acts increase the individual fitness of the recipient while decreasing the individual fitness of the donor. *Ch. 33*

alveolus, pl. alveoli (L. a small cavity) Microscopic air sacs in the lungs where oxygen enters the blood and carbon dioxide leaves. *Ch. 9*

ameba A protozoan phylum whose members have changing shapes brought about by cytoplasmic streaming, which forms cell extensions called pseudopodia. *Ch. 26*

amino acid A molecule containing an amino acid group ($-NH_2$), a carboxyl group ($-COOH$), a hydrogen atom, a carbon atom, and a functional group that differs among amino acids; an extremely diverse array of proteins is made from the 20 common amino acids. *Ch. 2*

amniocentesis (Gr. amnion, membrane around the fetus + centes, puncture) A prenatal diagnostic procedure in which a sampling of amniotic fluid is obtained by insertion of a needle into the amniotic cavity and withdrawn into a syringe; the removed fetal cells are then grown in tissue culture and tests are performed to determine if genetic abnormalities are present. *Ch. 20*

amnion (Gr. membrane around the fetus) A thin, protective membrane that grows around the embryo during the third and fourth weeks. Fluid fills the cavity between the amnion and the embryo or fetus. *Ch. 22*

amniotic egg (Gr. membrane around the fetus) A round or oval body covered with a shell or membrane, produced by the female of reptiles, birds, and monotremes, which protects the embryo from drying out, nourishes it, and enables it to develop outside of water. *Ch. 24*

amniotic fluid (Gr. membrane around the fetus) A fluid in which the fetus floats and moves; it also helps keep the temperature constant for fetal development. *Ch. 22*

amphibian (Gr. amphibios, double life) An animal capable of living on land and in the water. *Ch. 30*

amylase (Gr. amylon, starch + asis, colloid enzyme) An enzyme that breaks down starches and glycogen to sugars. *Ch. 8*

anabolic reactions Chemical reactions that use energy to build complex molecules from simpler molecules. *Ch. 4*

anaerobic (Gr. an, without + aer, air + bios, life) Literally, without oxygen; any process that can occur without oxygen; includes glycolysis and fermentation. *Ch. 5*

analagous (Gr. proportionate) Of differing evolutionary origin now similar in form and function. *Ch. 23*

anaphase (Gr. ana, up + phasis, form) The stage of mitosis characterized by the physical separation of sister chromatids and their movement to opposite poles of the cell. *Ch. 18*

angiosperm (Gr. angeion, vessel + sperma, seed) A vascular, flowering plant with protected seeds. *Ch. 27*

animal behavior A scientific discipline that was formed when the fields of ethology and behavioral psychology merged. The field of animal behavior applies both the physiological perspective of the ethologist and the psychological perspective of the behaviorist to animal study. *Ch. 32*

annual ring A growth ring in a tree, the result of cambium cells dividing; one ring equals one year's growth and can thus be used to calculate the age of a tree. *Ch. 28*

antagonists (Gr. antagonizesthai, to struggle against) An opposing pair of skeletal muscles. *Ch. 16*

anther (Gr. anthos, flower) The compartmentalized structure in the male stamen of a flower where haploid pollen grains are produced. *Ch. 27*

antheridium, pl. antheridia (Gr. anthos, flower) The structure where sperm is produced in certain algae, the bryophytes, and several divisions of vascular plants. *Ch. 27*

anthropoid (Gr. anthropos, man + eidos, form) A suborder of mammals that includes monkeys, apes, gorillas, chimpanzees, and humans; they differ from prosimians in the structure of the teeth, brain, skull, and limbs; they are diurnal, live in groups with complex social interactions, and care for their young for prolonged periods. *Ch. 24*

antibody (Gr. anti, against) A protein produced in the blood by a B lymphocyte that specifically binds circulating antigen and marks cells or viruses bearing antigens for destruction. *Ch. 11*

antibody-mediated immune response One of the two branches of the immune response; it is initiated by helper T cells that have been activated by interleukin-1 and the presence of antigens, and results in B cells producing antibodies. The antibodies bind to the antigens they encounter and mark them for destruction. *Ch. 11*

anticodon (Gr. anti, against + L. code) A portion of a tRNA molecule with a sequence of three base pairs complementary to a specific mRNA codon. *Ch. 18*

antidiuretic hormone (ADH) (Gr. anti, against + dia, intensive + ouresis, urination) A hormone produced by the hypothalamus but stored and released by the posterior lobe of the pituitary that helps control the volume of the blood by regulating the amount of water reabsorbed by the kidneys. *Ch. 12*

antigen (Gr. anti, against + genos, origin) A foreign substance inducing the formation of antibodies that specifically bind to the foreign substance, marking it for destruction. *Ch. 11*

anus The opening of the rectum for the elimination of feces. *Ch. 8*

aorta (Gr. aeirein, to lift) The largest artery in the body, it carries the blood from the left side of the heart throughout the body except the lungs. *Ch. 10*

aortic semilunar valve A one-way valve that permits blood flow from the left side of the heart and then snaps shut, preventing a backflow from the aorta to the heart. *Ch. 10*

apical meristem (L. apex, top + Gr. meristos, divided) A growth tissue at the tips of roots and the tips of shoots in a plant that allows a plant to grow taller and the roots to grow deeper. *Ch. 28*

appendicular skeleton (L. appendicula, a small appendage) The portion of the human skeleton, containing 126 bones, that consists of the bones of the appendages (arms and legs) and the bones that help attach the appendages to the axial skeleton. *Ch. 16*

appendix (L. appendere, to hang) A pouch that hangs from the beginning of the large intestine and serves no essential purpose in humans. *Ch. 8*

aqueous humor (L. aqua, water + humor, fluid) A watery fluid filling a chamber behind the cornea that nourishes the cornea and the lens and, together with vitreous, creates a pressure within the eyeball maintaining the eyeball's shape. *Ch. 15*

archebacteria (Gr. arche, beginning + Gr. bakterion, rod) A taxonomic section consisting of bacteria that are chemically different in certain structures and metabolic processes from all other bacteria. *Ch. 24*

archegonium, pl. archegonia (Gr. archegonos, first of a race) A structure in which eggs are formed in certain algae, the bryophytes, and several divisions of vascular plants. *Ch. 27*

arteriole A smaller artery that leads from an artery to a capillary. *Ch. 10*

arteriosclerosis A thickening and hardening of the walls of arteries. *Ch. 10*

artery, pl. arteries (Gr. arteria, artery) A blood vessel that carries blood away from the heart. *Ch. 10*

articulation *See* joint. *Ch. 16*

artificial selection The process of selecting organisms for a desirable characteristic and then breeding that organism with another organism exhibiting the same trait to propagate organisms with the desired characteristic. *Ch. 23*

asci Saclike structures that enclose the sexual spores of sac fungi. *Ch. 26*

asexual reproduction A type of reproduction in which a parent organism divides by mitosis and produces two identical organisms; does not introduce variation. *Ch. 18*

association area The area in the brain that connects all parts of the cerebral cortex and appears to be the site of higher cognitive activities such as memory, reasoning, intelligence, and personality. *Ch. 14*

aster In animal mitosis, an array of microtubules that radiates outward from the centrioles when the centrioles reach the poles of the cell. *Ch. 18*

atherosclerosis (Gr. athere, porridge + sklerosis, hardness) A disease in which the inner walls of the arteries accumulate fat deposits, narrowing the passageways and leading to elevated systolic blood pressure. *Ch. 10*

atom (Gr. atomos, indivisible) A core (nucleus) of protons and neutrons surrounded by shells containing electrons. The electrons largely determine the chemical properties of an atom. *Ch. 2*

atomic mass The combined mass of all the protons and neutrons of an atom without regard to its electrons. *Ch. 2*

atomic number The number of protons in an atom; the atomic number is also the same as the number of electrons in that atom. *Ch. 2*

ATP *See* adenosine triphosphate. *Ch. 4*

atrioventricular (AV) node A group of specialized cardiac muscle cells that receives the impulse initiated by the sinoatrial node and conducts it by way of the bundle of His. *Ch. 10*

atrium, pl. atria (L. main room) The upper chamber of each half of the heart; the right atrium receives deoxygenated blood from the body (except the lungs), and the left atrium receives oxygenated blood from the lungs through the pulmonary veins. *Ch. 10*

auditory canal (L. audire, to hear) A 1-inch long canal that receives sound waves and leads directly to the eardrum. *Ch. 15*

Australopithecus afarensis (L. australis, southern + Gr. pithekos, ape; L. Afar region of Ethiopia) A species of australopithecines considered to be the first hominids, although not human (members of the genus *Homo*); its oldest fossil skeleton is 3.5 million years old. *Ch. 24*

autonomic nervous system (Gr. autos, self + nomos, law) A branch of the peripheral nervous system consisting of motor neurons that control the involuntary and automatic responses of the glands and the nonskeletal muscles of the body. *Ch. 14*

autosomes (Gr. autos, self + soma, body) Chromosomes that carry the majority of the genetic information except information determining gender. *Ch. 20*

autotroph (Gr. autos, self + trophos, feeder) Self-feeder; an organism that produces its own food by photosynthesis or other chemical means. *Ch. 6*

auxin (Gr. auxein, to increase) A hormone that promotes growth through cell elongation and controls phototropism and geotropism in plants. *Ch. 27*

AV node *See* atrioventricular (AV) node. *Ch. 10*

axial skeleton The central axis of the human skeleton, consisting of 80 bones, that includes the skull, vertebral column, and rib cage. *Ch. 16*

axon (Gr. axle) A single projection extending from a neuron cell body that conducts impulses away from the cell body. *Ch. 13*

B lymphocyte (B cell) A type of lymphocyte that matures in human bone marrow and secretes antibodies during the antibody-mediated immune response; named after a digestive organ in birds (the bursa of Fabricius) in which these lymphocytes were discovered. *Ch. 11*

bacteriophage (phage) (Gr. bakterion, little rod + phagein, to eat) A virus that infects bacteria. *Ch. 25*

bacterium, pl. bacteria (Gr. bakterion, a little rod) The oldest, simplest, and most abundant organism; it is the only organism with a prokaryotic cellular organization. *Ch. 3*

ball-and-socket joint A kind of joint that allows rotation, or the movement of a bone around an axis. *Ch. 16*

bark The outer protective covering of a woody plant or tree that consists of the cork, the cork cambium, and the parenchyma cells. *Ch. 28*

basal body A structure composed of microtubules that serves to anchor a cilium or flagellum to the cell. *Ch. 3*

basal ganglion, pl. ganglia or ganglions Groups of nerve cell bodies located at the base of the cerebrum that control large, subconscious movements of the skeletal muscles. *Ch. 14*

base Any substance that combines with H^+ ions; a substance that has a pH value higher than water's neutral value of 7. *Ch. 2*

basidium, pl. basidia (Gr. basis, base) In a club fungus, a club-shaped structure from which its sexual spores hang. *Ch. 26*

basophil (Gr. basis, base + philein, to love) A type of granulocyte containing granules that rupture and release chemicals enhancing the body's response to injury or infection; they also play a role in causing allergic responses. *Ch. 10*

behavior The patterns of movement, sounds (vocalization), and body positions (posture) exhibited by an animal; behaviors also include any type of change in an animal, such as a change in coloration or the releasing of a scent, that can trigger certain behaviors in another animal. *Ch. 32*

behaviorism A school of thought that suggests that learning takes place in a stimulus/response fashion, possibly reinforced by some type of reward that may or may not be readily apparent. *Ch. 32*

benign tumors Growths or masses of cells that are made up of partially transformed cells, are confined to one location and are encapsulated, shielding them from surrounding tissues. *Ch. 20*

bicuspid valve A one-way valve in the heart through which blood flows from the left atrium to the lower, adjoining left ventricle. *Ch. 10*

bilateral symmetry (L. bi, two + lateris, side; Gr. symmetria, symmetry) Describing an object or an organism whose right side is a mirror image of its left side. *Ch. 29*

bile (L. bilis) A collection of molecules secreted by the liver that helps in the digestion of lipids. *Ch. 8*

bile pigments Substances produced by the liver from the breakdown of old, worn-out red blood cells; bile pigments enter the small intestine with the bile and are the cause of the characteristic color of the feces. *Ch. 12*

binary fission (L. binarius, consisting of two things or parts + fissus, split) Asexual reproduction in which one cell divides into two with no exchange of genetic material among cells; bacteria use this process. *Ch. 25*

binomial nomenclature (L. bi, twice, + Gr. nomos, law; L. nomen, name + calare, to call) The system of using the last two categories in the hierarchy of classification, genus and species, to provide the scientific name of an organism, usually written in Latin. *Ch. 1*

bioenergy The use of living plants to produce energy; the burning of wood to produce fuel is an example of bioenergy. *Ch. 38*

biological concentration A process by which some organisms accumulate certain harmful or deadly chemicals within their bodies. *Ch. 38*

biological diversity The richness and variety of species on Earth. *Ch. 38*

biological magnification A process by which toxins accumulate in organisms high in the food chain when they consume tainted prey lower in the food chain. *Ch. 38*

biome (Gr. bios, life + -oma, group) A distinctive, broad, terrestrial ecosystem of plants and animals that occurs over wide areas within specific climatic regions. *Ch. 37*

biosphere (Gr. bios, life + sphaira, ball) The global ecosystem of life on Earth that extends from the tops of the tallest mountains to the depths of the deepest seas. *Ch. 38*

biotechnology The use of scientific and engineering principles to manipulate organisms resulting in organisms with specific characteristics, organisms that synthesize particular products, and information about organisms that could otherwise not be known. *Ch. 31*

biotic Living factors within the environment, such as plants, animals, and microorganisms. *Ch. 35*

birth canal The vagina, through which the fetus passes during birth. *Ch. 22*

birth control pills Pills that contain estrogen and protesterone, which shut down the production of the pituitary hormones FSH and LH, preventing the maturation of the secondary oocyte and ovulation. *Ch. 21*

blade The flattened portion of a leaf. *Ch. 26*

blastocyst (Gr. blastos, germ + kytos, cell) The stage of development following the morula in which the embryo is a hollow ball of cells. *Ch. 22*

blastula (Gr. a little sprout) The general term used to describe the saclike blastocyst of mammals and, in other animals, the stage that develops a similar fluid-filled cavity. *Ch. 22*

blood pressure The pressure, determined indirectly, existing in the large arteries at the height of the pulse wave; the systolic intraarterial pressure; normal blood pressures values are 70 to 90 (mm of Hg) diastolic and 110 to 130 systolic. *Ch. 10*

bone (A.S. ban, bone) A hard material that forms the vertebrate skeleton; composed of collagen fibers that contribute flexibility and needle-shaped crystals of calcium that impart rigidity. *Ch. 16*

bony fish A kind of fish that lives in both salt and fresh water and derives its name from its bony internal skeleton. *Ch. 30*

Bowman's capsule (after Sir William Bowman, British physician) An apparatus at the front end of each nephron tube in a kidney that functions as a filter in the formation of urine. *Ch. 12*

brain That part of the central nervous system comprised of four main parts: the cerebrum, the cerebellum, the diencephalon, and the brainstem. *Ch. 13*

brainstem The part of the brain consisting of the midbrain, pons, and medulla that brings messages to and from the spinal cord and controls important body reflexes such as the rhythm of the heartbeat and rate of breathing. *Ch. 14*

breathing The movement of air into and out of the lungs. *Ch. 9*

bronchiole (L. bronchiolus, air passage) The last division of bronchi whose walls have clusters of tiny pouches, or alveoli. *Ch. 9*

bronchus, pl. bronchi (Gr. bronchos, windpipe) One of a pair of airway structures that branches from the lower end of the trachea into each lung. *Ch. 9*

brown algae Large, multicellular algae found predominantly on northern, rocky shores, some growing to enormous sizes; kelp is a brown algae. *Ch. 26*

bryophyte (Gr. bryon, moss + phyton, plant) A single division of small, low-growing plants that are commonly found in moist places, including the mosses, liverworts, and hornworts. *Ch. 27*

bulbourethral glands (L. bulbus, bulbous root + Gr. ourethra, urethra) A set of tiny glands lying beneath the prostate that secretes an alkaline fluid into the semen. *Ch. 21*

bundle of His (after Wilhelm His, Jr, German physician) A strand of impulse-conducing muscle that conducts heartbeat impulses from the right atrium to the ventricles of the heart. *Ch. 10*

C_3 photosynthesis A process that uses the normal Calvin cycle process to fix carbon. *Ch. 6*

C_4 photosynthesis In hot climates, a process by which plants deal with the problem of CO_2 being released by photorespiration by concentrating CO_2 within the cells where carbon fixation takes place. *Ch. 6*

calcitonin (CT) (L. calcem, lime) A hormone secreted by the thyroid that works with parathyroid hormone to regulate the concentration of calcium in the bloodstream. *Ch. 17*

calorie (L. calor, heat) The unit measurement of energy in food. *Ch. 8*

Calvin cycle (after Calvin, an American chemist) *See* light-independent reactions. *Ch. 6*

cambium (L. combiare, to exchange) Lateral meristem tissue in a dicot with a woody stem or trunk that consists of vascular cambium and cork cambium. *Ch. 28*

Cambrian period The oldest period within the Paleozoic era; the Cambrian period ended roughly 505 million years ago, and represents an important point in the evolution of life. All of the main phyla and divisions of organisms that exist today (except for the chordates and the land plants) had evolved by the end of the Cambrian period. *Ch. 24*

capillary (L. capillaris, hairlike) A microscopic blood vessel with a wall only one cell thick that connects the end of an arteriole with the beginning of a venule; the site where gases are exchanged between the blood and tissues, nutrients delivered, and wastes picked up. *Ch. 10*

capillary beds Networks of capillaries. *Ch. 10*

capsid (L. capsa, box) A protein that covers the nucleic acid core of a virus. *Ch. 25*

carbohydrate (L. carbo, charcoal + hydro, water) A molecule that contains carbon, hydrogen, and oxygen, with the concentration of hydrogen and oxygen atoms in a 2:1 ratio. *Ch. 2*

carbon fixation A process by which organisms use the ATP and NADPH produced by photosynthetic reactions to build organic molecules from atmospheric carbon dioxide. *Ch. 6*

carboxypeptidase An enzyme produced by the pancreas, which, together with trypsin and chymotrypsin, completes the digestion of proteins in the small intestine. *Ch. 8*

carcinogen (Gr. karkinos, cancer + gen) Any cancer-causing agent. *Ch. 20*

cardiac muscle A type of striated muscle fiber arranged in a special way critical for cardiac function; cardiac muscle makes up the heart, acting as a pump for the circulatory system. *Ch. 7*

cardiovascular system The heart and blood vessels, the "plumbing" of the circulatory system. *Ch. 10*

carnivore (L. carnivorous, flesh eating) An animal that eats meat. *Ch. 8*

carotenoids Pigments that absorb photons of green, blue, and violet wavelengths and reflect red, yellow, and orange; they are second only to the chlorophylls in importance in photosynthesis. *Ch. 6*

carpal (Gr. karpos, wrist) Any of the eight short bones, lined up in two rows of four, that make up the wrist. *Ch. 16*

carrying capacity The number of individuals within a population that can be supported within a particular environment for an indefinite period. *Ch. 34*

cartilage (L. cartilago, gristle) A specialized connective tissue that is hard and strong; composed of chondrocytes that secrete a matrix consisting of a semisolid gel and fibers. Laid down in long parallel arrays, cartilage is a firm and flexible tissue that does not stretch. *Ch. 7*

cartilaginous fish (L. cartilago, gristle) A marine fish such as a shark, skate, or ray having denticles that cover their skin. *Ch. 30*

catabolic (Gr. katabole, throwing down) Referring to a process in which complex molecules are broken down into simpler ones. *Ch. 4*

catalyst (Gr. kata, down + lysis, a loosening) A substance that increases the rate of a chemical reaction but is not chemically changed by the reaction; an enzyme is a catalyst. *Ch. 4*

cell (L. cella, a chamber or small room) A membrane-bounded unit containing hereditary material, cytoplasm, and organelles; a cell can release energy from fuel and use that energy to grow and reproduce. *Ch. 3*

cell body The body of a nerve cell or neuron, which contains a nucleus and other cell organelles and two types of cellular projections, axons and dendrites. *Ch. 13*

cell culture techniques Methods that enable scientists to remove cells from a parent plant and grow them into new plants. *Ch. 27*

cell cycle The life of a cell; a major portion is concerned with the activities of interphase and a minor portion with mitosis. *Ch. 18*

cell-mediated immune response A chain of events unleashed by T cells during which cytotoxic T cells attack body cells infected with viruses. *Ch. 11*

cell membrane A bilayer of phospholipid molecules studded with proteins to which carbohydrates are attached. The membrane proteins allow the cell to interact with the environment and perform various cellular functions. *Ch. 3*

cell plate During cytokinesis in plants and some algae, the formation of a partition. The partition begins as the plants manufacture new membrane sections that then accumulate at the metaphase plate and fuse. *Ch. 18*

cell theory A statement first formulated in the mid-1800s, which states that all living things are made up of cells, that cells are the smallest living units of life, and that all cells arise from preexisting cells. *Ch. 3*

cell wall The structure that surrounds the plasma membrane in cells. In many plants, algae, and fungi the wall is rigid and composed of cellulose and imparts a stiffness to the tissues; in single-celled organisms, cell walls give shape to the organisms and help protect them. *Ch. 3*

cellular respiration The complex series of chemical reactions—glycolysis, the Krebs cycle, and the electron transport chain—by which cells break down fuel molecules to release energy, using oxygen and producing carbon dioxide. *Ch. 5*

cellular slime molds A phyla of fungus like protists that look and behave like amebas during much of their life cycle, but can group together and form dormant, cystlike spores when food is scarce. These spores then revert to the amebalike form when conditions are favorable again. *Ch. 26*

cellulose (L. cellula, little cell) A polysaccharide that is the chief component of plant cell walls; because cellulose cannot readily be broken down, it works well as a biological structural material and occurs widely in plants. *Ch. 2*

central nervous system The site of information processing within the nervous system, comprising the brain and spinal cord. *Ch. 14*

centriole (Gr. kentron, center of a circle + L. olus, little one) An organelle surrounded by microtubule-organizing centers that play a key role in the mitotic process; most plants and fungi lack centrioles. *Ch. 18*

centromere (Gr. kentron, center + meros, a part) A constricted region where two identical chromosomes (sister chromatids) are joined. *Ch. 18*

cerebellum (L. little brain) The part of the brain located below the occipital lobes of the cerebrum that coordinates subconscious movements of the skeletal muscles.. *Ch. 14*

cerebral cortex The thin layer of tissue, gray matter, that forms the outer layer of the cerebrum; the major site of higher cognitive processes such as sense perception, thinking, learning, and memory. *Ch. 14*

cerebrospinal fluid A liquid that cushions the brain and the spinal cord, acting like a shock absorber. *Ch. 14*

cerebrum (L. brain) The largest and most dominant part of the human brain, divided into two hemispheres connected by the corpus callosum. *Ch. 14*

cervix (L. neck) The narrower bottom part of the uterus that opens into the vagina. *Ch. 21*

chemical bonds Forces that hold atoms together. *Ch. 2*

chemoautotroph (Gr. chemeia, chemistry + auto, self + trophos feeder) Organisms that make their own food by deriving energy from inorganic molecules. *Ch. 25*

chiasma, pl. chiasmata (Gr. a cross) In meiosis, the point of crossing-over where parts of chromosomes have been exchanged during synapsis; under a light microscope, a chiasma appears as an X-shaped structure. *Ch. 18*

chitin (Gr. chiton, tunic) A modified form of cellulose, relatively indigestible, that is a structural material in insects and many fungi. *Ch. 2*

chlorophyll (Gr. chloros, green + phyllon, leaf) The pigment that absorb photons of green, blue, and violet wavelengths; chlorophyll is the primary light gatherer in all plants and algae and in almost all photosynthetic bacteria. *Ch. 6*

chloroplast (Gr. chloros, green + plastos, molded) An energy-producing organelle found in the cells of plants and algae; the site of photosynthesis in plants. *Ch. 3*

cholecystokinin (CCK) (Gr. chole, bile + kystis, bladder + kinein, to move) A hormone that helps control digestion in the small intestine. *Ch. 8*

chondrocyte (Gr. chondros, cartilage + kytos, cell) A cell that produces cartilage, a specialized connective tissue that is hard and strong. *Ch. 7*

chordate (Gr. chorde, cord) An organism distinguished by three principal features: a nerve cord, a notochord, and pharyngeal slits; includes fishes, amphibians, reptiles, birds, mammals, and humans. *Ch. 24*

chorion An extraembryonic membrane that facilitates the transfer of nutrients, gases, and wastes between the embryo and the mother's body. It is a primary part of the placenta. *Ch. 22*

choroid (Gr. chorioeides, skinlike) A thin, dark-brown membrane that lines the sclera of the eye. The choroid contains blood-carrying vessels that nourish the retina, and a dark pigment that absorbs light rays so that they will not be reflected within the eyeball. *Ch. 15*

chromatin (Gr. chroma, color) The complex of proteins and the hereditary material, DNA, which make up the chromosomes of eukaryotes. *Ch. 18*

chromosome (Gr. chroma, color + soma, body) A discrete, threadlike body that forms as the nuclear material condenses during cell division and that carries genetic information. *Ch. 3*

chronic bronchitis An inflammation of the bronchi and bronchioles that lasts for at least 3 months each year for 2 consecutive years with no accompanying disease as a cause; one of the disorders commonly included in chronic obstructive pulmonary disease. *Ch. 9*

chronic obstructive pulmonary disease (COPD) A term used to refer to disorders that block the airways and impair breathing. *Ch. 9*

chyme (Gr. chymos, juice) The mixture of partly digested food, gastric juice, and mucus, having the consistency of pea soup, found in the stomach during digestion. *Ch. 8*

chymotrypsin An enzyme produced by the pancreas, which, together with trypsin and carboxypeptidase, completes the digestion of proteins in the small intestine. *Ch. 8*

ciliary muscle (L. ciliaris, pert. to eyelid) A tiny circular muscle that slightly changes the shape of the lens by contracting or relaxing. *Ch. 15*

ciliate (L. cilium, eyelash) A protozoan characterized by fine, short, hairline cellular extensions called cilia. *Ch. 26*

cilium, pl. cilia (L. eyelash) A short whiplike organelle of motility that protrudes from some eukaryotic cells. *Ch. 3*

citric acid cycle *See* Krebs cycle. *Ch. 5*

class A taxonomic subcategory of divisions and phyla; as organisms appear to be increasingly related to each other, they are more and more narrowly classified within taxonomic subcategories. *Ch. 1*

classical biotechnology The use of scientific and engineering principles to manipulate organisms, relying on the traditional techniques of selection, mutation, and hybridization. *Ch. 31*

classical conditioning A form of learning in which an animal is taught to associate a new stimulus with a natural stimulus that normally evokes a response in the animal. *Ch. 32*

cleavage The process of cell division that occurs without cell growth. *Ch. 22*

cleavage furrow During cytokinesis in animal cells, the area where a belt of microfilaments pinches the cell at the metaphase plate. *Ch. 18*

climax community A community in which the mix of plants and animals becomes stable; the last stage of succession. *Ch. 35*

clitellum, pl. clitella (L. clitellae, a packsaddle) A thickened band on an earthworm's body that secretes a mucus holding the worms together as they exchange sperm. *Ch. 29*

clitoris (Gr. kleitoris) A small mass of erectile and nervous tissue in the female genitalia that responds to sexual stimulation; it is homologous to the penis in males. *Ch. 21*

cloaca (L. sewer) In an aquatic animal, the terminal part of the gut into which ducts from the kidney and reproductive systems open. *Ch. 30*

clone (Gr. klon, a cutting used for propagation) A group of identical cells that arises by repeated mitotic divisions from one original cell. *Ch. 11*

closed circulatory system A body system in which blood is enclosed within vessels as it travels throughout the body of the organism. *Ch. 29*

club fungi Fungi that have club-shaped structures from which unenclosed spores are produced. Mushrooms, puffballs, and shelf fungi are all types of club fungi. *Ch. 26*

cnidarian (Gr. knide, sea nettle) A member of the phylum Cnidaria, which includes jellyfish, hydra, sea anemones, and corals; they have radially symmetrical bodies. *Ch. 29*

cochlea (Gr. kokhlos, land snail) A winding, cone-shaped tube forming a portion of the inner ear that contains the organ of Corti, the organ of hearing. *Ch. 15*

cochlear duct The inner tube of the cochlea containing specialized cells that are the receptors of hearing. *Ch. 15*

codominant Refers to traits in which the alternative forms of an allele are both dominant. *Ch. 20*

codon (L. code) A sequence of three nucleotide bases in transcribed mRNA that code for an amino acid, which is the building block of a polypeptide. *Ch. 18*

coelom (Gr. koilos, a hollow) A body cavity that is a fluid-filled enclosure within a bilaterally symmetrical organism. *Ch. 7, 29*

coelomates Organisms that have a fluid-filled body cavity. *Ch. 29*

coenzyme (L. co-, together + Gr. en, in + zyme, leaven) A cofactor that is a nonprotein organic (carbon-containing) molecule which helps an enzyme catalyze a chemical reaction. *Ch. 4*

coevolution (L. co-, together + e-, out + volvere, to fill) Interactions that involve the long-term, mutual evolutionary adjustment of the characteristics of the members of biological communities in relation to one another. *Ch. 24*

cofactor A special nonprotein molecule that helps an enzyme catalyze a chemical reaction. *Ch. 4*

cognitivism A school of thought that suggests that individuals acquire and then store information in memory; learning takes place as new information builds and merges with the old, leading to changes in behavior. *Ch. 32*

collagen (Gr. kolla, glue + gennan, to produce) The most abundant protein in the human body, whose fibers are strong and wavy; found in the connective tissue of skin, bone, and cartilage. *Ch. 2, 7*

collar cells Specialized, flagellated cells that line the inside cavity of the body of a sponge. *Ch. 29*

collecting duct A small duct that receives urine from the kidney nephrons. *Ch. 12*

colon (Gr. kolon) The large intestine; its function is to absorb sodium and water, to eliminate wastes, and to provide a home for friendly bacteria. *Ch. 8*

commensalism (L. cum, together with + mensa, table) A type of relationship between two organisms in which one species benefits and the other species neither benefits nor is harmed. *Ch. 35*

community (L. communitas, community, fellowship) The various populations of organisms that live together in a particular place. *Ch. 35*

compact bone A type of bone in the human skeleton that runs the length of long bones and has no spaces within its structure visible to the naked eye; it is hard and dense, which gives the bone the strength to withstand mechanical stress. *Ch. 7, 16*

companion cell A specialized cell in a vascular plant that actively transports fluid to and from a phloem cell. *Ch. 28*

competition The striving by organisms of different species that live near one another to obtain the same limited resources. *Ch. 35*

competitive behavior A type of behavior that occurs when two or more animals strive to obtain the same needed resource, such as food, water, nesting sites, or mates. *Ch. 33*

competitive exclusion A principle that states that when two species are competing with one another for the same limited resource in a specific location, the species able to use that resource most efficiently will eventually eliminate the other species in that location. *Ch. 34*

complement A group of proteins that kill foreign cells by creating holes in their membranes. *Ch. 11*

complementary DNA (cDNA) cloning The process of synthesizing DNA from mRNA and then making copies of the DNA by means of yeast or bacterial cells. *Ch. 31*

compound A molecule made up of the atoms of two or more elements. *Ch. 2*

compound eye A structure that is composed of many independent visual units, each containing a lens; an important structure of many anthropods such as bees, flies, moths, and grasshoppers. *Ch. 29*

condom A sheath for the penis, constructed of thin rubber or other natural or synthetic materials, that is designed to prevent sperm from reaching the uterine tubes. Helps reduce the risk of contracting sexually transmitted diseases. *Ch. 21*

cone A light receptor located within the retina at the back of the eye that functions in bright light and detects color. *Ch. 15*

conjugation (L. conjugare, to yoke together) A method of genetic recombination in bacteria during which a donor and a recipient cell make contact and the DNA from the donor is injected into the recipient cell. *Ch. 31*

connective tissue A collection of tissues and its cells that provides a framework for the body, joins its tissues, helps defend it from foreign invaders, and acts as storage sites for specific substances. *Ch. 7*

consumer An organism in an ecosystem that feeds on producers and other consumers, passing energy along that was once captured from the sun. *Ch. 36*

continental shelf The margin of land that extends out from the intertidal zone usually 50 to 100 kilometers (30 to 60 miles) and slopes to a depth of about 200 meters (approximately 650 feet) beneath the sea. *Ch. 37*

control In a scientific experiment, a standard against which observations or conclusions may be checked to establish their validity. *Ch. 1*

conus arteriosus A chamber in the heart of cartilaginous and bony fishes that pumps blood to the gills. *Ch. 30*

convergent evolution Change over time among different species of organisms having different ancestors that results in similar structures and adaptations. *Ch. 23*

coral A general term that refers to a variety of invertebrate animals of the phylum Cnidaria. *Ch. 29*

coral reef A community of organisms that live on masses of rocks in the ocean lying at or near the surface of the water. Coral reefs are partially made up of marine animals called corals (phylum *Cnidaria*) that secrete calcium carbonate, a hard shelllike substance. Other marine organisms also grow on the rocks and make up part of the coral reef. *Ch. 35*

core Viral nucleic acid. *Ch. 25*

cork cambium A lateral meristem tissue whose outermost cells produce cork cells that are waterproof and decay resistant; on maturity, the cork cells lose their cytoplasm and become hardened. *Ch. 28*

cornea (L. corneus, horny) The transparent portion of the eye's outer layer that permits light to enter the eye. *Ch. 15*

corpus callosum (N.L. callous body) The single, thick bundle of nerve fibers that connects the two hemispheres of the cerebrum in humans and primates. *Ch. 14*

corpus luteum (N.L. yellow body) A structure that emerges from a ruptured follicle in the ovary after ovulation; it secretes estrogen and progesterone, preparing the endometrium for the implantation of the fertilized egg. *Ch. 21*

cortex (L. rind, bark) *(1)* The outer layer of an organ as distinguished from the inner, as in the cerebral cortex. *Ch. 12,14*
(2) The outer superficial portion of the root of a vascular plant. *Ch. 28*

corticosteroids A group of hormones, secreted by the adrenal cortex in response to ACTH, that act on the nuclei of target cells, causing the cells' hereditary material to produce certain proteins. *Ch. 17*

cotyledon (Gr. kotyledon, a cup-shaped hollow) A seed leaf; in a dicot, the place where most of the food is stored for the developing embryo. *Ch. 27*

coupled channel A type of channel that has binding sites on one membrane transport protein for two types of molecules which are transported across the cell membrane in opposite directions. *Ch. 3*

coupled reaction Chemical reactions (changes involving the rearrangement of atoms or molecules), that occur in conjunction with one another. *Ch. 4*

courtship Behavior patterns that lead to mating. *Ch. 33*

courtship ritual Patterns of behavior, usually consisting of a series of fixed action patterns of movement, designed to lead to mating. Courtship rituals are unique to each species and occur after the formation of a male-female pair. *Ch. 33*

covalent bond (L. co-, together + valare, to be strong) A chemical bond created by atoms sharing one or more pairs of electrons. *Ch. 2*

cranial nerves (Gr. kranion, skull) Any of the twelve pairs of nerves that enter the brain through holes in the skull. *Ch. 14*

creatinine (Gr. kreas, flesh) A nitrogenous waste found in urine, derived primarily from creatine found in muscle cells. *Ch. 12*

Cretaceous period The third and last period of the Mesozoic era, in which flowering plants (or angiosperms) began to appear. *Ch. 24*

cri du chat syndrome A congenital disorder characterized by severe mental retardation and a moon face; caused by a deletion located on one of the number 5 chromosomes. *Ch. 20*

crista, pl. cristae (L. crest) The infoldings of the inner membrane of a mitochondrion. *Ch. 3*

Cro-Magnon (after a cave in southwestern France) An early member of *H. sapiens sapiens* whose anatomical features were similar to modern humans; they used sophisticated tools, hunted, and made elaborate cave paintings of animals and hunt scenes. *Ch. 24*

crossing-over An essential element of meiosis, which produces sister chromatids that are not identical with each other, resulting in new combinations of genes. *Ch. 18*

cuboidal (Gr. kubos, cube, + eidos, form) One of the main shapes of epithelial cells. Cuboidal cells have complex shapes but look like cubes when the tissue is cut at right angles to the surface; found lining tubules in the kidney and the ducts of glands. *Ch. 7*

cyanobacteria (Gr. kyanos, dark-blue + bakterion, dim. of baktron, a staff) Formerly blue-green algae; a group of photosynthetic bacteria that played a key role in the evolution of life by gradually oxygenating the atmosphere and the oceans around 2 billion years ago. *Ch. 24*

cyclic adenosine monophosphate A molecule that triggers enzymes that cause a cell to alter its functioning in response to a hormone. *Ch. 17*

cyclic electron flow A system that harvests energy in a process involving Photosystem I working independently of Photosystem II. This method of generating energy produces ATP but does not directly produce NADPH as noncyclic electron flow does. *Ch. 6*

cystic fibrosis (Gr. kystis, bladder + L. fibra, fiber + osis, condition) The most common fatal genetic disease of Caucasians in which affected individuals secrete a thick mucus that clogs the airways of the lungs and the passages of the pancreas and liver. *Ch. 20*

cytokinesis (Gr. kytos, hollow vessel + kinesis, movement) The physical division of the cytoplasm of a eukaryotic cell into two daughter cells. *Ch. 18*

cytoplasm (Gr. kytos, hollow vessel + plasma, anything molded) The viscous or gellike fluid within a cell that contains storage substances, a network of interconnected filaments and fibers, and cell organelles. *Ch. 3*

cytoskeleton (Gr. kytos, hollow vessel + skeleton, a dried body) A network of filaments and fibers within the cytoplasm that helps maintain the shape of the cell, move substances within cells, and anchor various structures in place. *Ch. 3*

cytotoxic T cell (Gr. kytos, cell + toxikon, poison) A type of T cell that breaks apart cells infected by viruses and foreign cells such as incompatible organ transplants. *Ch. 11*

deamination The process in which an amino compound loses an -NH_2 group. *Ch. 12*

deciduous (L. decidere, to fall off) A tree that drops its leaves and remains dormant throughout the winter. *Ch. 37*

decomposer An organism that obtains its energy by breaking down the organic material of dead organisms. Decomposers contribute to the recycling of nutrients in the environment. *Ch. 36*

deductive reasoning A pattern of thought in which a person begins with generalizations and proceeds to specific instances. *Ch. 1*

deep zone The area of pond or lake water into which light does not penetrate. *Ch. 36*

dehydration synthesis The process by which monomers are put together to form polymers. *Ch. 2*

deletion Refers to an abnormally short chromosome that has lost a chromosomal section. *Ch. 20*

deltas Fan-shaped areas of accumulated sediment deposited by a stream or river as it enters an open body of water. *Ch. 36*

dendrite (Gr. dendron, tree) A projection extending from a nerve cell body that acts as an antenna for the reception of nerve impulses and conducts these impulses toward the cell body. *Ch. 13*

density The number of organisms or individuals in a population per unit of area. *Ch. 34*

density-dependent limiting factors Environmental factors that result from the growth of a population but act to limit its subsequent growth. Density-dependent limiting factors increase in effectiveness as population density increases. *Ch. 34*

density-independent limiting factors Environmental factors that operate to limit a population's growth, regardless of its density. *Ch. 34*

denticle (L. denticulus, dim. of dentem, tooth) A small, pointed, toothlike scale that covers the skin of cartilaginous fishes and gives the skin a sandpaperlike texture. *Ch. 30*

deoxyribonucleic acid (DNA) A nucleic acid present in the chromosomes that is the chemical basis of heredity and the carrier of genetic information; arranged as two long chains that twist around each other to form a double helix. *Ch. 3*

dependent variable The factor within a controlled experiment that varies in response to changes in the independent variable. *Ch. 1*

depolarization The change in electrical potential of a receptor cell or nerve cell membrane. *Ch. 13*

depressant Drugs that slow down the activity of the central nervous system. *Ch. 13*

dermal tissue The outer protective covering of all plants with the exception of woody shrubs and trees, which have bark. *Ch. 28*

detritus (L. worn down) The refuse or waste material of an ecosystem. *Ch. 36*

deuterostome (Gr. deutero, second + stoma, mouth) A member of a branch of coelomate animals that during the gastrula stage of development, gives rise to a two-layered embryo; the first indentation becomes the anus and the second indentation becomes the mouth. *Ch. 29*

development The events between fertilization and maturity in an organism's life cycle. *Ch. 27*

diabetes (Gr. passing through) A set of disorders characterized by a high level of glucose in the blood, the underlying cause being the lack or partial lack of insulin; in type I or juvenile onset diabetes, a person has no effective insulin; in type II or maturity onset diabetes, a person has some effective insulin but not enough to meet body needs. *Ch. 17*

diaccharidases Enzymes, produced by specialized epithelial cells of the small intestine, that break down the disaccharides maltose, sucrose, and lactose to the monosaccharides glucose, fructose, and galactose. *Ch. 8*

dialysis (Gr. dia, through + lysis, dissolution) The filtering of blood through a selectively permeable membrane to remove toxic wastes, regulate blood pH, and regulate ion concentration; used in renal failure. *Ch. 12*

diaphragm A sheet of muscle that forms the horizontal partition between the thoracic cavity and the abdominal cavity. *Ch. 9*

diastolic period (Gr. diastole, expansion) The span of time during the first part of a heartbeat when the atria are filling; at this time the pressure in the arteries leading from the left side of the heart to the tissues of the body decreases slightly as the blood moves out of the arteries, through the vascular system, and into the atria. *Ch. 10*

diatom (Gr. diatemnein, to cut through) A type of microscopic golden algae that possesses a siliceous or calcium-containing cell wall. *Ch. 26*

diatomic molecule A molecule composed of two atoms of the same type. *Ch. 2*

dicot One of two groups of angiosperms in which the food reserves of its seeds is stored in the cotyledons, or seed leaves; a dicot usually has netlike veins in its leaves and its flower parts are in fours or fives. *Ch. 27*

diencephalon (Gr. dia, between + enkephalon, within the head) The part of the brain consisting of the thalamus and hypothalmus. *Ch. 14*

diffusion (L. diffundere, to pour out) The net movement of molecules from a region of higher concentration to a region of lower concentration, eventually resulting in a uniform distribution of the molecules. This movement is the result of random, spontaneous molecular motions. *Ch. 3*

digestion A process in which food particles are broken down into small molecules that can be absorbed by the body. *Ch. 8*

dihybrid (Gr. dis, twice + L. hybrida, mongrel) The product of two plants that differ from one another in two traits. *Ch. 19*

dinoflagellate (Gr. dinos, rotation + N.L. flagellum, whip) A type of flagellated unicellular red algae, characterized by stiff outer coverings; its flagella beat in two grooves, one encircling the cell like a belt and the other perpendicular to it. *Ch. 26*

diploid (Gr. diploos, double + eidos, form) A cell that contains double the haploid amount; a double set of the genetic information. *Ch. 18*

disaccharide (Gr. dis, twice + sakcharon, sugar) Two monosaccharides linked together; sucrose (table sugar) is a disaccharide formed by linking a molecule of glucose to a molecule of fructose. *Ch. 2*

dispersion The way in which the individuals of a population are arranged within their environment. *Ch. 34*

distal convoluted tubule The portion of the nephron tubule in which selective reabsorption takes place. *Ch. 12*

divisions Subdivisions of the kingdoms Monera, Plantae and Fungi, in which taxonomists group organisms that have similar evolutionary histories; taxonomically equivalent to phyla. *Ch. 1*

DNA *See* deoxyribonucleic acid. *Ch. 18*

dominant The form of a trait that will be expressed in a hybrid offspring. *Ch. 19*

Down syndrome (after J. Langdon Down, British physician) A genetic disorder produced when an individual receives three (instead of two) 21 chromosomes; also called trisomy 21. *Ch. 20*

drug addiction A compulsive urge to continue using a psychoactive drug, physical and/or psychological dependence on the drug, and a tendency to increase the dosage of the drug. *Ch. 13*

drug tolerance A decrease in the effects of the same dosage of a drug in a person who takes the drug over time. *Ch. 13*

duodenum (L. duodeni, twelve each; with reference to its length, about twelve finger breadths) The initial, short segment of the small intestine that is involved in digestion and absorption. *Ch. 8*

duplication A chromosomal abnormality in which a section of a chromosome has been replicated; in a karyotype, the chromosome carrying the duplication appears longer than its homologue. *Ch. 20*

ecology (Gr. oikos, house + logos, word) The field of biology that studies the interactions among organisms and between organisms and their environments in the laboratory and in nature. *Ch. 37*

ecosystem (Gr. oikos, house + systema, that which is put together) A community of plants, animals, and microorganisms that interacts with one another and with their environments and is interdependent on one another for survival. *Ch. 35*

ectoderm (Gr. ectos, outside + derma, skin) The outer layer of cells formed during the development of embryos; forms the outer layer of skin, the nervous system, and portions of the sense organs. *Ch. 22*

ectothermic (Gr. ectos, outside + therme, heat) Referring to animals like reptiles, amphibians, and fishes that regulate body temperature by taking in heat from the environment. *Ch. 30*

effector A muscle or gland that effects (or causes) responses when stimulated by nerves. *Ch. 13*

efferent neuron *See* motor neuron. *Ch. 13*

electrocardiogram (ECG) (Gr. elektron, amber + kardia, heart + -gram, written) A graphical representation of the electrical activity of the heart. *Ch. 10*

electron A subatomic particle, having very little mass and carrying a negative charge, that moves within a shell outside the nucleus of an atom. *Ch. 2*

electron transport chain A series of electron carriers located in the mitochondria. By passing electrons and protons in a series of redox reactions, the molecules in this chain create a flow of energy that works in special ways to produce most of the ATP of cellular respiration. *Ch. 5*

element A pure substance that is made up of a single kind of atom and that cannot be separated into different substances by ordinary chemical methods. *Ch. 2*

elimination A process that takes place as digestive wastes leave the body during defecation. *Ch. 8*

embolus A traveling blood clot. *Ch. 10*

embryo (Gr. embryon, something that swells in the body) The early stage of development in humans from the third to eighth weeks. *Ch. 22*

emigration (L. emigrare, to migrate) The movement of organisms out of a population. *Ch. 34*

emphysema A chronic obstructive pulmonary disease in which mucus plugs various bronchioles, trapping air within alveoli and often causing them to rupture. In this condition, the lungs lose their elasticity and the ability to recoil during exhalation, and instead stay filled with air. *Ch. 9*

endergonic (Gr. endon, within + ergos, work) Used to describe a reaction in which the products of the reaction contain more energy than the reactants, so energy must be supplied for the reaction to proceed. *Ch. 4*

endocrine gland (Gr. endon, within + krinein, to separate) A ductless gland that secretes hormones and spills these chemicals directly into the bloodstream. *Ch. 17*

endocrine system (Gr. endon, within + krinein, to separate) The collective term for the 10 different endocrine glands that secrete 30 different hormones. *Ch. 17*

endocytosis (Gr. endon, within + kytos, cell) A process in which cells engulf large molecules or particles and bring these substances into the cell packaged within vesicles. *Ch. 3*

endoderm (Gr. endon, within + derma, skin) The inner layer of cells formed during the early development of embryos; it gives rise to the digestive tract lining, the digestive organs, the respiratory tract, the lungs, the urinary bladder, and the urethra. *Ch. 22*

endodermis (Gr. endon, within + derma, skin) In the root of a vascular plant, the innermost layer of the cortex, which consists of specialized cells that regulate the flow of water between the vascular tissues and the outer portion of the root. *Ch. 28*

endometrium (Gr. endon, within + metrios, of the womb) The inner lining of the uterus. *Ch. 21*

endoplasmic reticulum (Gr. endon, within + plasma, from cytoplasm; L. reticulum, network) An extensive system of membranes that divides the interior of eukaryotic cells into compartments and channels. *Ch. 3*

endosperm (Gr. endon, within + sperma, seed) An extraembryonic tissue in a monocot where most of the food is stored for the developing embryo. *Ch. 27*

endosymbiont (Gr. endon, within + bios, life) An organism that lives within another in a mutualistic relationship; the major endosymbionts that occur in eukaryotic cells are mitochondria and chloroplasts. *Ch. 3*

endosymbiotic theory (Gr. endon, within + bios, life) The most widely-accepted theory regarding how eukaryotes arose. According to this theory, bacteria were engulfed by host prokaryotic cells. *Ch. 24*

endothermic (Gr. endon, within + therme, heat) Referring to organisms such as birds and mammals that regulate body temperature internally. These organisms maintain a constant, high internal body temperature. *Ch. 30*

end-product inhibition The process in which the enzyme catalyzing the first step in a series of chemical reactions has an inhibitor binding site to which the end product of the pathway binds. As the concentration of the end product builds up in the cell, it begins to bind to the first enzyme in the metabolic pathway, shutting off that enzyme. In this way, the end product feeds information back to the first enzyme in the pathway, shutting the pathway down when additional end product is not needed. *Ch. 4*

energy level An electron shell that surrounds the nucleus of an atom. *Ch. 2*

entropy (Gr. en, in + tropos, change in manner) The energy lost to disorder; it is a measure of the disorder of a system. *Ch. 4*

envelope A chemical layer over the capsid of many viruses that is rich in proteins, lipids, and carbohydrate molecules. *Ch. 25*

environment (M.E. envirounen, encircle) A general term for the biosphere; the land, air, water, and every living thing on the Earth. *Ch. 38*

enzyme (Gr. enzymos, leavened, from en, in + zyme, leaven) A protein that lowers the free energy of activation, thus allowing chemical reactions to take place. *Ch. 4*

eosinophil (Gr. eos, dawn (rose-colored) + philein, to love) A kind of granulocyte believed to be involved in allergic reactions; they also act against certain parasitic worms. *Ch. 10*

epidermis (Gr. epi, upon + derma, skin) In the root of a vascular plant, the outer layer of the cortex, which absorbs water and minerals. *Ch. 28*

epididymis (Gr. epi, upon + didymos, testis) A long, coiled tube that sits on the back side of the testes where sperm undergo further development after their formation within the testes. *Ch. 21*

epiglottis (Gr. epi, upon + glotta, tongue) A flap of tissue that folds back over the opening to the larynx, thus preventing food or liquids from entering the airway. *Ch. 8*

epinephrine (Gr. epi, on, over + nephros, kidney) A hormone produced by the adrenal medulla that readies the body to react to stress. *Ch. 17*

epiphyte (Gr. epi, on, over + phyton, plant) A plant of the tropical rain forest that grows on trees or other plants for support but draws its nourishment from rain water. *Ch. 37*

epithelial tissue A collection of tissues that cover and line internal and external surfaces of the body and compose the glands. *Ch. 7*

epithelium (Gr. epi, on + thele, nipple) The collective term for all epithelial cells, which have six different functions in the body: protection, absorption, sensation, secretion, excretion, and surface transport. *Ch. 7*

epoch (Gr. epechein, to hold on, check) A subdivision of a period. *Ch. 24*

era One of the five major periods of geological time since the formation of the Earth until the present day. *Ch. 24*

erythrocyte (Gr. erythros, red + kytos, hollow vessel) A red blood cell, packed with hemoglobin, which acts as a mobile transport unit, picking up and delivering gases primarily oxygen. *Ch. 7*

esophagus (Gr. oiso, carry + phagein, to eat) The food tube that connects the pharynx to the stomach. *Ch. 8*

essential amino acids The eight amino acids that humans cannot manufacture and therefore must be obtained from proteins in the food they eat. *Ch. 8*

estrogen (Gr. oistros, frenzy + genos, origin) Any of various hormones that develop and maintain the female reproductive structures such as the ovarian follicles, the lining of the uterus, and the breasts. *Ch. 21*

estuary (L. aestus, tide) A place where the fresh water of rivers and streams meets the salt water of oceans. *Ch. 37*

ethology (Gr. ethos, habit or custom + logos, discourse) The study of animal behavior in the natural environment, which examines the biological basis of the patterns of movement, sounds, and body positions of animals. *Ch. 21*

eubacteria (Gr. eus, good + bakterion, rod) The so-called true bacteria, which include all bacteria except the archaebacteria. *Ch. 25*

euglenoid (Gr. eus, good + glene, eyeball) A flagellate member of the genus *Euglena,* having chloroplasts and making its own food by photosynthesis. *Ch. 26*

eukaryotic cell (Gr. eu, good + karyon, kernel) A type of cell more complex than a prokaryote that makes up the bodies of plants, animals, protists, and fungi; a plasma membrane encloses the cytoplasm, which contains organelles and the nucleus. *Ch. 3*

eustachian tube (after Bartolomeo Eustachio, Italian anatomist) A structure that connects the middle ear with the nasopharynx; it equalizes air pressure on both sides of the eardrum when the outside air pressure is not the same. *Ch. 15*

eutrophication (Gr. eutrophic, thriving) The accumulation of inorganic nutrients in a lake. *Ch. 36*

evolution (L. evolvere, to unfold) The process of change over time by which existing populations of organisms develop from ancestral forms through modification of their characteristics. *Ch. 1, 23*

evolutionary tree A representation of the pattern of relationships among major groups of organisms. *Ch. 23*

excitatory synapse A type of synapse in which a neurotransmitter depolarizes the postsynaptic membrane, resulting in the continuation of the nerve impulse. *Ch. 13*

excretion A process whereby metabolic wastes, excess water, and excess salts are removed from the blood and passed out of the body. *Ch. 12*

exergonic (L. ex, out + Gr. ergon, work) Used to describe a reaction in which the products contain less energy than the reactants and the excess energy is released; exergonic reactions take place spontaneously. *Ch. 4*

exocrine gland (Gr. exo, outside + krinein, to separate) A term applied to a gland whose secretion reaches its destination by means of ducts. *Ch. 17*

exocytosis (Gr. ex, out of + kytos, cell) The reverse of endocytosis; the discharge of material by a cell by packaging it in a vesicle and moving the vesicle to the cell surface. *Ch. 3*

exon (Gr. exo, outside) A nucleotide sequence that encodes the amino acid sequence of a polypeptide. *Ch. 18*

exoskeleton A rigid outer covering of an organism that maintains its shape and supports structures of the body. *Ch. 29*

expiration (Gr. ex, out + L. spirare, to breathe) The expelling of air from the lungs in breathing; it occurs when the volume of the thoracic cavity is decreased and the resulting increased pressure forces air out of the lungs. *Ch. 9*

exponential growth A period of rapid increase in the number of organisms in a population. *Ch. 25*

external fertilization A union of eggs and sperm that takes place outside the body of the female. *Ch. 30*

external genitals Sexual organs, such as the penis, that are located on the outside of the body. *Ch. 21*

external respiration The exchange of carbon dioxide and oxygen gases by the red blood cells and alveoli in the lungs. *Ch. 9*

extinction (L. exstinguere, to quench) A dying out of species to the point where they no longer exist. *Ch. 24*

extraembryonic membranes Structure that form from the trophoblast and provide nourishment and protection to the developing embryo and fetus; so named because they is not a part of the embryo. *Ch. 22*

F plasmid A plasmid carrying a fertility factor, which consists of several genes that promote the transfer of the plasmid to other cells. *Ch. 25*

facilitated diffusion The movement of selected molecules across the cell membrane by specific transport proteins along the concentration gradient and without an expenditure of energy. *Ch. 3*

family A taxonomic subcategory of order. *Ch. 1*

fats Large molecules made up of carbon, hydrogen, and oxygen, with a hydrogen-to-oxygen ratio higher than 2:1. *Ch. 2*

fatty acid A long hydrocarbon chain ending in a carboxyl (-COOH) group; a fatty acid can be saturated, unsaturated, or polyunsaturated. *Ch. 2*

feces (L. faeces) Body waste discharged by way of the anus. *Ch. 8*

feedback loop A mechanism by which information regarding the status of a physiological situation or system is fed back to the system so that appropriate adjustments can be made. *Ch. 7*

female condom A birth control device that is shaped somewhat like a male condom, but has a ring at each end. One ring fits over the cervix and the other hangs outside the vagina. *Ch. 21*

femur The thigh bone in the appendicular skeleton of a human. *Ch. 16*

fermentation (L. fermentum, ferment) An anaerobic process by which certain organisms make ATP. *Ch. 5*

fertilization (L. ferre, to bear) The union of a male gamete (sperm) and a female gamete (egg). *Ch. 22*

fetus The term used to describe the stage of prenatal development after 8 weeks and until birth. *Ch. 22*

fibroblast (L. fibra, fiber + Gr. blastos, sprout) The most numerous of the connective tissue cells; they are flat, irregular, branching cells that secrete fibers into the matrix between them. *Ch. 7*

fibrocartilage (L. fibra, fiber + cartilago, gristle) A type of cartilage that has collagen fibers embedded in its matrix; used by the body as a "shock absorber" in the knee joint and as disks between the vertebrae. *Ch. 7*

filtrate The water and dissolved substances that are first filtered out of the blood during the formation of urine. *Ch. 12*

filtration The process by which the blood is passed through nephron membranes that separate blood cells and proteins from the water and small molecules of the blood. *Ch. 12*

first filial (F_1) generation The hybrid offspring of the parental (P) generation. *Ch. 19*

first law of thermodynamics A law stating that energy cannot be created or destroyed; it can only be changed from one form or state to another. *Ch. 4*

first trimester The first three months of pregnancy. *Ch. 22*

fixed action patterns Sequences of innate behaviors in which the actions follow an unchanging order of muscular movements. *Ch. 32*

flagellum, pl. flagella (L. flagellum, whip) A long, whiplike organelle of motility that protrudes from some cells. *Ch. 3*

fluid mosaic model The model of the cell membrane that describes the fluid nature of a lipid bilayer studded with a mosaic of proteins. *Ch. 3*

follicle (L. folliculus, little bag) The potential egg surrounded by its supporting cells while it matures in the ovary.

follicle-stimulating hormone (FSH) A gonadotropic hormone secreted by the anterior pituitary that triggers the maturation of one egg each month in females; it triggers sperm production in males. *Ch. 17*

food chain A series of organisms, from each trophic level, that feed on one another. *Ch. 36*

food web The interwoven food chain of an ecosystem; a diagram of who eats whom. *Ch. 36*

foramen, pl. foramina A hole in a bone. *Ch. 14*

foraminifer, pl. foraminifera (L. forare, to bore) A type of ameba that secretes beautifully sculpted shells made out of calcium carbonate; also called a foram. *Ch. 26*

formed elements The solid portion of blood plasma composed principally of erythrocytes, leukocytes, and platelets. *Ch. 10*

fossil (L. fodere, to dig) Any record of a dead organism; any trace or impression of an animal or plant that has been preserved in the Earth's crust. *Ch. 23*

fossil fuels Substances such as coal, oil, and natural gas that are formed over time (acted on by heat and pressure) from the undecomposed carbon compounds of organisms that died millions of years ago. *Ch. 38*

fovea (L. a pit) The area of sharpest vision within the retina due to its high concentration of cones. *Ch. 15*

free energy of activation The energy needed to initiate a chemical reaction. *Ch. 4*

free radical A charged molecule fragment with an unpaired electron that is highly reactive. *Ch. 20*

frond (L. foliage) The leaf of a fern. *Ch. 27*

frontal lobe A section on both hemispheres of the cerebral cortex dealing with motor activity. *Ch. 14*

fruit A vessel containing seeds produced by an angiosperm. *Ch. 27*

functional group Special groups of atoms attached to an organic molecule; important because most chemical reactions that occur within organisms involve the transfer of a functional group from one molecule to another. *Ch. 2*

fundamental niche The role that an organism might play in an ecosystem if competitors were not present. *Ch. 35*

fungus, pl. fungi (L. mushroom) A multicellular, eukaryotic organisms that feeds on dead or decaying organic material. *Ch. 26*

gall bladder In human beings, a sac attached to the underside of the liver, where excess bile is stored and concentrated. *Ch. 8*

gamete (Gr. wife) A sex cell; the female gamete is the egg, and the male gamete is the sperm. *Ch. 18*

gametophyte (Gr. gamete, wife + phyton, plant) The haploid phase of a plant life cycle that alternates with the diploid phase. *Ch. 27*

gametophyte (gamete-plant) generation The haploid phase of a plant life cycle; gametophytes produce haploid gametes by mitosis. *Ch. 27*

ganglion, pl. ganglia A group of nerve cell bodies located within the peripheral nervous system. *Ch. 14*

gastric glands Glands dotting the inner surface of the stomach that secrete a gastric juice of hydrochloric acid and pepsinogen. *Ch. 8*

gastrin A digestive hormone of the stomach that controls the production of acid. *Ch. 8*

gastrulation (L. little belly) The process by which various cell groups of the inner cell mass, migrate, divide, and differentiate resulting in a three layer embryo. *Ch. 22*

gene (Gr. genos, birth, race) A sequence of nucleotides in the hereditary material that codes for a particular polypeptide. *Ch. 18*

gene families Multiple copies of genes in eukaryotes, which are derived from a common ancestral gene and are a reflection of the evolutionary process. *Ch. 18*

gene mutation A change in the genetic message of a chromosome due to alterations of molecules within the structure of the chromosomal DNA. *Ch. 20*

gene synthesis cloning The process of synthesizing DNA in the laboratory based on knowledge of its nucleotide sequence and then making copies of the DNA by means of yeast or bacterial cells. *Ch. 31*

gene therapy The technique in which scientists try to cure inherited genetic disorders by inserting genes with the proper genetic message into patients having defective genes. *Ch. 18*

general adaptation syndrome The three stages of reaction by the body to stress: the alarm reaction, resistance, and exhaustion. *Ch. 17*

genetic counseling The process in which geneticists identify couples at risk of having children with genetic defects and help them have healthy children. *Ch. 20*

genetic engineering (recombinant DNA technology) New techniques of molecular biology that involve the purposeful manipulation of genes within organisms. *Ch. 25*

genetics (Gr. genos, birth, race) The branch of biology dealing with the principles of heredity and variation in organisms. *Ch. 19*

genital herpes A sexually transmitted disease, caused most often by the herpes simplex virus (HSV-2), that produces blisterlike sores on the genitals. *Ch. 25*

genital warts A sexually transmitted disease caused by the human papillomavirus (HPV). The warts are soft, pink, flat or raised growths that appear singly or in clusters on the external genitals and rectum. *Ch. 25*

genome An organism's total complement of genetic material. *Ch. 31*

genotype (Gr. genos, offspring + typos, form) The total set of genes that constitutes an organism's genetic makup. *Ch. 19*

genus A taxonomic subcategory of family. *Ch. 1*

geographic isolation A term describing organisms that live in sharply discontinuous habitats from one another. *Ch. 23*

geothermal energy The use of heat deep within the Earth for heating purposes or as part of a process to produce electricity. *Ch. 38*

germination (L. germinare, to sprout) The sprouting of a seed that begins when it receives water and has appropriate growing conditions. *Ch. 27*

gestation (L. gestare, to bear) The time of human development from conception until birth, approximately 8 $^1/_2$ months. *Ch. 22*

gibberellin (*Gibberella,* a genus of fungi) A hormone that together with auxin, promotes growth in plants through cell elongation. *Ch. 27*

gill An organ of respiration in certain animals such as fishes. *Ch. 29*

gizzard The digestive organ of a bird, often filled with grit, that grinds food. *Ch. 30*

gland (L. glans, acorn) An epithelial cell specialized to produce and discharge substances. *Ch. 7*

glial cell (Gr. glia, glue) A supporting nerve cell of the brain and spinal cord. *Ch. 7*

global warming A worldwide temperature increase that could result from the greenhouse effect. *Ch. 38*

glomerulus (L. a little ball) A tuft of capillaries surrounded by Bowman's capsule that acts as a filtration device in the formation of urine. *Ch. 12*

glottis The opening to the larynx and trachea. *Ch. 8*

glucagon A hormone produced in the islets of Langerhans that raises blood glucose level by converting glycogen into glucose. *Ch. 17*

glycogen (Gr. glykys, sweet + gen, of a kind) Animal starch; a storage form of glucose within animals. *Ch. 2*

glycolysis (Gr. glykys, sweet + lyein, to loosen) The first of three series of chemical reactions in cellular respiration, which results in the formation of two ATP molecules and two molecules of pyruvate from one molecule of glucose. *Ch. 5*

golden algae A phyla of unicellular algae that have gold-green photosynthetic pigments and store food as oil. *Ch. 26*

golden-brown algae A type of golden algae. *Ch. 26*

Golgi complex (after Camillo Golgi, Italian physician) The delivery system of the eukaryotic cell; it collects, modifies, packages, and distributes molecules that are made within the cell. *Ch. 3*

gonad (Gr. gone, seed) A male or female reproductive sex organ that produces sex cells, or gametes. *Ch. 21*

gonadotropins Two of the four tropic hormones; gonadotropins affect the male and female sex organs. *Ch. 17*

gonorrhea A sexually transmitted disease, caused by the bacterium *Neisseria gonorrhoeae,* that results in a primary infection and inflammation of the urethra (in men and women) and the vagina and cervix (in women). *Ch. 25*

gradient Differences in concentration, pressure, or electrical charge, which often results in a net movement of molecules in a particular direction in response to these differences. *Ch. 3*

grana, sing. granum (L. grain or seed) Within chloroplasts, stacks of thylakoids. *Ch. 6*

granulocyte (L. granulum, little grain + Gr. kytos, cell) A circulating leukocyte that gets its name from the tiny granules in its cytoplasm; granulocytes are classified into three groups by their staining properties. *Ch. 10*

greenhouse effect The selective energy absorption by carbon dioxide in the atmosphere, which allows heat to enter the Earth's atmosphere but prevents it from leaving; may be a contributing factor to global warming. *Ch. 38*

green algae One of the three phyla of multicellular algae; green algae include both unicellular and multicellular forms. Most forms of green algae are aquatic, but some species live in moist places on land. *Ch. 26*

ground tissue Plant tissue that stores the carbohydrates the plant produces. It forms the substance of the plant and is the tissue in which vascular tissue is embedded. *Ch. 28*

growth hormone (GH) A hormone secreted by the anterior pituitary that works with the thyroid hormones to control normal growth. *Ch. 17*

growth rate A measure of increase of a population. The growth rate of a population is determined by subtracting its death rate from its birth rate. *Ch. 34*

guard cells A pair of cells that brackets a stoma and regulates its opening and closing. *Ch. 28*

gymnosperm (Gr. gymnos, naked + sperma, seed) A vascular plant whose seeds are not completely enclosedby the tissues of the parent at the time it is pollinated; the most familiar gymnosperms are the conifers. *Ch. 27*

habitat (L. habitare, to inhabit) An area or space within an ecosystem, including the factors within it, in which each living thing resides. *Ch. 35*

habituation The ability of animals to "get used to" certain types of stimuli that they perceive as non-threatening. *Ch. 32*

half-life The length of time it takes for half the radioactive material in a sample to decay. *Ch. 23*

hallucinogens Psychedelic drugs that cause sensory perceptions that have no external stimuli; they cause a person to see, hear, smell, or feel things that do not exist. *Ch. 13*

haploid (Gr. haploos, single + eidos, form) A sex cell that contains half the amount of hereditary material of the original plant cell. *Ch. 18*

Haversian canal (after Clopton Havers, British physician and anatomist) A narrow channel that runs parallel to the length of the bone and that contains blood vessels and nerve cells. *Ch. 16*

heart The muscular pump that is the center of the cardiovascular or circulatory system. *Ch. 10*

helper T cell A kind of T lymphocyte that initiates the immune response by identifying foreign invaders and stimulating the production of other cells to fight them. *Ch. 11*

hemoglobin (Gr. haima, blood + L. globus, a ball) The iron-containing pigment that imparts the color to red blood cells; hemoglobin is produced within the red bone marrow and carries oxygen in the blood. *Ch. 9*

hemophilia (Gr. haima, blood + philein, to love) A hereditary condition in which the blood is slow to clot or does not clot at all. *Ch. 20*

herbivore (L. herba, grass + vorare, to eat) An animal that feeds on plants. *Ch. 8*

hermaphrodite (Gr. Hermaphroditos, son of Hermes and Aphrodite, who was man and woman combined) An animal or plant having both male and female reproductive organs. *Ch. 29*

heterotroph (Gr. heteros, other + trophos, feeder) An organism that cannot produce its own food; they obtain energy by breaking down organic material. *Ch. 6*

hetereozygous (Gr. heteros, other + zygotos, a pair) An individual having two different alleles for a trait. *Ch. 19*

hinge joint A kind of joint that allows movement in one plane only. *Ch. 16*

homeostasis (Gr. homeos, similar + stasis, standing) The maintenance of a stable internal environment in spite of a possibly very different external environment. *Ch. 7*

hominid (L. homo, man) The family of hominoids consisting of human beings; the only living hominid is *Homo sapiens sapiens. Ch. 24*

hominoid (L. homo, man) One of four superfamilies of the anthropoid suborder that includes apes and humans. *Ch. 24*

Homo erectus (L. homo, man + erectus, upright) An extinct species of hominids whose fossil record dates back to 1.6 million years ago; *H. erectus* was fully adapted to upright walking, made sophisticated tools, built shelters, used fire, and probably communicated with language. *Ch. 24*

Homo habilis (L. homo, man + habilis, skillful) An extinct species of hominid whose fossil record dates back about 2 million years; considered human because they exhibited a far greater intelligence than their ancestors by making tools and clothing. *Ch. 24*

Homo sapiens (L. homo, man + sapiens, wise) A hominid whose fossil record dates back about 200,000 years and most likely evolved from *H. erectus;* their anatomy featured larger brains, flatter heads, more sloping foreheads, and more protruding brow ridges than modern humans. *Ch. 24*

Homo sapiens sapiens (L. homo, man + sapiens, wise) Modern man; the subspecies of hominids who made their appearance 10,000 years before the Neanderthal subspecies died out and whose early members are called Cro-Magnons. *Ch. 24*

homologous (Gr. homologia, agreement) Of the same evolutionary origin, now differing in structure and function. *Ch. 23*

homozygous (Gr. homos, same or similar + zygotos, a pair) An individual having two identical alleles for a trait. *Ch. 19*

hormone (Gr. hormaien, to excite) A chemical messenger secreted an endocrine gland and sent to other cells of the body. *Ch. 17*

human chorionic gonadotropic (HCG) (Gr. chorion, chorion + gonos, genitals + trope, turning) A hormone secreted by embryonic placental tissue that maintains the corpus luteum so that it will continue to secrete progesterone and estrogen; the detection of this hormone in the urine is the basis for home pregnancy tests. *Ch. 21*

human genome project A worldwide scientific project, the goal of which is to decipher the DNA code of all 46 human chromosomes. *Ch. 18*

human immunodeficiency virus (HIV) The virus that causes AIDS, especially deadly because it destroys the ability of the immune system to mount a defense against any infection because it attacks and destroys helper T cells. *Ch. 11*

humerus (L. upper arm) The upper arm bone in the appendicular skeleton of a human. *Ch. 16*

Huntington's disease (after G. Huntington, U.S. physician) A fatal genetic disorder caused by a mutant dominant allele that results in a progressive deterioration of brain cells. *Ch. 20*

hyaline cartilage (Gr. hyalos, glass + L. cartilago, gristle) A type of cartilage that has very fine collagen fibers in its matrix; it is found on the ends of long bones, ringing the windpipe, and in the ribs and nose. *Ch. 7*

hybrid (L. hybrida, mongrel) The offspring of the cross between two different varieties or species. *Ch. 19*

hydrogen bond The attraction between the partial negative charge of one polar molecule and the partial positive charge of another polar molecule. *Ch. 2*

hydrolysis (Gr. hydro, water + lyse, to break) The process by which a polymer is disassembled by adding molecules of water. *Ch. 2*

hydrolyzing enzyme (Gr. hydro, water + lyse, to break) An biological catalyst that breaks down substances by adding molecule(s) of water. *Ch. 8*

hydrophilic (Gr. hydor, water + philos, loving) Referring to the phosphate functional groups in polar molecules, which form hydrogen bonds with water and, consequently, are soluble or dissolve in water. *Ch. 3*

hydrophobic (Gr. hydor, water + phobos, hating) Refers to nonpolar molecules like oil that cannot form hydrogen bonds with water; this is why oil and water do not mix. *Ch. 3*

hyperpolarization (Gr. hyper, above + polaris, pole) The resulting state in the interior of a rod cell when it becomes even more negatively charged than before because sodium channels have closed because of stimulation. *Ch. 15*

hypertonic (Gr. hyper, above + tonos, tension) Refers to a solution with a solute concentration higher than that of another fluid. *Ch. 3*

hypha, pl. hyphae (Gr. hyphe, web) A slender filament of a fungus barely visible to the naked eye. Hyphae grow into a weblike mass of intertwined filaments. *Ch. 26*

hypothalamus (Gr. hypo. under + thalamos, inner room) A mass of gray matter lying at the base of the cerebrum (beneath the thalamus) that controls the activities of various body organs. *Ch. 14, 17*

hypothesis (Gr. hypo, under + tithenai, to put) A plausible answer to a scientific question, based on available knowledge and generalizations made from observations. *Ch. 1*

hypotonic (Gr. hypo, under + tonos, tension) Refers to a solution with a solute concentration lower than that of another fluid. *Ch. 3*

ileum (L. groin, flank) The third and final part of the small intestine, following the jejunum. *Ch. 8*

immigration (L. immigrare, to migrate) The movement of organisms into a population. *Ch. 34*

immune response The body's specific response to foreign molecules, such as the production of antibodies directed against a particular antigen. The immune response is a result of the complex activities of recognition and defense carried out by the immune system. *Ch. 11*

immune system The collective term for populations of white blood cells that resist disease. *Ch. 11*

immunity (L. immunitas, safe) The specific defense of a human body that consists of cellular and molecular responses to particular foreign invaders. *Ch. 11*

imperfect fungi A division of the kingdom Fungi (organisms that live on dead material). The imperfect fungi have no sexual stage of reproduction and reproduce asexually by spores. Some members of this division are sources of antibiotics or are important in food production, but may cause disease in humans. *Ch. 26*

implantation (L. in, into + plantare, to plant) The embedding of the developing blastocyst into the posterior wall of the uterus approximately 1 week after fertilization. *Ch. 22*

imprinting A rapid and irreversible type of learning that takes place during an early developmental stage of some animals. *Ch. 32*

incomplete dominance A situation in which neither member of a pair of alleles exhibits dominance over the other. *Ch. 20*

incus One of the three bones of the middle ear that amplifies sound vibrations and carries them from the outer ear to the inner ear. The incus is also known as the *anvil*. *Ch. 15*

independent variable The factor that is manipulated during a controlled experiment. *Ch. 1*

induction (L. inductio, leading in) During prenatal development, the interaction of cytoplasmic stimulators and inhibitors to direct the process of cell differentiation. *Ch. 22*

inductive reasoning A pattern of thought in which a person develops generalizations from specific instances. *Ch. 1*

inferior vena cava A large vein that drains blood from the lower body and returns it to the right atrium of the heart. *Ch. 10*

inflammation (L. inflammare, to flame) A reaction when cells are damaged by microbes, chemicals, or physical substances. *Ch. 11*

inhibitory synapse A type of neurotransmitter that reduces the ability of the postsynaptic membrane to depolarize. *Ch. 13*

innate behaviors Patterns of movement, sounds, and body positions resulting from genetically determined neural programs that are part of the nervous system at the time of birth or develop at an appropriate point in maturation. *Ch. 32*

inner ear A complex of fluid-filled canals in the ear; receptor cells of this organ change the mechanical "sound" energy into nerve impulses. *Ch. 15*

insect societies Social groups that are formed by many species of insects, particularly bees and ants, and are characterized by a division of labor. *Ch. 33*

insertion The attachment of one end of a skeletal muscle to a bone that will move. *Ch. 16*

insight (reasoning) The capability of recognizing a problem and solving it mentally before ever trying out a solution. *Ch. 32*

inspiration (L. in, in + spirare, to breathe) Drawing air into the lungs; it occurs when the volume of the thoracic cavity is increased and the resulting negative pressure causes air to be sucked into the lungs; opposite of expiration. *Ch. 9*

insulin (L. insula, island) A hormone produced in the islets of Langerhans of the pancreas that promotes the regulation of glucose metabolism by the cells. *Ch. 17*

integration A function of the central nervous system in which the brain and spinal cord make sense of incoming sensory information and then produce appropriate outgoing motor impulses. *Ch. 14*

integument (L. integumentum, covering) The skin, hair, and nails of the human body. *Ch. 16*

integumentary system Another name for the vertebrate body's integument: the skin, hair, and nails. *Ch. 16*

intercostal muscles Literally, "between-the-rib muscles"; muscles that extend from rib to rib and assist the diaphragm in the breathing process. *Ch. 9*

internal respiration The exchange of oxygen and carbon dioxide between the blood and the tissue fluid. *Ch. 9*

interneuron (L. inter, between + Gr. neuron, nerve) A type of nerve cell found in the spinal cord and brain that receives incoming messages and sends outgoing messages in response. *Ch. 13*

internodes The portions of a plant's stem that lie between the nodes. *Ch. 28*

interphase The portion of the cell cycle preceding mitosis in which the cell grows and carries out normal life functions. During this time the cell also produces an exact copy of the hereditary material, DNA, as it prepares for cell division. *Ch. 18*

intertidal zone In a marine environment, the area between the highest tides and the lowest tides. *Ch. 37*

intervertebral disks Disks of fibrocartilage, positioned between all vertebrae except the sacrum and coccyx, that act as shock absorbers, provide the means of attachment between one vertebra and the next, and permit movement of the vertebral column. *Ch. 16*

intron (L. intra, within) A segment of DNA transcribed into mRNA but removed before translation. *Ch. 18*

inversion The process in which a broken piece of chromosome reattaches to the same chromosome but in a reversed direction. *Ch. 20*

ion (Gr. going) An electrically charged particle. *Ch. 2*

ionic bond An attraction between charged particles having opposite charges. *Ch. 2*

ionization A process of spontaneous formation of charged particles very typical of water molecules, caused by the breaking of the molecules' bonds. *Ch. 2*

ionizing radiation A form of energy, also known as electromagnetic energy, that can cause chromosomes to break or can cause changes in the nucleotide structure of DNA. X-rays and nuclear radiation are kinds of ionizing radiation. *Ch. 20*

iris (L. rainbow) A diaphragm lying between the cornea and the lens that controls the amount of light entering the eye. *Ch. 15*

islets of Langerhans (after Paul Langerhans, German anatomist) The separate types of cells within the exocrine cells of the pancreas that produce the hormones insulin and glucagon. *Ch. 17*

isomer (Gr. isos, equal + meros, part) One of two or more molecules that have the same molecular formula but are arranged slightly differently; fructose is an isomer of glucose. *Ch. 24*

isotonic (Gr. isos, equal + tonos, tension) Refers to solutions having equal solute concentrations to one another. *Ch. 3*

isotope (Gr. isos, equal, + topos, place) An atom of an element that has the same number of protons but different numbers of neutrons in its nucleus than another atom of that element. *Ch. 2*

jawless fish A tubular, scaleless, jawless organism that lives in the sea or brackish water. *Ch. 30*

jejunum (L. empty) The second portion of the small intestine extending from the duodenum to the ileum. *Ch. 8*

joint An articulation; a place where bone or bones and cartilage come together. *Ch. 16*

Jurassic period The second period of the Mesozoic era, in which large dinosaurs dominated the Earth and birds first appeared. *Ch. 24*

karyotype (Gr. karyon, kernel + typos, stamp or print) The particular array of chromosomes that belongs to an individual as shown by specially treating a cell just as it is about to divide. *Ch. 18*

kidney The organ in all vertebrates that carries out the processes of filtration, reabsorption, and excretion. *Ch. 12*

kidney stone Crystals of certain salts that develop in the kidney and block urine flow. *Ch. 12*

kinesis The change in the speed of the random, nondirected movements of an animal with respect to changes in certain environmental stimuli. *Ch. 32*

kinetic energy The energy of motion. *Ch. 4*

kingdoms Broad categories in which taxonomists group all living things; the taxonomic system used in this book recognizes five kingdoms of life: *Monera, Protista, Plantae, Fungi,* and *Animalia. Ch. 1*

Klinefelter syndrome (after Harry F. Klinefelter, Jr, U.S. physician) A genetic condition resulting from an XXY zygote that develops into a human male who is sterile, has many female body characteristics, and in some cases, has diminished mental capacity. *Ch. 20*

Krebs cycle (after Hans A. Krebs, German-born English biochemist) The series of eight reactions during which pyruvate, the end product of glycolysis, enters the cycle to form citric acid and is finally oxidized to carbon dioxide. Also called citric acid cycle. *Ch. 5*

labia majora Two longitudinal folds of skin that run posteriorly from the mons in the exterior genitals of the female. *Ch. 21*

labia minora Folds of skin covered by the labia majora in the exterior genitals of the female. *Ch. 21*

labor The sequence of events that leads to birth. *Ch. 22*

lacteal (L. lacteus, of milk) A lymphatic vessel within a villus in the small intestine that helps pass nutrients into the lymph and blood during absorption. *Ch. 8*

lacuna, pl. lacunae (L. a pit) The tiny chambers in which cartilage and bone cells lie. *Ch. 7*

laguno A fine body hair that appears over the body of a fetus toward the end of the third month of pregnancy but that is lost before birth. *Ch. 22*

lancelet A scaleless, fishlike marine chordate. *Ch. 30*

larynx (Gr.) The voice box located at the upper end of the human windpipe. *Ch. 9*

lateral buds Tiny undeveloped side shoots that develop at the angles between a plant's leaves and its stem. *Ch. 28*

lateral line system A complex system of mechanoreceptors possessed by all fishes and amphibians that detects mechanical stimuli such as sound, pressure, and movement. *Ch. 30*

laws of thermodynamics Two laws that govern all of the changes in energy that take place in the universe. *Ch. 4*

learning An alteration in behavior based on experience. *Ch. 32*

leaves The parts of a plant's structure where most photosynthesis takes place. *Ch. 28*

left atrium The upper left chamber of the heart, into which oxygenated blood enters from the lungs. *Ch. 10*

left ventricle A chamber in the heart into which blood flows from the left atrium. The left ventricle then pumps blood through the aorta into the arteries. *Ch. 10*

lens (L. lentil) A body in the eye lying just behind the aqueous humor that plays a major role in focusing the light entering the eye. *Ch. 15*

leukocyte (Gr. leukos, white + kytos, hollow vessel) Any of several kinds of white blood cells including macrophages and lymphocytes, all functioning to defend the body against invading microorganisms and foreign substances. *Ch. 7*

life cycle The progression of stages an organism passes through from its conception until it conceives another similar organism. *Ch. 18*

ligament (L. ligare, to bind) A bundle or strip of dense connective tissue that holds a bone to a bone. *Ch. 16*

light-dependent reactions The reactions of photosynthesis that produce ATP and NADPH; so called because they take place only in the presence of light. *Ch. 6*

ligase A sealing enzyme that helps reform the bonds between pieces of DNA that have been cut with the same restriction enzyme. *Ch. 31*

light-independent reactions The reactions of photosynthesis that use ATP and NADPH to drive the formation of sucrose and other organic compounds from carbon dioxide; so called because they do not need light to occur as long as ATP is available; also called the Calvin cycle. *Ch. 6*

limb bud The appearance of arms and legs that appear as microscopic flippers during the fourth week in an embryo. *Ch. 22*

limbic system (L. limbus, border) A network of neurons, which together with the hypothalamus, forms a ringlike border around the top of the brainstem; responsible for many of the most deep-seated drives and emotions of vertebrates. *Ch. 14*

lipase (Gr. lipos, fat + -ase, enzyme) An enzyme that breaks down the triglycerides in lipids to fatty acids and glycerol. *Ch. 8*

lipid (Gr. lipos, fat) Any of a wide variety of molecules, all of which are soluble in oil but insoluble in water;

important categories of lipids are oils, fats, and waxes; phospholipids; and steroids. *Ch. 2*

lipid bilayer The basic foundation of biological membranes; it forms a fluid, flexible covering for a cell and keeps the watery contents of the cell on one side of the membrane and the water environment on the other. *Ch. 3*

liver A large, complex organ weighing over 3 pounds, lying just under the diaphram, that performs over 500 functions in the body, including aiding in the digestion of lipids. *Ch. 8*

loop of Henle (after F.G.J. Henle, German anatomist) The descending and ascending loops of the renal tubule. *Ch. 12*

lumen, pl. lumina (L. light) The hollow core in the three layers of tissue that makes up the walls of the arteries through which blood flows. *Ch. 10*

luteinizing hormone (LH) A gonadotropic hormone secreted by the anterior pituitary that stimulates the release of an egg in females and stimulating the production of testosterone in males. *Ch. 17*

lymph (L. lympha, clear water) Tissue fluid. *Ch. 10*

lymphocytes A type of agranulocyte that recognizes and reacts to substances that are foreign to the body, sometimes producing a protective immunity to disease. Lymphocytes are cells that make up the immune system, the specific resistance to disease. *Ch. 10*

lymph nodes Small, ovoid "spongy" structures located in various places of the body along the route of the lymphatic vessels, filtering the lymph as it passes through. *Ch. 10*

lymphatic system The system of one-way, blind-ended vessels that collects and returns to the blood the approximately 10% of the fluid that does not return to the blood directly. *Ch. 10*

lysogenic cycle (Gr. lysis, dissolution + gennan, to produce) A pattern of viral replication in which a virus integrates its genetic material with that of a host and replicates each time the host cell replicates. *Ch. 25*

lysosome (Gr. lysis, a loosening + soma, a body) A membrane-bounded vesicle containing digestive enzymes that break down old cell parts or materials brought into the cell from the environment. *Ch. 3*

lytic cycle (Gr. lysis, dissolution) A pattern of viral replication in which a virus enters a cell, replicates, and then causes the cell to burst, releasing new viruses. *Ch. 25*

macromolecule (Gr. makros, large + L. moliculus, a little mass) A large organic molecule having many functional groups. *Ch. 2*

macrophage (Gr. makros, long + -phage, eat) A phagocytic cell in the bloodstream that engulfs foreign bacteria and antibody-coated cells or particles in the process of phagocytosis; macrophages act as the body's scavengers. *Ch. 7*

malignant A term describing cancerous tumors that have the ability to invade and kill other tissues and move to other areas of the body. *Ch. 20*

malleus A very small bone, connected to the internal side of the eardrum, that works with two other small bones, the incus and stapes, to amplify sound vibrations and carry them from the outer ear to the inner ear. The malleus is also known as the *hammer*. *Ch. 15*

Malpighian tubule (after Marcello Malpighi, Italian anatomist) A slender projection from the digestive tract in the body of a terrestrial anthropod that serves as an excretory system. *Ch. 29*

mammal (L. mamma, breast) A warm-blooded vertebrate that has hair; female mammals secrete milk from mammary glands to feed young. *Ch. 24*

mammary glands Milk-producing glands that lie over the chest muscles in females. *Ch. 21*

mantle In the body of a mollusk, a soft epithelium that arises from the dorsal body wall and encloses a cavity between itself and the visceral mass. *Ch. 29*

marsupial (L. marsupium, pouch) A subclass of mammals that give birth to immature young, which are then carried in a pouch. *Ch. 24*

mating Male-female behaviors that result in fertilization regardless of whether copulation occurs. *Ch. 33*

medulla (L. marrow) The lowest portion of the brainstem, continuous with the spinal cord below; the site of neuron tracts, which cross over one another delivering sensory information from the right side of the body to the left side of the brain and vice versa. *Ch. 12*

medusa, pl. medusae A cnidarian that is a free-floating, umbrella-shaped animal with the mouth usually located on the underside of the umbrella-shaped and tentacles hanging down around the umbrella edge. *Ch. 29*

megakaryocyte (Gr. megas, large + karyon, nucleus + kytos, cell) A large bone marrow cell that pinches off bits of its cytoplasm, resulting in a cell fragment, or platelet. *Ch. 10*

meiosis (Gr. meioun, to make smaller) The two-staged process of nuclear division in which the number of chromosomes in cells is halved during gamete formation. *Ch. 18*

Mendel's law of independent assortment The concept that the distribution of alleles for one trait into the gametes does not affect the distribution of alleles for other traits. *Ch. 19*

Mendel's law of segregation The concept that each gamete receives only one of an organism's pair of alleles. Chance determines which member of a pair of alleles becomes included in a gamete. *Ch. 19*

meninges, sing. meninx (Gr. membrane) The three layers of membranes covering both the brain and spinal cord. *Ch. 14*

menopause (L. mens, month + pausis, cessation) Permanent cessation of menstrual activity in a woman, usually between the ages of 50 and 55; the end of the menses. *Ch. 21*

menstruation (L. mens, month) The monthly sloughing of the blood-enriched lining of the uterus when pregnancy does not occur; the lining degenerates and causes a flow of blood, tissue, and mucus from the uterus out the vagina. *Ch. 21*

meristematic tissue (Gr. merizein, to divide) An undifferentiated type of tissue in a vascular plant that produces new plant cells during growth; also called meristem. *Ch. 28*

mesoderm (Gr. mesos, middle + derma, skin) The layer of cells in the developing embryo that differentiates into the skeleton, muscles, blood, reproductive organs, connective tissue, and the innermost layer of the skin. *Ch. 29*

mesopelagic A term referring to the middle ocean zone; little light penetrates this ocean zone from 200 to 1000 meters (650 to 3250 feet) *Ch. 37*

mesophyll (Gr. mesos, middle + phyllon, leaf) The parenchymal layer of a leaf, bounded by epidermis, where photosynthesis takes place. *Ch. 28*

messenger RNA (mRNA) A type of RNA that brings information from the DNA within the nucleus to the ribosomes in the cytoplasm to direct which polypeptide is assembled. *Ch. 18*

metabolism (Gr. metabole, change) All the chemical reactions that take place within a living organism. *Ch. 5*

metaphase (Gr. meta, middle + phasis, form) The stage of mitosis characterized by the alignment of the chromosomes equidistant from the two poles of the cell. *Ch. 18*

metastasis One of the characteristics of cancer cells; the ability to spread to multiple sites throughout the body. *Ch. 20*

microfilament (Gr. mikros, small + L. filum, a thread) A thin, twisted double-chain fiber of protein within the cytoskeleton that helps support and shape eukaroytic cells. *Ch. 3*

microtubule (Gr. mikros, small + tubulus, little pipe) A spiral array of protein subunits forming microscopic tubes within the cytoskeleton that provide intracellular support in the nondividing cell. *Ch. 3*

microvillus, pl. microvilli (Gr. mikros, small + L. villus, tuft of hair) Microscopic, cytoplasmic projections that cover the epithelial cells of the villi of the small intestine on their exposed surfaces. *Ch. 8*

midbrain The top part of the brainstem; it contains nerve tracts connecting the upper and lower parts of the brain, and nuclei that act as reflex centers for movement. *Ch. 14*

middle ear The middle portion of the ear containing three bones that act together like an amplifier to increase the force of sound vibrations. *Ch. 15*

migration Long-range, two-way movements by animals, often occurring yearly with the change of seasons. *Ch. 32*

minerals (L. minerale) Inorganic substances transported in organisms as ions dissolved in blood and other fluids in both plants and animals; a variety of minerals perform a variety of functions in the human body. *Ch. 8*

mitochondrion, pl. mitochondria (Gr. mitos, thread + chondrion, small grain) An oval, sausage-shaped, or threadlike organelle about the size of a bacterium, bounded by a double membrane whose function is to break down fuel molecules, thus releasing energy for cell work. *Ch. 3*

mitosis (Gr. mitos, thread) A process of cell division that produces two identical cells from an original parent cell. *Ch. 18*

molecule (L. molecula, little mass) A combination of tightly bound covalently-bonded atoms. *Ch. 2, 18*

Monera (Gr. moneres, individual) The kingdom that consists of the bacteria. *Ch. 25*

monocot One of the two groups of angiosperms in which the food reserves within its seeds are stored in extraembryonic tissue called endosperm; a monocot usually has parallel veins in its leaves and its flower parts are often in threes. *Ch. 27*

monocytes A group of agranulocytes that circulate as the granulocytes do; monocytes are attracted to the sites of injury or infection where they mature into macrophages and engulf any bacteria or dead cells that neutrophils may have left behind. *Ch. 10*

monohybrid (Gr. monos, single + L. hybrida, mongrel) The progeny or product of two plants that differ from one another in a single trait. *Ch. 19*

monosaccharide (Gr. monos, one + sakcharon, sugar) A simple sugar. *Ch. 2*

monotreme (Gr. mono, single + treme, hole) A mammal that lays eggs having leathery shells similar to those of a reptile. *Ch. 24*

mons pubis The mound of fatty tissue that lies over the place of attachment of the two pubic bones. *Ch. 21*

morphogenesis (Gr. morphe, form + genesis, origin) The early stage of development in a vertebrate when cells begin to move, or migrate, thus shaping the new individual. *Ch. 22*

mortality The death rate of a population. *Ch. 34*

morula A stage of development in which the pre-embryo consists of about 16 densely clustered cells and is still the same size as a newly-fertilized ovum. *Ch. 22*

motor area The part of the brain straddling the rearmost portion of the frontal lobe that sends messages to move the skeletal muscles. *Ch. 14*

motor neurons Nerve cells of the peripheral nervous system that transmit impulses away from the central nervous system to the muscles and glands. *Ch. 13*

multiple alleles A system of alleles in which a gene is represented by more than two alleles within a population. *Ch. 20*

muscle fiber A long, multinucleated cell packed with organized arrangements of microfilaments capable of contraction. *Ch. 16*

muscle tissue Any of three different kinds of muscle cells—smooth, skeletal, or cardiac—that are the workhorses of the body; characterized by an abundance of special thick and thin microfilaments. *Ch. 7*

mutation (L. mutare, to change) A permanent change in the genetic material. *Ch. 35*

mutualism (L. mutuus, lent, borrowed) A type of symbiosis in which both participating species benefit. *Ch. 35*

mycelium, pl. mycelia (Gr. mykes, fungus) In fungi, a mass of hyphae. *Ch. 26*

myelin sheath (Gr. myelinos, full of marrow) The fatty wrapping created by multiple layers of Schwann cell membranes; the myelin sheath insulates the axon of a nerve cell. *Ch. 13*

myofibril (Gr. myos, muscle + L. fibrilla, little fiber) A cylindrical, organized arrangement of special thick and thin microfilaments capable of shortening a muscle fiber. *Ch. 16*

myosin (Gr. mys, muscle + in, belonging to) One of the two protein components of myofilaments in a muscle fiber; actin is the other. *Ch. 16*

nasal cavities Two hollow areas, located above the oral cavity and behind the nose, that are bordered by projections of bone covered with moist epithelial tissue. *Ch. 9*

natural selection The process in which organisms having adaptive traits survive in greater numbers than organisms without such traits. *Ch. 23*

Neanderthal (Neander, valley in western Germany) A subspecies of *H. sapiens* that lived from about 125,000 to 35,000 years ago in Europe and the Middle East; Neanderthals were short and powerfully built, with large brains; they made diverse tools, took care of the sick and injured, and buried their dead. *Ch. 24*

negative feedback The process by which enzyme activity is regulated by inhibitors. *Ch. 4*

negative feedback loop A feedback loop in which the response of the regulating mechanism is negative with respect to the output. Most of the body's regulatory mechanisms work by means of negative feedback loops. *Ch. 7*

nephridium, pl. nephridia (Gr. nephros, kidney) A tubular structure through which wastes are removed in certain organisms. *Ch. 29*

nephron (Gr. nephros, kidney) Any of the millions of microscopic tubular units of the kidney where urine is formed. *Ch. 12*

neritic zone The area of shallow waters along the coasts of the continents, which extends from the low tide mark to waters as deep as 200 meters. Because light reaches the waters of most of this zone, it supports an abundant array of plant and animal life, including coral reefs. *Ch. 37*

nerve Bundles of neurons surrounded by numerous supporting cells. *Ch. 14*

nerve cord A single, hollow cord along the back that is a principal feature of chordates; in vertebrates, the nerve cord differentiates into a brain and spinal cord. *Ch. 24*

nerve impulse A rapid electrical/chemical signal of a neuron that reports information and/or initiates a quick response in specific tissues. *Ch. 13*

nervous system The body's complex network of neurons, which both gathers information about the body's internal and external environments, and processes and responds to that information. *Ch. 13*

nervous tissue Groups of similar cells that are specialized to conduct electrical/chemical impulses, and other supporting cells. *Ch. 7*

neuromuscular junction A synapse between a neuron and a skeletal muscle cell. *Ch. 13*

neuron (Gr. nerve) A nerve cell specialized to conduct an electrical/chemical current. *Ch. 7*

neurotransmitter (Gr. neuron, nerve + L. trans, across + mitere, to send) A chemical released when a nerve impulse reaches the axon tip of a nerve cell. *Ch. 13*

neurulation (Gr. neuron, nerve) The development of a hollow nerve cord. *Ch. 22*

neutron (L. neuter, neither) A subatomic particle found at the nucleus of atom, similar to a proton in mass but neutral and carrying no charge. *Ch. 2*

neutrophil (L. neuter, neither + Gr. philein, to move) A type of granulocyte that migrates to the site of an injury and sticks to the interior walls of blood vessels, where it forms projections and phagocytizes microorganisms and other foreign particles. *Ch. 10*

niche (L. nidus, nest) The special role each organism plays within an ecosystem; an organism's use of factors such as space, food, temperature, mating conditions, and moisture within an ecosystem. *Ch. 35*

nitrifying bacteria (Gr. nitron, salt) Bacteria that live in nodules in the roots of legumes and cycle nitrogen in ecosystems by converting ammonia to nitrates, a form of nitrogen used by plants. *Ch. 25*

nitrogenous wastes Nitrogen-containing molecules that are produced as waste products from the body's breakdown of proteins and nucleic acids. *Ch. 12*

noble gases Atoms of elements that have equal numbers of protons and electrons and have full outer-electron energy levels and thus do not react readily with other elements. *Ch. 2*

node of Ranvier (after L.A. Ranvier, French histologist) An uninsulated spot between two Schwann cells on a myelinated neuron. *Ch. 13*

nodes Locations on the stem of a plant where leaves form. *Ch. 28*

noncyclic electron flow A system that harvests energy in a process that involves photosystems I and II working together to produce NAPH and ATP; these two products power the generation of sucrose and other organic compounds from carbon dioxide in light-independent reactions. *Ch. 6*

nondisjunction The failure of homologous chromosomes to separate after synapsis, resulting in gamates with abnormal numbers of chromosomes. *Ch. 20*

nongonococcal urethritis A sexually transmitted disease caused by the bacterium *Chlamydia trachomatis,* that has gonorrhealike symptoms. *Ch. 25*

nonspecific defense A set of defenses that the body uses to act against foreign invaders in general; they include the skin and mucous membranes, chemicals that kill bacteria, and the inflammatory process. *Ch. 11*

nonrenewable resources Resources that are formed at a rate much slower than their consumption; they are thus finite in supply. Fossil fuels and minerals are nonrenewable resources. *Ch. 38*

nonvascular plants Plants that lack specialized transport tissues. *Ch. 27*

notochord (Gr. noto, back + L. chorda, cord) A structure that forms the midline axis along which the vertebral column (backbone) develops in all vertebrate animals. *Ch. 22*

nuclear envelope The outer, double membrane surrounding the nucleus of a eukaryotic cell. *Ch. 3*

nucleic acid A long polymer of repeating subunits called nucleotides which are each composed of a 5-carbon sugar, a phosphate group, and a nitrogen-containing base. The two types of nucleic acid within cells are deoxyribonucleic acid (DNA) and ribonucleic acid (RNA). *Ch. 2*

nucleolus, pl. nucleoli (L. a small nucleus) The site within the nucleus of ribosomal RNA synthesis; consists of ribosomal RNA plus some ribosomal proteins. *Ch. 3*

nucleotide A single unit of nucleic acid consisting of a five-carbon sugar, a phosphate group, and an organic nitrogen-containing molecule, or base. *Ch. 2*

nuclear power An energy source derived from nuclear fission reactions—the splitting of atoms of radioactive material. *Ch. 38*

nucleus (L. a kernal, dim. fr. nux, nut) (1) The central core of an atom containing protons and neutrons. (2) The double membrane vesicle of a eukaryotic cell that contains the hereditary material, or DNA. *Ch. 2, 3*

nutrient (L. nutritio, nourish) (1) A raw material of food; the six classes of nutrients are carbohydrates, fats, proteins, vitamins, minerals, and water. (2) Inorganic materials a plant needs to grow, such as hydrogen, oxygen, and nitrogen. *Ch. 8, 28*

occipital lobe The section of the cerebral cortex in each hemisphere of the brain having to do with vision, with different sites corresponding to different positions on the retina. *Ch. 14*

octet rule One of the three factors that influence whether an atom will interact with other atoms; the octet rule states that an atom with an unfilled outer shell has a tendency to interact with another atom or atoms in ways that will complete this outer shell. *Ch. 2*

olfactory receptors Neurons whose cell bodies are embedded in the nasal epithelium; they detect smells when different airborne chemicals bind with receptor chemicals in their ciliated dendrite endings. *Ch. 15*

omnivore (L. omnis, all + vorare, to eat) An organism that eats both plant and animal foods; human beings are omnivores. *Ch. 8*

oncogenes Cancer-causing genes. *Ch. 20*

one gene-one enzyme theory An explanation regarding hereditary material that states that the production of a given enzyme is under the control of a specific gene. If the gene mutates, the enzyme will not be synthesized properly or will not be made at all. Therefore, the reaction it catalyzes will not take place, and the product of the reaction will not be produced. *Ch. 18*

oogenesis (Gr. oon, egg + genesis, generation, birth) The process of meiosis and development that produces mature female sex cells, or eggs. *Ch. 21*

open circulatory system A body system in which blood flows in vessels leading to and from the heart but through irregular channels called blood sinuses in many parts of the body. *Ch. 29*

open-sea zone The great expanse of the open ocean that lies beyond the continental shelf. The open-sea zone is home to many diverse forms of life. *Ch. 37*

open-water zone The main body of pond or lake water through which light penetrates. *Ch. 37*

operant conditioning A form of learning in which an animal associates something that it does with a reward or punishment. *Ch. 32*

operculum (L. operire, to shut, cover) A flap in a bony fish that protects the gills and enhances water flow over the gills, thus bringing more oxygen in contact with their gas-exchanging surfaces and allowing a fish to breathe while stationary. *Ch. 30*

opiates Compounds derived from the milky juice of the poppy plant that act as narcotic analgesics, which stop or reduce pain without causing a person to lose consciousness. *Ch. 13*

optic nerve The nerve carrying impulses from the eye to the brain for the sense of sight. *Ch. 15*

order A taxonomic subcategory of class. *Ch. 1*

Ordovician period A period during the Paleozoic era, 505 to 430 million years ago during which worm-like aquatic animals began to evolve from ancient flatworms. *Ch. 24*

organ (L. organon, tool) Grouped tissues that form a structural and functional unit. *Ch. 7*

organic compounds The carbon-containing molecules that make up living things. *Ch. 2*

organisms Living things. Organisms are composed of cells and can be either multicellular or unicellular. *Ch. 1*

organ of Corti (after Alfonso Corti, Italian anatomist) The organ of hearing; the collective term for the hair cells, the supporting cells of the basilar membrane, and the overhanging tectorial membrane. *Ch. 15*

organ system A group of organs that function together to carry out the principal activities of the organism. *Ch. 7*

organelle (Gr. organella, little tool) Any of a number of highly specialized, intracellular structures that perform specific functions; a feature common to most eukaryotes but lacking in bacteria. *Ch. 1*

origin (L. oriri, to arise) With respect to muscles, the attachment of one end of a skeletal muscle to a stationary bone. *Ch. 16*

osmoregulation The control of water movement into and out of an organism. *Ch. 30*

osmosis (Gr. osmos, impulse + osis, condition) A special form of diffusion in which water molecules move from an area of higher concentration to an area of lower concentration across a differentially permeable membrane. *Ch. 3*

osmotic pressure The force that water molecules exert on a cell as they diffuse into a cell. *Ch. 3*

ossification The process of bone formation, which in humans begins approximately at the eighth week of development and continues beyond birth to the age of 18 or 19 years. *Ch. 22*

osteoblast (Gr. osteon, bone + blastos, cell) A cell that forms bone. *Ch. 16*

osteocyte (Gr. osteon, bone + kytos, hollow vessel) A cell that produces bone. *Ch. 7*

otolith (Gr. otos, ear + lithos, stone) Small pebbles of calcium carbonate embedded in a layer of jellylike material that is spread over the surface of ciliated and nonciliated cells within the saccule of the inner ear. Otoliths play a role in the determination of position with respect to gravity. *Ch. 15*

outer ear The part of the ear that funnels sound waves in toward the eardrum; includes the flaps of skin on the outside of the head called ears. *Ch. 15*

oval window The entrance to the inner ear. *Ch. 15*

ovary (L. ovum, egg) *(1)* A female gonad, located in the pelvic cavity of an animal, where egg production occurs. *(2)* In flowering plants, a chamber at the base of the female pistil that completely encloses and protects the ovules. *Ch. 21, 27*

oviparous (L. ovum, egg + parere, to bring forth) A method of reproduction in which the female lays eggs and the young hatch outside of the mother; birds and most reptiles are oviparous. *Ch. 30*

ovoviviparous (L. ovum, egg + vivus, alive + parere, to bring forth) A method of reproduction in which the female retains fertilized eggs within her oviducts until the young hatch; the young receive nutrition from the egg yolk and not from the mother. *Ch. 30*

ovulation (L. ovulum, little egg) The monthly process by which an egg is released from the ovary. *Ch. 21*

ovule (L. ovulum, little egg) A protective structure in which egg cells grow in a naked seed plant. *Ch. 27*

ovum, pl. ova (L. egg) A mature egg cell. *Ch. 21, 22*

oxidation (Fr. oxider, to oxidize) The loss of an electron by an atom or a molecule. *Ch. 5*

oxidation-reduction A chemical reaction in which one atom or molecule gains an electron while the other atom or molecule involved in the reaction loses an electron. *Ch. 5*

oxygen debt A term used to describe the oxygen needed to break down the lactic acid in the liver, delivered by the bloodstream from the muscles during strenuous exercise. *Ch. 5*

oxytocin (Gr. oxys, sharp + tokos, birth) A hormone produced by the hypothalamus but stored and released in the posterior lobe of the pituitary that affects the contraction of the uterus during childbirth and stimulates the mammary glands, allowing a new mother to nurse her child. *Ch. 17*

ozone (Gr. ozein, to smell) A principal chemical air pollutant, (O_3), formed by photochemical reactions on hydrocarbons and nitrogen oxides in the air, that is extremely irritating to the eyes and upper respiratory tract. *Ch. 38*

Paleocene epoch The earliest epoch of the Cretaceous period, in which the small animals that survived the extinctions of the Mesozoic underwent adaptive radiation and quickly filled the habitats vacated by the dinosaurs. *Ch. 24*

pancreas (Gr. pan, all + kreas, flesh) A long gland that lies beneath the stomach and is surrounded on one side by the curve of the duodenum; it secretes a number of digestive enzymes and the hormones insulin and glucagon. *Ch. 8*

pancreatic amylase A digestive enzyme secreted by the pancreas that works within the small intestine to break down starch and glycogen to maltose. *Ch. 8*

pancreatic lipase A digestive enzyme secreted by the pancreas that works in the small intestine to break down triglycerides to fatty acids and glycerol. *Ch. 8*

parasite (Gr. para, beside + sitos, food) An organism that lives in or on another. *Ch. 26, 35*

parasitism (Gr. para, beside + sitos, food) A relationship between organisms in which one benefits but the other (the host) is harmed. *Ch. 35*

parasympathetic system (Gr. para, beside + syn, with + pathos, feeling) A subdivision of the autonomic nervous system that generally stimulates the activities of normal internal body functions and inhibits alarm responses; opposite of sympathetic nervous system. *Ch. 14*

parathyroid gland (Gr. para, beside + thyreos, shield + eidos, form, shape) One of the four small glands embedded in the posterior side of the thyroid that produces parathyroid hormone, which regulates the concentration of calcium in the bloodstream. *Ch. 17*

parathyroid hormone (PTH) A hormone secreted by the parathyroid glands that works antagonistically to calcitonin to help maintain the proper blood levels of various ions, primarily calcium. *Ch. 17*

parenchyma cells (Gr. para, beside + en, in + chein, to pour) Cells in the ground tissue of vascular plants that function in photosynthesis and storage; the most common of all plant cell types. *Ch. 28*

parental (P) generation Plants having contrasting forms of single traits, artificially fertilized with one another, to produce hybrid offspring referred to as the first filial (F_1) generation. *Ch. 19*

parietal lobe The section of the cerebral cortex of each hemisphere of the brain containing sensory receptors from different parts of the body. *Ch. 14*

passive immunity The type of immunity produced by injection of antibodies into the subject to be protected or acquired by the fetus through the placenta. *Ch. 11*

passive transport Molecular movement down a gradient and across a cell membrane; the three types of passive transport are diffusion, osmosis, and facilitated diffusion. *Ch. 3*

peck order In chickens, another term for the rank order; chickens establish this social hierarchy by pecking at one another. *Ch. 33*

pectoral girdle (L. pectus, chest) The part of the appendicular skeleton made up of two pairs of bones: the clavicles, or collarbones, and the scapulae, or shoulder blades. *Ch. 16*

pedigree A diagram of the genetic relationships among family members over several generations. *Ch. 20*

pelagic zone The open ocean. *Ch. 37*

pelvic girdle (L. pelvis, basin) The part of the appendicular skeleton made up of the two bones called coxal bones, pelvic bones, or hip bones. *Ch. 16*

penis A cylindrical organ that transfers sperm from the male reproductive tract to the female reproductive tract; the male urinary organ. *Ch. 21*

pepsin (Gr. pepsis, digestion) An enzyme of the stomach that digests proteins, breaking them down into short peptides. *Ch. 8*

peptidases Enzymes produced by cells in the intestinal epithelium that work as a team with trypsin, chymotrypsin, and carboxypeptidase to break down polypeptides into shorter chains and then to amino acids. *Ch. 8*

peptide bond A covalent bond that links two amino acids formed during dehydration synthesis when the amino group at one end and the carboxyl group at the other end lose a molecule of water between them. *Ch. 2*

peptide hormones One of the two main classes of endocrine hormones. Peptide hormones are made of amino acids and are unable to pass through cell membranes; instead, they bind to receptor molecules embedded in the membranes of target cells. *Ch. 17*

peptides Short proteins. *Ch. 2*

pericycle (Gr. peri, around + kykos, circle) In the roots of vascular plants, a cylinder of cells made up of parenchymal cells that are able to undergo cell division to produce either branch roots or new vascular tissue. *Ch. 28*

peripheral nervous system (Gr. peripherein, to carry around) The part of the nervous system made up of the nerves of the body that bring messages to and from the brain and spinal cord. *Ch. 14*

peristalsis (Gr. peri, around + stellein, to wrap) The rhythmic wave of contractions by the muscles of the esophagus that moves food toward the stomach and of the small intestine that moves food being digested along its length, to the large intestine. *Ch. 8*

permafrost (Perma[nent] + frost) A layer of permanently frozen subsoil, found about 1 meter down from the surface, that is common throughout most of the Arctic regions. *Ch. 37*

petal (Gr. petalon, leaf) A part of a flower, frequently prominent and usually colorful, that serves to attract pollinators, such as birds and insects. *Ch. 27*

petiole (L. petiolus, a little foot) The slender stalk of a leaf. *Ch. 28*

pH scale A series of regularly-spaced numbers (0 to 14) that indicates the relative concentration of H^+ ions in a solution. Low pH values indicate high concentrations of H^+ ions (acids), and high pH values indicate low concentrations. *Ch. 2*

phagocytes White blood cells that patrol the body and destroy foreign cells by engulfing and ingesting them. *Ch. 11*

phagocytosis (Gr. phagein, to eat + kytos, hollow vessel) A type of endocytosis in which a cell ingests an organism or some other fragment of organic matter; macrophages and neutrophils are phagocytes. *Ch. 3*

pharyngeal (gill) slit (Gr. pharynx, gullet) A principal feature of chordate embryos that develops into the gill structure of a fish and into the ear, jaw, and throat structures of a terrestrial vertebrate. *Ch. 24*

pharynx (Gr. gullet) The upper part of the throat that extends from behind the nasal cavities to the openings of the esophagus and larynx. *Ch. 8*

phenotype (Gr. phainein, to show + typos, print) The outward appearance or expression of an organism's genes. *Ch. 19*

pheromone A chemical produced by one individual that alters the physiology or behavior of other individuals of the same species. *Ch. 33*

phloem (Gr. phloos, bark) A type of vascular tissue that conducts carbohydrates a plant uses as food as well as other needed substances. *Ch. 28*

phospholipid (Gr. phosphoros, light-bearer + lipos, fat) A molecule made up of a portion of a fat molecule with a phosphate functional group attached. Because these molecules have polar and nonpolar parts, they form a double layer of molecules (a bilayer) when in a watery environment. A phospholipid bilayer is the foundation of cell membranes. *Ch. 3*

photic zone An area of the open ocean where light is available to organisms from the water's surface to an approximate depth of 200 meters (650 feet). *Ch. 37*

photoautotroph (Gr. photos, light + auto, self + trophos, feeder) An organism that makes its own food by photosynthesis using the energy of the sun. *Ch. 25*

photocenter An array of pigment molecules within the thylakoid membranes of a chloroplast. It acts like a light antenna, capturing and directing photon energy toward a single molecule of a chlorophyll *a* that will participate in photosynthesis. *Ch. 6*

photochemical smog A type of air pollution caused by reactions between hydrocarbons and nitrogen oxides taking place in the presence of sunlight. *Ch. 38*

photon (Gr. photos, light) A discrete packet of energy from sunlight. *Ch. 6*

photosynthesis (Gr. photos, light -syn, together + tithenai, to place) The process whereby energy from the sun is captured by living organisms and used to produce molecules of food. *Ch. 6*

Photosystem I One of two photosynthetic pathways used by plants and algae that captures energy from the sun and directs this energy toward a molecule of chlorophyll *a* that will participate in photosynthesis. The photosystems are numbered to show the order in which they evolved in photosynthetic organisms. *Ch. 6*

Photosystem II One of two photosynthetic pathways used by plants and algae that captures energy from the sun and directs this energy toward a molecule of chlorophyll *a* that will participate in photosynthesis. The photosystems are numbered to show the order in which they evolved in photosynthetic organisms. *Ch. 6*

phyletic gradualism A theory that states that new species develop slowly and gradually as an entire species changes over time. *Ch. 23*

phylum, pl. phyla (Gr. phylon, race, tribe) A major taxonomic group that is a division of kingdom. *Ch. 1*

pigments Molecules that absorb some visible wavelengths of light and transmit or reflect others. *Ch. 6*

pineal gland (L. pinus, pine tree) A tiny gland lying deep within the brain whose exact function remains a mystery; it is the possible site of an individual's biological clock. *Ch. 17*

pinna (L. feather) The projected part of the outer ear, the ear flap. *Ch. 15*

pinocytosis (Gr. pinein, to drink + kytos, vessel) A type of endocytosis in which a cell ingests liquid material containing dissolved molecules. *Ch. 3*

pistil (L. pistillum, pestle) The female sex organ at the center of a flower. *Ch. 27*

pith A parenchymal (storage) tissue located in the center of the roots of a monocot. *Ch. 28*

pituitary A tiny gland hanging from the underside of the brain, under the control of the hypothalamus, that secretes nine different major hormones. *Ch. 17*

pivot joint A kind of joint that allows side-to-side movement. *Ch. 16*

placenta (L. a flat cake) A flat disk of tissue that grows into the uterine wall during the development of placental mammals through which the mother supplies the offspring with food, water, and oxygen and through which she removes wastes. *Ch. 22*

placental mammal A mammal (animal that is warm-blooded, has hair, and [if female] secretes milk from glands to feed its young) that nourishes its developing embryo within the body of the mother until maturity. *Ch. 24*

plankton (Gr. planktos, wandering) Organisms that drift in the water. *Ch. 36*

plaques Masses of cholesterol and other lipids that can build up within the walls of large and medium-sized arteries in a disease called atherosclerosis; the accumulation of such masses impairs the arteries' proper functioning. *Ch. 10*

plasma The fluid intercellular matrix within which blood cells float; contains practically every substance used and discarded by cells as well as nutrients, hormones, proteins, salts, ions, and albumin. *Ch. 7*

plasma cell Cells that differentiate from B cells of the immune system. They secrete antibodies, which kill or inactivate foreign antigens. *Ch. 11*

plasma membrane A thin, nonrigid structure that encloses the cell and regulates interactions between the cell and its environment. *Ch. 3*

plasmid A small fragment of DNA in a bacterium that replicates independently of the main chromosome. *Ch. 3, 31*

plasmodial (acellular) slime molds A phyla of funguslike protists that have an ameboid stage to their life cycle; they spend much of their time as non-walled, multinucleate masses of cytoplasm called plasmodia, but can form spores when food or moisture is in short supply. *Ch. 26*

plasmodium, pl. plasmodia (P.L. plasma, form + Gr. eidos, shape) A non-walled, multinucleate mass of cytoplasm distinctive of a particular stage in the life cycle of a slime mold. Plasmodial slime molds spend much of their life cycles in this form. *Ch. 26*

platelets Cell fragments present in blood that play an important role in blood clotting. *Ch. 10*

plate tectonics theory A scientific explanation of how and why the position and sizes of land masses have changed over time. *Ch. 24*

platelet A cell fragment present in blood that plays an important role in the clotting of blood. *Ch. 10*

pleura, pl, pleurae (Gr. side) A thin, delicate, sheetlike membrane that lines the interior walls of the thoracic cavity and folds back on itself to cover each lung; covered by a thin film of liquid, the pleura reduces friction during respiratory movements of the lungs. *Ch. 9*

point mutation (gene mutation) A change in the genetic message of a chromosome caused by alterations of molecules within the structure of the chromosomal DNA. *Ch. 20*

polar molecules Molecules such as water that have opposite partial charges at different ends of the molecule. *Ch. 2*

pollen grain (L. fine dust) A male gametophyte enclosed within a protective outer covering. Pollen grains produce sperm by mitosis. *Ch. 27*

pollination The transfer of pollen from the anther to the stigma in flowering plants usually accomplished by the wind, insects, or birds. *Ch. 27*

pollution (L. polluere, to pollute) Contamination of the water sources and air of the Earth, causing physical and chemical changes that harm living and nonliving things. *Ch. 38*

polymer (Gr. polus, many + meris, part) A giant molecule formed of long chains of similar molecules. *Ch. 2*

polyp (Gr. polypous, many feet) A cnidarian that exists as an aquatic, cylindrical animal with a mouth at one end ringed with tentacles; it anchors itself to rocks. *Ch. 29*

polypeptide (Gr. polys, many + peptein, to digest) A long chain of amino acids linked end to end by peptide bonds; proteins are long, complex polypeptides. *Ch. 2*

polysaccharide (Gr. polys, many + sakcharon, sugar) A long polymer composed of simple sugars. *Ch. 2*

polyunsaturated Referring to a fat composed of fatty acids that have more than one double bond; polyunsaturated fats have low melting points and are therefore liquid fats, or oils. *Ch. 2*

pons (L. bridge) The part of the brainstem consisting of bands of nerve fibers that act as "bridges" and connect various parts of the brain to one another; it also brings messages to and from the spinal cord. *Ch. 14*

population (L. populus, the people) A group that consists of the individuals of a given species that occur together at one place and at one time. *Ch. 35*

population ecologists Scientists who study how populations grow and interact. *Ch. 34*

population pyramid A bar graph that shows the composition of a population by age and sex. *Ch. 34*

positive feedback loop A physiological stimulus/response system in which the response of the regulating mechanism is positive with respect to the stimulus. *Ch. 7*

positive phototaxis An innate movement of an organism toward a source of light. *Ch. 26*

postsynaptic membrane The membrane of the target cell of the synaptic cleft. *Ch. 13*

potential energy Stored energy; energy not actively doing work but having the capacity to do so. *Ch. 4*

Precambrian A term referring to geological strata and fossil organisms the predate Cambrian time. *Ch. 24*

predation (L. praeda, prey) The killing and eating of an organism of one species by an organism of another species; an animal that kills and eats members of its own species is a cannibal. *Ch. 35*

predator (L. praeda, prey) An organism of one species that kills and eats organisms of another. *Ch. 35*

preembryo A term referring to the developing cell mass formed by the zygote as it begins to divide by mitotic cell division during the first two weeks of development. *Ch. 22*

prenatal development Life before birth; the gradual growth and progressive changes in a developing human from conception until the time the fetus leaves the mother's womb. *Ch. 22*

prey (L. prehendere, to grasp or seize) An organism killed and eaten by a predator. *Ch. 35*

primary bronchi The two airways branching immediately from the trachea into each lung. *Ch. 9*

primary consumer A herbivore, which feeds directly on green plants. *Ch. 36*

primary germ layers Three layers of cells that develop from the inner cell mass of the blastocyst and from which all the organs and tissues of the body develop. *Ch. 22*

primary growth A type of plant growth that occurs mainly at the tips of the roots and shoots. During primary growth the plant grows taller and its roots grow deeper in the ground. *Ch. 28*

primate (L. primus, first) A mammal that has characteristics reflecting a tree-dwelling lifestyle, such as hands and feet able to grasp things, flexible limbs, and a flexible spine. *Ch. 24*

producer An organism capable of capturing energy from the sun and converting it to chemical energy usable to themselves and consumers. *Ch. 1*

probe A molecule that binds to a specific gene or nucleotide sequence. *Ch. 31*

products Substances obtained as the result of chemical reaction; substrates that have undergone a chemical change. *Ch. 4*

prokaryotic cell (Gr. pro, before + karyon, kernel) A cell smaller than a eukaryote; it has a simple interior organization with a single, circular strand of hereditary material that is not enclosed by a membrane; bacteria are prokaryotes. *Ch. 3*

prolactin (L. pro, before + lac, milk) A hormone secreted by the anterior pituitary that in association with estrogen, progesterone, and other hormones, stimulates the mammary glands in the breasts to secrete milk after a woman has given birth to a child. *Ch. 17*

prophase (Gr. pro, before + phasis, form) The stage of mitosis characterized by the appearance of visible chromosomes. *Ch. 18*

proprioceptor (L. proprius, one's own + ceptor, a receiver) A type of nerve ending located within skeletal muscles, tendons, and the inner ear that gives the body information about the position of its parts relative to each other and to the pull of gravity. *Ch. 15*

prosimian (Gr. pro, before + L. simia, an ape) A suborder of primates that includes lemurs, indris, aye-ayes, and lorises; they are small animals, mostly nocturnal, with large ears and eyes, and enlongated snouts and rear limbs. *Ch. 24*

prostaglandins Hormones secreted by cells throughout the body that stimulate smooth muscle contraction and the dilation and constriction of blood vessels. *Ch. 17*

prostate (Gr. prostates, one standing in front) A gland surrounding the male urethra that adds a milky alkaline fluid to semen that neutralizes the acidity of the vagina. *Ch. 21*

proteases Digestive enzymes that break down proteins to smaller polypeptides, and polypeptides to amino acids. *Ch. 8*

protein (Gr. proteios, primary) A linear polymer of amino acids. *Ch. 2*

protists (Gr. protos, first) Organisms that comprise the kingdom Protista, which includes unicellular eukaryotes as well as some multicellular forms. *Ch. 26*

proton A subatomic particle in the nucleus of an atom that has mass and carries a positive charge. *Ch. 2*

protostomes (Gr. proto, first + stoma, mouth) One of two distinct evolutionary lines of coelomates that include the mollusks, annelids, and arthropods. Their embryological development is characterized by the mouth developing from the first indentation of the gastrula and a spiral pattern of cleavage. *Ch. 29*

protozoan (Gr. protos, first + zoon, animal) An animallike heterotrophic protist that eats organic matter for energy. *Ch. 26*

proximal convoluted tubule A coiled portion of the nephron closest to Bowman's capsule lying in the cortex of the kidney, the site where reabsorption begins as the filtrate passes through the proximal tubule. *Ch. 12*

pseudocoelomate (Gr. pseudos, false + koiloma, a hollow) An organism, such as a roundworm, that has a false body cavity (fluid-filled enclosure lined with connective tissue). *Ch. 29*

pseudopod, pl. pseudopodia (Gr. pseudos, false + pous, foot) A temporary, pushed out cell part of an ameba used for food procurement and locomotion. *Ch. 26*

pseudostratified (Gr. pseudos, false + L. stratificare, to arrange in layers) Refers to a type of epithelial tissue that gives the false appearance of being layered. *Ch. 7*

pulmonary artery An artery that extends from the right ventricle of the heart; it branches into two smaller arteries that carry deoxygenated blood to the lungs. *Ch. 10*

pulmonary circulation The part of the human circulatory system that brings blood to and from the lungs. *Ch. 10*

pulmonary semilunar valve A one-way valve that permits blood flow from the right ventricle of the heart to the pulmonary artery. *Ch. 10*

pulmonary veins Large blood vessels that carry oxygenated blood from the lungs to the left atrium of the heart. *Ch. 10*

punctuated equilibrium A theory regarding the mechanism of evolution that states that new species arise suddenly and rapidly as small subpopulations of a species split from the populations of which they were a part. *Ch. 23*

Punnett square (after Reginald C. Punnett, English geneticist) A diagram that visualizes the genotypes of progeny in simple Mendelian crosses and illustrates their expected ratios. *Ch. 19*

pupil The opening in the center of the iris through which light passes. *Ch. 15*

purines Double-ring compounds that are components of nucleotides. Both DNA and RNA contain the purines adenine and guanine. *Ch. 18*

Purkinje fibers Conducting fibers within the heart that branch from the bundle of His and initiate the almost simultaneous contraction of all the cells of the right and left ventricles. *Ch. 10*

pyrimidines Single-ring compounds that are components of nucleotides. Both DNA and RNA contain the pyrimidine cytosine; in addition, DNA contains the pyrimidine thymine, and RNA contains the pyrimidine uracil. *Ch. 18*

pyruvate The three-carbon molecule produced by glycolysis and the beginning substrate of the citric acid cycle. *Ch. 5*

Quaternary period The period from the end of the Tertiary to the present time. *Ch. 24*

radial symmetry (L. radius, a spoke of a wheel; Gr. symmetros, symmetry) Referring to organisms that have body parts that merge, or radiate, from a central point, much like spokes on a wheel. *Ch. 29*

radicle (L. radicula, root) The young root that is usually the first portion of the embryo to emerge from the germinating seed. *Ch. 27*

radiolarian (L. radius, ray) A kind of ameba that secretes a shell made of silica that is glasslike and delicate. *Ch. 26*

rank order A social hierarchy that exists in many groups of fishes, reptiles, birds, and mammals. *Ch. 33*

realized niche The role an organism actually plays in the ecosystem. *Ch. 35*

receptor A specialized cell component that detects stimuli; it provides the body with information about the internal environment, the position in space, and the external environment. *Ch. 13*

recessive The form of a trait that recedes or disappears entirely in a hybrid offspring. *Ch. 19*

recognition sites Certain nucleotide (base) sequences in DNA molecules recognized by restriction enzymes. Scientists use restriction enzymes to cut large pieces of DNA into smaller fragments at recognition sites. *Ch. 31*

rectum (L. straight) The lower part of the large intestine, which terminates at the anus. *Ch. 8*

red algae One of the three phyla of multicellular algae; red algae play an important role in the formation of coral reefs and produce gluelike substances that make them commercially useful. *Ch. 26*

red blood cell *See* erythrocyte.

red bone marrow The soft tissue that fills the spaces within the bony latticework of spongy bone; it is the place where most of the body's blood cells are formed. *Ch. 16*

redox reaction A term used to describe the occurrence of oxidation and reduction reactions, which always happen together. *Ch. 3*

reduction (L. reduction, a bringing back; originally "bringing back" a metal from its oxide) The gain of an electron by an atom or a molecule. *Ch. 5*

reflex (L. reflectare, to bend back) An automatic response to a nerve stimulation. *Ch. 14*

reflex arc The pathway of nervous activity in an automatic response to nerve stimulation. *Ch. 14*

refractory period The recovery period after membrane depolarization during which the membrane is unable to respond to additional stimulation. *Ch. 13*

regulatory sites In eukaryotic cells, specific nucleotide sequences that control the transcription of genes. *Ch. 18*

releasing hormones Hormones produced by the hypothalamus that affect the secretion of specific hormones from the anterior pituitary. *Ch. 17*

renal failure A disease that occurs when the filtration of the blood at the glomerulus either slows or stops; in acute renal failure, filtration stops suddenly, while in chronic renal failure, the filtration of blood at the glomerulus slows gradually. *Ch. 12*

renewable resources Resources that are produced by natural systems that replace themselves quickly enough to keep pace with consumption; wind power and solar power are renewable resources. *Ch. 38*

reptiles A class of vertebrates characterized by dry skins covered with scales that help retard water loss. *Ch. 30*

replication fork The split in a DNA molecule where the double-stranded DNA molecule separates during DNA replication. *Ch. 18*

respiration (L. respirare, to breathe) The uptake of oxygen and the release of carbon dioxide by the body. Cellular, internal, and external respiration are all part of the general process of respiration. *Ch. 9*

respiratory assembly The electron transport chain. The third series of reactions of cellular respiration during which 32 ATPs are produced. *Ch. 5*

respiratory bronchioles The last division of bronchi. Bronchioles are tiny air passageways whose walls have clusters of microscopic airsacs, or alveoli. *Ch. 9*

response (L. respondere, to reply) The reply or reaction of an organism to a stimulus; it can be an innate or inborn reflex or a learned response such as by operant or classical conditioning. *Ch. 13, 32*

resting potential An electrical potential difference, or electrical charge, along the membrane of the resting neuron. *Ch. 13*

restriction enzyme An enzyme that recognizes certain base sequences in a DNA strand and breaks the bonds between them at a particular point. *Ch. 31*

restriction fragments Pieces of DNA that have been cut from larger pieces of DNA by restriction enzymes. *Ch. 31*

reticular formation A complex network of neurons woven throughout the entire length of the brainstem, but concentrated in the medulla; the reticular formation monitors information concerning incoming stimuli and identifies important ones. *Ch. 14*

retina (L. a small net) Tissue at the back wall of the eye, composed of rod and cone cells, that is sensitive to light. *Ch. 15*

rhizoid (Gr. rhiza, root) In a bryophyte, a slender, rootlike projection that anchors the plant to its substrate; unlike roots, however, it consists of only a few cells and does not play a major role in the absorption of water or minerals. *Ch. 28*

rhizome (Gr. rhizoma, mass of roots) An underground stem of a plant that can produce a new plant by the process of vegetative propagation. *Ch. 27*

rhodopsin (Gr. rhodon, rose + opsis, vision) A molecular complex in the retina formed by the coupling of retinal and opsin. Rhodopsin absorbs photons of light and is integral to the physiology of vision. *Ch. 15*

rhythm method A form of birth control that involves avoiding sexual relations for three days preceding and one day following ovulation. *Ch. 21*

ribonucleic acid (RNA) One of two types of nucleic acid found in cells; differs from DNA in that its sugar is ribose and uracil is present rather than thymine. *Ch. 18*

ribosomal RNA (rRNA) The type of RNA found in ribosomes that plays a role in the manufacture of polypeptides. *Ch. 3*

ribosome A minute, round structure found on endoplasmic reticulum or free in the cytoplasm; ribosomes are the places where proteins are manufactured. *Ch. 3*

ribulose biphosphate (RuBP) A molecule that joins with carbon dioxide in the first stage of the Calvin cycle and results in the process of carbon fixation. *Ch. 6*

RNA polymerase The special enzyme that transcribes RNA from DNA. *Ch. 18*

rod A light receptor located within the retina at the back of the eye that functions in dim light and detects white light only. *Ch. 15*

root The part of a plant usually found below ground; it absorbs water and minerals and anchors the plant. *Ch. 28*

root cap A thimblelike mass of cells that covers and protects the root's apical meristem as it grows through the soil. *Ch. 28*

root pressure The force produced as water continually enters the root by osmosis as a result of special cellular pumps in the root cells, maintaining a higher concentration of dissolved minerals in root cells than the concentration of minerals in the water of the soil. *Ch. 28*

rough endoplasmic reticulum (Gr. endon, within + plasma, from cytoplasm; L. reticulum, network) The ribosome-studded intracellular membrane channels system that makes and transports proteins destined to leave the cell. *Ch. 3*

round window A membrane-covered hole at the wider end of the cochlea located in the inner ear. *Ch. 15*

SA node *See* sinoatrial (SA) node.

saccule (N.L. sacculus, small bag) A sac inside a bulge in the vestibule of the inner ear containing both ciliated and nonciliated cells, jellylike material, and pieces of calcium carbonate. *Ch. 15*

sac fungi One of the three divisions of fungi. Sac fungi live in both aquatic and terrestrial environments and are characterized by sexual spores borne in saclike structures. Familiar examples of this group are cup fungi and yeasts. *Ch. 26*

saliva (L. spittle) The secretion of the salivary glands; a solution consisting primarily of water, mucus, and the digestive enzyme salivary amylase. *Ch. 8*

salivary amylase (L. salivarius, slimy + G. amylon, starch -asis, coloid enzyme) A digestive enzyme that breaks down starch into molecules of the disaccharide maltose. *Ch. 8*

salivary glands (L. salivarius, slimy + glans, acorn) The paired glands of the mouth that secrete saliva. *Ch. 8*

saltatory conduction (L. saltatio, leaping) A very fast form of nerve impulse conduction in which impulses "jump" along myelinated neurons. *Ch. 13*

sanitary landfill An enormous depression in the ground where trash and garbage is dumped, compacted, and then covered with dirt. *Ch. 38*

saprophytic (Gr. sapros, putrid + phyton, plant) Feeding on dead or decaying organic material; fungi are saprophytic. *Ch. 26*

sarcomere (Gr. sarx, flesh + meris, part of) The repeating bands of actin and myosin myofilaments that appear between two Z lines in a muscle fiber. *Ch. 16*

sarcoplasmic reticulum (Gr. sarx, flesh + plassein, to form, mold; L. reticulum, network) A tubular, branching latticework of endoplasmic reticulum that wraps around each myofibril like a sleeve. *Ch. 16*

Schwann cells (after Theodor Schwann, German anatomist) Fatty cells that are wrapped around many of the long cell processes of sensory and motor neurons of the peripheral nervous system. *Ch. 13*

scientific laws Descriptions of natural phenomena. *Ch. 1*

scientific methods Various procedures used to answer scientific questions. *Ch. 1*

scientific theory A synthesis of hypotheses that helps scientists make dependable predictions about the world. Explanations of natural phenomena. *Ch. 1*

sclera (Gr. skleros, hard) The tough outer layer of connective tissue that covers and protects the eye. *Ch. 15*

sclerenchyma cell (Gr. skleros, hard + en, in + chymein, to pour) A dead, hollow cell with a strong wall found in the ground tissue of a vascular plant that helps support and strengthen the ground tissue. *Ch. 28*

scrotum (L. a bag) A sac of skin, located outside the lower pelvic area of the male, which houses the testicles, or testes. *Ch. 21*

sea squirt The common name given to most species of tunicates because of its habit of forcefully squirting water out of its excurrent siphons when disturbed. *Ch. 30*

secondary bronchi Outbranchings of each of the two primary bronchi. Three secondary bronchi serve the three right lobes of the lungs, whereas two secondary bronchi serve the two left lobes of the lungs. *Ch. 9*

secondary consumer A meat eater, or carnivore, that feeds on a herbivore. *Ch. 36*

secondary growth A type of plant growth in which the plant grows thicker. This type of growth occurs in all woody trees and shrubs such as pines, oaks, and rhododendrons. *Ch. 28*

second filial (F_2) generation The offspring of the F_1 generation, which are hybrid offspring produced from parent plants having contrasting forms of single traits.. *Ch. 19*

second law of thermodynamics A description of a natural phenomenon that states that disorder in the universe constantly increases. *Ch. 4*

second trimester The second three-month period of pregnancy. *Ch. 22*

secretin One of the hormones that controls digestion in the small intestine. *Ch. 8*

sedative-hypnotics Central nervous system depressants that induce sleep (sedatives) and reduce anxiety (hypnotics). *Ch. 13*

seed A structure from which a new plant grows. *Ch. 27*

seedless vascular plants Plants that reproduce by means of spores rather than seeds; a familiar member of this group is the fern. *Ch. 27*

selective reabsorption The process in which the kidneys return specific substances to the blood according to the body's needs. *Ch. 12*

semen (L. seed) Fluid produced by the accessory glands of the male reproductive system combined with sperm. *Ch. 21*

semicircular canal Any of three fluid-filled canals in the inner ear that detect the direction of an individual's movement. *Ch. 15*

seminal vesicle One of two accessory glands that secrete a thick, clear fluid, primarily composed of fructose, forming a part of the semen. *Ch. 21*

seminiferous tubule (L. semen, seed + ferre, to produce) A tightly coiled tube within a testis where sperm cells develop. *Ch. 21*

sensory area The area of the brain, located directly behind the motor area on the leading edge of the parietal lobe, that deals with information regarding stimuli; each point on the surface of the sensory area represents receptors from a different part of the body. *Ch. 14*

sensory neuron A nerve cell of the peripheral nervous system that transmits information to the central nervous system. *Ch. 13, 14*

sepal (L. sepalum, a covering) A member of the outermost whorl, or ring, of modified leaves that enclose and protect a growing flower bud. *Ch. 27*

septum, pl. septa (L. saeptum, a fence) (1) The tissue that separates the two sides of the heart. (2) In a fungus, a cross wall that divides a hypha into cells. (3) In an annelid, an internal partition that divides the segments. *Ch. 10, 26, 29*

seta, pl. setae (L. bristle) Bristles that protrude from the body of an earthworm that help anchor the worm during locomotion or when it is in its burrow. *Ch. 29*

sex chromosome Chromosomes that determine the gender of an individual as well as certain other characteristics. *Ch. 20*

sexual reproduction The type of generation of offspring that involves the fusion of gametes to produce the first cell of a new individual. *Ch. 18*

shell A term for the volume of space around an atom's nucleus where an electron is most likely to be found. *Ch. 2*

shoot The part of a vascular plant that exists above the ground. *Ch. 28*

shore zone The shallow water near the edges of a lake or pond in which plants with roots, such as cattails and water lilies, may grow. *Ch. 37*

shotgun cloning The process of synthesizing DNA by cutting the DNA of an entire genome into pieces, isolating and purifying the genomic DNA, and then making copies of this DNA by means of yeast or bacterial cells. *Ch. 31*

sickle-cell anemia (A.S. sicol, curved + L. cella, chamber + Gr. a, not + haima, blood) A recessive genetic disorder common to African blacks and their descendants in which affected individuals cannot transport oxygen to their tissues properly because the molecules within the red blood cells that carry oxygen, hemoglobin proteins, are defective. *Ch. 20*

sigmoid growth curve The exponential growth of a population and its subsequent stabilization at the level of its environment's carrying capacity. *Ch. 34*

simple eye A structure composed of a single visual unit having one lens; found in some arthropods and some flying insects. *Ch. 29*

sinoatrial (SA) node A small cluster of specialized cardiac muscle cells embedded in the upper wall of the right atrium of the heart that automatically and rhythmically sends out impulses initiating each heartbeat. *Ch. 10*

sinus venosus A chamber in the hearts of cartilaginous and bony fishes that along with the atrium collects blood from the organs. *Ch. 30*

sister chromatid Either of two identical structures held together at the centromere, composed of chromatin material that coils and condenses just before and during cell division to form chromosomes. *Ch. 18*

skeletal muscle A type of muscle that is voluntary; skeletal muscle is connected to bones and allows for body movement. *Ch. 7*

skeleton The collective term for the body's 206 bones. The skeleton provides support for the body, helps the body move (along with the muscles), and protects delicate internal structures. *Ch. 16*

skeletal system Another term for the skeleton. *Ch. 16*

skull The framework of the head, which in humans consists of 8 cranial and 14 facial bones. *Ch. 16*

slime mold A protist that is funguslike in one phase of its life cycle and amebalike in another phase of its life cycle. *Ch. 26*

small intestine The tubelike portion of the digestive tract that begins at the pyloric sphincter and ends at its T-shaped junction with the large intestine. *Ch. 8*

smog A type of air pollution that results from the burning of fossil fuels. *Ch. 38*

smooth endoplasmic reticulum (Gr. endon, within + plasma, from cytoplasm; L. reticulum, network) The type of membranous channel of eukaryotic cells that helps build carbohydrates and lipids within the cytoplasm; it does not have ribosomes attached to its surface and does not manufacture proteins. *Ch. 3*

smooth muscle A type of contractile tissue that contracts involuntarily and is located in the walls of certain internal structures such as blood vessels and the stomach. *Ch. 7*

social behaviors Activities of animals that help members of the same species communicate and interact with one another, each responding to stimuli from others. *Ch. 33*

sociobiology The biology of social behavior that applies the knowledge of evolution to the study of animal behavior. *Ch. 33*

sodium-potassium pump The term given to the coupled channel that uses energy to move sodium (Na^+) and potassium (K^+) ions across the cell membrane in opposite directions. *Ch. 3*

soft palate (L. palatum, palate) The tissue at the back of the roof of the mouth. *Ch. 8*

solar power The use of the sun for heating or electricity. *Ch. 38*

solid waste Garbage and trash. *Ch. 38*

soluble Able to be dissolved in water. *Ch. 2*

solute The molecules dissolved in water. *See* solution, solvent. *Ch. 3*

solution A mixture of molecules and ions dissolved in water. *Ch. 3*

solvent The most common of the molecules in a solution, usually water. *Ch. 3*

somatic nervous system (Gr. soma, body) The branch of the peripheral nervous system consisting of motor neurons that send messages to the skeletal muscles and control voluntary responses. *Ch. 14*

somite (Gr. soma, body) A chunk of mesoderm that gives rise to most of the axial skeleton and most of the dermis of the body. *Ch. 22*

spawn (M.E. spawnen, to spread out) The depositing of sperm and eggs directly into water by males and females; a method of reproduction used by most fishes, amphibians, and shellfish. *Ch. 30*

speciation The process by which new species are formed during the process of evolution. *Ch. 23*

species A population of organisms that interbreeds freely in their natural settings and does not interbreed with other populations. *Ch. 1, 2, 3*

specific defense The immune response of the body. *Ch. 11*

spermatid A haploid cell in the testes arising from a diploid cell called a spermatogonium. Spermatids develop and mature into spermatozoa. *Ch. 21*

spermatogenesis (Gr. sperma, sperm, seed + gignesthai, to be born) The development of sperm cells (spermatozoa) within the coiled tubules of the testis. *Ch. 21*

spermatozoon (sperm) A male sex cell. *Ch. 21*

spinal cord The part of the central nervous system that runs down the neck and back, receives information from the body, carries this information to the brain, and sends information from the brain to the body. *Ch. 14*

spinal nerve Bundles of motor and sensory neurons that bring information to and from the spinal cord. *Ch. 14*

spindle fibers Special microtubules that extend from centrioles (in animal cells) centrioles during the prophase stage of mitosis. *Ch. 18*

spiracle (L. spirare, to breathe) A specialized opening in the body of most terrestrial arthropods that allows air to pass into the trachea. *Ch. 29*

spleen An organ of the lymphatic system that stores an emergency blood supply and also contains white blood cells. *Ch. 10*

spongy bone A type of bone in the human skeleton that is composed of an open latticework of thin plates of bone; spongy bone is found at the ends of long bones and within short, flat, and irregularly shaped bones. *Ch. 7*

spore (Gr. spora, seed) A reproductive body produced by certain fungi and plants, and which is formed by meiosis in a diploid parent and by mitosis in a haploid parent; a spore is always haploid. *Ch. 26*

sporophyte (Gr. spora, seed + phyton, plant) In the life cycle of a plant, the diploid (spore-plant) generation that alternates with the haploid (gamete-plant) phase. *Ch. 27*

sporozoan (Gr. spora, seed + zoon, animal) A nonmotile, spore-forming, protozoan phylum that is a parasite of vertebrates, including humans; they are passed from host to host by various insect species. *Ch. 26*

squamous (L. squama, scale) One of the main shapes of epithelial cells. Squamous cells are thin and flat and are found in the air sacs of the lungs, the lining of blood vessels, and the skin. *Ch. 7*

stable population A population whose size remains the same through time. *Ch. 34*

stamen (L. thread) The male sex organ of a flower. A stamen consists of an anther, the structure where pollen grains are produced, supported by a filament. *Ch. 27*

stapes One of the three bones of the middle ear that amplify sound vibrations and carry them from the outer ear to the inner ear. The stapes is also known as the *stirrup*. *Ch. 15*

starch (A.S. stercan) Stored energy in plants formed by using glucose to produce polysaccharides. *Ch. 2*

stem The framework of a plant that supports and positions the leaves where most photosynthesis occurs. *Ch. 28*

stereoscopic vision (Gr. stereos, solid + skopein, to view) Vision having depth perception created by two eyes focusing on the same object from slightly different angles. *Ch. 24*

steroid hormones One of the two main classes of endocrine hormones. Steroid hormones are made of cholesterol and are able to pass through the cell membrane. Once inside the cell, these hormones bind to receptor molecules located within the cytoplasm of target cells *Ch. 17*

stimulants Drugs that enhance the activity of two neurotransmitters: norepinephrine and dopamine, resulting in an alerting and euphoric effect. *Ch. 13*

stoma, pl. stomata (Gr. mouth) An opening in the epidermis of a leaf bordered by guard cells. *Ch. 28*

stomach (Gr. stomachos, mouth) A muscular sac in which food is collected and partially digested by hydrochloric acid and proteases. *Ch. 8*

stratified (L. stratificare, to arrange in layers) Refers to epithelium that is made up of two or more layers. *Ch. 7*

stretch receptor Sensory neurons wrapped around specialized muscle spindles. When a muscle is stretched, the muscle spindle gets longer and stretches the nerve ending, repeatedly stimulating it to fire. *Ch. 15*

stroma, pl. stromata (Gr. anything spread out) The fluid that surrounds the thylakoids inside chloroplasts. *Ch. 6*

style (L. stilus, stylus) A narrow stalk arising from the top of the ovary that bears the stigma in a flowering plant. *Ch. 27*

submissive behavior Behavior that animals often exhibit in response to a threat display in order to avoid fighting. Submissive behavior may include making the body appear smaller or withdrawing fangs or claws. *Ch. 33*

substrates Substances entering into a chemical reaction. *Ch. 4*

succession The dynamic process of change, during which a sequence of communities replaces one another in an orderly and predictable way. *Ch. 35*

superior vena cava A large vein that drains blood from the upper body and returns it to the right atrium of the heart. *Ch. 10*

suppressor T cell A type of T cell that limits the immune response. *Ch. 11*

survivorship The proportion of an original population that lives to a certain age. *Ch. 34*

suture A type of immovable joint between two bones. *Ch. 16*

symbiosis (Gr. a living together) The living together of two or more organisms in a close association. *Ch. 3*

sympathetic nervous system A subdivision of the autonomic nervous system that generally mobilizes the body for greater activity; produces responses that are the opposite of the parasympathetic nervous system. *Ch. 14*

syphilis A sexually transmitted disease, caused by the bacterium *Treponema pallidum,* that produces three stages of infection, from localized to widespread. *Ch. 25*

synapse (Gr. synapsis, a union) A junction between an axon tip of a neuron and the dendrite of another, usually including a narrow gap separating the two neurons. *Ch. 13*

synapsis (Gr. union) The lining up of homologous chromosomes during prophase I of meiosis, initiating the process of crossing-over. *Ch. 18*

synaptic cleft The space or gap between two adjacent neurons. *Ch. 13*

synovial joint (L. synovia, joint fluid) A freely movable joint in which a space exists between articulating bones. *Ch. 16*

systemic circulation The pathway of blood vessels to the body regions and organs other than the lungs. *Ch. 10*

systolic period (Gr. systole, contraction) The pushing period of heart contraction, which ends with the closing of the aortic valve, during which a pulse of blood is forced into the systemic arterial system, immediately raising the blood pressure within these vessels. *Ch. 10*

T cell (T lymphocyte) A type of lymphocyte that carries out the cell-mediated immune response. *Ch. 11*

T lymphocyte *See* T cell.

taiga (Russ.) The northern coniferous forest extending across vast areas of Eurasia and North America; it is a marshy region south of the tundra that hs long, cold winters. *Ch. 37*

tarsal (Gr. tarsos, a broad, flat surface) Any of the seven short bones that make up the ankle in the appendicular skeleton of a human. *Ch. 16*

taste bud A microscopic chemoreceptor embedded within the papillae of the tongue that works with the olfactory receptors to produce the taste sensation. *Ch. 15*

taxis A directed movement by an animal toward or away from a stimulus, such as light, chemicals, or heat. *Ch. 32*

taxonomy (Gr. taxis, arrangement + nomos, law) The classification of the diverse array of species by categorizing organisms based on their common ancestry. *Ch. 1*

Tay-Sachs disease (after Warren Tay, British physician, and Bernard Sachs, U.S. neurologist) An incurable, fatal, recessive, hereditary disorder in which brain deterioration causes death by the age of 5 years; has a high incidence of occurrence among Jews of Eastern and Central Europe and among American Jews. *Ch. 20*

tectorial membrane (L. tectum, roof) The membrane that covers the hairs that stick up from the cochlear duct. *Ch. 15*

telophase (Gr. telos, end + phasis, form) The stage of mitosis during which the mitotic apparatus assembled during prophase is disassembled, the nuclear envelope is reestablished, and the normal use of the genes present in the chromosomes is reinitiated. *Ch. 18*

template A pattern used as a guide to duplicate a shape or structure; refers to the ability of any polynucleotide, such as a RNA molecule, that can act as a guide for the synthesis of a second polynucleotide based on the complementarity of its bases. *Ch. 24*

temporal lobe (L. temporalis, the temples) The section of the cerebral cortex in each hemisphere of the brain dealing with hearing; different surface areas correspond to different tones and rhythms. *Ch. 14*

tendon (Gr. tenon, stretch) A tissue that connects muscles to bones. *Ch. 16*

teratogen (Gr. teratos, monster + gennan, to produce) An agent, such as alcohol, that can induce malformations in the rapidly developing tissues and organs of an embryo or fetus. *Ch. 22*

tertiary period The first period of the Cenozoic era. *Ch. 24*

territorial behaviors Behaviors an animal may exhibit that involve marking off an area as its own and defending it against the same sex members of its species. *Ch. 33*

testcross A cross between a phenotypically dominant test plant with a known homozygous recessive plant; devised by Mendel to further test his conclusions. *Ch. 19*

testis, pl. testes (L. witness) The male gonads where sperm production occurs. *Ch. 21*

testosterone (Gr. testis, testicle + steiras, barren) A sex hormone secreted by the testes responsible for the development and maintenance of male secondary sexual characteristics. *Ch. 17*

tetrad (Gr. tetras, four) During prophase I of meiosis I, the name given to the paired homologous chromosomes that together have four chromatids. *Ch. 18*

tetrapod (Gr. tetrapodos, four-footed) A land-living four-limbed vertebrate. *Ch. 30*

thalamus (Gr. thalamos, chamber) A mass of gray matter lying at the base of the cerebrum that receives sensory stimuli, interprets some of these stimuli, and sends the remaining sensory messages to appropriate locations in the cerebrum. *Ch. 14*

thermal pollution A change in water temperature that occurs when heated water is released into rivers after using the water for cooling purposes. *Ch. 38*

third trimester Third three-month period of pregnancy. *Ch. 22*

thoracic cavity The chest cavity in human beings, located within the trunk of the body and extending above the diaphragm and below the neck. *Ch. 9*

thoroughfare channels Capillaries within capillary beds that connect arterioles and venules directly. *Ch. 10*

threat or **intimidation displays** A form of aggressive behavior that animals may exhibit during a competitive situation. The purpose of these displays is to scare other animals away or cause them to "back down" before fighting takes place. *Ch. 33*

threshold potential A level of nerve membrane depolarization in which sodium-specific channels in the membrane open, allowing sodium ions to diffuse into the cell. The level at which a stimulus results in an action potential along the nerve cell membrane. *Ch. 13*

thrombus (Gr. clot) A blood clot that forms in a blood vessel and interferes with the flow of blood. *Ch. 10*

thylakoid (Gr. thylakos, sac + -oides, like) A flat, saclike membrane in the chloroplast of a eukaryote; stacks of thylakoids are called grana. *Ch. 3*

thymus (Gr. thymos) A gland located in the neck (until puberty when it begins to degenerate) that plays an important role in the maturation of certain lymphocytes called T cells, which are an essential part of the immune system. *Ch. 10*

thyroid gland An important gland located in the neck just below the voice box that produces hormones regulating the body's metabolism, growth, and development. *Ch. 17*

thyroid-stimulating hormone (TSH) A tropic hormone produced by the anterior pituitary that triggers the thyroid gland to produce the three thyroid hormones. *Ch. 17*

thyroxine (T_4) A hormone produced by the thyroid gland that helps regulate the body's metabolism. *Ch. 17*

tidal volume The amount or volume of air inspired or expired with each breath. *Ch. 9*

tissue (L. texere, to weave) A group of similar cells that work together to perform a function. *Ch. 1*

tissue differentiation A prenatal developmental process in which groups of cells become distinguished from other groups of cells by the jobs they will perform in the body. *Ch. 22*

trachea (L. windpipe) The windpipe; the air passageway that runs down the neck in front of the esophagus and brings air to the lungs. *Ch. 9*

tracheid (Gr. tracheia, rough) A type of xylem made up of stacks of cells with tapering ends that has holes along the length of its walls; it carries water and dissolved minerals through a plant and also provides support. *Ch. 28*

trait An inherited feature or characteristic of a plant or person. *Ch. 19*

transcription (L. trans, across + scribere, to write) The first step in the process of polypeptide synthesis in which a gene is copied into a strand of messenger RNA. *Ch. 18*

transduction One of the ways in which bacteria transfer genetic material. Transduction occurs as DNA from a donor bacterium is transferred to a recipient bacterium by a virus. *Ch. 31*

transfer RNA (tRNA) A type of RNA that transports amino acids, used to build polypeptides, to the ribosomes; tRNA also aligns each amino acid at the correct place on the elongating polypeptide chain. *Ch. 18*

transformation One of the ways in which bacteria transfer genetic material. Transformation occurs as genetic material from a lysed bacteria cell is released into the surrounding medium and is taken up by a competent cell. *Ch. 31*

transgenic plants Plants that are genetically altered using the techniques of genetic engineering. *Ch. 31*

translation (L. trans, across + locare, to put or place) The second step of gene expression in which mRNA, using its copied DNA code, directs the synthesis of a polypeptide. *Ch. 18*

translocation (L. trans, across + locare, to put or place) A situation in which a section of a chromosome breaks off and then reattaches to another chromosome, producing an abnormally long chromosome. *Ch. 20*

transpiration (L. trans, across + spirare, to breathe) The process by which water vapor leaves a leaf. *Ch. 28*

Triassic period The first period of the Mesozoic era, in which small dinosaurs and primitive mammals appeared. *Ch. 24*

tricuspid valve A one-way valve within the heart, through which blood passes from the right atrium to the right ventricle. *Ch. 10*

triglyceride A fat molecule in which three fatty acids are attached to each of the three carbons of a glycerol molecule. *Ch. 2*

triiodothyronine (T_3) A hormone produced by the thyroid gland that helps regulate the body's metabolism.

trophic level (Gr. trophos, feeder) A feeding level in an ecosystem. *Ch. 36*

tropical rain forest deforestation The loss of the plants and animals that live in the tropical rain forest through cut-and-burn agriculture and logging. *Ch. 38*

tropic hormones Four of the seven hormones produced by the anterior pituitary gland; the tropic hormones turn on, or stimulate, other endocrine glands. *Ch. 17*

tropomyosin (Gr. tropos, turn + myos, muscle) A protein involved in muscle contraction. *Ch. 16*

troponin (Gr. tropos, turn) A protein that attaches to both actin and tropomyosin during muscle contraction; it is involved with calcium binding and cross-bridge formation. *Ch. 16*

true breeding Said of plants that produce offspring consistently identical to the parent with respect to certain defined characteristics after generations of self-fertilization. *Ch. 19*

trypsin (Gr. tripsis, friction) An enzyme produced by the pancreas, which together with chymotrypsin and carboxypeptidase, completes the digestion of proteins. *Ch. 8*

tubal ligation In females, an operation that involves the removal of a section of each of the two uterine tubes, through which the oocyte travels to the uterus, in order to bring about sterilization. *Ch. 21*

tubular secretion The process by which the kidneys excrete a variety of potentially harmful substances from the blood. *Ch. 12*

tumor (L. swollen) An uncontrolled growth of a large mass of cells. *Ch. 20*

tundra (Russ.) A vast, uniform, virtually treeless region encircling the top of the world across North America and Eurasia that covers one-fifth of the Earth's land surface; it has long and cold winters, and the ground below 1 meter is frozen even in summer. *Ch. 37*

Turner syndrome (after H.H. Turner, U.S. physician) A genetic condition resulting from an XO zygote that develops into a human female who is sterile, of short stature, with a webbed neck, low-set ears, a broad chest, infantile sex organs, and low-normal mental abilities. *Ch. 20*

tympanic membrane (Gr. tympanon, drum) The thin piece of fibrous connective tissue stretched over the opening to the middle ear; the eardrum. *Ch. 15*

ultrasound A noninvasive procedure that uses sound waves to produce an image of the fetus but that harms neither the mother nor the fetus; allows the fetus to be examined for major abnormalities. Ultrasound is used for other diagnostic purposes also. *Ch. 20*

umbilical artery and vein (L. umbilicus, navel) The blood lines of the umbilical cord; the artery brings blood to the placenta, which provides nourishment and removes wastes from the blood; the fetal blood then travels through the umbilical vein back to the fetus. *Ch. 22*

umbilical cord (L. umbilicus, navel) The attachment connecting the fetus with the placenta. *Ch. 22*

unsaturated Referring to a fat composed of fatty acids having many double bonds in their fatty acid chains. *Ch. 2*

urea (Gr. ouron, urine) The primary excretion product from the deamination of amino acids. *Ch. 12*

ureter (Gr. oureter) One of a set of tubes that carries urine from the kidney to the bladder. *Ch. 12*

urethra (Gr. ourein, to urinate) A muscular tube that brings urine from the urinary bladder to the outside; in men, the urethra also carries semen to the outside of the body during ejaculation. *Ch. 21*

uric acid A nitrogenous waste in the urine formed from the breakdown of nucleic acids (DNA and RNA) found in the cells of ingested food and from the metabolic turnover of bodily nucleic acids and ATP. *Ch. 12*

urinary bladder A hollow muscular organ that acts as a storage pouch for urine. *Ch. 12*

urinary system A set of interconnected organs that not only remove wastes, excess water, and excess ions from the blood but store this fluid until it can be expelled from the body. *Ch. 12*

urine (Gr. ouron, urine) The fluid produced by the kidneys made up of water and dissolved waste products. *Ch. 12*

urochordate (Gr. oura, tail + chorde, cord) A subphylum of tunicates that has a larval notochord in its tail. *Ch. 30*

uterine tube The passageway from an ovary to the uterus; commonly called a fallopian tube. *Ch. 21*

uterus (L. womb) The organ in females in which a fertilized ovum can develop; the womb. *Ch. 21*

utricle The larger of the two membranous sacs within the vestibule, containing both ciliated and nonciliated cells, that functions in the maintenance of bodily equilibrium and coordination. *Ch. 15*

vaccination (L. vaccinus, pert. to cows) An injection with disease-causing microbes or toxins that have been killed or changed in some way so as to be harmless; it causes the body to build up antibodies against a particular disease. *Ch. 11*

vacuole (L. vacuus, empty) A membrane-bounded storage sac within which such substances as water, food, and wastes can be found; most often found in plant cells, vacuoles play a major role in helping plant tissues stay rigid. *Ch. 3*

vagina (L. sheath) An organ in the body of a female whose muscular, tubelike passageway to the exterior has three functions; it accepts the penis during intercourse, it is the lower portion of the birth canal, and it provides an exit for the menstrual flow. *Ch. 21*

valve (L. valva, leaf of a folding door) Any of the one-way valves found in the heart and blood vessels, similarly constructed, that prevent the backflow of blood. *Ch. 10*

vascular bundle A cylinder of tissue made up of strands of xylem and phloem tissue positioned next to each other that constitute the transportation system of vascular plants. *Ch. 28*

vascular cambium A growth tissue in the woody stems of vascular plants that increases the diameter of a stem or root and gives rise to secondary xylem and secondary phloem. *Ch. 28*

vascular plant A plant having a system of specialized vessels that transports water and nutrients. *Ch. 27*

vascular tissue A differentiated tissue in vascular plants that conducts water, minerals, carbohydrates, and other substances throughout the plant; the principal types are xylem and phloem. *Ch. 28*

vas deferens In males, paired tubes that ascend from the epididymis into the pelvic cavity, looping over the side of the urinary bladder and eventually joining at the urethra. The vas deferens carry sperm during ejaculation. *Ch. 21*

vasectomy In males, an operation that involves the removal of a portion of the vas deferens, the tube through which sperm travels to the penis, in order to bring about sterilization. *Ch. 21*

vegetative propagation An asexual reproductive process in which a new plant develops from a portion of a parent plant. *Ch. 27*

vein (L. vena, a blood vessel) A blood vessel that brings blood to the heart. *Ch. 10*

ventricle (L. ventriculus. belly) Either of two lower chambers of the heart; the right ventricle pumps blood into the pulmonary artery and then the lungs, whereas the left ventricle pumps blood through the aorta into the arteries. *Ch. 10*

venule (L. vena, vein) A small vein that connects capillaries with larger veins. *Ch. 10*

vertebral column The collection of 26 bones in the middle of the back, stacked one on top of the other, that acts like a strong, flexible rod and supports the head in a human skeleton; the spine or backbone. *Ch. 16*

vertebrate (L. vertebra, vertebra) A member of a subphylum of chordates characterized by having a vertebral column surrounding a dorsal nerve cord. *Ch. 30*

vestibule A structure within the inner ear that detects the effects of gravity on the body. *Ch. 15*

villus, pl. villi (L. a tuft of hair) Fine, fingerlike projections that increase the surface absorption capability of the small intestine. *Ch. 8*

viruses Nonliving infectious agents that enter living organisms and cause disease. *Ch. 25*

visceral mass In a mollusk, a group of organs consisting of the digestive, excretory, and reproductive organs. *Ch. 29*

vitamin (L. vita, life + amine, of chemical origin) An organic molecule that performs functions such as helping the body use the energy of carbohydrates, fats, and proteins; 13 different vitamins play a vital role in the human body. *Ch. 8*

viviparous (L. vivus, alive + parere, to bring forth) A method of reproduction in which the young is born alive and the developing embryo derives primary nourishment from the mother and not the egg. *Ch. 30*

vocal cord Two pieces of elastic tissue covered with a mucous membrane stretched across the larynx that are involved in the production of sound. *Ch. 9*

vulva (L. covering) The collective term for the external genitals of a female. *Ch. 21*

water mold A mold that thrives in moist places and aquatic environments, parasitizing plants and animals. *Ch. 26*

white blood cell *See* leukocyte.

wind power The use of the wind to generate electricity, usually by means of windmills. *Ch. 38*

xylem (Gr. xylon, wood) A principal type of vascular tissue that conducts water and dissolved minerals in a plant. *Ch. 28*

yellow-green algae A type of golden algae. *Ch. 26*

yellow marrow A soft, fatty connective tissue that fills the hollow cylindrical core of long bones. *Ch. 16*

yolk (O.E. geolu, yellow) The nutrient material of an ovum that the developing organism can live on until nutrients can be derived from the mother. *Ch. 22*

yolk sac An extraembryonic membrane that produces blood for the embryo until its liver becomes functional. In addition, part of the yolk sac becomes the lining of the developing digestive tract. *Ch. 22*

zona pellucida (L. zona, girdle + pellucidus, transparent) A jellylike covering of the ovum. *Ch. 22*

zygospores Sexual spores formed by the zygote-forming fungi. *Ch. 26*

zygote (Gr. zygotos, paired together) A cell produced by the fusion of the haploid nuclei of the sperm and egg; the fertilized ovum. *Ch. 21*

zygote-forming fungi One of the three divisions of fungi. Zygote-forming fungi have a distinct sexual phase of reproduction that is characterized by the formation of sexual spores called zygospores. *Ch. 26*

CREDITS

Chapter 1

Opener, James King-Holmes/Science Photo Library/Photo Researcher, Inc. (top); Miki Koren Courtesy of Sigma Chemical Co. (middle); Laura J. Edwards (bottom)
Fig. 1-1, A-B, Stewart Halperin
Fig. 1-2, A-B, Richard Gross/Biological Photography; C, Ken Lucas/Biological Photo Service; D-F, Stewart Halperin
Fig. 1-3, A-B, Richard Gross/Biological Photography; C, Stewart Halperin (photo), E. Rohne Rudder (illustration); D, E. Rohne Rudder; E-H, Stewart Halperin; I, NASA
Fig. 1-4, Richard Gross/Biological Photography (photos)
Fig. 1-5, Stewart Halperin
Fig. 1-6, Richard Gross/Biological Photography
Fig. 1-7, Raychel Ciemma
Box (pg. 18), Stewart Halperin

Chapter 2

Opener, Stewart Halperin
Fig. 2-3, C, Michael Gadomski/Tom Stack & Associates
Fig. 2-5, Michael & Patricia Figden
Fig. 2-7, A, William Ober; B, George Bernard/Animals Animals; C, Eastcott/Momatiuk/The Image Works
Fig. 2-9, Lilli Robins
Fig. 2-14, B, Manfred Kagel/Peter Arnold, Inc.
Fig. 2-15, A, J.D. Litvay/Visuals Unlimited
Box (pg. 24), Simon Fraser/Medical Phisics, RVI, Newcastle/Science Photo Library/Custom Medical Stock Photo, Inc.
Box (pg. 30), Stewart Halperin

Chapter 3

Opener, David M. Phillips/Visuals Unlimited
Fig. 3-1, A,B,D, Richard Gross/Biological Photography; C, Lennart Nilsson, *Behold Man,* Little, Brown and Co.; E, Carolina Biological Supply Co.
Fig. 3-2, L.L. Sims/Visuals Unlimited
Fig. 3-4, A, David M. Phillips/Visuals Unlimited (photo); B, Bill Ober (art)
Fig. 3-7, B, J. David Roberson; C-F, Nadine Sokol
Fig. 3-10, A, J.V. Small; reprinted with permission from *Electron Micro. Rev.* 1:155, copyright 1988, with kind permission from Elsevier Science Ltd, The Boulevard, Langford Lane, Kidlington OX5 1GB, UK
Fig. 3-11, C, Richard Rodewald, University of Virginia
Fig. 3-14, B, K.G. Murti/Visuals Unlimited
Fig. 3-15, C, Dr. Thomas W. Tillack; D, Charles J. Flickinger
Fig. 3-16, Richard Gross/Biological Photography
Fig. 3-17, B, Richard Rodewald, University of Virginia/Biological Photo Service
Fig. 3-18, C, Brenda R. Eisenberg, Ph.D., University of Illinois Chicago
Fig. 3-19, C, Kenneth R. Miller
Fig. 3-20, Ellen Dirkson/Visuals Unlimited
Fig. 3-21, Paul W. Johnson/Biological Photo Service
Fig. 3-22, Barbara Cousins
Fig. 3-23, Ronald J. Erwin
Fig. 3-26, C, Charles L. Sanders/Biological Photo Service
Fig. 3-27, B, Dr. Birgit H. Satir
Box (pg. 52, 53), 3-A, Leonard Lessin/Peter Arnold, Inc.; 3-B(1),(2), Bruce Iverson; 3-C, Robert Brons/Biological Photo Service; 3-D, Kennedy/Biological Photo Service; 3-E,(1), T.D. Pugh and E.H. Newcomb, University of Wisconsin; 3-E(2), T.D. Pugh and E.H. Newcomb, University of Wisconsin/Biological Photo Service
Box (pg. 68), Stewart Halperin

Chapter 4

Opener, Stewart Halperin
Fig. 4-3, Barbara Cousins
Fig. 4-6, Stewart Halperin
Fig. 4-9, Barbara Cousins
Fig. 4-12, Barbara Cousins
Box (pg. 96), Stewart Halperin

Chapter 5

Opener, Stewart Halperin
Fig. 5-1, Nadine Sokol
Fig. 5-2, Stewart Halperin
Fig. 5-4, George Klart
Fig. 5-5, Nadine Sokol
Fig. 5-10, A, Custom Medical Stock Photo
Fig. 5-12, A, Richard Gross/Biological Photography; B, Stewart Halperin
Box (pg. 105), Stewart Halperin

Chapter 6

Opener, Stewart Halperin
Fig. 6-1, A-C, Stewart Halperin
Fig. 6-3, Nadine Sokol
Fig. 6-6, A, Richard Gross/Biological Photography
Fig. 6-7, Raychel Ciemma
Box (pg. 128), Stewart Halperin

Chapter 7

Opener, Lennart Nilsson, *Behold Man,* Little, Brown and Co.
Fig. 7-1, Nadine Sokol
Fig. 7-2, Christine Oleksyk (art); Tom Tracy/Photographic Resources
Fig. 7-4, Emma Shelton
Fig. 7-5, J.V. Small & F. Rinnerthaler
Fig. 7-6, *St. Louis Globe-Democrat*
Fig. 7-7, A, David J. Mascaro and Associates; B-C, John Hagen
Fig. 7-8, Lennart Nilsson, *Behold Man,* Little, Brown and Co.
Fig. 7-9, David M. Phillips/Visuals Unlimited
Fig. 7-10, Bill Ober

Fig. 7-11, Lennart Nilsson, *Behold Man,* Little, Brown and Co.
Fig. 7-12, Cynthia Turner Alexander/Terry Cockerham, Synapse Media Production/Christine Oleksyk
Fig. 7-13, Nadine Sokol
Table 7-2, Richard Gross/Biological Photography
Table 7-3, (1-4), Ed Reschke; (5-10), Richard Gross/Biological Photography
Table 7-4, Richard Gross/Biological Photography
Box (pg. 141), Stewart Halperin

Chapter 8

Opener, Stewart Halperin
Fig. 8-1, Nadine Sokol
Fig. 8-2, A, Nadine Sokol; B, G. David Brown
Fig. 8-3, Barbara Cousins
Fig. 8-4, Cynthia Turner Alexander/Terry Cockerham, Synapse Media Productions
Fig. 8-5, Nadine Sokol
Fig. 8-6, G. David Brown
Fig. 8-7, Bill Ober
Fig. 8-9, A, Barbara Cousins; B, David M. Phillips/Visuals Unlimited
Fig. 8-10, Nadine Sokol
Fig. 8-13, A-C, Stewart Halperin
Box (pg. 164), Stewart Halperin
Box (pg. 176), U.S. Department of Agriculture/U.S. Department of Health and Human Services; Aug., 1992

Chapter 9

Opener (top), American Cancer Society
Opener (bottom), James Stevenson/SPL/Photo Researchers, Inc.
Fig. 9-1, A, Cynthia Turner Alexander/Terry Cockerham, Synapse Media Productions
Fig. 9-2, Courtesy of AT&T Archives
Fig. 9-3, Ellen Dirkson/Visuals Unlimited
Fig. 9-4, Lennart Nilsson, *Behold Man,* Little, Brown and Co.
Fig. 9-5, Art Siegel, University of Pennsylvania
Fig. 9-6, Nadine Sokol
Fig. 9-9, Barbara Cousins
Fig. 9-10, Nadine Sokol
Fig. 9-11, A,C Stewart Halperin
Fig. 9-12, Moore/Visuals Unlimited
Box (pg. 190), Stewart Halperin

Chapter 10

Opener, Custom Medical Stock Photo
Fig. 10-1, Ronald J. Ervin
Fig. 10-2, Cynthia Turner Alexander/Terry Cockerham, Synapse Media Productions
Fig. 10-3, Bill Ober
Fig. 10-4, Ed Reschke
Fig. 10-5, Bill Ober
Fig. 10-6, D.W. Fawcett-T. Kuwabava/Visuals Unlimited
Fig. 10-7, Ed Reschke
Fig. 10-8, John D. Cunningham/Visuals Unlimited
Fig. 10-9, Christine Oleksyk
Fig. 10-10, Barbara Cousins
Fig. 10-11, Joan M. Beck
Fig. 10-12, Lisa Shoemaker/Joan M. Beck
Fig. 10-14, Mako Murayama/BPS/Tom Stack
Fig. 10-15, Raychel Ciemma
Fig. 10-16, Manfred Kage/Peter Arnold, Inc.
Fig. 10-17, B, Cynthia Turner Alexander/Terry Cockerham, Synapse Media Productions
Fig. 10-18, Ed Reschke
Fig. 10-19, Harry Ransom Humanities Research Center
Box (pg. 212), Stewart Halperin
Box (pg. 218), Stewart Halperin

Chapter 11

Opener, Lennart Nilsson
Fig. 11-1, Raychel Ciemma
Fig. 11-2, The Bettmann Archive
Fig. 11-7, Dr. A. Liepins/SPL/Photo Researchers, Inc.
Fig. 11-9, Barbara Cousins
Fig. 11-10, Secchi, LeCaque, Roussel Uclaf, CNRI/Science Source/Photo Researchers, Inc.
Fig. 11-11, Manfred Kage/Peter Arnold, Inc.
Fig. 11-15, Barbara Cousins
Fig. 11-16, Larry G. Arlain, Wright State University
Box (pg. 240), Stewart Halperin

Chapter 12

Opener, Stewart Halperin
Fig. 12-2, Stewart Halperin
Fig. 12-3, Larry Brock/Tom Stack & Associates
Fig. 12-4, Barbara Cousins
Fig. 12-5, Barbara Cousins
Fig. 12-6, Barbara Cousins
Fig. 12-13, Raychel Ciemma
Fig. 12-14, Barbara Cousins
Fig. 12-16, Stewart Halperin
Box (pg. 252), © 1991 Ted Horowitz/The Stock Market

Chapter 13

Opener, T. McCarthy/Custom Medical Stock Photo
Fig. 13-1, Peter Cohen/Custom Medical Stock Photo
Fig. 13-3, C.S. Raines/Visuals Unlimited
Fig. 13-4, A, David M. Phillips/Visuals Unlimited; C, ©Don W. Fawcett/Visuals Unlimited
Fig. 13-11, Stewart Halperin
Fig. 13-12, B, Heimer L. *Human Brain and Spinal Cord,* Springer-Verlag
Fig. 13-15, Dr. Michael J. Kuhar, NIDA Addiction Research Center
Box (pg. 276), Stewart Halperin

Chapter 14

Opener, William Gage/Custom Medical Stock Photo
Fig. 14-1, Cynthia Turner Alexander/Terry Cockerham, Synapse Media Productions
Fig. 14-3, Nadine Sokol
Fig. 14-5, Marcus Raichle, M.D./Washington University Medical School
Fig. 14-6, Karen Waldo
Fig. 14-7, Barbara Cousins
Fig. 14-8, Michael P. Schenk
Fig. 14-9, Nadine Sokol
Fig. 14-10, Michael P. Schenk
Fig. 14-11, Barbara Cousins
Fig. 14-12, Barbara Cousins
Fig. 14-13, Raychel Ciemma
Box (pg. 292), Stewart Halperin
Box (pg. 301), Stewart Halperin

Chapter 15

Opener, Nathan Benn/Woodfin Camp
Fig. 15-1, A-C, Stewart Halperin
Fig. 15-3, Barbara Cousins
Fig. 15-4, A, Marsha A. Dohrmann; B, Christine Oleksyk
Fig. 15-5, Raychel Ciemma
Fig. 15-7, Marsha J. Dohrmann
Fig. 15-8, G. David Brown
Fig. 15-9, A, Christine Oleksyk; B, Marsha J. Dohrmann; C, Scott Mittman
Fig. 15-10, Marsha J. Dohrmann
Fig. 15-12, A, Marsha J. Dohrmann; B, Kathy Mitchell Grey; C-D, G. David Brown
Box (pg. 313), E. Rohne Rudder

Chapter 16

Opener, Vince Rodriquez
Fig. 16-1, Ronald J. Ervin
Fig. 16-2, Laurie O'Keefe/John Daugherty
Fig. 16-3, Kate Sweeney
Fig. 16-4, Nadine Sokol
Fig. 16-5, Barbara Cousins
Fig. 16-6, Barbara Cousins
Fig. 16-7, Barbara Cousins
Fig. 16-8, Scott Bodell
Fig. 16-9, Kate Sweeney
Fig. 16-10, Nadine Sokol
Fig. 16-11, Barbara Cousins
Fig. 16-12, Richard Rodewald, University of Virginia
Fig. 16-14, John D. Cunningham/Visuals Unlimited
Fig. 16-15, Barbara Cousins

Table 16-1, Nadine Sokol after Rusty Jones
Box (pg. 331), Stewart Halperin
Box (pg. 339), Stewart Halperin

Chapter 17

Opener, Stewart Halperin
Fig. 17-2, Barbara Cousins
Fig. 17-3, Barbara Cousins
Fig. 17-4, Nadine Sokol
Fig. 17-5, Nadine Sokol
Fig. 17-6, Bettina Cirone/Photo Researchers, Inc.
Fig. 17-7, *American Journal of Medicine* 20(1956) 133
Fig. 17-8, NMSB/Custom Medical Stock Photo
Fig. 17-9, Barbara Cousins
Fig. 17-10, Coustom Medical Stock Photo
Fig. 17-11, Raychel Ciemma
Fig. 17-12, Raychel Ciemma
Fig. 17-13, Nadine Sokol
Fig. 17-14, A,B,D, Stewart Halperin
Fig. 17-15, Ed Reschke
Box (pg. 359), Stewart Halperin

Chapter 18

Opener, CNRI/SPL/Science Source
Fig. 18-1, U.K. Laemmli & J.R. Paulson
Fig. 18-2, Raychel Ciemma (art); Ada L. Olins/Biological Photo Service (photomicrograph)
Fig. 18-6, Cold Springs Harbor Laboratory Archives
Fig. 18-7, Cold Springs Harbor Laboratory Archives
Fig. 18-9, Christine Oleksyk
Fig. 18-12, Nadine Sokol
Fig. 18-14, Molly Babich
Fig. 18-16, Carlyn Iversen
Fig. 18-17, Molly Babich
Fig. 18-18, Nadine Sokol (art); 1982 C. Franke, J.E. Edstrom, A.W. McDowall & O.L. Miller, Jr. (photo)
Fig. 18-19, Barbara Cousins
Fig. 18-20, Carlyn Iversen
Fig. 18-21, Richard Gross/Biological Photography
Fig. 18-23, Barbara Cousins (art); Richard Gross/Biological Photography (photos)
Fig. 18-24, Dr. A.S. Bajer (photo)
Fig. 18-25, B.A. Palevitz & E.H. Newcomb/BPS/Tom Stack & Associates
Fig. 18-28, James Kezer, University of Oregon (photo)
Fig. 18-29, Barbara Cousins (art); C.A. Hasenkampf, University of Toronto/Biological Photo Service
Box (pg. 372), Nadine Sokol
Box (pg. 387), Stewart Halperin

Chapter 19

Opener, Stewart Halperin
Fig. 19-1, E. Rohne Rudder
Fig. 19-2, Nadine Sokol after Bill Ober
Fig. 19-3, Nadine Sokol after Bill Ober
Fig. 19-4, Nadine Sokol
Fig. 19-5, Nadine Sokol
Fig. 19-6, Nadine Sokol
Fig. 19-8, Nadine Sokol
Fig. 19-11, Carolina Biological Supply Company
Box (pg. 413), Stewart Halperin

Chapter 20

Opener, Culver Pictures, Inc.
Fig. 20-1, CNRI/SPL/Science Source/Photo Researchers, Inc.
Fig. 20-2, A, The Children's Hospital, Denver, Cytogenics Laboratory; B, R. Hutchings/Photo Researcher, Inc.
Fig. 20-3, Barbara Cousins
Fig. 20-5, Earl Plunkett; from Valentine, 1986
Fig. 20-6, Earl Plunkett; from Valentine, 1986
Fig. 20-8, Barbara Cousins
Fig. 20-9, VU/Cabisco
Fig. 20-12, Adam Hart-Davis/SPL/Custom Medical Stock Photo
Fig. 20-15, The Bettmann Archive
Fig. 20-16, M. Mrayama/Biological Photo Service
Fig. 20-17, Stewart Halperin (photos)
Fig. 20-19, Washington University School of Medicine
Box (pg. 430, 431), A-B, Jack Tandy

Chapter 21

Opener, Lennart Nilsson, *Behold Man,* Little, Brown and Co.
Fig. 21-1, Lennart Nilsson, *Behold Man,* Little, Brown and Co.
Fig. 21-2, Kate Sweeney
Fig. 21-3, Bill Ober
Fig. 21-4, Barbara Cousins
Fig. 21-6, Kate Sweeney
Fig. 21-8, Ed Reschke
Fig. 21-9, Kevin Somerville
Fig. 21-10, David J. Mascaro & Associates
Fig. 21-12, A-D, Stewart Halperin
Fig. 21-13, Laura J. Edwards
Fig. 21-14, Ronald J. Ervin
Box (pg. 460), Stewart Halperin

Chapter 22

Opener, Stewart Halperin
Fig. 22-1, A, Barbara Cousins; B, David M. Phillips/Visuals Unlimited
Fig. 22-2, A-B, Lennart Nilsson, *A Child is Born*
Fig. 22-3, A-B, Lennart Nilsson, *A Child is Born*
Fig. 22-4, A-B, Lennart Nilsson, *A Child is Born* (photos); Raychel Ciemma (art)
Fig. 22-5, Lennart Nilsson, *A Child is Born*
Fig. 22-6, Raychel Ciemma
Fig. 22-7, Scott Bodell
Fig. 22-8, Kate Sweeney
Fig. 22-9, Raychel Ciemma
Fig. 22-10, Barbara Cousins
Fig. 22-11, Kevin Somerville after Bill Ober
Fig. 22-12, Lennart Nilsson, *A Child is Born*
Fig. 22-13, Lennart Nilsson, *A Child is Born*
Fig. 22-14, Barbara Cousins
Fig. 22-15, Lennart Nilsson, *A Child is Born*
Fig. 22-16, Lennart Nilsson, *A Child is Born*
Fig. 22-17, Lennart Nilsson, *A Child is Born*
Fig. 22-18, Lennart Nilsson, *A Child is Born*
Fig. 22-20, Scott Bodell (art); Martin/Custom Medical Stock Photo
Box (pg. 476), Stewart Halperin

Chapter 23

Opener, Stewart Halperin
Fig. 23-3, A, Frank B. Gill/VIREO; B, John S. Dunning/VIREO
Fig. 23-4, A-C, Louise Van der Meid
Fig. 23-5, A, Ralph A. Reinhold©/Animals Animals; B, ©Gerard Lacz/Animals Animals
Fig. 23-8, The National Portrait Gallery, London
Fig. 23-9, Frank S. Balthis
Fig. 23-10, Molly Babich
Fig. 23-11, John D. Cunningham/Visuals Unlimited
Fig. 23-12, Richard Gross/Biological Photography
Fig. 23-14, Nadine Sokol after Bill Ober
Fig. 23-15, Don & Pat Valenti/Tom Stack & Associates
Fig. 23-16, A, George H.H. Huey/Earth Scenes; B, Arthur Gloor/Earth Scenes
Table 23-1, Alters, B.J. & McComas, W.F. (1994). Punctuated equilibrium: The missing link in evolution education. *The American Biology Teacher, 56,*(6), 334-339.
Box (pg. 501), Stewart Halperin

Chapter 24

Opener, NASA
Fig. 24-1, B, Kevin Walsh, University of California, San Diego
Fig. 24-2, Dudley Foster/Woods Hole Oceanographic Institution
Fig. 24-3, A, M.R. Walter, Macquarie University, Australia; C, Paul F. Hoffman
Fig. 24-4, Andrew H. Knoll, Botanical Museum of Harvard University
Fig. 24-6, A,D, David Bruton; B,C,E, Simon Conway Morris, University of Cambridge

Fig. 24-7, A, Richard Gross/Biological Photography; B, Frans Lanting/Minden Pictures
Fig. 24-8, A, Smithsonian Institution photo no. USNM 57628 Walcott, 1911; B, Heather Angel/Biofotos
Fig. 24-11, Steve Martin/Tom Stack & Associates
Fig. 24-12, Gary Milburn/Tom Stack & Associates
Fig. 24-13, Bill Ober
Fig. 24-14, A, John D. Cunningham/Visuals Unlimited; B, Barbara Laing/Black Star
Fig. 24-15, Stouffer Productions Ltd./Animals Animals
Fig. 24-18, Gerard Lacz/Animals Animals
Fig. 24-19, A, Doug Wechsler/Animals Animals/B, C.C. Lockwood/Animals Animals
Fig. 24-20, Russell A. Mittermeier
Fig. 24-21, Alan Nelson/Animals Animals
Fig. 24-23, Alan E. Mann
Fig. 24-25, John Reader
Fig. 25-26, A, The Living World, St. Louis Zoo, St. Louis, MO.; B, John Reader
Fig. 24-27, Allen Carroll and Jay Maternes/National Geographic, November 1985
Fig. 24-28, Douglas Waugh/Peter Arnold, Inc.
Fig. 24-29, SCODE/Overseas
Box (pg. 514), Stewart Halperin

Chapter 25

Opener, Science Source
Fig. 25-1, A, K.G. Murti/Visuals Unlimited; B, K. Namba and D.L.D. Caspar
Fig. 25-3, A, Science VU/Visuals Unlimited; B, Carlyn Iverson
Fig. 25-4, Nadine Sokol
Fig. 25-10, Nadine Sokol
Fig. 25-11, Elizabeth Gentt/Visuals Unlimited
Fig. 25-12, Runk/Schoenberger/Grant Heilman/Photography, Inc.
Fig. 25-13, Jo Handelsman and Steven A. Vicen
Fig. 25-14, Photo Researchers/CDC
Fig. 25-15, A.M. Siegleman/Visuals Unlimited; Science Source/Photo Researchers, Inc.
Fig. 25-16, Biophoto Associates/ Photo Researchers, Inc.; Visuals Unlimited
Box (pg. 552), Stewart Halperin
Box (pg. 558), S. Benjamin/Custom Medical Stock Photo, Inc.
Table 25-1, National Cancer Institute; Science Photo/Photo Researchers; VU/© Veronika Burmeister; Science VU/Visuals Unlimited; Howard Sochurek/The Stock market

Chapter 26

Opener, Stewart Halperin
Fig. 26-1, Courtesy of Stanley Erlandson
Fig. 26-2, A, Dr. E.W. Daniels, Argonne National Laboratory, Argonne, IL; and University of Illinois College of Medicine, Dept. of Anatomy & Cell Biology, Chicago; B-C, Richard Gross/Biological Photography
Fig. 26-3, A, M. Abbey/Visuals Unlimited; B, Carlyn Iversen
Fig. 26-4, Richard Gross/Biological Photography
Fig. 26-5, A, Richard Gross/Biological Photography; B, Manfred Kage/Peter Arnold, Inc.
Fig. 26-6, Courtesy of Diana Laulaien-Schein
Fig. 26-7, Bill Ober
Fig. 26-8, David M. Phillips/Visuals Unlimited
Fig. 26-10, K.G. Murti/Visuals Unlimited
Fig. 26-11, Nadine Sokol after Bill Ober
Fig. 26-12, A-B, Richard Gross/Biological Photography
Fig. 26-13, Bill Ober
Fig. 26-14, Sanford Berry/Visuals Unlimited
Fig. 26-15, A, David M. Phillips/Visuals Unlimited; B, Richard Gross/Biological Photography
Fig. 26-16, A, Philip Sze/Visuals Unlimited; B, William C. Jorgensen/Visuals Unlimited; C, Bob Evans/Peter Arnold, Inc.
Fig. 26-17, A, BioPhoto Associates/Photo Reseachers, Inc.; B-C, John D. Cunningham/Visuals Unlimited; D, Richard Gross/Biological Photography
Fig. 26-18, A, William C. Jorgensen/Visuals Unlimited; B, Gary K. Robinson/Visuals Unlimited; C, Brian Parker/Tom Stack & Associates
Fig. 26-19, A, Cabisco/Visuals Unlimited; B-F, Higuchi Bioscience Laboratory
Fig. 26-20, E.S. Ross
Fig. 26-21, A-C, Richard Gross/Biological Photography
Fig. 26-22, Dwight Kuhn
Fig. 26-23, Carlyn Iversen
Fig. 26-24, A, Stewart Halperin; B, Jeremy Burgess/Science Photo Library/Photo Researchers, Inc.; C, Carlyn Iversen
Fig. 26-25, A, Bob Evans/Peter Arnold, Inc.; B, Carlyn Iversen
Fig. 26-26, Gordon Langsbury/Bruce Coleman Ltd.
Fig. 26-28, A-C, Richard Gross/Biological Photography
Fig. 26-29, Carlyn Iversen
Fig. 26-30, A, David M. Phillips/Visuals Unlimited; B, BioPhoto Associates /Photo Researchers, Inc.; C, Richard Gross/Biological Photography
Fig. 26-31, Ken Greer/Visuals Unlimited
Fig. 26-32, A-C, Richard Gross/Biological Photography
Fig. 26-33, Ed Reschke
Box (pg. 588), Stewart Halperin

Chapter 27

Opener, Richard Gross/Biological Photography
Fig. 27-1, Barbara Cousins
Fig. 27-2, A, Kirtley-Perkins/Visuals Unlimited; B, Ken Davis/Tom Stack & Associates
Fig. 27-3, A, Stewart Halperin; B, John D. Cunningham/Visuals Unlimited
Fig. 27-4, Nadine Sokol after Bill Ober
Fig. 27-5, Bill Ober
Fig. 27-6, Nadine Sokol after Bill Ober
Fig. 27-7, Carlyn Iversen
Fig. 27-8, A, Whit Bronaugh; B, E.S. Ross; C, Michael & Patricia Fogden
Fig. 27-9, Carlyn Iversen
Fig. 27-10, A, Tom J. Ulrich/Visuals Unlimited; B, Ron Spomer/Visuals Unlimited; C-D, Stewart Halperin
Fig. 27-11, Carlyn Iversen
Fig. 27-12, A-E, Richard Gross/Biological Photography; F, Whit Bronaugh
Fig. 27-13, Stewart Halperin
Fig. 27-14, Jack M. Bostrack/Visuals Unlimited
Fig. 27-15, E. Webber/Visuals Unlimited
Fig. 27-16, Sylvan H. Wittwer, Michigan State University
Fig. 27-17, John D. Cunningham/Visuals Unlimited
Box (pg. 609), Stewart Halperin

Chapter 28

Opener, Ernst van Jaarsueld/Kirstenbosch National Botanical Garden/South Africa
Fig. 28-1, Carlyn Iversen
Fig. 28-2, A, Bill Ober; C, Randy Moore/Visuals Unlimited
Fig. 28-3, A, George J. Wilder/Visuals Unlimited; B, Richard Gross/Biological Photography
Fig. 28-4, A, Cabisco/Visuals Unlimited; B, Ed Reschke/Peter Arnold, Inc.; C, Fred E. Hossler/Visuals Unlimited
Fig. 28-5, Carlyn Iverson (art); Richard Gross/Biological Photography
Fig. 28-6, Carlyn Iverson (art); Richard Gross/Biological Photography
Fig. 28-7, Kent Wood
Fig. 28-8, A, John D. Cunningham/Visuals Unlimited; B-C, Richard Gross/Biological Photography
Fig. 28-9, A-B, Richard Gross/Biological Photography
Fig. 28-10, Carlyn Iverson
Fig. 28-11, Carlyn Iverson (art); Ray F. Evert (photo)

Fig. 28-12, Richard Gross/Biological Photography
Fig. 28-13, Carlyn Iverson
Fig. 28-14, A, Richard Gross/Biological Photography; B, Stewart Halperin
Fig. 28-16, Bill Ober (art); Ed Reschke/Peter Arnold, Inc. (photo)
Fig. 28-18, Carlyn Iverson
Box (pg. 631), Stewart Halperin

Chapter 29

Opener, Alex Kerstich
Fig. 29-2, Bill Ober
Fig. 29-3, Barbara Cousins
Fig. 29-4, Scott Bodell after Bill Ober
Fig. 29-5, A, Nadine Sokol after Bill Ober; B-C, Stephen Fink
Fig. 29-6, A-C, Stephen Frink
Fig. 29-7, E. Jeff Rottman
Fig. 29-8, Nadine Sokol after Bill Ober
Fig. 29-9, Nadine Sokol after Bill Ober
Fig. 29-10, A, Jim & Cathy Church; B, E.J. Cable/Tom Stack & Associates
Fig. 29-11, Nadine Sokol after Bill Ober
Fig. 29-12, V.R. Ferris
Fig. 29-13, M.G. Schultz from H. Zaimen (ed.) pictorial presentation of parasites
Fig. 29-14, Nadine Sokol
Fig. 29-15, Bill Ober (art); A, Milton Rand/Tom Stack & Associates; B, Stephen Frink; C, Fred Bavendam; D, Kjell Sandved
Fig. 29-16, A, Kjell Sandved; B, David Dennis/Tom Stack & Associates; C, Gwen Fidler/Tom Stack & Associates
Fig. 29-17, Nadine Sokol
Fig. 29-18, A-B, Stewart Halperin; C, Kjell Sandved
Fig. 29-19, A, Kjell Sandved; B, Cleveland P. Hickman, Jr.
Fig. 29-20, Bill Ober
Fig. 29-21, Kjell Sandved
Fig. 29-22, Bill Ober
Fig. 29-25, A,D, Stephen Frink; B, Carl Roessler/Tom Stack & Associates; C, David L. Pawson/Harbor Branch Oceanographic Institution, Inc.; E, Jeff Rotman
Fig. 29-26, Bill Ober
Box (pg. 644), Stewart Halperin

Chapter 30

Opener, Stewart Halperin
Fig. 30-1, Bill Ober
Fig. 30-2, A, Rick Harbo; B, Jim & Cathy Church; C, Bill Ober
Fig. 30-3, Richard Olson
Fig. 30-4, Nadine Sokol after Bill Ober
Fig. 30-5, A, Heather Angel; B, Bill Ober
Fig. 30-6, A-B, Marty Snyderman; C, Carl Roessler/Tom Stack & Associates
Fig. 30-9, A-B, Innerspace Visions
Fig. 30-10, A-C, Richard Gross/Biological Photography
Fig. 30-11, A-B, Michael Fogden
Fig. 30-12, A,C, Stewart Halperin; B,D, Richard Gross/Biological Photography
Fig. 30-13, Bill Ober
Fig. 30-14, Bill Ober
Fig. 30-15, A-C, Stewart Halperin; D, John D. Cunningham/Visuals Unlimited; E, William J. Weber/Visuals Unlimited
Fig. 30-16, Raychel Ciemma
Fig. 30-17, A, J.E. Wapstra/NPIAW; B, J. Alcock/Visuals Unlimited
Fig. 30-18, A-B, Stewart Halperin
Fig. 30-19, A-C, Stewart Halperin; E, Mary Snyderman
Box (pg. 677), Stewart Halperin

Chapter 31

Opener, Miki Koren, Courtesy of Sigma Chemical Co.
Fig. 31-1, Stewart Halperin
Fig. 31-3, John Durham/Photo Researchers
Fig. 31-5, Jack Griffith
Fig. 31-6, David P. Allison, Oak Ridge National Laboratory/Biological Photo Service
Fig. 31-7, Visuals Unlimited/Science Vu-URSCIM
Fig. 31-11, From Baron EJ, Peterson LR, Finegold SM: *Bailey & Scott's Diagnostic Microbiology,* ed 9, St. Louis, Mosby, 1994
Fig. 31-14, Wedgworth/Custom Medical Stock Photo
Fig. 31-15, M. English/Custom Medical Stock Photo
Fig. 31-16, Matt Meadows/Peter Arnold, Inc. (photo)
Fig. 31-17, M. Baret/Science Source/Photo Researchers
Box (pg. 696), Stewart Halperin

Chapter 32

Opener, Paul Lemmons/Frank Lane Picture Agency
Fig. 32-1, Jeff Foott/Bruce Coleman, Inc.
Fig. 32-2, Don & Pat Valenti/Tom Stack & Associates
Fig. 32-3, Raychel Ciemma
Fig. 32-4, Stoufer Productions, Ltd./Animals Animals
Fig. 32-5, E. Rohne Rudder
Fig. 32-6, E.S. Ross
Fig. 32-7, Thomas McAvoy, Life Magazine, 1955, Time Warner, Inc.
Fig. 32-8, FPG International Corp.
Fig. 32-9, Raychel Ciemma
Fig. 32-10, Omikron/Photo Researchers, Inc.
Fig. 32-11, Lee Boltin
Fig. 32-12, Babkin Papers, Osler Library of the History of Medicine, McGill University, Montreal, Canada

Chapter 33

Opener, Norbert Wu
Fig. 33-1, Jane Burton/Bruce Coleman Ltd.
Fig. 33-2, Bill Eades
Fig. 33-4, Tui De Roy
Fig. 33-5, Dale Sarver/Animals Animals
Fig. 33-6, Stewart Halperin
Fig. 33-7, C.W. Schwartz/Animals Animals
Fig. 33-8, Bill Ober (art); Oxford Scientific Films/Animals Animals (photo)
Fig. 33-9, Nadine Sokol
Fig. 33-10, Stewart Halperin
Fig. 33-11, Fred Bruemmer
Fig. 33-12, A, John Stern/Animals Animals; B, J.C. Stevenson/Animals Animals
Fig. 33-13, E.S. Ross
Fig. 33-14, E.S. Ross
Fig. 33-16, Mark Moffett
Fig. 33-17, Tom Stack & Associates
Box (pg. 731), Stewart Halperin
Box (pg. 733), Mike McKarett/Bruce Coleman Ltd.

Chapter 34

Opener, Thomas D. Mangelsen/Images of Nature
Fig. 34-2, Alan Fletcher Research Station, Queensland Department of Lands, Queensland, Australia
Fig. 34-7, Stewart Halperin
Fig. 34-9, A-B, Stewart Halperin (photos)
Box (pg. 754), Cameramann Int'l, LTD.
Box (pg. 756), Stewart Halperin

Chapter 35

Opener, Stephen Fink
Fig. 35-1, David Swanlung, Save-the-Redwoods League
Fig. 35-2, E.S. Ross
Fig. 35-3, A, Fred Whitehead/Animals Animals—Earth Scenes; B, Stewart Halperin; C, Jeri Clark
Fig. 35-4, Roger del Moral, University of Washington
Fig. 35-5, D.P. Wilson/Eric & David Hosking
Fig. 35-6, Barbara Cousins
Fig. 35-7, A, George H. Harrison; B, M.P. Kahl/National Geographic Society; C, Stewart Halperin
Fig. 35-8, M. Abbey/Visuals Unlimited
Fig. 35-10, Joe McDonald/Animals Animals
Fig. 35-11, Calgene, Inc.
Fig. 35-12, Stewart Halperin (top); E.S. Ross (bottom)

Fig. 35-13, A, James L. Castner; B, William J. Wesler/Visuals Unlimited; C, Alex Kerstich; D, E.S. Ross
Fig. 35-14, A, Patty Murray/Animals Animals; B, Dwight Kuhn; C, Paul A. Opler
Fig. 35-15, E.S. Ross
Fig. 35-16, A,D, Richard Gross/Biological Photography; B, E.S. Ross; C,E, James L. Castner
Fig. 35-17, Richard Gross/Biological Photography
Fig. 35-18, A, Kennan Ward; B, Frank S. Balthis
Fig. 35-19, Bill Wood/Bruce Coleman, Inc.
Fig. 35-20, Cabisco/Visuals Unlimited
Fig. 35-21, E.S. Ross
Fig. 35-22, Dwight Kuhn
Fig. 35-23, E.S. Ross
Fig. 35-24, Gary Braasch
Box (pg. 769), Times Mirror

Chapter 36

Opener, A.J. Stevens/Bruce Coleman Ltd.
Fig. 36-1, Nadine Sokol
Fig. 36-2, E.S. Ross
Fig. 36-3, Richard Gross/Biological Photography
Fig. 36-4, Raychel Ciemma
Fig. 36-5, Raychel Ciemma
Fig. 36-6, Lilli Robins
Fig. 36-9, Nadine Sokol
Fig. 36-10, Nadine Sokol
Fig. 36-12, Nadine Sokol
Fig. 36-13, A, Gerald A. Peters, Virginia Commonwealth University; B, Tom Stack & Associates; C, Stewart Halperin
Fig. 36-14, Nadine Sokol
Fig. 36-15, Norbert Wu
Box (pg. 790), John D. Cunningham/Visuals Unlimited

Chapter 37

Opener, James L. Castner
Fig. 37-1, Bill Ober
Fig. 37-4, Bill Ober
Fig. 37-7, Bill Ober
Fig. 37-8, A-B, Richard Gross/Biological Photography
Fig. 37-9, A, Richard Gross/Biological Photography; B, Doug Wechsler/VIREO; C-D, Michael & Patricia Fogden; E, E.S. Ross
Fig. 37-10, Stewart Halperin
Fig. 37-11, W. Perry Conway/Tom Stack & Associates
Fig. 37-12, Stewart Halperin
Fig. 37-13, A, Richard Kolar/Earth Scenes; B, Breck P. Kent/Earth Scenes; C, Arthur Gloor/Earth Scenes
Fig. 37-14, A, John Cancalosi/Tom Stack & Associates; B, William J. Hamilton III
Fig. 37-15, W.J. Weber/Visuals Unlimited
Fig. 37-16, Richard Gross/Biological Photography
Fig. 37-17, Lee Casebere
Fig. 37-18, Cleveland P. Hickman, Jr.
Fig. 37-19, Bill McRae
Fig. 37-20, Tom Stack & Associates
Fig. 37-21, Raychel Ciemma
Fig. 37-22, B-D, Chesapeake Bay Foundation
Fig. 37-23, Dan Gotshell
Fig. 37-24, Jeff Foott/Tom Stack & Associates
Fig. 37-25, Stewart Halperin
Fig. 37-26, A, Norbert Wu; B, C, Dale Glantz
Fig. 37-27, Raychel Ciemma
Fig. 37-28, A, Stewart Halperin; B, Doc White/Images Unlimited; C, Jeff Rotman Photography; D, Norbert Wu; E, Dudley Foster/Woods Hole Oceanographic Institution
Box (pg. 804), Stewart Halperin

Chapter 38

Opener, Chet Hanchet/Photographic Resources
Fig. 38-1, Nadine Sokol
Fig. 38-2, A, Thomas A. Schneider; B, Grant Heilman/Grant Heilman Photography; C, Keven Schafer/Peter Arnold, Inc.; D, Stan W. Elms/Visuals Unlimited
Fig. 38-4, Frans Lanting/Minden Pictures
Fig. 38-6, J.M. Rankin/Worldwide Fund for Nature
Fig. 38-7, Agricultural Research Service, USDA
Fig. 38-8, Stewart Halperin
Fig. 38-9, Ray Pfortner/Peter Arnold, Inc.
Fig. 38-10, Ray Pfortner/Peter Arnold, Inc.
Fig. 38-11, Raychel Ciemma
Fig. 38-13, Gamma Liaison Network
Fig. 38-14, Thomas A. Scheider
Fig. 38-16, A, Frank S. Balthis; B, Bruno J. Zehnder/Peter Arnold
Box (pg. 835), Biosphere 2

*I*NDEX

B

D

F

G

H

Q

R

T

W